HANDBOOK OF THE BEHAVIORAL NEUROBIOLOGY OF SEROTONIN

HANDBOOK OF THE BEHAVIORAL NEUROBIOLOGY OF SEROTONIN

Edited by

CHRISTIAN P. MÜLLER AND BARRY L. JACOBS

ELSEVIER

AMSTERDAM • BOSTON • HEIDELBERG • LONDON
NEW YORK • OXFORD • PARIS • SAN DIEGO
SAN FRANCISCO • SINGAPORE • SYDNEY • TOKYO

Academic Press is an imprint of Elsevier

Academic Press is an imprint of Elsevier
32 Jamestown Road, London NW1 7BY, UK
30 Corporate Drive, Suite 400, Burlington, MA 01803, USA
525 B Street, Suite 1900, San Diego, CA 92101-4495, USA

First edition 2010

British Library Cataloguing-in-Publication Data
A catalogue record for this book is available from the British Library

Library of Congress Cataloging-in-Publication Data
A catalog record for this book is available from the Library of Congress

ISBN : 978-0-12-374634-4

For information on all Academic Press publications
visit our website at www.elsevierdirect.com

Typeset by Macmillan Publishing Solutions
www.macmillansolutions.com

Printed and bound by CPI Group (UK) Ltd, Croydon, CR0 4YY
Transferred to Digital Print 2011

To our loved ones

C.P.M.
Bine & Merlin

B.L.J.
Susyn Berger

Contents

Section 3: Serotonin and Behavioral Control

Section 4: Serotonin in Disease Conditions

Preface

After more than 60 years of intensive research, the serotonin system has lost nothing of its attraction to researchers and people interested in the brain mechanisms of subjective experience and behavior. In fact, the more that becomes known about serotonin, the more it uncovers a brilliant complexity, which raises even more questions. Our understanding of the brain serotonergic system has been largely driven by pharmacological research. This is not surprising, given that the serotonin system, with its many specific genes and a steadily increasing number of proteins, offers a huge variety of possible pharmacological targets. Many of them proved to be effective in treating psychiatric diseases, such as depression and anxiety disorders. By the way, pharmacological progress was also shaping the theories about the origin of the disease and serotonin function in general. The parallel development of a behavioral neuroscience of serotonin always made use of the pharmacological tools. However, it is more focused on a general idea of the role of serotonin in behavior, a master plan, a predominant role, something like a brand identity. Needless to say, a full understanding of the role of a neurotransmitter in disease conditions is inconceivable without having a concept of how the transmitter system works in normal behavior. Despite many previous and still ongoing attempts to reduce serotonin to one key role, it has never been proved to fulfill any narrow claim. What we know for sure is only that there is no behavior and plasticity without it, while too much of it is equally detrimental. Its fate in behavioral neuroscience is still its omnipresent involvement in virtually every behavior, which is best reflected on the physiological level by the fact that a relatively small number of serotonin neurons modulate, by their projections, virtually all other regions of the central nervous system. To date the serotonergic system successfully resists a reductionist approach, but precisely that may finally be its beauty.

For this book we have invited well-known experts in many different fields of serotonin neuroscience to contribute their findings and views. The result is a compilation that presents a broad picture of how serotonin in the brain influences behavior and subjective experiences (including the coordinated role of physiology). The book is divided into four sections. In Section 1, state of the art knowledge of the functional anatomy of the serotonin system is presented. It starts with a view on the evolution of the serotonergic system. The genetic organization of the system is then outlined, where the genes that are unique for serotonergic signaling and their characteristics are discussed. This is followed by chapters that summarize the gross anatomy and ultrastructure of the serotonergic neurons, projections, and synapses in the brain. Serotonin receptors are the essential proteins that 'translate' the extracellular serotonin signal into intracellular signals. The distribution of the many different serotonin receptors and their particular signaling characteristics are presented in subsequent chapters. A further chapter reviews the ways in which serotonin activity can be detected in the brains of animals and humans during behavior.

Section 2 provides an overview of the neurophysiology of the serotonin system, which is the basis for an understanding of normal behavioral processes and pathological conditions. The activity of serotonergic neurons is discussed extensively, followed by an overview of the electrophysiology of the serotonin receptors. Essential for serotonin signaling are its synthesis and later inactivation metabolism. Both are discussed in detail, with the focus on the enzymes tryptophane hydroxylase and monoamine oxidase. The behavioral effects of serotonin precursor over- and undersupply are also critically evaluated. Once serotonin is released in the brain, it exerts effects on postsynaptic cells and subsequently other transmitter systems. These effects are reviewed in two chapters. Postsynaptic effects need to be in a certain 'activation window' that on the one hand assures appropriate levels of plasticity, and on the other prevents dangerous overactivation. This is achieved by a number of feedback control systems, which are discussed in the next chapter. As has become evident in recent years, serotonin also plays a major role in the generation of rhythmic activity in the brain and in CO_2 chemoreception; these aspects are summarized in separate chapters.

Section 3 deals with the role of serotonin in behavioral control, focusing on normal behavior. It starts with an in-depth review of how serotonin guides brain development, and how disturbances of the serotonergic system during prenatal to adolescent development affect adult behavior. The role of serotonin in a number of behaviors and subjective states is then discussed in detail in separate chapters focusing on motor control, appetite and ingestion, sexual behavior, mood and emotion, anxious states, reward and reinforcement, impulsivity, learning and memory, social behavior, and pain control.

In Section 4 the role of the serotonergic system in pathological conditions is extensively reviewed, discussing psychiatric disorders and serotonin-based pharmacotherapies. As such, separate chapters are devoted to the role of serotonin in stress, depression, drug addiction, hallucinogenic drug action, obsessive compulsive disorders, attention-deficit hyperactivity disorder, schizophrenia, hallucinogenic drug action, autism, panic and anxiety disorders, aggression, and anorexia and bulimia. Two further chapters focus on how genetic, epigenetic, and environmental factors influence psychiatric disorders by their interaction with the serotonergic system in humans and in animals, respectively. A final chapter discusses the pharmacogenetics of serotonin receptors in psychiatric drug action.

While the contribution of serotonin to many behaviors is evident, the claim that 'it is involved in everything, but responsible for nothing' might still hold true. However, we are now at the edge of a system's role in behavior. Further success can only be achieved with a thorough understanding of single transmitter systems and how they influence others. As such, this book is an attempt to present the current state of knowledge on the serotonin system in the control of behavior. We hope that it may serve as reference and starting point for future systems approaches incorporating the serotonergic system into a larger view of how subjective experience and behavior is organized by the brain.

Christian P. Müller and Barry L. Jacobs

List of Contributors

Albert Adell, Department of Neurochemistry and Neuropharmacology, Instituto de Investigaciones Biomédicas de Barcelona, CSIC (IDIBAPS), Centro de Investigación Biomédica en Red de Salud Mental (CIBERSAM), Barcelona, Spain

Rodrigo Andrade, Department of Pharmacology, Wayne State University School of Medicine, Detroit, USA

Francesc Artigas, Department of Neurochemistry and Neuropharmacology, Instituto de Investigaciones Biomédicas de Barcelona, CSIC (IDIBAPS); Centro de Investigación Biomédica en Red de Salud Mental (CIBERSAM), Barcelona, Spain

Efrain C. Azmitia, Professor of Biology, Department of Biology and Psychiatry, Center for Neuroscience, New York University, 100 Washington Square East, NY, USA

Ursula F. Bailer, University of California San Diego, School of Medicine, Department Psychiatry, La Jolla, CA, USA; Medical University of Vienna, Department of Psychiatry and Psychotherapy, Division of Biological Psychiatry, Vienna, Austria

Sheryl D. Beck, Abramson Research Center, Children's Hospital of Philadelphia and University of Pennsylvania School of Medicine, Philadelphia, PA, USA

Mary E. Blue, Hugo W. Moser Research Institute at Kennedy Krieger, Inc., Johns Hopkins University School of Medicine, Baltimore, MD, USA

Joël Bockaert, Institut de Génomique Fonctionnelle, Université de Montpellier, CNRS, and INSERM, Montpellier, France

Marco Bortolato, Department of Pharmacology and Pharmaceutical Sciences, School of Pharmacy, University of Southern California, Los Angeles, CA, USA

Analía Bortolozzi, Department of Neurochemistry and Neuropharmacology, Instituto de Investigaciones Biomédicas de Barcelona, CSIC (IDIBAPS); Centro de Investigación Biomédica en Red de Salud Mental (CIBERSAM), Barcelona, Spain

Turhan Canli, Department of Psychology; Graduate Program in Genetics, Stony Brook University, Stony Brook, USA

Robert J. Carey, VA Medical Center; SUNY Upstate Medical University, Syracuse, USA

Gregory V. Carr, University of Pennsylvania, School of Medicine, Philadelphia, PA, USA

Jean-Christophe Cassel, Laboratoire d'Imagerie et de Neurosciences Cognitives, Université de Strasbourg-CNRS, F-67000 Strasbourg, France

Pau Celada, Department of Neurochemistry and Neuropharmacology, Instituto de Investigaciones Biomédicas de Barcelona, CSIC (IDIBAPS); Centro de Investigación Biomédica en Red de Salud Mental (CIBERSAM), Barcelona, Spain

Kevin Chen, Department of Pharmacology and Pharmaceutical Sciences, School of Pharmacy, University of Southern California, Los Angeles, CA, USA

Sylvie Claeysen, Institut de Génomique Fonctionnelle, Université de Montpellier, CNRS, and INSERM, Montpellier, France

Peter G. Clifton, Department of Psychology, University of Sussex, Brighton, UK

Roser Cortés, Departamento Neuroquímica y Neurofarmacología, IIBB/CSIC-IDIBAPS, CIBERNED, Barcelona, Spain

Ian W. Craig, MRC Social, Genetic and Developmental Psychiatry (SGDP) Centre, Institute of Psychiatry, King's College London, UK

Molly J. Crockett, Department of Experimental Psychology and Behavioural and Clinical Neuroscience Institute, University of Cambridge, Cambridge, UK

John F. Cryan, School of Pharmacy, Department of Pharmacology and Therapeutics and Alimentary Pharmabiotic Centre, University College Cork, Ireland

Ursula M. D'Souza, MRC Social, Genetic and Developmental Psychiatry (SGDP) Centre, Institute of Psychiatry, King's College London, UK

Cristina Marta Del Ben, Department of Neuroscience and Behavioral Sciences, Hospital das Clínicas da Faculdade de Medicina de Ribeirão Preto, Universidade de São Paulo, Ribeirão, Preto, Brazil

Damiaan Denys, Department of Psychiatry, Academic Medical Center (AMC), University of Amsterdam; Institute for Neuroscience, an institute of the Royal Netherlands Academy of Arts and Sciences, Amsterdam, The Netherlands

Laurent Descarries, Departments of Pathology and Cell Biology and of Physiology, and Groupe de recherche sur le système nerveux central (GRSNC), Faculty of Medicine, Université de Montréal, Montreal, QC, Canada
Llorenç Díaz-Mataix, Department of Neurochemistry and Neuropharmacology, Instituto de Investigaciones Biomédicas de Barcelona, CSIC (IDIBAPS), Barcelona, Spain
Carolin I. Dohle, Department of Neurology, Yale University School of Medicine, New Haven, CT, USA
Elif Aysimi Duman, Department of Psychology, Stony Brook University, Stony Brook, USA
Aline Dumuis, Institut de Génomique Fonctionnelle, Université de Montpellier, CNRS, and INSERM, Montpellier, France
Vikas Duvvuri, University of California San Diego, School of Medicine, Department Psychiatry, La Jolla, CA, USA
Matthijs G.P. Feenstra, Netherlands Institute for Neuroscience, an institute of the Royal Netherlands Academy of Arts and Sciences, Amsterdam, The Netherlands
Casimir A. Fornal, Program in Neuroscience, Princeton University, Princeton, NJ, USA
Eva Friedel, Department of Psychiatry and Psychotherapy, Charité – University Medicine Berlin, CCM, Germany
Mark A. Geyer, Department of Psychiatry, University of California at San Diego, CA, USA
Frederico Guilherme Graeff, Department of Neuroscience and Behavioral Sciences, Faculdade de Medicina de Ribeirão Preto, Universidade de São Paulo, Brazil
Francisco Silveira Guimarães, Department of Pharmacology, Faculdade de Medicina de Ribeirão Preto, Universidade de São Paulo, Brazil
Jutatip Guptarak, Department of Biology, Texas Woman's University, Denton, TX, USA
Adam L. Halberstadt, Department of Psychiatry, University of California at San Diego, La Jolla, CA, USA
Matthew W. Hale, University of Colorado at Boulder, Department of Integrative Physiology, Boulder, CO, USA
Hiroyuki Hasegawa, Department of Biosciences, Teikyo University of Science and Technology, Uenohara, Japan
Andreas Heinz, Department of Psychiatry and Psychotherapy, Charité – University Medical Center Berlin, CCM, Germany
Julie G. Hensler, Department of Pharmacology, University of Texas Health Science Center, San Antonio, USA
Christine F. Hohmann, Department of Biology, Morgan State University, Baltimore, MD, USA
Jean-Pierre Hornung, Département de Biology Cellulaire et de Morphologie, Faculty of Biology and Medicine, University of Lausanne, Rue du Bugnon 9, 1005 Lausanne, Switzerland
Daniel Hoyer, Psychiatry/Neuroscience Research, Novartis Institutes for Biomedical Research, Basel, Switzerland
Joseph P. Huston, Institute of Physiological Psychology and Center for Biological and Medical Research, University of Düsseldorf, Germany
Barry L. Jacobs, Program in Neuroscience, Princeton University, Princeton, NJ, USA
Walter H. Kaye, University of California San Diego, School of Medicine, Department Psychiatry, La Jolla, CA, USA
Andre Klompmakers, Department of Psychiatry, Academic Medical Center (AMC), University of Amsterdam, The Netherlands
Michelle D. Lee, Department of Psychology, Swansea University, UK
Astrid C.E. Linthorst, Henry Wellcome Laboratories for Integrative Neuroscience and Endocrinology (HW-LINE), University of Bristol, Bristol, UK
Christopher A. Lowry, University of Colorado at Boulder, Department of Integrative Physiology, Boulder, USA; University of Bristol, Henry Wellcome Laboratories for Integrative Neuroscience and Endocrinology, Bristol, UK
Irwin Lucki, Departments of Psychiatry and Pharmacology, University of Pennsylvania, School of Medicine, Philadelphia, PA, USA
Gerard J. Marek, Abbott Laboratories, Neuroscience Development, GPRD, Abbott Park, IL, USA
Philippe Marin, Institut de Génomique Fonctionnelle, Université de Montpellier, CNRS, and INSERM, Montpellier, France
Charles A. Marsden, School of Biomedical Sciences, University of Nottingham Medical School, Nottingham, UK
William J. McBride, Institute of Psychiatric Research, Department of Psychiatry, Indiana University School of Medicine, Indianapolis, USA
Guadalupe Mengod, Departamento Neuroquímica y Neurofarmacología, IIBB/CSIC-IDIBAPS, CIBERNED, Barcelona, Spain
Klaus A. Miczek, Departments of Psychology, Psychiatry, Pharmacology and Neuroscience, Tufts University, Medford and Boston, MA, USA
Christian P. Müller, MRC SGDP-Center and Interdisciplinary Research Group Addiction, Institute of Psychiatry, King's College London, UK

Kazuhiro Nakamura, Department of Pathology, Juntendo University School of Medicine, Tokyo, Japan
David E. Nichols, Department of Medicinal Chemistry and Molecular Pharmacology, School of Pharmacy and Phamaceutical Sciences – RHPH, Purdue University, West Lafayette, IN, USA
Olivia F. O'Leary, School of Pharmacy, and Department of Pharmacology and Therapeutics, University College Cork, Ireland
Robert D. Oades, Clinic for Child and Adolescent Psychiatry and Psychotherapy, University of Duisburg-Essen, Essen, Germany
Martin Parent, Department of Pathology and Cell Biology, and Groupe de recherche sur le système nerveux central (GRSNC), Faculty of Medicine, Université de Montréal, Montreal, QC, Canada
Martin E. Pum, Institute of Physiological Psychology and Center for Biological and Medical Research, University of Düsseldorf, Germany
Isabel M. Quadros, Center for Neuroscience Research, Department of Neuroscience, Tufts University, Medford and Boston, MA, USA
Boris B. Quednow, Department of Experimental and Clinical Pharmacopsychology, University Hospital of Psychiatry, University of Zurich, Switzerland
Johannes M.H.M. Reul, Henry Wellcome Laboratories for Integrative Neuroscience and Endocrinology (HW-LINE), University of Bristol, Bristol, UK
Gavin P. Reynolds, Department of Psychiatry, Queen's University Belfast, Belfast, UK
Mustapha Riad, Department of Pathology and Cell Biology, Faculty of Medicine, Université de Montréal, Montreal, QC, Canada
George B. Richerson, Departments of Neurology and Cellular and Molecular Physiology, Yale University School of Medicine; Veterans Affairs Medical Center, New Haven, CT, USA
Trevor W. Robbins, Department of Experimental Psychology and Behavioural and Clinical Neuroscience Institute, University of Cambridge, Cambridge, UK
Noemí Santana, Department of Neurochemistry and Neuropharmacology, Instituto de Investigaciones Biomédicas de Barcelona, CSIC (IDIBAPS); Centro de Investigación Biomédica en Red de Salud Mental (CIBERSAM), Barcelona, Spain
Gunter Schumann, MRC SGDP-Center and Interdisciplinary Research Group Addiction, Institute of Psychiatry, King's College London, UK
Trevor Sharp, University Department of Pharmacology, Mansfield Road, Oxford, UK
Jean C. Shih, Department of Pharmacology and Pharmaceutical Sciences, School of Pharmacy; Department of Cell and Neurobiology, Keck School of Medicine; University of Southern California, Los Angeles, CA, USA
Claudia Sommer, Department of Neurology, University of Würzburg, Germany
Aki Takahashi, Department of Psychology, Tufts University, Medford and Boston, MA, USA
Lynda Uphouse, Department of Biology, Texas Woman's University, Denton, TX, USA
Geoffrey van der Plasse, Department of Psychiatry, Rudolf Magnus Institute of Neuroscience, Utrecht University Medical Centre, Utrecht, The Netherlands
Addy van Dijk, Department of Psychiatry, Academic Medical Center (AMC), University of Amsterdam, Amsterdam, The Netherlands
Robert P. Vertes, Center for Complex Systems and Brain Sciences, Florida Atlantic University, Boca Raton, USA
M. Teresa Vilaró, Departamento Neuroquímica y Neurofarmacología, IIBB/CSIC-IDIBAPS, CIBERNED, Barcelona, Spain
Patricia M. Whitaker-Azmitia, Department of Psychology, Stony Brook University, Stony Brook, USA
Olga O. Yevtushenko, Department of Neuropharmacology, Institute of Pharmacology and Toxicology AMS Ukraine, Kyiv, Ukraine
Hélio Zangrossi Jr, Department of Pharmacology, Faculdade de Medicina de Ribeirão Preto, Universidade de São Paulo, Brazil

SECTION 1

Functional Anatomy of the Serotonergic System

CHAPTER 1.1

Evolution of Serotonin: Sunlight to Suicide

Efrain C. Azmitia*

Department of Biology and Psychiatry, Center for Neuroscience, New York University, Washington Square East, New York

Abstract: Serotonin is involved in many of the behaviors and biological systems that are central to human life, extending from early developmental events related to neurogenesis and maturation, to apoptosis and neurodegeneration that underlie dementia and death. How can a single chemical be so powerful in determining the quality and quantity of human life? In this chapter, the evolution of serotonin and its biosynthetic pathways from tryptophan are examined. The essential components of the serotonin biosynthetic pathway are highly conserved. Tryptophan-based chemicals, including serotonin, melatonin and auxin, have important action in the differentiation, mitosis and survival of single cell organisms. As the complexity of life evolved into multicellular organisms, especially plants, serotonin levels rose dramatically. The importance of tryptophan, serotonin and auxin is evident in photosynthesis and plant growth. When the animal kingdom began, the ability to synthesize tryptophan was lost, and serotonin levels dropped accordingly. Animals had to develop many special mechanisms to secure tryptophan from their diets, and to carefully conserve its integrity during circulation throughout the body. There was the emergence of multiple receptors and reuptake proteins that permit serotonin actions without utilization of serotonin itself. Serotonin receptors appear as early as the blastula and gastrula stages of embryonic development, and continually monitor and regulate ontogenic changes as it serves phylogenic evolution.

The most amazing concept to emerge from an analysis of serotonin evolution is its relation to light. Beginning with the light-absorption properties of the indole ring of tryptophan, a direct path can be drawn to the effects of sunlight on photosynthesis and serotonin levels in plants. Progressing further in phylogeny, the effects of sunlight are seen on serotonin levels and on mood, sleep and suicide ideation in humans. The focus on the evolution of serotonin leads from an awareness of the beginning of life to the current human struggle to enjoy our dominant position on earth.

Keywords: seasonal affective disorder (SAD), tryptophan, auxin, hallucinogens, 5-HT$_{1A}$ receptor, photosynthesis, homeostasis indole, chloroplast, fungus, metazoa.

Introduction

It has always been an enigma how serotonin, a monoamine neurotransmitter, can have such diverse and important functions in the human brain. The question is further complicated when it is recognized that serotonin exists in all the organs of the body (e.g., the skin, gut, lung, kidney, liver, and testis) and in nearly every living organism on Earth (e.g., fungi, plants, and animals) (Azmitia, 1999). Serotonin is phylogentically ancient, and evolved prior to the appearance of neurons. Whatever function serotonin has in the brain, it should be consistent with its evolutionary history. However, little attention has been given to the biological emergence of serotonin. Most neuroscience studies focus on serotonin in mediating particular behaviors (e.g., feeding, sex, sleep, and learning) or its involvement in specific brain disorders (e.g., depression, Alzheimer's disease and autism). A broader view has been rarely raised. Over the past 50 years, several scientists have proposed general organistic roles for brain serotonin (Brodie and Shore, 1957; Woolley, 1961; Scheibel *et al.*, 1975). A more comprehensive theory that includes expanded functions proposes that serotonin acts as a homeostatic regulator which integrates mind and body with the outside world (Azmitia, 2001, 2007). In reading this chapter, it is hoped that the reader will appreciate why serotonin came to be so important in the mental health of humans.

The evolution proposed for serotonin begins with a discussion of its precursor tryptophan and its metabolites in unicellular organisms nearly 3 billion years ago. To make serotonin from tryptophan, oxygen is needed, and in the earliest geological times the Earth's atmosphere had little oxygen. Thus, serotonin is made specifically

*E-mail: eca1@nyu.edu

Christian Müller & Barry Jacobs (Eds.)
Handbook of Behavioral Neurobiology of Serotonin
ISBN 978-0-12-374634-4

DOI: 10.1016/B978-0-12-374634-4.00001-0

Table 1 Light spectrum of wavelengths that reach the Earth

Type	Wavelength (nm)
Infrared	>1000
Red	700–630
Orange	630–590
Yellow	590–560
Green	560–490
Blue	490–450
Violet	450–400
UVA	400–315
UVB	315–380
UVC	280–100

in unicellular systems capable of photosynthesis and the cellular production of oxygen. The conserved serotonin biosynthetic pathway began in the unicellular systems of cyanobacteria, green algae and fungi, and continually evolved to its current position in the human brain.

Photosynthesis is dependent on the energy derived from sunlight. The amount of light reaching Earth has seasonal variation and, because of the Earth's tilting, is most noticeable at the polar extremes. Light can be roughly divided into ultraviolet (UV) and visible light (Table 1). In humans, UVB radiation produces sunburn and some forms of skin cancer, while UVA (black light) will produce skin discoloration and is necessary for vitamin D production. Glass and plastic can block UVB rays. The two types of light mainly absorbed during photosynthesis in plants are blue light and red light. The light most effective in alleviating the depression associated with seasonal affective disorder is blue light.

Tryptophan

Light capturing

Indole

Indole is an organic compound consisting of a benzene ring and a pyrrole ring (Figure 1). The origin of the first photosynthetic pathway is proposed to be the photo-oxidation of uroporphyrinogen (a tetrapyrrole) by UVC, and this oxidation is accompanied by the release of molecular hydrogen. The oxidation of uroporphyrinogen to uroporphyrin, the first biogenetic porphyrin, could have occurred anaerobically and abiologically on the primordial Earth 3 billion years ago (Mercer-Smith *et al.*, 1985). The indole structure can be found in many organic compounds, like the amino acid tryptophan, and in tryptophan-containing enzymes and receptors, alkaloids and pigments. The indole ring is electron rich and will lose an

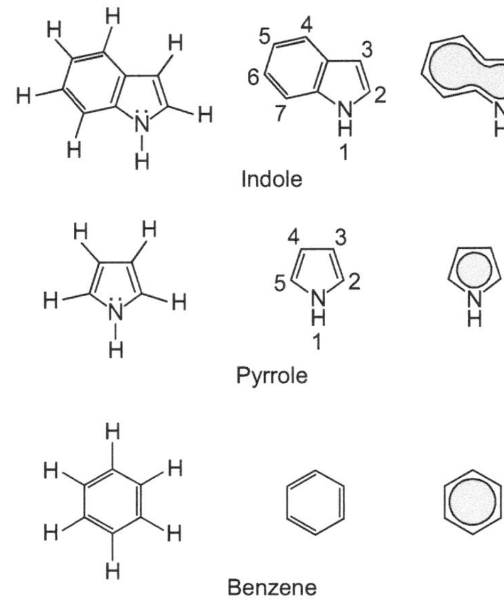

Figure 1 The indole structure consists of aromatic pyrrole and benzene rings. In the indole ring, the electrons are merged and a large pool of electrons becomes available for redox reactions.

electron (oxidized) to an electrophilic compound, like a heavy metal, to decrease its oxidation number. The most reactive position on the indole structure for electrophilic aromatic substitution is C-3, which is 10^{13} times more reactive than benzene. Presently, cationic porphyrins are known to bind to the tryptophan moiety of proteins (Zhou *et al.*, 2008). A similar situation to tryptophan occurs with chlorophyll, the light-capturing molecule of plants.

Photosynthesis

Most proteins are endowed with an intrinsic UV fluorescence because they contain aromatic amino acids, particularly phenylalanine, tyrosine and tryptophan. Comparing the three aromatic amino acids, tryptophan has the highest fluorescence quantum yield, overshadowing markedly the emissions of the other two. Generally, a distinction is made between tryptophan and non-tryptophan fluorescence. The tryptophan amino acid contains an indole backbone and absorbs light. Free tryptophan has characteristic fluorescence absorption at UVB (Borkman and Lerman, 1978; Lin and Sakmar, 1996), and the fluorescence emission is in the range of UVA-blue light (Du *et al.*, 1998). This amino acid in proteins is needed for solar energy to drive photosynthesis in cyanobacteria, algae and plants (Vavilin *et al.*, 1999). Solar energy conversion in photosynthesis involves electron transfer between an excited donor molecule and an acceptor molecule that are contained in the reaction center, an intrinsic membrane protein pigment complex (Wang *et al.*, 2007). In the photosynthetic reaction centers of *Rhodobacter sphaeroides*, there are 39 tryptophan residues. Initiation of the electron transfer reaction by excitation results in a transient change in the absorbance at UVB, near the peak of the tryptophan absorbance band. According to Wang *et al.* (2007):

> Given the similarity of the core features of the photosynthetic complexes from bacteria and plants, it is very likely that this same framework holds true for the initial electron transfer reaction of photosynthesis in general and possibly for other protein-mediated electron transfer reactions on similar time scales.

Absorption of blue light waves in chloroplasts leads to excitation of the indole structure of tryptophan so that it loses one of the electrons from its indole ring structure – it becomes oxidized. The electron that is lost from the indole ring passes through the intermediary of heavy metals to make its way down the electron chain. This process leads to the production of reducing cofactors such as nicotinamide adenine dinucleotide (NADH), or reduction of H_2O to O_2. In bacterial ferric cytochrome P-450, the initial event of photoreduction is the photo-ionization of tryptophan in the active site of P-450 (Pierre *et al.*, 1982). A laser flash at 265 nm, near the UVB range, triggers the nanosecond event of a protein structural change coincident with an ejection of electrons, and is followed by the photoreduction of the heme moiety (Bazin *et al.*, 1982). The transfer of electrons between tryptophan and the heavy metals occurs at extremely rapid rates of several nanoseconds (Shih *et al.*, 2008). This process takes place during photosynthesis, and is the procedure that converts solar energy into biological energy. CO_2 and water are converted to O_2 and glucose. This is the most important biochemical process on Earth. The Earth's atmosphere contains 20 percent oxygen, and drives the biological evolution of aerobic organisms. Photosynthesis occurs in certain bacteria (e.g., cyanobacteria), algae (e.g., green algae) and all plants (Bryant and Frigaard, 2006).

Chloroplasts

Plants evolved a specialized intracellular organelle, the chloroplast, not only to capture light, but also as the source of tryptophan synthesis. This organelle may have evolved from captured cyanobacteria, and contains several hundred copies of chlorophyll. Chlorophyll became the universal transformer of solar energy in bioenergetics, and is sensitive to both blue and red light. In addition to its role in photosynthesis, blue light absorbance has been implicated as an early step in certain blue light-mediated morphogenic events (Rubinstein and Stern, 1991). The spectra for blue light-stimulated stomatal opening and phototropism are specialized for sensory transduction in the chloroplast (Quiñones *et al.*, 1998). Thus blue light in plants not only underlies photosynthesis, it also produces morphological plasticity in the plant leaf to promote sensory transduction of light energy into bioenergy. Structural and functional advantages of the chlorophyll molecule may have determined its selection in evolution (Mauzerall, 1973). Chloroplast have a high need for tryptophan because of its light-absorption functions.

Amino acid

Tryptophan is the largest and most hydrophobic amino acid, and provides important folding signals to large proteins (Aoyagi *et al.*, 2001). Tryptophan is the least used amino acid in the composition of protein molecules, generally comprising around 1–2 percent of the protein weight. In the basic genetic code, only one code is used for tryptophan, 'UGG', and this is flanked by stop codons on UAA, UAG, and UGA. Yet many important molecules are derived from tryptophan, including the nucleic acids adenosine and thymidine in DNA. Furthermore, a novel nucleolar protein, WDR55, carries a tryptophan–aspartate repeat motif and is involved in the production of ribosomal RNA (rRNA) (Iwanami *et al.*, 2008). These findings suggest that WDR55 is a nuclear modulator of rRNA synthesis, cell cycle progression, and embryonic organogenesis. Thus, despite its limited abundance, tryptophan and its associated molecules are involved in all cellular aspects of the organism's life, and serve key regulatory roles in mitosis (Humphrey and Enoch, 1998), cell movement (Efimenko *et al.*, 2006) and maturation (Cooke *et al.*, 2002).

The synthesis of tryptophan takes several key enzymes: anthranilate synthase (EC 4.1.3.27), anthranilate phosphoribosyl-transferase (EC 2.4.2.18), anthranilate phosphoribosyl-isomerase (EC 5.3.1.24), indole-3-glycerol-phosphate synthase (EC 4.1.1.4) and tryptophan synthase (EC 4.2.1.20) (Figure 2). These enzymes are found in primitive unicellular organisms and in all plant systems. In many organisms there are isoforms of these that provide redundancy in tryptophan synthesis. In plants, tryptophan biosynthetic enzymes are synthesized as higher molecular weight precursors and then imported into chloroplasts and processed into their mature active forms (Zhao and Last, 1995). Chloroplast organelles contain the genes and enzymes, similar to those seen in cyanobacteria, required for tryptophan synthesis. Despite the abundant amount of tryptophan made by lower organisms, animals are tryptophan autotrophs, which lack the genes necessary for tryptophan synthesis. Tryptophan for serotonin synthesis is obtained from food. Often the main source of food is proteins from other animals; despite the fact that the amount of tryptophan in protein from animal sources is the least compared to the other amino acids.

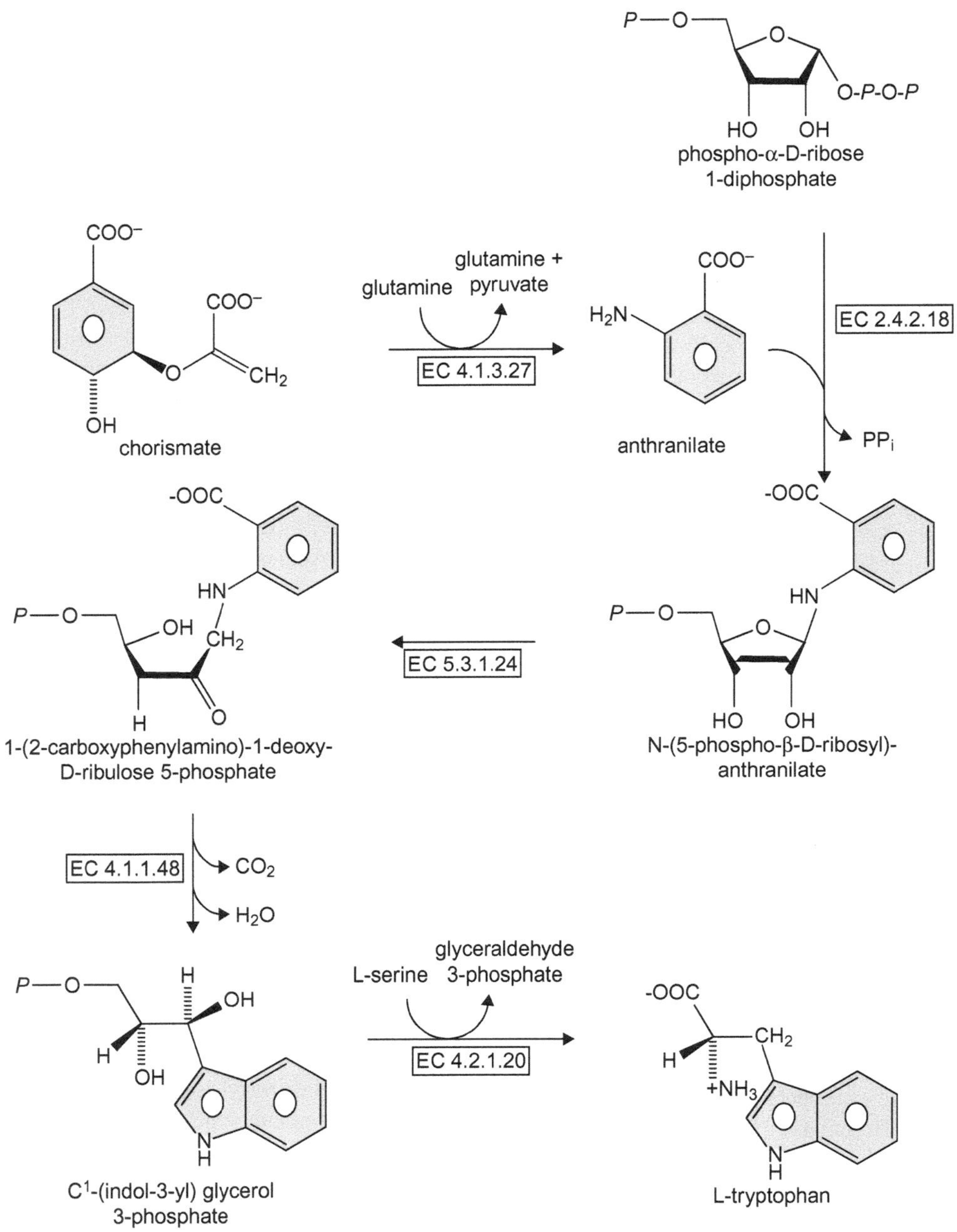

Figure 2 The five enzymatic steps in the synthesis of tryptophan from chorismate; the names of each are provided in the text. In plants, these enzymes and the chloroplast genes for them are located within the chloroplast. In animals, the genes are inactive and all tryptophan is obtained from external sources.

Tryptophan levels are much higher in fruits, vegetables and nuts. Once digested, tryptophan exists in a free and bound form in the plasma. Only the free form is available for transport into the brain, and its transport must compete with the uptake of all the other, more abundant, aromatic amino acids (e.g., tyrosine, phenylalanine, leucine, isoleucine, valine and methionine) (Curzon *et al.*, 1973). The synthesis of serotonin in humans appears to be restricted to a few cell types (e.g., mast cells and neurons). Selected cells in all organs have specific uptake proteins for capturing serotonin from the plasma where it is stored. The importance of this will be discussed later.

The capture of light by tryptophan is used at the active site by nearly all proteins (e.g., chlorophyll, rhodopsin and skin pigment cells) which capture light (Angiolillo and Vanderkooi, 1996). A long-range transfer of electrons to cytochromes from tryptophan can be demonstrated with a number of proteins, including paralbumin, aldolase and liver alcohol dehydrogenase (Dadak *et al.*, 1992). Thus, tryptophan is very sensitive to light in a variety of protein environments, and readily releases its electrons from the excited triplet state upon light exposure. This ability to transfer electrons is not only utilized in the initial events of photosynthesis. In the light receptor rhodopsin, there is rotation of a single tryptophan molecule when meta I is converted to meta II after illumination (Chabre and Breton, 1979). Photoactivation of rhodopsin involves a change in the relative disposition of transmembrane helices 3 and 6, which contain Trp126 and Trp265 respectively, within the receptor (Lin and Sakmar, 1996). These studies show that tryptophan serves key photo-signaling properties in light receptor rhodopsin, probably the most primitive of all G-protein receptors.

Tryptophan's ability to absorb UV light can be put to advantage. The bacterium *Bacillus subtilis* is protected from the lethal actions of ultraviolet radiation by tryptophan (Hunt, 1964). UV and X-ray light produce luminescence of tryptophan (Steen, 1967). This absorption provides protection by filtering harmful UV light before it can reach the underlying DNA and photoreceptors. The UV absorbance in chicken eye aqueous is partially accounted for by the presence of tryptophan (Ringvold *et al.*, 2000).

Tryptophan has many functions in the cells, not only as an essential amino acid and chromophore but also because it makes niacin and all its precursors. NAD+ and NADP+ cofactors are involved in nearly all aspects of cell metabolism (Cox *et al.*, 2000). *De novo* synthesis of nicotinamide adenine dinucleotides from tryptophan is a more important source of these coenzymes than is the utilization of dietary nicotinamide or nicotinic acid (Bender and Olufunwa, 1988). In the nucleus, niacin is important for DNA repair, and tryptophan capture of light appears to be responsible for the DNA photo-damage associated with mutation and cell death in the absence of repair (Friedberg *et al.*, 1995).

Tryptophan derivatives

Photosynthesis is a major risk to cells as well as nutritional benefits. Photosynthesis can be perturbed by upsetting the balance between the rates of light collection and light use, resulting in the production of reactive oxygen species (ROS) (Asada, 1996). Left unchecked, ROS are damaging to protein function and membrane integrity, and pose a serious threat to photosynthetic organisms. UV light reaction with tryptophan generates many photoproducts having deleterious actions. UVB radiation of N-formylkynurenine generates free radicals such as singlet oxygen and superoxide (Grossweiner, 1984). Tryptophan also generates fluorescent photoproducts which inhibit the growth and differentiation of cultured fertilized sea urchin eggs and mouse fibroblasts (Zigman and Hare, 1976). Thus a wide range of cells, from bacteria to mammals, are harmed by photo-oxidized tryptophan. The damage done to the human skin by sunlight can be largely attributed to the actions of tryptophan metabolism via the pyrrolase, kynureninine and niacin pathways (Binazzi and Calandra, 1975). Ultraviolet light sensitizes tryptophan to enhance the photoxidation of tyrosine to dopachrome, a precursor of melanin (Badu and Joshi, 1992).

The critical problem in oxygen generating cells was what to do with all the reactive oxidizing chemicals produced during the generation of oxygen. Many of the derivatives of tryptophan evolved to function as antioxidant molecules in simple cells long before they assumed more complex functions in animals and humans. Substituted indoles are derivatives of the tryptophan-based tryptamine alkaloids, such as serotonin, melatonin and auxin; the hallucinogens psilocybin, DMT, 5-MeO-DMT; and the ergots, such as ergotamine and LSD (Schultes and Hofmann, 1973). N-formylkynurenine, 5-methoxytryptamine, auxin and melatonin in prokaryotes function as photosensors, antioxidants and pattern generators for flagella movement, and in defense mechanisms. The defense mechanism may involve attack of plants by fungi, and a role for the phytohormone indole-3-acetic acid or a structurally related compound. Many of these products also have morphogenic actions and play a role in both apoptosis and mitosis. Many of the alkaloids produced from tryptophan have powerful actions on the human brain by acting mainly through the serotonin receptors. The structures of some of these compounds are shown in Figure 3.

Melatonin is a powerful antioxidant (Hardeland, 2005). Melatonin is found in all plants and animals, including algae. Melatonin has been shown to inhibit microtubule function in flagella and in mitosis. Melatonin synthesis has a diurnal rhythm in many protists and plants. It functions as an antioxidant in plants. Seasonal animals use melatonin as a biological signal for the organization of day-length dependent (photoperiodic) annual functions such as reproduction, behavior, coat growth and camouflage coloring (Chaturvedi, 1984). Melatonin is also related to the mechanism by which some amphibians and reptiles change the color of their skin, and, indeed, it was in this connection that the substance was first discovered (Sugden *et al.*, 2004).

Auxin (hydroxyindole acetic acid) is a protective, antioxidant compound found in unicellular organisms (Cooke *et al.*, 2002). Auxin is an important phototrophic hormone in plants, is present in bacteria, and functions to stimulate growth and survival. It has many different effects, such as inducing cell elongation and cell division, with

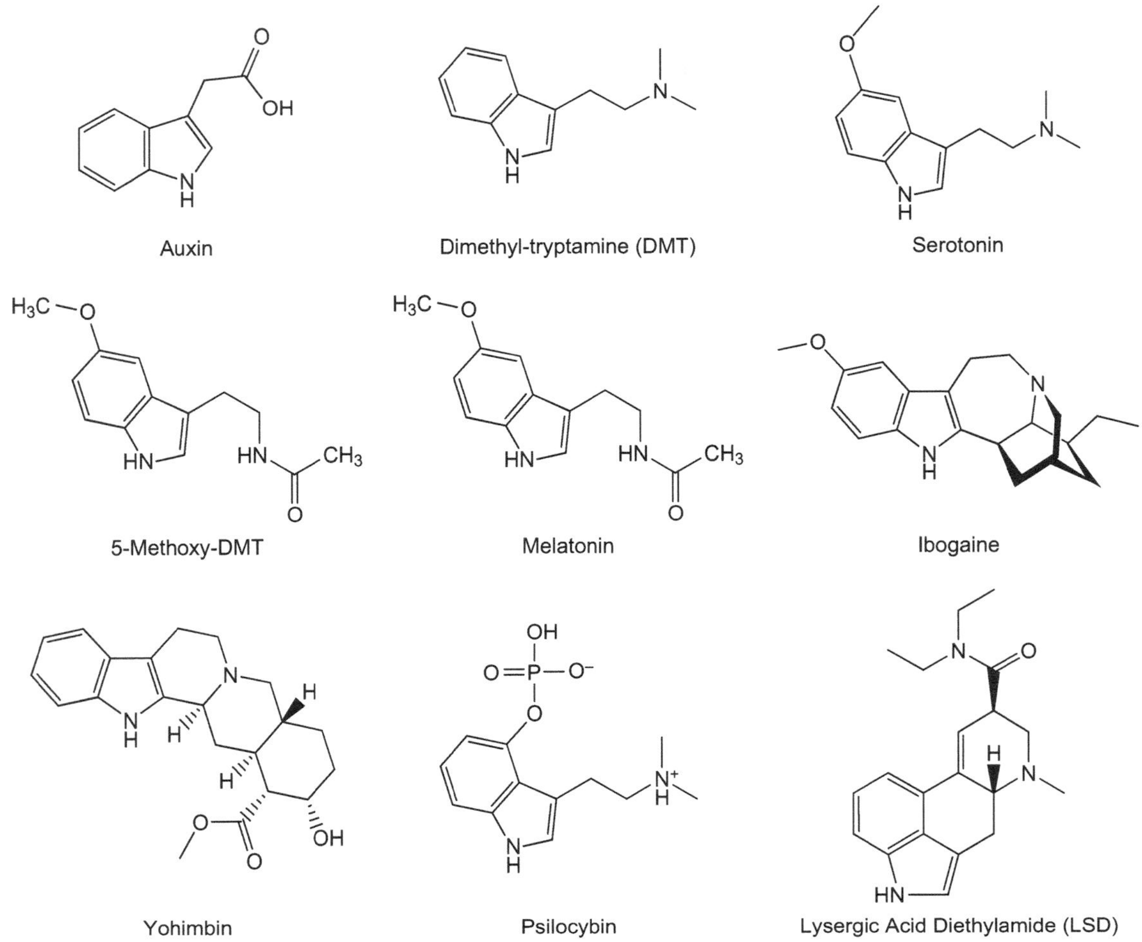

Figure 3 Many of the alkaloid compounds derived from the tryptophan. Some of these compounds, such as auxin, DMT and melatonin, appear to have been made in protista, occurring before plants. Some of these compounds are known to be potent hallucinogens when ingested by humans (see text). It is reported that serotonin is found in the Fungi Kingdom in yeast cells.

all the subsequent results for plant growth and development (Stern *et al.*, 2002; Mauseth, 2008). Auxin has been reported in several bacteria, and the key enzyme in the auxin synthetic pathway was found in the genome sequence database of *Paenibacillus polymyxa* (Phi *et al.*, 2008). On the cellular level, auxin is essential for cell growth, affecting both cell division and cellular expansion (Perrot-Rechenmann *et al.*, 2002). Depending on the specific tissue, auxin may promote axial elongation (as in shoots), lateral expansion (as in root swelling) or isodiametric expansion (as in fruit growth). Growth and division of plant cells together result in growth of tissue, and specific tissue growth contributes to the development of plant organs. Growth of cells contributes to the plant's size, but uneven localized growth produces bending, turning and orientation of organs – for example, stems turning toward light sources (phototropism), roots growing in response to gravity (gravitropism), and other tropisms. Auxin controls the orientation of cortical microtubules in maize, perhaps caused by a reduced microtubule turnover (Wiesler *et al.*, 2002). Actin transport is essential for its uneven distribution in plants. Auxin transport inhibitors impair vesicle motility and actin cytoskeleton dynamics in yeast, plants and animals (Dhonukshe *et al.*, 2008). In plants, auxin is synthesized in specialized cells and is transported throughout the plant to influence leaf and root growth. Its specific function in leaves is to move the entire photo-generating organ towards the source of solar energy.

N-formylkynurenine functions as a photosensitizer, and is made in mitochondria by the oxidation of tryptophan (Møller and Kristensen, 2006). 5-Methoxytryptamine is also found in nearly all living organism, and is a key intermediate in the pineal synthesis of melatonin in vertebrates. The production of all other 5-methoxyindoles in the pineal gland as well as in the retina is always larger than that of melatonin. In the pineal gland, 5-methoxytryptophan, for example, is synthesized in a quantity which is 60 to 170

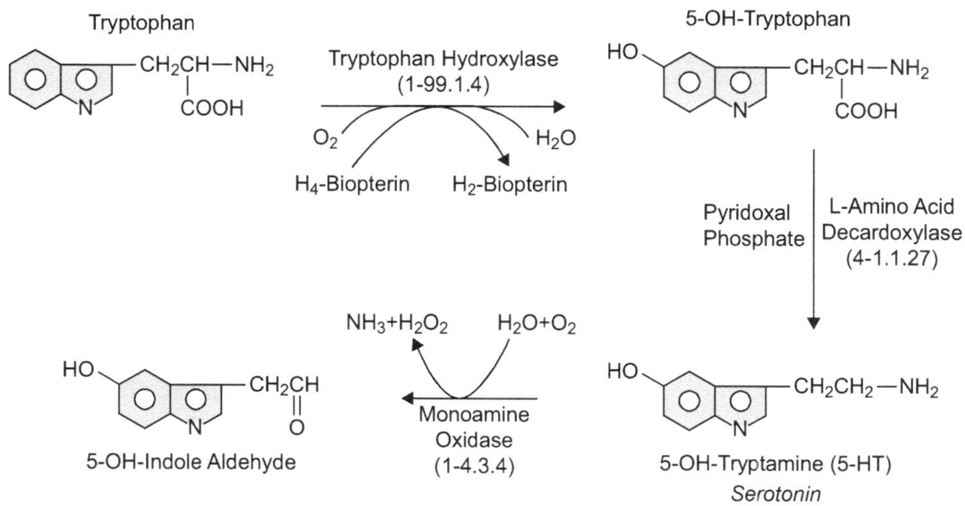

Figure 4 The synthesis of serotonin from tryptophan. The actions of tryptophan hydroxylase and the general L-amino acid decarboxylase are shown along with the necessary cofactors. The metabolism of serotonin to 5-OH-indole aldehyde is similar to auxin (indole aldehyde), a generated hydrogen peroxide (H_2O_2). Serotonin is also a precursor for melatonin synthesis in vertebrates.

times greater than that of melatonin, while in the retina the synthesized amount of 5-methoxytryptophan is even 60 to 1000 times greater than that of melatonin (Pévet *et al.*, 1981). This would provide a pathway for the synthesis of melatonin independent of one using 5-hydroxytryptamine, or serotonin.

Below is a list of some of the natural psychoactive drugs synthesized from tryptophan. All these compounds can function as antioxidants in single-cell organisms, and many have been used by man as entactogens and etheogens for thousands of years.

1. Psilocybin mushrooms (also called psilocybian mushrooms): these are fungi. Kingdom Fungi includes yeasts, rusts, smuts, molds, mushrooms and mildews. Though formerly classified as plants, fungi lack chlorophyll and the organized plant structures of stems, roots and leaves.
2. Dimethyltryptamine (DMT): this occurs in many species of plants, and is commonly used as an hallucinogen via drinking or smoking.
3. Bufotenine: this molecule is found in mushrooms, plants, and the skin and venom of toads (*Bufo* genus).
4. 5-MeO-DMT: this molecule is widely distributed in plants and toads.
5. Ergot, ergotamine: the parent compound of the major hallucinogen LSD has long been known to be produced by a fungus, genus *Claviceps*. The hallucinogen was synthesized by the addition of diethylamide to ergotamine by the chemist Albert Hofmann at the Sandoz Laboratory. LSD is believed to be one of the most potent mind-altering compounds discovered to date.
6. Ibogaine: this is found in a number of plants, principally in a member of the dogbane family known as iboga (*Tabernanthe iboga*).
7. Yohimbine: this is the principal alkaloid of the bark of the West-African evergreen *Pausinystalia yohimbe* Pierre. In Africa, yohimbine has traditionally been used as an aphrodisiac.

Serotonin

Biosynthesis

Serotonin is synthesized from tryptophan by two enzymes: tryptophan hydroxylase, which requires molecular oxygen; and tetrahydro-biopterin and aromatic amino-acid decarboxylase, which requires pyridoxal phosphate (Figure 4). It is principally metabolized by monoamine oxidase (A & B) to generate H_2O_2 and 5-hydroxyindole acetic acid. MAO A & B have been detected in Fungi (Sablin *et al.*, 1998). In the pineal gland of vertebrates, serotonin is methylated to produce melatonin. Precursors of these four enzymes (hydroxylase, decarboxylase, oxidase and methylase) are found in four major kingdoms of life (Bacteria, Fungi, Planta, and Animal).

Primitive hydroxylase enzymes

One of the solutions for removing excess O_2, and the potent threat of excess oxidation in cells, is to use the biological machinery that evolved to deal with CO_2. Anaerobic cells had developed a variety of enzymes for converting CO_2 into biological energy in the form of glucose, with O_2 as a byproduct. The most common process was to produce the sugar, glyceraldehyde-3 phosphate, a three-carbon sugar produced by three molecules of CO_2. The first enzymatic step in this reaction involves the attachment of a molecule of CO_2 to the five-carbon sugar, ribulose bisphosphate (RuBP). This step is emphasized because the enzyme that catalyzes this initial reaction, and possibly the most abundant enzyme on Earth, is RuBP carboxylase, also known as rubisco. In the very early stages of life on Earth, the carboxylase primarily attached to CO_2; however, as O_2 levels increased it was shown that this compound could react more favorably with oxygen (Smith, 1976). The oxygen-based reaction is energy inefficient, but serves to remove excess oxygen during periods of high light flux. This may have been the first enzyme to attach oxygen to a substrate (such as tryptophan to produce 5-hydroxytryptophan (5-HTP)). This enzyme is called a hydroxylase because only a single oxygen is used and the other forms water. RuBP carboxylase has the same phosphate-binding site sequence found in tryptophan biosynthetic enzymes (Wilmanns *et al.*, 1991).

Phenylalanine biosynthesis evolved prior to the addition of branches leading to tyrosine and tryptophan. An evolutionary scenario has been developed that begins with non-enzymatic reactions which may have operated in primitive systems, followed by the evolution of an enzymatic system that pre-dated the divergence of major lineages of modern bacteria (Gram-positive bacteria, Gram-negative purple bacteria, and cyanobacteria) (Ahmad and Jensen, 1998). The bacteria (*Chromobacterium violaceum*) enzyme phenylalanine hydroxylase contains copper ion, and if the enzyme is oxidized it is inactive (electron acceptor, gives up a hydrogen molecule); a single hydrogen ion (H^+) is sufficient to reduce the enzyme to a catalytically active state (Pember *et al.*, 1986). H_4-biopterin reductively activates the enzyme (donates electron). Bacteria (*Pseudomonas fluorescens*) contain a primitive biopterin hydroxylase, cyanide oxygenase, which can convert KCN to CO_2 (Kunz *et al.*, 2001). Many of these enzymatic properties and the use of biopterin cofactor are shared by the mammalian tryptophan hydroxylase, which is the rate-limiting step for serotonin biosynthesis.

The substrates for the primitive hydroxylase enzyme were tryptophan, tyrosine and phenylalanine, all of which can capture light (Grenett *et al.*, 1987; Boularand *et al.*, 1998; Wiens *et al.*, 1998). The hydroxylase enzyme gave rise to a very large number of complex alkaloids in plants, all of which are potent antioxidants in their own right. As we now know, cellular oxidation is important for cell maturation and division, but excess oxidation results in cell death. The synthesis of pharmaceutically important indoles involves the hydroxylase as well as decarboxylase enzymes (Facchini *et al.*, 2000). 5-HTP, the immediate precursor of serotonin, is formed from tryptophan hydroxylase. This molecule is rapidly converted to serotonin by the ubiquitous carboxylase working in reverse to function as a decarboxylase. Thus, serotonin is produced from tryptophan by enzymes commonly used in anaerobic organisms before O_2 was formed inside cells. Besides the algae, fungi and molds, the most efficient generators of O_2 and serotonin are plants.

Decarboxylase

The second enzyme in serotonin biosynthesis is tryptophan decarboxylase. The bacteria, *Enterobacteria cloacae* strains that are normally associated with plant roots, produce auxin by using the enzyme indolepyruvate decarboxylase (Zimmer *et al.*, 1994). Aromatic L-amino acid decarboxylase occurs in plants (Facchini *et al.*, 2000). For example, tryptophan decarboxylase in plants leads to the biosynthesis of pharmaceutically active indole alkaloids. These compounds are abundant throughout the plant kingdom, and are known to produce strong psychic or hallucinogenic responses in humans (see Figure 3). In animals, this enzymes losses its specificity for tryptophan and becomes a general aromatic acid decarboxylase. It is likely that as animal cells had a marked reduction in tryptophan due to loss of tryptophan synthesizing enzymes, there was less evolutionary need for a specific tryptophan decarboxylase enzyme.

Cofactors

Tryptophan is necessary for the synthesis of a variety of cofactors, including biopterin, NADH and pyridoxal phosphate (Nelson and Cox, 2008). Biopterin is an essential cofactor in hydroxylase enzymes, and is necessary for the synthesis of serotonin. NADH and NADPH are involved in nearly every redox reaction in all living cells. They act as both oxidizing and reducing agents. Pyridoxal-5′-phosphate is a catalysis (redox) partner in a variety of reactions involving amino acids (Alexander *et al.*, 1994). Pyridoxal phosphate, the decarboxylase cofactor, in the form of an adduct absorbing at 330–340 nm, is suggested as a candidate for photoactivation of the 5-hydroxytryptophan decarboxylation for the role of the photoactive chromophore of decarboxylase. The cofactor pyridoxal-5-phosphate appears to have emerged very early in biological evolution; conceivably, organic cofactors and metal ions were the first biological catalysts (Mehta and Christen, 2005).

Presence of serotonin

In prokaryotes and algae living under anaerobic or aerobic conditions, the evidence for the presence of serotonin is weak, although all serotonin biosynthetic enzymes and cofactors are present. The enzymes aromatic amino acid (tryptophan) hydroxylase, tryptophan (aromatic amino-acid) decarboxylase, monoamine oxidase and aromatic hydroxylase are conserved in unicellular organisms. The compounds produced by these enzymes all have roles in light-induced reactions involved in regulating the oxidative reactions in cells. In nearly every unicellular organism, tryptophan and its metabolites (auxin, melatonin and 5-methoytryptamine) function as light sensors (needed for diurnal rhythms) and as antioxidants. Melatonin and 5-methoxytryptamine appear to be ubiquitous substances found in nearly every living cell (Hardeland, 1999). Auxin is also found in many unicellular organisms (Overbeek, 1940; Jacobs *et al.*, 1985). In fungi, *Candida guilliermondii*, serotonin (and other hydroxylated tryptamines) and ergot derivatives provide protection from UV toxicity and stimulate cell proliferation (Strakhovskaia *et al.*, 1983; Belenikina *et al.*, 1991). *Candida* cannot make oxygen, but depends on oxygen for energy production (obligate aerobes).

The Plant Kingdom

In plants, the synthesis of serotonin is much higher than in mammals (Garattini and Valzelli, 1965; Smith, 1971; Sparks and Slevin, 1985). The explanation for the high levels of serotonin is abundance of intracellular oxygen, tryptophan, and the enzymes and cofactors necessary for the production of 5-hydroxytryptophan. The levels of serotonin inside plants far exceed those seen in the animal brain, by approximately 100-fold – for example, banana skin has a level of 40 mg/g, while rat hippocampus has a level of 0.4 mg/g. Interestingly, the immediate precursor of 5-HT, 5-HTP, accounts for 20 percent of the total fresh weight in seeds from *Grivonia simplicifolia*, a tropical shrub of west Africa, which has potent medicinal properties (Lemaire and Adosraku, 2002). This seed has 500 times the serotonin levels measured in mammalian brains, where 5-HTP levels are almost undetectable.

The hydroxylase enzyme is of double significance in plants, since not only is molecular oxygen removed from the cell but also a very reactive antioxidant molecule is produced. Serotonin also provides a source for auxin and other important alkaloid production in plants. Despite the high levels of serotonin in plants, most general botany textbooks say nothing on this topic.

The Animal Kingdom (Table 2)

Animal cells lack chloroplasts, which are organelles central to plant photosynthesis and tryptophan synthesis (Table 2). This lack of a key evolutionary mechanism of life led animals to develop a number of traits in order to survive. Animals acquire photosynthesis-produced carbon by forming symbioses with algae and cyanobacteria. These associations are widespread in the phyla Porifera (e.g., sponges) and Cnidaria (e.g., corals, hydra), but are otherwise uncommon or absent from animal phyla (Venn *et al.*, 2008). The sponges (Porifera), the most primitive animal form, evolved about 600 million years ago (Valentine, 2004). Animals had to move to capture organisms that contained tryptophan, and to breathe to extract O_2 from the atmosphere. 5-HT- and 5-hydroxytryptamine-derived alkaloids are found in sponges (Salmoun *et al.*, 2002). This species does not have a nervous system, and feeds by filtration. In hydra, the most primitive animal with a specialized nervous and motor system, 5-HT appears to be localized to sensory cells scattered along the epithelium of the organism. When a distinct nervous system is seen, such as in the flatworm (*Stenostomum leucops*), serotonin neurons are localized there (Wikgren and Reuter, 1985). Serotonin in animals is produced in very low quantities because of the scarcity of tryptophan, and this may explain the few cells that contain 5-HT. As seen in hydra, these specialized 5-HT cells are nevertheless ideally localized, and have pronounced actions on the life of the organism (Figure 5).

Table 2 The Metazoa (Animal Kingdom), a major subdivision of life that includes all animals[1]

Phylum	Age (Mya)[2]	Class, genus, species, common names
Porifera	600	Sponge
Cnidaria	600	Colenterates; coral, hydra
Annelida	525	Earthworm
Echinodermata	500	Sea urchins
Molluska	450	Gastropoda; aplysia
		Cephalopodia; squid
Anthropoda	400	Crustacean; lobster
Arthropoda	350	Insecta; drosophila
Nematoda	325	*C. elegans*
Chordata		Vertebrate: subphylum
	400	*Fish*
	300	*Reptiles*
	250	*Amphibians, Newt*
	200	*Birds*
	65	*Mammal*
	2.5	*Homo*
	0.2	Homo sapiens

Source: University of California Museum of Paleontology (ucmp.berkeley.edu/index.php).

[1]Fossil records indicate that most metazoan phyla developed during the Cambrian explosion (542–488 million years ago). Many of the classes of vertebrates have fossil records indicating their appearance after this time.

[2]Mya, million years ago.

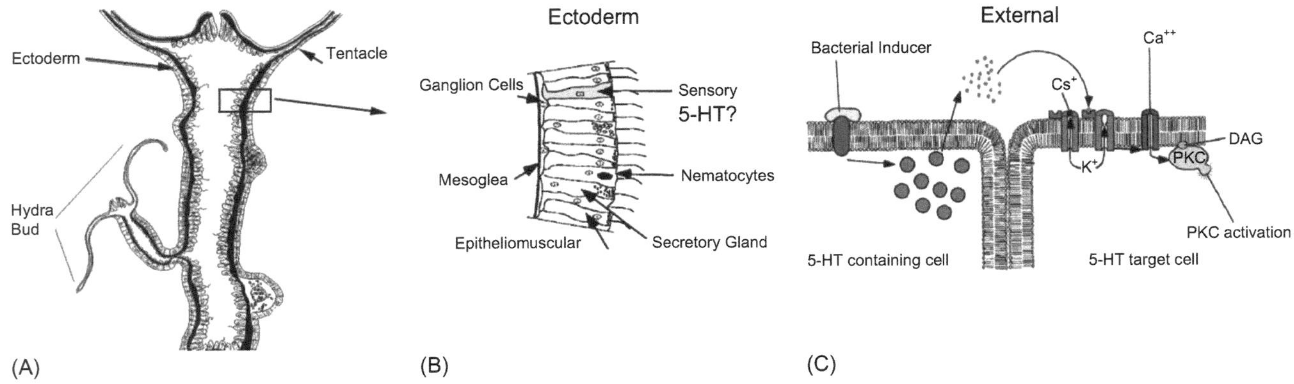

Figure 5 Proposed 5-HT action in hydra nervous system (modified from McCauley, 1997). The 5-HT content is produced and released from a sensory cell and interacts with specific receptors located on neighboring cells. A. An outline of the body plan of hydra shows a very primitive plan, with ectodermal cells lining the outside of the organism. B. A close-up view of the ectodermal layer shows occasional sensory neurons (5-HT?) scattered among non-sensory cells. C. Schematic diagram proposed by McCauley to demonstrate 5-HT release and interaction with a neighboring receptor that activates protein kinase C (PKC).

As these organisms have evolved a neuronal center for responding to their complex environment, serotonin continues to position itself to interact maximally with organizing centers. In aplysia there is a primitive brain, and from the very beginning serotonin is present. According to Marois and Carew (1997):

> The results indicate that the first serotonergic cells emerge at mid-embryogenesis and that a total of five cells make up the entire serotonergic system by hatching. The primitive serotonergic cells in aplysia are part of a newly discovered ganglion in aplysia, called the apical ganglion.

The 5-HT neuronal distribution is described in adult aplysia, and the vast serotonergic neuronal distribution through the nervous system, gut and periphery is notable (Goldstein *et al.*, 1984):

> Many neuronal cell bodies are stained in addition to the giant cerebral neuron of the cerebral ganglion and cells in the RB cluster of the abdominal ganglia which previously had been characterized biochemically and pharmacologically as being serotonergic. Neuronal cell bodies, both in central ganglia and in the wall of the gut, are encircled by plexuses of serotonergic varicosities. The neuropil of ganglia and the eye also contain fine, immunoreactive axons bearing varicosities. Intraganglionic connectives and nerves contain many stout fluorescent axons. Serotonergic varicosities are also observed in the connective tissue sheath surrounding central ganglia and nerves, as well as in heart and body muscle, blood vessels and gut.

The extensive distribution of serotonergic neurons in lower animals besides the brain and gut is not seen in mammals. The serotonergic neurons are located in the midbrain reticular formation, and have extensive axonal branching throughout the neuroaxis (Azmitia and Segal, 1978; Azmitia and Gannon, 1983) and the gut enterochromaffin cells and myenteric plexus (Gershon and Tack, 2007). Serotonergic axons from the myenteric plexus can innervate the nearby pancreas (Kirchgessner and Gershon, 1990). Mast cells located throughout the body and lying near blood vessels are the principal source of tryptophan hydroxylase in most peripheral tissues. These neuroendocrine mast cells contain serotonin at early gestational stages, and are positioned to serve a trophic function throughout the body as described for auxin in plants (Cutz, 1982).

In the brain of animals, the serotonin system has undergone considerable modification to keep abreast of the increased size and complexity of the brain (Figure 6). The neurons in the brains which contain serotonin actually decrease in relative size as one ascends from aplysia to primates. In lower organisms single serotonin neurons can regulate particular circuits, while in rodents the serotonergic neurons appear to act as large clusters of neurons acting together to influence general brain circuits. Finally, in primates the distribution of neurons in the raphe midbrain becomes more clustered, and these smaller groups send axons, often myelinated, to specific regions of cortex. This final arrangement allows serotonin to participate in multiple, discrete cortical functions than are not possible in lower mammals. The morphological organization appears to be continually evolving, and may enable more precise distribution of serotonin within the brain of humans.

Receptors

Capturing light is one of the most primitive functions of a receptor. Primitive rhodopsin GTP-linked receptors, which

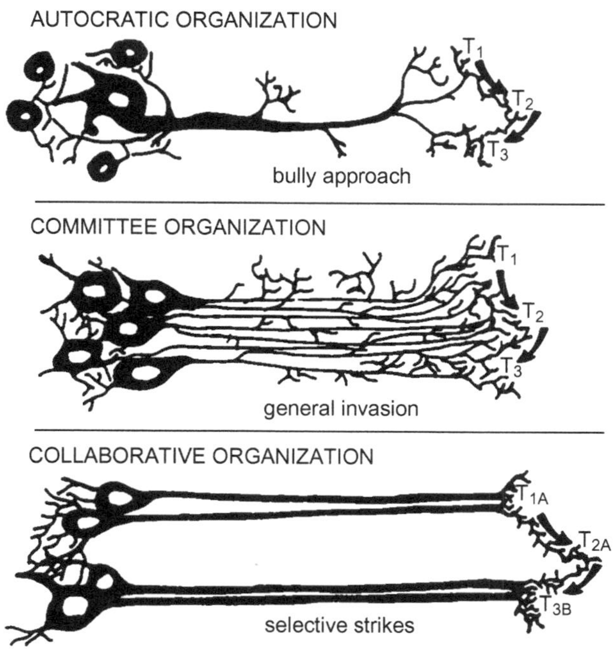

Figure 6 The progression of 5-HT neurons from aplysia to humans. In the aplysia nervous system there are only a few neurons which contain serotonin, and these neurons are large, with extensive connections. In the rodent brain, the 5-HT neurons are arranged in large grouping along the midline of the mescencephalon. The axons from these neurons ascend towards the forebrain in large bundles using mainly the ancient medial forebrain bundle. In primates, the distribution of serotonergic neurons in the mescencepahlon is into smaller clusters of neurons. In addition, many of the axons from these neurons are now myelinated. This new arrangement facilitated more precise and rapid delivery of sero-tonin to forebrain targets (Azmitia, 1987).

function in the human eye in rod cells used for night vision, are present in cyanobacteria (Ruiz-González and Marín, 2004). Cyanobacteria diverged about 2.5 billion years ago. The active site for solar capture in this receptor protein is tryptophan (Chabre and Breton, 1979; Hoersch *et al.*, 2008). The rhodopsin receptor and the 5-HT_{1A} receptor show substantial homology (Nowak *et al.*, 2006). In plants (*Arabidopsis thaliana*), a gene sequence was identified with a region of a putative receptor that is similar to sequences of serotonin receptors and other receptors of the so-called rhodopsin receptor family using a G-protein linked mechanism (Josefsson and Rask, 1997). From the gene diversity for serotonin receptors, the 5-HT_{1A} receptor is estimated to have evolved 750 million to 1 billion years ago, before the muscarinic, dopaminergic and adrenergic receptor systems (Peroutka and Howell, 1994). This age estimate would indicate that the receptor existed before the evolution of the most primitive animal form, sponges (Porifera), which evolved some 600 million years ago. If rhodopsin is considered the prototype of the 5-HT_{1A} receptor, the emergence of serotonin receptors occurred 3.5 to 2.5 million years ago in cyanobacteria.

Life began in sea water, where Na^+ and Cl^- ions are highly concentrated. Cells evolved a mechanism to exclude these ions in order to maintain a stable membrane potential, and 'neurotransmitters' evolved the ability to regulate these specific ion channels to rapidly manipulate the membrane potential. Second messengers, e.g., G proteins, c-AMP, and phospholipase C systems, appeared early in evolution and occur in all phyla that have been investigated. With the possible exception of the Porifera and Cnidaria, all the classical 'neurotransmitter' receptor subtypes identified in mammals occur throughout the animal phyla (see Walker *et al.*, 1996). Many of the serotonin receptors are seen in the embryonic stage – for example, H^3–5-HT binding is seen in the blastula and gastrula of sea urchins (Brown and Shaver, 1989). A gene from the sea urchin encoding the serotonin receptor (5-HT–hpr) was identified and showed sequence homology with the aplysia 5-HT_2 receptor (Katow *et al.*, 2004). Cells expressing the 5-HT receptor appeared near the tip of the archenteron in 33-h post-fertilization larvae. The serotonergic receptor cells developed 7 cellular tracts by 48 hours, and extended short fibers to the larval body surface through the ectoderm. These serotonergic receptor cells are a mesencephalic cell lineage, which appear to transmit serotonin signals to ectodermal cells at the start of gastrulation in sea urchins. In humans, the 5-HT_{1A} receptors are at their highest levels before birth (Bar-Peled *et al.*, 1991). In rats, the receptors for serotonin are not only present in the fetus, but can also be modified by injections of agonist (Whitaker-Azmitia *et al.*, 1987). All the invertebrate receptors so far cloned show homologies with mammalian receptors. This indicates that many of the basic serotonin receptor subtypes evolved during early geological periods and appeared at early ontogenic times. As the saying goes, 'Ontogeny recapitulates Phylogeny'.

Tremendous diversity has occurred in receptors in mammals. There are hundreds of serotonin receptor clones, and the human brain has at least 20 separate neuronal transcripts of 5-HT receptors (Moroz *et al.*, 2006). Serotonin is specifically bound to at least 16 specific receptor proteins in the human brain which regulate ion channels, c-AMP levels and kinase activity in neurons. The 5-HT receptors are found in every cell of the body. Why so many, and why such a large distribution? It can be speculated that the difficulty in making and obtaining tryptophan in animals results in low serotonin availability. The function of a receptor is to alert a cell that a chemical is present in the environment, without removing or altering the chemical. Thus, if a chemical is in short supply, the appearance of receptor molecules permits its actions to be transmitted throughout the organism. In order for this to be maximally effective, an efficient mechanism for the distribution of serotonin is required. Animals have specific tryptophan and serotonin binding proteins in their blood to help transport these molecules to specific target areas, such as the brain. Glial cells at the junction of the blood–brain barrier have special transport proteins for concentrating tryptophan and delivering it to the serotonin neurons (Bachmann, 2002; O'Kane and Hawkins, 2003). Serotonergic neurons developed long, unmyelinated axons that can take up tryptophan and utilize enzymes required for serotonin synthesis throughout the brain and gut. In summary, loss of tryptophan has promoted a highly branched, unmyelinated neural network, and a plethora of specific receptors to maximize serotonin's actions.

5-HT function

Plants do not have neurons or muscles, but they are nevertheless capable of limited movement by rotating their leaves towards the light and sending their roots deep into the soil to capture H_2O and nitrogen. In multicellular plant organisms, growth and mitosis are modulated by tryptophan-derived molecules, including serotonin. Furthermore, with the increase in organistic complexity, these molecules, in particular auxin, evolved homeostatic functions for effectively capturing and transporting solar energy and integrating plant rhythms and organism physiology. The plants developed effective transport systems for delivering tryptophan-based molecules from one region to another (e.g., leaf to root), depending on need (Mauseth, 2008). These two methods of producing movement, mitosis and maturation of plant cells, are similar to that seen in unicellular organisms and fungi (Eckert *et al.*, 1999). Auxin and other tryptophan regulate the rapid tracking of leaves toward the shifting source of light. The movements of both leaves and roots depend on compounds similar to serotonin, such as auxin (Ivanchenko *et al.*, 2008). The turning of the leaf to its source of energy depends on the rearrangement of the cell's cytoskeleton inside the leaf cells. In the root, the emersion into the soil is produced by regulating cell division and maturation. Serotonin, auxin and melatonin are involved in ion signaling in the dedifferentiation and differentiation of plant cells (Jones *et al.*, 2007). Auxin is the single most important trophic factor in plants (Perrot-Rechenmann *et al.*, 2002). Serotonin and melatonin have diurnal and seasonal rhythms in their synthesis and function in plants based on availability of solar energy. Thus, many of the transport and receptor mechanisms for serotonin and related molecules are established prior to the appearance of neurons. The actions of serotonin on the cell cytoskeleton and differentiation forecast the actions of serotonin in neuronal development and adult neuroplasticity in mammals (Azmitia, 1999).

Serotonin acquired many new functions in animals as the animals developed more complex processes. Serotonin-producing cells served a defense mechanism (stinging) in Cnidaria (coral) and in many Arthropoda (insects) (Horen, 1972; Weiger, 1997). In lower animals, serotonin neurons are primarily sensory neurons (activated by external stimuli), and influence food intake, defense withdrawal, and complex locomotor actions such as swimming (e.g., in sea urchins, Echinodermata) (Yaguchi and Katow, 2003). In the worm ganglia (Annelids), serotonin is first found in interneurons, which permits better regulation of complex behaviors such as swimming (Kristan and Nusbaum, 1982) and possibly learning and memory (Moss *et al.*, 2005). In *Caenorhabditis elegans* (Nematodes), 5-HT is involved in modulating feeding behavior by rapidly altering a chemosensory circuit (Chao *et al.*, 2004). The involvement of serotonin is also directed at neurons. The serotonin released from an apical ganglion interacts with specific neuronal receptors to increase or decrease the firing rate of its target cells involved in sensory and motor processing (Marois and Carew, 1997). Actions of serotonin on sexual activity and reproduction are evident (Boyle and Yoshino, 2005). In addition, serotonin changes cAMP and Ca^{2+} levels in its target neurons, influences their transcription rate and modifies cell morphology (Pettigrew *et al.*, 2005).

The actions of serotonin thus extend from that of antioxidant through morphogenesis and ascend to being involved in complex behaviors such as an organism's position in a social hierarchy. Serotonin in lobsters (Arthropods) regulates socially relevant behaviors such as dominance-type posture, offensive tail flicks, and escape responses (Kravitz, 2000). This action of serotonin may be through the 5-HT_{1A} receptor (Sosa *et al.*, 2004). 5-HT-regulated social and mental behaviors increased in number and complexity as these functions became more advanced and complicated. The many reports of increased social dominance in primates (Edwards and Kravitz, 1997) and improved mood and confidence in social interactions in humans after using drugs which increase serotonin levels are well documented (Kramer, 1993; Young and Leyton, 2002). In these higher animals, 5-HT continues in its role of a homeostatic regulator in adjusting the dynamic interactions of these many functions within the organism, and how the organism interacts with the outside world.

Trophic

The actions of serotonin in Metazoa begin very early in development. They are seen at both the blastula and gastrula stages, as noted by the appearance of serotonin receptors in the blastula stages. In Mollusca, serotonin is involved in the determination of the animal pole during early blastula stages (Buznikov *et al.*, 2003). Application of para-chlorophenylalanine (PCPA, a tryptophan hydroxylase inhibitor) interferes in morphogenesis by arresting gastrulation, which results in the disintegration of embryos. At lower concentrations of PCPA, retarded morphogenetic movements were observed that resulted in malformations in the anterior parts of the embryos and yolk granule degradation in the notochord (Hämäläinen and Kohonen, 1989). In mammals, the actions of serotonin on the developing fetus are felt from the time of conception due to the circulating serotonin in the plasma of the mother (Côté *et al.*, 2007). Serotonin neurons are formed very early in gestation in vertebrates (Lidov and Molliver, 1982; Wallace and Lauder, 1983; Okado *et al.*, 1989). Peripheral tissue also expresses cells which contain serotonin. For example, the mast cells (neuroendocrine) of the lung contain serotonin in the fetus (Kushnir-Sukhov *et al.*, 2006). It is logical to assume that if serotonin made a very early phylogenic appearance, then it should also make a very early ontogenic appearance. Plant seeds and animal embryos have the highest levels of serotonin.

By increasing cAMP and P-CREB, serotonin mediates a trophic response that may underlie both maturation and memory formation in aplysia (Glanzman *et al.*, 1990). Thus, in much the same way as serotonin and its derivatives influence the process and organelles of photosynthesis to move in order to track the source of light, in animals serotonin influences the morphology of sensory and motor neurons involved in neuronal networking in order to track the source of relevant stimuli. The changes in neuronal morphology are particularly intriguing, because they affect neuronal connectivity in much the same way as has been proposed for vertebrates. Even a relatively brief removal of serotonin from the brain of vertebrates results in loss of spine, dendritic profiles and synapses (Yan *et al.*, 1997; Okado *et al.*, 2001). This topic of neuroplasticity has been extensively reviewed by the current authors (Azmitia, 2001, 2007; Azmitia and Whitaker-Azmitia, 1991, 1997; Jacobs and Azmitia, 1992).

In mammals, serotonin has evolved a trophic relationship with glial cells. High-affinity receptors have been identified on astrocytes and Schwann cells from rodents and primates (Hertz *et al.*, 1984; Whitaker-Azmitia and Azmitia, 1986; Gaietta *et al.*, 2003). One function is for astrocytes to provide serotonergic neurons with tryptophan (Pow and Cook, 1997). The serotonin receptors on astrocytes can also release the neurite extension factor S100B, and glucose (Azmitia, 2001). Serotonin application induces glial-derived neurotrophic factor (GDNF) mRNA expression via the activation of fibroblast growth factor receptor 2 (FGFR2) (Tsuchioka *et al.*, 2008). Activation of serotonin receptors also promotes the development of glial cells in the brainstem of rats (Tajuddin *et al.*, 2003).

The recruitment of secondary cells to amplify serotonin's trophic actions emerged in animals. Astrocytes are found in vertebrates, and in *C. elegans* and drosophila. These supportive cells may have appeared even earlier. Using antibodies against a myelin marker and an astrocytic marker, evidence for glial cells was found in moths (Arthropoda) and aplysia (Roots, 1981). The emergence of these cells as targets for serotonin is what Brodie and Shore envisioned in 1957 when they termed the serotonin system the 'trophotrophic system'. Serotonin has trophic functions (mitosis, apoptosis, differentiation and metabolism), directly and through astrocytes, the other major cellular system in the brain, by receptor-mediated changes in glucose availability and trophic factor release. These actions are considered to be significant in the development and aging of the brain (Azmitia, 1999). Serotonin can in fact be considered to be important for the development and maturation of the entire organism, since serotonin and its receptors are found throughout the body – see Hansen and Witte, 2008 (gut), Wasserman, 1980 (lung), Raymond *et al.*, 1993 (kidney), Hagmann *et al.*, 1992 (liver) and Nordlind *et al.*, 2008 (skin).

The idea that serotonin functions as a trophic factor in vertebrate brains requires a new concept for how serotonin can be most efficiently distributed from axons. Traditionally, neurotransmitters are transported by the axon to a specific synaptic site where the neurotransmitter is released. This is seen for a proportion of the serotonin axons (Muller *et al.*, 2007). There were major discussions regarding whether serotonin could also function by diffusion from unmyelinated axonal varicosities on to non-synaptic sites (Beaudet and Descarries, 1978), and the controversy was settled by accepting the idea of diffuse release, as well as acknowledging that many of the receptor targets of serotonin are on non-neuronal cells. For example, in the rat brain, serotonin axons course through the lateral and III ventricles along ependymal cells (Møllgård and Wiklund, 1979). Serotonin fibers can be considered to be a 'drip irrigation system' for the brain. As long as the axons are intact, serotonin is efficiently released throughout the brain. In old age and in neurodegenerative diseases, serotonin axons in the human brain degenerate (Azmitia and Nixon, 2008).

Seasonal affective disorder and suicides

Dysfunctions in brain serotonin are implicated in many mental disorders, such as autism, Down's syndrome, anorexia nervosa, anxiety and depression. There are nearly 10,000 papers dealing with serotonin and various diseases ranging from alcohol addiction (Martinez *et al.*, 2008) to herpes zoster (Ohyama *et al.*, 2004). The relationship between serotonin and depression is cited in over 13,000 papers, with 1750 citations since 2007. Furthermore, a strong correlation exists between brain serotonin levels, depression and suicide, with the first paper in this area written over 40 years ago (Shaw *et al.*, 1967). Those attempting suicide had significantly lower levels of 5-HIAA in the CSF compared to controls (Mann *et al.*, 1996). PET studies indicate that the 5-HT_{2A} receptor is altered in depressed suicide attempters (Audenaert *et al.*, 2006). A decrease in serotonin has serious consequences on normal brain homeostasis, both structural and functional., and influences a person's desire to continue living. It is surprising to learn that sunlight has dramatic actions on the brain serotonin system of humans.

A seasonal variation in affective disorders was reported several decades ago (Videbech, 1975), and has certainly been noted from the earliest times of recorded history. All Northern hemispheric cultures since the Mesopotamians have developed special holidays to mark the nadir of light on Earth, and celebrations to counter winter's gloom (for example, Makar, Sankranti, Saturnalia, and Dong Zhi) (Count and Count, 2000). Seasonal affective disorder (SAD) consists of recurrent major depressive episodes in the fall/winter with remissions in spring/summer, and is effectively treated with serotonin drugs and/or light therapy (Westrin and Lam, 2007). Treatment with light therapy or antidepressant medication is associated with equivalent marked improvement in the assessment of psychosocial functioning and life quality. There was no significant difference in measures in 96 SAD patients receiving 8 weeks of treatment with either (1) 10,000-lux light treatment and a placebo capsule, or (2) 100-lux light treatment (placebo light) and 20 mg fluoxetine (Michalak *et al.*, 2007). Several studies have confirmed that patients respond favorable to light therapy (Yerevanian *et al.*, 1986; Stewart *et al.*, 1991; Rao *et al.*, 1992).

Light therapy has effects on serotonin parameters in humans. It has been shown that blood serotonin increases in healthy subjects and patients with non-seasonal depression after repeated visible light exposure. Blood samples from jugular veins in 101 healthy men showed that turnover of serotonin by the brain was lowest in winter, and directly related to the prevailing duration of bright sunlight (Lambert *et al.*, 2002). The production of serotonin measured by this procedure increased rapidly with exposure to increased luminosity. Serotonin levels were higher on bright days no matter what the time of year, and the amount of serotonin present reflected the hours of sun exposure on a particular day – conditions the day before had no effect. In a group of patients with a history of SAD, significantly lower plasma biopterin and tryptophan levels were measured that increased after light therapy (Hoekstra *et al.*, 2003).

A seasonal rhythm in plasma serotonin transporter in normal subjects was first demonstrated over 25 years ago (Whitaker *et al.*, 1984). There was a significant reduction in the Hamilton Depression Rating Scale (HAMD21) score after therapy vs before treatment, and the K_d for citalopram binding was significantly higher after phototherapy than before treatment (Swiecicki *et al.*, 2005). In agreement with the previous work, binding studies of 5-HTT show that this protein is in a sensitized state during depression in SAD, and normalizes after light therapy and in natural summer remission (Willeit *et al.*, 2008). Phototherapy had a significant influence on both the measured serotonin transport parameters (B_{max} and K_d).

This suggests that daily light therapy has a sound basis in biology and makes evolutionary sense.

Blue light is effective at increasing tryptophan absorption during photosynthesis in chloroplasts, and this light is efficient at treating patients suffering from SAD. As mentioned with the plant chloroplast system, it appears in human studies that blue light might be the most effective. Blue light can suppress melatonin levels and aid in circadian phase shift. Light therapy is effective at significantly reducing Hamilton Depression Rating Scale (SAD Version) when a narrow band of blue light (468 nm) is used (Glickman *et al.*, 2006). The UV-A spectrum does not increase the antidepressant response of light therapy, and clinical application of light therapy should use light sources that have the UV spectrum filtered (Lam *et al.*, 1992). Light therapy relieves suicidal ideation in patients with SAD consistent with overall clinical improvement. Emergence of suicidal ideas or behaviors is very uncommon with light therapy (Lam *et al.*, 2000). It has been proposed that the lighting standards in the home and workplace should be re-evaluated on the basis of new knowledge regarding the neurobiological effects of light (Jacobsen *et al.*, 1987). This might be considered one of the first steps taken by a society to achieve conditions conducive to enhancing serotonin function in the general population, and an acknowledgement of serotonin's special relationship with sunlight that began to emerge at the beginning of life on Earth (Figure 7).

Summary

Sunlight has beneficial effects on the serotonin system and the mood and stability of humans. This is consistent with the idea that serotonin is involved in homeostasis in humans (Azmitia, 2001) and contributes to the emergence of mind (Azmitia, 2007). What is surprising is the consistency of serotonin's function throughout evolution. The indole ring of tryptophan was the first and principal

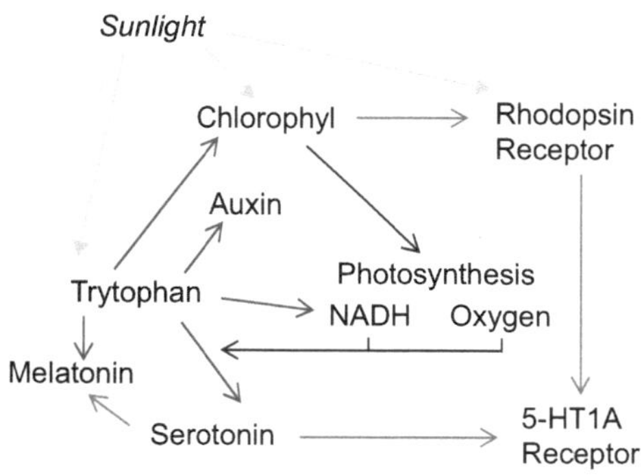

Figure 7 The evolutionary relationship between sunlight and the amino-acid tryptophan and its major derivatives: serotonin, melatonin, auxin and NADH. In addition, the connections of certain tryptophan containing proteins are illustrated: chlorophyll, rhodopsin and 5-HT_{1A} receptor. These relationships began in the earliest eukaryotic cells and continue in humans.

means of converting sunlight into biological energy. In the simplest cells, in the presence of sunlight, serotonin and related tryptophan derivates function as a powerful antioxidant and serve to help organisms to maintain homeostasis, regulate differentiation and promote cell division. The actions of serotonin have consistently been beneficial for the survival of the species throughout early periods of evolution, and for the growth of the organism from the earliest times of ontogeny. This action is maintained even though the availability of tryptophan, needed to produce serotonin, was shut down in animals. The main strategy used by animals is to become voracious consumers of tryptophan-containing plants and animals. In vertebrates, a sophisticated delivery system in the blood was developed to deliver tryptophan throughout the body so serotonin could be synthesized in sufficient quantities to meet its needs. Serotonin is distributed from plasma, neighboring mast cells, and specialized serotonin neurons. In mammals, the expression of serotonin receptors occurs at the earliest stages of ontogeny and is activated by circulating plasma serotonin from the mother. In the early stages of brain development, serotonergic neurons formed in the midbrain and immediately sent out extensive fibers to the forebrain. Serotonin-producing neurons grew complex axonal plexuses of unmyelinated axons dispensing serotonin by a 'drip irrigation' method. Serotonin-specific receptors proliferated in animals, with over a dozen types distributed in every cell and in every organ.

Serotonin, an important tryptophan derivative, has continued to maintain a central position in the evolution (phylogeny) and development (ontogeny) of organisms, from fungi to humans. It is important to recognize that loss of serotonin in the human brain, a major factor in suicides, means loss of synapses and retraction of dendrites and spines (Azmitia, 2001). Seasonal affective depression and suicidal ideation can be treated with light therapy, which is intended to mimic the beneficial effects of natural sunlight. This chapter has followed an evolutionary path from tryptophan absorption of light in photosynthesis to serotonin's actions in treating seasonal affective disorder. All cells and organs of the body, and especially of the brain, are affected by the serotonin system. The actions of sunlight may be the magic elixir to help maintain homeostasis between body and mind, improve social interactions, and create harmony among the phyla.

Acknowledgements

I am indebted to Dr. Patricia Whitaker-Azmitia for her comments during the preparation of this manuscript. I also acknowledge the many serotonin researchers whose work has inspired me to focus my career on a single brain chemical. Funding for this research was provided by a New York University Challenge grant.

References

Ahmad, S. and Jensen, R.A. (1998) New prospects for deducing the evolutionary history of metabolic pathways in prokaryotes: aromatic biosynthesis as a case-in-point. Orig. Life Evol. Biosph., 18: 41–57.

Alexander, F.W., Sandmeier, E., Mehta, P.K. and Christen, P. (1994) Evolutionary relationships among pyridoxal-5′-phosphate-dependent enzymes. Regio-specific alpha, beta and gamma families. Eur. J. Biochem., 219: 953–960.

Angiolillo, P.J. and Vanderkooi, J.M. (1996) Hydrogen atoms are produced when tryptophan within a protein is irradiated with ultraviolet light. Photochem. Photobiol., 64: 492–495.

Aoyagi, M., Arvai, A.S., Ghosh, S., Stuehr, D.J., Tainer, J.A. and Getzoff, E.D. (2001) Structures of tetrahydrobiopterin binding-site mutants of inducible nitric oxide synthase oxygenase dimer and implicated roles of Trp457. Biochemistry, 40: 12826–12832.

Asada, K. (1996) Radical production and scavenging in the chloroplast. In: Baker, N. (Ed.), Photosynthesis and the Environment. Kluwer Academic Publishers, Dordrecht, pp. 123–150.

Audenaert, K., Peremans, K., Goethals, I. and van Heeringen, C. (2006) Functional imaging, serotonin and the suicidal brain. Acta Neurol. Belg., 106: 125–131.

Azmitia, E.C. (1987) The primate serotonergic system: progression towards a collaborative organization. In: Meltzer, H. (Ed.), Psychopharmacology, Third Generation of Progress. Plenum Press, New York, NY, pp. 61–74.

Azmitia, E.C. (1999) Serotonin neurons, neuroplasticity, and homeostasis of neural tissue. Neuropsychopharmacology, 21(2 Suppl.): 33S–45S.

Azmitia, E.C. (2001) Modern views on an ancient chemical: serotonin effects on cell proliferation, maturation, and apoptosis. Brain Res. Bull., 56: 413–424.

Azmitia, E.C. (2007) Cajal and brain plasticity: insights relevant to emerging concepts of mind. Brain Res. Rev., 55: 395–405.

Azmitia, E.C. and Gannon, P.J. (1983) The ultrastructural localization of serotonin immunoreactivity in myelinated and unmyelinated axons within the medial forebrain bundle of rat and monkey. J. Neuroscience, 3: 2083–2090.

Azmitia, E.C. and Nixon, R. (2008) Dystrophic serotonergic axons in neurodegenerative diseases. Brain Res., 1217: 185–194.

Azmitia, E.C. and Segal, M. (1978) An autoradiographic analysis of the differential ascending projections of the dorsal and median raphe nuclei in the rat. J. Comp. Neurol., 179: 641–646.

Azmitia, E.C. and Whitaker-Azmitia, P.M. (1991) Awaking the sleeping giant: anatomy and plasticity of the brain serotonergic system. J. Clin. Psychiatry., 52: 4–16.

Azmitia, E.C. and Whitaker-Azmitia, P.M. (1997) Handbook of Experimental Pharmacology: Serotonergic Neurons and 5-HT Receptors in the CNS. Baumgarten, H.G. and Gothert, M. (Eds.), Berlin: Springer, pp. 1–39.

Bachmann, C. (2002) Mechanisms of hyperammonemia. Clin. Chem. Lab. Med., 40: 653–662.

Badu, V. and Joshi, P.C. (1992) Tryptophan as an endogenous photosensitizer to elicit harmful effects of ultraviolet B. Indian J. Biochem. Biophys., 29: 296–298.

Bar-Peled, O., Gross-Isseroff, R., Ben-Hur, H., Hoskins, I., Groner, Y. and Biegon, A. (1991) Fetal human brain exhibits a prenatal peak in the density of serotonin 5-HT1A receptors. Neurosci. Lett., 127: 173–176.

Bazin, M., Pierre, J., Debey, P. and Santus, R. (1982) One-electron photoreduction of bacterial cytochrome P-450 by ultraviolet light. II. A study using laser flash photolysis of the dynamics of the reduction of bacterial cytochrome P-450 in the presence of carbon monoxide. The role of the protein conformation. Eur. J. Biochem., 124: 539–544.

Beaudet, A. and Descarries, L. (1978) The monoamine innervation of rat cerebral cortex: synaptic and nonsynaptic axon terminals. Neuroscience, 3(10): 851–860.

Belenikina, N.S., Strakhovskaya, M.G. and Fraikin, G.Y. (1991) Near-UV activation of yeast growth. J. Photochem. Photobiol. B, 10: 51–55.

Bender, D.A. and Olufunwa, R. (1988) Utilization of tryptophan, nicotinamide and nicotinic acid as precursors for nicotinamide nucleotide synthesis in isolated rat liver cells. Br. J. Nutrition, 59: 279–287.

Binazzi, M. and Calandra, P. (1975) Tryptophan metabolism in skin diseases. Acta Vitaminol. Enzymol., 29: 161–163.

Borkman, R.F. and Lerman, S. (1978) Fluorescence spectra of tryptophan residues in human and bovine lens proteins. Exp. Eye Res., 26: 705–713.

Boularand, S., Biguet, N.F., Vidal, B., Veron, M., Mallet, J., Vincent, J.D., Dufour, S. and Vernier, P. (1998) Tyrosine hydroxylase in the European eel (*Anguilla anguilla*): cDNA cloning, brain distribution, and phylogenetic analysis. J. Neurochem., 71: 460–470.

Boyle, J.P. and Yoshino, T.P. (2005) Serotonin-induced muscular activity in *Schistosoma mansoni* larval stages: importance of 5-HT transport and role in daughter sporocyst production. J. Parasitol., 91: 542–550.

Brodie, B.B. and Shore, P.A. (1957) A concept for a role of serotonin and norepinephrine as chemical mediators in the brain. Ann. NY Acad. Sci., 66: 631–642.

Brown, K.M. and Shaver, J.R. (1989) [3H]serotonin binding to blastula, gastrula, prism, and pluteus sea urchin embryo cells. Comp. Biochem. Physiol. C, 93: 281–285.

Bryant, D.A. and Frigaard, N.-U. (2006) Prokaryotic photosynthesis and phototrophy illuminated. Trends Microbiol., 14: 488–496.

Buznikov, G.A., Nikitina, L.A., Voronezhskaya, E.E., Bezuglov, V.V., Dennis Willows, A.O. and Nezlin, L.P. (2003) Localization of serotonin and its possible role in early embryos of *Tritonia diomedea* (Mollusca: Nudibranchia). Cell Tissue Res., 311: 259–266.

Chabre, M. and Breton, J. (1979) Orientation of aromatic residues in rhodopsin. Rotation of one tryptophan upon the meta I to meta II transition after illumination. Photochem. Photobiol., 30: 295–299.

Chao, M.Y., Komatsu, H., Fukuto, H.S., Dionne, H.M. and Hart, A.C. (2004) Feeding status and serotonin rapidly and reversibly modulate a *Caenorhabditis elegans* chemosensory circuit. Proc. Natl. Acad. Sci. USA, 101: 15512–15517.

Chaturvedi, C.M. (1984) Effect of melatonin on the adrenal and gonad of the common mynah *Acridtheres tristis*. Austral. J. Zool., 32: 803–809.

Cooke, T.J., Poli, D., Sztein, A.E. and Cohen, J.D. (2002) Evolutionary patterns in auxin action. Plant Mol. Biol., 49: 319–338.

Côté, F., Fligny, C., Bayard, E., Launay, J.M., Gershon, M.D., Mallet, J. and Vodjdani, G. (2007) Maternal serotonin is crucial for murine embryonic development. Proc. Natl. Acad. Sci. USA, 104: 329–334.

Count, E.W. and Count, A.L. (2000) 4,000 Years of Christmas. Ulysses Press, Berkeley, CA.

Cox, M., Lehninger, A.L. and Nelson, D.R. (2000) Lehninger Principles of Biochemistry. Worth Publisher, New York, NY.

Curzon, G., Friedel, J. and Knott, P.J. (1973) The effect of fatty acids on the binding of tryptophan to plasma protein. Nature, 242: 198–200.

Cutz, E. (1982) Neuroendocrine cells of the lung. An overview of morphologic characteristics and development. Exp. Lung Res., 3: 185–208.

Dadak, V., Vanderkooi, J.M. and Wright, W.W. (1992) Electron transfer from excited tryptophan to cytochrome c: mechanism of phosphorescence quenching? Biochim. Biophys. Acta, 1100: 33–39.

Dhonukshe, P., Grigoriev, I., Fischer, R., Tominaga, M., Robinson, D.G., Hasek, J., Paciorek, T., Petrásek, J., Seifertová, D., Tejos, R., Meisel, L.A., Zazímalová, E., Gadella, T.W. Jr., Stierhof, Y.D., Ueda, T., Oiwa, K., Akhmanova, A., Brock, R., Spang, A. and Friml, J. (2008) Auxin transport inhibitors impair vesicle motility and actin cytoskeleton dynamics in diverse eukaryotes. Proc. Natl. Acad. Sci. USA, 105(11): 4489–4494.

Du, H., Fuh, R.A., Corkan, J., Li, A. and Lindsey, J.S. (1998) PhotochemCAD: a computer-aided design and research tool in photochemistry. Photochem. Photobiol., 68: 141–142.

Eckert, S.E., Hoffmann, B., Wanke, C. and Braus, G.H. (1999) Sexual development of *Aspergillus nidulans* in tryptophan auxotrophic strains. Arch. Microbiol., 172: 157–166.

Edwards, D.H. and Kravitz, E.A. (1997) Serotonin, social status and aggression. Curr. Opin. Neurobiol., 7: 812–819.

Efimenko, E., Blacque, O.E., Ou, G., Haycraft, C.J., Yoder, B.K., Scholey, J.M., Leroux, M.R. and Swoboda, P. (2006) Caenorhabditis elegans DYF-2, an orthologue of human WDR19, is a component of the intraflagellar transport machinery in sensory cilia. Mol. Biol. Cell, 17: 4801–4811.

Facchini, P.J., Huber-Allanach, K.L. and Tari, L.W. (2000) Plant aromatic L-amino acid decarboxylases: evolution,

biochemistry, regulation, and metabolic engineering applications. Phytochemistry, 54: 121–138.
Friedberg, E.C., Walker, G.C. and Siede, W. (1995) DNA Repair and Mutagenesis. ASM Press, Washington, DC.
Gaietta, G.M., Yoder, E.J., Deerinck, T., Kinder, K., Hanono, A., Han, A., Wu, C. and Ellisman, M.H. (2003) 5-HT2a receptors in rat sciatic nerves and Schwann cell cultures. J. Neurocytol., 32: 373–380.
Garattini, S. and Valzelli, L. (1965) Serotonin. Elsevier Publishing Co., Amsterdam.
Gershon, M.D. and Tack, J. (2007) The serotonin signaling system: from basic understanding to drug development for functional GI disorders. Gastroenterology, 132: 397–414.
Glanzman, D.L., Kandel, E.R. and Schacher, S. (1990) Target-dependent structural changes accompanying long-term synaptic facilitation in Aplysia neurons. Science, 249: 799–802.
Glickman, G., Byrne, B., Pineda, C., Hauck, W.W. and Brainard, G.C. (2006) Light therapy for seasonal affective disorder with blue narrow-band light-emitting diodes (LEDs). Biol. Psychiatry, 59: 502–507.
Goldstein, R., Kistler, H.B. Jr., Steinbusch, H.W. and Schwartz, J.H. (1984) Distribution of serotonin-immunoreactivity in juvenile Aplysia. Neuroscience, 11: 535–547.
Grenett, H.E., Ledley, F.D., Reed, L.L. and Woo, S.L. (1987) Full-length cDNA for rabbit tryptophan hydroxylase: functional domains and evolution of aromatic amino acid hydroxylases. Proc. Natl. Acad. Sci. USA, 84: 5530–5534.
Grossweiner, L.I. (1984) Photochemistry of proteins: a review. Curr. Eye Res., 3: 137–144.
Hagmann, W., Hacker, H.J. and Buchholz, U. (1992) Resident mast cells are the main initiators of anaphylactic leukotriene production in the liver. Hepatology, 16: 1477–1484.
Hämäläinen, M. and Kohonen, J. (1989) Studies on the effect of monoamine antagonists on the morphogenesis of the newt. Intl. J. Dev. Biol., 33: 157–163.
Hansen, M.B. and Witte, A.B. (2008) The role of serotonin in intestinal luminal sensing and secretion. Acta Physiol. (Oxf), 193: 311–323.
Hardeland, R. (1999) Melatonin and 5-methoxytryptamine in non-metazoans. Reprod. Nutr. Dev., 39: 399–408.
Hardeland, R. (2005) Antioxidative protection by melatonin: multiplicity of mechanisms from radical detoxification to radical avoidance. Endocrine, 27: 119–130.
Hertz, L., Schousboe, I., Hertz, L. and Schousboe, A. (1984) Receptor expression in primary cultures of neurons or astrocytes. Prog. Neuropsychopharmacol. Biol. Psychiatry, 8: 521–527.
Hoekstra, R., Fekkes, D., van de Wetering, B.J., Pepplinkhuizen, L. and Verhoeven, W.M. (2003) Effect of light therapy on biopterin, neopterin and tryptophan in patients with seasonal affective disorder. Psychiatry Res., 120: 37–42.
Hoersch, D., Otto, H., Wallat, I. and Heyn, M.P. (2008) Monitoring the conformational changes of photoactivated rhodopsin from microseconds to seconds by transient fluorescence spectroscopy. Biochemistry, 47: 11518–11527.
Horen, W.P. (1972) Insect and scorpion sting. J. Am. Med. Assoc., 221: 894–898.
Humphrey, T. and Enoch, T. (1998) Sum1, a highly conserved WD-repeat protein, suppresses S-M checkpoint mutants and inhibits the osmotic stress cell cycle response in fission yeast. Genetics, 148(4): 1731–1742.
Hunt, D.E. (1964) Amino acids as protective compounds in ultraviolet-irradiated *Hansenula wingei*. J. Bacteriol., 88: 893–895.
Ivanchenko, M.G., Muday, G.K. and Dubrovsky, J.G. (2008) Ethylene–auxin interactions regulate lateral root initiation and emergence in *Arabidopsis thaliana*. Plant J., 55: 335–347.
Iwanami, N., Higuchi, T., Sasano, Y., Fujiwara, T., Hoa, V.Q., Okada, M., Talukder, S.R., Kunimatsu, S., Li, J., Saito, F., Bhattacharya, C., Matin, A., Sasaki, T., Shimizu, N., Mitani, H., Himmelbauer, H., Momoi, A., Kondoh, H., Furutani-Seiki, M. and Takahama, Y. (2008) WDR55 is a nucleolar modulator of ribosomal RNA synthesis, cell cycle progression, and teleost organ development. PLoS Genet., 4(8): e1000171.
Jacobs, B.L. and Azmitia, E.C. (1992) Structure and function of the brain serotonin system. Physiol. Rev., 72: 165–229.
Jacobs, W.P., Falkenstein, K. and Hamilton, R.H. (1985) Nature and amount of auxin in algae: IAA from extracts of Caulerpa paspaloides (Siphonales). Plant Physiol., 78: 844–848.
Jacobsen, F.M., Wehr, T.A., Sack, D.A., James, S.P. and Rosenthal, N.E. (1987) Seasonal affective disorder: a review of the syndrome and its public health implications. Am. J. Public Health, 77: 57–60.
Jones, M.P., Cao, J., O'Brien, R., Murch, S.J. and Saxena, P.K. (2007) The mode of action of thidiazuron: auxins, indoleamines, and ion channels in the regeneration of *Echinacea purpurea* L. Plant Cell Rep., 26: 1481–1490.
Josefsson, L.G. and Rask, L. (1997) Cloning of a putative G-protein-coupled receptor from *Arabidopsis thaliana*. Eur. J. Biochem., 249: 415–420.
Katow, H., Yaguchi, S., Kiyomoto, M. and Washio, M. (2004) The 5-HT receptor cell is a new member of secondary mesenchyme cell descendants and forms a major blastocoelar network in sea urchin larvae. Mech. Dev., 121(4): 325–337.
Kirchgessner, A.L. and Gershon, M.D. (1990) Innervation of the pancreas by neurons in the gut. J. Neurosci., 10(5): 1626–1642.
Kramer, P. (1993) Listening to Prozac. Penguin, New York.
Kravitz, E.A. (2000) Serotonin and aggression: insights gained from a lobster model system and speculations on the role of amine neurons in a complex behavior. J. Comp. Physiol., 186: 221–238.
Kristan, W.B. Jr. and Nusbaum, M.P. (1982) The dual role of serotonin in leech swimming. J. Physiol. (Paris), 78: 743–747.
Kunz, D.A., Fernandez, R.F. and Parab, P. (2001) Evidence that bacterial cyanide oxygenase is a pterin-dependent hydroxylase. Biochem. Biophys. Res. Commun., 287: 514–518.
Kushnir-Sukhov, N.M., Gilfillan, A.M., Coleman, J.W., Brown, J.M., Bruening, S., Toth, M. and Metcalfe, D.D. (2006) 5-hydroxytryptamine induces mast cell adhesion and migration. J. Immunol., 177: 6422–6432.
Lam, R.W., Buchanan, A., Mador, J.A., Corral, M.R. and Remick, R.A. (1992) The effects of ultraviolet-A wavelengths in light therapy for seasonal depression. J. Affect Disord., 24: 237–243.
Lam, R.W., Tam, E.M., Shiah, I.S., Yatham, L.N. and Zis, A.P. (2000) Effects of light therapy on suicidal ideation in patients with winter depression. J. Clin. Psychiatry, 61: 30–32.
Lambert, G.W., Reid, C., Kaye, D.M., Jennings, G.L. and Esler, M.D. (2002) Effect of sunlight and season on serotonin turnover in the brain. Lancet, 360: 1840–1842.
Lemaire, P.A. and Adosraku, R.K. (2002) An HPLC method for the direct assay of the serotonin precursor, 5-hydroxytrophan, in seeds of *Grivonia simplicifolia*. Phytochem. Anal., 13: 333–337.
Lidov, H.G. and Molliver, M.E. (1982) An immunohistochemical study of serotonin neuron development in the

rat: ascending pathways and terminal fields. Brain Res. Bull., 8: 389–430.
Lin, S.W. and Sakmar, T.P. (1996) Specific tryptophan UV-absorbance changes are probes of the transition of rhodopsin to its active state. Biochemistry, 35: 11149–11159.
Mann, J.J., Malone, K.M., Psych, M.R., Sweeney, J.A., Brown, R.P., Linnoila, M., Stanley, B. and Stanley, M. (1996) Attempted suicide characteristics and cerebrospinal fluid amine metabolites in depressed inpatients. Neuropsychopharmacology, 15(6): 576–586.
Marois, R. and Carew, T.J. (1997) Ontogeny of serotonergic neurons in *Aplysia californica*. J. Comp. Neurol., 386: 477–490.
Martinez, D., Slifstein, M., Gil, R., Hwang, D.R., Huang, Y., Perez, A., Frankle, W.G., Laruelle, M., Krystal, J., Abi-Dargham, A., 2008. Positron emission tomography imaging of the serotonin transporter and 5-HT(1A) receptor in alcohol dependence. Biol Psychiatry, in press.
Mauseth, J.D. (2008) Botany: An Introduction to Plant Biology. 4th edn. Jones & Bartlett Publishers, Sudbury: MA.
Mauzerall, D.C. (1973) Why chlorophyll? Ann. NY Acad. Sci., 206: 483.
McCauley, D.W. (1997) Serotonin plays an early role in the metamorphosis of the hydrozoan *Phialidium gregarium*. Dev. Biol., 190: 229–240.
Mehta, P.K. and Christen, P. (2005) The molecular evolution of pyridoxal-5′-phosphate-dependent enzymes. Adv. Enzymol. Relat. Areas Mol. Biol., 74: 129–184.
Mercer-Smith, A., Raudino, A. and Mauzerall, D.C. (1985) A model for the origin of photosynthesis – III. The ultraviolet photochemistry of uroporphyrinogen. Photochem. Photobiol., 42: 239.
Michalak, E.E., Murray, G., Levitt, A.J., Levitan, R.D., Enns, M.W., Morehouse, R., Tam, E.M., Cheung, A. and Lam, R.W. (2007) Quality of life as an outcome indicator in patients with seasonal affective disorder: results from the Can-SAD study. Psychol. Med., 37(5): 727–736.
Møller, I.M. and Kristensen, B.K. (2006) Protein oxidation in plant mitochondria detected as oxidized tryptophan. Free Radic. Biol. Med., 40: 430–435.
Møllgård, K. and Wiklund, L. (1979) Serotoninergic synapses on ependymal and hypendymal cells of the rat subcommissural organ. J. Neurocytol., 8: 445–467.
Moroz, L.L., Edwards, J.R., Puthanveettil, S.V., Kohn, A.B., Ha, T., Heyland, A., Knudsen, B., Sahni, A., Yu, F., Liu, L., Jezzini, S., Lovell, P., Iannucculli, W., Chen, M., Nguyen, T., Sheng, H., Shaw, R., Kalachikov, S., Panchin, Y.V., Farmerie, W., Russo, J.J., Ju, J. and Kandel, E.R. (2006) Neuronal transcriptome of aplysia: neuronal compartments and circuitry. Cell, 127: 1453–1467.
Moss, B.L., Fuller, A.D., Sahley, C.L. and Burrell, B.D. (2005) Serotonin modulates axo-axonal coupling between neurons critical for learning in the leech. J. Neurophysiol., 94: 2575–2589.
Muller, J.F., Mascagni, F. and McDonald, A.J. (2007) Serotonin-immunoreactive axon terminals innervate pyramidal cells and interneurons in the rat basolateral amygdala. J. Comp. Neurol., 505: 314–335.
Nelson, D.L. and Cox, M.M. (2008) Lehninger Principles of Biochemistry. 5th edn. W.H. Freeman, New York, NY.
Nordlind, K., Azmitia, E.C. and Slominski, A. (2008) The skin as a mirror of the soul: exploring the possible roles of serotonin. Exp. Dermatol., 17: 301–311.
Nowak, M., Kolaczkowski, M., Pawlowski, M. and Bojarski, A.J. (2006) Homology modeling of the serotonin 5-HT1A receptor using automated docking of bioactive compounds with defined geometry. J. Med. Chem., 49: 205–214.
Ohyama, S., Kuniyoshi, M., Nishi, S. and Inanaga, K. (2004) [Postherpetic neuralgia alleviated by an SSRI fluvoxamine: two cases of PHN accompanied with depression were treated with fluvoxamine]. Nihon Shinkei Seishin Yakurigaku Zasshi, 24: 79–81.
Okado, N., Shibanoki, S., Ishikawa, K. and Sako, H. (1989) Developmental changes in serotonin levels in the chick spinal cord and brain. Brain Res. Dev. Brain Res., 50: 217–223.
Okado, N., Narita, M. and Narita, N. (2001) A biogenic amine-synapse mechanism for mental retardation and developmental disabilities. Brain Dev., 23(Suppl. 1): S11–S15.
O'Kane, R.L. and Hawkins, R.A. (2003) Na+-dependent transport of large neutral amino acids occurs at the abluminal membrane of the blood–brain barrier. Am. J. Physiol. Endocrinol. Metab., 285: E1167–E1173.
Overbeek, J.V. (1940) Auxin in marine algae. Plant Physiol., 15: 291–299.
Pember, S.O., Villafranca, J.J. and Benkovic, S.J. (1986) Phenylalanine hydroxylase from Chromobacterium violaceum is a copper-containing monooxygenase. Kinetics of the reductive activation of the enzyme. Biochemistry, 25: 6611–6619.
Peroutka, S.J. and Howell, T.A. (1994) The molecular evolution of G-protein-coupled receptors: focus on 5-hydroxytryptamine receptors. Neuropharmacology, 33: 319–324.
Perrot-Rechenmann, C., Hagen, G. and Perrot-Rechenmann, C. (2002) Auxin Molecular Biology. Springer Publisher, New York, NY.
Pettigrew, D.B., Smolen, P., Baxter, D.A. and Byrne, J.H. (2005) Dynamic properties of regulatory motifs associated with induction of three temporal domains of memory in aplysia. J. Comput. Neurosci., 18: 163–181.
Pévet, P., Balemans, M.G. and de Reuver, G.F. (1981) The pineal gland of the mole (*Talpa europaea* L.). VII. Activity of hydroxyindole-O-methyltransferase (HIOMT) in the formation of 5-methoxytryptophan, 5-methoxytryptamine, 5-methoxyindole-3-acetic acid, 5-methoxytryptophol and melantonin in the eyes and the pineal gland. J. Neural. Transm., 51: 271–282.
Phi, Q.T., Park, Y.M., Ryu, C.M., Park, S.H. and Ghim, S.Y. (2008) Functional identification and expression of indole-3-pyruvate decarboxylase from *Paenibacillus polymyxa* E681. J. Microbiol. Biotechnol., 18: 1235–1244.
Pierre, J., Bazin, M., Debey, P. and Santus, R. (1982) One-electron photoreduction of bacterial cytochrome P-450 by ultraviolet light. I. Steady-state irradiations. Eur. J. Biochem., 124: 533–537.
Pow, D.V. and Cook, D.K. (1997) Tryptophan is present in glial cells and photoreceptors in the chicken retina. Neuroreport, 8: 1767–1770.
Quiñones, M.A., Lu, Z. and Zeiger, E. (1998) Close correspondence between the action spectra for the blue light responses of the guard cell and coleoptile chloroplasts, and the spectra for blue light-dependent stomatal opening and coleoptile phototropism. Proc. Natl. Acad. Sci. USA, 93: 2224–2228.
Rao, M.L., Müller-Oerlinghausen, B., Mackert, A., Strebel, B., Stieglitz, R.D. and Volz, H.P. (1992) Blood serotonin, serum melatonin and light therapy in healthy subjects and in patients with nonseasonal depression. Acta Psychiatr. Scand., 86: 127–132.
Raymond, J.R., Kim, J., Beach, R.E. and Tisher, C.C. (1993) Immunohistochemical mapping of cellular and subcellular

distribution of 5-HT1A receptors in rat and human kidneys. Am. J. Physiol., 264: F9–F19.

Ringvold, A., Anderssen, E. and Kjønniksen, I. (2000) UV absorption by uric acid in diurnal bird aqueous humor. Invest. Ophthalmol. Vis. Sci., 41: 2067–2069.

Roots, B.I. (1981) Comparative studies on glial markers. J. Exp. Biol., 95: 167–180.

Rubinstein, B. and Stern, A.I. (1991) The role of plasma membrane redox activity in light effects in plants. J. Bioenerg. Biomembr., 23: 393–408.

Ruiz-González, M.X. and Marín, I. (2004) New insights into the evolutionary history of type 1 rhodopsins. J. Mol. Evol., 58: 348–358.

Sablin, S.O., Yankovskaya, V., Bernard, S., Cronin, C.N. and Singer, T.P. (1998) Isolation and characterization of an evolutionary precursor of human monoamine oxidases A and B. Eur. J. Biochem., 253: 270–279.

Salmoun, M., Devijver, C., Daloze, D., Braekman, J.C. and van Soest, R.W. (2002) 5-Hydroxytryptamine-derived alkaloids from two marine sponges of the genus Hyrtios. J. Nat. Prod., 65: 1173–1176.

Scheibel, M.S., Tomiyasu, U. and Scheibek, A.B. (1975) Do raphe nuclei of the reticular formation have a neurosecretory or vascular sensor function? Exp. Neurol., 47: 316–329.

Schultes, R.E. and Hofmann, A. (1973) The Botany and Chemistry of Hallucinogens. Charles C. Thomas Publisher, Springfield, IL.

Shaw, D.M., Camps, F.E. and Eccleston, E.G. (1967) 5-Hydroxytryptamine in the hind-brain of depressive suicides. Br. J. Psychiatry, 113: 1407–1411.

Shih, C., Museth, A.K., Abrahamsson, M., Blanco-Rodriguez, A.M., Di Bilio, A.J., Sudhamsu, J., Crane, B.R., Ronayne, K.L., Towrie, M., Vlcek, A. Jr., Richards, J.H., Winkler, J.R. and Gray, H.B. (2008) Tryptophan-accelerated electron flow through proteins. Science, 320: 1760–1762.

Smith, B.N. (1976) Evolution of C4 photosynthesis in response to changes in carbon and oxygen concentrations in the atmosphere through time. Biosystems, 8: 24–32.

Smith, T.A. (1971) The occurrence, metabolism and functions of amines in plants. Biol. Rev. Camb. Philos. Soc., 46: 201–241.

Sosa, M.A., Spitzer, N., Edwards, D.H. and Baro, D.J. (2004) A crustacean serotonin receptor: cloning and distribution in the thoracic ganglia of crayfish and freshwater prawn. J. Comp. Neurol., 473: 526–537.

Sparks, D.L. and Slevin, J.T. (1985) Determination of tyrosine, tryptophan and their metabolic derivatives by liquid chromatography-electrochemical detection: application to post mortem samples from patients with Parkinson's and Alzheimer's disease. Life Sci., 36: 449–457.

Steen, H.B. (1967) On the luminescence of L-tryptophane and L-tyrosine in aqueous solution at 77 degrees K induced by x-rays and UV light. Photochem. Photobiol., 6: 805–816.

Stern, K.R., Jansky, S., Bidlack, J.E. and Stern, K. (2002) Introductory Plant Biology. 9th edn. McGraw-Hill, New York, NY.

Stewart, K.T., Gaddy, J.R., Byrne, B., Miller, S. and Brainard, G.C. (1991) Effects of green or white light for treatment of seasonal depression. Psychiatry Res., 38: 261–270.

Strakhovskaia, M.G., Serdalina, A.M. and Fraikin, GIa (1983) [Effect of the photo-induced synthesis of serotonin on the photoreactivation of *Saccharomyces cerevisiae* yeasts]. Nauchnye Doki Vyss Shkoly Biol Nauki, 3: 25–28.

Sugden, D., Davidson, K., Hough, K. and The, M. (2004) Melatonin, melatonin receptors and melanophores: a moving story. Pigment Cell Res., 17(5): 454–460.

Swiecicki, L., Bidzinski, A. and Tonderska, A. (2005) [Platelet serotonin transport in the group of outpatients with seasonal affective disorder before and after light treatment, and in remission (in the summer)]. Psychiatr. Pol., 39(3): 459–468.

Tajuddin, N.F., Orrico, L.A., Eriksen, J.L. and Druse, M.J. (2003) Effects of ethanol and ipsapirone on the development of midline raphe glial cells and astrocytes. Alcohol, 29: 157–164.

Tsuchioka, M., Takebayashi, M., Hisaoka, K., Maeda, N. and Nakata, Y. (2008) Serotonin (5-HT) induces glial cell line-derived neurotrophic factor (GDNF) mRNA expression via the transactivation of fibroblast growth factor receptor 2 (FGFR2) in rat C6 glioma cells. J. Neurochem., 106: 244–257.

Valentine, J.W. (2004) On the Origin of Phyla. University of Chicago Press, Chicago, IL.

Vavilin, D.V., Ermakova-Gerdes, S.Y., Keilty, A.T. and Vermaas, W.F. (1999) Tryptophan at position 181 of the D2 protein of photosystem II confers quenching of variable fluorescence of chlorophyll: implications for the mechanism of energy-dependent quenching. Biochemistry, 38: 14690–14696.

Venn, A.A., Loram, J.E. and Douglas, A.E. (2008) Photosynthetic symbioses in animals. J. Exp. Bot., 59: 1069–1080.

Videbech, T. (1975) The psychopathology of anancastic endogenous depression. Acta Psychiatr. Scand., 52: 336–373.

Walker, R.J., Brooks, H.L. and Holden-Dye, L. (1996) Evolution and overview of classical transmitter molecules and their receptors. Parasitology, 113(Suppl.): S3–S33.

Wallace, J.A. and Lauder, J.M. (1983) Development of the serotonergic system in the rat embryo: an immunocytochemical study. Brain Res. Bull., 10: 459–479.

Wang, H., Lin, S., Allen, J.P., Williams, J.C., Blankert, S., Laser, C. and Woodbury, N.W. (2007) Protein dynamics control the kinetics of initial electron transfer in photosynthesis. Science, 316: 747–750.

Wasserman, S.I. (1980) The lung mast cell: its physiology and potential relevance to defense of the lung. Environ. Health Perspect., 35: 153–164.

Weiger, W.A. (1997) Serotonergic modulation of behaviour: a phylogenetic overview. Biol. Rev. Camb. Philos. Soc., 72: 61–95.

Westrin, A. and Lam, R.W. (2007) Seasonal affective disorder: a clinical update. Ann. Clin. Psychiatry, 19: 239–246.

Whitaker, P.M., Warsh, J.J., Stancer, H.C., Persad, E. and Vint, C.K. (1984) Seasonal variation in platelet 3H-imipramine binding: comparable values in control and depressed populations. Psychiatry Res., 11: 127–131.

Whitaker-Azmitia, P.M. and Azmitia, E.C. (1986) [3H]5-hydroxytryptamine binding to brain astroglial cells: differences between intact and homogenized preparations and mature and immature cultures. J. Neurochem., 46: 1186–1189.

Whitaker-Azmitia, P.M., Lauder, J.M., Shemmer, A. and Azmitia, E.C. (1987) Postnatal changes in serotonin receptors following prenatal alterations in serotonin levels: further evidence for functional fetal serotonin receptors. Brain Res., 430: 285–289.

Wiens, M., Koziol, C., Batel, R. and Muller, W.E. (1998) Phenylalanine hydroxylase from the sponge *Geodia cydonium*: implication for allorecognition and evolution of aromatic amino acid hydroxylases. Dev. Comp. Immunol., 22: 469–478.

Wiesler, B., Wang, Q.Y. and Nick, P. (2002) The stability of cortical microtubules depends on their orientation. Plant J., 32: 1023–1032.

Wikgren, M.C. and Reuter, M. (1985) Neuropeptides in a microturbellarian – whole mount immunocytochemistry. Peptides, 6(Suppl. 3): 471–475.

Willeit, M., Sitte, H.H., Thierry, N., Michalek, K., Praschak-Rieder, N., Zill, P., Winkler, D., Brannath, W., Fischer, M.B., Bondy, B., Kasper, S. and Singer, E.A. (2008) Enhanced serotonin transporter function during depression in seasonal affective disorder. Neuropsychopharmacology, 33: 1503–1513.

Wilmanns, M., Hyde, C.C., Davies, D.R., Kirschner, K. and Jansonius, J.N. (1991) Structural conservation in parallel beta/alpha-barrel enzymes that catalyze three sequential reactions in the pathway of tryptophan biosynthesis. Biochemistry, 30: 9161–9169.

Woolley, D.W. (1961) The Biochemical Basis of Psychosis. John Wiley and Sons, Inc, New York, NY.

Yaguchi, S. and Katow, H. (2003) Expression of tryptophan 5-hydroxylase gene during sea urchin neurogenesis and role of serotonergic nervous system in larval behavior. J. Comp. Neurol., 466: 219–229.

Yan, W., Wilson, C.C. and Haring, J.H. (1997) Effects of neonatal serotonin depletion on the development of rat dentate granule cells. Brain Res. Dev. Brain Res., 98: 177–184.

Yerevanian, B.I., Anderson, J.L., Grota, L.J. and Bray, M. (1986) Effects of bright incandescent light on seasonal and nonseasonal major depressive disorder. Psychiatry Res., 18: 355–364.

Young, S.N. and Leyton, M. (2002) The role of serotonin in human mood and social interaction. Insight from altered tryptophan levels. Pharmacol. Biochem. Behav., 71: 857–865.

Zhao, J. and Last, R.L. (1995) Immunological characterization and chloroplast localization of the tryptophan biosynthetic enzymes of the flowering plant Arabidopsis thaliana. J. Biol. Chem., 270(11): 6081–6087.

Zhou, B., Zhang, Z., Zhang, Y., Li, R., Xiao, Q., Liu, Y., Li, Z., 2008. Binding of cationic porphyrin to human serum albumin studied using comprehensive spectroscopic methods. J. Pharm. Sci., in press.

Zigman, S. and Hare, J.D. (1976) Inhibition of cell growth by near ultraviolet light photoproducts of tryptophan. Mol. Cell Biochem., 10: 131–135.

Zimmer, W., Hundeshagen, B. and Niederau, E. (1994) Demonstration of the indolepyruvate decarboxylase gene homologue in different auxin-producing species of the Enterobacteriaceae. Can. J. Microbiol., 40: 1072–1076.

CHAPTER 1.2

Genetic Organization of the Serotonergic System

Ursula M. D'Souza* and Ian W. Craig

MRC Social, Genetic and Developmental Psychiatry (SGDP) Centre, Institute of Psychiatry, King's College London, London, UK

Abstract: The serotonergic system consists of a number of genes and includes seven subfamilies of serotonin receptors (*5-HT$_{1-7}$*), with each subfamily revealing several subtypes. The system also includes tryptophan hydroxylase (*TPH-1* and *TPH-2*) and monoamine oxidase (*MAO*) enzymes which are involved in synthesis and degradation of serotonin, respectively. Furthermore, serotonin transporter (*SERT, 5-HTT*) is an important protein in the scheme as it controls the amount of serotonin in the synapse. Brain-derived neurotrophic factor (*BDNF*) is included in this system, as it is known to modulate serotonin transporter function. Each of the genes and/or gene family is described in terms of their exon–intron organization, chromosomal localization, any alternative splice variants, and 5′-flanking and promoter regions. Interestingly, most of these genes have been implicated in neuropsychiatric disorders; therefore, a discussion has been provided on significant functional polymorphisms that are associated with the phenotypes of interest. In general, all genes of the serotonergic system are organized into exons and introns except for the 5-HT$_1$ family of serotonin receptors, which are intronless and encoded by a single exon. Alternative splice forms have been identified in the *5-HT$_{2C}$*, *5-HT$_{3A}$*, *5-HT$_4$*, *5-HT$_7$*, *TPH-1*, *5-HTT* and *BDNF* genes, whereas RNA editing isoforms are only observed with the *5-HT$_{2C}$* receptor gene. The promoter and 5′-flanking regions have been characterized for the *5-HT$_{1A}$*, *5-HT$_{2A}$*, *5-HT$_{2C}$*, *5-HT$_{3B}$*, *5-HT$_4$*, *5-HT$_7$*, *TPH-1*, *TPH-2*, *5-HTT*, *MAOA* and *BDNF* genes, with transcription factors (DNA binding proteins) identified in some of the genes to further elucidate transcription regulation. Evidently, the diverse and complex genetic structure/organization and regulation of the serotonin system is associated with the key actions as an important monoamine neurotransmitter in the central nervous system.

Keywords: serotonin, gene structure, transcriptional, gene expression, alternative splice variants.

Introduction

Serotonin (5-hydroxytryptamine; 5-HT) is a biogenic monoamine and an important neurotransmitter/neuromodulator in the central and peripheral nervous system, where it plays a role in several behavioral functions, including mood, sleep cycles, aggression, appetite and learning (Veenstra-VanderWeele *et al.*, 2000; Mohammad-Zadeh *et al.*, 2008). It has been implicated in a wide range of neuropsychiatric conditions, including depression, anxiety disorders, obsessive-compulsive disorders, psychosis, eating disorders, and substance abuse and dependence (Lucki, 1998; Lesch, 2001). Serotonin is produced in two steps; first, the essential amino acid tryptophan is hydroxlyated to 5-hydroxytryptophan (5-HTP) by tryptophan hydroxylase (TPH). In the second step, 5-HTP is decarboxylated to form 5-HT by the enzyme L-aromatic amino acid decarboxylase. The primary catabolic pathway for serotonin is by the enzyme monoamine oxidase (MAO). This is a ubiquitous enzyme that exists in two major forms, MAO-A and MAO-B. Both are present in the brain, and the latter preferentially in serotonergic neurons; however, MAO-A has a higher affinity for serotonin as substrate (see, for example, Shih and Thompson 1999; see also Chapter 2.4 of this volume). Furthermore, the serotonin transporter (5-HTT) which is located on the presynaptic membrane is important in controlling the availability of serotonin in the synaptic cleft.

The diverse effects of serotonin are mediated by several 5-HT receptors. It was originally proposed by Gaddum and Picarelli that these receptors were broadly divided into 'D' and 'M' groups, based on the differential sensitivity of guinea-pig ileum 5-HT-induced contraction to some drugs. However, a new classification was later derived based on radio-ligand binding studies using agonists and antagonists (Bradley *et al.*, 1986; Peroutka, 1998). They proposed that there are three main groups, namely 5-HT$_1$, 5-HT$_2$ and 5-HT$_3$, each recognizing different synthetic

*Corresponding author
E-mail: spjgumd@iop.kcl.ac.uk

ISBN 978-0-12-374634-4
DOI: 10.1016/B978-0-12-374634-4.00002-2

ligands. However, with the discovery of molecular biology approaches, the nomenclature has evolved into an accepted classification which indicates seven subfamilies of serotonin receptors (5-HT_{1-7}), based on their structural and functional properties, with each subfamily comprising specific subtypes (Hoyer *et al.*, 1994, 2002; see also Chapter 1.6 of this volume). The complicated history of the discovery of genes and identification of ligand-binding sites of some of these subtypes resulted in a cumbersome and confusing nomenclature where different researchers used different terminologies. Thus the serotonin Club Nomenclature Committee revised the nomenclature to enable consistency and consolidation. To this effect, the 5-$HT_{1D\alpha}$ receptor subtype was renamed the 5-HT_{1D} receptor and the 5-$HT_{1D\beta}$ receptor renamed the 5-$HT_{1\beta}$ receptor. Additionally, the 5-HT_{1C} receptor was renamed 5-HT_{2C}, the 'classical' 5-HT_2 receptor known as the 5-HT_{2A} receptor, and the 5-HT_{2F} receptor called the 5-HT_{2B} receptor. For the sake of clarity, the chapter will use the current names of these particular receptor subtypes.

This chapter aims to describe the organization of the genes mentioned above that encompass the serotonergic system. Each gene and gene family will be described in separate sections detailing its gene structure (exon–intron organization), chromosomal localization, splice variants, 5′-flanking and promoter regions, and any significant functional polymorphisms that are associated with psychiatric disorders. Additionally, information on the distribution and second messenger responses of the serotonergic genes will only be covered in brief, as these topics have been reviewed extensively elsewhere (Peroutka, 1998; Barnes and Sharp, 1999; Veenstra-VanderWeele *et al.*, 2000; Lesch, 2001; Hoyer *et al.*, 2002; Cowen, 2007).

In general, the exon (coding regions) and intron (noncoding DNA sequences) boundaries of a gene are elucidated using molecular biology techniques involving cDNA and DNA cloning, and readers are referred to standard textbooks on this topic (Brown, 1999). Following the isolation of a gene, it is necessary to characterize its 5′-flanking and promoter region(s) to determine the transcription initiation site and the location of any exons in the untranslated region, and to identify any upstream regulatory sequence. Several molecular biological methods are used for these analyses and include 5′ Rapid Amplification of cDNA ends (5′ RACE), genomic walking, primer extension, RNase protection and S1 nuclease assays (Carey and Smale, 2000). Strategies for elucidating control regions involve the generation of serial 5′ deletion plasmid constructs fused with a reporter gene such as luciferase or chloramphenicol acetyl transferase (CAT). These constructs are transiently transfected into mammalian cell lines to determine transcriptional activity of the deletion fragments *in vitro*, or they can be used to generate transgenic mice and determine *in vivo* effects at a brain and behavioral level. Furthermore, electrophoretic mobility assays, DNase I footprinting and yeast one-hybrid studies have focused on identifying which DNA binding proteins (transcription factors) interact within the regulatory domains of several of the target genes.

The topics on gene structure, transcriptional gene regulation and gene expression have been reported comprehensively, and thus are not described here (Beebee and Burke, 1992; Mouradian *et al.*, 1994; Latchman, 1998; Carey and Smale, 2000; Brown and Brown, 2002; D'Souza *et al.*, 2003). Recent discoveries on other levels of gene regulation have been published, and include RNA interference, which is involved in gene silencing and noncoding RNA (ncRNA), or small RNAs that play a role in several pathways that also silence genes (Mattick and Makunin, 2006; Chu and Rana, 2007). Moreover, as DNA is packaged into a nucleoprotein complex known as chromatin, understanding this structure in addition to histone acetylation and DNA methylation as epigenetic modifications associated with heritable changes in gene expression has become a significant topic of current research (Vaissiere *et al.*, 2008).

Subfamilies of 5-HT receptors

5-HT_1 receptor genes

The 5-HT_1 class of serotonin receptors consists of five receptor types with 5-HT_{1A}, 5-HT_{1B} and 5-HT_{1D} characterized by a high affinity for 5-carboxamido-tryptamine, whereas 5-HT_{1E} and 5-HT_{1F} are characterized by a low affinity for this synthetic agonist (Lanfumey and Hamon, 2004). All five subtypes have nanomolar affinity for the endogenous ligand indolamine and couple preferentially, but not exclusively, to $G_{i/o}$ to inhibit adenylate cyclase (Hoyer *et al.*, 2002). The genes encoding the 5-HT_1 receptors have been cloned in both rodents and humans, and found to belong to the superfamily of G-protein coupled receptors having the characteristic seven transmembrane domains. The *5-HT_{1A}* receptor gene was one of the first of the family to be cloned in rat (Albert *et al.*, 1990; Fujiwara *et al.*, 1990) and in human (Kobilka *et al.*, 1987; Fargin *et al.*, 1988; Stam *et al.*, 1992; Chanda *et al.*, 1993). It was found to be intronless, and encoded by a single exon. The receptor gene is located on chromosome 5q11.2-q13 in humans (Kobilka *et al.*, 1987; Hoyer *et al.*, 1994) and on chromosome 13 in mice (Oakey *et al.*, 1991). The human and rat gene encodes predicted proteins of 421 and 422 amino acids, respectively, which share 89 percent homology at the amino acid level (Kobilka *et al.*, 1987; Fargin *et al.*, 1988; Albert *et al.*, 1990). The rat *5-HT$_{1A}$* receptor mRNA distribution in rat brain was comparable with the distribution pattern of the receptor's ligand binding sites, with highest mRNA expression in septum and hippocampus (Albert *et al.*, 1990). In general the 5-HT$_{1A}$ receptors are widely distributed in the brain, and in the raphe nuclei they act as autoreceptors to inhibit cell firing, whereas in limbic structures, especially the hippocampus, they serve as postsynaptic receptors (reviewed in Hoyer *et al.*, 2002; see also Chapter 1.6 of this volume). The primary signal transduction pathway of the 5-HT$_{1A}$ receptor is via inhibition of adenylyl cyclase (reviewed in Raymond *et al.*, 1999; Lesch, 2001). However, there is evidence showing that activation of the receptor leads to increased potassium conductance, and in some cell types to an increase in intracellular calcium (reviewed in Cowen, 2007).

The 5′-flanking and promoter regions of the mouse and human *5-HT$_{1A}$* receptor gene have been found to lack TATA box elements, but are rich in guanine and cytosine residues (Parks and Shenk, 1996). These 5′-flanking regions have 63 percent sequence identity, and consist of a guanine-cytosine-rich DNA sequence motif that interacts with the MAZ (Myc associated zinc finger) (Pur-1, Zif87) protein at four sites in HeLa (human cervical adenocarcinoma) cell nuclear extracts. However, three of the four MAZ binding sites were also shown to interact with the transcription factor Sp1, suggesting that MAZ and Sp1 both participate in regulating expression from the *5-HT$_{1A}$* receptor gene promoter (Parks and Shenk, 1996). Further analysis of the murine *5-HT$_{1A}$* receptor gene promoter *in vitro* and *in vivo* was undertaken in which upstream sequences extending to −4.5, −5.5 and −20 kb were examined for activity in cell culture and transgenic animals (Ansorge *et al.*, 2004). These promoter fragments were found to be active in *5-HT$_{1A}$* receptor mRNA-positive cells (neuronal cell line, mouse neuroblastoma/rat glioma hydrid NG108-15) and inactive in *5-HT$_{1A}$* receptor mRNA-negative cells (neuronal-like cell line, PC12 (rat adrenal pheochromocytoma) and non-neuronal cell line, COS-7 (African green monkey kidney, SV40 transformed). In adult mice these promoter fragments directed expression of the gene in specific brain regions and in cells with endogenous *5-HT$_{1A}$* receptor gene expression.

Genetic studies have focused on identifying a polymorphism, which is the existence of two or more alleles at significant frequencies (at least 1 percent) in the population and/or disease phenotypes, where an allele is one of several alternative forms of a gene or DNA sequence at a specific chromosomal location (locus) (see Strachan and Read, 1996; Plomin *et al.*, 2000). A functional SNP (single nucleotide polymorphism) (C-1019G) in the promoter region of the *5-HT$_{1A}$* receptor gene has been found to be associated with depression, suicide and panic disorder (Arias *et al.*, 2002; Serretti *et al.*, 2004), and recently with higher geriatric depression scale (GDS) scores (Lenze *et al.*, 2008). In serotonergic raphe RN46A cells, the human transcription factor Deaf-1 (nuclear deformed epidermal autoregulatory factor-1) repressed *5-HT$_{1A}$* receptor gene expression at the −1019C allele but not the −1019G allele. However, in non-serotonergic cells (septal SN48, neuroblastoma SKN-SH and neuroblastoma/glioma NG108-15 cell lines) that express *5-HT$_{1A}$* receptors Deaf-1 enhanced *5-HT$_{1A}$* promoter activity at the −1019C allele but not the −1019G allele (Czesak *et al.*, 2006). These results suggest that cell-specific regulation by Deaf-1 could underlie region-specific alterations in *5-HT$_{1A}$* receptor expression in different mood disorders.

Transcriptional initiation and regulation of the rat serotonin *5-HT$_{1A}$* receptor gene was characterized revealing a major site of transcriptional initiation located 58 bp downstream from a consensus TATA element (Storring *et al.*, 1999). This study also showed, using *5-HT$_{1A}$*-negative (pituitary GH4C1, L6 myoblast and C6 glioma) and *5-HT$_{1A}$*-positive (murine SN-48 and rat raphe RN46A) cell lines, that TATA-driven expression of the *5-HT$_{1A}$* receptor gene is regulated by a novel proximal tissue-specific enhancer region, a non-selective promoter and a distinct upstream repressor region. Furthermore, a 14-bp element was identified in the 5′-flanking region of the *5-HT$_{1A}$* receptor gene that mediates transcriptional repression in raphe cells (5-HT$_{1A}$ receptor expressing cells), and an additional 12-bp sequence identified in *5-HT$_{1A}$* receptor-negative cells that represses the expression of the gene (Ou *et al.*, 2000). The novel DNA binding protein Freud-1 (5′ repressor element under dual repression binding protein-1) was found to bind to the 14-bp 5′ repressor element of the *5-HT$_{1A}$* receptor gene and regulate the gene in neurons in a calcium-dependent manner (Ou *et al.*, 2003). More recently, there is evidence demonstrating that the long isoform of Freud-1 is the major isoform that regulates the human *5-HT$_{1A}$* receptor gene (Rogaeva and Albert, 2007). Furthermore, previous studies have shown that the rat *5-HT$_{1A}$* receptor gene is regulated by corticosteroids (Meijer *et al.*, 2000; Wissink *et al.*, 2000).

The classification of the 5-HT$_{1B}$ and 5-HT$_{1D}$ receptor subtypes was necessary following the discovery of related genes within the 5-HT$_{1}$ receptor family, and reveals a complicated history (Hartig *et al.*, 1996). This was due to the identification of species homologs of a gene (the equivalent gene in different species) which showed high homology at the amino acid level but displayed different pharmacological properties. On the contrary, there was further complexity when two different genes in the same species (intraspecies subtypes) showed only moderate homology at the amino acid level but displayed indistinguishable pharmacological properties (Hartig *et al.*, 1992). These receptors will be described under the subsequent section here on the 5-HT_{1B} receptors, and also under the subdivision reporting on the 5-HT_{1D} receptors below. The *$5\text{-}HT_{1B}$* receptor gene was cloned and characterized from rat brain, revealing a cDNA that contains an 1158-bp open reading frame which encodes a predicted protein having a molecular weight of 43,162 (Voigt *et al.*, 1991). The primary sequence of the gene indicated that it belongs to the 5-HT_1 receptor family, which was confirmed by pharmacological studies. The nucleotide sequence of the rat *$5\text{-}HT_{1B}$* receptor gene had 45 percent overall identity with the rat *$5\text{-}HT_{1A}$* receptor gene, and up to 60 percent identity when only the transmembrane domains are considered. *In situ* hybridization studies revealed mRNA expression of the *$5\text{-}HT_{1B}$* receptor in cells of the dorsal and median raphe nuclei, in agreement with previous findings indicating that this receptor acts as an autoreceptor on 5-HT terminals (Voigt *et al.*, 1991; Hoyer *et al.*, 2002). Interestingly, mRNA expression of the gene in hippocampus, striatum, cortex and cerebellum suggests that the 5-HT_{1B} receptor may also have a postsynaptic role (Voigt *et al.*, 1991; Barnes and Sharp, 1999). However, additionally, there is the suggestion that the 5-HT_{1B} receptor acts as a heteroreceptor, i.e., a modulatory receptor located on non-5HT terminals (reviewed in Barnes and Sharp, 1999).

The rat *$5\text{-}HT_{1B}$* receptor gene was also isolated through homology to a human *$5\text{-}HT_{1D\beta}$* reeceptor clone, and comparison of the amino acid sequence of the two genes showed a 93 percent overall identity and a 96 percent identity in the transmembrane regions, as well as a strong conservation in the intracellular and extracellular loops (Adham *et al.*, 1991). Additionally, this rat clone encoded a receptor with pharmacological properties matching the human *$5\text{-}HT_{1D}$* receptor gene, suggesting the *$5\text{-}HT_{1B}$* receptor gene is the rat homolog of the human *$5\text{-}HT_{1D\beta}$* receptor gene (see below under the section regarding *$5\text{-}HT_{1D}$* receptor genes). Cloning of the human *$5\text{-}HT_{1B}$* receptor gene revealed that it has characteristics of a G-protein coupled receptor and is localized on chromosome 6 at 6q13 (Jin *et al.*, 1992). The gene contains an intronless open reading frame of 1170 nucleotides which encodes a 390 amino acid protein. The isolation of another genomic clone of the human *$5\text{-}HT_{1B}$* receptor gene revealed it is a homolog of the rat 5-HT_{1B} receptor with pharmacological properties of 5-HT_{1D} receptors (Hamblin *et al.*, 1992). Thus a new nomenclature was proposed by the International Union of Pharmacology Committee on Receptor Nomenclature and Drug Classification (IUPHAR) and the previously known *$5\text{-}HT_{1D\beta}$* receptor was renamed the human 5-HT_{1B} receptor (Hartig *et al.*, 1996). The 5-HT_{1B} receptor negatively couples to adenylyl cyclase under forskolin-stimulated conditions in cell transfection studies (reviewed in Barnes and Sharp, 1999).

Evidence demonstrates that polymorphisms in the 5′ untranslated region of the human *$5\text{-}HT_{1B}$* receptor gene affects gene expression, and that some of these genetic variants are associated with several psychiatric conditions (Duan *et al.*, 2003). Recently, an A/G SNP polymorphism in the 3′UTR of the human *$5\text{-}HT_{1B}$* receptor gene was found to differentially alter transcriptional activity of a luciferase construct with the A allele being repressed by a microRNA (miR-96) in HeLa cells (Jensen *et al.*, 2008). It also appears to be associated with aggressive phenotypes (Jensen *et al.*, 2008).

The cloning of the originally named *$5\text{-}HT_{1C}$* receptor gene cDNA revealed that it belonged to the family of G-protein coupled receptors and has a distinct pharmacological profile (Lubbert *et al.*, 1987; Julius *et al.*, 1988). However, following the cloning of the 5-HT_2 receptor family the *$5\text{-}HT_{1C}$* receptor gene was reclassified as *$5\text{-}HT_{2C}$* (Humphrey *et al.*, 1993). Thus the *$5\text{-}HT_{1C}$* nomenclature is no longer used in the literature, and description of this particular gene is provided in the relevant section regarding *$5\text{-}HT_2$* receptor genes.

The cloning of the human *$5\text{-}HT_{1D}$* receptor gene showed that it contains an intronless open reading frame encoding a 377 amino acid polypeptide with characteristics of G-protein coupled receptors (Hamblin and Metcalf, 1991). The *$5\text{-}HT_{1D}$* receptor gene was localized to chromosome 1 by southern blot analysis of DNA from a rodent hybrid cell panel (Jin *et al.*, 1992). This first human *$5\text{-}HT_{1D}$* receptor gene was designated *$5\text{-}HT_{1D\alpha}$*, which was followed by the subsequent cloning of a second distinct gene *$5\text{-}HT_{1D\beta}$* receptor gene which was also intronless (Levy *et al.*, 1992a; Weinshank *et al.*, 1992). However, the human $5\text{-HT}_{1D\beta}$ receptor was redefined as the species homolog of the 5-HT_{1B} receptor, as its distribution and sequence were similar to the rat 5-HT_{1B} receptor. Subsequently, a rat gene was isolated which was homologous to the human $5\text{-HT}_{1D\alpha}$ receptor, having distinct pharmacological properties for a 5-HT_{1D} receptor, and thus the $5\text{-HT}_{1D\beta}$ receptor was renamed 5-HT_{1D} receptor (Hartig *et al.*, 1996). Both genes show strong similarity in sequence, pharmacological properities, inhibition of adenylate cyclase activity and their mRNA expressed in the cerebral cortex. Interestingly, both the receptors bind to sumatriptan (an anti-migraine drug) with high affinity (Weinshank *et al.*, 1992); however, genetic association studies do not provide evidence for the role of both genes in familial migraine (Racchi *et al.*, 2004) or clinical response to sumatriptan (Mehrotra *et al.*, 2007). Like the 5-HT_{1B} receptor, the 5-HT_{1D} receptor is located presynaptically on both serotonin and non-serotonin neurons (reviewed in Barnes and Sharp, 1999).

The human *5-HT* receptor gene was isolated in several studies (Levy *et al.*, 1992b; Zgombick *et al.*, 1992; Lovenberg *et al.*, 1993). The first isolated the gene from a human placental genomic library by using oligonucleotide probes derived from transmembrane regions of the cloned human $5\text{-}HT_{1D\beta}$ receptor (Zgombick *et al.*, 1992). The deduced amino acid of the isolated genomic clone was identical to the previously isolated but not characterized novel serotonin receptor gene S31 (Levy *et al.*, 1992b). The gene locates to human chromosome 6q14-q15 by *in situ* hybridization (Levy *et al.*, 1994). The nucleotide sequence of human $5\text{-}HT_{1E}$ receptor gene revealed 64 percent homology to the $5\text{-}HT_{1D\alpha}$ and $5\text{-}HT_{1D\beta}$ receptor subtypes in the transmembrane domain regions (Zgombick *et al.*, 1992). The rat $5\text{-}HT_{1E}$ receptor gene was cloned, revealing an intronless open reading frame encoding a 366 amino acid seven-transmembrane domain protein (Lovenberg *et al.*, 1993). Further characterization of the human homolog indicated its encoded protein to have 93 percent overall amino acid identity with the rat sequence. Within the transmembrane domain regions, the human gene showed 52, 59, 65 and 68 percent amino acid identity with the rat 5-HT_{1A}, rat 5-HT_{1B}, rat 5-HT_{1D} and human 5-HT_{1E} receptors, respectively. $5\text{-}HT_{1E}$ receptor mRNA is expressed in cortical regions and caudate and putamen, with lower levels detected in the amygdala and hypothalamic regions, similar to the receptors distribution in the brain, suggesting that the $5\text{-}HT_{1E}$ receptor is located postsynaptically where it mediates inhibits forskolin-stimulated adenylyl cyclase (reviewed in Barnes and Sharp, 1999).

The $5\text{-}HT_{1F}$ receptor gene was originally cloned in the mouse and expressed predominantly in hippocampus, with homology to the 5-HT_{1B}, 5-HT_{1D} and 5-HT_{1E} receptor subtypes (Amlaiky *et al.*, 1992). This was followed by cloning of the human gene, which was found to be intronless and encoded a protein 366 amino acids in length, having a relative molecular weight of 42,000, with distinct pharmacological properties compared with the other 5-HT_1 receptors. Its protein sequence had 70 percent homology with the $5\text{-}HT_{1E}$ receptor gene, 63 percent homology with the $5\text{-HT}_{1D\alpha}$ and 60 percent homology with the $5\text{-HT}_{1D\beta}$ receptors (Adham *et al.*, 1993). The mRNA expression of the gene was found in human brain but not present in other tissues such as kidney, liver, spleen, heart, pancreas and testes, and its specific mRNA distribution in guinea-pig coronal brain sections revealed expression in hippocampal pyramidal cells, in several thalamic nuclei and in the dorsal raphe nuclei, the latter results suggesting its possible role as an autoreceptor (Adham *et al.*, 1993; Hoyer *et al.*, 2002). The gene was localized to chromosome 3p12 using a monochromosomal mapping panel, followed by radiation-reduced hydrid mapping and fluorescent *in situ* hybridization (VanDenBrink *et al.*, 1998). Studies in cultured cells showed that the 5-HT_{1F} receptor couples to the inhibition of forskolin-stimulated adenylate cyclase (reviewed in Barnes and Sharp, 1999).

5-HT$_2$ receptor genes

The 5-HT_2 receptors belong to the superfamily of G-protein coupled receptors, and consists of three subtypes: 5-HT_{2A}, 5-HT_{2B} and 5-HT_{2C} (formerly known as the 5-HT_2, 5-HT_{2F} and 5-HT_{1C}, respectively) receptors with related molecular structure, amino acid sequence and signaling properties (Hoyer *et al.*, 1994; Leysen, 2004). They couple preferentially to $G_{q/11}$ to increase the hydrolysis of inositol phosphates and elevate intracellular calcium (Barnes and Sharp, 1999; Hoyer *et al.*, 2002). The $5\text{-}HT_{2A}$ receptor gene was cloned in rat (Prichett *et al.*, 1988; Julius *et al.*, 1990), hamster (Chambard *et al.*, 1990), mouse (Yang *et al.*, 1992) and human (Saltzman *et al.*, 1991; Stam *et al.*, 1992), and encodes a protein of 471 amino acids. The human $5\text{-}HT_{2A}$ gene is located on chromosome 13q14-q21 (Sparkes *et al.*, 1991), spans over 20 kb and consists of three exons separated by two introns (Chen *et al.*, 1992); it is 87 percent homologous with its rat counterpart. 5-HT_{2A} receptor is expressed in several regions in the brain, including cortical areas, caudate nucleus, nucleus accumbens, olfactory tubercle and hippocampus; these areas also correlate with its mRNA distribution, suggesting postsynaptic localization of the receptor in 5-HT neurons (reviewed in Barnes and Sharp, 1999).

The characterization of the rat $5\text{-}HT_{2A}$ promoter region revealed no TATA boxes or CCAAT sequences (Ding *et al.*, 1993; Du *et al.*, 1994) and the 5′-flanking region of human $5\text{-}HT_{2A}$ receptor gene showed multiple transcription initiation sites (Zhu *et al.*, 1995; Shih *et al.*, 1996) with similar initiation sites found in the mouse and rat genes (Chen *et al.*, 1992; Ding *et al.*, 1993, Du *et al.*, 1994). The promoter activity in the human gene was observed in a fragment that encompassed the transcription initiation sites when transfected into 5-HT_{2A} positive cell lines (SH-SY5Y (human neuroblastoma) and HeLa). Transcription factors Sp1, PEA3 (polyomavirus enhancer activator 3) and E-box binding proteins and two novel transcription factors were found to interact with this DNA fragment (Zhu *et al.*, 1995). PEA3 is the founding member of a subfamily of *ets* oncogenes, whereas the E-box proteins belong to a family of transcription factors that contain a basic helix–loop–helix (bHLH) domain.

Furthermore, two alternative promoters were present with a silencer region just downstream of the second promoter (Zhu *et al.*, 1995).

Polymorphisms in the *5-HT$_{2A}$* receptor gene have been found to be associated with schizophrenia and impulsive behavior (reviewed previously D'Souza and Craig, 2008; Nomura and Nomura, 2006). One of the gene variants is a SNP (A-1438G) located close to the promoter region of the gene, and this was investigated further for functional effects in our previous study (Parsons *et al.*, 2004). The transcriptional activity of the G allele was significantly lower than the A allele when the more downstream promoter was cloned into a reporter gene construct in the presence of an enhancer region (Parsons *et al.*, 2004). The effect was only seen in endogenous *5-HT$_{2A}$* receptor gene expressing SH-SY5Y and HeLa cells. However, no differences were found in allele-specific mRNA expression human cortical tissue (Bray *et al.*, 2004). Furthermore, even though the *5-HT$_{2A}$* receptor gene was found to be differentially expressed in several human *post-mortem* brain regions, the dose-dependent effect on gene expression in relation to the A-1438G SNP did not reach significance (Sugden *et al.*, 2009).

The *5-HT$_{2B}$* receptor gene was first cloned from stomach fundus in mouse (Foguet *et al.*, 1992a) and in rat (Foguet *et al.*, 1992b), the latter gene encoding a protein of 465 amino acids. This was followed by the cloning and characterization of the human *5-HT$_{2B}$* receptor gene (Kursar *et al.*, 1994; Schmuck *et al.*, 1994). The human *5-HT$_{2B}$* receptor cDNA isolated from a library derived from SH-SY5Y cells consisted of an open reading frame encoding 483 amino acids. It had an overall sequence homology of 80 percent and 90 percent within the membrane spanning regions when compared with the sequence of the rat gene (Schmuck *et al.*, 1994). Two introns were found in the human *5-HT$_{2B}$* gene in the same locations of the introns in the mouse *5-HT$_{2B}$* receptor subtypes (Foguet *et al.*, 1992a). However, the human *5-HT$_{2B}$* receptor gene isolated from a human placental genomic library had a single open reading frame encoding a protein of 481 amino acids showing 82 percent overall identity and 91.5 percent identity within the transmembrane domains at the amino acid level compared with the rat 5-HT$_{2B}$ receptor (Kursar *et al.*, 1994). The human *5-HT$_{2B}$* receptor gene was located to chromosome 2q36.3–2q37.1 (Horton *et al.*, 1996; Barnes and Sharp, 1999). The mRNA and protein of the *5-HT$_{2B}$* gene is detected at low levels in the brain compared with that of the 5-HT$_{2A}$ and 5-HT$_{2C}$ receptors; however, its immunostaining is restricted to cerebellum, lateral septum, dorsal hypothalamus and medial amygdala, suggesting a potential functional significance (reviewed in Barnes and Sharp, 1999).

As mentioned in the previous section, 5-HT$_{2C}$ was originally 5-HT$_{1C}$ in the older nomenclature, as they displayed similar pharmacological properties and second messenger systems, indicating that they were structurally related (Lubbert *et al.*, 1987). The *5-HT$_{2C}$* receptor gene was partially cloned in mice (Lubbert *et al.*, 1987), which subsequently led to isolation and sequencing of the full-length clone in rat (Julius *et al.*, 1988), in mouse (Yu *et al.*, 1991) and in human (Saltzman *et al.*, 1991). *5-HT$_{2C}$* receptor gene is different among the members of the 5-HT$_2$ receptor family, as it has a different genomic organization with three introns which separate the coding sequence into four exons. The first two introns are at equivalent positions as compared to the introns previously found in the *5-HT$_{2A}$* receptor gene, suggesting a close relationship between the two genes (Stam *et al.*, 1994). The 5-HT$_{2C}$ receptors are widely distributed in cortex, the limbic system and basal ganglia, where they are located postsynaptically; however, there are suggestions it may have a role presynaptically in projections of the habenula (reviewed in Barnes and Sharp, 1999). The *5-HT$_{2C}$* receptor gene was found to map to the human Xq24 chromosome and to mouse chromosome X D-X (reviewed in Barnes and Sharp, 1999). Interestingly, an alternatively splice variant of the 5-HT$_{2C}$ receptor has been identified in the rat, mouse and human, and termed 5-HT$_{2C-tr}$ (Canton *et al.*, 1996; Xie *et al.*, 1996). The sequence of this alternatively spliced form revealed a deletion of the region that codes for the putative second intracellular loop and the fourth transmembrane domain of the 5-HT$_{2C}$ receptor. This deletion results in a frameshift and premature termination, generating a protein of 248 amino acids. The mRNA of the short form is expressed in the same brain regions as the *5-HT$_{2C}$* receptor gene, including the striatum, hippocampus, hypothalamus and olfactory tubercle (Canton *et al.*, 1996); however, its expression is higher in choroid plexus tumors than in normal brain tissue (Xie *et al.*, 1996). Additionally, the *5-HT$_{2C}$* receptor gene is subject to other post-transcriptional modifications such as RNA editing, in which up to five adenosine residues are converted to inosines by double-stranded RNA adenosine deaminase(s) in the region of the receptor that codes for the second intracellular loop to produce up to 24 isoforms in brain with distinct structures and functions (Burns *et al.*, 1997; Fitzgerald *et al.*, 1999; Niswender *et al.*, 1999; Wang *et al.*, 2000; Werry *et al.*, 2008). Interestingly, evidence has shown that editing and splicing processes are correlated, and strongly suggests that RNA editing modulates alternative splicing in cell culture and human brain tissue (Flomen *et al.*, 2004).

Transcriptional regulation of the human *5-HT$_{2C}$* receptor gene has revealed its promoter region to be TATA-less with multiple initiation sites and a 7.3-kb 5′-flanking region which demonstrates reporter gene transcription activity in SK-N-SH and IMR32 neuroblastoma cells (Xie *et al.*, 1996). The *5-HT$_{2C}$* receptor gene has been associated with obesity (Yuan *et al.*, 2000) and reduced weight gain in subjects treated with antipsychotics (Reynolds *et al.*, 2002) in genetic studies where polymorphisms within the promoter region of the gene have been identified, including the −995G/A, −759C/T, −697G/C and GTn microsatellite (dinucleotide polymorphism) at −1027 bp (Yuan *et al.*, 2000; Reynolds *et al.*, 2002). Two of the haplotypes each consisting of a particular allele of the upstream GTn microsatellite (Z-6) have higher expression relative to the wild type in a mouse embryonal carcinoma (P19) cell line (Yuan *et al.*, 2000). Additionally, haplotypes containing the −759C allele showed reduced transcriptional activity compared with haplotypes containing the −759T allele in HEK293 (human embryonic kidney) and TE671 (human medulloblastoma) cell lines (Buckland *et al.*, 2005). Furthermore, haplotypes with either −759T or −697C alleles reduced promoter activity of the gene in SH-SY5Y human neuroblastoma cells (Hill and Reynolds, 2007). However, more recently the GTn microsatellite polymorphism was found not to affect mRNA expression of the *5-HT$_{2C}$* receptor in *post-mortem* brain tissue (Sugden *et al.*, 2009). Other studies have shown that the *5-HT$_{2C}$* mRNA is significantly decreased in *post-mortem* brain from schizophrenics (Castensson *et al.*, 2003), and this change in expression is not related to promoter polymorphisms or neuroleptic drug treatment (Castensson *et al.*, 2005).

5-HT$_3$ receptor genes

The 5-HT$_3$ receptors belong to the cys-loop family of pentameric ligand-gated ion channels, whereas all other 5HT receptors are G-protein coupled receptors. These receptors share characteristic features with other members of the ion channel family, including a large extracellular domain containing a conserved cysteine loop, four hydrophobic transmembrane domains, a large intracellular loop between the third and fourth transmembrane regions, and an extracellular C-terminus (Reeves and Lummis, 2002). The ion channel itself is an oligomeric complex consisting of five subunits, and each subunit has a protein structure with four transmembrane domains and a long N-terminal extracellular loop (Niesler *et al.*, 2003; Tzvetkov *et al.*, 2007). To date, five homologous genes coding for the 5-HT$_3$ receptor subunits (*5-HT$_{3A}$*, *5-HT$_{3B}$*, *5-HT$_{3C}$*, *5-HT$_{3D}$* and *5-HT$_{3E}$*) have been identified in humans (Tzvetkov *et al.*, 2007). However, some of these genes have also been cloned in other species, including rodents. Heteromeric receptors composed of 5-HT$_{3A}$ and 5-HT$_{3B}$ subunits have been experimentally shown to form functional receptors, whereas the role of the C, D and E subunits remains unclear. The 5-HT$_3$ receptors are involved in the regulation of gastrointestinal tract motility in the periphery. In the brain, they play an important role in regulation of nausea, vomiting and mental functions, with 5-HT$_3$ receptor antagonists used in drug abuse and the treatment of withdrawal symptoms (Costall and Naylor, 2004)

The cloning of the murine *5-HT$_{3A}$* receptor gene revealed its organization in 9 exons separated by 8 introns spanning over 12 kb of DNA sequence (Uetz *et al.*, 1994; Werner *et al.*, 1994). Interestingly, the mouse *5-HT$_{3A}$* gene intron–exon organization demonstrated similarities to muscular and neuronal nicotinic acetylcholine receptor genes (Wada *et al.*, 1988). Additionally, an alternative use of exon 9 splice acceptor sites results in the deletion of 6 amino acids in the large intracellular loop between transmembrane 3 and transmembrane 4, therefore generating a short (m5-HT$_{3A(a)}$ receptor) and a long (m5-HT$_{3A(b)}$ receptor) isoform of the murine m5-HT$_{3A}$ receptor (Hope *et al.*, 1993; Uetz *et al.*, 1994; Werner *et al.*, 1994). The *m5-HT$_{3A(a)}$* mRNA expression is about five times higher than the *m5-HT$_{3A(b)}$* mRNA in both neuroblastoma and neuronal tissues (Werner *et al.*, 1994). The long and short isoforms of the *5-HT$_{3A}$* receptor gene were also identified in the rat (Isenberg *et al.*, 1993) and in the guinea pig (Lankiewicz *et al.*, 1998). The coding sequence of the human *5-HT$_{3A}$* receptor gene was initially identified by screening a neuroblastoma library (Maricq *et al.*, 1991), and subsequently its cDNA sequence was further determined in three studies which revealed differences in four amino acid residues and in the 5′ and 3′ non-coding regions (Belelli *et al.*, 1995; Miyake *et al.*, 1995; Lankiewicz *et al.*, 1998). Additionally, about 84 percent amino acid sequence identity was observed with the mouse *5-HT$_{3A}$* receptor gene (Belelli *et al.*, 1995; Miyake *et al.*, 1995). Further elucidation of the exon–intron organization of the human *5-HT$_{3A}$* receptor gene revealed the gene spans 14.5 kb, and the coding region are separated by 8 introns at identical positions to the murine *5-HT$_{3A}$* receptor gene (Bruss *et al.*, 2000). This confirmed the previously published human *5-HT$_{3A}$* receptor gene cDNA sequence (Miyake *et al.*, 1995). However, the alternative splice acceptor in intron 8 of the murine gene was not present in the human homolog, and all human introns except for intron 5 were longer than their murine counterparts (Bruss *et al.*, 2000). Other alternative splice variants have been identified in humans, revealing that a short h-HT$_{3A}$ form is generated through skipping of exon 6, whereas a long form is produced through continuation of intron 7 (Bruss *et al.*, 1988). The human homolog of the mouse *5-HT$_{3A}$* receptor gene was assigned to human chromosome 11 by hybridization to rodent–human hybrid cell lines (Uetz *et al.*, 1994) and regionally localized to chromosome 11q23.1–q23.2 using fluorescence *in situ* hybridization (Weiss *et al.*, 1995).

The human *5-HT$_{3B}$* receptor gene was cloned using fetal brain and kidney cDNA libraries and found to encode a protein of 441 amino acid residues showing 41 percent amino acid identity with the 5-HT$_{3A}$ receptor subunit (Davies *et al.*, 1999). The human *5-HT$_{3B}$* receptor gene was also cloned from the small intestine cDNA library and the nucleotide sequence revealed a single large open reading frame of 1326 bp with 44 percent amino acid identity to the 5-HT$_{3B}$ receptor (Dubin *et al.*, 1999). Interestingly, evidence for the heteromeric assemblies of 5-HT$_{3A}$ and 5-HT$_{3B}$ subunits showed distinct pharmacological properties of the 5-HT$_3$ receptor which would have an effect on 5-HT activation of neuronal excitability and transmitter release (Davies *et al.*, 1999; Dubin *et al.*, 1999). Northern blot analysis and *in situ* hybridization studies demonstrated that the human *5-HT$_{3B}$* mRNA is co-expressed with the *m5-HT$_{3A}$* transcripts in the amygdala, caudate, hippocampus and cerebral cortical areas of the brain, which are also enriched with 5-HT$_3$ receptor proteins (Davies *et al.*, 1999; Dubin *et al.*, 1999). The human *5-HT$_{3B}$* receptor gene was found to localize to chromosome 11q23, which is also the position of the *5-HT$_{3A}$* receptor gene (Weiss *et al.*, 1995), and suggests that the homologs arose from a local gene duplication event (Davies *et al.*, 1999).

The *5-HT$_{3B}$* receptor gene was found to have two alternative promoter regions (which lack TATA boxes and CpG island motifs) that control the expression of different *5-HT$_{3B}$* transcripts with variations in the 5′ coding sequence in the peripheral and central nervous system. The transcription start sites in the brain were localized 4000 bp downstream of the first ATG start codon, and significant promoter activity was located from 1560 bp upstream to 93 bp downstream of the brain specific transcription start sites (Tzvetkov *et al.*, 2007). Examination of a deletion–insertion polymorphism (–100_–102delAAG) in the promoter region of the *5-HT$_{3B}$* receptor gene showed that the deletion allele increases promoter activity *in vitro* with 25 percent and 40 percent higher expression in PC-12 and HEK293 (human kidney transformed) cells, respectively, compared with the insertion construct (Meineke *et al.*, 2008).

The cloned human *5-HT$_{3C}$*, *5-HT$_{3D}$* and *5-HT$_{3E}$* receptor genes were found to be clustered in a subinterval of 100 kb on chromosome 3q27; they have restricted expression in the kidney, colon and liver and are widely expressed in the brain (Niesler *et al.*, 2003). The identification of these genes that encode different subunits provides insight to the complexity of the physiological activity of the 5-HT$_3$ receptor system. Furthermore, there is recent evidence of an association of a functional variant in the 3′UTR of the *5-HT$_{3E}$* receptor gene with diarrhea predominant irritable bowel syndrome which was affected by binding of microRNA-510 (Kapeller *et al.*, 2008).

5-HT$_4$ receptor genes

The 5-HT$_4$ receptors belong to the superfamily of G-protein coupled receptors, and couple positively to G$_s$ to stimulate adenylyl cyclase. They are encoded by a complex gene that spans 700 kb and consists of 38 exons which generate 7 carboxyl-terminal variants, termed a, b, c, d, e, f and g, whose sequences differ after the amino acid leucine in position 358 (Leu358). Another variant, referred to as the internal splice variant (h), is characterized by insertion of 14 amino acids into the second extracellular loop of the receptor (Bender *et al.*, 2000; Bockaert *et al.*, 2004). Furthermore, a novel 5-HT$_4$ receptor splice variant, 5-HT$_{4(i)}$ has also been identified (Brattelid *et al.*, 2004). The 5-HT$_4$ receptors play a role in regulating gastrointestinal tract motility and bladder contraction, and in the central nervous system it is involved in anxiety, memory and cognitive processing (Bender *et al.*, 2000; Maillet *et al.*, 2005).

The *5-HT$_{4A}$* receptor gene was cloned in the rat, showing a distinct pharmacological profile and that it belongs to the superfamily of G-protein coupled receptors (Gerald *et al.*, 1995). It was found to have two splice variants termed r5-HT$_{4(aL)}$ and r5-HT$_{4(aS)}$ for the long and short forms of the receptor, respectively, which differ in the length and sequence of their carboxyl terminus and gene expression levels in the brain (Gerald *et al.*, 1995). These alternatively spliced variants have been renamed r5-HT$_{4(a)}$ and r5-HT$_{4(b)}$ for the short and long forms of the receptor, respectively, following recommendations from the IUPHAR receptor nomenclature committee (Hoyer and Martin, 1997). The mouse *5-HT$_{4(a)}$* receptor gene has high homology with the previously cloned rat *5-HT$_{4(a)}$* receptor gene, and it has two splice variants which were expressed in most regions of mouse brain (Claeysen *et al.*, 1996). This is in contrast to the findings in rat, which showed that the long form is widely expressed in brain but the short form is restricted to the striatum (Gerald *et al.*, 1995). Subsequently, the cloning of the human *5-HT$_{4(a)}$* receptor gene from a human brain cDNA library showed that its sequence had strong similarity to the rat *5-HT$_{4(aL)}$* receptor cDNA (Van den Wyngaert *et al.*, 1997). Moreover the cloning of the human *5-HT$_{4a}$* receptor gene from cardiac tissue revealed the h5-HT$_{4(a)}$ receptor had 93 percent protein identity with the short form of the rat 5-HT$_{4(a)}$ receptor (Blondel *et al.*, 1997). Three new splice variants of the human 5-HT$_4$ receptor were subsequently identified, which were named h5-$HT_{4(b)}$, h5-$HT_{4(c)}$ and h5-$HT_{4(d)}$ receptors (Blondel *et al.*, 1998). These variants were also generated by alternative splicing events that occur in the carboxyl terminus of the human h5-HT_4 receptor, just after the amino acid leucine at position 358. The human *5-HT₄* receptor gene was localized to chromosome 5 bands 5q31–5q33 by *in situ* hybridization (Claeysen *et al.*, 1997). Sequence analysis indicates that the h5-$HT_{4(a)}$ and h5-$HT_{4(b)}$ receptors are human counterparts of the rat 5-$HT_{4(a)}$ and 5-$HT_{4(b)}$ receptor isoforms, respectively, whereas h5-$HT_{4(c)}$ and h5-$HT_{4(d)}$ are novel isoforms (Blondel *et al.*, 1998). The carboxyl terminus of h5-$HT_{4(c)}$ reveals a high number of phosphorylation sites and the h5-$HT_{4(d)}$ isoform produces an ultrashort form of the receptor with a truncation of two amino acids after the splicing site; data showed that the splice variants may differ in signal transduction following receptor activation (Blondel *et al.*, 1998). Similar splice variants in mouse, rat and human brain have been cloned and shown to have different coupling efficiency (Claeysen *et al.*, 1998). Three additional splice variants with shorter carboxyl-terminal sequences have been cloned from different species; these are termed rat 5-$HT_{4(e)}$, mouse 5-$HT_{4(f)}$ and human 5-$HT_{4(g)}$. These demonstrate higher capacity for isomerization from the inactive to active conformation of the receptor (Claeysen *et al.*, 1999). Furthermore, the human 5-$HT_{4(e)}$ (Mialet *et al.*, 2000) from heart and the human 5-$HT_{4(a\text{-}g)}$ spliced variants and a novel internal splice variant 5-$HT_{4(hb)}$ were cloned from a genomic library, revealing more information on the structure of the human *5-HT_4* gene (Bender *et al.*, 2000). The gene was found to span more than 130 kb and consists of 10 exons, with exons 2–5 encoding the most common part shared by all splice variants up to Leu358. The other exons downstream of exon 5 have been named alphabetically to correspond to the previous nomenclature given to the variants obtained by cDNA cloning. The novel splice variant results in the insertion of 14 amino acids in the second extracellular loop of the receptor and has a C-terminal b variant, thus it was termed 5-$HT_{4(hb)}$, which has a different second messenger activation from the other 5-HT_4 receptor isoforms (Bender *et al.*, 2000). Other human 5-HT_4 splice variants have been cloned and characterized, and include the 5-$HT_{4(n)}$ from the hippocampus, which contains none of the C-terminal exons (Vilaro *et al.*, 2002), and the 5-$HT_{4(i)}$ from pancreas, whose C-terminal does not influence coupling to adenylate cyclase (Brattelid *et al.*, 2004).

The 5′-flanking region and promoter of the *5-HT_4* gene from human placenta was characterized revealing a long (more than 5.1 kb) 5′ untranslated region and cell-specific gene expression within −210 to −105 bp in human choriocarcinoma T3M-3 (5-HT_4 receptor positive) and human hepatocellular carcinoma HepG2 (5-HT_4 receptor negative) cell lines (Hiroi *et al.*, 2001). Additionally, the untranslated exon 1 contains both negative (+112 to +182) and positive (+1 to +11) modulators, suggesting that exon 1 plays a role in the transcriptional regulation of the *5-HT_4* gene. Another study characterized the 5′UTR in human heart and found it to be 3.1 kb in length, containing a novel exon (Maillet *et al.*, 2005). The promoter region of this gene was 1.2 kb in length (−4298 to −3050) in human HeLa and IMR32 (neuroblastoma) cell lines but silent in monkey COS-7 cells, and lacked TATA and CAAT boxes. The enhancer region (−220/−61) in the long 5′UTR region was found to interact with the transcription factor Nkx2.5 (NK2-transcription factor-related 5, also called Csx), which is a member of the NK class of homeo-domain proteins (Maillet *et al.*, 2005).

5-HT_5 receptor genes

The 5-HT_5 receptor family consists of two subtypes, known as the 5-HT_{5A} and 5-HT_{5B} receptors, which belong to the superfamily of G-protein coupled receptors (see Nelson, 2004, for review). This family of 5-HT receptors is the least well understood amongst the serotonin receptor classes. The cDNA sequence of the *5-HT_{5A}* receptor gene was initially generated from a mouse-brain library with a pharmacological profile similar to the 5-HT_{1D} receptor but with distinct intracellular signaling properties and expression pattern (Plassat *et al.*, 1992). Subsequently, the gene was cloned in the rat (Erhlander *et al.*, 1993) and in human (Rees *et al.*, 1994). The human gene, similar to the mouse and rat *5-HT_{5A}* receptor genes, consists of two coding exons separated by a large intron (positioned in the middle of the third intracellular loop of the receptor) and codes for a protein of 357 amino acid residues. The human 5-HT_{5A} receptor has 93 percent nucleotide identity and 84 percent amino acid identity to the cloned mouse 5-HT_{5A} receptor, and is expressed in most regions of the brain with little or no expression in peripheral tissues. *In situ* hybridization studies in rat and mouse brain were in agreement regarding the widespread distribution of the gene; however, the transcript was present in the cerebral cortex, the dentate gyrus, the pyramidal cell layer within the hippocampus (CA1–3), the granule cell layer of the cerebellum and the tufted cells of the olfactory bulb (Plassat *et al.*, 1992). The *5-HT_{5A}* receptor gene was located on mouse chromosome 5 (position 5B) and human chromosome 7q36 (Matthes *et al.*, 1993).

The *5-HT_{5B}* receptor gene has been cloned in the rat (Erhlander *et al.*, 1993; Wisden *et al.*, 1993) and in the mouse (Matthes *et al.*, 1993), with a distinct ligand-binding profile and specific expression in medial habenula and hippocampal CA1 cells in adult brain. The gene was located on mouse chromosome 1 (position IF) and human chromosome 2q11-13 (Matthes *et al.*, 1993).

5-HT$_6$ *receptor genes*

The *5-HT*$_6$ receptor gene is the most recent addition to the 5-HT receptor family, and belongs to the superfamily of G-protein coupled receptors (as reviewed in Woolley *et al.*, 2004). The gene has been cloned in the rat (Monsma *et al.*, 1993; Ruat *et al.*, 1993a), human (Kohen *et al.*, 1996) and mouse (Kohen *et al.*, 2001). The receptor is positively coupled to adenylyl cyclase via G_s proteins, is expressed abundantly in the caudate nucleus of the brain and has distinct pharmacological properties, including high affinity for atypical antipsychotics such as clozapine. The human *5-HT*$_6$ cDNA has an open reading frame of 1320 bp, which is interrupted by two introns in positions corresponding to the third cytoplasmic loop and third extracellular loop, resulting in an apparent frame shift compared to the rat *5-HT*$_6$ receptor gene. It encodes a protein of 440 amino acids with a molecular size of 47 kDa. The rat *5-HT*$_6$ receptor gene was re-sequenced using the same clone used by Monsma *et al.* (1993), and the amino acid sequence of the human 5-HT$_6$ receptor was 89 percent similar to the corrected rat sequence (Kohen *et al.*, 1996). The human *5-HT*$_6$ receptor gene was found to map to 1p35-p36, which overlaps the assignment for the *5-HT*$_{1D\alpha}$ receptor gene, suggesting a close link between these genes (Kohen *et al.*, 1996). Similar to the human *5-HT*$_6$ receptor gene, the mouse *5-HT*$_6$ receptor gene is 1320 bp in length and codes for a protein of 440 amino acid residues and a molecular weight of 47 kDa (Kohen *et al.*, 2001). The mRNA expression of the gene is largely present in the central nervous system, and low levels are detected in the stomach and adrenal glands. In the brain the transcripts are present in the striatum, nucleus accumbens, hippocampus, cortex and olfactory tubercle by *in situ* hybridization and northern blot analyses (reviewed in Barnes and Sharp, 1999).

5-HT$_7$ *receptor genes*

The 5-HT$_7$ receptor family consists of several alternative splice variants, known as 5-HT$_{7(a)}$, 5-HT$_{7(b)}$, HT$_{7(c)}$, HT$_{7(d)}$ and HT$_{7(e)}$. The rat *5-HT*$_{7(a)}$ receptor gene was isolated from kidney proximal tubule (Shen *et al.*, 1993) and brain (Ruat *et al.*, 1993b), and expressed highly in the hypothalamus, hippocampus and brainstem. In the human, the *5-HT*$_{7(a)}$ receptor gene was initially identified as the long form of the *5-HT*$_7$ gene (Bard *et al.*, 1993) and later cloned in brain (Stam *et al.*, 1997). The human *5-HT*$_{7(b)}$ receptor gene, which has a 13 amino acid truncation that results from a 5-bp insertion that introduces a stop codon, was also isolated (Heidmann *et al.*, 1997; Jasper *et al.*, 1997; Stam *et al.*, 1997). The rat *5-HT*$_{7(b)}$ receptor cDNA was isolated from a brain library and was found to contain a 5-bp insertion (GTAAG) at an intron splice site which generated an 'in frame' stop codon which shortened the carboxy tail by 13 amino acids (Lovenberg *et al.*, 1993). Further isoforms have been identified and designated as the *5-HT*$_{7(c)}$ receptor gene in rat, *5-HT*$_{7(d)}$ receptor gene in human and *5-HT*$_{7(e)}$ receptor gene in rat (Heidmann *et al.*, 1997; Liu *et al.*, 2001). Both the 5-HT$_{7(c)}$ and 5-HT$_{7(d)}$ receptor isoforms are generated from the use of two distinct exon cassettes, located within intron 2 of the genes, which result in proteins differing in their carboxyl-terminal ends. The human *5-HT*$_7$ receptor gene was cloned and mapped to chromosome 10q21–q24 (Gelernter *et al.*, 1995) and the identification of a human *5-HT*$_7$ receptor pseudogene has also been reported (Olsen *et al.*, 1997; Qian *et al.*, 1998; Olsen and Schechter, 1999). In mouse, the *5-HT*$_{7(a)}$ receptor gene was previously cloned (Plassat *et al.*, 1993) and the splice variants m5-HT$_{7(b)}$ and m5-HT$_{7(c)}$ have been recently identified (Gellynck *et al.*, 2008).

The promoter region of the human *5-HT*$_7$ receptor gene lacks any TATA or CAAT boxes, but is GC rich, regulated by the transcription factors Sp1 and Sp3, and a region 653 bp upstream of the ATG start codon demonstrates highest transcriptional activity in both NS20Y (mouse neuroblastoma) and HEK293 cells (Laenen *et al.*, 2007).

Tryptophan hydroxylase genes

Tryptophan hydroxylase (also known as TrpOHase, or L-tryptophan tetrahydropteridine oxygen oxidoreductase) is the rate-limiting enzyme in the biosynthesis of serotonin that catalyzes the hydroxylation of the essential amino acid tryptophan to 5-hydroxytryptophan. Together with phenyalanine hydroxylase and tyrosine hydroxylase, trypophan hdroxylase is a member of a family of aromatic amino acid hydroxylases that utilize tetrahydropterins as substrates (Cooper *et al.*, 1996; Fitzpatrick, 1999). 5-HT synthesis in the brain is found mainly in serotonergic neurons of the dorsal raphe nucleus and additionally synthesized in the pineal gland, where it is a precursor in the synthetic pathway of melatonin. Interestingly, there are two genes that code for tryptophan hydroxylase; these are known as *TPH-1* and *TPH-2*. Generally, TPH-1 mainly catalyzes serotonin synthesis in peripheral organs, and the newly identified tryptophan hydroxylase-2

(TPH-2) is responsible for the synthesis of serotonin in the central nervous system (reviewed in Invernizzi, 2007).

Tryptophan hydroxylase 1 gene (TPH-1)

The *TPH-1* cDNA was first isolated from the rat (Darmon *et al.*, 1986, 1988) and rabbit pineal glands (Grenett *et al.*, 1987). Interestingly, several *TPH-1* mRNA transcripts have been revealed in rats that differ in the lengths of their 3′ (Darmon *et al.*, 1988) and 5′ untranslated regions, suggesting different mechanisms of *TPH-1* gene regulation in the pineal gland and brainstem (Delort *et al.*, 1989; Chamas and Sabban, 2002). The two rat mRNA transcripts generated by diversity at the 5′UTR have been designated *TPH-α* and *TPH-β* (having a longer 5′UTR), with expression of *TPH-α* significantly higher than that of *TPH-β* in the pineal gland and raphe nuclei in the brainstem (Delort *et al.*, 1989; Chamas and Sabban, 2002). *TPH-β* has a greater effect than *TPH-α* on translation of the heterologous gene luciferase in PC12 (rat pheochromocytoma) cells, indicating selective regulation of translation efficieny (Chamas and Sabban, 2002). Additionally, the rat gene was cloned from serotonergic neurons in the dorsal raphe nucleus (Kim *et al.*, 1991). In mouse, the gene was cloned from P815 mastocytoma cells, spans 21 kb, consists of 11 exons, and has been localized to the proximal half of chromosome 7 (Stoll *et al.*, 1990; Stoll and Goldman, 1991). Similar to the rat *TPH-1* gene, the coding sequence of the gene in human carcinoid tumor cells was 1332 bp and encodes a protein of 444 amino acids of a predicted molecular weight of 50,952 daltons (Boularand *et al.*, 1990). The nucleotide sequence identity was 86.8 percent and 91.2 percent between the human and rat genes (Boularand *et al.*, 1990). The human *TPH-1* gene was found to localize to chromosome 11 by somatic cell hybrid mapping (Ledley *et al.*, 1987), and specifically to chromosome 11p15.3-p14 by *in situ* hybridization (Craig *et al.*, 1991). The isolation and organization of the human *TPH-1* gene showed that it spans 29 kb and contains 11 exons (Boularand *et al.*, 1995a). Additionally, four mRNA species were observed in pineal gland and in carcinoid tumor resulting from differential splicing of three intron-like regions and of three exons located in the 5′UTR of the gene (Boularand *et al.*, 1995a). Furthermore, alternative splicing at the 3′ end of the cDNA sequence of the human gene generates two human TPH-1 isoforms with different carboxy termini that are expressed in both brain and pineal gland. However, only the spliced mRNA species was detected in most brain regions, including brainstem, midbrain and cerebellum, whereas the non-spliced transcript expression was limited to only some brain regions (Wang *et al.*, 1998).

The 5′UTR of the rat *TPH-1* gene has revealed the presence of two promoter regions, with the proximal one exhibiting two 'CCAAT homologies' (Delort *et al.*, 1989). In the mouse the promoter region was suggested to be upstream from the transcription initiation site, where a CAAT and a TATA box were present 55 bp and 25 bp, respectively (Stoll and Goldman, 1991). However, the transcriptional regulation of the mouse *TPH-1* gene was further elucidated, revealing the proximal promoter between −77 and −46 nucleotides and an activator region between nucleotides −343 and −21 bp in P815-HTR mastocytoma cells (Reed *et al.*, 1995). Furthermore, the transcription factor NF-Y (also known as CP-1 and CBP, and which interacts with CCAAT boxes found in Y boxes) was found to bind to the GGCCAAT motif in the mouse *TPH-1* promoter and activates transcription (Reed *et al.*, 1995). In humans, a minimal promoter region was active in pinealocyte cultures and PC12 cells between −73 and −51 nucleotides that drove cAMP-dependent transcription. It has an inverted CCAAT box rather than a cAMP-responsive element (Boularand *et al.*, 1995b). An additional inverted CCAAT box and a GC-rich region in the optimal promoter (−252/+29) of the human *TPH-1* gene interacted with NF-Y and Sp1 binding proteins in the basal and cAMP-stimulated transcriptional activation in pinealocytes (Cote *et al.*, 2002). Interestingly, other studies have shown that the transcription factor CCAAT box binding factor (CBF)/nuclear factor-Y (NF-Y) activates the transcription of the human *TPH-1* gene (Teerawatanasuk and Carr, 1998), whereas CCAAT displacement protein (CDP/Cut) interacts with the negative regulatory element of the gene (Teerawatanasuk *et al.*, 1999).

There is evidence showing that the TGT haplotype of 3 SNPs (T-1607C, G-1066A, T-346G) within the 5′ regulatory region of the *TPH-1* gene repressed transcriptional activity in human choriocarcinoma and colon adenocarcinoma cell lines; however, only the T-346G marker showed association with alcohol dependence in a Taiwanese group (Sun *et al.*, 2005).

Tryptophan hydroxylase 2 gene (TPH-2)

Gene targeting studies in mice led to the discovery of the second *TPH* gene, termed as *TPH-2*, with homologs of the gene additionally cloned and sequenced in humans and rats (Walther *et al.*, 2003). TPH-1 and TPH-2 show 71 percent amino acid identity in humans. In mice the *TPH-2* mRNA is primarily detected in the brain, whereas the *TPH-1* gene is expressed in the periphery, including the gut, pineal gland, spleen and thymus (Walther *et al.*, 2003; Cote *et al.*, 2003; Walther and Bader, 2003).

In rats, studies also revealed specific and high *TPH-2* mRNA expression in midbrain and brainstem raphe nuclei, with the *TPH-1* gene expression detected predominantly in the pineal gland (Patel *et al.*, 2004). In humans, *TPH-2* gene expression levels were highest in frontal cortex and thalamus followed by hypothalamus, lowest in the hippocampus and amygdala, and barely detectable in peripheral tissues (Zill *et al.*, 2004a). However, the first study demonstrated *TPH-1* and *TPH-2* mRNA expression in different regions of human *post-mortem* brain (Zill *et al.*, 2007). Interestingly, recent data from our group have revealed that mRNA expression of *TPH-1* in most human *post-mortem* brain regions is generally comparable or higher than that of *TPH-2* gene expression (Sugden *et al.*, 2009). These data confirm previous results where the *TPH-1* expression was higher than *TPH-2* in human cortex and thalamus and these differences were significant in hypothalamus and amygdala (Zill *et al.*, 2007). However, *TPH-1* gene expression levels were significantly higher than *TPH-2* mRNA expression in the striatum and cerebellum but not in the hippocampus, which showed higher *TPH-2* gene expression than *TPH-1* (Sugden *et al.*, 2009). This inconsistency could be related to factors such as the differing mean ages of samples analyzed in our study and that by Zill and colleagues (2007). Lately, a regulatory domain of 41 amino acids in the N terminus was found in TPH-2 and not in TPH-1, which controls enzyme protein expression and phosphorylation of serine 19 (a protein kinase A consensus site) located in this region, resulting in increased TPH-2 stability and enzyme activity in cell culture systems (Murphy *et al.*, 2008).

Several studies have identified functional polymorphisms in the *TPH-2* gene (reviewed in Zhang *et al.*, 2006). These include C1473G SNP, which reduced 5-HT levels in murine frontal cortex and striatum, indicating that TPH-2 controls brain serotonin synthesis (Zhang *et al.*, 2004). Interestingly, a functional SNP G1463A which results in an amino acid change from arginine to histidine in the human *TPH-2* gene resulted in 80 percent loss of function in serotonin synthesis when expressed in PC12 cells, and is more frequent in unipolar depression than bipolar depression (Zhang *et al.*, 2005). A previous human study showed a significant association with an SNP and a haplotype of SNPs between exons 5 and 7 of the *TPH-2* gene with major depression (Zill *et al.*, 2004b). More recently, two marker SNPs located in exons 7 and 9 showed differential expression of *TPH-2* mRNA levels in the pons region of human *post-mortem* brain. These SNPs are closely linked to *TPH-2* SNPs previously found to be associated with major depression and suicide (Lim *et al.*, 2007). Additionally, a novel short isoform of the *TPH-2* gene was identified with a truncated catalytic domain which was expressed in human brainstem, prefrontal cortex, hippocampus and amygdala, and an SNP (A22879G) in exon 6 of this short isoform was associated with major depression (Haghighi *et al.*, 2008). Nonetheless, higher levels of *TPH-2* mRNA were found in the dorsolateral prefrontal cortex in *post-mortem* brains from patients with bipolar disorder, suggesting a potential role of the gene in the pathophysiology of this disorder (De Luca *et al.*, 2005). Studies in rhesus monkey revealed two nonsynonymous SNPs in a haplotype with significantly higher expression of the *TPH-2* gene and protein, higher 5-HT production and lower mRNA stability than the wild type when stably expressed in PC12 and HEK293 cells (Chen and Miller, 2008).

Recent studies have indicated that the core promoter region of the human *TPH-2* gene is localized between −107 bp and +7 bp and in the region of +8 bp to +53 bp in the 5′UTR negatively regulated gene expression at transcriptional and post-transcriptional levels (Chen *et al.*, 2008). However, fMRI studies demonstrate that an SNP (rs570625; G-703T) in the upstream regulatory region of the gene increases amygdala reactivity to emotional stimuli (Brown *et al.*, 2005; Canli *et al.*, 2005) and to angry faces in patients with social anxiety disorder in a recent positron emission tomography (PET) study (Furmark *et al.*, 2009). Another T-473A (rs11178997) polymorphism in the 5′ regulatory region of the *TPH-2* gene was associated with unipolar and bipolar depression (Van Den Bogaert *et al.*, 2006), significantly reduced transcriptional activity in primary serotonergic neurons from rat raphe nuclei and in human small lung carcinoma SHP-77 cells, and reduced binding to the transcription factor POU3F2 (also known as Brn2, n-Oct-3) to the A-allele (Scheuch *et al.*, 2007). POU domain factors are transcriptional regulators characterized by a highly conserved unique bipartite DNA-binding domain referred to as the POU domain. The acronym POU (pronounced 'pow') is derived from the names of three mammalian transcription factors: the pituitary-specific Pit-1, the octamer-binding proteins Oct-1 and Oct-2, and the neural Unc-86 from *Caenorhabditis elegans*. Another recent study examined the functional effects of the haplotype of the two polymorphisms (G-703T and T-473A) in the 5′-flanking region of the human *TPH-2* gene mentioned above together with a third common variant A90G in the 5′UTR of the gene (Chen *et al.*, 2008). In both RN46A (a serotonergic cell line derived from embryonic day 13 rat medullary raphe nuclei) and HEK293 cells it was demonstrated that the haplotype (G-703T, T-473A and A90G) resulted in reduced transcriptional activity, and that all three polymorphisms potentially alter DNA–protein interactions (Chen *et al.*, 2008). Interestingly, further investigations on the regulation of the gene in mouse were reported where a novel

NRSE (neuron restrictive silencing element) motif interacts with the transcription factor REST/NRSF (RE-1 silencer of transcription/neuron restrictive silencer factor in C6 glioma cells) (Patel *et al.*, 2007). Finally, studies in rhesus monkeys demonstrated that 3′UTR polymorphisms and haplotypes modulate HPA axis function by altering the expression levels of the gene (Chen *et al.*, 2006).

Serotonin transporter gene (*SERT, 5-HTT, SLC6A4*)

The *SERT* gene is located on chromosome 17, and contains 14 exons spanning 31 kb (Lesch *et al.*, 1994). Its organization bears striking similarity to that of other members of the Na^+/Cl^- neurotransmitter transporter family. Bradely and Blakely (1997) used 5′ RACE to examine the pattern of transcripts and discovered alternative first exons (1A and 1B). Transcripts containing either one or other of these were identified in the human placental JAR cell line, in which interleukin-1 beta is a potent regulator of the *SERT* locus. Exon lB is not preceded by a TATA element, as suggested for the promoter upstream of exon lA. Exon 2 contains the translation initiation site. The wide range of transcripts sizes between 2 and 4 kb is partly explained by differential splicing, but also probably by the presence of two rare polyadenylation sites at the 3′ end of the gene (5′-AATGAA-3′ and 5′-AATATA-3′) and of a third polyadenylation signal similar to that found in rat and mouse *SERT* genes, thereby providing opportunity for variable polyadenylation (see Bradely and Blakely, 1997; Battersby *et al.*, 1999).

The most studied polymorphism of the *5-HTT* locus is located 1 kb upstream from the transcription start site and was identified initially by Heils *et al.* (1995, 1996). The basic polymorphism (*5-HTTLPR*) comprises a 44-bp insertion or deletion, creating either a 'long' 16-repeat allele or a 'short' 14-repeat allele, together with other low-frequency alleles including 15, 18, 19, 20 and 22 repeats, and there are various additional SNPs distinguishing some repeats (Delbruck *et al.*, 1997; Kunugi *et al.*, 1997; Michaelovsky *et al.*, 1999; Nakamura *et al.*, 2000). Flattem and Blakely (2000) noted a potential unstable sequence of 381 bp located between the *5-HTTLPR* and the transcription start site which is present in some but not other cloned sequences. They speculate that this may be detected in some individuals.

Classic studies have suggested that the short allele is associated with both lower transcriptional potential and anxiety-related personality traits, particularly neuroticism and affective disorders (reviewed in D'Souza and Craig, 2008). This review also summarizes the functional studies of the polymorphism in mammalian cell lines, lymphocytes, blood platelets and brain, and thus these are only briefly described here. The short variant of the *5-HTTLPR* had lower transcriptional activity, *5-HTT* expression and 5-HT uptake than the long variant in human placental choriocarcinoma (JAR) and peripheral lymphoblast cells (Heils *et al.*, 1996; Lesch *et al.*, 1996). This pattern has also been supported in part by neuroimaging studies (see D'Souza and Craig, 2008). A variety of subsequent investigations have generally supported the higher transcriptional levels observed for the long allele, but the picture is complicated by the existence of SNPs both within the promoter repeats and elsewhere in the gene (see Nakamura *et al.*, 2000; Hu *et al.*, 2006; Martin *et al.*, 2007). Very recent studies have investigated associations based on haplotypes which combine variants providing information throughout the locus and potentially tracking additional functional variants (see Mrazek *et al.*, 2008).

The importance of the *5-HTTLPR* region is highlighted in a study of gene-by-environment interaction (G $\times$ E), which showed that individuals having one or two copies of the short allele of the *5-HTTLPR* polymorphism exhibited more depressive symptoms, diagnosable depression, and suicidality in relation to stressful life events than individuals homozygous for the long allele (Caspi *et al.*, 2003). A second G $\times$ E study demonstrated that individuals with adolescent depression had a significant interaction between environmental risk and short alleles of *5-HTTLPR*, with the effect being observed in females only (Eley *et al.*, 2004). Additionally, independent groups have replicated the main findings of Caspi *et al.* (2003) (Kaufman *et al.*, 2004; Kendler *et al.*, 2005; Wilhelm *et al.*, 2006), although Gillespie *et al.* (2005) were unable to reproduce the gene by environment effect. In addition, Chorbov and colleagues (2007) described a G $\times$ E trauma interaction in the predisposition to major depressive disorder observed with high expression (long alleles – classified according to Hu *et al.*, 2006). Further replication is provided by studies on the Spanish PREDICT cohort, where both the SS homozygous genotype and increased life-threatening events enhanced the probability of depression, particularly after adjusting for gender age and family history of psychological problems (Cervilla *et al.*, 2007). Providing a possible insight into the mechanisms involved, it was found that girls homozygous for the S allele were apparently more reactive to stress measured by cortisol production than were those with an L allele (Gotlib *et al.*, 2008).

A sex-specific effect has been reported in which, for females, the S allele, combined with care-giving stress or low childhood socioeconomic status, was associated with higher depression scores as compared to participants in the non-stressor group and those with the long (L) allele. In contrast, for males, the L allele, combined with a stressor, was associated with higher depression scores as compared to those in the non-stressor group and those with the S allele (Brummett *et al.*, 2008).

Detailed examination of markers throughout the gene, including four haplotype tagging SNPs (rs2020942, rs140700, rs3798908 and rs1042173), found that, in addition to influences of the *5-HTTLPR*, the central region of the locus was also implicated in explanations of G × G×E interactions for mood (Lazary *et al.*, 2008). Overall, the original observations of Caspi *et al.* (2003) appear to have been replicated robustly. Most recently, other associated phenotypes also appear to fit the pattern of adversity interactions with genotype; specifically, S/S individuals with higher levels of maltreatment had significantly higher levels of anxiety susceptibility (although not neuroticism) than subjects in other groups (Stein *et al.*, 2008).

As reviewed previously (D'Souza and Craig, 2008), in addition to the promoter repeat, another example of a functional VNTR polymorphism in the serotonin transporter gene is located in intron 2 (termed as Stin2 VNTR). It comprises 9, 10 or 12 copies of a 16/17-bp element. Functional studies in differentiating embryonic stem cells *in vitro* and in transgenic mice *in vivo* revealed the Stin2 VNTR acts as a transcriptional regulator in an allele-dependent manner, with the 12-repeat allele demonstrating stronger enhancer activity than the 10-repeat allele. Additional investigations suggest that not only the number but also the primary structure of the repeats could affect the transcription of the gene (Lovejoy *et al.*, 2003). Furthermore, two transcription factors, YB-1 (Y box binding protein 1) and CTCF (CCTC-binding factor) were found to be responsible for the regulation of the Stin2 VNTR function as a transcriptional regulatory domain (Klenova *et al.*, 2004).

Indeed, given the possibility that the *5-HTT* locus may be a target for the therapeutic effects of lithium (LiCl) in treatment of depression, Roberts *et al.* (2007) demonstrated that LiCl modified the levels of CTCF and YB-1 mRNA and protein and that both CTCF and YB-1 showed differential binding to the allelic variants of the Stin2 VNTR and may result in alterations of *SLC6A4* expression. The complexity of the interactions is further increased by the potential interaction of LiCl with the *5-HTTLPR*, which contains domains potentially capable of interaction with CTCF.

There have been reported associations of Stin2 VNTR with behavioral disorders, including strong evidence for increased frequency of allele 12 of the VNTR with subjects having bipolar disorder (Collier *et al.*, 1996). Additionally, significant differences exist between patients with unipolar disorder and controls in the proportion of individuals having allele 9 (Battersby *et al.*, 1996; Ogilvie *et al.*, 1996). Moreover, an increased frequency of the 10 allele of this intron polymorphism together with the long allele of the promoter polymorphism of the serotonin transporter gene was observed in a suicide cohort having Slavic ethnicity (Hranilovic *et al.*, 2003). Furthermore, an independent report showed that the combined genotype at both *5-HTTLPR* and Stin2 locus ('low expressing' genotypes containing at least one short allele) lowered *5HTT* gene expression levels in lymphoblastoid cell lines from subjects having schizophrenia or schizoaffective disorder using a real-time PCR method (Hranilovic *et al.*, 2004).

Finally, epigenetic modification may provide an additional feature capable of modifying genotype effects in regulating *SERT* expression in a sex-specific manner. A study on lymphoblastoid lines indicated that the CpG island overlapping the 5′ region of the locus is more heavily methylated in females, resulting in reduced transcription (Philibert *et al.*, 2007, 2008)

Brain-derived neurotrophic factor gene (*BDNF*)

BDNF (brain-derived neurotrophic factor) has been shown to modulate serotonin transporter function in lymphoblast cells (Mossner *et al.*, 2000), and variants in this locus are highly relevant to the functionality of the monoamine pathways. The gene is located on 11p14 and codes for the precursor peptide (proBDNF) that is cleaved proteolytically to form the mature protein. BDNF is known to modulate hippocampal plasticity and hippocampal-dependent memory in cell models and animals. It plays an important role in regulating transmitter systems, neuronal survival and regeneration, and consequently is a key candidate for implication in behavioral disorders. The *BDNF* locus was originally reported to have a minimum of five 5′ non-coding exons (Aoyama *et al.*, 2001; Marini *et al.*, 2004); however, nine transcripts have been identified which differ in their use of alternative promoters, alternative splice donor and acceptor sites, and alternative polyadenylation sites (Liu *et al.*, 2006). Most transcripts encode proBDNF; however, two transcripts contain additional in-frame methionines that may result in proteins with longer N termini. These authors also identified non-coding RNAs transcribed from a gene that overlaps with BDNF on the opposite strand (BDNFOS) (Pruunsild *et al.*, 2007). In a detailed dissection of the *BDNF* locus, Pruunsild *et al.* (2007) show that the gene has 11 exons and 9 functional promoters which operate in a tissue- and brain-region-specific manner. A significant role may also be attributed to a range of differentially spliced non-coding antisense transcripts which are produced from an anti-BDNF locus (BDNFOS), which form double-stranded duplices in the brain *in vivo*, suggesting an important role for anti-BDNF in regulating BDNF expression (Pruunsild *et al.*, 2007). Furthermore, they identified non-coding natural antisense RNAs that display complex splicing and expression patterns which are transcribed from the BDNF gene locus. A total of at least 34 different transcripts can be produced, although their relative physiological significance remains to be established. Most recently, Baj and Tongiorgi (2009) have investigated the *BDNF* transcripts responsible for cell survival of neuroblastoma cell lines SH-SY-5Y and SK-N-BE in a model system employing small inhibitor (si) RNAs. The evidence suggests that mRNA isoforms coupling exons 4 or 6 or 9a to the coding exon 9 are necessary for survival.

Various *trans* acting factors may serve to modulate the expression of the locus, including the DNA binding protein Pitx3, a homeodomain transcription factor required for the development of midbrain dopaminergic neurons (Peng *et al.*, 2007). Additional evidence suggests that estrogen may also be important in the regulation of the *BDNF* gene (Scharfman and MacLusky, 2006; Sohrabji and Lewis, 2006). Also of interest is that the *BDNF* locus is one of the few well-defined targets of the methyl-binding protein *MECP2* gene, mutations in which are responsible for Rett syndrome (Chen *et al.*, 2003). More recently, microRNA, miR-30a-5p, has been found to act as a post-transcriptional inhibitor of BDNF in the prefrontal cortex (Mellios *et al.*, 2008).

A SNP has been identified in the 5′ pro-BDNF sequence at nucleotide 196 (G/A), producing an amino acid substitution (valine to methionine) at codon 66 (Val66Met). Genetic association studies and functional studies of this particular polymorphism and other variants in the gene have been reviewed recently and thus are only briefly summarized here (D'Souza and Craig, 2008). The Met allele has been found to be associated with bipolar disorder in adults (Neves-Pereira *et al.*, 2002; Sklar *et al.*, 2002; Lohoff *et al.*, 2005), although a very large study showed significant association only with a rapidly-cycling subset (Green *et al.*, 2006). Several studies have examined the functional impact of this polymorphism, revealing that Met allele exhibited abnormal intracellular trafficking, impaired secretion and storage of BDNF protein compared with the Val allele in neuronal cells (Egan *et al.*, 2003; Chen *et al.*, 2004) and with diminished hippocampal engagement using BOLD (blood oxygen level-dependent) fMRI in comparison with Val homozygotes (Hariri *et al.*, 2003).

Investigation of a complex structure embracing three types of dinucleotide repeats which is located about 1.0 kb from the translation initiation site has been determined. Insertion/deletion and nucleotide substitutions within the repeat motifs resulted in the identification of 23 alleles, and this BDNF-linked complex polymorphic region is known as *BDNF-LCPR*. One major allele was associated with bipolar disorder, and this allele showed reduced transcription in a luciferase coupled assay (Okada *et al.*, 2006). A synonymous second SNP (rs988748) has been described, but there are no significant functional or disease associations reported. Interestingly, the picture emerges of the *BDNF* locus with excellent credentials as a candidate gene in behavioral disorders, but also of one whose complexity of regulation and response to external and *trans* acting factors complicate simplistic interpretations.

Monoamine oxidase genes *(MAO)*

The monoamine oxidases A and B are the products of two abutting X-linked genes, and play an important role in the metabolism of biogenic amines in the central nervous system and in the periphery. Monoamine oxidases are oxygen oxidoreductase (deaminating) flavin-containing enzymes, and the two forms of the enzyme, MAO-A and MAO-B, differ in specificity for substrates, sensitivity to inhibitors, tissue distribution and immunological properties. In general, MAO-A preferentially oxidizes the biogenic amines such as serotonin, norephinephrine and epinephrine, whereas MOA-B catalyzes phenylethylamine and benzylamine. However, dopamine, tyramine and tryptamine are common substrates for both forms of the enzyme.

The cDNA nucleotide sequences of MAO-A and MAO-B genes were determined from human liver (Bach *et al.*, 1988). Studies on the genomic organization of the two loci indicate that they have considerable homology (73 percent at the amino acid level), and each comprises 15 exons with the genes arranged in a tail-to-tail configuration separated by about 40–45 kb (Chen *et al.*, 1991; Grimsby *et al.*, 1991). Furthermore, these studies showed that exon 12, which codes for the covalent FAD-binding site, is highly conserved (about 92 percent peptide identity) in both genes. The human *MAO-A* and *MAO-B* genes span 100 kb, and the exon/intron boundaries in each gene occur at homologous positions in the protein sequence, suggesting they originated by gene duplication (Grimsby *et al.*, 1991). The human *MAO-A* and *MAO-B* genes were assigned to the X chromosome using rodent–human hybrid somatic cell lines (Pintar *et al.*, 1981; Kochersperger *et al.*, 1986) and specifically mapped to Xp11.23-11.4 by *in situ* hybridization (Ozelius *et al.*, 1988; Lan *et al.*, 1989; Levy *et al.*, 1989).

MAO-A is expressed in the outer mitochondrial membrane of specific cells in various brain and peripheral tissues. In human liver *MAO-A* transcript is 5 kb in length;

however, in small intestine and placenta an additional 2-kb mRNA species is present resulting from alternative polyadenylation sites (Grimsby *et al.*, 1990; Chen *et al.*, 1991). *MAO-B* transcript is 3 kb in human liver and is also expressed widely, including in the brain, and the enzyme is also found in platelets. The highest expression of *MAO-A* and *MAO-B* genes was found in the frontal cortex and locus coeruleus brain regions (Grimsby *et al.*, 1990).

A detailed investigation employing transfection and sequencing techniques revealed maximal promoter-like function. This was observed in conserved high CpG regions directly flanking the transcriptional start sites. Additional regulatory sites involving direct repeats at approximately 1.3 kb and 0.2 kb for the A and B loci respectively were noted (Zhu *et al.*, 1992). Although additional upstream initiation sites have been described, Denney *et al.* (1994) observed that most transcripts for *MAO-A* initiate between −30 and −40, with minor transcription initiating at −95 and −136. Unlike the *MAO-B* promoter, the *MAOA* promoter region of two 90-bp repeats lacks a TATA box and CACCC element; however, the core promoter regions of both genes consist of two sets of overlapping Sp1 sites (Shih *et al.*, 1993, 1994). The significant differences in their promoter regions are presumed to explain the tissue- and cell-specific expression profiles of the two forms. Upstream repeating units were found to downregulate the promoter region of the human *MAO-A* gene independent of an initiator-like sequence in human glioma (1242 MG) and HeLa cells. Several transcription factors are known to regulate the human *MAO-A* promoter and include Sp1 (Zhu *et al.*, 1994), whereas R1 (RAM2/CDCA7L/JPO2) inhibits promoter and enzymatic activity in human neuroblastoma and glioblastoma cells (Chen *et al.*, 2005; Ou *et al.*, 2006). Comprehensive studies show that activation of the *MAO-A* promoter region by glucocorticoid and androgen is regulated differently by R1 and Sp1 (Ou *et al.*, 2006). The human *MAO-B* promoter region interacts with Sp1 and Sp3 (Wong *et al.*, 2001).

There is strong evidence that MAO-A plays an important role in human and animal behavior, and regulation of the *MAO-A* locus by an upstream VNTR has been the focus of many disease-association studies. These findings have recently been reviewed (D'Souza and Craig, 2008), and therefore only key results will be highlighted here, with a focus on updated information. First, a nonsense mutation (in exon 8 of the gene that leads to the production of a truncated, non-functional, protein) was found to be associated with impulsive aggressive behavior in affected males in a single large family studied in Holland (Brunner *et al.*, 1993). Increased aggression also characterized adult male transgenic mice having a deletion in the *MAO-A* gene (Cases *et al.*, 1995). A VNTR polymorphism is located 1.2 kb upstream of the coding region and consists of a 30-bp repeated sequence present in 3, 3.5, 4 or 5 copies with functional effects determined in human neuroblastoma and placental choriocarcinoma cell lines and also in *post-mortem* brain tissue (Sabol *et al.*, 1998; Deckert *et al.*, 1999; Balciuniene *et al.*, 2002).

Strong evidence, however, has been observed regarding the effects of the environment on the behavioral outcomes of different *MAO-A* promoter variants. In a gene–environment study, maltreated children were less likely to develop antisocial problems when they had the genotype in the 5′ regulatory region that confers high levels of *MAO-A* expression (Caspi *et al.*, 2002). Subsequent studies attempted to replicate these findings, demonstrating a significant role for *MAOA* variants interacting with stressful upbringing in predisposing to ASB (antisocial behavior) (Foley *et al.*, 2004; Nilsson *et al.*, 2006; Widom and Brzustowicz, 2006) and as predictors of destructive behavior during male adolescent alcohol consumption (Nilsson *et al.*, 2007).

There are far fewer data relating to *MAO-A* functional variants and females, possibly resulting from the complication of its X-localization and the perceived potential confound of its undetermined level of escape from inactivation; however, in a recent study, a gene by environment interaction was observed suggesting that girls with high-rather than low-activity alleles may be at increased risk of criminal behavior in the presence of psychosocial risk (Sjoberg *et al.*, 2007). A study of allelic expression patterns of *MAO-A* in brain has concluded that there was no evidence for skewing in normal individuals (Pinsonneault *et al.*, 2006). The interrelationship between the functional variants in the *MAO-A* promoter and HPA axis stress response that may underpin the G $\times$ E interaction has been recently reviewed (Craig, 2007).

Finally, the link between genotype and brain has been approached at the neuroimaging level. Neuroimaging studies by Meyer-Lindenberg *et al.* (2006) indicated that low-activity genotype in healthy males predicted pronounced reductions in volume that include virtually the entire cingulate gyrus and bilateral amygdalae. There were also genotype-dependent differences in amygdala activation during emotional arousal. Passamonti *et al.* (2006) employed blood oxygenation level-dependent (BOLD) fMRI to examine a genotype effect on a response inhibition task in healthy males. A greater response was observed in the Brodman's area in high-activity genotypes, whereas a greater response was observed in the right superior parietal cortex and bilateral extrastriate cortex in low-activity genotypes. In a first attempt to correlate brain MAO-A activity with genotype measured by positron emission tomography with the labeled ligand ^{11}chlorgyline, no significant differences in MAO-A activity were observed between the high- and low-activity genotypes (although a trend for higher activity was observed in the predicted direction for the visual cortex) (Fowler *et al.*, 2007). In a subsequent publication by the same group (Alia-Klein *et al.*, 2008), it was shown that lower brain MAO-A activity in cortical and subcortical brain regions correlated with higher self-reported trait aggression (observed in both high and low genotype groups). It seems, therefore, that genotype–brain and genotype–behavior relationships may be developmentally complex.

Other detailed functional, structural and connectivity investigations have suggested that the low-activity *MAO-A* allele adversely prejudices information processing within the amygdala, rostral cingulate and medial prefrontal cortex. High serotonin levels resulting from lower rates of metabolism may labilize such circuits required in social and emotional adjustment. This in turn may predicate an amplification of adverse early life experiences in genotype-vulnerable individuals (Buckholtz and Meyer-Lindenberg, 2008).

Conclusions

The serotonergic gene system consists of several genes, including a number of receptor subtypes with different pharmacological/signaling properties, anabolic and catabolic enzymes and the high-affinity sodium-chloride dependent transporter which is regulated by a neurotrophic factor. The genomic organization of these genes is diverse, with a complex transcriptional regulation, and is summarized in Table 1. Furthermore, RNA splicing and post-transcriptional medications such as RNA editing within some serotonergic genes generates a mechanism whereby single genes produce multiple proteins with different structures and functions. The plethora of gene products is important in the physiological and behavioral effects of serotonin, not only in the brain but throughout the whole organism, and they serve as targets for drug development/therapeutic strategies. Moreover, the functional polymorphisms in genes within this system have provided an understanding of the plausible molecular mechanisms underlying neurobehavioral disorders.

Acknowledgements

We would like to acknowledge the Medical Research Council (MRC), which supported research on a component grant entitled 'Functional analysis of mutations in genes of monoamine neurotransmitters metabolism that are associated with behavioral disorders' (Grant

Table 1 Summary of gene structure and regulation of the serotonergic genes as reviewed in the text

	Gene organization (coding and 5′ regulatory region)	Features in promoter region	Regulation by transcription factors
5-HT_{1A}	Intronless (single exon)	No TATA box and GC rich (human); TATA box (rat)	MAZ (Pur-1, Zif87), Sp1, Deaf-1, Freud-1, corticosteroids
5$HT_{1B/1D/1E/1F}$	Intronless (single exon)	—	—
5-HT_{2A}	3 exons separated by 2 introns	No TATA or CAAT boxes	Sp1, PEA3, E-box proteins
5-HT_{2B}	3 exons separated by 2 introns	—	—
5-HT_{2C}	4 exons separated by 3 introns	No TATA box	—
5-HT_{3A}	9 exons separated by 8 introns	—	—
5-HT_{3B}	—	Two promoter regions. No TATA or CpG islands	—
5-HT_4	10 exons, 1 untranslated exon	No TATA or CAAT boxes	Nkx2.5
5-HT_5	2 exons separated by 1 intron	—	—
5-HT_6	3 exons separated by 2 introns	—	—
5-HT_7	5 exons separated by 3 introns	No TATA or CAAT boxes, GC rich	Sp1, Sp3
TPH-1	11 exons	Two promoters in rat and in mouse TATA and CAAT boxes present	NF-Y, Sp1, CBF/NF-Y, CDP/Cut
TPH-2	Refer to cited articles	1 promoter region	POU3F2 (Brn2), REST/NRSF
5-HTT	14 exons		YB-1, CTCF
BDNF	11 exons	9 promoter regions	Pitx3, oestrogen
MAOA	15 exons	No TATA or CACC boxes and GC rich	Sp1, R1

Reference No: #G0000197). We also thank Dr. Andrew Makoff for constructive comments and Kelly Halton for secretarial support, especially with formatting of the text and references.

References

Adham, N., Romanienko, P., Hartig, P., Weinshank, R. and Branchek, T. (1991) The rat 5-hydroxytryptamine$_{1B}$ receptor is the species homologue of the human 5-hydroxytryptamine$_{1D\beta}$ receptor. Mol. Pharmacol., 41: 1–7.

Adham, N., Kao, H.-T., Schechter, L.E., Bard, J., Olsen, M., Urquhart, D., Durkin, M., Hartig, P.R., Weinshank, R.L. and Branchek, T.A. (1993) Cloning of another human serotonin receptor (5-HT$_{1F}$): a fifth 5-HT$_1$ receptor subtype coupled to the inhibition of adenylate cyclase. Proc. Natl. Acad. Sci. USA, 90: 408–412.

Albert, P.R., Zhou, Q.-Y., Van Tol, H.H.M., Bunzow, J.R. and Civelli, O. (1990) Cloning, functional expression and mRNA tissue distribution of the rat 5-hydroxytryptamine$_{1A}$ receptor gene. J. Biol. Chem., 265: 5825–5832.

Alia-Klein, N., Goldstein, R.Z., Kriplani, A., Logan, J., Tomasi, D., Williams, B., Telang, F., Shumay, E., Biegon, A., Craig, I.W., Henn, F., Wang, G.-J., Volkow, N.D. and Fowler, J.S. (2008) Brain monoamine oxidase A activity predicts trait aggression. J. Neurosci., 28: 5099–5104.

Amlaiky, N., Ramboz, S., Boschert, U., Plassat, J.-L. and Hen, R. (1992) Isolation of a mouse '5-HT1E-like' serotonin receptor expressed predominantly in hippocampus. J. Biol. Chem., 267: 19761–19764.

Ansorge, M., Tanneberger, C., Davies, B., Theuring, F. and Kusserow, H. (2004) Analysis of the murine 5-HT$_{1A}$ receptor gene promoter *in vitro* and *in vivo*. Eur. J. Neurosci., 20: 363–374.

Aoyama, M., Asai, K., Shishikura, T., Kawamoto, T., Miyachi, T., Yokoi, T., Togari, H., Wada, Y., Kato, T. and Nakagawara, A. (2001) Human neuroblastomas with unfavorable biologies express high levels of brain-derived neurotrophic factor mRNA and a variety of its variants. Cancer Lett., 164: 51–60.

Arias, B., Arranz, M.J., Gasto, C., Catalan, R., Pintor, L., Gutierrez, B., Kerwin, R.W. and Fananas, L. (2002) Analysis of structural polymorphisms and C-1018G promoter variant of the 5-HT$_{1A}$ receptor gene as putative risk factors in major depression. Mol. Psychiatry, 7: 930–932.

Bach, A.W.J., Lan, N.C., Johnson, D.L., Abell, C.W., Bembenek, M.E., Kwan, S.-W., Seeburg, P.H. and Shih, J.C. (1988) cDNA cloning of human liver monoamine oxidase A and B: molecular basis of differences in enzymatic properties. Proc. Natl. Acad. Sci. USA, 85: 4934–4938.

Baj, G. and Tongiorgi, E. (2009) BDNF splice variants from the second promoter cluster support cel survival of differentiated neuroblastoma upon cytotoxic stress. J. Cell. Sci., 122: 36–43.

Balciuniene, J., Emilson, L., Oreland, L., Pettersson, U. and Jazin, E.E. (2002) Investigation of the functional effect of monoamine oxidase polymorphisms in human brain. Hum. Genet., 110: 1–7.

Bard, J.A., Zgombick, J., Adham, N., Vaysse, P., Branchek, T.A. and Weinshank, R.L. (1993) Cloning of a novel human serotonin receptor (5-HT$_7$) positively linked to adenylate cyclase. J. Biol. Chem., 268: 23422–23426.

Barnes, N.M. and Sharp, T. (1999) A review of central 5-HT receptors and their function. Neuropharmacology, 38: 1083–1152.

Battersby, S., Ogilvie, A.D., Smith, C.A., Blackwood, D.H., Muir, W.J., Quinn, J.P., Fink, G., Goodwin, G.M. and Harmar, A.J. (1996) Structure of a variable number tandem repeat of the serotonin transporter gene and association with affective disorder. Psychiatr. Genet., 6: 177–181.

Battersby, S., Alan, D., Ogilvie, D.H.R., Blackwood, S.S., Miratul, M.K., Muqit, W.J., Muir, P.T., Goodwin, G.M. and Harmar, A.J. (1999) Presence of multiple functional polyadenylation signals and a single nucleotide polymorphism in the 3′ untranslated region of the human serotonin transporter gene. J. Neurochem., 72: 1384–1388.

Beebee, T. and Burke, J. (1992) Gene Structure and Transcription. IRL Press, Oxford University Press, Oxford.

Belelli, D., Balcarek, J.M., Hope, A.G., Peters, J.A., Lambert, J.J. and Blackburn, T.P. (1995) Cloning and functional expression of a human 5-hydroxytryptamine type 3A$_S$ receptor subunit. Mol. Pharmacol., 48: 1054–1062.

Bender, E., Pindon, A., van Oers, I., Zhang, Y.-B., Gommeren, W., Verhasselt, P., Jurzak, M., Leysen, J. and Luyten, W. (2000) Structure of the human serotonin 5-HT$_4$ receptor gene and cloning of a novel 5-HT$_4$ splice variant. J. Neurochem., 74: 478–489.

Blondel, O., Vandecasteele, G., Gastineau, M., Leclerc, S., Dahmoune, Y., Langlois, M. and Fischmeister, R. (1997) Molecular and functional characterisation of a 5-HT$_4$ receptor cloned from human atrium. FEBS Lett., 412: 465–474.

Blondel, O., Gastineau, M., Dahmoune, Y., Langlois, M. and Fischmeister, R. (1998) Cloning, expression and pharmacology of four human 5-hydroxytryptamine$_4$ receptor isoforms produced by alternative splicing in the carboxyl terminus. J. Neurochem., 70: 2252–2261.

Bockaert, J., Claeysen, S., Compan, V. and Dumuis, A. (2004) 5-HT$_4$ receptors. Curr. Drug Targets-CNS Neurol. Disord., 3: 39–51.

Boularand, S., Darmon, M.C., Ganem, Y., Launay, J.-M. and Mallet, J. (1990) Complete coding sequence of human tryptophan hydroxylase. Nucleic. Acids Res., 18: 4257.

Boularand, S., Darmon, M.C., Ravassard, P. and Mallet, J. (1995a) The human tryptophan hydroxylase gene. An unusual splicing complexity in the 5′ untranslated region. J. Biol. Chem., 270: 3748–3756.

Boularand, S., Darmon, M.C., Ravassard, P. and Mallet, J. (1995b) Characterisation of the human tryptophan hydroxylase gene promoter. J. Biol. Chem., 270: 3757–3764.

Bradely, C.C. and Blakely, R.D. (1997) Alternative splicing of the human serotonin transporter gene. J. Neurochem., 69: 1356–1367.

Bradley, P.B., Engel, G., Feniuk, W., Fozard, J.R., Humphrey, P.P., Middlemiss, D.N., Mylecharane, E.J., Richardson, B.P. and Saxena, P.R. (1986) Proposals for the classification and nomenclature of functional receptors for 5-hydroxytryptamine. Neuropharmacology, 25: 563–576.

Brattelid, T., Kvingedal, A.M., Krobert, K.A., Andressen, K., Bach, T., Hystad, M.E., Kaumann, A.J. and Levy, F.O. (2004) Cloning, pharmacological characterisation and tissue distribution of a novel 5-HT$_4$ receptor splice variant, 5-HT$_{4(i)}$. Naunyn-Schmiedeberg's Arch. Pharmacol., 369: 616–628.

Bray, N.J., Buckland, P.R., Hall, H., Owen, M.J. and O'Donovan, M.C. (2004) The serotonin-2A receptor gene locus does not contain common polymorphism affecting mRNA levels in adult brain. Mol. Psychiatry, 9: 109–114.

Brown, S.M., Peet, E., Manuck, S.B., Williamson, D.E., Dahl, R.E., Ferrell, R.E. and Hariri, A.R. (2005) A regulatory variant of the human tryptophan hydroxylase-2 gene biases amygdale reactivity. Mol. Psychiatry, 10: 884–888.

Brown, T. (1999) Essential Molecular Biology. Oxford University Press, Oxford.

Brown, W.M. and Brown, P.M. (2002) Transcription. Taylor & Francis, London.

Brummett, B.H., Boyle, S.H., Siegler, I.C., Kuhn, C.M., Ashley-Koch, A., Jonassaint, C.R., Züchner, S., Collins, A. and Williams, R.B. (2008) Effects of environmental stress and gender on associations among symptoms of depression and the serotonin transporter gene linked polymorphic region (5-HTTLPR). Behav. Genet., 38: 34–43.

Brunner, H.G., Nelen, M., Breakefield, X.O., Ropers, H.H. and van Oost, B.A. (1993) Abnormal behaviour associated with a point mutation in the structural gene for monoamine oxidase A. Science, 262: 578–580.

Bruss, M., Gothert, M., Hayer, M. and Bonisch, H. (1988) Molecular cloning of alternatively spliced human 5-HT_3 receptor Cdna. Ann. NY Acad. Sci., 861: 234–235.

Bruss, M., Eucker, T., Gothert, M. and Bonisch, H. (2000) Exon–intron organisation of the human 5-HT_{3A} receptor gene. Neuropharmacology, 39: 308–315.

Buckholtz, J.W. and Meyer-Lindenberg, A. (2008) MAOA and the neurogenetic architecture of human aggression. Trends Neurosci., 31: 120–129.

Buckland, P.R., Hoogendoorn, B., Guy, C.A., Smith, S.K., Coleman, S. and O'Donovan, M.C. (2005) Low gene expression conferred by association of an allele of the 5-HT_{2C} receptor gene with antipsychotic-induced weight gain. Am. J. Psychiatry, 162: 613–615.

Burns, C.M., Chu, H., Rueter, S.M., Hutchinson, L.K., Canton, H., Sanders-Bush, E. and Emeson, R.B. (1997) Regulation of serotonin-2C receptor G-protein coupling by RNA editing. Nature, 387: 303–308.

Canli, T., Congdon, E., Gutknecht, L., Constable, R.T. and Lesch, K.P. (2005) Amygdala responsiveness is modulated by tryptophan hydroxylase-2 gene variation. J. Neural. Transm., 112: 1479–1485.

Canton, H., Emeson, R.B., Barker, E.L., Backstrom, J.R., Lu, J.T., Chang, M.S. and Sanders-Bush, E. (1996) Identification, molecular cloning, and distribution of a short variant of the $\text{5-hydroxytryptamine}_{2C}$ receptor produced by alternative splicing. Mol. Pharmacol., 50: 799–807.

Carey, M. and Smale, S.T. (2000) Transcriptional Regulation in Eukaryotes: Concepts, Strategies and Techniques. Cold Spring Harbor Laboratory Press, Cold Spring Harbor, NY.

Cases, O., Seif, I., Grimsby, J., Gaspar, P., Chen, K., Pournin, S., Muller, U., Aguet, M., Babinet, C., Shih, J.C. et al. (1995) Aggressive behaviour and altered amounts of brain serotonin and norepinephrine in mice lacking MAOA. Science, 270: 362–364.

Caspi, A., McClay, J., Moffitt, T.E., Mill, J., Martin, J., Craig, I.W., Taylor, A. and Poulton, R. (2002) Role of genotype in the cycle of violence in maltreated children. Science, 297: 851–854.

Caspi, A., Sugden, K., Moffitt, T.E., Taylor, A., Craig, I.W., Harrington, H., McClay, J., Mill, J., Martin, J., Braithewaite, A. and Poulton, R. (2003) Influence of life stress on depression: moderation by a polymorphism in the 5-HTT gene. Science, 301: 386–389.

Castensson, A., Emilson, L., Sundberg, R. and Jazin, E. (2003) Decrease of serotonin receptor 2C in schizophrenia brains identified by high-resolution mRNA expression analysis. Biol. Psychiatry, 54: 1212–1221.

Castensson, A., Aberg, K., McCarthy, S., Saetre, P., Andersson, B. and Jazin, E. (2005) Serotonin receptor 2C (HTR2C) and schizophrenia: examination of possible medication and genetic influences on expression levels. Am. J. Med. Genet. B Neuropsychiatr. Genet., 134: 84–89.

Cervilla, J.A., Molina, E., Rivera, M., Torres-González, F., Bellón, J.A., Moreno, B., Luna, J.D., Lorente, J.A., Mayoral, F., King, M., Nazareth, I. and Gutiérrez, B. PREDICT Study Core Group (2007) The risk for depression conferred by stressful life events is modified by variation at the serotonin transporter 5HTTLPR genotype: evidence from the Spanish PREDICT-Gene cohort. Mol. Psychiatry, 8: 748–755.

Chamas, F. and Sabban, E.L. (2002) Role of the 5′ untranslated region (UTR) in the tissue-specific regulation of rat tryptophan hydroxylase gene expression by stress. J. Neurochem., 82: 645–654.

Chambard, J.C., van Obberghen-Schilling, E., Haslam, R.J., Vouret, V. and Pouyssegur, J. (1990) Chinese hamster serotonin (5HT) type 2 receptor cDNA sequence. Nucleic Acids Res., 18: 5282.

Chanda, P.K., Minchin, M.C.W., Davis, A.R., Greenberg, L., Reilly, Y., McGregor, W.H., Bhat, R., Lubeck, M.D., Mizutani, S. and Hung, P.P. (1993) Identification of residues important for ligand binding to the human $\text{5-hydroxytryptamine}_{1A}$ serotonin receptor. Mol. Pharmacol., 43: 516–520.

Chen, G.-L. and Miller, G.M. (2008) Rhesus monkey tryptophan hydroxylase-2 coding region haplotypes affect mRNA stability. Neuroscience, 155: 485–491.

Chen, Z.Y., Hotamisligil, G.S., Huang, J.-K., Wen, L., Exxeddine, D., Aydin-Muderrisoglu, N., Powell, J.F., Huang, R.H., Breakefield, X.O., Craig, I. and Hsu, Y.-P.P. (1991) Structure of the human gene for monoamine oxidase type A. Nucleic Acids Res., 19: 4537–4541.

Chen, K., Yang, W., Grimsby, J. and Shih, J.C. (1992) The human 5-HT_2 receptor is encoded by a multiple intron–exon gene. Mol. Brain. Res., 14: 20–26.

Chen, W.G., Chang, Q., Lin, Y., Meissner, A., West, A.E., Griffith, E.C., Jaenisch, R. and Greenberg, M.E. (2003) Derepression of BDNF transcription involves calcium-dependent phosphorylation of MeCP2. Science, 302: 885–889.

Chen, Z.-Y., Patel, P.D., Sant, G., Meng, C.-X., Teng, K.K., Hempstead, B.L. and Lee, F.S. (2004) Variant brain-derived neurotrophic factor (BDNF) (Met66) alters the intracellular trafficking and activity-dependent secretion of wild-type BDNF in neuroscretory cells and cortical neurons. J. Neurosci., 24: 4401–4411.

Chen, K., Ou, X.-M., Chen, G., Choi, S.H. and Shih, J.C. (2005) R1, a novel repressor of the human monoamine oxidase A. J. Biol. Chem., 280: 11552–11559.

Chen, G.-L., Novak, M.A., Hakim, S., Xie, Z. and Miller, G.M. (2006) Tryptophan hydroxylase-2 gene polymorphisms in rhesus monkeys: association with hypothalamic-pituitary-adrenal axis function and *in vitro* gene expression. Mol. Psychiatry, 11: 914–928.

Chen, G.-L., Vallender, E.J. and Miller, G.M. (2008) Functional characterisation of the human *TPH2* 5′ regulatory

region: untranslated region and polymorphisms modulate gene expression *in vitro*. Hum. Genet., 122: 645–657.
Chorbov, V.M., Lobos, E.A., Todorov, A.A., Heath, A.C., Botteron, K.N. and Todd, R.D. (2007) Relationship of 5-HTTLPR genotypes and depression risk in the presence of trauma in a female twin sample. Am. J. Med. Genet. B Neuropsychiatr. Genet., 144: 830–833.
Chu, C.-Y. and Rana, T.M. (2007) Small RNAs: regulators and guardians of the genome. J. Cell. Physiol., 213: 412–419.
Claeysen, S., Sebben, M., Journot, L., Bockaert, J. and Dumuis, A. (1996) Cloning, expression and pharmacology of the mouse 5-HT_{4L} receptor. FEBS Lett., 398: 19–25.
Claeysen, S., Faye, P., Sebben, M., Lemaire, S., Bockaert, J., Dumuis, A. and Taviaux, S. (1997) Assignment of 5-hydroxytryptamine receptor (HTR4) to human chromosome 5 bands q31–q33 by *in situ* hybridisation. Cytogenet. Cell. Genet., 78: 133–134.
Claeysen, S., Faye, P., Sebben, M., Taviaux, S., Bockaert, J. and Dumuis, A. (1998) 5-HT_4 receptors: cloning and expression of new splice variants. Ann. NY Acad. Sci., 861: 49–56.
Claeysen, S., Sebben, M., Becamel, C., Bockaert, J. and Dumuis, A. (1999) Novel brain-specific 5-HT_4 receptor splice variants show marked constitutive activity: role of the C-terminal intracellular domain. Mol. Pharmacol., 55: 910–920.
Collier, D.A., Arranz, M.J., Sham, P., Battersby, S., Vallada, H., Gill, P., Aitchison, K.J., Sodhi, M., Li, T., Roberts, G.W., Smith, B., Morton, J., Murray, R.M., Smith, D. and Kirov, G. (1996) The serotonin transporter is a potential susceptibility factor for bipolar affective disorder. Neuroreport, 7: 1675–1679.
Cooper, J.R., Bloom, F.E. and Roth, R.H. (1996) Serotonin (5-hydroxytryptamine) and histamine. In: Cooper, J.R., Bloom, F.E. and Roth, R.H. (Eds.), The Biochemical Basis of Neuropharmacology. Oxford University Press, New York, NY, pp. 352–409.
Costall, B. and Naylor, R.J. (2004) 5-HT3 receptors. Curr. Drug Targets-CNS Neurol. Disord., 3: 27–37.
Cote, F., Schussler, S., Boularand, S., Peirotes, A., Thevenot, E., Mallet, J. and Vodjdani, G. (2002) Involvement of NF-Y and Sp1 in basal and cAMP-stimulated transcriptional activation of the tryptophan hydroxylase (TPH) gene in the pineal gland. J. Neurochem., 81: 673–685.
Cote, F., Thevenot, E., Fligny, C., Fromes, Y., Darmon, M., Ripoche, M.-A., Bayard, E., Hanoun, N., Saurini, F., Lechat, P., Dandolo, L., Hamon, M., Mallet, J. and Vodjdani, G. (2003) Disruption of the nonneuronal *tph1* gene demonstrates the importance of peripheral serotonin in cardiac function. Proc. Natl. Acad. Sci. USA, 100: 13525–13530.
Cowen, D.S. (2007) Serotonin and neuronal growth factors – a convergence of signalling pathways. J. Neurochem., 101: 1161–1171.
Craig, I.W. (2007) The importance of stress and genetic variation in human aggression. Bioessays, 29: 227–236.
Craig, S.P., Boularand, S., Darmon, M.C., Mallet, J. and Craig, I.W. (1991) Localisation of human tryptophan hydroxylase (TPH) to chromosome 11p15.3-p14 by *in situ* hybridisation. Cytogenet. Cell. Genet., 56: 157–159.
Czesak, M., Lemonde, S., Peterson, E.A., Rogaeva, A. and Albert, P.R. (2006) Cell-specific repressor or enhancer activities of Deaf-1 at a serotonin 1A receptor gene polymorphism. J. Neurosci., 26: 1864–1871.
Darmon, M.C., Grima, B., Cash, C.D., Maitre, M. and Mallet, J. (1986) Isolation of a rat pineal gland cDNA clone homologous to tyrosine and phenylalanine hydroxylases. FEBS Lett., 206: 43–46.
Darmon, M.C., Guibert, B., Leviel, V., Ehret, M., Maitre, M. and Mallet, J. (1988) Sequence of two mRNAs encoding active rat tryptophan hydroxylase. J. Neurochem., 51: 312–316.
Davies, P.A., Pistis, M., Hanna, M.C., Peters, J.A., Lambert, J.J., Hales, T.G. and Kirkness, E.F. (1999) The 5-HT_{3B} subunit is a major determinant of serotonin-receptor function. Nature, 397: 359–363.
Deckert, J., Catalano, M., Syagailo, Y.V., Bosi, M., Okladnova, O., Di Bella, D., Nothen, M.M., Maffei, P., Franke, P., Fritze, J., Maier, W., Propping, P., Beckmann, H., Bellodi, L. and Lesch, K.-P. (1999) Excess of high activity monoamine oxidase A gene promoter alleles in female patients with panic disorder. Hum. Mol. Genet., 8: 621–624.
Delbruck, S.J., Wendel, B., Grunewald, I., Sander, T., Morris-Rosendahl, D., Crocq, M.A., Berrettini, W.H. and Hoehe, M.R. (1997) A novel allelic variant of the human serotonin transporter gene regulatory polymorphism. Cytogenet. Cell Genet., 79: 214–220.
Delort, J., Dumas, J.B., Darmon, M.C. and Mallet, J. (1989) An efficient strategy for cloning 5′ extremities of rare transcripts permits isolation of multiple 5′-untranslated regions of rat tryptophan hydroxylase mRNA. Nucleic Acids Res., 17: 6439–6447.
De Luca, V., Likhodi, O., Van Tol., H.H.M., Kennedy, J.L. and Wong, A.H.C. (2005) Tryptophan hydroxylase 2 gene expression and promoter polymorphisms in bipolar disorder and schizophrenia. Psychopharmacology, 183: 378–382.
Denney, R.M., Sharma, A., Dave, S.K. and Waguespack, A. (1994) A new look at the promoter of the human monoamine oxidase A gene: mapping transcription initiation sites and capacity to drive luciferase expression. J. Neurochem., 63: 843–856.
Ding, G., Toth, M., Zhou, Y., Parks, C., Hoffman, B.J. and Shenk, T. (1993) Glial cell-specific expression of the serotonin 2 receptor gene: selective reactivation of a repressed promoter. Mol. Brain Res., 20: 181–191.
D'Souza, U.M., Kel, A. and Sluyter, F. (2003) From transcriptional regulation to aggressive behavior. Behav. Genet., 33: 549–562.
D'Souza, U.M. and Craig, I.W. (2008) Functional genetic polymorphisms in serotonin and dopamine gene systems and their significance in behavioural disorders. Prog. Brain Res., 172: 73–98.
Du, Y.-L., Wilcox, B.D., Teitler, M. and Jeffrey, J.J. (1994) Isolation and characterisation of the rat 5-hydroxytryptamine type 2 receptor promoter: constitutive and inducible activity in myometrial smooth muscle cells. Mol. Pharmacol., 45: 1125–1131.
Duan, J., Sanders, A.R., Vander Molen, J.E., Martinolich, L., Mowry, B.J., Levinson, D.F., Crowe, R.R., Silverman, J.M. and Gejman, P.V. (2003) Polymorphisms in the 5′ untranslated region of the human serotonin receptor 1B (HTR1B) gene affect gene expression. Mol. Psychiatry, 8: 901–910.
Dubin, A.E., Huvar, R., D'Andrea, M.R., Pyati, J., Zhu, J.Y., Joy, K.C., Wilson, S.J., Galindo, J.E., Glass, C.A., Luo, L., Jackson, M.R., Lovenberg, T.W. and Erlander, M.G. (1999) The pharmacological and functional characterisation of the serotonin 5-HT_{3A} receptor are specifically modified by a 5-HT_{3B} receptor subunit. J. Biol. Chem., 274: 30799–30810.
Egan, M.F., Kojima, M., Callicott, J.H., Goldberg, T.E., Kolachana, B.S., Bertolino, A., Zaitsev, E., Gold, B.,

Goldman, D., Dean, M., Lu, B. and Weinberger, D.R. (2003) The BDNF val66met polymorphism affects activity-dependent secretion of BDNF and human memory and hippocampal function. Cell, 112: 257–269.

Eley, T.C., Sugden, K., Corsico, A., Gregory, A.M., Sham, P., McGuffin, P., Plomin, R. and Craig, I.W. (2004) Gene-environment interaction analysis of serotonin system markers with adolescent depression. Mol. Psychiatry, 9: 908–915.

Erlander, M.G., Lovenberg, T.W., Baron, B.M., De Lecea, L., Danielson, P.E., Racke, M., Slone, A.L., Siegel, B.W., Foye, P.E., Cannon, K., Burns, J.E. and Sutcliffe, J.G. (1993) Two members of a distinct subfamily of 5-hydroxytryptamine receptors differentially expressed in rat brain. Proc. Natl. Acad. Sci. USA, 90: 3452–3456.

Fargin, A., Raymond, J.R., Lohse, M.J., Kobilka, B.K., Caron, M.G. and Lefkowitz, R.J. (1988) The genomic clone G-21 which resembles a β-adrenergic receptor sequence encodes the 5-HT_{1A} receptor. Nature, 335: 358–360.

Fitzgerald, L.W., Iyer, G., Conklin, D.S., Krause, C.M., Marshall, A., Patterson, J.P., Tran, D.P., Jonak, G.J. and Hartig, P.R. (1999) Messenger RNA editing of the human serotonin 5-HT_{2C} receptor. Neuropsychopharmacology, 21: 82S–90S.

Fitzpatrick, P.F. (1999) Tetrahydropterin-dependent amino acid hydroxylases. Annu. Rev. Biochem., 68: 355–381.

Flattem, N.L. and Blakely, R.D. (2000) Modified structure of the human serotonin transporter promoter. Mol. Psychiatry, 5: 110–115.

Flomen, R., Knight, J., Sham, P., Kerwin, R. and Makoff, A. (2004) Evidence that RNA editing modulates splice site selection in the 5-HT2C receptor gene. Nucleic Acids Res., 32: 2113–2122.

Foguet, M., Nguyen, H., Le, H. and Lubbert, H. (1992a) Structure of the mouse 5-HT1C, 5-HT2 and stomach fundus serotonin receptor genes. Neuroreport, 3: 345–348.

Foguet, M., Hoyer, D., Pardo, L.A., Parekh, A., Kluxen, F.W., Kalkman, H.O., Stuhmer, W. and Lubbert, H. (1992b) Cloning and functional characterisation of the rat stomach fundus serotonin receptor. EMBO J., 11: 3481–3487.

Foley, D.L., Eaves, L.J., Wormley, B., Silberg, J.L., Maes, H.H., Kuhu, J. and Riley, B. (2004) Childhood adversity, monoamine oxidase A genotype, and risk for conduct disorder. Arch. Gen. Psychiatry, 61: 738–744.

Fowler, J.S., Alia-Klein, N., Kriplani, A., Logan, J., Williams, B., Zhu, W., Craig, I.W., Telang, F., Goldstein, R., Volkow, N.D., Vaska, P. and Wang, G.-W. (2007) Evidence that brain MAO A activity does not correspond to MAO A genotype in healthy male subjects. Biol. Psychiatry, 62: 355–358.

Fujiwara, Y., Nelson, D.L., Kashihara, K., Varga, E., Roeske, W.R. and Yamamura, H.I. (1990) The cloning and sequence analysis of the rat serotonin-1A receptor gene. Life Sci., 47: 127–132.

Furmark, T., Henningsson, S., Appel, L., Ahs, F., Linnman, C., Pissiota, A., Faria, V., Oreland, L., Bani, M., Pich, E.M., Eriksson, E. and Fredrikson, M. (2009) Genotype over diagnosis in amygdale responsiveness: affective processing in social anxiety disorder. J. Psychiatry Neurosci., 34: 30–40.

Gelernter, J., Rao, P.A., Pauls, D.L., Hamblin, M.W., Sibley, D.R. and Kidd, K.K. (1995) Assignment of the 5HT7 receptor gene (HTR7) to chromosome 10q and exclusion of genetic linkage with Tourette syndrome. Genomics, 26: 207–209.

Gellynck, E., Laenen, K., Andressen, K.W., Lintermans, B., De Martelaere, K., Matthys, A., Levy, F.O., Haegeman, G., Vanhoenacker, P. and van Craenenbroeck, K. (2008) Cloning, genomic organisation and functionality of 5-HT_7 receptor splice variants from mouse brain. Gene, 426: 23–31.

Gerald, C., Adham, N., Kao, H.-T., Olsen, M.A., Laz, T.M., Schechter, L.E., Bard, J.A., Vaysse, P.J.J., Hartig, P.R., Branchek, T.A. and Weinshank, R.L. (1995) The 5-HT_4 receptor: molecular cloning and pharmacological characterisation of two splice variants. EMBO J., 14: 2806–2815.

Gillespie, N.A., Whitfield, J.B., Williams, B., Heath, A.C. and Martin, N.G. (2005) The relationship between stressful life events, the serotonin transporter (5-HTTLPR) genotype and major depression. Psychol. Med., 35: 101–111.

Gotlib, I.H., Joormann, J., Minor, K.L. and Hallmayer, J. (2008) HPA axis reactivity: a mechanism underlying the association among 5-HTTLPR, stress and depression. Biol. Psychiatry, 63: 847–851.

Green, E.K., Raybould, R., Macgregor, S., Hyde, S., Young, A.H., O'Donovan, M.C., Owen, M.J., Kirov, G., Jones, L., Jones, I. and Craddock, N. (2006) Genetic variation of brain-derived neurotrophic factor (BDNF) in bipolar disorder: case-control study of over 3000 individuals from the UK. Br. J. Psychiatry, 188: 21–25.

Grenett, H.E., Ledley, F.D., Reed, L.L. and Woo, S.L.C. (1987) Full-length cDNA for rabbit tryptophan hydroxylase: functional domains and evolution of aromatic amino acid hydroxylases. Proc. Natl. Acad. Sci. USA, 84: 5530–5534.

Grimsby, J., Lan, N.C., Neve, R., Chen, K. and Shih, J.C. (1990) Tissue distribution of human monoamine oxidase A and B mRNA. J. Neurochem., 55: 1166–1169.

Grimsby, J., Chen, K., Wang, L.-J., Lan, N.C. and Shih, J.C. (1991) Human monoamine oxidase A and B genes exhibit identitical exon–intron organisation. Proc. Natl. Acad. Sci. USA, 88: 3637–3641.

Haghighi, F., Bach-Mizrachi, H., Huang, Y.Y., Arango, V., Shi, S., Dwork, A.J., Rosoklija, G., Sheng, H.T., Morozova, I., Ju, J., Russo, J.J. and Mann, J.J. (2008) Genetic architecture of the human tryptophan hydroxylase 2 gene: existence of neural isoforms and relevance for major depression. Mol. Psychiatry, 13: 813–820.

Hamblin, M.W. and Metcalf, M.A. (1991) Primary structure and functional characterization of a human 5-HT_{1D} type serotonin receptor. Mol. Pharmacol., 40: 143–148.

Hamblin, M.W., Metcalf, M.A., McGuffin, R.W. and Karpells, S. (1992) Molecular cloning and functional characterization of the human 5-HT_{1B} serotonin receptor: a homologue of the rat 5-HT_{1B} receptor with 5-HT_{1D}-like pharmacological specificity. Biochem. Biophys. Res. Comm., 184: 752–759.

Hariri, A.R., Goldberg, T.E., Mattay, V.S., Kolachana, B.S., Callicott, J.H., Egan, M.F. and Weinberger, D.R. (2003) Brain-derived neurotrophic factor val^{66}met polymorphism affects human memory-related hippocampal activity and predicts memory performance. J. Neurosci., 23: 6690–6694.

Hartig, P.R., Branchek, T.A. and Weinshank, R.L. (1992) A subfamily of 5-HT1D receptor genes. Trends Pharmacol. Sci., 13: 152–159.

Hartig, P.R., Hoyer, D., Humphrey, P.P.A. and Martin, G.R. (1996) Alignment of receptor nomenclature with the human genome: classification of the 5-HT_{1B} and 5-HT_{1D} receptor subtypes. Trends Pharmacol. Sci., 17: 103–105.

Heidmann, D.E.A., Metcalf, M.A., Kohen, R. and Hamblin, M.W. (1997) Four 5-hydroxytryptamine$_7$ (5-HT_7) receptor isoforms in human and rat produced by alternative

splicing: species differences due to altered intron–exon organisation. J. Neurochem., 68: 1372–1381.
Heils, A., Teufel, A., Petri, S., Seemann, M., Bengel, D., Balling, U., Riederer, P. and Lesch, K.-P. (1995) Functional promoter and polyadenylation site mapping of the human serotonin (5-HT) transporter gene. J. Neural Transm., 102: 247–254.
Heils, A., Teufel, A., Petri, S., Stober, G., Riederer, P., Bengel, D. and Lesch, K.P. (1996) Allelic variation of human serotonin transporter. J. Neurochem., 66: 2621–2624.
Hill, M.J. and Reynolds, G.P. (2007) 5-HT_{2C} receptor gene polymorphisms associated with antipsychotic drug action alter promoter activity. Brain Res., 1149: 14–17.
Hiroi, T., Hayashi-Kobayashi, N., Nagumo, S., Ino, M., Okawa, Y., Aoba, A. and Matsui, H. (2001) Identification and characterisation of the human serotonin-4 receptor gene promoter. Biochem. Biophys. Res. Comm., 289: 337–344.
Hope, A.G., Downie, D.L., Sutherland, L., Lambert, J.J., Peters, J.A. and Burchell, B. (1993) Cloning and functional expression of an apparent splice variant of the murine 5-HT3 receptor A subunit. Eur. J. Pharmacol., 245: 187–192.
Horton, Y.M., Lubbert, H. and Houslay, M.D. (1996) Localisation of the gene for the human serotonin 5-HT(2B) receptor to chromosome 2. Mol. Membr. Biol., 13: 29–31.
Hoyer, D. and Martin, G. (1997) 5-HT receptor classification and nomenclature: towards a harmonization with the human genome. Neuropharmacology, 36: 419–428.
Hoyer, D., Clarke, D.E., Fozard, J.R., Hartig, P.R., Martin, G.R., Mylecharane, E.J., Saxena, P.R. and Humphrey, P.P.A. (1994) International Union of Pharmacology Classification of Receptors for 5-hydroxytryptamine (serotonin). Pharmacol. Rev., 46: 157–203.
Hoyer, D., Hannon, J.P. and Martin, G.R. (2002) Molecular, pharmacological and functional diversity of 5-HT receptors. Pharmacol. Biochem. Behav.: 533–554.
Hranilovic, D., Stefulj, J., Furac, I., Kubat, M., Balija, M. and Jernej, B. (2003) Serotonin transporter gene promoter (5-HTTLPR) and intron 2 (VNTR) polymorphism in Croatian suicide victims. Biol. Psychiatry, 54: 884–889.
Hranilovic, D., Stefulj, J., Schwab, S., Borrmann-Hassenbach, M., Albus, M., Jernej, B. and Wildenauer, D. (2004) Serotonin transporter promoter and intron 2 polymorphisms: relationship between allelic variants and gene expression. Biol. Psychiatry, 55: 1090–1094.
Hu, X., Lipsky, R.H., Zhu, G., Akhtar, L.A., Taubman, J., Greenberg, B.D., Xu, K., Arnold, P.D., Richter, M.A., Kennedy, J.L., Murphy, D.L. and Goldman, D. (2006) Serotonin transporter promoter gain-of-function genotypes are linked to obsessive-compulsive disorder. Am. J. Hum. Genet., 78: 815–826.
Humphrey, P.P.A., Hartig, P. and Hoyer, D. (1993) A proposed new nomenclature for 5-HT receptors. Trends Pharmacol. Sci., 14: 233–237.
Invernizzi, R.W. (2007) Role of TPH-2 in brain function: news from behavioural and pharmacologic studies. J. Neurosci. Res., 85: 3030–3035.
Isenberg, K.E., Ukhun, I.A., Holstad, S.G., Jafri, S., Uchida, U., Zorumski, C.F. and Yang, J. (1993) Partial cDNA cloning and NGF regulation of a rat 5-HT3 receptor subunit. Neuroreport, 5: 121–124.
Jasper, J.R., Kosaka, Z.P., Chang, T.D.J. and Eglen, R.M. (1997) Cloning, expression and pharmacology of a truncated splice variant of the human 5-HT_7 receptor (h5-$HT_{7(b)}$). Br. J. Pharmacol., 122: 126–132.
Jensen, K.P., Covault, J., Conner, T.S., Tennen, H., Kranzler, H.R. and Furneaux, H.M. (2008) A common polymorphism in serotonin receptor 1B mRNA moderates regulation by miR-96 and associates with aggressive human behaviours. Mol. Psychiatry, 19: 1–9.
Jin, H., Oksenberg, D., Ashkenazi, A., Peroutka, S.J., Duncan, A.M.V., Rozmahel, R., Yang, Y., Mengod, G., Palacios, J.M. and O'Dowd, B.F. (1992) Characterisation of the human 5-hydroxytryptamine$_{1B}$ receptor. J. Biol. Chem., 267: 5735–5738.
Julius, D., MacDermott, A.B., Axel, R. and Jessell, T.M. (1988) Molecular characterization of a functional cDNA encoding the serotonin 1c receptor. Science, 241: 558–564.
Julius, D., Huang, K.N., Livelli, T., Axel, R. and Jessell, T.M. (1990) The 5-HT_2 receptor defines a family of structurally distinct but functionally conserved serotonin receptors. Proc. Natl. Acad. Sci. USA, 87: 928–932.
Kapeller, J., Houghton, L., Monnikes, H., Walstab, J., Moller, D., Bonisch, H., Burwinkel, B., Autschbach, F., Funke, B., Lasitschka, F., Gassler, N., Fischer, C., Whorwell, P.J., Atkinson, W., Fell, C., Buchner, K.J., Schmidtmann, M., van der Voort, I., Wisser, A.-S., Berg, T., Rappold, G. and Niesler, B. (2008) First evidence for an association of a functional variant in the microRNA-510 target site of the serotonin receptor-type 3E gene with diarrhea predominant irritable bowel syndrome. Hum. Mol. Genet., 17: 2967–2977.
Kaufman, J., Yang, B.-Z., Douglas-Palumberi, H., Houshyar, S., Lipschitz, D., Krystal, J.H. and Gerlenter, J. (2004) Social supports and serotonin transporter gene moderate depression in maltreated children. Proc. Natl. Acad. Sci. USA, 101: 17316–17321.
Kendler, K.S., Kuhn, J.W., Vittum, J., Prescott, C.A. and Riley, B. (2005) The interaction of stressful life events and a serotonin transporter polymorphism in the prediction of episodes of major depression: a replication. Arch. Gen. Psychiatry, 62: 529–535.
Kim, K.S., Wessel, T.C., Stone, D.M., Carver, C.H., Joh, T.H. and Park, D.H. (1991) Molecular cloning and characterisation of cDNA encoding tryptophan hydroxylase from rat central serotonergic neurons. Brain Res. Mol. Brain Res., 9: 277–283.
Klenova, E., Scott, A.C., Roberts, J., Shamsuddin, S., Lovejoy, E.A., Bergmann, S., Bubb, V.J., Royer, H.-D. and Quinn, J.P. (2004) YB-1 and CTCF differentially regulate the 5-HTT polymorphic intron 2 enhancer which predisposes to a variety of neurological disorders. J. Neurosci., 24: 5966–5973.
Kobilka, B.K., Frielle, T., Collins, S., Yang-Feng, T., Kobilka, T.S., Francke, U., Lefkowitz, R.J. and Caron, M.G. (1987) An intronless gene encoding a potential member of the family of receptors coupled to guanine nucleotide regulatory proteins. Nature, 329: 75–79.
Kochersperger, L.M., Parker, E.L., Siciliano, M., Darlington, G.J. and Denney, R.M. (1986) Assignment of genes for human monoamine oxidases A and B to the X chromosome. J. Neurosci. Res., 16: 601–616.
Kohen, R., Metcalf, M.A., Khan, N., Druck, T., Huebner, K., Lachowicz, J.E., Meltzer, H.Y., Sibley, D.R., Roth, B.L. and Hamblin, M.W. (1996) Cloning, characterisation and chromosomal localisation of a human 5-HT_6 serotonin receptor. J. Neurochem., 66: 47–56.
Kohen, R., Fashingbauer, L.A., Heidmann, D.E.A., Guthrie, C.R. and Hamblin, M.W. (2001) Cloning of the mouse 5-HT_6 serotonin receptor and mutagenesis studies of the third cytoplasmic loop. Mol. Brain Res., 90: 110–117.

Kunugi, H., Hattori, M., Kato, T., Tatsumi, M., Sakai, T., Sasaki, T., Hirose, T. and Nanko, S. (1997) Serotonin transporter gene polymorphisms: ethnic difference and possible association with bipolar affective disorder. Mol. Psychiatry, 2: 457–462.

Kursar, J.D., Nelson, D.L., Wainscott, D.B. and Baez, M. (1994) Molecular cloning, functional expression, and mRNA tissue distribution of the human 5-hydroxytryptamine$_{2B}$ receptor. Mol. Pharmacol., 46: 227–234.

Laenen, K., Haegeman, G. and Vanhoenacker, P. (2007) Structure of the human 5-HT$_7$ receptor gene and characterisation of its promoter region. Gene, 391: 252–263.

Lan, N.C., Heinzmann, C., Gal, A., Klisak, I., Orth, U., Lai, E., Grimsby, J., Sparkes, R.S., Mohandas, T. and Shih, J.C. (1989) Human monoamine oxidase A and B genes map to Xp 11.23 and are deleted in a patient with Norrie disease. Genomics, 4: 552–559.

Lanfumey, L. and Hamon, M. (2004) 5-HT1 receptors. Curr. Drug Targets-CNS Neurol. Disord., 3: 1–10.

Lankiewicz, S., Lobitz, N., Wetzel, C.H.R., Rupprecht, R., Gisselmann, G. and Hatt, H. (1998) Molecular cloning, functional expression and pharmacological characterisation of 5-hydroxytryptamine$_3$ receptor cDNA and its splice variants from guinea pig. Mol. Pharmacol., 53: 202–212.

Latchman, D. (1998) Gene Regulation: A Eukaryotic Perspective. 3rd edn. Stanley Thornes, Ltd., Cheltenham.

Lazary, J., Lazary, A., Gonda, X., Benko, A., Molnar, E., Juhasz, G. and Bagdy, G. (2008) New evidence for the association of the serotonin transporter gene (SLC6A4) haplotypes, threatening life events, and depressive phenotype. Biol. Psychiatry, 64: 498–504.

Ledley, F.D., Van Tuinen, P., Ledbetter, D., Gerhardt, T., Jones, C. and Woo, S.L.C. (1987) Assignment of tryptophan hydroxylase to human chromosome 11p. Duplication and rearrangement in aromatic acid hydroxylase evolution. Human Gene Mapping 9, Cytogenet. Cell Genet., 46: 646.

Lenze, E.J., Shardell, M., Ferrell, R.E., Orwig, D., Yu-Yahiro, J., Hawkes, W., Fredman, L., Miller, R. and Magaziner, J. (2008) Association of serotonin-1A and 2A receptor promoter polymorphisms with depressive symptoms and functional recovery in elderly persons after hip fracture. J. Affect. Disord., 111: 61–66.

Lesch, K.P. (2001) Serotonergic gene expression and depression: implications for developing novel antipdepressants. J. Affect. Disord., 62: 57–76.

Lesch, K.P., Balling, U., Gross, J., Strauss, K., Wolozin, B.L. and Murphy, D.L. (1994) Organisation of the human serotonin transporter gene. J. Neural Trans., 95: 157–162.

Lesch, K.P., Bengel, D., Heils, A., Sabol, S.Z., Greenberg, B.D., Petri, S., Benjamin, J., Muller, C.R., Hamer, D.H. and Murphy, D.L. (1996) Association of anxiety-related traits with a polymorphism in the serotonin transporter gene regulatory region. Science, 274: 1527–1531.

Levy, E.R., Powell, J.F., Buckle, V.J., Hsu, Y.P., Breakfield, X.O. and Craig, I.W. (1989) Localisation of human monoamine oxidase-A gene to Xp11.23–11.4 by *in situ* hybridisation: implications for Norrie disease. Genomics, 5: 368–370.

Levy, F.O., Gudermann, T., Perez-Reyes, E., Birnbaumer, M., Kaumann, A.J. and Birnbaumer, L. (1992a) Molecular cloning of a human serotonin receptor (S12) with a pharmacological profile resembling that of the 5-HT$_{1D}$ subtype. J. Biol. Chem., 267: 7553–7562.

Levy, F.O., Gudermann, T., Birnbaumer, M., Kaumann, A. and Birnbaumer, L. (1992b) Molecular cloning of a human gene (S31) encoding a novel serotonin receptor mediating inhibition of adenylyl cyclase. Fed. Eur. Biochem. Sci., 296: 201–206.

Levy, F.O., Holtgreve-Grez, H., Tasken, K., Solberg, R., Reid, T. and Gudermann, T. (1994) Assignment of the gene encoding the 5-HT$_{1E}$ serotonin receptor (S31) (locus HTR1E) to human chromosome 6q14-q15. Genomics, 22: 637–640.

Leysen, J.E. (2004) 5-HT2 receptors. Curr. Drug Targets-CNS Neurol. Disord., 3: 11–26.

Lim, J.-E., Pinsonneault, J., Sadee, W. and Saffen, D. (2007) Tryptophan hydroxylase 2 (*TPH2*) haplotypes predict levels of *TPH2* mRNA expression in human pons. Mol. Psychiatry, 12: 491–501.

Liu, H., Irving, H.R. and Coupar, I.M. (2001) Expression patterns of 5-HT7 receptor isoforms in the rat digestive tract. Life Sci., 69: 2467–2475.

Liu, Q.-R., Lu, L., Zhu, X.-G., Gong, J.-P., Shaham, Y. and Uhl, G.R. (2006) Rodent BDNF genes, novel promoters, novel splice variants, and regulation by cocaine. Brain Res., 1067: 1–12.

Lohoff, F.W., Sander, T., Ferraro, T.N., Dahl, J.P. and Berrettini, W.H. (2005) Confirmation of association between the Val66Met polymorphism in the brain derived neurotrophic factor (BDNF) gene and bipolar 1 Disorder. Am. J. Med. Genet. B Neuropsychiatr. Genet., 139: 51–53.

Lovejoy, E.A., Scott, A.C., Fiskerstrand, C.E., Bubb, V.J. and Quinn, J.P. (2003) The serotonin transporter intronic VNTR enhancer correlated with a predisposition to affective disorders has distinct regulatory elements within the domain based on the primary DNA sequence of the repeat unit. Eur. J. Neurosci., 17: 417–420.

Lovenberg, T.W., Erlander, M.G., Baron, B.M., Racke, M., Slone, A.L., Siegel, B.W., Craft, C.M., Burns, J.E., Danielson, P.E. and Sutcliffe, J.G. (1993) Molecular cloning and functional expression of the 5-HT$_{1E}$-like rat and human 5-hydroxytryptamine receptor genes. Proc. Natl. Acad. Sci. USA, 90: 2184–2188.

Lubbert, H., Hoffman, B.J., Snutch, T.P., van Dyke, T., Levine, A.J., Hartig, P.R., Lester, H.A. and Davidson, N. (1987) cDNA cloning of a serotonin 5-HT$_{1C}$ receptor by electrophysiological assays of mRNA-injected Xenopus oocytes. Proc. Natl. Acad. Sci. USA, 84: 4332–4336.

Lucki, I. (1998) The spectrum of behaviours influenced by serotonin. Biol. Psychiatry, 44: 151–162.

Maillet, M., Gastineau, M., Bochet, P., Asselin-Labat, M.-L., Morel, E., Laverriere, J.-N., Lompre, A.-M., Fischmeister, R. and Lezoualc'h, F. (2005) Functional studies of the 5′-untranslated region of human 5-HT$_4$ receptor mRNA. Biochem. J., 387: 463–471.

Maricq, A.V., Peterson, A.S., Brake, A.J., Myers, R.M. and Julius, D. (1991) Primary structure and functional expression of the 5-HT3 receptor, a serotonin-gated ion channel. Science, 254: 432–437.

Marini, A.M., Jiang, X., Wu, X., Tian, F., Zhu, D., Okagaki, P. and Lipsky, R.H. (2004) Role of brain-derived neurotrophic factor and NF-kappaB in neuronal plasticity and survival: from genes to phenotype. Restor. Neurol. Neurosci., 22: 121–130.

Martin, J., Cleak, J., Willis-Owen, S.A., Flint, J. and Shifman, S. (2007) Mapping regulatory variants for the serotonin transporter gene based on allelic expression imbalance. Mol. Psychiatry, 12: 421–422.

Matthes, H., Boschert, U., Amlaiky, N., Grailhe, R., Plassat, J.-L., Muscatelli, F., Mattei, M.-G. and Hen, R. (1993) Mouse

5-hydroxytryptamine$_{5A}$ and 5-hydroxytryptamine$_{5B}$ receptors define a new family of serotonin receptors: cloning, functional expression, and chromosomal localisation. Mol. Pharmacol., 43: 313–319.

Mattick, J.S. and Makunin I.V. (2006). Non-coding RNA. Hum. Mol. Genet., Spec. No. 1: R17-R29.

Mehrotra, S., Vanmolkot, K.R.J., Frants, R.R., van den Maagdenberg, A.M.J.M., Ferrari, M.D. and VanDenBrink, A.M. (2007) The Phe-124-Cys and A-161T variants of the human 5-HT$_{1B}$ receptor gene are not major determinants of the clinical response to sumatriptan. Headache, 47: 711–716.

Meijer, O.C., Williamson, A., Dallman, M.F. and Pearce, D. (2000) Transcriptional repression of the 5-HT1A receptor promoter by corticosterone via mineralocorticoid receptors depends on the cellular context. J. Neuroendocrinology, 12: 245–254.

Meineke, C., Tzvetkov, M.V., Bokelmann, K., Oetjen, E., Hirsch-Ernst, K., Kaiser, R. and Brockmoller, J. (2008) Functional characterisation of a −100_−102delAAG deletion–insertion polymorphism in the promoter region of the *HTR3B* gene. Pharmacogenetics Genomics, 18: 219–230.

Mellios, N., Huang, H.-S., Grigorenko, A., Rogaev, E. and Akbarian, S. (2008) A set of differentially expressed miRNAs, including miR-30a-5p, act as post-transcriptional inhibitors of BDNF in prefrontal cortex. Hum. Mol. Genet., 17: 3030–3042.

Meyer-Lindenberg, A., Buckholtz, J.W., Kolachana, B., Hariri, A.R., Pezawas, L., Blasi, G., Wabnitz, A., Honea, R., Verchinski, B., Callicott, J.H., Egan, M., Mattay, V. and Weinberger, D.R. (2006) Neural mechanisms of genetic risk for impulsivity and violence in humans. Proc. Natl. Acad. Sci. USA, 103: 6269–6274.

Mialet, J., Berque-Bestel, I., Eftekhari, P., Gastineau, M., Giner, M., Dahmoune, Y., Donzeau-Gouge, P., Hoebeke, J., Langlois, M., Sicsic, S., Fischmeister, R. and Lezoualc'h, F. (2000) Isolation of the serotonergic 5-HT$_{4(e)}$ receptor from human heart and comparative analysis of its pharmacological profile in C6-glial and CHO cell lines. Br. J. Pharmacol., 129: 771–781.

Michaelovsky, E., Frisch, A., Rockah, R., Peleg, L., Magal, N., Shohat, M. and Weizman, R. (1999) A novel allele in the promoter region of the human serotonin transporter gene. Mol. Psychiatry, 4: 97–99.

Miyake, A., Mochizuki, S., Takemoto, Y. and Akuzawa, S. (1995) Molecular cloning of human 5-hydroxytryptamine$_3$ receptor: heterogeneity in distribution and function among species. Mol. Pharmacol., 48: 407–416.

Mohammad-Zadeh, L.F., Moses, L. and Gwaltney-Brant, S.M. (2008) Serotonin: a review. J. Vet. Pharmacol. Ther., 31: 187–199.

Monsma, F.J. Jr., Shen, Y., Ward, R.P., Hamblin, M.W. and Sibley, D.R. (1993) Cloning and expression of a novel serotonin receptor with high affinity for tricyclic psychotropic drugs. Mol. Pharmacol., 43: 320–327.

Mossner, R., Daniel, S., Albert, D., Heils, A., Okladnova, O., Schmitt, A. and Lesch, K.-P. (2000) Serotonin transporter function is modulated by brain-derived neurotrophic factor (BDNF) but not nerve growth factor (NGF). Neurochem. Intl., 36: 197–202.

Mouradian, M.M., Minowa, M.T. and Minowa, T. (1994) Dopamine receptor genes: promoters and transcriptional activation. In: Niznik, H.B. (Ed.), Dopamine Receptors and Transporters: Pharmacology, Structure and Function. Marcel Dekker, New York, NY, pp. 205–235.

Mrazek, D.A., Rush, A.J., Biernacka, J.M., O'Kane, D.J., Cunningham, J.M., Wieben, E.D., Schaid, D.J., Drews, M.S., Courson, V.L., Snyder, K.A., Black, J.L. III and Weinshilboum, R.M. (2008). SLC6A4 variation and citalopram response. Am. J. Med. Genet. B Neuropsychiatr. Genet., 2008 Jul 10 (Epub ahead of print).

Murphy, K.L., Zhang, X., Gainetdinov, R.R., Beaulieu, J.-M. and Caron, M.G. (2008) A regulatory domain in the N terminus of tryptophan hydroxylase 2 controls enzyme expression. J. Biol. Chem., 283: 13216–13224.

Nakamura, M., Ueno, S., Sano, A. and Tanabe, H. (2000) The human serotonin transporter gene linked polymorphism (5-HTTLPR) shows ten novel allelic variants. Mol. Psychiatry, 5: 32–38.

Nelson, D.L. (2004) 5-HT$_5$ receptors. Curr. Drug Targets-CNS Neurol. Disord., 3: 53–58.

Neves-Pereira, M., Mundo, E., Muglia, P., King, N., Macciardi, F. and Kennedy, J.L. (2002) The brain-derived neurotrophic factor gene confers susceptibility to bipolar disorder: evidence from a family-based association study. Am. J. Hum. Genet., 71: 651–655.

Niesler, B., Frank, B., Kapeller, J. and Rappold, G.A. (2003) Cloning, physical mapping and expression analysis of the human 5-HT$_3$ serotonin receptor-like genes *HTR3C, HTR3D* and *HTR3E*. Gene, 310: 101–111.

Nilsson, K.W., Sjoberg, R.L., Damberg, M., Leppert, J., Ohrvik, J., Alm, P.O., Lindstrom, L. and Oreland, L. (2006) Role of monoamine oxidase A genotype and psychosocial factors in male adolescent criminal activity. Biol. Psychiatry, 59: 121–127.

Nilsson, K.W., Sjoberg, R.L., Wargelius, H.L., Leppert, J., Lindstrom, L. and Oreland, L. (2007) The monoamine oxidase A (MAO-A) gene, family function and maltreatment as predictors of destructive behaviour during male adolescent alcohol consumption. Addiction, 102: 389–398.

Niswender, C.M., Copeland, S.C., Herrick-Davis, K., Emeson, R.B. and Sanders-Bush, E. (1999) RNA editing of the human serotonin 5-hydroxytryptamine 2C receptor silences constitutive activity. J. Biol. Chem., 274: 9472–9478.

Nomura, M. and Nomura, Y. (2006) Psychological, neuroimaging and biochemical studies on functional association between impulsive behaviour and the 5-HT$_{2A}$ receptor gene polymorphisms in humans. Ann. NY Acad. Sci., 1086: 134–143.

Oakey, R.J., Caron, M.G., Lefkowitz, R.J. and Seldin, M.F. (1991) Genomic organization of adrenergic and serotonin receptors in the mouse, Linkage mapping of sequence-related genes provides a method for examining mammalian chromosome evolution. Genomics, 10: 338–344.

Ogilvie, A.D., Battersby, S., Bubb, V.J., Fink, G., Harmar, A.J., Goodwin, G. and Smith, C.A. (1996) Polymorphism in serotonin transporter gene associated with susceptibility to major depression. Lancet, 347: 731–733.

Okada, T., Hashimoto, R., Numakawa, T., Iijima, Y., Kosuga, A., Tatsumi, M., Kamijima, K., Kato, T. and Kunugi, H. (2006) A complex polymorphic region in the brain-derived neurotrophic factor (BDNF) gene confers susceptibility to bipolar disorder and affects transcriptional activity. Mol. Psychiatry, 11: 695–703.

Olsen, M.A. and Schechter, L.E. (1999) Cloning mRNA localisation and evolutionary conservation of a human 5-HT$_7$ receptor pseudogene. Gene, 227: 63–69.

Olsen, M.A., Nawoschik, S.P., Schurman, B.R., Smith, D.L., Burno, M. and Schechter, L.E. (1997) Identification of a human 5-HT_7 receptor pseudogene. Soc. Neurosci. Abstr., 23: 127.
Ou, X.-M., Jafar-Nejad, H., Storring, J.M., Meng, J.-H., Lemonde, S. and Albert, P.R. (2000) Novel dual repressor elements for neuronal cell-specific transcription of the rat 5-HT1A receptor gene. J. Biol. Chem., 275: 8161–8168.
Ou, X.-M., Lemonde, S., Jafar-Nejad, H., Bown, C.D., Goto, A., Rogaeva, A. and Albert, P.R. (2003) Freud-1: a neuronal calcium-regulated repressor of the 5-HT1A receptor gene. J. Neurosci., 23: 7415–7425.
Ou, X.M., Chen, K. and Shih, J.C. (2006) Glucocorticoid and androgen activation of monoamine oxidase A is regulated differently by R1 and Sp1. J. Biol. Chem., 281: 21512–21525.
Ozelius, L., Hsu, Y.P., Bruns, G., Powell, J.F., Chen, S., Weyler, W., Utterback, M., Zucker, D., Haines, J., Trofatter, J.A. et al. (1988) Human monoamine oxidase gene (MAOA): chromosome position (Xp21-p11) and DNA polymorphism. Genomics, 3: 53–58.
Parks, C.L. and Shenk, T. (1996) The serotonin 1a receptor gene contains a TATA-less promoter that responds to MAZ and Sp1. J. Biol. Chem., 271: 4417–4430.
Parsons, M.J., D'Souza, U.M., Arranz, M.J., Kerwin, R.W. and Makoff, A.J. (2004) The –1438A/G polymorphism in the 5-hydroxytryptamine type 2A receptor gene affects promoter activity. Biol. Psychiatry, 56: 406–410.
Passamonti, L., Fera, F., Magariello, A., Cerasa, A., Gioia, M.C., Muglia, M., Nicoletti, G., Gallo, O., Provinciali, L. and Quattrone, A. (2006) Monoamine oxidase – a genetic variation influence brain activity associated with inhibitory control: new insight into the neural correlates of impulsivity. Biol. Psychiatry, 59: 334–340.
Patel, P.D., Pontrello, C. and Burke, S. (2004) Robust and tissue-specific expression of TPH2 versus TPH1 in rat raphe and pineal gland. Biol. Psychiatry, 55: 428–433.
Patel, P.D., Bochar, D.A., Turner, D.L., Meng, F., Mueller, H.M. and Pontrello, C.G. (2007) Regulation of tryptophan hydroxylase-2 gene expression by a bipartite RE-1 silencer of transcription/neuron restrictive silencing factor (REST/NRSF) binding motif. J. Biol. Chem., 282: 26717–26724.
Peng, C., Fan, S., Li, X., Fan, X., Ming, M., Sun, Z. and Le, W. (2007) Overexpression of pitx3 upregulates expression of BDNF and GDNF in SH-SY5Y cells and primary ventral mesencephalic cultures. FEBS Lett., 581: 1357–1361.
Peroutka, S.J. (1998) 5-hydroxytryptamine receptor subtypes. Ann. Rev. Neurosci., 11: 45–60.
Philibert, R., Madan, A., Andersen, A., Cadoret, R., Packer, H. and Sandhu, H. (2007) Serotonin transporter mRNA levels are associated with methylation of an upstream CpG island. Am. J. Med. Genet. B Neuropsychiatr. Genet., 144: 101–105.
Philibert, R.A., Sandhu, H., Hollenbeck, N., Gunter, T., Adams, W. and Madan, A. (2008) The relationship of 5HTT (SLC6A4) methylation and genotype on mRNA expression and liability to major depression and alcohol dependence in subjects from the Iowa Adoption Studies. Am. J. Med. Genet. B Neuropsychiatr. Genet., 147: 543–549.
Pinsonneault, J.K., Papp, A.C. and Sandee, W. (2006) Allelic mRNA expression of X-linked monoamine oxidase (MAOA) in human brain: dissection of epigenetic and genetic factors. Hum. Mol. Genet., 15: 2636–2649.
Pintar, J.E., Barbosa, J., Francke, U., Castiglione, C.M., Hawkins, M. Jr. and Breakfield, X.O. (1981) Gene for monoamine oxidase type A assigned to the human X chromosome. J. Neurosci., 1: 166–175.
Plassat, J.-L., Boschert, U., Amlaiky, N. and Hen, R. (1992) The mouse 5HT5 receptor reveals a remarkable heterogeneity within the 5HT1D receptor family. EMBO J., 11: 4779–4786.
Plassat, J.-L., Amlaiky, N. and Hen, R. (1993) Molecular cloning of a mammalian serotonin receptor that activates adenylate cyclase. Mol. Pharmacol., 44: 229–236.
Plomin, R., DeFries, J.C., McClearn, G.E. and McGuffin, P. (2000) DNA: the basis of heredity. In: Plomin, R., DeFries, J.C., McClearn, G.E. and McGuffin, P. (Eds.), Behavioral Genetics. Worth Publishers, New York, NY, pp. 41–59.
Prichett, D.B., Bach, A.W.J., Wozny, M., Taleb, O., Dal Toso, R., Shih, J.C. and Seeburg, P.H. (1988) Structure and functional expression of cloned rat serotonin 5-HT2 receptor. EMBO J., 7: 4135–4140.
Pruunsild, P., Kazantseva, A., Aid, T., Palm, K. and Timmusk, T. (2007) Dissecting the human BDNF locus: bidirectional transcription, complex splicing, and multiple promoters. Genomics, 90: 397–406.
Qian, I.H., Kusumi, I., Ulpian, C., Tallerico, T., Nam, D., Liu, I.S., Seeman, M.V. and Seeman, P. (1998) A human serotonin-7 receptor pseudogene. Brain Res. Mol. Brain Res., 53: 339–343.
Racchi, M., Leone, M., Porrello, E., Rigamonti, A., Govoni, S., Sironi, M., Montomoli, C. and Bussone, G. (2004) Familial migraine with aura: association study with 5-$HT_{1B/1D}$, 5-HT_{2C}, and hSERT polymorphisms. Headache, 44: 311–317.
Raymond, J.R., Mukhin, Y.V., Gettys, T.W. and Garnovskaya, M.N. (1999) The recombinant 5-HT_{1A} receptor: G-protein coupling and signaling pathways. Br. J. Pharmacol., 127: 1751–1764.
Rees, S., den Daas, I., Foord, S., Goodson, S., Bull, D., Kilpatrick, G. and Lee, M. (1994) Cloning and characterisation of the human 5-HT_{5A} serotonin receptor. FEBS Lett., 355: 242–246.
Reed, G.E., Kirchner, J.E. and Carr, L.G. (1995) NF-Y activates mouse tryptophan hydroxylase transcription. Brain Res., 682: 1–12.
Reeves, D.C. and Lummis, S.C. (2002) The molecular basis of the structure and function of the 5-HT3 receptor: a model ligand-gated ion channel (review). Mol. Membr. Biol., 19: 11–26.
Reynolds, G.P., Zhang, Z.-J. and Zhang, X.-B. (2002) Association of antipsychotic drug-induced weight gain with a 5-HT2C receptor gene polymorphism. Lancet, 359: 2086–2087.
Roberts, J., Scott, A.C., Howard, M.R., Breen, G., Bubb, V.J., Klenova, E. and Quinn, J.P. (2007) Differential regulation of the serotonin transporter gene by lithium is mediated by transcription factors, CCCTC binding protein and Y-box binding protein 1, through the polymorphic intron 2 variable number tandem repeat. J. Neurosci., 27: 2793–2801.
Rogaeva, A. and Albert, P.R. (2007) The mental retardation gene CC2D1A/Freud-1 encodes a long isoform that binds conserved DNA elements to repress gene transcription. Eur. J. Neurosci., 26: 965–974.
Ruat, M., Traiffort, E., Arrang, J.-M., Tardivel-Lacombe, J., Diaz, J., Leurs, R. and Schwartz, J.-C. (1993a) A novel rat serotonin (5-HT_6) receptor: molecular cloning, localisation and stimulation of cAMP accumulation. Biochem. Biophys. Res. Comm., 193: 268–276.
Ruat, M., Traiffort, E., Leurs, R., Tardivel-Lacombe, J., Diaz, J., Arrang, J.-M. and Schwartz, J.-C. (1993b) Molecular cloning,

characterisation, and localisation of a high-affinity serotonin receptor (5-HT_7) activating cAMP formation. Proc. Natl. Acad. Sci. USA, 90: 8547–8551.

Sabol, S.Z., Hu, S. and Hamer, D. (1998) A functional polymorphism in the monoamine oxidase A gene promoter. Hum. Genet., 103: 273–279.

Saltzman, A.G., Morse, B., Whiteman, M.M., Ivanshchenko, Y., Jaye, M. and Felder, S. (1991) Cloning of the human serotonin 5-HT_2 and 5-HT_{1C} receptor subtypes. Biochem. Biophys. Res. Commun., 181: 1469–1478.

Scharfman, H.E. and MacLusky, N.J. (2006) Estrogen and brain-derived neurotrophic factor (BDNF) in hippocampus: complexity of steroid hormone-growth factor interactions in the adult CNS. Front. Neuroendocrinol., 27: 415–435.

Scheuch, K., Lautenschlager, M., Grohmann, M., Stahlberg, S., Kirchheiner, J., Zill, P., Heinz, A., Walther, D.J. and Priller, J. (2007) Characterisation of a functional promoter polymorphism of the human tryptophan hydroxylase 2 gene in serotonergic raphe neurons. Biol. Psychiatry, 62: 1288–1294.

Schmuck, K., Ullmer, C., Engels, P. and Lubbert, H. (1994) Cloning and functional characterisation of the human 5-HT_{2B} serotonin receptor. FEBS Lett., 342: 85–90.

Serretti, A., Artioli, P., Lorenzi, C., Pirovano, A., Tubazio, V. and Zanardi, R. (2004) The C(−1019)G polymorphism of the 5-HT1A gene promoter and antidepressant response in mood disorders: preliminary findings. Intl. J. Neuropsychopharmacology, 7: 453–460.

Shen, Y., Monsma, F.J. Jr., Metcalf, M.A., Jose, P.A., Hamblin, M.W. and Sibley, D.R. (1993) Molecular cloning and expression of a 5-$hydroxytryptamine_7$ serotonin receptor subtype. J. Biol. Chem., 268: 18200–18204.

Shih, J.C. and Thompson, R.F. (1999) Psychiatric Genetics '99. Monoamine oxidase in neuropsychiatry and behaviour. Am. J. Hum. Genet., 65: 593–598.

Shih, J.C., Grimsby, J., Chen, K. and Zhu, Q.-S. (1993) Structure and promoter organisation of the human monoamine oxidase A and B genes. J. Psychiatr. Neurosci., 18: 25–32.

Shih, J.C., Zhu, Q.-S., Grimbsy, J. and Chen, K. (1994) Identification of human monoamine oxidase (MAO) A and B gene promoters. J. Neural. Transm., 41: 27–33.

Shih, J.C., Zhu, Q.-S. and Chen, K. (1996) Determination of transcription initiation sites and promoter activity of the human 5-HT_{2A} receptor gene. Behav. Brain Res., 73: 59–62.

Sjoberg, R.L., Nilsson, K.W., Wargelius, H.-L., Leppert, J., Lindstrom, L. and Oreland, L. (2007) Adolescent girls and criminal activity: role of MAOA-LPR genotype and psychosocial factors. Am. J. Med. Genet. B Neuropsychiatr. Genet., 144: 159–164.

Sklar, P., Gabriel, S.B., McInnis, M.G., Bennett, P., Lim, Y.-M., Tsan, G., Schaffner, S., Kirov, G., Jones, I., Owen, M., Craddock, N., DePaulo, J.R. and Lander, E.S. (2002) Family-based association study of 76 candidate genes in bipolar disorder: BDNF is a potential risk locus. Mol. Psychiatry, 7: 579–593.

Sohrabji, F. and Lewis, D.K. (2006) Estrogen–BDNF interactions: implications for neurodegenerative diseases. Front. Neuroendocrinol., 27: 404–414.

Sparkes, R.S., Lan, N., Klisak, I., Mohandas, T., Diep, A., Kojis, T., Heinzmann, C. and Shih, J.C. (1991) Assignment of a serotonin 5HT-2 receptor gene (HTR2) to human chromosome 13q14–q21 and mouse chromosome 14. Genomics, 9: 461–465.

Stam, N.J., Van Huizen, F., Van Alebeek, C., Brands, J., Dijkema, R., Tonnaer, J.A. and Olijve, W. (1992) Genomic organization, coding sequence and functional expression of human 5-HT2 and 5-HT1A receptor genes. Eur. J. Pharmacol., 227: 153–162.

Stam, N.J., Vanderheyden, P., van Alebeek, C., Klomp, J., de Boer, T., van Delft, A.M. and Olijve, W. (1994) Genomic organization and functional expression of the gene encoding the human serotonin 5-HT2C receptor. Eur. J. Pharmacol., 269: 339–348.

Stam, N.J., Roesink, C., Dijcks, F., Garritsen, A., van Herpen, A. and Olijve, W. (1997) Human serotonin 5-HT_7 receptor: cloning and pharmacological characterisation of two receptor variants. FEBS Lett., 413: 489–494.

Stein, M.B., Schork, N.J. and Gelernter, J. (2008) Gene-by-environment (serotonin transporter and childhood maltreatment) interaction for anxiety sensitivity, an intermediate phenotype for anxiety disorders. Neuropsychopharmacology, 33: 312–319.

Stoll, J. and Goldman, D. (1991) Isolation and structural characterisation of the murine tryptophan hydroxylase gene. J. Neurosci. Res., 28: 457–465.

Stoll, J., Kozak, C.A. and Goldman, D. (1990) Characterisation and chromosomal mapping of a cDNA encoding tryptophan hydroxylase from a mouse mastocytoma cell line. Genomics, 7: 88–96.

Storring, J.M., Charest, A., Cheng, P. and Albert, P.R. (1999) TATA-driven transcriptional initiation and regulation of the rat serotonin 5-HT_{1A} receptor gene. J. Neurochem., 72: 2238–2247.

Strachan, T. and Read, A.P. (1996) Mutation and instability of human DNA. In: Strachan, T. and Read, A.P. (Eds.), Human Molecular Genetics. BIOS Scientific Publishers Limited, Oxford, pp. 241–272.

Sugden, K., Tichopad, A., Khan, N., Craig, I.W. and D'Souza, U.M. (2009) Gene within the serotonin system are differentially expressed in human brain. BMC Neuroscience, in press.

Sun, H.S., Fann, C.S., Lane, H.-Y., Chang, Y.-T., Chang, C.-J., Liu, Y.-L. and Cheng, T.-A. (2005) A functional polymorphism in the promoter region of the tryptophan hydroxylase gene is associated with alcohol dependence in one aboriginal group in Taiwan. Alcohol Clin. Exp. Res., 29: 1–7.

Teerawatanasuk, N. and Carr, L.G. (1998) CBF/NF-Y activates transcription of the human tryptophan hydroxylase gene through an inverted CCAAT box. Mol. Brain Res., 55: 61–70.

Teerawatanasuk, N., Skalnik, D.G. and Carr, L.G. (1999) CCAAT displacement protein (CDP/Cut) binds a negative regulatory element in the human tryptophan hydroxylase gene. J. Neurochem., 72: 29–39.

Tzvetkov, M.V., Meineke, C., Oetjen, E., Hirsch-Ernst, K. and Brockmoller, J. (2007) Tissue-specific alternative promoters of the serotonin receptor gene *HTR3B* in human brain and intestine. Gene, 386: 52–62.

Uetz, P., Abdelatty, F., Villarroel, A., Rappold, G., Weiss, B. and Koenen, M. (1994) Organisation of the murine 5-HT_3 receptor gene and assignment to human chromosome 11. FEBS Lett., 339: 302–306.

Vaissiere, T., Sawan, C. and Herceg, Z. (2008) Epigenetic interplay between histone modifications and DNA methylation in gene silencing. Mutat. Res., 659: 40–48.

Van Den Bogaert, K., Sleegers, K., De Zutter, S., Heyrman, L., Norrback, K.F., Adolfsson, R., Van Broeckhoven, C. and Del-Favero, J. (2006) Association of brain-specific tryptophan hydroxylase, *TPH2*, with unipolar and bipolar disorder in a northern Swedish, isolated population. Arch. Gen. Psychiatry, 63: 1103–1110.

VanDenBrink, A.M., Vergouwe, M.N., Ophoff, R.A., Naylor, S.L., Dauwerse, H.G., Saxena, P.R., Ferrari, M.D. and Frants, R.R. (1998) Chromosomal localisation of the 5-HT_{1F} receptor gene: no evidence for involvement in response to sumatriptan in migraine patients. Am. J. Med. Genet. A, 77: 415–420.

Van den Wyngaert, I., Gommeren, W., Verhasselt, P., Jurzak, M., Leysen, J., Luyten, W. and Bender, E. (1997) Cloning and expression of a human serotonin 5-HT_4 receptor Cdna. J. Neurochem., 69: 1810–1819.

Veenstra-VanderWeele, J., Anderson, G.M. and Cook, E.H. Jr. (2000) Pharmacogenetics and the serotonin system: initial studies and future directions. Eur. J. Pharmacol., 410: 165–181.

Vilaro, M.T., Domenech, T., Palacios, J.M. and Mengod, G. (2002) Cloning and characterisation of anovel human 5-HT_4 receptor variant that lacks the alternatively spliced carboxy terminal exon. RT-PCR distribution in human brain and periphery of multiple 5-HT_4 receptor variants. Neuropharmacology, 42: 60–73.

Voigt, M.M., Laurie, D.J., Seeburg, P.H. and Bach, A. (1991) Molecular cloning and characterisation of a rat brain cDNA encoding a 5-hydroxytryptamine$_{1B}$ receptor. EMBO J., 10: 4017–4023.

Wada, K., Ballivet, M., Boulter, J., Connolly, J., Wada, E., Deneris, E.S., Swanson, L.W., Heinemann, S. and Patrick, J. (1988) Functional expression of a new pharmacological subtype of brain nicotinic acetylcholine receptor. Science, 240: 330–334.

Walther, D.J. and Bader, M. (2003) A unique central tryptophan hydroxylase isoform. Biochem. Pharmacol., 66: 1673–1680.

Walther, D.J., Peter, J.-U., Bashammakh, S., Hortnagl, H., Voits, M., Fink, H. and Bader, M. (2003) Synthesis of serotonin by a second tryptophan hydroxylase isoform. Science: 299–376.

Wang, G.-A., Coon, S.L. and Kaufman, S. (1998) Alternative splicing at the 3′-cDNA of human tryptophan hydroxylase. J. Neurochem.: 1769–1772.

Wang, Q., O'Brien, P.J., Chen, C.-X., Cho, D.-S.C., Murray, J.M. and Nishikura, K. (2000) Altered G protein-coupling functions of RNA editing isoform and splicing variant serotonin$_{2C}$ receptors. J. Neurochem., 74: 1290–1300.

Weinshank, R.L., Zgombick, J.M., Macchi, M.J., Branchek, T.A. and Hartig, P.R. (1992) Human serotonin 1D receptor is encoded by a superfamily of two distinct genes: 5-$HT_{1D\alpha}$ and 5-$HT_{1D\beta}$. Proc. Natl. Acad. Sci. USA, 89: 3630–3634.

Weiss, B., Mertz, A., Schrock, E., Koenen, M. and Rappold, G. (1995) Assignment of a human homolog of the mouse Htr3 receptor gene to chromosome 11q23.1–q23.2. Genomics, 29: 304–305.

Werner, P., Kawashima, E., Reid, J., Hussy, N., Lundstrom, K., Buell, G., Humbert, Y. and Jones, K.A. (1994) Organisation of the mouse 5-HT_3 receptor gene and functional expression of two splice variants. Brain Res. Mol. Brain Res., 26: 233–241.

Werry, T.D., Loiacono, R., Sexton, P.M. and Christopoulos, A. (2008) RNA editing of the serotonin $5HT_{2C}$ receptor and its effects on cell signalling, pharmacology and brain function. Pharmacol. Ther., 119: 7–23.

Widom, C.S. and Brzustowicz, L.M. (2006) MAOA and the 'cycle of violence': childhood abuse and neglect, MAOA genotype, and risk for violent and antisocial behaviour. Biol. Psychiatry, 60: 684–689.

Wilhelm, K., Mitchell, P.B., Niven, H., Finch, A., Wedgwood, L., Scimone, A., Blair, I.P., Parker, G. and Schofield, P.R. (2006) Life events, first depression onset and the serotonin transporter gene. Br. J. Psychiatry, 188: 210–215.

Wisden, W., Parker, E.M., Mahle, C.D., Grisel, D.A., Nowak, H.P., Yocca, F.D., Felder, C.C., Seeburg, P.H. and Voigt, M.M. (1993) Cloning and characterisation of the rat 5-HT_{5B} receptor. FEBS Lett., 333: 25–31.

Wissink, S., Meijer, O., Pearce, D., van der Burg, B. and van der Saag, P.T. (2000) Regulation of the rat serotonin-1A receptor gene by corticosteroids. J. Biol. Chem., 275: 1321–1326.

Wong, W.K., Chen, K. and Shih, J.C. (2001) Regulation of human monoamine oxidase B gene by Sp1 and Sp3. Mol. Pharmacol., 59: 852–859.

Woolley, M.L., Marsden, C.A. and Fone, K.C. (2004) 5-HT_6 receptors. Curr. Drug Targets-CNS Neurol. Disord., 3: 59–79.

Xie, E., Zhu, L., Zhag, L. and Chang, L.-S. (1996) The human serotonin 5-HT_{2C} receptor: complete cDNA, genomic structure and alternatively spliced variant. Genomics, 35: 551–561.

Yang, W., Chen, K., Lan, N.C., Gallaher, T.K. and Shih, J.C. (1992) Gene structure and expression of the mouse 5-HT2 receptor. J. Neurosci. Res., 33: 196–204.

Yu, L., Nguyen, H., Le, H., Bloem, L.J., Kozak, C.A., Hoffman, B.J., Snutch, T.P., Lester, H.A., Davidson, N. and Lubbert, H. (1991) The mouse 5-HT1C receptor contains eight hydrophobic domains and is X-linked. Brain Res. Mol. Brain Res., 11: 143–149.

Yuan, X., Yamada, K., Ishiyama-Shigemoto, S., Koyama, W. and Nonaka, K. (2000) Identification of polymorphic loci in the promoter region of the serotonin 5-HT_{2C} receptor gene and their association with obesity and Type II diabetes. Diabetologia, 43: 373–376.

Zgombick, J.M., Schechter, L.E., Macchi, M., Hartig, P.R., Branchek, T.A. and Weinshank, R.L. (1992) Human gene S31 encodes the pharmacologically defined serotonin 5-hydroxytryptamine$_{1E}$ receptor. Mol. Pharmacol., 42: 180–185.

Zhang, X., Beaulieu, J.-M., Sotnikova, T.D., Gainetdinov, R.R. and Caron, M.G. (2004) Tryptophan hydroxylase-2 controls brain serotonin synthesis. Science, 305: 217.

Zhang, X., Gainetdinov, R.R., Beaulieu, J.M., Sotnikova, T.D., Burch, L.H., Williams, R.B., Schwartz, D.A., Krishnan, K.R. and Caron, M.G. (2005) Loss-of-function mutation in tryptophan hydroxylase-2 identified in unipolar major depression. Neuron, 45: 11–16.

Zhang, X., Beaulieu, J.-M., Gainetdinov, R.R. and Caron, M.G. (2006) Functional polymorphisms of the brain serotonin synthesizing enzyme tryptophan hydroxylase-2. Cell Mol. Life Sci., 63: 6–11.

Zhu, Q.-S., Grimsby, J., Chen, K. and Shih, J.C. (1992) Promoter organisation and activity of human monoamine oxidase (MAO) A and B genes. J. Neurosci., 12: 4437–4446.

Zhu, Q.-S., Chen, K. and Shih, J.C. (1994) Bidirectional promoter of human monoamine oxidase A (MAO A) controlled by transcription factor Sp1. J. Neurosci., 14: 7393–7403.

Zhu, Q.-S., Chen, K. and Shih, J.C. (1995) Characterisation of the human 5-HT_{2A} receptor gene promoter. J. Neurosci., 15: 4885–4895.

Zill, P., Buttner, A., Eisenmenger, W., Bondy, B. and Ackenheil, M. (2004a) Regional mRNA expression of a second tryptophan hydroxylase isoform in post-mortem tissue samples of two human brains. Eur. Neuropsychopharmacol., 14: 282–284.

Zill, P., Baghai, T.C., Zwanzger, P., Schule, C., Esser, D., Rupprecht, R., Moller, H.-J., Bondy, B. and Ackenheil, M. (2004b) SNP and haplotype analysis of a novel tryptophan hydroxylase isoform (TPH2) gene provide evidence for association with major depression. Mol. Psychiatry, 9: 1030–1036.

Zill, P., Buttner, A., Eisenmenger, W., Moller, H.-J., Ackenheil, M. and Bondy, B. (2007) Analysis of tryptophan hydroxylase I and II mRNA expression in the human brain: a *post-mortem* study. J. Psychiatr. Res., 41: 168–173.

CHAPTER 1.3

The Neuronatomy of the Serotonergic System

Jean-Pierre Hornung*

Département de Biologie Cellulaire et de Morphologie, Faculty of Biology and Medicine, University of Lansanne, Rue du Bugnon 9, 1005, Lausanne, Switzerland

Abstract: Neurons using serotonin as a neurotransmitter in the central nervous system are restricted to the brainstem, centered on the midline raphe nuclei, extending into the subnuclei of the lateral reticular formation. These neurons divide into two populations. A rostral group resides in the midbrain and rostral pons (caudal linear, dorsal and median raphe nuclei), with a major projection targeted to the forebrain. A caudal group located in the caudal pons and the medulla (raphe magnus, obscurus and pallidus nuclei, lateral medullary reticular formation) has a major projection targeted to the spinal cord. Both groups have projections to the brainstem. Subnuclei of the serotonergic system have distinctive afferent and efferent connections, with some entities more specifically linked to the sensory, motor or limbic systems. Serotonergic neurons segregate, based on their axonal morphology, into neurons with large varicose or small varicose axons, the latter being selectively sensitive to neurotoxins derived from amphetamines. This is the basis for a dual serotonergic projection system, which partially overlaps in certain structures such as the cerebral cortex. Small varicose axons form wide axonal arborization of various densities of synaptic boutons, sustaining a system of volume transmission. Large varicose axons form pericellular arrays of synaptic terminals surrounding the soma and proximal dendrites of selected neurons. Median raphe, raphe obscurus and raphe pallidus nuclei are the sources for serotonergic neurons with large varicose axons. As one of the diffuse aminergic systems of the brain, the serotonergic system segregates clearly into different entities, with subnuclei holding their own efferent and afferent connections, serotonergic neurons having different axonal morphologies and synaptic connections, and, finally, numerous serotonin receptor subtypes having different expression patterns across the brain.

Keywords: raphe nuclei, rostral group, caudal group, large varicose axons, small varicose axons.

Introduction

Serotonin was first isolated from enterochromaffin cells (and originally named 'enteramine' by Erspamer and Viali in 1937, and later identified as serotonin (Erspamer and Asero, 1952)). Serotonin synthesis involves the specific rate-limiting enzyme tryptophan hydroxylase (TPH), encoded in two distinct genes: TPH-1 in peripheral tissue, and TPH-2 in the central nervous system (Walther *et al.*, 2003). Serotonin-synthesizing neurons of the CNS express additional proteins for the regulation of serotonin neurotransmission, in particular the serotonin transporter (SERT) and the vesicular monoamine transporter 2 (VMAT2), which are essential for the sequestration and vesicular release of serotonin (Gaspar *et al.*, 2003; Murphy and Lesch, 2008). Although brain serotonin was detected biochemically (Twarog and Page, 1953; Bogdanski *et al.*, 1956) and its effects on behavior and mood hypothesized (Woolley and Shaw, 1954) during the 1950s, it was only in the 1960s that serotonergic neurons could be directly visualized for the first time using the histochemical fluorescent method (Dahlström and Fuxe, 1964). More accurate and detailed morphological analysis of the serotonergic neurons was later achieved using immunocytochemical techniques, with antibodies against TPH (Pickel *et al.*, 1977), serotonin itself (Steinbusch, 1981) and the serotonin transporter (Ovalle *et al.*, 1995; Sur *et al.*, 1996; Zhou *et al.*, 1996).

The serotonin-containing neurons were characterized by a distribution restricted to the brainstem, and particularly to clusters of cells named B1 to B9 (Dahlström and Fuxe, 1964) which were mainly restricted to the midline raphe nuclei. In total, a few tens (in rodents) to a few hundreds (in human) of thousands of serotonergic neurons, segregated into a rostral and a caudal group, give rise to a dense axonal innervation to nearly all divisions of the CNS. This feature, shared with the other central amine systems, led to the concept of a volume transmission

*E-mail: jean-pierre.hornung@unil.ch

ISBN 978-0-12-374634-4
DOI: 10.1016/B978-0-12-374634-4.00003-4

mechanism modulating the function of specific local circuits across systems (Fuxe *et al.*, 2007). However, parallel studies using various tracing and mapping methods outlined subdivisions and specific connection between groups of serotonergic neurons (see subsequent sections of this chapter). The specificity of connections within the serotonergic system relies on several parameters: (1) the subdivision of raphe nuclei into spatially and morphologically distinct neuronal clusters; (2) a contrasting density of the serotonergic axonal arborizations between cytoarchitectonic divisions of the brain; (3) a distinct axonal morphology and distribution originating from different raphe nuclei; and (4) a large repertoire of serotonin receptors with specific cellular and regional pattern of expression (see Chapter 1.6).

The lineage of the serotonin-synthesizing neurons associates the coordination of several transcription factors which specify their location caudal to the midbrain–hindbrain organizer in two cellular clusters in rhombomers 1–3 and 5–7 (Alenina *et al.*, 2006). The original anatomical subdivision of the serotonergic nuclei is well supported by the current molecular characterization of these neurons. Among them, Pet-1 is a serotonin specific transcription factor expressed in embryonic and adult serotonergic neurons (Hendricks *et al.*, 1999). The ongoing studies on the contribution of each of these factors to the establishment of the serotonergic system will further consolidate our knowledge of its functional organization (see Chapter 1.2).

Finally, it should be recognized that the serotonergic system is generated early during embryogenesis, and that serotonin transmission sustained by early serotonergic neurons, and by non-serotonergic neurons expressing transiently protein-regulating serotonin transmission (transporters, receptors), contributes to proper brain development (see Chapter 3.1).

Divisions of the raphe and associated serotonin-containing nuclei

Using the histochemical fluorescent method, the distribution of serotonergic neurons was first described as nine groups in the brainstem, labeled caudally to rostrally as B1 to B9 (Dahlström and Fuxe, 1964). Since a large proportion of serotonin-containing neurons reside at the midline, there has been an extension of the use of the name *raphe nuclei* to refer to the source of serotonergic innervation of the brain. The raphe nuclei belong to the midline division of the reticular formation of the brainstem, extending from the midbrain rostrally to the decussation of the pyramidal tract caudally (Brodal, 1981). Except for the dorsal raphe nucleus, which is located in the ventral periaqueductal gray, the raphe nuclei are positioned ventral to the medial longitudinal fasciculus. Histofluorescence and immmunocytochemical studies in various mammalian species have established that most serotonin-containing neurons are located in the raphe nuclei, with a population extending laterally into the reticular formation. The higher proportion of serotonergic neurons positioned in the mesopontic reticular formation is a characteristic feature of the primate brain (Nobin and Björklund, 1973; Hubbard and Di Carlo, 1974; Schofield and Everitt, 1981; Schofield and Dixson, 1982; Felten and Sladek, 1983; Azmitia *et al.*, 1986; Takahashi *et al.*, 1986; Hornung and Fritschy, 1988).

The cell density and the cytoarchitectonic boundaries of the raphe nuclei vary, the dorsal raphe (DR) nucleus having the sharpest limits, while delineation of the caudal raphe nuclei is poor. The proportion of serotonin-containing neurons in a given cytoarchitectonic subdivision varies from between 80 and 100 percent in regions of the dorsal raphe, to 10 percent in the lateral reticular formation. The raphe nuclei and the serotonergic system of common laboratory animals have been described in detail elsewhere (Jacobowitz and MacLean, 1978; Wiklund *et al.*, 1981; Pecci-Saavadra *et al.*, 1983; Steinbusch *et al.*, 1983; Jacobs *et al.*, 1984; Azmitia *et al.*, 1986; Hornung and Fritschy, 1988; Törk, 1990; Bjarkam *et al.*, 1997; Léger *et al.*, 1998; VanderHorst and Ulfhake, 2006).

This chapter describes systematically the cyto- and chemoarchitecture of the raphe nuclei along the rostrocaudal axis of the brainstem, focusing on the median (in the raphe nuclei) and lateral (in the lateral reticular formation) distribution of serotonergic neurons. The following section describes the connectivity of the raphe nuclei.

The rostral group

Three raphe nuclei constitute the rostral serotonergic group extending from the caudal mesencephalon to the mid pons – the caudal linear nucleus (CLi), the DR, and the median raphe nucleus (MnR) – as well as a large population of serotonergic neurons in the paramedian and lateral mesopontic reticular formation, mostly associated with the median nucleus.

Caudal linear nucleus

The most rostral population of serotonergic neurons is located in the caudal mesencephalon. It represents the rostral population of the B8 group (Dahlström and Fuxe, 1964). The CLi belongs to the ventral mesencephalic tegmentum (Halliday and Törk, 1986). It is located at the midline, surrounded ventrally by the interpeduncular nucleus and laterally on both sides by the red nuclei. Posteriorly, it is adjacent to the superior cerebellar peduncle decussation. Dorsally, at the level of the medial longitudinal fasciculi, CLi merges with the rostral-most division of the dorsal raphe at the level of the interfascicular division (see below). The serotonergic neurons of the CLi have an elongated dendritic arborization, with its main axis oriented in the rostro-caudal direction, parallel to the midline plane. In the primate brain there is an additional population of serotonergic neurons in the rostral division of the interpeduncular nucleus, with neurons displaying a distinctive multipolar dendritic morphology (Azmitia *et al.*, 1986; Hornung and Fritschy, 1988; Törk and Hornung, 1990).

Besides serotonergic neurons, the heterogeneous population of CLi neurons includes tyrosin hydroxylase (TH)-immunoreactive dopaminergic neurons, and substance P (SP)-containing neurons (Halliday *et al.*, 1990).

Dorsal raphe nucleus

The DR is the largest collection of serotonergic neurons, counting for about one third of all serotonergic neurons of the brain. It represents the population of the B7 (rostral) and B6 (caudal) groups, divided in five subnuclei located between the mesencephalic aqueduct and the medial longitudinal fasciculus (Dahlström and Fuxe, 1964), while all the other serotonergic nuclei are located ventral to this tract. The dorsal raphe spreads in the anteroposterior axis between the oculomotor and the abducens nuclei, with the widest expansion at the level of the trochlear nucleus. The dorsal division of the dorsal raphe (DRD) abuts the aqueduct and splits the periaqueductal gray. It contains medium multipolar serotonergic neurons. At the pontomesencephalic transition, a population of large multipolar neurons in the ventrolateral division of the dorsal raphe (DRVL) spreads dorsally and laterally to the eye muscle motoneurons. Small multipolar serotonergic neurons are restricted to the midline in the ventral division of the dorsal raphe (DRV), between the nuclei of the ocular motoneurons. The ventral-most serotonergic neurons of the dorsal raphe lie in the intermediate division of the dorsal raphe (DRI) between the two parasagittal tracts of the medial longitudinal fasciculus, with elongated dendritic arborization parallel to the sagittal plane, and contiguous to the sagittal division of the median raphe located ventrally and to the CLi located rostrally. In the pons, the dorsal raphe extends in two parasagittal columns of serotonergic neurons, the caudal division of the dorsal raphe (DRD), made of small neurons with dendrites, often parallel to the floor of the overlaying fourth ventricle.

All neurons of the DRI are serotonergic, and the proportion declines to 80 percent in the DRV and 60 percent in all other subdivisions. Several neurotransmitters have been identified among the DR neurons. Very few GABA-containing neurons in the DR synthesize 5-HT (Wang and Nakai, 1993; Stamp and Semba, 1995; Charara and Parent, 1998). The distribution of SP and catecholamine is restricted to the rostral DR. There is a high proportion of large to medium-sized SP-immunoreactive neurons with morphologies similar to the serotonergic neurons, and it is estimated that up to 40 percent of 5-HT neurons express SP (Baker *et al.*, 1991a). There are only few catecholaminergic neurons, of variable morphologies, scattered in the rostral DR. There are also a few neurons containing the neuropeptides dynorphin, angiotensin and enkephalin (Björklund and Hökfelt, 1985; Björklund *et al.*, 1990; Okere and Waterhouse, 2006). In the rostral DR (B7), there are nitric oxide (NO) neurons detected by nitric oxide synthase (NOS) immnunocytochemistry or NADPH-diaphorase histochemistry. NOS is present in 5-HT neurons of the midline dorsal and ventral division, while NOS and 5-HT are mostly found in distinct neuronal populations in the lateral division of the dorsal raphe (DRVL) (Léger *et al.*, 1998; Tagliaferro *et al.*, 2001; Okere and Waterhouse, 2006). In DRC, there is a dense and restricted population of neurotensin-containing neurons (Jennes *et al.*, 1982).

Median raphe nucleus (MnR) and pontomesencephalic reticular formation

(a) The median raphe nucleus

The median raphe nucleus (MnR, corresponding to the B8 and B5 groups) is made of a midline and a paramedian division. They extend from the caudal end of the decussation of the superior cerebellar peduncles rostrally, to the middle of the metencephalon, below the medial longitudinal fasciculus dorsally to the ventral border of the tegmentum. The midline group is contiguous dorsally with the DRI without clear segregation, and the neurons of the midline MnR have a similar elongated dendritic arborization parallel to the sagittal plane. There are two paramedian streams of neurons flanking the midline group, with medium-sized multipolar neurons (Törk, 1990).

(b) The serotonergic neurons of the pontomesencephalic reticular formation

In the pontomesencephalic region, numerous serotonergic neurons form two distinct clusters in the lateral reticular formation of the rostral half of the pons: one in the oral pontine nucleus (PnO), in the center of the brainstem, and the second more ventrally in the supralemniscal nucleus (SuL). The number of serotonergic neurons in the lateral versus medial/midline reticular formation varies between species, being lower in rodents and high in the rabbit

(Bjarkam *et al.*, 1997) and primates (Azmitia *et al.*, 1986), approaching even numbers in the latter species (Baker *et al.*, 1991b). In the PnO, serotonergic neurons are large with radiating dendrites, while they are medium-sized with preferentially horizontally oriented dendrites in the supralemniscal nucleus, parallel to the dorsal limit of the medial lemniscus (corresponding to the B9 group).

In total, the MnR and pontomesencephalic reticular formation contain the second largest group of serotonergic neurons, amounting to 20–25 percent of the total serotonergic population. In the MnR, about 80 percent of the neurons are serotonergic in the midline subdivision, a proportion reduced to 30 percent in the paramedian region and very low in the lateral reticular formation. In the MnR there are a small number of peptidergic neurons, containing SP (Baker *et al.*, 1991a), cholecystokinin (Schiffmann and Vanderhaeghen, 1991), dynorphin, enkephalin and neurotensin (Björklund and Hökfelt, 1985; Björklund *et al.*, 1990).

The caudal group

The caudal serotonergic group extends from the caudal metencephalon to the decussation of the pyramidal tract, at the brainstem–spinal cord junction. Three raphe nuclei – the raphe magnus nucleus (RMg), the raphe obscurus nucleus (ROb) and the raphe pallidus nucleus (RPa) – as well as a large population of serotonergic neurons in the lateral medullary reticular formation constitute the caudal group, in continuation caudally with the lateral wings of the raphe magnus. There is a narrow serotonin-free transversal segment at the middle of the pons that separates the rostral and the caudal groups.

Raphe magnus nucleus

The raphe magnus nucleus (RMg) is the rostral-most nucleus of this group centered at the junction between the pons and the medulla, at the level of the facial nucleus (corresponds to the B3 group). It is restricted to the midline, overlying the medial lemniscus. It is located in the ventral tegmentum, extending caudally up to the anterior limit of the inferior olive nucleus. At the caudal level, serotoninergic neurons extend more laterally in continuity with a column of serotonergic neurons in the lateral reticular medullary formation (lateral paragigantocellularis nucleus). At the midline serotonergic neurons are mainly bipolar (parallel to the sagittal plane), while they are multipolar in the paramedian and lateral reticular formation (Törk and Hornung, 1990).

Substance P has been shown to co-localize in RMg serotonergic neurons, as well as thyrotropin-releasing hormones (Halliday *et al.*, 1988a; Rikard-Bell *et al.*, 1990; Poulat *et al.*, 1992; Wu *et al.*, 1993), as well as other neuropeptides such as enkephalin, somatostatin and cholecystokinin (Björklund and Hökfelt, 1985; Björklund *et al.*, 1990). A population of NPY-containing neurons has been described in the human RMg (Halliday *et al.*, 1988b).

Raphe obscurus nucleus

The raphe obscurus nucleus (ROb) is a long stream of neurons along the midline, extending rostro-caudally from the level of the facial nucleus to the decussation of the pyramidal tract (corresponds to the B2 group). In the rostral medulla, ROb is located dorsally to the RMg. ROb serotonergic neurons are small to medium-sized, with dendrites oriented dorsoventrally. Numerous SP-IR neurons (Del Fiacco *et al.*, 1984; Halliday *et al.*, 1988a; Rikard-Bell *et al.*, 1990), and scattered galanin-containing neurons (Blessing and Gai, 1997), have been described in the ROb.

Raphe pallidus nucleus

The raphe pallidus nucleus (RPa) is the smallest raphe nucleus (corresponds to the B1 group). It is a narrow collection of cells along the midline at the dorsal edge of the pyramidal tract, flanked laterally by the inferior olivary nuclei, starting rostrally at the level of the caudal end of the RMg. Serotonergic neurons of the RPa are slender, elongated neurons with few dendritic processes oriented predominantly dorsoventrally (Loewy, 1981; Nakamura *et al.*, 2002).

Serotonergic neurons in the lateral medullary reticular formation

Serotonergic neurons are in the lateral reticular formation at all levels of the medulla, in continuity rostrally with the lateral extensions of the RMg. This column of neurons shifts laterally, in the caudal direction, into the lateral paragigantocellular reticular nucleus (LPGi). Caudally, serotonergic neurons are situated close to the medullary surface (Chan-Palay, 1977; Steinbusch, 1981; Jacobs *et al.*, 1984; Azmitia *et al.*, 1986).

Neurons of the lateral reticular formation are multipolar, and those near the ventrolateral border of the medulla have a bipolar morphology. In this region, there are serotonergic and catecholaminergic neurons which constitute two separate populations (Halliday *et al.*, 1988c).

Connectivity

The entire network of the serotonergic projection is readily visualized with immunocytochemical detection of tryptophan hydroxylase, serotonin or SERT, which are abundant and selectively expressed by serotonergic neurons (Pickel *et al.*, 1977; Steinbusch, 1981; Nielsen *et al.*, 2006; Raghanti *et al.*, 2008). It reveals a dense and diverging projection of these brainstem neurons to all divisions of the adult brain. Anterograde and retrograde tracing techniques have revealed that serotonergic and non-serotonergic neurons of the raphe nuciei have overlapping projections. Subdivisions of the raphe nuclei have distinct projections predominantly focalized on restricted regions. Both immunocytochemistry and anterograde tracing techniques revealed that serotonergic axons have distinct morphologies (with thin or thick varicosities) (Kosofsky and Molliver, 1987; Törk, 1990; Ridet *et al.*, 1994). These two morphological types of serotoninergic axons originate in separate raphe nuclei, and, due to the differential sensitivity of these two populations of serotonergic neurons to the specific neurotoxic agent MDMA, the preferential degeneration of serotoninergic thin varicose axons reveals the arborization of the thick varicose axons (Kosofsky and Molliver, 1987; Ridet *et al.*, 1994). There is also a difference in the functional ultrastructure of the serotonergic varicosities; the majority (small size varicosities) are devoid of associated postsynaptic specializations, while some (large varicosities) form classical chemical synapses with postsynaptic neurons bearing a discrete postsynaptic differentiation (Törk, 1990; Ridet *et al.*, 1994; Smiley and Goldman-Rakic, 1996; see also Chapter 1.4 of this volume). Therefore, the relative density of serotonergic axons or varicosities defines either the functional organization of the serotonergic projection, if one considers a volume neurotransmission system of thin varicose axons, or the nature of the postsynaptic neurons in the case of large varicose axon projections.

The raphe nuclei of the rostral group have a projection primarily directed towards the forebrain, and to a lesser extent to the brainstem, while those of the caudal group primarily send projections to the lower brainstem and spinal cord (see below for a summary of the large body of literature dedicated to this topic). These two regions of the raphe system follow distinctive differentiation processes throughout their development (Lidov and Molliver, 1982; Aitken and Törk, 1988).

The rostral group

Efferent projections from the rostral group

The rostral group has a major projection to the forebrain, ascending in two parallel pathways. The dorsal pathway is parallel to the medial longitudinal fasciculus, and collects the projections of the DR and the large multipolar serotonergic neurons of the MnR in the dorsolateral PnO, located near the DRVL (Azmitia *et al.*, 1986). The ventral pathway is located on each side of the midline MnR, starting in the pons and reaching the ventral tegmental area, where it continues more rostrally into the medial forebrain bundle. The ventral pathway distributes principally to the basal forebrain and the medial cortex; the dorsal pathway approaches the medial forebrain bundle at the midbrain–forebrain junction, where it exchanges fibers with the ventral serotonergic pathway. At this level, a number of serotonergic axons enter the internal capsule to innervate the lateral cortex. In addition, the rostral group projects to the brainstem and to the spinal cord.

The anterior dorsal raphe nucleus projects principally to the cerebral cortex, the neostriatum, the amygdala and the substantia nigra (Steinbusch *et al.*, 1980; van der Kooy and Hattori, 1980; Waterhouse *et al.*, 1986; Vertes, 1991; Corvaja *et al.*, 1993). The caudal division of the dorsal raphe (DRC), extending into the pons, projects to the hippocampus, the entorhinal cortex and the locus coeruleus (Kohler and Steinbusch, 1982; Imai *et al.*, 1986; Datiche *et al.*, 1995; Vertes and Crane, 1997). There is some topographic organization of the serotonergic innervation to the cerebral cortex, mainly in the dorsal raphe, with a rostral to caudal gradient of the neurons relative to the innervation of the frontal to occipital cortex, respectively (Waterhouse *et al.*, 1986; Wilson and Molliver, 1991). The dorsal raphe also projects to the rostral cholinergic neurons (Gasbarri *et al.*, 1999), to the lateral septum or the striatum (Waselus *et al.*, 2006), to the thalamus (Westlund *et al.*, 1990; Gonzalo-Ruiz *et al.*, 1995), to the ependyma and to the subcommissural organ (Mikkelsen *et al.*, 1997; Simpson *et al.*, 1998). In the brainstem, it is also connected to cholinergic and non-cholinergic neurons of the lateral tegmental area (Steininger *et al.*, 1997), to the superior colliculus (Parsons *et al.*, 2001), to somatosensory, auditory, or vestibular nuclei (Thompson *et al.*, 1995; Simpson *et al.*, 2003; Halberstadt and Balaban, 2006; Lee *et al.*, 2008), as well as to motor nuclei of the brainstem (Li *et al.*, 1993; Lee *et al.*, 2008), to the cerebellum and to the spinal cord (Kazakov *et al.*, 1993; Li *et al.*, 1993; Petit *et al.*, 1995). A projection from DR and CLi to the bed nucleus of the stria terminalis is made by neurons containing the vasoactive intestinal neuropeptide (Petit *et al.*, 1995).

Multiple retrograde labeling revealed several collateralized projections involving the prefrontal cortex and the nucleus accumbens (Van Bockstaele *et al.*, 1993), the lateral geniculate nucleus and the superior colliculus (Villar *et al.*, 1988), the trigeminal nucleus and the central amygdaloid nucleus (Halberstadt and Balaban, 2006), the trigeminal nucleus and the facial nucleus (Lee *et al.*, 2008), the trigeminal nucleus and the spinal cord (Li *et al.*, 1993), or even the cerebral cortex and the spinal cord (Kazakov *et al.*, 1993). Finally, in parallel to the projection of serotonergic neurons, there is a contribution of non-serotonergic neurons to the projection to cortical and subcortical regions originating from the dorsal raphe nucleus (Datiche *et al.*, 1995; Aznar *et al.*, 2004; Halberstadt and Balaban, 2006, 2007, 2008).

The MnR is divided into a median and paramedian division with an ascending projection centered predominantly on the basal forebrain, the septal region, the ventral tegmental area, the hypothalamus, the midline thalamic nuclei and the hippocampus (Azmitia and Segal, 1978; Kohler *et al.*, 1982; Vertes and Martin, 1988; Vertes *et al.*, 1999). There is a projection of the MnR to the cerebral cortex, but with a sparse distribution as compared to the partially overlapping and complementary serotonergic projection of the dorsal raphe nucleus (Kievit and Kuypers, 1975; van der Kooy and Kuypers, 1979; Kohler and Steinbusch, 1982; Porrino and Goldman-Rakic, 1982; Tigges *et al.*, 1982; Datiche *et al.*, 1995; Vertes *et al.*, 1999). The projection of MnR on the rostral midline structure is supplemented with a projection on the raphe nuclei of the caudal and rostral groups, as well as the laterodoral tegmental nucleus (Vertes *et al.*, 1999) and the cerebellum (Bishop and Ho, 1985; Kerr and Bishop, 1991). The ascending projection of MnR could modulate, directly and indirectly, the motor system and the basal ganglia by direct striatal projection and indirect nigral and intralaminar thalamic projection. Similarly, a direct hippocampal and indirect septal, mamillary and cingulate MnR projection could strongly modulate the limbic system. The supralemniscal division of MnR contributes, together with serotonergic neurons located more dorsally in the MnR and DR, to the cerebellar innervation (Vertes and Crane, 1997).

The cerebral cortex, including the hippocampus, receives a strong serotonergic innervation from axons with various morphologies and specific laminar and areal distribution (Kievit and Kuypers, 1975; Kohler and Steinbusch, 1982; Porrino and Goldman-Rakic, 1982; Tigges *et al.*, 1982; O'Hearn and Molliver, 1984; Waterhouse *et al.*, 1986; Kosofsky and Molliver, 1987; Törk, 1990). The areal differentiation of the serotonergic innervation between cortical areas is particularly differentiated in the primate species (Morrison *et al.*, 1982; Takeuchi and Sano, 1983; Campbell *et al.*, 1987; Berger *et al.*, 1988; DeFelipe and Jones, 1988; Hornung *et al.*, 1990; Wilson and Molliver, 1991). Based on their morphology, serotonergic axons could be segregated into small varicose axons (less than 1 micron) and large varicose axons. Small varicose axons are widely distributed in all cortical areas, whereas large varicose axons are more restricted to layers in the hippocampus and the medial and frontal neocortex. The application of amphetamine-derived neurotoxins (p-chloroamphetamine (PCA), 3,4-methylenedioxymethamphetamine (MDMA, ecstasy)) preferentially affects small varicose axons while leaving large varicose axons intact (Kosofsky and Molliver, 1987; O'Hearn *et al.*, 1988; Mamounas *et al.*, 1991). All cortical layers contain small varicose fibers, with distinctive patterns between areas. Characteristically, the laminar distribution of the serotonergic axons is highest in layer IV. The laminar distribution changes sharply between areas, and is most obvious at the transition between primary sensory and associative areas. The laminar pattern of serotonergic innervation of the prefrontal cortex increases in complexity and density in the human and great ape as compared to the innervation in the corresponding areas of the macaque monkey cortex, a differentiation which is not observed when comparing the serotonergic innervation of the motor cortex of the same species (Raghanti *et al.*, 2008). The laminar profile of the serotonergic axons is, for instance, already detectable 2 weeks after birth in the marmoset monkey, a New World primate, and will later sharpen into the adult laminar pattern (Figure 1). It suggests that the organization of the serotonergic innervation of the cerebral cortex is determined early during postnatal development but will further differentiate with the maturation of the cortical connectivity.

Large varicose axons form complex pericellular arrays closely surrounding neurons with the morphology of interneurons, and are shown to express the calcium binding protein calbindin or calretinin, but not parvalbumin, in the hippocampus of rodent and primates (Halasy *et al.*, 1992; Hornung and Celio, 1992; Miettinen and Freund, 1992; Acsady *et al.*, 1993). These large varicosities form classical chemical synapses onto the soma and dendrites of these interneurones. Calretinin- and calbindin-containing interneurons express the ionotrope 5-HT_3 receptor (Morales and Bloom, 1997), and are likely to mediate the serotonergic activation of GABA neurotransmission in the hippocampus (Ropert and Guy, 1991; Maeda *et al.*, 1994; McMahon and Kauer, 1997). Pericellular arrays have been visualized in the neocortex of cat (Mulligan and Törk, 1987), non-human primates (Hornung *et al.*, 1990; Wilson and Molliver, 1991; Jakab and Goldman-Rakic, 2000) and human (Kosofsky and Kowall, 1989; Hornung and De Tribolet, 1995; Trottier *et al.*, 1996). Electron microscopic studies of serotonergic axons have demonstrated that large varicosities in pericellular arrays indeed repetitively contact the same postsynaptic cell in forming classical chemical synaptic differentiation (Figure 2). Similar pericellular arrays of large varicose axons from serotonergic axons have been described in the lateral septum surrounding parvalbumin-containing neurons (Leranth and Vertes, 1999).

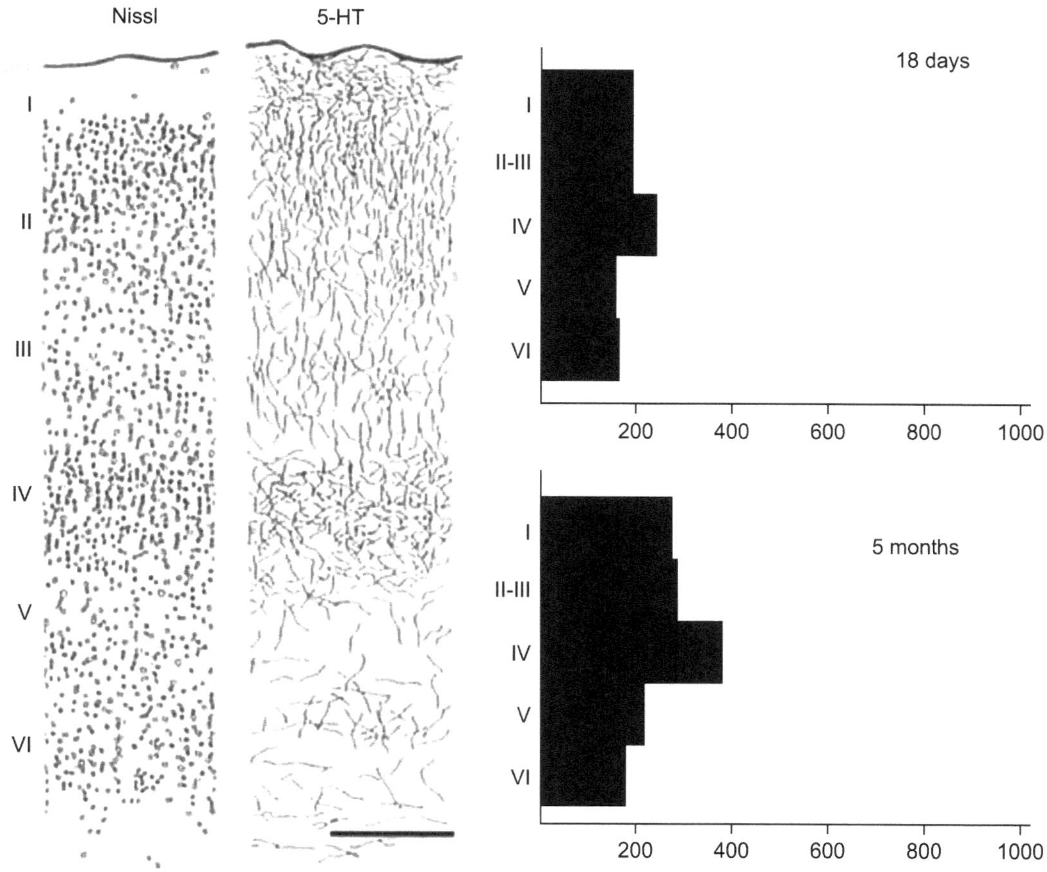

Figure 1 Left: drawing of the 5-HT axons and the outline of cells (Nissl stain) defining the laminar boundaries in a coronal section through the superior temporal cortex of the marmoset monkey. Notice a salient granular layer (layer IV) corresponding to the densest meshwork of serotonergic axons. On the right, the profile of axonal density is quantified under microscope visualization, measured as the number of intersections by immunoreactive fibers on a grid placed in the ocular. In a young marmoset a characteristic profile is already distinctive from other neighboring cortical areas (not shown). A similar analysis in the 5-month-old marmoset shows clearly a maturation and augmentation of the fiber density, but still with a profile comparable to the one in neonatal animals. Scale: 150 microns.

Afferent projections to the rostral group

Although it is likely that the raphe nuclei share the basic architecture of the projections between mammalian species, most of our knowledge is based on tract-tracing studies made in rodents. In primates, the increased complexity of the raphe system projecting to the forebrain (see above) may lead to further differentiation in the organization of their afferents. The lateral habenula is a major source of afferents to DR and MnR (Behzadi *et al.*, 1990; Peyron *et al.*, 1998), a pathway using glutamate as neurotransmitter (Behzadi *et al.*, 1990). Both raphe nuclei also receive an input from the medial septum (Kalén and Wiklund, 1989) and the diagonal band of Broca and the ventral pallidum (Behzadi *et al.*, 1990; Peyron *et al.*, 1998). Several hypothalamic projections targeted to different divisions of the DR and MnR originate from nuclei in the medial and lateral pre-optic areas, the lateral, dorsal, ventral and posterior divisions of the hypothalamus (Behzadi *et al.*, 1990; Peyron *et al.*, 1998; Lee *et al.*, 2003, 2005a, 2005b). There is an abundant GABAergic innervation of DR neurons (Wang *et al.*, 1992), and glutamatergic projections from the lateral habenula, the interpeduncular nucleus, several hypothalamic nuclei, the ventral tegmental areas, the laterodorsal tegmentum nuclei, and the cingulate cortex (Behzadi *et al.*, 1990). Stimulation of the medial prefrontal cortex inhibits the bursting of serotonergic neurons in the DR and MnR (Hajos *et al.*, 1998). The central nucleus of the amygdala projects to the DR (Peyron *et al.*, 1998; Lee *et al.*, 2007). There are, in addition, ascending projections to MnR originating from the RMg and RPa nuclei and from the nucleus prepositus hypoglossi (Behzadi *et al.*, 1990). Glycinergic projections to DR originate from the ventral and ventrolateral periaqueductal gray neurons and from neurons in the

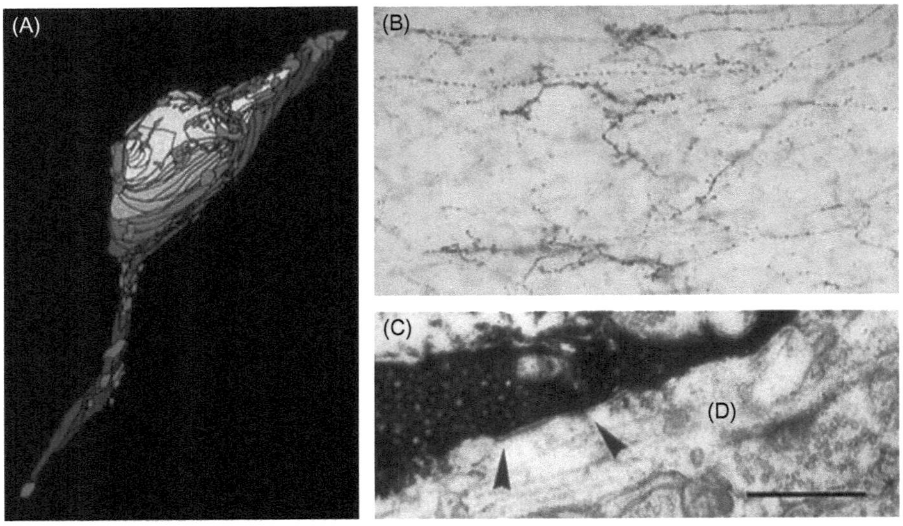

Figure 2 (A) Three-dimensional reconstruction of a cortical interneuron in layer II of the cat primary auditory cortex (green) contacted by varicose 5-HT-immunoreactive axons (red) at the surface of the soma and primary dendrites. (B) Micrograph of a coronal section in the marmoset frontal cortex containing 5-HT-immunoreactive thin and varicose axons, the latter forming conspicuous pericellular arrays surrounding tighly the cell body and proximal dendrites of individual neurons. (C) Electron micrograph of an unstained postsynaptic dendrite of a cat auditory cortex interneuron (D) forming a synaptic contact with a serotoninergic large varicose immunoreactive axon terminal (arrowheads). Scale bar: for (A), 8 microns; for (B), 40 microns; for (C), 1 micron. To see the full color version of this figure please refer to the color plate at the back of the book. Copies produced via our print on demand service do not contain color plates; if your copy does not have the color plate, please go to this webiste to view the figure in color www.elsevierdirect.com/companions/9780123746344

medullary rostral paragigantocellularis and rostral ventromedial reticular nuclei (Rampon *et al.*, 1999). Afferents from the medial nucleus of the solitary tract forward autonomic inputs to DR (Herbert, 1992).

The caudal group

Efferent projections from the caudal group

The medullary raphe nuclei have major projections directed towards the caudal brainstem and spinal cord (Brodal *et al.*, 1960; Carlsson *et al.*, 1964; Skagerberg and Björklund, 1985). Retrograde labeling combined with 5-HT immunocytochemistry revealed that half of the raphe nuclei projecting to the spinal cord synthesize serotonin, a proportion that falls to a quarter in the lateral medullary neuron (Jones and Light, 1992).

Two populations of serotonergic axons project into the spinal cord. Axons bearing small varicosities are predominant in the dorsal division, and axons bearing large varicosities in the intermediate and ventral divisions of the spinal cord. Large varicose axons form classical chemical synapses onto the motoneurons (Ridet *et al.*, 1993). The ROb and the RPa provide the primary serotonergic input to the brainstem somatic motor nuclei, such as the motor trigeminal nucleus (Li *et al.*, 1993; Ribeiro-do-Valle, 1997), the facial nucleus (Arita *et al.*, 1993, 1995; Li *et al.*, 1993), the retrofacial and the ambiguus nuclei (Arita *et al.*, 1993, 1995). In the spinal cord, it provides the innervation of the phrenic nucleus (Holtman *et al.*, 1990; Hosogai *et al.*, 1998) and the lumbar motoneurons (Tanaka *et al.*, 2006). There is also a projection to the autonomic motoneurons of the parasympathetic (dorsal vagal nucleus) and sympathetic (thoracic intermediolateral column, IML) systems (Loewy and McKellar, 1981; Poulat *et al.*, 1992; Yang *et al.*, 1994; Manaker and Fogarty, 1995). Serotonergic neurons projecting to the autonomic motoneurons co-localize the neuropeptide thyrotropin-releasing hormone (TRH) and SP (Poulat *et al.*, 1992; Wu *et al.*, 1993); those projecting to the somatic motoneurons lack SP (Wu *et al.*, 1993).

The raphe projections to the dorsal horn and dorsal column nuclei originated primarily from the RMg, and its serotonergic axons are devoid of TRH and SP (Wu and Wessendorf, 1992). In the sensory relays, there is a dense serotonergic innervation of laminae I and II of the spinal trigeminal nucleus (Costa *et al.*, 1994) spinal cord (Ridet *et al.*, 1993). The p-chloroamphetamine treatment has a similar selective neurotoxic effect as in the forebrain: the fine varicose serotonergic innervation disappears in the dorsal horn while sparing the large varicose serotonergic innervation in all divisions of the spinal cord (Ridet *et al.*, 1994). Inferring from the distribution of the serotonergic fibers in the spinal cord, the fine varicose fibers originate primarily in RMg and the large varicose ones in ROb and RPa. The descending projections in the spinal cord follow two pathways: one lateral (and dorsal, depending on the authors) originates in the RMg and terminates in the dorsal horn, and one medial (and ventral) originates in the ROb and the RPa and terminates in the ventral horn and IML (Felten and Sladek, 1983; Azmitia *et al.*, 1986). There is a mixture of unmyelinated and partially myelinated descending serotonergic axons in the dorsolateral tract (Westlund *et al.*, 1992; Leanza *et al.*, 1995). Although most raphe projections point toward the caudal divisions of the brain, serotonergic and non-serotonergic neurons of the caudal raphe nuclei send axon collaterals to both the lumbar intermediate and ventral gray matter and the medial pre-optic area of the hypothalamus (Leanza *et al.*, 1995). The RMg and RPa have ascending projections, in the rostral serotonin group, in the RMn (Behzadi *et al.*, 1990).

Afferent projections to the caudal group

Afferents to the anterior part of the caudal group (RMg, rostral RPa and Gia) originate in several hypothalamic nuclei, the dorsolateral periaqueductal gray, the central nucleus of the amygdala, the bed nucleus of the stria terminalis, and the medullary reticular formation (Zagon, 1993; Hermann *et al.*, 1997). Spinally projecting RMg and rostral RPa neurons receive direct catecholaminergic synaptic inputs (Tanaka *et al.*, 1994). The pneumotaxic area in the medial parabrachial area and the retrofacial nuclei has convergent inputs to the RMg (Gang *et al.*, 1993). There are also converging inputs onto RMg, ROb and RPa midline raphe neurons from visceral sensory afferents and ventrolateral periaqueductal gray matter neurons (Snowball *et al.*, 1997). The DR projects caudally on the RMg and Rob and LPGi (Vertes and Kocsis, 1994).

Conclusions

The serotonergic system is a collection of neurons and nuclei distributed along the brainstem in two groups sending projections in many directions, the rostral mainly ascending in the forebrain, the caudal with a major projection into the spinal cord. Each group subdivides further in subnuclei holding distinct sets of restricted afferent and efferent connections. Thus, serotonergic subnuclei are integrated components of, for instance, the cardiorespiratory system, the pain system, the motor system or the limbic system. This principle of organization accounts for the many functions with which serotonergic transmission is associated, and which are described in detail in the following chapters of this book.

In each of the regions or systems in which the serotonergic system is involved, the spatial pattern of the serotonergic afferents (regional and laminar specificity) positions the serotonergic input into the regulation of the system. The type of neurons connected (cellular specificity) and the type of serotonergic receptors associated (molecular specificity) contribute to the variety of cellular mechanisms modulated by serotonergic fibers.

The organization principles of the serotonergic system associate several parameters, such as the architecture of the connectivity, the expression pattern of serotonergic receptors and second messenger effectors in the various neuronal types of the target structure, with functional modulation of a system which will greatly vary from one area to the next. The following chapters will widely document these features in the various developmental, structural, physiological and pathological processes in which serotonin transmission is involved.

References

Acsady, L., Halasy, K. and Freund, T.F. (1993) Calretinin is present in non-pyramidal cells of the rat hippocampus-III. Their inputs from the median raphe and medial septal nuclei. Neuroscience, 52: 829–841.

Aitken, A.R. and Törk, I. (1988) Early development of serotonin-containing neurons and pathways as seen in whole-mount preparations of the fetal rat brain. J. Comp. Neurol., 274: 32–47.

Alenina, N., Bashammakh, S. and Bader, M. (2006) Specification and differentiation of serotonergic neurons. Stem Cell Rev., 2: 5–10.

Arita, H., Sakamoto, M., Hirokawa, Y. and Okado, N. (1993) Serotonin innervation patterns differ among the various medullary motoneuronal groups involved in upper airway control. Exp. Brain Res., 95: 100–110.

Arita, H., Ichikawa, K. and Sakamoto, M. (1995) Serotonergic cells in nucleus raphe pallidus provide tonic drive to posterior cricoarytenoid motoneurons via 5-hydroxytryptamine 2 receptors in cats. Neurosci. Lett., 197: 113–116.

Azmitia, E.C. and Segal, M. (1978) An autoradiographic analysis of the differential ascending projections of the dorsal and median raphe nuclei in the rat. J. Comp. Neurol., 179: 641–668.

Azmitia, E.C., Gannon, P.J. and Fahn, S. (1986) The primate serotonergic system: a review of human and animal studies and a report on Macaca fascicularis. In: Fahn, S., Marsden, C.D. and van Woert, M.H. (Eds.), Advances in Neurology, Myoclonus. Raven Press, New York, NY, Vol. 43, pp. 407–468.

Aznar, S., Qian, Z.X. and Knudsen, G.M. (2004) Non-serotonergic dorsal and median raphe projection onto parvalbumin- and calbindin-containing neurons in hippocampus and septum. Neuroscience, 124: 573–581.

Baker, K.G., Halliday, G.M., Hornung, J.P., Geffen, L.B., Cotton, R.G. and Törk, I. (1991a) Distribution, morphology and number of monoamine-synthesizing and substance P-containing neurons in the human dorsal raphe nucleus. Neuroscience, 42: 757–775.

Baker, K.G., Halliday, G.M., Halasz, P., Hornung, J.P., Geffen, L.B., Cotton, R.G. and Törk, I. (1991b) Cytoarchitecture of serotonin-synthesizing neurons in the pontine tegmentum of the human brain. Synapse, 7: 301–320.

Behzadi, G., Kalen, P., Parvopassu, F. and Wiklund, L. (1990) Afferents to the median raphe nucleus of the rat: retrograde cholera toxin and wheat germ conjugated horseradish peroxidase tracing, and selective D-[3H]aspartate labelling of possible excitatory amino acid inputs. Neuroscience, 37: 77–1002.

Berger, B., Trottier, S., Verney, C., Gaspar, P. and Alvarez, C. (1988) Regional and laminar distribution of the dopamine and serotonin innervation in the macaque cerebral cortex: a radioautographic study. J. Comp. Neurol., 273: 99–119.

Bishop, G.A. and Ho, R.H. (1985) The distribution and origin of serotonin immunoreactivity in the rat cerebellum. Brain Res., 331: 195–207.

Bjarkam, C.R., Sorensen, J.C. and Geneser, F.A. (1997) Distribution and morphology of serotonin-immunoreactive neurons in the brainstem of the New Zealand white rabbit. J. Comp. Neurol., 380: 507–519.

Björklund, A. and Hökfelt, T. (1985) GABA and Neuropeptides in the CNS, I. Elsevier, Amsterdam.

Björklund, A., Hökfelt, T. and Kuhar, M.J. (1990) Neuropeptides in the CNS, Part II. Elsevier, Amsterdam.

Blessing, W.W. and Gai, W.P. (1997) Caudal pons and medulla oblongata. In: Bloom, F.E., Björklund, A. and Hökfelt, T. (Eds.), Handbook of Chemical Neuroanatomy. Elsevier, Amsterdam, pp. 139–215.

Bogdanski, D.F., Pletscher, A., Brodie, B.B. and Undenfriend, S. (1956) Identification and assay of serotonin in brain. J. Pharmacol. Exp. Ther., 117: 82–88.

Brodal, A. (1981) The reticular formation. In: *Neurological Anatomy*, pp. 411–447. Oxford University Press, Oxford.

Brodal, A., Taber, E. and Walberg, F. (1960) The raphe nuclei of the brain stem in the cat. II. Efferent connections. J. Comp. Neurol., 114: 239–259.

Campbell, M.J., Lewis, D.A., Foote, S.L. and Morrison, J.H. (1987) Distribution of choline acetyltransferase-, serotonin-, dopamine- á-hydroxylase-, tyrosine hydroxylase-immunoreactive fibers in monkey primary auditory cortex. J. Comp. Neurol., 261: 209–220.

Carlsson, A., Falck, B., Fuxe, K. and Hillarp, N.A. (1964) Cellular localization of monoamines in the spinal cord. Acta Physiol. Scand., 60: 112–119.

Chan-Palay, V. (1977) Indoleamine neurons and their processes in the normal rat brain and in chronic diet-induced thiamine deficiency demonstrated by uptake of 3 H-serotonin. J. Comp. Neurol., 176: 467–494.

Charara, A. and Parent, A. (1998) Chemoarchitecture of the primate dorsal raphe nucleus. J. Chem. Neuroanat., 15: 111–127.

Corvaja, N., Doucet, G. and Bolam, J.P. (1993) Ultrastructure and synaptic targets of the raphe-nigral projection in the rat. Neuroscience, 55: 417–427.

Costa, J.J.L., Averill, S., Saavedra, J.P. and Priestley, J.V. (1994) Serotonin innervation of enkephalin containing neurones in the rat spinal trigeminal nucleus. Neurosci. Lett., 168: 167–171.

Dahlström, A. and Fuxe, K. (1964) Localization of monoamines in the lower brain stem. Experientia, 20: 398–399.

Datiche, F., Luppi, P.H. and Cattarelli, M. (1995) Serotonergic and non-serotonergic projections from the raphe nuclei to the piriform cortex in the rat: a cholera toxin B subunit (CTb) and 5-HT immunohistochemical study. Brain Res., 671: 27–37.

DeFelipe, J. and Jones, E.G. (1988) A light and electron microscopic study of serotonin-immunoreactive fibers and terminals in the monkey sensory-motor cortex. Exp. Brain Res., 71: 171–182.

Del Fiacco, M., Dessi, M.L. and Levanti, M.C. (1984) Topographical localization of substance P in the human postmortem brainstem. An immunohistochemical study in the newborn and adult tissue. Neuroscience, 12: 591–611.

Erspamer, V. and Asero, B. (1952) Identification of enteramine, the specific hormone of the enterochromaffin cell system, as 5-hydroxytryptamine. Nature, 169: 800–801.

Felten, D.L. and Sladek, J.R. Jr. (1983) Monoamine distribution in primate brain V. Monoaminergic nuclei: anatomy, pathways and local organization. Brain Res. Bull., 10: 171–284.

Fuxe, K., Dahlstrom, A., Hoistad, M., Marcellino, D., Jansson, A., Rivera, A., Diaz-Cabiale, Z., Jacobsen, K., Tinner-Staines, B., Hagman, B., Leo, G., Staines, W., Guidolin, D., Kehr, J., Genedani, S., Belluardo, N. and Agnati, L.F. (2007) From the Golgi-Cajal mapping to the transmitter-based characterization of the neuronal networks leading to two modes of brain communication: wiring and volume transmission. Brain Res. Rev., 55: 17–54.

Gang, S., Nakazono, Y., Aoki, M. and Aoki, N. (1993) Differential projections to the raphe nuclei from the medial parabrachial–Kolliker-Fuse (NPBM–KF) nuclear complex and the retrofacial nucleus in cats: retrograde WGA-HRP tracing. J. Auton. Nerv. Syst., 45: 241–244.

Gasbarri, A., Sulli, A., Pacitti, C. and McGaugh, J.L. (1999) Serotonergic input to cholinergic neurons in the substantia innominata and nucleus basalis magnocellularis in the rat. Neuroscience, 91: 1129–1142.

Gaspar, P., Cases, O. and Maroteaux, L. (2003) The developmental role of serotonin: news from mouse molecular genetics. Nat. Rev. Neurosci., 4: 1002–1012.

Gonzalo-Ruiz, A., Lieberman, A.R. and Sanz-Anquela, J.M. (1995) Organization of serotoninergic projections from the raphe, nuclei to the anterior thalamic nuclei in the rat: a combined retrograde tracing and 5-HT immunohistochemical study. J. Chem. Neuroanat., 8: 103–115.

Hajos, M., Richards, C.D., Szekely, A.D. and Sharp, T. (1998) An electrophysiological and neuroanatomical study of the medial prefrontal cortical projection to the midbrain raphe nuclei in the rat. Neuroscience, 87: 95–108.

Halasy, K., Miettinen, R., Szabat, E. and Freund, T.F. (1992) GABAergic interneurons are the major postsynaptic targets of median raphe afferents in the rat dentate gyrus. Eur. J. Neurosci., 4: 144–153.

Halberstadt, A.L. and Balaban, C.D. (2006) Serotonergic and nonserotonergic neurons in the dorsal raphe nucleus send collateralized projections to both the vestibular nuclei and the central amygdaloid nucleus. Neuroscience, 140: 1067–1077.

Halberstadt, A.L. and Balaban, C.D. (2007) Selective anterograde tracing of the individual serotonergic and nonserotonergic components of the dorsal raphe nucleus projection to the vestibular nuclei. Neuroscience, 147: 207–223.

Halberstadt, A.L. and Balaban, C.D. (2008) Selective anterograde tracing of nonserotonergic projections from dorsal raphe nucleus to the basal forebrain and extended amygdala. J. Chem. Neuroanat., 35: 317–325.

Halliday, G.M. and Törk, I. (1986) Comparative anatomy of the ventromedial mesencephalic tegmentum in the rat, cat, monkey and human. J. Comp. Neurol., 252: 423–445.

Halliday, G.M., Li, Y.W., Joh, T.H., Cotton, R.G.H., Howe, P.R.C., Geffen, L.B. and Blessing, W.W. (1988a) Distribution of substance P-like immunoreactive neurons in the human medulla oblongata: co-localization with monoamine-synthesizing neurons. Synapse, 2: 353–370.

Halliday, G.M., Li, Y.W., Oliver, J.R., Joh, T.H., Howe, P.R.C., Geffen, L.B. and Blessing, W.W. (1988b) The distribution of neuropeptide Y-like immunoreactive neurons in the human medulla. Neuroscience, 26: 179–191.

Halliday, G.M., Li, Y.W., Joh, T.H., Cotton, R.G.H., Howe, P.R.C., Geffen, L.B. and Blessing, W.W. (1988c) Distribution of monoamine-synthesizing neurons in the human medulla oblongata. J. Comp. Neurol., 273: 301–317.

Halliday, G.M., Gai, W.P., Blessing, W.W. and Geffen, L.B. (1990) Substance P-containing neurons in the pontomesencephalic tegmentum of the human brain. Neuroscience, 39: 81–96.

Hendricks, T., Francis, N., Fyodorov, D. and Deneris, E.S. (1999) The ETS domain factor Pet-1 is an early and precise marker of central serotonin neurons and interacts with a conserved element in serotonergic genes. J. Neurosci., 19: 10348–10356.

Herbert, H. (1992) Evidence for projections from medullary nuclei onto serotonergic and dopaminergic neurons in the midbrain dorsal raphe nucleus of the rat. Cell Tissue Res., 270: 149–156.

Hermann, D.M., Luppi, P.H., Peyron, C., Hinckel, P. and Jouvet, M. (1997) Afferent projections to the rat nuclei raphe magnus, raphe pallidus and reticularis gigantocellularis pars alpha demonstrated by iontophoretic application of choleratoxin (subunit b). J. Chem. Neuroanat., 13: 1–21.

Holtman, J.R. Jr., Vascik, D.S. and Maley, B.E. (1990) Ultrastructural evidence for serotonin-immunoreactive terminals contacting phrenic motoneurons in the cat. Exp. Neurol., 109: 269–272.

Hornung, J.P. and Celio, M.R. (1992) The selective innervation by serotonergic axons of calbindin-containing interneurons in the neocortex and hippocampus of the marmoset. J. Comp. Neurol., 320: 457–467.

Hornung, J.P. and De Tribolet, N. (1995) Chemical anatomy of the human cerebral cortex. In: Tracey, D., Paxinos, G. and Stone, J. (Eds.), Advances in Behavioral Biology, Vol. 43, Neurotransmitters in the Human Brain. Plenum, New York, NY, pp. 41–60.

Hornung, J.P. and Fritschy, J.M. (1988) Serotoninergic system in the brainstem of the marmoset: a combined immunocytochemical and three-dimensional reconstruction study. J. Comp. Neurol., 270: 471–487.

Hornung, J.P., Fritschy, J.M. and Törk, I. (1990) Distribution of two morphologically distinct subsets of serotoninergic axons in the cerebral cortex of the marmoset. J. Comp. Neurol., 297: 165–181.

Hosogai, M., Matsuo, S., Sibahara, T. and Kawai, Y. (1998) Projection of respiratory neurons in rat medullary raphe nuclei to the phrenic nucleus. Respir. Physiol., 112: 37–50.

Hubbard, J.B. and Di Carlo, V. (1974) Fluorescence histochemistry of monoamine-containing cell bodies in the brainstem of the squirrel monkey (*Saimiri sciureus*) III. Serotonin-containing groups. J. Comp. Neurol., 153: 385–396.

Imai, H., Steindler, D.A. and Kitai, S.T. (1986) The organization of divergent axonal projections from the midbrain raphe nuclei in the rat. J. Comp. Neurol., 243: 363–380.

Jacobowitz, D.M. and MacLean, P.D. (1978) A brainstem atlas of catecholaminergic neurons and serotonergic perikarya in a pygmy primate (*Cebuella pygmaea*). J. Comp. Neurol., 177: 397–416.

Jacobs, B., Gannon, P.J. and Azmitia, E.C. (1984) Atlas of serotonergic cell bodies in the cat brainstem: an immunohistochemical study. Brain Res. Bull., 13: 1–31.

Jakab, R.L. and Goldman-Rakic, P.S. (2000) Segregation of serotonin 5-HT2A and 5-HT3 receptors in inhibitory circuits of the primate cerebral cortex. J. Comp. Neurol., 417: 337–348.

Jennes, L., Stumpf, W.E. and Kalivas, P.W. (1982) Neurotensin: topographical distribution in rat brain by immunohistochemistry. J. Comp. Neurol., 210: 211–224.

Jones, S.L. and Light, A.R. (1992) Serotoninergic medullary raphespinal projection to the lumbar spinal cord in the rat: a retrograde immunohistochemical study. J. Comp. Neurol., 322: 599–610.

Kalén, P. and Wiklund, L. (1989) Projections from the medial septum and diagonal band of Broca to the dorsal and central superior raphe nuclei: a non-cholinergic pathway. Exp. Brain Res., 75: 401–416.

Kazakov, V.N., Kravtsov, P.Y., Krakhotkina, E.D. and Maisky, V.A. (1993) Sources of cortical, hypothalamic and spinal serotonergic projections: topical organization within the nucleus raphe dorsalis. Neuroscience, 56: 157–164.

Kerr, C.W. and Bishop, G.A. (1991) Topographical organization in the origin of serotoninergic projections to different regions of the cat cerebellar cortex. J. Comp. Neurol., 304: 502–515.

Kievit, J. and Kuypers, H.G.J.M. (1975) Subcortical afferents to the frontal lobe in the Rhesus monkey studied by means of retrograde horseradish peroxidase transport. Brain Res., 85: 261–266.

Kohler, C. and Steinbusch, H. (1982) Identification of serotonin and non-serotonin-containing neurons of the mid-brain raphe projecting to the entorhinal area and the hippocampal formation. A combined immunohistochemical and fluorescent retrograde tracing study in the rat brain. Neuroscience, 7: 951–975.

Kohler, C., Chan-Palay, V. and Steinbusch, H. (1982) The distribution and origin of serotonin-containing fibers in the septal area: a combined immunohistochemical and fluorescent retrograde tracing study in the rat. J. Comp. Neurol., 209: 91–111.

Kosofsky, B.E. and Kowall, N.W. (1989) Demonstration of two morphologic classes of serotonergic fibers in human cortex. Soc. Neurosci. Abstr., 15: 5.

Kosofsky, B.E. and Molliver, M.E. (1987) The serotoninergic innervation of cerebral cortex: different classes of axon terminals arise from dorsal and median raphe nuclei. Synapse, 1: 153–168.

Leanza, G., Perez, S., Pellitteri, R., Russo, A. and Stanzani, S. (1995) Branching serotonergic and non-serotonergic projections from caudal brainstem to the medial preoptic area and the lumbar spinal cord, in the rat. Neurosci. Lett., 200: 5–8.

Lee, H.S., Kim, M.A., Valentino, R.J. and Waterhouse, B.D. (2003) Glutamatergic afferent projections to the dorsal raphe nucleus of the rat. Brain Res., 963: 57–71.

Lee, H.S., Kim, M.A. and Waterhouse, B.D. (2005a) Retrograde double-labeling study of common afferent projections to the dorsal raphe and the nuclear core of the locus coeruleus in the rat. J. Comp. Neurol., 481: 179–193.

Lee, H.S., Lee, B.Y. and Waterhouse, B.D. (2005b) Retrograde study of projections from the tuberomammillary nucleus to the dorsal raphe and the locus coeruleus in the rat. Brain Res., 1043: 65–75.

Lee, H.S., Eum, Y.J., Jo, S.M. and Waterhouse, B.D. (2007) Projection patterns from the amygdaloid nuclear complex to subdivisions of the dorsal raphe nucleus in the rat. Brain Res., 1143: 116–125.

Lee, S.B., Lee, H.S. and Waterhouse, B.D. (2008) The collateral projection from the dorsal raphe nucleus to whisker-related, trigeminal sensory and facial motor systems in the rat. Brain Res., 1214: 11–22.

Léger, L., Charnay, Y., Burlet, S., Gay, N., Schaad, N., Bouras, C. and Cespuglio, R. (1998) Comparative distribution of nitric oxide synthase- and serotonin-containing neurons in the raphe nuclei of four mammalian species. Histochem. Cell Biol., 110: 517–525.

Leranth, C. and Vertes, R.P. (1999) Median raphe serotonergic innervation of medial septum/diagonal band of broca (MSDB) parvalbumin-containing neurons: possible involvement of the MSDB in the desynchronization of the hippocampal EEG. J. Comp. Neurol., 410: 586–598.

Li, Y.Q., Takada, M. and Mizuno, N. (1993) The sites of origin of serotoninergic afferent fibers in the trigeminal motor, facial, and hypoglossal nuclei in the rat. Neurosci. Res., 17: 307–313.

Lidov, H.G.W. and Molliver, M.E. (1982) Immunohistochemical study of the development of serotonergic neurons in the rat CNS. Brain Res., 78: 45–46.

Loewy, A.D. (1981) Raphe pallidus and raphe obscurus projections to the intermediolateral cell column in the rat. Brain Res., 222: 129–133.

Loewy, A.D. and McKellar, S. (1981) Serotonergic projections from the ventral medulla to the intermediolateral cell column in the rat. Brain Res., 211: 146–152.

Maeda, T., Kaneko, S. and Satoh, M. (1994) Inhibitory influence via 5-HT3 receptors on the induction of LTP in mossy fiber-CA3 system of guinea-pig hippocampal slices. Neurosci. Res., 18: 277–282.

Mamounas, L.A., Mullen, C.A., O'Hearn, E. and Molliver, M.E. (1991) Dual serotoninergic projections to forebrain in the rat: morphologically distinct 5-HT axon terminals exhibit differential vulnerability to neurotoxic amphetamine derivatives. J. Comp. Neurol., 314: 558–586.

Manaker, S. and Fogarty, P.F. (1995) Raphespinal and reticulospinal neurons project to the dorsal vagal complex in the rat. Exp. Brain Res., 106: 79–92.

McMahon, L.L. and Kauer, J.A. (1997) Hippocampal interneurons are excited via serotonin-gated ion channels. J. Neurophysiol., 78: 2493–2502.

Miettinen, R. and Freund, T.F. (1992) Convergence and segregation of septal and median raphe inputs onto different subsets of hippocampal inhibitory interneurons. Brain Res., 594: 263–272.

Mikkelsen, J.D., Hay-Schmidt, A. and Larsen, P.J. (1997) Central innervation of the rat ependyma and subcommissural organ with special reference to ascending serotoninergic projections from the raphe nuclei. J. Comp. Neurol., 384: 556–568.

Morales, M. and Bloom, F.E. (1997) The 5-HT 3 receptor is present in different subpopulations of GABAergic neurons in the rat telencephalon. J. Neurosci., 17: 3157–3167.

Morrison, J.H., Foote, S.L., Molliver, M.E., Bloom, F.E. and Lidov, H.G.W. (1982) Noradrenergic and serotonergic fibers innervate complementary layers in monkey primary visual cortex: an immunohistochemical study. Proc. Natl. Acad. Sci. USA, 79: 2401–2405.

Mulligan, K.A. and Törk, I. (1987) Serotonergic axons form basket-like terminals in cerebral cortex. Neurosci. Lett., 81: 7–12.

Murphy, D.L. and Lesch, K.P. (2008) Targeting the murine serotonin transporter: insights into human neurobiology. Nat. Rev. Neurosci., 9: 85–96.

Nakamura, K., Matsumura, K., Kaneko, T., Kobayashi, S., Katoh, H. and Negishi, M. (2002) The rostral raphe pallidus nucleus mediates pyrogenic transmission from the preoptic area. J. Neurosci., 22: 4600–4610.

Nielsen, K., Brask, D., Knudsen, G.M. and Aznar, S. (2006) Immunodetection of the serotonin transporter protein is a more valid marker for serotonergic fibers than serotonin. Synapse, 59: 270–276.

Nobin, A. and Björklund, A. (1973) Topography of the monoamine neuron systems in the human brain as revealed in fetuses. Acta Physiol. Scand., S388: 1–40.

O'Hearn, E. and Molliver, M.E. (1984) Organization of raphe-cortical projections in rat: a quantitative retrograde study. Brain Res. Bull., 13: 709–726.

O'Hearn, E., Battaglia, G., De Souza, E.B., Kuhar, M.J. and Molliver, M.E. (1988) Methylenedioxyamphetamine (MDA) and methylenedioxymethamphetamine (MDMA) cause selective ablation of serotonergic axon terminals in forebrain: immunocytochemical evidence for neurotoxicity. J. Neurosci., 8: 2788–2803.

Okere, C.O. and Waterhouse, B.D. (2006) Activity-dependent heterogeneous populations of nitric oxide synthase neurons in the rat dorsal raphe nucleus. Brain Res., 1086: 117–132.

Ovalle, S., Casanova, E., Garate, C., Alonso-Llamazares, A., Chinchetru, M.A. and Calvo, P. (1995) Immunodetection of serotonin transporter from mouse brain. NeuroReport, 6: 2353–2356.

Parsons, M.J., Benca, R.M., Brownfield, M.S. and Behan, M. (2001) Age-associated changes in the serotonergic system in rat superior colliculus and pretectum. Brain Res. Bull., 55: 435–444.

Pecci-Saavadra, J., Pasik, T., Pasik, P., Caputto, R. and Ajmone-Marsan, C. (1983) Immunocytochemistry of serotonergic neurons in the central nervous system of monkeys. In: Neurotransmission. Learning and Memory, pp. 81–96. Raven Press, New York.

Petit, J.M., Luppi, P.H., Peyron, C., Rampon, C. and Jouvet, M. (1995) VIP-like immunoreactive projections from the dorsal raphe and caudal linear raphe nuclei to the bed nucleus of the stria terminalis demonstrated by a double immunohistochemical method in the rat. Neurosci. Lett., 193: 77–80.

Peyron, C., Petit, J.M., Rampon, C., Jouvet, M. and Luppi, P.H. (1998) Forebrain afferents to the rat dorsal raphe nucleus demonstrated by retrograde and anterograde tracing methods. Neuroscience, 82: 443–468.

Pickel, V.M., Joh, T.H. and Reis, D.J. (1977) A serotonergic innervation of noradrenergic neurons in nucleus locus coeruleus: demonstration by immunocytochemical localization of the transmitter specific enzymes tyrosine and tryptophan hydroxylase. Brain Res., 131: 197–214.

Porrino, L.J. and Goldman-Rakic, P.S. (1982) Brainstem innervation of prefrontal and anterior cingulate cortex in the rhesus monkey revealed by retrograde transport of HRP. J. Comp. Neurol., 205: 63–76.

Poulat, P., Marlier, L., Rajaofetra, N. and Privat, A. (1992) 5-Hydroxytryptamine, substance P and thyrotropin-releasing hormone synapses in the intermediolateral cell column of the rat thoracic spinal cord. Neurosci Lett., 136: 19–22.

Raghanti, M.A., Stimpson, C.D., Marcinkiewicz, J.L., Erwin, J.M., Hof, P.R. and Sherwood, C.C. (2008) Differences in cortical serotonergic innervation among humans, chimpanzees, and macaque monkeys: a comparative study. Cereb. Cortex, 18: 584–597.

Rampon, C., Peyron, C., Gervasoni, D., Pow, D.V., Luppi, P.H. and Fort, P. (1999) Origins of the glycinergic inputs to the rat locus coeruleus and dorsal raphe nuclei: a study combining retrograde tracing with glycine immunohistochemistry. Eur. J. Neurosci., 11: 1058–1066.

Ribeiro-do-Valle, L.E. (1997) Serotonergic neurons in the caudal raphe nuclei discharge in association with activity of masticatory muscles. Braz. J. Med. Biol. Res., 30: 79–83.

Ridet, J.L., Rajaofetra, N., Teilhac, J.R., Geffard, M. and Privat, A. (1993) Evidence for nonsynaptic serotonergic and noradrenergic innervation of the rat dorsal horn and possible involvement of neuron-glia interactions. Neuroscience, 52: 143–157.

Ridet, J.L., Geffard, M. and Privat, A. (1994) Light and electron microscopic studies of the effects of p-chloroamphetamine on the monoaminergic innervation of the rat spinal cord. J. Comp. Neurol., 343: 281–296.

Rikard-Bell, G.C., Törk, I., Sullivan, C. and Scheibner, T. (1990) Distribution of substance P-like immunoreactive fibres and terminals in the medulla oblongata of the human infant. Neuroscience, 34: 133–148.

Ropert, N. and Guy, N. (1991) Serotonin facilitates GABAergic transmission in the CA1 region of rat hippocampus *in vitro*. J. Physiol., 441: 121–136.

Schiffmann, S.N. and Vanderhaeghen, J.J. (1991) Distribution of cells containing mRNA encoding cholecystokinin in the rat central nervous system. J. Comp. Neurol., 304: 219–233.

Schofield, S.P.M. and Dixson, A.F. (1982) Distribution of catecholamine and indoleamine neurons in the brain of the common marmoset (*Callithrix jacchus*). J. Anat., 134: 315–338.

Schofield, S.P.M. and Everitt, B.J. (1981) The organization of indoleamine neurons in the brain of the rhesus monkey (*Macaca mulatta*). J. Comp. Neurol., 197: 369–383.

Simpson, K.L., Fisher, T.M., Waterhouse, B.D. and Lin, R.C. (1998) Projection patterns from the raphe nuclear complex to the ependymal wall of the ventricular system in the rat. J. Comp. Neurol., 399: 61–72.

Simpson, K.L., Waterhouse, B.D. and Lin, R.C. (2003) Differential expression of nitric oxide in serotonergic projection neurons: neurochemical identification of dorsal raphe inputs to rodent trigeminal somatosensory targets. J. Comp. Neurol., 466: 495–512.

Skagerberg, G. and Björklund, A. (1985) Topographic principles in the spinal projections of serotonergic and non-serotonergic brainstem neurons in the rat. Neuroscience, 15: 445–480.

Smiley, J.F. and Goldman-Rakic, P.S. (1996) Serotonergic axons in monkey prefrontal cerebral cortex synapse predominantly on interneurons as demonstrated by serial section electron microscopy. J. Comp. Neurol., 367: 431–443.

Snowball, R.K., Dampney, R.A. and Lumb, B.M. (1997) Responses of neurones in the medullary raphe nuclei to inputs from visceral nociceptors and the ventrolateral periaqueductal grey in the rat. Exp. Physiol., 82: 485–500.

Stamp, J.A. and Semba, K. (1995) Extent of colocalization of serotonin and GABA in the neurons of the rat raphe nuclei. Brain Res., 677: 39–49.

Steinbusch, H.W., van der Kooy, D., Verhofstad, A.A. and Pellegrino, A. (1980) Serotonergic and non-serotonergic projections from the nucleus raphe dorsalis to the caudate-putamen complex in the rat, studied by a combined immunofluorescence and fluorescent retrograde axonal labeling technique. Neurosci. Lett., 19: 137–142.

Steinbusch, H.W.M. (1981) Distribution of serotonin-immunoreactivity in the central nervous system of the rat-cell bodies and terminals. Neuroscience, 6: 557–618.

Steinbusch, H.W.M., Nieuwenhuys, R. and Emson, P.C. (1983) The raphe nuclei of the rat brainstem: a cytoarchitectonic and immunohistochemical study. In: Chemical Neuroanatomy, pp. 131–207. Raven Press, New York, NY.

Steininger, T.L., Wainer, B.H., Blakely, R.D. and Rye, D.B. (1997) Serotonergic dorsal raphe nucleus projections to the cholinergic and noncholinergic neurons of the pedunculopontine tegmental region: a light and electron microscopic anterograde tracing and immunohistochemical study. J. Comp. Neurol., 382: 302–322.

Sur, C., Betz, H. and Schloss, P. (1996) Immunocytochemical detection of the serotonin transporter in rat brain. Neuroscience, 73: 217–231.

Tagliaferro, P., Ramos, A.J., Lopez-Costa, J.J., Lopez, E.M., Saavedra, J.P. and Brusco, A. (2001) Increased nitric oxide synthase activity in a model of serotonin depletion. Brain Res. Bull., 54: 199–205.

Takahashi, H., Nakashima, S., Ohama, E., Takeda, S. and Ikuta, F. (1986) Distribution of serotonin-containing bodies in the brainstem of the human fetus determined with immunohistochemistry using antiserotonin serum. Brain Dev., 8: 355–365.

Takeuchi, Y. and Sano, Y. (1983) Immunohistochemical demonstration of serotonin-containing nerve fibers in the inferior olivary complex of the rat, cat and monkey. Cell Tissue Res., 231: 17–28.

Tanaka, H., Amamiya, S., Miura, N., Araki, A., Ohinata, J. and Fujieda, K. (2006) Postnatal development of brainstem serotonin-containing neurons projecting to lumbar spinal cord in rats. Brain Dev., 28: 586–591.

Tanaka, M., Okamura, H., Tamada, Y., Nagatsu, I., Tanaka, Y. and Ibata, Y. (1994) Catecholaminergic input to spinally projecting serotonin neurons in the rostral ventromedial medulla oblongata of the rat. Brain Res. Bull., 35: 23–30.

Thompson, A.M., Moore, K.R. and Thompson, G.C. (1995) Distribution and origin of serotoninergic afferents to guinea pig cochlear nucleus. J. Comp. Neurol., 351: 104–116.

Tigges, J., Tigges, M., Cross, N.A., McBride, R.L., Letbetter, W.D. and Anschel, S. (1982) Subcortical structures projecting to visual cortical areas in squirrel monkey. J. Comp. Neurol., 209: 29–40.

Törk, I. (1990) Anatomy of the serotonergic system. Ann. NY Acad. Sci., 600: 9–34, discussion 34–35.

Törk, I. and Hornung, J.P. (1990) Raphe nuclei and the 5-HT system. In: Paxinos, G. (Ed.), The Human Nervous System. Academic Press, San Diego, CA, pp. 1001–1022.

Trottier, S., Evrard, B., Vignal, J.P., Scarabin, J.M. and Chauvel, P. (1996) The serotonergic innervation of the cerebral cortex in man and its changes in focal cortical dysplasia. Epilepsy Res., 25: 79–106.

Twarog, B.M. and Page, I.H. (1953) Serotonin content of some mammalian tissues and urine and a method for its determination. Am. J. Physiol., 175: 157–161.

Van Bockstaele, E.J., Biswas, A. and Pickel, V.M. (1993) Topography of serotonin neurons in the dorsal raphe nucleus that send axon collaterals to the rat prefrontal cortex and nucleus accumbens. Brain Res., 624: 188–198.

VanderHorst, V.G. and Ulfhake, B. (2006) The organization of the brainstem and spinal cord of the mouse: relationships between monoaminergic, cholinergic, and spinal projection systems. J. Chem. Neuroanat., 31: 2–36.

van der Kooy, D. and Hattori, T. (1980) Dorsal raphe cells with collateral projections to the caudate-putamen and substantia nigra: a fluorescent retrograde stud double labeling study in the rat. Brain Res., 186: 1–7.

van der Kooy, D. and Kuypers, H.G.J.M. (1979) Fluorescent retrograde double labeling: axonal branching in the ascending raphe and nigral projections. Science, 204: 873–875.

Vertes, R.P. (1991) A PHA-L analysis of ascending projections of the dorsal raphe nucleus in the rat. J. Comp. Neurol., 313: 643–668.

Vertes, R.P. and Crane, A.M. (1997) Distribution, quantification, and morphological characteristics of serotonin-immunoreactive cells of the supralemniscal nucleus (B9) and pontomesencephalic reticular formation in the rat. J. Comp. Neurol., 378: 411–424.

Vertes, R.P. and Kocsis, B. (1994) Projections of the dorsal raphe nucleus to the brainstem: PHA-L analysis in the rat. J. Comp. Neurol., 340: 11–26.

Vertes, R.P. and Martin, G.F. (1988) Autoradiographic analysis of ascending projections from the pontine and mesencephalic reticular formation and the median raphe nucleus in the rat. J. Comp. Neurol., 275: 511–541.

Vertes, R.P., Fortin, W.J. and Crane, A.M. (1999) Projections of the median raphe nucleus in the rat. J. Comp. Neurol., 407: 555–582.

Villar, M.J., Vitale, M.L., Hökfelt, T. and Verhofstad, A.A.J. (1988) Dorsal raphe serotoninergic branching neurons projecting both to the lateral geniculate body and superior colliculus: a combined retrograde tracing-immunohistochemical study in the rat. J. Comp. Neurol., 277: 126–140.

Walther, D.J., Peter, J.U., Bashammakh, S., Hortnagl, H., Voits, M., Fink, H. and Bader, M. (2003) Synthesis of serotonin by a second tryptophan hydroxylase isoform. Science, 299: 76.

Wang, Q.P. and Nakai, Y. (1993) Enkephalinergic innervation of GABAergic neurons in the dorsal raphe nucleus of the rat. Brain Res. Bull., 32: 315–320.

Wang, Q.P., Ochiai, H. and Nakai, Y. (1992) GABAergic innervation of serotonergic neurons in the dorsal raphe nucleus of the rat studied by electron microscopy double immunostaining. Brain Res. Bull., 29: 943–948.

Waselus, M., Galvez, J.P., Valentino, R.J. and Van Bockstaele, E.J. (2006) Differential projections of dorsal raphe nucleus neurons to the lateral septum and striatum. J. Chem. Neuroanat., 31: 233–242.

Waterhouse, B.D., Mihailoff, G.A., Baack, J.C. and Woodward, D.J. (1986) Topographical distribution of dorsal and median raphe neurons projecting to motor, sensorimotor, and visual cortical areas in the rat. J. Comp. Neurol., 249: 460–476.

Westlund, K.N., Sorkin, L.S., Ferrington, D.G., Carlton, S.M., Willcockson, H.H. and Willis, W.D. (1990) Serotoninergic and noradrenergic projections to the ventral posterolateral nucleus of the monkey thalamus. J. Comp. Neurol., 295: 197–207.

Westlund, K.N., Lu, Y., Coggeshall, R.E. and Willis, W.D. (1992) Serotonin is found in myelinated axons of the dorsolateral funiculus in monkeys. Neurosci. Lett., 141: 35–38.

Wiklund, L., Léger, L. and Persson, M. (1981) Monoamine cell distribution in the cat brain stem. A fluorescence histochemical study with quantification of indolaminergic and locus coeruleus cell groups. J. Comp. Neurol., 203: 613–647.

Wilson, M.A. and Molliver, M.E. (1991) The organization of serotonergic projections to cerebral cortex in primates: retrograde transport studies. Neuroscience, 44: 555–570.

Woolley, D.W. and Shaw, E. (1954) A biochemical and pharmacological suggestion about certain mental disorders. Proc. Natl. Acad. Sci. USA, 40: 228–231.

Wu, W. and Wessendorf, M.W. (1992) Organization of the serotonergic innervation of spinal neurons in rats – I. Neuropeptide coexistence in varicosities innervating some spinothalamic tract neurons but not in those innervating postsynaptic dorsal column neurons. Neuroscience, 50: 885–898.

Wu, W., Elde, R. and Wessendorf, M.W. (1993) Organization of the serotonergic innervation of spinal neurons in rats – III. Differential serotonergic innervation of somatic and parasympathetic preganglionic motoneurons as determined by patterns of co-existing peptides. Neuroscience, 55: 223–233.

Yang, H., Wu, S.V., Ishikawa, T. and Tache, Y. (1994) Cold exposure elevates thyrotropin-releasing hormone gene expression in medullary raphe nuclei: relationship with vagally mediated gastric erosions. Neuroscience, 61: 655–663.

Zagon, A. (1993) Innervation of serotonergic medullary raphe neurons from cells of the rostral ventrolateral medulla in rats. Neuroscience, 55: 849–867.

Zhou, F.C., Xu, Y., Bledsoe, S., Lin, R. and Kelley, M.R. (1996) Serotonin transporter antibodies: production, characterization, and localization in the brain. Mol. Brain Res., 43: 267–278.

CHAPTER 1.4

Ultrastructure of the Serotonin Innervation in the Mammalian Central Nervous System

Laurent Descarries*[,1,2,3], Mustapha Riad[1] and Martin Parent[1,3]

[1]*Department of Pathology & Cell Biology, Université de Montréal, Montreal, Quebec, Canada*
[2]*Department of Physiology, Université de Montréal, Montreal, Quebec, Canada*
[3]*Groupe de recherche sur le système nerveux central (GRSNC), Faculty of Medicine, Université de Montréal, Montreal, Quebec, Canada*

Abstract: In the past 40 years, cytochemical, radioautographic and imunocytochemical methods have been applied to identify serotonin (5-hydroxytryptamine, 5-HT) neurons at the electron-microscopic level, providing a wealth of information on their ultrastructural features in many parts of the mammalian central nervous system. This chapter summarizes much of these data, with an emphasis on the fine structural characteristics of 5-HT nerve terminals, particularly when they have been described in relation to chemically, physiologically or hodologically identified targets. Results obtained in various experimental models and during postnatal development have been included, as they provide insights into the morphological plasticity of the 5-HT system.

Keywords: axon terminals, cytochemistry, radioautography, immunocytochemistry, electron microscopy.

Introduction

The serotonin (5-hydroxytryptamine, 5-HT) system is one of the most widespread transmitter-defined neuronal systems in the mammalian central nervous system (CNS). It is now clear that there is no single region of the CNS without 5-HT innervation, including the cerebro-ventricular cavity. Following the discovery of 5-HT neurons and the first anatomical description of their distribution as fluorescent microscopical silhouettes throughout the rat CNS (Dahlström and Fuxe, 1964; Fuxe, 1965), it did not take long before electron-microscopic methods were developed for their specific identification and characterization at the ultrastructural level (for review, see Beaudet and Descarries, 1987). Cytochemical, radioautographic and immunocytochemical approaches were then applied in that order, yielding a wealth of information on the distribution, cytological features and ultrastructural relationships of 5-HT neurons in most parts of the CNS (Table 1).

The cytochemical technique was the first to demonstrate the 'axonal bouton' nature of the minute, evanescent, yellow fluorescent dots visualized with the histochemical fluorescence technique[1] (Hökfelt, 1967). However, it usually required immersion fixation procedures and/or the use of chemical markers of questionable specificity, and was soon replaced by the radioautographic technique after administration of tritiated 5-HT ([^{3}H]5-HT) *in vivo*, which was compatible with perfusion fixation procedures for electron microscopy and yielded results of unsurpassed ultrastructural quality (reviewed in Descarries and Beaudet, 1983). [^{3}H]5-HT uptake radioautography could also be used on brain slices to quantify 5-HT innervation density (Doucet *et al.*, 1988). However, radioautography after [^{3}H]5-HT uptake *in vivo* was technically demanding, and, because of the blood–brain barrier, its use was limited to brain regions accessible by cerebroventricular administration, topical application or local microinstillation of [^{3}H]5-HT. In this context, the advent of immunocytochemistry in the late 1970s represented a major breakthrough, allowing for ultrastructural investigations of the 5-HT system in any CNS region, after immunolabeling with antibodies against

[1]Note: Throughout this chapter, the words 'nerve ending', 'terminal', 'varicosity', 'bouton' are used as synonyms to designate axonal enlargements containing aggregated small vesicles and endowed or not with the morphologically defined membrane specializations (junctional complex) that are the hallmark of synapses.

*Corresponding author
E-mail: laurent.descarries@umontreal.ca

Christian Müller & Barry Jacobs (Eds.)
Handbook of Behavioral Neurobiology of Serotonin
ISBN 978-0-12-374634-4

DOI: 10.1016/B978-0-12-374634-4.00004-6

Table 1 Ultrastructural features of 5-HT axon terminals (varicosities) in mammalian CNS (as of December 2008)

Region	Species	Method	Synaptic incidence	References
Forebrain				
Cerebral cortex				
Fronto-parietal	Rat	[^{3}H]5-HT radioautography		Descarries *et al.*, 1975; Beaudet and Descarries, 1978, 1981
		SERT immunocytochemistry		Yamamoto *et al.*, 1998
Frontal	Rat	5-HT immunocytochemistry	32%	Séguéla *et al.*, 1989[a]; Descarries *et al.*, 1991, Descarries and Umbriaco, 1995
		TPH immunocytochemistry	22%	Cohen *et al.*, 1995[a]
Pre-frontal	Rat	SERT immunocytochemistry		Miner *et al.*, 2000
Parietal	Rat	5-HT immunocytochemistry	46%	Molliver *et al.*, 1982; Séguéla *et al.*, 1989[a]
Occipital	Rat	5-HT immunocytochemistry	37%	Papadopoulos *et al.*, 1987a, 1987b; Séguéla *et al.*, 1989[a]; Dori *et al.*, 1996; Paspalas and Papadopoulos, 2001[b]
Entorhinal	Rat	TPH immunocytochemistry	18%	Cohen *et al.*, 1995[a]
Auditory	Cat	5-HT immunocytochemistry	3%	De Felipe *et al.*, 1991[a]
Sensorimotor	Monkey	5-HT immunocytochemistry	2–3%	De Felipe and Jones, 1988[a]
Pre-frontal	Monkey	5-HT immunocytochemistry	23%	Smiley and Goldman-Rakic, 1996[a]
Visual	Monkey	5-HT immunocytochemistry		De Lima *et al.*, 1988
	Human	[^{3}H]5-HT radioautography		Calas *et al.*, 1981
Hippocampus				
CA1	Rat	5-HT immunocytochemistry	21%	Freund *et al.*, 1990[b]; Umbriaco *et al.*, 1995[a]
		TPH immunocytochemistry	12%	Cohen *et al.*, 1995[a]
		SERT immunocytochemistry		Zhou *et al.*, 1998
CA3	Rat	5-HT immunocytochemistry	19%; <20%	Descarries *et al.*, 1990; Freund *et al.*, 1990[b]; Oleskevich *et al.*, 1991[a]; Daszuta *et al.*, 1991[a]
Dentate gyrus	Rat	5-HT immunocytochemistry	24%; <20%	Anderson *et al.*, 1986; Descarries *et al.*, 1990; Oleskevich *et al.*, 1991[a]; Daszuta *et al.*, 1991[a]
Olfactory bulb	Rat	[^{3}H]5-HT radioautography		Halász *et al.*, 1978
Basal forebrain				
Diagonal band of Broca	Rat	5-HT immunocytochemistry	46%	Dinopoulos *et al.*, 1997[a]
Nucleus basalis	Monkey	5-HT immunocytochemistry		Smiley *et al.*, 1999
Septum	Rat	5-HT immunocytochemistry		Dinopoulos *et al.*, 1993; Milner and Veznedaroglu, 1993[b]; Léránth and Vertes, 1999[b]
Bed nucleus of the stria terminalis	Rat	5-HT immunocytochemistry		Phelix *et al.*, 1992
Amygdala (basolateral)	Rat	5-HT immunocytochemistry	76%	Muller *et al.*, 2007[a,b]
Neostriatum	Rat	[^{3}H]5-HT radioautography	13%	Arluison and De La Manche, 1980; Soghomonian *et al.*, 1989[a]
		5-HT immunocytochemistry	10%	Soghomonian *et al.*, 1989[a]
			6–8%	Van Luijtelaar *et al.*, 1991; Descarries *et al.*, 1992[a];

	Cat	[^{3}H]5-HT radioautography		Calas and Ségu, 1976; Calas *et al.*, 1976
	Monkey	5-HT immunocytochemistry		Pasik and Pasik, 1982; Pasik *et al.*, 1982
Nucleus accumbens	Rat	SERT immunocytochemistry		Van Bockstaele and Pickel, 1993[b]; Van Bockstaele *et al.*, 1996[b]; Pickel and Chan, 1999
Globus pallidus	Monkey	5-HT immunocytochemistry		Pasik *et al.*, 1984
Thalamus				
Ventrobasal complex	Rat	5-HT immunocytochemistry		Nothias *et al.*, 1988
Ventral lateral nucleus	Rat	5-HT immunocytochemistry	62%	Dori *et al.*, 1998[a]
Ventral posterior nucleus	Cat	5-HT immunocytochemistry	9%	Liu and Jones, 1991[a]
	Monkey	5-HT immunocytochemistry	10%	Liu and Jones, 1991[a]
Lateral geniculate nucleus	Rat	5-HT immunocytochemistry	42%	Papadopoulos and Parnavelas, 1990; Dinopoulos *et al.*, 1995[a]
	Cat	5-HT immunocytochemistry		De Lima and Singer, 1987a, 1987b
	Monkey	5-HT immunocytochemistry	1%	Pasik *et al.*, 1988; Wilson and Hendrickson, 1988[a]
Perigeniculate nucleus	Cat	5-HT immunocytochemistry		De Lima and Singer, 1987a
Zona incerta	Rat	[^{3}H]5-HT radioautography		Bosler *et al.*, 1984b[b]; Frankfurt and Beaudet, 1988[b]
Subthalamic nucleus	Monkey	SERT immunocytochemistry		Wallman *et al.*, 2007
Hypothalamus				
Pre-optic area	Rat	[^{3}H]5-HT radioautography		Kiss and Halász, 1985[b]
Suprachiasmatic nucleus	Rat	cytochemistry		Ajika and Ochi, 1975; Nojyo and Sano, 1978
		[^{3}H]5-HT radioautography		Kiss *et al.*, 1984[b]; Bosler and Beaudet, 1985a, 1985b[b]; Bosler *et al.*, 1986[b]
		5-HT immunocytochemistry	48%	Boulaïch *et al.*, 1994[a]; Arai *et al.*, 2002
N. dorsomedialis hypothalami	Rat	[^{3}H]5-HT radioautography		Descarries and Beaudet, 1978; Beaudet and Descarries, 1979, 1981; Frankfurt and Beaudet, 1987
Paraventricular nucleus	Rat	5-HT immunocytochemistry		Liposits *et al.*, 1987[b]
Nucleus arcuatus	Rat	[^{3}H]5-HT radioautography		Bosler *et al.*, 1984b[b], Bosler *et al.*, 1986[b]; Kiss *et al.*, 1984[b]; Bosler and Beaudet, 1985a[b]; Bosler *et al.*, 1986[b]
Supra-optic nucleus	Rat	5-HT immunocytochemistry		Boulaïch *et al.*, 1994[a]
Circumventricular organs				
Median eminence	Rat	5,7-DHT degeneration		Baumgarten and Lachenmayer, 1974
			[3H]5-HT radioautography	Calas *et al.*, 1974, 1978; Calas, 1977; Bosler, 1980; Nakai *et al.*, 1983[b]; Bouchaud and Bosler, 1986
Org. vasc. laminae term.	Rat	[^{3}H]5-HT radioautography		Bosler, 1978, 1980; Calas *et al.*, 1978; Bouchaud and Bosler, 1986; Bosler and Descarries, 1988; Saïdi and Bosler, 1990

(*Continued*)

Table 1 *(Continued)*

Region	**Species**	**Method**	**Synaptic incidence**	**References**
Subcommissural organ	Rat	[^{3}H]5-HT radioautography		Bouchaud and Arluison 1977; Bouchaud, 1979; Calas *et al.*, 1978; Møllgärd and Wiklund, 1979; Bosler, 1980; Bouchaud and Bosler, 1986
Subfornical organ	Rat	[^{3}H]5-HT radioautography		Calas *et al.*, 1978; Bouchaud and Bosler, 1986
	Monkey	[^{3}H]5-HT radioautography		Calas *et al.*, 1974
Area postrema	Rat	[^{3}H]5-HT radioautography		Armstrong *et al.*, 1984[b]; Pickel and Armstrong, 1984[b]; Pickel *et al.*, 1984[b]; Bouchaud and Bosler, 1986; Bosler *et al.*, 1986[b]
Supra-ependymal plexus	Rat	cytochemistry		Lorez and Richards, 1973, 1975; Richards *et al.*, 1973
		[^{3}H]5-HT radioautography		Chan-Palay 1976; Richards, 1977; Calas *et al.*, 1978; Bosler, 1980; De La Manche *et al.*, 1981
		5-HT immunocytochemistry		Dinopoulos and Dori, 1995
	Monkey	[^{3}H]5-HT radioautography		Chan-Palay, 1976
	Human	scanning electron microscopy		Richards *et al.*, 1980, 1981
Brainstem				
Substantia nigra	Rat	[^{3}H]5-HT radioautography		Parizek *et al.*, 1971
		5-HT immunocytochemistry		Mori *et al.*, 1987; Corvaja *et al.*, 1993
Reticulata			100%	Moukhles *et al.*, 1997[a,b]
Compacta			50%	Moukhles *et al.*, 1997[a,b]
	Guinea pig	5-HT immunocytochemistry		Nedergaard *et al.*, 1988[b]
Reticulata	Monkey	5-HT immunocytochemistry		Pasik *et al.*, 1984
Ventral tegmental area	Rat	[^{3}H]5-HT radioautography	50%	Bosler *et al.*, 1986[b]; Hervé *et al.*, 1987[a,b]; Van Bockstaele *et al.*, 1994[b]
Nucleus interpeduncularis	Rat	cytochemistry		Nojyo and Sano, 1978
Red nucleus	Rat	5-HT immunocytochemistry		André *et al.*, 1987[b]
	Cat	[^{3}H]5-HT radioautography		Bosler *et al.*, 1983
Oculomotor nucleus	Rat	[^{3}H]5-HT radioautography		Soghomonian *et al.*, 1986
Superior colliculus	Rat	5-HT immunocytochemistry		Dori *et al.*, 1998[a]
Superficial layers			55%	
Deep layers			38%	
	Cat	5-HT immunocytochemistry		Mize and Horner, 1989
	Hamster	5-HT immunocytochemistry		Crissman *et al.*, 1993
			6%	Arce *et al.*, 1992[a], Arce *et al.*, 1995
Dorsal periaqueductal gray	Rat	5-HT immunocytochemistry	23%	Lovick *et al.*, 2000[a]
Central gray	Rat	[^{3}H]5-HT radioautography		Aghajanian and Bloom, 1967
		5-HT immunocytochemistry		Clements *et al.*, 1985
Nucleus raphe dorsalis	Rat	[^{3}H]5-HT radioautography		Beaudet and Descarries, 1981; Chan-Palay, 1982; Descarries *et al.*, 1982; Brusco *et al.*, 1983

	Cat	5-HT immunocytochemistry		Chazal and Ralston, 1987
	Monkey	[^{3}H]5-HT radioautography		Chan-Palay, 1982
		5-HT immunocytochemistry		Kapadia *et al.*, 1985
Nucleus raphe magnus	Rat	5-HT immunocytochemistry		Chazal and Ma, 1989
Locus coeruleus	Rat	TPH immunocytochemistry		Pickel *et al.*, 1977
		[^{3}H]5-HT radioautography		Léger and Descarries, 1978
Mesopontine tegmentum	Rat	5-HT immunocytochemistry		Honda and Semba, 1994; Wang *et al.*, 2000[b]
	Rat	SERT immunocytochemistry		Steininger *et al.*, 1997
N. paragigantocellularis lateralis	Rat	[^{3}H]5-HT radioautography		Chan-Palay, 1978b
Nucleus paratrigeminalis	Rat	[^{3}H]5-HT radioautography		Chan-Palay, 1978a
	Monkey	[^{3}H]5-HT radioautography		Chan-Palay, 1978a
Mesencephalic trigeminal n.	Rat	5-HT immunocytochemistry		Liem *et al.*, 1993; Liem and Copray, 1996
	Cat	5-HT immunocytochemistry		Lazarov and Chouchkov, 1995
Spinal trigeminal nucleus	Rat	5-HT immunocytochemistry		López Costa *et al.*, 1994; Li *et al.*, 1997[b]
Trigeminal motor nucleus	Rat	[^{3}H]5-HT radioautography		Schaffar *et al.*, 1984
		5-HT immunocytochemistry		Saha *et al.*, 1991
	Cat	5-HT immunocytochemistry	27%	Nagase *et al.*, 1997[b]
Facial motor nucleus	Rat	[^{3}H]5-HT radioautography		Aghajanian and McCall, 1980
	Monkey	5-HT immunocytochemistry		Takeuchi *et al.*, 1983
Nucleus ambiguus	Monkey	5-HT immunocytochemistry		Takeuchi *et al.*, 1983
Nucleus tractus solitarius	Rat	[^{3}H]5-HT radioautography		Pickel *et al.*, 1984[b]; Bosler *et al.*, 1986[b]; Huang and Pickel, 2002
		SERT immunocytochemistry		Huang *et al.*, 2004
	Cat	5-HT immunocytochemistry		Elde *et al.*, 1982; Maley and Elde, 1982; Maley *et al.*, 1990
Inferior olivary complex	Rat	[^{3}H]5-HT radioautography		Wiklund *et al.*, 1981
	Opossum	[^{3}H]5-HT radioautography		King *et al.*, 1984
		5-HT immunocytochemistry		King *et al.*, 1984
Dorsal column nuclei	Cat	5-HT immunocytochemistry		Blomqvist and Broman, 1993
Cerebellum				
Cortex	Rat	[^{3}H]5-HT radioautography		Bloom *et al.*, 1972; Chan-Palay, 1975, 1977; Palay and Chan-Palay, 1976; Sotelo and Beaudet 1979; Beaudet and Descarries, 1981; Beaudet and Sotelo, 1981; Beaudet, 1982
	Monkey	[^{3}H]5-HT radioautography		Chan-Palay, 1975, 1977
Deep nuclei	Rat	[^{3}H]5-HT radioautography		Chan-Palay, 1975, 1977; Palay and Chan-Palay, 1976
	Monkey	[^{3}H]5-HT radioautography		Chan-Palay, 1975, 1977; Palay and Chan-Palay, 1976
Spinal cord				
Dorsal horn	Rat	5-HT immunocytochemistry	37%	Maxwell *et al.*, 1983, 1985; Mizukawa *et al.*, 1986; Marlier *et al.*, 1991, 1992; Ridet *et al.*, 1993[a]; Ridet, 1994

(*Continued*)

Table 1 *(Continued)*

Region	Species	Method	Synaptic incidence	References
	Cat	[^{3}H]5-HT radioautography		Ségu and Calas, 1978; Ruda and Gobel, 1980; Ruda *et al.*, 1981a, 1981b; Glazer and Basbaum, 1984[b]
		5-HT immunocytochemistry		Ruda *et al.*, 1982; Hoffert *et al.*, 1983; Light *et al.*, 1983; Nishikawa *et al.*, 1983; Miletic *et al.*, 1984; Maxwell and Jankowska, 1996[b]; Jankowska *et al.*, 1997[b]
	Monkey	5-HT immunocytochemistry		Nishikawa *et al.*, 1983; LaMotte and De Lanerolle, 1983; Rajaofetra *et al.*, 1992
Intermediolateral cell column	Rat	5-HT immunocytochemistry	100%	Mizukawa *et al.*, 1986; Vera *et al.*, 1990[b]; Poulat *et al.*, 1992[a]; Ranson *et al.*, 2006[b]
	Guinea pig	5-HT immunocytochemistry		Chiba and Masuko, 1987[b]
	Rabbit	5-HT immunocytochemistry		Jensen *et al.*, 1995[b]
Ventral horn	Rat	5-HT immunocytochemistry	95%	Pelletier *et al.*, 1981; Mizukawa *et al.*, 1986; Ridet, 1994[a]; Tai *et al.*, 1997
	Cat	5-HT immunocytochemistry		Ulfhake *et al.*, 1987; Holtman *et al.*, 1990; Pilowsky *et al.*, 1990; Alvarez *et al.*, 1998
	Monkey	5-HT immunocytochemistry		Pecci-Saavedra *et al.*, 1983

Review articles, theses and abstracts excluded, unless reporting data not published elsewhere.

[a]Studies providing quantitative estimates of the synaptic incidence of 5-HT terminals (%), as directly observed in serial thin sections or extrapolated stereologically from single thin sections.

[b]Studies combining techniques to identify targets of 5-HT axon terminals at the electron microscopic level.

the biosynthetic enzyme tryptophan hydoxylase (TPH) (Pickel *et al.*, 1976), 5-HT itself (Steinbusch, 1981), and more recently, the plasma membrane transporter of 5-HT (SERT) (Zhou *et al.*, 1998). Moreover, computer-assisted stereological techniques were developed for the unbiased count of neuronal cell bodies and axon terminals, which may now yield reliable estimates of 5-HT cell populations and density of 5-HT innervation in various parts of the CNS.

Looking back over the past forty years, it would seem that a huge amount of information has thus been gathered on the ultrastructure of the 5-HT innervation in the mammalian CNS (Table 1). Yet, it must be emphasized that most of these data come from rat, less frequently from cat and monkey, and that almost nothing is known for humans, because of technical difficulties in achieving adequate preservation of the fine structure of *post-mortem* brain tissue. Moreover, even in rat, many CNS regions have not been examined properly at the electron-microscopic level. It should also be noted that no ultrastructural studies of the 5-HT system have yet been carried out in the mouse, in spite of its increasing use in genetic, neurobiological, behavioral, pathological and pharmacological investigations of the 5-HT system. Another caveat is that many ultrastructural studies have relied on a very restricted sampling and/or have merely sought to provide (illustrate) examples of morphological configurations that were expected in support of hypothetical connectional schemes.

Information on the ultrastructure of 5-HT neurons in mammalian brain has already been reviewed in some detail toward the end of the 1980s (Beaudet and Descarries, 1987; Soghomonian *et al.*, 1988; see also Maley *et al.*, 1990). For this reason, the present chapter will insist on data acquired since that time. Moreover, in summarizing these 'new' data, more attention will be paid to the relational than intrinsic (internal) morphological features of 5-HT axon terminals. The existence of axon terminals of different morphology in terms of size, shape and content of their small vesicles, and presence or absence of larger, granular (dense-core) vesicles, is no longer viewed as reliable evidence for a distinct chemical identity or cellular origin of the corresponding neurons. Some 5-HT neurons have been demonstrated to send widely divergent projections to anatomical regions as distant from one another as the cerebral cortex and the spinal cord (Fallon and Loughlin, 1982; Kazakov *et al.*, 1993), in which axon terminals, albeit chemically defined, can hardly be expected to have the same morphology. Furthermore, progress in double-labeling techniques applicable at the electron-microscopic level has allowed for the combined demonstration of other transmitters and neuroactive molecules within 5-HT terminals, and for the physiological or hodological identification of neurons with which these are juxtaposed or in synaptic contact. This chapter will also pay particular attention to results obtained in various experimental models and during postnatal development, which provide insights on the morphological plasticity of the 5-HT system.

5-HT cell bodies and dendrites

Even if the common acceptance of the word innervation implies the release of transmitter by axonal nerve endings, there is strong evidence for a somatic and dendritic release of 5-HT by midbrain 5-HT neurons (Héry *et al.*, 1982), as initially demonstrated in the case of the dopamine neurons in ventral mesencephalon (Björklund and Lindvall, 1975; Cheramy *et al.*, 1981). This has prompted the search for organelles and/or membrane specializations equivalent to those observed at nerve terminals in ultrastructural studies of 5-HT mesencephalic nuclei. In rat, cat and monkey, 5-HT containing perikarya and dendrites of raphe dorsalis nucleus (DRN), whether identified by [^{3}H]5-HT uptake radioautography (Beaudet and Descarries, 1981; Descarries *et al.*, 1982; Chan-Palay, 1982) or by 5-HT immunocytochemistry (Brusco *et al.*, 1983; Kapadia *et al.*, 1985; Pecci-Saavedra *et al.*, 1986; Chazal and Ralston, 1987) were often observed to be directly apposed to other neuronal somata or dendrites. These contacts are solely characterized by an interruption of the interneuronal glial sheath and juxtaposition of the plasma membranes, but, in cat, Chazal and Ralston have observed clusters of vesicles near the apposed plasma membranes of 5-HT immunoreactive dendrites (Chazal and O'Hara, 1986; Chazal and Ralston, 1987). Interestingly, in a subsequent study in the rat, Chazal and Ma (1989) did not find vesicular aggregates or membrane differentiations suggestive of synaptic specializations in 5-HT-immunolabeled dendrites of nucleus raphe magnus.

In rat, it was also observed that most of the somato-dendritic elements directly apposed to 5-HT perikarya or dendrites belong to non-5-HT neurons. In the rostral pole of the DRN, some of these apposed cells have been identified as dopamine neurons by combined radioautography and tyrosine hydroxylase (TH) immunocytochemistry (Beaudet, personal communication). Differentiated synaptic contacts between adjacent 5-HT dendrites have, however, been described in monkey as well as in cat (Kapadia *et al.*, 1985; Chazal and Ralston, 1987). Direct juxtapositions of 5-HT dendrites were also observed in rat, but these appositions never showed junctional (synaptic) specializations (Descarries *et al.*, 1982). As the axonal 5-HT input to rat DRN is scarce (Descarries *et al.*, 1982) and 5-HT$_{1A}$ autoreceptors, known to control the firing and release of 5-HT neurons, are essentially somato-dendritic (Riad *et al.*, 2000), dendro-dendritic juxtapositions as well as junctions could represent the morphological substrate for this powerful auto- and mutual inhibition (Aghajanian *et al.*, 1972; Wang and Aghajanian, 1978, 1982).

5-HT innervation of the forebrain

Cerebral cortex

The 5-HT innervation of the cerebral cortex originates from the midbrain nuclei raphe dorsalis (DRN) and medianus (MRN), and spreads to every allo- and iso-cortical area (Andén *et al.*, 1966; Conrad *et al.*, 1974; Bobillier *et al.*, 1976; Beaudet and Descarries, 1976; Azmitia and Segal, 1978; Moore *et al.*, 1978; Lidov *et al.*, 1980; Steinbusch, 1981, 1984; Morrison *et al.*, 1982, 1984; Lindvall and Björklund, 1984). Areal and laminar differences in 5-HT innervation density have been documented in a quantitative radioautographic study of the anterior half of the rat cerebral cortex after [^{3}H]5-HT uptake in slices (Audet *et al.*, 1989). The mean regional density of the cortical 5-HT innervation was then estimated at 5.8×10^6 varicosities per mm^3 of tissue, with some variations between areas and layers. The highest laminar density was always that of layer I, except in pyriform cortex, and each region showed a distinct laminar pattern of 5-HT innervation. Such regional and laminar differences have been documented by histochemical fluorescence and 5-HT immunohistochemical studies in various species (see Chapter 1.3 of this volume). They appear to be even greater in the monkey than in the rat or cat cortex (Morrison *et al.*, 1982; Morrison and Foote, 1986; Campbell *et al.*, 1987), and need to be further documented in human brain (e.g., Trottier *et al.*, 1996; Austin *et al.*, 2002).

As shown in Table 1, both radioautography after [^{3}H]5-HT uptake *in vivo* and immunocytochemistry with antibodies against 5-HT itself, TPH or SERT have been used to examine the fine structural features of these nerve terminals in various cortical areas of rats and Old

World monkeys. Electron-microscopic pictures of SERT-labeled axon terminals in the rat cerebral cortex have also been published by Yamamoto *et al.* (1998), and an ultrastructural 5-HT immunocytochemical study has also been carried out in cat auditory cortex (De Felipe *et al.*, 1991). However, throughout the literature, only one electron micrograph has been published of an axon terminal labeled with [^{3}H]5-HT, in a pre-rolandic biopsy specimen of human cortex (Calas *et al.*, 1981).

The first electron-microscopic description of cortical 5-HT axon varicosities, identified in rat fronto-parietal cortex by radioautography after [^{3}H]5-HT uptake *in vivo* (Descarries *et al.*, 1975; see also Beaudet and Descarries, 1978, 1981, 1984), revealed that, much like axon terminals of the peripheral autonomic nervous system, many of these CNS terminals lacked the localized plasma membrane specializations (junctional complex) that are the hallmark of synapses (Peters and Palay, 1996). Except for a few discordant reports, based on preliminary results (Molliver *et al.*, 1982) and early immunocytochemical data from the visual cortex of rat and monkey (Parnavelas *et al.*, 1985; Papadopoulos *et al.*, 1987a, 1987b; De Lima *et al.*, 1988), every subsequent study in which the synaptic incidence of cortical 5-HT varicosities was determined confirmed the partly or largely asynaptic character of this innervation, whether in rat, cat or monkey. In particular, the systematic 5-HT immunocytochemical survey of Séguéla *et al.* (1989), carried out in three regions of adult rat neocortex, evaluated the synaptic incidence of 5-HT varicosities at 36 percent and 28 percent in the superficial and deep layers of primary motor cortex respectively, 46 percent in the primary somato-sensory cortex, and 37 percent in the primary visual cortex (e.g., Figure 1A). The contacts formed by synaptic 5-HT varicosities were asymmetrical, and more frequently found on dendritic spines than dendritic branches, at least within the deep frontal and the upper occipital layers. Similar proportions of 5-HT varicosities (22 percent and 18 percent) were later reported to be synaptic in rat frontoparietal and entorhinal cortex after TPH immunolabeling (Cohen *et al.*, 1995). It should also be noted that, in the upper layers of all three cortical areas examined by Séguéla *et al.* (1989), unlabeled axonal varicosities, presumably non-5-HT, were frequently encountered juxtaposed to the 5-HT immunoreactive terminals. Such appositions between axon terminals could well provide for a distal presynaptic control of the release of other transmitters by 5-HT and, conversely, for a local control of 5-HT release by other cortical afferents. However, in a subsequent double-immunolabeling study by Miner *et al.* (2000), dopamine (TH immunoreactive) and 5-HT (5-HTT immunoreactive) axon terminals were rarely found to be juxtaposed, and were typically separated by an average distance greater than 1 μm in adult rat, prelimbic prefrontal cortex.

In cat auditory cortex, De Felipe *et al.* (1991) reported the existence of small, dense clusters of 5-HT immunoreactive varicosities, first described as baskets (Mulligan and Törk, 1987, 1988), around the somata and primary dendrites of cortical neurons, many of which were also shown to be immunoreactive to GABA. However, of 135 5-HT boutons examined in serial ultrathin sections, 110 of which belonged to these basket formations, only 4 were found making a synaptic contact, all asymmetrical. In cynomolgus monkey sensory-motor cortex, the same authors observed only 5 of 191 5-HT immunoreactive boutons reconstructed from serial sections showing a synaptic specialization; all of these were again asymmetrical (De Felipe and Jones, 1988). In the prefrontal cortex of rhesus monkey, Smiley and Goldman-Rakic (1996) found only 23 percent of 213 5-HT-immunostained varicosities completely viewed in serial sections making identifiable synapses. All contacts established by these synaptic 5-HT varicosities were again asymmetrical, and mostly formed with dendritic branches resembling those of interneurons.

In a more recent electron-microscopic immunocytochemical study of the rat visual cortex, Paspalas and Papadopoulos (2001) combined labeling with antibodies against 5-HT and against either neuropeptide Y (NPY), somatostatin (SRIF) or vasoactive intestinal polypeptide (VIP). Synaptic contacts were then demonstrated between some 5-HT terminals and the perikarya and proximal dendrites of NPY and SRIF neurons, as well as with the distal dendrites of SRIF neurons, but never with VIP elements. However, in another study in which the relationships between 5-HT afferents and various peptidergic-GABAergic interneurons of the rat cerebral cortex were examined by dual immunofluorescence confocal microscopy, many of the VIPs as well as SRIF and NPY neurons appeared to be innervated by 5-HT afferents (Cauli *et al.*, 2004).

Together with similar data from other parts of brain, the largely asynaptic character of the cortical monoamine and acetylcholine innervations may be viewed as crucial evidence for the existence of a diffuse mode of transmission in addition to synaptic transmission by these neuronal systems (for reviews, see Descarries *et al.*, 1991; Descarries and Umbriaco, 1995; Descarries and Mechawar, 2000, 2008). Currently available data on the subcellular localization of the different 5-HT receptors in mammalian CNS are consistent with this view (Descarries *et al.*, 1986). In all regions thus far examined, and irrespective of their subtype, some of the 5-HT receptors have been visualized on the plasma membrane of somata, dendrites, axonal fibers or axon terminals, but never at sites of junctional specialization.

Hippocampus

The hippocampus has been shown by both radioautographic and immunocytochemical methods to receive a relatively dense 5-HT innervation (Azmitia and Marovitz, 1980; Lidov *et al.*, 1980, Köhler, 1984). In the dorsal hippocampus of rat, the regional and laminar distribution of this innervation has been quantified by radioautography after [^{3}H]5-HT uptake in slices (Oleskevich and Descarries, 1990). The average density of 5-HT innervation in subiculum, Ammon's horn and dentate gyrus was then measured to be in the order of 2.7×10^6 varicosities per mm^3, but with values in the subiculum > in Ammon's horn > in dentate gyrus, and a marked heterogeneity in laminar distribution in each of these regions.

Few electron-microscopic investigations of the fine structure and cellular relationships of 5-HT nerve terminals in hippocampus have thus far been carried out, all of them in rat and with 5-HT immunocytochemistry, except for Cohen *et al.* (1995), who used TPH antibodies. In their initial report, Anderson *et al.* (1986) identified 5-HT axon terminals in the dentate gyrus (DG) of normal rats, but also in rats having received fetal raphe transplants after neonatal lesion of the fimbria-fornix and entorhinal cortex. Two categories of immunoreactive profiles were then observed: large ones, which never appeared to form synaptic junctions; and smaller ones, which occasionally showed asymmetrical synapses upon dendritic spines or small dendritic branches. These ultrastructural characteristics were found in both normal and grafted animals.

Oleskevich *et al.* (1991) provided a detailed description of the ultrastructural features of the 5-HT terminals in two major sectors of adult rat hippocampus, CA3 of Ammon's horn and the crest of DG; their description was completed by Umbriaco *et al.* (1995) with similar data from the stratum radiatum of CA1. The hippocampal 5-HT varicosities were then shown to be of comparable size in the three anatomical sectors and five neuropil layers examined, and ranged in synaptic incidence from 19 to 24 percent, whether extrapolated stereologically to the whole volume of varicosities from the synaptic frequency observed in single ultrathin sections, or directly observed in serial sections across whole varicosities (e.g., Figure 1B). A similarly low proportion of synaptic 5-HT varicosities (12 percent) was reported in the TPH-immunocytochemical study of Cohen *et al.* (1995), carried out in the CA1 region. As in the cerebral cortex, most if not all synaptic 5-HT varicosities formed asymmetrical junctions. In CA3, only dendritic shafts were contacted, but in CA1 and DG there were also a few axo-spinous 5-HT synapses. In both CA3 and DG, as in the cerebral cortex, axonal varicosities were the most abundant neuronal element juxtaposed to the 5-HT-immunostained varicosities. Dendritic shafts were also relatively numerous, but dendritic spines infrequent. It is also noteworthy that in the rat hippocampus, as in the cerebral cortex of rat, cat or monkey, some GABAergic interneurons could be shown to be synaptically contacted by 5-HT varicosities in double-immunolabeling experiments (Freund *et al.*, 1990).

The ultrastructure of hippocampal 5-HT terminals has also been examined in a state of stabilized 5-HT hyperinnervation produced by a graft of fetal mesencephalic raphe neurons after previous cytotoxic destruction of the endogenous 5-HT innervation (Daszuta *et al.*, 1991). As in the previous study of Anderson *et al.* (1986), 5-HT terminals generated by the grafted 5-HT neurons were not found to differ significantly from normal, either in size or in synaptic characteristics. Despite the fact that they had grown in abnormal host tissue and in excessive number, they were apparently committed to express the same set of intrinsic and relational features as their normal phenotype.

Olfactory bulb

In spite of its strategic location as initial relay in olfactory pathways and known innervation by 5-HT fibers originating from the midbrain raphe nuclei (Dahlström *et al.*, 1965; Araneda *et al.*, 1980), there has been only one ultrastructural study of the 5-HT innervation of olfactory bulb, carried out 30 years ago in the rat by Halász *et al.* (1978), using radioautography after *in vivo* labeling by local injection of [^{3}H]5-HT. It was then shown that [^{3}H]5-HT-labeled boutons in the glomerular layer formed axo-dendritic synaptic contacts with periglomerular cells, whereas in the external plexiform and internal granular layers they synapsed with dendritic shafts and spines of granule cells. Synapses on mitral cells were not demonstrated. This suggested that 5-HT effects in the olfactory bulb are exerted via interneurons. Whether some of the periglomerular cells contacted by 5-HT afferents belong to the distinct subpopulation synthetizing dopamine (Halász *et al.*, 1977; Ljungdahl *et al.*, 1977) remains to be demonstrated.

Basal forebrain

The basal forebrain comprises the medial and lateral septum, the diagonal band of Broca, including the nuclei of its horizontal and vertical limbs, and the nucleus basalis. These brain regions contain cholinergic neurons, and have been shown to receive a prominent 5-HT innervation from the raphe medianus

(Moore and Halaris, 1975; Azmitia and Segal, 1978; Vertes and Martin, 1988).

An electron-microscopic study conducted in the rat medial septum and the diagonal band of Broca has shown symmetrical synaptic contacts between 5-HT immunoreactive axon varicosities and the soma-dendrites of neurons retrogradely labeled by axonal transport from the hippocampus of wheatgerm agglutinated apo-horseradish peroxidase conjugated to colloidal gold particles (Milner and Veznedaroglu, 1993). In accordance with this study, double-labeling experiments in these same brain areas have shown symmetric and asymmetric synaptic contacts made by 5-HT terminals with parvalbumin-containing GABAergic neurons (Léránth and Vertes, 1999) known to innervate the hippocampus (Naumann *et al.*, 1992; Peterson *et al.*, 1992) and to express 5-HT_{2A} receptors (Léránth and Vertes, 1999). This circuitry is presumably involved in the well-known desynchronizing effects of the raphe medianus on the hippocampal electroencephalogram, thought to be important for mnemonic functions (for review, see Vertes and Kocsis, 1997).

Dinopoulos *et al.* (1993, 1997) have also used 5-HT immunocytochemistry to examine the ultrastructural features of the 5-HT innervation in rat lateral and medial septum, as well as in the diagonal band of Broca, with emphasis on their postnatal development. They observed the 5-HT varicosities in single thin sections, but took care to extrapolate the synaptic incidence for whole varicosities at various postnatal ages, from birth to adult. In the lateral septum, they thus demonstrated that, irrespective of age, virtually all 5-HT axon varicosities made symmetrical synapses, mostly with somata and proximal dendrites (Dinopoulos *et al.*, 1993). In the dorsal portion of the lateral septum, characteristic basket-like, pericellular arrangements were observed, as previously described for dopamine terminals (Descarries and Beaudet, 1983). In contrast, in the developing horizontal limb of the diagonal band of Broca, the synaptic incidence of 5-HT varicosities followed a biphasic temporal pattern: it increased gradually from birth, reached 43 percent at the end of the second postnatal week, declined markedly in the following week, and increased again to its adult value of 46 percent. At all postnatal ages, both symmetrical and asymmetrical contacts were established, mostly with a dendritic shaft (Dinopoulos *et al.*, 1997).

To our knowledge, there has been only one ultrastructural study attempting to characterize the 5-HT innervation of the nucleus basalis, where 5-HT is likely to modulate the function of cholinergic neurons projecting to the cerebral cortex (Smiley *et al.*, 1999). A synaptic contact made by a 5-HT axon varicosity was then observed in the nucleus basalis of Old World monkeys, but relatively poor ultrastructural preservation of the tissue precluded a systematic characterization of this innervation.

Bed nucleus of the stria terminalis

As part of the limbic–hypothalamic–pituitary–adrenal system, the bed nucleus of the stria terminalis (BST) is tightly implicated in various visceral responses to stress and related behaviors. 5-HT innervation has been demonstrated by light- and electron-microscopic 5-HT immunocytochemistry in this nucleus, and shown to be denser in its medial than lateral subdivision (Phelix *et al.*, 1992). 5-HT varicosities in two size groups (large and small) were observed in the ventral BST, approximately 10 percent of which showed a synaptic specialization in single thin sections. These synaptic contacts were symmetrical or asymmetrical, and always made with dendritic branches. Even though some 5-HT varicosities were seen in close proximity to neuronal cell bodies, neither juxtapositions nor axo-somatic synapses were observed.

Amygdala

All subdivisions of the amygdalar complex receive a dense 5-HT innervation from the raphe dorsalis (Moore *et al.*, 1978; Steinbusch, 1981; Sadikot and Parent, 1990; Bauman and Amaral, 2005). In the basolateral amygdala of the rat, a double-labeling, immuno-electron microscopic study has shown that more than 75 percent of 5-HT immunoreactive axon varicosities formed synaptic junctions, which were mostly symmetrical, with dendritic spines and distal dendrites of pyramidal cells (Muller *et al.*, 2007). Interestingly, synaptic contacts were also observed with interneurons (Muller *et al.*, 2007), which have been reported to express 5-HT_{2A} and 5-HT_{3A} receptors in this part of the amygdala (Morales and Bloom, 1997; Mascagni and McDonald, 2007; McDonald and Mascagni, 2007). This could account for an indirect inhibition of pyramidal cells (Rainnie, 1999; Stutzmann and LeDoux, 1999) and a modulation of the activity of the basolateral amygdala in the generation of fast rhythmic oscillations during emotional arousal (Paré *et al.*, 2002; Muller *et al.*, 2005; Woodruff and Sah, 2007).

Neostriatum

The 5-HT innervation of mammalian neostriatum, first demonstrated in the rat by Fuxe and Jonnson (1967) using the histochemical fluorescence technique, was later shown to arise mainly from the raphe dorsalis and, to a lesser extent, from the raphe medianus (Jacobs *et al.*, 1974, 1978; Lorens and Guldberg, 1974; Bobillier *et al.*, 1975, 1976; Kellar *et al.*, 1977; Azmitia and Segal, 1978; Moore *et al.*, 1978). This innervation appears rather uniformly distributed, without any suggestion of a patch and matrix organizational pattern (Soghomonian *et al.*, 1987). Its density increases from rostral to caudal, and is always higher ventrally than dorsally, ranging from 4.8×10^6 varicosities per mm^3 rostrally to 6.3×10^6 caudally – an average of 5.6×10^6 (Mrini *et al.*, 1995), almost the same as that in cerebral cortex.

Early electron-microscopic observations made in cat and rat after radioautographic labeling by topical and intraventricular administration of [^{3}H]5-HT (Calas and Ségu, 1976; Calas *et al.*, 1976; Arluison and De La Manche, 1980), or later in cynomolgus monkey (*Macaca fascicularis*) after 5-HT immunocytochemistry (Pasik and Pasik, 1982; Pasik *et al.*, 1982, 1984), have suggested that neostriatal 5-HT terminals often lacked synaptic specialization in all three species. This was confirmed for rat in a detailed study where both methods were used in parallel to examine a significant number of [^{3}H]5-HT-labeled varicosities from a paraventricular sector of neostriatum in single thin sections, and numerous 5-HT-immunostained varicosities from the same paraventricular sector plus a dorsal neostriatal sector in serial thin sections (Soghomonian *et al.*, 1989). The proportion of junctional 5-HT varicosities was then found to be very low, whether extrapolated stereologically from single sections (10 percent) or directly observed in serial sections (13 percent) (Figure 1C). The rare 5-HT synapses were equally distributed between dendritic spines and dendritic branches, and, as in the cortex, almost always asymmetrical. Whether synaptic or not, 5-HT varicosities were directly apposed to a variety of structures comprising mostly other axon terminals, dendritic spines and branches, but rarely neuronal somata. These appositional and synaptic relationships were similar in both the paraventricular and dorsal neostriatum, suggesting that throughout this brain region 5-HT transmission could be diffuse as well as synaptic, and might exert a distal control on the release of various transmitters through axo-axonic appositions.

Interestingly, as observed in the hippocampus after 5-HT grafts, no changes in either the intrinsic or relational features of 5-HT nerve terminals could be demonstrated in a state of stabilized 5-HT hyperinnervation induced by neonatal destruction of the nigrostriatal dopamine innervation with 6-hydroxydopamine (Descarries *et al.*, 1992), in spite of a twofold increase in the number of 5-HT terminals in the neostriatum of these rats (Mrini *et al.*, 1995). Also noteworthy was the report of Van Luijtelaar *et al.* (1991), who, in the sole ultrastructural study of a 5-HT innervation in aged rats, confirmed the above characteristics for a majority of normal looking, thin varicose 5-HT immunoreactive fibers innervating the rat neostriatum, but also observed swollen 5-HT immunoreactive fibers containing distorted mitochondria and membrane-filled vacuole-like structures, suggestive of a degenerative process.

Nucleus accumbens

Three immuno-electron microscopy studies have been published regarding the 5-HT innervation in nucleus accumbens. In the first study, by Van Bockstaele and Pickel (1993), care was taken to distinguish between the core of the nucleus, which is related to the caudate-putamen and involved in motor functions, and the shell region, which is associated with the limbic system and implicated in motivational processes. In the core, 5-HT immunoreactive varicosities appeared generally smaller than in the shell, and they contained fewer large dense core vesicles, assumed to transport biosynthetic enzymes and neuropeptides. The synaptic incidence of these terminals amounted to 39 percent and 46 percent in single thin sections of the core and the shell, respectively, which, unless due to a sampling bias, suggests that most if not all were synaptic. In both core and shell, the postsynaptic targets of 5-HT terminals were mostly large to medium-sized dendritic branches. Less than 2 percent were dendritic spines. In the core, synaptic contacts were mostly asymmetrical, whereas they were mostly symmetrical in the shell. Appositions with unlabeled terminals were frequent, and some could be shown to be TH immunoreactive (i.e., presumably dopaminergic) in double-immunolabeling experiments.

A subsequent double-immunolabeling study with anti-5-HT and anti-GABA antibodies also identified appositions and symmetrical synaptic contacts between 5-HT terminals and GABA-labeled and unlabeled dendrites in the shell region of the rat nucleus accumbens (Van Bockstaele *et al.*, 1996). Numerous 5-HT and GABA immunoreactive axon terminals were then observed in direct apposition to one another, suggesting distal interactions between the two systems.

5-HT terminals in the rat nucleus accumbens have also been identified in another immuno-electron microscopy study carried out with antibodies against the plasma membrane 5-HT transporter (SERT) (Pickel and Chan, 1999). It was then verified that the SERT-labeled axons and terminals did contain 5-HT in dual-immunolabeling experiments. As in the first immuno-electron microscopy study ever carried out with a SERT antibody (Zhou *et al.*, 1998), gold immunolabeling confirmed the predominant localization of the transporter to the plasma membrane of fine unmyelinated axons and their varicosities, away from synaptic specializations, which indicated that the SERT may

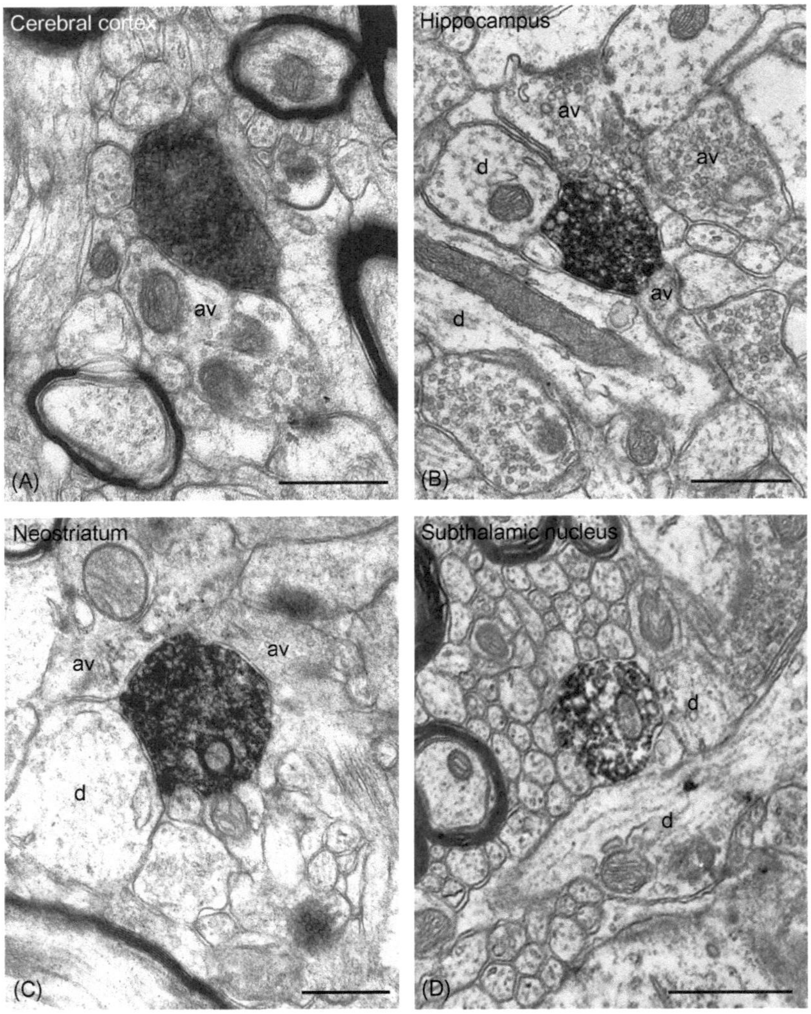

Figure 1 Examples of immunoreactive 5-HT axon varicosities (terminals) in different anatomical regions of rat brain. Immunoperoxidase-DAB technique, with primary antibodies against 5-HT itself, in (A)–(C), and against the plasma membrane 5-HT transporter (SERT) in (D). (A) Cerebral cortex. This varicosity from layer VI of the frontal cortex is heavily labeled. Its roundish small vesicles, outlined by DAB precipitate, are barely discernible. It is directly apposed to another axon varicosity (av), unlabeled, and several small axonal and dendritic profiles. No synaptic junction was visible in four adjacent thin sections across this varicosity. From Séguéla *et al.* (1989). (B) Hippocampus. This presumed non-junctional 5-HT varicosity is from the stratum radiatum of CA1. It is juxtaposed to three unlabeled axon varicosities (av) and two dendritic branches (d). Its small vesicles are indistinguishable from those in other immunolabeled or unlabeled axon varicosities. From Umbriaco *et al.* (1995). (C) Neostriatum. Typical 5-HT varicosity from the middle third of neostriatum. Most neostriatal 5-HT varicosities are non-junctional. This one appears juxtaposed to a dendritic branch (d) and unlabeled varicosities (av). From Descarries, unpublished. (D) Subthalamic nucleus. This relatively small 5-HT varicosity, partly surrounded by a bundle of small, unmyelinated axons, is also juxtaposed to two dendritic branches (d). Scale bars: 0.5 μm. From Parent *et al.*, unpublished.

not only play a role in terminating 5-HT synaptic transmission, but also regulate diffusely released and/or an ambient level of 5-HT (see also Tao-Cheng and Zhou, 1999).

Globus pallidus

Even though the globus pallidus is known to receive a relatively dense 5-HT input, estimated at 4.8×10^6 varicosities per mm^3 of tissue in the rat (Soghomonian *et al.*, 1987), little is known of the fine structural features of this innervation, which has only been examined in the monkey. In adolescent cynomolgus monkeys, Pasik *et al.* (1984) reported that 5-HT-immunolabeled fibers from this brain region formed large boutons, up to 1.5 μm in diameter, making asymmetrical synaptic contact with dendritic shafts. These authors also observed crest synapses comprising an unlabeled terminal facing the 5-HT immunoreactive bouton contacting the same dendritic spines.

Thalamus

The 5-HT innervation of the thalamus arises essentially from the raphe dorsalis, with minor contributions from the raphe medianus and the controlateral periaqueductal gray (Azmitia and Segal, 1978; Pasquier and Villar, 1982; De Lima and Singer, 1987b; Westlund *et al.*, 1990). The 5-HT fibers *en route* to the thalamus form a bundle that sweeps dorsoventrally within the midbrain tegmentum, and courses beneath the thalamus along its entire caudo-rostral extent. Several fiber fascicles break off from this main bundle at different levels, and ascend dorsally to pervade all thalamic nuclei (Lavoie and Parent, 1991). Within the thalamus, 5-HT axons are very thin, unmyelinated, and bear axon varicosities distributed at irregular intervals along their length. The regional density of 5-HT innervation varies markedly between thalamic nuclei (Cropper *et al.*, 1984). In the lateral geniculate nucleus (LGN) of the monkey (*Macaca nemestrina*), the density of 5-HT axon varicosities has been estimated at 1.4×10^6 varicosities per mm^3 of tissue (Wilson and Hendrickson, 1988).

Most ultrastructural studies of the 5-HT innervation in thalamus have focused on the LGN, whether in the rat (Papadopoulos and Parnavelas, 1990; Dinopoulos *et al.*, 1995; Dori *et al.*, 1998), cat (De Lima and Singer, 1987a) or monkey (Pasik *et al.*, 1988; Wilson and Hendrickson, 1988). The ventrobasal complex (Nothias *et al.*, 1988), the ventral lateral nucleus (Dori *et al.*, 1998) and the ventral posterior nucleus (Liu and Jones, 1991) have also been examined. In specific relay thalamic nuclei, varying proportions of 5-HT axon varicosities appeared to be endowed with a synaptic junction (Table 1). Except for one study, which claimed that all 5-HT varicosities in the rat LGN were synaptic when examined in serial sections (Papadopoulos and Parnavelas, 1990), all other studies carried out in single or serial thin sections reported partial values, notably that of Dinopoulos *et al.* (1995), who found a 42 percent incidence as stereologically extrapolated from single sections to the whole volume of varicosities (Dinopoulos *et al.*, 1995). The synaptic contacts established by thalamic 5-HT axon have been described as mainly asymmetrical and made with dendritic branches (De Lima and Singer, 1987a; Nothias *et al.*, 1988; Papadopoulos and Parnavelas, 1990; Liu and Jones, 1991; Dinopoulos *et al.*, 1995; Dori *et al.*, 1998).

In the study by Dinopoulos *et al.* (1995), the ultrastructural features of 5-HT terminals in the rat LGN were also examined during the postnatal period. Whereas only few axo-somatic symmetrical synaptic contacts had been observed in the adult rat LGN (Papadopoulos and Parnavelas, 1990), such contacts were frequently observed at the time of birth. As already described in the septum and the diagonal band of Broca, the synaptic incidence of LGN 5-HT varicosities appeared to increase gradually from birth to reach a peak at the end of the second postnatal week (51 percent), and then declined markedly in the following week before increasing again to reach its adult value of 42 percent. This led to the speculation that synapses formed in the first 2 weeks could be primarily involved in morphogenetic effects of 5-HT, and those formed later with 5-HT neurotransmission. However, a comparable study carried out in the ventral lateral thalamic nucleus of the rat (Dori *et al.*, 1998) showed that in this nucleus, the proportion of 5-HT varicosities engaged in synaptic contact increased gradually from birth to adulthood, and that the length of synaptic junctions, the type of synaptic contacts and the nature of the postsynaptic elements remained unchanged throughout the development (Dori *et al.*, 1998).

It needs to be pointed out that some neurons, including thalamocortical neurons, which do not synthetize 5-HT, are able to take up and store 5-HT during the postnatal period in rat (Lebrand *et al.*, 1996, 1998; Cases *et al.*, 1998; Hansson *et al.*, 1998). Data based on the 5-HT or SERT immunocytochemical identification of terminals during postnatal development must therefore be regarded with caution. Some of the axon terminals which are then observed to be 5-HT or SERT immunoreactive might not be those of neurons that will later be part of the 5-HT system.

Zona incerta

Two studies combining [^{3}H]5-HT uptake radioautography and tyrosine hydroxylase immunocytochemistry have demonstrated the existence of direct appositions between 5-HT axon varicosities and dopamine cell bodies and dendrites in the medial zona incerta of rat. In both studies, carried out in single thin sections, only a few of the [^{3}H]5-HT labeled varicosities were seen in synaptic junctions, which were all asymmetrical. Such observations were made in normal rats (Bosler *et al.*, 1983), but also in rats that had been treated with 5,7-dihydroxytryptamine 50 days earlier, and in which the examined terminals were those of regenerated axons (Frankfurt and Beaudet, 1988). This suggested the recapitulation of an innate program by 5-HT neurons, this time after partial lesion.

Subthalamic nucleus

The 5-HT innervation of the subthalamic nucleus (STN) arises mainly from the raphe dorsalis (Lavoie and Parent, 1990). Although many electrophysiological studies have underlined the influence of 5-HT on the activity of STN neurons (Stanford *et al.*, 2005; Xiang *et al.*, 2005; Shen and Johnson, 2008), there are not yet any data in the literature regarding the fine ultrastructural features of 5-HT axon varicosities in this key component of the basal ganglia (but see Wallman *et al.*, 2007). A preliminary study in our laboratory indicates that SERT immunoreactive axon varicosities are smaller in the rat than in the squirrel monkey (*Saimiri sciureus*) STN, but make synapses in a similar proportion (Figure 1D). The synaptic incidence extrapolated stereologically to the whole volume of varicosities is about 50 percent in both species. In monkeys, all junctions formed by the synaptic 5-HT varicosities are asymmetrical, compared to 77 percent in rats. In both species, synaptic junctions are almost equally shared between dendritic branches and spines. It is noteworthy that, despite a significantly higher density of 5-HT innervation in the anterior compared to the posterior half of monkey STN (1.7 vs 1.3×10^6 varicosities per mm^3 of tissue, according to unbiased count under the light microscope), there are no significant differences between the ultrastructural features of these 5-HT varicosities, including synaptic incidence, in the two halves of STN. These similar findings in rat and monkey strongly suggest that the raphe dorsalis exerts its mixed electrophysiological effects on STN neurons through both synaptic delivery and diffusion of 5-HT, which may then reach 5-HT receptors on neighboring axonal afferents and provide for some local presynaptic inhibition.

Hypothalamus

The hypothalamus receives a 5-HT innervation of variable density (Fuxe, 1965; Descarries and Beaudet, 1978; Steinbusch and Nieuwenhuys, 1981), which mainly originates from the midbrain and medial lemniscus 5-HT cell groups (group B9), as demonstrated after electrolytic lesioning of the nucleus raphe dorsalis and/or medianus (Aghajanian *et al.*, 1969; Kuhar *et al.*, 1972; Van de Kar and Lorens, 1979), or combined fluorescence 5-HT immunohistochemistry and retrograde transport labeling (Willoughby and Blessing, 1987). Thus far, all ultrastructural studies of hypothalamic 5-HT axon terminals have been carried out in the rat.

Rostrally, the *medial preoptic area* displays a moderate number of 5-HT terminals (Descarries and Beaudet, 1978; Steinbusch, 1984). In an electron-microscopic examination of this region after [^{3}H]5-HT uptake radioautography, approximately one-third of 5-HT varicosity profiles displayed a synaptic specialization in single thin sections (Kiss and Halász, 1985), which suggests that most of them, if not all, were actually synaptic. After concomitant immunolabeling with specific antibodies against the luteinizing hormone releasing hormone (LHRH), approximately 5 percent of [^{3}H]5-HT-labeled terminals were found in contact, synaptic or not, with LHRH-immunostained dendrites, indicating that, in this hypothalamic area, 5-HT might directly act on LHRH neurons to regulate pituitary gonadotrophin secretion.

The *suprachiasmatic nucleus* (SCN) receives one of the densest 5-HT inputs in rat brain (Descarries and Beaudet, 1978; Steinbusch, 1981), particularly in its ventral part (see Descarries and Beaudet, 1978; Bosler, 1980). This innervation is believed to arise primarily from the raphe medianus (Van de Kar and Lorens, 1979), but also partly from the raphe dorsalis (Kawano *et al.*, 1996). Cytochemical approaches were first used for the electron-microscopic identification of 5-HT terminals in this nucleus (Ajika and Ochi, 1975; Nojyo and Sano, 1978). [^{3}H]5-HT radioautographic studies followed, all of which combined the 5-HT labeling with that of another transmitter or neurohormone (Kiss *et al.*, 1984; Bosler and Beaudet, 1985a, 1985b; Bosler *et al.*, 1986). The 5-HT innervation of the rat SCN is only partly synaptic, as only 10 percent of axonal profiles labeled with [^{3}H]5-HT in this nucleus displayed a synaptic junction in single thin section, suggesting a synaptic incidence of at least 40 percent for the whole volume of these varicosities (Bosler and Beaudet, 1985a). This has been confirmed in a subsequent study of the synaptic connectivity of 5-HT graft efferents in the SCN as well as the supraoptic nucleus (SON) (Boulaïch *et al.*, 1994). Forty-eight percent of 5-HT immunoreactive terminals examined in serial sections were then found to make synapses in the SCN of normal rats, most of them with dendritic branches, and either symmetrical or asymmetrical. Interestingly, it has been reported that 5-HT axon terminals of the rat SCN lack monoamine oxidase B immunoreactivity, which is the type of monoamine oxidase found in 5-HT somadendrites (Arai *et al.*, 2002).

By combining [^{3}H]5-HT labeling with vasoactive intestinal peptide (VIP) immunocytochemistry, it was shown that almost one-third of SCN 5-HT terminals were either directly apposed to or in synaptic contact with VIP immunoreactive neurons (Kiss *et al.*, 1984; Bosler and Beaudet, 1985a, 1985b; see also Bosler *et al.*, 1986). More than 60 percent of all synaptic contacts established by 5-HT axons were made with VIP neurons. These contacts were mostly found on dendrites, but a few were also observed on neuronal somata. 5-HT axons within the SCN were also shown by combined radioautography and immunocytochemistry to be directly apposed to glutamic acid decarboxylase (GAD) immunoreactive axon terminals (Bosler *et al.*, 1984a). Although deprived of synaptic specialization, such appositional contacts were suggestive of 5-HT/GABA interactions in the SCN, where the 5-HT innervation is likely to be involved in a circadian regulation of gonadotrophin and prolactin secretion. Lastly, as illustrated in Figure 2A, it should be noted that in addition to synapses between 5-HT terminals and vasopressin immunoreactive dendrites, axo-axonic synapses between 5-HT and vasopressin axon terminals might be present in the rat SCN (Bosler, personal communication).

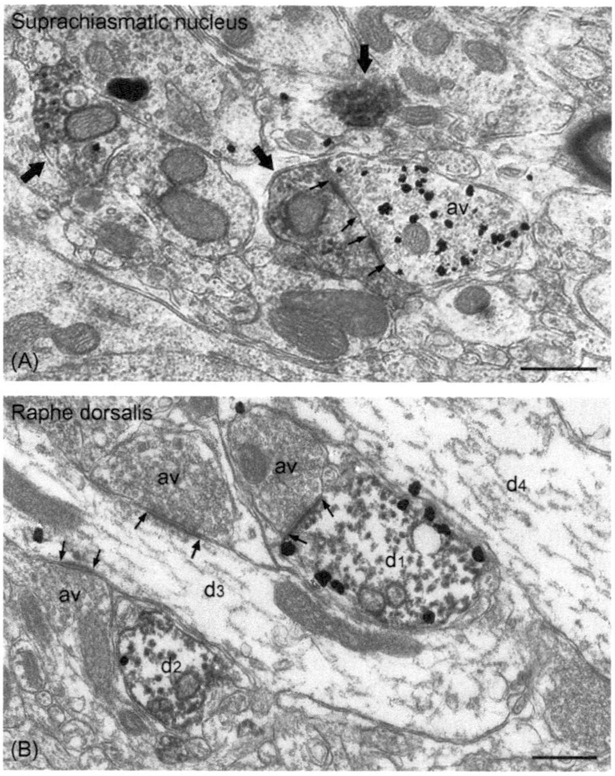

Figure 2 Suprachiasmatic nucleus and raphe dorsalis nucleus. (A) Suprachiasmatic nucleus. This 5-HT immunoreactive axon terminal, labeled with the silver-intensified immunogold technique, is seen in synaptic contact (between small arrows) with an immunoperoxidase-DAB-labeled vasopressin immunoreactive profile (arrows) that appears to be axonal also. Note that these synaptic partners are almost totally enwrapped by glial leaflets. Two other, vasopressin immunoreactive, axon-like processes are also visible (arrows), one of which (left) contains several large dense-cored vesicles. Reproduced courtesy of Olivier Bosler. (B) Raphe dorsalis nucleus. Two dendrites (d1 and d2), dually immunolabeled with primary antibodies against the 5-HT_{1A} receptor (immunoperoxidase-DAB technique) and against the plasma membrane 5-HT transporter (SERT, immunogold technique), run normal to longitudinally sectioned unlabeled dendrites (d3 and d4). In d1, the plasma membrane localization of the SERT is obvious after its labeling with the immunogold technique. In contrast, the labeling of 5-HT_{1A} receptors appears to be cytoplasmic, even if these receptors are known to be membrane bound, as visualized with the immunogold technique (Riad et al., 2000). Note the synaptic contacts (between small arrows) made by unlabeled terminals (av) onto the lower, unlabeled dendritic branch (d3). These unlabeled terminals are morphologically similar to the one which contacts the dually labeled, 5-HT dendrite. Scale bars: 0.5 μm. From Riad *et al.* (unpublished).

An ultrastructural study of the 5-HT innervation has also been carried out in the *hypothalamic paraventricular nucleus*, which is moderately innervated by 5-HT fibers, particularly in its parvocellular subdivision (Descarries and Beaudet, 1978; Steinbusch and Nieuwenhuys, 1981). Using the immunoperoxidase-DAB technique with two chromogens to identify 5-HT terminals in conjunction with corticotropin releasing factor (CRF) elements, Liposits *et al.* (1987) described 5-HT immunoreactive boutons which formed synapses 'en passant' with dendrites and somata of parvocellular neurons. Moreover, some of these terminals made axo-dendritic and axo-somatic synapses with CRF immunoreactive neurons.

The *arcuate nucleus* also receives a moderately dense 5-HT innervation, which originates from both the raphe dorsalis and medianus (Van de Kar and Lorens, 1979; Willoughby and Blessing, 1987). In this nucleus, few of the varicosities labeled by [^{3}H]5-HT uptake *in vivo* were found to establish synaptic contacts, which again were mostly made with dendritic shafts (Kiss and Halász, 1986). Combined radioautographic and 5-HT immunocytochemical examination demonstrated the existence of direct appositions between [^{3}H]5-HT-labeled terminals and nerve cell bodies and/or dendrites of neurons immunoreactive for TH and therefore belonging to the A12 and A13 dopamine cell groups (Bosler *et al.*, 1984b, 1986; Bosler and Beaudet, 1985a; Kiss and Halász, 1986). Thus, 5-HT may be implicated in a direct, proximal control of dopamine neuron function within the arcuate nucleus, and potentially affect the dopamine regulation of prolactin, gonadotrophin and somatotrophin secretion. [^{3}H]5-HT-labeled varicosities have also been observed next to adreno-corticotropic hormone ($ACTH_{1-24}$) immunoreactive perikarya and dendrites within the rat arcuate nucleus (Kiss *et al.*, 1984; Bosler and Beaudet, 1985b). Most of these 5-HT varicosities were merely juxtaposed to the $ACTH_{1-24}$ containing neurons, but a few were seen in synaptic contact with their dendritic shafts.

The only study in which 5-HT terminals of the *supraoptic nucleus* (SON) have been investigated at the ultrastructural level is that of Boulaïch *et al.* (1994), who focused on eventual changes in the intrinsic and relational features of these terminals in the SCN and SON after 5-HT reinnervation and/or hyperinnervation by grafts of fetal neurons following an initial 5,7-dihydroxytryptamine lesion. Thirty-eight percent of serially sectioned 5-HT immunoreactive terminals in the SON of normal rats were then found to make synapses, almost always on dendritic shafts as opposed to spines, and more frequently symmetrical than asymmetrical. Interestingly, as in the SCN and irrespective of the density of 5-HT re- or hyperinnervation, the synaptic incidence of the 5-HT terminals was remarkably conserved after grafting, as previously observed in the hippocampus under similar conditions (Daszuta *et al.*, 1991), or in the 5-HT hyperinnervated neostriatum after 6-OHDA destruction of nigrostriatal DA neurons (Descarries *et al.*, 1992). 5-HT synapses were mostly re-established with dendritic shafts, as in the normal innervation, but there was a tendency toward an increased number of symmetrical versus asymmetrical contacts in both the SON and SCN.

In an early histochemical fluorescence study after intraventricular administration of exogenous 5-HT (Fuxe and Ungerstedt, 1968) and subsequent radioautographic studies after intraventricular administration of [^{3}H]5-HT (Descarries and Beaudet, 1978; Beaudet and Descarries, 1979; see also Beaudet and Descarries, 1981), a small population of neurons endowed with the capacity to take up exogenous 5-HT or [^{3}H]5-HT was observed in the ventral part of the *dorsomedial nucleus* of adult rat hypothalamus, next to a particularly dense zone of innervation by [^{3}H]5-HT-labeled terminals in the radioautographic study. However, these cell bodies had not been seen in earlier fluorescence histochemical studies of untreated rats; neither were they shown to contain endogenous 5-HT in subsequent 5-HT immunocytochemical studies (e.g., Steinbusch and Nieuwenhuys, 1981), except in rats treated with a monoamine oxidase inhibitor and the 5-HT precursor, L-tryptophan (Frankfurt *et al.*, 1981). It has not been determined what proportion of the [^{3}H]5-HT-labeled terminals in the dense zone of presumed 5-HT innervation which was then observed at the electron-microscopic level might actually arise from the [^{3}H]5-HT-labeled cells. The morphology of these terminals was not different from that in other hypothalamic regions. Some were occasionally observed to make symmetrical or asymmetrical synapses with unlabeled nerve cell bodies, but synaptic incidence was not evaluated (Beaudet and Descarries, 1979).

Frankfurt and Beaudet (1987) have examined the ultrastructural features of [^{3}H]5-HT-labeled terminals in the dorsal part of the dorsomedial nucleus of hypothalamus at two survival times after a 5-HT denervation by intraventricular 5,7-dihydroxytryptamine – i.e., in a state of 5-HT reinnervation (30 days) and 5-HT hyperinnervation (50 days). At both survival times, as well as in controls, they found only a few synaptic contacts made by 5-HT varicosities, all axodendritic and asymmetrical, but an increase in the number of terminals abutting perikarya compared to the sham-treated rats. They concluded from these results that the 5-HT innervation in hypothalamus may regenerate with a great deal of cellular specificity.

Circumventricular organs

The circumventricular organs are a series of midline structures located around the ventricular system and open to neuro-hemal exchanges. They comprise the median eminence, organum vasculosum laminae terminalis, subfornical organ, subcommissural organ, pineal gland and area postrema (see Calas *et al.*, 1978; Bouchaud and Bosler, 1986).

First demonstrated in the rat as degenerating nerve terminals after intraventricular administration of the selective neurotoxin 5,7-dihydroxytryptamine (Baumgarten and Lachenmayer, 1974), the 5-HT innervation of the *median eminence* (ME) predominates in its outer layer, the inner layer showing only scattered terminals (Calas, 1977; Jennes *et al.*, 1982). In rat and monkey, the 5-HT terminals in the outer layer were shown by electron-microscopic radioautography to contact the basement membrane of portal capillaries (Calas *et al.*, 1974; Calas, 1977; Nakai *et al.*, 1983), or to directly abut on tanycytic processes, but were never found to make a synaptic junction. Many were, however, directly juxtaposed to other axonal varicosities, some of which were later identified as containing thyrotropin-releasing hormone (TRH) (Nakai *et al.*, 1983), tyrosine hydroxylase or $ACTH_{1-24}$ (Bosler and Beaudet, 1985a), in combined radioautographic and immunocytochemical experiments. A light-microscopic study by Jennes *et al.* (1982) emphasized the considerable overlap between 5-HT and LHRH immunoreactive axon terminals in the rat ME, but axo-axonal synaptic contacts have never been demonstrated in the many electron-microscopic studies of this brain region. As suggested by Calas more than 30 years ago (1977), these observations suggested diverse modes of actions for 5-HT in the ME: (1) direct release in the portal system; (2) effects on the permeability of capillaries to various neurosecretory factors; (3) actions on tanycytic function; and (4) regulation of the release of hypophysiotrophic hormones or other transmitters through axo-axonic appositions.

Relatively dense 5-HT innervation has also been discovered by fluorescence histochemistry in the rat *organum vasculosum laminae terminalis* (OVLT) (Moore, 1977). In subsequent [^{3}H]5-HT uptake radioautographic studies at the electron-microscopic level (Bosler, 1978, 1980; Calas *et al.*, 1978; Bouchaud and Bosler, 1986; Bosler and Descarries, 1988; Saïdi and Bosler, 1990), [^{3}H] 5-HT-labeled terminals were found to represent almost 10 percent of all axon terminals in this brain region. They were most abundant in the rostrocaudal and ventrodorsal portions of the juxtaventricular zone, as well as in the dorsal aspect of the juxtavascular zone, but absent from the vascular core. In the juxtaventricular zone, a significant proportion made symmetrical or asymmetrical synapses with dendritic shafts or spines, but none were ever seen in synaptic contact in the juxtavascular zone. In this latter zone, the 5-HT endings established close relationships with neurosecretory-like axons, and with astrocytic or tanycytic processes, with which they occasionally formed 'synaptoid' contacts. A few of these varicosities were also seen to abut directly on the outer basement membrane of the perivascular space. It is therefore likely that, in OVLT, 5-HT influences non-neural as well as neural elements. At a proximal level of regulation (juxtaventricular zone), it could act both synaptically and non-synaptically as interneuronal transmitter or modulator, whereas distally (juxtavascular zone) it could be released as a neurohormone in addition to modulating neurosecretion.

Interestingly, the 5-HT innervation of the OVLT has been shown to regenerate within several months after total destruction of its nerve terminals by cerebroventricular administration of 5,7-dihydroxytryptamine (Saïdi and Bosler, 1990). Sixteen months after the lesion, the regenerated terminals displayed a normal pattern of distribution and their ultrastructural features were similar to those of aged-matched controls, including their synapsing with dendritic shafts and spines in the juxtaventricular zone.

The rat *subcommissural organ* (SCO) also receives a relatively dense 5-HT innervation, arising mainly from the midbrain nuclei raphe dorsalis and medianus, with some contribution from the raphe pontis (Léger *et al.*, 1983). Within the SCO, these 5-HT terminals have been shown to establish asymmetrical synaptic contacts with both epen- and hypendymocytes (Bouchaud and Arluison, 1977; Bouchaud, 1979; Møllgärd and Wiklund, 1979). More than 75 percent of axon terminals synapsing with SCO ependymocytes could be labeled with [^{3}H]5-HT, suggesting that 5-HT plays a major role in regulating the secretory activity of these cells. According to Léger *et al.* (1983), the 5-HT afferents from the raphe dorsalis would inhibit the secretory activity of SCO ependymocytes, whereas those from the raphe medianus would exert the opposite effect. GABA uptake has also been demonstrated in some 5-HT fibers of the SCO (Gamrani *et al.*, 1981), but the functional significance of this co-localization remains to be determined.

The *subfornical organ* (SFO) appears to be sparsely 5-HT innervated in both radioautographic and immunohistochemical studies (Calas *et al.*, 1978; Takeuchi and Sano, 1983a; Tramu *et al.*, 1983). In the rat, [^{3}H]5-HT-labeled boutons were mainly found beneath the ependyma and within the vascular region. In this zone, they approach fenestrated capillaries (Calas *et al.*, 1978). According to this author, some of the perivascular 5-HT fibers exhibit a discontinuous thickening of their membrane in front of the basal membrane surrounding the pericapillary space, whereas intraparenchymal 5-HT fibers occasionally display synaptic-like differentiations (see also Bouchaud and Bosler, 1986). 5-HT terminals have also been identified by [^{3}H]5-HT radioautography in the SFO of the monkey (Calas *et al.*, 1974).

In an early histochemical fluorescence study, Fuxe and Owman (1965) described a small group of indolaminergic, presumably 5-HT neurons in the rat *area postrema* (AP), located ventrally to the so-called A2 cell group of catecholaminergic neurons. Since then, the presence of these 5-HT cells has been confirmed by immunohistochemistry (Steinbusch, 1981; Tramu *et al.*, 1983), and the ultrastructural features of 5-HT axonal varicosities described by radioautography after intraventricular administration of [^{3}H]5-HT, within the AP parenchyma or the pericapillary spaces (Pickel and Armstrong, 1984). In the parenchyma, these 5-HT varicosities frequently showed axo-dendritic synaptic contacts, whereas in the ventricular spaces they were described as free endings. Pickel *et al.* (1984) combined TH immunocytochemistry with the [^{3}H]5-HT labeling to demonstrate that some of the axodendritic synapses in the parenchyma were made with catecholaminergic as well as non-catecholaminergic neurons. After combined radioautographic and immunocytochemical labeling, Armstrong *et al.* (1984) also observed the presence of enkephalin-like immunoreactivity in some of the 5-HT varicosities in AP.

Supra-ependymal plexus

An extensive network of 5-HT axons forms a supra-ependymal plexus along the walls of the cerebral ventricles (Richards *et al.*, 1973; Lorez and Richards, 1973, 1975; Chan-Palay, 1976; Richards, 1977; Bosler, 1980; De La Manche *et al.*, 1981). These 5-HT fibers, which mostly originate from midbrain nuclei raphe dorsalis and medianus (Aghajanian and Gallager, 1975), pervade most of the ventricular cavities, including the surface of the OVLT and of the SFO (Takeuchi and Sano, 1983a). They are absent beneath the SCO, but the ones seen along the lateral edge of the organ appear to be continuous with the dense plexus of hypendymal fibers (Møllgärd and Wiklund, 1979; see also Bosler, 1980, and Takeuchi and Sano, 1983a). Electron-microscopic examination of supra-ependymal 5-HT varicosities after radioautographic [^{3}H]5-HT uptake labeling in rat and monkey (Chan-Palay 1976; Richards, 1977; Calas *et al.*, 1978; Bosler, 1980; De La Manche *et al.*, 1981) has revealed that these varicosities are secured to the ependyma by small desmosomal-like attachment plaques, and devoid of synaptic specializations. The supra-ependymal 5-HT fibers in the rat have also been visualized with 5-HT immunohistochemistry (Richards *et al.*, 1981), and shown to take up

[^{3}H]GABA (Gamrani *et al.*, 1984) and to be GABA immunoreactive (Harandi *et al.*, 1986). Striking pictures of these 5-HT fibers have also been obtained from *post-mortem* human brain by scanning electron microscopy (Richards *et al.*, 1980, 1981; see also Lorez and Richards, 1982).

Dinopoulos and Dori (1995) have also carried out a light- and electron-microscopic immunocytochemical study of the postnatal development of the supra-ependymal plexus in the lateral ventricles of the rat. They found this system to look morphologically mature by the end of the second postnatal week, and confirmed that its 5-HT varicosities, confined to the ventricular surface of the ependymal lining, never make synapses with the ependymal cells.

As initially described by Fuxe and Owman (1965) with the histochemical fluorescence technique, and later confirmed by 5-HT immunohistochemistry, there is a small group of several hundred 5-HT neuronal cell bodies, located ventrally to the A2 noradrenaline cell group, in the *area postrema* (AP) of the rat. 5-HT axon varicosities, labeled by [^{3}H]5-HT uptake, have also been identified in the rat AP (Armstrong *et al.*, 1984; Pickel and Armstrong, 1984; Bouchaud and Bosler, 1986). These are located either in the parenchyma, where they frequently show axo-dendritic synaptic differentiations, or within the pericapillary and the ventricular spaces, in the form of free nerve endings. After combined radioautographic and immunocytochemical labeling, Pickel *et al.* (1984) were able to demonstrate the existence of 5-HT terminals making synapses on dendrites of TH-immunoreactive (presumably noradrenergic) as well as non-catecholaminergic neurons in the rat AP, whereas Armstrong *et al.* (1984) reported that enkephalin-like immunoreactivity was occasionally present in some of these [^{3}H]5-HT labeled terminals (see also Bosler *et al.*, 1986).

5-HT innervaton of the brainstem

Substantia nigra

The substantia nigra (SN) receives some of the densest 5-HT innervation in the brain (Fuxe, 1965; Steinbusch, 1981) arising primarily from the mesencephalic raphe nuclei (Bobillier *et al.*, 1976; Fibiger and Miller, 1977; Azmitia and Segal, 1978; Parent *et al.*, 1981; Lavoie and Parent, 1990; Vertes, 1991). In monkey as well as rat, its pars reticulata is more densely innervated than its pars compacta (Pasik *et al.*, 1984). In the rat, Moukhles *et al.* (1997) have quantified the density of this innervation at 9×10^6 and 6×10^6 varicosities per mm^3 of tissue in reticulata and compacta, respectively.

Following an early radioautographic study demonstrating the existence of labeled terminals forming axo-dendritic synaptic contacts in the substantia nigra of rats administered [^{3}H]5-HT in a lateral ventricle of the brain (Parizek *et al.*, 1971), the ultrastructure of these 5-HT terminals has been examined by 5-HT immunocytochemistry in the rat and guinea pig as well as in the monkey (Table 1). In both pars reticulata and compacta of rat SN, Mori *et al.* (1987) and Corvaja *et al.* (1993) noted the presence of many 5-HT varicosities which made synaptic contact mostly with dendritic shafts and occasionally with dendritic spines (Corvaja *et al.*, 1993), and similar observations were made in the pars reticulata of the monkey's SN (Pasik *et al.*, 1984). Subsequently, the study of Moukhles *et al.* (1997) revealed a striking difference between the 5-HT varicosities in the two subdivisions of rat SN. As extrapolated stereologically to the whole volume of these varicosities, virtually all 5-HT varicosities in the pars reticulata, but only 50 percent of those in the pars compacta, formed synapses, essentially with dendritic branches. By combining the immunogold labeling of tyrosine hydroxylase (TH) with 5-HT immunoperoxidase labeling, these authors also confirmed an earlier observation of Nedergaard *et al.* (1988) of synaptic contacts between 5-HT terminals and dopamine (TH immunoreactive) dendrites in the guinea pig. In rat, they found that 20 percent of the synaptic 5-HT terminals in the pars reticulata contacted dopamine dendrites, and that a smaller percentage (6 percent) was in contact with unlabeled dendrites characterized by an unusually high cytoplasmic content in neurotubules.

Interestingly, Bunin and Wightman (1998) then used fast-scan voltammetry to compare 5-HT uptake and release in the pars reticulata of substantia nigra, in which all 5-HT terminals appear to be synaptic, to that in the raphe dorsalis, a somatodendritic region with rare synaptic 5-HT varicosities. The conclusion from their study was that diffuse 5-HT transmission occurred in both regions, suggesting that this mode of transmission might apply to synaptic as well as non-synaptic 5-HT terminals.

Ventral tegmental area

The ventral tegmental area (VTA), which contains the largest collection of dopamine-containing neurons in the brain (group A10, according to the nomenclature of Dahlström and Fuxe, 1964), also receives relatively dense 5-HT innervation from the raphe dorsalis and medianus. Two radioautographic studies after intraventricular administration of [^{3}H]5-HT *in vivo* have described the ultrastructural features of this 5-HT innervation in the rat (Hervé *et al.*, 1987; Van Bockstaele *et al.*, 1994; see also Bosler *et al.*, 1986). Hervé *et al.* (1987) demonstrated that more than 18 percent of these terminals showed a synaptic specialization in single thin sections, allowing the extrapolation that 50 percent were actually synaptic. Most of these synapses were asymmetrical, and formed with dendritic shafts. After concomitant immunolabeling for TH, it was apparent that some were made with the dendrites of dopamine neurons and others with non-dopamine dendrites. Similar observations were made by Van Bockstaele *et al.* (1994), who showed, in addition, that many of the neurons contacted by 5-HT immunoreactive terminals were labeled by retrograde axonal transport of gold-conjugated wheatgerm agglutinin injected into the nucleus accumbens. Thus, within the VTA, synaptic as well as non-synaptic release of 5-HT might modulate mesolimbic as well as mesocortical efferent systems.

Nucleus interpeduncularis

Presumed 5-HT axon terminals have also been visualized in the interpeduncular nucleus of the rat by Nojyo and Sano (1978) in their cytochemical study of the suprachiasmatic nucleus after loading with 5,6-dihydroxytryptamine. These authors described these 5-HT terminals as less frequently synaptic than in the suprachiasmatic nucleus.

Red nucleus

Even though 5-HT exerts various electrophysiological effects when applied iontophoretically into the red nucleus of cat or rat (Davis and Vaughan, 1969; Schmied *et al.*, 1991; Licata *et al.*, 1998), the density of 5-HT innervation appears to be moderate in this brain region (Steinbusch, 1981). Bosler *et al.* (1983) have described the ultrastructural features of this 5-HT innervation in the cat, after its radioautographic labeling by [^{3}H]5-HT uptake *in vitro*. Symmetrical synaptic junctions were then reported to be made by 5-HT axon varicosities, with small dendritic processes and with the cell bodies and large dendritic trunks of magnocellular neurons. In contrast, in a subsequent 5-HT immunocytochemical study of the rat red nucleus, 5-HT axon varicosities were not found to establish any clearly defined synaptic junction, indicating a very low synaptic incidence (André *et al.*, 1987). These authors also combined 5-HT with GABA immunolabeling to demonstrate occasional contacts between the 5-HT endings and dendrites of large-sized neurons receiving a GABA input, as well as direct appositions between 5-HT and GABA positive terminals, suggesting that red nucleus output may be dually regulated by 5-HT and GABA afferents and that distal axonal interactions might also play a role in red nucleus function.

Oculomotor nucleus

The density of 5-HT innervation in the rat oculomotor nucleus has been rated as low to medium, with a rostrocaudal decreasing gradient of distribution (Steinbusch, 1981). In the only electron-microscope radioautographic study of this nucleus, following local administration of [^{3}H]5-HT *in vivo*, Soghomonian *et al.* (1986) observed that about 50 percent of its 5-HT axon varicosities profiles exhibited a synaptic membrane differentiation in single thin sections, suggesting an entirely synaptic innervation and the frequent occurrence of more than one junction per varicosity. The synapses were predominantly asymmetrical when formed with small to large dendritic shafts, and exclusively symmetrical onto dendritic spines. A few 5-HT axon varicosities in synaptic contact with unlabeled cell body were also observed (Soghomonian *et al.*, 1986). This study suggests that, as in other motor nuclei (see below), 5-HT transmission is mostly synaptic within the oculomotor nucleus.

Superior colliculus

The mammalian superior colliculus receives multiple inputs that include projections from the retina, the visual cortex and the lateral geniculate nucleus. This laminated structure is well known to play a crucial role in the processing of information directing gaze toward objects of interest (McPeek and Keller, 2004). Its 5-HT innervation, which is particularly dense in the superficial layers (Fuxe, 1965), comes mainly from the raphe dorsalis, but also from the raphe medianus, raphe pontis and controlateral periaqueductal gray (Beitz *et al.*, 1986; Villar *et al.*, 1988; Mize and Horner, 1989).

Immuno-electron microscopy experiments conducted in the superior colliculus of the rat (Dori *et al.*, 1998), cat (Mize and Horner, 1989) and hamster (Arce *et al.*, 1992, 1995; Crissman *et al.*, 1993) have revealed that many of the 5-HT axon varicosities in this nucleus do not display synaptic junctions. As measured in single thin sections, the reported synaptic incidences for these terminals ranged from 4 percent in the hamster (Arce *et al.*, 1992, 1995; Crissman *et al.*, 1993) to 22 percent in the rat (Dori *et al.*, 1998). Only 3 of 50 5-HT axon varicosities examined in serial thin sections in the superficial layers of the hamster's superior colliculus were synaptic (Arce *et al.*, 1992), but, in the rat, stereological extrapolation to the whole volume of varicosities indicated a synaptic incidence of 55 percent (Dori *et al.*, 1998). In both superficial and deep layers of the rat superior colliculus, the majority of these synapses were symmetrical and made with dendritic shafts. Symmetrical or asymmetrical synapses on spines were also occasionally observed, but the rare synapses on cell bodies were exclusively symmetrical (Dori *et al.*, 1998). In the rat (Dori *et al.*, 1998), as well as the cat (Mize and Horner, 1989) and hamster (Arce *et al.*, 1995), 5-HT axon varicosities were often seen juxtaposed to unlabeled axonal profiles, and some of these axo-axonal contacts were described as synaptic (Arce *et al.*, 1995; Dori *et al.*, 1998). 5-HT_{1B} receptor binding sites have been demonstrated on retinocollicular afferents (Ségu *et al.*, 1986; Boulenguez *et al.*, 1993, 1996; Sari *et al.*, 1999), supporting the possibility of a distal, inhibitory gating of visual inputs by 5-HT.

As in the septum, the diagonal band of Broca and lateral geniculate, Dori *et al.* (1998) reported that the synaptic incidence of 5-HT axon terminals in the superficial layers of the developing superior colliculus of the rat increases markedly from birth to the end of the first postnatal week, and then decreases for 2 weeks prior to a further increase to its adult value. The first phase of synapse formation would coincide with the arrival of cortical and retinal afferents in the superior colliculus, which supports an influence of 5-HT in the development of these afferent projections (Rhoades *et al.*, 1993; Ke *et al.*, 1999). In contrast, the earlier studies conducted in the hamster superior colliculus had rather suggested a gradual age-related increase in the density of 5-HT innervation with a gradual decrease in the synaptic incidence measured in single thin sections, from 37 percent in the newborn to 4 percent in the adult (Crissman *et al.*, 1993). Rather than a progressive loss of synaptic contacts, this latter result could point to a progressive dilution of the synaptic 5-HT innervation by later-arriving 5-HT afferents that do not make synapses. The idea of a competition for synaptic sites between 5-HT axon varicosities and retinal afferent projections has been tested by postnatal enucleation in adult and newborn hamsters. Removal of retinal afferent projections at birth resulted in an increased density of 5-HT innervation and of their synaptic incidence, as measured in single thin sections, from 4 percent to 18 percent in adult hamsters (Arce *et al.*, 1995).

Dorsal periaqueductal and central gray

It is in the central gray of rat brainstem that axon terminals having taken up [^{3}H]5-HT were visualized for the first time by high-resolution radioautography (Aghajanian and Bloom, 1967). Almost 20 years later, a detailed light- and electron-microscopic 5-HT immunocytochemical study of the rat midbrain periaqueductal gray (PAG) was carried out by Clements *et al.* (1985). In this study, synaptic contacts established by 5-HT axon varicosities were rare. Similar observations were subsequently made by Lovick *et al.* (2000), who focused on the dorsolateral PAG and examined the ultrastructural relationships of 5-HT terminals in brain slices in which some PAG neurons had been filled with biocytin, but also in specimens from perfusion-fixed whole brain. In serial sections across 21 5-HT terminals apposed to biocytin-filled dendrites, these authors found only 4 displaying synaptic specialization, and they reported a similar synaptic incidence of 23 percent in the better-preserved material from whole brain.

Nucleus raphe dorsalis

In all raphe nuclei containing 5-HT neurons, the latter are found among numerous non-5-HT cell bodies. For instance, in the rat raphe dorsalis, 5-HT cells account for approximately one-third of a total population of 36,000 neurons (Descarries *et al.*, 1982), some of which are now known to be GABAergic or glutamatergic (Commons and Valentino, 2002; Commons *et al.*, 2005). Even though not numerous, 5-HT axon terminals have been identified in the rat, cat and monkey raphe dorsalis, either with [^{3}H]5-HT uptake radioautography or with 5-HT immunocytochemistry (Chan-Palay, 1982; Descarries *et al.*, 1982; Chazal and Ralston, 1987). As interconnections have been demonstrated between the different raphe nuclei (Bobillier *et al.*, 1976), some of these terminals could be those of 5-HT afferents from other 5-HT cell groups, and others of recurrent collaterals issued by intrinsic 5-HT neurons (Mosko *et al.*, 1977; Kalén *et al.*, 1985). Whatever their origin, 5-HT terminals in the raphe dorsalis make only rare synaptic specializations, which also seems to be the case in rat raphe magnus (Chazal and Ma, 1989). In both nuclei occasional synaptic contacts have, however, been observed between 5-HT terminals and unidentified dendritic branches, as well as with tyrosine hydroxylase immunoreactive dendrites within raphe dorsalis (see Soghomonian *et al.*, 1988), which presumably belong to the small population of dopamine neurons residing in the rostral pole of this nucleus (Descarries *et al.*, 1986).

In keeping with earlier radioautographic observations of [^{3}H]5-HT uptake by 5-HT neurons in the raphe dorsalis and the notion of a somatodendritic release of the transmitter, 5-HT transporter (SERT) has been shown to be present all along the plasma membrane of these neurons, including the soma-dendrites (Figure 3A), and not only on axons and axon terminals in territories of innervation. By dual-immunocytochemical labeling, it has also been possible to demonstrate the presence of 5-HT_{1A} autoreceptors

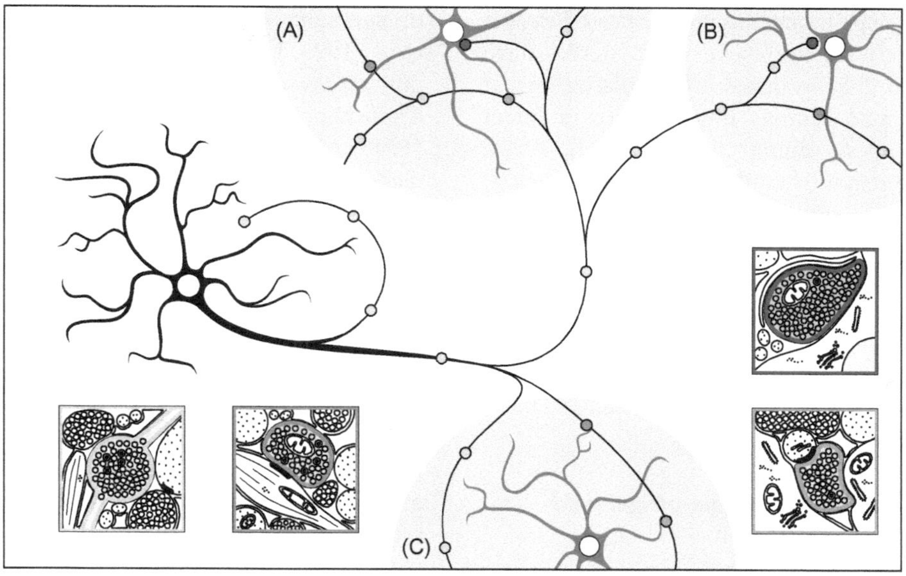

Figure 3 Schematic representation of a 5-HT neuron. As depicted here, a single 5-HT neuron may innervate different anatomical regions (A, B, C), including its nucleus of origin (through recurrent collaterals of its axon). In the territories of innervation, many of the 5-HT varicosities (small colored dots), distributed as rosary beads on distal branches of the axon, do not make synaptic contact (yellow dots). Others are endowed with junctional complexes, the small zone of membrane specialization, either symmetrical or asymmetrical, that is the hallmark of synapses, and make such a contact with dendritic branches (green dots), and, more rarely, dendritic spines (blue dots). Juxtapositions to cell bodies are also seen (red dots). These various configurations, schematized in the boxes, may be found within a given anatomical territory, and probably on a single axonal branch. The vesicular content of 5-HT varicosities may also vary from one anatomical region to another, and in some regions, such as the cerebellum with its mossy 5-HT fibers, or the locus coeruleus and its 5-HT varicosities filled with microvesicles and canaliculi, 5-HT terminals of a unique morphology may be found. It has been calculated that some 5-HT neurons may possess as many as hundreds of thousands axon varicosities (Descarries *et al.*, 1990). The demonstration of the existence of asynaptic 5-HT varicosities in mammalian CNS was at the origin of the concept of diffuse (volume) transmission (for review, see Descarries and Mechawar, 2000). To see the full color version of this figure please refer to the color plate at the back of the book. Copies produced via our print on demand service do not contain color plates; if your copy does not have the color plate, please go to this website to view the figure in color www.elsevierdirect.com/companions/9780123746344

on the soma-dendrites of the raphe dorsalis 5-HT neurons decorated with SERT (Figure 2B).

Locus coeruleus

The locus coeruleus is the major group of noradrenaline neurons in mammalian brain (A6, in the nomenclature of Dahlström and Fuxe, 1964). It receives a dense 5-HT innervation, which appears to originate from three raphe nuclei: the raphe dorsalis, raphe medianus and raphe pontis (Léger *et al.*, 1980). This 5-HT innervation has been characterized at the electron-microscopic level in the rat, using both TPH immunocytochemistry (Pickel *et al.*, 1977) and radioautography after *in vivo* uptake of [^{3}H] 5-HT (Léger and Descarries, 1978; see also Descarries and Léger, 1978). Its varicosities are often asynaptic, but some do establish typical asymmetrical synapses with dendritic branches and spines presumably belonging to noradrenaline neurons. Thus, both diffuse and synaptic 5-HT transmission might contribute to the modulatory effects of 5-HT on noradrenaline neurons (see, for example, Guiard *et al.*, 2008).

Mesopontine tegmentum

The 5-HT neurons located in the raphe dorsalis are also known to innervate the pedunculopontine and laterodorsal nuclei, and double-immunolabeling for electron microscopy has allowed for the demonstration of synaptic contacts between 5-HT axon varicosities and the cholinergic neurons in these two nuclei (Honda and Semba, 1994; Steininger *et al.*, 1997; Wang *et al.*, 2000). Both symmetrical and asymmetrical synaptic contacts were observed, which were mainly located on proximal dendrites and cell bodies (Honda and Semba, 1994; Steininger *et al.*, 1997). These observations were consistent with electrophysiological studies showing that 5-HT-induced hyperpolarization of cholinergic laterodorsal neurons involves a direct action of 5-HT (Luebke *et al.*, 1992), and with the suggested a role of the raphe dorsalis in the modulation of rapid eye movement (REM) sleep through this direct action (Steriade and McCarley, 2005). Electrophysiological studies have also suggested that 5-HT-induced hyperpolarization of cholinergic laterodorsal neurons might involve a direct action of 5-HT on these neurons (Luebke *et al.*, 1992). Interestingly, in the rostral laterodorsal tegmental nucleus a few 5-HT cell bodies and dendrites are present (Honda and Semba, 1994), and some of the 5-HT dendritic shafts contain aggregated synaptic vesicles, as already observed in the raphe nucleus of the cat (Chazal and O'Hara, 1986; Chazal and Ralston, 1987) and monkey (Kapadia *et al.*, 1985). It is therefore likely that somatodendritic as well as axonal release of 5-HT takes place in this nucleus.

Paragigantocellularis lateralis and paratrigeminal nuclei

The nucleus *paragigantocellularis lateralis* of the medullary reticular formation contains 5-HT neurons that were first identified as the B3 group by Dahlström and Fuxe (1964). The fine ultrastructural features of these neurons and axon terminals in this nucleus have been described in a radioautographic study conducted by Chan-Palay (1978a) in rhesus monkeys. Prominent axo-axonic as well as symmetrical and asymmetrical axo-dendritic synaptic contacts were then observed to be made by these 5-HT varicosities, whose synaptic incidence was reported to be as high as 84 percent in two adjacent thin sections, suggesting the existence of multiple synaptic junctions on many of these varicosities.

In another radioautographic study following injection of tritiated 5-HT in the *paratrigeminalis nucleus* of the rat and monkey, Chan-Palay (1978b) described the 5-HT varicosities of this nucleus as making not only axo-dendritic and axo-spinous but also axo-somatic synaptic contacts. She then estimated at 60 percent the synaptic incidence of these 5-HT varicosities in adjacent thin sections.

Trigeminal nuclei

The trigeminal nuclear complex extends throughout the entire brainstem and merges caudally with the dorsal horn of the spinal cord. Its three nuclei, the mesencephalic, spinal and motor trigeminal, all receive a dense 5-HT innervation mainly from the raphe dorsalis (Tashiro *et al.*, 1989; Copray *et al.*, 1991).

In the *mesencephalic trigeminal nucleus*, the synaptic incidence of 5-HT varicosities appears to differ between rat and cat, as observed by 5-HT immunocytochemistry in single thin sections. In the rat, the vast majority of 5-HT axon varicosities appear to make asymmetrical or symmetrical contacts mainly with dendrites (Liem *et al.*, 1993; Liem and Copray, 1996), whereas in the cat, only few would show identifiable junctions (Lazarov and Chouchkov, 1995). It is noteworthy that Liem and Copray (1996) were able to use an immunogold postembedding technique to label some of these 5-HT terminals. They could thus demonstrate the storage of 5-HT in large granular dense-core vesicles, as well as in small round or pleomorphic vesicles, as previously inferred from cytochemical, radioautographic and immunocytochemical observations (reviewed in Beaudet and Descarries, 1987).

In the *spinal trigeminal nucleus caudalis*, which corresponds to the dorsal horn of the cervical cord, the densest 5-HT innervation is that of lamina I (Li *et al.*, 1997; López Costa *et al.*, 1994), in which nociceptive primary afferents make synapses. As in the dorsal horn of spinal cord (see below), a majority of 5-HT varicosities in this region do not make synapses, as visualized with 5-HT immuno-electron microscopy in single thin sections (Li *et al.*, 1997; López Costa *et al.*, 1994). Some, however, display symmetrical junctions onto neuronal somadendrites that may be retrogradely labeled with lectin wheatgerm agglutinin conjugated to horseradish peroxidase (WGA-HRP) from the parabrachial or thalamic region (Li *et al.*, 1997). These synaptic contacts are then mostly found on dendritic profiles, occasionally on somata and rarely on dendritic spines. Because 5-HT_1 and 5-HT_3 receptors are known to be present on primary afferent axons of the dorsal horn (Daval *et al.*, 1987, Hamon *et al.*, 1989), but no axo-axonal synaptic contacts are made by its 5-HT afferents, it is likely that diffuse 5-HT transmission controls the release of neurotransmitters from these as well as the spinal cord nociceptive afferents (Li *et al.*, 1997).

In the rat *motor trigeminal nucleus*, the density of 5-HT innervation has been estimated at 1.8×10^6 varicosities per mm^3 of tissue (Schaffar *et al.*, 1984). At the electron-microscopic level, it has been suggested that, in both rat and cat, 5-HT neurotransmission might be only partially synaptic, as many [^{3}H]5-HT-labeled or 5-HT immunoreactive axon varicosities failed to show a synaptic junction in single thin sections (Schaffar *et al.*, 1984; Nagase *et al.*, 1997). However, in a discordant report, Saha *et al.* (1991) claimed that virtually all 5-HT axon terminals in the motor trigeminal nucleus of rat were endowed with a synaptic junction. Both symmetrical and asymmetrical synaptic contacts, mostly made with dendritic shafts, were observed in each of these studies.

Facial motor nucleus

The facial motor nucleus receives a prominent 5-HT innervation, as demonstrated by fluorescence histochemistry

(Fuxe, 1965) and biochemical measures (Palkovits *et al.*, 1974). In a radioautographic study of this nucleus after [^{3}H]5-HT uptake *in vivo*, Aghajanian and McCall (1980) demonstrated the existence of axo-somatic and axo-dendritic synapses formed by 5-HT axon varicosities and reported a synaptic incidence of 19 percent in single thin sections. This suggested that many of these terminals were synaptic, providing a morphological substrate for the facilitatory effects of 5-HT on the activity of motoneurons (McCall and Aghajanian, 1979).

Ambiguous nucleus

In a 5-HT immunocytochemical study, the ambiguous nucleus of *Macaca fuscata* was examined at the electron-microscopic level. Takeuchi *et al.* (1983) confirmed the existence of synapses, both symmetrical and asymmetrical, made by 5-HT terminals on cell bodies and dendrites of motoneurons in this nucleus, as well as in the motor trigeminal, facial and accessory nerves, and hypoglossal nuclei.

Nucleus of the solitary tract

The nucleus of the solitary tract is an important relay in processing respiratory and cardiovascular inputs to the central nervous system. This region contains a high level of endogenous 5-HT (Palkovits *et al.*, 1974), and its 5-HT innervation has been reported to be moderately dense, as visualized by fluorescence or light microscopy after immunocytochemical labeling (Fuxe, 1965; Steinbusch, 1981; Maley and Elde, 1982; Pickel *et al.*, 1984). Electron-microscopic studies in this nucleus have been conducted in the rat (Pickel *et al.*, 1984; Huang and Pickel, 2002; Huang *et al.*, 2004) and cat (Elde *et al.*, 1982; Maley and Elde, 1982; Maley *et al.*, 1990). After [^{3}H]5-HT uptake in rat, 30 percent of the single thin section profiles of labeled varicosities displayed a synaptic specialization. Symmetrical as well as asymmetrical synapses on either dendritic branches or spines were also observed (Pickel *et al.*, 1984). Combined radioautography and immunocytochemistry allowed the visualization of synaptic contacts between 5-HT axon varicosities and tyrosine hydroxylase-immunolabeled dendrites (Pickel *et al.*, 1984), suggesting a prominent role of 5-HT in cardiovascular regulation through synaptic control of noradrenergic neurons innervating the hypothalamus.

Studies of the subcellular distribution of 5-HT$_{2A}$ and 5-HT$_{3A}$ receptors in relation to 5-HTT-immunolabeled terminals have, however, underlined the importance of diffuse transmission in the rat nucleus of the solitary tract to activate receptors at a distance from 5-HT uptake sites (Huang and Pickel, 2002; Huang *et al.*, 2004). In the cat, serial sections analysis has confirmed the lack of synaptic specialization on a vast majority of 5-HT axon varicosities in this nucleus (Maley and Elde, 1982). The rare synaptic contacts that were then observed were asymmetrical and, as in rats, mostly made with dendritic branches or spines.

Inferior olivary complex

All subdivisions of the inferior olivary complex have been shown to receive a 5-HT input, with some degree of interspecies variability (Wiklund *et al.*, 1977; Takeuchi and Sano, 1983b; King *et al.*, 1984). In the rat, the source of this 5-HT innervation has been identified as the nucleus paragigantocellularis, which contains a part of the B3 5-HT group, as already mentioned (Bishop and Ho, 1986). There is strong immunohistochemical evidence for the coexistence of 5-HT and substance P in at least some of these olivary afferents (Paré *et al.*, 1987). In the dorsal accessory olive of rat, 5-HT axon terminals display a very low incidence of synaptic contacts (Wiklund *et al.*, 1981). Less than 5 percent of 750 [^{3}H]5-HT labeled profiles examined in single thin sections appeared to be synaptic. In the opossum, a similarly low incidence (2 percent) was observed using either [^{3}H]5-HT radioautography or 5-HT immunocytochemistry (King *et al.*, 1984). In both species, 5-HT synapses were mostly asymmetrical and established with dendritic branches. The non-synaptic 5-HT terminals were found in close apposition with dendritic shafts, somatic appendages or spines. In the caudal medial accessory nucleus of the opossum, a substantial number of 5-HT varicosities were observed in close apposition with nerve cell bodies or axon terminals (King *et al.*, 1984).

Dorsal column nuclei

According to the 5-HT immunocytochemical study of Blomqvist and Broman (1993), the 5-HT innervation of the dorsal column nuclei (DCN) is heterogeneously distributed in a pattern faithfully related to the cytoarchitectonic subdivisions of these nuclei in both cats and owl monkeys. In the cat, electron-microscopic examination showed that these 5-HT fibers were thin and unmyelinated, and studded with small varicosities juxtaposed to dendrites, other terminals and cell bodies, but rarely made a synaptic contact.

5-HT innervation of the cerebellum

The existence of a 5-HT innervation in the cerebellum was first demonstrated in the rat, with the histochemical fluorescence technique (Hökfelt and Fuxe, 1969). Subsequent 5-HT immunohistochemical studies showed this innervation to be sparse and heterogeneously distributed (Takeuchi *et al.*, 1982). The ultrastructure of this 5-HT innervation has been examined in both the cerebellar cortex and cerebellar nuclei of rat and monkey.

As demonstrated with [^{3}H]5-HT uptake radioautography in the *cerebellar cortex* of both species (Chan-Palay, 1975), a small contingent of the cerebellar 5-HT afferents corresponds to mossy fibers confined to the granular layer, whereas the majority belong to a diffuse projection system which arborizes in all three cortical layers (Chan-Palay, 1975; Beaudet and Sotelo, 1981), presumably arising from the brainstem as well as midbrain raphe cell groups (Chan-Palay, 1975). The large mossy fiber 5-HT varicosities typically form numerous asymmetrical synaptic contacts on the dendritic shafts of granule cells (Chan-Palay, 1975; Palay and Chan-Palay, 1976; Beaudet and Descarries, 1981; Beaudet and Sotelo, 1981). A few also make synapses 'en marron' with the dendrites of Golgi cells (Chan-Palay, 1975). In contrast, a very small proportion (<9 percent) of the smaller varicosities of the diffuse system make synaptic junctions (Beaudet and Sotelo, 1981). In single sections from the granular layer, these authors observed only 2 of 75 such terminals in synaptic contact with dendrites of Golgi cells, whereas in the molecular layer they found 2 of 50 such varicosity profiles in synaptic contact with Purkinje cell branchlet spines. In the *cerebellar nuclei*, the 5-HT axons are scarce and their varicosities have been reported to make synaptic contact with the soma of interneurons and the dendrites of large and small projection neurons (Chan-Palay, 1975; see also Chan-Palay, 1977).

Sotelo and Beaudet (1979) examined the 5-HT innervation in the agranular cerebellum of rat after neonatal X-irradiation (see also Beaudet and Sotelo, 1981). Interestingly, in this modified environment only the diffuse 5-HT system developed, and in the adult its varicosities were all found to make synapses with either the branchlet spines of Purkinje cell dendrites or the dendritic shafts of Golgi cells. This finding was interpreted as indicating that the territory of innervation can exert a determinant influence on the synaptic modeling of incoming 5-HT afferents.

5-HT innervation of the spinal cord

5-HT axon terminals are present in all laminae of the gray matter of the spinal cord (Fuxe, 1965; Ségu and Calas, 1978; Steinbusch, 1981). These axons mainly originate from the brainstem, namely from the medullary raphe complex and from 5-HT neurons located in the central gray and reticular nuclei of the midbrain (Bowker *et al.*, 1981, 1987). In rat and primates, 5-HT neurons have been visualized within the spinal cord itself (LaMotte *et al.*, 1982; Bowker, 1986a; Newton *et al.*, 1986), and could give rise to a small proportion of spinal 5-HT axon varicosities.

The *dorsal horn* (Rexed's laminae I–IV), and particularly its superficial layers I and II, receive the densest 5-HT innervation (Steinbusch, 1981; LaMotte and De Lanerolle, 1983). The ultrastructural features of these axon terminals have been examined in rat, cat and monkey, using the radioautographic or immunocytochemical approach (Table 1). In both cat and monkey, including the dorsal horn of caudal medulla in cat (Ruda *et al.*, 1981b), most of these 5-HT varicosities were shown to establish symmetrical synaptic contacts with the shaft of either proximal or distal dendrites (Ruda and Gobel, 1980; Ruda *et al.*, 1981a, 1982; LaMotte and De Lanerolle, 1983; Light *et al.*, 1983). Fewer 5-HT varicosities were found to make synapses on neuronal perikarya, but examples of somata contacted by multiple 5-HT endings (pericellular arrangements) were found in cat lamina I (Ruda *et al.*, 1982).

In cat, the morphological heterogeneity of dendrites synaptically contacted by 5-HT terminals led to the suggestion that several types of intrinsic neurons in the dorsal horn received a 5-HT input (Ruda *et al.*, 1982). Within layers I and II of the caudal medulla, membranous appositions and synaptic contacts between [^{3}H]5-HT-labeled axons and enkephalin immunoreactive dendrites could be demonstrated, indicating that enkephalin-containing neurons were one of the recipient neuronal types (Glazer and Basbaum, 1984). A number of light-microscopical observations were also made, suggesting that some 5-HT axon varicosities terminate on physiologically identified nociceptive specific cells as well as wide dynamic range neurons projecting to the mesencephalon and thalamus (Hoffert *et al.*, 1983; Miletic *et al.*, 1984; Hylden *et al.*, 1986). In a study of laminae III and IV combining retrograde tracing and 5-HT immunocytochemistry at light- and electron-microscopic levels, nearly all retrogradely labeled neurons which project to the dorsal column nuclei displayed 5-HT terminals apposed to their perikarya or proximal dendrites in both cat and monkey (Nishikawa *et al.*, 1983). Some of these terminals were observed to make symmetrical synaptic contacts, indicating that both wide dynamic range and low threshold mechanoreceptive neurons received 5-HT afferents. Subsequently, in cat, 5-HT terminals were also found on dorsal horn spino-cerebellar tract neurons and dorsal horn interneurons in pathways from muscle afferents identified according to electrophysiological criteria, as well as on interneurons labeled intracellularly with horseradish peroxidase (Maxwell and Jankowska, 1996; Jankowska *et al.*, 1997). Again, small symmetrical synaptic contacts were infrequently seen on these cell bodies, but more commonly on proximal and intermediate parts of their dendrites.

In the rat, as opposed to cat and monkey, few synaptic differentiations are made by 5-HT endings on either neuronal cell bodies or dendrites of the dorsal horn (Maxwell *et al.*, 1983, 1985; Mizukawa *et al.*, 1986; Marlier *et al.*, 1991, 1992; Ridet *et al.*, 1993; Ridet, 1994). Ridet *et al.* (1993) did examine a large population of these 5-HT terminals in serial thin sections, and estimated their synaptic incidence at 37 percent. In rat, as well as in cat and monkey dorsal horn, it was also noted that 5-HT terminals are often juxtaposed to other nerve endings (Maxwell *et al.*, 1983, 1985; Ruda *et al.*, 1982; LaMotte and De Lanerolle, 1983), suggesting a distal control by 5-HT of the release of other transmitters.

The ultrastructure of 5-HT axon terminals has also been characterized by immuno-electron microscopy in the *intermediolateral cell column* of rat, guinea pig and rabbit (Mizukawa *et al.*, 1986; Chiba and Masuko, 1987; Poulat *et al.*, 1992; Jensen *et al.*, 1995). Some of these nerve endings were then shown to make appositional or synaptic contacts on dendrites and neuronal cell bodies, which were later identified as those of preganglionic sympathetic neurons projecting to the superior cervical ganglion or the adrenal medulla by combining 5-HT immunocytochemistry with retrograde axonal tracing (Vera *et al.*, 1990; Jensen *et al.*, 1995). In rat, 5-HT terminals were also found in close apposition with cell bodies and proximal dendrites of preganglionic sympathetic neurons in the intermediolateral cell column or parasympathetic neurons in sacral nuclei that project to the major pelvic ganglion (Ranson *et al.*, 2006). Synaptic junctions were apparently not made by these 5-HT varicosities, whereas 5-HT varicosities in regions surrounding the labeled preganglionic cells could be seen in synaptic contact with unlabeled somata and dendritic processes of presumed interneurons. This was taken as an indication that in rat, the 5-HT modulation of pelvic preganglionic neurons primarily involved indirect connections via local interneurons. It should be noted that coexistence of 5-HT with substance P (SP) and with thyrotropin-releasing hormone (TRH) had been demonstrated by confocal immunofluorescence microscopy in fibers and terminals apposing identified preganglionic sympathetic neurons in the intermediolateral cell column of rat (Appel *et al.*, 1986, 1987). Consistent with these observations, the axon terminals visualized by Poulat and colleagues in the intermediolateral cell column of rat thoracic spinal cord after single immunocytochemical labeling for either 5-HT, SP or TRH displayed identical intrinsic and relational ultrastructural features and formed symmetrical and asymmetrical synaptic junctions with dendritic branches in identical proportion, as if belonging to the same neurons.

Since the first fluorescence histochemical demonstration of their existence in rat spinal cord (Fuxe, 1965), several studies have described the ultrastructural characteristics of 5-HT terminals around the large neurons in the *ventral horn* of rat, cat or monkey (Table 1). In all three species, 5-HT immuno-electron microscopic investigations showed that these 5-HT axon terminals synaptically contacted the soma or dendrites of motoneurons (Mizukawa and Takeuchi, 1982; Pecci-Saavedra *et al.*, 1983; Mizukawa *et al.*, 1986). Ridet (1994) subsequently demonstrated in serial thin sections that all 5-HT terminals in this part of the rat spinal cord were synaptic, accounting for the direct excitatory effects of 5-HT on ventral horn neurons described in micro-iontophoretic experiments (White and Neuman, 1980).

In the monkey ventral horn, a multiple fluorescence immunocytochemical study demonstrated that a subpopulation of 5-HT endings was also immunoreactive for substance P or TRH (Bowker, 1986b; Harkness and Brownfield, 1986), suggesting that the three molecules could be released from the same fibers to regulate the activity of motoneurons. An electron-microscopic study of axon terminals singly immunolabeled for 5-HT, SP and THR, carried out in the motor nucleus of spinal cord segments L7–S1 in the adult cat (Ulfhake *et al.*, 1987), described the immunoreactive boutons as preferentially distributed to the proximal dendritic domain of motoneurons, but rarely making a synaptic specialization in single thin sections, again suggesting that many of them might in fact contain the three transmitters. This was also in keeping with an earlier demonstration by Pelletier *et al.* (1981) that, in adjacent thin sections, both 5-HT and substance P immunoreactivity could be found in the same large granular (dense-core) vesicles of some anterior horn terminals.

Immuno-electron microscopic studies of the 5-HT innervation of intracellularly labeled neurons have also been carried out in the phrenic nucleus of cat at the C5 segment level (Holtman *et al.*, 1990; Pilowsky *et al.*, 1990). 5-HT immunoreactive boutons were then frequently found in close apposition or making symmetrical and asymmetrical synaptic contact with phrenic motoneurons, which were more common on dendrites than cell bodies. The ultrastructure of 5-HT immunoreactive terminals in the phrenic nucleus has also been described in some detail by Tai *et al.* (1997), who examined the plasticity of this innervation in the rat 30 days after hemisection at the C2 segment level. It was then shown that a majority of the 5-HT immunoreactive terminals formed axodendritic contacts, with either asymmetrical or symmetrical junctions onto retrogradely labeled phrenic motoneurons, but axosomatic synapses were also seen. After the hemisection, there was an increased number of 5-HT terminals in the phrenic nucleus, and the total number and length of asymmetrical and symmetrical junctional complexes per 5-HT terminal were significantly greater, as well as the proportion of 5-HT varicosities displaying multiple synapses.

Lastly, in a detailed immunocytochemical study of the 5-HT innervation of alpha-motoneurons in the lumbar spinal cord of adult cats, Alvarez *et al.* (1998) reported that the spatial distribution of 5-HT terminals observed in close contact with proximal and distal dendritic branches at the light-microscopic level essentially matched the distribution of the surface membrane area of these postsynaptic neurons. Moreover, in serial sections for electron microscopy, these authors observed many of these 5-HT varicosities making a synaptic junction, confirming that the bulbospinal 5-HT system does provide a significant direct synaptic input to spinal motoneurons innervating skeletal muscles.

Concluding remarks

Over the past 40 years, the various methods applicable to the ultrastructural identification of 5-HT neurons in mammalian CNS have yielded an image of these neurons quite distinct from the usual textbook schematization (Figure 3). Whereas some 5-HT axons have been demonstrated to be myelinated in projection pathways, particularly in the monkey (Azmitia and Gannon, 1983, 1986), widespread innervation by fine, unmyelinated axons whose multiple branches bear innumerable varicosities (presumptive release sites) appears to be the rule for most if not all central 5-HT neurons (Descarries *et al.*, 1990). It may also be assumed that such terminal arbors, bearing varicosities that often lack a junctional specialization and therefore lie free in the neuropil, are highly plastic, not only in a functional sense but also in their morphological configuration and relationship with their microenvironment (Descarries *et al.*, 1975; Beaudet and Descarries, 1978). This has already been considered as a determinant of the remarkable capacities of 5-HT neurons for regeneration and re-establishment of function, in contrast to those of more hard-wired CNS systems. The development of cell-imaging techniques for the examination of transmitter-identified axons in live brain should eventually provide definitive evidence of such a plasticity and capacities.

Wherever 5-HT endings make a synapse, double-immunolabeling methods and the combination of immunocytochemistry with other neurocytological methods already allow for identification of some of the cellular targets of 5-HT neurons in terms of transmitter(s) content, territory of projection, anatomical origin and/or functional properties. However, much remains to be learned about the role and mode of action of 5-HT released from non-synaptic axon varicosities, and which might then reach distant targets by diffusion in the extracellular space. In regions of high innervation density, the existence of a low ambient concentration of 5-HT must also be considered, which could permanently regulate the expression and or functional state of high-affinity receptor subtypes for 5-HT and for other transmitters located either on the 5-HT neurons themselves or on other neuronal types, as well as on glial cells and microvessels (for further discussion, see Descarries and Mechawar, 2000).

Current information on the ultrastructural localization of the 14 subtypes of 5-HT receptors thus far characterized in the mammalian CNS is still fragmentary (for review, see Descarries *et al.*, 2006). When the electron-microscopic visualization of one or the other of these receptors will be combined with that of 5-HT terminals, the morphological observations at the ultrastructural level will assume even greater functional significance. Further knowledge of the regional density and ultrastructural features of the 5-HT innervation and of the cellular and subcellular localization of 5-HT receptors should also help in achieving a more reliable interpretation of brain-imaging data soon to come regarding the regional distribution of the SERT and of the various 5-HT receptors in live animal or human brain, and health, disease or therapeutic states.

Acknowledgements

Personal studies referred to in the text were funded in large part by grants to Laurent Descarries from the former Medical Research Council of Canada and current Canadian Institutes of Health Research. Martin Parent was the recipient of a Herbert H. Jasper Fellowship from the Groupe de recherche sur le système nerveux central (GRSNC), and of a postdoctoral fellowship from the Fonds de la recherche en santé du Québec (FRSQ). The FRSQ also provided infrastructure support (GRSNC) to the authors' work. The authors also thank Claude Gauthier for his help in the preparation of Figure 3.

References

Aghajanian, G.K. and Bloom, F.E. (1967) Localization of tritiated serotonin in rat brain by electron microscopic radioautography. J. Pharmacol. Exp. Ther., 156: 23–30.

Aghajanian, G.K. and Gallager, D.W. (1975) Raphe origin of serotonergic nerves terminating in the cerebral ventricles. Brain Res., 88: 221–231.

Aghajanian, G.K. and McCall, R.B. (1980) Serotonergic synaptic input to facial motoneurons: localization by electron-microscopic radioautography. Neuroscience, 5: 2155–2161.

Aghajanian, B.K., Bloom, F.E. and Sheard, M.H. (1969) Electron microscopy of degeneration within the serotonin pathway of rat brain. Brain Res., 13: 266–273.

Aghajanian, G.K., Haigler, H.J. and Bloom, F.E. (1972) Lysergic acid diethylamine and serotonin direct action on serotonin containing neurons. Life Sci., 11: 615–622.

Ajika, K. and Ochi, J. (1975). Serotonin fibers in the rat hypothalamus, as revealed with pharmacological tools. In: Yamada, E. (Ed.), Proc. Xth Intl Congr. Anatomists, Science Council of Japan, Tokyo, p. 257.

Alvarez, F.J., Pearson, J.C., Harrington, D., Dewey, D., Torbeck, L. and Fyffe, R.E.W. (1998) Distribution of 5-hydroxytryptamine-immunoreactive boutons on α-motoneurons in the lumbar spinal cord of adult cats. J. Comp. Neurol., 393: 69–83.

Andén, N.-E., Dahlström, A., Fuxe, K., Larsson, K., Olson, L. and Ungerstedt, U. (1966) Ascending monoamine neurons to the telencephalon and diencepohalon. Acta Physiol. Scand., 67: 313–326.

Anderson, K.J., Holets, V.R., Mazur, P.C., Lasher, R.S. and Cotman, C.W. (1986) Immunocytochemical localization of serotonin in the rat dentate gyrus following raphe transplants. Brain Res., 369: 21–28.

André, D., Vuillon-Cacciuttolo, G. and Bosler, O. (1987) GABA nerve endings in the rat red nucleus: combined detection with serotonin terminals using dual immunocytochemistry. Neuroscience, 23: 1095–1102.

Appel, N.M., Wessendorf, M.W. and Elde, R.P. (1986) Coexistence of serotonin and substance P-like immunoreactivity in nerve fibers apposing identified sympathoadrenal preganglionic neurons in rat intermediolateral cell column. Neurosci. Lett., 65: 241–246.

Appel, N.M., Wessendorf, M.W. and Elde, R.P. (1987) Thyrotropin-releasing hormone in spinal cord: coexistence with serotonin and with substance P in fibers and terminals apposing identified preganglionic sympathetic neurons. Brain Res., 415: 137–143.

Arai, R., Karasawa, N., Kurokawa, K., Kanai, H., Horiike, K. and Ito, A. (2002) Differential subcellular location of mitochondria in rat serotonergic neurons depends of the presence and the absence of monoamine oxidase type B. Neuroscience, 114: 825–835.

Araneda, S., Gamrani, H., Font, C., Calas, A., Pujol, J.-F. and Bobillier, P. (1980) Retrograde axonal transport following injection of [^{3}H]-serotonin into the olfactory bulb. II. Radioautographic study. Brain Res., 196: 417–427.

Arce, E.A., Bennett-Clarke, C.A., Mooney, R.D. and Rhoades, R.W. (1992) Synaptic organization of the serotoninergic input to the superficial gray layer of the hamster's superior colliculus. Synapse, 11: 67–75.

Arce, E.A., Rhoades, R.W. and Mooney, R.D. (1995) Effects of neonatal and adult enucleation on the synaptic organization of the serotoninergic projection to the superficial gray layer of the hamster's superior colliculus. Brain Res. Dev. Brain Res., 90: 168–173.

Arluison, M. and De La Manche, I.S. (1980) High-resolution radioautographic styudy of the serotonin innervation of the rat corpus striatum after intraventricular administration of [^{3}H]5-hydroxytryptamine. Neuroscience, 5: 229–240.

Armstrong, D.M., Miller, R.J., Beaudet, A. and Pickel, V.M. (1984) Enkephalin-like immunoreactivity in rat area postrema: ultrastructural localization and coexistence with serotonin. Brain Res., 310: 269–278.

Audet, M.A., Descarries, L. and Doucet, G. (1989) Quantified regional and laminar distribution of the serotonin innervation in the anterior half of adult rat cerebral cortex. J. Chem. Neuroanat., 2: 29–44.

Austin, M.C., Whitehead, R.E., Edgar, C.L., Janovsky, J.E. and Lewis, D.A. (2002) Localized decrease in serotonin transporter-immunoreactive axons in the prefrontal cortex of depressed subjects committing suicide. Neuroscience, 114: 807–815.

Azmitia, E. and Gannon, P.J. (1983) The ultrastructural localization of serotonin immunoreactivity in myelinated and unmyelinated axons within the medial forebrain bundle of rat and monkey. J. Neurosci., 3: 2083–2090.

Azmitia, E. and Gannon, P.J. (1986) The primate serotonergic system: a review of human and animal studies and a report on *Macaca fascicularis*. In: Fahn, S., Marsden, C.D. and van Woert, M.H. (Eds.), Myoclonus, Vol. 43, Advances in Neurology. Raven Press, New York, NY, pp. 407–468.

Azmitia, E.C. and Marovitz, W.F. (1980) *In vitro* hippocampal uptake of tritiated serotonin (^{3}H-5-HT): a morphological, biochemical and pharmacological approach to specificity. J. Histochem. Cytochem., 28: 636–644.

Azmitia, E.C. and Segal, M.J. (1978) An radioautographic analysis of the differential ascending projections of the dorsal and median raphe nuclei in the rat. J. Comp. Neurol., 179: 641–667.

Bauman, M.D. and Amaral, D.G. (2005) The distribution of serotonergic fibers in the macaque monkey amygdala: an immunohistochemical study using antisera to 5-hydroxytryptamine. Neuroscience, 136: 193–203.

Baumgarten, H.G. and Lachenmayer, L. (1974) Indoleamine-containing nerve terminals in the rat median eminence. Z. Zellforsch., 147: 285–292.

Beaudet, A. (1982) High resolution radioautography of central 5-hydroxytryptamine (5-HT) neurons. J. Histochem. Cytochem., 30: 765–768.

Beaudet, A. and Descarries, L. (1976) Quantitative data on serotonin nerve terminals in adult rat neocortex. Brain Res., 111: 301–309.

Beaudet, A. and Descarries, L. (1978) The monoamine innervation of rat cerebral cortex: synaptic and non-synaptic relationships. Neuroscience, 3: 851–860.

Beaudet, A. and Descarries, L. (1979) Radioautographic characterization of a serotonin-accumulating nerve cell group in adult rat hypothalamus. Brain Res., 160: 231–243.

Beaudet, A. and Descarries, L. (1981) The fine structure of central serotonin neurons. J. Physiol. (Paris), 77: 193–203.

Beaudet, A. and Descarries, L. (1984) Fine structure of monoamine terminals in cerebral cortex. Neurol. Neurobiol., 10: 77–93.

Beaudet, A. and Descarries, L. (1987) Ultrastructural identification of serotonin neurons. In: Steinbusch, H.W.M. (Ed.), Methods in Neuroscience, Monoaminergic Neurons: Light Microscopy and Ultrastructure. John Wiley & Sons, Chichester, pp. 265–313.

Beaudet, A. and Sotelo, C. (1981) Synaptic remodeling of serotonin axon terminals in rat agranular cerebellum. Brain Res., 206: 305–329.

Beitz, A.J., Clements, J.R., Mullett, M.A. and Ecklund, L.J. (1986) Differential origin of brainstem serotoninergic projections to the midbrain periaqueductal gray and superior colliculus of the rat. J. Comp. Neurol., 250: 498–509.
Bishop, G.A. and Ho, R.H. (1986) Cell bodies of origin of serotonin-immunoreactive afferents to the inferior olivary complex of the rat. Brain Res., 399: 369–373.
Björklund, A. and Lindvall, O. (1975) Dopamine in dendrites of substantia nigra neurons: suggestions for a role in dendritic terminals. Brain Res., 83: 531–537.
Blomqvist, A. and Broman, J. (1993) Serotoninergic innervation of the dorsal column nuclei and its relation to cytoarchitectonic subdivisions: an immunohistochemical study in cats and monkeys (*Aotus trivirgatus*). J. Comp. Neurol., 327: 584–596.
Bloom, F.E., Hoffer, B.J., Siggins, G.R., Barker, J.L. and Nicoll, R.A. (1972) Effects of serotonin on central neurons: microiontophoretic administration. Fed. Proc., 31: 97–106.
Bobillier, P., Petitjean, F., Salvert, D., Ligier, M. and Seguin, S. (1975) Differential projections of the raphe nucleus, raphe dorsalis and nucleus raphe centralis as revealed by radioautography. Brain Res., 85: 205–210.
Bobillier, P., Seguin, S., Petitjean, F., Salvert, D., Touret, M. and Jouvet, M. (1976) The raphe nuclei of the cat brain stem: a topographical atlas of their efferent projections as revealed by radioautography. Brain Res., 113: 449–486.
Bosler, O. (1978) Radioautographic identification of serotonin axon terminals in the rat organum vasculosum laminae terminalis. Brain Res., 150: 177–181.
Bosler, O. (1980) Radioautographie des systèmes monoaminergiques centraux. Nouvelle possibilité d'investigation des neurones dopaminergiques. *Les Entretiens du Carla*, pp. 31–45. Centre de recherche Pierre Fabre, Castres.
Bosler, O. and Beaudet, A. (1985a) Relations ultrastructurales entre systèmes monoaminergiques et peptidergiques dans l'hypothalamus. Ann. Endocrinol. (Paris), 46: 19–26.
Bosler, O. and Beaudet, A. (1985b) VIP neurons as prime synaptic targets for serotonin afferents in rat suprachiasmatic nucleus: a combined radioautographic and immunocytochemical study. J. Neurocytol., 14: 749–763.
Bosler, O. and Descarries, L. (1988) Monoamine innervation of the organum vasculosum laminae terminalis (OVLT): a high resolution radioautographic study in the rat. J. Comp. Neurol., 272: 545–561.
Bosler, O., Nieoullon, A., Onteniente, B. and Dusticier, N. (1983) *In vitro* radioautographic study of the monoaminergic innervation of cat red nucleus. Identification of serotoninergic terminals. Brain Res., 259: 288–292.
Bosler, O., Beaudet, A. and Tappaz, M. (1984a) Morphological evidence for cellular interactions between serotoninergic terminals and VIP and GABAergic elements in rat suprachiasmatic nucleus. Excerp. Med. 7th Intl Congr. Endocrinol., 652: 486.
Bosler, O., Joh, T.H. and Beaudet, A. (1984b) Ultrastructural relationships between serotonin and dopamine neurons in the rat arcuate nucleus and medial zona incerta: a combined radioautographic and immunocytochemical study. Neurosci. Lett., 48: 279–285.
Bosler, O., Beaudet, A. and Pickel, V.M. (1986) Characterization of chemically defined neurons and their cellular relationships by combined immunocytochemistry and radioautographic localization of transmitter uptake sites. J. Electr. Microsc. Techn., 4: 21–39.
Bouchaud, C. (1979) Evidence for a multiple innervation of subcommissural ependymocytes in the rat. Neurosci. Lett., 12: 253–258.
Bouchaud, C. and Arluison, M. (1977) Serotoninergic innervation of ependymal cells in the rat subcommissural organ. A fluorescence, electron microscopic and radioautographic study. Biol. Cell., 30: 65–72.
Bouchaud, C. and Bosler, O. (1986) The circumventricular organs of the mammalian brain with special reference to monoaminergic innervation. Intl. Rev. Cytol., 105: 283–327.
Boulaïch, S., Daszuta, A., Geffard, M. and Bosler, O. (1994) Synaptic connectivity of serotonin graft efferents in the suprachiasmatic and supraoptic nuclei of the hypothalamus. Exp. Brain Res., 101: 353–364.
Boulenguez, P., Abdelkefi, J., Pinard, R., Christolomme, A. and Segu, L.J. (1993) Effects of retinal deafferentation on serotonin receptor types in the superficial gray layer of the superior colliculus of the rat. Chem. Neuroanat., 6: 167–175.
Boulenguez, P., Pinard, R. and Segu, L. (1996) Subcellular localization of 5-HT1B binding sites in the stratum griseum superficiale of the rat superior colliculus: an electron microscopic quantitative autoradiographic study. Synapse, 24: 203–212.
Bowker, R.M. (1986a) Intrinsic 5-HT-immunoreactive neurons in the spinal cord of the fetal non-human primate. Develop. Brain Res., 28: 137–143.
Bowker, R.M. (1986b) Serotonergic and peptidergic inputs to the primate ventral spinal cord as visualized with multiple chromagens on the same tissue section. Brain Res., 375: 345–350.
Bowker, R.M., Westlund, K.N. and Coulter, J.D. (1981) Origins of serotonergic projections to the spinal cord in rat: an immunocytochemical-retrograde transport study. Brain Res., 226: 187–199.
Bowker, R.M., Reddy, V.K., Fung, S.J., Chan, J.Y.H. and Barnes, C.D. (1987) Serotonergic and non-serotonergic raphe neurons projecting to the feline lumbar and cervical spinal cord: a quantitative horseradish peroxidase-immunocytochemical study. Neurosci. Lett., 75: 31–37.
Brusco, A., Peressini, S. and Pecci Saavedra, J. (1983) Serotonin-like immunoreactivity and anti-5-hydroxytryptamine (5-HT) antiobodies: ultrastructural application in the central nervous system. J. Histochem. Cytochem., 31: 524–530.
Bunin, M.A. and Wightman, R.M. (1998) Quantitative evaluation of 5-hydroxytryptamine (serotonin) release and uptake: an investigation of extrasynaptic transmission. J. Neurosci., 18: 4854–4860.
Calas, A. (1977) Radioautographic studies of aminergic neurons terminating in the median eminence. In: Costa, E. and Gessa, G.L. (Eds.), Advances in Biochemical Pharmacology. Raven Press, New York, NY, pp. 79–88.
Calas, A. and Ségu, L. (1976) Radioautographic localization and identification of monoaminergic neurons in the central nervous system. J. Microsc. Biol. Cell., 27: 249–252.
Calas, A., Alonso, G., Arnauld, E. and Vincent, J.D. (1974) Demonstration of indolaminergic fibres in the median eminence of the duck, rat and monkey. Nature, 250: 241–243.
Calas, A., Besson, M.J., Gauchy, C., Alonso, G., Glowinski, J. and Cheramy, A. (1976) Radioautographic study of *in vivo* incorporation of [^{3}H]monoamines in the rat caudate nucleus: identification of serotoninergic fibers. Brain Res., 118: 1–13.
Calas, A., Bosler, O., Arluison, M. and Bouchaud, C. (1978) Serotonin as a neurohormone in circumventricular organs

and supraependymal fibers. In: Scott, D.E., Kozlowski, G.P., Weindl, A. (Eds.), Brain–Endocrine Interactions III. Neural Hormones and Reproduction. 3rd Intl Symp. Würzburg. S. Karger A.G., Basel, pp. 238–250.

Calas, A., Dupuy, J.-J., Gamrani, H., Gonella, J., Mourre, C., Condamin, M., Pellissier, J.-F. and Van den Bosch, P. (1981) Radioautographic identification of serotonin cells. Adv. Exp. Med. Biol., 133: 51–66.

Campbell, M.J., Lewis, D.A., Foote, S.L. and Morrison, J.H. (1987) Distribution of choline acetyltransferase-, serotonin, dopamine-beta-hydroxylase-, and tyrosine hydroxylase-immunoreactive fibers in monkey primary auditory cortex. J. Comp. Neurol., 261: 209–220.

Cases, O., Lebrand, C., Giros, B., Vitalis, T., De Maeyer, E., Caron, M.G., Price, D.J., Gaspar, P. and Seif, I. (1998) Plasma membrane transporters of serotonin, dopamine, and norepinephrine mediate serotonin accumulation in atypical locations in the developing brain of monoamine oxidase A knock-outs. J. Neurosci., 18: 6914–6927.

Cauli, B., Tong, X.-K., Rancillac, A., Serluca, N., Lambolez, B., Rossier, J. and Hamel, E. (2004) Cortical GABA interneurons in neurovascular coupling: relays for subcortical vasoactive pathways. J. Neurosci., 24: 8940–8949.

Chan-Palay, V. (1975) Fine structure of labelled axons in the cerebellar cortex and nuclei of rodents and primates after intraventricular infusions with tritiated serotonin. Anat. Embryol., 148: 235–265.

Chan-Palay, V. (1976) Serotonin axons in the supra- and subependymal plexuses and in the leptomeninges: their roles in local alterations of cerebrospinal fluid and vasomotor activity. Brain Res., 102: 103–130.

Chan-Palay, V. (1977) Cerebellar Dentate Nucleus: Organization, Cytology and Transmitters. Springer-Verlag, Berlin.

Chan-Palay, V. (1978a) The paratrigeminal nucleus. II. Identification and inter-relations of catecholamine axons, indoleamine axons, and substance P immunoreactive cells in the neuropil. J. Neurocytol., 7: 419–442.

Chan-Palay, V. (1978b) Morphological correlates for transmitter synthesis, transport, release, uptake and catabolism: a study of serotonin neurons in the nucleus paragigantocellularis lateralis. In: Fonnum, F. (Ed.), Amino Acids as Chemical Transmitters. Plenum, New York, NY, pp. 1–30.

Chan-Palay, V. (1982) Serotonin neurons and their axons in the raphe dorsalis of rat and rhesus monkey: demonstration by high resolution radioautography with ^{3}H-serotonin. Neurol. Neurobiol., 1: 357–386.

Chazal, G. and Ma, W. (1989) A ultrastructural analysis of serotoninergic neurons in the nucleus raphe magnus of the rat. Neuroscience, 33: 301–310.

Chazal, G. and O'Hara, P.T. (1986) Vesicle-containing dendrites in the nucleus raphe dorsalis of the cat. A serial electron microscopic analysis. J. Neurocytol., 15: 777–787.

Chazal, G. and Ralston, H.J. III (1987) Serotonin-containing structures in the nucleus raphe dorsalis of the cat: an ultrastructural analysis of dendrites, presynaptic dendrites, and axon terminals. J. Comp. Neurol., 259: 317–329.

Cheramy, A., Leviel, V. and Glowinski, J. (1981) Dendritic release of dopamine in the substantia nigra. Nature, 289: 537–542.

Chiba, T. and Masuko, S. (1987) Synaptic structure of the monoamine and peptide nerve terminals in the intermediolateral nucleus of the guinea pig thoracic spinal cord. J. Comp. Neurol., 262: 242–255.

Clements, J.R., Beitz, A.J., Fletcher, T.F. and Mullett, M.A. (1985) Immunocytochemical localization of serotonin in the rat periaqueductal gray: a quantitative light and electron microscopic study. J. Comp. Neurol., 236: 60–70.

Cohen, Z., Ehret, M., Maître, M. and Hamel, E. (1995) Ultrastructural analysis of tryptophane hydroxylase immunoreactive nerve terminals in the rat cerebral cortex and hippocampus: their associations with local blood vessels. Neuroscience, 66: 555–569.

Commons, K.G. and Valentino, R.J. (2002) Cellular basis for the effects of substance P in the periaqueductal gray and raphe dorsalis. J. Comp. Neurol., 447: 82–97.

Commons, K.G., Beck, S.G. and Bey, V.W. (2005) Two populations of glutamatergic axons in the rat raphe dorsalis defined by the vesicular glutamate transporters 1 and 2. Eur. J. Neurosci., 21: 1577–1586.

Conrad, L.C.A., Leonard, C.M. and Pfaff, D.W. (1974) Connections of the median and dorsal raphe nuclei in the rat: an radioautographic and degeneration study. J. Comp. Neurol., 156: 179–206.

Copray, J.C., Liem, R.S., Ter Horst, G.J. and van Willigen, J.D. (1991) Origin, distribution and morphology of serotonergic afferents to the mesencephalic trigeminal nucleus of the rat. Neurosci. Lett., 121: 97–101.

Corvaja, N., Doucet, G. and Bolam, J.P. (1993) Ultrastructure and synaptic targets of the raphe-nigral projection in the rat. Neuroscience, 55: 417–427.

Crissman, R.S., Arce, E.A., Bennett-Clarke, C.A., Mooney, R.D. and Rhoades, R.W. (1993) Reduction in the percentage of serotoninergic axons making synapses during the development of the superficial layers of the hamster's superior colliculus. Brain Res. Dev. Brain Res., 75: 131–135.

Cropper, E.C., Eisenman, J.S. and Azmitia, E.C. (1984) An immunocytochemical study of the serotonergic innervation of the thalamus of the rat. J. Comp. Neurol., 224: 38–50.

Dahlström, A. and Fuxe, K. (1964) Evidence for the existence of monoamine neurons in the central nervous system. IV. Demonstration of monoamines in the cell body of brain stem neurons. Acta Physiol. Scand., 62(Suppl. 232): 1–55.

Dahlström, A., Fuxe, K., Olson, L. and Ungerstedt, U. (1965) On the distribution and possible function of monoamine nerve terminals in the olfactory bulb of the rabbit. Life Sci., 4: 2071–2074.

Daszuta, A., Chazal, G., Garcia, S., Oleskevich, S. and Descarries, L. (1991) Ultrastructural features of serotonin neurons grafted to adult rat hippocampus. An immunocytochemical analysis of their cell bodies and axon terminals. Neuroscience, 42: 793–811.

Daval, G., Vergé, D., Basbaum, A.I., Bourgoin, S. and Hamon, M. (1987) Radioautographic evidence of serotonin1 binding sites on primary afferent fibres in the dorsal horn of the rat spinal cord. Neurosci. Lett., 83: 71–76.

Davis, R. and Vaughan, P.C. (1969) Pharmacological properties of feline red nucleus. Int. J. Neuropharmacol., 8: 475–488.

De Felipe, J. and Jones, E.G. (1988) A light and electron microscopic study of serotonin-immunoreactive fibers and terminals in the monkey sensori-motor cortex. Exp. Brain Res., 71: 171–182.

De Felipe, J., Hendry, S.H.C., Hashikawa, T. and Jones, E.G. (1991) Synaptic relationships of serotonin-immunoreactive terminal baskets of GABA neurons in the cat auditory cortex. Cereb. Cortex, 1: 117–133.

De La Manche, I.S., Arluison, M. and Bouchaud, C. (1981) Experimental ultrastructural modifications of the content

of serotoninergic supra-ependymal nerve fibres in the rat. J. Physiol. (Paris), 77: 225–231.
De Lima, A.D. and Singer, W. (1987a) The serotoninergic fibers in the dorsal geniculate nucleus of the cat: distribution and synaptic connections demonstrated with immunocytochemistry. J. Comp. Neurol., 258: 339–351.
De Lima, A.D. and Singer, W. (1987b) The brainstem projection to the lateral geniculate nucleus in the cat: distribution and synaptic connections demonstrated with immunocytochemistry. J. Comp. Neurol., 259: 92–127.
De Lima, A.D., Bloom, F.E. and Morrison, J.H. (1988) Synaptic organization of serotonin–immunoreactive fibers in primary visual cortex of the Macaque monkey. J. Comp. Neurol., 274: 280–294.
Descarries, L. and Beaudet, A. (1978) The serotonin innervation of adult rat hypothalamus. In: Vincent, J.D. and Kordon, C. (Eds.), Biologie Cellulaire des Processus Neurosécrétoires, Coll. Internat. CNRS, Ed. CNRS, Paris, pp. 135–153.
Descarries, L. and Beaudet, A. (1983) The use of radioautography for investigating transmitter-specific neurons. In: Björklund, A. and Hökfelt, T. (Eds.), Handbook of Chemical Neuroanatomy, Vol. I, Methods. Elsevier/North-Holland, Amsterdam, pp. 286–364.
Descarries, L. and Léger, L. (1978) Serotonin nerve terminals in the locus coeruleus of the adult rat. In: Garattini, S., Pujol, J.F. and Samanin, R. (Eds.), Interactions Between Putative Neurotransmitters. Raven Press, New York, NY, pp. 355–367.
Descarries, L. and Mechawar, N. (2000) Ultrastructural evidence for diffuse transmission by monoamine and acetylcholine neurons of the central nervous system. Prog. Brain Res., 125: 27–47.
Descarries, L. and Mechawar, N. (2008) Structural organization of monoamine and acetylcholine neuron systems in the rat CNS. In: Lajtha, A. (Ed.) and Vizi, E.S. (Vol. ed.), Handbook of Neurochemistry and Molecular Neurobiology, 3rd edn., Neurotransmitter Systems, Springer, New York, NY, pp. 1–20.
Descarries, L. and Umbriaco, D. (1995) Ultrastructural basis of monoamine and acetylcholine function in CNS. Seminars Neurosci., 7: 309–318.
Descarries, L., Beaudet, A. and Watkins, K.C. (1975) Serotonin nerve terminals in adult rat neocortex. Brain Res., 100: 563–588.
Descarries, L., Watkins, K.C., Garcia, S. and Beaudet, A. (1982) The serotonin neurons in nucleus raphe dorsalis of adult rat: a light and electron microscope radioautographic study. J. Comp. Neurol., 207: 239–254.
Descarries, L., Berthelet, F., Garcia, S. and Beaudet, A. (1986) Dopaminergic projection from nucleus raphe dorsalis to neostriatum in the rat. J. Comp. Neurol., 249: 511–520.
Descarries, L., Audet, M.A., Doucet, G., Garcia, S., Oleskevich, S., Séguéla, P., Soghomonian, J.-J. and Watkins, K.C. (1990) Morphology of central serotonin neurons. Ann. NY Acad. Sci., 600: 81–92.
Descarries, L., Séguéla, P. and Watkins, K.C. (1991) Nonjunctional relationships of monoamine axon terminals in the cerebral cortex of adult rat. Adv. Neurosci., 1: 53–62.
Descarries, L., Soghomonian, J.-J., Garcia, S., Doucet, G. and Bruno, J.P. (1992) Ultrastructural analysis of the serotonin hyperinnervation in adult rat neostriatum following neonatal dopamine denervation with 6-hydroxydopamine. Brain Res., 569: 1–13.
Descarries, L., Cornea-Hébert, V. and Riad, M. (2006) Cellular and subcellular localization of serotonin receptors in the central nervous system. In: Roth, B.L. (Ed.), The Serotonin Receptors: From Molecular Pharmacology to Human Therapeutics. Humana Press, Totowa, NJ, pp. 277–317.
Dinopoulos, A. and Dori, I. (1995) The development of the serotonergic fiber network of the lateral ventricles of the rat brain: a light and electron microscopic immunocytochemical analysis. Exp. Neurol., 133: 73–84.
Dinopoulos, A., Dori, I. and Parnavelas, J.G. (1993) Serotonergic innervation of the mature and developing lateral septum of the rat: a light and electron microscopic immunocytochemical analysis. Neuroscience, 55: 209–222.
Dinopoulos, A., Dori, I. and Parnavelas, J.G. (1995) Serotonergic innervation of the lateral geniculate nucleus of the rat during postnatal development: a light and electron microscopic immunocytochemical analysis. J. Comp. Neurol., 363: 532–544.
Dinopoulos, A., Dori, I. and Parnavelas, J.G. (1997) The serotonin innervation of the basal forebrain shows a transient phase during development. Brain Res. Dev. Brain Res., 99: 38–52.
Dori, I.E., Dinopoulos, A. and Parnavelas, J.G. (1996) Regional differences in the ontogeny of the serotonergic preojection to the cerebral cortex. Exp. Neurol., 138: 1–14.
Dori, I.E., Dinopoulos, A., Blue, M.E. and Parnavelas, J.G. (1998) The development of the synaptic organization of the serotonergic system differs in brain areas with different functions. Exp. Neurol., 154: 113–125.
Doucet, G., Descarries, L., Audet, M.A., Garcia, S. and Berger, B. (1988) Radioautographic method for quantifying regional monoamine innervations in the rat brain. Application to the cerebral cortex. Brain Res., 441: 233–259.
Elde, R., Maley, B., Holets, V. and Seybold, V. (1982) Peptidergic and serotinergic neurons in neuroendocrine and central autonomic circuits. Neurol. Neurobiol., 1: 331–340.
Fallon, J.H. and Loughlin, S.E. (1982) Monoamine innervation of the forebrain. Collateralization. Brain Res. Bull., 9: 295–307.
Fibiger, H.C. and Miller, J.J. (1977) An anatomical and electrophysiological investigation of the serotonergic projections from the raphe dorsalis to the substantia nigra in the rat. Neuroscience, 2: 975–987.
Frankfurt, M. and Beaudet, A. (1987) Ultrastructural organization of regenerated serotonin axons in the dorsomedial hypothalamus of the adult rat. J. Neurocytol., 16: 799–809.
Frankfurt, M. and Beaudet, A. (1988) Reinnervation of dopamine neurons by regenerating serotonin axons in the rat medial zona incerta. Exp. Brain Res., 72: 473–480.
Frankfurt, M., Lauder, J.M. and Azmitia, E.C. (1981) The immunocytochemical localization of serotonergic neurons in the rat hypothalamus. Neurosci. Lett., 24: 227–232.
Freund, T.F., Gulyás, A.L., Acsády, L., Görcs, T. and Tóth, K. (1990) Serotonergic control of the hippocampus via local inhibitory interneurons. Proc. Natl. Acad. Sci. USA, 87: 8501–8505.
Fuxe, K. (1965) Evidence for the existence of monoamine neurons in the central nervous system. IV. Distribution of monoamine nerve terminals in the central nervous system. Acta Physiol. Scand., 65(Suppl. 247): 37–85.
Fuxe, K. and Jonnson, G. (1967) A modification of the histochemical fluorescence method for the improved localization of 5-hydroxytryptamine. Histochemie, 11: 161–166.
Fuxe, K. and Owman, C. (1965) Cellular localization of monoamines in the area postrema of certain mammals. J. Comp. Neurol., 125: 337–353.

Fuxe, K. and Ungerstedt, U. (1968) Histochemical studies on the distribution of catecholamines and 5-hydroxytryptamine after intraventricular injections. Histochemie, 13: 16–28.

Gamrani, H., Belin, M.F., Aguera, M., Calas, A. and Pujol, J.F. (1981) Radioautographic evidence for an innervation of the subcommissural organ by GABA-containing nerve fibres. J. Neurocytol., 10: 411–424.

Gamrani, H., Harandi, M., Bélin, M.F., Dubois, M.P. and Calas, A. (1984) Direct electron microscopic evidence for the coexistence of GABA uptake and endogenous serotonin in the same rat central neurons by coupled radioautographic and immunocytochemical procedures. Neurosci. Lett., 48: 25–30.

Glazer, E. and Basbaum, A.I. (1984) Axons which take up [^{3}H]serotonin are presynaptic to enkephalin immunoreactive neurons in cat dorsal horn. Brain Res., 298: 386–391.

Guiard, B.P., El Mansari, M., Merali, Z. and Blier, P. (2008) Functional interactions between dopamine, serotonin and norepinephrine neurons: an in-vivo electrophysiological study in rats with monoamine lesions. Intl. J. Neuropsychopharmacol., 11: 625–639.

Halász, N., Ljungdahl, A., Hökfelt, T., Johansson, O., Goldstein, M., Park, D. and Biberfeld, P. (1977) Transmitter histochemistry of the rat olfactory bulb. I. Immunohistochemical localization of monoamine synthesizing enzymes. Support for intrabulbar, periglomerular dopamine neurons. Brain Res., 126: 455–474.

Halász, N., Ljungdahl, A. and Hökfelt, T. (1978) Transmitter histochemistry of the rat olfactory bulb. II. Fluorescence histochemical, radioautographic and electron microscopic localization of monoamines. Brain Res., 154: 253–271.

Hamon, M., Gallissot, M.C., Menard, F., Gozlan, H., Bourgoin, S. and Vergé, D. (1989) 5-HT$_3$ receptor binding sites are on capsaicin-sensitive fibres in the rat spinal cord. Eur. J. Pharmacol., 164: 315–322.

Hansson, S.R., Mezey, E. and Hoffman, B.J. (1998) Serotonin transporter messenger RNA in the developing rat brain: early expression in serotonergic neurons and transient expression in non-serotonergic neurons. Neuroscience, 83: 1185–1201.

Harandi, M., Didier, M., Aguera, M., Calas, A. and Belin, M.F. (1986) GABA and serotonin (5-HT) pattern in the supraependymal fibers of the rat epithalamus: combined radioautographic and immunocytochemical studies. Effect of 5-HT content on [^{3}H]GABA accumulation. Brain Res., 370: 241–249.

Harkness, D.H. and Brownfield, M.S. (1986) A thyrotropin-releasing hormone-containing system in the rat dorsal horn separate from serotonin. Brain Res., 384: 323–333.

Hervé, D., Pickel, V.M., Joh, T.H. and Beaudet, A. (1987) Serotonin axon terminals in the ventral tegmental area of the rat: fine structure and synaptic input to dopaminergic neurons. Brain Res., 435: 71–83.

Héry, F., Faudon, M. and Ternaux, J.P. (1982) *In vivo* release of serotonin in two raphe nuclei (raphe dorsalis and magnus) of the cat. Brain Res. Bull., 8: 123–129.

Hoffert, M.J., Miletic, V., Ruda, M.A. and Dubner, R. (1983) Immunocytochemical identification of serotonin axonal contacts on characterized neurons in laminae I and II of the cat dorsal horn. Brain Res., 267: 361–364.

Hökfelt, T. (1967) Electron microscopic studies on brain slices from regions rich in catecholamine nerve terminals. Acta Physiol. Scand., 69: 119–120.

Hökfelt, T. and Fuxe, K. (1969) Cerebellar monoamine nerve terminals, a new type of afferent fibers in the cortex cerebelli. Exp. Brain Res., 9: 63–72.

Holtman, J.R. Jr., Vascik, D.S. and Maley, B.E. (1990) Ultrastructural evidence for serotonin-immunoreactive terminals contacting phrenic motoneurons in the cat. Exp. Neurol., 109: 269–272.

Honda, T. and Semba, K. (1994) Serotonergic input to cholinergic neurons in the rat mesopontine tegmentum. Brain Res., 647: 299–306.

Huang, J. and Pickel, V.M. (2002) Serotonin transporters (SERTs) within the nucleus of the solitary tract: subcellular distribution and relation to 5-HT2A receptors. J. Neurocytol., 31: 667–679.

Huang, J., Spier, A.D. and Pickel, V.M. (2004) 5-HT3 receptor subunits in the rat medial nucleus of the solitary tract: subcellular distribution and relation to the serotonin transporter. Brain Res., 1028: 156–169.

Hylden, J.L.K., Hayashi, H., Ruda, M.A. and Dubner, R. (1986) Serotonin innervation of physiologically identified lamina I projection neurons. Brain Res., 370: 401–404.

Jacobs, B.L., Wise, W.L. and Taylor, K.M. (1974) Differential behavioral and neurochemical effects following lesions of the dorsal or median raphe nuclei in rats. Brain Res., 79: 353–361.

Jacobs, B.L., Foote, S.L., Rajaofetra, N. et al. (1978) Differential projections of neurons within the raphe dorsalis of the rat: a horseradish peroxidase (HRP) study. Brain Res., 147: 149–153.

Jankowska, E., Maxwell, D.J., Dolk, S. and Dahlström, A. (1997) A confocal and electron microscopic study of contacts between 5-HT fibres and feline dorsal horn interneurons in pathways from muscle afferents. J. Comp. Neurol., 387: 430–438.

Jennes, L., Beckman, W.C., Stumpf, W.E. and Grzanna, R. (1982) Anatomical relationships of serotoninergic and noradrenalinergic projections with the GnRH system in septum and hypothalamus. Exp. Brain Res., 46: 331–338.

Jensen, I., Llewellyn-Smith, I.J., Pilowsky, P., Minson, J.B. and Chalmers, J. (1995) Serotonin inputs to rabbit sympathetic preganglionic neurons projecting to the superior cervical ganglion or adrenal medulla. J. Comp. Neurol., 353: 427–438.

Kalén, P., Karlson, M. and Wiklund, L. (1985) Possible excitatory amino acid afferents to nucleus raphe dorsalis of the rat investigated with retrograde wheat germ agglutinin and D-[^{3}H]aspartate tracing. Brain Res., 360: 285–297.

Kapadia, S.E., De Lanerolle, N.C. and LaMotte, C.C. (1985) Immunocytochemical and electron microscopic study of serotonin neuronal organization in the raphe dorsalis of the monkey. Neuroscience, 15: 729–746.

Kawano, H., Decker, K. and Reuss, S. (1996) Is there a direct retina–raphe–suprachiasmatic nucleus pathway in the rat?. Neurosci. Lett., 212: 143–146.

Kazakov, V.N., Kravtsov, P.Ya., Krakhotkina, E.D. and Maisky, V.A. (1993) Sources of cortical, hypothalamic and spinal serotonergic projections: topical organization within the nucleus raphe dorsalis. Neuroscience, 56: 157–164.

Ke, M., Mooney, R.D. and Rhoades, R.W. (1999) Increased serotonin in the developing superior colliculus affects receptive-field size of retinotectal afferents but not that of postsynaptic neurons. Vis. Neurosci., 16: 121–130.

Kellar, K.J., Brown, P.A., Madrid, J., Bernstein, M., Vernikos Danellis, J. and Melher, W.R. (1977) Origins of serotonin innervation of forebrain structure. Exp. Neurol., 56: 52–62.

King, J.S., Ho, R.H. and Burry, R.W. (1984) The distribution and synaptic organization of serotoninergic elements in the inferior olivary complex of the opossum. J. Comp. Neurol., 227: 357–368.

Kiss, J. and Halász, B. (1985) Demonstration of serotoninergic axons terminating on luteinizing hormone-releasing hormone neurons in the preoptic area of the rat using a combination of immunocytochemistry and high resolution radioautography. Neuroscience, 14: 69–78.

Kiss, J. and Halász, B. (1986) Synaptic connections between serotoninergic axon terminals and tyrosine hydroxylase-immunoreactive neurons in the arcuate nucleus of the rat hypothalamus. A combination of electron microscopic radioautography and immunocytochemistry. Brain Res., 364: 284–294.

Kiss, J., Léránth, C. and Halász, B. (1984) Serotoninergic endings on VIP neurons in the suprachiasmatic nucleus and on ACTH-neurons in the arcuate nucleus of the rat hypothalamus. A combination of high resolution radioautography and electron microscopic immunocytochemistry. Neurosci. Lett., 44: 119–124.

Köhler, C. (1984) The distribution of serotonin binding sites in the hippocampal region of the rat brain. An autoradiographic study. Neuroscience, 13: 667–680.

Kuhar, M.J., Aghajanian, B.K. and Roth, R.H. (1972) Tryptophan hydroxylase activity and synaptosomal uptake of serotonin in discrete brain regions after midbrain raphe lesions: correlations with serotonin levels and histochemical fluorescence. Brain Res., 44: 165–176.

LaMotte, C.C. and De Lanerolle, N.C. (1983) Ultrastructure of chemically defined neuron systems in the dorsal horn of the monkey. III. Serotonin immunoreactivity. Brain Res., 274: 65–77.

LaMotte, C.C., Johns, D.R. and de Lanerolle, N.C. (1982) Immunohistochemical evidence of indolamine neurons in monkey spinal cord. J. Comp. Neurol., 206: 359–370.

Lavoie, B. and Parent, A. (1990) Immunohistochemical study of the serotoninergic innervation of the basal ganglia in the squirrel monkey. J. Comp. Neurol., 299: 1–16.

Lavoie, B. and Parent, A. (1991) Serotoninergic innervation of the thalamus in the primate: an immunohistochemical study. J. Comp. Neurol., 312: 1–18.

Lazarov, N.E. and Chouchkov, C.N. (1995) Serotonin-containing projections to the mesencephalic trigeminal nucleus of the cat. Anat. Rec., 241: 136–142.

Lebrand, C., Cases, O., Adelbrecht, C., Doye, A., Alvarez, C., El Mestikawy, S., Seif, I. and Gaspar, P. (1996) Transient uptake and storage of serotonin in developing thalamic neurons. Neuron, 17: 823–835.

Lebrand, C., Cases, O., Wehrlé, R., Blakely, R.D., Edwards, R.H. and Gaspar, P. (1998) Transient developmental expression of monoamine transporters in the rodent forebrain. J. Comp. Neurol., 401: 506–524.

Léger, L. and Descarries, L. (1978) Serotonin nerve terminals in the locus coeruleus of adult rat: a radioautographic study. Brain Res., 145: 1–13.

Léger, L., Degueurce, A.M. and Pujol, J.F. (1980) Origin of the serotoninergic innervation of the rat locus coeruleus. C. R. Acad. Sci. Paris, série D., 290: 807–810.

Léger, L., Degueurce, A., Lundberg, J.J., Pujol, J.F. and Møllgård, K. (1983) Origin and influence of the serotoninergic innervation of the subcommissural organ in the rat. Neuroscience, 10: 411–423.

Léránth, C. and Vertes, R.P. (1999) Median raphe serotonergic innervation of medial septum/diagonal band of Broca (MSDB) parvalbumin-containing neurons: possible involvement of the MSDB in the desynchronization of the hippocampal EEG. J. Comp. Neurol., 410: 586–598.

Li, J.-L., Kaneko, T., Nomura, S., Li, Y.-Q. and Mizuno, N. (1997) Association of serotonin-like immunoreactive axons with nociceptive projection neurons in the caudal spinal trigeminal nucleus of the rat. J. Comp. Neurol., 384: 127–141.

Licata, F., Li Volsi, G., Ciranna, L., Maugeri, G. and Santangelo, F. (1998) 5-Hydroxytryptamine modifies neuronal responses to glutamate in the red nucleus of the rat. Exp. Brain Res., 118: 61–70.

Lidov, H.G.W., Grzanna, R. and Molliver, M.E. (1980) The serotonin innervation of the cerebral cortex in the rat: an immunohistochemical analysis. Neuroscience, 5: 207–227.

Liem, R.S.B. and Copray, J.C.V.M. (1996) Immunogold localization of serotonin within synaptic terminals in the rat mesencephalic trigeminal nucleus. Acta Anat. (Basel), 155: 50–56.

Liem, R.S.B., Copray, J.C.V.M. and van Willigen, J.D. (1993) Serotonin-immunoreactive terminals in the mesencephalic trigeminal nucleus of the rat: an electron microscopic immunocytochemical study. Acta Anat. (Basel), 148: 34–41.

Light, A.R., Kavoodjian, A.M. and Petrusz, P. (1983) The ultrastructure and synaptic connections of serotonin-immunoreactive terminals in spinal laminae I and II. Somatosensory Res., 1: 33–50.

Lindvall, O. and Björklund, A. (1984) General organization of cortical monoamine systems. Neurol. Neurobiol., 1: 9–40.

Liposits, Z., Phelix, C. and Paull, W.K. (1987) Synaptic interaction of serotonergic axons and corticotropin releasing factor (CRF) synthesizing neurons in the hypothalamic paraventricular nucleus of the rat. A light and electron microscopic immunocytochemical study. Histochemistry, 86: 541–549.

Liu, X.B. and Jones, E.G. (1991) The fine structure of serotonin and tyrosine hydroxylase immunoreactive terminals in the ventral posterior thalamic nucleus of cat and monkey. Exp. Brain Res., 85: 507–518.

Ljungdahl, A., Hökfelt, T., Halasz, N., Johansson, O. and Goldstein, M. (1977) Olfactory bulb dopamine neurons – the A15 catecholamine cell group. Acta Physiol. Scand., 452(Suppl.): 31–34.

López Costa, J.J., Averill, S., Pecci Saavedra, J.P. and Priestley, J.V. (1994) Serotonin innervation of enkephalin containing neurones in the rat spinal trigeminal nucleus. Neurosci. Lett., 168: 167–171.

Lorens, S.A. and Guldberg, H.C. (1974) Regional 5-hydroxytryptamine following selective midbrain raphe lesions in the rat. Brain Res., 78: 45–56.

Lorez, H.P. and Richards, J.G. (1973) Distribution of indolealkylamine nerve terminals in the ventricles of the rat brain. Z. Zellforsch., 144: 511–522.

Lorez, H.P. and Richards, J.G. (1975) 5-HT nerve terminals in the fourth ventricule of the rat brain: their identification and distribution studied by fluorescence histochemistry and electron microscopy. Cell Tissue Res., 165: 37–48.

Lorez, H.P. and Richards, J.G. (1982) Supra-ependymal serotoninergic nerves in mammalian brain: morphological, pharmacological and functional studies. Brain Res. Bull., 9: 727–741.

Lovick, T.A., Parry, D.M., Stezhka, V.V. and Lumb, B.M. (2000) Serotonergic transmission in the periaqueductal gray matter in relation to aversive behaviour: morphological evidence for direct modulatory effects on identified output neurons. Neuroscience, 95: 763–772.

Luebke, J.I., Greene, R.W., Semba, K., Kamondi, A., McCarley, R.W. and Reiner, P.B. (1992) Serotonin hyperpolarizes cholinergic low-threshold burst neurons in the rat laterodorsal tegmental nucleus in vitro. Proc. Natl. Acad. Sci. USA, 89: 743–747.

Maley, B. and Elde, R. (1982) The ultrastructural localization of serotonin immunoreactivity within the nucleus of the solitary tract of the cat. J. Neurosci., 2: 1499–1506.

Maley, B.E., Engle, M.G., Humphreys, S., Vascik, D.A., Howes, K.A., Newton, B.W. and Elde, R.P. (1990) Monoamine synaptic structure and localization in the central nervous system. J. Electr. Microsc. Techn., 15: 20–33.

Marlier, L., Sandillon, F., Poulat, P., Rajaofetra, N., Geffard, M. and Privat, A. (1991) Serotonergic innervation of the dorsal horn of rat spinal cord: light and electron microscopic immunocytochemical study. J. Neurocytol., 20: 310–322.

Marlier, L., Poulat, P., Rajaofetra, N., Sandillon, F. and Privat, A. (1992) Plasticity of serotonergic innervation of the dorsal horn of the rat spinal cord following neonatal capsaicin treatment. J. Neurosci. Res., 31: 346–358.

Mascagni, F. and McDonald, A.J. (2007) A novel subpopulation of 5-HT type 3A receptor subunit immunoreactive interneurons in the rat basolateral amygdala. Neuroscience, 144: 1015–1024.

Maxwell, D.J. and Jankowska, E. (1996) Synaptic relationships between serotonin-immunoreactive axons and dorsal horn spinocerebellar tract cells in the cat spinal cord. Neuroscience, 70: 247–253.

Maxwell, D.J., Léránth, Cs. and Verhofstad, A.A.J. (1983) Fine structure of serotonin-containing axons in the marginal zone of the rat spinal cord. Brain Res., 266: 253–259.

Maxwell, D.J., Léránth, C. and Verhofstad, A.A.J. (1985) Synaptic arrangements formed by serotonin-immunoreactive axons in the substantia gelatinosa of the rat spinal cord. Q. J. Exp. Physiol., 70: 377–388.

McCall, R. and Aghajanian, G.K. (1979) Serotonergic facilitation of facial motoneuron excitation. Brain Res., 169: 11–27.

McDonald, A.J. and Mascagni, F. (2007) Neuronal localization of 5-HT type 2A receptor immunoreactivity in the rat basolateral amygdala. Neuroscience, 146: 306–320.

McPeek, R.M. and Keller, E.L. (2004) Deficits in saccade target selection after inactivation of superior colliculus. Nat. Neurosci., 7: 757–763.

Miletic, V., Hoffert, M.J., Ruda, M.A., Dubner, R. and Shigenaga, Y. (1984) Serotinergic axonal contacts on identified car spinal dorsal horn neurons and their correlation with nucleus raphe magnus stimulation. J. Comp. Neurol., 228: 129–141.

Milner, T.A. and Veznedaroglu, E. (1993) Serotonin-containing terminals synapse on septohippocampal neurons in the rat. J. Neurosci. Res., 36: 260–271.

Miner, L.H., Schroeter, S., Blakely, R.D. and Sesack, S.R. (2000) Ultrastructural localization of the serotonin transporter in superficial and deep layers of the rat prelimbic prefrontal cortex and its spatial relationships to dopamine terminals. J. Comp. Neurol., 427: 220–234.

Mize, R.R. and Horner, L.H. (1989) Origin, distribution, and morphology of serotonergic afferents to the cat superior colliculus: a light and electron microscope immunocytochemistry study. Exp. Brain Res., 75: 83–98.

Mizukawa, K. and Takeuchi, Y. (1982) Electron microscopic investigation of monoaminergic terminals to alpha-motoneurons in the anterior horn of the cat spinal cord. Acta Med. Okayama, 36: 85–93.

Mizukawa, K., Otsuka, N. and Hattori, T. (1986) Serotonin-containing nerve fibers in the rat spinal cord: electron microscopic immunohistochemistry. Acta Med. Okayama, 40: 1–10.

Møllgärd, K. and Wiklund, L. (1979) Serotoninergic synapses on ependymal and hypendymal cells of the rat subcommissural organ. J. Neurocytol., 8: 445–467.

Molliver, M.E., Grzanna, R., Lidov, H.G.W., Morrisson, J.H. and Olschowka, J.A. (1982) Monoamine systems in the cerebral cortex. Neurol. Neurobiol., 1: 255–278.

Moore, R.Y. (1977) Organum vasculosum laminae terminalis: innervation by serotonin neurons of the midbrain raphe. Neurosci. Lett., 5: 297–302.

Moore, R.Y. and Halaris, A.E. (1975) Hippocampal innervation by serotonin neurons of the midbrain raphe in the rat. J. Comp. Neurol., 164: 171–183.

Moore, R.Y., Halaris, A.E. and Jones, B.E. (1978) Serotonin neurons of the midbrain raphe: ascending projections. J. Comp. Neurol., 180: 417–438.

Morales, M. and Bloom, F.E. (1997) The 5-HT_3 receptor is present in different subpopulations of GABAergic neurons in the rat telencephalon. J. Neurosci., 17: 3157–3167.

Mori, S., Matsuura, T., Takino, T. and Sano, Y. (1987) Light and electron microscopic immunohistochemical studies of serotonin nerve fibers in the substantia nigra of the rat, cat and monkey. Anat. Embryol., 176: 13–18.

Morrison, J.H. and Foote, S.L. (1986) Noradrenergic and serotoninergic innervation of cortical, thalamic, and tectal visual structures in Old and New World monkeys. J. Comp. Neurol., 243: 117–138.

Morrison, J.H., Foote, S.L., Molliver, M.E., Bloom, F.E. and Lidov, H.G.W. (1982) Noradrenergic and serotoninergic fibers innervate complementary layers in monkey primary visual cortex: an immunohistochemical study. Proc. Natl. Acad. Sci. USA, 79: 2401–2405.

Morrison, J.H., Foote, S.L. and Bloom, F.E. (1984) Regional, laminar, developmental and functional characteristics of noradrenaline and serotonin innervation patterns on monkey cortex. Neurol. Neurobiol., 10: 61–75.

Mosko, S.S., Haubrich, D. and Jacobs, B.L. (1977) Serotonergic afferents to the raphe dorsalis: evdience from HRP and synaptosomal uptake studies. Brain Res., 119: 269–290.

Moukhles, H., Bosler, O., Bolam, J.P., Vallée, A., Umbriaco, D., Geffard, M. and Doucet, G. (1997) Quantitative and morphometric data indicate precise cellular interactions between serotonin terminals and postsynaptic targets in rat substantia nigra. Neuroscience, 76: 1159–1171.

Mrini, A., Soucy, J.-P., Lafaille, F., Lemoine, P. and Descarries, L. (1995) Quantification of the serotonin hyperinnervation in adult rat neostriatum after neonatal 6-hydroxydopamine lesion of nigral dopamine neurons. Brain Res., 669: 303–308.

Muller, J.F., Mascagni, F. and McDonald, A.J. (2005) Coupled networks of parvalbumin-immunoreactive interneurons in the rat basolateral amygdala. J. Neurosci., 25: 7366–7376.

Muller, J.F., Mascagni, F. and McDonald, A.J. (2007) Serotonin-immunoreactive axon terminals innervate pyramidal cells and interneurons in the rat basolateral amygdala. J. Comp. Neurol., 505: 314–335.

Mulligan, K.A. and Törk, I. (1987) Serotonergic axons form basket-like terminals in cerebral cortex. Neurosci. Lett., 81: 7–12.

Mulligan, K.A. and Törk, I. (1988) Serotoninergic innervation of the cat cerebral cortex. J. Comp. Neurol., 270: 86–110.

Nagase, Y., Moritani, M., Nakagawa, S., Yoshida, A., Takemura, M., Zhang, L.-F., Kida, H. and Shigenaga, Y.

(1997) Serotonergic axonal contacts on identified cat trigeminal motoneurons and their correlation with medullary raphe nucleus stimulation. J. Comp. Neurol., 384: 443–455.
Nakai, Y., Shioda, S., Ochiai, H.J., Kudo, J. and Hashimoto, A. (1983) Ultrastructural relationship between monoamine- and TRH-containing axons in the rat median eminence, as revealed by combined radioautography and immunocytochemistry in the same tissue sections. Cell Tissue Res., 230: 1–14.
Naumann, T., Peterson, G.M. and Frotscher, M. (1992) Fine structure of rat septohippocampal neurons: II. A time course analysis following axotomy. J. Comp. Neurol., 325: 219–242.
Nedergaard, S., Bolam, J.P. and Greenfield, S.A. (1988) Facilitation of a dendritic calcium conductance by 5-hydroxytryptamine in the substantia nigra. Nature, 333: 174–177.
Newton, B.W., Maley, B.E. and Hamill, R.W. (1986) Immunohistochemical demonstration of serotonin neurons in autonomic regions of the rat spinal cord. Brain Res., 376: 155–163.
Nishikawa, N., Bennett, G.J., Ruda, M.A., Lu, G.W. and Dubner, R. (1983) Immunocytochemical evidence for a serotoninergic innervation of dorsal column postsynaptic neurons in cat and monkey: light and electron microscopic observations. Neuroscience, 10: 1333–1340.
Nojyo, Y. and Sano, Y. (1978) Ultrastructure of the serotonergic nerve terminals in the suprachiasmatic and interpeduncular nuclei of rat brains. Brain Res., 149: 482–488.
Nothias, F., Onteniente, B., Roudier, F. and Peschanksi, M. (1988) Immunocytochemical study of serotoninergic and noradrenergic innervation of the ventrobasal complex of the rat thalamus. Neurosci. Lett., 95: 59–63.
Oleskevich, S. and Descarries, L. (1990) Quantified distribution of the serotonin innervation in adult rat hippocampus. Neuroscience, 34: 19–33.
Oleskevich, S., Descarries, L., Watkins, K.C., Séguéla, P. and Daszuta, A. (1991) Ultrastructural features of the serotonin innervation in adult rat hippocampus. An immunocytochemical description in single and serial thin sections. Neuroscience, 42: 777–791.
Palay, S.L. and Chan-Palay, V. (1976) A guide to the synaptic analysis of the neuropil. Cold Spring Harb. Symp. Quant. Biol., XL: 1–16.
Palkovits, M., Brownstein, M. and Saavedra, J.M. (1974) Serotonin content of the brain stem nuclei in the rat. Brain Res., 80: 237–249.
Papadopoulos, G.C. and Parnavelas, J.G. (1990) Distribution and synaptic organization of serotoninergic and noradrenergic axons in the lateral geniculate nucleus of the rat. J. Comp. Neurol., 294: 345–355.
Papadopoulos, G.G., Parnavelas, J.G. and Buijs, R. (1987a) Monoaminergic fibers form conventional synapses in the cerebral cortex. Neurosci. Lett., 76: 275–279.
Papadopoulos, G.G., Parnavelas, J.G. and Buijs, R. (1987b) Light and electron microscopic immunocytochemical analysis of the serotonin innervation of the rat visual cortex. J. Neurocytol., 26: 883–892.
Paré, M., Descarries, L. and Wiklund, L. (1987) Innervation and reinnervation of rat inferior olive by neurons containing serotonin and substance P: an immunohistochemical study after 5,6-dihydroxytryptamine lesioning. J. Neurocytol., 16: 155–167.
Paré, D., Collins, D.R. and Pelletier, J.G. (2002) Amygdala oscillations and the consolidation of emotional memories. Trends Cogn. Sci., 6: 306–314.
Parent, A., Descarries, L. and Beaudet, A. (1981) Organization of ascending serotonin systems in the adult rat brain. A radioautographic study after intraventricular administration of [^{3}H]5-hydroxytryptamine. Neuroscience, 6: 115–138.
Parizek, J., Hassler, R. and Bak, I.J. (1971) Light and electron microscopic radioautography of substantia nigra of rat after intraventricular administration of tritium labelled norepinephrine, dopamine, serotonin and the precursors. Z. Zellforsch., 115: 137–148.
Parnavelas, J.G., Moises, H.C. and Speciale, S.G. (1985) The monoaminergic innervation of the rat visual cortex. Proc. R. Soc. Lond. B., 223: 319–329.
Pasik, T. and Pasik, P. (1982) Serotoninergic afferents in the monkey neostriatum. Acta Biol. Acad. Sci. Hung., 33: 277–288.
Pasik, P., Pasik, T. and Pecci Saavedra, J. (1982) Immunocytochemical localization of serotonin at the ultrastructural level. J. Histochem. Cytochem., 30: 760–764.
Pasik, P., Pasik, T., Holstein, G.R. and Pecci Saavedra, J. (1984) Serotoninergic innervation of the monkey basal ganglia: an immunocytochemical light and electron microscopic study. In: McKenzie, J.S., Kenn, R.E. and Wilcock, L.N. (Eds.), The Basal Ganglia: Structure and Function. Plenum Press, New York, NY, pp. 115–129.
Pasik, P., Pasik, T. and Holstein, G.R. (1988) Serotonin-immunoreactivity in the monkey lateral geniculate nucleus. Exp. Brain Res., 69: 662–666.
Paspalas, C.D. and Papadopoulos, G.C. (2001) Serotoninergic afferents preferentially innervate distinct subclasses of peptidergic interneurons in the rat visual cortex. Brain Res., 891: 158–167.
Pasquier, D.A. and Villar, M.J. (1982) Specific serotonergic projections to the lateral geniculate body from the lateral groups of the raphe dorsalis. Brain Res., 249: 142–146.
Pecci-Saavedra, J.P., Pasik, T. and Pasik, P. (1983) Immunocytochemistry of serotoninergic neurons in the central nervous system of monkeys. In: Caputto, R. and Ajmone Marsan, C. (Eds.), Neural Transmission, Learning and Memory. Raven Press, New York, NY, pp. 81–96.
Pecci-Saavedra, J.P., Brusco, A., Peressini, S. and Oliva, D. (1986) A new case for a presynaptic role of dendrites: an immunocytochemical study of the n. raphe dorsalis. Neurochem. Res., 11: 997–1009.
Pelletier, G., Steinbusch, H.W.M. and Verhofstad, A.A. (1981) Immunoreactive substance P and serotonin present in the same dense-core vesicles. Nature, 293: 71–72.
Peters, A. and Palay, S.L. (1996) The morphology of synapses. J. Neurocytol., 25: 687–700.
Peterson, G.M., Naumann, T. and Frotscher, M. (1992) Identified septohippocampal neurons survive axotomy: a fine-structural analysis in the rat. Neurosci. Lett., 138: 81–85.
Phelix, C.F., Liposits, Z. and Paull, W.K. (1992) Monoamine innervation of bed nucleus of stria terminalis: an electron microscopic investigation. Brain Res. Bull., 28: 949–965.
Pickel, V.M. and Armstrong, D.M. (1984) Ultrastructural localization of monoamines and peptides in rat area postrema. Fed. Proc., 43: 2949–2951.
Pickel, V.M. and Chan, J. (1999) Ultrastructural localization of the serotonin transporter in limbic and motor compartments of the nucleus accumbens. J. Neurosci., 19: 7356–7366.
Pickel, V.M., Joh, T.H. and Reis, D. (1976) Monoamine synthesizing enzymes in central dopaminergic, noradrenergic and serotonergic neurons. Immunocytochemical localization by light and electron microscopy. J. Histochem. Cytochem., 24: 792–806.

Pickel, V.M., Joh, T.H. and Reis, D. (1977) A serotonergic innervation of noradrenergic neurons in nucleus locus coeruleus: demonstration by immunocytochemical localization of the transmitter specific enzymes tyrosine and tryptophan hydroxylase. Brain Res., 131: 197–214.

Pickel, V.M., Joh, T.H., Chan, J. and Beaudet, A. (1984) Serotoninergic terminals: ultrastructure and synaptic interaction with catecholamine-containing neurons in the medial nuclei of the solitary tract. J. Comp. Neurol., 225: 291–301.

Pilowsky, P.M., de Castro, D., Llewellyn-Smith, I., Lipski, J. and Voss, M.D. (1990) Serotonin immunoreactive boutons make synapses with feline phrenic motoneurons. J. Neurosci., 10: 1091–1098.

Poulat, P., Marlier, L., Rajaofetra, N. and Privat, A. (1992) 5-hydroxytryptamine, substance P and thyrotropin-releasing hormone synapses in the intermediolateral cell column of the rat thoracic spinal cord. Neurosci. Lett., 136: 19–22.

Rainnie, D.G. (1999) Serotonergic modulation of neurotransmission in the rat basolateral amygdala. J. Neurophysiol., 82: 69–85.

Rajaofetra, N., Poulat, P., Marlier, L., Sandillon, F., Drian, M.J., König, N., Famose, F., Verschuere, B., Gouy, D., Geffard, M. and Privat, A. (1992) Transplantation of embryonic serotonin immunoreactive neurons into the transected spinal cord of adult monkey (*Macaca fascicularis*). Brain Res., 572: 329–334.

Ranson, R.N., Santer, R.M. and Watson, A.H.D. (2006) The relationship between serotonin, dopamine beta hydroxylase and GABA immunoreactive inputs and spinal preganglionic neurones projecting to the major pelvic ganglion of Wistar rats. Neuroscience, 141: 1935–1949.

Rhoades, R.W., Bennett-Clarke, C.A., Lane, R.D., Leslie, M.J. and Mooney, R.D. (1993) Increased serotoninergic innervation of the hamster's superior colliculus alters retinotectal projections. J. Comp. Neurol., 334: 397–409.

Riad, M., Garcia, S., Watkins, K.C., Jodoin, N., Doucet, E., Langlois, X., El Mestikawy, S., Hamon, M. and Descarries, L. (2000) Somatodendritic localization of 5-HT_{1A} and preterminal axonal localization of 5-HT_{1B} serotonin receptors in adult rat brain. J. Comp. Neurol., 417: 181–194.

Richards, J.G. (1977) Radioautographic evidence for the selective accumulation of [^{3}H]5-HT by supra-ependymal nerve terminals. Brain Res., 134: 151–157.

Richards, J.G., Lorez, H.P. and Tranzer, J.P. (1973) Indolealkylamine nerve terminals in cerebral ventricles: identification by electron microscopy and fluorescence histochemistry. Brain Res., 57: 277–288.

Richards, J.G., Lorez, H.P., Colombo, V.E., Guggenheim, R. and Kiss, D. (1980) Supraependymal nerve fibres in human brain: correlative transmission and scanning electron microscopical and fluorescence histochemical studies. Neuroscience, 5: 1489–1502.

Richards, J.G., Lorez, H.P., Colombo, V.E., Guggenheim, R., Kiss, D. and Wu, J.Y. (1981) Demonstration of supraependymal nerve fibres in human brain and their immunohistochemical identification in rat brain. J. Physiol. (Paris), 77: 219–224.

Ridet, J.-L. (1994) Organisation ultrastructurale et plasticité des systèmes monoaminergiques dans la moelle épinière du rat. Thèse de doctorat. Université de Montpellier II, Montpellier, p. 34.

Ridet, J.-L., Rajaofetra, N., Teilhac, J.R., Geffard, M. and Privat, A. (1993) Evidence for non-synaptic serotonergic and noradrenergic innervation of the rat dorsal horn and possible involvement of neuron-glia interactions. Neuroscience, 52: 143–157.

Ruda, M.A. and Gobel, S. (1980) Ultrastructural characterization of axonal endings in the substantia gelatinosa which take up [^{3}H]serotonin. Brain Res., 184: 57–83.

Ruda, M.A., Allen, B. and Gobel, S. (1981a) Ultrastructure of descending serotoninergic axonal endings in layers I and II of the dorsal horn. J. Physiol. (Paris), 77: 205–209.

Ruda, M.A., Allen, B. and Gobel, S. (1981b) Ultrastructural analysis of medial brain stem afferents to the superficial dorsal horn. Brain Res., 205: 175–180.

Ruda, M.A., Coffield, J. and Steinbusch, H.W.M. (1982) Immunocytochemical analysis of serotonergic axons in laminae I and II of the lumbar spinal cord of the cat. J. Neurosci., 2: 1660–1671.

Sadikot, A.F. and Parent, A. (1990) The monoaminergic innervation of the amygdala in the squirrel monkey: an immunohistochemical study. Neuroscience, 36: 431–447.

Saha, S., Appenteng, K. and Batten, T.F.C. (1991) Light and electron microscopical localisation of 5-HT immunoreactive boutons in the rat trigeminal motor nucleus. Brain Res., 559: 145–148.

Saïdi, H. and Bosler, O. (1990) Serotonin reinnervation of the rat organum vasculosum laminae terminalis (OVLT) after 5,7-dihydroxytryptamine deafferentation. Brain Res., 530: 151–155.

Sari, Y., Miquel, M.C., Brisorgueil, M.J., Ruiz, G., Doucet, E., Hamon, M. and Vergé, D. (1999) Cellular and subcellular localization of 5-hydroxytryptamine1B receptors in the rat central nervous system: immunocytochemical, autoradiographic and lesion studies. Neuroscience, 88: 899–915.

Schaffar, N., Jean, A. and Calas, A. (1984) Radioautographic study of serotoninergic axon terminals in the rat trigeminal motor nucleus. Neurosci. Lett., 44: 31–36.

Schmied, A., Farin, D., Amalric, M. and Dormont, J.F. (1991) Changes in motor performance and rubral single unit activity in cats after microinjections of serotonin into the red nucleus area. Brain Res., 567: 91–100.

Ségu, L. and Calas, A. (1978) The topographical distribution of serotoninergic terminals in the spinal cord of the cat: quantitative radioautographic studies. Brain Res., 153: 449–464.

Ségu, L., Abdelkefi, J., Dusticier, G. and Lanoir, J. (1986) High-affinity serotonin binding sites: autoradiographic evidence for their location on retinal afferents in the rat superior colliculus. Brain Res., 384: 205–217.

Séguéla, P., Watkins, K.C. and Descarries, L. (1989) Ultrastructural relationships of serotonin axon terminals in the cerebral cortex of the adult rat. J. Comp. Neurol., 289: 129–142.

Shen, K.Z. and Johnson, S.W. (2008) 5-HT inhibits synaptic transmission in rat subthalamic nucleus neurons in vitro. Neuroscience, 151: 1029–1033.

Smiley, J.F. and Goldman-Rakic, P.S. (1996) Serotonergic axons in monkey prefrontal cerebral cortex synapse predominantly on interneurons as demonstrated by serial section electron microscopy. J. Comp. Neurol., 367: 431–443.

Smiley, J.F., Subramanian, M. and Mesulam, M.M. (1999) Monoaminergic–cholinergic interactions in the primate basal forebrain. Neuroscience, 93: 817–829.

Soghomonian, J.-J., Descarries, L. and Lanoir, J. (1986) Monoamine innervation of the oculomotor nucleus in the rat. A radioautographic study. Neuroscience, 17: 1147–1157.

Soghomonian, J.-J., Doucet, G. and Descarries, L. (1987) Serotonin innervation in adult rat neostriatum. I. Quantified regional distribution. Brain Res., 425: 85–100.

Soghomonian, J.-J., Beaudet, A. and Descarries, L. (1988) Ultrastructural relationships of central serotonin neurons. In: Osborne, N.N. and Hamon, M. (Eds.), Neuronal Serotonin. John Wiley & Sons, Chichester, pp. 57–92.

Soghomonian, J.-J., Descarries, L. and Watkins, K.C. (1989) Serotonin innervation in adult rat neostriatum. II. Ultrastructural features: a radioautographic and immunocytochemical study. Brain Res., 481: 67–86.

Sotelo, C. and Beaudet, A. (1979) Influence of experimentally induced agranularity on the synaptogenesis of serotonin nerve terminals in rat cerebral cortex. Proc. R. Soc. Lond. B, 206: 133–138.

Stanford, I.M., Kantaria, M.A., Chahal, H.S., Loucif, K.C. and Wilson, C.L. (2005) 5-Hydroxytryptamine induced excitation and inhibition in the subthalamic nucleus: action at 5-HT(2C), 5-HT(4) and 5-HT(1A) receptors. Neuropharmacology, 49: 1228–1234.

Steinbusch, H.W.M. (1981) Distribution of serotonin immunoreactivity in the central nervous system of the rat. Cell bodies and terminals. Neuroscience, 6: 557–618.

Steinbusch, H.V.M. (1984) Serotonin-immunoreactive neurons and their projections in the CNS. In: Björklund, A., Hökfelt, T. and Kuhar, M.J. (Eds.), Handbook of Chemical Neuroanatomy. Vol. 3, Classical Transmitters and Transmitter Receptors in the CNS, Part II. Elsevier, Amsterdam, pp. 68–125.

Steinbusch, H.V.M. and Nieuwenhuys, R. (1981) Localization of serotonin-like immunoreactivity in the central nervous system and pituitary of the rat, with special references to the innervation of the hypothalamus. Adv. Exp. Med. Biol., 133: 7–35.

Steininger, T.L., Wainer, B.H., Blakely, R.D. and Rye, D.B. (1997) Serotonergic raphe dorsalis projections to the cholinergic and noncholinergic neurons of the pedonculopontine tegmental region: a light and electron microscopic anterograde tracing and immunocytochemical study. J. Comp. Neurol., 382: 302–322.

Steriade, M. and McCarley, R.W. (2005) Brain Control of Sleep and Wakefulness. Kluwer Academic Press, New York, NY.

Stutzmann, G.E. and LeDoux, J.E. (1999) GABAergic antagonists block the inhibitory effects of serotonin in the lateral amygdala: a mechanism for modulation of sensory inputs related to fear conditioning. J. Neurosci., 19: RC8.

Tai, Q., Palazzolo, K.L. and Goshgarian, G. (1997) Synaptic plasticity of 5-hydroxytryptamine-immunoreactive terminals in the phrenic nucleus following spinal cord injury: a quantitative electron microscopic analysis. J. Comp. Neurol., 386: 613–624.

Takeuchi, Y. and Sano, Y. (1983a) Serotonin distribution in the circumventricular organs of the rat. An immunohistochemical study. Anat. Embryol. (Berl.), 167: 311–319.

Takeuchi, Y. and Sano, Y. (1983b) Immunohistochemical demonstration of serotonin-containing nerve fibers in the inferior olivary complex of the rat, cat and monkey. Cell Tissue Res., 231: 17–28.

Takeuchi, Y., Kimura, H. and Sano, Y. (1982) Immunohistochemical demonstration of serotonin-containing nerve fibers in the cerebellum. Cell Tissue Res., 226: 1–12.

Takeuchi, Y., Kojima, M., Matsuura, T. and Sano, Y. (1983) Serotonergic innervation on the motoneurons in the mammalian brainstem. Light and electron microscopic immunohistochemistry. Anat. Embryol. (Berl.), 167: 321–333.

Tao-Cheng, J.-H. and Zhou, F.C. (1999) Differential polarization of serotonin transporters in axons versus soma-dendrites: an immunogold electron microscopy study. Neuroscience, 94: 821–830.

Tashiro, T., Satoda, T., Matsushima, R. and Mizuno, N. (1989) Enkephalin-, substance P- and serotonin-like immunoreactive axonal varicosities in close apposition to perikarya of mesencephalic trigeminal nucleus neurons in the cat. Brain Res., 494: 162–167.

Tramu, G., Pillez, A. and Leonardelli, J. (1983) Serotonin axons of the ependyma and circumventricular organs in the forebrain of the guinea pig. An immunohistochemical study. Cell Tissue Res., 228: 297–311.

Trottier, S., Evrard, B., Vignal, J.P., Scarabin, J.M. and Chauvel, P. (1996) The serotonergic innervation of the cerebral cortex in man and its changes in focal cortical dysplasia. Epilepsy Res., 25: 79–106.

Ulfhake, B., Arvidsson, U., Cullheim, S., Hökfelt, T., Brodin, E., Verhofstad, A. and Visser, T. (1987) An ultrastructural study of 5-hydroxytryptamine-, thyrotropin-releasing hormone- and substance P-immunoreactive axonal boutons in the motor nucleus of spinal cord segments L7–S1 in the adult cat. Neuroscience, 23: 917–929.

Umbriaco, D., Garcia, S., Beaulieu, C. and Descarries, L. (1995) Relational features of acetylcholine, noradrenaline, serotonin and GABA axon terminals in the stratum radiatum of adult rat hippocampus (CA1). Hippocampus, 5: 605–620.

Van de Kar, L.D. and Lorens, S.A. (1979) Differential serotonergic innervation of individual hypothalamic nuclei and other forebrain regions by the dorsal and median midbrain raphe nuclei. Brain Res., 162: 45–54.

Van Bockstaele, E.J. and Pickel, V.M. (1993) Ultrastructure of serotonin-immunoreactive terminals in the core and shell of the rat nucleus accumbens: cellular substrates for interactions with catecholamine afferents. J. Comp. Neurol., 334: 603–617.

Van Bockstaele, E.J., Cestari, D.M. and Pickel, V.M. (1994) Synaptic structure and connectivity of serotonin terminals in the ventral tegmental area: potential sites for modulation of mesolimbic dopamine neurons. Brain Res., 647: 307–322.

Van Bockstaele, E.J., Chan, J. and Pickel, V.M. (1996) Pre- and postsynaptic sites for serotonin modulation of GABA-containing neurons in the shell region of the rat nucleus accumbens. J. Comp. Neurol., 370: 116–128.

Van Luijtelaar, M.G., Wouterlood, F.G., Tonnaer, J.A. and Steinbusch, H.W.M. (1991) Ultrastructure of aberrant serotonin-immunoreactive fibers in the caudate putamen complex of the aged rat. Synapse, 8: 162–168.

Vera, P.L., Holets, V.R. and Miller, K.E. (1990) Ultrastructural evidence of synaptic contacts between substance P-, enkephalin, and serotonin-immunoreactive terminals and retrogradely labeled sympathetic preganglionic neurons in the rat: a study using a double-peroxidase procedure. Synapse, 6: 221–229.

Vertes, R.P. (1991) A PHA-L analysis of ascending projections of the raphe dorsalis in the rat. J. Comp. Neurol., 313: 643–668.

Vertes, R.P. and Kocsis, B. (1997) Brainstem-diencephalo-septohippocampal systems controlling the theta rhythm of the hippocampus. Neuroscience, 81: 893–926.

Vertes, R.P. and Martin, G.F. (1988) Radioautographic analysis of ascending projections from the pontine and mesencephalic reticular formation and the raphe medianus in the rat. J. Comp. Neurol., 275: 511–541.

Villar, M.J., Vitale, M.L., Hökfelt, T. and Verhofstad, A.A. (1988) Dorsal raphe serotoninergic branching neurons projecting both to the lateral geniculate body and superior colliculus: a combined retrograde tracing-immunohistochemical study in the rat. J. Comp. Neurol., 277: 126–140.

Wallman, M.-J., Parent, M., Descarries, L. and Parent, A. (2007). Serotonin innervation of the subthalamic nucleus in monkey and human. Program No. 513.7. *2007. Abstract Viewer/ Itinerary Planner.* Washington, DC: *Society for Neuroscience.* Online.

Wang, Q.-P., Guan, J.-L. and Shioda, S. (2000) Synaptic contacts between serotonergic and cholinergic neurons in the rat raphe dorsalis and laterodorsal tegmental nucleus. Neuroscience, 97: 553–563.

Wang, R.Y. and Aghajanian, G.K. (1978) Collateral inhibition of serotonergic neurones in the rat raphe dorsalis: pharmacological evidence. Neuropharmacology, 17: 819–825.

Wang, R.Y. and Aghajanian, G.K. (1982) Correlative firing patterns of serotonergic neurons in rat raphe dorsalis. J. Neurosci., 2: 11–16.

Westlund, K.N., Sorkin, L.S., Ferrington, D.G., Carlton, S.M., Willcockson, H.H. and Willis, W.D. (1990) Serotoninergic and noradrenergic projections to the ventral posterolateral nucleus of the monkey thalamus. J. Comp. Neurol., 295: 197–207.

White, S.R. and Neuman, R.S. (1980) Faciliatation of spinal motoneurone excitability by 5-hydroxytryptamine and noradrenaline. Brain Res., 188: 119–127.

Wiklund, L., Björklund, A. and Sjölund, B. (1977) The indolaminergic innervation of the inferior olive. 1. Convergence with the direct spinal afferents in the areas projecting to the cerebellar anterior lobe. Brain Res., 131: 1–21.

Wiklund, L., Descarries, L. and Møllgärd, K. (1981) Serotoninergic axon terminals in the rat dorsal accessory olive: normal ultrastructure and light microscopic demonstration of regeneration after 5,6-dihydroxytryptamine lesioning. J. Neurocytol., 10: 1009–1027.

Willoughby, J.O. and Blessing, W.W. (1987) Origin of serotonin innervation of the arcuate and ventromedial hypothalamic region. Brain Res., 418: 170–173.

Wilson, J.R. and Hendrickson, A.E. (1988) Serotonergic axons in the monkey's lateral geniculate nucleus. Vis. Neurosci., 1: 125–133.

Woodruff, A.R. and Sah, P. (2007) Inhibition and synchronization of basal amygdala principal neuron spiking by parvalbumin-positive interneurons. J. Neurophysiol., 98: 2956–2961.

Xiang, Z., Wang, L. and Kitai, S.T. (2005) Modulation of spontaneous firing in rat subthalamic neurons by 5-HT receptor subtypes. J. Neurophysiol., 93: 1145–1157.

Yamamoto, H., Fujimiya, M., Shirai, Y., Nakashita, M., Oyasu, M. and Saito, N. (1998) Immunohistochemical localization of serotonin transporter in normal and colchicine treated rat brain. Neurosci. Res., 32: 305–312.

Zhou, F.C., Tao-Cheng, J.-H., Segu, L., Patel, T. and Wang, Y. (1998) Serotonin transporters are located on the axons beyond the synaptic junctions: anatomical and functional evidence. Brain Res., 805: 241–254.

CHAPTER 1.5

Classification and Signaling Characteristics of 5-HT Receptors

Joël Bockaert*, Sylvie Claeysen, Aline Dumuis and Philippe Marin

Institut de Génomique Fonctionnelle, Université de Montpellier, CNRS UMR 5203, INSERM U661, Montpellier, France

Abstract: Serotonin (5-hydroxytryptamine; 5-HT), along with its receptors, is one of the oldest molecular devices assuming intercellular communication, being present as soon as simple nervous systems appeared during evolution. 5-HT receptors (7 classes, 5-HT_1 to 5-HT_7, coded by 17 different genes) are G protein-coupled receptors (GPCR) except for 5-HT_3 receptors, which are ionic channels. Splicing and editing add to the diversity of 5-HT receptors. They were first defined on a pharmacological basis and then on a gene structural basis, but the concordance between these two classifications is remarkable. The complexity of signaling events, including G protein-independent signaling of the GPCR classes, is not yet fully understood, and underscores the physiological importance of 5-HT. 5-HT receptors are associated with a large number of proteins involved in their localization, traffic and the fine-tuning signaling. Homodimerization and heterodimerization of 5-HT receptors with other GPCRs but also ligand-directed signaling (signaling dependent on the ligand and not only on the receptor) add to the complexity. Finally, 5-HT receptors are the targets of many therapeutic drugs on the market, in development and yet to be developed.

Keywords: serotonin, 5-hydroxytryptamine, 5-HT, G protein-coupled receptors, GPCR, signaling, G protein, G protein-associated proteins, GIPs, pharmacology.

Introduction

G protein-coupled receptors (GPCRs) are certainly among the oldest devices devoted to signal transduction, being present before plants, fungi and animal. Fungi (for example, *Saccharomyces cerevisiae*) express pheromone- and glucose-sensing GPCRs (for further reading, see Bockaert and Pin, 1999; Fredriksson *et al.*, 2003; Bockaert *et al.*, 2005). Although serotonin (5-hydroxytryptamine, 5-HT) is also one of the oldest cell–cell communication signaling molecules, it seems that both serotonin and associated receptors did not appear before the emergence of excitable cells and simple nervous systems such as those of coelenterates (Cnidaria) (Hajj-Ali and Anctil, 1997), tunicates, nematodes, mollusks and insects (Kroeze and Roth, 2006). Once present, 5-HT receptors were 'tinkered' with to generate a large diversity of molecules, probably to ensure a fine control of intracellular signaling and cellular functions. Indeed, there are at least seven subclasses of 5-HT receptors in the nematode *Caenorhabditis elegans* and four in *Aplysia*, *Drosophila*, and *Anopheles* mosquitoes (Kroeze and Roth, 2006).

The classification of 5-HT receptors can be based on both their primary sequence dendrograms (trees based on sequence alignments) and/or their pharmacology. The pharmacology of invertebrate 5-HT receptors is rather poor for obvious reasons (compounds are not developed by the pharmaceutical industry), and only sequence characteristics can be used for classification. The dendrograms show few discrete groupings for invertebrate GPCRs, apart from one for 5-HT_7-like receptors (Kroeze and Roth, 2006). Thus, the classification of invertebrate 5-HT receptors is almost non-existent.

In vertebrates, the classification has followed a slow evolution over the past 50 years. Two main periods can be distinguished: a pharmacological period, during which the development of specific 5-HT ligands permitted a fairly good classification, and a molecular period in which the cloning of 5-HT receptors completed the pharmacological classification. It is interesting to note that an excellent match was found between the two (Figure 1). The pharmacological period defined four classes (5-HT_{1-4}), and the molecular cloning period three more (5-HT_{5-7}). Some

*Corresponding author
E-mail: joel.bockaert@igf.cnrs.fr

ISBN 978-0-12-374634-4
DOI: 10.1016/B978-0-12-374634-4.00005-8

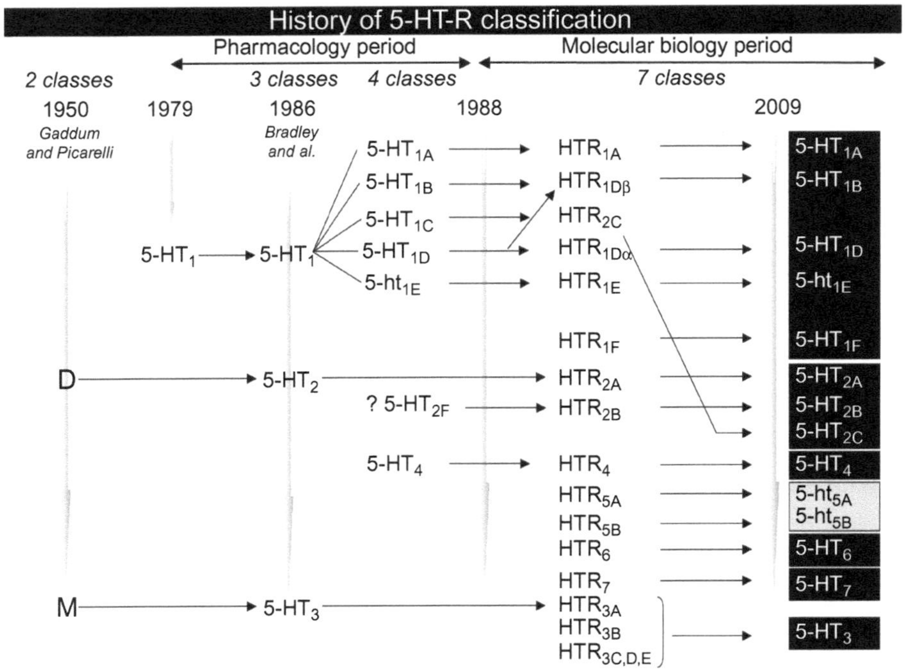

Figure 1 Historical perspective of 5-HT-R classification. Scientists interested in 5-HT receptors know that Gaddum and Picarelli proposed the first classification. Using binding studies and functional studies (generally cAMP production), 8 receptors were 'pharmacologically' characterized. Finally, during the 'molecular biology' period 13 different genes coding 5-HT receptors were cloned. There is a remarkable concordance between the pharmacological classification and the classification based on gene identification. The only major discrepancy was the 5-HT_{2C} receptor, which was named 5-HT_{1C} during the pharmacological period.

subclasses have also been defined by molecular biology. The actual dendrogram of vertebrate 5-HT receptors shows groupings that fit with the classification defined by pharmacology (binding site structure). The historical evolution of the 5-HT receptor classification will be reviewed here, and we will also show that not only the coding sequences but also the intron/exon structure of the 5-HT receptor genes can be classified into seven classes.

It is evident that the point of having so many receptors is *not* just to allow development of specific and therapeutic drugs by the pharmaceutical industry. The point, in a physiological and evolutionary perspective, is that these receptors have different cellular expression and control differently, via different fine signaling events, cell physiology. Indeed, each receptor subclass has a different signaling repertoire. The classical signaling events associated with most of these receptors have been reviewed (Bockaert *et al.*, 2006; Hannon and Hoyer, 2008). We will update these data, insisting on the new concepts and data showing that knowledge of the complexity of signaling cascades of 5-HT receptor classes and subclasses is still in its infancy.

The different drugs and radioligands characterizing each receptor are not presented systematically here, since this has recently been well done (Hannon and Hoyer, 2008).

Historical perspective of 5-HT receptor classification: a ping-pong game between pharmacology and molecular biology

As already stated, there have been two main periods of classification; pharmacological and molecular biology.

The four pharmacologically defined 5-HT receptor classes

The pharmacological period obviously started in 1950 when Gaddum and colleagues (Gaddum and Picarelli, 1957) found that, in the guinea pig ileum, the 5-HT-induced contracting effect could be blocked in part by dibenziline (the D receptor, which was finally characterized as a postsynaptic 5-HT_2 receptor) and in part by morphine (the M receptor, which was finally characterized as a presynaptic 5-HT_3 receptor).

Between 1975 and 1980, great efforts were made by Fillion in France (Fillion *et al.*, 1976) and Peroutka and Snyder in the USA (Peroutka and Snyder, 1979) to use radiolabeled [^{3}H]-5-HT, [^{3}H]-LSD and [^{3}H]-spiperone a dopamine/5-HT antagonist to label brain 5-HT receptors. Two types of receptors were pharmacologically characterized with these binding studies; 5-HT_1 (labeled with

[^{3}H]-5-HT and [^{3}H]-LSD) and 5-HT_2 receptors (labeled with [^{3}H]-spiperone and [^{3}H]-LSD). Based on both data from Gaddum and colleagues and binding data, Bradley and co-workers (Bradley *et al.*, 1986) proposed the existence of three 5-HT receptor families, 5-HT_1, 5-HT_2 and 5-HT_3, the latter being only based, at that time, on the existence of the Gaddum 'M' response. It is remarkable that even after 20 years of intensive discovery of new 5-HT receptor ligands and intensive molecular cloning, the 3 classes remain pharmacologically the same.

In addition to binding studies and organ responses (such as guinea-pig contractions) another tool was introduced in the 1980s: the use of second messenger production in intact cells or membranes following 5-HT stimulation. This tool was superb for studying structure–response relationships, and had the advantage of clearly distinguishing agonists from antagonists. It was at this time that we discovered the 5-HT_4 receptor which was able to stimulate cAMP production in colliculi neurons (Dumuis *et al.*, 1988a; Bockaert *et al.*, 2008).

Pharmacologically defined 5-HT_1 and 5-HT_2 sub-classes

Before molecular cloning began, binding studies (using more and more selective radio-ligands) and second messenger studies were used to characterize the subtypes of 5-HT_1 receptors.

5-HT_{1A} was first characterized using [^{3}H]-8-OH-DPAT (Gozlan *et al.*, 1983), and corresponds to the spiperone sensitive [^{3}H]-5-HT binding sites (Pedigo *et al.*, 1981). A few years later this receptor was found, in neurons, to be negatively coupled to adenylyl-cyclase (De Vivo and Maayani, 1986, Weiss *et al.*, 1986).

5-HT_{1B} was first characterized as the part of the [^{3}H]-5-HT binding sites insensitive to spiperone and 8-OH-DPAT, and then as binding sites labeled with [^{125}I]-CYP (cyanopindolol). We found that this presynaptic receptor was negatively coupled to adenylyl cyclase in substantia nigra (Bouhelal *et al.*, 1988).

5-HT_{1C} had a very transient existence in the 5-HT_1 class. This receptor, found at high density in the choroid plexus and labeled with [^{3}H]-5-HT (as with other 5-HT_1 receptors) but coupled to inositol phosphate production and Ca^{2+} signaling (as with 5-HT_2 receptors), definitively joined the 5-HT_2 family following its early cloning, its sequence homology with other 5-HT_2 receptors and an obvious 5-HT_2 pharmacology (Hannon and Hoyer, 2008) (Figure 1).

The history of 5-HT_{1D} is even more complex. It was first observed that the [^{125}I]-CYP ligand, which binds to 5-HT_{1B} in rodents, fish and opossum, was not labeling such homolog receptors in guinea-pig, pig and human. However, a functionally similar receptor (inhibition of 5-HT release; inhibition of adenylyl-cyclase) was present in those species and thus was called 5-HT_{1D}. It was the cloning period that shed some light. Indeed, in humans, two genes coding for receptors having a 5-HT_{1D} pharmacology were cloned (5-$HT_{1D\alpha}$ and the 5-$HT_{1D\beta}$), having a very slight difference in pharmacology. In rats, too, two genes are present; α and β. Thus, the 5-$HT_{1D\beta}$ gene in both rodent and human codes for the now termed 5-HT_{1B} receptor (Figure 1). The binding of [^{125}I]-CYP ligand to the rodent receptor and not to the human receptor and, more generally, the low affinity of the human receptor for β-blockers is due to a single residue difference in TM7 (an asparagine, N355, in human, and a threonine in rodent) (Adham *et al.*, 1994).

5-$HT_{1D\alpha}$ codes for 5-HT_{1D} receptor in both rodents and humans. Selective ligands are now available to distinguish between 5-HT_{1B} and $_{1D}$ receptors (Hannon and Hoyer, 2008). All active anti-migraine drugs (triptans) act on 5-HT_{1B} receptors. Whether or not some may be active via a specific action on 5-HT_{1D} receptors is still a matter of debate (Hannon and Hoyer, 2008).

The putative 5-ht_{1E} receptor was identified in the human frontal cortex as a residual [^{3}H]-5-HT binding when all others 5-HT_1 receptors were blocked. It has not been found in rodents, and no specific drugs are available. A gene has been cloned which codes for a receptor that could correspond to the pharmacologically 5-ht_{1E} binding site. Its exact signaling properties remain to be discovered (Hannon and Hoyer, 2008).

When examining the status of 5-HT_2 receptors during the pharmacological period, the actual 5-HT_{2A} and 5-HT_{2C} receptors were clearly defined. 5-HT_{2A} was indeed the D receptor of Gaddum (Gaddum and Picarelli, 1957) as well as the 5-HT_2 receptor found in brain by Peroutka and Snyder using binding studies (Peroutka *et al.*, 1981). [^{3}H]-Ketanserin rapidly became the reference ligand for 5-HT_{2A}. 5-HT_{2C} was, in fact, defined as 5-HT_{1C} during this period (see above). Both were found to be coupled to phospholipase C. The existence of the actually characterized 5-HT_{2B} was 'present' at a subluminal stage during this period. Indeed, a 5-HT_2 receptor was present in the fundus (thus called 5-HT_{2F}), in which it induced contraction, but no convincing pharmacological tools were available at that time for its definitive characterization (Hannon and Hoyer, 2008).

The molecular cloning period

G-protein-coupled receptors

The molecular cloning period started with the cloning of the 5-HT_{1A} receptor. An orphan receptor (G21) isolated by cross-hybridization at reduced stringency with a full-length β_2-adrenergic receptor probe appeared to code for the 5-HT_{1A} receptor (Fargin *et al.*, 1988). Most of the pharmacologically characterized 5-HT receptors were rapidly cloned (Hannon and Hoyer, 2008), including the fundus receptor first called 5-HT_{2F} (for fundus) (Foguet *et al.*, 1992) but later named 5-HT_{2B} in view of the existence, at that time, of 5-HT_{2A} and 5-HT_{2C} receptors.

In addition, 'new' genes coding for 5-HT receptors were cloned which were not yet pharmacologically characterized. These included 5-HT_{1F}, 5-ht_{5A}, 5-ht_{5B}, 5-HT_6 and 5-HT_7.

5-HT_{1F} was cloned on the basis of its sequence homology with the 5-$HT_{1B/1D}$ receptors (Adham *et al.*, 1993). It is expressed in many brain areas, and has a high affinity for triptans and LSD. It may be a better target for anti-migraine drugs than 5-HT_{1B} receptors, since it could be devoid of vasomotor effects. It is negatively coupled to adenylyl-cyclase.

The cloned 5-ht_{5A} and 5-ht_{5B} proteins (Matthes *et al.*, 1993) have not yet reached the status of 'receptors', since neither their signal transduction nor their physiological function are clearly established.

The 5-HT_6 receptor was suspected to be present in the striatum and some glioblastomes such as NCB20. It was positively coupled with adenylyl cyclase and sensitive to antipsychotic drugs. The cloning was realized in 1993 in the rat (Ruat *et al.*, 1993a). It is essentially present in brain, and is indeed positively coupled to adenylyl cyclase in striatal neurons (Sebben *et al.*, 1994).

Another 5-HT receptor positively coupled with adenylyl cyclase was discovered by homology cloning: the 5-HT_7 receptor (Ruat *et al.*, 1993b). Interestingly, this receptor shares low homology with other members of the 5-HT receptor family ($<$50 percent) and high homology with a drosophila 5-HT receptor positively coupled with an adenylate cyclase ($>$94 percent) (Kroeze and Roth, 2006).

Ionotropic 5-HT_3 receptors

This 5-HT receptor is the only one that is not a metabotropic receptor but a ligand-gated ion channel. It is interesting to note that the main neurotransmitters (glutamate, GABA, acetylcholine, 5-HT) have managed, during evolution, to use both rapid (milliseconds) ion-gated channel-mediated and slow (seconds) GPCR-mediated transmission.

Two genes coding for 5-HT_3 receptors have been cloned, 5-HT_{3A} (Maricq *et al.*, 1991) and 5-HT_{3B} (Dubin *et al.*, 1999; Maricq *et al.*, 1991). They have a high level of identity with other members of the Cys–Cys loop ligand-gated channel superfamily (e.g., nicotinic, GABA-A, glycine receptors), especially in the channel domain area.

The native receptor, as revealed by electron microscopy, is pentameric (Boess *et al.*, 1995). 5-HT_{3B} is unable to form a functional receptor on its own; however, it can co-assemble with 5-HT_{3A} to form functional 5-$HT_{3A/B}$ channels with properties similar to those of native 5-HT_3 receptors. Other human (but not rodent) sequences have been cloned for the 5-$HT_{3C–E}$ receptors, but their role as 5-HT_3 receptors is obscured although they can have a modulating role associated with 5-HT_{3A} (Niesler *et al.*, 2007; Jensen *et al.*, 2008). Since 5-HT_{3A} and 5-HT_{3B} receptors are co-distributed in the brain as well as in dorsal root ganglia and trigeminal ganglia, hetero-pentameric structures may exist.

Evolutionary consideration

As seen in Figure 1, the classification proposed during the pharmacological period was, with few exceptions, as discussed in the previous paragraph, confirmed during the molecular cloning period. This indicates that the pharmacologists did a good job in their classification based on structure–function relationships, binding data, and the few elements of signal transduction that were available at that time. In addition, this indicates that the binding sites of all these different receptors co-evolved with the whole primary structure of their corresponding receptors. Indeed, there is a perfect coincidence between the grouping of vertebrate 5-HT receptors as revealed by the dendogram based on sequence alignments, and classes and sub-classes defined by the pharmacology (Kroeze and Roth, 2006). This was not necessarily expected. Why have such different binding sites to recognize a common neurotransmitter? With the increasing complexity of the neuronal network it was certainly interesting to have an increased diversity of 5-HT receptor signaling (and thus of the receptor domains involved in signaling), as well as a diversity in cellular expression. However, their binding site could have remained similar. Note, however, that the evolution of the binding sites had allowed receptors with different affinity for 5-HT.

Finally, the 5-HT receptor dendrogram shows that the nature of the primary coupling with the G protein is not an element of clustering when considering sequence homologies. The 5-HT_6 receptor coupled to Gs, for example, is more similar to 5-HT_2 receptors coupled to Gq than to 5-HT_7 receptors coupled to Gs.

Genomic, post-genomic characteristics of 5-HT receptors

5-HTRs are generated from 17 different genes in the human genome (NCBI, Map Viewer at http://www.ncbi.

nlm.nih.gov/mapview/); 12 genes encoding metabotropic 5-HT receptors and 5 genes encoding ionotropic 5-HT_3 receptors. There is one pseudogene: 5-HT_{5B}.

Metabotropic 5-HT receptor genes

G-protein-coupled receptor genes diverged from a common ancestor, and their structure reflects their evolutionary relationship (Barnes and Sharp, 1999). 5-HT_1 receptors are coded by intronless genes, and clearly constitute a subgroup of genes that arose by duplication events implicating several chromosomes (1, 3, 5 and 6) (Figure 2). 5-HT_2 receptor genes are also closely related, presenting two similar introns located at the end of transmembrane domain 2 (TM2) and within TM4-coding sequences. The 5-HT_{2C} receptor gene presents an additional intron in the extracellular domain of the coding sequences, and extends over more than 183 kb. Alternative splicing of the 5-HT_2 receptor gene does not generate any functional splice variant (Xie *et al.*, 1996). The 5-HT_4 receptor gene is far more complex. The coding sequence extends over more than 185 kb and includes at least 14 exons, leading to multiple splicing combinations. To date, 10 functional splice variants have been described that can be generated by alternative splicing of the human transcript (Bender *et al.*, 2000; Hiroi *et al.*, 2001). For nine of these variants, splicing gives rise to alternate C-terminal domains (Figure 2). As some exons (such as the h or the i exons) do not present any in-frame stop codons, many additional combinations probably remain to be discovered (Brattelid *et al.*, 2004). Indeed, a recent publication described 10 novel splice variants of the porcine 5-HT_4 receptor, including a functional homofusion variant (De Maeyer *et al.*, 2008). The existence of these variants, some of them arising from duplication events, remains to be demonstrated in the human genome. The relevance and the functional role of such a multiplicity of isoforms also remain to be elucidated. The finding that some less well represented 5-HT_4 splice variants (e.g., $5\text{-HT}_{4(a)}$ receptor and $5\text{-HT}_{4(e)}$ receptor) interact with specific intracellular proteins supports the view that these isoforms are in charge of the fine regulation of signal transduction and are not the result of transcriptional leakage (Joubert *et al.*, 2004). Two 5-HT_5 receptor genes have been described in the rodent, but only the 5-HT_{5A} receptor can be generated in humans (Rees *et al.*, 1994). A 5-HT_{5B} receptor pseudogene had been localized on chromosome 2 in Build 35.1 of the human genome; however, this annotation has been permanently suppressed in the actual build (36.3) because of insufficient evidence that this locus could be transcribed. The 5-HT_6 receptor gene presents two splicing sites: the first lies in the middle of the intracellular loop 3 and is similar to the splicing site found in the 5-HT_{5A} receptor gene; the second separates exons encoding TM6 and TM7 (Figure 2). A non-functional splice variant has been described for this receptor (Olsen *et al.*, 1999). The human 5-HT_7 receptor gene is more complex: different acceptor and donor sites can be used to generate three receptors that differ in the composition and length of their C-terminal domains ($5\text{-HT}_{7(a)}$, $5\text{-HT}_{7(b)}$, and $5\text{-HT}_{7(d)}$). Because of its altered intron–exon organization, the rat gene generates the $5\text{-HT}_{7(c)}$ but not the $5\text{-HT}_{7(d)}$ isoform (Heidmann *et al.*, 1997). This splicing position giving rise to alternate C-terminuses is similar to that found in the 5-HT_4R gene. The different C-terminal domains of the 5-HT_7 receptor variants can probably recruit specific sets of intracellular proteins that might modulate their function as demonstrated for the 5-HT_4 receptor isoforms.

Ionotropic 5-HT₃ receptor genes

The subunits of the pentameric 5-HT_3 ion channel are encoded by five different genes (Barnes *et al.*, 2009; see also Figure 2). HTR3A and HT3B genes arose by duplication in the same locus (11q23.1). Other duplication events generated multiple copies of the 5-HT_3 receptor gene in the 3q27.1 locus, generating the HTR3C, HTR3D and HTR3E genes, and even a pseudogene (htr3c2). With the exception of HTR3D, all genes presented a common structure of eight to nine exons, the TM regions being encoded by the two to three last exons. An alterative splicing site in the HTR3A gene generates two splice variants of 510 and 478 amino acids (annotated respectively $5\text{-HT}_{3A(a)}$ and $5\text{-HT}_{3A(b)}$ in the NCBI Gene Database) which differ in TM3 and intracellular loop 3 (Belelli *et al.*, 1995; Miyake *et al.*, 1995; see also Figure 2). The number of 5-HT_{3A} receptor isoforms is not fixed to date, as additional splice variants have been reported (Bruss *et al.*, 2000). The HTR3D gene has been described as encoding a 4TM-subunit of 279 residues, lacking the major part of the large N-terminal domain (Niesler *et al.*, 2007; see also Figure 2). However, two alternative transcripts of this gene have been reported that could produce proteins of 454 or 233 amino acids (see accession numbers Q70Z44-1 and Q70Z44-3 in the Uniprot Database, http://www.uniprot.org). Further work remains to be done to conclude the relevance of this observation.

Editing

RNA editing is a dynamic mechanism that generates molecular and functional diversity. The most common type of RNA editing involves nucleotide substitution that consists of either cytidine to uridine (C to U) or, most frequently,

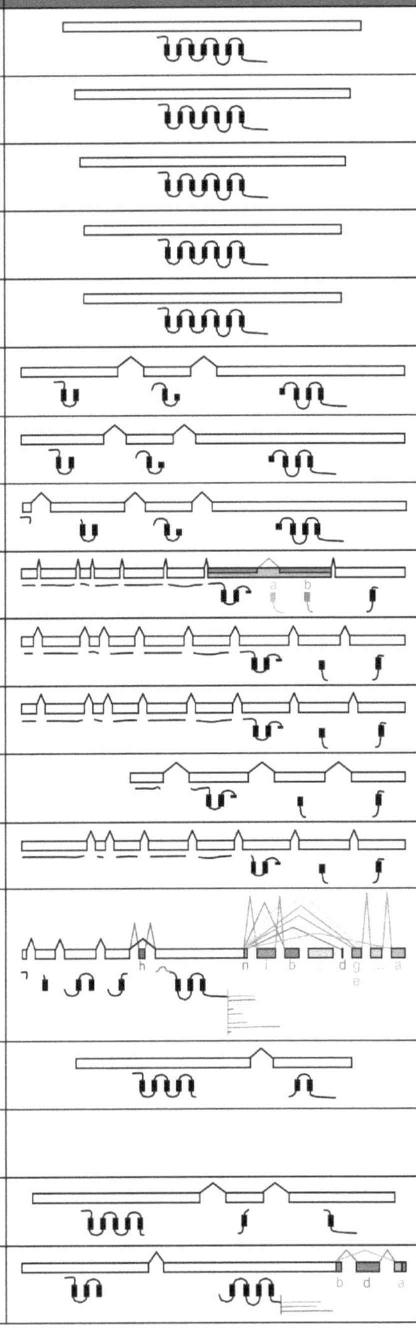

Gene	Gene structure	Gene products	Cds
HTR1A 5q11.2-q13		5-HT_{1A}: 422aa	1269b
HTR1B 6q13		5-HT_{1B}: 390aa	1173b
HTR1D 1p36.3-p34.3		5-HT_{1D}: 377aa	1134b
HTR1E 6q14-q15		5-HT_{1E}: 365aa	1098b
HTR1F 3p12		5-HT_{1F}: 366aa	1101b
HTR2A 6q14-q21		5-HT_{2A}: 471aa	>60kb
HTR2B 2q36.3-q37.1		5-HT_{2B}: 481aa	>15kb
HTR2C Xq24		5-HT_{2C}: 458aa	>183kb
HTR3A 11q23.1		$5\text{-HT}_{3A(a)}$: 510aa $5\text{-HT}_{3A(b)}$: 478aa	>15kb
HTR3B 11q23.1		5-HT_{3B}: 441aa	>41kb
HTR3C 3q27.1		5-HT_{3C}: 447aa	>7kb
HTR3D 3q27.1		5-HT_{3D}: 279aa	>7kb
HTR3E 3q27.1		5-HT_{3E}: 471aa	>6kb
HTR4 5q31-q33		$5\text{-HT}_{4(a)}$: 387aa $_{(b)}$: 388aa, $_{(c)}$: 411aa, $_{(d)}$: 360aa, $_{(e)}$: 371aa, $_{(f)}$: 363aa, $_{(g)}$: 378aa, $_{(hb)}$: 402aa, $_{(i)}$: 428aa, $_{(n)}$: 359aa,	>185kb
HTR5A 7q36.1		5-HT_{5A} : 357aa	>14kb
HTR5B 2q14.1		Pseudogene	none
HTR6 1p36-p35		5-HT_{6} : 440aa	>15kb
HTR7 10q21-q24		$5\text{-HT}_{7(a)}$: 445aa $5\text{-HT}_{7(b)}$: 432aa $5\text{-HT}_{7(d)}$: 479aa	>116kb

Figure 2 Human gene structures of 5-HT receptor coding sequences. This figure was generated using released data of the human genome project available on the NCBI site (http://www.ncbi.nlm.nih.gov) using Map Viewer interface, Gene Database and Consensus CDS Database. Builds 36.2–36.3 were used to localize the exons, except for HTR5B which was only annotated on build 35. The first column presents gene names and cytogenetic localizations. The second column schematizes the exon composition of the coding sequences and the part of the receptor encoded by each exon: open boxes, exons, bridges, splicing events, black boxes, transmembrane domains, curved lines, loops or N- or C-terminal domains. Colors indicate alternative exons and the receptor part encoded in the splice variants. The third column indicates the name and size of the gene products and splice variants (aa, amino acids). The fourth column presents size of the complete coding sequence on the human genome (b, bases; kb, kilobases). To see the full color version of this figure please refer to the color plate at the back of the book. Copies produced via our print on demand service do not contain color plates; if your copy does not have the color plate, please go to this website to view the figure in color www.elsevierdirect.com/companions/9780123746344

adenosine to inosine (A to I). Inosine is read as guanosine by the translational machinery of the cell (Higuchi *et al.*, 1993). The $5HT_{2C}$ receptor is the only known GPCR to be regulated by editing. Editing takes place at four major positions on exon V (corresponding to the second exon in the coding sequence), termed sites A, B, C and D, and one minor site, termed E. 5-HT_{2C} receptor mRNA editing has been identified as A to I discrepancies between genomic and cDNA sequences in the region encoding the second intracellular loop of the receptor. This A to I substitution alters the coding potentials of three codons. Theoretically, 5-HT_{2C} receptor editing can generate up to 32 different mRNAs that encode 24 receptor isoforms from the non-edited form (INI) to the fully edited form (VGV) (Fitzgerald *et al.*, 1999). The editing of 5-HT_{2C} transcripts is tissue-specific. Seven major 5-HT_{2C} receptor isoforms, encoded by 11 distinct RNA species, are expressed in rat brain, compared with 14 receptor isoforms encoded by 23 trancripts in human brain. The most prevalent isoform in human brain is 5-HT_{2C}-VSV, whereas 5-HT_{2C}-VNV is the most abundant in the rat. Editing is regulated by 5-HT itself. Depletion of 5-HT increases the expression of forms having a high affinity for 5-HT, whereas stimulation of 5-HT_{2C} receptors has the opposite effect (Gurevich *et al.*, 2002a). Altered RNA editing of the 5-HT_{2C} receptor may play a role in depression. 5-HT_{2C} receptor mRNA isoforms expressed in human prefrontal cortex differ in tissues obtained from depressed suicide victims compared with healthy controls. In suicide patients, three abundant isoforms, resulting from an increase at the E and C sites and a decrease at the D site and encoding receptors, are expressed. These isoforms exhibit decreased basal activity and decreased agonist affinity and potency (Gurevich *et al.*, 2002b).

Recently, it has been reported that the small nucleolar RNA (snoRNA) HBII-52, which exhibits sequence complementarity to the alternative spliced exon V encoding for a functional 5-HT_{2C} receptor, regulates receptor mRNA splicing. Notably, Prader-Willy syndrome patients (characterized by neonatal muscular hypotonia, early childhood obesity, hypogonadism and mental retardation) who do not express HBII-52 snoRNA exhibit abnormally low levels of the non-edited 5-HT_{2C} receptor.

G protein-mediated signaling

The pleiotropic functions of 5-HT are not only exerted by concerted actions of multiple receptor subtypes, splicing and editing, but also by multiple coupling of each receptor species. Since we have written two recent reviews (Bockaert *et al.*, 2006; Millan *et al.*, 2008) on the subject, we will focus here on the more recent and original aspects of this important question. In addition, we will try to make the distinction between signaling events established in native cells versus those observed when the receptor is expressed in cell lines.

5-HT_{1A} receptor

This is an inhibitory receptor coupled to Gi/Go inhibiting the adenylyl-cyclase but also opening G-protein-gated inwardly rectifying gK^+ (GIRK) and inhibiting gCa^{2+} (Figure 3A). Interestingly, 5-HT_{1A} receptors have different properties depending on their localization on 5-HT neurons (auto-receptors) versus on postsynaptic heterologous neurons. In particular, drugs have different efficacies, and the auto-receptors have a much greater ability to desensitize. The reasons are still obscure, but one possibility may be that autoreceptors are preferentially coupled to Gi_3 and postsynaptic receptors to Go (Mannoury la Cour *et al.*, 2006). Other signaling events mediated through activation of Gi/Go in native tissues (and in some cases *in vivo*) include: (1) activation or inhibition of ERK (extracellular signal-regulated kinase), depending on the brain area (Crane *et al.*, 2007); (2) activation of the Akt pathway (Cowen *et al.*, 2005); and (3) inhibition of NMDA transmission in prefrontal cortex, an effect potentiated by dampening RGS4 function (Gu *et al.*, 2007). RGS4 is an RGS gene associated with schizophrenia (Gu *et al.*, 2007). In COS7 cells, 5-HT_{1A} receptors also stimulate both an ERK-dependent anti-apoptotic pathway and a Janus-kinase 2 (Jak2)-dependent pro-apoptotic pathway as well as c-jun N-terminal kinase (Turner *et al.*, 2007a, 2007b).

5-$HT_{1B/1D/1E/1F}$ receptors

5-HT_{1B} receptors are negatively coupled to adenylyl-cyclase via Gi/Go in substantia nigra (Bouhelal *et al.*, 1988; Hamblin and Metcalf, 1991). In transfected cells, 5-HT_{1B} receptors regulate PLC, PLD, Akt kinase, neuronal nitric oxide synthase (nNOS), and the ERK pathway. They also activate K^+ channels and inhibit Ca^{2+} channels. The negative coupling of 5-$HT_{1D/1E/1F}$ receptors to adenylyl-cyclase has only been demonstrated in transfected cells, but likely occurs *in vivo* (Bockaert *et al.*, 2006).

5-HT_{2A} receptors

5-HT_{2A} receptors activate PLC, PLA2 and the ERK pathway (downstream of PLC) in neurons (Bockaert *et al.*, 2006; Figure 3B). Recently, an original signaling pathway has been associated with 5-HT_{2A} receptor stimulation

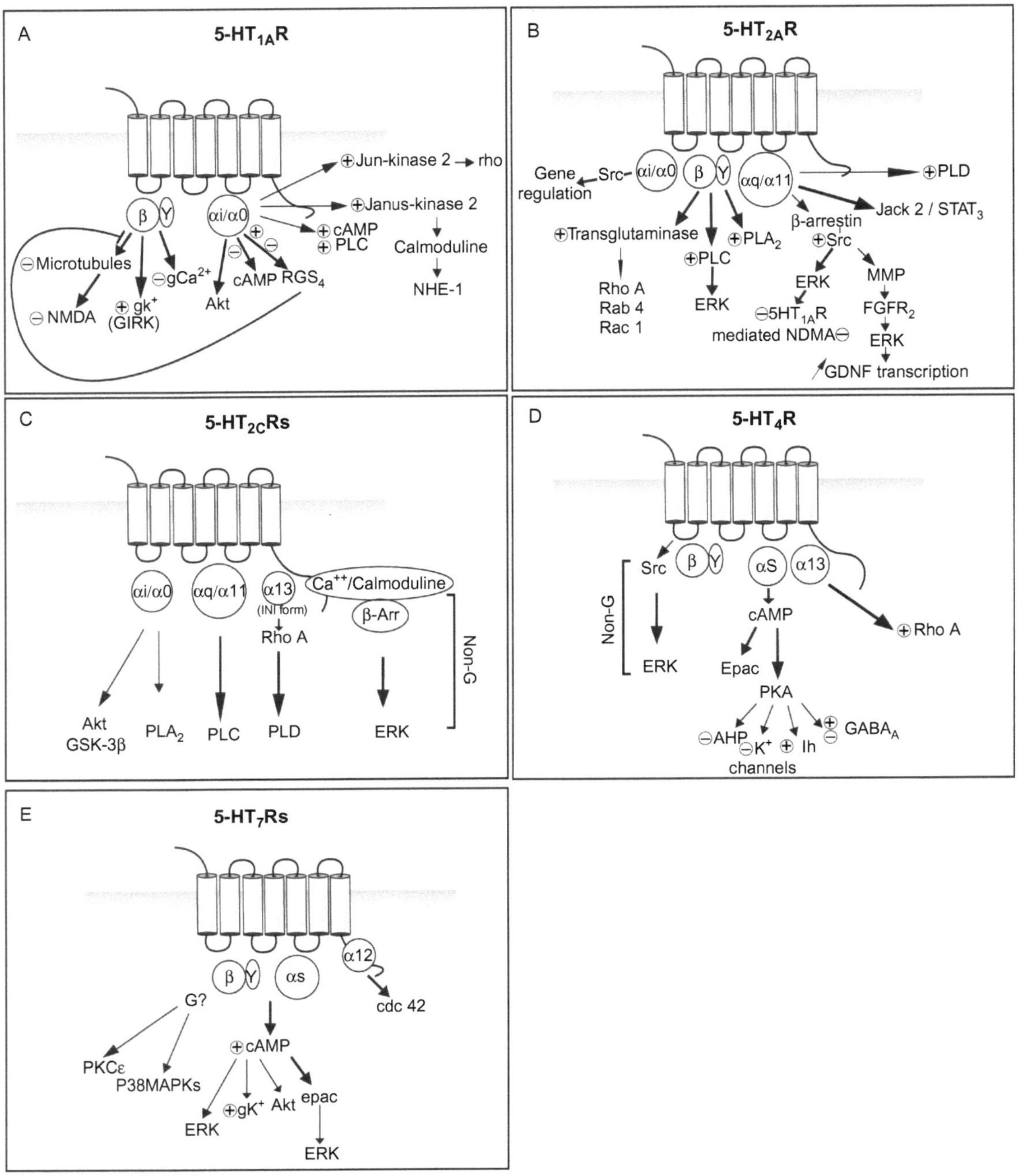

Figure 3 Signaling pathways of major 5-HT receptors. Heavy arrows indicate the signaling pathways demonstrated in native cells, the others being described in heterologous cells.

in platelets and a rat cortical cell line: the covalent link of serotonin to a glutamine residue of small G proteins such as RhoA and Rab4 as well as Rac1, a process which renders them constitutively active (Walther *et al.*, 2003; Dai *et al.*, 2008). Physical interaction of 5-HT-bound, activated Rab4 (Rab4-GTP) with the serotonin transporter (SERT) has been found to 'paralyze' its translocation from intracellular compartments to the plasma membrane (Ahmed *et al.*, 2008). 5-HT_{2A} receptors inhibit the 5-HT_{1A}-mediated modulation of NMDA current in pyramidal neurons via an arrestin/Src/ERK pathway (Yuen *et al.*, 2008). In C6 glioma cells, 5-HT_{2A} activates matrix metalloproteinases (MMPs) via the Src pathway, the release of FGF receptor 2 ligand, the activation of FGFR2 receptor, the activation of downstream ERK and transcription of GDNF (Tsuchioka *et al.*, 2008). Thus, 5-HT_{2A} receptors activate ERK through several pathways. They also activate the Janus kinase-2 (Jak2/STAT) transcription pathway in skeletal and smooth muscles (Guillet-Deniau *et al.*, 1997; Banes *et al.*, 2005) and PLD in non-neuronal cells.

5-HT$_{2B}$ receptors

Similarly to 5-HT$_{2A}$ receptors, 5-HT$_{2B}$ receptors are coupled to Gq/G11 proteins and activate PLC. However, to date, no study has reported such a coupling within neurons. 5-HT$_{2B}$ receptors have also been reported to stimulate Ca^{2+} mobilization in astrocytes derived from various rat cerebral regions. Additional 5-HT$_{2B}$ receptor-mediated signaling pathways have been described in non-neuronal cells expressing endogenous receptors. Stimulation of 5-HT$_{2B}$ receptors is associated with an increase in cyclic GMP through the dual activation of constitutive and inducible NOS. Gq/11-coupled 5-HT$_{2B}$ receptors also activate PI3K/Akt and ERK1/2 signaling pathways (Bockaert *et al.*, 2006).

5-HT$_{2C}$ receptors

5-HT$_{2C}$ receptors have been reported to activate PLC via Gq/11 in many brain regions, including the choroid plexus (Sanders-Bush and Breeding, 1988; Figure 3C). The non-edited form (5-HT$_{2CINI}$) is more potent in stimulating PLC, and exhibits higher constitutive activity than the edited forms (Niswender *et al.*, 1999). Moreover, 5-HT$_{2CINI}$ receptors are capable of coupling to the G13 protein, activating PLD signaling pathway and increasing Rho GTPase activity (McGrew *et al.*, 2002), whereas the fully edited isoform (5-HT$_{2CVGV}$) fails to activate G13. RNA editing has an important implication for signal transduction *in vivo* (Price et al., 2001).

In CHO cells, 5-HT$_{2C}$ receptors also activate Gi/Go, modulating PLA2 and Akt/GSK3-β (Cussac *et al.*, 2002).

5-HT$_4$ receptors

Since 5-HT$_4$ receptor was discovered from its capacity to activate adenylyl cyclase, the 5-HT$_4$Rs mediated Gs/cAMP/PKA signaling pathway is obviously the best characterized (Figure 3D). In transfected cell lines other couplings have been found, including coupling to Gq and G13 (Bockaert *et al.*, 2006, 2008). Following PKA activation, a series of ionic currents are modulated. These include a long-lasting inhibition of K^+ currents (mediated by a PKA-dependent inhibition of phosphatases), which enhances neuronal excitability, and a decrease in spike accommodation (Ansanay *et al.*, 1995), inhibition of AHP, activation of I_H, and activation or inhibition of GABA$_A$ receptor (Bockaert *et al.*, 2006, 2008). cAMP produced upon 5-HT$_4$ receptor activation also stimulates the exchange factor Epac, which activates α-secretase and the release of sAPPα (amyloid precursor protein) via an Epac/Rap1/Ras pathway (Lezoualc'h and Robert, 2003).

5-HT$_5$ receptor

The signaling pathways associated with 5-ht$_5$ putative receptors are still obscure. No data are available in native cells (Hannon and Hoyer, 2008).

5-HT$_6$ receptors

Like 5-HT$_4$ receptors and 5-HT$_7$ receptors, 5-HT$_6$ receptors are positively coupled to adenylyl-cyclase in neurons (Sebben *et al.*, 1994). 5-HT$_6$ receptors also activate the ERK pathway in CHO cells via a mechanism partially dependent on PKA and requiring the tyrosine kinase Fyn (Yun *et al.*, 2007).

5-HT$_7$ receptors

The coupling of 5-HT$_7$ receptors is more documented than that of the 5-HT$_6$ receptor (Figure 3E). The 5-HT$_7$ receptor has been recognized as the receptor coupled to adenylyl cyclase exhibiting a high affinity for 5-carboxamidotryptamine (5-CT) in the guinea-pig hippocampus (Shenker *et al.*, 1987; Dumuis *et al.*, 1988b). 5-HT$_7$ receptors also engage the ERK1/2 pathway in hippocampal neurons via activation of Epac (Lin *et al.*, 2003). Activation of ERK via PKA is also possible. 5-HT$_7$ receptor activation increases neurite length through a Gα12/RhoA/Cdc42 signaling pathway in hippocampal neurons (Lin *et al.*, 2003; Kvachnina *et al.*, 2005). In astrocytoma cells, 5-HT$_7$ receptors activate p38 and PKCε. The implication of G proteins in this pathway is not demonstrated (Lieb *et al.*, 2005).

Non-G-protein dependent signaling

Recently, many GPCRs have also been shown to signal without activation of G proteins. This non-G-protein mediated signaling generally implicates interactions with GPCR interacting proteins (Bockaert *et al.*, 2004). A classic non-G signaling pathway, which implicates β-arrestins, is the ERK pathway. Long-lasting activation (>5 minutes up to several hours) of ERK induced by many GPCRs, including β-adrenergic receptors or AT$_{1A}$

angiotensin II receptors, is due to their co-internalization with β-arrestin after their uncoupling from G proteins, and takes place in the cytosol. Internalized GPCR-β-arrestin complex recruits many proteins, including the Src–Raf–MEK–ERK module. This contrasts with the rapid (<5 min) activation of ERK evoked by GPCRs, which is generally G-protein-dependent and leads to nuclear translocation of ERK. However, several GPCRs, including 5-HT receptors, differ from this general scheme. For instance, both early and sustained phases of ERK activation mediated by the 5-HT_{2C} receptor are independent of the receptor's cognate G proteins (Gq, G13 and Gi/o) and require physical interaction of calmodulin with the receptor C-terminal domain, and β-arrestin1 and 2 recruitment by the receptor (Labasque *et al.*, 2008) (Figure 3C). This non-G signaling pathway, established not only in transfected HEK-293 cells but also in cortical neurons and choroid plexus epithelial cells expressing native 5-HT_{2C} receptor, might be involved in neurogenesis induced by chronic treatment with 5-HT_{2C} receptor agonists and their antidepressant-like activity. Activation of 5-HT_4 receptors likewise induces a sustained ERK phosphorylation in both transfected HEK-293 cells and cultured colliculi neurons (Barthet *et al.*, 2007). Again, this activation was mostly Gs/cAMP/PKA-independent and did not involve any other G proteins (Gq, Gi, Go) and associated downstream messengers. Interestingly, 5-HT_4 receptor-mediated ERK activation did not involve β-arrestins but was dependent on Src tyrosine kinase. This receptor-operated pathway might be important for 5-HT_4 receptor-mediated long-term potentiation (LTP) (Huang and Kandel, 2007). Collectively, these findings demonstrated that multiple non-G mechanisms, either requiring or not requiring recruitment of β-arrestins and additional partners, can contribute to the activation of the ERK pathway by 5-HT receptors.

Constitutive activity

Some GPCRs are capable of activating G proteins in the absence of agonists. This intrinsic activity was mainly established for heterogeneously expressed GPCRs, but also in authentic cellular contexts for some GPCRs. Some antagonists can inhibit these 'constitutive' activities and thus behave as 'inverse agonists'. Several 5-HT receptors exhibit constitutive activity in neurons. Constitutive activity at postsynaptic 5-HT_{1A} receptors expressed in hippocampus has been described. Such an intrinsic activity has not been yet demonstrated for 5-HT_{1A} autoreceptors (Martel *et al.*, 2007). As previously mentioned, constitutive activity of 5-HT_{2C} receptors is highly dependent on the degree of mRNA edition: non-edited or partially edited versions of the receptor exhibit the highest levels of constitutive activity, whereas the fully edited one is devoid of constitutive activity (Berg *et al.*, 2005; Chanrion *et al.*, 2008). Constitutive activation of non-edited or partially edited isoforms is associated with constitutive internalization (Marion *et al.*, 2004). Physiological constitutive activity of 5-HT_{2A} receptors might explain impairment of associative learning by inverse agonists (Berg *et al.*, 2005). 5-HT_4 receptors display a particularly high constitutive activity in transfected cells, depending of the splice variant analyzed. The shorter ones exhibit a higher constitutive activity than the longer ones (Bockaert *et al.*, 2008). Unfortunately, no data are available concerning a possible *in vivo* existence of constitutive activity. Constitutive activity of native 5-HT_7 receptors also remains to be demonstrated (Millan *et al.*, 2008).

Agonist-directed signaling

It has been recently proposed that a given GPCR can be stabilized under different active conformational states depending of the nature of the agonist – a phenomenon called 'ligand-directed signaling' (LDS) (Kenakin, 2007). LDS has been demonstrated for several 5-HT receptor subtypes. 5-HT_{1A} receptor agonists show differential efficacies to stimulate various G proteins and to activate different channels such as GIRKs versus smooth inward-current channels (I_{smooth}). The different efficacies of agonists at autoreceptors versus postsynaptic receptors might reflect their coupling to Gi_3 versus Go, respectively (Millan *et al.*, 2008). The rank order of efficacy for recruiting human 5-HT_{2C} receptors differs for PLC versus PLA2 and for Gq/11 versus Gi_3. Similarly, contrasting agonist efficacies at PLC versus PLA2 have been observed for human 5-HT_{2A} receptors (Millan *et al.*, 2008). A particularly interesting example of LDS at 5-HT receptors is the different signaling induced by psychedelic hallucinogens (such as LSD, DOI and (−)DOB) versus non-hallucinogen agonists (such as lisuride) of 5-HT_{2A} receptors. Both ligand categories engage the Gq/PLC pathway, leading to *c-fos* induction in the somato-sensory cortex and the medial-prefrontal cortex (Figure 4) (Benneyworth *et al.*, 2007; Gonzalez-Maeso *et al.*, 2007). In contrast, only hallucinogenic drugs activate a Gi/o-Src pathway, leading to induction of *egr-2* expression in somato-sensory cortex and cortical neurons in primary culture (Gonzalez-Maeso *et al.*, 2007; Benneyworth *et al.*, 2007). Recent reports add to the complexity and interest of this hallucinogenic ligand signaling. A physiological antagonism exists between 5-HT_{2A} and $mGlu_{2/3}$ receptors in these cortical areas (Marek *et al.*, 2000). Activation of $mGlu_{2/3}$ receptors attenuates behavioral effects of hallucinogens known to

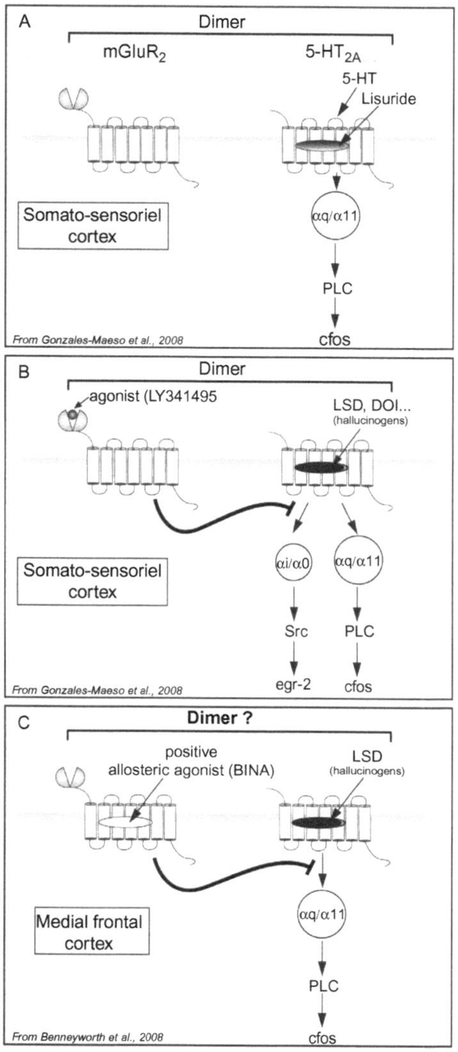

Figure 4 Hallucinogens effects of some 5-HT_{2A} agonists such as LSD and DOI: implication of 'ligand directed signaling' of 5-HT_{2A} receptors and inhibition of hallucinogens effects by metabotropic glutamate receptors. (A) In somato-sensory cortex, Gonzalez-Maeso *et al.* (2008) showed a heterodimerization between mGluR2 and 5-HT_{2A} receptors. In cortical cultures and somato-sensory cortex, stimulation by non-hallucinogenic such as lisuride, 5-HT itself, and hallucinogenic drugs such as LSD and DOI engaged the Gq/G_{11}/PLC/c-fos pathway (Gonzalez-Maeso *et al.*, 2007, 2008). (B) In cortical cultures (Gonzalez-Maeso *et al.*, 2007) and somato-sensory cortex (Gonzalez-Maeso *et al.*, 2007, 2008), hallucinogenic drugs such as LSD and DOI, but not non-hallucinogenic drugs, activated, in addition to the Gq/G_{11}/PLC/c-fos pathway, the Gi/Go/Src/erg-2 pathway. In the somato-sensory cortex, activation of mGluR2 with LY 341495 inhibited the Gi/Go/Src/erg-2 pathway but not the Gq/G_{11}/PLC/c-fos pathway. LY 341495, an mGluR2 othosteric agonist, inhibited the stereotypical head-twitch behavior induced by hallucinogens (Gonzalez-Maeso *et al.*, 2008). (C) In the medial frontal cortex, the allosteric agonist BINA inhibited the stereotypical head-twitch behavior and c-fos induction of hallucinogens. Whether or not there is a regional difference (somato-sensory versus medial frontal cortex) between the control triggered by mGluR2 on the c-fos induction by hallucinogens remains to be seen (Benneyworth *et al.*, 2007).

produce a state of drug intoxication that resembles many of the symptoms of acute schizophrenia (Marek *et al.*, 2000). Consistent with this hypothesis, data from a recent phase II clinical trial show that an $mGlu_{2/3}$ receptor agonist is an effective antipsychotic therapy (Aghajanian and Marek, 1999; Schoepp, 2006). The mechanism of $mGlu_{2/3}$ receptor agonist effect is controversial. $mGlu_{2/3}$ receptors are supposed to be localized on glutamatergic terminals connecting pyramidal neurons. 5-HT_{2A} receptors are postsynaptically localized on layer V pyramidal neurons. Since activation of $mGlu_{2/3}$ receptors inhibits the 5-HT_{2A} receptor-induced excitatory postsynaptic currents, the following hypothesis has been proposed to explain the effects of $mGlu_{2/3}$ receptors: (1) activation of postsynaptic layer V pyramidal neurons mediates induction of glutamate release from presynaptic nerve terminals connecting these neurons via a retrograde messenger; (2) glutamate induces postsynaptic excitatory currents in layer V pyramidal neurons, and it is this release which is inhibited following activation of presynaptic $mGlu_{2/3}$ receptors (Aghajanian and Marek, 1999; Benneyworth *et al.*, 2007). However, this hypothesis has been recently challenged by a report showing that $mGlu_2$ receptors can form heterodimers with 5-HT_{2A} receptors in postsynaptic pyramidal neurons (Gonzalez-Maeso *et al.*, 2008). Activation of $mGlu_2$ receptors will 'directly' inhibit the hallucinogen-induced activation of the Gi/o–Src pathway, preventing induction of *egr-2* (Figure 4), without affecting the Gq–PLC pathway and *c-fos* induction (Gonzalez-Maeso *et al.*, 2008). This absence of inhibition of hallucinogen-induced Gq/*c–fos* pathway by mGlu2 receptors observed in the somato-sensory cortex by Gonzales-Maseo and colleagues contrasts with the observation made in the prefrontal cortex by Benneyworth and colleagues (2007). Further work is obviously needed to resolve this important paradox.

Toward the notion of a 5-HT receptor complex

An increasing body of evidence indicates that 5-HT receptors are not 'isolated islands' floating within the plasma membrane, but that they are part of multiprotein complexes that incorporate a dimer (or an oligomer) of receptors associated with numerous intracellular proteins in addition to G proteins and the ubiquitous GPCR signaling modulators β-arrestins. To date, 5-HT receptors are among the GPCRs for which interacting proteins have been the most extensively characterized (Table 1), often by means of yeast two-hybrid screens or proteomic strategies. The list is still growing, and will continue to grow in the future, as the majority of interactions certainly remain to be discovered. Although the functional consequences of most of them remain to be established

Table 1 Proteins identified as specific partners of 5-HT receptors and their role at the cellular level and *in vivo* (only receptor subtypes for which interacting proteins have been unambiguously identified are mentioned)

		Interaction domain			
Receptor subtype	**Interacting protein**	**Intracellular loops**	**C-terminus**	**Function of identified interactions**	**References**
5-HT$_{1A}$	Calmodulin	+		Signaling, trafficking	Turner *et al.* (2004), Della Rocca *et al.* (1999)
	Yif1B		+	Dendritic targeting	Carrel *et al.* (2008)
5-HT$_{1B}$	P11	+		Trafficking, signaling	Svenningsson *et al.* (2006)
5-HT$_{2A}$	PDZ proteins*		+	Dendritic targeting	Xia *et al.* (2003b)
	PSD-95	+	+	Trafficking, signaling	Xia *et al.* (2003a), Becamel *et al.* (2004)
	MUPP1	+	+	Trafficking, targeting	Becamel *et al.* (2001)
	SAP97		+	Signaling	Becamel *et al.* (2004)
	MAGI2		+	Targeting	Becamel *et al.* (2004)
	MPP3		+	Signaling	Becamel *et al.* (2004)
	CIPP	+	+	Phosphorylation,	Becamel *et al.* (2004)
	Caveolin 1,2		+	desensitization	Bhatnagar *et al.* (2004)
	ARF1-6		+	Signaling	Robertson *et al.* (2003)
	MAP1A		+	Signaling	Cornea-Hebert *et al.* (2002)
	JAK2/STAT3			(NO synthesis)	Guillet-Deniau *et al.* (1997)
	RSK2		+	Desensitization	Sheffler *et al.* (2006), Strachen *et al.* (2008)
	Calmodulin		+	Receptor phosphorylation	Turner and Raymond (2005)
5-HT$_{2B}$	PDZ proteins*		+	Desensitization, trafficking	Manivet *et al.* (2000)
	MUPP1		+	Desensitization, trafficking	Becamel *et al.* (2001)
5-HT$_{2C}$	PDZ proteins*		+	Signaling	Backstrom *et al.* (2000)
	MUPP1		+	Receptor dephosphorylation, addiction	Ullmer *et al.* (1998), Becamel *et al.* (2001), Parker *et al.* (2003)
	PSD-95		+	Trafficking, signaling, antidepressant response to receptor stimulation	Becamel *et al.* (2002), Becamel *et al.* (2004), Gavarini *et al.* (2006)
	SAP97		+	Targeting	Becamel *et al.* (2004)
	SAP102		+	Targeting	Becamel *et al.* (2004)
	MAGI-2		+	Signaling	Becamel *et al.* (2004)
	MPP3		+		Becamel *et al.* (2002), Becamel *et al.* (2004), Gavarini *et al.* (2006)
	Veli3/CASK/Mint1		+		Becamel *et al.* (2002), Becamel *et al.* (2004)
	Calmodulin		+	Trafficking, signaling	Becamel *et al.* (2002), Labasque *et al.* (2008)
	PICOT				Becamel *et al.* (2002)
	PTD4		+		Becamel *et al.* (2002)
	PTEN		+		Ji *et al.* (2006)
			+		Warner-Schmidt *et al.* (2009)

Receptor subtype	Interacting protein	Interaction domain		Function of identified interactions	References
		Intracellular loops	C-terminus		
5-HT_4	P11[$]		+		Joubert *et al.* (2004)
5-$HT_{4(a)}$	SNX27		+		Joubert *et al.* (2004)
	NHERF-1 (EBP50)		+		Joubert *et al.* (2004)
	MAGI2				Joubert *et al.* (2004)
	MPP3		+		Joubert *et al.* (2004)
	Veli 1-3		+		Joubert *et al.* (2004)
	CRMP2		+		Barthet *et al.* (2007)
	Src		+		
5-$HT_{4(e)}$	CIPP				Joubert *et al.* (2004)
	nNOS				Joubert *et al.* (2004)
	Sec23				Joubert *et al.* (2004)
5-HT_6	Fyn				Yun *et al.* (2007)

[*]Refers to experiments using receptors mutated on their PDZ binding motif.
[$]P11 binds to the third intracellular loop of 5-HT_4 receptors and therefore might equally interact with all receptor splice variants, including 5-HT_{4a} and 5-HT_{4e} receptors.

in authentic cellular contexts, a set of recent studies highlights their impact in CNS physiology and suggests their possible therapeutic exploitation for the treatment of CNS disorders. We do not provide an extensive description of 5-HT receptor interacting proteins here (see Table 1), as the majority of these interactions have been reviewed elsewhere (Bockaert *et al.*, 2004, 2006), but rather illustrate, using selected examples, their influence on receptor signal transduction and subcellular compartmentation and their possible implications in psychiatric disorders such as depression and addiction.

5-HT receptor-interacting proteins can modulate G protein-operated signaling by influencing localization of receptors at the plasma membrane, their desensitization and their coupling to G proteins. 5-$HT_{2A/2C}$ receptors express PDZ binding motifs at their C-terminal extremity, which recruit specific sets of PDZ domain-containing proteins (PDZ proteins) (Becamel *et al.*, 2002, 2004; Gavarini *et al.*, 2004). These interactions govern receptor trafficking in and out of the plasma membrane, and receptor desensitization. These effects depend on both the nature of the PDZ protein bound to the receptor and the receptor subtype. Association of the 5-HT_{2A} receptor with PSD-95 stabilizes receptor at the plasma membrane (Xia *et al.*, 2003a). Accordingly, this interaction amplifies receptor-operated signals. In contrast, interaction of PSD-95 with 5-HT_{2C} receptors increases spontaneous and agonist-dependent receptor internalization, whereas association of the receptor with another PDZ protein, MAGUK p55 subfamily member 3 (MPP3), prevents both receptor desensitization and internalization (Gavarini *et al.*, 2006). Overall, disruption of 5-HT_{2C} receptor/PDZ protein interactions in cortical neurons increases receptor desensitization, indicating that PDZ proteins predominantly inhibit receptor desensitization and internalization in neurons. Besides PDZ proteins, p11 (also called S100A10 or calpactin I light chain or annexin II light chain) is a member of the S100 Ca^{2+}-binding protein superfamily that binds to several 5-HT receptor subtypes and affects receptor trafficking. Moreover, p11 is unique among the S100 family members for its capacity to interact with 5-HT receptors (Warner-Schmidt *et al.*, 2009). P11 was first identified, using yeast two-hybrid screening, as a 5-HT_{1B} receptor interacting protein (Svenningsson *et al.*, 2006). P11/5-HT_{1B} receptor interaction promotes receptor expression at the plasma membrane, and thereby enhances receptor signal transduction efficacy (Svenningsson *et al.*, 2006). More recently, a physical interaction between p11 and 5-HT_4 receptors has been described (Warner-Schmidt *et al.*, 2009). This interaction likewise increases 5-HT_4 receptor cell surface expression and downstream signaling.

Interaction of 5-HT receptors with accessory proteins also affects receptor phosphorylation and thereby receptor desensitization. 5-HT_{2A} receptors (third intracellular loop) bind to p90 ribosomal S6 kinase 2 (RSK2), which induces receptor phosphorylation at Ser^{314} located in the second intracellular loop (Sheffler *et al.*, 2006; Strachen *et al.*, 2008). Ser^{314} phosphorylation leads to a tonic inhibition of receptor-operated signaling. Moreover, 5-HT_{2A} receptor phosphorylation by RSK2 is required for epidermal growth factor-mediated heterologous desensitization of the 5-HT_{2A} receptor. Another partner that might influence receptor phosphorylation and desensitization is calmodulin (CaM). The CaM binding site on the 5-HT_{2A} receptor C-terminus overlaps with a putative PKC phosphorylation site. In fact, CaM binding to the receptor C-terminus and receptor phosphorylation by PKC were found to be antagonistic, suggesting a role of CaM in the regulation of receptor desensitization (Turner and Raymond, 2005). CaM association with 5-HT$_{1A}$ receptor likewise interferes with receptor phosphorylation by PKC (Turner *et al.*, 2004).

Interacting proteins might also directly influence coupling of 5-HT receptors to G proteins. For instance, interaction of CaM with 5-HT$_{2A}$ receptors (second intracellular loop) prevents their coupling to Gq (Turner and Raymond, 2005), whereas receptor association with caveolins facilitates coupling to Gq (Bhatnagar *et al.*, 2004).

Beyond G protein-dependent signal transduction, interaction of 5-HT receptors with accessory proteins might promote non-G signaling. As already mentioned, constitutive interaction of 5-HT$_4$ receptor with Src is essential for receptor-operated ERK1,2 signaling, which is G-protein- and arrestin-independent (Barthet *et al.*, 2007). Activation of the ERK pathway by 5-HT$_{2C}$ receptor is likewise independent of the G proteins known to be activated by the receptor, but is intimately related to arrestin recruitment – a process facilitated by CaM bound to the receptor C-terminal domain (Labasque *et al.*, 2008). Thus, this non-G signaling seems to involve a tripartite 'signalosome' including the 5-HT$_{2C}$ receptor, CaM and β-arrestin. Finally, activation of the ERK pathway by the 5-HT$_6$ receptor, which is largely insensitive to PKA inhibition, is mediated by a direct interaction of Fyn (*via* its SH3 domain) with the receptor C-terminal domain (Yun *et al.*, 2007).

5-HT receptor-interacting proteins are not only important for signal transduction; they are also key determinants of subcellular distribution of 5-HT receptors. The 5-HT$_{2A}$ receptor PDZ binding motif is essential for the preferential receptor targeting to the dendritic compartment of pyramidal cortical neurons (Xia *et al.*, 2003b). The nature of the PDZ protein(s) that govern dendritic targeting of the receptor remains to be elucidated. PSD-95, the major receptor PDZ partner, is a possible candidate. In dendrites, the receptor is mainly confined in intracellular compartments. This intracellular localization seems to be related to its ability to interact with the microtubule-associated protein MAP1A (Cornea-Hebert *et al.*, 2002). Interaction of 5-HT$_{1A}$ receptor C-terminus with Yif1b, a protein of the endoplasmic reticulum/Golgi trafficking machinery, is essential for receptor targeting in the somatodendritic compartment (Carrel *et al.*, 2008).

The association of 5-HT$_{2A}$ receptor with caveolin-1, detected in both C6 glioma cells and rat brain synaptic membrane preparations, is essential for targeting the receptor to lipid rafts and, as mentioned above, for receptor-mediated signal transduction (Bhatnagar *et al.*, 2004). In pulmonary artery smooth muscle cells, the 5-HT$_{2A}$ receptor forms a ternary complex with caveolin-1 and voltage-gated K^+ channels (K_v1.5), which was found to be critical for 5-HT$_{2A}$ receptor-mediated inhibition of K_v1.5 channels (Cogolludo *et al.*, 2006).

The essential role of 5-HT receptor-interacting proteins in modulating signal transduction suggests that these interactions might be physiologically relevant or contribute to pathological situations. Interaction of p11 with both 5-HT$_{1B}$ and 5-HT$_4$ receptors seems to be required for antidepressant actions of 5-HT$_{1B}$ and 5-HT$_4$ agonists (Svenningsson *et al.*, 2006; Warner-Schmidt *et al.*, 2009). Indeed, p11 knock-out mice exhibit a depression-like phenotype and reduced responsiveness to 5-HT$_{1B}$ and 5-HT$_4$ agonists in rodent behavioral models of depression. Further supporting a role of p11 in depression-like states, p11 brain expression is increased by antidepressant treatments, whereas the level of the protein is reduced in an animal model of depression as well as in the brain of patients who had suffered from unipolar major depression (Svenningsson *et al.*, 2006). Moreover, overexpression of p11 in neurons produces behavioral features reminiscent of those observed after antidepressant treatment. Collectively, these data suggest that alteration of 5-HT$_{1B}$ receptor/p11 and 5-HT$_4$ receptor/p11 interactions might contribute, at least in part, to the etiology of depressed states. 5-HT$_{2C}$ receptors interact with the tumor suppressor PTEN, an enzyme exhibiting both lipid and protein phosphatase activities (Ji *et al.*, 2006). This interaction was detected in dopaminergic neurons of the ventral tegmental area (VTA) innervating the accumbens nucleus, which are tonically inhibited by 5-HT$_{2C}$ receptors, and prevented agonist-induced receptor phosphorylation (Ji *et al.*, 2006). Disruption of 5-HT$_{2C}$ receptor/PTEN interaction in the VTA reproduced the inhibitory effect of 5-HT$_{2C}$ agonists on VTA dopaminergic neuron firing, and prevented the rewarding effects of cannabinoids that were mediated by increased activity of VTA dopaminergic neurons (Ji *et al.*, 2006). Importantly, disconnecting 5-HT$_{2C}$ receptors from PTEN did not reproduce the side effects evoked by 5-HT$_{2C}$ agonists, such as anxiety, penile erection, hypophagia, and suppression of locomotor activity. These data suggest that 5-HT$_{2C}$ receptor/PTEN interaction might contribute to the reinforcing role of drugs, and thereby constitutes a potential target for the treatment of drug addiction-related behaviors.

5-HT$_3$ receptor signaling

The signaling events associated with 5-HT$_3$ receptors are, in principle, very simple. Being cationic-channel permeable Na^+, Ca^{2+} and K^+, activation of 5-HT$_3$ receptors results in cell depolarization. 5-HT_3 receptors mediate fast synaptic responses postsynaptically, and stimulate the release of 5-HT and some other neurotransmitters presynaptically. There are discrepancies between the biophysical properties of the homo-pentameric 5-HT_{3A} receptors and native 5-HT_3 receptors. Following the cloning of 5-HT_{3B} receptors, these discrepancies have been, at least in part, solved by the observation that the 5-HT_{3AB} receptors have properties more similar to native receptors. The complexity of native channels is likely to be much greater owing to the existence of splice variants of 5-HT_{3A} and 5-HT_{3B} receptors and the possible formation of hetero-pentamers between 5-HT_{3A} and the newly cloned 5-HT_{3C-E} forms (Jensen *et al.*, 2008).

References

Adham, N., Kao, H.T., Schecter, L.E., Bard, J., Olsen, M., Urquhart, D., Durkin, M., Hartig, P.R., Weinshank, R.L. and Branchek, T.A. (1993) Cloning of another human serotonin receptor (5-HT_{1F}): a fifth 5-HT_1 receptor subtype coupled to the inhibition of adenylate cyclase. Proc. Natl. Acad. Sci. USA, 90: 408–412.

Adham, N., Tamm, J.A., Salon, J.A., Vaysse, P.J., Weinshank, R.L. and Branchek, T.A. (1994) A single point mutation increases the affinity of serotonin 5-HT_{1D} alpha, 5-HT_{1D} beta, 5-HT_{1E} and 5-HT_{1F} receptors for beta-adrenergic antagonists. Neuropharmacology, 33: 387–391.

Aghajanian, G.K. and Marek, G.J. (1999) Serotonin, via 5-HT_{2A} receptors, increases EPSCs in layer V pyramidal cells of prefrontal cortex by an asynchronous mode of glutamate release. Brain Res., 825: 161–171.

Ahmed, B.A., Jeffus, B.C., Bukhari, S.I., Harney, J.T., Unal, R., Lupashin, V.V., Van Der Sluijs, P. and Kilic, F. (2008) Serotonin transamidates Rab4 and facilitates its binding to the C terminus of serotonin transporter. J. Biol. Chem., 283: 9388–9398.

Ansanay, H., Dumuis, A., Sebben, M., Bockaert, J. and Fagni, L. (1995) A cyclic AMP-dependent, long-lasting inhibition of a K^+ current in mammalian neurons. Proc. Natl. Acad. Sci. USA, 92: 6635–6639.

Backstrom, J.R., Price, R.D., Reasoner, D.T. and Sanders-Bush, E. (2000) Deletion of the serotonin 5-HT_{2C} receptor PDZ recognition motif prevents receptor phosphorylation and delays resensitization of receptor responses. J. Biol. Chem., 275: 23620–23626.

Banes, A.K., Shaw, S.M., Tawfik, A., Patel, B.P., Ogbi, S., Fulton, D. and Marrero, M.B. (2005) Activation of the JAK/STAT pathway in vascular smooth muscle by serotonin. Am. J. Physiol. Cell Physiol., 288: C805–C812.

Barnes, N.M. and Sharp, T. (1999) A review of central 5-HT receptors and their function. Neuropharmacology, 38: 1083–1152.

Barnes, N.M., Hales, T.G., Lummis, S.C. and Peters, J.A. (2009) The 5-HT(3) receptor – the relationship between structure and function. Neuropharmacology, 56: 273–284.

Barthet, G., Framery, B., Gaven, F., Pellissier, L., Reiter, E., Claeysen, S., Bockaert, J. and Dumuis, A. (2007) 5-hydroxytryptamine 4 receptor activation of the extracellular signal-regulated kinase pathway depends on Src activation but not on G protein or beta-arrestin signaling. Mol. Biol. Cell., 18: 1979–1991.

Becamel, C., Figge, A., Poliak, S., Dumuis, A., Peles, E., Bockaert, J., Lubbert, H. and Ullmer, C. (2001) Interaction of serotonin 5-hydroxytryptamine type 2C receptors with PDZ10 of the multi-PDZ domain protein MUPP1. J. Biol. Chem., 276: 12974–12982.

Becamel, C., Alonso, G., Galeotti, N., Demey, E., Jouin, P., Ullmer, C., Dumuis, A., Bockaert, J. and Marin, P. (2002) Synaptic multiprotein complexes associated with 5-HT(2C) receptors: a proteomic approach. EMBO J., 21: 2332–2342.

Becamel, C., Gavarini, S., Chanrion, B., Alonso, G., Galeotti, N., Dumuis, A., Bockaert, J. and Marin, P. (2004) The serotonin 5-HT_{2A} and 5-HT_{2C} receptors interact with specific sets of PDZ proteins. J. Biol. Chem., 279: 20257–20266.

Belelli, D., Balcarek, J.M., Hope, A.G., Peters, J.A., Lambert, J.J. and Blackburn, T.P. (1995) Cloning and functional expression of a human 5-hydroxytryptamine type 3AS receptor subunit. Mol. Pharmacol., 48: 1054–1062.

Bender, E., Pindon, A., Van Oers, I., Zhang, Y.B., Gommeren, W., Verhasselt, P., Jurzak, M., Leysen, J. and Luyten, W. (2000) Structure of the human serotonin 5-HT_4 receptor gene and cloning of a novel 5-HT_4 splice variant. J. Neurochem., 74: 478–489.

Benneyworth, M.A., Xiang, Z., Smith, R.L., Garcia, E.E., Conn, P.J. and Sanders-Bush, E. (2007) A selective positive allosteric modulator of metabotropic glutamate receptor subtype 2 blocks a hallucinogenic drug model of psychosis. Mol. Pharmacol., 72: 477–484.

Berg, K.A., Harvey, J.A., Spampinato, U. and Clarke, W.P. (2005) Physiological relevance of constitutive activity of 5-HT_{2A} and 5-HT_{2C} receptors. Trends Pharmacol. Sci., 26: 625–630.

Bhatnagar, A., Sheffler, D.J., Kroeze, W.K., Compton-Toth, B. and Roth, B.L. (2004) Caveolin-1 interacts with 5-HT_{2A} serotonin receptors and profoundly modulates the signaling of selected Galphaq-coupled protein receptors. J. Biol. Chem., 279: 34614–34623.

Bockaert, J. and Pin, J.P. (1999) Molecular tinkering of G protein-coupled receptors: an evolutionary success. EMBO J., 18: 1723–1729.

Bockaert, J., Roussignol, G., Becamel, C., Gavarini, S., Joubert, L., Dumuis, A., Fagni, L. and Marin, P. (2004) GPCR-interacting proteins (GIPs): nature and functions. Biochem. Soc. Trans., 32: 851–855.

Bockaert, J., Claeysen, S., Joubert, L., Fagni, L. and Dumuis, A. (2005) Molecular and Cellular Determinants of GPCR Splice Variant Constitutive Activity. Wiley-VCH, Weinheim.

Bockaert, J., Claeysen, S., Becamel, C., Dumuis, A. and Marin, P. (2006) Neuronal 5-HT metabotropic receptors: fine-tuning of their structure, signaling, and roles in synaptic modulation. Cell Tissue Res., 326: 553–572.

Bockaert, J., Claeysen, S., Compan, V. and Dumuis, A. (2008) 5-HT(4) receptors: history, molecular pharmacology and brain functions. Neuropharmacology, 55: 922–931.

Boess, F.G., Beroukhim, R. and Martin, I.L. (1995) Ultrastructure of the 5-hydroxytryptamine3 receptor. J. Neurochem., 64: 1401–1405.

Bouhelal, R., Smounya, L. and Bockaert, J. (1988) 5-HT_{1B} receptors are negatively coupled with adenylate cyclase in rat substantia nigra. Eur. J. Pharmacol., 151: 189–196.

Bradley, P.B., Engel, G., Feniuk, W., Fozard, J.R., Humphrey, P.P., Middlemiss, D.N., Mylecharane, E.J., Richardson, B.P. and Saxena, P.R. (1986) Proposals for the classification and nomenclature of functional receptors for 5-hydroxytryptamine. Neuropharmacology, 25: 563–576.

Brattelid, T., Kvingedal, A.M., Krobert, K.A., Andressen, K.W., Bach, T., Hystad, M.E., Kaumann, A.J. and Levy, F.O. (2004) Cloning, pharmacological characterisation and tissue distribution of a novel 5-HT4 receptor splice variant, 5-HT4(i). Naunyn Schmiedebergs Arch. Pharmacol., 369: 616–628.

Bruss, M., Barann, M., Hayer-Zillgen, M., Eucker, T., Gothert, M. and Bonisch, H. (2000) Modified 5-HT_{3A} receptor function by co-expression of alternatively spliced human 5-HT_{3A} receptor isoforms. Naunyn Schmiedebergs Arch. Pharmacol., 362: 392–401.

Carrel, D., Masson, J., Al Awabdh, S., Capra, C.B., Lenkei, Z., Hamon, M., Emerit, M.B. and Darmon, M. (2008) Targeting of the 5-HT_{1A} serotonin receptor to neuronal dendrites is mediated by Yif1B. J. Neurosci., 28: 8063–8073.

Chanrion, B., Mannoury La Cour, C., Gavarini, S., Seimandi, M., Vincent, L., Pujol, J.F., Bockaert, J., Marin, P. and Millan, M.J. (2008) Inverse agonist and neutral antagonist actions of antidepressants at recombinant and native 5-hydroxytryptamine2C receptors: differential modulation of cell surface expression and signal transduction. Mol. Pharmacol., 73: 748–757.

Cogolludo, A., Moreno, L., Lodi, F., Frazziano, G., Cobeno, L., Tamargo, J. and Perez-Vizcaino, F. (2006) Serotonin inhibits voltage-gated K^+ currents in pulmonary artery smooth muscle cells. Role of 5-HT_{2A} receptors, caveolin-1, and KV1.5 channel internalization. Circ. Res., 98: 931–938.

Cornea-Hebert, V., Watkins, K.C., Roth, B.L., Kroeze, W.K., Gaudreau, P., Leclerc, N. and Descarries, L. (2002) Similar ultrastructural distribution of the 5-HT(2A) serotonin receptor and microtubule-associated protein MAP1A in cortical dendrites of adult rat. Neuroscience, 113: 23–35.

Cowen, D.S., Johnson-Farley, N.N. and Travkina, T. (2005) 5-HT receptors couple to activation of Akt, but not extracellular-regulated kinase (ERK), in cultured hippocampal neurons. J. Neurochem., 93: 910–917.

Crane, J.W., Shimizu, K., Carrasco, G.A., Garcia, F., Jia, C., Sullivan, N.R., D'Souza, D.N., Zhang, Y., Van De Kar, L.D., Muma, N.A. and Battaglia, G. (2007) 5-HT_{1A} receptors mediate (+)8-OH-DPAT-stimulation of extracellular signal-regulated kinase (MAP kinase) *in vivo* in rat hypothalamus: time dependence and regional differences. Brain Res., 1183: 51–59.

Cussac, D., Newman-Tancredi, A., Duqueyroix, D., Pasteau, V. and Millan, M.J. (2002) Differential activation of Gq/11 and Gi(3) proteins at 5-hydroxytryptamine(2C) receptors revealed by antibody capture assays: influence of receptor reserve and relationship to agonist-directed trafficking. Mol. Pharmacol., 62: 578–589.

Dai, Y., Dudek, N.L., Patel, T.B. and Muma, N.A. (2008) Transglutaminase-catalyzed transamidation: a novel mechanism for Rac1 activation by 5-hydroxytryptamine2A receptor stimulation. J. Pharmacol. Exp. Ther., 326: 153–162.

De Maeyer, J.H., Aerssens, J., Verhasselt, P. and Lefebvre, R.A. (2008) Alternative splicing and exon duplication generates 10 unique porcine 5-HT_4 receptor splice variants including a functional homofusion variant. Physiol. Genomics., 34: 22–33.

Della Rocca, G.J., Mukhin, Y.V., Garnovskaya, M.N., Daaka, Y., Clark, G.J., Luttrell, L.M., Lefkowitz, R.J. and Raymond, J.R. (1999) Serotonin 5-HT_{1A} receptor-mediated Erk activation requires calcium/calmodulin-dependent receptor endocytosis. J. Biol. Chem., 274: 4749–4753.

De Vivo, M. and Maayani, S. (1986) Characterization of the 5-hydroxytryptamine$_{1A}$ receptor-mediated inhibition of forskolin-stimulated adenylate cyclase activity in guinea pig and rat hippocampal membranes. J. Pharmacol. Exp. Ther., 238: 248–253.

Dubin, A.E., Huvar, R., D'Andrea, M.R., Pyati, J., Zhu, J.Y., Joy, K.C., Wilson, S.J., Galindo, J.E., Glass, C.A., Luo, L., Jackson, M.R., Lovenberg, T.W. and Erlander, M.G. (1999) The pharmacological and functional characteristics of the serotonin 5-HT(3A) receptor are specifically modified by a 5-HT(3B) receptor subunit. J. Biol. Chem., 274: 30799–30810.

Dumuis, A., Bouhelal, R., Sebben, M., Cory, R. and Bockaert, J. (1988a) A nonclassical 5-hydroxytryptamine receptor positively coupled with adenylate cyclase in the central nervous system. Mol. Pharmacol., 34: 880–887.

Dumuis, A., Bouhelal, R., Sebben, M. and Bockaert, J. (1988b) A 5-HT receptor in the central nervous system, positively coupled with adenylate cyclase, is antagonized by ICS 205 930. Eur. J. Pharmacol., 146: 187–188.

Fargin, A., Raymond, J.R., Lohse, M.J., Kobilka, B.K., Caron, M.G. and Lefkowitz, R.J. (1988) The genomic clone G-21 which resembles a beta-adrenergic receptor sequence encodes the 5-HT_{1A} receptor. Nature, 335: 358–360.

Fillion, G., Fillion, M.P., Jacob, J. and Rousselle, J.C. (1976) 5-HT and LSD high affinity binding sites to brain synaptosomal membranes [proceedings]. Br. J. Pharmacol., 58: 425P–426P.

Fitzgerald, L.W., Iyer, G., Conklin, D.S., Krause, C.M., Marshall, A., Patterson, J.P., Tran, D.P., Jonak, G.J. and Hartig, P.R. (1999) Messenger RNA editing of the human serotonin 5-HT_{2C} receptor. Neuropsychopharmacology, 21: 82S–90S.

Foguet, M., Hoyer, D., Pardo, L.A., Parekh, A., Kluxen, F.W., Kalkman, H.O., Stuhmer, W. and Lubbert, H. (1992) Cloning and functional characterization of the rat stomach fundus serotonin receptor. EMBO J., 11: 3481–3487.

Fredriksson, R., Lagerstrom, M.C., Lundin, L.G. and Schioth, H.B. (2003) The G-protein-coupled receptors in the human genome form five main families. Phylogenetic analysis, paralogon groups, and fingerprints. Mol. Pharmacol., 63: 1256–1272.

Gaddum, J.H. and Picarelli, Z.P. (1957) Two kinds of tryptamine receptors. Br. J. Pharmacol., 12: 323–328.

Gavarini, S., Becamel, C., Chanrion, B., Bockaert, J. and Marin, P. (2004) Molecular and functional characterization of proteins interacting with the C-terminal domains of 5-HT_2 receptors: emergence of 5-HT_2 'receptosomes'. Biol. Cell, 96: 373–381.

Gavarini, S., Becamel, C., Altier, C., Lory, P., Poncet, J., Wijnholds, J., Bockaert, J. and Marin, P. (2006) Opposite effects of PSD-95 and MPP3 PDZ proteins on serotonin 5-hydroxytryptamine2C receptor desensitization and membrane stability. Mol. Biol. Cell, 17: 4619–4631.

Gonzalez-Maeso, J., Weisstaub, N.V., Zhou, M., Chan, P., Ivic, L., Ang, R., Lira, A., Bradley-Moore, M., Ge, Y., Zhou, Q., Sealfon, S.C. and Gingrich, J.A. (2007) Hallucinogens recruit specific cortical 5-HT(2A) receptor-mediated signaling pathways to affect behavior. Neuron, 53: 439–452.

Gonzalez-Maeso, J., Ang, R.L., Yuen, T., Chan, P., Weisstaub, N.V., Lopez-Gimenez, J.F., Zhou, M., Okawa, Y., Callado, L.F., Milligan, G., Gingrich, J.A., Filizola, M., Meana, J.J. and Sealfon, S.C. (2008) Identification of a serotonin/glutamate receptor complex implicated in psychosis. Nature, 452: 93–97.

Gozlan, H., El Mestikawy, S., Pichat, L., Glowinski, J. and Hamon, M. (1983) Identification of presynaptic serotonin autoreceptors using a new ligand: 3H-PAT. Nature, 305: 140–142.

Gu, Z., Jiang, Q. and Yan, Z. (2007) RGS4 modulates serotonin signaling in prefrontal cortex and links to serotonin dysfunction in a rat model of schizophrenia. Mol. Pharmacol., 71: 1030–1039.

Guillet-Deniau, I., Burnol, A.F. and Girard, J. (1997) Identification and localization of a skeletal muscle secrotonin 5-HT_{2A} receptor coupled to the Jak/STAT pathway. J. Biol. Chem., 272: 14825–14829.

Gurevich, I., Englander, M.T., Adlersberg, M., Siegal, N.B. and Schmauss, C. (2002a) Modulation of serotonin 2C receptor editing by sustained changes in serotonergic neurotransmission. J. Neurosci., 22: 10529–10532.

Gurevich, I., Tamir, H., Arango, V., Dwork, A.J., Mann, J.J. and Schmauss, C. (2002b) Altered editing of serotonin 2C receptor pre-mRNA in the prefrontal cortex of depressed suicide victims. Neuron, 34: 349–356.

Hajj-Ali, I. and Anctil, M. (1997) Characterization of a serotonin receptor in the cnidarian Renilla koellikeri: a radiobinding analysis. Neurochem. Intl., 31: 83–93.

Hamblin, M.W. and Metcalf, M.A. (1991) Primary structure and functional characterization of a human 5-HT_{1D}-type serotonin receptor. Mol. Pharmacol., 40: 143–148.

Hannon, J. and Hoyer, D. (2008) Molecular biology of 5-HT receptors. Behav. Brain. Res., 195: 198–213.

Heidmann, D.E., Metcalf, M.A., Kohen, R. and Hamblin, M.W. (1997) Four 5-hydroxytryptamine7 (5-HT_7) receptor isoforms in human and rat produced by alternative splicing: species differences due to altered intron–exon organization. J. Neurochem., 68: 1372–1381.

Higuchi, M., Single, F.N., Kohler, M., Sommer, B., Sprengel, R. and Seeburg, P.H. (1993) RNA editing of AMPA receptor subunit GluR-B: a base-paired intron–exon structure determines position and efficiency. Cell, 75: 1361–1370.

Hiroi, T., Hayashi-Kobayashi, N., Nagumo, S., Ino, M., Okawa, Y., Aoba, A. and Matsui, H. (2001) Identification and characterization of the human serotonin-4 receptor gene promoter. Biochem. Biophys. Res. Commun., 289: 337–344.

Huang, Y.Y. and Kandel, E.R. (2007) 5-Hydroxytryptamine induces a protein kinase A/mitogen-activated protein kinase-mediated and macromolecular synthesis-dependent late phase of long-term potentiation in the amygdala. J. Neurosci., 27: 3111–3119.

Jensen, A.A., Davies, P.A., Brauner-Osborne, H. and Krzywkowski, K. (2008) 3B but which 3B? And that's just one of the questions: the heterogeneity of human 5-HT(3) receptors. Trends Pharmacol. Sci., 29: 437–444.

Ji, S.P., Zhang, Y., Van Cleemput, J., Jiang, W., Liao, M., Li, L., Wan, Q., Backstrom, J.R. and Zhang, X. (2006) Disruption of PTEN coupling with 5-HT_{2C} receptors suppresses behavioral responses induced by drugs of abuse. Nat. Med., 12: 324–329.

Joubert, L., Hanson, B., Barthet, G., Sebben, M., Claeysen, S., Hong, W., Marin, P., Dumuis, A. and Bockaert, J. (2004) New sorting nexin (SNX27) and NHERF specifically interact with the 5-HT_{4A} receptor splice variant: roles in receptor targeting. J. Cell. Sci., 117: 5367–5379.

Kenakin, T. (2007) Functional selectivity through protean and biased agonism: who steers the ship?. Mol. Pharmacol., 72: 1393–1401.

Kroeze, W.K. and Roth, B.L. (2006) Molecular Biology and Genomic Organization of G Protein-coupled Serotonin Receptors. Humana Press, Totowa, NJ.

Kvachnina, E., Liu, G., Dityatev, A., Renner, U., Dumuis, A., Richter, D.W., Dityateva, G., Schachner, M., Voyno-Yasenetskaya, T.A. and Ponimaskin, E.G. (2005) 5-HT_7 receptor is coupled to G alpha subunits of heterotrimeric G12-protein to regulate gene transcription and neuronal morphology. J. Neurosci., 25: 7821–7830.

Labasque, M., Reiter, E., Becamel, C., Bockaert, J. and Marin, P. (2008) Physical interaction of calmodulin with the 5-hydroxytryptamine2C receptor C-terminus is essential for G protein-independent, arrestin-dependent receptor signaling. Mol. Biol. Cell., 19: 4640–4650.

Lezoualc'h, F. and Robert, S.J. (2003) The serotonin 5-HT_4 receptor and the amyloid precursor protein processing. Exp. Gerontol., 38: 159–166.

Lieb, K., Biersack, L., Waschbisch, A., Orlikowski, S., Akundi, R.S., Candelario-Jalil, E., Hull, M. and Fiebich, B.L. (2005) Serotonin via 5-HT_7 receptors activates p38 mitogen-activated protein kinase and protein kinase C epsilon resulting in interleukin-6 synthesis in human U373 MG astrocytoma cells. J. Neurochem., 93: 549–559.

Lin, S.L., Johnson-Farley, N.N., Lubinsky, D.R. and Cowen, D.S. (2003) Coupling of neuronal 5-HT_7 receptors to activation of extracellular-regulated kinase through a protein kinase A-independent pathway that can utilize Epac. J. Neurochem., 87: 1076–1085.

Manivet, P., Mouillet-Richard, S., Callebert, J., Nebigil, C.G., Maroteaux, L., Hosoda, S., Kellermann, O. and Launay, J.M. (2000) PDZ-dependent activation of nitric-oxide synthases by the serotonin 2B receptor. J. Biol. Chem., 275: 9324–9331.

Mannoury La Cour, C., El Mestikawy, S., Hanoun, N., Hamon, M. and Lanfumey, L. (2006) Regional differences in the coupling of 5-hydroxytryptamine-1A receptors to G proteins in the rat brain. Mol. Pharmacol., 70: 1013–1021.

Marek, G.J., Wright, R.A., Schoepp, D.D., Monn, J.A. and Aghajanian, G.K. (2000) Physiological antagonism between 5-hydroxytryptamine(2A) and group II metabotropic glutamate receptors in prefrontal cortex. J. Pharmacol. Exp. Ther., 292: 76–87.

Maricq, A.V., Peterson, A.S., Brake, A.J., Myers, R.M. and Julius, D. (1991) Primary structure and functional expression of the 5HT3 receptor, a serotonin-gated ion channel. Science, 254: 432–437.

Marion, S., Weiner, D.M. and Caron, M.G. (2004) RNA editing induces variation in desensitization and trafficking of 5-hydroxytryptamine 2c receptor isoforms. J. Biol. Chem., 279: 2945–2954.

Martel, J.C., Ormiere, A.M., Leduc, N., Assie, M.B., Cussac, D. and Newman-Tancredi, A. (2007) Native rat hippocampal 5-HT_{1A} receptors show constitutive activity. Mol. Pharmacol., 71: 638–643.

Matthes, H., Boschert, U., Amlaiky, N., Grailhe, R., Plassat, J.L., Muscatelli, F., Mattei, M.G. and Hen, R. (1993) Mouse 5-hydroxytryptamine5A and 5-hydroxytryptamine5B receptors define a new family of serotonin receptors: cloning, functional expression, and chromosomal localization. Mol. Pharmacol., 43: 313–319.

McGrew, L., Chang, M.S. and Sanders-Bush, E. (2002) Phospholipase D activation by endogenous 5-hydroxytryptamine 2C receptors is mediated by Galpha13 and pertussis toxin-insensitive Gbetagamma subunits. Mol. Pharmacol., 62: 1339–1343.

Millan, M.J., Marin, P., Bockaert, J. and Mannoury La Cour, C. (2008) Signaling at G-protein-coupled serotonin receptors:

recent advances and future research directions. Trends Pharmacol. Sci., 29: 454–464.
Miyake, A., Mochizuki, S., Takemoto, Y. and Akuzawa, S. (1995) Molecular cloning of human 5-hydroxytryptamine3 receptor: heterogeneity in distribution and function among species. Mol. Pharmacol., 48: 407–416.
Niesler, B., Walstab, J., Combrink, S., Moller, D., Kapeller, J., Rietdorf, J., Bonisch, H., Gothert, M., Rappold, G. and Bruss, M. (2007) Characterization of the novel human serotonin receptor subunits 5-HT_{3C}, 5-HT_{3D}, and 5-HT_{3E}. Mol. Pharmacol., 72: 8–17.
Niswender, C.M., Copeland, S.C., Herrick-Davis, K., Emeson, R.B. and Sanders-Bush, E. (1999) RNA editing of the human serotonin 5-hydroxytryptamine 2C receptor silences constitutive activity. J. Biol. Chem., 274: 9472–9478.
Olsen, M.A., Nawoschik, S.P., Schurman, B.R., Schmitt, H.L., Burno, M., Smith, D.L. and Schechter, L.E. (1999) Identification of a human 5-HT6 receptor variant produced by alternative splicing. Brain Res. Mol. Brain Res., 64: 255–263.
Parker, L.L., Backstrom, J.R., Sanders-Bush, E. and Shieh, B.H. (2003) Agonist-induced phosphorylation of the serotonin 5-HT_{2C} receptor regulates its interaction with multiple PDZ protein 1. J. Biol. Chem., 278: 21576–21583.
Pedigo, N.W., Yamamura, H.I. and Nelson, D.L. (1981) Discrimination of multiple [3H]5-hydroxytryptamine binding sites by the neuroleptic spiperone in rat brain. J. Neurochem., 36: 220–226.
Peroutka, S.J. and Snyder, S.H. (1979) Multiple serotonin receptors: differential binding of [3H]5-hydroxytryptamine, [3H]lysergic acid diethylamide and [3H]spiroperidol. Mol. Pharmacol., 16: 687–699.
Peroutka, S.J., Lebovitz, R.M. and Snyder, S.H. (1981) Two distinct central serotonin receptors with different physiological functions. Science, 212: 827–829.
Price, R.D., Weiner, D.M., Chang, M.S. and Sanders-Bush, E. (2001) RNA editing of the human serotonin 5-HT_{2C} receptor alters receptor-mediated activation of G13 protein. J. Biol. Chem., 276: 44663–44668.
Rees, S., Den Daas, I., Foord, S., Goodson, S., Bull, D., Kilpatrick, G. and Lee, M. (1994) Cloning and characterisation of the human 5-HT_{5A} serotonin receptor. FEBS Lett., 355: 242–246.
Robertson, D.N., Johnson, M.S., Moggach, L.O., Holland, P.J., Lutz, E.M. and Mitchell, R. (2003) Selective interaction of ARF1 with the carboxy-terminal tail domain of the 5-HT_{2A} receptor. Mol. Pharmacol., 64: 1239–1250.
Ruat, M., Traiffort, E., Arrang, J.M., Tardivel-Lacombe, J., Diaz, J., Leurs, R. and Schwartz, J.C. (1993a) A novel rat serotonin (5-HT_6) receptor: molecular cloning, localization and stimulation of cAMP accumulation. Biochem. Biophys. Res. Commun., 193: 268–276.
Ruat, M., Traiffort, E., Leurs, R., Tardivel-Lacombe, J., Diaz, J., Arrang, J.M. and Schwartz, J.C. (1993b) Molecular cloning, characterization, and localization of a high-affinity serotonin receptor (5-HT_7) activating cAMP formation. Proc. Natl. Acad. Sci. USA, 90: 8547–8551.
Sanders-Bush, E. and Breeding, M. (1988) Putative selective 5-HT-2 antagonists block serotonin 5-HT-1C receptors in the choroid plexus. J. Pharmacol. Exp. Ther., 247: 169–173.
Schoepp, D. (2006) New directions in the treatment of schizophrenia: modulators of mGluR2 and or mGluR3 receptors. Neuropsychopharmacology, 32: S25.
Sebben, M., Ansanay, H., Bockaert, J. and Dumuis, A. (1994) 5-HT_6 receptors positively coupled to adenylyl cyclase in striatal neurones in culture. Neuroreport, 5: 2553–2557.
Sheffler, D.J., Kroeze, W.K., Garcia, B.G., Deutch, A.Y., Hufeisen, S.J., Leahy, P., Bruning, J.C. and Roth, B.L. (2006) p90 ribosomal S6 kinase 2 exerts a tonic brake on G protein-coupled receptor signaling. Proc. Natl. Acad. Sci. USA, 103: 4717–4722.
Shenker, A., Maayani, S., Weinstein, H. and Green, J.P. (1987) Pharmacological characterization of two 5-hydroxytryptamine receptors coupled to adenylate cyclase in guinea pig hippocampal membranes. Mol. Pharmacol., 31: 357–367.
Strachen, R.T., Sheffler, D.J., Willard, B., Kinter, M., Kiselar, J.G. and Roth, B.L. (2008) Ribosomal S6 kinase 2 directly phosphorylates the 5-HT_{2A} serotonin receptor thereby modulating 5-HT_{2A} signaling. J. Biol. Chem., 284: 5557–5573.
Svenningsson, P., Chergui, K., Rachleff, I., Flajolet, M., Zhang, X., Yacoubi, M.E., Vaugeois, J.M., Nomikos, G.G. and Greengard, P. (2006) Alterations in 5-HT_{1B} receptor function by p11 in depression-like states. Science, 311: 77–80.
Tsuchioka, M., Takebayashi, M., Hisaoka, K., Maeda, N. and Nakata, Y. (2008) Serotonin (5-HT) induces glial cell line-derived neurotrophic factor (GDNF) mRNA expression via the transactivation of fibroblast growth factor receptor 2 (FGFR2) in rat C6 glioma cells. J. Neurochem., 106: 244–257.
Turner, J.H. and Raymond, J.R. (2005) Interaction of calmodulin with the serotonin 5-hydroxytryptamine2A receptor. A putative regulator of G protein coupling and receptor phosphorylation by protein kinase C. J. Biol. Chem., 280: 30741–30750.
Turner, J.H., Gelasco, A.K. and Raymond, J.R. (2004) Calmodulin interacts with the third intracellular loop of the serotonin 5-hydroxytryptamine1A receptor at two distinct sites: putative role in receptor phosphorylation by protein kinase C. J. Biol. Chem., 279: 17027–17037.
Turner, J.H., Garnovskaya, M.N., Coaxum, S.D., Vlasova, T.M., Yakutovich, M., Lefler, D.M. and Raymond, J.R. (2007a) Ca^{2+}-calmodulin and janus kinase 2 are required for activation of sodium-proton exchange by the Gi-coupled 5-hydroxytryptamine 1a receptor. J. Pharmacol. Exp. Ther., 320: 314–322.
Turner, J.H., Garnovskaya, M.N. and Raymond, J.R. (2007b) Serotonin 5-HT_{1A} receptor stimulates c-Jun N-terminal kinase and induces apoptosis in Chinese hamster ovary fibroblasts. Biochim. Biophys. Acta, 1773: 391–399.
Ullmer, C., Schmuck, K., Figge, A. and Lubbert, H. (1998) Cloning and characterization of MUPP1, a novel PDZ domain protein. FEBS Lett., 424: 63–68.
Walther, D.J., Peter, J.U., Winter, S., Holtje, M., Paulmann, N., Grohmann, M., Vowinckel, J., Alamo-Bethencourt, V., Wilhelm, C.S., Ahnert-Hilger, G. and Bader, M. (2003) Serotonylation of small GTPases is a signal transduction pathway that triggers platelet alpha-granule release. Cell, 115: 851–862.
Warner-Schmidt, J.L., Flajolet, M., Maller, A., Chen, E.Y., Qi, H., Svenningsson, P. and Greengard, P. (2009) Role of p11 in cellular and behavioral effects of 5-HT_4 receptor stimulation. J. Neurosci., 29: 1937–1946.
Weiss, S., Sebben, M., Kemp, D.E. and Bockaert, J. (1986) Serotonin 5-HT_1 receptors mediate inhibition of cyclic AMP production in neurons. Eur. J. Pharmacol., 120: 227–230.
Xia, Z., Gray, J.A., Compton-Toth, B.A. and Roth, B.L. (2003a) A direct interaction of PSD95 with 5-HT_{2A} serotonin receptors regulates receptor trafficking and signal transduction. J. Biol. Chem., 278: 21901–21908.

Xia, Z., Hufeisen, S.J., Gray, J.A. and Roth, B.L. (2003b) The PDZ-binding domain is essential for the dendritic targeting of 5-HT_{2A} serotonin receptors in cortical pyramidal neurons *in vitro*. Neuroscience, 122: 907–920.

Xie, E., Zhu, L., Zhao, L. and Chang, L.S. (1996) The human serotonin 5-HT_{2C} receptor: complete cDNA, genomic structure, and alternatively spliced variant. Genomics, 35: 551–561.

Yuen, E.Y., Jiang, Q., Chen, P., Feng, J. and Yan, Z. (2008) Activation of 5-$HT_{2A/C}$ receptors counteracts 5-HT_{1A} regulation of n-methyl-D-aspartate receptor channels in pyramidal neurons of prefrontal cortex. J. Biol. Chem., 283: 17194–17204.

Yun, H.M., Kim, S., Kim, H.J., Kostenis, E., Kim, J.I., Seong, J.Y., Baik, J.H. and Rhim, H. (2007) The novel cellular mechanism of human 5-HT_6 receptor through an interaction with Fyn. J. Biol. Chem., 282: 5496–5505.

CHAPTER 1.6

Distribution of 5-HT Receptors in the Central Nervous System

Guadalupe Mengod[1], Roser Cortés[1], M. Teresa Vilaró[1] and Daniel Hoyer[*2]

[1]*Departamento Neuroquímica y Neurofarmacología, IIBB/CSIC-IDIBAPS,CIBERNED, Barcelona, Spain*
[2]*Psychiatry/Neuroscience Research, Novartis Institutes for Biomedical Research Basel, Switzerland*

Abstract: The almost incredible complexity of the 5-HT receptor family translates into the diversity of the brain distribution of these receptors, as briefly illustrated in this chapter. However, the picture is by no means complete, due either to the lack of tools for some of these receptors or their components (selective ligands or good antibodies that label exclusively a single receptor are still rare) and/or to the genomic complexity of some of these receptors; thus 5-HT_4 and 5-HT_7 receptors come in numerous splice variants, the 5-HT_{2C} receptor has even more potential editing variants, whereas the 5-HT_3 receptor has multiple potential subunits. There is debate as to which of the latter are co-expressed and form functional receptors, or which splice/editing variants of the former are expressed and where. In addition, one tends to be misled by density – for example, the 5-HT_{2C} receptor is not only a choroid plexus receptor, the dorsal raphe nucleus expresses more than just 5-HT_{1A} receptors, and it may be premature to suggest that the low density of 5-HT_3 receptors outside of the brainstem has low functional relevance. The 5-HT_{1e} receptor is not expressed in rodents, and 5-HT_{5b} receptors do not exist in man; in addition, both receptors share an almost total lack of adequate tools for their identification. On the other hand, we have tried to illustrate the fact that initial receptor distribution studies have been crucial in the definition of new 5-HT receptors, especially in the days preceding molecular biology/receptor cloning, and that the knowledge of receptor distribution in health and disease is essential for a (patho)physiological understanding of their function.

Keywords: 5-HT receptor families and subtypes, autoradiography, *in situ* hybridization, immunocytochemistry, brain distribution, species differences.

Introduction

The serotonin (5-HT, 5-hydroxytryptamine) system is one of the oldest neurotransmitter/hormone systems in evolution, probably as old as 800 million years; this may explain why 5-HT interacts with such a diversity of receptors of the G-protein-coupled family (GPCRs) and the ligand-gated ion channel family. This is similar to other 'old' neurotransmitters, such as acetylcholine, GABA or glutamate, which also have high levels of complexity, although the latter neurotransmitters show a high diversity at the level of their ligand-gated channel receptors, whereas 5-HT's complexity lies primarily at the levels of its many GPCRs. 5-HT was discovered in the gut in the 1930s and called enteramine, then rediscovered in the 1940s in the blood and called serotonin, based on its vasoconstrictor features. Although the major production of 5-HT is in the periphery, the brain has its share of 5-HT and related receptors and transporters, which in a number of cases appear to be involved in neurological or neuropsychiatric pathologies, such as mood disorders, anxiety and depression, schizophrenia and psychosis, addiction, memory impairment, pain, migraine, and chemotherapy- or surgical-induced vomiting where the trigger zones are central. It is therefore not surprising that 5-HT, tryptophan hydroxylases, the 5-HT transporter and a great variety of 5-HT receptors are to be found in the brain, some receptors having highly localized patterns, whereas others may be rather widespread. We will keep to the nomenclature that has been adopted and refined by the serotonin nomenclature committee (Hartig *et al.*, 1993; Hoyer *et al.*, 1994, 2002).

5-HT_{1A} receptors

The 5-HT_{1A} receptor was the first of the 5-HT family to be cloned (Kobilka *et al.*, 1987; Fargin *et al.*, 1988); it

*Corresponding author
E-mail: daniel.hoyer@novartis.com

Christian Müller & Barry Jacobs (Eds.)
Handbook of Behavioral Neurobiology of Serotonin
ISBN 978-0-12-374634-4

DOI: 10.1016/B978-0-12-374634-4.00006-X

was also the first for which really selective ligands were described, such as 8-OH-DPAT (Gozlan *et al.*, 1983; Middlemiss and Fozard, 1983). This had a tremendous impact on the pharmacological characterization of 5-HT receptors in general and that of 5-HT_{1A} in particular, including the distribution of 5-HT_{1A} receptors in the brain, since [^{3}H]-8-OH-DPAT was an excellent ligand for membrane binding (Gozlan *et al.*, 1988, 1995; Hoyer *et al.*, 1985a), but also in the performance of autoradiography in tissue slices of various species (Pazos and Palacios, 1985; Pazos *et al.*, 1987a). 8-OH-DPAT also allowed the definition of non-5-HT_{1A} sites, which developed rather rapidly in the mid-1980s (Hoyer *et al.*, 1986a; Bruinvels *et al.*, 1992; Waeber and Palacios, 1992), in both radioligand binding and autoradiographic studies. These allowed comparison of brain distribution patterns in various species, which meant that receptors could be distinguished or regrouped even when the pharmacological signature was different across species (Hoyer and Middlemiss, 1989). Although we now know that 8-OH-DPAT has some affinity for 5-HT_7 receptors (Tsou *et al.*, 1994), the conditions under which autoradiography was performed excluded significant binding to 5-HT_7 sites.

Thus, the brain distribution of 5-HT_{1A} receptors has been studied extensively by receptor autoradiography using a range of agonists, including initially [^{3}H]-5-HT, [^{3}H]-8-OH-DPAT, [^{3}H]-ipsapirone, [^{125}I]-BH-8-MeO-N-PAT (Pazos and Palacios, 1985; Palacios *et al.*, 1987; Ponchant *et al.*, 1988) and more recently a few antagonists [^{125}I]-p-MPPI and [^{3}H]-WAY 100635 (Laporte *et al.*, 1994) which allow *in vivo* labeling of 5-HT_{1A} receptors. PET studies have used [^{11}C]-WAY 100635 and related compounds (Sandell *et al.*, 2001) to image 5-HT_{1A} receptors in the living human brain.

The density of 5-HT_{1A} binding sites is high in limbic brain areas, notably hippocampus (Figure 1B), lateral septum, cortical areas (particularly cingulate and entorhinal cortex), and the mesencephalic raphe nuclei (both dorsal and median raphe nuclei) (Figure 1D). In contrast, levels of 5-HT_{1A} binding sites in the basal ganglia and cerebellum are barely detectable. 5-HT_{1A} receptors are located both postsynaptic to 5-HT neurons (in forebrain regions) and also on the 5-HT neurons themselves at the level of the soma and dendrites in the mesencephalic and medullary raphe nuclei. This is evident from studies on the effects of neuronal lesions on 5-HT_{1A} binding sites and mRNA, and studies of the cellular localization of the 5-HT_{1A} receptor using immunocytochemistry (Kia *et al.*, 1996a). The distribution of mRNA encoding the 5-HT_{1A} receptor is almost identical to that of the 5-HT_{1A} binding site (Figure 1A–D) (Pompeiano *et al.*, 1992). Antibodies have been generated to study 5-HT_{1A} receptor distribution in the brain. In general, there is good agreement between the distribution of 5-HT_{1A} binding sites and signals obtained using immunocytochemistry. At the cellular level, *in situ* hybridization and immunocytochemical studies demonstrate the presence of 5-HT_{1A} receptors in cortical pyramidal neurons as well as pyramidal and granular neurons of the hippocampus (Figure 1A). The 5-HT_{1A} receptor is expressed in 5-HT-containing neurons in the raphe nuclei (Figure 1C), cholinergic neurons in the septum, and (probably) glutamatergic (pyramidal) neurons in the cortex and hippocampus (DeFelipe *et al.*, 2001; Santana *et al.*, 2004; de Almeida and Mengod, 2008). They are co-expressed with the 5-HT_{2A} receptors in the pyramidal cells of the rat prefrontal cortex (Amargós-Bosch *et al.*, 2004). The ultrastructural location of the 5-HT_{1A} receptor identifies synaptic membranes, as well as extrasynaptic structures (Kia *et al.*, 1996b). There are reports of 5-HT_{1A} receptors in brain glial cells, but this has not been confirmed. In the human brain, similar findings have been reported whether the older or more recent tools have been used (Pazos *et al.*, 1987a; Varnas *et al.*, 2004a). The pattern of 5-HT_{1A} receptor distribution is similar across species, although the laminar organization of the 5-HT_{1A} receptor in cortical and hippocampal areas of humans differs somewhat from that in the rodent (Santana *et al.*, 2004; de Almeida and Mengod, 2008).

5-HT_{1B} receptors

The 5-HT_{1B} receptor was initially rather loosely defined as non-5-HT_{1A}; in other words, [^{3}H]-5-HT sites which were not sensitive to 8-OH-DPAT (Middlemiss and Fozard, 1983; Pazos and Palacios, 1985), had high affinity for some beta-blocking ligands such as [^{125}I]-cyanopindolol in rodents (Hoyer *et al.*, 1985b), but much less so in non-rodents and primates (Hoyer *et al.*, 1986a). This is when the difficulty started, as the rodent receptor was then defined as 5-HT_{1B} (also in opossum) and the non-rodent receptor (pig, calf, monkey, human and other species) as 5-HT_{1D}, since no pharmacologically equivalent site could be described in non-rodent species (Hoyer *et al.*, 1986a). However, based on similarities in coupling, function and distribution (for example, inhibition of adenylate cyclase activity, autoreceptor activity, and marked expression in striatum/substantia nigra), we proposed these two apparently pharmacologically distinct receptors (5-HT_{1B} and 5-HT_{1D}) would be species homologs (Hoyer and Middlemiss, 1989). This was then confirmed with the cloning of all members of the 5-HT_1 receptor family and with the availability of much more selective tools – for example, 5-HT_{1B}-selective and 5-HT_{1D}-selective ligands such as SB 224289 and BRL15572, respectively. The situation is still somewhat confused, since the 5-HT_{1D} and

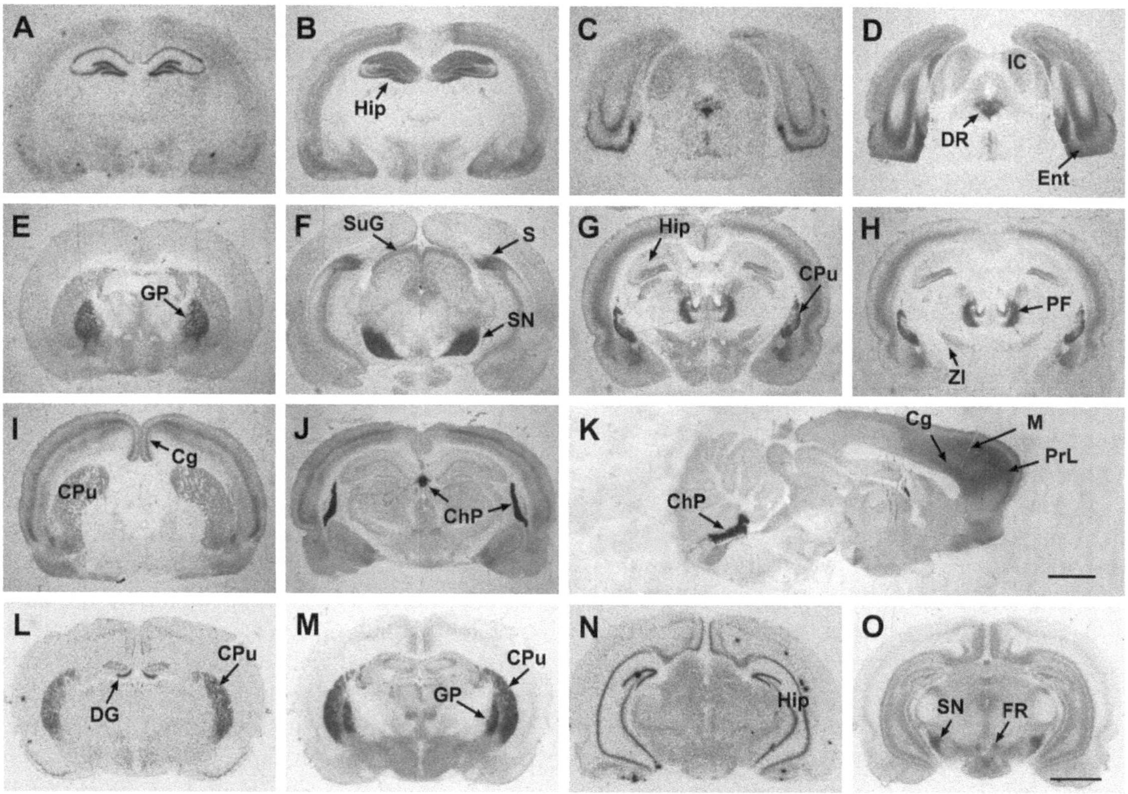

Figure 1 Autoradiographic distribution of radioligand binding sites and mRNA for several 5-HT receptor subtypes in the rat brain. Images are photomicrographs from film autoradiograms where dark areas correspond to regions rich in binding sites or mRNA. (A, C) Distribution of mRNA coding for 5-HT_{1A} receptors detected with ^{33}P-labeled oligonucleotide probes at two coronal levels of the rat brain. (B, D) Distribution of 5-HT_{1A} receptors labeled by [^{3}H]8-OH-DPAT (1 nM) in close sections. (E, F) Distribution of 5-$HT_{1B/1D}$ receptors labeled by [^{3}H]GR 125743 (1 nM). (G, H) Distribution of binding sites labeled by [^{3}H]-Sumatriptan (5 nM) alone, which labels 5-HT_{1B}, 5-HT_{1D}, and 5-HT_{1F} receptors (G), or in the presence of 10-nM 5-CT (5-carboxamidotryptamine), which results in the preferential visualization of 5-HT_{1F} receptors (H). (I) Distribution of 5-HT_{2A} receptors labeled by [^{3}H]MDL 100,907 (0.4 nM). (J, K) Distribution of sites labeled by [^{3}H]Mesulergine (5 nM) in coronal (J) and sagittal (K) sections. Note the very high densities of labeled sites in the choroid plexus that correspond to 5-HT_{2C} receptors. The autoradiographic signal in cortical regions represents mainly binding of the radioligand to 5-HT_{2A} receptors. (L, N) Distribution of mRNA coding for 5-HT_4 receptors detected with ^{33}P-labeled oligonucleotide probes that recognize all known splice variants of this receptor subtype. (M, O) Distribution of 5-HT_4 receptors labeled by [^{125}I]SB 207710 (0.02 nM) in close sections. Scale bars: 3 mm. The bar in O applies to all the coronal sections (Cg, cingulate cortex; ChP, choroid plexus; CPu, caudate putamen; DG, dentate gyrus; DR, dorsal raphe nucleus; Ent, entorhinal cortex; FR, fasciculus retroflexus; GP, globus pallidus; Hip, hippocampus; IC, inferior colliculus; M, motor cortex; PF, parafascicular thalamic nucleus; PrL, prelimbic cortex; S, subiculum; SN, substantia nigra; SuG, superficial gray layer of the superior colliculus; ZI, zona incerta).

5-HT_{1B} receptors remain very similar, in pharmacological terms especially, in non-rodents, and coexist in most species. They even seem to be similarly distributed, but it is very difficult to demonstrate the presence of significant amounts of 5-HT_{1D} protein or mRNA in the brain, whereas 5-HT_{1B} is strongly expressed across species. From this discussion, it is evident that brain receptor distribution studies have been an integral part of the process of 5-HT receptor discovery and characterization.

Autoradiographic studies using [^{3}H]-5-HT (in the presence of 8-OH-DPAT) (Pazos and Palacios, 1985), [^{125}I]-cyanopindolol (in the presence of isoprenaline) (Hoyer *et al.*, 1985b) or [^{125}I]-GTI (serotonin-5-O-carboxymethyl-glycyl-[^{125}I]tyrosinamide) (Waeber *et al.*, 1990) demonstrate a high density of 5-HT_{1B} sites in the rat basal ganglia, particularly the substantia nigra, globus pallidus, ventral pallidum and entopeduncular nucleus, but also many other regions. With appropriate displacing agents, both [^{125}I]-cyanopindolol and [^{125}I]-GTI allow discrimination of 5-HT_{1B} binding sites from 5-HT_{1D} binding sites in rodents, but currently there are no selective radioligands that allow this in non-rodent species (Waeber *et al.*, 1988a). The discrimination of 5-HT_{1B} and 5-HT_{1D} receptors in both rodent and non-rodent species has become more straightforward with the availability of a new 5-HT_{1B}/$_{1D}$ radioligand, [^{3}H]-GR125743 (Figure 1E, F), as well as cold ligands which discriminate 5-HT_{1B} and 5-HT_{1D} receptors. However, [^{3}H]-GR125743,

[^{3}H]-L-694,247 and [^{3}H]-Sumatriptan share these features, the latter also being able to label 5-HT_{1F} receptors (Figure 1G, H). Evidence from radioligand binding experiments using 5-HT neuronal lesions is equivocal regarding the synaptic location of the rat 5-HT_{1B} receptor, with some studies finding that the lesion causes upregulation of 5-HT_{1B} binding sites and others finding downregulation in the same areas. However, *in situ* hybridization studies have located mRNA encoding the 5-HT_{1B} receptor in the dorsal and median raphe nuclei. Furthermore, 5-HT_{1B} receptor mRNA in the raphe nuclei is markedly reduced by a 5-HT neuronal lesion. Together, these data suggest that 5-HT_{1B} receptors are located both presynaptically and postsynaptically relative to 5-HT neurons. It is speculated that in some brain areas (including substantia nigra and globus pallidus) 5-HT_{1B} binding sites may be located on non-5-HT nerve terminals, having been synthesized and then transported from cell bodies in other regions. Overall, the anatomical location of the 5-HT_{1B} receptor provides strong evidence to support the idea that the 5-HT_{1B} receptor has a role as both a 5-HT autoreceptor and a 5-HT heteroreceptor – i.e., controlling transmitter release. The receptors are also found on cerebral arteries and other vascular tissues, but no binding studies have confirmed this. Further, it seems that the receptor is 'silent' and may become responsive in conditions such as atherosclerosis, or at least needs some trigger. Peripheral effects have been described, such as inhibition of noradrenaline release in the vena cava and inhibition of plasma extravasation produced by trigeminal ganglion stimulation in guinea pigs and rats. 5-HT_{1B} receptors mediate contraction of rat caudal arteries. In non-rodents, they exhibit the 5-HT_{1D} 'pharmacology'.

Some forebrain areas with high levels of 5-HT_{1B} binding sites (e.g., striatum) also express 5-HT_{1B} receptor mRNA. However, other areas with high levels of 5-HT_{1B} binding sites have little detectable mRNA (e.g., substantia nigra, globus pallidus and entopeduncular nucleus). Similar mismatches between brain distribution of 5-HT_{1B} receptor mRNA and binding sites have been found in the primate and human brain (Varnas *et al.*, 2004a, 2004b). At the cellular level, *in situ* hybridization studies have localized 5-HT_{1B} receptor mRNA to granule and pyramidal cells within the hippocampus, and medium spiny neurons of the caudate putamen which are probably GABAergic. Immunocytochemical studies are now necessary to reveal the synaptic location of the receptors. mRNA has been located to some blood vessels. Antibodies are available, and have been used to label 5-HT_{1B} receptors in the brain of rodents.

Very recently, [^{3}H]AZ10419369 and its [^{11}C] analog have been described in autoradiographic studies in both macaque and human brain. Altogether, the data suggest that [^{11}C]AZ10419369 is a suitable radioligand for PET studies on 5-HT_{1B} receptor distribution and occupancy *in vivo* (Pierson *et al.*, 2008).

5-HT_{1D} receptors

5-HT_{1D} receptor has been cloned, and functionally expressed in recombinant systems; ligands can label it specifically (Zgombick *et al.*, 1995, 1996, 1997), although most of them will also label 5-HT_{1B} receptors (see above). 5-HT_{1D} selective antagonists have been developed for migraine, with the hope of avoiding cardiovascular side effects, but with less success than anticipated (see, for example, PNU-142633; McCall *et al.*, 2002), especially since both 5-HT_{1B} and 5-HT_{1D} receptors are present in the trigeminal ganglion (Hou *et al.*, 2001). It may well be that the 5-HT_{1B} component in the triptans is essential for their therapeutic effects. However, to determine the true contribution of 5-HT_{1D} in brain binding has proven difficult because of very low 5-HT_{1D} expression levels and the fact that, when present, 5-HT_{1D} mRNA is expressed in regions where much higher levels of 5-HT_{1B} mRNA have been found, and presumably the same is true for the corresponding proteins (Bruinvels *et al.*, 1993, 1994).

In situ hybridization allows detection of 5-HT_{1D} mRNA in rat brain regions, including the caudate putamen, nucleus accumbens, olfactory cortex, dorsal raphe nucleus and locus coeruleus. mRNA had low abundance in all regions but, interestingly, was undetectable in certain regions, including the globus pallidus, ventral pallidum and substantia nigra, where 5-$HT_{1B/1D}$ binding sites appear to be present. Together, these data are reminiscent of the findings with the 5-HT_{1B} receptor, and indicative of the 5-HT_{1D} receptor being located predominantly on axon terminals of both 5-HT and non-5-HT neurons.

It has been difficult to determine the distribution of 5-HT_{1D} receptors (see above) because protein levels appear to be low and there is a lack of radioligands able to discriminate 5-HT_{1D} from 5-HT_{1B} receptors (Schlicker *et al.*, 1997a, 1997b). Autoradiographic studies in rat utilizing [^{125}I]-GTI (serotonin-5-O-carboxymethyl-glycyl-[^{125}I]tyrosinamide) in the presence of CP 93129 to mask the rat 5-HT_{1B} binding site suggest that the 5-HT_{1D} site is present in various regions, but especially the basal ganglia (particularly the globus pallidus, substantia nigra and caudate putamen) and also the hippocampus and cortex. The distribution of 5-HT_{1D} receptors in human brain, as defined by the ketanserin-sensitive component of the [^{3}H]-Sumatriptan binding site, indicated their presence in the basal ganglia (globus pallidus and substantia nigra) as well as specific regions of the midbrain (periaqueductal gray) and spinal cord (Varnas *et al.*, 2001).

5-ht$_{1E}$ receptors

The 5-ht$_{1E}$ receptor was first detected in human brain membranes (this is crucial, see below) using radioligand binding studies that found that [^{3}H]5-HT, in the presence of blocking agents for other 5-HT$_1$ subtypes that were known at that time (5-HT$_{1A}$, 5-HT$_{1B}$, 5-HT$_{1C/2C}$), demonstrated a biphasic competition curve to 5-CT (Leonhardt *et al.*, 1989; Miller and Teitler, 1992; Bruinvels *et al.*, 1993; Barone *et al.*, 1994). The site with high affinity for 5-CT was thought to represent the 5-HT$_{1D}$ receptor. The low-affinity site had an unknown pharmacology, and was suggested to be a novel 5-HT receptor (5-ht$_{1E}$). Although 5-ht$_{1E}$ binding sites have been described in rodents, the relevance of these sites remains enigmatic as the receptor has not been cloned from rats or mice (Hoyer *et al.*, 1994), presumably because the gene is non-functional. Thus, the so-called rodent 5-HT$_{1E}$ sites that have been reported cannot be 5-HT$_{1E}$ receptors.

The 5-CT-insensitive [^{3}H]5-HT binding site was found in the cortex and caudate membranes of human, but also in guinea pig, rabbit and dog. Although we now know that other 5-HT receptor subtypes also have high affinity for [^{3}H]5-HT, are 5-CT insensitive (5-HT$_{1F}$, 5-HT$_6$) and have probably contributed to the initially described 5-ht$_{1E}$ binding, a human gene encoding for a receptor with 5-ht$_{1E}$ pharmacology (and structural features typical of a 5-HT$_1$ receptor) was subsequently isolated (McAllister *et al.*, 1992; Adham *et al.*, 1994a). In spite of the absence of selective radioligands for the 5-ht$_{1E}$ receptor, autoradiographic studies have provided a picture of the distribution of non-5-HT$_{1A/1B/1D/2C}$ [^{3}H]5-HT binding sites in human, rat, mouse and guinea-pig brain using rather complex technical conditions with co-incubation of ligands to block binding to the other known 5-HT$_1$ receptors. These studies indicate that, in all species, higher levels of these binding sites are present in the cortex (particularly entorhinal cortex), caudate putamen and claustrum, with detectable levels in other areas, such as the hippocampus (subiculum) and amygdala.

In hindsight, these receptor autoradiography studies may have been detecting a combination of 5-ht$_{1E}$ and 5-HT$_{1F}$ binding sites, and they were rather predictive. Thus, in the human and monkey brain, 5-ht$_{1E}$ mRNA is present in cortical areas (including entorhinal cortex) and the caudate and putamen, with lower levels in the amygdala and hypothalamus (Bruinvels *et al.*, 1994; Mengod *et al.*, 2006). 5-ht$_{1E}$ mRNA appears to be postsynaptic, consistent with receptor autoradiography studies finding no change in levels of the 5-ht$_{1E}$ binding site in rat forebrain following 5-HT neuronal lesions, but these studies were performed in rats.

The human (Lovenberg *et al.*, 1993) and the guinea-pig receptors have been cloned (Bai *et al.*, 2004), but remain enigmatic in both function and distribution; related to this, there are no selective ligands that allow a clear-cut functional or anatomical characterization of that receptor. Since the receptor is not expressed in the most commonly used laboratory animals, rat and mouse, the usual approach to create knock-out (KO) animals to study a phenotype that may be linked to lack of receptor expression or even the receptor KO/beta galactosidase knock-in is not very helpful for the study of either the distribution or the function. In addition, the guinea pig is not a favored species in behavioral studies, especially when selective tools are absent.

5-HT$_{1F}$ receptors

The 5-HT$_{1F}$ receptor could be described as a pure product of molecular biology: it was identified by homology screening starting from the existing known 5-HT$_1$ receptor sequences (Adham *et al.*, 1996, 1997). The receptor is still relatively poorly characterized; it is pharmacologically and structurally close to 5-ht$_{1E}$, but has been the subject of intense research (Adham *et al.*, 1994b; Phebus *et al.*, 1996) since Sumatriptan, the 5-HT$_{1B/1D}$ agonist, also has affinity for 5-HT$_{1F}$ receptors. Thus it was hypothesized that 5-HT$_{1F}$ receptor modulation may be a new approach in migraine intervention without the cardiovascular side effects of the triptans, since there is no evidence for functional vascular 5-HT$_{1F}$ receptors (Johnson *et al.*, 1997; Mitsikostas *et al.*, 1999).

Initial studies located 5-HT$_{1F}$ mRNA in the mouse and guinea-pig brain using *in situ* hybridization. 5-HT$_{1F}$ mRNA was found in hippocampus (CA1–CA3 cell layers), cortex (particularly cingulate and entorhinal cortices), and dorsal raphe nucleus (Bruinvels *et al.*, 1994). These results were confirmed in a subsequent more detailed mapping in the guinea-pig brain (Adham *et al.*, 1997), although levels of 5-HT$_{1F}$ mRNA in the raphe nuclei appeared to be much less abundant than in the initial report.

Brain regions containing 5-HT$_{1F}$ mRNA also display 5-CT-insensitive 5-HT$_1$, but non-5-HT$_{1A/1B/2C/1D}$ sites as detected in autoradiography studies. In autoradiography studies, [^{3}H]-Sumatriptan was used in the presence of 5-CT to label 5-HT$_{1F}$ binding sites in the guinea pig, rat (Figure 1G, H) and human brain (Lovenberg *et al.*, 1993; Waeber and Moskowitz, 1995; Mengod *et al.*, 1996; Palacios *et al.*, 1996). The distribution of 5-CT-insensitive [^{3}H]-Sumatriptan binding sites demonstrates a very good correlation with that of 5-HT$_{1F}$ mRNA (in the guinea pig) with the highest levels of binding in cortical and hippocampal areas, the claustrum and the caudate nucleus. Although the receptor is located in various parts of the basal ganglia, in contrast to 5-HT_{1B} and 5-HT_{1D} binding sites, 5-HT_{1F} binding sites appear to be barely detectable in the substantia nigra.

The brain distribution of 5-HT_{1F} sites became more precise (or more convincing) as labeled by the selective 5-HT_{1F} radioligand, [^{3}H]LY334370 (Lucaites *et al.*, 2005). In rat brain, specific 5-HT_{1F} binding was found in layers IV–V of cortical regions, the olfactory bulb and tubercle, nucleus accumbens, caudate putamen, parafascicular nucleus of the thalamus, medial mammillary nucleus, CA3 region of the hippocampus, subiculum, and amygdaloid nuclei. Interestingly, rat-brain autoradiography with [^{3}H]LY334370 and [^{3}H]-Sumatriptan showed labeling in the same brain regions when performed side by side. Some species differences in the distribution of the 5-HT_{1F} receptor were noted between rat, guinea pig, monkey and human brain. 5-HT_{1F} receptors have not been found in blood vessels, which is interesting in the context of migraine.

5-HT_{2A} receptors

The 5-HT_{2A} receptor is the classical 5-HT_2 or 5-HT_D receptor which was originally identified in the vascular (most blood vessels) and non-vascular smooth muscle (e.g., guinea-pig ileum). The receptor is also present in platelets and is rather prominently expressed in the brain. We were in the fortunate situation very early on to have adequate radioligands with good selectivity to label 5-HT_{2A} receptors – for example, [^{3}H] ketanserin or [^{3}H] spiperone (under appropriate conditions, since ketanserin can label alpha$_1$ adrenoceptors and spiperone has also high affinity for D_2 receptors) (Hoyer *et al.*, 1987). These ligands allowed good definition of the distribution of 5-HT_{2A} receptors even before it became apparent that subtypes of 5-HT_2 receptors existed (5-HT_{2B} and 5-HT_{2C}) (Hoyer *et al.*, 1986b). Such information regarding the receptor distribution was crucial, when the list of 5-HT_2 receptors expanded. Thus, it was found that [^{3}H]5-HT labeled sites in the choroid plexus were not identified by [^{3}H]ketanserin but by [^{3}H]mesulergine, a ligand known to bind dopamine D_2 and 5-HT_2 receptors (Pazos *et al.*, 1984a). An extensive pharmacological and distribution characterization (Pazos *et al.*, 1984b) led to the conclusion that the new sites could not be the classical 5-HT_2 receptors, and we proposed naming it 5-HT_{1C} (because of its high affinity for 5-HT), to be renamed later 5-HT_{2C} due to clear structural and transductional homology with the 5-HT_2 class.

More recently, [^{3}H]MDL 100,907 (also known as M 100,907) (Figure 1I) became another very selective tool to study 5-HT_{2A} receptors *in vitro*, and [^{11}C] analogs can be used *in vivo* for PET studies (Ito *et al.*, 1998). LSD and its derivatives do label 5-HT_2 receptors, but are not selective, similar to the agonist DOI. High levels of 5-HT_{2A}-binding sites have been found in cortical areas, the caudate, nucleus accumbens, olfactory tubercule and hippocampus (Pazos *et al.*, 1985, 1987b; López-Giménez *et al.*, 1997, 2001a). Generally, there is close overlap between the distribution of 5-HT_{2A}-binding sites, immunoreactivity and mRNA, suggesting that this subtype has a postsynaptic location (López-Giménez *et al.*, 1997, 1998, 1999, 2001a, 2001b). In areas such as the cortex, 5-HT_{2A} receptors have been localized on GABAergic interneurons and also on glutamatergic projection neurons (Santana *et al.*, 2004; de Almeida and Mengod, 2007). The distribution of human 5-HT_{2A} receptors has been successfully mapped using various tools, and altogether the data were consistent with rodent data (Pazos *et al.*, 1987b; Varnas *et al.*, 2004a). A peculiarity of the human 5-HT_{2A} receptor is its location in striosomes (Waeber and Palacios, 1994; López-Giménez *et al.*, 1999).

5-HT_{2B} receptors

The receptor was first identified in the rat fundus, but is also present in a number of vessels and the heart (Ullmer *et al.*, 1995). The receptor has been cloned (Foguet *et al.*, 1992; Schmuck *et al.*, 1994; Wainscott *et al.*, 1996), and received a great deal of attention more recently when it was established that fenfluramine caused valvulopathies by indirect activation of the 5-HT_{2B} receptor (Fitzgerald *et al.*, 2000). Initially its presence in the brain was largely neglected, since *in situ* hybridization studies did not provide positive data. However, the presence of 5-HT_{2B} receptor-like immunoreactivity was eventually reported in rat brain, although the immunostaining is restricted to a few regions, particularly the cerebellum, lateral septum, dorsal hypothalamus and medial amygdala (Duxon *et al.*, 1997). The cells expressing 5-HT_{2B} receptor-like immunoreactivity have a neuronal rather than astrocytic morphology. The brain 5-HT_{2B} receptor's function remains to be described.

5-HT_{2C} receptors

The 5-HT_{2C} was discovered by the combined use of membrane receptor binding/pharmacology and brain-slice autoradiography, where an atypical 5-HT binding site was identified initially. We had noticed the presence of a 5-HT site in the choroid plexus which was not 5-HT_{1A}, 5-HT_{1B} or 5-HT_{2A}, and was then named 5-HT_{1C}. The pharmacology and distribution of the new site was different from what was known (Hoyer *et al.*, 1985a; Pazos *et al.*, 1984a, 1984b), and this putative 5-HT receptor was found across species in the choroid plexus and other brain regions. Radioligands that label 5-HT_{2C} receptors are [^{3}H]mesulergine (in the presence of a selective 5-HT_{2A} antagonist), [^{3}H]5-HT (with adequate protection using a cocktail of 5-HT_1 ligands), [^{125}I]SCH23982 (also dopamine D1), [^{125}I]LSD (in the presence of adequate 5-HT_{2A} selective drugs) and the 5-HT_2 receptor agonist [^{125}I]DOI (also in the presence of selective 5-HT_{2A}-blocking drug).

5-HT_{2C} receptor localization is restricted to the CNS, unlike that of 5-HT_{2A} and 5-HT_{2B} receptors. Autoradiographic studies have identified this receptor in the choroid plexus, cortex, nucleus accumbens, hippocampus, amygdala, caudate and substantia nigra in rat brain. The 5-HT_{2C} receptors, in addition to having a postsynaptic location, may also be presynaptic. In the mouse, [^{3}H]mesulergine binding sites (in the presence of spiperone to block binding to 5-HT_{2A} receptors; Mengod *et al.*, 1990) throughout the wild-type brain show strong labeling in choroid plexus, which is predominant, although the presence of low/very low specific [^{3}H]mesulergine signals are detected in the nucleus accumbens, patches of the caudate putamen, the olfactory tubercle, claustrum, septum, cingular cortex, amygdala, dentate gyrus, periaqueductal gray, entorhinal cortex, and several brainstem motor nuclei. This binding is no longer detected in the 5-HT_{2C} receptor knock-out mouse brain (López-Giménez *et al.*, 2002), making the point that 5-HT_{2C} receptors are indeed present in the brain, although at much lower densities than 5-HT_{2A} receptors (with the remarkable exception of the choroid plexus). The distribution of mRNA is very similar to that of protein or binding sites, except for high levels in the habenular nucleus, where binding site levels are very low (Mengod *et al.*, 1990; López-Giménez *et al.*, 2001c). There are multiple splice and editing variants of 5-HT_{2C} receptors (Niswender *et al.*, 1998; Fitzgerald *et al.*, 1999) which are beyond the scope of this chapter; they are not discriminated as far as is known by the antagonist radioligands used in autoradiographic studies.

In monkey, 5-HT_{2C} mRNA is present in the choroid plexus, in layer V of most cortical regions, in the nucleus accumbens, ventral anterior caudate and putamen, septal nuclei, diagonal band, ventral striatum and extended amygdala (López-Giménez *et al.*, 2001c). Several thalamic, midbrain and brainstem nuclei also express 5-HT_{2C} mRNA. [^{3}H]mesulergine binding and mRNA showed a good correlation across the brain, supporting a predominant somatodendritic localization of 5-HT_{2C} receptors. However, in a few instances a lack of correlation between both patterns of signal suggests a possible location on axon terminals. Examples of poor correlation are the septal nuclei and horizontal limb of the diagonal band (presence of mRNA with apparent absence of binding sites) and the interpeduncular nucleus (presence of binding sites with apparent absence of mRNA).

The cellular localization of 5-HT_{2C} receptor mRNA in relation to serotonergic and GABAergic neurons has been studied in the anterior raphe nuclei of the rat (Serrats *et al.*, 2005). In the dorsal and median raphe nuclei, 5-HT_{2C} receptor mRNA was not detected in serotonergic cells identified as those expressing 5-HT transporter mRNA. In contrast, 5-HT_{2C} receptor mRNA was found in most GABAergic cells, recognized by the presence of glutamic acid decarboxylase mRNA. Such 5-HT_{2C} receptor-positive GABAergic neurons were mainly located in the intermediolateral and lateral portions of the dorsal raphe and lateral part of the median raphe. The present data give anatomical support to a previous hypothesis that proposed a negative-feedback loop involving reciprocal connections between GABAergic interneurons bearing 5-$HT_{2A/2C}$ receptors and 5-HT neurons in the dorsal raphe and surrounding areas. According to this model, the excitation of GABAergic interneurons through these 5-HT_{2C} (and 5-HT_{2A}) receptors would result in the suppression of 5-HT cell firing.

5-HT_3 receptors

The 5-HT_3 or 5-HT-M receptor was first identified in the guinea-pig ileum, and then more widely in the peripheral nervous system (PNS) (Gaddum and Picarelli, 1957). The presence of 5-HT_3 receptors in the brain has been a matter of controversy. With the development of adequate ligands this issue has been settled, as both membrane binding and autoradiography were successfully performed in brain tissue with compounds such as [^{3}H]ICS-205930 (Waeber *et al.*, 1988b), [^{3}H]ondansetron (Kilpatrick *et al.*, 1987), [^{3}H]zacopride (Parker *et al.*, 1996) and its iodinated analog [^{125}I]DAIZAC (Hewlett *et al.*, 1999) following initial successful binding in neuroblastoma glioma cells (Hoyer and Neijt, 1987) that suggested indeed the presence of 5-HT_3 receptors in the CNS. There are currently five 5-HT_3 receptor subunits that have been cloned (Maricq *et al.*, 1991; Belelli *et al.*, 1995; Davies *et al.*, 1999; Dubin *et al.*, 1999; Hanna *et al.*, 2000), and the receptor is a pentamer like, for example, nicotinic acetylcholine receptors. We will limit ourselves to the 5-HT_{3A} and 5-HT_{3B} subunits, which appear to form the basis of the functional heteromeric channel (Davies *et al.*, 1999). The radioligands used do not distinguish the subunits. The function of the latter three subunits, 5-$HT_{3C,\ 3D,\ 3E}$, is still being debated (Niesler *et al.*, 2003, 2007). On their own they have no channel activity, but are able to heteromerize with

5-HT_{3A}, and form functional channels, as is known for 5-$HT_{3A/3B}$. On the other hand, these three subunits seem to be absent from the rodent genome. Since 5-HT_{3B} appears to be sparse in the rodent brain (although see below), whereas 5-HT_{3C} is possibly present in the brain, and since splice variants of the 5-HT_{3A} exist, the situation is far from simple (van Hooft and Yakel, 2003). In addition, 5-HT_{3A} and nicotinic receptor subunits can co-assemble and form channels in cell systems, but this has not been demonstrated in the brain (see Chameau and van Hooft, 2006 for discussion and further references).

The highest levels of 5-HT_3 receptor binding sites are within the dorsal vagal complex in the brainstem (Kilpatrick *et al.*, 1989; Pratt *et al.*, 1990). This region comprises the nucleus tractus solitarius, area postrema, and dorsal motor nucleus of the vagus nerve, which are intimately involved in the initiation and coordination of the vomiting reflex; antagonism of 5-HT_3 receptors in these nuclei is therefore likely to contribute to the anti-emetic action of 5-HT_3 receptor antagonists. 5-HT_3 receptor expression in the forebrain is low, in most cases less than 40 fmol/mg or even lower. The highest levels outside the brainstem are expressed in regions such as the hippocampus, amygdala and superficial layers of the cerebral cortex. The distribution of 5-HT_3 receptor recognition sites within the forebrain displays species variations. For example, in humans, relatively high levels of 5-HT_3 receptor recognition sites have been located within the caudate nucleus and putamen (Abi-Dargham *et al.*, 1993) whereas low levels are detected within cortical regions (Barnes *et al.*, 1989; Waeber *et al.*, 1989; Abi-Dargham *et al.*, 1993). The majority of species investigated so far, however, express high levels of 5-HT_3 receptors within the hippocampus relative to other forebrain regions (e.g., mouse, rat, man) (Parker *et al.*, 1996).

5-HT_{3A} receptor mRNA transcripts are similarly distributed, in the rodent brain, to 5-HT_3 autoradiographic binding (e.g., the piriform cortex, entorhinal cortex, hippocampus) (Tecott *et al.*, 1993). In the hippocampus and prefrontal cortex, mRNA is present in interneurons (Tecott *et al.*, 1993; Puig *et al.*, 2004); this distribution indicates that the 5-HT_3 receptor may mediate the indirect inhibition of excitatory pyramidal neurons via activation of GABAergic interneurons. 5-HT_3 receptor-like immunoreactivity is primarily associated with GABA-containing neurons in the cerebral cortex and hippocampus that often co-localize with CCK (but not somatostatin) (Morales *et al.*, 1996a, 1996b; Morales and Bloom, 1997) or calbindin (but not parvalbumin) in the CA1/CA3 fields (Morales and Bloom, 1997).

Attempts to define the cellular location of the 5-HT_3 receptor expressed in the human basal ganglia indicate that they are not principally located on dopaminergic neurons, since their density is not influenced by the neurodegeneration within this region associated with Parkinson's disease. However, a significant population of the 5-HT_3 receptors in this region is associated with neurons that degenerate in Huntington's disease (Steward *et al.*, 1993). Huntington's is neuropathologically characterized by the degeneration of neurons that have their cell bodies within the caudate-putamen, which include the GABAergic projection neurons. 5-HT_{3B} immunohistochemical labeling has been reported with a selective 5-HT_{3B} antibody in the hippocampus (Monk *et al.*, 2001); however, the density of signals is very low and this has not been generally replicated, raising the possibility that 5-HT_{3A} and 5-HT_{3B} do not necessarily co-localize in the brain, and therefore the molecular nature of the brain 5-HT_3 receptor is still in debate. What is not debated, though, is that brain 5-HT_3 receptors can be both postsynaptic (e.g., in the interneurons) and as presynaptic: 5-HT_3 receptor binding in the nucleus of the solitary tract and the dorsal motor nucleus of the vagus is lost upon deafferentiation by nodose ganglionectomy or vagotomy; in addition, these nuclei show no 5-HT_3 mRNA expression, supporting the idea that 5-HT_3 receptors present in these nuclei are presynaptic. It has been reported that 5-HT_3 receptor activation modulates dopamine release, although this action may be indirect, and, directly, GABA release, as documented electrophysiologically in the hippocampus and amygdala. Thus, although 5-HT_3 receptors can be convincingly demonstrated in various brain regions, there is still work to be done to reconcile function and localization as well as subunit composition and co-expression. However, more recently a new antibody has been reported to selectively label 5-HT_{3B} subunits in various species, especially mouse and rat (Doucet *et al.*, 2007): in contrast to previous reports, intense labeling was seen in peripheral ganglia (trigeminal, nodose, superior cervical and dorsal root ganglia), where almost 100 percent of neurons expressing 5-HT_{3A} subunits were also labeled by anti-5-HT_{3B} antibody. 5-HT_{3B} immunoreactivity was also found in the rat hippocampus (DG and CA1 layer) and in isolated cortical neurons. Interestingly, both 5-HT_{3A} and 5-HT_{3B} immunoreactivities were increased following nerve lesion, confirming co-localization of both subunits. These data suggest that 5-HT_3 subunit *in situ* hybridization does not inform about heteromeric 5-$HT_{3A/B}$ as compared to homomeric 5-HT_{3A} receptors in the PNS or CNS of rodents.

5-HT_4 receptors

The 5-HT_4 receptor has been known for some time in the gastrointestinal tract (GIT), heart and brain (where it stimulates cAMP production in the mouse colliculi)

(Dumuis *et al.*, 1988). However, it took the development of selective compounds and, eventually, the cloning of the receptor to make its mark (Dumuis *et al.*, 1989; Villalon *et al.*, 1990; Bockaert *et al.*, 1992; Gerald *et al.*, 1995). The 5-HT_4 receptor exists in a number of splice variants (at least 10, and the number is still increasing) which are not equal across species. When referring to autoradiography, we assume that the ligands will bind rather indiscriminately to all splice variants; *in situ* hybridization and RT-PCR have been performed with a number of selective probes in the brain in various species, but a detailed account would be too complex for the current scope. Suffice it to say that 5-HT_4 receptors are prominently expressed in the brain and that, depending on the species studied, various splice variants can be found, although not all are present in brain.

Derivatives of a number of 5-HT_4 receptor ligands have made useful radioligands to map and pharmacologically characterize the 5-HT_4 receptor (e.g., [^{3}H]GR113808, [^{3}H]RS 57639, [^{3}H]BIMU1, [^{3}H]prucalopride and especially [^{125}I]SB207710). A consistent finding across the species investigated is the presence of relatively high levels of the 5-HT_4 receptor in the nigrostriatal and mesolimbic systems of rat, guinea pig, pig, cow, monkey and man (Grossman *et al.*, 1993; Waeber *et al.*, 1993; Jakeman *et al.*, 1994; Vilaró *et al.*, 1996, 2002, 2005). In rat brain, [^{3}H]GR113808 presented prominent binding in the limbic system (islands of Calleja, hippocampus, ventral pallidum, fundus striate, olfactory tubercle, septum and amygdala) and the striato-nigro-tectal pathway. In human brain (Varnas *et al.*, 2003), [^{125}I]SB207710 labeled the basal ganglia (caudate nucleus, putamen, nucleus accumbens, globus pallidus and substantia nigra) and the hippocampal formation (CA1 and subiculum) (Figure 1M,O). In the neocortex, high levels were seen in superficial layers, and lower levels in deep cortical layers. 5-HT_4 receptors appear to be largely localized postsynaptically. However, presynaptic localization on terminals of GABAergic (dentate gyrus), dopaminergic and serotonergic neurons is also likely, since the release of these neurotransmitters is modulated by 5-HT_4 agonists.

Vilaró and colleagues (2005) have combined several approaches in rat and guinea-pig brain to study the three variants 5-$HT_{4(a)}$, 5-$HT_{4(b)}$, and 5-$HT_{4(e)}$. They compared the distribution of 5-HT_4 receptors using [^{125}I]-SB 207710, as well as that of 5-$HT_{4(a)}$, 5-$HT_{4(b)}$ and 5-$HT_{4(e)}$ mRNA visualized by *in situ* hybridization (ISHH) (Figure 1L,N) and RT-PCR. In several projection systems (striato-nigral and striato-pallidal pathways, in the hippocampus from dentate granule cells to CA3 habenulo-interpeduncular pathway), 5-HT_4 receptors may be located both somatodendritically and axonally. RT-PCR showed all three variants to be rather widely distributed (olfactory tubercle, striatum, hippocampus, inferior colliculus, substantia nigra, parietal cortex). 5-$HT_{4(b)}$ mRNA was present in all regions examined, whereas 5-$HT_{4(a)}$ and 5-$HT_{4(e)}$ distribution was somewhat more restricted. In other regions (periaqueductal gray, reticular formation, medial septum, diagonal band), faint ISHH signals are observed for 5-$HT_{4(a)/4(e)}$ mRNAs, whereas 5-$HT_{4(b)}$ mRNA signals are almost undetectable. Finally, neurotoxic lesions of basal ganglia components in guinea pig (Vilaró *et al.*, 2005) also indicate a location of these receptors on terminals of striatal projection neurons.

Altogether, the brain distribution of 5-HT_4 receptors is compatible with a role for these receptors in memory and emotion (Compan *et al.*, 2004; Lucas *et al.*, 2007).

5-ht_{5a} receptors

The 5-ht_{5a} and 5-ht_{5b} receptors are still considered to be 'orphans', since there is no clear function attached to these receptors and 5-ht_{5b} does not exist in humans (Grailhe *et al.*, 2001), because the human gene is interrupted by stop codons. Further, even the coupling of the receptors is not firmly established.

The distribution of 5-ht_{5a} receptors was studied using *in situ* hybridization (Pasqualetti *et al.*, 1998), specific antibodies and receptor autoradiography, although no selective and adequate radioligand is available. Carson and colleagues (1996), using 5-ht_{5A}-selective antibodies, found immunoreactivity in astrocytes, and reported increased 5-ht_{5A} mRNA levels in reactive gliosis. However, more recent studies have not found 5-ht_{5A} receptors in astrocytes. Oliver and co-workers (Oliver *et al.*, 2000) and others (Doly *et al.*, 2004) reported on the distribution of 5-ht_{5a} immunoreactivity in brain and spinal cord. 5-ht_{5A}-ir was expressed in various layers of the cerebral cortex, in the hippocampus, septum, amygdala and hypothalamus, mostly in axon terminal fields from the median raphe. Furthermore, the suprachiasmatic nucleus, basal ganglia, diagonal band nuclei, amygdala (especially central, basolateral and lateral nuclei), ventral pallidum, entorhinal and cingulate cortices and paraventricular hypothalamus were also identified.

Radioligand binding was performed on WT and 5-ht_{5A}-knock-out mouse brains (Waeber *et al.*, 1998; Grailhe *et al.*, 1999). [^{125}I]LSD, a high-affinity ligand for the 5-ht_{5A} receptor, was used in the presence of adequate spiperone and clozapine concentrations to block 5-HT_{2A}, 5-HT_{2C}, 5-HT_6, 5-HT_7, and dopamine receptor binding. High levels of [^{125}I]LSD binding sites were found in the olfactory bulb and medial habenula of WT mice, with lower densities in the neocortex, hippocampus and caudate putamen. In contrast, 5-ht_{5A}-knock-out mice displayed little specific binding except in the medial habenula (possibly 5-ht_{5B} binding). In spite of these data, the function of the 5-ht_5 receptors remains elusive in the absence of selective tools or genetic links to disease. Functional studies performed with the only selective 5-ht_{5A} antagonist, SB-699551-A, suggest an autoreceptor role for the 5-ht_{5A} receptor in guinea-pig raphe nucleus (Thomas *et al.*, 2006). Furthermore, Serrats *et al.* (2004) reported that 5-ht_{5b} receptor mRNA was co-localized with that of the 5-HT transporter primarily in the dorsal raphe, suggesting a possible autoreceptor role for 5-ht_{5B} as well.

5-HT_6 receptors

The existence of the 5-HT_6 receptor has been suspected in striatum and in neuroblastoma cells, where a positively cAMP coupled 5-HT receptor with atypical pharmacology was described and confirmed when the 5-HT_6 receptor was cloned. It seems that this receptor is brain-specific.

5-HT_6 receptor mRNA is largely confined to the central nervous system, although low levels may be found in the stomach and adrenal glands (Ruat *et al.*, 1993; but see also Monsma *et al.*, 1993; Mengod *et al.*, 2006). High levels of 5-HT_6 receptor mRNA are consistently detected within the striatum (caudate nucleus) of rat, guinea pig and human (Ruat *et al.*, 1993; Mengod *et al.*, 2006). Relatively high levels are also present in the olfactory tubercle, nucleus accumbens and hippocampus.

Subsequently, receptor distribution was studied using the selective antagonist [^{125}I]SB-258585 (Roberts *et al.*, 2002), which shows high specific binding when used in autoradiographic studies in rat: very prominent signals light up in the caudate putamen, nucleus accumbens, islands of Calleja, olfactory tubercle and choroid plexus. Moderate levels were present in the hippocampus, cerebral cortex, thalamus, hypothalamus and substantia nigra; and low levels in the globus pallidus, cerebellum, other mesencephalic regions and the rhombencephalon.

Immunocytochemistry was also used to map 5-HT_6 receptor distribution (Gerard *et al.*, 1997; Hamon *et al.*, 1999): dense signals were found in the frontal, entorhinal and piriform cortices, nucleus accumbens, cerebellum, caudate-putamen, hippocampus (dentate gyrus and CA1), olfactory tubercle and islands of Calleja. Moderate staining was detected in other cortical zones, the taenia tecta, substantia nigra, oculomotor nucleus, red nucleus, motor trigeminal nucleus and facial nucleus. The septum, globus pallidus, hypothalamus, colliculi and raphe nuclei were negative. These immunohistochemical studies, including electron-microscopic investigations, have shown prominent dendritic localization of the 5-HT_6 receptor. The striatal binding sites may be on intrinsic GABAergic or cholinergic neurons, or on terminals of projection neurons from the thalamus or cerebral cortex.

Given the prominent expression in the striatum, it was suspected that 5-HT_6 receptors interact with dopamine transmission. A number of antipsychotics have high affinity for 5-HT_6 receptors. However, 6-OHDA lesions in the substantia nigra and striatum produced no significant changes in [^{125}I]SB-258585 binding in any brain region examined, suggesting that 5-HT_6 receptors are not located on dendritic, somatic or terminal elements of dopaminergic neurons. Similarly, the depletion of serotoninergic innervation via 5,7-dihydroxytryptamine did not affect 5-HT_6 receptor expression, suggesting that 5-HT_6 receptors are not presynaptic autoreceptors (Bourson *et al.*, 1998).

5-HT_7 receptors

The 5-HT_7 receptor has been known for some time in the cardiovascular or GIT system, and was originally termed 5-HT_1-like, since it has rather high affinity for 5-CT and, to some extent, for 8-OH-DPAT (Hoyer *et al.*, 1994). When cloned, it was realized that although the pharmacology is very similar to that of the so-called 5-HT_1-like receptor, it is obvious that 5-HT_7 and any of the 5-HT_1 receptor subtypes have very little in common structurally; in addition, 5-HT_7 receptors essentially mediate cAMP stimulation, in contrast to 5-HT_1 receptors (Hoyer *et al.*, 1994).

The 5-HT_7 receptor has a number of splice variants, and attempts have been made to quantify the expression of these using *in situ* hybridization or PCR (To *et al.*, 1995). Initially, based on the high affinity of 5-HT_7 receptors for 5-CT, autoradiographic studies were performed using [^{3}H]5-CT in the presence of blocking concentrations of ligands interacting with 5-HT_1 receptors (Vanhoenacker *et al.*, 2000). More recently, [^{3}H]SB-269970 has been described as a highly selective radioligand for 5-HT_7 receptors, and autoradiographic studies could be performed (Varnas *et al.*, 2004b). In various species, including human brain, 5-HT_7 binding is found in anterior thalamus and the hippocampus (dentate gyrus) at high levels. Other regions expressing intermediate levels are the septum and hypothalamus (suprachiasmatic nucleus SCN), other hippocampal regions (CA1, CA2), the anterior cingulate and other cerebral cortical areas (in pyramidal cells and on GABAergic neurons), some amygdala and brainstem nuclei and basal ganglia, and Purkinje cells in the cerebellum. Similar findings have been reported in rat and guinea pig (To *et al.*, 1995; Gustafson *et al.*, 1996), and are in agreement with mRNA distribution (Vanhoenacker *et al.*, 2000). The receptor is also found, as expected, in the GIT and vascular smooth muscle, and a few sympathetic ganglia, as well as in the cerebral and meningeal circulation. The localization in the SCN may relate to sleep-modulating effects of 5-HT$_7$ ligands, and it has become evident that a number of antipsychotics act as inverse agonists at 5-HT$_7$ receptors. However, it is not established that a 5-HT$_7$ antagonist may be an antipsychotic in its own right.

With respect to the distribution of the isoforms of the 5-HT$_7$ receptor, large tissue-specific differences in the splicing of pre-mRNA within a species are not apparent (Heidmann *et al.*, 1997, 1998). However, the relative abundance of the 5-HT$_{7(b)}$ receptor isoform displays marked differences between rat (low) and human (high) tissues. Altogether, 5-HT$_7$ receptors are expressed in the brain, and protein and mRNA distribution appear to correlate. Immunocytochemistry has also been used to map 5-HT$_7$ receptor distribution (Moyer and Kennaway, 1999; Geurts *et al.*, 2002; Doly *et al.*, 2005) with good results.

Conclusion

5-HT receptors (and the 5-HT transporters as well as tryptophan hydroxylase, not addressed here) are present in the brain, in general, in rather similar manners across species. 5-HT receptors are clearly main players in various physiologic and pathophysiologic situations either directly or indirectly following long-term effects of selective serotonin or noradrenalin reuptake inhibitors on their transporters. Not all 5-HT receptors have found their niche, and there are odd exceptions, such as 5-ht$_{5B}$, which does not exist in human, or 5-HT$_{1e}$, whose functional protein is not produced in rodents. There are still question marks regarding other receptors such as 5-HT$_{2B}$, whose presence and function in the brain is not resolved. Additional complexity relates to apparent existence of the numerous editing variants of the 5-HT$_{2C}$ receptors, or the many splice variants of 5-HT$_4$ or 5-HT$_7$ receptors whose distribution and specific function cannot be studied easily due to the lack of tools. There is also evidence that mRNA and protein levels do not need to match, and even their distribution can be rather different between cell bodies and terminals. The bigger challenge appears to be 5-HT$_3$ receptor subunits, for which mRNA may not be easily found in the brain or even peripheral nervous system, whereas antibodies may detect significant levels of protein. Another challenge is cellular co-localization, for which the tools must be exquisitely selective and sensitive before conclusions can be drawn. Finally, receptor distribution studies need to reach the subcellular level if we are to understand whether receptor heteromers are produced *in situ*, as has been suggested in recombinant systems, being of the same family (e.g., 5-HT$_{1B/1D}$), or more complex situations such as dopamine/5-HT receptor heteromers, or channels that combine 5-HT$_3$ and nicotinic receptor subunits. The diversity of combinations appears endless, and we have not addressed the function of accessory proteins, which may well explain a number of pathologic conditions depending on whether receptor addressing may function normally or not.

References

Abi-Dargham, A., Laruelle, M., Wong, D.T., Robertson, D.W., Weinberger, D.R. and Kleinman, J.E. (1993) Pharmacological and regional characterization of [3H]LY278584 binding sites in human brain. J. Neurochem., 60: 730–737.

Adham, N., Vaysse, P.J., Weinshank, R.L. and Branchek, T.A. (1994a) The cloned human 5-HT1E receptor couples to inhibition and activation of adenylyl cyclase via two distinct pathways in transfected BS-C-1 cells. Neuropharmacology, 33: 403–410.

Adham, N., Tamm, J.A., Salon, J.A., Vaysse, P.J., Weinshank, R.L. and Branchek, T.A. (1994b) A single point mutation increases the affinity of serotonin 5-HT1D alpha, 5-HT1D beta, 5-HT1E and 5-HT1F receptors for beta-adrenergic antagonists. Neuropharmacology, 33: 387–391.

Adham, N., Gerald, C., Schechter, L., Vaysse, P., Weinshank, R. and Branchek, T. (1996) [^{3}H]5-Hydroxytryptamine labels the agonist high affinity state of the cloned rat 5-HT$_4$ receptor. Eur. J. Pharmacol., 304: 231–235.

Adham, N., Bard, J.A., Zgombick, J.M., Durkin, M.M., Kucharewicz, S., Weinshank, R.L. and Branchek, T.A. (1997) Cloning and characterization of the guinea pig 5-HT$_{1F}$ receptor subtype: a comparison of the pharmacological profile to the human species homolog. Neuropharmacology, 36: 569–576.

Amargós-Bosch, M., Bortolozzi, A., Puig, M.V., Serrats, J., Adell, A., Celada, P., Toth, M., Mengod, G. and Artigas, F. (2004) Co-expression and *in vivo* interaction of serotonin1A and serotonin2A receptors in pyramidal neurons of prefrontal cortex. Cereb. Cortex, 14: 281–299.

Bai, F., Yin, T., Johnstone, E.M., Su, C., Varga, G., Little, S.P. and Nelson, D.L. (2004) Molecular cloning and pharmacological characterization of the guinea pig 5-HT$_{1E}$ receptor. Eur. J. Pharmacol., 484: 127–139.

Barnes, J.M., Barnes, N.M., Costall, B., Ironside, J.W. and Naylor, R.J. (1989) Identification and characterisation of 5-hydroxytryptamine 3 recognition sites in human brain tissue. J. Neurochem., 53: 1787–1793.

Barone, P., Jordan, D., Atger, F., Kopp, N. and Fillion, G. (1994) Quantitative autoradiography of 5-HT1D and 5-HT1E binding sites labelled by [3H]5-HT, in frontal cortex and the hippocampal region of the human brain. Brain Res., 638: 85–94.

Belelli, D., Balcarek, J.M., Hope, A.G., Peters, J.A., Lambert, J.J. and Blackburn, T.P. (1995) Cloning and functional expression of a human 5-hydroxytryptamine type 3A(S) receptor subunit. Mol. Pharmacol., 48: 1054–1062.

Bockaert, J., Fozard, J.R., Dumuis, A. and Clarke, D.E. (1992) The 5-HT4 receptor: a place in the sun. Trends Pharmacol. Sci., 13: 141–145.

Bourson, A., Boess, F.G., Bos, M. and Sleight, A.J. (1998) Involvement of 5-HT_6 receptors in nigro-striatal function in rodents. Br. J. Pharmacol., 125: 1562–1566.

Bruinvels, A.T., Lery, H., Nozulak, J., Palacios, J.M. and Hoyer, D. (1992) 5-HT_{1D} binding sites in various species: similar pharmacological profile in dog, monkey, calf, guinea-pig and human brain membranes. Naunyn Schmiedebergs Arch. Pharmacol., 346: 243–248.

Bruinvels, A.T., Palacios, J.M. and Hoyer, D. (1993) Autoradiographic characterisation and localisation of 5-HT1D compared to 5-HT1B binding sites in rat brain. Naunyn Schmiedebergs Arch. Pharmacol., 347: 569–582.

Bruinvels, A.T., Landwehrmeyer, B., Gustafson, E.L., Durkin, M.M., Mengod, G., Branchek, T.A., Hoyer, D. and Palacios, J.M. (1994) Localization of 5-HT1B, 5-HT1D alpha, 5-HT1E and 5-HT1F receptor messenger RNA in rodent and primate brain. Neuropharmacology, 33: 367–386.

Carson, M.J., Thomas, E.A., Danielson, P.E. and Sutcliffe, J.G. (1996) The 5-HT_{5A} serotonin receptor is expressed predominantly by astrocytes in which it inhibits cAMP accumulation: a mechanism for neuronal suppression of reactive astrocytes. Glia, 17: 317–326.

Chameau, P. and van Hooft, J.A. (2006) Serotonin 5-HT(3) receptors in the central nervous system. Cell Tissue Res., 326: 573–581.

Compan, V., Zhou, M., Grailhe, R., Gazzara, R.A., Martin, R., Gingrich, J., Dumuis, A., Brunner, D., Bockaert, J. and Hen, R. (2004) Attenuated response to stress and novelty and hypersensitivity to seizures in 5-HT4 receptor knock-out mice. J.Neurosci., 24: 412–419.

Davies, P.A., Pistis, M., Hanna, M.C., Peters, J.A., Lambert, J.J., Hales, T.G. and Kirkness, E.F. (1999) The 5-HT3B subunit is a major determinant of serotonin-receptor function. Nature, 397: 359–363.

de Almeida, J. and Mengod, G. (2007) Quantitative analysis of glutamatergic and GABAergic neurons expressing 5-HT(2A) receptors in human and monkey prefrontal cortex. J. Neurochem., 103: 475–486.

de Almeida, J. and Mengod, G. (2008) Serotonin 1A receptors in human and monkey prefrontal cortex are mainly expressed in pyramidal neurons and in a GABAergic interneuron subpopulation: implications for schizophrenia and its treatment. J. Neurochem., 107: 488–496.

DeFelipe, J., Arellano, J.I., Gomez, A., Azmitia, E.C. and Munoz, A. (2001) Pyramidal cell axons show a local specialization for GABA and 5-HT inputs in monkey and human cerebral cortex. J. Comp. Neurol., 433: 148–155.

Doly, S., Fischer, J., Brisorgueil, M.J., Verge, D. and Conrath, M. (2004) 5-HT5A receptor localization in the rat spinal cord suggests a role in nociception and control of pelvic floor musculature. J. Comp. Neurol., 476: 316–329.

Doly, S., Fischer, J., Brisorgueil, M.J., Verge, D. and Conrath, M. (2005) Pre- and postsynaptic localization of the 5-HT7 receptor in rat dorsal spinal cord: immunocytochemical evidence. J. Comp. Neurol., 490: 256–269.

Doucet, E., Latremoliere, A., Darmon, M., Hamon, M. and Emerit, M.B. (2007) Immunolabelling of the 5-HT 3B receptor subunit in the central and peripheral nervous systems in rodents. Eur. J. Neurosci., 26: 355–366.

Dubin, A.E., Huvar, R., D'Andrea, M.R., Pyati, J., Zhu, J.Y., Joy, K.C., Wilson, S.J., Galindo, J.E., Glass, C.A., Luo, L., Jackson, M.R., Lovenberg, T.W. and Erlander, M.G. (1999) The pharmacological and functional characteristics of the serotonin 5-HT(3A) receptor are specifically modified by a 5-HT(3B) receptor subunit. J. Biol. Chem., 274: 30799–30810.

Dumuis, A., Bouhelal, R., Sebben, M., Cory, R. and Bockaert, J. (1988) A nonclassical 5-hydroxytryptamine receptor positively coupled with adenylate cyclase in the central nervous system. Mol. Pharmacol., 34: 880–887.

Dumuis, A., Sebben, M. and Bockaert, J. (1989) The gastrointestinal prokinetic benzamide derivatives are agonists at the non-classical 5-HT receptor (5-HT_4) positively coupled to adenylate cyclase in neurons. Naunyn Schmiedebergs Arch. Pharmacol., 340: 403–410.

Duxon, M.S., Flanigan, T.P., Reavley, A.C., Baxter, G.S., Blackburn, T.P. and Fone, K.C. (1997) Evidence for expression of the 5-hydroxytryptamine2B receptor protein in the rat central nervous system. Neuroscience, 76: 323–329.

Fargin, A., Raymond, J.R., Lohse, M.J., Kobilka, B.K., Caron, M.G. and Lefkowitz, R.J. (1988) The genomic clone G-21 which resembles a beta-adrenergic receptor sequence encodes the 5-HT_{1A} receptor. Nature, 335: 358–360.

Fitzgerald, L.W., Iyer, G., Conklin, D.S., Krause, C.M., Marshall, A., Patterson, J.P., Tran, D.P., Jonak, G.J. and Hartig, P.R. (1999) Messenger RNA editing of the human serotonin 5-HT2C receptor. Neuropsychopharmacology, 21: 82S–90S.

Fitzgerald, L.W., Burn, T.C., Brown, B.S., Patterson, J. P., Corjay, M.H., Valentine, P.A., Sun, J.H., Link, J.R., Abbaszade, I., Hollis, J.M., Largent, B.L., Hartig, P.R., Hollis, G.F., Meunier, P.C., Robichaud, A.J. and Robertson, D.W. (2000) Possible role of valvular serotonin 5-HT(2B) receptors in the cardiopathy associated with fenfluramine. Mol. Pharmacol., 57: 75–81.

Foguet, M., Hoyer, D., Pardo, L.A., Parekh, A., Kluxen, F.W., Kalkman, H.O., Stuhmer, W. and Lubbert, H. (1992) Cloning and functional characterization of the rat stomach fundus serotonin receptor. EMBO J., 11: 3481–3487.

Gaddum, J.H. and Picarelli, Z.P. (1957) Two kinds of tryptamine receptor. Br. J. Pharmacol. Chemother., 12: 323–328.

Gerald, C., Adham, N., Kao, H.T., Olsen, M.A., Laz, T.M., Schechter, L.E., Bard, J.A., Vaysse, P.J., Hartig, P.R. and Branchek, T.A. (1995) The 5-HT4 receptor: molecular cloning and pharmacological characterization of two splice variants. EMBO J., 14: 2806–2815.

Gerard, C., Martres, M.P., Lefevre, K., Miquel, M.C., Verge, D., Lanfumey, L., Doucet, E., Hamon, M. and el Mestikawy, S. (1997) Immuno-localization of serotonin 5-HT_6 receptor-like material in the rat central nervous system. Brain Res., 746: 207–219.

Geurts, F.J., De Schutter, E. and Timmermans, J.P. (2002) Localization of 5-HT_{2A}, 5-HT_3, 5-HT_{5A} and 5-HT_7 receptor-like immunoreactivity in the rat cerebellum. J. Chem. Neuroanat., 24: 65–74.

Gozlan, H., el Mestikawy, S., Pichat, L., Glowinski, J. and Hamon, M. (1983) Identification of presynaptic serotonin autoreceptors using a new ligand: 3H-PAT. Nature, 305: 140–142.

Gozlan, H., Ponchant, M., Daval, G., Verge, D., Menard, F., Vanhove, A., Beaucourt, J.P. and Hamon, M. (1988) 125I-Bolton-Hunter-8-methoxy-2-[N-propyl-N-propylamino]tetralin as a new selective radioligand of 5-HT_{1A} sites in the rat brain. *In vitro* binding and autoradiographic studies. J. Pharmacol. Exp. Ther., 244: 751–759.

Gozlan, H., Thibault, S., Laporte, A.M., Lima, L. and Hamon, M. (1995) The selective 5-HT1A antagonist radioligand

[3H]WAY 100635 labels both G-protein-coupled and free 5-HT1A receptors in rat brain membranes. Eur. J. Pharmacol., 288: 173–186.

Grailhe, R., Waeber, C., Dulawa, S.C., Hornung, J.P., Zhuang, X.X., Brunner, D., Geyer, M.A. and Hen, R. (1999) Increased exploratory activity and altered response to LSD in mice lacking the 5-HT5A receptor. Neuron, 22: 581–591.

Grailhe, R., Grabtree, G.W. and Hen, R. (2001) Human 5-HT_5 receptors: the 5-HT_{5A} receptor is functional but the 5-HT_{5B} receptor was lost during mammalian evolution. Eur. J. Pharmacol., 418: 157–167.

Grossman, C.J., Kilpatrick, G.J. and Bunce, K.T. (1993) Development of a radioligand binding assay for 5-HT4 receptors in guinea-pig and rat brain. Br. J. Pharmacol., 109: 618–624.

Gustafson, E.L., Durkin, M.M., Bard, J.A., Zgombick, J. and Branchek, T.A. (1996) A receptor autoradiographic and *in situ* hybridization analysis of the distribution of the 5-ht_7 receptor in rat brain. Br. J. Pharmacol., 117: 657–666.

Hamon, M., Doucet, E., Lefèvre, K., Miquel, M.C., Lanfumey, L., Insausti, R., Frechilla, D., Delrio, J. and Vergé, D. (1999) Antibodies and antisense oligonucleotide for probing the distribution and putative functions of central 5-HT_6 receptors. Neuropsychopharmacology, 21: S68–S76.

Hanna, M.C., Davies, P.A., Hales, T.G. and Kirkness, E.F. (2000) Evidence for expression of heteromeric serotonin 5-HT(3) receptors in rodents. J. Neurochem., 75: 240–247.

Hartig, P.R., Adham, N., Zgombick, J., Macchi, M., Kao, H.T., Schechter, L., Branchek, T. and Weinshank, R. (1993) Serotonin receptor subtypes. Psychopharmacol. Ser., 10: 15–25.

Heidmann, D.E.A., Metcalf, M.A., Kohen, R. and Hamblin, M.W. (1997) Four 5-hydroxytryptamine(7) (5-HT_7) receptor isoforms in human and rat produced by alternative splicing: species differences due to altered intron–exon organization. J. Neurochem., 68: 1372–1381.

Heidmann, D.E.A., Szot, P., Kohen, R. and Hamblin, M.W. (1998) Function and distribution of three rat 5-hydroxytryptamine$_7$ (5-HT_7) receptor isoforms produced by alternative splicing. Neuropharmacology, 37: 1621–1632.

Hewlett, W.A., Trivedi, B.L., Zhang, Z.J., De Paulis, T., Schmidt, D.E., Lovinger, D.M., Ansari, M.S. and Ebert, M.H. (1999) Characterization of (S)-des-4-amino-3-[125I]iodozacopride ([125I]DAIZAC), a selective high-affinity radioligand for 5-hydroxytryptamine3 receptors. J. Pharmacol. Exp. Ther., 288: 221–231.

Hou, M., Kanje, M., Longmore, J., Tajti, J., Uddman, R. and Edvinsson, L. (2001) 5-HT(1B) and 5-HT(1D) receptors in the human trigeminal ganglion: co-localization with calcitonin gene-related peptide, substance P and nitric oxide synthase. Brain Res., 909: 112–120.

Hoyer, D. and Middlemiss, D.N. (1989) Species differences in the pharmacology of terminal 5-HT autoreceptors in mammalian brain. Trends Pharmacol. Sci., 10: 130–132.

Hoyer, D. and Neijt, H.C. (1987) Identification of serotonin 5-HT3 recognition sites by radioligand binding in NG108-15 neuroblastoma-glioma cells. Eur. J. Pharmacol., 143: 291–292.

Hoyer, D., Engel, G. and Kalkman, H.O. (1985a) Molecular pharmacology of 5-HT_1 and 5-HT_2 recognition sites in rat and pig brain membranes: radioligand binding studies with [^{3}H]5-HT, [^{3}H]8-OH-DPAT, (−)[^{125}I]iodocyanopindolol, [^{3}H]mesulergine and [^{3}H]ketanserin. Eur. J. Pharmacol., 118: 13–23.

Hoyer, D., Engel, G. and Kalkman, H.O. (1985b) Characterization of the 5-HT_{1B} recognition site in rat brain: binding studies with ^{125}I-iodocyanopindolol. Eur. J. Pharmacol., 118: 1–12.

Hoyer, D., Clarke, D.E., Fozard, J.R., Hartig, P.R., Martin, G.R., Mylecharane, E.J., Saxena, P.R. and Humphrey, P.P. (1994) International Union of Pharmacology classification of receptors for 5-hydroxytryptamine (serotonin). Pharmacol. Rev., 46: 157–203.

Hoyer, D., Pazos, A., Probst, A. and Palacios, J.M. (1986a) Serotonin receptors in the human brain. I. Characterization and autoradiographic localization of 5-HT_{1A} recognition sites. Apparent absence of 5-HT_{1B} recognition sites. Brain Res., 376: 85–96.

Hoyer, D., Pazos, A., Probst, A. and Palacios, J.M. (1986b) Serotonin receptors in the human brain. II. Characterization and autoradiographic localization of 5-HT_{1C} and 5-HT_2 recognition sites. Brain Res., 376: 97–107.

Hoyer, D., Vos, P., Closse, A., Pazos, A., Palacios, J.M. and Davies, H. (1987) [3H]ketanserin labels 5-HT2 receptors and alpha 1-adrenoceptors in human and pig brain membranes. Naunyn Schmiedebergs Arch. Pharmacol., 335: 226–230.

Hoyer, D., Hannon, J.P. and Martin, G.R. (2002) Molecular, pharmacological and functional diversity of 5-HT receptors. Pharmacol. Biochem. Behav., 71: 533–554.

Ito, H., Nyberg, S., Halldin, C., Lundkvist, C. and Farde, L. (1998) PET imaging of central 5-HT2A receptors with carbon-11-MDL 100,907. J. Nucl. Med., 39: 208–214.

Jakeman, L.B., To, Z.P., Eglen, R.M., Wong, E.H. and Bonhaus, D.W. (1994) Quantitative autoradiography of 5-HT_4 receptors in brains of three species using two structurally distinct radioligands, [3H]GR113808 and [3H]BIMU-1. Neuropharmacology, 33: 1027–1038.

Johnson, K.W., Schaus, J.M., Durkin, M.M., Audia, J.E., Kaldor, S.W., Flaugh, M.E., Adham, N., Zgombick, J., Cohen, M.L., Branchek, T.A. and Phebus, L.A. (1997) 5-HT_{1F} receptor agonists inhibit neurogenic dural inflammation in guinea pigs. Neuroreport, 8: 2237–2240.

Kia, H.K., Miquel, M.C., Brisorgueil, M.J., Daval, G., Riad, M., El Mestikawy, S., Hamon, M. and Vergé, D. (1996a) Immunocytochemical localization of serotonin(1A) receptors in the rat central nervous system. J. Comp. Neurol., 365: 289–305.

Kia, H.K., Brisorgueil, M.J., Hamon, M., Calas, A. and Verge, D. (1996b) Ultrastructural localization of 5-hydroxytryptamine1A receptors in the rat brain. J. Neurosci. Res., 46: 697–708.

Kilpatrick, G.J., Jones, B.J. and Tyers, M.B. (1987) Identification and distribution of 5-HT_3 receptors in rat brain using radioligand binding. Nature, 330: 746–748.

Kilpatrick, G.J., Jones, B.J. and Tyers, M.B. (1989) Binding of the 5-HT3 ligand, [3H]GR65630, to rat area postrema, vagus nerve and the brains of several species. Eur. J. Pharmacol., 159: 157–164.

Kobilka, B.K., Frielle, T., Collins, S., Yang-Feng, T., Kobilka, T.S., Francke, U., Lefkowitz, R.J. and Caron, M.G. (1987) An intronless gene encoding a potential member of the family of receptors coupled to guanine nucleotide regulatory proteins. Nature, 329: 75–79.

Laporte, A.-M., Lima, L., Gozlan, H. and Hamon, M. (1994) Selective *in vivo* labelling of brain 5-HT_{1A} receptors by [^{3}H]WAY 100635 in the mouse. Eur. J. Pharmacol., 271: 505–514.

Leonhardt, S., Herrick-Davis, K. and Titeler, M. (1989) Detection of a novel serotonin receptor subtype (5-HT_{1E}) in human brain: interaction with a GTP-binding protein. J. Neurochem., 53: 465–471.

López-Giménez, J.F., Mengod, G., Palacios, J.M. and Vilaró, M.T. (1997) Selective visualization of rat brain 5-HT_{2A} receptors by autoradiography with [3H]MDL 100,907. Naunyn Schmiedebergs Arch. Pharmacol., 356: 446–454.

López-Giménez, J.F., Vilaró, M.T., Palacios, J.M. and Mengod, G. (1998) [3H]MDL 100,907 labels 5-HT_{2A} serotonin receptors selectively in primate brain. Neuropharmacology, 37: 1147–1158.

López-Giménez, J.F., Mengod, G., Palacios, J.M. and Vilaró, M.T. (1999) Human striosomes are enriched in 5-HT2A receptors: autoradiographical visualization with [3H]MDL100,907,[125I](+/−)DOI and [3H]ketanserin. Eur. J. Neurosci., 11: 3761–3765.

López-Giménez, J.F., Villazon, M., Brea, J., Loza, M.I., Palacios, J.M., Mengod, G. and Vilaró, M.T. (2001a) Multiple conformations of native and recombinant human 5-hydroxytryptamine(2a) receptors are labelled by agonists and discriminated by antagonists. Mol. Pharmacol., 60: 690–699.

López-Giménez, J.F., Vilaró, M.T., Palacios, J.M. and Mengod, G. (2001b) Mapping of 5-HT2A receptors and their mRNA in monkey brain: [3H]MDL100,907 autoradiography and *in situ* hybridization studies. J. Comp. Neurol., 429: 571–589.

López-Giménez, J.F., Mengod, G., Palacios, J.M. and Vilaró, M.T. (2001c) Regional distribution and cellular localization of 5-HT_{2C} receptor mRNA in monkey brain: comparison with [3H]mesulergine binding sites and choline acetyltransferase mRNA. Synapse, 42: 12–26.

López-Giménez, J.F., Tecott, L.H., Palacios, J.M., Mengod, G. and Vilaró, M.T. (2002) Serotonin 5-HT (2C) receptor knockout mice: autoradiographic analysis of multiple serotonin receptors. J. Neurosci. Res., 67: 69–85.

Lovenberg, T.W., Erlander, M.G., Baron, B.M., Racke, M., Slone, A.L., Siegel, B.W., Craft, C.M., Burns, J.E., Danielson, P.E. and Sutcliffe, J.G. (1993) Molecular cloning and functional expression of 5-HT_{1E}-like rat and human 5-hydroxytryptamine receptor genes. Proc. Natl. Acad. Sci. USA, 90: 2184–2188.

Lucaites, V.L., Krushinski, J.H., Schaus, J.M., Audia, J.E. and Nelson, D.L. (2005) [3H]LY334370, a novel radioligand for the 5-HT_{1F} receptor. II. Autoradiographic localization in rat, guinea pig, monkey and human brain. Naunyn Schmiedebergs Arch. Pharmacol., 371: 178–184.

Lucas, G., Rymar, V.V., Du, J., Mnie-Filali, O., Bisgaard, C., Manta, S., Lambas-Senas, L., Wiborg, O., Haddjeri, N., Pineyro, G., Sadikot, A.F. and Debonnel, G. (2007) Serotonin(4) (5-HT(4)) receptor agonists are putative antidepressants with a rapid onset of action. Neuron, 55: 712–725.

Maricq, A.V., Peterson, A.S., Brake, A.J., Myers, R.M. and Julius, D. (1991) Primary structure and functional expression of the 5-HT_3 receptor, a serotonin-gated ion chanel. Science, 254: 432–436.

McAllister, G., Charlesworth, A., Snodin, C., Beer, M.S., Noble, A.J., Middlemiss, D.N., Iversen, L.L. and Whiting, P. (1992) Molecular cloning of a serotonin receptor from human brain (5-HT_{1E}): a fifth 5-HT_1-like subtype. Proc. Natl. Acad. Sci. USA, 89: 5517–5521.

McCall, R.B., Huff, R., Chio, C.L., TenBrink, R., Bergh, C.L., Ennis, M.D., Ghazal, N.B., Hoffman, R.L., Meisheri, K., Higdon, N.R. and Hall, E. (2002) Preclinical studies characterizing the anti-migraine and cardiovascular effects of the selective 5-HT_{1D} receptor agonist PNU-142633. Cephalalgia, 22: 799–806.

Mengod, G., Nguyen, H., Le, H., Waeber, C., Lubbert, H. and Palacios, J.M. (1990) The distribution and cellular localization of the serotonin 1C receptor mRNA in the rodent brain examined by *in situ* hybridization histochemistry. Comparison with receptor binding distribution. Neuroscience, 35: 577–591.

Mengod, G., Vilaró, M.T., Raurich, A., López-Giménez, J.F., Cortés, R. and Palacios, J.M. (1996) 5-HT receptors in mammalian brain: receptor autoradiography and *in situ* hybridization studies of new ligands and newly identified receptors. Histochem. J., 28: 747–758.

Mengod, G., Vilaró, M.T., Cortés, R., López-Giménez, J.F., Raurich, A. and Palacios, J.M. (2006) Chemical neuroanatomy of 5-HT receptor subtypes in the mammalian brain. In: Roth, B.L. (Ed.), The Serotonin Receptors: From Mol. Pharmacol. to Human Therapeutics. Humana Press, Totowa, NJ, pp. 319–364.

Middlemiss, D.N. and Fozard, J.R. (1983) 8-Hydroxy-2-(di-n-propylamino)-tetralin discriminates between subtypes of the 5-HT_1 recognition site. Eur. J. Pharmacol., 90: 151–153.

Miller, K.J. and Teitler, M. (1992) Quantitative autoradiography of 5-CT-sensitive (5-HT_{1D}) and 5-CT-insensitive (5-HT_{1E}) serotonin receptors in human brain. Neurosci. Lett., 136: 223–226.

Mitsikostas, D.D., Delrio, M.S., Moskowitz, M.A. and Waeber, C. (1999) Both 5-HT_{1B} and 5-HT_{1F} receptors modulate c-fos expression within rat trigeminal nucleus caudalis. Eur. J. Pharmacol., 369: 271–277.

Monk, S.A., Desai, K., Brady, C.A., Williams, J.M., Lin, L., Princivalle, A., Hope, A.G. and Barnes, N.M. (2001) Generation of a selective 5-HT_{3B} subunit-recognising polyclonal antibody; identification of immunoreactive cells in rat hippocampus. Neuropharmacology, 41: 1013–1016.

Monsma, F.J.J., Shen, Y., Ward, R.P., Hamblin, M.W. and Sibley, D.R. (1993) Cloning and expression of a novel serotonin receptor with high affinity for tricyclic psychotropic drugs. Mol. Pharmacol., 43: 320–327.

Morales, M. and Bloom, F.E. (1997) The 5-HT_3 receptor is present in different subpopulations of GABAergic neurons in the rat telencephalon. J. Neurosci., 17: 3157–3167.

Morales, M., Battenberg, E., de Lecea, L. and Bloom, F.E. (1996a) The type 3 serotonin receptor is expressed in a subpopulation of GABAergic neurons in the rat neocortex and hippocampus. Brain Res., 731: 199–202.

Morales, M., Battenberg, E., Delecea, L., Sanna, P.P. and Bloom, F.E. (1996b) Cellular and subcellular immunolocalization of the type 3 serotonin receptor in the rat central nervous system. Mol. Brain Res., 36: 251–260.

Moyer, R.W. and Kennaway, D.J. (1999) Immunohistochemical localization of serotonin receptors in the rat suprachiasmatic nucleus. Neurosci. Lett., 271: 147–150.

Niesler, B., Frank, B., Kapeller, J. and Rappold, G.A. (2003) Cloning, physical mapping and expression analysis of the human 5-HT3 serotonin receptor-like genes HTR3C, HTR3D and HTR3E. Gene, 310: 101–111.

Niesler, B., Walstab, J., Combrink, S., Moller, D., Kapeller, J., Rietdorf, J., Bonisch, H., Gothert, M., Rappold, G. and Bruss, M. (2007) Characterization of the novel human serotonin receptor subunits 5-HT_{3C}, 5-HT_{3D}, and 5-HT_{3E}. Mol. Pharmacol., 72: 8–17.

Niswender, C.M., Sanders-Bush, E. and Emeson, R.B. (1998) Identification and characterization of RNA editing events within the 5-HT_{2C} receptor. Ann. NY Acad. Sci., 861: 38–48.

Oliver, K.R., Kinsey, A.M., Wainwright, A. and Sirinathsinghji, D.J. (2000) Localization of 5-ht(5A) receptor-like immunoreactivity in the rat brain. Brain Res., 867: 131–142.

Palacios, J.M., Pazos, A. and Hoyer, D. (1987) Characterization and mapping of 5-HT_{1A} sites in the brain of animals and man. In: Dourish, C.T., Ahlenius, S. and Hutson, P.H. (Eds.), Brain 5-HT1A Receptors. Ellis Horwood, Chichester, pp. 67–81.

Palacios, J.M., Raurich, A., Mengod, G., Hurt, S.D. and Cortés, R. (1996) Autoradiographic analysis of 5-HT receptor subtypes labelled by [3H]5-CT ([3H]5-carboxamidotryptamine). Behav. Brain Res., 73: 239–243.

Parker, R.M., Barnes, J.M., Ge, J., Barber, P.C. and Barnes, N.M. (1996) Autoradiographic distribution of [3H]-(S)-zacopride-labelled 5-HT_3 receptors in human brain. J. Neurol. Sci., 144: 119–127.

Pasqualetti, M., Ori, M., Nardi, I., Castagna, M., Cassano, G.B. and Marazziti, D. (1998) Distribution of the 5-HT5A serotonin receptor mRNA in the human brain. Brain Res. Mol. Brain Res., 56: 1–8.

Pazos, A. and Palacios, J.M. (1985) Quantitative autoradiographic mapping of serotonin receptors in the rat brain. I. Serotonin-1 receptors. Brain Res., 346: 205–230.

Pazos, A., Hoyer, D. and Palacios, J.M. (1984a) The binding of serotonergic ligands to the porcine choroid plexus: characterization of a new type of serotonin recognition site. Eur. J. Pharmacol., 106: 539–546.

Pazos, A., Hoyer, D. and Palacios, J.M. (1984b) Mesulergine, a selective serotonin-2 ligand in the rat cortex, does not label these receptors in porcine and human cortex: evidence for species differences in brain serotonin-2 receptors. Eur. J. Pharmacol., 106: 531–538.

Pazos, A., Cortés, R. and Palacios, J.M. (1985) Quantitative autoradiographic mapping of serotonin receptors in the rat brain. II. Serotonin-2 receptors. Brain Res., 346: 231–249.

Pazos, A., Probst, A. and Palacios, J.M. (1987a) Serotonin receptors in the human brain – III. Autoradiographic mapping of serotonin-1 receptors. Neuroscience, 21: 97–122.

Pazos, A., Probst, A. and Palacios, J.M. (1987b) Serotonin receptors in the human brain – IV. Autoradiographic mapping of serotonin-2 receptors. Neuroscience, 21: 123–139.

Phebus, L.A., Johnson, K.W., Audia, J.E., Cohen, M.L., Dressman, B.A., Fritz, J.E., Kaldor, S.W., Krushinski, J.H., Schenck, K.W., Zgombick, J.M., Branchek, T.A., Adham, N. and Schaus, J.M. (1996) Characterization of LY334370, a potent and selective 5-HT_{1F} receptor agonist, in the neurogenic dural inflammation model of migraine pain. Soc. Neurosci. Abstr., 22: 1331.

Pierson, M.E., Andersson, J., Nyberg, S., McCarthy, D.J., Finnema, S.J., Varnas, K., Takano, A., Karlsson, P., Gulyas, B., Medd, A.M., Lee, C.M., Powell, M.E., Heys, J.R., Potts, W., Seneca, N., Mrzljak, L., Farde, L. and Halldin, C. (2008) [11C]AZ10419369: a selective 5-HT1B receptor radioligand suitable for positron emission tomography (PET). Characterization in the primate brain. Neuroimage, 41: 1075–1085.

Pompeiano, M., Palacios, J.M. and Mengod, G. (1992) Distribution and cellular localization of mRNA coding for 5-HT1A receptor in the rat brain: correlation with receptor binding. J. Neurosci., 12: 440–453.

Ponchant, M., Beaucourt, J.P., Vanhove, A., Daval, G., Verge, D., Hamon, M. and Gozlan, H. (1988) 125I-BH-8-MeO-N-PAT, a new ligand for the study of 5-HT1A receptors in the central nervous system. C. R. Acad. Sci. III, 306: 147–152.

Pratt, G.D., Bowery, N.G., Kilpatrick, G.J., Leslie, R.A., Barnes, N.M., Naylor, R.J., Jones, B.J., Nelson, D.R., Palacios, J.M. and Slater, P. (1990) Consensus meeting agrees distribution of 5-HT_3 receptors in mammalian hindbrain. Trends Pharmacol. Sci., 11: 135–137.

Puig, M.V., Santana, N., Celada, P., Mengod, G. and Artigas, F. (2004) *In vivo* excitation of GABA interneurons in the medial prefrontal cortex through 5-HT_3 receptors. Cereb. Cortex, 14: 1365–1375.

Roberts, J.C., Reavill, C., East, S.Z., Harrison, P.J., Patel, S., Routledge, C. and Leslie, R.A. (2002) The distribution of 5-HT_6 receptors in rat brain: an autoradiographic binding study using the radiolabelled 5-HT_6 receptor antagonist [^{125}I]SB-258585. Brain Res., 934: 49–57.

Ruat, M., Traiffort, E., Arrang, J.M., Tardivel-Lacombe, J., Diaz, J., Leurs, R. and Schwartz, J.C. (1993) A novel rat serotonin (5-HT6) receptor: molecular cloning, localization and stimulation of cAMP accumulation. Biochem. Biophys. Res. Commun., 193: 268–276.

Sandell, J., Halldin, C., Pike, V.W., Chou, Y.H., Varnas, K., Hall, H., Marchais, S., Nowicki, B., Wikstrom, H.V., Swahn, C.G. and Farde, L. (2001) New halogenated [11C]WAY analogues, [11C]6FPWAY and [11C]6BPWAY – radiosynthesis and assessment as radioligands for the study of brain 5-HT1A receptors in living monkey. Nucl. Med. Biol., 28: 177–185.

Santana, N., Bortolozzi, A., Serrats, J., Mengod, G. and Artigas, F. (2004) Expression of serotonin1A and serotonin2A receptors in pyramidal and GABAergic neurons of the rat prefrontal cortex. Cereb. Cortex, 14: 1100–1109.

Schlicker, E., Fink, K., Molderings, G.J., Price, G.W., Duckworth, M., Gaster, L., Middlemiss, D.N., Zentner, J., Likungu, J. and Gothert, M. (1997a) Effects of selective h5-HT1B (SB-216641) and h5-HT1D (BRL-15572) receptor ligands on guinea-pig and human 5-HT auto- and heteroreceptors. Naunyn Schmiedebergs Arch. Pharmacol., 356: 321–327.

Schlicker, E., Fink, K., Molderings, G.J., Price, G.W., Middlemiss, D.N., Zentner, J., Likungu, J. and Gothert, M. (1997b) Effects of SB 216641 and BRL 15572 (selective h5-HT_{1B} and h5-HT_{1D} receptor antagonists, respectively) on guinea-pig and human 5-HT auto- and hetero-receptors. Br. J. Pharmacol., 120(Suppl.): 143P.

Schmuck, K., Ullmer, C., Engels, P. and Lubbert, H. (1994) Cloning and functional characterization of the human 5-HT_{2B} serotonin receptor. FEBS Lett., 342: 85–90.

Serrats, J., Raurich, A., Vilaró, M.T., Mengod, G. and Cortés, R. (2004) 5-ht_{5B} receptor mRNA in the raphe nuclei: coexpression with serotonin transporter. Synapse, 51: 102–111.

Serrats, J., Mengod, G. and Cortes, R. (2005) Expression of serotonin 5-HT2C receptors in GABAergic cells of the anterior raphe nuclei. J. Chem. Neuroanat., 29: 83–91.

Steward, L.J., Bufton, K.E., Hopkins, P.C., Davies, W.E. and Barnes, N.M. (1993) Reduced levels of 5-HT3 receptor recognition sites in the putamen of patients with Huntington's disease. Eur. J. Pharmacol., 242: 137–143.

Tecott, L.H., Maricq, A.V. and Julius, D. (1993) Nervous system distribution of the serotonin 5-HT_3 receptor mRNA. Proc. Natl. Acad. Sci. USA, 90: 1430–1434.

Thomas, D.R., Soffin, E.M., Roberts, C., Kew, J.N., de la Flor, R.M., Dawson, L.A., Fry, V.A., Coggon, S.A., Faedo, S.,

Hayes, P.D., Corbett, D.F., Davies, C.H. and Hagan, J.J. (2006) SB-699551-A (3-cyclopentyl-N-[2-(dimethylamino)ethyl]-N-[(4′-{[(2-phenylethyl)amino]methyl}-4-biphenylyl)methyl]propanamide dihydrochloride), a novel 5-ht$_{5A}$ receptor-selective antagonist, enhances 5-HT neuronal function: evidence for an autoreceptor role for the 5-ht$_{5A}$ receptor in guinea pig brain. Neuropharmacology, 51: 566–577.

To, Z.P., Bonhaus, D.W., Eglen, R.M. and Jakeman, L.B. (1995) Characterization and distribution of putative 5-ht$_7$ receptors in guinea-pig brain. Br. J. Pharmacol., 115: 107–116.

Tsou, A.P., Kosaka, A., Bach, C., Zuppan, P., Yee, C., Tom, L., Alvarez, R., Ramsey, S., Bonhaus, D.W. and Stefanich, E. (1994) Cloning and expression of a 5-hydroxytryptamine7 receptor positively coupled to adenylyl cyclase. J. Neurochem., 63: 456–464.

Ullmer, C., Schmuck, K., Kalkman, H.O. and Lubbert, H. (1995) Expression of serotonin receptor mRNAs in blood vessels. FEBS Lett., 370: 215–221.

Vanhoenacker, I., Haegeman, I. and Leysen, J.E. (2000) 5-HT$_7$ receptors: current knowledge and future prospects. Trends Pharmacol. Sci., 21: 70–77.

van Hooft, J.A. and Yakel, J.L. (2003) 5-HT3 receptors in the CNS: 3B or not 3B? Trends Pharmacol. Sci., 24: 157–160.

Varnas, K., Hall, H., Bonaventure, P. and Sedvall, G. (2001) Autoradiographic mapping of 5-HT(1B) and 5-HT(1D) receptors in the post mortem human brain using [(3)H]GR 125743. Brain Res., 915: 47–57.

Varnas, K., Halldin, C., Pike, V.W. and Hall, H. (2003) Distribution of 5-HT4 receptors in the postmortem human brain – an autoradiographic study using [125I]SB 207710. Eur. Neuropsychopharmacol., 13: 228–234.

Varnas, K., Halldin, C. and Hall, H. (2004a) Autoradiographic distribution of serotonin transporters and receptor subtypes in human brain. Hum. Brain Mapp., 22: 246–260.

Varnas, K., Thomas, D.R., Tupala, E., Tiihonen, J. and Hall, H. (2004b) Distribution of 5-HT$_7$ receptors in the human brain: a preliminary autoradiographic study using [3H]SB-269970. Neurosci. Lett., 367: 313–316.

Vilaró, M.T., Cortés, R., Gerald, C., Branchek, T.A., Palacios, J.M. and Mengod, G. (1996) Localization of 5-HT4 receptor mRNA in rat brain by *in situ* hybridization histochemistry. Mol. Brain Res., 43: 356–360.

Vilaró, M.T., Domènech, T., Palacios, J.M. and Mengod, G. (2002) Cloning and characterization of a novel human 5-HT$_4$ receptor variant that lacks the alternatively spliced carboxy terminal exon. RT-PCR distribution in human brain and periphery of multiple 5-HT$_4$ receptor variants. Neuropharmacology, 42: 60–73.

Vilaró, M.T., Cortés, R. and Mengod, G. (2005) Serotonin 5-HT4 receptors and their mRNAs in rat and guinea pig brain: distribution and effects of neurotoxic lesions. J. Comp. Neurol., 484: 418–439.

Villalon, C.M., den Boer, M.O., Heiligers, J.P. and Saxena, P.R. (1990) Mediation of 5-hydroxytryptamine-induced tachycardia in the pig by the putative 5-HT$_4$ receptor. Br. J. Pharmacol., 100: 665–667.

Waeber, C. and Moskowitz, M.A. (1995) Autoradiographic visualisation of [^{3}H]5-carboxamidotryptamine binding sites in the guinea pig and rat brain. Eur. J. Pharmacol., 283: 31–46.

Waeber, C. and Palacios, J.M. (1992) Non 5-HT$_{1A}$/5-HT$_{1C}$ [^{3}H]5-HT binding sites in the hamster, opossum and rabbit brain show similar regional distribution but different sensitivity to beta-adrenoceptor antagonists. Synapse, 12: 261–270.

Waeber, C. and Palacios, J.M. (1994) Binding sites for 5-hydroxytryptamine-2 receptor agonists are predominantly located in striosomes in the human basal ganglia. Mol. Brain Res., 24: 199–209.

Waeber, C., Dietl, M.M., Hoyer, D., Probst, A. and Palacios, J.M. (1988a) Visualization of a novel serotonin recognition site (5-HT$_{1D}$) in the human brain by autoradiography. Neurosci. Lett., 88: 11–16.

Waeber, C., Dixon, K., Hoyer, D. and Palacios, J.M. (1988b) Localisation by autoradiography of neuronal 5-HT$_3$ receptors in the mouse CNS. Eur. J. Pharmacol., 151: 351–352.

Waeber, C., Hoyer, D. and Palacios, J.M. (1989) 5-Hydroxytryptamine$_3$ receptors in the human brain: autoradiographic visualization using [^{3}H]ICS 205-930. Neuroscience, 31: 393–400.

Waeber, C., Zhang, L. and Palacios, J.M. (1990) 5-HT$_{1D}$ receptors in the guinea-pig brain: pre- and postsynaptic localizations in the striato-nigral pathway. Brain Res., 528: 197–206.

Waeber, C., Sebben, M., Grossman, C., Javoy-Agid, F., Bockaert, J. and Dumuis, A. (1993) [3H]-GR113808 labels 5-HT4 receptors in the human and guinea-pig brain. Neuroreport, 4: 1239–1242.

Waeber, C., Grailhe, R., Yu, X.J., Hen, R. and Moskowitz, M.A. (1998) Putative 5-ht$_5$ receptors: localization in the mouse CNS and lack of effect in the inhibition of dural protein extravasation. Ann. NY Acad. Sci., 861: 85–90.

Wainscott, D.B., Lucaites, V.L., Kursar, J.D., Baez, M. and Nelson, D.L. (1996) Pharmacologic characterization of the human 5-hydroxytryptamine2B receptor: evidence for species differences. J. Pharmacol. Exp. Ther., 276: 720–727.

Zgombick, J.M., Schechter, L.E., Kucharewicz, S.A., Weinshank, R.L. and Branchek, T.A. (1995) Ketanserin and ritanserin discriminate between recombinant human 5-HT$_{1D\alpha}$ and 5-HT$_{1D\beta}$ receptor subtypes. Eur. J. Pharmacol., 291: 9–15.

Zgombick, J.M., Schechter, L.E., Adham, N., Kucharewicz, S.A., Weinshank, R.L. and Branchek, T.A. (1996) Pharmacological characterizations of recombinant human 5-HT$_{1D\alpha}$ and 5-HT$_{1D\beta}$ receptor subtypes coupled to adenylate cyclase inhibition in clonal cell lines: apparent differences in drug intrinsic efficacies between human 5 HT$_{1D}$ subtypes. Naunyn-Schmiedeberg's Arch. Pharmacol., 354: 226–236.

Zgombick, J.M., Bard, J.A., Kucharewicz, S.A., Urquhart, D.A., Weinshank, R.L. and Branchek, T.A. (1997) Molecular cloning and pharmacological characterization of guinea pig 5-HT$_{1B}$ and 5-HT$_{1D}$ receptors. Neuropharmacology, 36: 513–524.

CHAPTER 1.7

Measuring Serotonin Activity *in vivo* in the Brains of Animals and Humans

Charles A. Marsden*

School of Biomedical Sciences, University of Nottingham Medical School, Nottingham, UK

Abstract: Serotonin research has resulted in the identification of up to 14 receptors in the brain. A real challenge is to develop experimental strategies and methods that can be used to investigate the role of these receptors in the integrated function of the serotonergic systems in the animal and human brain *in vivo*. There are now methods available that can measure *in vivo* functional changes in extracellular serotonin (microdialysis, voltammetry). However, with microdialysis temporal and spatial resolution is limited, and while voltammetry has the capability to improve on both these limitations, development of suitable readily usable electrodes is still required. The serotonin transporter and some serotonin receptors can be monitored using positron emission tomography (PET), but work is still required to develop new and better ligands for other receptors. Tryptophan depletion studies can be effectively used to investigate altered serotonergic function in subjects with an underlying trait dysfunction, though its ability to be used to investigate the role of serotonin in normal function remains debatable. Magnetic resonance imaging techniques, in particular fMRI combined with selective pharmacological interventions, offer an exciting new approach to investigating the interactions between serotonin pathways and other neurotransmitters in integrated behavioral responses.

Keywords: microdialysis, voltammetry, positron emission tomography, PET, tryptophan depletion, functional magnetic resonance imaging, fMRI, pharmacological MRI, phMRI.

Introduction

The advances made over the past 20 years with the identification of a multiplicity of serotonin receptors (presently numbering 14) found in the brain associated with the wide and diverse serotonergic innervation, derived from a few midbrain neurones, provides a particular challenge to those neuroscientists and pharmacologists interested in receptor function. This interest is not just confined to the identification of the function of individual receptors, though this offers considerable possibilities in terms of drug development, but also to their role in the integrated function of the serotonergic systems in the brain. Examples of the latter include the complex interaction between the $5HT_{1A}$ and $5HT_{2C}$ receptors in the pathways responsible for psychosis (Geyer and Vollenweider, 2008) or aversive behaviors (Millan, 2006), or the role of $5HT_{2C}$ in feeding and neurodendocrine function (Lucki, 1998). Such studies are now becoming possible with the development of a range of methodologies that have allowed investigation of the mechanisms that regulate serotonergic function *in vivo* and during behavior. With the current emphasis on so-called translational research, it is important that we have methodologies that can be applied to both animal and human studies.

This short chapter will consider the principles underlying the methods and provide selected examples of their application from a historical viewpoint. The major *in vivo* methods to be considered are those that measure serotonin release (microdialysis and voltammetry), receptors and transporters (positron emission tomography (PET) imaging) and integrated neuronal function in response to pharmacological and behavioral interventions (pharmacological and functional magnetic resonance imaging (phMRI, fMRI)). Other methods that allow manipulation of serotonin levels in the brain (tryptophan depletion studies) or indirect measurement of receptor function in the brain (neuroendocrine markers) will also be briefly discussed. Emphasis will be placed on the way these methods can be used in the non-anesthetized and preferably freely moving animal (or human).

*E-mail: charles.marsden@nottingham.ac.uk

Christian Müller & Barry Jacobs (Eds.)
Handbook of Behavioral Neurobiology of Serotonin
ISBN 978-0-12-374634-4

DOI: 10.1016/B978-0-12-374634-4.00007-1

Measurement of serotonin release (efflux) *in vivo*

Microdialysis

The quest to develop a method by which transmitter release could be measured *in vivo* has been active for many years, with the original approaches based on the push–pull cannula (Myers and Beleslin, 1970), but it was the development of the hollow fiber microdialysis probe (Ungerstedt and Pycock, 1974) linked to a highly selective high performance liquid chromatography (HPLC) separation combined with the excellent sensitivity of the electrochemical detector (ECD) (Keller *et al.*, 1976) that made expectation a reality. While the initial microdialysis studies concentrated on the measurement of striatal dopamine due to the abundance of dopamine in the striatum and the relative ease at targeting that brain area (Zetterstrom *et al.*, 1983), it was not long before the method was applied to serotonin. Early studies concentrated on drug-induced changes and, importantly, studies showing that a selective agonist of the 5-HT_{1A} receptor reduced serotonin release (Marsden, 1985; Sharp *et al.*, 1989) – a finding supported by complementary *in vitro* studies (Middlemiss and Hutson 1990), and one that resulted in the identification of the 5-HT_{1A} receptor as the somatodendritic autoreceptor (Sprouse and Aghajanian, 1987). With improvements in technology the microdialysis method was soon adapted for use in freely moving animals, and consequently to behavioral situations. One of the early behavioral microdialysis studies demonstrated increased hippocampal serotonin release when rats were exposed to an aversive environment (elevated plus maze) (Wright *et al.*, 1992), thus providing *in vivo* functional evidence for the importance of serotonin in the acute response to aversion.

Microdialysis is a catheter-based sampling technique in which the catheter essentially mimics the function of a capillary blood vessel (for review, see Li *et al.*, 2006). The tip of the catheter (probe) is formed by a dialysis membrane of known molecular size (normally 15 kDa cut-off) that is continuously perfused with artificial cerebrospinal fluid (CSF) and when implanted in a brain area allows unbound molecules to transfer from the extracellular space around the probe into the artificial CSF, which is then collected and analyzed. As mentioned already, serotonin in the perfusate can then be measured by HPLC-ECD or alternatively by mass spectrometry (MS). The method is not just restricted to the measurement of neurotransmitters (amines and amino acids) and can be applied widely to other molecules important in brain/neuronal function (glucose, lactate), as well as sampling brain levels of administered drugs. Importantly, microdialysis can be used in humans as well as animals, though human brain studies are obviously limited to very special circumstances; however, in contrast, it has been widely used in humans to monitor blood metabolites. In one such study changes in intravenous serotonin were measured in cancer patients in response to cisplatin, and comparisons made between the effects of two 5-HT_3 antagonists on the cisplatin response (Castejon *et al.*, 1999). In another recent microdialysis study skin serotonin was shown to increase in patients with burn injuries, indicating that it may have a role in local vascular control and formation of edema (Samuelsson *et al.*, 2008).

It is important to understand that microdialysis does not truly measure neurotransmitter release, as the diameter of the probe (0.2–0.5 mm) is far greater than the size of a synapse. What is really measured is transmitter overflow or efflux into the extracellular space, which occurs following release and subsequent reuptake or metabolism, and a more precise description of what microdialysis measures are changes in extracellular levels rather than transmitter release.

Microdialysis offers many advantages, including relative ease of use in freely moving animals, a high level of specificity and a wide range of substances that can be successfully detected. There are, however, negative aspects to the method. One important factor is the size of the probe, in terms of both its external diameter (0.2–0.5 mm) and the length of dialysis membrane (normally 3–4 mm) required to obtain sufficient serotonin for detection by the assay methods available. The main limitation, however, is the time resolution of the method, which is measured in minutes per sample and generally ranges from 5 to 30 minutes, depending on the sensitivity of the assay for the transmitter under investigation and the brain area being sampled. For serotonin, the time resolution is generally in the 15- to 30-minute range rather than the 5- to 15-minute range. This contrasts with the time resolution of the changes in neuronal activity and transmitter release observed following neuronal activation by physiological or behavioral stimuli, which will be in the range of milliseconds to seconds. While this problem is not too serious when using microdialysis to look at the time effects of drugs on serotonin release, which occur over a comparable period of time to the temporal resolution of the probe – and indeed the effects of drugs are often cumulative – it is more of an issue in behavioral studies aimed at identifying key serotonergic triggers in a behavioral sequence. The issue of time resolution has been one of the driving forces in the attempts to develop voltammetric probes suitable for the detection of serotonin.

Microdialysis has been used in a wide range of pharmacological and behavioral situations to gain understanding of the alterations in release of serotonin. It is beyond the scope of this review to discuss these studies in detail, as they will be covered elsewhere in this volume. It is clear, however, that serotonin release is influenced by a wide range of external stimuli and, depending on the nature of the stimulus, the effect may be stimulus-specific as well as regional-specific (Lucki, 1998). From a brief look at some of the early publications using microdialysis to measure extracellular serotonin, it is clear that changes are observed which result from non-specific behavioral activation (Reuter *et al.*, 1997), but there are also numerous studies showing increases in limbic areas (hippocampus, prefrontal cortex) in response to environmental stressors such as exposure to the elevated plus maze (Wright *et al.*, 1992), footshock (Fulford and Marsden, 1998), conditioned emotional response (Muchimapura *et al.*, 2002; Figure 1) and contextual fear conditioning (Wilkinson *et al.*, 1996). In the isolated reared rat, which shows hyper-responsiveness to novelty and stress, the increase in serotonin in the hippocampus in response to footshock or fear conditioning fails to occur (Muchimapura *et al.*, 2002; Figure 1), while the serotonergic response in the nucleus accumbens is greatly enhanced (Fulford and Marsden, 1998). These results suggest that the forebrain changes in extracellular serotonin in response to external stressors are associated with coping with the stressor, thus supporting the view that drugs such as the serotonin selective reuptake inhibitors (SSRIs) act in anxious states by enhancing serotonergic inhibition of the response to stress. It is also interesting to note that while acute stressors, as described above, increase extracellular serotonin in limbic areas, more prolonged stress – such as forced swimming (Kirby *et al.*, 1995), aggression and drug withdrawal states (Parsons *et al.*, 1995) – reduce extracellular serotonin. In a recent study we have shown that short-term (10-minute) restraint stress increases hippocampal extracellular serotonin, while 1-hour restraint stress produces no change or a decrease in the extracellular levels, indicating that the serotonergic response to stress in limbic areas is time-dependent.

More recent interest in serotonin has focused on the role of specific (5-HT_{1A}, 5-HT_4 and 5-HT_6) receptors in cognition (for review, see King *et al.*, 2008). For example, 5-HT_6 receptor antagonists have been shown to prolong retention and enhance consolidation of information using a range of tests. Microdialysis studies that have measured not only extracellular serotonin but also glutamate and GABA have helped to understand the serotonergic mechanisms involved in the pro-cognitive effects of these drugs. Administration of the 5-HT_6 antagonist SB-271046 increased extracellular glutamate levels in the frontal cortex (Dawson *et al.*, 2000), while agonists such as WAY-181187 and WAY-208466 increased extracellular GABA in the frontal cortex as well as the amygdala, hippocampus and striatum. Immunohistochemical studies have located the 5-HT_6 receptors on GABA neurons, and this, combined, with the glutamate and GABA information, indicates that activation of 5HT_6 receptors increases GABA release which then inhibits glutamate release, so administration of the antagonists prevents the increase

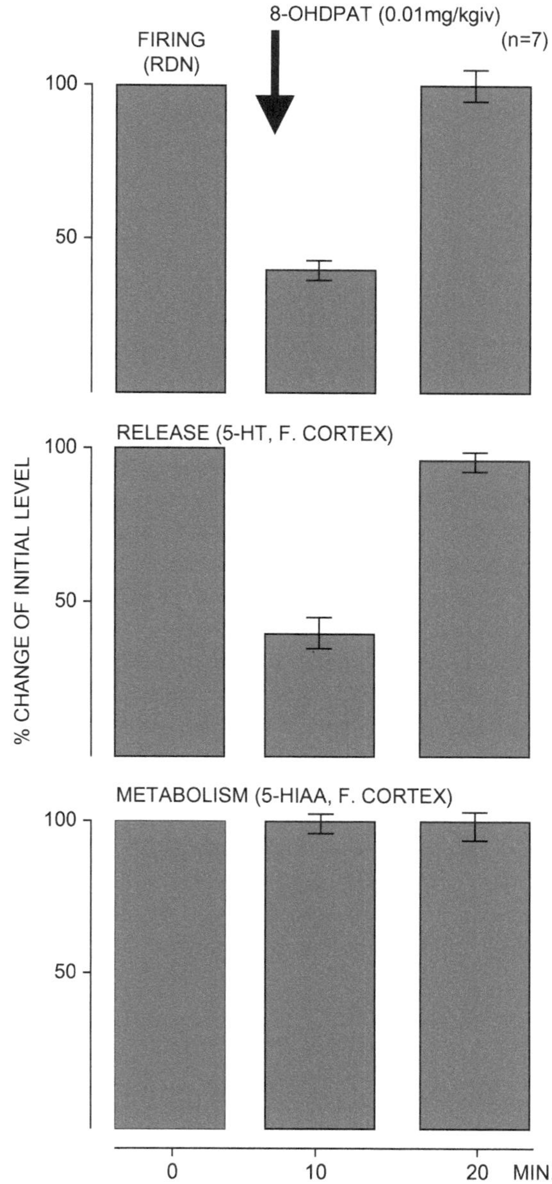

Figure 1 Graph showing the effect of administration of the 5-HT_{1A} agonist 8-OHDPAT on the electrophysiological recordings from dorsal raphe neurons of the rat while measuring extracellular serotonin and 5-HIAA in the frontal cortex with a Nafion®-coated electrochemical electrode. Note the decrease in firing due to activation of the somatodendritic autoreceptors in the dorsal raphe and the associated decrease in serotonin release but no change in 5-HIAA, indicating that measurement of the metabolite is not a good index of release. Data modified from Crespi *et al.* (1990).

in inhibitory GABA release, leading to improved cognitive function. Increasingly we need to look at the integrated picture of the interactions between different neurotransmitters to gain understanding, rather than just concentrating on changes in one transmitter. The use of multiple-probe microdialysis offers the opportunity to do this type of study in the individual animal, particularly when combined with suitable behavioral paradigms.

Voltammetry

The idea of developing fine voltammetric probes that could be implanted into specific brain areas and used to measure electroactive neurotransmitters such as dopamine, noradrenaline and serotonin (Figure 2) came from Ralph Adam's inventive laboratory at the University of Kansas (Adams, 1976). The concept is to scale down the electrochemical detector used with HPLC and then apply a suitable potential to catalyze the oxidation of serotonin before measuring the current produced (Figure 2). However, unlike the HPLC-ECD methodology there is no separation of the individual amines prior to their oxidation, and as they oxidize at rather similar potentials to each other, and most importantly also at a similar potential to ascorbic acid, this raised important issues about the specificity of the signals detected. Ascorbic acid is abundant in the brain, and its release from neuronal and non-neuronal cells in the brain is associated with transmitter release, and it soon became apparent that the signals detected in the early voltammetric studies and attributed to either dopamine or serotonin (Conti *et al.*, 1978) were contaminated by the oxidation of ascorbate (Gonon *et al.*, 1980).

Since those early days with voltammetry, several important advances have been made to the methods of scanning (fast scan cyclic voltammetry, chronoamperometry, constant potential amperometry) as well as development of suitable electrodes and biosensors (for review, see Robinson *et al.*, 2008) such that it is now possible to measure dopamine release in real time (Phillips *et al.*, 2003) using electrodes (tip diameters of 10 μm or less) that can simultaneously make electrophysiological recordings while also monitoring behavior (Cheer *et al.*, 2007). Advances in the area of serotonin have not been so rapid due to its rather more complex oxidation pattern, which results in a build-up of oxidation products on the electrode surface, leading to a loss of the functional properties of the electrode. Secondly, the oxidation profile is similar to that of dopamine and ascorbate, again creating difficulties with selectivity. Two approaches to this problem have been adopted. The first approach was to coat the electrodes with Nafion®, which allows penetration of the amines, but not their acid metabolites or ascorbate, to the electrode surface (Gerhardt *et al.*, 1984). Early studies indicated that this could be a way of developing an electrode with sufficient sensitive and selectivity to measure serotonin (Crespi *et al.*, 1988). Studies with these electrodes have demonstrated that serotonin measured using voltammetry relates to changes in serotonergic neuronal function, but this is not the case with the

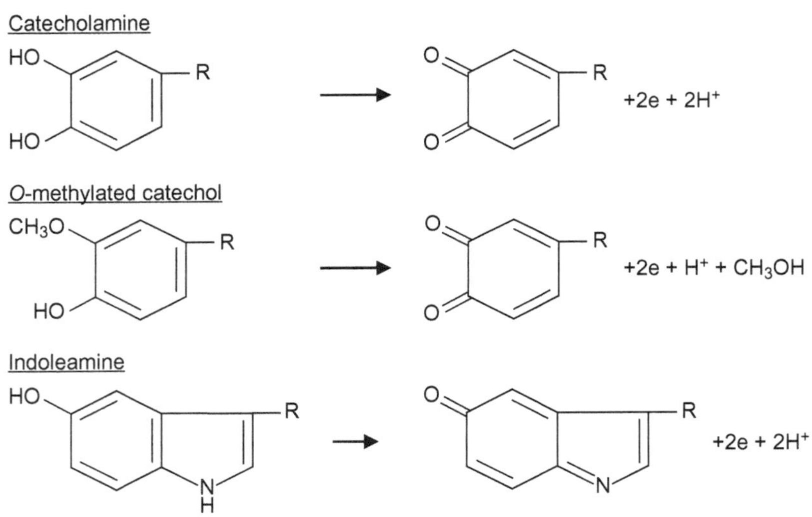

Figure 2 Electrochemistry of catechols and indoles. When this is performed at the surface of a suitable electrode so that it acts as the oxidizing agent, the release of electrons is measured in the form of current. The amount of current produced is directly proportional to the number of molecules oxidized. The problem is to separate the different molecules that can be oxides, including abundant levels of ascorbate. In recent years this process has been achieved by improved scanning methodology and the development of biosensors using Nafion®. Full details of the electrochemistry of all molecules that might be detected using electrochemical electrodes are given in Robinson *et al.* (2008).

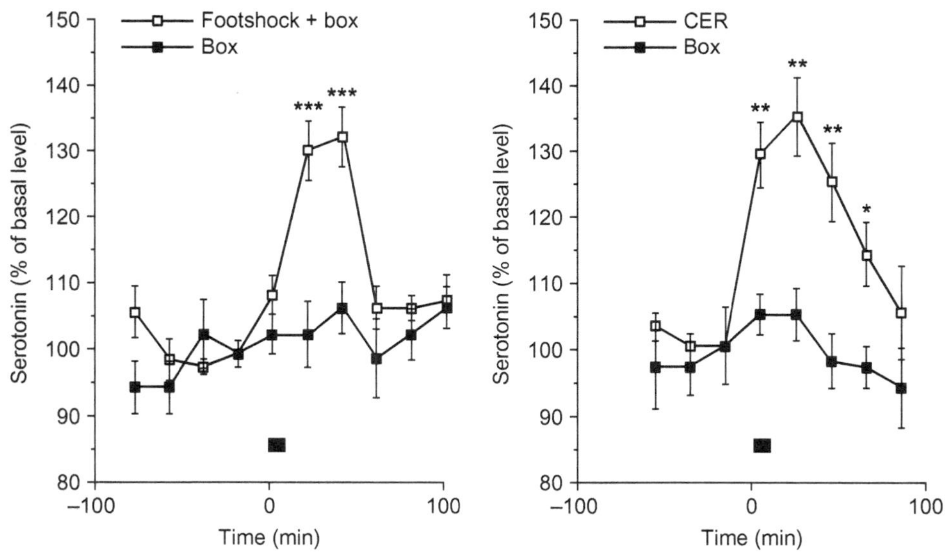

Figure 3 The effect of mild footshock and conditioned emotional response (CER) on hippocampal serotonin measured in the dorsal hippocampus of the rat. Mild footshock (0.4 mA) (n = 6/group) was given for 1 second every minute for 10 mins using a conditioning box. For the CER study, the footshock was given as described on day 1 in the conditioning box and then the rat was returned to its home cage; 24 hours later the rat was placed back in the conditioning box but without shock being applied so that the response observed is to the contextual cue. The point at which mild footshock or CER was performed is indicated by the black square at the base of the graph. Microdialysis samples were collected using concentric probes at 20-min intervals from the CA1 region of the hippocampus. The levels of serotonin in the dialysates were measured using HPLC with electrochemical detection. Note the marked increase in serotonin in response to both mild footshock and CER compared to the controls that were just exposed to the conditioning box without footshock. The data have been adapted from Muchimapura *et al.* (2002).

major serotonin metabolite 5-hydroxyindoleacetic acid (5-HIAA) (Crespi *et al.*, 1990; Figure 3). Another approach has been to maximize the fast scan cyclic voltammetric approach, developed to measure dopamine, to meet the requirements of serotonin. This has been done successfully to measure serotonin release *in vitro* from brain slices (Bunin *et al.*, 1999), and very recently for *in vivo* measurement of serotonin release in the Red Nucleus of the rat (Hashemi and Wightman, 2008).

The development of selective and sensitive electrodes or biosensors based on electrochemical properties needs to be advanced, as they offer real advantages over the microdialysis method in terms of understanding the precise role of serotonin in specific behavioral events. While advances have been made, we are not yet at the point where voltammetric electrodes are widely available for chronic behavioral studies. Voltammetric electrodes have the advantage of real-time resolution combined with improved tip diameter. Comparing the scale of an implanted microdialysis probe and a voltammetric electrode is like comparing a tree trunk with a twig.

Measurement of receptors and transporters in vivo

SPECT and PET are the only methods available that allow visualization of receptors and transporters in the living animal or human and which offer the possibility of investigating changes associated with physiological, pharmacological and pathological events. The key to successful PET imaging is the identification of suitable receptor and transporter ligands, and it is really only in the past 10 years or so that suitable ligands for the serotonin transporter (SERT) and now some serotonin receptors (e.g., 5-HT_{1A} and 5-HT_{2A}) have been identified. Suitable ligands must be able to enter the brain, have very high selectivity for the specific site targeted, show low non-specific binding (otherwise this can result in false positive information) and show an appropriate rate of release from the specific binding sites.

SERT tracer ligands and PET applications in animals and humans

Early studies employed [^{123}I]-βCIT with single photon emission computed tomography (SPECT), but there were issues with its specificity as it had a similar affinity for both the dopamine and serotonin transporters and it also had relatively low sensitivity, which prevented location of SERT in some brain regions of interest – the anterior cingulate cortex, for example (Bhagwagar *et al.*, 2007). The original PET SERT ligand was *trans*-1,2,3,5,6,

10-β-hexahydro-6-[4-(methylthio)phenyl[pyrrolo-[2,1-a]isoquinoline (+)-^{11}C-McN5652 ([^{11}C]McN5652), but this too has issues associated with its use due to the level of non-specific binding and the kinetics of the compound (Szabo *et al.*, 1999; Buck *et al.*, 2000). A more recent additionis3-^{11}C-amino-4-(2-dimethylaminomethylphenyl-sulfanyl)benzonitrile (^{11}C-DASB) (Wilson *et al.*, 2000), which appears to have both high affinity and selectivity for SERT *in vitro* and *in vivo*.

Szabo *et al.* (2002) and Frankle *et al.* (2004) have compared the two ligands, [^{11}C]McN5652 and ^{11}C-DASB, using PET, in both the baboon, under various experimental conditions (control, ecstasy (MDMA) treatment and paroxetine treatment), and humans. The results found a high correlation between the binding of the ligands and the known regional levels of SERT, with preferential binding in areas known to have high SERT. The binding of both ligands was reduced by MDMA treatment in a comparable manner, and paroxetine treatment altered both the kinetics and metabolism of the two ligands, with marked displacement from areas with high endogenous SERT binding. There were, however, differences between the two tracers that indicated preference for ^{11}C-DASB, as it has higher brain activity, a faster removal rate and provided greater contrast between subcortical and cortical regions.

PET studies on SERT have provided important information regarding the long-running issue as to whether MDMA produces long-term loss of serotonergic function in the human in a similar manner to that reported in animals. Studies using [^{11}C] McN5652 demonstrated a loss of serotonergic terminals (McCann *et al.*, 1998; Buchert *et al.*, 2003), but these studies had various limitations in terms of methodology – for example, the Buchert study used the cerebellum as the reference region to control for non-specific binding, though it is now known that SERT is present there. More recently McCann and co-workers (McCann *et al.*, 2005) repeated their study using both [^{11}C] McN5652 and ^{11}C-DASB, and again found generalized loss of SERT in the MDMA users, as well as a good correlation between the two tracers. The question remains as to whether these results, based on 23 subjects, all of whom reported using MDMA on at least 25 occasions but not within 2 weeks of scanning, can be generalized to all MDMA users. One problem, common to all drug-abuse studies, is that MDMA users do not necessarily restrict their drug use to that single drug, and interactions between drugs may occur in terms of serotonergic effects. Furthermore, it is not clear what the clinical significance is of the loss in SERT. The assumption is that loss of SERT reflects a loss of serotonergic nerve terminals and consequently a decrease in serotonergic function, a view that supports animal studies showing loss of serotonergic terminals after MDMA treatment. There is an alternative explanation, as an increase in the number of SERT binding sites, as reported to be associated with seasonal affective disorder, may indicate loss of serotonin in the synaptic cleft due to enhanced reuptake, and so also decreased serotonergic function. Conversely, decreased SERT binding may reflect less reuptake and so increased function when this occurs in the absence of a concomitant loss of terminals. For example, there is debate about what happens to SERT binding, measured by either SPECT or PET, in depressed patients. A recent study, using ^{11}C-DASB as the tracer, found no change in SERT in brain areas associated with depression (amygdala, anterior cingulate cortex, caudate nucleus, frontal cortex, hippocampus and dorsal raphe nucleus) in recovered male depressed patients, indicating that recurrent depression may not be associated with persistent deficits in SERT. This does not mean that depression is not associated with altered serotonergic function, as there are reported differences in cortical postsynaptic receptors (Bhagwagar *et al.*, 2004, 2006). Clearly changes in SERT measured by SPECT or PET need to be interpreted with care, and should be looked at together with other markers of serotonergic function before reaching a final conclusion.

Ligands for serotonin receptors for use with PET

A key area for future development of tracer ligands for use with PET will center on serotonin receptors and, while the present discussion will be restricted to 5-HT$_{1A}$ and 5-HT$_{2A}$ receptors, other receptors such as the 5-HT$_6$ receptor are obvious future targets.

The availability of selective antagonists for the 5-HT$_{1A}$ receptor had a substantial impact on research into the function of this receptor (Forster *et al.*, 1995) which was then translated into the development of new radiotracers for this receptor. The first was [18F]-FCWAY (Neumeister *et al.*, 2004a), and although this compound gave reliable signals in subcortical areas it was less successful in cortical areas, as bone retained the fluoride signal (Carson *et al.*, 2003). A ligand with excellent delineation and precision for PET use is [^{11}C] WAY-100635 (Pike *et al.*, 1996). This radiotracer has been successfully used to show persistent reductions in 5-HT$_{1A}$ postsynaptic receptors in specific brain areas in recovered depressed patients (Bhagwagar *et al.*, 2004) and both pre- and postsynaptic receptors in patients with panic disorder (Nash *et al.*, 2008). This radiotracer offers real promise for future PET studies in both animals and humans.

The 5-HT$_{2A}$ receptor has generated interest in terms of its possible importance in depression, as well as in the mode of action of antipsychotic drugs and hallucinogenic and psychotic behavior (Geyer and Vollenweider, 2008). Two radiotracer ligands that target this receptor are ^{18}F-Setoperone and [^{11}C] MDL 100,907. The latter compound appears to have suitable properties, as it shows high brain uptake, good contrast of the PET images, and does not produce radiolabeled metabolites. It demonstrates good binding potential to the receptor, and is recommended as the best PET ligand for the 5-HT_{2A} receptor at present (Hinz *et al.*, 2007). Interestingly, PET studies in both animals (Hirani *et al.*, 2003) and humans (Talbot *et al.*, 2004) have shown that binding of the ligand to the 5-HT_{2A} receptor is independent of changes in synaptic serotonin concentrations, which suggests it can be used to monitor 5-HT_{2A} function without having to consider the effect of changes in serotonin levels.

Tryptophan depletion and the measurement of serotonin function

The special nature of the synthesis of serotonin in the brain offers a unique way to alter serotonergic function in both animals and humans. The enzyme tryptophan hydroxylase, which converts tryptophan to 5-hydroxytryptophan, is unsaturated at normal brain tryptophan concentrations; thus the synthesis of serotonin in the brain is dependent upon tryptophan availability and its uptake by the brain. Tryptophan uptake into the brain is by the large neutral amino acid transport system for which tryptophan competes with other essential amino acids for entry into the brain. It is possible to deplete plasma tryptophan and reduce its uptake by the brain by administering to subjects a tryptophan-free amino acid mixture of essential amino acids that compete with tryptophan for transport into the brain across the blood–brain barrier (Figure 4). Preclinical studies have shown that this results in decreased plasma tryptophan levels and decreases in brain tryptophan, serotonin and its metabolite 5-hydroxyindole acetic acid (5-HIAA) in the rat (Gessa *et al.*, 1974; Curzon, 1979). The same procedure in the human has been shown to decrease plasma tryptophan and the levels of this amino acid as well as those of 5-HIAA in cerebrospinal fluid (CSF) (Williams *et al.*, 1999). These data clearly point to tryptophan depletion reducing brain serotonergic function; however, a note of caution is needed, as a recent microdialysis study in the rat found that while a tryptophan-free amino acid mixture reduced plasma tryptophan, there was no change in the extracellular levels of either serotonin or dopamine in the prefrontal cortex, indicating no change in serotonergic function and thus that reduced serotonergic function may not fully explain the behavioral effects measured both in animals and humans with this method (Van der Plaase *et al.*, 2007).

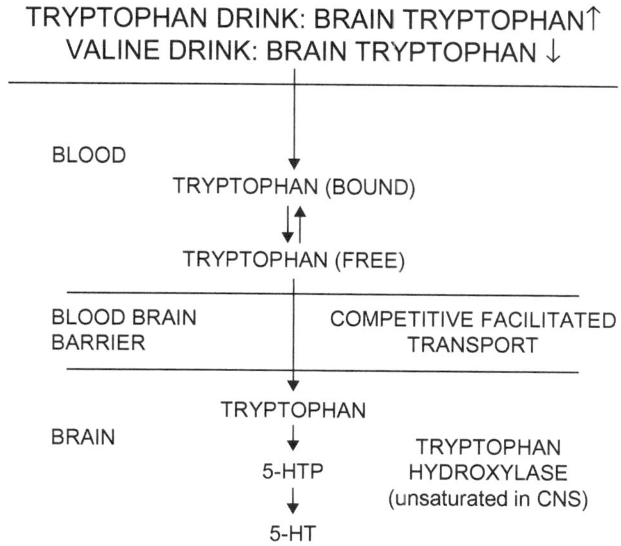

Figure 4 Diagram showing the relationship between the plasma tryptophan and brain serotonin (5HT) levels. Plasma tryptophan can bind to plasma protein so only unbound (free) tryptophan is available to the competitive facilitated transporter that transports the essential (large neutral) amino acids across the blood–brain barrier (BBB). Drinks containing no tryptophan and high levels of another large neutral amino acid which compete for the transporter, such as valine, will reduce levels of tryptophan in the brain and thus reduce the synthesis of serotonin in the brain.

One potential explanation is that serotonergic function is only reduced by tryptophan depletion when there is an underlying trait abnormality of serotonergic function, but not when serotonergic function is normal (as in the experiment described above). Indeed, the most striking results with tryptophan depletion in humans have been found in remitted depressed patients, when tryptophan depletion results in a transient return of the symptoms during the period of tryptophan depletion (Smith *et al.*, 1997; Neumeister *et al.*, 2004b), but not in control subjects with no familial risk of depression (Moreno *et al.*, 1999). These findings support the idea of a trait serotonergic abnormality in depressed patients, and it has been suggested that tryptophan depletion may be clinically useful in identifying individuals at risk of developing major depression (Moreno *et al.*, 2000). Similar conclusions appear to prevail with social anxiety disorder, as patients with this disorder are improved by SSRI treatment but this effect was reversed by tryptophan depletion (Argyropoulos *et al.*, 2004).

There are, however, studies showing changes in cognitive function in healthy young volunteers following tryptophan depletion (Rogers *et al.*, 1999; Schmitt *et al.*, 2000; Finger *et al.*, 2007), as well as activity in the prefrontal cortex measured by fMRI (Rubia *et al.*, 2005).

It remains to be determined whether these effects, in normal subjects, on cognitive functioning can be fully explained by altered serotonergic activity or by some other neural mechanism influenced by tryptophan depletion. Overall, tryptophan depletion studies offer an excellent non-invasive method for identifying disorders with underlying serotonergic dysfunction, but its role in the study of normal serotonergic function remains debatable.

Neuroendocrine markers and serotonergic function

The importance of various serotonergic receptors in the regulation of pituitary neuroendocrine function is well established, particularly in relation to the HPA axis and the release of cortisol in humans and corticosterone in rats involving 5-HT_{1A}, 5-HT_{2A} and 5-HT_{2C} receptors (for review, see Fuller, 1996). The same receptors in the hypothalamic paraventricular nucleus are involved with the regulation of prolactin and renin release (Lucki, 1998). This has led to the use of a range of pharmacological agents (SSRIs, 5-HTP, agonists and antagonists, and serotonin releasers as well as tryptophan depletion) with relative selectivity for these receptors to investigate their functional state in various psychiatric disorders by measurement of changes in the peripheral release of the hormones (for review, see Murphy *et al.*, 1996). With the development of improved PET ligands for these receptors, it is becoming increasingly possible to make direct brain measurements rather than indirectly using hormonal markers.

Measurement of neuronal activity – neural pathways

Electrophysiological techniques have played a key role in understanding the detailed mechanisms involved in serotonin receptor function and their signal transduction pathways, but will not be discussed here because this information in covered elsewhere in this book. The emphasis of this discussion will be on approaches available to look not specifically at serotonergic function, but rather at how manipulation of serotonergic function (for example, using drugs) can lead to changes in brain region activity and neural connections.

Regional metabolic activity as an index of neuronal activity has been very successfully measured using labeled 2-deoxyglucose in conjunction with PET for many years in both controls and patients (Neumeister *et al.*, 2004a). More recently this approach has been superseded by functional magnetic resonance imaging (fMRI), which is non-invasive and provides great potential, as it can be readily linked with cognitive and emotional testing.

It is not the role of this chapter to discuss in detail the potential of fMRI to help our understanding of the ways in which serotonergic neurones interact with other pathways in both normal and abnormal behavior. However, fMRI offers as translational approach, as it can be used both in animals (Steward *et al.*, 2005) and in humans (Northoff *et al.*, 2002; Kemp *et al.*, 2004) to investigate brain regional patterns of neuronal activation and deactivation following drug administration using methods such as Blood Oxygen Level Dependence (BOLD). These drug studies are often referred to as pharmacological MRI (phMRI), to distinguish them from studies (fMRI) using behavioral or cognitive interventions and monitoring the subsequent changes in regional neuronal function. These studies may also incorporate serotonin-selective drugs, for example SSRIs, into their design. An interesting example of this approach is provided by the studies looking at the neural responses of normal subjects and patients to fearful facial expressions of emotion, and how these neural responses are modified by drug treatment (Sheline *et al.*, 2001; Kemp *et al.*, 2004; Chan *et al.*, 2009). These studies have not only identified the brain areas associated with the interpretation of visual emotional stimuli, but also shown greater neural responses to fearful stimuli in depressives and those at risk for depression (Chan *et al.*, 2009). They showed too that SSRI treatment enhanced the neural response to pleasant stimuli and reduced the response to fearful stimulus (Kemp *et al.*, 2004). Neuroimaging of neural activity can play a major role in improved understanding of the pathways involved in the complex interactions between serotonin and other neurotransmitters that drive cognitive and emotional behaviors.

Conclusions

Considerable effort has been put into the development of methodologies that can be used to monitor serotonergic function during behavior. The results obtained have improved understanding of the function of specific receptors or specific pathways, and benefited the quest for new approaches to the treatment of certain psychiatric disorders in particular. The real challenge is to understand the role of serotonin in the integrated neural pathways that drive behavior, and to do this the methods still require improvement. Measurement of transmitter release requires better time and spatial resolution, while PET methodology needs new ligands for serotonin receptors as yet not detectable by the method. Magnetic resonance imaging will play a central role in the future investigation of integrated function, as it makes it possible to monitor activity within neural pathways involved in cognitive and emotional behavior combined with an ability to modify responses using serotonergic selective drugs. Such an approach is essential for successful translational studies using both animals and humans.

Acknowledgements

Studies at Nottingham described in this chapter have been supported by the MRC, BBSRC and Wellcome Trust. We would also like to thank GSK and Lundbeck a/s for financial support.

References

Adams, R.N. (1976) Probing brain chemistry with electroanalytical techniques. Anal. Chem., 48: 1126A–1138A.

Argyropoulos, S.P., Hood, S.D., Adrover, M., Bell, C.J., Rich, A.S., Nash, J.R., Rich, N.C., Witchel, H.J. and Nutt, D.J. (2004) Tryptophan depletion reverses the therapeutic effect of selective serotonin reuptake inhibitors in social anxiety disorder. Biol. Psychiatry, 56: 503–509.

Bhagwagar, Z., Rabiner, E.A., Sargent, P.A., Grasby, P.M. and Cowen, P.J. (2004) Persistent reduction in brain serotonin 1A receptor binding in recovered depressed men measured by positron emission tomography with [^{11}C]WAY-100635. Mol. Psychiatry, 9: 386–392.

Bhagwagar, Z., Hinz, R., Taylor, M., Fancy, S., Cowen, P. and Grasby, P. (2006) Increased 5-HT$_{2A}$ receptor binding in euthymic, medication-free patients recovered from depression. A positron emission study with [^{11}C]MDL 100,907. Am. J. Psychiatry, 163: 1580–1587.

Bhagwagar, Z., Murthy, N., Selaraj, S., Hinz, R., Taylor, M., Fancy, S., Grasby, P. and Cowen, P. (2007) 5-HTT binding in recovered depressed patients and healthy volunteers: a positron emission tomography study with [^{11}C]DASB. Am. J. Psychiatry, 164: 1858–1865.

Buchert, R., Thomasius, R., Nebeling, B., Petersen, K., Obrocki, J., Jenicke, L., Wilke, F., Wartberg, L., Zapletalova, P. and Clausen, M. (2003) Long-term effects of 'ectasy' use on serotonin transporters of the brain investigated by PET. J. Nucl. Med., 44: 375–384.

Buck, A., Gucker, P.M., Schonbachler, R.D., Arigoni, M., Kneifel, S., Vollenweider, X.F., Ametamey, S.M. and Burger, C. (2000) Evaluation of serotonergic transporters using PET and [11C](+)McN-5652: assessment of methods. J. Cereb. Blood Flow Metab., 20: 244–252.

Bunin, M.A., Prioleau, C., Mailman, R.B. and Wightman, R.M. (1999) Release and uptake rates of 5-hydroxytryptamine in the dorsal raphe and substantia nigra reticulate of the rat brain. J. Neurochem., 70: 1077–1087.

Carson, R.E., Wu, Y., Lang, L., Ma, Y., Der, M.G., Herscovitch, P. and Eckelman, W.C. (2003) Brain uptake of the acid metabolites of F-18 labelled WAY 100635 analogs. J. Cereb. Blood Flow Metab., 23: 249–260.

Castejon, A.M., Paez, X., Hernandez, L. and Cubeddu, L.X. (1999) Use of intravenous microdialysis to monitor changes in serotonin release and metabolism induced by cisplatin in cancer patients: comparative effects of Granisetron and Ondansetron. J. Pharmacol. Exp. Therap., 291: 960–966.

Chan, S.W.Y., Norbury, R., Goodwin, G. and Harmar, C.J. (2009) Risk for depression and neural responses to fearful facial expressions of emotion. Br. J. Psych., 194: 139–145.

Cheer, J., Aragona, B., Heien, M., Seipel, A., Carelli, R. and Wightman, R. (2007) Coordinated accumbal dopamine release and neural activity drive goal-directed behavior. Neuron, 54: 237–244.

Conti, J.C., Strope, E., Adams, R.N. and Marsden, C.A. (1978) Voltammetry in brain tissue chronic recording of stimulated dopamine and 5-hydroxytryptamine release. Life Sci., 23: 2705–2716.

Crespi, F., Martin, K.F. and Marsden, C.A. (1988) Measurement of extracellular basal levels of 5-HT *in vivo* using Nafion-coated carbon fibre electrodes combined with DPV. Neuroscience, 27: 885–896.

Crespi, F., Garratt, J.C., Sleight, A.J. and Marsden, C.A. (1990) *In vivo* evidence that 5-hydroxytryptamine (5-HT) neuronal firing and release are not necessarily correlated with metabolism. Neuroscience, 35: 139–144.

Curzon, G. (1979) Relationships between plasma, CSF and brain tryptophan. J. Neural. Transm., 15(Suppl.): 81–92.

Dawson, L.A., Nguyen, H.Q. and Li, P. (2000) *In vivo* effects of the 5-HT6 antagonist SB-271046 on striatal and frontal cortex extracellular concentrations of noradrenaline, dopamine, 5-HT, glutamate and aspartate. Br. J. Pharmacol., 130: 23–26.

Finger, E.C., Marsh, A.A., Buzas, B., Kamel, N., Rhodes, R., Vythilingham, M., Pine, D.S., Goldman, D. and Blair, J.R. (2007) The impact of tryptophan depletion and 5-HTTLRP genotype on passive avoidance and response reversal instrumental learning tasks. Neuropsychopharmacology, 32: 206–215.

Forster, E.A., Cliffe, I.A., Bill, D.J. et al. (1995) A pharmacological profile of the selective silent 5-HT$_{1A}$ receptor antagonist, WAY 100635. Eur. J. Pharmacol., 281: 81–88.

Frankle, W.G., Huang, Y., Hwang, D.R., Talbot, P.S., Slifstein, M., Van Heertum, R. et al. (2004) Comparative evaluation of serotonin transporters radioligands 11C-DASB and 11C-McN 5652 in healthy humans. J. Nucl. Med., 45: 682–694.

Fulford, A.J. and Marsden, C.A. (1998) Conditioned release of 5-hydroxytryptamine in vivo in the nucleus accumbens following isolation-rearing in the rat. Neuroscience, 83: 481–487.

Fuller, R.W. (1996) Serotonin receptors involved in regulation of pituitary-adrenocortical function in rats. Behav. Brain. Res., 73: 215–219.

Gerhardt, G.A., Oke, A.F., Nagy, G., Moghaddam, B. and Adams, R.N. (1984) Nafion-coated electrodes with high selectivity for CNS electrochemistry. Brain Res., 290: 390–395.

Gessa, G.L., Biggio, G., Fadda, F., Corsini, G.U. and Tagliamonte, A. (1974) Effect of the oral administration of tryptophan-free amino acid mixtures on serum tryptophan, brain tyrptophan and serotonin metabolism. J. Neurochem., 22: 869–870.

Geyer, M.A. and Vollenweider, F.X. (2008) Serotonin research: contributions to understanding psychosis. Trends Pharmacol. Sci., 29: 445–453.

Gonon, F., Buda, M., Cespuglio, R., Jouvet, M. and Pujol, J.F. (1980) *In vivo* electrochemical detection of catechols in the striatum of anesthetized rats: Dopamine or DOPAC? Nature, 286: 902–904.

Hashemi, P. and Wightman, R.M. (2008) Application of fast scan cyclic voltammetry to in vivo -5-HT monitoring. Proceedings of the 12th International Conference on *in vivo*

Methods, University of British Columbia 72–74. ISBN: 978-0-9310624-0-2.

Hinz, R., Bhagwagar, Z., Cowen, P.J., Cunningham, V.J. and Grasby, P.M. (2007) Validation of a tracer kinetic model for the quantification of 5-HT_{2A} receptors in human brain with [^{11}C]MDL 100,907. J. Cereb. Blood Flow Metab., 27: 161–172.

Hirani, E., Sharp, T., Sprakes, M., Grasby, P. and Hume, S. (2003) Fenfluramine evokes 5-HT_{2A} receptor-mediated responses but does not displace [^{11}C]MDL 100907: small animal PET and gene expression studies. Synapse, 50: 251–260.

Keller, R., Oke, A., Mefford, I. and Adams, R.N. (1976) Liquid chromatographic analysis of catecholamines: routine assay for regional brain mapping. Life Sci., 19: 995–1004.

Kemp, A.H., Gray, M.A., Silberstein, R.B., Armstrong, S.M. and Nathan, P.J. (2004) Augmentation of serotonin enhances pleasant and suppresses unpleasant cortical electrophysiological responses to visual emotional stimuli in humans. NeuroImage, 22: 1084–1096.

King, M.V., Marsden, C.A. and Fone, K.C.F. (2008) A role for the 5-HT_{1A}, 5-HT_4 and 5-HT_6 receptors in learning and memory. Trends Pharmacol. Sci., 29: 482–492.

Kirby, L.G., Allen, A.R. and Lucki, I. (1995) Regional differences in the effects of forced swimming on extracellular levels of 5-hydroxytryptamine and 5-hydroxyindoleacetic acid. Brain Res., 682: 189–196.

Li, Y., Peris, J., Li, Z. and Derendorf, H. (2006) Microdialysis as a tool in local pharmacodynamics. AAPS J., 8: E222–E235.

Lucki, I. (1998) The spectrum of behaviors influenced by serotonin. Biol. Psychiatry, 44: 151–162.

Marsden, C.A. (1985) *In vivo* monitoring of pharmacological and physiological changes in endogenous serotonin release and metabolism. In: Green, A.R. (Ed.), Neuropharmacology of Serotonin. Oxford University Press, Oxford, pp. 218–252.

McCann, U.D., Szabo, Z., Scheffel, U., Dannals, R.F. and Ricaurte, G.A. (1998) Positron emission tomographic evidence of toxic effect of MDMA (Ecstasy) on brain serotonin neurons in human beings. Lancet, 352: 1433–1437.

McCann, U.D., Szabo, Z., Seckin, E., Rosenblatt, P., Mathews, W.B., Ravert, H.T., Dannals, R.F. and Ricaurte, G.A. (2005) Quantitative PET studies of the serotonin transporter in MDMA users and controls using [^{11}C]McN5652 and [^{11}C]DASB. Neuropsychopharmacology, 30: 1741–1750.

Middlemiss, D.N. and Hutson, P.H. (1990) Measurement of the *in vitro* release of endogenous monoamine neurotransmitters as a means of identification of prejunctional receptors. J. Neurosci. Methods, 34: 23–28.

Millan, M.J. (2006) Multi-target strategies for the improved treatment of depressive states; conceptual foundations and neuronal substrates, drug discovery and therapeutic application. Pharmacol. Ther., 110: 135–370.

Moreno, F.A., Gelenberg, A.J., Heninger, G.R., McKnight, K.M., Allen, J., Phillips, A.P. and Delgado, P.L. (1999) Tyrptophan depletion and depressive vulnerability. Biol. Psychiatry, 46: 498–505.

Moreno, F.A., Heninger, G.R., McGahuey, C.A. and Delgado, P.L. (2000) Tryptophan depletion and risk of depression relapse: a prospective study of tryptophan depletion as a potential predictor of depressive episodes. Biol. Psychiatry, 48: 327–329.

Muchimapura, S., Fulford, A.J., Mason, R. and Marsden, C.A. (2002) Isolation rearing in the rat disrupts the hippocampal response to stress. Neuroscience, 112: 697–705.

Murphy, D.L., Aulakh, C., Mazzola-Pomietto, P. and Briggs, N.C. (1996) Neuroendocrine responses to serotonergic agonists as indices of the functional status of central serotonin neurotransmission in humans: a preliminary comparative analysis of neuroendocrine endpoints versus other endpoint measures. Behav. Brain. Res., 73: 209–214.

Myers, R.D. and Beleslin, D.B. (1970) The spontaneous release of 5-hydroxytryptamine and acetylcholine within the diencephalon of the unanaesthetized rhesus monkey. Exp. Brain Res., 11: 539–552.

Nash, J.R., Sargent, P.A., Rabiner, E.A., Hood, S.D., Argyropoulos, S.V., Potokae, J.P., Grasby, P.M. and Nutt, D.J. (2008) Serotonin 5-HT1A receptor binding in people with panic disorder: positron emission tomography study. Br. J. Psychiatry, 193: 229–234.

Neumeister, A., Bain, E., Nugent, A.C., Carson, R.E., Bonne, O., Luckenbaugh, D.A., Eckelman, W., Herscovitch, P., Charney, D. and Drevets, W.C. (2004a) Reduced serotonin type 1A binding in panic disorder. J. Neurosci., 24: 589–591.

Neumeister, A., Nugent, A.C., Waldeck, T., Geraci, M., Schwarz, M., Bonne, O., Bain, E.E., Luckenbaugh, D.A., Herscovitch, P., Charney, D.S. and Drevets, W.C. (2004b) Neural and behavioral responses to tryptophan depletion in unmedicated patients with remitted major depressive disorder and controls. Arch. Gen. Psychiatry, 61: 765–773.

Northoff, G.W., Thomas, R.A., Gessner, M., Schlagenhauf, F., Fell, J., Baumgart, F. et al. (2002) GABA-ergic modulation of prefrontal spatio-temporal activation pattern during emotional processing: a combined fMRI/MEG study with placebo and lorazepam. J. Cogn. Neurosci., 14: 348–370.

Parsons, L.H., Koob, G.F. and Weiss, F. (1995) Serotonin dysfunction in the nucleus accumbens of rats during withdrawal after unlimited access to intravenous cocaine. J. Pharmacol. Exp. Ther., 274: 1182–1191.

Phillips, P.E., Stuber, G.D., Heien, M.L., Wightman, R.M. and Carelli, R.M. (2003) Subsecond dopamine release promotes cocaine seeking. Nature, 422: 614–618.

Pike, V.W., McCarron, J.A., Lammertsma, A.A., Osman, S., Hume, S.P., Sargent, P.A., Bench, C.J., Cliffe, I.A., Fletcher, A. and Grasby, P.M. (1996) Exquisite delineation of 5-HT1A receptors in human brain with PET and [carbonyl-11C]WAY-100635. Eur. J. Pharmacol., 301: R5–R7.

Reuter, L.E., Fornal, C.A. and Jacobs, B.L. (1997) A critical review of 5-HT microdialysis and behavior. Rev. Neurosci., 8: 117–137.

Robinson, D.L., Hermans, A., Seipel, A.T. and Wightman, R.M. (2008) Monitoring rapid chemical communication in the brain. Chem. Rev., 108: 2554–2584.

Rogers, R.D., Blackshaw, A.J., Middleton, H.C., Matthews, K., Hatwin, K., Crowley, C., Hopwood, A., Wallace, C., Deakin, J.F., Sahakian, B.J. and Robbins, T.W. (1999) Tryptophan depletions impairs stimulus–reward learning while methylphenidate disrupts attentional control in healthy young adults: implications for the monoaminergic basis of impulse behavior. Psychopharmacology, 146: 482–491.

Rubia, K., Lee, F., Cleare, A.J., Tunstall, N., Fu, C.H.Y., Brammer, M. and McGuire, P. (2005) Tryptophan depletion reduces right inferior prefrontal activation during response inhibition in fast, event-related fMRI. Psychopharmacology, 179: 791–803.

Samuelsson, A., Abdiu, A., Wackenfors, A. and Sjoberg, F. (2008) Serotonin kinetics in patients with burn injuries: a comparison between the local and systemic responses measured by microdialysis – a pilot study. Burns, 34: 617–622.

Schmitt, J.A., Jorissen, B.L., Sobczak, S., van Boxtel, M.P., Hogervorst, E., Deutz, N.E. and Reidel, W.J. (2000) Tryptophan depletion impairs memory consolidation but improves focused attention in healthy young volunteers. J. Psychopharmacol., 14: 21–29.

Sharp, T., Bramwell, S.R. and Grahame-Smith, D.G. (1989) 5-HT_1 agonists reduce 5-hydroxytryptamine release in rat hippocampus *in vivo* as determined by brain microdialysis. Br. J. Pharmacol., 96: 283–290.

Sheline, Y.I., Barch, D.M., Donnelly, J.M., Ollinger, J.M., Snyder, A.Z. and Mintun, M.A. (2001) Increased amygdale responses to masked emotional faces in depressed subjects resolves with antidepressant treatment: an fMRI study. Biol. Psychiatry, 50: 651–658.

Smith, K.A., Fairburn, G.G. and Cowen, P.J. (1997) Relapse of depression after rapid depletion of tryptophan. Lancet, 349: 915–919.

Sprouse, J.S. and Aghajanian, G.K. (1987) Electrophysiological responses of serotoninergic dorsal raphe neurons to 5-HT1A and 5-HT1B agonists. Synapse, 1: 3–9.

Steward, C.A., Marsden, C.A., Prior, M.J., Morris, P.G. and Chapman, V. (2005) Methodological considerations in rat brain BOLD contrast pharmacological MRI. Psychopharmacology, 180: 687–704.

Szabo, Z., Scheffel, U., Mathews, W.B., Ravert, H., Szabo, K., Kraut, M., Palmon, S. et al. (1999) Kinetic analysis of [11C] McN5652: a serotonin transporter ligand. J. Cereb. Blood Flow Metab., 19: 967–981.

Szabo, Z., McCann, U.D., Wilson, A.A., Scheffel, U., Owonikoko, T., Mathews, W.B., Ravert, H.T., Hilton, J., Dannals, R.F. and Ricaurte, G.A. (2002) Comparison of (+)-^{11}C-McN5652 and ^{11}C-DASB as serotonin transporter radioligands under various experimental conditions. J. Nucl. Med., 43: 678–692.

Talbot, P.S., Frankle, W.G., Hwang, D., Huang, Y., Abi-Dargham, A. and Laruelle, M. (2004) Effects of transient reduction of endogenous 5-HT levels on brain 5-HT transporter and 5-HT2A receptor availability: a PET study using [11C]DASB, [11C]MDL 100907 and the rapid tryptophan depletion method in healthy humans. Biol. Psychiatry, 55: 63S–64S.

Ungerstedt, U. and Pycock, Ch. (1974) Functional correlates of dopamine neurotransmission. Bull. Schweiz. Akad. Med. Wiss., 30: 44–55.

Van der Plaase, G., Meerkerk, T.J., Lieben, C.K.J., Blokland, A. and Feenstra, M.G.P. (2007) Lack of evidence for reduced prefrontal cortical serotonin and dopamine efflux after acute tryptophan depletion. Psychopharmacology, 195: 377–385.

Wilkinson, L.S., Humby, T., Killcross, S., Robbins, T.W. and Everitt, B.J. (1996) Dissociations in hippocampal 5-hydroxytryptamine release in the rat following Pavlovian aversive conditioning to discrete and contextual stimuli. Eur. J. Neurosci., 8: 1479–1487.

Williams, W.A., Shoaf, S.E., Hommer, D., Rawlings, R. and Linnoila, M. (1999) Effects of acute tryptophan depletion on plasma and cerebrospinal fluid tryptophan and 5-hydroxyindoleacteic acid in normal volunteers. J. Neurochem., 72: 1641–1647.

Wilson, A.A., Ginovart, N., Schmidt, M., Meyer, J.H., Threlkeld, P.G. and Houle, S. (2000) Novel radiotracers for imaging the serotonin transporter by positron emission tomography: synthesis, radiosynthesis and in vitro and ex vivo evaluation of (11)C-labeled 2-(phenylthio)araalkylamine. J. Med. Chem., 43: 3103–3110.

Wright, I.K., Upton, N. and Marsden, C.A. (1992) Effect of established and putative anxiolytics on extracellular 5-HT and 5-HIAA in the ventral hippocampus of rats during behavior on the elevated plus maze. Psychopharmacology, 109: 338–346.

Zetterstrom, T., Sharp, T., Marsden, C.A. and Ungerstedt, U. (1983) *In vivo* measurement of dopamine and its metabolites by intracerebral dialysis: changes after d-amphetamine. J. Neurochem., 41: 1769–1773.

SECTION 2

The Neurophysiology of Serotonin

CHAPTER 2.1

Activity of Brain Serotonergic Neurons in Relation to Physiology and Behavior

Barry L. Jacobs* and Casimir A. Fornal

Program in Neuroscience, Princeton University, Princeton, New Jersey

Abstract: We have explored the activity of brain serotonin (5-hydroxytryptamine; 5-HT) neurons in behaving cats exposed to a wide variety of behavioral and physiological conditions. In general, the activity of these neurons is strongly related to spontaneous changes in behavioral state (being highest in active waking and lowest during REM sleep). Across a wide variety of experimental manipulations, the activity of serotonergic neurons was relatively unchanged. This includes dramatic alterations in general physiology. However, one condition, motor activity, strongly affected neuronal activity. A general relationship obtains between level of tonic motor activity and serotonergic neuronal activity across all major groups of serotonergic neurons. Superimposed upon this in some neurons is an additional relationship where a further, often dramatic, activation is seen in association with repetitive, central pattern generator (CPG)-mediated behaviors. The exact nature of this relationship varies both with the specific serotonergic neuronal group and within a particular group. We hypothesize that the primary function of this increased serotonergic neuronal activity is to facilitate behavioral output by coordinating autonomic and neuroendocrine function in association with motor demand, and by a concomitant general suppression of afferent input from sensory processing channels.

Keywords: electrophysiological recordings, sleep–wake cycle, stressors, autonomic function, motor activity, hypercapnia, respiration, fatigue.

Abbreviations: 5-HT, 5-hydroxytryptamine (serotonin); CCK, cholecystokinin; CNS, central nervous system; CPG, central pattern generator; DRN, dorsal raphe nucleus; EEG, electroencephalogram; EOG, electro-oculogram; EMG, electromyogram; GABA, γ-aminobutyric acid; NCS, nucleus centralis superior; NRM, nucleus raphe magnus; NRO, nucleus raphe obscurus; NRP, nucleus raphe pallidus; REM, rapid eye movement

Introduction

As the chapters in this volume amply demonstrate, a large variety of approaches have been used to investigate serotonin's functional role(s) in the CNS. Over the past several decades, the confluence of results derived from these varied perspectives (anatomical, physiological, genetic, clinical, behavioral, etc.) has shown the breadth of serotonin's influence on physiology and behavior. Although our laboratory at Princeton University has employed a number of these methodologies over the years, our primary focus has been neurophysiological in nature; specifically, single neuron recordings of brain serotonergic neurons in behaving animals. This chapter represents an overview of this work, with an emphasis on our more recent studies. (Some of our previous review chapters can be consulted for detailed descriptions of our earlier work in this area – Jacobs and Fornal, 1997a, 1998.)

The overarching goal of our research has been the examination of the behavioral/physiological role of the mammalian CNS serotonin system. Does it have a specific role or roles, or is serotonin a ubiquitous neurotransmitter, like GABA or glutamate, exerting effects on virtually all brain processes, and thus serving no specifiable behavioral or physiological function? Is there a general theory of CNS serotonin function which could account, at least in part, for the broad array of data on this topic?

The primary tool that we employed in our work was measuring the electrical activity of individual serotonergic neurons ('single unit recordings') in unanesthetized, freely behaving domestic cats (*F. catus*). Our philosophy was to follow leads provided by observing the variations

*Corresponding author
E-mail: barryj@princeton.edu

ISBN 978-0-12-374634-4

DOI: 10.1016/B978-0-12-374634-4.00001-0

in neuronal activity that emerged under a diversity of behavioral and physiological conditions.

At the outset, our thinking was guided by certain fundamental facts about brain serotonin. First, the basic plan for the cell bodies of these neurons is conserved throughout the vertebrates. This implies that its function must, to some degree, be common in the physiology and behavior of fish, amphibians and reptiles as well as mammals, including primates. Second, virtually all of the cell bodies of serotonergic neurons are found in the brainstem, on or near the midline. This suggests a somewhat stronger involvement in basic processes, especially those associated with axial functions, such as controlling the trunk and proximal limb muscles. Finally, these neurons and their processes are among the first to develop during ontogeny. All of this suggests that the central serotonin system subserves basic or fundamental functions within the CNS.

In an attempt to establish generalizations about serotonin's role, we examined the activity of serotonergic neurons throughout the neuraxis. The clusters of brain serotonergic neurons can be divided into two major groups. The rostral or mesencephalic group contains two primary nuclei which supply most of the serotonin to the forebrain: the nucleus centralis superior (NCS) and the dorsal raphe nucleus (DRN). The caudal or medullary group contains three primary nuclei which supply most of the serotonin to the brainstem and spinal cord: the nucleus raphe magnus (NRM), the nucleus raphe obscurus (NRO), and the nucleus raphe pallidus (NRP).

Basic neuronal characteristics

Our recording technique employing chronically implanted microwires allows us to record the activity of the same neuron over long periods of time and despite active, gross movements. Many presumed 5-HT neurons have a distinctive 'neuronal signature' that makes them unique in the CNS and allows them confidently to be identified online as serotonergic. They discharge in a slow (typically 1–6 Hz) and highly regular manner. This clock-like activity is the manifestation of an endogenous pacemaker. These neurons also possess an unusually long-duration action potential (2–3 ms). These neuronal characteristics appear to be somewhat universal since they are: (1) observed in *in vitro* preparations, under anesthesia, and in behaving animals; (2) seen in the several species in which this has been examined; and (3) found in the various brainstem groups of 5-HT neurons (Jacobs and Azmitia, 1992).

Although the preponderance of serotonergic neurons manifest these characteristics ('typical'), brain serotonergic neurons with other electrophysiological features ('atypical') have also been identified: see work from the laboratories of Sakai (Sakai and Crochet, 2001), Beck (Beck *et al.*, 2004), Sharp (Hajos *et al.*, 1995) and Urbain (Urbain *et al.*, 2006). Unfortunately, there has been little attempt to examine the response of these 'atypical' serotonergic neurons across varied behavioral and physiological conditions, other than the sleep–wake–arousal cycle. Furthermore, studies of these neurons have been limited, almost exclusively, to the DRN. Finally, whereas our studies have been carried out in domestic cats, with the exception of the studies of Sakai the rest of this work has been conducted in laboratory rats.

Sleep–wake–arousal cycle

In the quiet waking state in the cat, the activity of DRN 5-HT neurons is slow (~3 spikes/s) and regular, characteristic of the unmodulated operation of an endogenous pacemaker (Jacobs and Fornal, 1991). During an aroused state or in response to sensory stimuli, neuronal activity can be increased 30–50 percent above the quiet waking level. The most dramatic changes in the activity of 5-HT neurons occur during sleep. As animals become drowsy and enter slow-wave sleep, neuronal activity slows to approximately 50 percent of the quiet waking level and loses its regularity. Finally, during REM sleep, the activity of most brain 5-HT neurons declines dramatically, in many cases falling to zero. This general pattern of activity across the sleep–wake–arousal cycle displayed by 5-HT neurons in the DRN is also seen, with some variation, in the other major groups of brainstem 5-HT neurons: NCS, NRM, NRO and NRP (Figure 1) (Jacobs and Fornal, 1991).

Response to stressors

As a starting point for our research program, we reasoned that it would be informative to examine the activity of brain 5-HT neurons under extreme and varied conditions. Thus, we studied neuronal responses in cats exposed to a number of strong environmental and physiological stimuli. Accordingly, the activity of 5-HT neurons in the DRN, NCS or NRM was examined while cats were exposed to: a heated environment or a pyrogen; drug-induced increases or decreases in blood pressure; insulin-induced hypoglycemia; phasic or tonic painful stimuli; systemic injections of morphine; loud noise; physical restraint; or a natural enemy (dog) (see Jacobs and Fornal (1991) for details). Despite the fact that all of these conditions evoked strong behavioral responses and/or physiological changes indicative of sympathetic activation, none of them significantly activated 5-HT neuronal activity in the DRN, NCS or NRM beyond the level typically seen

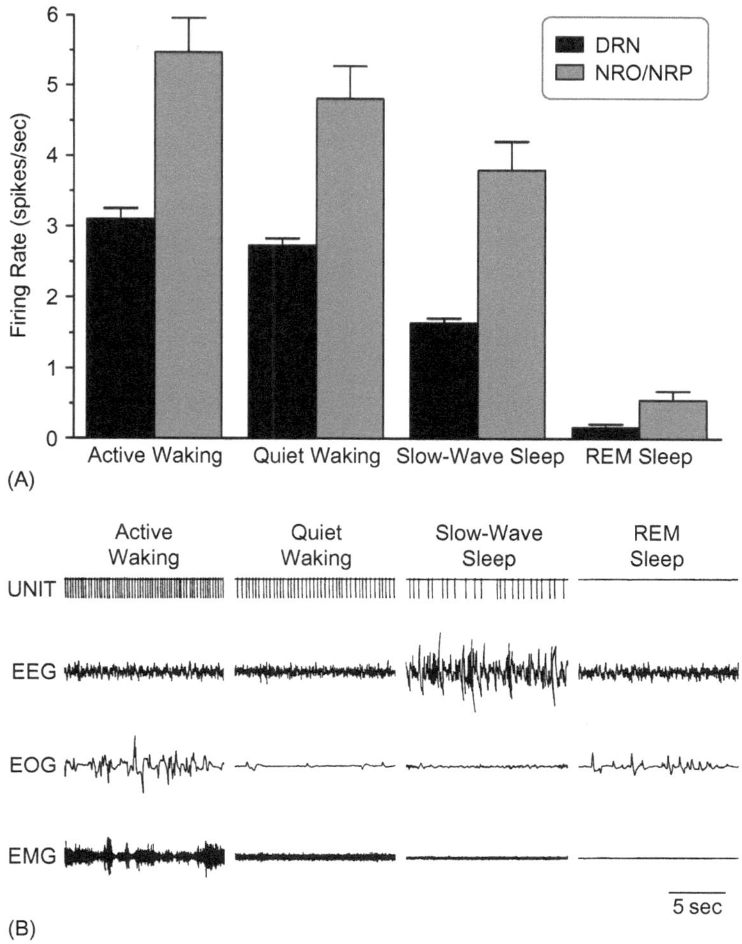

Figure 1 Activity of 5-HT neurons across the sleep–wake–arousal cycle. Panel A shows mean firing rates for a group of DRN and NRO/NRP neurons. In general, the spontaneous firing rate of medullary raphe neurons is higher than that of mesencephalic raphe neurons during comparable behavioral states. Panel B shows the activity of a typical medullary 5-HT neuron across the sleep–wake–arousal cycle. Note the positive relationship between neuronal activity and the level of tonic motor activity/muscle tone in the nuchal EMG. EEG, electroencephalogram; EOG, electro-oculogram; EMG, electromyogram.

during an undisturbed active waking state (studies of the response of caudal medullary 5-HT neurons – in NRO and NRP – to autonomic nervous system perturbations are described in the following section).

A couple of examples from this series of studies illustrate this point. The activity of DRN 5-HT neurons was examined in response to both increased ambient temperature and pyrogen-induced fever, stimuli eliciting opposite thermoregulatory responses (Fornal *et al.*, 1987). Neuronal activity remained unaffected as ambient temperature was increased from 25°C to 43°C. Following prolonged heat exposure, cats displayed intense continuous panting, relaxation of posture, and a progressive rise in body/brain temperature (range 0.5–2.0°C), yet no change in 5-HT neuronal activity occurred. In a parallel study, a synthetic pyrogen (muramyl dipeptide) was administered systemically, resulting in increased body/brain temperature within 30 minutes and lasting for approximately 6 hours. The peak elevation of body temperature was typically 1.5–2.5°C, yet, once again, no change in neuronal activity was observed.

A large and relatively consistent body of evidence implicates central 5-HT in analgesia, especially those 5-HT neurons localized in the NRM and projecting to the dorsal horn. Accordingly, we examined the activity of NRM 5-HT neurons in behaving cats exposed to a variety of phasic or tonic painful stimuli. No change in neuronal activity was produced by these stimuli relative to the discharge rate during an undisturbed active waking baseline (Auerbach *et al.*, 1985). There was also no change in 5-HT neuronal activity in response to the systemic administration of morphine, in a dose that produced analgesia. These results have been confirmed in studies reporting that identified serotonergic NRM neurons in the rat were not activated by painful stimuli eliciting the withdrawal reflex or by analgesic doses of morphine (Potrebic *et al.*, 1994; Gao *et al.*, 1998).

In summary, these data indicate that mild to relatively strong stressors, drawn from a number of different environmental and physiological categories, do not significantly perturb brainstem serotonergic neurons. It is important, however, to emphasize that stressors do activate the serotonergic system, but that they do so no more than other non-stressful, activating conditions. One way of reconciling this with the behavioral and physiological processes with which 5-HT has been implicated is to invoke the concept of conjunctive activation. For example, 5-HT released during feeding may exert a role in satiety only in the presence of increased release of cholecystokinin (CCK). Employing this line of reasoning, 5-HT released in reaction to exposure to stressors would exert a different effect than it does during feeding because it now interacts with corticosterone rather than CCK. In addition, this acknowledges that such interactions may occur only in specific brain areas (e.g., hypothalamus versus hippocampus). In an important demonstration of this principle, Stutzman and colleagues (Sutzmann *et al.*, 1998) found that the electrophysiological effect of 5-HT on amygdala neurons in rats was abolished in the absence of corticosterone.

Autonomic function

The activity of medullary 5-HT neurons (NRO and NRP) and their relation to autonomic function is of special interest because these neurons have extensive projections to the intermediolateral cell column of the thoracic spinal cord, where they make connections with sympathetic preganglionic neurons, including those that innervate the adrenal medulla (Bacon *et al.*, 1990). Reciprocally, 5-HT medullary neurons receive direct synaptic inputs from the rostral ventrolateral medulla (Zagon, 1993), the primary source of sympathetic activity. A number of studies have shown that 5-HT exerts predominantly an excitatory influence on sympathetic outflow (McCall, 1984; Huangfu *et al.*, 1994). Therefore, medullary 5-HT neurons may serve a homeostatic role in regulating arterial blood pressure and heart rate, body temperature, and blood glucose concentrations.

In this context, we examined the activity of medullary 5-HT neurons during manipulations of the cardiovascular system by producing transient alterations in arterial blood pressure (and sympathetic activity) and by intravenous administration of vasoactive drugs (Martin-Cora *et al.*, 2005). The activity of 5-HT NRO/NRP neurons was not significantly altered in response to phenylephrine (a vasoconstrictor) or sodium nitroprusside (a vasodilator), at doses which elicited marked reflex bradycardia and tachycardia, respectively. Furthermore, no significant changes in neuronal activity were observed after systemic administration of hydralazine, a direct long-acting vasodilator, despite prolonged reflex activation of the sympathetic nervous system, as indicated by sustained increases in heart rate and plasma norepinephrine levels. Acute venous hemorrhage, up to 22.5 percent of estimated total blood volume, also had no effect on the activity of NRO/NRP neurons, despite the fact that plasma catecholamine levels were significantly elevated by this manipulation. Overall, these results do not support a direct role of 5-HT NRO/NRP neurons in the reflex alterations in sympathetic (and parasympathetic) outflow evoked by these cardiovascular manipulations. This suggests that any involvement of 5-HT medullary neurons in cardiovascular regulation is mediated independently of cardiovascular afferent activity.

We have also used insulin-induced hypoglycemia to preferentially activate the sympathoadrenal system (Martin-Cora *et al.*, 2002). Administration of insulin (2–4 IU/kg, i.v.) produced a 50 percent decrease in blood glucose and a near eight-fold increase in plasma epinephrine levels. Surprisingly, the discharge rate of 5-HT NRO/NRP neurons was *reduced* by approximately 40 percent after insulin administration. The subsequent administration of glucose reversed the effect of insulin on blood glucose, plasma epinephrine and neuronal activity. Thus, firing rate was directly correlated with blood glucose and, more importantly, inversely correlated with plasma epinephrine levels. Since medullary 5-HT neurons are thought to subserve a sympathoexcitatory role, these results contradict what would be predicted for these cells. However, animals treated with insulin also displayed diminished muscle tone and signs of muscle weakness, which closely paralleled the decreases in neuronal activity observed after insulin administration. Thus, the effect of systemic insulin on 5-HT neuronal activity may be related to changes in motor output rather than to changes in sympathetic outflow (see below).

Medullary 5-HT neurons may also participate in the central control of respiration through connections with respiratory-related structures in the pons/medulla and phrenic motoneurons (see Chapter 2.10 in this volume). Recent electrophysiological studies have shown that rat medullary 5-HT neurons recorded *in vitro* are highly sensitive to changes in CO_2 and/or pH, suggesting that they may function as 'central chemoreceptors' to modulate respiratory output (Wang *et al.*, 2001). To determine whether medullary 5-HT neurons are chemosensitive *in vivo*, we examined the responses of these neurons to hypercapnia (Veasey *et al.*, 1995), a potent respiratory stimulus which augments ventilatory output by activating primarily central, and to a lesser extent peripheral, chemoreceptors. In support of a role in respiratory modulation, a subset (22 percent) of NRO/NRP neurons was activated in association with increased respiration induced by inhalation of carbon dioxide (Figure 2). The magnitude of the neuronal response was correlated positively with the fraction of inspired CO_2 and was related to ventilatory motor output, specifically inspiratory amplitude. The threshold level of CO_2 for producing a significant increase in firing rate was about 3 percent, suggesting relatively high chemosensitivity. During sleep, the responsiveness of these neurons to systemic CO_2 stimulation was greatly reduced or abolished, and paralleled the diminished ventilatory sensitivity to CO_2. These data suggest that diminished activation of 5-HT neurons during sleep may contribute to the reduced motor responses to hypercapnia and therefore may play an important role in sleep-related breathing disorders, such as sleep apnea and sudden infant death syndrome (SIDS). Additionally, a small subset (25 percent) of 5-HT neurons in the DRN was also activated by hypercapnia (Veasey *et al.*, 1997). Some of these cells, which are known to project to the cervical spinal cord, may directly influence respiratory motor output. Although these studies provide evidence that 5-HT neurons in both the rostral and caudal raphe nuclei are sensitive to hypercarbia under physiological conditions, it is not clear whether the neuronal responses were intrinsic or synaptically mediated by other cells involved in respiratory control, as this would require blockade of afferent inputs to address. Furthermore, although hypercapnia can induce profound effects on cerebral blood flow, cardiac output and arterial pressure, 5-HT neurons in general are unresponsive to alterations in cardiovascular activity, as discussed above. Therefore, it is unlikely that the observed neuronal responses to increased CO_2 are mediated by secondary changes in cardiovascular function induced by hypercarbia. Consistent with our findings in awake animals that some 5-HT neurons are activated by hypercapnia, systemic CO_2 stimulation in cats has been reported to induce *c-fos* expression (a marker of neuronal activation) in the caudal raphe (Larnicol *et al.*, 1994); however, double-labeling experiments are needed to positively identify these cells as serotonergic.

Motor activity

Initial clues that there might be an important relationship between 5-HT neuronal activity and motor activity

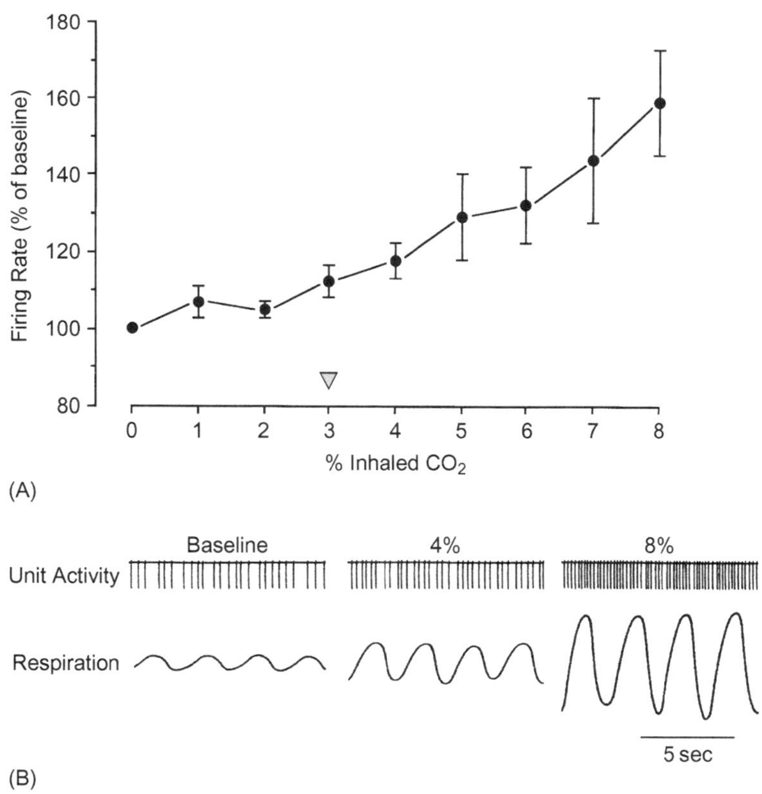

Figure 2 Activation of medullary 5-HT neurons during increased respiration induced by inhalation of carbon dioxide (CO_2). Panel A shows the concentration–response curve to CO_2 for a group of responsive cells. The arrowhead denotes the threshold concentration of CO_2 for producing a significant increase in mean firing rate. Panel B shows the activity of one of these cells, along with respiration, at baseline, when the cat was breathing room air, and then at 4 and 8 percent CO_2. Note the progressive increase in respiratory depth and firing rate of this cell. Adapted from Veasey *et al.* (1995).

came from our early experiments where motor output was drastically altered by using pontine lesions to block the atonia of REM sleep (Trulson *et al.*, 1981) and by microinjecting carbachol into the pons to produce atonia during waking (Steinfels *et al.*, 1983). These data indicated that the activity of rostral 5-HT neurons was more closely related to the level of tonic motor activity than to the behavioral 'state' of the animal.

In recent experiments, we have observed more specific relationships between 5-HT neuronal activity and *phasic* motor function. When cats engage in a variety of types of CPG-mediated oral–buccal activities, such as chewing/biting, licking or grooming, approximately one-fourth of DRN 5-HT neurons increase their activity by as much as two- to five-fold (Figure 3) (Fornal *et al.*, 1996). In contrast, the rest of the 5-HT neurons in this nucleus maintain their slow and rhythmic activity. These increases in neuronal activity often *precede* the initiation of movement by several seconds, but they invariably terminate coincident with the end of the behavioral sequence. Equally impressive is the fact that even brief (1–5 s) spontaneous pauses in these behaviors are accompanied by an immediate decrease in neuronal activity back to baseline levels, or below. During a variety of other non-rhythmic episodic or purposive movements, even those involving oral–buccal responses such as yawning, no increase, or even a decrease, in neuronal activity is seen. During dramatic attentional shifts, such as those occurring during orienting movements in response to imperative stimuli, the activity of DRN 5-HT neurons may fall silent for several seconds (Fornal *et al.*, 1996). This occurs in association with large eye movements, turning the head toward the stimulus, and suppression of any other ongoing behaviors.

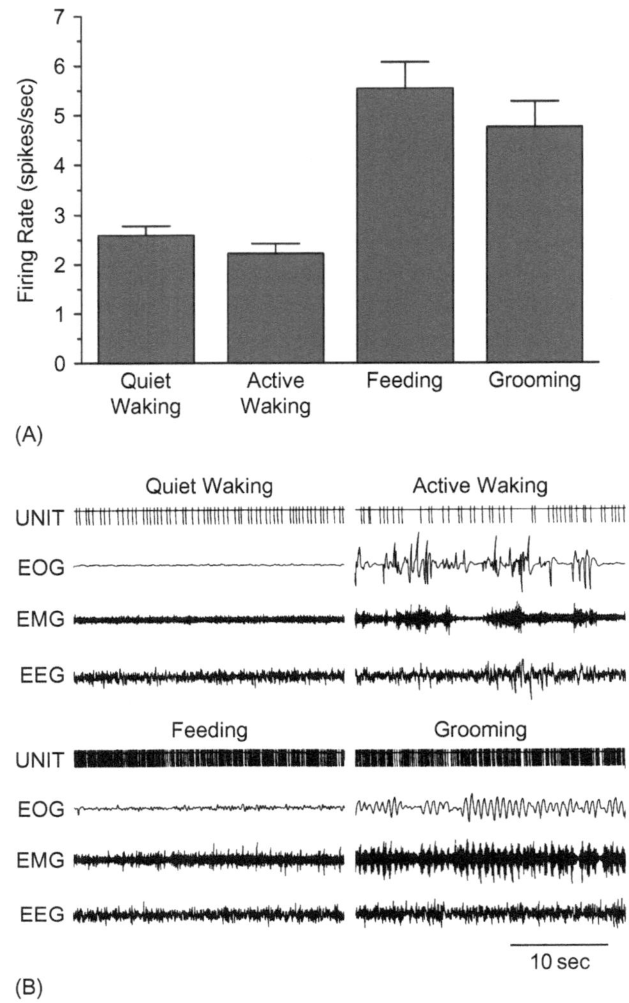

Figure 3 Activation of DRN 5-HT neurons during spontaneous feeding and grooming behavior. Panel A shows mean firing rates for a group of oral–buccal related cells during feeding and grooming in relation to active and quiet waking baseline firing rates. Panel B shows the activity of one of these cells across all four behavioral conditions. The firing rate of this cell was increased approximately three-fold during feeding and grooming, as compared to waking. The rhythmic nature of the grooming can be seen in the EOG and nuchal EMG traces. Also, note the decrease in unit discharge rate and regularity during active waking (cat moving) as compared to quiet waking. EEG, electroencephalogram; EOG, electro-oculogram; EMG, electromyogram. Adapted from Fornal *et al.* (1996).

Based on these data demonstrating a strong relationship between DRN 5-HT neurons (those that innervate forebrain) and motor activity, we explored these same issues for NRO and NRP neurons (those that innervate the spinal cord). In contrast to mesencephalic 5-HT neurons, where only a subgroup is activated during CPG-mediated behaviors, virtually all medullary serotonergic neurons are activated under at least some of these conditions (Veasey *et al.*, 1995). The magnitude of activation, however, is much less (i.e., 50–100 percent above baseline versus 100–400 percent above baseline for the DRN). In this context, it may be important to note that the basal, quiet waking discharge rate of medullary serotonergic neurons is approximately twice that of mesencephalic serotonergic neurons (5–6 vs 2–3 spikes/s). There also appears to be at least some degree of response specificity for these neurons. Thus, virtually all medullary serotonergic neurons are activated during treadmill-induced locomotion (Figure 4), but only subgroups are activated during increased respiratory output induced by hypercapnea, or during chewing/licking. Many of these individual neurons are activated in association with more than one of these motor activities. Initial indications are that, like the rostral 5-HT neurons, at least some of these caudal 5-HT neurons can be activated by somotosensory stimulation. It is still not clear whether there is a preferential body area (e.g., trunk or limbs versus head) where such stimulation is most effective. Finally, *unlike* DRN 5-HT neurons, NRO and NRP

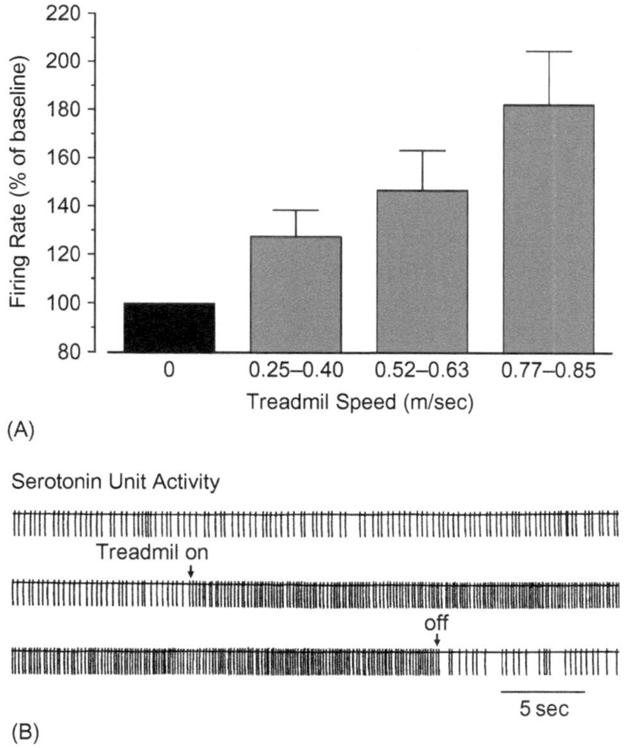

Figure 4 Activation of medullary 5-HT neurons during treadmill locomotion. Panel A shows the relationship between firing rate and speed of locomotion for a group of locomotor-related NRO/NRP cells. Panel B shows the activity of one of these cells during a locomotor trial. This cell increased its firing rate immediately at the onset of running on the treadmill (speed = 0.4 m/s). Neuronal activity returned immediately back to, or just below, baseline levels when the treadmill was stopped. Adapted from Veasey *et al.* (1995).

neuronal activity is not suppressed during orienting responses. This may reflect a lack of involvement of the spinal cord-projecting neurons in functions associated with cognition, such as attention.

During treadmill-induced locomotion the magnitude of increased neuronal activity showed a strong positive correlation with the speed of running (from 0.25 to 0.85 m/s) (Veasey *et al.*, 1995). A few cells even displayed an increased discharge that was in phase with the step cycle. Finally, the excitatory neuronal response during treadmill locomotion was precisely coincident with the onset of the treadmill, and remained stable for a given speed. When the treadmill was stopped, unit activity returned immediately to, or just below, baseline. This suggests that the firing rate of these neurons is tightly coupled to motor activation rather than autonomic activity, such as heart rate and blood pressure, which remain elevated after locomotor activity has stopped.

During feeding, two distinct patterns of neuronal activation were observed for NRO/NRP neurons (Veasey *et al.*, 1995). For most cells the increase occurred precisely at feeding onset and was maintained throughout the feeding period. Unit activity abruptly returned to baseline levels with the cessation of eating. Thus, the activity of these neurons appears to be temporally correlated with ingestive behaviors such as chewing, licking or swallowing. In contrast, a few cells demonstrated gradual increases in unit activity during feeding, reaching a plateau 60–80 s after feeding onset, and then only gradually returned to baseline minutes after the termination of feeding. The elevated activity of this subset of neurons is more temporally related to ongoing digestive processes, such as gastric secretion and motility. We hypothesize that these neurons play a role in gastrointestinal function via projections to the parasympathetic preganglionic neurons in the dorsal vagal complex.

Finally, when DRN neurons were examined under identical conditions, none were activated during treadmill-induced locomotion or cold exposure (Veasey *et al.*, 1997; Martin-Cora *et al.*, 2000), but some were activated during carbon dioxide-induced hyperpnea (Veasey *et al.*, 1997).

Fatigue

Physical exercise cannot be maintained indefinitely, and eventually fatigue will terminate motor output. Traditionally, peripheral factors, such as intramuscular depletion of energy substrates (e.g., glycogen) or accumulation of metabolic byproducts (e.g., lactic acid), have been thought to be responsible for fatigue, although there is now clear evidence that changes within the central nervous system can contribute to fatigue (Davis and Bailey, 1997). Besides its direct role in motor output, it is becoming increasingly clear that the debilitating fatigue associated with some chronic diseases (e.g., chronic fatigue syndrome and multiple sclerosis) also appears to be mainly of central origin (Davis and Bailey, 1997; Swain, 2000), and is moreover an important factor in clinical depression (Swain, 2000).

Central fatigue consists principally of a failure to maintain full motor unit recruitment and/or optimal motor unit discharge for muscle force production (Sale, 1987), and has been linked to an inadequate descending drive from the motor cortex (Gandevia *et al.*, 1996). Serotonin is the neurotransmitter most often implicated in human forms of fatigue and in animal experiments examining exercise-induced fatigue (Newsholme and Blomstrand, 1995; Davis and Bailey, 1997; Davis *et al.*, 2000). Recent work in our laboratory suggests that 5-HT neurons in the NRO and NRP may play an important role in various aspects of central fatigue (Fornal *et al.*, 2006). These neurons are known to facilitate both the excitability and the recruitment of α-motoneurons (Jacobs and Fornal, 1997b). As described above, in the awake cat the activity of these 5-HT neurons is suppressed in relation to diminished motor capacity (i.e., experimentally induced muscle hypotonia/atonia, and muscle weakness). These findings have led us to hypothesize that diminished activity of the descending medullary 5-HT system may be an important component of central fatigue. During prolonged treadmill locomotion (0.4 m/s for 30–60 min), we have found that the activity of some NRO/NRP 5-HT neurons progressively decreases from baseline levels (Fornal *et al.*, 2006). An additional more rapid decline is often observed immediately after the treadmill is turned off, at the time the cat can no longer keep pace – i.e., 'fatigue'. Neuronal activity in these animals gradually returns to baseline within 30–60 min. During this recovery period, activity can be reactivated momentarily by turning on the treadmill, indicating that increased motor output is still capable of driving these cells, albeit at a reduced rate. Preliminary evidence suggests that these effects of sustained locomotor activity may be unique to NRO/NRP 5-HT neurons, inasmuch as DRN 5-HT neurons do not show this decline. We hypothesize that during prolonged locomotor activity a central mechanism is activated inhibiting the output of medullary 5-HT neurons, which in turn leads to a disfacilitation of α-motoneuron excitability (and autonomic dysregulation), thus contributing to the development of fatigue.

Some of the factors that might be responsible for this fatigue-related decrease in 5-HT neuronal activity and, thereby, contribute to the development of fatigue are: (1) whole body carbohydrate stores which become depleted during prolonged exercise (glucose administration to animals exhibiting signs of fatigue may reverse these effects); (2) a progressive increase in core body temperature, which is thought to limit physical output (cooling or antipyretic drugs may therefore reverse these alterations in 5-HT neuronal activity); and (3) hypoxia, which may be important in the decreased 5-HT neuronal activity seen during fatigue.

Discussion

A general relationship exists between level of tonic motor activity and 5-HT neuronal activity across all groups of 5-HT neurons that we have studied. Superimposed upon this in some neurons is an additional relationship in which a further, often dramatic, activation is seen in association with repetitive, CPG-mediated behaviors. The exact nature of this relationship varies both with 5-HT grouping (e.g., locomotor-related NRP neuronal activity versus grooming-related DRN neuronal activity) and within a particular group (e.g., respiratory-related and feeding-related NRP neuronal activity). The primary function of this increased 5-HT neuronal activity in association with tonic and repetitive motor output is to coordinate autonomic and neuroendocrine function in association with the existing motor demand, and to suppress activity in most sensory information processing channels (Prochazka, 1989; Nelson, 1996).

There are several reasons for believing that the facilitation of motor output by 5-HT is general in nature. First, 5-HT neuronal activity is generally tonically elevated during the execution of a particular behavior, suggesting that 5-HT is facilitating the 'behavior' rather than a particular motor act or muscle group that is a component of the behavior. Secondly, the distribution of 5-HT axon terminal innervation of the brainstem and spinal cord is consistent with the involvement of 5-HT in patterned movement employing gross skeletal muscles rather than movements using finer or more discrete muscles, such as those controlling the eyes or the digits. Thus, in the spinal cord there is a denser input to the medial portion of the ventral horn, where axial motoneurons serving the trunk and limbs are found, compared to the lateral portions, where distal motoneurons serving paws and digits are found (Steinbusch, 1981). Thirdly, there is evidence that spinally projecting medullary 5-HT neurons collateralize at various levels of the cord, suggesting a broad relationship to motor activity (Allen and Cechetto, 1994). Finally, the fact that single medullary 5-HT neurons may project to both the intermediolateral column of the spinal cord as well as to the ventral horn suggests an integrative function involving motor and autonomic control (Allen and Cechetto, 1994).

Several important functions may be served by these 5-HT inputs to motor structures. They may smooth motor outputs and may also obviate the need for continuous repetitive excitatory inputs to maintain a continuous output in motor systems. By augmenting weak or polysynaptic inputs, 5-HT may also bring motoneurons to their firing threshold. The anticipation of motor activity by 5-HT neurons suggests that they may serve a 'priming', and possibly a conditioning, function for motor output. (It is not yet clear if NRO/NRP 5-HT neurons manifest this anticipatory response, as do DRN 5-HT neurons.) The simultaneous inhibition of irrelevant sensory information processing acts to suppress inputs that might disrupt motor output. Finally, 5-HT's involvement in autonomic and neuroendocrine regulation serves a support function for the demands associated with changes in the level of motor output, such as increased oxygenation of the blood and redistribution of blood flow to skeletal muscles, or increased glycogenolysis for maintaining a stable glucose supply to the brain.

One of the most impressive aspects of 5-HT's involvement in motor control is its generality across phylogeny.

There is evidence in a variety of invertebrate species that 5-HT may play a broad, and even integrative, role in such diverse functions as postural control, swimming, and feeding. As discussed above with regards to mammals, there is evidence that in invertebrates 5-HT exerts its effects on behavior at multiple levels: on motoneurons, muscles, CPGs, cardiovascular system, etc. (Kupfermann and Weiss, 1981; Lent, 1985).

In sum, we hypothesize that the primary role of the 5-HT system is to provide coordination of motor, autonomic and sensory processes, which are activated with central motor commands. Furthermore, we believe that *all other* behavioral and physiological functions that have been associated with 5-HT in the brain and spinal cord are derivative of this fundamental relationship.

Acknowledgement

The authors' research described in this chapter was supported primarily by research grants from the USPHS (NIH – NIMH). The authors wish to thank the many contributions of research assistants, graduate students and postdoctoral fellows who have worked in our laboratory over the years.

References

Allen, G.V. and Cechetto, D.F. (1994) Serotoninergic and non-serotoninergic neurons in the medullary raphe system have axon collateral projections to autonomic and somatic cell groups in the medulla and spinal cord. J. Comp. Neurol., 350: 357–366.

Auerbach, S., Fornal, C.A. and Jacobs, B.L. (1985) Response of serotonin-containing neurons in nucleus raphe magnus to morphine, noxious stimuli, and periaqueductal gray stimulation in freely moving cats. Exp. Neurol., 88: 609–628.

Bacon, S.J., Zagon, A. and Smith, A.D. (1990) Electron microscopic evidence of a monosynaptic pathway between cells in the caudal raphé nuclei and sympathetic preganglionic neurons in the rat spinal cord. Exp. Brain Res., 79: 589–602.

Beck, S.G., Pan, Y.Z., Akanwa, A.C. and Kirby, L.G. (2004) Median and dorsal raphe neurons are not electrophysiologically identical. J. Neurophysiol., 91: 994–1005.

Davis, J.M., Alderson, N.L. and Welsh, R.S. (2000) Serotonin and central nervous system fatigue: nutritional considerations. Am. J. Clin. Nutr., 72: 573S–578S.

Davis, J.M. and Bailey, S.P. (1997) Possible mechanisms of central nervous system fatigue during exercise. Med. Sci. Sports Exerc., 29: 45–57.

Fornal, C.A., Litto, W.J., Morilak, D.A. and Jacobs, B.L. (1987) Single-unit responses of serotonergic dorsal raphe nucleus neurons to environmental heating and pyrogen administration in freely moving cats. Exp. Neurol., 98: 338–403.

Fornal, C.A., Metzler, C.W., Marrosu, F., Ribeiro-do-Valle, L.E. and Jacobs, B.L. (1996) A subgroup of dorsal raphe serotonergic neurons in the cat is strongly activated during oral–buccal movements. Brain Res., 716: 123–133.

Fornal, C.A., Martin-Cora, F.J. and Jacobs, B.L. (2006) 'Fatigue' of medullary but not mesencephalic raphe serotonergic neurons during locomotion in cats. Brain Res., 1072: 55–61.

Gandevia, S.C., Allen, G.M., Butler, J.E. and Taylor, J.L. (1996) Supraspinal factors in human muscle fatigue: evidence for suboptimal output from the motor cortex. J. Physiol., 490: 529–536.

Gao, K., Chen, D.O., Genzen, J.R. and Mason, P. (1998) Activation of serotonergic neurons in the raphe magnus is not necessary for morphine analgesia. J. Neurosci., 18: 1860–1868.

Hajos, M., Gartside, S.E., Villa, A.D. and Sharp, T. (1995) Evidence for a repetitive (burst) firing pattern in a sub-population of 5-hydroxytryptamine neurons in the dorsal and median raphe nuclei of the rat. Neuroscience, 69: 189–197.

Huangfu, D., Hwang, L.J., Riley, T.A. and Guyenet, P.G. (1994) Role of serotonin and catecholamines in sympathetic responses evoked by stimulation of rostral medulla. Am. J. Physiol., 226: R338–R352.

Jacobs, B.L. and Azmitia, E.C. (1992) Structure and function of the brain serotonin system. Physiol. Rev., 72: 165–229.

Jacobs, B.L. and Fornal, C.A. (1991) Activity of serotonergic neurons in the behaving animal. Pharmacol. Rev., 43: 563–578.

Jacobs, B.L. and Fornal, C.A. (1997a) Physiology and pharmacology of brain serotonergic neurons. In: Baumgarten, H.G. and Göthert, M. (Eds.), Handbook of Experimental Pharmacology, Serotonergic Neurons and 5-HT Receptors in the CNS. Springer-Verlag, Berlin, Vol. 129, pp. 91–116.

Jacobs, B.L. and Fornal, C.A. (1997b) Serotonin and motor activity. Curr. Opin. Neurobiol., 7: 820–825.

Jacobs, B.L. and Fornal, C.A. (1998) An integrative role for serotonin in the central nervous system. In: Lydic, R.B. and Baghdoyan, H.A. (Eds.), Handbook of Behavioral State Control: Cellular and Molecular Mechanisms. CRC Press, Boca Raton, FL, pp. 181–193.

Kupfermann, I. and Weiss, K.R. (1981) The role of serotonin in arousal of feeding behavior in *Aplysia*. In: Jacobs, B.L. and Gelperin, A. (Eds.), Serotonin Neurotransmission and Behavior. MIT Press, Cambridge, MA, pp. 255–287.

Larnicol, N., Wallois, F., Berquin, P., Gros, F. and Rose, D. (1994) c-*fos*-like immunoreactivity in the cat's neuraxis following moderate hypoxia or hypercapnia. J. Physiol., 88: 81–88.

Lent, C.M. (1985) Serotonergic modulation of the feeding behavior of the medicinal leech. Brain Res. Bull., 14: 643–655.

Martin-Cora, F.J., Fornal, C.A., Metzler, C.W. and Jacobs, B.L. (2000) Single-unit responses of serotonergic medullary and pontine raphe neurons to environmental cooling in freely moving cats. Neuroscience, 98: 301–309.

Martin-Cora, F.J., Fornal, C.A., Metzler, C.W. and Jacobs, B.L. (2002) Insulin-induced hypoglycemia decreases single-unit activity of serotonergic medullary raphe neurons in freely moving cats: relationship to sympathetic and motor output. Eur. J. Neurosci., 16: 722–734.

Martin-Cora, F.J., Fornal, C.A. and Jacobs, B.L. (2005) Single-unit responses of serotonergic medullary raphe neurons to cardiovascular challenges in freely moving cats. Eur. J. Neurosci., 22: 3195–3204.

McCall, R.B. (1984) Evidence for a serotonergically mediated sympathoexcitatory response to stimulation of medullary raphe nuclei. Brain Res., 311: 131–139.

Nelson, R.J. (1996) Interactions between motor commands and somatic perception in sensorimotor cortex. Curr. Opin. Neurobiol., 6: 801–810.

Newsholme, E.A. and Blomstrand, E. (1995) Tryptophan, 5-hydroxytryptamine and a possible explanation for central fatigue. Adv. Exp. Med. Biol., 384: 315–320.

Potrebic, S.B., Field, H.L. and Mason, P. (1994) Serotonin immunoreactivity is contained in one physiological cell class in the rat rostral ventromedial medulla. J. Neurosci., 14: 1655–1665.

Prochazka, A. (1989) Sensorimotor gain control: a basic strategy of motor systems? Prog. Neurobiol., 33: 281–307.

Sakai, K. and Crochet, S. (2001) Differentiation of presumed serotonergic dorsal raphe neurons in relation to behavior and wake–sleep states. Neuroscience, 104: 1141–1155.

Sale, D.G. (1987) Influence of exercise and training on motor unit activation. Exerc. Sport Sci. Rev., 15: 95–151.

Steinbusch, H.W. (1981) Distribution of serotonin-immunoreactivity in the central nervous system of the rat-cell bodies and terminals. Neuroscience, 6: 557–618.

Steinfels, G.F., Heym, J., Strecker, R.E. and Jacobs, B.L. (1983) Raphe unit activity in freely moving cats is altered by manipulations of central but not peripheral motor systems. Brain Res., 279: 77–84.

Sutzmann, G.E., LeDoux, L.E., Magarinos, A.M. and McEwen, B.S. (1998) Serotonin modulation of sensory inputs to the lateral amygdala: dependency on corticosterone. J. Neurosci., 18: 9529–9538.

Swain, M.G. (2000) Fatigue in chronic disease. Clin. Sci., 99: 1–8.

Trulson, M.E., Jacobs, B.L. and Morrison, A.R. (1981) Raphe unit activity during REM sleep in normal cats and in pontine lesioned cats displaying REM sleep without atonia. Brain Res., 226: 75–91.

Urbain, N., Creamer, K. and Debonnel, G. (2006) Electrophysiological diversity of the dorsal raphe cells across the sleep–wake cycle of the rat. J. Physiol., 573: 679–695.

Veasey, S.C., Fornal, C.A., Metzler, C.W. and Jacobs, B.L. (1995) Response of serotonergic caudal raphe neurons in relation to specific motor activities in freely moving cats. J. Neurosci., 15: 5346–5359.

Veasey, S.C., Fornal, C.A., Metzler, C.W. and Jacobs, B.L. (1997) Single-unit responses of serotonergic dorsal raphe neurons to specific motor challenges in freely moving cats. Neuroscience, 79: 161–169.

Wang, W., Tiwari, J.K., Bradley, S.R., Zaykin, R.V. and Richerson, G.B. (2001) Acidosis-stimulated neurons of the medullary raphe are serotonergic. J. Neurophysiol., 85: 2224–2235.

Zagon, A. (1993) Innervation of serotonergic medullary raphe neurons from cells of the rostral ventrolateral medulla in rats. Neuroscience, 55: 849–867.

CHAPTER 2.2

Electrophysiology of Serotonin Receptors

Gerard J. Marek*

Abbott Laboratories, Neuroscience Development, Abbott Park, Illinois

Abstract: The 14 different serotonin or 5-hydroxytryptamine (5-HT) receptors are widely distributed throughout the central nervous system. Of these 14 serotonin receptors, 13 mediate the slow modulatory action of 5-HT upon fast excitatory or inhibitory effects of glutamate/aspartate and γ-aminobutyric acid (GABA)/glycine, respectively. The presence of different 5-HT receptors in discrete cellular compartments through the neuroaxis allows for an integrated effect of serotonergic neurotransmission on sleep and arousal processes, executive functions of the prefrontal cortex, affective behavior, aggression, learning and memory, and modulation of pain. This is accomplished though specific 5-HT receptors controlling the pacemaker functions of midbrain and brainstem 5-HT-containing neurons which project throughout the neuroaxis, as discussed elsewhere. Different 5-HT receptors are localized on local or projection GABAergic neurons in addition to the different compartments (axons, axon terminals, dendrites, cell bodies) of glutamatergic local or output cells. 5-HT receptors are also present either on other biogenic amine-containing cells or on cells modulating the other biogenic amine-containing cells to modulate the electrophysiological action of norepinephrine, dopamine and acetylcholine throughout the neuroaxis.

Keywords: 5-HT receptors, raphe nuclei, prefrontal cortex, hippocampus, striatum, amygdala, autoreceptors, heteroceptors.

Abbreviations: BLA, basolateral amygdala; BNST, bed nucleus of the stria terminalis; EPSC/EPSP, excitatory postsynaptic current/potential; 5-HT, 5-hydroxytryptamine; GABA, γ-aminobutyric acid; IPSC/IPSP, inhibitory postsynaptic current/potential; mPFC, medial prefrontal cortex; mGlu, metabotropic glutamate receptors; MAOIs, monoamine oxidase inhibitors (MAOIs); PFC, prefrontal cortex; SSRIs, selective serotonin reuptake inhibitors; SNRIs, serotonin-norepinephrine reuptake inhibitors; SNpr, substantia nigra pars reticulate; SNpc, substantial nigra pars compacta; STN, subthalamic nucleus; TTX, tetrodotoxin; TCAs, tricyclic antidepressants; VTA, ventral tegmental area; VPM, ventroposterior medial nucleus; VPL, ventroposterior lateral nucleus.

Introduction

The primary focus of this chapter will be on identifying those 5-hydroxytryptamine (5-HT) receptor subtypes that appear to have the greatest relevance in mediating salient behavioral effects of psychotherapeutic drugs or mediating profound physiological/behavioral changes. A secondary and related focus will be to understand the interplay of different 5-HT receptors within both microcircuits and macrocircuits where multiple 5-HT receptors either interfere with the function of other 5-HT receptors or add to/synergize with the effects of other 5-HT receptors. This chapter will concentrate on the best-characterized 5-HT receptor subtypes, but will also point out questions and issues which remain unanswered or not addressed.

The transduction pathways used by the fourteen 5-HT receptors is discussed elsewhere. However, there are five 5-HT receptor subtypes that tend to be coupled to G_i/G_o proteins and are negatively coupled to adenylyl cyclase and the production of cAMP (the $5HT_{1A}$, $5\text{-}HT_{1B}$, $5\text{-}HT_{1D}$, $5\text{-}HT_{1E}$ and $5\text{-}HT_{1F}$ receptors). The $5\text{-}HT_5$ receptor family may also share these transduction pathways. These receptors tend to have direct inhibitory effects. The $5\text{-}HT_3$ receptor is the only ionotropic 5-HT receptor, and has direct excitatory effects due to net inward flux of sodium. The $5\text{-}HT_2$ receptor family members ($5\text{-}HT_{2A}$, $5\text{-}HT_{2B}$, $5\text{-}HT_{2C}$) are coupled to G_q/G_{11} proteins often linked to phospholipase C. These receptors tend to depolarize cells and/or increase the excitability of the cellular compartments in which they are found. The remaining

*E-mail: gerard.marek@abbott.com

Christian Müller & Barry Jacobs (Eds.)
Handbook of Behavioral Neurobiology of Serotonin
ISBN 978-0-12-374634-4

DOI: 10.1016/B978-0-12-374634-4.00001-0

5-HT receptors ($5\text{-}HT_4$, $5\text{-}HT_6$, and $5\text{-}HT_7$) are coupled to G_s proteins and a positive modulation of adenylyl cyclase that usually results in direct excitatory effects.

$5\text{-}HT_{1A}$ receptors

The $5\text{-}HT_{1A}$ receptor has been one of the most studied members of the serotonin receptor family. This receptor plays a profound role as a somatodendritic autoreceptor on serotonin-containing neurons throughout the midbrain and brainstem. These $5\text{-}HT_{1A}$ autoreceptors modulate the pacemaker activity of these neurons in a negative feedback fashion, and are discussed elsewhere (see Chapter 2.1 of this volume). In other regions of the brain, such as the prefrontal cortex, neocortex, hippocampal formation and amygdala, $5\text{-}HT_{1A}$ receptors exert direct inhibitory effects. In these different brain regions, the $5\text{-}HT_{1A}$ receptors usually are expressed in somatodendritic compartments of principal cells. In the neocortex, $5\text{-}HT_{1A}$ receptors are localized to the initial axon segment (or axon hillock) of layer III and layer V pyramidal cells (Azmitia *et al.*, 1996). In the hippocampus, activation of $5\text{-}HT_{1A}$ receptors is known to suppress the activity of CA3 pyramidal cells, and this is an effect that is enhanced by many antidepressant drugs (Haddjeri *et al.*, 1998). At a macro-level, activation of $5\text{-}HT_{1A}$ receptors reduces the frequency of hippocampal rhythmic slow activity, analogous to the actions of anxiolytic drugs such as benzodiazepines.

Hippocampal $5\text{-}HT_{1A}$ receptors: modulation by antidepressant drugs

Differential changes in presynaptic and postsynaptic $5\text{-}HT_{1A}$ receptors appear relevant for understanding how serotonergic function may be modified by antidepressant drugs. The desensitization of somatodendritic $5\text{-}HT_{1A}$ autoreceptors was postulated to explain the delayed therapeutic effects of selective serotonin reuptake inhibitors (SSRIs), serotonin-norepinephrine reuptake inhibitors (SNRIs) and monoamine oxidase (MAO) inhibitors (Blier *et al.*, 1986; Chaput *et al.*, 1986; Rueter *et al.*, 1998; Pineyro and Blier, 1999). The activation of somatodendritic $5\text{-}HT_{1A}$ receptors in serotonin-containing nuclei decreased the basal firing rate of these neurons and attenuated the enhancement of ambient synaptic 5-HT concentrations throughout the forebrain. However, the enhancement of the inhibitory effects of $5\text{-}HT_{1A}$ receptor activation has been found to be a general feature following the subchronic administration (2–4 weeks) of most antidepressant drugs, including tricyclic antidepressants (de Montigny and Aghajanian, 1978), SSRIs (Blier and de Montigny, 1983; Blier and Bouchard, 1992), SNRIs (Rueter *et al.*, 1998), MAO inhibitors, atypical antidepressants which block $5\text{-}HT_3$ and $5\text{-}HT_{2A/2C}$ receptors (Blier *et al.*, 1984; Haddjeri *et al.*, 1997), electroconvulsive shock (Blier and Bouchard, 1992), lithium (Haddjeri *et al.*, 2000) and $5\text{-}HT_{1A}$ receptor partial agonists (Blier and de Montigny, 1987).

Several antidepressant drug combination treatments that have been investigated in clinical practice have also been examined to see whether the drug combination speeds up or increases the maximal enhancement of hippocampal $5\text{-}HT_{1A}$ receptor function. The co-administration of mirtazapine and the SSRI paroxetine appeared to accelerate the enhancement of $5\text{-}HT_{1A}$ receptor function in CA3 pyramidal cells, developing over a period of 3–4 weeks, in addition to being more effective than either drug alone (Besson *et al.*, 2000). This type of improved therapeutic response in the clinic for the treatment of depressed patients with the combination of either mirtazapine or mianserin has been successful in double-blind, placebo-controlled studies as described elsewhere (Marek *et al.*, 2003). Another medication well known to enhance the clinical effect of antidepressant drugs, especially in treatment-refractory patients, is the addition of lithium carbonate to ongoing treatment with a variety of antidepressant drugs. The addition of lithium to the tricyclic antidepressant (TCA) imipramine, the SSRI paroxetine and the monoamine oxidase inhibitor (MAOI) tranylcypromine has also been found to increase the disinhibition of dorsal hippocampal CA3 pyramidal neurons over that seen with either treatment alone (Haddjeri *et al.*, 2000). Addition of the β-adrenergic receptor antagonist and $5\text{-}HT_{1A}$ receptor antagonist pindolol to SSRIs results in greater increases in synaptic 5-HT during the early stages of treatment, and can accelerate the development of an enhanced inhibitory response in CA3 pyramidal cells to $5\text{-}HT_{1A}$ receptor activation. Some of the clinical trials suggest that this combination of SSRIs and pindolol for depressed patients may also accelerate the development of therapeutic responses, though this is not without controversy (Artigas *et al.*, 1996). In addition, a $5\text{-}HT_{1A}$ receptor agonist and $5\text{-}HT_{2A}$ receptor antagonist flibanserin appeared to accelerate both the recovery of firing rate for dorsal raphe neurons with chronic flibanserin administration and the enhancement of the $5\text{-}HT_{1A}$ response in CA3 neurons of the rat hippocampus (Rueter and Blier, 1999). Clinical results testing this hypothesis have not been reported.

$5\text{-}HT_{1A}$ receptors induce postsynaptic hyperpolarizations throughout the limbic forebrain. In CA1 pyramidal cells of the hippocampus, one of the most conspicuous effects of $5\text{-}HT_{1A}$ receptor activation is to produce a hyperpolarization by a postsynaptic mechanism. This 5-HT-induced hyperpolarization frequently is followed by a depolarizing response mediated by 5-HT_4 receptor activation (Andrade *et al.*, 1986; Andrade and Nicoll, 1987; Andrade and Chaput, 1991a; Torres *et al.*, 1996).

Medial prefrontal cortex (mPFC) 5-HT_{1A} receptors

The 5-HT_{1A} receptor message ribonucleic acid (mRNA) and binding sites are well distributed throughout the prefrontal cortex and the neocortex (Pompeiano *et al.*, 1992). Estimates from mRNA *in situ* hybridization studies are that ~60 percent of glutamatergic cells and ~25 percent of GABAergic cells express 5-HT_{1A} receptor mRNA in the rat prefrontal cortex (Santana *et al.*, 2004). Single-cell polymerase chain reaction (PCR) studies suggested that nearly all prefrontal cortical pyramidal cells (92 percent of 26 cells) expressed 5-HT_{1A} receptors (Feng *et al.*, 2001). Activation of 5-HT_{1A} receptors results in a hyperpolarization of the cell (Araneda and Andrade, 1991; Ashby *et al.*, 1994; Tanaka and North, 1994). In a sample of layer V mPFC cells, 22 of 90 exhibited a pure hyperpolarizing response while 26 of 90 cells were first hyperpolarized, followed by a depolarization dependent on 5-HT_{2A} receptor activation (Araneda and Andrade, 1991). In supragranular cortical layers (layers II–III), 50 percent of 28 sampled pyramidal cells were hyperpolarized by 5-HT with pharmacological characteristics consistent with 5-HT_{1A} receptors. With respect to supragranular interneurons, 33 percent of 54 cells were hyperpolarized by 5-HT. About 63 percent of the interneurons with axons restricted to layer I were hyperpolarized by 5-HT. Approximately 50 percent of the 54 cells sampled were depolarized by 5-HT_{2A} receptor activation. Thus, 5-HT_{1A} receptors exert an important physiological role in prefrontal cortical and neocortical functioning. However, the hyperpolarizing effects of 5-HT_{1A} receptor activation appear to work in tandem with the increases in cellular excitability induced by 5-HT_{2A} receptors. In this context, behavioral effects such as the suppression by 5-HT_{1A} receptor agonists of head shakes induced by serotonergic hallucinogens (which all share in common activation of 5-HT_{2A} receptors) are not surprising (Darmani *et al.*, 1990).

5-HT_{1A} receptors in the extended amygdala

In addition to a potential role of prefrontal cortical 5-HT_{1A} receptors in mediating behavior relevant to fear, anxiety and aggression, 5-HT_{1A} receptors in the extended amygdala may contribute to 'emotional' behavior. A number of studies suggest that activation of 5-HT_{1A} receptors suppress neurotransmitter release without altering the postsynaptic cell when recording from the basolateral nucleus (BLA). One study using a slice preparation found that 8-OH-DPAT and non-selective 5-HT_{1A} receptor agonists suppressed excitatory postsynaptic potentials (EPSPs) without altering the resting membrane potential and neuronal resistence (Cheng *et al.*, 1998). These same investigators demonstrated that 5-HT_{1A} receptor activation could block the potentiation of excitatory synaptic transmission induced by β-adrenergic receptor activation at a site downstream from cAMP formation (Wang *et al.*, 1999). A somewhat different picture emerged from studies using mechanically dissociated BLA neurons which preserve native presynaptic nerve terminals. 5-HT_{1A} receptor activation was found to suppress the frequency of $GABA_A$-mediated inhibitory postsynaptic currents (IPSCs) without altering the miniature IPSC amplitude (Kishimoto *et al.*, 2000). Consistent with presynaptic control of neurotransmitter release, activation of 5-HT_{1A} receptors has been found to inhibit voltage-dependent calcium currents of the N- and P/Q-types rather than L-type (Lin *et al.*, 2001). A convergence of 5-HT_3 and 5-HT_{1A} receptor function was found for mechanically dissociated BLA neurons as the 5-HT_{1A} receptor agonist 8-OH-DPAT was able to suppress the transient facilitation of miniature IPSC frequency induced by the 5-HT3 receptor agonist mCPBG via an apparent membrane-delimited mechanism (Koyama *et al.*, 2002a).

Not all 5-HT_{1A} receptor-mediated responses in the extended amygdala involve inhibition of transmitter release. The anterolateral bed nucleus of the stria terminalis (BNST AL) receives a relatively heavy serotonin projection. Experiments with brain slices suggested that the main action of 5-HT and 5-CT in this area was evoking a hyperpolarization without an alteration in input resistence (Levita *et al.*, 2004). This action of 5-HT and 5-CT was blocked by the relatively selective 5-HT_{1A} receptor antagonist WAY100635. Curiously, though, 8-OH-DPAT did not mimic the effects of 5-HT and 5-CT. The importance of these data from a translation to animal behavior is that infusion of 5-CT into the BNST did reduce the acoustic startle response without affecting the animals' general activity level.

5-HT_{1A} receptors and the basal ganglia

5-HT_{1A} receptors are also present in motor circuitry. The subthalamic nucleus (STN) is an important structure, since high-frequency stimulation of this brain region is useful in patients with Parkinson's disease. While 5-HT_{2C} and 5-HT_7 receptors appear to depolarize or induce inward currents in STN neurons (Xiang *et al.*, 2005), activation of 5-HT_{1A} receptors induces hyperpolarization or outward currents (Stanford *et al.*, 2005; Shen *et al.*, 2007). The globus pallidus (as well as the substantia nigra pars reticulata, SNpr) are innervated by fibers originating from the subthalamic nucleus. While globus pallidal neurons have recently been shown to be depolarized by 5-HT_7 receptor activation, other neurons are hyperpolarized by 5-HT_{1A} receptor action through a postsynaptic action (Chen *et al.*, 2008; Hashimoto and Kita, 2008).

5-HT_{1B} receptors

5-HT_{1B} receptors ($5\text{-HT}_{1D\beta}$ gene product) play a major role throughout the brain, acting as autoreceptors in the terminals of serotonin-containing cells originating from the brainstem raphe nuclei (see Chapter 2.1). Interestingly, 5-HT_{1B} receptors localized at postsynaptic sites relative to serotonin-containing nerve terminals appear to make up the vast majority of 5-HT_{1B} receptors throughout the brain, since lesions of raphe neurons do not decrease the number of 5-HT_{1B} receptor binding sites. 5-HT_{1B} receptors also are used as heteroceptors in many cells throughout the brain to inhibit the release of neurotransmitters other than serotonin. The striking dissociation between localization of the mRNA and receptor binding for rodent 5-HT_{1B} receptors, together with results of lesion studies, have established that this 5-HT receptor is predominantly localized to axon terminals (Boschert *et al.*, 1994).

In the anterior cingulate cortex, 5-HT suppresses electrically-stimulated EPSPs/excitatory postsynaptic currents (EPSCs) recorded from layer V pyramidal cells, and does not have an effect on evoked inhibitory postsynaptic potentials (IPSPs)/IPSCs when stimulating either in the white matter deep to the cortex or in layer V (Tanaka and North, 1993). This 5-HT response was mediated by 5-HT_{1B} receptors based on the effects of 5-HT_{1B} receptor agonists (TFMPP and CGS12066B) and antagonists (cyanopindolol) as well as agonists and antagonists for other 5-HT and β-adrenergic receptors. This 5-HT_{1B} receptor-mediated inhibition of EPSPs was also independent of membrane potential. The absence of direct hyperpolarizing effects of 5-HT_{1B} receptor activation coupled with the observation that ~58 percent of pyramidal cells contain 5-HT_{1B} receptor mRNA would suggest that cortico-cortical excitatory fibers are modulated by 5-HT_{1B} heteroceptors. 5-HT also suppressed EPSPs evoked by corpus callosum stimulation when recording from layer V pyramidal cells of the mPFC. The pharmacology of these responses appeared to agree with mediation by 5-HT_{1B} rather than 5-HT_{1A} receptors (Read *et al.*, 1994).

The 5-HT_{1B} receptor also plays a physiological developmental role for sensory cortical areas. The 5-HT_{1B} receptor is transiently expressed along with the 5-HT transporter in thalamocortical fibers originating in the ventrobasal thalamus and terminating in layer IV. Activation of these 5-HT_{1B} receptors reduces the monosynaptic thalamocortical EPSCs evoked by low-frequency stimulation of the internal capsule (Laurent *et al.*, 2002). A clear developmental function for this receptor is suggested by studies of the barrelless monoamine oxidase knockout mice, where inactivation of the 5-HT_{1B} receptor then restores the barrels. The expression of the 5-HT_{1B} receptor in the thalamocortical pathway of the ventrobasal complex (ventral posterior medial nucleus (VPM)/ventral posterior lateral nucleus (VPL)) and the dorsal lateral geniculate disappears by the end of the second week.

A number of pathways exist where significant mismatches between 5-HT_{1B} mRNA and receptor binding are found for the cells of origin and their respective terminal fields. Lesion studies complement the electrophysiological data, suggesting the presence of presynaptic 5-HT_{1B} heteroceptors in these pathways. A variety of extracellular and intracellular studies demonstrate that 5-HT_{1B} receptor activation decreases GABA release from striatopallidal terminals and leads to excitation of globus pallidum neurons, consistent with results from striatal kainic lesions (Sari *et al.*, 1999; Querejeta *et al.*, 2005; Hashimoto and Kita, 2008; Rav-Acha *et al.*, 2008). Similarly, a whole-cell patch-clamp recording from the substantia nigra pars reticulata found that 5-HT_{1B} receptor activation suppressed evoked GABA_A receptor-mediated IPSCs (Stanford and Lacey, 1996). Consistent with this electrophysiological study, an antibody recognizing the 5-HT_{1B} receptor at the electron microscopic level found immunoperoxidase staining in fine unmyelinated axons and nerve terminals (Sari *et al.*, 1997). Accordingly, neostriatal kainic acid lesions markedly decreased 5-HT_{1B} receptor-like immunoreactivity (Sari *et al.*, 1999). Intracellular recordings from the superior colliculus combined with the loss of collicular 5-HT_{1B} receptors supported the working hypothesis that 5-HT_{1B} receptor activation suppresses glutamate release in the retinocollicular pathway (Mooney *et al.*, 1994, 1996). Activation of 5-HT_{1B} receptors also suppresses polysynaptic EPSCs modeling the network activity of CA1/CA1 synapses evoked by electrical stimulation of the stratum radiatum (Mlinar *et al.*, 2003). The CA1 pyramidal neurons project both to other CA1 pyramidal cells in the stratum radiatum and to the subiculum. Knife-cut lesions disrupting the CA1/subicular synapse and CA1 chemical lesions causing a deafferentation of both the CA1/subicular and the CA1/CA1 synapse decrease 5-HT_{1B} receptor binding (Amara *et al.*, 1995, 2001). Another region with relatively weak 5-HT_{1B} receptor mRNA expression yet strong 5-HT_{1B} receptor binding is the subthalamic nucleus, where both

IPSCs and EPSCs are inhibited via activation of 5-HT_{1B} receptors.

5-HT_{1D} receptors

There is relatively little variation in the binding of most ligands to the human gene products of the 5-$HT_{1D\alpha}$ and the 5-$HT_{1D\beta}$ receptor genes. In contrast, a mutation in the rodent 5-$HT_{1D\beta}$ receptor gene (5-HT_{1B} receptors) provides pharmacological differentiation for a number of ligands compared to the gene product for the 5-$HT_{1D\alpha}$ receptor or 5-HT_{1D} receptors. In the rodent, there is a relatively sparse expression of 5-HT_{1D} mRNA (5-$HT_{1D\alpha}$ gene product) compared to 5-$HT_{1D\beta}$ mRNA (Bruinvels *et al.*, 1994). Since the triptans used for migraine treatments activate 5-HT_{1D}, 5-HT_{1B} and 5-HT_{1F} receptors, and 5-HT_{1D} receptors tend to have a relatively restricted CNS distribution, most electrophysiology relevant to 5-HT_{1D} receptors has concentrated on trigeminal neurons or the nucleus tractus solitarius, and will not be dealt with in detail here (Goadsby and Knight, 1997; Cumberbatch *et al.*, 1999; Hoskin *et al.*, 2004; Jeggo *et al.*, 2007).

5-HT_{1E} receptors

While the 5-HT_{1E} receptor was first cloned in 1992, advancing knowledge for this receptor has been hampered by lack of selective radiotracers and adequate blocking compounds. The mRNA for this receptor is clearly present in the human striatum and layer IV of the visual cortex (Bruinvels *et al.*, 1994). Similarly, quantitative real-time polymerase chain reaction (qRT-PCR) analysis demonstrated 5-HT_{1E} mRNA in the guinea-pig brain regions in, in descending order, the hippocampus, olfactory bulb, cortex, thalamus, pons, hypothalamus, midbrain, striatum and cerebellum (Bai *et al.*, 2004). Further advancements in understanding this receptor will require selective receptor agonists, receptor antagonists or small interfering RNA (siRNA).

5-HT_{1F} receptors

While the 5-HT_{1F} receptor is activated by the triptan compounds that have been approved for the treatment of migraine (Ramadan *et al.*, 2003), little work has suggested applications outside of this indication. A novel radioligand, [^{3}H]LY334370, has been described for the 5-HT_{1F} receptor. Autoradiographic studies demonstrate that this receptor is localized to a number of limbic forebrain-related regions such as layers IV–V of all the neocortical areas examined, the olfactory bulb, nucleus accumbens, caudate-putamen, parafascicular nucleus of the thalamus, medial mammillary nucleus, the CA3 hippocampal region, subiculum and several amygdaloid nuclei (Lucaites *et al.*, 2005). This localization would suggest that electrophysiological and behavioral studies would be warranted with selective 5-HT_{1F} agonists.

5-HT_{2A} receptors

5-HT_{2A} receptors have been implicated in the psychotomimetic effects of serotonergic hallucinogens (Titeler *et al.*, 1988; Aghajanian and Marek, 1999a; Nichols, 2004). Cortical 5-HT_{2A} receptors have been a natural focus to explain the psychotomimetic effects of lysergic acid diethylamide (LSD) or mescaline, since the cortex has one of the highest expressions of 5-HT_{2A} receptor protein in the rodent and human brain. Secondly, one of the more popular models for studying hallucinogenic effects in rodents is the head shakes that are induced by all three major classes of serotonergic hallucinogens (phenethylamine, indoleamine and ergot hallucinogens). These head shakes can be induced by infusion of phenethylamine hallucinogens directly into the prelimbic region of the rat mPFC (Willins and Meltzer, 1997). An understanding of 5-HT_{2A} receptor function in the prefrontal cortex is dependent upon understanding the electrophysiological actions of 5-HT_{2A} receptors in all of the cellular compartments for which they are found. Cellular compartments where 5-HT_{2A} receptors are localized to include postsynaptic sites on layer II, III, and V pyramidal cells, postsynaptic sites on certain GABAergic interneurons, glia, and a more controversial presence in axons projecting to layer V pyramidal cells.

In the prefrontal cortex, 5-HT_{2A} receptors are known to have excitatory actions in many pyramidal cells, especially those in layer V, which are the principal cortical output cells (Araneda and Andrade, 1991; Amargos-Bosch *et al.*, 2004). In layer V pyramidal cells, direct activation of 5-HT_{2A} receptor and 5-HT_{1A} receptors mediates excitatory inward currents/depolarizing potentials and inhibitory outward currents/hyperpolarizing potentials (Figure 1), respectively (Araneda and Andrade, 1991; Tanaka and North, 1993). In fact, a majority of the electrophysiological effects of 5-HT upon pyramidal cells in the PFC and neocortex are direct inhibitory effects mediated by non-5-HT_{2A} (effects such as suppressing electrically-evoked EPSPs) receptors, while the 5-HT_{2A} receptor provides the major direct excitatory drive onto different cellular compartments such as glutamatergic pyramidal cells and GABAergic interneurons (Tanaka and North, 1993; Read *et al.*, 1994; Aghajanian and Marek, 1999b; Zhou and Hablitz, 1999).

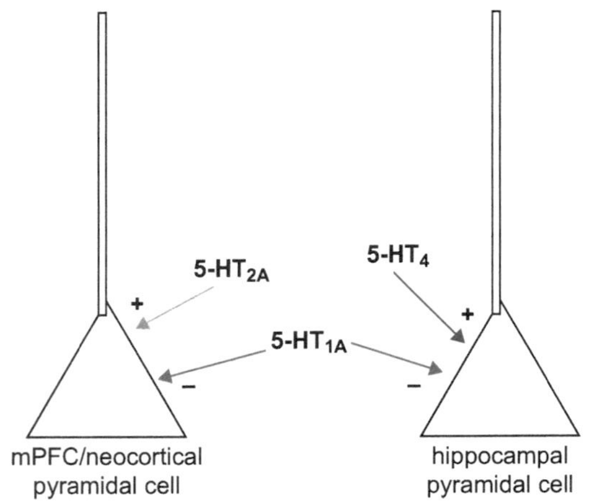

Figure 1 Opposing effects of postsynaptic excitatory 5-HT receptors and postsynaptic inhibitory 5-HT_{1A} receptors on pyramidal cells in the medial prefrontal cortex/neocortex and the hippocampus. The 5-HT_{2A} receptor and the 5-HT_4 receptor play similar roles with respect to physiological effects on pyramidal cells of the prefrontal cortex (PFC)/neocortex and the hippocampus, respectively, with respect to the reduction of the slow afterhyperpolarization (sAHP) and a slow depolarization. In both types of cortical regions, activation of 5-HT_{1A} receptor exerts a slow hyperpolarization which acts in an opposite direction from the influence of 5-HT_{2A} or 5-HT_4 receptor activation. This arrangement is highlighted here as the layer V pyramidal cells of the PFC/neocortex are the output cells of the allocortex and reflect integration of activity within cortical columns. Thus, antagonism of key 5-HT receptors at critical sites within microcircuits/macrocircuits could augment or attenuate the effects of therapeutic treatments such as serotonin reuptake inhibitors, which enhance activity at all 5-HT receptors.

5-HT_{2A} receptors: excitation of cortical interneurons

5-HT_{2A} receptors are known to be localized in both large and small GABAergic interneurons, including both parvalbumin- and calbindin-containing interneurons in the PFC and neocortex (Willins *et al.*, 1997; Jakab and Goldman-Rakic, 1998, 2000). In layer III pyramidal cells of the mPFC the influence of IPSCs from GABAergic interneurons via 5-HT_{2A} receptor activation appears to be greater than in layer V pyramidal cells, where only about ~15 percent of the synaptic currents are IPSPs/IPSCs (Aghajanian and Marek, 1997; Zhou and Hablitz, 1999). In the hippocampus, the 5-HT_{2A} and 5-HT_3 receptors appear to act together to excite GABAergic interneurons that induce IPSPs in the dentate gyrus and the CA1 pyramidal cells, respectively (Araneda and Andrade, 1991; Piguet and Galvan, 1994; Tanaka and North, 1993; Shen and Andrade, 1998). In the piriform cortex, a subpopulation of GABAergic cells are excited to fire action potentials in slices by 5-HT acting at 5-HT_{2A} receptors (Sheldon and Aghajanian, 1990, 1991; Gellman and Aghajanian, 1994; Marek and Aghajanian, 1994). In these cells, both the indoleamine hallucinogen LSD and the phenethylamine hallucinogen DOI act as 5-HT_{2A} receptor partial agonists (Marek and Aghajanian, 1996).

5-HT_{2A} receptors enhance mPFC glutamate release

An electrophysiological action of 5-HT_{2A} receptor activation first explicitly described in the mid-1990s may be especially relevant to understanding the important role this receptor plays in mediating the modulating or mediating salient effects of antidepressant, antipsychotic and hallucinogenic drugs. 5-HT_{2A} receptor activation has been found to increase the frequency of spontaneous (not electrically-evoked) excitatory postsynaptic currents/potentials (EPSCs/EPSPs) when recording during slice preparations from the primary output cell of the PFC, the layer V pyramidal cells (Aghajanian and Marek, 1997). A survey of different cortical layers found that the frequency of excitatory synaptic currents induced in layer V pyramidal cells was at least four-fold greater than in layer II/III pyramidal cells. In contrast to layer V pyramidal cells, the layer VI pyramidal cells demonstrated no significant change in either the frequency or amplitude of EPSCs following 5-HT bath application (Lambe *et al.*, 2000). These excitatory synaptic currents induced by 5-HT_{2A} receptor action are blocked by AMPA receptor activation in a manner consistent with a postsynaptic effect (Aghajanian and Marek, 1997; Zhang and Marek, 2008). In keeping with the hypothesis that these 5-HT-induced excitatory synaptic currents are linked to wider behavioral effects such as DOI-induced head shakes, AMPA receptor antagonists suppress DOI-induced head shakes (Zhang and Marek, 2008).

Several pharmacological aspects of suppressing these 5-HT-induced EPSCs suggest that a presynaptic mechanism may be involved. First, activation and blockade of mGlu2/3 receptors suppresses or enhances the frequency of 5-HT-induced EPSCs relative to changes in the amplitude of EPSCs consistent with a presynaptic influence of mGlu2 receptors on excitatory amino acid release (Marek *et al.*, 2000). Activation of mGlu2 receptors and blockade of mGlu2 receptors suppresses or enhances, respectively, the frequency of head shakes induced by phenethylamine hallucinogens (Benvenga *et al.*, 2006; Benneyworth *et al.*, 2007). This is of interest, not only from an antipsychotic drug development standpoint (Marek and Aghajanian, 1995; Patil *et al.*, 2007; Rorick-Kehn *et al.*, 2007a, 2007b), but is also intriguing given the primary physiological function of mGlu2 receptors as presynaptic autoreceptors present in the preterminal axonal compartments (Schoepp, 2001).

Other heteroceptors interacting with 5-HT_{2A} receptors both at the level of 5-HT-induced EPSCs in the mPFC

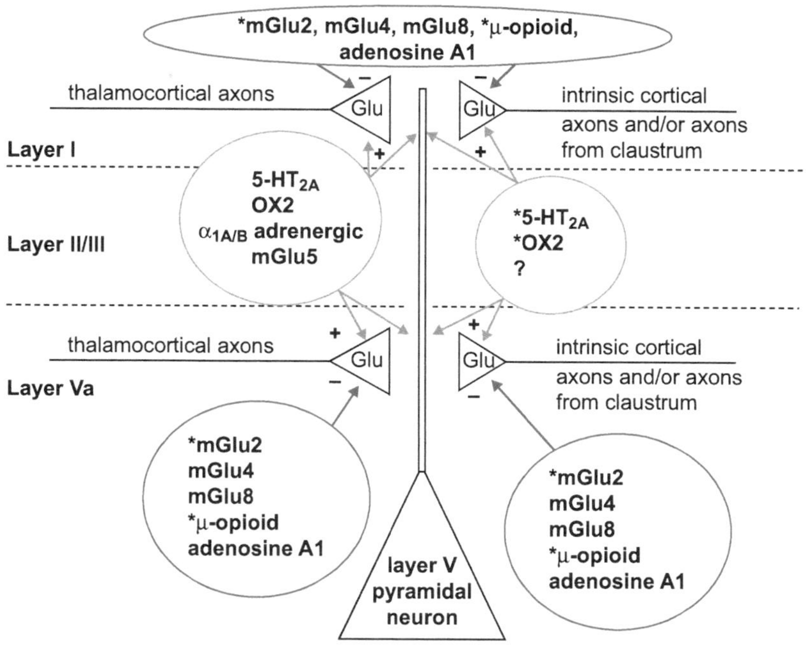

Figure 2 5-HT_{2A} receptors work in tandem with other monoaminergic ($\alpha_{1A/1B}$-adrenergic), glutamatergic (mGlu5), and peptidergic (orexin OX2) receptors to enhance glutamate release (spontaneous excitatory postsynaptic potentials (EPSPs)/excitatory postsynaptic currents (EPSCs)) onto the apical dendritic tree of layer V pyramidal cells (layers I and Va) from both thalamocortical and claustral/intrinsic cortical afferents. In the thalamocortical and claustral/intrinsic cortical afferents, metabotropic glutamate (mGlu2, mGlu4 and mGlu8) autoreceptors, μ-opioid heteroceptors, and adenosine A1 heteroceptors act to attenuate glutamate release. Evidence supporting the existence of thalamocortical mGlu2 autoreceptors and thalamocortical μ-opioid heteroceptors also includes data from lesion studies (*). These relationships emphasize the manner in which single 5-HT receptors may work in tandem with a variety of other aminergic, glutamatergic and peptidergic receptors. Challenges remain in understanding the nature of the relationship between modulation of synaptic inputs to layer V pyramidal cells versus postsynaptic modulation of the passive properties of these same layer V pyramidal cells shown in Figure 1. In addition, where both positive and negative presynaptic modulation of excitatory neurotransmission occurs, challenges also remain in addressing whether this modulations occurs in preterminal axons or in axon terminals or in both. An unresolved issue at the present time is whether the non-thalamic component of the 5-HT-induced EPSCs is mediated through afferents from intrinsic cortical neurons vs. afferents from a structure such as the claustrum.

and also with respect to DOI-induced head shakes are μ-opioid receptors. Activation of μ-opioid receptors, but not κ- or δ-opioid receptors, suppresses 5-HT-induced EPSCs with little, if any, evidence for simultaneous postsynaptic effects in the layer V pyramidal cells (Marek and Aghajanian, 1998). Similarly, activation of μ-opioid receptors, but not κ- or δ-opioid receptors, suppresses DOI-induced head shakes (Marek, 2003). The importance of this interaction is the widespread physiological role played by inhibition of neurotransmitter release for all of the opioid receptors.

A third heteroceptor interaction with 5-HT_{2A} receptors both at the level of 5-HT-induced EPSCs in the mPFC and also with respect to DOI-induced head shakes is that of adenosine A1 receptors. Activation of adenosine A1 receptors selectively suppressed 5-HT-induced EPSCs recorded from layer V pyramidal cells, but did not suppress fast EPSCs evoked by stimulation of the white matter (Stutzman *et al.*, 2001). Activation and blockade of adenosine A1 receptors also suppressed and enhanced, respectively, DOI-induced head shakes (Marek, 2009). Again, the importance of these observations is in the widely accepted physiological heteroceptor role played by adenosine A1 receptors in glutamatergic neuron axons and the extensive overlap between mGlu2 and adenosine A1 receptors (Figure 2).

A presynaptic role for 5-HT_{2A} receptors in mediating this increase in the frequency of EPSCs was not supported by the experiments with tetrodotoxin (TTX) and the subsequent block of impulse flow. Usually spontaneous neurotransmitter release induced by a presynaptic heteroceptor is not suppressed by TTX administration. However, the 5-HT-induced EPSCs are totally suppressed by application of TTX onto the slice (Aghajanian and Marek, 1997). Since mGlu2 receptors generally are not in the presynaptic terminal and are often located in a preterminal region of the axon, it would also seem consistent that 5-HT_{2A} receptors inducing glutamate release might be proximal in the axonal arbor to the mGlu2 receptors. In fact, some axonal arbors such as thalamocortical axons are known to support local or ectopic impulse flow (Pinault, 1995). Experiments utilizing transgenic strategies such as deletion of 5-HT_{2A} receptors and then local rescue of 5-HT_{2A} receptors in the cortex have suggested that the source of glutamate being released by 5-HT_{2A} receptor activation is on afferents to layer V pyramidal cells rather than retrograde release phenomena such as occur with cannabinoids and arachidonic acid (Beique *et al.*, 2007).

Thalamocortical pathways and 5-HT_{2A} receptors

Lesion studies directed at determining the source of a subcortical glutamatergic afferent have suggested that the thalamus contributes to a substantial portion (50–60 percent) of the 5-HT-induced EPSCs. Both fiber-sparing NMDA-induced lesions and electrolytic lesions reduced the frequency of 5-HT-induced EPSCs by ~50–60 percent (Lambe and Aghajanian, 2001; Marek *et al.*, 2001). The importance of this finding is the excellent match between the layers of the PFC and neocortex where 5-HT_{2A} receptors are most densely expressed (layers I and Va) and the layers where local application of 5-HT from a glass micropipette induces the excitatory current within seconds despite the absence of cells in these layers which fire in response to 5-HT (Aghajanian and Marek, 1997). Furthermore, the intralaminar and midline thalamic nuclei terminate most heavily in layers I and Va, which are the layers where 5-HT-induced EPSCs are supported by local application. Accordingly, three GPCRs (mGlu2 autoreceptors, μ-opioid heteroceptors and adenosine A1 receptors) whose activation suppresses 5-HT-induced EPSCs have mRNA expressed in the midline and thalamic intralaminar nuclei. These three presynaptic or preterminal receptors are expressed in a laminar pattern in the cortex (heaviest in layers I and Va) consistent with autoreceptor and heteroceptor function, respectively, from these thalamic nuclei intimately involved in arousal and attention. Furthermore, the binding of agonist radiotracers for mGlu2 and μ-opioid receptors were decreased by the same thalamic lesions which suppressed by ~50–60 percent, but did not completely eliminate, the 5-HT-induced EPSCs (Marek *et al.*, 2001).

Other observations besides the regulation of DOI-induced head shakes support functional relationships in thalamocortical pathways. For example, the DOI-induced increase in c-fos expression in the somatosensory cortex is attenuated by lesions which destroy the posterior thalamic nucleus and VPM/VPL (Scruggs *et al.*, 2000). Thus, the intralaminar and midline thalamic nuclei may be one of several sources of afferents upon which 5-HT_{2A} receptor activation induces glutamate release into the apical dendrites (layers I and Va) of the layer V pyramidal cells.

Challenges to presynaptic thalamocortical 5-HT_{2A} receptors

A series of studies employing mice with a deletion of 5-HT_{2A} receptors and then rescue strategies to recover 5-HT_{2A} receptors either in the prefrontal cortex/neocortex and the thalamus has challenged the hypothesis that 5-HT_{2A} receptors are acting in a presynaptic manner. These studies have demonstrated a loss of 5-HT-induced EPSCs, DOI-induced immediate early gene expression, and DOI-induced head shakes in transgenic mice lacking 5-HT_{2A} receptors (Weisstaub *et al.*, 2006; Gonzalez-Maeso *et al.*, 2007). These electrophysiological, biochemical and behavioral effects have been rescued in transgenic mice with a rescue of cortical 5-HT_{2A} receptors, but not in mice with a rescue of thalamic 5-HT_{2A} receptors. However, both of these genetic rescue strategies are compromised by serious confounds. The use of the gene *Emx1* to drive cortical 5-HT_{2A} expression in the *htr2a*$^{-/-}$ mice also led to expression of 5-HT_{2A} receptors in the claustrum. The claustrum is an additional candidate area, besides the midline and intralaminar thalamic nuclei, which also sends cortical projections to layers I and V of the prefrontal cortex and neocortex, and is activated by 5-HT_{2A} receptor expression as judged by immediate early gene expression. While not well understood, and infrequently studied, the claustrum is known to be involved in the multi-modal sensory processing throughout the cortex. Thus, afferents from the association cortex or neocortex form only one of two known candidate regions (the other being the claustrum) where 5-HT_{2A} receptors are rescued in the *htr2a*$^{-/-}$ x *Emx1-Cre* mice.

Confounds also compromise the use of a rescue strategy for putative thalamic 5-HT_{2A} receptors. The thalamic rescue strategy introduced the 5-HT_{2A} receptor gene into a region controlled by a promoter for the transient expression of the serotonin transporter in all of the 'first-order' or sensory thalamic nuclei, such as the dorsal lateral geniculate, medial geniculate nucleus, ventroposterior medial and ventroposterior lateral nuclei, and the ventral lateral and ventral anterior nuclei (Lebrand *et al.*, 1996; Persico *et al.*, 2001). These first-order nuclei project primarily to the mid-layer of the cortex or layer IV, and tend to receive cortical afferents (modulatory inputs) from layer VI pyramidal cells rather than layer V pyramidal cells. These first-order thalamic nuclei do not send a significant projection to layer I of the prefrontal cortex and neocortex, as is the case for the midline and intralaminar thalamic nuclei. The midline and intralaminar thalamic neurons also receive 'driver' inputs from layer V pyramidal cells (Sherman and Guillery, 1996, 1998). Given the lack of available evidence that the midline and intralaminar thalamic nuclei possess transient expression of the serotonin transporter, as do the first-order thalamic nuclei, the use of *ht2r*$^{-/-}$ x *Htt-Cre* mice might have limited 5-HT_{2A} receptor expression to thalamic neurons (e.g., sensory thalamic nuclei) which are not the candidate thalamic nuclei hypothesized to mediate 50–60 percent of the 5-HT-induced EPSCs in rats. It should be pointed out, however, that other evidence from *in vivo* electrophysiological experiments with anesthetized rats and *in vivo* dialysis has not supported the hypothesis that 5-HT_{2A} receptors induce local cortical effects via releasing glutamate from thalamocortical neurons (Puig *et al.*, 2003). Understanding the importance of 5-HT_{2A} receptor activation in modulating afferent axons to the layer V pyramidal cells as opposed to the simple postsynaptic effects of 5-HT_{2A} receptors will require further correlation to higher-order behavioral functions, such as different executive functions of the PFC.

5-HT_{2A} receptors: antidepressant-mediated down-regulation

A substantial body of literature has grown describing the down-regulation/desensitization of 5-HT_{2A} receptor binding/functional 5-HT_{2A} receptor mediated responses in the mPFC following the chronic administration of most antidepressants other than SSRIs and serotonin-norepinephrine reuptake inhibitors (SNRIs). There has been relatively little study of the modulation of 5-HT_{2A} receptor-mediated responses at the single-cell level. A subchronic duration of imipramine treatment (15 mg/kg $\times$ 21 days) has been found to down-regulate the maximal response observed both for 5-HT-induced EPSCs and the 5-HT-induced excitation of piriform cortical interneurons (Marek, 2008a). These results agree with a plethora of studies showing down-regulation of 5-HT_{2A} receptor binding by imipramine and other TCAs. A dissociation was observed for the effects of 10 daily electroconvulsive treatments on cortical 5-HT_{2A} receptors. This subchronic electroconvulsive shock (ECS) treatment desensitized the 5-HT-induced EPSCs while increasing the maximal response of cortical GABAergic interneurons to 5-HT (Marek, 2008a). Thus, both subchronic imipramine and ECS treatment down-regulated/desensitized glutamatergic compartments to the effects of 5-HT_{2A} receptor activation. In contrast, the concentration–response curve for 5-HT-induced EPSCs was similar in PFC slices from both subchronic fluoxetine (10 mg/kg $\times$ 21 days) and vehicle-treated rats (Marek, 2008b). In extracellular recordings examining the effects of 5-HT receptor agonists on spontaneous epileptiform discharges induced in an Mg^{2+-} free slice perfused with picrotoxin, the ability of 8-OH-DPAT to suppress epileptiform activity was enhanced while the ability of DOI to enhance these discharges was attenuated (Bobula *et al.*, 2003). These findings are in general agreement with the opposing action of cortical 5-HT_{1A} and 5-HT_{2A} receptors, and that the balance of activity with chronic antidepressant treatment is toward inhibitory actions of 5-HT and away from the excitatory actions of 5-HT_{2A} receptor activation.

Other important roles exist for 5-HT_{2A} receptors such as the modulation of locus coeruleus activity (Rasmussen and Aghajanian, 1986, 1988). The 5-HT_{2A} receptor appears to specifically regulate dopaminergic VTA neurons (see work by E. Pehek and B. Roth); this relationship is relevant to the involvement of these pathways in responses to environmental stress.

5-HT_{2B} receptors

5-HT_{2B} mRNA has been detected at low levels in the brain of cat and monkey, but not rat (Helton *et al.*, 1994; Schmuck *et al.*, 1994). This receptor is present in highest abundance in the liver and kidney. Agonist action at this receptor has also been suggested to be the mechanism underlying the valvular heart disease associated with fenfluramine (Evrard *et al.*, 1999; Rothman *et al.*, 2000), precluding significant interest by the pharmaceutical industry in this target with limited brain distribution.

5-HT_{2C} receptors

Unlike 5-HT_{2A} receptors, which are preferentially distributed to the prefrontal cortex and neocortex, 5-HT_{2C} receptors are preferentially distributed to subcortical regions, including the basal ganglia and related structures (Mengod *et al.*, 1990; Wright *et al.*, 1995; Lopez-Gimenez *et al.*, 1997, 1998, 2001). 5-HT_{2C} receptors play a role in mediating effects in a number of basal ganglia-related structures ranging from the substantia nigra pars reticulata (SNpr), substantia nigra pars compacta (SNpc), ventral tegmental area (VTA), striatum and globus pallidus to the subthalamic nucleus. 5-HT_{2C} receptor activation induced inward currents in SNpr neurons in addition to increasing the frequency of $GABA_A$ receptor-mediated IPSCs, which were probably due to excitation of other GABAergic SNpr neurons (Stanford and Lacey, 1996). The 5-$HT_{2C/1B}$ receptor agonist m-chlorophenylpiperazine (mCPP) excited all non-dopaminergic neurons in the VTA while differentially increasing the firing rate of non-dopaminergic SNpr cells depending on whether these cells also responded to a peripheral footpinch stimulus (Di Giovanni *et al.*, 2001). A number of 5-HT_{2C} receptor agonists (RO 60-0175, mCPP, MK-801) increased the firing frequency of dopaminergic VTA cells, often with an increase in burst frequency or spike frequency within bursts that typically result in supralinear dopamine release in terminal fields (Di Matteo *et al.*, 1999, 2000, 2001; Di Giovanni *et al.*, 2000, 2001; Gobert *et al.*, 2000). A number of 5-HT_{2C} receptor antagonists either blocked the effects of the 5-HT_{2C} agonist or decreased the firing rate of VTA cells. In contrast to the VTA cells, 5-HT_{2C} receptor agonists and antagonists had little, if any, effect on dopaminergic SNpc neurons (Di Matteo *et al.*, 1999; Di Giovanni *et al.*, 2000). Another differential effect of 5-$HT_{2A/2C}$ receptor agonism was seen in the interaction with a dopamine D2 receptor-mediated inhibition of the firing rate. Both a selective 5-HT_{2A} receptor antagonist and a selective 5-HT_{2C} receptor antagonist blocked the DOI-induced enhancement of the quinpirole (D2 receptor agonist)-induced reduction of VTA cells. DOI did not enhance the quinpirole-induced reduction in the firing rate of SNpc (A9) cells (Olijslagers *et al.*, 2004).

5-HT_{2C} receptors exert direct excitatory effects on neurons from regions of the basal ganglia outside of the brainstem VTA and the substantia nigra. Striatal cholinergic interneurons contain 5-HT_{2C}, 5-HT_6 and 5-HT_7 receptor mRNA. Depolarization or inward currents can be induced by 5-HT, and is mediated by all three of these 5-HT receptor subtypes (Bonsi *et al.*, 2007). In the subthalamic nucleus, 5-HT_{2C} and 5-HT_4 receptor activation induces direct postsynaptic excitatory effects while activation of the 5-HT_{1A} receptor mediates a postsynaptic inhibitory effect (Stanford *et al.*, 2005). Thus, 5-HT_{2C} receptors are densely expressed in the basal ganglia-related structures and exert direct excitatory effects in a variety of neuron types throughout this motor circuitry. Given the preclinical antipsychotic-like profile exerted by 5-HT_{2C} receptor agonists, the results of phase II clinical testing for the Wyeth 5-HT_{2C} receptor agonist vabicaserin in schizophrenia patients should be informative.

5-HT_3 receptors

The 5-HT_3 receptors form the only 5-HT receptor subtype which functions as a ligand-gated ion channel (i.e., an ionotropic receptor). This receptor was, in the original classification of 5-HT receptors by Gaddum and Picarelli, known as the M-receptor. Two different subunits, 5-HT_{3A} and 5-HT_{3B}, are present in peripheral sites, whereas only the 5-HT_{3A} receptor subunit is present in central neurons (Morales and Wang, 2002). While a high density of 5-HT_3 receptors is found in the hindbrain in areas thought to play a role in anti-emetic uses of 5-HT_3 antagonists (e.g., ondansetron and metoclopramide), a moderate density of 5-HT_3 receptors is present in the prefrontal cortex, neocortex, hippocampus and amygdala. In these forebrain regions, 5-HT_3 receptors are localized in many of the calbindin- and CCK-containing interneurons (Morales and Bloom, 1997). 5-HT_3 receptors also are localized to some calretinin-containing interneurons, especially in the agranular insular cortex and the piriform cortex.

Hippocampal 5-HT_3 receptors in interneurons

Initial experiments with central neurons found that 5-HT or 2-methyl-5-HT are agonists that induce rapidly densensitizing responses; the excitatory responses in cultured hippocampal neurons were blocked by ICS-205-930, curare or metoclopramide (Yakel and Jackson, 1988). Recordings from the dentate gyrus of hippocampal slices revealed that both GABAergic basket cells and the principal excitatory cell dentate cells responded to 5-HT with spontaneous IPSCs which were blocked by the $GABA_A$ receptor antagonist bicuculline (Kawa, 1994). In addition to this apparent excitatory action of 5-HT_3 receptors on GABA release, activation of 5-HT_3 receptors also induced inward currents in the basket cells. Hippocampal GABAergic interneurons located in the stratum radiatum of the CA1 field, which receives a strong projection from the median raphe nuclei, are directly excited by postsynaptic 5-HT_3 receptors (McMahon and Kauer, 1997). These interneurons in CA1 send axonal projections to the dentate gyrus, as well as extensive ramifications within CA1. Another study using single-cell RT-PCR found that only the short splice variant of the 5-HT_{3A} receptor subunit was expressed in hippocampal interneurons of CA1 (Sudweeks *et al.*, 2002). Furthermore, half the interneurons containing 5-HT_3 receptors appeared to assemble with $\alpha4\beta2$ nicotine receptor subunits to form functional heteromeric receptors. More recent work using single rat hippocampal neuron cultures has confirmed that 5-HT_3 receptors control GABA, but not glutamate, release (Dorostkar and Boehm, 2007). Further, activation of 5-HT_3 receptors mediates opposite effects on spontaneous as compared to action-dependent release. Finally, the presynaptic action of 5-HT_3 receptor activation to enhance GABA release onto CA3 pyramidal cells was found to gradually increase with development (Choi *et al.*, 2007).

Multiple 5-HT receptors may act together or oppose each other in different hippocampal cellular compartments. For example, in the dentate gyrus granule cells, a postsynaptic effect of 5-HT_{1A} receptor activation leads to hyperpolarization (Piguet and Galvan, 1994). In contrast, very brief local 5-HT applications induce a hyperpolarization and a transient burst of spontaneous IPSPs that are blocked by tetrodotoxin or 5-HT_3 receptor antagonists (Piguet and Galvan, 1994). More prolonged increases in spontaneous IPSPs induced by 5-HT were blocked by the 5-HT_{2A} receptor antagonist M100907 (Piguet and Galvan, 1994). Using a 'synaptic bouton preparation' by acutely isolating CA1 neurons allowed for studying the presence of multiple 5-HT receptors in GABAergic nerve terminals (Katsurabayashi *et al.*, 2003). The GABAergic interneuron terminals within the CA1 hippocampal field have also been shown to contain 5-HT_{1A} heteroceptors. All of the terminals were found to contain 5-HT_{1A} receptors, while only a subset contained both inhibitory 5-HT_{1A} receptors and excitatory 5-HT_3 receptors (Katsurabayashi *et al.*, 2003). Accordingly, a 5-HT_{1A} receptor antagonist WAY100635 and a 5-HT_3 receptor antagonist Y-25130 had opposing influences on hippocampal high-frequency (200-Hz) or 'ripple' oscillations in the CA1 area of awake, behaving rats (Ponomarenko *et al.*, 2003).

Amygdaloid 5-HT_3 receptors

5-HT_3 receptors play a physiological role in both the lateral amygdala and the basolateral amygdala. In the lateral amygdala brain slices, 5-HT_3 receptor activation was found to increase excitatory synaptic potentials (Sugita *et al.*, 1992). In contrast, recordings made from mechanically dissociated BLA neurons with preserved presynaptic nerve terminals found that 5-HT_3 receptor activation had only a presynaptic effect to increase GABA release from the terminals of interneurons (Koyama *et al.*, 2000). Somewhat analogous to the interactions of 5-HT_3 and 5-HT_{1A} receptors in GABAergic terminals in the CA1 hippocampus, activation of heteroceptors for both of these 5-HT receptors on GABAergic nerve terminals of the BLA had opposing effects on the frequency of miniature IPSCs (Koyama *et al.*, 2002b).

Prefrontal cortical/neocortical 5-HT_3 receptors

An important role for 5-HT_3 receptors in the prefrontal cortex and neocortex has been identified based on the observation that over 90 percent of the cortical and hippocampal neurons expressing 5-HT_3 receptors are GABAergic interneurons (Morales *et al.*, 1996). With respect to calcium binding proteins, the 5-HT_3 receptors are not found in parvalbumin-containing interneurons, but are expressed in either calbindin- or calretinin-containing interneurons (Morales and Bloom, 1997). With respect to peptides, the 5-HT_3 receptors are present in 35–66 percent of CCK-containing interneurons in different cortical regions (Morales and Bloom, 1997). However, 5-HT_3 receptors are not co-localized in somatostatin-expressing interneurons. An important role for 5-HT_3 receptors expressed in cortical interneurons is also present for primates. 5-HT_3 receptors are present in interneurons targeting pyramidal cells dendrites, small GABA-, substance P receptor-, and calbindin-containing interneurons, and medium-size calretinin-containing interneurons in the macaque cerebral cortex (Jakab and Goldman-Rakic, 2000). Electrophysiological studies have found that 5-HT_3 receptor activation can suppress spontaneously active cells in the rat mPFC (Ashby *et al.*, 1989). The suggestion that this is a direct effect, though, appears to be in contradiction to the overwhelming expression of 5-HT_3 receptors in interneurons. A large, desensitizing inward current and a transient enhancement of spontaneous IPSCs can also be seen in a minority of interneurons in the rat PFC (Zhou and Hablitz, 1999). However, 5-HT_{2A} receptor-mediated changes in spontaneous IPSCs were much more frequent than 5-HT_3 receptor effects. Intracellular recordings from fast-spiking (FS) interneurons, low threshold spike (LTS) interneurons, and pyramidal cells of the visual cortex revealed opposing effects of 5-HT_3 and 5-HT_{1A} receptors somewhat analogous to findings in the hippocampus and amygdala (Xiang and Prince, 2003). Approximately 75 percent of the FS and LTS interneurons responded to 5-HT. 5-HT induced outward currents in 50 percent of 5-HT-responsive FS interneurons; these inward currents were blocked by a 5-HT_3 receptor antagonist tropisetron. 5-HT induced outward currents in 41 percent of the responsive cells' FS interneurons; these outward currents were blocked by the 5-HT_{1A} receptor antagonist NAN-190. Outward currents blocked by NAN-190 were preferentially induced in LTS cells (81 percent of 5-HT-responsive cells). In contrast, 5-HT-induced inward currents (blocked by tropisetron) in only 15 percent of 5-HT-responsive LTS cells.

5-HT_3 receptors and dopamine-containing midbrain cells

The ability of 5-HT_3 receptor antagonists to block dopamine-mediated effects, especially using local administration of dopamine agonists in the nucleus accumbens, prompted intense interest in the potential antipsychotic action for this class of compounds (Costall and Naylor, 1994). Differences in the ability of antipsychotic drugs to decrease the spontaneous activity of neurons in the ventral tegmental area (VTA; A10) compared to the substantia nigra pars compacta (SNpc; A9) has been proposed and studied as a means to predict atypical antipsychotic action (Grace *et al.*, 1997). Emerging dopamine-cell depolarization block with continued administration of antipsychotic drugs has been suggested to mediate these effects. The emergence of the depolarization block can be empirically demonstrated by apomorphine reversing the chronic effects of the antipsychotic drug upon the firing rate of the dopamine-containing cells.

The effects of 5-HT_3 receptor antagonists in the dopamine-cell depolarization block paradigm have been inconsistent. For example, MDL 73,147EF (palonosetron), unlike haloperidol, did not acutely increase the number of active dopamine neurons in A9. However, when given subchronically, MDL 73,147EF (like haloperidol) decreased the number of active A10 (VTA) and A9 (SNpc) cells (Sorensen *et al.*, 1989). Chronic administration of another 5-HT_3 receptor antagonist, BRL 43694, failed to alter the number of active dopamine neurons encountered in either A10 (VTA) or A9 (SNpc) as seen with known antipsychotic drugs (Ashby *et al.*, 1990). The effects of another 5-HT_3 receptor antagonist were different from those of these other 5-HT_3 receptor antagonists and antipsychotic drugs. Acute and chronic administration of zatosetron (LY277359) decreased the number of spontaneously active A10 (VTA) cells without altering A9 (SNpc) cells (Rasmussen *et al.*, 1991). While chronic zatosetron did decrease the spontaneous firing rate of A10 cells similar to atypical antipsychotic drugs, this effect was not reversed by apomorphine as seen for atypical antipsychotic drug-induced depolarization block of dopamine-containing cells. However, the data from a double-blind, placebo-controlled trial of ondansetron in schizophrenic patients was generally negative (Costall and Naylor, 1994). The clinical evidence for anxiolytic activity with 5-HT_3 receptor antagonists is also far from compelling (Greenshaw and Silverstone, 1997). This raises a fundamental question as to whether this lack of translation from the preclinical arena to the clinic is because modulating only a few subpopulations of GABAergic interneuron is not enough for a robust therapeutic effect, or because the model systems used to increase enthusiasm in this target are limited compared to those used to drive the successful translation of mGlu2/3 receptor agonists (e.g., positive phase II or phase III data for schizophrenia and GAD). Alternatively, clinical studies with 5-HT_3 receptor antagonists were conducted in the absence of a radiotracer for receptor occupancy or other robust biomarker strategies so that the optimal dose may not have been tested.

An additional aspect of 5-HT_3 receptors not discussed here is their sensitivity to alcohol. The reader is referred to a number of manuscripts by David Lovinger and colleagues which discuss this functional relationship.

5-HT_4 receptors

An extensive number of studies have looked at depolarizing effects of 5-HT in the hippocampus that are mediated by activation of 5-HT_4 receptors. In addition to hyperpolarizing effects of 5-HT_{1A} receptor activation, 5-HT exerts several other prominent postsynaptic effects on hippocampal CA1 pyramidal neurons, including a slow membrane depolarization and a reduction of the after-hyperpolarization potential or AHP (Andrade and Nicoll, 1987; Colino and Halliwell, 1987; Chaput *et al.*, 1990). Further pharmacological experiments characterized these responses as due to 5-HT_4 receptor activation (Andrade and Chaput, 1991b; Beck, 1992; Torres *et al.*, 1994; Birnstiel and Beck, 1995). In fact, Andrade and Chaput noted that a number of the cellular effects mediated by activation of 5-HT_{2A} receptors in the prefrontal cortex and neocortex are mediated by 5-HT_4 receptor activation (see Figure 1) in the hippocampus (Andrade and Chaput, 1991b).

5-HT_4 receptor activation, via a presynaptic action, has also been found to potentiate the synapse from mossy fibers (originating from dentate granule cells) terminating onto CA3 pyramidal cells (Kobayashi *et al.*, 2008) in several different strains of mice. 5-HT_{1A} receptor action depressed this same synapse. Interestingly, chronic fluoxetine (10 mg/kg per day) enhanced the 5-HT_4 receptor-mediated potentiation of the mossy fiber-CA3 synapse at low 5-HT concentrations, and attenuated the 5-HT-induced potentiation at high 5-HT concentrations. Chronic oral fluoxetine treatment did not alter light/dark behavior, elevated plus maze behavior, and immobility in the forced swim test. This treatment did, however, reduce the activity of these mice in a novel environment.

As discussed previously, activation of 5-HT_4 receptors exerts depolarizing effects in several movement-related regions with the 5-HT_{2C} receptor. These regions include the subthalamic nucleus (Stanford *et al.*, 2005; Xiang *et al.*, 2005) and the globus pallidus (Shen *et al.*, 2007; Chen *et al.*, 2008).

5-HT_4 receptor stimulation has been found to potentiate the dendritic population EPSP evoked in the stratum radiatum of the CA1 hippocampal field by stimulation of the Schaffer collateral pathway (Tokarski and Bijak, 1996; Bijak, 1997; Matsumoto *et al.*, 2001). Chronic administration of antidepressants (the TCA imipramine and three SSRIs) was found to attenuate effects of 5-HT_4 receptor-induced potentiation of the Schaffer collateral stimulation (Tokarski and Bijak, 1996; Bijak, 1997). Given that 5-HT_{1A} receptor-mediated inhibitory responses tend to be enhanced, antidepressant-induced regulation of 5-HT responses in the hippocampal formation appears to shift the balance of activity towards inhibitory effects and away from the 5-HT_4 receptor-mediated excitatory effects.

5-HT$_5$ receptors

The 5-HT$_{5A}$ (expressed in human, rat and mouse) and 5-HT$_{5B}$ (expressed in rat and mouse, but not human) receptors were recently reviewed (Nelson, 2004). This receptor also has been plagued by a relative lack of specific reagents. However, the localization of 5-HT$_{5A}$ receptor mRNA in the mouse or rat cerebral cortex, hippocampal CA1-3 fields, and habenula is quite intriguing. The 5-HT$_{5A}$ receptor mRNA in the habenula coupled with immunohistochemical evidence of 5-HT$_{5A}$ receptor expression on dorsal raphe serotonergic cell bodies and axons raises interesting functional questions. What is the nature of serotonergic modulation of long-loop feedback from forebrain regions (excluding the prefrontal cortex) to the raphe nuclei? Do 5-HT$_{5A}$ receptors play an autoreceptor function in the raphe nuclei themselves?

5-HT$_6$ receptors

A 5-HT$_6$ receptor antibody showed the highest immunoreactivity in the plexiform layer of the olfactory tubercle, the frontal cortex and entorhinal cortex, nucleus accumbens, striatum, hippocampus (stratum oriens and radiatum of the CA1 field, molecular layer of dentate gyrus), and the molecular layer of the cerebellum. A moderate intensity of specific immunolabeling was seen in the thalamus, substantia nigra, superficial layer of the superior colluculus, motor trigeminal and facial nucleus. The distribution of immunoreactivity for the 5-HT$_6$ receptor was similar to the 5-HT$_6$ receptor mRNA distribution (Gerard *et al.*, 1996, 1997). Frankly, much of the interest in this receptor as a target for cognition impairment associated with schizophrenia (CIAS) has been driven by the relatively high affinity of clozapine and olanzapine for the 5-HT$_6$ receptor. These data have been complemented by the effects of 5-HT$_6$ receptor antagonists in increasing extracellular levels of acetylcholine and glutamate in the forebrain, in addition to improving cognition in rodents.

5-HT$_6$ receptor antagonists generally do not have effects in behavioral or electrophysiological screens for antipsychotic drugs. The repeated oral administration of the 5-HT$_6$ receptor antagonist SB-271046 (1–10 mg/kg × 21 days) did not alter the firing pattern of dopaminergic neurons in the VTA compared to vehicle-treated animals. The two higher doses did increase the number of spontaneously active SNC dopaminergic neurons compared to control (Gerard *et al.*, 1996, 1997; Minabe *et al.*, 2004). This finding would not necessarily contradict the hypothesis that 5-HT$_6$ receptors might be targets for treating the cognitive impairment associated with schizophrenia (CIAS). However, despite a press release in December 2005 citing positive effects of the 5-HT$_6$ receptor antagonist SGS518 or LY483518 for the treatment of CIAS, there are no studies for this putative medication currently in progress cited at the clinical trial registry (www.clinicaltrials.gov).

There is relatively little in the way of other electrophysiological studies addressing 5-HT$_6$ receptors. In addition to a positive role for 5-HT$_{2C}$ and 5-HT$_7$ receptors, activation of 5-HT$_6$ receptors appears to depolarize and induce inward currents in cholinergic interneurons of the striatum (Bonsi *et al.*, 2007). While optimal agonist reagents are not available, a number of antagonist reagents are available, which can be synthesized and used in experiments to better understand the physiological role played by this 5-HT receptor.

5-HT$_7$ receptors

Amongst the regions with the highest concentration of 5-HT$_7$ receptors in the human brain, by [^{3}H]SB-269970 autoradiography, are the anterior thalamus, mediodorsal thalamus, dentate gyrus, pulvinar, substantia nigra, VTA, dorsal raphe nuclei, hippocampus, hypothalamus and anterior cingulate cortex (Varnas *et al.*, 2004). The 5-HT$_7$ receptor is known to be coupled to G$_s$ proteins, leading to an increase in cAMP concentration that frequently results in a direct depolarizing or excitatory effect.

Hippocampal 5-HT$_7$ receptors

One of the initial pharmacological characterizations of a 5-HT$_7$ receptor-mediated electrophysiological response in the forebrain was in the hippocampus (Bacon and Beck, 2000). Activation of 5-HT$_7$ receptors increased the excitability of CA3 pyramidal cells in brain slice experiments as 5-HT was found to decrease the slow afterhyperpolarization (sAHP) in a concentration-dependent manner. The rank order of potency of other 5-HT agonists (5-carboxyamidotryptamine [5-CT] > 5-HT > 5-methoxytryptamine (5-MeOT) to decrease sAHP agreed well with their affinity for the 5-HT$_7$ receptor during experiments where WAY-100635 and GR-113808 were used to block 5-HT$_{1A}$ and 5-HT$_4$ receptors, respectively. Furthermore, the 5-HT$_{2A/2C}$, 5-HT$_3$, 5-HT$_{1B}$, 5-HT$_{1D}$ and 5-HT$_6$ receptors were ruled out as agonists for this electrophysiological response, as agonists for these 5-HT receptors (DOI, 2-methyl 5-HT, α-methyl 5-HT, sumatriptan and clozapine) did not mimic the effect of 5-HT at 10-μM concentrations. Formal analysis of 5-HT receptor antagonists with Schild analyses using the relatively selective 5-HT$_7$

receptor antagonist SB-269770, mesulergine and ritanserin was also consistent with the involvement of the 5-HT_7 receptor in this response. In contrast, antagonists for the 5-$HT_{2A/2C}$, 5-HT_6 receptors (clozapine) and 5-HT_3 receptors (ICS-205,930) had no effect on the 5-HT-induced decrease in sAHP. Interestingly, clozapine had little effect either as an agonist or an antagonist for the 5-HT-induced suppression of the sAHP in hippocampal CA3 pyramidal cells. A 5-HT-induced decrease in sAHP and frequency adaptation of action potential firing that is mediated via 5-HT_7 receptor activation has also been found when recording from pyramidal cells of the CA1 area during rat hippocampal slice experiments (Tokarski *et al.*, 2003). One other electrophysiological effect identified for 5-HT_7 receptors was a weak but significant effect of 5-HT_7 and 5-HT_2 receptor activation at suppressing the perforant path input to CA1 pyramidal cells (Otmakhova *et al.*, 2005). 5-HT did not affect paired-pulse facilitation of this pathway, suggesting that the weak effect of 5-HT on suppressing the perforant pathway input is a postsynaptic rather than a presynaptic effect. The mechanism underlying this effect remains to be determined, as other receptors that act similarly on this pathway generally are coupled to G_i/G_o proteins such as mGlu2 and α_2-adrenergic receptors. However, the general effect of 5-HT_7 receptor activation in the hippocampus appears to be an increase in the excitability of CA1 and CA3 pyramidal cells.

Thalamic 5-HT_7 receptors

Another region where an electrophysiological effect of the 5-HT_7 receptor is well-characterized from both pharmacological and physiological standpoints is the anterodorsal nucleus (ADn) of the thalamus (Chapin and Andrade, 2001a, 2001b). 5-HT_7 receptor agonism exerts an excitatory effect (depolarization or inward current associated with modulation of an I_h current) on the glutamatergic cells of the ADn of the thalamus. Pharmacological characterization was aided by a greater potency (18-fold) of 5-CT than 5-HT and blockade by both mesulergine (using Schild analysis) and the potent, selective 5-HT_7 receptor antagonist SB-269770. Other 5-HT agonists (e.g., DOI for 5-$HT_{2A/2C}$ receptors) did not induce an inward current, similar to 5-HT. Furthermore, 8-OH-DPAT induced weak induction of an inward current as well as blocking the effect of 5-HT, consistent with a partial agonist action for this compound at 5-HT_7 receptors. These findings build on the importance of this receptor for controlling hippocampal physiology, as these ADn neurons project to the retrosplenial granular cortex and receive afferents from the subiculum (including the pre- and parasubiculum) (Steriade *et al.*, 1997).

Antidepressants modulate hippocampal 5-HT_7 receptors

A differential action of antidepressant drugs and electroconvulsive shock (ECS) has been found for the excitatory 5-HT_7 responses in the rat hippocampal CA3 field. Repeated daily treatment for 2 weeks with either the SSRI citalopram or the tricyclic antidepressant imipramine attenuated the 5-CT-induced increase in bursting frequency of CA3 pyramidal cells recorded from hippocampal slices (Toharski *et al.*, 2005). In contrast to the monoamine reuptake inhibitors, electroconvulsive shock enhanced the 5-HT_7 receptor-mediated enhanced epileptiform activity from extracellular recordings of CA3 pyramidal cells (Pitra *et al.*, 2007).

Other 5-HT_7 receptor mediated responses

Another region where 5-HT_7 receptors appear to play an important role is in the suprachiasmatic nucleus, an important regulatory center of circadian physiology. For example, activation of 5-HT_7 receptors by 5-HT for cultured suprachiasmatic neurons is able to inhibit GABA-activated currents (Kawahara *et al.*, 1994). This response is also modulated by cAMP, as expected for a 5-HT_7 receptor. However, other 5-HT receptors are also involved in altering suprachiasmatic neuron physiology, including 5-$HT_{2A/2C}$ receptors.

Activation of both 5-HT_7 receptors and other 5-HT receptors appears to contribute to depolarizing effects in a number of different cellular compartments throughout the brain. For example, 5-HT_7 responses in the subthalamic nucleus and the globus pallidus were discussed above in the section on 5-HT_{1A} receptors. The presence of 5-HT_7 receptors on striatal cholinergic interneurons is also discussed above in the section on 5-HT_6 receptors.

Summary

One of the most remarkable features of psychopharmacology over the past 25 years has been the failure of selective 5-HT receptor agonists and antagonists to take their place in the armentarium of psychotherapeutic drugs. We continue to have little understanding regarding which of the 5-HT receptors play a critical positive and negative role in mediating the effects of SSRIs and SNRIs in mood and anxiety disorders. Beyond a contribution for 5-HT_{2A} receptors in mediating some of the actions of atypical antipsychotics, the contributions played by other 5-HT receptors in treating psychotic disorders (5-HT_{2C}, 5-HT_6, and 5-HT_7) remains uncertain. Certainly, the efficacy of selective 5-HT_{2A} receptor antagonists for schizophrenia (activity of M100907 and eplivanserin intermediate between placebo and haloperidol) and 5-HT_{1A} receptor agonists for mood and anxiety disorders remains limited. The relative lack of success in converting agonists or antagonists highly selective for individual 5-HT receptor subtypes into effective novel therapeutic agents increases the need for translating the interactions between different 5-HT receptors and the serotonin transporter or between different 5-HT receptors and glutamate and GABA receptor subtypes to paradigms exploring the complexity of these interactions at higher levels of organization, such as oscillatory behavior of circuits, sleep and arousal processes, and behaviors closely linked to limbic-related macrocircuits.

References

Aghajanian, G.K. and Marek, G.J. (1997) Serotonin induces excitatory postsynaptic potentials in apical dendrites of neocortical pyramidal cells. Neuropharmacology, 36: 589–599.

Aghajanian, G.K. and Marek, G.J. (1999a) Serotonin and hallucinogens. Neuropsychopharmacology, 21: 16S–23S.

Aghajanian, G.K. and Marek, G.J. (1999b) Serotonin, via 5-HT_{2A} receptors, increases EPSCs in layer V pyramidal cells of prefrontal cortex by an asynchronous mode of glutamate release. Brain Res., 825: 161–171.

Amara, D.A., Segu, L., Naili, S. and Buhot, M.C. (1995) Serotonin 1B receptor regulation after dorsal subiculum deafferentation. Brain Res. Bull., 38: 17–23.

Amara, D.A., Segu, L. and Buhot, M.C. (2001) Region-specific decrease in 5-HT1A and 5-HT1B binding sites after intra-hippocampal ibotenic acid injections in the rat. Neurosci. Lett., 310: 25–28.

Amargos-Bosch, M., Bortolozzi, A., Puig, M.V., Serrats, J., Adell, A., Celada, P., Toth, M., Mengod, G. and Artigas, F. (2004) Co-expression and *in vivo* interaction of $serotonin_{1a}$ and $serotonin_{2a}$ receptors in pyramidal neurons of prefrontal cortex. Cereb. Cortex, 14: 281–299.

Andrade, R. and Chaput, Y. (1991a) 5-HT_4-like receptors mediate the slow excitatory response to serotonin in the rat hippocampus. J. Pharmacol. Exp. Ther., 257: 930–937.

Andrade, R. and Chaput, Y. (1991b) 5-Hydroxytryptamine4-like receptors mediate the slow excitatory response to serotonin in the rat hippocampus. J. Pharmacol. Exp. Ther., 257: 930–937.

Andrade, R. and Nicoll, R.A. (1987) Pharmacologically distinct actions of serotonin on single pyramidal neurones of the rat hippocampus recorded *in vitro*. J. Physiol., 394: 99–124.

Andrade, R., Malenka, R.C. and Nicoll, R.A. (1986) A G protein couples serotonin and $GABA_B$ receptors to the same channels in hippocampus. Science, 234: 1261–1265.

Araneda, R. and Andrade, R. (1991) $5\text{-Hydroxytryptamine}_2$ and $5\text{-hydroxytryptamine}_{1A}$ receptors mediate opposing responses on membrane excitability in rat association cortex. Neuroscience, 40: 399–412.

Artigas, F., Romero, L., de Montigny, C. and Blier, P. (1996) Acceleration of the effect of selected antidepressant drugs in major depression by 5-HT1A antagonists. Trends Neurosci., 19: 378–383.

Ashby, C.R., Edwards, E., Harkins, K. and Wang, R.Y. (1989) Characterization of 5-hydroxytryptamine-3 receptors in the medial prefrontal cortex: a microiontophoretic study. Eur. J. Pharmacol., 173: 193–196.

Ashby, C.R., Jiang, L.H. and Wang, R.Y. (1990) Chronic BRL 43694, a selective 5-HT3 receptor antagonist, fails to alter the number of spontaneously active midbrain dopamine neurons. Eur. J. Pharmacol., 175: 347–350.

Ashby, C.R., Edwards, E. and Wang, R.Y. (1994) Electrophysiological evidence for a functional interaction between 5-HT_{1A} and 5-HT_{2A} receptors in the rat medial prefrontal cortex: an iontophoretic study. Synapse, 17: 173–181.

Azmitia, E.C., Gannon, P.J., Kheck, N.M. and Whitaker-Azmitia, P.M. (1996) Cellular localization of the 5-HT1A receptor in primate brain neurons and glial cells. Neuropsychopharmacology, 14: 35–46.

Bacon, W.L. and Beck, S.G. (2000) 5-Hydroxytryptamine(7) receptor activation decreases slow afterhyperpolarization amplitude in CA3 hippocampal pyramidal cells. J. Pharmacol. Exp. Ther., 294: 672–679.

Bai, F., Yin, T., Johnson, E.M., Su, C., Varga, G., Little, S.P. and Nelson, D.L. (2004) Molecular cloning and pharmacological characterization of the guinea pig 5-HT1E receptor. Eur. J. Pharmacol., 484: 127–139.

Beck, S.G. (1992) 5-Hydroxytryptamine increases excitability of CA1 hippocampal pyramidal cells. Synapse, 10: 334–340.

Beique, J.-C., Imad, M., Mladenovic, L., Gingrich, J.A. and Andrade, R. (2007) Mechanism of the $5\text{-hydroxytryptamine}_{2A}$ receptor-mediated facilitation of synaptic activity in prefrontal cortex. Proc. Natl. Acad. Sci. USA, 104: 9870–9875.

Benneyworth, M.A., Xiang, Z., Smith, R.L., Garcia, E.E., Conn, P.J. and Sanders-Bush, E. (2007) A selective positive allosteric modulator of metabotropic glutamate receptor subtype 2 blocks a hallucinogenic drug model of psychosis. Mol. Pharmacol., 72: 477–484.

Benvenga, M.J., Chaney, S.F., Baez, M., Britton, T.C., Hornback, W.J., Monn, J.A. and Marek, G.J. (2006) Attenuation of DOI-induced head twitches in mGluR2 KO mice. FASEB J., 20: A246.

Besson, A., Haddjeri, N., Blier, P. and de Montigny, C. (2000) Effects of the co-administration of mirtazapine and paroxetine on serotonergic neurotransmission in the rat brain. Eur. Neuropsychopharmacol., 10: 177–188.

Bijak, M. (1997) Imipramine-induced subsensitivity to the 5-HT4 receptor activation, a possible mediation via an alteration in the postreceptor transduction mechanism involving adenylate cyclase. Pol. J. Pharmacol., 49: 345–350.

Birnstiel, S. and Beck, S.G. (1995) Modulation of the 5-hydroxytryptamine4 receptor-mediated response by short-term and long-term administration of corticosterone in rat CA1 hippocampal pyramidal neurons. J. Pharmacol. Exp. Ther., 273: 1132–1138.

Blier, P. and Bouchard, C. (1992) Effect of repeated electroconvulsive shocks on serotonergic neurons. Eur. J. Pharmacol., 211: 365–373.

Blier, P. and de Montigny, C. (1983) Electrophysiological investigations on the effect of repeated zimelidine administration on serotonergic neurotransmission in the rat. J. Neurosci., 3: 1270–1278.

Blier, P. and de Montigny, C. (1987) Modification of 5-HT neuron properties by sustained administration of the 5-HT1A agonist gepirone: electrophysiological studies in the rat brain. Synapse, 1: 470–480.

Blier, P., de Montigny, C. and Tardif, D. (1984) Effects of the two antidepressant drugs mianserin and indalpine on the serotonergic system: single-cell studies in the rat. Psychopharmacology, 84: 242–249.

Blier, P., de Montigny, C. and Azzaro, A.J. (1986) Modification of serotonergic and noradrenergic neurotransmissions by repeated administration of monoamine oxidase inhibitors: electrophysiological studies in the rat central nervous system. J. Pharmacol. Exp. Ther., 237: 987–994.

Bobula, G., Tokarski, K., Zahorodna, A. and Hess, G. (2003) Adaptive changes in the reactivity of 5-HT_{1A} and 5-HT_2 receptors induced in rat frontal cortex by repeated imipramine and citalopram. Naunyn Schmiedebergs Arch. Pharmacol., 367: 444–450.

Bonsi, P., Cuomo, D., Ding, J., Sciamanna, G., Ulrich, S., Tscherter, A., Bernardi, G., Surmeier, D.J. and Pisani, A. (2007) Endogenous serotonin excites striatal cholingergic interneurons via the activation of 5-HT2C, 5-HT6, and 5-HT7 serotonin receptors: implications for extrapyramidal side effects of serotonin reuptake inhibitors. Neuropsychopharmacology, 32: 1840–1854.

Boschert, U., Ait, D., Amara, D.A., Segu, L. and Hen, R. (1994) The mouse 5-$hydroxytryptamine_{1B}$ receptor is localized predominantly on axon terminals. Neuroscience, 58: 167–182.

Bruinvels, A.T., Landwehrmeyer, B., Gustafson, E.L., Durkin, M.M., Mengod, G., Branchek, T.A., Hoyer, D. and Palacios, J.M. (1994) Localization of 5-HT_{1B}, 5-HT_{1Da}, 5-HT_{1E} and 5-HT_{1F} receptor messenger RNA in rodent and primate brain. Neuropharmacology, 33: 367–386.

Chapin, E.M. and Andrade, R. (2001a) A 5-HT(7) receptor-mediated depolarization in the anterodorsal thalamus. I. Pharmacological characterization. J. Pharmacol. Exp. Ther., 297: 395–402.

Chapin, E.M. and Andrade, R. (2001b) A 5-HT(7) receptor-mediated depolarization in the anterodorsal thalamus. II. Involvement of the hyperpolarization-activated current I(h). J. Pharmacol. Exp. Ther., 297: 403–409.

Chaput, Y., de Montigny, C. and Blier, P. (1986) Effects of a selective 5-HT reputake blocker, citalopram, on the sensitivity of 5-HT autoreceptors: electrophysiological studies in the rat brain. Naunyn-Schmiedebergs Arch. Pharmacol., 333: 342–348.

Chaput, Y., Araneda, R.C. and Andrade, R. (1990) Pharmacological and functional analysis of a novel serotonin receptor in the rat hippocampus. Eur. J. Pharmacol., 182: 441–456.

Chen, L., Yung, K.K., Chan, Y.S. and Yung, W.H. (2008) 5-HT excites globus pallidus neurons by multiple receptor mechanisms. Neuroscience, 151: 439–451.

Cheng, L.L., Wang, S.J. and Gean, P.W. (1998) Serotonin depresses excitatory synaptic transmission and depolarization-evoked $Ca2^+$ influx in rat basolateral amygdala via 5-HT1A receptors. Eur. J. Neurosci., 10: 2163–2172.

Choi, I.S., Cho, J.H., Kim, J.T., Park, E.J., Lee, M.G., Shin, H.I., Choi, B.J. and Jang, I.S. (2007) Serotonergic modulation of GABAergic synaptic transmission in developing rat CA3 pyramidal neurons. J. Neurochem., 103: 2342–2353.

Colino, A. and Halliwell, J.V. (1987) Differential modulation of three separate K-conductances in hippocampal CA1 neurons by serotonin. Nature, 328: 73–77.

Costall, B. and Naylor, R.J. (1994) 5-HT3 Receptor Antagonism and Schizophrenia. In: King, F.D., Jones, B.J. and Sanger, G.J. (Eds.), 5-Hydroxytryptamine-3 Receptor Antagonists. CRC Press, Inc., Boca Raton, FL, pp. 183–201.

Cumberbatch, M.J., Williamson, D.J., Mason, G.S., Hill, R.G. and Hargreaves, R.J. (1999) Dural vasodilation causes a sensitization of rat caudal trigeminal neurones *in vivo* that is blocked by a 5-HT1B/1D agonist. Br. J. Pharmacol., 126: 1478–1486.

Darmani, N.A., Martin, B.R., Pandey, U. and Glennon, R.A. (1990) Do functional relationships exist between 5-HT_{1A} and 5-HT_2 receptors? Pharmacol. Behav. Biochem., 36: 901–906.

De Montigny, C. and Aghajanian, G.K. (1978) Tricyclic antidepressants: long-term treatment increases responsivity of rat forebrain neurons to serotonin. Science, 202: 1303–1306.

Di Giovanni, G., Di Matteo, V., Di Mascio, M. and Esposito, E. (2000) Preferential modulation of mesolimbic vs nigrostiatal dopaminergic function by serotonin(2C/2B) receptor agonists: a combined *in vivo* electrophysiological and microdialysis study. Synapse, 35: 53–61.

Di Giovanni, G., Di Matteo, V., La Grutta, V. and Esposito, E. (2001) m-Chlorophenylpiperazine excites non-dopaminergic neurons in the rat substantia nigra and ventral tegmental area by activating serotonin-2C receptors. Neuroscience, 103: 111–116.

Di Matteo, V., Di Giovanni, G., Di Mascio, M. and Esposito, E. (1999) SB 242084, a selective serotonin2C receptor antagonist, increases dopaminergic transmission in the mesolimbic system. Neuropharmacology, 38: 1195–1205.

Di Matteo, V., Di Giovanni, G., Di Mascio, M. and Esposito, E. (2000) Biochemical and electrophysiological evidence that RO 60-0175 inhibits mesolimbic dopaminergic function through serotonin(2C) receptors. Brain Res., 865: 85–90.

Di Matteo, V., De Blasi, A., Di Giulio, C. and Esposito, E. (2001) Role of 5-HT_{2C} receptor in the control of central dopamine function. TIPS, 22: 229–232.

Dorostkar, M.M. and Boehm, S. (2007) Opposite effects of presynaptic 5-HT3 receptor activation on spontaneous and action potential-evoked GABA release at hippocampal synapses. J. Neurochem., 100: 395–405.

Evrard, A., Laporte, A.M., Chastanet, M., Hen, R., Hamon, M. and Adrien, J. (1999) 5-HT1A and 5-HT1B receptors control the firing of serotoninergic neurons in the dorsal raphe nucleus of the mouse: studies in 5-HT1B knockout mice. Eur. J. Neurosci., 11: 3823–3831.

Feng, J., Cai, X., Zhao, J. and Yan, Z. (2001) Serotonin receptors modulate GABAA receptor channels through activation of anchored protein kinase C in prefrontal cortical neurons. J. Neurosci., 21: 6502–6511.

Gellman, R.L. and Aghajanian, G.K. (1994) $Serotonin_2$ receptor-mediated excitation of interneurons in pirform cortex: antagonism by atypical antipsychotic drugs. Neuroscience, 58: 515–525.

Gerard, C., El Mestikawy, S., Lebrand, C., Adrien, J., Ruat, M., Traiffort, E., Hamon, M. and Martres, M.P. (1996) Quantitative RT-PCR distribution of serotonin 5-HT6 receptor mRNA in the central nervous system of control or 5,7-dihydroxytryptamine-treated rats. Synapse, 23: 164–173.

Gerard, C., Martres, M.P., Lefevre, K., Miquel, M.C., Verge, D., Lanfumey, L., Doucet, E., Hamon, M. and el Mestikawy, S. (1997) Immuno-localization of serotonin 5-HT6 receptor-like material in the rat central nervous system. Brain Res., 746: 207–219.

Goadsby, P.J. and Knight, Y. (1997) Inhibition of trigeminal neurones after intravenous administration of naratriptan through an action at 5-hydroxy-tryptamine (5-HT(1B/1D)) receptors. Br. J. Pharmacol., 122: 918–922.

Gobert, A., Rivet, J.M., Lejeune, F., Newman-Tancredi, A., Adhumeau-Auclair, A., Nicolas, J.P., Cistarelli, L., Melon, C. and Millan, M.J. (2000) Serotonin(2C) receptors tonically suppress the activity of mesocortical dopaminergic and adrenergic, but not serotonergic, pathwsay: a combined dialysis and electrophysiological analysis in the rat. Synapse, 36: 205–221.

Gonzalez-Maeso, J., Weisstaub, N.V., Zhou, M., Chan, P., Ivic, L., Ang, R., Lira, A., Bradley-Moore, M., Ge, Y., Zhou, Q., Sealfon, S.C. and Gingrich, J.A. (2007) Hallucinogens recruit specific cortical 5-HT_{2A} receptor-mediated signalling pathways to affect behavior. Neuron, 53: 439–452.

Grace, A.A., Bunney, B.S., Moore, H. and Todd, C.L. (1997) Dopamine-cell depolarization block as a model for the therapeutic actions of antipsychotic drugs. Trends Neurosci., 20: 31–37.

Greenshaw, A.J. and Silverstone, P.H. (1997) The non-antiemetic uses of serotonin 5-HT_3 receptor antagonists. Drugs, 53: 20–39.

Haddjeri, N., Blier, P. and de Montigny, C. (1997) Effects of long-term treatment with the a_2-adrenoceptor antagonist mirtazepine on 5-HT neurotransmission. Naunyn-Schmiedeberg's Arch. Pharmacol., 355: 20–29.

Haddjeri, N., Blier, P. and deMontigny, C. (1998) Long-term antidepressant treatments result in a tonic activation of forebrain 5-HT_{1A} receptors. J. Neurosci., 18: 10150–10156.

Haddjeri, N., Szabo, S.T., de Montigny, C. and Blier, P. (2000) Increased tonic activation of rat forebrain 5-HT1A receptors by lithium addition to antidepressant treatments. Neuropsychopharmacology, 22: 346–356.

Hashimoto, K. and Kita, H. (2008) Serotonin activates presynaptic and postsynaptic receptors in rat globus pallidus. J. Neurophysiol., 99: 1723–1732.

Helton, L.A., Thor, K.B. and Baez, M. (1994) 5-hydroxytryptamine2A, 5-hydroxytryptamine2B, and 5-hydroxytryptamine2C receptor mRNA expression in the spinal cortd of rat, cat, monkey and human. Neuroreport, 5: 2617–2620.

Hoskin, K.L., Lambert, G.A., Donaldson, C. and Zagami, A.S. (2004) The 5-hydroxytryptamine1B/1D/1F receptor agonists eletriptan and naratriptan inhibit trigeminovascular input to the nucleus tractus solitarius in the cat. Brain Res., 998: 91–99.

Jakab, R.L. and Goldman-Rakic, P.S. (1998) 5-HT2A serotonin receptors in the primate cerebral cortex: possible site of action of hallucinogens in pyramidal cell apical dendrites. Proc. Natl. Acad. Sci. USA, 95: 735–740.

Jakab, R.L. and Goldman-Rakic, P.S. (2000) Segregation of serotonin 5-HT2A and 5-HT3 receptors in inhibitory circuits of the primate cerebral cortex. J. Comp. Neurol., 417: 337–348.

Jeggo, R.D., Wang, Y., Jordan, D. and Ramage, A.G. (2007) Activation of 5-HT1B and 5-HT1D receptors in the rat nucleus tractus solitarius: opposing action on neurones that receive an excitatory vagal C-fibre afferent input. Br. J. Pharmacol., 150: 987–995.

Katsurabayashi, S., Kubota, H., Tokutomi, N. and Akaike, N. (2003) A distinct distribution of functional presynaptic 5-HT receptor subtypes on GABAergic nerve terminals projecting to single hippocampal CA1 pyramidal neurons. Neuropharmacology, 44: 1022–1030.

Kawa, K. (1994) Distribution and functional properties of 5-HT3 receptors in the rat hippocampus dentate gyrus: a patch-clamp study. J. Neurophysiol., 71: 1935–1947.

Kawahara, F., Saito, H. and Katsuki, H. (1994) Inhibition of 5-HT7 receptor stimulation of GABAA receptor-activated current in cultured rat suprachiasmatic neurons. J. Physiol., 478(Pt 1): 67–73.

Kishimoto, K., Koyama, S. and Akaike, N. (2000) Presynaptic modulation of synaptic gamma-aminobutyric acid transmission by tandospirone in rat basolateral amygdala. Eur. J. Pharmacol., 407: 257–265.

Kobayashi, K., Ikeda, Y., Haneda, E. and Suzuki, H. (2008) Chronic fluoxetine bidirectionally modulates potentiating effects of serotonin on the hippocampal mossy fiber synaptic transmission. J. Neurosci., 28: 6272–6280.

Koyama, S., Matsumoto, N., Kubo, C. and Akaike, N. (2000) Presynaptic 5-HT3 receptor-mediated modulation of synaptic GABA release in the mechanically dissociated rat amygdala neurons. J. Physiol., 529: 373–383.

Koyama, S., Matsumoto, N., Murakami, N., Kubo, C., Nabekura, J. and Akaike, N. (2002a) Role of presynaptic 5-HT1A and 5-HT3 receptors in modulation of synaptic GABA transmission in dissociated rat basolateral amygdala neurons. Life Sci., 72: 375–387.

Koyama, S., Matsumoto, N., Murakami, K., Kubo, C., Nabekura, J. and Akaike, N. (2002b) Role of presynaptic 5-HT1A and 5-HT3 receptors in modulation of synaptic GABA transmission in dissociated rat basolateral amygdala neurons. Life Sci., 72: 375–387.

Lambe, E.K. and Aghajanian, G.K. (2001) The role of Kv1.2-containing potassium channels in serotonin-induced glutamate release from thalamocortical terminals in rat frontal cortex. J. Neurosci., 21: 9955–9963.

Lambe, E.K., Goldman-Rakic, P.S. and Aghajanian, G.K. (2000) Serotonin induces EPSCs preferentially in layer V pyramidal neurons of the frontal cortex in the rat. Cereb. Cortex, 10: 974–980.

Laurent, A., Goaillard, J.-M., Cases, O., Lebrand, C., Gaspar, P. and Ropert, N. (2002) Activity-dependent presynaptic effect of serotonin 1B receptors on the somatosensory thalamocortical transmission in neonatal mice. J. Neurosci., 22: 886–900.

Lebrand, C., Cases, O., Adelbrecht, C., Doye, A., Alvarez, C., El Mestikawy, S., Seif, I. and Gaspar, P. (1996) Transient uptake nad storage of serotonin in developing thalamic neurons. Neuron, 17: 991–1003.

Levita, L., Hammack, S.E., Mania, I., Li, X.Y., Davis, M. and Rainnie, D.G. (2004) 5-Hydroxytryptamine1A-like receptor activation in the bed nucleus of the stria terminalis: electrophysiological and behavioral studies. Neuroscience, 128: 583–596.

Lin, C.H., Huang, Y.C., Tsai, J.J. and Gean, P.W. (2001) Modulation of voltage-dependent calcium currents by serotonin in acutely dissociated isolated rat amygdala neurons. Synapse, 41: 351–359.

Lopez-Gimenez, J., Vilaro, M.T., Palacios, J.M. and Mengod, G. (1998) [^{3}H]MDL 100,907 labels 5-HT_{2A} serotonin receptors selectively in primate brain. Neuropharmacology, 37: 1147–1158.

Lopez-Gimenez, J.F., Mengod, G., Palacios, J.M. and Vilaro, M.T. (1997) Selective visualization of rat brain 5-HT2A receptors

by autoradiography with [3H]MDL 100,907. Naunyn-Schmiedeberg's Arch. Pharmacol., 356: 446–454.
Lopez-Gimenez, J.F., Mengod, G., Palacios, J.M. and Vilaro, M.T. (2001) Regional distribution and cellular localization of 5-HT2C receptor mRNA in monkey brain: comparison with [3H]mesulergine binding sites and choline acetyltransferase mRNA. Synapse, 42: 12–26.
Lucaites, V.L., Krushinski, J.H., Schaus, J.M., Audia, J.E. and Nelson, D.L. (2005) [3H]LY334370, a novel radioligand for the 5-HT1F receptor. II. Autoradiographic localization in rat, guinea pig, monkey and human brain. Naunyn-Schmiedeberg's Arch. Pharmacol., 371: 178–184.
Marek, G.J. (2003) Behavioral evidence for m-opioid and 5-HT_{2A} receptor interactions. Eur. J. Pharmacol., 474: 77–83.
Marek, G.J. (2008a) Regulation of rat cortical 5-hydroxytryptamine$_{2A}$-receptor mediated electrophysiological responses by repeated daily treatment with electroconvulsive shock or imipramine. Eur. Neuropsychopharmacol., 18: 498–507.
Marek, G.J. (2008b) Cortical 5-hydroxytryptamine$_{2A}$-receptor mediated excitatory synaptic currents in the rat following repeated daily fluoxetine administration. Neurosci. Lett., 438: 312–316.
Marek, G.J. (2009) Activation of adenosine A1(A1) receptors suppreseses head shakes induced by serotonergic hallucinogenin rats. Neuropharmacol., 56: 1082–1087.
Marek, G.J. and Aghajanian, G.K. (1994) Excitation of interneurons in piriform cortex by 5-hydroxytryptamine: blockade by MDL 100,907, a highly selective 5-HT_{2A} receptor antagonist. Eur. J. Pharmacol., 259: 137–141.
Marek, G.J. and Aghajanian, G.K. (1995) Protein kinase C inhibitors enhance the 5-HT_{2A} receptor-mediated excitatory effects of serotonin on interneurons in rat piriform cortex. Synapse, 21: 123–130.
Marek, G.J. and Aghajanian, G.K. (1996) LSD and the phenethylamine hallucinogen DOI are potent partial agonists at 5-HT_{2A} receptors on neurons in the rat piriform cortex. J. Pharmacol. Exp. Ther., 278: 1373–1382.
Marek, G.J. and Aghajanian, G.K. (1998) 5-HT-induced EPSCs in neocortical layer V pyramidal cells: suppression by μ-opiate receptor activation. Neurosci, 86: 485–497.
Marek, G.J., Wright, R.A., Schoepp, D.D., Monn, J.A. and Aghajanian, G.K. (2000) Physiological antagonism between 5-hydroxytryptamine$_{2A}$ and group II metabotropic glutamate receptors in prefrontal cortex. J. Pharmacol. Exp. Ther., 292: 76–87.
Marek, G.J., Wright, R.A., Gewirtz, J.C. and Schoepp, D.D. (2001) A major role for thalamocortical afferents in serotonergic hallucinogen receptor function in the rat neocortex. Neuroscience, 105: 379–392.
Marek, G.J., Carpenter, L.L., McDougle, C.J. and Price, L.H. (2003) Synergistic action of 5-HT_{2A} antagonists and selective serotonin reuptake inhibitors in neuropsychiatric disorders. Neuropsychopharmacology, 28: 402–412.
Matsumoto, M., Togashi, H., Mori, K., Ueno, K., Ohashi, S., Kojima, T. and Yoshioka, M. (2001) Evidence for involvement of central 5-HT(4) receptors in cholinergic function associated with cognitive processes: behavioral, electrophysiological, and neurochemical studies. J. Pharmacol. Exp. Ther., 296: 676–682.
McMahon, L.L. and Kauer, J.A. (1997) Hippocampal interneurons are excited via serotonin-gated ion channels. J. Neurophysiol., 78: 2493–2502.
Mengod, G., Nguyen, H., Lee, H., Waeber, C., Lubbert, H. and Palacios, J.M. (1990) The distribution and cellular localization of 5-HT_{1C} receptor mRNA in the rodent brain examined by *in situ* hybridization histochemistry. Comparison with receptor binding distribution. Neuroscience, 35: 577–592.
Minabe, Y., Shirayama, Y., Hashimoto, K., Routledge, C., Hagan, J.J. and Ashby, C.R. Jr. (2004) Effect of the acute and chronic administration of the selective 5-HT6 receptor antagonist SB-271046 on the activity of midbrain dopamine neurons in rats: an in vivo electrophysiological study. Synapse, 52: 20–28.
Mlinar, B., Falsini, C. and Corradetti, R. (2003) Pharmacological characterization of 5-HT_{1B} receptor-mediated inhibition of local excitatory synaptic transmission in the CA1 region of the rat hippocampus. Br. J. Pharmacol., 138: 71–80.
Mooney, R.D., Shi, M.Y. and Rhoades, R.W. (1994) Modulation of retinotectal transmission by presynaptic 5-HT1B receptors in the superior colliculus of the adult hamster. J. Neurophysiol., 72: 3–13.
Mooney, R.D., Huang, X., Shi, M.Y., Bennett-Clarke, C.A. and Rhoades, R.W. (1996) Serotonin modulates retinotectal and corticotectal convergence in the superior colliculus. Prog. Brain Res., 112: 57–69.
Morales, M.E. and Bloom, F.E. (1997) The 5-HT_3 receptor is present in different subpopulations of GABAergic neurons in the rat telencephalon. J. Neurosci., 17: 3157–3167.
Morales, M.E. and Wang, S.-D. (2002) Differential composition of 5-hydroxytryptamine$_3$ receptors synthesized in the rat CNS and peripheral nervous system. J. Neurosci., 22: 6732–6741.
Morales, M., Battenberg, E., de Lecea, L. and Bloom, F.E. (1996) The type 3 serotonin receptor is expressed in a subpopulation of GABAergic neurons in the rat neocortex and hippocampus. Brain Res., 731: 199–202.
Nelson, D.L. (2004) 5-HT_5 receptors. Curr. Drug Targets CNS Neurol. Disord., 3: 53–58.
Nichols, D.E. (2004) Hallucinogens. Pharmacol. Ther., 101: 131–181.
Olijslagers, J.E., Werkman, T.R., McCreary, A.C., Siarey, R., Kruse, C.G. and Wadman, W.J. (2004) 5-HT2 receptors differentially modulate dopamine-mediated auto-inhibition in A9 and A10 midbrain areas of the rat. Neuropharmacology, 46: 504–510.
Otmakhova, N.A., Lewey, J., Asrican, B. and Lisman, J.E. (2005) Inhibition of perforant path input to the CA1 region by serotonin and noradrenaline. J. Neurophysiol., 94: 1413–1422.
Patil, S.T., Zhang, L., Martenyi, F., Lowe, S.L., Jackson, K.A., Andreev, B.V., Avedisova, A.S., Bardenstein, L.M., Gurovich, I.Y., Morozova, M.A., Mosolov, S.N., Neznanov, N.G., Reznik, A.M., Smulevich, A.B., Tochilov, V.A., Johnson, B.G., Monn, J.A. and Schoepp, D.D. (2007) Activation of mGlu2/3 receptors as a new approach to treat schizophrenia: a randomized Phase 2 clinical trial. Nat. Med., 13: 1103–1108.
Persico, A.M., Mengual, E., Moessner, R., Hall, S.F., Revay, R.S., Sora, I., Arellano, J., DeFelipe, J., Gimenez-Amaya, J.M., Conciatori, M., Marino, R., Baldi, A., Cabib, S., Pascucci, T., Uhl, G.R., Murphy, D.L., Lesch, K.P. and Keller, F. (2001) Barrell pattern formation requires serotonin uptake by thalamocortical afferents, and not vesicular monoamine release. J. Neurosci., 21: 6862–6873.
Piguet, P. and Galvan, M. (1994) Transient and long-lasting actions of 5-HT on rat dentate gyrus neurones *in vitro*. J. Physiol. (Lond.), 481: 629–639.

Pinault, D. (1995) Backpropagation of action potentials generated at ectopic axonal loci: hypothesis that axon terminals integrate local environmental signals. Brain Res. Rev., 21: 42–92.

Pineyro, G. and Blier, P. (1999) Autoregulation of serotonin neurons: role in antidepressant drug action. Pharmacol. Rev., 51: 533–591.

Pitra, P., Tokarski, K., Grzegorzewska, M. and Hess, G. (2007) Effects of repetitive administration of tianeptine, zinc hydroaspartate and electroconvulsive shock on the reactivity of 5-HT(7) receptors in rat hippocampus. Pharmacol. Rep., 59: 627–635.

Pompeiano, M., Palacios, J.M. and Mengod, G. (1992) Distribution and cellular localization of mRNA coding for 5-HT$_{1A}$ receptor in the rat: correlation with receptor binding. J. Neurosci., 12: 440–453.

Ponomarenko, A.A., Knoche, A., Korotkova, T.M. and Haas, H.L. (2003) Aminergic control of high-frequency (approximately 200 Hz) network oscillations in the hippocampus of the behaving rat. Neurosci. Lett., 348: 101–104.

Puig, M.V., Celada, P., Diaz-Mataix, L. and Artigas, F. (2003) In vivo modulation of the activity of pyramidal neurons in the rat medial prefrontal cortex by 5-HT$_{2A}$ receptors. Relationship to thalamocortical afferents. Cereb. Cortex, 13: 870–882.

Querejeta, E., Oviedo-Chavez, A., Araujo-Alvarez, J.M., Quinones-Cardenas, A.R. and Delgado, A. (2005) In vivo effects of local activation and blockade of 5-HT1B receptors on globus pallidus neuronal spiking. Brain Res., 1043: 186–194.

Ramadan, N.M., Skljarevski, V., Phebus, L.A. and Johnson, K.W. (2003) 5-HT1F receptor agonists in acute migraine treatment: a hypothesis. Cephalalgia, 23: 776–785.

Rasmussen, K. and Aghajanian, G.K. (1986) Effects of hallucinogens on spontaneous and sensory-evoked locus coeruleus unit activity in the rat: reversal by selective 5-HT$_2$ antagonists. Brain Res., 385: 395–400.

Rasmussen, K. and Aghajanian, G.K. (1988) Potency of antipsychotics in reversing the effects of a hallucinogenic drug on locus coeruleus neurons correlates with 5-HT$_2$ binding affinity. Neuropsychopharmacology, 1: 101–107.

Rasmussen, K., Stockton, M.E. and Czachura, J.F. (1991) The 5-HT3 receptor antagonist zatosetron decreases the number of spontaneously active A10 dopamine neurons. Eur. J. Pharmacol., 205: 113–116.

Rav-Acha, M., Bergman, H. and Yarom, Y. (2008) Pre- and postsynaptic serotonergic excitation of globus pallidus neurons. J. Neurophysiol., 100: 1053–1066.

Read, H.L., Beck, S.G. and Dun, N.J. (1994) Serotonergic suppression of interhemispheric cortical synaptic potentials. Brain Res., 643: 17–28.

Rorick-Kehn, L.M., Johnson, B.G., Burkey, J.L., Wright, R.A., Calligaro, D.O., Marek, G.J., Nisenbaum, E.S., Catlow, J.T., Kingston, A.E., Giera, D., Herin, M.F., Monn, J.A., McKinzie, D.L. and Schoepp, D.D. (2007a) Pharmacological and pharmacokinetic properties of a structurally novel, potent, and selective metabotropic glutamate 2/3 receptor agonist: in vitro characterization of agonist (−)-(1R,4S,5S,6S)-4-amino-2-sulfonylbicyclo[3.1.0]-hexane-4,6-dicarboxylic acid (LY404039). J. Pharmacol. Exp. Ther., 321: 308–317.

Rorick-Kehn, L.M., Johnson, B.G., Knitowski, K.M., Salhoff, C.R., Witkin, J.M., Perry, K.W., Griffey, K.I., Tizzano, J.P., Monn, J.A., McKinzie, D.L. and Schoepp, D.D. (2007b) *In vivo* pharmacological characterization of the structurally novel, potent, selective mGlu2/3 receptor agonist LY404039 in animal models of psychiatric disorders. Psychopharmacol, 193: 121–136.

Rothman, R.B., Baumann, M.H., Savage, J.E., Rauser, L., McBride, A., Hufeisen, S.J. and Roth, B.L. (2000) Evidence for possible involvement of 5-HT(2B) receptors in the cardiac valvulopathy associated with fenfluramine and other serotonergic medications. Circulation, 102: 2836–2841.

Rueter, L.A., Montigny, C. and Blier, P. (1998) Electrophysiological characterization of the effect of long-term duloxetine administration on the rat serotonergic and noradrenergic systems. J. Pharmacol. Exp. Ther., 285: 404–412.

Rueter, L.E. and Blier, P. (1999) Electrophysiological examination of the effects of sustained flibanserin administration on serotonin receptors in rat brain. Br. J. Pharmacol., 126: 627–638.

Santana, N., Bortolozzi, A., Serrats, J., Mengod, G. and Artigas, F. (2004) Expression of serotonin$_{1A}$ and serotonin$_{2A}$ receptors in pyramidal and GABAergic neurons of the rat prefrontal cortex. Cereb. Cortex, 14: 1100–1109.

Sari, Y., Lefevre, K., Bancila, M., Quignon, M., Miguel, M.C., Langlois, X., Hamon, M. and Verge, D. (1997) Light and electron microscopic immunocytochemical visualization of 5-HT1B receptors in the rat brain. Brain Res., 760: 281–286.

Sari, Y., Miquel, M.C., Brisorgueil, M.J., Ruiz, G., Doucet, E., Hamon, M. and Verge, D. (1999) Cellular and subcellular localization of 5-hydroxytryptamine1B receptors in the rat central nervous system: immunocytochemical, autoradiographic and lesion studies. Neuroscience, 88: 899–915.

Schmuck, K., Ullmer, C., Engels, P. and Lubbert, H. (1994) Cloning and functional characterization of the human 5-HT2B serotonin receptor. FEBS Lett., 342: 85–90.

Schoepp, D.D. (2001) Unveiling the functions of presynaptic metabotropic glutamate receptors in the central nervous system. J. Pharmacol. Exp. Ther., 299: 12–20.

Scruggs, J.L., Patel, S., Bubser, M. and Deutch, A.Y. (2000) DOI-induced activation of the cortex: dependence upon 5-HT$_{2A}$ heteroceptors on thalamocortical glutamatergic neurons. J. Neurosci., 20: 8846–8852.

Sheldon, P.W. and Aghajanian, G.K. (1990) Serotonin (5-HT) induces IPSPs in pyramidal layer cells of rat piriform cortex: evidence for the involvement of a 5-HT$_2$-activated interneuron. Brain Res., 506: 62–69.

Sheldon, P.W. and Aghajanian, G.K. (1991) Excitatory responses to serotonin (5-HT) in neurons of the rat piriform cortex: evidence for mediation by 5-HT$_{1C}$ receptors in pyramidal cells and 5-HT$_2$ receptors in interneurons. Synapse, 9: 208–218.

Shen, K.Z., Kozell, L.B. and Johnson, S.W. (2007) Multiple conductances are modulated by 5-HT receptor subtypes in rat subthalamic nucleus neurons. Neuroscience, 148: 996–1003.

Shen, R.Y. and Andrade, R. (1998) 5-Hydroxytryptamine2 receptor facilitates GABAergic neurotransmission in rat hippocampus. J. Pharmacol. Exp. Ther., 285: 805–812.

Sherman, S.M. and Guillery, R.W. (1996) Functional organization of thalamocortical relays. J. Neurophysiol., 76: 1367–1395.

Sherman, S.M. and Guillery, R.W. (1998) On the actions that one nerve cell can have on another: distinguishing 'drivers' from 'modulators'. Proc. Natl. Acad. Sci. USA, 95: 7121–7126.

Sorensen, S.M., Humphreys, T.M. and Palfreyman, M.G. (1989) Effect of acute and chronic MDL 73,147EF, a 5-HT3 receptor

antagonist, on A9 and A10 dopamine neurons. Eur. J. Pharmacol., 163: 115–118.
Stanford, I.M. and Lacey, M.G. (1996) Differential actions of serotonin, mediated by 5-HT1B and 5-HT2C receptors, on GABA-mediated synaptic input to rat substantia nigra pars reticulata neurons *in vitro*. J Neurosci., 16: 7566–7573.
Stanford, I.M., Kantaria, M.A., Chahal, H.S., Loucif, K.C. and Wilson, C.L. (2005) 5-Hydroxytryptamine induced excitation and inhibition in the subthalamic nucleus: action at 5-HT(2C), 5-HT(4), 5-HT(1A) receptors. Neuropharmacology, 49: 1228–1234.
Steriade, M., Jones, E.G. and McCormick, D.A. (1997) Thalamus: Organization and Function. Elsevier, New York, NY.
Stutzman, G.E., Marek, G.J. and Aghajanian, G.K. (2001) Adenosine preferentially suppresses serotonin2A receptor-enhanced excitatory postsynaptic currents in layer V neurons of the rat medial prefrontal cortex. Neuroscience, 105: 55–69.
Sudweeks, S.N., Hooft, J.A. and Yakel, J.L. (2002) Serotonin 5-HT(3) receptors in rat CA1 hippocampal interneurons: functional and molecular characterization. J. Physiol., 544: 715–726.
Sugita, S., Shen, K.Z. and North, R.A. (1992) 5-Hydroxytryptamine is a fast excitatory transmitter at 5-HT_3 receptors in rat amygdala. Neuron, 8: 199–203.
Tanaka, E. and North, R.A. (1993) Actions of 5-hydroxytryptamine on neurons of the rat cingulate cortex. J. Neurophysiol., 69: 1749–1757.
Tanaka, E. and North, R.A. (1994) Opioid actions on rat anterior cingulate cortex neurons *in vitro*. J. Neurosci., 14: 1106–1113.
Titeler, M., Lyon, R.A. and Glennon, R.A. (1988) Radioligand binding evidence implicates the brain 5-HT_2 receptor as a site of action for LSD and phenethylamine hallucinogens. Psychopharmacology, 94: 213–216.
Toharski, K., Zahorodna, A., Bobula, B., Grzegorzewska, M., Pitra, P. and Hess, G. (2005) Repeated administration of citalopram and imipramine alters the responsiveness of rat hippocampal circuitry to the activation of 5-HT7 receptors. Eur. J. Pharmacol., 524: 60–66.
Tokarski, K. and Bijak, M. (1996) Antidepressant-induced adaptive changes in the effects of 5-HT, 5-HT1A and 5-HT4 agonists on the population spike recorded in hippocampal CA1 cells do not involve presynaptic effects on excitatory synaptic transmission. Pol. J. Pharmacol., 48: 565–573.
Tokarski, K., Zahorodna, A., Bobula, B. and Hess, G. (2003) 5-HT7 receptors increase the excitability of rat hippocampal CA1 pyramidal neurons. Brain Res., 993: 230–234.
Torres, G.E., Holt, I.L. and Andrade, R. (1994) Antagonists of 5-HT4 receptor-mediated responses in adult hippocampal neurons. J. Pharmacol. Exp. Ther., 271: 255–261.
Torres, G.E., Arfken, C.L. and Andrade, R. (1996) 5-hydroxytryptamine$_4$ receptors reduce afterhyperpolarization in hippocampus by inhibiting calcium-induced calcium release. Mol. Pharmacol., 50: 1316–1322.
Varnas, K., Thomas, D.R., Tupala, E., Tiihonen, J. and Hall, H. (2004) Distribution of 5-HT7 receptors in the human brain: a preliminary autoradiographic study using [3H]SB-269970. Neurosci. Lett., 367: 313–316.
Wang, S.J., Cheng, L.L. and Gean, P.W. (1999) Cross-modulation of synaptic plasticity by beta-adrenergic and 5-HT1A receptors in the rat basolateral amygdala. J. Neurosci., 19: 570–577.
Weisstaub, N.V., Zhou, M., Lira, A., Lambe, E.K., Gonzalez-Maeso, J., Hornung, J.-P., Sibille, E., Underwood, M.D., Itohara, S., Dauer, W.T., Ansorge, M.S., Morelli, E., Mann, J.J., Toth, M., Aghajanian, G.K., Sealfon, S.C., Hen, R. and Gingrich, J.A. (2006) Cortical 5-HT_{2A} receptor signalling modulates anxiety-like behaviors in mice. Science, 313: 536–540.
Willins, D.L. and Meltzer, H.Y. (1997) Direct injection of 5-HT2A receptor agonists into the medial prefrontal cortex produces a head-twitch response in rats. J. Pharmacol. Exp. Ther., 282: 699–706.
Willins, D.L., Deutch, A.Y. and Roth, B.L. (1997) Serotonin 5-HT2A receptors are expressed on pyramidal cells and interneurons in the rat cortex. Synapse, 27: 79–82.
Wright, D.E., Seroogy, K.B., Lundgren, K.H., Davis, B.M. and Jennes, L. (1995) Comparative localization of serotonin$_{1A,\ 1C,}$ and $_2$ receptor subtype mRNAs in rat brain. J. Comp. Neurol., 351: 357–373.
Xiang, Z. and Prince, D.A. (2003) Heterogenous actions of serotonin on interneurons in rat visual cortex. J. Neurophysiol., 89: 1278–1287.
Xiang, Z., Wang, L. and Kitai, S.T. (2005) Modulation of spontaneous firing in rat subthalamic neurons by 5-HT receptor subtypes. J. Neurophysiol., 93: 1145–1157.
Yakel, J.L. and Jackson, M.B. (1988) 5-HT_3 receptors mediate rapid responses in cultured hippocampus and a clonal cell line. Neuron, 1: 615–621.
Zhang, C. and Marek, G.J. (2008) AMPA receptors involvement in 5-hydroxytryptamine$_{2A}$ receptor-mediated prefrontal cortical excitatory synaptic currents and DOI-induced head shakes. Prog. Neuropsychopharmacol. Biol. Psychiatry, 32: 62–71.
Zhou, F.-M. and Hablitz, J.J. (1999) Activation of serotonin receptors modulates synaptic transmission in rat cerebral cortex. J. Neurophysiol., 82: 2989–2999.

CHAPTER 2.3

Tryptophan Hydroxylase and Serotonin Synthesis Regulation

Hiroyuki Hasegawa*,[1] and Kazuhiro Nakamura[2]

[1]*Department of Biosciences, Teikyo University of Science and Technology, Uenohara, Japan*
[2]*Department of Pathology, Juntendo University School of Medicine, Tokyo, Japan*

Abstract: Potential determinants of existing tryptophan hydroxylase (TPH) activity *in vivo* are (1) a supply of tryptophan via the tryptophan transporter, (2) *de novo* synthesis of tetrahydrobiopterin and efficient regeneration of this coenzyme, and (3) iron metabolism. TPH is an iron protein. Iron binding is relatively strong in TPH2, while it is very weak in TPH1. Owing to the poorly liganded iron of TPH, various complexities arise. Irreversible inactivation of TPH and neural deterioration seem to be caused by an H_2O_2 reaction with non-liganded or poorly liganded Fe^{2+}. Within the cell, the amount of ferrous iron available to TPH is below saturation. The metabolism of iron may control serotonin production by attenuating actual TPH activity in the cell interior. Phosphorylation of TPH was studied in terms of its functional aspects with the use of brain TPH, while the identification of phosphorylation sites was mainly conducted on the basis of the amino acid sequence of TPH1. Protein turnover is cell-type specific in terms of its machinery, and its regulation is one of the major post-translational modifications of a protein. TPH degradation is likely linked to protein phosphorylation. Functionally, TPH, the rate-limiting enzyme in serotonin biosynthesis, is involved in neural development, which likely affects animal behaviors in the adult stages. Future studies will define the sensitive period during development when TPH activity is required for the fine-tuning of neuronal organization that is responsible for each behavior in the adult.

Keywords: iron protein, tetrahydrobiopterin, proteasome, hydroxyl radical, chelatable iron, iron deficiency, neural development, behavior, psychiatric disorder.

Abbreviations: TPH, tryptophan hydroxylase; 5-HT, 5-hydroxytryptamine (serotonin); 5-HTP, 5-hydroxy-L-tryptophan; AADC, aromatic L-amino acid decarboxylase; DHPR, dihydropteridine reductase; DHFR, dihydrofolate reductase; TH, tyrosine hydroxylase; PAH, phenylalanine hydroxylase; DTT, dithiothreitol; MDMA, 3,4-methylenedioxymethamphetamine; PCPA, para-chlorophenylalanine; PPI, prepulse inhibition.

Serotonin biosynthesis

Serotonin (5-HT) is synthesized from L-tryptophan by a two-step enzyme reaction (Figure 1):

$$\text{L-Tryptophan} \rightarrow \text{5-hydroxy-L-tryptophan (5-HTP)} \quad (1)$$

$$\text{5-Hydroxy-L-tryptophan} \rightarrow \text{5-hydroxytryptamine (5-HT)} \quad (2)$$

Reaction (1) is catalyzed by tryptophan hydroxylase (TPH; 1.14.16.4), and reaction (2) by aromatic-L-amino acid decarboxylase (AADC; EC 4.1.1.28). Among the serotonin-producing cells, tryptophan hydroxylase is expressed by such cells as serotonin-secreting neurons in the central nervous system and enteric neural plexus, enterochromaffin cells of the gastrointestinal mucosa, mast cells of various connective tissues, and pinealocytes of the pineal gland. A high concentration of 5-HT in these tissues is not always an indication of a strong expression of TPH. Platelets are an extreme example of this. They are the largest 5-HT carrier in the body, but hardly synthesize 5-HT at all. Instead, they take it up from the blood stream. Most 5-HT in the blood is secreted by the enterochromaffin cells of the gastrointestinal tract. Since the tissue level of 5-HT is maintained by a counterbalance between its production and release, it cannot be used to determine how rapidly 5-HT is regenerated unless the steady state is interrupted. In a classic challenge to this line of thought (Koe and Weissman, 1966), it was reported that the half-decay time of 5-HT in rat brain was about 12 hours

*Corresponding author
E-mail: hsgw-bh4@asahi-net.email.ne.jp

Christian Müller & Barry Jacobs (Eds.)
Handbook of Behavioral Neurobiology of Serotonin
ISBN 978-0-12-374634-4

DOI: 10.1016/B978-0-12-374634-4.00010-1

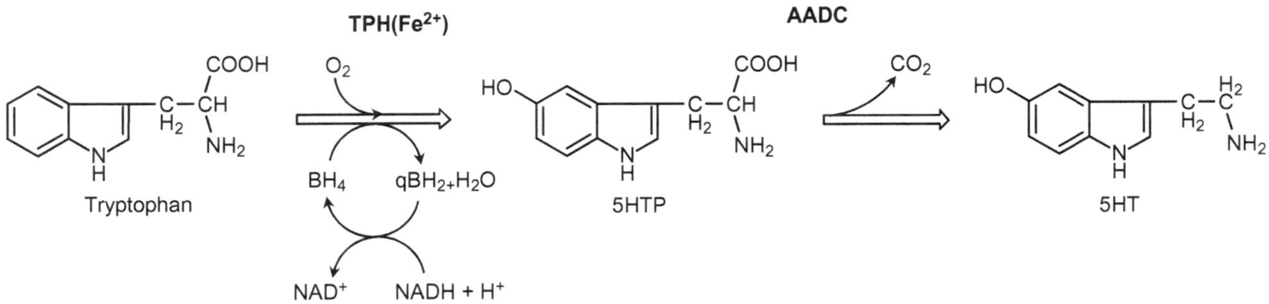

Figure 1 Biosynthesis of serotonin from tryptophan. Serotonin (5-HT, 5-hydroxytryptamine) is synthesized from L-tryptophan by a two-step enzyme reaction. Tryptophan hydroxylase (TPH) requires molecular oxygen (O_2) and tetrahydrobiopterin (BH_4) as a co-substrate. The enzyme has an absolute requirement for ferrous iron (Fe^{2+}). The reaction product, 5-hydroxy-L-tryptophan (5-HTP), is converted to serotonin by aromatic L-amino acid decarboxylase (AADC).

while that in the blood was about 3 days, the authors having arrived at their determinations by arresting the TPH reaction with high doses of p-chlorophenylalanine (PCPA), a pseudo-irreversible inhibitor of TPH. Neural 5-HT levels are relatively low compared with levels in peripheral organs such as the pineal gland and gastrointestinal mucosa, but production and utilization of 5-HT in the brain is much more rapid than all such tissues in the periphery.

AADC is the same entity as 'dopa decarboxylase', which catalyzes decarboxylation of dihydroxyphenylalanine (DOPA) in the production of catecholamines such as dopamine, noradrenaline and adrenaline (Lovenberg *et al.*, 1962; Christenson *et al.*, 1972). The enzyme is a homo-dimer containing one catalytically active pyridoxal 5′-phosphate per subunit (PALP-enzyme, or B6-enzyme) and catalyzes the decarboxylation of 5-HTP, yielding 5-HT in serotonin-producing cells (Voltattorni *et al.*, 1971). However, the enzyme AADC is not responsible for histamine production, which is catalyzed by the other PALP enzyme, histidine decarboxylase (HDC; EC 4.1.1.22). Distribution of AADC has been known to be much wider than that of tyrosine hydroxylase or TPH. In histochemical studies, many secretory cells derived from embryonic neural crest cells have been recognized as APUDs (members of the 'amine-precursor uptake and decarboxylation series'), which are characterized by monoamine accumulation when the precursor, dopa or 5-HTP, is supplied, and it is possible that these cells might express AADC. One idea that has been gaining support is that AADC is not only distributed non-specifically but also expressed in abundance. However, cellular levels of 5-HTP, the AADC substrate, are roughly 25–30 percent those of 5-HT on a molar basis in whole brains of mice, suggesting that this enzyme is not always furnished in excess. It is also noteworthy that 5-HT is condensed in secretory (particle) vesicles in 5-HT-secreting cells. These cells are usually furnished with the uptake transporter VMAT which is not bidirectional – i.e., not capable of outward movement. Since 5-HTP is bidirectionally relocated across the cell membrane, it is probably delivered via the amino acid transporter for tryptophan.

Background of tryptophan hydroxylase

Factors responsible for TPH activity in vivo

Tryptophan hydroxylase is a biopterin-dependent monooxygenase (Kaufman, 1962). Monooxygenases catalyze reactions by oxygenating their substrate with one atom of oxygen following cleavage of the double bond of dioxygen with two electrons donated by a co-substrate. These reactions require assistance from transient metals such as iron. In the TPH reaction, one oxygen atom is inserted into the indole ring at C(5) of tryptophan and the other oxygen is accepted by BH_4 (Figures 1 and 2) yielding 4a-hydroxy-BH_4. The hydroxylase engages in three enzyme reactions all connected to the biopterin cofactor:

$$\text{L-Tryptophan} + O_2 + BH_4 \rightarrow \text{5-HTP} + \text{4a-hydroxy} - BH_4 \quad (1)$$

$$\text{4a-hydroxy} - BH_4 \rightarrow qBH_2 + H_2O \quad (2)$$

$$qBH_2 + NAD(P)H + H^+ \rightarrow BH_4 + NAD(P)^+ \quad (3)$$

TPH reaction (1) produces 5-hydroxy-L-tryptophan (5-HTP) and 4a-hydroxy-BH_4 (pterin-4a-calbinolamine derivative of BH_4. Reaction (2) proceeds spontaneously or can be catalyzed by a dehydrating enzyme,[1] and reaction (3) proceeds by dihydropteridine reductase (DHPR; EC 1.6.99.7) (Kaufman, 1964). The total reaction process is as follows,

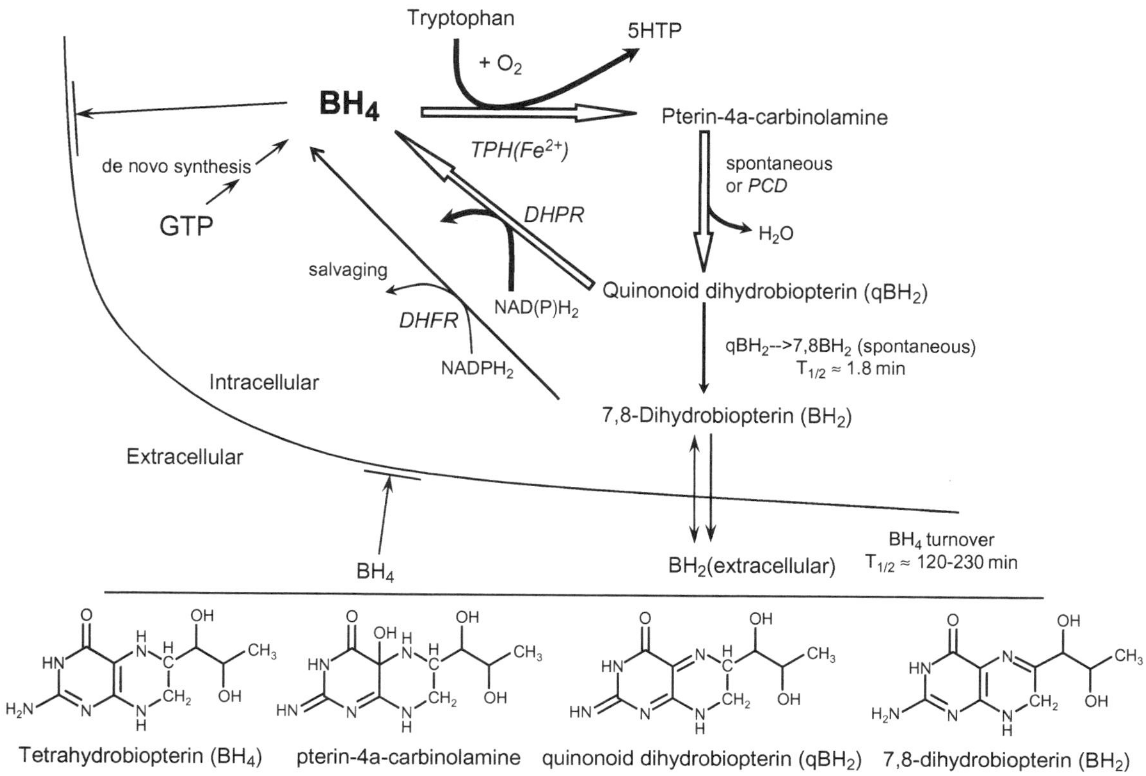

Figure 2 Recycling of BH4, the essential coenzyme. BH_4 works as a two-electron donor in the TPH reaction. The reaction product is 4a-hydroxy-BH_4 (pterin-4a-carbinolamine) which liberates H_2O, yielding quinonoid dihydrobiopterin (qBH_2). qBH_2 is reduced back to BH_4 by the action of dihydropteridine reductase (DHPR). qBH_2 spontaneously converts to 7,8-dihydrobiopterin (BH_2). BH_2 can be recycled by dihydrofolate reductase (DHFR) to BH_4, but DHFR is not always present in a large enough amount to convert BH_2 in 5-HT-producing cells. Hence, conversion of qBH_2 to BH_2 is virtually synonymous with inactivation of BH_4.

where a catalytic amount of BH_4 functions as an essential coenzyme for the TPH reaction in the cell:

$$\text{L-Tryptophan} + O_2 + \text{NAD(P)H} + H^+ \rightarrow \text{5-HTP} + H_2O + \text{NAD(P)}^+$$

Hence, TPH requires three substrates for its hydroxylation: (1) dioxygen, (2) tryptophan, and (3) tetrahydrobiopterin (BH_4). Dissociable ferrous iron is also a limiting component of the enzyme reaction under the actual intracellular conditions.

1. O_2 is most actively reduced in mitochondria to H_2O; however, a considerable portion of O_2 is incompletely reduced, yielding $O_2^{-\bullet}$ and H_2O_2. The reduced oxygen may be benign as long as ferrous iron is absent. However, TPH is highly susceptible to H_2O_2, presumably due to its poor liganding of Fe^{2+} (Kell, 2009), as discussed below. This might be the key to understanding various problems related to TPH, including cytotoxic events caused by drugs targeting 5-HT-producing cells.
2. L-Tryptophan, an essential amino acid for humans, is supplied to the brain through the blood circulation. Supplementation with tryptophan increases production of serotonin in the brain,[2] suggesting an insufficient supply of tryptophan as far as the central nervous system (CNS) is concerned (Koe and Weissman, 1966; Fernstrom and Wurtman, 1971).
3. BH_4 in the brain is believed to be synthesized in the CNS. The concentration of BH_4 as a TPH substrate is not at saturation level in the CNS (Miwa *et al.*, 1985), hence the BH_4 level would be directly responsible for 5-HT production.

Post-translational modification and TPH degradation would control 5-HT biosynthesis as well as TPH gene expression. The following sections will focus on cellular mechanisms for maintaining cofactor levels, as well as on the involvement of (1) BH4 and (2) ferrous iron. Then, the focus will shift to (3) enzyme stability, which is closely related to incomplete liganding with iron, followed by a discussion of (4) phosphorylation and, finally, (5) TPH degradation.

Two isoforms of tryptophan hydroxylase

Tryptophan hydroxylase exists as two isoforms, TPH1 and TPH2, which are encoded on two independent genes. The molecular identity of the brain type of TPH was unknown for the 15 years until the discovery of the second gene encoding TPH2.

In the late 1960s, tryptophan hydroxylase was first characterized in extracts from a neonatal tissue of peripheral origin (Hosoda and Glick, 1965; Lovenberg *et al.*, 1965) and from brain tissue (Grahame, 1967; Lovenberg *et al.*, 1967). The peripheral TPH and the brain enzyme were distinguished as different entities based on details of kinetic parameters and physicochemical properties especially reflecting its stability, as well as by immunoreactivity (Kuhn *et al.*, 1980a; Nakata and Fujisawa, 1982a, 1982b; Hasegawa *et al.*, 1987). However, the molecular basis of their separate status – i.e., the difference in their amino acid sequences – was not clearly recognized until 2003, when the second gene was found that represented the major portion of the brain enzyme (Cote *et al.*, 2003; Walther *et al.*, 2003; Zhang *et al.*, 2004).

Before gene cloning came into common use, an enzyme was characterized using the purified protein. In the case of TPH, however, the enzymes from both the peripheral organs and the brain were very difficult to purify, largely due to their[3] unusual instability (Ichiyama *et al.*, 1974; Kuhn *et al.*, 1980b; Cash, 1998). Even when the purification method was established in the early 1980s (Nakata and Fujisawa, 1982a, 1982b; Cash *et al.*, 1985), this extreme instability hampered any detailed characterization of the enzyme. The cloning of the cDNA could have facilitated characterization of the relevant enzyme. TPH cDNA was first cloned in 1987 from a rabbit peripheral organ, the pineal gland (Grenett *et al.*, 1987). Since then, many cDNAs have been cloned from various libraries of other sources including the brain (Darmon *et al.*, 1988; Boularand *et al.*, 1990; Stoll *et al.*, 1990; Kim *et al.*, 1991; D'Sa *et al.*, 1996; Florez *et al.*, 1996; Kowlessur and Kaufman, 1999). The full-length amino acid sequence of TPH has not been determined with purified enzyme, and hence has never been compared with the sequence deduced from the cDNA. Although multiple splicing was noted in the non-coding regions (Darmon *et al.*, 1988; Boularand *et al.*, 1995), they all included a homologous ORF, and the authors did not explain the known difference between the brain TPH and the peripheral enzyme. It was speculated that the difference arose from region-specific post-translational modification (Kim *et al.*, 1991), but no conclusive evidence was provided. Many researchers investigated 5-HT synthesis in the brain using probes based on the commonly cloned cDNA, which we now know to be TPH1, a minor portion of TPH in the brain but a major protein in the periphery. The molecular identity of the brain counterpart was established in 2003, and the new isoform was termed TPH2 (Cote *et al.*, 2003; Walther *et al.*, 2003; Zhang *et al.*, 2004). Thus, research on 5-HT-production in the brain had a very arduous 15 years in terms of the molecular aspects of neuronal TPH.

The amino acid sequence of TPH1 (444 amino acids) (Grenett *et al.*, 1987; Darmon *et al.*, 1988) is highly preserved in the amino acid sequence of TPH2 except for an additional 41 amino acid peptide at the N-terminal (485 amino acids (Walther *et al.*, 2003)), as depicted in Figure 3. They are both composed of three major domains; a C-terminal tetramerization sequence, a central catalytic body, and an N-terminal regulatory peptide. In addition, both are essentially soluble enzymes and behave as a tetramer (Nakata and Fujisawa, 1981, 1982a). The expression of TPH2 is predominant in the neurons and that of TPH1 in the peripheral organs, including the pineal gland. However, the minor expression of either enzymes in the other's favored location has also been shown (Nakamura *et al.*, 2008). TPH1 is the primary form and expresses earlier in neural development (Nakamura *et al.*, 2006a; Nakamura and Hasegawa, 2007). Studies on the local and temporal distribution of the two isoforms would lead to further understanding of their roles in particular functions as well as in neural development. In this context, the use of mono-specific antibodies raised against a distinctive region(s) of the respective TPH (Sakowski *et al.*, 2006; Nakamura *et al.*, 2008) (Figure 3) would allow researchers to distinguish between the two isoforms, and would be a powerful tool for further investigation.

BH_4 requirement

BH_4 is an essential coenzyme for both tryptophan hydroxylase and tyrosine hydroxylase. BH_4 regulates the rate of 5-HT production.

Cells engaged in production of 5-HT contain a catalytic amount of BH_4. Since the intracellular concentration of BH_4 is not likely to be at saturation level (Miwa *et al.*, 1985), which is around or lower than the K_m of TPH (roughly 20–30 μM), TPH exerts only a fraction of its potential activity (V_{max}). When BH_4 is delivered directly to the brain, by microdialysis infusion (Miwa *et al.*, 1985), by direct injection into the brain (Kapatos *et al.*, 1982), or by immersion of brain slices into 6RBH_4-containing

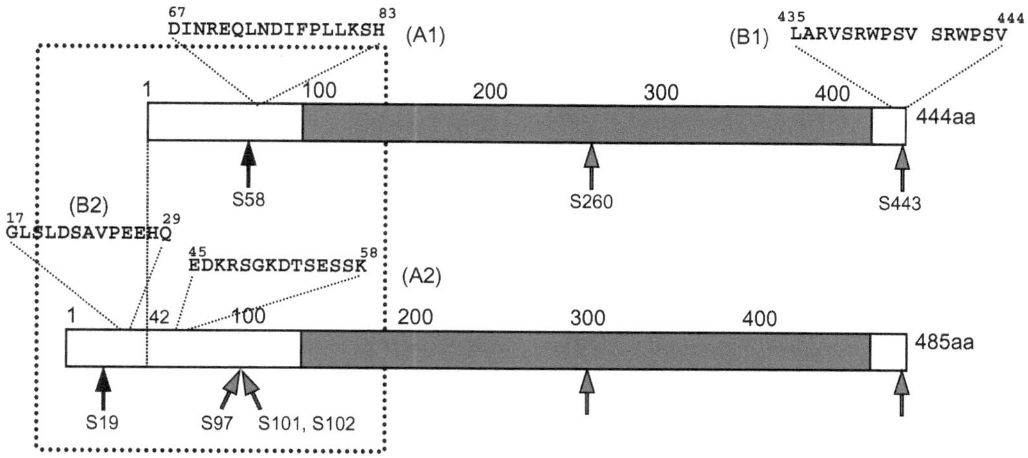


```
rTPH1    1:-----------------------------------------MIEDNKENKDHSSERGRV 18
rTPH2    1:MQPAMMMFSSKYWARRGLSLDSAVPEEHQILGGLTQNKATASKSEDKRSGKD-TSESSKT 59
                                                    **     **   **
rTPH1   19:TLIFSLKNEVGGLIKALKIFQENHVNLLHIESRKSKRRNSEFEIFVDCDINREQLNDIFP 78
rTPH2   60:AVVFSLKNEVGGLVRALRLFQEKHVNMLHIESRRSRRRSSEVEIFVDCECGKTEFNELIQ 119
              **********  **  *** *** ****** * ** ** ******         *
rTPH1   79:LLKSHTTVLSVDSPDQLPEKEDVMETVPWFPKKISDLDFCANRVLLYGSELDADHPGFKD 138
rTPH2  120:LLKFQTTIVTLNPPDNIWTEEEELEDVPWFPRKISELDRCSHRVLMYGTELDADHPGFKD 179
           ***  **      **     *   * ***** *** ** *  *** ** ***********
```

Figure 3 Schematic diagram of determined or putative phosphorylation sites of tryptophan hydroxylase TPH1 and TPH2. In TPH, three domains from the N-terminal adjacent to the C-terminal are known to be regulatory, catalytic (gray) and tetramer-forming, with amino acid sequences that are well conserved (marked with an asterisk*), while 41 additional peptides are unique to TPH2. A1 and A2 denote peptides employed in raising antisera that exclusively distinguish TPH1 and TPH2, respectively (bold underlined) (Nakamura *et al.*, 2008). B1 and B2 similarly represent mono-specific antisera (bold italics) (Sakowski *et al.*, 2006). Numbered serine residues denote determined or putative phosphorylation sites (references in text).

saline (Sawada *et al.*, 1986), TPH activity in the treated rat brain or tissues significantly increased, which was determined by a metabolic assay using the decarboxylase inhibitor NSD-1015 or NSD-1055. Similar enhancement was described in a report on RBL2H3 cells under monolayer culture expressing the peripheral type of TPH; cellular TPH activity was stimulated two-fold when the level of intracellular BH_4 was increased by means of sepiapterin administration[4] (Hasegawa *et al.*, 1999). Hence, serotonin-producing cells do not use up their full potential TPH activity (for review, see Sawada and Nagatsu, 1988). The concentration of the active co-substrate, BH_4, is maintained mainly through recycling (Figure 2). As an analogy to the phenylalanine hydroxylase reaction in the liver, the primary product of BH_4 in the TPH reaction might be an unstable 4a-hydroxy-BH_4, pterin-4a-carbinolamine (Curtius *et al.*, 1993). Pterin-4a-carbinolamine could be dehydrated by an enzyme, pterin-4a-carbinolamine dehydratase (PCD; EC 4.2.1.96), but this protein is not likely expressed in the brain (Davis *et al.*, 1992; Lei and Kaufman, 1999). PCD/DCoH is a bifunctional protein, the DCoH portion serving as a cofactor regulating dimerization of the hepatic nuclear factor, NHF1, a homeodomain transcription factor (Citron *et al.*, 1992). However, this pterin intermediate is quite unstable at a neutral pH and dehydration of pterin-4a-carbinolamine might take place, spontaneously producing quinonoid dihydrobiopterin (qBH_2) (Kaufman, 1976). In the recycling system, the actual concentration of BH_4 is determined by a balance between reducing DHPR activity and BH_4-oxidizing activity. Unless sufficient DHPR is furnished within the cell, conversion of qBH_2 to the more stable 7,8-BH_2 will occur spontaneously; the half decay time of qBH_2 was estimated to be 1.8 minutes (Watabe, 1978). The enzyme DHPR prefers NADH to NADPH as its co-substrate; however, there is a ready availability of NADPH in the cytosol of living cells, which could be much more abundant than that of NADH.[5] Hence, an insufficient supply of NADPH might allow qBH_2 to convert to the inactive 7,8-BH_2, and 7,8-BH_2 could then be salvaged by reaction (4) catalyzed by dihydrofolate reductase (DHFR; EC 1.5.1.3) (Kaufman, 1967; Nichol *et al.*, 1983):

$$7,8\text{-}BH_2 + NADPH + H^+ \rightarrow BH_4 + NADP^+ \quad (4)$$

The enzyme catalyzes reduction of the 7,8-dihydroform of both conjugated and unconjugated pterin[6] using NADPH as the coenzyme. This enzyme is widely distributed in most tissues and hence the amount of DHFR in the body as a whole is sufficient for reducing 7,8-BH_2 back to active BH_4 for economical recycling of BH_4 (Sawabe *et al.*, 2004); however, the level of DHFR is not high enough to replace DHPR in supporting local monoamine production.[7] DHFR was reported not to distribute to neurons in a high enough amount to salvage *in situ* oxidized BH_4 for the local production of monoamines especially in the CNS (Bonafe *et al.*, 2001). The key to explaining these contradictory observations could be the short life of qBH_2 as mentioned above ($T_{1/2}$ = 1.8 min). When the quantity of DHPR is insufficient for reducing qBH_2 before qBH_2 conversion to 7,8-BH_2 within a short period, as determined by the short half-life, the formed 7,8-BH_2 exits from the cell into the body circulation.[8] BH_4 is virtually impermeable across the plasma membrane and there are no known transporters to date, but its oxidized form, 7,8-dihydrobiopterin (7,8-BH_2), is somewhat more permeable. In this context, the conversion of qBH_2 to 7,8-BH_2 might be the major route by which BH_4 is withdrawn from the cell. Hence, the production of 7,8-BH_2 in the cell could be accounted for by the inactivation of BH_4 – that is, a local loss of active BH_4 in the cell. DHPR is therefore fully responsible for the intracellular recycling of BH_4 through the reduction of qBH_2.

As mentioned above, BH_4 levels in the cell determine the rate of 5-HT production. However, the mechanism by which BH_4 is maintained in the cell at a given level which is obviously higher than the extracellular concentration remains unclear. High-affinity uptake was suggested (Hasegawa *et al.*, 2007), but the evidence is not yet conclusive. The biosynthesis rate is counterbalanced with the loss of BH_4, essentially by conversion of qBH_2 to 7,8-BH_2 and the leaking out of the latter. The turnover rate expressed by the half-life of BH_4 was 120 to 230 minutes in monoamine-producing cells under monolayer culture (Hasegawa *et al.*, 2005). The rate was estimated by the initial decrease in BH_4 after the addition of DAHP (2,4-diamino-6-hydroxypyrimidine), an inhibitor of GPT-cyclohydrolase I. BH_4 supplementation can be achieved by administering 7,8-BH_2 or, preferably, sepiapterin (Sawabe *et al.*, 2008). If the deficiency is in the CNS, however, peripheral administration of BH_4 is very inefficient due to the presence of the blood–brain barrier (BBB). The mechanism for rejection of BH_4 at the BBB is not known to date. BH_4 supplementation has only ineffectively raised 5-HT production (Brand *et al.*, 1996). Moreover, it has not come into common use owing to the rapid renal clearance[9] that takes place when it is exogenously supplied, in addition to its poor ability to cross the BBB. Oral supplementation with BH_4 has been prescribed only for BH_4-deficient patients whose conditions would otherwise be very serious. If there were means to overcome these obstacles, BH_4 supplementation could be a potent therapy in the future for stimulating 5-HT synthesis.

TPH as an iron enzyme

TPH is an iron protein. Iron binding is relatively strong in TPH2 while it is very weak in TPH1.

The biopterin-dependent aromatic amino acid hydroxylases, phenylalanine-, tyrosine- and tryptophan-hydroxylase, share many structural homologies, and they all require bound iron for their catalytic functions. Hence, a brief comparison of these three hydroxylases would help us to understand their intrinsic natures. Phenylalanine hydroxylase (PAH; EC1.14.3.1) carries the most tightly bound iron, and the enzyme is active *in vitro* when purified without iron enrichment (Fisher *et al.*, 1972; Gottschall *et al.*, 1982). However, according to Shiman and Jefferson (1982), a large proportion of rat liver PAH (20–45 percent) is in an inactive state due to an iron vacancy, which represents a state of dynamic equilibrium controlled by the metabolism of hepatic iron. Tyrosine hydroxylase (TH; EC1.14.16.2) requires iron enrichment for an appreciable reaction *in vitro*. TH was found to contain iron at a less than stoichiometric concentration when purified to homogeneity from bovine adrenal medulla (Haavik *et al.*, 1988). Recombinant human TH isoforms, the alternative mRNA splicing products TH1–TH4, each expressed in *E. coli* and purified without iron enrichment in the presence of 1 μM EDTA, contained little iron, amounting to less than 0.1 atom/subunit. These isoforms were almost inactive, and required Fe^{2+} plus mercaptan for their enzyme activity (Haavik *et al.*, 1991). The status of iron binding is distinctive in tryptophan hydroxylase isoforms TPH1 and TPH2. Using rat brainstem, Nakata and Fujisawa purified TPH to around 5000-fold greater than the amount in crude extracts by maintaining a concentration of 50-μM EDTA almost throughout the entire purification. The purified enzyme was still active without iron enrichment, although it was 'activated' several fold by incubation with Fe^{2+} (Nakata and Fujisawa, 1982a). The stable iron-binding of the brain TPH was also observed in another laboratory; in these assays, the enzyme did not lose its activity even after a 17-h incubation with 100-μM EDTA at 4°C (Cash *et al.*, 1985). In contrast, the peripheral type of TPH, TPH1, obtained from mouse mastocytoma P-815, contained less than a stoichiometric amount of iron (0.05 atom/subunit). In addition, it was virtually inactive unless Fe^{2+} was added together with the appropriate mercaptan as described below (Hasegawa and Ichiyama, 2005). Iron-free TPH1 was available just by mixing with EDTA (100 μM) and successively removing the chelator by gel filtration.

In this context, TPH2 appeared to have bound iron with relatively high affinity, while TPH1 was the least stable in terms of iron-binding among the aromatic amino acid hydroxylases so far examined. In any case, TPH has an iron-binding site composed of neighboring His, and Glu at the bottom of the tryptophan-binding groove in the 3D-structure, which was illustrated using recombinant TPH1 crystallized as a stable but inactive complex in the presence of Fe^{3+} and 7,8-BH_2 (Wang *et al.*, 2002).

Activation of TPH, or restoration of the lost activity of TPH

Owing to the poorly liganded iron of TPH, various complexities arose: reversible inactivation of TPH seems to be brought about by oxidation of Fe^{2+} to Fe^{3+}.

TPH is quite unstable, whether in tissue extracts or as a purified preparation, and undergoes a serious loss in its activity under atmospheric conditions. A portion of the lost activity is restorable by an 'activation' procedure. The TPH reaction was carried out *in vitro* in the presence of Fe^{2+} and appropriate thiols such as 2-mercaptoethanol or DTT in an empirical effort to preserve the tetrahydropterin cofactor. Ichiyama's group developed a TPH assay that included a two-step incubation, anaerobic and aerobic, to 'activate' the enzyme and the subsequent TPH reaction, respectively (Ichiyama *et al.*, 1974). The procedure was very effective with most TPH preparations of peripheral origin, i.e., largely containing TPH1.[10] Anaerobicity was viewed as essential at that time, but it is actually unnecessary when more than 2 mg/ml of catalase are included in the activation mixture. 'Activation' of the enzyme in the former incubation with ferrous iron (50 μM) and DTT (30 mM) had to be separated from the enzyme reaction, they claimed, because the activation was a time-dependent process and its progress was interrupted when tryptophan was added. The activation process was completed within about 15 minutes. Several-fold higher activity was obtained with freshly prepared tissue extracts (Hasegawa *et al.*, 1987), and even more activity was achieved with purified enzyme which had lost activity during the purification procedure. A similar protocol was used for 'activation' of the brain enzyme. Activation (or recovery of lost activity) with the brain TPH required a prolonged incubation, 20–24 h, under similar anaerobic conditions (Hamon *et al.*, 1978; Kuhn *et al.*, 1980b; Stone *et al.*, 1989a). These features of TPH activation (or restoration) suggested a homology between TPH1 and TPH2, but at the same time we noted a great difference in the rate of the process, presumably due to their respective iron-binding affinities, which differed by more than two orders of magnitude. Since the activation did not alter kinetic parameters of TPH of either the brain or peripheral enzymes except for the V_{max}, it is considered that the procedure increased the number of active TPH molecules rather than altered the functional properties of the enzyme. The same activation procedure, i.e., treatment of the enzyme with ferrous iron and a thiol under anaerobic conditions, was employed in the restoration of the homologous hydroxylases TH (Okuno and Fujisawa, 1981; Haavik *et al.*, 1988; Ramsey *et al.*, 1995) and PAH (Fisher *et al.*, 1972) after *in vitro* iron deprivation. Furthermore, PAH inactivated by removal of iron from the liver under perfusion was also restored with a similar treatment *in vitro* (Shiman and Jefferson, 1982). However, the restoration conditions were similar to those employed in the restoration or activation of other non-heme iron proteins such as aconitase (Kennedy *et al.*, 1983) and phosphoenolpyruvate carboxykinase (Bentle and Lardy, 1977), suggesting common iron liganding behavior.

Since TPH is very fragile *in vitro*, it was uncertain for long time whether the inactivation, or activation, occurred only under cell-free conditions as an artifact or even whether it occurred *in vivo* as a physiological event. In this regard, an acute depression in TPH activity in rat brain after administration of 3,4-methylenedioxymethamphetamine (MDMA, single dose of 4–10 mg/kg, i.p.) (Schmidt and Taylor, 1987) was reported as evidence that the 'inactivation' did indeed take place *in vivo*. Subsequently, it was demonstrated that the enzyme activity was recovered *in vitro* as well as *in vivo* (Stone *et al.*, 1989b). The *in vivo* recovery was not likely brought about by protein synthesis, because it took place even in rats whose protein synthesis was blocked by cycloheximide (1 mg/kg, i.p.), suggesting that the re-activation occurred through a molecular modification. The *in vitro* reactivation of TPH obtained from rats treated with MDMA or its metabolites was achieved by anaerobic incubation with Fe^{2+}/DTT (Elayan *et al.*, 1993). Thus, this iron-related inactivation and reactivation is considered to take place *in vivo*. *In vitro* inactivation of TPH (brain enzyme) was demonstrated by nitric oxide (NO) exposure (Kuhn and Arthur, 1996). NO-driven inactivation *in vitro* was analyzed using recombinant TPH1, and the inactivated enzyme was similarly recovered by anaerobic incubation with Fe^{2+} and a thiol compound such as DTT (Kuhn and Arthur, 1997). Since the *in vitro* inactivation was highly antagonized by thiols, it was considered to be caused by oxidation of protein sulfhydryl (Stone *et al.*, 1989a; Kuhn and Arthur, 1997). The reason for the need for a high concentration of thiols has been controversial (Hasegawa and Ichiyama, 2005). Nonetheless, the involvement of nitric oxide (NO) was evidenced by the *in vivo* depletion of TPH induced by MDMA treatment (Zheng and Laverty, 1998; Darvesh *et al.*, 2005). NO-mediated inactivation of TPH might be a key cause of the neurotoxicity of such drugs as MDMA or their metabolites.

Unusual instability of reversible or irreversible TPH inactivation

Irreversible inactivation of TPH and neural deterioration seem to be caused by an H_2O_2 reaction with non-liganded or poorly liganded Fe^{2+}.

The peripheral TPH, TPH1, was purified as an inactive but restorable form from bovine pineal gland (Ichiyama *et al.*, 1976) and mouse mastocytoma P815 (Nakata and Fujisawa, 1982b). In the case of brain TPH, its unusual instability has long hindered the progress in TPH research as a means of understanding the neural regulation of 5-HT production. The inactivation of TPH2 was slower than that of TPH1 but it was still very fast, which made purification difficult. Purification of the brain TPH was achieved by adding catalase to the enzyme solution (the enzyme to be purified had to be mixed with catalase for purification) (Nakata and Fujisawa, 1982a; Cash *et al.*, 1985). The instability was enhanced by the addition of Fe^{2+} and a reducing compound, such as ascorbate or mercaptan, in the presence of oxygen (Kuhn *et al.*, 1980b). The protective effect of catalase was remarkable,[11] suggesting involvement of H_2O_2. These features can be accounted for by the Fenton Reaction (Stadtman and Berlett, 1991; Wardman and Candeias, 1996). The key reactions in the TPH-inactivating system including Fe^{2+}, DTT, BH_4, and O_2 might be the reduction of O_2 to H_2O_2 and of Fe^{3+} to Fe^{2+} and the subsequent formation of an activated oxygen (hydroxy radical $HO^{\bullet}$ or ferryl ion FeO^{2+}):

$$H_2O_2 + Fe^{2+} \rightarrow Fe^{3+} + HO^{\bullet} + HO^{-}, \text{ or } FeO^{2+} + H_2O.$$

In the case of the reaction of free Fe^{2+} or its aquo-ion with H_2O_2, impairment of an amino acid by the activated oxygen would take place at random – i.e., non-specific with regard to the chemical modification site, comparable to the exposure of proteins to ionizing radiation (Stadtman and Berlett, 1991). Taking the short range of $HO^{\bullet}$ free movement into account, the Fenton reaction with bound iron takes place in a site-specific manner (Stadtman and Oliver, 1991). The inactivation of TPH was efficiently prevented by the substrate, tryptophan, in a saturable manner (unpublished data). The allowance of H_2O_2 access suggested that the Fe^{2+} was poorly liganded at the bottom of the active site as proposed by the crystallographic 3D structure (Wang *et al.*, 2002). This fact was consistent with the lack of protection by chemical $HO^{\bullet}$ scavengers (Kuhn *et al.*, 1980b; Stadtman and Oliver, 1991; Cash, 1998). The oxidation of Fe^{2+} by H_2O_2 providing Fe^{3+} results in TPH inactivation, as illustrated in Figure 4A. While oxidation of Fe^{2+} with H_2O_2 efficiently inactivates TPH, this inactivation can be reactivated by removing Fe^{3+} and supplying ferrous iron Fe^{2+}. Simultaneously generated $HO^{\bullet}$ attacks the amino acid at the active site, although at less than 100 percent probability, leading to fatal inactivation of TPH, which is irretrievable on anaerobic incubation with Fe^{2+}/DTT. H_2O_2 plays an essential role in both the reversible and irreversible inactivation of TPH; therefore, the scavenging of H_2O_2 by catalase or glutathione peroxidase prevents this inactivation. In living cells, however, H_2O_2 or its precursor $O_2^{-\bullet}$ is continuously produced in normal mitochondria, accompanied by the reduction of O_2 to water at a considerable rate (*ca.* 1–4 percent of O_2 reduced). To add to our confusion, catalase resides in the peroxisome and not in the cytosolic space accommodating TPH. Considering that TPH levels are maintained in the cell, the following possibilities are suggested:

1. TPH is protected by an unknown enzyme(s) scavenging H_2O_2 or $HO^{\bullet}$
2. Fe^{2+} of native TPH *in vivo* is somehow in a more completely liganded form so that H_2O_2 does not react to cause site-specific modification
3. The actual concentration of Fe^{2+} is much lower than under our experimental conditions ($>10^{-5}$ M)
4. TPH is actually inactivated by $HO^{\bullet}$ at a fast rate and turns over rapidly.

At present, we do not have any supporting evidence. With aconitase, another H_2O_2-sensitive and 'Fe^{2+}/DTT-activated' non-heme iron protein, the degradation rate was reported to be enhanced by chemical modification caused by reactive oxygen radicals (Grune *et al.*, 1998), although the cytosolic concentration of chelatable Fe^{2+} might have been lower than 10^{-6} M, as discussed below.

Unique iron requirement of TPH

TPH1 activity has an absolute requirement for Fe^{2+}. Fe^{3+} antagonizes Fe^{2+}, and Fe^{3+}-bound TPH is inactive.

The method for TPH activation, anaerobic incubation with Fe^{2+} and DTT, was empirically established, but it proved to be useful in a wide range of applications involving pterin-dependent hydroxylases and other non-heme iron proteins, as mentioned above. However, the duration required for its completion differed with each application (probably due to differences in liganding stability). The activation mechanism for one enzyme may often serve as a guide for understanding others. With the use of TPH1 purified from mouse mastocytoma P815, 'DTT-activation' was examined and a mechanism was proposed (Hasegawa and Ichiyama, 2005). Since TPH1 and TPH2 have not been compared in the literature as yet, the core of the work with TPH1 described below could represent a sort of case study. The iron content of purified TPH1 was far less than the expected stoichiometry (0.05 atm/subunit), even though

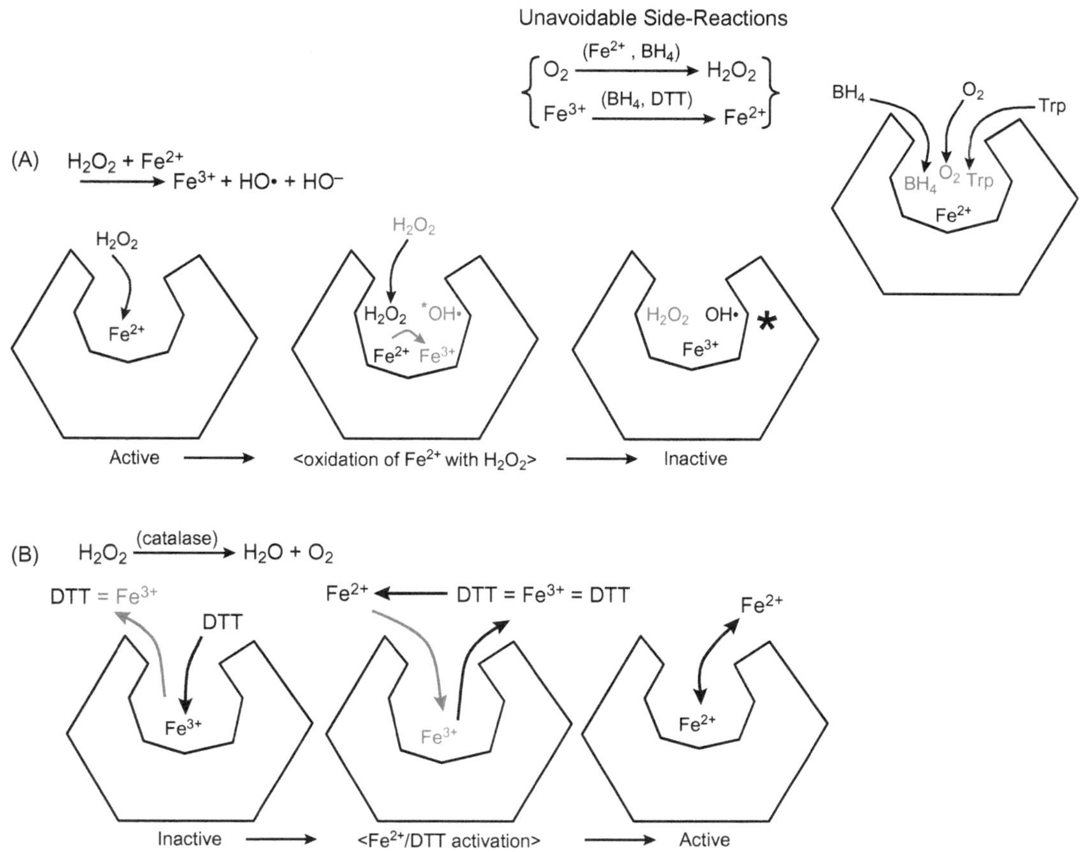

Figure 4 Schematic representation of mechanisms for inactivation of tryptophan hydroxylase and its reactivation by DTT. In enzyme activation and subsequent reaction, or even during storage, TPH is unavoidably exposed to Fe^{2+} and dioxygen in the presence of high concentrations of reducing compounds – conditions favorable for the Fenton reaction. (A) The Fenton reaction oxidizes Fe^{2+} to Fe^{3+} and generates HO$^{\bullet}$. TPH is inactivated irreversibly from an occasional site-directed attack by OH$^{\bullet}$ at the active-site amino acid residue. (B) DTT, an Fe^{3+}-chelating thiol, masks Fe^{3+} and removes it from the inactive TPH resulting in the enzyme reactivation. The presence of H_2O_2 is key, but H_2O_2 can be scavenged by a high concentration of catalase of probably more than 4 mg/ml.

the purification was achieved by one-step affinity chromatography with a yield of 50–70 percent. It appeared as if iron was not essential for the enzyme activity, because the enzyme was fully active as long as it was incubated with DTT but without the addition of exogenous iron. However, to confuse matters, the iron-free apo-enzyme assayed had an absolute requirement for Fe^{2+} which was not replaceable even with Fe^{3+} and even in the absence of DTT (Hasegawa and Ichiyama, 1986). Special care was taken to prevent contamination by ambient iron. The point of confusion was explained by a 'shared use of exiting iron'[12] based on the rapid exchange of dissociable Fe^{2+} at the effective concentration required by the enzyme. The minimum effective concentration of Fe^{2+} on apo-TPH for the enzyme to be active was around 10^{-10}M (Hasegawa and Ichiyama, 2005), which was estimated using an Fe^{2+}-ion buffer system[13] TPH had no activity in the presence of Fe^{2+} at an estimated concentration of 1.8×10^{-12}M (metal buffer: 0.1 mM Fe^{2+} + 1 mM EDTA at pH 6.7). On the other hand, TPH showed full activity (5-HTP formation: 5 μmol/min per mg) in the presence of 8.7×10^{-9} M Fe^{2+} (metal buffer: 0.5 mM Fe^{2+} + 1 mM EGTA at pH 6.7). The stability constant (LogK_{ML}, L denotes TPH) falls between 8 and 12. Fe^{3+} -removal from inactive TPH was achieved by incubation with metal-chelating sulfhydryl compounds with a common structure, HSCCOH or HSCCNH, and either was replaceable with EGTA. 8-Mercaptoquinoline, an Fe^{2+}-chelating thiol, did not function in this case, and potential SS breakers such as S^{2-} or tri-n-butylphosphine failed to activate the enzyme. The concentration of the reagent required for effective activation of TPH was much higher than that expected for disulfide bond cleavage, suggesting that DTT was unlikely to be a disulfide bond breaker for which a 30-fold molar excess was thought to be sufficient. Thus, the 'Fe^{2+}/DTT activation' can be understood as illustrated in Figure 4B: DTT preferably binds to free Fe^{3+} and prevents it from binding to TPH, and the removal of Fe^{3+} from TPH by a high concentration of DTT takes place simultaneously with the regeneration of Fe^{2+}. In practice, an *in vitro* reaction mixture for the TPH assay requires an antioxidant as a preservative for BH_4. Avoidance of the site-specific Fenton Reaction by the addition of catalase (>4 mg/ml) were essential for the reaction to proceed quantitatively under aerobic conditions.

Availability of ferrous iron in the cell

The metabolism of iron may control 5-HT production by attenuating actual TPH activity in the cell interior.

In the previous section, the distinctive requirement that TPH has for iron was described in terms of the relatively rapid equilibrium of TPH with free Fe^{2+}, $8 < \text{Log}\, K_{ML} < 12$ (M and L denote Fe^{2+} and TPH, respectively). This feature was demonstrated by delivering ferrous iron to cells with hemin or via an iron shuttle with cell-permeable chelators. Hemin is Fe^{2+}-bound porphyrin, which integrates the iron catalytically through action of ferrochelatase (EC 4.99.1.1) and liberates Fe^{2+} by complex degradation catalyzed by heme-oxygenase (EC 1.14.99.3.) in the cell. When hemin was administered to cells in culture, it was taken up and the iron delivered, resulting in an elevation of the 'labile' or 'chelatable' iron pool (Rouault *et al.*, 1985). Hemin administration to RBL2H3 cells, a peripheral 5-HT-producing cell line, strongly elevated intracellular TPH activity in a steep and concentration-dependent manner (Hasegawa *et al.*, 1999). The hemin-induced elevation of TPH activity was completely inhibited by administration of the cell-permeable chelator, salicylaldehyde isonicotinoyl hydrazone (SIH) or desferrioxamine (DFO), suggesting that hemin had liberated Fe^{2+} into the cytosolic pool of chelatable iron. Fe^{3+} complex with SIH also brought iron into the cell, resulting in only a two-fold elevation of TPH activity. The presumed equilibrium between free SIH and SIH-Fe^{3+} and also between free and Fe^{2+}-bound SIH both in and out of cells was established based on the fact that SIH was not metabolized but rather shuttled iron across the cell membrane (Ponka *et al.*, 1984). When administered to cells, free SIH suppressed TPH activity, as did DFO, and it may be that these compounds harbor chelated iron ions, as seen in Figure 5. TPH activity remained constant before and after treatment with hemin, SIH or DFO, as measured by a cell-free TPH assay, suggesting that the amount of TPH had not increased or decreased. The intracellular chelatable iron pool might be composed primarily of Fe^{2+}, but also contains multiple ligands, including peptides, nucleotides, and amino acids. This meant that a large proportion of TPH was in an inactive form due to a low concentration of available ferrous iron. In the erythroleukemia cell line K562, the chelatable iron pool was estimated *in situ* to be between 2 and 5×10^{-7} M using the fluorescent chelator calcein[14] (Breuer *et al.*, 1995). Furthermore, the concentration of chelatable iron in isolated hepatocytes under primary

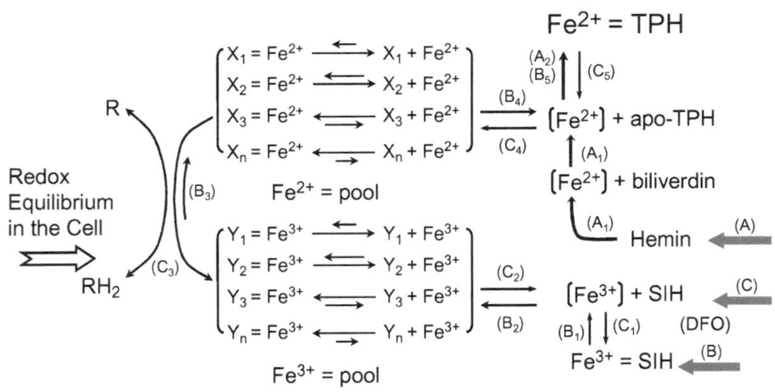

Figure 5 Chelatable iron pool. Extracellular administration (thick gray arrows) of (A) hemin, (B) Fe^{3+}-bound salicylaldehyde isonicotinoyl hydrazone (SIH), or the free cell-permeable chelators SIH or desferrioxamine (DFO). Internalized hemin is degraded to release Fe^{2+} irreversibly through the action of heme-oxygenase. The liberated Fe^{2+} pushes a series of reactions otherwise equilibrated along lines A1 and A2, resulting in a continuous increase in Fe^{2+}-TPH. Internalized Fe^{3+}-bound SIH (B) dissociates in part to liberate Fe^{3+}. This pushes iron ion pools along lines B2 up to B5 to increase free Fe^{2+}, also resulting in the increase in Fe^{2+}-TPH. However, the supply of dissociated Fe^{2+} is restrained, presumably due to an equilibrium between free and bound SIH as if it were a metal buffer. Addition of SIH (C) pushes a reaction to reserve Fe^{3+}-SIH (C1), consuming Fe^{3+} in the process, and this pulls Fe^{2+} through reactions along lines C2 down from C5, resulting in a decrease in Fe^{2+} = TPH (where ' = ' denotes liganding). RH_2 represents the intracellular reducing pressure provided by energy metabolism. X and Y denote possible ligands with various affinities to iron ions.

culture was estimated to be $5.4 \pm 1.3 \times 10^{-6}$M using newly synthesized fluorescent chelators (Ma *et al.*, 2006). Fe^{3+} may be a minor component of the chelatable iron pool owing to the great capacity of cytosolic ferritin to bind under physiological conditions. Hence, Fe^{3+} might not be competitive with Fe^{2+} in the cell, although TPH is very sensitive to added or even ambient Fe^{3+} under cell-free experimental conditions *in vitro*. As mentioned in the above section on the 'Unique iron requirement of TPH', the apo-enzyme of TPH1 was fully active when the enzyme reaction was performed in 8.7×10^{-9}M Fe^{2+}-EGTA buffer; however, TPH was in fact in a 'hungry' state within the cell. In other words, the dissociable Fe^{2+} in living cells did not support the full potential activity of the enzyme, suggesting that the 'chelatable' Fe^{2+} pool that was available for TPH was smaller than that provided by the Fe^{2+}-EGTA buffer. Since the chelatable iron pool might be composed of ligands of varying affinity, and since measurement with a chelating probe is based on competition with intracellular ligands, estimated values might vary depending on the target cells and on the strength of the chelator. It is therefore considered that a large portion of the 'chelatable' Fe^{2+}, which was estimated using relatively strong chelators as the probe, might have not been available to TPH for ligand competition in the actual cell interior. A similar dissociation/association of iron ion was reported *in vivo* and *in vitro* with rat-liver PAH, a homologous pterin-dependent monooxygenase (Shiman and Jefferson, 1982). PAH activity was also manipulated in the liver by shuttling iron with transferrin under perfusion. As mentioned earlier, the stability of iron binding with PAH seems much higher than TPH1. PAH altered iron occupancy within 30 minutes, which was surprisingly rapid. As yet, TPH2 has not been compared with TPH1 with respect to its Fe^{2+} -hunger in the CNS. The size of the chelatable pool of Fe^{2+} or the concentration of Fe^{2+} available for TPH might be closely related to iron metabolism in the body as a whole. A latent iron deficiency in weaning rats gave rise to a 5-HT decrease in the CNS (Shukla *et al.*, 1989). Taken together, iron deficiency and 5-HT-related behavior appears to be an interesting subject for future exploration.

Phosphorylation

Phosphorylation of TPH was studied in terms of its functional aspects with the use of brain TPH while identification of phosphorylation sites was mainly conducted on the basis of the amino acid sequence of TPH1.

Phosphorylation of TPH has drawn intense interest as a potential regulatory mechanism. In general, phosphorylation is a post-translational modification which changes an enzyme's characteristics and can be reversed by dephosphorylation or by enzyme degradation. An approximately two-fold, *in vitro* activation with native brain TPH under Ca^{2+}-dependent phosphorylating conditions was reported in the late 1970s, and its responsible kinase was identified in 1983 to be Ca^{2+}/calmodulin-dependent protein kinase II (CaMK II; EC 2.7.11.17) (Yamauchi and Fujisawa, 1983). Phosphorylation of brain TPH *in vivo* by cAMP-dependent protein kinase (PKA; EC 2.7.11.11) was also suggested (Garber and Makman, 1987), and *in vitro* reconstitution of phosphorylation was carried out with brain extracts (Johansen *et al.*, 1995, 1996); however, enzyme activation *in vitro* was ambiguous. In both cases, CaMK II and PKA, and a third protein factor which could be present in incompletely purified enzyme preparations, were strongly implicated as mediators of the brain TPH activation. The enzyme activity of carefully purified native TPH was very similarly augmented by CaMK II or PKA, and each caused about a two-fold activation in the presence of 14-3-3 protein (Yamauchi *et al.*, 1981; Furukawa *et al.*, 1993; Winge *et al.*, 2008), suggesting that this protein was required for consolidating the activation. It was also suggested that phosphorylation with CaMK II attenuates proteolytic degradation in the presence of 14-3-3 protein (Murphy *et al.*, 2008; Winge *et al.*, 2008). The increase in catalytic activity with enzyme activation has been studied extensively with the brain enzyme, mainly consisting of TPH2. In contrast, phosphorylation of TPH1 triggered rapid degradation of the protein, as discussed in the next section (Iida *et al.*, 2002).

Phosphorylation sites on cDNA-deduced amino acid sequences of TPH1 and TPH2 are depicted in Figure 3. Phosphorylation at Ser58 in the regulatory domain of TPH1 has been well studied (for reviews, see Hufton *et al.*, 1995; Fitzpatrick, 1999; Mockus and Vrana, 1998), but most sequence related work on 'brain TPH' prior to 2003 would have involved TPH1. CaMK II was first discovered as a protein responsible for activation of brain TPH, as mentioned above. Its phosphorylation site has recently been identified to be Ser19 in the N-terminal region, a site which can be phosphorylated with PKA as well (Kuhn *et al.*, 2007; Murphy *et al.*, 2008; Winge *et al.*, 2008). The homologous downstream TPH2 region includes phosphorylation sequences similar to those of TPH1; however, to date, native TPH2 eludes phosphorylation of these sites.

TPH protein turnover

The machinery of protein turnover is cell-type specific. TPH degradation is likely linked to protein phosphorylation.

Steady state levels of an enzyme are maintained as a counterbalance between its biosynthesis and degradation.

However, the enzyme level of TPH in a given cell or tissue generally does not correlate with its turnover rate. At any point in time, the amount measurable represents the steady state level; however, the turnover rate predicts how quickly the current level can increase or decrease on demand, usually as a response to stimulation (a good description can be found in Alberts *et al.*, 2002). A conventional but powerful method for estimating the turnover rate might be chasing the initial decrease rate after quickly stopping its biosynthesis. The decrease in the rate usually obeys first-order kinetics unless there is a particular underlying process which is driven by a limited number of high-affinity action sites. In this case, a linear decrease may be observed. In the case of the proteasome, which is a major and ubiquitous cytosolic apparatus, a virtually unlimited number of action sites may contribute. The observed initial rate of decrease would otherwise be undetectable due to compensation via biosynthesis. In a classic work by Meek and Neff (Meek and Neff, 1972), the half-life of the brainstem TPH was 2.5 days or longer as measured from the decline in the slope of TPH activity after administering a rat model a high dose of para-chlorophenylalanine (PCPA). A similar half-life was observed in MDMA-treated rat brain (single dose, 10 mg/kg, s.c.) using a metabolic assay which measured cortical 5-HTP accumulation after arresting AADC action with NSD1015 (Stone *et al.*, 1989b). When TPH levels in the brain are maintained under slow turnover conditions, a rapid stimulus response could be achieved by certain mechanisms other than quick degradation.

As for TPH1, its turnover rate was measured by administering cycloheximide (1 μg/ml) to cells in culture. The turnover rate expressed by the half-life ($T_{1/2}$) was 15–40 minutes, depending on the cell conditions or on the characteristics of the 5-HT-producing cell lines; rodent mast cell lines RBL2H3, P-815 and FMA3.[15] The degradation was driven by the ubiquitin-proteasome pathway (Hasegawa *et al.*, 1995; Kojima *et al.*, 2000). It was demonstrated that phosphorylation of TPH functioned as a tag for TPH1 degradation by the ubiquitin–proteasome pathway (Iida *et al.*, 2002). The proteolytic degradation was ATP-dependent, and was prevented with proteasome-specific inhibitors such as lactacystin, as well as other proteasome inhibitors such as MG132 (carbobenzyl-Leu-Leu-Leu-H), but was not inhibited with E-64, a calcium-dependent inhibitor of calpains, or with non-lysosomal cysteine proteases also expressed ubiquitously in mammalian cells. The responsible protein kinase has not clearly been determined yet; *in vitro* phosphorylation by CaM/kinase II partially functioned to trigger MG-132-sensitive proteolysis, but PKA did not. TPH degradation by the proteasome through the phosphorylation-ubiquitination axis could be a major stimulus response affecting the TPH level, at least in the mast cell system. The intracellular environment rather than the TPH protein itself might be responsible for the stimulus–response control of TPH turnover. The ubiquitin–proteasome system also determines properties of specifically differentiated cells. In this context, studies on regulation of TPH turnover should be restricted to particular cell systems operative in 5-HT-synthesizing cells.

Tryptophan hydroxylase in behavioral neuroscience

Involvement of TPH in neural development

The early expression of neurotransmitters and receptors in the developing brain has drawn attention to their potential contribution in modulating neuronal developmental processes. Monoamines are among the first neurotransmitters to develop during embryogenesis. Serotonergic neurons are generated on embryonic days 10 (E10) to E12 in the mouse (Levitt and Rakic, 1982). One day after their generation, raphe neurons can synthesize 5-HT.

The possible involvement of TPH at several stages of neural development was suggested from observations using a TPH blocker, *p*-chlorophenylalanine (PCPA). PCPA proved to specifically retard the onset of neuronal differentiation in brain regions known to contain 5-HT terminals or to have a high 5-HT content in the adult (5-HT target cells) in a study using long-survival 3H-thymidine autoradiography in rats (Lauder and Krebs, 1978). This finding indicates that TPH might have a role in neuronal differentiation, an early stage of neural development.

Neuronal differentiation is followed by the extension of axons and dendrites and the formation of the neural network. 5-HT is essential for 5-HT neurons to develop in an autocrine manner (Gaspar *et al.*, 2003), in addition to its action as a crucial neurotransmitter in various regions of the adult brain. Changes in the brain 5-HT content during development could disturb the development of 5-HT neurons. The consequence of TPH blockade on these developmental stages was further assessed using serotonergic RN46A cells under conditions where 5-HT is produced and released – that is, in the presence of BDNF and a high K^+ (White *et al.*, 1994). Targeted deletions of molecules involved in the metabolism or transport of 5-HT in mice affected dendritic maturation during development (Gaspar *et al.*, 2003). However, gross abnormalities were not found in these mice. Since actin contributes to the fine-tuning of neural circuits during development (Luo, 2002), actin may affect maturation of 5-HT neurons. An 8-day blockade of TPH using PCPA resulted in a marked accumulation of F-actin just beneath the membranes in both the soma and the proximal and distal neuritic segments

(Nakamura *et al.*, 2006b), suggesting that TPH activity might affect maturation of 5-HT neurons via appropriate actin polymerization.

There is also evidence that TPH activity is required for the maturation of non-5-HT neurons. Since the cerebral cortex is widely innervated by 5-HT-containing axons originating from neurons in the raphe nuclei, an appropriate amount of 5-HT during development likely underlies the development of cortical neurons and dendritic arborization (Levitt *et al.*, 1997; Vitalis and Parnavelas, 2003). Major events in corticogenesis take place at the E12-17 stage of embryonic development in rats. E12-17 PCPA treatment altered the maturation of pyramidal neurons of layers III and V of the somatosensory cortex in rats, with these cells displaying reduced dendritic arborization and complexity (Vitalis *et al.*, 2007). In addition, the PCPA treatment transiently altered the incorporation in the cortical plate of interneurons derived from the caudal ganglionic eminence, and persistently affected the differentiation of a subpopulation expressing calretinin and/or cholecystokinin. Thus, TPH activity could contribute to multiple stages of neuronal development of multiple cell types.

Effect of TPH on behavioral traits

5-HT neurons project to several areas, such as the cortex and limbic system. These regions are responsible for the generation and expression of many brain functions such as sensorimotor functions and motivational and affective states. Thus, 5-HT is believed to be important through its neuromodulatory role in these regions. Accordingly, TPH, the rate-limiting enzyme in 5-HT biosynthesis, has the potential to affect multiple brain functions in the adult through regulation of brain 5-HT contents in these regions. For instance, repeated treatment with PCPA caused attenuation of prepulse inhibition (PPI) (Fletcher *et al.*, 2001; Prinssen *et al.*, 2002), the suppression of the startle response by a relatively weak earlier stimulus, which theoretically reflects sensorimotor gating (Braff and Geyer, 1990; Graham, 1975). As for aggression, PCPA increased social aggression in birds (Buchanan *et al.*, 1994) and induced muricidal aggression in male and female laboratory rats (Valzelli *et al.*, 1981), but inhibited fighting behavior of postpartum mice toward unfamiliar male intruder mice (Ieni and Thurmond, 1985).

In other studies, 5-HT was shown to be essential for 5-HT neurons to form in an autocrine manner during development (Gaspar *et al.*, 2003). The formation of proper neural circuits is sensitive to activities that occur during well-defined periods of postnatal development (Berardi *et al.*, 2000). Proper TPH activity is likely required for the maturation of some brain functions within a narrow timeframe during development, and if this activity is compromised, it can adversely affect brain functions in the adult.

Pharmacological experiments were performed to define the sensitive period in the development of PPI. The blockade of TPH using PCPA from postnatal day 21 (P21) to P24, but not at earlier or later time-points, perturbed the PPI of adult mice. Reduced PPI in adulthood was also observed when PCPA was injected into rats during P21–P24. When the 5-HT precursor 5-HTP was given together with PCPA from P21 to P24, the attenuation of PPI was rescued in these rats. Therefore, in rodents, the sensitive period when proper TPH activity is required for PPI in adulthood proved to be from P21 to P24 (Nakamura *et al.*, 2006b). In addition, blockade of TPH during the embryonic period resulted in alterations in motor activity and learning in the adult. In rats, the prenatal PCPA increased activity in an open field in offspring treated on either gestational days (GD) 8–11 or 14–17. The highest levels of activity were noted in male and female offspring treated on GD 14–17. Similarly, the offspring treated on GD 14–17 showed a significant learning impairment in the Morris water maze (Vataeva *et al.*, 2007).

Since TPH2 mRNA is preferentially expressed in the brain (Walther *et al.*, 2003) and in peripheral myenteric neurons in the gut (Cote *et al.*, 2003), TPH2 might be involved in animal behavior by regulating 5-HT contents in the brain. A few reports support this notion. First, TPH2 mRNA in caudal dorsal raphe nucleus (DRN) was associated with lower anxiety-like behavior, whereas TPH2 mRNA in rostral dorsomedial DRN was associated with increased anxiety-like behavior (Hiroi *et al.*, 2006). Second, fluoxetine treatment for 4 and 8 weeks significantly increased basal TPH2 mRNA levels in the midbrain, an effect that was correlated with the appearance of antidepressant-like effects in the forced swim test, and a significant induction of TPH2 and 5-HT transporter mRNAs was detected in the midbrain of untreated rats 24 hours after the swim test (Shishkina *et al.*, 2007).

In contrast, TPH1 mRNA is mainly expressed in the pineal gland and the periphery (Walther *et al.*, 2003). However, the expression of TPH1 is up-regulated in the brain under some circumstances. Estrogen increased TPH1 mRNA expression in DRN of wild-type and ER-alpha knockout mice but not ER-beta knockouts, indicating that ER-beta is responsible for mediating estrogen-regulated TPH1 expression in the murine DRN (Gundlah *et al.*, 2005). Furthermore, a quantitative real-time PCR showed that TPH1 mRNA from male Wistar rats was up-regulated 2.5-fold in DRN by imposing daily restraint stress for

1 week. However, the expression of TPH2 was not affected by the stress in DRN (Abumaria *et al.*, 2008). These data raise the attractive hypothesis that TPH1 in serotonergic neurons of the DRN might be involved in stress-induced psychopathologies.

TPH in psychiatric conditions

The human TPH1 gene represents a candidate for genes involved in psychiatric conditions. The polymorphism A779C in intron 7 of TPH1 has been reported to be associated with alcoholic offenders (Nielsen *et al.*, 1998). In addition, the association of TPH1 with suicidal behavior has been well studied. As one example, A779C has been reported to be associated with suicidal behavior in depressed patients (Mann *et al.*, 1997). Another study reported that the association between TPH1 and suicidal behavior is stronger with violent suicidal behavior (Abbar *et al.*, 2001). Several recent meta-analyses have concluded that there is an association between A218C and a history of suicide attempts in Caucasians (Rujescu *et al.*, 2003; Bellivier *et al.*, 2004). Suicidal behavior has been implicated in the relationship between TPH1 and mood disorder. Bellivier and colleagues (1998) initially reported an association between TPH1 and bipolar disorder. However, in their study, the frequency of the allele was significantly associated with a history of suicidal behavior in bipolar subjects. Therefore, the reported association with bipolar disorder might be linked to suicidal behavior. Four subsequent studies found no association between TPH1 and bipolar disorder, and three of these studies, in which the population of bipolar patients was stratified according to their history of suicidal behavior, found no association with suicidal behavior in bipolar subjects (Furlong *et al.*, 1998; Kirov *et al.*, 1999; McQuillin *et al.*, 1999; Tsai *et al.*, 1999). Likewise, Zaboli *et al.* (2006) proposed that TPH1 is associated with borderline personality disorder in suicidal women. Collectively, TPH1 may be involved in susceptibility to suicidal behavior in some populations. Courtet *et al.* (2005) suggested that TPH may be a quantitative risk factor where the more pronounced the serotonergic dysfunction, the higher the levels of anger and the more severe the suicidal act (involving lethality, violence, repeated suicide attempts, completed suicide).

As for TPH2, there are reports suggesting an association between TPH2 and mood disorders. In addition, TPH2 gene variation is suggested to influence personality traits and disorders related to emotional dysregulation (Gutknecht *et al.*, 2007; Reuter *et al.*, 2007). Further studies will confirm the involvement of TPH2 in specific psychiatric conditions.

Notes

1. The responsible enzyme, PCD/DCoH, is a bifunctional protein which is not likely to be expressed in the CNS (Davis and Kaufman, 1989). This intermediate dehydrates spontaneously at a neutral pH.
2. Tryptophan was formerly a popular 'home remedy'. Since very small amounts of impurities were formed during its mass production, the supplement was judged to be neurotoxic and disappeared from the market. The impurity might have been generated during the drying process under air.
3. The major reason for the instability seems to be related to the nature of ferrous iron, whose behavior is difficult to control under aerobic conditions *in vitro*, and is dissociable from the enzyme. See the section on 'Iron requirement'.
4. Sepiapterin administration is so far the best method for increasing the BH_4 concentration inside many types of the cells, including RBL2H3, PC12 and primary cultures of hepatocytes. Sepiapterin is incorporated by a transporter(s) and converted to BH_4 by the salvage pathway, while BH4 is rejected by these cells by incorporation into the cytosolic compartment followed by immediate oxidation and release (Hasegawa *et al.*, 2005; Sawabe *et al.*, 2008).
5. With purified preparations, the K_m values of DHPR, as determined by the cytochrome-c method, were reported to be 1.5 and 74 μM for NADH and NDDPH (V_{max}, 8.5 μmol/min per mg), respectively. Enzymic data of DHPR from the livers of sheep, cows, and humans were concisely summarized in *Methods in Enzymology*, Vol. 142 (1987). As an indication of available reduced pyridine nucleotides, the ratios of reduced and oxidized forms (NAD^+/NADH and $NADP^+$/NADPH) within the cytosolic compartment of liver cells were 1.16 and 0.014, respectively (Krebs and Veech, 1969).
6. Folic acid is a 'conjugated pterin' and biopterin is an 'unconjugated pterin'. Owing to the enormous role of DHFR in reducing 7,8-dihydrofolate to tetrahydrofolate, which functions as the essential carrier of 1-carbon in DNA synthesis, its inhibitors such as methotrexate MTX have been employed as classic anti-tumor drugs. In animal experiments, tissue BH_4 levels were significantly decreased or those of BH_2 were elevated by administration of MTX (Sawabe *et al.*, 2004).
7. Patients with a hereditary DHPR deficiency secrete large amounts of 7,8-BH2 in the urine, but suffer from a severe deficiency of its reduced form (BH_4) which supports monoamine biosynthesis.

8. Circulating BH_2 is quickly taken up by many cells, including erythrocytes, where it is reduced back to BH_4 (Hasegawa *et al.*, 2005; Sawabe *et al.*, 2008). No large amounts of BH_2 appear in the urine; BH_4, the fully reduced form, appears instead.
9. Since there is no major degradation pathway of BH_4 as far as we know, renal excretion essentially reflects the amount of BH_4 biosynthesis. Daily renal excretion of a healthy rat was around 0.4 mg/kg (adult male, SD, unpublished). Healthy daily excretion of BH_4 into the human adult urine amounts to roughly 20 mg a day. The daily dose required to treat a BH_4 deficiency (5–20 mg/kg per day) is 15- to 60-fold greater than the putative amount synthesized in the body.
10. The activation procedure had the great advantage of retrieving the full potential activity of the enzyme. However, the activation required strictly anaerobic conditions, and this was achieved in a Thunberg tube connected to a vacuum line switchable to inert gas (argon or nitrogen). In addition, the enzyme reaction had to be performed within a short period to assess the linear progress against inactivation under air. These could be the reasons why the method did not become widely used.
11. In the reaction for catalase scavenging of H_2O_2, the reaction rate is determined by the rate of collision between the enzyme and H_2O_2, similar to the non-enzymic chemical reaction. H_2O_2 is generated by an oxygen molecule coupled with the reaction, Fe^{2+} to Fe^{3+} and/or XH2 to X, where XH2 stands for the reducing reagent. The H_2O_2 concentration is quite low, and in this case, the catalase concentration should be high enough for effective scavenging. Note that a 'high concentration' of catalase, such as 1 mg/ml, is merely a catalase-subunit concentration of 17 μM; not a very high concentration. With many published methods, a catalase concentration below 2 mg/ml may not be sufficient for this purpose, depending on how urgently the scavenging is required.
12. Under typical assay conditions, the TPH concentration (0.2 μg pure enzyme in a 150-μL reaction mixture) was calculated to contain around 2.67×10^{-8} mol/l of TPH molecules (subunit mw: 53,000). If 5 percent of subunits carried bound Fe^{3+}, the iron might have been converted by DTT to 1.33×10^{-9} M Fe^{2+}, which was close to the Fe^{2+} concentration in the 8.7×10^{-9} M Fe^{2+}-EGTA buffer mentioned in the text. As long as the bound iron was dissociable, this concentration was high enough for the enzyme to exert virtually full activity.
13. The stability constants of representative chelators EGTA and EDTA with ferrous iron Fe^{2+} were found to be 11.9 and 14.3 ($\log K_{ML}$), respectively, in log expression. An Fe^{2+}-EGTA buffer, 100 μM $Fe(NH_4)2(SO_4)2$, 1 mM EGTA, might have yielded 8.7×10^{-9} M Fe^{2+} aquo ions at pH 6.7, or by replacing EGTA with EDTA, 1.8×10^{-12} M Fe^{2+}. Since the stability constants of these chelators to Fe^{3+} ($\log K_{ML}$ for Fe^{3+}) are much higher, 20.5 and 25.1, respectively, the free Fe^{3+} ion was selectively masked off the system (Hasegawa and Ichiyama, 1986, 2005). The ion concentration was calculated using the PC application 'EQCAL' (Backman, 1988) based on an algorithm developed by Eriksson (1979).
14. Calcein, the fluorescent probe employed in this estimation, is a strong ligand of iron ion, comparable with EDTA (Breuer *et al.*, 1995). The stability constants ($LogK_{ML}$) of EDTA to Fe^{3+} and Fe^{2+} are 25.1 and 14.33, respectively. Those of EGTA are 20.5 and 11.92, respectively. As an Fe^{2+} ligand, EGTA was not competitive with TPH1, but at the same time, EGTA could remove the bound Fe^{3+} from inactive TPH1 (Hasegawa and Ichiyama, 2005).
15. These mast cells respond to a stem cell factor, SCF; a ligand for c-kit, while P-815 and FMA3 carry a mutation for constitutive activation at the c-kit, KIT ligand receptor tyrosine kinase (Tsujimura *et al.*, 1996). P-815 and FMA3 continuously express high levels of TPH1. RBL2H3, a rat basophilic leukemia derived cell line, expresses relatively low levels of TPH1 while its expression is controlled by antigen-IgE stimulation (Hasegawa *et al.*, 1996).

References

Abbar, M., Courtet, P., Bellivier, F., Leboyer, M., Boulenger, J.P., Castelhau, D., Ferreira, M., Lambercy, C., Mouthon, D., Paoloni-Giacobino, A., Vessaz, M., Malafosse, A. and Buresi, C. (2001) Suicide attempts and the tryptophan hydroxylase gene. Mol. Psychiatry, 6: 268–273.

Abumaria, N., Ribic, A., Anacker, C., Fuchs, E. and Flugge, G. (2008) Stress upregulates TPH1 but not TPH2 mRNA in the rat dorsal raphe nucleus: identification of two TPH2 mRNA splice variants. Cell Mol. Neurobiol., 28: 331–342.

Alberts, B., Bray, D., Lewis, J., Raff, M., Roberts, K. and Watson, J.D. (2002) Molecular Biology of the Cell, 3rd Edition, pp. 728, Fig. 715–710. Garland Publishing, Inc., New York.

Backman, L. (1988) EQCAL. Biosoft, Cambridge.

Bellivier, F., Leboyer, M., Courtet, P., Buresi, C., Beaufils, B., Samolyk, D., Allilaire, J.F., Feingold, J., Mallet, J. and Malafosse, A. (1998) Association between the tryptophan hydroxylase gene and manic-depressive illness. Arch. Gen. Psychiatry, 55: 33–37.

Bellivier, F., Chaste, P. and Malafosse, A. (2004) Association between the TPH gene A218C polymorphism and suicidal behavior: a meta-analysis. Am. J. Med. Genet. B Neuropsychiatr. Genet., 124B: 87–91.

Bentle, L.A. and Lardy, H.A. (1977) P-enolpyruvate carboxykinase ferroactivator. Purification and some properties. J. Biol. Chem., 252: 1431–1440.

Berardi, N., Pizzorusso, T. and Maffei, L. (2000) Critical periods during sensory development. Curr. Opin. Neurobiol., 10: 138–145.

Bonafe, L., Thony, B., Penzien, J.M., Czarnecki, B. and Blau, N. (2001) Mutations in the sepiapterin reductase gene cause a novel tetrahydrobiopterin-dependent monoamine-neurotransmitter deficiency without hyperphenylalaninemia. Am. J. Hum. Genet., 69: 269–277.

Boularand, S., Darmon, M.C., Ganem, Y., Launay, J.M. and Mallet, J. (1990) Complete coding sequence of human tryptophan hydroxylase. Nucleic Acids Res., 18: 4257.

Boularand, S., Darmon, M.C. and Mallet, J. (1995) The human tryptophan hydroxylase gene. An unusual splicing complexity in the 5'-untranslated region. J. Biol. Chem., 270: 3748–3756.

Braff, D.L. and Geyer, M.A. (1990) Sensorimotor gating and schizophrenia. Human and animal model studies. Arch. Gen. Psychiatry, 47: 181–188.

Brand, M.P., Hyland, K., Engle, T., Smith, I. and Heales, S.J. (1996) Neurochemical effects following peripheral administration of tetrahydropterin derivatives to the hph-1 mouse. J. Neurochem., 66: 1150–1156.

Breuer, W., Epsztejn, S. and Cabantchik, Z.I. (1995) Iron acquired from transferrin by K562 cells is delivered into a cytoplasmic pool of chelatable iron(II). J. Biol. Chem., 270: 24209–24215.

Buchanan, C.P., Shrier, E.M. and Hill, W.L. (1994) Time-dependent effects of PCPA on social aggression in chicks. Pharmacol. Biochem. Behav., 49: 483–488.

Cash, C.D. (1998) Why tryptophan hydroxylase is difficult to purify: a reactive oxygen-derived species-mediated phenomenon that may be implicated in human pathology. Gen. Pharmacol., 30: 569–574.

Cash, C.D., Vayer, P., Mandel, P. and Maitre, M. (1985) Tryptophan 5-hydroxylase. Rapid purification from whole rat brain and production of a specific antiserum. Eur. J. Biochem., 149: 239–245.

Christenson, J.G., Dairman, W. and Udenfriend, S. (1972) On the identity of DOPA decarboxylase and 5-hydroxytryptophan decarboxylase (immunological titration-aromatic L-amino acid decarboxylase-serotonin-dopamine-norepinephrine). Proc. Natl. Acad. Sci. USA, 69: 343–347.

Citron, B.A., Davis, M.D., Milstien, S., Gutierrez, J., Mendel, D.B., Crabtree, G.R. and Kaufman, S. (1992) Identity of 4a-carbinolamine dehydratase, a component of the phenylalanine hydroxylation system, and DCoH, a transregulator of homeodomain proteins. Proc. Natl. Acad. Sci. USA, 89: 11891–11894.

Cote, F., Thevenot, E., Fligny, C., Fromes, Y., Darmon, M., Ripoche, M.A., Bayard, E., Hanoun, N., Saurini, F., Lechat, P., Dandolo, L., Hamon, M., Mallet, J. and Vodjdani, G. (2003) Disruption of the nonneuronal tph1 gene demonstrates the importance of peripheral serotonin in cardiac function. Proc. Natl. Acad. Sci. USA, 100: 13525–13530.

Courtet, P., Jollant, F., Castelnau, D., Buresi, C. and Malafosse, A. (2005) Suicidal behavior: relationship between phenotype and serotonergic genotype. Am. J. Med. Genet. C Semin Med. Genet., 133C: 25–33.

Curtius, H.-C., Ghisla, S., Hasegawa, H., Blau, N. and Rebrin, I. (1993) Progress in the study of biosynthesis and role of 7-substituted pterins: function of pterin-4a-carbinolamine dehydratase. Adv. Exp. Med. Biol., 338: 107–110.

Darmon, M.C., Guibert, B., Leviel, V., Ehret, M., Maitre, M. and Mallet, J. (1988) Sequence of two mRNAs encoding active rat tryptophan hydroxylase. J. Neurochem., 51: 312–316.

Darvesh, A.S., Yamamoto, B.K. and Gudelsky, G.A. (2005) Evidence for the involvement of nitric oxide in 3,4-methylenedioxymethamphetamine-induced serotonin depletion in the rat brain. J. Pharmacol. Exp. Ther., 312: 694–701.

Davis, M.D. and Kaufman, S. (1989) Evidence for the formation of the 4a-carbinolamine during the tyrosine-dependent oxidation of tetrahydrobiopterin by rat liver phenylalanine hydroxylase. J. Biol. Chem., 264: 8585–8596.

Davis, M.D., Kaufman, S. and Milstien, S. (1992) Distribution of 4a-hydroxytetrahydropterin dehydratase in rat tissues. Comparison with the aromatic amino acid hydroxylases. FEBS Lett., 302: 73–76.

D'Sa, C.M., Arthur, R.E. Jr., States, J.C. and Kuhn, D.M. (1996) Tryptophan hydroxylase: cloning and expression of the rat brain enzyme in mammalian cells. J. Neurochem., 67: 900–906.

Elayan, I., Gibb, J.W., Hanson, G.R., Lim, H.K., Foltz, R.L. and Johnson, M. (1993) Short-term effects of 2,4,5-trihydroxyamphetamine, 2,4,5-trihydroxymethamphetamine and 3,4-dihydroxymethamphetamine on central tryptophan hydroxylase activity. J. Pharmacol. Exp. Ther., 265: 813–818.

Eriksson, G. (1979) An algorithm for the computation of aqueous multi-component, multiphase equilibria. Analytica. Chimica. Acta, 112: 375–383.

Fernstrom, J.D. and Wurtman, R.J. (1971) Brain serotonin content: physiological dependence on plasma tryptophan levels. Science, 173: 149–152.

Fisher, D.B., Kirkwood, R. and Kaufman, S. (1972) Rat liver phenylalanine hydroxylase, an iron enzyme. J. Biol. Chem., 247: 5161–5167.

Fitzpatrick, P.F. (1999) Tetrahydropterin-dependent amino acid hydroxylases. Annu. Rev. Biochem., 68: 355–381.

Fletcher, P.J., Selhi, Z.F., Azampanah, A. and Sills, T.L. (2001) Reduced brain serotonin activity disrupts prepulse inhibition of the acoustic startle reflex. Effects of 5,7-dihydroxytryptamine and p-chlorophenylalanine. Neuropsychopharmacology, 24: 399–409.

Florez, J.C., Seidenman, K.J., Barrett, R.K., Sangoram, A.M. and Takahashi, J.S. (1996) Molecular cloning of chick pineal tryptophan hydroxylase and circadian oscillation of its mRNA levels. Brain Res. Mol. Brain Res., 42: 25–30.

Furlong, R.A., Ho, L., Rubinsztein, J.S., Walsh, C., Paykel, E.S. and Rubinsztein, D.C. (1998) No association of the tryptophan hydroxylase gene with bipolar affective disorder, unipolar affective disorder, or suicidal behaviour in major affective disorder. Am. J. Med. Genet., 81: 245–247.

Furukawa, Y., Ikuta, N., Omata, S., Yamauchi, T., Isobe, T. and Ichimura, T. (1993) Demonstration of the phosphorylation-dependent interaction of tryptophan hydroxylase with the 14-3-3 protein. Biochem. Biophys. Res. Commun., 194: 144–149.

Garber, S.L. and Makman, M.H. (1987) Regulation of tryptophan hydroxylase activity by a cyclic AMP-dependent mechanism in rat striatum. Brain Res., 427: 1–10.

Gaspar, P., Cases, O. and Maroteaux, L. (2003) The developmental role of serotonin: news from mouse molecular genetics. Nat. Rev. Neurosci., 4: 1002–1012.

Gottschall, D.W., Dietrich, R.F., Benkovic, S.J. and Shiman, R. (1982) Phenylalanine hydroxylase. Correlation of the iron

content with activity and the preparation and reconstitution of the apoenzyme. J. Biol. Chem., 257: 845–849.
Graham, F.K. (1975) Presidential Address, 1974. The more or less startling effects of weak prestimulation. Psychophysiology, 12: 238–248.
Grahame, S.D. (1967) The biosynthesis of 5-hydroxytryptamine in brain. Biochem. J., 105: 351–360.
Grenett, H.E., Ledley, F.D., Reed, L.L. and Woo, S.L. (1987) Full-length cDNA for rabbit tryptophan hydroxylase: functional domains and evolution of aromatic amino acid hydroxylases. Proc. Natl. Acad. Sci. USA, 84: 5530–5534.
Grune, T., Blasig, I.E., Sitte, N., Roloff, B., Haseloff, R. and Davies, K.J.A. (1998) Peroxynitrite increases the degradation of aconitase and other cellular proteins by proteasome. J. Biol. Chem., 273: 10857–10862.
Gundlah, C., Alves, S.E., Clark, J.A., Pai, L.Y., Schaeffer, J.M. and Rohrer, S.P. (2005) Estrogen receptor-beta regulates tryptophan hydroxylase-1 expression in the murine midbrain raphe. Biol. Psychiatry, 57: 938–942.
Gutknecht, L., Jacob, C., Strobel, A., Kriegebaum, C., Muller, J., Zeng, Y., Markert, C., Escher, A., Wendland, J., Reif, A., Mossner, R., Gross, C., Brocke, B. and Lesch, K.P. (2007) Tryptophan hydroxylase-2 gene variation influences personality traits and disorders related to emotional dysregulation. Int. J. Neuropsychopharmacol., 10: 309–320.
Haavik, J., Andersson, K.K., Petersson, L. and Flatmark, T. (1988) Soluble tyrosine hydroxylase (tyrosine 3-monooxygenase) from bovine adrenal medulla: large-scale purification and physicochemical properties. Biochim. Biophys. Acta, 953: 142–156.
Haavik, J., Le, B.B., Martinez, A., Flatmark, T. and Mallet, J. (1991) Recombinant human tyrosine hydroxylase isozymes – reconstitution with iron and inhibitory effect of other metal ions. Eur. J. Biochem., 199: 371–378.
Hamon, M., Bourgoin, S., Hery, F. and Simonnet, G. (1978) Characteristics of the activation by dithiothreitol and Fe(2+) of tryptophan hydroxylase from the rat brain. Neurochem. Res., 3: 585–598.
Hasegawa, H. and Ichiyama, A. (1986) Strict requirement of Fe^{2+} for tryptophan 5-monooxygenase from mouse mastocytoma, P-815. In: Cooper, B.A. and Whitehead, V.M. (Eds.), Chemistry and Biology of Pteridines 1986. Walter de Gruyter, Berlin, pp. 369–372.
Hasegawa, H. and Ichiyama, A. (2005) Distinctive iron requirement of tryptophan 5-monooxygenase: TPH1 requires dissociable ferrous iron. Biochem. Biophys. Res. Commun., 338: 277–284.
Hasegawa, H., Yanagisawa, M., Inoue, F., Yanaihara, N. and Ichiyama, A. (1987) Demonstration of non-neural tryptophan 5-mono-oxygenase in mouse intestinal mucosa. Biochem. J., 248: 501–509.
Hasegawa, H., Kojima, M., Oguro, K. and Nakanishi, N. (1995) Rapid turnover of tryptophan hydroxylase in serotonin producing cells: demonstration of ATP-dependent proteolytic degradation. FEBS Lett., 368: 151–154.
Hasegawa, H., Kojima, M., Iida, Y., Oguro, K. and Nakanishi, N. (1996) Stimulation of tryptophan hydroxylase production in a serotonin-producing cell line (RBL2H3) by intracellular calcium mobilizing reagents. FEBS Lett., 392: 289–292.
Hasegawa, H., Oguro, K., Naito, Y. and Ichiyama, A. (1999) Iron dependence of tryptophan hydroxylase activity in RBL2H3 cells and its manipulation by chelators. Eur. J. Biochem., 261: 734–739.
Hasegawa, H., Sawabe, K., Nakanishi, N. and Wakasugi, O.K. (2005) Delivery of exogenous tetrahydrobiopterin (BH4) to cells of target organs: role of salvage pathway and uptake of its precursor in effective elevation of tissue BH4. Mol. Genet. Metab., 86(Suppl. 1): S2–S10.
Hasegawa, H., Sawabe, K., Sugawara, Y., Maruyama, S., Matsuoka, H., Nakanishi, N. and Wakasugi, O.K. (2007) Recent studies on sepiapterin and biopterin transport in mammalian cells. In: Blau, N. and Beat, T. (Eds.), Proceedings of 13th International Symposium on Chemistry & Biology of Pteridines & Folates. SPS Publications, Eustis, FL, pp. 254–268.
Hiroi, R., McDevitt, R.A. and Neumaier, J.F. (2006) Estrogen selectively increases tryptophan hydroxylase-2 mRNA expression in distinct subregions of rat midbrain raphe nucleus: association between gene expression and anxiety behavior in the open field. Biol. Psychiatry, 60: 288–295.
Hosoda, S. and Glick, D. (1965) Biosynthesis of 5-hydroxytryptophan and 5-hydroxytryptamine from tryptophan by neoplastic mouse mast cells. Biochim. Biophys. Acta, 111: 67–78.
Hufton, S.E., Jennings, I.G. and Cotton, R.G. (1995) Structure and function of the aromatic amino acid hydroxylases. Biochem. J., 311(Pt 2): 353–366.
Ichiyama, A., Hori, S., Mashimo, Y., Nukiwa, T. and Makuuchi, H. (1974) The activation of bovine pineal tryptophan 5-monooxygenase. FEBS Lett., 40: 88–91.
Ichiyama, A., Hasegawa, H., Tohyama, C., Dohmoto, C. and Kataoka, T. (1976) Some properties of bovine pineal tryptophan hydroxylase. Adv. Exp. Med. Biol., 74: 103–117.
Ieni, J.R. and Thurmond, J.B. (1985) Maternal aggression in mice: effects of treatments with PCPA, 5-HTP and 5-HT receptor antagonists. Eur. J. Pharmacol., 111: 211–220.
Iida, Y., Sawabe, K., Kojima, M., Oguro, K., Nakanishi, N. and Hasegawa, H. (2002) Proteasome-driven turnover of tryptophan hydroxylase is triggered by phosphorylation in RBL2H3 cells, a serotonin producing mast cell line. Eur. J. Biochem., 269: 4780–4788.
Johansen, P.A., Jennings, I., Cotton, R.G.H. and Kuhn, D.M. (1995) Tryptophan hydroxylase is phosphorylated by protein kinase A. J. Neurochem., 65: 882–888.
Johansen, P.A., Jennings, I., Cotton, R.G.H. and Kuhn, D.M. (1996) Phosphorylation and activation of tryptophan hydroxylase by exogenous protein kinase A. J. Neurochem., 66: 817–823.
Kapatos, G., Katoh, S. and Kaufman, S. (1982) Biosynthesis of biopterin by rat brain. J. Neurochem., 39: 1152–1162.
Kaufman, S. (1962) Aromatic Hydroxylations. In: Hayaishi, O. (Ed.), Oxygenases. Academic Press, New York, NY, pp. 129–180.
Kaufman, S. (1964) Studies on the structure of the primary oxidation product formed from tetrahydropteridines during phenylalamine hydroxylation. J. Biol. Chem., 239: 332–338.
Kaufman, S. (1967) Metabolism of the phenylalanine hydroxylation cofactor. J. Biol. Chem., 242: 3934–3943.
Kaufman, S. (1976) On the nature of an intermediate that is formed during the enzymatic conversion of phenylalanine to tyrosine. In: Yasunobu, K.T., Mower, H.F. and Hayaishi, O. (Eds.), Iron and Copper Proteins, Vol. 74 in Advances in Experimental Medicine and Biology. Plenum Publishing, New York, NY, pp. 91–102.
Kell, D.B. (2009) Iron behaving badly: inappropriate iron chelation as a major contributor to the aetiology of vascular and

other progressive inflammatory and degenerative diseases. BMC Med. Genomics, 2: 2.

Kennedy, M.C., Emptage, M.H., Dreyer, J.L. and Beinert, H. (1983) The role of iron in the activation–inactivation of aconitase. J. Biol. Chem., 258: 11098–11105.

Kim, K.S., Wessel, T.C., Stone, D.M., Carver, C.H., Joh, T.H. and Park, D.H. (1991) Molecular cloning and characterization of cDNA encoding tryptophan hydroxylase from rat central serotonergic neurons. Brain Res. Mol. Brain Res., 9: 277–283.

Kirov, G., Owen, M.J., Jones, I., McCandless, F. and Craddock, N. (1999) Tryptophan hydroxylase gene and manic-depressive illness. Arch. Gen. Psychiatry, 56: 98–99.

Koe, B.K. and Weissman, A. (1966) p-Chlorophenylalanine: a specific depletor of brain serotonin. J. Pharmacol. Exp. Ther., 154: 499–516.

Kojima, M., Oguro, K., Sawabe, K., Iida, Y., Ikeda, R., Yamashita, A., Nakanishi, N. and Hasegawa, H. (2000) Rapid turnover of tryptophan hydroxylase is driven by proteasomes in RBL2H3 cells, a serotonin producing mast cell line. J. Biochem. (Tokyo), 127: 121–127.

Kowlessur, D. and Kaufman, S. (1999) Cloning and expression of recombinant human pineal tryptophan hydroxylase in Escherichia coli: purification and characterization of the cloned enzyme. Biochim. Biophys. Acta, 1434: 317–330.

Krebs, H.A. and Veech, R.L. (1969) Equilibrium relations between pyridine nucleotides and adenine nucleotides and their roles in the regulation of metabolic processes. Adv. Enzyme Regul., 7: 397–413.

Kuhn, D.M. and Arthur, R.E. Jr. (1996) Inactivation of brain tryptophan hydroxylase by nitric oxide. J. Neurochem., 67: 1072–1077.

Kuhn, D.M. and Arthur, R. Jr. (1997) Molecular mechanism of the inactivation of tryptophan hydroxylase by nitric oxide: attack on critical sulfhydryls that spare the enzyme iron center. J. Neurosci., 17: 7245–7251.

Kuhn, D.M., Meyer, M.A. and Lovenberg, W. (1980a) Comparisons of tryptophan hydroxylase from a malignant murine mast cell tumor and rat mesencephalic tegmentum. Arch. Biochem. Biophys., 199: 355–361.

Kuhn, D.M., Ruskin, B. and Lovenberg, W. (1980b) Tryptophan hydroxylase. The role of oxygen, iron, and sulfhydryl groups as determinants of stability and catalytic activity. J. Biol. Chem., 255: 4137–4143.

Kuhn, D.M., Sakowski, S.A., Geddes, T.J., Wilkerson, C. and Haycock, J.W. (2007) Phosphorylation and activation of tryptophan hydroxylase 2: identification of serine-19 as the substrate site for calcium, calmodulin-dependent protein kinase II. J. Neurochem., 103: 1567–1573.

Lauder, J.M. and Krebs, H. (1978) Serotonin as a differentiation signal in early neurogenesis. Dev. Neurosci., 1: 15–30.

Lei, X.D. and Kaufman, S. (1999) Characterization of expression of the gene for human pterin carbinolamine dehydratase/dimerization cofactor of HNF1. DNA Cell Biol., 18: 243–252.

Levitt, P. and Rakic, P. (1982) The time of genesis, embryonic origin and differentiation of the brain stem monoamine neurons in the rhesus monkey. Brain Res., 256: 35–57.

Levitt, P., Harvey, J.A., Friedman, E., Simansky, K. and Murphy, E.H. (1997) New evidence for neurotransmitter influences on brain development. Trends Neurosci., 20: 269–274.

Lovenberg, M., Weissbach, H. and Udenfriend, S. (1962) Aromatic L-amino acid decarboxylase. J. Biol. Chem., 237: 89–93.

Lovenberg, W., Levine, R.J. and Sjoerdsma, A. (1965) A tryptophan hydroxylase in cell-free extracts of malignant mouse mast cells. Biochem. Pharmacol., 14: 887–889.

Lovenberg, W., Jequier, E. and Sjoerdsma, A. (1967) Tryptophan hydroxylation: measurement in pineal gland, brainstem, and carcinoid tumor. Science, 155: 217–219.

Luo, L. (2002) Actin cytoskeleton regulation in neuronal morphogenesis and structural plasticity. Annu. Rev. Cell Dev. Biol., 18: 601–635.

Ma, Y., de Groot, H., Liu, Z., Hider, R.C. and Petrat, F. (2006) Chelation and determination of labile iron in primary hepatocytes by pyridinone fluorescent probes. Biochem. J., 395: 49–55.

Mann, J.J., Malone, K.M., Nielsen, D.A., Goldman, D., Erdos, J. and Gelernter, J. (1997) Possible association of a polymorphism of the tryptophan hydroxylase gene with suicidal behavior in depressed patients. Am. J. Psychiatry, 154: 1451–1453.

McQuillin, A., Lawrence, J., Kalsi, G., Chen, A., Gurling, H. and Curtis, D. (1999) No allelic association between bipolar affective disorder and the tryptophan hydroxylase gene. Arch. Gen. Psychiatry, 56: 99–101.

Meek, J.L. and Neff, N.H. (1972) Tryptophan 5-hydroxylase: approximation of half-life and rate of axonal transport. J. Neurochem., 19: 1519–1525.

Miwa, S., Watanabe, Y. and Hayaishi, O. (1985) 6R-L-erythro-5,6,7,8-tetrahydrobiopterin as a regulator of dopamine and serotonin biosynthesis in the rat brain. Arch. Biochem. Biophys., 239: 234–241.

Mockus, S.M. and Vrana, K.E. (1998) Advances in the molecular characterization of tryptophan hydroxylase. J. Mol. Neurosci., 10: 163–179.

Murphy, K.L., Zhang, X., Gainetdinov, R.R., Beaulieu, J.M. and Caron, M.G. (2008) A regulatory domain in the N terminus of tryptophan hydroxylase 2 controls enzyme expression. J. Biol. Chem., 283: 13216–13224.

Nakamura, K. and Hasegawa, H. (2007) Developmental role of tryptophan hydroxylase in the nervous system. Mol. Neurobiol., 35: 45–54.

Nakamura, K., Sugawara, Y., Sawabe, K., Ohashi, A., Tsurui, H., Xiu, Y., Ohtsuji, M., Lin, Q.S., Nishimura, H., Hasegawa, H. and Hirose, S. (2006a) Late developmental stage-specific role of tryptophan hydroxylase 1 in brain serotonin levels. J. Neurosci., 26: 530–534.

Nakamura, K., Koyama, Y., Takahashi, K., Tsurui, H., Xiu, Y., Ohtsuji, M., Lin, Q.S., Sugawara, Y., Sawabe, K., Ohashi, A., Ohnuma, T., Arai, H., Nishimura, H., Hasegawa, H. and Hirose, S. (2006b) Requirement of tryptophan hydroxylase during development for maturation of sensorimotor gating. J. Mol. Biol., 363: 345–354.

Nakamura, K., Sato, T., Ohashi, A., Tsurui, H. and Hasegawa, H. (2008) Role of a serotonin precursor in development of gut microvilli. Am. J. Pathol., 172: 333–344.

Nakata, H. and Fujisawa, H. (1981) Simple and rapid purification of tryptophan 5-monooxygenase from rabbit brain by affinity chromatography. J. Biochem. (Tokyo), 90: 567–569.

Nakata, H. and Fujisawa, H. (1982a) Purification and properties of tryptophan 5-monooxygenase from rat brain-stem. Eur. J. Biochem., 122: 41–47.

Nakata, H. and Fujisawa, H. (1982b) Tryptophan 5-monooxygenase from mouse mastocytoma P815. A simple purification and general properties. Eur. J. Biochem., 124: 595–601.

Nichol, C.A., Lee, C.L., Edelstein, M.P., Chao, J.Y. and Duch, D.S. (1983) Biosynthesis of tetrahydrobiopterin by de novo and salvage pathways in adrenal medulla extracts, mammalian cell cultures, and rat brain *in vivo*. Proc. Natl. Acad. Sci. USA, 80: 1546–1550.

Nielsen, D.A., Virkkunen, M., Lappalainen, J., Eggert, M., Brown, G.L., Long, J.C., Goldman, D. and Linnoila, M. (1998) A tryptophan hydroxylase gene marker for suicidality and alcoholism. Arch. Gen. Psychiatry, 55: 593–602.

Okuno, S. and Fujisawa, H. (1981) Inactivation of tyrosine 3-monooxygenase by acetone precipitation and its restoration by incubation with a sulfhydryl agent and iron. Biochim. Biophys. Acta, 658: 327–333.

Ponka, P., Grady, R.W., Wilczynska, A. and Schulman, H.M. (1984) The effect of various chelating agents on the mobilization of iron from reticulocytes in the presence and absence of pyridoxal isonicotinoyl hydrazone. Biochim. Biophys. Acta, 802: 477–489.

Prinssen, E.P., Assie, M.B., Koek, W. and Kleven, M.S. (2002) Depletion of 5-HT disrupts prepulse inhibition in rats: dependence on the magnitude of depletion, and reversal by a 5-HT precursor. Neuropsychopharmacology, 26: 340–347.

Ramsey, A.J., Daubner, S.C., Ehrlich, J.I. and Fitzpatrick, P.F. (1995) Identification of iron ligands in tyrosine hydroxylase by mutagenesis of conserved histidinyl residues. Protein Sci., 4: 2082–2086.

Reuter, M., Kuepper, Y. and Hennig, J. (2007) Association between a polymorphism in the promoter region of the TPH2 gene and the personality trait of harm avoidance. Int. J. Neuropsychopharmacol., 10: 401–404.

Rouault, T., Rao, K., Harford, J., Mattia, E. and Klausner, R.D. (1985) Hemin, chelatable iron, and the regulation of transferrin receptor biosynthesis. J. Biol. Chem., 260: 14862–14866.

Rujescu, D., Giegling, I., Sato, T., Hartmann, A.M. and Moller, H.J. (2003) Genetic variations in tryptophan hydroxylase in suicidal behavior: analysis and meta-analysis. Biol. Psychiatry, 54: 465–473.

Sakowski, S.A., Geddes, T.J., Thomas, D.M., Levi, E., Hatfield, J.S. and Kuhn, D.M. (2006) Differential tissue distribution of tryptophan hydroxylase isoforms 1 and 2 as revealed with monospecific antibodies. Brain Res., 1085: 11–18.

Sawabe, K., Wakasugi, K.O. and Hasegawa, H. (2004) Tetrahydrobiopterin uptake in supplemental administration: elevation of tissue tetrahydrobiopterin in mice following uptake of the exogenously oxidized product 7,8-dihydrobiopterin and subsequent reduction by an anti-folate-sensitive process. J. Pharmacol. Sci., 96: 124–133.

Sawabe, K., Yamamoto, K., Harada, Y., Ohashi, A., Sugawara, Y., Matsuoka, H. and Hasegawa, H. (2008) Cellular uptake of sepiapterin and push–pull accumulation of tetrahydrobiopterin. Mol. Genet. Metab., 94: 410–416.

Sawada, M. and Nagatsu, T. (1988) Tryptophan hydroxylase activity in brain slices. Int. J. Biochem., 20: 1033–1038.

Sawada, M., Sugimoto, T., Matsuura, S. and Nagatsu, T. (1986) (6R)-tetrahydrobiopterin increases the activity of tryptophan hydroxylase in rat raphe slices. J. Neurochem., 47: 1544–1547.

Schmidt, C.J. and Taylor, V.L. (1987) Depression of rat brain tryptophan hydroxylase activity following the acute administration of methylenedioxymethamphetamine. Biochem. Pharmacol., 36: 4095–4102.

Shiman, R. and Jefferson, L.S. (1982) Iron-dependent regulation of rat liver phenylalanine hydroxylase activity in vivo, in vitro, and in perfused liver. J. Biol. Chem., 257: 839–844.

Shishkina, G.T., Kalinina, T.S. and Dygalo, N.N. (2007) Up-regulation of tryptophan hydroxylase-2 mRNA in the rat brain by chronic fluoxetine treatment correlates with its antidepressant effect. Neuroscience, 150: 404–412.

Shukla, A., Agarwal, K.N., Chansuria, J.P. and Taneja, V. (1989) Effect of latent iron deficiency on 5-hydroxytryptamine metabolism in rat brain [see comments]. J. Neurochem., 52: 730–735.

Stadtman, E.R. and Berlett, B.S. (1991) Fenton chemistry. Amino acid oxidation. J. Biol. Chem., 266: 17201–17211.

Stadtman, E.R. and Oliver, C.N. (1991) Metal-catalyzed oxidation of proteins. Physiological consequences. J. Biol. Chem., 266: 2005–2008.

Stoll, J., Kozak, C.A. and Goldman, D. (1990) Characterization and chromosomal mapping of a cDNA encoding tryptophan hydroxylase from a mouse mastocytoma cell line. Genomics, 7: 88–96.

Stone, D.M., Johnson, M., Hanson, G.R. and Gibb, J.W. (1989a) Acute inactivation of tryptophan hydroxylase by amphetamine analogs involves the oxidation of sulfhydryl sites. Eur. J. Pharmacol., 172: 93–97.

Stone, D.M., Hanson, G.R. and Gibb, J.W. (1989b) *In vitro* reactivation of rat cortical tryptophan hydroxylase following in vivo inactivation by methylenedioxymethamphetamine. J. Neurochem., 53: 572–581.

Tsai, S.J., Hong, C.J. and Wang, Y.C. (1999) Tryptophan hydroxylase gene polymorphism (A218C) and suicidal behaviors. Neuroreport, 10: 3773–3775.

Tsujimura, T., Morii, E., Nozaki, M., Hashimoto, K., Moriyama, Y., Takebayashi, K., Kondo, T., Kanakura, Y. and Kitamura, Y. (1996) Involvement of transcription factor encoded by the mi locus in the expression of c-kit receptor tyrosine kinase in cultured mast cells of mice. Blood, 88: 1225–1233.

Valzelli, L., Bernasconi, S. and Garattini, S. (1981) p-Chlorophenylalanine-induced muricidal aggression in male and female laboratory rats. Neuropsychobiology, 7: 315–320.

Vataeva, L.A., Kudrin, V.S., Vershinina, E.A., Mosin, V.M., Tiul'kova, E.I. and Otellin, V.A. (2007) Behavioral alteration in the adult rats prenatally exposed to para-chlorophenylalanine. Brain Res., 1169: 9–16.

Vitalis, T. and Parnavelas, J.G. (2003) The role of serotonin in early cortical development. Dev. Neurosci., 25: 245–256.

Vitalis, T., Cases, O., Passemard, S., Callebert, J. and Parnavelas, J.G. (2007) Embryonic depletion of serotonin affects cortical development. Eur. J. Neurosci., 26: 331–344.

Voltattorni, C.B., Minelli, A. and Turano, C. (1971) Spectral properties of the coenzyme bound to DOPA decarboxylase from pig kidney. FEBS Lett., 17: 231–235.

Walther, D.J., Peter, J.U., Bashammakh, S., Hortnagl, H., Voits, M., Fink, H. and Bader, M. (2003) Synthesis of serotonin by a second tryptophan hydroxylase isoform. Science, 299: 76.

Wang, L., Erlandsen, H., Haavik, J., Knappskog, P.M. and Stevens, R.C. (2002) Three-dimensional structure of human tryptophan hydroxylase and its implications for the biosynthesis of the neurotransmitters serotonin and melatonin. Biochemistry, 41: 12569–12574.

Wardman, P. and Candeias, L.P. (1996) Fenton chemistry: an introduction. Radiat. Res., 145: 523–531.

Watabe, S. (1978) Purification and characterization of tetrahydrofolate protein complex in bovine liver. J. Biol. Chem., 253: 6673–6679.

White, L.A., Eaton, M.J., Castro, M.C., Klose, K.J., Globus, M.Y., Shaw, G. and Whittemore, S.R. (1994) Distinct regulatory pathways control neurofilament expression and neurotransmitter synthesis in immortalized serotonergic neurons. J. Neurosci., 14: 6744–6753.

Winge, I., McKinney, J.A., Ying, M., D'Santos, C.S., Kleppe, R., Knappskog, P.M. and Haavik, J. (2008) Activation and stabilization of human tryptophan hydroxylase 2 by phosphorylation and 14-3-3 binding. Biochem. J., 410: 195–204.

Yamauchi, T. and Fujisawa, H. (1983) Purification and characterization of the brain calmodulin-dependent protein kinase (kinase II), which is involved in the activation of tryptophan 5-monooxygenase. Eur. J. Biochem., 132: 15–21.

Yamauchi, T., Nakata, H. and Fujisawa, H. (1981) A new activator protein that activates tryptophan 5-monooxygenase and tyrosine 3-monooxygenase in the presence of Ca^{2+}-, calmodulin-dependent protein kinase. Purification and characterization. J. Biol. Chem., 256: 5404–5409.

Zaboli, G., Gizatullin, R., Nilsonne, A., Wilczek, A., Jonsson, E.G., Ahnemark, E., Asberg, M. and Leopardi, R. (2006) Tryptophan hydroxylase-1 gene variants associate with a group of suicidal borderline women. Neuropsychopharmacology, 31: 1982–1990.

Zhang, X., Beaulieu, J.M., Sotnikova, T.D., Gainetdinov, R.R. and Caron, M.G. (2004) Tryptophan hydroxylase-2 controls brain serotonin synthesis. Science, 305: 217.

Zheng, Y. and Laverty, R. (1998) Role of brain nitric oxide in (+/−)3,4-methylenedioxymethamphetamine (MDMA)-induced neurotoxicity in rats. Brain Res., 795: 257–263.

CHAPTER 2.4

The Degradation of Serotonin: Role of MAO

Marco Bortolato[1], Kevin Chen[1] and Jean C. Shih[*,1,2]

[1]*Department of Pharmacology and Pharmaceutical Sciences, School of Pharmacy, University of Southern California, Los Angeles, California*
[2]*Department of Cell and Neurobiology, Keck School of Medicine, University of Southern California, Los Angeles, California*

Abstract: The enzymatic degradation of serotonin (5-hydroxytryptamine, 5-HT) is mainly served by monoamine oxidase (MAO), a mitochondrial-bound enzyme requiring flavin adenine dinucleotide as a cofactor. MAO catalyzes the oxidative deamination of 5-HT by converting it into 5-hydroxy-3-indolacetaldehyde (5-HIAL), which is further processed into 5-hydroxy-3-indolacetic acid (5-HIAA) by aldehyde dehydrogenase. MAO also serves the degradation of other monoamine neurotransmitters, such as norepinephrine, dopamine and trace amines. Two MAO isoenzymes, named A and B, have been identified, which differ in genetic bases, transcriptional regulation, substrate preference, inhibitor affinity and regional distribution. Although MAO A exhibits the highest preference for 5-HT and catalyzes its metabolism under physiological conditions, MAO B is the only isoenzyme that has been identified in 5-HTergic neurons. This chapter reviews the evidence on the biochemistry, genetics and neurobiology of MAOs, in relationship to their functions in the regulation of 5-HTergic neurotransmission, as assessed by studies on the outcomes of their pharmacological and genetic inactivation.

Keywords: monoamine oxidase, 5-HT, catabolism.

Introduction: pathways of serotonin degradation

One of the key functional premises for the proper regulation of serotonin (5-hydroxytryptamine, 5-HT) neurotransmission is its rapid reuptake and enzymatic degradation. The main pathway of 5-HT metabolism, in the brain and peripheral tissues, is mediated by the enzyme monoamine oxidase (MAO), and consists of its deaminative oxidation to 5-hydroxy-3-indolacetaldehyde (5-HIAL).

This compound is further processed by an NAD^+-dependent aldehyde dehydrogenase (ALDH) into 5-hydroxy-3-indolacetic acid (5-HIAA). Small amounts of 5-HIAL have been shown to be converted into 5-hydroxytryptophol (5-HTOL) by either an NADH-dependent aldehyde reductase (ALR) or an NADPH-dependent alcohol-dehydrogenase (ADH) (Feldstein and Williamson, 1968; Consalvi *et al.*, 1986; Svensson *et al.*, 1999) (Figure 1). While the physiological brain concentration of 5-HTOL is only 1–5% of the 5-HIAA levels (Beck *et al.*, 1984), intake of ethanol has been shown to produce a dramatic increase of 5-HTOL production (Helander *et al.*, 1993), probably as a result of competitive inhibition of ALDH, as well as an increased NADH/NAD ratio, which enhances ADH activity (Davis *et al.*, 1967; Lahti and Majchrowicz, 1974; Beck *et al.*, 1982). Interestingly, this observation has led some authors to propose 5-HTOL/5-HIAA ratio as a sensitive marker of recent alcohol drinking (Voltaire *et al.*, 1992; Helander *et al.*, 1996).

5-HIAA is rapidly eliminated by diffusion, transported into the bloodstream and excreted through the kidneys (Udenfriend *et al.*, 1956) through a combined mechanism of glomerular filtration and active tubular excretion (Despopoulos and Weissbach, 1957). Urinary 5-HIAA concentrations are considered a dependable index of 5-HT peripheral concentrations (or MAO activity), and are commonly used to monitor the effects of MAO pharmacological inhibition or to screen for carcinoid syndrome, a disorder characterized by increased 5-HT release by some gastrointestinal endocrine tumors.

Similar considerations have also prompted researchers to use 5-HIAA concentrations in the cerebrospinal fluid (CSF) as an indicator of 5-HT turnover (Jimerson, 1985). Low CSF 5-HIAA levels have been repeatedly associated with certain psychiatric conditions, such as impulsivity, attempted suicide, antisocial behavior (Asberg *et al.*, 1976; van Praag, 1983; Mehlman *et al.*, 1994; Virkkunen *et al.*, 1995).

[*]Corresponding author
E-mail: jcshih@usc.edu

Christian Müller & Barry Jacobs (Eds.)
Handbook of Behavioral Neurobiology of Serotonin
ISBN 978-0-12-374634-4

DOI: 10.1016/B978-0-12-374634-4.00001-0

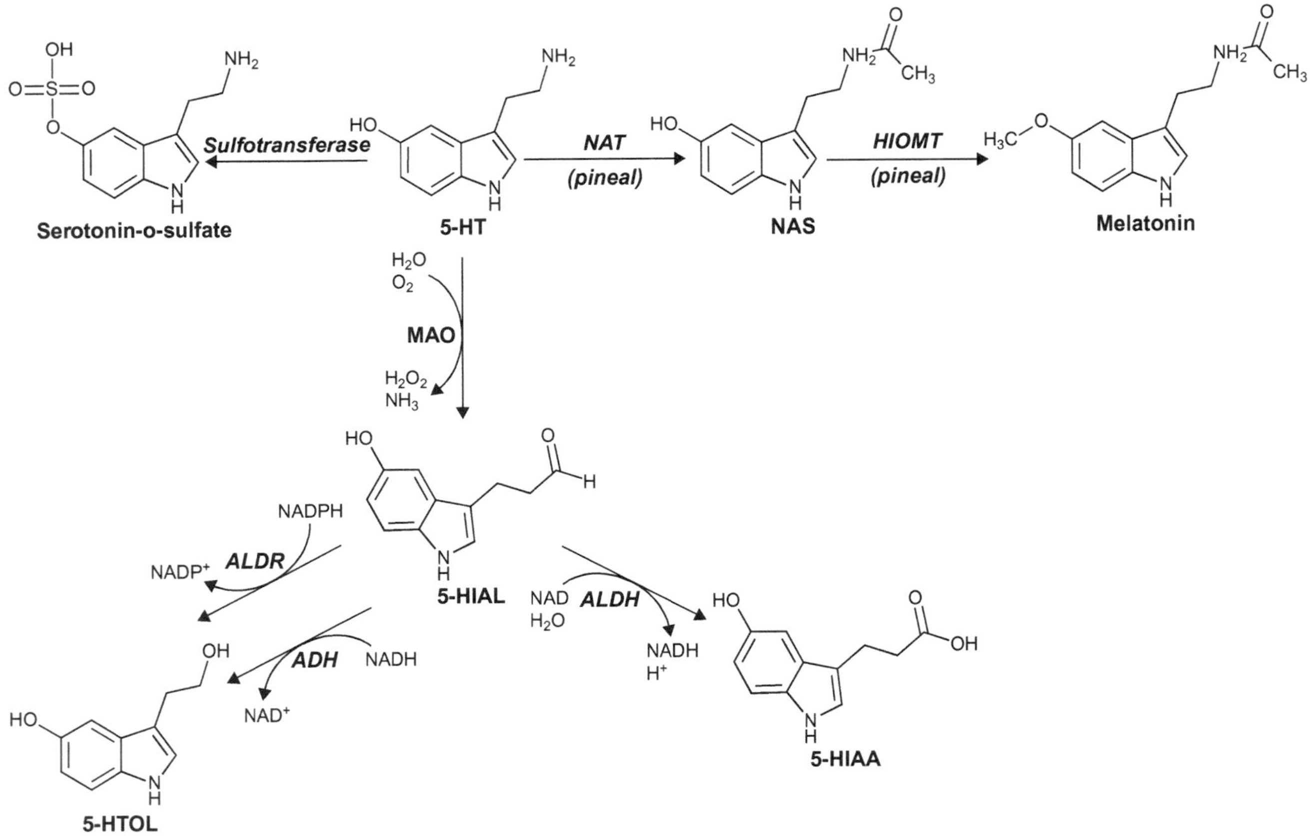

Figure 1 Metabolic pathways of 5-HT. MAO catalyzes the conversion of 5-HT into 5-hydroxy-3-indolacetaldehyde (5-HIAL), which is then further degraded to 5-hydroxy-3-indolacetic acid (5-HIAA). Alternatively, 5-HIAL can be converted into 5-hydroxytryptophol (5-HTOL) by aldehyde reductase (ALDR) or alcohol dehydrogenase (ADH). 5-HT can also be conjugated to sulfate or glucuronyl (not shown) groups by transferases. In the pineal gland, 5-HT is converted to *N*-acetylserotonin (NAS) by N-acetyl transferase. NAS is then transformed into melatonin by hydroxyindole-O-methyltransferase. For further details, see text.

Besides MAO-mediated degradation, 5-HT can undergo other secondary bioconversion pathways, such as conjugation to sulfate by means of a sulfotransferase (Hammond *et al.*, 1981; Hidaka *et al.*, 1969; Elchisak, 1983; Stuart *et al.*, 2003) (Figure 1), or glucuronylation by means of a glucuronyl-transferase (Tyce *et al.*, 1968). 5-HT is also converted to melatonin by acetylation (catalyzed by N-acetyl transferase) and subsequent methylation by hydroxyindole-O-methyltransferase (Namboodiri *et al.*, 1987; Sugden *et al.*, 1987) (Figure 1). This pathway has been shown to occur mainly in the pineal gland, although it is present also in other tissues, such as retina, gastrointestinal tract and thyroid (Kvetnoy, 1999). Other metabolites of 5-HT, such as 5-hydroxyindole thiazoladine carboxylic acid (5-HITCA), have been recently identified (Squires *et al.*, 2006). Nevertheless, these secondary pathways seem to play only a minor role in the metabolism and regulation of 5-HT.

In consideration of the predominant role of MAO in 5-HT metabolism, this chapter will provide a synoptic view on the state of the art of MAO biochemistry and genetics, and will illustrate the available evidence on the key functional role played by this enzyme in the regulation of 5-HT signaling and its functional correlates in behavioral organization.

MAO: general characteristics

MAO (amine: oxygen oxidoreductase (deaminating) (flavin-containing); MAO; E.C. 1.4.3.4) is a mitochondrial-bound flavoprotein, catalyzing the oxidative deamination of 5-HT and other monoamine neurotransmitters, such as norepinephrine, dopamine, epinephrine and trace amines (β-phenylethylamine, tryptamine, tyramine, etc.). Furthermore, MAO also serves a cytoprotective role by degrading exogenous amines, which exert their toxicity by affecting cardiovascular and endocrine homeostasis.

The enzyme was first isolated from extracts of rabbit liver in 1928, by Mary Hare, who named it tyramine oxidase in reference to its ability to catalyze the conversion of tyramine into *p*-hydroxyphenylacetic acid (Hare, 1928). The current nomenclature was actually proposed 10 years later, following the identification of the full spectrum of the monoaminergic substrates served by the enzyme (Zeller, 1938).

The chemical reaction catalyzed by MAO requires flavin adenine dinucleotide (FAD) as a redox cofactor is performed on monoamine substrates in their neutral, deprotonated forms (Edmondson *et al.*, 1993; Miller and Edmondson, 1999) and consists of the conversion of monoamines into the corresponding aldehydes.

Research has shown that the catalytic action of MAO starts with the cleavage of the C_a–H bond of the deprotonated form of monoamine substrates (Edmondson *et al.*, 1993; Miller and Edmondson, 1999). Specifically, FAD accepts two hydrogen atoms and is reduced to $FADH_2$, while the amine is converted into the correspondent imine:

$$RCH_2NH_2 + FAD \rightarrow RCH = NH + FADH_2.$$

Although the details of this process are still unclear, several theories have been proposed to explain them, including: an addition–elimination mechanism (Kim *et al.*, 1993) with formation of flavin-amine adducts; a proton abstraction reaction with a carbanion intermediate (Walker and Edmondson, 1994); and the formation of amine radical cations by single electron transfer to FAD (Silverman and Hawe, 1995). This last reaction, however, is considered unlikely, as it is not favored thermodynamically (Edmondson *et al.*, 2007).

In most cases, the second step of MAO reaction consists of the re-oxidization of $FADH_2$, with the conversion of oxygen into hydrogen peroxide:

$$\begin{aligned} RCH &= NH + FADH_2 + O_2 \rightarrow RCH \\ &= NH + FAD + H_2O_2. \end{aligned}$$

Finally, the imine is hydrolyzed into the corresponding aldehyde and ammonia:

$$RCH = NH + H_2O_2 \rightarrow RCHO + NH_3.$$

As mentioned above, aldehydes are mainly oxidized into acids by ALDH, but can also be converted into alcohols or glycols by ADH or ALR. A typical example of this latter mechanism is the peripheral metabolism of norepinephrine, which is converted into 3,4-dihyroxyphenylglycol (DHPG) and then further degraded by COMT into 3-methoxy-4-hydroxyphenylglycol (MHPG).

The byproducts of MAO-catalyzed reactions include potentially neurotoxic species, such as hydrogen peroxide and ammonia. In particular, hydrogen peroxide can trigger the production of reactive oxygen species (ROS) and induce mitochondrial damage and neuronal apoptosis. The oxidative stress induced by MAO activation may be critical in the pathogenetic mechanisms of several disorders in which this enzyme has been implicated, such as Parkinson's disease or depression (Bortolato *et al.*, 2008).

MAO isoenzymes

In 1968, two separate MAO isoenzymes – respectively named A and B – were postulated based on the differential selectivity MAO inhibitors. MAO A was defined as the isoenzyme inhibited by low doses of clorgyline, while MAO B was blocked by nanomolar concentrations of (R)-deprenyl (Johnston, 1968; Squires, 1968; Knoll and Magyar, 1972). This distinction was also paralleled by findings on the spectrum of affinity for different monoamine substrates. Although the two isoenzymes do not have absolute substrate specificity, 5-HT is mainly degraded by MAO A, with affinity 120-fold higher than MAO B. MAO A also prefers norepinephrine, while MAO B is the main enzyme for β-phenylethylamine catabolism.

The metabolism of dopamine and tyramine is generally served by both isoenzymes, albeit with considerable differences across species. For instance, while the metabolism of dopamine is mainly served by glial MAO B in humans, primates and other mammals (Garrick and Murphy, 1980), it is only metabolized by MAO A under basal conditions and by both isoenzymes at high concentrations in mice (Cases *et al.*, 1995; Fornai *et al.*, 1999). Finally, dopamine is always metabolized by MAO A in rats, regardless of the concentrations (Fornai *et al.*, 1999).

The different molecular nature of the two MAO isoenzymes was unequivocally demonstrated in 1988, with the cloning of the cDNA of *Maoa* and *Maob* genes from humans (Bach *et al.*, 1988; Hsu *et al.*, 1988). The comparison of the deduced amino acid sequences of the two isoenzymes indicates that they share 70 percent sequence identity, consist of 527 and 520 amino acids, and have molecular weights of 59.7 kDa and 58.8 kDa, respectively. The cloning of bovine and rat MAO A and MAO B (Ito *et al.*, 1988; Powell *et al.*, 1989; Kuwahara *et al.*, 1990; Kwan and Abell, 1992) revealed high levels of sequence identity across different mammalian species.

The availability of MAO cDNA clones made possible the isolation and characterization of the human *Maoa* (Chen *et al.*, 1991; Grimsby *et al.*, 1991; Kwan *et al.*, 1992) and *Maob* (Grimsby *et al.*, 1991; Kwan *et al.*, 1992) gene structure and localization.

Both genes are located on the X chromosome (Xp11.23 in humans), in opposite directions with tail-to-tail orientation (Lan *et al.*, 1989a; Levy *et al.*, 1989; Ozelius *et al.*, 1988), and are composed of 15 exons and span over 60 kb. The observation that MAO A and MAO B share identical exon–intron organization suggests their origin from a common ancestral gene (Grimsby *et al.*, 1991). The occurrence of two MAO isoenzymes can be traced back to the anuran amphibians (Kobayashi *et al.*, 1981): in these vertebrates, MAO A is notably predominant in the tadpole stage, while MAO B expression increases through metamorphosis (Nicotra and Senatori, 1988). In consideration of the different roles of various monoamines in the regulation of blood pressure and heart functioning, the development of target-specific MAO isoenzymes was probably instrumental to the new challenges posed by terrestrial life to the cardiovascular system, in parallel to other adaptive neuroendocrine mechanisms (Takei *et al.*, 2007).

The availability of MAO A and B clones allowed the selective transfection of their cDNAs in mammalian cells, leading to the discovery that the expressed enzymes were catalytically active and displayed substrate and inhibitor specificities similar to those of the endogenous enzymes (Lan *et al.*, 1989b) (Figure 2). This important finding demonstrated that only one polypeptide was required for the catalytically active form of MAO A or MAO B.

Expression of site-directed MAO mutants in COS cells confirmed that covalent attachment of FAD to cystein residues in both isoenzymes (Cys-406 in MAO A and Cys-397 in MAO B) (Wu *et al.*, 1993) is critical for their enzymatic activity. Additional site-directed mutagenesis studies determined that the cysteine residues that play a critical role for enzymatic activity are in position 374 for MAO A and positions 156 and 365 for MAO B (Wu *et al.*, 1993). We also demonstrated that substrate and inhibitor specificities in human MAO isoenzymes are conferred by an isoleucine residue in position 335 and a tyrosine residue in position 326 in human MAO A and B, respectively (Geha *et al.*, 2001).

Studies on MAO A/B chimeric enzymes ascertained that the difference in substrate preference and inhibitor specificity was conferred by an internal segment between amino acids 161 and 375 for MAO A and amino acids 152 and 366 for MAO B, but not by the N- or C-terminal regions (Gottowik *et al.*, 1993; Chen *et al.*, 1996; Shih *et al.*, 1998; Geha *et al.*, 2000). Furthermore, it was discovered that FAD was covalently bound – through a thioester linkage – to the cystein residue of a pentapeptidic sequence (Ser–Gly–Gly–Cys–Tyr) located between aa 403 and 407 in MAO A and aa 394 and 398 in MAO B (Kearney *et al.*, 1971; Bach *et al.*, 1988; Hsu *et al.*, 1988).

The determination of MAO 3D structure with X-crystallographic studies (Binda *et al.*, 2002, 2003; Ma *et al.*, 2004; De Colibus *et al.*, 2005) revealed that human MAO A displays a monomeric organization (Andrés *et al.*, 2004; De Colibus *et al.*, 2005), with a monoamine-binding cavity size of 550 $Å^3$ and a FAD site opposite to the entrance. Two tyrosine residues are predicted to facilitate substrate placement for the catalytic activity (Edmondson *et al.*, 2007). In contrast, human MAO B crystallizes as a dimer, and displays two cavities separated by a isoleucine gate, which can exist in either an open or a closed

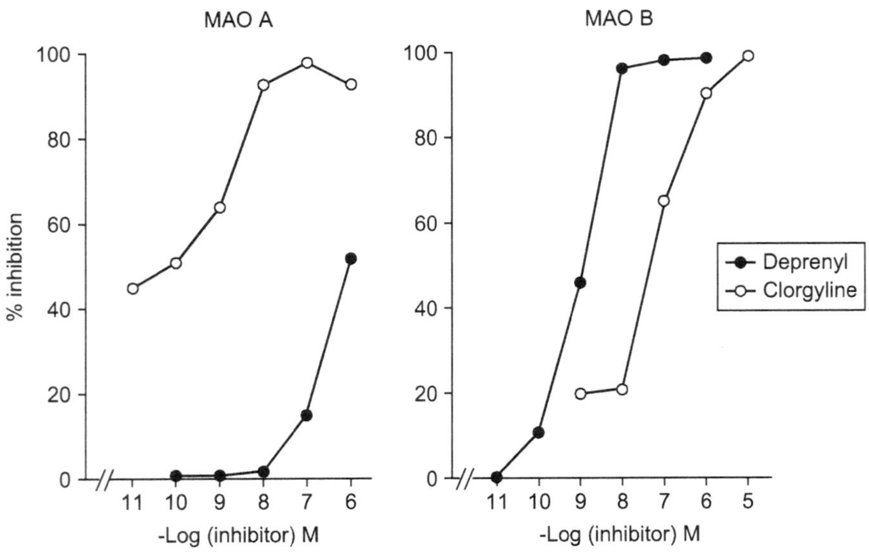

Figure 2 Inhibition of MAO A and MAO B expressed in COS cells by deprenyl and clorgyline. MAO A and MAO B activity was assessed using 5-HT and β-phenylethylamine, respectively, as substrates in the presence of various concentrations of inhibitors. Modified from Lan *et al.* (1989a).

conformation depending on the size of the substrate (Hubalek *et al.*, 2005). The FAD binding site is similar to that identified for MAO A (Edmondson *et al.*, 2007). Overall, these results are in substantial agreement with earlier findings of mutagenesis studies (Geha *et al.*, 2001, 2002).

Transcription and translation of MAO

Over the past 20 years, our group has afforded significant contributions in the understanding of the transcriptional regulation of *Maoa* and *Maob* genes. While the detailed illustration of these findings is beyond the scope of the present chapter (see Chen, 2004; Shih and Chen, 2004), it is worth noting that the promoters of both genes, as well as their mechanisms of activation and repression, are remarkably different (Shih *et al.*, 1994). Human MAO B expression, for example, is activated by protein kinase C and MAPK signaling transduction pathways (Wong *et al.*, 2002), while it is decreased by methylation (Wong *et al.*, 2003). A key role in the regulation of the expression of both genes is played by the family of the transcription factor Sp1 (Zhu *et al.*, 1994; Wong *et al.*, 2001). We also identified R1, a novel repressor of *Maoa* gene expression (Chen *et al.*, 2005). Notably, both R1 and Sp1 activities are modulated by glucocorticoids and androgens (Ou *et al.*, 2006).

Several studies have indicated that the transcriptional activity of human *Maoa* gene may be affected by different variants of an extensive repeat structure located in the promoter, 1.2 kb upstream of the MAO A coding sequence. The functional significance of this motif and its role in behavioral regulation will be discussed below.

Like other mitochondrial proteins, MAOs are synthesized by free ribosomes in the cytosol (Sagara and Ito, 1982). Neither isoenzyme, however, features the signal sequence in the N-terminal domain typical of mitochondrial protein precursors (Roise and Schatz, 1988). Indeed, it has been established that the C-terminus of MAO B contains the sequence for targeting the mitochondrial outer membrane (Mitoma and Ito, 1992) and attaching to it (Rebrin *et al.*, 2001). Insertion of both MAO A and MAO B into the mitochondrial outer membrane requires ubiquitin and ATP (Zhuang *et al.*, 1988, 1992; Zhuang and McCauley, 1989). The specific mechanism of incorporation into the outer mitochondrial membrane, however, is still poorly understood.

Localization and distribution of MAO

The localization of both MAO A and MAO B in the central nervous system and across different peripheral tissues has been studied in different species using numerous complementary approaches, such as the use of enzyme inhibitors, immunohistochemistry, autoradiography and *in situ* hybridization.

As MAO is bound to the outer mitochondrial membrane (Greenawalt and Schnaitman, 1970), it is present in most cells, with the obvious exception of red blood cells. Nevertheless, the pattern of distribution of MAO A and MAO B is remarkably distinct.

In humans, MAO A is highly expressed in the liver, lung, placenta and small intestine, while its lowest levels are found in the spleen and brain microvessels. Furthermore, MAO A is not found in platelets and lymphocytes (Bond and Cundall, 1977; Donnelly and Murphy, 1977). In contrast, the levels of MAO B expression are highest in the small intestine, liver and brain, and poor in the pancreas, spleen, lung, and skin fibroblasts. Interestingly, this isoenzyme is absent in placental tissue (Dahlstroem and Fuxe, 1964; Grimsby *et al.*, 1990).

Although the brain exhibits high concentrations of MAO A and MAO B, their regional and cell-specific localization is also strikingly different. Specifically, MAO A is predominantly found in catecholaminergic neurons (and in particular in the locus coeruleus, nucleus accumbens, hypothalamus and mammillary complex). In contrast, MAO B is found in 5-HTergic neurons, but also in histaminergic cells and in astrocytes (Westlund *et al.*, 1988; Saura *et al.*, 1994; Luque *et al.*, 1995; Jahng *et al.*, 1997).

The functional significance of the pattern of distribution of MAO isoenzymes in the brain is still partially unclear, in consideration of their substrate specificity. For instance, the localization of MAO B in the 5-HTergic neurons is seemingly at variance with rich evidence showing that 5-HT is preferentially degraded by MAO A (see above) and that inactivation of MAO A, but not MAO B, results in 5-HT level enhancement (Mousseau and Greenshaw, 1989; Cases *et al.*, 1995; Grimsby *et al.*, 1997). Indeed, MAO B is the only isoenzyme identified in the cell bodies and dendrites of 5-HTergic neurons in the raphe nuclei (Levitt *et al.*, 1982). Nevertheless, MAO A mRNA has been repeatedly observed in these cells (Luque *et al.*, 1995; Jahng *et al.*, 1997; Filipenko *et al.*, 2002).

An intriguing hypothesis that may account for these apparent discrepancies is that, in 5-HTergic neurons, MAO A protein may actually be synthesized in the cell body (or axon hillock), anchored to select mitochondria and transported by microtubules or other mechanisms to the axon terminals located in different brain areas, such as the cortex, amygdala, hippocampus, etc. (Jacobs and Azmitia, 1992) (Figure 3). This possibility is indirectly supported by the finding that MAO B expression is abundant in the mitochondria located in the somata of

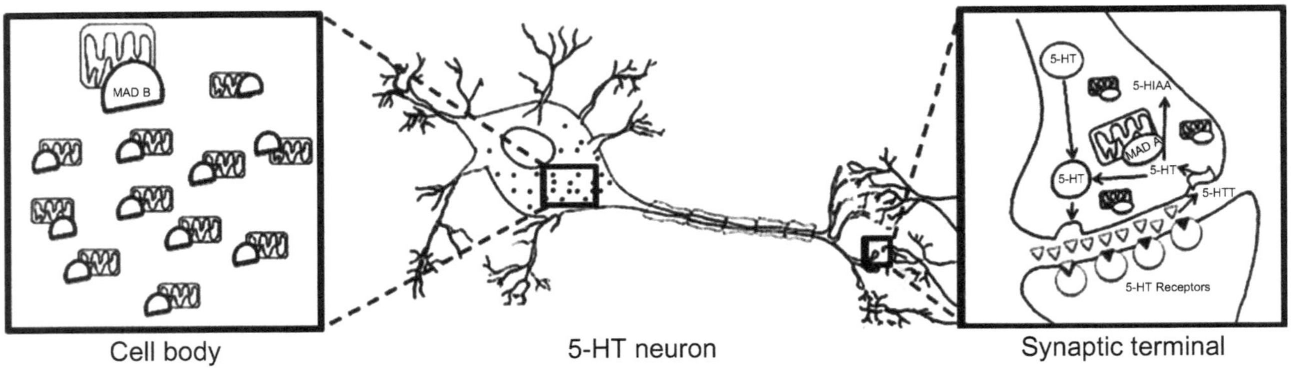

Figure 3 Hypothetical compartmentalization of MAO A and MAO B in 5-HTergic neurons. We propose that the localization of MAO B (represented as a semi-oval shape in the left box) is limited to the mitochondria of the cell body and dendrites of the 5-Htergic neuron. Conversely, MAO A (schematized as an elliptic shape on the mitochondria in the right box) may be mainly distributed in the synaptic terminals, where it would degrade the 5-HT after its reuptake operated by 5-HT transporter (5-HTT). For more details, see text.

5-HTergic neurons, but completely absent from the mitochondria in their axon terminals (Arai *et al.*, 2002). The differential subcellular location of MAO B-containing and MAO B-free mitochondria may reflect their different interactions with kinesins, the motor proteins in charge of anterograde mitochondrial transport through microtubules (Goldstein, 2001; Hollenbeck and Saxton, 2005).

Interestingly, the identification of MAO A-containing mitochondria in axon terminals (Westlund *et al.*, 1993) suggests that they are efficiently transported to this compartment by microtubules. Further investigations regarding the possible interactions between MAO isoenzymes and kinesin receptors are warranted to understand the molecular mechanisms underpinning these differences.

The proposed compartmentalization of MAO isoenzymes in 5-HTergic neurons might respond to specific physiological requirements for the proper processing of 5-HT synthesis, vesicular uptake and metabolism. Indeed, several lines of evidence point to an independent regulation of formation and degradation of 5-HT in the cell body and axon terminals of 5-HTergic neurons (Neckers, 1982; Daszuta *et al.*, 1984a, 1984b).

Although 5-HT synthesis can occur in the axon terminal, it is physiologically more abundant in the cell body (Daszuta *et al.*, 1984a; Pivac *et al.*, 2003). Thus, in consideration of the high affinity of MAO A for 5-HT, the localization of this enzyme in the soma may arguably compete with the gradient-based mechanism of 5-HT vesicular uptake. On the other hand, the presence of MAO A in the synaptic terminal may facilitate the degradation of 5-HT after its reuptake mediated by 5-HT transporter (5-HTT). Conversely, the presence of MAO B in the body of 5-HTergic neurons may prevent the uptake of wrong amine neurotransmitters in the vesicles. Furthermore, this isoenzyme may catalyze the degradation of trace amines such as β-phenylethylamine and tryptamine, which have been suggested to deplete the content of 5-HT in the vesicles by virtue of an amphetamine-like mechanism (Burchett and Hicks, 2006).

Role of MAO in behavioral regulation: pharmacological inhibition

The development and design of MAO inhibitors was originally prompted by the fortuitous discovery that iproniazid – a derivative of isoniazid originally designed to improve the antitubercular properties of this agent – exerted remarkable mood-enhancing properties (Fox and Gibas, 1953). Capitalizing on the finding that this compound was also a potent, irreversible MAO blocker (Zeller and Barsky, 1952), several researchers postulated a link between MAO, its substrates and mood control. This theory was substantiated by the antidepressant activity of novel irreversible MAO inhibitors, such as nialamide (Rowe *et al.*, 1959) and phenelzine (Saunders *et al.*, 1959). The widespread success and use of MAO inhibitors in clinical practice was paralleled by multiple reports of their side effects. In particular, serious concerns stemmed from the description of the 'cheese reaction', an often lethal hypertensive crisis due to the absorption of tyramine or other sympathomimetic amines contained in fermented food, such as cheese, wine, etc. (Anderson *et al.*, 1993). Due to the inactivation of intestinal MAO, high concentrations of these dietary amines could circulate in the bloodstream and cause excessive sympathetic activation and systemic vasoconstriction.

The hazards of MAO inhibitors propelled the development of novel categories of antidepressant drugs, such as tricyclics and selective serotonin reuptake inhibitors (SSRIs), leading to the characterization of selective antidepressant agents with better tolerability which gradually supplanted MAO inhibitors. Only the recent production of reversible MAO A inhibitors (which can be replaced by dietary amines and do not induce the cheese effect) has rekindled interest for this category of drugs.

There is a general consensus that the antidepressant properties of MAO inhibitors are actually to be ascribed to the blockade of MAO A, particularly in view of its action on 5-HT metabolism and the consequent ability to counter the decline in 5-HT observed in depression. Indeed, MAO A inhibition has been shown to increase 5-HT concentrations in the synaptic cleft (Sharp *et al.*, 1997) and to affect firing of 5-HTergic neurons in the dorsal raphe nucleus (Aghajanian *et al.*, 1970; Blier and de Montigny, 1985), and these effects have been suggested to underpin the antidepressant effects of MAO inhibitors. The link between the antidepressant effects of MAO inhibitors and 5-HTergic system is also supported by several lines of evidence, showing that the neurochemical, neurophysiological and behavioral outcomes of acute MAO blockade are reversed by 5-HT_{1A} receptor antagonists (Verge *et al.*, 1986; Sotelo *et al.*, 1990; Sharp *et al.*, 1997; Moser and Sanger, 1999). Nonetheless, it is possible that the actions of MAO inhibitors on norepinephrine (and, to a more limited extent, dopamine) signaling may contribute to their high efficacy in depression-related symptoms which tend to respond poorly to SSRIs (Amsterdam and Shults, 2005), such as dysthymia (Vallejo *et al.*, 1987), bulimia and hypersomnolence (Thase *et al.*, 1992; Carruba *et al.*, 2001).

MAO inhibitors are also indicated for a number of anxiety-spectrum disorders which have been shown to depend on alterations of 5-HTergic signaling, including social phobia, agoraphobia and panic disorder, post-traumatic stress disorder (Cyr and Farrar, 2000) and obsessive-compulsive disorder (OCD) (Jenike *et al.*, 1983).

The involvement of 5-HT in the antidepressant effects of MAO inhibitors is also indirectly confirmed by the notion that MAO B selective inhibitors, such as deprenyl, induce only modest mood-enhancing effects, which may also be due to the action of this isoenzyme on oxidative stress (Bortolato *et al.*, 2008).

Role of MAO in behavioral regulation: human evidence

Deletions of MAO A and MAO B in Norrie disease

The first case of human MAO deficiency was originally reported in some patients afflicted by Norrie disease (ND). This genetic disorder is caused by a mutation of the NDP (Norrie disease pseudoglioma) gene, located in Xp 11.4, and is characterized by congenital blindness, progressive hearing loss, and developmental delays in motor and cognitive skills (Warburg, 1975). Given the close proximity of NDP to MAO A and MAO B, some individuals afflicted by ND were found to carry deletions of MAO A and B (Sims *et al.*, 1989; Chen *et al.*, 1995). The loss of MAO function in ND results in high urinary concentrations of β-phenylethylamine and tyramine metabolites (Murphy *et al.*, 1990, 1991; Collins *et al.*, 1992). This alteration is accompanied by an array of atypical symptoms, such as growth failure, severe mental retardation, autistic-like behaviors, autonomic dysfunction and altered sleep patterns (Sims *et al.*, 1989; Murphy *et al.*, 1990; Collins *et al.*, 1992). Nevertheless, the predominance of the sensory deprivation in ND patients made it difficult to define the phenotypical outcomes of congenital MAO ablation.

MAO A total deficiency: Brunner syndrome

In 1993, Brunner and colleagues described for the first time a genetic disorder caused by a point nonsense mutation in exon 8 of *Maoa* gene, which induced total deficiency of MAO A (Brunner syndrome) in several male subjects within a Dutch pedigree (Brunner *et al.*, 1993a, 1993b). The nosographic characteristics of this complex behavioral syndrome included mild mental retardation, impulsive aggressiveness, and violent responses to environmental stressors (Brunner *et al.*, 1993a, 1993b). The genetic defect in MAO A was also associated with a five-fold increase in urinary levels of 5-HT and O-metabolites of dopamine and norepinephrine.

MAO polymorphisms

Several polymorphisms of the *Maoa* gene have been identified (Ozelius *et al.*, 1988; Black *et al.*, 1991; Hotamisligil and Breakefield, 1991; Hinds *et al.*, 1992) and employed in association studies for different psychiatric alterations (Shih and Thompson, 1999). The first report of an association between MAO A polymorphism and psychiatric disorder concerned the correlation between a long dinucleotide (CA)n repeat in intron 2 (Black *et al.*, 1991) and susceptibility to alcoholism/substance abuse in males (Vanyukov *et al.*, 1995).

Research has shown that some of the different polymorphisms of the *Maoa* gene may be associated with differences in enzymatic activity (Hotamisligil and Breakefield, 1991). In particular, the functional polymorphism of MAO A that has been characterized most extensively is a variable-number tandem repeat (VNTR) polymorphism, located 1.2 kb upstream of the transcription initiation site

(Sabol *et al.*, 1998). Six different allelic variants have been characterized in the human population (Huang *et al.*, 2004; Sabol *et al.*, 1998), the most common of which presents three repeats (3R) and four repeats (4R) (Sabol *et al.*, 1998; Deckert *et al.*, 1999; Jonsson *et al.*, 2000). The first studies on VNTR variants in MAO A promoter consistently documented higher MAO A gene transcription and enzyme activity in association with the 4R allelic variant in transfected cells (Sabol *et al.*, 1998) and in cultures of human male skin fibroblasts (Denney *et al.*, 1999). In agreement with this finding, human studies have shown that the 3R variant is associated with behavioral features linked to low MAO A activity, such as impulsive aggressiveness and antisocial personality (Oreland *et al.*, 2007; Buckholtz and Meyer-Lindenberg, 2008). Conversely, 4R carriers display higher levels of 5-hydroxyindoleacetic acid (5-HIAA) in the cerebrospinal fluid (Jonsson *et al.*, 2000).

Furthermore, functional magnetic resonance imaging (fMRI) studies show a link between MAO A promoter VNTR variants and structural and functional differences in the brain (Meyer-Lindenberg *et al.*, 2006). Mutations in this promoter region have also been associated with a number of psychiatric disorders (Shih and Thompson, 1999; Chen, 2004).

Nevertheless, recent findings questioned the strength of the relation between different VNTR variants and MAO A activity. For example, *post-mortem* studies found a tendency for higher MAO A enzyme activity in brain samples from 4R adult carriers as compared to 3R counterparts, but this difference was not significant (Balciuniene *et al.*, 2002). Likewise, significant differences in MAO A activity (calculated as positron emission tomography (PET) time-activity data for [^{11}C]clorgyline) between 3R and 4R adult subjects were found only in the primary visual cortex, but not in other brain regions (Alia-Klein *et al.*, 2008). An interesting possibility is that the 4R variant may actually be associated with a higher MAO A promoter than 3R in an age-dependent fashion, and the correlation between VNTR variants and MAO A activity may vary over time due to the exposure to a number of environmental factors that affect MAO A levels and activity, such as stress (Maura and Vaccari, 1975), diet changes (Jahng *et al.*, 1998), tobacco smoking (Fowler *et al.*, 1996), physical exercise (Morishima *et al.*, 2006), social environment (Filipenko *et al.*, 2002), and aging itself (Saura *et al.*, 1994). This hypothesis is corroborated by the finding that 3R-carriers with a history of early stress are more likely to commit violent crimes and to be diagnosed with conduct disorders (Caspi *et al.*, 2002).

Irrespective of the functional significance of MAO A polymorphic variants, these data point to a link between low-activity polymorphic variants and antisocial traits (Manuck *et al.*, 2000). Indeed, MAO A activity in cortical and subcortical brain regions has been recently shown to be inversely correlated with the levels of self-reported aggression in men (Alia-Klein *et al.*, 2008). In line with this perspective, Brunner syndrome may be regarded as the end of a broad spectrum of antisocial manifestations correlated with low MAO A activity.

Role of MAO in behavioral regulation: animal evidence

Our group and others have developed several lines of transgenic mice with nonsense mutations for each MAO isoenzyme. The characterization of the phenotypes of these animals has afforded substantial contributions to our understanding of the role of MAO in the regulation of behavior and brain functions and complemented the information from human studies.

MAO A knockout mice

The first line of MAO A knockout (KO) mice was generated through the serendipitous insertion of an interferon-β cassette in exons 2 and 3 of the *Maoa* gene of C3H/HeJ mice (Cases *et al.*, 1995). The genetic defect followed an X-linked recessive line of transmission, and resulted in the total ablation of MAO A enzyme, leading to a complex array of abnormal morphological and behavioral phenotypes. MAO A KO pups display locomotor alterations, such as intense head bobbing, trembling, prolonged righting, and delay in the use of hindlimbs in swimming (Cases *et al.*, 1995; Cazalets *et al.*, 2000). This retarded maturation of the locomotor network is accompanied by alterations in the respiratory (Bou-Flores *et al.*, 2000) and sensory functions. Between days 11 and 16, MAO A KO mice exhibit locomotor hyperactivity, jumping and abnormal postures (Cases *et al.*, 1995). Adult MAO A KO mice display heightened levels of impulsive aggressiveness in the resident–intruder task and in other encounters with cage mates (Cases *et al.*, 1995; Chen *et al.*, 1999). This abnormal behavior is accompanied by marked reductions in exploratory activity and startle response, increased grasping during courtship, as well as enhanced motility in the forced swim test (Cases *et al.*, 1995). MAO A KO mice also show a number of alterations of emotional behaviors, such as increased tendency to stay in the center of an open arena (Cases *et al.*, 1995; Agatsuma *et al.*, 2006), higher levels of retention of aversive memories (Kim *et al.*, 1997) or retention of conditioned passive avoidance (Dubrovina and Zinov'ev, 2007). Nevertheless, they do not show overt alterations of anxiety-like behaviors in the elevated plus maze (Popova *et al.*, 2001) and have low endocrine reaction to major stressors, such as physical restraint, water deprivation, cold temperature and chronic variable stress (Popova *et al.*, 2006). Since adult C3H mice are blind, some of these alterations have been questioned to depend on the specific characteristic of the strain. Nevertheless, the recent identification of a novel line of 129/Sv MAO A KO mice, harboring a nonsense point mutation of the exon 8 (in a position close to that altered in Brunner syndrome) substantially confirmed most of these alterations and strengthened their link with MAO A function (Bortolato *et al.*, 2008; Scott *et al.*, 2008). Taken together, the behavioral abnormalities observed in MAO A KO mice suggest a general impairment in environmental reactivity, with aberrant regulation of adaptive responses to stress, in a fashion reminiscent of that reported in Brunner syndrome patients.

The morphological analysis of neuronal patterns in the brain of MAO A KO mice indicates the possibility that their behavioral abnormalities may depend on the sensory deficits induced by the deficiency of this enzyme. Indeed, MAO A KO mice display alterations in the auditory and visual pathways (Upton *et al.*, 1999; Thompson, 2008), as well as the dysmorphogenesis of barrel fields in layer IV of the somatosensory cortex (S1) (Cases *et al.*, 1995). Barrel fields are the cortical representations of the mystacial vibrissae in the rodent snout, and their formation is dependent on the thalamocortical projections arising from the ventrobasal thalamic nuclei (Erzurumlu and Jhaveri, 1990). As whiskers play a key role in the regulation of exploratory behavior and thigmotaxis in rodents (Vincent, 1912; Luhmann *et al.*, 2005), the alterations of barrel fields may underpin the abnormal responses in the open field in MAO A KO mice.

The neurochemical underpinnings of the phenotypical alterations in MAO A KO mice are still elusive. Interestingly, most behavioral and neuromorphological alterations observed in MAO A KO mice can be partially reproduced by chronic treatment with MAO A inhibitors during the early developmental stages, but not in adulthood. For instance, while pharmacological blockade of MAO A in adults fails to induce aggression, the same regimen in perinatal stages has been shown to produce a decrease in latency to attack in adult rats (Mejia *et al.*, 2002). Furthermore, early treatment with clorgyline induces behavioral alterations and impairments in thalamocortical development that resemble those observed in MAO A KO mice (Cases *et al.*, 1995; Boylan *et al.*, 2000). Conversely, we have shown that genetic reinstatement of MAO A in the forebrain of MAO A KO mice rescues their behavioral and morphogenetic abnormalities (Chen *et al.*, 2007).

Converging lines of evidence support a key role of 5-HT in the morphological and functional alterations in MAO A KO mice. In fact, early administration of p-chlorophenylalanine (PCPA), an inhibitor of 5-HT synthesis, reverses most aberrations in MAO A KO mice (Cases *et al.*, 1995; Upton *et al.*, 1999; Bou-Flores *et al.*, 2000). Indeed, hemizygous males displayed high levels of 5-HT and norepinephrine and modest increases in dopamine (Figure 4). Notably, the differences in 5-HT concentrations between MAO A KO mice and their wild-type (WT) counterparts are particularly elevated (up to 10-fold) during the early postnatal stages (Cases *et al.*, 1995; Lajard *et al.*, 1999).

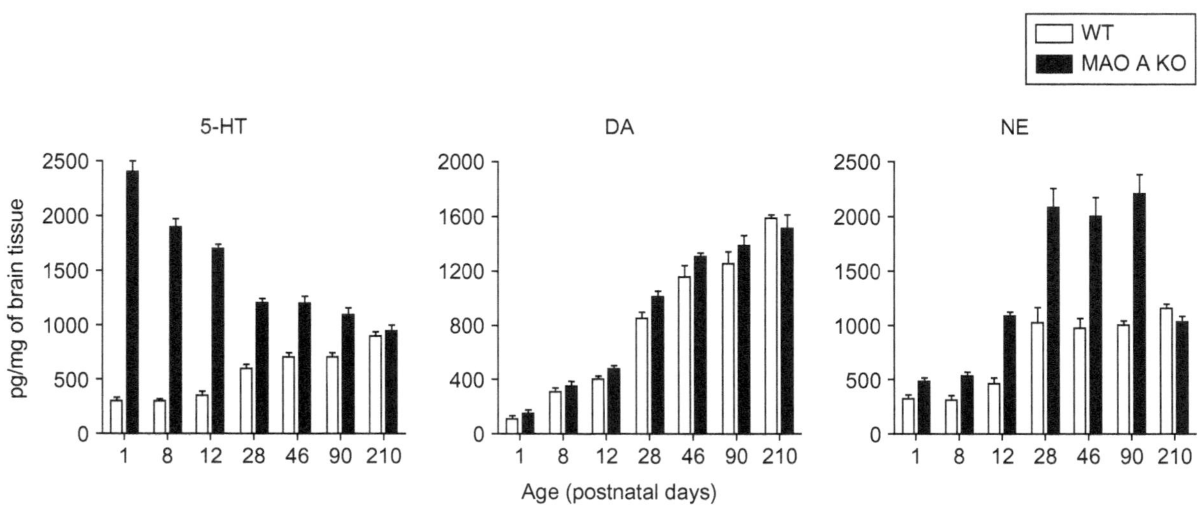

Figure 4 Levels of 5-HT, dopamine (DA), and norepinephrine (NE) are significantly higher in MAO A KO mice than WT conspecifics in an age-related fashion. Values from HPLC assays are expressed in pg/mg of wet brain and represent the mean ± SEM. Modified from Cases *et al.* (12195).

The key role of 5-HT has been inequivocably shown in the dysmorphogenetic phenomena observed in MAO A KO mice. This notion is in agreement with extensive evidence showing that 5-HT has a trophic effect on neuronal progenitors (Lieske *et al.*, 1999; Lotto *et al.*, 1999) and on the morphogenesis of the cortex (Vitalis and Parnavelas, 2003) as well as sensory and locomotor networks (Nakajima *et al.*, 1998; Lebrand *et al.*, 2006), probably through activation of different sets of receptors. Accordingly, the abnormal formation of barrels of MAO A KO mice has been shown to be underpinned by excessive levels of 5-HT in the first week of postnatal life (Vitalis *et al.*, 1998), through activation of 5-HT_{1B} receptors (Salichon *et al.*, 2001). In contrast, the morphological alterations of the respiratory centers in the medulla and the cervical phrenic motoneurons in MAO A KO are reversed by blockade of 5-HT_{2A} receptors (Bou-Flores *et al.*, 2000).

The effects of MAO A inactivation on the 5-HTergic system have been analyzed in some investigations. In comparison with WT controls, MAO A KO mice show a reduced electric activity of 5-HTergic neurons, as well as down-regulation of 5-HTT, 5-HT_{1A}, 5-HT_{2A} and 5-HT_{2C} receptors (Shih *et al.*, 2000; Evrard *et al.*, 2002; Lanoir *et al.*, 2006).

The association of high levels of 5-HT and marked impulsive aggressiveness in MAO A KO mice is an interesting exception to the inverse relationship generally observed between these two phenomena (Coccaro, 1989). Indeed, our group found that aggression in MAO A KO mice is suppressed by ketanserin, the prototypical 5-HT_2 receptor antagonist, suggesting a key role of this receptor in the behavioral abnormalities caused by MAO A deficiency (Shih and Chen, 1999). 5-HT plays a pivotal role in the regulation of threat detection (Cools *et al.*, 2008) and in the pathophysiology of antisocial behavior (Deakin, 2003). Thus, high levels of 5-HT may set the emotional threshold for the enactment of aggressive responses, which on the other hand may reflect complex interactions with other neurotransmitters. Accordingly, both norepinephrine and dopamine have been extensively implicated in the regulation of impulsivity and aggression (Haller *et al.*, 1998; de Almeida *et al.*, 2005).

MAO B knockout mice

In agreement with the low affinity of MAO B for 5-HT, mice deficient in this isoenzyme display only significant changes in brain levels of β-phenylethylamine, which are accompanied by behavioral disinhibition and alterations in stress-related and anxiety-like responses (Cases *et al.*, 1995; Bortolato *et al.*, 2009). The phenotypical characteristics of MAO B KO mice have been described elsewhere (Bortolato *et al.*, 2008, 2009).

MAO A/B knockout mice

We have recently characterized and developed a novel MAO AB KO line, harboring a spontaneous point mutation of MAO A and ablation of MAO B (Chen *et al.*, 2004). These mice display an array of unique phenotypes, which cannot be recapitulated by a summation of MAO A KO and MAO B KO. 5-HT, norepinephrine, dopamine and β-phenylethylamine levels are highly increased in comparison to either the MAO A or the MAO B KO mice. Furthermore, the behavioral phenotype in these mutant animals is characterized by low novelty-induced locomotion, high levels of anxiety-like behaviors in the elevated plus maze, and low latency to attack in the resident-intruder paradigm, which may be likened to an increase in impulsivity (Chen *et al.*, 2004). The role of 5-HT in these alterations of emotional behaviors, however, is still poorly understood.

Conclusions

MAO is the key enzyme in 5-HT degradation, and its regulation is fundamental in the management of 5-HT signaling. By means of several complementary approaches, research has provided a great deal of knowledge about the structure and function of both MAO isoenzymes and their key role in the regulation of brain architecture and behavior. These studies have set the stage for future investigations which will further our understanding of the role of these enzyme in the pathophysiology of several mental disorders and examine the specific role of 5-HT in the morphological and behavioral outcomes of MAO inactivation.

Acknowledgements

This work was supported by NIMH grants, R01MH67968, R37MH39085 (MERIT Award), and the Boyd and Elsie Welin Professorship. We thank Lauren Burgeno and Sean Godar for their insightful comments and valuable contribution in the preparation of the figures.

References

Agatsuma, S., Lee, M., Zhu, H., Chen, K., Shih, J.C., Seif, I. and Hiroi, N. (2006) Monoamine oxidase A knockout mice exhibit impaired nicotine preference but normal responses to novel stimuli. Hum. Mol. Genet., 15: 2721–2731.

Aghajanian, G.K., Graham, A.W. and Sheard, M.H. (1970) Serotonin-containing neurons in brain: depression of firing by monoamine oxidase inhibitors. Science, 169: 1100–1102.

Alia-Klein, N., Goldstein, R.Z., Kriplani, A., Logan, J., Tomasi, D., Williams, B., Telang, F., Shumay, E., Biegon, A., Craig, I.W., Henn, F., Wang, G.J., Volkow, N.D. and Fowler, J.S. (2008) Brain monoamine oxidase A activity predicts trait aggression. J. Neurosci., 28: 5099–5104.

Amsterdam, J.D. and Shults, J. (2005) MAOI efficacy and safety in advanced stage treatment-resistant depression: a retrospective study. J. Affect Disord., 89: 183–188.

Anderson, M.C., Hasan, F., McCrodden, J.M. and Tipton, K.F. (1993) Monoamine oxidase inhibitors and the cheese effect. Neurochem. Res., 18: 1145–1149.

Andrés, A.M., Soldevila, M., Navarro, A., Kidd, K.K., Oliva, B. and Bertranpetit, J. (2004) Positive selection in MAOA gene is human exclusive: determination of the putative amino acid change selected in the human lineage. Hum. Genet., 115: 377–386.

Arai, R., Karasawa, N., Kurokawa, K., Kanai, H., Horiike, K. and Ito, A. (2002) Differential subcellular location of mitochondria in rat serotonergic neurons depends on the presence and the absence of monoamine oxidase type B. Neuroscience, 114: 825–835.

Asberg, M., Traskman, L. and Thoren, P. (1976) 5-HIAA in the cerebrospinal fluid. A biochemical suicide predictor? Arch. Gen. Psychiatry, 33: 1193–1197.

Bach, A.W., Lan, N.C., Johnson, D.L., Abell, C.W., Bembenek, M.E., Kwan, S.W., Seeburg, P.H. and Shih, J.C. (1988) cDNA cloning of human liver monoamine oxidase A and B: molecular basis of differences in enzymatic properties. Proc. Natl. Acad. Sci. USA, 85: 4934–4938.

Balciuniene, J., Emilsson, L., Oreland, L., Petterson, U. and Jazin, E.E. (2002) Investigation of the functional effect of monoamine oxidase polymorphisms in human brain. Hum. Genet., 110: 1–7.

Beck, O., Borg, S., Edman, G., Fyro, B., Oxenstierna, G. and Sedvall, G. (1984) 5-hydroxytryptophol in human cerebrospinal fluid: conjugation, concentration gradient, relationship to 5-hydroxyindoleacetic acid, and influence of hereditary factors. J. Neurochem., 43: 58–61.

Beck, O., Borg, S., Eriksson, L. and Lundman, A. (1982) 5-hydroxytryptophol in the cerebrospinal fluid and urine of alcoholics and healthy subjects. Naunyn Schmiedebergs. Arch. Pharmacol., 321: 293–297.

Binda, C., Li, M., Hubalek, F., Restelli, N., Edmondson, D.E. and Mattevi, A. (2003) Insights into the mode of inhibition of human mitochondrial monoamine oxidase B from high-resolution crystal structures. Proc. Natl. Acad. Sci. USA, 100: 9750–9755.

Binda, C., Newton-Vinson, P., Hubálek, F., Edmondson, D.E. and Mattevi, A. (2002) Structure of human monoamine oxidase, B., a drug target for the treatment of neurological disorders. Nat. Struct. Biol., 9: 22–26.

Black, G.C., Chen, Z.Y., Craig, I.W. and Powell, J.F. (1991) Dinucleotide repeat polymorphism at the MAOA locus. Nucleic Acids Res., 19: 689.

Blier, P. and de Montigny, C. (1985) Serotoninergic but not noradrenergic neurons in rat central nervous system adapt to long-term treatment with monoamine oxidase inhibitors. Neuroscience, 16: 949–955.

Bond, P.A. and Cundall, R.L. (1977) Properties of monoamine oxidase (MAO) in human blood platelets, plasma. Clin. Chim Acta, 80: 317–326.

Bortolato, M., Chen, K. and Shih, J.C. (2008) Monoamine oxidase inactivation: from pathophysiology to therapeutics. Adv. Drug Deliv. Rev., 60: 1527–1533.

Bortolato, M., Godar, S.C., Davarian, S., Chen, K. and Shih, J.C. (2009) Behavioral disinhibition and reduced anxiety-like behaviors in monoamine oxidase B-deficient mice. Neuropsychopharmacology, in press.

Bou-Flores, C., Lajard, A.M., Monteau, R., De Maeyer, E., Seif, I., Lanoir, J. and Hilaire, G. (2000) Abnormal phrenic motoneuron activity and morphology in neonatal monoamine oxidase A-deficient transgenic mice: possible role of a serotonin excess. J. Neurosci., 20: 4646–4656.

Boylan, C.B., Bennett-Clarke, C.A., Crissman, R.S., Mooney, R.D. and Rhoades, R.W. (2000) Clorgyline treatment elevates cortical serotonin and temporarily disrupts the vibrissae-related pattern in rat somatosensory cortex. J. Comp. Neurol., 427: 139–149.

Brunner, H.G., Nelen, M., Breakefield, X.O., Ropers, H.H. and van Oost, B.A. (1993) Abnormal behavior associated with a point mutation in the structural gene for monoamine oxidase A. Science, 262: 578–580.

Brunner, H.G., Nelen, M.R., van Zandvoort, P., Abeling, N.G., van Gennip, A.H., Wolters, E.C., Kuiper, M.A., Ropers, H.H. and van Oost, B.A. (1993) X-linked borderline mental retardation with prominent behavioral disturbance: phenotype, genetic localization, and evidence for disturbed monoamine metabolism. Am. J. Hum. Genet., 52: 1032–1039.

Buckholtz, J.W. and Meyer-Lindenbery, A. (2008) MAO A and the neurogenetic architecture of human aggression. Trends Neurosci., 31: 120–129.

Burchett, S.A. and Hicks, T.P. (2006) The mysterious trace amines: protean neuromodulators of synaptic transmission in mammalian brain. Prog. Neurobiol., 79: 223–246.

Carruba, M.O., Cuzzolaro, M., Riva, L., Bosello, O., Liberti, S., Castra, R., Dalle Grave, R., Santonastaso, P., Garosi, V. and Nisoli, E. (2001) Efficacy and tolerability of moclobemide in bulimia nervosa: a placebo-controlled trial. Intl. Clin. Psychopharmacol., 16: 27–32.

Cases, O., Seif, I., Grimsby, J., Gaspar, P., Chen, K., Pournin, S., Muller, U., Aguet, M., Babinet, C. and Shih, J.C. (1995) Aggressive behavior and altered amounts of brain serotonin and norepinephrine in mice lacking MAOA. Science, 268: 1763–1766.

Caspi, A., McClay, J., Moffitt, T.E., Mill, J., Martin, J., Craig, I.W., Taylor, A. and Poulton, R. (2002) Role of genotype in the cycle of violence in maltreated children. Science, 297: 851–854.

Cazalets, J.R., Gardette, M. and Hilaire, G. (2000) Locomotor network maturation is transiently delayed in the MAOA-deficient mouse. J. Neurophysiol., 83: 2468–2470.

Chen, K. (2004) Organization of MAO A and MAO B promoters and regulation of gene expression. Neurotoxicology, 25: 31–36.

Chen, K., Cases, O., Rebrin, I., Wu, W., Gallaher, T.K., Seif, I. and Shih, J.C. (2007) Forebrain-specific expression of monoamine oxidase A reduces neurotransmitter levels, restores the brain structure, and rescues aggressive behavior in monoamine oxidase A-deficient mice. J. Biol. Chem., 282: 115–123.

Chen, Z.Y., Denney, R.M. and Breakefield, X.O. (1995) Norrie disease and MAO genes: nearest neighbors. Hum. Mol. Genet., 4 Spec No: 1729–1737.

Chen, L., He, M., Sibille, E., Thompson, A., Sarnyai, Z., Baker, H., Shippenberg, T. and Toth, M. (1999) Adaptive changes in postsynaptic dopamine receptors despite unaltered dopamine dynamics in mice lacking monoamine oxidase B. J. Neurochem., 73: 647–655.

Chen, K., Holschneider, D.P., Wu, W., Rebrin, I. and Shih, J.C. (2004) A spontaneous point mutation produces monoamine oxidase A/B knockout mice with greatly elevated monoamines and anxiety-like behavior. J. Biol. Chem., 279: 39645–39652.

Chen, Z.Y., Hotamisligil, G.S., Huang, J.K., Wen, L., Ezzeddine, D., Aydin-Muderrisoglu, N., Powell, J.F., Huang, R.H., Breakefield, X.O., Craig, I. et al. (1991) Structure of the human gene for monoamine oxidase type A. Nucleic Acids Res., 19: 4537–4541.

Chen, K., Ou, X.-M., Chen, G., Choi, S.H. and Shih, J.C. (2005) R1, a novel repressor of the human monoamine oxidase A. J. Biol. Chem., 280: 11552–11559.

Chen, K., Wu, H.F. and Shih, J.C. (1996) Influence of C terminus on monoamine oxidase A and B catalytic activity. J. Neurochem., 66: 797–803.

Coccaro, E.F. (1989) Central serotonin and impulsive aggression. Br. J. Psychiatry Suppl.: 52–62.

Collins, F.A., Murphy, D.L., Reiss, A.L., Sims, K.B., Lewis, J.G., Freund, L., Karoum, F., Zhu, D., Maumenee, I.H. and Antonarakis, S.E. (1992) Clinical, biochemical, and neuropsychiatric evaluation of a patient with a contiguous gene syndrome due to a microdeletion Xp11.3 including the Norrie disease locus and monoamine oxidase (MAOA and MAOB) genes. Am. J. Med. Genet., 42: 127–134.

Consalvi, V., Mardh, G. and Vallee, B.L. (1986) Human alcohol dehydrogenases and serotonin metabolism. Biochem. Biophys. Res. Commun., 139: 1009–1016.

Cools, R., Roberts, A.C. and Robbins, T.W. (2008) Serotoninergic regulation of emotional and behavioural control processes. Trends Cogn. Sci., 12: 31–40.

Cyr, M. and Farrar, M.K. (2000) Treatment for posttraumatic stress disorder. Ann. Pharmacother, 34: 366–376.

Dahlstroem, A. and Fuxe, K. (1964) Evidence for the existence of monoamine-containing neurons in the central nervous system. I. Demonstration of monoamines in the cell bodies of brain stem neurons. Acta Physiol. Scand. Suppl., 232: 1–55.

Daszuta, A., Faudon, M. and Hery, F. (1984a) *In vitro* 3H-serotonin (5-HT) synthesis and release in BALBc and C57BL mice. II. Cell body areas. Brain Res. Bull., 12: 565–570.

Daszuta, A., Hery, F. and Faudon, M. (1984b) *In vitro* 3H-serotonin (5-HT) synthesis and release in BALBc and C57BL mice. I. Terminal areas. Brain Res. Bull., 12: 559–563.

Davis, V.E., Brown, H., Huff, J.A. and Cashaw, J.L. (1967) The alteration of serotonin metabolism to 5-hydroxytryptophol by ethanol ingestion in man. J. Lab. Clin. Med., 69: 132–140.

Deakin, J.F. (2003) Depression and antisocial personality disorder: two contrasting disorders of 5HT function. J. Neural Transm. Suppl.: 79–93.

de Almeida, R.M., Ferrari, P.F., Parmigiani, S. and Miczek, K.A. (2005) Escalated aggressive behavior: dopamine, serotonin and GABA. Eur. J. Pharmacol., 526: 51–64.

Deckert, J., Catalano, M., Syagailo, Y.V., Bosi, M., Okladnova, O., Di Bella, D., Nöthen, M.M., Maffei, P., Franke, P., Fritze, J., Maier, W., Propping, P., Beckmann, H., Bellodi, L. and Lesch, K.-P. (1999) Excess of high activity monoamine oxidase A gene promoter alleles in female patients with panic disorder. Hum. Mol. Genet., 8: 621–624.

De Colibus, L., Li, M., Binda, C., Lustig, A., Edmondson, D.E. and Mattevi, A. (2005) Three-dimensional structure of human monoamine oxidase A (MAO A): relation to the structures of rat MAO A and human MAO B. Proc. Natl. Acad. Sci. USA, 102: 12684–12689.

Denney, R.M., Koch, H. and Craig, I.W. (1999) Association between monoamine oxidase A activity in human male skin fibroblasts and genotype of the MAO-A promoter-associated variable number tandem repeat. Hum. Genet., 105: 542–551.

Despopoulos, A. and Weissbach, H. (1957) Renal metabolism of 5-hydroxyindolacetic acid. Am. J. Physiol., 189: 548–550.

Donnelly, C.H. and Murphy, D.L. (1977) Substrate- and inhibitor-related characteristics of human platelet. Biochem. Pharmacol., 26: 853–858.

Dubrovina, N.I. and Zinov'ev, D.R. (2007) [Contribution of GABA receptors in extinction of memory trace in norm and depressive-like state]. Ross Fiziol Zh Im I M Sechenova, 93: 1285–1291.

Edmondson, D.E., Bhattacharyya, A.K. and Walker, M.C. (1993) Spectral and kinetic studies of imine product formation in the oxidation of p-(N,N-dimethylamino)benzylamine analogues by monoamine oxidase B. Biochemistry, 32: 5196–5202.

Edmondson, D.E., Binda, C. and Mattevi, A. (2007) Structural insights into the mechanism of amine oxidation by monoamine oxidases A and B. Arch. Biochem. Biophys., 464: 269–276.

Elchisak, M.A. (1983) Determination of dopamine-O-sulfate and norepinephrine-O-sulfate isomers and serotonin-O-sulfate by high-performance liquid chromatography using dual-electrode electrochemical detection. J. Chromatogr., 264: 119–127.

Erzurumlu, R.S. and Jhaveri, S. (1990) Thalamic axons confer a blueprint of the sensory periphery onto the developing rat somatosensory cortex. Brain Res. Dev. Brain Res., 56: 229–234.

Evrard, A., Malagie, I., Laporte, A.M., Boni, C., Hanoun, N., Trillat, A.C., Seif, I., De Maeyer, E., Gardier, A., Hamon, M. and Adrien, J. (2002) Altered regulation of the 5-HT system in the brain of MAO-A knockout mice. Eur. J. Neurosci., 15: 841–851.

Feldstein, A. and Williamson, O. (1968) 5-Hydroxytryptamine metabolism in rat brain and liver homogenates. Br. J. Pharmacol., 34: 38–42.

Filipenko, M.L., Beilina, A.G., Alekseyenko, O.V., Dolgov, V.V. and Kudryavtseva, N.N. (2002) Repeated experience of social defeats increases serotonin transporter and monoamine oxidase A mRNA levels in raphe nuclei of male mice. Neurosci. Lett., 321: 25–28.

Fornai, F., Chen, K., Giorgi, F.S., Gesi, M., Alessandri, M.G. and Shih, J.C. (1999) Striatal dopamine metabolism in monoamine oxidase B-deficient mice: a brain dialysis study. J. Neurochem., 73: 2434–2440.

Fowler, J.S., Volkow, N.D., Wang, G.-J., Pappas, N., Logan, J., Shea, C., Alexoff, D., MacGregor, R.R., Schlyer, D.J., Zezulkova, I. and Wolf, A.P. (1996) Brain monoamine oxidase A inhibition in cigarette smokers. Proc. Natl. Acad. Sci. USA, 93: 14065–14069.

Fox, H.H. and Gibas, J.T. (1953) Synthetic tuberculostats. V. Alkylidene derivatives of isonicotinylhydrazine. J. Org. Chem., 18: 983–989.

Garrick, N.A. and Murphy, D.L. (1980) Species differences in the deamination of dopamine and other substrates for monoamine oxidase in brain. Psychopharmacology (Berl.), 72: 27–33.

Geha, R.M., Chen, K. and Shih, J.C. (2000) Phe(208) and Ile(199) in human monoamine oxidase A and B do not determine substrate and inhibitor specificities as in rat. J. Neurochem., 75: 1304–1309.

Geha, R.M., Chen, K., Wouters, J., Ooms, F. and Shih, J.C. (2002) Analysis of conserved active site residues in monoamine oxidase A and B and their three-dimensional molecular modeling. J. Biol. Chem., 277: 17209–17216.

Geha, R.M., Rebrin, I., Chen, K. and Shih, J.C. (2001) Substrate and inhibitor specificities for human monoamine oxidase A and B are influenced by a single amino acid. J. Biol. Chem., 276: 9877–9882.

Goldstein, L.S. (2001) Kinesin molecular motors: transport pathways, receptors, and human disease. Proc. Natl. Acad. Sci. USA, 98: 6999–7003.

Gottowik, J., Cesura, A.M., Malherbe, P., Lang, G. and Da Prada, M. (1993) Characterisation of wild-type and mutant forms of human monoamine oxidase A and B expressed in a mammalian cell line. FEBS Lett., 317: 152–156.

Greenawalt, J.W. and Schnaitman, C. (1970) An appraisal of the use of monoamine oxidase as an enzyme marker for the outer membrane of rat liver mitochondria. J. Cell Biol., 46: 173–179.

Grimsby, J., Chen, K., Wang, L.J., Lan, N.C. and Shih, J.C. (1991) Human monoamine oxidase A and B genes exhibit identical exon–intron organization. Proc. Natl. Acad. Sci. USA, 88: 3637–3641.

Grimsby, J., Lan, N.C., Neve, R., Chen, K. and Shih, J.C. (1990) Tissue distribution of human monoamine oxidase A and B mRNA. J. Neurochem., 55: 1166–1169.

Grimsby, J., Toth, M., Chen, K., Kumazawa, T., Klaidman, L., Adams, J.D., Karoum, F., Gal, J. and Shih, J.C. (1997) Increased stress response and b-phenylethylamine in *MAOB*-deficient mice. Nat. Genet., 17: 206–210.

Haller, J., Makara, G.B. and Kruk, M.R. (1998) Catecholaminergic involvement in the control of aggression: hormones, the peripheral sympathetic, and central noradrenergic systems. Neurosci. Biobehav. Rev., 22: 85–97.

Hammond, D.L., Yaksh, T.L. and Tyce, G.M. (1981) Presence of conjugated 5-hydroxytryptamine in in vivo superfusates of rat spinal cord. J. Neurochem., 37: 1068–1071.

Hare, M.L. (1928) Tyramine oxidase: a new enzyme system in liver. Biochem. J., 22: 968–979.

Helander, A., Beck, O., Jacobsson, G., Lowenmo, C. and Wikstrom, T. (1993) Time course of ethanol-induced changes in serotonin metabolism. Life Sci., 53: 847–855.

Helander, A., Beck, O. and Jones, A.W. (1996) Laboratory testing for recent alcohol consumption: comparison of ethanol, methanol, and 5-hydroxytryptophol. Clin. Chem., 42: 618–624.

Hidaka, H., Nagatsu, T. and Yagi, K. (1969) Occurrence of a serotonin sulphotransferase in the brain. J. Neurochem., 16: 783–785.

Hinds, H.L., Hendriks, R.W., Craig, I.W. and Chen, Z.Y. (1992) Characterization of a highly polymorphic region near the first exon of the human MAOA gene containing a GT dinucleotide and a novel VNTR motif. Genomics, 13: 896–897.

Hollenbeck, P.J. and Saxton, W.M. (2005) The axonal transport of mitochondria. J. Cell Sci., 118: 5411–5419.

Hotamisligil, G.S. and Breakefield, X.O. (1991) Human monoamine oxidase A gene determines levels of enzyme activity. Am. J. Hum. Genet., 49: 383–392.

Hsu, Y.P., Weyler, W., Chen, S., Sims, K.B., Rinehart, W.B., Utterback, M.C., Powell, J.F. and Breakefield, X.O. (1988) Structural features of human monoamine oxidase A elucidated from cDNA and peptide sequences. J. Neurochem., 51: 1321–1324.

Huang, Y.-Y., Cate, S.P., Battistuzzi, C., Oquendo, M.A., Brent, D. and Mann, J.J. (2004) An association between a functional polymorphism in the monoamine oxidase A gene promoter, impulsive traits and early abuse experiences. Neuropsychopharmacology, 29: 1498–1505.

Hubalek, F., Binda, C., Khalil, A., Li, M., Mattevi, A., Castagnoli, N. and Edmondson, D.E. (2005) Demonstration of isoleucine 199 as a structural determinant for the selective inhibition of human monoamine oxidase B by specific reversible inhibitors. J. Biol. Chem., 280: 15761–15766.

Ito, A., Kuwahara, T., Inadome, S. and Sagara, Y. (1988) Molecular cloning of a cDNA for rat liver monoamine oxidase B. Biochem. Biophys. Res. Commun., 157: 970–976.

Jacobs, B.L. and Azmitia, E.C. (1992) Structure and function of the brain serotonin system. Physiol. Rev., 72: 165–229.

Jahng, J.W., Houpt, T.A., Joh, T.H. and Son, J.H. (1998) Differential expression of monoamine oxidase, A., serotonin transporter, tyrosine hydroxylase and norepinephrine transporter mRNA by anorexia mutation and food deprivation. Dev. Brain Res., 107: 241–246.

Jahng, J.W., Houpt, T.A., Wessel, T.C., Chen, K., Shih, J.C. and Joh, T.H. (1997) Localization of monoamine oxidase A and B mRNA in the rat brain by in situ hybridization. Synapse, 25: 30–36.

Jenike, M.A., Surman, O.S., Cassem, N.H., Zusky, P. and Anderson, W.H. (1983) Monoamine oxidase inhibitors in obsessive-compulsive disorder. J. Clin. Psychiatry, 44: 131–132.

Jimerson, D.C.B.W. (1985) Cerebrospinal fluid amine metabolite studies in depression: research update. In: Beckman, H. and Riederer, P. (Eds.), Pathochemical Markers in Major Psychosis. Springer Verlag, Berlin, pp. 128–143.

Johnston, J.P. (1968) Some observations upon a new inhibitor of monoamine oxidase in brain tissue. Biochem. Pharmacol., 17: 1285–1297.

Jonsson, E.G., Norton, N., Gustavsson, J.P., Oreland, L., Owen, M.J. and Sedvall, G.C. (2000) A promoter polymorphism in the monoamine oxidase A gene and its relationships to monoamine metabolite concentrations in CSF of healthy volunteers. J. Psychiatr Res., 34: 239–244.

Kearney, E.B., Salach, J.I., Walker, W.H., Seng, R.L., Kenney, W., Zeszotek, E. and Singer, T.P. (1971) The covalently-bound flavin of hepatic monoamine oxidase. 1. Isolation and sequence of a flavin peptide and evidence for binding at the 8alpha position. Eur. J. Biochem., 24: 321–327.

Kim, J.J., Shih, J.C., Chen, K., Chen, L., Bao, S., Maren, S., Anagnostaras, S.G., Fanselow, M.S., De Maeyer, E., Seif, I. and Thompson, R.F. (1997) Selective enhancement of emotional, but not motor, learning in monoamine oxidase A-deficient mice. Proc. Natl. Acad. Sci. USA, 94: 5929–5933.

Kim, J.M., Bogdan, M.A. and Mariano, P.S. (1993) Mechanistic analysis of the 3-methyllumiflavin-promoted oxidative deamination of benzylamine. A potential model for monoamine oxidase catalysis. J. Am. Chem. Soc., 115: 10591.

Knoll, J. and Magyar, K. (1972) Some puzzling pharmacological effects of monoamine oxidase inhibitors. Adv. Biochem. Psychopharmacol., 5: 393–408.

Kobayashi, S., Takahara, K. and Kamijo, K. (1981) Monoamine oxidase in frog liver and brain. Comp. Biochem. Physiol. C, 69: 179–183.

Kuwahara, T., Takamoto, S. and Ito, A. (1990) Primary structure of rat monoamine oxidase A deduced from cDNA and its expression in rat tissues. Agric. Biol. Chem., 54: 253–257.

Kvetnoy, I.M. (1999) Extrapineal melatonin: location and role within diffuse neuroendocrine system. Histochem. J., 31: 1–12.

Kwan, S.W. and Abell, C.W. (1992) cDNA cloning and sequencing of rat monoamine oxidase A: comparison with the human and bovine enzymes. Comp. Biochem. Physiol. B, 102: 143–147.

Kwan, S.W., Bergeron, J.M. and Abell, C.W. (1992) Molecular properties of monoamine oxidases A and B. Psychopharmacology (Berl.), 106(Suppl.): S1–S5.

Lajard, A.M., Bou, C., Monteau, R. and Hilaire, G. (1999) Serotonin levels are abnormally elevated in the fetus of the monoamine oxidase-A-deficient transgenic mouse. Neurosci. Lett., 261: 41–44.

Lahti, R.A. and Majchrowicz, E. (1974) Ethanol and acetaldehyde effects on metabolism and binding of biogenic amines. Q. J. Stud. Alcohol., 35: 1–14.

Lan, N.C., Chen, C.H. and Shih, J.C. (1989a) Expression of functional human monoamine oxidase A and B cDNAs in mammalian cells. J. Neurochem., 52: 1652–1654.

Lan, N.C., Heinzmann, C., Gal, A., Klisak, I., Orth, U., Lai, E., Grimsby, J., Sparkes, R.S., Mohandas, T. and Shih, J.C. (1989b) Human monoamine oxidase A and B genes map to Xp 11.23 and are deleted in a patient with Norrie disease. Genomics, 4: 552–559.

Lanoir, J., Hilaire, G. and Seif, I. (2006) Reduced density of functional 5-HT1A receptors in the brain, medulla and spinal cord of monoamine oxidase-A knockout mouse neonates. J. Comp. Neurol., 495: 607–623.

Lebrand, C., Gaspar, P., Nicolas, D. and Hornung, J.P. (2006) Transitory uptake of serotonin in the developing sensory pathways of the common marmoset. J. Comp. Neurol., 499: 677–689.

Levitt, P., Pintar, J.E. and Breakefield, X.O. (1982) Immunocytochemical demonstration of monoamine oxidase B in brain astrocytes and serotonergic neurons. Proc. Natl. Acad. Sci. USA, 79: 6385–6389.

Levy, E.R., Powell, J.F., Buckle, V.J., Hsu, Y.P., Breakefield, X.O. and Craig, I.W. (1989) Localization of human monoamine oxidase-A gene to Xp11.23-11.4 by *in situ* hybridization: implications for Norrie disease. Genomics, 5: 368–370.

Lieske, V., Bennett-Clarke, C.A. and Rhoades, R.W. (1999) Effects of serotonin on neurite outgrowth from thalamic neurons in vitro. Neuroscience, 90: 964–967.

Lotto, B., Upton, L., Price, D.J. and Gaspar, P. (1999) Serotonin receptor activation enhances neurite outgrowth of thalamic neurones in rodents. Neurosci. Lett., 269: 87–90.

Luhmann, H.J., Huston, J.P. and Hasenohrl, R.U. (2005) Contralateral increase in thigmotactic scanning following unilateral barrel-cortex lesion in mice. Behav. Brain Res., 157: 39–43.

Luque, J.M., Kwan, S.W., Abell, C.W., Da Prada, M. and Richards, J.G. (1995) Cellular expression of mRNAs encoding monoamine oxidases A and B in the rat central nervous system. J. Comp. Neurol., 363: 665–680.

Ma, J., Yoshimura, M., Yamashita, E., Nakagawa, A., Ito, A. and Tsukihara, T. (2004) Structure of rat monoamine oxidase A and its specific recognitions for substrates and inhibitors. J. Mol. Biol., 338(1): 103–114.

Manuck, S.B., Flory, J.D., Ferrel, R.E., Mann, J.J. and Muldoon, M.F. (2000) A regulatory polymorphism of monoamine oxidase-A gene may be associated with variability in aggression, impulsivity, and central nervous system serotonergic responsivity. Psychiat Res., 95: 9–23.

Maura, G. and Vaccari, A. (1975) Relationships between age of submission to environmental stress, and monoamine oxidase activity in rats. Experientia, 31: 191–193.

Mehlman, P.T., Higley, J.D., Faucher, I., Lilly, A.A., Taub, D.M., Vickers, J., Suomi, S.J. and Linnoila, M. (1994) Low CSF 5-HIAA concentrations and severe aggression and impaired impulse control in nonhuman primates. Am. J. Psychiatry, 151: 1485–1491.

Mejia, J.M., Ervin, F.R., Baker, G.B. and Palmour, R.M. (2002) Monoamine oxidase inhibition during brain development induces pathological aggressive behavior in mice. Biol. Psychiatry, 52: 811–821.

Meyer-Lindenberg, A., Buckholtz, J.W., Kolachana, B.R., Hariri, A.R., Pezawas, L., Blasi, G., Wabnitz, A., Honea, R., Verchinski, B., Callicott, J.H., Egan, M., Matty, V. and Weinberger, D.R. (2006) Neural mechanisms of genetic risk for impulsivity and violence in humans. Proc. Natl. Acad. Sci. USA, 103: 6269–6274.

Miller, J.R. and Edmondson, D.E. (1999) Influence of flavin analogue structure on the catalytic activities and flavinylation reactions of recombinant human liver monoamine oxidases A and B. J. Biol. Chem., 274: 23515–23525.

Mitoma, J. and Ito, A. (1992) Mitochondrial targeting signal of rat liver monoamine oxidase B is located at its carboxy terminus. J. Biochem., 111: 20–24.

Morishima, M., Harada, N., Hara, S., Sano, A., Seno, H., Takahashi, A., Morita, Y. and Nakaya, Y. (2006) Monoamine oxidase A activity and norepinephrine level in hippocampus determine hyperwheel running in SPORTS rats. Neuropsychopharmacology, 31: 2627–2638.

Moser, P.C. and Sanger, D.J. (1999) 5-HT1A receptor antagonists neither potentiate nor inhibit the effects of fluoxetine and befloxatone in the forced swim test in rats. Eur. J. Pharmacol., 372: 127–134.

Mousseau, D.D. and Greenshaw, A.J. (1989) Chronic effects of clomipramine and clorgyline on regional levels of brain amines and acid metabolites in rats. J. Neural. Transm., 75: 73–79.

Murphy, D.L., Sims, K.B., Karoum, F., de la Chapelle, A., Norio, R., Sankila, E.M. and Breakefield, X.O. (1990) Marked amine and amine metabolite changes in Norrie disease patients with an X-chromosomal deletion affecting monoamine oxidase. J. Neurochem., 54: 242–247.

Murphy, D.L., Sims, K.B., Karoum, F., Garrick, N.A., de la Chapelle, A., Sankila, E.M., Norio, R. and Breakefield, X.O. (1991) Plasma amine oxidase activities in Norrie disease patients with an X-chromosomal deletion affecting monoamine oxidase. J. Neural Transm. Gen. Sect., 83: 1–12.

Nakajima, K., Matsuyama, K. and Mori, S. (1998) Prenatal administration of para-chlorophenylalanine results in suppression of serotonergic system and disturbance of swimming movements in newborn rats. Neurosci. Res., 31: 155–169.

Namboodiri, M.A., Dubbels, R. and Klein, D.C. (1987) Arylalkylamine N-acetyltransferase from mammalian pineal gland. Methods Enzymol., 142: 583–590.

Neckers, L.M. (1982) Serotonin turnover and regulation. In: Osborne, N.N. (Ed.), Biology of Serotonergic Transmission. Wiley, New York, NY, pp. 139–158.

Nicotra, A. and Senatori, O. (1988) Changes in monoamine oxidase activity by mitochondria isolated from late embryos of *Bufo bufo*. Comp. Biochem. Physiol. C, 89: 5–9.

Oreland, L., Nilsson, K., Damberg, M. and Hallman, J. (2007) Monoamine oxidases: activities, genotypes and the shaping of behaviour. J. Neural. Transm., 114: 817–822.

Ou, X.-M., Chen, K. and Shih, J.C. (2006) Glucocorticoid and androgen activation of monoamine oxidase A are regulated differently by R1 and Sp1. J. Biol. Chem., 281: 21512–21525.

Ozelius, L., Hsu, Y.P., Bruns, G., Powell, J.F., Chen, S., Weyler, W., Utterback, M., Zucker, D., Haines, J., Trofatter, J.A. et al. (1988) Human monoamine oxidase gene (MAOA): chromosome position (Xp21-p11) and DNA polymorphism. Genomics, 3: 53–58.

Pivac, N., Muck-Seler, D. and Diksic, M. (2003) Effects of desipramine on regional serotonin synthesis in the rat brain: acute and chronic autoradiographic studies. Neurochem. Intl., 43: 611–619.

Popova, N.K., Maslova, L.N., Morosova, E.A., Bulygina, V.V. and Seif, I. (2006) MAO A knockout attenuates adrenocortical response to various kinds of stress. Psychoneuroendocrinology, 31: 179–186.

Popova, N.K., Skrinskaya, Y.A., Amstislavskaya, T.G., Vishnivetskaya, G.B., Seif, I. and de Meier, E. (2001) Behavioral characteristics of mice with genetic knockout of monoamine oxidase type A. Neurosci. Behav. Physiol., 31: 597–602.

Powell, J.F., Hsu, Y.P., Weyler, W., Chen, S.A., Salach, J., Andrikopoulos, K., Mallet, J. and Breakefield, X.O. (1989) The primary structure of bovine monoamine oxidase type A. Comparison with peptide sequences of bovine monoamine oxidase type B and other flavoenzymes. Biochem. J., 259: 407–413.

Rebrin, I., Geha, R.M., Chen, K. and Shih, J.C. (2001) Effects of carboxyl-terminal truncations on the activity and solubility of human monoamine oxidase B. J. Biol. Chem., 276: 29499–29506.

Roise, D. and Schatz, G. (1988) Mitochondrial presequences. J. Biol. Chem., 263: 4509–4511.

Rowe, R.P., Bloom, B.M., Pan, S.Y. and Finger, K.F. (1959) Pharmacological studies with nialamide, a new antidepressant agent. Proc. Soc. Exp. Biol. Med., 101: 832–836.

Sabol, S.Z., Hu, S. and Hamer, D. (1998) A functional polymorphism in the monoamine oxidase A gene promoter. Hum. Genet., 103: 273–279.

Sagara, Y. and Ito, A. (1982) In vitro synthesis of monoamine oxidase of rat liver outer mitochondrial membrane. Biochem. Biophys. Res. Commun., 109: 1102–1107.

Salichon, N., Gaspar, P., Upton, A.L., Picaud, S., Hanoun, N., Hamon, M., De Maeyer, E., Murphy, D.L., Mossner, R., Lesch, K.P., Hen, R. and Seif, I. (2001) Excessive activation of serotonin (5-HT) 1B receptors disrupts the formation of sensory maps in monoamine oxidase A and 5-HT transporter knockout mice. J. Neurosci., 21: 884–896.

Saunders, J.C., Roukema, R.W., Kline, N.S. and Bailey, S.D. (1959) Clinical results with phenelzine. Am. J. Psychiatry, 116: 71–72.

Saura, J., Richards, J.G. and Mahy, N. (1994) Age-related changes on MAO in Bl/C57 mouse tissues: a quantitative radioautographic study. J. Neural. Transm. Suppl., 41: 89–94.

Scott, A.L., Bortolato, M., Chen, K. and Shih, J.C. (2008) Novel monoamine oxidase A knock out mice with human-like spontaneous mutation. Neuroreport, 19: 739–743.

Sharp, T., Gartside, S.E. and Umbers, V. (1997) Effects of co-administration of a monoamine oxidase inhibitor and a 5-HT1A receptor antagonist on 5-hydroxytryptamine cell firing and release. Eur. J. Pharmacol., 320: 15–19.

Shih, J.C. and Chen, K. (1999) MAO-A and -B gene knockout mice exhibit distinctly different behavior. Neurobiology (Bp), 7: 235–246.

Shih, J.C. and Chen, K. (2004) Regulation of MAO-A and MAO-B gene expression. Curr. Med. Chem., 11: 1995–2005.

Shih, J.C., Chen, K. and Geha, R.M. (1998) Determination of regions important for monoamine oxidase (MAO) A and B substrate and inhibitor selectivities. J. Neural Transm. Suppl., 52: 1–8.

Shih, J.C., Chen, K., Ridd, M.J. and Seif, I. (2000) Ginkgo biloba abolishes aggression in mice lacking MAO A. Antioxid. Redox. Signal., 2: 467–471.

Shih, J.C. and Thompson, R.F. (1999) Monoamine oxidase in neuropsychiatry and behavior. Am. J. Hum. Genet., 65: 593–598.

Shih, J.C., Zhu, Q.S., Grimsby, J. and Chen, K. (1994) Identification of human monoamine oxidase (MAO) A and B gene promoters. J. Neural. Transm. Suppl., 41: 27–33.

Silverman, R.B. and Hawe, W.P. (1995) SAR studies of fluorine-substituted benzylamines and substituted 2-phenylethylamines as substrates and inactivators of monoamine oxidase B. J. Enzyme Inhib., 9: 203–215.

Sims, K.B., Ozelius, L., Corey, T., Rinehart, W.B., Liberfarb, R., Haines, J., Chen, W.J., Norio, R., Sankila, E., de la Chapelle, A. et al. (1989) Norrie disease gene is distinct from the monoamine oxidase genes. Am. J. Hum. Genet., 45: 424–434.

Sotelo, C., Cholley, B., El Mestikawy, S., Gozlan, H. and Hamon, M. (1990) Direct Immunohistochemical evidence of the existence of 5-HT1A autoreceptors on serotoninergic neurons in the midbrain raphe nuclei. Eur. J. Neurosci., 2: 1144–1154.

Squires, L.N., Jakubowski, J.A., Stuart, J.N., Rubakhin, S.S., Hatcher, N.G., Kim, W.S., Chen, K., Shih, J.C., Seif, I. and Sweedler, J.V. (2006) Serotonin catabolism and the formation and fate of 5-hydroxyindole thiazolidine carboxylic acid. J. Biol. Chem., 281: 13463–13470.

Squires, R.F. (1968) Additional evidence for the existence of several forms of mitochondrial monoamine oxidase in the mouse. Biochem. Pharmacol., 17: 1401–1409.

Stuart, J.N., Zhang, X., Jakubowski, J.A., Romanova, E.V. and Sweedler, J.V. (2003) Serotonin catabolism depends upon location of release: characterization of sulfated and gamma-glutamylated serotonin metabolites in Aplysia californica. J. Neurochem., 84: 1358–1366.

Sugden, D., Cena, V. and Klein, D.C. (1987) Hydroxyindole O-methyltransferase. Methods Enzymol., 142: 590–596.

Svensson, S., Some, M., Lundsjo, A., Helander, A., Cronholm, T. and Hoog, J.O. (1999) Activities of human alcohol dehydrogenases in the metabolic pathways of ethanol and serotonin. Eur. J. Biochem., 262: 324–329.

Takei, Y., Ogoshi, M. and Inoue, K. (2007) A 'reverse' phylogenetic approach for identification of novel osmoregulatory and cardiovascular hormones in vertebrates. Front Neuroendocrinol., 28: 143–160.

Thase, M.E., Frank, E., Mallinger, A.G., Hamer, T. and Kupfer, D.J. (1992) Treatment of imipramine-resistant recurrent depression, III: Efficacy of monoamine oxidase inhibitors. J. Clin. Psychiatry, 53: 5–11.

Thompson, A.M. (2008) Serotonin immunoreactivity in auditory brainstem neurons of the postnatal monoamine oxidase-A knockout mouse. Brain Res., 1228: 58–67.

Tyce, G.M., Flock, E.V. and Owen, C.A. Jr. (1968) Metabolism of serotonin by the isolated perfused rat liver – effect of glucuronyl transferase deficiency or monoamine oxidase inhibition. Biochem. Pharmacol., 17: 1543–1552.

Udenfriend, S., Titus, E., Weissbach, H. and Peterson, R.E. (1956) Biogenesis and metabolism of 5-hydroxyindole compounds. J. Biol. Chem., 219: 335–344.

Upton, A.L., Salichon, N., Lebrand, C., Ravary, A., Blakely, R., Seif, I. and Gaspar, P. (1999) Excess of serotonin (5-HT) alters the segregation of ispilateral and contralateral retinal projections in monoamine oxidase A knockout mice: possible role of 5-HT uptake in retinal ganglion cells during development. J. Neurosci., 19: 7007–7024.

Vallejo, J., Gasto, C., Catalan, R. and Salamero, M. (1987) Double-blind study of imipramine versus phenelzine in melancholias and dysthymic disorders. Br. J. Psychiatry, 151: 639–642.

van Praag, H.M. (1983) CSF 5-HIAA and suicide in non-depressed schizophrenics. Lancet, 2: 977–978.

Vanyukov, M.M., Moss, H.B., Yu, L.M., Tarter, R.E. and Deka, R. (1995) Preliminary evidence for an association of a dinucleotide repeat polymorphism at the MAOA gene with early onset alcoholism/substance abuse. Am. J. Med. Genet., 60: 122–126.

Verge, D., Daval, G., Marcinkiewicz, M., Patey, A., El Mestikawy, S., Gozlan, H. and Hamon, M. (1986) Quantitative autoradiography of multiple 5-HT1 receptor subtypes in the brain of control or 5,7-dihydroxytryptamine-treated rats. J. Neurosci., 6: 3474–3482.

Vincent, S.B. (1912) The function of vibrissae in the behavior of the white rat. Behav. Monogr., 1: 1–82.

Virkkunen, M., Goldman, D., Nielsen, D.A. and Linnoila, M. (1995) Low brain serotonin turnover rate (low CSF 5-HIAA) and impulsive violence. J. Psychiatry Neurosci., 20: 271–275.

Vitalis, T., Cases, O., Callebert, J., Launay, J.M., Price, D.J., Seif, I. and Gaspar, P. (1998) Effects of monoamine oxidase A inhibition on barrel formation in the mouse somatosensory cortex: determination of a sensitive developmental period. J. Comp. Neurol., 393: 169–184.

Vitalis, T. and Parnavelas, J.G. (2003) The role of serotonin in early cortical development. Dev. Neurosci., 25: 245–256.

Voltaire, A., Beck, O. and Borg, S. (1992) Urinary 5-hydroxytryptophol: a possible marker of recent alcohol consumption. Alcohol. Clin. Exp. Res., 16: 281–285.

Walker, M.C. and Edmondson, D.E. (1994) Structure–activity relationships in the oxidation of benzylamine analogues by bovine liver mitochondrial monoamine oxidase B. Biochemistry, 33: 7088–7098.

Warburg, M. (1975) Norrie's disease –differential diagnosis and treatment. Acta Ophthalmol. (Copenh.), 53: 217–236.

Westlund, K.N., Denney, R.M., Rose, R.M. and Abell, C.W. (1988) Localization of distinct monoamine oxidase A and monoamine oxidase B cell populations in human brainstem. Neuroscience, 25: 439–456.

Westlund, K.N., Krakower, T.J., Kwan, S.W. and Abell, C.W. (1993) Intracellular distribution of monoamine oxidase A in selected regions of rat and monkey brain and spinal cord. Brain Res., 612: 221–230.

Wong, W.K., Chen, K. and Shih, J.C. (2001) Regulation of human monoamine oxidase B gene by Sp1 and Sp3. Mol. Pharmacol., 59: 852–859.

Wong, W.K., Ou, X.M., Chen, K. and Shih, J.C. (2002) Activation of human monoamine oxidase B gene expression by a protein kinase CMAPK signal transduction pathway involves c-Jun and Egr-1. J. Biol. Chem., 277: 22222–22230.

Wong, W.K., Chen, K. and Shih, J.C. (2003) Decreased methylation and transcription repressor Sp3 up-regulated human monoamine oxidase (MAO) B expression during Caco-2 differentiation. J. Biol. Chem., 278: 36227–36235.

Wu, H.F., Chen, K. and Shih, J.C. (1993) Site-directed mutagenesis of monoamine oxidase A and B: role of cysteines. Mol. Pharmacol., 43: 888–893.

Zeller, E.A. (1938) Ueber den enzymathischen Abbau von Histamin and Diaminen. Helv. Chem. Acta, 21: 881–890.

Zeller, E.A. and Barsky, J. (1952) *In vivo* inhibition of liver and brain monoamine oxidase by 1-isonicotinyl-2-isopropyl hydrazine. Proc. Soc. Exp. Biol. NY, 81: 459–461.

Zhu, Q.-S., Chen, K. and Shih, J.C. (1994) Bidirectional promoter of human monoamine oxidase A (MAO A) controlled by transcription factor Sp1. J. Neurosci., 14: 7393–7403.

Zhuang, Z., Hogan, M. and McCauley, R. (1988) The *in vitro* insertion of monoamine oxidase B into mitochondrial outer membranes. FEBS Lett., 238: 185–190.

Zhuang, Z.P., Marks, B. and McCauley, R.B. (1992) The insertion of monoamine oxidase A into the outer membrane of rat liver mitochondria. J. Biol. Chem., 267: 591–596.

Zhuang, Z.P. and McCauley, R. (1989) Ubiquitin is involved in the in vitro insertion of monoamine oxidase B into mitochondrial outer membranes. J. Biol. Chem., 264: 14594–14596.

CHAPTER 2.5

Cellular Effects of Serotonin in the CNS

Rodrigo Andrade[*,1] and Sheryl G. Beck[2]

[1]*Department of Pharmacology, Wayne State University School of Medicine, E. Canfield, Detroit, Michigan*
[2]*Abramson Research Center, Children's Hospital of Philadelphia, University of Pennsylvania, Philadelphia, Pennsylvania*

Abstract: Serotonergic neurons innervate the entire neuroaxis, thus allowing serotonin to regulate the function of most CNS neurons. Work during the past two decades has made clear that the specific form taken by this regulation is dependent on the complement of serotonin receptors expressed by each cell. This general principle is now being extended to different cell types within defined neuronal networks to understand how serotonin regulates network function. In this chapter we review progress along this front in two intensely studied areas, the hippocampus and the cerebral cortex, and highlight areas of progress while noting outstanding issues that await elucidation.

Keywords: serotonin, hippocampus, cortex, 5-HT_{1A}, 5-HT_{2A}, 5-HT_3, 5-HT_4, 5-HT_7, pyramidal neurons, GABAergic interneurons.

Abbreviations: 1AKO, 5-HT_{1A} receptor knockout mouse; AMPA, alpha-amino-3-hydroxy-5-methyl-4-isoxazolepropionic acid; CA, cornu ammonis; CB, calbindin; CCK, choleocystokinin; CNS, central nervous system; CNQX, 6-cyano-7-nitroquinoxaline-2,3-dione; DNQX, 6,7-dinitroquinoxaline-2,3-dione; eEPSC, evoked excitatory postsynaptic current; Gα, G-protein alpha subunit; GABA, gamma aminobutyric acid; GIRK, G-protein-mediated inward rectifying potassium channel; GPCR, G-protein-coupled receptors; 5-HT, 5-hydroxytryptamine, serotonin; Kir3.x, inward rectifying potassium channel; Kv1.2, delayed rectifier potassium channel; NMDA, N-methyl-D-aspartate; PV, parvalbumin; sIPSC, spontaneous inhibitory postsynaptic current; SLM, stratum lacunosum moleculare; SO, stratum oriens; SP, stratum pyramidale; SR, stratum radiatum; TTX, tetrodotoxin; WT, wild type.

Introduction

What does serotonin do? We hear that question often, from colleagues, from students, from neighbors, even from fellow passengers in airplanes. And although the question has different meanings depending on who asks it, we would like to argue that ultimately the answers collapse around the central issue of what serotonin does to the functioning of central neuronal networks.

This idea, which admittedly appears quite myopic on first encounter, is grounded in one of the central concepts underlying modern neurobiology: that there is a direct correspondence between the coordinated activity of central neuronal networks and behavior – in other words, that behavior is the macroscopic embodiment of brain activity. Seen in this manner, to ask about the function of serotonin in health and disease is also to ask how serotonin functions within the context of neuronal networks. We believe this holds true even if at the present time our understanding of the relationship between neuronal activity and behavior remains primitive. Thus, in this chapter we would like to argue that it is not that serotonin regulates sleep, hunger, movement or anxiety, to name just a few. Rather, what serotonin does is help code for the specific functional states of central networks that create the substrate upon which those specific behaviors take place. Admittedly, in this formulation, the function of serotonin does not lend itself well to descriptors arising from our everyday experience. However, we believe that such a formulation provides a fruitful framework for understanding how serotonin may be involved in almost everything.

This formulation emerges from our own work on the cellular electrophysiology of serotonin. However, its core is grounded on three fundamental features of serotonergic systems that we believe frame any attempt to understand what serotonin does in the brain.

The first feature that is singularly striking about serotonin as a neurotransmitter in the mammalian brain is that a relatively small number of serotonin-synthesizing neurons

*Corresponding author
E-mail: randrade@med.wayne.edu

Christian Müller & Barry Jacobs (Eds.)
Handbook of Behavioral Neurobiology of Serotonin
ISBN 978-0-12-374634-4

DOI: 10.1016/B978-0-12-374634-4.00001-0

in the brainstem innervate nearly the entire neuroaxis (Dahlstrom and Fuxe, 1964; Steinbusch, 1981; Jacobs and Azmitia, 1992). Of course this is not a unique feature of the serotonergic system; one only needs to consider the noradrenergic innervation of the brain originating from locus coeruleus and other adrenergic nuclei to see the pattern replicated. Nevertheless, this is a central feature of the organization of serotonergic systems in the brain. This implies, as pointed out by others (such as Jacobs and Fornal, 1999), that whatever serotonin does must be quite general.

The second feature is that the activity of serotonergic neurons is associated with the behavioral state of the animal (Jacobs and Azmitia, 1992; Jacobs and Fornal, 1999; Jacobs *et al.*, 2002). This key observation, which dates back to the seminal studies of Jacobs and associates in the late 1970s, indicates that whatever general function serotonin plays in the brain is somehow related to the global behavioral state of the animal, although a more specific association with motor activity has also been suggested (Jacobs and Fornal, 1999; Jacobs *et al.*, 2002).

The third telling feature of serotonergic neurotransmission is that serotonin acts in the brain predominantly, albeit not exclusively, through G-protein-coupled receptors (GPCRs; the exception is the 5-HT_3 receptor). GPCRs are not well suited to carry millisecond to millisecond synaptic operations, the elementary building blocks of computational operations. Rather, these receptors regulate the way neurons, the basic computational units of neuronal networks, function. This realization, codified in the 1980s through groundbreaking work on invertebrate experimental preparations (reviewed by Harris-Warrick and Marder, 1991), brought to close the decades-long controversy that asked whether serotonin (or dopamine, etc.) was excitatory or inhibitory. These insights ushered a reformulation of how we think about neurotransmitters that act predominantly through GPCRs, and brought forth the idea that by modifying the properties of neurons, neurotransmitters acting through GPCRs allow neuronal networks to produce varying outputs.

Together, these ideas led to a reconceptualization of what serotonin does in the brain. For, if behavioral states represent the organism-level counterparts of functional states of central neuronal circuits, then the function of serotonin in the brain is probably best understood as contributing to the coding for these functional states. In other words, what serotonin does in the brain is to help orchestrate neuronal function such that neuronal networks operate in a manner that executes behavior appropriate to the circumstances. As such, serotonin forms part of an ensemble of neurotransmitters specifying the functional states of the central neuronal networks that define the universe of behavioral states available to an organism.

But how can serotonin accomplish this task? Studies examining the electrophysiological effects of serotonin in central neurons suggest that this is made possible by the existence of multiple serotonin receptor subtypes and the ability of neurons to co-express them. The co-expressed receptors act as translators that link localized rises in serotonin concentration to the activation of clusters of cellular processes defined by the specific combination of serotonin receptor subtypes present at a given site. This allows serotonin to regulate central neurons in a manner that is specific and flexible, as the effects in each cell reflect the combinatorial action of multiple serotonin receptors each signaling a qualitatively invariant response. And, of course, within specific neuronal networks, each cell type expresses a unique amalgamation of receptors. We believe this is how a highly divergent neuronal system that releases serotonin onto a large fraction of the neurons in the brain can attain the flexibility needed to tune the function of neuronal circuits as needed to help code the behavioral state of the organism.

But of course these are all theoretical considerations. How are these principles implemented in real neuronal networks? In this chapter we provide some examples of how this may happen in two areas of the brain, the hippocampus and the cerebral cortex. We do so, fully cognizant that there is a vast and extraordinary literature on the effects of serotonin on many different neural networks, and also painfully aware that in writing this review we are forced to leave most of it out. We ask for the forgiveness of our colleagues. We focus on these two areas not only because we work in them, but also because a strong case can be made that they represent the best extant example we have of how serotonin, through the use of different receptor subtypes, can regulate central neuronal networks.

The hippocampus

It is probably not surprising that the hippocampus is the brain area where the effects of serotonin have been most thoroughly examined. This reflects the simple fact that the 'light is better' in the hippocampus. Among the principal advantages of working in the hippocampus is the fact that the basic circuitry of this region is organized along parallel lamellae, each containing the basic functional elements of the trisynaptic hippocampal circuit including the granule cells of the dentate gyrus, the pyramidal cells of the CA3 and CA1 regions and their synaptic connections. Equally important from an experimental point of view, the cell bodies of the principal cells, the granule cells of the dentate gyrus and the pyramidal cells of the CA fields, are arranged in well-defined layers and give rise to consistently oriented apical and basal dendrites, while afferent and efferent pathways are generally segregated within specific layers. This precise cytoarchitecture makes it easy for electrophysiologists to stimulate and record from identified cells or even isolated dendrites, and study the properties of select synapses.

From a functional standpoint, all the principal neurons of the hippocampal trisynaptic circuit use glutamate as a neurotransmitter and thus are excitatory. As such, this circuit can be understood as a three-stage processing network, with a main input comprising the enthorhinal afferents to the dentate gyrus and a main output comprising the CA1 projection to the subiculum. However, it is important to understand that the function of this circuit is dramatically shaped by GABAergic interneurons. These cells are highly heterogeneous in terms of their soma location, axonal and dendritic arborizations, neurochemical content (neuropeptides, calcium-binding proteins) and electrophysiological characteristics, and this heterogeneity leads to major differences in the functionality of these interneurons (reviewed by Ascoli *et al.*, 2008; Klausberger and Somogyi, 2008). Although at the present time it is not known precisely how many different types of interneurons are present in the hippocampus, it is clear that these cells shape information flow through the trisynaptic hippocampal circuit (Ascoli *et al.*, 2008; Klausberger and Somogyi, 2008).

Pyramidal cells

Anatomical studies suggest strongly that most cell types in the hippocampus, including the pyramidal cells of the CA fields as well as the granule cells of the dentate gyrus, may be regulated by serotonin. Thus, the hippocampus receives a robust serotonergic innervation that originates primarily from the median raphe, although both the median and the dorsal raphe contribute fibers to the innervation of the ventral hippocampus. Serotonergic fibers are found throughout all fields of the hippocampus, but are most numerous in the stratum radiatum and stratum lacunosum moleculare of the CA1 and CA3 regions (Freund *et al.*, 1990; Gulyas *et al.*, 1999). Interestingly, 5-HT fibers appear to make very few synaptic specializations with the pyramidal cells on either soma or dendrites (Gulyas *et al.*, 1999), although there are 5-HT receptors on the pyramidal cell processes. This indicates that volume or paracrine transmission may be the predominant mode of transmission between the serotonergic fibers and serotonin receptors expressed on hippocampal pyramidal cells and at least some types of interneurons (Figure 1).

Seeking to understand how serotonin regulates hippocampal function, initial work focused on pyramidal cells of the CA1 region and characterized a serotonin-induced hyperpolarization that resulted in inhibition of the population spike (Segal, 1980; Beck *et al.*, 1985; Andrade and Nicoll, 1987). Subsequent work showed that this effect was mediated by 5-HT$_{1A}$ receptors (Segal, 1980; Beck *et al.*, 1985; Andrade and Nicoll, 1987) acting through G$\alpha_{i/o}$ and leading to the activation of a potassium current carried through GIRK (Kir3.x) potassium channels (Andrade and Nicoll, 1987; Beck *et al.*, 1992) (Figure 2C). A similar 5-HT$_{1A}$ receptor-mediated hyperpolarizing effect is seen in granule cells of the dentate gyrus (Piguet and Galvan, 1994) and pyramidal neurons of the CA3 region. However, in this latter area the 5-HT$_{1A}$ response is less potent than the response in CA1, but greater in magnitude, even though the density of the receptors appears to be less (Beck *et al.*, 1992). Differences notwithstanding, these results showed that 5-HT$_{1A}$ receptors signal reductions in cellular excitability in each of the principal cells of the hippocampal trisynaptic circuit (Figure 2C).

Subsequent studies indicated that 5-HT$_{1A}$ receptors are not the sole serotonin receptors expressed on pyramidal cells (Andrade and Nicoll, 1987; Colino and Halliwell, 1987) and CA1 pyramidal neurons were shown to also express 5-HT$_4$ receptors (Andrade and Nicoll, 1987; Andrade and Chaput, 1991; Torres *et al.*, 1994) while CA3 pyramidal neurons were shown to express 5-HT$_4$ and 5-HT$_7$ receptors (Bacon and Beck, 2000). Both of these receptors couple to Gα_s, activate adenylate cyclase and protein kinase A, and signal an inhibition of the calcium-activated potassium current that mediates the slow afterhyperpolarization present in these cells (Figure 2B). Since this current acts as a brake on the ability of the pyramidal cells to fire repetitively, its suppression by 5-HT$_4$ and 5-HT$_7$ receptors results in a large facilitation in the ability of strong stimuli to drive cell firing. As such, 5-HT$_4$ and 5-HT$_7$ receptors could be considered as signaling 'excitatory' responses to serotonin, an observation that initially seemed at odds with the observation that these same cells also expressed 'inhibitory' 5-HT$_{1A}$ receptors. This paradox was resolved with the demonstration that the coactivation of 5-HT$_{1A}$ and 5-HT$_4$ receptors resulted in a large change in neuronal gain, a measure of how neurons encode excitatory input into firing output, such that weak excitatory stimuli are suppressed while strong stimuli are facilitated (Andrade and Nicoll, 1987).

Finally, pyramidal neurons of the CA1, but not the CA3, field express 5-HT$_{1B}$ receptors that are targeted to their axonal projection terminals located in the subiculum (Ait *et al.*, 1995). Activation of these receptors inhibits glutamate-mediated synaptic transmission in this region (Boeijinga and Boddeke, 1993) by reducing the probability of glutamate release at individual synaptic terminals (Figure 2E). One interesting feature of this presynaptic

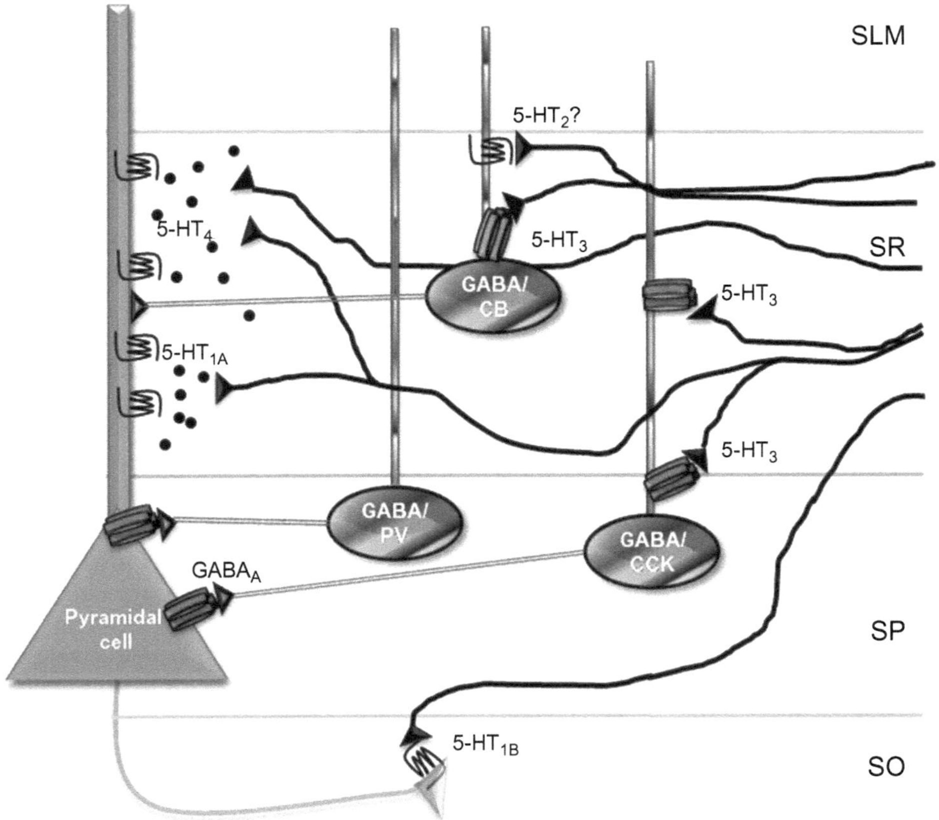

Figure 1 Cartoon depicting the known interactions of serotonin with components of the CA1 hippocampal subfield. The pyramidal cell, in gray, is the principal neuron surrounded by three different types of GABAergic interneurons (in gradient gray). The 5-HT innervation, depicted in black, is found in all the layers, but is most numerous in the stratum radiatum (SR) and stratum lacunosum molecular (SLM). Current evidence suggests that a large fraction of serotonin fibers do not make specialized synaptic contact with the pyramidal or the parvalbumin (PV)-expressing GABAergic interneurons. What is depicted, therefore, is release of 5-HT from axonal terminals into the extrasynaptic space. On the pyramidal cells there are 5-HT$_{1A}$ and 5-HT$_4$ receptors. On the GABA/CCK and GABA/CB interneurons serotonin does make synaptic contacts that are thought to activate the 5-HT$_3$ receptors. 5-HT$_2$ receptors are known to increase GABA synaptic activity, but it is not known exactly on which GABA interneuron type they are located; in this illustration we have placed it on a calbindin interneuron but such location is conjectural. 5-HT$_{1B}$ receptors are located in the subiculum on synaptic terminals originating from CA1 pyramidal cells.

inhibition is that it is frequency dependent, such that high-frequency volleys can be expected to be relatively unaffected by serotonin (Boeijinga and Boddeke, 1996). This feature dovetails with the serotonin-induced change in gain at the level of the soma, and suggests a general principle for the effects of serotonin on the hippocampal trisynaptic circuit – namely, that serotonergic activity results in a change in the functioning of this circuit such that transmission of weak/low-frequency stimuli is suppressed by serotonin acting on 5-HT$_{1A}$ receptors located on the dendrites of the principal neurons, and on 5-HT$_{1B}$ receptors located on the CA1 terminals in subiculum, while strong/high-frequency stimuli are facilitated in the CA1–CA3 subfield through the activation of 5-HT$_4$ and 5-HT$_7$ receptors. And of course the frequency dependence of the 5-HT$_{1B}$ receptor-mediated presynaptic inhibition allows unfettered transmission of such strong/high-frequency action potential firing at the output stage in subiculum. In other words, serotonin acting through multiple receptors in the different hippocampal fields appears to shape the functional mode of the trisynaptic circuit.

GABAergic interneurons

Unfortunately much less is known about the effects of serotonin on the different types of GABAergic interneurons, although it is clear that serotonergic fibers show remarkable selectivity in their innervation. Thus, for example, serotonergic fibers often follow the dendrites and cell bodies of calbindin neurons in stratum radiatum and stratum oriens (Freund *et al.*, 1990; Gulyas *et al.*, 1999). Cholecystokinin- and parvalbumin-expressing interneurons both make perisomatic contacts

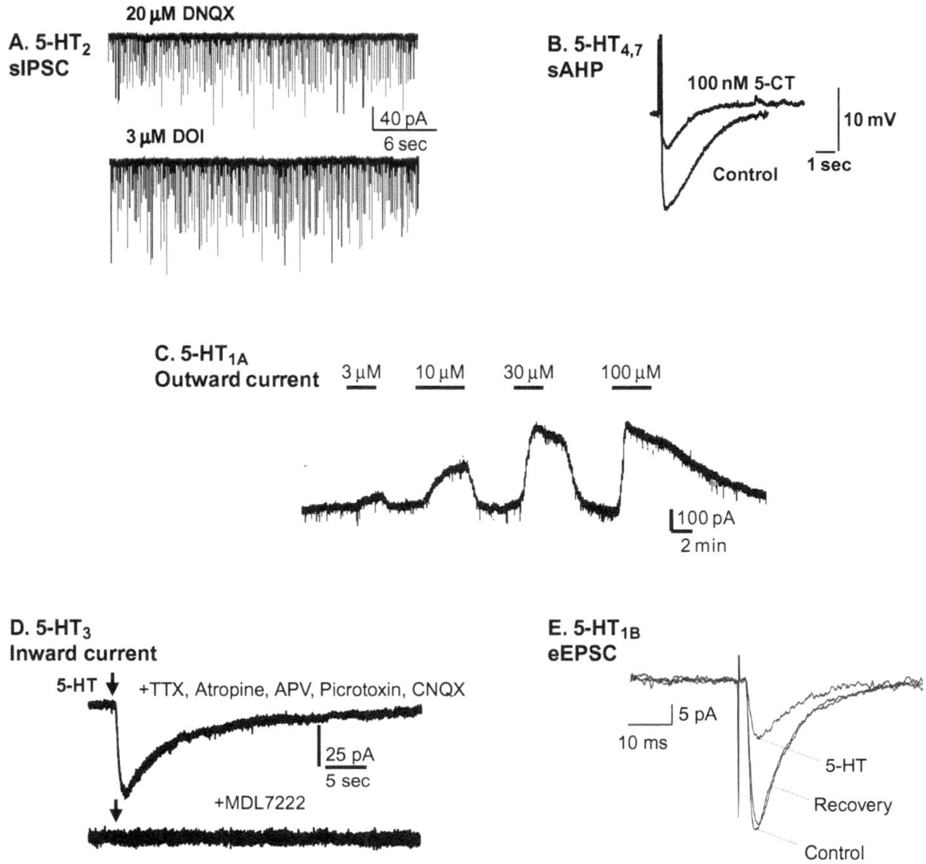

Figure 2 Electrophysiological effects elicited by activation of different 5-HT receptors in the CA1 subfield of the hippocampus. (A) Activation of 5-HT_2 receptors elicits an increase in GABAergic synaptic activity onto CA1 pyramidal neurons. This cell was voltage clamped at -60 mV, and spontaneous GABAergic inhibitory synaptic currents (sIPSCs, downward deflections) were isolated by the administration of the AMPA receptor antagonist DNQX. Administration of the 5-HT_2 agonist DOI increases the frequency and amplitude of the sIPSCs. (B) Activation of the 5-$HT_{4,7}$ receptors produces a decrease in the amplitude of the slow afterhyperpolarization (sAHP) that follows a train of action potentials elicited by a depolarizing current pulse. (C) Activation of 5-HT_{1A} receptors elicits a concentration-dependent outward current mediated by activation of GIRK (Kir3.x) potassium channels. (D) Activation of 5-HT_3 receptors in the presence of the action potential blocker TTX, and antagonists for muscarinic cholinergic (atropine), NMDA (APV), AMPA (CNQX), and $GABA_A$ (picrotoxin) receptors, elicits an inward current that can be blocked by the 5-HT_3 antagonist MDL7222. (E) Activation of 5-HT_{1B} receptors located on the presynaptic terminals of the CA1 pyramidal cells inhibits evoked excitatory synaptic transmission (eEPSC) onto pyramidal neurons in the subiculum.

with pyramidal cells, but only cholecystokinin-expressing interneurons receive 5-HT synaptic contacts (Freund *et al.*, 1990; Gulyas *et al.*, 1999). This anatomical specificity hints at a precise regulation of different interneuron types by serotonin.

From an electrophysiological standpoint, serotonin elicits highly variable effects on over 64 percent of GABAergic interneurons (Parra *et al.*, 1998), a variability that is likely to reflect the functional heterogeneity of these cells. The clearest findings concern parvalbumin- and cholecystokinin-expressing interneurons, which play a significant role in controlling gamma and theta oscillations (Freund, 2003; Klausberger *et al.*, 2003; Freund and Katona, 2007) through a strong perisomatic innervation of the pyramidal cells. Interestingly, while both of these cells types make anatomically similar connections with pyramidal cells, they differ in function. Thus parvalbumin-expressing interneurons are proposed to operate in a precise clockwork manner maintaining the constant theta activity, in a non-plastic manner with little interference from afferent input. In contrast, the cholecystokinin-expressing interneurons are susceptible to subcortical pathway input regarding the motivational, the emotional and the general physiological state of the animal. Remarkably, cholecystokinin- but not parvalbumin-expressing interneurons express 5-HT_3 receptors (Morales and Backman, 2002). Of course, activation of the 5-HT_3 receptors elicits a fast inward current (Figure 2D) that depolarizes these GABAergic interneurons (McMahon and Kauer, 1997) and thus can be expected to increase inhibitory input to the soma of the pyramidal cell, reducing the likelihood of action potential initiation and thus directly influencing gamma and theta activity. This suggests a mechanism through which serotonin could modify the function of the perisomatic network.

Other classes of interneurons are also likely to be regulated by serotonin acting on other receptor subtypes, thus allowing for their modulation. For example, 5-HT_2 receptors increase GABAergic synaptic activity, but it is unknown on which interneurons they are located (Figure 2A). Anatomical studies indicate very precise or selective innervation of only the calbindin- and cholecystokinin-expressing interneurons, but other physiological studies indicate a more global influence of serotonin on interneurons (Parra *et al.*, 1998), perhaps through volume transmission. Thus clearly one of the key tasks at hand is to understand if and how serotonin regulates the function of the other types of GABAergic interneurons in the hippocampus. This knowledge, coupled with a better understanding of the role of interneurons in shaping hippocampal function, should allow us to begin the development of a mechanistic picture of how serotonin regulates hippocampal circuits and hippocampal function.

Development

Serotonin fibers innervate the hippocampus during gestation, and serotonin receptors, most notably those of the 5-HT_{1A} subtype, are detectable at birth and increase thereafter (Patel and Zhou, 2005; Deng *et al.*, 2007). A series of recent studies using genetically engineered mice has revealed a remarkable role for this serotonergic receptor in terms of influencing the development of the hippocampus. 5-HT_{1A} receptor knockout (1AKO) mice were generated by targeted gene disruption in three different laboratories (Heisler *et al.*, 1998; Parks *et al.*, 1998; Ramboz *et al.*, 1998). In parallel to the complete lack of 5-HT_{1A} receptors demonstrated through receptor binding in 1AKO mice, all these mouse lines showed elevated levels of anxiety in several different animal models (Heisler *et al.*, 1998; Ramboz *et al.*, 1998; Parks *et al.*, 1998). To better understand the basis for this phenotype, Gross and colleagues (2002) used gene targeting technology to generate mice with conditional expression of the 5-HT_{1A} receptor in forebrain regions, including the hippocampus, amygdala and frontal cortex. This approach allowed for the rescue of functional 5-HT_{1A} receptors (i.e., the rescue of the 5-HT_{1A}-receptor induced hyperpolarization) in the forebrain. Remarkably, these 5-HT_{1A} 'rescued' mice behaved like WT mice in the behavioral tests for anxiety (Gross *et al.*, 2002). If the receptor was rescued when the mouse was an adult or after postnatal day 21, the anxiety phenotype was still present, indicating that the 5-HT_{1A} receptor needs to be present during postnatal days 0–21 to relieve the anxiety phenotype. These results point to a key role for 5-HT_{1A} receptors during development, most likely at the level of the hippocampal circuitry, and suggest that disruption of serotonergic function in this area during development can lead to long-lasting behavioral changes. Future studies will have to be conducted to determine exactly what changes occur during this critical time period when 5-HT is not present that lead to the anxious phenotype.

The cerebral cortex

In comparison to the hippocampus, with its diaphanous cytoarchitecture, the cerebral cortex presents a much more complex target for understanding the effects of serotonin. As such, this section represents an even more tentative progress report.

Like the hippocampus, the cerebral cortex contains two main neuronal cell types, glutamate-releasing excitatory pyramidal neurons and GABA-releasing inhibitory interneurons. Pyramidal neurons are the predominant cell type, and constitute the only projection cell of the cerebral cortex. Histologically, pyramidal neurons are arranged in multiple cell layers, generally six, with each layer having a preferential projection target. Thus, in general, layer VI projects to thalamus, layer V projects predominantly to subcortical structures, and the more superficial layers II–III to ipsi- and contralateral cortex. Layer IV, when present, receives thalamic input. Recent work, however, suggests that within this broad pyramidal cell classification lies considerable molecular and functional heterogeneity (Molnar and Cheung, 2006; Sugino *et al.*, 2006; Molyneaux *et al.*, 2007; Leone *et al.*, 2008). Furthermore, as in hippocampus (Klausberger and Somogyi, 2008), GABAergic interneurons fall into several distinct anatomical and functional classes.

In general, anatomical studies have shown that the serotonergic innervation of the cerebral cortex is extensive, and fairly homogeneous (Lidov *et al.*, 1980). However, it is important to note that the expression of different serotonin receptor subtypes displays considerably variation from cortical area to cortical area. Thus, for example, in adult rats 5-HT_{2A} receptors show a steep anteroposterior expression gradient (Lopez-Gimenez *et al.*, 1997), while 5-HT_7 receptors are essentially absent from the prefrontal cortex in adult rodents but are reasonably well expressed at more posterior levels of the cingulate (Gustafson *et al.*, 1996). This heterogeneity is not surprising given the functional heterogeneity of the cortex. However, it does make it difficult to provide a single answer to the question of what serotonin does to cortical networks, as the answer will clearly depend on the area of cortex in question.

Pyramidal cells

These limitations notwithstanding, considerable progress has been made elucidating the actions of serotonin in particular areas of the cerebral cortex. Several groups have focused on pyramidal cells of the prefrontal, somatosensory and visual cortices, and in this section we emphasize the work in these areas. The prefrontal cortex in particular has emerged in the last decade as a key brain region in the organization of behavior (Buckner *et al.*, 2008; Tanji and Hoshi, 2008), and has been implicated in a variety of pathophysiological processes ranging from drug abuse to autism and schizophrenia (see, for example, Bachevalier and Loveland, 2006; Tan *et al.*, 2007; Buckner *et al.*, 2008; Clay *et al.*, 2008; Tanji and Hoshi, 2008). As such, it has received considerable attention from serotonin physiologists working predominantly in rodent model systems. This has resulted in substantial strides in our understanding of the effects of serotonin in this region. However, it is important to realize that while there are important similarities in terms of connectivity and function between the prefrontal cortex of rodents and of primates, there are also some important differences (Uylings *et al.*, 2003). Furthermore, there are significant cytoarchitectonic differences – most notably, the absence of a well-developed granular layer IV in rodent prefrontal cortex (van Eden and Uylings, 1985). It will be important to keep these differences in mind as we seek to use the knowledge gained in rodent model systems to understand the pathophysiology of the prefrontal cortex and its relationship to serotonergic mechanisms.

The effects of serotonin on cortical pyramidal cells, and especially those of layer V, have been assessed by several groups during the past two decades. There is a general consensus that administration of serotonin results in a large increase in spontaneous synaptic activity that is accompanied by variable changes in membrane potential and a large modulation of neuronal gain, a measure of how effectively neurons transform excitatory drive into firing (Araneda and Andrade, 1991; Spain, 1994; Aghajanian and Marek, 1997; Zhou and Hablitz, 1999; Foehring *et al.*, 2002; Beique *et al.*, 2004a, 2004b, 2007; Villalobos *et al.*, 2005; Zhang and Arsenault, 2005; Higgs *et al.*, 2006). The simplest of these effects, the changes in membrane potential and gain, have been extensively studied, and it is now apparent that they can be accounted for, at least in adult animals, by the co-expression of 5-HT$_{1A}$ and 5-HT$_{2A}$ receptors in pyramidal neurons. A large fraction of layer V pyramidal neurons in the prefrontal cortex express 5-HT$_{1A}$ receptors that signal a membrane hyperpolarization that can be expected to suppress excitability (Araneda and Andrade, 1991; Martin-Ruiz *et al.*, 2001; del Burgo *et al.*, 2008). However, most pyramidal neurons also co-express 5-HT$_{2A}$ receptors (Araneda and Andrade, 1991; Vysokanov *et al.*, 1998; Martin-Ruiz *et al.*, 2001), which generally signal a variable membrane depolarization and inhibit a calcium-activated potassium current (Araneda and Andrade, 1991; Villalobos *et al.*, 2005; Zhang and Arsenault, 2005). As in the hippocampus, this co-expression of receptors signaling opposite effects on excitability was initially puzzling. However, because the slow calcium-activated potassium current limits the ability of a neuron to transform a sustained excitatory drive into spiking, its inhibition by serotonin acting on 5-HT$_{2A}$ receptors effectively increases a neuron's gain (Araneda and Andrade, 1991; Zhang and Arsenault, 2005). Furthermore, the balance between hyperpolarizing and depolarizing influences can effectively adjust the excitatory input range over which neurons can encode excitatory input into firing activity. This suggests that the co-expression of 5-HT$_{1A}$ and 5-HT$_{2A}$ in pyramidal cells allows serotonin to regulate how cells transform excitatory input into firing in a manner that is easily tunable by changing receptor expression levels (Araneda and Andrade, 1991).

Interestingly, previous studies have shown that uncoordinated synaptic noise can effectively increase neuronal gain (Higgs *et al.*, 2006, and references therein). Since serotonin very strongly facilitates spontaneous synaptic activity, this effect alone would be expected to increase gain in pyramidal cells irrespective of any changes in neuronal properties. It is tempting to speculate that perhaps these two mechanisms could synergize in pyramidal cells, although experimental confirmation of this conjecture is missing at this time (W.J. Spain, personal communication).

While the direct effects of serotonin on membrane excitability outlined above have been understood for well over a decade, the very robust increase in spontaneous glutamate receptor-mediated synaptic activity elicited by serotonin in prefrontal cortex has proven much more contentious. Many studies have shown that serotonin induces this increase in glutamate release through the activation of 5-HT$_{2A}$ receptors (Aghajanian and Marek, 1999; Marek and Aghajanian, 1999; Zhou and Hablitz, 1999; Foehring *et al.*, 2002; Beique *et al.*, 2004b, 2007). However, the simplest explanation for such an effect, namely that serotonin excited pyramidal cells, was thought unlikely given the small 5-HT$_{2A}$ receptor-induced depolarization generally seen in pyramidal neurons. Instead it was proposed that the activation of 5-HT$_{2A}$ receptors on pyramidal cells led to the release of a retrograde message that excited thalamocortical afferents, by inhibiting Kv1.2 channels (Zhou and Hablitz, 1999; Lambe and Aghajanian, 2001). Recent work, however, has shown this possibility to be untenable (Beique *et al.*, 2007); instead, it has proposed that 5-HT_{2A} receptors excite a population of pyramidal cells in cortex that had previously gone unsampled, and that it is the excitation of this population of pyramidal cells that accounts for the increase in glutamate release widely observed in this region (Beique *et al.*, 2007) (Figure 3). These results emphasize the importance of addressing the functional heterogeneity of pyramidal cells when considering the effects of serotonin in cortex.

In addition to these effects on pyramidal cells, administration of 5-HT to cortical slices also inhibits evoked excitatory synaptic transmission onto pyramidal neurons (Tanaka and North, 1993; Rhoades *et al.*, 1994; Murakoshi *et al.*, 2001; Laurent *et al.*, 2002; Torres-Escalante *et al.*, 2004). This effect reflects a presynaptic inhibitory effect of serotonin on glutamate release that is mediated by 5-$HT_{1B/D}$ receptors (Tanaka and North, 1993; Rhoades *et al.*, 1994; Laurent *et al.*, 2002). Unfortunately, the exact identity of the afferents expressing the 5-HT_{1B} receptors and thus inhibited by serotonin is unclear. In young animals, it has been suggested that serotonin acting via 5-HT_{1B} receptors regulates the activity of thalamocortical axons innervating somatosensory cortex (Rhoades *et al.*, 1994; Laurent *et al.*, 2002). However, since current technical limitations preclude the selective stimulation of different cortical afferents, it is not known whether these findings can be generalized to different ages and to different cortical areas. In any case, the observation that administration of serotonin increases spontaneous glutamate synaptic activity in prefrontal and somatosensory cortices suggests that this presynaptic inhibitory effect in cortex must exhibit at least some selectivity for specific populations of afferents.

Interneurons

Cortical interneurons constitute a diverse group of GABAergic neurons whose function is to sculpt the functional properties of pyramidal cell networks (Bacci *et al.*, 2005; Klausberger and Somogyi, 2008). While the exact number of interneuron classes is still uncertain, it is now clear that different cortical interneurons identified on morphological grounds or based upon the expression of molecular markers also represent distinct functional classes. Thus it will come as no surprise that different classes of cortical interneurons also express different serotonin receptors or possibly combinations of serotonin receptors.

Our understanding of the effects of serotonin on different classes of interneurons is still in its infancy, but has

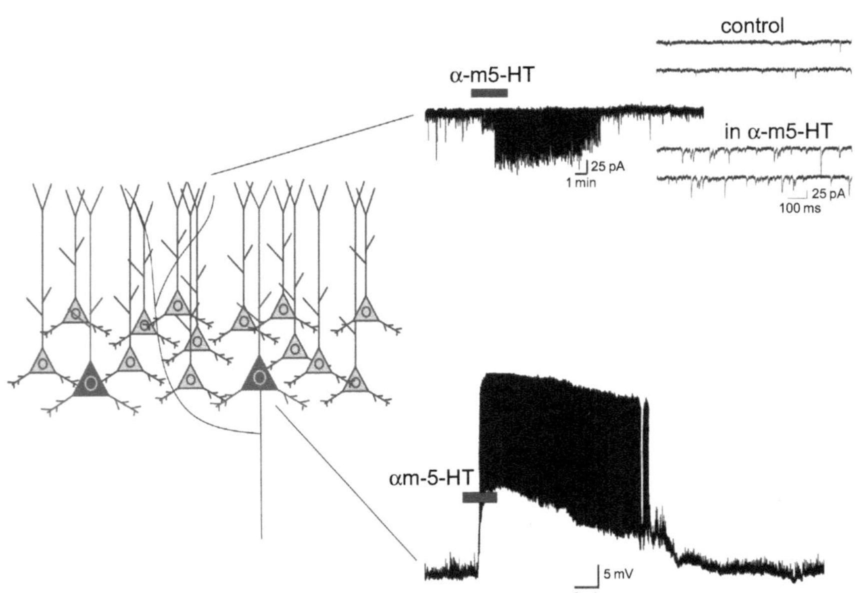

Figure 3 Serotonin increases glutamate release in the prefrontal cortex by exciting a subpopulation of deep layer V pyramidal neurons. Administration of the 5-HT_{2A} agonist alpha-methyl serotonin depolarizes and excites the large pyramidal cells of deep layer V in the rat prefrontal cortex. In contrast, administration of this agonist elicits only a modest depolarization/inward current in the smaller pyramidal neurons of upper layer V but elicits a large increase in spontaneous glutamate-mediated synaptic activity onto these cells. This effect is thought to reflect the excitation of the deep layer V cells that synapse onto the more superficial pyramidal neurons in this region.

been greatly helped by the identification of molecular markers that identify subsets of interneurons (Bacci *et al.*, 2005). For example, one class of large cortical interneurons is identifiable by the expression of parvalbumin. These cells correspond to the classic chandelier and basket cells innervating the soma and initial segment of pyramidal cells, which allows them to function as key regulators of the output (firing) of pyramidal cells. Anatomical studies conducted in the prefrontal cortex have shown that these cells robustly express 5-HT_{2A} receptors and 5-HT_{2C} receptors (Willins *et al.*, 1997; Jakab and Goldman-Rakic, 2000; de Almeida and Mengod, 2007; Liu *et al.*, 2007) but express 5-HT_{1A} receptors only infrequently (de Almeida and Mengod, 2008). In contrast, in this same region, a different class of interneurons identifiable by the expression of calbindin expresses 5-HT_{1A}, 5-HT_{2A} and 5-HT_3 receptors (Jakab and Goldman-Rakic, 2000; de Almeida and Mengod, 2007, 2008).

Electrophysiological studies show patterns of response that are consistent with the presence of all these receptor subtypes in cortical interneurons (Zhou and Hablitz, 1999; Foehring *et al.*, 2002; Xiang and Prince, 2003). However, because the electrophysiological criteria generally used in these studies are insufficient to identify specific interneuron subtypes, these results are difficult to place in a functional context. Thus clearly additional studies are needed to elucidate how serotonin regulates cortical interneurons. One important question that will need to be addressed is whether serotonin regulates different interneuron types in a uniform manner across different layers and cortical regions. The growing availability of transgenic mice expressing fluorescent proteins in selected subpopulations of cells, including interneurons (Ma *et al.*, 2006), should greatly facilitate such studies.

Serotonergic regulation of cortical synaptic plasticity

In addition to regulating neuronal excitability, serotonin also plays an important role in regulating synaptic plasticity in cortex. Since synapses strengthen or weaken as a function of activity levels, serotonin-induced changes in cellular excitability alone can be expected to regulate synaptic plasticity. However, it is now widely believed that neurotransmitters acting via GPCRs, and serotonin in particular, may in fact have a much more pervasive effect, functioning as gatekeepers for synaptic plasticity – a process often called metaplasticity.

The visual cortex provides what is probably the clearest example of a serotonin regulation of synaptic plasticity. It has been known for many years that synaptic inputs to the visual cortex during development can express activity-dependent plasticity during a transient phase known as the critical period. Work in the mid-1990s showed that at least some forms of synaptic plasticity expressed during this period were dependent on 5-HT_2 serotonin receptor activation (Gu and Singer, 1995; Kojic *et al.*, 1997). Subsequent work showed that expression of this serotonin-dependent synaptic plasticity coincided with the transient expression of 5-HT_2 receptor patches in visual cortex, and that activation of 5-HT_2 receptors facilitated the expression of long-term potentiation in neurons of layer IV (Kojic *et al.*, 2000, 2001). Interestingly, a recent study has reported that administration of fluoxetine to adult rats restores visual plasticity lost after the critical period, although the relationship of these observations to the earlier studies remains unspecified (Maya Vetencourt *et al.*, 2008). These limitations notwithstanding, the studies provide compelling support for the idea that serotonin could regulate synaptic plasticity in cortex.

These studies, while tantalizing, could not address potential cellular and molecular mechanisms underlying this phenomenon. A recent study from the Kirkwood laboratory (Seol *et al.*, 2007), however, while not directly addressing serotonergic mechanisms, has shown that neurotransmitter receptors coupling to $G\alpha_s$ and $G\alpha_{q-11}$ in the visual cortex promote long-term potentiation and long-term depression in pyramidal cells through the phosphorylation of glutamate receptors. These are exciting findings that link signaling steps downstream from GPCRs to our rapidly growing understanding of the molecular mechanisms underlying synaptic plasticity. However, it is noteworthy that the polarity of the 5-HT_2 receptor-mediated metaplasticity observed in the earlier studies in visual cortex are the reverse of those predicted by these more recent findings (Choi *et al.*, 2005; Seol *et al.*, 2007). This is a puzzling discrepancy that may reflect the significant experimental differences between the two sets of studies. In any case, direct support for the idea that $G\alpha_{q-11}$ coupled 5-HT_2 receptors may promote long-term depression in cerebral cortex has recently been obtained in prefrontal cortex (Zhong *et al.*, 2008), although the mechanism proposed to underlie this effect appears different from that expected from the work by the Kirkwood lab.

Clearly our understanding of the mechanism by which serotonin receptors regulate synaptic plasticity is in its early infancy. There are multiple forms of synaptic plasticity, both long-term potentiation and long-term depression, and it seems highly unlikely that different synapses in cortex would uniformly express a single type of either of these. Furthermore, different synapses in cortex may be differentially regulated by serotonin depending on the specific form(s) of synaptic plasticity they express and the serotonin receptor subtypes expressed in the presynaptic and postsynaptic elements. Given the technical difficulties inherent in studying homogeneous populations of synapses in cortex, it will take clever, technically innovative approaches to address these issues.

Cortical serotonin during development

The expression of serotonin receptor subtypes in the brain, and the cerebral cortex in particular, is under strong developmental control (see, for example, Beique *et al.*, 2004a; Bonnin *et al.*, 2006; Waeber *et al.*, 1994; Vizuete *et al.*, 1997). Changes in receptor expression levels appear particularly prominent during the prenatal and early postnatal periods, which represent key times for synaptogenesis and synaptic refinement (Zhang, 2006). Recent studies have shown that these changes in receptor levels can fundamentally reshape what serotonin does in the cerebral cortex (Zhang, 2003; Beique *et al.*, 2004a, 2004b). For example, during the early postnatal period, pyramidal cells of layer V of the prefrontal cortex are profoundly depolarized by serotonin, an effect that is mediated by the activation of 5-HT_{2A} and also by 5-HT_7 receptors that are transiently expressed in this area at this age (Zhang, 2003; Beique *et al.*, 2004a). By the third postnatal week, however, 5-HT_7 receptors are no longer expressed on pyramidal cells of the prefrontal cortex, and 5-HT_{2A} receptors have become largely uncoupled from the ion channels mediating the membrane depolarization (Beique *et al.*, 2004a). *Pari passu* with these changes, 5-HT_{1A} receptors begin to appear, signaling the characteristic hyperpolarization that is the hallmark effect of these receptors in adult animals (Beique *et al.*, 2004a). These changes in excitability have been hypothesized to play a role in setting the rules for the synaptic refinement that takes place during this developmental epoch. However, how specifically these effects contribute to the refinement of prefrontal cortical circuits, if at all, remains to be determined.

Perhaps the best understood effect of serotonin on the development of cortical circuits is the 5-HT_{1B} receptor modulation of the thalamic innervation of somatosensory cortex. 5-$HT_{1B/D}$ receptors are transiently expressed in the thalamic fibers innervating the barrel cortex, and either depletion of serotonin or inhibition of 5-$HT_{1B/D}$ receptor function results in impairment of the formation of vibrissae-related patches in layer IV (Young-Davies *et al.*, 2000, and references therein). The mechanism underlying this phenomenon remained mysterious for many years. However, recently Bonnin and associates, in a remarkable technical *tour de force*, have shown that serotonin acting on 5-$HT_{1B/D}$ receptors regulates the guidance clues transmitted through netrin-1 onto thalamocortical axons (Bonnin *et al.*, 2007). In the absence of serotonin, or 5-$HT_{1B/D}$ receptors, the resulting deficit in pathfinding is thought to account for the disruption of the sensory map.

The road ahead

Over 25 years ago the authors of this review started to work on serotonin, fascinated by the possibility of making sense of its bewildering electrophysiology in central neurons through an understanding of serotonin receptor subtypes. Thanks to the work of many labs – many more than can be acknowledged here – a good portion of what we wanted to know, namely understanding what each receptor subtype signals in central neurons, has been clarified. Yet, as is patently clear in this chapter, this knowledge has proven grossly insufficient as a way to understand how serotonin regulates neuronal function in the brain.

A number of critical challenges remain. First, going in the systems neuroscience direction, its is now clear that, to understand what serotonin does, we need to understand what it does in the context of specific neuronal circuits. That is the fundamental message from this chapter. But beyond that, we need to understand how serotonin receptors are activated by both volume transmission and synaptically released serotonin under physiological conditions. For example, does a single serotonergic cell, or group of serotonergic neurons, control all the different serotonin receptors on a given postsynaptic neuron, or only some in specific locations, or some of a particular subtype? These, and many more similar questions, are critical to our understanding of the physiology of serotonin; yet at the present time we cannot provide answers.

Second, and now moving in the direction of cellular neuroscience, it is now clear that GPCRs do not signal solely via G proteins. Rather, they function as ligand-activated signaling platforms capable of involving a variety of molecular partners. This is likely to shape serotonergic signaling mediated through the G protein and to add additional complexities to what is being signaled. Integrating these advances in our understanding of serotonin receptors' function into the physiological context of central neurons is likely to consume the next 25 years and beyond.

Acknowledgements

Work in the authors' laboratories has been supported by NIH grants MH43987 and the State of Michigan (to Rodrigo Andrade) and NIH grants MH075047, MH39004, MH63078 and NS28512 (to Sheryl G. Beck). The authors thank Dr Jerrel Yakel for the 5-HT_3 trace used in Figure 2.

References

Aghajanian, G.K. and Marek, G.J. (1997) Serotonin induces excitatory postsynaptic potentials in apical dendrites of neocortical pyramidal cells. Neuropharmacology, 36: 589–599.

Aghajanian, G.K. and Marek, G.J. (1999) Serotonin, via 5-HT2A receptors, increases EPSCs in layer V pyramidal cells of prefrontal cortex by an asynchronous mode of glutamate release. Brain Res., 825: 161–171.

Ait, A.D., Segu, L., Naili, S. and Buhot, M.C. (1995) Serotonin 1B receptor regulation after dorsal subiculum deafferentation. Brain Res. Bull., 38: 17–23.

Andrade, R. and Chaput, Y. (1991) 5-Hydroxytryptamine$_4$-like receptors mediate the slow excitatory response to serotonin in the rat hippocampus. J. Pharmacol. Exp. Ther., 257: 930–937.

Andrade, R. and Nicoll, R.A. (1987) Pharmacologically distinct actions of serotonin on single pyramidal neurones of the rat hippocampus recorded *in vitro*. J. Physiol., 394: 99–124.

Araneda, R. and Andrade, R. (1991) 5-Hydroxytryptamine2 and 5-hydroxytryptamine 1A receptors mediate opposing responses on membrane excitability in rat association cortex. Neuroscience, 40: 399–412.

Ascoli, G.A., Onso-Nanclares, L., Anderson, S.A., Barrionuevo, G., Avides-Piccione, R., Burkhalter, A., Buzsaki, G., Cauli, B., Defelipe, J., Fairen, A., Feldmeyer, D., Fishell, G., Fregnac, Y., Freund, T.F., Gardner, D., Gardner, E.P., Goldberg, J.H., Helmstaedter, M., Hestrin, S., Karube, F., Kisvarday, Z.F., Lambolez, B., Lewis, D.A., Marin, O., Markram, H., Munoz, A., Packer, A., Petersen, C.C., Rockland, K.S., Rossier, J., Rudy, B., Somogyi, P., Staiger, J.F., Tamas, G., Thomson, A.M., Toledo-Rodriguez, M., Wang, Y., West, D.C. and Yuste, R. (2008) Petilla terminology: nomenclature of features of GABAergic interneurons of the cerebral cortex. Nat. Rev. Neurosci., 9: 557–568.

Bacci, A., Huguenard, J.R. and Prince, D.A. (2005) Modulation of neocortical interneurons: extrinsic influences and exercises in self-control. Trends Neurosci., 28: 602–610.

Bachevalier, J. and Loveland, K.A. (2006) The orbitofrontal-amygdala circuit and self-regulation of social-emotional behavior in autism. Neurosci. Biobehav. Rev., 30: 97–117.

Bacon, W.L. and Beck, S.G. (2000) 5-Hydroxytryptamine(7) receptor activation decreases slow afterhyperpolarization amplitude in CA3 hippocampal pyramidal cells. J. Pharmacol. Exp. Ther., 294: 672–679.

Beck, S.G., Choi, K.C. and List, T.J. (1992) Comparison of 5-hydroxytryptamine$_{1A}$-mediated hyperpolarization in CA1 and CA3 hippocampal pyramidal cells. J. Pharmacol. Exp. Ther., 263: 350–359.

Beck, S.G., Clarke, W.P. and Goldfarb, J. (1985) Spiperone differentiates multiple 5-hydroxytryptamine responses in rat hippocampal slices *in vitro*. Eur. J. Pharmacol., 116: 195–197.

Beique, J.C., Campbell, B., Perring, P., Hamblin, M.W., Walker, P., Mladenovic, L. and Andrade, R. (2004) Serotonergic regulation of membrane potential in developing rat prefrontal cortex: coordinated expression of 5-hydroxytryptamine (5-HT)1A, 5-HT2A, and 5-HT7 receptors. J. Neurosci., 24: 4807–4817.

Beique, J.C., Chapin-Penick, E.M., Mladenovic, L. and Andrade, R. (2004) Serotonergic facilitation of synaptic activity in the developing rat prefrontal cortex. J. Physiol., 556: 739–754.

Beique, J.C., Imad, M., Mladenovic, L., Gingrich, J.A. and Andrade, R. (2007) Mechanism of the 5-hydroxytryptamine 2A receptor-mediated facilitation of synaptic activity in prefrontal cortex. Proc. Natl Acad. Sci. USA, 104: 9870–9875.

Boeijinga, P.H. and Boddeke, H.W. (1993) Serotonergic modulation of neurotransmission in the rat subicular cortex *in vitro*: a role for 5-HT1B receptors. Naunyn Schmiedebergs Arch. Pharmacol., 348: 553–557.

Boeijinga, P.H. and Boddeke, H.W. (1996) Activation of 5-HT1B receptors suppresses low but not high frequency synaptic transmission in the rat subicular cortex *in vitro*. Brain Res., 721: 59–65.

Bonnin, A., Peng, W., Hewlett, W. and Levitt, P. (2006) Expression mapping of 5-HT1 serotonin receptor subtypes during fetal and early postnatal mouse forebrain development. Neuroscience, 141: 781–794.

Bonnin, A., Torii, M., Wang, L., Rakic, P. and Levitt, P. (2007) Serotonin modulates the response of embryonic thalamocortical axons to netrin-1. Nat. Neurosci., 10: 588–597.

Buckner, R.L., Andrews-Hanna, J.R. and Schacter, D.L. (2008) The brain's default network: anatomy, function, and relevance to disease. Ann. NY Acad. Sci., 1124: 1–38.

Choi, S.Y., Chang, J., Jiang, B., Seol, G.H., Min, S.S., Han, J.S., Shin, H.S., Gallagher, M. and Kirkwood, A. (2005) Multiple receptors coupled to phospholipase C gate long-term depression in visual cortex. J. Neurosci., 25: 11433–11443.

Clay, S.W., Allen, J. and Parran, T. (2008) A review of addiction. Postgrad. Med., 120: E01–E07.

Colino, A. and Halliwell, J.V. (1987) Differential modulation of three separate K-conductances in hippocampal CA1 neurons by serotonin. Nature, 328: 73–77.

Dahlstrom, A. and Fuxe, K. (1964). Evidence for the existence of monoamine-containing neurons in the central nervous system: I. Demonstration of monoamines in cell bodies of brain neurons. Acta Physiol. Scand. Suppl, 232: 1–55.

de Almeida, A.J. and Mengod, G. (2007) Quantitative analysis of glutamatergic and GABAergic neurons expressing 5-HT(2A) receptors in human and monkey prefrontal cortex. J. Neurochem., 103: 475–486.

de Almeida, A.J. and Mengod, G. (2008). Serotonin 1A receptors in human and monkey prefrontal cortex are mainly expressed in pyramidal neurons and in a GABAergic interneuron subpopulation: implications for schizophrenia and its treatment. J. Neurochem., in press.

del Burgo, L.S., Cortes, R., Mengod, G., Zarate, J., Echevarria, E. and Salles, J. (2008) Distribution and neurochemical characterization of neurons expressing GIRK channels in the rat brain. J. Comp Neurol., 510: 581–606.

Deng, D.R., Djalali, S., Holtje, M., Grosse, G., Stroh, T., Voigt, I., Kusserow, H., Theuring, F., Ahnert-Hilger, G. and Hortnagl, H. (2007) Embryonic and postnatal development of the serotonergic raphe system and its target regions in 5-HT1A receptor deletion or overexpressing mouse mutants. Neuroscience, 147: 388–402.

Foehring, R.C., van Brederode, J.F., Kinney, G.A. and Spain, W.J. (2002) Serotonergic modulation of supragranular neurons in rat sensorimotor cortex. J. Neurosci., 22: 8238–8250.

Freund, T.F. (2003) Interneuron Diversity series: rhythm and mood in perisomatic inhibition. Trends Neurosci., 26: 489–495.

Freund, T.F., Gulyas, A.I., Acsady, L., Gorcs, T. and Toth, K. (1990) Serotonergic control of the hippocampus via local

inhibitory interneurons. Proc. Natl Acad. Sci. USA, 87: 8501–8505.
Freund, T.F. and Katona, I. (2007) Perisomatic inhibition. Neuron, 56: 33–42.
Gross, C., Zhuang, X., Stark, K., Ramboz, S., Oosting, R., Kirby, L., Santarelli, L., Beck, S. and Hen, R. (2002) Serotonin1A receptor acts during development to establish normal anxiety-like behaviour in the adult. Nature, 416: 396–400.
Gu, Q. and Singer, W. (1995) Involvement of serotonin in developmental plasticity of kitten visual cortex. Eur. J. Neurosci., 7: 1146–1153.
Gulyas, A.I., Acsady, L. and Freund, T.F. (1999) Structural basis of the cholinergic and serotonergic modulation of GABAergic neurons in the hippocampus. Neurochem. Int., 34: 359–372.
Gustafson, E.L., Durkin, M.M., Bard, J.A., Zgombick, J. and Branchek, T.A. (1996) A receptor autoradiographic and in situ hybridization analysis of the distribution of the 5-ht7 receptor in rat brain. Br. J. Pharmacol., 117: 657–666.
Harris-Warrick, R.M. and Marder, E. (1991) Modulation of neural networks for behavior. Annu. Rev. Neurosci., 14: 39–57.
Heisler, L.K., Chu, H.M., Brennan, T.J., Danao, J.A., Bajwa, P., Parsons, L.H. and Tecott, L.H. (1998) Elevated anxiety and antidepressant-like responses in serotonin 5-HT1A receptor mutant mice. Proc. Natl Acad. Sci. USA, 95: 15049–15054.
Higgs, M.H., Slee, S.J. and Spain, W.J. (2006) Diversity of gain modulation by noise in neocortical neurons: regulation by the slow afterhyperpolarization conductance. J. Neurosci., 26: 8787–8799.
Jacobs, B.L. and Azmitia, E.C. (1992) Structure and function of the brain serotonin system. Physiol. Rev., 72: 165–229.
Jacobs, B.L. and Fornal, C.A. (1999) Activity of serotonergic neurons in behaving animals. Neuropsychopharmacology, 21: 9S–15S.
Jacobs, B.L., Martin-Cora, F.J. and Fornal, C.A. (2002) Activity of medullary serotonergic neurons in freely moving animals. Brain Res. Brain Res. Rev., 40: 45–52.
Jakab, R.L. and Goldman-Rakic, P.S. (2000) Segregation of serotonin 5-HT2A and 5-HT3 receptors in inhibitory circuits of the primate cerebral cortex. J. Comp. Neurol., 417: 337–348.
Klausberger, T., Magill, P.J., Marton, L.F., Roberts, J.D., Cobden, P.M., Buzsaki, G. and Somogyi, P. (2003) Brain-state- and cell-type-specific firing of hippocampal interneurons *in vivo*. Nature, 421: 844–848.
Klausberger, T. and Somogyi, P. (2008) Neuronal diversity and temporal dynamics: the unity of hippocampal circuit operations. Science, 321: 53–57.
Kojic, L., Dyck, R.H., Gu, Q., Douglas, R.M., Matsubara, J. and Cynader, M.S. (2000) Columnar distribution of serotonin-dependent plasticity within kitten striate cortex. Proc. Natl. Acad. Sci. USA, 97: 1841–1844.
Kojic, L., Gu, Q., Douglas, R.M. and Cynader, M.S. (2001) Laminar distribution of cholinergic- and serotonergic-dependent plasticity within kitten visual cortex. Brain Res. Dev. Brain Res., 126: 157–162.
Kojic, L., Gu, Q., Douglas, R.M. and Cynader, M.S. (1997) Serotonin facilitates synaptic plasticity in kitten visual cortex: an *in vitro* study. Brain Res. Dev. Brain Res., 101: 299–304.
Lambe, E.K. and Aghajanian, G.K. (2001) The role of Kv1.2-containing potassium channels in serotonin-induced glutamate release from thalamocortical terminals in rat frontal cortex. J. Neurosci., 21: 9955–9963.
Laurent, A., Goaillard, J.M., Cases, O., Lebrand, C., Gaspar, P. and Ropert, N. (2002) Activity-dependent presynaptic effect of serotonin 1B receptors on the somatosensory thalamocortical transmission in neonatal mice. J. Neurosci., 22: 886–900.
Leone, D.P., Srinivasan, K., Chen, B., Alcamo, E. and McConnell, S.K. (2008) The determination of projection neuron identity in the developing cerebral cortex. Curr. Opin. Neurobiol., 18: 28–35.
Lidov, H.G., Grzanna, R. and Molliver, M.E. (1980) The serotonin innervation of the cerebral cortex in the rat – an immunohistochemical analysis. Neuroscience, 5: 207–227.
Liu, S., Bubar, M.J., Lanfranco, M.F., Hillman, G.R. and Cunningham, K.A. (2007) Serotonin2C receptor localization in GABA neurons of the rat medial prefrontal cortex: implications for understanding the neurobiology of addiction. Neuroscience, 146: 1677–1688.
Lopez-Gimenez, J.F., Mengod, G., Palacios, J.M. and Vilaro, M.T. (1997) Selective visualization of rat brain 5-HT2A receptors by autoradiography with [3H]MDL 100,907. Naunyn Schmiedebergs Arch. Pharmacol., 356: 446–454.
Ma, Y., Hu, H., Berrebi, A.S., Mathers, P.H. and Agmon, A. (2006) Distinct subtypes of somatostatin-containing neocortical interneurons revealed in transgenic mice. J. Neurosci., 26: 5069–5082.
Marek, G.J. and Aghajanian, G.K. (1999) 5-HT2A receptor or alpha1-adrenoceptor activation induces excitatory postsynaptic currents in layer V pyramidal cells of the medial prefrontal cortex. Eur. J. Pharmacol., 367: 197–206.
Martin-Ruiz, R., Puig, M.V., Celada, P., Shapiro, D.A., Roth, B.L., Mengod, G. and Artigas, F. (2001) Control of serotonergic function in medial prefrontal cortex by serotonin-2A receptors through a glutamate-dependent mechanism. J. Neurosci., 21: 9856–9866.
Maya Vetencourt, J.F., Sale, A., Viegi, A., Baroncelli, L., De, P.R., O'Leary, O.F., Castren, E. and Maffei, L. (2008) The antidepressant fluoxetine restores plasticity in the adult visual cortex. Science, 320: 385–388.
McMahon, L.L. and Kauer, J.A. (1997) Hippocampal interneurons are excited via serotonin-gated ion channels. J. Neurophysiol., 78: 2493–2502.
Molnar, Z. and Cheung, A.F. (2006) Towards the classification of subpopulations of layer V pyramidal projection neurons. Neurosci. Res., 55: 105–115.
Molyneaux, B.J., Arlotta, P., Menezes, J.R. and Macklis, J.D. (2007) Neuronal subtype specification in the cerebral cortex. Nat. Rev. Neurosci., 8: 427–437.
Morales, M. and Backman, C. (2002) Coexistence of serotonin 3 (5-HT3) and CB1 cannabinoid receptors in interneurons of hippocampus and dentate gyrus. Hippocampus, 12: 756–764.
Murakoshi, T., Song, S.Y., Konishi, S. and Tanabe, T. (2001) Multiple G-protein-coupled receptors mediate presynaptic inhibition at single excitatory synapses in the rat visual cortex. Neurosci. Lett., 309: 117–120.
Parks, C.L., Robinson, P.S., Sibille, E., Shenk, T. and Toth, M. (1998) Increased anxiety of mice lacking the serotonin1A receptor. Proc. Natl. Acad. Sci. USA, 95: 10734–10739.
Parra, P., Gulyas, A.I. and Miles, R. (1998) How many subtypes of inhibitory cells in the hippocampus? Neuron, 20: 983–993.
Patel, T.D. and Zhou, F.C. (2005) Ontogeny of 5-HT1A receptor expression in the developing hippocampus. Brain Res. Dev. Brain Res., 157: 42–57.

Piguet, P. and Galvan, M. (1994) Transient and long-lasting actions of 5-HT on rat dentate gyrus neurones *in vitro*. J. Physiol., 481(Pt 3): 629–639.

Ramboz, S., Oosting, R., Amara, D.A., Kung, H.F., Blier, P., Mendelsohn, M., Mann, J.J., Brunner, D. and Hen, R. (1998) Serotonin receptor 1A knockout: an animal model of anxiety-related disorder. Proc. Natl. Acad. Sci. USA, 95: 14476–14481.

Rhoades, R.W., Benett-Clarke, C.A., Shi, M.Y. and Mooney, R.D. (1994) Effects of 5-HT on thalamocortical synaptic transmission in the developing rat. J. Neurophysiol., 72: 2438–2450.

Segal, M. (1980) The action of serotonin in the rat hippocampal slice preparation. J. Physiol., 303: 423–439.

Seol, G.H., Ziburkus, J., Huang, S., Song, L., Kim, I.T., Takamiya, K., Huganir, R.L., Lee, H.K. and Kirkwood, A. (2007) Neuromodulators control the polarity of spike-timing-dependent synaptic plasticity. Neuron, 55: 919–929.

Spain, W.J. (1994) Serotonin has different effects on two classes of Betz cells from the cat. J. Neurophysiol., 72: 1925–1937.

Steinbusch, H.W. (1981) Distribution of serotonin-immunoreactivity in the central nervous system of the rat-cell bodies and terminals. Neuroscience, 6: 557–618.

Sugino, K., Hempel, C.M., Miller, M.N., Hattox, A.M., Shapiro, P., Wu, C., Huang, Z.J. and Nelson, S.B. (2006) Molecular taxonomy of major neuronal classes in the adult mouse forebrain. Nat. Neurosci., 9: 99–107.

Tan, H.Y., Callicott, J.H. and Weinberger, D.R. (2007) Dysfunctional and compensatory prefrontal cortical systems, genes and the pathogenesis of schizophrenia. Cereb. Cortex, 17(Suppl. 1): i171–i181.

Tanaka, E. and North, R.A. (1993) Actions of 5-hydroxytryptamine on neurons of the rat cingulate cortex. J. Neurophysiol., 69: 1749–1757.

Tanji, J. and Hoshi, E. (2008) Role of the lateral prefrontal cortex in executive behavioral control. Physiol. Rev., 88: 37–57.

Torres, G.E., Holt, I.L. and Andrade, R. (1994) Antagonists of 5-HT4 receptor-mediated responses in adult hippocampal neurons. J. Pharmacol. Exp. Ther., 271: 255–261.

Torres-Escalante, J.L., Barral, J.A., Ibarra-Villa, M.D., Perez-Burgos, A., Gongora-Alfaro, J.L. and Pineda, J.C. (2004) 5-HT1A, 5-HT2, and GABAB receptors interact to modulate neurotransmitter release probability in layer 2/3 somatosensory rat cortex as evaluated by the paired pulse protocol. J. Neurosci. Res., 78: 268–278.

Uylings, H.B., Groenewegen, H.J. and Kolb, B. (2003) Do rats have a prefrontal cortex? Behav. Brain Res., 146: 3–17.

van Eden, C.G. and Uylings, H.B. (1985) Cytoarchitectonic development of the prefrontal cortex in the rat. J. Comp. Neurol., 241: 253–267.

Villalobos, C., Beique, J.C., Gingrich, J.A. and Andrade, R. (2005) Serotonergic regulation of calcium-activated potassium currents in rodent prefrontal cortex. Eur. J. Neurosci., 22: 1120–1126.

Vizuete, M.L., Venero, J.L., Traiffort, E., Vargas, C., Machado, A. and Cano, J. (1997) Expression of 5-HT7 receptor mRNA in rat brain during postnatal development. Neurosci. Lett., 227: 53–56.

Vysokanov, A., Flores-Hernandez, J. and Surmeier, D.J. (1998) mRNAs for clozapine-sensitive receptors co-localize in rat prefrontal cortex neurons. Neurosci. Lett., 258: 179–182.

Waeber, C., Sebben, M., Nieoullon, A., Bockaert, J. and Dumuis, A. (1994) Regional distribution and ontogeny of 5-HT4 binding sites in rodent brain. Neuropharmacology, 33: 527–541.

Willins, D.L., Deutch, A.Y. and Roth, B.L. (1997) Serotonin 5-HT2A receptors are expressed on pyramidal cells and interneurons in the rat cortex. Synapse, 27: 79–82.

Xiang, Z. and Prince, D.A. (2003) Heterogeneous actions of serotonin on interneurons in rat visual cortex. J. Neurophysiol., 89: 1278–1287.

Young-Davies, C.L., Benett-Clarke, C.A., Lane, R.D. and Rhoades, R.W. (2000) Selective facilitation of the serotonin(1B) receptor causes disorganization of thalamic afferents and barrels in somatosensory cortex of rat. J. Comp. Neurol., 425: 130–138.

Zhang, Z.W. (2003) Serotonin induces tonic firing in layer V pyramidal neurons of rat prefrontal cortex during postnatal development. J. Neurosci., 23: 3373–3384.

Zhang, Z.W. (2006) Canadian Association of Neurosciences review: postnatal development of the mammalian neocortex: role of activity revisited. Can. J. Neurol. Sci., 33: 158–169.

Zhang, Z.W. and Arsenault, D. (2005) Gain modulation by serotonin in pyramidal neurones of the rat prefrontal cortex. J. Physiol., 566: 379–394.

Zhou, F.M. and Hablitz, J.J. (1999) Activation of serotonin receptors modulates synaptic transmission in rat cerebral cortex. J. Neurophysiol., 82: 2989–2999.

Zhong, P., Liu, W., Gu, Z. and Yan, Z. (2008) Serotonin facilitates long-term depression induction in prefrontal cortex via p38 MAPK/Rab5-mediated enhancement of AMPA receptor internalization. J. Physiol., 586: 4465–4479.

CHAPTER 2.6

Serotonergic Feedback Control

Trevor Sharp*

University Department of Pharmacology, Oxford, UK

Abstract: This chapter discusses current knowledge of the pharmacology and physiology of the feedback mechanisms that control serotonin (5-hydroxytryptamine; 5-HT) neurons, and considers their possible contribution to the pathophysiology and treatment of psychiatric disorders, especially major depression. A key mechanism involved in the control of 5-HT neurons is feedback inhibition by presynaptic 5-HT autoreceptors. These autoreceptors are implicated in depression and anxiety neurobiology, and form the basis of various ongoing 5-HT drug discovery efforts in these therapeutic areas. Recent experiments have discovered further complexity of 5-HT neuron control, specifically in the form of postsynaptic feedback mechanisms. These mechanisms have the physiological effects of 5-HT autoreceptors but use additional 5-HT receptor subtypes, and operate via neural inputs to 5-HT neurons. There is also recent evidence for postsynaptic feedback systems that excite 5-HT neurons. The postsynaptic feedback mechanisms are also providing a new source of drug targets for therapeutic application. Overall, current data suggest the presence of a previously unsuspected, complex arrangement of pre- and postsynaptic 5-HT receptor-mediated feedback mechanisms that control 5-HT neuron function.

Keywords: 5-HT autoreceptor, 5-HT$_{1A}$, 5-HT$_{1B}$, 5-HT$_2$, 5-HT$_4$, dorsal raphe nucleus, electrophysiology, depression, antidepressant.

Introduction

Serotonin (5-hydroxytryptamine; 5-HT) is implicated in numerous important brain functions, ranging from regulation of circadian changes in the sleep–wake cycle to processing of complex information linked to emotion and cognition. Moreover, changes in 5-HT have been linked to both the pathophysiology and successful treatment of many common psychiatric disorders, including depression, anxiety and schizophrenia.

The 5-HT innervation of the forebrain derives from clusters of 5-HT neurons in the midbrain raphe nuclei, especially the dorsal raphe nucleus (DRN). Across a range of species, many (but not all) DRN 5-HT neurons are characterized by a spontaneous, slow and regular firing pattern (Allers and Sharp, 2003), suggesting the presence of strict feedback control mechanisms. This idea arose from earlier observations that 5-HT receptor agonists inhibited the firing of 5-HT neurons *in vivo* (Aghajanian *et al.*, 1972). In addition, application of 5-HT or 5-HT receptor agonists to *in vitro* brain preparations was found to inhibit the release of preloaded radiolabeled 5-HT (Farnebo and Hamberger, 1971; Gothert and Weinheimer, 1979). Together these findings suggested the presence of presynaptic '5-HT autoreceptors' which acted at the somatodendrites and nerve terminals of 5-HT neurons to inhibit transmission. Subsequent pharmacological and genetic analysis of these effects, combined with new information on the diversity of 5-HT receptor subtypes and their cellular localization, led to the current 5-HT autoreceptor model which invokes a key role for the 5-HT$_1$ receptor family.

5-HT autoreceptors

The current 5-HT autoreceptor model proposes a population of 5-HT autoreceptors (5-HT$_{1A}$) located on the soma and dendrites of 5-HT neurons which open potassium channels (and possibly reduce Ca^{2+} conductance) via G-protein coupling to inhibit 5-HT cell firing (Innis *et al.*, 1988; Barnes and Sharp, 1999). Another population of 5-HT autoreceptors (5-HT$_{1B}$) is located on nerve terminals which G-protein couple to inhibit adenylate cyclase and/or open potassium channels to directly inhibit 5-HT release (Middlemiss and Hutson, 1990; Barnes and Sharp, 1999). The presence of additional populations of 5-HT autoreceptors (5-HT$_{1D}$ and 5-HT$_{5A}$), which modulate 5-HT release at the level of the soma, dendrites and possibly

*E-mail: trevor.sharp@pharm.ox.ac.uk

Christian Müller & Barry Jacobs (Eds.)
Handbook of Behavioral Neurobiology of Serotonin
ISBN 978-0-12-374634-4

DOI: 10.1016/B978-0-12-374634-4.00013-7

nerve terminals, is also proposed (Piñeyro *et al.*, 1995a; Stamford *et al.*, 2000; Thomas *et al.*, 2006).

5-HT$_{1A}$ autoreceptors

In vivo electrophysiological studies in the 1970s found that systemically and locally applied LSD inhibited 5-HT cell firing in the rat DRN (Aghajanian *et al.*, 1972; Haigler and Aghajanian, 1974). Subsequent detailed pharmacological analysis of this phenomenon identified the role of 5-HT$_{1A}$ receptors. For example, electrophysiological studies found that a wide range of 5-HT$_{1A}$ receptor agonists (including 8-OH-DPAT) had the same effect, and that this action was blocked by non-selective (spiperone) and selective (WAY 100635) 5-HT$_{1A}$ receptor antagonists (Sprouse and Aghajanian, 1988; Craven *et al.*, 1994; Corradetti *et al.*, 1996). Partial 5-HT$_{1A}$ agonists such as NAN-190 and SDZ 216,525 were also found to inhibit 5-HT cell firing via 5-HT$_{1A}$ receptor activation (Lanfumey *et al.*, 1993; Fornal *et al.*, 1994).

These electrophysiological data were strongly supported by findings from *in vivo* microdialysis measurements of extracellular 5-HT. For example, such studies demonstrated that administration of 5-HT$_{1A}$ agonists (8-OH-DPAT) or partial agonists (e.g., NAN-190, SDZ 216,525) decreased extracellular 5-HT in rat forebrain regions via 5-HT$_{1A}$ receptor activation (Sharp *et al.*, 1989a, 1996; Sharp and Hjorth, 1990). More recent electrophysiological and microdialysis experiments using 5-HT$_{1A}$ knockout mice confirm the pharmacological identity of the 5-HT$_{1A}$ autoreceptor (Knobelman *et al.*, 2001; Gross *et al.*, 2002; Guilloux *et al.*, 2006).

Curiously, the non-selective 5-HT receptor antagonists methiothepin, metergoline and methysergide were found to inhibit 5-HT cell firing (Haigler and Aghajanian, 1974) and release (Sharp *et al.*, 1989b), which may be due to 5-HT$_{1A}$ receptor activation, although α_1-adrenoceptor or indeed partial 5-HT$_2$ receptor agonist activity may be contributing factors (see later). In fact, it was only very recently that the inhibitory effect of LSD on 5-HT cell firing was shown to be mediated by 5-HT$_{1A}$ receptors and not other 5-HT receptor subtypes subsequently found to inhibit 5-HT cell firing (Quérée *et al.*, 2009).

Findings regarding 5-HT$_{1A}$ receptor localization support electrophysiological evidence for the presence of 5-HT$_{1A}$ receptors on midbrain 5-HT neurons. For example, radioligand binding and immunocytochemical studies in rodent and human tissue demonstrated the presence of high densities of 5-HT$_{1A}$ receptors localized in the DRN (Verge *et al.*, 1986; Sotelo *et al.*, 1990; Castro *et al.*, 2000). Moreover, 5-HT$_{1A}$ receptor immunolabeling coexisted with 5-HT neuron markers, and 5-HT$_{1A}$ binding sites and mRNA in the DRN were diminished by 5-HT neuron lesions (Miquel *et al.*, 1992; Radja *et al.*, 1992; Kia *et al.*, 1996).

Although the inhibition of 5-HT cell firing *in vivo* by 5-HT$_{1A}$ agonist administration was thought to be mediated via 5-HT$_{1A}$ receptors located on midbrain 5-HT neurons, subsequent data raised the possibility of a contribution from postsynaptic 5-HT$_{1A}$ receptors in the forebrain (see later).

5-HT$_{1B}$ autoreceptors

There is convincing evidence that the 5-HT$_{1B}$ receptor functions as a 5-HT autoreceptor at the nerve terminal. For example, *in vitro* brain slice studies demonstrate a strong correlation between the potency with which 5-HT receptor agonists and antagonists inhibit and enhance (respectively) release of preloaded radiolabeled 5-HT, and their affinity for the rat 5-HT$_{1B}$ binding site (Middlemiss, 1984; Engel *et al.*, 1986; Limberger *et al.*, 1991; Buhlen *et al.*, 1996). Work using the same model demonstrated the presence of the 5-HT$_{1B}$ autoreceptor (5-HT$_{1D\beta}$ in old nomenclature) in human brain (Galzin *et al.*, 1992; Maura *et al.*, 1993; Schlicker *et al.*, 1997). Such *in vitro* findings are supported by the effects of selective 5-HT$_{1B}$ receptor agonists and antagonists on extracellular 5-HT in brain microdialysis studies (Sharp *et al.*, 1989a; Hjorth and Tao, 1991; Knobelman *et al.*, 2000). Moreover, both *in vivo* microdialysis (Trillat *et al.*, 1997; Knobelman *et al.*, 2001) and *in vitro* brain slice studies (Piñeyro *et al.*, 1995a; Rutz *et al.*, 2006) demonstrated that 5-HT$_{1B}$ receptor agonists do not inhibit the release of 5-HT in 5-HT$_{1B}$ receptor knockout mice. While the latter studies tend to rule out a significant role for the 5-HT$_{1D}$ receptor in feedback control, *in vitro* data suggest that 5-HT$_{1D}$ receptors inhibit 5-HT efflux specifically in the DRN (Piñeyro *et al.*, 1995b; Stamford *et al.*, 2000). However, the precise molecular and cellular basis of this mechanism remains obscure.

Findings on 5-HT$_{1B}$ receptor localization are fully consistent with a nerve terminal 5-HT$_{1B}$ autoreceptor (Sari, 2004). Specifically, in the midbrain raphe nuclei 5-HT$_{1B}$ receptor mRNA is present in 5-HT cell bodies (Doucet *et al.*, 1995), and much evidence suggests that 5-HT$_{1B}$ receptor protein is synthesized at the soma and then transported to the nerve terminals. For example, studies using cell cultures transfected with 5-HT$_{1B}$ receptor constructs demonstrated that the protein is preferentially addressed away from the soma to axons (Darmon *et al.*, 1998; Ghavami *et al.*, 1999). Moreover, ultra-structural studies found that 5-HT nerve terminals autographically labeled with radioactive 5-HT, co-localize 5-HT$_{1B}$ receptor immunoreactivity (Sari, 2004).

Antidepressant augmentation using 5-HT autoreceptor antagonists

A shortcoming of even the most modern antidepressants including selective 5-HT (serotonin) reuptake inhibitors

(SSRIs) is that they are slow to act, often poorly tolerated, and fail to produce a satisfactory response in a high proportion (>30 percent) of patients (Artigas *et al.*, 2006). Current pharmacological strategies aimed to augment the therapeutic effect of SSRIs include combination with a 5-HT_1 autoreceptor antagonist.

The rationale for these strategies is evidence that when administered acutely, SSRIs indirectly activate 5-HT_{1A} and 5-HT_{1B} autoreceptors, which limits the ability of these drugs to enhance 5-HT transmission. Evidence suggests that 5-HT autoreceptors desensitize after a course (weeks) of SSRI treatment (Blier *et al.*, 1988; Blier and Bouchard, 1994; Newman *et al.*, 2004) leading to an increase in 5-HT function, which triggers the critical downstream events (e.g., increase synaptic plasticity, etc.) that are thought to underpin the relief of depression. The idea is that the use of autoreceptor antagonists would circumvent the delay in time needed for the desensitization process to come about.

In support of this thinking, 5-HT_{1A} receptor antagonists were shown to prevent the inhibitory effect of the SSRIs on 5-HT cell firing in different species (Gartside *et al.*, 1995; Bjorvatn *et al.*, 2000). Moreover, 5-HT_{1A} and 5-HT_{1B} receptor antagonists were found to augment the effect of SSRIs on 5-HT function. For instance, microdialysis studies showed that 5-HT_{1A} receptor antagonists facilitated the effect of SSRIs (as well as monoamine oxidase inhibitors and certain tricyclic antidepressants) on extracellular 5-HT (Gartside *et al.*, 1995; Gobert *et al.*, 1997; Invernizzi *et al.*, 1997). Also, 5-HT_{1B} antagonists augmented SSRI effects in some (but not all) microdialysis experiments (Rollema *et al.*, 1996; Gobert *et al.*, 1997), and potentiated the effects of SSRI/5-HT_{1A} receptor antagonist combinations (Gobert *et al.*, 1997; Sharp *et al.*, 1997; Dawson and Nguyen, 2000). More recently, co-administration of a 5-HT_{1A} receptor antagonist with a SSRI resulted in enhanced responses in some behavioral models (e.g., Duxon *et al.*, 2000; Muraki *et al.*, 2008) but not in others (Cryan *et al.*, 1999), and enhanced 5-HT mediated expression of activity-dependent genes (Castro *et al.*, 2003; Tordera *et al.*, 2003).

Such studies have lead to the clinical investigation of the utility of the non-selective 5-HT_{1A} ligand pindolol as an SSRI augmenting agent. Some trials found that pindolol/SSRI combinations were beneficial, but data are complicated by a number of issues, including differences in the demographic characteristics of the patients studied (Ballesteros and Callado, 2004; Segrave and Nathan, 2005). A recent meta-analysis of 11 randomized trials showed a greater response with pindolol augmentation versus placebo and an overall beneficial clinical effect, most clearly up to 4 weeks of treatment (Whale *et al.*, 2008). Unfortunately, the progression of such studies is hampered by the non-selectivity and partial agonist properties of pindolol (Clifford *et al.*, 1998), and the difficulty of finding a tolerable dose that achieves high levels of 5-HT_{1A} receptor occupancy (Rabiner *et al.*, 2000).

Current research is focused on the development of selective 5-HT_{1A} and 5-HT_{1B} receptor antagonists and inverse agonists, and 'dual' and 'triple' action drugs which have combined 5-HT transporter/5-HT_{1A}/5-HT_{1B} blocking properties (Millan, 2009). Preliminary data on a number of 'dual-activity' 5-HT transporter/5-HT_{1A} receptor ligands have been reported, including vilazodone (Heinrich *et al.*, 2004; Hughes *et al.*, 2005), VN2222 (Romero *et al.*, 2003), BMS-296859 (Taber *et al.*, 2003), WAY-163426 (Hornby *et al.*, 2004), Lu 36-274 (Mørk and Hogg, 2003) and 4-aryl-pyrido[1,2-c]pyrimidine (Herold *et al.*, 2009). Whilst analysis of the actions of these agents remains incomplete, both vilazodone and VN2222 appear to demonstrate evidence of partial 5-HT_{1A} agonist properties. Recently, data on the first reported triple action SSRI/5-HT_{1A}/5-HT_{1B} antagonist, SB-649915-B were described (Hughes *et al.*, 2007). In electrophysiological and microdialysis experiments, SB-649915 did not alter 5-HT cell firing *per se* but blocked the inhibitory effect of 8-OH-DPAT, and caused a greater increase of extracellular 5-HT in rat cortex than an SSRI (Figure 1). Importantly, in an anxiety model SB-649915 was anxiolytic, with a faster onset of effect compared to an SSRI (Starr *et al.*, 2007). Clearly, it is of great interest to see how drugs of this class fare in clinical trials.

Physiological and pathophysiological effects of 5-HT autoreceptors

Earlier studies using animal models link the decrease in 5-HT transmission evoked by 5-HT_1 autoreceptor activation to a range of behavioral and physiological effects, including anxiolysis, hypolocomotion, hyperphagia, and hypothermia (De Vry, 1995; Barnes and Sharp, 1999). More recently, 5-HT_1 autoreceptor activation and the consequent inhibition of 5-HT cell firing and release has been associated with reduced aggression (de Boer and Koolhaas, 2005) and enhancement of psychostimulant-induced hyperactivity and sensitization (Carey *et al.*, 2008).

5-HT_1 autoreceptors are implicated in both the cause and treatment of a variety of psychiatric illnesses, but in particular depression and other stress-related disorders. For example, abnormal 5-HT_{1A} autoreceptor function and consequent changes in 5-HT release are thought to contribute to maladaptive responses to stress in animal models of depression as well as patients with unipolar or bipolar depression (Lanfumey *et al.*, 1999; Lesch and Gutknecht, 2004; Cowen, 2008; Sullivan *et al.*, 2009). Moreover, in human volunteers, natural variation in

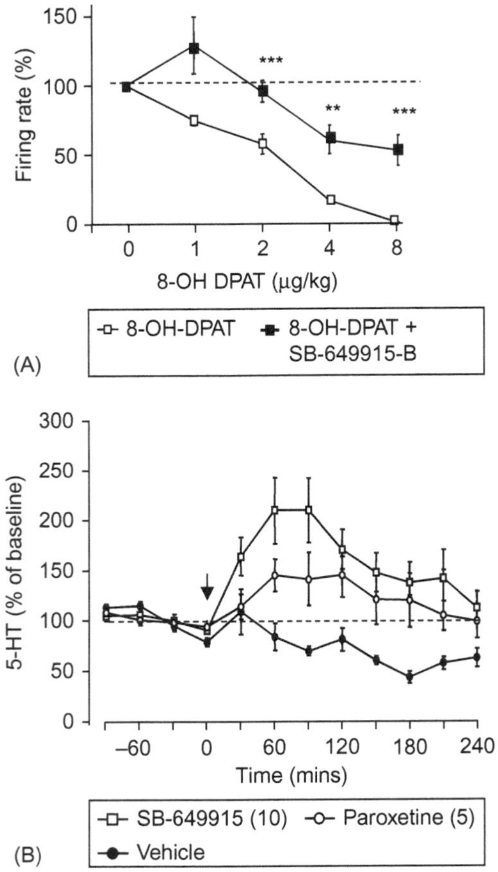

Figure 1 Effects of the SSRI/5-HT_{1A}/5-HT_{1B} receptor antagonist SB-649915-B on (A) 5-HT cell firing in the DRN and (B) cortical extracellular 5-HT, in the rat. In (A), note the blockade of the inhibitory effect of the 5-HT_{1A} agonist 8-OH-DPAT by SB-649915-B (3 mg/kg i.v.). $^{**}P < 0.01$, $^{***}P < 0.001$ versus vehicle. In (B), both the SSRI paroxetine and SB-649915-B have a significant effect of 5-HT compared to controls (doses are in mg/kg as indicated). Data taken in part from Hughes *et al.* (2007). Mean ± SEM values shown (n = 6–8 per group).

levels of 5-HT_{1A} autoreceptor density, possibly due to polymorphic variation in the 5-HT_{1A} receptor gene, have been associated with neuroimaging correlates of anxiety (Fisher *et al.*, 2006; Fakra *et al.*, 2009).

In an exciting recent development, altered 5-HT homeostasis brought about through abnormalities in 5-HT_{1A} autoreceptor have been linked to the catastrophic autonomic sequelae underlying sudden infant death (Audero *et al.*, 2008). Specifically, it was found that mice overexpressing the 5-HT_{1A} autoreceptor died early in life due to failure of necessary autonomic changes likely to compensate for changes in environment and stress.

Given that 5-HT_{1A} receptor activation has such powerful effects on 5-HT transmission, and the above evidence that changes in 5-HT_{1A} autoreceptor function can have important pathophysiological outcomes, it is surprising that evidence for the presence of 5-HT_{1A} autoreceptor tone in *in vivo* models is controversial. For example, reports on the effects of 5-HT_{1A} receptor antagonists administered systemically alone on 5-HT cell firing *in vivo* are inconsistent, although slight increases have been detected in anesthetized animals in some studies (Gartside *et al.*, 1995; Hajós *et al.*, 2001). Similarly, in *in vivo* microdialysis studies, neither selective 5-HT_{1A} nor 5-HT_{1B} receptor antagonists by themselves consistently increased extracellular 5-HT in brain regions of either anesthetized or awake animals (Gobert *et al.*, 1997; Sharp *et al.*, 1997). In support of these findings, it is reported that brain extracellular levels of 5-HT in brain regions of 5-HT_{1A} and 5-HT_{1B} receptor knockout mice are not different from those in wild-type controls (He *et al.*, 2001; Gardier *et al.*, 2003; Guilloux *et al.*, 2006).

This said, the 5-HT_{1A} receptor antagonist WAY 100635 had a clear-cut stimulatory effect on 5-HT cell firing in cats in the active-awake state (Fornal *et al.*, 1996). Moreover, the latter drug also increased 5-HT cell firing in anesthetized rats with noradrenaline depletions, suggesting that 5-HT tone at the 5-HT_{1A} autoreceptor may be masked by modulatory influences of noradrenaline (Haddjeri *et al.*, 2004). Thus, these data can be taken as evidence that 5-HT_{1A} autoreceptors are under strong physiological tone *in vivo*, at least in some conditions.

Postsynaptic 5-HT feedback control mechanisms

Whilst the importance of 5-HT_1 autoreceptors is clear, emerging findings indicate additional complexity of 5-HT feedback control, specifically in the form of feedback mechanisms involving postsynaptic 5-HT receptors (i.e., 5-HT receptors located on non-5-HT neurons). This idea derives in part from *in vivo* electrophysiological studies showing that not only 5-HT_{1A} receptor agonists but also agonists for a range of other 5-HT receptor subtypes alter 5-HT cell firing and release (Figure 2; Table 1).

Feedback involving postsynaptic 5-HT_{1A} receptors

Evidence from animal studies indicates that 5-HT_{1A} receptor agonist-induced inhibition of 5-HT cell firing can be dissociated from an action on somatodendritic 5-HT_{1A} autoreceptors. For example, transection of the rostral forebrain attenuated the inhibition of 5-HT cell firing induced by (systemically administered) 5-HT_{1A} receptor agonists even though the sensitivity of somatodendritic 5-HT_{1A} receptors in the DRN was not changed (Ceci *et al.*, 1994; Hajós *et al.*, 1999). Also, local injection of a

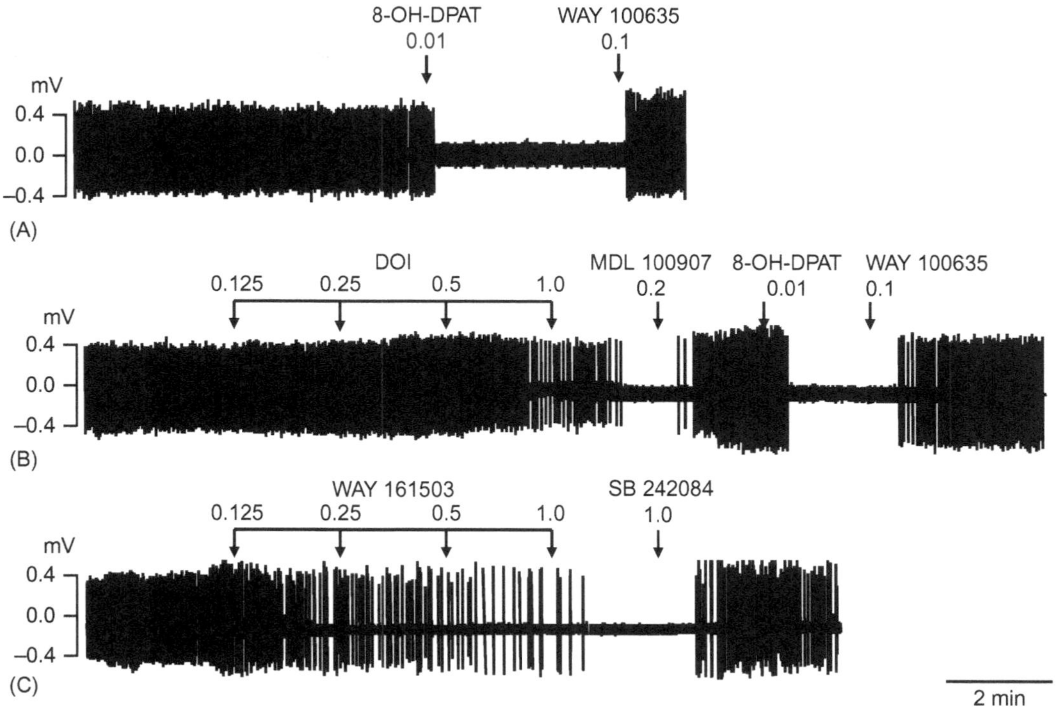

Figure 2 Electrophysiological evidence for the presence of 5-HT_{1A}, 5-HT_{2A} and 5-HT_{2C} feedback control of midbrain 5-HT neurons. Traces are spike trains from recordings of individual 5-HT neurons in the DRN of anesthetized rats, injected systemically with specific 5-HT receptor agonists. (A) Inhibitory effect of the 5-HT_{1A} receptor agonist 8-OH-DPAT and reversal by the 5-HT_{1A} receptor antagonist WAY 100635. (B) Inhibitory effect of the 5-HT_2 receptor agonist DOI and reversal by the 5-HT_{2A} receptor antagonist MDL 100907. Note that the same neuron is also inhibited by 8-OH-DPAT, providing evidence for control of this neuron by both 5-HT_{2A} and 5-HT_{1A} receptors. (C) Inhibitory effect of the 5-HT_{2C} receptor agonist WAY 161503 and reversal by the 5-HT_{2C} receptor antagonist SB 242084. Drug administration (mg/kg i.v.) as indicated by arrows.

5-HT_{1A} receptor agonist into regions outside of the DRN decreased 5-HT cell firing and 5-HT release (Hajós *et al.*, 1999; Celada *et al.*, 2001). Importantly, 5-HT_{1A} agonist-induced inhibition of 5-HT cell firing and release were not markedly influenced by localized pharmacological blockade or inactivation of 5-HT_{1A} receptors in the DRN (Romero *et al.*, 1994; Martín-Ruiz and Ugedo, 2001). Collectively, these findings suggest that a population of 5-HT_{1A} receptors outside the DRN contributes to 5-HT_{1A} agonist-induced inhibition of 5-HT cell firing and release – i.e., that both pre- and postsynaptic 5-HT_{1A} receptors are involved.

Feedback involving postsynaptic 5-HT_2 receptors

Accumulating evidence suggests that postsynaptic 5-HT_2 receptors also regulate 5-HT cell firing. In particular, when administered systemically, phenethylamine derivatives, such as DOM (1-(2,5-di-methoxy-4-methylphenyl)-2-aminopropane) and DOI (2,5-dimethoxy-4-iodo-amphetamine), which have 5-HT_2 agonist properties in a range of models (Barnes and Sharp, 1999), decreased both 5-HT cell firing and brain extracellular 5-HT (Wright *et al.*, 1990; Boothman *et al.*, 2003). Recent experiments confirm the involvement of 5-HT_2 receptors in these effects. For example, the inhibition of 5-HT cell firing induced by DOI was blocked by the 5-HT_2 receptor antagonist, ritanserin, as well as the 5-HT_{2A} receptor antagonist MDL 100907, suggesting the involvement of the 5-HT_{2A} receptor subtype (Figure 2; Martín-Ruiz *et al.*, 2001; Boothman *et al.*, 2003).

Interestingly, the 5-$HT_{2B/C}$ receptor antagonist SB 206553 partially attenuated DOI-induced inhibition of 5-HT cell firing (Boothman *et al.*, 2003), implicating a role for 5-$HT_{2B/C}$ receptors. This was confirmed by recent studies demonstrating that the putative 5-HT_{2C} receptor agonists WAY 161503, Ro 600175 and mCPP inhibited 5-HT cell firing, and in each case the effect was reversed by the selective 5-HT_{2C} receptor antagonist SB 242084 (Boothman *et al.*, 2006b; Quérée and Sharp, 2006; Quérée *et al.*, 2009).

Collectively, these data support a role for both 5-HT_{2A} and 5-HT_{2C} receptor subtypes in the inhibitory feedback control of 5-HT cell firing. Since neither 5-HT_{2A} nor 5-HT_{2C} receptors are expressed by 5-HT neurons

Table 1 Some evidence implicating specific postsynaptic 5-HT receptor subtypes in the feedback control of midbrain 5-HT neurons

5-HT receptor subtype	Evidence	Comment	Reference
5-HT_{1A}	See main text		
5-HT_{1B}	Systemic administration of 5-HT_{1B} agonists (RU 24969, CP 94253) increases the firing of 5-HT neurons in rat DRN *in vivo*.	Effect blocked by 5-HT_{1B} antagonist (GR 127935). Data do not exclude 'disinhibition' of 5-HT neurones through activation of presynaptic 5HT_{1B} autoreceptor.	Evrard *et al.*, 1999; Adell *et al.*, 2001
5-HT_{2A}	See main text		
5-HT_{2C}	See main text		
5-HT_3	Local application of 5-HT_3 agonists (2-methyl-5-HT) increased 5-HT efflux in rat and guinea pig brain regions (hippocampus, cortex) *in vitro* and *in vivo*.	Effect blocked by 5-HT_3 antagonist (e.g. MDL 72222).	
5-HT_4	See main text		
5-HT_{5A}	Local application of a 5-HT_{5A} antagonist (SB-699551-A) attenuated the 5-CT-induced depression in 5-HT cell firing in guinea pig DRN *in vitro*.	Effect may be mediated by 5-HT_{5A} receptors located presynaptically on 5-HT neurons.	Thomas *et al.*, 2005
5-HT_6	Systemic administration of 5-HT_6 agonist (WAY-181187) decreased 5-HT efflux in rat cortex *in vivo*.	Effect blocked by 5-HT_6 antagonist (SB-271046). Effect region-specific (cortex but not 5 other regions).	Schechter *et al.*, 2008
5-HT_7	Local application of 5-HT_7 antagonist (SB269970-A) inhibited 5-HT efflux in guinea pig DRN *in vitro*.		Roberts *et al.*, 2004

(Pazos *et al.*, 1985; Cornea-Hebert *et al.*, 1999), this form of feedback control likely involves postsynaptic mechanisms.

Postsynaptic feedback – beyond 5-HT_{1A} and 5-HT_2 receptor subtypes

In addition to 5-HT_{1A}, 5-HT_{2A} and 5-HT_{2C} receptors, other 5-HT receptor subtypes are implicated in post-synaptic 5-HT feedback control, and these include 5-HT_3, 5-HT_4, 5-HT_{5A}, 5-HT_6 and 5-HT_7 (Table 1). The most investigated candidate is the 5-HT_4 receptor, which, interestingly, may mediate positive feedback control of 5-HT neurons. Specifically, Debonnel and colleagues (Lucas and Debonnel, 2002; Lucas *et al.*, 2005) reported that systemic administration of various 5-HT_4 receptor agonists, including RS 67333, increased 5-HT cell firing in the rat DRN. Moreover, this effect was attenuated by the 5-HT_4 receptor antagonist GR 125487, which by itself decreased 5-HT cell firing (Lucas and Debonnel, 2002). In keeping with these data, basal DRN 5-HT cell firing was reduced in 5-HT_4 receptor knockout mice, and increased in mice with virally-induced forebrain 5-HT_4 receptor overexpression (Lucas *et al.*, 2005; Conductier *et al.*, 2006).

These electrophysiological data accord with earlier microdialysis experiments (Ge and Barnes, 1996) showing that 5-HT_4 receptor agonists increased, whereas 5-HT_4 receptor antagonists decreased, extracellular 5-HT in rat hippocampus. Collectively, these data suggest that 5-HT_4 receptors exert a tonic, excitatory effect on 5-HT cell firing and release. This effect is likely mediated by 5-HT_4 receptors located postsynaptically, since 5-HT_4 receptors are abundant in the forebrain but lack measurable expression in the midbrain (Waeber *et al.*, 1994; Vilaro *et al.*, 1996).

Neural pathways involved in postsynaptic 5-HT feedback

The neural substrates of postsynaptic 5-HT feedback mechanisms are likely to lie amongst the many inputs to midbrain 5-HT neurons. Inputs that are already emerging as important include projections from the medial prefrontal cortex (mPFC) and lateral habenula nucleus (LHb) to the DRN, as well as DRN GABA neurons.

The mPFC–raphe pathway comprises a cortical glutamate projection that preferentially targets DRN GABA neurons to inhibit 5-HT cell firing (Hajós *et al.*, 1998; Varga *et al.*, 2001; Jankowski and Sesack, 2004). A minor glutamate projection from the mPFC that directly excites 5-HT cell firing in the DRN is also reported (Hajós *et al.*,

1998; Jankowski and Sesack, 2004). The LHb–DRN pathway also comprises an excitatory (principally glutamate) projection that targets DRN GABA neurons but there is also evidence of a direct inhibitory LHb–raphe GABA projection (Ferraro *et al.*, 1996; Varga *et al.*, 2003).

Electrophysiological data suggest that both mPFC–DRN and LHb–DRN projections influence large numbers of 5-HT neurons, probably through targeting the extensive network of GABA neurons in the DRN (Hajós *et al.*, 1998; Varga *et al.*, 2003). Indeed, morphological analysis of juxtacellular-labeled, single DRN GABA neurons revealed dendrites and varicose axon collaterals that spanned across DRN sub-regions and even across the midline to the contralateral side, suggesting an interaction with potentially large parts of the DRN 5-HT system (Allers and Sharp, 2003).

mPFC-DRN pathways and postsynaptic 5-HT feedback

The mPFC–DRN pathway is a putative neural substrate of postsynaptic 5-HT_{1A} receptor feedback. The mPFC is rich in 5-HT_{1A} receptors, and local application of 5-HT_{1A} receptor agonists into this region decreased 5-HT cell firing in the DRN (Hajós *et al.*, 1999). Also, either lesion of the mPFC or intra-mPFC administration of a 5-HT_{1A} receptor antagonist attenuated the inhibition of 5-HT cell firing and/or release induced by systemic 5-HT_{1A} receptor agonist administration (Casanovas *et al.*, 1999; Hajós *et al.*, 1999). In support of these findings, 5-HT_{1A} receptor agonists increased mPFC and DRN GABA neuron activity in electrophysiological and *c-fos* (a molecular marker of neural activity) studies (Borsini *et al.*, 1995; Hajós *et al.*, 1999; Raley *et al.*, 2004). Although 5-HT_{1A} receptors are inhibitory, it is possible that activation of 5-HT_{1A} receptors present on GABA interneurons in the mPFC (Aznar *et al.*, 2003; Santana *et al.*, 2004) results in disinhibition of the output glutamate neurons. Therefore, a potential scenario is that 5-HT_{1A} receptors in the mPFC excite the mPFC–raphe glutamate projection to activate DRN GABA neurons, which then inhibit 5-HT cell firing (Figure 3).

Interestingly, the mPFC–DRN pathway is a putative substrate of 5-HT_{2A} receptor feedback. For instance, the mPFC is rich in (excitatory) 5-HT_{2A} receptors located on glutamate neurons (Cornea-Hebert *et al.*, 1999), and *c-fos* expression studies demonstrate that 5-HT_{2A} receptor-mediated activation of both mPFC neurons (Raley *et al.*, 2004; Boothman and Sharp, 2005) and excitotoxic lesions of the mPFC attenuate the inhibitory effect of DOI on 5-HT cell firing (this laboratory, in preparation). DOI also evoked a 5-HT_{2A} receptor-mediated increase in *c-fos* expression in DRN GABA neurons (Boothman and Sharp, 2005). Thus, 5-HT_{2A} receptor activation may stimulate the mPFC–raphe glutamate projection to excite DRN GABA neurons, which then inhibit 5-HT cell firing (Figure 4).

Interestingly, evidence indicates that postsynaptic 5-HT_4 receptor feedback may also operate via the mPFC–DRN projection. Specifically, intra-mPFC injection of a viral construct designed to over-express 5-HT_4 receptors increased 5-HT cell firing in the DRN (Lucas *et al.*, 2005). Thus, activation of (excitatory) 5-HT_4 receptors in the mPFC may stimulate the excitatory mPFC–DRN projection to directly excite 5-HT neurons (Figure 3).

LHb–DRN pathways and postsynaptic 5-HT feedback

The LHb–DRN rather than the mPFC–DRN pathway is a putative substrate of the 5-HT_{2C} feedback mechanism. The LHb is abundant in 5-HT_{2C} receptor mRNA

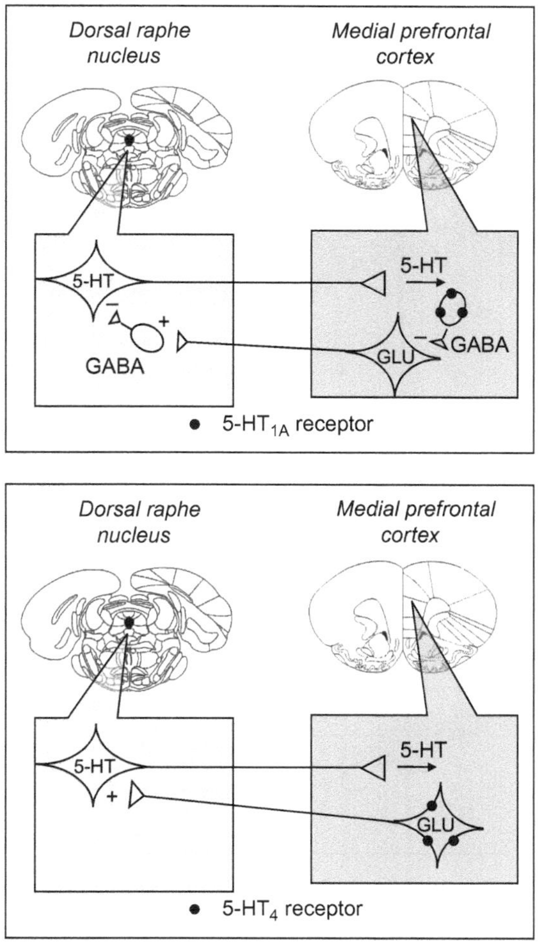

Figure 3 Diagram illustrating the theoretical neural pathways underlying feedback control of midbrain 5-HT neurons by postsynaptic 5-HT_{1A} and 5-HT_4 receptors located on non-5-HT neurons in the mPFC. See text for details.

(Pompeiano *et al.*, 1994), and systemic administration of 5-HT_{2C} receptor agonists including WAY 161503 elicited a striking, 5-HT_{2C} receptor-mediated increase in *c-fos* expression in the LHb (Quérée *et al*, unpublished data). 5-HT_{2C} receptor agonist administration also increased *c-fos* expression in DRN GABA neurons (Boothman *et al.*, 2006b; Quérée *et al.*, 2009). Importantly, in preliminary experiments, excitotoxic lesions of the LHb markedly reduce 5-HT_{2C} agonist-induced inhibition of 5-HT cell firing (Quérée *et al.*, unpublished observation). Thus, 5-HT_{2C} receptor stimulation in the LHb may activate the LHb–DRN pathway to trigger the activity of DRN GABA neurons, which inhibits the firing of 5-HT neurons (Figure 4).

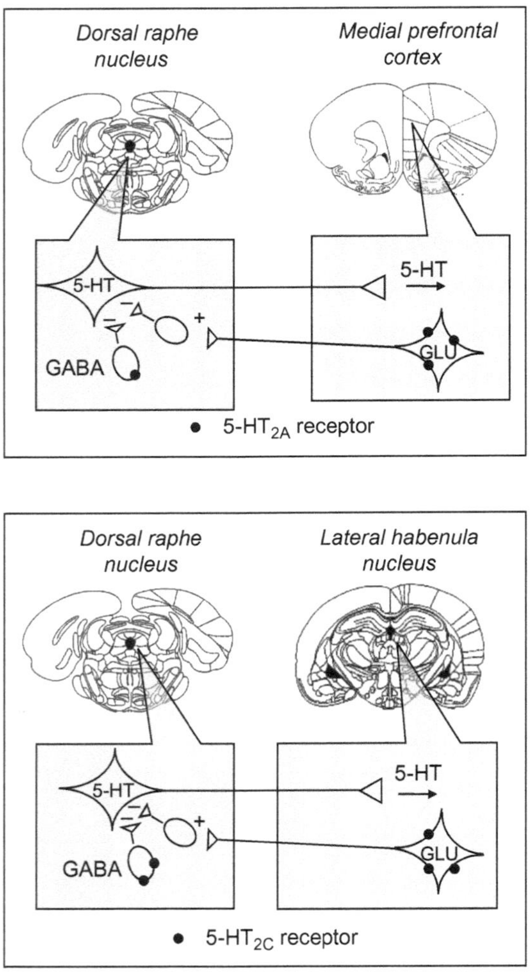

Figure 4 Diagram illustrating the theoretical neural pathways underlying feedback control of midbrain 5-HT neurons by postsynaptic 5-HT_{2A} and 5-HT_{2C} receptors located on non-5-HT neurons in the mPFC or LHb. See text for details.

DRN GABA neurons and postsynaptic 5-HT feedback

DRN GABA neurons themselves may form a local postsynaptic feedback circuit that inhibits the firing of neighboring 5-HT neurons. Anatomical studies demonstrate reciprocal synaptic interactions between 5-HT and GABA neurons in the DRN, and in *in vitro* electrophysiological locally applied 5-HT evoked GABA-mediated inhibitory currents in DRN 5-HT neurons, this effect involving both 5-HT_{2A} and 5-HT_{2C} receptor subtypes (Liu *et al.*, 2000). In accord with these data, 5-HT_2 receptor agonist administration evoked a 5-HT_{2A} receptor-mediated increase in *c-fos* expression in DRN GABA neurons (Figure 5; Boothman and Sharp, 2005; Boothman *et al.*, 2006b), and 5-HT_{2C} receptor agonists have a similar effect (Boothman *et al.*, 2006b; Quérée *et al.*, 2009). Whilst activated mPFC/LHb inputs might contribute to the latter effects, both 5-HT_{2A} and 5-HT_{2C} receptors directly activate DRN GABA neurons *in vitro* (Liu *et al.*, 2000), and immunocytochemical data suggest that DRN GABA neurons express 5-HT_{2C} receptors (Serrats *et al.*, 2005; Boothman *et al.*, 2006b).

Possible physiological roles of postsynaptic 5-HT feedback

To date, the evidence for the presence of postsynaptic 5-HT feedback derives largely from the use of 5-HT agonist probes, and it is not known whether these systems have important roles in physiology and pathophysiology, and if so which roles they play. It certainly appears to be the case that postsynaptic 5-HT feedback control is a common property of 5-HT neurons. In recent studies, almost all 5-HT neurons tested were responsive to $5\text{-HT}_{2A/C}$ receptor agonist administration (Boothman *et al.*, 2003; Boothman *et al.*, 2006b). Also, 5-HT_4 receptor agonists were found to increase the firing rate in approximately 50 percent of DRN 5-HT neurons examined (Lucas and Debonnel, 2002), which is similar to the estimated number of DRN 5-HT neurons subject to feedback control by postsynaptic 5-HT_{1A} receptors (Martín-Ruiz and Ugedo, 2001). Thus, significant numbers of 5-HT neurons appear responsive to postsynaptic feedback control. Moreover, since numerous individual 5-HT neurons were found to respond to both 5-HT_{1A} and $5\text{-HT}_{2A/C}$ receptor agonists (Boothman *et al.*, 2003; Boothman *et al.*, 2006b), it is likely that single 5-HT neurons are influenced by more than one feedback pathway.

Clues regarding the physiological and pathophysiological importance of 5-HT_{1A} autoreceptors have arisen in part from studies of the behavioral effects of 5-HT_{1A}

receptor agonists and antagonists, aided by evidence that these receptors (at least in some conditions) are tonically active (see above). In the case of the postsynaptic 5-HT_2 feedback systems, there are many behavioral effects associated with 5-HT_{2A} and 5-HT_{2C} receptor activation (Barnes and Sharp, 1999), but the possibility that these effects link in some way to 5-HT feedback mechanisms has not yet been considered. Data so far suggest that the 5-$HT_{2A/C}$ receptor feedback mechanisms are not tonically active. That is, in electrophysiological and microdialysis studies to date 5-$HT_{2A/C}$ receptor antagonists did not increase 5-HT cell firing or release. However, demonstration of tone at 5-HT_{1A} autoreceptors has been problematic, and required specific experimental conditions (see above). It is theoretically possible that tone on any one feedback mechanism may be difficult to reveal using antagonists, because blockade of one mechanism may result in compensatory changes in another. Against this argument, evidence suggests that the 5-HT_4 feedback is under the tone of endogenous 5-HT, since 5-HT_4 receptor antagonists alone decreased 5-HT cell firing (Lucas and Debonnel, 2002; Lucas *et al.*, 2005).

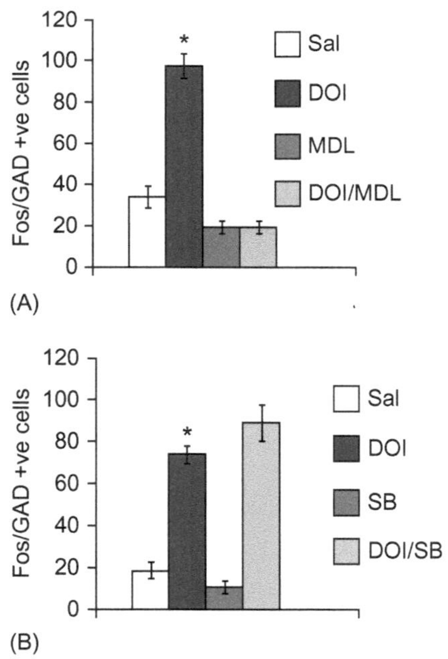

Figure 5 Effect of the 5-HT_2 agonist DOI on Fos immunoreactivity in DRN neurons immunopositive for the GABA marker, glutamate decarboxylase (GAD). In (A), DOI (2 mg/kg i.p.) was administered in combination with the 5-HT_{2A} selective antagonist MDL 100907 (0.3 mg/kg i.p.). In (B), DOI (2 mg/kg i.p.) was administered in combination with the 5-HT_{2C} selective antagonist SB 242084 (0.5 mg/kg i.p.). Note the increase in Fos/GAD double-labeled neurons by DOI and selective blockade by MDL 100907. *$P < 0.001$ versus vehicle controls. Data taken from Raley *et al.* (2004).

From a physiological perspective, it may be relevant that the mPFC–DRN and LHb–DRN pathways are emerging as key neuroanatomical substrates for postsynaptic 5-HT feedback control, because both pathways are thought to play important roles in mediating the impact of stress. In particular, it is proposed that the mPFC–DRN pathway protects DRN 5-HT neurons from excessive activation by prolonged stress that results in behaviors characteristic of depression (learned helplessness) (Amat *et al.*, 2005). Thus, in animals, chemical inactivation of the mPFC–DRN pathway led to amplification of the effect of stress, such that controllable stress assumed the deleterious impact of uncontrollable stress (Amat *et al.*, 2005). In this respect, there is evidence that the LHb–DRN pathway may play a similar role to the mPFC–DRN pathway (Amat *et al.*, 2001).

Therefore, one possible physiological function of certain postsynaptic 5-HT feedback mechanisms is to sense and control stress-evoked changes in 5-HT availability. Thus, postsynaptic feedback systems utilizing 5-HT receptors in the mPFC (5-HT_{1A}, 5-HT_{2A}, 5-HT_4) and LHb (5-HT_{2C}) might act in unison with presynaptic 5-HT autoreceptors (5-HT_{1A}, 5-HT_{1B}) to provide the necessary balancing influence of inhibitory and excitatory feedback control on 5-HT neurons and mitigate against adverse environmental stimuli such as excessive stress. Recently genetically engineered mice with forebrain and even cortex-specific modulations of 5-HT receptor expression have been developed (Gross *et al.*, 2002; Weisstaub *et al.*, 2006). This technology may allow further testing of the importance of postsynaptic 5-HT receptor-mediated feedback in stress control mechanisms.

Novel therapeutic strategies exploiting postsynaptic feedback

As noted above, much effort has been invested in translating knowledge of 5-HT_1 autoreceptor pharmacology to the development of ligands for therapeutic use, particularly in the areas of depression and anxiety. In the same way, knowledge of the pharmacology of postsynaptic 5-HT feedback might open up interesting therapeutic possibilities. Indeed, recent preclinical data show that 5-HT_{2C} receptor antagonists enhance the effect of 5-HT reuptake inhibitors on 5-HT function (Cremers *et al.*, 2004), and 5-HT_{2A} receptor antagonists may also have this effect (Boothman *et al.*, 2006a). This 5-HT enhancing property of 5-HT_{2C} and 5-HT_{2A} receptor antagonists has been associated with antidepressant augmenting effects of these agents in animal models (Marek *et al.*, 2003; Cremers *et al.*, 2004). The latter observation is consistent with clinical evidence that drugs with 5-HT_2 receptor antagonist properties (e.g., risperidone, olanzapine) enhance the therapeutic effect of SSRIs in treatment-resistant depression (Marek *et al.*, 2003). Interestingly, 5-HT_4 receptor agonists were recently found to have antidepressant effects in preclinical models (Lucas *et al.*, 2007), which might link to the excitatory 5-HT_4 feedback mechanism causing increased 5-HT output.

The above data clearly advocate the further testing of antagonists that block 5-HT_1 and 5-HT_2 negative feedback mechanisms, and agonists that activate the 5-HT_4 positive feedback mechanism, in antidepressant combinations (or dual action drugs) in clinical trials. The possibility of synergistic effects of blocking multiple 5-HT feedback systems is another avenue of clinical (and preclinical) research, although the safety of this approach would need to be proven. Continued investigation of other 5-HT feedback mechanisms (5-HT_6, 5-HT_7) might also encourage the identification of novel pharmacological strategies for translating to humans.

Conclusion

Feedback control appears to be a key mechanism maintaining 5-HT neurons under homeostatic control. Evidence for feedback inhibition by presynaptic 5-HT_{1A} and 5-HT_{1B} autoreceptors is well established. These autoreceptors are implicated in the treatment of depression and anxiety, and form the basis of various 5-HT drug discovery efforts in these therapeutic areas. Recent findings have discovered that feedback control of 5-HT neurons also involves 5-HT receptors located on postsynaptic targets. These postsynaptic feedback mechanisms utilize multiple 5-HT receptor subtypes (5-HT_{1A}, 5-HT_{2A}, 5-HT_{2C} and 5-HT_4 receptors, although there may be more) and appear to operate via neural pathways that input to 5-HT neurons. In this respect, inputs from the mPFC and LHb are emerging as particularly important. The physiological role of these postsynaptic feedback mechanisms is unknown, but it is argued herein that they act in unison with 5-HT autoreceptors to sense and control extreme changes in 5-HT availability such as might be evoked by high levels of stress. New findings suggest that drugs targeting these feedback systems may have therapeutic utility in depression and other psychiatric disorders.

Acknowledgements

This work was supported by a grant from the European Community (FP6 Integrated Network; NEWMOOD, LMSH-CT-2004-503474), and studentships from Wellcome Trust, MRC and the Government of Jersey. The excellent scientific and technical contributions to this work of Laura Boothman, Josie Raley and Philip Quérée are gratefully acknowledged.

References

Adell, A., Celada, P. and Artigas, F. (2001) The role of 5-HT1B receptors in the regulation of serotonin cell firing and release in the rat brain. J. Neurochem., 79: 172–182.

Aghajanian, G.K., Haigler, H.J. and Bloom, F.E. (1972) Lysergic acid diethylamide and serotonin: direct actions on serotonin-containing neurons in rat brain. Life Sci. I, 11: 615–622.

Allers, K.A. and Sharp, T. (2003) Neurochemical and anatomical identification of fast- and slow-firing neurones in the rat dorsal raphe nucleus using juxtacellular labelling methods *in vivo*. Neuroscience, 122: 193–204.

Amat, J., Baratta, M.V., Paul, E., Bland, S.T., Watkins, L.R. and Maier, S.F. (2005) Medial prefrontal cortex determines how stressor controllability affects behavior and dorsal raphe nucleus. Nat. Neurosci., 8: 365–371.

Amat, J., Sparks, P.D., Matus-Amat, P., Griggs, J., Watkins, L.R. and Maier, S.F. (2001) The role of the habenular complex in the elevation of dorsal raphe nucleus serotonin and the changes in the behavioral responses produced by uncontrollable stress. Brain Res., 917: 118–126.

Artigas, F., Adell, A. and Celada, P. (2006) Pindolol augmentation of antidepressant response. Curr. Drug Targets, 7: 139–147.

Audero, E., Coppi, E., Mlinar, B., Rossetti, T., Caprioli, A., Banchaabouchi, M.A., Corradetti, R. and Gross, C. (2008) Sporadic autonomic dysregulation and death associated with excessive serotonin autoinhibition. Science, 321: 130–133.

Aznar, S., Qian, Z., Shah, R., Rahbek, B. and Knudsen, G.M. (2003) The 5-HT1A serotonin receptor is located on calbindin- and parvalbumin-containing neurons in the rat brain. Brain Res., 959: 58–67.

Ballesteros, J. and Callado, L.F. (2004) Effectiveness of pindolol plus serotonin uptake inhibitors in depression: a meta-analysis of early and late outcomes from randomised controlled trials. J. Affect. Disord., 79: 137–147.

Barnes, N.M. and Sharp, T. (1999) A review of central 5-HT receptors and their function. Neuropharmacology, 38: 1083–1152.

Bjorvatn, B., Fornal, C.A., Martin, F.J., Metzler, C.W. and Jacobs, B.L. (2000) Venlafaxine and its interaction with WAY 100635: effects on serotonergic unit activity and behavior in cats. Eur. J. Pharmacol., 404: 121–132.

Blier, P. and Bouchard, C. (1994) Modulation of 5-HT release in the guinea-pig brain following long-term administration of antidepressant drugs. Br. J. Pharmacol., 113: 485–495.

Blier, P., De Montigny, C. and Chaput, Y. (1988) Electrophysiological assessment of the effects of antidepressant treatments on the efficacy of 5-HT neurotransmission. Clin. Neuropharmacol., 11(Suppl. 2): S1–S10.

Boothman, L.J., Allers, K.A., Rasmussen, K. and Sharp, T. (2003) Evidence that central 5-HT2A and 5-HT2B/C receptors regulate 5-HT cell firing in the dorsal raphe nucleus of the anaesthetised rat. Br. J. Pharmacol., 139: 998–1004.

Boothman, L.J., Mitchell, S.N. and Sharp, T. (2006a) Investigation of the SSRI augmentation properties of 5-HT(2) receptor antagonists using *in vivo* microdialysis. Neuropharmacology, 50: 726–732.

Boothman, L., Raley, J., Denk, F., Hirani, E. and Sharp, T. (2006b) *In vivo* evidence that 5-HT(2C) receptors inhibit 5-HT neuronal activity via a GABAergic mechanism. Br. J. Pharmacol., 149: 861–869.

Boothman, L.J. and Sharp, T. (2005) A role for midbrain raphe gamma aminobutyric acid neurons in 5-hydroxytryptamine feedback control. Neuroreport, 16: 891–896.

Borsini, F., Ceci, A., Bietti, G. and Donetti, A. (1995) BIMT 17, a 5-HT1A receptor agonist/5-HT2A receptor antagonist, directly activates postsynaptic 5-HT inhibitory responses in the rat cerebral cortex. Naunyn Schmiedebergs Arch. Pharmacol., 352: 283–290.

Buhlen, M., Fink, K., Boing, C. and Gothert, M. (1996) Evidence for presynaptic location of inhibitory 5-HT1D beta-like autoreceptors in the guinea-pig brain cortex. Naunyn Schmiedebergs Arch. Pharmacol., 353: 281–289.

Carey, R.J., Huston, J.P. and Muller, C.P. (2008) Pharmacological inhibition of dopamine and serotonin activity blocks spontaneous and cocaine-activated behaviour. Prog. Brain Res., 172: 347–360.

Casanovas, J.M., Hervas, I. and Artigas, F. (1999) Postsynaptic 5-HT1A receptors control 5-HT release in the rat medial prefrontal cortex. Neuroreport, 10: 1441–1445.

Castro, M.E., Harrison, P.J., Pazos, A. and Sharp, T. (2000) Affinity of (+/−)-pindolol, (−)-penbutolol, and (−)-tertatolol for pre- and postsynaptic serotonin 5-HT(1A) receptors in human and rat brain. J. Neurochem., 75: 755–762.

Castro, E., Tordera, R.M., Hughes, Z.A., Pei, Q. and Sharp, T. (2003) Use of Arc expression as a molecular marker of increased postsynaptic 5-HT function after SSRI/5-HT1A receptor antagonist co-administration. J. Neurochem., 85: 1480–1487.

Ceci, A., Baschirotto, A. and Borsini, F. (1994) The inhibitory effect of 8-OH-DPAT on the firing activity of dorsal raphe serotoninergic neurons in rats is attenuated by lesion of the frontal cortex. Neuropharmacology, 33: 709–713.

Celada, P., Puig, M.V., Casanovas, J.M., Guillazo, G. and Artigas, F. (2001) Control of dorsal raphe serotonergic neurons by the medial prefrontal cortex: involvement of serotonin-1A, GABA(A), and glutamate receptors. J. Neurosci., 21: 9917–9929.

Clifford, E.M., Gartside, S.E., Umbers, V., Cowen, P.J., Hajós, M. and Sharp, T. (1998) Electrophysiological and neurochemical evidence that pindolol has agonist properties at the 5-HT1A autoreceptor *in vivo*. Br. J. Pharmacol., 124: 206–212.

Conductier, G., Dusticier, N., Lucas, G., Cote, F., Debonnel, G., Daszuta, A., Dumuis, A., Nieoullon, A., Hen, R., Bockaert, J. and Compan, V. (2006) Adaptive changes in serotonin neurons of the raphe nuclei in 5-HT(4) receptor knockout mouse. Eur. J. Neurosci., 24: 1053–1062.

Cornea-Hebert, V., Riad, M., Wu, C., Singh, S.K. and Descarries, L. (1999) Cellular and subcellular distribution of the serotonin 5-HT2A receptor in the central nervous system of adult rat. J. Comp. Neurol., 409: 187–209.

Corradetti, R., Pugliese, A.M., Le Poul, E., Laaris, N., Hamon, M. and Lanfumey, L. (1996) Electrophysiological effects of WAY 100635, a new 5-HT1A receptor antagonist, on dorsal raphe nucleus serotoninergic neurones and CA1 pyramidal cells *in vitro*. Acta Physiol. Hung, 84: 407–409.

Cowen, P.J. (2008) Serotonin and depression: pathophysiological mechanism or marketing myth? Trends Pharmacol. Sci., 29: 433–436.

Craven, R., Grahame-Smith, D. and Newberry, N. (1994) WAY-100635 and GR127935: effects on 5-hydroxytryptamine-containing neurones. Eur. J. Pharmacol., 271: R1–R3.

Cremers, T.I., Giorgetti, M., Bosker, F.J., Hogg, S., Arnt, J., Mork, A., Honig, G., Bogeso, K.P., Westerink, B.H., Den Boer, H., Wikstrom, H.V. and Tecott, L.H. (2004) Inactivation of 5-HT(2C) receptors potentiates consequences of serotonin reuptake blockade. Neuropsychopharmacology, 29: 1782–1789.

Cryan, J.F., Mcgrath, C., Leonard, B.E. and Norman, T.R. (1999) Onset of the effects of the 5-HT1A antagonist, WAY-100635, alone, and in combination with paroxetine, on olfactory bulbectomy and 8-OH-DPAT-induced changes in the rat. Pharmacol. Biochem. Behav., 63: 333–338.

Darmon, M., Langlois, X., Suffisseau, L., Fattaccini, C.M. and Hamon, M. (1998) Differential membrane targeting and pharmacological characterization of chimeras of rat serotonin 5-HT1A and 5-HT1B receptors expressed in epithelial LLC-PK1 cells. J. Neurochem., 71: 2294–2303.

Dawson, L.A. and Nguyen, H.Q. (2000) The role of 5-HT(1A) and 5-HT(1B/1D) receptors on the modulation of acute fluoxetine-induced changes in extracellular 5-HT: the mechanism of action of (+/−)pindolol. Neuropharmacology, 39: 1044–1052.

De Boer, S.F. and Koolhaas, J.M. (2005) 5-HT1A and 5-HT1B receptor agonists and aggression: a pharmacological challenge of the serotonin deficiency hypothesis. Eur. J. Pharmacol., 526: 125–139.

De Vry, J. (1995) 5-HT1A receptor agonists: recent developments and controversial issues. Psychopharmacology (Berl.), 121: 1–26.

Doucet, E., Pohl, M., Fattaccini, C.M., Adrien, J., Mestikawy, S.E. and Hamon, M. (1995) In situ hybridization evidence for the synthesis of 5-HT1B receptor in serotoninergic neurons of anterior raphe nuclei in the rat brain. Synapse, 19: 18–28.

Duxon, M.S., Starr, K.R. and Upton, N. (2000) Latency to paroxetine-induced anxiolysis in the rat is reduced by co-administration of the 5-HT(1A) receptor antagonist WAY100635. Br. J. Pharmacol., 130: 1713–1719.

Engel, G., Gothert, M., Hoyer, D., Schlicker, E. and Hillenbrand, K. (1986) Identity of inhibitory presynaptic 5-hydroxytryptamine (5-HT) autoreceptors in the rat brain cortex with 5-HT1B binding sites. Naunyn Schmiedebergs Arch. Pharmacol., 332: 1–7.

Evrard, A., Laporte, A.M., Chastanet, M., Hen, R., Hamon, M. and Adrien, J. (1999) 5-HT1A and 5-HT1B receptors control the firing of serotoninergic neurons in the dorsal raphe nucleus of the mouse: studies in 5-HT1B knockout mice. Eur. J. Neurosci., 11: 3823–3831.

Fakra, E., Hyde, L.W., Gorka, A., Fisher, P.M., Munoz, K.E., Kimak, M., Halder, I., Ferrell, R.E., Manuck, S.B. and Hariri, A.R. (2009) Effects of HTR1A C(-1019)G on amygdala reactivity and trait anxiety. Arch. Gen. Psychiatry, 66: 33–40.

Farnebo, L.O. and Hamberger, B. (1971) Drug-induced changes in the release of 3 H-monoamines from field stimulated rat brain slices. Acta Physiol. Scand. Suppl., 371: 35–44.

Ferraro, G., Montalbano, M.E., Sardo, P. and La Grutta, V. (1996) Lateral habenular influence on dorsal raphe neurons. Brain Res. Bull., 41: 47–52.

Fisher, P.M., Meltzer, C.C., Ziolko, S.K., Price, J.C., Moses-Kolko, E.L., Berga, S.L. and Hariri, A.R. (2006) Capacity

for 5-HT1A-mediated autoregulation predicts amygdala reactivity. Nat. Neurosci., 9: 1362–1363.
Fornal, C.A., Marrosu, F., Metzler, C.W., Tada, K. and Jacobs, B.L. (1994) Effects of the putative 5-hydroxytryptamine1A antagonists BMY 7378, NAN 190 and (–)-propranolol on serotonergic dorsal raphe unit activity in behaving cats. J. Pharmacol. Exp. Ther., 270: 1359–1366.
Fornal, C.A., Metzler, C.W., Gallegos, R.A., Veasey, S.C., Mccreary, A.C. and Jacobs, B.L. (1996) WAY-100635, a potent and selective 5-hydroxytryptamine1A antagonist, increases serotonergic neuronal activity in behaving cats: comparison with (S)-WAY-100135. J. Pharmacol. Exp. Ther., 278: 752–762.
Galzin, A.M., Poirier, M.F., Lista, A., Chodkiewicz, J.P., Blier, P., Ramdine, R., Loo, H., Roux, F.X., Redondo, A. and Langer, S.Z. (1992) Characterization of the 5-hydroxytryptamine receptor modulating the release of 5-[3H]hydroxytryptamine in slices of the human neocortex. J. Neurochem., 59: 1293–1301.
Gardier, A.M., David, D.J., Jego, G., Przybylski, C., Jacquot, C., Durier, S., Gruwez, B., Douvier, E., Beauverie, P., Poisson, N., Hen, R. and Bourin, M. (2003) Effects of chronic paroxetine treatment on dialysate serotonin in 5-HT1B receptor knockout mice. J. Neurochem., 86: 13–24.
Gartside, S.E., Umbers, V., Hajós, M. and Sharp, T. (1995) Interaction between a selective 5-HT1A receptor antagonist and an SSRI *in vivo*: effects on 5-HT cell firing and extracellular 5-HT. Br. J. Pharmacol., 115: 1064–1070.
Ge, J. and Barnes, N.M. (1996) 5-HT4 receptor-mediated modulation of 5-HT release in the rat hippocampus *in vivo*. Br. J. Pharmacol., 117: 1475–1480.
Ghavami, A., Stark, K.L., Jareb, M., Ramboz, S., Segu, L. and Hen, R. (1999) Differential addressing of 5-HT1A and 5-HT1B receptors in epithelial cells and neurons. J. Cell Sci., 112(Pt 6): 967–976.
Gobert, A., Rivet, J.M., Cistarelli, L. and Millan, M.J. (1997) Potentiation of the fluoxetine-induced increase in dialysate levels of serotonin (5-HT) in the frontal cortex of freely moving rats by combined blockade of 5-HT1A and 5-HT1B receptors with WAY 100,635 and GR 127,935. J. Neurochem., 68: 1159–1163.
Gothert, M. and Weinheimer, G. (1979) Extracellular 5-hydroxytryptamine inhibits 5-hydroxytryptamine release from rat brain cortex slices. Naunyn Schmiedebergs Arch. Pharmacol., 310: 93–96.
Gross, C., Zhuang, X., Stark, K., Ramboz, S., Oosting, R., Kirby, L., Santarelli, L., Beck, S. and Hen, R. (2002) Serotonin1A receptor acts during development to establish normal anxiety-like behaviour in the adult. Nature, 416: 396–400.
Guilloux, J.P., David, D.J., Guiard, B.P., Chenu, F., Reperant, C., Toth, M., Bourin, M. and Gardier, A.M. (2006) Blockade of 5-HT1A receptors by (+/−)-pindolol potentiates cortical 5-HT outflow, but not antidepressant-like activity of paroxetine: microdialysis and behavioral approaches in 5-HT1A receptor knockout mice. Neuropsychopharmacology, 31: 2162–2172.
Haddjeri, N., Lavoie, N. and Blier, P. (2004) Electrophysiological evidence for the tonic activation of 5-HT(1A) autoreceptors in the rat dorsal raphe nucleus. Neuropsychopharmacology, 29: 1800–1806.
Haigler, H.J. and Aghajanian, G.K. (1974) Lysergic acid diethylamide and serotonin: a comparison of effects on serotonergic neurons and neurons receiving a serotonergic input. J. Pharmacol. Exp. Ther., 188: 688–699.
Hajós, M., Hajós-Korcsok, E. and Sharp, T. (1999) Role of the medial prefrontal cortex in 5-HT1A receptor-induced inhibition of 5-HT neuronal activity in the rat. Br. J. Pharmacol., 126: 1741–1750.
Hajós, M., Hoffmann, W.E., Tetko, I.V., Hyland, B., Sharp, T. and Villa, A.E. (2001) Different tonic regulation of neuronal activity in the rat dorsal raphe and medial prefrontal cortex via 5-HT(1A) receptors. Neurosci. Lett., 304: 129–132.
Hajós, M., Richards, C.D., Szekely, A.D. and Sharp, T. (1998) An electrophysiological and neuroanatomical study of the medial prefrontal cortical projection to the midbrain raphe nuclei in the rat. Neuroscience, 87: 95–108.
He, M., Sibille, E., Benjamin, D., Toth, M. and Shippenberg, T. (2001) Differential effects of 5-HT1A receptor deletion upon basal and fluoxetine-evoked 5-HT concentrations as revealed by *in vivo* microdialysis. Brain Res., 902: 11–17.
Heinrich, T., Bottcher, H., Schiemann, K., Holzemann, G., Schwarz, M., Bartoszyk, G.D., Van Amsterdam, C., Greiner, H.E. and Seyfried, C.A. (2004) Dual 5-HT1A agonists and 5-HT reuptake inhibitors by combination of indole-butyl-amine and chromenonyl-piperazine structural elements in a single molecular entity. Bioorg. Med. Chem., 12: 4843–4852.
Herold, F., Chodkowski, A., Izbicki, L., Krol, M., Kleps, J., Turlo, J., Nowak, G., Stachowicz, K., Dybala, M. and Siwek, A. (2009) Novel 4-aryl-pyrido[1,2-c]pyrimidines with dual SSRI and 5-HT1A activity, part 1. Eur. J. Med. Chem., 44: 1710–1717.
Hjorth, S. and Tao, R. (1991) The putative 5-HT1B receptor agonist CP-93,129 suppresses rat hippocampal 5-HT release *in vivo*: comparison with RU 24969. Eur. J. Pharmacol., 209: 249–252.
Hornby, G.A., Stack, G., Dawson, L.A., Nguyen, H., Spangler, T., Smith, D.L., Mazandarani, H., Tran, M., Kalgaonkar, S., Zhang, G., Schechter, L. and Andree, T. (2004) SSRI/5-HT1A antagonists as future antidepressant agents: a neurochemical assessment of WAY-426. Soc. Neurosci. Abstr., 1024: 14.
Hughes, Z.A., Starr, K.R., Langmead, C.J., Hill, M., Bartoszyk, G.D., Hagan, J.J., Middlemiss, D.N. and Dawson, L.A. (2005) Neurochemical evaluation of the novel 5-HT1A receptor partial agonist/serotonin reuptake inhibitor, vilazodone. Eur. J. Pharmacol., 510: 49–57.
Hughes, Z.A., Starr, K.R., Scott, C.M., Newson, M.J., Sharp, T., Watson, J.M., Hagan, J.J. and Dawson, L.A. (2007) Simultaneous blockade of 5-HT1A/B receptors and 5-HT transporters results in acute increases in extracellular 5-HT in both rats and guinea pigs: *in vivo* characterization of the novel 5-HT1A/B receptor antagonist/5-HT transport inhibitor SB-649915-B. Psychopharmacology (Berl.), 192: 121–133.
Innis, R.B., Nestler, E.J. and Aghajanian, G.K. (1988) Evidence for G protein mediation of serotonin- and GABAB-induced hyperpolarization of rat dorsal raphe neurons. Brain Res., 459: 27–36.
Invernizzi, R., Velasco, C., Bramante, M., Longo, A. and Samanin, R. (1997) Effect of 5-HT1A receptor antagonists on citalopram-induced increase in extracellular serotonin in the frontal cortex, striatum and dorsal hippocampus. Neuropharmacology, 36: 467–473.
Jankowski, M.P. and Sesack, S.R. (2004) Prefrontal cortical projections to the rat dorsal raphe nucleus: ultrastructural features and associations with serotonin and gamma-aminobutyric acid neurons. J. Comp. Neurol., 468: 518–529.
Kia, H.K., Miquel, M.C., Brisorgueil, M.J., Daval, G., Riad, M., El Mestikawy, S., Hamon, M. and Verge, D. (1996)

Immunocytochemical localization of serotonin1A receptors in the rat central nervous system. J. Comp. Neurol., 365: 289–305.

Knobelman, D.A., Hen, R., Blendy, J.A. and Lucki, I. (2001) Regional patterns of compensation following genetic deletion of either 5-hydroxytryptamine(1A) or 5-hydroxytryptamine(1B) receptor in the mouse. J. Pharmacol. Exp. Ther., 298: 1092–1100.

Knobelman, D.A., Kung, H.F. and Lucki, I. (2000) Regulation of extracellular concentrations of 5-hydroxytryptamine (5-HT) in mouse striatum by 5-HT(1A) and 5-HT(1B) receptors. J. Pharmacol. Exp. Ther., 292: 1111–1117.

Lanfumey, L., Haj-Dahmane, S. and Hamon, M. (1993) Further assessment of the antagonist properties of the novel and selective 5-HT1A receptor ligands (+)-WAY 100 135 and SDZ 216-525. Eur. J. Pharmacol., 249: 25–35.

Lanfumey, L., Pardon, M.C., Laaris, N., Joubert, C., Hanoun, N., Hamon, M. and Cohen-Salmon, C. (1999) 5-HT1A autoreceptor desensitization by chronic ultramild stress in mice. Neuroreport, 10: 3369–3374.

Lesch, K.P. and Gutknecht, L. (2004) Focus on the 5-HT1A receptor: emerging role of a gene regulatory variant in psychopathology and pharmacogenetics. Intl. J. Neuropsychopharmacol., 7: 381–385.

Limberger, N., Deicher, R. and Starke, K. (1991) Species differences in presynaptic serotonin autoreceptors: mainly 5-HT1B but possibly in addition 5-HT1D in the rat, 5-HT1D in the rabbit and guinea-pig brain cortex. Naunyn Schmiedebergs Arch. Pharmacol., 343: 353–364.

Liu, R., Jolas, T. and Aghajanian, G. (2000) Serotonin 5-HT(2) receptors activate local GABA inhibitory inputs to serotonergic neurons of the dorsal raphe nucleus. Brain Res., 873: 34–45.

Lucas, G., Compan, V., Charnay, Y., Neve, R.L., Nestler, E.J., Bockaert, J., Barrot, M. and Debonnel, G. (2005) Frontocortical 5-HT4 receptors exert positive feedback on serotonergic activity: viral transfections, subacute and chronic treatments with 5-HT4 agonists. Biol. Psychiatry, 57: 918–925.

Lucas, G. and Debonnel, G. (2002) 5-HT4 receptors exert a frequency-related facilitatory control on dorsal raphe nucleus 5-HT neuronal activity. Eur. J. Neurosci., 16: 817–822.

Lucas, G., Rymar, V.V., Du, J., Mnie-Filali, O., Bisgaard, C., Manta, S., Lambas-Senas, L., Wiborg, O., Haddjeri, N., Piñeyro, G., Sadikot, A.F. and Debonnel, G. (2007) Serotonin(4) (5-HT(4)) receptor agonists are putative antidepressants with a rapid onset of action. Neuron, 55: 712–725.

Marek, G.J., Carpenter, L.L., Mcdougle, C.J. and Price, L.H. (2003) Synergistic action of 5-HT2A antagonists and selective serotonin reuptake inhibitors in neuropsychiatric disorders. Neuropsychopharmacology, 28: 402–412.

Martín-Ruiz, R., Puig, M.V., Celada, P., Shapiro, D.A., Roth, B.L., Mengod, G. and Artigas, F. (2001) Control of serotonergic function in medial prefrontal cortex by serotonin-2A receptors through a glutamate-dependent mechanism. J. Neurosci., 21: 9856–9866.

Martín-Ruiz, R. and Ugedo, L. (2001) Electrophysiological evidence for postsynaptic 5-HT(1A) receptor control of dorsal raphe 5-HT neurones. Neuropharmacology, 41: 72–78.

Maura, G., Thellung, S., Andrioli, G.C., Ruelle, A. and Raiteri, M. (1993) Release-regulating serotonin 5-HT1D autoreceptors in human cerebral cortex. J. Neurochem., 60: 1179–1182.

Middlemiss, D.N. (1984) Stereoselective blockade at [3H]5-HT binding sites and at the 5-HT autoreceptor by propranolol. Eur. J. Pharmacol., 101: 289–293.

Middlemiss, D.N. and Hutson, P.H. (1990) The 5-HT1B receptors. Ann. NY Acad. Sci., 600: 132–147, discussion 347–48

Millan, M.J. (2009) Dual- and triple-acting agents for treating core and co-morbid symptoms of major depression: novel concepts, new drugs. Neurotherapeutics, 6: 53–77.

Miquel, M.C., Doucet, E., Riad, M., Adrien, J., Verge, D. and Hamon, M. (1992) Effect of the selective lesion of serotoninergic neurons on the regional distribution of 5-HT1A receptor mRNA in the rat brain. Brain Res. Mol. Brain Res., 14: 357–362.

Mørk, A. and Hogg, S. (2003) *In vivo* profile of Lu 36-274: a compound with combined 5-HTT/5-HT1A activity and putative fast onset of antidepressant action. Soc. Neurosci. Abstr.: 958.17.

Muraki, I., Inoue, T. and Koyama, T. (2008) Effect of co-administration of the selective 5-HT1A receptor antagonist WAY 100,635 and selective 5-HT1B/1D receptor antagonist GR 127,935 on anxiolytic effect of citalopram in conditioned fear stress in the rat. Eur. J. Pharmacol., 586: 171–178.

Newman, M.E., Shalom, G., Ran, A., Gur, E. and Van De Kar, L.D. (2004) Chronic fluoxetine-induced desensitization of 5-HT1A and 5-HT1B autoreceptors: regional differences and effects of WAY-100635. Eur. J. Pharmacol., 486: 25–30.

Pazos, A., Cortes, R. and Palacios, J.M. (1985) Quantitative autoradiographic mapping of serotonin receptors in the rat brain. II. Serotonin-2 receptors. Brain Res., 346: 231–249.

Piñeyro, G., Castanon, N., Hen, R. and Blier, P. (1995a) Regulation of [3H]5-HT release in raphe, frontal cortex and hippocampus of 5-HT1B knockout mice. Neuroreport, 7: 353–359.

Piñeyro, G., De Montigny, C. and Blier, P. (1995b) 5-HT1D receptors regulate 5-HT release in the rat raphe nuclei. *In vivo* voltammetry and *in vitro* superfusion studies. Neuropsychopharmacology, 13: 249–260.

Pompeiano, M., Palacios, J.M. and Mengod, G. (1994) Distribution of the serotonin 5-HT2 receptor family mRNAs: comparison between 5-HT2A and 5-HT2C receptors. Brain Res. Mol. Brain Res., 23: 163–178.

Quérée, P., Peters, S. and Sharp, T. (2009) Further pharmacological characterisation of 5-HT2C receptor agonist-induced inhibition of 5 HT neuronal activity in the dorsal raphe nucleus *in vivo*. Br. J. Pharmacol., in press.

Quérée, P. and Sharp, T. (2006). Electrophysiological evidence for 5-HT2C receptor-mediated control of 5-HT cell firing in the anaesthetised rat. Proceedings of the British Pharmacological Society, available online at: http://www.pA2online.org/abstracts/Vol4Issue2abst050P.

Rabiner, E.A., Gunn, R.N., Castro, M.E., Sargent, P.A., Cowen, P.J., Koepp, M.J., Meyer, J.H., Bench, C.J., Harrison, P.J., Pazos, A., Sharp, T. and Grasby, P.M. (2000) beta-blocker binding to human 5-HT(1A) receptors *in vivo* and *in vitro*: implications for antidepressant therapy. Neuropsychopharmacology, 23: 285–293.

Radja, F., Daval, G., Hamon, M. and Verge, D. (1992) Pharmacological and physicochemical properties of preversus postsynaptic 5-hydroxytryptamine1A receptor binding sites in the rat brain: a quantitative autoradiographic study. J. Neurochem., 58: 1338–1346.

Raley, J., Boothman, L. and Sharp, T. (2004) Immunohistochemical evidence that 5-HT1A and 5-HT2 receptor agonists activate GABA neurones in the dorsal raphe nucleus. Proceedings of the British Pharmacological Society, available online at: http://www.pA2online.org/Vol2Issue4abst066P.

Rollema, H., Clarke, T., Sprouse, J.S. and Schulz, D.W. (1996) Combined administration of a 5-hydroxytryptamine (5-HT)1D antagonist and a 5-HT reuptake inhibitor synergistically increases 5-HT release in guinea pig hypothalamus *in vivo*. J. Neurochem., 67: 2204–2207.

Romero, L., Celada, P. and Artigas, F. (1994) Reduction of *in vivo* striatal 5-hydroxytryptamine release by 8-OH-DPAT after inactivation of Gi/G(o) proteins in dorsal raphe nucleus. Eur. J. Pharmacol., 265: 103–106.

Romero, L., Celada, P., Martín-Ruiz, R., Díaz-Mataix, L., Mourelle, M., Delgadillo, J., Hervas, I. and Artigas, F. (2003) Modulation of serotonergic function in rat brain by VN2222, a serotonin reuptake inhibitor and 5-HT1A receptor agonist. Neuropsychopharmacology, 28: 445–456.

Rutz, S., Riegert, C., Rothmaier, A.K., Buhot, M.C., Cassel, J.C. and Jackisch, R. (2006) Presynaptic serotonergic modulation of 5-HT and acetylcholine release in the hippocampus and the cortex of 5-HT1B-receptor knockout mice. Brain Res. Bull., 70: 81–93.

Santana, N., Bortolozzi, A., Serrats, J., Mengod, G. and Artigas, F. (2004) Expression of serotonin1A and serotonin2A receptors in pyramidal and GABAergic neurons of the rat prefrontal cortex. Cereb. Cortex., 14: 1100–1109.

Sari, Y. (2004) Serotonin1B receptors: from protein to physiological function and behavior. Neurosci. Biobehav. Rev., 28: 565–582.

Schechter, L.E., Lin, Q., Smith, D.L., Zhang, G., Shan, Q., Platt, B., Brandt, M.R., Dawson, L.A., Cole, D., Bernotas, R., Robichaud, A., Rosenzweig-Lipson, S. and Beyer, C.E. (2008) Neuropharmacological profile of novel and selective 5-HT(6) receptor agonists: WAY-181187 and WAY-208466. Neuropsychopharmacology, 33: 1323–1335.

Schlicker, E., Fink, K., Molderings, G.J., Price, G.W., Duckworth, M., Gaster, L., Middlemiss, D.N., Zentner, J., Likungu, J. and Gothert, M. (1997) Effects of selective h5-HT1B (SB-216641) and h5-HT1D (BRL-15572) receptor ligands on guinea-pig and human 5-HT auto- and heteroreceptors. Naunyn Schmiedebergs Arch. Pharmacol., 356: 321–327.

Segrave, R. and Nathan, P.J. (2005) Pindolol augmentation of selective serotonin reuptake inhibitors: accounting for the variability of results of placebo-controlled double-blind studies in patients with major depression. Hum. Psychopharmacol., 20: 163–174.

Serrats, J., Mengod, G. and Cortes, R. (2005) Expression of serotonin 5-HT2C receptors in GABAergic cells of the anterior raphe nuclei. J. Chem. Neuroanat., 29: 83–91.

Sharp, T., Bramwell, S.R. and Grahame-Smith, D.G. (1989a) 5-HT1 agonists reduce 5-hydroxytryptamine release in rat hippocampus *in vivo* as determined by brain microdialysis. Br. J. Pharmacol., 96: 283–290.

Sharp, T., Bramwell, S.R., Hjorth, S. and Grahame-Smith, D.G. (1989b) Pharmacological characterization of 8-OH-DPAT-induced inhibition of rat hippocampal 5-HT release *in vivo* as measured by microdialysis. Br. J. Pharmacol., 98: 989–997.

Sharp, T. and Hjorth, S. (1990) Application of brain microdialysis to study the pharmacology of the 5-HT1A autoreceptor. J. Neurosci. Methods, 34: 83–90.

Sharp, T., Umbers, V. and Hjorth, S. (1996) The role of 5-HT1A autoreceptors and alpha 1-adrenoceptors in the inhibition of 5-HT release – II NAN-190 and SDZ 216-525. Neuropharmacology, 35: 735–741.

Sharp, T., Umbers, V. and Gartside, S.E. (1997) Effect of a selective 5-HT reuptake inhibitor in combination with 5-HT1A and 5-HT1B receptor antagonists on extracellular 5-HT in rat frontal cortex *in vivo*. Br. J. Pharmacol., 121: 941–946.

Sotelo, C., Cholley, B., El Mestikawy, S., Gozlan, H. and Hamon, M. (1990) Direct immunohistochemical evidence of the existence of 5-HT1A autoreceptors on serotoninergic neurons in the midbrain raphe nuclei. Eur. J. Neurosci., 2: 1144–1154.

Sprouse, J.S. and Aghajanian, G.K. (1988) Responses of hippocampal pyramidal cells to putative serotonin 5-HT1A and 5-HT1B agonists: a comparative study with dorsal raphe neurons. Neuropharmacology, 27: 707–715.

Stamford, J.A., Davidson, C., Mclaughlin, D.P. and Hopwood, S.E. (2000) Control of dorsal raphe 5-HT function by multiple 5-HT(1) autoreceptors: parallel purposes or pointless plurality? Trends Neurosci., 23: 459–465.

Starr, K.R., Price, G.W., Watson, J.M., Atkinson, P.J., Arban, R., Melotto, S., Dawson, L.A., Hagan, J.J., Upton, N. and Duxon, M.S. (2007) SB-649915-B, a novel 5-HT1A/B autoreceptor antagonist and serotonin reuptake inhibitor, is anxiolytic and displays fast onset activity in the rat high light social interaction test. Neuropsychopharmacology, 32: 2163–2172.

Sullivan, G.M., Ogden, R.T., Oquendo, M.A., Kumar, J.S., Simpson, N., Huang, Y.Y., Mann, J.J. and Parsey, R.V. (2009) Positron emission tomography quantification of serotonin-1A receptor binding in medication-free bipolar depression. Biol. Psychiatry, in press.

Taber, M.T., Torrente, J.R., Palmer, J., Poss, M.A., Tortolandi, D.R., Yevich, Y.P., Stark, A.D. and Mattson, R.J. (2003) Pharmacological characterisation of BMS-196859, a novel serotonin uptake inhibitor and serotonin 5-HT1A receptor antagonist. Soc. Neurosci. Abstr.: 958.15.

Thomas, D.R., Soffin, E.M., Roberts, C., Kew, J.N., de la Flor, R.M., Dawson, L.A., Fry, V.A., Coggon, S.A., Faedo, S., Hayes, P.D., Corbett, D.F., Davies, C.H. and Hagan, J.J. (2006) SB-699551-A (3-cyclopentyl-N-[2-(dimethylamino)ethyl]-N-[(4′-{[(2-phenylethyl)amino]methyl}-4-biphenylyl)methyl] propanamide dihydrochloride), a novel 5-ht5A receptor-selective antagonist, enhances 5-HT neuronal function: evidence for an autoreceptor role for the 5-ht5A receptor in guinea pig brain. Neuropharmacology, 51: 566–577.

Tordera, R., Pei, Q., Newson, M., Gray, K., Sprakes, M. and Sharp, T. (2003) Effect of different 5-HT1A receptor antagonists in combination with paroxetine on expression of the immediate-early gene Arc in rat brain. Neuropharmacology, 44: 893–902.

Trillat, A.C., Malagie, I., Scearce, K., Pons, D., Anmella, M.C., Jacquot, C., Hen, R. and Gardier, A.M. (1997) Regulation of serotonin release in the frontal cortex and ventral hippocampus of homozygous mice lacking 5-HT1B receptors: *in vivo* microdialysis studies. J. Neurochem., 69: 2019–2025.

Varga, V., Kocsis, B. and Sharp, T. (2003) Electrophysiological evidence for convergence of inputs from the medial prefrontal cortex and lateral habenula on single neurons in the dorsal raphe nucleus. Eur. J. Neurosci., 17: 280–286.

Varga, V., Szekely, A.D., Csillag, A., Sharp, T. and Hajós, M. (2001) Evidence for a role of GABA interneurones in the cortical modulation of midbrain 5-hydroxytryptamine neurones. Neuroscience, 106: 783–792.

Verge, D., Daval, G., Marcinkiewicz, M., Patey, A., El Mestikawy, S., Gozlan, H. and Hamon, M. (1986) Quantitative autoradiography of multiple 5-HT1 receptor subtypes in the brain of control or 5,7-dihydroxytryptamine-treated rats. J. Neurosci., 6: 3474–3482.

Vilaro, M.T., Cortes, R., Gerald, C., Branchek, T.A., Palacios, J.M. and Mengod, G. (1996) Localization of 5-HT4 receptor mRNA in rat brain by in situ hybridization histochemistry. Brain Res. Mol. Brain Res., 43: 356–360.

Waeber, C., Sebben, M., Nieoullon, A., Bockaert, J. and Dumuis, A. (1994) Regional distribution and ontogeny of 5-HT4 binding sites in rodent brain. Neuropharmacology, 33: 527–541.

Weisstaub, N.V., Zhou, M., Lira, A., Lambe, E., Gonzalez-Maeso, J., Hornung, J.P., Sibille, E., Underwood, M., Itohara, S., Dauer, W.T., Ansorge, M.S., Morelli, E., Mann, J.J., Toth, M., Aghajanian, G., Sealfon, S.C., Hen, R. and Gingrich, J.A. (2006) Cortical 5-HT2A receptor signaling modulates anxiety-like behaviors in mice. Science, 313: 536–540.

Whale, R., Terao, T., Cowen, P., Freemantle, N. and Geddes, J. (2008) Pindolol augmentation of serotonin reuptake inhibitors for the treatment of depressive disorder: a systematic review. J. Psychopharmacol., in press.

Wright, I.K., Garratt, J.C. and Marsden, C.A. (1990) Effects of a selective 5-HT2 agonist, DOI, on 5-HT neuronal firing in the dorsal raphe nucleus and 5-HT release and metabolism in the frontal cortex. Br. J. Pharmacol., 99: 221–222.

CHAPTER 2.7

Tryptophan Depletion and Serotonin Release – A Critical Reappraisal

Matthijs G.P. Feenstra[*,1] and Geoffrey van der Plasse[2]

[1]*Netherlands Institute for Neuroscience, An Institute of the Royal Netherlands Academy of Arts and Sciences, Amsterdam, The Netherlands*
[2]*Department of Psychiatry, Rudolf Magnus Institute of Neuroscience, Utrecht University Medical Centre, Utrecht, The Netherlands*

Abstract: Tryptophan depletion is often assumed to lead to a decrease in neuronal serotonin release in the brain. Here we review the literature and show that only in animal studies in which either serotonin synthesis rate is already decreased or serotonin utilization is increased has an effect of tryptophan depletion on serotonin release been demonstrated. In the absence of convincing evidence for reduced central serotonin release, the possibility that other mechanisms are involved cannot be discarded. We therefore conclude that one should be careful to interpret tryptophan depletion-related effects as reflecting a widespread reduction of central serotonin release.

Keywords: tryptophan depletion, serotonin release, serotonin synthesis, serotonin receptors, microdialysis, tryptophan-free diet, 5-hydroxyindoleacetic acid, cerebrospinal fluid.

Introduction

Tryptophan depletion was introduced about 30 years ago as an experimental method to obtain a rapid and reversible decrease of the concentration of serotonin in the brain (Biggio *et al.*, 1974; Gessa *et al.*, 1974). By ingestion of a mixture of amino acids without the serotonin precursor tryptophan, the rate of serotonin synthesis in the brain is strongly decreased (Nishizawa *et al.*, 1997). Since its introduction in human psychopharmacology (Young *et al.*, 1985), this method has been used increasingly as a relatively simple and safe method to assess serotonergic functions. A PubMed search delivers about 50 articles per year for 'tryptophan depletion', and the popularity of the method appears still to be increasing.

Most authors tacitly assume that tryptophan depletion will reliably and reproducibly decrease serotonin transmission because of its effect on synthesis. However, this is only possible if serotonin release and binding to postsynaptic receptors is affected, as well. While some authors indeed mention this explicitly, the actual evidence is extremely sparse and often misquoted. Therefore, we will review what evidence is available to support that tryptophan depletion affects serotonin release and serotonin receptor binding.

*Corresponding author
E-mail: m.feenstra@nin.knaw.nl

Tryptophan depletion

The rate-limiting step in the synthesis of serotonin is the hydroxylation of tryptophan to 5-hydroxytryptophan by tryptophan hydroxylase, the enzymatic marker of serotonin neurons. This means that synthesis is directly dependent on the supply of the precursor, tryptophan. Free plasma concentrations of this essential amino acid are determined by its intake, plasma protein binding and utilization in a number of physiological processes, of which serotonin synthesis is only one. Tryptophan can enter the brain through active uptake by a carrier shared with other large, neutral amino acids. Intake of a tryptophan-free diet (Biggio *et al.*, 1974) or a mixture of essential amino acids (Gessa *et al.*, 1974) leads to increased demand for tryptophan for protein synthesis and decreased brain uptake. Blocking protein synthesis indeed prevents the effect on brain uptake (Moja *et al.*, 1991). Thus, plasma and brain concentrations of tryptophan are strongly decreased, and this leads to decreased concentrations of serotonin and its metabolite 5-hydroxyindoleacetic acid (5-HIAA) in the brain tissue (Biggio *et al.*, 1974; Gessa *et al.*, 1974). Whether this decrease will have consequences for serotonergic neurotransmission depends on the utilization (i.e., the net loss) of the transmitter because of activity-dependent release and metabolism. If utilization is too high in comparison to the reduced synthesis, releasable pools of

ISBN 978-0-12-374634-4
DOI: 10.1016/B978-0-12-374634-4.00014-9

serotonin will be exhausted and serotonergic function will be affected.

Numerous physiological, pharmacological and behavioral experiments have been carried out after tryptophan depletion, and these have been reviewed regularly (Delgado *et al.*, 1989; Young and Teff, 1989; Reilly *et al.*, 1997; Fadda, 2000; Moore *et al.*, 2000; Bel *et al.*, 2001; van der Does, 2001; Riedel *et al.*, 2002; Booij *et al.*, 2003; Fusar-Poli *et al.*, 2006; Evers *et al.*, 2007; Ruhé *et al.*, 2007). An important issue is that tryptophan depletion is often used to prove that a function is dependent on serotonergic activity or not – i.e., as an 'acid test' for serotonergic involvement. We feel that the evidence for this is not strong enough. For a reliable use of the test, unequivocal evidence should be available that tryptophan depletion reproducibly decreases serotonin release and serotonergic transmission. We will argue that this is not the case, and that the effects strongly depend on the precise experimental conditions. We will review the evidence that tryptophan depletion affects serotonergic neuronal activity, serotonin release and serotonin receptor activation in the following sections.

Effects of tryptophan depletion on 5-HT release and neuronal activity

The firing activity of serotonergic neurons in the raphe nuclei is subject to serotonergic control through autoreceptors. Increases or decreases in extracellular serotonin from dendritic or collateral release are reflected in decreased or increased neuronal activity, respectively. Early studies showed that the former effect, decreased neuronal activity, was indeed consistently observed after tryptophan administration (Aghajanian, 1972; Trulson and Jacobs, 1975). A tryptophan-free diet, however, did not significantly change firing in cats, even though brain tryptophan, serotonin and 5-HIAA concentrations were decreased (Trulson, 1985).

The same author showed that serotonin release measured by push–pull perfusion was not affected in this condition, although the concentrations of 5-HIAA in the perfusate were decreased. Later studies of extracellular serotonin concentrations used microdialysis sampling, which is now the method of choice to determine serotonin neuronal release (Westerink, 1995). These studies are summarized in Table 1. Our conclusion is that serotonin release is only reduced by tryptophan depletion when the diet is chronically administered, when a serotonin-reuptake inhibitor is used, or when serotonin release is strongly stimulated. The effect of chronic administration of the diet is clearly illustrated by the results of Fadda and colleagues (Fadda, 2000; Fadda *et al.*, 2000a, 2000b), who showed that administration of the diet for 3–5 days leads to steadily decreasing extracellular levels of serotonin in rat hippocampus and frontal cortex. Chronic administration of such diets is also known to affect body weight (see, for example, D'Souza *et al.*, 2004; Cahir *et al.*, 2007) and plasma corticosterone levels (D'Souza *et al.*, 2004). It cannot be expected that the physiological or neurochemical alterations after long-term diet administration are exclusively dependent on serotonergic mechanisms.

Addition of serotonin reuptake blockers to the dialysis fluid was often applied in older studies to increase the extracellular concentrations, and therefore the detectability of serotonin in the dialysates. However, it has long been known that reuptake blockade leads to decreased serotonin synthesis rates (Carlsson and Lindqvist, 1978), and this will increase the chances that synthesis cannot keep up with utilization when the tryptophan availability is reduced. The effect of uptake inhibition is clearly demonstrated in the set of experiments reported by Bel and Artigas (1996), who studied the effects of peripheral administration of the uptake blocker fluvoxamine. Only after chronic fluvoxamine did tryptophan depletion lead to decreased serotonin release in the dorsal raphe nucleus or frontal cortex; after saline treatment, no effect was observed. Fasting or feeding the rats before the experiment did not alter the fluvoxamine effect.

Gartside *et al.* (1992b) also reported no effect on basal release of tryptophan depletion by treatment with one of the essential amino acids that competes with tryptophan, i.e. valine. However, they observed a reduced effect of electrical stimulation of the raphe nucleus or administration of fenfluramine, resulting in two-fold and six-fold increases of release in control animals, respectively (Gartside *et al.*, 1992a, 1992b). It would be interesting to test whether normal, physiological activation of serotonergic activity would also uncover the effects of tryptophan depletion on *in vivo* serotonin release. Van der Plasse *et al.* (2007) exposed rats to a novel environment and measured release after administration of the tryptophan-free diet. However, this stimulus was apparently too weak to activate serotonin release. More conclusive evidence might be obtained by combining tryptophan depletion with studies of *in vivo* release under conditions of more consistently increased serotonergic activity, such as motor activity (Jacobs and Fornal, 1997) or stress (Amat *et al.*, 2005).

A further issue that has not been taken into account is the differential innervation of brain areas by the median and dorsal raphe nuclei. These form two distinct serotonergic systems, with different morphological and pharmacological properties (Molliver, 1987). As differences have been shown to exist in stress-induced serotonin release and the effects of chronic uptake inhibition in these systems (Auerbach and

Table 1 Effects on serotonin release

Release	Measurement	Subjects	Method	Biochemical effect	Effect on 5-HT release	Reference
Push–pull perfusion (^{3}H-trp)	Lateral ventricle	Cat	Fast; trp+ or − diet	−49% trp; −27% 5-HT tissue	No signif effect	Trulson, 1985
In vitro slice (fluoxetine)	Hypothalamus	Rat	trp+ or − superfusion	+ or − tissue trp, 5-HT	+ or − basal and evoked	Schaechter and Wyrtman, 1990
Microdialysis (citalopram)	Frontal cortex	Rat; anesth	AA-mix oral	−61% trp; −9% 5-HT tissue	−30% basal	Heslop *et al.*, 1991
Microdialysis (citalopram)	Ventral hippocampus	Rat	AA-mix ip inj	Not reported	−32% raphe stimulation −20% fenfluramine effect	Gartside *et al.*, 1992a
Microdialysis (citalopram)	Ventral hippocampus	Rat; anesth	Valine ip inj	Not reported	−40% fenfluramine effect	Gartside *et al.*, 1992a
Microdialysis (citalopram)	Ventral hippocampus	Rat; anesth	valine ip inj	Not reported	no effect basal −53% raphe stimulation	Gartside *et al.*, 1992b
Microdialysis (chron fluvox sc)	Frontal cortex, d-raphe	Rat	Fast; AA-mix oral	Not reported	−50% basal d-raphe −39% basal frontal cortex	Bel and Artigas, 1996
Microdialysis	Frontal cortex, d-raphe	Rat	AA-mix oral	Not reported	no signif effect	Bel and Artigas, 1996
Microdialysis (chron fluvox sc)	Frontal cortex, d-raphe	Rat	AA-mix oral	Not reported	−68% basal frontal cortex	Bel and Artigas, 1996
Microdialysis (citalopram)	Dorsal hippocampus	Rat	Fast; AA-mix oral	Not reported	−35% basal −62% fenfluramine effect	Stancampiano et al., 1997a
Microdialysis (citalopram)	Cortex, d. hippocampus	Rat	Fast; trp-free diet	Not reported	−30% cortex −50% hippocampus	Stancampiano *et al.*, 1997b
Microdialysis (citalopram)	Frontal cortex	Rat	Fast; trp-free diet	Not reported	No effect (day 1); −75% (day 3)	Fadda *et al.*, 2000a
Microdialysis (alaproclate)	Frontal cortex	Rat	Fast; trp-free diet	Not reported	−40% (day 1); −75% (day 3)	Fadda *et al.*, 2000b
Microdialysis	Dorsal hippocampus	Rat	Low or high trp diet	−20%; no effect	−44% (day 4); +92% (day 7)	van der Stelt *et al.*, 2004
Microdialysis	Medial PFC	Rat	Fast; gelatin oral	−71% plasma trp	No signif effect	van der Plasse *et al.*, 2007

Abbreviations: AA, amino acid; anesth, anesthesized; chron, chronic; d, dorsal; fluvox, fluvoxamine; ip, intraperitoneal; sc, subcutaneous; signif, significant; trp, tryptophan.

Hjorth, 1995; Storey *et al.*, 2006), it would be interesting to study their response to tryptophan depletion.

In conclusion, only in conditions of enhanced serotonin utilization or reduced serotonin synthesis has an effect of tryptophan depletion on serotonin release been demonstrated.

Direct measurements of serotonin extracellular concentrations in primate brain tissue have not been reported, but a few studies were performed using cerebrospinal fluid (CSF), which is in direct contact with the extracellular fluid of the brain and spinal cord. All studies showed profound decreases of tryptophan concentrations in plasma, CSF or brain tissue, and furthermore reported 5-HIAA decreases of 30 percent (Table 2). No study, however, reported on serotonin concentrations. The value of 5-HIAA measurements is limited, as extracellular concentrations of 5-HIAA cannot be expected to be a reliable indicator of neuronal serotonin release (Crespi, 1990).

Deaminated monoamine acidic metabolites are not stored in the neuron, like the parent transmitter, but freely exchange between intra- and extracellular spaces. Their concentration is determined by the metabolism of released, but also of newly synthesized, transmitter, and by active transport out of the brain by the probenecid-sensitive carrier (Westerink, 1995). While extracellular 5-HIAA may show changes similar to those of serotonin, this is not necessarily the case, and differential effects may be expected when either uptake or synthesis is altered (Stenfors and Ross, 2004; van der Stelt *et al.*, 2005). In summary, alterations in CSF 5-HIAA levels cannot be taken as proof of altered serotonin release.

Another outcome of all these studies is that a profound reduction of plasma tryptophan, the usual measure of the effectiveness of tryptophan depletion in human studies, is not necessarily associated with decreased *in vivo* serotonin release in the brain. A clear example is provided by van der Plasse *et al.* (2007), who showed strongly decreased plasma tryptophan in the absence of a significant alteration of rat-brain serotonin release. While this should be confirmed in other studies, it supports the conclusion that a central imbalance between serotonin synthesis and utilization is needed to observe an effect of tryptophan depletion.

Effects of tryptophan depletion on serotonin receptors

An alternative approach to study the effects of tryptophan depletion on serotonin release is to determine *in vivo* occupation of serotonin receptors or serotonin receptor-mediated effects or *ex vivo* receptor numbers. An overview of the few studies that have been performed is presented in Table 3. Decreased serotonin release should be reflected as an increase in radioligand binding potential. This has not been reported. Human PET studies of radioligand binding to 5-HT_{1A} receptors after tryptophan depletion did not detect any alteration, which is remarkable, as a study of the effects of inhibition of tryptophan hydroxylase on 5-HT_{1A} binding in rats using the same radioligand (^{18}F-MPPF) did report the expected clear increase in binding (Zimmer *et al.*, 2003). The only study of 5-HT_2 receptors did not detect an increase, but rather a decrease, in binding potential (Yatham *et al.*, 2001).

Animal studies on receptor-occupancy or receptor-mediated effects are sparse, too (Table 3). Moreover, most of the results were obtained after chronic tryptophan-free or low-tryptophan diets (and not all these experiments are reported here). The only acute effect is a decrease in 5-HT_{1A} receptor numbers in the raphe nucleus (Cahir *et al.*, 2007), which again cannot be explained by a reduction in occupancy by serotonin. Instead, the authors suggest a homeostatic response to reduce autoreceptor feedback, and they discuss the possibility that alterations in corticosterone release might mediate this effect. Others, however, conclude that serotonin is only involved in cortisol/corticosterone release in response to stress, but not under basal conditions (Porter *et al.*, 2007).

Thus, none of these studies provides any evidence of decreased serotonin release after tryptophan depletion.

Other effects of tryptophan depletion – alternative mechanisms

As was mentioned previously, tryptophan is used in many physiological processes, both peripherally and centrally.

Table 2 Measurements in primates

Release, synthesis	Measurement	Subjects	Method	Biochemical effect	Effect on 5-HT	Reference
CSF concentrate	Cisternal CSF	Vervet monkey	Fast; AA mixture	CSF −61% trp; −34% 5-HIAA	Not measured	Young *et al.*, 1989
5-HT synthesis	Human PET; ^{11}C-Me-L-Trp	Volunteers	Diet; AA drink	Plasma −72% trp; synth rate −87% male; plasma −89% trp; synth rate −97% female	Not measured	Nishizawa *et al.*, 1997
CSF concentrate	Lumbar CSF	Volunteers	Diet; AA drink	Plasma −89% trp; CSF −92% trp; −31% 5-HIAA	Not measured	Carpenter *et al.*, 1998
CSF concentrate	Lumbar CSF	Volunteers	Diet; AA drink	Plasma −86% trp; CSF −92% trp; −33% 5-HIAA	Not measured	Williams *et al.*, 1999

Abbreviations: AA, amino acid; CSF, cerebrospinal fluid; Me-L-trp, α-methyl-L-tryptophan; PET, positron emission tomography; synth, synthesis.

Table 3 Receptor measurements

Receptor	Receptor measurement	Subjects	Method	Biochemical effect	Effect on receptor	Reference
5-HT_{1A}	Human PET; ^{18}F-MPPF	Volunteers	Diet; AA drink	−62% plasma trp	No signif eff	Udo de Haes *et al.*, 2002
5-HT_{1A}	Human PET; ^{18}F-MPPF	Remitted patients	AA drink	−85% plasma trp	No signif eff	Prashak-Rieder *et al.*, 2004
5-HT_{1A}	Human PET; 18F-FCWAY	Volunteers	Not reported	Not reported	No signif eff	Williams *et al.*, 2002
5-HT_2	Human PET; ^{18}F-setoperone	Volunteers	AA drink	−71.5% plasma trp	−7.9% BP cortical areas	Yatham *et al.*, 2001
5-HT_{1A}	Blood hormones	Rat	3-week trp-free diet	−41% 5-HT FCo	−40% OXT resp agon −69% ACTH resp agon −100% corticost resp agon	D'Souza *et al.*, 2004
5-HT_{1A}	Rec binding brain	Rat	Acute or chron diet	−80% plasma trp; −30–80% brain 5-HT	Acute: −14% raphe rec number	Cahir *et al.*, 2007
5-HT_{1A}	Prolactin resp; rec binding brain	Rat	1-week low-trp diet	−30–40% plasma trp −20% cortex 5-HTP	+ 200% 2 min resp agon no signif eff rec number	Franklin *et al.*, 1999
5-HT_{2A}	Rec binding brain	Rat	Acute or chron diet	−80% plasma trp; −30–80% brain 5-HT	3 weeks: +46% cortex rec number	Cahir *et al.*, 2007
5-HT_{2C}	Prolactin resp; rec binding brain	Rat	6-week low-trp diet	−40–60% plasma trp −43% cortex 5-HTP	+36% AUC resp agon no signif eff rec number	Franklin *et al.*, 1999

Abbreviations: AA, amino acid; ACTH, adrenocorticotroph hormone; agon, agonist; AUC, area-under-the-curve; BP, binding potential; chron, chronic; corticost, corticosterone; FCo, frontal cortex; FCWAY, Trans-4-Fluoro-N-(2-[4-(2-methoxyphenyl)piperazin-1-yl]ethyl)-N-(2-pyridyl) cyclohexanecarboxamide; MPPF, (4-(2′-methoxyphenyl)-1-[2′-(N-2″-pyridinyl)-p-fluorobenzamido]-ethyl piperazine); OXT, oxytocine; PET, positron emission tomography; rec, receptor; resp, response; signif eff, significant effect; trp, tryptophan.

While the possible consequences of a reduction of central serotonin synthesis receive most attention, it cannot be excluded that other processes are involved in the physiological or behavioral effects that have been reported after tryptophan depletion. Therefore, we need to consider first whether serotonergic or non-serotonergic peripheral processes might be involved in these effects, and second, whether non-serotonergic central processes might play a role.

Peripheral processes

Despite the fact that only 5 percent of the total amount of serotonin present in the body is found in the brain, and that serotonin receptors are distributed throughout the body (see, for example, Ramage and Villalón, 2008), relatively little is know about consequences of tryptophan depletion on serotonergic functions outside the brain. Serotonin has been implicated as an important modulator of cardiovascular and gastrointestinal activity. Although peripheral mechanisms can be expected to predominate, these processes may also partly depend on central serotonin availability. It is thus not surprising that tryptophan has been suggested to link somatic and psychiatric illnesses (Russo *et al.*, 2003). The fact that most serotonin is synthesized and utilized peripherally supports such a notion.

Peripheral serotonin is synthesized in, and released from, the enterochromaffin cells of the intestinal system. It is known to be involved in the regulation of gastrointestinal activity (Kilkens *et al.*, 2004), and tryptophan depletion affects gastric emptying in healthy female controls (van Nieuwenhoven *et al.*, 2004). A potentially important observation is that tryptophan depletion in a control group and in patients with irritable bowel syndrome (IBS) not only impaired affective memory performance, but also heightened visceral perception. Shufflebotham *et al.* (2006) replicated these findings in patients with IBS, but not in healthy controls.

Central as well as peripheral actions of serotonin on cardiovascular output have been well described. In short, serotonin acts both directly via the raphe nucleus on (para)sympathetic innervation, and as a modulator on serotonergic receptors on the smooth muscle tissue of the heart and vascular system. As such, serotonin receptor activation has been shown to increase, as well as decrease, heart rate (Jonnakuty and Gragnoli, 2008), and to regulate blood pressure (Nilsson *et al.*, 1999). The effect of tryptophan depletion on these variables remains, however, unclear. A 2-day dietary depletion of tryptophan enhanced stress-induced increases in blood pressure in recovered anxiety patients treated with serotonin-reuptake inhibitors (Davies *et al.*, 2006), but Roiser *et al.* (2008) did not observe any change in blood pressure or heart rate after acute tryptophan depletion in healthy controls. A study by van der Veen *et al.* (2008) has, to the best of our knowledge, been the only one to find a correlation between tryptophan depletion and cardiac slowing, although the authors attributed this to an effect of task rather than a depletion of serotonin levels.

These examples are illustrative of the idea that perception of bodily functions (e.g., increased heart rate or gastrointestinal activation) affects cognitive appraisal – an idea that is certainly not new. The influential concept was originally posed in the nineteenth century by William James and Carl Lange, and later took the form of the somatic-marker hypothesis posed by Damasio (1996). This model, in essence, entails the notion that affective signals from the body are used to guide decision-making, and as such play an important role in the expression of behavior.

Although the exact nature of peripheral effects induced by (acute) tryptophan depletion is poorly understood, there are positive indications that tryptophan depletion does affect functions outside the brain that deserve consideration when interpreting the central effects of tryptophan depletion.

Central processes

In addition to possible effects of tryptophan depletion on peripheral processes, an important question is the extent to which tryptophan depletion might directly exert its effects through modulation of other central monoamines. Effects of tryptophan depletion might be expected, as the catecholamine precursor tyrosine is one of the amino acids that are administered and that compete for transport over the blood–brain barrier (Badawy, 2005). However, *in vivo* microdialysis studies showed that neither noradrenaline nor dopamine release was altered after tryptophan depletion in rats (Fadda *et al.*, 2000b; van der Plasse *et al.*, 2007). In addition, brain concentrations of these neurotransmitters and their metabolites were not affected (Lieben *et al.*, 2004a; Ardis *et al.*, 2008). Even tyrosine concentrations were not consistently altered (Lieben *et al.*, 2004a; Jans *et al.*, 2007a).

Measurements in CSF indicated that tyrosine concentrations were not changed (Young *et al.*, 1989) or increased (Carpenter *et al.*, 1998) while, again, dopamine and noradrenaline metabolites were not altered. Therefore, there is no indication that tryptophan depletion leads to alterations in brain catecholamines that might explain the physiological and behavioral effects.

Effects of tryptophan depletion on non-serotonergic processes that use tryptophan have not been studied in great detail. However, Lieben *et al.* (2004b) performed a pilot study on the effect of tryptophan depletion on kynurenine-related metabolites in serum, liver and brain. An interesting finding was that tryptophan was decreased in serum and brain, but not in liver. In liver, only the concentration of quinolinic acid was decreased, but the brain concentrations of neither kynurenine, nor kynurenic acid or quinolinic acid were decreased (Lieben *et al.*, 2004b), making it unlikely that tryptophan depletion affects central processes through this pathway.

Consequences for the interpretation of tryptophan-depletion induced effects

The greatest impact that the tryptophan-depletion method has made is undoubtedly in the field of psychiatry. Since the reports that tryptophan depletion may lower mood in normal males (Young *et al.*, 1985) and in remitted depressed patients (Delgado *et al.*, 1990), this method has been used to study serotonergic involvement in psychiatric disorders (Reilly *et al.*, 1997; Bel *et al.*, 2001; van der Does, 2001; Booij *et al.*, 2003). Recently, attention has increasingly been directed to the use of tryptophan depletion in cognitive research, as well (Park *et al.*, 1994; Riedel *et al.*, 1999; Riedel, 2004; Fusar-Poli *et al.*, 2007). Tryptophan-depletion studies now form the basis for novel theories about the role of serotonin in behavior and cognition (Evers *et al.*, 2007; Cools *et al.*, 2008).

However, before we accept the apparent use of tryptophan depletion as an 'acid test' for serotonergic involvement in human clinical and cognitive research, we have to return to our earlier conclusion that tryptophan depletion only decreases serotonin release and function in specific conditions – i.e., treatment with serotonin reuptake blockers, reduced serotonin synthesis and increased serotonin utilization. Recent clinical literature indeed shows that the reported effects of tryptophan depletion are not always reproducible. Indications that effects are observed only in special experimental conditions or only in subpopulations of either patients or normal controls abound in the literature. Thus, the effectiveness of tryptophan depletion as a method for studying serotonin involvement in mood is restricted to subpopulations that show a vulnerability to serotonin dysregulation (Fadda, 2000; Booij *et al.*, 2003; Fusar-Poli *et al.*, 2006; Jans *et al.*, 2007b; Ruhé *et al.*, 2007). These subgroups include remitted depressed patients (generally those treated with selective serotonin reuptake inhibitors; see Bremner *et al.*, 1997; Young and Leyton, 2002; Neumeister *et al.*, 2004), females (who show lower synthesis rates; see Ellenbogen *et al.*, 1996; Nishizawa *et al.*, 1997; Schmitt *et al.*, 2000; Booij *et al.*, 2002) and subjects with certain genetic variations in the serotonin transporter gene (associated with reduced sero tonin reuptake; see Roiser *et al.*, 2006; Finger *et al.*, 2007). Based on the animal data, one might add that the level of activation due to different task demands and its interaction with the biological vulnerability will prove to be an important factor, as well.

The examples mentioned in the section on peripheral processes also suggest that the effects of tryptophan depletion probably depend on the individual vulnerability of the participants. The same should apply to the studies of animal and human cognition. Regarding the human studies, the issue is further complicated by the fact that many studies using tryptophan depletion report changes in brain activation measured by functional MRI, in the absence of effects on performance or mood (Fusar-Poli *et al.*, 2007; Anderson *et al.*, 2008). Given the uncertainty regarding underlying mechanisms, the inferences for serotonergic functions of these results remain tentative.

The possibility that individual differences in emotional processing and its underlying biological processes are a major factor in the variability of tryptophan-depletion induced effects on cognition has been raised (Evers *et al.*, 2007; Harmer, 2008). Supportive evidence was recently provided by (1) the work of Sambeth *et al.* (2007), who found that impairments in memory recall after tryptophan depletion were stronger in females than in males, and (2) the work of Olivier *et al.* (2008), who tested object memory in rats with genetic deletions of the serotonin transporter. Tryptophan depletion led to similar effects on plasma tryptophan in all genotypes, but differential effects on brain serotonin concentrations and object recognition.

Conclusion

The evidence that tryptophan depletion leads to a reduction of serotonin release and transmission is sparse, and is restricted to animal studies in which either serotonin synthesis rate is already decreased or serotonin utilization is increased. The variability in the results of human tryptophan depletion studies can tentatively be explained by taking the biological vulnerability and the task demands into account. However, in the absence of convincing evidence for reduced central serotonin release, the possibility that other mechanisms are involved cannot be discarded. We conclude that one should be careful to interpret tryptophan depletion-related effects as reflecting a widespread reduction of central serotonin release.

Acknowledgement

This study was funded by the ZonMW TOP grant 912-02-050 from the Dutch Organization for Scientific Research (NWO).

References

Aghajanian, G.K. (1972) Chemical-feedback regulation of serotonin-containing neurons in brain. Ann. NY Acad. Sci., 193: 86–94.

Amat, J., Baratta, M.V., Paul, E., Bland, S.T., Watkins, L.R. and Maier, S.F. (2005) Medial prefrontal cortex determines how stressor controllability affects behavior and dorsal raphe nucleus. Nat. Neurosci., 8: 365–371.

Anderson, I.M., McKie, S., Elliott, R., Williams, S.R. and Deakin, J.F.W. (2008) Assessing human 5-HT function *in vivo* with pharmacoMRI. Neuropharmacology, 55: 1029–1037.

Ardis, T.C., Cahir, M., Elliott, J.J., Bell, R., Reynolds, G.P. and Cooper, S.J. (2008). Effect of acute tryptophan depletion on noradrenaline and dopamine in the rat brain. J. Psychopharmacol., in press

Auerbach, S.B. and Hjorth, S. (1995) Effect of chronic administration of the selective serotonin (5-HT) uptake inhibitor citalopram on extracellular 5-HT and apparent autoreceptor sensitivity in rat forebrain *in vivo*. Naunyn Schmiedeberg's Arch. Pharmacol., 352: 597–606.

Badawy, A.A.-B. (2005) Acute tryptophan or tyrosine depletion test: time for reappraisal? J. Psychopharmacol., 19: 429–430.

Bel, C., Abrams, J. and Nutt, D. (2001) Tryptophan depletion and its implications for psychiatry. Br. J. Psychiatry, 178: 399–405.

Bel, N. and Artigas, F. (1996) Reduction of serotonergic function in rat brain by tryptophan depletion: effects in control and fluvoxamine-treated rats. J. Neurochem., 67: 669–676.

Biggio, G., Fadda, F., Fanni, P., Tagliamonte, A. and Gessa, G.L. (1974) Rapid depletion of serum tryptophan, serotonin and 5-hydroxyindoleacetic acid by a tryptophan-free diet. Life Sci., 14: 1321–1329.

Booij, L., van der Does, W., Benkelfat, C., Bremner, J.D., Cowen, P.J., Fava, M., Gillin, C., Leyton, M., Moore, P.,

Smith, K.A. and Van der Kloot, W.A. (2002) Predictors of mood response to acute tryptophan depletion: a reanalysis. Neuropsychopharmacology, 27: 852–861.

Booij, L., van der Does, A.J.W. and Riedel, W.J. (2003) Monoamine depletion in psychiatric and healthy populations: review. Mol. Psychiatry, 8: 951–973.

Bremner, J.D., Innis, R.B., Salomon, R.M., Staib, L.H., Ng, C.K., Miller, H.L., Bronen, R.A., Krystal, J.H., Duncan, J., Rich, D., Price, L.H., Malison, R., Dey, H., Soufer, R. and Charney, D.S. (1997) Positron emission tomography measurement of cerebral metabolic correlates of tryptophan depletion-induced depressive relapse. Arch. Gen. Psychiatry, 54: 364–374.

Cahir, M., Ardis, T., Reynolds, G.P. and Cooper, S.J. (2007) Acute and chronic tryptophan depletion differentially regulate central 5-HT_{1A} and 5-HT_{2A} receptor binding in the rat. Psychopharmacology, 190: 497–506.

Carlsson, A. and Lindqvist, M. (1978) Effects of antidepressant agents on the synthesis of brain monoamines. J. Neural. Transm., 43: 73–91.

Carpenter, L.L., Anderson, G.M., Pelton, G.H., Gudin, J.A., Kirwin, P.D.S., Price, L.H., Heninger, G.R. and McDougle, C.J. (1998) Tryptophan depletion during continuous CSF sampling in healthy human subjects. Neuropsychopharmacology, 19: 26–35.

Cools, R., Roberts, A.C. and Robbins, T.W. (2008) Serotonergic regulation of emotional and behavioural control processes. Trends Cogn. Sci., 12: 31–40.

Crespi, F. (1990) *In vivo* voltammetry with micro-biosensors for analysis of neurotransmitter release and metabolism. J. Neurosci. Meth., 34: 53–65.

Damasio, A.R. (1996) The somatic marker hypothesis and the possible functions of the prefrontal cortex. Philos. Trans. R Soc. Lond. B Biol. Sci., 351: 1413–1420.

Davies, S.J.C., Hood, S.D., Argyropoulos, S.V., Morris, K., Bell, C., Witchel, H.J., Jackson, P.R., Nutt, D.J. and Potokar, J.P. (2006) Depleting serotonin enhances both cardiovascular and psychological stress reactivity in recovered patients with anxiety disorder. J. Clin Psychopharmacol., 26: 414–418.

Delgado, P.L., Charney, D.S., Price, L.H., Aghajanian, G.K., Landis, H. and Heninger, G.R. (1990) Serotonin function and the mechanism of antidepressant action. Reversal of antidepressant-induced remission by rapid depletion of plasma tryptophan. Arch. Gen. Psych., 47: 411–418.

Delgado, P.L., Charney, D.S., Price, L.H., Landis, H. and Heninger, G.R. (1989) Neuroendocrine and behavioral effects of dietary tryptophan restriction in healthy volunteers. Life Sci., 45: 2323–2332.

D'Souza, D.N., Zhang, Y., Garcia, F., Battaglia, G. and van de Kar, L.D. (2004) Fluoxetine-induced changes in body weight and 5-HT_{1A} receptor-mediated hormone secretion in rats on a tryptophan-deficient diet. Am. J. Physiol. Regul. Integr. Comp. Physiol., 286: R390–R397.

Ellenbogen, M.A., Young, S.N., Dean, P., Palmour, R.M. and Benkelfat, C. (1996) Mood response to acute tryptophan depletion in healthy volunteers: sex differences and temporal stability. Neuropsychopharmacology, 15: 465–474.

Evers, E.A.T., van der Veen, F.M., Fekkes, D. and Jolles, J. (2007) Serotonin and cognitive flexibility: neuroimaging studies into the effect of acute tryptophan depletion in healthy volunteers. Curr. Med. Chem. Rev., 14: 2989–2995.

Fadda, F. (2000) Tryptophan-free diets: a physiological tool to study brain serotonin function. News Physiol. Sci., 15: 260–264.

Fadda, F., Cocco, S., Rossetti, Z.L., Melis, G. and Stancampiano, R. (2000a) A tryptophan-free diet markedly reduces frontocortical 5-HT release, but fails to modify ethanol preference in alcohol-preferring (sP) and non-preferring (sNP) rats. Behav. Brain. Res., 108: 127–132.

Fadda, F., Cocco, S. and Stancampiano, R. (2000b) A physiological method to selectively decrease brain serotonin release. Brain Res. Brain Res. Protoc., 5: 219–222.

Finger, E.C., Marsh, A.A., Buzas, B., Kamel, N., Rhodes, R., Vythilingham, M., Pine, D.S., Goldman, D. and Blair, J.R. (2007) The impact of tryptophan depletion and 5-HTTLPR genotype on passive avoidance and response reversal instrumental learning tasks. Neuropsychopharmacology, 32: 206–215.

Franklin, M., Craven, R.D., Dowling, B., Campling, G., Elliott, J.M. and Cowen, P.J. (1999) Effect of a long-term low tryptophan diet on the prolactin responses to the 5-HT_{1A} and 5-HT_{2C} agonists, 8-OH-DPAT and mCPP in the male rat. J. Psychopharmacol., 13: 58–63.

Fusar-Poli, P., Allen, P., McGuire, P., Placentino, A., Cortesi, M. and Perez, J. (2006) Neuroimaging and electrophysiological studies of the effects of acute tryptophan depletion: a systematic review of the literature. Psychopharmacology, 188: 131–143.

Fusar-Poli, P., Allen, P. and Cortesi, M. (2007) Can acute tryptophan depletion modulate brain function in absence of behavioural effects? Med. Hypoth., 68: 722.

Gartside, S.E., Cowen, P.J. and Sharp, T. (1992a) Effect of amino acid loads on hippocampal 5-HT release *in vivo* evoked by electrical stimulation of the dorsal raphe nucleus and d-fenfluramine administration. Br. J. Pharmacol., 107: 448P.

Gartside, S.E., Cowen, P.J. and Sharp, T. (1992b) Evidence that the large neutral amino acid L-valine decreases electrically-evoked release of 5-HT in rat hippocampus *in vivo*. Psychopharmacology, 109: 251–253.

Gessa, G.L., Biggio, G., Fadda, F., Corsini, G.U. and Tagliamonte, A. (1974) Effect of the oral administration of tryptophan-free amino acid mixtures on serum tryptophan, brain tryptophan and serotonin metabolism. J. Neurochem., 22: 869–870.

Harmer, C.J. (2008) Serotonin and emotional processing: does it help to explain antidepressant drug action? Neuropharmacology, 55: 1023–1028.

Heslop, K., Portas, C.M. and Curzon, G. (1991) Effect of altered tryptophan availability on tissue and extracellular serotonin in the rat. Fifth International Conference on *In vivo* Methods, pp. 259–261. University of Groningen Press, Groningen.

Jacobs, B.L. and Fornal, C.A. (1997) Serotonin and motor activity. Curr. Opin. Neurobiol., 7: 820–825.

Jans, L.A.W., Lieben, C.K.J. and Blokland, A. (2007a) Influence of sex and estrous cycle on the effects of acute tryptophan depletion induced by a gelatin-based mixture in adult Wistar rats. Neuroscience, 147: 304–317.

Jans, L.A.W., Riedel, W.J., Markus, C.R. and Blokland, A. (2007b) Serotonergic vulnerability and depression: assumptions, experimental evidence and implications. Mol. Psychiatry, 12: 522–543.

Jonnakuty, C. and Gragnoli, C. (2008) What do we know about serotonin? J. Cell Phys., 217: 301–306.

Kilkens, T.O.C., Honig, A., van Nieuwenhoven, M.A., Riedel, W.J. and Brummer, R.-J.M. (2004) Acute tryptophan depletion affects brain-gut responses in irritable bowel syndrome patients and controls. Gut, 53: 1794–1800.

Lieben, C.K.J., Blokland, A., Westerink, B. and Deutz, N.E.P. (2004a) Acute tryptophan and serotonin depletion using an optimized tryptophan-free protein–carbohydrate mixture in the adult rat. Neurochem. Intl., 44: 9–16.

Lieben, C.K.J., Schwarcz, R., Guidetti, P. and Blokland, A. (2004b) Effects of an acute tryptophan depletion on the kynurenine pathway. In: Lieben, C.K.J., Serotonin and behavior: animal studies applying the method of acute tryptophan depletion. PhD Thesis, University of Maastricht, pp. 204–213.

Moja, E.A., Restani, P., Corsini, E., Stacchezzini, M.C., Assereto, R. and Galli, C.L. (1991) Cycloheximide blocks the fall of plasma and tissue tryptophan levels after tryptophan-free amino acid mixtures. Life Sci., 49: 1121–1128.

Molliver, M.E. (1987) Serotonergic neuronal systems: what their anatomic organization tells us about function. J. Clin. Psychopharmacol., 7(Suppl.): 3S–23S.

Moore, P., Landolt, H.P., Seifritz, E., Clark, C., Bhatti, T., Kelsoe, J., Rapaport, M. and Gillin, J.C. (2000) Clinical and physiological consequences of rapid tryptophan depletion. Neuropsychopharmacology, 23: 601–622.

Neumeister, A., Nugent, A.C., Waldeck, T., Geraci, M., Schwarz, M., Bonne, O., Bain, E.E., Luckenbaugh, D.A., Herscovitch, P., Charney, D.S. and Drevets, W.C. (2004) Neural and behavioral responses to tryptophan depletion in unmedicated patients with remitted major depressive disorder and controls. Arch. Gen. Psychiatry, 61: 765–773.

Nilsson, T., Longmore, J., Shaw, D., Pantev, E., Bard, J.A., Branchek, T. and Edvinsson, L. (1999) Characterisation of 5-HT receptors in human coronary arteries by molecular and pharmacological techniques. Eur. J. Pharmacol., 372: 49–56.

Nishizawa, S., Benkelfat, C., Young, S.N., Leyton, M., Mzengeza, S., de Montigny, C., Blier, P. and Diksic, M. (1997) Differences between males and females in rates of serotonin synthesis in human brain. Proc. Natl. Acad. Sci. USA, 94: 5308–5313.

Olivier, J.D.A., Jans, L.A.W., Korte-Bouws, G.A.H., Korte, S.M., Deen, P.M.T., Cools, A.R., Ellenbroek, B.A. and Blokland, A. (2008) Acute tryptophan depletion dose dependently impairs object memory in serotonin transporter knockout rats. Neuroscience, 200: 243–254.

Park, S.B., Coull, J.T., Young, A.T., Sahakian, B.J., Robbins, T.W. and Cowen, P.J. (1994) Tryptophan depletion in normal volunteers produces selective impairments in learning and memory. Neuropharmacology, 33: 578–588.

Porter, R.J., Gallagher, P. and O'Brien, J.T. (2007) Effects of rapid tryptophan depletion on salivary cortisol in older people recovered from depression, and the healthy elderly. J. Psychopharmacol., 21: 71–75.

Prashak-Rieder, N., Hussey, D., Wilzon, A.A., Carella, A., Lee, M., Dunn, E., Willeit, M., Bagby, R.M., Houle, S. and Meyer, J.H. (2004) Tryptophan depletion and serotonin loss in selective serotonin reuptake inhibitor-treated depression: an [^{18}F]MPPF positron emission tomography study. Biol. Psychiatry, 56: 587–591.

Ramage, A.G. and Villalón, C.M. (2008) 5-Hydroxytryptamine and cardiovascular regulation. Trends Pharmacol. Sci., 29: 472–481.

Reilly, J.G., McTavish, S.F.B. and Young, A.H. (1997) Rapid depletion of plasma tryptophan: a review of studies and experimental methodology. J. Psychpharmacol., 11: 381–392.

Riedel, W.J. (2004) Cognitive changes after acute tryptophan depletion: what can they tell us? Psychol. Med., 34: 3–8.

Riedel, W.J., Klaassen, T., Deutz, N.E.P., van Someren, A. and van Praag, H.M. (1999) Tryptophan depletion in normal volunteers produces selective impairment in memory consolidation. Psychopharmacology, 141: 362–369.

Riedel, W.J., Klaassen, T. and Schmitt, J.A.J. (2002) Tryptophan, mood, and cognitive function. Brain Behav. Immun., 16: 581–589.

Roiser, J.P., Blackwell, A.D., Cools, R., Clark, L., Rubinsztein, D.C., Robbins, T.W. and Sahakian, B.J. (2006) Serotonin transporter polymorphism mediates vulnerability to loss of incentive motivation following acute tryptophan depletion. Neuropsychopharmacology, 31: 2264–2272.

Roiser, J.P., Levy, J., Fromm, S.F., Wang, H., Hasler, G., Sahakian, B.J. and Drevets, W.C. (2008) The effect of acute tryptophan depletion on the neural correlates of emotional processing in healthy volunteers. Neuropsychopharmacology, 33: 1992–2006.

Ruhé, H.G., Mason, N.S. and Schene, A.H. (2007) Mood is indirectly related to serotonin, norepinephrine and dopamine levels in humans: a meta-analysis of monoamine depletion studies. Mol. Psychiatry, 12: 331–359.

Russo, S., Kema, I.P., Fokkema, R., Boon, J.C., Willemse, P.H.B., de Vries, E.G.E., den Boer, J.A. and Korf, J. (2003) Tryptophan as a link between psychopathology and somatic states. Psychosomatic Med., 65: 665–671.

Sambeth, A., Blokland, A., Harmer, C.J., Kilkens, T.O.C., Nathan, P.J., Porter, R.J., Schmitt, J.A.J., Scholtissen, B., Sobczak, S., Young, A.H. and Riedel, W.J. (2007) Sex differences in the efffect of acute tryptophan depletion on declarative episodic memory: a pooled analysis of nine studies. Neurosci. Biobehav. Rev., 31: 516–529.

Schaechter, J.D. and Wyrtman, R.J. (1990) Serotonin release varies with brain tryptophan levels. Brain Res., 532: 203–210.

Schmitt, J.A., Jorissen, B.L., Sobczak, S., van Boxtel, M.P., Hogervorst, E., Deutz, N.E. and Riedel, W.J. (2000) Tryptophan depletion impairs memory consolidation but improves focussed attention in healthy young volunteers. J. Psychopharmacol., 14: 21–29.

Shufflebotham, J., Hood, S., Hendry, J., Hince, D.A., Morris, K., Nutt, D., Probert, C. and Potokar, J. (2006) Acute tryptophan depletion alters gastrointestinal and anxiety symptoms in irritable bowel syndrome. Am. J. Gastroenterol., 101: 2582–2587.

Stancampiano, R., Melis, F., Sarais, L., Cocco, S., Cugusi, C. and Fadda, F. (1997a) Acute administration of a tryptophan-free amino acid mixture decreases 5-HT release in rat hippocampus *in vivo*. Am. J. Physiol., 272: R991–R994.

Stancampiano, R., Cocco, S., Melis, F., Cugusi, C., Sarais, L. and Fadda, F. (1997b) The decrease of serotonin release induced by a tryptophan-free amino acid diet does not affect spatial and passive avoidance learning. Brain Res., 762: 269–274.

Stenfors, C. and Ross, S.B. (2004) Changes in extracellular 5-HIAA concentrations as measured by *in vivo* microdialysis technique in relation to changes in 5-HT release. Psychopharmacology, 172: 119–128.

Storey, J.D., Robertson, D.A.F., Beattie, J.E., Reid, I.C., Mitchell, S.N. and Balfour, D.J.K. (2006) Behavioural and neurochemical responses by repeated exposure to an elevated open platform. Behav. Brain Res., 166: 220–229.

Trulson, M.E. (1985) Dietary tryptophan does not alter the function of brain serotonin neurons. Life Sci., 37: 1067–1072.

Trulson, M.E. and Jacobs, B.J. (1975) Raphe neurons: depression of activity by L-5-hydroxytryptophan. Brain Res., 97: 350–355.

Udo de Haes, J.I., Bosker, F.J., van Waarde, A., Pruim, J., Willemsen, A.T.M., Vaalburg, W. and den Boer, J.A. (2002) 5-HT_{1A} receptor imaging in the human brain: effect of tryptophan depletion and infusion of [^{18}F]MPPF binding. Synapse, 46: 108–115.

van der Does, A.J.W. (2001) The effects of tryptophan depletion on mood and psychiatric symptoms. J. Affect. Disord., 64: 107–119.

van der Plasse, G., Meerkerk, T.J., Lieben, C.K.J., Blokland, A. and Feenstra, M.G.P. (2007) Lack of evidence for reduced prefrontal cortical serotonin and dopamine efflux after acute tryptophan depletion. Psychopharmacology, 195: 377–385.

van der Stelt, H.M., Broersen, L.M., Olivier, B. and Westenberg, H.G.M. (2004) Effects of dietary tryptophan variations on extracellular serotonin in the dorsal hippocampus of rats. Psychopharmacology, 172: 137–144.

van der Stelt, H.M., Breuer, M.E., Olivier, B. and Westenberg, H.G.M. (2005) Permanent deficits in serotonergic functioning of olfactory bulbectomized rats: an *in vivo* microdialysis study. Biol. Psychiatry, 57: 1061–1067.

van der Veen, F.M., Mies, G.W., van der Molen, M.W. and Evers, E.A. (2008) Acute tryptophan depletion in healthy males attenuates phasic cardiac slowing but does not affect electrocortical response to negative feedback. Psychopharmacology, 199: 255–263.

van Nieuwenhoven, M.A., Valks, S.D.M., Sobczak, S., Riedel, W.J. and Brummer, R.-J.M. (2004) Acute tryptophan depletion slows gastric emptying in females. Br. J. Nutr., 91: 351–355.

Westerink, B.H.C. (1995) Brain microdialysis and its application for the study of animal behaviour. Behav. Brain Res., 70: 103–124.

Williams, W.A., Shoaf, S.E., Hommer, D., Rawlings, R. and Linnoila, M. (1999) Effects of acute tryptophan depletion on plasma and cerebrospinal fluid tryptophan and 5-hydroxyindoleacetic acid in normal volunteers. J. Neurochem., 72: 1641–1647.

Williams, W.A., Carson, R.E., Momeman, R., Lang, L., Heinz, A., Bjork, J., Geyer, C., Rawlings, R., Phillips, M., Johnson, B., Excleman, W.C., Hommer, D. and Herscovitch, P. (2002) Effects of acute plasma tryptophan depletion on serotonin receptor occupancy using PET in healthy human subjects. Program No 826.4, Society for Neuroscience.

Yatham, L.N., Liddle, P.F., Shiah, I.-S., Lam, R.W., Adam, M.J., Zis, A.P. and Ruth, T.J. (2001) Effects of rapid tryptophan depletion on brain 5-HT_2 receptors: a PET study. Br. J. Psychiatry, 178: 448–453.

Young, S.N. and Leyton, M. (2002) The role of serotonin in human mood and social interaction. Insight from altered tryptophan levels. Pharmacol. Biochem. Behav., 71: 857–865.

Young, S.N. and Teff, K.L. (1989) Tryptophan availability, 5HT synthesis and 5HT function. Prog. Neuro-Psychopharmacol. Biol. Psychiatry, 13: 373–379.

Young, S.N., Smith, S.E., Pihl, R.O. and Ervin, F.R. (1985) Tryptophan depletion causes a rapid lowering of mood in normal males. Psychopharmacology, 87: 173–177.

Young, S.N., Ervin, F.R., Pihl, R.O. and Finn, P. (1989) Biochemical aspects of tryptophan depletion in primates. Psychopharmacology, 98: 508–511.

Zimmer, L., Rbah, L., Giacomelli, F., Le Bars, D. and Renaud, B. (2003) A reduced extracellular serotonin level increases the 5-HT_{1A} PET ligand ^{18}F-MPPF binding in the rat hippocampus. J. Nucl. Med., 44: 1495–1501.

CHAPTER 2.8

Serotonin Interaction with Other Transmitter Systems

Albert Adell[*,1,2], Analía Bortolozzi[1,2,3], Llorenç Díaz-Mataix[1], Noemí Santana[1,2], Pau Celada[1,2,3] and Francesc Artigas[1,2]

[1]*Department of Neurochemistry and Neuropharmacology, Instituto de Investigaciones Biomédicas de Barcelona, Barcelona, Spain*
[2]*Centro de Investigación Biomédica en Red de Salud Mental (CIBERSAM), Barcelona, Spain*
[3]*Institut d'Investigacions Biomèdiques August Pi i Sunyer, Barcelona, Spain*

Abstract: Serotonergic fibers in the brain originate from a small number of cell bodies confined to the raphe nuclei of the brainstem and spread profusely, giving rise to an extensive arborization of axons in terminal fields. For this reason, it is not surprising that the brain serotonergic system is involved in several physiological as well as pathological conditions, and the regulation of serotonergic transmission at the level of raphe nuclei is of crucial importance in terms of serotonin (5-HT) action in nerve endings. The mechanisms that are responsible for the regulation of serotonergic transmission can be divided into intrinsic, i.e. those circumscribed to the internal function of 5-HT cells, and extrinsic, i.e. those that depend fundamentally on the interaction between other transmitter systems. The afferent connections of the raphe nuclei provide the neuroanatomical substrate for the interaction between different brain areas and serotonergic neurons. However, it is worth noting that these interactions are reciprocal. In this chapter we summarize the mutual interactions between serotonin and noradrenaline, dopamine, acetylcholine, GABA and glutamate. The considerable amount of interest in those interactions is due to the critical role they appear to play in mental functions. Special attention has been paid to the reciprocal control of the prefrontal cortex (PFC) and the midbrain raphe nuclei, which are involved in psychiatric diseases like major depression and schizophrenia.

Keywords: serotonin, noradrenaline, dopamine, acetylcholine, GABA, glutamate, raphe, prefrontal cortex.

Abbreviations: 1S,3S-ACPD, 1S,3S-aminocyclopentane-1,3-dicarboxylic acid; 5-HT, serotonin; 8-OH-DPAT, 8-hydroxy-2-(di-*n*-propylamino)tetralin; AMPA, α-amino-3-hydroxy-5-methyl-4-isoxazole-4-propionic acid; BAY x 3702, *R*-(–)-2-{4-[(chroman-2-ylmethyl)-amino]-butyl}-1,1-dioxo-benzo[d]isothiazolone; DNQX, 6,7-dinitroquinoxaline-2,3-dione; DOI, 1-[2,5-dimethoxy-4-iodophenyl-2-aminopropane; DR, dorsal raphe nucleus; GABA, γ-aminobutyric acid; IGluRS, ionotropic glutamate receptors; LC, locus coeruleus; LDT, laterodorsal tegmentum; LSD, lysergic acid diethylamide; M100907, (R)-(+)-(2,3-dimethoxyphenyl)-1-[2-(4-fluorophenyl)ethyl]-4-piperidine methanol; mGluRs, metabotropic glutamate receptors; MK-801, dizocilpine; MnR, median raphe nucleus; nAChRs, nicotinic acetylcholine receptors; NBQX, 2,3-dioxo-6-nitro-1,2,3,4-tetrahydrobenzo[f]quinoxaline-7-sulfonamide; NMDA, N-methyl-D-aspartate; PFC, prefrontal cortex; PPTg, pedunculopontine tegmentum; REM, rapid eye movement; SCH 23390, (R)-(+)-7-chloro-8-hydroxy-3-methyl-1-phenyl-2,3,4,5-tetrahydro-1*H*-3-benzazepine; SKF 38393, (±)-1-phenyl-2,3,4,5-tetrahydro-(1*H*)-3-benzazepine-7,8-diol; TTX, tetrodotoxin; VP, ventral pallidum; VTA, ventral tegmental area; WAY 100635, N-[2-[4-(2-methoxyphenyl)-1-piperazinyl]ethyl]-N-2-pyridinylcyclohexanecarboxamide.

Introduction

Most of the neurons synthesizing serotonin (5-HT) in the central nervous system are located in the raphe nuclei of the brainstem, and serotonergic terminals can be found in virtually every brain region (Jacobs and Azmitia, 1992; Halliday *et al.*, 1995; Leger *et al.*, 2001). With this widespread distribution, it is not surprising that serotonin has a role in many physiological functions as well as pathological states. Interactions between serotonergic and other transmitter systems can be found at the level of 5-HT cell bodies of the raphe nuclei and the level of serotonergic terminals in a great variety of brain areas. In both locations these interactions can be bidirectional – i.e., release of 5-HT may influence other transmitter systems, which in turn can also have an impact on serotonergic

*Corresponding author
E-mail: aacnqi@iibb.csic.es

Christian Müller & Barry Jacobs (Eds.)
Handbook of Behavioral Neurobiology of Serotonin
ISBN 978-0-12-374634-4

DOI: 10.1016/B978-0-12-374634-4.00015-0

transmission. The knowledge of these interactions in complex neuronal circuits may pave the way to a powerful pharmacological approach for the development of therapeutic strategies directed to severe psychiatric conditions such as depression and schizophrenia. The neuromodulatory effects of 5-HT on diverse transmitter systems can be achieved through multiple postsynaptic 5-HT receptors, although the precise molecular mechanisms by which this synaptic efficacy is affected remain to be fully elucidated. Several papers have reviewed the main features of transmitter receptors that can regulate 5-HT firing and release at the level of raphe nuclei (Jacobs and Azmitia, 1992; Piñeyro and Blier, 1999; Adell *et al.*, 2002). In this chapter, we will update previous data on this matter and also focus on the interaction of 5-HT with other transmitter systems at the level of the prefrontal cortex (PFC). The PFC is involved in many higher brain functions, including mnemonic processes, executive functions, goal-directed behavior, working memory and control of emotional signals (Fuster, 2001; Miller and Cohen, 2001; Romo and Salinas, 2003). To this end, PFC functions rely on its extensive and usually reciprocal connections with a vast array of cortical and subcortical structures (Groenewegen and Uylings, 2000; Fuster, 2001; Miller and Cohen, 2001). The PFC, as the rest of cortical areas, is formed predominantly by excitatory (glutamatergic) and inhibitory (GABAergic) neurons. Glutamatergic (pyramidal) neurons project to cortical and subcortical areas and GABAergic interneurons tightly control the activity of pyramidal neurons, thus providing the level of inhibition necessary for the regular functioning of cortical operation. Pyramidal neurons located in layer V of the medial PFC (mPFC) project heavily to the dorsal raphe nuclei (DR) (Hajós *et al.*, 1998; Peyron *et al.*, 1998; Vertes, 2004; Gabbott *et al.*, 2005), which in turn send serotonergic innervation back to the mPFC. This serotonergic projection can impinge upon both pyramidal and GABAergic neurons and exert excitatory (through 5-HT_{2A}, 5-HT_3, 5-HT_4, 5-HT_6, and 5-HT_7 receptors) and inhibitory (mainly through 5-HT_{1A} receptors) actions on either neuronal type. The global influence of the serotonergic system on the PFC will thus depend directly on the final balance of the action of the transmitter upon all of its receptors. However, serotonergic neurons of the DR also send projections to other brainstem monoaminergic nuclei, such as the nucleus locus coeruleus (LC) and ventral tegmental area (VTA), which may exert an important regulation of cortical activity through their respective projections. In this chapter, the functional interactions between 5-HT and other transmitter systems, such as noradrenaline, dopamine, acetylcholine, GABA and glutamate, are discussed with the focus on microdialysis and electrophysiological studies in the rodent brain.

Noradrenaline

Influence of 5-HT receptors on brain noradrenaline

There is a serotonergic projection from the DR to the LC (Vertes and Kocsis, 1994; Luppi *et al.*, 1995; Kaehler *et al.*, 1999), which appears to exert an inhibitory role, since the lesion of serotonergic neurons enhanced the firing activity of the noradrenergic cells of the LC (Haddjeri *et al.*, 1997). The 5-HT_{1A} receptor agonist, 8-OH-DPAT, and the 5-HT_{1A} receptor antagonist, WAY 100635, enhanced and suppressed, respectively, the firing activity of LC noradrenergic neurons (Szabo and Blier, 2001). These actions were entirely dependent on intact serotonergic neurons. The inhibitory effects of WAY 100635 were antagonized by M100907, which suggests an involvement of 5-HT_{2A} receptors. The absence of 5-HT_{2A}-immunolabeling in the LC (Fay and Kubin, 2000), however, seems to indicate that this population of 5-HT_{2A} receptors is located elsewhere in the brain. It is possible that 5-HT_{2A} located in the mPFC may play a role in these effects through its excitatory influence on the LC (Arnsten and Goldman-Rakic, 1984; Jodo *et al.*, 1998; Vertes, 2004). In fact, the intra-mPFC perfusion of the 5-$HT_{2A/2C}$ agonist DOI elicits an increase in the extracellular concentration of noradrenaline in this region in an impulse-dependent manner (Figure 1). On the other hand, there is electrophysiological and neurochemical evidence that 5-HT_{2C} receptors exert a tonic suppressant influence on the activity of LC noradrenergic transmission, since 5-HT_{2C} agonists decrease the firing of LC noradrenergic cells and 5-HT_{2C} antagonists increased the release of noradrenaline in LC and mPFC (Gobert *et al.*, 2000; Millan *et al.*, 2000). From a physiological point of

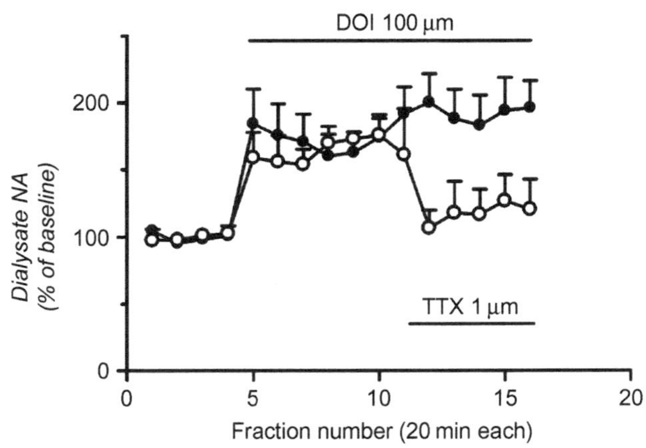

Figure 1 The local perfusion of the 5-$HT_{2A/2C}$ receptor agonist DOI (100µM) increased the release of noradrenaline in the mPFC. This effect was dependent on nerve impulse because it was abolished by the co-perfusion of tetrodotoxin (TTX; 1µM). $P < 0.05$ versus DOI alone.

view, it has to be taken into consideration that the mutual influence of noradrenergic and serotonergic neurons may vary across the sleep–wake cycle since both neuronal groups are REM-off. Thus, the α_1-adrenoceptor agonist phenylephrine, applied in the DR, reversed the cessation of serotonergic discharge occurring during REM sleep in cats (Sakai and Crochet, 2000).

Influence of noradrenergic receptors on brain serotonin

It is well known that noradrenergic afferents from the LC innervate the DR (Baraban and Aghajanian, 1980; Peyron *et al.*, 1996), which has been corroborated by the presence of noradrenaline and its uptake sites in the DR (Tejani-Butt, 1992; Ordway *et al.*, 1997; Strazielle *et al.*, 1999). Noradrenaline exert a tonic facilitation of serotonergic transmission through α_1-adrenoceptors (Baraban and Aghajanian, 1980; Marwaha and Aghajanian, 1982; Vandermaelen and Aghajanian, 1983; Pan *et al.*, 1994; Adell and Artigas, 1999; Bortolozzi and Artigas, 2003; Pudovkina *et al.*, 2003) located in the DRN (Pieribone *et al.*, 1994; Day *et al.*, 1997), possibly in the cell bodies' of serotonergic neurons (Gallager and Aghajanian, 1976a). Consistent with their action on the firing rate of serotonergic neurons, α_1-adrenoceptor antagonists decrease the release of 5-HT in projection areas (Rouquier *et al.*, 1994; Hjorth *et al.*, 1995; de Boer *et al.*, 1996; Pudovkina *et al.*, 2003). In addition to their localization in the DR, a notable amount of α_1-adrenoceptors is distributed along the PFC (Palacios *et al.*, 1987; McCune *et al.*, 1993; Pieribone *et al.*, 1994; Day *et al.*, 1997; Domyancic and Morilak, 1997; Cahir *et al.*, 2004) and expressed both in pyramidal cells and GABAergic interneurons (Figure 2). The activation of these cortical α_1-adrenoceptors also elicits a facilitatory action on 5-HT neurons (Figure 3). Thus, their stimulation by the α_1-adrenoceptor agonist cirazoline increases the release of 5-HT not only locally in the PFC but also distally in the DRN (Amargós-Bosch *et al.*, 2003). This latter effect is likely accounted for by a stimulation of pyramidal cells projecting to the DRN (Araneda and Andrade, 1991; Marek and Aghajanian, 1999) because it is blocked by the cortical perfusion of the α-amino-3-hydroxy-5-methyl-4-isoxazole-4-propionic acid (AMPA)/kainate receptor antagonist, NBQX, and the non-selective metabotropic glutamate receptor (mGluR) II/III agonist 1S,3S-ACPD (Amargós-Bosch *et al.*, 2003). Furthermore, there may be differences with regard to the role of α_1-adrenoceptor subtypes that stimulate serotonergic transmission in the DR and PFC because the predominant subtypes in the PFC are α_{1A} and α_{1D} (Figure 2), whereas the α_{1B}-subtype prevails in the DR (Pieribone *et al.*, 1994; Day *et al.*, 1997).

In addition to the α_1-adrenoceptor, the DR also bears one of the highest densities of α_2-adrenoceptors in the brain (Rosin *et al.*, 1993; Strazielle *et al.*, 1999). However, whereas the α_1-adrenoceptor has an excitatory role, the α_2-adrenoceptor is inhibitory. Thus, local application of the α_2-adrenoceptor agonist clonidine *in vivo* reduces 5-HT release in the DR and, consequently, in the PFC (Bortolozzi and Artigas, 2003; Pudovkina *et al.*, 2003). This effect is likely due to a decrease in noradrenergic transmission, because it can be mimicked by perfusing clonidine into the LC (Pudovkina *et al.*, 2002). The inhibitory action of the α_2-adrenoceptors on serotonergic transmission seems to possess a tonic nature, since the administration of α_2-adrenoceptor antagonists enhances 5-HT release in the DR (Hopwood and Stamford, 2001; Bortolozzi and Artigas, 2003; Pudovkina *et al.*, 2003). It is possible that the endpoint effect of noradrenaline on α_2-adrenoceptors results from a balance between its actions on α_2-autoreceptors located in noradrenergic terminals and α_2-heteroreceptors located on serotonergic neurons. The elevated density of α_{2A}-adrenoceptor mRNA in DR (Scheinin *et al.*, 1994) suggests that this subtype is the dominant player in this nucleus. The results obtained with microdialysis are in good concordance with electrophysiological work showing that the firing rate of 5-HT cells in the DR is suppressed by α_2-adrenoceptor agonists (Svensson *et al.*, 1975; Marwaha and Aghajanian, 1982) and increased by α_2-adrenoceptor antagonists (Freedman and Aghajanian, 1984; Garratt *et al.*, 1991; Haddjeri *et al.*, 1996). Since α_1-adrenoceptors and α_2-adrenoceptors possess an opposite impact on serotonergic transmission, it is therefore possible that the actions of noradrenaline on α_1-adrenoceptors are cancelled out by those on α_2-adrenoceptors.

Dopamine

Influence of 5-HT receptors on brain dopamine

Much attention has been paid to the interaction between dopaminergic and serotonergic systems of the brain, as both systems are involved in the control of complex behaviors and implicated in the actions of drugs used to treat psychiatric illnesses. Dopamine patterns of distribution in the brain are not as diffuse as those of serotonin or noradrenaline. There are three major dopaminergic systems in the brain: the nigrostriatal system, originating in the substantia nigra pars compacta; the mesolimbic system, originating in the ventral tegmental area (VTA) and terminating in the nucleus accumbens (NAC); and the mesocortical system, originating in the VTA, but innervating the PFC. Unlike serotonin, the dopamine system

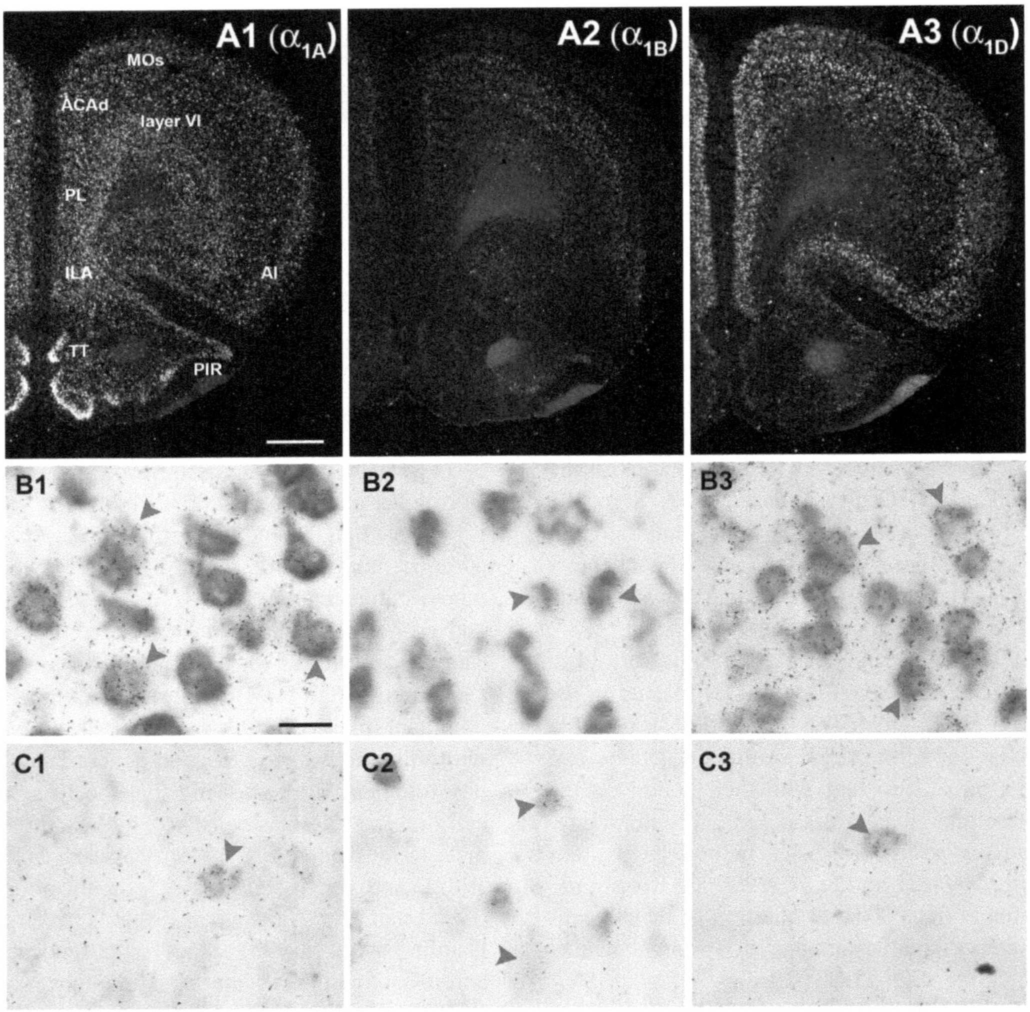

Figure 2 (A) Low-magnification dark-field photomicrographs showing the localization of α_{1A}-adrenoceptor (A1), α_{1B}-adrenoceptor (A2) and α_{1D}-adrenoceptor (A3) mRNAs in the rat prefrontal cortex using *in situ* hybridization histochemistry. The sections correspond approximately to AP + 3.7 mm (Paxinos and Watson, 2005). Each receptor transcript was detected with ^{33}P-labeled oligonucleotides. (B) High-magnification photomicrographs showing the presence of: α_{1A}-adrenoceptor (B1), α_{1B}-adrenoceptor (B2) and α_{1D}-adrenoceptor (B3) mRNA (^{33}P-labeled oligonucleotides, seen as silver grains) in pyramidal cells of the prelimbic area. Pyramidal neurons were identified by the presence of vGluT1 mRNA (Dig-labeled oligonucleotides). (C) High-magnification photomicrographs showing the presence of: α_{1A}-adrenoceptor (C1), α_{1B}-adrenoceptor (C2) and α_{1D}-adrenoceptor (C3) mRNA (^{33}P-labeled oligonucleotides, seen as silver grains) in GABAergic cells of the prelimbic area. GABAergic neurons were identified by the presence of $GAD_{65/67}$ mRNA (Dig-labeled oligonucleotides). Red arrowheads mark cells positive for both transcripts, blue arrowheads mark cells only positive for the receptor mRNA. For the sake of simplicity, only a few cells of each type are marked. Abbreviations: ACAd, dorsal anterior cingulate cortex; PL, prelimbic area; ILA; infralimbic area; TT, tenia tecta; PIR, piriform cortex; AI, agranular insular cortex; MOs, secondary motor area (nomenclature from Swanson, 1998). Bar size: 1 mm (A1), 20 μm (B1). To see the full color version of this figure please refer to the color plate at the back of the book. Copies produced via our print on demand service do not contain color plates. If your copy does not have the color plate, please go to this website to view the figure in color www.elsevierdirect.com/companions/9780123746344

does not undergo major cyclic shifts in activity levels during the course of the sleep–wake cycle. The cell bodies and terminal regions of all three dopaminergic pathways are innervated by serotonergic neurons originating in the DR (Beart and McDonald, 1982; Hervé *et al.*, 1987; Nedergaard *et al.*, 1988; Van Bockstaele *et al.*, 1994; Broderick and Phelix, 1997). Local application of 5-HT or pharmacological or electrical stimulation of the DR evoked excitations and/or inhibitions of VTA dopaminergic neurons (Cameron *et al.*, 1997; Gervais and Rouillard, 2000). The excitatory or inhibitory nature of this regulation may thus depend on the particular 5-HT receptor involved and the cellular type where it is expressed. Therefore, alterations in serotonergic activity could be responsible for dysregulation of dopaminergic function as inferred to occur in schizophrenia.

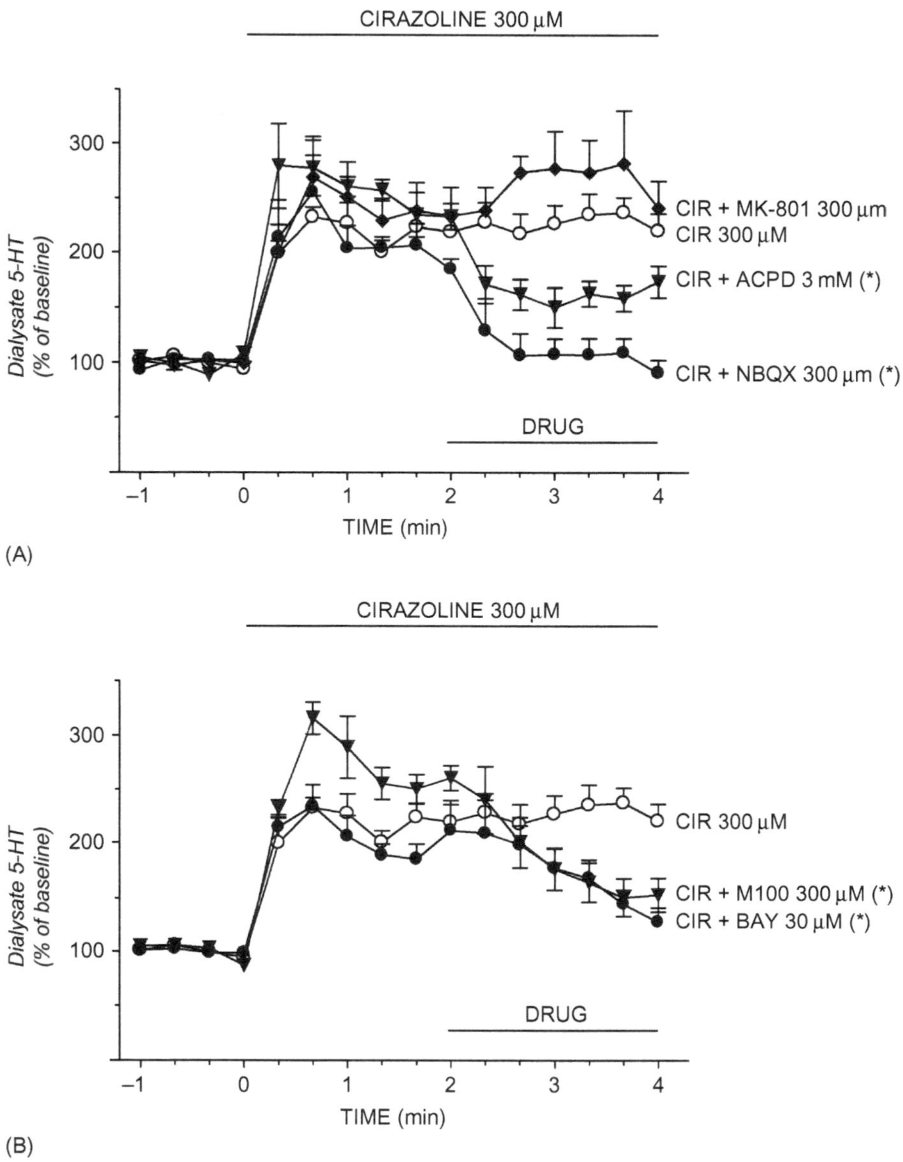

Figure 3 The local perfusion of the α_1-adrenoceptor agonist cirazoline (CIR, 300 μM) increased the 5-HT release in the mPFC. (A) The effect of cirazoline was reverted by the co-perfusion of the AMPA receptor antagonist NBQX (300 μM) and by the non-selective mGluR II/III agonist 1*S*,3*S*-ACPD (3 mM), but not by the NMDA receptor antagonist MK-801 (300 μM). (B) The effect of cirazoline was reverted by the co-perfusion of the 5-HT_{1A} receptor agonist BAY x 3702 (30 μM) and by the 5-HT_{2A} receptor antagonist M100907 (300 μM). (*) $P < 0.05$ versus cirazoline alone. Data taken and redrawn from Amargós-Bosch *et al.* (2003), with permission.

Several studies have described the presence of 5-HT_{1A} receptors in the VTA (Pazos and Palacios, 1985; Burnet *et al.*, 1995; Wright *et al.*, 1995), localized to dopaminergic as well as non-dopaminergic (Doherty and Pickel, 2001), possibly GABAergic cells (Gronier, 2008). The systemic administration of 5-HT_{1A} receptor agonists evoked biphasic actions on VTA dopaminergic neurons. Thus, low doses increased the firing rate and the number of bursts, whereas high doses resulted in an eventual decrease of firing (Arborelius *et al.*, 1993a; Prisco *et al.*, 1994; Lejeune *et al.*, 1997; Lejeune and Millan, 1998; Díaz-Mataix *et al.*, 2005). Consequently, low doses of 5-HT_{1A} receptor agonists increase dopamine release in the VTA (Chen and Reith, 1995) and the prefrontal cortex (Arborelius *et al.*, 1993b; Rasmusson *et al.*, 1994; Tanda *et al.*, 1994; Gobert *et al.*, 1998; Sakaue *et al.*, 2000), but sparing the nigrostriatal system. In line with these findings, the administration of selective 5-HT_{1A} agonists (8-OHDPAT and BAY x 3702) increased significantly the output of dopamine in the mPFC of wild-type, but not of 5-HT_{1A}

receptor knockout mice (Bortolozzi *et al.*, 2007), which further strengthens the role of 5-HT_{1A} receptors in controlling dopaminergic transmission. This effect is dependent on cortical 5-HT_{1A} receptors, since it vanished in rats subjected to PFC transection (Figure 4 Díaz-Mataix *et al.*, 2005). It is thus conceivable that low concentrations of 5-HT_{1A} receptor agonists preferentially activate 5-HT_{1A} receptors in GABAergic interneurons of the PFC. The presence of 5-HT_{1A} receptors in mPFC GABAergic interneurons (Santana *et al.*, 2004), together with the reversal of the stimulatory effect of BAY x 3702 by bicuculline, suggests that low concentrations of 5-HT_{1A} receptor agonists preferentially activate 5-HT_{1A} receptors in GABA interneurons. This may eventually result in disinhibition of pyramidal neurons projecting to the VTA. A higher concentration of such compounds may overcome this effect, activating directly pyramidal 5-HT_{1A} receptors and reducing the prefrontal excitatory output to dopaminergic neurons. Although such cellular difference in 5-HT_{1A} receptor sensitivity remains to be established, previous reports are consistent with this possibility (Sprouse and Aghajanian, 1986, 1987; Beck *et al.*, 1992), perhaps because of a different expression level in different neuronal populations (Hoyer and Boddeke, 1993).

Recently, 5-HT_{2A} receptors have been localized to both dopaminergic (about 20–40 percent of the population, according to different authors) and non-dopaminergic (putatively GABA-containing) neurons in the VTA (Cornea-Hébert *et al.*, 1999; Doherty and Pickel, 2000; Ikemoto *et al.*, 2000; Nocjar *et al.*, 2002). The 5-HT_{2A} receptor subtype in the VTA may be of particular relevance to some psychiatric illnesses, in particular schizophrenia, since 5-HT_{2A} receptor agonists such as LSD are potent hallucinogens, whereas atypical antipsychotic drugs are 5-HT_{2A} receptor antagonists (Schmidt *et al.*, 1995; Meltzer *et al.*, 2003). From a functional point of view, it has been observed that the administration of 5-HT_{2A} receptor ligands modulates dopaminergic transmission (Gobert and Millan 1999; Ichikawa *et al.*, 2001; Minabe *et al.*, 2001; Pehek *et al.*, 2001, 2002) and

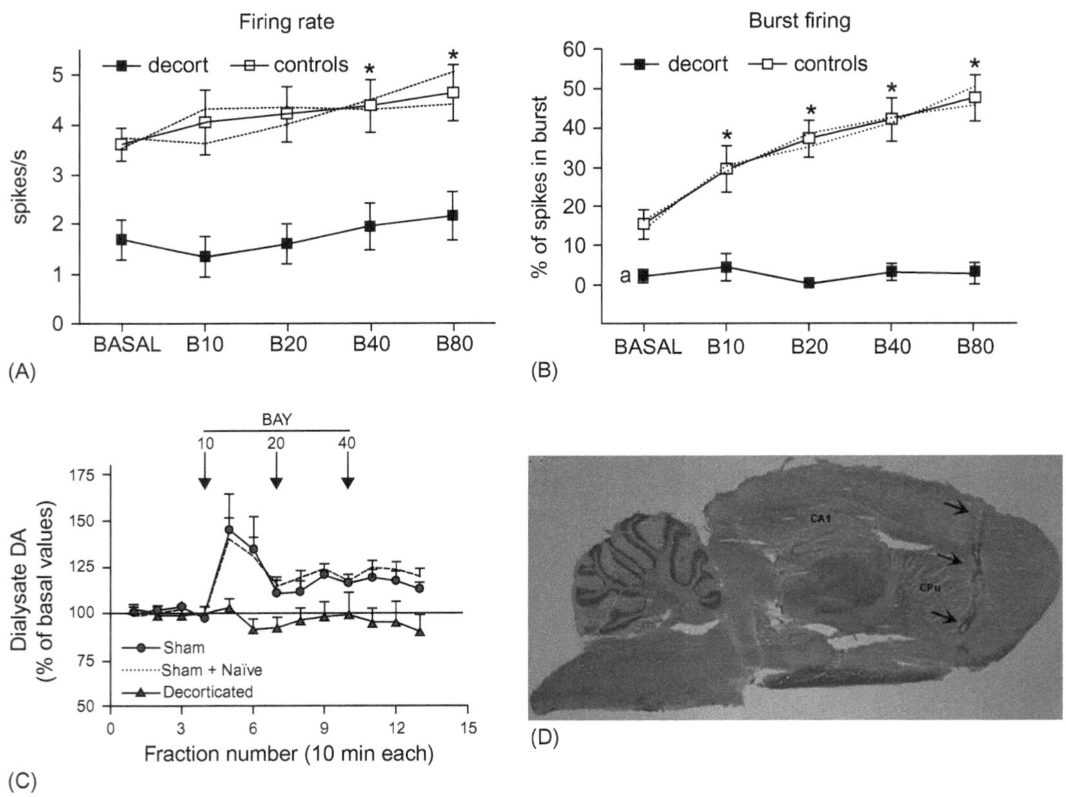

Figure 4 Dose–response curves of the 5-HT_{1A} receptor agonist BAY x 3702 (10–80µg/kg, i.v.) on the firing rate (A) and burst firing (B) of VTA dopaminergic neurons in sham-operated and cortically transected rats. (C) The intravenous administration of BAY x 3702 (10, 20 and 40µg/kg) elevated the extracellular concentration of dopamine in the VTA of sham-operated rats. The dotted line shows the effect in naïve and sham rats. This effect was totally abolished by cortical transection. (D) Sagittal section of a rat brain at the approximate lateral coordinate 3.4, taken from midline (Paxinos and Watson, 2005). The transection lesion is shown by arrows. $^{*}P < 0.05$ versus baseline; $^{a}P < 0.05$ versus controls. Data taken and redrawn from Díaz-Mataix *et al.* (2005), with permission.

recent results from our lab indicate that the activity of dopaminergic neurons of the VTA is under the excitatory control of 5-HT_{2A} receptors located in the PFC (Figure 5; Bortolozzi *et al.*, 2005).

Besides 5-HT_{2A} receptors, 5-HT_{2C} receptors play also a prominent role in the regulation of dopaminergic function. Thus, 5-HT_{2C} receptor agonists decrease the firing of VTA dopaminergic neurons (Prisco *et al.*, 1994; Di Giovanni *et al.*, 2000; Di Matteo *et al.*, 2000; Gobert *et al.*, 2000) and the release of dopamine in corticolimbic areas (Millan *et al.*, 1998; Di Giovanni *et al.*, 2000; Di Matteo *et al.*, 2000; Gobert *et al.*, 2000) through the activation of inhibitory, presumably GABAergic, interneurons (Di Giovanni *et al.*, 2001). In accord with this, 5-HT_{2C} receptors have been found in GABAergic neurons within the VTA (Pompeiano *et al.*, 1994; Eberle-Wang *et al.*, 1997; Bubar and Cunningham, 2007). It appears that 5-HT_{2C} receptors may exert a tonic inhibitory control of the activity of ascending dopamine neurons (Millan *et al.*, 1998; Di Matteo *et al.*, 1998, 1999; De Deurwaerdère and Spampinato, 1999; Di Giovanni *et al.*, 1999; Gobert *et al.*, 2000; Pozzi *et al.*, 2002), although this effect may depend on the activity of dopaminergic neurons (Pozzi *et al.*, 2002).

In addition to receptors belonging to the 5-HT_1 and 5-HT_2 families, a role for other 5-HT receptors in the control of the activity of dopamine neurons of the VTA has been suggested. Thus, the somatodendritic release of dopamine in the VTA is markedly increased by a 5-HT_3 receptor agonist, and slightly reduced by a 5-HT_3 receptor antagonist (Campbell *et al.*, 1996; Liu *et al.*, 2006; Rodd *et al.*, 2007). This is consistent with the finding that 5-HT_3 receptor antagonists reduce the number of spontaneously active VTA dopaminergic cells in rodents (Minabe *et al.*, 1991; Rasmussen *et al.*, 1991), and suggests that the somatodendritic release of dopamine in the VTA may be regulated by a local tonic excitatory action of 5-HT acting on 5-HT_3 receptors.

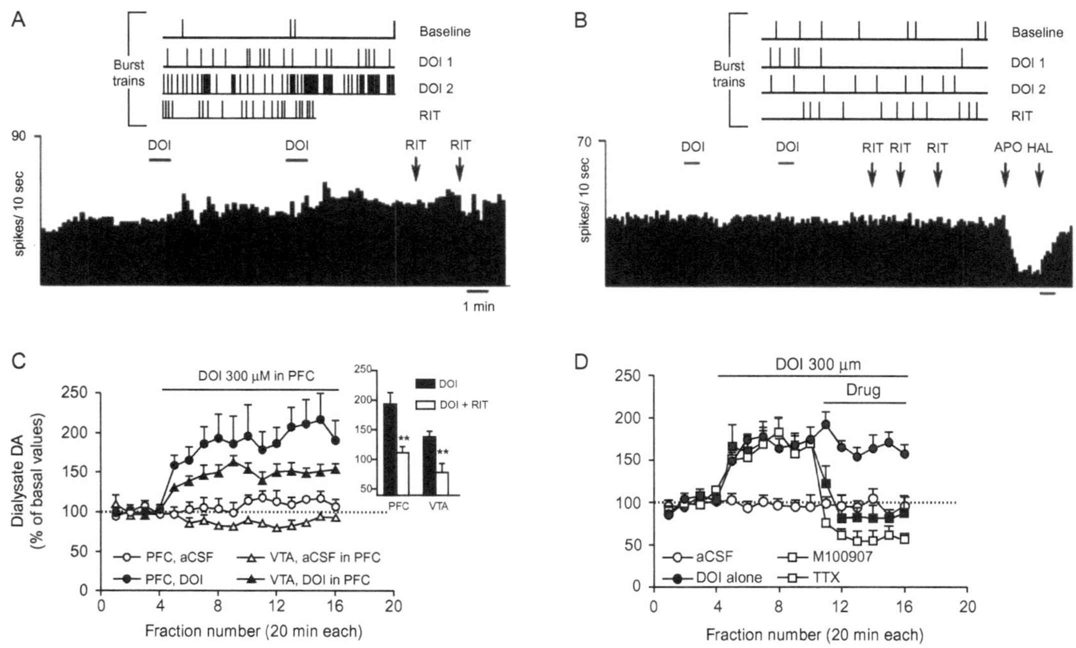

Figure 5 (A) Representative example of a ventral tegmental area (VTA) dopaminergic neuron whose burst firing was increased by the local application of the $5\text{-HT}_{2A/2C}$ receptor agonist DOI in the mPFC. Shown is the integrated firing rate histogram (abscissa, spikes/10s; ordinate, min). The four upper traces show burst trains in baseline conditions, after cortical DOI application (first and second doses) and after systemic ritanserin (RIT) administration. This unit had a baseline 1% of burst firing. The application of DOI (0.2 μl, 40 pmol, denoted by one 1-min horizontal bar) in the mPFC increased burst firing to 9 and 35% (first and second applications). The administration of ritanserin (1 + 1 mg/kg, i.v.) reduced burst firing to 14%. (B) Example of a VTA dopaminergic neuron insensitive to the application of DOI in the mPFC. The unit activity was suppressed by the administration of apomorphine (APO; 36 μg/kg, i.v.) and reversed by the subsequent administration of haloperidol (HAL; 0.4 mg/kg, i.v.). (C) In dual-probe microdialysis experiments, application of DOI in the mPFC increased dopamine (DA) release in the mPFC and VTA. Controls were perfused with artificial CSF (aCSF) for the entire experiment. Inset: reversal by ritanserin (300 μM, applied in the mPFC) of the elevation of dopamine release in the mPFC and VTA produced by the application of DOI (300 μM) in the mPFC. (D) In single probe experiments, the application of 300 μM DOI in the mPFC increased the local dopamine release. This effect was reversed by co-perfusion of 1 μM tetrodotoxin (TTX) and the selective 5-HT_{2A} receptor antagonist M100907. Controls were perfused with aCSF for the entire experiment. Horizontal bars show the time of drug application. $^{**}P < 0.001$ versus DOI alone. Data reproduced from Bortolozzi *et al.* (2005), with permission.

Influence of dopamine receptors on brain serotonin

In a reciprocal way, dopaminergic cells also control the activity of 5-HT neurons through projections from dopaminergic nuclei to the DR (Sakai *et al.*, 1977; Beckstead *et al.*, 1979; Kalén *et al.*, 1988; Peyron *et al.*, 1995; Kitahama *et al.*, 2000). In addition, a population of dopamine-containing neurons has been found within the boundaries of the DR (Hökfelt *et al.*, 1976; Descarries *et al.*, 1986; Geffard *et al.*, 1987; Kalén *et al.*, 1988; Hasue and Shammah-Lagnado, 2002; Lu *et al.*, 2006).

The activation of local dopamine D1 receptors with SKF 38393 does not alter 5-HT release in raphe nuclei (Ferré *et al.*, 1994; Adell and Artigas, 1999), which is consistent with the virtual absence of such receptors in these nuclei (Dawson *et al.*, 1988; Wamsley *et al.*, 1989). However, the systemic administration of the non-selective dopamine receptor agonist apomorphine enhances serotonergic cell firing and release in the DR, an effect fully antagonized by the D1 receptor antagonist SCH 23390 (Ferré *et al.*, 1994; Martín-Ruiz *et al.*, 2001a). This suggests that the activity of serotonergic neurons of the DR is partly under the control of D1 receptors located outside this nucleus. Furthermore, the effect of apomorphine is also attenuated by the D2/D3 receptor antagonist raclopride (Martín-Ruiz *et al.*, 2001a), which indicates an involvement of dopamine D2 receptors as well. The systemic administration of the dopamine D2 receptor agonist quinpirole slightly increases 5-HT cell firing and release in the DR (Martín-Ruiz *et al.*, 2001a). Nevertheless, the local application of quinpirole failed to alter the release of 5-HT in the DR and median raphe nucleus (MnR) (Adell and Artigas, 1999; Martín-Ruiz *et al.*, 2001a) or DR serotonergic cell firing (Martín-Ruiz *et al.*, 2001a). In summary, dopamine D1 and D2 receptors exert a facilitatory influence on the activity of serotonergic neurons in the DR. The differences between the results of local and systemic administration of the dopaminergic compounds seem to favor the view that these receptors are located outside the raphe nuclei. A possible location for such dopamine receptors is the substantia nigra, since the local perfusion of apomorphine in this region enhances the release of 5-HT in the DR (Martín-Ruiz *et al.*, 2001a).

Acetylcholine

The existence of an interaction between serotonin and acetylcholine at the level of DR is based on several facts. First, there is anatomical evidence that the DR receives a cholinergic input from the adjacent laterodorsal tegmentum (LDTg) and the pedunculopontine tegmentum (PPTg) (Woolf and Butcher, 1989; Wang *et al.*, 2000; Galindo-Charles *et al.*, 2008). Second, immunocytochemical studies have demonstrated the presence of postsynaptic $\alpha7$- and $\alpha4\beta2$-containing nicotinic acetylcholine receptors (nAChRs) in serotonergic and non-serotonergic neurons of the DR (Seguela *et al.*, 1993; Bitner *et al.*, 2000; Bitner and Nikkel, 2002; Galindo-Charles *et al.*, 2008). Third; electrophysiological studies using *in vitro* preparations have also shown that nicotine stimulated DR 5-HT release (Mihailescu *et al.*, 2002). Fourth, the systemic administration of nicotine increased the concentration of serotonin not only in the DR but also in terminal areas of the brain, such as the cortex, striatum, hippocampus and hypothalamus (Seth *et al.*, 2002). Fifth, postsynaptic $\alpha7$- and $\alpha4\beta2$-containing nAChRs in serotonergic and non-serotonergic neurons of the DR are fully functional as evidenced by electrophysiological studies using selective antagonists (Galindo-Charles *et al.*, 2008).

The activities of both cholinergic and serotonergic systems are elevated during arousal and wake states. However, in contrast to serotonergic cells of the DR, the majority of LDT and PPTg cholinergic neurons are highly active during both wakefulness and REM sleep (Steriade *et al.*, 1990a, 1990b). Furthermore, cholinergic LDT and PPTg neurons are tonically inhibited by 5-HT_{1A} receptor agonists during the waking state, but not during REM sleep state (Thakkar *et al.*, 1998). This distinct interaction may have a fundamental physiological implication in terms of the maintenance of the waking state by the cholinergic input to the DR (Hobson *et al.*, 2000).

GABA

Several experiments evidenced that serotonergic projection neurons and GABA-containing neurons present in the raphe nuclei mutually regulate each other's activity (Gallager and Aghajanian, 1976b; Forchetti and Meek, 1981). Serotonergic neurons of the raphe nuclei are under an inhibitory GABAergic influence not only from intrinsic GABA-containing cells (Belin *et al.*, 1979; Harandi *et al.*, 1987; Wang *et al.*, 1992) but also from the periaqueductal gray (Jolas and Aghajanian, 1997) and multiple distant sources (Gervasoni *et al.*, 2000). This is supported by a great number of GABAergic perikarya and nerve endings present in the DR and its vicinity (Nanopoulos *et al.*, 1982; Belin *et al.*, 1983; Harandi *et al.*, 1987; Stamp and Semba, 1995; Abellán *et al.*, 2000a).

Two are the GABA receptors involved in the regulation of serotonergic transmission – $GABA_A$ and $GABA_B$ receptors – and both subtypes have been found in the raphe nuclei (Bowery *et al.*, 1987; Chu *et al.*, 1990; Bischoff *et al.*, 1999; Lu *et al.*, 1999; Abellán *et al.*, 2000a, 2000b). $GABA_A$ receptors are expressed in 5-HT cells (Mennini *et al.*, 1986; Gao *et al.*, 1993), whereas $GABA_B$ receptors are present both in serotonergic neurons (Wirtshafter and Sheppard, 2001; Varga *et al.*, 2002; Serrats *et al.*, 2003) and, as autoreceptors, in GABA-containing neurons (Waldmeier *et al.*, 1988; Bowery, 1993; Serrats *et al.*, 2003). From a physiological viewpoint, it is well known that GABA inhibits DR 5-HT cell firing (Gallager and Aghajanian, 1976b; Wang and Aghajanian, 1977; Levine and Jacobs, 1992; Gervasoni *et al.*, 2000) as well as 5-HT synthesis and turnover (Forchetti and Meek, 1981; Nishikawa and Scatton, 1985), which results in a suppression of 5-HT release in forebrain (Becquet *et al.*, 1990; Abellán *et al.*, 2000b). In microdialysis studies, the local application of GABA did not alter 5-HT release in the DR (Abellán *et al.*, 2000b). However, the $GABA_A$ receptor agonist muscimol decreases the release of 5-HT in the DR (Tao *et al.*, 1996). Different influence of the GABAergic input in the raphe nuclei might explain such a discrepancy. The GABAergic tone on 5-HT neurons and the release of GABA in the DR changes across the sleep–wake cycle, being minimal during light conditions and increasing with sleep (Jacobs and Fornal, 1991; Nitz and Siegel, 1997; Gervasoni *et al.*, 2000). Thus, the different waking state of the animals, such as light conditions (Abellán *et al.*, 2000a, 2000b) or darkness (Tao *et al.*, 1996), could account for these inconsistent observations. The local application of the $GABA_A$ receptor antagonist bicuculline elevates 5-HT release in the DR (Tao *et al.*, 1996), which would suggest that GABAergic transmission exerts a tonic inhibition on 5-HT release in the DR through $GABA_A$ receptors. However, the fact that GABA inhibits 5-HT cell firing but not release suggests that at least a portion of 5-HT release in the DR does not depend on serotonergic cell firing.

Conflicting results have also been reported with respect to the influence of $GABA_B$ receptors on serotonergic transmission. Thus, both systemic and local administration of baclofen (a selective $GABA_B$ receptor agonist) were shown to increase (Abellán *et al.*, 2000a, 2000b) or decrease (Tao *et al.*, 1996) 5-HT release in the raphe nuclei. On the other hand, it seems that baclofen activates preferentially presynaptic $GABA_B$ receptors in the DR, as indicated by a pre- versus postsynaptic EC_{50} ratio of ~20 (Abellán *et al.*, 2000b). This is in good concordance with its actions in terminal areas (Howe *et al.*, 1987; Davies *et al.*, 1990). Hence, in the presence of an active GABAergic input onto 5-HT neurons (light conditions, see above), baclofen would remove the tonic $GABA_A$ input through its predominant activation of presynaptic $GABA_B$ receptors. In this manner, it might increase 5-HT release through an effect resembling that of bicuculline – i.e., $GABA_A$ disinhibition. In contrast, during darkness (a condition associated with a low GABAergic input onto 5-HT neurons) a direct inhibitory effect of baclofen on 5-HT neurons would prevail, thus resulting in a reduction of 5-HT release (Tao *et al.*, 1996). Furthermore, it is worth noting that $GABA_B$ receptor antagonists are without effect in either light or dark conditions (Tao *et al.*, 1996; Abellán *et al.*, 2000a). Therefore, it appears that $GABA_B$ receptors do not exert a tonic control on the release of 5-HT in the raphe nuclei.

The inhibitory GABAergic control of serotonergic neurons of the raphe nuclei is also under the influence of 5-HT receptors. Thus, in addition to the autoreceptor role of $5\text{-}HT_{1A}$ and $5\text{-}HT_{1B}$ receptor subtypes localized to serotonergic neurons, these receptors are also expressed in GABAergic interneurons within the raphe nuclei (Serrats, Mengod and Cortés, unpublished results). Consequent with the inhibitory function of $5\text{-}HT_{1A}$ and $5\text{-}HT_{1B}$ receptors, it has been shown that their stimulation by serotonin reduced the release of GABA *in vitro* (Bagdy *et al.*, 2000; see Harsing, 2006 for review).

Electrophysiological studies first suggested the presence of $5\text{-}HT_{2C}$ receptors on GABA-containing neurons, within or in the vicinity of the DR (Liu *et al.*, 2000). This has been recently supported by the observation that $5\text{-}HT_{2C}$ receptor mRNA existed in most GABAergic cells, recognized by the presence of glutamic acid decarboxylase mRNA, but was not detected in serotonin transporter mRNA-containing cells in the DR (Serrats *et al.*, 2005). Such $5\text{-}HT_{2C}$ receptor-positive GABAergic neurons were mainly located in the intermediolateral and lateral portions of the DR and provide an anatomical support to the proposed negative-feedback loop involving reciprocal connections between GABAergic interneurons bearing $5\text{-}HT_{2C}$ receptors and 5-HT neurons in the DR (Liu *et al.*, 2000; Boothman *et al.*, 2003; Boothman and Sharp, 2005). A possible role of DR $5\text{-}HT_{2A}$ receptors in this negative-feedback loop has also been suggested (Boothman *et al.*, 2003). Nevertheless, although $5\text{-}HT_{2A}$ receptor messenger RNA was detected in the DR, $5\text{-}HT_{2A}$ receptor immunoreactivity in this nucleus was not (Wright *et al.*, 1995; Cornea-Hébert *et al.*, 1999; Fay and Kubin, 2000). For this reason, the influence of $5\text{-}HT_{2A}$ receptor activation on DR serotonergic neurons may be mediated by an action on $5\text{-}HT_{2A}$ receptors located on distant sites, most likely the medial prefrontal cortex (Martín-Ruiz *et al.*, 2001b).

Glutamate

Numerous electrophysiological and biochemical studies have evidenced a strong glutamatergic input to the DR (Pan and Williams, 1989; Tao and Auerbach, 1996; Tao *et al.*, 1997; Celada *et al.*, 2001; Puig *et al.*, 2003;

Amargós-Bosch *et al.*, 2004; Gabbott *et al.*, 2005), which originates in the lateral habenula, periaqueductal gray, lateral hypothalamus and mPFC (Aghajanian and Wang, 1977; Kalén *et al.*, 1985, 1986; Hajós *et al.*, 1998; Peyron *et al.*, 1998; Vertes, 2004). Ionotropic glutamate receptors (iGluR) can be divided in three subtypes, namely AMPA, kainate and *N*-methyl-D-aspartate (NMDA) receptors, and each of them appears to be responsible for the glutamatergic stimulation of serotonergic cells in the DR. Thus, intracellular recordings in midbrain slice preparations have identified AMPA/kainate- and NMDA-mediated currents in DR 5-HT neurons (Pan and Williams, 1989). Accordingly, the activation of these neurons *in vivo* elicited by electrical stimulation of the mPFC is abolished by AMPA/kainate and NMDA receptor antagonists (Celada *et al.*, 2001). Likewise, *in vivo* microdialysis studies gives support to the involvement of the three iGluRs in the control of 5-HT release in the DR. Thus, kainate, which binds to both AMPA and kainate receptors (Hall *et al.*, 1994), elevates 5-HT release in DR in an impulse-dependent manner (Tao *et al.*, 1997; Tao and Auerbach, 2000). AMPA can also increase extracellular 5-HT in DR, but only if the rapid desensitization of the receptor is prevented (Tao *et al.*, 1997). The effects of AMPA/kainate receptor agonists in DR can be blocked by kynurenate or DNQX (Tao *et al.*, 1997), which by themselves fail to alter 5-HT release (Tao *et al.*, 1997). Similarly to AMPA/kainate receptor agonists, NMDA evokes an increase of 5-HT release in the DR *in vivo* (Tao and Auerbach, 1996) and in midbrain primary cultures (Becquet *et al.*, 1993). The effect of NMDA can be antagonized by simultaneous perfusion of competitive or non-competitive NMDA receptor antagonists (Tao and Auerbach, 1996, 2000). In contrast to iGluRs, there is a lack of information about the role of mGluRs in the control of the activity of serotonergic cells within the DR, probably because the scanty presence of these receptors in the raphe nuclei (Shigemoto and Mizuno, 2000).

Projection from prefrontal cortex to raphe nuclei: crossroad of multiple receptors and transmitters

Special attention must be paid to the projection from the mPFC to the DR (Hajós *et al.*, 1998; Peyron *et al.*, 1998; Vertes, 2004). The axons of 5 percent of pyramidal neurons located in layer V of mPFC project to DR (Gabbott *et al.*, 2005) and tightly control the activity of serotonergic cells of this nucleus (Hajós *et al.*, 1998; Celada *et al.*, 2001; Martín-Ruiz *et al.*, 2001b). Furthermore, this influence is dependent on the array of different receptors of multiple transmitters present in pyramidal neurons as well as GABAergic interneurons that control cortical circuits. Due to the connectivity of the mPFC with brainstem monoaminergic nuclei (Figure 6), this brain region constitutes a crossroads at which several transmitters, such as glutamate, GABA, 5-HT, dopamine and noradrenaline, interact. Therefore, the main excitatory afferents to the mPFC (mediodorsal nucleus of the thalamus and hippocampus) as well as the cortical output can be modulated in turn by serotonergic pathways arising from the midbrain raphe nuclei. These multiple interactions integrate excitatory and inhibitory inputs, and are of paramount importance for a correct operation of higher brain functions. As a rule, the stimulation of excitatory and inhibitory receptors located in pyramidal neurons will lead, respectively, to an increase or decrease of 5-HT cell firing and/or release. However, a great number of receptors have a dual location in both pyramidal cells and GABAergic interneurons, and the final response of serotonergic cells will depend on a balance of the actions of different transmitters on their respective receptors present in both cellular types. Therefore, the extracellular concentration of 5-HT in the mPFC can be regarded as an *in vivo* neurochemical index of the pyramidal influence on ascending midbrain 5-HT neurons. This experimental approach is based on several observations. First, as mentioned above, anatomical, electrophysiological and neurochemical studies indicate the presence of a very close reciprocal relationship between the mPFC and the midbrain raphe nuclei. Thus, the electrical stimulation of the mPFC elicited profound changes in most DR 5-HT neurons and *vice versa* (Hajós *et al.*, 1998; Celada *et al.*, 2001; Puig *et al.*, 2005). As a consequence, the electrical stimulation of the mPFC increased 5-HT release in the DR (Celada *et al.*, 2001). Second, the activation of excitatory (5-HT_{2A}, α_1-adrenergic, AMPA) or inhibitory (5-HT_{1A}, μ-opioid, mGluR2/3) receptors in mPFC increased and decreased, respectively, the local 5-HT release and – when examined – the firing of 5-HT neurons and 5-HT release in the DR (Celada *et al.*, 2001; Martín-Ruiz *et al.*, 2001b; Amargós-Bosch *et al.*, 2003, 2004; Bortolozzi *et al.*, 2003; Puig *et al.*, 2003). Similar results have been obtained by stimulating mPFC pyramidal neurons through a disinhibition of the thalamocortical pathway (Puig *et al.*, 2003; Amargós-Bosch *et al.*, 2007). Another 5-HT receptor, namely 5-HT_4 receptor, has been recently shown to exert a facilitatory influence on a subpopulation of DR 5-HT neurons (Lucas and Debonnel, 2002). Furthermore, this 5-HT_4-dependent facilitatory control on DR 5-HT neurons stems from the mPFC (Lucas *et al.*, 2005) and appears to be accounted for by a desensitization of 5-HT_{1A} autoreceptors in the DR (Lucas *et al.*, 2007). Most importantly, such an effect is achieved after only a 3-day treatment with a 5-HT_4 receptor agonist. Thus, these findings suggest that 5-HT_4 receptor agonists might be regarded as a new class of

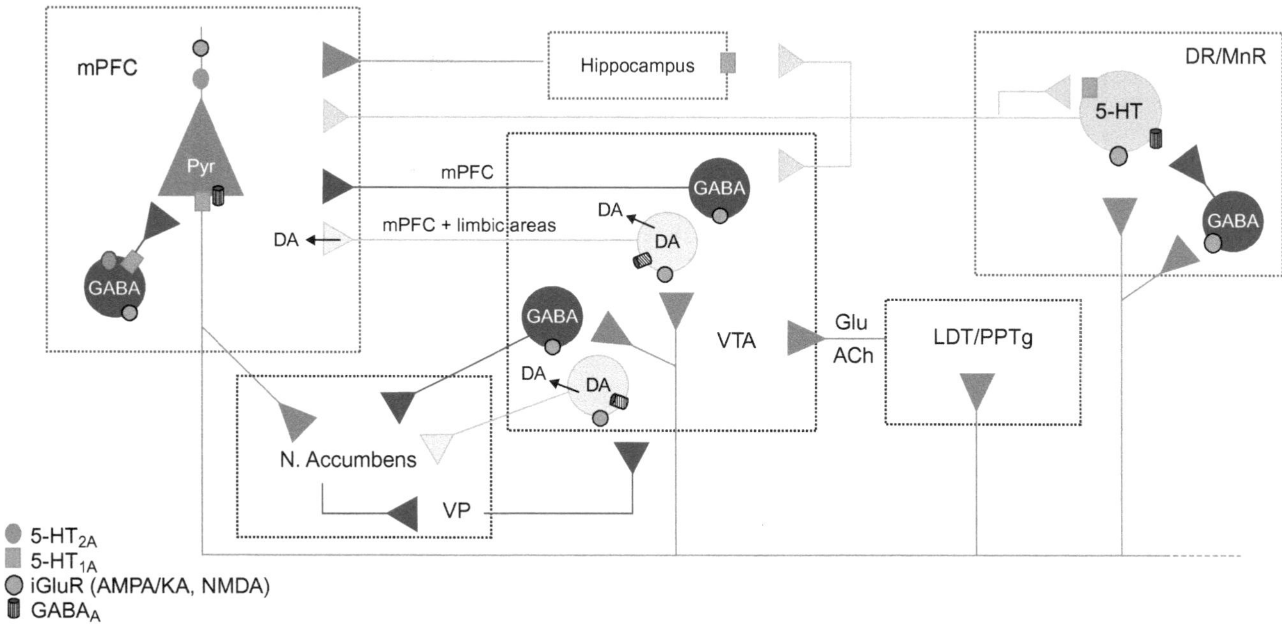

Figure 6 Anatomical and functional relationship between the mPFC, the VTA, and the dorsal and median raphe nuclei (DR/MnR). Pyramidal neurons of the mPFC project directly to mesocortical (but not mesoaccumbal) dopamine (DA) neurons, closing an mPFC-VTA circuit. GABAergic cells in the VTA project to mPFC and limbic areas as well. The mPFC may also modulate the activity of VTA neurons indirectly, through the basal ganglia circuit (e.g., mPFC nucleus accumbens ventral pallidum (VP) pathway) or through afferents to the LDT/PPTg. Likewise, the mPFC is reciprocally connected with the DR/MnR. Pyramidal neurons of the mPFC project to raphe 5-HT and GABA neurons and modulate their activity. In turn, 5-HT neurons modulate the activity of pyramidal cells in the mPFC through various receptors, in particular 5-HT_{1A} and 5-HT_{2A} receptors, which are expressed by pyramidal and GABAergic neurons. The pharmacological activation of these receptors in the mPFC has been shown to modulate 5-HT neuron activity and terminal 5-HT release via descending afferents to the midbrain. In addition, to the mPFC, 5-HT_{1A} receptors are densely expressed in the DR/MnR (autoreceptors) and in areas projecting to the mPFC, such as the hippocampal formation. Direct 5-HT afferents to the VTA may also be involved in the control of DA neurons by other 5-HT receptors, notably the 5-HT_{2C} subtype, present in GABAergic neurons of the VTA. Abbreviations: Glu, glutamate; Pyr, pyramidal neuron; Ach, acetylcholine; N. Accumbens, nucleus accumbens; iGluR, ionotropic glutamate receptor; KA, kainic acid. To see the full color version of this figure please refer to the color plate at the back of the book. Copies produced via our print on demand service do not contain color plates. If your copy does not have the color plate, please go to this website to view the figure in color www.elsevierdirect.com/companions/9780123746344

antidepressant drugs with a rapid onset of action. Taken together, all these effects are likely driven by changes in the excitatory output of the mPFC to DR and the subsequent change in serotonergic activity (Celada *et al.*, 2001; Martín-Ruiz *et al.*, 2001b).

The projections of the mPFC to other brainstem nuclei such as the VTA and the locus coeruleus underscore the importance of the modulation of the activity of the pyramidal cell in the mPFC by 5-HT receptors. For example, our research group has shown that 5-HT_{1A} receptors in the mPFC are deeply involved in the modulation of dopaminergic activity because the 5-HT_{1A} agonist, BAY x 3702, failed to alter dopaminergic cell activity or dopamine release in the VTA in cortically transected rats (Díaz-Mataix *et al.*, 2005). This complex interplay confers a crucial importance for the interactions between glutamate and other transmitter systems at the level of mPFC, and may help to elucidate the mechanisms involved in the elevation of mesocortical dopamine release produced by atypical antipsychotics.

Acknowledgements

This work was supported by the Spanish Ministry of Health (FIS Grant PI070111 to Albert Adell and FIS Grant PI060264 to Pau Celada), the Spanish Ministry of Education and Science (Grant SAF 2007-62378 to Francesc Artigas), and the Generalitat de Catalunya (SGR2005/00758). Analía Bortolozzi is supported by a Ramón y Cajal Research contract from the Ministry of Education and Science. Pau Celada is supported by the Researcher Stabilization Program of the Health Department of the Generalitat de Catalunya.

References

Abellán, M.T., Adell, A., Honrubia, M.A., Mengod, G. and Artigas, F. (2000a) $GABA_B$-RI receptors in serotonergic neurons: effects of baclofen on 5-HT output in rat brain. Neuroreport, 11: 941–945.

Abellán, M.T., Jolas, T., Aghajanian, G.K. and Artigas, F. (2000b) Dual control of dorsal raphe serotonergic neurons by $GABA_B$ receptors. Electrophysiological and microdialysis studies. Synapse, 36: 21–34.

Adell, A. and Artigas, F. (1999) Regulation of the release of 5-hydroxytryptamine in the median raphe nucleus of the rat by catecholaminergic afferents. Eur. J. Neurosci., 11: 2305–2311.

Adell, A., Celada, P., Abellán, M.T. and Artigas, F. (2002) Origin and functional role of the extracellular serotonin in the midbrain raphe nuclei. Brain Res. Brain Res. Rev., 39: 154–180.

Aghajanian, G.K. and Wang, R.Y. (1977) Habenular and other midbrain raphe afferents demonstrated by a modified retrograde tracing technique. Brain Res., 122: 229–242.

Amargós-Bosch, M., Adell, A., Bortolozzi, A. and Artigas, F. (2003) Stimulation of α_1-adrenoceptors in the rat medial prefrontal cortex increases the local *in vivo* 5-hydroxytryptamine release: reversal by antipsychotic drugs. J. Neurochem., 87: 831–842.

Amargós-Bosch, M., Bortolozzi, A., Puig, M.V., Serrats, J., Adell, A., Celada, P., Toth, M., Mengod, G. and Artigas, F. (2004) Co-expression and *in vivo* interaction of $serotonin_{1A}$ and $serotonin_{2A}$ receptors in pyramidal neurons of prefrontal cortex. Cereb. Cortex, 14: 281–299.

Amargós-Bosch, M., Adell, A. and Artigas, F. (2007) Antipsychotic drugs reverse the AMPA receptor-stimulated release of 5-HT in the medial prefrontal cortex. J. Neurochem., 102: 550–561.

Araneda, R. and Andrade, R. (1991) $5\text{-Hydroxytryptamine}_2$ and $5\text{-hydroxytryptamine}_{1A}$ receptors mediate opposing responses on membrane excitability in rat association cortex. Neuroscience, 40: 399–412.

Arborelius, L., Chergui, K., Murase, S., Nomikos, G.G., Hook, B.B., Chouvet, G., Hacksell, U. and Svensson, T.H. (1993a) The 5-HT1A receptor selective ligands, (R)-8-OH-DPAT and (S)-UH-301, differentially affect the activity of midbrain dopamine neurons. Naunyn Schmiedebergs Arch. Pharmacol., 347: 353–362.

Arborelius, L., Nomikos, G.G., Hacksell, U. and Svensson, T.H. (1993b) (R)-8-OH-DPAT preferentially increases dopamine release in rat medial prefrontal cortex. Acta Physiol. Scand., 148: 465–466.

Arnsten, A.F. and Goldman-Rakic, P.S. (1984) Selective prefrontal cortical projections to the region of the locus coeruleus and raphe nuclei in the rhesus monkey. Brain Res., 306: 9–18.

Bagdy, E., Kiraly, I. and Harsing, L.G. Jr. (2000) Reciprocal innervation between serotonergic and GABAergic neurons in raphe nuclei of the rat. Neurochem. Res., 25: 1465–1473.

Baraban, J.M. and Aghajanian, G.K. (1980) Suppression of firing activity of 5-HT neurons in the dorsal raphe by alpha-adrenoceptor antagonists. Neuropharmacology, 19: 355–363.

Beart, P.M. and McDonald, D. (1982) 5-Hydroxytryptamine and 5-hydroxytryptaminergic-dopaminergic interactions in the ventral tegmental area of rat brain. J. Pharm. Pharmacol., 34: 591–593.

Beck, S.G., Choi, K.C. and List, T.J. (1992) Comparison of $5\text{-hydroxytryptamine}_{1A}$-mediated hyperpolarization in CA1 and CA3 hippocampal pyramidal cells. J. Pharmacol. Exp. Ther., 263: 350–359.

Beckstead, R.M., Domesick, V.B. and Nauta, W.J. (1979) Efferent connections of the substantia nigra and ventral tegmental area in the rat. Brain Res., 175: 191–217.

Becquet, D., Faudon, M. and Hery, F. (1990) The role of serotonin release and autoreceptors in the dorsalis raphe nucleus in the control of serotonin release in the cat caudate nucleus. Neuroscience, 39: 639–647.

Becquet, D., Hery, M., Deprez, P., Faudon, M., Fache, M.P., Giraud, P. and Hery, F. (1993) *N*-methyl-D-aspartic acid/glycine interactions on the control of 5-hydroxytryptamine release in raphe primary cultures. J. Neurochem., 61: 1692–1697.

Belin, M.F., Aguera, M., Tappaz, M., McRae-Degueurce, A., Bobillier, P. and Pujol, J.F. (1979) GABA-accumulating neurons in the nucleus raphe dorsalis and periaqueductal gray in the rat: a biochemical and radioautographic study. Brain Res., 170: 279–297.

Belin, M.F., Nanopoulos, D., Didier, M., Aguera, M., Steinbusch, H., Verhofstad, A., Maitre, M. and Pujol, J.F. (1983) Immunohistochemical evidence for the presence of gamma-aminobutyric acid and serotonin in one nerve cell. A study on the raphe nuclei of the rat using antibodies to glutamate decarboxylase and serotonin. Brain Res., 275: 329–339.

Bischoff, S., Leonhard, S., Reymann, N., Schuler, V., Shigemoto, R., Kaupmann, K. and Bettler, B. (1999) Spatial distribution of $GABA_BR1$ receptor mRNA and binding sites in the rat brain. J. Comp. Neurol., 412: 1–16.

Bitner, R.S. and Nikkel, A.L. (2002) Alpha-7 nicotinic receptor expression by two distinct cell types in the dorsal raphe nucleus and locus coeruleus of rat. Brain Res., 938: 45–54.

Bitner, R.S., Nikkel, A.L., Curzon, P., Donnelly-Roberts, D.L., Puttfarcken, P.S., Namovic, M., Jacobs, I.C., Meyer, M.D. and Decker, M.W. (2000) Reduced nicotinic receptor-mediated antinociception following *in vivo* antisense knock-down in rat. Brain Res., 871: 66–74.

Boothman, L.J. and Sharp, T. (2005) A role for midbrain raphe γ aminobutyric acid neurons in 5-hydroxytryptamine feedback control. Neuroreport, 16: 891–896.

Boothman, L.J., Allers, K.A., Rasmussen, K. and Sharp, T. (2003) Evidence that central 5-HT_{2A} and $5\text{-HT}_{2B/C}$ receptors regulate 5-HT cell firing in the dorsal raphe nucleus of the anaesthetised rat. Br. J. Pharmacol., 139: 998–1004.

Bortolozzi, A. and Artigas, F. (2003) Control of 5-hydroxytryptamine release in the dorsal raphe nucleus by the noradrenergic system in rat brain. Role of α-adrenoceptors. Neuropsychopharmacology, 28: 421–434.

Bortolozzi, A., Díaz-Mataix, L., Scorza, M.C., Celada, P. and Artigas, F. (2005) The activation of 5-HT_{2A} receptors in prefrontal cortex enhances dopaminergic activity. J. Neurochem., 95: 1597–1607.

Bortolozzi, A., Díaz-Mataix, L., Toth, M., Celada, P. and Artigas, F. (2007) *In vivo* actions of aripiprazole on serotonergic and dopaminergic systems in rodent brain. Psychopharmacology, 191: 745–758.

Bowery, N.G. (1993) $GABA_B$ receptor pharmacology. Annu. Rev. Pharmacol. Toxicol., 33: 109–147.

Bowery, N.G., Hudson, A.L. and Price, G.W. (1987) $GABA_A$ and $GABA_B$ receptor site distribution in the rat central nervous system. Neuroscience, 20: 365–383.

Broderick, P.A. and Phelix, C.F. (1997) I. Serotonin (5-HT) within dopamine reward circuits signals open-field behavior. II. Basis for 5-HT–DA interaction in cocaine dysfunctional behavior. Neurosci. Biobehav. Rev., 21: 227–260.

Bubar, M.J. and Cunningham, K.A. (2007) Distribution of serotonin 5-HT_{2C} receptors in the ventral tegmental area. Neuroscience, 146: 286–297.

Burnet, P.W., Eastwood, S.L., Lacey, K. and Harrison, P.J. (1995) The distribution of 5-HT_{1A} and 5-HT_{2A} receptor mRNA in human brain. Brain Res., 676: 157–168.

Cahir, M., Mawhinney, T. and King, D.J. (2004) Differential region-specific regulation of central α_1-adrenoceptor binding following chronic haloperidol and clozapine administration in the rat. Psychopharmacology, 172: 196–201.

Cameron, D.L., Wessendorf, M.W. and Williams, J.T. (1997) A subset of ventral tegmental area neurons is inhibited by dopamine, 5-hydroxytryptamine and opioids. Neuroscience, 77: 155–166.

Campbell, A.D., Kohl, R.R. and McBride, W.J. (1996) Serotonin-3 receptor and ethanol-stimulated somatodendritic dopamine release. Alcohol, 13: 569–574.

Celada, P., Puig, M.V., Casanovas, J.M., Guillazo, G. and Artigas, F. (2001) Control of dorsal raphe serotonergic neurons by the medial prefrontal cortex: involvement of serotonin-1A, $GABA_A$, and glutamate receptors. J. Neurosci., 21: 9917–9929.

Chen, N.H. and Reith, M.E. (1995) Monoamine interactions measured by microdialysis in the ventral tegmental area of rats treated systemically with (±)-8-hydroxy-2-(di-n-propylamino)tetralin. J. Neurochem., 64: 1585–1597.

Chu, D.C., Albin, R.L., Young, A.B. and Penney, J.B. (1990) Distribution and kinetics of $GABA_B$ binding sites in rat central nervous system: a quantitative autoradiographic study. Neuroscience, 34: 341–357.

Cornea-Hébert, V., Riad, M., Wu, C., Singh, S.K. and Descarries, L. (1999) Cellular and subcellular distribution of the serotonin 5-HT2A receptor in the central nervous system of adult rat. J. Comp. Neurol., 409: 187–209.

Davies, C.H., Davies, S.N. and Collingridge, G.L. (1990) Paired-pulse depression of monosynaptic GABA-mediated inhibitory postsynaptic responses in rat hippocampus. J. Physiol., 424: 513–531.

Dawson, T.M., Barone, P., Sidhu, A., Wamsley, J.K. and Chase, T.N. (1988) The D1 dopamine receptor in the rat brain: quantitative autoradiographic localization using an iodinated ligand. Neuroscience, 26: 83–100.

Day, H.E., Campeau, S., Watson, S.J. Jr. and Akil, H. (1997) Distribution of α_{1a}-, α_{1b}- and α_{1d}-adrenergic receptor mRNA in the rat brain and spinal cord. J. Chem. Neuroanat., 13: 115–139.

de Boer, T.H., Nefkens, F., van Helvoirt, A. and van Delft, A.M. (1996) Differences in modulation of noradrenergic and serotonergic transmission by the alpha-2 adrenoceptor antagonists, mirtazapine, mianserin and idazoxan. J. Pharmacol. Exp. Ther., 277: 852–860.

De Deurwaerdère, P. and Spampinato, U. (1999) Role of serotonin$_{2A}$ and serotonin$_{2B/2C}$ receptor subtypes in the control of accumbal and striatal dopamine release elicited *in vivo* by dorsal raphe nucleus electrical stimulation. J. Neurochem., 73: 1033–1042.

Descarries, L., Berthelet, F., Garcia, S. and Beaudet, A. (1986) Dopaminergic projection from nucleus raphe dorsalis to neostriatum in the rat. J. Comp. Neurol., 249: 511–515.

Díaz-Mataix, L., Scorza, M.C., Bortolozzi, A., Toth, M., Celada, P. and Artigas, F. (2005) Involvement of 5-HT_{1A} receptors in prefrontal cortex in the modulation of dopaminergic activity: role in atypical antipsychotic action. J. Neurosci., 25: 10831–10843.

Di Giovanni, G., De Deurwaerdère, P., Di Mascio, M., Di Matteo, V., Esposito, E. and Spampinato, U. (1999) Selective blockade of serotonin-2C/2B receptors enhances mesolimbic and mesostriatal dopaminergic function: a combined *in vivo* electrophysiological and microdialysis study. Neuroscience, 91: 587–597.

Di Giovanni, G., Di Matteo, V., Di Mascio, M. and Esposito, E. (2000) Preferential modulation of mesolimbic vs. nigrostriatal dopaminergic function by serotonin$_{2C/2B}$ receptor agonists: a combined *in vivo* electrophysiological and microdialysis study. Synapse, 35: 53–61.

Di Giovanni, G., Di Matteo, V., La Grutta, V. and Esposito, E. (2001) *m*-Chlorophenylpiperazine excites non-dopaminergic neurons in the rat substantia nigra and ventral tegmental area by activating serotonin-2C receptors. Neuroscience, 103: 111–116.

Di Matteo, V., Di Giovanni, G., Di Mascio, M. and Esposito, E. (1998) Selective blockade of serotonin$_{2C/2B}$ receptors enhances dopamine release in the rat nucleus accumbens. Neuropharmacology, 37: 265–272.

Di Matteo, V., Di Giovanni, G., Di Mascio, M. and Esposito, E. (1999) SB 242084, a selective serotonin$_{2C}$ receptor antagonist, increases dopaminergic transmission in the mesolimbic system. Neuropharmacology, 38: 1195–1205.

Di Matteo, V., Di Giovanni, G., Di Mascio, M. and Esposito, E. (2000) Biochemical and electrophysiological evidence that Ro 60-0175 inhibits mesolimbic dopaminergic function through serotonin$_{2C}$ receptors. Brain Res., 865: 85–90.

Doherty, M.D. and Pickel, V.M. (2000) Ultrastructural localization of the serotonin 2A receptor in dopaminergic neurons in the ventral tegmental area. Brain Res., 864: 176–185.

Doherty, M.D. and Pickel, V.M. (2001) Targeting of serotonin 1A receptors to dopaminergic neurons within the parabrachial subdivision of the ventral tegmental area in rat brain. J. Comp. Neurol., 433: 390–400.

Domyancic, A.V. and Morilak, D.A. (1997) Distribution of α_{1A} adrenergic receptor mRNA in the rat brain visualized by *in situ* hybridization. J. Comp. Neurol., 386: 358–378.

Eberle-Wang, K., Mikeladze, Z., Uryu, K. and Chesselet, M.F. (1997) Pattern of expression of the serotonin$_{2C}$ receptor messenger RNA in the basal ganglia of adult rats. J. Comp. Neurol., 384: 233–247.

Fay, R. and Kubin, L. (2000) Pontomedullary distribution of 5-HT2A receptor-like protein in the rat. J. Comp. Neurol., 418: 323–345.

Ferré, S., Cortés, R. and Artigas, F. (1994) Dopaminergic regulation of the serotonergic raphe-striatal pathway: microdialysis studies in freely moving rats. J. Neurosci., 14: 4839–4846.

Forchetti, C.M. and Meek, J.L. (1981) Evidence for a tonic GABAergic control of serotonin neurons in the median raphe nucleus. Brain Res., 206: 208–212.

Freedman, J.E. and Aghajanian, G.K. (1984) Idazoxan (RX 781094. selectively antagonizes α_2-adrenoceptors on rat central neurons. Eur. J. Pharmacol., 105: 265–272.

Fuster, J.M. (2001) The prefrontal cortex – an update: time is of the essence. Neuron, 30: 319–333.

Gabbott, P.L., Warner, T.A., Jays, P.R., Salway, P. and Busby, S.J. (2005) Prefrontal cortex in the rat: projections to subcortical autonomic, motor, and limbic centers. J. Comp. Neurol., 492: 145–177.

Galindo-Charles, L., Hernandez-Lopez, S., Galarraga, E., Tapia, D., Bargas, J., Garduno, J., Frias-Dominguez, C., Drucker-Colin, R. and Mihailescu, S. (2008) Serotoninergic dorsal raphe neurons possess functional postsynaptic nicotinic acetylcholine receptors. Synapse, 62: 601–615.

Gallager, D.W. and Aghajanian, G.K. (1976a) Effect of antipsychotic drugs on the firing of dorsal raphe cells. I. Role of adrenergic system. Eur. J. Pharmacol., 39: 341–355.

Gallager, D.W. and Aghajanian, G.K. (1976b) Effect of antipsychotic drugs on the firing of dorsal raphe cells. II. Reversal by picrotoxin. Eur. J. Pharmacol., 39: 357–364.

Gao, B., Fritschy, J.M., Benke, D. and Mohler, H. (1993) Neuron-specific expression of $GABA_A$-receptor subtypes: differential association of the α1- and α3-subunits with serotonergic and GABAergic neurons. Neuroscience, 54: 881–892.

Garratt, J.C., Crespi, F., Mason, R. and Marsden, C.A. (1991) Effects of idazoxan on dorsal raphe 5-hydroxytryptamine neuronal function. Eur. J. Pharmacol., 193: 87–93.

Geffard, M., Tuffet, S., Mons, N. and Chagnaud, J.L. (1987) Simultaneous detection of indoleamines and dopamine in rat dorsal raphe nuclei using specific antibodies. Histochemistry, 88: 61–64.

Gervais, J. and Rouillard, C. (2000) Dorsal raphe stimulation differentially modulates dopaminergic neurons in the ventral tegmental area and substantia nigra. Synapse, 35: 281–291.

Gervasoni, D., Peyron, C., Rampon, C., Barbagli, B., Chouvet, G., Urbain, N., Fort, P. and Luppi, P.H. (2000) Role and origin of the GABAergic innervation of dorsal raphe serotonergic neurons. J. Neurosci., 20: 4217–4225.

Gobert, A. and Millan, M.J. (1999) Serotonin $(5\text{-HT})_{2A}$ receptor activation enhances dialysate levels of dopamine and noradrenaline, but not 5-HT, in the frontal cortex of freely-moving rats. Neuropharmacology, 38: 315–317.

Gobert, A., Rivet, J.M., Audinot, V., Newman-Tancredi, A., Cistarelli, L. and Millan, M.J. (1998) Simultaneous quantification of serotonin, dopamine and noradrenaline levels in single frontal cortex dialysates of freely-moving rats reveals a complex pattern of reciprocal auto- and heteroreceptor-mediated control of release. Neuroscience, 84: 413–429.

Gobert, A., Rivet, J.M., Lejeune, F., Newman-Tancredi, A., Adhumeau-Auclair, A., Nicolas, J.P., Cistarelli, L., Melon, C. and Millan, M.J. (2000) $Serotonin_{2C}$ receptors tonically suppress the activity of mesocortical dopaminergic and adrenergic, but not serotonergic, pathways: a combined dialysis and electrophysiological analysis in the rat. Synapse, 36: 205–221.

Groenewegen, H.J. and Uylings, H.B. (2000) The prefrontal cortex and the integration of sensory, limbic and autonomic information. Prog. Brain Res., 126: 3–28.

Gronier, B. (2008) Involvement of glutamate neurotransmission and *N*-methyl-d-aspartate receptor in the activation of midbrain dopamine neurons by 5-HT1A receptor agonists: an electrophysiological study in the rat. Neuroscience, 156: 995–1004.

Haddjeri, N., Blier, P. and de Montigny, C. (1996) Effect of the alpha-2-adrenoceptor antagonist mirtazapine on the 5-hydroxytryptamine system in the rat brain. J. Pharmacol. Exp. Ther., 277: 861–871.

Haddjeri, N., de Montigny, C. and Blier, P. (1997) Modulation of the firing activity of noradrenergic neurones in the rat locus coeruleus by the 5-hydroxtryptamine system. Br. J. Pharmacol., 120: 865–875.

Hajós, M., Richards, C.D., Szekely, A.D. and Sharp, T. (1998) An electrophysiological and neuroanatomical study of the medial prefrontal cortical projection to the midbrain raphe nuclei in the rat. Neuroscience, 87: 95–108.

Hall, R.A., Kessler, M. and Lynch, G. (1994) Kainate binding to the AMPA receptor in rat brain. Neurochem. Res., 19: 777–782.

Halliday, G., Harding, A. and Paxinos, G. (1995) Serotonin and tachykynin systems. In: Paxinos, G. (Ed.), The Rat Nervous System. Academic Press, San Diego, CA, pp. 929–974.

Harandi, M., Aguera, M., Gamrani, H., Didier, M., Maitre, M., Calas, A. and Belin, M.F. (1987) γ-Aminobutyric acid and 5-hydroxytryptamine interrelationship in the rat nucleus raphe dorsalis: combination of radioautographic and immunocytochemical techniques at light and electron microscopy levels. Neuroscience, 21: 237–251.

Harsing, L.G. (2006) The pharmacology of the neurochemical transmission in the midbrain raphe nuclei of the rat. Curr. Neuropharmacol., 4: 313–339.

Hasue, R.H. and Shammah-Lagnado, S.J. (2002) Origin of the dopaminergic innervation of the central extended amygdala and accumbens shell: a combined retrograde tracing and immunohistochemical study in the rat. J. Comp. Neurol., 454: 15–33.

Hervé, D., Pickel, V.M., Joh, T.H. and Beaudet, A. (1987) Serotonin axon terminals in the ventral tegmental area of the rat: fine structure and synaptic input to dopaminergic neurons. Brain Res., 435: 71–83.

Hjorth, S., Bengtsson, H.J., Milano, S., Lundberg, J.F. and Sharp, T. (1995) Studies on the role of 5-HT_{1A} autoreceptors and α_1-adrenoceptors in the inhibition of 5-HT release – I. BMY7378 and prazosin. Neuropharmacology, 34: 615–620.

Hobson, J.A., Pace-Schott, E.F. and Stickgold, R. (2000) Dreaming and the brain: toward a cognitive neuroscience of conscious states. Behav. Brain Sci., 23: 793–842.

Hökfelt, T., Johansson, O., Fuxe, K., Goldstein, M. and Park, D. (1976) Immunohistochemical studies on the localization and distribution of monoamine neuron systems in the rat brain. I. Tyrosine hydroxylase in the mes- and diencephalon. Med. Biol., 54: 427–453.

Hopwood, S.E. and Stamford, J.A. (2001) Noradrenergic modulation of serotonin release in rat dorsal and median raphe nuclei via α_1 and α_{2A} adrenoceptors. Neuropharmacology, 41: 433–442.

Howe, J.R., Sutor, B. and Zieglgansberger, W. (1987) Baclofen reduces post-synaptic potentials of rat cortical neurones by an action other than its hyperpolarizing action. J. Physiol., 384: 539–569.

Hoyer, D. and Boddeke, H.W. (1993) Partial agonists, full agonists, antagonists: dilemmas of definition. Trends Pharmacol. Sci., 14: 270–275.

Ichikawa, J., Dai, J. and Meltzer, H.Y. (2001) DOI, a $5\text{-HT}_{2A/2C}$ receptor agonist, attenuates clozapine-induced cortical dopamine release. Brain Res., 907: 151–155.

Ikemoto, K., Nishimura, A., Okado, N., Mikuni, M., Nishi, K. and Nagatsu, I. (2000) Human midbrain dopamine neurons express serotonin 2A receptor: an immunohistochemical demonstration. Brain Res., 853: 377–380.

Jacobs, B.L. and Azmitia, E.C. (1992) Structure and function of the brain serotonin system. Physiol. Rev., 72: 165–229.

Jacobs, B.L. and Fornal, C.A. (1991) Activity of brain serotonergic neurons in the behaving animal. Pharmacol. Rev., 43: 563–578.

Jodo, E., Chiang, C. and Aston-Jones, G. (1998) Potent excitatory influence of prefrontal cortex activity on noradrenergic locus coeruleus neurons. Neuroscience, 83: 63–79.

Jolas, T. and Aghajanian, G.K. (1997) Opioids suppress spontaneous and NMDA-induced inhibitory postsynaptic currents in the dorsal raphe nucleus of the rat *in vitro*. Brain Res., 755: 229–245.

Kaehler, S.T., Singewald, N. and Philippu, A. (1999) Dependence of serotonin release in the locus coeruleus on dorsal raphe neuronal activity. Naunyn Schmiedebergs Arch. Pharmacol., 359: 386–393.

Kalén, P., Karlson, M. and Wiklund, L. (1985) Possible excitatory amino acid afferents to nucleus raphe dorsalis of the rat investigated with retrograde wheat germ agglutinin and D-[^{3}H]aspartate tracing. Brain Res., 360: 285–297.

Kalén, P., Pritzel, M., Nieoullon, A. and Wiklund, L. (1986) Further evidence for excitatory amino acid transmission in the lateral habenular projection to the rostral raphe nuclei: lesion-induced decrease of high affinity glutamate uptake. Neurosci. Lett., 68: 35–40.

Kalén, P., Skagerberg, G. and Lindvall, O. (1988) Projections from the ventral tegmental area and mesencephalic raphe to the dorsal raphe nucleus in the rat. Evidence for a minor dopaminergic component. Exp. Brain Res., 73: 69–77.

Kitahama, K., Nagatsu, I., Geffard, M. and Maeda, T. (2000) Distribution of dopamine-immunoreactive fibers in the rat brainstem. J. Chem. Neuroanat., 18: 1–9.

Leger, L., Charnay, Y., Hof, P.R., Bouras, C. and Cespuglio, R. (2001) Anatomical distribution of serotonin-containing neurons and axons in the central nervous system of the cat. J. Comp. Neurol., 433: 157–182.

Lejeune, F. and Millan, M.J. (1998) Induction of burst firing in ventral tegmental area dopaminergic neurons by activation of serotonin (5-HT)1A receptors: WAY 100,635-reversible actions of the highly selective ligands, flesinoxan and S 15535. Synapse, 30: 172–180.

Lejeune, F., Newman-Tancredi, A., Audinot, V. and Millan, M.J. (1997) Interactions of (+)- and (−)-8- and 7-hydroxy-2-(di-n-propylamino)tetralin at human (h)D3, hD2 and h serotonin$_{1A}$ receptors and their modulation of the activity of serotoninergic and dopaminergic neurones in rats. J. Pharmacol. Exp. Ther., 280: 1241–1249.

Levine, E.S. and Jacobs, B.L. (1992) Neurochemical afferents controlling the activity of serotonergic neurons in the dorsal raphe nucleus: microiontophoretic studies in the awake cat. J. Neurosci., 12: 4037–4044.

Liu, R., Jolas, T. and Aghajanian, G. (2000) Serotonin 5-HT$_2$ receptors activate local GABA inhibitory inputs to serotonergic neurons of the dorsal raphe nucleus. Brain Res., 873: 34–45.

Liu, W., Thielen, R.J., Rodd, Z.A. and McBride, W.J. (2006) Activation of serotonin-3 receptors increases dopamine release within the ventral tegmental area of Wistar and alcohol-preferring (P) rats. Alcohol, 40: 167–176.

Lu, J., Jhou, T.C. and Saper, C.B. (2006) Identification of wake-active dopaminergic neurons in the ventral periaqueductal gray matter. J. Neurosci., 26: 193–202.

Lu, X.Y., Ghasemzadeh, M.B. and Kalivas, P.W. (1999) Regional distribution and cellular localization of γ-aminobutyric acid subtype 1 receptor mRNA in the rat brain. J. Comp. Neurol., 407: 166–182.

Lucas, G. and Debonnel, G. (2002) 5-HT$_4$ receptors exert a frequency-related facilitatory control on dorsal raphé nucleus 5-HT neuronal activity. Eur. J. Neurosci., 16: 817–822.

Lucas, G., Compan, V., Charnay, Y., Neve, R.L., Nestler, E.J., Bockaert, J., Barrot, M. and Debonnel, G. (2005) Frontocortical 5-HT$_4$ receptors exert positive feedback on serotonergic activity: viral transfections, subacute and chronic treatments with 5-HT$_4$ agonists. Biol. Psychiatry, 57: 918–925.

Lucas, G., Rymar, V.V., Du, J., Mnie-Filali, O., Bisgaard, C., Manta, S., Lambas-Senas, L., Wiborg, O., Haddjeri, N., Piñeyro, G., Sadikot, A.F. and Debonnel, G. (2007) Serotonin$_4$ (5-HT$_4$) receptor agonists are putative antidepressants with a rapid onset of action. Neuron, 55: 712–725.

Luppi, P.H., Aston-Jones, G., Akaoka, H., Chouvet, G. and Jouvet, M. (1995) Afferent projections to the rat locus coeruleus demonstrated by retrograde and anterograde tracing with cholera-toxin B subunit and *Phaseolus vulgaris* leucoagglutinin. Neuroscience, 65: 119–160.

Marek, G.J. and Aghajanian, G.K. (1999) 5-HT$_{2A}$ receptor or α_1-adrenoceptor activation induces excitatory postsynaptic currents in layer V pyramidal cells of the medial prefrontal cortex. Eur. J. Pharmacol., 367: 197–206.

Martín-Ruiz, R., Ugedo, L., Honrubia, M.A., Mengod, G. and Artigas, F. (2001a) Control of serotonergic neurons in rat brain by dopaminergic receptors outside the dorsal raphe nucleus. J. Neurochem., 77: 762–775.

Martín-Ruiz, R., Puig, M.V., Celada, P., Shapiro, D.A., Roth, B.L., Mengod, G. and Artigas, F. (2001b) Control of serotonergic function in medial prefrontal cortex by serotonin-2A receptors through a glutamate-dependent mechanism. J. Neurosci., 21: 9856–9866.

Marwaha, J. and Aghajanian, G.K. (1982) Relative potencies of alpha-1 and alpha-2 antagonists in the locus ceruleus, dorsal raphe and dorsal lateral geniculate nuclei: an electrophysiological study. J. Pharmacol. Exp. Ther., 222: 287–293.

McCune, S.K., Voigt, M.M. and Hill, J.M. (1993) Expression of multiple alpha adrenergic receptor subtype messenger RNAs in the adult rat brain. Neuroscience, 57: 143–151.

Meltzer, H.Y., Li, Z., Kaneda, Y. and Ichikawa, J. (2003) Serotonin receptors: their key role in drugs to treat schizophrenia. Prog. Neuropsychopharmacol. Biol. Psychiatry, 27: 1159–1172.

Mennini, T., Gobbi, M. and Romandini, S. (1986) Localization of GABA$_A$ and GABA$_B$ receptor subtypes on serotonergic neurons. Brain Res., 371: 372–375.

Mihailescu, S., Guzmán-Marín, R., Dominguez, M.C. and Drucker-Colín, R. (2002) Mechanisms of nicotine actions on dorsal raphe serotoninergic neurons. Eur. J. Pharmacol., 452: 77–82.

Millan, M.J., Dekeyne, A. and Gobert, A. (1998) Serotonin (5-HT)$_{2C}$ receptors tonically inhibit dopamine (DA) and noradrenaline (NA), but not 5-HT, release in the frontal cortex *in vivo*. Neuropharmacology, 37: 953–955.

Millan, M.J., Lejeune, F. and Gobert, A. (2000) Reciprocal autoreceptor and heteroreceptor control of serotonergic, dopaminergic and noradrenergic transmission in the frontal cortex: relevance to the actions of antidepressant agents. J. Psychopharmacol., 14: 114–138.

Miller, E.K. and Cohen, J.D. (2001) An integrative theory of prefrontal cortex function. Annu. Rev. Neurosci., 24: 167–202.

Minabe, Y., Ashby, C.R. Jr. and Wang, R.Y. (1991) The effect of acute and chronic LY 277359, a selective 5-HT3 receptor antagonist, on the number of spontaneously active midbrain dopamine neurons. Eur. J. Pharmacol., 209: 151–156.

Minabe, Y., Hashimoto, K., Watanabe, K.I. and Ashby, C.R. Jr. (2001) Acute and repeated administration of the selective 5-HT$_{2A}$ receptor antagonist M100907 significantly alters the activity of midbrain dopamine neurons: an *in vivo* electrophysiological study. Synapse, 40: 102–112.

Nanopoulos, D., Belin, M.F., Maitre, M., Vincendon, G. and Pujol, J.F. (1982) Immunocytochemical evidence for the existence of GABAergic neurons in the nucleus raphe dorsalis. Possible existence of neurons containing serotonin and GABA. Brain Res., 232: 375–389.

Nedergaard, S., Bolam, J.P. and Greenfield, S.A. (1988) Facilitation of a dendritic calcium conductance by 5-hydroxytryptamine in the substantia nigra. Nature, 333: 174–177.

Nishikawa, T. and Scatton, B. (1985) Inhibitory influence of GABA on central serotonergic transmission. Raphe nuclei as the neuroanatomical site of the GABAergic inhibition of cerebral serotonergic neurons. Brain Res., 331: 91–103.

Nitz, D. and Siegel, J. (1997) GABA release in the dorsal raphe nucleus: role in the control of REM sleep. Am. J. Physiol., 273: R451–R455.

Nocjar, C., Roth, B.L. and Pehek, E.A. (2002) Localization of 5-HT_{2A} receptors on dopamine cells in subnuclei of the midbrain A10 cell group. Neuroscience, 111: 163–176.

Ordway, G.A., Stockmeier, C.A., Cason, G.W. and Klimek, V. (1997) Pharmacology and distribution of norepinephrine transporters in the human locus coeruleus and raphe nuclei. J. Neurosci., 17: 1710–1719.

Palacios, J.M., Hoyer, D. and Cortés, R. (1987) α_1-Adrenoceptors in the mammalian brain: similar pharmacology but different distribution in rodents and primates. Brain Res., 419: 65–75.

Pan, Z.Z. and Williams, J.T. (1989) GABA- and glutamate-mediated synaptic potentials in rat dorsal raphe neurons *in vitro*. J. Neurophysiol., 61: 719–726.

Pan, Z.Z., Grudt, T.J. and Williams, J.T. (1994) α_1-Adrenoceptors in rat dorsal raphe neurons: regulation of two potassium conductances. J. Physiol., 478(Pt 3): 437–447.

Paxinos, G. and Watson, C. (2005) The Rat Brain in Stereotaxic Coordinates. Elsevier/Academic Press, Amsterdam.

Pazos, A. and Palacios, J.M. (1985) Quantitative autoradiographic mapping of serotonin receptors in the rat brain. I. Serotonin-1 receptors. Brain Res., 346: 205–230.

Pehek, E.A., McFarlane, H.G., Maguschak, K., Price, B. and Pluto, C.P. (2001) M100,907, a selective 5-HT_{2A} antagonist, attenuates dopamine release in the rat medial prefrontal cortex. Brain Res., 888: 51–59.

Peyron, C., Luppi, P.H., Kitahama, K., Fort, P., Hermann, D.M. and Jouvet, M. (1995) Origin of the dopaminergic innervation of the rat dorsal raphe nucleus. Neuroreport, 6: 2527–2531.

Peyron, C., Luppi, P.H., Fort, P., Rampon, C. and Jouvet, M. (1996) Lower brainstem catecholamine afferents to the rat dorsal raphe nucleus. J. Comp. Neurol., 364: 402–413.

Peyron, C., Petit, J.M., Rampon, C., Jouvet, M. and Luppi, P.H. (1998) Forebrain afferents to the rat dorsal raphe nucleus demonstrated by retrograde and anterograde tracing methods. Neuroscience, 82: 443–468.

Pieribone, V.A., Nicholas, A.P., Dagerlind, A. and Hokfelt, T. (1994) Distribution of α_1 adrenoceptors in rat brain revealed by *in situ* hybridization experiments utilizing subtype-specific probes. J. Neurosci., 14: 4252–4268.

Piñeyro, G. and Blier, P. (1999) Autoregulation of serotonin neurons: role in antidepressant drug action. Pharmacol. Rev., 51: 533–591.

Pompeiano, M., Palacios, J.M. and Mengod, G. (1994) Distribution of the serotonin 5-HT_2 receptor family mRNAs: comparison between 5-HT_{2A} and 5-HT_{2C} receptors. Brain Res. Mol. Brain Res., 23: 163–178.

Porras, G., Di, M.V., Fracasso, C., Lucas, G., De Deurwaerdere, P., Caccia, S., Esposito, E. and Spampinato, U. (2002) 5-HT_{2A} and $5\text{-HT}_{2C/2B}$ receptor subtypes modulate dopamine release induced *in vivo* by amphetamine and morphine in both the rat nucleus accumbens and striatum. Neuropsychopharmacology, 26: 311–324.

Pozzi, L., Acconcia, S., Ceglia, I., Invernizzi, R.W. and Samanin, R. (2002) Stimulation of 5-hydroxytryptamine (5-HT_{2C}) receptors in the ventrotegmental area inhibits stress-induced but not basal dopamine release in the rat prefrontal cortex. J. Neurochem., 82: 93–100.

Prisco, S., Pagannone, S. and Esposito, E. (1994) Serotonin–dopamine interaction in the rat ventral tegmental area: an electrophysiological study *in vivo*. J. Pharmacol. Exp. Ther., 271: 83–90.

Pudovkina, O.L., Cremers, T.I. and Westerink, B.H. (2002) The interaction between the locus coeruleus and dorsal raphe nucleus studied with dual-probe microdialysis. Eur. J. Pharmacol., 445: 37–42.

Pudovkina, O.L., Cremers, T.I. and Westerink, B.H. (2003) Regulation of the release of serotonin in the dorsal raphe nucleus by α_1 and α_2 adrenoceptors. Synapse, 50: 77–82.

Puig, M.V., Celada, P., Díaz-Mataix, L. and Artigas, F. (2003) *In vivo* modulation of the activity of pyramidal neurons in the rat medial prefrontal cortex by 5-HT_{2A} receptors: relationship to thalamocortical afferents. Cereb. Cortex, 13: 870–882.

Puig, M.V., Artigas, F. and Celada, P. (2005) Modulation of the activity of pyramidal neurons in rat prefrontal cortex by raphe stimulation *in vivo*: involvement of serotonin and GABA. Cereb. Cortex, 15: 1–14.

Rasmussen, K., Stockton, M.E. and Czachura, J.F. (1991) The 5-HT3 receptor antagonist zatosetron decreases the number of spontaneously active A10 dopamine neurons. Eur. J. Pharmacol., 205: 113–116.

Rasmusson, A.M., Goldstein, L.E., Deutch, A.Y., Bunney, B.S. and Roth, R.H. (1994) 5-HT_{1A} agonist (±)-8-OH-DPAT modulates basal and stress-induced changes in medial prefrontal cortical dopamine. Synapse, 18: 218–224.

Rodd, Z.A., Gryszowka, V.E., Toalston, J.E., Oster, S.M., Ji, D., Bell, R.L. and McBride, W.J. (2007) The reinforcing actions of a serotonin-3 receptor agonist within the ventral tegmental area: evidence for subregional and genetic differences and involvement of dopamine neurons. J. Pharmacol. Exp. Ther., 321: 1003–1012.

Romo, R. and Salinas, E. (2003) Flutter discrimination: neural codes, perception, memory and decision making. Nat. Rev. Neurosci., 4: 203–218.

Rosin, D.L., Zeng, D., Stornetta, R.L., Norton, F.R., Riley, T., Okusa, M.D., Guyenet, P.G. and Lynch, K.R. (1993) Immunohistochemical localization of α_{2A}-adrenergic receptors in catecholaminergic and other brainstem neurons in the rat. Neuroscience, 56: 139–155.

Rouquier, L., Claustre, Y. and Benavides, J. (1994) α_1-Adrenoceptor antagonists differentially control serotonin release in the hippocampus and striatum: a microdialysis study. Eur. J. Pharmacol., 261: 59–64.

Sakai, K. and Crochet, S. (2000) Serotonergic dorsal raphe neurons cease firing by disfacilitation during paradoxical sleep. Neuroreport, 11: 3237–3241.

Sakai, K., Salvert, D., Touret, M. and Jouvet, M. (1977) Afferent connections of the nucleus raphe dorsalis in the cat as visualized by the horseradish peroxidase technique. Brain Res., 137: 11–35.

Sakaue, M., Somboonthum, P., Nishihara, B., Koyama, Y., Hashimoto, H., Baba, A. and Matsuda, T. (2000) Postsynaptic

5-hydroxytryptamine$_{1A}$ receptor activation increases *in vivo* dopamine release in rat prefrontal cortex. Br. J. Pharmacol., 129: 1028–1034.

Santana, N., Bortolozzi, A., Serrats, J., Mengod, G. and Artigas, F. (2004) Expression of serotonin$_{1A}$ and serotonin$_{2A}$ receptors in pyramidal and GABAergic neurons of the rat prefrontal cortex. Cereb. Cortex, 14: 1100–1109.

Scheinin, M., Lomasney, J.W., Hayden-Hixson, D.M., Schambra, U.B., Caron, M.G., Lefkowitz, R.J. and Fremeau, R.T. Jr. (1994) Distribution of alpha 2-adrenergic receptor subtype gene expression in rat brain. Brain Res. Mol. Brain Res., 21: 133–149.

Schmidt, C.J., Sorensen, S.M., Kehne, J.H., Carr, A.A. and Palfreyman, M.G. (1995) The role of 5-HT$_{2A}$ receptors in antipsychotic activity. Life Sci., 56: 2209–2222.

Seguela, P., Wadiche, J., Dineley-Miller, K., Dani, J.A. and Patrick, J.W. (1993) Molecular cloning, functional properties, and distribution of rat brain α_7: a nicotinic cation channel highly permeable to calcium. J. Neurosci., 13: 596–604.

Serrats, J., Artigas, F., Mengod, G. and Cortés, R. (2003) GABA$_B$ receptor mRNA in the raphe nuclei: co-expression with serotonin transporter and glutamic acid decarboxylase. J. Neurochem., 84: 743–752.

Serrats, J., Mengod, G. and Cortés, R. (2005) Expression of serotonin 5-HT$_{2C}$ receptors in GABAergic cells of the anterior raphe nuclei. J. Chem. Neuroanat., 29: 83–91.

Seth, P., Cheeta, S., Tucci, S. and File, S.E. (2002) Nicotinic–serotonergic interactions in brain and behaviour. Pharmacol. Biochem. Behav., 71: 795–805.

Shigemoto, R. and Mizuno, N. (2000) Metabotropic glutamate receptors – immunocytochemical and *in situ* hybridization analyses. In: Ottersen, O.P. and Storm-Mathisen, J. (Eds.), Handbook of Chemical Neuroanatomy. Elsevier, Amsterdam, Vol. 18, pp. 63–98.

Sprouse, J.S. and Aghajanian, G.K. (1986) (−)-Propranolol blocks the inhibition of serotonergic dorsal raphe cell firing by 5-HT$_{1A}$ selective agonists. Eur. J. Pharmacol., 128: 295–298.

Sprouse, J.S. and Aghajanian, G.K. (1987) Electrophysiological responses of serotoninergic dorsal raphe neurons to 5-HT$_{1A}$ and 5-HT$_{1B}$ agonists. Synapse, 1: 3–9.

Stamp, J.A. and Semba, K. (1995) Extent of colocalization of serotonin and GABA in the neurons of the rat raphe nuclei. Brain Res., 677: 39–49.

Steriade, M., Datta, S., Pare, D., Oakson, G. and Curro Dossi, R.C. (1990a) Neuronal activities in brain-stem cholinergic nuclei related to tonic activation processes in thalamocortical systems. J. Neurosci., 10: 2541–2559.

Steriade, M., Pare, D., Datta, S., Oakson, G. and Curro, D.R. (1990b) Different cellular types in mesopontine cholinergic nuclei related to ponto-geniculo-occipital waves. J. Neurosci., 10: 2560–2579.

Strazielle, C., Lalonde, R., Hebert, C. and Reader, T.A. (1999) Regional brain distribution of noradrenaline uptake sites, and of α_1-, α_2- and β-adrenergic receptors in PCD mutant mice: a quantitative autoradiographic study. Neuroscience, 94: 287–304.

Svensson, T.H., Bunney, B.S. and Aghajanian, G.K. (1975) Inhibition of both noradrenergic and serotonergic neurons in brain by the alpha-adrenergic agonist clonidine. Brain Res., 92: 291–306.

Swanson, L.W. (1998) Brain Maps: Structure of the Rat Brain. Elsevier, Amsterdam.

Szabo, S.T. and Blier, P. (2001) Functional and pharmacological characterization of the modulatory role of serotonin on the firing activity of locus coeruleus norepinephrine neurons. Brain Res., 922: 9–20.

Tanda, G., Carboni, E., Frau, R. and Di Chiara, G. (1994) Increase of extracellular dopamine in the prefrontal cortex: a trait of drugs with antidepressant potential? Psychopharmacology, 115: 285–288.

Tao, R., Ma, Z. and Auerbach, S.B. (1996) Differential regulation of 5-hydroxytryptamine release by GABA$_A$ and GABA$_B$ receptors in midbrain raphe nuclei and forebrain of rats. Br. J. Pharmacol., 119: 1375–1384.

Tao, R. and Auerbach, S.B. (1996) Differential effect of NMDA on extracellular serotonin in rat midbrain raphe and forebrain sites. J. Neurochem., 66: 1067–1075.

Tao, R. and Auerbach, S.B. (2000) Regulation of serotonin release by GABA and excitatory amino acids. J. Psychopharmacol., 14: 100–113.

Tao, R., Ma, Z. and Auerbach, S.B. (1997) Influence of AMPA/kainate receptors on extracellular 5-hydroxytryptamine in rat midbrain raphe and forebrain. Br. J. Pharmacol., 121: 1707–1715.

Tejani-Butt, S.M. (1992) [^{3}H]Nisoxetine: a radioligand for quantitation of norepinephrine uptake sites by autoradiography or by homogenate binding. J. Pharmacol. Exp. Ther., 260: 427–436.

Thakkar, M.M., Strecker, R.E. and McCarley, R.W. (1998) Behavioral state control through differential serotonergic inhibition in the mesopontine cholinergic nuclei: a simultaneous unit recording and microdialysis study. J. Neurosci., 18: 5490–5497.

Van Bockstaele, E.J., Cestari, D.M. and Pickel, V.M. (1994) Synaptic structure and connectivity of serotonin terminals in the ventral tegmental area: potential sites for modulation of mesolimbic dopamine neurons. Brain Res., 647: 307–322.

Vandermaelen, C.P. and Aghajanian, G.K. (1983) Electrophysiological and pharmacological characterization of serotonergic dorsal raphe neurons recorded extracellularly and intracellularly in rat brain slices. Brain Res., 289: 109–119.

Varga, V., Sik, A., Freund, T.F. and Kocsis, B. (2002) GABA$_B$ receptors in the median raphe nucleus: distribution and role in the serotonergic control of hippocampal activity. Neuroscience, 109: 119–132.

Vertes, R.P. (2004) Differential projections of the infralimbic and prelimbic cortex in the rat. Synapse, 51: 32–58.

Vertes, R.P. and Kocsis, B. (1994) Projections of the dorsal raphe nucleus to the brainstem: PHA-L analysis in the rat. J. Comp. Neurol., 340: 11–26.

Waldmeier, P.C., Wicki, P., Feldtrauer, J.J. and Baumann, P. A. (1988) Potential involvement of a baclofen-sensitive autoreceptor in the modulation of the release of endogenous GABA from rat brain slices *in vitro*. Naunyn Schmiedebergs Arch. Pharmacol., 337: 289–295.

Wamsley, J.K., Gehlert, D.R., Filloux, F.M. and Dawson, T.M. (1989) Comparison of the distribution of D-1 and D-2 dopamine receptors in the rat brain. J. Chem. Neuroanat., 2: 119–137.

Wang, Q.P., Ochiai, H. and Nakai, Y. (1992) GABAergic innervation of serotonergic neurons in the dorsal raphe nucleus of the rat studied by electron microscopy double immunostaining. Brain Res. Bull., 29: 943–948.

Wang, Q.P., Guan, J.L. and Shioda, S. (2000) Synaptic contacts between serotonergic and cholinergic neurons in the

rat dorsal raphe nucleus and laterodorsal tegmental nucleus. Neuroscience, 97: 553–563.
Wang, R.Y. and Aghajanian, G.K. (1977) Physiological evidence for habenula as major link between forebrain and midbrain raphe. Science, 197: 89–91.
Wirtshafter, D. and Sheppard, A.C. (2001) Localization of $GABA_B$ receptors in midbrain monoamine containing neurons in the rat. Brain Res. Bull., 56: 1–5.
Woolf, N.J. and Butcher, L.L. (1989) Cholinergic systems in the rat brain: IV. Descending projections of the pontomesencephalic tegmentum. Brain Res. Bull., 23: 519–540.
Wright, D.E., Seroogy, K.B., Lundgren, K.H., Davis, B.M. and Jennes, L. (1995) Comparative localization of serotonin1A, 1C, and 2 receptor subtype mRNAs in rat brain. J. Comp. Neurol., 351: 357–373.

CHAPTER 2.9

Serotonergic Regulation of Rhythmical Activity of the Brain, Concentrating on the Hippocampus

Robert P. Vertes*

Center for Complex Systems and Brain Sciences, Florida Atlantic University, Boca Raton, Florida

Abstract: The dorsal (DRN) and median raphe (MRN) nuclei are the major serotonin-containing nuclei of the brainstem, with extensive projections to the forebrain. With a few exceptions, DRN and MRN give rise to separate, non-overlapping sets of projections to the forebrain. The MRN is mainly a midline/paramidline system with fibers primarily distributing to medial parts of the hypothalamus and thalamus, the medial basal forebrain, the septum and the hippocampus. By contrast, DRN fibers distribute to most of the remaining regions of the forebrain, including the substantia nigra, the lateral hypothalamus, the amygdala, and widespread regions of the basal forebrain and cortex. DRN and MRN essentially only send common projections to the midline and intralaminar thalamus. These differential projections undoubtedly reflect functional differences of the two systems. There are two major states of the hippocampal electroencephalogram (EEG), the theta and non-theta states, and the non-theta state has been further divided into large and small irregular activity, LIA and SIA, respectively. Non-theta states of the hippocampal EEG appear to be directly controlled by serotonergic neurons of MRN that exert desynchronizing actions on the hippocampal EEG (or blockade of theta) by disrupting the discharge of theta rhythmically firing neurons of the septum and/or the hippocampus. MRN stimulation desynchronizes the hippocampal EEG, MRN lesions produce constant, uninterrupted theta, and a network of neurons has been identified in MRN with discharge characteristics related to the hippocampal EEG. This includes populations of 5-HT projection and non-projection neurons and GABAergic interneurons that interact to modulate the hippocampal EEG. The theta rhythm serves a well-recognized role in memory-processing functions of the hippocampus. Accordingly, it seems that by blocking theta (or desynchronizing the hippocampal EEG) 5-HT MRN cells may suppress or temporarily suspend memory-processing functions of the hippocampus. The MRN may be an important part of a system of connections that directs the hippocampus to essentially disregard insignificant environmental events. The DRN contains relatively significant numbers of serotonergic and GABAergic cells that fire rhythmically phase-coupled to theta. These DRN cells may provide theta synchronous input to several DRN targets known to exhibit theta rhythmicity (anterior thalamus, amygdala, striatum and prefrontal cortex), possibly to coordinate their activity during theta states or conditions.

Keywords: dorsal and median raphe nuclei, septum, hippocampus, theta rhythm, electroencephalogram, memory.

Serotonergic systems of the brainstem

It is well established that the brainstem contains discrete groups of serotonin-containing (5-hydroxytryptamine, 5-HT) neurons, extending from the caudal medulla to the rostral midbrain. It has further been shown that the axonal processes of these cells are extensive, distributing to structures throughout the neuroaxis. As would be expected from this widespread distribution, 5-HT cells participate in a range of functions which may, in part, be mediated through a serotonergic modulation of rhythmical activities.

In their original report in the rat, Dalhstrom and Fuxe (1964) identified nine serotonin-containing cell groups of the brainstem which they termed B1–B9. With the exception of B9, these 5-HT containing nuclei are located on the midline and, as such, were designated raphe nuclei; *raphe* meaning seam, or in this context the line of

*E-mail: vertes@ccs.fau.edu

Christian Müller & Barry Jacobs (Eds.)
Handbook of Behavioral Neurobiology of Serotonin
ISBN 978-0-12-374634-4

DOI: 10.1016/B978-0-12-374634-4.00016-2

junction (midline) of the two halves of the brainstem. With the possible exclusion of B9, the alpha-numerical designations for the raphe nuclei have been replaced by names which reflect their anatomical location or characteristics, such as B1, raphe pallidus; B2, raphe obscurus; B3, raphe magnus; B5, raphe pontis; B7, dorsal raphe nucleus (DRN); and B8, median raphe nucleus (MRN) (Halliday *et al.*, 1995; Harding *et al.*, 2004).

In addition to these cell groups, recently developed histochemical procedures (Arita *et al.*, 1993; Vertes and Crane, 1997) have revealed the presence of relatively significant numbers of 5-HT neurons lateral to the midline within the medulla and the pons/midbrain. Those of the medulla primarily extend laterally from nucleus raphe magnus along the floor of the rostral medulla, while 5-HT cells of the pons/midbrain (outside of B9) stretch laterally from the DRN across the dorsal pontine tegmentum and ventrolaterally within the pontomesencephalic reticular formation (RF) (Vertes and Crane, 1997). For the most part, the caudal raphe nuclei (B1, B2, B3) give rise to descending projections to the lower brainstem and spinal cord, while the rostral raphe nuclei, particularly DRN and MRN, project widely throughout the upper brainstem and forebrain (Vertes and Martin, 1988; Vertes, 1991; Jacobs and Azmitia, 1992; Morin and Meyer-Bernstein, 1999, Vertes *et al.*, 1999; Aznar *et al.*, 2004).

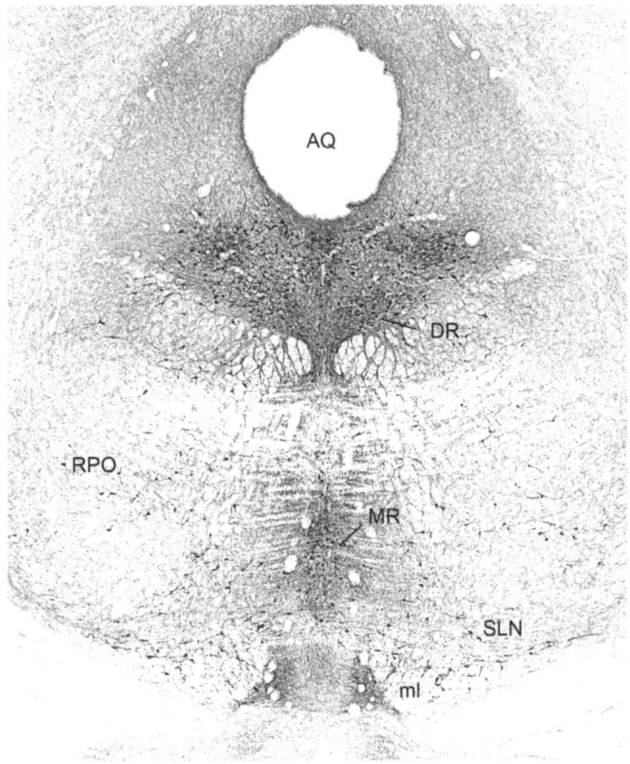

Figure 1 Lightfield photomicrograph of a transverse section through the upper brainstem showing 5-HT immunostained cells in the dorsal raphe nucleus (DR) (including the lateral wings of DR), the median raphe nucleus (MR) and the supralemniscal nucleus (SLN/B9). Abbreviations: AQ, cerebral aqueduct; ml, medial lemniscus; RPO, nucleus pontis oralis.

Serotonergic neurons of the dorsal and median raphe nuclei

As is well recognized, the dorsal and median raphe are densely populated with serotonergic neurons. At the midbrain, 5-HT cells of DRN are fairly concentrated along the midline, whereas further caudally they remain densely packed medially, but also extend laterally from the core of DRN to its lateral 'wings' (Figure 1). Serotonergic cells of DRN are typically large (30–40 μm), fusiform in shape, stain darkly for 5-HT and contain about four or five primary dendrites radiating from the cell body.

By comparison with DRN, there are considerably fewer 5-HT containing neurons in MRN. As shown in Figure 1, serotonergic MRN cells are dispersed fairly evenly throughout the nucleus. Those located dorsally in MRN are predominantly small (10–12 μm), oval, and stain lightly for 5-HT, whereas 5-HT cells of the ventral MRN are medium-sized (15–22 μm) and either oval or spindle-shaped. The dendrites of the oval cells are short (10–20 μm) and coarse, while those of the spindle-shaped neurons are generally long (40–150 μm).

Surprisingly, with the degree of interest in 5-HT systems, few studies have quantified numbers of 5-HT cells in raphe nuclei (Lowry *et al.*, 2008). Early studies estimated the number of 5-HT cells in MRN at ~1100 cells (Descarries *et al.*, 1982) and in DRN at 11,500 cells (Moore *et al.*, 1978). More recently, we reported the following values for the rat: 5-HT cells in DRN = 15,191; in MRN = 4114; in the supralemniscal nucleus (or B9) = 4571 and in the pontomesencephalic reticular formation = 1948 (Vertes and Crane, 1997; Vertes and Linley, 2008).

Ascending projections of the dorsal raphe nucleus

It is well documented that DRN fibers distribute widely throughout the neuroaxis (Vertes, 1991; Jacobs and Azmitia, 1992; Halliday *et al.*, 1995; Harding *et al.*, 2004). With some differences among reports, several early studies, using older tracing techniques, showed that DRN strongly innervates several forebrain sites, including the lateral hypothalamus, the midline and intralaminar nuclei of the thalamus, parts of the amygdala, the dorsal and ventral striatum, the septum, the hippocampus, and widespread regions of the cortex (Azmitia and Segal, 1978; Moore *et al.*, 1978). More recent examinations of DRN projections (Vertes, 1991; Vertes and Kocsis, 1994; Morin and Meyer-Bernstein, 1999; Waselus *et al.*, 2006; Kanno *et al.*, 2008), primarily using the anterograde tracer *Phaseolus vulgaris* leuccoagglutinin (PHA-L), confirmed some earlier findings and described additional ones as follows.

Specifically, it has been shown that the main subcortical targets of DRN are the midbrain/diencephalic central gray, the ventral tegmental area (VTA), the substantia nigra-pars compacta (SNc), and the lateral (LHy) and supramammillary (SUM) nuclei of hypothalamus; the anterior (anteroventral and anteromedial), lateral (lateral dorsal and lateral posterior), mediodorsal, midline and intralaminar, and lateral geniculate nuclei (LGN) of the thalamus; the central, lateral, basolateral, basomedial and amygdalo-piriform transition zone of the amygdala; and the nucleus accumbens (ACC), dorsal striatum (caudate-putamen, CP), bed nucleus of stria terminalis (BST), lateral septum (LS), lateral preoptic area (LPO), substantia innominata (SI), magnocellular preoptic area (MgPO), ventral pallidum (VP), endopiriform nucleus (EN) and claustrum (CLA) of the basal forebrain (Vertes, 1988, 1991; Morin and Meyer-Bernstein, 1999).

The main cortical targets of DRN are the entorhinal (EC), piriform, agranular insular (dorsal, ventral and posterior divisions), lateral agranular (frontal) cortex, medial orbital (MO) and medial prefrontal cortex (mPFC) (Vertes, 1991; Vertes and Linley, 2008). The latter includes the medial agranular (AGm), anterior cingulate (AC), prelimbic (PL) and infralimbic (IL) cortices. DRN distributes moderately to the hippocampal formation (HF).

DRN projections to the dopaminergic (DA) system: projections to dopaminergic cells and to their sites of termination

It is well recognized that the serotonergic system interacts with and influences dopaminergic systems of the brain (Kapur and Remington, 1996; Di Matteo *et al.*, 2002). DRN fibers distribute significantly to major DA cell groups of the midbrain – the substantia nigra, pars compacta (SNc) and ventral tegmental area (VTA). In a complementary manner, DRN projects to main SNc and VTA targets; namely, to the basal nuclei of the amygdala, dorsal and ventral (nucleus accumbens) striatum and BST, and widely throughout the frontal cortex to the agranular insular, mPFC, and orbital cortices. In effect, DRN is positioned to influence the origins and terminal destinations of the nigrostriatal (from SNc) and mesolimbic (from VTA) systems.

DRN projections to the thalamus

The primary destinations of DRN fibers to the thalamus are the anteromedial, anteroventral, mediodorsal (MD), LGN and the midline and intralaminar nuclei of the thalamus (Pechanski and Besson, 1984; Vertes, 1991; Morin and Meyer-Bernstein, 1999; Krout *et al.*, 2002; McKenna and Vertes, 2004). The latter group mainly includes the reuniens (RE), rhomboid (RH), interanteromedial (IAM), paratenial (PT) and paraventricular (PV) nuclei of the midline thalamus, and the central medial (CEM), central lateral (CL), paracentral (PC) and parafascicular nuclei of the intralaminar thalamus. Accordingly, with the possible exception of the lateral geniculate nucleus (LGN), DRN projects heavily to a subset of thalamic nuclei that have been characterized as the 'limbic thalamus', based on their diverse and widespread afferents from the brainstem/diencephalon and selective projections to the limbic forebrain (Vertes, 2006; Vertes *et al.*, 2006; Vertes and Hoover, 2008). DRN/serotonergic input to the 'limbic thalamus' would appear to be important for such processes as arousal, attention and affect (Bentivoglio *et al.*, 1991; Van der Werf *et al.*, 2002; Kirouac *et al.*, 2006; Vertes, 2002, 2004, 2006; Hsu and Price, 2009).

DRN projections to the basal forebrain

DRN fibers distribute widely over the basal forebrain (BF) to the BST, LS, and dorsal and ventral striatum, and along the floor of the basal forebrain to the substantia innominata (SI), ventral pallidum (VP), and lateral preoptic area (Vertes, 1991; Morin and Meyer-Bernstein, 1999). As will subsequently be discussed in greater detail, both DRN and MRN project strongly to the septum, but to separate regions thereof. DRN distributes substantially to regions of the basal forebrain that contain dense populations of cholinergic (ACh) neurons: the magnocellular preoptic nucleus (MgPO), SI and VP (Vertes, 1988; Gasbarri *et al.*, 1999; Zaborszky, 2002). As is well-established, ACh-containing fibers of the basal forebrain project widely throughout the cortex (Rye *et al.*, 1984; Woolf *et al.*, 1984; Luiten *et al.*, 1987) and reportedly serve an important role in behavioral/EEG arousal (Zaborszky, 2002; Jones, 2004; Sarter *et al.*, 2005) as well as in cognitive functions (Cassel and Jeltsch, 1995; Jeltsch-David *et al.*, 2008), DRN projections to these sites (as well as those to the limbic thalamus and cortex) would appear to modulate processes of arousal/attention.

DRN projections to the cortex

Similar to the thalamus, the foremost targets of DRN fibers to the cortex are to parts of the 'limbic cortex', including the medial orbital, agranular insular, piriform, mPFC (AC, PL, IL) and entorhinal cortices. DRN distributes much less abundantly to associational, somatosensory, and special sensory areas of the cortex. The marked DRN projections to 'limbic cortices' are consistent with global 5-HT actions on the forebrain, including a significant influence on affect and mood (Hensler, 2006; and this volume).

Ascending projections of the median raphe nucleus

Unlike the widespread distribution of DRN fibers to the forebrain, MRN projects to a much more restricted group of forebrain structures (Vertes and Martin, 1988; Morin and Meyer-Bernstein, 1999; Vertes *et al.*, 1999; Kanno *et al.*, 2008). MRN fibers predominantly distribute to midline structures of the forebrain – including medial regions of the hypothalamus, thalamus and basal forebrain – as well as to the hippocampus. In effect, the MRN represents a midline/paramidline system of connections. Primary forebrain targets of MRN are the interpeduncular nucleus (IP), VTA, medial mammillary nucleus, supramammillary nucleus (SUM), posterior nucleus of the hypothalamus (PH), the dopamine-cell containing A13 region of medial zona incerta (ZI), the parafascicular, reuniens, mediodorsal, central medial, paracentral and central lateral nuclei of the midline/intralaminar thalamus, the lateral habenula, the suprachiasmatic nucleus (SCN), the vertical and horizontal limbs of diagonal band nuclei, the medial septum (MS) and the hippocampus. Overall, MRN projections to the (neo) cortex are light, and essentially restricted to the perirhinal, entorhinal and parts of the prefrontal cortices.

In contrast to DRN, MRN sends some fibers to VTA, but mainly targets the A13 group of the medial zona incerta. Among putative functions, DA cells of the medial ZI have been implicated in the release of gonadotropins; MRN stimulation enhances the secretion of gonadotropins (James *et al.*, 1987; Morello *et al.*, 1989).

Although there is some overlap of MRN and DRN projections to the hippocampus, MRN distributes much more heavily than DRN to the hippocampus. MRN fibers permeate all parts of the dorsal and ventral HF, terminating densely in the outer molecular layer (stratum lacunosum-moleculare) of CA1 and CA3 of Ammon's horn and within the inner molecular and granule cell layers of the dentate gyrus (DG).

The pronounced MRN projections to the septum and hippocampus suggest that MRN may distribute to both sites via axon collaterals originating from a common group of cells. We recently examined this, using fluorescent retrograde tracers (McKenna and Vertes, 2001), and found that: (1) approximately 8–12 percent of MRN neurons were double-labeled (collateral projections to medial septum and hippocampus); (2) double-labeled cells were predominately located in the rostral MRN; and (3) considerably more double-labeled cells were present in MRN than DRN. This suggests that a discrete population of MRN neurons (mainly of the rostral MRN) may simultaneously influence the septum and hippocampus.

Differential DRN and MRN projections

As may be apparent from the foregoing, with few exceptions, the dorsal and median raphe nuclei distribute to separate, essentially non-overlapping, sites in the forebrain. As described, MRN is predominately a midline/para-midline system of projections, with pronounced projections to the medial hypothalamus and thalamus, and to the septum and hippocampus. With the exception of modest projections to the perirhinal and entorhinal cortices, MRN sends few projections to other (lateral) parts of the forebrain.

By contrast, DRN fibers distribute to most of the remaining regions of the forebrain (not served by MRN), primarily including the substantia nigra, the lateral hypothalamus, the amygdala, and widespread regions of the basal forebrain and cortex. DRN and MRN essentially only send common (overlapping) projections to the midline and intralaminar thalamus. Both nuclei distribute to the septum and the nucleus accumbens, but to different (non-overlapping) parts of these structures. In effect, then, the raphe/5-HT innervation of the forebrain is shared by MRN and DRN. These differential projections undoubtedly reflect functional differences of the two systems.

Differential *serotonergic* DRN and MRN projections to the septum and hippocampus

We recently examined the possible differential distribution of *5-HT* DRN and MRN fibers to the forebrain, concentrating on the septum and hippocampus, by initially determining the overall patterns of distribution of 5-HT fibers to the forebrain and then comparing these patterns to those seen following selective 5-HT lesions of MRN (with 5,7-dihydroxytryptamine (5,7-DHT) or DRN (with

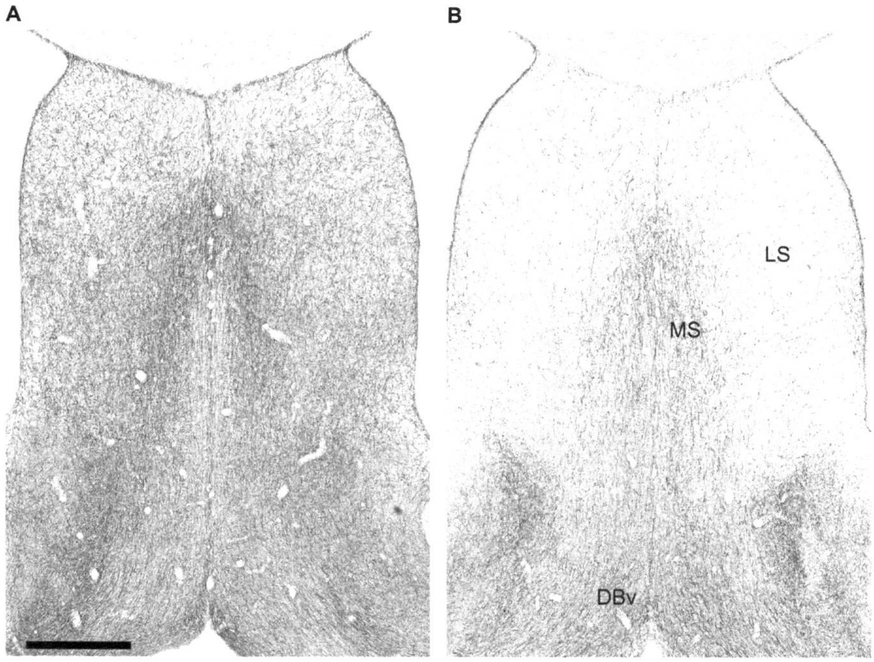

Figure 2 Lightfield photomicrographs of transverse sections through the rostral forebrain showing the distribution of 5-HT fibers within the septal region visualized with an antisera for the serotonin transporter protein (SERT), in (A) a naive rat, and (B) a PCA lesioned rat. Note a complete loss of 5-HT axons throughout the dorsal and ventral lateral septum (LS) in the PCA lesioned rat, while fibers remain intact in the medial septum (MS) and the nucleus of diagonal band, vertical division (DBv). Scale bar: 500 μm.

p-choloroamphetamine (PCA)) (Vertes *et al.*, 2007). In the following, we show representative samples of normal patterns of 5-HT immunolabeling in the septum and hippocampus, and patterns observed following the selective destruction of 5-HT MRN or DRN cells.

Figure 2A shows the normal pattern of 5-HT innervation of the septum, and the pattern seen following peripheral injections of PCA (Figure 2B). As depicted in Figure 2A, 5-HT immunolabeled fibers fill the lateral and medial septum (MS) with greatest density in MS and the vertical limb of the diagonal band nuclei (DBv). Following PCA treatment which selectively destroys 5-HT DRN fibers (Mamounas and Molliver, 1988; Mamounas *et al.*, 1991), there is a complete loss of 5-HT axons dorsoventrally throughout the lateral septum, while those of MS and DBv remain intact (Figure 2B). This is consistent with

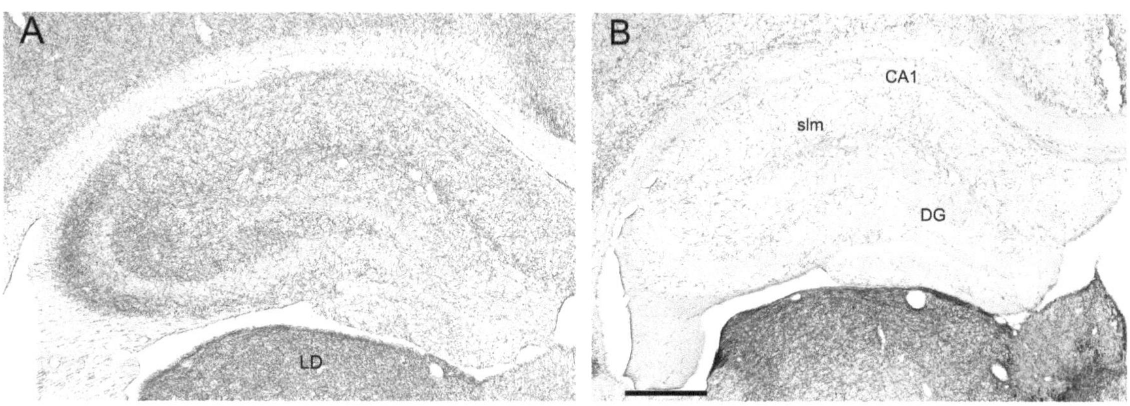

Figure 3 Lightfield photomicrographs of transverse sections through the dorsal hippocampus showing the distribution of 5-HT fibers within the hippocampus visualized with an antisera for the serotonin transporter protein (SERT), in (A) a naive rat, and (B) a 5,7-DHT MR lesioned rat. Note the heavy concentration of 5-HT fibers throughout the hippocampus in the naive rat (A) and a dramatic loss of 5-HT fibers following an injection of 5,7-DHT in MR (B). Abbreviations: CA1, CA1 field of Ammon's horn of the dorsal hippocampus; DG, dentate gyrus of hippocampus; LD, laterodorsal nucleus of the thalamus; slm, stratum lacunosum-moleculare. Scale bar: for (A), 500 μm; for (B), 625 μm.

selective MRN projections to MS and DRN projections to the intermediate LS.

Figure 3A shows the normal pattern of distribution of 5-HT fibers to the hippocampal formation (HF), and the patterns seen following 5,7-DHT injections into MRN (Figure 3B). As depicted in Figure 3A, 5-HT fibers distribute heavily throughout the hippocampus, with a particularly dense concentration within the stratum lacunosum-moleculare (slm) of CA1, strata oriens, pyramidal, and radiatum of CA2/CA3, and the polymorphic layer of the upper blade of the dentate gyrus. As demonstrated in Figure 3B, there is a dramatic loss of 5-HT fibers throughout the hippocampus following a 5,7-DHT injection in MRN. This is consistent with the demonstration that HF receives its main raphe (and 5-HT) input from MRN.

Role of the median raphe (MRN) nucleus in the control of non-theta states of the hippocampal electroencephalogram (EEG) – or hippocampal EEG desynchronization

An extensive body of evidence indicates that the MRN serves a direct role in controlling the non-theta (or desynchronized) state of the hippocampal EEG. It was demonstrated early on that MRN stimulation desynchronized the hippocampal EEG (Macadar *et al.*, 1974; Assaf and Miller, 1978; Yamamoto *et al.*, 1979; Vertes, 1981), and that MRN lesions produced uninterrupted theta, independent of behavior (Maru *et al.*, 1979; Yamamoto *et al.*, 1979). It was further shown that the desynchronizing effects of MRN on the hippocampal EEG were mediated by serotonergic MRN neurons. Specifically, Assaf and Miller (1978) demonstrated that the disruptive effect of MRN stimulation on septal pacemaking cells and the hippocampal EEG was blocked by pre-treatment with the 5-HT synthesis inhibitor p-chlorophenylalanine (PCPA), which produced a 60–80 percent depletion of forebrain serotonin, while Yamamoto *et al.* (1979) reported that ongoing theta produced by MRN lesions could be temporarily disrupted by injections of the serotonin precursor L-5-hydroxytryptophan (L-5-HTP). More recently, we demonstrated that injections of various substances into MRN that either suppressed 5-HT MRN neurons (5-HT_{1A} autoreceptor agonists or GABA agonists) or reduced excitatory drive to them (excitatory amino acid antagonists) produced theta at short latencies (1–2 min) and for long durations (20–80 min) (Kinney *et al.*, 1994, 1995, 1996; Vertes *et al.*, 1994). In a similar manner, Varga *et al.* (2002) identified $GABA_B$ receptors on serotonergic neurons of MRN, and further reported that infusions of the $GABA_B$ agonist, baclofen, into MRN produced long-lasting theta. They concluded (Varga *et al.*, 2002) that the effects of baclofen on theta 'resulted from suppression of the serotonergic output from the median raphe'.

In examinations of the effects of manipulations of MRN on the hippocampal EEG in awake rabbits, Vinogradova and colleagues (Kitchigina *et al.*, 1999; Vinogradova *et al.*, 1999) showed that low-amplitude MRN stimulation disrupted the bursting discharge of medial septal cells and abolished theta in the hippocampus, and that the reversible suppression of MRN with local injections of lidocaine increased the frequency and regularity of discharge of theta-bursting neurons of the septum and hippocampus and produced continuous theta in the hippocampus. Kitchigina *et al.* (1999) concluded that 'the median raphe nucleus can be regarded as a functional antagonist of the reticular formation, powerfully suppressing theta bursts of the medial septal area neurons and the hippocampal theta rhythm'. In a follow-up report, Kudina *et al.* (2004) demonstrated that single injections of the serotonin reuptake inhibitor, fluoxetine, into the lateral ventricles (intracerebroventricular) produced a greater than 50 percent reduction in the amplitude of theta that persisted for periods of 60–90 minutes.

Recent examination of the effects of MRN stimulation on the hippocampal EEG and hippocampal unit activity

It has traditionally been thought that the hippocampal EEG consists of two states: large-amplitude irregular activity (LIA), and theta rhythm (Bland, 1986; Bland and Colom, 1993; Vertes and Kocsis, 1997). Although some early reports identified a third state of the hippocampal EEG, termed small irregular activity (SIA) (Vanderwolf, 1971; Whishaw, 1972), SIA has only recently received significant attention (Jarosiewicz *et al.*, 2002; Jarosiewicz and Skaggs, 2004a, 2004b). SIA appears to be much less prevalent than LIA and, according to recent reports (Jarosiewicz and Skaggs, 2004a, 2004b) is predominantly seen in transitions between stages of sleep, from sleep to waking, or during waking, with the presentation of novel or unexpected stimuli that are not accompanied by orienting movements. Jarosiewicz and Skaggs (2004b) have suggested that SIA represents a level of arousal intermediate between that of sleep and waking.

Early studies that examined the effects of MRN stimulation on the hippocampal EEG (Assaf and Miller, 1978; Yamamoto *et al.* 1979; Vertes, 1981) generally failed to define the specific properties of the EEG produced by such stimulation, commonly referring to it as hippocampal desynchronization, non-theta, or perhaps LIA. In contrast to this, Bland and colleagues (Jackson *et al.*, 2008) recently reported that MRN stimulation produced small-amplitude irregular activity (SIA). For instance, they showed that high-amplitude (>600 μA) stimulation of MRN gave rise to very low-amplitude (~25–50 μV) EEG activity in the hippocampus, or, according to them, 'a near flattening of the electroencephalogram'. Jackson *et al.* (2008) stated that MRN stimulation 'does not elicit hippocampal field activity typical of the non-theta (LIA, sharp wave, slow oscillation) but rather produces an activated neocortex and small amplitude irregular hippocampal activity'. Their findings seemed to indicate, then, that the MRN is responsible for producing a third state of the hippocampal EEG (or SIA), during the unique conditions in which it is present in rats.

Although this remains a possibility, it may be the case that the very low-amplitude EEG (flattening) produced by MRN stimulation represents a nearly complete suppression of all hippocampal EEG activity (SIA, LIA and theta) rather than the generation of a third pattern of activity – or SIA. Even if this activity is not SIA, Jackson *et al.* (2008) nonetheless drew (needed) attention to the fact that MRN stimulation does not (as generally believed) give rise to LIA, but rather produces a marked overall reduction in the amplitude of the hippocampal EEG – or profound blockade of theta. Supporting this (blocking theta): (1) Jackson *et al.* (2008) showed that low- (as opposed to high-) amplitude MRN stimulation produced less of a reduction in EEG amplitude, and partially blocked theta but did not elicit LIA; (2) previous reports have similarly demonstrated that MRN stimulation virtually flattened the hippocampal EEG (Vertes, 1981); and (3) pharmacological suppression of MRN generates long trains of uninterrupted theta (Kinney *et al.*, 1994, 1995, 1996; Vertes *et al.*, 1994; Varga *et al.*, 2002)

Regarding possible mechanisms for the suppression of theta with MRN activation, Jackson *et al.* (2008) found that MRN stimulation significantly inhibited the activity of several types of cells of the hippocampus, including phasic theta-on, tonic theta-on, theta-off and non-theta related cells. Specifically, MRN stimulation completely inhibited the activity of 40 percent (16/40) of phasic theta-on cells, produced a marked decrease in firing with a loss of rhythmicity in 53 percent (21/40) of these cells, and no change in discharge rates but a loss of rhythmicity in 7 percent of the cells. In a somewhat similar manner, MRN stimulation produced: (1) a reduction in firing in 68 percent (13/19) of tonic theta-on cells; (2) an increase in discharge in 16 percent (3/19) of them; and (3) no change in activity in another 16 percent of cells. All six of the theta-off cells (those firing at higher rates during non-theta than during theta) showed reduced rates of discharge with MRN stimulation, generally delivered during non-theta conditions. The authors proposed that the latter finding (i.e., theta-off cell suppression) supports their position that MRN stimulation produces a third state of the hippocampal EEG which is different from theta and LIA – that is, MRN stimulation suppresses the activity of theta-off cells during LIA, and it might be expected to enhance it, if MRN stimulation produced LIA (Jackson *et al.*, 2008).

In sum, Jackson *et al.* (2008) showed that MRN stimulation inhibited the activity of the large majority of both theta-on (phasic and tonic) and theta-off cells in the hippocampus, which undoubtedly contributes to the marked suppression (flattening) of theta/hippocampal EEG with MRN stimulation.

Unit activity in the median raphe nucleus in relation to the hippocampal EEG – serotonergic and non-serotonergic neurons

In an initial examination of the discharge properties of DRN and MRN neurons in urethane anesthetized rats, we showed that a relatively large percentage of cells of both nuclei fired synchronously with the theta rhythm (Kocsis and Vertes, 1996). For instance, with respect to MRN, 5 of 10 MRN cells exhibited theta rhythmical patterns of discharge and high unit-theta coherence values (0.47 to 0.89), indicating strong theta rhythmicity. Interestingly, the 'theta-rhythmic' MRN cells fired at high mean rates of activity (21–42 Hz), suggesting that they were not serotonergic neurons – or the classically defined population of 5-HT raphe cells with slow (1- to 4-Hz), regular rates of discharge. Accordingly, it was tentatively suggested that these MRN cells were GABAergic neurons.

In a subsequent comprehensive analysis of the firing characteristics of MRN neurons with respect to theta in anesthetized rats, we demonstrated that 145/181 MRN cells (or ~80 percent) showed theta-related changes in activity; that is, they fired at significantly higher (theta-on cells) or lower (theta-off cells) rates of activity during theta compared to non-theta states (Viana Di Prisco and Vertes, 2006). The theta-on (68 percent) and theta-off (32 percent) MRN neurons were further divided into slow- (~1–3 Hz), moderate- (5–11 Hz) and fast-firing (>12 Hz) cells. Unlike the previous report, which involved a much smaller sample of MRN neurons (Kocsis and Vertes, 1996), approximately one-third (49/145) of MRN cells discharged at slow regular rates of activity similar to 'classic' serotonin-containing neurons of the raphe and were presumed to be 5-HT cells (but see below). About 30 percent of the slow-firing (SF) neurons were theta-off cells, while the remaining (or the majority) were theta-on cells.

Figure 4 depicts discharge profiles of a slow-firing theta-off (Figure 4A, B) and SF theta-on (Figure 4C, D)

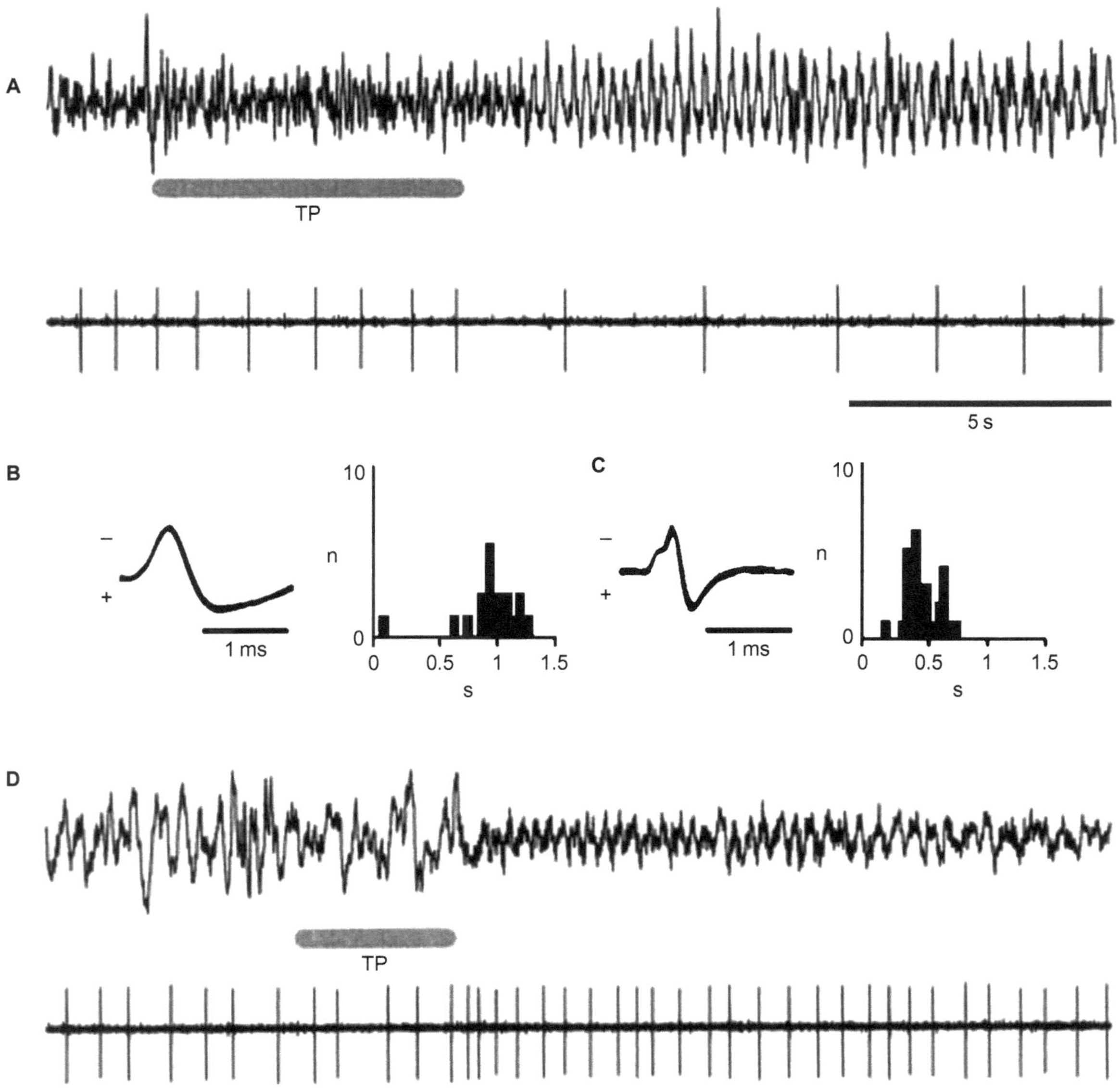

Figure 4 (A) The discharge characteristics of a slow firing (SF) cell of the median raphe nucleus that decreased in discharge from non-theta (left side) to theta elicited with tail pinch (TP, horizontal gray bar). (B) Superimposed actions potentials of the theta-off cell (A) showing a wide spike (~2 ms) and interspike interval histogram (ISIH) demonstrating the slow rate of firing of the cell (~1.2 Hz) (0.8-s peak in ISIH) during non-theta conditions. (C) Superimposed actions potentials of the SF theta-on cell of (D) showing a wide spike (~2 ms) and the ISIH demonstrating enhanced firing of the cell to 2.4 Hz (0.4-s peak in ISIH) in the 10-s period following TP-elicited theta. (D) The discharge properties of the SF theta-on cell of C showing increased rates of firing during TP-elicited theta.

MRN neuron. As illustrated, both the SF theta-on and the theta-off cells had wide spikes (~2 ms) (Figure 4B, C) and fired at significantly decreased (theta-off) or increased (theta-on) rates of activity during tail pinch (TP) elicited theta (Figure 4A, D).

As will be discussed in greater detail below, several recent reports, using various methods to identity 5-HT raphe neurons (mainly of the DRN), have shown that serotonergic cells are not a homogeneus population of neurons as originally described (slow, regular rates of discharge and wide action potentials) (Aghajanian *et al.*, 1978; Vandermaelen and Aghajanian, 1983; Jacobs and Fornal, 1991; Jacobs and Azmitia, 1992), but fire over a range of frequencies up to >20 Hz (Allers and Sharp, 2003; Kocsis *et al.*, 2006; Urbain *et al.*, 2006). In this regard, we identified a second population of MRN neurons, termed moderately-firing (MF) cells, that discharged at mean baseline rates (non-theta) of ~7 Hz (range 5–11 Hz) (Viana Di Prisco *et al.*, 2002). These cells constituted about 37 percent (53/145) of MRN neurons and, unlike SF neurons, more were theta-off (60 percent) than theta-on (40 percent) cells. The theta-off MF cells showed a dramatic decrease in firing rate from non-theta to theta from mean rates about 7.0 Hz to rates of 2.0 Hz. An interesting subset of theta-off MF cells discharge at very regular rates similar to classic slow-firing MRN neurons; the main difference between the two was rates of discharge during baseline conditions (SF cells ~1–3 Hz and MF cells ~5–7 Hz). Although not systematically examined, a few regular-firing MF theta-off cells were strongly inhibited by systemic injections of the 5-HT_{1A} agonist, 8-hydroxy-2(di-*n*-propylamino)tetralin (8-OH-DPAT), indicating that they were serotonergic neurons.

Finally, a third population of MRN neurons (37/145) discharged at high baseline rates of activity (12–22 Hz) and exhibited further increases from non-theta to theta states. All fast-firing (FF) MRN neurons had narrow action potentials, were theta-on cells (no theta-off FF MRN neurons) and were subdivided into tonic and phasically discharging cells. The phasic FF MRN neurons not only showed pronounced increases in rates of discharge with theta (from mean rates of 12.0 to 20.7 Hz), but also fired rhythmically, synchronous with theta. Figure 5 depicts one such cell. As illustrated (Figure 5A), the cell fired arrhythmically in the absence of theta and rhythmically at increased rates with TP-elicited theta. The

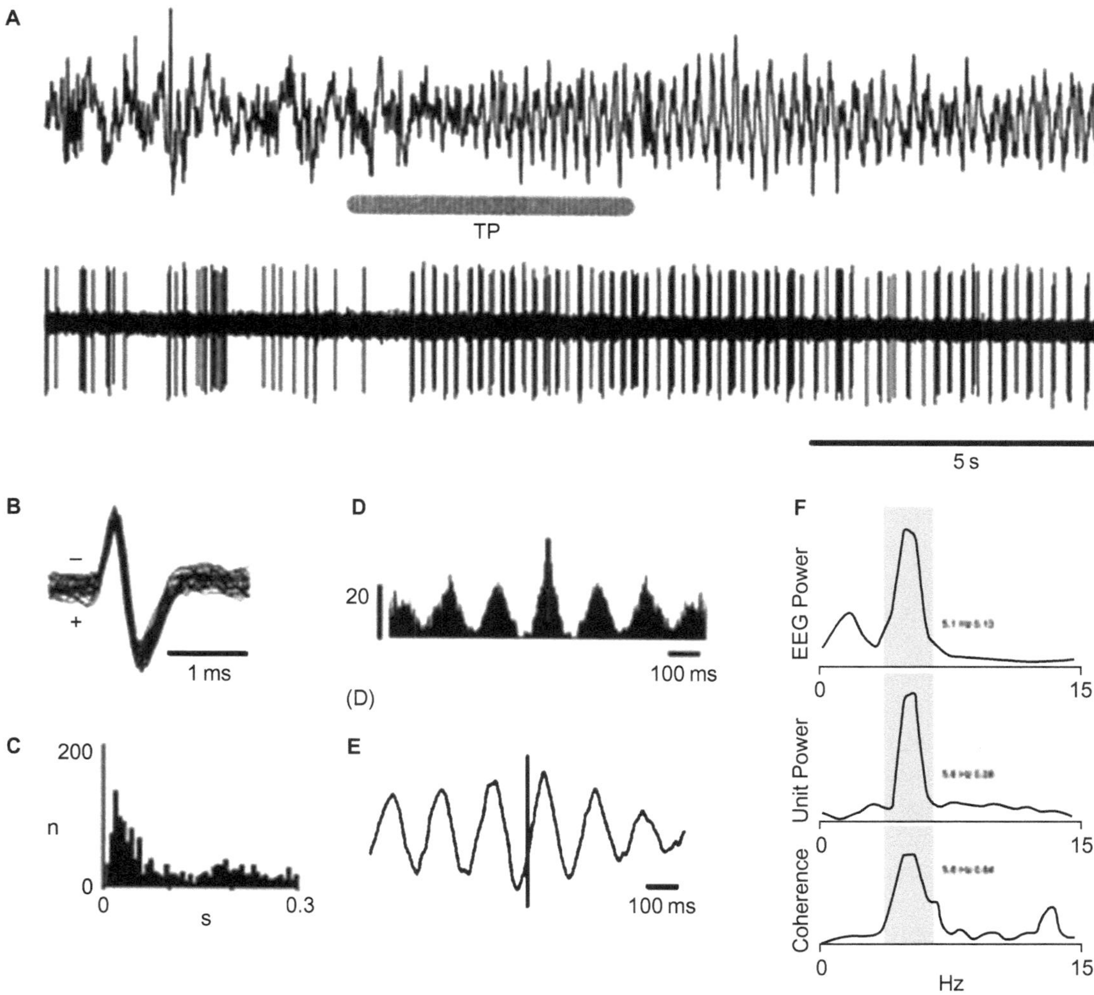

Figure 5 (A) The discharge characteristics of a fast-firing median raphe nucleus (MRN) cell that increased in rate of discharge and fired rhythmically to tail pinch (TP)-elicited theta. (B) Superimposed APs of the cell showing a narrow spike width of ~1 ms. (C) ISI histogram showing clustering at two intervals (~20 ms and ~200 ms) reflecting inter- and intra-burst frequencies during theta. (D, E) Autocorrelogram and crosscorrelogram depicting the rhythmical discharge of the cell (D) locked to theta (E) during theta conditions. (F) spectral and coherence plots showing peaks in the EEG and unit signals at theta frequency (~5.6 Hz) and significant coherence (0.64) between EEG and unit signals at theta frequency.

rhythmical discharge of the cell is shown by: (1) the second peak in the ISI histogram at approximately 200 ms (5 Hz) (Figure 5C); (2) rhythmical peaks at theta frequency in the autocorrelogram (Figure 5D); (3) unit-EEG locked oscillations in the cross-correlogram (Figure 5E); (4) peaks in the unit and EEG autospectra at approximately 5 Hz (Figure 5F); and (5) pronounced coherence between unit and EEG signals at theta frequency (Figure 5F). Rhythmic FF cells fired at higher rates and showed greater coherences with theta during TP-elicited than during spontaneous theta; that is, rates and coherences of 20.7 Hz and 0.67 with TP theta compared to 16.2 Hz and 0.56 with spontaneous theta.

The discharge characteristics of these neurons were very similar to those of the previously described subpopulation of cells that fired at high rates (20–40 Hz) phase-locked to theta (Kocsis and Vertes, 1996). Based on this property and others, including narrow APs, these cells were considered to be GABAergic neurons.

In this regard, in addition to 5-HT cells, the median raphe nucleus contains significant numbers of GABAergic neurons (Mugnaini and Oertel, 1985; Jacobs and Azmitia, 1992; Maloney *et al.*, 1999) which have been shown to contact and inhibit 5-HT MRN cells (Forchetti and Meek, 1981; Nishikawa and Scatton, 1985a, 1985b). As discussed, injections of $GABA_A$ (Kinney *et al.*, 1995) or $GABA_B$ agonists (Varga *et al.*, 2002) into MRN generate persistent theta. This suggests a GABAergic MRN influence on 5-HT cells of MRN in the modulation of the hippocampal EEG.

Supporting earlier findings, Li *et al.* (2005) recently showed, using reverse dialysis, that infusions of the $GABA_A$ agonist muscimol, or the $GABA_B$ agonist baclofen, into MRN generated prolonged theta, presumably through the inhibition of 5-HT cells of MRN, whereas infusions of the $GABA_A$ antagonist bicuculline into MRN eliminated theta, presumably through the disinhibition of 5-HT MRN cells.

Interestingly, however, bicuculline had complex effects that were partially attributed to the presence of $GABA_A$ receptors on both 5-HT cells and GABAergic interneurons of the MRN. Specifically, at low concentrations bicuculline either suppressed theta or gave rise to a short period of theta before desynchronizing the hippocampal EEG, while at high concentrations bicuculline produced a 5- to 10-minute period of desynchronization followed by long-lasting theta. Presumably, these effects largely reflect a delicate balance between sites of actions of bicuculline; that is, either a disinhibition of GABAergic interneurons resulting in stronger inhibitory actions on 5-HT MRN neurons and theta, or a direct disinhibition of 5-HT MRN cells and hippocampal EEG desynchronization.

Based on the foregoing, we suggested the following for various types of MRN cells (Viana Di Prisco *et al.*, 2002): (1) the slow-firing cells (theta-on and theta-off) and a subset of the moderately discharging cells were serotonergic neurons, and the phasic and tonic fast-firing cells were mainly GABAergic neurons; (2) the 5-HT theta-off (or desynchronization-on) cells were projection neurons, and the 5-HT theta-on and GABAergic cells were primarily interneurons; and (3) these populations of cells mutually interact in the modulation of the hippocampal EEG. In effect, the activation of local 5-HT theta-on cells as well as the GABAergic theta-on cells would inhibit 5-HT theta-off projection cells to release or generate theta, whereas suppression of 5-HT theta-on and/or GABAergic theta-on activity would disinhibit 5-HT theta-off (desynchronization-on) cells, resulting in a blockade of theta or a desynchronization of the hippocampal EEG.

A recent examination (Kocsis *et al.*, 2006) of the discharge properties of MRN and DRN neurons, combined with procedures for transmitter identification (juxtacellular labeling), largely supports the foregoing scheme – with some additional complexity. Specifically, Kocsis *et al.* (2006) described the discharge properties of ten serotonergic and five non-serotonergic cells of the DRN/MRN. Of the ten 5-HT cells, the activity of four corresponded to classic 5-HT neurons – i.e., broad spikes and, as indicated by the authors, 'a slow metronome-like firing mode'. Three of four of the slow firing neurons were theta-off cells and the other one was a theta-on cell. Interestingly, three of the ten serotonergic neurons fired at high rates of activity (>8 Hz) during theta and non-theta states, and synchronously (phase-locked) to hippocampal theta. The discharge characteristics of these cells were very similar to subsets of fast-firing cells described previously (Kocsis and Vertes, 1996; Viana Di Prisco *et al.*, 2002), which were tentatively identified as GABAergic neurons. Additional studies using larger samples of cells (with transmitter identification) will be needed to determine the relative proportion of FF 'theta rhythmic' cells of DRN/MRN that are GABAergic or serotonergic neurons. If it proves to be the case that a significant percentage of these cells are serotonergic, it would not alter the foregoing model for MRN involvement in the hippocampal EEG. Specifically, the fast-firing 'theta rhythmic' 5-HT MRN neurons could serve a role similar to that proposed for the slow- and moderately-firing 5-HT theta-on cells – i.e., 5-HT interneurons that modulate the activity of 5-HT MRN projection cells in the control of the hippocampal EEG. This was also indicated by Kocsis *et al.* (2006), who stated that 'Rhythmically firing serotonergic neurons have local axon collaterals and can therefore participate in suppressing the activity of slow firing neurons either through $5\text{-}HT_{1A}$ autoreceptors or through a local $5\text{-}HT_2$ receptor-activated GABAergic circuit.'

In one of the few studies examining the activity of MRN cells in behaving animals, Jacobs and colleagues

(Marrosu *et al.*, 1996) showed in awake cats that putative 5-HT cells of MRN exhibited properties indicative of a role in the desynchronization of the hippocampal EEG. These cells fire at highest rates during 'automatic' behaviors of waking and slow-wave sleep (SWS) (desynchronized states of the hippocampus), and at lowest rates during the exploration of waking and REM sleep (theta states) (Jacobs and Azmitia, 1992; Marrosu *et al.*, 1996).

Site(s) of action of MRN desynchronizing effects on the hippocampal EEG

It is now well established that hippocampal theta is generated by a network of structures extending from the brainstem to the hippocampus (Bland and Colom, 1993; Kocsis and Vertes, 1994; Vertes and Kocsis, 1997; Vertes *et al.*, 2004). Specifically, during theta, tonically firing cells of nucleus pontis oralis (RPO) of the rostral pons activate putative glutamatergic neurons of the supramammillary nucleus (SUM) of the hypothalamus, which convert this steady barrage into a rhythmical pattern of discharge which is relayed to 'pacemaking' cells of the medial septum and then to the hippocampus (HF) to generate theta. Accordingly, MRN could modulate the hippocampal EEG via actions on one or more structures of the theta-generating system; that is, RPO, SUM, the medial septum/diagonal band nuclei (MS/DBv) or the hippocampus. As described, MRN distributes significantly to each of these sites: RPO, SUM, MS/DBv and HF (Vertes and Martin, 1988; Morin and Meyer-Bernstein, 1999; Vertes *et al.*, 1999).

It has been shown that MS/DBv lesions completely abolish theta (Vertes and Kocsis, 1997; Vertes *et al.*, 2004); this suggests that the medial septum is critical for the control of the hippocampal EEG – theta and non-theta. Accordingly, the desynchronizing actions of MRN on the hippocampal EEG would appear to be primarily mediated through the MS/DBv. In this regard, MRN stimulation disrupts the rhythmical discharge of septal pacemaking cells and abolishes theta (Assaf and Miller, 1978; Kitchigina *et al.*, 1999), while the suppression of MRN in awake rabbits (with injections of lidocaine) was found to increase the frequency and regularity of discharge of septal bursting neurons and produce continuous theta in the hippocampus. In addition, Kinney *et al.* (1996) showed that injections of the 5-$HT1_A$ agonist, 8-OH-DPAT, into MRN (which inhibits 5-HT MRN neurons) rhythmically activated septal pacemaking cells and generated theta.

As discussed, Bland and colleagues (Jackson *et al.*, 2008) recently demonstrated that high-frequency MRN stimulation both desynchronized the hippocampal EEG and inhibited the activity of a large percentage of phasic and tonic theta-on cells of the hippocampus. While the authors acknowledge that it is difficult to determine whether MRN-elicited effects are routed through the septum and directly affect the hippocampus, their findings favor a direct action on the hippocampus. It is unclear, however, whether the suppressive effects of MRN stimulation on unit activity in the hippocampus indirectly resulted from the disruption of bursting activity of MS/DBv neurons – which, as discussed above, occurs with MRN stimulation.

In summary, the MS/DBv is the nodal point for converting ascending influences into patterns of hippocampal EEG activity, and as such the activation or disruption of a common pool of MS/DBv cells would appear responsible for controlling theta and non-theta states, respectively, of the hippocampal EEG.

Functional significance of theta and by inference the functional role of MRN-elicited blockade of theta- or the desynchronization (non-theta) of the hippocampal EEG

Although the theta rhythm has been implicated in several functions including 'arousal' (Green and Arduini, 1954) and, recently, sensorimotor integration (Bland and Colom, 1993; Bland and Oddie, 2001), the prevailing view is that theta serves a critical role in mnemonic processes of the hippocampus (Vertes and Kocsis, 1997; Buzsáki, 2002; Kirk and Mackay, 2003; Vertes *et al.*, 2004; Hasselmo, 2005; Vertes, 2005). Several lines of evidence indicate a direct involvement of theta in memory. This includes the demonstration that: (1) discrete MS/DBv lesions that abolish theta produce severe learning/memory deficits in rats; (2) LTP is optimally elicited in the hippocampus with stimulation at theta frequency (i.e., theta burst or primed burst potentiation); (3) stimulation delivered in the presence of theta, and on its positive phase, significantly potentiates population responses in the hippocampus; (4) the loss of LTP with primed burst (theta) stimulation in mutant mice is accompanied by a pronounced disruption of place cell activity and spatial memory; and (5) theta is present in humans during the encoding and retrieval stages of various tasks (for review, see Vertes *et al.*, 2004).

We recently advanced the theory that theta serves as a 'tag' for short-term memory in the hippocampus (Vertes, 2005). In short, we proposed that the hippocampus receives two main types of input; theta from ascending brainstem-diencephalo-septal systems, and 'information bearing' mainly from thalamocortical and cortical systems. The temporal convergence of activity of these two systems results in the encoding of information in the hippocampus, primarily reaching it via cortical routes. By analogy to processes associated with long-term potentiation (LTP), we suggest that theta represents a strong depolarizing influence on NMDA receptor-containing cells of the hippocampus. The temporal coupling of a theta-induced depolarization and the release of glutamate to these cells from intra- and extra-hippocampal sources activates them. This, in turn, initiates processes leading to a (short-term) strengthening of connections between presynaptic ('information-bearing') and postsynaptic neurons of the hippocampus. Theta is selectively present in the rat during active exploratory movements. During exploration, a rat continually gathers and updates information about its environment. If this information is temporally coupled to theta (as with the case of locomotion), it becomes temporarily stored in the hippocampus by mechanisms similar to the early phase of LTP (E-LTP). If the exploratory behavior of the rat goes unreinforced, these relatively short lasting traces (1–3 hours) gradually weaken and eventually fade, to be reupdated. On the other hand, if the explorations of the rat lead to rewards (or punishments), additional modulatory inputs to the hippocampus become activated and convert short-term, theta-dependent memories into long-term stores.

If, as indicated, theta serves a critical role in memory, the question arises regarding the functional significance of the serotonergic MRN system that actively blocks theta. A likely possibility may be that the 5-HT MRN system suppresses or suspends memory. In this regard, several reports have shown that serotonergic agents block LTP and that 5-HT antagonists (mainly 5-HT_3 antagonists) enhance LTP and/or memory (Corradetti *et al.*, 1992; Staubli and Otaky, 1994; Staubli and Xu, 1995; Buhot, 1997). MRN may be an important part of a system of connections that directs the hippocampus to essentially disregard insignificant environmental events. In a variant of this, Kitchigina *et al.* (1999) suggested that the MRN may serve to terminate attention or shift attention to new stimuli, thereby stabilizing previously learned information.

In addition to its effects on the hippocampal EEG, the MRN serves a well-documented role in the control of locomotion. MRN stimulation suppresses locomotor behavior (Graeff *et al.*, 1980; Peck and Vanderwolf, 1991), while various manipulations that inhibit the activity of MRN cells produce locomotion (Paris and Lorens, 1987; Hillegaart and Hjorth, 1989; Wirtshafter *et al.*, 1993; Shim *et al.*, 1997). Further, Sinnamon *et al.* (2000) demonstrated that injections of GABA (and agonists) into MRN enhanced hypothalamic-induced stepping behavior (locomotion) and increased the power of theta in the 4- to 5-Hz band. The possible dual MRN control of locomotion and states of the hippocampal EEG may have important functional consequences. It is obviously critical for a rat (or other species) when exploring its environment to commit relevant aspects of it to memory, hence the coupling of locomotion with theta to encode information during exploratory behaviors. On the other hand, there may be less of a demand to commit information to memory when an animal is stationary or engaged in automatic behaviors such as grooming or consumatory acts (non-theta states); hence the absence of theta during these conditions.

Unit activity in the dorsal raphe nucleus (DRN) in relation to the hippocampal EEG (serotonergic and non-serotonergic neurons) and possible functional roles

In an early examination of the discharge properties of DRN neurons in the behaving rat, we (Kocsis and Vertes, 1992) showed that activity of 17 of 30 DRN cells was correlated with the hippocampal theta rhythm; the remaining 13 cells showed no relationship to theta. It was generally the case that theta-related cells fired irregularly during quiet waking and slow wave sleep (non-theta states), and changed to a regular bursting pattern synchronous with theta during waking with movements and REM sleep. Units were divided into slow-firing, fast-firing (11–40 Hz) and very fast-firing (55–70 Hz) cells, and subpopulations of each fired rhythmically with theta. In a follow-up report in the urethane anesthetized rat (Kocsis and Vertes, 1996), a slightly higher percentage of DRN neurons (7/10) was correlated with theta and discharge rates varied over a wide range (5–42 Hz) to include slow- and fast-firing DRN neurons.

As discussed with MRN, it was thought that the slow-firing neurons of these reports (Kocsis and Vertes, 1992, 1996) were serotonergic cells and the fast-firing neurons were non-5-HT or mainly GABAergic cells. While this division may still largely hold, several recent reports have indicated that slow-firing DRN cells are not exclusively serotonergic and subsets of fast-discharging DRN units are 5-HT neurons. For instance, partially consistent with classic descriptions, Allers and Sharp (2003), using juxtacellular techniques, reported that 8 of 10 DRN cells with narrow spikes and high discharge rates (>12 Hz) were GABAergic neurons. With respect, however, to slow- and regular-firing cells with broad spikes, about half (12/24) were monoaminergic neurons (serotonin or dopamine), but the other half were not – indicating that some neurons with these characteristics are not serotonergic cells. In like manner, Kocsis *et al.* (2006) recently demonstrated that 7 of 10 slow-firing neurons of DRN/MRN were 5-HT cells, while the remaining three fired at high baseline rates (>8 Hz) phase-coupled to theta. Of the five non-serotonergic cells, 2/5 discharged at low rates (similar to the majority of 5-HT cells); while 3/5 fired at high rates (~10 Hz), particularly in the presence of theta. The activity of one of these cells was phase-coupled to theta. The foregoing indicates, then, that most DRN cells exhibit classic characteristics; that is, slow-firing cells with broad APs are 5-HT neurons and fast-firing cells with narrow spikes are non-5-HT or mainly GABAergic neurons. Despite this, it seems that relatively significant numbers of DRN neurons may not fit these categories.

With regard to a DRN influence on the hippocampal EEG (theta or non-theta), it appears unlikely that DRN has a direct role in the generation of theta or desynchronized states of the hippocampal EEG – as demonstrated for the MRN. For instance, electrical stimulation of the DRN has no effect on the hippocampal EEG (Vertes, 1981). The DRN receives descending theta-related input from forebrain structures such as the supramammillary nucleus (Vertes, 1992), which may rhythmically drive DRN neurons at theta frequencies (Kocsis and Vertes, 1992, 1996; Kocsis *et al.*, 2006). The theta rhythmic cells of DRN could, in turn, exert effects on the MRN in the modulation of the hippocampal EEG, in the manner in which local 5-HT and GABAergic interneurons of the MRN modulate 5-HT MRN projection cells in the blockade of theta or the desynchronization of the hippocampal EEG. The DRN projects to (Vertes, 1991; Vertes and Kocsis, 1994; Morin and Meyer-Bernstein, 1999; Tischler and Morin, 2003) and functionally affects the MRN (Mokler *et al.*, 2009). In addition, through its widespread forebrain projections (Vertes, 1991; Morin and Meyer-Bernstein, 1999; Waselus *et al.*, 2006; Kanno *et al.*, 2008) the DRN may provide theta synchronous input to several forebrain structures known to exhibit theta rhythmicity, which include the anterior thalamus, amygdala, striatum and prefrontal cortex (Vertes *et al.*, 2001; Albo *et al.*, 2003; Pape *et al.*, 2005; DeCoteau *et al.*, 2007; Mitchell *et al.*, 2008; Young and McNaughton, 2009). The DRN may serve a critical role in coordinating the activity of these various structures during theta states or behaviors, possibly to bind emotional and sensorimotor aspects of behavior.

Acknowledgement

This work was supported by the National Institute of Mental Health, Grant number: MH63519, and the National Science Foundation, Grant number IOS-082639.

References

Aghajanian, G.K., Wang, R.Y. and Baraban, J. (1978) Serotonergic and non-serotonergic neurons of the dorsal raphe: reciprocal changes in firing induced by peripheral nerve stimulation. Brain Res., 153: 169–175.

Albo, Z., Viana Di Prisco, G. and Vertes, R.P. (2003) Anterior thalamic unit discharge profiles and coherence with hippocampal theta rhythm. Thal. Rel. Syst., 2: 133–144.

Allers, K.A. and Sharp, T. (2003) Neurochemical and anatomical identification of fast- and slow-firing neurones in the rat dorsal raphe nucleus using juxtacellular labelling methods *in vivo*. Neuroscience, 122: 193–204.

Arita, H., Sakamoto, M., Hirokawa, Y. and Okado, N. (1993) Serotonin innervation patterns differ among the various medullary motoneuronal groups involved in upper airway control. Exp. Brain Res., 95: 100–110.

Assaf, S.Y. and Miller, J.J. (1978) Role of a raphe serotonin system in control of septal unit activity and hippocampal desynchronization. Neuroscience, 3: 539–550.

Azmitia, E.C. and Segal, M. (1978) An autoradiographic analysis of the differential ascending projections of the dorsal and median raphe nuclei in the rat. J. Comp. Neurol., 179: 641–667.

Aznar, S., Qian, Z.X. and Knudsen, G.M. (2004) Non-serotonergic dorsal and median raphe projection onto parvalbumin- and calbindin-containing neurons in hippocampus and septum. Neuroscience, 124: 573–581.

Bentivoglio, M., Balercia, G. and Kruger, L. (1991) The specificity of the nonspecific thalamus: the midline nuclei. Prog. Brain Res., 87: 53–80.

Bland, B.H. (1986) The physiology and pharmacology of hippocampal formation theta rhythms. Prog. Neurobiol., 26: 1–54.

Bland, B.H. and Colom, L.V. (1993) Extrinsic and intrinsic properties underlying oscillation and synchrony in limbic cortex. Prog. Neurobiol., 41: 157–208.

Bland, B.H. and Oddie, S.D. (2001) Theta band oscillation and synchrony in the hippocampal formation and associated structures: the case for its role in sensorimotor integration. Behav. Brain Res., 127: 119–136.

Buhot, M.C. (1997) Serotonin receptors in cognitive behaviors. Curr. Opin. Neurobiol., 7: 243–254.

Buzsáki, G. (2002) Theta oscillations in the hippocampus. Neuron, 33: 325–340.

Cassel, J.C. and Jeltsch, H. (1995) Serotonergic modulation of cholinergic function in the central nervous system: cognitive implications. Neuroscience, 69: 1–41.

Corradetti, R., Ballerini, L., Pugliese, A.M. and Pepeu, G. (1992) Serotonin blocks the long-term potentiation induced by primed burst stimulation in the CA1 region of rat hippocampal slices. Neuroscience, 46: 511–518.

Dalhstrom, A. and Fuxe, K. (1964) Evidence for the existence of monoamine containing neurons in the central nervous system: I. Demonstrations of monoamines in cell bodies of brainstem neurons. Acta Physiol. Scand., 232: 1–55.

DeCoteau, W.E., Thorn, C., Gibson, D.J., Courtemanche, R., Mitra, P., Kubota, Y. and Graybiel, A.M. (2007) Learning-related coordination of striatal and hippocampal theta rhythms during acquisition of a procedural maze task. Proc. Natl. Acad. Sci. USA, 104: 5644–5649.

Descarries, L., Watkins, K.C., Garcia, S. and Beaudet, A. (1982) The serotonin neurons in nucleus raphe dorsalis of adult rat: a light and electron microscopic radioautographic study. J. Comp. Neurol., 207: 239–254.

Di Matteo, V., Cacchio, M., Di Giulio, C. and Esposito, E. (2002) Role of serotonin (2C) receptors in the control of brain

dopaminergic function. Pharmacol. Biochem. Behav., 71: 727–734.
Forchetti, C.M. and Meek, J.L. (1981) Evidence for a tonic GABAergic control of serotonin neurons in the median raphe nucleus. Brain Res., 206: 208–212.
Gasbarri, A., Sulli, A., Pacitti, C. and McGaugh, J.L. (1999) Serotonergic input to cholinergic neurons in the substantia innominata and nucleus basalis magnocellularis in the rat. Neuroscience, 91: 1129–1142.
Graeff, F.G., Quintero, S. and Gray, J.A. (1980) Median raphe stimulation, hippocampal theta rhythm and threat-induced behavioral inhibition. Physiol. Behav., 25: 253–261.
Green, J.D. and Arduini, A.A. (1954) Hippocampal electrical activity in arousal. J. Neurophysiol., 17: 533–557.
Halliday, G., Harding, A. and Paxinos, G. (1995) Serotonin and tachykinin systems. In: Paxinos, G. (Ed.), The Rat Nervous System. Academic Press, New York, NY, pp. 929–974.
Harding, A., Paxinos, G. and Halliday, G. (2004) The serotonin and tachykinin systems. In: Paxinos, G. (Ed.), The Rat Nervous System. Academic Press, New York, NY, pp. 1205–1256.
Hasselmo, M.E. (2005) What is the function of hippocampal theta rhythm? Linking behavioral data to phasic properties of field potential and unit recording data. Hippocampus, 15: 936–949.
Hensler, J.G. (2006) Serotonergic modulation of the limbic system. Neurosci. Biobehav. Rev., 30: 203–214.
Hillegaart, V. and Hjorth, S. (1989) Median raphe, but not dorsal raphe, application of the 5-HT1A agonist 8-OH-DPAT stimulates rat motor activity. Eur. J. Pharmacol., 160: 303–307.
Hsu, D.T. and Price, J.L. (2009) Paraventricular thalamic nucleus: subcortical connections and innervation by serotonin, orexin, and corticotropin-releasing hormone in macaque monkeys. J. Comp. Neurol., 512: 825–848.
Jackson, J., Dickson, C.T. and Bland, B.H. (2008) Median raphe stimulation disrupts hippocampal theta via rapid inhibition and state-dependent phase reset of theta-related neural circuitry. J. Neurophysiol., 99: 3009–3026.
Jacobs, B.L. and Azmitia, E.C. (1992) Structure and function of the brain serotonin system. Physiol. Rev., 72: 165–229.
Jacobs, B.L. and Fornal, C.A. (1991) Activity of brain serotonergic neurons in the behaving animal. Pharmacol. Rev., 43: 563–578.
James, M.D., MacKenzie, E.J., Tuohy-Jones, P.A. and Wilson, C.A. (1987) Dopaminergic neurones in the zona incerta exert a stimulatory control on gonadotrophin release via D1 dopamine receptors. Neuroendocrinology, 45: 348–355.
Jarosiewicz, B. and Skaggs, W.E. (2004a) Hippocampal place cells are not controlled by visual input during the small irregular activity state in the rat. J. Neurosci., 24: 5070–5077.
Jarosiewicz, B. and Skaggs, W.E. (2004b) Level of arousal during the small irregular activity state in the rat hippocampal EEG. J. Neurophysiol., 91: 2649–2657.
Jarosiewicz, B., McNaughton, B.L. and Skaggs, W.E. (2002) Hippocampal population activity during the small-amplitude irregular activity state in the rat. J. Neurosci., 22: 1373–1384.
Jeltsch-David, H., Koenig, J. and Cassel, J.C. (2008) Modulation of cholinergic functions by serotonin and possible implications in memory: general data and focus on 5-HT(1A) receptors of the medial septum. Behav. Brain Res., 195: 86–97.
Jones, B.E. (2004) Activity, modulation and role of basal forebrain cholinergic neurons innervating the cerebral cortex. Prog. Brain Res., 145: 157–169.
Kanno, K., Shima, S., Ishida, Y. and Yamanouchi, K. (2008) Ipsilateral and contralateral serotonergic projections from dorsal and median raphe nuclei to the forebrain in rats: immunofluorescence quantitative analysis. Neurosci. Res., 61: 207–218.
Kapur, S. and Remington, G. (1996) Serotonin-dopamine interaction and its relevance to schizophrenia. Am. J. Psychiatry, 153: 466–476.
Kinney, G.G., Kocsis, B. and Vertes, R.P. (1994) Injections of excitatory amino acid antagonists into the median raphe nucleus produce hippocampal theta rhythm in the urethane anesthetized rat. Brain Res., 654: 96–104.
Kinney, G.G., Kocsis, B. and Vertes, R.P. (1995) Injections of muscimol into the median raphe nucleus produce hippocampal theta rhythm in the rat. Psychopharmacology, 120: 244–248.
Kinney, G.G., Kocsis, B. and Vertes, R.P. (1996) Medial septal unit firing characteristics following injections of 8-OH-DPAT into the median raphe nucleus. Brain Res., 708: 116–122.
Kirk, I.J. and Mackay, J.C. (2003) The role of theta-range oscillations in synchronising and integrating activity in distributed mnemonic networks. Cortex, 39: 993–1008.
Kirouac, G.J., Parsons, M.P. and Li, S. (2006) Innervation of the paraventricular nucleus of the thalamus from cocaine- and amphetamine-regulated transcript (CART) containing neurons of the hypothalamus. J. Comp. Neurol., 497: 155–165.
Kitchigina, V.F., Kudina, T.A., Kutyreva, E.V. and Vinogradova, O.S. (1999) Neuronal activity of the septal pacemaker of theta rhythm under the influence of stimulation and blockade of the median raphe nucleus in the awake rabbit. Neuroscience, 94: 453–463.
Kocsis, B. and Vertes, R.P. (1992) Dorsal raphe neurons: synchronous discharge with the theta rhythm of the hippocampus in the freely behaving rat. J. Neurophysiol., 68: 1463–1467.
Kocsis, B. and Vertes, R.P. (1994) Characterization of neurons of the supramammillary nucleus and mammillary body that discharge rhythmically with the hippocampal theta rhythm in the rat. J Neurosci., 14: 7040–7052.
Kocsis, B. and Vertes, R.P. (1996) Midbrain raphe cell firing and hippocampal theta rhythm in urethane-anesthetized rats. Neuroreport, 7: 2867–2872.
Kocsis, B., Varga, V., Dahan, L. and Sik, A. (2006) Serotonergic neuron diversity: identification of raphe neurons with discharges time-locked to the hippocampal theta rhythm. Proc. Natl. Acad. Sci. USA, 103: 1059–1064.
Krout, K.E., Belzer, R.E. and Loewy, A.D. (2002) Brainstem projections to midline and intralaminar thalamic nuclei of the rat. J. Comp. Neurol., 448: 53–101.
Kudina, T.A., Sudnitsyn, V.V., Kutyreva, E.V. and Kichigina, V.F. (2004) The serotonin reuptake inhibitor fluoxetine suppresses theta oscillations in the electroencephalogram of the rabbit hippocampus. Neurosci. Behav. Physiol., 34: 929–933.
Li, S., Varga, V., Sik, A. and Kocsis, B. (2005) GABAergic control of the ascending input from the median raphe nucleus to the limbic system. J. Neurophysiol., 94: 2561–2574.
Lowry, C.A., Evans, A.K., Gasser, P.J., Hale, M.H., Staub, D.R. and Shekhar, A. (2008) Topographic organization and chemoarchitecture of the doral raphe nucleus and the median raphe nucleus. In: Monti, J.A., Pandi-Perumal, S.R., Jacobs, B.L. and Nutt, D.J. (Eds.), Serotonin and Sleep: Functional and Clinical Aspects. Birkhäuser Verlag, Switzerland, pp. 25–67.
Luiten, P.G., Gaykema, R.P., Traber, J. and Spencer, D.G. Jr. (1987) Cortical projection patterns of magnocellular

basal nucleus subdivisions as revealed by anterogradely transported Phaseolus vulgaris leucoagglutinin. Brain Res., 413: 229–250.

Macadar, A.W., Chalupa, L.M. and Lindsley, D.B. (1974) Differentiation of brain stem loci which affect hippocampal and neocortical electrical activity. Exp. Neurol., 43: 499–514.

Maloney, K.J., Mainville, L. and Jones, B.E. (1999) Differential c-Fos expression in cholinergic, monoaminergic, and GABAergic cell groups of the pontomesencephalic tegmentum after paradoxical sleep deprivation and recovery. J. Neurosci., 19: 3057–3072.

Mamounas, L.A. and Molliver, M.E. (1988) Evidence for dual serotonergic projections to neocortex: axons from the dorsal and median raphe nuclei are differentially vulnerable to the neurotoxin p-chloroamphetamine (PCA). Exp. Neurol., 102: 23–36.

Mamounas, L.A., Mullen, C.A., O'Hearn, E. and Molliver, M.E. (1991) Dual serotoninergic projections to forebrain in the rat: morphologically distinct 5-HT axon terminals exhibit differential vulnerability to neurotoxic amphetamine derivatives. J. Comp. Neurol., 314: 558–586.

Maru, E., Takahashi, L.K. and Iwahara, S. (1979) Effects of median raphe nucleus lesions on hippocampal EEG in the freely moving rat. Brain Res., 163: 223–234.

Marrosu, F., Fornal, C.A., Metzler, C.W. and Jacobs, B.L. (1996) 5-HT_{1A} agonists induce hippocampal theta activity in freely moving cats: role of presynaptic 5-HT_{1A} receptors. Brain Res., 739: 192–200.

McKenna, J.T. and Vertes, R.P. (2001) Collateral projections from the median raphe nucleus to the medial septum and hippocampus. Brain Res. Bull., 54: 619–630.

McKenna, J.T. and Vertes, R.P. (2004) Afferent projections to nucleus reuniens of the thalamus. J. Comp. Neurol., 480: 115–142.

Mitchell, D.J., McNaughton, N., Flanagan, D. and Kirk, I.J. (2008) Frontal-midline theta from the perspective of hippocampal 'theta.' Prog. Neurobiol., 86: 156–185.

Mokler, D.J., Dugal, J.R., Hoffman, J.M. and Morgane, P.J. (2009) Functional interrelations between nucleus raphé dorsalis and nucleus raphé medianus: a dual probe microdialysis study of glutamate-stimulated serotonin release. Brain Res. Bull., 78: 132–138.

Moore, R.Y., Halaris, A.E. and Jones, B.E. (1978) Serotonin neurons of the midbrain raphe: ascending projections. J. Comp. Neurol., 180: 417–438.

Morello, H., Caligaris, L., Haymal, B. and Taleisnik, S. (1989) Inhibition of proestrous LH surge and ovulation in rats evoked by stimulation of the medial raphe nucleus involves a GABA-mediated mechanism. Neuroendocrinology, 50: 81–87.

Morin, L.P. and Meyer-Bernstein, E.L. (1999) The ascending serotonergic system in the hamster: comparison with projections of the dorsal and median raphe nuclei. Neuroscience, 91: 81–105.

Mugnaini, E. and Oertel, W.H. (1985) An atlas of the distribution of GABAergic neurons and terminals in the rat CNS as revealed by GAD immunohistochemistry. In: Bjorklund, A. and Hökfelt, T. (Eds.), Handbook of Chemical Neuroanatomy: GABA and Neuropeptides in the CNS. Elsevier, New York, NY, pp. 436–608.

Nishikawa, T. and Scatton, B. (1985a) Inhibitory influence of GABA on central serotonergic transmission. Involvement of the habenulo-raphe pathways in the GABAergic inhibition of ascending cerebral serotonergic neurons. Brain Res., 331: 81–90.

Nishikawa, T. and Scatton, B. (1985b) Inhibitory influence of GABA on central serotonergic transmission. Raphe nuclei as the neuroanatomical site of the GABAergic inhibition of cerebral serotonergic neurons. Brain Res., 331: 91–103.

Pape, H.C., Narayanan, R.T., Smid, J., Stork, O. and Seidenbecher, T. (2005) Theta activity in neurons and networks of the amygdala related to long-term fear memory. Hippocampus, 15: 874–880.

Paris, J.M. and Lorens, S.A. (1987) Intra-median raphe infusions of muscimol and the substance P analogue DiMe-C7 produce hyperactivity: role of serotonin neurons. Behav. Brain Res., 26: 139–151.

Pechanski, M. and Besson, J.M. (1984) Diencephalic connections of the raphe nuclei of the rat brainstem: an anatomical study with reference to the somatosensory system. J. Comp. Neurol., 224: 509–534.

Peck, B.K. and Vanderwolf, C.H. (1991) Effects of raphe stimulation on hippocampal and neocortical activity and behaviour. Brain Res., 568: 244–252.

Rye, D.B., Wainer, B.H., Mesulam, M.M., Mufson, E.J. and Saper, C.B. (1984) Cortical projections arising from the basal forebrain: a study of cholinergic and noncholinergic components employing combined retrograde tracing and immunohistochemical localization of choline acetyltransferase. Neuroscience, 13: 627–643.

Sarter, M., Hasselmo, M.E., Bruno, J.P. and Givens, B. (2005) Unraveling the attentional functions of cortical cholinergic inputs: interactions between signal-driven and cognitive modulation of signal detection. Brain Res. Rev., 48: 98–111.

Shim, I., Javaid, J. and Wirtshafter, D. (1997) Dissociation of hippocampal serotonin release and locomotor activity following pharmacological manipulations of the median raphe nucleus. Behav. Brain Res., 89: 191–198.

Sinnamon, H.M., Jassen, A.K. and Ilch, C. (2000) Hippocampal theta activity and facilitated locomotor stepping produced by GABA injections in the midbrain raphe region. Behav. Brain Res., 107: 93–103.

Staubli, U. and Otaky, N. (1994) Serotonin controls the magnitude of LTP induced by theta bursts via an action on NMDA-receptor-mediated responses. Brain Res., 643: 10–16.

Staubli, U. and Xu, F. (1995) Effects of 5-HT3 receptor antagonism on hippocampal theta rhythm, memory, and LTP induction in the freely moving rat. J. Neurosci., 15: 2445–2452.

Tischler, R.C. and Morin, L.P. (2003) Reciprocal serotonergic connections between the hamster median and dorsal raphe nuclei. Brain Res., 981: 126–132.

Urbain, N., Creamer, K. and Debonnel, G. (2006) Electrophysiological diversity of the dorsal raphe cells across the sleep–wake cycle of the rat. J. Physiol., 573: 679–695.

Vandermaelen, C.P. and Aghajanian, G.K. (1983) Electrophysiological and pharmacological characterization of serotonergic dorsal raphe neurons recorded extracellularly and intracellularly in rat brain slices. Brain Res., 289: 109–119.

Van der Werf, Y.D., Witter, M.P. and Groenewegen, H.J. (2002) The intralaminar and midline nuclei of the thalamus. Anatomical and functional evidence for participation in processes of arousal and awareness. Brain Res. Rev., 39: 107–140.

Vanderwolf, C.H. (1971) Limbic-diencephalic mechanisms of voluntary movement. Psychol. Rev., 78: 83–113.

Varga, V., Sik, A., Freund, T.F. and Kocsis, B. (2002) GABA (B) receptors in the median raphe nucleus: distribution and role in the serotonergic control of hippocampal activity. Neuroscience, 109: 119–132.
Vertes, R.P. (1981) An analysis of ascending brain stem systems involved in hippocampal synchronization and desynchronization. J. Neurophysiol., 46: 1140–1159.
Vertes, R.P. (1988) Brainstem afferents to the basal forebrain in the rat. Neuroscience, 24: 907–935.
Vertes, R.P. (1991) A PHA-L analysis of ascending projections of the dorsal raphe nucleus in the rat. J. Comp. Neurol., 313: 643–668.
Vertes, R.P. (1992) PHA-L analysis of projections from the supramammillary nucleus in the rat. J. Comp. Neurol., 326: 595–622.
Vertes, R.P. (2002) Analysis of projections from the medial prefrontal cortex to the thalamus in the rat, with emphasis on nucleus reuniens. J. Comp. Neurol., 442: 163–187.
Vertes, R.P. (2004) Differential projections of the infralimbic and prelimbic cortex in the rat. Synapse, 51: 32–58.
Vertes, R.P. (2005) Hippocampal theta rhythm: a tag for short-term memory. Hippocampus, 15: 923–935.
Vertes, R.P. (2006) Interactions among the medial prefrontal cortex, hippocampus and midline thalamus in emotional and cognitive processing in the rat. Neuroscience, 142: 1–20.
Vertes, R.P. and Kocsis, B. (1994) Projections of the dorsal raphe nucleus to the brainstem: PHA-L analysis in the rat. J. Comp. Neurol., 340: 11–26.
Vertes, R.P. and Crane, A.M. (1997) Distribution, quantification, and morphological characteristics of serotonin-immunoreactive cells of the supralemniscal nucleus (B9) and pontomesencephalic reticular formation in the rat. J. Comp. Neurol., 378: 411–424.
Vertes, R.P. and Hoover, W.B. (2008) Efferent projections of the paraventricular and paratenial nuclei of the dorsal midline thalamus in the rat. J. Comp. Neurol., 508: 212–237.
Vertes, R.P. and Kocsis, B. (1997) Brainstem-diencephalo-septohippocampal systems controlling the theta rhythm of the hippocampus. Neuroscience, 81: 893–926.
Vertes, R.P. and Linley, S.B. (2008) Efferent and afferent connections of the dorsal and median raphe nuclei in the rat. In: Monti, J.A., Pandi-Perumal, S.R., Jacobs, B.L. and Nutt, D.J. (Eds.), Serotonin and Sleep: Functional and Clinical Aspects. Birkhäuser Verlag, Switzerland, pp. 69–102.
Vertes, R.P. and Martin, G.F. (1988) Autoradiographic analysis of ascending projections from the pontine and mesencephalic reticular formation and the median raphe nucleus in the rat. J. Comp. Neurol., 275: 511–541.
Vertes, R.P., Kinney, G.G., Kocsis, B. and Fortin, W.J. (1994) Pharmacological suppression of the median raphe nucleus with serotonin1A agonists, 8-OH-DPAT and buspirone, produces hippocampal theta rhythm in the rat. Neuroscience, 60: 441–451.
Vertes, R.P., Fortin, W.J. and Crane, A.M. (1999) Projections of the median raphe nucleus in the rat. J. Comp. Neurol., 407: 555–582.
Vertes, R.P., Albo, Z. and Viana Di Prisco, G. (2001) Theta-rhythmically firing neurons in the anterior thalamus: implications for mnemonic functions of Papez's circuit. Neuroscience, 104: 619–625.
Vertes, R.P., Hoover, W.B. and Viana Di Prisco, G. (2004) Theta rhythm of the hippocampus: subcortical control and functional significance. Behav. Cogn. Neurosci. Rev., 3: 173–200.
Vertes, R.P., Hoover, W.B., Do Valle, A.C., Sherman, A. and Rodriguez, J.J. (2006) Efferent projections of reuniens and rhomboid nuclei of the thalamus in the rat. J. Comp. Neurol., 499: 768–796.
Vertes, R.P., Linley, S.B., Hoover, W.B. and Hughes, K. (2007) Differential serotonergic innervation of the forebrain by the dorsal (DR) and median (MR) raphe nuclei as revealed by SERT staining and selective 5-HT DR and MR lesions. Soc. Neurosci. Abstr., 780: 17.
Viana Di Prisco, G. and Vertes, R.P. (2006) Excitatory actions of the ventral midline thalamus (rhomboid/reuniens) on the medial prefrontal cortex in the rat. Synapse, 60: 45–55.
Vinogradova, O.S., Kitchigina, V.F., Kudina, T.A. and Zenchenko, K.I. (1999) Spontaneous activity and sensory responses of hippocampal neurons during persistent theta-rhythm evoked by median raphe nucleus blockade in the rabbit. Neuroscience, 94: 745–753.
Waselus, M., Galvez, J.P., Valentino, R.J. and Van Bockstaele, E.J. (2006) Differential projections of dorsal raphe nucleus neurons to the lateral septum and striatum. J. Chem. Neuroanat., 31: 233–242.
Whishaw, I.Q. (1972) Hippocampal electroencephalographic activity in the Mongolian gerbil during natural behaviours and wheel running and in the rat during wheel running and conditioned immobility. Can. J. Psychol., 26: 219–239.
Wirtshafter, D., Stratford, T.R. and Pitzer, M.R. (1993) Studies on the behavioral activation produced by stimulation of GABA-B receptors in the median raphe nucleus. Behav. Brain Res., 59: 83–93.
Woolf, N.J., Eckenstein, F. and Butcher, L.L. (1984) Cholinergic systems in the rat brain: I. Projections to the limbic telencephalon. Brain Res. Bull., 13: 751–784.
Yamamoto, T., Watanabe, S., Oishi, R. and Ueki, S. (1979) Effects of midbrain raphe stimulation and lesion on EEG activity in rats. Brain Res. Bull., 4: 491–495.
Young, C.K. and McNaughton, N. (2009) Coupling of theta oscillations between anterior and posterior midline cortex and with the hippocampus in freely behaving rats. Cereb. Cortex, 19: 24–40.
Zaborszky, L. (2002) The modular organization of brain systems. Basal forebrain: the last frontier. Prog. Brain Res., 136: 359–372.

CHAPTER 2.10

5-HT Neurons and Central CO_2 Chemoreception

Carolin I. Dohle[1] and George B. Richerson*[,2]

[1]Department of Neurology, Yale University School of Medicine, New Haven, Connecticut
[2]Departments of Neurology and Cellular & Molecular Physiology, Yale University School of Medicine, and Veterans Affairs Medical Center, New Haven, Connecticut

Abstract: Many neurons within the medullary raphe project to nuclei that control respiratory motor output. Through release of serotonin (5-HT) and the co-localized neuropeptides substance P and thyrotropin-releasing hormone, these neurons provide an essential tonic drive that maintains ventilation. Many 5-HT neurons are closely associated with large arteries of the brainstem, where they respond to changes in arterial CO_2 levels by increasing their firing rate. These properties allow them to play an important role as central respiratory chemoreceptors, causing an increase in ventilation in response to increased CO_2 levels and returning CO_2 and pH back to normal. Some 5-HT neurons in the midbrain may play a similar role in sensing changes in blood CO_2, and may induce arousal when CO_2 levels increase during sleep. Defects in 5-HT neurons occur in a variety of human diseases and may contribute to abnormalities of respiratory control and blood gas homeostasis.

Introduction

The neural control of breathing and its role in human disease has been the subject of intensive research. Much of this effort has been aimed at identifying neurons that act as central respiratory chemoreceptors (CRCs) – neurons that sense changes in CO_2 and initiate ventilatory changes that help maintain pH homeostasis. In this chapter, we discuss evidence that 5-HT neurons are CRCs. We review their role in the control of breathing and the ventilatory response to hypercapnia, and discuss data showing they have intrinsic chemosensitivity. We also review data from mice in which 5-HT neurons are genetically deleted, and in which there is a large decrease in the response to increased CO_2. Finally, we discuss the implications for human disease.

The 5-HT system stimulates respiratory output

Substantial evidence obtained from *in vitro* and *in vivo* work indicates that 5-HT neurons in the medullary raphe cause an increase in respiratory output via a neuromodulatory effect on brainstem neurons (Richerson, 2004; Hodges and Richerson, 2008a). Their ability to influence breathing is based in part on their extensive connections to multiple elements of the respiratory control network throughout the brainstem, including the pre-Bötzinger Complex (pre-BötC), nucleus tractus solitarius (NTS), retrotrapezoid nucleus, and phrenic and hypoglossal motor nuclei.

Many studies have examined the influence of the 5-HT system on respiratory output *in vitro* and *in vivo*. For example, an excitatory effect of 5-HT receptor activation has been shown using *in vitro* slice preparations containing the pre-BötC, a region of the medulla that is involved in respiratory rhythm generation (Smith *et al.*, 1991). Blockade of 5-HT_{2A} receptors leads to disruption of rhythm generation in slices containing the pre-BötC (Pena and Ramirez, 2002; Ptak *et al.*, 2009). In addition, the neuropeptides substance P (SP) and thyrotropin-releasing hormone (TRH), which are both co-localized in 5-HT neurons of the medulla, have strong stimulatory effects on respiratory activity in slices (Dekin *et al.*, 1985; Funk *et al.*, 1994; Greer *et al.*, 1996; Pena and Ramirez, 2004). The effects of 5-HT, TRH and SP occur in part due to depolarization of motor neurons (Bayliss *et al.*, 1997; Rekling *et al.*, 2000) and other respiratory neurons (Rekling *et al.*, 1996; Pena and Ramirez, 2004). In addition, TRH induces bursting pacemaker activity in the NTS (Dekin *et al.*, 1985), and 5-HT has the same effect on respiratory neurons in the pre-BötC (Ptak *et al.*, 2009). This increased bursting would enhance respiratory output. 5-HT, TRH and SP also each stimulate respiratory output in the *en bloc* brainstem-spinal cord preparation of the newborn rat (Morin *et al.*, 1990; Ptak and Hilaire, 1999; Richerson, 2004). Similarly, there is a large body of data demonstrating that raphe neurons, as well as exogenous 5-HT and TRH receptor agonists, stimulate

*Corresponding author
E-mail: george.richerson@yale.edu

ISBN 978-0-12-374634-4
DOI: 10.1016/B978-0-12-374634-4.00017-4

breathing *in vivo* (Richerson, 2004). For example, direct electrical stimulation of neurons in the medullary raphe in anesthetized, paralyzed and ventilated cats *in vivo* leads to an increase in respiratory frequency and phrenic nerve activity (Holtman *et al.*, 1986).

Although there is evidence that 5-HT, TRH, and SP each have a predominantly stimulatory effect on breathing *in vivo* (Manzke *et al.*, 2003; Richerson, 2004; Hodges and Richerson, 2008a; Hodges *et al.*, 2008), there are also contradictory data. For example, some older studies have concluded that 5-HT neurons have an inhibitory effect on respiration *in vivo* (Olson *et al.*, 1979; Lalley *et al.*, 1997). However, there are reasons to believe that the data that led to this conclusion are subject to other interpretations (Richerson, 2004; Hodges and Richerson, 2008a), and the bulk of the data support the conclusion that 5-HT neurons provide a tonic excitatory drive to the respiratory network *in vivo*. Since 5-HT neurons are tonically active during wakefulness, their effect on breathing would be expected to be greatest during the awake state. Thus, a reduction in firing of 5-HT neurons during sleep (Jacobs *et al.*, 1990, 2002; Fenik *et al.*, 2005) may contribute to the decrease in ventilation that occurs during sleep.

The 5-HT system is involved in the response to hypercapnia

There are relatively few conditions other than changes in the state of arousal that alter the firing of 5-HT neurons *in vivo* (Jacobs and Azmitia, 1992; Jacobs and Fornal, 1999). One is a change in the level of inspired CO_2. In unanesthetized behaving cats, a subset of physiologically-identified 5-HT neurons in the raphe obscurus and in the dorsal raphe respond to hypercapnia with an increase in firing rate (Veasey *et al.*, 1995, 1997). Interestingly, some of these neurons respond to CO_2 less during sleep (Veasey *et al.*, 1995), consistent with the observation that there is a reduction in the ventilatory response to CO_2 during sleep. These data are supported by other *in vivo* experiments in unanesthetized animals showing that hypercapnia activates c-fos in neurons of the medullary raphe, in some cases defined as serotonergic (see Figure 2A, below) (Larnicol *et al.*, 1994; Haxhiu *et al.*, 2001; Okada *et al.*, 2002; Pete *et al.*, 2002; Johnson *et al.*, 2005). The conclusion that 5-HT neurons are chemosensitive *in vivo* is also supported by experiments using microdialysis in the hypoglossal nucleus of unanesthetized mice, which showed that hypercapnia (mean inhaled CO_2 = 7%) causes a 2.5-fold increase in extracellular 5-HT levels (Kanamaru and Homma, 2007). The response to hypercapnia of 5-HT neurons in the ventrolateral medulla is much smaller in halothane anesthetized rats (Mulkey *et al.*, 2004), probably due to a depressant effect of halothane on respiratory chemoreception.

In vivo experiments have shown that microinjection of acetazolamide into the medullary raphe, which causes focal acidosis, stimulates respiratory output (Bernard *et al.*, 1996). Similarly, focal CO_2 microdialysis in the

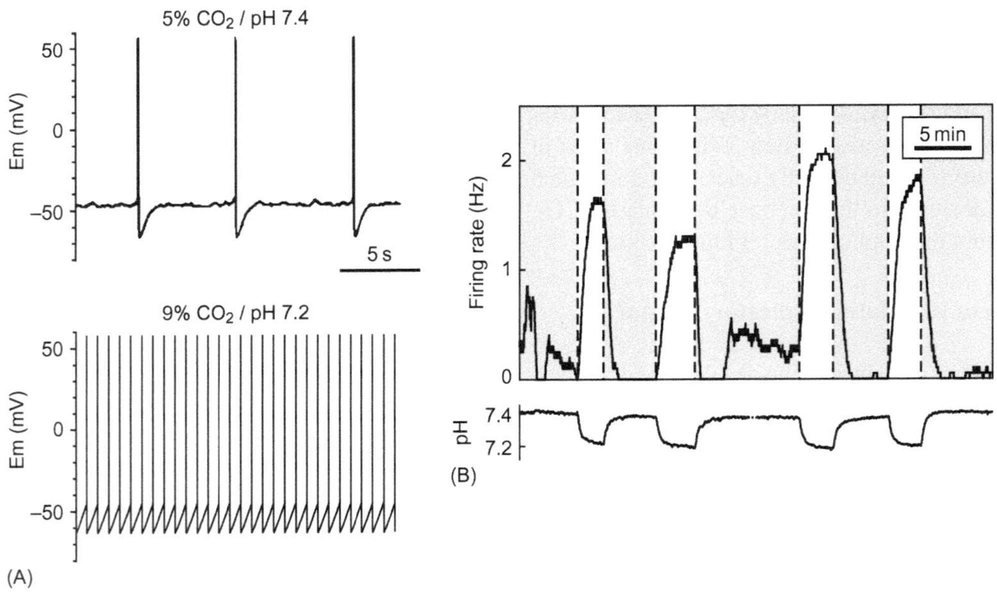

Figure 1 5-HT neurons increase their firing rate *in vitro* in response to an increase in CO_2 or decrease in pH. (A) Membrane potential (V_m) of a medullary raphe neuron in culture under baseline conditions and when exposed to hypercapnia. (B) The firing rate of 5-HT neurons *in vitro* increases significantly in response to a small decrease in pH. Shown is the firing rate of a cultured midbrain raphe neuron (upper panel) in response to repeated changes in pH from 7.4 to 7.2 (lower panel). (A) Reproduced with permission from Wiley-Blackwell: *Journal of Physiology* (Wang *et al.*, 2002) 540:951–70, ©2002. (B) Reproduced with permission from Nature Publishing Group: *Nature Neuroscience* (Severson *et al.*, 2003), 6: 1139–40, ©2003.

medullary raphe of the rat increases ventilation in sleep, but not in the awake state (Nattie and Li, 2001). In contrast, the same approach does increase ventilation in awake goats (Hodges *et al.*, 2004a, 2004b).

5-HT neurons have cellular properties consistent with a role as CRCs

A subset of neurons from the ventromedial medulla increase their firing rate in brain slices and primary tissue culture when extracellular CO_2 is increased from 5 percent to 9 percent (Figure 1) (Richerson, 1995; Wang *et al.*, 1998). These neurons are stimulated to the same degree by a similar drop in pH at constant PCO_2 (isocapnic acidosis), indicating that the primary stimulus is pH and not CO_2 *per se*. These neurons respond maximally to changes in pH between 7.2 and 7.6, which is relatively close to the range of pH changes that occur under normal physiological conditions.

Neurons from the ventromedial medulla that are stimulated by acidosis have electrophysiological properties characteristic of 5-HT neurons, with a highly regular firing rate, a wide action potential and a prominent afterhyperpolarization (Jacobs and Azmitia, 1992; Mason, 1997; Wang *et al.*, 1998). Studies in both brain slices and culture using immunohistochemistry for tryptophan hydroxylase have shown that these acidosis-stimulated neurons are serotonergic (Wang *et al.*, 2001; Bradley *et al.*, 2002). Chemosensitive 5-HT neurons in tissue culture and in brain slices continue to respond to hypercapnia after blockade of fast synaptic transmission either by antagonists of glutamate and GABA receptors or by high calcium/low magnesium solution (Richerson, 1995; Wang *et al.*, 2001; Bradley *et al.*, 2002). This does not prove that their ability to sense CO_2 is an intrinsic property, but that has recently been verified using acute dissociation, after which medullary 5-HT neurons from mice continue to be robustly chemosensitive (Wu *et al.*, 2008).

When mature, 5-HT neurons increase their firing rate to an average of 300 percent of their baseline level in response to a decrease in pH from 7.4 to 7.2 (Wang *et al.*, 1998, 2001, 2002). This is a large response compared to neurons studied in other proposed sites of CRCs (Putnam *et al.*, 2004). Interestingly, the response of 5HT neurons is not fully mature until several weeks after birth. In rat brain slices, there are few 5-HT raphe neurons that are chemosensitive until animals are older than postnatal day 12 (Wang and Richerson, 1999). In cell culture of rat and mouse raphe neurons there is a similar developmental time course (Wang and Richerson, 1999; Wu *et al.*, 2008), taking 10–11 days for 5-HT neurons to develop chemosensitivity. After that time the size of the response increases progressively until 4 weeks of age. This developmental maturation of chemosensitivity parallels development of the hypercapnic ventilatory response (HCVR) in rats *in vivo*. Several studies have shown that the HCVR of rats is not mature at birth, and undergoes postnatal

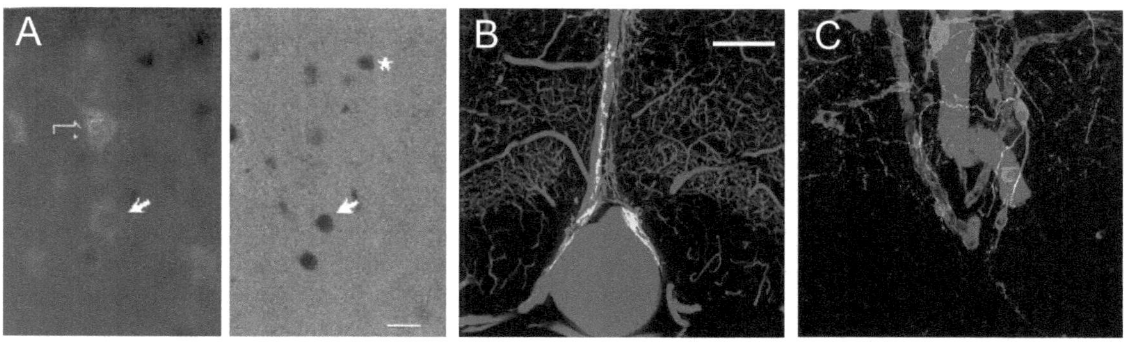

Figure 2 5-HT neurons are chemosensitive *in vivo* and are closely associated with large blood vessels. (A) A subset of neurons in the raphe pallidus express c-Fos when exposed to hypercapnia (right panel). Some of these neurons are immunoreactive for 5-HT (left panel). Open arrow: 5-HT neuron that does not express c-Fos. Closed arrows: 5-HT neuron that is immunoreactive for c-Fos. Asterisk: C-Fos positive neuron that does not express 5-HT. Scale bar: 20 μm. (B) Medullary raphe neurons are closely related to the basilar artery and its large penetrating branches. Shown is a transverse section of the medulla. 5-HT neurons are immunostained with an antibody against tryptophan hydroxylase (green). Blood vessels are filled with fluorescent albumin (red). Scale bar: 200 μm. (C). Shown is a neuron in the medullary raphe that was filled with biocytin (green) after patch-clamp recording. This neuron was stimulated by acidosis and was immunoreactive for tryptophan hydroxylase (blue). It had processes that were closely associated with large blood vessels (red). (A) Reproduced with permission from Elsevier: *Respiration Physiology* (Haxhiu *et al.*, 2001) 129: 191–209, ©2001. (B, C) Reproduced with permission from Nature Publishing Group: *Nature Neuroscience* (Bradley *et al.*, 2002) 5: 401–2, ©2002. To see the full color version of this figure please refer to the color plate at the back of the book. Copies produced via our print on demand service do not contain color plates; if your copy does not have the color plate, please go to this website to view the figure in color www.elsevierdirect.com/companions/9780123746344

maturation after approximately P12 (Serra *et al.*, 2001; Stunden *et al.*, 2001; Davis *et al.*, 2006). It is possible that the developmental increase in chemosensitivity of 5-HT neurons is responsible in part for the developmental increase in the HCVR of rats.

5-HT neurons in both the medullary and midbrain raphe have an anatomical relationship with large penetrating arteries of the brainstem that is consistent with their playing a role in sensing blood CO_2 (Bradley *et al.*, 2002; Severson *et al.*, 2003). The highest concentration of 5-HT neurons in the medulla is immediately adjacent to the basilar artery and its large midline branches. Neurons in brain slices with this anatomical arrangement are stimulated by acidosis, and electron microscopic images show that processes of serotonergic neurons are located directly adjacent to the vessel walls (Figure 2B, C) (Bradley *et al.*, 2002; Severson *et al.*, 2003). The medullary raphe is traversed primarily by large arteries and is relatively devoid of veins, which would lead to a local tissue PCO_2 that more accurately reflects the PCO_2 of arterial blood rather than mixed venous blood. The fact that brainstem 5-HT neurons are closely related to blood vessels has also been shown for neurons in the nucleus raphe pontis and in the nucleus linealis rostralis which enwrap large penetrating arteries with their processes (Scheibel *et al.*, 1975). This anatomical specialization of 5-HT neurons would allow them to closely monitor blood CO_2 levels in a manner analogous to the way peripheral chemoreceptors monitor CO_2 in the carotid arteries and aorta. The level of PCO_2 in these large arteries more accurately parallels changes in lung ventilation than PCO_2 in more distal vessels such as capillaries.

Respiratory output and chemoreception are altered by interfering with the 5-HT system

Supporting the hypothesis that 5-HT neurons are important for the ventilatory response to hypercapnia, disrupting the function of the medullary raphe system has an inhibitory effect on respiratory chemoreception. Reversible silencing with lidocaine or permanent lesions with ibotenic acid of the piglet medullary raphe both lead to a decrease in phrenic and hypoglossal nerve activity and the response to hypercapnia (Dreshaj *et al.*, 1998). After selective lesions of medullary 5-HT neurons with the toxin saporin conjugated to an antibody to the 5-HT transporter, there is a decrease in the ventilatory response to hypercapnia in both the awake state and non-REM sleep (Nattie *et al.*, 2004). Similarly, intraventricular injection of 5,7-dihydroxytryptamine lesions 5-HT neurons and causes hypoventilation (Olson *et al.*, 1979; Mueller *et al.*, 1984). Microdialysis of the medullary raphe with 8-hydroxy-2-di-n-propylaminotetralin (8-OH-DPAT) to cause inhibition of 5-HT neurons via 5-HT_{1A} autoreceptors also significantly decreases the ventilatory response to hypercapnia in piglets (Messier *et al.*, 2004). Thus, there are data from a variety of approaches that implicate 5-HT neurons as important CRCs *in vivo*.

To better evaluate the role of 5-HT neurons in control of breathing, new genetic approaches have been developed. These mouse models have taken advantage of knowledge about the transcription factors that are required for development of 5-HT neurons. The transcription factor Pet1 is found exclusively in 5-HT neurons (Hendricks *et al.*, 1999). In Pet-1 knockout (KO) mice, the number of 5-HT neurons is decreased by approximately 70 percent, with a concomitant decrease in 5-HT levels to 10–15 percent compared to wild-type animals (Hendricks *et al.*, 2003). Plethysmographic studies in these mice show a decrease in breathing frequency by about 30 percent at postnatal day 4.5 (Erickson *et al.*, 2007). At this time, there is also increased variability in the breathing pattern. This variability decreases with maturation of Pet-1 KO mice, with improved breathing at P 9.5, indicating a crucial role of 5-HT in the early maturation of respiration. At P4.5, Pet-1 KO mice respond normally to hypercapnic challenges with 5 percent CO_2, raising the possibility that either 5-HT neurons do not play an important role as chemoreceptors until they are older and develop chemosensitivity (see above), or that the remaining 30 percent of 5-HT neurons were able to compensate for this loss (Blier, 2008). As adults, there is a decrease in the hypercapnic ventilatory response in male Pet-1 KO mice, but not in females (Hodges *et al.*, 2005), pointing to an important effect of gender in the ability to compensate for the decrease in 5-HT neurons.

Lmx1b is a homeobox gene that is crucial for the development of the brainstem 5-HT system, but is also involved in development of other neuronal types (Ding *et al.*, 2003). Lmx1b KO mice lack all 5-HT neurons, but also lack numerous other cell types and do not survive the perinatal period (Zhao *et al.*, 2006). Therefore, an alternative method has been developed. For this purpose, loxP sites were inserted around exons 4, 5, and 6 of the Lmx1b gene (Zhao *et al.*, 2006). These floxed *Lmx1b* mice were bred with mice that expressed Cre recombinase downstream of the enhancer region for Pet-1 (Scott *et al.*, 2005) to generate progeny in which Lmx1b is specifically deleted only in precursors of 5-HT neurons. These $Lmx1b^{f/f/p}$ mice are born essentially devoid of any central 5-HT neurons (Zhao *et al.*, 2006).

Respiratory frequency is abnormally low in adult $\text{Lmx1b}^{f/f/p}$ mice under resting conditions (Figure 3A), but ventilation is normal (Hodges *et al.*, 2008). Thus, it is possible for nearly normal baseline breathing to occur in adults if 5-HT neurons are absent from embryonic life. However, baseline ventilation and the respiratory pattern are severely disrupted in $Lmx1b^{f/f/p}$ mice during the neonatal period, with an average of seven apneas per minute. This normalizes during the first 2 weeks of life (Richerson *et al.*, 2008). Interpretation of these results is complicated by the plasticity of the neonatal brain. The presence of near-normal breathing in adult $Lmx1b^{f/f/p}$ mice does not rule out a critical role of 5-HT neurons for normal respiratory output in adults. The respiratory network may simply be able to compensate for the absence of 5-HT neurons by undergoing a process of homeostatic plasticity during development (Turrigiano and Nelson, 2004). Future work will be required to determine the effect on breathing of acute deletion of all 5-HT neurons in adult animals.

The response to hypoxia is normal in adult $Lmx1b^{f/f/p}$ mice. However, the ventilatory response to 5 percent and 7 percent CO_2 is reduced by about 50 percent (Figure 3B) (Hodges *et al.*, 2008). When $Lmx1b^{f/f/p}$ mice are exposed to 10 percent CO_2, their HCVR is comparable to wild-type mice (Hodges and Richerson, 2008b), suggesting that 5-HT neurons are critical for the response to small changes in CO_2, but the response to very high levels of CO_2 may be mediated by non-5-HT neurons (Nattie, 1999). This is consistent with *in vitro* results demonstrating that 5-HT neurons have a large response to small changes in CO_2/pH, whereas other CRC candidates are less sensitive (Putnam *et al.*, 2004).

$Lmx1b^{f/f/p}$ mice also have decreased ability to maintain core body temperature when exposed to cold. This failure in thermoregulation is due to a decrease in shivering and a decrease in metabolism of brown adipose tissue. It is unclear how this defect in thermoregulation relates to the defect in chemoreception, but it is possible that they are linked by metabolic rate, which is influenced by body temperature and influences blood gases. 5-HT neurons would be well situated to play a role in coordinating these functions.

Role of 5-HT neurons in forebrain effects of hypercapnia

There is no evidence that 5-HT neurons in the midbrain raphe play a major role in control of breathing, and yet they are also chemosensitive to CO_2 *in vivo* (Veasey *et al.*, 1997)

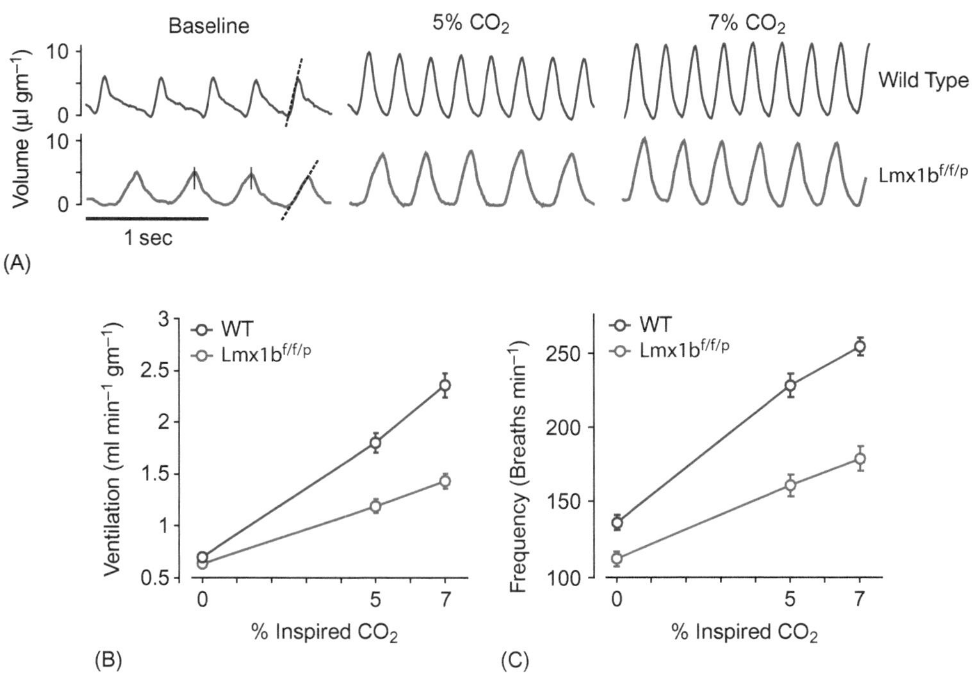

Figure 3 Mice with genetic deletion of 5-HT neurons have a defect in central respiratory chemoreception. (A) Plethysmograph recordings from WT (blue) and $Lmx1b^{f/f/p}$ (red) mice. $Lmx1b^{f/f/p}$ breathe with a slower frequency at baseline, and do not have as large a response to 5% and 7% CO_2. (B, C) $Lmx1b^{f/f/p}$ have a smaller response to an increase in ambient CO_2 and this is due to a smaller increase in breathing frequency. Reproduced with permission from the Society for Neuroscience: *Journal of Neuroscience* (Hodges *et al.*, 2008) 28: 2495–505, ©2008. To see the full color version of this figure please refer to the color plate at the back of the book. Copies produced via our print on demand service do not contain color plates; if your copy does not have the color plate, please go to this website to view the figure in color www.elsevierdirect.com/companions/9780123746344

and *in vitro* (Severson *et al.*, 2003). The question arises whether this chemosensitivity plays any role in normal brain function. Since 5-HT neurons in the midbrain project to the forebrain, it would be expected that any effect they have would be seen as a change in forebrain function. Forebrain projections of 5-HT neurons are particularly abundant to the limbic system and the thalamus, where they play important roles in anxiety and arousal. As it turns out, hypercapnia has particularly strong effects in inducing dyspnea, a form of anxiety interpreted as suffocation, and arousal from sleep.

Intense dyspnea is induced by an increase in arterial PCO_2 (Chonan *et al.*, 1990; American Thoracic Society, 1999). The dyspnea induced by hypercapnia occurs in part independent of any indirect effects on changes in respiratory motor output. For example, quadriplegic patients experience dyspnea when inspired CO_2 is increased (Opie *et al.*, 1959; Banzett *et al.*, 1989). Similarly healthy, mechanically ventilated human subjects paralyzed with vecuronium develop dyspnea when exposed to an increase in inspired CO_2 (Banzett *et al.*, 1990). There is other evidence indicating that changes in CO_2 can induce dyspnea independent of changes in respiratory effort (Buchanan *et al.*, 2008), suggesting that there are direct effects of hypercapnia, presumably mediated by stimulation of CO_2 chemoreceptors.

Functional imaging studies with positron emission tomography and functional magnetic resonance imaging have shown that hypercapnia activates multiple limbic regions of the brain (Buchanan *et al.*, 2008). Furthermore, SSRIs reduce anxiety and panic thought to be due to limbic activation (Hensler, 2006). Thus, the presence of diffuse projections from midbrain raphe nuclei to forebrain and limbic structures involved in dyspnea, and the influence of 5-HT on limbic function, supports the notion that direct projections from 5-HT neurons could mediate the sensation of dyspnea in response to hypercapnia.

Hypercapnia is also a potent arousal stimulus. In adult humans, hypercapnia induces arousal from both non-REM and REM sleep, but with a lower threshold for REM compared to non-REM (Berthon-Jones and Sullivan, 1984). In dogs, hypercapnia induces arousals from both non-REM and REM but does so more quickly and with a lower threshold in non-REM than in REM (Phillipson *et al.*, 1977). In lambs, repeated 60-second challenges with 8 percent CO_2 greatly increase arousal probability compared with control conditions (Johnston *et al.*, 2007). It has previously been proposed that 5-HT neurons could contribute to this arousal response to hypercapnia (Washburn *et al.*, 2002; Severson *et al.*, 2003) since 5-HT neurons of the midbrain are chemosensitive and 5-HT causes conversion of thalamocortical activity from a sleep-like pattern to a waking pattern (Steriade *et al.*, 1993). Consistent with this possibility, preliminary data from *Lmx1b*$^{f/f/p}$ mice indicate that 5-HT neurons are required for the normally robust electroencephalographic and behavioral changes associated with arousal in response to hypercapnia (Buchanan *et al.*, 2007). It is possible that midbrain 5-HT neurons induce arousal in response to low levels of CO_2, whereas the same neurons induce anxiety and dyspnea at higher CO_2 levels.

The 5-HT system, ventilation and human disease

A variety of human diseases have been linked to abnormalities of the 5-HT system. Some of these are also associated with abnormalities of breathing and/or the response to hypercapnia. These include sudden infant death syndrome (SIDS), sudden unexplained death in epilepsy (SUDEP), sleep apnea, panic disorder and Parkinson's disease.

SIDS

SIDS is the leading cause of mortality in infants between 1 month and 1 year of age in the developed world with an incidence of around 0.6 per 1000 (Kinney *et al.*, 2009), accounting for 6 deaths in the US every day. It is defined as 'the sudden and unexpected death of an infant under 12 months of age that is usually associated with a sleep period and remains unexplained after a complete autopsy, death scene investigation, and review of the clinical history' (Krous *et al.*, 2004). Based on the evidence at the time, Hunt and Brouillette proposed in 1987 that 'the most compelling hypothesis continues to be that SIDS is related to a brainstem abnormality in the neuroregulation of cardiorespiratory control' (Hunt and Brouillette, 1987). It is now known that a small subset of SIDS is due to mutations of cardiac ion channels that lead to long QT syndrome (Weese-Mayer *et al.*, 2007). However, the majority of cases are thought to be due to abnormal brainstem regulation of cardiac or respiratory function (Kinney *et al.*, 2001, 2005, 2009). Recognition that the majority of cases are found in the prone position, and the hypothesis that this causes airway obstruction, resulted in the 'Back to Sleep' campaign, which reduced the incidence of SIDS by 50 percent. Insight into the mechanisms of SIDS has also recently been advanced by the demonstration of abnormalities in the 5-HT system in four independent datasets of pathology material from SIDS cases (Kinney *et al.*, 2009). For example, there is a decrease in binding of lysergic acid diethylamide (LSD), a ligand for 5 $HT_{1A–D}$ and $5HT_2$ receptors, in the medullary arcuate nucleus of SIDS infants compared to control cases. This nucleus is of interest because it is homologous to chemosensitive fields of the ventrolateral medulla of cats (Filiano *et al.*, 1990), and because it contains cell bodies of 5-HT neurons (Figure 4A). It has also been found to be hypoplastic in a small number of cases of SIDS. Other regions containing 5-HT neurons also have decreased LSD binding in SIDS victims (Figure 4B) (Kinney *et al.*, 2009). It is now known that this decrease in 5-HT receptor binding is due to a decrease in 5-HT_{1A} autoreceptors present on 5-HT neurons themselves, because there is also a decrease in binding of 8-OH-DPAT (Paterson *et al.*, 2006).

Recently it has been found that there is an increase in number of 5-HT neurons in the medulla of SIDS cases (Figure 4C), but that the majority of these neurons have an 'immature' morphology (Paterson *et al.*, 2006). It is not yet clear how these pathological changes relate to the function of the 5-HT system, but one possibility is that there is a delay in development of 5-HT neurons, possibly induced by environmental exposures or genetic factors (Weese-Mayer *et al.*, 2007; Kinney *et al.*, 2009). Environmental factors that have been identified include exposure to cigarette smoke and maternal use of alcohol (Moon *et al.*, 2007). Genetic factors include allelic variants of the promoter region of the gene for the 5-HT transporter (Weese-Mayer *et al.*, 2007). There may also be effects of sex hormones, since there is a higher incidence of SIDS in males compared to females, as well as differences in the pathology between genders in the raphe obscurus (Paterson *et al.*, 2006). Therefore, the working hypothesis is that the pathological changes in SIDS reflect a decrease in function of the 5-HT system, which causes abnormal respiratory output. If that is the case, then infants at risk may have abnormalities of breathing, cardiac function and thermoregulation similar to (but less severe than) the mouse models described above.

SUDEP

Some epilepsy patients die unexpectedly without apparent cause. The incidence of this SUDEP has been estimated at 0.09 per 1000 person-years for a general population of epilepsy patients, and at 9.3 per 1000 person-years in high-risk epilepsy surgery candidates (Tomson *et al.*, 2008). That translates to 9.3 percent of the latter population that will die of SUDEP over 10 years. The epidemiology of SUDEP has many parallels with SIDS. For example, a subset of SUDEP cases also has mutations of long QT syndrome genes, but the majority are thought to be due to abnormalities of cardiorespiratory control (Ackerman, 2005; Tomson *et al.*, 2008; Johnson *et al.*, 2009). As with SIDS, patients are often found in bed,

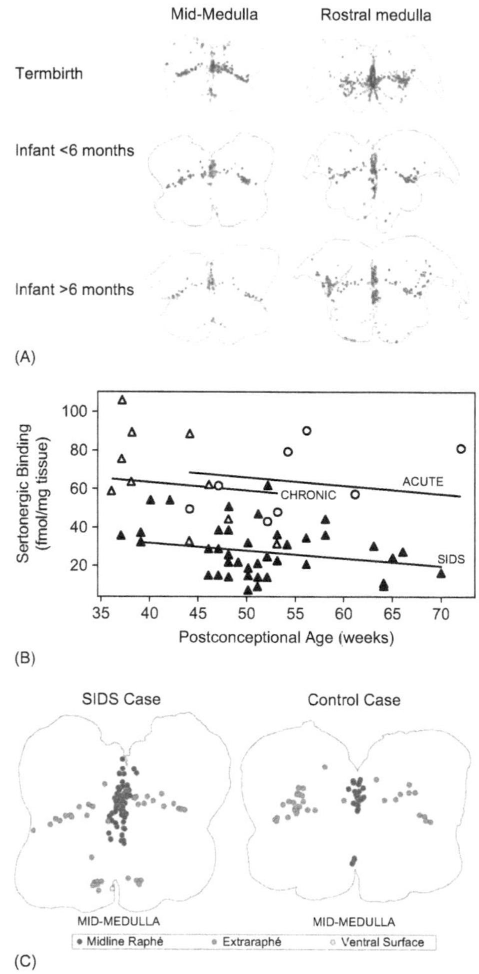

Figure 4 Sudden infant death syndrome (SIDS) is associated with abnormalities of the 5-HT system. (A) Developmental changes in the human medullary raphe nuclei at the mid- and rostral medullary levels. Blue dots, raphe 5-HT neurons; green dots, extra-raphe 5-HT neurons; red dots, 5-HT neurons on the ventral surface including the arcuate nucleus. Scale bar: 1 cm. (B) Tritiated LSD binding to 5-HT receptors is decreased in the raphe obscurus of infants that have died of SIDS. The scatter plot shows ^{3}H-LSD binding as measured with quantitative tissue receptor autoradiograhy in the raphe obscurus relative to postconceptional age for SIDS cases (solid triangles) compared to control infants that died of acute causes (open circles) or diseases that caused chronic hypoxia (closed circles). (C) Comparison of distribution of 5-HT neurons of a SIDS case and a control case. In the midline raphe of the SIDS infant there is an increased number of 5-HT neurons. (A) Reproduced with permission from Elsevier: *Autonomic Neuroscience* (Kinney, 2007) 132: 81–102, ©2007. (B) Reproduced with permission from Lippincott, Williams & Wilkins, Inc.: *Journal of Neuropathology and Experimental Neurology* (Panigrahy *et al.*, 2000) 59: 377–387. (C) Reproduced with permission from the American Medical Association: *Journal of the American Medical Association* (Paterson et *al.*, 2006) 296: 2124–2132 ©2006. To see the full color version of this figure please refer to the color plate at the back of the book. Copies produced via our print on demand service do not contain color plates; if your copy does not have the color plate, please go to this website to view the figure in color www.elsevierdirect.com/companions/9780123746344

and 71 percent are found in the prone position (Kloster and Engelskjon, 1999). It is also more common in males. Recently it has been found that there is severe postictal respiratory depression in many epilepsy patients, with oxygen saturations dropping below 90 percent after 33 percent of seizures (Bateman *et al.*, 2008). This occurs even in complex partial seizures without generalization, so it is not due to loss of motor function (Figure 5A). Changes in respiratory function have also been seen in animal models of seizures (Johnston *et al.*, 1997; St-John *et al.*, 2006). SUDEP has been associated with abnormalities of 5-HT function (So, 2008). For example, in the DBA/2 mouse model of seizures, there is a high incidence

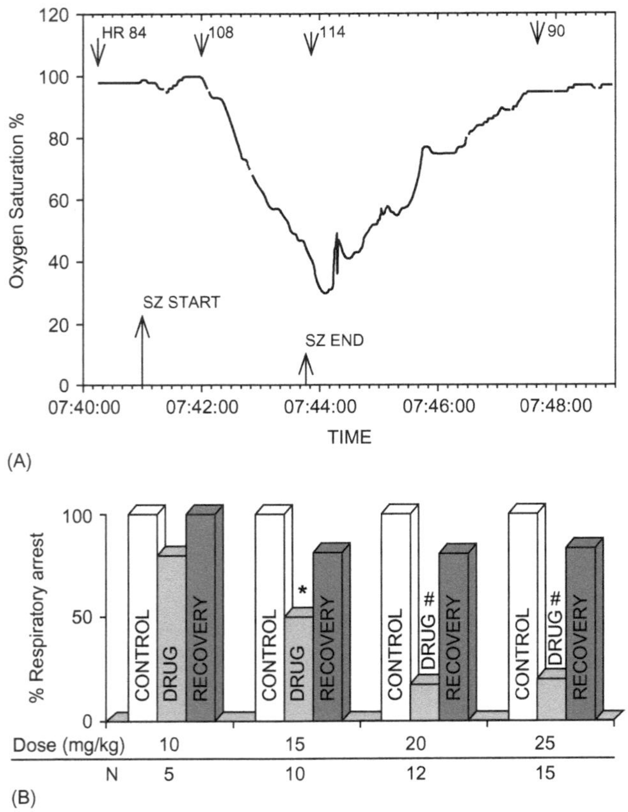

Figure 5 Some cases of sudden unexpected death in epilepsy (SUDEP) may involve postictal blunting of serotonergic respiratory drive. (A) Recording of oxygen saturation in a patient with a complex partial left temporal onset seizure. Despite the lack of secondary generalization there is severe and prolonged desaturation to 30 percent. The patient remained sitting the entire seizure. (B) Fluoxetine reduces postictal respiratory arrest in DBA/2 mice. All mice had respiratory arrest after audiogenic seizures. This was significantly reduced by fluoxetine treatment in a dose dependent manner. (A) Reproduced with permission from the Oxford University Press: *Brain* (Bateman *et al.*, 2008) 131 (Pt 12) 3239–45, ©2008. (B) Reproduced with permission from the Oxford University Press: *Epilepsia* (Tupal and Faingold, 2006) 47: 21–6, ©2006.

of postictal apnea and death, both of which are reduced after treatment with fluoxetine, a 5-HT reuptake inhibitor (SSRI) (Figure 5B) (Tupal and Faingold, 2006). Thus, it is reasonable to conclude that seizures may cause temporary dysfunction of 5-HT neurons due to invasion of the brainstem by electrical discharges, and this could cause respiratory arrest.

Sleep apnea

The involvement of 5-HT neurons in control of breathing and the decrease in firing rate of 5-HT neurons during sleep (Jacobs *et al.*, 1990) suggest that airway obstruction in sleep apnea could be worsened in patients with defects in the 5-HT system. It is clear that motor neurons that innervate the airway receive serotonergic projections (Brandes *et al.*, 2006). There is evidence that a decrease in 5-HT release contributes to the decrease in airway tone during sleep in animals (Fenik *et al.*, 2005), although there is also evidence against this (Horner, 2008). Nonetheless, a subset of patients with sleep apnea improve with treatment by SSRIs (Hanzel *et al.*, 1991). It is possible that patients whose sleep apnea is due to a decrease in function of the 5-HT system may be more likely to respond to treatment with serotonergic drugs.

Panic disorder

It has been proposed that panic attacks are due to false activation of a 'suffocation alarm' that is present in everyone and is mediated by neural circuitry that initiates a protective response to conditions that prevent adequate ventilation of the lungs (Klein, 1993; Papp *et al.*, 1993). Patients with panic disorder experience spontaneous episodes of severe dyspnea, hyperventilation, palpitations and anxiety. These panic attacks can be induced in many patients with panic disorder by exposure to elevated ambient CO_2 (Gorman *et al.*, 1994). For example, two breaths of 35 percent CO_2 is adequate to induce a panic attack in many panic disorder patients (Fyer *et al.*, 1987). There is no animal model of panic disorder, and the neural mechanisms of panic attacks in humans are poorly understood, but it is widely believed that defects in the 5-HT system contribute (Coplan *et al.*, 1992). For example, SSRIs are the first-line treatment for panic disorder patients. Thus it is reasonable to conclude that patients with panic disorder have an electrophysiological defect in 5-HT neurons that causes them to develop paroxysmal firing in response to levels of CO_2 that are not high enough to be activating in normal humans. This in turn would cause inappropriate activation of many neural circuits that are involved in the

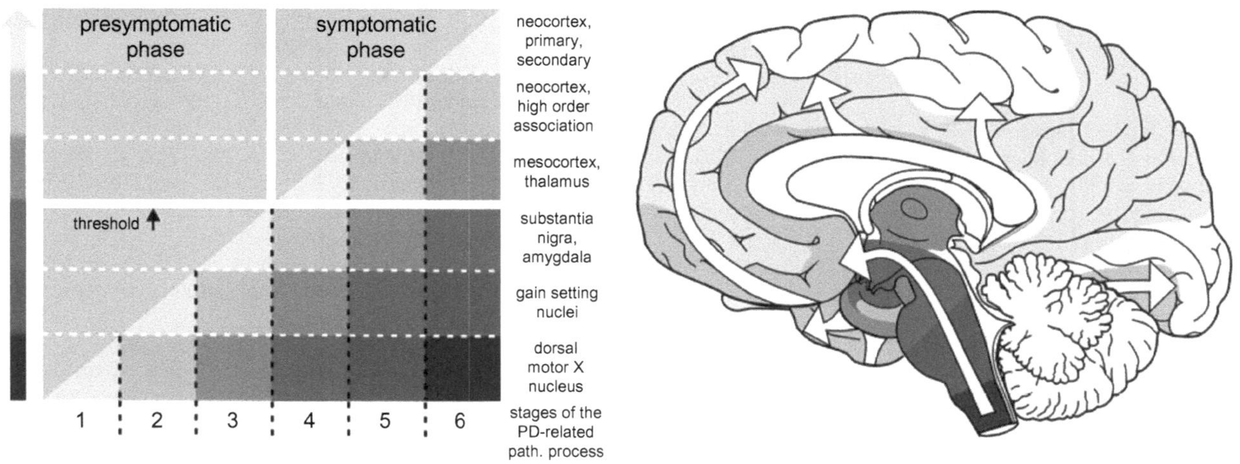

(A) (B)

Figure 6 Parkinson's disease involves the 5-HT system early and many patients have abnormalities of breathing and chemoreception. (A) Braak staging of neurodegeneration in idiopathic PD. Note that 'gain-setting nuclei', which includes the raphe nuclei, are involved earlier in the disease process than dopaminergic neurons of the sustantia nigra. (B) Progression of disease in PD begins in the lower brainstem and progresses rostrally. (A) Reproduced with permission from Springer: *Cell Tissue Research* (Braak *et al.*, 2004) 318: 121–34, ©2004. To see the full color version of this figure please refer to the color plate in the back of the book. Copies produced via our print on demand service do not contain color plates. If your copy does not have the color plate, please go to this website to view the figure in color www.elsevierdirect.com/companions/9780123746344

response to hypercapnia, including respiratory, cardiac and limbic circuits (Buchanan *et al.*, 2008).

Parkinson's disease

It has long been thought that Parkinson's disease (PD) is due to specific neurodegeneration of dopaminergic neurons of the substantia nigra, and that the symptoms are relatively limited to defects in motor control. However, that view has recently been drastically changed by the realization that most PD patients have a large constellation of non-motor symptoms that precede motor symptoms by many years (Lim *et al.*, 2009; Tolosa *et al.*, 2009). These symptoms include many forms of autonomic dysfunction, pain/sensory disturbances, sleep and mood disorders, and olfactory deficits. Consistent with these wide-ranging symptoms, it has recently been discovered that there are many pathological changes that precede degeneration of the substantia nigra, including in the enteric nervous system and medullary autonomic nuclei such as the dorsal motor nucleus of the vagus (Figure 6) (Braak *et al.*, 2003, 2004). The raphe nuclei are one of the earliest neuronal groups that degenerate (Braak *et al.*, 2004). This is relevant to the above discussion because a large percentage of PD patients complain of problems with breathing, including dyspnea in 40 percent of patients (Witjas *et al.*, 2002), and respiratory dyskinesia in many others. There is also blunting of the hypoxic ventilatory response in many PD patients (Onodera *et al.*, 2000), and the leading cause of death is pneumonia. It is possible that defects in 5-HT neurons could contribute to some of these symptoms, and indeed there is some evidence that PD patients may benefit from adjunctive treatment with serotonergic drugs (Di Matteo *et al.*, 2008).

Each of these examples illustrates how understanding the role of 5-HT neurons in control of breathing and CO_2 sensation may lead to unexpected insights into the pathophysiology of a variety of human diseases that are not necessarily associated with breathing. There are additional examples for which a link may also exist, but where the exact relationship remains unclear, such as migraine headaches, hypercapnia-induced changes in cerebral blood flow, and hypocapnia-induced absence seizures. Defining these relationships through a better understanding of the role of 5-HT neurons in chemoreception will be an important goal for future work in these areas.

Acknowledgements

This work was supported by NIH P01HD36379, NIH HD052772, the Bumpus Foundation and the VAMC.

References

Ackerman, M.J. (2005) Cardiac causes of sudden unexpected death in children and their relationship to seizures and syncope: genetic testing for cardiac electropathies. Semin. Pediatr. Neurol., 12: 52–58.

American Thoracic Society (1999) Dyspnea. Mechanisms, assessment and management: a consensus statement. Am. J. Respir. Crit. Care Med., 159: 321–340.

Banzett, R.B., Lansing, R.W., Reid, M.B., Adams, L. and Brown, R. (1989) 'Air hunger' arising from increased PCO_2 in mechanically ventilated quadriplegics. Respir. Physiol., 76: 53–67.

Banzett, R.B., Lansing, R.W., Brown, R., Topulos, G.P., Yager, D., Steele, S.M., Londono, B., Loring, S.H., Reid, M.B. and Adams, L. (1990) 'Air hunger' from increased PCO_2 persists after complete neuromuscular block in humans. Respir. Physiol., 81: 1–17.

Bateman, L.M., Li, C.S. and Seyal, M. (2008) Ictal hypoxemia in localization-related epilepsy: analysis of incidence, severity and risk factors. Brain, 131: 3239–3245.

Bayliss, D.A., Viana, F., Talley, E.M. and Berger, A.J. (1997) Neuromodulation of hypoglossal motoneurons: cellular and developmental mechanisms. Resp. Physiol., 110: 139–150.

Bernard, D.G., Li, A.H. and Nattie, E.E. (1996) Evidence for central chemoreception in the midline raphe. J. Appl. Physiol., 80: 108–115.

Berthon-Jones, M. and Sullivan, C.E. (1984) Ventilation and arousal responses to hypercapnia in normal sleeping humans. J. Appl. Physiol., 57: 59–67.

Blier, P. (2008) Resiliency of monoaminergic systems: the 80% rule and its relevance to drug development. J. Psychopharmacol., 22: 587–589.

Braak, H., Del, T.K., Rub, U., de Vos, R.A., Jansen Steur, E.N. and Braak, E. (2003) Staging of brain pathology related to sporadic Parkinson's disease. Neurobiol. Aging, 24: 197–211.

Braak, H., Ghebremedhin, E., Rub, U., Bratzke, H. and Del, T.K. (2004) Stages in the development of Parkinson's disease-related pathology. Cell. Tissue Res., 318: 121–134.

Bradley, S.R., Pieribone, V.A., Wang, W., Severson, C.A., Jacobs, R.A. and Richerson, G.B. (2002) Chemosensitive serotonergic neurons are intimately associated with large arteries of the ventral medulla. Nat. Neurosci., 5: 401–402.

Brandes, I.F., Zuperku, E.J., Stucke, A.G., Jakovcevic, D., Hopp, F.A. and Stuth, E.A. (2006) Serotonergic modulation of inspiratory hypoglossal motoneurons in decerebrate dogs. J. Neurophysiol., 95: 3449–3459.

Buchanan, G.F., Hodges, M.R. and Richerson, G.B. (2007) Deficiency in hypercapnia-induced arousal in mice with genetic deletion of 5-HT neurons. Soc. Neurosci. Abst., 33: 73612.

Buchanan, G.F., Hodges, M.R. and Richerson, G.B. (2008) Contribution of chemosensitive serotonergic neurons to interactions between the sleep–wake cycle and respiratory control. In: Monti, J.M., Pandi-Perumal, S.R., Jacobs, B.L. and Nutt, D.J. (Eds.), Serotonin and Sleep: Molecular, Functional and Clinical Aspects. Birkhauser Verlag, Basel, pp. 529–554.

Chonan, T., Mulholland, M.B., Leitner, J., Altose, M.D. and Cherniack, N.S. (1990) Sensation of dyspnea during hypercapnia, exercise, and voluntary hyperventilation. J. Appl. Physiol., 68: 2100–2106.

Coplan, J.D., Gorman, J.M. and Klein, D.F. (1992) Serotonin related functions in panic-anxiety: a critical overview. Neuropsychopharmacology, 6: 189–200.

Davis, S.E., Solhied, G., Castillo, M., Dwinell, M., Brozoski, D. and Forster, H.V. (2006) Postnatal developmental changes in CO_2 sensitivity in rats. J. Appl. Physiol., 101: 1097–1103.

Dekin, M.S., Richerson, G.B. and Getting, P.A. (1985) Thyrotropin-releasing hormone induces rhythmic bursting in neurons of the nucleus tractus solitarius. Science, 229: 67–69.

Di Matteo, V., Pierucci, M., Esposito, E., Crescimanno, G., Benigno, A. and Di, G.G. (2008) Serotonin modulation of the basal ganglia circuitry: therapeutic implication for Parkinson's disease and other motor disorders. Prog. Brain Res., 172: 423–463.

Ding, Y.Q., Marklund, U., Yuan, W., Yin, J., Wegman, L., Ericson, J., Deneris, E., Johnson, R.L. and Chen, Z.F. (2003) Lmx1b is essential for the development of serotonergic neurons. Nat. Neurosci., 6: 933–938.

Dreshaj, I.A., Haxhiu, M.A. and Martin, R.J. (1998) Role of the medullary raphe nuclei in the respiratory response to CO_2. Resp. Physiol., 111: 15–23.

Erickson, J.T., Shafer, G., Rossetti, M.D., Wilson, C.G. and Deneris, E.S. (2007) Arrest of 5HT neuron differentiation delays respiratory maturation and impairs neonatal homeostatic responses to environmental challenges. Respir. Physiol. Neurobiol., 159: 85–101.

Fenik, V.B., Davies, R.O. and Kubin, L. (2005) Noradrenergic, serotonergic and GABAergic antagonists injected together into the XII nucleus abolish the REM sleep-like depression of hypoglossal motoneuronal activity. J. Sleep Res., 14: 419–429.

Filiano, J.J., Choi, J.C. and Kinney, H.C. (1990) Candidate cell populations for respiratory chemosensitive fields in the human infant medulla. J. Comp. Neurol., 293: 448–465.

Funk, G.D., Smith, J.C. and Feldman, J.L. (1994) Development of thyrotropin-releasing-hormone and norepinephrine potentiation of inspiratory-related hypoglossal motoneuron discharge in neonatal and juvenile mice *in vitro*. J. Neurophysiol., 72: 2538–2541.

Fyer, M.R., Uy, J., Martinez, J., Goetz, R., Klein, D.F., Fyer, A., Liebowitz, M.R. and Gorman, J. (1987) CO2 challenge of patients with panic disorder. Am. J. Psych., 144: 1080–1082.

Gorman, J.M., Papp, L.A., Coplan, J.D., Martinez, J.M., Lennon, S., Goetz, R.R., Ross, D. and Klein, D.F. (1994) Anxiogenic effects of CO2 and hyperventilation in patients with panic disorder. Am. J. Psych., 151: 547–553.

Greer, J.J., al Zubaidy, Z. and Carter, J.E. (1996) Thyrotropin-releasing hormone stimulates perinatal rat respiration *in vitro*. Am. J. Physiol., 271: R1160–R1164.

Hanzel, D.A., Proia, N.G. and Hudgel, D.W. (1991) Response of obstructive sleep apnea to fluoxetine and protriptyline. Chest, 100: 416–421.

Haxhiu, M.A., Tolentino-Silva, F., Pete, G., Ke, P. and Mack, S.O. (2001) Monoaminergic neurons, chemosensation and arousal. Resp. Physiol., 129: 191–209.

Hendricks, T., Francis, N., Fyodorov, D. and Deneris, E.S. (1999) The ETS domain factor Pet-1 is an early and precise marker of central serotonin neurons and interacts with a conserved element in serotonergic genes. J. Neurosci., 19: 10348–10356.

Hendricks, T.J., Fyodorov, D.V., Wegman, L.J., Lelutiu, N.B., Pehek, E.A., Yamamoto, B., Silver, J., Weeber, E.J., Sweatt, J.D. and Deneris, E.S. (2003) Pet-1 ETS gene plays a critical role in 5-HT neuron development and is required for normal anxiety-like and aggressive behavior. Neuron, 37: 233–247.

Hensler, J.G. (2006) Serotonergic modulation of the limbic system. Neurosci. Biobehav. Rev., 30: 203–214.

Hodges, M.R. and Richerson, G.B. (2008a) Contributions of 5-HT neurons to respiratory control: neuromodulatory and trophic effects. Respir. Physiol. Neurobiol., 164: 222–232.

Hodges, M.R. and Richerson, G.B. (2008b) Interaction between defects in ventilatory and thermoregulatory control in mice lacking 5-HT neurons. Respir. Physiol. Neurobiol., 164: 350–357.

Hodges, M.R., Klum, L., Leekley, T., Brozoski, D.T., Bastasic, J., Davis, S., Wenninger, J.M., Feroah, T.R., Pan, L.G. and Forster, H.V. (2004a) Effects on breathing in awake and sleeping goats of focal acidosis in the medullary raphe. J. Appl. Physiol., 96: 1815–1824.

Hodges, M.R., Martino, P., Davis, S., Opansky, C., Pan, L.G. and Forster, H.V. (2004b) Effects on breathing of focal acidosis at multiple medullary raphe sites in awake goats. J. Appl. Physiol., 97: 2303–2309.

Hodges, M.R., Best, S., Deneris, E.S. and Richerson, G.B. (2005) Adult Pet-1 knockout mice exhibit an attenuated hypercapnic ventilatory response. Program No. 352.4. *2005 Abstract Viewer/Itinerary Planner* [http://sfn.scholarone.com/itin2005/index.html]. Washington, DC: Society for Neuroscience.

Hodges, M.R., Tattersall, G.J., Harris, M.B., McEvoy, S.D., Richerson, D.N., Deneris, E.S., Johnson, R.L., Chen, Z.F. and Richerson, G.B. (2008) Defects in breathing and thermoregulation in mice with near-complete absence of central serotonin neurons. J. Neurosci., 28: 2495–2505.

Holtman, J.R. Jr., Anastasi, N.C., Norman, W.P. and Dretchen, K.L. (1986) Effect of electrical and chemical stimulation of the raphe obscurus on phrenic nerve activity in the cat. Brain Res., 362: 214–220.

Horner, R.L. (2008) Neuromodulation of hypoglossal motoneurons during sleep. Respir. Physiol. Neurobiol., 164: 179–196.

Hunt, C.E. and Brouillette, R.T. (1987) Sudden infant death syndrome: 1987 perspective. J. Pediatr., 110: 669–678.

Jacobs, B.L. and Azmitia, E.C. (1992) Structure and function of the brain serotonin system. Physiol. Rev., 72: 165–229.

Jacobs, B.L. and Fornal, C.A. (1999) Activity of serotonergic neurons in behaving animals. Neuropsychopharmacology, 21: 9S–15S.

Jacobs, B.L., Wilkinson, L.O. and Fornal, C.A. (1990) The role of brain serotonin. A neurophysiologic perspective. Neuropsychopharmacology, 3: 473–479.

Jacobs, B.L., Martin-Cora, F.J. and Fornal, C.A. (2002) Activity of medullary serotonergic neurons in freely moving animals. Brain Res. Brain Res. Rev., 40: 45–52.

Johnson, J.N., Hofman, N., Haglund, C.M., Cascino, G.D., Wilde, A.A. and Ackerman, M.J. (2009) Identification of a possible pathogenic link between congenital long QT syndrome and epilepsy. Neurology, 72: 224–231.

Johnson, P.L., Hollis, J.H., Moratalla, R., Lightman, S.L. and Lowry, C.A. (2005) Acute hypercarbic gas exposure reveals functionally distinct subpopulations of serotonergic neurons in rats. J. Psychopharmacology, 19: 327–341.

Johnston, R.V., Grant, D.A., Wilkinson, M.H. and Walker, A.M. (2007) The effects of repeated exposure to hypercapnia on arousal and cardiorespiratory responses during sleep in lambs. J. Physiol., 582: 369–378.

Johnston, S.C., Siedenberg, R., Min, J.K., Jerome, E.H. and Laxer, K.D. (1997) Central apnea and acute cardiac ischemia in a sheep model of epileptic sudden death. Ann. Neurol., 42: 588–594.

Kanamaru, M. and Homma, I. (2007) Compensatory airway dilation and additive ventilatory augmentation mediated by dorsomedial medullary 5-hydroxytryptamine 2 receptor activity and hypercapnia. Am. J. Physiol. Regul. Integr. Comp. Physiol., 293: R854–R860.

Kinney, H.C. (2007) The development of the medullary serotonergic system in early human life. Auton. Neurosci., 132: 81–102.

Kinney, H.C., Filiano, J.J. and White, W.F. (2001) Medullary serotonergic network deficiency in the sudden infant death syndrome: review of a 15-year study of a single dataset. J. Neuropathol. Exp. Neurol., 60: 228–247.

Kinney, H.C., Myers, M.M., Belliveau, R.A., Randall, L.L., Trachtenberg, F.L., Ten Fingers, S., Youngman, M., Habbe, D. and Fifer, W.P. (2005) Subtle autonomic and respiratory dysfunction in the sudden infant death syndrome associated with serotonergic brainstem abnormalities: a case report. J. Neuropathol. Exp. Neurol., 64: 689–694.

Kinney, H.C., Richerson, G.B., Dymecki, S.M., Darnall, R.A. and Nattie, E.E. (2009) Serotonin and the brainstem in the sudden infant death syndrome: a review. Annu. Rev. Pathol., in press.

Klein, D.F. (1993) False suffocation alarms, spontaneous panics, and related conditions. An integrative hypothesis. Archiv. Gen. Psych., 50: 306–317.

Kloster, R. and Engelskjon, T. (1999) Sudden unexpected death in epilepsy (SUDEP): a clinical perspective and a search for risk factors. J. Neurol. Neurosurg. Psychiatry, 67: 439–444.

Krous, H.F., Beckwith, J.B., Byard, R.W., Rognum, T.O., Bajanowski, T., Corey, T., Cutz, E., Hanzlick, R., Keens, T.G. and Mitchell, E.A. (2004) Sudden infant death syndrome and unclassified sudden infant deaths: a definitional and diagnostic approach. Pediatrics, 114: 234–238.

Lalley, P.M., Benacka, R., Bischoff, A.M. and Richter, D.W. (1997) Nucleus raphe obscurus evokes 5-HT-1A receptor-mediated modulation of respiratory neurons. Brain Res., 747: 156–159.

Larnicol, N., Wallois, F., Berquin, P., Gros, F. and Rose, D. (1994) c-fos-like immunoreactivity in the cat's neuraxis following moderate hypoxia or hypercapnia. J. Physiol. (Paris), 88: 81–88.

Lim, S.Y., Fox, S.H. and Lang, A.E. (2009) Overview of the extranigral aspects of Parkinson disease. Arch. Neurol., 66: 167–172.

Manzke, T., Guenther, U., Ponimaskin, E.G., Haller, M., Dutschmann, M., Schwarzacher, S. and Richter, D.W. (2003) 5-HT_{4a} receptors avert opioid-induced breathing depression without loss of analgesia. Science, 301: 226–229.

Mason, P. (1997) Physiological identification of pontomedullary serotonergic neurons in the rat. J. Neurophysiol., 77: 1087–1098.

Messier, M.L., Li, A. and Nattie, E.E. (2004) Inhibition of medullary raphe serotonergic neurons has age dependent effects on the CO_2 response in newborn piglets. J. Appl. Physiol., 96: 1909–1919.

Moon, R.Y., Horne, R.S. and Hauck, F.R. (2007) Sudden infant death syndrome. Lancet, 370: 1578–1587.

Morin, D., Hennequin, S., Monteau, R. and Hilaire, G. (1990) Serotonergic influences on central respiratory activity: an *in vitro* study in the newborn rat. Brain Res., 535: 281–287.

Mueller, R.A., Towle, A.C. and Breese, G.R. (1984) Supersensitivity to the respiratory stimulatory effect of TRH in 5,7-dihydroxytryptamine-treated rats. Brain Res., 298: 370–373.

Mulkey, D.K., Stornetta, R.L., Weston, M.C., Simmons, J. R., Parker, A., Bayliss, D.A. and Guyenet, P.G. (2004) Respiratory control by ventral surface chemoreceptor neurons in rats. Nat. Neurosci., 7: 1360–1369.

Nattie, E.E. (1999) CO_2, brainstem chemoreceptors and breathing. Prog. Neurobiol., 59: 299–331.

Nattie, E.E. and Li, A. (2001) CO_2 dialysis in the medullary raphe of the rat increases ventilation in sleep. J. Appl. Physiol., 90: 1247–1257.

Nattie, E.E., Li, A., Richerson, G.B. and Lappi, D.A. (2004) Medullary serotonergic neurons and adjacent neurons that express neurokinin-1 receptors are both involved in chemoreception *in vivo*. J. Physiol., 556: 235–253.

Okada, Y., Chen, Z., Jiang, W., Kuwana, S. and Eldridge, F.L. (2002) Anatomical arrangement of hypercapnia-activated cells in the superficial ventral medulla of rats. J. Appl. Physiol., 93: 427–439.

Olson, E.B.J., Dempsey, J.A. and McCrimmon, D.R. (1979) Serotonin and the control of ventilation in awake rats. J. Clin. Inv., 64: 689–693.

Onodera, H., Okabe, S., Kikuchi, Y., Tsuda, T. and Itoyama, Y. (2000) Impaired chemosensitivity and perception of dyspnoea in Parkinson's disease. Lancet, 356: 739–740.

Opie, L.H., Smith, A.C. and Spalding, J.M. (1959) Conscious appreciation of the effects produced by independent changes of ventilation volume and of end-tidal PCO_2 in paralysed patients. J. Physiol., 149: 494–499.

Panigrahy, A., Filiano, J., Sleeper, L.A. et al. (2000) Decreased serotonergic receptor binding in rhombic lip-derived regions of the medulla oblongata in the sudden infant death syndrome. J. Neuropathol. Exp. Neurol., 59: 377–387.

Papp, L.A., Klein, D.F. and Gorman, J.M. (1993) Carbon dioxide hypersensitivity, hyperventilation, and panic disorder. Am. J. Psych., 150: 1149–1157.

Paterson, D.S., Trachtenberg, F.L., Thompson, E.G., Belliveau, R.A., Beggs, A.H., Darnall, R., Chadwick, A.E., Krous, H.F. and Kinney, H.C. (2006) Multiple serotonergic brainstem abnormalities in sudden infant death syndrome. J. Am. Med. AssocA, 296: 2124–2132.

Pena, F. and Ramirez, J.M. (2002) Endogenous activation of serotonin-2A receptors is required for respiratory rhythm generation *in vitro*. J. Neurosci., 22: 11055–11064.

Pena, F. and Ramirez, J.M. (2004) Substance P-mediated modulation of pacemaker properties in the mammalian respiratory network. J. Neurosci., 24: 7549–7556.

Pete, G., Mack, S.O., Haxhiu, M.A., Walbaum, S. and Gauda, E.B. (2002) CO2-induced c-Fos expression in brainstem preprotachykinin mRNA containing neurons. Resp. Physiol. Neurobiol., 130: 265–274.

Phillipson, E.A., Kozar, L.F., Rebuck, A.S. and Murphy, E. (1977) Ventilatory and waking responses to CO2 in sleeping dogs. Am. Rev. Respir. Dis., 115: 251–259.

Ptak, K. and Hilaire, G. (1999) Central respiratory effects of substance P in neonatal mice: an *in vitro* study. Neurosci. Lett., 266: 189–192.

Ptak, K., Yamanishi, T., Aungst, J., Milescu, L.S., Zhang, R., Richerson, G.B., Smith, J.C. (2009) Raphé neurons stimulate respiratory circuit activity by multiple mechanisms via endogenously released serotonin and substance P. J. Neurosci., 29(12):3720–3737.

Putnam, R.W., Filosa, J.A. and Ritucci, N.A. (2004) Cellular mechanisms involved in CO_2 and acid signaling in chemosensitive neurons. Am. J. Physiol. Cell Physiol., 287: C1493–C1526.

Rekling, J.C., Champagnat, J. and Denavit-Saubie, M. (1996) Thyrotropin-releasing hormone (TRH) depolarizes a subset of inspiratory neurons in the newborn mouse brain stem *in vitro*. J. Neurophysiol., 75: 811–819.

Rekling, J.C., Funk, G.D., Bayliss, D.A., Dong, X.W. and Feldman, J.L. (2000) Synaptic control of motoneuronal excitability. Physiol. Rev., 80: 767–852.

Richerson, G.B. (1995) Response to CO_2 of neurons in the rostral ventral medulla *in vitro*. J. Neurophysiol., 73: 933–944.

Richerson, G.B. (2004) Serotonin neurons as CO_2 sensors that maintain pH homeostasis. Nat. Rev. Neurosci., 5: 449–461.

Richerson, G.B., Wehner, M.R., Hodges, M.R., Ptak, K., Aungst, J. and Smith, J.C. (2008) Neonatal mice lacking central 5-HT neurons have severe apnea and a high mortality. Soc. Neurosci. Abstr., 34.

Scheibel, M.E., Tomiyasu, U. and Scheibel, A.B. (1975) Do raphe nuclei of the reticular formation have a neurosecretory or vascular sensor function? Exp. Neurol., 47: 316–329.

Scott, M.M., Wylie, C.J., Lerch, J.K., Murphy, R., Lobur, K., Herlitze, S., Jiang, W., Conlon, R.A., Strowbridge, B.W. and Deneris, E.S. (2005) A genetic approach to access serotonin neurons for *in vivo* and *in vitro* studies. Proc. Natl. Acad. Sci. USA, 102: 16472–16477.

Serra, A., Brozoski, D., Hedin, N., Franciosi, R. and Forster, H.V. (2001) Mortality after carotid body denervation in rats. J. Appl. Physiol., 91: 1298–1306.

Severson, C.A., Wang, W.G., Pieribone, V.A., Dohle, C.I. and Richerson, G.B. (2003) Midbrain serotonergic neurons are central pH chemoreceptors. Nat. Neurosci., 6: 1139–1140.

Smith, J.C., Ellenberger, H.H., Ballanyi, K., Richter, D.W. and Feldman, J.L. (1991) Pre-Botzinger complex: a brainstem region that may generate respiratory rhythm in mammals. Science, 254: 726–729.

So, E.L. (2008) What is known about the mechanisms underlying SUDEP? Epilepsia, 49(Suppl. 9): 93–98.

Steriade, M., McCormick, D.A. and Sejnowski, T.J. (1993) Thalamocortical oscillations in the sleeping and aroused brain. Science, 262: 679–685.

St-John, W.M., Rudkin, A.H., Homes, G.L. and Leiter, J.C. (2006) Changes in respiratory-modulated neural activities, consistent with obstructive and central apnea, during fictive seizures in an *in situ* anaesthetized rat preparation. Epilepsy Res., 70: 218–228.

Stunden, C.E., Filosa, J.A., Garcia, A.J., Dean, J.B. and Putnam, R.W. (2001) Development of *in vivo* ventilatory and single chemosensitive neuron responses to hypercapnia in rats. Resp. Physiol., 127: 135–155.

Tolosa, E., Gaig, C., Santamaria, J. and Compta, Y. (2009) Diagnosis and the premotor phase of Parkinson disease. Neurology, 72: S12–S20.

Tomson, T., Nashef, L. and Ryvlin, P. (2008) Sudden unexpected death in epilepsy: current knowledge and future directions. Lancet Neurol., 7: 1021–1031.

Tupal, S. and Faingold, C.L. (2006) Evidence supporting a role of serotonin in modulation of sudden death induced by seizures in DBA/2 mice. Epilepsia, 47: 21–26.

Turrigiano, G.G. and Nelson, S.B. (2004) Homeostatic plasticity in the developing nervous system. Nat. Rev. Neurosci., 5: 97–107.

Veasey, S.C., Fornal, C.A., Metzler, C.W. and Jacobs, B.L. (1995) Response of serotonergic caudal raphe neurons in relation to specific motor activities in freely moving cats. J. Neurosci., 15: 5346–5359.

Veasey, S.C., Fornal, C.A., Metzler, C.W. and Jacobs, B.L. (1997) Single-unit responses of serotonergic dorsal raphe neurons to specific motor challenges in freely moving cats. Neuroscience, 79: 161–169.

Wang, W. and Richerson, G.B. (1999) Development of chemosensitivity of rat medullary raphe neurons. Neuroscience, 90: 1001–1011.

Wang, W., Pizzonia, J.H. and Richerson, G.B. (1998) Chemosensitivity of rat medullary raphe neurones in primary tissue culture. J. Physiol., 511: 433–450.

Wang, W., Zaykin, A.V., Tiwari, J.K., Bradley, S.R. and Richerson, G.B. (2001) Acidosis-stimulated neurons of the medullary raphe are serotonergic. J. Neurophysiol., 85: 2224–2235.

Wang, W., Bradley, S.R. and Richerson, G.B. (2002) Quantification of the response of rat medullary raphe neurones to independent changes in pH_o and PCO_2. J. Physiology, 540: 951–970.

Washburn, C.P., Sirois, J.E., Talley, E.M., Guyenet, P.G. and Bayliss, D.A. (2002) Serotonergic raphe neurons express TASK channel transcripts and a TASK-like pH- and halothane-sensitive K+ conductance. J. Neurosci., 22: 1256–1265.

Weese-Mayer, D.E., Ackerman, M.J., Marazita, M.L. and Berry-Kravis, E.M. (2007) Sudden Infant Death Syndrome: review of implicated genetic factors. Am. J. Med. Genet. A, 143A: 771–788.

Witjas, T., Kaphan, E., Azulay, J.P., Blin, O., Ceccaldi, M., Pouget, J., Poncet, M. and Cherif, A.A. (2002) Nonmotor fluctuations in Parkinson's disease: frequent and disabling. Neurology, 59: 408–413.

Wu, Y., Hodges, M.R. and Richerson, G.B. (2008) Stimulation by hypercapnic acidosis in mouse 5-HT neurons *in vitro* is enhanced by age and increased temperature. Soc. Neurosci. Abstr., 34.

Zhao, Z.Q., Scott, M., Chiechio, S., Wang, J.S., Renner, K.J., Gereau, R.W., Johnson, R.L., Deneris, E.S. and Chen, Z.F. (2006) Lmx1b is required for maintenance of central serotonergic neurons and mice lacking central serotonergic system exhibit normal locomotor activity. J. Neurosci., 26: 12781–12788.

SECTION 3

Serotonin and Behavioral Control

CHAPTER 3.1

Serotonin and Development[1]

Patricia M. Whitaker-Azmitia*

Department of Psychology, Stony Brook University, Stony Brook, New York

Abstract: Serotonin is a key neurotransmitter in the adult brain, as well as a developmental signal in the developing brain. Serotonin itself is expressed very early in both human and rodent brain, and its receptors, enzymes and transporter sites are all overdeveloped initially, peaking in some cases in the first 2 years of life and in other cases in early adolescence. In either case, the serotonin system begins to decline by early adulthood. This early peak in activity reflects the developmental function of serotonin; however, even in the adult brain, serotonin is likely to play a role in maintenance and plasticity. As a developmental signal, serotonin acts in a variety of ways, depending on the developmental stage. Earliest on, serotonin acts in an autoregulatory manner, influencing the development of its own neurons. As brain development continues, serotonin begins to act in all subsequent stages – neurogenesis/neuroprotection, gliogenesis, migration and finally differentiation and maturation. Any factor which influences the amount of serotonin during development may result in enduring changes in the brain. In this regard, concerns have been raised regarding the effects of maternal stress or drug use during pregnancy. Understanding the role of serotonin in development has made significant contributions to the study of developmental disabilities, including fetal alcohol syndrome and autism.

Keywords: neonatal, fetal, neurogenesis, gliogenesis, synaptogenesis, ontogeny, neuroprotection, serotonin, development.

Development of the serotonergic system

Although the number of serotonin neurons is very small in all species studied, once born, the neurons develop into the most widely spread neurotransmitter system in the brain. Serotonin is also one of the earliest developing neurotransmitter systems in the mammalian brain (Aitken and Tork, 1988; Lauder, 1990). During brain development, serotonin itself, as well as its associated enzymes, uptake sites and receptors, all reach peak levels before declining into adult values. This peak in serotonergic activity early in brain development reflects the dual role of serotonin in the brain – as a developmental signal and as a neurotransmitter. In the immature brain, significantly higher levels of serotonin are required in order to serve the developmental role, and any disruption can have serious consequences to the final mature brain. Thus, knowing the timing and means by which the serotonin system itself develops can be key to understanding much of adult brain function.

[1]This chapter is dedicated to Dr Jean Lauder, on the occasion of her retirement.

*Corresponding author
E-mail: pwhitaker@notes.cc.sunysb.edu.

Serotonin cell determination

Serotonin neurons appear early in development, mostly in two groups of nuclei in the anterior and posterior medulla which later become the well-recognized nine raphe nuclei, nuclei B5–9 (which includes the largest, the dorsal raphe) arise from the anterior group and B1–4 from the posterior group. Consequently, B5–9 project largely to cortical and some cerebellar targets, while B1–4 principally project to the spinal cord. Serotonin cell development, like all cells, is determined by both the cell's environment and its genome.

There appear to be distinct differences in the genetics of early development in the two clusters, and the anterior group forms slightly earlier than the posterior (Cordes, 2005). Serotonergic induction occurs in cell populations defined by Sonic hedgehog (Shh) (Hynes and Rosenthal, 1999), along the dorsal–ventral axes and FGF 4 in the antero-posterior axis in the posterior groups. The anterior group induction takes places through the positioning actions not only of Shh and FGF4, but also of FGF 8 (Ye *et al.*, 1998). Further, FGF 8 levels determine formation of dorsal or medial raphe nuclei, in the anterior group. The transcription factor Nkx2.2 is also involved in the generation of the posterior groups (Briscoe *et al.*, 1999).

Christian Müller & Barry Jacobs (Eds.)
Handbook of Behavioral Neurobiology of Serotonin
ISBN 978-0-12-374634-4

DOI: 10.1016/B978-0-12-374634-4.00018-6

Once the position of serotonin precursor cells is determined, other transcription factors are necessary to establish the actual serotonergic phenotype in the post-mitotic cells (Gaspar *et al.*, 2003). The principal transcription factor required to establish the neurochemical specificity of serotonin neurons (i.e., tryptophan hydroxylase and the serotonin transporter) is the ETS family factor Pet-1 (Hendricks *et al.*, 1999). Pet-1 is found in the entire extent of the raphe nuclei, but localized only to those cells which will express tryptophan hydroxylase. Conserved Pet-1 binding sites are found in the promoter regions of the specific serotonin neuronal markers – 5-HT1A receptor, the serotonin transporter (SERT), tryptophan hydroxylase and aromatic-L-amino acid decarboxylase (Hendricks *et al.*, 1999). Other important factors, including the homeobox gene Lmx1b, Mash1, Gata2, Gata3 and Phox2b, form a 'transcriptional network' which further specifies differentiation (Alenina *et al.*, 2006).

Development of the serotonin system in human brain

In the human brain, serotonergic neurons are first evident in the brainstem by 5 weeks of gestation (Sundstrom *et al.*, 1993), and increase dramatically by the tenth week of gestation (Shen *et al.*, 1989; Kontur *et al.*, 1993; Levallois *et al.*, 1993), where they are found predominantly in the dorsal raphe nuclei (Shen *et al.*, 1989). Very early on, serotonin neurons are distributed in association with cranial nerve nuclei and reticular formation nuclei associated with respiration and blood pressure regulation (Kinney *et al.*, 2007). By at least 20 weeks of gestation (Kinney *et al.*, 2007) or as early as 15 weeks (Takahashi *et al.*, 1986), the typical organization of serotonin cell bodies into the raphe nuclei can be seen. The number of serotonergic cells shows a reduction as fetal development continues (Shen *et al.*, 1989). By birth, the highly regulated and important diurnal rhythm of serotonin is already present (Attanasio *et al.*, 1986).

The serotonin fibers reach the cortical anlage at 8 weeks of gestation, the subplate at 10 weeks of gestation and the cortical plate at 13 weeks of gestation (Verney *et al.*, 2002). At the same time, the serotonin transporter can be visualized in non-serotonergic fibres in thalamocortical processes (Verney *et al.*, 2002) in the interior capsule (between putamen and caudate), suggesting that serotonin is being produced and must be regulated in order for development to proceed. Synthesis of serotonin in the brain continues to increase throughout gestation and into the first 2 years of life, peaking at levels approximately twice that found in the adult brain at 5 years of age (Chugani *et al.*, 1999) and then declining to adult levels (Hedner *et al.*, 1986; Toth and Fekete, 1986). Imaging studies of the serotonin transporter in developing neurologically typical children show a steady increase from 3 years of age to 18 years (Makkonen *et al.*, 2008), but by 20 years of age PET studies have shown that the serotonin transporter begins to decrease and continues to do so at the approximate rate of 10 percent per decade up until the eighth decade (Yamamoto *et al.*, 2002). A similar age-related loss of serotonergic tone, using endocrine challenge tests, has been reviewed and reported (Lerer *et al.*, 1999).

5-HT1A serotonin receptors are highest in the developing brain, estimated to be at gestational weeks 16–22 (Bar-Peled *et al.*, 1991; del Olmo and Pazos, 2001), and levels fall to that found in the young adult by birth (del Olmo *et al.*, 1998). There is a transient appearance of this receptor in the fetal and neonatal cerebellum, a region from which it is absent in the adult (del Olmo *et al.*, 1994). After the age of 20, PET studies show a decrease of 3.6–10 percent per decade of the 5-HT1A receptor in cortical regions; however, there is no loss in the raphe autoreceptor (Tauscher *et al.*, 2001), with the greater decline in females (Costes *et al.*, 2005). This pattern in the 5-HT1A receptor is therefore similar to the patterns found for serotonin terminals and serotonin levels. This is not surprising, since this is the receptor most closely associated with the developmental effects of serotonin. Moreover, this receptor has been localized in high numbers to immature astroglial cells, and much of serotonin's actions on neuronal maturation may be through release of growth factors from astroglial cells.

5-HT2 receptors also show peak levels prenatally, but later than the 5-HT1A receptor, the 5-HT2A showing the highest level in raphe nuclei at midgestation (Zec *et al.*, 1996; Paterson *et al.*, 2004). The 5-HT2A receptor first appears in insular and temporal cortex by 25 weeks of gestation but does not peak in number until 31/32-gestation-week specimens, relatively late in the development of cortex (Wai *et al.*, 2008). In human PET studies, the number of 5-HT2A receptors dramatically decreases beginning at the age of 20 (Sheline *et al.*, 2002).

Thus, serotonin fibers are growing into the human cortex, as migration and differentiation are taking place. Moreover, the amount of serotonin begins to decrease by age 20, when the final stages of brain maturation are thought to have occurred.

Development of the serotonergic system in rodent models

In rodent brain, a very similar pattern of development is seen – that is, an early and robust development followed by a decline as the animal reaches adolescence.

Serotonergic cells are first identifiable at gestational day 12 (GD 12), as two groups of cells in the midline of the rhombencephalon (Hendricks *et al.*, 2003), with the rostral group evident earlier (Zhou *et al.*, 2000). Genesis of serotonin cells continues until GD 18 (Hansson *et al.*, 1998). The caudal group (designated B1–5) becomes the descending serotonin system of the raphe pallidus, raphe magnus, raphe pontis and raphe obscurus. The rostral group (B6–8) becomes the ascending serotonin system of the dorsal and median raphe nuclei (Lauder *et al.*, 1982; Lauder, 1985, 1990).

Initial axonal elongation is evident as early as GD 12 and continues until GD 16 (Hansson *et al.*, 1998; Zhou *et al.*, 2000) when axons enter the medial forebrain bundle and reach the basal forebrain and hippocampus (Hansson *et al.*, 1998), stria terminalis, hypothalamus, dorsal thalamus, cingulate cortex, superior olivary nucleus and cochlear nucleus (Narboux-Neme *et al.*, 2008). Serotonin projections continue to grow, traveling along pre-existing fiber tracts on GD 16–19 and arriving in the frontal cortex by GD 20 (Lidov and Molliver, 1982; Wallace and Lauder, 1983; Zhou *et al.*, 2000) in rat. Development in mice is essentially the same as in rats (Brüning *et al.*, 1997). Although serotonin axons reach target areas by birth, the final innervation arborization occurs robustly up until postnatal day (PND 21). In many cases, the terminal density decreases after that point (Hansson *et al.*, 1998). In somatosensory cortex the peak is even earlier, and is decreased to adult levels by 21 days, suggesting a very important role for serotonin in development of this region (D'Amato *et al.*, 1987). It is during this time of terminal arborization that serotonin also reaches its highest brain levels in rodents, much like the first 2–5 years of human life, and also at the time when synaptogenesis is at its peak. As in human brain, there is a transient expression of the serotonin transporter in cells which are not serotonergic. In particular, this has been observed in thalamocortical projections to sensory cortex (Brüning and Liangos, 1997), where serotonin is thought to play a role in barrel field formation (see below).

Similar to human brain, amongst the serotonin receptors examined, the 5-HT1A appears to be the receptor which shows the earliest appearance in development and can be transiently expressed in brain regions from which it is later absent, including the cerebellum (Daval *et al.*, 1987). The receptor is found densely by GD 15 in mice, in the thalamus, hippocampus and in cortex, in a medial to lateral gradient as development takes place (Bonnin *et al.*, 2006). In the developing hippocampus of rats, the 5-HT1A receptor is found in virtually all neurons once neurogenesis is complete (Patel and Zhou, 2005). Astroglial cells transiently express high amounts of this receptor postnatally, further suggesting that the role of serotonin in development may be in part mediated through effects on the astroglial cell (Whitaker-Azmitia *et al.*, 1990a; Patel and Zhou, 2005). Early work showed that these receptors could undergo plastic changes if animals were treated prenatally with serotonin receptor agonists (Whitaker-Azmitia *et al.*, 1987).

5-HT2 receptors also show a pattern similar to human development. The receptor peaks later than the 5-HT1A, reaching adult levels postnatally (Basura and Walker, 2000; Basura *et al.*, 2008).

Role of serotonin in brain development

Over the course of development, the actions of serotonin clearly change as well. Early on serotonin's role will be involvement in cell survival, and later, as the brain matures, serotonin will play a role in migration and differentiation. It is also likely that, even in the adult brain, serotonin will play a role in synaptic maintenance and plasticity, in many cases by the same mechanisms used in development. The following sections are arranged by developmental properties over time, and combine both human and animal studies.

Autocrine/autoregulatory effects

In animal studies, one of the earliest observed effects of serotonin on brain development was that it could inhibit the outgrowth of serotonin-producing neurons themselves (Whitaker-Azmitia and Azmitia, 1986a; Haydon *et al.*, 1987; Diefenbach *et al.*, 1995; Goldberg, 1998). Thus, developmental treatment with a variety of drugs which increase serotonin during development, including agonists (Shemer *et al.*, 1988, 1991), releasers (Akbari *et al.*, 1992), MDMA (Koprich *et al.*, 2003), monoamine-oxidase inhibitors (MAO-I) (Whitaker-Azmitia *et al.*, 1994) and uptake inhibitors (Cabrera-Vera *et al.*, 1997; Cabrera-Vera and Battaglia, 1998), have all been shown to decrease the serotonin activity in the adult brain.

Cell cultures of serotonin neurons derived from embryonic raphe neurons from mice lacking 5-HT1A or 5-HT1B receptor show that both of these receptors, but particularly the 5-HT1A receptor, show tonic inhibition of serotonin neuronal survival. Specifically, inhibition of cAMP by the autoreceptor 5-HT1A appears to be involved in inhibition of development (Rumajogee *et al.*, 2004).

Two recent studies using serotonin reuptake inhibitors during development show that there are behavioral changes associated with the decreased serotonergic outgrowth (Ansorge *et al.*, 2008; Popa *et al.*, 2008). Studies in 5-HTT$^{-/-}$ mice show a similar effect, with a 50 percent loss of serotonin neurons overall in the knockout animals (Lira *et al.*, 2003) and resultant behavioral changes such as increased measures of depression (Lira *et al.*, 2003) and anxiety (Holmes *et al.*, 2003). These behavioral changes in the 5-HTT$^{-/-}$ mice are suggested to be due to the effects of increased 'serotonin tone' in early development, and if mice are treated with PCPA or a 5-HT1A antagonist from PND 5-28, the depression and anxiety improve. Again, this suggests a role for the 5-HT1A receptor in inhibition of serotonergic development (Alexandre *et al.*, 2006).

This autoregulatory developmental effect of serotonin may play a role in the symptoms of autism, as children with autism often have an increased blood level of serotonin. This could access the brain early in development, before the blood–brain barrier is complete, and cause loss of serotonin terminals (Whitaker-Azmitia, 2005). Recent findings show that there is indeed a loss of serotonin terminals in autism (Chugani *et al.*, 1997, 1999; Makkonen *et al.*, 2008).

Neurogenesis/neuroprotection

Although serotonin fibers are present in cortex during late stages of neurogenesis in both humans (Sidman and Rakic, 1973) and rodents, there is very little clear evidence that serotonin plays a direct role in increasing cell proliferation. Rather, the earliest work by Jean Lauder, using parachlorophenylalanine (PCPA)-depletion of serotonin and thymidine incorporation to determine the effects of serotonin on development, found that serotonin actually ceased cell division by influencing the onset of differentiation (Lauder and Krebs, 1978; Lauder *et al.*, 1981; Lauder, 1983).

Similar work by Vitalis *et al.* (2007) treated developing rat fetuses from GD 12 to 17 with PCPA and found no change in BrdU incorporation or changes in cell cycle markers at GD 16.5 or 17.5, indicating no direct effect on cell proliferation. However, these authors have introduced the possibility that any decrease in cell number that may be observed could be due to the neuroprotectant role of serotonin, and thus any apparent loss in cell number in a serotonin-depleted animal may actually be due to increased apoptosis (see below). Indeed, in a similar PCPA-depletion study by Khozhai and Otellin (2006), the authors found that as the animals aged, there was a greater number of cells dying.

The case in the hippocampus may be different, where there is evidence that serotonin may play a limited role in neurogenesis of granule cells in the adult brain. Brezun and Daszuta (1999) found that depleting serotonin with the toxin 5,7 dihydroxytryptamine in adult rats decreased the number of BrdU-labeled granule cells in the dentate gyrus, and that transplanting fetal raphe cells into the hippocampus restored neurogenesis (Brezun and Daszuta, 2000). It has been suggested that the 5-HT1A receptor plays a role in neurogenesis in the hippocampus (Gould, 1999).

As a result of these studies on adult hippocampal neurogenesis, it had been suggested that serotonergic drugs such as the selective serotonin reuptake inhibitors (SSRIs) also increase neurogenesis in the adult brain, and that this mechanism might explain antidepressant therapeutic effects (Santarelli *et al.*, 2003; Marcussen *et al.*, 2008). However, recently studies have failed to support this observation (Cowen *et al.*, 2008; Huang *et al.*, 2008; Navailles *et al.*, 2008).

Gliogenesis and maturation

Serotonin plays a role in generating and maturing astroglial cells. One important aspect of this function is the release of the calcium-binding protein S100B from astroglial cells into the neuropil. Once released, S100B may actually be the mediator of many of the effects of serotonin on later brain development, such as differentiation and maturation of neurons.

The earliest studies of Lauder *et al.* (1982) combining thymidine incorporation and serotonin immunochemistry found that developing serotonin axons were closely associated with proliferating astroglial cells, and it was suggested that growing serotonin axons release trophic factors from these cells as they grow, which in turn matures neighboring neurons (Lauder, 1985). When rat pups are depleted of serotonin prenatally (with PCPA) there are fewer GFAP-positive astroglial cells as adults (Khozhai, 2006), while prenatal treatment with a 5-HT1A agonist increases glial number (Tajuddin *et al.*, 2003). 5-HT1A overexpressing mice show increased S100B positive cells in the cortex, striatum and hippocampus at PND 15.5 (Deng *et al.*, 2007). In adult rat, SSRI-treatment prevents stress-induced loss of astroglial cells (Czéh *et al.*, 2006) and may promote gliogenesis (Manev *et al.*, 2003). These animal findings have led to the suggestion that SSRIs may play a role in improving outcomes in neurological disorders, including epilepsy, stroke and Alzheimer's disease (Mostert *et al.*, 2008).

Astroglial cells express a number of serotonin receptors as well as the serotonin transporter (Whitaker and Warsh, 1983; Whitaker-Azmitia and Azmitia, 1986b). The 5-HT2 receptor is expressed, and, through this receptor, serotonin plays a role in release of Glial-cell line Derived Neurotrophic protein (GDNF) – a protein which serves as a neuroprotectant and neuronal survival factor (Tsuchioka *et al.*, 2008).

The high-affinity 5-HT1A receptor is also expressed on astroglial cells, in higher numbers in the immature cell and lower numbers as the cell matures. This may account for many of the transiently expressed serotonin receptors in whole brain. Some studies have suggested that SSRIs also influence release of S100B, and thereby exert their therapeutic effect (Tramontina *et al.*, 2008). Moreover, the 5-HT1A receptor is responsible for regulating the release of the very important cytokine, S100B.

In cell culture studies, treatment with a 5-HT1A agonist increases maturation of astroglial cells, taking on a stellate mature phenotype while releasing the neurotrophic factor, S100B (Azmitia *et al.*, 1990; Whitaker-Azmitia *et al.*, 1990a). S100B is a member of the EF-hand family of calcium-binding proteins, and plays a role in many cells by virtue of its ability to regulate both intra- and extracellular calcium. S100B also influences cellular function by binding to the cell surface receptor RAGE (receptor for advanced glycation endproducts). Amongst its known functions, S100B induces neurite outgrowth and prevents apoptosis. However, at higher concentrations and after brain trauma S100B can lead to increased cell death and advanced aging (Donato, 2003, 2007; Rothermundt *et al.*, 2003). Thus, serotonin has been suggested to play a role in both development and plasticity through regulation of S100B (Whitaker-Azmitia and Azmitia, 1994), and serotonergic regulation of this factor can be key to normal brain function.

Apoptosis

Apoptosis (programmed cell death) is an important part of development, as the maturing brain overproduces neurons and many must be removed as a normal course of development. In the human brain, approximately 50 percent of neurons are removed immediately prior to birth (reviewed in Andersen, 2003). Serotonin reduces apoptosis in the developing brain. For example, in serotonin transporter knockout mice, the higher levels of serotonin lead to decreased cell loss in the striatum, thalamus, hypothalamus, cortex and hippocampus (Persico *et al.*, 2003). Neuronal cell lines transfected with the 5-HT1A receptor show greater survival (Fricker *et al.*, 2005).

Apoptosis can be induced in developing cells in a variety of ways. In cell culture, 5-HT1A receptor agonists inhibit apoptosis induced by ethanol (Lee *et al.*, 2008) or serum deprivation (Ahlemeyer *et al.*, 1999). Serotonin may also protect against apoptosis in the mature brain, and similar findings can be observed in animal models, where a 5-HT1A agonist is neuroprotective against ischemia (Bode-Greuel *et al.*, 1990; Bielenberg and Burkhardt, 1990) and NMDA-induced excitotoxicity (Oosterink *et al.*, 1998; Madhavan *et al.*, 2003). Removal of serotonin prior to the ischemia model increases damage (Nakata *et al.*, 1997). In an animal model of myocardial infarction, an SSRI inhibited apoptosis (Wann *et al.*, 2008). In some cases, the neuroprotective effect of serotonin is not directly on the neuron, but rather through the effects of serotonin 5-HT1A receptors releasing the astroglial protein S100B (Ahlemeyer *et al.*, 2000; Brewton *et al.*, 2001; Ramos *et al.*, 2004)

Migration

Once cells are born, they must move (or migrate) to their final position in the brain before differentiation can take place. In human cortex, migration is complete by 26 weeks of gestation and the cortical layers are identifiable (Sidman and Rakic, 1973). At this time serotonin fibers are already in place, and serotonin may pay a role in cortical layering. In mouse tissue culture models, serotonin inhibits interneuron migration, an effect blocked by a 5-HT6 receptor antagonist. These same neurons are also disrupted in the 5-HTT knockout mouse (Riccio *et al.*, 2008). Treatment with the non-receptor selective serotonin agonist 5-methoxytryptamine (5-MT) during gestation causes a decrease in the amount of reelin produced by Cajal-Retzius cells of the cortex. Reelin, a glycoprotein associated with extracellular matrix, is involved in cortical column formation (Quattrocchi *et al.*, 2002), and the 5-MT-treated animals also showed pronounced disturbances in columnar formation (Janusonis *et al.*, 2005). Interestingly, deficits in reelin have been reported in autism (Fatemi *et al.*, 2005), which is often thought to be related to both cortical disturbed development and serotonin dysregulation.

Axonal elongation

Once in place, neurons undergo the final steps in differentiation, axonal elongation, and synaptogenesis and dendritic elaboration. In rodent models, much of the axon elongation influenced by serotonin appears to be in sensory systems, and occurs in the first week of neonatal life. In humans, this corresponds to the final trimester of gestation.

For axonal elongation, the most striking role of serotonin is in the development of thalamocortical efferents – cells that do not express serotonin, but which express, transiently, the serotonin transporter. Thus, these axons appear to grow to specific sites in the cortex by responding to and controlling levels of serotonin in the neuropil of already developed cells (Cases *et al.*, 1996). Most of the work in this area suggests that serotonin's largest influence is at PND 0–8 in the rat, a time when serotonin would be innervating virtually all of its targets, and a great deal of serotonin could be released from axonal varicosities by volume transmission (Zhou *et al.*, 2005). Excess serotonin leads to loss of the typical barrel-like patterns of cells in somatosensory cortex (Cases *et al.*, 1996; Lieske *et al.*, 1999), and a greater percent of the cortex is taken up by thalamic projections (Lane *et al.*, 2006). 5-HT1B receptors appear to be the predominant receptor involved in the control of thalamocortical outgrowth (Lotto *et al.*, 1999; Bou-Florés *et al.*, 2000; Salichon *et al.*, 2001). Developing neonatal rat pups depleted of serotonin by tryptophan depletion show a delay in removing excess or misplaced connections in retinotectal projections (González *et al.*, 2008), and a similar role for serotonin has recently been described for ascending auditory pathways (Thompson, 2008). Thus, these studies all show a role for serotonin in the fine tuning of tactile, auditory and visual connections.

Terminal elaboration of serotonin neurons has an important relationship with developing dopamine terminals. It has been shown in animal studies that depletion of dopamine early in development leads to hyperinnervation by serotonin terminals (Basura and Walker, 1999, 2000), while developmental treatment with a dopamine D1 agonist decreases serotonin terminal growth (Whitaker-Azmitia *et al.*, 1990b). This interaction between monoamines has been suggested to play a highly signficiant role in 'miswiring' of the cortex, leading to psychopathologies (Benes *et al.*, 2000).

Synaptogenesis and dendritic elaboration

The final step in development in which serotonin plays a role is in synaptogenesis and dendritic elaboration. Since this is the developmental process that actually takes place throughout life, serotonin's effects can last lifelong, and indeed there are animal studies showing that serotonin is an important synaptic maintenance factor, and indications that serotonergic drugs such as SSRIs may exert their pharmacological action by improving synaptic density.

In humans, synaptogenesis begins in midgestation (Huttenlocher *et al.*, 1983) and peaks at approximately 2 years of age (Chugani *et al.*, 1987). At 2 years, synaptogenesis and dendritic elaboration still continue at a high rate, but begin to slow down into adolescence and adulthood. This time-course coincides with what is known about the levels and activity of the serotonin system. Serotonin may influence synaptogenesis, directly through stimulation of 5-HT1A receptors, or through release of the trophic factor S100B (Azmitia, 2001).

The studies described below have shown effects for serotonin at four ages – fetal rat, which approximates midgestation in humans; neonatal rat, which represents the final trimester of gestation in humans; more than 1 week of age but less than 3 in rats, representing the first two years of human life; and, finally, adult studies. The pattern of effects on synaptogenesis varies with the critical time of depletion, with the earliest time showing the most long-lasting effects.

Fetal rat, midgestation human

PCPA depletion of serotonin from GD12–17 produces fewer dendrites in pyramidal neurons in layers III and V of somatosensory cortex which lasts into adulthood (Vitalis *et al.*, 2007). It appears that the 5-HT1A receptor is important, since in cultures of cerebeller Purkinje cells (Kondoh *et al.*, 2004) or hippocampal cells (Nishi *et al.*, 1996), synaptogenesis is increased by a 5-HT1A agonist.

Neonatal rat, third trimester human

Transient depletion of serotonin (by PCA or the selective 5-HT1A antagonist NAN-190) beginning at PND 3 shows that the 5-HT1A receptor plays a role in promoting synaptogenesis in dentate gyrus granule cells, particularly synaptic connections to dendritic spines. However, in this case the normal synaptic density is recovered by PND 60, suggesting it is delay in maturation rather than the permanent loss seen in animals depleted earlier in development. If serotonin is removed permanently beginning at PND 3 with the selective neurotoxin 5,7-DHT, dendritic development does not recover (Faber and Haring, 1999). Infusing the animals at this age with an antibody against S100B also shows a loss of synapses, which is not additive to the effects of a 5-HT1A antagonist (Wilson *et al.*, 1998). All of this shows that serotonin 5-HT1A-mediated release of S100B beginning at PND 3 (or the last trimester of human pregnancy) promotes synaptogenesis in the hippocampus. If serotonin levels are temporarily decreased, there is later recovery. If serotonin does not recover, there is also no recovery in synaptic density.

1–3 weeks rat: first 2 years of human life

Depletion of serotonin with PCPA from PND 10–20 leads to loss of presynaptic (synaptophysin) and dendritic markers (MAP-2) in the hippocampus of adolescent animals. As the animals age, there is recovery of presynaptic markers, but no improvement of the dendritic density. In the aged animal there is overgrowth of presynaptic elements, suggesting that a reactive synaptogenesis takes place over time (Mazer *et al.* 1997).

Adult

Studies in adult animals show that the role serotonin plays in synaptogenesis continues in the form of synaptic maintenance into adulthood. PCPA depletion of serotonin in adult animals causes loss of synapses in all CA1, CA3 and granule layer cells (Matsukawa *et al.*, 1997), a loss of the dendritic marker MAP-2 throughout the hippocampus (Whitaker-Azmitia *et al.*, 1995), and a loss of neurofilaments of axons within the hippocampus (Ramos *et al.*, 2000). In adult female rats, PCPA decreases dendritic spine density in the hippocampus, independent of hormonal effects (Alves *et al.*, 2002). In the cortex, this depletion causes an estimated 39–50 percent loss of synapses (Chen *et al.*, 1994). In studies using PCA depletion of serotonin, loss of dendritic and synaptic markers can be seen not only in the hippocampus, but also in the hypothalamus and cortex (Azmitia *et al.*, 1995). Synaptic maintenance is through the 5-HT1A-mediated release of S100B. In 5-HT-depleted animals S100B can be seen to accumulate within the astroglial cell rather than being released into the neuropil (Ramos *et al.*, 2000), and treatment with a 5-HT1A agonist can reverse the synaptic loss (Azmitia *et al.*, 1995).

Environmental effects on the development of serotonin

Many of the animal studies discussed above show that changing the levels of serotonin during development changes not only brain morphology but also behavior – in particular, lifelong alterations in behaviors modeling human anxiety (Borella *et al.*, 1997; Holmes *et al.*, 2003; Hohmann *et al.*, 2007; Cannizzaro *et al.*, 2008), autism (Kahne *et al.*, 2002; McNamara *et al.*, 2008), depression (Lira *et al.*, 2003; Vataeva *et al.*, 2007) and aggression (Farabollini *et al.*, 1988; Whitaker-Azmitia and Azmatia, 1994). These observations make it very important to know what variables during development affect serotonin, as these may be predisposing factors to psychopathology in later life.

Many external factors have been shown to interfere with the development of the serotonin system, including hypoxia (Shiraishi *et al.*, 2008; Strackx *et al.*, 2008), diet (Gonzalez-Burgos *et al.*, 1996; Feria-Velasco *et al.*, 2002; Himpel *et al.*, 2006; O'Reilly *et al.*, 2007; Pôrto *et al.*, 2008; Serfaty *et al.*, 2008), infections (Winter *et al.*, 2008), and environmental toxins such as pesticides (Garcia *et al.*, 2001; Aldridge *et al.*, 2005) and heavy metals. However, the most urgent concerns regarding potential damage to the serotonin system and long-term psychopathology are those centered on the effects of stress and of prenatal drug exposure.

Stress

Stress-induced chronic activation of the hypothalamic–pituitary–adrenal axis has effects on the serotonergic system of the mature brain (Lanfumey *et al.*, 2008). For example, restraint models of stress raise corticosterone levels and desensitize the 5-HT1A receptor (Mendelson and McEwen, 1991), and decrease serotonin release. Some of the loss of the 5-HT1A receptor may be on astroglial cells, decreasing S100B support for other cells (Nishi and Azmitia, 1996). The effects of glucocorticoids on the serotonin system have been proposed to be the basis of stress-induced depression in the mature brain. The developing brain may be even more vulnerable to stress-induced damage, and the consequences longer lasting (McEwen, 2003).

Prenatal stress or treatment with dexamethasone from GD 15–21 causes decreased synaptic number and decreased serotonin in the hippocampus into adulthood (Muneoka *et al.*, 1997), and studies looking at various times of stress suggest that this is the time of greatest vulnerability prenatally in rodents (Peters, 1986). Rats born following *in utero* stress also have decreased 5-HT1A receptors in the hippocampus (Van den Hove *et al.*, 2006a). Pups born to adrenalectomized mothers show increased serotonin in the cortex and hippocampus at PND 1 and 8, but a decrease at PND 22 (Leret *et al.*, 2004). This could reflect an overgrowth of serotonin terminals in the absence of glucocorticoids, followed by the autoregulatory removal of excess serotonin terminals. Prenatal stress also decreases hippocampal content of S100B, suggesting that increased levels of corticosterone decrease S100B, resulting in lower levels of serotonin (Van den Hove *et al.*, 2006b).

Later in development, removal from the dam (Matthews *et al.*, 2001; Veenema *et al.*, 2006) or post-weaning social isolation both lead to a permanent loss of serotonin terminals in the hippocampus (Whitaker-Azmitia *et al.*, 2000). In the rodent *Octodon degus*, early deprivation leads to a loss of serotonin fibers in the hippocampus, but an apparent increase in the amygdala. This shows the importance of emotional experience in setting up serotonergic networks (Gos *et al.*, 2006)

In human infants, prenatal maternal levels of cortisol are correlated with lower levels of serotonin in the infant (Field *et al.*, 2003), and maternal depression during pregnancy also leads to loss of serotonin in the infant (Field *et al.*, 2004). However, at later times in development, massaging or stroking the infant or adolescent human increases blood serotonin (Field, 2002), suggesting that some of the loss of serotonin in rodent isolation studies may be due to loss of physical contact with parents or peers.

Drugs

Many drugs exert their effects through the serotonergic system and thus would be expected to have an effect on the developing neurons themselves, at least in part through the autoregulatory/autocrine effects described above. For example, cocaine and MDMA would be expected to have serious effect on brain development by virtue of their effect on serotonin, and in animal studies, they do. However, greater concern should be placed on those substances more commonly taken during pregnancy – alcohol and the SSRI antidepressants. As well, since human brain development spans a long period of time – at least up to 20 years of age – the use of these substances even into adolescence should be considered carefully.

The most common non-genetic form of mental retardation continues to be that associated with fetal alcohol syndrome (FAS), and both animal and human studies show a role for alcohol-induced deficits in the serotonin system, and thus of serotonin regulation of brain development, in this condition. Treatment of animals at a very young fetal age (before serotonin neurons are born) with alcohol leads to a loss of both ascending and descending serotonin neurons (Sari and Zhou, 2004). This early exposure may reflect the increased incidence of sudden infant death (SID) in FAS (Kinney *et al.*, 2003; Zhou *et al.*, 2008). At later times of treatment (E11–15), although serotonin cell number is not affected, the number of terminals is decreased in the adult. This loss of serotonin fibers also results in the expected concomitant loss of maturation of regions along the pathway of the developing serotonin axons (Zhou *et al.*, 2005). This loss of neurons can be prevented by concomitant treatment with S100B or a 5-HT1A receptor agonist (Druse *et al.*, 2007), which blocks the alcohol-induced apoptosis of neurons (Druse *et al.*, 2006).

In human *post-mortem* studies, using tissue from children dying of SIDS, a significant loss of serotonin functioning in medullary nuclei was found in those children exposed to alcohol *in utero* (Kinney *et al.*, 2003). SPECT image analysis of the serotonin transporter in children with FAS show decreased transporter, suggesting loss of serotonin terminals (Riikonen *et al.*, 2005).

The use of antidepressants during pregnancy is an ongoing controversy, but epidemiological studies show that as many as 5 percent of women do take an antidepressant during pregnancy (Oberlander *et al.*, 2006). Antidepressants do not cause typical teratogenic effects on organ formation, and many women do need continual treatment during pregnancy. However, the long-term outcomes in terms of behavioral and neurochemical teratogenicity are not known, and animal studies suggest there may be persisting effects.

Many animal studies have looked at the effects of developmental exposure to antidepressants on the serotonin system in adults. In most cases, these studies have shown a reduction of serotonin (Hilakivi *et al.*, 1987; Feenstra *et al.*, 1996; Vijayakumar and Meti, 1999) or serotonin transporter (Hansen and Mikkelsen, 1998; Whitaker-Azmitia *et al.*, 2004; Maciag *et al.*, 2006) in the adult brain. This loss of terminals is what would be expected, as increased serotonin *in utero* could lead to the autoregulatory loss of serotonin terminals. Loss of serotonin in the mature brain following neonatal exposure to antidepressants has the expected behavioral changes – principally depressive-like symptoms (Hilakivi and Hilakivi, 1987; Fernandez-Pardal and Hilakivi, 1989; Vogel *et al.*, 1990; Velazquez-Moctezuma and Ruiz, 1992; Hansen *et al.*, 1997)

Thus, animal studies would predict that exposure to antidepressants during development would lead to loss of serotonin as adults and the behavioral phenotype of depression.

In the first few weeks of life, children exposed *in utero* to antidepressants do show a specific behavioral syndrome which is thought to be related to antidepressant withdrawal. In the early neonatal period, the exposed infants show irritability, tremor, restlessness and rigidity (Nordeng *et al.*, 2001; Zeskind and Stephens, 2004; Sanz *et al.*, 2005). Some studies show respiratory distress (Oberlander *et al.*, 2004). At 3 months of age, there is an altered HPA stress response (Oberlander *et al.*, 2008). The severity of this syndrome has been inversely correlated with decreased 5-HIAA in cord blood, suggesting that the exposed infants have decreased serotonin (Laine *et al.*, 2003). A study in newborn blood also indicates a loss of serotonin (Anderson *et al.*, 2004). Unfortunately there are no published studies on children older than 4 years, who do have subtle motor changes (Casper *et al.*, 2003), making it difficult to know whether or not the children will develop psychopathology related to low serotonin.

Summary

Serotonin plays an important role in development of the human brain. Unfortunately, there are many factors which can alter the levels of serotonin during development. These changes in serotonin at critical times can lead to lifelong changes in mental functioning and may underlie serious psycholopathology, including depression, anxiety and autism. However, knowing the role serotonin plays at crucial periods may also enable the therapeutic use of serotonergic agents, to restore appropriate development.

References

Ahlemeyer, B., Beier, H., Semkova, I., Schaper, C. and Krieglstein, J. (2000) S-100beta protects cultured neurons against glutamate- and staurosporine-induced damage and is involved in the antiapoptotic action of the 5 HT(1A)-receptor agonist, Bay x 3702. Brain Res., 858: 121–128.

Aitken, A.R. and Tork, I. (1988) Early development of serotonin-containing neurons and pathways as seen in wholemount preparations of the fetal rat brain. J. Comp. Neurol., 274: 32–47.

Akbari, H.M., Kramer, H.K., Whitaker-Azmitia, P.M., Spear, L.P. and Azmitia, E.C. (1992) Prenatal cocaine exposure disrupts the development of the serotonergic system. Brain Res., 572(1–2): 57–63.

Alenina, N., Bashammakh, S. and Bader, M. (2006) Specification and differentiation of serotonergic neurons. Stem Cell Rev., 2(1): 5–10.

Alexandre, C., Popa, D., Fabre, V., Bouali, S., Venault, P., Lesch, K.P., Hamon, M. and Adrien, J. (2006) Early life blockade of 5-hydroxytryptamine 1A receptors normalizes sleep and depression-like behavior in adult knockout mice lacking the serotonin transporter. J. Neurosci., 26(20): 5554–5564.

Alves, S.E., Hoskin, E., Lee, S.J., Brake, W.G., Ferguson, D., Luine, V., Allen, P.B., Greengard, P. and McEwen, B.S. (2002) Serotonin mediates CA1 spine density but is not crucial for ovarian steroid regulation of synaptic plasticity in the adult rat dorsal hippocampus. Synapse, 45: 143–151.

Andersen, S.L. (2003) Trajectories of brain development: point of vulnerability or window of opportunity? Neurosci. Biobehav. Rev., 27: 3–18.

Anderson, G.M., Czarkowski, K., Ravski, N. and Epperson, C.N. (2004) Platelet serotonin in newborns and infants: ontogeny, heritability, and effect of *in utero* exposure to selective serotonin reuptake inhibitors. Pediatr. Res., 56: 418–422.

Ansorge, M.S., Morelli, E. and Gingrich, J.A. (2008) Inhibition of serotonin but not norepinephrine transport during development produces delayed, persistent perturbations of emotional behaviors in mice. J. Neurosci., 28(1): 199–207.

Attanasio, A., Rager, K. and Gupta, D. (1986) Ontogeny of circadian rhythmicity for melatonin, serotonin, and N-acetylserotonin in humans. J. Pineal. Res., 3(3): 251–256.

Azmitia, E.C. (2001) Neuronal instability: implications for Rett's syndrome. Brain Dev., 23(Suppl. 1): S1–S10.

Azmitia, E.C., Dolan, K. and Whitaker-Azmitia, P.M. (1990) S-100B but not NGF, EGF, insulin or calmodulin is a CNS serotonergic growth factor. Brain Res., 516(2): 354–356.

Azmitia, E.C., Rubinstein, V.J., Strafaci, J.A., Rios, J.C. and Whitaker-Azmitia, P.M. (1995) 5-HT1A agonist and dexamethasone reversal of para-chloroamphetamine induced loss of MAP-2 and synaptophysin immunoreactivity in adult rat brain. Brain Res., 677(2): 181–192.

Bar-Peled, O., Gross-Isseroff, R., Ben-Hur, H., Hoskins, I., Groner, Y. and Biegon, A. (1991) Fetal human brain exhibits a prenatal peak in the density of serotonin 5-HT1A receptors. Neurosci. Lett., 127(2): 173–176.

Basura, G.J. and Walker, P.D. (1999) Suppression of serotonin hyperinnervation does not alter the dysregulatory influences of dopamine depletion on striatal neuropeptide gene expression in rodent neonates. Neurosci. Lett., 274(1): 9–12.

Basura, G.J. and Walker, P.D. (2000) Stimulated serotonin release from hyperinnervated terminals subsequent to neonatal dopamine depletion regulates striatal tachykinin, but not enkephalin gene expression. Brain Res., 81(1–2): 80–91.

Basura, G.J., Abbas, A.I., O'Donohue, H., Lauder, J.M., Roth, B.L., Walker, P.D. and Manis, P.B. (2008) Ontogeny of serotonin and serotonin2A receptors in rat auditory cortex. Hear Res., 244(1–2): 45–50.

Benes, F.M., Taylor, J.B. and Cunningham, M.C. (2000) Convergence and plasticity of monoaminergic systems in the medial prefrontal cortex during the postnatal period: implications for the development of psychopathology. Cereb. Cortex, 10(10): 1014–1027.

Bielenberg, G.W. and Burkhardt, M. (1990) 5-hydroxytryptamine1A agonists. A new therapeutic principle for stroke treatment. Stroke, 21(Suppl. 12): IV161–IV163.

Bode-Greuel, K.M., Klisch, J., Horváth, E., Glaser, T. and Traber, J. (1990) Effects of 5-hydroxytryptamine1A-receptor agonists on hippocampal damage after transient forebrain ischemia in the Mongolian gerbil. Stroke, 21(Suppl. 12): IV164–IV166.

Bonnin, A., Peng, W., Hewlett, W. and Levitt, P. (2006) Expression mapping of 5-HT1 serotonin receptor subtypes during fetal and early postnatal mouse forebrain development. Neuroscience, 141(2): 781–794.

Borella, A., Bindra, M. and Whitaker-Azmitia, P.M. (1997) Role of the 5-HT1A receptor in development of the neonatal rat brain: preliminary behavioral studies. Neuropharmacology, 36(4–5): 445–450.

Bou-Florés, C., Lajard, A.-M., Monteau, R., De Maeyer, E., Seif, I., Lanoir, J. and Hilaire, G. (2000) Abnormal phrenic motoneuron activity and morphology in neonatal monoamine oxidase A-deficient transgenic mice: possible role of a serotonin excess. J. Neurosci., 20: 4646–4656.

Brewton, L.S., Haddad, L. and Azmitia, E.C. (2001) Colchicine-induced cytoskeletal collapse and apoptosis in N-18 neuroblastoma cultures is rapidly reversed by applied S-100β. Brain Res., 912: 9–16.

Brezun, J.M. and Daszuta, A. (1999) Depletion in serotonin decreases neurogenesis in the dentate gyrus and the subventricular zone of adult rats. Neuroscience, 89(4): 999–1002.

Brezun, J.M. and Daszuta, A. (2000) Serotonin may stimulate granule cell proliferation in the adult hippocampus, as observed in rats grafted with foetal raphe neurons. Eur. J. Neurosci., 12(1): 391–396.

Briscoe, J., Sussel, L., Serup, P., Hartigan-O'Connor, D., Jessell, T.M., Rubenstein, J.L. and Ericson, J. (1999) Homeobox gene Nkx2.2 and specification of neuronal identity by graded Sonic hedgehog signalling. Nature, 398(6728): 622–627.

Brüning, G. and Liangos, O. (1997) Transient expression of the serotonin transporter in the developing mouse thalamocortical system. Acta Histochem., 99(1): 117–121.

Brüning, G., Liangos, O. and Baumgarten, H.G. (1997) Prenatal development of the serotonin transporter in mouse brain. Cell Tissue Res., 289(2): 211–221.

Cabrera-Vera, T.M. and Battaglia, G. (1998) Prenatal exposure to fluoxetine (Prozac) produces site-specific and age-dependent alterations in brain serotonin transporters in rat progeny: evidence from autoradiographic studies. J. Pharmacol. Exp. Ther., 286(3): 1474–1481.

Cabrera-Vera, T.M., Garcia, F., Pinto, W. and Battaglia, G. (1997) Effect of prenatal fluoxetine (Prozac) exposure on brain serotonin neurons in prepubescent and adult male rat offspring. J. Pharmacol. Exp. Ther., 280(1): 138–145.

Cannizzaro, C., Plescia, F., Gagliano, M., Cannizzaro, G., Mantia, G., La Barbera, M., Provenzano, G. and Cannizzaro, E. (2008) Perinatal exposure to 5-methoxytryptamine, behavioural-stress

reactivity and functional response of 5-HT1A receptors in the adolescent rat. Behav. Brain Res., 186(1): 98–106.
Cases, O., Vitalis, T., Seif, I., De Maeyer, E., Sotelo, C. and Gaspar, P. (1996) Lack of barrels in the somatosensory cortex of monoamine oxidase A-deficient mice: role of a serotonin excess during the critical period. Neuron, 16(2): 297–307.
Casper, R.C., Fleisher, B.E., Lee-Ancajas, J.C., Gilles, A., Gaylor, E., De Battista, A. and Hoyme, H.E. (2003) Follow-up of children of depressed mothers exposed or not exposed to antidepressant drugs during pregnancy. J. Pediatr., 142: 402–408.
Chen, L., Hamaguchi, K., Ogawa, M., Hamada, S. and Okado, N. (1994) PCPA reduces both monoaminergic afferents and nonmonoaminergic synapses in the cerebral cortex. Neurosci. Res., 19: 111–115.
Chugani, D.C., Muzik, O., Rothermel, R., Behen, M., Chakraborty, P., Mangner, T., da Silva, E.A. and Chugani, H.T. (1997) Altered serotonin synthesis in the dentatothalamocortical pathway in autistic boys. Ann. Neurol., 42(4): 666–669.
Chugani, D.C., Muzik, O., Behen, M., Rothermel, R., Janisse, J.J., Lee, J. and Chugani, H.T. (1999) Developmental changes in brain serotonin synthesis capacity in autistic and nonautistic children. Ann. Neurol., 45(3): 287–295.
Chugani, H.T., Phelps, M.E. and Mazziotta, J.C. (1987) Positron emission tomography study of human brain functional development. Ann. Neurol., 22(4): 487–497.
Cordes, S.P. (2005) Molecular genetics of the early development of hindbrain serotonergic neurons. Clin. Genet., 68(6): 487–494.
Costes, N., Merlet, I., Ostrowsky, K., Faillenot, I., Lavenne, F., Zimmer, L., Ryvlin, P. and Le Bars, D. (2005) A 18F-MPPF PET normative database of 5-HT1A receptor binding in men and women over aging. J. Nucl. Med., 46(12): 1980–1989.
Cowen, D.S., Takase, L.F., Fornal, C.A. and Jacobs, B.L. (2008) Age-dependent decline in hippocampal neurogenesis is not altered by chronic treatment with fluoxetine. Brain Res., 1228: 14–19.
Czéh, B., Simon, M., Schmelting, B., Hiemke, C. and Fuchs, E. (2006) Astroglial plasticity in the hippocampus is affected by chronic psychosocial stress and concomitant fluoxetine treatment. Neuropsychopharmacology, 31(8): 1616–1626.
D'Amato, R.J., Blue, M.E., Largent, B.L., Ledbetter, D.J., Molliver, M.E. and Snyder, S.H. (1987) Ontogeny of the serotonergic projection to rat neocortex: transient expression of a dense innervation to primary sensory areas. Proc. Natl. Acad. Sci. USA, 84: 4322–4326.
Daval, D., Verge, A., Becerril, H., Gozlan, U., Spampinato, J. and Hamon, M. (1987) Transient expression of 5-HT1A receptors binding sites in some areas of the rat CNS during postnatal development. Intl. J. Dev. Neurosci., 5: 171–189.
del Olmo, E. and Pazos, A. (2001) Aminergic receptors during the development of the human brain: the contribution of *in vitro* imaging techniques. J. Chem. Neuroanat., 22(1–2): 101–114.
del Olmo, E., Díaz, A., Guirao-Piñeyro, M., del Arco, C., Pascual, J. and Pazos, A. (1994) Transient localization of 5-HT1A receptors in human cerebellum during development. Neurosci. Lett., 166(2): 149–152.
del Olmo, E., López-Giménez, J.F., Vilaró, M.T., Mengod, G., Palacios, J.M. and Pazos, A. (1998) Early localization of mRNA coding for 5-HT1A receptors in human brain during development. Brain Res. Mol. Brain Res., 60(1): 123–126.
Deng, D.R., Djalali, S., Holtje, M., Grosse, G., Stroh, T., Voigt, I., Kusserow, H., Theuring, F., Ahnert-Hilger, G. and Hortnagi, H. (2007) Embryonic and postnatal development of the serotonergic raphe system and its target regions in 5-HT1A receptor deletion or overexpressing mouse mutants. Neuroscience, 147(2): 338–402.
Diefenbach, T.J., Sloley, B.D. and Goldberg, J.I. (1995) Neurite branch development of an identified serotonergic neuron from embryonic *Helisoma*: evidence for autoregulation by serotonin. Dev. Biol., 167(1): 282–293.
Donato, R. (2003) Intracellular and extracellular roles of S100 proteins. Microsc. Res. Tech., 60(6): 540–551.
Donato, R. (2007) RAGE: a single receptor for several ligands and different cellular responses: the case of certain S100 proteins. Curr. Mol. Med., 7(8): 711–724.
Druse, M.J., Tajuddin, N.F., Gillespie, R.A. and Le, P. (2006) The effects of ethanol and the serotonin(1A) agonist ipsapirone on the expression of the serotonin(1A) receptor and several antiapoptotic proteins in fetal rhombencephalic neurons. Brain Res., 1092(1): 79–86.
Druse, M.J., Gillespie, R.A., Tajuddin, N.F. and Rich, M. (2007) S100B-mediated protection against the pro-apoptotic effects of ethanol on fetal rhombencephalic neurons. Brain Res., 1150: 46–54.
Faber, K.M. and Haring, J.H. (1999) Synaptogenesis in the postnatal rat fascia dentata is influenced by 5-HT1a receptor activation. Dev. Brain Res., 114: 245–252.
Farabollini, F., Hole, D.R. and Wilson, C.A. (1988) Behavioral effects in adulthood of serotonin depletion by P-chlorophenylalanine given neonatally to male rats. Intl. J. Neurosci., 41(3-4): 187–199.
Fatemi, S.H., Snow, A.V., Stary, J.M., Araghi-Niknam, M., Reutiman, T.J., Lee, S., Brooks, A.I. and Pearce, D.A. (2005) Reelin signaling is impaired in autism. Biol. Psychiatry, 57(7): 777–787.
Feenstra, M.G., van Galen, H., Te Riele, P.J., Botterblom, M.H. and Mirmiran, M. (1996) Decreased hypothalamic serotonin levels in adult rats treated neonatally with clomipramine. Pharmacol. Biochem. Behav., 55: 647–652.
Feria-Velasco, A., del Angel, A.R. and Gonzalez-Burgos, I. (2002) Modification of dendritic development. Prog. Brain Res., 136: 135–143.
Fernandez-Pardal, J. and Hilakivi, L.A. (1989) Effect of low alcohol dose on behavioral despair in rats neonatally treated with antidepressant drugs. Alcohol, 6: 93–95.
Field, T. (2002) Violence and touch deprivation in adolescents. Adolescence, 37(148): 735–749.
Field, T., Diego, M., Hernandez-Reif, M., Schanberg, S., Kuhn, C., Yando, R. and Bendell, D. (2003) Pregnancy anxiety and comorbid depression and anger: effects on the fetus and neonate. Depress. Anxiety, 17(3): 140–151.
Field, T., Diego, M., Dieter, J., Hernandez-Reif, M., Schanberg, S. and Kuhn, C. (2004) Prenatal depression effects on the fetus and the newborn. Infant Behav. Dev., 27: 216–229.
Fricker, A.D., Rios, C., Devi, L.A. and Gomes, I. (2005) Serotonin receptor activation leads to neurite outgrowth and neuronal survival. Brain Res. Mol. Brain Res., 138(2): 228–235.
Garcia, G., Tagliaferro, P., Bortolozzi, A., Madariaga, M.J., Brusco, A., Evangelista de Duffard, A.M., Duffard, R. and Saavedra, J.P. (2001) Morphological study of 5-HT neurons and astroglial cells on brain of adult rats perinatal or chronically exposed to 2,4-dichlorophenoxyacetic acid. Neurotoxicology, 22(6): 733–741.

Gaspar, P., Cases, O. and Maroteaux, L. (2003) The developmental role of serotonin: news from mouse molecular genetics. Nat. Rev. Neurosci., 4(12): 1002–1012.

Goldberg, J.I. (1998) Serotonin regulation of neurite outgrowth in identified neurons from mature and embryonic Helisoma trivolvis. Perspect. Dev. Neurobiol., 5(4): 373–387.

González, E.M., Penedo, L.A., Oliveira-Silva, P., Campello-Costa, P., Guedes, R.C. and Serfaty, C.A. (2008) Neonatal tryptophan dietary restriction alters development of retinotectal projections in rats. Exp. Neurol., 211(2): 441–448.

Gonzalez-Burgos, I., del Angel-Meza, A.R., Barajas-López, G. and Feria-Velasco, A. (1996) Tryptophan restriction causes long-term plastic changes in corticofrontal pyramidal neurons. Intl. J. Dev. Neurosci., 14(5): 673–679.

Gos, T., Becker, K., Bock, J., Malecki, U., Bogerts, B., Poeggel, G. and Braun, K. (2006) Early neonatal and postweaning social emotional deprivation interferes with the maturation of serotonergic and tyrosine hydroxylase-immunoreactive afferent fiber systems in the rodent nucleus accumbens, hippocampus and amygdala. Neuroscience, 140(3): 811–821.

Gould, E. (1999) Serotonin and hippocampal neurogenesis. Neuropsychopharmacology, 21(Suppl. 2): 46S–51S.

Hansen, H., Sanchez, C. and Meier, E. (1997) Neonatal administration of the selective serotonin reuptake inhibitor Lu 10-134-C increases forced swimming-induced immobility of adult rats: a putative animal model of depression? J. Pharm. Exp. Ther., 283: 1333–1341.

Hansen, H.H. and Mikkelsen, J.D. (1998) Long-term effects on serotonin transporter mRNA expression of chronic neonatal exposure to a serotonin reuptake inhibitor. Eur. J. Pharmacol., 352: 307–315.

Hansson, S.R., Mezey, E. and Hoffman, B.J. (1998) Serotonin transporter messenger RNA in the developing rat brain: early expression in serotonergic neurons and transient expression in non-serotonergic neurons. Neuroscience, 83: 1185–1201.

Haydon, P.G., McCobb, D.P. and Kater, S.B. (1987) The regulation of neurite outgrowth, growth cone motility, and electrical synaptogenesis by serotonin. J. Neurobiol., 18(2): 197–215.

Hedner, J., Lundell, K.H., Breese, G.R., Mueller, R.A. and Hedner, T. (1986) Developmental variations in CSF monoamine metabolites during childhood. Biol. Neonate., 49(4): 190–197.

Hendricks, T., Francis, N., Fyodorov, D. and Deneris, E.S. (1999) The ETS domain factor Pet-1 is an early and precise marker of central serotonin neurons and interacts with a conserved element in serotonergic genes. J. Neurosci., 19(23): 10348–10356.

Hendricks, T.J., Fyodorov, D.V., Wegman, L.J., Lelutiu, N.B., Pehek, E.A., Yamamoto, B., Silver, J., Weeber, E.J., Sweatt, J.D. and Deneris, E.S. (2003) Pet-1 ETS gene plays a critical role in 5-HT neuron development and is required for normal anxiety-like and aggressive behavior. Neuron, 37(2): 233–247.

Hilakivi, L.A. and Hilakivi, I. (1987) Increased adult behavioral 'despair' in rats neonatally exposed to desipramine or zimeldine: an animal model of depression? Pharmacol. Biochem. Behav., 28: 367–369.

Himpel, S., Bartels, J., Zimdars, K., Huether, G., Adler, L., Dawirs, R.R. and Moll, G.H. (2006) Association between body weight of newborn rats and density of serotonin transporters in the frontal cortex at adulthood. J. Neural Transm., 113(3): 295–302.

Hohmann, C.F., Walker, E.M., Boylan, C.B. and Blue, M.E. (2007) Neonatal serotonin depletion alters behavioral responses to spatial change and novelty. Brain Res., 1139: 163–177.

Holmes, A., Lit, Q., Murphy, D.L., Gold, E. and Crawley, J.N. (2003) Abnormal anxiety-related behavior in serotonin transporter null mutant mice: the influence of genetic background. Genes Brain Behav., 2(6): 365–380.

Huang, G.J., Bannerman, D. and Flint, J. (2008) Chronic fluoxetine treatment alters behavior, but not adult hippocampal neurogenesis, in BALB/cJ mice. Mol. Psychiatry, 13(2): 119–121.

Huttenlocher, P.R., De Courten, C., Garey, L.J. and Van der Loos, H. (1983) Synaptic development in human cerebral cortex. Intl. J. Neurol., 16–17: 144–154.

Hynes, M. and Rosenthal, A. (1999) Specification of dopaminergic and serotonergic neurons in the vertebrate CNS. Curr. Opin. Neurobiol., 9(1): 26–36.

Janusonis, S., Gluncic, V. and Rakic, P. (2005) Early serotonergic projections to Cajal-Retzius cells: relevance for cortical development. J. Neurosci., 24(7): 1652–1659.

Kahne, D., Tudorica, A., Borella, A., Shapiro, L., Johnstone, F., Huang, W. and Whitaker-Azmitia, P.M. (2002) Behavioral and magnetic resonance spectroscopic studies in the rat hyperserotonemic model of autism. Physiol. Behav., 75(3): 403–410.

Khozhai, L.I. (2006) Formation of the astroglia in the mouse neocortex after temporary prenatal blockade of serotonin synthesis. Neurosci. Behav. Physiol., 36(3): 275–278.

Khozhai, L.I. and Otellin, V.A. (2006) Formation of the neocortex in mice developing in conditions of prenatal serotonin deficiency. Neurosci. Behav. Physiol., 36(5): 513–517.

Kinney, H.C., Randall, L.L., Sleeper, L.A., Willinger, M., Belliveau, R.A., Zec, N., Rava, L.A., Dominici, L., Iyasu, S., Randall, B., Habbe, D., Wilson, H., Mandell, F., McClain, M. and Welty, T.K. (2003) Serotonergic brainstem abnormalities in Northern Plains Indians with the sudden infant death syndrome. J. Neuropathol. Exp. Neurol., 62(11): 1178–1191.

Kinney, H.C., Belliveau, R.A., Trachtenberg, F.L., Rava, L.A. and Paterson, D.S. (2007) The development of the medullary serotonergic system in early human life. Auton. Neurosci., 132(1–2): 81–102.

Kondoh, M., Shiga, T. and Okado, N. (2004) Regulation of dendrite formation of Purkinje cells by serotonin through serotonin(1A) and serotonin(2A) receptors in culture. Neurosci. Res., 48: 101–109.

Kontur, P.J., Leranth, C., Redmond, D.E. Jr., Roth, R.H. and Robbins, R.J. (1993) Tyrosine hydroxylase immunoreactivity and monoamine and metabolite levels in cryopreserved human fetal ventral mesencephalon. Exp. Neurol., 121(2): 172–180.

Koprich, J.B., Chen, E.Y., Kanaan, N.M., Campbell, N.G., Kordower, J.H. and Lipton, J.W. (2003) Prenatal 3,4-methylenedioxymethamphetamine (ecstasy) alters exploratory behavior, reduces monoamine metabolism, and increases forebrain tyrosine hydroxylase fiber density of juvenile rats. Neurotoxicol. Teratol., 25(5): 509–517.

Laine, K., Heikkinen, T., Ekblad, U. and Kero, P. (2003) Effects of exposure to selective serotonin reuptake inhibitors during pregnancy on serotonergic symptoms in newborns and cord blood monoamine and prolactin concentrations. Arch. Gen. Psychiatry, 60(7): 720–726.

Lane, R.D., Chiaia, N.L., Kesterson, K.L., Rhoades, R.W. and Mooney, R.D. (2006) Boundary-limited serotonergic influences on pattern organization in rat sensory cortex. Neurosci. Lett., 395(2): 165–169.

Lanfumey, L., Mongeau, R., Cohen-Salmon, C. and Hamon, M. (2008) Corticosteroid-serotonin interactions in the neurobiological mechanisms of stress-related disorders. Neurosci. Biobehav. Rev., 32(6): 1174–1184.

Lauder, J.M. (1983) Hormonal and humoral influences on brain development. Psychoneuroendocrinology, 8(2): 121–155.
Lauder, J.M. (1985) Roles for neurotransmitters in development: possible interaction with drugs during the fetal and neonatal periods. Prog. Clin. Biol. Res., 163C: 375–380.
Lauder, J.M. (1990) Ontogeny of the serotonergic system in the rat: serotonin as a developmental signal. Ann. NY Acad. Sci., 600: 297–313.
Lauder, J.M. and Krebs, H. (1978) Serotonin as a differentiation signal in early neurogenesis. Dev. Neurosci., 1(1): 15–30.
Lauder, J.M., Wallace, J.A. and Krebs, H. (1981) Roles for serotonin in neuroembryogenesis. Adv. Exp. Med. Biol., 133: 477–506.
Lauder, J.M., Petrusz, P., Wallace, J.A., Dinome, A., Wilkie, M.B. and McCarthy, K. (1982) Combined serotonin immunocytochemistry and 3H-thymidine autoradiography: *in vivo* and *in vitro* methods. J. Histochem. Cytochem., 30(8): 788–793.
Lee, J.H., Tajuddin, N.F., Druse, M.J. (2008) Effects of ethanol and ipsapirone on the expression of genes encoding anti-apoptotic proteins and an antioxidant enzyme in ethanol-treated neurons. *Brain Res* 1 November [Epub ahead of print].
Lerer, B., Gelfin, Y. and Shapira, B. (1999) Neuroendocrine evidence for age-related decline in central serotonergic function. Neuropsychopharmacology, 21(2): 321–322.
Leret, M.L., Peinado, V., Suárez, L.M., Tecedor, L., Gamallo, A. and González, J.C. (2004) Role of maternal adrenal glands on the developing serotoninergic and aminoacidergic systems of the postnatal rat brain. Intl. J. Dev. Neurosci., 22(2): 87–93.
Levallois, C., Calvet, M.C. and Petite, D. (1993) Immunocytochemical demonstration of monoamine-containing cells in human fetal brain stem dissociated cultures. Dev. Brain Res., 75(1): 141–145.
Lidov, H.G. and Molliver, M.E. (1982) An immunohistochemical study of serotonin neuron development in the rat: ascending pathways and terminal fields. Brain Res. Bull., 8(4): 389–430.
Lieske, V., Bennett-Clarke, C.A. and Rhoades, R.W. (1999) Effects of serotonin on neurite outgrowth from thalamic neurons *in vitro*. Neuroscience, 90(3): 967–974.
Lira, A., Zhou, M., Castanon, N., Ansorge, M.S., Gordon, J.A., Francis, J.H., Bradley-Moore, M., Lira, J., Underwood, M.D., Arango, V., Kung, H.F., Hofer, M.A., Hen, R. and Gingrich, J.A. (2003) Altered depression-related behaviors and functional changes in the dorsal raphe nucleus of serotonin transporter-deficient mice. Biol. Psychiatry, 54(10): 960–971.
Lotto, B., Upton, L., Price, D.J. and Gaspar, P. (1999) Serotonin receptor activation enhances neurite outgrowth of thalamic neurones in rodents. Neurosci. Lett., 269: 87–90.
Maciag, D., Williams, L., Coppinger, D. and Paul, I.A. (2006) Neonatal citalopram exposure produces lasting changes in behavior which are reversed by adult imipramine treatment. Eur. J. Pharmacol., 532(3): 265–269.
Madhavan, L., Freed, W.J., Anantharam, V. and Kanthasamy, A. G. (2003) 5-hydroxytryptamine1A receptor activation protects against N-methyl-Daspartate-induced apoptotic cell death in striatal and mesencephalic cultures. J. Pharmacol. Exp. Ther., 304: 913–923.
Makkonen, I., Riikonen, R., Kokki, H., Airaksinen, M.M. and Kuikka, J.T. (2008) Serotonin and dopamine transporter binding in children with autism determined by SPECT. Dev. Med. Child Neurol., 50(8): 593–597.
Manev, H., Uz, T. and Manev, R. (2003) Glia as a putative target for antidepressant treatments. J. Affect Disord., 75(1): 59–64.
Marcussen, A.B., Flagstad, P., Kristjansen, P.E., Johansen, F.F. and Englund, U. (2008) Increase in neurogenesis and behavioural benefit after chronic fluoxetine treatment in Wistar rats. Acta Neurol. Scand., 117(2): 94–100.
Matsukawa, M., Ogawa, M., Nakadate, K., Maeshima, T., Ichitani, Y., Kawai, N. and Okado, N. (1997) Serotonin and acetylcholine are crucial to maintain hippocampal synapses and memory acquisition in rats. Neurosci. Lett., 230: 13–16.
Matthews, K., Dalley, J.W., Matthews, C., Tsai, T.H. and Robbins, T.W. (2001) Periodic maternal separation of neonatal rats produces region- and gender-specific effects on biogenic amine content in postmortem adult brain. Synapse, 40(1): 1–10.
Mazer, C., Muneyyirci, J., Taheny, K., Raio, N., Borella, A. and Whitaker-Azmitia, P.M. (1997) Serotonin depletion during synaptogenesis leads to decreased synaptic density and learning deficits in the adult rat: a possible model of neurodevelopmental disorders with cognitive deficits. Brain Res., 760(1–2): 68–73.
McEwen, B.S. (2003) Early life influences on life-long patterns of behavior and health. Ment. Retard. Dev. Disabil. Res. Rev., 9(3): 149–154.
McNamara, I., Borella, A., Bialowas, L. and Whitaker-Azmitia, P.M. (2008) Further studies in the developmental hyperserotonemia model (DHS) of autism: social, behavioral and peptide changes. Brain Res., 1189: 204–213.
Mendelson, S.D. and McEwen, B.S. (1991) Autoradiographic analyses of the effects of restraint-induced stress on 5-HT1A, 5-HT1C and 5-HT2 receptors in the dorsal hippocampus of male and female rats. Neuroendocrinology, 54: 454–461.
Mostert, J.P., Koch, M.W., Heerings, M., Heersema, D.J. and De Keyser, J. (2008) Therapeutic potential of fluoxetine in neurological disorders. CNS Neurosci. Ther., 14(2): 153–164.
Muneoka, K., Mikuni, M., Ogawa, T., Kitera, K., Kamei, K., Takigawa, M. and Takahashi, K. (1997) Prenatal dexamethasone exposure alters brain monoamine metabolism and adrenocortical response in rat offspring. Am. J. Physiol., 273(5 Pt 2): R1669–R1675.
Nakata, N., Suda, H., Izumi, J., Tanaka, Y., Ikeda, Y., Kato, H., Itoyama, Y. and Kogure, K. (1997) Role of hippocampal serotonergic neurons in ischemic neuronal death. Behav. Brain Res., 83(1–2): 217–220.
Narboux-Neme, N., Pavone, L.M., Avallone, L., Zhuang, X. and Gaspar, P. (2008) Serotonin transporter transgenic (SERT(cre)) mouse line reveals developmental targets of serotonin specific reuptake inhibitors (SSRIs). Neuropharmacology, 55(6): 994–1005.
Navailles, S., Hof, P.R. and Schmauss, C. (2008) Antidepressant drug-induced stimulation of mouse hippocampal neurogenesis is age-dependent and altered by early life stress. J. Comp. Neurol., 509(4): 372–381.
Nishi, M., Whitaker-Azmitia, P.M. and Azmitia, E.C. (1996) Enhanced synaptophysin immunoreactivity in rat hippocampal culture by 5-HT 1A agonist, S100b, and corticosteroid receptor agonists. Synapse, 23(1): 1–9.
Nordeng, H., Lindemann, R., Perminov, K.V. and Reikvam, A. (2001) Neonatal withdrawal syndrome after *in utero* exposure to selective serotonin reuptake inhibitors. Acta Paediatr., 90(3): 288–291.
Oberlander, T.F., Misri, S., Fitzgerald, C.E., Kostaras, X., Rurak, D. and Riggs, W. (2004) Pharmacologic factors associated with transient neonatal symptoms following prenatal psychotropic medication exposure. J. Clin. Psychiatry, 65(2): 230–237.

Oberlander, T.F., Warburton, W., Misri, S., Aghajanian, J. and Hertzman, C. (2006) Neonatal outcomes after prenatal exposure to selective serotonin reuptake inhibitor antidepressants and maternal depression using population-based linked health data. Arch. Gen. Psychiatry, 63(8): 898–906.

Oberlander, T.F., Grunau, R., Mayes, L., Riggs, W., Rurak, D., Papsdorf, M., Misri, S. and Weinberg, J. (2008) Hypothalamic–pituitary–adrenal (HPA) axis function in 3-month old infants with prenatal selective serotonin reuptake inhibitor (SSRI) antidepressant exposure. Early Hum. Dev., 84(10): 689–697.

Oosterink, B.J., Korte, S.M., Nyakas, C., Korf, J. and Luiten, P.G. (1998) Neuroprotection against N-methyl-D-aspartate-induced excitotoxicity in rat magnocellular nucleus basalis by the 5-HT1A receptor agonist 8-OH-DPAT. Eur. J. Pharmacol., 358: 147–152.

O'Reilly, K.C., Trent, S., Bailey, S.J. and Lane, M.A. (2007) 13-cis-Retinoic acid alters intracellular serotonin, increases 5-HT1A receptor, and serotonin reuptake transporter levels *in vitro*. Exp. Biol. Med., 232(9): 1195–1203.

Patel, T.D. and Zhou, F.C. (2005) Ontogeny of 5-HT1A receptor expression in the developing hippocampus. Dev. Brain Res., 157(1): 42–57.

Paterson, D.S., Belliveau, R.A., Trachtenberg, F. and Kinney, H.C. (2004) Differential development of 5-HT receptor and the serotonin transporter binding in the human infant medulla. J. Comp. Neurol., 472(2): 221–231.

Persico, A.M., Baldi, A., Dell'Acqua, M.L., Moessner, R., Murphy, D.L., Lesch, K.P. and Keller, F. (2003) Reduced programmed cell death in brains of serotonin transporter knockout mice. Neuroreport, 14: 341–344.

Peters, D.A. (1986) Prenatal stress: effect on development of rat brain serotonergic neurons. Pharmacol. Biochem. Behav., 24(5): 1377–1382.

Popa, D., Léna, C., Alexandre, C. and Adrien, J. (2008) Lasting syndrome of depression produced by reduction in serotonin uptake during postnatal development: evidence from sleep, stress, and behavior. J. Neurosci., 28(14): 3546–3554.

Pôrto, L.C., Sardinha, F.L., Telles, M.M., Guimarães, R.B., Albuquerque, K.T., Andrade, I.S., Oyama, L.M., Nascimento, C.M., Santos, O.F., Ribeiro, E.B. (2008) Impairment of the serotonergic control of feeding in adult female rats exposed to intra-uterine malnutrition. *Br J Nutr* [Epub ahead of print].

Quattrocchi, C.C., Wannenes, F., Persico, A.M., Ciafré, S.A., D'Arcangelo, G., Farace, M.G. and Keller, F.J. (2002) Reelin is a serine protease of the extracellular matrix. Biol. Chem., 277(13): 11616.

Ramos, A.J., Tagliaferro, P., Lopez, E.M., Pecci Saavedra, J. and Brusco, A. (2000) Neuroglial interactions in a model of parachlorophenylalanine-induced serotonin depletion. Brain Res., 883: 1–14.

Ramos, A.J., Rubio, M.D., Defagot, C., Hischberg, L., Villar, M. J. and Brusco, A. (2004) The 5HT1A receptor agonist, 8-OH-DPAT, protects neurons and reduces astroglial reaction after ischemic damage caused by cortical devascularization. Brain Res., 1030(2): 201–220.

Riccio, O., Potter, G., Walzer, C., Vallet, P., Szabó, G., Vutskits, L., Kiss, J.Z., Dayer, A.G. (2008) Excess of serotonin affects embryonic interneuron migration through activation of the serotonin receptor 6. *Mol Psychiatry* [Epub ahead of print].

Riikonen, R.S., Nokelainen, P., Valkonen, K., Kolehmainen, A.I., Kumpulainen, K.I., Könönen, M., Vanninen, R.L. and Kuikka, J.T. (2005) Deep serotonergic and dopaminergic structures in fetal alcoholic syndrome: a study with nor-beta-CIT-single-photon emission computed tomography and magnetic resonance imaging volumetry. Biol. Psychiatry, 57(12): 1565–1572.

Rothermundt, M., Peters, M., Prehn, J.H. and Arolt, V. (2003) S100B in brain damage and neurodegeneration. Microsc. Res. Tech., 60(6): 614–632.

Rumajogee, P., Vergé, D., Hanoun, N., Brisorgueil, M.J., Hen, R., Lesch, K.P., Hamon, M. and Miquel, M.C. (2004) Adaption of the serotoninergic neuronal phenotype in the absence of 5-HT autoreceptors or the 5-HT transporter: involvement of BDNF and cAMP. Eur. J. Neurosci., 19(4): 937–944.

Salichon, N., Gaspar, P., Upton, A.L., Picaud, S., Hanoun, N., Hamon, M., De Maeyer, E., Murphy, D.L., Mössner, R., Lesh, K.P., Hen, R. and Seif, I. (2001) Excessive activation of serotonin (5-HT) 1B receptors disrupts the formation of sensory maps in monoamine oxidase A and 5-HT transporter knockout mice. J. Neurosci., 21: 884–896.

Santarelli, L., Saxe, M., Gross, C., Surget, A., Battaglia, F., Dulawa, S., Weisstaub, N., Lee, J., Duman, R., Arancio, O., Belzung, C. and Hen, R. (2003) Requirement of hippocampal neurogenesis for the behavioral effects of antidepressants. Science, 301(5634): 805–809.

Sanz, E.J., De-las-Cuevas, C., Kiuru, A., Bate, A. and Edwards, R. (2005) Selective serotonin reuptake inhibitors in pregnant women and neonatal withdrawal syndrome: a database analysis. Lancet, 365(9458): 482–487.

Sari, Y. and Zhou, F.C. (2004) Prenatal alcohol exposure causes long-term serotonin neuron deficit in mice. Alcohol. Clin. Exp. Res., 28: 941–948.

Serfaty, C.A., Oliveira-Silva, P., Faria Melibeu, A.D. and Campello-Costa, P. (2008) Nutritional Tryptophan Restriction and the Role of Serotonin in Development and Plasticity of Central Visual Connections. Neuroimmunomodulation, 15(3): 170–175.

Sheline, Y.I., Mintun, M.A., Moerlein, S.M. and Snyder, A.Z. (2002) Greater loss of 5-HT(2A) receptors in midlife than in late life. Am. J. Psychiatry, 159(3): 430–435.

Shemer, A., Whitaker-Azmitia, P.M. and Azmitia, E.C. (1988) Effects of prenatal 5-methoxytryptamine and parachlorophenylalanine on serotonergic uptake and behavior in the neonatal rat. Pharmacol. Biochem. Behav., 30(4): 847–851.

Shemer, A.V., Azmitia, E.C. and Whitaker-Azmitia, P.M. (1991) Dose-related effects of prenatal 5-methoxytryptamine (5-MT) on development of serotonin terminal density and behavior. Dev. Brain Res., 59: 59–65.

Shen, W.Z., Luo, Z.B., Zheng, D.R. and Yew, D.T. (1989) Immunohistochemical studies on the development of 5-HT (serotonin) neurons in the nuclei of the reticular formations of human fetuses. Pediatr. Neurosci., 15(6): 291–295.

Shiraishi, M., Takizawa, Y., Ide, S., Obonai, T., Goto, Y. and Itoh, M. (2008) Brainstem monoamine pathology of neonatal hypoxic-ischemic brain damage: a model of acute stage of neonatal asphyxia. Brain Res., 1213: 120–126.

Sidman, R.L. and Rakic, P. (1973) Neuronal migration, with special reference to developing human brain: a review. Brain Res., 62(1): 1–35.

Strackx, E., van den Hove, D.L., Steinbusch, H.P., Prickaerts, J., Vles, J.S., Blanco, C.E., Steinbusch, H.W. and Gavilanes, A.W. (2008) Fetal asphyxia leads to a decrease in dorsal raphe serotonergic neurons. Dev. Neurosci., 30: 358–366.

Sundstrom, E., Kolare, S., Souverbie, F., Samuelsson, E. B., Pschera, H., Lunell, N.O. and Seiger, A. (1993) Neurochemical differentiation of human bulbospinal monoaminergic neurons during the first trimester. Dev. Brain Res., 75(1): 1–12.

Tajuddin, N.F., Orrico, L.A., Eriksen, J.L. and Druse, M.J. (2003) Effects of ethanol and ipsapirone on the development of midline raphe glial cells and astrocytes. Alcohol, 29(3): 157–164.

Takahashi, H., Nakashima, S., Ohama, E., Takeda, S. and Ikuta, F. (1986) Distribution of serotonin-containing cell bodies in the brainstem of the human fetus determined with immunohistochemistry using antiserotonin serum. Brain Dev., 8(4): 355–365.

Tauscher, J., Verhoeff, N.P., Christensen, B.K., Hussey, D., Meyer, J.H., Kecojevic, A., Javanmard, M., Kasper, S. and Kapur, S. (2001) Serotonin 5-HT1A receptor binding potential declines with age as measured by [11C]WAY-100635 and PET. Neuropsychopharmacology, 24(5): 522–530.

Thompson, A.M. (2008) Serotonin immunoreactivity in auditory brainstem neurons of the postnatal monoamine oxidase-A knockout mouse. Brain Res., 1228: 58–67.

Toth, G. and Fekete, M. (1986) 5-Hydroxyindole acetic excretion in newborns, infants and children. Acta Paediatr. Hung., 27(3): 221–226.

Tramontina, A.C., Tramontina, F., Bobermin, L.D., Zanotto, C., Souza, D.F., Leite, M.C., Nardin, P., Gottfried, C. and Gonçalves, C.A. (2008) Secretion of S100B, an astrocyte-derived neurotrophic protein, is stimulated by fluoxetine via a mechanism independent of serotonin. Prog. Neuropsychopharmacol. Biol. Psychiatry, 32(6): 1580–1583.

Tsuchioka, M., Takebayashi, M., Hisaoka, K., Maeda, N. and Nakata, Y. (2008) Serotonin (5-HT) induces glial cell line-derived neurotrophic factor (GDNF) mRNA expression via the transactivation of fibroblast growth factor receptor 2 (FGFR2) in rat C6 glioma cells. J. Neurochem., 106(1): 244–257.

Van den Hove, D.L., Lauder, J.M., Scheepens, A., Prickaerts, J., Blanco, C.E. and Steinbusch, H.W. (2006a) Prenatal stress in the rat alters 5-HT1A receptor binding in the ventral hippocampus. Brain Res., 1090(1): 29–34.

Van den Hove, D.L., Steinbusch, H.W., Bruschettini, M., Gazzolo, D., Frulio, R., Scheepens, A., Prickaerts, J. and Blanco, C.E. (2006b) Prenatal stress reduces S100B in the neonatal rat hippocampus. Neuroreport, 17(10): 1077–1080.

Vataeva, L.A., Kudrin, V.S., Vershinina, E.A., Mosin, V.M., Tiul'kova, E.I. and Otellin, V.A. (2007) Behavioral alteration in the adult rats prenatally exposed to para-chlorophenylalanine. Brain Res., 1169: 9–16.

Veenema, A.H., Blume, A., Niederle, D., Buwalda, B. and Neumann, I.D. (2006) Effects of early life stress on adult male aggression and hypothalamic vasopressin and serotonin. Eur. J. Neurosci., 24(6): 1711–1720.

Velazquez-Moctezuma, J. and Ruiz, O.D. (1992) Neonatal treatment with clomipramine increased immobility in forced swim test: an attribute of animal models of depression. Pharm. Biochem. Behav., 42: 737–739.

Verney, C., Lebrand, C. and Gaspar, P. (2002) Changing distribution of monoaminergic markers in the developing human cerebral cortex with special emphasis on the serotonin transporter. Anat. Rec., 267(2): 87–93.

Vijayakumar, M. and Meti, B.L. (1999) Alterations in the levels of monoamines in discrete brain regions of clomipramine-induced animal model of endogenous depression. Neurochem. Res., 24(3): 345–349.

Vitalis, T., Cases, O., Passemard, S., Callebert, J. and Parnavelas, J.G. (2007) Embryonic depletion of serotonin affects cortical development. Eur. J. Neurosci., 26(2): 331–344.

Vogel, G., Neill, D., Hagler, M. and Kors, D. (1990) A new animal model of endogenous depression: a summary of present findings. Neurosci. Biobehav. Rev., 14: 85–91.

Wai, M.S., Shi, C., Kwong, W.H., Zhang, L., Lam, W.P. and Yew, D.T. (2008) Development of the human insular cortex: differentiation, proliferation, cell death, and appearance of 5HT-2A receptors. Histochem. Cell Biol., 130: 1199–1204.

Wallace, J.A. and Lauder, J.M. (1983) Development of the serotonergic system in the rat embryo: an immunocytochemical study. Brain Res. Bull., 10(4): 459–479.

Wann, B.P., Bah, T.M., Kaloustian, S., Boucher, M., Dufort, A.M., Le Marec, N., Godbout, R., Rousseau, G. (2008) Behavioural signs of depression and apoptosis in the limbic system following myocardial infarction: effects of sertraline. J. Psychopharmacol. [Epub ahead of print].

Whitaker-Azmitia, P., Zhou, F., Hobin, J. and Borella, A. (2000) Isolation-rearing of rats produces deficits as adults in the serotonergic innervation of hippocampus. Peptides, 21(11): 1755–1759.

Whitaker, P.M. and Warsh, J.J. (1983) Specific binding of radiolabelled antidepressants to brain astroglial cells. Adv. Biochem. Psychopharmacol., 37: 477–485.

Whitaker-Azmitia, P.M. (2005) Behavioral and cellular consequences of increasing serotonergic activity during brain development: a role in autism? Intl. J. Dev. Neurosci., 23(1): 75–83.

Whitaker-Azmitia, P.M. and Azmitia, E.C. (1986a) Autoregulation of fetal serotonergic neuronal development: Role of high affinity serotonin receptors. Neurosci. Lett., 67: 307–312.

Whitaker-Azmitia, P.M. and Azmitia, E.C. (1986b) [3H]5-hydroxytryptamine binding to brain astroglial cells: differences between intact and homogenized preparations and mature and immature cultures. J. Neurochem., 46(4): 1186–1189.

Whitaker-Azmitia, P.M. and Azmitia, E.C. (1994) Astroglial 5-HT1a receptors and S-100 beta in development and plasticity. Perspect. Dev. Neurobiol., 2(3): 233–238.

Whitaker-Azmitia, P.M., Lauder, J.M., Shemmer, A. and Azmitia, E.C. (1987) Postnatal changes in serotonin receptors following prenatal alterations in serotonin levels: further evidence for functional fetal serotonin receptors. Brain Res., 430(2): 285–289.

Whitaker-Azmitia, P.M., Murphy, R. and Azmitia, E.C. (1990a) Stimulation of astroglial 5-HT1A receptors releases the serotonergic growth factor, protein S-100, and alters astroglial morphology. Brain Res., 528(1): 155–158.

Whitaker-Azmitia, P.M., Quartermain, D. and Shemer, A.V. (1990b) Prenatal treatment with a selective D1 receptor agonist (SKF 38393) alters adult [3H]paroxetine binding and dopamine and serotonin behavioral sensitivity. Brain Res., 57(2): 181–185.

Whitaker-Azmitia, P.M., Zhang, X. and Clarke, C. (1994) Effects of gestational exposure to monoamine oxidase inhibitors in rats: preliminary behavioral and neurochemical studies. Neuropsychopharmacology, 11(2): 125–132.

Whitaker-Azmitia, P.M., Borella, A. and Raio, N. (1995) Serotonin depletion in the adult rat causes loss of the dendritic marker MAP-2. A new animal model of schizophrenia? Neuropsychopharmacology, 12(3): 269–272.

Wilson, C.C., Faber, K.M. and Haring, J.H. (1998) Serotonin regulates synaptic connections in the dentate molecular layer of adult rats via 5-HT1a receptors: evidence for a glial mechanism. Brain Res., 782(1–2): 235–239.

Winter, C., Reutiman, T.J., Folsom, T.D., Sohr, R., Wolf, R. J., Juckel, G. and Fatemi, S.H. (2008) Dopamine and serotonin levels following prenatal viral infection in mouse – implications for psychiatric disorders such as schizophrenia and autism. Eur. Neuropsychopharmacol., 18(10): 712–716.
Yamamoto, M., Suhara, T., Okubo, Y., Ichimiya, T., Sudo, Y., Inoue, M., Takano, A., Yasuno, F., Yoshikawa, K. and Tanada, S. (2002) Age-related decline of serotonin transporters in living human brain of healthy males. Life Sci., 71(7): 751–757.
Ye, W., Shimamura, K., Rubenstein, J.L., Hynes, M.A. and Rosenthal, A. (1998) FGF and Shh signals control dopaminergic and serotonergic cell fate in the anterior neural plate. Cell, 93(5): 755–766.
Zec, N., Filiano, J.J., Panigrahy, A., White, W.F. and Kinney, H. C. (1996) Developmental changes in [3H]lysergic acid diethylamide ([3H]LSD) binding to serotonin receptors in the human brainstem. J. Neuropathol. Exp. Neurol., 55(1): 114–126.
Zeskind, P.S. and Stephens, L.E. (2004) Maternal selective serotonin reuptake inhibitor use during pregnancy and newborn neurobehavior. Pediatrics, 113(2): 368–375.
Zhou, F.C., Sari, Y. and Zhang, J.K. (2000) Expression of serotonin transporter protein in developing rat brain. Dev. Brain Res., 119(1): 33–45.
Zhou, F.C., Sari, Y. and Powrozek, T.A. (2005) Fetal alcohol exposure reduces serotonin innervation and compromises development of the forebrain along the serotonergic pathway. Alcohol. Clin. Exp. Res., 29(1): 141–149.
Zhou, F.C., Fang, Y. and Goodlett, C. (2008) Peptidergic agonists of activity-dependent neurotrophic factor protect against prenatal alcohol-induced neural tube defects and serotonin neuron loss. Alcohol. Clin. Exp. Res., 32(8): 1361–1371.

CHAPTER 3.2

Serotonin and Basal Sensory–Motor Control

Robert J. Carey*

VA Medical Center, and SUNY Upstate Medical University, Syracuse, New York

Abstract: This chapter provides a brief synopsis of the essential role of brain serotonergic neurons in the integration of sensory input and motoric output. Although serotonin neurons are proportionally small numerically in the brain, the serotonin neurons have anatomical communication with virtually all neurons in the brain. For sensory–motoric function, the serotonin neurons have a particular prominence in that they densely project to primary sensory cortices and to cranial and spinal motor neurons. By virtue of this anatomical position at the entrance to and exit from brain and spinal cord, the serotonergic system is uniquely situated to modulate behavior across the full spectrum of the sleep–wakefulness cycle.

Introduction

Serotonin-containing neurons represent a phylogenetically very old neurotransmitter system. The serotonin (5-HT) neurons are located in clusters of raphe nuclei which encompass areas from the midbrain to the medulla. While the total number of serotonin (5-HT) neurons is small relative to total neuronal cells in the brain (e.g., approximately 20,000 in rat brain), the axons of these cells have a massive number of collaterals so that virtually the entire brain and spinal cord receive 5-HT innervation (Steinbusch, 1981; Jacobs and Azmitia, 1992). Consistent with the rostral/caudal distribution of 5-HT raphe nuclei, the caudal pontine and medullary clusters send projections to the cerebellum, medulla and spinal cord, whereas the midbrain raphe nuclei innervate the forebrain. In keeping with the descending projections to the medulla, the 5-HT neurons have a substantial impact on basic functions related to respiration, thermal regulation and cardiovascular function. Projections from 5-HT nucleus raphe obscurus and ventral lateral medullary neurons project extensively to the ventral horn of the spinal cord and profusely innervate α-motor neurons, thereby impacting substantially upon motoric function (Skagerberg and Bjorklund, 1985). More elusive in terms of function are the forebrain projections of the dorsal and median raphe nuclei situated in the midbrain reticular formation. These 5-HT axons project to neo- and paleocortex as well as subcortical striatal and limbic nuclei (Fallon and Loughlin, 1987). Not surprisingly, perturbations in the forebrain 5-HT have been implicated in several diverse functions, including cognition, memory, aggression, feeding and reproduction (Brown and Linnola, 1990; Spoont, 1992; Lucki, 1998; Vanderwolf, 1998; Birger *et al.*, 2003). Such complex behavioral processes encompass interactions among many neuronal systems involving other neurotransmitters as well as peptides. Clearly, 5-HT systems are contributory to a variety of important behavioral functions. In linking serotonin and the serotonergic neural network to behavior, it is necessary to recognize that the underlying foundation of behavior is sensory–motor integration and that the serotonergic system is anatomically organized to play a major integrative role. In an effort to better define this serotonergic role in sensory–motor integration, we will initially consider some methodological problems and then focus upon several key observations which offer a potential unifying conceptual framework for the role of serotonin in sensory–motor function.

Methodological considerations in brain/behavior analyses

A basic approach to a functional analysis of neurotransmitter systems is to perform experimental manipulation that selectively impairs neuronal activity in a single neurotransmitter system and assess the impact upon behavior. We will briefly examine basic issues that need to be considered in terms of linking 5-HT alterations to behavior. For example, the recent utilization of 'knockout' mouse preparations has provided one approach. While an animal model lacking 5-HT neurons is non-viable, it is possible to create viable knockouts lacking in particular 5-HT receptor subtypes (e.g., the 5-H_{1B} receptor) (Rocha *et al.*, 1998; Sora *et al.*, 2001). Although behavioral changes can be observed in these preparations, the impact cannot be easily ascribed to the loss of a specific receptor

*E-mail: careyr@cnyrc.org

Christian Müller & Barry Jacobs (Eds.)
Handbook of Behavioral Neurobiology of Serotonin
ISBN 978-0-12-374634-4

DOI: 10.1016/B978-0-12-374634-4.00019-8

subtype in that many compensatory changes in 5-HT neurons and other transmitter systems occur which make it difficult to connect any of the observed behavioral changes to the absence of the receptor subtype vs the compensatory changes to this missing receptor subtype. Furthermore, this type of genetic subtraction manipulation affects the expression of the 5-HT receptors throughout the brain. In order to assess 5-HT involvement in complex behavioral effects, it is necessary to selectively manipulate 5-HT neurons which project to the forebrain – namely, the dorsal and median raphe nuclei. Indeed, there have been vast numbers of studies conducted in which dorsal and median raphe nuclei have been damaged by various methods, including electrolytic- and neurotoxin-induced lesions. In view of the relatively inaccessible location and the elongated topography of the raphe nuclei, it is difficult to produce virtually complete lesions of these nuclei using non-specific lesion techniques such as electrolytic and radio-frequency lesion procedures without also damaging surrounding reticular formation and other neuronal tissue. Even more selective lesion methods involving intracerebral injections of neurotoxins such as 5,7-dihydroxy tryptamine infusions (5,7-DHT) can produce asymmetric and incomplete lesions of 5-HT neurons, and can also include non-5HT neuronal damage. In other neurotransmitter systems in which there is an extensive axonal collateralization (e.g., dopamine), it is well documented that neuronal cell loss in the rat must be severe (i.e., $\geq$90 percent; Fornaguera *et al.*, 1994) to produce persistent behavioral deficits. Additionally, it is well-known that in the case of human degenerative diseases such as Parkinson's disease (dopamine neurons) and amyotrophic lateral sclerosis (ALS) (ventral horn motor neurons) that observable behavior deficits tend to occur when cell losses are in the order of 60–80 percent (Agid *et al.*, 1990; Nirmalananthan and Greensmith, 2005). The difficult requirement to produce severe cell loss restricted exclusively to dorsal and median raphe nuclei has diminished the utility of lesion methods; in addition, there is no human neurodegenerative disorder which is specifically linked to the dorsal and median raphe serotonin neurons and that has a specific neurological deficit profile.

Coupled to the intrinsic experimental difficulties of creating an animal model preparation in which the dorsal and median raphe nuclei are selectively destroyed is the related problem of development of appropriate behavioral methods to detect and measure the effects of such manipulations. In rodent models of learning and memory, it is critical to differentiate between learning and performance variables. For example, in a reward-based learning task, a manipulation which alters the reward value of a reinforcement would clearly have a performance effect in that the capacity to acquire the behavioral response may be intact but the incentive to perform would be changed. Consequently, if an experimental treatment alters the reinforcement value of a stimulus used to motivate learning, then retardation in the acquisition of the learned behavior may occur. Unless a change in the reward value of the motivational stimulus is identified, a deficit in a reward learning task could be mislabeled as a learning/memory deficit. There are also widely used aversive learning tasks in which painful stimuli such as footshock are used to motivate learning. The most commonly used tasks include both passive avoidance tasks, in which the behavior learned is response inhibition of spontaneous behavior that leads to footshock, and, conversely, active avoidance tasks, in which the learned behavior is the acquisition of an active response to avoid a footshock. Clearly, 5HT manipulations that alter sensitivity to the aversive stimulus and/or response output impact upon such learning tests. A reduction in response output would facilitate passive avoidance learning but impair active avoidance; and a 5HT manipulation that increases response output would have the opposite effect. A modification in reactivity to the aversive stimulus used to motivate behavior could enhance/impair learning in both types of tasks. Again, effects upon the motivational properties of the reward/aversive stimulus used in a learning task would be performance effects, rather than effects upon associative capacity. Needless to say, profound effects upon response output are obvious; however, if a serotonergic manipulation leaves an animal seemingly intact motorically, then subtle changes such as the tendency, rather than the capacity, to respond are more difficult to detect. What is needed is a simple measure of the spontaneous baseline activity level of an animal. In fact, a wide variety of behavioral measurement methods have been developed in an effort to make the basic assessment of baseline response output. One of the most widely used methods in rodents is to place a test animal in an open arena and record movement during a fixed time interval. While automated systems incorporating photobeam interruptions have been widely used, this procedure provides only an indirect measure and does not capture the full range of behavioral movement (e.g., rearing, grooming, locomotion). Direct observations in scoring of responses combined with videotape scoring, while labor intensive, can provide a reliable and comprehensive capture of behavioral activity. Detailed behavioral recording is necessary but, nonetheless, is not sufficient in that many environmental variables need to be considered (e.g., arena size, lighting, and test duration and frequency). Lighting conditions are important factors in that white light in rodents is aversive, so that a 5-HT manipulation could impact upon activity indirectly by diminishing or enhancing the aversive quality of the white light. While the diurnal cycle is a substantial variable affecting rodent activity, for acute testing of activity level dark or red light conditions favor behavioral activation, and therefore may provide less problematic testing conditions for rodents (Nasello *et al.*, 1998). Besides behavioral recording methods and lighting, it is necessary to consider the testing protocol as well.

A major determinant of behavioral activity in a test arena is the novelty/familiarity of the test environment (Cerbone and Sadile, 1994). The greater the novelty of the environment, the greater the behavioral activity in the control treatment groups. Typically, treatment manipulations designed to modify brain serotonin do not use a novel environment to test change in response output. A consideration relevant to the use of a novel environment is that the control group activity level is high, so it becomes more difficult to detect an increase in activity. In several studies, however, with stimulant drugs (i.e., cocaine), we have shown that cocaine at a modest dose level that does not induce behavioral abnormalities or stereotypes increases locomotor activity in a novel environment (Carey and Damianopoulos, 2006). This result indicates that initial activity in a novel environment does not create a response-output ceiling effect, and could be an effective technique to assess 5-HT manipulations on behavioral activation capacity. On the other hand, changes in overall motility can be misleading in that activity changes can occur within a test session (i.e., habituation). 5-HT treatments can modulate locomotor activation in all these forms of expression. In intact control animals, locomotor activation is highest in the initial phase of testing in a novel environment but then progressively declines as the animal acquires familiarity with the test environment. A 5-HT manipulation, therefore, could increase locomotor activation by interfering with the capacity to adapt and acquire familiarity with the test environment, Thus, an overall increase in activity level by itself is not a sufficient assessment of response-output effects of treatment variables, since it may indicate an inability to acquire information about the environment rather than an overactive response system. In order to avoid this possibility, another approach is to place the animal in a test environment for a substantial time period prior to performing a manipulation upon the 5-HT system. In that the pre-exposure lowers the response output of the control treatment to a stable level of response output, this protocol can be seen as an effective way to detect a change in response output. An increase in behavioral activity in such a testing protocol, however, does not necessarily imply direct activation of response systems, in that the increase could also be attributed to an effect of the 5-HT manipulation upon the acquired habituation/familiarity to the test environment so that the environment now becomes more novel and in this way evokes an increase in response output. Thus, the increase in response output could be a direct motoric effect or it could be indirect by virtue of an impact on stimulus processing or on memory due to a change in context familiarity/habituation. Altogether, these considerations indicate that the seemingly simple measurement of response output can be a daunting task. Alternatively, important information regarding the involvement of the serotonergic system in sensory–motor function has been obtained in studies that have monitored 5-HT activity during the sleep–wakefulness cycle and in response to well-defined stimuli.

Serotonin and motoric function

The most critical source of information about the between activity in serotonergic neurons and motor function has been provided by studies involving serotonergic hyper- and hypoactivity. A number of investigations have shown that pharmacological over-stimulation of serotonergic system can generate a hypermotility syndrome (Jacobs, 1976) and, in addition, reverse neuroleptic induced hypomotility (Carey *et al.*, 2000). In that excessive serotonergic activity generates a behavioral state of over-activation, the expectation would be that inactivation of serotonin would be linked to behavioral inactivation. Given the difficulty of severely reducing 5-HT activity alluded to earlier, the support for serotonergic inactivity being linked to motoric inactivation has been provided by an unusual source. Studies using unit activity recordings (Trulson and Jacobs, 1979; Jacobs *et al.*, 1981; Trulson and Trulson, 1982) in the serotonergic neurons during the sleep–wakefulness cycle revealed a positive correlation between serotonergic activity and wakefulness. A surprising and startling observation provided by this research was the finding that serotonergic unit activity became essentially silent during rapid eye movement sleep (REM). This is of central importance with regard to the role of serotonin in motoric function in that during REM sleep, when there is intense brain activity there is a movement paralysis (Stock, 1982; Lai and Siegel, 1988). This naturalistic observational research uncovered a vital function of serotonin in permitting REM sleep without the hazard of sleep-walking/running. In addition, these and other findings indicate that serotonergic activity appears to be critical for movement (Harris-Warrick and Cohen 1982; Jacobs *et al.*, 2000). Although 5-HT activity may be necessary for movement, other sources of evidence have shown that serotonergic activity, by itself, is insufficient to activate lower motor neuron activity (Myslinsky and Anderson, 1978; Takahashi and Berger, 1990). This necessary but not sufficient role for a neurotransmitter is certainly not unique to serotonin, but rather has pointed-up the critical contribution of serotonin as an enabler of movement through interaction with other neurotransmitter systems (McCall and Aghajanian, 1979; White and Neuman, 1980). A number of investigations have identified the importance of the interaction between the excitatory amino acid (glutamate) and serotonin in the modulation of lower motor neuron activity (Singer *et al.*, 1996; Kodama *et al.*, 1998; Datta *et al.*, 2003). Importantly, this serotonin–glutamate linkage points to new directions in the understanding and possible treatment of the devastating motor neuron disease of amyotrophic lateral sclerosis (ALS) (Steinke and Tyler, 1964; Bertel *et al.*, 1991; Sandyk, 2006).

Serotonin and sensory input interactions

While the critical contribution of serotonin neurons to movement in the 'final common motoric neuron pathway' is well established, the contribution and interaction of seronergic neurons to higher brain function and to more global activity processes labeled as behavior, needless to say, are less evident. A vast literature has been developed involving a variety of pharmacological, genetic knockouts and lesion manipulations. As mentioned in the initial comments of this chapter, independent variables (i.e., lesions) and dependent variables (i.e., measures of behavior) used in these studies raise many questions and issues regarding the extant literature. Some recent reports, however, appear to suggest an important and new way to understand the critical contribution of the serotonergic system to behavior. A key finding is in the reports of Müller *et al.* (2007) and Pum *et al.* (2007, 2008). Analogous to the electrophysiolgical findings of Jacobs and Fornal (1999), these investigations employed an *in vivo* monitoring of serotonergic activity using microdialysis in sensory cortex to monitor serotonin release in the awake behaving rat. Critically, this *in vivo* recording technique as with electrophysiological unit recordings involved monitoring of 5-HT systems, and therefore did not disrupt normative ongoing 5-HT activity. These microdialysis investigations showed that drugs (e.g., cocaine) which increase serotonin extracellularly increase serotonin in primary sensory areas of the neocortex, such as the visual cortex. Additionally, these investigations showed that a behaviorally significant visual stimulus (i.e., flashing light) also increased serotonin release in the visual cortex, and this release was accompanied by increased behavioral activation. Importantly, these microdialysis findings have provided a new way to conceptualize the role of serotonin in sensory–motor function by showing that activation of the serotonergic system can directly modulate sensory information at the level of the sensory areas of the neocortex, and in this way modify the salience or significance of environmental stimuli. Clearly, if the salience of the stimulus is enhanced then the behavioral response will also be amplified. By having a modulatory influence upon direct primary sensory stimuli at the cortical level and on the response output systems, serotonin is situated to have a pivotal role in reactivity to external environmental stimuli by modulation of sensory information in sensory cortices as well as efferent motor neurons in the spinal cord; thereby promoting action, emotion and memory storage.

A study we conducted recently (Carey *et al.*, 2008) is pertinent to the issue of the role of serotonin in processing exteroceptive stimuli and consequent modulation of relevant behavior. We manipulated serotonin availability by the use of low autoreceptor preferring dose levels of the selective 5-HT_{1A} agonist (8-OHDPAT) and antagonist (WAY-100635) in a testing paradigm with rats in which we measured the behavioral responsiveness to a small novel object placed at the center of an open field. At dose levels which did not alter overall motility we found that the 5-HT_{1A} agonist 8-OHDPAT, at an autoreceptor dose level (0.025 mg/kg), which inhibits activity in serotonin neurons, diminished behavioral responses to the novel object visual stimulus, whereas a selective autoreceptor antagonist, WAY-100635, which can enhance ongoing serotonin activity (Müller *et al.*, 2002), increased responsiveness to the novel object. As the dose level of these drugs was increased to 0.05 mg/kg the effect became exaggerated; and in the case of the agonist 8-OHDPAT it virtually eliminated responsivity to the novel object as well as motor activity. Consistent with a stimulus salience interpretation, these findings can be related to the motor paralysis during REM sleep when the serotonergic system becomes inactive. That is, the absence of serotonergic activity during REM sleep would not only impede activity in lower motor neurons, but seemingly would also blunt the influence of external stimuli by severely diminishing their salience and in this way complement the movement paralysis to allow the intense internal brain activity of REM sleep.

Serotonin and sensory–motor reactivity

On a less dramatic level in the sleep–wakefulness cycle it can be argued that modest changes in serotonergic activity can alter behavioral reactivity to stimuli and response readiness. This conceptualization of the role of serotonin fits well with a framework implicating decreased 5-HT activation in behavioral disorders such as depression. Seemingly, reduced 5-HT activity would contribute to a diminished interest in external environmental stimuli and events by not providing the critical amplification of stimulus salience necessary for eliciting effective and appropriate responses. Alternatively, a hyperactivity of the 5-HT system may contribute to hypomania and drug-seeking activity by exaggeration of the salience of reward-related stimuli. Clearly, serotonergic systems have a vital modulatory role in sensory–motor function, and the use of exploratory behavior elicited by stimulus salience variation is a new and important direction for understanding the contribution of the serotonergic brain systems in cognition and behavior. Overall, the anatomy of the serotonergic system provides the basis for the critical role of the serotonergic systems in sensory–motor function. In the primary sensory neocortex the serotonergic neurons have dense arborization (Azmitia and Segal, 1978); similarly, in the final common path at the level of the α-motor neurons, the serotonertic arborization is also dense (Takeuchi *et al.*, 1983; Doly *et al.*, 2004). Thus, at two key loci for the control of behavior – one for the stimulus (sensory cortices) and one for the response (α-motor neurons) – the serotonin projections are anatomically positioned to profoundly modulate behavior.

References

Agid, Y., Ruberg, M., Reisman, R., Hirsch, E. and Javoy-Agid, F. (1990) The biochemistry of Parkinson's disease. In: Stern, G.M. (Ed.), Parkinson's Disease. Chapman and Hall Medical, London, pp. 99–125.

Azmitia, E.C. and Segal, M. (1978) An autoradiographic analysis of the differential ascending projections of the dorsal and median raphe nuclei in the rat. J. Comp. Neurol., 179: 641–667.

Bertel, G., Malessa, S., Siuga, E. and Hornykiewicz, G. (1991) Amytrophic lateral sclerosis: changes of noradrenergic and serotonergic transmitter systems in the spinal cord. Brain. Res., 566: 54–60.

Birger, M., Swartz, M., Cohen, D., Alesh, Y., Grishpan, C. and Koteir, M. (2003) Aggression; the testosterone–serotonin link. Isr. Med. Assoc. J., 5: 653–658.

Brown, G.I. and Linnola, M.I. (1990) CSF serotonin metabolite (5-HIAA) studies in depression, impulsivity, and violence. J. Clin. Psychiatry., 51(Suppl.): 31–41.

Carey, R.J. and Damianopoulos, E.N. (2006) Cocaine conditioning and sensitization: The habituation factor. Pharmacol. Biochem. Behav., 84: 128–133.

Carey, R.J., Damianopoulos, E.N. and DePalma, G. (2000) 8-OH DPAT can restore locomotor stimulant effects of cocaine blocked by haloperidol. Pharamacol. Biochem. Behav., 66: 863–872.

Carey, R.J., Shanahan, A., Damianopoulos, E.N., Müller, C.P. and Huston, J.P. (2008) Behavior selectively elicited by novel stimuli: modulation by the 5-HT1A agonist 8-OHDPAT and antagonist WAY-100635. Behav. Pharmacol., 19: 361–364.

Cerbone, A. and Sadile, A.G. (1994) Behavioral habituation to spatial novelty: interference and noninterference studies. Neurosci. Biobehav. Rev., 18: 497–518.

Datta, S., Spoley, E.E. and Patterson, R.H. (2003) Microinjection of glutamate into the pedunculopontine tegmentum induces REM sleep and wakefulness in the rat. Am. J. Physiol. Regul. Integr. Comp. Physiol., 280: R752–R759.

Doly, S., Madeira, A., Fischer, J., Brisorgueil, M.J., Daval, J., Bernard, R., Verge, D. and Conrath, M. (2004) The 5-HT2A receptor is widely distributed in the rat spinal cord and mainly localized at the plasma membrane of postsynaptic receptors. J. Comp. Neurol., 472: 496–511.

Fallon, J.H. and Loughlin, S.E. (1987) Monoamine innervation of cerebral cortex and a theory of the role of monoamines in the cerebral cortex and basal ganglia. In: Jones, E.G. and Peters, A. (Eds.), Cerebral cortex: Further Aspects of Cortical Function Including Hippocampus. Plenum, New York, NY, pp. 41–127.

Fornaguera, J., Carey, R.J., Huston, J.P. and Schwarting, R.K.W. (1994) Behavioral asymmetries and recovery in rats with different degrees of unilateral striatal dopamine depletion. Brain Res., 664: 178–188.

Harris-Warrick, R.M. and Cohen, A.H. (1982) Serotonin modulates the central pattern generator for locomotion in the isolated lamprey spinal cord. J. Exp. Biol., 116: 27–46.

Jacobs, B.L. (1976) An animal behavioral model for studying central serotonergic synapses. Life Sci., 19: 777–785.

Jacobs, B.L. and Azmitia, E.C. (1992) Structure and function of the brain serotonergic system. Physiol. Rev., 72: 165–229.

Jacobs, B.L. and Fornal, C.A. (1999) Activity of serotonergic neurons in behaving animals. Neuropsychopharmacology, 21(2 Suppl.): 98–138.

Jacobs, B.L., Heym, J. and Trulson, M.E. (1981) Behavioral and physiological correlates of brain serotonergic unit activity. J. Physiol. (Paris), 77L: 431–436.

Jacobs, B.L., Martin-Cora, P.M. and Fornal, C.A. (2000) Activity of medullary serotonergic neurons in freely moving rats. Brain Res. Brain Res. Rev., 40: 45–52.

Kodama, T., Lai, Y.Y. and Siegel, J.M. (1998) Enhanced glutamatergic release during REM sleep in the rostromedulla as measured by *in vivo* microdialysis. Brain Res., 780: 176–179.

Lai, Y.Y. and Siegel, J.M. (1988) Medullary regions mediating atonia. J. Neurosci., 8: 4790–4796.

Lucki, I. (1998) The spectrum of behaviors influenced by serotonin. Biol. Psychiat., 44: 151–162.

McCall, R.B. and Aghajanian, G.K. (1979) Serotonergic facilitation of facial motoneuron excitation. Brain Res., 169: 13–27.

Müller, C.P., Carey, R.J., de Souza Silva, M.A., Jochan, G. and Huston, J.P. (2002) Cocaine increases serotonergic activity in the hippocampus and nucleus accumbens *in vivo* 5-HT1A-receptor antagonism blocks behavioral but potentiates serotonergic activation. Synapse, 45: 66–67.

Müller, C.P., de Souza Silva, M.A. and Huston, J.P. (2007) Double dissociating effects of sensory stimulation and cocaine on serotonin activity in the occipital and temporal cortex. Neuropharmacology, 52: 854–862.

Myslinsky, N.R. and Anderson, E.G. (1978) The effect of serotonin precursors on alpha and gamma motoneuron activity. J. Pharmacol. Exp. Ther., 204: 19–26.

Nasello, A.G., Machado, C., Bastos, J.F. and Felicio, L.F. (1998) Sudden darkness induces a high activity–low anxiety state in male and female rats. Physiol. Rev., 63: 451–454.

Nirmalananthan, N. and Greensmith, L. (2005) Amyotrophic lateral sclerosis: recent advances and future therapies. Curr. Opin. Neurol., 18: 712–719.

Pum, M., Carey, R.J., Huston, J.P. and Müller, C.P. (2007) Dissociating effects of cocaine and d-amphetamine on dopaminergic and serotonergic activity in entorhinal, perirhinal and prefrontal cortex: *in vivo* microdialysis study. Psychopharmacology, 193(3): 375–390.

Pum, M., Huston, J.P., de Souza Silva, M.A. and Müller, C.P. (2008) Visual sensory motor-gating by serotonin-activation in the medial prefrontal and occipital, but not in the rhinal cortices in rats. Neuroscience, 153: 361–372.

Rocha, B.A., Fumagali, F., Gainetdinov, R.R., Jones, S.R., Ator, Giros, R., Miller, G.W. and Caron, M.G. (1998) Cocaine self-administration in dopamine-transporter knockout mice. Nat. Neurosci., 1: 132–137.

Sandyk, R. (2006) Serotonergic mechanisms in amyotrophic lateral sclerosis. Intl. J. Neurosci., 116: 775–826.

Singer, J.H., Bellingham, M.C. and Berger, A.J. (1996) Presynaptic inhibition of glutamatergic synaptic transmission in rat motoneurones by serotonin. Neurophysiology, 76: 797–807.

Skagerberg, G. and Bjorklund, A. (1985) Topographic principles in the spinal projections of serotonergic and non-serotonergic brain stem neurons in the rat. Neuroscience, 15: 445–480.

Sora, I., Hall, F.S., Andrews, A.M., Itokawa, M., Li, S.-F., Wei, H.-B., Wichems, C., Lesch, K.-P., Murphy, D.L. and Uhl, O.R. (2001) Molecular mechanisms of cocaine reward. Combined dopamine and serotonin transporter knockouts eliminate cocaine preference. Proc. Natl. Acad. Sci. USA, 98: 5300–5305.

Spoont, M.R. (1992) Modulatory role of serotonin in neural information processing: implications for human psychopathology. Psychol. Bull., 112: 330–350.

Steinbusch, H.W.M. (1981) Distribution of immunoreactivity in the central system of the rat-cell bodies and terminals. Neuroscience, 4: 557–618.

Steinke, I. and Tyler, H.R. (1964) The association of amyotrophic lateral sclerosis (motor neuron disease) and carbohydrate intolerance, a clinical study. Metabolism, 13: 1376–1381.

Stock, G. (1982) Neurobiology of REM sleep. In: Ganten, D. and Pfaff, D. (Eds.), Sleep: Clinical and Experimental Aspects. Springer Verlag, Berlin, pp. 1–36.

Takahashi, T. and Berger, A.G. (1990) Direct excitation of rat spinal motoneurones by serotonin. J. Physiol., 423: 63–76.

Takeuchi, Y., Kojima, M., Matsoura, T. and Sano, Y. (1983) Serotonergic innervation on the motoneurons in the mammalian brain stem. Light and electron microscopic immunohistochemistry. Anatomy. Embryol. (Berl.), 167: 321–333.

Trulson, M.E. and Jacobs, B.L. (1979) Raphe unit activity in freely-moving rats: correlation with level of behavioral arousal. Brain Res., 163: 135–150.

Trulson, M.E. and Trulson, V.M. (1982) Activity of nucleus raphe pallidus neurons across the sleep–waking cycle in freely moving rats. Brain Res., 237: 232–237.

Vanderwolf, C.H. (1998) A general role for serotonin in the control of behavior: studies with intracerebral 5,7, dihydroxytryptamine. Brain Res., 504: 198.

White, S.R. and Neuman, R.S. (1980) Facilitation of spinal motoneuron excitability by 5-hydroxytryptamine and noradrenaline. Brain Res., 188: 119–127.

CHAPTER 3.3

Role of the Serotonergic System in Appetite and Ingestion Control

Michelle D. Lee[1] and Peter G. Clifton[*,2]

[1]*Department of Psychology, Swansea University, Swansea, UK*
[2]*Department of Psychology, University of Sussex, Brighton, UK*

Abstract: Three decades on from Blundell's suggestion that serotonin serves to orchestrate endogenous satiety, we present an overview of the behavioral, pharmacological and neurological evidence that underpins our understanding of the role of serotonin in the control of appetite. Early human and rodent studies, many using fenfluramine, provided initial support for the hypothesis. The more recent characterization of serotonin receptor subtypes, the development of selective pharmacological probes, and the use of transgenic mice unable to express certain 5-HT receptor subtypes has identified specific roles for the 5-HT_{2C}, 5-HT_{1B} and 5-HT_6 receptor subtypes in feeding behavior. It has also provided support for the development of 5-HT_{2C} receptor agonists as a treatment for obesity. Serotonin acts at multiple brain sites to enhance satiety; these include the nucleus of the solitary tract and lateral parabrachial nucleus in the brainstem and hypothalamic sub-regions such as the arcuate and paraventricular nuclei. There are important interactions between serotonin- and hypothalamic peptide-containing cells, especially those expressing α-melanocyte stimulating hormone and Neuropeptide Y. Serotonin also has important interactions with cholecystokinin at the brainstem level. The multiplicity of these interactions has been reflected in the realization that serotonin has multiple influences at the behavioral level, which range from the mediation of short-term satiation in response to gastric distension and the presence of gut nutrients, to modulation of the responses evoked by conditioned cues that lead to the initiation of feeding.

Keywords: serotonin, satiety, appetite, 5-HT, 5-HT2C, 5-HT1B, 5-HT6, obesity, feeding, body weight, fenfluramine

Abbreviations: αMSH, α-melanocyte stimulating hormone; 5-HTP, 5-hydroxy-L-tryptophan; 8-OH DPAT, 8-hydroxy-2-di-n-propylamino-tetralin; AGRP, Agouti-related peptide; CCK, cholecystokinin; GABA, gamma-amino butyric acid; KO, knockout; LH, lateral hypothalamus; MC, melanocortin; mCPP, m-chlorophenylpiperazine; NPY, Neuropeptide Y; NTS, nucleus of the solitary tract; PBN, parabrachial nucleus of the pons; POMC, pro-opiomelanocortin; *p*CPA, para-chlorophenylalanine; TFMPP, N-[3-trifloromethyl]piperazine

Introduction

Evidence for the involvement of serotonin in the neurobiology of appetite first began to accumulate in the early 1970s. Amongst the factors that stimulated these early studies were an interest in the role of the hypothalamic neurotransmitter systems in the regulation of feeding and drinking (Grossmann, 1960), and the discovery of fenfluramine, a non-stimulatory derivative of amphetamine (van Rossum and Simons, 1969) that quickly replaced amphetamine itself in the clinical treatment of obesity. Samanin's group was the first to report that lesions of the raphe nucleus attenuated the hypohagic effect of fenfluramine but had no effect on reductions of food intake induced by amphetamine (Samanin *et al.*, 1972) and, shortly afterwards, showed that depletion of central dopamine, using intraventricular 6-hydroxydopamine with pargyline pretreatment, attenuated the hypophagic effects of amphetamine and several other psychostimulants (Samanin *et al.*, 1975). The clear implication was that fenfluramine-induced hypophagia depended on enhanced brain serotonergic activity whereas amphetamine probably achieved its effect through a dopaminergic mechanism. However, Samanin did not suggest, at this point, that serotonin might play a role in the endogenous modulation of feeding behavior. Early studies by Blundell demonstrated further dissociations in the effects of fenfluramine

*Corresponding author
E-mail: pete@sussex.ac.uk

Christian Müller & Barry Jacobs (Eds.)
Handbook of Behavioral Neurobiology of Serotonin
ISBN 978-0-12-374634-4

DOI: 10.1016/B978-0-12-374634-4.00020-4

and amphetamine. The hypophagic effect of amphetamine was reduced in rats with lateral hypothalamic (LH) lesions, whereas the effect of fenfluramine was enhanced (Blundell and Leshem, 1974). An early microstructural study demonstrated that fenfluramine reduced meal size whereas amphetamine increased intermeal intervals (Blundell *et al.*, 1976). In addition, para-chlorophenylalanine (*p*CPA), which depletes serotonin, was shown to increase food intake and result in obesity (Breisch *et al.*, 1976). Blundell's seminal review (Blundell, 1977), which became a citation classic, synthesized these early studies and proposed that endogenous serotonin inhibits food consumption by augmenting short-term satiety processes both during a meal, and the subsequent post-ingestive period.

Serotonin and feeding behavior: early studies with human participants

Much of our understanding of the role of the serotonergic system in the control of appetite and ingestive behavior has developed from studies carried out in the 1980s and early 1990s using the serotonin releaser and reuptake inhibitor fenfluramine (Garattini *et al.*, 1986, 1987), initially as the racemic dl-fenfluramine and later the more potent isomer d-fenfluramine. In fact, until 1997, when fenfluramine was withdrawn as an anti-obesity treatment due to its association with valvular heart disease (Connolly *et al.*, 1997), it was the prototypical compound for investigating serotonergic mechanisms underpinning food intake and satiety.

Goodhall and Silverstone characterized both the behavioral and pharmacological profile of fenfluramine-induced anorexia in a series of studies in lean human volunteers (Goodall and Silverstone, 1988a,b; McGuirk *et al.*, 1991; Goodall *et al.*, 1992, 1993). In all of the studies participants were allowed to choose an *ad libitum* lunch through the use of an adapted self-service food dispenser that recorded both the timing and choice of single food items. These food items were varied to give a choice of macronutrients and also both sweet and non-sweet foods. The food types that were offered, such as a chicken sandwich or a portion of crackers and cheese, were high in the proportion of energy derived from either carbohydrate or fat, whereas protein content was kept relatively stable. The initial studies (Goodall and Silverstone, 1988a,b) indicated that the d-fenfluramine reduced subjective ratings of hunger as well as intake of non-sweet carbohydrates and fat, but spared protein intake. Furthermore, administration of the non-selective 5-$HT_{1/2}$ receptor antagonist metergoline partially reversed the reduction in non-sweet food caused by d-fenfluramine. Similar effects of d-fenfluramine have been shown in an obese population (Wurtman *et al.*, 1985).

In a more detailed microstructural analysis in humans, McGuirk *et al.* (1991) found that d-fenfluramine decreased the rate of eating without altering the duration of the meal. This pattern of eating was distinct from that of d-amphetamine, which reduced intake by shortening meal duration, but did not affect the rate of eating. This finding was consistent with an earlier report by Rogers and Blundell (1979), where dl-fenfluramine slowed within-meal rate of eating but did not alter latency to initiate eating, unlike d-amphetamine administration, which delayed the start of eating and increased eating rate.

Goodall *et al.* (1993) found that the 5-$HT_{2A/2C}$ receptor antagonist ritanserin blocked the preferential anorexigenic action of d-fenfluramine on cumulative fat intake during a self-service lunch. Moreover, ritanserin alone did not increase food intake compared to placebo, consistent with a true antagonist effect. Interestingly, the effect of d-fenfluramine on cumulative fat intake became more pronounced as the meal progressed, suggesting a drug-induced enhancement of satiation during the meal.

The findings of Silverstone and colleagues were significant in establishing the pharmacological profile of d-fenfluramine-induced hypophagia in humans, as well as demonstrating the importance of behavioral measures of food intake. Their data also suggested that the inhibitory effect of d-fenfluramine on food consumption was mediated by 5-HT_1 and/or 5-HT_2 receptor subtypes. This was consistent with contemporary data obtained from studies using rats; Neill and Cooper (1989) utilized the most selective pharmacological tools available at the time in an attempt to elucidate the receptor subtypes through which d-fenfluramine exerted its anorexigenic effect. The effect of d-fenfluramine was reversed by the 5-$HT_{2A/2C}$ antagonist ritanserin and the 5-$HT_{1A/1B}$ antagonist cyanopindolol, as well as the non-selective 5-HT_1 antagonist metergoline, but not by the peripheral antagonist xylamidine, or by the 5-HT_2 antagonists ketanserin and ICI 169 369 (both with low affinity for 5-HT_{2C} receptors).

It was widely assumed that rather than directly stimulating postsynaptic receptors, d-fenfluramine reduced food consumption indirectly by increasing 5-HT availability at these receptors. However, Curzon and colleagues (Gibson *et al.*, 1993; Oluyomi *et al.*, 1994; Curzon *et al.*, 1997) demonstrated, in rats, that d-fenfluramine was still able to reduce food intake after a dose of *p*CPA that significantly depleted extra-raphe 5-HT levels and blocked the ability of d-fenfluramine to increase 5-HT transmission in medial hypothalamic areas. These findings clearly pointed to a direct effect of d-fenfluramine at 5-HT receptors. In addition, Gibson *et al.* (1993) confirmed that the active metabolite d-norfenfluramine stimulated 5-HT_{2C} receptors by showing that the anorexigenic effect of this metabolite, but not the parent compound, was reversed by the 5-HT_{2C}

antagonist 1-naphthylpiperazine. As more selective antagonists have become available it has been possible to confirm that the direct and indirect effects of d-fenfluramine on food intake are mediated by 5-HT_{2C} receptors (Vickers *et al.*, 2001). Further support has come from studies using transgenic mouse models of 5-HT function where mice lacking functional 5-HT_{2C} receptors have been shown to be obese (Tecott *et al.*, 1995), and resistant to d-fenfluramine-induced hypophagia.

Since the withdrawal of d-fenfluramine as an anti-obesity treatment, sibutramine is the only serotonergic anti-obesity drug currently available in the clinic (Heal *et al.*, 1998). Nevertheless, this extensive body of work centered on-fenfluramine has served to focus attention on the role of specific 5-HT receptors in the inhibitory control of food intake. What is more, it has provided much of the impetus for the development 5-HT_{2C} receptors agonists, such as lorcaserin, as potential anti-obesity agents (Thomsen *et al.*, 2008).

Serotonin and satiety: the role of serotonin receptor subtypes

5-HT_{1B} and 5-HT_{2C} receptors and feeding

Satiation, which is the process that leads to the termination of a meal, is mediated by a cascade of cognitive and pre-ingestive cues. Satiety, a state associated with a greatly lowered probability of feeding, is maintained by post-ingestive cues that have resulted from the digestion and absorption of nutrients gained from previous meals. The initiation of feeding depends on both interoceptive cues that arise from gradually decreasing energy availability, as well as response to incentive-related cues and to more general signals such as time of day. Blundell (1977) explicitly postulated a role for serotonin in satiation and satiety. Since then it has become clear that serotonin and its receptors may modulate both the termination of a meal, and the initiation of the next meal, through actions at multiple sites within the nervous system. These actions are likely to be triggered as a consequence of effects on gut functioning, the processing of both interoceptive and exteroceptive stimuli relevant to ingestion, as well as effects on motivational systems. Subsequently, Simansky (1996) presented a robust argument in favor of a pivotal role for serotonin in regulating satiety that was reinforced by pharmacological evidence. The case for serotonin as a CNS satiety factor was based on accumulating evidence from rodent studies using techniques that afforded a detailed analysis of the organization of feeding responses both within and between meals and over longer time periods. Two such microstructural paradigms that have been particularly useful in determining the contribution of serotonergic mechanisms to ingestive behavior are the behavioral satiety sequence and meal-pattern analysis in free-feeding rodents and humans.

The behavioral satiety sequence (for an extensive review see Halford *et al.*, 1998) describes a repertoire of behaviors in rodents that occur in an orderly progression during a feeding episode of between 30 and 60 minutes. When food is first presented, feeding (manipulating or chewing food) immediately becomes the dominant behavior, which then gradually gives way to a period of activity (rearing, sniffing, locomotion) followed by grooming, rest and sleep. The satiety sequence is absent during sham feeding (Young *et al.*, 1974), but is advanced in time with a more rapid transition to rest, by pre-feeding a small amount of food prior to the test meal (Cooper and Francis, 1993; Kitchener and Dourish, 1994). The critical measure of augmentation of satiety processes, therefore, is that a compound or manipulation should reduce food intake and hasten the transition through the sequence, whilst keeping the overall pattern of behavior intact. Perhaps the original example of the application of the satiety sequence is the pioneering work of Smith and Gibbs, who first described the gastrointestinal peptide cholecystokinin (CCK) as a satiety factor on the basis that it curtailed sham feeding and elicited the complete satiety sequence (Antin *et al.*, 1975).

Microstructural analyses of meal patterns (for review see Clifton, 2000) typically involve apparatus that allows each rat to access 45-mg food pellets from a food magazine in such a way that each pellet removed can be time-logged precisely. To ensure an accurate record of intake the cage environment must be adjusted to prevent hoarding, so that each pellet is eaten before another can be taken. In addition, water intake is monitored in fine detail by logging each tongue contact with a drinking spout and delivering water through activation of a pump. In general, rodents can be housed in this meal-pattern apparatus for several weeks at a time, allowing for continuous records of feeding and drinking responses to be collected before, during and after drug treatment. Consequently, analysis of feeding records can provide measures of latency to feed, meal size and duration, meal frequency, and rate of eating, with similar data for drinking records. From these various inter-meal and intra-meal parameters, it is possible to tease out any satiety-enhancing consequences of an experimental manipulation. Thus, a decrease in meal size without a delay in beginning the meal is a behavioral correlate of within-meal satiation (the process which terminates a meal), whilst an increase in inter-meal interval suggests enhanced post-ingestive satiety. Simansky (1996) suggested that drug-induced alterations of feeding rate require careful interpretation, as slowing of feeding rate could indicate non-specific motor effects. However, there is evidence, in both humans (Rogers and Blundell, 1979) and rats (Clifton, 2000), demonstrating that the rate of eating decreases towards the end of a meal as satiation becomes advanced.

These techniques have been applied to studies of drugs that stimulate serotonin availability, direct 5-HT agonists, selective 5-HT receptor antagonists, and 5-HT receptor knockout mice, leading to a clear consensus that much of the inhibitory effect of serotonergic agents on food consumption arises through the potentiation of endogenous satiety. Confirmation of earlier work (Blundell *et al.*, 1976) suggesting that fenfluramine potentiated satiety processes was provided by Burton *et al.* (1981), who showed that fenfluramine reduced meal size in a dose-dependent manner, whilst meal duration and inter-meal interval were unchanged. These early findings were expanded by meal-patterning studies using the serotonin precursor 5-hydroxy-L-tryptophan (5-HTP) (Fletcher, 1986) and the serotonin reuptake inhibitors fluoxetine (Clifton, 1989; Lee and Clifton, 1992) and sertraline (Simansky and Vaidya, 1990). These studies confirmed a fenfluramine-like reduction in meal size and rate of eating without any alteration of meal frequency when serotonin levels were raised by compounds structurally distinct from fenfluramine.

Meal-pattern analyses of a range of 5-HT agonists lend support to the view that serotonin serves to inhibit ingestion by modulating satiety, and have also provided a more accurate picture of the role of specific subtypes of 5-HT receptors. The 5-$HT_{1B/2C}$ agonist mCPP has been shown to decrease meal size and rate of eating (Simansky and Vaidya, 1990; Clifton *et al.*, 1993), but also had marked effects on water intake (Clifton *et al.*, 1993). These studies pinpointed a role for 5-HT_{1B} and 5-HT_{2C} receptors in controlling meal-related satiety consistent with Vickers and colleagues (Vickers *et al.*, 2001), who reversed the hypophagic effect of d-fenfluramine with the selective 5-HT_{2C} antagonist SB-242084 but not the 5-$HT_{1B/1D}$ antagonist GR-127935 or the 5-HT_{1B} antagonist SB-224289. The preferential 5-HT_{2C} agonist Ro 60-0175 also proved a useful probe of the role of this receptor in ingestive behavior. Clifton *et al.* (2000) compared Ro 60-0175 and d-fenfluramine in a meal-pattern experiment. Ro 60-0175 and d-fenfluramine reduced meal size, without changing meal frequency. However, d-fenfluramine but not Ro 60-0175, except at the highest dose, slowed the rate of eating. Moreover, the selective effect of the agonist on meal size was completely reversed by the 5-HT_{2C} antagonist SB-242084. The 5-HT_{2C} antagonist also reversed the decrease in feeding rate caused by d-fenfluramine, strongly supporting a role for 5-HT_{2C} receptors in the control of eating rate despite the lack of effect of Ro 60-0175 on this parameter. Together these findings suggested that stimulation of role 5-HT_{2C} receptors selectively inhibits food consumption by augmenting within-meal satiation.

Meal-pattern analyses of compounds acting at 5-HT_{1B} receptors have shown that there are subtle differences in the behavioral profile of drugs that act at 5-HT_{1B} receptors compared to 5-HT_{2C} receptors. Early investigations of the role of 5-HT_{1B} receptors in ingestion control depended on the administration of RU 24969, which has high affinity for both 5-HT_{1A} and 5-HT_{1B} receptors. However, the decrease in feeding seemed to be attributable to 5-HT_{1B} receptors, as the selective 5-HT_{1B} receptor antagonists cyanopindolol and pindolol attenuated RU 24949-induced hypophagia (Kennett *et al.*, 1987; Kennett and Curzon, 1988). Most accounts of the affect of RU 24969 on feeding only provide short-term intake data rather than meal patterns, but Simansky and Vaidya (1990) reported that an ED_{50} dose of RU 24969 reduced meal duration, but had no direct affect on the rate of eating. The selective 5-HT_{1B} agonist CP-94,253 (Koe *et al.*, 1992) proved a more reliable probe of 5-HT_{1B} receptor function in ingestive behavior. A meal-pattern analysis using CP-94,253 was conducted by Lee *et al.* (2002), demonstrating a dose-related reduction in meal size and duration without any concomitant slowing of the rate of eating, consistent with the profile of RU 24969 (Simansky and Vaidya, 1990). However, the authors also measured the pattern of water consumption and found that CP-94,253-treatment caused a significant hypodipsia which outlasted the effect of this agent on food intake – an effect which had not previously been seen after 5-HT_{2C} agonist administration (Clifton *et al.*, 2000). Lee *et al.* (2002) concluded that it was likely that 5-HT_{1B} and 5-HT_{2C} receptor subtypes controlled separate elements of satiety, with 5-HT_{1B} receptors responsible for meal size and 5-HT_{2C} receptors regulating rate of eating. This conclusion was consistent with the findings of Grignaschi and Samanin (1992), who demonstrated that the meal-size effect of d-fenfluramine was blocked by pretreatment with the 5-$HT_{1A/1B}$ antagonist (±)-cyanopindolol, whereas the effect on feeding rate was blocked by the 5-$HT_{2A/2C}$ antagonist ritanserin. The action of d-fenfluramine on both meal size and rate of eating was blocked by the 5-$HT_{1/2}$ antagonist metergoline (Grignaschi and Samanin, 1992). Further evidence that recruitment of both receptor subtypes is necessary for a complete effect on satiety came from meal pattern analyses of the 5-$HT_{1B/2C}$ agonist mCPP, which decreased meal size *and* reduced feeding rate (Simansky and Vaidya, 1990; Clifton *et al.*, 1993).

The consensus view that the effects of 5-HT_{1B} and 5-HT_{2C} receptors on satiety process can be dissociated has been reinforced by findings from studies employing the behavioral satiety sequence. These findings are, in general, consistent across several laboratories utilizing a range of test meals (palatable wet mash, standard rodent chow, sweet solutions) and recording methods (time-sampled observations, analysis of video-tape, and continuous recording) in both rats and mice. A number of authors have demonstrated that raising levels of available 5-HT using serotonin reuptake inhibitors leads to a more rapid transition through the behavioral satiety sequence whilst maintaining the structure of the sequence, consistent with a potentiation of satiety (Clifton *et al.*, 1989; Simansky and Vaidya, 1990; McGuirk *et al.*, 1992). Perhaps unsurprisingly, given its more complex mechanism of action, and the use of different isomers, the effect of fenfluramine on the satiety sequence has been more variable. Some early reports in rats suggested that both the racemate and the more potent *d* isomer of fenfluramine inhibited the resting phase of the sequence, and should therefore not be considered to be a satiety-enhancing agent when given acutely (Montgomery and Willner, 1988; Willner *et al.*, 1990), although this was not consistent with earlier work by Blundell and McArthur (1981). For a more detailed discussion of this issue, see Halford *et al.* (1998). More recently, Vickers *et al.* (1999) and Hewitt *et al.* (2002) have demonstrated a selective effect of d-fenfluramine in advancing the satiety sequence, including an enhancement of rest, in C57BL/6 mice.

Comparisons of the behavioral profile of a range of selective agonists have proved fruitful in elucidating the contribution of 5-HT receptor subtypes to endogenous satiety. Thus, Kitchener and Dourish found that 5-$HT_{1B/2C}$ agonists mCPP and TFMPP elicited the complete satiety sequence advanced in time, whereas the 5-$HT_{1A/1B}$ agonist RU 24969 and the 5-$HT_{2A/2C}$ agonist DOI disrupted the satiety sequence, both by increasing active behaviors and by retarding the transition into rest (Kitchener and Dourish, 1994). It is likely that the disruption to the satiety sequence caused by RU 24969 is due to its action at 5-HT_{1A} receptors, as later studies using the more 5-HT_{1B} agonist CP-94,253, which has a 45-fold greater affinity for 5-HT_{1B} over 5-HT_{1A} receptors (Lee and Simansky, 1997; Lee *et al.*, 2002), showed an early termination of feeding, without disrupting the sequence. Furthermore, resting behavior was increased in the final phase of the sequence when a palatable wet mash was used as the test meal (Lee *et al.*, 2002). This finding is consistent with those of O'Neill and Parameswaran (1997), who found dual stimulation of both 5-HT_{1A} and 5-HT_{1B} receptors was necessary before increases in locomotor activity were noticeable. The characterization of the preferential 5-HT_{2C} agonist Ro 60-0175 on indices of satiety consolidated earlier data collected using d-fenfluramine (Vickers *et al.*, 1999). Thus, Hewitt and colleagues compared two doses of Ro 60-0175 against a behaviorally selective dose of d-fenfluramine in the mouse, and found that this agonist produced the expected leftwards shift in the sequence of behaviors, and also increased rest at the end of the sequence (Hewitt *et al.*, 2002). Perhaps the most selective 5-HT_{2C} agonist characterized so far is VER23779 (Richter *et al.*, 2006). This highly selective compound, consistent with earlier reports of the effects of Ro 60-0175, reduced food intake by enhancing satiety in a behaviorally selective manner (Somerville *et al.*, 2007).

Converging evidence in support of a critical role for 5-HT_{2C} receptors in ingestion control has also come from work in receptor knockout mice. More specifically, mice lacking functional 5-HT_{2C} receptors eat more and are heavier compared to lean wild-type littermates (Tecott *et al.*, 1995; Heisler *et al.*, 1998), consistent with tonic inhibitory control of intake by serotonergic systems. Vickers *et al.* (1999) analyzed the effect of d-fenfluramine on the behavioral satiety sequence in 5-HT_{2C} receptor knockout mice and wild-type controls. In wild-type mice consuming a palatable wet mash meal, d-fenfluramine reduced food intake and produced a temporally advanced, yet intact, satiety sequence. This satiety-like effect of d-fenfluramine was blunted in mice lacking the 5-HT_{2C} receptor. A later study (Clifton, 2000) showed that the d-fenfluramine-induced reduction in meal size, measured in a mouse meal-pattern paradigm, was absent when 5-HT_{2C} knockout mice were treated with the same dose of d-fenfluramine.

It has proved more difficult, using mouse genetic models, to further clarify the role of 5-HT_{1B} receptors in the serotonergic control of appetite. 5-HT_{1B} knockout mice do display a hyperphagic phenotype and increased body weight compared to wild-type mice (Bouwknecht *et al.*, 2001). These animals are also insensitive to the effects of dl-fenfluramine both on feeding behavior and on c-*fos* induction in hypothalamic and amygdala sub-regions (Lucas *et al.*, 1998). However, the behavioral data are in striking contrast to those collected using selective 5-HT_{1B} antagonists. Lee *et al.* (2004) reported that 5-HT_{1B} knockout mice showed reduced sensitivity to the hypophagic and satiety-enhancing profile of d-fenfluramine. However, in wild-type mice d-fenfluramine-induced anorexia, and changes in the satiety sequence were not reversed by either the 5-$HT_{1B/1D}$ antagonist GR-127935 or the selective 5-HT_{1B} antagonist SB-224289. Clifton *et al.* (2003) demonstrated that 5-HT_{1B} knockout mice were less responsive to an active dose of the 5-HT_{2C} receptor agonist Ro 60-0175, which significantly reduced food intake in wild-types. Moreover, the effect of the same dose of Ro 60-0175 was not reversed by the selective 5-HT_{1B} antagonist SB224289. The reduced responsiveness to probes of 5-HT_{2C} function also generalized to measures of locomotor activity, and to the induction of c-*fos* immunoreacitivity in several brain regions. These findings prompted Clifton *et al.* (2003) to conclude that compensatory changes in 5-HT_{2C} receptor function following loss of 5-HT_{1B} receptors limited the conclusions that could be drawn from this knockout model in relation to the serotonergic control of feeding.

One potentially important difference between the effects of 5-HT_{1B} and 5-HT_{2C} receptor activation on ingestive behavior has emerged from the use of modified second-order schedules of reinforcement in mice. In such a schedule, mice are trained to respond for food during 30-minute sessions in which, during the first 5 minutes, responding is maintained by the presentation of a conditioned stimulus associated with food delivery rather than by food itself (Somerville *et al.*, 2007). Since the schedule is relatively lean even after this initial period, food consumption typically starts about 10 minutes into the session. Thus, each test session can be considered as comprising an initial appetitive component followed by a mixed appetitive/consummatory component. The 5-HT_{1B} receptor agonist CP94,253 led to similar reductions in responding during both the appetitive and mixed components, whereas the selective 5-HT_{2C} receptor agonist VER23779 produced a much greater relative reduction in appetitive responding (Somerville *et al.*, 2007). This result implies that 5-HT_{2C} receptor activation may play a particular role in modulating response to conditioned cues associated with food. These behavioral differences were associated with differences in regional induction of c-*fos* immunoreactivity by the two drugs. In particular, VER23779 produced marked activation of the basolateral amygdale, which was absent with CP94,253-treatment.

5-HT_{1A} receptors and feeding

The 5-HT_{1A} receptor has a well-defined inhibitory action in the raphe nucleus as a somatodendritic autoreceptor. There are also important postsynaptic populations of 5-HT_{1A} receptors in hippocampal sub-regions. In rodents, the prototypical 5-HT_{1A} receptor agonist 8-hydroxy-2-di-n-propylamino-tetralin (8-OH DPAT) has a biphasic action on feeding. Low doses (typically less than 0.5 mg/kg) stimulate food intake, whereas higher doses reduce food intake (Dourish *et al.*, 1985). Food intake is also stimulated when 8-OH-DPAT is directly injected into the median raphe nucleus (Currie and Coscina, 1993), suggesting that this effect arises from inhibition of serotonergic raphe cell firing due to the activation of 5-HT_{1A} autoreceptors. Higher doses that reduce feeding are likely to act at postsynaptic 5-HT_{1A} receptors, and their effects are completely reversed by the pretreatment with the 5-HT_{1A} receptor antagonist WAY-100635, although the antagonist had no intrinsic action on feeding in the test paradigm (Vickers *et al.*, 1996). Such doses of 8-OH-DPAT induce a characteristic behavioral syndrome (Tricklebank *et al.*, 1984), which may be partly responsible for the reduction in food intake. The enhanced feeding induced by low doses of 8-OH-DPAT is a variable phenomenon. For example, doses that stimulate feeding in sated animals may depress food intake in food-deprived animals (Dourish *et al.*, 1985; Ebenezer, 1992). In addition, a dose of 8-OH-DPAT that stimulates feeding in normal rats has no effect on feeding in older obese Zucker rats, and inhibits feeding in younger obese animals (Voigt *et al.*, 2002). The same shift in the effect of 8-OH-DPAT is observed in rats that have consumed a sugar-rich diet and become obese (Inam *et al.*, 2008). Both studies suggest that as obesity develops, there may be changes in the sensitivity of raphe serotonergic neurons to autoreceptor feedback.

Animals over-expressing the 5-HT_{1A} receptor do not show changes in food intake or body-weight regulation (Kusserow *et al.*, 2004). The same is true of 5-HT_{1A} receptor null mutant mice (Zhuang *et al.*, 1999). However, Lucki and colleagues have recently demonstrated that female, but not male, 5-HT_{1A} receptor knockout mice over-consume a sucrose solution (Bechtholt *et al.*, 2008). There are other indications that 5-HT_{1A} receptors may be involved in sex differences in feeding. For example, Currie *et al.* (2004) demonstrated that, although intra-raphe infusion of 8-OH-DPAT could reverse the hypophagic effect of the selective serotonin reuptake inhibitor fluoxetine in male rats, the same treatment was ineffective in female rats unless they had been ovariectomized.

5-HT_6 receptors and feeding

The 5-HT_6 receptor was only characterized in the 1990s and, as with the majority of other 5-HT receptor subtypes, is a G protein positively linked to adenyl cyclase (Barnes and Sharp, 1999). 5-HT_6 receptor mRNA is found at high levels in the striatum and hypothalamus in several mammalian species (Kohen *et al.*, 1996). Rat autoradiographic studies, using the specific 5-HT_6 radioligand, [^{125}I]-SB-258585, demonstrated a high density of binding in discrete hypothalamic nuclei, including the arcuate, lateral, anterior and dorsomedial hypothalamus (East *et al.*, 2002).

The first significant indication that 5-HT_6 receptors might play an important role in feeding and body-weight regulation came from a study in which the primary focus was on cognitive performance. Woolley and colleagues used two techniques to reduce 5-HT_6 receptor function. These were intracerebroventricular (ICV) administration of an antisense oligonucleotide that was complementary to the gene for the rat 5-HT_6 receptor, and chronic systemic administration of the putative 5-HT_6 receptor antagonist Ro 04-6790 (Woolley *et al.*, 2001). In addition to enhancing retention for spatial position in the Morris water maze, both manipulations reduced food intake and body weight. These findings are complemented by studies of 5-HT_6 gene manipulations in mice. Null mutation of the 5-HT_6 receptor gene had no effect on food intake and body weight when mice were fed a normal diet. However, when these animals were given access to a high fat diet over an 11-week period, food intake was reduced by 8 percent and body weight decreased by 35 percent relative to wild-type controls (Frassetto *et al.*, 2008). As these data suggest, the relatively large reduction did not solely depend on lower food intake, but was amplified by a reduction in the efficiency with which food was transformed into body mass. Data obtained using more potent 5-HT_6 antagonists, such as E-6837, have confirmed that chronic treatment will reduce food intake and body weight (Fisas *et al.*, 2006), but more detailed behavioral analyses are not yet available. Heal *et al.* (2008) have reviewed the clinical implications of these data in detail.

Other 5-HT receptor subtypes and feeding

Although at least 14 different serotonin receptor subtypes have been identified (Barnes and Sharp, 1999), there is a relative lack of evidence to suggest specific or selective involvement of 5-HT_{2A}, 5-HT_{2B}, 5-HT_4, 5-HT_5 and 5-HT_7 receptors in feeding (Clifton and Kennett, 2006), and they are not discussed further here. There is also relatively little evidence to suggest an *independent* role for the ligand-gated, ionotrophic 5-HT_3 receptor in feeding, although there are important interactions between this receptor and CCK in mediating effects on feeding behavior; the relevant data are discussed below.

Neural mechanisms underpinning serotonergic modulation of feeding behavior

Taste-responsive neurons project initially to rostral areas of the nucleus of the solitary tract (NTS) (Norgren *et al.*, 2006). The onward projection is to the parabrachial nucleus of the pons (PBN), and then via two routes to the forebrain. One component projects to the thalamus and thence to gustatory cortex; the second projects to limbic areas, including the central gray, amygdala, hypothalamus and bed nucleus of the stria terminalis. This latter pathway is also likely to be carrying non-gustatory information. Visceral information relating to, for example, gastric stretch also projects to the nucleus of the solitary tract, parabrachial nucleus and then on to these limbic structures.

Brainstem central pattern generators are responsible for the patterning of chewing and other components of ingestive behavior (Lund and Kolta, 2006). They may be directly activated by taste and other stimuli associated with food. Such responses remain intact in the absence of forebrain inputs, and are likely to result from direct excitation by sensory cells within the nucleus of the solitary tract and other brainstem areas. These motor responses are also likely to be under tonic inhibitory control by the basal ganglia, and specifically from GABAergic outputs of the pallidum (Grillner *et al.*, 2005). It is likely that hypothalamic influences on feeding behavior are expressed both through this mechanism and also by more direct connections from the lateral hypothalamus to the brainstem (Goto *et al.*, 2005). The motor components of ingestive behavior are also under additional forebrain control from the cortical masticatory area (Lund and Kolta, 2006).

Serotonin agonists might reduce food intake through their action at several sites, given the widespread distribution of serotonergic neurons and serotonin receptors (Palacios *et al.*, 1990; Törk, 1990) in the nervous system. There is direct evidence to suggest an effect in the brainstem. d-Fenfluramine reduces food intake when infused into the fourth ventricle (Grill *et al.*, 1997). The 5-$HT_{2C/1B}$ receptor agonist mCPP also reduces short-term feeding responses when infused into the fourth ventricle (Kaplan *et al.*, 1998). In both cases, the response is seen in neurally intact rats and is also observed in rats in which the neural connections from brainstem to forebrain have been disrupted by a supracollicular section. Such animals have to be maintained by tube feeding, and their feeding responses can only be monitored by infusing a palatable solution into the mouth via an indwelling cannula. However, the reduction of ingestive responses induced by either d-fenfluramine or mCPP suggests an important role for this caudal brainstem structure. The hypophagic effect of d-fenfluramine infused into the fourth ventricle is partially reversed by systemic administration of the 5-$HT_{1/2}$ antagonist metergoline (Grill *et al.*, 1997). More significantly, fourth-ventricle infusions of the 5-$HT_{2C/2A}$ antagonist mesulergine can abolish the hypophagic effect of systemically administered mCPP (Kaplan *et al.*, 1998), strongly supporting these authors' contention that caudal brainstem serotonin receptors are both '*necessary and sufficient*' for the effects of mCPP on feeding.

Candidate structures for these effects of mCPP and d-fenfluramine include the NTS and the lateral PBN

(Palacios *et al.*, 1990; Wright *et al.*, 1995). There is evidence to support the involvement of both of these structures in the serotonergic modulation of feeding behavior. Cells in both areas show neural activation, as revealed by activation of immediate early genes such as c-*fos* and c-*jun* by fenfluramine (Li and Rowland, 1996). Selective serotonin agonists also activate both areas, though with some sub-regional differences. The selective 5-HT_{1B} receptor agonist CP94,253 produces activation in the NTS, though not in the adjacent dorsal vagal and hypoglossal nuclei (Lee *et al.*, 2004). The selective 5-HT_{2C} agonist VER23779, by comparison, produced a more restricted activation in the posterior part of the NTS, at the level of the dorsal motor nucleus of the vagus (Somerville *et al.*, 2007). It has also been reported that a selectively hypophagic dose of mCPP activates a specific population of cells within the NTS (Lam *et al.*, 2009). These cells are tyrosine hydroxlase-containing, and are most likely a population of noradrenergic cells which relay visceroceptive information from the brain stem to forebrain areas (Sawchenko and Swanson, 1982). Relevant and important targets of this projection are the periventricular and lateral hypothalamic areas, both of which are implicated in the control of feeding.

The lateral parabrachial nucleus of the pons is another plausible brainstem target for the action of serotonin agonists on feeding behavior. The selective 5-HT_{1B} receptor agonist CP94,25 produces an especially marked activation of the external lateral and dorsal regions of the lateral PBN, although the effects of the selective 5-HT_{2C} receptor agonist VER23799 were much less marked (Somerville *et al.*, 2007). Other studies indicate a functional role for both 5-HT_{1B} and 5-HT_{2C} receptors in this area. Axon-sparing lesions of the lateral PBN markedly attenuated the hypophagic effects of systemically administered d-fenfluramine in one study (Li *et al.*, 1994). In addition, this study showed that the lesion leads to a marked attenuation of d-fenfluramine-induced immediate early gene immunoreactivity in the central nucleus and bed nucleus of the stria terminalis. This is consistent with the suggestion that this pathway may mediate some serotonergically-mediated peripheral effects of the ingestion of food on feeding behavior. It also demonstrates that the neural activation of these areas by d-fenfluramine is unlikely to depend on serotonin receptors in those areas, but is indirect, or transynaptic. However, in a second study (Trifunovic and Reilly, 2001), lesions of the lateral PBN had no effect on d-fenfluramine-induced hypophagia. In this study, lesions of the more medial, gustatory area of the PBN enhanced the reduction of feeding behavior induced by d-fenfluramine. One possibility is that this area may contribute to the hedonic evaluation of foods, and thus to the activation of feeding behavior. It is therefore interesting that lesions of the LH area also enhance the hypophagic action of dl-fenfluramine (Blundell and Leshem, 1974).

Direct infusion of the 5-HT_{1B} receptor agonist CP93,129 into the lateral PBN produces a marked reduction in feeding at a low nanomolar dose (Lee *et al.*, 1998). No effects on drinking, grooming or locomotor activity were observed at this dose. In an earlier analysis of the effects of systemically administered 5-HT_{1B} receptor agonist CP94,253, Lee and Simansky (1997) established that the changes in drinking microstructure in rats presented with a palatable sucrose solution resembled those associated with enhanced satiety rather than decreased hedonic impact. This observation was consistent with the advancement of the satiety sequence observed by Halford and Blundell (Halford *et al.*, 1998); however, it is clear that both the lateral PBN and the NTS are involved in other regulatory functions, including sodium intake. In this case, activation of serotonergic neurons in the dorsal raphe that project to the lateral PBN is associated with reductions in the intake of a hypertonic sodium solution (Godino *et al.*, 2007).

The hypothalamus remains the pre-eminent area associated with the control of food intake and body weight. There is a substantial body of evidence to suggest that serotonin functions as a key neurotransmitter component of the hypothalmic feeding network. Leibowitz and co-workers, in an extensive series of experiments reported during the 1980s (reviewed by Leibowitz and Alexander, 1998), demonstrated that medial hypothalamic administration of serotonin could reduce food intake. More detailed behavioral studies indicated that this effect resulted from a decrease in meal size while latency to feed and meal frequency were unaffected. These data were consistent with studies that had examined the effects of systemically administered serotonergic agents, and were interpreted in terms of a role for serotonin in satiation. In addition to these effects on food intake, additional studies showed more selective effects when rats were presented with pure macronutrients. A typical result was that the consumption of carbohydrate, presented as polycose, was reduced, whereas intake of fat (typically lard) or protein (typically casein powder) was unchanged. A number of possible interpretations exist for these data. Liebowitz, following the hypothesis initially outlined by Wurtman *et al.* (1986), suggested that the protein-sparing component of the response to serotonergic agents will lead to relatively high circulating levels of tryptophan, the amino acid precursor for serotonin synthesis. This, in turn, will raise the availability of tryptophan in the brain, and increase serotonin synthesis. The net result will be that a high carbohydrate meal leads to increased serotonergic function that will tend to promote protein consumption at the expense of carbohydrate – a classic negative feedback system. There is a substantial amount of evidence to support different elements of this hypothesis.

However, there are also some important caveats that should be recognized in relation to the experimental data supporting this hypothesis. The macronutrient sources that are used in diet-selection experiments differ in several respects, in addition to their nutrient composition. These include taste, texture and water content. All of these factors may influence feeding behavior directed towards a particular dietary component. In a detailed exploration of this issue, Lawton and Blundell demonstrated that selective suppression of carbohydrate by prior administration of d-fenfluramine could only be obtained under a restricted set of conditions (Lawton and Blundell, 1992). Selective suppression of carbohydrate intake was only observed when hydrated chow (a semi-liquid wet mash) was pitted against dry polycose powder in a choice test. Non-selective suppression of intake was observed in several other conditions. These data must lead to some reservations regarding the elegant theoretical arguments outlined in the previous paragraph.

Serotonin and downstream peptidergic mechanisms in the hypothalamus

At least since the mid-1990s, it has been clear that there are important interactions between amine and peptide neurotransmitters in the hypothalamus. The arcuate nucleus plays a crucial role in receiving humoral and other inputs that initiate hypothalamic contributions to feeding. Two distinct populations of cells are involved at this level. The first of these express Pro-opiomelanocortin (POMC), a precursor of αMSH which, when released at terminal endings in the paraventricular (PVN) or LH, acts at MC_4 receptors, leading to a reduction in feeding behavior (Heisler *et al.*, 2003). Disruption of the genes for either POMC or the MC_4 receptor leads to a severe obesity syndrome in both rodents and humans (Heisler *et al.*, 2003). A second population of cells co-express Neuropeptide Y and the endogenous MC_4 antagonist Agouti-related peptide (AGRP). Animals that over-express AGRP show the expected obesity phenotype (Adan *et al.*, 2003). Although constitutive knockout of the NPY gene has no effect on body weight (Erickson *et al.*, 1996), recent studies have demonstrated that targeted ablation of AGRP-containing cells, manipulated so that they also express a diptheria toxin receptor, leads to reductions in feeding and body weight (Gropp *et al.*, 2005). This suggests that loss of NPY receptors early in development is subject to a compensatory process that renders them less important in feeding regulation during adult life.

More recently, studies by Heisler and colleagues have provided some very interesting insights into the action of serotonergic drugs that reduce feeding behavior on both 5-HT_{2C} and 5-HT_{1B} receptors expressed by different cell groups in the hypothalamus. In an initial study (Heisler *et al.*, 2002), they determined a threshold dose for the reduction of feeding by d-fenfluramine in rats. This dose was then used in immunohistochemical studies which identified POMC-containing cells activated by fenfluramine. Finally, in an electrophysiological study, it was shown that fenfluramine could activate these POMC-containing cells. More recent work (Heisler *et al.*, 2006) has identified a specific role for 5-HT_{1B} receptor activation. First, it was demonstrated that 5-HT_{1B} receptors are localized on AGRP containing neurons in the arcuate nucleus. Subsequent experiments demonstrated that the effects of fenfluramine, and of the selective 5-HT_{1B} receptor agonist CP94,253, were blunted in mice that either over-expressed AGRP or had non-functional MC_4 receptors. This led these authors to propose a model in which 5-HT_{1B} receptors are expressed by NPY/AGRP-containing neurons which, when activated, hyperpolarize them. This would lead to both a reduction in the release of AGRP and NPY, which would lower the activation of downstream feeding centers in the lateral hypothalamus, and a reduction in the secretion of GABA by collaterals of these same neurons which normally inhibit αMSH-containing neurons in the arcuate nucleus (Cowley *et al.*, 2001). αMSH-containing cells express 5-HT_{2C} receptors (Heisler *et al.*, 2002). Fenfluramine, either through its indirect releasing effects on serotonin, or from metabolites such as *nor*-fenfluramine (which are themselves 5-HT_{2C} receptor agonists), depolarizes these cells and increases their firing rate (Heisler *et al.*, 2002). The resultant increase in release of αMSH would also be expected to decrease the activation of downstream feeding centers. Lam and colleagues used the selective 5-HT_{2C} agonist BVT.X and demonstrated that its anorexigenic action was absent in MC_4 knockout mice, again confirming this receptor as a downstream mediator of the effects of 5-HT_{2C} receptor stimulation on feeding (Lam *et al.*, 2008).

Serotonin–CCK interactions

There are important interactions between serotonin and other neurotransmitter systems involved in the mediation of hunger and satiety. The best-described relationship is between CCK and serotonin. Stallone *et al.* (1989) first demonstrated that the non-selective serotonin receptor antagonist metergoline was able to attenuate the hypophagic effect of CCK. A central site of action seemed likely, since xylamidine, another non-selective antagonist which is not centrally available after systemic administration, was without effect. Subsequently Poeschla *et al.* (1993) showed a similar effect using the relatively selective 5-HT_{2C} receptor antagonist mianserin. By contrast, antagonists at the 5-HT_{2A} or 5-HT_3 receptor were ineffective. Asarian (2009) has shown that these effects are very likely to involve hindbrain 5-HT_{2C} receptors. First, she demonstrated that the hypophagic effect of CCK is not observed in 5-HT_{2C} KO mice. In addition, although the induction of c-*fos* immunoreactivity by CCK was normal in the paraventricular and arcuate nuclei of 5-HT_{2C} KO mice, CCK failed to evoke the normal c-*fos* response in the nucleus of the solitary tract of 5-HT_{2C} KO mice. One likely account of this interaction would be that the satiating effect of CCK, perhaps acting at peripheral vagal afferents, is dependent on 5-HT_{2C} receptors located in the nucleus of the solitary tract.

Other evidence suggests a complementary interaction between CCK and 5-HT neurotransmission in the modulation of feeding. Cooper and colleagues were able to show that reductions in feeding behavior induced by fenfluramine were attenuated by pretreatment with the CCK_A receptor antagonist devazapide (Cooper *et al.*, 1992). Clifton and Cooper (1992) reported that this effect resulted from a specific reversal in the reduction of meal size that was induced by fenfluramine, such that short-term food intake was restored to normal, whereas the subsequent intermeal interval tended to be longer. This suggested that, while serotonergically mediated intra-meal satiation depended on a CCK_A mechanism, the maintenance of inter-meal satiety did not depend on CCK neurotransmission.

The 5-HT_3 receptor, unlike any other so far discussed, is a ligand-gated ion channel receptor. Although these receptors are quite widely distributed within the nervous system (Morales *et al.*, 1998), they are found in locations that would permit a role in the regulation of feeding, and especially in satiety responses. For example, they are expressed on vagal afferents and in several hindbrain nuclei. The evidence for any direct involvement in the modulation of feeding is contradictory. Several reports suggest no effect on food intake of normally fed or of or food-deprived rats (Sugimoto *et al.*, 1996). The selective 5-HT_3 antagonist ondansetron had paradigm-specific effects in another study (Cooper and Francis, 1993). Intake of a sweetened palatable mash was increased, whereas consumption of a palatable sucrose solution was reduced. In a subsequent study from the same group (van der Hoek and Cooper, 1994), the effects of ondansetron on consumption of sweetened mash were examined in greater detail. On this occasion the treatment led to significant *decreases* in intake, which resulted from a decrease in the duration of feeding rather than from any decrease in feeding rate. High doses of the 5-HT_3 antagonist mCPBG have been reported to reduce food intake, although it was not possible to demonstrate this effect actually involved activation of 5-HT_3 receptors since it was not reversed by the 5-HT_3 receptor antagonist MDL-72222 (Mazzola-Pomietto *et al.*, 1995).

There is also some evidence to suggest that 5-HT_3 receptors may mediate the acquisition of the aversion that develops to a diet unbalanced in amino acids, since it is attenuated by the 5-HT_3 antagonist tropisetron (Aja *et al.*, 1999). The hypophagic response itself may be mediated by CCK_A receptors, as the effect of tropisetron is blocked by the CCK_A receptor antagonist devazepide (Aja *et al.*, 1999). 5-HT_3 receptors may play a particular role in the conditioning process that associates the sensory characteristics of the diet with the consequences of amino acid imbalance (Bellinger *et al.*, 2005). 5-HT_3 receptors, again in combination with CCK_A receptors, may also modulate satiety responses to fat solutions. In one reported paradigm, food was made available to rats for 6 hours each day and the animals then received a duodenal infusion of a fat solution. This manipulation reduced meal size and increased inter-meal interval. Tropisetron, a 5-$HT_{3/4}$ receptor antagonist, normalized the inter-meal interval, but not meal size, whereas combined treatment with tropisetron and devazepide resulted in a normalization of both meal size and intermeal interval (Burton-Freeman *et al.*, 1999).

In a recent group of reports, Hayes and colleagues have considerably extended our understanding of the role of 5-HT_3 receptors in mediating satiety-like responses resulting both from gastric distension and from intestinal nutrient content at the level of the hindbrain. For example, in one study using c-*fos* induction as a measure of neuronal activation, they were able to show that CCK-dependent activation of neurons by gastric distension in the dorsal vagal complex, and especially in the nucleus of the solitary tract, depended on activation of 5-HT_3 receptors (Hayes and Covasa, 2006). Thus, this c-*fos* response was attentuated by the 5-HT_3 receptor antagonist ondansetron. In a related study, they examined the role of 5-HT_3 receptors in mediating the supression of feeding induced by the presence of lipids in the duodenum. Intralipid (a proprietary stable fat emulsion) was infused into the duodenum and resulted in reduced consumption of either a palatable sucrose solution or of chow; these reductions were attenuated by pretreatment with odansetron, which, in contrast to tropisetron, is selective to the 5-HT_3 receptors (Savastano *et al.*, 2007). In summary, recent evidence suggests that 5-HT_3 receptors, interacting with CCK, may play an important role in mediating short-term satiation that results from either gastric distension or the presence of nutrients, especially fat, in the upper intestinal tract.

This section has emphasized the contribution of different serotonin receptor subtypes, especially 5-HT_{2C} and 5-HT_{1B}, to the modulation of feeding. It has also reviewed the interactions between serotonin and hypothalamic peptides, and between serotonin and CCK. However, it is likely that additional roles will come to wider attention. Two examples will suffice. Asarian (2008), in addition to demonstrating the absence of hypophagia induced by CCK in 5-HT_{2C} knockout mice, also showed that these animals were insensitive to the effects of glucagon-like peptide (GLP-1) in reducing feeding behavior. Interactions between serotonin and dopamine have been well studied in relation to unconditioned locomotor activity and the effects of drugs of abuse. However, it is also clear that there may be important dopamine–serotonin interactions in the modulation of feeding. Thus, Fletcher *et al.* (2004) demonstrated that infusion of Ro 60-0175 into the ventral tegmental area (VTA) of the rat not only reduced responding for a drug reward but also reduced responding for food.

Concluding remarks

The potential utility of serotonergic agents in the treatment of obesity is covered in Chapter 4.11 of this volume. The most advanced candidates at present are 5-HT_{2c} receptor agonists, although the 5-HT_6 receptor is also a plausible target. It will of course be crucial for any 5-HT_{2C} agonist to be sufficiently selective for 5-HT_{2C} over 5-HT_{2B} receptors, as the valvulopathies associated with d-fenfluramine are due to expression of 5-HT_{2B} receptors in cardiac tissue (Fitzgerald *et al.*, 2000). Activation of 5-HT_{1B} receptors is also contraindicated because of the potential association with pulmonary arterial hypertension (Dempsie and MacLean, 2008). Although a number of 5-HT_{2C} agonists have been developed and progressed to at least Phase I trials, only lorcaserin (Thomsen *et al.*, 2008) is known to be in large-scale Phase III trials.

In summary, our understanding of the role of serotonin in the regulation of feeding behavior has greatly advanced since the original elaboration of the hypothesis that this neurotransmitter enhances satiation and satiety (Blundell, 1977). It is now clear that serotonin plays a broad role in this regard, through its actions at multiple sites in the nervous system, and by modulating a number of distinct behavioral processes. These range from the mediation of short-term satiation in response to gastric distension and the presence of nutrients in the intestinal tract, to modulation of the responses evoked by conditioned cues that lead to the initiation of a feeding bout.

References

Adan, R.A., Hillebrand, J.J., De Rijke, C., Nijenhuis, W., Vink, T., Garner, K.M. and Kas, M.J. (2003) Melanocortin system and eating disorders. Ann. N.Y. Acad. Sci., 994: 267–274.

Aja, S.M., Barrett, J.A. and Gietzen, D.W. (1999) CCK(A) and 5-HT3 receptors interact in anorectic responses to amino acid deficiency. Pharmacol. Biochem. Behav., 62: 487–491.

Antin, J., Gibbs, J., Holt, J., Young, R.C. and Smith, G.P. (1975) Cholecystokinin elicits the complete behavioral sequence of satiety in rats. J. Comp. Physiol. Psychol., 89: 784–790.

Asarian, L. (2009) Loss of cholecystokinin and glucagon-like peptide-1-induced satiation in mice lacking serotonin 2C receptors. Am. J. Physiol. Regul. Integr. Comp. Physiol., 296: R51–R56.

Barnes, N.M. and Sharp, T. (1999) A review of central 5-HT receptors and their function. Neuropharmacology, 38: 1083–1152.

Bechtholt, A.J., Smith, K., Gaughan, S. and Lucki, I. (2008) Sucrose intake and fasting glucose levels in 5-HT(1A) and 5-HT(1B) receptor mutant mice. Physiol. Behav., 93: 659–665.

Bellinger, L.L., Williams, F.E., Lucente, J., Pavelka, J., Dixon, K.D. and Gietzen, D.W. (2005) Autonomic efferents affect intake of imbalanced amino acid diets by rats. Pharmacol. Biochem. Behav., 81: 24–31.

Blundell, J.E. (1977) Is there a role for serotonin (5-hydroxytryptamine) in feeding? Intl. J. Obesity, 1: 15–42.

Blundell, J.E. and Leshem, M.B. (1974) Central action of anorexic agents: effects of amphetamine and fenfluramine in rats with lateral hypothalamic lesions. Eur. J. Pharmacol., 28: 81–88.

Blundell, J.E. and McArthur, R.A. (1981) Behavioural flux and feeding: continuous monitoring of food intake and food selection, and the video-recording of appetitive and satiety sequences for the analysis of drug action. In: Samanin, R. and Garittini, S. (Eds.), Anorectic Agents: Mechanisms of Action and Tolerance. Raven Press, New York, NY, pp. 19–43.

Blundell, J.E., Latham, C.J. and Leshem, M.B. (1976) Differences between the anorexic actions of amphetamine and fenfluramine – possible effects on hunger and satiety. J. Pharm. Pharmacol., 28: 471–477.

Bouwknecht, J.A., van der Gugten, J., Hijzen, T.H., Maes, R.A., Hen, R. and Olivier, B. (2001) Male and female 5-HT(1B) receptor knockout mice have higher body weights than wildtypes. Physiol. Behav., 74: 507–516.

Breisch, S.T., Zemlan, F.P. and Hoebel, B.G. (1976) Hyperphagia and obesity following serotonin depletion by intraventricular p-chlorophenylalanine. Science, 192: 382–385.

Burton, M.J., Cooper, S.J. and Popplewell, D.A. (1981) The effect of fenfluramine on the microstructure of feeding and drinking in the rat. Br. J. Pharmacol., 72: 621–633.

Burton-Freeman, B., Gietzen, D.W. and Schneeman, B.O. (1999) Cholecystokinin and serotonin receptors in the regulation of fat-induced satiety in rats. Am. J. Physiol., 276: R429–R434.

Clifton, P., Lee, M., Somerville, E.M., Kennett, G. and Dourish, C. (2003) 5-HT1B receptor knockout mice show a compensatory reduction in 5-HT2C receptor function. Eur. J. Neurosci., 17: 185–190.

Clifton, P.G. (2000) Meal patterning in rodents: psychopharmacological and neuroanatomical studies. Neurosci. Biobehav. Rev., 24: 213–222.

Clifton, P.G. and Cooper, S.J. (1992) 5-HT/CCK interactions influence meal size in free feeding rats. In: Dourish, C.T.

(Ed.), Multiple Cholecystokinin Receptors: Progress towards CNS Therapeutic Targets. Oxford University Press, Oxford.

Clifton, P.G. and Kennett, G.A. (2006) Monoamine receptors in the regulation of feeding behavior and energy balance. CNS Neurol. Dis. Drug. Targets, 5: 293–312.

Clifton, P.G., Barnfield, A.M. and Philcox, L. (1989) A behavioral profile of fluoxetine-induced anorexia. Psychopharmacology, 97: 89–95.

Clifton, P.G., Barnfield, A.M. and Curzon, G. (1993) Effects of food deprivation and mCPP treatment on the microstructure of ingestive behaviour of male and female rats. J. Psychopharmacol., 7: 257–264.

Clifton, P.G., Lee, M.D. and Dourish, C.T. (2000) Similarities in the action of Ro 60-0175, a 5-HT2C receptor agonist and d-fenfluramine on feeding patterns in the rat. Psychopharmacol., 152: 256–267.

Connolly, H.M., Crary, J.L., McGoon, M.D., Hensrud, D.D., Edwards, B.S., Edwards, W.D. and Schaff, H.V. (1997) Valvular heart disease associated with fenfluramine-phentermine. N. Engl. J. Med., 337: 581–588.

Cooper, S.J., Dourish, C.T. and Clifton, P.G. (1992) CCK antagonists and CCK-monoamine interactions in the control of satiety. Am. J. Clin. Nutr., 55: 291S–295S.

Cooper, S.J. and Francis, J. (1993) A microstructural analysis of the effects of presatiation on feeding behavior in the rat. Physiol. Behav., 53: 413–416.

Cowley, M.A., Smart, J.L., Rubinstein, M., Cerdán, M.G., Diano, S., Horvath, T.L., Cone, R.D. and Low, M.J. (2001) Leptin activates anorexigenic POMC neurons through a neural network in the arcuate nucleus. Nature, 411: 480–484.

Currie, P.J., Braver, M., Mirza, A. and Sricharoon, K. (2004) Sex differences in the reversal of fluoxetine-induced anorexia following raphe injections of 8-OH-DPAT. Psychopharmacol., 172: 359–364.

Currie, P.J. and Coscina, D.V. (1993) Diurnal variations in the feeding response to 8-OH-DPAT injected into the dorsal or median raphe. Neuroreport, 4: 1105–1107.

Curzon, G., Gibson, E.L. and Oluyomi, A.O. (1997) Appetite suppression by commonly used drugs depends on 5-HT receptors but not on 5-HT availability. Trends Pharmacol. Sci., 18: 21–25.

Dempsie, Y. and MacLean, M.R. (2008) Pulmonary hypertension: therapeutic targets within the serotonin system. Br. J. Pharmacol., 155: 455–462.

Dourish, C.T., Hutson, P.H. and Curzon, G. (1985) Characteristics of feeding induced by the serotonin agonist 8-hydroxy-2-(di-n-propylamino) tetralin (8-OH-DPAT). Brain Res. Bull., 15: 377–384.

East, S.Z., Burnet, P.W., Leslie, R.A., Roberts, J.C. and Harrison, P.J. (2002) 5-HT6 receptor binding sites in schizophrenia and following antipsychotic drug administration: autoradiographic studies with [^{125}I]SB-258585. Synapse, 45: 191–199.

Ebenezer, I.S. (1992) Effects of the 5-HT1A agonist 8-OH-DPAT on food intake in food-deprived rats. Neuroreport, 3: 1019–1022.

Erickson, J.C., Clegg, K.E. and Palmiter, R.D. (1996) Sensitivity to leptin and susceptibility to seizures of mice lacking neuropeptide Y. Nature, 381: 415–421.

Fisas, A., Codony, X., Romero, G., Dordal, A., Giraldo, J., Mercé, R., Holenz, J., Vrang, N., Sørensen, R.V., Heal, D., Buschmann, H. and Pauwels, P.J. (2006) Chronic 5-HT6 receptor modulation by E-6837 induces hypophagia and sustained weight loss in diet-induced obese rats. Br. J. Pharmacol., 148: 973–983.

Fitzgerald, L.W., Burn, T.C., Brown, B.S., Patterson, J.P., Corjay, M.H., Valentine, P.A., Sun, J.H., Link, J.R., Abbaszade, I., Hollis, J.M., Largent, B.L., Hartig, P.R., Hollis, G.F., Meunier, P.C., Robichaud, A.J. and Robertson, D.W. (2000) Possible role of valvular serotonin 5-HT(2B) receptors in the cardiopathy associated with fenfluramine. Mol. Pharmacol., 57: 75–81.

Fletcher, P., Chintoh, A., Sinyard, J. and Higgins, G. (2004) Injection of the 5-HT2C receptor agonist Ro60-0175 into the ventral tegmental area reduces cocaine-induced locomotor activity and cocaine self-administration. Neuropsychopharmacology, 29: 308–318.

Frassetto, A., Zhang, J., Lao, J.Z., White, A., Metzger, J.M., Fong, T.M. and Chen, R.Z. (2008) Reduced sensitivity to diet-induced obesity in mice carrying a mutant 5-HT6 receptor. Brain Res., 1236: 140–144.

Garattini, S., Mennini, T., Bendotti, C., Invernizzi, R. and Samanin, R. (1986) Neurochemical mechanism of action of drugs which modify feeding via the serotoninergic system. Appetite, 7(Suppl): 15–38.

Garattini, S., Mennini, T. and Samanin, R. (1987) From fenfluramine racemate to d-fenfluramine. Specificity and potency of the effects on the serotoninergic system and food intake. Ann. N.Y. Acad. Sci., 499: 156–166.

Gibson, E.L., Kennedy, A.J. and Curzon, G. (1993) d-Fenfluramine- and d-norfenfluramine-induced hypophagia: differential mechanisms and involvement of postsynaptic 5-HT receptors. Eur. J. Pharmacol., 242: 83–90.

Godino, A., De Luca, L.A., Antunes-Rodrigues, J. and Vivas, L. (2007) Oxytocinergic and serotonergic systems involvement in sodium intake regulation: satiety or hypertonicity markers? Am. J. Physiol. Regul. Integr. Comp. Physiol., 293: R1027–1036.

Goodall, E.M., Cowen, P.J., Franklin, M. and Silverstone, T. (1993) Ritanserin attenuates anorectic, endocrine and thermic responses to d-fenfluramine in human volunteers. Psychopharmacology, 112: 461–466.

Goodall, E., Feeney, S., McGuirk, J. and Silverstone, T. (1992) A comparison of the effects of d- and l-fenfluramine and d-amphetamine on energy and macronutrient intake in human subjects. Psychopharmacology, 106: 221–227.

Goodall, E. and Silverstone, T. (1988a) The interaction of metergoline, a 5-HT receptor blocker, and dexfenfluramine in human feeding. Clin. Neuropharmacol., 11(Suppl. 1): 135–138.

Goodall, E. and Silverstone, T. (1988b) Differential effect of d-fenfluramine and metergoline on food intake in human subjects. Appetite, 11: 215–228.

Goto, M., Canteras, N., Burns, G. and Swanson, L. (2005) Projections from the subfornical region of the lateral hypothalamic area. J. Comp. Neurol., 493: 412–438.

Grignaschi, G. and Samanin, R. (1992) Role of 5-HT receptors in the effect of d-fenfluramine on feeding patterns in the rat. Eur. J. Pharmacol., 212: 287–289.

Grill, H.J., Donahey, J.C., King, L. and Kaplan, J.M. (1997) Contribution of caudal brainstem to d-fenfluramine anorexia. Psychopharmacology, 130: 375–381.

Grillner, S., Hellgren, J., Ménard, A., Saitoh, K. and Wikström, M.A. (2005) Mechanisms for selection of basic motor programs – roles for the striatum and pallidum. Trends Neurosci., 28: 364–370.

Gropp, E., Shanabrough, M., Borok, E., Xu, A., Janoschek, R., Buch, T., Plum, L., Balthasar, N., Hampel, B., Waisman, A.,

Barsh, G., Horvath, T. and Brüning, J. (2005) Agouti-related peptide-expressing neurons are mandatory for feeding. Nat. Neurosci., 8: 1289–1291.

Grossmann, S.P. (1960) Eating or drinking elicited by direct adrenergic or cholinergic stimulation of hypothalamus. Science, 132: 301–302.

Halford, J.C., Wanninayake, S.C. and Blundell, J.E. (1998) Behavioral satiety sequence (BSS) for the diagnosis of drug action on food intake. Pharmacol. Biochem. Behav., 61: 159–168.

Hayes, M.R. and Covasa, M. (2006) Gastric distension enhances CCK-induced Fos-like immunoreactivity in the dorsal hindbrain by activating 5-HT3 receptors. Brain Res., 1088: 120–130.

Heal, D.J., Cheetham, S.C., Prow, M.R., Martin, K.F. and Buckett, W.R. (1998) A comparison of the effects on central 5-HT function of sibutramine hydrochloride and other weight-modifying agents. Br. J. Pharmacol., 125: 301–308.

Heal, D.J., Smith, S.L., Fisas, A., Codony, X. and Buschmann, H. (2008) Selective 5-HT6 receptor ligands: progress in the development of a novel pharmacological approach to the treatment of obesity and related metabolic disorders. Pharmacol. Ther., 117: 207–231.

Heisler, L.K., Chu, H.M. and Tecott, L.H. (1998) Epilepsy and obesity in serotonin 5-HT2C receptor mutant mice. Ann. N.Y. Acad. Sci., 861: 74–78.

Heisler, L.K., Cowley, M.A., Kishi, T., Tecott, L., Fan, W., Low, M.J., Smart, J.L., Rubinstein, M., Tatro, J.B., Zigman, J.M., Cone, R. and Elmquist, J. (2003) Central serotonin and melanocortin pathways regulating energy homeostasis. Ann. N.Y. Acad. Sci., 994: 169–174.

Heisler, L.K., Cowley, M.A., Tecott, L., Fan, W., Low, M.J., Smart, J.L., Rubinstein, M., Tatro, J.B., Marcus, J.N., Holstege, H., Lee, C.E., Cone, R. and Elmquist, J. (2002) Activation of central melanocortin pathways by fenfluramine. Science, 297: 609–611.

Heisler, L., Jobst, E., Sutton, G., Zhou, L., Borok, E., Thorntonjones, Z., Liu, H., Zigman, J., Balthasar, N. and Kishi, T. (2006) Serotonin reciprocally regulates melanocortin neurons to modulate food intake. Neuron, 51: 239–249.

Hewitt, K.N., Lee, M., Dourish, C. and Clifton, P. (2002) Serotonin 2C receptor agonists and the behavioral satiety sequence in mice. Pharmacol. Biochem. Behav., 71: 691–700.

Inam, Q.U., Jabeen, B., Haleem, M.A. and Haleem, D.J. (2008) Long-term consumption of sugar-rich diet decreases the effectiveness of somatodendritic serotonin-1A receptors. Nutr. Neurosci., 11: 277–282.

Kaplan, J.M., Song, S. and Grill, H.J. (1998) Serotonin receptors in the caudal brainstem are necessary and sufficient for the anorectic effect of peripherally administered mCPP. Psychopharmacology, 137: 43–49.

Kennett, G.A. and Curzon, G. (1988) Evidence that hypophagia induced by mCPP and TFMPP requires 5-HT1C and 5-HT1B receptors; hypophagia induced by RU 24969 only requires 5-HT1B receptors. Psychopharmacology, 96: 93–100.

Kennett, G.A., Dourish, C.T. and Curzon, G. (1987) 5-HT1B agonists induce anorexia at a postsynaptic site. Eur. J. Pharmacol., 141: 429–435.

Kitchener, S.J. and Dourish, C.T. (1994) An examination of the behavioral specificity of hypophagia induced by 5-HT1B, 5-HT1C and 5-HT2 receptor agonists using the post-prandial satiety sequence in rats. Psychopharmacology (Berl), 113: 369–377.

Koe, B.K., Lebel, L.A., Fox, C.B. and Macor, J.E. (1992) Characterization of [3H]CP-96,501 as a selective radioligand for the serotonin 5-HT1B receptor: binding studies in rat brain membranes. J. Neurochem., 58: 1268–1276.

Kohen, R., Metcalf, M.A., Khan, N., Druck, T., Huebner, K., Lachowicz, J.E., Meltzer, H.Y., Sibley, D.R., Roth, B.L. and Hamblin, M.W. (1996) Cloning, characterization, and chromosomal localization of a human 5-HT6 serotonin receptor. J. Neurochem., 66: 47–56.

Kusserow, H., Davies, B., Hörtnagl, H., Voigt, I., Stroh, T., Bert, B., Deng, D.R., Fink, H., Veh, R.W. and Theuring, F. (2004) Reduced anxiety-related behavior in transgenic mice overexpressing serotonin 1A receptors. Brain Res. Mol. Brain Res., 129: 104–116.

Lam, D.D., Przydzial, M.J., Ridley, S.H., Yeo, G.S., Rochford, J.J., O'Rahilly, S. and Heisler, L.K. (2008) Serotonin 5-HT2C receptor agonist promotes hypophagia via downstream activation of melanocortin 4 receptors. Endocrinology, 149: 1323–1328.

Lam, D.D., Zhou, L., Vegge, A., Xiu, P.Y., Christensen, B.T., Osundiji, M.A., Yueh, C.Y., Evans, M.L. and Heisler, L.K. (2009) Distribution and neurochemical characterization of neurons within the nucleus of the solitary tract responsive to serotonin agonist-induced hypophagia. Behav. Brain Res., 196: 139–143.

Lawton, C.L. and Blundell, J.E. (1992) The effect of d-fenfluramine on intake of carbohydrate supplements is influenced by the hydration of the test diets. Behav. Pharmacol., 3: 517–523.

Lee, M.D., Aloyo, V.J., Fluharty, S.J. and Simansky, K.J. (1998) Infusion of the serotonin1B (5-HT1B) agonist CP-93,129 into the parabrachial nucleus potently and selectively reduces food intake in rats. Psychopharmacology, 136: 304–307.

Lee, M.D. and Clifton, P.G. (1992) Partial reversal of fluoxetine anorexia by the 5-HT antagonist metergoline. Psychopharmacology, 107: 359–364.

Lee, M.D., Kennett, G.A., Dourish, C.T. and Clifton, P.G. (2002) 5-HT1B receptors modulate components of satiety in the rat: behavioural and pharmacological analyses of the selective serotonin1B agonist CP-94,253. Psychopharmacology, 164: 49–60.

Lee, M.D. and Simansky, K.J. (1997) CP-94, 253: a selective serotonin1B (5-HT1B) agonist that promotes satiety. Psychopharmacology, 131: 264–270.

Lee, M., Somerville, E.M., Kennett, G., Dourish, C. and Clifton, P. (2004) Tonic regulation of satiety by 5-HT receptors in the mouse: converging evidence from behavioral and c-fos immunoreactivity studies? Eur. J. Neurosci., 19: 3017–3025.

Leibowitz, S.F. and Alexander, J.T. (1998) Hypothalamic serotonin in control of eating behavior, meal size, and body weight. Biol. Psychiatry, 44: 851–864.

Li, B.H. and Rowland, N.E. (1996) Effect of chronic dexfenfluramine on Fos in rat brain. Brain Res., 728: 188–192.

Li, B.H., Spector, A.C. and Rowland, N.E. (1994) Reversal of dexfenfluramine-induced anorexia and c-Fos/c-Jun expression by lesion in the lateral parabrachial nucleus. Brain Res., 640: 255–267.

Lucas, J.J., Yamamoto, A., Scearce-Levie, K., Saudou, F. and Hen, R. (1998) Absence of fenfluramine-induced anorexia and reduced c-Fos induction in the hypothalamus and central amygdaloid complex of serotonin 1B receptor knockout mice. J. Neurosci., 18: 5537–5544.

Lund, J.P. and Kolta, A. (2006) Generation of the central masticatory pattern and its modification by sensory feedback. Dysphagia, 21: 167–174.

Mazzola-Pomietto, P., Aulakh, C.S. and Murphy, D.L. (1995) Temperature, food intake, and locomotor activity effects of a 5-HT3 receptor agonist and two 5-HT3 receptor antagonists in rats. Psychopharmacology, 121: 488–493.

McGuirk, J., Goodall, E., Silverstone, T. and Willner, P. (1991) Differential effects of d-fenfluramine, l-fenfluramine and d-amphetamine on the microstructure of human eating behavior. Behav. Pharmacol., 2: 113–119.

McGuirk, J., Muscat, R. and Willner, P. (1992) Effects of the 5-HT uptake inhibitors, femoxetine and paroxetine, and a 5-HT1A/B agonist, eltoprazine, on the behavioral satiety sequence. Pharmacol. Biochem. Behav., 41: 801–805.

Montgomery, A.M. and Willner, P. (1988) Fenfluramine disrupts the behavioral satiety sequence in rats. Psychopharmacology, 94: 397–401.

Morales, M., Battenberg, E. and Bloom, F.E. (1998) Distribution of neurons expressing immunoreactivity for the 5HT3 receptor subtype in the rat brain and spinal cord. J. Comp. Neurol., 402: 385–401.

Neill, J.C. and Cooper, S.J. (1989) Evidence that d-fenfluramine anorexia is mediated by 5-HT1 receptors. Psychopharmacology, 97: 213–218.

Norgren, R., Hajnal, A. and Mungarndee, S.S. (2006) Gustatory reward and the nucleus accumbens. Physiol. Behav., 89: 531–535.

O'Neill, M.F. and Parameswaran, T. (1997) RU24969-induced behavioral syndrome requires activation of both 5HT1A and 5HT1B receptors. Psychopharmacology, 132: 255–260.

Oluyomi, A.O., Gibson, E.L., Barnfield, A.M. and Curzon, G. (1994) d-Fenfluramine and d-norfenfluramine hypophagias do not require increased hypothalamic 5-hydroxytryptamine release. Eur. J. Pharmacol., 264: 111–115.

Palacios, J.M., Waeber, C., Hoyer, D. and Mengod, G. (1990) Distribution of serotonin receptors. Ann. N.Y. Acad. Sci., 600: 36–52.

Poeschla, B., Gibbs, J., Simansky, K.J., Greenberg, D. and Smith, G.P. (1993) Cholecystokinin-induced satiety depends on activation of 5-HT1C receptors. Am. J. Physiol., 264: R62–R64.

Richter, H.G., Adams, D.R., Benardeau, A., Bickerdike, M. J., Bentley, J.M., Blench, T.J., Cliffe, I.A., Dourish, C., Hebeisen, P., Kennett, G.A., Knight, A.R., Malcolm, C.S., Mattei, P., Misra, A., Mizrahi, J., Monck, N.J., Plancher, J.M., Roever, S., Roffey, J.R., Taylor, S. and Vickers, S.P. (2006) Synthesis and biological evaluation of novel hexahydro-pyrido[3′,2′:4,5]pyrrolo[1,2-a]pyrazines as potent and selective 5-HT(2C) receptor agonists. Bioorg. Med. Chem. Lett., 16: 1207–1211.

Rogers, P.J. and Blundell, J.E. (1979) Effect of anorexic drugs on food intake and the micro-structure of eating in human subjects. Psychopharmacology, 66: 159–165.

Samanin, R., Bernasconi, S. and Garattini, S. (1975) The effect of selective lesioning of brain catecholamine-containing neurons on the activity of various anorectics in thr rat. Eur. J. Pharmacol., 34: 373–375.

Samanin, R., Ghezzi, D., Valzelli, L. and Garattini, S. (1972) The effects of selective lesioning of brain serotonin or catecholamine containing neurones on the anorectic activity of fenfluramine and amphetamine. Eur. J. Pharmacol., 19: 318–322.

Savastano, D.M., Hayes, M.R. and Covasa, M. (2007) Serotonin-type 3 receptors mediate intestinal lipid-induced satiation and Fos-like immunoreactivity in the dorsal hindbrain. Am. J. Physiol. Regul. Integr. Comp. Physiol., 292: R1063–R1070.

Sawchenko, P.E. and Swanson, L.W. (1982) The organization of noradrenergic pathways from the brainstem to the paraventricular and supraoptic nuclei in the rat. Brain Res., 257: 275–325.

Simansky, K.J. (1996) Serotonergic control of the organization of feeding and satiety. Behav. Brain Res., 73: 37–42.

Simansky, K.J. and Vaidya, A.H. (1990) Behavioral mechanisms for the anorectic action of the serotonin (5-HT) uptake inhibitor sertraline in rats: comparison with directly acting 5-HT agonists. Brain Res. Bull., 25: 953–960.

Somerville, E.M., Horwood, J., Lee, M., Kennett, G. and Clifton, P. (2007) 5-HT(2C) receptor activation inhibits appetitive and consummatory components of feeding and increases brain c-fos immunoreactivity in mice. Eur. J. Neurosci., 25: 3115–3124.

Stallone, D., Nicolaïdis, S. and Gibbs, J. (1989) Cholecystokinin-induced anorexia depends on serotoninergic function. Am. J. Physiol., 256: R1138–R1141.

Sugimoto, Y., Yamada, J., Yoshikawa, T., Noma, T. and Horisaka, K. (1996) Effects of peripheral 5-HT2 and 5-HT3 receptor agonists on food intake in food-deprived and 2-deoxy-D-glucose-treated rats. Eur. J. Pharmacol., 316: 15–21.

Tecott, L.H., Sun, L.M., Akana, S.F., Strack, A.M., Lowenstein, D.H., Dallman, M.F. and Julius, D. (1995) Eating disorder and epilepsy in mice lacking 5-HT2c serotonin receptors. Nature, 374: 542–546.

Thomsen, W.J., Grottick, A.J., Menzaghi, F., Reyes-Saldana, H., Espitia, S., Yuskin, D., Whelan, K., Martin, M., Morgan, M., Chen, W., Al-Shamma, H., Smith, B., Chalmers, D. and Behan, D. (2008) Lorcaserin, a novel selective human 5-hydroxytryptamine2C agonist: *in vitro* and *in vivo* pharmacological characterization. J. Pharmacol. Exp. Ther., 325: 577–587.

Tricklebank, M.D., Forler, C. and Fozard, J.R. (1984) The involvement of subtypes of the 5-HT1 receptor and of catecholaminergic systems in the behavioral response to 8-hydroxy-2-(di-n-propylamino)tetralin in the rat. Eur. J. Pharmacol., 106: 271–282.

Trifunovic, R. and Reilly, S. (2001) Medial versus lateral parabrachial nucleus lesions in the rat: effects on cholecystokinin- and D-fenfluramine-induced anorexia. Brain Res., 894: 288–296.

Törk, I. (1990) Anatomy of the serotonergic system. Ann. N.Y. Acad. Sci., 600: 9–34, discussion 34–35

van der Hoek, G.A. and Cooper, S.J. (1994) Ondansetron, a selective 5-HT3 receptor antagonist, reduces palatable food consumption in the nondeprived rat. Neuropharmacology, 33: 805–811.

van Rossum, J.M. and Simons, F. (1969) Locomotor activity and anorexogenic action. Psychopharmacologia, 14: 248–254.

Vickers, S.P., Clifton, P.G. and Dourish, C.T. (1996) Behavioral evidence that d-fenfluramine-induced anorexia in the rat is not mediated by the 5-HT1A receptor subtype. Psychopharmacology, 125: 168–175.

Vickers, S.P., Clifton, P.G., Dourish, C.T. and Tecott, L.H. (1999) Reduced satiating effect of d-fenfluramine in serotonin 5-HT(2C) receptor mutant mice. Psychopharmacology, 143: 309–314.

Vickers, S.P., Dourish, C.T. and Kennett, G.A. (2001) Evidence that hypophagia induced by d-fenfluramine and d-norfenfluramine in the rat is mediated by 5-HT2C receptors. Neuropharmacology, 41: 200–209.

Voigt, J.P., Schade, R., Fink, H. and Hörtnagl, H. (2002) Role of 5-HT1A receptors in the control of food intake in obese Zucker rats of different ages. Pharmacol. Biochem. Behav., 72: 403–409.

Willner, P., McGuirk, J., Phillips, G. and Muscat, R. (1990) Behavioral analysis of the anorectic effects of fluoxetine and fenfluramine. Psychopharmacology, 102: 273–277.

Woolley, M.L., Bentley, J.C., Sleight, A.J., Marsden, C.A. and Fone, K.C. (2001) A role for 5-ht6 receptors in retention of spatial learning in the Morris water maze. Neuropharmacology, 41: 210–219.

Wright, D.E., Seroogy, K.B., Lundgren, K.H., Davis, B.M. and Jennes, L. (1995) Comparative localization of serotonin1A, 1C, and 2 receptor subtype mRNAs in rat brain. J. Comp. Neurol., 351: 357–373.

Wurtman, J., Wurtman, R., Mark, S., Tsay, R., Gilbert, W. and Growdon, J. (1985) d-Fenfluramine selectively suppresses carbohydrate snacking by obese subjects. Intl. J. Eating Disord., 4: 89–99.

Young, R.C., Gibbs, J., Antin, J., Holt, J. and Smith, G.P. (1974) Absence of satiety during sham feeding in the rat. J. Comp. Physiol. Psychol., 87: 795–800.

Zhuang, X., Gross, C., Santarelli, L., Compan, V., Trillat, A.C. and Hen, R. (1999) Altered emotional states in knockout mice lacking 5-HT1A or 5-HT1B receptors. Neuropsychopharmacology, 21: 52S–60S.

CHAPTER 3.4

Serotonin and Sexual Behavior

Lynda Uphouse* and Jutatip Guptarak

Department of Biology, Texas Woman's University, Denton, Texas

Abstract: Serotonin's (5-HT) classical role in regulation of sexual behavior is to inhibit copulation. However, the neurotransmitter exerts a more complex influence on the behavior dependent on the particular aspect of the behavior examined. In this chapter, the historical basis for consideration of 5-HT as a major player in sexual behavior is discussed. Current information about the role of 5-HT in the sexual dysfunction that is associated with the use of selective serotonin reuptake inhibitors is reviewed. Information on specific 5-HT receptor subtypes, though far from complete, implicates 5-HT_{1A}, 5-HT_{1B}, 5-HT_{2A}, 5-HT_{2C}, 5-HT_3 and 5-HT_7 receptors as potential mediators of 5-HT's influence on sexual behavior. Evidence for the involvement of each of these receptor subtypes is reviewed, and their different effects in males and females examined. Finally, the possible biological significance of serotonin's contribution to sexual behavior is examined.

Keywords: lordosis, ejaculation, erection, sexual motivation, serotonin receptors, SSRIs, premature ejaculation, appetitive and consummatory sexual behavior.

Abbreviations: 5-CT, 5-carboxyamidetryptamine – non-selective, but 5-HT_7 receptor agonist; 5-HT, serotonin; 5-HTP, 5-hydroxytryptophan; 5,7-DHT, 5,7-dihydroxytryptamine – serotonin neurotoxin; 8-OH-DPAT, 8-hydroxy-2-(di-N-propylamino) tetralin – 5-HT_{1A} receptor agonist; AH, anterior hypothalamus; CNS, central nervous system; CGS-12066A, 7-trifluoromethyl-4(4-methyl-1-piperazinyl)-pirrolo[1,2a]quinoxaline maleate – 5-HT_{1B} receptor agonist; DHT, dihydrotestosterone; DOI, (+/−)-1-(2,5-dimethoxy-4-iodophenyl)-2-aminopropane HCl – 5-$HT_{2A/2C}$ receptor agonist; DRN, dorsal raphe nucleus; GR 127935, N-[4-methoxy-3-(4-methyl-1-piperazinyl)phenyl]-2′-methyl-4′-(5-methyl-1,2,4-oxadiazol-3-yl)-1-1′-biphenyl-4-carboxamide hydrochloride – 5-$HT_{1B/1D}$ receptor antagonist; KOIB, 5-HT_{1B} receptor knockout; L/M, lordosis to mount ratio; MBH, mediobasal hypothalamus; mCPP, 1-(m-chlorophenyl) piperazine – non-selective, but 5-HT_{1B} and 5-HT_{2C} receptor agonist; MDL 100907, (±)-2,3-dimethoxyphenyl)-1-[2-(4-piperdine)-methanol] – 5-HT_{2A} receptor antagonist; MDMA, 3,4 methylenedioxymethamphetamine, 'ecstasy' – 5-HT depleting agent; mPOA, medial preoptic area; mPOA-AH, medial preoptic area-anterior hypothalamus; nPGi, nucleus paragigantocellularis; MRN, median raphe nucleus; POA, preoptic area; PAG, periaqueductal gray; PCPA, parachlorophenylalanine – tryptophan hydroxylase inhibitor; PVN, paraventricular nucleus of the hypothalamus; RS 102221, 8-[5-(2,4-dimethoxy-5-(4-trifluoro methylphenylsulfonamido)phenyl-5-oxopentyl)]-1,3,8-triazaspiro[4.5]decane-2,4-dione hydrochloride – 5-HT_{2C} receptor antagonist; SB206553, N-3-pyridinyl-3,5-dihydro-5-methyl-benzo(1,2-b:4,5-b′)dipyrrole-(2 H)carboxamide – 5-$HT_{2B/2C}$ receptor antagonist; SB 269970A, 2-(2-(4-methyl-piperidin-l-yl)-pyrrolidine-1-sulfonyl)-phenol – selective 5-HT_7 receptor antagonist; SERT, serotonin transporter; SSRIs, selective serotonin reuptake inhibitors; TFMPP, 1-[m-trifluoro methylphenyl]piperazine – non-selective, but 5-HT_{1B} and 5-HT_{2C} receptor agonist; VMN, ventromedial nucleus of the hypothalamus; WAY100635, N-[2[4-(2-methoxyphenyl)-1-piperazinyl]ethyl]N-(2-pyridinyl)cyclohexanecarboxamide trihydrochloride – 5-HT_{1A} receptor antagonist; WAY100135, chiral N-t-butyl-3-(1-(4-(2-methoxy) phenyl)piperazinyl)-1-phenylpropionamide dihydrochloride, quarter hydrate – 5-HT_{1A} receptor antagonist, but partial agonist.

Introduction

Serotonin (5-HT) generally exerts inhibitory control over sexual behavior, but, depending on the 5-HT receptor subtype involved, gender, and the behavioral parameter investigated, 5-HT's role is considerably more complex. Global increases in 5-HT inhibit male or female sexual behavior while reductions in 5-HT increase sexual behavior, but effects of 5-HT receptor selective agents reveal both inhibitory and facilitatory effects (for review, see Mendelson, 1992; Uphouse, 2000; Hull *et al.*, 2004;

*Corresponding author
E-mail: luphouse@mail.twu.edu

ISBN 978-0-12-374634-4
DOI: 10.1016/B978-0-12-374634-4.00021-6

de Jong *et al.*, 2006). Furthermore, effects of 5-HT vary depending on the specific aspect of sexual behavior examined. More emphasis has been placed on consummatory behavior than on sexual motivation. In the following discussion, 5-HT's contribution to male and female sexual behavior will be examined from a historical perspective, leading to current understanding of the subject. Males and females will be discussed separately, followed by an overview of the biological significance of 5-HT's role in sexual behavior and possible directions for future research.

Female sexual behavior

Description of female sexual behavior

Estrogen and progesterone coordinate the female reproductive cycle so that mating and ovulation are synchronized to optimize fertilization of the ruptured ova (Sodersten, 1981; Pfaff and Modianos, 1985). The complete repertoire of female sexual behavior includes solicitation, proceptivity (including sexual motivation) and receptivity (Sodersten, 1981; Erskine, 1989). Proceptive and solicitous behaviors advertise the female's readiness to mate. Receptivity refers to the female's consummatory response (the lordosis reflex in rodents) required for successful mating to occur. The lordosis quotient (or lordosis to mount (L/M) ratio × 100) has been the most commonly assessed parameter of female sexual behavior. Solicitation and proceptivity have been measured by hopping/darting and ear-wiggling in rodents, or by the female's pacing of her mating activity (Erskine, 1989). Motivated behavior has been studied primarily in the partner-preference paradigm or pacing paradigm.

The lordosis reflex is a stereotyped posture made by the female in response to displacement of mechanoreceptors in the flank, rump, tail base and perineum (Pfaff and Modianos, 1985). In non-primate mammals, hormonal priming is required for female sexual behavior to occur, and ovariectomy leads to abrupt cessation of the behavior (Sodersten, 1981; Pfaff and Modianos, 1985). In a naturally cycling female rat, both estrogen and progesterone contribute to the elicitation of sexual receptivity (Sodersten, 1981). In ovariectomized rats, estrogen, alone, facilitates lordosis behavior, but progesterone reduces the dose of estrogen priming that is required and is essential for the full repertoire of female sexual behavior (Sodersten, 1981).

The basic neuronal circuitry for lordosis is contained in the spinal cord. An abbreviated lordosis response can be obtained in a spinal preparation (McKenna *et al.*, 1991a), but forebrain areas are required for complete expression of the reflex (Pfaff and Modianos, 1985). Many brain areas contribute to the complete behavioral repertoire of the sexually receptive female (Cottingham and Pfaff, 1986; Daniels *et al.*, 1999; Flanagan-Cato *et al.*, 2006), but diencephalic sites (including the preoptic area, POA, and ventromedial nucleus of the hypothalamus, VMN), midbrain and brainstem sites are crucial for hormonal control of lordosis (Pfaff and Modianos, 1985; Daniels and Flanagan-Cato, 2000; Flanagan-Cato *et al.*, 2006). Input from multiple brain areas converges in the VMN (Cottingham and Pfaff, 1986; Flanagan-Cato *et al.*, 2006), which typically has been viewed as a site where hormonal priming acts to increase the lordosis reflex (Pfaff and Modianos, 1985). Pathways from the VMN descend to the midbrain (especially to the periaqueductal gray, PAG) where integration of the sensory–motor reflex occurs (Canteras *et al.*, 1994; Daniels *et al.*, 1999).

Implications for a role of 5-HT in female sexual behavior

Early investigators noted that the administration of drugs such as reserpine, tetrabenazine or parachlorophenylalanine (PCPA), which reduce the overall concentration of monoamines, activated lordosis in estrogen-primed, ovariectomized females as effectively as did progesterone. Moreover, a variety of non-selective 5-HT receptor antagonists were reported to increase consummatory, but not solicitous, behavior. A thorough review of these early studies can be found in Mendelson (1992). In general, any treatment which reduced 5-HT activity or which decreased overall monoamine concentrations increased lordosis, while treatments that increased 5-HT activity reduced the lordosis reflex. However, drug effects interact with the nature and degree of hormonal priming, with both estrogen and progesterone offering some protection against the effects of serotonergic agents (Trevino *et al.*, 1999; Jackson and Etgen, 2001; Sinclair-Worley and Uphouse, 2004).

Direct infusion of 5-HT or 5-HT-active compounds into various brain areas implicated the mediobasal hypothalamus (MBH) (especially the VMN) as a site where 5-HT-mediated inhibition occurred (Foreman and Moss, 1978; Allen *et al.*, 1993; Frankfurt *et al.*, 1994), and it was suggested that 5-HT levels in the MBH were inversely correlated with lordosis (Luine and Paden, 1982; Allen *et al.*, 1993; Farmer *et al.*, 1996). Consistent with this suggestion, a reduction of 5-HT input to the MBH by lesioning of 5-HT neurons with 5,7-dihydroxytryptamine (5,7-DHT) reduced the dose of estrogen that was required to facilitate the lordosis reflex (Luine *et al.*, 1984; Frankfurt *et al.*, 1985), and this enhanced response to estrogen was lost when 5-HT axons reinnervated the hypothalamus (Frankfurt *et al.*, 1994). Moreover, the concentration of

5-HT available to activate postsynaptic receptors in the MBH declines near the time of sexual receptivity (Farmer *et al.*, 1996; Maswood *et al.*, 1999). Progesterone, after estrogen priming, rather than estrogen alone, may be responsible for the 5-HT decline (Gereau *et al.*, 1993; Farmer *et al.*, 1996; Maswood *et al.*, 1999; Lu and Bethea, 2002). However, estrogen may influence the temporal availability of 5-HT by modulation of 5-HT reuptake into nerve terminals (McQueen *et al.*, 1997; Bethea *et al.*, 2002; Bertrand *et al.*, 2005; Benmansour *et al.*, 2008).

If an increase in 5-HT decreases sexual behavior, then blocking the serotonin transporter (SERT) should reduce sexual behavior. This was appreciated in early studies with non-specific monoamine reuptake inhibitors, and later with selective serotonin reuptake inhibitors (SSRIs). Due to the sexual dysfunction associated with their use, there has been resurgence in interest regarding SSRIs, which are used to treat a variety of mood disorders especially prevalent in women (Gregorian *et al.*, 2002; Montgomery *et al.*, 2002). In animal studies, both acute and chronic treatment with SSRIs reduces female rat lordosis behavior but there is considerable variation in the timing and magnitude of the response (Matuszczyk *et al.*, 1998; Frye *et al.*, 2003; Sarkar *et al.*, 2008a). For example, Matuszczk *et al.* (1998) treated ovariectomized rats daily with 10 mg/kg fluoxetine, and tested sexual behavior after weekly priming with estrogen and progesterone. Effects on sexual receptivity were small, and did not develop until approximately 21 days of treatment. In contrast, a single treatment with fluoxetine reduced lordosis behavior in hormonally-primed, ovariectomized female rats (Frye *et al.*, 2003; Sarkar *et al.*, 2008a) and lateral displacement in female hamsters (Frye and Rhodes, 2003). After longer treatment in rats, fluoxetine's inhibitory effects were attenuated (Frye *et al.*, 2003; Sarkar *et al.*, 2008a). These findings differ from the human literature, where sexual side effects are often reported to continue through the duration of SSRI treatment. However, few studies have been designed to evaluate human female sexual activity prior to the onset of SSRI treatment. Some investigators report improvement after chronic fluoxetine treatment, while others report continued sexual dysfunction.

The degree to which SSRI effects on the neuroendocrine system contribute to their effects on sexual behavior in rodent models is still uncertain, and little information is available in the human population. In animal models, some investigators have reported normal estrous cyclicity when female rats were treated daily with fluoxetine (Matuszczyk *et al.*, 1998; Heisler *et al.*, 1999; Van de Kar *et al.*, 2002), while disruption of estrous cyclicity and sexual behavior was reported in Fischer female rats, with the implication that the SSRI's effect on sexual behavior was not independent from the estrous cycle disruption (Uphouse *et al.*, 2006; Sarkar *et al.*, 2008b). However, strain differences may account for the variable findings (Maswood *et al.*, 2008). Neural sites responsible for SSRI-induced female sexual behavior have not been identified, but effects have been attributed to CNS and/or peripheral effects of the SSRI-induced elevation in 5-HT (Clayton, 2002; Frohlich and Meston, 2005). Additional studies will be required to determine if sexual behavioral effects are the direct result of elevations in CNS 5-HT, or consequences of endocrine or other peripheral or neural modifications that occur following treatment with SSRIs.

5-HT receptors and female consummatory behaviors

The 5-HT_{1A}, 5-HT_{1B}, 5-$HT_{2A/2C}$, 5-HT_3 and 5-HT_7 receptors have been implicated in the modulation of lordosis behavior. The 5-HT_1 and 5-HT_2 families have received the most attention, leading to the conclusion that activation of 5-HT_{1A} receptors inhibits lordosis behavior while activation of 5-HT_2 receptors facilitates the reflex (for review, see Mendelson, 1992; Uphouse, 2000). Similar to 5-HT_{1A} receptors, 5-HT_7 receptors may also inhibit lordosis behavior (Siddiqui and Shaharyar, 2007; Siddiqui *et al.*, 2007). Activation of the ligand-gated 5-HT_3 receptor in the MBH may facilitate lordosis behavior (Maswood *et al.*, 1997, 1998), but systemic treatments with 5-HT_3 receptors agonists have generally been ineffective in altering the lordosis response (Tanco *et al.*, 1993, 1994).

5-HT_{1A} receptors

Several investigators have reported that systemic administration of 5-HT_{1A} receptor agonists reduced lordosis behavior of ovariectomized females, hormonally primed with estrogen or with estrogen and progesterone, and effects were attenuated by a variety of 5-HT_{1A} receptor antagonists (for review, see Mendelson, 1992; Uphouse, 2000). Since systemic administration of 5-HT_{1A} receptor agonists activates 5-HT_{1A} receptors that are postsynaptic to 5-HT terminals as well as 5-HT_{1A} somatodendritic autoreceptors, localization of the effective sites is important in understanding the role of 5-HT in the modulation of lordosis behavior. At the postsynaptic target, 5-HT_{1A} receptor agonists simulate effects of increased 5-HT release from nerve terminals. However, at soma and dendrites of 5-HT neurons, activation of 5-HT_{1A} autoreceptors decreases firing of 5-HT neurons and decreases release of 5-HT from synaptic terminals (Sprouse and Aghajanian, 1987; Hjorth and Sharp, 1991; Adell *et al.*, 1993). Low doses of ipsapirone or gespirone, but not other 5-HT_{1A} receptor agonists, were reported to facilitate lordosis behavior of ovariectomized rats primed only with estrogen, but not in rats primed with estrogen and progesterone (Mendelson and Gorzalka, 1986a,b). Because somatodendritic 5-HT_{1A} autoreceptors may be especially sensitive to 5-HT_{1A} receptor agonists (Kennett *et al.*, 1987; Beer *et al.*, 1990), facilitation of lordosis behavior after a low dose of these drugs might reflect activation of somatodendritic autoreceptors and a consequent reduction of 5-HT release in terminal fields. To our knowledge, direct application of a low dose of 5-HT_{1A} receptor agonists to the somatodendritic autoreceptors has not been examined in a suboptimally hormone-primed animal, so the importance of these autoreceptors to potential facilitation of lordosis behavior has yet to be determined.

Direct infusion of the 5-HT_{1A} receptor agonist 8-OH-DPAT (8-hydroxy-2-(di-N-propylamino) tetralin) into the dorsal raphe nucleus (DRN) had no effects on lordosis behavior of sexually receptive female rats (Uphouse *et al.*, 1992). However, some decline was evident following infusion of 8-OH-DPAT into the median raphe nucleus (MRN) (Uphouse *et al.*, 1994a). In contrast, infusion of 8-OH-DPAT into regions terminal to 5-HT neurons produced robust inhibition of lordosis behavior (Gonzalez *et al.*, 1997; Uphouse, 2000). Moreover, inhibition was observed in rats with intrahypothalamic 5,7-dihydroxytryptamine (5,7-DHT) lesions (Aiello-Zaldivar *et al.*, 1992) or in rats with lesions of the DRN (Kishitake and Yamanouchi, 2005). Effective sites for 5-HT_{1A} receptor agonists include the medial preoptic area (mPOA), VMN, and ventrolateral PAG (Gonzalez *et al.*, 1997; Uphouse, 2000; Caldwell and Albers, 2002), while infusion into the dorsal or anterior hypothalamus (AH) is relatively ineffective (Gonzalez *et al.*, 1997; Trevino *et al.*, 1999; Guptarak *et al.*, 2004). Of terminal areas, the VMN is the brain region in the rat where the lowest dose of 8-OH-DPAT inhibits lordosis behavior. For this brain area, Uphouse and colleagues have argued that the 5-HT_{1A} receptor provides a brake on sexual receptivity under conditions of high risk to the female (Uphouse, 2000). However, in female hamsters, infusion of 8-OH-DPAT into the mPOA-AH reduced mean lordosis duration but increased the number of lordosis episodes during the observation session (Caldwell and Albers, 2002). These authors suggested a role for mPOA 5-HT_{1A} receptors in the regulation of the temporal pacing of lordosis.

If activation of 5-HT_{1A} receptors tonically decreases lordosis behavior, treatment with 5-HT_{1A} receptor antagonists might increase lordosis. The selective 5-HT_{1A} receptor antagonist WAY100635 (N-[2[4-(2-methoxyphenyl)-1-piperazinyl]ethyl]N-(2-pyridinyl)cyclohexanecarboxamide trihydrochloride) did increase lordosis in ovariectomized rats that had a low initial sexual receptivity (Kishitake and Yamanouchi, 2004), but when suboptimally hormonally-primed, ovariectomized rats received VMN infusion with WAY100635, lordosis behavior was not facilitated (Uphouse and Wolf, 2004). However, some facilitation was observed following infusion with the partial agonist WAY100135 (chiral N-t-butyl-3-(1-(4-(2-methoxy) phenyl)piperazinyl)-1-phenylpropionamide dihydrochloride, quarter hydrate) (Uphouse *et al.*, 1996). The effects of a 5-HT_{1A} receptor antagonist may vary as a function of the female's initial level of receptivity (e.g., if receptivity is too low, facilitation may not be possible; and if receptivity is high, a further increase is not possible). Nevertheless, 5-HT_{1A} receptors do not appear to exert tonic inhibition of lordosis behavior but may increase the lordosis threshold. In that case, a 5-HT_{1A} receptor antagonist could reduce the amount of estrogen and progesterone priming required to facilitate the reflex. Such a change in sensitivity to estrogen priming is thought to be responsible for the enhanced lordosis seen after 5,7-DHT lesion (Frankfurt *et al.*, 1985; Luine *et al.*, 1987). However, this hypothesis has not been tested directly.

The potency of 5-HT_{1A} receptor agonists for inhibiting sexual behavior does, however, decrease with an increase in hormonal priming, and both estrogen and progesterone appear to be contributors (Jackson and Uphouse, 1998; Jackson and Etgen, 2001; Truitt *et al.*, 2003). Although estrogen does not alter the density of 5-HT_{1A} receptors (Jackson and Etgen, 2001; Landry and Di Paolo, 2003), it may reduce receptor coupling to effector G-proteins (Raap *et al.*, 2000; Mize *et al.*, 2003) and thereby alter the effectiveness of the 5-HT_{1A} receptor agonists.

While there is general agreement about the inhibitory effects of 8-OH-DPAT on female sexual behavior in a variety of species, facilitation of sexual receptivity was reported in the female ferret (Paredes *et al.*, 1994). The opposite effect of the drug in ferrets and rats may reflect a species difference in the response to the drug (e.g., male ferrets also respond to 8-OH-DPAT in a manner opposite to that of rats; see male sexual behavior section below). Whether this reflects species differences in the behavioral pattern typical of sexual behavior, a difference in neuronal circuitry underlying the control of sexual behavior, or different pharmacokinetics of the drug is uncertain.

Since 8-OH-DPAT also has affinity for 5-HT_7 receptors (Cornfield *et al.*, 1991), it is difficult to exclude 5-HT_7 receptors in all the reported inhibitory effects of 8-OH-DPAT. However, pindolol blocked the effects of systemic 8-OH-DPAT treatment, and in the VMN, 5-HT_7 receptors probably make minimal contribution to the effects of 8-OH-DPAT. In this brain area, both WAY100635 and pindolol, neither of which are active at 5-HT_7 receptors (Lovenberg *et al.*, 1993; Fletcher *et al.*, 1996), attenuated effects of 8-OH-DPAT (Uphouse *et al.*, 1996; Uphouse and Wolf, 2004).

In summary, most evidence is consistent with an inhibitory role of 5-HT_{1A} receptors in female sexual behavior, and receptors postsynaptic to 5-HT terminals appear to mediate the inhibitory effect.

5-HT_{1B} receptors

Since terminal 5-HT_{1B} receptors function as autoreceptors to control 5-HT release (Sari, 2004), they might be expected to play a positive role in control of lordosis behavior. In early experiments with non-selective 5-HT_{1B} receptor agonists such as mCPP (1-(m-chlorophenyl) piperazine) or TFMPP (1-[m-trifluoromethyl-phenyl]piperazine), increases in lordosis behavior were reported (for review, see Mendelson, 1992). In rats with intrahypothalamic 5,7-DHT lesions, TFMPP facilitated lordosis behavior (Aiello-Zaldivar *et al.*, 1992) and infusion of TFMPP into the VMN of suboptimally-hormone primed rats increased lordosis behavior (Wolf *et al.*, 1998). However, Wolf *et al.* (1998) attributed these effects to actions of TFMPP at 5-HT_2 receptors. Since most of the drugs originally implicating 5-HT_{1B} receptors in lordosis behavior also share agonist action at multiple 5-HT receptors (Sari, 2004), the role of 5-HT_{1B} receptors has been difficult to identify.

However, recently the 5-$HT_{1B/1D}$ receptor antagonist N-[4-methoxy-3-(4-methyl-1-piperazinyl)phenyl]-2′-methyl-4′-(5-methyl-1,2,4-oxadiazol-3-yl)-1-1′-biphenyl-4-carboxamide hydrochloride (GR 127935) was found to inhibit lordosis behavior of ovariectomized rats, hormonally primed only with estrogen (Uphouse *et al.*, 2008). No effects were present in rats hormonally primed with estrogen and progesterone. Although these findings are consistent with earlier speculations for a facilitatory effect of 5-HT_{1B} receptors (Mendelson, 1992), additional work is required before such a role can be definitively identified. In particular, since GR 127935 can act as a partial agonist under some conditions (Beer *et al.*, 1998; de Groote *et al.*, 2003), it is imperative that additional selective 5-HT_{1B} receptor drugs be examined.

5-HT_2 receptors

A possible lordosis-facilitatory role for 5-$HT_{2A/2C}$ receptors stemmed from findings that systemic treatment with non-selective 5-$HT_{2A/2C}$ receptor antagonists inhibited lordosis in ovariectomized rats, hormonally primed with estrogen or with estrogen and progesterone, and that 5-$HT_{2A/2C}$ receptor agonists facilitated lordosis in ovariectomized rats primed only with estrogen (for review, see Mendelson, 1992). Intracranial studies with the 5-$HT_{2A/2C}$ receptor selective agonist DOI ((+/−)-1-(2,5-dimethoxy-4-iodophenyl)-2-aminopropane HCl) support such a facilitative role (Gonzalez *et al.*, 1997; Wolf *et al.*, 1998). However, in females that were already sexually receptive, DOI did not alter lordosis behavior but increased hopping and darting behavior (Rossler *et al.*, 2006). Since females showed high L/M ratios (0.7–0.8) prior to treatment with DOI, additional facilitation may not have been observable.

The VMN (Wolf *et al.*, 1998) and mPOA (Gonzalez *et al.*, 1997) have been implicated as neural sites where 5-HT_2 receptor-mediated facilitation of lordosis occurs, but it remains unclear if 5-HT_{2A} or 5-HT_{2C} receptors are responsible. The 5-HT_{2A} receptor antagonist MDL 100907 ((±)-2,3-dimethoxyphenyl)-1-[2-(4-piperdine) methanol) did not alter the effect of DOI following infusion into the VMN, but the selective 5-HT_{2C} receptor antagonist SB206553 (N-3-pyridinyl-3,5-dihydro-5-methyl-benzo (1,2-b:4,5-b′)dipyrrole-(2 H)carboxamide) reduced the facilitation (Wolf *et al.*, 1999). However, in the mPOA, facilitatory effects of DOI were attributed to activation of 5-HT_{2A}, rather than 5-HT_{2C}, receptors (Gonzalez *et al.*, 1997). In female hamsters, infusion of DOI into the mPOA-AH increased mean lordosis duration but reduced the number of lordosis episodes, and there was no effect of mCPP. Since mCPP has agonist action at 5-HT_{2C} receptors, these authors concluded that it was the 5-HT_{2A}, and not the 5-HT_{2C}, receptor that was critical to the mPOA facilitation in hamsters (Caldwell and Albers, 2002). Given the potential role of the mPOA in female rat sexual motivation (Guarraci *et al.*, 2004) and the established role for the VMN in lordosis behavior (Pfaff and Modianos, 1985), it is possible that 5-HT_{2A} receptors in the mPOA are important for sexual motivation while 5-HT_{2C} receptors in the VMN facilitate lordosis behavior. Additional studies are required to pursue these possibilities.

5-HT_3 receptors

5-HT_3 receptor antagonists were initially reported to have no effect on sexual behavior of sexually receptive females, but to increase lordosis in non-receptive rats (discussed in Tanco *et al.*, 1993). In an attempt to replicate these initial findings, no effects of 5-HT_3 receptor active compounds were found (Tanco *et al.*, 1993, 1994). In contrast, a substantial decline of lordosis behavior and lordosis quality was found in hormonally primed ovariectomized rats after VMN infusion of the 5-HT_3 receptor antagonist tropisetron, and effects were attenuated by co-infusion with the 5-HT_3 receptor agonist m-chlorophenylbiguanide, but in proestrous rats, the inhibitory effects of tropisetron were not dose-dependent and showed a considerably lower potency (Maswood *et al.*, 1997, 1998). However, a facilitatory role for the 5-HT_3 receptor in the periphery has been suggested (McKenna *et al.*, 1991b). These authors argued that the 5-HT_3 receptor might sensitize sensory fibers and thereby reduce the amount of stimulation required to elicit the reflex (McKenna *et al.*, 1991b).

In summary, in comparison to the 5-HT_1 and 5-HT_2 family of receptors, the ligand-gated 5-HT_3 receptor may play a minor role in female sexual behavior. Nevertheless, given the paucity of attempts to identify any potential role for $5\text{-}HT_3$ receptors in female sexual behavior, their apparent lack of importance to the behavior should be evaluated cautiously.

$5\text{-}HT_7$ receptors

To our knowledge, only two studies have been designed to evaluate effects of $5\text{-}HT_7$ receptors on lordosis behavior. In each, the authors based their conclusions upon a comparison of effects of 8-OH-DPAT and 5-CT (5-carboxyamidetryptamine) (Siddiqui and Shaharyar, 2007; Siddiqui *et al.*, 2007). Both drugs reduced lordosis behavior, and the $5\text{-}HT_{1A}$ receptor antagonist (partial agonist) WAY100135 attenuated the decline. A $5\text{-}HT_7$ receptorantagonist,SB269970A(2-(2-(4-methyl-piperidin-1-yl)-pyrrolidine-1-sulfonyl)-phenol), attenuated effects of 5-CT but not of 8-OH-DPAT. In non-receptive rats, both WAY100135 and SB 269970A produced a slight facilitation of lordosis behavior. These investigators concluded that activation of the $5\text{-}HT_7$ receptor may inhibit female sexual behavior like $5\text{-}HT_{1A}$ receptors, and, similar to the effects of estrogen on $5\text{-}HT_{1A}$ receptors, the effects of 5-CT were partially attenuated by a high dose of estrogen (Siddiqui and Shaharyar, 2007; Siddiqui *et al.*, 2007).

Receptor interaction

Although it has been convenient to attempt to dissect the influence of the various 5-HT receptors to the neurotransmitter's modulation of female sexual behavior, it is, at best, an imprecise view of the endogenous role played by 5-HT. Endogenous 5-HT acts upon all available 5-HT receptors so that the net behavioral outcome is a composite of all the information received from receptor activation. For example, while 8-OH-DPAT inhibits lordosis behavior following infusion into the VMN, co-infusion with either $5\text{-}HT_2$ receptor agonists or a $5\text{-}HT_3$ receptor agonist can prevent the lordosis inhibition (Uphouse *et al.*, 1994b; Maswood *et al.*, 1998). Similarly, the $5\text{-}HT_{2A/2C}$ receptor agonist DOI can attenuate the reduction in lordosis behavior initiated by the $5\text{-}HT_3$ receptor antagonist, tropisetron (Maswood *et al.*, 1998). Therefore, composite effects of the individual receptors are likely to provide important clues as to how the endogenous system is actually functioning.

It is also increasingly clear that peripheral effects of 5-HT contribute to the neurotransmitter's modulation of female sexual behavior (Frohlich and Meston, 2000). Emphasis on central mechanisms has been important in identifying sites and mechanisms of action responsible for CNS mediation of the behavioral change in response to 5-HT, but in the human population, pharmaceutical agents modifying 5-HT function are administered systemically; therefore, the contribution of peripheral events is equally important (Frohlich and Meston, 2000, 2005).

5-HT and female sexual motivation

There have been surprisingly few studies designed to examine effects of 5-HT on female sexual motivation. In a pacing paradigm, when the latency for females to return to the male after an ejaculation was examined, PCPA increased the return duration, interpreted as a reduction in sexual motivation (Emery and Moss, 1984). However, given the potential effects of 5-HT on vaginal sensitivity and/or response to stimuli (Frohlich and Meston, 2000), it is difficult to conclude that control and experimental females received equivalent sensory stimuli.

In a partner-preference paradigm, however, the SSRI fluoxetine was reported to reduce the female's preference for spending time with the male (Matuszczyk *et al.*, 1998). The difference was present only at the first test, after 7 days of fluoxetine treatment; thereafter, male preference scores for the control females declined to the same level as the fluoxetine-treated group and remained relatively constant for the duration of treatment. A fluoxetine-induced decline in lordosis behavior did not develop for at least 7 days after the first male preference test.

In contrast to the study by Matuszczyk *et al.* (1998), Sarkar *et al.* (2008a) found no decline in male preference scores when fluoxetine-treated, ovariectomized rats were hormonally primed with estrogen and progesterone. The differences in these findings may reflect the use of intact rats by Matuszczyk *et al.* (1998) and ovariectomized rats by Sarkar *et al.* (2008a). If, as previously suggested (Uphouse *et al.*, 2006), fluoxetine has the potential to disrupt the normal hormonal events that drive the reproductive cycle, estrogen and progesterone priming may have overridden the effects of fluoxetine on male preference. However, Maswood *et al.* (2008) failed to find a change in male preference after 10 days of fluoxetine treatment in intact female rats.

Given the current interest in drug-induced disruption of female sexual desire and satisfaction, additional studies are warranted in order to understand the potential involvement of serotonergic mechanisms to female sexual motivation.

Summary

It is now well established that female consummatory sexual behavior is reduced by activation of $5\text{-}HT_{1A}$ receptors and possibly increased by activation of $5\text{-}HT_{2A/2C}$

receptors. The definitive role for 5-HT_{1B} and 5-HT_3 receptors remains to be determined. Female sexual motivation has received far less attention, and should receive greater emphasis in future studies. Finally, there is no information about any potential role of the 5-HT_4, 5-HT_5 or 5-HT_6 receptors in female sexual activity.

Male sexual behavior

Description of male sexual behavior

Male sexual behavior, including appetitive (sexual motivation; recognition and approach to a sexually relevant partner) and consummatory (copulatory) activities, is controlled by both central and peripheral mechanisms (Cruz *et al.*, 1999; Giuliano and Rampin, 2004; de Jong *et al.*, 2006; Hull and Dominguez, 2007). Mating culminates with ejaculation, which is followed by a post-ejaculatory interval during which the male shows little interest in his sexual partner. The lumbosacral spinal cord houses the sensory–motor circuitry required for ejaculation (Truitt and Coolen, 2002; Allard *et al.*, 2005), while supraspinal control is inhibitory to the reflex (Coolen *et al.*, 2004; Allard *et al.*, 2005; de Jong *et al.*, 2006). Information from forebrain areas is transmitted to the spinal cord via the PAG and then to the nucleus paragigantocellularis (nPGi) of the medulla (Rizvi *et al.*, 1996; Marson and Gravitt, 2004). In spinal-cord transected, anesthetized rats, a 'urethrogenital reflex' (muscle contraction and urethral expulsion) can be elicited (McKenna *et al.*, 1991b). However, the reflex can also be elicited in rats with lesion of the nPGi (Marson and McKenna, 1990) or by stimulation of the mPOA (Marson and McKenna, 1994). Therefore, supraspinal information, probably derived from the nPGi, tonically inhibits the ejaculation reflex (de Jong *et al.*, 2006) while the mPOA, via a relay to the PAG, reduces this tonic inhibition and thereby reduces the ejaculatory threshold, facilitating ejaculation (de Jong *et al.*, 2006). Critical sites for descending control of ejaculation include the mPOA, paraventricular nucleus (PVN), PAG and nPGi (Giuliano, 2007). The mPOA, likely via projections to the PAG, influences descending information to the spinal cord and is a critical site for integrating the complex sensory and motor information involved in copulation (Mas, 1995; Hull and Dominguez, 2007). Appetitive behavior is influenced by the mPOA, nucleus accumbens and additional brain areas (Pfaus and Phillips, 1991; Paredes *et al.*, 1993, 1998; Mas, 1995; Pfaus *et al.*, 1995). The anterior lateral hypothalamus may be important in the cessation of sexual behavior that occurs during the post-ejaculatory interval (Mas, 1995; Hull *et al.*, 1999).

Male sexual behavior requires a minimal amount of testosterone priming. Adult castration leads to a decline in testosterone over a few days, while sexual behavior declines at a slower rate, depending on prior sexual experience (Hull and Dominguez, 2007). Sexual behavior can be restored by hormonal priming with testosterone (McGinnis *et al.*, 1989), which is metabolized via aromatization to estrogen, or via 5-alpha reductase to dihydrotestosterone (DHT) (Wilson *et al.*, 1993; Wilson, 1999). Both metabolites are required to produce the full repertoire of male sexual behavior (McGinnis and Dreifuss, 1989; Vagell and McGinnis, 1997; Putnam *et al.*, 2003).

Implications for a role of 5-HT in male sexual behavior

An inhibitory effect of serotonin on male sexual behavior has been appreciated since the 1970s, with observations that various treatments which increased serotonin led to a reduction in copulatory behavior. Excellent prior reviews are available (Hull *et al.*, 2004; de Jong *et al.*, 2006). Like female sexual behavior, evaluation of the consummatory response has dominated the study of male sexual behavior. When male sexual motivation has been the focus of interest, investigators have used the partner-preference paradigm, conditioned place preference, or sexual motivation has been inferred from the latency to ejaculate.

Although different in their time-course and effectiveness, most agents that elevate 5-HT reduce ejaculation efficiency (i.e., increase the number of intromissions preceding ejaculation) and/or increase ejaculation latency (i.e., delay ejaculation), while reductions in 5-HT generally facilitate sexual behavior (Hull *et al.*, 2004; de Jong *et al.*, 2006). However, in anesthetized and spinally-transected rats, 5-HT may exert a facilitatory influence on penile reflexes such as erection and ejaculation (Rehman *et al.*, 1999; Yonezawa *et al.*, 2000) and may even elicit erection in intact, non-anesthetized models (Millan and Perrin-Monneyron, 1997).

5-HT enhancing agents have included the 5-HT precursor 5-hydroxytryptophan (5-HTP), parachloramphetamine or fenfluramine, which lead to release of 5-HT from nerve terminals, or SSRIs, which increase 5-HT by blocking reuptake into the nerve terminal. Procedures to reduce 5-HT activity have included lesioning of 5-HT neurons with 5,7, DHT, inhibition of 5-HT synthesis, or a variety of 5-HT receptor antagonists (for review, see Hull *et al.*, 2004; de Jong *et al.*, 2006). In the anterior lateral hypothalamic area, 5-HT increases following ejaculation and may contribute to the reduced sexual behavior during the post-ejaculatory interval (Lorrain *et al.*, 1997).

Serotonin's inhibitory effects on copulation occur throughout the central nervous system, because direct application of 5-HT to a variety of areas, including the nucleus accumbens (Hillegaart *et al.*, 1991), mPOA (Verma *et al.*, 1989), DRN or MRN (Hillegaart *et al.*, 1989) and spinal cord (Svensson and Hansen, 1984; Mas *et al.*, 1985), inhibit copulation. However, facilitatory effects of a low concentration of 5-HT in DRN or MRN reduced ejaculation latency and number of intromissions to ejaculation (Hillegaart *et al.*, 1989), and an increase in copulation was reported after 5-HT application to the MRN but not DRN (Larsson and Ahlenius, 1999). Facilitatory effects of 5-HT in raphe regions may reflect autoinhibitory effects of 5-HT and consequent reduction of 5-HT's release in terminal areas. Evidence that lesions of the MRN significantly enhanced male sexual behavior is consistent with this possibility (reviewed in Larsson and Ahlenius, 1999).

Of the agents that modify 5-HT, SSRIs have been of special interest because of their effectiveness in the treatment of premature ejaculation in humans (Waldinger *et al.*, 2005). Individuals being treated with SSRIs for depression often develop sexual dysfunction, evidenced as a delay in or prevention of ejaculation (Waldinger and Olivier, 1998; Gregorian *et al.*, 2002; Montgomery *et al.*, 2002; Nurnberg, 2008). This undesirable side effect is of therapeutic advantage for individuals with premature ejaculation (Waldinger and Olivier, 2004; Waldinger, 2005; Mercado *et al.*, 2008). However, the mechanisms responsible for these effects of SSRIs remain unclear. In spite of comparable enhancement of extracellular 5-HT, SSRIs vary in their ability to delay ejaculation (Waldinger *et al.*, 1998; Waldinger and Olivier, 2004). Furthermore, clinical improvement of premature ejaculation requires weeks of treatment with SSRIs (Waldinger *et al.*, 1998; Waldinger *et al.*, 2005), while an increase in extracellular 5-HT is more rapid in onset (Gobert *et al.*, 1997; Tao *et al.*, 2000; Felton *et al.*, 2003).

In animal studies, an ejaculatory delay after acute treatment with SSRIs has been reported by some investigators (Yells *et al.*, 1994, 1995; Lorrain *et al.*, 1997), but most have not seen such acute effects (Waldinger *et al.*, 2002; de Jong *et al.*, 2005a,b). This may be a function of the use of different SSRIs and/or differences in the precise experimental procedures and/or route of administration of the SSRIs. Nevertheless, repeated treatment with SSRIs is clearly more effective than a single treatment in delaying ejaculation. Because chronic treatment with SSRIs leads to downregulation/desensitization of 5-HT_{1A} and 5-HT_{1B} receptors (Hensler, 2003; Newman *et al.*, 2004), some investigators have argued that the ability of SSRIs to delay ejaculation requires the desensitization of these receptors (de Jong *et al.*, 2006). This argument is particular appealing, given the facilitatory role of 5-HT_{1A} receptors and putative inhibitory role of 5-HT_{1B} receptors in male sexual behavior (see below). Evidence that chronic treatment with the SSRI paroxetine attenuated some, but not all, effects of a 5-HT_{1A} receptor agonist on sexual behavior provides some evidence in support of this idea (de Jong *et al.*, 2005c). However, with newer, short-acting reuptake inhibitors such as dapoxetine, effects in patients suffering from premature ejaculation may be seen within hours (Clement *et al.*, 2007; Giuliano, 2007; Kendirci *et al.*, 2007; Wang *et al.*, 2007). Effects of SSRIs on male ejaculation probably result primarily from actions of the drugs in the CNS (de Jong *et al.*, 2005b), but peripheral effects cannot be ruled out (Medina *et al.*, 2000; Giuliano, 2007).

Effects of SSRIs on non-contact penile erections have been varied, with some investigators reporting fluoxetine-induced increases (Simon *et al.*, 1993; Millan and Perrin-Monneyron, 1997) and others reporting decreases after treatment (Sukoff Rizzo *et al.*, 2008). While the explanation for these differences is unclear, it appears that when the level of spontaneous erections is low, erections are increased by fluoxetine. If spontaneous erections are high, they may be reduced by fluoxetine.

Receptors and male sexual behavior

With the available collection of pharmaceutical agents and molecular tools, the most coherent story is that 5-HT_{1A} receptor agonists can inhibit or facilitate copulation in a species-dependent manner (e.g., facilitation in rats and inhibition in mice); the 5-HT_{1B} receptor inhibits copulation but may facilitate penile erections. The 5-HT_2 receptors reduce ejaculation but increase (via 5-HT_{2C} receptor) penile erection. The 5-HT_3 receptor plays a minor, if any, role in male sexual behavior.

5-HT_{1A} receptors

Activation of 5-HT_{1A} receptors increases ejaculation in rats but decreases it in mice. In rats, 5-HT_{1A} receptors facilitate male sexual behavior by delaying ejaculation latency, but reduce spontaneous penile erections (Schnur *et al.*, 1989; Looney *et al.*, 2005; de Jong *et al.*, 2006). The first report of facilitation in rats was in 1981 (Ahlenius *et al.*, 1981), and was rapidly followed by additional studies indicating that a variety of 5-HT_{1A} receptor agonists reduced ejaculatory latency in rats but did not enhance motivational indices such as mounting or intromission latency (for review, see Hull *et al.*, 1999, 2004; de Jong *et al.*, 2006). Facilitatory effects were blocked with 5-HT_{1A} receptor antagonists (Ahlenius and Larsson, 1991; Hillegaart and Ahlenius, 1998), including the selective 5-HT_{1A} receptor antagonist WAY100635 (Ahlenius and Larsson, 1998). Alone, WAY100635 did not alter ejaculatory latency but did enhance the inhibitory effects of agents (e.g., 5-HTP or SSRIs) that increase 5-HT (Ahlenius and Larsson, 1998, 1999; de Jong *et al.*, 2005a; Looney *et al.*, 2005).

Similarly, WAY100635, alone, had only minor effects on penile erection, but the combined blockage of 5-HT_{1A} and 5-HT_{1B} receptors led to a five-fold enhancement of the acute erectogenic effects of fluoxetine (Millan and Perrin-Monneyron, 1997). De Jong *et al.* (2006) argued that 5-HT_{1A} receptors might become important only when 5-HT levels are increased. Such a possibility is interesting, given the variety of environmental challenges that can elevate 5-HT.

The most extensively studied 5-HT_{1A} receptor agonist has been 8-OH-DPAT. Systemic 8-OH-DPAT facilitated ejaculation in intact or hormone-primed males (Mos *et al.*, 1991; Pomerantz *et al.*, 1993a; Fernandez-Guasti and Rodriguez-Manzo, 1997), antagonized the effect of 5-HTP on ejaculation (Ahlenius and Larsson, 1985) and reduced the amount of testosterone required for mating to occur (Rowland and Houtsmuller, 1998), but inhibited penile erections in dogs (Yonezawa *et al.*, 2004), rats (Schnur *et al.*, 1989; Rehman *et al.*, 1999) and monkeys (Pomerantz *et al.*, 1993b). Intrathecal administration of 8-OH-DPAT reduced penile erections and ejaculation in *ex copula* models (Schnur *et al.*, 1989; Lee *et al.*, 1990). Doses of 8-OH-DPAT that facilitated copulatory behavior reduced the weight of seminal emissions and led to the conclusion that effects of 8-OH-DPAT were not equivalent to normal copulatory/ejaculatory function (Schnur *et al.*, 1989).

In contrast to 8-OH-DPAT's facilitation of copulation in rats, 8-OH-DPAT inhibited male sexual behavior in ferrets (Paredes *et al.*, 1994), mice (Rodriguez-Manzo *et al.*, 2002) and rabbits (Paredes *et al.*, 2000). In dogs, 8-OH-DPAT reduced ejaculatory semen and increased the latency to ejaculate in response to continuous manual stimulation (Yonezawa *et al.*, 2004); in monkeys, 8-OH-DPAT's facilitation was dose-dependent, occurring only at a low dose of the drug (Pomerantz *et al.*, 1993a). At higher doses, monkeys did not initiate copulation. Explanations for these species differences remain unclear, but may reflect differences in the male's mating pattern, sensitivity to stimuli, or the relative balance between inhibitory and facilitatory 5-HT receptor families. Unfortunately, little attention has been given to the examination of these species differences.

A facilitatory effect of 5-HT_{1A} receptor agonists on sexual behavior in rats could result from an autoreceptor-mediated reduction in extracellular 5-HT. The observation that a low dose of 8-OH-DPAT facilitated sexual behavior following infusion into the MRN provides some support for this possibility (Hillegaart *et al.*, 1991). However, no effects of 8-OH-DPAT were found following infusion into the DRN (Hillegaart *et al.*, 1991; Fernandez-Guasti *et al.*, 1992). Moreover, inhibition of 5-HT synthesis with PCPA or destruction of 5-HT cell bodies with 5,7-DHT did not block facilitatory effects of 8-OH-DPAT (Fernandez-Guasti and Escalante, 1991). Finally, direct infusion of 8-OH-DPAT into multiple brain areas that are targeted by 5-HT neurons increased copulatory behavior (Fernandez-Guasti *et al.*, 1992; Hull *et al.*, 1999, 2004; de Castilhos *et al.*, 2006; de Jong *et al.*, 2006). Consequently, postsynaptic 5-HT_{1A} receptors are clearly included in the facilitatory effects of 5-HT_{1A} receptor agonists on male sexual behavior in the rat.

However, in many studies with local application of 8-OH-DPAT, specificity for the 5-HT_{1A} receptor has not been confirmed. While relatively selective for 5-HT_{1A} receptors, 8-OH-DPAT also has affinity for other receptors. Effects of systemic 8-OH-DPAT treatment can be attenuated by 5-HT_{1A} receptor antagonists, but comparable experiments have not been done for many intracranial studies. Confirmation that it is the 5-HT_{1A} receptor subtype that mediates effects of direct application is badly needed, as illustrated by the finding that facilitatory effects of 8-OH-DPAT following infusion into the mPOA may not reside in its serotonergic effects (Lorrain *et al.*, 1998). Facilitatory effects of the 8-OH-DPAT infusion to the mPOA were not blocked by a 5-HT_{1A} receptor antagonist but were blocked by a dopamine2 receptor antagonist (Lorrain *et al.*, 1998; Hull *et al.*, 1999), so that Hull and coworkers have suggested that 8-OH-DPAT's facilitatory effect in the mPOA is mediated by its effect on dopamine (Hull *et al.*, 1999, 2004).

Therefore, there is a need for greater clarification of the sites of action and receptor specificity of 5-HT_{1A} receptor agonists on male copulatory behavior. A clear dissociation between peripheral and central effects of the drug is also needed. Effects of systemic 8-OH-DPAT on ejaculatory latency could include an enhancement of the effects of genital stimulation, since the drug can compensate for bilateral transection of the dorsal penile nerve (Dahlof *et al.*, 1988). Moreover, 5-HT_{1A} receptor agonists, as reported for gespirone and flexinoxan, may increase the male's attractiveness to the female and thereby influence mating behavior (Mos *et al.*, 1991). These effects of 5-HT_{1A} receptor agonists require further examination.

Consequently, in spite of general agreement about the copulation-enhancing effects of 5-HT_{1A} receptors in some species, there is still information to be gained about the mechanisms whereby 5-HT_{1A} receptors influence male sexual behavior.

5-HT_{1B} receptors

A variety of 5-HT_{1B} receptor agonists have been reported to mimic the effects of 5-HT by reducing male copulatory behavior. Many 5-HT_{1B} receptor agonists have been investigated, but a majority of data in support of the inhibitory role for 5-HT_{1B} receptors has been generated with TFMPP or mCPP (for review, see Hull *et al.*, 2004; de Jong *et al.*, 2006). In addition to agonist action at 5-HT_{1B} receptors, TFMPP has agonist action at 5-HT_{1A} and 5-HT_2 receptors, and mCPP is also a 5-HT_{2C} receptor agonist, so the specificity of their effect at 5-HT_{1B} receptors is not clear. Nevertheless, it is unlikely that 5-HT_{1A} receptors are responsible for the inhibitory effects of TFMPP on male sexual behavior, since 5-HT_{1A} receptor agonists facilitate sexual behavior (at least in rats). Moreover, under comparable experimental conditions, following infusion into the nucleus accumbens, the 5-HT_{1A} receptor agonist 8-OH-DPAT reduced ejaculatory latency while it was increased by TFMPP (Fernandez-Guasti *et al.*, 1992). Additionally, the effect of a 5-HT_{1B} receptor agonist, anpirtoline, on male sexual behavior was antagonized by a 5-HT_{1B} receptor antagonist, but not by a 5-HT_{1A} receptor antagonist (Hillegaart and Ahlenius, 1998). Furthermore, 8-OH-DPAT reduces penile erections, while 5-HT_{1B} and 5-HT_{2C} receptor agonists increase penile erections (Berendsen *et al.*, 1990). Finally, sexual behavior of 5-HT_{1B} receptor knockout (KO1B) mice is enhanced relative to wild-type mice and is not inhibited by either 5-HTP or TFMPP (Rodriguez-Manzo *et al.*, 2002), suggesting that these agents require the 5-HT_{1B} receptor for their inhibition. However, 8-OH-DPAT also failed to inhibit ejaculation in KO1B mice (Rodriguez-Manzo *et al.*, 2002), raising the question of the relationship between 5-HT_{1A} and 5-HT_{1B} receptors to the inhibitory effects of 5-HT in male mice. Therefore, in spite of the limited selectivity of the available 5-HT_{1B} receptor compounds, the weight of evidence indicates that it is their activation of 5-HT_{1B} receptors that leads to inhibition of sexual behavior. Nevertheless, data with more selective compounds are needed.

Intracranial effects of 5-HT_{1B} receptor compounds have received less emphasis than their 5-HT_{1A} receptor counterparts. Intracranial treatment with TFMPP in nucleus accumbens or mPOA reduced ejaculation latency (Fernandez-Guasti *et al.*, 1992), but, to our knowledge, male sexual behavior following intracranial infusion of more selective 5-HT_{1B} receptor compounds has not been reported.

5-HT_2 receptors

5-HT_2 receptors were initially implicated as facilitatory to sexual behavior when the 5-HT_2 receptor antagonists pirenperone and ketanserin were reported to reduce the proportion of males exhibiting sexual behavior (Mendelson and Gorzalka, 1985). However, studies with the 5-$HT_{2A/2C}$ selective receptor agonist DOI indicated inhibitory effects of 5-HT_2 receptors on male sexual behavior. DOI increased ejaculation latency and reduced the number of rats achieving ejaculation, and inhibitory effects of DOI were antagonized by 5-HT_2 receptor antagonists (for review, see Hull *et al.*, 2004; de Jong *et al.*, 2006). Inhibitory effects of mCPP on ejaculation, originally attributed to 5-HT_{1B} receptor activation, are now viewed as the result of 5-HT_{2C} receptor activation (Mendelson and Gorzalka, 1990; Pomerantz *et al.*, 1993b).

Although effects of various 5-HT releasing compounds are reversed by 5-HT_2 receptor antagonists (Millan *et al.*, 1997; de Jong *et al.*, 2006), inhibitory effects of 5-HTP are not (Ahlenius and Larsson, 1998), allowing the suggestion that 5-HT_2 receptor subtypes may differentially influence components of the male sexual response. In particular, 5-HT_{2A} and 5-HT_{2C} receptors may exert opposite roles in the control of copulation and penile reflexes (Lau *et al.*, 2007; Yonezawa *et al.*, 2008). For example, mCPP was recently reported to exert a biphasic effect on ejaculation and penile erections (Yonezawa *et al.*, 2008). At low doses, mCPP increased penile erection, the incidence of ejaculation, and the amount of seminal emissions. However, at higher doses, the drug inhibited both erection and ejaculation. Pro-ejaculatory effects of mCPP were attenuated by both a 5-HT_{2C} and a 5-HT_{2A} receptor antagonist, but only the 5-HT_{2C} receptor antagonist reduced the effect of mCPP on penile erection (Yonezawa *et al.*, 2008).

The 5-HT_2 receptor subtype responsible for all the effects of 5-HT_2 receptor agonists is not clear, but their erectogenic effect probably relies upon activation of 5-HT_{2C} receptors. Earlier reported effects of the erectogenic effects of mCPP have now been attributed to effects at 5-HT_{2C} receptors (Berendsen and Broekkamp, 1990; Bagdy *et al.*, 1992), and more selective 5-HT_{2C} receptor agents reinforce the facilitatory role of 5-HT_{2C} receptors in penile erection (Yonezawa *et al.*, 2008). An increase in firing of nerves innervating the penis and an increased intracavernous pressure in anesthetized rats was seen following mCPP treatment (Steers and de Groat, 1989). In addition, several investigators have considered that SSRI effects on ejaculation and erection may be via increased activation of 5-HT_{2C} receptors (Pomerantz *et al.*, 1993a; Waldinger and Olivier, 1998). A selective 5-HT_{2C} receptor antagonist, SB 206553, abolished penile erections elicited by fluoxetine (Millan *et al.*, 1997).

The neural sites outside the spinal cord where 5-HT_2 receptors agonists are involved in inhibition of copulatory behavior have received minimal attention, but infusion of DOI into the amygdala increased the latency to the first ejaculation and reduced the frequency of ejaculation

(Drago *et al.*, 1999). A role for the 5-HT_{2A} receptor in peripheral serotonergic pathways has been suggested (Lau *et al.*, 2007). Clearly, additional studies (and more selective receptor agonists and antagonists) are needed to dissect the relative contribution of the 5-HT_2 receptor subtypes to the modulation of male sexual behavior.

5-HT_3 receptors

In few studies has the potential involvement of 5-HT_3 receptors in male sexual behavior been examined. Lau *et al.* (2007) suggested a role for peripheral 5-HT_3 receptors in the erectile process, but systemic treatment with 5-HT_3 receptor antagonists had no effect on male sexual behavior (Tanco *et al.*, 1993). Administration of a 5-HT_3 receptor agonist, l-phenylbiguanide, into the lateral ventricle reduced ejaculatory latency without significant effects on other components of the behavior (Tanco *et al.*, 1994).

Measures of male sexual motivation

Compared to the study of copulation and penile reflexes, less emphasis has been placed on serotonin's contribution to male sexual motivation. However, under conditions where 8-OH-DPAT reduced ejaculatory latency in rats, the drug did not alter conditioned place preference following the reward of ejaculation (Camacho *et al.*, 2007). Therefore, these authors concluded that 8-OH-DPAT's facilitation of sexual behavior in male rats was not associated with an enhancement of the rewarding properties of sex.

In mice, where 8-OH-DPAT produces an inhibition of sexual behavior, the drug also produced a decline in sexual motivation, as measured by the time the male spent near a sexually receptive female (Popova and Amstislavskaya, 2002a). From a series of studies designed to examine the effects of 5-HT receptors on preference time for the female, Popova and Amstislavskaya (2002a) have concluded that, in mice, 5-HT_{1A} and possibly 5-HT_{1B} receptors block sexual motivation, while 5-HT_{2A} and 5-HT_{2C} receptors increase sexual motivation, and at least part of the effect may be mediated by alterations in the male's testosterone response to the presence of a receptive female. In the non-contact, partner-preference paradigm, the 5-HT_{1A} receptor agonist, 8-OH-DPAT, and the 5-HT_{1B} receptor agonist, CGS-12066A (7-trifluoromethyl-4 (4-methyl-1-piperazinyl)-pirrolo[1,2a]quinoxaline maleate), reduced the time male mice spent near the female (Popova and Amstislavskaya, 2002a). Only 8-OH-DPAT reduced the testosterone increase that usually occurs when males are exposed to females. However, using a similar paradigm, the 5-HT_{2C} receptor antagonist, RS 102221 (8-[5-(2,4-dimethoxy-5-(4-trifluoromethylphenylsulfonamido)phenyl-5-oxopentyl)]-1,3,8-triazaspiro[4.5]decane-2,4-dione hydrochloride) increased the male's time near the female and increased the testosterone response (Popova and Amstislavskaya, 2002b). Ketanserin and cyproheptadine, 5-HT_{2A} receptor-preferring antagonists, produced the reciprocal response (i.e., less time with the female and reduced testosterone response). In general agreement with these findings, DOI increased interest (i.e., increased pursuit and thrusting or increased genital sniffing) in sexual partners in prepubertal and young adult males (Padoin and Lucion, 1995).

These studies are consistent with a possible facilitatory effect of 5-HT_{2C} receptors on sexual motivation, and also with species differences in the response to 5-HT_{1A} receptor agonists. They further illustrate that copulatory behavior and sexual motivation are distinct behaviors, and may be independently modulated by neurally active compounds. Given the emerging interest in the sexual dysfunction resulting from treatment with a variety of therapeutic pharmaceuticals, there is an increasing need for the evaluation of serotonergic compounds on male sexual motivation.

In the few studies in which sexual motivation has been assessed during treatment with SSRIs, the consequences for sexual motivation are not consistent. When conditioned level changing in a bilevel chamber was used as the measure of sexual motivation, 10 mg/kg fluoxetine reduced conditioned level changing and reduced ejaculatory responses, leading the authors to conclude that fluoxetine reduced both appetitive and consummatory responses (Cantor *et al.*, 1999).

However, daily treatment for at least 4 weeks with a low dose of fluoxetine (0.75 mg/kg) failed to alter time spent with an estrous female or the number of urinary markings near the female (Taylor *et al.*, 1996). In contrast, copulatory effects became evident as early as the first week of treatment and progressed throughout the experiment. Similarly, 10 mg/kg fluoxetine failed to alter the number of crossings of an electrified grill that males made to reach a receptive female, and comparable findings were obtained with trimipramine (Benelli *et al.*, 2004). In contrast, a daily dose of fluoxetine (10 mg/kg) increased ejaculation latency after 9 days of treatment; preference for time spent with the female was reduced by 14 days of treatment and was eliminated by 3 weeks of treatment (Vega Matuszcyk *et al.*, 1998). However, the reduced female preference may have reflected an effect of prior experience, since rats were tested repeatedly. It was suggested that the males had developed an association between the female and the failure to ejaculate, and that this association may explain the reduction in preference for the female (Frank *et al.*, 2000). Similar to the findings by Taylor and colleagues, Frank *et al.* (2000) reported that males chronically treated with fluoxetine continued to pursue receptive females, and therefore appeared to exhibit continued sexual motivation.

In one study with a 5-HT-depleting agent (3,4 methylenedioxymethamphetamine, MDMA, 'ecstasy'), conditioned place preference was examined (Straiko *et al.*, 2007). Rats treated with MDMA failed to form conditioned place preference for sexual reward.

Summary

Effects of 5-HT on male sexual behavior have been studied most extensively in rats, and ejaculation and penile erection have been the primary measures of sexual activity. Evidence for an effect of 5-HT_{1A} receptors is clearly established, but species differences are present in the direction of the effect. Identification of the mechanisms responsible for these species differences would be of value in further understanding of the serotonergic mechanisms involved in control of sexual behavior. The inhibitory effect of 5-HT_{1B} receptors on copulation and the possible erectile effect of 5-HT_{2C} receptors may be present across species, but the absence of substantial data with selective agonists limits understanding of the receptor subtypes involved.

Overall summary and future directions

An inhibitory effect of 5-HT on both male and female consummatory behavior is now well established. From animal studies, as well as clinical findings, it appears that 5-HT also contributes to sexual motivation; but considerably less attention has been given to the neurotransmitter's motivational role. Furthermore, the mechanisms involved in sexual effects of 5-HT continue to be elusive, and the complete story regarding how different 5-HT receptor subtypes contribute to the modulation of sexual behavior remains to be written. In the absence of receptor-selective agonists and antagonists for many 5-HT receptors, it is still difficult to precisely define which receptor(s) mediate the divergent effects of 5-HT's effect on sexual behavior; and a majority of 5-HT receptors remain to be investigated.

Since 5-HT does not tonically inhibit most elements of sexual behavior in males or females, sexual interest does not require that 5-HT function be removed. In fact, under some conditions, activation of select 5-HT receptors may even facilitate sexual behavior. What, then, is the biological significance of 5-HT's modulation of sexual behavior? Clues to 5-HT's role can be derived from the diversity of actions of the neurotransmitter. 5-HT exerts influence on behaviors that permit individual survival in a complex and potentially dangerous environment. The neurotransmitter contributes to behaviors such as fear, anxiety and arousal (Blier and Abbott, 2001; Carrasco and Van de Kar, 2003; Beekman *et al.*, 2005; Kitson, 2007), which allow individuals successfully to escape or adapt to threatening situations. In addition, 5-HT is involved in control of sleep, nutrient intake, and immune responses (Kaye, 2008; Nonogaki, 2008; Sanger, 2008; Vinkers *et al.*, 2008), which facilitate the organism's ability to cope with internal challenge. The neurotransmitter's modulation of sexual behavior needs to be viewed within the context of this homeostatic regulation. Rather than controlling sexual behavior *per se*, 5-HT coordinates behavioral outcomes that enhance the probability of individual survival. For example, if individuals are given the choice between satisfaction of hunger and sexual activity, sexual activity may decline. If forced to choose between escape from predation and sexual activity, individuals will choose to escape. The contribution of 5-HT to sexual behavior will therefore vary as a function of changes in internal and external demands.

For example, we have previously argued that 5-HT_{1A} receptors reduce sexual activity under conditions of potential threat (Uphouse, 2000; Uphouse *et al.*, 2003). That 5-HT_{1A} receptors facilitate copulatory behavior in male rats, while inhibiting sexual behavior in female rats, seems, at first, to contradict this suggestion. However, rapid ejaculation in the presence of a potential threat would maximize the rat's probability both of escaping predation and of producing offspring. To delay ejaculation in the presence of a threatening environment reduces the rat's probability of survival. With this perspective, the occurrence of species differences in the response to a drug such as 8-OH-DPAT (which is facilitatory in male rats but inhibitory in male mice) begins to coalesce in that the drug primarily functions to inhibit sexual activity. Therefore, from this perspective, 5-HT_{1A} receptor activation produces a comparable sexual consequence (e.g., acts as a brake on sexual activity) in males and females. Such a perspective is also consistent with evidence that 5-HT_{1A} receptors antagonists, alone, do not seem to increase sexual behavior, but have the ability to attenuate inhibition induced by 5-HT releasing agents or by stressful events (Uphouse *et al.*, 2007).

Given the lack of a large amount of data for other receptor subtypes, it may be premature to attempt to identify additional unifying roles. However, data for 5-HT_2 receptors fit well within the above theme. 5-HT_2 receptors are implicated in the facilitation of female sexual behavior, as well as in male sexual motivation and erection. Of interest, then, is the reciprocal role played by 5-HT_{2C}

receptors in sexual and eating behavior, increasing sexual behavior (Nedergaard *et al.*, 2004; Giuliano, 2007) but reducing food intake (Langhans, 2007; Somerville *et al.*, 2007). Moreover, 5-HT_{1A} and 5-HT_2 receptors seem to function together to fine-tune the behavioral outcome (Berendsen and Broekkamp, 1990; Krebs-Thomson and Geyer, 1998; Mikkelsen *et al.*, 2004).

Thus, 5-HT's role in modulation of sexual behavior is likely to be far more complex than any identification of its facilitatory or inhibitory components. Instead, 5-HT's widespread distribution in the CNS and the periphery enables the neurotransmitter to facilitate selection of one set of behavioral outcomes at the expense of another. In many cases, this behavioral competition sacrifices sexual activity while facilitating behaviors that may be more conducive to individual survival.

Acknowledgements

The authors acknowledge support of grants NIH HD28419 and GM 55380. Appreciation is expressed to Dr Sharmin Maswood for reading an earlier version of the chapter.

References

Adell, A., Carceller, A. and Artigas, F. (1993) *In vivo* brain dialysis study of the somatodendritic release of serotonin in the Raphe nuclei of the rat: effects of 8-hydroxy-2-(di-n-propylamino)tetralin. J. Neurochem., 60: 1673–1681.

Ahlenius, S. and Larsson, K. (1985) Antagonism by lisuride and 8-OH-DPAT of 5-HTP-induced prolongation of the performance of male rat sexual behavior. Eur. J. Pharmacol., 110: 379–381.

Ahlenius, S. and Larsson, K. (1991) Opposite effects of 5-methoxy-N,N-di-methyl-tryptamine and 5-hydroxytryptophan on male rat sexual behavior. Pharmacol. Biochem. Behav., 38: 201–205.

Ahlenius, S. and Larsson, K. (1998) Evidence for an involvement of 5-HT1B receptors in the inhibition of male rat ejaculatory behavior produced by 5-HTP. Psychopharmacology (Berl.), 137: 374–382.

Ahlenius, S. and Larsson, K. (1999) Synergistic actions of the 5-HT1A receptor antagonist WAY-100635 and citalopram on male rat ejaculatory behavior. Eur. J. Pharmacol., 379: 1–6.

Ahlenius, S., Larsson, K., Svensson, L., Hjorth, S., Carlsson, A., Lindberg, P., Wikstrom, H., Sanchez, D., Arvidsson, L.E., Hacksell, U. and Nilsson, J.L. (1981) Effects of a new type of 5-HT receptor agonist on male rat sexual behavior. Pharmacol. Biochem. Behav., 15: 785–792.

Aiello-Zaldivar, M., Luine, V. and Frankfurt, M. (1992) 5,7-DHT facilitated lordosis: effects of 5-HT agonists. Neuroreport, 3: 542–544.

Allard, J., Truitt, W.A., McKenna, K.E. and Coolen, L.M. (2005) Spinal cord control of ejaculation. World J. Urol., 23: 119–126.

Allen, D.L., Renner, K.J. and Luine, V.N. (1993) Pargyline-induced increase in serotonin levels: correlation with inhibition of lordosis in rats. Pharmacol. Biochem. Behav., 45: 837–841.

Bagdy, G., Kalogeras, K.T. and Szemeredi, K. (1992) Effect of 5-HT1C and 5-HT2 receptor stimulation on excessive grooming, penile erection and plasma oxytocin concentrations. Eur. J. Pharmacol., 229: 9–14.

Beekman, M., Flachskamm, C. and Linthorst, A.C. (2005) Effects of exposure to a predator on behaviour and serotonergic neurotransmission in different brain regions of C57bl/6N mice. Eur. J. Neurosci., 21: 2825–2836.

Beer, M., Kennett, G.A. and Curzon, G. (1990) A single dose of 8-OH-DPAT reduces raphe binding of [3H]8-OH-DPAT and increases the effect of raphe stimulation on 5-HT metabolism. Eur. J. Pharmacol., 178: 179–187.

Beer, M.S., Heald, M.A., McAllister, G. and Stanton, J.A. (1998) Pharmacological characterisation of a cloned dog 5-HT1B receptor cell line. Eur. J. Pharmacol., 360: 117–121.

Benelli, A., Frigeri, C., Bertolini, A. and Genedani, S. (2004) Influence of mirtazapine on the sexual behavior of male rats. Psychopharmacology (Berl.), 171: 250–258.

Benmansour, S., Piotrowski, J.P., Altamirano, A.V. and Frazer, A. (2008) Impact of ovarian hormones on the modulation of the serotonin transporter by fluvoxamine. *Neuropsychopharmacology*, in press.

Berendsen, H.H. and Broekkamp, C.L. (1990) Behavioural evidence for functional interactions between 5-HT-receptor subtypes in rats and mice. Br. J. Pharmacol., 101: 667–673.

Berendsen, H.H., Jenck, F. and Broekkamp, C.L. (1990) Involvement of 5-HT1C-receptors in drug-induced penile erections in rats. Psychopharmacology (Berl.), 101: 57–61.

Bertrand, P.P., Paranavitane, U.T., Chavez, C., Gogos, A., Jones, M. and van den Buuse, M. (2005) The effect of low estrogen state on serotonin transporter function in mouse hippocampus: a behavioral and electrochemical study. Brain Res., 1064: 10–20.

Bethea, C.L., Lu, N.Z., Gundlah, C. and Streicher, J.M. (2002) Diverse actions of ovarian steroids in the serotonin neural system. Front Neuroendocrinol., 23: 41–100.

Blier, P. and Abbott, F.V. (2001) Putative mechanisms of action of antidepressant drugs in affective and anxiety disorders and pain. J. Psychiatry Neurosci., 26: 37–43.

Caldwell, H.K. and Albers, H.E. (2002) The effects of serotonin agonists on the hypothalamic regulation of sexual receptivity in Syrian hamsters. Horm. Behav., 42: 78–84.

Camacho, F.J., Castro, M., Hernandez, V. and Paredes, R.G. (2007) Facilitation of ejaculation induced by 8-OH-DPAT does not produce conditioned place preference in male rats. Behav. Neurosci., 121: 579–585.

Canteras, N.S., Simerly, R.B. and Swanson, L.W. (1994) Organization of projections from the ventromedial nucleus of the hypothalamus: a *Phaseolus vulgaris*–leucoagglutinin study in the rat. J. Comp. Neurol., 348: 41–79.

Cantor, J.M., Binik, Y.M. and Pfaus, J.G. (1999) Chronic fluoxetine inhibits sexual behavior in the male rat: reversal with oxytocin. Psychopharmacology (Berl.), 144: 355–362.

Carrasco, G.A. and Van de Kar, L.D. (2003) Neuroendocrine pharmacology of stress. Eur. J. Pharmacol., 463: 235–272.

Clayton, A.H. (2002) Female sexual dysfunction related to depression and antidepressant medications. Curr. Women's Health Rep., 2: 182–187.

Clement, P., Bernabe, J., Gengo, P., Denys, P., Laurin, M., Alexandre, L. and Giuliano, F. (2007) Supraspinal site of action for the inhibition of ejaculatory reflex by dapoxetine. Eur. Urol., 51: 825–832.

Coolen, L.M., Allard, J., Truitt, W.A. and McKenna, K.E. (2004) Central regulation of ejaculation. Physiol. Behav., 83: 203–215.

Cornfield, L.J., Lambert, G., Arvidsson, L.E., Mellin, C., Vallgarda, J., Hacksell, U. and Nelson, D.L. (1991) Intrinsic activity of enantiomers of 8-hydroxy-2-(di-n-propylamino)tetralin and its analogs at 5-hydroxytryptamine1A receptors that are negatively coupled to adenylate cyclase. Mol. Pharmacol., 39: 780–787.

Cottingham, S.L. and Pfaff, D. (1986) Interconnectedness of steroid hormone-binding neurons: existence and implications. In: Ganten, D. and Pfaff, D. (Eds.), Curr. Topics Neuroendocrinol. Springer, Berlin, Vol. 7, pp. 223–249.

Cruz, M.R., Liu, Y.C., Manzo, J., Pacheco, P. and Sachs, B.D. (1999) Peripheral nerves mediating penile erection in the rat. J. Auton. Nerv. Syst., 76: 15–27.

Dahlof, L.G., Ahlenius, S. and Larsson, K. (1988) Copulatory performance of penile desensitized male rats following the administration of 8-OH-DPAT. Physiol. Behav., 43: 841–843.

Daniels, D. and Flanagan-Cato, L.M. (2000) Functionally-defined compartments of the lordosis neural circuit in the ventromedial hypothalamus in female rats. J. Neurobiol., 45: 1–13.

Daniels, D., Miselis, R.R. and Flanagan-Cato, L.M. (1999) Central neuronal circuit innervating the lordosis-producing muscles defined by transneuronal transport of pseudorabies virus. J. Neurosci., 19: 2823–2833.

de Castilhos, J., Marcuzzo, S., Forti, C.D., Frey, R.M., Stein, D., Achaval, M. and Rasia-Filho, A.A. (2006) Further studies on the rat posterodorsal medial amygdala: dendritic spine density and effect of 8-OH-DPAT microinjection on male sexual behavior. Brain. Res. Bull., 69: 131–139.

de Groote, L., Klompmakers, A.A., Olivier, B. and Westenberg, H.G. (2003) An evaluation of the effect of NAS-181, a new selective 5-HT(1B) receptor antagonist, on extracellular 5-HT levels in rat frontal cortex. Naunyn Schmiedebergs Arch. Pharmacol., 367: 89–94.

de Jong, T.R., Pattij, T., Veening, J.G., Dederen, P.J., Waldinger, M.D., Cools, A.R. and Olivier, B. (2005a) Citalopram combined with WAY 100635 inhibits ejaculation and ejaculation-related Fos immunoreactivity. Eur. J. Pharmacol., 509: 49–59.

de Jong, T.R., Pattij, T., Veening, J.G., Dederen, P.J., Waldinger, M.D., Cools, A.R. and Olivier, B. (2005b) Effects of chronic paroxetine pretreatment on (+/−)-8-hydroxy-2-(di-n-propyl-amino)tetralin induced c-fos expression following sexual behavior. Neuroscience, 134: 1351–1361.

de Jong, T.R., Pattij, T., Veening, J.G., Waldinger, M.D., Cools, A.R. and Olivier, B. (2005c) Effects of chronic selective serotonin reuptake inhibitors on 8-OH-DPAT-induced facilitation of ejaculation in rats: comparison of fluvoxamine and paroxetine. Psychopharmacology (Berl.), 179: 509–515.

de Jong, T.R., Veening, J.G., Waldinger, M.D., Cools, A.R. and Olivier, B. (2006) Serotonin and the neurobiology of the ejaculatory threshold. Neurosci. Biobehav. Rev., 30: 893–907.

Drago, F., Busa, L., Benelli, A. and Bertolini, A. (1999) Acute low doses of melatonin stimulate rat sex behavior: the role of serotonin neurotransmission. Eur. J. Pharmacol., 385: 1–6.

Emery, D.E. and Moss, R.L. (1984) p-Chlorophenylalanine alters pacing of copulation in female rats. Pharmacol. Biochem. Behav., 20: 337–341.

Erskine, M.S. (1989) Solicitation behavior in the estrous female rat: a review. Horm. Behav., 23: 473–502.

Farmer, C.J., Isakson, T.R., Coy, D.J. and Renner, K.J. (1996) *In vivo* evidence for progesterone dependent decreases in serotonin release in the hypothalamus and midbrain central grey: relation to the induction of lordosis. Brain Res., 711: 84–92.

Felton, T.M., Kang, T.B., Hjorth, S. and Auerbach, S.B. (2003) Effects of selective serotonin and serotonin/noradrenaline reuptake inhibitors on extracellular serotonin in rat diencephalon and frontal cortex. Naunyn Schmiedebergs Arch. Pharmacol., 367: 297–305.

Fernandez-Guasti, A. and Escalante, A. (1991) Role of presynaptic serotonergic receptors on the mechanism of action of 5-HT1A and 5-HT1B agonists on masculine sexual behaviour: physiological and pharmacological implications. J. Neural. Transm. Gen. Sect., 85: 95–107.

Fernandez-Guasti, A., Escalante, A.L., Ahlenius, S., Hillegaart, V. and Larsson, K. (1992) Stimulation of 5-HT1A and 5-HT1B receptors in brain regions and its effects on male rat sexual behaviour. Eur. J. Pharmacol., 210: 121–129.

Fernandez-Guasti, A. and Rodriguez-Manzo, G. (1997) 8-OH-DPAT and male rat sexual behavior: partial blockade by noradrenergic lesion and sexual exhaustion. Pharmacol. Biochem. Behav., 56: 111–116.

Flanagan-Cato, L.M., Lee, B.J. and Calizo, L.H. (2006) Co-localization of midbrain projections, progestin receptors, and mating-induced fos in the hypothalamic ventromedial nucleus of the female rat. Horm. Behav., 50: 52–60.

Fletcher, A., Forster, E.A., Bill, D.J., Brown, G., Cliffe, I.A., Hartley, J.E., Jones, D.E., McLenachan, A., Stanhope, K.J., Critchley, D.J., Childs, K.J., Middlefell, V.C., Lanfumey, L., Corradetti, R., Laporte, A.M., Gozlan, H., Hamon, M. and Dourish, C.T. (1996) Electrophysiological, biochemical, neurohormonal and behavioural studies with WAY-100635, a potent, selective and silent 5-HT1A receptor antagonist. Behav. Brain Res., 73: 337–353.

Foreman, M.M. and Moss, R.L. (1978) Role of hypothalamic serotonergic receptors in the control of lordosis behavior in the female rat. Horm. Behav., 10: 97–106.

Frank, J.L., Hendricks, S.E. and Olson, C.H. (2000) Multiple ejaculations and chronic fluoxetine: effects on male rat copulatory behavior. Pharmacol. Biochem. Behav., 66: 337–342.

Frankfurt, M., McKittrick, C.R., Mendelson, S.D. and McEwen, B.S. (1994) Effect of 5,7-dihydroxytryptamine, ovariectomy and gonadal steroids on serotonin receptor binding in rat brain. Neuroendocrinology, 59: 245–250.

Frankfurt, M., Renner, K., Azmitia, E. and Luine, V. (1985) Intrahypothalamic 5,7-dihydroxytryptamine: temporal analysis of effects on 5-hydroxytryptamine content in brain nuclei and on facilitated lordosis behavior. Brain Res., 340: 127–133.

Frohlich, P.F. and Meston, C.M. (2000) Evidence that serotonin affects female sexual functioning via peripheral mechanisms. Physiol. Behav., 71: 383–393.

Frohlich, P. and Meston, C.M. (2005) Fluoxetine-induced changes in tactile sensation and sexual functioning among clinically depressed women. J. Sex Marital Ther., 31: 113–128.

Frye, C.A., Petralia, S.M., Rhodes, M.E. and Stein, B. (2003) Fluoxetine may influence lordosis of rats through effects on midbrain 3 alpha,5 alpha-THP concentrations. Ann. N.Y. Acad. Sci., 1007: 37–41.

Frye, C.A. and Rhodes, M.E. (2003) Zaprinast, a phosphodiesterase 5 inhibitor, overcomes sexual dysfunction produced by fluoxetine, a selective serotonin reuptake inhibitor in hamsters. Neuropsychopharmacology, 28: 310–316.

Gereau, R.W., Kedzie, K.A. and Renner, K.J. (1993) Effect of progesterone on serotonin turnover in rats primed with estrogen implants into the ventromedial hypothalamus. Brain. Res. Bull., 32: 293–300.
Giuliano, F. (2007) 5-Hydroxytryptamine in premature ejaculation: opportunities for therapeutic intervention. Trends Neurosci., 30: 79–84.
Giuliano, F. and Rampin, O. (2004) Neural control of erection. Physiol. Behav., 83: 189–201.
Gobert, A., Rivet, J.M., Cistarelli, L. and Millan, M.J. (1997) Potentiation of the fluoxetine-induced increase in dialysate levels of serotonin (5-HT) in the frontal cortex of freely moving rats by combined blockade of 5-HT1A and 5-HT1B receptors with WAY 100,635 and GR 127,935. J. Neurochem., 68: 1159–1163.
Gonzalez, M.I., Greengrass, P., Russell, M. and Wilson, C.A. (1997) Comparison of serotonin receptor numbers and activity in specific hypothalamic areas of sexually active and inactive female rats. Neuroendocrinology, 66: 384–392.
Gregorian, R.S., Golden, K.A., Bahce, A., Goodman, C., Kwong, W.J. and Khan, Z.M. (2002) Antidepressant-induced sexual dysfunction. Ann. Pharmacother., 36: 1577–1589.
Guarraci, F.A., Megroz, A.B. and Clark, A.S. (2004) Paced mating behavior in the female rat following lesions of three regions responsive to vaginocervical stimulation. Brain Res., 999: 40–52.
Guptarak, J., Selvamani, A. and Uphouse, L. (2004) GABAA-5-HT1A receptor interaction in the mediobasal hypothalamus. Brain Res., 1027: 144–150.
Heisler, L.K., Kanarek, R.B. and Homoleski, B. (1999) Reduction of fat and protein intakes but not carbohydrate intake following acute and chronic fluoxetine in female rats. Pharmacol. Biochem. Behav., 63: 377–385.
Hensler, J.G. (2003) Regulation of 5-HT1A receptor function in brain following agonist or antidepressant administration. Life Sci., 72: 1665–1682.
Hillegaart, V. and Ahlenius, S. (1998) Facilitation and inhibition of male rat ejaculatory behaviour by the respective 5-HT1A and 5-HT1B receptor agonists 8-OH-DPAT and anpirtoline, as evidenced by use of the corresponding new and selective receptor antagonists NAD-299 and NAS-181. Br. J. Pharmacol., 125: 1733–1743.
Hillegaart, V., Ahlenius, S. and Larsson, K. (1989) Effects of local application of 5-HT into the median and dorsal raphe nuclei on male rat sexual and motor behavior. Behav. Brain Res., 33: 279–286.
Hillegaart, V., Ahlenius, S. and Larsson, K. (1991) Region-selective inhibition of male rat sexual behavior and motor performance by localized forebrain 5-HT injections: a comparison with effects produced by 8-OH-DPAT. Behav. Brain Res., 42: 169–180.
Hjorth, S. and Sharp, T. (1991) Effect of the 5-HT1A receptor agonist 8-OH-DPAT on the release of 5-HT in dorsal and median raphe-innervated rat brain regions as measured by *in vivo* microdialysis. Life Sci., 48: 1779–1786.
Hull, E.M. and Dominguez, J.M. (2007) Sexual behavior in male rodents. Horm. Behav., 52: 45–55.
Hull, E.M., Lorrain, D.S., Du, J., Matuszewich, L., Lumley, L.A., Putnam, S.K. and Moses, J. (1999) Hormone–neurotransmitter interactions in the control of sexual behavior. Behav. Brain Res., 105: 105–116.
Hull, E.M., Muschamp, J.W. and Sato, S. (2004) Dopamine and serotonin: influences on male sexual behavior. Physiol. Behav., 83: 291–307.
Jackson, A. and Etgen, A.M. (2001) Estrogen modulates 5-HT(1A) agonist inhibition of lordosis behavior but not binding of [(3)H]-8-OH-DPAT. Pharmacol. Biochem. Behav., 68: 221–227.
Jackson, A. and Uphouse, L. (1998) Dose-dependent effects of estradiol benzoate on 5-HT1A receptor agonist action. Brain Res., 796: 299–302.
Kaye, W. (2008) Neurobiology of anorexia and bulimia nervosa. Physiol. Behav., 94: 121–135.
Kendirci, M., Salem, E. and Hellstrom, W.J. (2007) Dapoxetine, a novel selective serotonin transport inhibitor for the treatment of premature ejaculation. Ther Clin. Risk Manag., 3: 277–289.
Kennett, G.A., Marcou, M., Dourish, C.T. and Curzon, G. (1987) Single administration of 5-HT1A agonists decreases 5-HT1A presynaptic, but not postsynaptic receptor-mediated responses: relationship to antidepressant-like action. Eur. J. Pharmacol., 138: 53–60.
Kishitake, M. and Yamanouchi, K. (2004) Facilitatory effects of WAY-100635, a 5-HT1A receptor antagonist, on lordosis in female rats. Neurosci. Lett., 371: 147–151.
Kishitake, M. and Yamanouchi, K. (2005) Effects of 5-HT1A-receptor agonist, 8-OH-DPAT, and GABAB-receptor agonist, baclofen, on lordosis in female rats with lesions in either the dorsal raphe nucleus or septum. J. Pharmacol. Sci., 98: 419–424.
Kitson, S.L. (2007) 5-hydroxytryptamine (5-HT) receptor ligands. Curr. Pharm. Des., 13: 2621–2637.
Krebs-Thomson, K. and Geyer, M.A. (1998) Evidence for a functional interaction between 5-HT1A and 5-HT2 receptors in rats. Psychopharmacology (Berl.), 140: 69–74.
Landry, M. and Di Paolo, T. (2003) Effect of chronic estradiol, tamoxifen or raloxifene treatment on serotonin 5-HT1A receptor. Brain Res. Mol. Brain Res., 112: 82–89.
Langhans, W. (2007) Signals generating anorexia during acute illness. Proc. Nutr. Soc., 66: 321–330.
Larsson, K. and Ahlenius, S. (1999) Brain and sexual behavior. Ann. N.Y. Acad. Sci., 877: 292–308.
Lau, D.H., Thompson, C.S., Mumtaz, F.H., Morgan, R.J. and Mikhailidis, D.P. (2007) Serotonin induces a biphasic response in rabbit cavernosal smooth muscle: relevance to the erectile process. Urol. Int., 79: 255–261.
Lee, R.L., Smith, E.R., Mas, M. and Davidson, J.M. (1990) Effects of intrathecal administration of 8-OH-DPAT on genital reflexes and mating behavior in male rats. Physiol. Behav., 47: 665–669.
Looney, C., Thor, K.B., Ricca, D. and Marson, L. (2005) Differential effects of simultaneous or sequential administration of paroxetine and WAY-100,635 on ejaculatory behavior. Pharmacol. Biochem. Behav., 82: 427–433.
Lorrain, D.S., Matuszewich, L., Friedman, R.D. and Hull, E.M. (1997) Extracellular serotonin in the lateral hypothalamic area is increased during the postejaculatory interval and impairs copulation in male rats. J. Neurosci., 17: 9361–9366.
Lorrain, D.S., Matuszewich, L. and Hull, E.M. (1998) 8-OH-DPAT influences extracellular levels of serotonin and dopamine in the medial preoptic area of male rats. Brain Res., 790: 217–223.
Lovenberg, T.W., Baron, B.M., de Lecea, L., Miller, J.D., Prosser, R.A., Rea, M.A., Foye, P.E., Racke, M., Slone, A.L., Siegel, B.W., Danielson, P.E., Sutcliffe, J.G. and Erlander, M.G. (1993) A novel adenylyl cyclase-activating serotonin receptor (5-HT7) implicated in the regulation of mammalian circadian rhythms. Neuron, 11: 449–458.

Lu, N.Z. and Bethea, C.L. (2002) Ovarian steroid regulation of 5-HT1A receptor binding and G protein activation in female monkeys. Neuropsychopharmacology, 27: 12–24.
Luine, V.N. and Paden, C.M. (1982) Effects of monoamine oxidase inhibition on female sexual behavior, serotonin levels and type A and B monoamine oxidase activity. Neuroendocrinology, 34: 245–251.
Luine, V.N., Renner, K.J., Frankfurt, M. and Azmitia, E.C. (1984) Facilitated sexual behavior reversed and serotonin restored by raphe nuclei transplanted into denervated hypothalamus. Science, 226: 1436–1439.
Luine, V.N., Thornton, J.E., Frankfurt, M. and MacLusky, N.J. (1987) Effects of hypothalamic serotonin depletion on lordosis behavior and gonadal hormone receptors. Brain Res., 426: 47–54.
Marson, L. and Gravitt, K. (2004) Spinal neurons activated with the urethrogenital reflex in the male rat. Brain Res., 1026: 108–115.
Marson, L. and McKenna, K.E. (1990) The identification of a brainstem site controlling spinal sexual reflexes in male rats. Brain Res., 515: 303–308.
Marson, L. and McKenna, K.E. (1994) Stimulation of the hypothalamus initiates the urethrogenital reflex in male rats. Brain Res., 638: 103–108.
Mas, M. (1995) Neurobiological correlates of masculine sexual behavior. Neurosci. Biobehav. Rev., 19: 261–277.
Mas, M., Zahradnik, M.A., Martino, V. and Davidson, J.M. (1985) Stimulation of spinal serotonergic receptors facilitates seminal emission and suppresses penile erectile reflexes. Brain Res., 342: 128–134.
Maswood, N., Caldarola-Pastuszka, M. and Uphouse, L. (1997) 5-HT3 receptors in the ventromedial nucleus of the hypothalamus and female sexual behavior. Brain Res., 769: 13–20.
Maswood, N., Caldarola-Pastuszka, M. and Uphouse, L. (1998) Functional integration among 5-hydroxytryptamine receptor families in the control of female rat sexual behavior. Brain Res., 802: 98–103.
Maswood, N., Sarkar, J. and Uphouse, L. (2008) Modest effects of repeated fluoxetine on estrous cyclicity and sexual behavior of Sprague Dawley female rats. Brain Res., in press.
Maswood, S., Truitt, W., Hotema, M., Caldarola-Pastuszka, M. and Uphouse, L. (1999) Estrous cycle modulation of extracellular serotonin in mediobasal hypothalamus: role of the serotonin transporter and terminal autoreceptors. Brain Res., 831: 146–154.
Matuszczyk, J.V., Larsson, K. and Eriksson, E. (1998) Subchronic administration of fluoxetine impairs estrous behavior in intact female rats. Neuropsychopharmacology, 19: 492–498.
McGinnis, M.Y. and Dreifuss, R.M. (1989) Evidence for a role of testosterone–androgen receptor interactions in mediating masculine sexual behavior in male rats. Endocrinology, 124: 618–626.
McGinnis, M.Y., Mirth, M.C., Zebrowski, A.F. and Dreifuss, R.M. (1989) Critical exposure time for androgen activation of male sexual behavior in rats. Physiol. Behav., 46: 159–165.
McKenna, K.E., Chung, S.K. and McVary, K.T. (1991b) A model for the study of sexual function in anesthetized male and female rats. Am. J. Physiol., 261: R1276–R1285.
McKenna, K.E., Knight, K.C. and Mayers, R. (1991a) Modulation by peripheral serotonin of the threshold for sexual reflexes in female rats. Pharmacol. Biochem. Behav., 40: 151–156.
McQueen, J.K., Wilson, H. and Fink, G. (1997) Estradiol-17 beta increases serotonin transporter (SERT) mRNA levels and the density of SERT-binding sites in female rat brain. Brain Res. Mol. Brain Res., 45: 13–23.
Medina, P., Segarra, G., Ballester, R., Chuan, P., Domenech, C., Vila, J.M. and Lluch, S. (2000) Effects of antidepressants in adrenergic neurotransmission of human vas deferens. Urology, 55: 592–597.
Mendelson, S.D. (1992) A review and reevaluation of the role of serotonin in the modulation of lordosis behavior in the female rat. Neurosci. Biobehav. Rev., 16: 309–350.
Mendelson, S.D. and Gorzalka, B.B. (1985) A facilitatory role for serotonin in the sexual behavior of the female rat. Pharmacol. Biochem. Behav., 22: 1025–1033.
Mendelson, S.D. and Gorzalka, B.B. (1986a) 5-HT1A receptors: differential involvement in female and male sexual behavior in the rat. Physiol. Behav., 37: 345–351.
Mendelson, S.D. and Gorzalka, B.B. (1986b) Effects of 5-HT1A selective anxiolytics on lordosis behavior: interactions with progesterone. Eur. J. Pharmacol., 132: 323–326.
Mendelson, S.D. and Gorzalka, B.B. (1990) Sex differences in the effects of 1-(m-trifluoromethylphenyl) piperazine and 1-(m-chlorophenyl) piperazine on copulatory behavior in the rat. Neuropharmacology, 29: 783–786.
Mercado, M.G., Kimmer, S.L., Holman, J.R., Wanserski, G. and Lo, V. (2008) Clinical inquiries. What's the best drug treatment for premature ejaculation? J. Fam. Pract., 57: 192–194.
Mikkelsen, J.D., Hay-Schmidt, A. and Kiss, A. (2004) Serotonergic stimulation of the rat hypothalamo-pituitary–adrenal axis: interaction between 5-HT1A and 5-HT2A receptors. Ann. N.Y. Acad. Sci., 1018: 65–70.
Millan, M.J. and Perrin-Monneyron, S. (1997) Potentiation of fluoxetine-induced penile erections by combined blockade of 5-HT1A and 5-HT1B receptors. Eur. J. Pharmacol., 321: R11–R13.
Millan, M.J., Peglion, J.L., Lavielle, G. and Perrin-Monneyron, S. (1997) 5-HT2C receptors mediate penile erections in rats: actions of novel and selective agonists and antagonists. Eur. J. Pharmacol., 325: 9–12.
Mize, A.L., Young, L.J. and Alper, R.H. (2003) Uncoupling of 5-HT1A receptors in the brain by estrogens: regional variations in antagonism by ICI 182,780. Neuropharmacology, 44: 584–591.
Montgomery, S.A., Baldwin, D.S. and Riley, A. (2002) Antidepressant medications: a review of the evidence for drug-induced sexual dysfunction. J. Affect. Disord., 69: 119–140.
Mos, J., Van Logten, J., Bloetjes, K. and Olivier, B. (1991) The effects of idazoxan and 8-OH-DPAT on sexual behaviour and associated ultrasonic vocalizations in the rat. Neurosci. Biobehav. Rev., 15: 505–515.
Nedergaard, P., Sanchez, C. and Mellerup, E. (2004) Different roles of 5-HT2A and 5-HT2C receptors in regulation of female rat paced mating behaviour. Behav. Brain Res., 149: 151–157.
Newman, M.E., Shalom, G., Ran, A., Gur, E. and Van de Kar, L.D. (2004) Chronic fluoxetine-induced desensitization of 5-HT1A and 5-HT1B autoreceptors: regional differences and effects of WAY-100635. Eur. J. Pharmacol., 486: 25–30.
Nonogaki, K. (2008) Ghrelin and feedback systems. Vitam. Horm., 77: 149–170.
Nurnberg, H.G. (2008) An evidence-based review updating the various treatment and management approaches to serotonin

reuptake inhibitor-associated sexual dysfunction. Drugs Today (Barc.), 44: 147–168.
Padoin, M.J. and Lucion, A.B. (1995) The effect of testosterone and DOI (1-(2,5-dimethoxy-4-iodophenyl)-2-aminopropane) on male sexual behavior of rats. Eur. J. Pharmacol., 277: 1–6.
Paredes, R.G., Highland, L. and Karam, P. (1993) Socio-sexual behavior in male rats after lesions of the medial preoptic area: evidence for reduced sexual motivation. Brain Res., 618: 271–276.
Paredes, R.G., Kica, E. and Baum, M.J. (1994) Differential effects of the serotonin1A agonist, 8-OH-DPAT, on masculine and feminine sexual behavior of the ferret. Psychopharmacology (Berl.), 114: 591–596.
Paredes, R.G., Contreras, J.L. and Agmo, A. (2000) Serotonin and sexual behavior in the male rabbit. J. Neural. Transm., 107: 767–777.
Paredes, R.G., Tzschentke, T. and Nakach, N. (1998) Lesions of the medial preoptic area/anterior hypothalamus (MPOA/AH) modify partner preference in male rats. Brain Res., 813: 1–8.
Pfaff, D.W. and Modianos, D. (1985) Neural mechanisms of female reproductive behavior. In: Adler, D., Pfaff, D. and Goy, R.W. (Eds.), Handbook of Behavioral Neurobiology. Plenum Press, New York, NY, pp. 423–493.
Pfaus, J.G., Damsma, G., Wenkstern, D. and Fibiger, H.C. (1995) Sexual activity increases dopamine transmission in the nucleus accumbens and striatum of female rats. Brain Res., 693: 21–30.
Pfaus, J.G. and Phillips, A.G. (1991) Role of dopamine in anticipatory and consummatory aspects of sexual behavior in the male rat. Behav. Neurosci., 105: 727–743.
Pomerantz, S.M., Hepner, B.C. and Wertz, J.M. (1993a) Serotonergic influences on male sexual behavior of rhesus monkeys: effects of serotonin agonists. Psychopharmacology (Berl.), 111: 47–54.
Pomerantz, S.M., Hepner, B.C. and Wertz, J.M. (1993b) 5-HT1A and 5-HT1C/1D receptor agonists produce reciprocal effects on male sexual behavior of rhesus monkeys. Eur. J. Pharmacol., 243: 227–234.
Popova, N.K. and Amstislavskaya, T.G. (2002a) Involvement of the 5-HT(1A) and 5-HT(1B) serotonergic receptor subtypes in sexual arousal in male mice. Psychoneuroendocrinology, 27: 609–618.
Popova, N.K. and Amstislavskaya, T.G. (2002b) 5-HT2A and 5-HT2C serotonin receptors differentially modulate mouse sexual arousal and the hypothalamo-pituitary-testicular response to the presence of a female. Neuroendocrinology, 76: 28–34.
Putnam, S.K., Sato, S. and Hull, E.M. (2003) Effects of testosterone metabolites on copulation and medial preoptic dopamine release in castrated male rats. Horm. Behav., 44: 419–426.
Raap, D.K., DonCarlos, L., Garcia, F., Muma, N.A., Wolf, W.A., Battaglia, G. and Van de Kar, L.D. (2000) Estrogen desensitizes 5-HT(1A) receptors and reduces levels of G(z), G(i1) and G(i3) proteins in the hypothalamus. Neuropharmacology, 39: 1823–1832.
Rehman, J., Kaynan, A., Christ, G., Valcic, M., Maayani, S. and Melman, A. (1999) Modification of sexual behavior of Long-Evans male rats by drugs acting on the 5-HT1A receptor. Brain Res., 821: 414–425.
Rizvi, T.A., Murphy, A.Z., Ennis, M., Behbehani, M.M. and Shipley, M.T. (1996) Medial preoptic area afferents to periaqueductal gray medullo-output neurons: a combined Fos and tract tracing study. J. Neurosci., 16: 333–344.
Rodriguez-Manzo, G., Lopez-Rubalcava, C., Hen, R. and Fernandez-Guasti, A. (2002) Participation of 5-HT(1B) receptors in the inhibitory actions of serotonin on masculine sexual behaviour of mice: pharmacological analysis in 5-HT(1B) receptor knockout mice. Br. J. Pharmacol., 136: 1127–1134.
Rossler, A.S., Bernabe, J., Denys, P., Alexandre, L. and Giuliano, F. (2006) Effect of the 5-HT receptor agonist DOI on female rat sexual behavior. J. Sex Med., 3: 432–441.
Rowland, D.L. and Houtsmuller, E.J. (1998) 8-OH-DPAT interacts with sexual experience and testosterone to affect ejaculatory response in rats. Pharmacol. Biochem. Behav., 60: 143–149.
Sanger, G.J. (2008) 5-Hydroxytryptamine and the gastrointestinal tract: where next? Trends Pharmacol. Sci., in press.
Sari, Y. (2004) Serotonin1B receptors: from protein to physiological function and behavior. Neurosci. Biobehav. Rev., 28: 565–582.
Sarkar, J., Hiegel, C., Ginis, G.E., Hilbun, E. and Uphouse, L. (2008a) Subchronic treatment with fluoxetine attenuates effects of acute fluoxetine on female rat sexual behavior. Brain Res., 1190: 56–64.
Sarkar, J., Hiegel, C., Maswood, N. and Uphouse, L. (2008b) Daily male exposure attenuates estrous cycle disruption by fluoxetine. Behav. Brain Res., 189: 83–91.
Schnur, S.L., Smith, E.R., Lee, R.L., Mas, M. and Davidson, J.M. (1989) A component analysis of the effects of DPAT on male rat sexual behavior. Physiol. Behav., 45: 897–901.
Siddiqui, A., Niazi, A., Shaharyar, S. and Wilson, C.A. (2007) The 5HT(7) receptor subtype is involved in the regulation of female sexual behaviour in the rat. Pharmacol. Biochem. Behav., 87: 386–392.
Siddiqui, A. and Shaharyar, S. (2007) Attenuating effects of prior oestradiol benzoate priming on 5-HT-mediated lordosis behavior in rats are dose-dependent. Acta Physiol. Hung., 94: 209–221.
Simon, P., Bertrand, J. and Costentin, J. (1993) 5-HT1A receptor blockade increases penile erections induced by indirect serotonin agonists. Neuroreport, 5: 229–230.
Sinclair-Worley, L. and Uphouse, L. (2004) Effect of estrogen on the lordosis-inhibiting action of ketanserin and SB 206553. Behav. Brain Res., 152: 129–135.
Sodersten, P. (1981) Estradiol–progesterone interactions in the reproductive behavior of female rats. In: Ganten, D. and Pfaff, D. (Eds.), Current Topics in Neuroendocrinology: Actions of Progesterone on the Brain. Springer-Verlag, New York, NY, pp. 141–174.
Somerville, E.M., Horwood, J.M., Lee, M.D., Kennett, G.A. and Clifton, P.G. (2007) 5-HT(2C) receptor activation inhibits appetitive and consummatory components of feeding and increases brain c-fos immunoreactivity in mice. Eur. J. Neurosci., 25: 3115–3124.
Sprouse, J.S. and Aghajanian, G.K. (1987) Electrophysiological responses of serotoninergic dorsal raphe neurons to 5-HT1A and 5-HT1B agonists. Synapse, 1: 3–9.
Steers, W.D. and de Groat, W.C. (1989) Effects of m-chlorophenylpiperazine on penile and bladder function in rats. Am. J. Physiol., 257: R1441–R1449.
Straiko, M.M., Gudelsky, G.A. and Coolen, L.M. (2007) Treatment with a serotonin-depleting regimen of MDMA prevents conditioned place preference to sex in male rats. Behav. Neurosci., 121: 586–593.
Sukoff Rizzo, S.J., Schechter, L.E. and Rosenzweig-Lipson, S. (2008) A novel approach for predicting antidepressant-induced

sexual dysfunction in rats. Psychopharmacology (Berl.), 195: 459–467.
Svensson, L. and Hansen, S. (1984) Spinal monoaminergic modulation of masculine copulatory behavior in the rat. Brain Res., 302: 315–321.
Tanco, S.A., Watson, N.V. and Gorzalka, B.B. (1993) Lack of effects of 5-HT3 antagonists on normal and morphine-attenuated sexual behaviours in female and male rats. Experientia, 49: 238–241.
Tanco, S.A., Watson, N.V. and Gorzalka, B.B. (1994) Effects of 5-HT3 agonists on reproductive behaviors in rats. Psychopharmacology (Berl.), 115: 245–248.
Tao, R., Ma, Z. and Auerbach, S.B. (2000) Differential effect of local infusion of serotonin reuptake inhibitors in the raphe versus forebrain and the role of depolarization-induced release in increased extracellular serotonin. J. Pharmacol. Exp. Ther., 294: 571–579.
Taylor, G., Bardgett, M., Csernansky, J., Early, T., Haller, J., Scherrer, J. and Womack, S. (1996) Male reproductive systems under chronic fluoxetine or trimipramine treatment. Physiol. Behav., 59: 479–485.
Trevino, A., Wolf, A., Jackson, A., Price, T. and Uphouse, L. (1999) Reduced efficacy of 8-OH-DPAT's inhibition of lordosis behavior by prior estrogen treatment. Horm. Behav., 35: 215–223.
Truitt, W.A. and Coolen, L.M. (2002) Identification of a potential ejaculation generator in the spinal cord. Science, 297: 1566–1569.
Truitt, W., Harrison, L., Guptarak, J., White, S., Hiegel, C. and Uphouse, L. (2003) Progesterone attenuates the effect of the 5-HT1A receptor agonist, 8-OH-DPAT, and of mild restraint on lordosis behavior. Brain Res., 974: 202–211.
Uphouse, L. (2000) Female gonadal hormones, serotonin, and sexual receptivity. Brain Res. Brain Res. Rev, 33: 242–257.
Uphouse, L., Andrade, M., Caldarola-Pastuszka, M. and Jackson, A. (1996) 5-HT1A receptor antagonists and lordosis behavior. Neuropharmacology, 35: 489–495.
Uphouse, L., Andrade, M., Caldarola-Pastuszka, M. and Maswood, S. (1994b) Hypothalamic infusion of the 5-HT2/1C agonist, DOI, prevents the inhibitory actions of the 5-HT1A agonist, 8-OH-DPAT, on lordosis behavior. Pharmacol. Biochem. Behav., 47: 467–470.
Uphouse, L., Caldarola-Pastuszka, M. and Droge, M. (1992) 8-OH-DPAT in the midbrain central gray inhibits lordosis behavior. Pharmacol. Biochem. Behav., 43: 833–838.
Uphouse, L., Hensler, J.G., Sarkar, J. and Grossie, B. (2006) Fluoxetine disrupts food intake and estrous cyclicity in Fischer female rats. Brain Res., 1072: 79–90.
Uphouse, L., Hiegel, C., Perez, E. and Guptarak, J. (2007) Serotonin receptor involvement in effects of restraint on female rat lordosis behavior. Pharmacol. Biochem. Behav., 86: 631–636.
Uphouse, L., Hiegel, C., Guptarak, J. and Maswood, N. (2008) Progesterone reduces the effect of the serotonin 1B/1D receptor antagonist, GR 127935, on lordosis behavior. Horm. Behav., in press.
Uphouse, L., Maswood, S. and Caldarola-Pastuszka, M. (1994a) Agonist activation of 5-HT1A receptors in the median raphe nucleus and female rat lordosis behavior. Brain Res., 668: 271–275.
Uphouse, L. and Wolf, A. (2004) WAY100635 and female rat lordosis behavior. Brain Res., 1013: 260–263.
Uphouse, L., White, S., Harrison, L., Hiegel, C., Majumdar, D., Guptarak, J. and Truitt, W.A. (2003) Restraint accentuates the effects of 5-HT2 receptor antagonists and a 5-HT1A receptor agonist on lordosis behavior. Pharmacol. Biochem. Behav., 76: 63–73.
Vagell, M.E. and McGinnis, M.Y. (1997) The role of aromatization in the restoration of male rat reproductive behavior. J. Neuroendocrinol., 9: 415–421.
Van de Kar, L.D., Raap, D.K., Battaglia, G., Muma, N.A., Garcia, F. and DonCarlos, L.L. (2002) Treatment of cycling female rats with fluoxetine induces desensitization of hypothalamic 5-HT(1A) receptors with no change in 5-HT(2A) receptors. Neuropharmacology, 43: 45–54.
Vega Matuszcyk, J., Larsson, K. and Eriksson, E. (1998) The selective serotonin reuptake inhibitor fluoxetine reduces sexual motivation in male rats. Pharmacol. Biochem. Behav., 60: 527–532.
Verma, S., Chhina, G.S., Mohan Kumar, V. and Singh, B. (1989) Inhibition of male sexual behavior by serotonin application in the medial preoptic area. Physiol. Behav., 46: 327–330.
Vinkers, C.H., van Bogaert, M.J., Klanker, M., Korte, S.M., Oosting, R., Hanania, T., Hopkins, S.C., Olivier, B. and Groenink, L. (2008) Translational aspects of pharmacological research into anxiety disorders: the stress-induced hyperthermia (SIH) paradigm. Eur. J. Pharmacol., 585: 407–425.
Waldinger, M.D. (2005) Lifelong premature ejaculation: definition, serotonergic neurotransmission and drug treatment. World J. Urol., 23: 102–108.
Waldinger, M.D. and Olivier, B. (1998) Selective serotonin reuptake inhibitor-induced sexual dysfunction: clinical and research considerations. Intl. Clin. Psychopharmacol., 13(Suppl. 6): S27–S33.
Waldinger, M.D., Hengeveld, M.W., Zwinderman, A.H. and Olivier, B. (1998) Effect of SSRI antidepressants on ejaculation: a double-blind, randomized, placebo-controlled study with fluoxetine, fluvoxamine, paroxetine, and sertraline. J. Clin. Psychopharmacol., 18: 274–281.
Waldinger, M.D. and Olivier, B. (2004) Utility of selective serotonin reuptake inhibitors in premature ejaculation. Curr. Opin. Investig. Drugs, 5: 743–747.
Waldinger, M.D., Schweitzer, D.H. and Olivier, B. (2005) On-demand SSRI treatment of premature ejaculation: pharmacodynamic limitations for relevant ejaculation delay and consequent solutions. J. Sex Med., 2: 121–131.
Waldinger, M.D., van De Plas, A., Pattij, T., van Oorschot, R., Coolen, L.M., Veening, J.G. and Olivier, B. (2002) The selective serotonin re-uptake inhibitors fluvoxamine and paroxetine differ in sexual inhibitory effects after chronic treatment. Psychopharmacology (Berl.), 160: 283–289.
Wang, W.F., Chang, L., Minhas, S. and Ralph, D.J. (2007) Selective serotonin reuptake inhibitors in the treatment of premature ejaculation. Chin. Med. J. (Engl.), 120: 1000–1006.
Wilson, J.D. (1999) The role of androgens in male gender role behavior. Endocr. Rev., 20: 726–737.
Wilson, J.D., Griffin, J.E. and Russell, D.W. (1993) Steroid 5 alpha-reductase 2 deficiency. Endocr. Rev., 14: 577–593.
Wolf, A., Caldarola-Pastuszka, M., DeLashaw, M. and Uphouse, L. (1999) 5-HT2C receptor involvement in female rat lordosis behavior. Brain Res., 825: 146–151.
Wolf, A., Caldarola-Pastuszka, M. and Uphouse, L. (1998) Facilitation of female rat lordosis behavior by hypothalamic infusion of 5-HT(2A/2C) receptor agonists. Brain Res., 779: 84–95.
Yells, D.P., Prendergast, M.A., Hendricks, S.E. and Miller, M.E. (1995) Monoaminergic influences on temporal patterning of sexual behavior in male rats. Physiol. Behav., 58: 847–852.

Yells, D.P., Prendergast, M.A., Hendricks, S.E. and Nakamura, M. (1994) Fluoxetine-induced inhibition of male rat copulatory behavior: modification by lesions of the nucleus paragigantocellularis. Pharmacol. Biochem. Behav., 49: 121–127.

Yonezawa, A., Ando, R., Imai, M., Watanabe, C., Furuta, S., Kutsuwa, M., Kimura, Y. and Sakurada, S. (2004) Differential effects of yohimbine, naloxone and 8-OH-DPAT on ejaculatory response in male dogs. Methods Find Exp. Clin. Pharmacol., 26: 47–51.

Yonezawa, A., Watanabe, C., Ando, R., Furuta, S., Sakurada, S., Yoshimura, H., Iwanaga, T. and Kimura, Y. (2000) Characterization of p-chloroamphetamine-induced penile erection and ejaculation in anesthetized rats. Life Sci., 67: 3031–3039.

Yonezawa, A., Yoshizumi, M., Ebiko, M., Ise, S.N., Watanabe, C., Mizoguchi, H., Kimura, Y. and Sakurada, S. (2008) Ejaculatory response induced by a 5-HT2 receptor agonist m-CPP in rats: differential roles of 5-HT2 receptor subtypes. Pharmacol. Biochem. Behav., 88: 367–373.

CHAPTER 3.5

Serotonin in Mood and Emotion

Julie G. Hensler*

Department of Pharmacology, University of Texas Health Science Center-San Antonio, San Antonio, Texas

Abstract: The topographical organization of serotonergic innervation of the forebrain, as well as the paracrine nature of serotonergic neurotransmission in limbic structures, support the contention that serotonin plays a neuromodulatory role in the brain. The lack of synaptic specializations at serotonergic terminals and varicosities in many brain regions suggests that serotonergic innervation may be particularly pliable. As conscious feeling or the cognitive aspect of emotion involves learning and memory, the serotonergic system may be particularly well suited for modulating mood. Pharmacological manipulations of serotonergic neurotransmission alter emotional processing, attentional bias, emotional memory, dysfunctional attitudes and decision-making. In healthy human subjects, increases in serotonergic neurotransmission result in enhanced attention and recognition of positive emotional material. Contrary to expectation, acute increases in serotonergic neurotransmission also increase attentional bias towards negative or fearful stimuli. In general, decreased serotonergic neurotransmission results in impaired attention and recognition of positive emotional material, and increases the attentional bias towards negative stimuli in healthy subjects. A polymorphism in the serotonin transporter gene also alters emotional processing. Understanding not only the neurobiology of emotional responses, but also serotonergic modulation of emotional states, will be important as we attempt to elucidate the etiology of emotional disorders. Future research with agents specific to various serotonin receptors may identify important therapeutic targets for the treatment of cognitive and affective disorders.

Keywords: limbic system, serotonin receptors, tryptophan depletion, selective serotonin reuptake inhibitors, serotonin transporter, MDMA, fenfluramine.

Introduction

Webster's *New Collegiate Dictionary* defines mood as a conscious state of mind or predominant emotion. Emotion is defined as the affective aspect of consciousness. In addition to this conscious feeling or cognitive aspect of emotion, there is a visceral component, specifically the characteristic physical responses associated with emotional state. The present chapter focuses on the role of the neurotransmitter serotonin (5-hydroxytryptamine; 5-HT) in mood and emotion. A brief overview of the limbic system and innervation of these structures in brain, which are responsible for emotion, by serotonergic neurons is provided before discussing serotonergic modulation of emotion. In many instances, review articles are cited, in addition to original papers, in the hope that the interested reader will pursue these topics further. The role of serotonin, however, in psychopathologies or clinical disease states will not be specifically addressed in this chapter. The reader is referred to Chapters 4.2, 4.9 and 4.10 for the role of serotonin in major depressive disorder, in anxiety and panic disorders, and in aggression and violent behavior, respectively.

Neuroanatomy of emotion

The neural representation of feelings and emotion lies in the limbic system of the brain. The limbic system is composed of cortical as well as subcortical structures, which are intimately interconnected. The major structures of the limbic system include the prefrontal cortex, cingulate cortex, entorhinal cortex, hippocampus, nucleus accumbens (ventral striatum), ventral pallidum, amygdala, and anterior hypothalamus (Swanson and Petrovich, 1998; Iversen *et al.*, 2000; Heimer, 2003). Connections between these structures form complex circuits. For an excellent review of the organization of major neuronal circuits and neurochemical networks of the limbic system, the reader is referred to Morgane *et al.* (2005). As emphasized by Morgane *et al.* (2005), complex psychological states such as emotion and mood warrant consideration of the corre-

*E-mail: hensler@uthscsa.edu

Christian Müller & Barry Jacobs (Eds.)
Handbook of Behavioral Neurobiology of Serotonin
ISBN 978-0-12-374634-4

DOI: 10.1016/B978-0-12-374634-4.00022-8

sponding anatomical components as a whole, in that these psychological processes are not performed by single structures but by complexes of interacting systems in the brain.

The unconscious, visceral responses associated with emotion are mediated by subcortical areas. These systemic autonomic, endocrine and skeletal motor responses involve connections between the nuclei of the amygdala, the hypothalamus and the brainstem (Figure 1). The lower brainstem and spinal cord receive extensive connections from the limbic forebrain and limbic midbrain, and constitute a major output for the limbic system (Holstege, 1992; Holstege and Georgiadis, 2004). Cell body groups for various neurotransmitters are located in the midbrain – for example, serotonin (raphe nuclei), dopamine (substantia nigra and ventral tegmental area), acetylcholine (laterodorsal and pedunculopontine tegmental nuclei), norepinephrine (locus coeruleus) and epinephrine (nucleus of the solitary tract). The limbic midbrain can be divided into ventral and medial components. The medial tegmentum consists of the raphe nuclei, the central gray and the dorsal and ventral tegmental nuclei of Gudden. The lateral tegmentum is comprised of the ventral tegmental area, the

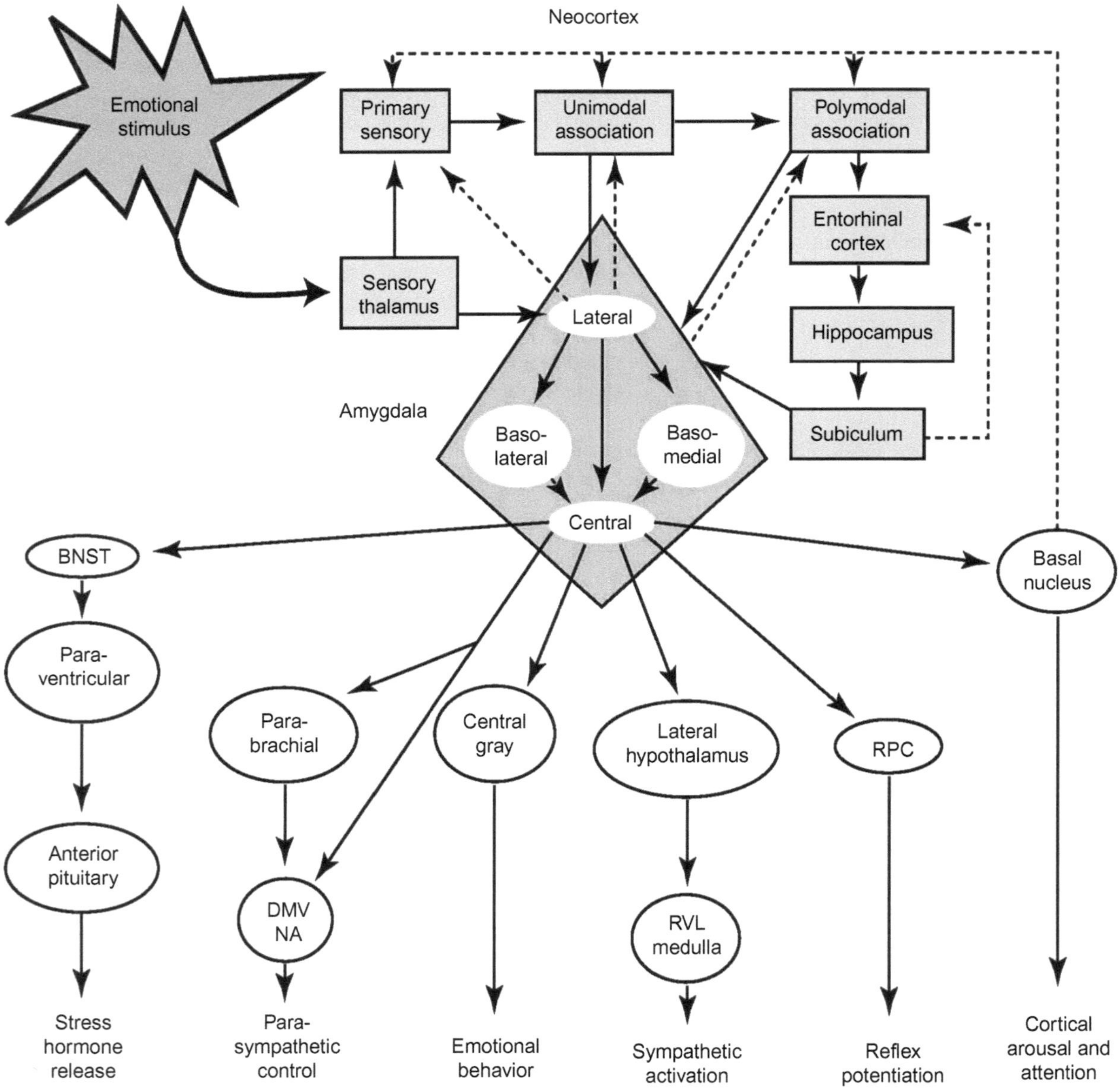

Figure 1 The amygdala serves as a hub, connecting the hypothalamus and brainstem, responsible for unconscious visceral responses associated with emotion, with cortical areas and hippocampus responsible for conscious feeling or cognitive aspects of emotion. BNST, bed nucleus of the stria terminalis; DMV, dorsal nucleus of the vagus; NA, nucleus ambiguus; RPC, nucleus reticularis pontis oralis; RVL, rostral ventral nucleus of the medulla. This figure was published in Eichenbaum, H.B, 'Learning and memory: brain systems', pp. 1299–1328, *Fundamental Neuroscience*, 2nd edition, edited by L.R. Squire *et al.* Copyright Elsevier (2003).

locus coeruleus, the substantia nigra, the nucleus ambiguus and the nucleus tractus solitarius. The hypothalamus is important in the regulation of homeostasis, and serves to integrate autonomic responses and endocrine function to maintain the constancy of the internal environment. The hypothalamus, which is intimately linked to the brainstem and limbic forebrain, coordinates the peripheral expression of emotion (Iversen *et al.*, 2000). For an excellent review and discussion of the limbic midbrain, and the involvement of the lower brainstem and spinal cord in unconscious visceral or somatic responses associated with emotion, the reader is referred to Morgane *et al.* (2005) and Morgane and Mokler (2006).

The amygdala is central to both cognitive and visceral aspects of emotion, in that it plays a critical role in coordinating the conscious experience of emotion and the peripheral expressions of emotion. The amygdala serves as a hub, connecting the hypothalamus and brainstem with cortical areas and hippocampus (Figure 1). Thus, the amygdala is not only essential for the expression of inherent or inborn emotional responses, but also plays a critical role in the acquisition and expression of learned emotional responses (see, for example, Armony and Le Doux, 1997; Hamann *et al.*, 1999; Baxter and Murray, 2002; Phelps and LeDoux, 2005). The reciprocal connections between the amygdala and the neocortex permit the incorporation of learning and experience into the cognitive aspects of emotion (Figure 1). For example, neuroimaging studies indicate that the amygdala and orbitofrontal cortex are common brain structures involved in the recognition and processing of facial expressions representing a variety of emotions (e.g., fear, disgust, sadness, happiness, and anger); both structures are activated by the perception of different emotional faces (Del-Ben *et al.*, 2008).

Conscious feeling or the cognitive aspect of emotion is believed to be mediated by the hippocampus and neocortex. Neuronal connections from hippocampus and neocortical areas to the amygdala make it possible for not just external stimuli but also memory and imagination to evoke emotional responses or feelings (Iversen *et al.*, 2000). The amygdala, in combination with the hippocampus and prefrontal cortex, plays an important role in the retrieval of memories for emotional events (Buchanan, 2007). As proposed by Armony and Le Doux (1997), the amygdala integrates various sensory and cognitive inputs from forebrain structures (i.e., the thalamus, hippocampus and neocortex). The thalamus provides information as to the basic nature or features of the stimulus, while cortical areas and the hippocampus provide additional 'higher-order' information about the complex nature of the stimulus. The hippocampus also plays a critical role in setting the emotional context (Williams *et al.*, 2005; Gentili *et al.*, 2008; Marschner *et al.*, 2008; Park *et al.*, 2008). The extinction of emotional responses is an active form of learning, and is mediated by a network of three structures: the amygdala, hippocampus and prefrontal cortex (see, for example, Phelps and LeDoux, 2005; Quirk and Mueller, 2008).

As discussed above, forebrain structures including orbitofrontal cortex, anterior cingulate cortex, insula and amygdala mediate higher-order emotional processing and learning. These involve the representation of stimulus values, the emotional enhancement of sensory processing, and predictions of values associated with actions in order to bias decision-making. Neuroimaging studies have provided substantial information about the neural basis of emotion, specifically the functional anatomy of emotion. The reader is referred to the following for recent reviews of the neuroanatomy of laughter (Meyer *et al.*, 2007), love (Zeki, 2007), regret (Coricelli *et al.*, 2007) and aversive emotional experiences (Phelps and LeDoux, 2005).

Serotonergic innervation of the limbic system

Especially high levels of serotonin are found in limbic forebrain structures. Serotonin projections to these forebrain structures arise from serotonergic cell body groups in the midbrain. Dahlström and Fuxe (1964), using the Falck-Hillarp technique of histofluorescence, observed that the majority of serotonergic soma were found in cell body groups previously designated by Taber *et al.* (1960) as the raphe nuclei based on cytoarchitectural criteria – i.e., on cell body structural characteristics and organization. Dahlström and Fuxe described nine groups of serotonin-containing cell bodies, designated B1–B9, which correspond for the most part with the raphe nuclei (Descarries *et al.*, 1982; Kohler and Steinbusch, 1982; Molliver, 1987; Törk, 1990).

Although serotonin-containing neuronal cell bodies are restricted to discrete groups of cells or nuclei located along the midline of the brainstem, their axonal projections innervate nearly every area of the central nervous system. Serotonergic cell body groups B1–B4 are more caudally situated (mid-pons to caudal medulla), and give rise to serotonergic axons that project within the brainstem and to the spinal cord. Ascending serotonergic projections innervating the forebrain, including limbic areas, arise primarily from the dorsal raphe (B6 and B7), median raphe (B8) and B9 cell group (Parent *et al.*, 1981; Kohler and Steinbusch, 1982). These serotonergic neurons form a dense plexus of axonal processes. The serotonergic projections, arising primarily from the dorsal and median raphe nuclei, compose two distinct serotonergic systems that differ in their electrophysiological characteristics (Kirby *et al.*, 2003; Beck *et al.*,

2004), topographic organization and morphology, as well as sensitivity to neurotoxins and perhaps psychoactive or therapeutic agents (Molliver, 1987). These differences may be extremely important for understanding the role of these two serotonergic systems in normal brain function and in mental illness. For a more detailed description of serotonergic innervation of limbic structures, the reader is referred to Hensler (2006).

In many areas of the limbic system, serotonergic neurotransmission can best be described as paracrine or volume transmission, which requires that the neurotransmitter diffuses to remote receptor sites. Extensive ultrastructural studies by Descarries and colleagues (Descarries and Mechawar, 2000; see also Chapter 1.4 of this volume) have shown that in the cerebral cortex and hippocampus of the rat, the majority of serotonergic varicosities are not associated with synaptic specializations. These findings indicate that the serotonergic innervation of cortex and hippocampus is predominantly non-junctional; specifically, high densities of small vesicles exist in varicosities lacking synaptic specializations and located far from any recognizable postsynaptic density. Additional factors determining the type of neurotransmission (i.e., restricted to the synaptic cleft or 'hard-wired' versus diffuse volume transmission) include the location of receptors with respect to release sites, the amount of neurotransmitter released, the rate of diffusion away from the release site, and the removal or reuptake of the neurotransmitter by its transporter. The extrasynaptic location of serotonin receptors and the serotonin transporter is further evidence for paracrine serotonergic neurotransmission in many areas of brain (Hensler, 2006).

It has been proposed by Descarries that neuronal varicosities deprived of synaptic attachment might undergo translocation and reshaping along their parent fiber, and that in the absence of synaptic junctions, the released neurotransmitter might diffuse in the intercellular space to act on relatively distant cellular targets (Descarries *et al.*, 1975; Beaudet and Descarries, 1976). This raises intriguing possibilities for synaptic plasticity, and specifically for the pliant or tractable nature of the functional architecture of serotonergic neurons innervating structures of the limbic system. As conscious feeling or the cognitive aspect of emotion involves learning and memory, the serotonergic innervation of the forebrain may be particularly well suited for modulating mood.

Serotonergic modulation of emotion

The topographical organization of serotonergic innervation of forebrain structures (Molliver, 1987) as well as the paracrine nature of serotonergic neurotransmission in limbic structures support the contention that serotonin plays a neuromodulatory role in the brain.

Not surprisingly, then, serotonin neurotransmission is important in determining emotional information processing and mood. Pharmacological manipulations of the serotonergic systems in the brain affect attentional bias, recognition of facial emotion, emotional memory, dysfunctional attitudes and decision-making. Negative biases in information processing may play a role in the maintenance of major depressive disorder. For excellent recent reviews evaluating the literature pertaining to serotonin, emotional information processing, cognition and depression, the reader is referred to Merens *et al.* (2007) and Harmer (2008).

Agents that increase serotonergic neurotransmission

Tryptophan is an essential amino acid found in dietary protein, and a precursor to serotonin in its synthesis. Supplementing the diet with tryptophan is believed to increase serotonin synthesis and therefore neurotransmission. In healthy volunteers, increased secretion of anterior pituitary hormones and a psychological state mimicking the effect of serotonergic antidepressants are taken as an indication that brain serotonergic function is increased (Cowen, 1993; Yatham and Steiner 1993; Attenburrow *et al.*, 2001; Harmer *et al.*, 2003a). Increased brain serotonergic function by acute dietary supplementation with tryptophan appears to have broad effects on cognitive performance in healthy subjects, and may impair affective working memory (Luciana *et al.*, 2001). As discussed below, increased serotonergic neurotransmission results in enhanced attention and recognition of positive emotional material. Contrary to expectation, acute increases in serotonergic neurotransmission also increase the attentional bias towards negative or fearful stimuli. In normal, healthy subjects, these cognitive changes, and changes in emotional processing, are not accompanied by changes in mood (Merens *et al.*, 2007; Del-Ben *et al.*, 2008; Harmer, 2008).

Identification or recognition of facial expressions representing basic emotions has been widely used to investigate the functional anatomy and neurochemistry of emotion. Acute tryptophan loading by administration of a single dose of nutritionally sourced tryptophan facilitates the perception of facial expressions of both happiness and fear in healthy female volunteers (Attenburrow *et al.*, 2003). This increased threat processing, as well as the bias in emotional processing towards positive emotional material, is also observed in healthy volunteers following a single dose of the selective serotonin reuptake inhibitor (SSRI) citalopram (Table 1). Acute administration of citalopram to healthy volunteers of both genders improves performance in recognizing happy facial expressions (Harmer *et al.*, 2003a) and shifts attentional bias towards positive words (Browning *et al.*, 2007). Acute citalopram administration has also been shown to improve or facilitate the recognition of fearful facial expressions, and to enhance fear-potentiated startle to predictable and unpredictable shocks (Harmer *et al.*, 2003a; Browning *et al.*, 2007; Grillon *et al.*, 2007). Taken together, these studies indicate that increases in serotonergic neurotransmission are not anxiogenic in the absence of threatening or fearful stimuli, but increase threat processing and exacerbate anxiety experienced in responses to threat. That acute increases in serotonergic neurotransmission increase recognition of and attention to both positive and negative emotional material suggests that serotonin may have a general effect on attention and perception.

Table 1 Serotonergic modulation of emotion in healthy human subjects

Agents that increase serotonergic neurotransmission				
Acute tryptophan loading	Females	↑	Recognition of happy facial expressions	Attenburrow *et al.*, 2003
		↑	Recognition of fearful facial expressions	
Acute SSRI administration	Both genders	↑	Recognition of happy facial expressions	Harmer *et al.*, 2003b; Browning *et al.*, 2007
		↑	Recognition of fearful facial expressions	
		↑	Attentional bias toward positive words	Browning *et al.*, 2007
		↑	Fear potentiated startle responses	Grillon *et al.*, 2007
Repeated tryptophan administration	Females	↑	Recognition of happy facial expressions	Murphy *et al.*, 2006
		↓	Recognition of disgusted facial expressions	
	Males	↔	Recognition of happy facial expressions	
Repeated SSRI administration	Both genders	↔	Recognition of happy facial expressions	Harmer *et al.*, 2004, 2006a
		↓	Recognition of fearful facial expressions	
		↑	Recall of words associated with positive personality traits	
Immediately after MDMA	Both genders	↑	Intense positive emotional state	Mørland, 2000; Baylen and Rosenburg, 2006
		↑	Recognition of fearful facial expressions	Hoshi *et al.*, 2004
Acute fenfluramine administration	Both genders	↓	Dysfunctional attitudes	Meyer *et al.*, 2003
Agents that decrease serotonergic neurotransmission				
Acute tryptophan depletion	Both genders	↓	Recall of neutral or positive words	Klaassen *et al.*, 2002
		↑	Emotional interference by negative words	Hayward *et al.*, 2005
		↑	Recognition of happy facial expressions	
	Females	↓	Speed in responding to happy words	Murphy *et al.*, 2002
	Both genders	↔	Speed in responding to happy words	Rubinsztein *et al.*, 2001
	Females	↓	Recognition of fearful facial expressions	Harmer *et al.*, 2003a
	Males	↔	Recognition of emotional facial expressions	
4 days after MDMA	Both genders	↓	Recognition of fearful facial expressions	Hoshi *et al.*, 2004

It might be expected that increased threat processing in healthy volunteers, as a result of acute increases in central serotonergic neurotransmission, would be accompanied by increases in the activation of the amygdala to fearful faces. However, neuroimaging studies have shown that increases in serotonergic function following acute administration of an SSRI attenuate amygdala activation (Moret and Briley, 1996; Attenburrow *et al.*, 2001; David *et al.*, 2003; Felton *et al.*, 2003; McKie *et al.*, 2005). Following acute citalopram administration, activation of the right orbitofrontal cortex (BA47) and amygdala in response to aversive or fearful faces is decreased (Del-Ben *et al.*, 2005; Anderson *et al.*, 2007). These studies of Del-Ben *et al.* (2005) and Anderson *et al.* (2007) were, however, conducted only in male subjects, and there is considerable evidence for sexual dimorphism in emotional processing (Cahill, 2006).

A longer period of dietary supplementation with tryptophan (i.e., for 14 days) has been found to facilitate the recognition of happy expressions and decrease the recognition of disgusted faces in healthy women; the perception of fearful faces is not changed (Murphy *et al.*, 2006). Interestingly, this positive bias in the processing of emotional material as a result of longer periods of tryptophan supplementation is not observed in male volunteers (Murphy *et al.*, 2006). A similar bias of emotional processing away from negative and towards positive emotional material is observed following administration of the SSRI citalopram to healthy volunteers of both genders for 7 days. While acute administration of an SSRI increases the processing of both happiness and fear in healthy individuals, repeated administration of an SSRI to healthy subjects results in a decreased recognition of anger and fear (Harmer *et al.*, 2004, 2006a). Repeated citalopram administration to healthy volunteers also increased memory for positive material, and specifically increased recall of words associated with positive personality traits (Harmer *et al.*, 2004) (Table 1). Differences between the effect of acute and of chronic increases in serotonergic neurotransmission on emotional processing may be related to compensatory, neuroadaptive changes that occur with prolonged increases in serotonergic neurotransmission – e.g., the down-regulation or desensitization of specific serotonin receptors as a result of increased extracellular serotonin. That repeated dietary supplementation with tryptophan, or repeated administration of SSRIs, increases recognition of happy emotional faces and recall of positive material, and decreases perception of fearful faces, suggests a divergence of emotional processing for positive and negatively valenced material.

As has been demonstrated for acute administration of SSRIs, neuroimaging studies indicate that administration of citalopram for 7 days to healthy volunteers of both genders decreases activation of the amygdala in response to fearful faces versus happy faces (Harmer *et al.*, 2006a). Administration of the SSRI escitalopram (the active, S-isomer of citalopram) to healthy volunteers for 21 days also attenuates activation of the amygdala in response to negatively, but not positively, valenced emotional faces (Arce *et al.*, 2008).

MDMA (3,4-methylendioxymethamphetamine) inhibits serotonin reuptake and stimulates the release of serotonin stored in presynaptic vesicles. MDMA also inhibits the catabolic enzyme monoamine oxidase (MAO), reducing the metabolism of serotonin within the nerve terminal and therefore contributing to the increased release of serotonin by MDMA (Green *et al.*, 2003). Immediately following an acute dose, MDMA elicits an intense positive emotional state accompanied by feelings of empathy, sociability and closeness; these are believed to be a result of increased serotonergic neurotransmission (Mørland, 2000; Baylen and Rosenberg, 2006), although MDMA also increases dopaminergic and noradrenergic neurotransmission through similar mechanisms (Green *et al.*, 2003). MDMA also facilitates the recognition of fearful faces after acute administration (Hoshi *et al.*, 2004). The increased accuracy of fear recognition may be due to increased serotonin release as a result of MDMA administration (Green *et al.*, 1995), consistent with enhanced threat processing observed following increased serotonergic neurotransmission as a result of acute administration of the SSRI citalopram or acute dietary supplementation with tryptophan (Table 1).

Fenfluramine is a substrate for the serotonin transporter, and is a potent releaser of serotonin from nerve terminals. It is believed that fenfluramine releases serotonin from neurons via a non-exocytotic, carrier-mediated exchange mechanism involving the serotonin transporter. The metabolite norfenfluramine is a potent serotonin 5-HT_{2B} and 5-HT_{2C} receptor agonist (Rothman and Baumann, 2002). Increasing serotonin function by administration of d-fenfluramine to healthy subjects results in a reduction in dysfunctional attitudes, specifically negatively biased views of oneself, the world and the future that are exaggerated in depression (Meyer *et al.*, 2003; Table 1).

Agents that decrease serotonergic neurotransmission

Intake of a mixture of essential amino acids free of tryptophan, the precursor of 5-HT synthesis, results in a reduction of brain serotonergic function. Acute tryptophan depletion is therefore a useful tool to investigate the effects of lowered serotonin function in humans. Acute tryptophan depletion results in a temporary depressive relapse in recovered depressed patients taking serotonergic antidepressants (Delgado *et al.*, 1990; Van der

Does, 2001). However, only small effects (i.e., lowered or decreased mood) are found in healthy subjects, and especially in those subjects with a family history of affective disorders (Ellenbogen *et al.*, 1999; Klaassen *et al.*, 2002). In general, as discussed below, decreased serotonergic neurotransmission results in impaired attention and recognition of positive emotional material, and increases the attentional bias towards negative stimuli in healthy subjects. The shift in attentional bias toward negative emotional material may be related, however, to the type of test or task performed.

Acute tryptophan depletion impairs recall of neutral and positive, but not negative, words in healthy volunteers (Klaassen *et al.*, 2002). Reaction time of healthy control subjects to emotionally charged stimuli compared with neutral stimuli is increased in the emotional Stroop task, an indication that healthy volunteers are more easily distracted by negative stimuli after acute tryptophan depletion (Hayward *et al.*, 2005). In healthy female volunteers, acute tryptophan depletion slows responses to happy, but not sad, target words in the affective go/no-go task (Murphy *et al.*, 2002); this, however, has not been a consistent observation (Rubinsztein *et al.*, 2001; Table 1).

In contrast to the observed shift in attentional bias to negative material in verbal tasks, acute tryptophan depletion has been found to impair the recognition of fearful facial expressions by healthy female volunteers (Harmer *et al.*, 2003b), and enhance the recognition of happy facial expressions in healthy volunteers of both genders (Hayward *et al.*, 2005). These results are in agreement with the observations that increased serotonin function, by acute administration of the SSRI citalopram or by acute dietary supplementation with tryptophan, results in increased threat processing and shifts attentional bias towards negative or fearful stimuli (Attenburrow *et al.*, 2003; Harmer *et al.*, 2003a; Browning *et al.*, 2007) (Table 1).

Neuroimaging studies indicate that the decreases in serotonin synthesis and function in brain as a result of tryptophan depletion in healthy men and women enhance amygdala activation in response to fearful faces in contrast to happy faces (Cools *et al.*, 2005; van der Veen *et al.*, 2007). This is in contrast to what has been shown following acute or repeated SSRI administration, as discussed above.

In addition to inducing an acute and rapid release of serotonin, MDMA inhibits the activity of tryptophan hydroxylase, the rate-limiting enzyme required for serotonin synthesis (Green *et al.*, 2003). Decreased serotonin synthesis, as a result of the inhibition of tryptophan hydroxylase, causes serotonin depletion. Serotonergic neurotransmission is therefore reduced during the days following MDMA administration. Four days after MDMA administration, the perception of fearful faces is impaired (Hoshi *et al.*, 2004). This decrease in the accuracy of fear recognition is believed to be due to the subsequent depletion of serotonin (Green *et al.*, 1995). Decreased serotonergic neurotransmission may also be responsible for undesirable emotional and cognitive effects experienced more intensely following MDMA use. These include negative mood, impaired concentration, lack of energy, irritability, agitation, depression and anxiety (Mørland, 2000; Green *et al.*, 2003; Baylen and Rosenberg, 2006). MDMA exposure may also result in long-term neurotoxic damage to serotonergic neurons. Administration of MDMA to rodents and non-human primates results in a long-term reduction in serotonin levels in several brain regions; biochemical and histological evidence indicates that this reflects degeneration of serotonergic terminals (Green *et al.*, 1995). Limited, indirect clinical evidence indicates that similar changes may occur in human brain with MDMA exposure, although whether long-term neurotoxic effects seen in animals occurs in humans has yet to be unequivocally demonstrated (Green *et al.*, 2003; Cowen, 2007).

Serotonin transporter gene polymorphism and emotional processing

The serotonin transporter is responsible for terminating the action of serotonin in the extracellular space through the energy dependent reuptake of serotonin into the presynaptic terminal. Thus, the serotonin transporter plays a key role in regulating serotonergic neurotransmission. The serotonin transporter gene (5-HTTLPR) has two frequent alleles, designated long (*l*) and short (*s*). The *s*-allele results in decreased serotonin transporter function and expression *in vitro* (Heils *et al.*, 1995, 1996; Lesch *et al.*, 1996). A functional, single nucleotide variant has been recently detected within the *l*-allele, designated l_A and l_G. l_A is associated with increased expression of the serotonin transporter *in vitro*, whereas l_G is similar to *s* in being associated with low expression (Nakamura *et al.*, 2000; Hu *et al.*, 2006). Imaging studies using the new and highly selective ligand [^{11}C]DASB have extended these *in vitro* studies, reporting that these alleles are indeed associated with differences in serotonin transporter expression *in vivo* (Praschak-Rieder *et al.*, 2007; Reimold *et al.*, 2007).

Emotional appraisal, or how emotions are interpreted and reflected upon, is one of the most important aspects of emotional regulation. A recent study has compared emotional appraisal profiles of negative (fear and sadness) and positive (joy) emotions as a function of serotonin transporter gene alleles. Healthy volunteers that carry the *s*-allele of the serotonin transporter gene score higher than non-carriers for unpleasantness and goal-hindrance, and scored lower for coping ability in the case of fear and sadness (Szily *et al.*, 2008) (Table 2). These findings suggest that the *s*-allele, which conveys decreased function and expression of the serotonin transporter, is associated with a vulnerable cognitive style related to the appraisal of negative emotions (Szily *et al.*, 2008). Indeed, the *s*-allele has been shown to be associated with major depressive disorder in those who have experienced traumatic or stressful life events (see, for example, Caspi *et al.*, 2003; Kendler *et al.*, 2005; Jacobs *et al.*, 2006) (Table 2).

The effect of acute tryptophan depletion on the processing of emotional facial expressions has also been shown to vary as a function of genotype at the 5-HTTLPR. Tryptophan depletion significantly impairs the recognition of fearful facial expressions in carriers of the *s*-allele (Marsh *et al.*, 2006). This impairment is specific to fear expressions. However, in healthy volunteers homozygous for the *l*-allele, acute tryptophan depletion did not alter the recognition of fearful faces (Marsh *et al.*, 2006). The findings of this study suggest that individuals homozygous for the *l*-allele are less sensitive to the effects of tryptophan depletion on emotional processing (Table 2).

Low expression and/or function of the serotonin transporter associated with the *s*-allele might be expected to result in increased serotonin in the extracellular space, and enhanced emotional processing of positively valenced material as a result of increased serotonergic neurotransmission. However, genetic determinants of serotonin transporter function most likely result in physiological and neuroadaptive changes during development, and may not necessarily result in increases in serotonergic neurotransmission in the long term. For example, serotonin transporter knockout mice exhibit a variety of neurological and physiological abnormalities, and exhibit enhanced fear, and anxiety-like and depression-like behaviors (for review, see Murphy and Lesch, 2008).

Table 2 Serotonin transporter gene polymorphism and emotional processing

s-allele carriers (*s/s, s/l*)		
↑	Emotional appraisal score for negative emotions	Szily *et al.*, 2008
↓	Recognition of fearful facial expressions with tryptophan depletion	Marsh *et al.*, 2006
↑	Activation of amygdala in response to fearful stimuli	Hariri *et al.*, 2002
l-allele homozygotes (*l/l*)		
↔	Recognition of emotional facial expressions with tryptophan depletion	Marsh *et al.*, 2006

Carriers of the *s*-allele of the serotonin transporter exhibit increased activation of the amygdala in response to fearful stimuli compared with individuals homozygous for the *l*-allele (Hariri *et al.*, 2002) (Table 2). These results indicate genotype-related differences in excitability of the amygdala to emotional stimuli, which may contribute to the increased fear and anxiety typically associated with the *s*-allele (Murphy and Lesch, 2008). Morphometrical analyses indicate genotype-related differences in anatomy as well. Healthy individuals who are carriers of the s-allele have reduced gray matter volume in limbic regions believed to be critical for the processing of negative emotion, specifically perigenual cingulate and amygdala (Pezawas *et al.*, 2005). Neuroimaging techniques show tight functional coupling of these brain regions during perceptual processing of fearful stimuli. The functional connectivity between these brain regions is critical for emotion regulation and for the extinction of negative affect and fear (see, for example, Pezawas *et al.*, 2005; Phelps and LeDoux, 2005; Quirk and Mueller, 2008). Most interestingly, *s*-allele carriers showed a relative uncoupling of this circuit, which inversely predicted almost 30 percent of variation in temperamental anxiety (Pezawas *et al.*, 2005). These functional and neuroanatomical studies provide evidence of a systems-level mechanism underlying normal emotional reactivity and genetic susceptibility for depression.

Agents acting directly on specific subtypes of serotonin receptor

A recent review by Engin and Treit (2008) provides an extensive overview of the effects of intracerebral infusion of a variety of agents on animals' unconditioned fear reactions. Of interest are 5-HT_{1A} and 5-HT_3 receptors. 5-HT_{1A} receptors are found in high density in the dorsal and median raphe, where they function as somatodendritic autoreceptors. These receptors, when activated by serotonin or agonist, decrease serotonergic neuronal firing. 5-HT_{1A} receptors are also located postsynaptically to serotonergic neurons, and are present in high density in cortical and limbic structures (see Hensler, 2003, and references therein). In general, 5-HT_{1A} receptor agonists produced anxiolytic-like behavioral effects in animal models when infused into the dorsal or median raphe nuclei. However, inconsistent effects have been reported when 5-HT_{1A} receptor agonists are injected into the amygdala, septum and hippocampus (Engin and Treit, 2008). Most of the experimental findings to date are in agreement with the hypothesis that increases and decreases in serotonergic output from the raphe translate into anxiogenic and anxiolytic effects, respectively. However, overall the data are not entirely consistent with this view. Inconsistencies in the literature may be due in part to the variety of behavioral tests used and the complexity of inhibitory feedback mechanisms, which include circuits involving frontal cortex and amygdala (Ceci *et al.*, 1994; Bosker *et al.*, 1997; Hajos *et al.*, 1999; Celada *et al.*, 2001).

The 5-HT_3 receptor is the only serotonin receptor that is a member of the ligand-gated ion channel superfamily of receptors. Specifically, this receptor is a cation channel that is gated by serotonin. In brain, 5-HT_3 receptors are found in cortical and limbic structures. This localization is consistent with behavioral studies in animals that suggest that 5-HT_3 receptor antagonists may have potential anxiolytic, antidepressant and cognitive effects (Barnes and Sharp, 1999; Costall and Naylor, 2004). 5-HT_3 receptor antagonists produce anxiolytic-like effects in the amygdala (Engin and Treit, 2008). The anxiety-related effects of ligands of other serotonin-receptor subtypes have been inconsistent after infusion into the amygdala. Other receptors of interest for future studies include 5-HT_2, 5-HT_6 and 5-HT_7 receptors in cortical areas.

Surprisingly little work has been done to investigate the effects of agents that act on specific subtypes of serotonin receptor on emotional processing in humans. Harmer *et al.* (2006b) have examined the effects of the 5-HT_3 receptor antagonist ondansetron on facial emotion recognition, emotional categorization and memory, and emotion-potentiated startle, in healthy volunteers. Although ondansetron did not affect subjective measures of mood, anxiety or wellbeing, ondanstron did reduce the emotion-potentiated startle, especially in response to negative pictures. Interestingly, facial expression recognition and emotional memory were not significantly altered, in contrast to the effects of acute SSRI administration on facial emotion recognition (Harmer *et al.*, 2006b). These findings are consistent with those of animal studies, and suggest a role for $5HT_3$ receptors in some elements of fear processing.

Conclusion

The effects of manipulations in serotonergic neurotransmission on cognition and emotion are complex. This is perhaps not unexpected, based on the dense plexus of innervation of forebrain limbic structures by serotonergic neurons arising from the dorsal and median raphe. The very nature of serotonergic neurotransmission in many brain regions is consistent with a neuromodulatory role of this neurotransmitter. Understanding not only the neurobiology of emotional responses, but also the serotonergic modulation of emotional states, will be important as we attempt to elucidate the etiology of emotional disorders. By examining how emotional responses can be manipulated, we may be able to change maladaptive emotional reactions. Future research with agents specific to various serotonin receptors may identify important therapeutic targets for the treatment of cognitive and affective disorders.

Acknowledgements

This work was supported in part by US PHS grant MH 52369.

References

Anderson, I.M., Del-Ben, C.M., Mckie, S., Richardson, P., Williams, S.R., Elliott, R. and Deakin, J.F. (2007) Citalopram modulation of neuronal responses to aversive face emotions: a functional MRI study. Neuroreport, 18: 1351–1355.

Arce, E., Simmons, A.N., Lovero, K.L., Stein, M.B. and Paulus, M.P. (2008) Escitalopram effects on insula and amygdala BOLD activation during emotional processing. Psychopharmacology (Berl.), 196: 661–672.

Armony, J.L. and Le Doux, J.E. (1997) How the brain processes emotional information. Ann. NY Acad. Sci., 821: 259–270.

Attenburrow, M.J., Mitter, P.R., Whale, R., Terao, T. and Cowen, P.J. (2001) Low-dose citalopram as a 5-HT neuroendocrine probe. Psychopharmacology, 155: 323–326.

Attenburrow, M.J., Williams, C., Odontiadis, J., Reed, A., Powell, J., Cowen, P.J. et al. (2003) Acute administration of nutritionally sourced tryptophan increases fear recognition. Psychopharmacology, 169: 104–107.

Barnes, N.M. and Sharp, T. (1999) A review of central 5-HT receptors and their function. Neuropharmacology, 38: 1083–1152.

Baxter, M.G. and Murray, E.A. (2002) The amygdala and reward. Nat. Rev. Neurosci., 3: 563–573.

Baylen, C.A. and Rosenberg, H. (2006) A review of the acute subjective effects of MDMA/ecstasy. Addiction, 101: 933–947.

Beaudet, A. and Descarries, L. (1976) Quantitative data on serotonin nerve terminals in adult rat neocortex. Brain Res., 11: 301–309.

Beck, S.G., Pan, Y.-S., Akanwa, A.C. and Kirby, L.G. (2004) Median and dorsal raphe neurons are not electrophysiologically identical. J. Neurophysiol., 91: 994–1005.

Bosker, F.J., Klompmakers, A.A. and Westenberg, H.G.M. (1997) Postsynaptic 5-HT_{1A} receptors mediate 5-hydroxytryptamine release in the amygdala through a feedback to the caudal linear raphe. Eur. J. Pharmacol., 333: 147–157.

Browning, M., Reid, C., Cowen, P.J., Goodwin, G.M. and Harmer, C.J. (2007) A single dose of citalopram increases fear recognition in healthy subjects. J. Psychopharmacol., 21: 684–690.

Buchanan, T.W. (2007) Retrieval of emotional memories. Psychol. Bull., 133: 761–779.

Cahill, L. (2006) Why sex matters for neuroscience. Nat. Rev. Neruosci., 7: 477–484.

Caspi, A., Sugden, K., Moffitt, T.E., Taylor, A., Craig, I.W., Harrington, H., McClay, J., Mill, J., Martin, J., Braithwaite, A. and Poulton, R. (2003) Influence of life stress on depression: moderation by a polymorphism in the 5-HTT gene. Science, 301: 386–389.

Ceci, A., Baschirotto, A. and Borsini, F. (1994) The inhibitory effect of 8-OH-DPAT on the firing activity of dorsal raphe serotoninergic neurons in rats is attenuated by lesion of the frontal cortex. Neuropharmacology, 33: 709–713.

Celada, P., Puig, M.V., Casanovas, J.M., Guillazo, G. and Artigas, F. (2001) Control of dorsal raphe serotonergic neurons by the medial prefrontal cortex: involvement of serotonin-1A, GABAA and glutamate receptors. J. Neurosci., 21: 9917–9929.

Cools, R., Calder, A.J., Lawrence, A.D., Clark, L., Bullmore, E. and Robbins, T.W. (2005) Individual differences in threat sensitivity predict serotonergic modulation of amygdala response to fearful faces. Psychopharmacology, 180: 670–679.

Coricelli, G., Dolan, R.J. and Sirigu, A. (2007) Brain, emotion and decision making: the paradigmatic example of regret. Trends Cogn. Sci., 11: 258–265.

Costall, B. and Naylor, R.J. (2004) 5-HT3 receptors. Curr. Drug Targets CNS Neurol. Disord., 3: 27–37.

Cowen, P.J. (1993) Serotonin receptor subtypes in depression: evidence from studies in neuroendocrine regulation. Clin. Neuropharmacol., 16: 287–296.

Cowen, R.L. (2007) Neuroimaging research in human MDMA users: a review. Psychopharmacology (Berl.)., 189: 539–556.

Dahlström, A. and Fuxe, K. (1964) Evidence for the existence of monoaminecontaining neurons in the central nervous system. I. Demonstration of monoamines in cell bodies of brain neurons. Act. Physiol. Scand, 62(Suppl. 232): 1–55.

David, D.J., Bourin, M., Jego, G., Przybylski, C., Jolliet, P. and Gardier, A.M. (2003) Effects of acute treatment with paroxetine, citalopram and venlafaxine *in vivo* on noradrenaline and serotonin outflow: a microdialysis study in Swiss mice. Br. J. Pharmacol., 140: 1128–1136.

Del-Ben, C.M., Deakin, J.F., McKie, S., Delvai, N.A., Williams, S.R., Elliott, R., Dolan, M. and Anderson, I.M. (2005) The effect of citalopram pretreatment on neuronal responses to neuropsychological tasks in normal volunteers: an FMRI study. Neuropsychopharmacology, 30: 1724–1734.

Del-Ben, C.M., Ferreira, C.A., Alves-Neto, W.C. and Graeff, F.G. (2008) Serotonergic modulation of face-emotion recognition. Braz. J. Med. Biol. Res., 41: 263–269.

Delgado, P.L., Charney, D.S., Price, L.H., Aghajanian, G.K., Lands, H. and Heninger, G.R. (1990) Serotonin function and the mechanism of antidepressant action. Reversal of antidepressant-induced remission by rapid depletion of plasma tryptophan. Arch. Gen. Psychiatry, 47: 411–418.

Descarries, L. and Mechawar, N. (2000) Ultrastructural evidence for diffuse transmission by monoamine and acetylcholine neurons of the central nervous system. Prog. Brain Res., 125: 27–47.

Descarries, L., Beaudet, A. and Watkins, K.C. (1975) Serotonin nerve terminals in adult rat neocortex. Brain Res., 100: 563–588.

Descarries, L., Watkins, K.C., Garcia, S. and Beaudet, A. (1982) The serotonin neurons in the nucleus raphe dorsalis of adult rat: a light and electron microscope radioautographic study. J. Comp. Neurol., 207: 239–254.

Eichenbaum, H.B. (2003) Learning and memory: Brain systems. In: Squire, L.R., Bloom, F.E., McConnell, S.K., Roberts, J.L., Spitzer, N.C. and Sigmond, M.J. (Eds.), Fundamental neuroscience. 2nd edn. Academic Press, San Diego, CA, pp. 1299–1328.

Ellenbogen, M.A., Young, S.N., Dean, P., Palmour, R.M. and Benkelfat, C. (1999) Acute tryptophan depletion in healthy young women with a family history of major affective disorder. Psychol. Med., 29: 35–46.

Engin, E. and Treit, D. (2008) The effects of intra-cerebral drug infusions on animals' unconditioned fear reactions: a systematic review. Prog. Neuropsychopharmacol. Biol. Psychiatry, 32: 1399–1419.

Felton, T.M., Kang, T.B., Hjorth, S. and Auerbach, S.B. (2003) Effects of selective serotonin and serotonin/noradrenaline reuptake inhibitors on extracellular serotonin in rat diencephalon and frontal cortex. Naunyn Schmiedebergs Arch. Pharmacol., 367: 297–305.

Gentili, C., Gobbini, M.I., Ricciardi, E., Vanello, N. Pietrini, P., Haxby, J.V. and Guazzelli,.M. (2008) Differential modulation of neural activity throughout the distributed neural system for face perception in patients with Social Phobia and healthy subjects. Brain Res. Bull., in press.

Green, A.R., Cross, A.J. and Goodwin, G.M. (1995) Review of the pharmacology and clinical pharmacology of 3,4-methylenedioxymethamphetamine (MDMA or 'Ecstasy'). Psychopharmacology, 119: 247–260.

Green, A.R., Mechan, A.O., Elliott, J.M., O'Shea, E. and Colado, M.I. (2003) The pharmacology and clinical pharmacology of 3,4-methylenedioxymethamphetamine (MDMA, 'ecstasy'). Pharmacol. Rev., 55: 463–508.

Grillon, C., Levenson, J. and Pine, D.S. (2007) A single dose of the selective serotonin reuptake inhibitor citalopram exacerbates anxiety in humans: a fear-potentiated startle study. Neuropsychopharmacology, 32: 225–231.

Hajos, M., Hajos-Korsok, E. and Sharp, T. (1999) Role of the medial prefrontal cortex in 5-HT1A receptor-induced inhibition of 5-HT neuronal activity in the rat. Br. J. Pharmacol., 126: 1741–1750.

Hamann, S.B., Ely, T.D., Grafton, S.T. and Kilts, C.D. (1999) Amygdala activity related to enhanced memory for pleasant and aversive stimuli. Nat. Neurosci., 2: 289–293.

Hariri, A.R., Mattay, V.S., Tessitore, A., Kolachana, B., Fera, F., Goldman, D., Egan, M.F. and Weinberger, D.R. (2002) Serotonin transporter genetic variation and the response of the human amygdala. Science, 297: 400–403.

Harmer, C.J. (2008) Serotonin and emotional processing: does it help explain antidepressant drug action? Neuropharmacology, 55: 1023–1028.

Harmer, C.J., Bhagwagar, Z., Perrett, D.I., Vollm, B.A., Cowen, P.J. and Goodwin, G.M. (2003a) Acute SSRI administration affects the processing of social cues in healthy volunteers. Neuropsychopharmacology, 28: 148–152.

Harmer, C.J., Rogers, R.D., Tunbridge, E., Cowen, P.J. and Goodwin, G.M. (2003b) Tryptophan depletion decreases the recognition of fear in female volunteers. Psychopharmaoclogy, 167: 411–417.

Harmer, C.J., Shelley, N.C., Cowen, P.J. and Goodwin, G.M. (2004) Increased positive versus negative affective perception and memory in healthy volunteers following selective serotonin and norepinephrine reuptake inhibition. Am. J. Psychiatry, 161: 1256–1263.

Harmer, C.J., Mackay, C.E., Reid, C.B., Cowen, P.J. and Goodwin, G.M. (2006a) Antidepressant drug treatment modifies the neural processing of nonconscious threat cues. Biol. Psychiatry, 59: 816–820.

Harmer, C.J., Reid, C.B., Ray, M.K., Goodwin, G.M. and Cowen, P.J. (2006b) 5-HT3 antagonisms abolishes the emotion potentiated startle effect in humans. Psychopharmacology, 186: 18–24.

Hayward, G., Goodwin, G.M., Cowen, P.J. and Harmer, C.J. (2005) Lowdose tryptophan depletion in recovered depressed patients induces changes in cognitive processing without depressive symptoms. Biol. Psychiatry, 57: 517–524.

Heils, A., Teufel, A., Petri, S., Seemann, M., Bengel, D., Balling, U., Riederer, P. and Lesch, K.P. (1995) Functional promoter and polyadenylation site mapping of the human serotonin (5-HT) transporter gene. J Neural. Transm. Gen. Sect., 102: 247–254.

Heils, A., Teufel, A., Petri, S., Stober, G., Riederer, P., Bengel, D. and Lesch, K.P. (1996) Allelic variation of human serotonin transporter gene expression. J. Neurochem., 66: 2621–2624.

Heimer, L. (2003) A new anatomical framework for neuropsychiatric disorders and drug abuse. Am. J. Psych., 160: 1726–1739.

Hensler, J.G. (2003) Regulation of 5-HT_{1A} receptor function in brain following agonist or antidepressant administration. Life Sci., 72: 1665–1682.

Hensler, J.G. (2006) Serotonergic modulation of the limbic system. Neurosci. Biobehavior. Revs., 30: 203–214.

Holstege, G. (1992) The emotional motor system. Eur. J. Morphol., 30: 67–79.

Holstege, G. and Georgiadis, J. (2004) The emotional brain; neural correlates of cat sexual behavior and human male ejaculation. Prog. Brain Res., 143: 39–45.

Hoshi, R., Bisla, J. and Curran, H.V. (2004) The acute and subacute effects of 'ecstasy' (MDMA) on processing of facial expressions: preliminary findings. Drug Alcohol. Depend, 76: 297–304.

Hu, X.Z., Lipsky, R.H., Zhu, G., Akhtar, L.A., Taubman, J., Greenberg, B.D., Xu, K., Arnold, P.D., Richter, M.A., Kennedy, J.L., Murphy, D.L. and Goldman, D. (2006) Serotonin transporter promoter gain-of-function genotypes are linked to obsessive-compulsive disorder. Am. J. Hum. Genet., 78: 815–826.

Iversen, S., Kupfermann, I. and Kandel, E.R. (2000) Emotional states and feelings. In: Kandel, E.R., Schwartz, J.H. and Jessel, T.M. (Eds.), Principles of Neural Science. 4th edn. McGraw Hill, New York, NY, pp. 982–997.

Jacobs, N., Kenis, G., Peeters, F., Derom, C., Vlietinck, R. and van Os, J. (2006) Stress-related negative affectivity and genetically altered serotonin transporter function: evidence of synergism in shaping risk of depression. Arch. Gen. Psychiatry, 63: 989–996.

Kendler, K.S., Kuhn, J.W., Vittum, J., Prescott, C.A. and Riley, B. (2005) The interaction of stressful life events and a serotonin transporter polymorphism in the prediction of episodes of major depression: a replication. Arch. Gen. Psychiatry, 62: 529–535.

Kirby, L.G., Pernar, L., Valentino, R.J. and Beck, S.G. (2003) Distinguishing characteristics of serotonin and non-serotonin-containing cells in the dorsal raphe nucleus: electrophysiological and immunohistochemical studies. Neuroscience, 116: 669–683.

Klaassen, T., Riedel, W.J., Deutz, N.E.P. and van Praag, H.M. (2002) Mood congruent memory bias induced by tryptophan depletion. Psychol. Med., 32: 167–172.

Kohler, C. and Steinbusch, H.W.M. (1982) Identification of serotonin and nonserotonin- containing neurons of the mid-brain raphe projecting to the entorhinal area and the hippocampal formation. A combined immunohistochemical and fluorescent retrograde tracing study in the rat. Neuroscience, 7: 951–959.

Lesch, K.P., Bengel, D., Heils, A., Sabol, S.Z., Greenberg, B.D., Petri, S., Benjamin, J., Müller, C.R., Hamer, D.H. and Murphy, D.L. (1996) Association of anxiety-related traits with a polymorphism in the serotonin transporter gene regulatory region. Science, 274: 1527–1531.

Luciana, M., Burgund, E.D., Berman, M. and Hanson, K. L. (2001) Effects of tryptophan loading on verbal, spatial and affective working memory functions in healthy adults. J. Psychopharmacol., 15: 219–230.

Marschner, A., Kalisch, R., Vervliet, B., Vansteenwegen, D. and Büchel, C. (2008) Dissociable roles for the hippocampus and the amygdala in human cued versus context fear conditioning. J. Neurosci., 28: 9030–9036.

Marsh, A.A., Finger, E.C., Buzas, B., Soliman, N., Richell, R.A., Vythilingham, M., Pine, D.S., Goldman, D. and Blair, R.J. (2006) Impaired recognition of fear facial expressions in 5-HTTLPR S-polymorphism carriers following tryptophan depletion. Psychopharmacology, 189: 387–394.

McKie, S., Del-Ben, C., Elliott, R., Williams, S., del Vai, N., Anderson, I. and Deakin, J.F. (2005) Neuronal effects of acute citalopram detected by pharmacoMRI. Psychopharmacology, 180: 680–686.

Merens, W., Willem Van der Does, A.J. and Spinhoven, P. (2007) The effects of serotonin manipulations on emotional information processing and mood. J. Affect. Disord., 103: 43–62.

Meyer, J.H., McMain, S., Kennedy, S.H., Korman, L., Brown, G.M., DaSilva, J.N., Wilson, A.A., Blak, T., Eynan-Harvey, R., Goulding, V.S., Houle, S. and Links, P. (2003) Dysfunctional attitudes and 5-HT2 receptors during depression and self-harm. Am. J. Psychiatry, 160: 90–99.

Meyer, M., Baumann, S., Wildgruber, D. and Alter, K. (2007) How the brain laughs. Comparative evidence from behavioral, electrophysiological and neuroimaging studies in human and monkey. Behav. Brain Res., 182: 245–260.

Molliver, M.E. (1987) Serotonergic neuronal systems: what their anatomic organization tells us about function. J. Clin. Psychopharmacol., 7: 3S–23S.

Moret, C. and Briley, M. (1996) Effects of acute and repeated administration of citalopram on extracellular levels of serotonin in rat brain. Eur. J. Pharmacol., 295: 189–197.

Morgane, P.J. and Mokler, D.J. (2006) The limbic brain: continuing resolution. Neurosci. Biobehav. Rev., 30: 119–125.

Morgane, P.J., Galler, J.R. and Mokler, D.J. (2005) A review of systems and networks of the limbic forebrain/limbic midbrain. Prog. Neurobiol., 75: 143–160.

Mørland, J. (2000) Toxicity of drug abuse – amphetamine designer drugs (ecstasy): mental effects and consequences of single dose use. Toxicol. Lett., 112–113: 147–152.

Murphy, D.L. and Lesch, K.P. (2008) Targeting the murine serotonin transporter: insights into human neurobiology. Nat. Rev. Neurosci., 9: 85–96.

Murphy, F.C., Smith, K.A., Cowen, P.J., Robbins, T.W. and Sahakian, B.J. (2002) The effects of tryptophan depletion on cognitive and affective processing in healthy volunteers. Psychopharmacology, 163: 42–53.

Murphy, S.E., Longhitano, C., Ayres, R.E., Cowen, P.J. and Harmer, C.J. (2006) Tryptophan supplementation induces a positive bias in the processing of emotional material in healthy female volunteers. Psychopharmacology, 187: 121–130.

Nakamura, M., Ueno, S., Sano, A. and Tanabe, H. (2000) The human serotonin transporter gene linked polymorphism (5-HTTLPR) shows ten novel allelic variants. Mol. Psychiatry, 5: 32–38.

Park, J.Y., Gu, B.M., Kang, D.H., Shin, Y.W., Choi C.H., Lee, J.M. and Kwon, J.S. (2008) Integration of cross-modal emotional information in the human brain: an fMRI study. Cortex, in press.

Parent, A., Descarries, L. and Beaudet, A. (1981) Organization of ascendingserotonin neurons in the adult rat brain. A radioautographic study after intraventricular administration of [3H]5-hydroxytryptamine. Neuroscience, 6: 115–138.

Pezawas, L., Meyer-Lindenberg, A., Drabant, E.M., Verchinski, B.A., Munoz, K.E., Kolachana, B.S., Egan, M.F., Mattay, V.S., Hariri, A.R. and Weinberger, D.R. (2005) 5-HTTLPR polymorphism impacts human cingulate-amygdala interactions: a genetic susceptibility mechanism for depression. Nat. Neurosci., 8: 828–834.

Phelps, E.A. and LeDoux, J.E. (2005) Contributions of the amygdala to emotion processing: from animal models to human behavior. Neuron, 48: 175–187.

Praschak-Rieder, N., Kennedy, J., Wilson, A.A., Hussey, D., Boovariwala, A., Willeit, M., Ginovart, N., Tharmalingam, S., Masellis, M., Houle, S. and Meyer, J.H. (2007) Novel 5-HTTLPR allele associates with higher serotonin transporter binding in putamen: a [(^{11})C] DASB positron emission tomography study. Biol. Psychiatry, 62: 327–331.

Quirk, G.J. and Mueller, D. (2008) Neural mechanisms of extinction learning and retrieval. Neuropsychopharmacology, 33: 56–72.

Reimold, M., Smolka, M.N., Schumann, G., Zimmer, A., Wrase, J., Mann, K., Hu, X.Z., Goldman, D., Reischl, G., Solbach, C., Machulla, H.J., Bares, R. and Heinz, A. (2007) Midbrain serotonin transporter binding potential measured with [(^{11})C]DASB is affected by serotonin transporter genotype. J. Neural. Transm., 114: 635–639.

Rothman, R.B. and Baumann, M.H. (2002) Serotonin releasing agents. Neurochemical, therapeutic and adverse effects. Pharmacol. Biochem. Behav., 71: 825–836.

Rubinsztein, J.S., Mehta, M.A., Robbins, T.W., Rogers, R.D., Riedel, W.J. and Sahakian, B.J. (2001) Acute dietary tryptophan depletion impairs maintenance of 'affective set' and delayed visual recognition in healthy volunteers. Psychopharmacology, 154: 319–326.

Swanson, L.W. and Petrovich, G.D. (1998) What is the amygdala? Trends Neurosci., 21: 323–331.

Szily, E., Bowen, J., Unoka, Z., Simon, L. and Keri, S. (2008) Emotion appraisal is modulated by the genetic polymorphism of the serotonin transporter. J. Neural. Transm., 115: 819–822.

Taber, E., Brodal, A. and Walberg, F. (1960) The raphe nuclei of the brain stem in the cat. I. Normal topography and cytoarchitecture and general discussion. J. Comp. Neurol., 114: 161–187.

Törk, I. (1990) Anatomy of the serotonergic system. Ann. N.Y. Acad. Sci., 600: 9–34.

Van der Does, A.J.W. (2001) The effects of tryptophan depletion on mood and psychiatric symptoms. J. Affect. Disord., 64: 107–119.

van der Veen, F.M., Evers, E.A., Deutz, N.E. and Schmitt, J.A. (2007) Effects of acute tryptophan depletion on mood and facial emotion perception related brain activation and performance in healthy women with and without a family history of depression. Neuropsychopharmacology, 32: 216–222.

Williams, M.A., McGlone, F., Abbott, D.F. and Mattingley, J.B. (2005) Differential amygdala responses to happy and fearful facial expressions depend on selective attention. Neuroimage, 24: 417–425.

Yatham, L.N. and Steiner, M. (1993) Neuroendocrine probes of serotonergic function: A critical review. Life Sci., 53: 447–463.

Zeki, S. (2007) The neurobiology of love. FEBS Lett., 581: 2575–2579.

CHAPTER 3.6

Serotonin and the Neurobiology of Anxious States

Christopher A. Lowry*[,1,2] and Matthew W. Hale[1]

[1]*University of Colorado at Boulder, Department of Integrative Physiology, Boulder, Colorado*
[2]*University of Bristol, Henry Wellcome Laboratories for Integrative Neuroscience and Endocrinology, Bristol, UK*

Abstract: Anxious states are emotional states associated with an increase in avoidance behavior, particularly in situations where a conflict between approach and avoidance occurs, as when both potential rewarding outcomes and potential aversive outcomes exist. An important component of anxious states is that they are trans-situational, transferring from one situation to another. Serotonergic systems play an important role in modulating anxious states. Serotonergic signaling can either facilitate or attenuate anxious states, depending on the site of action and the specific serotonin receptor subtype involved. In this chapter, we present an operational definition of anxious states, describe a number of stimuli that can induce anxious states with distinct time-courses, and consider the evidence that serotonergic systems play a role in modulation of these anxious states. We focus on a subpopulation of serotonergic neurons that projects to forebrain limbic structures implicated in the regulation of anxious states, including the ventral hippocampus/subiculum and the basolateral amygdaloid complex. We then compare and contrast this serotonergic system with another serotonergic system implicated in the inhibition of panic-like responses, a principal component of a subset of anxiety disorders in humans.

Keywords: 5-hydroxytryptamine, anxiety, basolateral amygdala, bed nucleus of the stria terminalis, circuit, hippocampus, raphe, serotonin.

Abbreviations: 5-HTT, 5-hydroxytryptamine transporter, serotonin transporter; 5-HTTLPR, 5-HTT-linked polymorphic region; BB_1, bombesin-like peptide type I receptor; BB_2, bombesin-like peptide type II receptor; CCK, cholecystokinin; CRF, corticotropin-releasing factor; CRF_1, corticotropin-releasing factor type I receptor; CRF_2, corticotropin-releasing factor type II receptor; dHPC, dorsal hippocampus; DMCM, methyl 6,7-dimethoxy-4-ethyl-beta-carboline-3-carboxylate; DPAG, dorsal periaqueductal gray; DRN, dorsal raphe nucleus, dorsal part; DRVL/VLPAG, dorsal raphe nucleus, ventrolateral part/ventrolateral part of the periaqueductal gray; FG-7142, N-methyl-β-carboline-3-carboxamide; fMRI, functional magnetic resonance imaging; GABA, γ-aminobutyric acid; MNR, median raphe nucleus; RVLM, rostral ventrolateral medulla; SERT, 5-hydroxytryptamine transporter, serotonin transporter; SSRI, selective serotonin reuptake inhibitor; TPH2, tryptophan hydroxylase 2; vHPC, ventral hippocampus.

Introduction

Aims and scope of the chapter

Anxious states and anxiety-related behaviors appear to be regulated by a distributed and interconnected network of neural circuits in the brain. Serotonin is an important modulator of these anxiety-related circuits, but the mechanisms through which serotonin modulates these circuits and the associated physiological and behavioral arousal are not well understood. The current chapter has three aims: (1) to present an operational definition of anxious states and anxiety-related behavior; (2) to briefly define neural systems involved in modulating anxiety-related behaviors; and (3) to consider how serotonergic systems specifically modulate these anxiety-related circuits.

Serotonergic systems are topographically organized. This means that serotonergic neurons located in different parts of the brainstem raphe complex give rise to projections to specific and different neural targets (Figure 1). In order to understand the role of serotonergic systems in the regulation of anxious states and anxiety-related behavior, it is important to characterize the anatomical and functional properties of those serotonergic neurons that project to neuronal structures that regulate anxiety. It is not possible within the constraints of this chapter to consider the effects of serotonergic

*Corresponding author
E-mail: christopher.lowry@colorado.edu

Christian Müller & Barry Jacobs (Eds.)
Handbook of Behavioral Neurobiology of Serotonin
ISBN 978-0-12-374634-4

DOI: 10.1016/B978-0-12-374634-4.00023-X

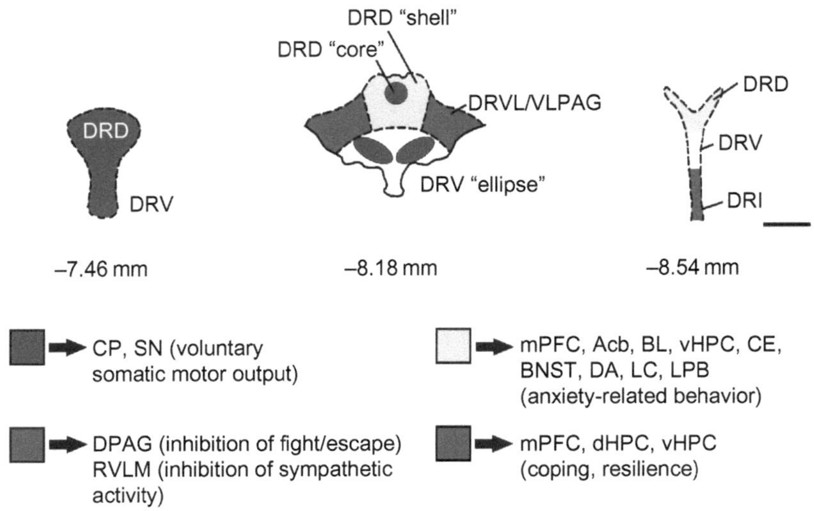

Figure 1 Diagrammatic illustration of the topographical distribution of serotonergic neurons projecting to functionally related forebrain and hindbrain targets in the rat brain. Serotonergic neurons projecting to the caudate putamen and substantia nigra (blue) are located predominantly in the rostral part of the DR and are concentrated within clusters in the DRD 'core' and DRV 'ellipse' regions of the mid-rostrocaudal parts of the DR. Serotonergic neurons projecting to brain structures implicated in the regulation of anxious states and anxiety-related behaviors (yellow) are concentrated in the mid-rostrocaudal DRD 'shell' region and caudal DR. Populations of serotonergic neurons in the DRVL/VLPAG region (red) project to the DPAG and RVLM and have been implicated in the inhibition of panic-like responses. Serotonergic neurons in the DRI (green) have projections to the prefrontal cortex, hippocampus, and midline thalamus, are thought to be activated by peripheral immune activation, and may play an important role in coping and resilience. Abbreviations: Acb, nucleus accumbens; BL, basolateral amygdala, BNST, bed nucleus of the stria terminalis; CE, central nucleus of the amygdala; CP, caudate putamen; DA, dorsal hypothalamic area; dHPC, dorsal hippocampus; DRD, dorsal raphe nucleus, dorsal part; DRI, dorsal raphe nucleus, interfascicular part, DRV, dorsal raphe nucleus, ventral part; DRVL/VLPAG, dorsal raphe nucleus, ventrolateral part/ventrolateral periaqueductal gray; DPAG, dorsal periaqueductal gray; LC, locus coeruleus; LPB, lateral parabrachial nucleus; mPFC, medial prefrontal cortex; RVLM, rostral ventrolateral medulla; SN, substantia nigra; vHPC, ventral hippocampus. Scale bar, 500 μm. Reproduced from *Stress: The Biology Of Stress*, 12 January 2005, Christopher A. Lowry, Philip L. Johnson, Anders Hay-Schmidt *et al.* 'Modulation of anxiety circuits by serotonergic systems' 8:4, pp. 233–246, reprinted by permission of Taylor & Francis Ltd, http://www.tandf.co.uk/journals. To see the full color version of this figure please refer to the color plate at the back of the book. Copies produced via our print on demand service do not contain color plates; if your copy does not have the color plate, please go to this website to view the figure in color www.elsevierdirect.com/companions/9780123746344

systems on all components of a distributed neuronal circuit regulating anxiety; instead we will focus on the ventral hippocampus/subiculum and the basolateral nucleus of the amygdala, two important components of the neural circuits modulating anxiety-related behavior in situations that involve conflict between approach and avoidance. We will also briefly discuss the dorsal periaqueductal gray (DPAG) and rostral ventrolateral medulla (RVLM), two important components of the neural circuits determining vulnerability to panic-like responses, a significant component of some anxiety disorders, such as panic disorder and post-traumatic stress disorder (PTSD) in humans.

Operational definition of anxious states and anxiety-related behavior

In this chapter, we will consider two different symptoms associated with anxiety disorders in humans that can be modeled in animal studies. The first is an *emotional state* of anxiety that involves a conflict when both potential rewarding and aversive outcomes are present; this emotional state leads to increased avoidance of potentially aversive outcomes. The second is a *vulnerability* to panic-like responses following exposure to normally innocuous stimuli, a component of some anxiety disorders including panic disorder and PTSD (American Psychiatric Association, 2000). The former, also referred to as conflict anxiety, is a complex emotional state associated with increased vigilance and arousal, and a sustained increase in avoidance behavior in environmental situations characterized by a level of uncertainty or unpredictability, including situations where a conflict between approach and avoidance behaviors exists (Figure 2). An important element of the emotional state of anxiety is that it is trans-situational. In other words, the anxiety state transfers to different environments or situations (Figure 2). Panic-like responses include increases in heart rate and blood pressure, and escape-like behaviors (Graeff *et al.*, 1996; Shekhar *et al.*, 2003). The neural circuits regulating anxious states and vulnerability to panic-like

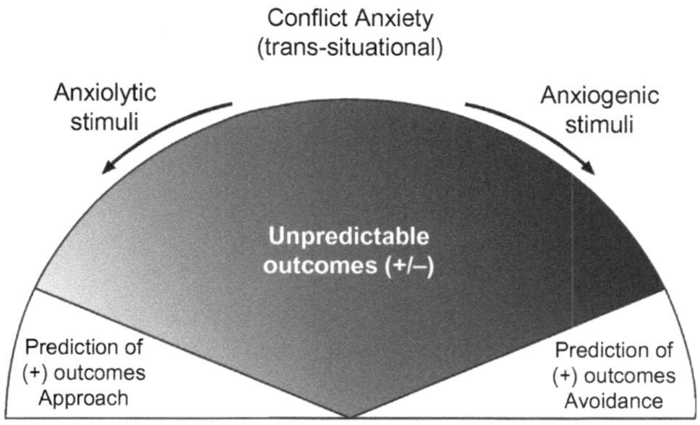

Figure 2 Diagrammatic illustration of a model of conflict anxiety. In a highly predictable environment the individual is faced with a straightforward behavioral strategy, approach-type behaviors in the case of a rewarding outcome and avoidance-type behaviors in the case of an aversive outcome. However, in a complex unpredictable environment the individual may perceive simultaneously the potential for both rewarding and aversive outcomes of specific behavioral strategies. This conflict, between approaching a potentially rewarding stimulus and avoiding a potentially aversive stimulus, is central to many anxiety-states. Anxiolytic drugs (benzodiazepines) or stimuli (safety signals) can shift the balance towards approach-type behaviors, while anxiogenic drugs (FG-7142, caffeine) or stimuli (bright light in rodents) can shift the balance towards avoidance-type behaviors. An important component of anxious states is that the associated increases in physiological and behavioral arousal are not tied to a specific context, but are trans-situational. Serotonin can have anxiolytic or anxiogenic effects depending on the site of action and the 5-HT receptor subtype involved.

responses are different, and therefore we would expect that the serotonergic systems modulating them would be different.

Models of anxious states with distinct temporal profiles

An emotional state of anxiety can be induced by a number of drugs or experiential factors. Anxious states can be transient, lasting a few minutes, a few hours, a few days, or even a lifetime, in which case the individual is said to have anxiety traits. Anxiety traits can be dependent on genetic influences or developmental factors during early life. Below, we consider different pharmacological, experiential or genetic influences that can increase anxious states or lead to development of persistent anxiety traits.

The most extensively studied model of anxious states is treatment with anxiogenic drugs, including β-carbolines, inverse partial agonists and agonists that act at the benzodiazepine allosteric site of the $GABA_A$ receptor, such as N-methyl-β-carboline-3-carboxamide (FG-7142) (Evans and Lowry, 2007) and methyl 6,7-dimethoxy-4-ethyl-beta-carboline-3-carboxylate (DMCM) (Cole *et al.*, 1995), and benzodiazepine antagonists such as flumazenil (File and Pellow, 1984). These drugs can act very quickly, within 5 minutes, to induce an anxiety state that, depending on the dose used, can last 30–90 minutes (Evans and Lowry, 2007). Other classes of anxiogenic drugs include adenosine receptor antagonists such as caffeine, $5\text{-HT}_{2A/2C}$ receptor agonists such as m-chlorophenyl piperazine (mCPP), and α_2-adrenoreceptor antagonists such as yohimbine (for references, see Singewald and Sharp, 2000; see Singewald *et al.*, 2003). These drugs appear to activate a common, distributed neural system in the brain, including the basolateral nucleus of the amgdala, bed nucleus of the stria terminalis, and brainstem monoaminergic systems including noradrenergic and serotonergic systems (Singewald and Sharp, 2000; Singewald *et al.*, 2003; Abrams *et al.*, 2005; Evans and Lowry, 2007). These common effects are observed despite the fact that the drugs have such diverse pharmacological specificity, suggesting that there are many nodes through which this anxiety circuitry can be activated. Importantly for this chapter, these drugs activate a subpopulation of serotonergic neurons within the dorsal raphe nucleus, discussed further below.

A number of endogenous anxiogenic neuropeptides and neurotransmitters are implicated in activation of anxiety circuits and induction of anxious states, including corticotropin-releasing factor (CRF) acting at CRF type I (CRF_1) receptors (Dunn and Berridge, 1987; Heinrichs *et al.*, 1997), cholecystokinin acting at CCK_A and CCK_B receptors (Harro, 2006), bombesin-like peptides acting at BB_1/BB_2 receptors (Merali *et al.*, 2006), noradrenaline acting at α_1 and β adrenoceptors (Morilak *et al.*, 2005), and serotonin acting at 5-HT_{2C} receptors (Anderson *et al.*, 2002; Campbell and Merchant, 2003). The anxiogenic effects of activation of these neuropeptide or neurotransmitter receptors are evident quickly, and, depending on the dose, can last up to several hours.

Environmental stimuli can also increase anxiety state. These stimuli tend to be diffuse, species-specific aversive cues, such as bright light in rodents (for review, see Walker *et al.*, 2003). Rodents avoid bright light and open spaces – conditions that increase the likelihood of aversive events, including predation. Effectively, any stimulus that increases the perceived threat or increases the perceived potential for aversive outcomes can increase anxiety state. Bright light induced anxiety in rodents has a rapid onset, evident within 5–20 minutes, and persists for approximately 20 minutes after the light is turned off (see, for example, Walker *et al.*, 2003; Bouwknecht *et al.*, 2007). Other aversive experiences can also increase anxiety state. An extensively studied model in rodents is the increase in anxiety state induced by exposure to immobilization stress. The anxiety state induced by immobilization stress has a rapid onset, occurring within 15 minutes, but is transient (Cecchi *et al.*, 2002a, 2002b). Other aversive experiences also induce a trans-situational anxiety state; these include shock (see below), predator odor (Adamec *et al.*, 2006), swim stress (Heinrichs *et al.*, 1994), restraint (Heinrichs *et al.*, 1994; Martijena *et al.*, 1997; Mendonca and Guimaraes, 1998), social defeat (Heinrichs *et al.*, 1992, 1994; Ruis *et al.*, 1999) and predator exposure (Adamec and Shallow, 1993; Adamec *et al.*, 2004; for review, see Korte and De Boer, 2003; Armario *et al.*, 2008).

If the aversive experience is intense and unmanageable, such as repeated, uncontrollable shock, it can induce a prolonged increase in anxiety state that lasts up to 24–72 hours (Gewirtz *et al.*, 1998; for review, see Walker *et al.*, 2003; Maier and Watkins, 2005). The extended time-course of increased anxiety state following intense, uncontrollable stress suggests that unique mechanisms are involved, relative to anxiety states with shorter time-courses. In studies of social avoidance, shock-induced increases in anxiety state can last up to 5–10 days (Leveleki *et al.*, 2006). Social defeat by a conspecific may be an ethologically relevant model of intense, uncontrollable stress that also leads to a prolonged increase in anxiety state, an emotional state that can last at least 2 weeks (Ruis *et al.*, 1999; Nakayasu and Ishii, 2008). Interestingly, the prolonged anxiety state following social defeat is abrogated if the animal is put in social housing conditions (Ruis *et al.*, 1999; Nakayasu and Ishii, 2008). Likewise, the prolonged anxiety state following repeated shock is abrogated if the animal has control over termination of the shock (Maier and Watkins, 2005). The effect of stressor controllability to prevent the development of subsequent anxious states is dependent on activation of the medial prefrontal cortex and the consequent inhibition of serotonergic neurons in the caudal dorsal raphe nucleus (Maier and Watkins, 2005).

Many of the pharmacological and experiential anxiogenic stimuli described above share in common that they activate a subset of serotonergic neurons within the mid-rostrocaudal and caudal dorsal raphe nucleus, a region that gives rise to the majority of serotonergic projections to forebrain structures controlling anxious states and anxiety-related behavior. This will be discussed further below. Perhaps more importantly, in many of the cases described above, the development of anxiety states and expression of anxiety-related behavior appears to be dependent on activation of serotonergic systems (Maier and Watkins, 2005; Merali *et al.*, 2006; Meloni *et al.*, 2007; Sziray *et al.*, 2007; Cooper *et al.*, 2008; Harada *et al.*, 2008; Strong *et al.*, 2008).

Models of anxious traits

Anxiety traits are stable individual differences, resulting in an increased tendency to respond with state anxiety to environmental situations. Genetic composition plays an important role in determining anxiety traits. In addition, a lifelong increase in anxiety trait can result from a number or adverse experiences, either during prenatal or neonatal development, or in adolescence.

Genetic models

Genetic influences play an important role in determining anxiety traits, including genes that control serotonergic neurotransmission. A key step in regulating 5-HT neurotransmission at the synapse is its rapid reuptake into the presynaptic terminal by the serotonin transporter (SERT; 5-HTT). Thus, blocking SERT with selective serotonin reuptake inhibitors (SSRIs) has become the most effective strategy to treat mood and anxiety disorders over the past three decades. This large-scale use of SSRIs in treating mood and anxiety disorders and the genetic heterogeneity of responses to SSRIs has also facilitated the demonstration that the SERT expression in the brain is a major determinant of not only the response to SSRIs, but also coping with stress and the risk of development of mood and anxiety disorders (Lesch *et al.*, 1996; Ogilvie *et al.*, 1996; Rotondo *et al.*, 2002; Van den Hove *et al.*, 2006). A common polymorphism consisting of short (*s*) versus long (*l*) forms of the promoter region of the SERT gene *slc6A4* (5-HTTLPR) has been associated with variability in rates of SERT transcription and serotonin reuptake (Lesch *et al.*, 1996). A number of recent reports have provided data supporting the hypothesis that genetic polymorphisms in 5-HTTLPR may determine the efficacy of (Smeraldi *et al.*, 1998; Pollock *et al.*, 2000; Rausch *et al.*, 2003) as well as the side effects (Perlis *et al.*, 2003) seen with antidepressants that are serotonin reuptake inhibitors – e.g., the SSRIs. Individuals with *s*/*s* (population frequency, 20 percent) and *s*/*l* genotypes (population frequency, 40 percent) have reduced SERT sites compared to individuals who are homozygous for the long allele (*l*/*l*) (population frequency, 40 percent) (Lotrich *et al.*, 2001). Individuals with *s*/*s* and *s*/*l* have been shown to have decreased antidepressant response and increased side effects compared to individuals with the *l*/*l* phenotype (Smeraldi *et al.*, 1998; Pollock *et al.*, 2000).

More importantly, recent studies in healthy subjects have reported that the 5-HTTLPR polymorphism regulating SERT expression also determines amygdala activation in response to negative stimuli as measured with functional magnetic resonance imaging (fMRI) (Hariri *et al.*, 2002, 2005). In these studies, subjects with one or more *s* allele were shown to have increased amygdala activation in response to a facial emotion recognition task. Genetic variation in SERT expression is an important factor in regulating neuroticism, anxiety response and amygdala activation in healthy volunteers (Heinz *et al.*, 2005; Brocke *et al.*, 2006; Canli *et al.*, 2006; Rhodes *et al.*, 2007; Smolka *et al.*, 2007), as well as development of anxiety disorders and depression (Munafo *et al.*, 2008). Lower SERT expression is a strong vulnerability factor for increased risk of anxiety disorders (Furmark *et al.*, 2004; Hariri and Holmes, 2006; Maron and Shlik, 2006; Xu *et al.*, 2006), perhaps due to developmental effects or to compensatory responses following chronic elevation of synaptic serotonin concentrations. Low availability of SERT measured *in vivo* is correlated with anxiety in patients with major depression (Reimold *et al.*, 2008). Therefore, low SERT expression appears to greatly facilitate amygdala excitation and vulnerability to anxiety and mood disorders by an as yet unknown mechanism.

Animal models with reduced SERT expression, similar to humans, likewise appear to have a vulnerability to development of anxious states. For example, SERT knockout mice, but not their heterozygous or wild-type counterparts, respond with a persistent increase in anxious state following exposure to predator odor (Adamec *et al.*, 2006, 2008). Mice and rats with inactivation of SERT expression demonstrate increases in anxiety traits (Kalueff *et al.*, 2007; Olivier *et al.*, 2008). Conversely, mice with overexpression of SERT have a low anxiety phenotype (Jennings *et al.*, 2006).

Evidence suggests that polymorphisms in other genes that are important determinants of serotonergic neurotransmission also affect anxiety traits. A variety of evidence supports a role for abnormal 5-HT_{1A} receptor function in panic disorder and social anxiety disorder. These receptors are found both postsynaptically and presynaptically, where they mediate somatodendritic negative feedback inhibition of serotonergic neuronal firing. Patients with panic disorder show subsensitivity of these receptors in laboratory challenge tests (Lesch *et al.*, 1992). Neumeister *et al.* (2004) reported reduced amounts of 5-HT_{1A} receptor binding using positron emission tomography (PET) in unmedicated patients with panic disorder compared to healthy controls, especially in the amygdala, cingulate cortex and midbrain raphe complex. Another study recently confirmed this by demonstrating reduced 5-HT_{1A} receptor binding in the amygdala and midbrain raphe complex of subjects with panic disorder using [^{11}C]WAY-100635, a more selective PET ligand for 5-HT_{1A} receptors (Nash *et al.*, 2008). Another PET study using [^{11}C]WAY-100635 as the ligand found reduced 5-HT_{1A} receptor binding in subjects with social anxiety disorder (Lanzenberger *et al.*, 2007). The G allele of the C(-1019)G variant of the 5-HT_{1A} receptor has been associated with depression, panic disorder, neuroticism, and reduced response to antidepressant drug treatment (Drago *et al.*, 2008; Le *et al.*, 2008).

Mice with genetic inactivation of the 5-HT_{1A} receptor display increased anxiety traits (Heisler *et al.*, 1998; Parks *et al.*, 1998; Ramboz *et al.*, 1998; Gross *et al.*, 2002). This phenotype can be reversed by selective expression of 5-HT_{1A} receptors within dentate granule cells in the hippocampus, including the ventral hippocampus (Tsetsenis *et al.*, 2007), suggesting an important role for 5-HT_{1A} receptors in hippocampal circuitry in the regulation of anxious states (see below).

Although less is known about associations between 5-HT_{2A} receptor function and anxiety disorders, at least one study has reported association of polymorphism in the 5-HT_{2A} receptor gene and panic disorder (Inada *et al.*, 2003), while recent independent studies have found that polymorphisms in the 5-HT_{2A} receptor are associated with symptom severity in panic disorder (Unschuld *et al.*, 2007; Yoon *et al.*, 2008). Thus, genetic studies support a role for both 5-HT_{1A} and 5-HT_{2A} receptors in the pathophysiology and symptom severity of anxiety disorders.

Mice with genetic inactivation of 5-HT_{2A} receptors have decreased anxiety traits as measured in conflict anxiety tests (Weisstaub *et al.*, 2006). Selective restoration of 5-HT_{2A} receptors, principally in layer V of the cortex, reversed the anxiety phenotype (Weisstaub *et al.*, 2006), suggesting that cortical 5-HT_{2A} receptors normally play a role in development of anxiety traits.

Individuals with a single nucleotide polymorphism C178T in the regulatory region of the serotonin receptor type 3 gene (*htr3A*) show altered amygdala activation during a face recognition task (Iidaka *et al.*, 2005). Individuals with the C/C phenotype had greater activity in the amygdala and dorsal and medial prefrontal cortices than those with the C/T phenotype. Mice with genetic deletion of the 5-HT_3 receptor show reduced anxiety traits (Kelley *et al.*, 2003; Bhatnagar *et al.*, 2004), consistent with the hypothesis that the 5-HT_3 receptor may play a role in development of anxiety traits. Altogether, these genetic studies in humans and rodents suggest that 5-HT_{1A} receptors (particularly hippocampal 5-HT_{1A} receptors) decrease anxiety traits, while 5-HT_{2A} receptors (particularly cortical 5-HT_{2A} receptors) and 5-HT_3 receptors play a role in facilitation of anxiety traits. Clarification of the role of these receptors in the modulation of anxious states and the development of anxiety traits, as well as the involvement of other serotonin receptor subtypes in development of anxiety traits, will require further study.

Adverse early life experience models

Maternal stress during pregnancy can result in long-term increases in anxiety traits (for review, see Weinstock, 2008). A variety of psychological and physical stressors during gestation, including restraint (Van den Hove *et al.*, 2005), footshock (Estanislau and Morato, 2005), exposure to the synthetic glucocorticoid dexamethasone (Nagano *et al.*, 2008) and exposure to a predator (Patin *et al.*, 2005), can result in increased anxiety traits in offspring. The long-term behavioral consequences of prenatal stress appear to be sex-specific (Weinstock, 2007). Restraint stress during the final 10 days of gestation results in increased anxiety-like behavior in the elevated plus maze test in male offspring, but a reduction of anxiety-like behavior in female offspring (Zuena *et al.*, 2008). As well as the long-term behavioral consequences, maternal stress also induces neurobiological changes in the offspring (Weinstock, 2008). Exposure to prenatal stress is associated with changes in cell morphology (Kraszpulski *et al.*, 2006), higher levels of CRF (Cratty *et al.*, 1995), and greater numbers of CRF receptors (Ward *et al.*, 2000) in the basolateral amygdala. Importantly in the context of this chapter, prenatal stress results in a decrease in 5-HT_{1A} receptor binding in male, but not female, pups in the ventral, but not dorsal, hippocampus (Van den Hove *et al.*, 2006). These changes in the neurobiology of the basolateral amygdala and ventral hippocampus may have important consequences in the development of anxiety traits in adult rats.

Adverse early life experience during the neonatal period can result in increased anxiety traits that persist throughout the lifetime of the organism. Exposure of rats to prolonged periods of neonatal maternal separation has been suggested as an animal model of vulnerability to development of anxiety states and a depression-like syndrome (Willner, 1990; Hall, 1998; Plotsky *et al.*, 1998; Ladd *et al.*, 2000). Although the mechanisms involved are not clear, this vulnerability is associated with increased CRF receptor expression in the dorsal raphe nucleus (Ladd *et al.*, 1996), and changes in serotonergic systems within the brain, including altered stress-induced serotonergic neurotransmission during adulthood (Gartside *et al.*, 2003; Daniels *et al.*, 2004; van Riel *et al.*, 2004). In contrast, exposure of rats to short periods of neonatal maternal separation has been suggested as an animal model of resilience to stress and stress-related pathology, effects that are dependent on epigenetic programming and altered limbic serotonergic function (Smythe *et al.*, 1994; Meaney and Szyf, 2005).

Social isolation during a critical phase of development in adolescence results in persistent, stable increases in anxiety traits in adult rats (for review, see Fone and Porkess, 2008). Social isolation during adolescence increases anxiety-like behavior in the open-field (Hall, 1998), elevated plus maze (Wright *et al.*, 1991) and social interaction tests (Lukkes *et al.*, 2008a). The increase in anxiety traits in social isolates is associated with an up-regulation of CRF_2 receptors in the dorsal raphe nucleus, and facilitation of CRF-induced serotonin release in the nucleus accumbens (Lukkes *et al.*, 2008b). Thus, adverse experience during prenatal, neonatal and adolescent periods can lead to development of anxiety traits in adulthood, possibly via alterations in serotonergic systems signaling to forebrain structures controlling anxious states and anxiety-related behavior.

Priming models

Repeated subthreshold disinhibition or activation of key nodes of the anxiety circuitry (a process referred to as priming) can result in a persistent, seemingly permanent increase in anxiety state, and thus would be considered an anxiety trait. Priming of the basolateral amygdala results in increases in anxious states, as measured in a number of tests including the social interaction test and elevated plus maze test, but also results in vulnerability to panic-like responses to normally innocuous agents, such as sodium lactate (Sanders *et al.*, 1995; Sajdyk and Shekhar, 2000; Shekhar *et al.*, 1999, 2003, 2005). On the other hand, priming of the bed nucleus of the stria terminalis results in stable increases in anxious states as measured in the social interaction test but not the elevated plus maze test (Lee *et al.*, 2008). Similarly, rats subjected to priming of the bed nucleus of the stria terminalis do not display vulnerability to panic-like responses to sodium lactate (Lee *et al.*, 2008). Therefore, it has been suggested that priming of the basolateral amygdala may model anxiety disorders that include vulnerability to panic attacks, including panic disorder and PTSD, while priming of the bed nucleus of the stria terminalis may model social anxiety disorder (Lee *et al.*, 2008). Persistent disinhibition of key nodes of the anxiety circuitry, such as the bed nucleus of the stria terminalis (Sajdyk *et al.*, 2008) and dorsomedial hypothalamus (Shekhar *et al.*, 1996), can also result in a stable increase in anxious state and (in the case of the dorsomedial hypothalamus) vulnerability to panic-like responses to sodium lactate.

Interactions among genetic influences, adverse early life experience, and adverse experience during adulthood

Interactions among the risk factors described above, such as genetic influences, adverse early life experience and stressful life experience during adulthood, are likely to play an important role in the development of anxiety traits in rodents and of anxiety and affective disorders in humans (see, for example, Caspi *et al.*, 2003).

Models of panic-like states

As mentioned briefly above, animal models of a stable vulnerability to panic-like responses have been developed. For example, priming of the basolateral amygdala (Sajdyk *et al.*, 1999; Sajdyk and Shekhar, 2000) or chronic disinhibition of the dorsomedial hypothalamus (Shekhar *et al.*, 1996) results in a persistent vulnerability to panic-like responses (tachycardia, hypertension) in response to i.v. infusions of sodium lactate.

Neural circuits modulating anxiety-related behavior and anxious states

Attempts have been made to understand the neural systems involved in the modulation of the conflict between approach and avoidance that is inherent in many forms of anxious states (Gray, 1982; Gray and McNaughton, 1987, 2000; Millan, 2003). Intuitively, it seems that neural systems mediating approach and avoidance behaviors must be involved, but that higher-order processes are likely to be involved in predicting outcomes of specific behavioral responses and resolving conflicts between two opposing behavioral strategies. For example, Gray has suggested that the septo-hippocampal system plays a critical role in the neurobiology of anxious states, and in particular that it plays a critical role in predicting the outcome of specific behavioral responses – an idea elaborated in what is called the 'comparator model of hippocampal function' (Gray, 1982; Gray and McNaughton, 2000; Vinogradova, 2001). In the context of conflict anxiety the hippocampus acts as a comparator (Gray, 1982), comparing expected outcomes versus actual outcomes. According to this model, the hippocampus and the subiculum sample multimodal sensory information via connections with the entorhinal cortex (Witter *et al.*, 1988; Van Groen and Wyss, 1990; Tamamaki and Nojyo, 1995; Vinogradova, 2001) in order to detect mismatch between expected and actual outcomes. In the case of a mismatch between the expected and actual outcomes (unpredictability), as might occur during exposure to a novel environment, the CA1 region of the ventral hippocampus and subiculum send input to the basolateral amygdala, which is a critical structure in the regulation of anxiety-related behavior following unpredictable stimuli (Herry *et al.*, 2007). The basolateral amygdala attaches emotional salience to the novel stimulus, and regulates the autonomic (via connections with the central nucleus of the amygdala) and behavioral (via connections with the bed nucleus of the stria terminalis) responses to the anxiety-related stimulus. These ideas are consistent with recent findings that hippocampal 'place cells' encode not only instantaneous place, but also information about past and future events (Ferbinteanu and Shapiro, 2003; Jeffery, 2004). For example, some hippocampal place cells are modulated (either switched 'on' or 'off ') depending on whether the rat is going to turn right or left in a maze to obtain a reward. This function of the hippocampus may be particularly relevant to risk assessment behavior, where predictions about the potential outcomes (rewarding/aversive) of different behavioral strategies (approach/avoidance) are required. If the outcome does not match the predicted outcome, the basolateral amygdala may be involved in assigning emotional salience to both potential rewarding and potential aversive outcomes. In support of a functional association between the hippocampus and the basolateral amygdala in the regulation of anxious states and anxiety-related behavior, we have recently reported that exposure to a mild anxiogenic stimulus (an open field arena in low-light conditions) activates basolateral amygdala-projecting neurons in the CA1 region of the ventral hippocampus, the entorhinal cortex and the subiculum, and that this activation is inversely correlated with habituation to the open field during the 15-minute test (Hale *et al.*, 2008a; Figure 3). That is, rats that did not habituate to the open field had a high level of activation of this anxiety circuit projecting to the basolateral amygdala.

The basolateral amygdala as a nodal structure in anxiety-related circuits

The basolateral amygdala is activated during anxious states, as determined using induction of immediate–early gene expression (Singewald *et al.*, 2003; Hale *et al.*, 2006). Consistent with a key role of the basolateral amygdala in the development of anxious states, inhibition of the basolateral amygdala following blockade of N-methyl-D-aspartate (NMDA) and non-NMDA receptors, or administration of the benzodiazepine receptor agonist midazolam, decreases anxiety-related behaviors (File and Gonzalez, 1996), while pharmacological activation of the basolateral amygdala with the $GABA_A$ receptor antagonist bicuculline methiodide, or the anxiety-related neuropeptides CRF or urocortin 1, increases anxiety-related behaviors (Sajdyk and Shekhar, 1997; Sajdyk *et al.*, 1999; Spiga *et al.*, 2006). As mentioned above, priming of the basolateral amygdala results in a persistent increase in anxious state. Imaging studies in humans, although they don't have the resolution to specifically identify the basolateral amygdala, have consistently shown that the amygdala is activated during anxious states, including following challenge with anxiogenic drugs such as the 5-$HT_{2A/2C}$ receptor agonist mCPP (Anderson *et al.*, 2002).

The ventral hippocampus/subiculum as a nodal structure in anxiety-related circuits

The hippocampus is an important component of the distributed neuronal network regulating anxious states.

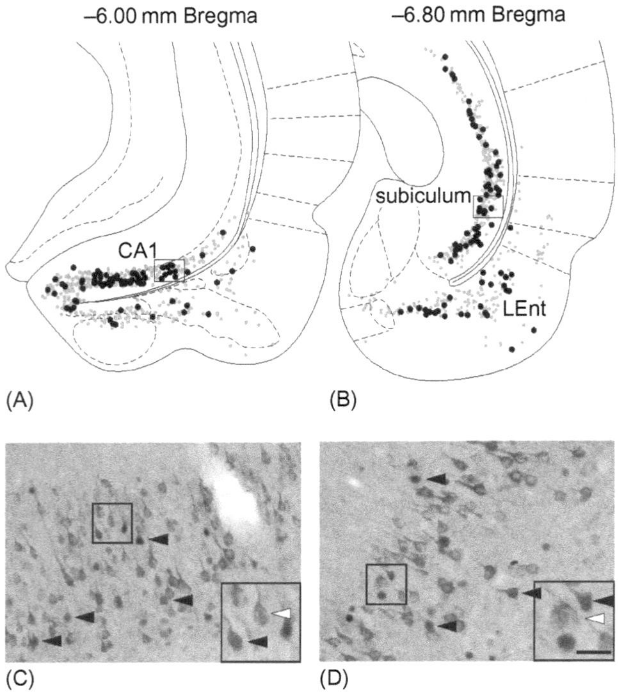

Figure 3 Neural circuits underlying conflict anxiety. Retrograde tracing combined with detection of immediate-early gene induction reveals that basolateral amygdala-projecting neurons in the CA1 region of the ventral hippocampus, subiculum and entorhinal cortex are activated following exposure to a mild anxiogenic stimulus (exposure to an open-field under unfamiliar, low-light conditions). (A, B) Diagrammatic illustrations of basolateral amygdala-projecting neurons in (A) the ventral hippocampus at −6.00 mm Bregma and (B) the subiculum and entorhinal cortex at −6.80 mm Bregma. Basolateral amygdala-projecting neurons in the dorsal raphe nucleus were also activated (not shown), suggesting that a dorsal raphe nucleus–basolateral amygdala circuit modulates this anxiety network during anxious states. Rats received unilateral injection of the retrograde tracer Cholera Toxin B-subunit (CTb) into the basolateral amygdala. Seven days later rats were exposed to an open-field arena in unfamiliar, low-light conditions. Brains were then processed for immunohistochemical detection of c-Fos, as a marker of neuronal activation (c-Fos-ir), and for CTb, as a marker of basolateral amygdala-projecting neurons (CTb-ir). Black circles illustrate c-Fos-ir/CTb-ir neurons and the gray circles illustrate c-Fos-immunonegative/CTb-ir neurons. Black boxes in (A) and (B) are shown as photomicrographs in (C) and (D) respectively. (C) Photomicrograph illustrating c-Fos-ir nuclei and CTb-ir neurons in the CA1. (D) Photomicrograph illustrating c-Fos-ir nuclei and CTb-ir neurons in the subiculum and lateral entorhinal cortex (LEnt). Black arrowheads indicate examples of c-Fos-ir/CTb-ir neurons and white arrowheads indicate examples of c-Fos-immunonegative/CTb-ir neurons. Black boxes in C and D are shown at higher magnification in insets in the lower right-hand corner of each image. Abbreviations: CA1, CA1 region of the ventral hippocampus; LEnt, lateral entorhinal cortex. Scale bar, 500 μm (A, B), 50 μm (C, D), 25 μm (insets).

The hippocampus is not a homogeneous structure, however, and convergent lines of evidence suggest that there is a functional dissociation across its dorsal–ventral axis (Bannerman *et al.*, 2004). Neurons in the ventral hippocampus that project to the basolateral amygdala are activated even following mild-anxiety challenges such as exposure to a novel open field under low-light conditions (Hale *et al.*, 2008a). Lesions to the ventral (vHPC), but not dorsal hippocampus (dHPC), reduce anxiety-related behavior in the social interaction test (McHugh *et al.*, 2004), the elevated plus maze (Kjelstrup *et al.*, 2002) and the novelty induced suppression of feeding test (McHugh *et al.*, 2004). Temporary inactivation of the vHPC with sodium channel blockers tetrodotoxin or lidocaine, or the NMDA-receptor antagonist (±)-2-amino5-phosphopentanoic acid (AP-5), decreases anxiety-related behavior in the elevated plus maze, while inactivation of the dHPC is without effect (Degroot and Treit, 2004; Bertoglio *et al.*, 2006; Nascimento Hackl and Carobrez, 2007). A proportion of neurons within the dorsal raphe nucleus that project to the basolateral amygdala also give rise to collateral (branched) projections to the vHPC (Imai *et al.*, 1986), suggesting that a subset of serotonergic neurons has coordinate control over these two components of anxiety circuitry.

Neural circuits modulating vulnerability to panic-like responses

Attempts have been made to understand the neural systems involved in the modulation of panic-like responses (Graeff, 1990, 1991, Graeff et al., 1996). As with neural circuits regulating anxious states, it seems that neural systems mediating fight-or-flight responses must be involved, but that higher-order modulatory processes are likely to be involved. This is supported by the finding that priming of the amygdala or chronic disinhibition of the dorsomedial hypothalamus can result in a chronic vulnerability to panic-like responses to normally innocuous agents such as sodium lactate (Shekhar *et al.*, 1996; Sajdyk and Shekhar, 2000). Two key structures involved in mediating panic-like responses are the dorsal periaqueductal gray (DPAG), which mediates escape responses (Graeff *et al.*, 1996), and the rostroventrolateral medulla (RVLM), which controls the sympathetic motor component of cardiovascular responses (Dampney, 1994; Saper, 1995) (Figure 4).

The dorsal periaqueductal gray as a nodal structure in mediating flight behavior

The DPAG is activated during panic- or escape-like responses, as determined using induction of immediate–early gene expression (Silveira *et al.*, 2001; Lamprea *et al.*, 2002), and is considered to be a model of panic anxiety (Graeff, 1994; Jenck *et al.*, 1995). Consistent with a key role of the DPAG in vulnerability to panic-like responses,

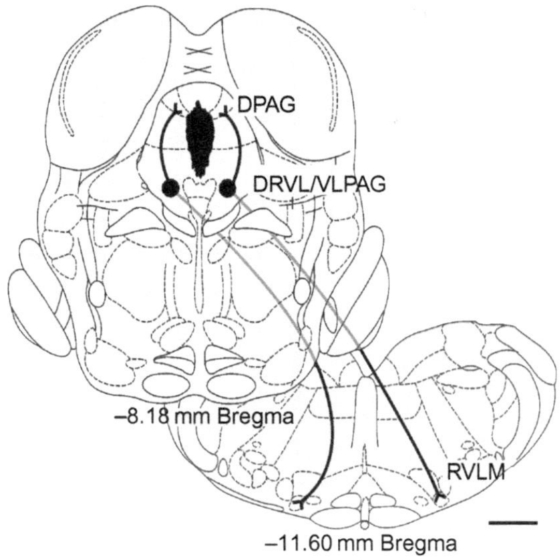

Figure 4 Schematic diagram of two coronal sections from a rat brain (−8.18 and −11.60 mm Bregma) illustrating known projections of serotonergic neurons within the 'lateral wings' or DRVL/VLPAG region of the dorsal raphe nucleus. Serotonergic neurons within the DRVL/VLPAG region project to the DPAG where, via actions on 5-HT_{1A} and 5-HT_2 receptors, serotonin evokes a decrease in the functional excitability of the DPAG and fight-or-flight or panic-like responses. Serotonergic neurons in the DRVL/VLPAG region also project to the RVLM, where, via actions on 5-HT_{1A} receptors, serotonin evokes sympathoinhibition and hypotension. Overall, activation of DRVL serotonergic neurons would be expected to promote a more passive emotional coping strategy. Abbreviations: DPAG, dorsal periaqueductal gray; DRVL/VLPAG, dorsal raphe nucleus, ventrolateral part/ventrolateral periaqueductal gray; RVLM, rostral ventrolateral medulla. Scale bar, 500 μm.

pharmacological inhibition of the DPAG using the benzodiazepine midazolam (de Bortoli *et al.*, 2008), or the $GABA_A$ receptor agonist muscimol (de Menezes *et al.*, 2008; Bueno *et al.*, 2005), or the $GABA_B$ receptor agonist baclofen (Bueno *et al.*, 2005) impairs panic-like and escape behaviors, while electrical stimulation of the DPAG (Jenck *et al.*, 1995), pharmacological activation of the DPAG with the neuropeptide cholecystokinin sulfated octapeptide (CCK-8s) (Zanoveli *et al.*, 2004), or disinhibition of the DPAG with the $GABA_A$ receptor antagonist bicuculline methiodide (Graeff *et al.*, 1986) increases escape responses.

The RVLM as a nodal structure mediating cardiovascular responses

The RVLM is an important component of the neural mechanisms regulating blood pressure under baseline and stimulated conditions (Madden and Sved, 2003), and an important component of the descending pathway mediating the sympathetic motor component of cardiovascular responses to stimulation of the dorsomedial hypothalamus (Horiuchi *et al.*, 2006). As such, it could play a role in dysregulation of panic-like responses that have been described following chronic disinhibition of the dorsomedial hypothalamus (Shekhar *et al.*, 1996; Johnson *et al.*, 2007).

Modulation of anxiety-related circuits by serotonin

Having described parts of the neural circuits regulating anxious states and panic-like responses, in this section we highlight how serotonin modulates anxious states via effects on the ventral hippocampus and basolateral amygdala, and modulates panic-like responses via effects on the DPAG and RVLM.

Effects of serotonin within the basolateral amygdala

The basolateral amygdala is innervated by both serotonergic and non-serotonergic neurons within the dorsal raphe nucleus (Hale *et al.*, 2008b). Although the basolateral amygdala-projecting neurons are widely distributed across the rostrocaudal extent of the raphe complex, the highest concentrations of basolateral amygdala-projecting neurons are located in the mid-rostrocaudal region of the dorsal part of the dorsal raphe nucleus (DRD) (Ottersen, 1981; Imai *et al.*, 1986; Abrams *et al.*, 2005). We have recently reported that the majority of basolateral amygdala-projecting neurons in the midbrain raphe complex are serotonergic (65.2 ± 3.0 percent; Hale *et al.*, 2008b). However, the numbers of basolateral amygdala-projecting neurons that are serotonergic vary depending on the anatomical subdivision of the midbrain raphe complex. Fewer than half of the basolateral amygdala-projecting neurons in the ventrolateral part of the dorsal raphe nucleus (DRVL) region (both at −8.00 and −8.18 mm Bregma) were found to be serotonergic, whereas basolateral amygdala-projecting neurons in the mid-rostrocaudal DRD were mostly serotonergic (73.1 ± 5.8 percent at −8.00 mm Bregma and 85.5 ± 7.2 percent at −8.18 mm Bregma; Figure 5). The neurochemical phenotype(s) of the non-serotonergic neurons projecting from the dorsal raphe nucleus to the basolateral amygdala is/are not known. The hypothesis that basolateral amygdala-projecting neurons in the dorsal raphe nucleus are involved in modulation of anxiety-related behavior is supported by the recent finding that the basolateral amygdala-projecting subset of neurons within the dorsal raphe nucleus is activated following exposure of rats to a mild anxiogenic stimulus – exposure to an open-field under unfamiliar, low-light conditions (Hale *et al.*, 2008b).

Serotonergic neurons in the DRD region have been associated with anxiety-like responses (for review, see Lowry *et al.*, 2005). The DRD receives dense projections from anxiety-related forebrain areas, including the bed nucleus of the stria terminalis and central amygdaloid nucleus, and gives rise to projections to anxiety-related forebrain areas, including the basolateral amygdala and the ventral hippocampus (Abrams *et al.*, 2005; Hale *et al.*, 2008b; for review, see Imai *et al.*, 1986; Lowry *et al.*, 2005). Serotonergic neurons within the DRD are activated following a number of stress- and anxiety-related stimuli, including uncontrollable stress (Grahn *et al.*, 1999), social defeat (Gardner *et al.*, 2005), anxiogenic drugs (Abrams *et al.*, 2005) and anxiety-related neuropeptides (Staub *et al.*, 2005, 2006). Injection of CRF into the caudal, but not rostral, dorsal raphe nucleus induces anxiety- or fear-related behaviors measured 24 hours later in a model of learned helplessness (Hammack *et al.*, 2002); an effect that is thought to be due to an overactivation of serotonergic neurons in the caudal dorsal raphe nucleus (Maier and Watkins, 2005). Thus, serotonergic neurons in the DRD are anatomically integrated with a distributed anxiety-related circuit, serotonergic neurons in this region are selectively activated by anxiogenic stimuli, and CRF-induced activation of this region induces behavior consistent with an increased anxious state.

Anxiogenic stimuli increase extracellular serotonin concentrations in the basolateral amygdala (Amat *et al.*, 1998a, 2004). Serotonin acting within the basolateral amygdala may have either anxiolytic or anxiogenic effects, depending on the cellular site of action and the serotonin receptor subtype involved. A role for serotonin in facilitating coping mechanisms in the amygdala is supported by findings that the overall effect of serotonin in the basolateral amygdala is a 5-HT_{2A} receptor-mediated enhancement of GABA-mediated inhibition of projection neurons (Rainnie, 1999). In addition, recent evidence from McDonald and coworkers (Muller *et al.*, 2007) suggests that serotonergic fibers preferentially target distinct subpopulations of interneurons within the basolateral

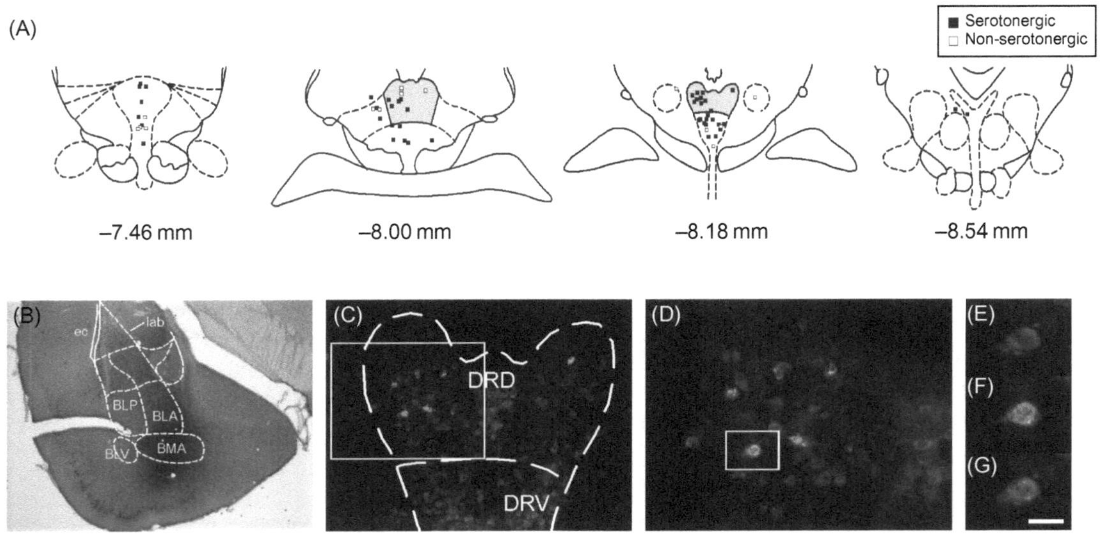

Figure 5 Illustrations and photomicrographs showing retrogradely labeled neurons in the dorsal raphe nucleus following injections of Cholera Toxin B-subunit (CTb) into the basolateral nucleus of the amygdala. (A) Illustrations of the distribution of basolateral amygdala-projecting serotonergic (black squares) and non-serotonergic (white squares) neurons within subdivisions of the dorsal raphe nucleus. Retrogradely labeled neurons are concentrated in the mid-rostrocaudal dorsal raphe nucleus (gray shaded areas at −8.00 mm and −8.18 mm Bregma) (Paxinos and Watson, 1998). Most but not all BL-projecting neurons in the dorsal raphe nucleus are serotonergic. (B) Photomicrograph illustrating the CTb injection site restricted to the BLA and BMA. (C) Double-labeled immunofluorescence photomicrograph showing the mid-rostrocaudal dorsal raphe nucleus (corresponding to −8.18 mm Bregma illustration in (A)). White box in (C) indicates region shown at higher magnification in (D). White box in (D) indicates region shown at higher magnification in (E), (F) and (G). Tryptophan hydroxylase-immunoreactive (TrpOH-ir; serotonergic) neurons appear red (E), while CTb-ir (BL-projecting) neuron appears green (F). Double TrpOH/CTb immunofluorescent neuron appears yellow. Abbreviations: BLA, basolateral nucleus of the amygdala, anterior part; BLP, basolateral nucleus of the amygdala, posterior part; BLV, basolateral nucleus of the amygdala, ventral part; BMA, basomedial nucleus of the amygdala, anterior part; DRD, dorsal raphe nucleus, dorsal part; DRV, dorsal raphe nucleus, ventral part; ec, external capsule; lab, longitudinal association bundle. Scale bar: 448 μm (A); 500 μm (B); 100 μm (C); 50 μm (D), 25 μm (E, F, G). To see the full color version of this figure please refer to the color plate at the back of the book. Copies produced via our print on demand service do not contain color plates; if your copy does not have the color plate, please go to this website to view the figure in color www.elsevier direct.com/companions/9780123746344

amygdala, in addition to targeting projection neurons. Thus, activation of dorsal raphe nucleus serotonergic systems would enhance inhibition in the basolateral amygdala and facilitate coping with the stressor. Significantly, activation of 5-HT_2 receptors in the basolateral amgydala reduces fear and anxiety in guinea pigs, as measured by a tonic immobility test (Leite-Panissi *et al.*, 2006). Imaging studies in humans support a potential role for serotonin within the amygdala in resilience (Anderson *et al.*, 2008).

Serotonin acting on presynaptic 5-HT_{1A} receptors within the basolateral amygdala inhibits both inhibitory (Kishimoto *et al.*, 2000) and excitatory (Cheng *et al.*, 1998) transmission. As might be expected based on these electrophysiological effects, the behavioral effects of 5-HT_{1A} receptor activation in the basolateral amygdala are complex, and appear to be dependent on the paradigm used to measure anxiety-related behavior (Zangrossi and Graeff, 1994; Gonzalez *et al.*, 1996; Zangrossi *et al.*, 1999).

In contrast to the anxiolytic or mixed effects of 5-HT_{2A} and 5-HT_{1A} receptor activation, respectively, in the basolateral amygdala, activation of 5-HT_{2C} receptors appears to be anxiogenic. The cellular mechanisms are not clear, but direct injection of 5-HT_{2C} receptor agonists into the basolateral amygdala increases anxiety-related behavior – an effect that can be reversed by 5-HT_{2C} receptor-selective antagonists (Campbell and Merchant, 2003). Thus, pharmacological studies support a role for serotonergic systems in both facilitating and attenuating anxiety-related behavior via actions on different presynaptic and postsynaptic signaling mechanisms. It remains to be determined if these opposing effects are associated with activation of different subsets of serotonergic neurons in the dorsal raphe complex.

Effects of serotonin within the ventral hippocampus/subiculum

Like the basolateral amygdala, the vHPC is an important site for the modulatory effects of serotonin on anxious states and anxiety-related behavior. The hippocampus receives dense projections from the midbrain raphe complex. The median raphe nucleus (MNR) sends projections predominantly to the dHPC, while the dorsal raphe nucleus sends projections predominantly to the vHPC (Azmitia and Segal, 1978; Imai *et al.*, 1986; Vertes, 1991). Consistent with this anatomical specificity, electrical stimulation of the MNR induces 5-HT release in the dHPC and the vHPC, whereas stimulation of the DR induces 5-HT release in the vHPC only (McQuade and Sharp, 1997). Ventral hippocampus-projecting neurons in the dorsal raphe nucleus are located in the caudal part of the nucleus (Imai *et al.*, 1986), a region known to contain a population of serotonergic neurons that are sensitive to anxiogenic drugs and anxiety-related neuropeptides (Abrams *et al.*, 2005; Staub *et al.*, 2005, 2006) and anxiogenic stimuli (Grahn *et al.*, 1999; Gardner *et al.*, 2005). A subset of neurons in the caudal dorsal raphe nucleus that projects to the vHPC also projects to the basolateral amygdala (Imai *et al.*, 1986), suggesting that the dorsal raphe nucleus exerts coordinate control over multiple nodes of anxiety circuits. Similar collateral projections of serotonergic neurons in the dorsal raphe nucleus to different nodes of anxiety circuitry have been described for the prefrontal cortex and nucleus accumbens (Van Bockstaele *et al.*, 1993), hippocampus and entorhinal cortex (Kohler and Steinbusch, 1982), entorhinal cortex and septum (Kohler *et al.*, 1982), hippocampus and locus coeruleus, and locus coeruleus and amygdala (Imai *et al.*, 1986). These findings suggest that a small subset of serotonergic neurons gives rise to collateral projections to an extensive, but anatomically distributed and functionally related, anxiety circuit (Figure 1).

Serotonin release in the vHPC is increased by anxiogenic stimuli (Amat *et al.*, 1998b; Kagamiishi *et al.*, 2003). Microdialysis studies suggest an increase in serotonin release in the vHPC following exposure to the elevated plus maze (Wright *et al.*, 1992; Voigt *et al.*, 1999; Rex *et al.*, 2005) an effect that can be blocked with anxiolytic drugs such as diazepam or the 5-HT_{1A} receptor agonist ipsapirone. It should be noted that while systemic administration of diazepam blocks the release of serotonin in the vHPC and decreases anxiety-like behavior in the elevated plus maze, ipsapirone blocks only the serotonin release, with no change in anxiety-like behavior. This may suggest a presynaptic effect of ipsapirone at the level of the dorsal raphe nucleus to inhibit serotonin release, and a postsynaptic effect, possibly in the vHPC, regulating behavior. Microinjection of the 5-HT_{1A} receptor antagonist WAY-100635 into the vHPC reduces anxiety-related behavior in the elevated plus maze in mice naïve to the apparatus, but has no effect in mice pre-exposed to the apparatus 24 hours prior to testing. Microinjection of WAY-100635 into the dHPC was without effect on anxiety-like behavior in the elevated plus maze (Nunes-de-Souza *et al.*, 2002).

Despite the evidence for a dissociation of function of the hippocampus along its dorsal–ventral axis and the possible involvement of serotonin within the vHPC in the modulation of anxious states and anxiety-related behavior, microinjection studies with 5-HT_{1A} agonists and antagonists, similar to studies of the basolateral amygdala, provide no clear consensus (for review, see Engin and Treit, 2007). Within the vHPC, agonists at the 5-HT_{1A} receptor have no effect in the social interaction test (Hogg *et al.*, 1994) or elevated plus maze (File and Gonzalez, 1996), and antagonists at the 5-HT_{1A} receptor have been shown to have no effect in the social interaction test (Hogg *et al.*, 1994). 5-HT_{1A} receptor antagonists have anxiolytic effects in the elevated plus maze (Nunes-de-Souza *et al.*, 2002). As in the basolateral amygdala, agonists at the 5-HT_{2C} receptor have anxiogenic effects in the vHPC, but not in the dHPC (Alves *et al.*, 2004), reinforcing the idea that within the hippocampus the vHPC plays a primary role in regulation of anxious states. Further studies are required to fully explore the anxiolytic and anxiogenic effects of serotonergic compounds in the ventral hippocampus, and the receptor systems involved.

Effects of serotonin in panic circuitry

Evidence suggests that there is an anatomical and functional topography of serotonergic neurons within the dorsal raphe nucleus (for review, see Lowry *et al.*, 2008a) (Figure 1). One of the most distinct anatomical and functional subsets of serotonergic neurons is that subset located within the 'lateral wings' of the dorsal raphe nucleus. These are large multipolar neurons that are scattered throughout the DRVL part of the dorsal raphe nucleus and the adjacent VLPAG. These cells are part of a sympathomotor command center, with multisynaptic connections to both hindlimb muscles and the adrenal medulla, and therefore are positioned to regulate multiple components of the 'fight-or-flight' or panic-like responses (Kerman *et al.*, 2006).

One site that may be involved in DRVL/VLPAG regulation of panic-like responses is the RVLM. DRVL/VLPAG serotonergic neurons project to the RVLM (Underwood *et al.*, 1999; Bago *et al.*, 2002), and chemical stimulation of the VLPAG region results in sympathoinhibition and hypotension. VLPAG stimulation-induced sympathoinhibition and hypotension can be blocked by microinjection of the 5-HT_{1A} receptor antagonist WAY-100635 into the RVLM (Bago and Dean, 2001). In approximately half of the animals, microinjection of WAY-100635 into the RVLM not only prevented the VLPAG-induced sympathoinhibition and hypotension, but also evoked sympathoexcitation and hypertension (Bago and Dean, 2001). DRVL/VLPAG neurons also project to the DPAG (Stezhka and Lovick, 1997). Activation of 5-HT_{1A} or 5-HT_2 receptors in the DPAG reduces the excitability of the DPAG and reduces panic-like behavioral responses (for review, see Graeff *et al.*, 1996). Thus, serotonergic projections arising from the DRVL/VLPAG region innervate a sympathomotor command center and appear to inhibit both autonomic and behavioral components of panic-like responses.

Functional anatomical studies support a role for the DRVL/VLPAG region in inhibition of panic-like responses. We have shown that a number of panicogenic stimuli, including sodium lactate infusion (Johnson *et al.*, 2008) and exposure to hypercapnia (elevated concentrations of carbon dioxide in inhaled air; Johnson *et al.*, 2005), selectively increase c-Fos expression in the DRVL/VLPAG region (for review, see Johnson *et al.*, 2004). Importantly, we have found that infusion of sodium lactate, while it increases c-Fos expression in serotonergic neurons within the DRVL/VLPAG region in normal rats that do not respond to sodium lactate, has no effect on c-Fos expression in serotonergic neurons in panic-prone rats that respond to sodium lactate infusion with tachycardia and hypertension (Johnson *et al.*, 2008). These data are consistent with the hypothesis that serotonergic neurons in the DRVL/VLPAG region normally play a role in suppression of autonomic responses to panicogenic stimuli, but that failure to activate this serotonergic system disinhibits panic responses.

Adverse early life experience may result in vulnerability to stress-induced effects on the DRVL/VLPAG group of serotonergic neurons. We have shown that exposure of rats to a social defeat paradigm increased the expression of TPH2 and SERT mRNA expression, but that this effect was only apparent in rats that had been exposed to adverse early life experience (maternal separation) (Gardner *et al.*, 2006; Lowry *et al.*, 2008b). In contrast, stress had no effect on TPH2 or SERT mRNA expression in other regions of the dorsal raphe nucleus, even in rats exposed to maternal separation. These site-specific effects may be due to activation of specific afferent inputs to this region (Peyron *et al.*, 1998). Interestingly, when rats were pretreated with a 5-HT_{1A} receptor antagonist, to prevent autoreceptor-mediated negative feedback to serotonergic neurons, exposure of rats to the forced swim test selectively increased c-Fos expression in serotonergic neurons in the DRVL/VLPAG region (Commons, 2008), consistent with the hypothesis that stimuli that evoke a passive emotional coping strategy (Keay and Bandler, 2001) activate DRVL/VLPAG serotonergic neurons.

Summary and future directions

Anxious states can be induced by a number of pharmacological approaches or by adverse experience, during either development or adult life. Genetic influences and adverse early life experiences can result in stable individual differences in anxiety traits. These anxious states and anxiety traits appear to be controlled by a distributed network of interconnected brain regions, including the brainstem noradrenergic and serotonergic sytems. We argue that in order to understand the role of serotonergic systems in the modulation of anxious states, it is important to identify key nodes of anxiety circuitry, and to understand the properties of the subpopulation(s) of serotonergic neurons that projects to anxiety circuits. Using this approach, it is clear that a subset of serotonergic neurons, predominantly located in the mid-rostrocaudal and caudal parts of the dorsal raphe nucleus, projecting to the basolateral amygdala and ventral hippocampus, modulates anxiety states, while a different subset of serotonergic neurons in the lateral wings of the dorsal raphe nucleus, projecting to the DPAG and RVLM, is part of a sympathomotor command system regulating the fight-or-flight or panic response.

Future studies should clarify the role of different serotonin receptor subtypes within these different anxiety circuits in the regulation of anxious states and panic-like responses, and under what conditions these serotonergic systems are activated. Selective lesion of different subsets of serotonergic neurons should play an important role in clarifying some of these issues. If different subsets of serotonergic neurons have different molecular phenotypes, as might be suspected based on their different afferent input, molecular profiling of subsets of serotonergic neurons should provide opportunities for selective pharmacological manipulation of subsets of serotonergic neurons involved in modulation of anxious states, a strategy that could limit side effects. Future studies should investigate the interactions among genetic vulnerability, adverse early life experience and adverse experience during adulthood on serotonergic systems, anxious states and the development of anxiety traits. Finally, strategies should be explored for reversal of these anxiety states and anxiety traits.

Acknowledgements

We are grateful to Dr. Anders Hay-Schmidt and Dr. Jens Mikkelsen for assistance with the retrograde tracing experiments. Christopher Lowry is a recipient of an NARSAD 2007 Young Investigator Award.

References

Abrams, J.K., Johnson, P.L., Hay-Schmidt, A., Mikkelsen, J.D., Shekhar, A. and Lowry, C.A. (2005) Serotonergic systems associated with arousal and vigilance behaviors following administration of anxiogenic drugs. Neuroscience, 133: 983–997.

Adamec, R., Burton, P., Blundell, J., Murphy, D.L. and Holmes, A. (2006) Vulnerability to mild predator stress in serotonin transporter knockout mice. Behav. Brain Res., 170: 126–140.

Adamec, R.E. and Shallow, T. (1993) Lasting effects on rodent anxiety of a single exposure to a cat. Physiol. Behav., 54: 101–109.

Adamec, R., Walling, S. and Burton, P. (2004) Long-lasting, selective, anxiogenic effects of feline predator stress in mice. Physiol. Behav., 83: 401–410.

Adamec, R., Holmes, A. and Blundell, J. (2008) Vulnerability to lasting anxiogenic effects of brief exposure to predator stimuli: sex, serotonin and other factors – relevance to PTSD. Neurosci. Biobehav. Rev., 32: 1287–1292.

Alves, S.H., Pinheiro, G., Motta, V., Landeira-Fernandez, J. and Cruz, A.P. (2004) Anxiogenic effects in the rat elevated plus-maze of 5-HT(2C) agonists into ventral but not dorsal hippocampus. Behav. Pharmacol., 15: 37–43.

Amat, J., Matus-Amat, P., Watkins, L.R. and Maier, S.F. (1998a) Escapable and inescapable stress differentially alter extracellular levels of 5-HT in the basolateral amygdala of the rat. Brain Res., 812: 113–120.

Amat, J., Matus-Amat, P., Watkins, L.R. and Maier, S.F. (1998b) Escapable and inescapable stress differentially and selectively alter extracellular levels of 5-HT in the ventral hippocampus and dorsal periaqueductal gray of the rat. Brain Res., 797: 12–22.

Amat, J., Tamblyn, J.P., Paul, E.D., Bland, S.T., Amat, P., Foster, A.C., Watkins, L.R. and Maier, S.F. (2004) Microinjection of urocortin 2 into the dorsal raphe nucleus activates serotonergic neurons and increases extracellular serotonin in the basolateral amygdala. Neuroscience, 129: 509–519.

American Psychiatric Association (2000) Diagnostic and Statistical Manual of Mental Disorders, Text Revision. 4th Ed. American Psychiatric Association, Washington, DC.

Anderson, I.M., Clark, L., Elliott, R., Kulkarni, B., Williams, S.R. and Deakin, J.F. (2002) 5-HT(2C) receptor activation by m-chlorophenylpiperazine detected in humans with fMRI. Neuroreport, 13: 1547–1551.

Anderson, I.M., McKie, S., Elliott, R., Williams, S.R. and Deakin, J.F. (2008) Assessing human 5-HT function *in vivo* with pharmacoMRI. Neuropharmacology, 55: 1029–1037.

Armario, A., Escorihuela, R.M. and Nadal, R. (2008) Long-term neuroendocrine and behavioural effects of a single exposure to stress in adult animals. Neurosci. Biobehav. Rev., 32: 1121–1135.

Azmitia, E.C. Jr. and Segal, M. (1978) An autoradiographic analysis of the differential ascending projection of the dorsal and median raphe nuclei in the rat. J. Comp. Neurol., 179: 651–668.

Bago, M. and Dean, C. (2001) Sympathoinhibition from ventrolateral periaqueductal gray mediated by 5-HT(1A) receptors in the RVLM. Am. J. Physiol. Regul. Integr. Comp. Physiol., 280: R976–R984.

Bago, M., Marson, L. and Dean, C. (2002) Serotonergic projections to the rostroventrolateral medulla from midbrain and raphe nuclei. Brain Res., 945: 249–258.

Bannerman, D.M., Rawlins, J.N., McHugh, S.B., Deacon, R.M., Yee, B.K., Bast, T., Zhang, W.N., Pothuizen, H.H. and Feldon, J. (2004) Regional dissociations within the hippocampus – memory and anxiety. Neurosci. Biobehav. Rev., 28: 273–283.

Bertoglio, L.J., Joca, S.R. and Guimaraes, F.S. (2006) Further evidence that anxiety and memory are regionally dissociated within the hippocampus. Behav. Brain Res., 175: 183–188.

Bhatnagar, S., Sun, L.M., Raber, J., Maren, S., Julius, D. and Dallman, M.F. (2004) Changes in anxiety-related behaviors and hypothalamic–pituitary–adrenal activity in mice lacking the 5-HT-3A receptor. Physiol. Behav., 81: 545–555.

Bouwknecht, J.A., Spiga, F., Staub, D.R., Hale, M.W., Shekhar, A. and Lowry, C.A. (2007) Differential effects of exposure to low-light or high-light open-field on anxiety-related behaviors: Relationship to c-Fos expression in serotonergic and non-serotonergic neurons in the dorsal raphe nucleus. Brain Res. Bull., 72: 32–43.

Brocke, B., Armbruster, D., Muller, J., Hensch, T., Jacob, C.P., Lesch, K.P., Kirschbaum, C. and Strobel, A. (2006) Serotonin transporter gene variation impacts innate fear processing: Acoustic startle response and emotional startle. Mol. Psychiatry, 11: 1106–1112.

Bueno, C.H., Zangrossi, H. Jr., Nogueira, R.L., Soares, V.P. and Viana, M.B. (2005) Panicolytic-like effect induced by the stimulation of GABAA and GABAB receptors in the dorsal periaqueductal grey of rats. Eur. J. Pharmacol., 516: 239–246.

Campbell, B.M. and Merchant, K.M. (2003) Serotonin 2C receptors within the basolateral amygdala induce acute fear-like responses in an open-field environment. Brain Res., 993: 1–9.

Canli, T., Qiu, M., Omura, K., Congdon, E., Haas, B.W., Amin, Z., Herrmann, M.J., Constable, R.T. and Lesch, K.P. (2006) Neural correlates of epigenesis. Proc. Natl. Acad. Sci. USA, 103: 16033–16038.

Caspi, A., Sugden, K., Moffitt, T.E., Taylor, A., Craig, I.W., Harrington, H., McClay, J., Mill, J., Martin, J., Braithwaite, A. and Poulton, R. (2003) Influence of life stress on depression: moderation by a polymorphism in the 5-HTT gene. Science, 301: 386–389.

Cecchi, M., Khoshbouei, H., Javors, M. and Morilak, D.A. (2002a) Modulatory effects of norepinephrine in the lateral bed nucleus of the stria terminalis on behavioral and neuroendocrine responses to acute stress. Neuroscience, 112: 13–21.

Cecchi, M., Khoshbouei, H. and Morilak, D.A. (2002b) Modulatory effects of norepinephrine, acting on alpha 1 receptors in the central nucleus of the amygdala, on behavioral and neuroendocrine responses to acute immobilization stress. Neuropharmacology, 43: 1139–1147.

Cheng, L.L., Wang, S.J. and Gean, P.W. (1998) Serotonin depresses excitatory synaptic transmission and depolarization-evoked $Ca2^{+}$ influx in rat basolateral amygdala via 5-HT1A receptors. Eur. J. Neurosci., 10: 2163–2172.

Cole, B.J., Hillmann, M., Seidelmann, D., Klewer, M. and Jones, G.H. (1995) Effects of benzodiazepine receptor partial inverse agonists in the elevated plus maze test of anxiety in the rat. Psychopharmacology (Berl.), 121: 118–126.

Commons, K.G. (2008) Evidence for topographically organized endogenous 5-HT-1A receptor-dependent feedback inhibition of the ascending serotonin system. Eur. J. Neurosci., 27: 2611–2618.

Cooper, M.A., McIntyre, K.E. and Huhman, K.L. (2008) Activation of 5-HT1A autoreceptors in the dorsal raphe nucleus reduces the behavioral consequences of social defeat. Psychoneuroendocrinology, 33: 1236–1247.

Cratty, M.S., Ward, H.E., Johnson, E.A., Azzaro, A.J. and Birkle, D.L. (1995) Prenatal stress increases corticotropin-releasing factor (CRF) content and release in rat amygdala minces. Brain Res., 675: 297–302.

Dampney, R.A.L. (1994) Functional organization of central pathways regulating the cardiovascular system. Physiol. Rev., 74: 323–364.

Daniels, W.M., Pietersen, C.Y., Carstens, M.E. and Stein, D.J. (2004) Maternal separation in rats leads to anxiety-like behavior and a blunted ACTH response and altered neurotransmitter levels in response to a subsequent stressor. Metab. Brain Dis., 19: 3–14.

de Bortoli, V., Nogueira, R.L. and Zangrossi, H. Jr. (2008) Alprazolam potentiates the antiaversive effect induced by the activation of 5-HT(1A) and 5-HT (2A) receptors in the rat dorsal periaqueductal gray. Psychopharmacology (Berl.), 198: 341–349.

de Menezes, R.C., Zaretsky, D.V., Sarkar, S., Fontes, M.A. and DiMicco, J.A. (2008) Microinjection of muscimol into the periaqueductal gray suppresses cardiovascular and neuroendocrine response to air jet stress in conscious rats. Am. J. Physiol. Regul. Integr. Comp. Physiol., 295: R881–R890.

Degroot, A. and Treit, D. (2004) Anxiety is functionally segregated within the septo-hippocampal system. Brain Res., 1001: 60–71.

Drago, A., Ronchi, D.D. and Serretti, A. (2008) 5-HT1A gene variants and psychiatric disorders: a review of current literature and selection of SNPs for future studies. Int. J. Neuropsychopharmacol., 11: 701–721.

Dunn, A.J. and Berridge, C.W. (1987) Corticotropin-releasing factor administration elicits a stress-like activation of cerebral catecholamine systems. Pharmacol. Biochem. Behav., 27: 685–691.

Engin, E. and Treit, D. (2007) The role of hippocampus in anxiety: intracerebral infusion studies. Behav. Pharmacol., 18: 365–374.

Estanislau, C. and Morato, S. (2005) Prenatal stress produces more behavioral alterations than maternal separation in the elevated plus maze and in the elevated T-maze. Behav. Brain Res., 163: 70–77.

Evans, A.K. and Lowry, C.A. (2007) Pharmacology of the beta-carboline FG-7142, a partial inverse agonist at the benzodiazepine allosteric site of the GABA A receptor: neurochemical, neurophysiological, and behavioral effects. CNS Drug. Rev., 13: 475–501.

Ferbinteanu, J. and Shapiro, M.L. (2003) Prospective and retrospective memory coding in the hippocampus. Neuron, 40: 1227–1239.

File, S.E. and Gonzalez, L.E. (1996) Anxiolytic effects in the plus-maze of 5-HT1A-receptor ligands in dorsal raphe and ventral hippocampus. Pharmacol. Biochem. Behav., 54: 123–128.

File, S.E. and Pellow, S. (1984) The anxiogenic action of Ro 15-1788 is reversed by chronic, but not by acute, treatment with chlordiazepoxide. Brain Res., 310: 154–156.

Fone, K.C. and Porkess, M.V. (2008) Behavioural and neurochemical effects of post-weaning social isolation in rodents – relevance to developmental neuropsychiatric disorders. Neurosci. Biobehav. Rev., 32: 1087–1102.

Furmark, T., Tillfors, M., Garpenstrand, H., Marteinsdottir, I., Langstrom, B., Oreland, L. and Fredrikson, M. (2004) Serotonin transporter polymorphism related to amygdala excitability and symptom severity in patients with social phobia. Neurosci. Lett., 362: 189–192.

Gardner, K.L., Thrivikraman, K.V., Lightman, S.L., Plotsky, P.M. and Lowry, C.A. (2005) Early life experience alters behavior during social defeat: focus on serotonergic systems. Neuroscience, 136: 181–191.

Gardner, K. L., Lightman, S. L., Plotsky, P. M., and Lowry, C. A. (2006) Early life experience programs gene expression in serotonin neurons. Neuroscience Meeting Planner. Atlanta, GA: Society for Neuroscience, 2006. Online. Program No. 562.15.

Gartside, S.E., Johnson, D.A., Leitch, M.M., Troakes, C. and Ingram, C.D. (2003) Early life adversity programs changes in central 5-HT neuronal function in adulthood. Eur. J. Neurosci., 17: 2401–2408.

Gewirtz, J.C., McNish, K.A. and Davis, M. (1998) Lesions of the bed nucleus of the stria terminalis block sensitization of

the acoustic startle reflex produced by repeated stress, but not fear-potentiated startle. Prog. Neuropsychopharmacol. Biol. Psychiatry., 22: 625–648.
Gonzalez, L.E., Andrews, N. and File, S.E. (1996) 5-HT1A and benzodiazepine receptors in the basolateral amygdala modulate anxiety in the social interaction test, but not in the elevated plus-maze. Brain Res., 732: 145–153.
Graeff, F.G. (1990) Brain defense systems and anxiety. In: Roth, M., Burrows, G.D. and Noyes, R. (Eds.), Handbook of Anxiety. Elsevier, Amsterdam, Vol. 3, pp. 307–354.
Graeff, F.G. (1991) Neurotransmitters in the dorsal periaqueductal gray and animal models of panic anxiety. In: Briley, M. and File, S.E. (Eds.), New Concepts in Anxiety. MacMillan Press, London, pp. 288–312.
Graeff, F.G. (1994) Neuroanatomy and neurotransmitter regulation of defensive behaviors and related emotions in mammals. Braz. J. Med. Biol. Res., 27: 811–829.
Graeff, F.G., Brandao, M.L., Audi, E.A. and Schutz, M.T. (1986) Modulation of the brain aversive system by GABAergic and serotonergic mechanisms. Behav. Brain. Res., 22: 173–180.
Graeff, F.G., Guimaraes, F.S., De Andrade, T.G. and Deakin, J.F. (1996) Role of 5-HT in stress, anxiety, and depression. Pharmacol. Biochem. Behav., 54: 129–141.
Grahn, R.E., Will, M.J., Hammack, S.E., Maswood, S., McQueen, M.B., Watkins, L.R. and Maier, S.F. (1999) Activation of serotonin-immunoreactive cells in the dorsal raphe nucleus in rats exposed to an uncontrollable stressor. Brain Res., 826: 35–43.
Gray, J.A. (1982) The Neuropsychology of Anxiety: An Enquiry into the Functions of the Septo-Hippocampal System. Clarendon Press, Oxford.
Gray, J.A. and McNaughton, N. (1987) The neuropsychology of anxiety: reprise. In: Dienstbier, R.A. and Hope, D.A. (Eds.), Perspectives on Anxiety, Panic, and Fear. University of Nebraska Press, Lincoln, Nebraska, pp. 61–134.
Gray, J.A. and McNaughton, N. (2000) The Neuropsychology of Anxiety. 2nd Ed. Oxford Medical Publications, Oxford.
Gross, C., Zhuang, X., Stark, K., Ramboz, S., Oosting, R., Kirby, L., Santarelli, L., Beck, S. and Hen, R. (2002) Serotonin1A receptor acts during development to establish normal anxiety-like behaviour in the adult. Nature, 416: 396–400.
Hale, M.W., Bouwknecht, J.A., Spiga, F., Shekhar, A. and Lowry, C.A. (2006) Exposure to high- and low-light conditions in an open-field test of anxiety increases c-Fos expression in specific subdivisions of the rat basolateral amygdaloid complex. Brain. Res. Bull., 71: 174–182.
Hale, M.W., Hay-Schmidt, A., Mikkelsen, J.D., Poulsen, B., Shekhar, A. and Lowry, C.A. (2008a) Exposure to an open-field arena increases c-Fos expression in a distributed anxiety-related system projecting to the basolateral amygdaloid complex. Neuroscience, 155: 659–672.
Hale, M.W., Hay-Schmidt, A., Mikkelsen, J.D., Poulsen, B., Bouwknecht, J.A., Evans, A.K., Stamper, C.E., Shekhar, A. and Lowry, C.A. (2008b) Exposure to an open-field arena increases c-Fos expression in a subpopulation of neurons in the dorsal raphe nucleus, including neurons projecting to the basolateral amygdaloid complex. Neuroscience, 157: 733–748.
Hall, F.S. (1998) Social deprivation of neonatal, adolescent, and adult rats has distinct neurochemical and behavioral consequences. Crit. Rev. Neurobiol., 12: 129–162.
Hammack, S.E., Richey, K.J., Schmid, M.J., LoPresti, M.L., Watkins, L.R. and Maier, S.F. (2002) The role of corticotropin-releasing hormone in the dorsal raphe nucleus in mediating the behavioral consequences of uncontrollable stress. J. Neurosci., 22: 1020–1026.
Harada, K., Yamaji, T. and Matsuoka, N. (2008) Activation of the serotonin 5-HT2C receptor is involved in the enhanced anxiety in rats after single-prolonged stress. Pharmacol. Biochem. Behav., 89: 11–16.
Hariri, A.R., Drabant, E.M., Munoz, K.E., Kolachana, B.S., Mattay, V.S., Egan, M.F. and Weinberger, D.R. (2005) A susceptibility gene for affective disorders and the response of the human amygdala. Arch. Gen. Psychiatr., 62: 146–152.
Hariri, A.R. and Holmes, A. (2006) Genetics of emotional regulation: the role of the serotonin transporter in neural function. Trends. Cogn. Sci., 10: 182–191.
Hariri, A.R., Mattay, V.S., Tessitore, A., Kolachana, B., Fera, F., Goldman, D., Egan, M.F. and Weinberger, D.R. (2002) Serotonin transporter genetic variation and the response of the human amygdala. Science, 297: 400–403.
Harro, J. (2006) CCK and NPY as anti-anxiety treatment targets: promises, pitfalls, and strategies. Amino Acids, 31: 215–230.
Heinrichs, S.C., Pich, E.M., Miczek, K.A., Britton, K.T. and Koob, G.F. (1992) Corticotropin-releasing factor antagonist reduces emotionality in socially defeated rats via direct neurotropic action. Brain Res., 581: 190–197.
Heinrichs, S.C., Menzaghi, F., Pich, E.M., Baldwin, H.A., Rassnick, S., Britton, K.T. and Koob, G.F. (1994) Anti-stress action of a corticotropin-releasing factor antagonist on behavioral reactivity to stressors of varying type and intensity. Neuropsychopharmacology, 11: 179–186.
Heinrichs, S.C., Lapsansky, J., Lovenberg, T.W., Desouza, E. B. and Chalmers, D.T. (1997) Corticotropin-releasing factor CRF_1, but not CRF_2, receptors mediate anxiogenic-like behavior. Regul. Pept., 71: 15–21.
Heinz, A., Braus, D.F., Smolka, M.N., Wrase, J., Puls, I., Hermann, D., Klein, S., Grusser, S.M., Flor, H., Schumann, G., Mann, K. and Buchel, C. (2005) Amygdala–prefrontal coupling depends on a genetic variation of the serotonin transporter. Nat. Neurosci., 8: 20–21.
Heisler, L.K., Chu, H.M., Brennan, T.J., Danao, J.A., Bajwa, P., Parsons, L.H. and Tecott, L.H. (1998) Elevated anxiety and antidepressant-like responses in serotonin 5-HT1A receptor mutant mice. Proc. Natl. Acad. Sci. USA, 95: 15049–15054.
Herry, C., Bach, D.R., Esposito, F., Di, S.F., Perrig, W.J., Scheffler, K., Luthi, A. and Seifritz, E. (2007) Processing of temporal unpredictability in human and animal amygdala. J. Neurosci., 27: 5958–5966.
Hogg, S., Andrews, N. and File, S.E. (1994) Contrasting behavioural effects of 8-OH DPAT in the dorsal raphe nucleus and ventral hippocampus. Neuropharmacology, 33: 343–348.
Horiuchi, J., McDowall, L.M. and Dampney, R.A. (2006) Differential control of cardiac and sympathetic vasomotor activity from the dorsomedial hypothalamus. Clin. Exp. Pharmacol. Physiol., 33: 1265–1268.
Iidaka, T., Ozaki, N., Matsumoto, A., Nogawa, J., Kinoshita, Y., Suzuki, T., Iwata, N., Yamamoto, Y., Okada, T. and Sadato, N. (2005) A variant C178T in the regulatory region of the serotonin receptor gene HTR3A modulates neural activation in the human amygdala. J. Neurosci., 25: 6460–6466.
Imai, H., Steindler, D.A. and Kitai, S.T. (1986) The organization of divergent axonal projections from the midbrain raphe nuclei in the rat. J. Comp. Neurol., 243: 363–380.
Inada, Y., Yoneda, H., Koh, J., Sakai, J., Himei, A., Kinoshita, Y., Akabame, K., Hiraoka, Y. and Sakai, T. (2003) Positive

association between panic disorder and polymorphism of the serotonin 2A receptor gene. Psychiatry. Res., 118: 25–31.

Jeffery, K.J. (2004) Remembrance of futures past. Trends. Cogn. Sci., 8: 197–199.

Jenck, F., Moreau, J.L. and Martin, J.R. (1995) Dorsal periaqueductal gray-induced aversion as a simulation of panic anxiety: elements of face and predictive validity. Psychiatry. Res., 57: 181–191.

Jennings, K.A., Loder, M.K., Sheward, W.J., Pei, Q., Deacon, R.M.J., Benson, M.A., Olverman, H.J., Hastie, N.D., Shen, S., Harmar, A.J. and Sharp, T. (2006) Increased expression of the 5-HT transporter confers a low-anxiety phenotype linked to decreased 5-HT transmission. J. Neurosci., 26: 8955–8964.

Johnson, P.L., Lightman, S.L. and Lowry, C.A. (2004) A functional subset of serotonergic neurons in the rat ventrolateral periaqueductal gray implicated in the inhibition of sympathoexcitation and panic. Ann. N.Y. Acad. Sci., 1018: 58–64.

Johnson, P.L., Hollis, J.H., Moratalla, R., Lightman, S.L. and Lowry, C.A. (2005) Acute hypercarbic gas exposure reveals functionally distinct subpopulations of serotonergic neurons in rats. J. Psychopharmacol., 19: 327–341.

Johnson, P.L., Truitt, W.A., Fitz, S.D., Lowry, C.A. and Shekhar, A. (2007) Neural pathways underlying lactate-induced panic. Neuropsychopharmacology, 33: 2093–2107.

Johnson, P.L., Lowry, C.A., Truitt, W. and Shekhar, A. (2008) Disruption of GABAergic tone in the dorsomedial hypothalamus attenuates responses in a subset of serotonergic neurons in the dorsal raphe nucleus following lactate-induced panic. J. Psychopharmacol., 22: 642–645.

Kagamiishi, Y., Yamamoto, T. and Watanabe, S. (2003) Hippocampal serotonergic system is involved in anxiety-like behavior induced by corticotropin-releasing factor. Brain Res., 991: 212–221.

Kalueff, A.V., Fox, M.A., Gallagher, P.S. and Murphy, D.L. (2007) Hypolocomotion, anxiety and serotonin syndrome-like behavior contribute to the complex phenotype of serotonin transporter knockout mice. Genes Brain Behav., 6: 389–400.

Keay, K.A. and Bandler, R. (2001) Parallel circuits mediating distinct emotional coping reactions to different types of stress. Neurosci. Biobehav. Rev., 25: 669–678.

Kelley, S.P., Bratt, A.M. and Hodge, C.W. (2003) Targeted gene deletion of the 5-HT3A receptor subunit produces an anxiolytic phenotype in mice. Eur. J. Pharmacol., 461: 19–25.

Kerman, I.A., Shabrang, C., Taylor, L., Akil, H. and Watson, S.J. (2006) Relationship of presympathetic-premotor neurons to the serotonergic transmitter system in the rat brainstem. J. Comp. Neurol., 499: 882–896.

Kishimoto, K., Koyama, S. and Akaike, N. (2000) Presynaptic modulation of synaptic gamma-aminobutyric acid transmission by tandospirone in rat basolateral amygdala. Eur. J. Pharmacol., 407: 257–265.

Kjelstrup, K.G., Tuvnes, F.A., Steffenach, H.A., Murison, R., Moser, E.I. and Moser, M.B. (2002) Reduced fear expression after lesions of the ventral hippocampus. Proc. Natl. Acad. Sci. USA, 99: 10825–10830.

Kohler, C. and Steinbusch, H. (1982) Identification of serotonin and non-serotonin-containing neurons of the mid-brain raphe projecting to the entorhinal area and the hippocampal formation. A combined immunohistochemical and fluorescent retrograde tracing study in the rat brain. Neuroscience, 7: 951–975.

Kohler, C., Chan-Palay, V. and Steinbusch, H. (1982) The distribution and origin of serotonin-containing fibers in the septal area: a combined immunohistochemical and fluorescent retrograde tracing study in the rat. J. Comp. Neurol., 209: 91–111.

Korte, S.M. and De Boer, S.F. (2003) A robust animal model of state anxiety: fear-potentiated behaviour in the elevated plus-maze. Eur. J. Pharmacol., 463: 163–175.

Kraszpulski, M., Dickerson, P.A. and Salm, A.K. (2006) Prenatal stress affects the developmental trajectory of the rat amygdala. Stress, 9: 85–95.

Ladd, C.O., Owens, M.J. and Nemeroff, C.B. (1996) Persistent changes in corticotropin-releasing factor neuronal systems induced by maternal deprivation. Endocrinology, 137: 1212–1218.

Ladd, C.O., Huot, R.L., Thrivikraman, K.V., Nemeroff, C.B., Meaney, M.J. and Plotsky, P.M. (2000) Long-term behavioral and neuroendocrine adaptations to adverse early experience. Prog. Brain Res., 122: 81–103.

Lamprea, M.R., Cardenas, F.P., Vianna, D.M., Castilho, V.M., Cruz-Morales, S.E. and Brandao, M.L. (2002) The distribution of fos immunoreactivity in rat brain following freezing and escape responses elicited by electrical stimulation of the inferior colliculus. Brain Res., 950: 186–194.

Lanzenberger, R.R., Mitterhauser, M., Spindelegger, C., Wadsak, W., Klein, N., Mien, L.K., Holik, A., Attarbaschi, T., Mossaheb, N., Sacher, J., Geiss-Granadia, T., Kletter, K., Kasper, S. and Tauscher, J. (2007) Reduced serotonin-1A receptor binding in social anxiety disorder. Biol. Psychiatry, 61: 1081–1089.

Le, F.B., Czesak, M., Steubl, D. and Albert, P.R. (2008) Transcriptional regulation at a HTR1A polymorphism associated with mental illness. Neuropharmacology, 55: 977–985.

Lee, Y., Fitz, S., Johnson, P.L. and Shekhar, A. (2008) Repeated stimulation of CRF receptors in the BNST of rats selectively induces social but not panic-like anxiety. Neuropsychopharmacology, in press

Leite-Panissi, C.R., Ferrarese, A.A., Terzian, A.L. and Menescal-de-Oliveira, L. (2006) Serotoninergic activation of the basolateral amygdala and modulation of tonic immobility in guinea pig. Brain Res. Bull., 69: 356–364.

Lesch, K.P., Wiesmann, M., Hoh, A., Muller, T., Disselkamp-Tietze, J., Osterheider, M. and Schulte, H.M. (1992) 5-HT1A receptor-effector system responsivity in panic disorder. Psychopharmacology (Berl.), 106: 111–117.

Lesch, K.P., Bengel, D., Heils, A., Sabol, S.Z., Greenberg, B.D., Petri, S., Benjamin, J., Muller, C.R., Hamer, D.H. and Murphy, D.L. (1996) Association of anxiety-related traits with a polymorphism in the serotonin transporter gene regulatory region. Science, 274: 1527–1531.

Leveleki, C., Sziray, N., Levay, G., Barsvari, B., Soproni, K., Mikics, E. and Haller, J. (2006) Pharmacological evaluation of the stress-induced social avoidance model of anxiety. Brain Res. Bull., 69: 153–160.

Lotrich, F.E., Pollock, B.G. and Ferrell, R.E. (2001) Polymorphism of the serotonin transporter: implications for the use of selective serotonin reuptake inhibitors. Am. J. Pharmacogenomics, 1: 153–164.

Lowry, C.A., Johnson, P.L., Hay-Schmidt, A., Mikkelsen, J. and Shekhar, A. (2005) Modulation of anxiety circuits by serotonergic systems. Stress, 8: 233–246.

Lowry, C.A., Evans, A.K., Gasser, P.J., Hale, M.W., Staub, D.R. and Shekhar, A. (2008a) Topographical organization and chemoarchitecture of the dorsal raphe nucleus and the median raphe nucleus. In: Monti, J.M., Pandi-Perumal, B.L., Jacobs, B.L. and Nutt, D.L. (Eds.), Serotonin and Sleep: Molecular, Functional and Clinical Aspects. Birkhauser, Basel, pp. 25–68.

Lowry, C. A., Gardner, K. L., Lightman, S. L. and Plotsky, P. M. (2008b) Adverse early life experience and adult social defeat interact to increase serotonin transporter mRNA expression in the dorsal raphe nucleus. 2008 Abstracts Viewer/Itinerary Planner. Washington, DC Society for Neuroscience 2008. CD-ROM, Program No. 281.5.

Lukkes, J.L., Mokin, M.V., Scholl, J.L. and Forster, G.L. (2008a) Adults rats exposed to early-life social isolation exhibit increased anxiety and conditioned fear behavior, and altered hormonal stress responses. Horm. Behav., in press, doi:10.1016/j.yhbeh.2008.10.014

Lukkes, J.L., Summers, C.H., Scholl, J.L., Renner, K.J. and Forster, G.L. (2008b) Early life social isolation alters corticotropin-releasing factor responses in adult rats. Neuroscience, in press, doi:10.1016/j.neuroscience.2008.10.036

Madden, C.J. and Sved, A.E. (2003) Rostral ventrolateral medulla C1 neurons and cardiovascular regulation. Cell. Mol. Neurobiol., 23: 739–749.

Maier, S.F. and Watkins, L.R. (2005) Stressor controllability and learned helplessness: the roles of the dorsal raphe nucleus, serotonin, and corticotropin-releasing factor. Neurosci. Biobehav. Rev., 29: 829–841.

Maron, E. and Shlik, J. (2006) Serotonin function in panic disorder: important, but why? Neuropsychopharmacology, 31: 1–11.

Martijena, I.D., Calvo, N., Volosin, M. and Molina, V.A. (1997) Prior exposure to a brief restraint session facilitates the occurrence of fear in response to a conflict situation: behavioral and neurochemical correlates. Brain Res., 752: 136–142.

McHugh, S.B., Deacon, R.M., Rawlins, J.N. and Bannerman, D.M. (2004) Amygdala and ventral hippocampus contribute differentially to mechanisms of fear and anxiety. Behav. Neurosci., 118: 63–78.

McQuade, R. and Sharp, T. (1997) Functional mapping of dorsal and median raphe 5-hydroxytryptamine pathways in forebrain of the rat using microdialysis. J. Neurochem., 69: 791–796.

Meaney, M.J. and Szyf, M. (2005) Maternal care as a model for experience-dependent chromatin plasticity? Trends Neurosci., 28: 456–463.

Meloni, E.G., Reedy, C.L., Cohen, B.M. and Carlezon, W.A. Jr. (2007) Activation of Raphe Efferents to the Medial Prefrontal Cortex by Corticotropin-Releasing Factor: Correlation with Anxiety-Like Behavior. Biol. Psychiatry, 63: 832–839.

Mendonca, F.H. and Guimaraes, F.S. (1998) Intra-hippocampal administration of cycloheximide attenuates the restraint-induced exploratory deficit of an elevated plus maze. Behav. Brain Res., 91: 207–211.

Merali, Z., Bedard, T., Andrews, N., Davis, B., McKnight, A.T., Gonzalez, M.I., Pritchard, M., Kent, P. and Anisman, H. (2006) Bombesin receptors as a novel anti-anxiety therapeutic target: BB1 receptor actions on anxiety through alterations of serotonin activity. J. Neurosci., 26: 10387–10396.

Millan, M.J. (2003) The neurobiology and control of anxious states. Prog. Neurobiol., 70: 83–244.

Morilak, D.A., Barrera, G., Echevarria, D.J., Garcia, A.S., Hernandez, A., Ma, S. and Petre, C.O. (2005) Role of brain norepinephrine in the behavioral response to stress. Prog. Neuropsychopharmacol. Biol. Psychiatry, 29: 1214–1224.

Muller, J.F., Mascagni, F. and McDonald, A.J. (2007) Serotonin-immunoreactive axon terminals innervate pyramidal cells and interneurons in the rat basolateral amygdala. J. Comp. Neurol., 505: 314–335.

Munafo, M.R., Freimer, N.B., Ng, W., Ophoff, R., Veijola, J., Miettunen, J., Jarvelin, M.R., Taanila, A. and Flint, J. (2008) 5-HTTLPR genotype and anxiety-related personality traits: A meta-analysis and new data. Am. J. Med. Genet. B Neuropsychiatr. Genet., doi:10.1002/ajmg.b.30808

Nagano, M., Ozawa, H. and Suzuki, H. (2008) Prenatal dexamethasone exposure affects anxiety-like behaviour and neuroendocrine systems in an age-dependent manner. Neurosci. Res., 60: 364–371.

Nakayasu, T. and Ishii, K. (2008) Effects of pair-housing after social defeat experience on elevated plus-maze behavior in rats. Behav. Processes, 78: 477–480.

Nascimento Hackl, L.P. and Carobrez, A.P. (2007) Distinct ventral and dorsal hippocampus AP5 anxiolytic effects revealed in the elevated plus-maze task in rats. Neurobiol. Learn. Mem., 88: 177–185.

Nash, J.R., Sargent, P.A., Rabiner, E.A., Hood, S.D., Argyropoulos, S.V., Potokar, J.P., Grasby, P.M. and Nutt, D.J. (2008) Serotonin 5-HT1A receptor binding in people with panic disorder: positron emission tomography study. Br. J. Psychiatry, 193: 229–234.

Neumeister, A., Bain, E., Nugent, A.C., Carson, R.E., Bonne, O., Luckenbaugh, D.A., Eckelman, W., Herscovitch, P., Charney, D.S. and Drevets, W.C. (2004) Reduced serotonin type 1A receptor binding in panic disorder. J. Neurosci., 24: 589–591.

Nunes-de-Souza, R.L., Canto-de-Souza, A. and Rodgers, R.J. (2002) Effects of intra-hippocampal infusion of WAY-100635 on plus-maze behavior in mice. Influence of site of injection and prior test experience. Brain Res., 927: 87–96.

Ogilvie, A.D., Battersby, S., Bubb, V.J., Fink, G., Harmar, A.J., Goodwim, G.M. and Smith, C.A. (1996) Polymorphism in serotonin transporter gene associated with susceptibility to major depression. Lancet, 347: 731–733.

Olivier, J.D., Van Der Hart, M.G., Van Swelm, R.P., Dederen, P.J., Homberg, J.R., Cremers, T., Deen, P.M., Cuppen, E., Cools, A.R. and Ellenbroek, B.A. (2008) A study in male and female 5-HT transporter knockout rats: an animal model for anxiety and depression disorders. Neuroscience, 152: 573–584.

Ottersen, O.P. (1981) Afferent connections of the amygdaloid complex of the rat with some observations in the cat. III. Afferents from the lower brain stem. J. Comp. Neurol., 202: 335–356.

Parks, C.L., Robinson, P.S., Sibille, E., Shenk, T. and Toth, M. (1998) Increased anxiety of mice lacking the serotonin1A receptor. Proc. Natl. Acad. Sci. USA, 95: 10734–10739.

Patin, V., Lordi, B., Vincent, A. and Caston, J. (2005) Effects of prenatal stress on anxiety and social interactions in adult rats. Dev. Brain Res., 160: 265–274.

Paxinos, G. and Watson, C. (1998) The Rat Brain in Stereotaxic Coordinates. 4th Ed.. Academic Press, San Diego, CA.

Perlis, R.H., Mischoulon, D., Smoller, J.W., Wan, Y.J., Lamon-Fava, S., Lin, K.M., Rosenbaum, J.F. and Fava, M. (2003) Serotonin transporter polymorphisms and adverse effects with fluoxetine treatment. Biol. Psychiatry, 54: 879–883.

Peyron, C., Petit, J.-M., Rampon, C., Jouvet, M. and Luppi, P.-H. (1998) Forebrain afferents to the rat dorsal raphe nucleus demonstrated by retrograde and anterograde tracing methods. Neuroscience, 82: 443–468.

Plotsky, P.M., Owens, M.J. and Nemeroff, C.B. (1998) Psychoneuroendocrinology of depression. Hypothalamic–pituitary–adrenal axis. Psychiatr. Clin. North Am., 21: 293–307.

Pollock, B.G., Ferrell, R.E., Mulsant, B.H., Mazumdar, S., Miller, M., Sweet, R.A., Davis, S., Kirshner, M.A., Houck, P.R., Stack, J.A., Reynolds, C.F. and Kupfer, D.J. (2000) Allelic variation in the serotonin transporter promoter affects onset of paroxetine treatment response in late-life depression. Neuropsychopharmacology, 23: 587–590.

Rainnie, D.G. (1999) Serotonergic modulation of neurotransmission in the rat basolateral amygdala. J. Neurophysiol., 82: 69–85.

Ramboz, S., Oosting, R., Amara, D.A., Kung, H.F., Blier, P., Mendelsohn, M., Mann, J.J., Brunner, D. and Hen, R. (1998) Serotonin receptor 1A knockout: an animal model of anxiety-related disorder. Proc. Natl. Acad. Sci. USA, 95: 14476–14481.

Rausch, J.L., Moeller, F.G. and Johnson, M.E. (2003) Initial platelet serotonin (5-HT) transport kinetics predict nortriptyline treatment outcome. J. Clin. Psychopharmacol., 23: 138–144.

Reimold, M., Batra, A., Knobel, A., Smolka, M.N., Zimmer, A., Mann, K., Solbach, C., Reischl, G., Schwarzler, F., Grunder, G., Machulla, H.J., Bares, R. and Heinz, A. (2008) Anxiety is associated with reduced central serotonin transporter availability in unmedicated patients with unipolar major depression: a [11C]DASB PET study. Mol. Psychiatry, 13: 606–613, 557

Rex, A., Voigt, J.P. and Fink, H. (2005) Anxiety but not arousal increases 5-hydroxytryptamine release in the rat ventral hippocampus in vivo. Eur. J. Neurosci., 22: 1185–1189.

Rhodes, R.A., Murthy, N.V., Dresner, M.A., Selvaraj, S., Stavrakakis, N., Babar, S., Cowen, P.J. and Grasby, P.M. (2007) Human 5-HT transporter availability predicts amygdala reactivity in vivo. J. Neurosci., 27: 9233–9237.

Rotondo, A., Mazzanti, C., Dell'Osso, L., Rucci, P., Sullivan, P., Bouanani, S., Gonnelli, C., Goldman, D. and Cassano, G.B. (2002) Catechol o-methyltransferase, serotonin transporter, and tryptophan hydroxylase gene polymorphisms in bipolar disorder patients with and without comorbid panic disorder. Am. J. Psychiatry, 159: 23–29.

Ruis, M.A., te Brake, J.H., Buwalda, B., De Boer, S.F., Meerlo, P., Korte, S.M., Blokhuis, H.J. and Koolhaas, J.M. (1999) Housing familiar male wildtype rats together reduces the long-term adverse behavioural and physiological effects of social defeat. Psychoneuroendocrinology, 24: 285–300.

Sajdyk, T.J. and Shekhar, A. (1997) Excitatory amino acid receptor antagonists block the cardiovascular and anxiety responses elicited by gamma-aminobutyric acidA receptor blockade in the basolateral amygdala of rats. J. Pharmacol. Exp. Ther., 283: 969–977.

Sajdyk, T.J. and Shekhar, A. (2000) Sodium lactate elicits anxiety in rats after repeated GABA receptor blockade in the basolateral amygdala. Eur. J. Pharmacol., 394: 265–273.

Sajdyk, T.J., Schober, D.A., Gehlert, D.R. and Shekhar, A. (1999) Role of corticotropin-releasing factor and urocortin within the basolateral amygdala of rats in anxiety and panic responses. Behav. Brain Res., 100: 207–215.

Sajdyk, T.J., Johnson, P.L., Fitz, S.D. and Shekhar, A. (2008) Chronic inhibition of GABA synthesis in the bed nucleus of the stria terminalis elicits anxiety-like behavior. J. Psychopharmacol., 22: 633–641.

Sanders, S.K., Morzorati, S.L. and Shekhar, A. (1995) Priming of experimental anxiety by repeated subthreshold GABA blockade in the rat amygdala. Brain Res., 699: 250–259.

Saper, C.B. (1995) Central autonomic system. In: Paxinos, G. (Ed.), The Rat Nervous System. Academic Press, San Diego, CA, pp. 107–135.

Shekhar, A., Keim, S.R., Simon, J.R. and McBride, W.J. (1996) Dorsomedial hypothalamic GABA dysfunction produces physiological arousal following sodium lactate infusions. Pharmacol. Biochem. Behav., 55: 249–256.

Shekhar, A., Sajdyk, T.S., Keim, S.R., Yoder, K.K. and Sanders, S.K. (1999) Role of the basolateral amygdala in panic disorder. Ann. NY Acad. Sci., 877: 747–750.

Shekhar, A., Sajdyk, T.J., Gehlert, D.R. and Rainnie, D.G. (2003) The amygdala, panic disorder, and cardiovascular responses. Ann. NY Acad. Sci., 985: 308–325.

Shekhar, A., Truitt, W., Rainnie, D. and Sajdyk, T. (2005) Role of stress, corticotrophin releasing factor (CRF) and amygdala plasticity in chronic anxiety. Stress, 8: 209–219.

Silveira, M.C., Zangrossi, H., de B, V., Silveira, R. and Graeff, F.G. (2001) Differential expression of Fos protein in the rat brain induced by performance of avoidance or escape in the elevated T-maze. Behav. Brain Res., 126: 13–21.

Singewald, N. and Sharp, T. (2000) Neuroanatomical targets of anxiogenic drugs in the hindbrain as revealed by Fos immunocytochemistry. Neuroscience, 98: 759–770.

Singewald, N., Salchner, P. and Sharp, T. (2003) Induction of c-Fos expression in specific areas of the fear circuitry in rat forebrain by anxiogenic drugs. Biol. Psychiatry, 53: 275–283.

Smeraldi, E., Zanardi, R., Benedetti, F., Di, B.D., Perez, J. and Catalano, M. (1998) Polymorphism within the promoter of the serotonin transporter gene and antidepressant efficacy of fluvoxamine. Mol. Psychiatry, 3: 508–511.

Smolka, M.N., Buhler, M., Schumann, G., Klein, S., Hu, X.Z., Moayer, M., Zimmer, A., Wrase, J., Flor, H., Mann, K., Braus, D.F., Goldman, D. and Heinz, A. (2007) Gene–gene effects on central processing of aversive stimuli. Mol. Psychiatry, 12: 307–317.

Smythe, J.W., Rowe, W.B. and Meaney, M.J. (1994) Neonatal handling alters serotonin (5-HT) turnover and 5-HT2 receptor binding in selected brain regions: relationship to the handling effect on glucocorticoid receptor expression. Dev. Brain Res., 80: 183–189.

Spiga, F., Lightman, S.L., Shekhar, A. and Lowry, C.A. (2006) Injections of urocortin 1 into the basolateral amygdala induce anxiety-like behavior and c-Fos expression in brainstem serotonergic neurons. Neuroscience, 138: 1265–1276.

Staub, D.R., Spiga, F. and Lowry, C.A. (2005) Urocortin 2 increases c-Fos expression in topographically organized subpopulations of serotonergic neurons in the rat dorsal raphe nucleus. Brain Res., 1044: 176–189.

Staub, D.R., Evans, A.K. and Lowry, C.A. (2006) Evidence supporting a role for corticotropin-releasing factor type 2 (CRF(2)) receptors in the regulation of subpopulations of serotonergic neurons. Brain Res., 1070: 77–89.

Stezhka, V.V. and Lovick, T.A. (1997) Projections from dorsal raphe nucleus to the periaqueductal grey matter: studies in slices of rat midbrain maintained in vitro. Neurosci. Lett., 230: 57–60.

Strong, P.V., Greenwood, B.N. and Fleshner, M. (2008) The effects of the selective 5-HT(2C) receptor antagonist SB 242084 on learned helplessness in male Fischer 344 rats. Psychopharmacology (Berl.), doi:10.1017/s00213-008-1413-3

Sziray, N., Leveleki, C., Levay, G., Marko, B., Harsing, L.G. Jr., Mikics, E., Barsy, B. and Haller, J. (2007) Mechanisms underlying the long-term behavioral effects of traumatic experience in rats: the role of serotonin/noradrenaline balance and NMDA receptors. Brain Res. Bull., 71: 376–385.

Tamamaki, N. and Nojyo, Y. (1995) Preservation of topography in the connections between the subiculum, field CA1, and the entorhinal cortex in rats. J. Comp. Neurol., 353: 379–390.

Tsetsenis, T., Ma, X.H., Lo, I.L., Beck, S.G. and Gross, C. (2007) Suppression of conditioning to ambiguous cues by pharmacogenetic inhibition of the dentate gyrus. Nat. Neurosci., 10: 896–902.

Underwood, M.D., Arango, V., Bakalian, M.J., Ruggiero, D.A. and Mann, J.J. (1999) Dorsal raphe nucleus serotonergic neurons innervate the rostral ventrolateral medulla in rat. Brain Res., 824: 45–55.

Unschuld, P.G., Ising, M., Erhardt, A., Lucae, S., Kloiber, S., Kohli, M., Salyakina, D., Welt, T., Kern, N., Lieb, R., Uhr, M., Binder, E.B., Muller-Myhsok, B., Holsboer, F. and Keck, M.E. (2007) Polymorphisms in the serotonin receptor gene HTR2A are associated with quantitative traits in panic disorder. Am. J. Med. Genet. B Neuropsychiatr. Genet., 144B: 424–429.

Van Bockstaele, E.J., Biswas, A. and Pickel, V.M. (1993) Topography of serotonin neurons in the dorsal raphe nucleus that send axon collaterals to the rat prefrontal cortex and nucleus accumbens. Brain Res., 624: 188–198.

Van den Hove, D.L., Blanco, C.E., Aendekerk, B., Desbonnet, L., Bruschettini, M., Steinbusch, H.P., Prickaerts, J. and Steinbusch, H.W. (2005) Prenatal restraint stress and long-term affective consequences. Dev. Neurosci., 27: 313–320.

Van den Hove, D.L., Lauder, J.M., Scheepens, A., Prickaerts, J., Blanco, C.E. and Steinbusch, H.W. (2006) Prenatal stress in the rat alters 5-HT1A receptor binding in the ventral hippocampus. Brain Res., 1090: 29–34.

Van Groen, T. and Wyss, J.M. (1990) Extrinsic projections from area CA1 of the rat hippocampus: olfactory, cortical, subcortical, and bilateral hippocampal formation projections. J. Comp. Neurol., 302: 515–528.

Van Riel, E., van Gemert, N.G., Meijer, O.C. and Joels, M. (2004) Effect of early life stress on serotonin responses in the hippocampus of young adult rats. Synapse, 53: 11–19.

Vertes, R.P. (1991) A PHA-L analysis of ascending projections of the dorsal raphe nucleus in the rat. J. Comp. Neurol., 313: 643–668.

Vinogradova, O.S. (2001) Hippocampus as comparator: role of the two input and two output systems of the hippocampus in selection and registration of information. Hippocampus, 11: 578–598.

Voigt, J.P., Rex, A., Sohr, R. and Fink, H. (1999) Hippocampal 5-HT and NE release in the transgenic rat TGR(mREN2)27 related to behavior on the elevated plus maze. Eur. Neuropsychopharmacol., 9: 279–285.

Walker, D.L., Toufexis, D.J. and Davis, M. (2003) Role of the bed nucleus of the stria terminalis versus the amygdala in fear, stress, and anxiety. Eur. J. Pharmacol., 463: 199–216.

Ward, H.E., Johnson, E.A., Salm, A.K. and Birkle, D.L. (2000) Effects of prenatal stress on defensive withdrawal behavior and corticotropin releasing factor systems in rat brain. Physiol. Behav., 70: 359–366.

Weinstock, M. (2007) Gender differences in the effects of prenatal stress on brain development and behaviour. Neurochem. Res., 32: 1730–1740.

Weinstock, M. (2008) The long-term behavioural consequences of prenatal stress. Neurosci. Biobehav. Rev., 32: 1073–1086.

Weisstaub, N.V., Zhou, M., Lira, A., Lambe, E., Gonzalez-Maeso, J., Hornung, J.P., Sibille, E., Underwood, M., Itohara, S., Dauer, W.T., Ansorge, M.S., Morelli, E., Mann, J.J., Toth, M., Aghajanian, G., Sealfon, S.C., Hen, R. and Gingrich, J.A. (2006) Cortical 5-HT2A receptor signaling modulates anxiety-like behaviors in mice. Science, 313: 536–540.

Willner, P. (1990) Animal models of depression: an overview. Pharmacol. Ther., 45: 425–455.

Witter, M.P., Griffioen, A.W., Jorritsma-Byham, B. and Krijnen, J.L. (1988) Entorhinal projections to the hippocampal CA1 region in the rat: an underestimated pathway. Neurosci. Lett., 85: 193–198.

Wright, I.K., Upton, N. and Marsden, C.A. (1991) Resocialisation of isolation-reared rats does not alter their anxiogenic profile on the elevated X-maze model of anxiety. Physiol. Behav., 50: 1129–1132.

Wright, I.K., Upton, N. and Marsden, C.A. (1992) Effect of established and putative anxiolytics on extracellular 5-HT and 5-HIAA in the ventral hippocampus of rats during behaviour on the elevated X-maze. Psychopharmacology (Berl.), 109: 338–346.

Xu, K., Ernst, M. and Goldman, D. (2006) Imaging genomics applied to anxiety, stress response, and resiliency. Neuroinformatics, 4: 51–64.

Yoon, H.K., Yang, J.C., Lee, H.J. and Kim, Y.K. (2008) The association between serotonin-related gene polymorphisms and panic disorder. J. Anxiety. Disord., 22: 1529–1534.

Zangrossi, J.H. and Graeff, F.G. (1994) Behavioral effects of intra-amygdala injections of GABA and 5-HT acting drugs in the elevated plus-maze. Braz. J. Med. Biol. Res., 27: 2453–2456.

Zangrossi, H. Jr., Viana, M.B. and Graeff, F.G. (1999) Anxiolytic effect of intra-amygdala injection of midazolam and 8-hydroxy-2-(di-n-propylamino)tetralin in the elevated T-maze. Eur. J. Pharmacol., 369: 267–270.

Zanoveli, J.M., Netto, C.F., Guimaraes, F.S. and Zangrossi, H. Jr. (2004) Systemic and intra-dorsal periaqueductal gray injections of cholecystokinin sulfated octapeptide (CCK-8s) induce a panic-like response in rats submitted to the elevated T-maze. Peptides, 25: 1935–1941.

Zuena, A.R., Mairesse, J., Casolini, P., Cinque, C., Alema, G.S., Morley-Fletcher, S., Chiodi, V., Spagnoli, L.G., Gradini, R., Catalani, A., Nicoletti, F. and Maccari, S. (2008) Prenatal restraint stress generates two distinct behavioral and neurochemical profiles in male and female rats. PLoS ONE, 3: e2170.

CHAPTER 3.7

Role of Serotonin in Brain Reward and Regulation of Alcohol Drinking Behavior

W.J. McBride*

Institute of Psychiatric Research, Department of Psychiatry, Indiana University School of Medicine, Indianapolis, Indiana

Abstract: This chapter will review the involvement of serotonin (5-HT) in mediating brain reward processes, and extend the review to the regulation of alcohol drinking by 5-HT. The first three sections of the chapter will focus on studies using intracranial self-stimulation, place conditioning, and conditioned reward and delay of reward procedures to examine the involvement of 5-HT in the regulation of brain reward systems. The fourth section will focus on studies of the involvement of 5-HT in regulating the rewarding properties of alcohol, as measured by alcohol drinking behavior of rodent models. Overall, the data suggest that 5-HT systems negatively modulate brain reward processes and alcohol drinking. However, some findings suggest that the effects of 5-HT on brain reward and alcohol drinking is dependent upon the subtype of 5-HT receptor.

Keywords: brain reward system; intracranial self-stimulation; place conditioning; conditioned reward; delay of reward; alcohol drinking behavior.

Abbreviations: AA, Alko-Alcohol; ACB, nucleus accumbens; CNS, central nervous system; CPA, conditioned place aversion; CPP, conditioned place preference; CR, conditioned reward; CS, conditioned stimulus; CTA, conditioned taste aversion; DA, dopamine; 5,7-DHT, 5,7-dihydroxytryptamine; DLR, delay of reward; DRN, dorsal raphe nucleus; EtOH, ethanol; FH, Fawn-Hooded; ICSA, intracranial self-administration; ICSS, intracranial self-stimulation; HAD, high-alcohol-drinking; 5-HT, serotonin; LAD, low-alcohol-drinking; LH, lateral hypothalamus; MFB, medial forebrain bundle; MPF, medial prefrontal cortex; MRN, median raphe nucleus; NP, alcohol non-preferring; 8-OH-DPAT, 8-hydroxy-2-(di-*n*-propylamino)-tetralin; P, alcohol preferring; PCPA, para-chlorophenyalanine; VP, ventral pallidum; VMT, ventral tegmentum; VTA, ventral tegmental area.

Introduction

Serotonin (5-HT) has been implicated in modulating many behaviors and in regulating the reinforcing effects of certain drugs of abuse, such as cocaine and alcohol. The main sources of 5-HT are the dorsal and median raphe nuclei (DRN and MRN, respectively) located in the brainstem. Fibers from these nuclei project to almost all parts of the brain, suggesting the involvement of the 5-HT system in modulating many different functions. The widespread innervations from the DRN and MRN suggest that 5-HT is involved in many behaviors. Moreover, there are a high number of 5-HT receptor subtypes, suggesting that 5-HT involvement in modulating behaviors is likely to be highly complex and difficult to define clearly.

Brain reward systems incorporate multiple limbic pathways, most notably the medial forebrain bundle (MFB) and the lateral hypothalamus (LH). Within these pathways, the ventral tegmental area (VTA) dopamine (DA) projections to several regions, for example, the nucleus accumbens (ACB), ventral pallidum (VP) and medial prefrontal cortex (MPF), play important roles in mediating the rewarding actions of drugs of abuse, including alcohol, and are considered central structures in brain reward systems.

Several experimental approaches were used to examine the involvement of 5-HT in brain reward. The intracranial self-stimulation (ICSS) technique defines discrete brain sites that support electrical self-stimulation. With this technique, stimulating electrodes are stereotaxically implanted into discrete brain sites; if the animal responds to receiving an electrical stimulation, then the site is supporting self-stimulation and the effects are considered reinforcing. This site may, therefore, be part of the brain

*E-mail: wmcbride@iupui.edu

Christian Müller & Barry Jacobs (Eds.)
Handbook of Behavioral Neurobiology of Serotonin
ISBN 978-0-12-374634-4

DOI: 10.1016/B978-0-12-374634-4.00024-1

reward system. Using the ICSS technique in combination with pharmacological treatments can provide insights into the involvement of 5-HT in brain reward mechanisms.

Another approach is the intracranial self-administration (ICSA) procedure. The ICSA technique is a cross between the ICSS and operant drug self-administration. The ICSA procedure has been used to identify neuroanatomical substrates mediating the reinforcing effects of various drugs of abuse (McBride *et al.*, 1999). With this procedure, the animal responds to receiving the site-specific microinjection of a given pharmacological agent (e.g., cocaine, morphine, ethanol, etc.). The ICSS and ICSA procedures have been used to map CNS sites, neuronal pathways and neurotransmitter receptors involved in brain reward.

The place conditioning technique can be used to examine the positive or negative conditioning effects of a given treatment on the subsequent preference or aversion of a compartment previously associated with the treatment, but tested in the absence of the treatment. This technique has been used extensively to test whether pharmacological agents are reinforcing or aversive. With this technique, the animal is administered the test compound and is then placed in one compartment of a two- or three-compartment chamber. After three or four conditioning/treatment trials, the animal is placed in the two- or three-compartment apparatus without any treatment and allowed access to all compartments. If the animal prefers the compartment associated with the treatments, then the agent is producing a positive or rewarding effect (conditioned place preference, CPP). In contrast, if the animal avoids the compartment associated with the treatments, then the agent is considered to be producing negative or aversive effects (conditioned place aversion, CPA).

In the conditioned reward (CR) procedure, the animal responds to cues associated with a reward in the absence of the reward. Usually, with this procedure the animal is conditioned to a cue (conditioned stimulus, CS+), such as a tone or light, prior to receiving a reward (e.g., food, water, drug, etc.). With this procedure, the CS+ takes on rewarding characteristics of its own, as exemplified by the animal responding for the CS+ in the absence of the reward.

The above behavioral techniques, in combination with site-specific microinjections, have the potential of delineating the involvement of certain 5-HT pathways and receptors in modulating brain reward processes. This chapter will review the results of several experimental approaches that examined the involvement of 5-HT in regulating brain reward function.

The involvement of 5-HT systems in regulating alcohol drinking behavior has been studied in a number of animal models (see reviews by LeMarquand *et al.*, 1994; Li and McBride, 1995; McBride and Li, 1998; Murphy *et al.*, 2002). The final section of this chapter will review key experiments on the involvement of 5-HT in regulating the reinforcing effects of alcohol and alcohol-drinking behavior.

ICSS and ICSA studies and the effects of 5-HT

Effects of 5-HT on electrical self-stimulation of the LH and MFB

Most ICSS studies use electrodes implanted in the LH or MFB, which activate DA pathways originating in the VTA (Rolls *et al.*, 1974; Gallistel *et al.*, 1982). Thus, studies designed to test 5-HT effects on ICSS with electrodes implanted in the LH or MFB are presumed to be testing effects on the activity of VTA DA pathways.

Studies on the effects of 5-HT on ICSS of the LH and MFB involved increasing or decreasing 5-HT synthesis, increasing extracellular 5-HT levels by increasing release or inhibiting reuptake, and administration of selective 5-HT receptor agonists or antagonists (Table 1). Increasing 5-HT synthesis with systemic administration of tryptophan or 5-hydroxytryptophan (5-HTP) reduced self-stimulation of the LH (Patkina and Lapin, 1976), whereas inhibiting 5-HT synthesis with para-chlorophenylalanine (PCPA) had no effect on ventral tegmentum (VMT) self-stimulation (Miliaressis, 1977) or self-stimulation of the LH (Van der Kooy *et al.*, 1977). These results suggest that 5-HT systems may have a role in modulating brain reward stimulation, although it is difficult to interpret the effects on 5-HT synthesis, because it is not known if these treatments significantly altered extracellular levels of 5-HT.

Several studies used pharmacological treatments with agents known to increase the extracellular levels of 5-HT, and tested the effects of these agents on brain electrical self-stimulation (Table 1). Dexfenfluramine and fenfluramine, which are known to increase the release of 5-HT, suppressed self-stimulation of the LH (Hoebel *et al.*, 1988) and MFB (Olds, 1995). In agreement with the results of the 5-HT releasing agents, the 5-HT reuptake inhibitor fluoxetine increased the threshold for electrical self-stimulation of the VTA or MFB (Lee and Kornetsky, 1998) and reduced the rates of self-stimulation of the MFB (Katz and Carroll, 1977). The combined results suggest that general activation of 5-HT pathways may have an inhibitory role on brain reward stimulation, involving the VTA DA system.

Several studies used 5-HT receptor agonists and antagonists to better understand the role of 5-HT in brain electrical self-stimulation (Table 1). Two of these studies examined the effects of systemic administration of 5-HT_{1A} receptor agonists on ICSS of the LH (Montgomery *et al.*,

Table 1 Involvement of 5-HT in brain reward as measured with intra-cranial self-stimulation and intra-cranial self-administration techniques

Treatment	CNS region	Results/conclusions	Reference
A. ICSS studies			
Tryptophan and 5-HTP; systemic; cats	LH	Increasing 5-HT synthesis reduced ICSS, suggesting that 5-HT exerts inhibitory effects on the brain reward system	Patkina and Lapin, 1976
PCPA; systemic	VMT or MRN	Inhibiting 5-HT synthesis reduced ICSS in MRN, but not in VMT; these results suggest that 5-HT neurons in the MRN may be involved in brain reward stimulation	Miliaressis, 1977
PCPA; systemic	HIP or LH	Inhibiting 5-HT synthesis reduced ICSS in HIP but not in LH; reward stimulation in the HIP mediated by 5-HT	Van der Kooy *et al.*, 1977
Dexfenfluramine; systemic	LH	Increasing 5-HT release inhibited ICSS; these results indicate that 5-HT exerts a negative effect on LH brain reward stimulation	Hoebel *et al.*, 1988
Fenfluramine, i.p., Sprague-Dawley rats	MFB	Fenfluramine suppressed ICSS, supporting the idea that 5-HT exerts an inhibitory effect on the brain reward system	Olds, 1995
Fluoxetine, i.p.	Caudal MFB	Inhibiting 5-HT reuptake reduced rates of ICSS; these results suggest that increasing extracellular levels of 5-HT will attenuate brain reward stimulation	Katz and Carroll, 1977
Fluoxetine, i.p.; Wistar rats	VTA or MFB	Inhibiting 5-HT reuptake increased the threshold for ICSS; these results indicate that increased extracellular levels of 5-HT can result in reduced sensitivity to brain reward stimulation	Lee and Kornetsky, 1998
8-OH-DPAT, buspirone; i.p.; Lister hooded rats	LH	Low-dose 8-OH-DPAT increased and high doses reduced ICSS; buspirone produced depression of ICSS; these results suggest that low 8-OH-DPAT doses reduce 5-HT activity via activation of 5-HT_{1A} autoreceptors, resulting in increased ICSS; activation of postsynaptic 5-HT_{1A} receptors suppresses ICSS	Montgomery *et al.*, 1991
8-OH-DPAT and p-MPPI; systemic	LH	The 5-HT_{1A} agonist reduced ICSS threshold at low doses, but increased the threshold at higher doses; the 5-HT_{1A} antagonist had no effect; these results suggest that LH stimulation does not involve activation of postsynaptic 5-HT_{1A} receptors, but activation of these 5-HT_{1A} receptors reduces brain reward stimulation	Harrison and Markou, 2001
RU24969 and GR127935; s.c.; Wistar rats	LH	The 5-$HT_{1A/1B}$ receptor agonist increased ICSS threshold, but the 5-$HT_{1B/1D}$ antagonist had no effect on ICSS; these results support the idea that LH stimulation does not involve activation of 5-HT_{1B} receptors, but activation of 5-HT_{1B} receptors can reduce brain reward stimulation	Harrison *et al.*, 1999
8-OH-DPAT, microinjection, MRN	LH	Microinjection of the 5-HT_{1A} agonist into MRN reduced ICSS threshold; these results indicate that activating 5-HT_{1A} receptors in MRN increases brain reward stimulation, suggesting that 5-HT projections from the MRN exert inhibitory influence on brain reward	Fletcher *et al.*, 1995
8-OH-DPAT; microinject, MRN, DRN	LH	The 5-HT_{1A} agonist injected into the MRN reduced the ICSS threshold, but had no effect when injected into the DRN; these results indicate that activating 5-HT_{1A} receptors in the MRN may be reducing 5-HT neuronal activity, suggesting 5-HT projections from MRN, but not DRN, are exerting inhibitory influence on brain reward stimulation	Harrison and Markou, 2001
M100907; systemic; Sprague-Dawley rats	LH	The 5-HT_{2A} antagonist did not alter ICSS, indicating that activation of 5-HT_{2A} receptors may not be involved in brain reward stimulation	Benaliouad *et al.*, 2007
Ondansetron, s.c.; rats	LH	The 5-HT_3 antagonist did not alter ICSS, indicating that activation of 5-HT_3 receptors may not be involved in brain reward stimulation	Herberg *et al.*, 1992
Ondansetron, s.c.	LH	The 5-HT_3 antagonist did not alter ICSS, supporting the idea that activation of 5-HT_3 receptors may not be involved in brain reward stimulation	Montgomery *et al.*, 1993

(*Continued*)

Table 1 *Continued*

Treatment	CNS region	Results/conclusions	Reference
SB-204070-A; systemic; rats	LH	The 5-HT_4 antagonist did not alter ICSS; indicating that activation of 5-HT_4 receptors may not be involved in brain reward stimulation	Reavill *et al.*, 1998
PCPA, i.g., rats	DRN or MRN	Administration of the 5-HT synthesis inhibitor reduced ICSS in the DRN and MRN, suggesting that activation of 5-HT neurons are involved in self-stimulation of the DRN and MRN	Van der Kooy *et al.*, 1978
Metergoline, cyproheptadine, PCA, 5,7-DHT	MRN	Non-selective 5-HT antagonists, depletion of 5-HT or destruction of 5-HT neurons had no effect on ICSS of MRN, suggesting that the rewarding effects of electrical stimulation in MRN do not appear to involve 5-HT neurons	Deakin, 1980
B. ICSA studies			
Muscimol, microinjection; rats	DRN, MRN	Muscimol was self-infused into the MRN and DRN, suggesting that activating $GABA_A$ receptors in the MRN and DRN is producing reinforcing effects, presumably thru inhibition of 5-HT neuronal activity	Liu and Ikemoto, 2007
1-(m-chloro-phenyl)-biguanide; microinjection; rats	VTA	The 5-HT_3 agonist is self-infused into the posterior VTA, suggesting that activation of 5-HT_3 receptors produces reinforcing effects	Rodd *et al.*, 2007

Abbreviations: ICSA, intracranial self-administration; ICSS, intracranial self-stimulation; LH, lateral hypothalamus; DRN, dorsal raphe nucleus; MRN, median raphe nucleus; PCA, para-chloro-amphetamine; PCPA, para-chlorophenylalanine; 5,7-DHT, 5,7-dihydroxytryptamine; 5-HTP, 5-hydroxytryptophan; 8-OH-DPAT, 8-hydroxy-2-(di-*n*-propylamino)-tetralin; HIP, hippocampus; VMT, ventral tegmentum; MFB, medial forebrain bundle; VTA, ventral tegmental area; s.c., subcutaneous; i.g., intragastric; i.p., intraperitoneal.

1991; Harrison and Markou, 2001). These studies reported that the 5-HT_{1A} agonist 8-hydroxy-2-(di-*n*-propylamino)-tetralin (8-OH-DPAT) produced a biphasic effect on ICSS of the LH, with low doses increasing responding and lowering the threshold for ICSS, and higher doses reducing responding and increasing the threshold for ICSS. Low doses of 8-OH-DPAT are thought to act at 5-HT_{1A} autoreceptors (Dourish *et al.*, 1988), whereas higher doses act at both presynaptic 5-HT_{1A} autoreceptors and postsynaptic 5-HT_{1A} receptors. Therefore, the results of the 5-HT_{1A} receptor agonist studies suggest that activating postsynaptic 5-HT_{1A} receptors reduces brain reward stimulation, whereas inhibiting 5-HT neuronal activity, via activation of cell body 5-HT_{1A} autoreceptors, enhances brain reward stimulation. Consistent with the postsynaptic effects of 8-OH-DPAT on 5-HT_{1A} receptors, activation of 5-HT_{1B} receptors, with systemic administration of a 5-HT_{1B} receptor agonist, increased the threshold for self-stimulation of the LH (Harrison *et al.*, 1999), suggesting that activation of 5-HT projections to postsynaptic 5-HT_{1B} receptors can also inhibit LH self-stimulation (Table 1).

To test the hypothesis that activating 5-HT_{1A} autoreceptors will enhance self-stimulation of the LH, Fletcher *et al.* (1995) and Harrison and Markou (2001) examined the direct effect of microinjecting 8-OH-DPAT into the MRN on ICSS of the LH (Table 1). Both studies reported that microinjecting 8-OH-DPAT into the MRN reduced the threshold for self-stimulation of the LH, suggesting that inhibiting 5-HT neuronal activity in the MRN, by activating 5-HT_{1A} autoreceptors with 8-OH-DPAT, enhances brain reward stimulation. On the other hand, microinjecting 8-OH-DPAT into the DRN had no significant effect on the ICSS of the LH (Harrison and Markou, 2001). The results observed with microinjecting 8-OH-DPAT into the MRN are compatible with the results reported above for 5-HT releasers and reuptake inhibitors, and support an interpretation that the 5-HT system from the MRN plays an inhibitory role in regulating the rewarding effects of stimulating the mesolimbic DA system.

Systemic administration of 5-HT_{1A} (Harrison and Markou, 2001), 5-HT_{1B} (Harrison *et al.*, 1999), 5-HT_{2A} (Benaliouad *et al.*, 2007), 5-HT_3 (Herberg *et al.*, 1992; Montgomery *et al.*, 1993) or 5-HT_4 (Reavill *et al.*, 1998) receptor antagonists had little effect on MFB or LH self-stimulation (Table 1). These results suggest that any tonic activation of these receptors within this part of the brain reward system is not having a significant effect on modulating electrical self-stimulation of the MFB or LH. It is possible that other subtypes of 5-HT receptors may be involved, and/or the effects of 5-HT may be on accumulated actions at multiple receptors, such that inhibiting only one receptor at a time may not be sufficient to alter electrical self-stimulation of the LH or MFB.

5-HT and electrical self-stimulation of the hippocampus, MRN and DRN

Numerous studies indicate that multiple brain regions can support reinforcement processes (reviewed in McBride *et al.*, 1999), including the hippocampus and MRN. Van der Kooy *et al.* (1977) reported that the hippocampus supported electrical self-stimulation, and that systemic administration of the 5-HT synthesis inhibitor PCPA reduced self-stimulation of this region (Table 1). Miliaressis (1977) reported that the MRN supported self-stimulation and that inhibition of 5-HT synthesis reduced self-stimulation of this region. In addition, Van der Kooy *et al.* (1978) reported that both the MRN and DRN supported electrical self-stimulation and that PCPA also reduced self-stimulation of these raphe nuclei (Table 1). The results of these studies suggest that activation of 5-HT systems promotes self-stimulation of the hippocampus and raphe nuclei. In contrast, another study (Deakin, 1980) reported that non-selective 5-HT receptor antagonists, depletion of 5-HT, or destruction of 5-HT neurons had little effect on electrical self-stimulation of the MRN (Table 1), suggesting that 5-HT neurons may not be involved. It is difficult to reconcile the differences between the studies; it is possible that differences in the placement of electrodes may account for the discrepancies between the studies. In the study of Deakin (1980), the electrodes may be stimulating fibers of passage, and therefore little effect was observed by destroying 5-HT neurons.

ICSA studies: 5-HT involvement in brain reward

Few studies have been carried out with the ICSA technique to study the involvement of 5-HT in brain reward mechanisms. Liu and Ikemoto (2007) reported that muscimol, a $GABA_A$ receptor agonist, was self-infused into the MRN and DRN of Wistar rats (Table 1). Activating $GABA_A$ receptors within the raphe nuclei will likely inhibit 5-HT neuronal activity. Therefore, if this is the case, then inhibiting 5-HT neuronal activity may remove tonic inhibition of the mesolimbic DA system; this disinhibition results in activation of brain reward processes. These results are compatible with the majority of studies cited above for an inhibitory effect of 5-HT on electrical self-stimulation of the LH and MFB. However, this ICSA study is at odds with the ICSS results indicating the MRN and DRN support electrical self-stimulation (Van der Kooy *et al.*, 1978; Miliaressis, 1977), unless electrical self-stimulation of the raphe nuclei is producing its effect via activation of GABA interneurons. However, electrical self-stimulation of raphe nuclei may also activate fibers of passage or adjacent structures, which could account of the results of electrical self-stimulation of the raphe nuclei (Miliaressis, 1977; Van der Kooy *et al.*, 1978).

Another ICSA study (Rodd *et al.*, 2007) indicated that a $5\text{-}HT_3$ agonist was self-infused into the posterior VTA (Table 1), suggesting that activation of local $5\text{-}HT_3$ receptors is producing reinforcing effects. Moreover, the results of this study suggest that local activation of DA neurons is involved in mediating the reinforcing effects of the $5\text{-}HT_3$ agonist. Although these results support an involvement of the 5-HT system in brain reward mechanisms, they are somewhat counter to the overall evidence that suggests activating the 5-HT system may inhibit brain reward stimulation. However, the influence of the 5-HT system on different processes is likely to be complex, and dependent in part on the subtype of 5-HT receptor and the CNS region.

In summary, the data support a mainly inhibitory role for 5-HT in brain reward stimulation. However, most studies supporting this view were conducted with local electrical self-stimulation of the mesolimbic DA system. On the other hand, the results of some studies suggest that the MRN and DRN can also support reinforcement processes and may be part of the brain reward system.

Place conditioning and effects of 5-HT

The rewarding and aversive effects of drugs of abuse and many other compounds have been studied using place conditioning. With the place conditioning paradigm, the intrinsic effects of the agent are paired with a previously neutral environment during the conditioning trials. If the effects of the compound are positive or rewarding, then the animal will spend more time in the environment previously associated with the drug effects, when tested in the absence of the drug. Conversely, if the agent produces a negative or aversive effect, then the animal will spend less time in the environment previously associated with drug administration, when tested in the absence of the drug. The place conditioning procedure has been used extensively to evaluate the positive and negative effects of a wide variety of compounds. Comprehensive reviews of the use of the place conditioning paradigm have been published (Tzschentke 1998, 2007).

Several studies were undertaken to assess the effects of increasing the extracellular levels of 5-HT on place conditioning (Table 2). In two studies (Davies and Parker, 1993; Rea *et al.*, 1998), the effects of the 5-HT releaser fenfluramine were tested in rats, and found to produce CPA, suggesting that 5-HT overflow is producing aversive effects. On the other hand, a third study (Meehan and Schechter, 1994) reported that fenfluramine failed to produce either CPP or CPA in Sprague-Dawley rats, although these investigators found that fenfluramine produced CPA in Fawn-Hooded (FH) rats. Differences in the experimental protocols may account in part for the disagreement between the studies with Sprague-Dawley rats.

Risinger (1997) tested the effects of the 5-HT reuptake inhibitor fluoxetine on place conditioning in mice, and

Table 2 Involvement of 5-HT in brain reward as measured with place conditioning or conditioned taste aversion

Treatment	Measurement	Results/conclusions	Reference
Fenfluramine; i.p.; Sprague-Dawley rats	CPA	Administration of fenfluramine produced CPA, suggesting that increasing extracellular 5-HT levels produced aversive effects	Davies and Parker, 1993
Fenfluramine, systemic; Sprague-Dawley rats	CPP, CPA	Fenfluramine failed to produce CPP or CPA, suggesting that increasing extracellular levels of 5-HT had little effect on the brain reward or aversion systems	Meehan and Schechter, 1994
Fenfluramine; systemic; rats	CPP, CPA	Fenfluramine produced dose-related CPA, indicating that increasing extracellular levels of 5-HT produces aversive effects	Rea *et al.*, 1998
Fluoxetine, systemic; mice	CPP, CTA	Administration of the 5-HT reuptake inhibitor had no effect on either measure; these results indicate that increasing the extracellular 5-HT levels has little effect on brain reward or aversion systems in mice	Risinger, 1997
Paroxetine, fluoxetine, sertraline; systemic; Wistar rats	CPP	All three 5-HT reuptake inhibitors produced significant CPP; these results suggest that elevating extracellular 5-HT can produce rewarding effects	Subhan *et al.*, 2000
8-OH-DPAT; systemic	CPP, CPA	Low doses of 8-OH-DPAT produced CPP; high doses produced CPA; low dose but not high dose effects blocked by PCPA and DA antagonists; supporting the idea that low dose 8-OH-DPAT acting thru 5-HT_{1A} autoreceptor inhibition of 5-HT activity, whereas high dose effects are at postsynaptic 5-HT_{1A} receptors	Papp and Willner, 1991
CP 94,253, GR127935, systemic	CPP, CPA	Administration of a 5-HT_{1B} agonist produced CPA, whereas a partial 5-HT_{1B} agonist had no effect; these results suggest that activating 5-HT_{1B} receptors produces aversive effects	Cervo *et al.*, 2002
TFMPP, WAY161503; s.c.; Sprague-Dawley rats	CPP	Neither the mixed 5-$HT_{1A/1B/2C}$ receptor agonist, TFMPP, nor the 5-HT_{2C} agonist, WAY161503, induced place conditioning; these results suggest that activating 5-HT_{2C} receptors is not producing rewarding effects	Mosher *et al.*, 2005
WAY161503, s.c.; Sprague-Dawley rats	CPA, CTA	Administration of the 5-HT_{2C} agonist induced CPA and CTA, suggesting activation of 5-HT_{2C} receptors produces aversion	Mosher *et al.*, 2006
Mianserine, eltoprazine, ketanserine, mCPP; i.p.; rats	CPP, CPA	The 5-HT_{1C} agonist had little effect on CPP, CPA, whereas 5-HT_{1C} antagonists produced CPA; the 5-HT_2 antagonist had little effect; these results indicate that inhibiting 5-HT_{1C} receptors produces aversive effects, suggesting that 5-HT_{1C} receptors may be having a positive tonic influence on brain reward processes	Rocha *et al.*, 1993a
Mianserine, i.p.; 5,7-DHT lesions; rats	CPA	Destruction of 5-HT neurons enhanced CPA produced by the 5-HT_{1C} antagonist; reduced activation of 5-HT_{1C} receptors enhances aversive effects	Rocha *et al.*, 1993b
8-OH-DPAT, microinjection; DRN, MRN	CPP	Microinjecting a 5-HT_{1A} agonist into DRN or MRN produces CPP; indicating that inhibition of 5-HT neuronal activity produces reinforcing effects; these results suggest that both 5-HT systems exert tonic inhibitory effects on the brain reward system	Fletcher *et al.*, 1993
Muscimol, microinjection, MRN, DRN, rats	CPP	Administration of the $GABA_A$ agonist into the MRN produced CPP, whereas injection into the DRN did not, suggesting that inhibiting MRN 5-HT neurons produces rewarding effects	Liu and Ikemoto, 2007
Muscimol, microinjection, MRN	CPP	Activating $GABA_A$ receptors in the MRN did not produce CPP; these results suggest that inhibiting 5-HT neurons in the MRN does not activate brain reward processes	Le *et al.*, 2008

Abbreviations: DRN, dorsal raphe nucleus; MRN, median raphe nucleus; CPP, conditioned place preference; CTA, conditioned taste aversion; CPA, conditioned place aversion; mCPP, 5-HT_{1C} agonist; PCPA, para-chlorophenylalanine; 8-OH-DPAT, 8-hydroxy-2-(di-*n*-propylamino)-tetralin; TFMPP, mixed 5-$HT_{1A/1B/2C}$ receptor agonist; i.p., intraperitoneal; s.c., subcutaneous.

found no effect on either CPP or CPA (Table 2). In contrast, Subhan *et al.* (2000) tested three 5-HT reuptake inhibitors (paroxetine, fluoxetine and sertraline) on place conditioning in Wistar rats, and reported that all three reuptake inhibitors produced CPP. These results suggest that activating 5-HT systems in the rat produces rewarding effects, but this may not be the case for mice.

The results with the 5-HT releaser fenfluramine and 5-HT reuptake inhibitors are not in agreement. Administration of a 5-HT releaser produced CPA (Davies and Parker, 1993; Rea *et al.*, 1998), whereas the 5-HT reuptake inhibitors (Subhan *et al.*, 2000) produced CPP (Table 2). The differences between the two treatments may reflect their relative specificities and/or modes of action. Fenfluramine blocks uptake but it also accelerates release, causing higher extracellular levels of 5-HT at both 'active' and 'inactive' synapses. Reuptake inhibitors have their most significant effects on extracellular levels of 5-HT at 'active' synapses. Therefore, the 5-HT reuptake inhibitors may be enhancing 5-HT effects at selected synapses, whereas fenfluramine is increasing 5-HT at most 5-HT synapses, including those that are producing aversive effects, which may mask any positive actions. Furthermore, non-specific effects of these compounds on other monoamine systems may be having a significant role in their actions.

Systemic administration of agonists for 5-HT_{1A}, 5-HT_{1B} and 5-HT_{2C} receptors was also tested in the place conditioning paradigm (Table 2). The 5-HT_{1A} agonist 8-OH-DPAT, at low doses, produced CPP, but at high doses produced CPA (Papp and Willner, 1991). The low-dose effect is postulated to result from selective activation of cell body 5-HT_{1A} autoreceptors, which reduces 5-HT neuronal activity. The high-dose effect is thought to result from the action of 8-OH-DPAT at postsynaptic 5-HT_{1A} receptors. These results are similar to the effects of 8-OH-DPAT on ICSS of the LH (Montgomery *et al.*, 1991; Harrison and Markou, 2001). Administration of a 5-HT_{1B} (Cervo *et al.*, 2002) or a 5-HT_{2C} (Mosher *et al.*, 2006) receptor agonist produced CPA (Table 2). In addition, a mixed 5-$HT_{1A,1B,2C}$ agonist did not produce place conditioning (Mosher *et al.*, 2005). Administration of a 5-HT_{1C} receptor agonist had little effect on CPP or CPA (Rocha *et al.*, 1993a), whereas inhibiting 5-HT_{1C} receptors produced CPA (Rocha *et al.*, 1993a, 1993b). The inhibition of 5-HT_2 receptors with non-selective antagonists had little effect on CPP or CPA (Rocha *et al.*, 1993a). Overall, these results suggest that activating postsynaptic 5-HT_{1A}, 5-HT_{1B} or 5-HT_{2C} receptors can produce aversive effects (Table 2). On the other hand, inhibiting 5-HT_{1C} receptors produces aversive effects, suggesting that this subtype of receptor may be actively involved in positive modulation of brain reward processes.

The results with the 5-HT agonists producing CPA (Cervo *et al.*, 2002; Mosher *et al.*, 2005, 2006), or with the low-dose effects of 8-OH-DPAT producing CPP (Papp and Willner, 1991), are not compatible with the actions of 5-HT reuptake inhibitors (Subhan *et al.*, 2000) producing CPP (Table 2). It is difficult to reconcile these differences, since the increased extracellular 5-HT produced by the reuptake inhibitors would be expected to occur at 5-HT_{1A}, 5-HT_{1B} and 5-HT_{2C} receptors. It is possible that the reuptake inhibitors may be having their main effects at certain active 5-HT synapses involving 5-HT_{1C} or other 5-HT receptor subtypes. Alternatively, the 5-HT reuptake inhibitors may be having their initial effects at cell body 5-HT_{1A} autoreceptors, thereby reducing 5-HT neuronal activity.

Several microinjection studies were undertaken to examine the effects of inhibiting 5-HT neuronal activity in the raphe nuclei on place conditioning (Table 2). Fletcher *et al.* (1993) reported that microinjection of the 5-HT_{1A} receptor agonist 8-OH-DPAT, into either the MRN or the DRN, produced CPP. Liu and Ikemoto (2007) reported that microinjection of the $GABA_A$ receptor agonist muscimol into the MRN produced CPP, whereas injection into the DRN did not. In contrast, Le *et al.* (2008) reported that microinjection of muscimol into the MRN did not produce CPP. The apparent discrepancy between the latter two studies may be due to differences in the placement of the injection cannula and the experimental paradigm. Microinjection of 8-OH-DPAT and muscimol into the raphe nuclei would be expected to reduce 5-HT neuronal activity. The results of the microinjection studies suggest that reducing 5-HT neuronal activity enhances rewarding effects. These findings are compatible with most of the ICSS studies and with the place conditioning studies, using systemic injections, and support a mainly inhibitory role for 5-HT regulation of brain reward.

Involvement of 5-HT in conditioned reward and delay of reward

Conditioned reward and 5-HT

Increased incentive motivation refers to the process by which appetitive behaviors are produced by rewarding stimuli (Beninger, 1983; Fibiger and Phillips, 1986). The conditioned reward (CR) paradigm has been used to study incentive motivation. In this paradigm, an initial neutral stimulus is paired with a primary reinforcer; after several pairings, the animals are given access to two levers, an inactive lever and a lever that delivers the CR. Incentive motivation is demonstrated when the responses on the CR lever are significantly higher than responses on the inactive lever. Several studies indicate a role for dopaminergic systems in mediating CR (see, for example, Robbins, 1978; Beninger, 1983; Robbins *et al.*, 1983; Fibiger and Phillips, 1986; Beninger and Ranaldi, 1992). Since changes in the 5-HT system modify reward-related behavioral responses to drugs acting on the DA system (e.g., Smith *et al.*, 1986, 1987; Carroll *et al.*, 1990), a logical next step would be to conduct studies examining the involvement of 5-HT in CR (Table 3).

Fletcher (1995) examined the effects of d-fenfluramine, a 5-HT releaser, on water-deprived Sprague-Dawley rats (Table 3). In this study, water-deprived rats were trained to associate a compound stimulus (light and tone) with water during the conditioning phase. During the test phase, water was not delivered, but the compound stimulus was delivered when the correct lever was pressed. Systemic administration of d-fenfluramine reduced responding for the CR (Fletcher, 1995), suggesting that increasing 5-HT release was attenuating the conditioned reinforcing effects of the compound stimulus.

Sasaki-Adams and Kelley (2001) examined the effects of chronic fluoxetine, a 5-HT reuptake inhibitor, in a CR paradigm, in which Sprague-Dawley rats were trained to associate a compound stimulus (light and click of food dispenser) with the delivery of a sugar pellet. During CR testing, the compound stimulus, but not the sugar pellet, was delivered when the correct lever was pressed. Administration of fluoxetine increased responding on the active lever for the conditioned stimulus (Table 3), which is not in agreement with the results with d-fenfluramine (Fletcher, 1995). Since both d-fenfluramine and fluoxetine increase the extracellular levels of 5-HT, their differential effects on CR may be due to a combination of factors – e.g., acute versus chronic drug administration, and/or 5-HT being released by fenfluramine at essentially all 5-HT synapses, whereas the greater effects of the 5-HT reuptake inhibitor are at 'active' 5-HT synapses.

In a follow-up study, Fletcher *et al.* (1999) examined the effects of selective 5,7-DHT lesions of both the MRN and DRN on responding in a CR paradigm (compound stimulus of light and tone associated with delivery of water in water-deprived Sprague-Dawley rats). In support of the findings of the d-fenfluramine study (Fletcher, 1995), the combined 5-HT lesions of the MRN and DRN significantly enhanced responding on the active lever in the CR test (Table 3). Overall, these results support the idea that general activation of 5-HT systems may decrease incentive motivation. Although some parts of the 5-HT system may be actively involved in a positive manner on CR, activation of the majority of 5-HT synapses may produce negative effects and counter any positive actions of 5-HT.

Fletcher and Higgins (1997) reported that administration of ondansetron, a 5-HT_3 receptor antagonist, did not alter learning a CR (Table 3), suggesting that activation of 5-HT_3 receptors is not involved in this process.

Delay of reward and 5-HT

When choosing between a larger but delayed reward and a smaller but more immediate reward, a comparison is made between the 'values' associated with each reward, and the reward associated with the larger value is often chosen (Platt, 2002). The possible involvement of 5-HT in the timescale of reward prediction has been briefly reviewed (Schweighofer *et al.*, 2007).

Fletcher (1993) determined the effects of microinjecting 8-OH-DPAT (to activate local cell body 5-HT_{1A} autoreceptors and reduce 5-HT neuronal activity) into the MRN and DRN of Sprague-Dawley rats on responding in two different operant paradigms – i.e., extinction responding and during omission training (delay of reward paradigm). Microinjection of 8-OH-DPAT into the DRN did not significantly alter responding in either behavioral paradigm, whereas microinjection of 8-OH-DPAT into the MRN increased extinction responding, enhanced operant performance during omission training, and slightly increased responding during periods of non-reward (Table 3). To provide additional data that the effects of 8-OH-DPAT were a result of inhibition of 5-HT neuronal activity, Fletcher (1993) reported that microinjection of muscimol (a $GABA_A$ agonist) into the MRN also increased extinction responding. These results could be interpreted to indicate that MRN 5-HT neurons, but not DRN 5-HT neurons, may be involved in controlling behavioral inhibition under conditions of non-reward.

In agreement with the above study, Wogar *et al.* (1993) reported that the combined 5,7-DHT lesions of the MRN and DRN shortened the delay to the larger reinforcer so that the short and delayed reinforcer were equally effective (Table 3). Similar findings were reported by Al-Ruwaitea *et al.* (1999) and Bizot *et al.* (1999), in which 5,7-DHT lesions of the raphe nuclei reduced the number of choices for the larger but delayed reward over the smaller but more immediate reward (Table 3). Taking the opposite approach, Bizot *et al.* (1999) found that increasing the extracellular levels of 5-HT by administration of fluoxetine increased preference for the larger but delayed reward (Table 3). These results suggest that increased activity of the 5-HT system may improve the effectiveness of a larger but delayed reward, possibly indicating more controlled and less impulsive behavior.

Inhibition of 5-HT neuronal activity and activation of postsynaptic 5-HT_{1A} receptors following systemic administration of 8-OH-DPAT decreased performance for long delays (6 and 12 s) and increased performance for a

Table 3 Modulation of conditioned reward and delay of reward by 5-HT

Treatment	**Procedure**	**Results/conclusions**	**Reference**
d-fenfluramine; i.p.; Sprague-Dawley rats	CR	Increased extracellular levels of 5-HT reduced responding for the CR, suggesting that 5-HT has an inhibitory action on brain reward circuitry	Fletcher, 1995
Fluoxetine, systemic, rats	CR	Increased extracellular levels of 5-HT enhanced CR responding; these results suggest that 5-HT facilitated reward-related behavior	Sasaki-Adams and Kelley, 2001
5,7-DHT lesions of both DRN, MRN; Sprague-Dawley rats	CR	Depletion of 5-HT innervation increased responding for CR, supporting the idea that 5-HT exerts an inhibitory influence on reward processes	Fletcher *et al.*, 1999
Ondansetron, systemic, rats	CR	Inhibiting 5-HT_3 receptors does not alter learning CR; these results suggest that 5-HT_3 receptors are not involved with learning a CR	Fletcher and Higgins, 1997
8-OH-DPAT, microinjection MRN, DRN; Sprague-Dawley rats	DLR	Inhibiting 5-HT activity in the MRN increased responding during extinction, during delay of reward, and for non-reward; these results indicate that the MRN 5-HT system may be involved in controlling behavioral inhibition by detecting signals of non-reward	Fletcher, 1993
5,7-DHT lesions, both MRN and DRN	DLR	Lesions shortened the delay to the larger reward that rendered the short and delayed reward equally effective; these results suggest that the 5-HT systems are involved in maintaining the effectiveness of delayed reinforcers	Wogar *et al.*, 1993
5,7-DHT, microinjection, both DRN, MRN	DLR	Reduction of 5-HT innervation increased the time within a trial at which proportional choice of the delayed reward attained a value of 50% of the shorter reward; these results indicate that 5-HT system exerts control (inhibition) of behavior of delayed reward	Al-Ruwaitea *et al.*, 1999
5,7-DHT in DRN; PCPA treatment; 8-OH-DPAT, buspirone, ipsapirone, MDL73005EF, fluoxetine; i.p.	DLR	Reduced 5-HT innervation from the DRN and reduced 5-HT synthesis caused a shift toward immediate reward; activating 5-HT_{1A} receptors and increasing extracellular 5-HT increased preference for the larger but delayed reward; blocking 5-HT_{1A} receptors had no effect on preference; these results suggest that the 5-HT system may be involved in modulating pathways that evaluate reward value	Bizot *et al.*, 1999
8-OH-DPAT, systemic, rats	DLR	Activating 5-HT_{1A} receptors increased lever press performance for a short delay; these results suggest that activating postsynaptic 5-HT_{1A} receptors and/or reducing 5-HT neuronal activity improves performance during a short delay for reward	Balleine *et al.*, 1996

Abbreviations: CR, conditioned reward; DLR, delay of reward; MRN, median raphe nucleus; DRN, dorsal raphe nucleus; 5,7-DHT, 5, 7-dihydroxytryptamine; PCPA, para-chlorophenylalanine; 8-OH-DPAT, 8-hydroxy-2-(di-*n*-propylamino)-tetralin; i.p., intraperitoneal.

short (3-s) delay (Balleine *et al.*, 1996). In contrast to the results of this latter study, Bizot *et al.* (1999) reported that systemic administration of 8-OH-DPAT increased preference for the larger but delayed (25 s) reward. The length of the delay may be the main factor that contributes to the apparent disagreement between the two studies. Bizot *et al.* (1999) reported that, with a 15-s delay, 8-OH-DPAT had a small effect in reducing the number of choices for the larger reward, but with a 25-s delay, 8-OH-DPAT significantly increased preference for the larger reward. The effects of partial 5-HT_{1A} receptor agonists were also dependent on the length of the delay (Bizot *et al.*, 1999). The differences in the effects of 8-OH-DPAT between the studies of Balleine *et al.* (1996) and Bizot *et al.* (1999) appear to be due to the differences in delay to the larger reward, with the longer delay being more sensitive to the effects of 8-OH-DPAT. These results suggest that the 5-HT system may have a role in evaluating the incentive value of a reward, and modulating control of behavior of immediate gratification. It is difficult to interpret the effects of 8-OH-DPAT on delay of reward, since systemic injection of this agonist will likely act at both cell body autoreceptors and postsynaptic 5-HT_{1A} receptors. Therefore, the combination of reducing 5-HT neuronal activity and activating postsynaptic 5-HT_{1A} receptors likely underlies the effects of 8-OH-DPAT on delay of reward.

Summary/conclusions – involvement of 5-HT in brain reward

Serotonin has a major role in modulating many different behaviors. It exerts significant modulation on the mesolimbic DA system, with evidence for both inhibitory and excitatory effects, depending upon the 5-HT receptor. The findings with electrical self-stimulation of the LH or MFB (which activates the mesolimbic DA system) indicate that reducing activity of MRN and DRN 5-HT neurons increases brain reward stimulation, suggesting that these 5-HT pathways exert a general inhibitory control on the mesolimbic reward system (Table 1). A similar conclusion of a general inhibitory effect of 5-HT on brain reward is supported by place conditioning studies. However, place conditioning studies have an advantage in that this procedure can measure both the rewarding and aversive effects of pharmacological agents. Reducing 5-HT neuronal activity (e.g., microinjecting 8-OH-DPAT into the raphe nuclei) promoted CPP, whereas some studies reported that increasing the extracellular levels of 5-HT (e.g., with fenfluramine) promoted CTA, although this was not observed in other cases (Table 2). In agreement with the findings reported for ICSS and place conditioning, studies with CR found that reduced 5-HT innervation increased responding for a CR (Fletcher *et al.*, 1999), whereas increased release of 5-HT (Fletcher, 1995) reduced responding for a CR (Table 3). The studies using DLR paradigms (Table 3) suggest that 5-HT systems are also involved in assessing the incentive value of a reward, with reduced 5-HT innervation decreasing the preference for longer delays with larger rewards, and increased extracellular levels of 5-HT increasing preference for the longer delays and larger rewards (Bizot *et al.*, 1999).

Involvement of 5-HT in regulation of alcohol drinking behavior in rodent models

In some rat lines selectively bred for high or low alcohol consumption, there appear to be innate differences in 5-HT systems (Table 4). Initial findings in the selectively bred alcohol-preferring (P) and alcohol-non-preferring (NP) lines indicated that there were reduced tissue levels of 5-HT in CNS regions of the P compared to NP rats (Murphy *et al.*, 1982, 1987). Similar differences in CNS tissue levels were also reported for high-alcohol-drinking (HAD) compared to low-alcohol-drinking (LAD) rats (Gongwer *et al.*, 1989), although in other selectively bred rat lines such widespread differences were not observed (reviewed in McBride and Li, 1998; Murphy *et al.*, 2002; see also Devoto *et al.*, 1998). The lower contents of 5-HT in the CNS of P compared to NP rats appear to be due to reduced 5-HT innervation to these regions, which was a result of fewer 5-HT raphe neurons in the P compared to NP rats (Zhou *et al.*, 1991a, 1991b, 1994), suggesting a general overall reduction in 5-HT modulation of CNS function in the P than NP rat. This reduced effect of 5-HT may promote higher ethanol (EtOH) intakes in the P and HAD lines of rats. In agreement with this interpretation, Yoshimoto and Komura (1987) reported a significant inverse correlation between whole brain levels of 5-HT and alcohol intake of inbred and congenic mouse strains (Table 4).

If there is an inverse relationship between CNS 5-HT function and alcohol intake, then increasing CNS 5-HT function should result in decreased EtOH intake. Several studies examined the effects of increasing extracellular levels of 5-HT or giving agonists selective for subtypes of 5-HT receptors on EtOH intake in different rodent alcohol-drinking models (Table 4). Administration of the 5-HT reuptake inhibitor fluoxetine reduced EtOH drinking of P and HAD rats (Murphy *et al.*, 1985, 1988; McBride *et al.*, 1990, 1992), suggesting that increasing 5-HT function in these 5-HT deficient rat lines reduces their high-alcohol-drinking characteristics. In agreement with the findings with the 5-HT reuptake inhibitor, several studies (Rowland and Morian, 1992; Rezvani and Grady, 1994; Yu *et al.*, 1997) reported that administration of the 5-HT releaser fenfluramine also reduced EtOH intake in P, Fawn-Hooded (FH) and alcohol-dependent rats (Table 4).

Conversely, if general CNS 5-HT function is reduced in a 'normal' population of rats, then the above hypothesis would predict an increase in EtOH intake. Two experimental approaches that would produce a general reduction in 5-HT function would be 5,7-DHT lesions of raphe nuclei, and inhibiting 5-HT neuronal activity within the raphe nuclei, using microinjections of 5-HT_{1A} or $GABA_A$ receptor agonists. Lesions of the MRN and DRN, which reduced 5-HT levels more than 90 percent, did not alter EtOH intake of Sprague-Dawley rats (Adell and Myers, 1995). In contrast to these latter findings, reducing 5-HT neuronal activity, by microinjecting a 5-HT_{1A} or $GABA_A$ receptor agonist into either the MRN or DRN, increased EtOH intake of Wistar rats during a 40-minute scheduled access period (Tomkins *et al.*, 1994a; Tomkins and Fletcher, 1996). The contradictory findings between the studies of Tomkins and co-workers and that of Adell and Myers (1995) may be a result of a combination of factors. For example, under 24-hour continuous access conditions,

Table 4 Serotonin modulation of alcohol drinking behavior in rodent models

Treatment	Measurement	Results/conclusions	Reference(s)
A. Innate 5-HT			
P-NP rats	CNS regional levels of 5-HT	Lower 5-HT tissue levels in limbic regions associated with high EtOH intake	Murphy *et al.*, 1982, 1987
P-NP rats	Immuno-cytochemistry 5-HT	Lower 5-HT innervation from raphe nuclei to limbic regions associated with high EtOH intake	Zhou *et al.*, 1991a, 1991b, 1994
HAD-LAD rats	CNS regional levels of 5-HT	Lower 5-HT tissue levels in limbic regions associated with high EtOH intake	Gongwer *et al.*, 1989
sP-sNP rats	CNS regional levels of 5-HT	Lower levels of 5-HT in frontal cortex associated with high EtOH intake	Devoto *et al.*, 1998
Inbred and congenic mouse strains	Whole brain 5-HT content	A significant inverse correlation between alcohol intakes and whole brain tissue levels of 5-HT	Yoshimoto and Komura, 1987
B. Extracellular 5-HT			
P and HAD rats; fluoxetine; i.p. or i.g.	EtOH drinking, 24-hr F-C, scheduled access; i.g. EtOH self-administration	Inhibiting 5-HT reuptake reduced oral and i.g. EtOH self-administration	Murphy *et al.*, 1985, 1988; McBride *et al.*, 1990, 1992
P, FH and alcohol-dependent rats; fenfluramine; systemic	EtOH drinking, 24-hr F-C, scheduled access	Increasing 5-HT release resulted in reduced EtOH intake	Rezvani and Grady, 1994; Rowland and Morian, 1992; Yu *et al.*, 1997
C. MRN, DRN			
Sprague-Dawley rats; microinject 5,7-DHT; MRN, DRN	EtOH drinking, 24-hr F-C	Combined lesions of MRN and DRN reduced CNS 5-HT tissue levels >90% but did not alter EtOH intake; maintenance of low EtOH intake in Sprague-Dawley rats not regulated by CNS 5-HT system	Adell and Myers, 1995
Wistar rats; 8-OH-DPAT, muscimol, microinjection; MRN, DRN	EtOH intake; 40-min scheduled access period	Reducing 5-HT neuronal activity in either RN by activating 5-HT_{1A} or $GABA_A$ receptors increased EtOH intake; suggest that 5-HT systems negatively regulate alcohol drinking	Tomkins *et al.*, 1994a; Tomkins and Fletcher, 1996
D. 5-HT Receptors			
5-HT_{1A}; agonists: ipsapirone, 8-OH-DPAT; antagonist: WAY100135; systemic; Sprague-Dawley, Wistar, Long-Evans, P and HAD rats	24-hr F-C; limited access; operant sessions	Low doses of 8-OH-DPAT increased EtOH intake under 24-hr F-C and limited access conditions but had no effect in operant paradigm; high doses of 5-HT_{1A} agonists reduced EtOH drinking under all conditions; blocking 5-HT_{1A} receptors had little effect on EtOH intake; overall, the results suggest that inhibiting 5-HT neuronal activity can increase EtOH intake, and that postsynaptic 5-HT_{1A} receptors do not appear to be actively involved in modulating EtOH intake	McBride *et al.*, 1990; Svensson *et al.*, 1993; Tomkins *et al.*, 1994b; Wilson *et al.*, 1996; Ciccocioppo *et al.*, 1997; Tomkins and O'Neill, 2000; McKenzie-Quirk and Miczek, 2003
5-HT_{1B} agents; Wistar rats; agonists: RU24969, CGS12066b; antagonist: GR127935	Operant EtOH self-administration	Activating 5-HT_{1B} receptors reduces EtOH self-administration; blocking 5-HT_{1B} receptors had little effect on EtOH intake; these results suggest that postsynaptic 5-HT_{1B} receptors can exert a negative effect on EtOH intake, but these receptors are not tonically activated under operant conditions	Tomkins and O'Neill, 2000
5-HT_{2A}; antagonists: s.c., FG5974; amperozide; P, AA, FH rats	24-hr F-C; 1-hr limited access	Amperozide more effective than FG5974 in reducing EtOH intake; amperozide also reduced total fluid and food intake; both antagonists produced conditioned taste aversion; these results suggest that 5-HT_{2A} receptors are involved in regulating consummatory behaviors in general	Overstreet *et al.*, 1997

(Continued)

Table 4 *Continued*

Treatment	Measurement	Results/conclusions	Reference(s)
$5\text{-}HT_{2C}$ agents; agonist: RO 60-0175, s.c.; antagonist: SB242,084, i.p.; Wistar rats	Operant FR-4; 12% EtOH	Activating $5\text{-}HT_{2C}$ receptors reduces EtOH responding, whereas inhibiting $5\text{-}HT_{2C}$ receptors increases responding; these results suggest that $5\text{-}HT_{2C}$ receptors may be involved in limiting EtOH drinking	Tomkins *et al.*, 2002
$5\text{-}HT_3$ antagonists: MDL72222; ICS 205-930; s.c. or i.p.; sP and P rats	24-hr F-C, acquisition, on-going, relapse; limited access; operant	(a) Blocking $5\text{-}HT_3$ receptors effectively reduced EtOH drinking under 24-hr F-C conditions during acquisition and on-going drinking; (b) $5\text{-}HT_3$ receptor antagonists are less effective with 24-hr F-C relapse drinking; (c) under scheduled limited access (1 or 2 hr) $5\text{-}HT_3$ antagonists were ineffective in reducing EtOH intake, unless the limited access period was randomly presented at different times during the night cycle; (d) the results indicate that $5\text{-}HT_3$ receptors have a significant role in regulating EtOH intake when certain anticipatory conditioning factors are absent	Fadda *et al.*, 1991; McKinzie *et al.*, 1998, 2000; Rodd-Henricks *et al.*, 2000
$5\text{-}HT_4$ antagonist: GR113808; s.c.; rats	2-hr limited access; 10% EtOH	Blocking $5\text{-}HT_4$ receptors dose-dependently reduced EtOH intake without altering water or food intakes, suggesting that $5\text{-}HT_4$ receptors have a significant role in maintaining alcohol drinking	Panocka *et al.*, 1995

Abbreviations: P, alcohol-preferring; NP, alcohol-non-preferring; HAD, high-alcohol-drinking; LAD, low-alcohol-drinking; 5,7-DHT, 5,7-dihydroxytryptamine; 8-OH-DPAT, 8-hydroxy-2-(di-*n*-propylamino)-tetralin; MRN, median raphe nucleus; DRN, dorsal raphe nucleus; RN, raphe nuclei; F-C, free-choice; FR, fixed ratio; EtOH, ethanol; s.c., subcutaneous; i.p. intraperitoneal; i.g., intragastric; AA, Alko Alcohol; FH, Fawn Hooded; sP, Sardinian alcohol-preferring; sNP, Sardinian alcohol-non-preferring.

Sprague-Dawley rats may not consume sufficient EtOH to produce measurable BACs; therefore, there may not be a CNS effect of EtOH that can be influenced by removing CNS 5-HT modulation. In addition, the effects of 5,7-DHT lesions occurs over several days, and there may be compensatory mechanism developing during this time. Although the intakes of Wistar rats are low (Tomkins *et al.*, 1994a, 1994b), these rats appear to consume sufficient EtOH in the limited access period to produce measurable BACs; therefore, in this case there may be a CNS effect of EtOH, which can be modulated by manipulating 5-HT neuronal activity.

Selective receptor agonists or antagonists have been used to explore the involvement of certain 5-HT receptor subtypes in regulating EtOH intake (Table 4). Systemic administration of low doses of 8-OH-DPAT, a $5\text{-}HT_{1A}$ receptor agonist, increased EtOH intake (Tomkins *et al.*, 1994b; McKenzie-Quirk and Miczek, 2003), presumably through its action at cell body $5\text{-}HT_{1A}$ autoreceptors. However, higher doses of the $5\text{-}HT_{1A}$ receptor agonist reduced EtOH intake (McBride *et al.*, 1990; Svensson *et al.*, 1993; Tomkins *et al.*, 1994b; Wilson *et al.*, 1996; McKenzie-Quirk and Miczek, 2003), presumably through its additional actions at postsynaptic $5\text{-}HT_{1A}$ receptors. In contrast to the effects of an agonist, systemic administration of a $5\text{-}HT_{1A}$ receptor antagonist had little effect on alcohol intake (Ciccocioppo *et al.*, 1997; Tomkins and O'Neill, 2000). Similarly, activating $5\text{-}HT_{1B}$ receptors reduces EtOH intake, whereas administration of a $5\text{-}HT_{1B}$ receptor antagonist had little effect on EtOH intake (Tomkins and O'Neill, 2000). Overall, the results support the idea that reduced 5-HT neuronal activity increased EtOH intake, and that activation of postsynaptic $5\text{-}HT_{1A}$ and $5\text{-}HT_{1B}$ receptors can inhibit EtOH intake, although these receptors do not appear to be actively involved in modulating ongoing EtOH drinking.

The involvement of $5\text{-}HT_{2A}$ and $5\text{-}HT_{2C}$ receptors in regulating alcohol drinking has also been studied (Table 4). Overstreet *et al.* (1997) examined the effects of $5\text{-}HT_{2A}$ receptor antagonists on EtOH, water and food intake of P, Alko-Alcohol (AA) and Fawn Hooded (FH) rats, all high-alcohol-drinking rats. The two $5\text{-}HT_{2A}$ receptor antagonists tested reduced EtOH, total fluid and food intakes, and produced conditioned taste aversion (Table 4), suggesting that these receptors are involved in modulating consummatory behaviors in general. The effects of $5\text{-}HT_{2C}$ receptors have been studied under operant conditions (Tomkins *et al.*, 2002); systemic administration of an agonist reduced responding, whereas injection of an antagonist increased responding (Table 4).

These results suggest that 5-HT_{2C} receptors can modulate EtOH drinking and be involved in a mechanism for limiting the amount of EtOH consumed.

The involvement of 5-HT_3 receptors in regulating alcohol drinking has received considerable attention (see overview by McBride *et al.*, 2004) because of the evidence that EtOH can potentiate the excitatory actions of 5-HT at this receptor (Lovinger and White, 1991; Lovinger *et al.*, 2000). However, despite the documented action of EtOH at the 5-HT_3 receptor, the effects of 5-HT_3 receptor antagonists depend upon the alcohol drinking conditions (Table 4). Fadda *et al.* (1991) demonstrated that i.p. administration of the 5-HT_3 receptor antagonist MDL 72222 suppressed 24-hourr free-choice drinking in the Sardinian alcohol-preferring (sP) rats. In agreement with this latter study, Rodd-Henricks *et al.* (2000) reported that systemic administration of 5-HT_3 antagonists reduced 24-hour free-choice EtOH intake of P rats under acquisition or maintenance EtOH drinking conditions. However, these latter investigators reported that these same 5-HT_3 receptor antagonists were much less effective under relapse conditions in reducing EtOH intake. McKinzie *et al.* (1998) reported that MDL72222 and ICS 205-930, well-established 5-HT_3 receptor antagonists, reduced EtOH intake of P rats under 24-hour free-choice conditions, but the same doses of the antagonists were not effective in reducing EtOH intake of P rats under 2- or 4-hour limited access conditions, unless the limited EtOH access (only the 2-hour period tested) period was randomly given. In a follow-up study, McKinzie *et al.* (2000), using operant scheduled access, reported that systemic administration of MDL 72222 had no effect on operant responding for EtOH by P rats when the operant session was scheduled at the same time each day, but the antagonist was effective at reducing responding on the EtOH lever if the 1-hour operant session was randomly given each day. The results of the latter two studies suggest that 5-HT_3 receptors are actively involved in regulating EtOH intake, but conditioning factors produce neuronal alterations that reduce the effectiveness of the 5-HT_3 receptor antagonists.

The effects of the s.c. administration of the 5-HT_4 antagonist GR113808 were tested under 2-hour limited access conditions (Table 4). The results of this study (Panocka *et al.*, 1995) indicate that inhibiting 5-HT_4 receptors dose-dependently reduced EtOH drinking without altering water or food intakes, suggesting that 5-HT_4 receptors have a significant role in maintaining alcohol drinking.

Overall, the results suggest that, in general, lower functioning of the 5-HT systems originating from the MRN and DRN is associated with higher alcohol drinking, and that activation of postsynaptic 5-HT_{1A}, 5-HT_{1B} and 5-HT_{2C} receptors has a negative effect on EtOH intake. On the other hand, there is evidence supporting a positive influence for activation of 5-HT_3 and 5-HT_4 receptors on EtOH intakes. Therefore, it is possible that activation of certain 5-HT pathways involving 5-HT_3 and 5-HT_4 synapses would promote alcohol intake, whereas activation of 5-HT pathways involving 5-HT_{1A}, 5-HT_{1B} and 5-HT_{2C} postsynaptic receptors would reduce EtOH intake.

Summary/conclusions

In general, 5-HT appears to have a negative modulatory effect on the brain reward system, in that general activation of 5-HT MRN and DRN systems reduces reward, whereas reducing the function of these 5-HT systems increases reward, as measured in studies of ICSS of the LH and MFB place conditioning experiments, and CR experiments. Similarly, low function of the 5-HT system appears to be associated with the rewarding effects of alcohol, as indicated by effects on alcohol drinking behavior. Although inhibitory effects appear to be the main effect of the 5-HT systems on reward, there is also evidence that some synaptic pathways involving certain subtypes of 5-HT receptors may promote reinforcing effects. In summary, the MRN and DRN 5-HT neuronal systems have important roles in modulating brain reward.

Acknowledgements

Portions of the author's work were supported by research grants from the National Institute of Alcohol Abuse and Alcoholism: AA07611, AA12262 and AA10721.

References

Adell, A. and Myers, R.D. (1995) Selective destruction of midbrain raphe nuclei by 5,7-DHT: is brain 5-HT involved in alcohol drinking in Sprague-Dawley rats?. Brain Res., 693: 70–79.

Al-Ruwaitea, A.S., Chiang, T.J., Al-Zahrani, S.S., Ho, M.Y., Bradshaw, C.M. and Szabadi, E. (1999) Effect of central 5-hydroxytryptamine depletion on tolerance of delay of reinforcement: evidence from performance in a discrete-trials 'time-left' procedure. Psychopharmacology (Berl.), 141: 22–29.

Balleine, B.W., Fletcher, N. and Dickinson, A. (1996) Effect of the 5-HT1A agonist, 8-OH-DPAT on instrumental performance in rats. Psychopharmacology (Berl.), 125: 79–88.

Benaliouad, F., Kapur, S. and Rompré, P.P. (2007) Blockade of 5-HT2a receptors reduces haloperidol-induced attenuation of reward. Neuropsychopharmacology, 32: 551–561.

Beninger, R.J. (1983) The role of dopamine in locomotor activity and learning. Brain Res. Rev., 6: 173–196.

Beninger, R.J. and Ranaldi, R. (1992) The effects of amphetamine, apomorphine, SKF38393, quinpirole and bromodriptine on

responding for conditioned reward in rats. Behav. Pharmacol., 3: 155–163.
Bizot, J., Le Bihan, C., Puech, A.J., Hamon, M. and Thiébot, M. (1999) Serotonin and tolerance to delay of reward in rats. Psychopharmacology (Berl.), 146: 400–412.
Carroll, M.E., Lac, S.T., Asencio, M. and Kragh, R. (1990) Fluoxetine reduces intravenous cocaine self-administration in rats. Pharmacol. Biochem. Behav., 35: 237–244.
Cervo, L., Rozio, M., Ekalle-Soppo, C.B., Carnovali, F., Santangelo, E. and Samanin, R. (2002) Stimulation of serotonin 1B receptors induces conditioned place aversion and facilitates cocaine place conditioning in rats. Psychopharmacology (Berl.), 163: 142–150.
Ciccocioppo, R., Panocka, I., Polidori, C., Dourish, C.T. and Massi, M. (1997) Blockade of pre- and post-synaptic 5-HT1A receptors does not modify the effect of fluoxetine or 5-hydroxytryptophan on ethanol and food intake in rats. Psychopharmacology (Berl.), 134: 55–63.
Davies, A.M. and Parker, L.A. (1993) Fenfluramine-induced place aversion in a three-choice apparatus. Pharmacol. Biochem. Behav., 44: 595–600.
Deakin, J.F. (1980) On the neurochemical basis of self-stimulation with midbrain raphe electrode placements. Pharmacol. Biochem. Behav., 13: 525–530.
Devoto, P., Colombo, G., Stefanini, E. and Gessa, G.L. (1998) Serotonin is reduced in the frontal cortex of Sardinian ethanol-preferring rats. Alcohol Alcohol., 33: 226–229.
Dourish, C.T., Cooper, S.J., Gilbert, F., Coughlin, J. and Iversen, S.D. (1988) The 5-HT1A agonist 8-OH-DPAT increases consumption of palatable wet mash and liquid diets in the rat. Psychopharmacology, 94: 58–63.
Fadda, F., Garau, B., Marchei, F., Colombo, G. and Gessa, G.L. (1991) MDL 72222, a selective 5-HT3 receptor antagonist, suppresses voluntary ethanol consumption in alcohol-preferring rats. Alcohol Alcohol., 26: 107–110.
Fibiger, H.C. and Phillips, A.G. (1986) Reward, motivation, cognition: psychobiology of mesotelencephalic dopamine systems. In: Mountcastle, V.B., Bloom, F.E. and Geiger, S.R. (Eds.), Handbook of Physiology: The Nervous System. American Physiological Society, Bethesda, MD, Vol. 4, pp. 647–675.
Fletcher, P.J. (1993) A comparison of the effects of dorsal or median raphe injections of 8-OH-DPAT in three operant tasks measuring response inhibition. Behav. Brain Res., 54: 187–197.
Fletcher, P.J. (1995) Effects of d-fenfluramine and metergoline on responding for conditioned reward and the response potentiating effect of nucleus accumbens d-amphetamine. Psychopharmacology (Berl.), 118: 155–163.
Fletcher, P.J. and Higgins, G.A. (1997) Differential effects of ondansetron and alpha-flupenthixol on responding for conditioned reward. Psychopharmacology (Berl.), 134: 64–72.
Fletcher, P.J., Ming, Z.H. and Higgins, G.A. (1993) Conditioned place preference induced by microinjection of 8-OH-DPAT into the dorsal or median raphe nucleus. Psychopharmacology (Berl.), 113: 31–36.
Fletcher, P.J., Tampakeras, M. and Yeomans, J.S. (1995) Median raphe injections of 8-OH-DPAT lower frequency thresholds for lateral hypothalamic self-stimulation. Pharmacol. Biochem. Behav., 52: 65–71.
Fletcher, P.J., Korth, K.M. and Chambers, J.W. (1999) Selective destruction of brain serotonin neurons by 5,7-dihydroxytryptamine increases responding for a conditioned reward. Psychopharmacology (Berl.), 147: 291–299.
Gallistel, C.R., Boytim, M., Gomita, Y. and Klebanoff, L. (1982) Does pimozide block the reinforcing effect of brain stimulation? Pharmacol. Biochem. Behav., 17: 769–781.
Gongwer, M.A., Murphy, J.M., McBride, W.J., Lumeng, L. and Li, T.-K. (1989) Regional brain contents of serotonin, dopamine and their metabolites in the selectively bred high- and low-alcohol drinking lines of rats. Alcohol, 6: 317–320.
Harrison, A.A. and Markou, A. (2001) Serotonergic manipulations both potentiate and reduce brain stimulation reward in rats: involvement of serotonin-1A receptors. J. Pharmacol. Exp. Ther., 297: 316–325.
Harrison, A.A., Parsons, L.H., Koob, G.F. and Markou, A. (1999) RU 24969, a 5-HT1A/1B agonist, elevates brain stimulation reward thresholds: an effect reversed by GR 127935, a 5-HT1B/1D antagonist. Psychopharmacology (Berl.), 141: 242–250.
Herberg, L.J., De Belleroche, J.S., Rose, I.C. and Montgomery, A.M. (1992) Effect of the 5-HT3 receptor antagonist ondansetron on hypothalamic self-stimulation in rats and its interaction with the CCK analogue caerulein. Neurosci. Lett., 140: 16–18.
Hoebel, B.G., Hernandez, L., McClelland, R.C. and Schwartz, D. (1988) Dexfenfluramine and feeding reward. Clin. Neuropharmacol., 11: S72–S85.
Katz, R.J. and Carroll, B.J. (1977) Intracranial reward after Lilly 110140 (fluoxetine HCl): evidence for an inhibitory role for serotonin. Psychopharmacology (Berl.), 51: 189–193.
Le, A.D., Funk, D., Harding, S., Juzytsch, W., Li, Z. and Fletcher, P.J. (2008) Intra-median raphe nucleus (MRN) infusions of muscimol, a GABA-A receptor agonist, reinstate alcohol-seeking in rats: role of impulsivity and reward. Psychopharmacology, 195: 605–615.
Lee, K. and Kornetsky, C. (1998) Acute and chronic fluoxetine treatment decreases the sensitivity of rats to rewarding brain stimulation. Pharmacol. Biochem. Behav., 60: 539–544.
LeMarquand, D., Pihl, R.O. and Benkelfat, C. (1994) Serotonin and alcohol intake, abuse, and dependence: findings of animal studies. Biol. Psychiatry, 36: 395–421.
Li, T.-K. and McBride, W.J. (1995) Pharmacogenetic models of alcoholism. Clin. Neurosci., 3: 182–188.
Liu, Z.H. and Ikemoto, S. (2007) The midbrain raphe nuclei mediate primary reinforcement via GABA-A receptors. Eur. J. Pharmacol., 25: 735–743.
Lovinger, D.M. and White, G. (1991) Ethanol potentiation of 5-hydroxytryptamine-3 receptor-mediated ion current in neuroblastoma cells and isolated adult mammalian neurons. Mol. Pharmacol., 40: 263–270.
Lovinger, D.M., Sung, K.W. and Zhou, Q. (2000) Ethanol and trichloroethanol alter gating of 5-HT3 receptor-channels in NCB-20 neuroblastoma cells. Neuropharmacology, 39: 561–570.
McBride, W.J. and Li, T.-K. (1998) Animal models of alcoholism: neurobiology of high alcohol-drinking behavior in rodents. Crit. Rev. Neurobiol., 12: 339–369.
McBride, W.J., Murphy, J.M., Lumeng, L. and Li, T.-K. (1990) Serotonin, dopamine and GABA involvement in alcohol drinking of selectively bred rats. Alcohol, 7: 199–205.
McBride, W.J., Murphy, J.M., Lumeng, L. and Li, T.-K. (1992) Serotonin and alcohol consumption. In: Naranjo, C.A. and Sellers, E.M. (Eds.), Novel Pharmacological Interventions for Alcoholism. Springer-Verlag, New York, NY, pp. 59–67.
McBride, W.J., Murphy, J.M. and Ikemoto, S. (1999) Localization of brain reinforcement mechanisms: intracranial

self-administration and intracranial place-conditioning studies. Behav. Brain Res., 101: 129–152.

McBride, W.J., Lovinger, D.M., Machu, T., Thielen, R.J., Rodd, Z.A., Murphy, J.M., Roache, J.D. and Johnson, B.A. (2004) Serotonin-3 receptors in the actions of alcohol, alcohol reinforcement, and alcoholism. Alcohol. Clin. Exp. Res., 28: 257–267.

McKenzie-Quirk, S.D. and Miczek, K.A. (2003) 5-HT1A agonists: alcohol drinking in rats and squirrel monkeys. Psychopharmacology (Berl.), 167: 145–152.

McKinzie, D.L., Eha, R., Cox, R., Stewart, R.B., Dyr, W., Murphy, J.M., McBride, W.J., Lumeng, L. and Li, T.-K. (1998) Serotonin-3 receptor antagonism of alcohol intake: effects of drinking conditions. Alcohol, 15: 291–298.

McKinzie, D.L., McBride, W.J., Murphy, J.M., Lumeng, L. and Li, T.-K. (2000) Effects of MDL 72222, a serotonin3 antagonist, on operant responding for ethanol by alcohol-preferring P rats. Alcohol. Clin. Exp. Res., 24: 1500–1504.

Meehan, S.M. and Schechter, M.D. (1994) Conditioned place preference/aversion to fenfluramine in Fawn Hooded and Sprague-Dawley rats. Prog. Neuropsychopharmacol. Biol. Psychiatry, 18: 575–584.

Miliaressis, E. (1977) Serotonergic basis of reward in median raphé of the rat. Pharmacol. Biochem. Behav., 7: 177–180.

Montgomery, A.M., Rose, I.C. and Herberg, L.J. (1991) 5-HT1A agonists and dopamine: the effects of 8-OH-DPAT and buspirone on brain-stimulation reward. J. Neural Transm. Gen. Sect., 83: 139–148.

Montgomery, A.M., Rose, I.C. and Herberg, L.J. (1993) The effect of a 5-HT3 receptor antagonist, ondansetron, on brain stimulation reward, and its interaction with direct and indirect stimulants of central dopaminergic transmission. J. Neural Transm. Gen. Sect., 91: 1–11.

Mosher, T.M., Hayes, D. and Greenshaw, A.J. (2005) Differential effects of 5-HT2C receptor ligands on place conditioning and locomotor activity in rats. Eur. J. Pharmacol., 515: 107–116.

Mosher, T.M., Smith, J.G. and Greenshaw, A.J. (2006) Aversive stimulus properties of the 5-HT2C receptor agonist WAY 161503 in rats. Neuropharmacology, 51: 641–650.

Murphy, J.M., McBride, W.J., Lumeng, L. and Li, T.-K. (1982) Regional brain levels of monoamines in alcohol-preferring and -nonpreferring rats. Pharmacol. Biochem. Behav., 16: 145–149.

Murphy, J.M., Waller, M.B., Gatto, G.J., McBride, W.J., Lumeng, L. and Li, T.-K. (1985) Monoamine uptake inhibitors attenuate ethanol intake in alcohol-preferring (P) rats. Alcohol, 2: 349–352.

Murphy, J.M., McBride, W.J., Lumeng, L. and Li, T.-K. (1987) Contents of monoamines in forebrain regions of alcohol-preferring (P) and -nonpreferring (NP) lines of rats. Pharmacol. Biochem. Behav., 26: 389–392.

Murphy, J.M., Waller, M.B., Gatto, G.J., McBride, W.J., Lumeng, L. and Li, T.-K. (1988) Effects of fluoxetine on the intragastric self-administration of ethanol in the alcohol preferring P line of rats. Alcohol, 5: 283–286.

Murphy, J.M., Stewart, R.B., Bell, R.L., Badia-Elder, N.E., Carr, L.G., McBride, W.J., Lumeng, L. and Li, T.-K. (2002) Phenotypic and genotypic characterization of the Indiana University rat lines selectively bred for high and low alcohol preference. Behav. Genet., 32: 363–388.

Olds, M.E. (1995) Dopamine agonists prevent or counteract the suppression of brain stimulation reward by fenfluramine. Pharmacol. Biochem. Behav., 50: 41–48.

Overstreet, D.H., McArthur, R.A., Rezvani, A.H. and Post, C. (1997) Selective inhibition of alcohol intake in diverse alcohol-preferring rat strains by the 5-HT2A antagonists amperozide and FG 5974. Alcohol. Clin. Exp. Res., 21: 1448–1454.

Panocka, I., Ciccocioppo, R., Polidori, C., Pompei, P. and Massi, M. (1995) The 5-HT4, receptor antagonist, GR113808, reduces ethanol intake in alcohol-preferring rats. Pharmacol. Biochem. Behav., 52: 255–259.

Papp, M. and Willner, P. (1991) 8-OH-DPAT-induced place preference and place aversion: effects of PCPA and dopamine antagonists. Psychopharmacology, 103: 99–102.

Patkina, N.A. and Lapin, I.P. (1976) Effect of serotoninergic drugs on positive and negative reinforcing systems in cats. Pharmacol. Biochem. Behav., 5: 241–245.

Platt, M.L. (2002) Neural correlates of decisions. Curr. Opin. Neurobiol., 12: 141–148.

Rea, W.P., Rothman, R.B. and Shippenberg, T.S. (1998) Evaluation of the conditioned reinforcing effects of phentermine and fenfluramine in the rat: concordance with clinical studies. Synapse, 30: 107–111.

Reavill, C., Hatcher, J.P., Lewis, V.A., Sanger, G.J. and Hagan, J. (1998) 5-HT4 receptor antagonism does not affect motor and reward mechanisms in the rat. Eur. J. Pharmacol., 357: 115–120.

Rezvani, A.H. and Grady, D.R. (1994) Suppression of alcohol consumption by fenfluramine in Fawn-Hooded rats with serotonin dysfunction. Pharmacol. Biochem. Behav., 48: 105–110.

Risinger, F.O. (1997) Fluoxetine's effects on ethanol's rewarding, aversive and stimulus properties. Life Sci., 61: 235–242.

Robbins, T.W. (1978) The acquisition of responding with conditioned reinforcement: Effects of pipradrol, methylphenidate, d-amphetamine and nomifensine. Psychopharmacology, 58: 79–87.

Robbins, T.W., Watson, B.A., Gaskin, M. and Ennis, C. (1983) Contrasting interactions of pipradrol, d-amphetamine, cocaine, cocaine analogues, apomorphine and other drugs with conditioned reinforcement. Psychopharmacology, 80: 113–119.

Rocha, B., Di Scala, G., Jenck, F., Moreau, J.L. and Sandner, G. (1993a) Conditioned place aversion induced by 5-HT1C receptor antagonists. Behav. Pharmacol., 4: 101–106.

Rocha, B., Di Scala, G., Riga, M., Hoyer, D. and Sandner, G. (1993b) Effect of 5,7-dihydroxytryptamine lesion on mianserin-induced conditioned place aversion and on 5-hydroxytryptamine 1C receptors in rat brain. Neuroscience, 56: 687–693.

Rodd, Z.A., Gryszowka, V.E., Toalston, J.E., Oster, S.M., Ji, D., Bell, R.L. and McBride, W.J. (2007) The reinforcing actions of a serotonin-3 receptor agonist within the ventral tegmental area: evidence for subregional and genetic differences and involvement of dopamine neurons. J. Pharmacol. Exp. Ther., 321: 1003–1012.

Rodd-Henricks, Z.A., McKinzie, D.L., Edmundson, V.E., Dagon, C.L., Murphy, J.M., McBride, W.J., Lumeng, L. and Li, T.-K. (2000) Effects of 5-HT(3) receptor antagonists on daily alcohol intake under acquisition, maintenance, and relapse conditions in alcohol-preferring (P) rats. Alcohol, 21: 73–85.

Rolls, E.T., Kelly, P.H. and Shaw, S.G. (1974) Noradrenaline, dopamine and brain-stimulation reward. Pharmacol. Biochem. Behav., 2: 735–740.

Rowland, N.E. and Morian, K.R. (1992) Effect of dexfenfluramine on alcohol intake in alcohol-preferring 'P' rats. Alcohol, 9: 559–561.

Sasaki-Adams, D.M. and Kelley, A.E. (2001) Serotonin–dopamine interactions in the control of conditioned reinforcement and motor behavior. Neuropsychopharmacology, 25: 440–452.

Schweighofer, N., Tanaka, S.C. and Doya, K. (2007) Serotonin and the evaluation of future rewards: Theory, experiments, and possible neural mechanisms. Ann. N.Y. Acad. Sci., 1104: 289–300.

Smith, F.L., Yu, D.S.L., Smith, D.G., Leccese, A.P. and Lyness, W.H. (1986) Dietary tryptophan supplements attenuate amphetamine self-administration in the rat. Pharmacol. Biochem. Behav., 25: 849–855.

Smith, J.E., Shultz, K., Co, C., Goeders, N.E. and Dworkin, S.I. (1987) Effects of 5,7-dihydroxytryptamine lesions of the nucleus accumbens on rat intravenous morphine self-administration. Pharmacol. Biochem. Behav., 26: 607–612.

Subhan, F., Deslandes, P.N., Pache, D.M. and Sewell, R.D. (2000) Do antidepressants affect motivation in conditioned place preference?. Eur. J. Pharmacol., 408: 257–263.

Svensson, L., Fahlke, C., Hard, E. and Engle, J.A. (1993) Involvement of the serotonergic system in ethanol intake in the rat. Alcohol, 10: 219–224.

Tomkins, D.M. and Fletcher, P.J. (1996) Evidence that GABA(A) but not GABA(B) receptor activation in the dorsal raphe nucleus modulates ethanol intake in Wistar rats. Behav. Pharmacol., 7: 85–93.

Tomkins, D.M. and O'Neill, M.F. (2000) Effect of 5-HT(1B) receptor ligands on self-administration of ethanol in an operant procedure in rats. Pharmacol. Biochem. Behav., 66: 129–136.

Tomkins, D.M., Sellers, E.M. and Fletcher, P.J. (1994a) Median and dorsal raphe injections of the 5-HT1A agonist, 8-OH-DPAT, and the GABA-A agonist, muscimol, increase voluntary ethanol intake in Wistar rats. Neuropharmacology, 33: 349–358.

Tomkins, D.M., Higgins, G.A. and Sellers, E.M. (1994b) Low doses of the 5-HT1A agonist 8-hydroxy-2-(di-n-propylamino)-tetralin (8-OH DPAT) increase ethanol intake. Psychopharmacology (Berl.), 115: 173–179.

Tomkins, D.M., Joharchi, N., Tampakeras, M., Martin, J.R., Wichmann, J. and Higgins, G.A. (2002) An investigation of the role of 5-HT(2C) receptors in modifying ethanol self-administration behaviour. Pharmacol. Biochem. Behav., 71: 735–744.

Tzschentke, T.M. (1998) Measuring reward with the conditioned place preference paradigm: a comprehensive review of drug effects, recent progress and new issues. Prog. Neurobiol., 56: 613–672.

Tzschentke, T.M. (2007) Measuring reward with the conditioned place preference (CPP) paradigm: update of the last decade. Addict. Biol., 12: 227–462.

Van Der Kooy, D., Fibiger, H.C. and Phillips, A.G. (1977) Monoamine involvement in hippocampal self-stimulation. Brain Res., 136: 119–130.

Van Der Kooy, D., Fibiger, H.C. and Phillips, A.G. (1978) An analysis of dorsal and median raphe self-stimulation: effects of para-chlorophenylalanine. Pharmacol. Biochem. Behav., 8: 441–445.

Wilson, A.W., Neill, J.C. and Costall, B. (1996) The 5-HT1A receptor agonist 8-OH-DPAT reduces ethanol intake and maintained behavior in female Sprague-Dawley rats. Alcohol, 13: 407–413.

Wogar, M.A., Bradshaw, C.M. and Szabadi, E. (1993) Effect of lesions of the ascending 5-hydroxytryptaminergic pathways on choice between delayed reinforcers. Psychopharmacology (Berl.), 111: 239–243.

Yoshimoto, K. and Komura, S. (1987) Reexamination of the relationship between alcohol preference and brain monoamines in inbred strains of mice including senescence-accelerated mice. Pharmacol. Biochem. Behav., 27: 317–322.

Yu, Y.L., Fisher, H., Sekowski, A. and Wagner, G.C. (1997) Amphetamine and fenfluramine suppress ethanol intake in ethanol-dependent rats. Alcohol, 14: 45–48.

Zhou, F.C., Bledsoe, S., Lumeng, L. and Li, T.-K. (1991a) Serotonergic immunostained terminal fibers are lower in selected forebrain regions of alcohol-preferring rats. Alcohol, 8: 1–7.

Zhou, F.C., Bledsoe, S., Lumeng, L. and Li, T.-K. (1991b) Immunostained serotonergic fibers are decreased in selected brain regions of alcohol-preferring rats. Alcohol, 8: 425–431.

Zhou, F.C., Bledsoe, S., Lumeng, L. and Li, T.-K. (1994) Reduced serotonergic immunoreactive fibers in the forebrain of alcohol-preferring rats. Alcohol. Clin. Exp. Res., 18: 571–579.

CHAPTER 3.8

Role of Central Serotonin in Impulsivity and Compulsivity: Comparative Studies in Experimental Animals and Humans

Trevor W. Robbins* and Molly J. Crockett*

Department of Experimental Psychology and Behavioural and Clinical Neuroscience Institute, University of Cambridge, Cambridge, UK

Abstract: The involvement of serotonin in impulsivity (the tendency to respond prematurely) and compulsivity (the tendency to perseverate) is reviewed from the joint perspective of animal and human studies. Evidence is provided to support a role for serotonin in some forms of impulsivity, but not others, and in compulsivity. However, it is difficult to accommodate these roles into a common scheme implicating behavioral inhibition. The implications for neuropsychiatric disorders such as obsessive-compulsive disorder are considered.

Keywords: impulsivity, compulsivity, behavioral inhibition, striatum, prefrontal cortex, ADHD, obsessive-compulsive disorder.

Introduction

Serotonin (or 5-hydroxytryptamine, 5-HT) has been implicated in many behavioral and physiological functions, which is not surprising, given the diverse ramification of central 5-HT neurons and the existence of at least 14 distinct 5-HT receptors. However, the unitary functions most often attributed to this system have been those of aversive motivation (Wise *et al.*, 1972; Deakin and Graeff, 1991) and behavioral inhibition (Gray, 1982; Soubrié, 1986). The latter is most relevant to theories of impulsivity and compulsivity, although 5-HT may also be linked with these constructs through its possible role in aversive motivation. This chapter will summarize the research especially relevant to the impulsivity/compulsivity theme, and will draw on work performed in this laboratory as well as others, in both experimental animals and human volunteers.

As Gray (1982) himself realized, there are clear reciprocal relationships between apparently opponent appetitive and aversive systems, whereby the omission of an expected positive reinforcer (or 'reward') has aversive consequences, and the omission or postponement of an aversive reinforcer is perceived in a positive light. Whether these two contingencies actually are equivalent to receiving an aversive or appetitive reinforcer, respectively, seems somewhat dubious, although they may share some generalized emotional congruence. There is a similar confound in terms of the relationship between aversive motivation and inhibition: a punishing contingency normally results in the suppression of appetitive behavior, whereas the unconditioned response for a rodent to a pavlovian aversive reinforcer is to freeze. So a manipulation of the 5-HT system which reduces the unconditioned response to an aversive reinforcer (e.g., freezing in rodents) generally *also* causes increases in overall motor output, or 'behavioral disinhibition'. The question is, which of these roles ascribed to 5-HT is primary: inhibition or aversiveness? A further complication has been the fact that different aspects of aversiveness also implicate 5-HT; thus Deakin and Graeff (1991) pointed out that increasing levels of 5-HT transmission are associated with anxiety, and decreasing levels with depression. Throughout this chapter we will observe how seemingly opposite functional effects can follow superficially similar manipulations of 5-HT neurotransmission, in a manner that almost appears bound to confound any attempt to assign a unitary function to this system. These paradoxes, however, serve to strengthen our understanding of the 5-HT system, and how its operation may depend on subtle contextual demands.

*Corresponding authors
E-mail: twrz@cam.ac.uk (T.W. Robbins);
mollycrockett@gmail.com (M.J. Crockett)

Christian Müller & Barry Jacobs (Eds.)
Handbook of Behavioral Neurobiology of Serotonin
ISBN 978-0-12-374634-4

DOI: 10.1016/B978-0-12-374634-4.00025-3

5-HT and impulsivity

Impulsivity is generally defined as the tendency to respond prematurely without adequate foresight, sometimes with adverse consequences. This definition conceals a number of possibly distinct aspects of impulsivity which may each implicate an inhibitory process – for example, inhibition of an action or response or inhibition of the salience of an appetitive goal. Moreover, an impulsive decision can result from an altered criterion in terms of perceptual evidence sampled, or from altered processing of reinforcement contingencies. In our work with rats we employ three tests of impulsivity that capture these distinct forms; premature responding in a five-choice serial reaction time task; stop-signal inhibition; and temporal discounting of reward. These tests reveal different neural and neurochemical substrates for these distinct forms of impulsivity, although with some overlap sufficient to encourage the notion of a unitary construct (Robinson *et al.*, 2008a; but see also Winstanley *et al.*, 2004a).

Impulsivity in the five-choice serial reaction time task (5-CSRTT)

The 5-CSRTT is a test paradigm in which rats are trained to detect brief visual targets presented randomly in five locations to earn food (Figure 1). Rats have a propensity to respond prematurely in this task, particularly as the duration of the visual target is reduced, although this behavior is punished by time-out from positive reinforcement. This tendency is quite variable between individuals, with a small proportion of rats exhibiting high degrees of impulsive responding, even after extended training.

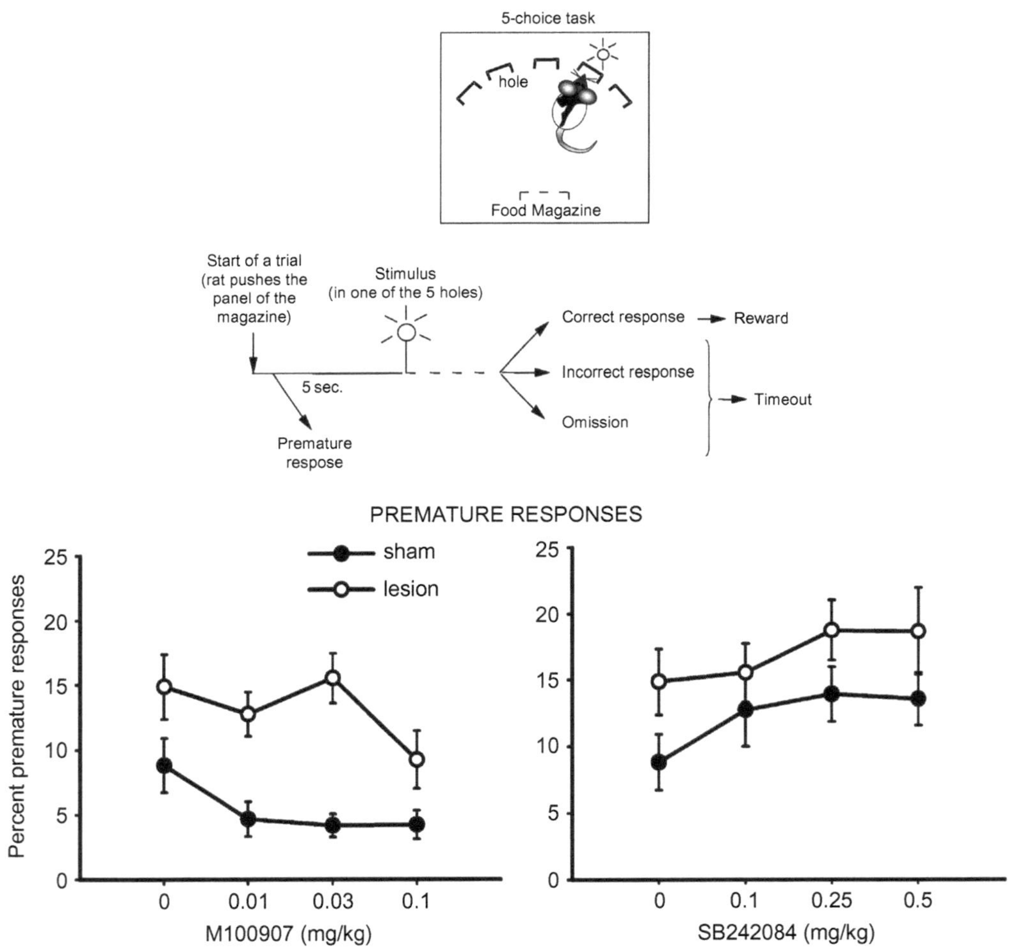

Figure 1 Contrasting effects of the 5-HT_{2A} receptor antagonist M100907 and the 5-HT_{2C} receptor antagonist SB242084 on premature responding in the 5-choice serial reaction time task in rats with either forebrain 5-HT depletion from the forebrain or the control treatment (sham surgery). The five-choice paradigm is also shown. The figures are reproduced from Winstanley *et al.* (2004b) and Robbins (2002), by permission of the publishers (Springer-Verlag).

An early experiment showed that profound (>85 percent loss in most forebrain regions assayed, including neocortex and striatum) depletion of forebrain 5-HT induced by intraventricular administration of the selective neurotoxin 5,7 dihydroxytryptamine (5,7-DHT) produced significant and essentially permanent increases in premature responding (Harrison *et al.*, 1997a). Further experiments targeting separately the dorsal and medial raphe nuclei indicated that lesions of the dorsal nucleus, which projects preferentially to the neocortex and striatum, produced higher levels of impulsive responding, sometimes in conjunction with *improved* accuracy (Harrison *et al.*, 1997b).

The increase in premature responses in the 5-CSRTT following intraventricular 5,7-DHT was replicated and extended by Winstanley *et al.* (2004b), who also demonstrated that the effect was ameliorated to some extent by systemic administration of M100907, the 5-HT_{2A} receptor antagonist. However, the suppression of premature responses was enhanced by lower doses of M100907 in sham-operated control rats (see Figure 1). By contrast, systemic administration of the 5-HT_{2C} receptor antagonist SB242084 exacerbated impulsivity produced by the 5-HT depletion, although again the exacerbation was obtained at lower doses in controls (Figure 1). This different pattern of effects of the 5-HT_{2A} and 5-HT_{2C} receptor antagonists (Figure 2) has since been reproduced following intracerebral administration into the nucleus accumbens (Robinson *et al.*, 2008a), a structure known to be implicated in impulsivity on the 5-CSRTT, and possibly reflecting an interaction between 5-HT and mesolimbic dopamine (DA) (Berg *et al.*, 2008). These contrasting effects of 5-HT_{2A} and 5-HT_{2C} agents are consistent with other data on effects of 5-HT_{2A} and 5-HT_{2C} agonists on impulsive responding when administered systemically (Koskinen *et al.*, 2000; Fletcher *et al.*, 2007; Navarra *et al.*, 2008). The opposite effects of the 5-HT_{2A} and 5-HT_{2C} antagonist are perhaps surprising in view of their structural similarities, and commonalities in their effects on secondary messengers (Conn and Sanders-Bush, 1987), but presumably reflect their differential involvement in neuronal circuitry modulated by serotonin (see, for example, Pompeiano *et al.*, 1994; Berg *et al.*, 2008; Calcagno *et al.*, 2009).

Contrasting effects of 5HT2A and 2C antagonists on measures of impulsive and compulsive responding in the rat

	IMPULSIVITY e.g. premature responses on 5CSRTT	COMPULSIVITY e.g. perseverative responses on discrimination reversal learning
5HT2A receptor antagonist (M100927)	Antagonise, remediate (ILC-PFC. N.Acc.)	Exacerbate (OFC-PFC)
5HT2C receptor antagonist (SB240804)	Exacerbate (N.Acc.)	Antagonise, remediate (OFC-PFC)

Figure 2 Contrasting effects of $5HT_{2A}$ and 5-HT_{2C} antagonists on measures of impulsive and compulsive responding in the rat. *Abbreviations:* OFC, orbitofrontal cortex; N. Acc., nucleus accumbens. Summary of data from Winstanley *et al.* (2003, 2004b) and Boulougouris *et al.* (2008).

Whereas ventral striatal 5-HT is implicated in impulsive responding induced on the 5-CSRTT, the involvement of cortical 5-HT is a little less clear. Previous work has established that the infralimbic (IL-) or ventromedial prefrontal cortex (PFC) is particularly implicated in the control of impulsivity, but not accuracy, of responding in the 5-CSRTT (Chudasama *et al.*, 2003; Murphy *et al.*, 2005). Dalley *et al.* (2002) actually found that the more highly impulsive rats tended to have *higher* levels of 5-HT in the medial prefrontal cortex (mPFC) as measured using *in vivo* microdialysis, seemingly in contradiction of a role for 5-HT in behavioral inhibition. This putative increase in mPFC 5-HT neurotransmission was apparently supported by findings that intra-mPFC infusions of ketanserin (a mixed 5-$HT_{2A/C}$ antagonist; Passetti *et al.*, 2003) and the more selective 5-HT_{2A} receptor antagonist M100907 (Winstanley *et al.*, 2003) reduced premature responses on the 5-CSRTT. However, these effects were only obtained in modified versions of the task when the duration of the visual targets was reduced. With standard task parameters, Robinson *et al.* (2008b) found no effect of either M100907 or the 5-HT_{2C} receptor antagonist SB242084 when infused intra-mPFC. However, Carli *et al.* (2006) found that intra-mPFC M100907 did antagonize the increased premature responses induced by intra-mPFC infusions of the NMDA receptor antagonist CPP, consistent with the possibility of the modulation of glutamate responses by the 5-HT_{2A} receptor in this region. This blockading effect is also consistent with the fact that premature responses had a high baseline in the study by Winstanley *et al.* (2003), but a low one in the experiment of Robinson *et al.* (2008b). Thus, it appears that the 5-HT_{2A} receptor does modulate impulsivity within the mPFC under some circumstances, consistent with the high density of 5-HT_{2A} receptors in this region (Santana *et al.*, 2004). Recent evidence suggests that there is also a balance between the influence on pyramidal cell output of 5-HT_{2A} and 5-HT_{2C} receptor activity in this region (Calcagno *et al.*, 2009). Thus, in parallel with the enhanced impulsivity elicited by intra-mPFC CPP and its reversal by M100907, the enhanced glutamate release following intra-mPFC CPP by M100907 was blocked by concurrent administration of a 5-HT_{2C} receptor antagonist. Furthermore, Calcagno *et al.* (2009) have found that the enhanced impulsivity produced by intra-IL-PFC CPP can be blocked by a 5-HT_{2C} agonist. The functional interaction between 5-HT_{2A} and 5-HT_{2C} receptors may be mediated by their neuronal effects at pyramidal cells, and on parvalbumin-containing GABA neurons, respectively (see Liu *et al.*, 2007; Calcagno *et al.*, 2009).

This series of studies has shown that the involvement of 5-HT in one form of impulsivity, premature responding in the 5-CSRTT, is quite complex, with behavioral disinhibition certainly being the outcome of global forebrain 5-HT loss. However, there is evidence that the effects of two subtypes of the 5-HT_2 receptor are reciprocally involved in the regulation of impulsivity. This reciprocity is also expressed within the nucleus accumbens and the mPFC, as well as other possible neural sites. A further question is, what form of impulsivity is captured by the premature responding measure in the 5-CSRTT? It quite possibly reflects perceptual and motivational, as well as motor, aspects.

Response inhibition: stop signal inhibition and Go/NoGo paradigms

As will be seen below, there are several ways of inferring effects of 5-HT manipulations on motor inhibition, which lead to a surprising, but informative, conclusion (Table 1). In this section, we also compare studies performed in the same paradigm with human as well as animal subjects. Although it is feasible to manipulate 5-HT function in humans, there is a relative dearth of means for doing this. Some of the most often used methods are shown in Box 1.

Stop-signal reaction time task (SSRTT)

Logan's stop-signal reaction time task was devised originally for human subjects with applications especially to patients with attention deficit hyperactivity disorder (ADHD) (Logan *et al.*, 1984). Simply speaking, the SSRTT measures the speed with which an initiated action can be cancelled or inhibited (see Eagle and Robbins, 2008a, for a full theoretical description). Recently, it has

Table 1 Summary of reviewed findings of serotoninergic manipulations on measures of response inhibition

Reference	Manipulation	Task	Is response inhibition before or after response selection?	Effect on response inhibition
Rats				
Harrison *et al.*, 1997b	5,7 DHT	5CSRTT	Before	Impaired
Harrison *et al.*, 1999	Dorsal raphe lesion	5CSRTT	Before	Impaired
Eagle *et al.*, 2008a	Acute SSRI	SSRTT	After	No effect
Bari *et al.*, 2009	Acute SSRI	SSRTT	After	No effect
Eagle *et al.*, 2008b	5,7 DHT	SSRTT	After	No effect
Eagle *et al.*, 2008b	5,7 DHT	SSRTT with waiting period	?	Impaired
Harrison et *al.*, 1999	5,7 DHT	GNG	Before	Impaired
Humans				
Chamberlain *et al.*, 2006	Acute SSRI	SSRTT	After	No effect
Clark *et al.*, 2005	ATD	SSRTT	After	No effect
Crean *et al.*, 2002	ATD	SSRTT	After	Impaired in FH+ males; improved in FH− males
Walderhaug *et al.*, 2002	ATD	CPT	Before	Impaired
Walderhaug *et al.*, 2007	ATD	CPT	Before	Impaired in males; improved in females
LeMarquand *et al.*, 1999	ATD	GNG	Before	Impaired in FH+ males; no effect in FH− males
Evers *et al.*, 2006	ATD	GNG	Before	No effect
Crockett *et al.*, 2008	ATD	GNG	Before	No effect
Scholes *et al.*, 2006	ATD	Stroop	Before	Enhanced
Evers *et al.*, 2006	ATD	Stroop	Before	Enhanced

Abbreviations: 5,7 DHT: neurotoxic lesions to 5-HT-releasing neurons; 5CSRTT: 5-choice serial reaction time task; SSRTT: Stop signal reaction time task; GNG: Go/NoGo task; ATD: Acute tryptophan depletion; FH+: family history of alcoholism; FH−: no family history of alcoholism; CPT: Continuous performance task.

Common methods for manipulating 5-HT in humans

Acute tryptophan depletion (ATD)

ATD induces a transient 5-HT deficit in the CNS (Carpenter *et al.*, 1998). In the ATD procedure, participants ingest an amino acid load via a drink that contains all of the large neutral amino acids (LNAAs) except for tryptophan, which is the chemical precursor of 5-HT. Tryptophan competes with other LNAAs for access via the blood–brain barrier through the LNAA transporter. Lowering the tryptophan:LNAA ratio almost completely halts tryptophan transport into the brain. Microdialysis in rats has shown that ATD attenuates 5-HT release in frontal cortex and midbrain raphe nuclei (Bel and Artigas, 1996). In human subjects, plasma and CSF tryptophan levels reach nadir 5 hours after the administration of the amino acid mixture (Carpenter *et al.*, 1998).

Tryptophan supplementation and loading

Tryptophan supplementation and loading are thought to enhance 5-HT function by increasing availability of tryptophan. Tryptophan loading involves a similar procedure to ATD, but the amino acid load contains a large amount of tryptophan. In the tryptophan supplementation procedure, participants take daily supplements of tryptophan in tablet form, often for several weeks. Both procedures increase tryptophan availability to levels that saturate tryptophan hydroxylase, the rate-limiting enzyme that converts tryptophan to 5-HT (Young and Gauthier, 1981).

Acute vs chronic SSRI administration

Selective serotonin reuptake inhibitors (SSRIs) increase the concentration of 5-HT in the synapse by blocking its presynaptic reuptake. They do so by blocking the presynaptic active transport mechanism in the 5-HT transporter (5-HTT). Consequently, the action of 5-HT on postsynaptic receptors is enhanced (Spinks and Spinks, 2002). The onset of an SSRI's clinical effects can be delayed 1 or 2 weeks from the start of treatment (Taylor *et al.*, 2006), despite the fact that these drugs inhibit 5-HT reuptake within hours of administration (Bymaster *et al.*, 2002). The underlying mechanisms of this effect are presumed to involve the presynaptic 5-HT1A autoreceptors. When released 5-HT activates these receptors, potassium channels are opened, leading to hyperpolarization of the cell and inhibition of cell firing, which results in decreased 5-HT release (Blier *et al.*, 1998). SSRI-induced increases in synaptic 5-HT are thus countered by negative feedback on serotonergic release caused by 5-HT binding to autoreceptors, which may explain why SSRIs do not show clinical effects immediately after administration. The effects of acute administration of SSRIs may therefore mirror those of serotonin depletions, such as ATD (Carpenter *et al.*, 19981). The lower acute doses used in research in human volunteers are particularly likely to have autoreceptor effects to decrease extracellular 5-HT levels. Chronic administration of SSRIs, however, is thought to enhance serotonergic neurotransmission, as autoreceptors desensitize with repeated drug administration and firing of 5-HT neurons is restored (Blier *et al.*, 1998).

Pharmacological fMRI

All of the above methods can be combined with functional magnetic resonance imaging (fMRI) to probe the effects of altered 5-HT neurotransmission on the neural correlates of behavior. Interpreting the effects of 5-HT manipulations on neural activity is not straightforward, however, and must be approached with caution when the manipulation does not produce effects on behavior. When 5-HT manipulation does affect fMRI results, it is not yet possible to determine the underlying mechanism; for example, enhanced neural activity could reflect increased neurotransmission due to greater efficiency, or the need to recruit more neurons due to impaired efficiency (Anderson *et al.*, 2008).

proven possible to configure this task appropriately for monkeys and rodents (Eagle and Robbins, 2003; Eagle *et al.*, 2008a) allowing for comparisons to be made across species. Performance in the SSRTT appears to involve structures other than the nucleus accumbens in the rat; in fact, an orbitofrontal cortex (OFC)–medial striatal–subthalamic nucleus system has been implicated, which may overlap with the neural network engaged during performance of the task in an fMRI paradigm in humans (Aron and Poldrack, 2006).

So far, manipulations of the central 5-HT system have shown few effects on performance of the SSRTT. The selective serotonin reuptake inhibitor (SSRI) citalopram has no effect on performance in rats (Eagle *et al.*, 2008a; Bari *et al.*, 2009) on acute administration. More strikingly, global forebrain 5-HT depletion in rats also has no effect on the SSRTT, which contrasts markedly with its effects to induce impulsive responding in the 5-CSRTT or in the Go/NoGo paradigm (Eagle *et al.*, 2008b). However, if the period for which a response has to be inhibited on the SSRTT is prolonged, then 5-HT depletion in the rat does produce behavioral disinhibition; there is a deficit in 'waiting' but not 'stopping' itself (Eagle *et al.*, 2008b).

The effects of 5-HT manipulations in humans on SSRTT performance for the most part parallel the findings in rats, with a few complications (Table 1). Acute tryptophan depletion (ATD) had no effect on response inhibition in the SSRTT (Clark *et al.*, 2005), and neither did acute administration of the SSRI citalopram (Chamberlain *et al.*, 2006). However, another study found that ATD impaired SSRTT performance in males with a family history of alcoholism, but *improved* performance in males without a family history of alcoholism (Crean *et al.*, 2002). Whether these more complex findings reflect the release by ATD of contrasting underlying individual tendencies, or differential individual responses to the ATD procedure, remains unclear.

Go/NoGo discrimination

Go/NoGo discrimination is another paradigm used to measure response inhibition. In this task, inhibition has to be exerted in the response selection phase, whereas in the SSRTT, response selection has already occurred but the performance of the response has to be curtailed. Considerable evidence links a reduction of 5-HT to behavioral disinhibition in Go/NoGo paradigms, including classical findings reviewed by Soubrié (1986). The interpretation of such data is generally along the lines that impaired 5-HT function results in a reduction of response inhibition, and hence the increase in non-reinforced or even punished NoGo responding. An alternative account is that the aversive consequences of responding on NoGo trials are reduced by 5-HT depletion. The latter hypothesis is difficult to refute, but we have attempted to address it by studying in parallel two similar tasks. In the first of these, acquiring a Go/NoGo discrimination depended on responding differentially to a fast-flashing or slow-flashing visual stimulus (counterbalanced across rats). Global 5-HT depletion actually completely prevented the acquisition of this discrimination, as the depleted rats failed to inhibit responding to the NoGo stimulus (regardless of whether it was fast or slow), although withholding the response, like performing the Go response, was symmetrically reinforced with food. If rats were pre-trained to solve the discrimination, 5-HT depletion impaired performance by inducing inappropriate responding on NoGo trials – though not to the extent of the complete behavioral disinhibition observed during acquisition (Harrison *et al.*, 1999).

In a complementary experiment, rats were trained in a conditional Go/Go discrimination based on similar fast and slow-flashing visual stimuli. In this situation, 5-HT depletion actually *improved* acquisition of the discrimination. Therefore, it is difficult to argue that 5-HT depletion reduced the impact of reward omission, as this would presumably also have impaired trial-and-error discrimination learning. The more parsimonious hypothesis is that it impaired response inhibition in the Go/NoGo procedure, but this deficit was largely irrelevant to the Go/Go task (Ward *et al.*, 1999). Note that these findings also approximate the effect of forebrain 5-HT loss on performance on the 5-CSRTT, where again premature responding was enhanced, but visual discrimination was unaffected or even enhanced (Harrison *et al.*, 1997a, 1997b).

In humans, the effects of 5-HT manipulations on Go/NoGo discrimination are more complex. Paralleling the findings in rats, Walderhaug *et al.* (2002) reported increased impulsivity or disinhibition following ATD in a continuous performance task requiring Go/NoGo discrimination. However, several other findings are less straightforward (Table 1). A later study using the same continuous performance task showed that ATD increased impulsivity in males, but *decreased* impulsivity in females (Walderhaug *et al.*, 2007). LeMarquand *et al.* (1999) found that ATD increased impulsive responses in the Go/NoGo task in males with a family history of alcoholism, but not in males without a family history of alcoholism. Two more recent studies found no effect of ATD on Go/NoGo discrimination in male and female healthy volunteers (Evers *et al.*, 2006; Crockett *et al.*, 2008). Furthermore, it could be argued that performance on the Stroop task involves inhibitory control in the response selection phase, and thus taps into the same mechanisms of the Go/NoGo task. However, two studies have found that ATD *enhances* performance on the Stroop task (Evers *et al.*, 2006; Scholes *et al.*, 2006). Thus, in humans the effects of ATD on Go/NoGo discrimination are not straightforward and appear to depend to a certain extent on individual vulnerability factors, such as gender or family history of psychopathology.

Researchers have used fMRI to investigate whether 5-HT manipulations modulate brain activity during Go/NoGo discrimination. One study found that ATD attenuated activity in the right inferior frontal gyrus (rIFG) during the Go/NoGo task (Rubia *et al.*, 2005); another study reported no effect of ATD on neural activity during the Go/NoGo task (Evers *et al.*, 2006). A third study found enhanced rIFG activity during the Go/NoGo task following acute pre-treatment with the SSRI citalopram (Del-Ben *et al.*, 2005). These findings provide some evidence for serotonergic modulation of the neural correlates of response inhibition, but it is important to note that none of these studies found effects of ATD on behavioral measures of response inhibition.

One potential explanation for these mixed results is that the behavioral tasks typically used to measure response inhibition in humans are non-reinforced: subjects do not receive rewarding or punishing feedback for their responses. But the 5-HT-inhibition hypothesis was based on animal work that employed reinforced tasks (Soubrié, 1986). This suggests that the results obtained in animal studies could be due to the involvement of 5-HT in reinforcement processing. 5-HT has indeed been linked with behavioral and neural indices of reinforcement processing in a variety of paradigms.

In summary, 5-HT manipulations have strikingly different effects on performance in two different tasks designed to measure response (including motor) inhibition. When response selection is involved prior to response initiation, 5-HT appears to have an important role in suppressing impulsive NoGo responding. This parallels the effects of 5-HT depletion on premature responding in the 5-CSRTT. In contrast, after a response has been initiated, 5-HT appears to no longer play a role in motor inhibition. Behavioral manipulations suggest that the predominant effect of 5-HT loss is to induce deficits in the capacity to wait rather than stop (Table 1). Clearly, this distinction goes beyond a simple theory of 5-HT involvement in behavioral inhibition; response inhibition may be involved in both circumstances, but the 5-HT system appears to be implicated only in one of them.

Temporal discounting of reward: impulsive choice

This refers to the choice between small immediate rewards and large delayed ones – an elemental form of decision-making. Other versions of temporal discounting or 'inter-temporal choice' may pit the choice between small immediate rewards and avoidance of large delayed punishment, as may occur in several decisions involving personal health choices, such as those involved in addiction. Previous work has suggested that down-regulating 5-HT function in rats by combined dorsal and median raphe 5,7-DHT increases impulsive choice (see, for example, Mobini *et al.*, 2000). However, using the same intraventricular treatment with 5,7-DHT as previously, we found no effect on temporal discounting in an operant procedure in which hungry rats responded for either immediate single food pellets, or for four pellets presented at varying delays culminating in 60 s. The resulting choice behavior could readily be fit by a hyperbolic discounting function, but there was no effect on temporal discounting following profound forebrain 5-HT loss. However, such 5-HT depletion did have effects on temporal discounting affected by systemic d-amphetamine. The latter treatment (repeated three times) dose-dependently *reduced* impulsive choice, inducing rats to respond on the lever associated with the large, delayed reward. This effect was only significant, however, for those rats with initially steep discounting functions – i.e., in rats showing impulsive choice. Intriguingly, given that d-amphetamine is an indirect catecholamine agonist, its effects to reduce impulsive choice were blocked by 5-HT depletion. The effects of d-amphetamine could also be blocked by the mixed D1/D2 receptor blocker α-flupenthixol, but only in 5-HT-depleted rats, suggesting that they were mediated by an *interaction* between DA and 5-HT systems. Further evidence for such an interaction comes from the fact that mesolimbic DA depletion achieved by intra-accumbens 6-hydroxydopamine failed by itself to block the anti-impulsive effect of d-amphetamine. Furthermore, the 5-HT_{1A} receptor antagonist WAY 100635 enhanced the action of d-amphetamine to reduce impulsive choice (Winstanley *et al.*, 2005).

The most likely site of action for these interactions is the nucleus accumbens, core region, as excitotoxic lesions of this structure have been found to cause impulsive choice in the same paradigm (Cardinal *et al.*, 2001). 5-HT_{1A} receptors are known to be located on DA cells in the ventral tegmental area that projects to the nucleus accumbens (as well as within the nucleus accumbens itself). Acute, systemic treatment with the 5-HT_{1A} agonist 8-OH-DPAT elicits strong impulsive choice, an effect which is blocked by a 5-HT_{1A} receptor antagonist, but which is not blocked by 5-HT depletion, suggesting that it is postsynaptic in origin. In contrast, such impulsivity is blocked by 6-OHDA lesions of the nucleus accumbens, which deplete dopamine there (Figure 3). At the dose used to induce impulsive choice, 8-OH-DPAT may thus induce impulsive responding by facilitating release of DA in the nucleus accumbens. These findings again highlight the intimate interactions of DA and 5-HT in the control of impulsive behavior.

An *in vivo* microdialysis study of delayed discounting by Winstanley *et al.* (2006) found elevated extracellular 5-HT within the mPFC specifically related to an instrumental choice condition, independent of the effects in separate groups of passive reward presentation, or of instrumental behavior without choice. This contrasted with findings for DA in mPFC which was elevated following passive reward presentation or during instrumental behavior, as well as during impulsive choice. Within the OFC, however, DA was more reliably associated with impulsive choice behavior.

The relationship between low or impaired 5-HT and 'impulsive' inter-temporal choice described above has been extended to humans. An early study found no effect of ATD on a temporal discounting of reward task (Crean *et al.*, 2002). The authors suggested that the

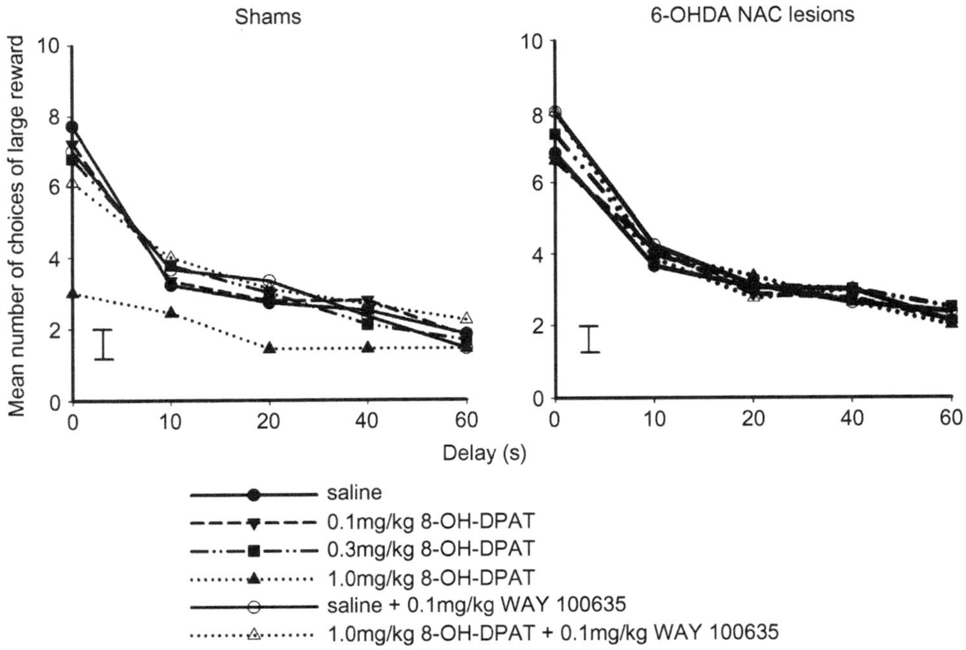

Figure 3 Hyperbolic temporal discounting in the rat as affected by an acute dose of the 5-HT1A receptor agonist 8-OHDPAT. The steeper discounting (i.e., impulsive responding) produced by this drug is abolished by prior dopamine depletion in the nucleus accumbens, which by itself has no effect. Reproduced from Winstanley *et al.* (2005).

inter-temporal choice task, which was questionnaire-based, was perhaps insufficiently sensitive to detect effects of altered 5-HT. A later study used an inter-temporal choice task with experiential delays (as are used in the animal studies reviewed above), and found that ATD increased choices for the smaller, sooner reward, paralleling at least some of the complex findings from rat studies (Schweighofer *et al.*, 2008). An fMRI study using the same task suggested that ATD increases impulsive inter-temporal choice by enhancing activity in the ventral striatum during short-term reward prediction (Tanaka *et al.*, 2007). In the same study, augmenting 5-HT function with tryptophan supplementation enhanced activity in the dorsal striatum during long-term reward prediction. It will be interesting to see if these data can be confirmed in animal studies.

In summary, as in the case of the premature responding in the 5-CSRTT, impulsive behavior was modulated by 5-HT mechanisms. Although the relevant brain regions still have not been fully elucidated, available data again point to a cortico-striatal mediation, as well as to complex interactions with dopamine, probably at both a cortical and striatal level. Evidence for a strong version of the 5-HT hypothesis of behavioral inhibition is lacking. Instead, the effects of 5-HT manipulations on impulsive choice appear to be more related to altered representations of instrumental outcomes.

5-HT and decision-making cognition

Representing reward and punishment contingencies of decision outcomes is of critical importance in gambling and risk-taking tasks, where subjects must weigh the relative potential costs and benefits of different response options. An early study with the Cambridge Gamble Task found that ATD reduced choices for the most probable outcome (i.e., the outcome with the highest expected value) (Rogers *et al.*, 1999a), suggesting that 5-HT is important for integrating information about the value and likelihood of choice outcomes. However, a subsequent study using the same task found the opposite effect; ATD *increased* choices for the most probable outcome (Talbot *et al.*, 2006). Either way, these results imply that 5-HT modulates the influence of reinforcement contingencies on choice. This process may be supported by a serotonergic modulation of the representation of reward magnitudes. Supporting this view, Rogers *et al.* (2003) reported that ATD reduced discrimination between magnitudes of expected rewards associated with different choice outcomes in a risk-taking task. Enhancing 5-HT function with tryptophan supplementation did not improve reward discrimination, however. Instead, tryptophan supplementation reduced framing effects in the loss domain – following tryptophan supplementation; subjects chose certain small losses over uncertain larger losses significantly more frequently than the placebo group (Murphy *et al.*, 2009). The authors interpreted this finding as evidence that tryptophan supplementation reduced loss aversion, which fits with other work linking reduced 5-HT to enhanced punishment processing. The neural correlates of these behavioral effects remain unclear, and it is still not well understood how 5-HT modulates the impact of expected rewards and punishments on decision-making cognition.

Social decision-making also involves weighing short-term desires against long-term goals (e.g., compliance with social norms). To investigate decision-making cognition in social contexts, researchers typically set up games with multiple players. In these games, subjects must decide whether or not to cooperate with, trust, or retaliate against their partner. Social decision-making has been linked to 5-HT function. Chronic (2-week) pre-treatment with the SSRI citalopram enhanced cooperative decisions in the Prisoner's Dilemma game (Tse and Bond, 2002); ATD had the opposite effect (Wood *et al.*, 2006). In the Ultimatum Game, ATD increased retaliation against unfair treatment (Crockett *et al.*, 2008). Notably, the effects of ATD on Ultimatum Game decision-making mirror the effects of lesions to the ventromedial prefrontal cortex (Koenigs and Tranel, 2007), suggesting a role for 5-HT in modulating decision-making cognition in this brain region.

5-HT and compulsivity

A clinical distinction is often made between impulsivity and compulsivity (Hollander and Rosen, 2000). In the latter case, we may define this as the inappropriate perseveration of responding, despite adverse consequences. Whereas ADHD is perhaps the most characteristic clinical example of impulsivity, obsessive-compulsive disorder is the paradigmatic example of compulsivity. Like impulsivity, compulsivity can be measured in a variety of ways and is probably not a unitary construct. For example, compulsivity can be measured in some sense by a failure to stop responding, and so the SSRTT may have an ambiguous position in this respect. However, perseveration in tasks in which altered reinforcement contingencies dictate that it is appropriate to switch responding are generally used to characterize compulsivity. The simplest version of this is extinction (reinforcement omission); increased resistance to extinction can be taken as an example of perseverative responding likely to involve inhibitory processes, given that modern conceptions of extinction suggest that it involves an active inhibitory process that suppresses stimulus–reward or action–outcome contingencies. Another measure is provided by discrimination reversal learning, in which previous associations between one stimulus and a rewarding outcome, and the other and reward omission or punishment, are reversed, so that the animal has to adapt appropriately to the altered environmental situation. Failure to reverse may in part occur because of a failure to learn the new stimulus–response associations; this learning is prevented if the animal persists in responding to the previously reinforced stimulus.

Reversal learning and 5-HT

Reversal learning is known to depend on an OFC-striatal system in monkeys and rats. Fiber-sparing lesions of the OFC impair reversal learning, whether between visual objects in rhesus monkeys, marmosets (Dias *et al.*, 1996) or rats (Chudasama and Robbins, 2003), or between spatial locations in the rat (Boulougouris *et al.*, 2007). Part of the deficit in each case appears to be a tendency to perseverate in the previously reinforced response. In the past few years, it has become apparent that the ascending 5-HT systems powerfully modulate reversal learning at the level of the OFC. Thus, in marmoset monkeys, depletion of 5-HT selectively from the PFC produces striking deficits in reversal learning, particularly in serial reversal, where the contingencies are continually reversed upon attainment of the learning criterion (Clarke *et al.*, 2004, 2005). Micro-analysis of responding during the reversal stage shows that the deficit is at least partly because of a perseveration of responding to the previously rewarded stimulus immediately following contingency reversal, implicating the failure of a response inhibitory control process. A further experiment has shown that the deficit is clearly perseverative in the sense that the 5-HT-depleted monkeys do not disengage from a previously reinforced stimulus; if it is removed and replaced by a novel stimulus, discrimination learning is normal (Clarke *et al.*, 2007). However, it is apparent that the 'perseverative' deficit again is more specific than would be expected of a global disinhibitory problem. First, performance on another type of shifting task, the extra-dimensional shift, which is more dependent on the lateral PFC, is *not* impaired by PFC 5-HT depletion (Clarke *et al.*, 2005). Second, an obvious way of explaining the perseveration is in terms of retarded resistance to extinction. However, a recent experiment has ruled out this mechanism, as OFC 5-HT depletion failed to affect extinction of a single (non-discriminated) rewarded response – although OFC DA loss did produce massive resistance to extinction (Walker *et al.*, 2008). A subsequent experiment on how PFC 5-HT loss affects the acquisition of control over behavior by conditioned stimuli (i.e., conditioned reinforcers) has indicated that the 5-HT loss appears to produce a 'bias' towards some aspects of the stimulus which can retard acquisition of new responding – this again contrasts with the effects of OFC DA loss which is associated with no such bias.

In more general terms, the link of 5-HT to reversal learning has been strikingly confirmed by other studies in primates – for example, linking the short-armed alleles of the 5-HT transporter polymorphism to inferior reversal learning (Izquierdo *et al.*, 2007). In humans, ATD slows response times (Murphy *et al.*, 2002) and increases errors to criterion during reversal learning (Park *et al.*, 1994; Rogers *et al.*, 1999b; Murphy *et al.*, 2002). Talbot *et al.* (2006) failed to find any significant effects of tryptophan depletion on reversal learning, but this may simply highlight possible heterogeneity of response due to genetic variation in response to low tryptophan.

One challenge that arises in studying reversal learning in humans is that simple versions of the task (such as those used with rats and marmosets) are typically not sufficiently difficult to be sensitive to drug manipulations. To address this issue, a probabilistic version of the reversal learning task was developed. In the probabilistic reversal task, subjects receive misleading feedback on 20 percent of trials. The behavioral and neural effects of 5-HT manipulations on probabilistic reversal learning in humans appear to be somewhat distinct from the effects of global prefrontal 5-HT depletion on classic reversal learning in rats and marmosets. A neuroimaging study of reversal learning in humans did not observe any effects of ATD on perseverative responding or OFC activity (Evers *et al.*, 2005). Instead, the authors found that ATD enhanced activity in the dorsomedial PFC in response to negative feedback throughout the task. Another study using the same task showed that acute SSRI (citalopram) administration impaired performance by increasing the tendency to switch following misleading negative feedback, which fits with the hypothesis that low 5-HT heightens responding to negative feedback (Chamberlain *et al.*, 2006). Similar behavioral and neural responses have been observed in depressed patients, and this mechanism is thought to underlie the 'catastrophic response to perceived failure' seen in depression (Taylor Tavares *et al.*, 2008).

In summary, 5-HT in the OFC appears responsible for biasing responding to particular aspects of the stimulus without necessarily being involved in control of the vigor of responding during extinction, whereas OFC DA has the opposite role. This dissociation may, of course, not hold for subcortical structures to which the OFC projects, such as the medial striatum, which is, however, also implicated in reversal learning (Clarke *et al.*, 2008).

Thus far, we have been unable to examine which 5-HT receptors are implicated in these behavioral effects in marmosets. However, we have begun such experiments in the rat, by using a simple spatial reversal paradigm which is also sensitive to OFC lesions (Boulougouris *et al.*, 2007). Systemic administration of the 5-HT_{2A} receptor antagonist M100907 impairs reversal on this task, whereas the 5-HT_{2C} receptor antagonist SB-242084 again had the opposite effect, in this case of facilitation of reversal learning – and in both cases, the effects appear to be mediated, at least in part, by perseverative responding (Boulougouris *et al.*, 2008). These findings may be theoretically important, as they appear to provide problems for the hypothesis that 5-HT is implicated in behavioral inhibition. As Figure 2 shows, whereas the 5-HT_{2A} receptor antagonist ameliorates impulsive behavior, it induces perseverative responding during reversal. By contrast, the 5-HT_{2C} receptor antagonist exacerbates impulsive responding, but facilitates reversal learning. It is difficult to see how a unitary response inhibition mechanism could account for this. Either there is a common effect on another mediating variable (e.g., anxiety or enhanced aversiveness) which accounts for both effects, or they are controlled by separate neuronal circuitries with differing involvement of 5-HT receptors. Given the involvement of an OFC-medial striatal circuitry in aspects of compulsive behavior, and the IL-PFC-nucleus accumbens in impulsive responding, this does appear plausible. A clear conclusion, however, is that impulsivity and compulsivity do appear to be distinct behavioral constructs, and probably have separate if interactive roles in the explanation of such syndromes as OCD and ADHD, and also substance abuse.

Conclusions

We have reviewed studies in experimental animals and humans, mainly from the perspective of our own studies, on the role of 5-HT in inhibitory control and reinforcement processing, relevant to such constructs as impulsivity, compulsivity and anxiety. Progress has clearly been made with respect to theories of central 5-HT function, although, as initially admitted, it is very difficult to arrive at a unitary role for such a complex system. Nevertheless, a few generalizations can be made. The functions of 5-HT in global behavioral inhibition are increasingly questioned; however, 5-HT depletion may especially 'disinhibit' responding in situations in which 'waiting' is required, perhaps through an exaggeration of the negative aversive qualities of the waiting period. A comparison of effects of 5-HT manipulations on paradigms tapping 'impulsivity' and 'compulsivity' suggests again that a behavioral construct such as 'inhibition' cannot adequately account for the findings. The effects of 5-HT loss on perseveration in monkeys also argue against a simple inhibitory account, in favor of one suggesting a modulation of the OFC in the control of stimulus preference. This theme that manipulations of 5-HT can bias the emotional processing of salient stimuli is consistent with many of the effects of 5-HT manipulations in humans, where, for example, mild and transient 5-HT loss leads to a bias to respond to negative feedback and to exhibit heightened aversive responses to emotional stimuli. We have also raised the possibility of specific effects of 5-HT manipulations on higher cognitive processes such as decision-making cognition, suggesting specific functions for this neurotransmitter in brain regions such as the prefrontal cortex. Here, the question is whether such effects ultimately derive from actions of 5-HT in lower level processes. The resolution of these paradoxes has been part and parcel of research on 5-HT functions in health and disease over the past 50 years or so. We assume that this will be possible as we understand the impact of current manipulations in more detail in the context of those neural systems on which they impinge, as new methods for studying the role of 5-HT emerge, and as knowledge about the precise functions of the different neural systems modulated by 5-HT accrues.

Acknowledgements

This work was mainly supported by a Wellcome Trust Programme Grant awarded to T.W. Robbins, B.J. Everitt, A.C. Roberts and B.J. Sahakian, and completed within the Behavioural and Clinical Neuroscience Institute supported by a joint grant from the MRC UK and the Wellcome Trust. Molly Crockett is a Gates Scholar.

References

Anderson, I.M., Mckie, S., Elliott, R., Williams, S.R. and Deakin, J.F.W. (2008) Assessing human 5-HT function *in vivo* with pharmacoMRI. Neuropharmacology, 55: 1029–1037.

Aron, A.R. and Poldrack, R.A. (2006) Cortical and subcortical contributions to Stop signal response inhibition: role of the subthalamic nucleus. J. Neurosci., 26: 2424–2433.

Bari, A., Eagle, D.M., Mar, A.C., Robinson, E.S.J. and Robbins, T.W. (2009) Dissociable effects of noadrenaline, dopamine and serotonin uptake blockade on stop task performance in rats. Psychopharmacology, 205; 273–283.

Bel, N. and Artigas, F. (1996) Reduction of serotonergic function in rat brain by tryptophan depletion: effects in control and fluvoxamine-treated rats. J. Neurochem., 67: 669–676.

Berg, K.A., Clarke, W.P. and Cunningham, K.A. (2008) Fine-tuning serotonin 2_C receptor function in the brain: molecular and functional implications. Neuropsychopharmacology, 55: 969–976.

Blier, P., Pineyro, G., El Mansari, M., Bergeron, R. and De Montigny, C. (1998) Role of somatodendritic 5-HT autoreceptors in modulating 5- HT neurotransmission. Ann. NY Acad. Sci., 861: 204–216.

Boulougouris, V., Dalley, J.W. and Robbins, T.W. (2007) Effects of orbitofrontal, infralimbic and prelimbic cortical lesions on serial spatial reversal learning in the rat. Behav. Brain Res., 179: 219–228.

Boulougouris, V., Glennon, J.C. and Robbins, T.W. (2008) Dissociable effect of selective 5-HT2A and 5-HT2C receptor antagonists on serial spatial reversal learning in rats. Neuropsychopharmacology, 33: 2007–2019.

Bymaster, F.P., Zhang, W., Carter, P.A., Shaw, J., Chernet, E., Phebus, L., Wong, D.T. and Perry, K.W. (2002) Fluoxetine, but not other selective serotonin uptake inhibitors, Increases norepinephrine and dopamine extracellular levels in prefrontal cortex. Psychopharmacology, 160: 353–361.

Calcagno, E., Carli, M., Baviera, M. and Invernizzi, R.W. (2009) Endogenous serotonin and serotonin-2_C receptors are involved in the ability of M100207 to suppress cortical glutamate release induced by NMDA receptor blockade. J. Neurochem., 108: 521–532.

Cardinal, R.N., Pennicott, C., Sugathapala, C.L., Robbins, T.W. and Everitt, B.J. (2001) Impulsive choice induced in rats by lesions of the nucleus accumbens core. Science, 292: 2499–2501.

Carli, M., Baviera, M., Invernizzi, R.W. and Balducci, C. (2006) Dissociable contribution of 5-HT1A and 5-HT2A receptors in the medial prefrontal cortex to different aspects of executive control such as impulsivity and compulsive perseveration in rats. Neuropsychopharmacology, 31: 757–767.

Carpenter, L.L., Anderson, G.M., Pelton, G.H., Gudin, J.A., Kirwin, P.D.S., Price, L.H., Heninger, G.R. and McDougle, C.J. (1998) Tryptophan depletion during continuous CSF sampling in healthy human subjects. Neuropsychopharmacology, 19: 26–35.

Chamberlain, S.R., Müller, U., Blackwell, A.D., Clark, L., Robbins, T.W. and Sahakian, B.J. (2006) Neurochemical modulation of response inhibition and probabilistic learning in humans. Science, 311: 861–863.

Chudasama, Y. and Robbins, T.W. (2003) Dissociable contributions of the orbitofrontal and infralimbic cortex in pavlovian autoshaping and discrimination reversal learning: Further evidence for the functional heterogeneity of the rodent frontal cortex. J. Neurosci., 23: 8771–8780.

Chudasama, Y., Passetti, F., Desai, A., Rhodes, S., Lopian, D. and Robbins, T.W. (2003) Dissociable aspects of performance on the 5 choice serial reaction time task following lesions of the dorsal anterior cingulate, infralimbic and orbitofrontal cortex in the rat: differential effects on selectivity, impulsivity and compulsivity. Behav. Brain Res., 146: 105–119.

Clark, L., Roisier, J.P., Cools, R., Runinzstein, D.C., Sahakian, B.J. and Robbins, T.W. (2005) Stop signal response inhibition is not modulated by tryptophan depletion or the serotonin transporter polymorphism in healthy volunteers: implications for the 5-HT theory of impulsivity. Psychopharmacology, 182: 570–578.

Clarke, H.F., Dalley, J.W., Crofts, H.S., Robbins, T.W. and Roberts, A.C. (2004) Cognitive inflexibility after prefrontal serotonin depletion. Science, 304: 878–880.

Clarke, H.F., Walker, S.C., Crofts, H.S., Dalley, J.W., Robbins, T.W. and Roberts, A.C. (2005) Prefrontal serotonin depletion affects reversal learning but not attentional set shifting. J. Neurosci., 12: 532–538.

Clarke, H.F., Walker, S.C., Dalley, J.W., Robbins, T.W. and Roberts, A.C. (2007) Cognitive inflexibility after prefrontal serotonin depletion is behaviourally and neurochemically specific. Cereb. Cortex, 17: 18–27.

Clarke, H.F., Robbins, T.W. and Roberts, A.C. (2008) Lesions of the medial striatum in monkeys produce perseverative impairments during reversal learning similar to those produced by lesions of the orbitofrontal cortex. J. Neurosci., 28: 10972–10982.

Conn, P.J. and Sanders-Bush, E. (1987) Relative efficacies of piperazines at the phosphoinositide hydrolysis-linked serotonergic (5-HT-2 and 5-HT-1C) receptors. J. Pharm. Exp. Ther., 142: 552–557.

Crean, J., Richards, J.B. and de Wit, H. (2002) Effect of tryptophan depletion on impulsive behavior in men with or without a family history of alcoholism. Behav. Brain Res., 136: 349–357.

Crockett, M.J., Clark, L., Tabibnia, G., Lieberman, M.D. and Robbins, T.W. (2008) Serotonin modulates behavioural reactions to unfairness. Science, 320: 1739.

Dalley, J.W., Theobald, D.E., Eagle, D.M., Passettii, F. and Robbins, T.W. (2002) Deficits in impulse control associated with tonically elevated serotonergic function in rat prefrontal cortex. Neuropsychopharmacology, 26: 716–728.

Deakin, J.F.W. and Graeff, F.G. (1991) 5-HT and mechanisms of defence. J. Psychopharmacol., 5: 305–315.

Del-Ben, C.M., Deakin, J.F.W., Mckie, S., Deivai, N.A., Williams, S.R., Elliott, , Dolan, M. and Anderson, I.M. (2005) The effect of citalopram pretreatment on neuronal responses to neuropsychological tasks in normal volunteers: an fMRI study. Neuropsychopharmacology, 30: 1724–1734.

Dias, R., Robbins, T.W. and Roberts, A.C. (1996) Dissociation in prefrontal cortex of affective and attentional shifts. Nature, 380: 69–72.

Eagle, D.M. and Robbins, T.W. (2003) Inhibitory control in rats performing on the stop-signal reaction time task: effects of lesions of the medial striatum and d-amphetamine. Behav. Neurosci., 117: 1302–1317.

Eagle, D.M., Bari, A. and Robbins, T.W. (2008a) The neuropsychopharmacology of action inhibition: cross-species translation of the stop-signal and go/no-go tasks. Psychopharmacology, 199: 439–456.

Eagle, D.M., Lehmann, O., Theobald, D.E.H., Pena, Y., Zakaria, R., Ghosh, R., Dalley, J.W. and Robbins, T.W. (2008b) Serotonin depletion impairs waiting but not stop-signal reaction time in rats: implications for theories of the role of 5-HT in behavioral inhibition. Neuropsychopharmacology. doi:10.1038/npp.2008.202

Evers, E.A.T., van der Veen, F.M., Jolles, J., Deutz, N.E.P. and Schmitt, J.A.J. (2006) Acute tryptophan depletion improves performance and modulates the BOLD response during a Stroop task in healthy females. Neuroimage, 32: 248–255.

Fletcher, P.J., Tampakeras, M., Sinyard, J. and Higgins, G.A. (2007) Opposing effects of 5-HT2_A and 5-HT2_C receptor antagonists in the rat and mouse on premature responding in the five-choice serial reaction time task. Psychopharmacology, 195: 223–234.

Gray, J. (1982) The Neuropsychology of Anxiety. Oxford University Press, London.

Harrison, A.A., Everitt, B.J. and Robbins, T.W. (1997a) Central 5-HT depletion enhances impulsive responding without affecting the accuracy of attentional performance: interactions with dopaminergic mechanisms. Psychopharmacology, 133: 329–342.

Harrison, A.A., Everitt, B.J. and Robbins, T.W. (1997b) Doubly dissociable effects of median- and dorsal-raphé lesions on the performance of the 5-choice serial reaction time test of attention in rats. Behav. Brain Res., 89: 135–149.

Harrison, A.A., Everitt, B.J. and Robbins, T.W. (1999) Central serotonin depletion impairs both the acquisition and performance of a symmetrically reinforced go/no-go conditional visual discrimination. Behav. Brain Res., 100: 99–112.

Hollander, E. and Rosen, J. (2000) Impulsivity. J. Psychopharmacol., 14: S39–S44.

Izquierdo, A., Newman, T.K., Higley, J.D. and Murray, E.A. (2007) Genetic modulation of cognitive flexibility and socioemtional behaviour in rhesus monkeys. Proc. Nat. Acad. Sci., 104: 14128–14133.

Koenigs, M. and Tranel, D. (2007) Irrational economic decision-making after ventromedial prefrontal damage: evidence from the Ultimatum Game. J. Neurosci., 27: 951–956.

Koskinen, T., Ruotsaleinen, S., Puumala, T., Lappalainen, R., Koivisto, E., Mannisto, P.T. and Sirvio, J. (2000) Activation of 5-HT2A receptors impairs response control of rats in a 5-choice serial reaction time task. Neuropsychopharmacology, 39: 471–481.

LeMarquand, D.G., Benkelfat, C., Pihl, R.O., Palmour, R.M. and Young, S.N. (1999) Behavioral disinhibition induced by tryptophan depletion in non-alcoholic young men with multigenerational family histories of paternal alcoholism. Am. J. Psychiatry, 156: 1771–1779.

Liu, S., Bubar, M.J., lanfranc, M.F., Hillman, G.R. and Cunningham, K.A. (2007) Serotonin 2_C receptor localisation in GABA neurons of the rat medial prefrontal cortex: implications for understanding the neurobiology of addiction. Neuroscience, 146: 1677–1688.

Logan, G.D., Cowan, W.B. and Davis, K.A. (1984): On the ability to inhibit simple and choice reaction time responses: a model and a method. J. Exp. Psycho. Hum. Percept. Perform., 10: 276–291.

Mobini, S., Chiang, T.J., Ho, M.Y., Bradshaw, C.M. and Szabadi, E. (2000) Effects of central 5-hydroxytryptamine depletion on sensitivity to delayed and probabilistic reinforcement. Psychopharmacology, 152: 390–397.

Murphy, E.R., Dalley, J.W. and Robbins, T.W. (2005) Local glutamate receptor antagonism in the rat prefrontal cortex disrupts response inhibition in a visuospatial attentional task. Psychopharmacology, 179: 99–107.

Murphy, F.C., Smith, K.A., Cowen, P.J., Robbins, T.W. and Sahakian, B.J. (2002) The effects of tryptophan depletion on cognitive and affective processing in healthy volunteers. Psychopharmacology, 163: 42–53.

Murphy, S.E., Longhitano, C., Ayres, R.E., Cowen, P.J., Harmer, C.J. and Rogers, R.D. (2009) The role of serotonin in nonnormative risky choice: The effects of tryptophan supplements on loss-aversion in healthy adult volunteers. J. Cogn. Neurosci., in press

Navarra, R., Comery, T.A., Graf, R., Rosenzweig-Lipson, S. and Day, M. (2008) The 5-HT_{2C} receptor agonist WAY-163909 decreases impulsivity in the 5-choice serial reaction time test. Behav. Brain Res., 188: 412–415.

Park, S.B., Coull, J.T., McShane, R.H., Young, A.H., Sahakian, B.J., Robbins, T.W. and Cowen, P.J. (1994) Tryptophan depletion in normal volunteers produces selective impairments in learning and memory. Neuropharmacology, 33: 575–588.

Passetti, F., Dalley, J.W. and Robbins, T.W. (2003) Double dissociation of serotoninergic and dopaminergic mechanisms in attentional performance using a rodent five choice reaction time task. Psychopharmacology, 165: 136–145.

Pompeiano, M., Palacios, J.M. and Mengod, G. (1994) Distribution of the serotonin receptor 5-HT2 receptor family mRNAs: comparison between 5-HT2A and 5-HT2C receptors. Mol. Brain Res., 23: 163–178.

Robbins, T.W. (2002) The five-choice serial reaction time task: behavioural pharmacology and functional neurochemistry. Psychopharmacology, 163: 362–380.

Robinson, E.S.J., Eagle, D.M., Mar, A.C., Bari, A., Banerjee, G., Jiang, J.S., Dalley, J.W. and Robbins, T.W. (2008a) Similar effects of the noradrenaline reuptake inhibitor atomoxetine on three distinct forms of impulsivity in the rat. Neuropsychopharmacology, 33: 1028–1037.

Robinson, E.S.J., Dalley, J.W., Theobald, D.E.H., Glennon, J.C., Pezze, M.A., Murphy, E.R. and Robbins, T.W. (2008b) Opposing roles for 5-HT2A and 5-HT2C receptors in the nucleus accumbens on inhibitory response control in the 5-choice serial reaction time task. Neuropsychopharmacology, 33: 2398–2406.

Rogers, R.D., Everitt, B.J., Baldacchino, A., Blackshaw, A.J., Swainson, R., Wynne, K., Baker, N.B., Hunter, J., Carthy, T., Booker, E., London, M., Deakin, J.F.W., Sahakian, B.J. and Robbins, T.W. (1999a) Dissociable deficits in the decision-making cognition of chronic amphetamine abusers, opiate abusers, patients with focal damage to prefrontal cortex, and tryptophan-depleted normal volunteers: evidence for monoaminergic mechanisms. Neuropsychopharmacology, 20: 322–339.

Rogers, R.D., Blackshaw, A.J., Middleton, H.C., Matthews, K., Hawtin, K., Crowley, C., Hopwood, A., Wallace, C., Deakin, J.F.W., Sahakian, B.J. and Robbins, T.W. (1999b) Tryptophan depletion impairs stimulus-reward learning while methylphenidate disrupts attentional control in healthy young adults: implications for the monoaminergic basis of impulsive behaviour. Psychopharmacology, 146: 482–491.

Rogers, R.D., Tunbridge, E.M., Bhagwagar, Z., Drevets, W.C., Sahakian, B.J. and Carter, C.S. (2003) Tryptophan depletion alters the decision-making of healthy volunteers through altered processing of reward cues. Neuropsychopharmacology, 28: 153–162.

Rubia, K., Lee, F., Cleare, A.J., Tunstall, N., Fu, C.H., Brammer, M. and McGuire, P. (2005) Tryptophan depletion reduces right inferior prefrontal activation during response inhibition in fast, event-related fMRI. Psychopharmacology, 179: 791–803.

Santana, N., Bortolozzi, A., Serrats, J., Mengod, G. and Artigas, F. (2004) Expression of serotonin1A and serotonin2A receptors in pyramidal and GABAergic neurons of the rat prefrontal cortex. Cereb. Cortex, 14: 1100–1109.

Scholes, K.E., Harrison, B.J., O'Neill, B.V., Leung, S., Croft, R.J., Pipingas, A., Phan, K.L. and Nathan, P.J. (2006) Acute serotonin and dopamine depletion improves attentional control: findings from the Stroop Task. Neuropsychopharmacology, 32: 1600–1610.

Schweighofer, N., Bertin, M., Shishida, K., Okamoto, Y., Tanaka, S.C., Yamawaki, S. and Doya, K. (2008) Low-serotonin levels increase delayed reward discounting in humans. J. Neurosci., 28: 4528–4532.

Soubrié, P. (1986) Reconciling the role of central serotonin neurones in human and animal behaviour. Behav. Brain Res., 9: 319–364.

Spinks, D. and Spinks, G. (2002) Serotonin reuptake inhibition: an update on current research strategies. Curr. Med. Chem., 9: 799–810.

Talbot, P.S., Watson, D.R., Barrett, S.L. and Cooper, S.J. (2006) Rapid tryptophan depletion improves decision-making cognition in healthy humans without affecting reversal learning or set shifting. Neuropsychopharmacology, 31: 1519–1525.

Tanaka, S.C., Schweighofer, N., Asahi, S., Shishida, K., Okamoto, Y., Yamawaki, S. and Doya, K. (2007) Serotonin differentially regulates short- and long-term prediction of rewards in the ventral and dorsal striatum. PLoS ONE, 2: e1333.

Taylor, M.J., Freemantle, N., Geddes, J.R. and Bhagwagar, Z. (2006) Early onset of selective serotonin reuptake inhibitor antidepressant action: systematic review and meta-analysis. Arch. Gen. Psychiatry, 63: 1217–1223.

Taylor Tavares, J.V., Clark, L., Furey, M.L., Williams, G.B., Sahakian, B.J. and Drevets, W.C. (2008) Neural basis of abnormal response to negative feedback in unmedicated mood disorders. Neuroimage, 42: 1118–1126.

Tse, W.S. and Bond, A.J. (2002) Serotonergic intervention affects both social dominance and affiliative behaviour. Psychopharmacology, 161: 324–330.

Walderhaug, E., Lunde, H., Nordvik, J.E., Landro, N.I., Refsum, H. and Magnusson, A. (2002) Lowering of serotonin by rapid tryptophan depletion increases impulsiveness in normal individuals. Psychopharmacology, 164: 385–391.

Walderhaug, E., Magnusson, A., Neumeister, A., Lappalainen, J., Lunde, H., Refsum, H. and Landro, N.I. (2007) Interactive effects of sex and 5-HTTLPR on mood and impulsivity during tryptophan depletion in healthy people. Biol. Psychiatry, 62: 593–599.

Walker, S.C., Robbins, T.W. and Roberts, A.C. (2008) Differential contributions of dopamine and serotonin to orbitofrontal cortex function in the marmoset. Cereb. Cortex. 19: 889–898.

Ward, B.O., Wilkinson, L.S., Robbins, T.W. and Everitt, B.J. (1999) Forebrain serotonin depletion facilitates the acquisition and performance of a conditional visual discrimination task in rats. Behav. Brain Res., 100: 51–65.

Winstanley, C.A., Dalley, J.W., Chudasama, Y., Theobald, D.E.H. and Robbins, T.W. (2003) Intra-prefrontal 8-OH-DPAT and M100907 improve visuospatial attention and decrease impulsivity on the five choice serial reaction time task in rats. Psychopharmacology, 167: 304–314.

Winstanley, C.A., Dalley, J.W., Theobald, D.E.H. and Robbins, T.W. (2004a) Fractionating impulsivity: contrasting effects of central 5-HT depletion on different measures of impulsive behaviour. Neuropsychopharmacology, 29: 1331–1343.

Winstanley, C.A., Theobald, D.E.H., Dalley, J.W., Glennon, J.C. and Robbins, T.W. (2004b) 5-HT2A and 5-HT2C receptor antagonists have opposing effects on a measure of impulsivity: interactions with global 5-HT depletion. Psychopharmacology, 176: 376–385.

Winstanley, C.A., Theobald, D.E., Dalley, J.W. and Robbins, T.W. (2005) Interactions between serotonin and dopamine in the control of impulsive choice in rats: therapeutic implications for impulse control disorders. Neuropsychopharmacology, 30: 669–682.

Winstanley, C.A., Theobald, D.E.H., Dalley, J.W., Cardinal, R.N. and Robbins, T.W. (2006) Double dissociation between serotonergic and dopaminergic modulation of medial prefrontal and orbitofrontal cortex during a test of impulsive choice. Cereb. Cortex, 16: 106–114.

Wise, C.D., Stein, L. and Berger, B.D. (1972) Benzodiazepines – reducing anxiety by reduction of serotonin turnover in brain. Science, 177: 180–183.

Wood, R.M., Rilling, J.K., Sanfey, A.G., Bhagwagar, Z. and Rogers, R.D. (2006) Effects of tryptophan depletion on the performance of an iterated Prisoner's Dilemma game in healthy adults. Neuropsychopharmacology, 31: 1075–1084.

Young, S.N. and Gauthier, S. (1981) Effect of tryptophan administration on tryptophan, 5-hydroxyindoleacetic acid and indoleacetic acid in human lumbar and cisternal cerebrospinal fluid. J. Neurol. Neurosurg. Psychiatry, 44: 323–328.

CHAPTER 3.9

Experimental Studies on the Role(s) of Serotonin in Learning and Memory Functions

Jean-Christophe Cassel*

Laboratoire d'Imagerie et de Neurosciences Cognitives, Université de Strasbourg FRE 3289 UdS-CNRS, IFR 37 des Neurosciences, Strasbourg, France

Abstract: In invertebrates, serotonin (5-HT) is involved in both non-associative and associative learning, and thus memory formation. Indeed, as shown in *Aplysia californica*, snails, leeches, etc., whereas the formation of memories is more difficult after 5-HT depletion, it is facilitated when serotonin functions are stimulated. Although it is generally considered that the evolution can be relatively conservative, the overall picture is much more complicated in higher vertebrates: when 5-HT systems are selectively damaged, the lesions have behavioral consequences, as can be evidenced, for instance, in rats or mice (e.g., on impulsivity), but learning is still possible, and most often memory capabilities appear normal. Serotonergic functions may also be enhanced by pharmacological tools, such as drugs inhibiting the reuptake of 5-HT (SSRIs). Although there is no consensus in the literature, SSRIs may have two types of effects according to whether they are given before or after training. When SSRIs are given before training, memory functions appear unchanged in some reports or altered in others; exceptionally, they can be improved. Given immediately after training, such drugs have no effect on memory or may improve it. In higher vertebrates, modifications of 5-HT functions have nevertheless consequences on the mnesic impact of treatments affecting other neurotransmitter functions, such as blockade of cholinergic receptors or cholinergic lesions. Therefore, with regard to learning and memory, one of the roles of the ascending serotonergic system taken as a whole could be to modulate more direct functional implications of other neurotransmitter systems. These modulations could operate in a region-specific and perhaps receptor-specific manner.

Keywords: 5-HT, invertebrates, learning, memory, serotonin, vertebrates.

Abbreviations: 5-HT, 5-hydroxytryptamine (or serotonin); 5,6-DHT, 5,6-dihydroxytryptamine; 5,7-DHT, 5,7-dihydroxytryptamine; CNS, central nervous system; cAMP, cyclic adenosine monophosphate; CREB, cAMP response element-binding; GABA, gamma-amino butyric acid; MAPK, mitogen-activated protein kinase; MDMA, 3,4-methylenedioxymethamphetamine (or 'ecstasy'); pCA, parachloroamphetamine; PCPA, paracholorophenylalanine; PKA, protein kinase A; SSRI(s), selective serotonin reuptake inhibitor(s); THA, tetrahydroaminoacridine; THC, Δ^9-tetrahydrocannabinol.

Introduction

Since (1) the discovery by Vittorio Erspamer and Maffo Vialli of the substance they named enteramine (Erspamer and Vialli, 1937); (2) the isolation and chemical characterization of this indole by Irvine Page, Arda Alden Green and Maurice Rapport (Rapport *et al.*, 1948a, 1948b; Rapport, 1949), who named it serotonin (5-HT); (3) its subsequent discovery as a neurotransmitter in the invertebrate smooth muscle by Betty Mack Twarog (Twarog and Page, 1953); and, finally, (4) the suspicion that it might be a neurotransmitter involved in mental illness by Dilworth Wayne Wooley (Wooley and Shaw, 1954), research on 5-HT has been the focus of sustained and widespread investigations extending ramifications in all possible corridors of basic and applied pharmacology (neuro)psycho pharmacology, psychiatry and neuroscience. The fact that virtually all regions of the mammalian brain receive serotonergic afferents arising from two relatively small nuclei of the midbrain, namely the dorsal and medial raphe – also called B6–B7 and B8–B9 groups in Dahlström and Fuxe's (1964) nomenclature – is probably not the least reason why 5-HT has generated such a fascination in the scientific community. Not only is it one of the most ancient signaling

*E-mail: jean-christophe.cassel@linc.u-strasbg.fr

ISBN 978-0-12-374634-4
DOI: 10.1016/B978-0-12-374634-4.00026-5

molecules (Azmitia, 2007; Nichols and Nichols, 2008); also, to exert its pharmacological actions, this neurotransmitter has 14 different receptor targets classified into 7 families denominated 5-HT_1–5-HT_7 (see, for example, Hoyer *et al.*, 2002; Nichols and Nichols, 2008). Most of these receptors have been described to participate, in one way or another, in a large variety of physiological and behavioral functions, including pain, feeding, sleep, sexual behavior, cardiac regulation, and both cognitive and emotional processes. Actually, it has become more and more apparent that where there is behavior, even at a very primary level in the evolutionary tree, there is a place for 5-HT.

The current chapter reviews some of the data from the literature indicating that mechanisms driven or modulated by 5-HT may participate in learning and memory functions. In no case, however, has the present chapter the pretension to provide an exhaustive overview of the vast literature on the implications of 5-HT in learning and memory functions. A Pubmed search performed on 3 September 2008 using '[(serotonin OR *5-HT*) and (*memory*)]' and '[(*serotonin* OR *5-HT*) and (*learning*)]' as combinations of keywords yielded 1660 and 4197 articles or reviews, respectively. Clearly, a list of 4197 references would probably cover at least 100 pages in a typical handbook, and reading all of them in order to compile their contribution into a short book chapter not exceeding 20 pages is probably anything but a realistic enterprise. Thus, rather than trying to achieve inaccessible completeness, the chapter will focus on a series of experiments in which serotonergic systems have been manipulated, mainly in terms of lesions or drug actions, in order to understand if and to what extent they contributed to learning and memory functions. Although mice and rats are very commonly used in such experiments, the chapter begins by considering a series of experiments carried out in animals having a much simpler nervous system – namely, invertebrates. Invertebrates, as will be illustrated, also encode, consolidate and retrieve information from their environment. Therefore, the role of 5-HT in memory functions is certainly easier to investigate and describe in such (sometimes very) simple organisms than in the higher vertebrates. Ideally, this chapter should have also covered the implications, as far as they are known, of the various 5-HT receptor types and subtypes in learning and memory functions. This task, though, would probably deserve one chapter per receptor family for which data are available in the literature.

Exploring the role of serotonin in memory using invertebrate system models

Serotonin and adaptation of the gill withdrawal reflex in Aplysia californica

In 1894, Santiago Ramon Y Cajal (see also Jones, 1994) proposed that the material substrate of memory formation could be the establishment of new connections between neurons. In 1949, Hebb proposed that learning and memory might rely upon experience-triggered functional reorganizations that can become lasting within neuronal networks (Hebb, 1949). About 20 years after this proposal, one of the three 2000 Nobel Prize Laureates in Physiology or Medicine, namely Eric Kandel, was seeking a simple animal model on which cellular and molecular studies of memory substrates would become possible. He started to work on an invertebrate named *Aplysia californica*, a giant marine snail (30 cm in length and 1 kg in weight for an adult specimen) which offers three major advantages: the number of neurons in its nervous system is relatively limited (about 20,000, to be compared with the 10^{12} neurons of the human brain); these neurons are large, even gigantic (sometimes 1000 μm in diameter); and many of them are uniquely identifiable (among other reasons, because of their particular pigmentation). It is perhaps worth remembering that neuronal gigantism had already been a wonderful and extremely precious partner in the discovery of the mechanisms underlying the generation of an action potential and its propagation along the axon (see, for example, Huxley, 2002). In addition, an Aplysia displays various forms of non-associative (habituation, sensitization, dishabituation, desensitization) or associative (classical or operant conditioning) learning capabilities. Kandel focused his interests on the so-called gill withdrawal reflex, which is a protection reflex: when the siphon is subjected to the application of a tactile stimulus, the snail retracts it and withdraws its gill under the mantle shelf (Figure 1). The functional characteristics of this reflex can be altered by three types of learning: habituation (the reflex undergoes weakening as a consequence of repeated stimulations), sensitization (the reflex becomes stronger as a result of an aversive shock), and classical conditioning (a neutral stimulus, which normally produces a weak or no reflex, is able to elicit this reflex as a result of its repeated association with an unconditioned stimulus invariably producing a marked reflex).

These forms of learning were found in higher vertebrates, including humans, and undergo a conversion from a short-term storage which is sensitive to interference to a long-term storage that depends on protein synthesis and is resistant to interference, as largely described in mammals. Schematically, in the Aplysia, a short-term memory is generated when stimulations eliciting the withdrawal reflex are weak or sparse. Conversely, a long-term memory establishes when the stimulations are more intense, numerous, repeated and/or grouped. In 1976, Castellucci and Kandel reported that following sensitization of the gill-withdrawal reflex, at precisely the synapse linking sensory and motor neurons, the amount of 5-HT released in response to the application of a tactile stimulation to the siphon was increased, and that this modification could

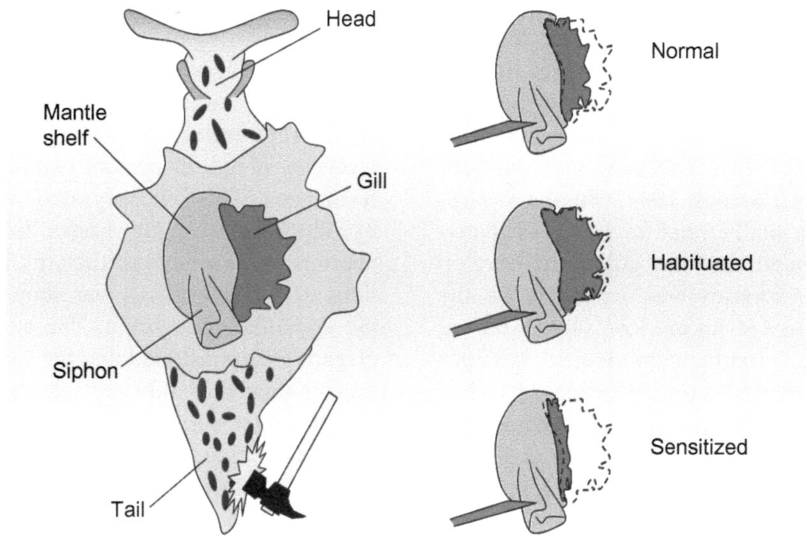

Figure 1 A dorsal view of an Aplysia showing the animal's respiratory organ: the gill. When the siphon is stimulated by the application of a tactile stimulus, the siphon is contracted and the gill is withdrawn under the mantle shelf (right, top). This withdrawal reflex is a protection reflex that undergoes habituation (right, middle) when the stimulation is repeated, or which may show sensitization after a noxious stimulus is applied to another part of the body (e.g., a hurt on the tail; right bottom). Dishabituation (not illustrated) is a term used to describe what happens once habituation has been established and a noxious stimulus is applied to another part of the body; in this case, the habituation of the withdrawal reflex is relieved. The scattered line indicates the limits of the non-retracted gill, as shown on the *Aplysia*, (left). To see the full color version of this figure please refer to the color plate in the back of the book. Copies produced via our print on demand service do not contain color plates. If your copy does not have the color plate, please go to this website to view the figure in color www.elsevierdirect.com/companions/9780123746344

last for days, and even weeks. Moreover, the sensitization response could be induced by application of 5-HT and blocked by treatment with the 5-HT_2 receptor antagonist cinanserin, or the mixed 5-HT_1/5-HT_2 antagonist cyproheptadine (Brunelli *et al.*, 1976; Mercer *et al.*, 1991). It is also noteworthy that lesions of the serotonergic neurons, which were produced with 5,7-dihydroxytryptamine (5,7-DHT; see below), reduced the synaptic facilitation response, as well as the dishabituation of the gill-withdrawal reflex (Glanzman *et al.*, 1989; see Figure 1). Although Glanzman *et al.* (1989) did not study the sensitization response in their lesion experiment, it is noteworthy that dishabituation can be induced in habituated animals using the same kind of manipulations as those leading to sensitization in non-habituated animals.

What, exactly, is the role of 5-HT in this type of learning? In fact, as concerns short-term memory, by acting on a variety of 5-HT receptor subtypes, of which four have been cloned (Barbas *et al.*, 2003), 5-HT is able to activate a second messenger system involving cAMP. 5-HT applications result in increased cAMP levels (Cedar and Schwartz, 1972), and direct injections of cAMP into presynaptic neurons is sufficient to induce presynaptic facilitation (Brunelli *et al.*, 1976). It was also shown that this facilitation involved PKA and the phosphorylation of K^+ channels (Klein and Kandel, 1980). Conversely to these 5-HT-triggered processes underlying short-term memory, the long-term memory processes, which make the memory available for days, weeks or longer, require protein synthesis – a major step of trace consolidation. In case of repeated stimulations, the 5-HT-triggered cAMP rise persists for minutes, which enables activation of PKA and, consequently, of MAPK, which are both translocated to the nucleus, where they activate a transcriptional cascade involving CREB-1. The injection of a phosphorylated form of Aplysia CREB-1a is able to initiate long-term memory (in fact, lasting synaptic facilitation) in the absence of any other manipulation (Bartsch *et al.*, 1998). Downstream to this process, the growth of new synaptic connections is initiated, and these connections may last as long as the memory is present (Bailey *et al.*, 1992a, 1992b). Thus, in the Aplysia, under the influence of an appropriate experience, a 5-HT-triggered mechanism is changing the signal transmission characteristics between sensory and motor neurons. This modification may be triggered by different molecular and cellular mechanisms which underlie short- and long-term memory of sensitization of the gill-withdrawal reflex. In both cases, there is one crucial conductor in this biochemical orchestra: 5-HT!

Serotonin and learning in other invertebrates

Interestingly, lesions of serotonergic neurons or other manipulations of serotonergic functions in other invertebrates (e.g., drosophilae, leeches, slugs, snails, worms) may also alter learning and memory capabilities (see, for example, Balaban *et al.*, 1987; Zakharov and Balaban, 1987; Ehrlich *et al.*, 1992; Sahley, 1994; Shirahata *et al.*, 2006). For instance, in a snail (*Helix lucorom*), Zakharov and Balaban (1987) found that the ability to develop avoidance-conditioned responses was dependent on the levels of 5-HT. When these levels are low, such as during the first post-hatching month, conditioning sessions do not result in memory formation. Conversely, when 5-HT levels are higher, as is the case later on in adults, the conditioning leads to memory formation. Interestingly, when 5-HT levels are depleted by 5,7-DHT treatment in adult snails, the latter again behave as do the immature ones: there is no memory formation. In a terrestrial slug (*Limax valentianus*), Shirahata *et al.* (2006) found that the memory for odor–taste associative learning could be disrupted by an about 55 percent reduction of 5-HT levels: after depletion, the slugs were unable to exhibit evidence for short-term memory (tested with a 24-hour post-acquisition delay), while their long-term memory (post-acquisition delay of 5 or 6 days) was normal. That short-term memory may be affected by 5-HT depletion whereas long-term memory is not might sound a bit surprising, if not enigmatic, especially when considering that before being consolidated for days or weeks, a memory must have existed in a kind of short-term buffer. The authors of this work nevertheless regard their findings as an indication that the mechanisms involved in each type of memory are different, one being 5-HT-dependent, the other not. These data also suggest that short- and long-term memories in slugs are processed separately from the beginning of the acquisition onwards. This interpretation appears compatible with findings showing that in Aplysia, although being both depending on 5-HT, short- and long-term memories have different processing pathways, the former being blocked by cyproheptadine, a 5-HT receptor antagonist, but not the latter (Emptage and Carew, 1993). However, this interpretation seems to ignore that short- and long-term memories are induced by different conditioning protocols (see above).

In a recent report, Sitaraman *et al.* (2008) demonstrated that 5-HT was critical for place memory formation in a fly, namely *Drosophila*. Briefly, flies were trained in a two-compartment heat box. When, during training, a fly crossed the midline in one direction, the box rapidly (within seconds) heated up to reach 41°C – a temperature the fly will spontaneously avoid when possible. When the fly again crossed the midline and returned to the first compartment, the box cooled down to 24°C. After efficient training, a normal fly learned to avoid the compartment associated with the onset of heating. Sitaraman *et al.* (2008) genetically manipulated the serotonergic and dopaminergic neurons, and found that flies in which 5-HT levels were decreased by about 80 percent and dopamine levels by about 75 percent exhibited weak memory performances. This memory deficit could be reproduced by blocking synaptic transmission from serotonergic neurons or, in wild-type flies, by pharmacological reductions of 5-HT levels. It was shown to be independent of the dopaminergic system. The authors of this work also elegantly demonstrated that the various genetic manipulations in their flies did not affect thermosensitivity, indicating that their nervous system was still able to 'process' the heated compartment as potentially hazardous.

A series of experiments has also been carried out in the leech (*Hirudo medicinalis*). For instance, after the report by Moss *et al.* (2005) showing that 5-HT modulated the axonal coupling between the neurons (the so-called S cells), which plays a critical role in the sensitization of the shortening reflex (a tactile stimulation induces whole-body shortening), Burrell and Sahley (2005) demonstrated that the 5-HT-induced potentiation of S-cell excitability could be blocked by a 5-HT receptor antagonist (methysergide), a G-protein inhibitor (GDP-β-S) and a PKA inhibitor (Rp-cAMP). They also found that the effects of 5-HT could be mimicked by a cAMP analog that activates PKA (db-cAMP). Methysergide and GDP-β-S prevented learning-induced potentiation of S-cell excitability, as well as the excitability change that accompanies sensitization. When the preparations were exposed to the 5-HT antagonist or the G-protein inhibitor, the sensitization of the shortening reflex could not be acquired. In this organism, Catarsi *et al.* (1990) had already detected that the efficiency of a simple form of non-associative learning followed the seasonal variation in 5-HT levels, being maximal when the 5-HT levels were highest.

5-HT seems to play a role in memory formation in even a much simpler invertebrate organism, namely a nematode. The nematode *Caenorhabditis elegans* has a very simple nervous system, which is made of only 302 neurons (more than 60 times less than in *Aplysia californica*). In a recent report, Zhang *et al.* (2005) have shown that such worms displayed a relatively complex learning faculty, as they were able to associate chemosensory stimuli with illness and subsequently to avoid the illness-associated stimuli when they were given an opportunity to make a choice. In their study, Zhang and colleagues bred their worms, which feed on bacteria, in the presence of pathogenic or non-pathogenic bacteria. Those bred on the pathogenic regimen were found later to significantly avoid it when given the choice between this regimen and a non-pathogenic one. Interestingly, 5-HT seems to play an important role in this learning. Indeed, worms raised on the pathogenic regimen showed increased 5-HT levels in several neurons, including chemosensory ones. Furthermore, worms that underwent a mutation of the gene coding for tryptophan hydroxylase, which resulted in 5-HT depletion, were unable to establish the association between the pathogenic bacteria regimen and illness. Although these data demonstrate a faculty to learn a relatively complex association, the findings reported by Zhang *et al.* (2005) do not unequivocally demonstrate that 5-HT is part of the mechanism of the memory formation process. They do, however, show that 5-HT is an important triggering signal in relation to the memory regarding the pathogenic bacteria.

Concluding remarks

In invertebrates, 5-HT appears to play a crucial role in the establishment of non-associative and associative learning experiences. In the case of functional disruption of the 5-HT system, memories are more difficult or impossible to establish. On the contrary, in some of these animals, an increase of 5-HT functions may facilitate learning. Given the principle of conservatism of adapted mechanisms along the phylum, it is reasonable, as stated by Kandel (2001), to consider that cellular and molecular processes used by the Aplysia and other invertebrates to store either short- or long-term memories may also be used in vertebrates, including mammals. Therefore, it makes sense also to tackle the role of 5-HT in memory processes in vertebrates, and perhaps, before anything else, in the most common laboratory animals – namely, mice and rats.

Exploring the role of serotonin in memory by alterations of the 5-HT system in higher vertebrates

In higher vertebrates, it is not possible to study memory processes at the level of a single particular neuron or synapse that would have its exact and easily identifiable equivalent within a particular brain structure of any subject of a given species. In approaches in higher vertebrates, the reasoning is usually focused on particular neuroanatomically defined serotonergic systems – in general, the dorsal and/or medial raphe. The functional outputs of these nuclei can be pharmacologically enhanced by administration of 5-HT-releasing substances or dietary tryptophan loading (rarely used), or, contrarily, depleted by the means of selective lesions, blockade of 5-HT synthesis or tryptophan-depleted food regimens (used more frequently than tryptophan loading, especially in human studies). The role of 5-HT can also be studied by direct intraparenchymal infusions of 5-HT, or of a variety of substances that more or less specifically activate or block the numerous 5-HT receptors (see Figure 2). As outlined hereafter, however, the approaches are much more complicated than in invertebrates, in terms of identifying the role(s) of 5-HT in memory functions.

Enhancing 5-HT functions

Inhibiting 5-HT reuptake

One of the most frequent methods used to boost the 5-HT systems is based on the administration of selective serotonin reuptake inhibitors (SSRIs) such as fluoxetine or citalopram. It must be kept in mind, however, that the administration regimen is a very important issue that must be considered in the interpretation of the drug effects on learning and memory performance. Indeed, when given acutely, SSRIs lead to the stimulation of 5-HT receptors at the postsynaptic level by increasing the availability of 5-HT in the synaptic cleft, but at the same time reduce (sometimes dramatically) the activity of the 5-HT system by a sustained action of 5-HT at the somato-dendritic inhibitory 5-HT_{1A} receptors, which respond in the nanomolar range. When SSRIs are given subchronically or chronically, however, these 5-HT_{1A} receptors undergo substantial down-regulation, whereby the net effects of such drugs become predominantly postsynaptic and correspond to a genuine activation of the 5-HT system.

An indication that SSRIs might interfere with memory formation can be derived from the report by Stewart and Reid (2000). Focusing on the hippocampal formation, these authors observed that repeated electroconvulsive stimulations, a treatment known to disrupt consolidation processes (see, for example, Sara and Hars, 2006), produced long-term potentiation in the perforant paths to granule-cell synapses that could be mimicked by fluoxetine treatment. While electroconvulsive stimulation disrupted spatial learning capabilities, however, fluoxetine left the latter intact. Keith *et al.* (2007) also reported that spatial learning was not affected by fluoxetine treatment of intact rats. However, when rats were subjected to lesions of the granule cells of the dentate gyrus, a hippocampal sub-region, such a treatment prevented recovery of function. Using other models of amnesia, which consisted of treating rats with a muscarinic receptor antagonist (scopolamine) or a cannabinoid receptor agonist (Δ^9-tetrahydrocannabinol, or THC), Egashira *et al.* (2006) found that a low dose of citalopram was able to attenuate the amnestic effects of scopolamine or THC in a spatial memory task (the radial maze). Using a water-maze task, Yau *et al.* (2002) found chronic fluoxetine treatment to have no effect on performance, whether in a population of

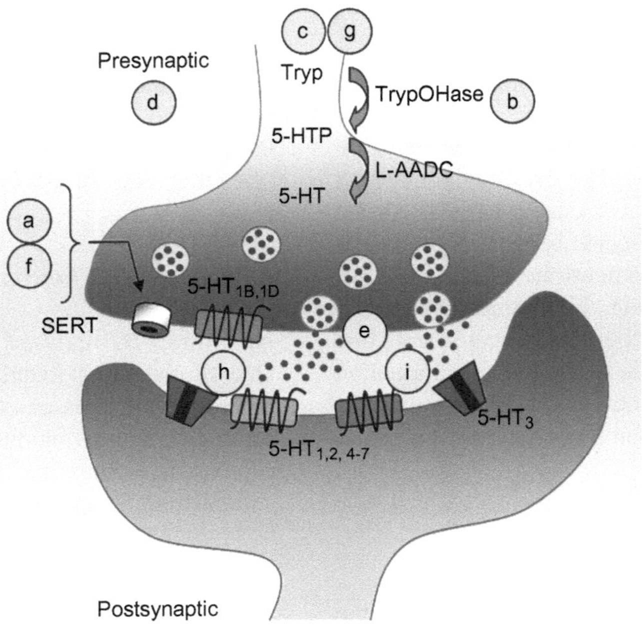

Figure 2 Alterations of the serotonergic system/synapses can be achieved by different methods. Depletion may be obtained by irreversible damage to the neurons by the means of intracerebroventricular 5,7-dihydroxytryptamine (5,7-DHT), which is taken up by the serotonin transporter (a), by pCPA-induced blockade of the 5-HT-synthesizing enzyme, tryptophane hydroxylase (b), by placing subjects under tryptophan-depleted food regimens (c), or by pCA- or MDMA-induced alterations of 5-HT terminals (d). Increased 5-HT functions can be obtained by acute pCA-induced release of 5-HT (e), by blockade of the serotonin transporter (SERT; f), or the administration of tryptophan-enriched food regimens (g). Another way to interfere with serotonergic function is based on the administration of ligands that may more or less selectively block (antagonists) or activate (agonists) the receptors (h), which are of the metabotropic (5-HT_1, 5-HT_2, 5-HT_4, 5-HT_7) or inotropic (5-HT_3) type; depending on the species, the autoreceptor is of the 5-HT_{1B} or 5-HT_{1D} subtype. As regards the receptors, it is also possible to genetically delete them in mice (i). In general, (d), (g) and (h) can be used both in experimental animals and human subjects, but all other approaches are restricted to animal models (most often to mice, rats, monkeys, less frequently to guinea pigs, as concerns the study of CNS functions). To see the full color version of this figure please refer to the color plate in the back of the book. Copies produced via our print on demand service do not contain color plates. If your copy does not have the color plate, please go to this website to view the figure in color www.elsevierdirect.com/companions/9780123746344

young adult rats or in a population of old rats exhibiting an age-related impairment. Valluzzi and Chan (2007) confirmed that fluoxetine (chronic treatment) had no effect on water-maze learning, but impaired hippocampal-independent learning in an object recognition task and an appetitive Pavlovian-conditioning task. Using citalopram or fluoxetine, Burghardt *et al.* (2007) recently reported that both drugs, when injected before the test following a drug-free acquisition, were able to exacerbate the expression of previously acquired conditioned auditory fear. This finding, however, does not demonstrate a facilitation of trace retrieval. Rather, it shows that SSRIs may produce symptoms of anxiety that contribute to the potentiation of the expression of fear. This observation is at variance with the report by Hashimoto *et al.* (1996), who found that the pre-testing administration of citalopram reduced freezing to a fear-conditioned context. Interestingly, when citalopram was given before the conditioning, acquisition of the auditory fear was ameliorated (Burghardt *et al*, 2004). However, there has been a report on weaker freezing in a fear-conditioned context after pre-conditioning citalopram treatment (Inoue *et al.*, 1996). Thus, it seems that, depending on whether conditioning is auditory or contextual, pre-conditioning or pre-testing administration of SSRIs has opposing effects. This can be linked to the fact that auditory conditioning relies upon amygdala-dependent processes, whereas contextual fear conditioning engages both the amygdala and the hippocampus. Interestingly enough, Burghardt *et al.* (2004) have reported a marked difference in the effects of citalopram on fear acquisition: while acute treatment facilitated auditory fear acquisition (probably because of a pharmacologically induced anxiety), chronic treatment in fact attenuated it. For more data on the effects of SSRIs on memory the reader is referred to the excellent review by Monleón and colleagues (2008), in which the authors point to a memory-impairing effect of SSRIs given before training, against an absence of impairing effects when such drugs were administered after training. In general, these authors conclude that the effects of SSRIs could be memory phase-dependent, being detrimental during encoding and early consolidation, and beneficial during later consolidation phases and retrieval.

Stimulating 5-HT release
Parachloroamphetamine (pCA) has also been used to boost 5-HT function. This may seem paradoxical, as this drug is often used to produce 5-HT depletion. In fact, this molecule has opposing effects depending on the post-administration delay. Shortly after administration it induces a massive release of 5-HT from the terminals, but then, in the long-term, destroys at least part of these terminals (Fuller, 1992). When used as a 5-HT-releasing agent, pCA treatment was found to impair acquisition of a passive avoidance response. Indeed, when injected before training, pCA produced a dose-dependent alteration of retention performances assessed with a 24-hour delay (Misane and Ögren, 2000); this alteration was not attributable to serotonergic toxicity. Similar findings were subsequently reported by Solana-Figueroa *et al.* (2002), who concomitantly demonstrated that this impairment could be attenuated by an increase of the foot-shock intensity, and thus more intense learning. In general, these observations are in agreement with the data reported earlier by Santucci *et al.* (1996), who also assessed the effects of acute pCA treatment on passive avoidance performance, and completed their approach on retention performance after eight-arm radial-maze training. In the passive avoidance task, Santucci *et al.* found a dose-dependent impairment of retention performance when pCA was injected 30 minutes before the training session. In rats that had achieved a good level of working- and reference-memory performance in a radial-maze task, treatment with pCA disrupted retrieval performance on the first day of administration, but not on two subsequent treatment days. All these observations are compatible with the idea that an increase of the serotonergic tonus can be detrimental to memory functions, either by altering consolidation processes or by interfering with recall.

Infusing 5-HT into brain regions
Intuitively, this is probably the most direct approach that comes to mind to study the role of a neurotransmitter in a particular function. In addition, intraparenchymal (directly into brain tissue) infusions enable an acceptable neuroanatomical specificity, as the neurotransmitter can be infused in virtually any structure or region of the brain, whether superficial or deep in the organ. A major drawback of this approach is the fact that all mechanisms ensuring the inactivation of the latter (i.e., enzymatic degradation or reuptake) are still operating and may therefore reduce the half-life of the infused substance dramatically. Another drawback is that acute flooding of all neurotransmitter-specific receptors within a given brain region or system is probably all but physiological, because *in situ*, the various regulations operating in this region or system certainly undergo a much finer multi-transmitter and receptor-specific neuropharmacological tuning than those mimicked by the receptor flooding. Nevertheless, there are a few studies in which 5-HT has been infused directly into regions such as the striatum, substantia nigra (which may participate in aspects of procedural memory), or amygdala and hippocampus (which may participate in aspects of 'declarative-like' memory). As regards the striatum, Prado-Alcalá *et al.* (2003) reported that an infusion of 5-HT into the posterior region of this structure after the acquisition of a step-through inhibitory avoidance task induced amnesia and thus disrupted consolidation processes. These observations are in line with the aforementioned acute effects of pCA administered into the striatum, a treatment resulting in the marked increase of 5-HT release from endogenous stores (see 'Stimulating 5-HT release', above). Díaz del Guante *et al.* (2004) also used an inhibitory avoidance task, but they infused 5-HT into the substantia nigra before the training session (*vs* cerebral cortex and zona incerta). Interestingly, the authors report a dose-dependent retention impairment, indicating that increased 5-HT activity in the substantia nigra may also produce anterograde amnesia. In earlier studies, 5-HT was infused into the amygdala (Plaznik *et al.*, 1985) or the hippocampus (Plaznik *et al.*, 1983) of rats before training, and was also found to impair the retention of a passive avoidance response.

Depleting 5-HT functions

For many years, experiments in behavioral neurobiology relied upon lesion techniques that selectively damaged well-delineated regions of the brain (nuclei or/and pathways to various target structures) in order to assess the behavioral outcomes of such damage. While very useful in studying and establishing possible structure–function relationships, such approaches were of limited interest regarding our understanding of the neuropharmacological regulations involved in the modulation of region- or system-specific functions. That kind of investigation requires pharmacological approaches or/and lesion techniques that can be selective for particular neurotransmitter systems. With regard to the serotonergic systems, an approach based on neurochemical selectivity of the lesions became possible with the emergence of a series of compounds such as 5,6-dihydroxytryptamine (5,6-DHT) and 5,7-dihydroxytryptamine (5,7-DHT), which were able to produce damage to serotonergic neurons that was selective when noradrenergic neurons were protected by a pharmacological inhibition of the noradrenaline reuptake carrier (Baumgarten *et al.*, 1971, 1973a; Baumgarten and Lachenmayer, 1972). Subsequently, 5,7-DHT, which proved to be more useful than 5,6-dihydroxytryptamine (the latter having a high non-specific toxicity and poor diffusion properties; Baumgarten *et al.*, 1973b), has been extensively used to damage serotonergic neurons in the rodent brain and to investigate the functional correlates of such damage at various levels of analysis, including memory. 5,7-DHT can be infused directly in specific brain regions such as the dorsal or/and median raphe. Its intraventricular administration to rats or mice, using a sufficient amount (generally about 150 μg), produces about 90 percent depletion of 5-HT in target structures such as the cortex and the hippocampus, with minimal or no damage to other monoaminergic systems. Moreover, this depletion is permanent. For the mechanisms of action of 5,7-DHT, the reader may refer to Sinhababu and Borchardt (1988) or Kawase *et al.* (1990). Alternatives to this lesion approach consist of treating the animals with parachlorophenylalanine (pCPA), which produces dramatic 5-HT depletion by inhibition of tryptophane hydroxylase, or with parachloroamphetamine (pCA; see above). Administration of 3,4-methylenedioxymethamphetamine (MDMA or 'ecstasy') at a high dose and under appropriate ambient temperature conditions may also produce serotonergic toxicity (see, for example, Green *et al.*, 2003). The potential problem with the latter compounds is that depletion, which may be massive with pCPA and pCA, is generally followed by protracted recovery, which can extend for over 1 year with MDMA (see, for example, Battaglia *et al.*, 1991). After pCPA treatment there is also a recovery, which is not detectable 3 days after administration but becomes significant after 8 days (Ben Hamida and Cassel, unpublished results). Furthermore, while MDMA affects serotonergic neurons in rats, it produces damage to dopaminergic neurons in mice (see, for example, Green *et al.*, 2003). In any case, lesions or depletion induced by one or the other of these compounds provide interesting data as to the role of 5-HT systems in learning and memory functions.

5,7-dihydroxytryptamine (5,7-DHT) lesions

There seems to be a relatively general consensus on the fact that 5,7-DHT-induced lesions do not alter learning and memory capabilities in a dramatic way. Early studies showed that electrolytic or ibotenate lesions of the median raphe impaired the acquisition of a delayed alternation task (Asin and Fibiger, 1984), and the same was true for the effect of an electrolytic lesion on performance in a radial-maze task (Asin *et al.*, 1985). In each of these reports, however, the authors also showed that when 5,7-DHT was injected into the median raphe, which induced 5-HT depletion as severe as following the electrolytic lesion, performance was comparable to that found in the controls, whatever the task (Asin and Fibiger, 1984; Asin *et al.*, 1985). As electrolytic or ibotenate lesions were not selective, these observations suggest that it was not the disruption of serotonergic neurons that accounted for the behavioral effects of both other less-selective lesions. Similar conclusions were drawn from studies in which 5,7-DHT, whether injected intracerebroventricularly or intraparenchymally, did not alter the capability to acquire a reference memory in a water-maze task and subsequently to retrieve it in a probe trial, or to perform a working memory task in the same test apparatus (Nilsson *et al.*, 1988; Richter-Levin and Segal, 1991; Murtha and Pappas, 1994; Lehmann *et al.*, 2000, 2002; Galani *et al.*, 2007). While most data in the literature seem to support such conclusions, some issues need to be addressed.

First, lesions of serotonergic neurons in neonatal rodents may result in an alteration of some cognitive functions or behaviors relevant to cognition (e.g., anxiety) when the animals are young adults (e.g., Hohmann *et al.*, 2007; McLean *et al.*, 1993); spatial memory capabilities, however, appear to be preserved (Volpe *et al.*, 1992). In the approaches showing cognitive deficits, it is nevertheless possible that the lesions contribute to altering aspects of CNS development which, in the end, affect functional dynamics within non-serotonergic neurotransmission systems more crucial to memory functions.

Second, whereas the effects of 5,7-DHT lesions on long-term memory seem weak, short-term (or working) memory appears more susceptible to such lesions, although not all of the few available studies agree on this point (e.g., Wirth *et al.*, 2000). Hritcu *et al.* (2007) recently published an article in which they showed that 5,7-DHT lesions disrupted short-term memory processes, as assessed in a Y-maze alternation task and an eight-arm radial-maze task, whereas long-term memory capabilities were preserved. Using a T-maze alternation task in rats, Ricaurte *et al.* (1993) found that 5,7-DHT lesions impaired choice accuracy. As this impairment was found with a 5-second inter-trial delay, and was still observed with up to a 3-minute one, it is compatible with an impairment of short-term memory processes. In another approach, Lieben *et al.* (2006) used an object recognition task in rats and found that serotonergic lesions altered the capability to identify a novel object as such (and thus to remember the familiar one) at a 1-hour delay between the exploration and testing sessions; unfortunately, longer delays were not tested. Along this line, it can be recalled that serotonergic lesions were also found to disrupt short-term memory in a terrestrial slug, while long-term memory was preserved (Shirahata *et al.*, 2006; see above). Finally, Cassaday *et al.* (2003) used a non-match to sample object recognition task in a Y-maze, and observed that rats with 5,7-DHT lesions were impaired in acquiring a working memory task but not in performing the task once acquired. These rats were also impaired during the early phase of a T-maze alternation task. However, using a short-term memory olfactory recognition task, Wirth *et al.* (2000) did not observe deficits after 5,7-DHT lesions.

Third, it cannot be ignored that a few reports even demonstrated facilitating effects of 5,7-DHT-induced serotonergic lesions on memory functions. For instance, using the Stone 14-unit T-maze, Altman *et al.* (1990) found that infusion of 5,7-DHT into the fimbria–fornix/cingular bundle led to lesioned rats requiring fewer trials to reach a performance criterion than their intact counterparts. In a more recent study, Anguiano-Rodríguez *et al.* (2007) tested rats in a water maze for their ability to use only proprioceptive information in an egocentric learning task. The authors found that 5-HT depletion confined to the striatum facilitated this kind of learning, most probably by modulating dopamine functions. Similar observations were made when 5,7-DHT was injected into the cerebral ventricles (Olvera-Cortés *et al.*, 2001). Using the same kind of lesions, Ward *et al.* (1999) reported on facilitated acquisition of an operant conditional discrimination task; these authors also observed that the performance of the lesioned rats was less sensitive to parametric manipulations (changing the cognitive, sensory or motivational constraints of the task) as compared to sham-operated controls.

Other approaches (MDMA, pCPA, pCA administration, dietary tryptophan restrictions)

There are also a few studies assessing the effects on memory functions of other 5-HT-depleting treatments. Using a regimen of MDMA treatment that was toxic for 5-HT neurons, Morley *et al.* (2001) found that adult rats exhibited impaired object recognition performances when tested 15 minutes after training. Likewise, in rats that had been given MDMA every fifth day between the ages of 35 and 60 days, and were then tested at the age of 65 days, object recognition performance was impaired (Piper and Meyer, 2004). Rats treated with a neurotoxic regimen of MDMA have also been tested in both the Cincinnati water maze (made of nine interconnected and flooded T-mazes in which rats have to learn the only possible pathway between a start and a goal location) and the more classical Morris water maze (Able *et al.*, 2006). In both tasks, the rats given MDMA and exhibiting depleted 5-HT levels showed modest but significant impairments. As was the case with 5,7-DHT, in the study by Skelton *et al.* (2006), the effects of MDMA have also been addressed in rats intoxicated as juveniles (between 1 and 10 or 11 and 20 postnatal days) and subsequently tested when young adults (at about 60 days). Whatever the age of intoxication, MDMA produced comparable 5-HT depletion (assessed at the age of 105 days). In both a sequential and a spatial reference-memory task in the water maze, rats given MDMA showed a slight but significant impairment, although only when the treatment was given between 11 and 20 days, which again raises the question of the exclusively serotonergic nature of the reported deficits. As already mentioned regarding the effects of neonatal administration of 5,7-DHT, developmental abnormalities may have contributed as well. In the study by Ricaurte *et al.* (1993), the neurotoxic regimen of MDMA had no effect on percent alternation in a T-maze. Finally, in a study by Robinson *et al.* (1993), MDMA-induced 5-HT depletion had no effect on a series of behavioral tests taxing the place–navigation learning set, skilled forelimb use, and the ability to make complex judgements regarding the stimulus properties.

In adult rats, pCPA also was found to alter spontaneous alternation in a T-maze, although it had no effect on passive avoidance tested with a post-acquisition delay of 24 hours (Hritcu *et al.*, 2007). This observation is compatible with the idea that 5-HT depletion induces short-term memory impairments. In most studies, however, pCPA had weak or no effects on a variety of learning and memory tasks (Richter-Levin and Segal, 1989; Vanderwolf *et al.*, 1990; Riekkinen and Riekkinen, 1994; Beiko *et al.*, 1997; Bertrand *et al.*, 2001).

Similar approaches were carried out in rats subjected to pCA-induced depletion of 5-HT. Some studies reported no effect on working memory performance (Sakurai and Wenk, 1990; Jäkälä *et al.*, 1993a), or even on spatial working memory (Altman *et al.*, 1989). Others found the treatment to improve learning capabilities in the Stone 14-unit T-maze (Altman *et al.*, 1989; Normile *et al.*, 1990). In a study assessing passive avoidance retention, Heinsbroek *et al.* (1988) found that the effects of pCA were dependent on the sex, with untreated males performing better than females during the retention test, but this difference was abolished after pCA treatment. The authors of the latter study nevertheless excluded that the effects of pCA were due to memory disruption.

Finally, using a tryptophan-restricted food diet in rats, González-Burgos *et al.* (1998) found that rats with brain 5-HT reduction (which was not measured in this study) were faster than their normally fed controls in learning a pathway between a start and a goal box in the Biel maze (a maze in which rats have to learn the only possible way between a start and a goal box).

Concluding remarks

Whereas among various invertebrates, such as the Aplysia, the leech, worms, etc., the role of 5-HT in the generation of associative memories appears relatively comparable, the outcome of studies aiming at boosting or depleting serotonergic functions in vertebrates, and especially in extremely common laboratory mammals such as mice and rats, is much more variable. Based on the latter studies, it is not possible to achieve a clear and consensual view regarding the role of 5-HT in learning and memory functions. For many learning and memory tasks, whether associative, spatial or other, it seems that general depletion of the serotonergic systems has no or rather positive consequences on memory function, whereas general boosting has usually no or rather negative consequences. It is noteworthy, however, that for each series of results that tend to support the latter assertion, whether in the positive or negative directions, there are several studies providing qualifying or even contradictory data. A relatively accurate way to characterize the involvement of the serotonergic systems in learning and memory functions in mammals would therefore consist of considering it as not absolutely crucial to such functions, while able to participate in their modulation by, for example, influencing the outputs from other systems which are perhaps more directly implicated in learning and memory processes. One such neurotransmitter system could be constituted by the cholinergic system of the basal forebrain, although the role of the septo-hippocampal cholinergic projections in memory has been subjected to discussion based on data obtained with highly selective and extensive cholinergic lesions in the septal nuclei (for review, see Parent and Baxter, 2004). The involvement of the cholinergic basalo-cortical pathways in memory is not crucial (Galani *et al.*, 2002), but its contribution to attention seems much better established (e.g., Harati *et al.*, 2008).

Manipulating other neurotransmitter systems in combination with alterations of 5-HT functions: the cholinergic system as an example

If 5-HT modulates the functional outputs of another system involved in memory, then damage to or functional alterations of serotonergic neurons and neurons of this system should demonstrate a synergistic interaction. In this case, the result of a double lesion or acute pharmacological alteration should be – in quality and/or magnitude – significantly more or significantly less than a simple addition of the effects of each single lesion. Over the past, several studies have provided some support to such a view, the earlier of which generally relied upon approaches lacking neuroanatomical and/or neurochemical selectivity. Indeed, while the neurotoxin 5,7-DHT has been used since the early 1970s (Baumgarten and Lachenmayer, 1972), a toxin showing an excellent and consensually acknowledged selectivity for the cholinergic system (of the basal forebrain), namely 192 IgG-saporin (for review, see, for example, Parent and Baxter, 2004), has been available only since the beginning of the 1990s (see, for example, Wiley *et al.*, 1991).

Cholinergic lesions in combination with 5-HT depletion

Non-selective cholinergic lesions

One of the first experiments designed to explore the effects on memory of combined cholinergic and serotonergic lesions as compared to those of either lesion alone was made by Nilsson *et al.* (1988). These authors performed radiofrequency lesions of the medial septum in combination with 5,7-DHT lesions (the toxin was administered intrecerebroventricularly) of the ascending serotonergic systems in the rat. Subsequently, they tested their animals using a classical Morris water-maze task that taxed spatial reference memory. In rats with only 5, 7-DHT lesions, there was no deficit. Conversely, rats with only a septal lesion showed an impaired ability to find the platform. In the animals bearing both lesions, however, the deficit was significantly larger than with septal lesions alone. Subsequently, Richter-Levin and Segal (1991) published a study in which they had combined partial electrolytic lesions of the septal region with 5,7-DHT lesions of the serotonergic neurons. Neither single lesion affected Morris water-maze performance in a significant way, but after double lesions both reference- and working-memory performances were dramatically impaired. Murtha and Pappas (1994) reported similar conclusions in rats subjected to intraseptal injections of a neurotoxic dose of N-methyl-D-aspartate and intrafornical infusions of 5, 7-DHT: only rats with double lesions were impaired in a water-maze task.

Serotonergic lesions were also combined with lesions of the nucleus basalis magnocellularis (NBM), another major cholinergic nucleus of the basal forebrain, which sends projections to the cortical mantle. For instance, Normile *et al.* (1990) used rats that sustained ibotenate-induced lesions of the NBM alone or in combination with systemic administration of pCA. Subsequently, these authors assessed the acquisition of a complex spatial discrimination task in a Stone 14-unit T-maze, in which the animals have to learn a route from a start to a goal box. The authors found that the rats given pCA alone showed better performance than their unlesioned counterparts. Although NBM lesions had no effect on spatial learning, they prevented the pCA-induced improvement. Riekkinen *et al.* (1990) compared the effects of separate or combined lesions of the dorsal raphe and the NBM on acquisition of a water-maze task in rats. These authors reported the dorsal raphe lesions to be ineffective, the NBM lesions to impair performance, and both lesions combined to induce a deficit greater than that due to NBM lesions alone. Using passive-avoidance and water-maze tasks, Jäkälä *et al.* (1993b) investigated the consequences of NBM lesions produced by intraparenchymal quisqualic acid injections, or inhibition of brain 5-HT synthesis performed by systemic pCPA treatment, or both types of lesions combined. They found that NBM lesions severely impaired passive-avoidance retention and only slightly disturbed water-maze performance. pCPA alone had no effect in either of these tasks, but potentiated the effects of NBM lesions. Markowska and Wenk (1991) also compared the behavioral effects of NBM lesions (injection of ibotenic acid) combined or not with 5-HT depletion induced by systemic pCA treatment, but they used a non-spatial memory task (cued delayed non-matching to sample task) and investigated more precisely both the acute and long-term effects of the treatments. They found no effect of 5-HT depletion, whether shortly after treatment (4 weeks) or later on when testing was prolonged. In contrast, NBM lesions resulted in a deficit in the non-spatial memory task. Combined pCA treatment and NBM lesions did not produce a larger deficit than NBM lesions alone. However, when testing was prolonged, rats with NBM lesions showed some improvement over time, whereas those given both NBM lesions and pCA treatment did not. Overall, the effects of NBM lesions combined with 5-HT depletion on a delayed non-matching to sample task appeared to be rather weak – a conclusion that was in line with another report by Sahgal and Keith (1993), who also used a delayed non-matching to sample paradigm in rats given NBM lesions combined with 5,7-DHT-induced 5-HT depletion.

Thus, this series of experiments suggests that the cognitive effects of cholinergic denervation of the hippocampus (septal lesions) or of the neocortex (nucleus basalis lesions) can be exacerbated by concomitant 5-HT depletion. It is noteworthy, however, that the techniques used to perform the cholinergic damage were all but selective. In 1997, Matsukawa and colleagues published the data of an experiment in which they combined a pCPA treatment with an AF64A-induced lesion (AF64A is a cholinergic neurotoxin) and, beside other results, found the combined lesions to alter performance during the acquisition of a water-maze task. However, the cholinergic specificity of AF64A has been questioned (see, for example, Eva *et al.*, 1987); the variable that was considered to account for learning in the Matsukawa study (i.e., platform latency) is sensitive to non-cognitive biases, the deficit reported was weak, and learning was not prevented in the double-lesioned group (unfortunately, the authors did not report probe-trial performance). As regards the cholinergic selectivity of the lesions, 192 IgG-saporin would probably have been a better choice.

Selective cholinergic lesions

To the best of our knowledge, we have been the first to use 192 IgG-saporin lesions in combination with another type of neurochemically selective damage, namely 5,7-DHT-induced 5-HT depletion (Lehmann *et al.*, 2000, 2002). In general, lesion experiments based on intracerebroventricular injections of 192 IgG-saporin and 5,7-DHT clearly suggest that the cognitive effects of a cholinergic denervation of both the hippocampus (septal lesions) and the neocortex (NBM) can be exacerbated by concomitant and widespread 5-HT depletion. Indeed, we found that 5,7-DHT, used in combination with 192 IgG-saporin, induced working-memory deficits in the water maze and in the radial maze, which neither toxin produced when injected alone. In that way, a study assessing behavioral (locomotor activity, forced T-maze alternation, beam-walking, Morris water maze and radial-maze) and neurochemical effects of intracerebroventricular injections of 192 IgG-saporin and of 5,7-DHT in adult Long-Evans female rats showed that cholinergic lesions, which reduced the concentration of acetylcholine by about 40 percent in the hippocampus but had no effect in the striatum, induced only severe motor deficits. Serotonergic lesions, which reduced the concentration of 5-HT by 80 percent in the hippocampus and the striatum, produced diurnal and nocturnal hyperactivity but no other behavioral effect. Finally, rats with combined lesions were more active than those with only serotonergic lesions, showed motor dysfunctions similar to those found in rats with cholinergic lesions alone, and exhibited impaired performance in the T-maze alternation test, the water-maze working-memory test and the radial-maze (Lehmann *et al.*, 2000). Another study conducted in our group showed that 5-HT depletion confined to the hippocampus – by injections of 5,7-DHT directly into the cingular bundle and the fimbria–fornix – could attenuate some of the behavioral deficits produced by a large dose of 192 IgG-saporin, which this time was injected directly into the medial septum and the diagonal band of Broca (Lehmann *et al.*, 2002). In this experiment, rats with single or combined damage were tested for locomotor activity, spontaneous T-maze alternation, sensorimotor, water-maze and radial-maze performance. The data showed that the cholinergic lesions, which decreased the hippocampal concentration of acetylcholine by about 65 percent (see also Chang and Gold, 2004), induced nocturnal hyperlocomotion, reduced T-maze alternation, impaired both reference-memory in the water maze and working-memory in the radial maze, but had no effect on sensorimotor performance and working-memory in the water maze. In the group of rats given combined lesions, all deficits produced by the cholinergic lesions were observed; surprisingly, however, the nocturnal hyperactivity and the working-memory deficits in the radial maze were significantly attenuated. These results suggest that reduction of the serotonergic tone in the hippocampus may compensate for some dysfunctions subsequent to a loss of cholinergic hippocampal inputs. We have reason to believe that part of the mechanism could be some facilitation of acetylcholine release (Birthelmer *et al.*, 2003).

Intrahippocampal grafts of cholinergic and serotonergic neurons

Although this is probably less directly related to the topic of this chapter, it is also noteworthy that after extensive hippocampal denervation by fimbria–fornix lesions, and thus lesions altering cholinergic, GABAergic, noradrenergic and serotonergic afferents of the hippocampus, intrahippocampal grafts rich in cholinergic and serotonergic neurons and providing a neurotransmitter-specific neurochemical restoration to the hippocampus (see, for example, Cassel *et al.*, 1992a) were found to produce the most efficient benefit as regards recovery of memory functions (see, for example, Nilsson *et al.*, 1990; Jeltsch *et al.*, 1994; Balse *et al.*, 1999; but see also Cassel *et al.*, 1992b or 1997 for review). For instance, in rats given radiofrequency (and thus unselective) lesions of the medial septum combined with intracerebroventricular injections of 5,7-DHT, Nilsson and his collaborators measured the behavioral effects of septal grafts alone (which are rich in cholinergic neurons but devoid of serotonergic ones), of raphe grafts alone (which are rich in serotonergic neurons but largely devoid of cholinergic ones), and of a combination of both types of grafts (Nilsson *et al.*, 1990). All these grafts were placed into the hippocampus as a cell suspension preparation, and cognitive function was assessed several months post-grafting in a Morris water maze according to a protocol taxing reference memory. Whereas neither type of single grafts produced beneficial effects on maze performance, the combined grafts were able to improve this performance at a post-grafting delay of 10 months. Similar observations were made after electrolytic lesions (Jeltsch *et al.*, 1994) or aspiration lesions (Balse *et al.*, 1999) of the fimbria and the dorsal fornix. Under some circumstances (i.e., serotonergic lesions combined with partial cholinergic lesions in the septal region), it seems that even the sole serotonergic reinnervation of the hippocampus may be sufficient to promote significant recovery of spatial reference as well as working-memory performance (Richter-Levin *et al.*, 1993). All these data demonstrate an implication of serotonergic and cholinergic processes in cognitive function, and may perhaps fit with the idea of serotonergic modulation of some cognitive abilities in which cholinergic mechanisms have been recognized to contribute a significant role. This view is further supported by experiments that combined pharmacological manipulations of the cholinergic system in 5-HT depleted animals.

Cholinergic drugs in combination with 5-HT depletion

Matsuno *et al.* (1993) found that pCA-induced impairment in passive-avoidance learning in mice could be attenuated by the administration of tetrahydroaminoacridine (efficient dose: 10 mg/kg, p.o.) or physostigmine (efficient doses: 0.25 and 0.5 mg/kg, i.p.), two acetylcholinesterase inhibitors. Thus, a deficit induced by serotonergic depletion can be partially reversed by increasing the half-life of acetylcholine. Acetylcholine binds to muscarinic and nicotinic receptors.

In one of the first studies assessing the effects of muscarinic blockade in 5-HT-depleted rats, Vanderwolf (1987) compared the passive-avoidance and water-maze learning performances of intact rats to that of pCPA-treated rats exposed to scopolamine (5 mg/kg, s.c.) or atropine (50 mg/kg, i.p.) – two antagonists of muscarinic receptors – during acquisition or retention trials. Scopolamine disrupted acquisition more than retention, an effect mimicked by atropine. pCPA was ineffective (i.e., did not alter learning/memory functions), but scopolamine produced a larger deficit in pCPA-treated rats than in intact rats. These data show that the cognitive effects of muscarinic blockade may be potentiated by serotonergic depletion. Similar findings were reported by Richter-Levin and Segal (1989), who administered atropine (20 mg/kg, i.p.) to pCPA-treated rats tested in the water-maze task. These authors found that this dose of atropine disrupted spatial learning performance only in pCPA-treated rats; pCPA had no proper effect. Riekkinen *et al.* (1991a) tested intact and 5,7-DHT-treated rats for water-maze performance after scopolamine (0.8 mg/kg, i.p.) or pilocarpine (muscarinic agonist; 4 and 10 mg/kg, i.p.) treatment. While 5,7-DHT lesions produced no deficit, scopolamine severely impaired spatial learning in both intact and lesioned rats. In intact rats, the scopolamine-induced deficit was abolished by pilocarpine treatment at a dose of 4 mg/kg (i.p.), whereas in the rats with 5,7-DHT lesions a dose of 10 mg/kg was necessary to counterbalance the scopolamine-induced effect. Thus, 5-HT depletion decreased the efficacy of pilocarpine in reversing a scopolamine-induced spatial reference memory deficit. In another study, Riekkinen *et al.* (1991b) have reported that scopolamine (0.8 mg/kg, i.p.) and the 5-HT reuptake blocker alaproclate (7.5 and 20 mg/kg, i.p.) were able to impair water-maze navigation performance. Furthermore, the combined injections of scopolamine (0.8 mg/kg) and alaproclate (20 mg/kg) produced greater impairment than after only scopolamine treatment. This effect could be partially reversed by administration of pilocarpine (6 mg/kg). A subamnestic dose of atropine has also been described to produce cognitive perturbations in rats given 5,7-DHT into the ventricles or directly into the fimbria–fornix, while sham-operated rats showed normal performance (Richter-Levin *et al.*, 1994).

Riekkinen *et al.* (1992) also demonstrated that nicotinic receptors may be involved in the cholinergic–serotonergic regulation of cognitive function. Indeed, this group has investigated the cognitive effects of mecamylamine (10 mg/kg, i.p.), a nicotinic antagonist, in rats subjected to massive 5-HT depletion (pCPA). The authors focused on passive-avoidance and water-maze performances. Neither performance was affected by pCPA treatment. Given alone in intact rats, mecamylamine delayed spatial learning in the water maze but did not affect passive avoidance. In pCPA-treated rats, however, mecamylamine induced a marked deficit in passive avoidance. Subsequently, Riekkinen *et al.* (1993) reported the results of a study in which they investigated the effects of mecamylamine (7.5 mg/kg, i.p.) in pCPA-treated rats on working memory in a water maze and on a passive-avoidance response. They confirmed that pCPA did not affect water-maze navigation or passive-avoidance performances, and showed that mecamylamine (slightly) impaired working memory and (substantially) altered passive-avoidance performances in intact rats. This effect was counterbalanced or attenuated by nicotine treatment (0.3 mg/kg, i.p.). In pCPA-treated rats, the mecamylamine-induced impairments were more pronounced and the preventive effect of nicotine was less marked than in intact rats. Riekkinen *et al.* (1994) reported that medial septal lesions impaired performance in both passive-avoidance and water-maze learning tasks. These deficits were aggravated following pCPA-induced 5-HT depletion, which, alone, left learning intact. Pre-training trial injections of tetrahydroaminoacridine (THA, 1, 3 and 5 mg/kg, i.p.), an acetylcholinesterase inhibitor, or nicotine (0.03, 0.1 and 0.3 mg/kg, i.p.) attenuated the deficit in rats with septal lesions at the respective doses of 3 mg/kg (THA), 0.1 and 0.3 mg/kg (nicotine) and in both tasks. In rats that had sustained both a septal lesion and 5-HT depletion, only THA was found to exert beneficial effects.

In some memory tasks, intact attention resources may be required to achieve optimal performance. Therefore, it is worth mentioning a possible implication of cholinergic–serotonergic interactions in attention processes. Using rats, Jäkälä *et al.* (1992) examined the effects of pCPA and/or scopolamine treatment on attentional processes measured in the five-choice serial reactional time test. This test measures the rats' ability to detect and respond to brief flashes of light presented in one of five food-rewarded locations. The authors found that scopolamine (0.2 mg/kg, i.p.) did not alter the discriminative accuracy (proportion of appropriate locations detected among all responses) in the rats that did not sustain pCPA-induced lesions. Under standard test conditions (intensity of visual stimulus and speed of presentation fixed), the discriminative accuracy of pCPA-treated rats tended to decrease. Addition of scopolamine treatment did not further impair pCPA-treated rats. Only when stimulus intensity was reduced or speed of presentation increased did pCPA-treated rats show impaired accuracy. These observations suggest that part of the synergistic effects of 5-HT depletion and cholinergic blockade on memory performance might involve difficulties in attentional function. A weak behavioral effect of concurrent serotonergic and cholinergic manipulations has also been reported by Robinson *et al.* (1993) in a water-maze learning task, and thus on spatial reference memory processes. These authors induced long-lasting serotonergic depletion by treating rats with high doses of MDMA, and found that 5-HT-depleted rats performed similarly to intact control rats. In addition, MDMA-treated rats showed a sensitivity to systemically-administered atropine (50 mg/kg, i.p.) that did not differ from that of intact rats. Robinson and colleagues interpreted these weak effects as being due to insufficient depletion of cortical 5-HT (−73 percent in the neocortex, −32 percent in the caudate nucleus).

Reversing the effects of cholinergic drugs by acting on 5-HT$_{1A}$ somatodendritic autoreceptors

Although this approach may at first appear to be outside the scope of this chapter, it is noteworthy that the activation of 5-HT$_{1A}$ receptors in the raphe nuclei produces a transient reduction of the overall serotonergic tonus in regions such as the cortical mantle and the hippocampus. Therefore, the acute functional consequences of activation of these receptors can be paralleled with 5-HT depleting treatments (e.g., Invernizzi *et al.* 1991), and their interaction with treatments altering cholinergic functions will be briefly reviewed. In 1995, Carli and colleagues published the result of an experiment in which rats were trained in a two-platform spatial discrimination task after intrahippocampal infusions of scopolamine (Carli *et al.*, 1995). Most interestingly, the systemic injection of a relatively selective 5-HT$_{1A}$ receptor antagonist, namely WAY 100135, prevented this impairment, although not through action on the 5TH$_{1A}$ receptors located in the hippocampus, as the directly intrahippocampal infusion of WAY 100135 had no effect on the scopolamine-induced impairment. Later on, these effects were confirmed with a highly selective 5-HT$_{1A}$ receptor antagonist (Carli *et al.*, 1997), WAY 100635. These studies did not resolve the question of which 5-HT_{1A} receptors accounted for the observed effects, those located postsynaptically, or the somatodendritic ones located on the raphe neurons. This is partly why Carli and colleagues undertook an experiment in which they infused a 5-HT_{1A} receptor agonist directly into the dorsal raphe (Carli *et al.*, 1998; see also Carli *et al.*, 2000). These authors elegantly demonstrated that the effects described in their 1995 and 1997 articles could be reproduced by activation of the somatodendritic 5-HT_{1A} receptors of the raphe serotonergic neurons, and thus by transient reduction of the serotonergic tonus in the target regions of these neurons.

Concluding remarks

Together, the results summarized in the third section of this chapter clearly show that 5-HT may modulate the efficacy of the basal forebrain cholinergic contribution to learning and memory functions. Indeed, acute (using drug treatments), semi-chronic (using pCA or pCPA) or permanent (using 5,7-DHT) alterations in serotonergic function may potentiate the effects on learning and memory of a drug- or lesion-induced cholinergic dysfunctioning within the septo-hippocampal or/and basalocortical systems. Here, the cholinergic system has been taken as one of the possible examples illustrating how changes in 5-HT levels may affect (reduce or enhance) the efficacy of other treatments targeting the cholinergic synapses or neurons. It is likely that interactions between the serotonergic system and other neurotransmitter systems might also have important implications in learning and memory functions. This could be the case, for instance, for GABAergic, glutamatergic, noradrenergic and dopaminergic systems (see, for example, Myhrer, 2003). Regarding the hippocampus, which plays a crucial role in cognitive functions, an idea of the complexity of the different modulating influences that may change serotonergic functions at simply a presynaptic level can be found in the excellent review by Vizi and Kiss (1998; see also Fink and Göthert, 2007).

General conclusions

In the nervous system of invertebrates such as Aplysia, snails, worms, leeches, etc., 5-HT participates in the establishment of non-associative and associative memories. When the 5-HT system is disrupted, the generation of these types of memories is more difficult, or simply impossible. Contrarily, in at least some of these animals, increased 5-HT function may facilitate learning.

With regard to higher vertebrates such as rodents, although there is no consensus in the literature, the implication of the ascending serotonergic systems in learning and memory functions might be summarized as follows: when these systems are damaged in a selective manner (such as is the case with 5,7-DHT), even when they are destroyed massively, this damage may have consequences on various aspects of a rat's or mouse's behavior (see, for example, Cools *et al.*, 2008), but learning and memory capabilities seem largely intact, except when lesions or pharmacological manipulations affecting other neurotransmitter systems are performed conjointly. This is true for most classical tests used in laboratories to explore the neurobiological and neuropharmacological substrates of learning and memory functions, and some results point to the fact that rats in which 5-HT has been depleted may show impairments in tasks which are taxing short-term memory. Serotonergic functions may also be pharmacologically enhanced – for instance, under the effects of MDMA or the influence of antidepressant drugs such as SSRIs. Based on a recent review that focused on the effects of antidepressant drugs on memory (Monleón *et al.*, 2008), SSRIs may have two types of effects depending on whether they are given before or after training. When administered before training, they leave memory functions either largely intact or altered, and may even improve them, but the latter case is very exceptional. When given after training, they have no effect on memory or they may improve it. Some of the variation among studies might be due to differences in the duration of the produced effects (acute, semi-chronic, permanent) and in the types of test and task used, as well as in the nature of serotonergic (for instance SSRIs may affect neurogenesis) or/and non-serotonergic (e.g., anticholinergic or 'pro'-noradrenergic) side effects of the various drugs tested so far. Furthermore, it cannot be excluded that factors such as species, strain and sex have also interfered with the outcome of the various experiments considered herein. From the experiments that have combined alterations of serotonergic functions with disruption of, for example, the cholinergic system – the example considered and documented to some extent in this chapter – it is relatively clear that the ascending serotonergic system cooperates with the cholinergic system in the modulation of learning and memory functions. Regarding the basal forebrain cholinergic system, the literature now provides a series of convincing arguments supporting this view, although the link between septo-hippocampal cholinergic functions and memory has been challenged over the past 15 years (for a review, see Parent and Baxter, 2004).

One of the aspects not considered in this chapter is the role in learning and memory of the 14 different types or subtypes of serotonergic receptors (Nichols and Nichols, 2008). Whereas in rats and mice overall functional modifications (depletion, stimulation) of the serotonergic system have generally weak effects on learning and memory, alterations that affect particular receptor types or subtypes, sometimes only in a particular and restricted region of the brain, have been found to disrupt or to enhance cognition (see, for example, Meneses, 2003; King *et al.*, 2008). This is not the place to review all these receptors and describe their transductional characteristics or other functional specificities/involvement, especially as some of this information is provided in several chapters of the current handbook. Moreover, the task of reviewing the literature regarding the effects on learning and memory of the activation, blockade or genetic manipulation of each of these receptors appears Herculean.

The weak consequences on learning and memory of massive 5-HT depletion vs the sometimes extensive impairments induced by the manipulation of a given receptor (see, for example, Koenig *et al.*, 2008) may seem surprising, if not paradoxical. However, in a recent review Azmitia (2007) has reminded that 5-HT influences almost all biological processes in vertebrates, in which many functions are acting in a combination described as holistic. Why, then, has an overall lesion of the serotonergic neurons so little impact on memory, whereas the blockade or activation of a particular receptor may have dramatic consequences? It is impossible to provide a clear response to this question. Given the wide distribution of serotonergic terminals, and of receptors, in the brain of vertebrates, especially mammalians, it is nevertheless tempting to speculate that one of the basic roles to which serotonergic neurons have evolved over the phylum is the maintenance of a given functional equilibrium among the different target structures of these neurons. Such a conception can be compared to a mechanism serving homeostasis, as stated by Azmitia (2007). Thus, overall functional adaptations of the serotonergic system could contribute to maintaining the brain's capability for elaborate normal operations despite variations of external or/and internal constraints (such as fatigue, stress, hunger, hyperthermia, effort, etc.) within a given range. Regarding learning and memory, such adaptations might occur conjointly in all serotonergic targets relevant to these functions. As long as the aforementioned overall equilibrium is maintained (which is the case when the serotonergic tonus is increased conjointly in all target structures, by, for example, a systemically given SSRI, or is depleted by, for example, a more or less extensive 5,7-DHT-induced lesion), most of the operations underlying cognition could remain possible. However, there might be at least two situations in which these operations may be compromised and thus memory dysfunctions emerge: when some of the other neurotransmitter systems undergo functional alterations in addition to the overall 5-HT dysfunction (such as after cholinergic lesions or muscarinic blockade, for instance), or when the overall serotonergic equilibrium is disrupted, which may be the case after a specific action on a specific serotonergic receptor. The latter point of view is totally speculative, and although it is not in contradiction to the role of the serotonergic system as a regulator of brain homeostasis, it is certainly no more than a possible support to designing working hypotheses driving future experiments.

Acknowledgements of Strasbourg

I would like to acknowledge the Centre National de la Recherche Scientifique, the University of strasbourg, the Alexander von Humboldt Stiftung and the National Institute for Drug Abuse for their support for my serotonin-related research. Thank you also to Dr Anne Pereira de Vasconcelos, Lucas Lecourtier and Bruno Will for their critical reading of a first draft of this chapter, and their helpful comments.

References

Able, J.A., Gudelsky, G.A., Vorhees, C.V. and Williams, M.T. (2006) 3,4-Methylenedioxymethamphetamine in adult rats produces deficits in path integration and spatial reference memory. Biol. Psychiatry, 59(12): 1219–1226.

Altman, H.J., Ogren, S.O., Berman, R.F. and Normile, H.J. (1989) The effects of p-chloroamphetamine, a depletor of brain serotonin, on the performance of rats in two types of positively reinforced complex spatial discrimination tasks. Behav. Neural. Biol., 52(2): 131–144.

Altman, H.J., Normile, H.J., Galloway, M.P., Ramirez, A. and Azmitia, E.C. (1990) Enhanced spatial discrimination learning in rats following 5,7-DHT-induced serotonergic deafferentation of the hippocampus. Brain Res., 518(1–2): 61–66.

Anguiano-Rodríguez, P.B., Gaytán-Tocavén, L. and Olvera-Cortés, M.E. (2007) Striatal serotonin depletion facilitates rat egocentric learning via dopamine modulation. Eur. J. Pharmacol., 556 (1–3): 91–98.

Asin, K.E. and Fibiger, H.C. (1984) Spontaneous and delayed spatial alternation following damage to specific neuronal elements within the nucleus medianus raphe. Behav. Brain Res., 13(3): 241–250.

Asin, K.E., Wirtshafter, D. and Fibiger, H.C. (1985) Electrolytic, but not 5,7-dihydroxytryptamine, lesions of the nucleus medianus raphe impair acquisition of a radial maze task. Behav. Neural. Biol., 44(3): 415–424.

Azmitia, E.C. (2007) Serotonin and brain: evolution, neuroplasticity and homeostasis. Intl. Rev. Neurobiol., 77: 31–56.

Bailey, C.H., Chen, M., Keller, F. and Kandel, E.R. (1992a) Serotonin-mediated endocytosis of apCAM: an early step of learning-related synaptic growth in Aplysia. Science, 256(5057): 645–649.

Bailey, C.H., Montarolo, P., Chen, M., Kandel, E.R. and Schacher, S. (1992b) Inhibitors of protein and RNA synthesis

block structural changes that accompany long-term heterosynaptic plasticity in Aplysia. Neuron, 9(4): 749–758.
Balaban, P.M., Vehovszky, A., Maximova, O.A. and Zakharov, I.S. (1987) Effect of 5,7-dihydroxytryptamine on the food-aversive conditioning in the snail *Helix lucorum* L. Brain Res., 404(1–2): 201–210.
Balse, E., Lazarus, C., Kelche, C., Jeltsch, H., Jackisch, R. and Cassel, J.C. (1999) Intrahippocampal grafts containing cholinergic and serotonergic fetal neurons ameliorate spatial reference but not working memory in rats with fimbria–fornix/cingular bundle lesions. Brain Res. Bull., 49(4): 263–272.
Barbas, D., DesGroseillers, L., Castellucci, V.F., Carew, T.J. and Marinesco, S. (2003) Multiple serotonergic mechanisms contributing to sensitization in aplysia: evidence of diverse serotonin receptor subtypes. Learn Mem., 10(5): 373–386.
Bartsch, D., Casadio, A., Karl, K.A., Serodio, P. and Kandel, E.R. (1998) CREB1 encodes a nuclear activator, a repressor, and a cytoplasmic modulator that form a regulatory unit critical for long-term facilitation. Cell, 95(2): 211–223.
Battaglia, G., Sharkey, J., Kuhar, M.J. and de Souza, E.B. (1991) Neuroanatomic specificity and time course of alterations in rat brain serotonergic pathways induced by MDMA (3,4-methylenedioxymethamphetamine): assessment using quantitative autoradiography. Synapse, 8(4): 249–260.
Baumgarten, H.G., Björklund, A., Lachenmayer, L., Nobin, A. and Stenevi, U. (1971) Long-lasting selective depletion of brain serotonin by 5,6-dihydroxytryptamine. Acta Physiol. Scand. Suppl., 373: 1–15.
Baumgarten, H.G. and Lachenmayer, L. (1972) 5,7-dihydroxytryptamine: improvement in chemical lesioning of indoleamine neurons in the mammalian brain. Z. Zellforsch. Mikrosk. Anat., 135(3): 399–414.
Baumgarten, H.G., Björklund, A., Lachenmayer, L. and Nobin, A. (1973a) Evaluation of the effects of 5,7-dihydroxytryptamine on serotonin and catecholamine neurons in the rat CNS. Acta Physiol. Scand. Suppl., 391: 1–19.
Baumgarten, H.G., Victor, S.J. and Lovenberg, W. (1973b) Effect of intraventricular injection of 5,7-dihydroxytryptamine on regional tryptophan hydroxylase of rat brain. J. Neurochem., 21: 251–253.
Beiko, J., Candusso, L. and Cain, D.P. (1997) The effect of non-spatial water maze pretraining in rats subjected to serotonin depletion and muscarinic receptor antagonism: a detailed behavioural assessment of spatial performance. Behav. Brain Res., 88(2): 201–211.
Bertrand, F., Lehmann, O., Galani, R., Lazarus, C., Jeltsch, H. and Cassel, J.C. (2001) Effects of MDL 73005 on water-maze performances and locomotor activity in scopolamine-treated rats. Pharmacol. Biochem. Behav., 68(4): 647–660.
Birthelmer, A., Schweizer, T., Jeltsch, H., Jackisch, R. and Cassel, J.C. (2003) 5,7-Dihydroxytryptamine lesions enhance and serotonergic grafts normalize the evoked overflow of acetylcholine in rat hippocampal slices. Eur. J. Neurosci., 16(10): 1839–1849.
Brunelli, M., Castellucci, V. and Kandel, E.R. (1976) Synaptic facilitation and behavioral sensitization in Aplysia: possible role of serotonin and cyclic AMP. Science, 194(4270): 1178–1181.
Burghardt, N.S., Sullivan, G.M., McEwen, B.S., Gorman, J.M. and LeDoux, J.E. (2004) The selective serotonin reuptake inhibitor citalopram increases fear after acute treatment but reduces fear with chronic treatment: a comparison with tianeptine. Biol. Psychiatry, 55(12): 1171–1178.
Burghardt, N.S., Bush, D.E., McEwen, B.S. and LeDoux, J.E.. (2007) Acute selective serotonin reuptake inhibitors increase conditioned fear expression: blockade with a 5-HT(2C) receptor antagonist. Biol. Psychiatry, 62(10): 1111–1118.
Burrell, B.D. and Sahley, C.L. (2005) Serotonin mediates learning-induced potentiation of excitability. J. Neurophysiol., 94(6): 4002–4010.
Cajal, S.R. (1894) The Croonian Lecture: La fine structure des centres nerveux. Proc. R. Soc. London, 55: 444–468.
Carli, M., Luschi, R. and Samanin, R. (1995) (S)-WAY 100135, a 5-HT1A receptor antagonist, prevents the impairment of spatial learning caused by intrahippocampal scopolamine. Eur. J. Pharmacol., 283(1–3): 133–139.
Carli, M., Bonalumi, P. and Samanin, R. (1997) WAY 100635, a 5-HT1A receptor antagonist, prevents the impairment of spatial learning caused by intrahippocampal administration of scopolamine or 7-chloro-kynurenic acid. Brain Res., 774 (1–2): 167–174.
Carli, M., Bonalumi, P. and Samanin, R. (1998) Stimulation of 5-HT1A receptors in the dorsal raphe reverses the impairment of spatial learning caused by intrahippocampal scopolamine in rats. Eur. J. Neurosci., 10(1): 221–230.
Carli, M., Balducci, C. and Samanin, R. (2000) Low doses of 8-OH-DPAT prevent the impairment of spatial learning caused by intrahippocampal scopolamine through 5-HT(1A) receptors in the dorsal raphe. Br. J. Pharmacol., 131(2): 375–381.
Cassaday, H.J., Norman, C., Shilliam, C.S., Vincent, C. and Marsden, C.A. (2003) Intraventricular 5,7-dihydroxytryptamine lesions disrupt acquisition of working memory task rules but not performance once learned. Prog. Neuropsychopharmacol. Biol. Psychiatry, 27(1): 147–156.
Cassel, J.C., Neufang, B., Kelche, C., Aiple, F., Will, B.E., Hertting, G. and Jackisch, R. (1992a) Effects of septal and/or raphe cell suspension grafts on hippocampal choline acetyltransferase activity, high affinity synaptosomal uptake of choline and serotonin, and behavior in rats with extensive septohippocampal lesions. Brain Res., 585(1–2): 243–254.
Cassel, J.C., Kelche, C., Majchrzak, M. and Will, B. (1992b) Factors influencing structure and function of intracerebral grafts in the mammalian brain: a review. Rest. Neurol. Neurosci., 4: 65–96.
Cassel, J.C., Duconseille, E., Jeltsch, H. and Will, B. (1997) The fimbria–fornix/cingular bundle pathways: a review of neurochemical and behavioural approaches using lesions and transplantation techniques. Prog. Neurobiol., 51(6): 663–716.
Castellucci, V. and Kandel, E.R. (1976) Presynaptic facilitation as a mechanism for behavioral sensitization in Aplysia. Science, 194(4270): 1176–1178.
Catarsi, S., Garcia-Gil, M., Traina, G. and Brunelli, M. (1990) Seasonal variation of serotonin content and nonassociative learning of swim induction in the leech *Hirudo medicinalis*. J. Comp. Physiol., 167(4): 469–474.
Cedar, H. and Schwartz, J.H. (1972) Cyclic adenosine monophosphate in the nervous system of Aplysia californica. II. Effect of serotonin and dopamine. J. Gen. Physiol., 60(5): 570–587.
Chang, Q. and Gold, P.E. (2004) Impaired and spared cholinergic functions in the hippocampus after lesions of the medial septum/vertical limb of the diagonal band with 192 IgG-saporin. Hippocampus, 14(2): 170–179.
Cools, R., Roberts, A.C. and Robbins, T.W. (2008) Serotoninergic regulation of emotional and behavioural control processes. Trends Cogn. Sci., 12(1): 31–40.

Dahlström, A. and Fuxe, K. (1964) Localization of monoamines in the lower brain stem. Experientia, 20(7): 398–399.

Díaz del Guante, M.A., Rivas, M., Prado-Alcalá, R.A. and Quirarte, G.L. (2004) Amnesia produced by pre-training infusion of serotonin into the substantia nigra. Neuroreport, 15(16): 2527–2529.

Egashira, N., Matsumoto, Y., Mishima, K., Iwasaki, K., Fujioka, M., Matsushita, M., Shoyama, Y., Nishimura, R. and Fujiwara, M. (2006) Low dose citalopram reverses memory impairment and electroconvulsive shock-induced immobilization. Pharmacol. Biochem. Behav., 83(1): 161–167.

Ehrlich, J.S., Boulis, N.M., Karrer, T. and Sahley, C.L. (1992) Differential effects of serotonin depletion on sensitization and dishabituation in the leech, *Hirudo medicinalis*. J. Neurobiol., 23(3): 270–279.

Emptage, N.J. and Carew, T.J. (1993) Long-term synaptic facilitation in the absence of short-term facilitation in Aplysia neurons. Science, 262(5131): 253–256.

Erspamer, V. and Vialli, M. (1937) Ricerche sul secreto delle cellule enterochromaffini. Boll. d Soc. Med.-chir Pavia, 51: 357–363.

Eva, C., Fabrazzo, M. and Costa, E. (1987) Changes of cholinergic, noradrenergic and serotonergic synaptic transmission indices elicited by ethylcholine aziridinium ion (AF64A) infused intraventricularly. J. Pharmacol. Exp. Ther., 241(1): 181–186.

Fink, K.B. and Göthert, M. (2007) 5-HT receptor regulation of neurotransmitter release. Pharmacol. Rev., 59(4): 360–417.

Fuller, R.W. (1992) Effects of p-chloroamphetamine on brain serotonin neurons. Neurochem. Res., 17(5): 449–456.

Galani, R., Lehmann, O., Bolmont, T., Aloy, E., Bertrand, F., Lazarus, C., Jeltsch, H. and Cassel, J.C. (2002) Selective immunolesions of CH4 cholinergic neurons do not disrupt spatial memory in rats. Physiol. Behav., 76(1): 75–90.

Galani, R., Berthel, M.C., Lazarus, C., Majchrzak, M., Barbelivien, A., Kelche, C. and Cassel, J.C. (2007) The behavioral effects of enriched housing are not altered by serotonin depletion but enrichment alters hippocampal neurochemistry. Neurobiol. Learn Mem., 88(1): 1–10.

Glanzman, D.L., Mackey, S.L., Hawkins, R.D., Dyke, A.M., Lloyd, P.E. and Kandel, E.R. (1989) Depletion of serotonin in the nervous system of Aplysia reduces the behavioural enhancement of gill withdrawal as well as the heterosynaptic facilitation produced by tail shock. J. Neurosci., 9(12): 4200–4213.

González-Burgos, I., Pérez-Vega, M.I., Del Angel-Meza, A.R. and Feria-Velasco, A. (1998) Effect of tryptophan restriction on short-term memory. Physiol. Behav., 63(2): 165–169.

Green, A.R., Mechan, A.O., Elliott, J.M., O'Shea, E. and Colado, M.I. (2003) The pharmacology and clinical pharmacology of 3,4-methylenedioxymethamphetamine (MDMA, 'ecstasy'). Pharmacol. Rev., 55(3): 463–508.

Harati, H., Barbelivien, A., Cosquer, B., Majchrzak, M. and Cassel, J.C. (2008) Selective cholinergic lesions in the rat nucleus basalis magnocellularis with limited damage in the medial septum specifically alter attention performance in the five-choice serial reaction time task. Neuroscience, 153(1): 72–83. 2008 Apr 22.

Hashimoto, S., Inoue, T. and Koyama, T. (1996) Serotonin reuptake inhibitors reduce conditioned fear stress-induced freezing behavior in rats. Psychopharmacology, 123(2): 182–186.

Hebb, D.O. (1949) The Organization of Behavior: A Neuropsychological Theory. Wiley, New York, NY.

Heinsbroek, R.P., Feenstra, M.G., Boon, P., Van Haaren, F. and Van de Poll, N.E. (1988) Sex differences in passive avoidance depend on the integrity of the central serotonergic system. Pharmacol. Biochem. Behav., 31(2): 499–503.

Hohmann, C.F., Walker, E.M., Boylan, C.B. and Blue, M.E. (2007) Neonatal serotonin depletion alters behavioral responses to spatial change and novelty. Brain Res., 1139: 163–177.

Hoyer, D., Hannon, J.P. and Martin, G.R. (2002) Molecular, pharmacological and functional diversity of 5-HT receptors. Pharmacol. Biochem. Behav., 71(4): 533–554.

Hritcu, L., Clicinschi, M. and Nabeshima, T. (2007) Brain serotonin depletion impairs short-term memory, but not long-term memory in rats. Physiol. Behav., 91(5): 652–657.

Huxley, A.F. (2002) Hodgkin and the action potential 1935–1952. J. Physiol., 538: 2.

Inoue, T., Hashimoto, S., Tsuchiya, K., Izumi, T., Ohmori, T. and Koyama, T. (1996) Effect of citalopram, a selective serotonin reuptake inhibitor, on the acquisition of conditioned freezing. Eur. J. Pharmacol., 311(1): 1–6.

Invernizzi, R., Carli, M., Di Clemente, A. and Samanin, R. (1991) Administration of 8-hydroxy-2-(di-n-propylamino) tetralin in raphe nuclei dorsalis and medianus reduces serotonin synthesis in the rat brain: differences in potency and regional sensitivity. J. Neurochem., 56(1): 243–247.

Jäkälä, P., Sirviö, J., Jolkkonen, J., Riekkinen, P. Jr., Acsady, L. and Riekkinen, P. (1992) The effects of p-chlorophenylalanine-induced serotonin synthesis inhibition and muscarinic blockade on the performance of rats in a 5-choice serial reaction time task. Behav. Brain Res., 51(1): 29–40.

Jäkälä, P., Sirviö, J., Riekkinen, P. Jr. and Riekkinen, P.J. Sr. (1993a) Effects of p-chlorophenylalanine and methysergide on the performance of a working memory task. Pharmacol. Biochem. Behav., 44(2): 411–418.

Jäkälä, P., Mazurkiewicz, M., Sirviö, J., Riekkinen, P. Sr. and Riekkinen, P. Jr. (1993b) The behavioral effects of serotonin synthesis inhibition and quisqualic acid induced lesions of the nucleus basalis magnocellularis in rats. Gen. Pharmacol., 24(5): 1141–1148.

Jeltsch, H., Cassel, J.C., Neufang, B., Kelche, C., Hertting, G., Jackisch, R. and Will, B. (1994) The effects of intrahippocampal raphe and/or septal grafts in rats with fimbria–fornix lesions depend on the origin of the grafted tissue and the behavioural task used. Neuroscience, 63(1): 19–39.

Jones, E.G. (1994) Santiago Ramón y Cajal and the Croonian lecture, March 1894. Trends Neurosci., 17: 190–192.

Kandel, E.R. (2001) The molecular biology of memory storage: a dialogue between genes and synapses. Science, 294(5544): 1030–1038.

Kawase, M., Sinhababu, A.K. and Borchardt, R.T. (1990) Mechanism of autoxidation of 5,7-dihydroxytryptamine: effect of fluorine substitution at positions 4 and/or 6. Chem. Pharm. Bull., 38(11): 2939–2946.

Keith, J.R., Wu, Y., Epp, J.R. and Sutherland, R.J. (2007) Fluoxetine and the dentate gyrus: memory, recovery of function, and electrophysiology. Behav. Pharmacol., 18(5–6): 521–531.

King, M.V., Marsden, C.A. and Fone, K.C. (2008) A role for the 5-HT(1A), 5-HT(4) and 5-HT(6) receptors in learning and memory. Trends Pharmacol. Sci., in press.

Klein, M. and Kandel, E.R. (1980) Mechanism of calcium current modulation underlying presynaptic facilitation and behavioral sensitization in Aplysia. Proc. Natl. Acad. Sci. USA, 77(11): 6912–6916.

Koenig, J., Cosquer, B. and Cassel, J.-C. (2008) Activation of septal 5-HT1A receptors alters spatial memory encoding, interferes with consolidation, but does not affect retrieval in rats subjected to a water-maze task. Hippocampus, 18(1): 99–118.

Lehmann, O., Jeltsch, H., Lehnardt, O., Pain, L., Lazarus, C. and Cassel, J.C. (2000) Combined lesions of cholinergic and serotonergic neurons in the rat brain using 192 IgG-saporin and 5,7-dihydroxytryptamine: neurochemical and behavioural characterization. Eur. J. Neurosci., 12(1): 67–79.

Lehmann, O., Bertrand, F., Jeltsch, H., Morer, M., Lazarus, C., Will, B. and Cassel, J.C. (2002) 5,7-DHT-induced hippocampal 5-HT depletion attenuates behavioural deficits produced by 192 IgG-saporin lesions of septal cholinergic neurons in the rat. Eur. J. Neurosci., 15(12): 1991–2006.

Lieben, C.K., Steinbusch, H.W. and Blokland, A. (2006) 5, 7-DHT lesion of the dorsal raphe nuclei impairs object recognition but not affective behavior and corticosterone response to stressor in the rat. Behav. Brain Res., 168(2): 197–207.

McLean, J.H., Darby-King, A., Sullivan, R.M. and King, S.R. (1993) Serotonergic influence on olfactory learning in the neonate rat. Behav. Neural. Biol., 60(2): 152–162.

Markowska, A.L. and Wenk, G.L. (1991) Serotonin influences the behavioral recovery of rats following nucleus basalis lesions. Pharmacol. Biochem. Behav., 38(4): 731–737.

Matsukawa, M., Ogawa, M., Nakadate, K., Maeshima, T., Ichitani, Y., Kawai, N. and Okado, N. (1997) Serotonin and acetylcholine are crucial to maintain hippocampal synapses and memory acquisition in rats. Neurosci. Lett., 230(1): 13–16.

Matsuno, K., Senda, T., Matsunaga, K., Mita, S. and Kaneto, H. (1993) Similar ameliorating effects of benzomorphans and 5-HT2 antagonists on drug-induced impairment of passive avoidance response in mice: comparison with acetylcholinesterase inhibitors. Psychopharmacology, 112(1): 134–141.

Mercer, A.R., Emptage, N.J. and Carew, T.J. (1991) Pharmacological dissociation of modulatory effects of serotonin in Aplysia sensory neurons. Science, 254(5039): 1811–1813.

Meneses, A. (2003) A pharmacological analysis of an associative learning task: 5-HT(1) to 5-HT(7)receptor subtypes function on a pavlovian/instrumental autoshaped memory. Learn Mem., 10(5): 363–372. 2003 Sep–Oct.

Misane, I. and Ögren, S.O. (2000) Multiple 5-HT receptors in passive avoidance: comparative studies of p-chloroamphetamine and 8-OH-DPAT. Neuropsychopharmacology, 22(2): 168–190.

Monleón, S., Vinader-Caerols, C., Arenas, M.C. and Parra, A. (2008) Antidepressant drugs and memory: insights from animal studies. Eur. Neuropsychopharmacol., 18(4): 235–248.

Morley, K.C., Gallate, J.E., Hunt, G.E., Mallet, P.E. and McGregor, I.S. (2001) Increased anxiety and impaired memory in rats 3 months after administration of 3,4-methylenedioxymethamphetamine ('ecstasy'). Eur. J. Pharmacol., 433(1): 91–99.

Moss, B.L., Fuller, A.D., Sahley, C.L. and Burrell, B.D. (2005) Serotonin modulates axo-axonal coupling between neurons critical for learning in the leech. J. Neurophysiol., 94(4): 2575–2589.

Murtha, S.J. and Pappas, B.A. (1994) Neurochemical, histopathological and mnemonic effects of combined lesions of the medial septal and serotonin afferents to the hippocampus. Brain Res., 651(1–2): 16–26.

Myhrer, T. (2003) Neurotransmitter systems involved in learning and memory in the rat: a meta-analysis based on studies of four behavioral tasks. Brain Res. Brain Res. Rev., 41(2–3): 268–287.

Nichols, D.E. and Nichols, C.D. (2008) Serotonin receptors. Chem. Rev., 108(5): 1614–1641.

Nilsson, O.G., Strecker, R.E., Daszuta, A. and Björklund, A. (1988) Combined cholinergic and serotonergic denervation of the forebrain produces severe deficits in a spatial learning task in the rat. Brain Res., 453(1–2): 235–246.

Nilsson, O.G., Brundin, P. and Björklund, A. (1990) Amelioration of spatial memory impairment by intrahippocampal grafts of mixed septal and raphe tissue in rats with combined cholinergic and serotonergic denervation of the forebrain. Brain Res., 515(1–2): 193–206.

Normile, H.J., Jenden, D.J., Kuhn, D.M., Wolf, W.A. and Altman, H.J. (1990) Effects of combined serotonin depletion and lesions of the nucleus basalis magnocellularis on acquisition of a complex spatial discrimination task in the rat. Brain Res., 536(1–2): 245–250.

Olvera-Cortés, E., Barajas-Pérez, M., Morales-Villagrán, A. and González-Burgos, I. (2001) Cerebral serotonin depletion induces egocentric learning improvement in developing rats. Neurosci. Lett., 313(1–2): 29–32.

Parent, M.B. and Baxter, M.G. (2004) Septohippocampal acetylcholine: involved in but not necessary for learning and memory? Learn Mem., 11(1): 9–20.

Piper, B.J. and Meyer, J.S. (2004) Memory deficit and reduced anxiety in young adult rats given repeated intermittent MDMA treatment during the periadolescent period. Pharmacol. Biochem. Behav., 79(4): 723–731.

Plaznik, A., Danysz, W. and Kostowski, W. (1983) Some behavioral effects of microinjections of noradrenaline and serotonin into the hippocampus of the rat. Physiol. Behav., 31(5): 625–631.

Plaznik, A., Danysz, W. and Kostowski, W. (1985) Some behavioral effects of microinjections of noradrenaline and serotonin into the amygdaloid body of the rat brain. Physiol. Behav., 34(4): 481–487.

Prado-Alcalá, R.A., Ruiloba, M.I., Rubio, L., Solana-Figueroa, R., Medina, C., Salado-Castillo, R. and Quirarte, G.L. (2003) Regional infusions of serotonin into the striatum and memory consolidation. Synapse, 47(3): 169–175.

Rapport, M.M. (1949) Serum vasoconstrictor (serotonin). V. The presence of creatinine in the complex: A proposed structure of the vasoconstrictor principle. J. Biol. Chem., 180: 961–969.

Rapport, M.M., Green, A.A. and Page, I.H. (1948a) Partial purification of the vasoconstrictor in beef serum. J. Biol. Chem., 174: 735–738.

Rapport, M.M., Green, A.A. and Page, I.H. (1948b) Crystalline serotonin. Science, 108: 329–330.

Ricaurte, G.A., Markowska, A.L., Wenk, G.L., Hatzidimitriou, G., Wlos, J. and Olton, D.S. (1993) 3,4-Methylenedioxymetha mphetamine, serotonin and memory. J. Pharmacol. Exp. Ther., 266(2): 1097–1105.

Richter-Levin, G. and Segal, M. (1989) Spatial performance is severely impaired in rats with combined reduction of serotonergic and cholinergic transmission. Brain Res., 477(1–2): 404–407.

Richter-Levin, G. and Segal, M. (1991) The effects of serotonin depletion and raphe grafts on hippocampal electrophysiology and behavior. J. Neurosci., 11(6): 1585–1596.

Richter-Levin, G., Greenberger, V. and Segal, M. (1993) Regional specificity of raphe graft-induced recovery of behavioral functions impaired by combined serotonergic/cholinergic lesions. Expl. Neurol., 121: 256–260.

Richter-Levin, G., Greenberger, V. and Segal, M. (1994) The effects of general and restricted serotonergic lesions on hippocampal electrophysiology and behavior. Brain Res., 642 (1–2): 111–116.

Riekkinen, M. and Riekkinen, P. Jr. (1994) Effects of THA and physostigmine on spatial navigation and avoidance performance in mecamylamine and PCPA-treated rats. Exp. Neurol., 125(1): 111–118.

Riekkinen, P. Jr., Sirviö, J. and Riekkinen, P. (1990) Interaction between raphe dorsalis and nucleus basalis magnocellularis in spatial learning. Brain Res., 527(2): 342–345.

Riekkinen, P. Jr., Sirviö, J., Valjakka, A., Miettinen, R. and Riekkinen, P. (1991a) Pharmacological consequences of cholinergic plus serotonergic manipulations. Brain Res., 552(1): 23–26.

Riekkinen, P. Jr., Jäkälä, P., Sirviö, J. and Riekkinen, P. (1991b) The effects of increased serotonergic and decreased cholinergic activities on spatial navigation performance in rats. Pharmacol. Biochem. Behav., 39(1): 25–29.

Riekkinen, P. Jr., Riekkinen, M., Sirviö, J. and Riekkinen, P. (1992) Effects of concurrent nicotinic antagonist and PCPA treatments on spatial and passive avoidance learning. Brain Res., 575(2): 247–250.

Riekkinen, M., Sirviö, J. and Riekkinen, P. Jr. (1993) Pharmacological consequences of nicotinergic plus serotonergic manipulations. Brain Res., 622(1–2): 139–146.

Riekkinen, P. Jr., Sirviö, J. and Riekkinen, M. (1994) Serotonin depletion decreases the therapeutic effect of nicotine, but not THA in medial septal-lesioned rats. Brain Res., 662(1–2): 95–102.

Robinson, T.E., Castañeda, E. and Whishaw, I.Q. (1993) Effects of cortical serotonin depletion induced by 3,4-methylenedioxymethamphetamine (MDMA) on behavior, before and after additional cholinergic blockade. Neuropsychopharmacology, 8(1): 77–85.

Sahgal, A. and Keith, A.B. (1993) Combined serotonergic–cholinergic lesions do not disrupt memory in rats. Pharmacol. Biochem. Behav., 45(4): 995–1001.

Sahley, C.L. (1994) Serotonin depletion impairs but does not eliminate classical conditioning in the leech Hirudo medicinalis. Behav. Neurosci., 108(6): 1043–1052.

Sakurai, Y. and Wenk, G.L. (1990) The interaction of acetylcholinergic and serotonergic neural systems on performance in a continuous non-matching to sample task. Brain Res., 519 (1–2): 118–121.

Santucci, A.C., Knott, P.J. and Haroutunian, V. (1996) Excessive serotonin release, not depletion, leads to memory impairments in rats. Eur. J. Pharmacol., 295(1): 7–17.

Sara, S.J. and Hars, B. (2006) In memory of consolidation. Learn Mem., 13(5): 515–521.

Shirahata, T., Tsunoda, M., Santa, T., Kirino, Y. and Watanabe, S. (2006) Depletion of serotonin selectively impairs short-term memory without affecting long-term memory in odor learning in the terrestrial slug Limax valentianus. Learn Mem., 13(3): 267–270.

Sinhababu, A.K. and Borchardt, R.T. (1988) Molecular mechanism of biological action of the serotonin neurotoxin 5,7-dihydroxytryptamine. Neurochem. Intl., 12: 273–284.

Sitaraman, D., Zars, M., Laferriere, H., Chen, Y.C., Sable-Smith, A., Kitamoto, T., Rottinghaus, G.E. and Zars, T. (2008) Serotonin is necessary for place memory in Drosophila. Proc. Natl. Acad. Sci. USA, 105(14): 5579–5584.

Skelton, M.R., Williams, M.T. and Vorhees, C.V. (2006) Treatment with MDMA from P11-20 disrupts spatial learning and path integration learning in adolescent rats but only spatial learning in older rats. Psychopharmacology, 189(3): 307–318.

Solana-Figueroa, R., Salado-Castillo, R., Quirarte, G.L., Galindo, L.E. and Prado-Alcalá, R.A. (2002) Enhanced inhibitory avoidance training protects against the amnesic effect of p-chloroamphetamine. Life Sci., 14;71(4): 391–399.

Stewart, C.A. and Reid, I.C. (2000) Repeated ECS and fluoxetine administration have equivalent effects on hippocampal synaptic plasticity. Psychopharmacology, 148(3): 217–223.

Twarog, B.M. and Page, I.H. (1953) Serotonin content of some mammalian tissues and urine and a method for its determination. Am. J. Physiol., 175: 157–161.

Valluzzi, J.A. and Chan, K. (2007) Effects of fluoxetine on hippocampal-dependent and hippocampal-independent learning tasks. Behav. Pharmacol., 18(5–6): 507–513.

Vanderwolf, C.H. (1987) Near-total loss of 'learning' and 'memory' as a result of combined cholinergic and serotonergic blockade in the rat. Behav. Brain Res., 23(1): 43–57.

Vanderwolf, C.H., Dickson, C.T. and Baker, G.B. (1990) Effects of p-chlorophenylalanine and scopolamine on retention of habits in rats. Pharmacol. Biochem. Behav., 35(4): 847–853.

Vizi, E.S. and Kiss, J.P. (1998) Neurochemistry and pharmacology of the major hippocampal transmitter systems: synaptic and nonsynaptic interactions. Hippocampus, 8(6): 566–607.

Volpe, B.T., Hendrix, C.S., Park, D.H., Towle, A.C. and Davis, H.P. (1992) Early post-natal administration of 5,7-dihydroxytryptamine destroys 5-HT neurons but does not affect spatial memory. Brain Res., 589(2): 262–267.

Ward, B.O., Wilkinson, L.S., Robbins, T.W. and Everitt, B.J. (1999) Forebrain serotonin depletion facilitates the acquisition and performance of a conditional visual discrimination task in rats. Behav. Brain Res., 100(1–2): 51–65.

Wiley, R.G., Oeltmann, T.N. and Lappi, D.A. (1991) Immunolesioning: selective destruction of neurons using immunotoxin to rat NGF receptor. Brain Res., 562(1): 149–153.

Wirth, S., Lehmann, O., Bertrand, F., Lazarus, C., Jeltsch, H. and Cassel, J.C. (2000) Preserved olfactory short-term memory after combined cholinergic and serotonergic lesions using 192 IgG-saporin and 5,7-dihydroxytryptamine in rats. Neuroreport, 11(2): 347–350.

Wooley, D.W. and Shaw, E. (1954) A biochemical and pharmacological suggestion about certain mental disorders. Proc. Natl. Acad. Sci. USA, 40: 228–231.

Yau, J.L., Hibberd, C., Noble, J. and Seckl, J.R. (2002) The effect of chronic fluoxetine treatment on brain corticosteroid receptor mRNA expression and spatial memory in young and aged rats. Brain Res. Mol. Brain Res., 106(1–2): 117–123.

Zakharov, I.S. and Balaban, P.M. (1987) Neural mechanisms of age-dependent changes in avoidance behaviour of the snail *Helix lucorum*. Neuroscience, 23(2): 721–729.

Zhang, Y., Lu, H. and Bargmann, C.I. (2005) Pathogenic bacteria induce aversive olfactory learning in Caenorhabditis elegans. Nature, 438(7065): 179–184.

CHAPTER 3.10

Social Behavior and Serotonin

Elif Aysimi Duman[1] and Turhan Canli[*,1,2]

[1]Department of Psychology, Stony Brook University, Stony Brook, New York
[2]Graduate Program in Genetics, Stony Brook University, Stony Brook, New York

Abstract: Serotonin (5-hydroxytryptamine, 5-HT) is implicated in the regulation of a wide range of social behaviors, including interpersonal aggression. Earlier studies investigating the role of 5-HT in aggression relied on the measurement of levels of the 5-HT metabolite 5-hydroxyindoleacetic acid (5-HIAA), and in later animal studies applied sophisticated pharmacological manipulations. With advances in molecular genetic techniques, the field has recently begun to focus on the genetic basis of individual differences in aggression. This chapter reviews studies of genetically modified animals and of gene polymorphisms in humans, with a particular focus on receptor genes, and gene polymorphisms located within the serotonin transporter (*5-HTT*), monoamine oxidase A (*MAOA*) and tryptophan hydroxylase 2 (*TPH2*) genes. We will also highlight recent imaging genetics studies and interactions with other genes and environmental factors, in order to point to promising future research approaches on the effects of serotonin on social behavior.

Keywords: serotonin, aggression, serotonin transporter (5-HTT), monoamine oxidase A (MAOA), tryptophan hydroxylase 2 (*TPH2*), gene polymorphism.

Introduction

Social neuroscience is one of the fastest growing areas of neuroscience, reflecting an increased interest in the neural underpinnings of complex behaviors of interacting individuals. These behaviors are shaped by both genetic and socio-environmental factors (Robinson *et al.*, 2008): genetic factors contribute to the structure and function of the brain, presumably including neural circuits that are relevant to social behavior, whereas socio-environmental factors shape an individual's behavior through encoded experiences and modifications in neural circuits. Through epigenetic and other mechanisms, socio–environmental factors can also regulate gene expression. Individual differences in social behavior may reflect the influence of common variations within genes that affect the neural circuitry underlying social behavior, as well as the influence of individuals' non-shared environment.

In this chapter, we will therefore focus on the role of genes and of common gene variants ('polymorphisms') associated with individual differences in social behavior. Because serotonin plays a central role in the regulation of social behavior, particularly aggression (see below), we will therefore focus on serotonergic gene variants and their role in interpersonal aggression (the discussion of self-directed aggression is beyond the scope of this chapter), which is of considerable societal importance (World Health Organization, 2007), and is a trait that shows both genetic and social environmental contributions (Seroczynski *et al.*, 1999).

Interpersonal aggression can be conceptualized in different ways. In many non-human animal studies, interpersonal aggression is conceptualized in terms of the stimuli that provoke the aggressive act, such as territorial aggression (Moyer, 1968). In many human studies, interpersonal aggression is conceptualized in terms of impulsive (reactive, uncontrolled) versus instrumental (goal-oriented, controlled) aggression, which are mediated by dissociable neural systems (Nelson and Trainor, 2007). Progress is made in our understanding of genetic and socio-environmental factors in these various forms of aggression. Much of this understanding rests on the neurobiology of the serotonergic system, as discussed below.

Serotonin and aggression

Serotonin (5-hydroxytryptamine, 5-HT) is a critical monoamine neurotransmitter of the central nervous system (CNS) that is produced mainly by the serotonergic

*Corresponding author
E-mail: turhan.canli@sunysb.edu

Christian Müller & Barry Jacobs (Eds.)
Handbook of Behavioral Neurobiology of Serotonin
ISBN 978-0-12-374634-4

DOI: 10.1016/B978-0-12-374634-4.00027-7

neurons found along the midline of the brainstem, from where it projects throughout the entire CNS (Jacobs and Azmitia, 1992). Raphe nuclei of the midbrain serve as a major source of serotonin in the forebrain, and send axons to various structures such as the hypothalamus, cortex, hippocampus, amygdala and striatum (Lucki, 1998). Due to its wide distribution throughout the CNS, serotonin is involved in the regulation of many physiological processes and behaviors, including aggression.

Evidence for a role of serotonin in aggression first came from studies measuring the main metabolite of serotonin, 5-hydroxyindoleacetic acid (5-HIAA). Most of these studies reported an inverse relationship between serotonergic activity and aggression, although this view has recently come under criticism (Olivier, 2004) because more recent data show that the relationship between serotonin levels and aggression varies with the type of aggression that is measured.

On the one hand, several studies have reported a negative correlation. For example, in mice, the turnover rate calculated from the 5-HT and 5-HIAA contents of *post-mortem* brain tissue is much lower in isolated aggressive individuals than it is in non-aggressive individuals (Giacalone *et al.*, 1968; Valzelli and Bernasconi, 1979). Human studies using peripheral measures of lumbar-punctured CSF 5-HIAA also suggest an inverse relationship between serotonin and aggression (Brown *et al.*, 1979; Kruesi *et al.*, 1990). These measures, however, were non-specific, as they were also correlated with other behavioral measures such as impulsivity, risk-taking behavior and suicidal tendencies (Kruesi *et al.*, 1990; Mann, 2003), suggesting that low levels of CSF 5-HIAA may be particularly relevant to impulsive forms of aggression. Indeed, this interpretation is supported by studies of dogs and non-human primates, which also reported negative correlations between CSF 5-HIAA levels and impulsive aggression (Higley *et al.*, 1996; Reisner *et al.*, 1996).

On the other hand, there is evidence for a *positive* correlation between serotonin and aggression, as has been reported between measures of CSF 5-HIAA levels and competitive aggression, which is used to attain higher social dominance (Higley *et al.*, 1996; Westergaard *et al.*, 1999). Thus, different forms of aggression may show opposite associations with serotonin levels.

Another important variable may be the state- versus trait-like properties of aggression. Van der Vegt and colleagues reported that, in rats, CSF 5-HIAA levels correlated positively with trait-like, but not with state-like, aggression (van der Vegt *et al.*, 2003). Of course, with regard to genetic mechanisms, one would search for genetic predictors of aggression particularly in trait-like phenotypes. Specific candidate genes that may impart individual differences in trait-like behavior will be discussed in the next section.

Serotonergic candidate genes

Serotonin function in the CNS is regulated by proteins that are involved in its synthesis, degradation, transport and reception. Therefore, the search for candidate genes involved in aggression begins with those that play a central role in coding for these proteins. For example, serotonin is synthesized from the amino acid tryptophan in two steps: the first rate-limiting step catalyzed by the tryptophan hydroxylase (TPH) and the second step catalyzed by the aromatic *l*-amino acid decarboxylase. TPH is produced in two isoforms, encoded by two genes: one is *TPH1*, which is primarily expressed in the periphery and the pineal gland; the other is *TPH2*, which is primarily expressed in the CNS (Walther and Bader, 2003). Another example concerns the degradation of serotonin by the enzymes monoamine oxidase A and B, which are encoded by the X-linked *MAOA* and *MAOB* genes, respectively. MAOA is especially important in the degradation of serotonin following its reuptake from the synaptic cleft by the serotonin transporter (Craig, 2007). There is only one mechanism for the reuptake of extracellular serotonin, regulating serotonergic transmission and tone, which is accomplished by the serotonin transporter, encoded by the serotonin transporter gene (*5-HTT, SERT*) (Gainetdinov and Caron, 2003). Postsynaptically, serotonin binds to any of 14 serotonin receptor subtypes: 5-HT_{1A}, 5-HT_{1B}, 5-HT_{1D}, 5-HT_{1E}, 5-HT_{1F}, 5-HT_{2A}, 5-HT_{2B}, 5-HT_{2C}, 5-HT_{3}, 5-HT_{4}, 5-HT_{5A}, 5-HT_{5B}, 5-HT_{6} and 5-HT_{7}. Several of these receptors have been shown to influence aggressive behavior, as discussed in the following section.

Serotonin receptors and aggression

The role of serotonin receptor subtypes in aggression has been mostly investigated in drug and genetic knockout mouse studies (Lesch and Merschdorf, 2000; Miczek *et al.*, 2002; Popova, 2006). One of these receptor subtypes is 5-HT_{1B}, which was studied extensively in mouse models (Olivier and van Oorschot, 2005). 5-HT_{1B} receptors are expressed in many brain regions, such as the raphe nuclei, periaquaductal gray, lateral septum, basal ganglia and hippocampus. 5-HT_{1B} receptors can function as autoreceptors that inhibit serotonergic neuron firing and thus serotonin release, or as heteroreceptors that release other neurotransmitters. In an earlier study, Saudou and colleagues reported that 5-HT_{1B} knockout mice exhibited more aggressive behaviors than controls (Saudou *et al.*, 1994), and this finding was also supported by pharmacological studies utilizing 5-HT_{1B} agonists (Olivier and van Oorschot, 2005).

A role for 5-HT_{1A} in aggression was documented in a study that compared 5-HT_{1A} receptor levels across two strains of mice: Korte and colleagues compared highly aggressive SAL male mice (who were genetically selected to exhibit short attack latencies to a non-aggressive opponent in a neutral cage) with low-aggression LAL mice, and reported that SAL mice were characterized by more postsynaptic 5-HT_{1A} receptors in limbic and cortical regions (Korte *et al.*, 1996). However, this phenotype may not be specific to aggression, because the SAL mice were also characterized by low corticosterone levels and a general pattern of active behavioral stress responsiveness (Korte *et al.*, 1996). On the other hand, data from humans suggest the opposite pattern between 5-HT_{1A} levels and aggression: evidence from pharmacological and positron emission tomography (PET) studies suggests an inverse relationship between questionnaire-assessed aggression and 5-HT_{1A} receptor function and binding in various regions of the brain (especially in the orbital prefrontal cortex) (Cleare and Bond, 2000; Parsey *et al.*, 2002).

Another serotonin receptor that is associated with aggression is the 5-HT_{2A} receptor, which is mainly found in the neocortex (Olivier, 2004). Antagonists of this receptor were shown to decrease aggressive behavior in rats and human clinical populations, although there are concerns that these effects are not aggression-specific, but rather act indirectly on aggressive behavior by altering locomotor activity (Miczek *et al.*, 2002). In humans, a recent human PET study (Meyer *et al.*, 2008) compared levels of 5-HT_{2A} binding potential (an index of receptor density) in the prefrontal cortex of violent aggressive individuals with healthy controls. It was reported that there is a negative correlation between aggression and prefrontal 5-HT_{2A} binding. Interestingly, this relation was only observed in younger subjects of the sample, and not in older subjects. Again, the association with aggression was confounded, because lower prefrontal binding was also associated with higher impulsivity scores.

Taken together, these studies suggest the involvement of 5-HT_{1A}, 5-HT_{1B} and 5-HT_{2A} receptor subtypes in aggression across both animal and human studies, although it should be noted that there may be other neurotransmitters systems acting upon the serotonergic neurons to elicit these changes (Siever, 2008). Moreover, the available data illustrate that future studies of the serotonergic regulation of aggressive behavior need to be designed to assess the role of potential confounding variables, such as age and trait impulsivity.

Serotonin transporter and aggression

The serotonin transporter, 5-HTT, enables the reuptake of excess serotonin from the synaptic cleft. The link to aggressive behavior has been demonstrated with genetically modified knockout *5-HTT* male mice, which were shown to have more extracellular 5-HT and exhibit less aggressive behavior, as measured by the latency and number of attacks towards an intruder, than did wild-type animals (Holmes *et al.*, 2002).

In non-human primates and in humans, a common polymorphism in the promoter region (*5-HTTLPR*) of the *5-HTT* gene has been the subject of many association studies investigating personality traits and psychiatric disorders (Lesch, 2007; Schinka *et al.*, 2004). This polymorphism exists in the form of two alleles (there is additional evidence for a single nucleotide polymorphism (SNP) within one of these alleles): a short (*s*) and a long (*l*) variant. The presence of one or two copies of the short variant (*s*-carriers) is associated with lower mRNA transcription (Lesch *et al.*, 1996). With respect to aggressive behavior, the *s*-variant is associated with violence in male adults (Liao *et al.*, 2004; Retz *et al.*, 2004), as well as aggressive behavior in children (Davidge *et al.*, 2004).

The *5-HTTLPR* polymorphism has also been shown to differentially moderate the effects of environmental factors, illustrating a gene-by-environment (G $\times$ E) interaction. For example, Barr and colleagues investigated the effects of early-life rearing conditions (mother- *vs* peer-reared) and *rh5-HTTLPR* genotype on aggression in juvenile macaque monkeys, as measured by bites, hairpulls, aggressive chases, hitting or slapping behaviors (Barr *et al.*, 2003). There was evidence for a G $\times$ E interaction: in *s/l* individuals peer-rearing was associated with much more aggressive behavior than was mother-rearing, but in *l/l* subjects rearing condition did not differentially affect aggressive behavior. A similar G $\times$ E interaction was observed in humans by Reif and colleagues, who reported that presence of the *5-HTTLPR* short allele moderates the effects of childhood maltreatment on later aggressive behavior (Reif *et al.*, 2007).

Using self-report measures of life stress history, we found evidence for a G $\times$ E interaction in the human brain (Canli *et al.*, 2006). High levels of life stress correlated positively with resting activation (as measured with perfusion imaging) in the amygdala and hippocampus in carriers of the *s*-allele, but correlated negatively in non-carriers. A similar pattern was recorded with fMRI during processing of emotional stimuli, and was observed in a wide range of brain regions involved in social behavior (Canli *et al.*, 2006; Canli and Lesch, 2007). We therefore speculated that *5-HTTLPR* genotype and a history of life stress interact to differentially shape neural circuits that generate social behavior (including aggression) (Canli and Lesch, 2007), but an explicit test of this hypothesis remains to be conducted.

Although an individual's genotype and life history are *ex post facto*, and thus represent a vulnerability that cannot be altered by therapeutic intervention, there is evidence that the current environment can ameliorate such vulnerability. For example, G × E vulnerability may be reduced through social support, as is suggested by Kaufman *et al.* (2004) in a study of maltreated children. These investigators showed that maltreated children with the *5-HTTLPR* short allele were more likely to become depressed than maltreated children who were homozygous for the long allele, but that this vulnerability could be significantly ameliorated if these children had social support in the form of a confidante. Future studies will hopefully address which kinds of social interventions are most successful in reducing the impact of adverse genetic and life history variables on later aggressive behavior. The discoveries from such studies would clearly be of broad scientific and public policy significance.

MAOA and aggression

Of the two forms of monoamine oxidases (MAOA and MAOB), most of the literature has focused on MAOA due to its role as the principal enzyme in 5-HT degradation, and its links with various types of aggressive behavior (Shih *et al.*, 1997; Popova, 2006). Disruption of the *MAOA* gene in mice is associated with increased aggressiveness, as measured by the intruder paradigm, as well as decreased firing in 5-HT neurons (Cases *et al.*, 1995; Popova *et al.*, 2001). Although the *MAOA* knockout mice showed increases in both serotonin and norepinephrine, the observed behavioral changes such as increased male aggression were linked to decreased serotonin degradation (Cases *et al.*, 1995). None of these effects were observed in *MAOB* knockout mice (Shih *et al.*, 1997, 1999). Very striking analogous human data come from a Dutch family study, in which a rare point mutation resulted in a stop codon in the *MAOA* gene, leading to a functional modification that is comparable to a 'knockout'. This mutation was associated with behavioral abnormalities like impulsive aggression in 14 male members of the family through four generations (Brunner *et al.*, 1993).

In addition to knockout studies, a variable number tandem repeat (VNTR) polymorphism in the promoter region of MAOA (*MAOA-LPR*) was shown to interact with both *5-HTTLPR* polymorphism and various environmental factors. This polymorphism results in 3, 3.5, 4 and 5 copies of this 30-bp sequence in the promoter region, leading to differential levels of expression of the gene and consequent differences in enzymatic activity. Accordingly, gene variants are classified as either low-activity (three- and five-repeat alleles) or high-activity alleles (Sabol *et al.*, 1998). The low-activity alleles were associated with reduced aggression in a community sample of males, as measured by the Aggression/Impulsivity (AI) index (Manuck *et al.*, 2000).

Non-human primate studies considering the orthologous *MAOA* VNTR polymorphism in rhesus monkeys (*rhMAOA-LPR*) have suggested that this polymorphism interacts with early life experiences (mother-reared or not) to regulate aggressive behavior. For example, Newman *et al.* (Newman *et al.*, 2005) investigated the role of *MAOA* genotype and rearing experience on aggressive behavior during competition for food and during social interactions in male rhesus monkeys. Whereas there was no main effect of either genotype or rearing condition on aggression, there was evidence for a G × E interaction: the most aggressive individuals were those who were reared by their mother and who also carried the low-activity *rhMAOA-LPR*. In contrast, animals with the same genotype who were reared by peers were less involved in aggressive behaviors, and did not differ significantly from mother-reared individuals with the high-activity allele (Newman *et al.*, 2005). This result is surprising, in light of data from *rh5-HTTLPR* and macaques (reviewed in the previous section) that suggest lower levels of aggression in genetically vulnerable individuals if they are reared by their mother. Newman and colleagues suggest that the higher level of aggression seen in mother-reared low-activity *rhMAOA-LPR*.monkeys may be adaptive, because the ability to show aggression in competitive social interactions is important for monkeys to obtain a higher rank in their social group. This argument may be valid, but could be equally applied to the *rh5-HTTLPR* data, and thus does not resolve the apparent contradiction. Yet the notion that aggression is not necessarily ill-adaptive is important, and future work could benefit from operationalizing the utility of aggression and measuring the influence of G × E interactions with respect to the adaptivity of aggressive behavior, rather than simply the extent of displayed aggressive behavior.

The first study to demonstrate any specific G × E interaction in humans was conducted by Caspi and colleagues, (Caspi *et al.*, 2002) in which the authors reported that the effect of childhood maltreatment on later development of antisocial behavior is moderated by the *MAOA-LPR* genotype. Maltreated children (all males in this study) with the low-activity variant of *MAOA* were more likely to show antisocial behavior than the ones with the high-activity variant. A replication study by Reif and colleagues also found an interaction between presence of the low-activity variant of *MAOA* and adverse childhood environment predicting violent behavior, but also identified significant main effects for both of these variables (Reif *et al.*, 2007). Other studies demonstrated that the interaction between *MAOA* genotype and childhood maltreatment influence the risk for both aggression and impulsive behavior in general (Caspi *et al.*, 2002; Huang *et al.*, 2004), an observation that was confirmed by a recent meta-analysis (Kim-Cohen *et al.*, 2006). These results suggest that the low-activity variants of the *MAOA* gene render individuals vulnerable towards an aggressive phenotype.

Another variable that may moderate the association between MAOA and aggression is the individual's sex. It was suggested that sex differences are not merely a result of *MAOA* being X-linked, but are also due to the different effects of sex hormones in males and females (Buckholtz and Meyer-Lindenberg, 2008). For example, it was reported that in rhesus monkeys, estrogen receptors are involved in regulating *MAOA* transcription in amygdala, anterior cingulate cortex (ACC) and orbitofrontal cortex (OFC), which are implicated in emotion regulation and partly in aggression (MacLusky *et al.*, 1986). Considering the significant sex differences in the expression of aggressive behaviors, the mechanisms by which sex hormones moderate the impact of *MAOA* on aggressive behavior will likely be a fruitful area of future work.

Genetic variation in *MAOA* and *5-HTT* may play an additive role in contributing to an aggressive phenotype. Wendland and colleagues noted that macaque species differ in their degree of tolerance with each other and with respect to hierarchical structures. Tolerant macaque societies were characterized by monomorphic *MAOA* and *5-HTTLPR* genotypes, whereas intolerant and hierarchical macaque societies were polymorphic at one or more of these loci (Wendland *et al.*, 2006). An explicit test of this hypothesis remains to be conducted.

The mechanism by which *MAOA* and *5-HTT* could both contribute to an aggressive phenotype may involve genetic regulation of the same neural circuitry. Indeed, recent imaging genetics studies (for a review and commentary, see Buckholtz and Meyer-Lindenberg, 2008) highlight the importance of *MAOA* and *5-HTT* on brain function. The low-activity variant of the *MAOA* was reported to increase amygdala and insula activation towards emotional facial stimuli, and decrease the involvement of PFC in emotion regulation (through OFC and ACC) (Meyer-Lindenberg *et al.*, 2006). Furthermore, impulsive aggressive individuals show decreased 5-HTT availability in the ACC, compared to controls (Frankle *et al.*, 2005), suggesting that modulation of serotonin levels in the ACC by either *MAOA* or *5-HTTLPR* variation may contribute to individual differences in ACC regulation of limbic regions, and potential dysregulation leading to impulsive aggression.

TPH2 and aggression

Tryptophan depletion studies have long pointed to a role for TPH in human aggressive behavior (Moeller *et al.*, 1996; LeMarquand *et al.*, 1998; Dougherty *et al.*, 1999; Marsh *et al.*, 2002). More recently, the *TPH2* gene was shown to contain various polymorphisms (Rotondo *et al.*, 1999), one of which (C779A) was associated with individual differences in CSF 5-HIAA levels (Nielsen *et al.*, 1994, 1998), which in turn have been linked to impulsivity and impulsive aggression (see above). More direct evidence for a role in aggression comes from an animal study in which an SNP in *Tph2* (C1473G) was shown to affect the enzymatic activity in mice and was associated with aggressive behavior. Mice that were homozygous for the C allele had higher *Tph2* activation in the midbrain than did mice that carried the G allele. The C/C mice also displayed higher levels of intermale aggression, as measured by the frequency of attacks towards an intruder, compared to mice that were homozygous for the G allele (Kulikov *et al.*, 2005).

Another SNP in the upstream regulatory sequence of *TPH2* (G703T) was recently associated with differential response in the amygdala towards both positive and negative emotional face stimuli (Brown *et al.*, 2005; Canli *et al.*, 2005). Using fMRI, our group reported that the presence of the T allele is associated with increased amygdala activation towards both happy *vs* neutral and fearful and sad *vs* neutral faces (Canli *et al.*, 2005). Using three different experimental paradigms (emotional picture perception, face perception, and an emotional word Stroop task) and two different brain activation measures (event-related potentials (ERPs) and fMRI), we and our collaborators reported additive effects of the *5-HTTLPR* polymorphism and *TPH2* G703T polymorphism in a sample of healthy volunteers (Herrmann *et al.*, 2007; Canli *et al.*, 2008). In both studies, subjects were divided into groups according to their *5-HTT* (*s*-carrier *vs* non-carrier) and *TPH2* (T-allele-carrier *vs* non-carrier) genotypes. The ERP study found that individuals who carried both of the gene variants that were associated with greater emotional reactivity (relative to a neutral control condition), i.e., the *s*-allele of *5-HTTLPR* and the T-allele of the *TPH2* polymorphism, showed the greatest neural activation of any group in response to emotional, compared to neutral, images (Herrmann *et al.*, 2007). The fMRI study found a similar pattern of activation in those who carried both the *s*-allele of *5-HTTLPR* and the T-allele of the *TPH2* polymorphisms, compared to all other individuals, when measuring brain activation in the putamen and amygdala in response to emotional faces and words (Canli *et al.*, 2008). These studies illustrate the additive effects of *5-HTT* and *TPH2* gene polymorphism in an individual's response towards emotional stimuli. At this time, an additive effect in impulsive aggression, or other forms of aggression, remains to be evaluated.

Concluding thoughts

In this chapter, we have reviewed the evidence for a role of the serotonergic system in different types of aggression. We integrated data from behavioral, genetic and neuroimaging studies, and highlighted the role for G × E interactions. To better understand the genetic contributions to individual differences in aggression, we highlighted studies that illustrate how polymorphisms within genes that play an important role in the synthesis, regulation or breakdown of serotonin have been associated with different forms of aggressive behavior. Moving towards more explicit mechanisms underlying G × E interactions, we discussed evidence for interactions between these gene polymorphisms and environmental factors, as in the case of *MAOA* and *5-HTT* interacting with adverse childhood environment. We discussed imaging studies that show associations between these genetic variations or G × E interactions and individual differences within neural circuits involved in the processing of emotional stimuli or involved in emotion regulation.

The reviewed literature illustrates how individual differences in social behavior can be traced to specific candidate genes, genetic variations, gene–gene additive effects, and G × E interactions, which are associated with individual differences in neural circuits that regulate social behavior. Future efforts will need to be directed at the underlying causal mechanisms by which genetic and environmental variables interact with each other, and how these interactions alter the structural and functional features of neural circuits. One likely candidate mechanism by which environmental factors interact with genes involves epigenetic alterations that modulate gene expression without altering the DNA sequence itself. The power of epigenetic regulation through social experience was first illustrated in the case of rat maternal care affecting both the offsprings' stress responses and the associated methylation of the glucocorticoid receptor gene (Weaver *et al.*, 2004). Intriguingly, a similar pattern has now been reported in human brains. McGowan and colleagues compared *post-mortem* brain tissue from suicide victims with a history of childhood abuse, no abuse, and controls, and found differential methylation in a promoter region of the glucocorticoid receptor gene as a function of history of childhood abuse (McGowan *et al.*, 2009).

As we advance our understanding of the mechanisms by which biological and environmental variables interact to shape social behavior, there will likely be strong interest in applying this knowledge to social policy. For example, reliable and accurate prediction of aggressive behavior would be relevant in the judicial system for parole decisions. Appropriate courses of punishment, treatment or interventions for juvenile delinquents or domestic abusers might be informed by a better understanding of any individual offender's genetic and life-history profile. Scientific discoveries could – indeed should – come to bear in these kinds of scenarios, with the goal of improving existing solutions. This is not to say that science can solve all social ills, or that scientific discoveries are around the corner that can deliver such accurate predictions. Indeed, one increasingly important duty of scientists should be to temper public expectations and a rush towards public policy applications based on preliminary findings. However, in times of social upheaval, understanding the mechanisms that explain individual differences in social behavior is a worthwhile effort with eventual significant public policy implications, if pursued with scientific rigor and public support.

Acknowledgement

This work was supported by the National Science Foundation, NSF # BCS-0224221, to Turhan Canli.

References

Barr, C.S., Newman, T.K., Becker, M.L., Parker, C.C., Champoux, M., Lesch, K.P., Goldman, D., Suomi, S.J. and Higley, J.D. (2003) The utility of the non-human primate; model for studying gene by environment interactions in behavioral research. Genes Brain Behav., 2: 336–340.

Brown, G.L., Goodwin, F.K., Ballenger, J.C., Goyer, P.F. and Major, L.F. (1979) Aggression in humans correlates with cerebrospinal fluid amine metabolites. Psychiatry Res., 1: 131–139.

Brown, S.M., Peet, E., Manuck, S.B., Williamson, D.E., Dahl, R.E., Ferrell, R.E. and Hariri, A.R. (2005) A regulatory variant of the human tryptophan hydroxylase-2 gene biases amygdala reactivity. Mol. Psychiatry, 10: 884–888.

Brunner, H.G., Nelen, M., Breakefield, X.O., Ropers, H.H. and van Oost, B.A. (1993) Abnormal behavior associated with a point mutation in the structural gene for monoamine oxidase A. Science, 262: 578–580.

Buckholtz, J.W. and Meyer-Lindenberg, A. (2008) MAOA and the neurogenetic architecture of human aggression. Trends Neurosci., 31: 120–129.

Canli, T. and Lesch, K.P. (2007) Long story short: the serotonin transporter in emotion regulation and social cognition. Nat. Neurosci., 10: 1103–1109.

Canli, T., Congdon, E., Gutknecht, L., Constable, R.T. and Lesch, K.P. (2005) Amygdala responsiveness is modulated by tryptophan hydroxylase-2 gene variation. J. Neural Transm., 112: 1479–1485.

Canli, T., Qiu, M., Omura, K., Congdon, E., Haas, B.W., Amin, Z., Herrmann, M.J., Constable, R.T. and Lesch, K.P. (2006) Neural correlates of epigenesis. Proc. Natl. Acad. Sci. USA, 103: 16033–16038.

Canli, T., Congdon, E., Constable, R.T. and Lesch, K.P. (2008) Additive effects of serotonin transporter and tryptophan hydroxylase-2 gene variation on neural correlates of affective processing. Biol. Psychol., 79: 118–125.

Cases, O., Seif, I., Grimsby, J., Gaspar, P., Chen, K., Pournin, S., Muller, U., Aguet, M., Babinet, C., Shih, J.C. et al. (1995) Aggressive behavior and altered amounts of brain serotonin and norepinephrine in mice lacking MAOA. Science, 268: 1763–1766.

Caspi, A., McClay, J., Moffitt, T.E., Mill, J., Martin, J., Craig, I.W., Taylor, A. and Poulton, R. (2002) Role of genotype in the cycle of violence in maltreated children. Science, 297: 851–854.

Cleare, A. and Bond, A. (2000) Ipsapirone challenge in aggressive men shows an inverse correlation between 5-HT 1A receptor function and aggression. Psychopharmacology, 148: 344–349.

Craig, I.W. (2007) The importance of stress and genetic variation in human aggression. Bioessays, 29: 227–236.

Davidge, K.M., Atkinson, L., Douglas, L., Lee, V., Shapiro, S., Kennedy, J.L. and Beitchman, J.H. (2004) Association of the serotonin transporter and 5HT1D [beta] receptor genes with extreme, persistent and pervasive aggressive behaviour in children. Psychiatr. Genet., 14: 143.

Dougherty, D.M., Bjork, J.M., Marsh, D.M. and Moeller, F.G. (1999) Influence of trait hostility on tryptophan depletion-induced laboratory aggression. Psychiatry Res., 88: 227–232.

Frankle, W.G., Lombardo, I., New, A.S., Goodman, M., Talbot, P.S., Huang, Y., Hwang, D.R., Slifstein, M., Curry, S. and Abi-Dargham, A. (2005) Brain serotonin transporter distribution in subjects with impulsive aggressivity: a positron emission study with [^{11}C] McN 5652. Am. J. Psychiatry, 162: 915.

Gainetdinov, R.R. and Caron, M.G. (2003) Monoamine transporters: from genes to behavior. Annu. Rev. Pharmacol. Toxicol., 43: 261–284.

Giacalone, E., Tansella, M., Valzelli, L. and Garattini, S. (1968) Brain serotonin metabolism in isolated aggressive mice. Biochem. Pharmacol., 17: 1315–1327.

Herrmann, M.J., Huter, T., Muller, F., Muhlberger, A., Pauli, P., Reif, A., Renner, T., Canli, T., Fallgatter, A.J. and Lesch, K.P. (2007) Additive effects of serotonin transporter and tryptophan hydroxylase-2 gene variation on emotional processing. Cereb. Cortex, 17: 1160–1163.

Higley, J.D., King Jr, S.T., Hasert, M.F., Champoux, M., Suomi, S.J. and Linnoila, M. (1996) Stability of interindividual differences in serotonin function and its relationship to severe aggression and competent social behavior in rhesus macaque females. Neuropsychopharmacology, 14: 67–76.

Holmes, A., Murphy, D.L. and Crawley, J.N. (2002) Reduced aggression in mice lacking the serotonin transporter. Psychopharmacology (Berl.), 161: 160–167.

Huang, Y.Y., Cate, S.P., Battistuzzi, C., Oquendo, M.A., Brent, D. and Mann, J.J. (2004) An association between a functional polymorphism in the monoamine oxidase a gene promoter, impulsive traits and early abuse experiences. Neuropsychopharmacology, 29: 1498–1505.

Jacobs, B.L. and Azmitia, E.C. (1992) Structure and function of the brain serotonin system. Am. Physiol. Soc., 72: 165–229.

Kaufman, J., Yang, B.Z., Douglas-Palumberi, H., Houshyar, S., Lipschitz, D., Krystal, J.H. and Gelernter, J. (2004) Social supports and serotonin transporter gene moderate depression in maltreated children. Proc. Natl. Acad. Sci. USA, 101: 17316–17321.

Kim-Cohen, J., Caspi, A., Taylor, A., Williams, B., Newcombe, R., Craig, I.W. and Moffitt, T.E. (2006) MAOA, maltreatment, and gene-environment interaction predicting children's mental health: new evidence and a meta-analysis. Mol. Psychiatry, 11: 903–913.

Korte, S.M., Meijer, O.C., de Kloet, E.R., Buwalda, B., Keijser, J., Sluyter, F., van Oortmerssen, G. and Bohus, B. (1996) Enhanced 5-HT1A receptor expression in forebrain regions of aggressive house mice. Brain Res., 736: 338–343.

Kruesi, M.J., Rapoport, J.L., Hamburger, S., Hibbs, E., Potter, W.Z., Lenane, M. and Brown, G.L. (1990) Cerebrospinal fluid monoamine metabolites, aggression, and impulsivity in disruptive behavior disorders of children and adolescents. Arch. Gen. Psychiatry, 47: 419–426.

Kulikov, A.V., Osipova, D.V., Naumenko, V.S. and Popova, N.K. (2005) Association between Tph2 gene polymorphism, brain tryptophan hydroxylase activity and aggressiveness in mouse strains. Genes Brain Behav., 4: 482–485.

LeMarquand, D.G., Pihl, R.O., Young, S.N., Tremblay, R.E., Seguin, J.R., Palmour, R.M. and Benkelfat, C. (1998) Tryptophan depletion, executive functions, and disinhibition in aggressive, adolescent males. Neuropsychopharmacology, 19: 333–341.

Lesch, K.P. (2007) Linking emotion to the social brain: the role of the serotonin transporter in human social behaviour. EMBO Reports 8 Special No: S24–S29.

Lesch, K.P. and Merschdorf, U. (2000) Impulsivity, aggression, and serotonin: a molecular psychobiological perspective. Behav. Sci. Law, 18: 581–604.

Lesch, K.P., Bengel, D., Heils, A., Sabol, S.Z., Greenberg, B.D., Petri, S., Benjamin, J., Muller, C.R., Hamer, D.H. and Murphy, D.L. (1996) Association of anxiety-related traits with a polymorphism in the serotonin transporter gene regulatory region. Science, 274: 1527–1531.

Liao, D.L., Hong, C.J., Shih, H.L. and Tsai, S.J. (2004) Possible association between serotonin transporter promoter region polymorphism and extremely violent crime in Chinese males. Neuropsychobiology, 50: 284–287.

MacLusky, N.J., Naftolin, F. and Goldman-Rakic, P.S. (1986) Estrogen formation and binding in the cerebral cortex of the developing rhesus monkey. Proc. Natl. Acad. Sci., 83: 513–516.

Mann, J.J. (2003) Neurobiology of suicidal behaviour. Nat. Rev. Neurosci., 4: 819–828.

Manuck, S.B., Flory, J.D., Ferrell, R.E., Mann, J.J. and Muldoon, M.F. (2000) A regulatory polymorphism of the monoamine oxidase-A gene may be associated with viability in aggression, impulsivity, and central nervous system serotonergic responsivity. Psychiatry Res., 95: 9–23.

Marsh, D.M., Dougherty, D.M., Moeller, F.G., Swann, A.C. and Spiga, R. (2002) Laboratory-measured aggressive behavior of women: acute tryptophan depletion and augmentation. Neuropsychopharmacology, 26: 660–671.

McGowan, P.O., Sasaki, A., D'Alessio, A.C., Dymov, S., Labonte, B., Szyf, M., Turecki, G. and Meaney, M.J. (2009) Epigenetic regulation of the glucocorticoid receptor in human brain associates with childhood abuse. Nat. Neurosci., 12: 342–348.

Meyer, J.H., Wilson, A.A., Rusjan, P., Clark, M., Houle, S., Woodside, S., Arrowood, J., Martin, K. and Colleton, M. (2008) Serotonin2A receptor binding potential in people with aggressive and violent behaviour. J. Psychiatry Neurosci., 33: 499.

Meyer-Lindenberg, A., Buckholtz, J.W., Kolachana, B., A, R.H., Pezawas, L., Blasi, G., Wabnitz, A., Honea, R., Verchinski, B.,

Callicott, J.H., Egan, M., Mattay, V. and Weinberger, D.R. (2006) Neural mechanisms of genetic risk for impulsivity and violence in humans. Proc. Natl. Acad. Sci. USA, 103: 6269–6274.

Miczek, K.A., Fish, E.W., de Bold, J.F. and de Almeida, R.M. (2002) Social and neural determinants of aggressive behavior: pharmacotherapeutic targets at serotonin, dopamine and γ-aminobutyric acid systems. Psychopharmacology, 163: 434–458.

Moeller, F.G., Dougherty, D.M., Swann, A.C., Collins, D., Davis, C.M. and Cherek, D.R. (1996) Tryptophan depletion and aggressive responding in healthy males. Psychopharmacology, 126: 97–103.

Moyer, K.E. (1968) Kinds of aggression and their physiological basis. Comm. Behav. Biol., 2: 65–87.

Nelson, R.J. and Trainor, B.C. (2007) Neural mechanisms of aggression. Nat. Rev. Neurosci., 8: 536–546.

Newman, T.K., Syagailo, Y.V., Barr, C.S., Wendland, J.R., Champoux, M., Graessle, M., Suomi, S.J., Higley, J.D. and Lesch, K.P. (2005) Monoamine oxidase A gene promoter variation and rearing experience influences aggressive behavior in rhesus monkeys. Biol. Psychiatry, 57: 167–172.

Nielsen, D.A., Goldman, D., Virkkunen, M., Tokola, R., Rawlings, R. and Linnoila, M. (1994) Suicidality and 5-hydroxyindoleacetic acid concentration associated with a tryptophan hydroxylase polymorphism. Arch. Gen. Psychiatry, 51: 34–38.

Nielsen, D.A., Virkkunen, M., Lappalainen, J., Eggert, M., Brown, G.L., Long, J.C., Goldman, D. and Linnoila, M. (1998) A tryptophan hydroxylase gene marker for suicidality and alcoholism. Arch. Gen. Psychiatry, 55: 593–602.

Olivier, B. (2004) Serotonin and aggression. Ann. NY. Acad. Sci., 1036: 382–392.

Olivier, B. and van Oorschot, R. (2005) 5-HT1B receptors and aggression: a review. Eur. J. Pharmacol., 526: 207–217.

Parsey, R.V., Oquendo, M.A., Simpson, N.R., Ogden, R.T., Van Heertum, R., Arango, V. and Mann, J.J. (2002) Effects of sex, age, and aggressive traits in man on brain serotonin 5-HT1A receptor binding potential measured by PET using [C-11]WAY-100635. Brain Res., 954: 173–182.

Popova, N.K. (2006) From genes to aggressive behavior: the role of serotonergic system. Bioessays, 28: 495–503.

Popova, N.K., Gilinsky, M.A., Amstislavskaya, T.G., Morosova, E.A., Seif, I. and De Maeyer, E. (2001) Regional serotonin metabolism in the brain of transgenic mice lacking monoamine oxidase A. J. Neurosci. Res., 66: 423–427.

Reif, A., Rösler, M., Freitag, C.M., Schneider, M., Eujen, A., Kissling, C., Wenzler, D., Jacob, C.P., Retz-Junginger, P. and Thome, J. (2007) Nature and nurture predispose to violent behavior: serotonergic genes and adverse childhood environment. Neuropsychopharmacology, 32: 2375–2383.

Reisner, I.R., Mann, J.J., Stanley, M., Huang, Y. and Houpt, K.A. (1996) Comparison of cerebrospinal fluid monoamine metabolite levels in dominant-aggressive and non-aggressive dogs. Brain Res., 714: 57–64.

Retz, W., Retz-Junginger, P., Supprian, T., Thome, J. and Rösler, M. (2004) Association of serotonin transporter promoter gene polymorphism with violence: relation with personality disorders, impulsivity, and childhood ADHD psychopathology. Behav. Sci. Law, 22: 415–425.

Robinson, G.E., Fernald, R.D. and Clayton, D.F. (2008) Genes and social behavior. Science, 322: 896–900.

Rotondo, A., Schuebel, K.E., Bergen, A.W., Aragon, R., Virkkunen, M., Linnoila, M., Goldman, D. and Nielsen, D.A. (1999) Identification of four variants in the tryptophan hydroxylase promoter and association to behavior. Mol. Psychiatry, 4: 360–368.

Sabol, S.Z., Hu, S. and Hamer, D. (1998) A functional polymorphism in the monoamine oxidase a gene promoter. Hum. Genet., 103: 273–279.

Saudou, F., Amara, D.A., Dierich, A., LeMeur, M., Ramboz, S., Segu, L., Buhot, M.C. and Hen, R. (1994) Enhanced aggressive behavior in mice lacking 5-HT1B receptor. Science, 265: 1875–1878.

Schinka, J.A., Busch, R.M. and Robichaux-Keene, N. (2004) A meta-analysis of the association between the serotonin transporter gene polymorphism (5-HTTLPR) and trait anxiety. Mol. Psychiatry, 9: 197–202.

Seroczynski, A.D., Bergeman, C.S. and Coccaro, E.F. (1999) Etiology of the impulsivity/aggression relationship: genes or environment? Psychiatry Res., 86: 41–57.

Shih, J.C., Grimsby, J. and Chen, K. (1997) Molecular biology of monoamine oxidase A and B: their role in the degradation of serotonin. Handb. Exp. Pharmacol., 129: 655–670.

Shih, J.C., Chen, K. and Ridd, M.J. (1999) Monoamine oxidase: from genes to behavior. Annu. Rev. Neurosci., 22: 197–217.

Siever, L.J. (2008) Neurobiology of aggression and violence. Am. J. Psychiatry, 165: 429–442.

Valzelli, L. and Bernasconi, S. (1979) . Aggressiveness by isolation and brain serotonin turnover changes in different strains of mice, Neuropsychobiology, 5: 129–135.

van der Vegt, B.J., Lieuwes, N., Cremers, T., de Boer, S.F. and Koolhaas, J.M. (2003) Cerebrospinal fluid monoamine and metabolite concentrations and aggression in rats. Horm. Behav., 44: 199–208.

Walther, D.J. and Bader, M. (2003) A unique central tryptophan hydroxylase isoform. Biochem. Pharmacol., 66: 1673–1680.

Weaver, I.C., Cervoni, N., Champagne, F.A., D'Alessio, A.C., Sharma, S., Seckl, J.R., Dymov, S., Szyf, M. and Meaney, M.J. (2004) Epigenetic programming by maternal behavior. Nat. Neurosci., 7: 847–854.

Wendland, J.R., Lesch, K.P., Newman, T.K., Timme, A., Gachot-Neveu, H., Thierry, B. and Suomi, S.J. (2006) Differential functional variability of serotonin transporter and monoamine oxidase a genes in macaque species displaying contrasting levels of aggression-related behavior. Behav. Genet., 36: 163–172.

Westergaard, G.C., Mehlman, P.T., Suomi, S.J. and Higley, J.D. (1999) CSF 5-HIAA and aggression in female macaque monkeys: species and interindividual differences. Psychopharmacology, 146: 440–446.

World Health Organization (2007) Third Milestones of a Global Campaign for Violence Prevention Report 2007: Scaling Up. WHO, Geneva.

CHAPTER 3.11

Serotonin in Pain and Pain Control

Claudia Sommer*

Department of Neurology, University of Würzburg, Germany

Abstract: In pain processing and modulation, serotonin (5-hydroxtryptamine, 5-HT) has excitatory (hyperalgesic) and inhibitory (analgesic) actions, depending on the site of action, the cell type and the type of receptor. In the periphery, 5-HT sensitizes afferent nerve fibers, thus contributing to inflammatory and neuropathic hyperalgesia. In the trigeminal system, agonism at $5\text{-HT}1_{B/D}$ receptors reduces neurotransmitter release, whereas actions through the 5-HT_{2A} receptor may underlie chronic headache. Genetic alterations in the 5-HT system may influence the susceptibility to migraine and to other pain disorders. 5-HT is involved in descending inhibitory pathways in the CNS, and modulation of this system is the most likely mechanism of action of antidepressant drugs in analgesia. Recently, facilitatory serotonergic pathways have been discovered, which may be functionally important.

Keywords: serotonin (5-HT), pain, hyperalgesia, neurotransmitters, pain pathways, inflammation.

Introduction

The effect mediated by 5-HT release is dependent on the cell and receptor type the neurotransmitter acts upon, and on the integration of the signal into the respective neuronal network. The classical understanding of serotonin is that of an analgesic action at spinal sites, which is considered part of the system of descending inhibition (Yaksh and Wilson, 1979). Strengthening this system, together with noradrenergic descending pathways, is the most likely mechanism of action of antidepressant drugs in pain control. Many experimental studies have further shown an interaction of the serotonergic system and morphine analgesia. Recently, facilitatory spinal serotonergic pathways have also been described (Suzuki *et al.*, 2004a). In peripheral tissues, 5-HT acts as a proalgesic inflammatory mediator (Moalem *et al.*, 2005). A specific role for 5-HT has been discussed in the pathophysiology of several types of headache (in particular migraine) and facial pain, of the fibromyalgia syndrome, and in neurogenic inflammation. For instance, 5-HT acting on $5\text{-HT}_{1B/D}$ receptors in the trigeminal system blocks the release of excitatory neurotransmitters and is analgesic. Down-regulation of 5-HT receptors is considered one of the pathomechanisms of medication-induced headache. This chapter will review the physiology of 5-HT in pain and analgesia, and summarize the evidence for a role of 5-HT in specific painful disorders.

*Corresponding author
E-mail: sommer@uni-wuerzburg.de

Basic physiology of serotonin in pain and pain control

5-HT in the periphery

Findings in animal models

The 5-HT tissue content increases rapidly in inflammation or injury. The main cellular sources of 5-HT in peripheral tissues are platelets and mast cells. After release of 5-HT, its action depends on the receptors present on afferent nerves in the vicinity (Figure 1). The mRNA for the 5-HT_{1B}, 5-HT_{1D}, $5\text{-HT}_{2A\text{-}C}$, 5-HT_3, 5-HT_4 and 5-HT_7 receptor subtypes has been detected in dorsal root ganglia (DRG) (Pierce *et al.*, 1996; Nicholson *et al.*, 2003), suggesting the presence of all these receptors on peripheral sensory nerves. However, which receptors are preferentially expressed may depend on various factors. For example, in isolated DRG neurons, stimulation of 5-HT_1 receptors had a hyperpolarizing effect, whereas stimulation 5-HT_2 receptors had a depolarizing effect on nociceptors (Todorovic and Anderson, 1992).

In most preparations, 5-HT was more potent in enhancing algesic effects of other mediators than in inducing pain by itself (Lang *et al.*, 1990; Abbott *et al.*, 1996). In rats, intradermally injected serotonin produces a dose-dependent hyperalgesia with a very short latency, indicating a direct excitatory effect on primary afferent neurons. This hyperalgesia can be mimicked by agonists for the 5-HT_{1A} receptor (Taiwo and Levine, 1992). Application

Christian Müller & Barry Jacobs (Eds.)
Handbook of Behavioral Neurobiology of Serotonin
ISBN 978-0-12-374634-4

DOI: 10.1016/B978-0-12-374634-4.00028-9

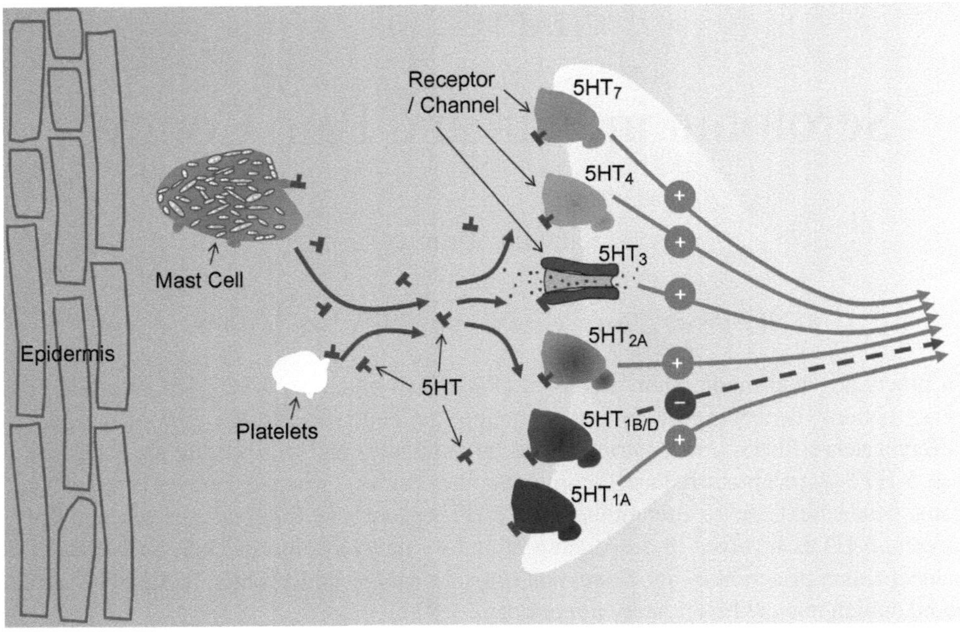

Figure 1 Simplified schematic drawing of 5-HT action in the periphery. To see the full color version of this figure please refer to the color plate in the back of the book. Copies produced via our print on demand service do not contein color plates. If your copy does not have the color plate, Please go to this website to view the figure in color www. elsevierdirect.com/companions/9780123746344

of 5-HT to nerve roots in rats induced short-term pain behavior similar to that with nucleus pulposus application, suggesting that 5-HT plays a role in the early phase of the pathogenesis of sciatica (Kato *et al.*, 2008). Thus, direct hyperalgesic effects as well as facilitatory effects of 5-HT for other peripheral mediators are possible.

In mice lacking the serotonin transporter 5-HTT, which is necessary for 5-HT reuptake into cells from the extracellular space (Bengel *et al.*, 1998), thermal hyperalgesia after nerve injury does not develop, most likely because of the reduced 5-HT levels in peripheral tissues of these mice (Vogel *et al.*, 2003). A similar reduction of thermal hyperalgesia was found in a model of inflammatory pain using complete Freund's adjuvant (CFA) injection into the mouse hindpaw (Palm *et al.*, 2008). 5-HT levels were increased after inflammation in the sciatic nerves of wild-type mice but not of 5-HTT knockout mice. The phenotype could be rescued by intraplantar injection of 5-HT, giving proof of the role of peripheral 5-HT in this model. In both models, chronic constriction injury (CCI) of the sciatic nerve and CFA-induced inflammation, the development of thermal hyperalgesia was dependent on an increase of 5-HT in the sciatic nerve. This finding may be of clinical relevance, since gene variants in humans modifying 5-HTT function might be one of the factors determining the extent of hyperalgesia in inflammation. Genetic variants are also known for guanosine triphosphate (GTP) cyclohydrolase, the rate-limiting enzyme for tetrahydrobiopterin synthesis, which is a cofactor for serotonin production, among others (Tegeder *et al.*, 2006). A haplotype leading to low tetrahydrobiopterin synthesis is associated with reduced postoperative and experimental pain.

The 5-HT_3 receptor, a ligand-gated cation channel, is the most widely studied 5-HT receptor in the concept of peripheral pain and analgesia (for review, see Sommer, 2004). Other investigators have identified 5-HT_1 and 5-HT_4 as the receptors involved in nociceptive behavior in formalin-induced paw inflammation in the rat (Eschalier *et al.*, 1989; Doak and Sawynok, 1997). 5-HT_3 receptor antagonists applied subcutaneously reduced formalin and complete Freund's adjuvant (CFA) induced pain (Giordano and Dyche, 1989; Giordano and Rogers, 1989). Pain but not edema in carrageenin-induced inflammation was blocked by preventive treatment with a 5-HT_3 antagonist (Eschalier *et al.*, 1989). Repeated injection of 5-HT into rat masticatory muscle evoked brief afferent fiber discharges, which could be significantly attenuated by the selective 5-HT_3 receptor antagonist tropisetron (Sung *et al.*, 2008). 5-HT, through activation of peripheral 5-HT_3 receptors, thus excites slowly conducting masticatory muscle afferent fibers. Similar mechanisms might contribute to 5-HT-evoked muscle pain in humans. In mice deficient in 5-HT_3 receptors, assays for acute pain, phase 1 of the formalin test, and tests for hyperalgesia and allodynia after partial nerve lesion gave normal results, but pain behavior in phase 2 of the formalin test was reduced, indicating a role of the 5-HT_3 receptor in tissue-injury-induced persistent nociception (Zeitz *et al.*, 2002). Furthermore, the visceral response to intraperitoneal

5-HT was reduced in the mutant mice. In the setting of CFA-induced hindpaw inflammation, the 5-HT_3 receptor was found to mediate activation of nociceptors but did not contribute to injury-associated edema. This result was explained by the distribution of 5-HT_3 receptor in DRG neurons in a minority of neurons of all sizes, the majority of which do not coexpress substance P or the vanilloid receptor TRPV1, thus defining a new subclass of DRG neurons (Zeitz *et al.*, 2002). Thus, the 5-HT_3 receptor is an important, but not the only, receptor mediating the peripheral algesic action of 5-HT.

A number of experimental studies point to an important role of the G-protein-coupled 5-HT_2 receptor in peripheral 5-HT-mediated pain. 5-HT_2 receptors are expressed in CGRP-containing small-diameter DRG neurons. They are upregulated in inflammatory pain, and a 5-HT_{2A} receptor antagonist reduces inflammatory hyperalgesia (Okamoto *et al.*, 2002). 5-HT_2 agonists injected into the plantar surface of the paw in rats induced lifting and licking, which was greatly enhanced in combination with prostaglandin E_2 and noradrenalin, and could be inhibited by the 5-$HT_{2A/2C}$ antagonist ketanserin (Abbott *et al.*, 1996). The density of 5-HT_{2A} receptor immunoreactive axon terminals in the superficial layers of the dorsal horn was increased in vincristine neuropathy. Furthermore, 5-HT_{2A} knockout mice did not develop neuropathic pain after vincristine administration, indicating that 5-HT_{2A} receptors are involved in peripheral sensitization and spinal nociceptive processing (Thibault *et al.*, 2008). Injection of 5-HT and of the 5-HT_{2A} receptor agonist alpha-methyl 5-HT reduced the paw-withdrawal latency to noxious heat in rats, indicating a role of the 5-HT_{2A} receptor in thermal hyperalgesia (Tokunaga *et al.*, 1998). Treatment with 5-HT_{2A} antagonists reduced the pain response in rats injected with formalin and carrageenan (Abbott *et al.*, 1997; Obata *et al.*, 2000; Wei *et al.*, 2005). Mechanistically, it has been suggested that activation of 5-HT_2 receptors depolarizes capsaicin-sensitive DRG neurons by reducing a resting potassium conductance, which might contribute to excitation and sensitization (McMahon *et al.*, 2006).

5-HT_4 receptors have been shown to increase tetrodotoxin-resistant sodium currents via stimulation of cyclic AMP production (Cardenas *et al.*, 1997, 2001). This would increase the probability of action potential firing, and thus be proalgesic. 5-HT_4 receptor agonists have been mostly discussed in the context of treating irritable bowel syndrome (Tonini and Pace, 2006) and in the prevention of opioid-induced respiratory depression (Manzke *et al.*, 2003; Lötsch *et al.*, 2005). The 5-HT_7 receptor has also been implied in peripheral hyperalgesia. Local administration of 5-HT_7 antagonist reduced formalin induced flinching in phase 1 and phase 2 (Rocha-Gonzalez *et al.*, 2005).

The activation of 5-$HT_{1B/D}$ receptors on trigeminal afferents inhibits adenylate cyclase via activation of $G_{i/o}$ proteins. 5-$HT_{1B/D}$ agonists (called triptans), such as sumatriptan and others, act on 5-$HT_{1B/D}$ receptors and inhibit neuropeptide release from the terminals of these afferents (Arvieu *et al.*, 1996). Sumatriptan also prevented capsaicin-induced hyperemia, a sign of neurogenic inflammation (Zochodne and Ho, 1994). Furthermore, sumatriptan reduced thermal hyperalgesia in mice with peripheral inflammation, but had no effect on nerve-injury-induced hyperalgesia (Bingham *et al.*, 2001). In summary, the majority of data point to a proalgesic action of 5-HT in the periphery at most receptors, with the exception of agonists at the 5-$HT_{1B/D}$ receptor, which may reduce neurogenic inflammation.

5-HT agonists and antagonists in human studies

In healthy volunteers, 5-HT infused into muscle did not induce pain or hyperalgesia by itself, but sensitized the tissue to bradykinin (Babenko *et al.*, 2000). 5-HT administered through intradermal microdialysis membranes produced pain and sometimes itching (Lischetzki *et al.*, 2001; Schmelz *et al.*, 2003). Injections of 5-HT into the masseter muscle in healthy human females induced pain and hyperalgesia which could be antagonized by a 5-HT_3-antagonist (Ernberg *et al.*, 2000, 2006). Topical application of the 5-HT_3 receptor antagonist ondansetron reduced capsaicin-induced pain and hyperalgesia in human volunteers (Giordano *et al.*, 1998). In contrast, the application of 5-HT_3 receptor antagonists locally in humans with fibromyalgia or similar pain states gave equivocal results (Ernberg *et al.*, 2003; Stratz and Müller, 2003).

Mechanisms of 5-HT actions on nociception in the periphery

The neuronal mechanism of this pain and hyperalgesia were investigated in *in vitro* preparations of peripheral nerves. In these preparations, 5-HT has an excitatory effect on peripheral nerve fibers. Some investigators used a mixture of inflammatory mediators ('inflammatory soup') (Kessler *et al.*, 1992), and found ectopic excitation and sensitization of acutely axotomized afferent nerve fibers in the rat (Michaelis *et al.*, 1997). In DRG neurons in culture, 5-HT alone had no major effect, but a proton-induced current was increased by the combination of 5-HT with other inflammatory mediators (Kress *et al.*, 1997).

Recently, the selective administration of 5-HT and a 5-HT_3 receptor agonist has also been shown to sensitize C fibers in isolated segments of rat sural nerve (Moalem *et al.*, 2005). The characteristics of this sensitization were not altered after previous nerve injury, which led the authors to conclude that constitutive rather than inducible expression of axonal 5-HT receptors is responsible for this effect. Others provided indirect evidence that 5-HT may have a stronger effect on lesioned or inflamed tissues and peripheral nerve fibers than on intact ones (Aley *et al.*, 2000; Song *et al.*, 2003).

Direct and indirect mechanisms of action have been postulated for the peripheral hyperalgesic action of 5-HT. 5-HT modulates tetrodotoxin-resistant (TTX-R) sodium currents, where it increases the magnitude of the current, shifts its conductance–voltage relationship to a hyperpolarized direction, and increases its rate of activation and inactivation (Gold *et al.*, 1996). The 5-HT_3 receptor, itself a ligand-gated ion channel, might directly enhance neuronal activity. Since 5-HT can sensitize nerve fibers to the actions of bradykinin, an effect on bradykinin receptors might be expected. Alternatively, downstream signaling of the mediators might converge and mutually enhance the effects. Protein kinase A (PKA) was identified as a factor in the signaling cascade of 5-HT after its intraplantar injection (Aley and Levine, 1999), and NO was suggested to facilitate this process, in analogy to findings with PGE_2 (Aley *et al.*, 1998). In an extensive pharmacological study, 5-HT-induced hyperalgesia was significantly reduced by local blockade of the 5-HT_3 receptor by tropisetron, by the non-specific selectin inhibitor fucoidan, by the cyclooxygenase inhibitor indomethacin, by guanethidine depletion of norepinephrine in the sympathetic terminals, and by local blockade of the $beta_1$- or $beta_2$-adrenergic receptors. The authors concluded that there is an indirect hyperalgesic action of 5-HT, mediated by a combination of mechanisms involved in inflammation such as neurophil migration and the local release of prostaglandin and norepinephrine (Oliveira *et al.*, 2007).

5-HT in the central nervous system

Data from animal experiments

Numerous studies have been performed to delineate the role of 5-HT in central pain processing. A variety of approaches were used, including methods to reduce 5-HT levels in the CNS (like dietary tryptophan depletion or inhibition of serotonin biosynthesis, to block or activate 5-HT receptors) and selective lesions of serotonergic projections. Through many experimental studies, 5-HT has been recognized, together with norepinephrine, as one of the main neurotransmitters involved in endogenous supraspinal pain-modulating systems (Basbaum and Fields, 1978). This is in accordance with the clinical use of drugs increasing the availability of 5-HT and norepinephrine, like antidepressants, in the treatment of chronic pain. However, excitatory effects of spinal 5-HT have repeatedly been described, and have to be integrated into our understanding of the serotonergic system (Figure 2).

Spinal 5-HT is largely derived from serotonergic neurons in the rostroventral medulla (RVM), in particular in the nucleus raphe magnus (NRM). Serotonergic tracts descend to the spinal cord through the dorsolateral funiculus and form synapses in laminae I, II, IV and V (Millan, 2002). The serotonergic RVM neurons are activated by noxious stimuli through opioid mechanisms (Zhang *et al.*, 2000). Their electrical stimulation evokes the release of 5-HT in the spinal cord cerebrospinal fluid, and 5-HT antagonists reduce the analgesia produced by this stimulation (Le Bars and Villanueva, 1988). In slice experiments containing RVM, some raphe neurons exhibit spontaneous firing, suggesting that 5-HT is released tonically onto the dorsal horn neurons (Yoshimura and Furue, 2006). Chronic pain states may reduce 5-HT in RVM neurons, thus reducing endogenous pain control and the effect of morphine (Sounvoravong *et al.*, 2004).

Stimulation of RVM or injections of 5-HT-receptor agonists into the spinal cord have shown both inhibitory and facilitatory effects on the pain behavior in animals. Electrophysiological studies showed an inhibition of the response of dorsal horn neurons to noxious stimulation when 5-HT was applied by iontophoreses (Duggan and Headley, 1977; Jordan *et al.*, 1978). Spinal administration of 5-HT mostly reduces pain behavior in animals, but results depend on the stimulus quality and the test used (Bardin *et al.*, 1997; Hains *et al.*, 2003). The antinociceptive activity of 5-HT is generally weak in models of neuropathic and inflammatory pain (Bardin *et al.*, 1997; Obata *et al.*, 2001). Excitatory effects have also been reported, mediated via 5-HT_7 receptors (Rocha-Gonzalez *et al.*, 2005), and possibly caused by increases in cyclic AMP levels, which shifts the voltage dependence of the cation current to more depolarized potentials and thus increases neuronal excitability (Cardenas *et al.*, 1999).

In the spinal cord, 5-HT receptors are present on terminals of primary afferent neurons, on projection neurons, and on excitatory and inhibitory interneurons, which makes the action of 5-HT very complex. Accordingly, blockade or depletion of spinal 5-HT can enhance nociception (Millan *et al.*, 1997; Pertovaara *et al.*, 2001). Excellent reviews on this subject have been published recently (Millan, 2002; Suzuki *et al.*, 2004b; Lopez-Garcia, 2006; Yoshimura and Furue, 2006). In particular, 5-HT_{2A} and 5-HT_3 receptors, among others (Hamon and Bourgoin, 1999; Nicholson *et al.*, 2003), are expressed by primary afferent fibers that convey nociceptive messages from the periphery to the CNS (Hamon *et al.*, 1989; Carlton and Coggeshall, 1997). 5-HT presynaptically

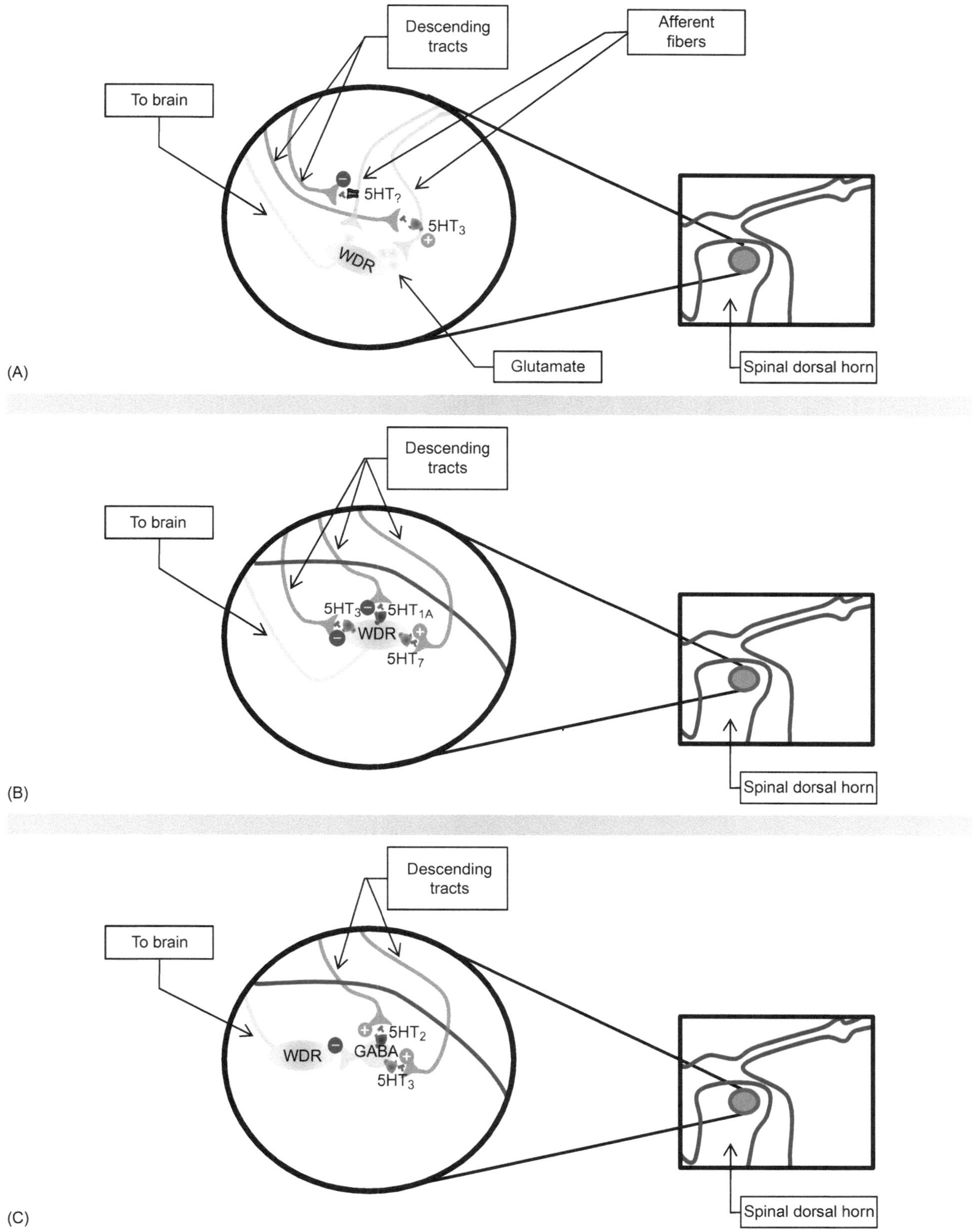

Figure 2 Simplified schematic drawing of the 5-HT descending system in the spinal cord. To see the full color version of this figure please refer to the color plate in the back of the book. Copies produced via our print on demand service do not contein color plates. If your copy does not have the color plate, Please go to this website to view the figure in color www. elsevierdirect.com/companions/9780123746344

inhibits the glutamate release from nociceptors, but the receptor responsible for this effect has not yet been identified (Ito *et al.*, 2000). 5-HT_{1A} and 5-HT_{1B} receptors are present at high densities in superficial dorsal horn neurons, for example on nociceptive-specific or on wide dynamic range (WDR) neurons. 5-HT_{1A} receptors, when activated, cause opening of K^+ and a closing of Ca^{2+} channels through coupling negatively to adenylate cyclase. Their activation should thus directly hyperpolarize substantia gelatinosa neurons and produce an inhibitory effect on the sensory transmission. However, application of 5-HT_{1A} and 5-HT_{1B} receptor agonist was also shown to increase responses of WDR neurons to peripheral stimulation (Zhang *et al.*, 2001a). This may be mediated by the presence of 5-HT_1 receptors on GABAergic interneurons, which would consequently be inhibited upon stimulation of the 5-HT_1 receptors (Millan *et al.*, 1996). 5-HT_2 receptors are positively coupled to phospholipase C, and thus exert excitatory influcences on neuronal activity. The antinociceptive effects attributed to the 5-HT_{2A} subtype (Courade *et al.*, 2001; Kjorsvik *et al.*, 2001; Radhakrishnan *et al.*, 2003; Sasaki *et al.*, 2003) would thus also have to be caused by the activation of inhibitory interneurons. Interestingly, 5-HT_{2A} receptors are scarce in normal spinal cord but are up-regulated after inflammation, indicating a state-specific action of 5-HT on these receptors (Zhang *et al.*, 2001b).

The ionotropic 5-HT_3 receptor increases neuronal excitability, and has been found to mediate descending facilitation through serotonergic pathways (Rahman *et al.*, 2004; Suzuki *et al.*, 2004a). A study with mice deficient of the 5-HT_3 receptor came to the conclusion that this receptor was proalgesic at peripheral and spinal sites (Zeitz *et al.*, 2002). If 5-HT acts on a presynaptic spinal 5-HT_3 receptor, excitatory neurotransmitter release is increased. Concordantly, ondansetron, a 5-HT_3 antagonist, reduces pain behavior upon mechanical stimuli after nerve injury (Suzuki *et al.*, 2004a) and mechanical allodynia after spinal cord injury in rats (Oatway *et al.*, 2004). Furthermore, from experiments ablating NK1-positive neurons in the spinal cord it was postulated that NK1 positive lamina I projection neurons form the origin of a spino-bulbo–spinal loop, which controls spinal excitability through the activation of a descending serotonergic pathway (Suzuki and Dickenson, 2005; Suzuki *et al.*, 2005). On the other hand, there is an inhibitory action of spinal 5-HT through the release of GABA from inhibitory interneurons via 5-HT_3 receptors (Alhaider *et al.*, 1991). This is in accordance with the recent finding of a co-localization of 5-HT_3 receptors and GABA in the spinal cord dorsal horn (Huang *et al.*, 2008).

In the formalin model, 5-HT_{1B}, 5-HT_{2C}, 5-HT_3 and 5-HT_4 receptors have been implied in 5-HT mediated analgesia (Jeong *et al.*, 2004). Kayser and colleagues investigated mice deficient in 5-HT_{1A}, 5-HT_{1B}, 5-HT_{2A} and 5-HT_{3A} in parallel with a panel of behavioral tests in the attempt to dissect out spinal, supraspinal and peripheral effects of 5-HT (Kayser *et al.*, 2007). 5-HT_{1A} knockout mice differed from wild types by higher thermal sensitivity, and 5-HT_{1B} knockout mice by higher thermal and formalin sensitivity. Both 5-HT_{2A} and 5-HT_{3A} knockout mice differed from wild types by a dramatic decrease in the formalin-induced nociceptive responses for phase 2. The authors concluded that 5-HT_{1B} and to a lesser degree 5-HT_{1A} receptors were responsible for endogenous inhibitory control of nociception by 5-HT, whereas 5-HT_{2A} and 5-HT_3 receptors had a role in peripheral hyperalgesia.

5-HT released from the brainstem nuclei was first shown to have an analgesic action at spinal sites as part of the system of descending inhibition (Yaksh and Wilson, 1979). Stimulation in the PAG and NRM releases 5-HT in the dorsal horn and excites 5-HT_2 or 5-HT_3 receptors on inhibitory interneurons, which in turn inhibit dorsal horn neurons (Peng *et al.*, 1996). However, more recently data have accumulated, first questioning a major role of 5-HT in descending analgesia, then showing even a facilitatory effect of 5-HT on nociceptive transmission. Depending upon the stimulus parameters, brainstem stimulation can cause hyperalgesia, and there is evidence that this is also mediated by descending serotonergic pathways (Zhuo and Gebhart, 1991). One reason underlying this two-fold action of 5-HT may lie in the fact that the result of RVM stimulation depends on whether 'on cells' or 'off cells' are activated (Fields *et al.*, 1983). Accordingly, 5-HT receptor antagonists can reduce both the pain-facilitating and pain-inhibiting effects of stimulation of the RVM. Some of the serotonergic RVM neurons appear to be 'on cells' themselves, because they are labeled with the neural activity marker c-fos after noxious stimulation of a rat hindpaw (Suzuki *et al.*, 2002). However, others dispute that either 'on' or 'off' cells are serotonergic (Fields, 2004), and regard the RVM-serotonergic system as a parallel but distinct pain-controlling system that may be under opioidergic control (Fields, 2004). In analogy to the opioidergic system of 'on' and 'off' cells, 5-HT is supposed to act in an activity-dependent mode, thus eliciting different behaviors at different times.

A very telling model to study the role of central 5-HT is provided by conditional knockout mice selectively deficient of 5-HT in central neurons. This was achieved by generating mice deficient of a transcription factor required for the differentiation of postmitotic 5-HT neurons (Ding *et al.*, 2003). These $Lmx1b^{f/f/p}$ mice are less sensitive to mechanical stimuli, but showed normal thermal and visceral pain responses. The mice had increased mechanical hypersensitivity after capsaicin injection, and increased pain behavior in the second phase of the formalin test (Zhao *et al.*, 2007a). The phenotype could be rescued by intrathecal injection of 5-HT. Antidepressants (fluoxetine, amitriptyline and duloxetine) had no effect on thermal pain in the Lmx1b$^{f/f/p}$ mice. Flinching in the second phase of the formalin test was attenuated by duloxetine, but not by the selective 5-HT reuptake inhibitor fluoxetine. Together, the findings from this study support the view that 5-HT is of fundamental importance for the analgesic effect of antidepressants (Zhao *et al.*, 2007a). Lmx1b$^{f/f/p}$ mice were also helpful in delineating the role of 5-HT in the action of opioids (Zhao *et al.*, 2007b).

5-HT is also involved in other complex regulatory circuits. For example, rapid eye movement (REM) sleep deprivation facilitates pain sensitivity. REM sleep deprivation in rats produced marked hypersensitivity to tactile stimuli via endogenous 5-HT acting on 5-HT$_{1A}$ and 5-HT$_{2C}$ receptors (Wei *et al.*, 2008). Thus, sustained activation of these receptors contributes to maintenance of cutaneous hypersensitivity in sleep-deprived animals.

5-HT receptor agonism and antagonism for pain relief

As mentioned above, the 5-HT$_{1B/D}$ agonist sumatriptan, 5-HT$_{2A}$ antagonists, and the 5-HT$_3$ receptor antagonist tropisetron have been shown to reduce pain behavior in certain animal models. Furthermore, a state-dependent analgesic effect of the 5-HT$_{1A}$ agonist F 13640 was found in rat models of acute and chronic nociceptive and neuropathic pain (Deseure *et al.*, 2007).

In contrast, the spinal excitatory action of 5-HT on 5-HT$_3$ receptors was shown to be necessary for the analgesic effect of gabapentin (Suzuki *et al.*, 2005). This finding may be of clinical relevance, since not all patients with neuropathic pain respond favorably to gabapentin. The 5HT$_3$-mediated spino-bulbo spinal loop, influenced by the affective state of the patient, might be part of determining this response (Suzuki *et al.*, 2005).

The greatly reduced analgesic effect of antidepressants in mice lacking central serotonergic neurons (Zhao *et al.*, 2007a) also indicates a dependence of their effect on 5-HT, even in dual antidepressants like duloxetine. By studying the expression of norepinephrine transporter, the authors could largely exclude a secondary effect through the norepinephrine system.

5-HT and the action of opioids

Various experimental studies have shown an interaction of the serotonergic system and morphine analgesia. Serotonergic pathways are thought to be functionally distinct from but to interact with the opioid-mediated pain-modulatory circuit (Fields, 2004). 5-HT in the NRM may be necessary for the action of morphine (Sounvoravong *et al.*, 2004). Morphine has direct actions at the 5-HT$_3$ receptor, suppressing 5-HT-induced currents in neurons in a competitive way (Wittmann *et al.*, 2006). Using the selective receptor antagonist SB-269970, the analgesic effects of systemic morphine in mice were shown to be dependent on the 5-HT$_7$ receptor (Dogrul and Seyrek, 2006). Interactions between the serotonergic, GABAergic and opioidergic system have been shown (Nemmani and Mogil, 2003).

Lmx1b$^{f/f/p}$ mice which lack central serotonergic neurons have no analgesia after administration of a kappa opioid receptor agonist and significantly reduced analgesic effects of μ- and δ-opioid receptor agonists at both spinal and supraspinal sites. In contrast, morphine tolerance and morphine reward behavior developed normally in these mice (Zhao *et al.*, 2007b). Thus, the central serotonergic system is a necessary component of the supraspinal pain modulatory circuitry mediating opioid analgesia. In contrast, morphine tolerance and morphine reward appear to be independent of the central serotonergic system. Interestingly, tramadol, a drug successfully used in pain management, acts through a combination of μ-opioid receptor and 5-HT$_{1A}$ receptors (Berrocoso *et al.*, 2007).

Data from human studies

Plasma 5-HT levels were found to be increased in several painful conditions (Table 1) – for example, in the complex regional pain syndrome (CRPS) (Wesseldijk *et al.*, 2008). Also, interstitial concentrations of 5-HT in the trapezius muscle were increased in patients with whiplash-associated disorders (Gerdle *et al.*, 2008) and with work-associated trapezius muscle pain (Larsson *et al.*, 2008). Both findings point to a role of 5-HT as a peripheral pain mediator, among others, in these conditions. In an earlier study, blood 5-HT levels correlated negatively with pain detection thresholds but not with pain tolerance, also indicating a peripheral nociceptive effect of 5-HT (Pickering *et al.*, 2003). As detailed above in the section on 5-HT agonists and antagonists, 5-HT injection in human volunteers induces sensitization to other mediators (Babenko *et al.*, 2000), or pain and itching (Lischetzki *et al.*, 2001; Schmelz *et al.*, 2003), or hyperalgesia (Ernberg *et al.*, 2000, 2006). An indirect indication for the role of 5-HT in experimental pain comes from the study on the GTP cyclohydrolase haplotype (Tegeder *et al.*, 2006), the rate-limiting enzyme for tetrahydrobiopterin synthesis, a cofactor for serotonin production. Low tetrahydrobiopterin (and thus presumably low 5-HT synthesis) was associated with reduced experimental pain.

Genetic variants in 5-HT receptors or in the 5-HT transporter have been hypothesized to underlie the variability

Table 1 List of painful disorders where a role of 5-HT has been suggested

Disorder	Finding	Reference
Migraine, between attacks	Lower plasma 5-HT, higher 5-HIAA	Ferrari *et al.*, 1989; Sicuteri *et al.*, 1961
Migraine, during attacks	Higher plasma 5-HT, lower 5-HIAA	Ferrari *et al.*, 1989
Medication overuse in migraine	Lower plasma 5-HT levels	Hering *et al.*, 1993
Migraine, between attacks	Increased availability of the 5-HTT in brainstem	Schuh-Hofer *et al.*, 2007
Migraine treatment	$5\text{-HT}_{1B/1D}$ receptor agonists abort migraine attacks	Ferrari *et al.*, 2002
Migraine with aura	Higher frequency of the short allele of the 5-HTTLPR	Borroni *et al.*, 2005; Marziniak *et al.*, 2005
Medication overuse headache	Platelet 5-HT uptake increased	Ayzenberg *et al.*, 2008
Complex regional pain syndrome (CRPS)	Plasma 5-HT levels increased	Wesseldijk *et al.*, 2008
Whiplash associated pain	Interstitial 5-HT concentrations in the trapezius muscle increased	Gerdle *et al.*, 2008
Work-associated trapezius muscle pain	Interstitial 5-HT concentrations in the trapezius muscle increased	Larsson *et al.*, 2008
Fibromyalgia syndrome	Reduced 5-HT levels in CSF	Houvenagel *et al.*, 1990; Russell *et al.*, 1992a
Fibromyalgia syndrome	Reduced serum 5-HT levels	Russell *et al.*, 1992b; Stratz *et al.*, 1993
Fibromyalgia syndrome	Higher frequency of the short allele of the 5-HTTLPR	Offenbaecher *et al.*, 1999; but see also Gürsoy *et al.*, 2002
Fibromyalgia syndrome	T102C polymorphism of the 5-HT_{2A}-receptor	Bondy *et al.*, 1999; Gürsoy *et al.*, 2001

Abbreviations: 5-HIAA, 5-hydroxyindole-acetic acid; 5-HTT, 5-HT transporter; 5-HTTLPR, 5-HTT-gene linked polymorphic region; CSF, cerebrospinal fluid.

in pain sensations in normal and diseased states. Recently, the intensity of cold pressure pain was investigated in 11 volunteers who had undergone positron emission tomography (PET) examination to quantify their 5-HT_{1A} binding potential. Cold pressure pain intensity inversely correlated with the 5-HT_{1A} binding potential in multiple cortical and subcortical areas. Subjects with high availability of 5-HT_{1A} receptors had low cold pressure pain intensity and a high capacity for central suppression of pain by a Valsalva maneuver (Martikainen *et al.*, 2007). This led to the conclusion that brain 5-HT_{1A} receptors influence pain thresholds and the capacity for analgesia in healthy subjects.

In a PET study using the 5-HT_{2A} receptor antagonist [(18)F]altanserin for imaging, positive correlations were found between tonic pain ratings and binding in orbitofrontal, medial inferior frontal, primary sensory-motor and posterior cingulate cortices (Kupers *et al.*, 2009). The authors concluded that 5-HT_{2A} receptor availability co-varies with responses to tonic pain, and that the correlation between 5-HT_{2A} receptor availability in prefrontal cortex and tonic pain suggests a possible role of this brain region in the modulation or cognitive appreciation of pain.

5-HT in headache

The study of 5-HT in headache, in particular in migraine, has a long history (Sicuteri *et al.*, 1961). Lower plasma 5-HT and relatively higher levels of 5-hydroxyindole-acetic acid, a metabolite of serotonin degradation, were reported in patients with migraine between attacks (Sicuteri *et al.*, 1961; Ferrari *et al.*, 1989). In contrast, during an attack increases in plasma 5-HT concomitant with decreases in 5-HIAA levels have been found (Ferrari *et al.*, 1989; for review, see Hamel, 2007). This led to the hypothesis that a chronic 5-HT deficit may be at the biochemical basis of migraine, while intermittent increases in 5-HT are triggers for migraine attacks. The 5-HT inhibitor methysergide was the first potent drug for migraine prophylaxis (Koehler and Tfelt-Hansen, 2008).

5-HT binds to the $5\text{-HT}_{1B/D}$ receptors with higher affinity than to 5-HT_{2A} receptors. Binding of 5-HT to the $5\text{-HT}_{1B/D}$ receptors, when present at low concentrations in the migraine-free period results in stabilization of the perivascular nociceptors. When released at higher concentrations, 5-HT will bind to the proalgesic 5-HT_{2A} receptors and may thus trigger a migraine attack. This is another example of the dual action of 5-HT in the nociceptive system depending on concentration, receptor availability and receptor affinity. Medication overuse further leads to reduced 5-HT levels in the blood of migraine patients (Hering *et al.*, 1993). This in turn may lead to upregulation of 5-HT_2 receptors, which when stimulated can increase the occurrence of migraine attacks, possibly via NO (Srikiatkhachorn *et al.*, 2002). These changes are reversible after medication withdrawal (Smith, 2004).

In a single photon emission computed tomography (SPECT) study using a selective 5-HTT ligand, increased availability of 5-HTT was found in the brainstem interictally in migraine patients (Schuh-Hofer *et al.*, 2007). This finding is compatible with decreased extracellular levels of 5-HT at the synaptic cleft and thus deregulation of the 5-HT system in the brainstem.

Physiological studies found that the amplitude of auditory- and visual-evoked potentials increased in migraineurs interictally, indicative of low 5-HT transmission. This was attributed to a deficit in cortical habituation (Coppola *et al.*, 2005). The triptans, $5\text{-HT}_{1B/1D}$ receptor agonists, activate inhibitory presynaptic $5\text{-HT}_{1B/1D}$ autoreceptors to decrease 5-HT synthesis. They increase the amplitude of auditory-evoked potentials in both normal subjects and migraine patients (Proietti-Cecchini *et al.*, 1997). Treatment with the β-blocker propranolol increases the rate of brain 5-HT synthesis, which may be the mechanism of action of β-blockers in migraine prophylaxis (Chugani *et al.*, 1999). There is some experimental evidence for a connection between 5-HT, cortical spreading depression (which is thought to underlie the migraine aura) and trigemino-vascular activation (which is thought to underlie the cephalalgia in migraine). Rats with low 5-HT had enhanced cortical spreading depression waves and an increased number of activated neurons in the trigeminal nucleus caudalis (Supornsilpchai *et al.*, 2006). Thus, chronically low brain 5-HT may facilitate trigeminal nociception through induction of cortical spreading depression, consistent with a deficit in the serotonergic descending pain inhibitory system (Panconesi, 2008).

The triptans, $5\text{-HT}_{1B/1D}$ receptor agonists, are the most successful drugs for acute migraine attacks (Ferrari *et al.*, 2002). It is assumed that their action via the $5\text{-HT}_{1B/D}$ receptors inhibits excitation of the perivascular trigeminal fibers, and thus blocks the release of vasoactive peptides (substance P, calcitonin gene-related peptide/CGRP) in the dura mater (Potrebic *et al.*, 2003). Triptans may exert part of their actions on trigeminal nuclei, inhibiting the activity trigeminal fibers centrally (Edvinsson and Uddman, 2005). Serotonergic RVM neurons project to cerebral vessels (Bradley *et al.*, 2002), where they regulate microcirculation (Cohen *et al.*, 1996). When raphe nuclei are activated, they can increase cerebral blood flow (Goadsby, 2005). Triptans also bind to receptors on the dorsal raphe and periaqueductal gray (Bonaventure *et al.*, 1998), indicating that, if they penetrate the blood–brain barrier, they may also act at these nuclei.

Genetic factors within the serotonergic pathways have been investigated for their role in migraine susceptibility. Most studies investigating 5-HT receptors, the TPH gene or MAOA, MAOB polymorphisms were negative. The most convincing evidence so far for a genetic association between the serotonergic system and migraine has been obtained for the association of polymorphisms in 5-HTT, which controls 5-HT brain homeostasis. We examined a functional deletion/insertion polymorphism in the transcriptional control region 1 kb upstream of the 5-HTT gene's transcription initiation site (5-HTT-linked polymorphic region, HTTLPR). *In vitro* and *in vivo* studies have demonstrated the functional relevance of this polymorphism with reduced transcriptional activity and low 5-HT reuptake activity determined by the short form of the allele (Lesch *et al.*, 1996). Patients with migraine with aura had a significantly higher frequency of the short allele of the 5-HTTLPR in comparison to a control population and to patients with migraine without aura (Marziniak *et al.*, 2005). These results were confirmed in an independent Italian population (Borroni *et al.*, 2005), while a Hungarian group found an association with migraine in general in women (Gonda *et al.*, 2007) and yet another group reported a link with increased frequency of migraine attacks (Kotani *et al.*, 2002). Interestingly, platelet 5-HT uptake is increased in patients with medication-overuse headache concomitantly with a transient increase in 5-HTT activity, and normalizes after withdrawal in parallel to the improvement of headache frequency (Ayzenberg *et al.*, 2008).

Fibromyalgia syndrome

Fibromyalgia syndrome (FMS) is a complex of symptoms associated with chronic widespread pain and tenderness upon pressure on at least 11 of 18 predefined anatomical points (Wolfe *et al.*, 1990). Additional symptoms are fatigue, sleep disturbance, depression, and gastrointestinal symptoms such as diarrhea and constipation. An effect of systemic 5-HT_3 antagonists was occasionally reported (Spath *et al.*, 2004). Together with the constellation of symptoms, this led to the hypothesis that 5-HT might be involved in the pathophysiology of FMS. Two studies investigating 5-HT levels in CSF found a reduction in patients with FMS compared to controls (Russell *et al.*, 1992a) or to reference data from the general population (Houvenagel *et al.*, 1990). In serum, reduced levels of serotonin were found in three studies (Russell *et al.*, 1992b; Stratz *et al.*, 1993; Wolfe *et al.*, 1997). Others investigated whether antibodies against serotonin or serotonin receptors were present in patients with FMS (Klein *et al.*, 1992; Werle *et al.*, 2001), and found these in a higher percentage than in controls. It is unclear if these antibodies play a pathophysiological role in FMS.

Of two studies investigating an association of FMS with polymorphisms in the 5-HTT promoter region, one found a higher percentage of the *S/S* genotype; these patients had higher scores in tests for depression and psychosocial impairment (Offenbaecher *et al.*, 1999). In the other study, however, no association was found (Gürsoy, 2002). Of the four studies that investigated serotonin receptors (Bondy *et al.*, 1999; Gürsoy *et al.*, 2001; Frank *et al.*, 2004; Tander *et al.*, 2008), two were positive, both of them for the T102C polymorphism of the 5-HT_{2A}-receptor (Bondy *et al.*, 1999; Gürsoy *et al.*, 2001). A study investigating other 5-HT_{2A} receptor gene polymorphisms (Tander *et al.*, 2008) and one studying variations in the 5-HT_{3A} and 5-HT_{3B} genes (Frank *et al.*, 2004) were negative.

5-HT_3 antagonists have been repeatedly studied in open trial in FMS. However, in contrast to chemotherapy-induced vomiting, where 5-HT_3 antagonists are extremely successful (Faerber *et al.*, 2007), in FMS only one randomized controlled trial has shown a moderate effect upon short-term treatment (Färber *et al.*, 2000).

In summary, a definite role of 5-HT in the pathogenesis of FMS is possible, but has not yet been proven.

Implications for the treatment of chronic pain with antidepressants

Metareviews indicate that combined serotonergic and noradrenergic reuptake inhibitors such as the tricyclic antidepressants (TCA) are efficient analgesics, in particular in neuropathic pain, while selective serotonin reuptake inhibitors (SSRIs) are less efficient (Saarto and Wiffen, 2005). Similarly, the effect of SSRIs in the prevention of migraine and tension-type headache is also poor (Moja *et al.*, 2005). Since relative selective noradrenalin reuptake inhibitors are also less effective in pain treatment, it has been suggested that a combined serotonergic and noradrenergic mechanism is needed. In view of the dual role of 5-HT in the pain-modulating system, it is not surprising that an increased availability of 5-HT at synapses may not in all instances lead to analgesia. Since the good pain control with TCA is set off by their side effects, combined serotonin–noradrenalin reuptake inhibitors (SNRIs) such as venlafaxine and duloxetine have been studied in clinical trials; these have fewer side effects and efficacy only slightly inferior to that of TCA (Sindrup *et al.*, 2005). In particular, in FMS, SNRIs appear to be effective in reducing pain levels (Clauw *et al.*, 2008; Üceyler *et al.*, 2008a, 2008b).

Conclusion

There is no simple answer regarding the question of the role of 5-HT in pain and in pain control. It may be most easily solved in the periphery, where 5-HT, released into inflamed or injured tissues, contributes to peripheral sensitization of nerve fibers (Figure 1). In the trigeminovascular system and in the CNS there are dual actions of 5-HT, depending on receptor availability and affinity, ligand concentration, and the neural network involved. The situation is even more complicated concerning the algesic or analgesic action of 5-HT in descending pathways and in the spinal cord (Figure 2). The concept of state-dependent actions, as recently proposed for the opioid system (Fields, 2004), may have to be adapted to the serotonergic system as well. Analgesics aiming at the serotonergic system have to take all these points into account.

Acknowledgements

The author's work cited in this review was supported by Volkswagen-Stiftung, by Bundesministerium für Bildung und Forschung (BMBF, Deutscher Forschungsverbund Neuropathischer Schmerz, DFNS) and by research funds of the University of Würzburg.

References

Abbott, F.V., Hong, Y. and Blier, P. (1996) Activation of 5-HT2A receptors potentiates pain produced by inflammatory mediators. Neuropharmacology, 35: 99–110.

Abbott, F.V., Hong, Y. and Blier, P. (1997) Persisting sensitization of the behavioural response to formalin-induced injury in the rat through activation of serotonin2A receptors. Neuroscience, 77: 575–584.

Aley, K.O. and Levine, J.D. (1999) Role of protein kinase A in the maintenance of inflammatory pain. J. Neurosci., 19: 2181–2186.

Aley, K.O., McCarter, G. and Levine, J.D. (1998) Nitric oxide signaling in pain and nociceptor sensitization in the rat. J. Neurosci., 18: 7008–7014.

Aley, K.O., Messing, R.O., Mochly-Rosen, D. and Levine, J.D. (2000) Chronic hypersensitivity for inflammatory nociceptor sensitization mediated by the epsilon isozyme of protein kinase C. J. Neurosci., 20: 4680–4685.

Alhaider, A.A., Lei, S.Z. and Wilcox, G.L. (1991) Spinal 5-HT3 receptor-mediated antinociception: possible release of GABA. J. Neurosci., 11: 1881–1888.

Arvieu, L., Mauborgne, A., Bourgoin, S., Oliver, C., Feltz, P., Hamon, M. and Cesselin, F. (1996) Sumatriptan inhibits the release of CGRP and substance P from the rat spinal cord. Neuroreport, 7: 1973–1976.

Ayzenberg, I., Obermann, M., Leineweber, K., Franke, L., Yoon, M.S., Diener, H.C. and Katsarava, Z. (2008) Increased activity of serotonin uptake in platelets in medication overuse headache following regular intake of analgesics and triptans. J. Headache Pain, 9: 109–112.

Babenko, V., Svensson, P., Graven-Nielsen, T., Drewes, A.M., Jensen, T.S. and Arendt-Nielsen, L. (2000) Duration and distribution of experimental muscle hyperalgesia in humans following combined infusions of serotonin and bradykinin. Brain Res., 853: 275–281.

Bardin, L., Bardin, M., Lavarenne, J. and Eschalier, A. (1997) Effect of intrathecal serotonin on nociception in rats: influence of the pain test used. Exp. Brain Res., 113: 81–87.

Basbaum, A.I. and Fields, H.L. (1978) Endogenous pain control mechanisms: review and hypothesis. Ann. Neurol., 4: 451–462.

Bengel, D., Murphy, D.L., Andrews, A.M., Wichems, C.H., Feltner, D., Heils, A., Mossner, R., Westphal, H. and Lesch, K.P. (1998) Altered brain serotonin homeostasis and locomotor insensitivity to 3, 4- methylenedioxymethamphetamine ('Ecstasy') in serotonin transporter-deficient mice. Mol. Pharmacol., 53: 649–655.

Berrocoso, E., De Benito, M.D. and Mico, J.A. (2007) Role of serotonin 5-HT1A and opioid receptors in the antiallodynic effect of tramadol in the chronic constriction injury model of neuropathic pain in rats. Psychopharmacology (Berl.), 193: 97–105.

Bingham, S., Davey, P.T., Sammons, M., Raval, P., Overend, P. and Parsons, A.A. (2001) Inhibition of inflammation-induced thermal hypersensitivity by sumatriptan through activation of 5-HT(1B/1D) receptors. Exp. Neurol., 167: 65–73.

Bonaventure, P., Langlois, X. and Leysen, J.E. (1998) Co-localization of 5-HT1B- and 5-HT1D receptor mRNA in serotonergic cell bodies in guinea pig dorsal raphe nucleus: a double labeling in situ hybridization histochemistry study. Neurosci. Lett., 254: 113–116.

Bondy, B., Spaeth, M., Offenbaecher, M., Glatzeder, K., Stratz, T., Schwarz, M., de Jonge, S., Kruger, M., Engel, R.R., Farber, L., Pongratz, D.E. and Ackenheil, M. (1999) The T102C polymorphism of the 5-HT2A-receptor gene in fibromyalgia. Neurobiol. Dis., 6: 433–439.

Borroni, B., Brambilla, C., Liberini, P., Rao, R., Archetti, S., Gipponi, S., Dalla Volta, G. and Padovani, A. (2005) Functional serotoin 5-HTTLPR polymorphism is a risk factor for migraine with aura. J. Headache Pain, 6: 182–184.

Bradley, S.R., Pieribone, V.A., Wang, W., Severson, C.A., Jacobs, R.A. and Richerson, G.B. (2002) Chemosensitive serotonergic neurons are closely associated with large medullary arteries. Nat. Neurosci., 5: 401–402.

Cardenas, C.G., Del Mar, L.P., Cooper, B.Y. and Scroggs, R.S. (1997) 5HT4 receptors couple positively to tetrodotoxin-insensitive sodium channels in a subpopulation of capsaicin-sensitive rat sensory neurons. J. Neurosci., 17: 7181–7189.

Cardenas, C.G., Mar, L.P., Vysokanov, A.V., Arnold, P.B., Cardenas, L.M., Surmeier, D.J. and Scroggs, R.S. (1999) Serotonergic modulation of hyperpolarization-activated current in acutely isolated rat dorsal root ganglion neurons. J. Physiol., 518(Pt 2): 507–523.

Cardenas, L.M., Cardenas, C.G. and Scroggs, R.S. (2001) 5HT increases excitability of nociceptor-like rat dorsal root ganglion neurons via cAMP-coupled TTX-resistant Na(+) channels. J. Neurophysiol., 86: 241–248.

Carlton, S.M. and Coggeshall, R.E. (1997) Immunohistochemical localization of 5-HT2A receptors in peripheral sensory axons in rat glabrous skin. Brain Res., 763: 271–275.

Chugani, D.C., Niimura, K., Chaturvedi, S., Muzik, O., Fakhouri, M., Lee, M.L. and Chugani, H.T. (1999) Increased brain serotonin synthesis in migraine. Neurology, 53: 1473–1479.

Clauw, D.J., Mease, P., Palmer, R.H., Gendreau, R.M. and Wang, Y. (2008) Milnacipran for the treatment of fibromyalgia in adults: a 15-week, multicenter, randomized, double-blind, placebo-controlled, multiple-dose clinical trial. Clin. Ther., 30: 1988–2004.

Cohen, Z., Bonvento, G., Lacombe, P. and Hamel, E. (1996) Serotonin in the regulation of brain microcirculation. Prog. Neurobiol., 50: 335–362.

Coppola, G., Vandenheede, M., Di Clemente, L., Ambrosini, A., Fumal, A., De Pasqua, V. and Schoenen, J. (2005) Somatosensory evoked high-frequency oscillations reflecting thalamo-cortical activity are decreased in migraine patients between attacks. Brain, 128: 98–103.

Courade, J.P., Chassaing, C., Bardin, L., Alloui, A. and Eschalier, A. (2001) 5-HT receptor subtypes involved in the spinal antinociceptive effect of acetaminophen in rats. Eur. J. Pharmacol., 432: 1–7.

Deseure, K., Breand, S. and Colpaert, F.C. (2007) Curative-like analgesia in a neuropathic pain model: parametric analysis of the dose and the duration of treatment with a high-efficacy 5-HT(1A) receptor agonist. Eur. J. Pharmacol., 568: 134–141.

Ding, Y.Q., Marklund, U., Yuan, W., Yin, J., Wegman, L., Ericson, J., Deneris, E., Johnson, R.L. and Chen, Z.F. (2003) Lmx1b is essential for the development of serotonergic neurons. Nat. Neurosci., 6: 933–938.

Doak, G.J. and Sawynok, J. (1997) Formalin-induced nociceptive behavior and edema: involvement of multiple peripheral 5-hydroxytryptamine receptor subtypes. Neuroscience, 80: 939–949.

Dogrul, A. and Seyrek, M. (2006) Systemic morphine produce antinociception mediated by spinal 5-HT7, but not 5-HT1A and 5-HT2 receptors in the spinal cord. Br. J. Pharmacol., 149: 498–505.

Duggan, A.W. and Headley, P.M. (1977) The effects of noradrenaline and 5-hydroxytryptamine on the responses of dorsal horn neurones to noxious and innocuous skin stimuli. Br. J. Pharmacol., 61: 149P.

Edvinsson, L. and Uddman, R. (2005) Neurobiology in primary headaches. Brain Res. Brain Res. Rev., 48: 438–456.

Ernberg, M., Lundeberg, T. and Kopp, S. (2000) Pain and allodynia/hyperalgesia induced by intramuscular injection of serotonin in patients with fibromyalgia and healthy individuals. Pain, 85: 31–39.

Ernberg, M., Lundeberg, T. and Kopp, S. (2003) Effects on muscle pain by intramuscular injection of granisetron in patients with fibromyalgia. Pain, 101: 275–282.

Ernberg, M., Hedenberg-Magnusson, B., Kurita, H. and Kopp, S. (2006) Effects of local serotonin administration on pain and microcirculation in the human masseter muscle. J. Orofac. Pain, 20: 241–248.

Eschalier, A., Kayser, V. and Guilbaud, G. (1989) Influence of a specific 5-HT3 antagonist on carrageenan-induced hyperalgesia in rats. Pain, 36: 249–255.

Faerber, L., Drechsler, S., Ladenburger, S., Gschaidmeier, H. and Fischer, W. (2007) The neuronal 5-HT3 receptor network after 20 years of research – evolving concepts in management of pain and inflammation. Eur. J. Pharmacol., 560: 1–8.

Färber, L., Stratz, T., Bückle, W., Späth, M., Pongratz, D., Lautenschlager, J., Kötter, I., Zöller, B., Peter, H.H., Neeck, G., Alten, R. and Müller, W. (2000) Efficacy and tolerability of tropisetron in primary fibromyalgia – a highly selective and competitive 5-HT3 receptor antagonist. German Fibromyalgia Study Group. Scand. J. Rheumatol. Suppl., 113: 49–54.

Ferrari, M.D., Odink, J., Tapparelli, C., Van Kempen, G.M., Pennings, E.J. and Bruyn, G.W. (1989) Serotonin metabolism in migraine. Neurology, 39: 1239–1242.

Ferrari, M.D., Goadsby, P.J., Roon, K.I. and Lipton, R.B. (2002) Triptans (serotonin, 5-HT1B/1D agonists. in migraine: detailed results and methods of a meta-analysis of 53 trials. Cephalalgia, 22: 633–658.

Fields, H. (2004) State-dependent opioid control of pain. Nat. Rev. Neurosci., 5: 565–575.

Fields, H.L., Bry, J., Hentall, I. and Zorman, G. (1983) The activity of neurons in the rostral medulla of the rat during withdrawal from noxious heat. J. Neurosci., 3: 2545–2552.

Frank, B., Niesler, B., Bondy, B., Spath, M., Pongratz, D.E., Ackenheil, M., Fischer, C. and Rappold, G. (2004) Mutational analysis of serotonin receptor genes: HTR3A and HTR3B in fibromyalgia patients. Clin. Rheumatol., 23: 338–344.

Gerdle, B., Lemming, D., Kristiansen, J., Larsson, B., Peolsson, M. and Rosendal, L. (2008) Biochemical alterations in the trapezius muscle of patients with chronic whiplash associated disorders (WAD) – a microdialysis study. Eur. J. Pain, 12: 82–93.

Giordano, J. and Dyche, J. (1989) Differential analgesic actions of serotonin 5-HT3 receptor antagonists in the mouse. Neuropharmacology, 28: 423–427.

Giordano, J. and Rogers, L.V. (1989) Peripherally administered serotonin 5-HT3 receptor antagonists reduce inflammatory pain in rats. Eur. J. Pharmacol., 170: 83–86.

Giordano, J., Daleo, C. and Sacks, S.M. (1998) Topical ondansetron attenuates nociceptive and inflammatory effects of intradermal capsaicin in humans. Eur. J. Pharmacol., 354: R13–R14.

Goadsby, P.J. (2005) Migraine pathophysiology. Headache, 45(Suppl. 1): S14–S24.

Gold, M.S., Reichling, D.B., Shuster, M.J. and Levine, J.D. (1996) Hyperalgesic agents increase a tetrodotoxin-resistant Na+ current in nociceptors. Proc. Natl. Acad. Sci. USA, 93: 1108–1112.

Gonda, X., Rihmer, Z., Juhasz, G., Zsombok, T. and Bagdy, G. (2007) High anxiety and migraine are associated with the s allele of the 5HTTLPR gene polymorphism. Psychiatry Res., 149: 261–266.

Gürsoy, S. (2002) Absence of association of the serotonin transporter gene polymorphism with the mentally healthy subset of fibromyalgia patients. Clin. Rheumatol., 21: 194–197.

Gürsoy, S., Erdal, E., Herken, H., Madenci, E. and Alasehirli, B. (2001) Association of T102C polymorphism of the 5-HT2A receptor gene with psychiatric status in fibromyalgia syndrome. Rheumatol. Intl., 21: 58–61.

Hains, B.C., Willis, W.D. and Hulsebosch, C.E. (2003) Serotonin receptors 5-HT1A and 5-HT3 reduce hyperexcitability of dorsal horn neurons after chronic spinal cord hemisection injury in rat. Exp. Brain Res., 149: 174–186.

Hamel, E. (2007) Serotonin and migraine: biology and clinical implications. Cephalalgia, 27: 1293–1300.

Hamon, M. and Bourgoin, S. (1999) Serotonin and its receptors in pain control. In: Sawynok, J. and Cowan, A. (Eds.), Novel Aspects of Pain Management: Opioids and Beyond. Wiley-Liss, Inc, New York, NY, pp. 203–228.

Hamon, M., Gallissot, M.C., Menard, F., Gozlan, H., Bourgoin, S. and Verge, D. (1989) 5-HT3 receptor binding sites are on capsaicin-sensitive fibres in the rat spinal cord. Eur. J. Pharmacol., 164: 315–322.

Hering, R., Glover, V., Pattichis, K., Catarci, T. and Steiner, T.J. (1993) 5HT in migraine patients with medication-induced headache. Cephalalgia, 13: 410–412.

Houvenagel, E., Forzy, G., Leloire, O., Gallois, P., Hary, S., Hautecoeur, P., Convain, L., Henniaux, M., Vincent, G. and Dhondt, J.L. (1990) [Cerebrospinal fluid monoamines in primary fibromyalgia]. Rev. Rhum. Mal. Osteoartic, 57: 21–23.

Huang, J., Wang, Y.Y., Wang, W., Li, Y.Q., Tamamaki, N. and Wu, S.X. (2008) 5-HT(3A) receptor subunit is expressed in a subpopulation of GABAergic and enkephalinergic neurons in the mouse dorsal spinal cord. Neurosci. Lett., 441: 1–6.

Ito, A., Kumamoto, E., Takeda, M., Shibata, K., Sagai, H. and Yoshimura, M. (2000) Mechanisms for ovariectomy-induced hyperalgesia and its relief by calcitonin: participation of 5-HT1A-like receptor on C-afferent terminals in substantia gelatinosa of the rat spinal cord. J. Neurosci., 20: 6302–6308.

Jeong, C.Y., Choi, J.I. and Yoon, M.H. (2004) Roles of serotonin receptor subtypes for the antinociception of 5-HT in the spinal cord of rats. Eur. J. Pharmacol., 502: 205–211.

Jordan, L.M., Kenshalo, D.R. Jr., Martin, R.F., Haber, L.H. and Willis, W.D. (1978) Depression of primate spinothalamic tract neurons by iontophoretic application of 5-hydroxytryptamine. Pain, 5: 135–142.

Kato, K., Kikuchi, S., Konno, S. and Sekiguchi, M. (2008) Participation of 5-hydroxytryptamine in pain-related behavior induced by nucleus pulposus applied on the nerve root in rats. Spine, 33: 1330–1336.

Kayser, V., Elfassi, I.E., Aubel, B., Melfort, M., Julius, D., Gingrich, J.A., Hamon, M. and Bourgoin, S. (2007) Mechanical, thermal and formalin-induced nociception is differentially altered in 5-HT1A−/−, 5-HT1B−/−, 5-HT2A−/−, 5-HT3A−/− and 5-HTT−/− knockout male mice. Pain, 130: 235–248.

Kessler, W., Kirchhoff, C., Reeh, P.W. and Handwerker, H.O. (1992) Excitation of cutaneous afferent nerve endings *in vitro* by a combination of inflammatory mediators and conditioning effect of substance P. Exp. Brain Res., 91: 467–476.

Kjorsvik, A., Tjolsen, A. and Hole, K. (2001) Activation of spinal serotonin(2A/2C) receptors augments nociceptive responses in the rat. Brain Res., 910: 179–181.

Klein, R., Bansch, M. and Berg, P.A. (1992) Clinical relevance of antibodies against serotonin and gangliosides in patients with primary fibromyalgia syndrome. Psychoneuroendocrino logy, 17: 593–598.

Koehler, P.J. and Tfelt-Hansen, P.C. (2008) History of methysergide in migraine. Cephalalgia, 28: 1126–1135.

Kotani, K., Shimomura, T., Shimomura, F., Ikawa, S. and Nanba, E. (2002) A polymorphism in the serotonin transporter gene regulatory region and frequency of migraine attacks. Headache, 42: 893–895.

Kress, M., Reeh, P.W. and Vyklicky, L. (1997) An interaction of inflammatory mediators and protons in small diameter dorsal root ganglion neurons of the rat. Neurosci. Lett., 224: 37–40.

Kupers, R., Frokjaer, V.G., Naert, A., Christensen, R., Budtz-Joergensen, E., Kehlet, H. and Knudsen, G.M. (2009) A PET [(18)F]altanserin study of 5-HT(2A) receptor binding in the human brain and responses to painful heat stimulation. Neuroimage, 44: 1001–1007.

Lang, E., Novak, A., Reeh, P.W. and Handwerker, H.O. (1990) Chemosensitivity of fine afferents from rat skin *in vitro*. J. Neurophysiol., 63: 887–901.

Larsson, B., Rosendal, L., Kristiansen, J., Sjogaard, G., Sogaard, K., Ghafouri, B., Abdiu, A., Kjaer, M. and Gerdle, B. (2008) Responses of algesic and metabolic substances to 8 h of repetitive manual work in myalgic human trapezius muscle. Pain, 140: 479–490.

Le Bars, D. and Villanueva, L. (1988) Electrophysiological evidence for the activation of descending inhibitory controls by nociceptive afferent pathways. Prog. Brain Res., 77: 275–299.

Lesch, K.P., Bengel, D., Heils, A., Sabol, S.Z., Greenberg, B. D., Petri, S., Benjamin, J., Muller, C.R., Hamer, D.H. and Murphy, D.L. (1996) Association of anxiety-related traits with a polymorphism in the serotonin transporter gene regulatory region [see comments]. Science, 274: 1527–1531.

Lischetzki, G., Rukwied, R., Handwerker, H.O. and Schmelz, M. (2001) Nociceptor activation and protein extravasation induced by inflammatory mediators in human skin. Eur. J. Pain, 5: 49–57.

Lopez-Garcia, J.A. (2006) Serotonergic modulation of spinal sensory circuits. Curr. Top Med. Chem., 6: 1987–1996.

Lötsch, J., Skarke, C., Schneider, A., Hummel, T. and Geisslinger, G. (2005) The 5-hydroxytryptamine 4 receptor agonist mosapride does not antagonize morphine-induced respiratory depression. Clin. Pharmacol. Ther., 78: 278–287.

Manzke, T., Guenther, U., Ponimaskin, E.G., Haller, M., Dutschmann, M., Schwarzacher, S. and Richter, D.W. (2003) 5-HT4(a) receptors avert opioid-induced breathing depression without loss of analgesia. Science, 301: 226–229.

Martikainen, I.K., Hirvonen, J., Kajander, J., Hagelberg, N., Mansikka, H., Nagren, K., Hietala, J. and Pertovaara, A. (2007) Correlation of human cold pressor pain responses with 5-HT(1A) receptor binding in the brain. Brain Res., 1172: 21–31.

Marziniak, M., Mossner, R., Schmitt, A., Lesch, K.P. and Sommer, C. (2005) A functional serotonin transporter gene polymorphism is associated with migraine with aura. Neurology, 64: 157–159.

McMahon, S.B., Bennett, D.L.H., B.I. and Bevan, S. (2006) Inflammatory mediators and modulators of pain. In: Textbook of Pain. pp. 49–72. Eds. S.B. McMahon, K.M. Elsevier: Amsterdam.

Michaelis, M., Vogel, C., Blenk, K.H. and Janig, W. (1997) Algesics excite axotomised afferent nerve fibres within the first hours following nerve transection in rats. Pain, 72: 347–354.

Millan, M.J. (2002) Descending control of pain. Prog. Neurobiol., 66: 355–474.

Millan, M.J., Seguin, L., Honore, P., Girardon, S. and Bervoets, K. (1996) Pro- and antinociceptive actions of serotonin (5-HT)1A agonists and antagonists in rodents: relationship to algesiometric paradigm. Behav. Brain Res., 73: 69–77.

Millan, M.J., Girardon, S. and Bervoets, K. (1997) 8-OH-DPAT-induced spontaneous tail-flicks in the rat are facilitated by the selective serotonin (5-HT)2C agonist, RO 60-0175: blockade of its actions by the novel 5-HT2C receptor antagonist SB 206,553. Neuropharmacology, 36: 743–745.

Moalem, G., Grafe, P. and Tracey, D.J. (2005) Chemical mediators enhance the excitability of unmyelinated sensory axons in normal and injured peripheral nerve of the rat. Neuroscience, 134: 1399–1411.

Moja, P., Cusi, C., Sterzi, R. and Canepari, C. (2005) Selective serotonin re-uptake inhibitors (SSRIs) for preventing migraine and tension-type headaches. Cochrane Database Syst. Rev.: CD002919.

Nemmani, K.V. and Mogil, J.S. (2003) Serotonin-GABA interactions in the modulation of mu- and kappa-opioid analgesia. Neuropharmacology, 44: 304–310.

Nicholson, R., Small, J., Dixon, A.K., Spanswick, D. and Lee, K. (2003) Serotonin receptor mRNA expression in rat dorsal root ganglion neurons. Neurosci. Lett., 337: 119–122.

Oatway, M.A., Chen, Y. and Weaver, L.C. (2004) The 5-HT3 receptor facilitates at-level mechanical allodynia following spinal cord injury. Pain, 110: 259–268.

Obata, H., Saito, S., Ishizaki, K. and Goto, F. (2000) Antinociception in rat by sarpogrelate, a selective 5-HT(2A) receptor antagonist, is peripheral. Eur. J. Pharmacol., 404: 95–102.

Obata, H., Saito, S., Sasaki, M., Ishizaki, K. and Goto, F. (2001) Antiallodynic effect of intrathecally administered 5-HT(2) agonists in rats with nerve ligation. Pain, 90: 173–179.

Offenbaecher, M., Bondy, B., de Jonge, S., Glatzeder, K., Kruger, M., Schoeps, P. and Ackenheil, M. (1999) Possible association of fibromyalgia with a polymorphism in the serotonin transporter gene regulatory region. Arthritis Rheum., 42: 2482–2488.

Okamoto, K., Imbe, H., Morikawa, Y., Itoh, M., Sekimoto, M., Nemoto, K. and Senba, E. (2002) 5-HT2A receptor subtype in the peripheral branch of sensory fibers is involved in the potentiation of inflammatory pain in rats. Pain, 99: 133–143.

Oliveira, M.C., Pelegrini-da-Silva, A., Parada, C.A. and Tambeli, C.H. (2007) 5-HT acts on nociceptive primary afferents through an indirect mechanism to induce hyperalgesia in the subcutaneous tissue. Neuroscience, 145: 708–714.

Palm, F., Mossner, R., Chen, Y., He, L., Gerlach, M., Bischofs, S., Riederer, P., Lesch, K.P. and Sommer, C. (2008) Reduced thermal hyperalgesia and enhanced peripheral nerve injury after hind paw inflammation in mice lacking the serotonin-transporter. Eur. J. Pain, 12: 790–797.

Panconesi, A. (2008) Serotonin and migraine: a reconsideration of the central theory. J. Headache Pain, 9: 267–276.

Peng, Y.B., Lin, Q. and Willis, W.D. (1996) The role of 5-HT3 receptors in periaqueductal gray-induced inhibition of nociceptive dorsal horn neurons in rats. J. Pharmacol. Exp. Ther., 276: 116–124.

Pertovaara, A., Keski-Vakkuri, U., Kalmari, J., Wei, H. and Panula, P. (2001) Response properties of neurons in the rostroventromedial medulla of neuropathic rats: attempted modulation of responses by [1DMe]NPYF, a neuropeptide FF analogue. Neuroscience, 105: 457–468.

Pickering, G., Januel, F., Dubray, C. and Eschalier, A. (2003) Serotonin and experimental pain in healthy young volunteers. Clin. J. Pain, 19: 276–279.

Pierce, P.A., Xie, G.X., Levine, J.D. and Peroutka, S.J. (1996) 5-Hydroxytryptamine receptor subtype messenger RNAs in rat peripheral sensory and sympathetic ganglia: a polymerase chain reaction study. Neuroscience, 70: 553–559.

Potrebic, S., Ahn, A.H., Skinner, K., Fields, H.L. and Basbaum, A.I. (2003) Peptidergic nociceptors of both trigeminal and dorsal root ganglia express serotonin 1D receptors: implications for the selective antimigraine action of triptans. J. Neurosci., 23: 10988–10997.

Proietti-Cecchini, A., Afra, J. and Schoenen, J. (1997) Intensity dependence of the cortical auditory evoked potentials as a surrogate marker of central nervous system serotonin transmission in man: demonstration of a central effect for the 5HT1B/1D agonist zolmitriptan (311C90, Zomig). Cephalalgia, 17: 849–854, discussion 799

Radhakrishnan, R., King, E.W., Dickman, J.K., Herold, C.A., Johnston, N.F., Spurgin, M.L. and Sluka, K.A. (2003) Spinal 5-HT(2) and 5-HT(3) receptors mediate low, but not high, frequency TENS-induced antihyperalgesia in rats. Pain, 105: 205–213.

Rahman, W., Suzuki, R., Rygh, L.J. and Dickenson, A.H. (2004) Descending serotonergic facilitation mediated through rat spinal 5HT3 receptors is unaltered following carrageenan inflammation. Neurosci. Lett., 361: 229–231.

Rocha-Gonzalez, H.I., Meneses, A., Carlton, S.M. and Granados-Soto, V. (2005) Pronociceptive role of peripheral and spinal 5-HT7 receptors in the formalin test. Pain, 117: 182–192.

Russell, I.J., Michalek, J.E., Vipraio, G.A., Fletcher, E.M., Javors, M.A. and Bowden, C.A. (1992b) Platelet 3H-imipramine uptake receptor density and serum serotonin levels in patients with fibromyalgia/fibrositis syndrome. J. Rheumatol., 19: 104–109.

Russell, I.J., Vaeroy, H., Javors, M. and Nyberg, F. (1992a) Cerebrospinal fluid biogenic amine metabolites in fibromyalgia/fibrositis syndrome and rheumatoid arthritis. Arthritis Rheum, 35: 550–556.

Saarto, T. and Wiffen, P. (2005) Antidepressants for neuropathic pain. Cochrane Database Syst. Rev.: CD005454.

Sasaki, M., Obata, H., Saito, S. and Goto, F. (2003) Antinociception with intrathecal alpha-methyl-5-hydroxytryptamine, a 5-hydroxytryptamine 2A/2C receptor agonist, in two rat models of sustained pain. Anesth. Analg., 96: 1072–1078, table of contents.

Schmelz, M., Schmidt, R., Weidner, C., Hilliges, M., Torebjork, H.E. and Handwerker, H.O. (2003) Chemical response pattern of different classes of C-nociceptors to pruritogens and algogens. J. Neurophysiol., 89: 2441–2448.

Schuh-Hofer, S., Richter, M., Geworski, L., Villringer, A., Israel, H., Wenzel, R., Munz, D.L. and Arnold, G. (2007) Increased serotonin transporter availability in the brainstem of migraineurs. J. Neurol., 254: 789–796.

Sicuteri, F., Testi, A. and Anselm, B. (1961) Biochemical investigations in headache: increase in hydroxyindoleacetic acid excretion during migraine attacks. Int. Arch. Allergy Appl. Immunol., 53: 537–542.

Sindrup, S.H., Otto, M., Finnerup, N.B. and Jensen, T.S. (2005) Antidepressants in the treatment of neuropathic pain. Basic Clin. Pharmacol. Toxicol., 96: 399–409.

Smith, N.L. (2004) Serotonin mechanisms in pain and functional syndromes: management implications in comorbid fibromyalgia, headache, and irritable bowl syndrome – case study and discussion. J. Pain Palliat Care Pharmacother, 18: 31–45.

Sommer, C. (2004) Serotonin in pain and analgesia: actions in the periphery. Mol. Neurobiol., 30: 117–125.

Song, X.J., Zhang, J.M., Hu, S.J. and LaMotte, R.H. (2003) Somata of nerve-injured sensory neurons exhibit enhanced responses to inflammatory mediators. Pain, 104: 701–709.

Sounvoravong, S., Nakashima, M.N., Wada, M. and Nakashima, K. (2004) Decrease in serotonin concentration in raphe magnus nucleus and attenuation of morphine analgesia in two mice models of neuropathic pain. Eur. J. Pharmacol., 484: 217–223.

Spath, M., Stratz, T., Neeck, G., Kotter, I., Hammel, B., Amberger, C.C., Haus, U., Farber, L., Pongratz, D. and Muller, W. (2004) Efficacy and tolerability of intravenous tropisetron in the treatment of fibromyalgia. Scand. J. Rheumatol., 33: 267–270.

Srikiatkhachorn, A., Suwattanasophon, C., Ruangpattanatawee, U. and Phansuwan-Pujito, P. (2002) 2002 Wolff Award. 5 -HT2A receptor activation and nitric oxide synthesis: a possible mechanism determining migraine attacks. Headache, 42: 566–574.

Stratz, T. and Müller, W. (2003) Lokalbehandlung rheumatischer Erkrankungen mit dem 5-HT3-Rezeptor-Antagonisten Tropisetron. Schmerz, 17: 200–203.

Stratz, T., Samborski, W., Hrycaj, P., Pap, T., Mackiewicz, S., Mennet, P. and Muller, W. (1993) Die Serotoninkonzentration im Serum bei Patienten mit generalisierter Tendomyopathie (Fibromyalgie) und chronischer Polyarthritis. Med. Klin., 88: 458–462.

Sung, D., Dong, X., Ernberg, M., Kumar, U. and Cairns, B.E. (2008) Serotonin (5-HT) excites rat masticatory muscle afferent fibers through activation of peripheral 5-HT3 receptors. Pain, 134: 41–50.

Supornsilpchai, W., Sanguanrangsirikul, S., Maneesri, S. and Srikiatkhachorn, A. (2006) Serotonin depletion, cortical spreading depression, and trigeminal nociception. Headache, 46: 34–39.

Suzuki, R. and Dickenson, A. (2005) Spinal and supraspinal contributions to central sensitization in peripheral neuropathy. Neurosignals, 14: 175–181.

Suzuki, R., Morcuende, S., Webber, M., Hunt, S.P. and Dickenson, A.H. (2002) Superficial NK1-expressing neurons control spinal excitability through activation of descending pathways. Nat. Neurosci., 5: 1319–1326.

Suzuki, R., Rahman, W., Hunt, S.P. and Dickenson, A.H. (2004a) Descending facilitatory control of mechanically evoked responses is enhanced in deep dorsal horn neurones following peripheral nerve injury. Brain Res., 1019: 68–76.

Suzuki, R., Rahman, W., Rygh, L.J., Webber, M., Hunt, S.P. and Dickenson, A.H. (2005) Spinal-supraspinal serotonergic circuits regulating neuropathic pain and its treatment with gabapentin. Pain, 117: 292–303.

Suzuki, R., Rygh, L.J. and Dickenson, A.H. (2004b) Bad news from the brain: descending 5-HT pathways that control spinal pain processing. Trends Pharmacol. Sci., 25: 613–617.

Taiwo, Y.O. and Levine, J.D. (1992) Serotonin is a directly-acting hyperalgesic agent in the rat. Neuroscience, 48: 485–490.

Tander, B., Gunes, S., Boke, O., Alayli, G., Kara, N., Bagci, H. and Canturk, F. (2008) Polymorphisms of the serotonin-2A receptor and catechol-O-methyltransferase genes: a study on fibromyalgia susceptibility. Rheumatol. Intl., 28: 685–691.

Tegeder, I., Costigan, M., Griffin, R.S., Abele, A., Belfer, I., Schmidt, H., Ehnert, C., Nejim, J., Marian, C., Scholz, J., Wu, T., Allchorne, A., Diatchenko, L., Binshtok, A.M., Goldman, D., Adolph, J., Sama, S., Atlas, S.J., Carlezon, W.A., Parsegian, A., Lotsch, J., Fillingim, R.B., Maixner, W., Geisslinger, G., Max, M.B. and Woolf, C.J. (2006) GTP cyclohydrolase and tetrahydrobiopterin regulate pain sensitivity and persistence. Nat. Med., 12: 1269–1277.

Thibault, K., Van Steenwinckel, J., Brisorgueil, M.J., Fischer, J., Hamon, M., Calvino, B. and Conrath, M. (2008) Serotonin 5-HT2A receptor involvement and Fos expression at the spinal level in vincristine-induced neuropathy in the rat. Pain, 140: 305–322.

Todorovic, S. and Anderson, E.G. (1992) Serotonin preferentially hyperpolarizes capsaicin-sensitive C type sensory neurons by activating 5-HT1A receptors. Brain Res., 585: 212–218.

Tokunaga, A., Saika, M. and Senba, E. (1998) 5-HT2A receptor subtype is involved in the thermal hyperalgesic mechanism of serotonin in the periphery. Pain, 76: 349–355.

Tonini, M. and Pace, F. (2006) Drugs acting on serotonin receptors for the treatment of functional GI disorders. Dig. Dis., 24: 59–69.

Üceyler, N., Hauser, W. and Sommer, C. (2008a) A systematic review on the effectiveness of treatment with antidepressants in fibromyalgia syndrome. Arthritis Rheum., 59: 1279–1298.

Üceyler, N., Offenbacher, M., Petzke, F., Hauser, W. and Sommer, C. (2008b) New treatment options for fibromyalgia: critical appraisal of duloxetine. Neuropsychiatr. Dis. Treat., 4: 525–529.

Vogel, C., Mossner, R., Gerlach, M., Heinemann, T., Murphy, D. L., Riederer, P., Lesch, K.P. and Sommer, C. (2003) Absence of thermal hyperalgesia in serotonin transporter-deficient mice. J. Neurosci., 23: 708–715.

Wei, H., Chen, Y. and Hong, Y. (2005) The contribution of peripheral 5-hydroxytryptamine2A receptor to carrageenan-evoked hyperalgesia, inflammation and spinal Fos protein expression in the rat. Neuroscience, 132: 1073–1082.

Wei, H., Ma, A., Wang, Y.X. and Pertovaara, A. (2008) Role of spinal 5-HT receptors in cutaneous hypersensitivity induced by REM sleep deprivation. Pharmacol. Res., 57: 469–475.

Werle, E., Fischer, H.P., Muller, A., Fiehn, W. and Eich, W. (2001) Antibodies against serotonin have no diagnostic relevance in patients with fibromyalgia syndrome. J. Rheumatol., 28: 595–600.

Wesseldijk, F., Fekkes, D., Huygen, F.J., Bogaerts-Taal, E. and Zijlstra, F.J. (2008) Increased plasma serotonin in complex regional pain syndrome type 1. Anesth. Analg., 106: 1862–1867.

Wittmann, M., Peters, I., Schaaf, T., Wartenberg, H.C., Wirz, S., Nadstawek, J., Urban, B.W. and Barann, M. (2006) The effects of morphine on human 5-HT3A receptors. Anesth. Analg., 103: 747–752.

Wolfe, F., Smythe, H.A., Yunus, M.B., Bennett, R.M., Bombardier, C., Goldenberg, D.L., Tugwell, P., Campbell, S.M., Abeles, M., Clark, P. et al. (1990) The American College of Rheumatology 1990 Criteria for the Classification of Fibromyalgia. Report of the Multicenter Criteria Committee. Arthritis Rheum., 33: 160–172.

Wolfe, F., Russell, I.J., Vipraio, G., Ross, K. and Anderson, J. (1997) Serotonin levels, pain threshold, and fibromyalgia symptoms in the general population. J. Rheumatol., 24: 555–559.

Yaksh, T.L. and Wilson, P.R. (1979) Spinal serotonin terminal system mediates antinociception. J. Pharmacol. Exp. Ther., 208: 446–453.

Yoshimura, M. and Furue, H. (2006) Mechanisms for the antinociceptive actions of the descending noradrenergic and serotonergic systems in the spinal cord. J. Pharmacol. Sci., 101: 107–117.

Zeitz, K.P., Guy, N., Malmberg, A.B., Dirajlal, S., Martin, W.J., Sun, L., Bonhaus, D.W., Stucky, C.L., Julius, D. and Basbaum, A.I. (2002) The 5-HT3 subtype of serotonin receptor contributes to nociceptive processing via a novel subset of myelinated and unmyelinated nociceptors. J. Neurosci., 22: 1010–1019.

Zhang, Y., Gao, X., Yang, Z.L., Huang, Y.L. and Wu, G.C. (2000) Expression of 5-HT(1A) receptor mRNA in rat nucleus raphe magnus neurons after peripheral inflammation. Brain Res., 887: 465–468.

Zhang, Y., Yang, Z., Gao, X. and Wu, G. (2001a) The role of 5-hydroxytryptamine1A and 5-hydroxytryptamine1B receptors in modulating spinal nociceptive transmission in normal and carrageenan- injected rats. Pain, 92: 201–211.

Zhang, Y.Q., Gao, X., Ji, G.C. and Wu, G.C. (2001b) Expression of 5-HT2A receptor mRNA in rat spinal dorsal horn and some nuclei of brainstem after peripheral inflammation. Brain Res., 900: 146–151.

Zhao, Z.Q., Chiechio, S., Sun, Y.G., Zhang, K.H., Zhao, C.S., Scott, M., Johnson, R.L., Deneris, E.S., Renner, K.J., Gereau, R.W.t. and Chen, Z.F. (2007a) Mice lacking central serotonergic neurons show enhanced inflammatory pain and an impaired analgesic response to antidepressant drugs. J. Neurosci., 27: 6045–6053.

Zhao, Z.Q., Gao, Y.J., Sun, Y.G., Zhao, C.S., Gereau, R.W. and Chen, Z.F. (2007b) Central serotonergic neurons are differentially required for opioid analgesia but not for morphine tolerance or morphine reward. Proc. Natl. Acad. Sci. USA, 104: 14519–14524.

Zhuo, M. and Gebhart, G.F. (1991) Spinal serotonin receptors mediate descending facilitation of a nociceptive reflex from the nuclei reticularis gigantocellularis and gigantocellularis pars alpha in the rat. Brain Res., 550: 35–48.

Zochodne, D.W. and Ho, L.T. (1994) Sumatriptan blocks neurogenic inflammation in the peripheral nerve trunk. Neurology, 44: 161–163.

SECTION 4

Serotonin in Disease Conditions

CHAPTER 4.1

The Impact of Stress on Serotonergic Neurotransmission

Astrid C.E. Linthorst* and Johannes M.H.M. Reul

Henry Wellcome Laboratories for Integrative Neuroscience and Endocrinology (HW-LINE), University of Bristol, Bristol, UK

Abstract: Levels of stress perceived by people in our modern society seem to have risen considerably, a phenomenon which may be related to the increased incidence of stress-related mental health problems such as depression and anxiety disorders. Whereas the body is well equipped to deal with single, acute stressful challenges, regulatory systems may derail when stress is of a repeated or prolonged nature. Two of the systems that appear to be malfunctioning in psychiatric disease are the hypothalamic–pituitary–adrenocortical (HPA) axis (i.e., the stress hormone axis) and the serotonin (5-HT) system. This chapter will focus on the effects of stress on serotonergic neurotransmission. We will first discuss how stress affects the activity of the serotonergic cell bodies in the rostral (midbrain/pontine) raphe nuclei (the dorsal and median raphe nucleus), followed by an overview of the effects of stress on 5-HT synthesis. The dorsal and median raphe nuclei project to major limbic forebrain structures involved in the coordination of the stress response, such as the frontal cortex, hippocampus, amygdala and hypothalamus. Therefore, we will subsequently highlight the effects of stress on the extracellular levels of 5-HT in these structures as assessed by *in vivo* microdialysis. Next, the (aberrant) functioning of serotonergic neurotransmission in animal models with changes in HPA axis regulation will be discussed, followed by an excursion into the beneficial effects of moderate exercise on stress coping mechanisms.

The reviewed data clearly demonstrate that the neurotransmitter 5-HT is not only highly responsive to acute stress; its responses are also intricately coordinated in a stressor-dependent and brain region-specific manner. This notion holds true for the regulation of the firing rate of serotonergic neurons, the activity and expression of tryptophan hydroxylase (the rate-limiting enzyme in the synthesis of 5-HT) and the extracellular levels of 5-HT in terminal regions. Furthermore, these data show that the 5-HT system can become compromised during situations of aberrant functioning of the HPA axis – for example, due to changes in the corticotropin-releasing factor system or in corticosteroid receptor function. Most data presented here are obtained during and following acute stressful challenges. This information is of great relevance for our understanding of the relationship between stress and 5-HT. However, there is unfortunately only very limited information available on the effects of repeated and chronic stress. With respect to mental health disease (and animal welfare), it will therefore be essential to make the next step and increase our knowledge on how prolonged stress may perturb serotonergic neurotransmission. Finally, as stress will not suddenly 'disappear' from our society, and the costs of stress-related disease are burdening the health systems of many countries, it is important to study the mechanisms underlying adequate coping with stress and stress resilience, and to find ways to improve and support such mechanisms – for example, by exercise.

Abbreviations: 5-HT, serotonin (5-hydroxytryptamine); 5-HIAA, 5-hydroxyindoleacetic acid; CRF, corticotropin-releasing factor; CRF1, corticotropin-releasing factor receptor type 1; CRF2, corticotropin-releasing factor receptor type 2; DRN, dorsal raphe nucleus; GR, glucocorticoid receptor; HPA, hypothalamic-pituitary-adrenocortical; MRN, median raphe nucleus; SSRI, selective serotonin reuptake inhibitor; TPH, tryptophan hydroxylase.

Introduction

In our modern society stress seems to be an omnipresent phenomenon. Many people complain about having to deal with stressful situations either at work or in the family, or are suffering from stress for an extended period of time. Even children will sometimes say that they feel 'stressed'. A detailed discussion of the reasons for the rise in the perceived levels of stress nowadays is beyond the scope of this chapter, but the increased speed of life in general and the concurrent floods of information to be processed may play a prominent role. Naturally, uncertainties regarding the economy, together with an unstable job and housing market, present major causes of stress in many people's lives.

*Corresponding author
E-mail: astrid.lindhorst@bristol.ac.uk

Christian Müller & Barry Jacobs (Eds.)
Handbook of Behavioral Neurobiology of Serotonin
ISBN 978-0-12-374634-4

DOI: 10.1016/B978-0-12-374634-4.00029-0

When combined with a sedentary lifestyle and unhealthy eating habits, stress coping mechanisms may fail and the damaging effects of chronic stress on physical and mental health will become evident.

In principal, the human body is well-equipped to deal with single stressful events, but when stress becomes chronic (for example, during care of ill or older relatives, prolonged financial problems, etc.) it may precipitate disease. It is now well accepted that stress – or maybe, more accurately phrased, mal-coping with stress – is involved in the etiology of certain psychiatric diseases such as major depression and anxiety disorders. In depressed patients, major aberrations have been found in the regulation of the stress hormone system, the hypothalamic–pituitary–adrenocortical (HPA) axis, under both baseline and challenged conditions (De Kloet *et al.*, 2005). Thus, major depression is, for instance, associated with a flattened circadian rhythm of cortisol and with impaired suppression in the dexamethasone suppression test – a neuroendocrine challenge test to assess negative feedback efficacy within the HPA axis via the glucocorticoid receptor (GR). Analyses of CSF samples and *post-mortem* tissues of depressed patients have provided evidence for alterations at the level of the corticotropin-releasing factor (CRF) system, which is one of the key regulators of the activity of the HPA axis and also plays an important role in stress-related behaviors such as fear and emotion (Nemeroff *et al.*, 1984, 1988; Raadsheer *et al.*, 1994, 1995; Arborelius *et al.*, 1999). Furthermore, modern molecular techniques have shown that alterations at the level of the genome – in genes involved in HPA axis regulation – may render individuals more susceptible to the development of a mood or anxiety disorder (Binder *et al.*, 2004a; Derijk *et al.*, 2006; van Rossum *et al.*, 2006).

As described in detail elsewhere in this volume, the neurotransmitter serotonin (5-HT) plays an important role in the etiology of mood and anxiety disorders (see also Maes and Meltzer, 1995; Mann, 1998; Ressler and Nemeroff, 2000). Many of the available pharmacological treatments for these illnesses involve manipulating the 5-HT system in one way or another (e.g., selective serotonin reuptake inhibitors (SSRIs) and buspirone, a 5-HT_{1A} receptor partial agonist). Importantly, it has been found that alterations in genes of the 5-HT system, such as polymorphism in the promoter region of the 5-HT transporter gene, can moderate the impact of stressful life events on depression (Caspi *et al.*, 2003). Thus, given the prominent role of both the HPA axis and 5-HT in stress-related psychiatric disease, the question arises as to whether and how stress influences the 5-HT system and, furthermore, whether stress exerts its effects on 5-HT via components of the stress hormone axis (i.e., involving glucocorticoids and neuropeptides of the CRF family). We endeavor to answer these questions in this chapter as follows. In the next section, we will first review evidence demonstrating the responsiveness of the rostral serotonergic raphe nuclei to stressful challenges; this in terms of c-Fos expression in 5-HT neurons and changes in the firing rate of these neurons. Next, the effects of stress on the synthesis of 5-HT and on the expression of the rate-limiting enzyme tryptophan hydroxylase (TPH) will be discussed. The subsequent section will be devoted to the effects of stress on extracellular levels of 5-HT – as assessed by *in vivo* microdialysis – in regions of the forebrain which are essential for the coordination of the various components of the stress response. Finally, the consequences of manipulating the HPA axis and the CRF system for serotonergic neurotransmission will be considered. The data discussed in this chapter underscore the concept that stress impacts on the 5-HT system in a stressor- and brain-region-dependent manner, involving, under certain circumstances, glucocorticoid hormones and CRF and related neuropeptides. Given that current pharmacological therapies for the treatment of stress-related mental health diseases are still unsatisfactory – especially because of profound side effects, delay in clinical efficacy and a significant proportion of treatment-resistant patients – the intricate interaction between stress and 5-HT should be taken as a stimulant to further investigate the role of and identify new drug targets in these systems in depression and anxiety disorders. As outlined towards the end of the chapter, it will furthermore be of great importance to study the impact of moderate exercise on the 5-HT system in more detail so as to increase our understanding of why exercise is beneficial in the treatment of mental health diseases, and why, putatively, it could be used for the prevention of the development of such illnesses.

Effects of stress on the raphe nuclei: c-Fos expression and neuronal firing

In this chapter we will focus on the rostral 5-HT system, with cell bodies located in the dorsal raphe nucleus (DRN), the caudal linear nucleus and the median raphe nucleus (MRN), and providing a dense innervation to the whole forebrain (for neuroanatomical details, see Chapter 1.3 of this volume).

The immediate early gene c-Fos is often used as a marker to identify neurons that are activated by various manipulations (Kovacs, 1998). Using this strategy, it has become increasingly clear that stress indeed activates neurons within the DRN and MRN. Increased expression of c-Fos mRNA has been demonstrated after exposure of animals to different psychological stress paradigms, such as inescapable tail-shock, elevated plus maze (a behavioral test for the assessment of anxiety levels in rodents), social defeat stress, immobilization stress, restraint and fear conditioning (Pezzone *et al.*, 1993; Silveira *et al.*, 1993; Watanabe *et al.*, 1994; Cullinan *et al.*, 1995; Beck and Fibiger, 1995; Martinez *et al.*, 1998; Grahn *et al.*, 1999; Ishida *et al.*, 2002; Rioja *et al.*, 2006; Lehner *et al.*, 2008). Forcing rats to swim – a stressor with not only psychological but also physical aspects (see below) – also induces c-Fos mRNA expression in both the DRN and MRN (Cullinan *et al.*, 1995). In contrast, activation of the immune system, which is regarded as a 'systemic' stressor representing an immediate physiological threat to the organism, seems to have no clear effect on c-Fos expression in the DRN and MRN in the rat (Ericsson *et al.*, 1994; Elmquist *et al.*, 1996; Laflamme *et al.*, 1999). However, in a recent study Hollis and colleagues demonstrated Fos-immunoreactivity in a specific subset of serotonergic neurons in the DRN (i.e., within the interfascicular and ventrolateral division of this raphe nucleus) after administration of bacterial endotoxin to mice (Hollis *et al.*, 2006). Interestingly, a topographically specific activation of neurons in the DRN has also been described after forced swim stress. Fifteen minutes of forced swimming in water at 25°C results in Fos-immunoreactivity in the dorsolateral DRN of rats (Roche *et al.*, 2003). Most of these Fos-positive neurons are, however, not 5-HT neurons but GABAergic interneurons, leading to the hypothesis that stress may regulate the neuronal activity of raphe 5-HT neurons (also) via its actions on the inhibitory neurotransmitter GABA within this nucleus. Given the observation that these Fos-positive neurons are enveloped by CRF fibers and contain CRF receptors (Roche *et al.*, 2003), it is of interest to note that intracerebroventricular administration of the CRF receptor type 1 (CRF1) ligands, CRF and urocortin 1, also causes a pronounced increase in Fos-immunoreactivity in the DRN (together with a moderate increase in the MRN) (Bittencourt and Sawchenko, 2000). In contrast, the CRF receptor type 2 (CRF2) ligand urocortin 2 has been found either to have no effect (Reyes *et al.*, 2001) or to increase the number of c-Fos-positive neurons mainly in the caudal regions of the DRN (Staub *et al.*, 2005). Administration of urocortin 2 directly into the caudal DRN results in an increased number of dual-labeled c-Fos/5-HT neurons in the caudal, middle and rostral DRN (Amat *et al.*, 2004). Interestingly, a recent study shows that in the caudal part of the DRN, levels of c-Fos expression correlate with the level of CRF-induced enhancement of the startle response in rats (Meloni *et al.*, 2008).

The above-described studies clearly show that stress (and members of the CRF neuropeptide family) results in a distinct and topographically specific neuronal activation of serotonergic (and other) neurons in the raphe nuclei. However, does this neuronal activation lead to changes in neuronal firing of 5-HT cell bodies in these nuclei? Unfortunately, data from studies aiming to answer this important question are so far inconclusive. Studies by Barry Jacobs and colleagues have shown, using an *in vivo* electrophysiology approach, that in the cat the firing pattern of 5-HT cell bodies remains unchanged (as compared to the firing rate during active waking) during various forms of stress, such as noise stress, pain, and thermoregulatory and glucoregulatory challenges (for references, see Jacobs and Azmitia, 1992). Also, predator exposure, a challenge which increases extracellular 5-HT levels in higher brain structures in rodents, was found to be without effect in the DRN of the cat (Wilkinson and Jacobs, 1988). In contrast, using a brain slice approach, Laaris and colleagues showed that a longer period of stress (30-minute restraint followed by 24-hour isolation or 16-hour exposure to novel uncontrolled environmental conditions), but not acute restraint stress, results in a decreased 5-HT_{1A} receptor-mediated inhibition of 5-HT cell firing rate in rats, which is dependent on corticosterone (Laaris *et al.*, 1997, 1999). Interestingly, early maternal separation (6 hours, 10 occasions, between postnatal days 5 and 20) has been found, *in vitro*, to decrease the sensitivity of DRN 5-HT neurons for $\alpha 1$-mediated excitation, but was without significant effect on the sensitivity of 5-HT_{1A} receptors. Furthermore, this manipulation did not change the proportion of classical and burst-firing 5-HT neurons, although the range of firing rates seems to be increased in the early-life stressed group (Gartside *et al.*, 2003). Subsequent work by the same group elegantly demonstrated that the balance between $\alpha 1$-mediated excitation and 5-HT_{1A}-mediated inhibition of the firing rate of DRN neurons is dependent on the level of corticosterone (Judge *et al.*, 2004). Thus, ADX rats implanted with a 2% corticosterone pellet show an enhanced $\alpha 1$-mediated excitation and a lower 5-HT_{1A}-mediated inhibition of DRN firing as compared to animals with a 70% corticosterone pellet (Judge *et al.*, 2004). By contrast, using an *in vivo* electrophysiology approach, it was found that prolonged (180-minute, postnatal days 2–14) maternal deprivation results in an increased 5-HT_{1A} autoreceptor-induced inhibition of raphe neuronal firing (Arborelius *et al.*, 2004). Studies on the effects of CRF on the firing rate of 5-HT neurons in the raphe nuclei have also provided evidence for a topographically specific influence. Thus, low doses of CRF were found to decrease the *in vivo* firing rate of 5-HT neurons in the rostral and medial parts of the DRN. Higher doses were without effect (Price *et al.*, 1998; Kirby *et al.*, 2000), whereas CRF increases the *in vitro* firing rate of a selective population of 5-HT neurons in the ventral and interfascicular region of the caudal DRN (Lowry *et al.*, 2000).

From the above, the picture is emerging that stress impacts on the activity of 5-HT neurons in the raphe nuclei, but that it probably does so in a stressor- and topographically-specific way. Clearly more research is needed, implementing other stressful (acute and chronic) challenges and the effects of stress on non-classical (i.e., burst-firing) 5-HT neurons. Also, the role of glucocorticoid hormones needs to be further elucidated.

Effect of stress on the synthesis of 5-HT

5-HT is synthesized from L-tryptophan, an essential amino acid which needs to be taken up via the diet and is transported into the brain via an L-type neutral amino acid transporter in the blood–brain barrier. The rate-limiting step in the synthesis of 5-HT is the conversion of L-tryptophan into 5-hydroxytryptophan (5-HTP), a step which is catalyzed by the enzyme tryptophan hydroxylase (TPH). TPH is present in enterochromaffin cells, myenteric neurons of the gut, the pineal gland, and the brain raphe nuclei. Next, 5-HTP is converted into 5-HT via decarboxylation by the enzyme aromatic L-amino acid decarboxylase (AADC). There is some evidence that, under certain circumstances, the availability of L-tryptophan to the brain can affect the level of synthesis of 5-HT (Fernstrom and Wurtman, 1971; Knott and Curzon, 1972; Boadle-Biber, 1993).

TPH is activated when stores of 5-HT need to be refilled, as evidenced by the rise in the synthesis of 5-HT from L-tryptophan during electrical stimulation of serotonergic cell bodies (Shields and Eccleston, 1972; Herr *et al.*, 1975; Boadle-Biber *et al.*, 1986). As discussed above, stress has distinct effects on neuronal activation and, putatively, consequently on the firing rate of 5-HT neurons in the raphe nuclei, leading to the obvious question of whether stress also influences TPH activity and expression. After the initial study of Azmitia and McEwen describing that prolonged cold and electric foot-shock stress increased TPH activity in the midbrain and forebrain of rats (Azmitia and McEwen, 1974), Margaret Boadle-Biber and colleagues have performed a large series of studies characterizing the effects of loud sound stress on 5-HT synthesis using an *ex vivo* TPH activity assay in rats. Applying this assay, it was found that one session of randomly presented sound stress causes a short-lasting activation of TPH in the cortex. This increase in TPH activity is persistent even 24 hours later when rats are exposed to repeated sound stress sessions (Boadle-Biber *et al.*, 1989). Importantly, as pointed out above, the enhanced TPH activity is related to increases in neuronal firing rate, as no effects were found when 5-HT neurons were silenced by pretreatment of the animals with the 5-HT_{1A} receptor agonist gepirone (Corley *et al.*, 1992). The effects of stress seem to be dependent on the availability of glucocorticoids, as adrenalectomy has been found to block the rise in TPH activity caused by sound stress (Singh *et al.*, 1990). The CRF system may also play a role in the regulation of TPH activity. Intracerebroventricular administration of CRF increases TPH activity, an effect in which glucocorticoid receptors play a permissive role (Singh *et al.*, 1992).

From the above studies, it has become clear that stress indeed impacts on the activity of TPH. However, the next question that arises is whether stress affects 5-HT synthesis in a neuroanatomically specific manner. Using an assay measuring the accumulation of 5-HTP after inhibition of AADC by m-hydroxylbenzylamine, it was found that sound stress, forced swimming and tail-shock increased TPH activity in the MRN only, and were without effect in the DRN (Dilts and Boadle-Biber, 1995; Corley *et al.*, 2002). Furthermore, whereas CRF was found to increase THP activity in the MRN (Singh *et al.*, 1992), this neuropeptide had no effect in the mediobasal hypothalamus (Van Loon *et al.*, 1982). A neuroanatomically specific effect of stress is further underscored by studies utilizing immobilization stress. Immobilization stress had no effect on TPH activity in the hypothalamus, hippocampus, amygdala and DRN after five sessions (Palkovits *et al.*, 1976), although it seemed to increase TPH activity in the locus coeruleus and to decrease activity in the suprachiasmatic nucleus after seven sessions (Culman *et al.*, 1984). Furthermore, using an *in vivo* microdialysis approach to measure the accumulation of 5-HTP, it was found that 1-hour immobilization results in increased 5-HT synthesis in the nucleus accumbens and the medial prefrontal cortex of the rat (Nakahara and Nakamura, 1999).

So far, we have reviewed the literature regarding the effects of stress on the activity of the TPH enzyme. However, much recent research has focused on the expression of TPH mRNA and on the levels of the TPH protein. In 2003, Walther and colleagues, in a seminal study published in *Science*, described for the first time the existence of a second TPH gene, *Tph2* (Walther *et al.*, 2003). The previously known gene is now called *Tph1*. Importantly, they demonstrated that *Tph2* is expressed exclusively in the brain (Walther *et al.*, 2003), and it was subsequently found that TPH2 controls the synthesis of 5-HT in the brain (Zhang *et al.*, 2004). The highest expression of TPH2 mRNA is found in the ventromedial DRN and MRN at the midrostral level (Clark *et al.*, 2006). Interestingly, TPH2 mRNA levels follow a circadian rhythm which is under the control of the suprachiasmatic nucleus (SCN) and is mediated by the circadian rhythm of corticosterone (Malek *et al.*, 2007). In contrast to TPH2, TPH1 mRNA is abundantly expressed in the pineal gland and the periphery. Although some studies have described the absence of *Tph1* in the raphe nuclei (Walther *et al.*, 2003; Cote *et al.*, 2003), the picture is now emerging that low levels of TPH1 mRNA are expressed in the brain serotonergic cell bodies (Patel *et al.*, 2004; Abumaria *et al.*, 2008) in rodents. In *post-mortem* human brain, low levels of TPH1 mRNA are found in the raphe nuclei but more substantial levels have been observed in several forebrain regions (e.g., the hypothalamus and amygdala) (Zill *et al.*, 2007). Furthermore, TPH1 may play a role in the late stage of development, as peak levels of TPH1 mRNA at P21 have been described recently (Nakamura *et al.*, 2006). The effects of stress on the expression of the two *Tph* genes have not been fully clarified yet. Although TPH2 seems to be the most important enzyme form for 5-HT synthesis in the brain, no effects of 1-week daily restraint stress on TPH2 mRNA levels were found (Abumaria *et al.*, 2008). Also, chronic social stress was found to have no effect on the expression of TPH2 mRNA (Abumaria *et al.*, 2006). Furthermore, stress during early gestation seems to only cause marginal (trend to significance) effects on TPH2 mRNA in the DRN of male offspring (Mueller and Bale, 2008), whereas chronic administration of ovine CRF or urocortin 2 had no effect on the absolute expression of TPH2 mRNA in the DRN (although changes in the relative expression between the midrostral ventromedial and core dorsomedial DRN were observed) (Clark *et al.*, 2007). Recently it has been described that daily subcutaneous administration of the glucocorticoid receptor agonist dexamethasone (0.3–3 mg/kg subcutaneously for 4 days) decreased TPH2 protein levels in the raphe nuclei of mice, but, given the poor penetration of dexamethasone into the brain, the underlying mechanism of this observation needs to be further elucidated (Clark *et al.*, 2008). With respect to the marginal effects of stress on TPH2 described so far, it is interesting that, despite its low levels of expression in the brain, TPH1 is more clearly affected by stress. Thus, repeated immobilization or restraint stress has been found to substantially (by 2- to 10-fold) increase TPH1 mRNA levels in the DRN and MRN (Chamas *et al.*, 1999, 2004; Abumaria *et al.*, 2008). However, the effect of repeated immobilization stress on TPH1 mRNA expression seems to be independent of circulating glucocorticoids, as similar effects of the stressor were observed in adrenalectomized rats (Chamas *et al.*, 2004). The above-described findings clearly demonstrate the need for further research into the effects of stress on the two forms of TPH (at the gene and protein level), taking into account the possible stressor dependency and neuroanatomical specificity of such effects. Such studies will be of great importance not only for our basic understanding of the regulation of serotonergic neurotransmission, but also, given the changes observed in *Tph2* expression in suicide victims (Bach-Mizrachi *et al.*, 2008) and the putative relationship between *Tph1* and *Tph2* gene variants and major depression (Zill *et al.*, 2004; Gizatullin *et al.*, 2006; Galfalvy *et al.*, 2008, but also see Gizatullin *et al.*, 2008), for further clarifying the role of 5-HT in stress-related psychiatric illnesses.

Effects of stress on the extracellular levels of 5-HT and 5-hydroxyindoleacetic acid

During the past 25 years, *in vivo* microdialysis has become the method of choice to study extracellular levels of neurotransmitters in the brain. Microdialysis is predominantly used in neuropharmacology to evaluate the effects of (newly developed) drugs on neurotransmitter systems, and to elucidate their mechanisms of action. However, microdialysis is increasingly becoming an important tool to study the role of neurotransmitter systems in physiology and behavior. As we have recently discussed (Linthorst and Reul, 2008), several refinements in the performance of the technique, including the use of gas anesthesia, guide cannulas and special freely-moving animal systems, together with improved analytical methods, have made microdialysis extremely suitable for the study of various aspects of the stress response in rats and (mutant) mice (Figure 1). Below we will discuss the effects of stress on the extracellular levels of 5-HT and its metabolite 5-hydroxyindoleacetic acid (5-HIAA) in (1) three higher brain regions important in the behavioral and psychological aspects of the stress responses (i.e., the hippocampus, the frontal cortex and the amygdala) and (2) the hypothalamus and pre-optic area, because of their key role in the regulation of the endocrine and autonomic responses to stress.

Hippocampus, prefrontal cortex and amygdala

During recent years it has become clear that stress affects hippocampal levels of 5-HT and its metabolite 5-HIAA in a stressor-dependent way. Submitting rodents to predator stress, an ethologically highly relevant psychological stressor, has been found to increase extracellular levels of 5-HT and 5-HIAA in the hippocampus of rats and mice (Rueter and Jacobs, 1996; Linthorst *et al.*, 2000). In a recent study, we used a rapid sampling design (5 minutes during and immediately after the stressful challenge) to study the effects of exposure to a rat on the serotonergic neurotransmission in the hippocampus of male C57BL/6N mice (Beekman *et al.*, 2005). For this stressor, a male Wistar rat was introduced into the home cage of

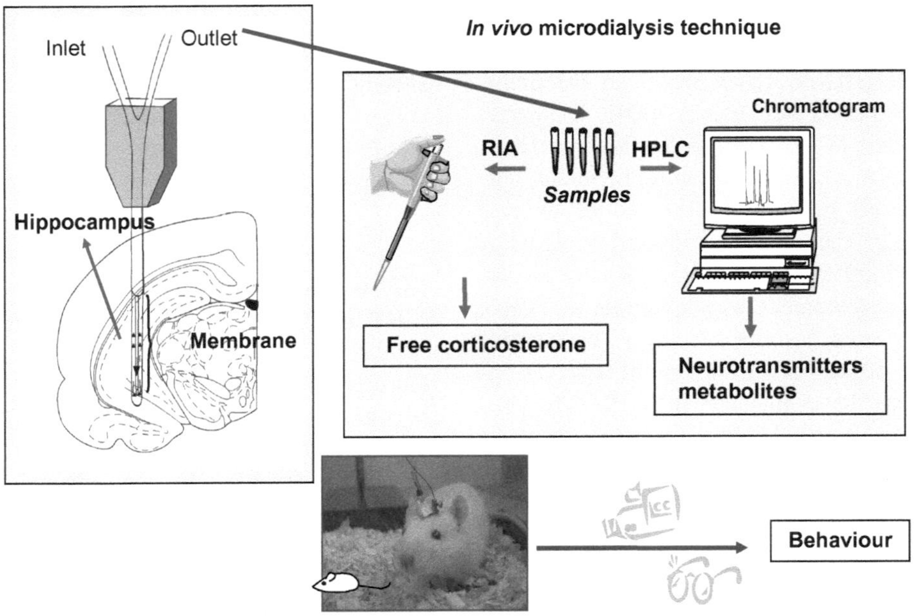

Figure 1 Schematic overview of the microdialysis technique as applied in our laboratories. A microdialysis probe is inserted through a guide cannula (not depicted, implanted 7–10 days earlier; the dialysis probe depicted is the CMA12 of CMA Microdialysis/AB, Stockholm, Sweden) exposing the dialysis membrane to the brain tissue of interest (shown here is the hippocampus). The microdialysis probe is perfused with Ringer solution via a microinfusion pump (not depicted). Drugs can be delivered to the target tissue by inclusion in the perfusion fluid (so-called retrodialysis or reverse dialysis). Animals are connected to a swivel system and counterbalance arm via a little peg anchored to the dental cement on their skull (see picture) giving them the opportunity to move freely in all directions. Samples are collected using automated microsamplers, and are subsequently used to measure various molecules in the dialysates – e.g., neurotransmitters and their metabolites – by high-pressure liquid chromatography (HPLC) with electrochemical detection or free corticosterone levels via radioimmunoassay (RIA). Behavior of the rats or mice is continuously scored during the microdialysis experiments by visual observation or is subsequently analyzed from videotape. Experiments are performed under baseline conditions, and during and following stressful challenges or behavioral testing. Note that this method allows the simultaneous collection of information on the status and responsivity of various neurotransmitter systems in the brain, on the activity of the HPA axis and on behavior, providing strong, integrative data and representing a clear example of refinement and reduction (Russell and Burch, 1959) in animal research. To see the full color version of this figure, please refer to the color plate in the back of the book. Copies produced via our print on demand service do not contain color plates. If your copy does not have the color plate, please go to this website to view the figure in color www.elsevierdirect.com/companions/9780123746344

the mouse but remained separated from the mouse by a Plexiglas wall with small holes. Even during the first 5 minutes of the total 30 minutes of rat exposure, 5-HT levels increased rapidly, reaching maximum levels of about 300 percent of baseline. 5-HT remained elevated during the next 10 minutes, but started to decrease in the second half of the rat exposure procedure. The levels of 5-HIAA also increased with the introduction of the rat, but this response was somewhat delayed as compared to the immediate 5-HT response. It is of interest to contrast the effects of the presence of the predator on hippocampal 5-HT with the behavioral changes concurrently taking place in the stressed mice. Immediately at the start of exposure to the rat, mice usually move to a corner of the cage and become highly vigilant. Next, various risk-assessment behaviors are initiated, such as sniffing the air, stretching towards and retracting from the separation wall (flat back posture) and then sniffing at the separation wall. During the second half of the predator exposure, mice were seen to be extensively sniffing the bedding of their cage (Beekman *et al.*, 2005). From these results we concluded that during psychological stress 5-HT in the hippocampus increases especially during the initial phase, supporting the assessment of a new and potentially life-threatening situation and assuring adequate coping responses regarding the challenge. Increases in extracellular concentrations in 5-HT and 5-HIAA were also observed in the frontal cortex, the amygdala and lateral septum (Rueter and Jacobs, 1996; Beekman *et al.*, 2005).

Whereas the above-discussed predator stress paradigm is an example of a psychological form of stress, most stressors used to investigate the effect of stress on hippocampal 5-HT have combined psychological and physical aspects. Tail pinch (attaching a paper clip to the tail of a rat or mouse) causes arousal and also activates pain sensory pathways and consummatory behaviors. Microdialysis studies have shown that this form of stress results in increases in extracellular 5-HT in the hippocampus, prefrontal cortex and amygdala, although in some studies this response was delayed until after termination of the stressor (Kalén *et al.*, 1989; Pei *et al.*, 1990; Vahabzadeh and Fillenz, 1994; Rueter and Jacobs, 1996; Fujino *et al.*, 2002). The delayed increase in 5-HT could well be caused by the increased consummatory and grooming behavior after the removal of the paper clip – a notion that may be underscored by the demonstration of a subset of 5-HT neurons which is highly active during oral–buccal movements in the cat (Fornal *et al.*, 1996), and by our own observation that grooming behavior in the home cage profoundly increases hippocampal 5-HT in rodents (Linthorst *et al.*, unpublished data). Electric foot- or tail-shock represents another example of a stressor with combined physical and psychological aspects. Surprisingly, there is not much information available regarding the effects of this stressor on extracellular levels of 5-HT in forebrain regions. *Post-mortem* tissue studies suggest that electrical foot-shock increases 5-HT turnover in the hippocampus, amygdala and prefrontal cortex (for references, see Linthorst, 2005). Interestingly, it seems that the exact response to electrical shock is dependent on the extent of control an animal can exert over the shock. Thus, an augmented turnover and extracellular response of 5-HT has been found in the frontal cortex and ventral hippocampus of rats for whom the shock was inescapable, as compared to their conspecifics that could escape the shock (Amat *et al.*, 1998; Heinsbroek *et al.*, 1991). Interestingly, the prefrontal cortex plays a key role in inhibiting the activation of serotonergic neurons in the DRN when rats are given the opportunity to escape shocks – i.e., when the stressor is controllable by the animal (Amat *et al.*, 2005). Conditioned fear stress also leads to an increase in hippocampal and prefrontal cortical extracellular levels of 5-HT, although the neurotransmitter response was not studied during the training phase when animals actually receive the electrical shocks (Yoshioka *et al.*, 1995; Wilkinson *et al.*, 1996; Hashimoto *et al.*, 1999). Importantly, Wilkinson and colleagues demonstrated that the 5-HT activation in the hippocampus induced by conditioned fear stress was due to the contextual aversive cues (i.e., the box in which the animal had received the shocks) but not to the conditioned discrete stimulus (Wilkinson *et al.*, 1996).

During the past years, we and others have extensively studied the effects of the combined psychological/physical stressor forced swimming on extracellular levels of 5-HT and 5-HIAA. Forced swim stress is performed by placing rats or mice in a glass beaker filled with water from which they cannot escape. Forced swim stress causes profound behavioral changes. During the swim session, animals will first try to escape by climbing against the wall and, in the case of rats, will explore possible escape routes under the water surface (diving). This period is followed by swimming in the container and by floating behavior. Although increased floating behavior/immobility is used as an index for despair or depressed behavior in the Porsolt swim test, it is now increasingly seen as an index of a learning process – i.e., the animal learns a behavioral coping strategy aimed to conserve energy when forced to swim on a second exposure (De Pablo *et al.*, 1989; Abel, 1993; Korte, 2001; Bilang-Bleuel *et al.*, 2005; Chandramohan *et al.*, 2008). However, apart from increased motor activity during parts of this challenge, the physiology of the animals is compromised by a rapid and extensive loss of body temperature (Stone, 1970; Linthorst *et al.*, 2007) (see below). Fifteen minutes of forced swim stress at 25°C results in an immediate increase in hippocampal levels of 5-HT. This increase is followed by a return to pre-stress levels when animals are returned to their home cage, and a subsequent second rise in the neurotransmitter levels (Figure 2C). Interestingly, whereas swimming in warm water (35°C) causes a rapid and prolonged rise in hippocampal extracellular 5-HT (Figure 2E), no effect on 5-HT was found when the animals were forced to swim in cold water (19°C), although levels increased 15–30 minutes after the termination of the stressor (Figure 2A). The effects of forced swim stress on extracellular levels of 5-HIAA were also highly water-temperature-dependent. Thus, forced swim stress in water at 35°C causes an immediate and sustained increase in hippocampal extracellular 5-HIAA (Figure 2F). However, the levels of this metabolite actually first decrease significantly in rats forced to swim in cold water at 19°C, after which they increase gradually (Figure 2B) (Linthorst *et al.*, 2007). Using an *in vivo* biotelemetry approach, it was found that 15 minutes of forced swimming had minor effects on the body temperature of the rats when water of 35°C was used (a decrease of about 1°C). However, at 25°C and 19°C, rats lost, respectively, approximately 8°C and 12°C of body temperature (Figure 2) (Linthorst *et al.*, 2007). These observations, together with the microdialysis findings described above, have led us to postulate that the final outcome of the neurotransmitter response to stress is shaped by the combination of the psychological aspects of the stressor and the impact of the concurrent physical changes. Recent investigations into the effects of forced swim stress on extracellular levels of GABA in the hippocampus have further substantiated this hypothesis. Thus, forced swim stress at 35°C increased hippocampal levels of GABA, whereas swimming at 25°C led to a profound decrease in extracellular neurotransmitter concentrations in this brain structure (De Groote and Linthorst, 2007). The physical impact of forced swim stress may also explain why there are some conflicting results reported in the literature with regard to the effects of this stressor on hippocampal 5-HT. Thus, an

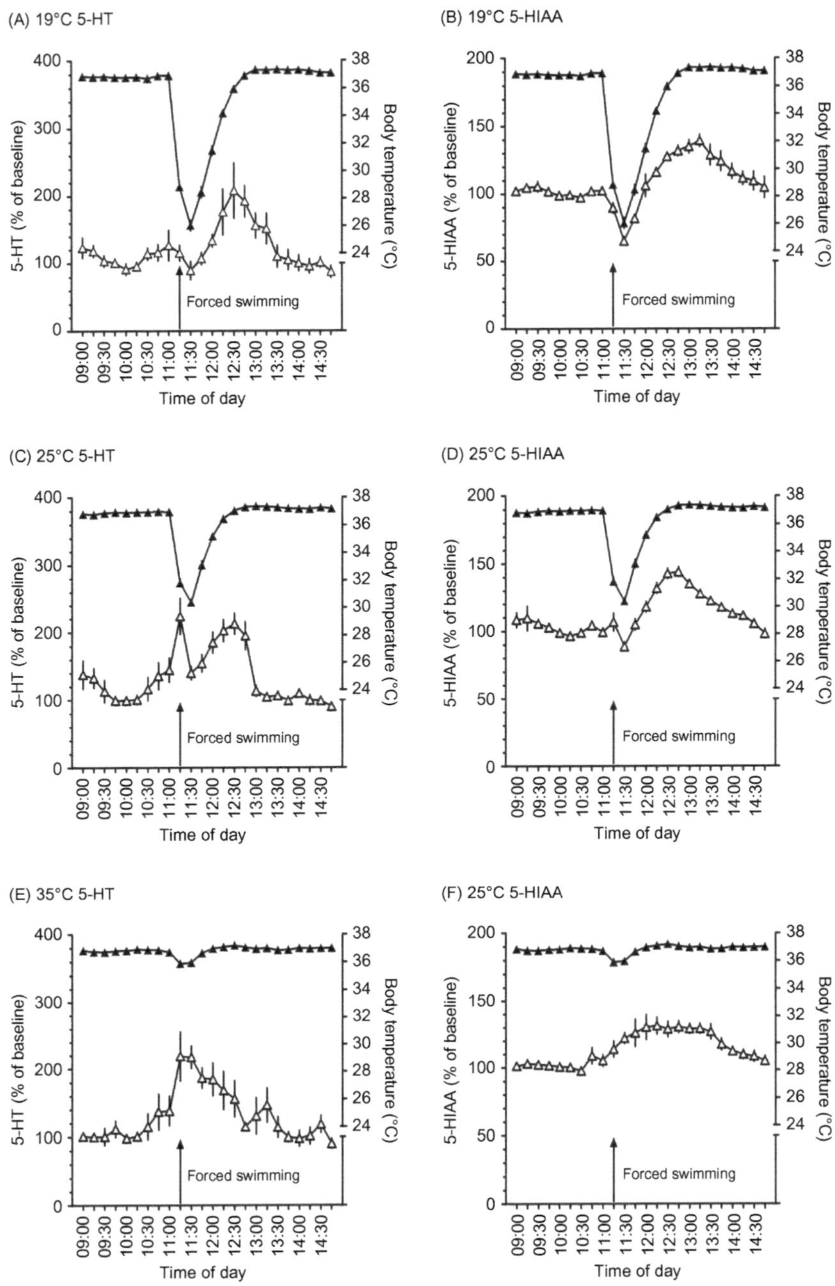

Figure 2 Comparison of the effects of forced swim stress at different water temperatures – (A, B) 19°C; (C, D) 25°C; (E, F) 35°C – on core body temperature (closed triangles) and hippocampal extracellular levels of 5-HT and 5-HIAA (open triangles) in male Wistar rats. 5-HT and 5-HIAA levels were assessed by *in vivo* microdialysis and body temperature (in separate animals) by *in vivo* biotelemetry. Levels of 5-HT and 5-HIAA are expressed as percentage of baseline. The time points on the x-axis indicate the time of the day at which collection of the respective sample was started. After collection of nine 15-min baseline samples, rats were placed in a beaker glass with water for 15 min (at 11:15 h, indicated by the arrow). After the forced swim session they were taken out of the water, dried with a towel and returned to their home cage, following which another fourteen 15-min dialysis samples were collected. 5-HT and 5-HIAA concentrations were analyzed by HPLC with electrochemical detection. Body temperature data were collected at 2-min intervals, but were averaged over 15-min bins to match the 15-min dialysate sample duration. Values are mean ± SEM. For 5-HT and 5-HIAA $n = 5$–7; for body temperature $n = 6$. Note that swimming for 15 min at 19 and 25°C produced a pronounced decrease in core body temperature and that the 5-HT and 5-HIAA responses in the hippocampus are dependent on the water temperature used. Comparison of the body temperature and neurotransmitter profiles suggests that two different processes may shape the ultimate 5-HT and 5-HIAA responses to swim stress. Thus, forced swimming causes an increase in extracellular levels of 5-HT and 5-HIAA, which is temporarily quenched or concentrations are transiently decreased when core body temperature falls below approximately 31°C. Reproduced from Linthorst *et al.* (2007), with permission of *Stress* (Informa Healthcare).

increase in hippocampal 5-HT was found by Rueter and Jacobs in rats that were forced to swim for 30 minutes in water of 30–35°C (Rueter and Jacobs, 1996), while Kirby and colleagues only observed a decrease in extracellular 5-HIAA (and no effect on 5-HT) when rats had to swim for 30 minutes in water of 21–22°C (Kirby *et al.*, 1995). Moreover, these authors found also disparate effects of swim stress in the frontal cortex and amygdala – i.e., increased levels or no effect/decreased levels of 5-HT and 5-HIAA (Kirby *et al.*, 1995; Rueter and Jacobs, 1996).

Immune stress is an example of a stressor with an immediate threat to the integrity and survival of an organism, and therefore is expected to merely be processed at the level of lower brain structures such as the hypothalamus and brainstem nuclei (Herman and Cullinan, 1997). However, we have provided extensive evidence that an immunological challenge impacts on serotonergic neurotransmission in the hippocampus of rats. Thus, systemic administration of bacterial endotoxin (lipopolysaccharide, LPS) results in a pronounced and prolonged ($>$6 hours) elevation of hippocampal extracellular levels of 5-HT and 5-HIAA (Linthorst *et al.*, 1995a). These effects can be mimicked by peripheral and intracerebroventricular administration of interleukin-1β and interleukin-2, but not by TNF-α (Linthorst *et al.*, 1995a; Merali *et al.*, 1997; Pauli *et al.*, 1998). Studies in rats treated chronically with CRF suggest that a full hippocampal 5-HT response is a prerequisite for the adequate development of sickness behavior during immune activation (Linthorst *et al.*, 1997). Based on these observations, we have concluded that immune stress also possesses a distinct psychological component probably involved in the mood changes occurring during infection.

Hypothalamus and pre-optic area

The hypothalamus and pre-optic area are key structures in the coordination of the HPA axis and the autonomic response to stress. Most studies using *post-mortem* tissues have demonstrated that stress increases or has no effect on the turnover of 5-HT in the different subnuclei of the hypothalamus (for references, see Linthorst, 2005). Surprisingly, there is little information available regarding the effects of stress on extracellular levels of 5-HT and 5-HIAA as assessed by *in vivo* microdialysis. We have studied the effects of immune stress on serotonergic neurotransmission in the pre-optic area of the rat. In contrast to the profound activation of 5-HT in the hippocampus as described above, no effect on extracellular 5-HT and only a minute rise in extracellular 5-HIAA was found in the pre-optic area (Linthorst *et al.*, 1995b). The absence of an effect of immune stress on pre-optic 5-HT is underscored by similar observations made in the cat (Wilkinson *et al.*, 1991). In contrast, immune system activation was found to increase both extracellular 5-HT and 5-HIAA levels in the hypothalamus of the rat (Mohankumar *et al.*, 1993; Shintani *et al.*, 1993; Lavicky and Dunn, 1995). With respect to psychological stress, we have recently demonstrated that exposure to a predator increases extracellular levels of 5-HT and its metabolite in the paraventricular nucleus and anterior hypothalamus of Balb/c mice (Beekman and Linthorst, unpublished observations). Another form of psychological stress, immobilization, was also found to increase markedly the extracellular levels of 5-HT in the hypothalamus (Shimizu *et al.*, 1992; Shintani *et al.*, 1995), although only a marginal effect on 5-HIAA was found. Interestingly, these effects could be blocked by infusion of an IL-1 receptor antagonist (Shintani *et al.*, 1995). From the above, the general picture is emerging that stress increases extracellular 5-HT in the hypothalamus (but not the pre-optic area), while it is at the same time clear that more research (implementing both acute and chronic stress paradigms) will be needed to further substantiate this notion.

Repeated and chronic stress

In previous sections of this chapter we have given examples of studies demonstrating that repeated stress and stressors of a longer duration may change the electrophysiological properties of raphe 5-HT neurons and the synthesis of this neurotransmitter. However, similarly to 5-HT firing rate and synthesis, there is, unfortunately, only limited information available regarding the effects of repeated or chronic stress on extracellular levels of 5-HT and 5-HIAA. This may have to do with the complicating factor that microdialysis is normally performed over a restricted number of days. However, there are a few interesting studies which have investigated the status and responsiveness of the 5-HT system at the end of a chronic stress paradigm.

Under baseline conditions, extracellular levels of 5-HT are closely correlated with the vigilance state of the animal. Thus, in various brain structures, among them the hippocampus, highest levels of 5-HT are found during active waking, while levels decrease when animals are entering slow-wave sleep (Houdouin *et al.*, 1991; Imeri *et al.*, 1994; Portas and McCarley, 1994; Portas *et al.*, 1998; Park *et al.*, 1999; Shouse *et al.*, 2000; Python *et al.*, 2001; Peñalva *et al.*, 2003). Interestingly, extracellular levels of 5-HT are lowest during rapid eye movement sleep, probably caused by the almost complete silencing of the firing rate of raphe 5-HT neurons during this sleep stage. In a combined *in vivo* microdialysis/sleep EEG study, we found that during a longer period of stress (i.e., a 4-hour period of sleep deprivation), hippocampal 5-HT levels are increased throughout the whole period (Peñalva *et al.*, 2003). However, after termination of the sleep-deprivation procedure, hippocampal levels of 5-HT decrease during the rebound phase and the normal 5-HT/sleep stage relationship is immediately re-established. Of interest in this respect is a recent study demonstrating that a chronic mild stress paradigm is able to disrupt exactly this tight relationship between hippocampal 5-HT levels and sleep stage by an as yet unknown mechanism (Gronli *et al.*, 2007).

Another example of the effects of chronic stress on 5-HT levels is the isolation-rearing model, as developed by Charles Marsden. These authors have studied the behavioral and neurochemical consequences of isolation rearing – i.e., single housing of rats for 6 weeks immediately following weaning – as compared to social, group housing. Isolation induces profound behavioral changes in the animals, but no changes in baseline 5-HT levels in the dorsal hippocampus were observed. However, the 5-HT system is clearly compromised in isolation-reared rats when challenged with a stress paradigm. Thus, footshock stress and stress caused by placing the animal on an elevated x-maze both result in a rise in hippocampal 5-HT in group-reared rats, but have no effect in isolated rats (Bickerdike *et al.*, 1993; Muchimapura *et al.*, 2002). There are some indications that the effects of chronic stress on extracellular 5-HT may be stressor-dependent. Repeated exposure of rats to an elevated open platform not only increased extracellular levels of 5-HT and 5-HIAA in the dorsal hippocampus under baseline conditions, but also resulted in elevated 5-HT levels when animals were tested on day 10 of the repeated platform exposure; this as compared to naïve animals which showed no changes in hippocampal 5-HT when placed on the elevated open platform for the first time (Storey *et al.*, 2006).

Effects of stress on serotonergic neurotransmission in animal models of aberrant HPA axis functioning

As discussed above, changes at the level of the genome may be implicated in the etiology of stress-related psychiatric disease. Mutant mouse strategies may therefore represent a useful tool to further study the regulation of the stress response and, importantly, to gain insight into (malfunctioning) stress coping mechanisms. Based on the putative involvement of the CRF system in depression and anxiety, and our observation that long-term elevation of brain CRF levels precipitate altered responsiveness of hippocampal 5-HT to stress (Linthorst *et al.*, 1997), we became interested in studying a genetic mouse model of CRF1 impairment (Timpl *et al.*, 1998). Performance of *in vivo* microdialysis in freely behaving CRF1 knockout mice over the diurnal cycle demonstrated that deletion of CRF1 results in significantly elevated levels of 5-HIAA, but not of 5-HT, in the hippocampus during both the light and dark periods (Peñalva *et al.*, 2002). There is substantial evidence that 5-HIAA levels may be an indicator for newly-synthesized 5-HT which has not been released from the presynaptic element (Grahame-Smith, 1974; Kuhn *et al.*, 1986; Peñalva *et al.*, 2002). Thus, the increased levels of 5-HIAA suggest that the synthesis of 5-HT is increased (in the hippocampus) in CRF1 knockout mice. Naturally, the question arises whether the deletion of CRF1 also has consequences for the response of serotonergic neurotransmission to stress. To address this question, CRF1 knockout and wild-type mice were subjected to a 10-minute session of forced swim stress (25°C). Interestingly, although the 5-HIAA response is similar in both genotypes, CRF1 knockout mice show an augmented 5-HT increase in the hippocampus when subjected to this form of stress (Peñalva *et al.*, 2002). To the best of our knowledge, this is the only information available on extracellular 5-HT and 5-HIAA in a mutant mouse model with alterations at the level of the CRF system. Together with the pronounced effects of CRF and the urocortins on hippocampal 5-HT and 5-HIAA levels demonstrated after intracerebroventricular administration (De Groote *et al.*, 2005), these data clearly show the intricate relationship between the CRF and 5-HT systems; aberrant functioning of this relationship may therefore underlie the development of certain psychiatric diseases.

Apart from changes in CRF, changes in GR and at the level of the mineralocorticoid receptor, both critical in the regulation of the HPA axis, may also play a role in stress-related mental health disorders. In an earlier study we investigated the stress responsiveness of GR-impaired mice with respect to behavior, HPA axis activity and hippocampal serotonergic neurotransmission (Linthorst *et al.*, 2000). Exposure of GR-impaired mice to a predator (rat; see above) resulted in an abnormal behavioral response to this potentially life-threatening stressor. GR-impaired mice showed abnormal risk assessment behaviors (i.e., an immediate inspection of the separation wall) which was accompanied by an absent HPA axis response to this challenge. Interestingly, during predator stress an augmented rise in 5-HT levels in the hippocampus in GR-impaired mice was found. Thus, although impairment of GR did not lead to changes in baseline levels of 5-HT and 5-HIAA, the response of this neurotransmitter to psychological stress was clearly exaggerated (Linthorst *et al.*, 2000). Alterations in serotonergic neurotransmission have also been found in other HPA axis-related mutant mice (Wei *et al.*, 2004; Rozeboom *et al.*, 2007), but unfortunately these studies were limited to the measurement of 5-HT_{1A}

receptor mRNA expression, and to date no studies on the extracellular levels of 5-HT and its metabolite in these mouse lines have been published.

Enhanced stress coping by exercise

As put forward in the introduction of this chapter, understanding how stress affects serotonergic neurotransmission is essential for increasing our insight into the mechanisms underlying stress-related psychiatric disease. However, equally important will be a further understanding of factors and mechanisms promoting adequate psychological and physiological coping with stress. There is now abundant evidence demonstrating that moderate exercise has beneficial effects on the cardiovascular system, on weight and food intake control and on sleep profiles (Lancel *et al.*, 2003). Moreover, during the past few years we have provided extensive evidence showing that voluntary exercise (i.e., wheel running in rats and mice that have access to a running wheel in their home cage) increases the flexibility of the HPA axis response to a stressful situation (Droste *et al.*, 2003, 2007). Thus, when corticosterone is needed to mobilize energy during a stressor with physical impact, such as during forced swimming, the plasma response of this stress hormone is markedly enhanced as compared to the response in sedentary animals. In contrast, during a mild psychological stressor, such as exposure to a novel environment, the plasma corticosterone response is significantly reduced in exercised rats and mice (Droste *et al.*, 2003, 2007). Interestingly, the lower plasma corticosterone response correlates with less anxious and risk-assessment behaviors during the period of exposure to novelty, suggesting a better behavioral and physiological coping with the stressful challenge in exercised animals (Droste *et al.*, 2007; Collins *et al.*, 2009). The notion that the dynamics of the HPA axis may be increased in exercising subjects is also underscored by an increased amplitude in the ultradian rhythm of free corticosterone (as assessed by *in vivo* microdialysis; see Droste *et al.*, 2008; Linthorst and Reul, 2008) in exercised rats as compared to sedentary control animals in the hours before the start of the active period (Droste et al., 2009). The above described evidence clearly supports a beneficial role of exercise in the regulation of physiological parameters, but, also for the purpose of this chapter, the logical next question would be whether exercise also supports mental health and, if so, which neurotransmitter pathways would be involved. Although exercise is increasingly being prescribed as a co-treatment for a number of psychiatric illnesses, its scientific underpinning needs to be strengthened. We have demonstrated that exercised mice show reduced anxiety and less impulsivity in a number of behavioral tests (Binder *et al.*, 2004b). Moreover, exercised rats show profound alterations in the expression of GABA-A receptor subunits in the brain (Hill *et al.*, 2008), which is of interest given that GABA-A receptors are an important drug target for the treatment of stress-related psychiatric disease. But what about 5-HT? Unfortunately, research on the effects of exercise on serotonergic neurotransmission in higher limbic brain regions has mainly been conducted using a forced running paradigm – i.e., forcing rodents to run on a treadmill. This paradigm causes stress in rodents, as it is usually performed during the light period – i.e., the period when the animals would normally rest and sleep. With this caveat in mind, treadmill experiments indicate that exercise may increase 5-HT and 5-HIAA levels in various brain structures (Meeusen *et al.*, 1996; Wilson and Marsden, 1996; Gomez-Merino *et al.*, 2001), which would fit with the observation that there is a close relationship between extracellular levels of 5-HT and vigilance state (with highest levels attained during active waking), as described above. Interestingly, recently increased expression of TPH2 mRNA in the DRN and MRN at zeitgeber time 10 (but not at zeitgeber time 2) has been described in rats after voluntary wheel running for 6 weeks (Malek *et al.*, 2007). Moreover, voluntary exercise significantly reduces the levels of 5-HT transporter mRNA in the DRN and MRN after 3 and 6 weeks (Greenwood *et al.*, 2005). However, from the limited available information, it is clear that, for a better understanding of the beneficial effects of exercise on mood, affect and anxiety, it will be of critical importance to further study the impact of exercise on the 5-HT system (and on other neurotransmitter systems including GABA) during voluntary exercise and using an *in vivo* microdialysis paradigm.

Concluding remarks and future

As reviewed in this chapter, the current state of our knowledge on the relationship between stress and the 5-HT system clearly shows that this neurotransmitter system is highly responsive to acute forms of stress, albeit in a stressor-dependent and brain region-specific manner. Without doubt, the available information on the consequences of chronic stress on serotonergic neurotransmission is limited, and no firm conclusions can be drawn. In our opinion it will be of great importance to further elucidate the effects of stress on the 5-HT system with more attention to the impact of chronic stress and to the mechanisms underlying putative alterations in serotonergic neurotransmission. Studies implementing the 5-HT system in exercising subjects will be needed to further increase our knowledge on how to improve stress coping mechanisms, and will contribute to our understanding of how to increase stress resilience in people suffering from stress in today's hectic and stressful world.

Acknowledgements

The authors acknowledge the excellent contributions to this research of many co-workers over the past 15 years. The authors are presently sponsored by The Wellcome Trust, Biological and Biosciences Research Council, Medical Research Council and Neuroendocrinology Charitable Trust.

References

Abel, E.L. (1993) Physiological correlates of the forced swim test in rats. Physiol. Behav., 54: 309–317.

Abumaria, N., Ribic, A., Anacker, C., Fuchs, E. and Flugge, G. (2008) Stress upregulates TPH1 but not TPH2 mRNA in the rat dorsal raphe nucleus: identification of two TPH2 mRNA splice variants. Cell Mol. Neurobiol., 28: 331–342.

Abumaria, N., Rygula, R., Havemann-Reinecke, U., Ruther, E., Bodemer, W., Roos, C. and Flugge, G. (2006) Identification of genes regulated by chronic social stress in the rat dorsal raphe nucleus. Cell Mol. Neurobiol., 26: 145–162.

Amat, J., Baratta, M.V., Paul, E., Bland, S.T., Watkins, L.R. and Maier, S.F. (2005) Medial prefrontal cortex determines how stressor controllability affects behavior and dorsal raphe nucleus. Nat. Neurosci., 8: 365–371.

Amat, J., Matus-Amat, P., Watkins, L.R. and Maier, S.F. (1998) Escapable and inescapable stress differentially and selectively alter extracellular levels of 5-HT in the ventral hippocampus and dorsal periaqueductal gray of the rat. Brain Res., 797: 12–22.

Amat, J., Tamblyn, J.P., Paul, E.D., Bland, S.T., Amat, P., Foster, A.C., Watkins, L.R. and Maier, S.F. (2004) Microinjection of urocortin 2 into the dorsal raphe nucleus activates serotonergic neurons and increases extracellular serotonin in the basolateral amygdala. Neuroscience, 129: 509–519.

Arborelius, L., Hawks, B.W., Owens, M.J., Plotsky, P.M. and Nemeroff, C.B. (2004) Increased responsiveness of presumed 5-HT cells to citalopram in adult rats subjected to prolonged maternal separation relative to brief separation. Psychopharmacology (Berl.), 176: 248–255.

Arborelius, L., Owens, M.J., Plotsky, P.M. and Nemeroff, C.B. (1999) The role of corticotropin-releasing factor in depression and anxiety disorders. J. Endocrinol., 160: 1–12.

Azmitia, E.C. Jr. and McEwen, B.S. (1974) Adrenalcortical influence on rat brain tryptophan hydroxylase activity. Brain Res., 78: 291–302.

Bach-Mizrachi, H., Underwood, M.D., Tin, A., Ellis, S.P., Mann, J.J. and Arango, V. (2008) Elevated expression of tryptophan hydroxylase-2 mRNA at the neuronal level in the dorsal and median raphe nuclei of depressed suicides. Mol. Psychiatry, 13: 507–513, 465

Beck, C.H. and Fibiger, H.C. (1995) Conditioned fear-induced changes in behavior and in the expression of the immediate early gene c-fos: with and without diazepam pretreatment. J. Neurosci., 15: 709–720.

Beekman, M., Flachskamm, C. and Linthorst, A.C.E. (2005) Effects of exposure to a predator on behavior and serotonergic neurotransmission in different brain regions of C57bl/6N mice. Eur. J. Neurosci., 21: 2825–2836.

Bickerdike, M.J., Wright, I.K. and Marsden, C.A. (1993) Social isolation attenuates rat forebrain 5-HT release induced by KCI stimulation and exposure to a novel environment. Behav. Pharmacol., 4: 231–236.

Bilang-Bleuel, A., Ulbricht, S., Chandramohan, Y., De Carli, S., Droste, S.K. and Reul, J.M.H.M. (2005) Psychological stress increases histone H3 phosphorylation in adult dentate gyrus granule neurons: involvement in a glucocorticoid receptor-dependent behavioral response. Eur. J. Neurosci., 22: 1691–1700.

Binder, E.B., Salyakina, D., Lichtner, P., Wochnik, G.M., Ising, M., Putz, B., Papiol, S., Seaman, S., Lucae, S., Kohli, M.A., Nickel, T., Kunzel, H.E., Fuchs, B., Majer, M., Pfennig, A., Kern, N., Brunner, J., Modell, S., Baghai, T., Deiml, T., Zill, P., Bondy, B., Rupprecht, R., Messer, T., Kohnlein, O., Dabitz, H., Bruckl, T., Muller, N., Pfister, H., Lieb, R., Mueller, J.C., Lohmussaar, E., Strom, T.M., Bettecken, T., Meitinger, T., Uhr, M., Rein, T., Holsboer, F. and Muller-Myhsok, B. (2004a) Polymorphisms in FKBP5 are associated with increased recurrence of depressive episodes and rapid response to antidepressant treatment. Nat. Genet., 36: 1319–1325.

Binder, E., Droste, S.K., Ohl, F. and Reul, J.M.H.M. (2004b) Regular voluntary exercise reduces anxiety-related behavior and impulsiveness in mice. Behav. Brain Res., 155: 197–206.

Bittencourt, J.C. and Sawchenko, P.E. (2000) Do centrally administered neuropeptides access cognate receptors? An analysis in the central corticotropin-releasing factor system. J. Neurosci., 20: 1142–1156.

Boadle-Biber, M.C. (1993) Regulation of serotonin synthesis. Prog. Biophys. Mol. Biol., 60: 1–15.

Boadle-Biber, M.C., Johannessen, J.N., Narasimhachari, N. and Phan, T.H. (1986) Tryptophan hydroxylase: increase in activity by electrical stimulation of serotonergic neurons. Neurochem. Intl., 8: 83–92.

Boadle-Biber, M.C., Corley, K.C., Graves, L., Phan, T.H. and Rosecrans, J. (1989) Increase in the activity of tryptophan hydroxylase from cortex and midbrain of male Fischer 344 rats in response to acute or repeated sound stress. Brain Res., 482: 306–316.

Caspi, A., Sugden, K., Moffitt, T.E., Taylor, A., Craig, I.W., Harrington, H., McClay, J., Mill, J., Martin, J., Braithwaite, A. and Poulton, R. (2003) Influence of life stress on depression: moderation by a polymorphism in the 5-HTT gene. Science, 301: 386–389.

Chamas, F., Serova, L. and Sabban, E.L. (1999) Tryptophan hydroxylase mRNA levels are elevated by repeated immobilization stress in rat raphe nuclei but not in pineal gland. Neurosci. Lett., 267: 157–160.

Chamas, F.M., Underwood, M.D., Arango, V., Serova, L., Kassir, S.A., Mann, J.J. and Sabban, E.L. (2004) Immobilization stress elevates tryptophan hydroxylase mRNA and protein in the rat raphe nuclei. Biol. Psychiatry, 55: 278–283.

Chandramohan, Y., Droste, S.K., Arthur, J.S. and Reul, J.M.H.M. (2008) The forced swimming-induced behavioral immobility response involves histone H3 phospho-acetylation and c-Fos induction in dentate gyrus granule neurons via activation of the N-methyl-D-aspartate/extracellular signal-regulated kinase/mitogen- and stress-activated kinase signalling pathway. Eur. J. Neurosci., 27: 2701–2713.

Clark, J.A., Flick, R.B., Pai, L.Y., Szalayova, I., Key, S., Conley, R.K., Deutch, A.Y., Hutson, P.H. and Mezey, E. (2008) Glucocorticoid modulation of tryptophan hydroxylase-2 protein in raphe nuclei and 5-hydroxytryptophan concentrations in frontal cortex of C57/Bl6 mice. Mol. Psychiatry, 13: 498–506.

Clark, M.S., McDevitt, R.A. and Neumaier, J.F. (2006) Quantitative mapping of tryptophan hydroxylase-2, 5-HT1A, 5-HT1B, and serotonin transporter expression across the anteroposterior axis of the rat dorsal and median raphe nuclei. J. Comp. Neurol., 498: 611–623.

Clark, M.S., McDevitt, R.A., Hoplight, B.J. and Neumaier, J.F. (2007) Chronic low dose ovine corticotropin releasing factor or urocortin II into the rostral dorsal raphe alters exploratory behavior and serotonergic gene expression in specific subregions of the dorsal raphe. Neuroscience, 146: 1888–1905.

Collins, A., Hill, L.E., Chandramohan, Y., Whitcomb, D., Droste, S.K. and Reul, J.M.H.M. (2009) Exercise improves cognitive responses to psychological stress through enhancement of epigenetic mechanisms and gene expression in the dentate gyrus. PLoS One, 4: e4330

Corley, K.C., Singh, V.B., Phan, T.H. and Boadle-Biber, M.C. (1992) Effect of gepirone on increases in tryptophan hydroxylase in response to sound stress. Eur. J. Pharmacol., 213: 417–425.

Corley, K.C., Phan, T.H., Daugherty, W.P. and Boadle-Biber, M.C. (2002) Stress-induced activation of median raphe serotonergic neurons in rats is potentiated by the neurotensin antagonist, SR 48692. Neurosci. Lett., 319: 1–4.

Cote, F., Thevenot, E., Fligny, C., Fromes, Y., Darmon, M., Ripoche, M.A., Bayard, E., Hanoun, N., Saurini, F., Lechat, P., Dandolo, L., Hamon, M., Mallet, J. and Vodjdani, G. (2003) Disruption of the nonneuronal tphl gene demonstrates the importance of peripheral serotonin in cardiac function. Proc. Natl. Acad. Sci. USA, 100: 13525–13530.

Cullinan, W.E., Herman, J.P., Battaglia, D.F., Akil, H. and Watson, S.J. (1995) Pattern and time course of immediate early gene expression in rat brain following acute stress. Neuroscience, 64: 477–505.

Culman, J., Kiss, A. and Kvetnansky, R. (1984) Serotonin and tryptophan hydroxylase in isolated hypothalamic and brain stem nuclei of rats exposed to acute and repeated immobilization stress. Exp. Clin. Endocrinol., 83: 28–36.

De Groote, L. and Linthorst, A.C.E. (2007) Exposure to novelty and forced swimming evoke stressor-dependent changes in extracellular GABA in the rat hippocampus. Neuroscience, 148: 794–805.

De Groote, L., Penalva, R.G., Flachskamm, C., Reul, J.M.H.M. and Linthorst, A.C.E. (2005) Differential monoaminergic, neuroendocrine and behavioral responses after central administration of corticotrophin-releasing factor receptor type 1 and type 2 agonists. J. Neurochem., 94: 45–56.

De Kloet, E.R., Joëls, M. and Holsboer, F. (2005) Stress and the brain: from adaptation to disease. Nat. Rev. Neurosci., 6: 463–475.

De Pablo, J.M., Parra, A., Segovia, S. and Guillamon, A. (1989) Learned immobility explains the behavior of rats in the forced swimming test. Physiol. Behav., 46: 229–237.

Derijk, R.H., Wust, S., Meijer, O.C., Zennaro, M.C., Federenko, I.S., Hellhammer, D.H., Giacchetti, G., Vreugdenhil, E., Zitman, F.G. and De Kloet, E.R. (2006) A common polymorphism in the mineralocorticoid receptor modulates stress responsiveness. J. Clin. Endocrinol. Metab., 91: 5083–5089.

Dilts, R.P. and Boadle-Biber, M.C. (1995) Differential activation of the 5-hydroxytryptamine-containing neurons of the midbrain raphe of the rat in response to randomly presented inescapable sound. Neurosci. Lett., 199: 78–80.

Droste, S.K., Gesing, A., Ulbricht, S., Muller, M.B., Linthorst, A.C.E. and Reul, J.M.H.M. (2003) Effects of long-term voluntary exercise on the mouse hypothalamic–pituitary–adrenocortical axis. Endocrinology, 144: 3012–3023.

Droste, S.K., Chandramohan, Y., Hill, L.E., Linthorst, A.C.E. and Reul, J.M.H.M. (2007) Voluntary exercise impacts on the rat hypothalamic–pituitary–adrenocortical axis mainly at the adrenal level. Neuroendocrinology, 86: 26–37.

Droste, S.K., Collins, A., Lightman, S.L., Linthorst, A.C.E. and Reul, J.M.H.M. (2009) Distinct, time-dependent effects of voluntary exercise on circadian and utradian rhythms and stress responses of free corticosterone in the rat hippocampus. Endocrinology, 150: 4170–4179.

Droste, S.K., De Groote, L., Atkinson, H.C., Lightman, S.L., Reul, J.M.H.M. and Linthorst, A.C.E. (2008) Corticosterone levels in the brain show a distinct ultradian rhythm but a delayed response to forced swim stress. Endocrinology, 149: 3244–3253.

Elmquist, J.K., Scammell, T.E., Jacobson, C.D. and Saper, C.B. (1996) Distribution of Fos-like immunoreactivity in the rat brain following intravenous lipopolysaccharide administration. J. Comp. Neurol., 371: 85–103.

Ericsson, A., Kovacs, K.J. and Sawchenko, P.E. (1994) A functional anatomical analysis of central pathways subserving the effects of interleukin-1 on stress-related neuroendocrine neurons. J. Neurosci., 14: 897–913.

Fernstrom, J.D. and Wurtman, R.J. (1971) Brain serotonin content: physiological dependence on plasma tryptophan levels. Science, 173: 149–152.

Fornal, C.A., Metzler, C.W., Marrosu, F., Ribiero-do-Valle, L.E. and Jacobs, B.L. (1996) A subgroup of dorsal raphe serotonergic neurons in the cat is strongly activated during oral–buccal movements. Brain Res., 716: 123–133.

Fujino, K., Yoshitake, T., Inoue, O., Ibii, N., Kehr, J., Ishida, J., Nohta, H. and Yamaguchi, M. (2002) Increased serotonin release in mice frontal cortex and hippocampus induced by acute physiological stressors. Neurosci. Lett., 320: 91–95.

Galfalvy, H., Huang, Y.Y., Oquendo, M.A., Currier, D. and Mann, J.J. (2008) Increased risk of suicide attempt in mood disorders and TPH1 genotype. J. Affect. Disord. in press, doi:10.1016/j.jad.2008.09.019.

Gartside, S.E., Johnson, D.A., Leitch, M.M., Troakes, C. and Ingram, C.D. (2003) Early life adversity programs changes in central 5-HT neuronal function in adulthood. Eur. J. Neurosci., 17: 2401–2408.

Gizatullin, R., Zaboli, G., Jonsson, E.G., Asberg, M. and Leopardi, R. (2006) Haplotype analysis reveals tryptophan hydroxylase (TPH) 1 gene variants associated with major depression. Biol. Psychiatry, 59: 295–300.

Gizatullin, R., Zaboli, G., Jonsson, E.G., Asberg, M. and Leopardi, R. (2008) The tryptophan hydroxylase (TPH) 2 gene unlike TPH-1 exhibits no association with stress-induced depression. J. Affect. Disord., 107: 175–179.

Gomez-Merino, D., Bequet, F., Berthelot, M., Chennaoui, M. and Guezennec, C.Y. (2001) Site-dependent effects of an acute intensive exercise on extracellular 5-HT and 5-HIAA levels in rat brain. Neurosci. Lett., 301: 143–146.

Grahame-Smith, D.G. (1974) How important is the synthesis of brain 5-hydroxytryptamine in the physiological control of its central function? Adv. Biochem. Psychopharmacol., 10: 83–91.

Grahn, R.E., Will, M.J., Hammack, S.E., Maswood, S., McQueen, M.B., Watkins, L.R. Maier, and SF. (1999) Activation of serotonin-immunoreactive cells in the dorsal raphe nucleus in rats exposed to an uncontrollable stressor. Brain Res., 826: 35–43.

Greenwood, B.N., Foley, T.E., Day, H.E., Burhans, D., Brooks, L., Campeau, S. and Fleshner, M. (2005) Wheel running alters serotonin (5-HT) transporter, 5-HT1A, 5-HT1B, and alpha 1b-adrenergic receptor mRNA in the rat raphe nuclei. Biol. Psychiatry, 57: 559–568.

Gronli, J., Fiske, E., Murison, R., Bjorvatn, B., Sorensen, E., Ursin, R. and Portas, C.M. (2007) Extracellular levels of serotonin and GABA in the hippocampus after chronic mild stress in rats. A microdialysis study in an animal model of depression. Behav. Brain Res., 181: 42–51.

Hashimoto, S., Inoue, T. and Koyama, T. (1999) Effects of conditioned fear stress on serotonin neurotransmission and freezing behavior in rats. Eur. J. Pharm., 378: 23–30.

Heinsbroek, R.P., van Haaren, F., Feenstra, M.G., Boon, P. and van de Poll, N.E. (1991) Controllable and uncontrollable footshock and monoaminergic activity in the frontal cortex of male and female rats. Brain Res., 551: 247–255.

Herman, J.P. and Cullinan, W.E. (1997) Neurocircuitry of stress: central control of the hypothalamo-pituitary–adrenocortical axis. Trends Neurosci., 20: 78–84.

Herr, B.E., Gallager, D.W. and Roth, R.H. (1975) Tryptophan hydroxylase: activation *in vivo* following stimulation of central serotonergic neurons. Biochem. Pharmacol., 24: 2019–2023.

Hill, L.E., Droste, S.K., Nutt, D.J., Linthorst, A.C.E. and Reul, J.M.H.M. (2008) Voluntary exercise alters GABAA receptor subunit and glutamic acid decarboxylase-67 gene expression in the rat forebrain. J. Psychopharmacol. in press, doi:10.1177/0269881108096983.

Hollis, J.H., Evans, A.K., Bruce, K.P., Lightman, S.L. and Lowry, C.A. (2006) Lipopolysaccharide has indomethacin-sensitive actions on Fos expression in topographically organized subpopulations of serotonergic neurons. Brain Behav. Immun., 20: 569–577.

Houdouin, F., Cespuglio, R., Gharib, A., Sarda, N. and Jouvet, M. (1991) Detection of the release of 5-hydroxyindole compounds in the hypothalamus and the n. raphe dorsalis throughout the sleep-waking cycle and during stressful situations in the rat: a polygraphic and voltammetric approach. Exp. Brain Res., 85: 153–162.

Imeri, L., De Simoni, M.G., Giglio, R., Clavenna, A. and Mancia, M. (1994) Changes in the serotonergic system during the sleep–wake cycle – simultaneous polygraphic and voltammetric recordings in hypothalamus using a telemetry system. Neuroscience, 58: 353–358.

Ishida, Y., Hashiguchi, H., Takeda, R., Ishizuka, Y., Mitsuyama, Y., Kannan, H., Nishimori, T. and Nakahara, D. (2002) Conditioned-fear stress increases Fos expression in monoaminergic and GABAergic neurons of the locus coeruleus and dorsal raphe nuclei. Synapse, 45: 46–51.

Jacobs, B.L. and Azmitia, E.C. (1992) Structure and function of the brain serotonin system. Physiol. Rev., 72: 165–229.

Judge, S.J., Ingram, C.D. and Gartside, S.E. (2004) Moderate differences in circulating corticosterone alter receptor-mediated regulation of 5-hydroxytryptamine neuronal activity. J. Psychopharmacol., 18: 475–483.

Kalén, P., Rosegren, E., Lindvall, O. and Björklund, A. (1989) Hippocampal noradrenaline and serotonin release over 24 hours as measured by the dialysis technique in freely moving rats: correlation to behavioral activity state, effect of handling and tail-pinch. Eur. J. Neurosci., 1: 181–188.

Kirby, L.G., Allen, A.R. and Lucki, I. (1995) Regional differences in the effects of forced swimming on extracellular levels of 5-hydroxytryptamine and 5-hydroxyindoleacetic acid. Brain Res., 682: 189–196.

Kirby, L.G., Rice, K.C. and Valentino, R.J. (2000) Effects of corticotropin-releasing factor on neuronal activity in the serotonergic dorsal raphe nucleus. Neuropsychopharmacology, 22: 148–162.

Knott, P.J. and Curzon, G. (1972) Free tryptophan in plasma and brain tryptophan metabolism. Nature, 239: 452–453.

Korte, S.M. (2001) Corticosteroids in relation to fear, anxiety and psychopathology. Neurosci. Biobehav. Rev., 25: 117–142.

Kovacs, K.J. (1998) C-Fos as a transcription factor - a stressful (re)view from a functional map. Neurochem. Intl., 33: 287–297.

Kuhn, D.M., Wolf, W.A. and Youdim, M.B. (1986) Serotonin neurochemistry revisited: a new look at some old axioms. Neurochem. Intl., 8: 141–154.

Laaris, N., Le Poul, E., Hamon, M. and Lanfumey, L. (1997) Stress-induced alterations of somatodendritic 5-HT1A autoreceptor sensitivity in the rat dorsal raphe nucleus--*in vitro* electrophysiological evidence. Fundam. Clin. Pharmacol., 11: 206–214.

Laaris, N., Le Poul, E., Laporte, A.M., Hamon, M. and Lanfumey, L. (1999) Differential effects of stress on presynaptic and postsynaptic 5-hydroxytryptamine-1A receptors in the rat brain: an *in vitro* electrophysiological study. Neuroscience, 91: 947–958.

Laflamme, N., Feuvrier, E., Richard, D. and Rivest, S. (1999) Involvement of serotonergic pathways in mediating the neuronal activity and genetic transcription of neuroendocrine corticotropin-releasing factor in the brain of systemically endotoxin-challenged rats. Neuroscience, 88: 223–240.

Lancel, M., Droste, S.K., Sommer, S. and Reul, J.M.H.M. (2003) Influence of regular voluntary exercise on spontaneous and social stress-affected sleep in mice. Eur. J. Neurosci., 17: 2171–2179.

Lavicky, J. and Dunn, A.J. (1995) Endotoxin administration stimulates cerebral catecholamine release in freely moving rats as assessed by microdialysis. J. Neurosci. Res., 40: 407–413.

Lehner, M., Taracha, E., Skorzewska, A., Turzynska, D., Sobolewska, A., Maciejak, P., Szyndler, J., Hamed, A., Bidzinski, A., Wislowska-Stanek, A. and Plaznik, A. (2008) Expression of c-Fos and CRF in the brains of rats differing in the strength of a fear response. Behav. Brain Res., 188: 154–167.

Linthorst, A.C.E. (2005) Stress, corticotropin-releasing factor and serotonergic neurotransmission. In: Steckler, T., Kalin, N.H. and Reul, J.M.H.M. (Eds.), Handbook of Stress and the Brain, Part 1: The Neurobiology of Stress. Elsevier, Amsterdam, Vol. 15, pp. 503–524.

Linthorst, A.C.E. and Reul, J.M.H.M. (2008) Stress and the brain: solving the puzzle using microdialysis. Pharmacol. Biochem. Behav., 90: 163–173.

Linthorst, A.C.E., Flachskamm, C., Müller-Preuss, P., Holsboer, F. and Reul, J.M.H.M. (1995a) Effect of bacterial endotoxin and interleukin-1 beta on hippocampal serotonergic neurotransmission, behavioral activity, and free corticosterone levels: an *in vivo* microdialysis study. J. Neurosci., 15: 2920–2934.

Linthorst, A.C.E., Flachskamm, C., Holsboer, F. and Reul, J.M.H.M. (1995b) Intraperitoneal administration of bacterial endotoxin enhances noradrenergic neurotransmission in the rat preoptic area: relationship with body temperature and hypothalamic–pituitary–adrenocortical axis activity. Eur. J. Neurosci., 7: 2418–2430.

Linthorst, A.C.E., Flachskamm, C., Hopkins, S.J., Hoadley, M.E., Labeur, M.S., Holsboer, F. and Reul, J.M.H.M. (1997) Long-term intracerebroventricular infusion of corticotropin-releasing hormone alters neuroendocrine, neurochemical, autonomic, behavioral, and cytokine responses to a systemic inflammatory challenge. J. Neurosci., 17: 4448–4460.

Linthorst, A.C.E., Flachskamm, C., Barden, N., Holsboer, F. and Reul, J.M.H.M. (2000) Glucocorticoid receptor impairment alters CNS responses to a psychological stressor: an *in vivo* microdialysis study in transgenic mice. Eur. J. Neurosci., 12: 283–291.

Linthorst, A.C.E., Flachskamm, C. and Reul, J.M.H.M. (2007) Water temperature determines neurochemical and behavioral responses to forced swim stress: an *in vivo* microdialysis and biotelemetry study in rats. Stress, 11: 88–100.

Lowry, C.A., Rodda, J.E., Lightman, S.L. and Ingram, C.D. (2000) Corticotropin-releasing factor increases *in vitro* firing rates of serotonergic neurons in the rat dorsal raphe nucleus: evidence for activation of a topographically organized mesolimbocortical serotonergic system. J. Neurosci., 20: 7728–7736.

Maes, M. and Meltzer, H.Y. (1995) The serotonin hypothesis of major depression. In: Bloom, F.E. and Kupfer, D.J. (Eds.), Psychopharmacology: The Fourth Generation of Progress. Raven, New York, NY, pp. 933–944.

Malek, Z.S., Sage, D., Pevet, P. and Raison, S. (2007) Daily rhythm of tryptophan hydroxylase-2 messenger ribonucleic acid within raphe neurons is induced by corticoid daily surge and modulated by enhanced locomotor activity. Endocrinology, 148: 5165–5172.

Mann, J.J. (1998) The neurobiology of suicide. Nat. Med., 4: 25–30.

Martinez, M., Phillips, P.J. and Herbert, J. (1998) Adaptation in patterns of c-fos expression in the brain associated with exposure to either single or repeated social stress in male rats. Eur. J. Neurosci., 10: 20–33.

Meeusen, R., Thorre, K., Chaouloff, F., Sarre, S., De Meirleir, K., Ebinger, G. and Michotte, Y. (1996) Effects of tryptophan and/or acute running on extracellular 5-HT and 5-HIAA levels in the hippocampus of food-deprived rats. Brain Res., 740: 245–252.

Meloni, E.G., Reedy, C.L., Cohen, B.M. and Carlezon, W.A. Jr. (2008) Activation of raphe efferents to the medial prefrontal cortex by corticotropin-releasing factor: correlation with anxiety-like behavior. Biol. Psychiatry, 63: 832–839.

Merali, Z., Lacosta, S. and Anisman, H. (1997) Effects of interleukin-1-beta and mild stress on alterations of norepinephrine, dopamine and serotonin neurotransmission – a regional microdialysis study. Brain Res., 761: 225–235.

Mohankumar, P.S., Thyagarajan, S. and Quadri, S.K. (1993) Interleukin-1 beta increases 5-hydroxyindoleacetic acid release in the hypothalamus *in vivo*. Brain Res. Bull., 31: 745–748.

Muchimapura, S., Fulford, A.J., Mason, R. and Marsden, C.A. (2002) Isolation rearing in the rat disrupts the hippocampal response to stress. Neuroscience, 112: 697–705.

Mueller, B.R. and Bale, T.L. (2008) Sex-specific programming of offspring emotionality after stress early in pregnancy. J. Neurosci., 28: 9055–9065.

Nakahara, D. and Nakamura, M. (1999) Differential effect of immobilization stress on *in vivo* synthesis rate of monoamines in medial prefrontal cortex and nucleus accumbens of conscious rats. Synapse, 32: 238–242.

Nakamura, K., Sugawara, Y., Sawabe, K., Ohashi, A., Tsurui, H., Xiu, Y., Ohtsuji, M., Lin, Q.S., Nishimura, H., Hasegawa, H. and Hirose, S. (2006) Late developmental stage-specific role of tryptophan hydroxylase 1 in brain serotonin levels. J. Neurosci., 26: 530–534.

Nemeroff, C.B., Owens, M.J., Bissette, G., Andorn, A.C. and Stanley, M. (1988) Reduced corticotropin releasing factor binding sites in the frontal cortex of suicide victims. Arch. Gen. Psychiatry, 45: 577–579.

Nemeroff, C.B., Widerlov, E., Bissette, G., Walleus, H., Karlsson, I., Eklund, K., Kilts, C.D., Loosen, P.T. and Vale, W. (1984) Elevated concentrations of CSF corticotropin-releasing factor-like immunoreactivity in depressed patients. Science, 226: 1342–1344.

Palkovits, M., Brownstein, M., Kizer, J.S., Saavedra, J.M. and Kopin, I.J. (1976) Effect of stress on serotonin concentration and tryptophan hydroxylase activity of brain nuclei. Neuroendocrinology, 22: 298–304.

Park, S.P., Lopez-Rodriguez, F., Wilson, C.L., Maidment, N., Matsumoto, Y. and Engel, J. (1999) *In vivo* microdialysis measures of extracellular serotonin in the rat hippocampus during sleep–wakefulness. Brain Res., 833: 291–296.

Patel, P.D., Pontrello, C. and Burke, S. (2004) Robust and tissue-specific expression of TPH2 versus TPH1 in rat raphe and pineal gland. Biol. Psychiatry, 55: 428–433.

Pauli, S., Linthorst, A.C.E. and Reul, J.M.H.M. (1998) Tumour necrosis factor-alpha and interleukin-2 differentially affect hippocampal serotonergic neurotransmission, behavioral activity, body temperature and hypothalamic–pituitary–adrenocortical axis activity in the rat. Eur. J. Neurosci., 10: 868–878.

Pei, Q., Zetterstrom, T. and Fillenz, M. (1990) Tail pinch-induced changes in the turnover and release of dopamine and 5-hydroxytryptamine in different brain regions of the rat. Neuroscience, 35: 133–138.

Peñalva, R.G., Flachskamm, C., Zimmermann, S., Wurst, W., Holsboer, F., Reul, J.M.H.M. and Linthorst, A.C.E. (2002) Corticotropin-releasing hormone receptor type 1-deficiency enhances hippocampal serotonergic neurotransmission: an *in vivo* microdialysis study in mutant mice. Neuroscience, 109: 253–266.

Peñalva, R.G., Lancel, M., Flachskamm, C., Reul, J.M.H.M., Holsboer, F. and Linthorst, A.C.E. (2003) Effect of sleep and sleep deprivation on serotonergic neurotransmission in the hippocampus: a combined *in vivo* microdialysis/EEG study in rats. Eur. J. Neurosci., 17: 1896–1906.

Pezzone, M.A., Lee, W.S., Hoffman, G.E., Pezzone, K.M. and Rabin, B.S. (1993) Activation of brainstem catecholaminergic neurons by conditioned and unconditioned aversive stimuli as revealed by c-Fos immunoreactivity. Brain Res., 608: 310–318.

Portas, C.M. and McCarley, R.W. (1994) Behavioral state-related changes of extracellular serotonin concentration in the dorsal raphe nucleus: a microdialysis study in the freely moving cat. Brain Res., 648: 306–312.

Portas, C.M., Bjorvatn, B., Fagerland, S., Gronli, J., Mundal, V., Sorensen, E. and Ursin, R. (1998) On-line detection of extracellular levels of serotonin in dorsal raphe nucleus and frontal cortex over the sleep/wake cycle in the freely moving rat. Neuroscience, 83: 807–814.

Price, M.L., Curtis, A.L., Kirby, L.G., Valentino, R.J. and Lucki, I. (1998) Effects of corticotropin-releasing factor on brain serotonergic activity. Neuropsychopharmacology, 18: 492–502.

Python, A., Steimer, T., de Saint, H.Z., Mikolajewski, R. and Nicolaidis, S. (2001) Extracellular serotonin variations during vigilance states in the preoptic area of rats: a microdialysis study. Brain Res., 910: 49–54.

Raadsheer, F.C., Hoogendijk, W.J.G., Stam, F.C., Tilders, F.J.H. and Swaab, D.F. (1994) Increased numbers of corticotropin-releasing hormone expressing neurons in the hypothalamic paraventricular nucleus of depressed patients. Neuroendocrinology, 60: 436–444.

Raadsheer, F.C., Van Heerikhuize, J.J., Lucassen, P.J., Hoogendijk, W.J.G., Tilders, F.J.H. and Swaab, D.F. (1995) Corticotropin-releasing hormone mRNA levels in the paraventricular nucleus of patients with Alzheimer's disease and depression. Am. J. Psychiatry, 152: 1372–1376.

Ressler, K.J. and Nemeroff, C.B. (2000) Role of serotonergic and noradrenergic systems in the pathophysiology of depression and anxiety disorders. Depress. Anxiety, 12: 2–19.

Reyes, T.M., Lewis, K., Perrin, M.H., Kunitake, K.S., Vaughan, J., Arias, C.A., Hogenesch, J.B., Gulyas, J., Rivier, J., Vale, W.W. and Sawchenko, P.E. (2001) Urocortin II: a member of the corticotropin-releasing factor (CRF) neuropeptide family that is selectively bound by type 2 CRF receptors. Proc. Natl. Acad. Sci. USA, 98: 2843–2848.

Rioja, J., Santin, L.J., Dona, A., de Pablos, L., Minano, F.J., Gonzalez-Baron, S. and Aguirre, J.A. (2006) 5-HT1A receptor activation counteracts c-Fos immunoreactivity induced in serotonin neurons of the raphe nuclei after immobilization stress in the male rat. Neurosci. Lett., 397: 190–195.

Roche, M., Commons, K.G., Peoples, A. and Valentino, R.J. (2003) Circuitry underlying regulation of the serotonergic system by swim stress. J. Neurosci., 23: 970–977.

Rozeboom, A.M., Akil, H. and Seasholtz, A.F. (2007) Mineralocorticoid receptor overexpression in forebrain decreases anxiety-like behavior and alters the stress response in mice. Proc. Natl. Acad. Sci. USA, 104: 4688–4693.

Rueter, L.E. and Jacobs, B.L. (1996) A microdialysis examination of serotonin release in the rat forebrain induced by behavioral environmental manipulations. Brain Res., 739: 57–69.

Russell, W.M.S. and Burch, R.L. (1959) The principles of humane experimental technique. Methuen, London.

Shields, P.J. and Eccleston, D. (1972) Effects of electrical stimulation of rat midbrain on 5-hydroxytryptamine synthesis as determined by a sensitive radioisotope method. J. Neurochem., 19: 265–272.

Shimizu, N., Take, S., Hori, T. and Oomura, Y. (1992) *In vivo* measurement of hypothalamic serotonin release by intracerebral microdialysis: significant enhancement by immobilization stress in rats. Brain Res. Bull., 28: 727–734.

Shintani, F., Kanba, S., Nakaki, T., Nibuya, M., Kinoshita, N., Suzuki, E., Yagi, G., Kato, R. and Asai, M. (1993) Interleukin-1 beta augments release of norepinephrine, dopamine, and serotonin in the rat anterior hypothalamus. J. Neurosci., 13: 3574–3581.

Shintani, F., Nakaki, T., Kanba, S., Sato, K., Yagi, G., Shiozawa, M., Aiso, S., Kato, R. and Asai, M. (1995) Involvement of interleukin-1 in immobilization stress-induced increase in plasma adrenocorticotropic hormone and in release of hypothalamic monoamines in the rat. J. Neurosci., 15: 1961–1970.

Shouse, M.N., Staba, R.J., Saquib, S.F. and Farber, P.R. (2000) Monoamines and sleep: microdialysis findings in pons and amygdala. Brain Res., 860: 181–189.

Silveira, M.C., Sandner, G. and Graeff, F.G. (1993) Induction of Fos immunoreactivity in the brain by exposure to the elevated plus-maze. Behav. Brain Res., 56: 115–118.

Singh, V.B., Corley, K.C., Phan, T.H. and Boadle-Biber, M.C. (1990) Increases in the activity of tryptophan hydroxylase from rat cortex and midbrain in response to acute or repeated sound stress are blocked by adrenalectomy and restored by dexamethasone treatment. Brain Res., 516: 66–76.

Singh, V.B., Hao-Phan, T., Corley, K.C. and Boadle-Biber, M.C. (1992) Increase in cortical and midbrain tryptophan hydroxylase activity by intracerebroventricular administration of corticotropin releasing factor: block by adrenalectomy, by RU 38486 and by bilateral lesions to the central nucleus of the amygdala. Neurochem. Intl., 20: 81–92.

Staub, D.R., Spiga, F. and Lowry, C.A. (2005) Urocortin 2 increases c-Fos expression in topographically organized subpopulations of serotonergic neurons in the rat dorsal raphe nucleus. Brain Res., 1044: 176–189.

Stone, E.A. (1970) Behavioral and neurochemical effects of acute swim stress are due to hypothermia. Life Sci. (Pt 1) Physiol. Pharmacol., 9: 877–888.

Storey, J.D., Robertson, D.A., Beattie, J.E., Reid, I.C., Mitchell, S.N. and Balfour, D.J. (2006) Behavioral and neurochemical responses evoked by repeated exposure to an elevated open platform. Behav. Brain Res., 166: 220–229.

Timpl, P., Spanagel, R., Sillaber, I., Kresse, A., Reul, J.M.H.M., Stalla, G.K., Blanquet, V., Steckler, T., Holsboer, F. and Wurst, W. (1998) Impaired stress response and reduced anxiety in mice lacking a functional corticotropin-releasing hormone receptor 1. Nat. Genet., 19: 162–166.

Vahabzadeh, A. and Fillenz, M. (1994) Comparison of stress-induced changes in noradrenergic and serotonergic neurons in the rat hippocampus using microdialysis. Eur. J. Neurosci., 6: 1205–1212.

Van Loon, G.R., Shum, A. and Ho, D. (1982) Lack of effect of corticotropin releasing factor on hypothalamic dopamine and serotonin synthesis turnover rates in rats. Peptides, 3: 799–803.

van Rossum, E.F., Binder, E.B., Majer, M., Koper, J.W., Ising, M., Modell, S., Salyakina, D., Lamberts, S.W. and Holsboer, F. (2006) Polymorphisms of the glucocorticoid receptor gene and major depression. Biol. Psychiatry, 59: 681–688.

Walther, D.J., Peter, J.U., Bashammakh, S., Hörtnagl, H., Voits, M., Fink, H. and Bader, M. (2003) Synthesis of serotonin by a second tryptophan hydroxylase isoform. Science, 299: 76.

Watanabe, Y., Stone, E. and McEwen, B.S. (1994) Induction and habituation of c-fos and zif/268 by acute and repeated stressors. Neuroreport, 5: 1321–1324.

Wei, Q., Lu, X.Y., Liu, L., Schafer, G., Shieh, K.R., Burke, S., Robinson, T.E., Watson, S.J., Seasholtz, A.F. and Akil, H. (2004) Glucocorticoid receptor overexpression in forebrain: a mouse model of increased emotional lability. Proc. Natl. Acad. Sci. USA, 101: 11851–11856.

Wilkinson, L.O. and Jacobs, B.L. (1988) Lack of response of serotonergic neurons in the dorsal raphe nucleus of freely moving cats to stressful stimuli. Exp. Neurol., 101: 445–457.

Wilkinson, L.O., Auerbach, S.B. and Jacobs, B.L. (1991) Extracellular serotonin levels change with behavioral state

but not with pyrogen-induced hyperthermia. J. Neurosci., 11: 2732–2741.
Wilkinson, L.S., Humby, T., Killcross, S., Robbins, T.W. and Everitt, B.J. (1996) Dissociations in hippocampal 5-hydroxytryptamine release in the rat following pavlovian aversive conditioning to discrete and contextual stimuli. Eur. J. Neurosci., 8: 1479–1487.
Wilson, W.M. and Marsden, C.A. (1996) *In vivo* measurement of extracellular serotonin in the ventral hippocampus during treadmill running. Behav. Pharmacol., 7: 101–104.
Yoshioka, M., Matsumoto, M., Togashi, H. and Saito, H. (1995) Effects of conditioned fear stress on 5-HT release in the rat prefrontal cortex. Pharmacol. Biochem. Behav., 51: 515–519.
Zhang, X., Beaulieu, J.M., Sotnikova, T.D., Gainetdinov, R.R. and Caron, M.G. (2004) Tryptophan hydroxylase-2 controls brain serotonin synthesis. Science, 305: 217.
Zill, P., Baghai, T.C., Zwanzger, P., Schule, C., Eser, D., Rupprecht, R., Moller, H.J., Bondy, B. and Ackenheil, M. (2004) SNP and haplotype analysis of a novel tryptophan hydroxylase isoform (TPH2) gene provide evidence for association with major depression. Mol. Psychiatry, 9: 1030–1036.
Zill, P., Buttner, A., Eisenmenger, W., Moller, H.J., Ackenheil, M. and Bondy, B. (2007) Analysis of tryptophan hydroxylase I and II mRNA expression in the human brain: a post-mortem study. J. Psychiatr. Res., 41: 168–173.

CHAPTER 4.2

The Role of Serotonin in Depression

Gregory V. Carr and Irwin Lucki*

Departments of Psychiatry and Pharmacology, University of Pennsylvania School of Medicine, Philadelphia, Pennsylvania

Abstract: A substantial body of research has developed concerning the role of serotonin (5-HT) in the etiology and treatment of major depressive disorder (MDD). The first section of this chapter summarizes the clinical evidence implicating alterations of the serotonergic system in the etiology of MDD. A general serotonin vulnerability has been proposed as a major risk factor in MDD, consisting of the composite risk factors from different components of 5-HT transmission. The effects of abnormal 5-HT synthesis, receptor function and genetic polymorphisms are discussed. The second section reviews some of the preclinical models and tests that are used to measure depressive behaviors and the efficacy of antidepressant drugs, such as selective serotonin reuptake inhibitors (SSRIs). This section is focused on models that are commonly used in rodent species. Models and tests that require either acute or chronic antidepressant administration to achieve behavioral alterations are covered. The third section reviews the preclinical (rodent) literature concerning the role of individual 5-HT receptors in the development and pharmacological treatment of depressive behaviors. The effects of 5-HT depletion and of manipulation of the 5-HT transporter are also discussed. Although identifying the 5-HT receptors that support the therapeutic effects of SSRIs may lead to the development of more effective antidepressant drugs with fewer side effects, it is unclear whether the complete antidepressant effect of SSRIs can be reproduced by a single selective 5-HT receptor agonist or antagonist.

Keywords: depression, antidepressants, SSRIs, models of depression, knockout mice, serotonin, rat, psychiatry.

Introduction

Since the identification of iproniazid, an early antidepressant, as a monoamine oxidase inhibitor (MAOI) that prevents the degradation of serotonin (5-HT), the role of the serotonergic system in the etiology and treatment of depression has been an intense area of study. In the years since this discovery, many advances have been made in our understanding of the components of the serotonergic system. At least 14 serotonin receptor subtypes belonging to seven major families have been identified in the brain, each associated with distinct topographical distributions and cell signaling mechanisms. The nuclei of serotonin-producing neurons in the hindbrain and midbrain have been identified, and their axonal projections have been mapped (Barnes and Sharp, 1999). In addition to detailed molecular and anatomical information, there is a large body of behavioral and pharmacological data that has served to solidify the link between different components of the 5-HT system and depressive behavior.

This chapter will consider in turn the roles for 5-HT in causing major depressive disorder (MDD) and the role of 5-HT mechanisms underlying therapeutic treatment. Efforts to attribute the cause of depression to deficient 5-HT transmission stem from a historical literature that measured changes in the metabolism of 5-HT in patients with MDD. More sophisticated techniques are now available for clinical studies that have allowed the association of MDD with specific genes related to a family of 5-HT-related targets, and the imaging of specific 5-HT receptors in patients. This literature will be reviewed in brief because it is too large to be contained in this chapter.

Among the frontline class of antidepressant treatments, selective serotonin reuptake inhibitors (SSRIs) exhibit their primary pharmacological effects through manipulation of the 5-HT system. SSRIs produce many adaptations that may be involved in their clinical efficacy. Preclinical studies have attempted to identify particular 5-HT receptor subtypes associated with antidepressant behavioral responses, with the goal of eventually providing targeted treatments. Genetically modified animals have been studied as models of developmental genetic vulnerability, and

*Corresponding author
E-mail: lucki@pharm.med.upenn.edu

ISBN 978-0-12-374634-4
DOI: 10.1016/B978-0-12-374634-4.00030-7

to elucidate the mechanism of action of SSRIs. This literature will be covered in the second part of this chapter.

Serotonin in the etiology of depression

A variety of evidence has accumulated suggesting that deficits in 5-HT neurotransmission can contribute to the development of MDD. The earliest evidence reported decreased levels of 5-HT metabolites in the cerebrospinal fluid (CSF) of depressed patients, but much of the association may actually have been based on suicidal behavior rather than depression (Asberg, 1997; Placidi *et al.*, 2001). Also, several reports found that the availability of L-tryptophan in plasma was significantly lower in subjects with MDD (Cowen *et al.*, 1989; Maes *et al.*, 1990), suggesting an overall reduction of 5-HT metabolism. Depletion of tryptophan, and lowering of brain 5-HT, can cause transient depressive symptoms in individuals that are vulnerable to depression, based on their personal or family history of depression (Moreno *et al.*, 1999; Neumeister *et al.*, 2004). In recently remitted depressed patients receiving SSRIs, acute tryptophan depletion led to a rapid clinically significant return of depressive symptoms (Delgado *et al.*, 1999). In contrast, remitted depressed patients maintained with tricyclic antidepressants were less prone to relapse following tryptophan depletion. Although these findings suggested that the synthesis of 5-HT is necessary for the maintenance of remission induced by SSRIs, they are also consistent with the evocation of transient withdrawal symptoms following chronic SSRI treatment.

Another strategy to assess serotonergic vulnerability has been used in studies that measured HPA-axis hormone secretion following the administration of 5-HT precursors, drugs that stimulate 5-HT release or directly activate 5-HT receptors (Maes and Meltzer, 1995). In particular, 5-HT_{1A}, 5-HT_{2A} and 5-HT_{2C} receptors stimulate cortisol and prolactin secretion in man. In patients with MDD, blunted prolactin responses appear to support evidence for decreased function of 5-HT_{1A} receptors and enhanced cortisol responses for enhanced function of 5-HT_{2A} and 5-HT_{2C} receptors.

Patients with MDD have been studied at autopsy using histological methods to measure changes in brain 5-HT receptors. Reliable assessment of neurochemicals after autopsy is subject to variability from the use of different laboratory techniques, the *post-mortem* interval, heterogeneity of diagnoses and the uncontrolled use of accompanying medications. A reduction of serotonin transporter sites has been reported in the prefrontal cortex of depressed patients, possibly representing widespread impairment of serotonergic function (Arango *et al.*, 2002). This finding may be accompanied by decreased 5-HT_{1A} and increased 5-HT_{2A} receptor binding sites found in deceased depressed patients (Stockmeier, 2003).

Although lacking the resolution of histology, imaging studies using positron emission tomography can be used to measure serotonergic targets in MDD (Stout *et al.*, 1996; Stockmeier, 2003; Meyer, 2007). These studies can also control patient selection and medication status. Binding potential for 5-HT transporters has been reported to be elevated in MDD (Cannon *et al.*, 2007; Meyer, 2007). However, there does not appear to be a longstanding change in 5-HT transporter binding potential in recovered medication-free MDD patients (Bhagwagar *et al.*, 2007), indicating normalization with resolution of MDD. Several studies have reported reduced binding potential for 5-HT_{1A} receptor imaging ligands in depressed patients (Drevets *et al.*, 2007). The detection of similar findings in unmedicated (Hirvonen *et al.*, 2008) and remitted depressed individuals (Bhagwagar *et al.*, 2004) support suggestions that reduced 5-HT_{1A} receptor activity is a trait marker associated with MDD. The binding potential for cortical 5-HT_{2A} receptors was reported decreased (Stockmeier, 2003) or increased in unmedicated depressed patients (Meyer *et al.*, 2003), depending on medication history, and remained elevated in unmedicated recovered patients when compared with controls (Bhagwagar *et al.*, 2006).

The family of genes that regulate 5-HT transmission or 5-HT receptors has been reported to be associated with MDD or with the therapeutic effects of SSRIs. Several genes for the 5-HT transporter or 5-HT receptors are distributed widely in more than one form across the human population, and they are associated with variations in 5-HT function. The most widely known polymorphism involves the insertion/deletion of either a 528-base pair sequence (long form) or a 484-base pair sequence (short form) within the promoter region of the 5-HT transporter (Lesch *et al.*, 1996). The long form is associated with greater transcription and function. There are many studies that have examined the occurrence of 5-HT transporter polymorphisms with depression and anxiety (Jans *et al.*, 2007), and the short-allele has been associated with depression, anxiety and aggression (Lesch *et al.*, 1996; Lesch, 2001). The 5-HT_{1A} receptor is a key regulator of 5-HT transmission because the presynaptic 5-HT_{1A} receptor regulates neuronal discharge and negative feedback while the postsynaptic 5-HT_{1A} receptor is located in key limbic regions involved in affective behavior. A common polymorphism in the human 5-HT_{1A} promoter region has been identified and was associated with depression and suicide (Le Francois *et al.*, 2008). A polymorphism was identified in the *htph2* gene that regulates the synthesis of serotonin and was associated with depression (Zhang *et al.*, 2005), although this was not replicated in a broader sample of depressed patients (Glatt *et al.*, 2005; Van Den Bogaert *et al.*, 2005; Zhou *et al.*, 2005). In addition to providing vulnerability for depression and anxiety disorders, several 5-HT genes have been associated with the therapeutic efficacy of SSRIs. Thus, 5-HT transporter polymorphic variations in alleles for the 5-HT transporter, 5-HT_{1A} receptor and 5-HT_{2A} receptors have been associated with rates of recovery for SSRIs in treating depression in some studies (Serretti and Artioli, 2004; McMahon *et al.*, 2006; Lekman *et al.*, 2008; Lin and Chen, 2008). Thus, the family of genes related to 5-HT transmission may serve either as endogenous markers of vulnerability for depression and anxiety or as pharmacogenetic indicators of therapeutic efficacy.

It is unlikely that any one patient would show dysfunction of all of the components of 5-HT transmission reported to be altered in MDD. Rather, a general serotonin vulnerability has been proposed as a major risk factor in MDD. Thus, patients may be vulnerable for developing MDD from a composite of individual risk factors associated with the many functional components of 5-HT transmission (Jans *et al.*, 2006). Similarly, therapeutic effectiveness may rely on a composite of 5-HT receptors distributed throughout the brain, rather than on the activity of a single 5-HT receptor in a single locus. The evidence indicates that more than a single type of 5-HT receptor is likely to be involved in producing the behavioral effects of SSRIs.

Behavioral pharmacology of antidepressant drugs

This section will focus on behavioral pharmacology studies in rodents that have helped to determine the role of components of the 5-HT system that are involved in treating depression. First, animal models and tests of depression and antidepressant activity that have been used to identify the neurobiological effects of clinically effective antidepressants will be reviewed. Second, the roles of serotonergic mechanisms, including the activation or blockade of individual 5-HT receptors, will be summarized in relation to the animal behavior tests. Over the years, a number of ligands for 5-HT receptors, both agonists and antagonists, have been developed as pharmacological tools in order to delineate the respective contributions of the receptor subtypes. Also, the incorporation of knockout (KO) mice into behavioral pharmacology has generated a wealth of insight into the roles of individual receptors in a number of different behaviors. To this end, numerous 5-HT receptor KO mice have been used to investigate the etiology of depression-like behavior. These studies will also be covered in this section. For a comprehensive review of 5-HT receptors and signaling mechanisms, see Barnes and Sharp (1999).

Animal tests and models of depression

This section will divide animal tests into two types, based on the time-course for the effects of antidepressant compounds in the respective tests and models. First, behavioral tests in which an antidepressant-like effect is evident after a single or small number of administrations will be discussed. The second part will discuss the tests and models that require chronic treatment (>2 weeks) with antidepressants in order to display efficacy.

Acute tests

Forced swim test (FST)

The FST was developed by Porsolt and colleagues as a way to measure the effects of antidepressant compounds in rats and mice (Porsolt *et al.*, 1977). The test involves placing the rodent in a container of water (usually cylindrical) from which the subject is unable to escape. After a period of time in which the subject attempts to escape from the container, the subject adopts a posture of immobility. Immobility is characterized by the lack of movement except that which is necessary to keep the subject's nose above the water level. In rats, the test consists of two swim exposures. The first is a 15-minute exposure. The second, conducted 24 hours later, is a 5-minute exposure. Immobility time is recorded during the second 5-minute test. In mice, the procedure consists of a single 6-minute test (Porsolt *et al.*, 1978). In the FST, antidepressants cause a decrease in immobility time at doses that do not increase general locomotor activity, unlike psychomotor stimulants (e.g., amphetamine, cocaine), which decrease immobility at doses that increase general locomotor activity.

Motivated by the inability of this test to measure the effects of SSRIs, changes in the procedure and scoring of the FST were introduced by Lucki and colleagues as the modified rat FST (Detke *et al.*, 1995a). This version of the FST introduced a set of changes that served to increase the utility of the test. The most important modifications were the switch to identification of component active behaviors along with immobility, the original FST scored only the total duration of immobile time. In the modified rat FST, the frequency of immobility, swimming and climbing is measured over the 5-minute test (Detke *et al.*, 1995a). Swimming is defined as horizontal movement throughout the chamber, and climbing is defined as vertical movement of the forepaws directed towards the sides of the chamber. Serotonergic antidepressants, including SSRIs, selectively increase swimming behavior. This effect was difficult to measure under the test conditions usually used for the original version of the FST.

In addition, the modified rat FST is able to differentiate between antidepressants that work through serotonergic mechanisms or noradrenergic mechanisms, as noradrenergic compounds selectively increase climbing behavior compared to the increase in swimming behavior seen with serotonergic compounds. Many laboratories have now adopted this scoring system (Cryan *et al.*, 2005).

Tail suspension test (TST)

The TST is a test of antidepressant activity in mice that shares its experimental rationale with the FST (Steru *et al.*, 1985). In the test, mice are suspended from a lever by their tails and their behavior is recorded over a 6-minute time period. As in the FST, mice struggle to escape for a period of time and then adopt a posture of immobility. The time spent immobile is the dependent measure. As in the FST, antidepressants decrease the amount of time spent immobile without increasing general locomotor activity.

Learned helplessness (LH)

LH is characterized by the inability of a subject to terminate a controllable stressor by performing a behavior after having been exposed to an uncontrollable stressor (usually foot-shock) (Seligman *et al.*, 1975). LH is thought to simulate many of the symptoms of depression (Seligman *et al.*, 1975). Treatment with antidepressants decreases the escape latency and escape failures in the test after exposure to LH (Leshner *et al.*, 1979; Sherman and Petty, 1980).

Chronic tests and models

The next set of tests and models requires chronic treatment with antidepressants in order to induce a change in behavior. These tests are of particular interest to researchers because SSRIs usually require chronic treatment before they produce their clinical effects in depression or anxiety disorders.

Novelty suppression of feeding (NSF)/Novelty-induced hypophagia (NIH)

The NSF and NIH tests are unique among anxiety tests in that they respond to treatment with antidepressants (Bodnoff *et al.*, 1988; Dulawa *et al.*, 2004). This fact makes them extremely valuable tools in drug discovery, because antidepressants are effective against, and widely prescribed for, anxiety disorders.

Both the NSF and NIH tests are conflict-based anxiety tests. They take advantage of rodent species' natural aversion to bright, open spaces. The measures of anxiety in these tests are the latency to consume food and the amount of food consumed in the novel arena. The differences between the two tests are that the NSF test uses standard chow and food deprivation in order to drive consumption in the novel arena, and the NIH test uses palatable foods (sweetened) consumed without food deprivation (Dulawa and Hen, 2005). This modification in the NIH procedure removes potential confounds caused by deprivation stress. Consumption in the novel arena is compared to the consumption levels in the home cage. The novel arena is brightly lit, and structurally different from the subject's home cage. Untreated subjects demonstrate a dramatic decrease in consumption in the novel arena compared to their home-cage consumption. Anxiolytics, such as benzodiazepines, decrease novel arena latency and increase novel arena food consumption without altering home-cage behavior. Antidepressant drugs, when administered chronically, are also able to produce an anxiolytic effect in these tests.

Chronic mild stress (CMS)

The rationale for the CMS model is that increased exposure to stressors, particularly unpredictable and uncontrollable stressors, increases the likelihood that a person will develop depression. In order to reproduce this in the laboratory, researchers typically use a set of mild stressors (e.g., strobe lighting, soiled cages, 24-hour lighting) administered in a semi-random fashion in order to mimic the stress of everyday life (Willner, 1997, 2005). A number of groups have been able to show that when administered over a period of weeks, the CMS procedures can lead to behavioral and endocrine changes that appear to be similar to human depression. The most common behavioral output measured during CMS experiments is the presence of anhedonia, represented as a decrease in consumption of a sucrose solution. However, changes in FST behavior, sleep alterations and locomotor activity suggest that the pathological changes resulting from CMS are more comprehensive than just alterations in a single endophenotype (Willner, 2005).

The alterations caused by CMS can be reversed by chronic treatment with a number of antidepressant drugs. Another feature that makes CMS an intriguing model is the fact that not all subjects exposed to CMS show a pathological change in behavior, and not all CMS-responsive subjects show a reversal after treatment with antidepressants (Jayatissa *et al.*, 2006). These features are present in human depression, and provide opportunities for research into the neural underpinnings of stress and treatment resistance. A major hindrance to the universal application of CMS in depression research is the difficulty with inter-lab reproducibility. A number of labs have been successful in the utilization of CMS, but many others have had difficulty implementing the procedure. This may be due to the lack of a standardized stress protocol, but it may also be due to unavoidable differences in animal husbandry (e.g., colony room space/size, ventilation) and other issues between institutions (Cryan and Slattery, 2007).

Olfactory bulbectomy (OB)
Bilateral removal of the olfactory bulbs causes severe behavioral and endocrine changes in rodents (Song and Leonard, 2005). The most common behavioral change monitored in the OB model is hyperactivity in an open-field apparatus. These changes can be reversed by chronic antidepressant drug treatments. These pathological changes are not due purely to the anosmia caused by removal of the olfactory bulbs, since peripheral blockade does not result in an analogous syndrome. There are morphological changes to areas of the limbic system, such as the piriform cortex, that may underlie the effects of the OB procedure (Jarosik *et al.*, 2007).

5-HT system and antidepressant responses

The current frontline treatments for depression in humans are the SSRIs. Consequently, these compounds have been tested extensively in rodents. This section will review not the vast number of studies utilizing these compounds alone, but rather the studies that attempt to identify the underlying components of the antidepressant response elicited by the SSRIs. The effects of 5-HT depletion on subsequent treatment effects, the role of individual 5-HT receptors and recent studies of rodents with targeted genetic mutations will be considered. These studies are summarized in Table 1.

5-HT depletion

Pharmacological depletion of 5-HT, by administration of the tryptophan hydroxylase (TPH) inhibitor parachlorophenylalanine (pCPA), does not produce effects on baseline behavior of antidepressant tests in rats or mice (Page *et al.*, 1999; Gavioli *et al.*, 2004; O'Leary *et al.*, 2007). pCPA did increase immobility of female 5-HT_{1B} knockout mice because of their high baseline levels of extracellular 5-HT (Jones and Lucki, 2005a). Mice generated with genetic deletion of the isoforms of TPH, either the major isoform in brain (TPH2) or a double knockout of both isoforms (TPH1 and TPH2), have been examined in tests for depression-related behavior (Savelieva *et al.*, 2008). Mice with complete deletion of TPH (TPH1 and TPH2) showed complete depletion of brain 5-HT from deletion of both TPH isoforms, and were significantly more immobile in the TST. TPH2 KO mice, with a substantial but incomplete depletion of 5-HT, also demonstrated increased TST immobility, but only after repeated testing. Although FST immobility was increased in both sets of mice, this was caused by their inability to swim. A TPH2 knock-in mouse line with reduced TPH2 activity and an 80 percent reduction in brain 5-HT was also reported to have significantly increased immobility times in the TST (Beaulieu *et al.*, 2008). The presence of a depressive behavioral phenotype in the genetic models may be due to the role of 5-HT in development. Considering the accumulating evidence for associating 5-HT insufficiency with vulnerability to depression in humans, these latter genetic preparations provide the first evidence in animals for a pathological role of 5-HT depletion in depressive behavior.

Depletion of 5-HT with pCPA blocks the effects of fluoxetine in the FST and the TST, while the effects of desipramine are unaffected by 5-HT depletion (Cesana *et al.*, 1993; Page *et al.*, 1999; Gavioli *et al.*, 2004; O'Leary *et al.*, 2007). These effects demonstrate that serotonergic mechanisms underlie the acute behavioral effects of SSRIs on tests of depressive behavior. These effects in rodents correspond with a similar pattern of relapse induction in clinical patients following the depletion of serotonin or catecholamines. Patients treated with SSRIs will relapse temporarily if 5-HT is depleted by restricting 5-HT synthesis, but not by catecholamine depletion; patients treated with tricyclics demonstrate the complementary pattern of response to relapse induction (Delgado *et al.*, 1991; Delgado, 2004). These studies affirm the importance of changes in 5-HT neurotransmission in contributing to the effects of antidepressant treatments.

5-HT_{1A} receptor

There is a large body of evidence that 5-HT_{1A} receptor agonists produce antidepressant-like effects (Table 1) across a number of different tests in animals (De Vry, 1995; Blier and Ward, 2003). A number of 5-HT_{1A} agonists decrease immobility in the rat FST (Kostowski *et al.*, 1992; Singh and Lucki, 1993; Lucki *et al.*, 1994). Also, the effects of 5-HT_{1A} agonists, along with the effects of the antidepressants buspirone and desipramine, in the FST are blocked by pretreatment with 5-HT_{1A} antagonists (Detke *et al.*, 1995b). The agonist MKC-242 produces antidepressant (AD)-like effects in the mouse FST even when 5-HT-producing neurons are destroyed or 5-HT synthesis is inhibited. This suggests that the AD-like effects of 5-HT_{1A} receptor agonists are due to activation of postsynaptic 5-HT_{1A} receptors (Matsuda *et al.*, 1995). These effects are not confined to the FST. Alnespirone and VN2222, a 5-HT_{1A} agonist and a mixed

Table 1 5-HT receptors and rodent behavioral tests of depression

Receptor	KO	Agonist	Antagonist	References
5-HT1A	AD-like	AD-like	AD-like	Kowtowski *et al.*, 1992
				MacSweeney *et al.*, 1998
	Augments SSRIs			Tordera *et al.*, 2000
				Cryan *et al.*, 1997
				Przegalinski *et al.*, 1995
				Hogg and Dalvi, 2004
				Jones and Lucki, 2005
				Parks *et al.*, 1998
				Heisler *et al.*, 1998
				Mayorga *et al.*, 2001
5-HT1B	AD-like (females)	Prodepressive	AD-like	Cervo *et al.*, 1989
				Dawson *et al.*, 2006
	Augments FLX			Hogg and Dalvi, 2004
				Mayorga *et al.*, 2001
				Jones and Lucki, 2005
5-HT2A	No report		AD-like	Albinsson *et al.*, 1994
				Patel *et al.*, 2004
			Augments FLX	Marek *et al.*, 2005
5-HT2C	No report	AD-like		Cryan and Lucki, 2000
				Rosenzweig-Lipson *et al.*, 2007
5-HT3	Prodepressive (females)		AD-like	Martin *et al.*, 1992
				Bhatnagar *et al.*, 2004
				Bravo and Maswood, 2006
5-HT4	No report	AD-like		Lucas *et al.*, 2007
5-HT6	No report	AD-like	AD-like	Svenningsson *et al.*, 2007
				Wesolowska and Nikiforuk, 2007
				Wesolowska *et al.*, 2007
5-HT7	AD-like		AD-like	Guscott *et al.*, 2005
			Augments citalopram	Hedlund *et al.*, 2005
				Wesolowska *et al.*, 2006
SERT	No effect (C57BL6)			Holmes *et al.*, 2002
	Mixed effect (129S6)			Lira *et al.*, 2003
	Blocked effects of DMI			

reuptake inhibitor and 5-HT_{1A} agonist respectively, have been shown to produce AD-like effects in the LH paradigm (MacSweeney *et al.*, 1998; Tordera *et al.*, 2002). Flesinoxan produced AD-like effects in the OB model (Cryan *et al.*, 1997). Also, chronic treatment with the partial agonist buspirone produced AD-like effects in the CMS paradigm (Przegalinski *et al.*, 1995). The 5-HT_{1A} receptor agonist 8-OH-DPAT also increases neurogenesis and survival of new neurons in the dentate gyrus after chronic treatment (Banasr *et al.*, 2004), effects produced by other clinically active ADs. Nevertheless, with the exception of buspirone and gepirone, the development of 5-HT_{1A} receptor agonists as antidepressants has not been clinically successful (Blier and Ward, 2003).

Clinically active AD treatments modify the response of 5-HT_{1A} receptors in a number of brain regions. It has been postulated that desensitization of 5-HT_{1A} autoreceptors after chronic treatment with antidepressant drugs is a key factor in antidepressant efficacy, and can account for the delay between the commencement of treatment and the alleviation of depressive symptoms (Piñeyro and Blier, 1999). In contrast, chronic administration of antidepressant treatments facilitates transmission mediated by 5-HT_{1A} receptors in postsynaptic regions, like the hippocampus (Burnet *et al.*, 1995; Haddjeri *et al.*, 1998; Shen *et al.*, 2002). Because antagonism of 5-HT_{1A} autoreceptors in the raphe nuclei could facilitate AD-like effects through disinhibition of 5-HT release, 5-HT_{1A} receptor antagonists were proposed as a promising adjunct to traditional AD treatment (Artigas *et al.*, 1996).

5-HT_{1A} KO mice have been generated and studied by multiple labs (Heisler *et al.*, 1998; Parks *et al.*, 1998). Loss of the 5-HT_{1A} receptor results in an antidepressant-like and anxiogenic phenotype (Heisler *et al.*, 1998; Parks *et al.*, 1998; Zhuang *et al.*, 1999; Jones and Lucki, 2005b). This lifelong phenotype is caused by the loss of the 5-HT_{1A} receptor during early development in the rodent (Gross *et al.*, 2002). Although 5-HT_{1A} KO mice exhibit normal basal and K^+-evoked release of 5-HT, the absence of 5-HT_{1A} autoreceptors causes an exaggerated increase of 5-HT levels in response to fluoxetine (Knobelman *et al.*, 2001). Nevertheless, 5-HT_{1A} KO mice demonstrate refractory behavioral responses to acute or chronic SSRI treatment, probably because of the absence of postsynaptic 5-HT_{1A} receptors (Mayorga *et al.*, 2001; Santarelli *et al.*, 2003). A polymorphic allele in the 5-HT_{1A} receptor promoter region in humans has been identified as a risk allele for depression and anxiety disorders, and resistance to the effects of SSRIs (Le Francois *et al.*, 2008).

5-HT$_{1B}$ receptor

The 5-HT_{1B} receptor is regulated by exposure to behavioral tests for antidepressants and following chronic antidepressant treatments. The stress of the LH procedure causes an up-regulation of the 5-HT_{1B} receptor in the cortex, hippocampus, septum and dorsal raphe (Edwards *et al.*, 1991; Neumaier *et al.*, 1997). Chronic AD treatment decreases the mRNA of the 5-HT_{1B} receptor in the dorsal raphe and decreases the efficacy of 5-HT_{1B} autoreceptors, which may lead to an increase in 5-HT release (Neumaier *et al.*, 1996; Piñeyro and Blier, 1999; Anthony and Sexton, 2000; Gur *et al.*, 2000). 5-HT_{1B} heteroreceptors in the dentate gyrus appear to contribute to increases of neurogenesis, a marker of AD treatment (Banasr *et al.*, 2004). Also, up-regulation of p11, a protein that increases 5-HT_{1B} receptor function in a number of brain regions, reduces depression-like symptoms (Svenningsson *et al.*, 2006).

The actions of ligands at the 5-HT_{1B} receptor produce results that correspond with the changes in the receptor seen after stress and AD treatment (Table 1). The agonist TFMPP, given systemically and locally within the ventral tegmental area (VTA), blocks the effects of desipramine in the rat FST (Cervo *et al.*, 1989). 5-HT_{1B} receptor antagonists have been shown to produce AD-like effects and augment the effects of traditional ADs in the FST (Hogg and Dalvi, 2004; Tatarczynska *et al.*, 2004; Dawson *et al.*, 2006).

A 5-HT_{1B} receptor knockout mouse has been generated that demonstrates increased aggression and reduced anxiety (Zhuang *et al.*, 1999). Although male 5-HT_{1B} KO male mice do not exhibit an AD-like baseline response in the TST, they do show exaggerated increases of 5-HT and augmented behavioral responses to fluoxetine (Knobelman *et al.*, 2001; Mayorga *et al.*, 2001), suggesting an important role for 5-HT_{1B} autoreceptors. 5-HT_{1B} KO females exhibit a decrease in baseline immobility in the TST and FST compared to KO males and WT mice because of a constitutive disinhibition of 5-HT release (Jones and Lucki, 2005b).

5-HT$_{2A}$ receptor

Much of the research on 5-HT_{2A} receptors has focused on their role in the modulation of dopamine and glutamate release in the medial prefrontal cortex. Antagonists at the receptor have been shown to inhibit dopamine release in the prefrontal cortex (Pehek *et al.*, 2006). Treatment with the $5\text{-HT}_{2A/2C}$ receptor agonist DOI causes an increase in glutamate release in the cortex that is blocked by pretreatment with a selective 5-HT_{2A} receptor antagonist (Scruggs *et al.*, 2003). Selective 5-HT_{2A} receptor antagonists may cause AD-like effects (Table 1) in a number of rodent tests (Albinsson *et al.*, 1994; Patel *et al.*, 2004), and co-administration of 5-HT_{2A} receptor antagonists can augment the antidepressant-like effects of SSRIs (Marek *et al.*, 2003, 2005).

5-HT$_{2c}$ receptor

Treatment with 5-HT_{2C} receptor agonists produces AD-like effects (Table 1) in the modified rat FST, resident–intruder social stress model, and OB models of depression (Cryan and Lucki, 2000; Rosenzweig-Lipson *et al.*, 2007). In addition, the selective 5-HT_{2C} receptor antagonist SB206533 blocks the behavioral effects of fluoxetine (Cryan and Lucki, 2000). Although the antidepressant mianserin is a 5-HT_{2C} receptor antagonist, it produces antidepressant-like effects by blocking α_2 noradrenergic receptors (Cryan and Lucki, 2000). In the mouse FST, the 5-HT_{2C} receptor agonist Ro 60-0175 synergized with subactive doses of imipramine, paroxetine, citalopram and fluvoxamine, but antagonized active doses of paroxetine and fluoxetine at higher doses (Clenet *et al.*, 2001). 5-HT_{2C} KO mice demonstrate an anxiolytic phenotype with a blunted

CRF response to stress, but their phenotype in depression tests was not reported (Heisler *et al.*, 2007). That 5-HT_{2C} KO mice are hyper-responsive to the effects of repeated stress would seem to support the findings that agonists can produce AD-like effects (Chou-Green *et al.*, 2003). On the other hand, some 5-HT_{2C} receptor antagonists, including the combined melatonin agonist/5-HT_{2C} antagonist agomelatine, have been reported to produce antidepressant-like effects (Bourin *et al.*, 2004; Dekeyne *et al.*, 2008).

5-HT_3 receptor

Antagonists at the 5-HT_3 receptor produce antidepressant-like effects (Table 1) in a number of tests and models (Martin *et al.*, 1992; Mahesh *et al.*, 2007) and co-treatment can augment the effects of sub-threshold doses of SSRIs (Redrobe and Bourin, 1997; Ramamoorthy *et al.*, 2008). On the other hand, 5-HT_3 receptor agonists were shown to attenuate the effects of antidepressant compounds in the rat FST (Nakagawa *et al.*, 1998). The effects of 5-HT_3 receptor antagonists may be due to a direct receptor interaction, because some clinically effective antidepressants have been reported to be functional antagonists at the 5-HT_3 receptor (Eisensamer *et al.*, 2003). Interestingly, female 5-HT_3 receptor KO mice show increased immobility in the FST compared to WT controls, suggesting that there is sex-specific regulation of the receptor (Martin *et al.*, 1992; Bhatnagar *et al.*, 2004; Bravo and Maswood, 2006).

5-HT_4 receptor

5-HT_4 receptor agonists exhibit antidepressant-like effects (Table 1) in rats in a number of tests (FST, OB, and CMS). Most interestingly, 5-HT_4 receptor agonists produce a rapid onset of a number of adaptations seen after chronic AD treatment. Only a 3-day treatment regimen was required to desensitize 5-HT_1 autoreceptors, increase hippocampal neurogenesis, decrease hyperlocomotion in the OB model, and normalize sucrose consumption in the CMS model (Lucas *et al.*, 2007). Pharmacological blockade or genetic deletion of the 5-HT_4 receptor appears to exert anxiolytic effects (Smriga and Torii, 2003; Compan *et al.*, 2004; Conductier *et al.*, 2006).

5-HT_6 receptor

Research has shown that similar effects are seen after treatment with both agonists and antagonists (Table 1). Both sets of compounds produce antidepressant-like and anxiolytic effects in a number of rodent tests (Svenningsson *et al.*, 2007; Wesolowska and Nikiforuk, 2007; Wesolowska *et al.*, 2007). There is also evidence that the effects of 5-HT_6 receptor agonists may involve modulation of GABAergic neurotransmission (Schechter *et al.*, 2008).

5-HT_7 receptor

Genetic deletion and treatment with 5-HT_7 receptor antagonists produce AD-like effects (Table 1) in the FST and TST (Guscott *et al.*, 2005; Hedlund *et al.*, 2005; Wesolowska *et al.*, 2006). Research also suggests that 5-HT_7 receptor antagonists have potential as adjunct AD treatments. A recent study showed that 5-HT_7 receptor antagonists increased the effects of citalopram on TST (mice) and REM sleep (rats) (Bonaventure *et al.*, 2007).

5-HT transporter (SERT) KO mice

The depression-related behavior of SERT KO mice appears to be affected by the background strain used to create the mouse. With a C57/BL6 background, there are no differences in baseline behavior (Table 1) in the FST or TST between KO and wild-type (WT) mice (Holmes *et al.*, 2002). However, on the 129S6 background, SERT KO mice exhibit an increase in immobility in the FST (pro-depressive) and a decrease in immobility in the TST (AD-like) response at baseline compared to WT controls (Holmes *et al.*, 2002; Lira *et al.*, 2003). On the C57/BL6 background, fluoxetine has no effect on behavior in the TST, while antidepressants that are not dependent on 5-HT to exert their effects (desipramine and imipramine) are still effective (Holmes *et al.*, 2002). It is unclear why deletion of the SERT protein does not produce the same behavioral effects as pharmacological blockade of the protein with SSRIs, but possible explanations include the developmental alteration of 5-HT receptor function and/or changes in the endogenous release of 5-HT (Fabre *et al.*, 2000; Gobbi *et al.*, 2001).

Conclusion

A large amount of research has elucidated the role of the 5-HT system in genetic and molecular mechanisms underlying the pathology and treatment of depression. Instead of focusing on one component of 5-HT transmission as the source for depression vulnerability, the current concept identifies different 5-HT genes and pathways as a family of risk factors that can contribute to the pathology of depression (Jans *et al.*, 2007). The genetic models of the deletion of TPH and 5-HT receptors emphasizes the critical role of 5-HT during development for the lifelong expression of normal affective behaviors. Although genetic deletion of individual 5-HT receptors has not yet reliably reproduced a depressive behavioral phenotype, the genetic deletion of TPH may reproduce a depressive-like behavioral phenotype.

Behavioral and neuropharmacology studies have attempted to identify the 5-HT receptors associated with the therapeutic effects of SSRIs. Identifying the 5-HT receptors that support the therapeutic effects of SSRIs may lead to the development of more effective antidepressant drugs with fewer side effects. However, the present findings indicate that multiple 5-HT receptors are likely to participate in antidepressant responses, and it is unclear that the entire antidepressant effect of SSRIs can be recapitulated by a selective 5-HT receptor agonist or antagonist. However, many of the 5-HT receptors have been discovered only recently. Therefore, selective agonists and antagonists have not yet been synthesized, and specific investigation of the effects mediated by these receptors has not been fully possible. Undoubtedly, behavioral testing in animals, particularly utilizing the models outlined in this chapter, will play an important role in the further exploration of the relationship between the 5-HT system and depression.

References

Albinsson, A., Bjork, A., Svartengren, J., Klint, T. and Andersson, G. (1994) Preclinical pharmacology of FG5893: a potential anxiolytic drug with high affinity for both 5-HT1A and 5-HT2A receptors. Eur. J. Pharmacol., 261: 285–294.

Anthony, J.P., Sexton, T.J. and Neumaier, J.F. (2000) Antidepressant-induced regulation of 5-HT(1b) mRNA in rat dorsal raphe nucleus reverses rapidly after drug discontinuation. J. Neurosci. Res., 61: 82–87.

Arango, V., Underwood, M.D. and Mann, J.J. (2002) Serotonin brain circuits involved in major depression and suicide. Prog. Brain Res., 136: 443–453.

Artigas, F., Romero, L., de Montigny, C. and Blier, P. (1996) Acceleration of the effect of selected antidepressant drugs in major depression by 5-HT1A antagonists. Trends Neurosci., 19: 378–383.

Asberg, M. (1997) Neurotransmitters and suicidal behavior. The evidence from cerebrospinal fluid studies. Ann. N. Y. Acad. Sci., 836: 158–181.

Banasr, M., Hery, M., Printemps, R. and Daszuta, A. (2004) Serotonin-induced increases in adult cell proliferation and neurogenesis are mediated through different and common 5-HT receptor subtypes in the dentate gyrus and the subventricular zone. Neuropsychopharmacology, 29: 450–460.

Barnes, N.M. and Sharp, T. (1999) A review of central 5-HT receptors and their function. Neuropharmacology, 38: 1083–1152.

Beaulieu, J.M., Zhang, X., Rodriguiz, R.M., Sotnikova, T.D., Cools, M.J., Wetsel, W.C., Gainetdinov, R.R. and Caron, M. G. (2008) Role of GSK3 beta in behavioral abnormalities induced by serotonin deficiency. Proc. Natl. Acad. Sci. USA, 105: 1333–1338.

Bhagwagar, Z., Rabiner, E.A., Sargent, P.A., Grasby, P.M. and Cowen, P.J. (2004) Persistent reduction in brain serotonin1A receptor binding in recovered depressed men measured by positron emission tomography with [11C]WAY-100635. Mol. Psychiatry, 9: 386–392.

Bhagwagar, Z., Hinz, R., Taylor, M., Fancy, S., Cowen, P. and Grasby, P. (2006) Increased 5-HT(2A) receptor binding in euthymic, medication-free patients recovered from depression: a positron emission study with [(11)C]MDL 100,907. Am. J. Psychiatry, 163: 1580–1587.

Bhagwagar, Z., Murthy, N., Selvaraj, S., Hinz, R., Taylor, M., Fancy, S., Grasby, P. and Cowen, P. (2007) 5-HTT binding in recovered depressed patients and healthy volunteers: a positron emission tomography study with [11C]DASB. Am. J. Psychiatry, 164: 1858–1865.

Bhatnagar, S., Nowak, N., Babich, L. and Bok, L. (2004) Deletion of the 5-HT3 receptor differentially affects behavior of males and females in the Porsolt forced swim and defensive withdrawal tests. Behav. Brain Res., 153: 527–535.

Blier, P. and Ward, N.M. (2003) Is there a role for 5-HT1A agonists in the treatment of depression? Biol. Psychiatry, 53: 193–203.

Bodnoff, S.R., Suranyi-Cadotte, B., Aitken, D.H., Quirion, R. and Meaney, M.J. (1988) The effects of chronic antidepressant treatment in an animal model of anxiety. Psychopharmacology (Berl.), 95: 298–302.

Bonaventure, P., Kelly, L., Aluisio, L., Shelton, J., Lord, B., Galici, R., Miller, K., Atack, J., Lovenberg, T.W. and Dugovic, C. (2007) Selective blockade of 5-hydroxytryptamine (5-HT)7 receptors enhances 5-HT transmission, antidepressant-like behavior, and rapid eye movement sleep suppression induced by citalopram in rodents. J. Pharmacol. Exp. Ther., 321: 690–698.

Bourin, M., Mocaer, E. and Porsolt, R. (2004) Antidepressant-like activity of S 20098 (agomelatine) in the forced swimming test in rodents: involvement of melatonin and serotonin receptors. J. Psychiatry Neurosci., 29: 126–133.

Bravo, G. and Maswood, S. (2006) Acute treatment with 5-HT3 receptor antagonist, tropisetron, reduces immobility in intact female rats exposed to the forced swim test. Pharmacol. Biochem. Behav., 85: 362–368.

Burnet, P.W., Mead, A., Eastwood, S.L., Lacey, K., Harrison, P.J. and Sharp, T. (1995) Repeated ECS differentially affects rat brain 5-HT1A and 5-HT2A receptor expression. Neuroreport, 6: 901–904.

Cannon, D.M., Ichise, M., Rollis, D., Klaver, J.M., Gandhi, S.K., Charney, D.S., Manji, H.K. and Drevets, W.C. (2007) Elevated serotonin transporter binding in major depressive disorder assessed using positron emission tomography and [11C]DASB; comparison with bipolar disorder. Biol. Psychiatry, 62: 870–877.

Cervo, L., Grignaschi, G., Nowakowska, E. and Samanin, R. (1989) 1-(3-Trifluoromethylphenyl) piperazine (TFMPP) in the ventral tegmental area reduces the effect of desipramine in the forced swimming test in rats: possible role of serotonin receptors. Eur. J. Pharmacol., 171: 119–125.

Cesana, R., Ceci, A., Ciprandi, C. and Borsini, F. (1993) Mesulergine antagonism towards the fluoxetine anti-immobility effect in the forced swimming test in mice. J. Pharm. Pharmacol., 45: 473–475.

Chou-Green, J.M., Holscher, T.D., Dallman, M.F. and Akana, S.F. (2003) Repeated stress in young and old 5-HT(2C) receptor knockout mice. Physiol. Behav., 79: 217–226.

Clenet, F., De Vos, A. and Bourin, M. (2001) Involvement of 5-HT(2C) receptors in the anti-immobility effects of antidepressants in the forced swimming test in mice. Eur. Neuropsychopharmacol., 11: 145–152.

Compan, V., Zhou, M., Grailhe, R., Gazzara, R.A., Martin, R., Gingrich, J., Dumuis, A., Brunner, D., Bockaert, J. and Hen, R. (2004) Attenuated response to stress and novelty and hypersensitivity to seizures in 5-HT4 receptor knockout mice. J. Neurosci., 24: 412–419.

Conductier, G., Dusticier, N., Lucas, G., Cote, F., Debonnel, G., Daszuta, A., Dumuis, A., Nieoullon, A., Hen, R., Bockaert, J. and Compan, V. (2006) Adaptive changes in serotonin neurons of the raphe nuclei in 5-HT(4) receptor knockout mouse. Eur. J. Neurosci., 24: 1053–1062.

Cowen, P.J., Parry-Billings, M. and Newsholme, E.A. (1989) Decreased plasma tryptophan levels in major depression. J. Affect. Disord., 16: 27–31.

Cryan, J.F. and Lucki, I. (2000) Antidepressant-like behavioral effects mediated by 5-Hydroxytryptamine(2C) receptors. J. Pharmacol. Exp. Ther., 295: 1120–1126.

Cryan, J.F. and Slattery, D.A. (2007) Animal models of mood disorders: recent developments. Curr. Opin. Psychiatry, 20: 1–7.

Cryan, J.F., Redmond, A.M., Kelly, J.P. and Leonard, B.E. (1997) The effects of the 5-HT1A agonist flesinoxan, in three paradigms for assessing antidepressant potential in the rat. Eur. Neuropsychopharmacol., 7: 109–114.

Cryan, J.F., Valentino, R.J. and Lucki, I. (2005) Assessing substrates underlying the behavioral effects of antidepressants using the modified rat forced swimming test. Neurosci. Biobehav. Rev., 29: 547–569.

Dawson, L.A., Hughes, Z.A., Starr, K.R., Storey, J.D., Bettelini, L., Bacchi, F., Arban, R., Poffe, A., Melotto, S., Hagan, J.J. and Price, G.W. (2006) Characterisation of the selective 5-HT1B receptor antagonist SB-616234-A (1-[6-(cis-3,5-dimethylpiperazin-1-yl)-2,3-dihydro-5-methoxyindol-1-yl]-1- [2'-methyl-4'-(5-methyl-1,2,4-oxadiazol-3-yl)biphenyl-4-yl]methanone hydrochloride): *in vivo* neurochemical and behavioural evidence of anxiolytic/antidepressant activity. Neuropharmacology, 50: 975–983.

Dekeyne, A., Mannoury la Cour, C., Gobert, A., Brocco, M., Lejeune, F., Serres, F., Sharp, T., Daszuta, A., Soumier, A., Papp, M., Rivet, J.M., Flik, G., Cremers, T.I., Muller, O., Lavielle, G. and Millan, M.J. (2008) S32006, a novel 5-HT2C receptor antagonist displaying broad-based antidepressant and anxiolytic properties in rodent models. Psychopharmacology (Berl.), 199: 549–568.

Delgado, P.L. (2004) How antidepressants help depression: mechanisms of action and clinical response. J. Clin. Psychiatry, 65(Suppl 4): 25–30.

Delgado, P.L., Price, L.H., Miller, H.L., Salomon, R.M., Licinio, J., Krystal, J.H., Heninger, G.R. and Charney, D.S. (1991) Rapid serotonin depletion as a provocative challenge test for patients with major depression: relevance to antidepressant action and the neurobiology of depression. Psychopharmacol. Bull., 27: 321–330.

Delgado, P.L., Miller, H.L., Salomon, R.M., Licinio, J., Krystal, J.H., Moreno, F.A., Heninger, G.R. and Charney, D.S. (1999) Tryptophan-depletion challenge in depressed patients treated with desipramine or fluoxetine: implications for the role of serotonin in the mechanism of antidepressant action. Biol. Psychiatry, 46: 212–220.

Detke, M.J., Rickels, M. and Lucki, I. (1995a) Active behaviors in the rat forced swimming test differentially produced by serotonergic and noradrenergic antidepressants. Psychopharmacology (Berl.), 121: 66–72.

Detke, M.J., Wieland, S. and Lucki, I. (1995b) Blockade of the antidepressant-like effects of 8-OH-DPAT, buspirone and desipramine in the rat forced swim test by 5HT1A receptor antagonists. Psychopharmacology (Berl.), 119: 47–54.

De Vry, J. (1995) 5-HT1A receptor agonists: recent developments and controversial issues. Psychopharmacology (Berl.), 121: 1–26.

Drevets, W.C., Thase, M.E., Moses-Kolko, E.L., Price, J., Frank, E., Kupfer, D.J. and Mathis, C. (2007) Serotonin-1A receptor imaging in recurrent depression: replication and literature review. Nucl. Med. Biol., 34: 865–877.

Dulawa, S.C., Holick, K.A., Gundersen, B. and Hen, R. (2004) Effects of chronic fluoxetine in animal models of anxiety and depression. Neuropsychopharmacology, 29: 1321–1330.

Dulawa, S.C. and Hen, R. (2005) Recent advances in animal models of chronic antidepressant effects: the novelty-induced hypophagia test. Neurosci. Biobehav. Rev., 29: 771–783.

Edwards, E., Harkins, K., Wright, G. and Henn, F.A. (1991) 5-HT1b receptors in an animal model of depression. Neuropharmacology, 30: 101–105.

Eisensamer, B., Rammes, G., Gimpl, G., Shapa, M., Ferrari, U., Hapfelmeier, G., Bondy, B., Parsons, C., Gilling, K., Zieglgansberger, W., Holsboer, F. and Rupprecht, R. (2003) Antidepressants are functional antagonists at the serotonin type 3 (5-HT3) receptor. Mol. Psychiatry, 8: 994–1007.

Fabre, V., Beaufour, C., Evrard, A., Rioux, A., Hanoun, N., Lesch, K.P., Murphy, D.L., Lanfumey, L., Hamon, M. and Martres, M.P. (2000) Altered expression and functions of serotonin 5-HT1A and 5-HT1B receptors in knockout mice lacking the 5-HT transporter. Eur. J. Neurosci., 12: 2299–2310.

Gavioli, E.C., Vaughan, C.W., Marzola, G., Guerrini, R., Mitchell, V.A., Zucchini, S., De Lima, T.C., Rae, G.A., Salvadori, S., Regoli, D. and Calo, G. (2004) Antidepressant-like effects of the nociceptin/orphanin FQ receptor antagonist UFP-101: new evidence from rats and mice. Naunyn. Schmiedebergs Arch. Pharmacol., 369: 547–553.

Glatt, C.E., Carlson, E., Taylor, T.R., Risch, N., Reus, V.I. and Schaefer, C.A. (2005) Response to Zhang *et al.* (2005): loss-of-function mutation in tryptophan hydroxylase-2 identified in unipolar major depression. Neuron, 45: 11–16. Neuron 48: 704–705; author reply 705–706.

Gobbi, G., Murphy, D.L., Lesch, K. and Blier, P. (2001) Modifications of the serotonergic system in mice lacking serotonin transporters: an *in vivo* electrophysiological study. J. Pharmacol. Exp. Ther., 296: 987–995.

Gross, C., Zhuang, X., Stark, K., Ramboz, S., Oosting, R., Kirby, L., Santarelli, L., Beck, S. and Hen, R. (2002) Serotonin1A receptor acts during development to establish normal anxiety-like behaviour in the adult. Nature, 416: 396–400.

Gur, E., Lerer, B., Dremencov, E. and Newman, M.E. (2000) Chronic repetitive transcranial magnetic stimulation induces subsensitivity of presynaptic serotonergic autoreceptor activity in rat brain. Neuroreport, 11: 2925–2929.

Guscott, M., Bristow, L.J., Hadingham, K., Rosahl, T.W., Beer, M.S., Stanton, J.A., Bromidge, F., Owens, A.P., Huscroft, I.,

Myers, J., Rupniak, N.M., Patel, S., Whiting, P.J., Hutson, P.H., Fone, K.C., Biello, S.M., Kulagowski, J.J. and McAllister, G. (2005) Genetic knockout and pharmacological blockade studies of the 5-HT7 receptor suggest therapeutic potential in depression. Neuropharmacology, 48: 492–502.

Haddjeri, N., Blier, P. and de Montigny, C. (1998) Long-term antidepressant treatments result in a tonic activation of forebrain 5-HT1A receptors. J. Neurosci., 18: 10150–10156.

Hedlund, P.B., Huitron-Resendiz, S., Henriksen, S.J. and Sutcliffe, J.G. (2005) 5-HT7 receptor inhibition and inactivation induce antidepressantlike behavior and sleep pattern. Biol. Psychiatry, 58: 831–837.

Heisler, L.K., Chu, H.M., Brennan, T.J., Danao, J.A., Bajwa, P., Parsons, L.H. and Tecott, L.H. (1998) Elevated anxiety and antidepressant-like responses in serotonin 5-HT1A receptor mutant mice. Proc. Natl. Acad. Sci. USA, 95: 15049–15054.

Heisler, L.K., Zhou, L., Bajwa, P., Hsu, J. and Tecott, L.H. (2007) Serotonin 5-HT(2C) receptors regulate anxiety-like behavior. Genes Brain Behav., 6: 491–496.

Hirvonen, J., Karlsson, H., Kajander, J., Lepola, A., Markkula, J., Rasi-Hakala, H., Nagren, K., Salminen, J.K. and Hietala, J. (2008) Decreased brain serotonin 5-HT1A receptor availability in medication-naive patients with major depressive disorder: an *in-vivo* imaging study using PET and [carbonyl-11C]WAY-100635. Int. J. Neuropsychopharmacol., 11: 465–476.

Hogg, S. and Dalvi, A. (2004) Acceleration of onset of action in schedule-induced polydipsia: combinations of SSRI and 5-HT1A and 5-HT1B receptor antagonists. Pharmacol. Biochem. Behav., 77: 69–75.

Holmes, A., Yang, R.J., Murphy, D.L. and Crawley, J.N. (2002) Evaluation of antidepressant-related behavioral responses in mice lacking the serotonin transporter. Neuropsychopharmacology, 27: 914–923.

Jans, L.A., Riedel, W.J., Markus, C.R. and Blokland, A. (2006) Serotonergic vulnerability and depression: assumptions, experimental evidence and implications. Mol. Psychiatry, 12: 522–543.

Jans, L.A., Riedel, W.J., Markus, C.R. and Blokland, A. (2007) Serotonergic vulnerability and depression: assumptions, experimental evidence and implications. Mol. Psychiatry, 12: 522–543.

Jarosik, J., Legutko, B., Unsicker, K., von Bohlen, and Und Halbach, O. (2007) Antidepressant-mediated reversal of abnormal behavior and neurodegeneration in mice following olfactory bulbectomy. Exp. Neurol., 204: 20–28.

Jayatissa, M.N., Bisgaard, C., Tingstrom, A., Papp, M. and Wiborg, O. (2006) Hippocampal cytogenesis correlates to escitalopram-mediated recovery in a chronic mild stress rat model of depression. Neuropsychopharmacology, 31: 2395–2404.

Jones, M.D. and Lucki, I. (2005a) Sex differences in the regulation of serotonergic transmission and behavior in 5-HT receptor knockout mice. Neuropsychopharmacology, 30: 1039–1047.

Jones, M.D. and Lucki, I. (2005b) Sex differences in the regulation of serotonergic transmission and behavior in 5-HT receptor knockout mice. Neuropsychopharmacology, 30: 1039–1047.

Knobelman, D.A., Hen, R. and Lucki, I. (2001) Genetic regulation of extracellular serotonin by 5-hydroxytryptamine(1A) and 5-hydroxytryptamine(1B) autoreceptors in different brain regions of the mouse. J. Pharmacol. Exp. Ther., 298: 1083–1091.

Kostowski, W., Dyr, W., Krzascik, P., Jarbe, T. and Archer, T. (1992) 5-Hydroxytryptamine1A receptor agonists in animal models of depression and anxiety. Pharmacol. Toxicol., 71: 24–30.

Le Francois, B., Czesak, M., Steubl, D. and Albert, P.R. (2008) Transcriptional regulation at a HTR1A polymorphism associated with mental illness. Neuropharmacology, 55: 977–985.

Lekman, M., Paddock, S. and McMahon, F.J. (2008) Pharmacogenetics of major depression: insights from level 1 of the Sequenced Treatment Alternatives to Relieve Depression (STAR*D) trial. Mol. Diagn. Ther., 12: 321–330.

Lesch, K.P. (2001) Serotonergic gene expression and depression: implications for developing novel antidepressants. J. Affect. Disord., 62: 57–76.

Lesch, K.P., Bengel, D., Heils, A., Sabol, S.Z., Greenberg, B.D., Petri, S., Benjamin, J., Muller, C.R., Hamer, D.H. and Murphy, D.L. (1996) Association of anxiety-related traits with a polymorphism in the serotonin transporter gene regulatory region. Science, 274: 1527–1531.

Leshner, A.I., Remler, H., Biegon, A. and Samuel, D. (1979) Desmethylimipramine (DMI) counteracts learned helplessness in rats. Psychopharmacology (Berl.), 66: 207–208.

Lin, E. and Chen, P.S. (2008) Pharmacogenomics with antidepressants in the STAR*D study. Pharmacogenomics, 9: 935–946.

Lira, A., Zhou, M., Castanon, N., Ansorge, M.S., Gordon, J.A., Francis, J.H., Bradley-Moore, M., Lira, J., Underwood, M.D., Arango, V., Kung, H.F., Hofer, M.A., Hen, R. and Gingrich, J.A. (2003) Altered depression-related behaviors and functional changes in the dorsal raphe nucleus of serotonin transporter-deficient mice. Biol. Psychiatry, 54: 960–971.

Lucas, G., Rymar, V.V., Du, J., Mnie-Filali, O., Bisgaard, C., Manta, S., Lambas-Senas, L., Wiborg, O., Haddjeri, N., Piñeyro, G., Sadikot, A.F. and Debonnel, G. (2007) Serotonin(4) (5-HT(4)) receptor agonists are putative antidepressants with a rapid onset of action. Neuron, 55: 712–725.

Lucki, I., Singh, A. and Kreiss, D.S. (1994) Antidepressant-like behavioral effects of serotonin receptor agonists. Neurosci. Biobehav. Rev., 18: 85–95.

MacSweeney, C.P., Lesourd, M. and Gandon, J.M. (1998) Antidepressant-like effects of alnespirone (S 20499) in the learned helplessness test in rats. Eur. J. Pharmacol., 345: 133–137.

Maes, M. and Meltzer, H.Y. (1995) The Serotonin Hypothesis of Major Depression. Raven Press, New York.

Maes, M., Jacobs, M.P., Suy, E., Minner, B., Leclercq, C., Christiaens, F. and Raus, J. (1990) Suppressant effects of dexamethasone on the availability of plasma L-tryptophan and tyrosine in healthy controls and in depressed patients. Acta Psychiatr. Scand., 81: 19–23.

Mahesh, R., Rajkumar, R., Minasri, B. and Venkatesha Perumal, R. (2007) Potential antidepressants: pharmacology of 2-methyl piperazin-1-yl)-1,8-naphthyridine-3-carbonitrile in rodent behavioural models. Pharmazie, 62: 919–924.

Marek, G.J., Carpenter, L.L., McDougle, C.J. and Price, L.H. (2003) Synergistic action of 5-HT2A antagonists and selective serotonin reuptake inhibitors in neuropsychiatric disorders. Neuropsychopharmacology, 28: 402–412.

Marek, G.J., Martin-Ruiz, R., Abo, A. and Artigas, F. (2005) The selective 5-HT2A receptor antagonist M100907 enhances antidepressant-like behavioral effects of the SSRI fluoxetine. Neuropsychopharmacology, 30: 2205–2215.

Martin, P., Gozlan, H. and Puech, A.J. (1992) 5-HT3 receptor antagonists reverse helpless behaviour in rats. Eur. J. Pharmacol., 212: 73–78.

Matsuda, T., Somboonthum, P., Suzuki, M., Asano, S. and Baba, A. (1995) Antidepressant-like effect by postsynaptic 5-HT1A receptor activation in mice. Eur. J. Pharmacol., 280: 235–238.

Mayorga, A.J., Dalvi, A., Page, M.E., Zimov-Levinson, S., Hen, R. and Lucki, I. (2001) Antidepressant-like behavioral effects in 5-hydroxytryptamine(1A) and 5-hydroxytryptamine(1B) receptor mutant mice. J. Pharmacol. Exp. Ther., 298: 1101–1107.

McMahon, F.J., Buervenich, S., Charney, D., Lipsky, R., Rush, A.J., Wilson, A.F., Sorant, A.J., Papanicolaou, G.J., Laje, G., Fava, M., Trivedi, M.H., Wisniewski, S.R. and Manji, H. (2006) Variation in the gene encoding the serotonin 2A receptor is associated with outcome of antidepressant treatment. Am. J. Hum. Genet., 78: 804–814.

Meyer, J.H. (2007) Imaging the serotonin transporter during major depressive disorder and antidepressant treatment. J. Psychiatry Neurosci., 32: 86–102.

Meyer, J.H., McMain, S., Kennedy, S.H., Korman, L., Brown, G.M., DaSilva, J.N., Wilson, A.A., Blak, T., Eynan-Harvey, R., Goulding, V.S., Houle, S. and Links, P. (2003) Dysfunctional attitudes and 5-HT2 receptors during depression and self-harm. Am. J. Psychiatry, 160: 90–99.

Moreno, F.A., Gelenberg, A.J., Heninger, G.R., Potter, R.L., McKnight, K.M., Allen, J., Phillips, A.P. and Delgado, P.L. (1999) Tryptophan depletion and depressive vulnerability. Biol. Psychiatry, 46: 498–505.

Nakagawa, Y., Ishima, T. and Takashima, T. (1998) The 5-HT3 receptor agonist attenuates the action of antidepressants in the forced swim test in rats. Brain Res., 786: 189–193.

Neumaier, J.F., Root, D.C. and Hamblin, M.W. (1996) Chronic fluoxetine reduces serotonin transporter mRNA and 5-HT1B mRNA in a sequential manner in the rat dorsal raphe nucleus. Neuropsychopharmacology, 15: 515–522.

Neumaier, J.F., Petty, F., Kramer, G.L., Szot, P. and Hamblin, M.W. (1997) Learned helplessness increases 5-hydroxytryptamine1B receptor mRNA levels in the rat dorsal raphe nucleus. Biol. Psychiatry, 41: 668–674.

Neumeister, A., Nugent, A.C., Waldeck, T., Geraci, M., Schwarz, M., Bonne, O., Bain, E.E., Luckenbaugh, D.A., Herscovitch, P., Charney, D.S. and Drevets, W.C. (2004) Neural and behavioral responses to tryptophan depletion in unmedicated patients with remitted major depressive disorder and controls. Arch. Gen. Psychiatry, 61: 765–773.

O'Leary, O.F., Bechtholt, A.J., Crowley, J.J., Hill, T.E., Page, M.E. and Lucki, I. (2007) Depletion of serotonin and catecholamines block the acute behavioral response to different classes of antidepressant drugs in the mouse tail suspension test. Psychopharmacology (Berl.), 192: 357–371.

Page, M.E., Detke, M.J., Dalvi, A., Kirby, L.G. and Lucki, I. (1999) Serotonergic mediation of the effects of fluoxetine, but not desipramine, in the rat forced swimming test. Psychopharmacology (Berl.), 147: 162–167.

Parks, C.L., Robinson, P.S., Sibille, E., Shenk, T. and Toth, M. (1998) Increased anxiety of mice lacking the serotonin1A receptor. Proc. Natl. Acad. Sci. USA, 95: 10734–10739.

Patel, J.G., Bartoszyk, G.D., Edwards, E. and Ashby, C.R. Jr. (2004) The highly selective 5-hydroxytryptamine (5-HT)2A receptor antagonist, EMD 281014, significantly increases swimming and decreases immobility in male congenital learned helpless rats in the forced swim test. Synapse, 52: 73–75.

Pehek, E.A., Nocjar, C., Roth, B.L., Byrd, T.A. and Mabrouk, O.S. (2006) Evidence for the preferential involvement of 5-HT2A serotonin receptors in stress- and drug-induced dopamine release in the rat medial prefrontal cortex. Neuropsychopharmacology, 31: 265–277.

Piñeyro, G. and Blier, P. (1999) Autoregulation of serotonin neurons: role in antidepressant drug action. Pharmacol. Rev., 51: 533–591.

Placidi, G.P., Oquendo, M.A., Malone, K.M., Huang, Y.Y., Ellis, S.P. and Mann, J.J. (2001) Aggressivity, suicide attempts, and depression: relationship to cerebrospinal fluid monoamine metabolite levels. Biol. Psychiatry, 50: 783–791.

Porsolt, R.D., Le Pichon, M. and Jalfre, M. (1977) Depression: a new animal model sensitive to antidepressant treatments. Nature, 266: 730–732.

Porsolt, R.D., Bertin, A. and Jalfre, M. (1978) 'Behavioural despair' in rats and mice: strain differences and the effects of imipramine. Eur. J. Pharmacol., 51: 291–294.

Przegalinski, E., Moryl, E. and Papp, M. (1995) The effect of 5-HT1A receptor ligands in a chronic mild stress model of depression. Neuropharmacology, 34: 1305–1310.

Ramamoorthy, R., Radhakrishnan, M. and Borah, M. (2008) Antidepressant-like effects of serotonin type-3 antagonist, ondansetron: an investigation in behaviour-based rodent models. Behav. Pharmacol., 19: 29–40.

Redrobe, J.P. and Bourin, M. (1997) Partial role of 5-HT2 and 5-HT3 receptors in the activity of antidepressants in the mouse forced swimming test. Eur. J. Pharmacol., 325: 129–135.

Rosenzweig-Lipson, S., Sabb, A., Stack, G., Mitchell, P., Lucki, I., Malberg, J.E., Grauer, S., Brennan, J., Cryan, J.F., Sukoff Rizzo, S.J., Dunlop, J., Barrett, J.E. and Marquis, K.L. (2007) Antidepressant-like effects of the novel, selective, 5-HT2C receptor agonist WAY-163909 in rodents. Psychopharmacology (Berl.), 192: 159–170.

Santarelli, L., Saxe, M., Gross, C., Surget, A., Battaglia, F., Dulawa, S., Weisstaub, N., Lee, J., Duman, R., Arancio, O., Belzung, C. and Hen, R. (2003) Requirement of hippocampal neurogenesis for the behavioral effects of antidepressants. Science, 301: 805–809.

Savelieva, K.V., Zhao, S., Pogorelov, V.M., Rajan, I., Yang, Q., Cullinan, E. and Lanthorn, T.H. (2008) Genetic disruption of both tryptophan hydroxylase genes dramatically reduces serotonin and affects behavior in models sensitive to antidepressants. PLoS ONE, 3: e3301.

Schechter, L.E., Lin, Q., Smith, D.L., Zhang, G., Shan, Q., Platt, B., Brandt, M.R., Dawson, L.A., Cole, D., Bernotas, R., Robichaud, A., Rosenzweig-Lipson, S. and Beyer, C.E. (2008) Neuropharmacological profile of novel and selective 5-HT6 receptor agonists: WAY-181187 and WAY-208466. Neuropsychopharmacology, 33: 1323–1335.

Scruggs, J.L., Schmidt, D. and Deutch, A.Y. (2003) The hallucinogen 1-[2,5-dimethoxy-4-iodophenyl]-2-aminopropane (DOI) increases cortical extracellular glutamate levels in rats. Neurosci. Lett., 346: 137–140.

Seligman, M.E., Rosellini, R.A. and Kozak, M.J. (1975) Learned helplessness in the rat: time course, immunization, and reversibility. J. Comp. Physiol. Psychol., 88: 542–547.

Serretti, A. and Artioli, P. (2004) The pharmacogenomics of selective serotonin reuptake inhibitors. Pharmacogenomics J., 4: 233–244.

Shen, C., Li, H. and Meller, E. (2002) Repeated treatment with antidepressants differentially alters 5-HT1A agonist-stimulated [35S]GTP gamma S binding in rat brain regions. Neuropharmacology, 42: 1031–1038.

Sherman, A.D. and Petty, F. (1980) Neurochemical basis of the action of antidepressants on learned helplessness. Behav. Neural Biol., 30: 119–134.

Singh, A. and Lucki, I. (1993) Antidepressant-like activity of compounds with varying efficacy at 5-HT1A receptors. Neuropharmacology, 32: 331–340.

Smriga, M. and Torii, K. (2003) L-Lysine acts like a partial serotonin receptor 4 antagonist and inhibits serotonin-mediated intestinal pathologies and anxiety in rats. Proc. Natl. Acad. Sci. USA, 100: 15370–15375.

Song, C. and Leonard, B.E. (2005) The olfactory bulbectomised rat as a model of depression. Neurosci. Biobehav. Rev., 29: 627–647.

Steru, L., Chermat, R., Thierry, B. and Simon, P. (1985) The tail suspension test: a new method for screening antidepressants in mice. Psychopharmacology (Berl.), 85: 367–370.

Stockmeier, C.A. (2003) Involvement of serotonin in depression: evidence from postmortem and imaging studies of serotonin receptors and the serotonin transporter. J. Psychiatr. Res., 37: 357–373.

Stout, S.C., Owens, M.J. and Nemeroff, C.B. (1996) CRF1 receptor binding, mRNA expression and signal transduction after chronic antidepressant treatment in rats. In: Society for Neuroscience Annual Meeting.

Svenningsson, P., Chergui, K., Rachleff, I., Flajolet, M., Zhang, X., El Yacoubi, M., Vaugeois, J.M., Nomikos, G.G. and Greengard, P. (2006) Alterations in 5-HT1B receptor function by p11 in depression-like states. Science, 311: 77–80.

Svenningsson, P., Tzavara, E.T., Qi, H., Carruthers, R., Witkin, J.M., Nomikos, G.G. and Greengard, P. (2007) Biochemical and behavioral evidence for antidepressant-like effects of 5-HT6 receptor stimulation. J. Neurosci., 27: 4201–4209.

Tatarczynska, E., Klodzinska, A., Stachowicz, K. and Chojnacka-Wojcik, E. (2004) Effect of combined administration of 5-HT1A or 5-HT1B/1D receptor antagonists and antidepressants in the forced swimming test. Eur. J. Pharmacol., 487: 133–142.

Tordera, R.M., Monge, A., Del Rio, J. and Lasheras, B. (2002) Antidepressant-like activity of VN2222, a serotonin reuptake inhibitor with high affinity at 5-HT1A receptors. Eur. J. Pharmacol., 442: 63–71.

Van Den Bogaert, A., De Zutter, S., Heyrman, L., Mendlewicz, J., Adolfsson, R., Van Broeckhoven, C. and Del-Favero, J. (2005) Response to Zhang *et al.* (2005): loss-of-function mutation in tryptophan hydroxylase-2 identified in unipolar major depression. Neuron, 45: 11–16. Neuron 48: 704; author reply 705–706

Wesolowska, A., Nikiforuk, A., Stachowicz, K. and Tatarczynska, E. (2006) Effect of the selective 5-HT7 receptor antagonist SB 269970 in animal models of anxiety and depression. Neuropharmacology, 51: 578–586.

Wesolowska, A. and Nikiforuk, A. (2007) Effects of the brain-penetrant and selective 5-HT6 receptor antagonist SB-399885 in animal models of anxiety and depression. Neuropharmacology, 52: 1274–1283.

Wesolowska, A., Nikiforuk, A. and Stachowicz, K. (2007) Anxiolytic-like and antidepressant-like effects produced by the selective 5-HT6 receptor antagonist SB-258585 after intrahippocampal administration to rats. Behav. Pharmacol., 18: 439–446.

Willner, P. (1997) Validity, reliability and utility of the chronic mild stress model of depression: a 10-year review and evaluation. Psychopharmacology (Berl.), 134: 319–329.

Willner, P. (2005) Chronic mild stress (CMS) revisited: consistency and behavioural–neurobiological concordance in the effects of CMS. Neuropsychobiology, 52: 90–110.

Zhang, X., Gainetdinov, R.R., Beaulieu, J.M., Sotnikova, T.D., Burch, L.H., Williams, R.B., Schwartz, D.A., Krishnan, K.R. and Caron, M.G. (2005) Loss-of-function mutation in tryptophan hydroxylase-2 identified in unipolar major depression. Neuron, 45: 11–16.

Zhou, Z., Peters, E.J., Hamilton, S.P., McMahon, F., Thomas, C., McGrath, P.J., Rush, J., Trivedi, M.H., Charney, D.S., Roy, A., Wisniewski, S., Lipsky, R. and Goldman, D. (2005) Response to Zhang *et al.* (2005): loss-of-function mutation in tryptophan hydroxylase-2 identified in unipolar major depression. Neuron, 45: 11–16. , Neuron 48: 702–703; author reply 705–706.

Zhuang, X., Gross, C., Santarelli, L., Compan, V., Trillat, A.C. and Hen, R. (1999) Altered emotional states in knockout mice lacking 5-HT1A or 5-HT1B receptors. Neuropsychopharmacology, 21: 52S–60S.

CHAPTER 4.3

The Role of Serotonin in Drug Addiction

Christian P. Müller*[,1], Martin E. Pum[2], Gunter Schumann[1] and Joseph P. Huston[2]

[1]*MRC SGDP-Center and Interdisciplinary Research Group Addiction, Institute of Psychiatry, King's College London, London, UK*
[2]*Institute of Physiological Psychology and Center for Biological and Medical Research, University of Düsseldorf, Düsseldorf, Germany*

Abstract: Drug addiction is a major psychiatric disease with, currently, no truly effective treatment available. Drug addiction involves many different behaviors. Given the involvement of serotonin (5-HT) in virtually all behaviors and brain plasticity at cellular, molecular and systems levels, one can readily assume a 5-HT contribution to the establishment and maintenance of addiction-related behaviors and their neuronal mechanisms. In this chapter we review the role of the 5-HT system in addiction-related behaviors for major addictive drugs, such as cocaine, amphetamine, methamphetamine, MDMA (ecstasy), morphine, heroin and nicotine. We discuss the impact of these drugs on 5-HT activity, and how this activity might translate into behavior by activating 5-HT receptors. The effects of serotonergic manipulations on addiction-related behaviors in animal models and in human therapy of addiction are reviewed. Intensive research has shown that drug-induced 5-HT activation does not have a uniform 'proactive effect' (i.e., neurochemical activity that triggers behavioral activation) on addiction-related behavior, but might often serve a 'counteractive function', thus limiting the expression of behavior which is driven by other neurochemical effects of the drug. 5-HT receptors can have very different effects on addiction-related behaviors depending on their localization in the brain, the cell type, and the synapse. Some receptor populations do appear to have proactive effects – i.e., translate the drug-induced 5-HT increase into addiction-related behaviors – while others counteract them, or are not involved. While acknowledging this complexity will be important for an understanding of the 5-HT role in drug addiction, it also provides a challenge to the use of 5-HT system components for therapy in humans.

Keywords: serotonin, addiction, animal models, clinical trials, cocaine, amphetamine, methamphetamine, MDMA, morphine, nicotine.

Introduction

Drug addiction in humans is a serious and complex psychiatric disease (*DSM-IV*, American Psychiatric Association, 1994), which not only affects the individual user but all aspects of social and economic life. Even after decades of intensive research in drug addiction, there is still only a limited understanding of the brain mechanisms involved. This is most obviously reflected in the lack of effective behavioral and/or pharmacotherapies for addiction. The quest to understand the development of drug abuse habits and drug addiction, therefore, is still a major challenge for behavioral neuroscience. Drug addiction involves different behaviors, which are in competition with non-drug-related behaviors. Most of the drug-related behaviors do not establish a new class of behavioral plasticity, but are essentially modifications of normal (i.e., non-drug-related) behaviors. They involve, for example, instrumental and Pavlovian conditioning. Accordingly, an understanding of the brain mechanisms of addiction would necessarily need to co-evolve with an understanding of how natural behavioral plasticity is mediated in the brain.

Addiction was seen as a malfunction of the brain's hedonic tone regulation (Wise, 1980, 2002), and later as malfunction of the motivational systems (Koob, 1992; Robinson and Berridge, 1993, 2003; Di Chiara, 1995) in that these systems were 'highjacked' by intense pharmacological reinforcers. While there are still problems to characterize brain mechanisms serving the psychological concepts of hedonia and motivation, it proved to be more successful to characterize mechanisms of reinforcement learning (Robbins and Everitt, 1996; Kelley and Berridge, 2002). With this background information, it is easier to

*Corresponding author
E-mail: christian.muller@kcl.ac.uk

Christian Müller & Barry Jacobs (Eds.)
Handbook of Behavioral Neurobiology of Serotonin
ISBN 978-0-12-374634-4

DOI: 10.1016/B978-0-12-374634-4.00031-9

understand drug effects within these systems and to view addiction as a disturbance of the brain reinforcement systems (Kalivas and Volkow, 2005; Koob and Le, 2008). However, reinforcement learning is just one type of learning and memory, leaving a number of important addiction-related behaviors out of the concept. This issue was better addressed when evidence accumulated that drugs of addiction cause molecular effects and structural and functional changes to the brain that are comparable with those of normal learning (Robinson and Kolb, 2004). Furthermore, when drugs induced this plasticity, it prevented normal stimulation and behavior from inducing it (Kolb *et al.*, 2003). In turn, natural stimuli (such as environmental enrichment and induced exploration behavior) were able to compete with reinstatement of drug-related behavior (Solinas *et al.*, 2008). These findings might highlight the competitive character of drug-related behaviors with non-drug-related behaviors. An alternate view on drug addiction is to look at it as a disturbance of the brain's memory systems (Nestler, 2002; Kelley, 2004; Hyman *et al.*, 2006). This conceptualization allows an incorporation of the previous models of hedonic, motivational and reinforcement learning into a broader view.

Serotonin (5-HT) was shown to be of essential importance not only to maintain synaptic plasticity throughout the whole lifespan (see Chapter 3.2), for the hedonic tone (see Chapter 4.13), in motivational processes (see Chapter 3.3) and in reinforcement processes (see Chapter 3.7), but also for learning and memory (see Chapter 3.9). Accordingly, it is not surprising that drugs of addiction, which induce profound changes in extracellular 5-HT activity and 5-HT receptor function, change the behavior-organizing circuitry of the brain either directly, by modulating the 5-HT system, or indirectly, by 5-HT effects on other transmitter systems (see Chapter 2.8). In this chapter we review the action of major addictive drugs, such as cocaine, amphetamine, methamphetamine, 3,4-methylenedioxymethamphetamine (MDMA, ecstasy), heroin, morphine and nicotine, on the 5-HT system, and how components of the 5-HT system contribute to the establishment and expression of addiction-related behaviors (for the role of 5-HT in alcohol addiction: see Chapter 3.7). We describe the influence of increased and decreased 5-HT activity on addiction-related behavior in animal models. It should be noted that drugs of addiction usually have many more effects on brain function and behavior than those related to addiction in general (see, for example, Kalivas, 2005; Müller *et al.*, 2007a); this, however, will not be discussed here. A major part of the chapter will focus on the contribution of single 5-HT receptor subtypes. In addition, we discuss how components of the 5-HT system have been tested as potential targets for a pharmacotherapy of drug addiction in human patients.

Measuring drug addiction in animal models

Many drugs of addiction evoke at low to medium doses, and parallel to pleasant feelings, an increase in locomotor activity. Acute hyperlocomotor effects, although not directly recognized as a harmful addiction-related behavior in humans and animals, are often investigated as a crude indicator for the subjective effects of the drug.

Due to the brain's function to maintain homeostasis, counter-regulatory processes are evoked by the strong neurochemical impact of a psychostimulant drug (Solomon, 1980; Koob and Le, 2008). Immediately after the intake of a psychostimulant drug, functional changes in the brain occur that may endure beyond the presence of the actual drug/metabolites in the brain (see, for example, Ungless *et al.*, 2001). These functional changes and corresponding homeostatic responses may be responsible for behavioral and subjective effects of the drug intake that outlast the presence of the drug in the brain. These effects, believed to be neuronal adaptations, are also the basis for a sensitization of the acute hyperlocomotor effects – i.e., when the same dose of the drug has increasing effects after only few administrations (Vanderschuren and Kalivas, 2000).

When the consumption of a drug is repeated, the frequency of intake may increase, and eventually a transition to a binge-taking pattern can take place (Kramer *et al.*, 1967; Gawin, 1991). This transition occurs because the euphoric effects of the drug develop tolerance – i.e., the same dose of a drug induces less euphoria with repeated administration. At this point, not only do the acute effects of the intake change, but gross behavioral changes may also appear. These new behaviors are related to drug-seeking and the self-administration of the drug. This behavioral pattern occurs despite the possible development of severe negative consequences of the drug consumption (Deroche Gamonet *et al.*, 2004; Vanderschuren and Everitt, 2004). The occurrence and frequency of such newly established behaviors is usually considered when drug use, abuse or addiction is diagnosed. Chronic drug use induces neuroadaptive changes, the behavioral manifestations of which can often be observed in the absence of the drug. In this case, drug-related behaviors are initiated in a drug-free state. The mechanisms behind these effects are not neurochemical rebound responses, but are based on morphological and functional long-term changes in the brain (Robinson and Kolb, 2004). Accordingly, brain substrates mediating the acute pharmacological drug effects on behavior are not necessarily the same as those underlying addiction-related behaviors in a drug-free state. Since the addiction-related behaviors promote continued drug taking, the resulting long-term neurophysiological alterations in brain function continue to be reinforced by the acute neuropharmacological effects of the drug.

In this chapter, we focus on the major paradigms to measure drug addiction-related behaviors in animal models. Basic experiments measure the effects of serotonergic manipulations on the acute drug-induced locomotor activity. When drugs are administered repeatedly, the acute behavioral effects can sensitize. Regarding the role of 5-HT in the sensitization of the hyperlocomotor effects, induction and expression of sensitization can be distinguished and independently modulated by various treatments. The discriminative stimulus effects of addictive drugs in animals serve as a model for the subjective drug effects in humans. In animal studies, discriminative stimulus properties are usually assessed by pairing a drug cue with the availability of a food or water reward when pressing one of two levers. If a certain drug can be readily differentiated from saline, responding takes place on the rewarded lever. This paradigm allows testing of the prerequisites for drug discrimination in the brain. It also investigates the ability of a pharmacological treatment to substitute for the drug as a pharmacological cue.

Chronic treatment with addictive drugs leads to the development of various addiction-related behaviors that are expressed in an undrugged state. When the effects of a pharmacological treatment are repeatedly paired with a certain place context (or cues), and in a free choice this place is preferred over another, originally equally preferred place, a conditioned place preference (CPP) is inferred. If the paired place is avoided in the test, a conditioned place aversion (CPA) is concluded (Bardo and Bevins, 2000; Tzschentke, 2007).

Self-administration of a drug is probably the most informative test for addiction related behaviors in animals (Spealman and Goldberg, 1978). In this paradigm, animals learn to perform an operant response to receive the drug, or a sensory stimulus (secondary reinforcer) which was previously associated with the delivery of the drug. Animals work to obtain both stimuli that are associated with the drug (conditioned reinforcers), as well the drug itself. The motivation to self-administer addictive drugs can be measured in fixed-ratio and progressive-ratio schedules of reinforcement. The response level at which the animal ceases to respond (the breakpoint) in a progressive-ratio schedule of reinforcement is an indicator of the motivational level (Depoortere *et al.*, 1993; Richardson and Roberts, 1996; Roberts *et al.*, 2002). If the behavioral activity that led to drug administration is no longer followed by the drug, the behavior undergoes extinction. Extinguished self-administration behavior can, however, be reinstated by stress, by cues signaling the availability of the drug, or by application of the drug itself (De Wit and Stewart, 1981; Shaham *et al.*, 2003). Of increasing importance in addiction research is the question of whether a pharmacological treatment can prevent the reinstatement of previously extinguished drug-seeking behavior.

When the well-established self-administration of certain drugs is abruptly discontinued in humans, a severe withdrawal syndrome, characterized by dysphoria, anhedonia depression, anxiety and psychomotor alterations, is observed. While dysphoria and depression may disappear after a few days, a sensitized hyperlocomotor response and sensitized behavioral stereotypies may still be observed after years of abstinence (Kramer *et al.*, 1967; Gawin, 1991). To target the aversive effects of withdrawal is an important component of a potential treatment strategy for drug addiction. The 5-HT system has been investigated for its role in the increased reward threshold during withdrawal, using intracranial self-stimulation (ICSS). In the ICSS paradigm, animals learn to perform an operant task reinforced by electrical stimulation of, for example, the lateral hypothalamus (Olds and Milner, 1954; Markou and Koob, 1992). An increase in the stimulation threshold that is required to maintain the ICSS behavior indicates a reduced reward value of the electrical stimulation, and may be seen as a reduced hedonic tone in the brain. For detailed reviews on the advantages and shortcomings of each animal model of drug addiction, see Sanchis-Segura and Spanagel (2006) and Olmstead (2006).

Cocaine

The effects of cocaine on 5-HT activity

Cocaine, a major alkaloid of the South American shrub *Erythroxylum coca*, is a powerful pharmacological reinforcer with a highly judged abuse potential (Gawin, 1991; Nutt *et al.*, 2007). A great number of studies reported increased extracellular 5-HT levels after acute cocaine administration. In the rat, acute cocaine administration leads to a temporally limited increase in extracellular 5-HT levels in various subcortical structures, such as the nucleus accumbens (Nac) (Broderick *et al.*, 1993; Parsons and Justice, 1993; Teneud *et al.*, 1996; Andrews and Lucki, 2001; Müller *et al.*, 2002a), dorsal striatum (Bradberry *et al.*, 1993), ventral pallidum (Sizemore *et al.*, 2000), hippocampus (Müller *et al.*, 2002a, 2004a), thalamus (Rutter *et al.*, 1998), hypothalamus (Shimizu *et al.*, 1992), ventral tegmental area (VTA) (Parsons and Justice, 1993; Chen and Reith, 1994; Reith *et al.*, 1997) and dorsal raphe nucleus (DRN) (Parsons and Justice, 1993). Cocaine also increased extracellular 5-HT levels in various neocortical areas, such as the prefrontal cortex (PFC) (Mangiavacchi *et al.*, 2001; Pum *et al.*, 2007), the occipital and temporal cortices (Müller *et al.*, 2007b), and the entorhinal and perirhinal cortices (Pum *et al.*, 2007, Figure 1). Small doses of cocaine, which had little effect on, for example, locomotor activity, led to a much more intense and prolonged increase in 5-HT than sensory stimulation (Müller *et al.*, 2007b; Pum *et al.*, 2007, 2008) or behavioral activity (Rueter *et al.*, 1997). When animals are allowed to self-administer cocaine with an unlimited access, 5-HT levels increase in the Nac and ventral pallidum, and then remain at a plateau (Parsons *et al.*, 1995; Sizemore *et al.*, 2000), possibly self-titrated by the self-administration behavior. The profound cocaine effects on extracellular 5-HT levels, which have occurred in virtually all brain areas that have been investigated so far, show a similar time-course to many acute behavioral and subjective effects, suggesting that 5-HT might contribute to, or at least modulate, the effects of cocaine.

When access to cocaine is stopped after prolonged drug self-administration, a withdrawal syndrome commences. This is characterized by dysphoria, anhedonia, depression and anxiety in humans. These behavioral effects appear to be associated with profound changes in 5-HT activity in rodents. It has been reported that extracellular 5-HT levels in the Nac and ventral pallidum, which were increased during cocaine self-administration, drop to below predrug baseline levels within 2 hours of withdrawal (Parsons *et al.*, 1995; Sizemore *et al.*, 2000). This reduction in 5-HT activity in the Nac could be observed for at least 12 hours (Parsons *et al.*, 1995). When cocaine was administered by the experimenter on 10 consecutive days, no change in basal 5-HT levels was found in the Nac, the VTA or the DRN 1 day after this treatment regimen in rats (Parsons and Justice, 1993). However, the chronic treatment led to a significantly higher percentage 5-HT increase after an acute cocaine challenge in all three areas, suggesting an augmented responsiveness of the 5-HT system after chronic cocaine treatment (Parsons and Justice, 1993). Given the similarities in the behavioral state between psychostimulant withdrawal and depression, it can be argued that reduced 5-HT activity in the limbic system may be a common neurochemical mechanism (Parsons *et al.*, 1995; Baumann and Rothman, 1998; Markou *et al.*, 1998).

Reducing and potentiating 5-HT activity

Studies on the role of the 5-HT system in cocaine's behavioral effects used selective neurotoxins to lesion the 5-HT neurons, or used tryptophan hydroxylase inhibitors to reduce 5-HT synthesis. Scheel-Krüger and colleagues (1977) used p-chlorphenylalanin (pCPA) as a 5-HT synthesis inhibitor in rats. They found that cocaine-induced hyperlocomotion was potentiated in rats after pCPA pretreatment. This finding was confirmed later by Svingos and Hitzemann (1992) and Herges and Taylor (1998) for several doses of cocaine. Also, potentiation of cocaine-induced hyperlocomotion was reported by Morrow and Roth (1996), who used 5,7-dihydroxytryptamine (5,7-DHT) to lesion serotonergic neurons. Local lesion studies, however, showed significant regional differences in the contribution of 5-HT to cocaine-induced hyperactivity. A local lesion of the mPFC with 5,7-DHT attenuated the hyperlocomotor effects of cocaine, while 5-HT depletion of the entorhinal or occipital cortices had no effect (Pum *et al.*, 2008). Together, several treatments that reduced the availability of 5-HT in the brain suggest an inhibitory role of the 5-HT system in cocaine-induced hyperlocomotion, with, however, regional differences in the brain. This view is further supported by studies that increased the availability of 5-HT in the brain by applying a 5-HT precursor, a selective serotonin reuptake inhibitor (SSRI) or a 5-HT releaser. In an early study, Pradhan *et al.* (1978) injected rats with 5-hydroxytryptophan, the direct 5-HT precursor, and found an attenuated locomotor response after cocaine. Molina and colleagues (2001) used L-tryptophan, the amino acid precursor of 5-HT, to increase 5-HT availability. They also reported a diminished locomotor response to cocaine afterwards. However, a potentiated locomotor response to cocaine was reported after the SSRIs fluoxetine and fluvoxamine (Herges and Taylor, 1998; Bubar *et al.*, 2003), but not after citalopram or sertraline (Fletcher *et al.*, 2004a). Interestingly, the fluoxetine-enhancing effects were also seen in animals with a 5,7-DHT-induced 5-HT depletion, which suggested the involvement of non-serotonergic effects of fluoxetine (Fletcher *et al.*, 2004a).

The SSRIs fluoxetine and citalopram did not substitute for cocaine in discriminative stimulus testing in rats. However, several SSRIs were shown to enhance the effects of, in particular, a sub-threshold dose of cocaine, resulting in a leftward shift of the dose–response curve in rats (Baker *et al.*, 1993; Callahan and Cunningham, 1997; Kleven and Koek, 1998; Klein and Gulley, 2009). In monkeys, in contrast, the SSRIs citalopram and fluoxetine attenuated the discriminative stimulus effects of cocaine (Spealman, 1993).

Cocaine induces a profound CPP in rats and mice (Tzschentke, 2007). A recent study in rats investigated the role of 5-HT in various cortical areas that showed an increase in the 5-HT response after acute cocaine treatment (Pum *et al.*, 2007; see also Figure 1). A depletion of 5-HT in the medial PFC (mPFC) by 90 percent and in entorhinal cortex by 61 percent significantly attenuated cocaine-induced CPP. In contrast, 78 percent depletion in the occipital cortex had no effect on CPP. This suggests multiple dissociations in the functional role of 5-HT in

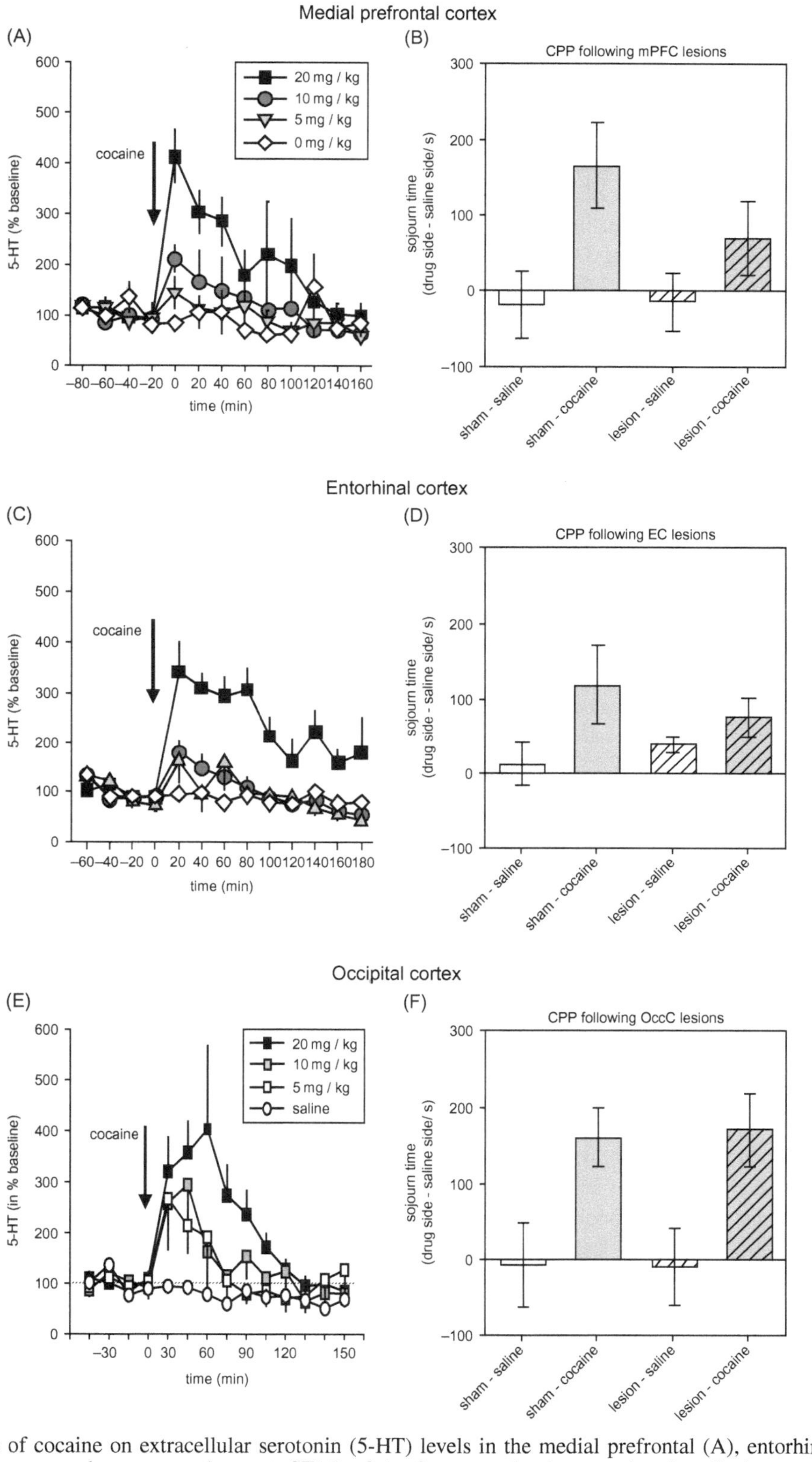

Figure 1 Acute effects of cocaine on extracellular serotonin (5-HT) levels in the medial prefrontal (A), entorhinal (C), and occipital cortex (E). Results are expressed as percent (mean ± SEM) of the four pre-stimulus samples (baseline) set as 100%. Place preference scores (time spent in conditioning compartment − time spent in pseudo-conditioning compartment; mean ± SEM) of rats with 5,7-DHT- or sham-lesions of the medial prefrontal (B), entorhinal (D), and occipital cortex (F). Figures modified from Müller *et al.* (2007a), with permission from Elsevier; and from Pum *et al.* (2007, 2008), with kind permission from Springer Science.

the cortex in cocaine-induced CPP and hyperlocomotor effects.

Global as well as forebrain- or amygdala-restricted 5-HT lesions, using 5,7-DHT, increased responding for cocaine in a progressive ratio schedule in rats (Loh and Roberts, 1990). A more complex role of 5-HT was suggested in studies by Tran-Nguyen *et al.* (1999, 2001), showing that global 5-HT reductions, using either pCPA or 5,7-DHT, attenuated cocaine-seeking during extinction of responding for cocaine, but increased the effectiveness of a cocaine priming injection in reinstating drug-seeking behavior. Complementary to this, the SSRI fluoxetine reduced the self-administration of low doses of cocaine in rats, but did not affect it at higher doses (Porrino *et al.*, 1989; Peltier and Schenk, 1993). Alaproclate, however, decreased responding for cocaine in a second-order schedule of drug delivery in squirrel monkeys (Czoty *et al.*, 2002). Pre-treatment with the 5-HT precursor L-tryptophan decreased the breakpoints in a progressive ratio schedule of intravenous cocaine self-administration in rats (McGregor *et al.*, 1993).

Inhibitory effects of 5-HT on the motivational/reinforcing effect of cocaine were confirmed in genetic approaches using either serotonin transporter (SERT) knockout (KO) mice or target-selected ENU-induced mutagenesis of the SERT in rats. SERT KO mice and rats showed an enhanced cocaine-induced CPP (Sora *et al.*, 2001; Homberg *et al.*, 2008). For the rats (Homberg *et al.*, 2008), but not for the mice (Thomsen *et al.*, 2009), enhanced cocaine self-administration was also reported. This would be largely in line with the studies showing that 5-HT activation exerts a rather inhibitory influence on major addiction-related behaviors.

5-HT receptors

5-HT$_{1A}$ receptor

A prominent, acute behavioral effect of psychostimulant drugs is an increase in locomotor activity at low to medium doses (e.g. Segal, 1975; Scheel-Krüger *et al.*, 1977; Miczek and Yoshimura, 1982). The hyperlocomotor response was shown to have a similar time course as the acute subjective effects and coincides with many neurochemical effects. A considerable number of studies have investigated the contribution of 5-HT$_{1A}$ receptors in the brain to the hyperlocomotor effects of cocaine. The hyperlocomotor effects of a low dose of cocaine (5 mg/kg) were blocked by pre-treatment with the 5-HT$_{1A}$ receptor agonist 8-OH-DPAT in rats (Przegalinski and Filip, 1997). Also, the partial 5-HT$_{1A}$ receptor agonist, S 16924, and the full agonist, osemozotan, blocked cocaine-induced hyperlocomotion in rats and mice (Millan *et al.*, 1998; Nakamura *et al.*, 2006). Other studies, in contrast, found that stimulation of the 5-HT$_{1A}$ receptor with 8-OH-DPAT potentiated cocaine-induced hyperlocomotion (De La Garza and Cunningham, 2000; Carey *et al.*, 2001, 2002, 2004a; Müller *et al.*, 2003). Since the effects of 8-OH-DPAT alone on locomotor activity depend on experimental parameters, such as the size of the arena, pre-test handling, and habituation procedures, which interfere with the emotional status of the animals, it may not be surprising that 5-HT$_{1A}$-receptor effects on cocaine-induced hyperlocomotion depend on these factors. The 5-HT$_{1A}$ receptor antagonist, NAN-190, blocked cocaine-induced behavioral activity in rats (King *et al.*, 1993). The selective 5-HT$_{1A}$ receptor antagonist, WAY 100135, however, failed to affect small dose cocaine-induced hyperlocomotion (Przegalinski and Filip, 1997), while the selective and silent 5-HT$_{1A}$ receptor antagonist, WAY 100635, clearly attenuated the acute hyperlocomotor effects of cocaine in rats (Carey *et al.*, 2000, 2001, 2002; Müller *et al.*, 2002a, b). In mice, however, pre-treatment with WAY100635 potentiated cocaine-induced hyperlocomotion (Nakamura *et al.*, 2006) or had no effect (Herges and Taylor (1998) when administered 1 hour before cocaine.

In order to determine the contribution of different local 5-HT$_{1A}$ receptor populations to the behavioral effects of psychostimulants, two different approaches have been used: (1) local application studies targeting 5-HT$_{1A}$ receptors in restricted brain areas, and (2) systemic application studies with very low doses of 5-HT$_{1A}$ receptor agonists or antagonists, which preferentially target the more sensitive 5-HT$_{1A}$ autoreceptors in the raphe nuclei. In the systemic approach with low doses it was shown that 8-OH-DPAT, which preferentially stimulate 5-HT$_{1A}$ autoreceptors, blocked the acute hyperlocomotor effects of cocaine (Carey *et al.*, 2004a, b, 2005a). Interestingly, the acute inhibitory effect of a low dose of 8-OH-DPAT was reversed by repeated treatment over 9 days (Carey *et al.*, 2005a); this was caused by a transformation of the 8-OH-DPAT drug cue into a cocaine-conditioned stimulus, which by itself elicits hyperlocomotor activity (Carey *et al.*, 2005b, c). Pre-treatment with the 5-HT$_{1A}$ receptor antagonist WAY 100635 at a low dose blocked the acute hyperlocomotor effects of cocaine (Carey *et al.*, 2004b). The studies using local application also mainly support an inhibitory role of 5-HT$_{1A}$ autoreceptors in cocaine-induced hyperlocomotion, but also showed that there are differences in the contribution of several 5-HT$_{1A}$ autoreceptor populations. Activating 5-HT$_{1A}$ autoreceptors in the median raphe nucleus (MRN) by local application of 8-OH-DPAT or blocking them with WAY 100635 did not affect cocaine-induced hyperlocomotion (Herges and Taylor, 1999; Szumlinski *et al.*, 2004). In contrast, the activation of 5-HT_{1A} autoreceptors in the DRN by 8-OH-DPAT potentiated cocaine-induced hyperlocomotion (Szumlinski *et al.*, 2004), suggesting a facilitatory role for these receptors. The application of WAY100635 into the DRN potentiated cocaine-induced hyperlocomotion (Herges and Taylor, 1999). Local application studies have also revealed multiple dissociations in the contribution of the several postsynaptic 5-HT_{1A} receptor subpopulations to the hyperlocomotor effects of cocaine. Application of 8-OH-DPAT into the ventral hippocampus of rats showed that local 5-HT_{1A} receptors suppress cocaine-induced hyperlocomotion (Müller *et al.*, 2004a), while application of 8-OH-DPAT into the ventral striatal region potentiated cocaine-induced hyperlocomotion (Müller *et al.*, 2004b). Overall, the available studies support the view that 5-HT_{1A} receptors are essentially involved in acute cocaine-induced hyperlocomotion. However, it appears that 5-HT_{1A} autoreceptors and postsynaptic 5-HT_{1A} receptors contribute differently (for review, see Müller and Huston, 2006; Müller *et al.*, 2007a).

A sensitization of the acute hyperlocomotor effects of cocaine can be observed after only a few days of repeated treatment (see, for example, Shuster *et al.*, 1977; Kalivas *et al.*, 1988; Yeh and Haertzen, 1991; Camp *et al.*, 1994). The induction of sensitization for the hyperlocomotor effects of cocaine was potentiated by 8-OH-DPT in rats (De La Garza and Cunningham, 2000; Carey *et al.*, 2001), while it was attenuated by antagonism of the 5-HT_{1A} receptor with WAY 100635 (Carey *et al.*, 2001). Furthermore, the expression of cocaine-induced locomotor sensitization was attenuated by the 5-HT_{1A} receptor antagonist NAN-190 (King *et al.*, 1993). In mice, however, the 5-HT_{1A} receptor agonist osemozotan blocked the establishment and expression, but not the maintenance, of cocaine-induced sensitization (Ago *et al.*, 2006b). The majority of the findings suggest a facilitatory role of 5-HT_{1A} receptors in the induction and expression of cocaine-induced sensitization, but no role in its maintenance.

Preferentially stimulating 5-HT_{1A} autoreceptors with the systemic application of low doses of 8-OH-DPAT (0.05 mg/kg, i.p.) initially blocked the acute hyperlocomotor effects of cocaine, but facilitated the induction of locomotor sensitization during repeated cocaine treatment (Carey *et al.*, 2005a). This effect, however, appeared not to be a true sensitization effect, but rather an 8-OH-DPAT drug-cue conditioned stimulus effect (Carey *et al.*, 2005b). A local application study targeted 5-HT_{1A} autoreceptors in the DRN and MRN with 8-OH-DPAT. 5-HT_{1A} receptors were found to have a facilitatory role in the MRN, but not in the DRN, in the induction of cocaine-induced locomotor sensitization (Szumlinski *et al.*, 2004). This finding supports the view that there are not only dissociable contributions of 5-HT_{1A} autoreceptors and postsynaptic receptors, but also dissociations within the population of the 5-HT_{1A} autoreceptors, with a dominant role for the MRN (Müller and Huston, 2006).

Cocaine can be discriminated from saline by several species. However, stimulation of 5-HT_{1A} receptors with 8-OH-DPAT, or the partial agonists buspirone and gepirone, does not substitute for cocaine in rats (Callahan and Cunningham, 1997). The discriminative stimulus properties of cocaine were also not modulated by medium doses of the partial 5-HT_{1A} receptor agonist buspirone, or the full agonist 8-OH-DPAT, in rats (Rapoza, 1993; Callahan and Cunningham, 1995, 1997; Przegalinski and Filip, 1997; Kleven and Koek, 1998). Buspirone and 8-OH-DPAT nevertheless reduced the discriminative stimulus properties of cocaine when administered at higher doses. At these doses, both agonists also suppressed general responding (Rapoza, 1993; Callahan and Cunningham, 1995, 1997), which suggests rather unspecific effects. The authors explained this finding, in part, by the prominent actions of buspirone, but not gepirone, at the D2 dopamine (DA) receptor (Van Wijngaarden *et al.*, 1990), which was shown to block the discriminative stimulus properties of the D2 DA receptor agonist apomorphine in rats and monkeys (Kamien and Woolverton, 1990; Rijnders and Slangen, 1993). A study in pigeons reported a partial substitution for cocaine by 8-OH-DPAT in two of four animals tested, and a partial blockade of the discriminative stimulus effects of cocaine by the 5-HT_{1A} receptor antagonist NAN-190 (Johanson and Barrett, 1993). Since only four animals were tested in this study and statistical results are somewhat unclear, conclusions must be limited. The 5-HT_{1A} receptor antagonist NAN-190 neither substituted for cocaine nor altered its discriminative stimulus properties in rats (Callahan and Cunningham, 1995). The local application of 8-OH-DPAT into the VTA or SN neither substituted for cocaine nor modulated its discriminative stimulus properties (De La Garza *et al.*, 1998). Altogether, most studies show that 5-HT_{1A} receptor agonists do not substitute for cocaine, and 5-HT_{1A} receptor antagonists do not modulate the discriminative stimulus properties. From these findings, it may be concluded that 5-HT_{1A} receptors do not play a major role in the discriminative stimulus properties of cocaine.

Cocaine induces a profound CPP in rodents (Spyraki *et al.*, 1987; Durazzo *et al.*, 1994; Bardo and Bevins, 2000; Tzschentke, 2007). The 5-HT_{1A} receptor agonist 8-OH-DPAT facilitates the expression of cocaine-induced CPP in mice (Ali and Kelly, 1996). However, the partial 5-HT_{1A} receptor agonist buspirone did not modify the establishment, as well as the expression, of cocaine-induced CPP (Ali and Kelly, 1997; Ettenberg and Bernardi, 2007). Nevertheless, evidence suggests that buspirone can attenuate the delayed aversive effects of cocaine resulting in CPA (Ettenberg and Bernardi, 2007). Although not much is known about the role of 5-HT_{1A} receptors in the establishment and expression of cocaine-induced CPP, the available data suggest no contribution to the establishment, but a possible facilitatory role in CPP expression and in subsequent aversive effects.

Several studies have shown drug-seeking and self-administration of cocaine in rodents (e.g., Arroyo *et al.*, 1998), dogs (Risner and Jones, 1980) and monkeys (e.g., Johanson *et al.*, 1976; Bedford *et al.*, 1980; Ritz and Kuhar, 1989) under different schedules of reinforcement. An inverted U-shaped dose–response function was established for infusion of cocaine under fixed-ratio schedules (Spealman, 1993; Arroyo *et al.*, 1998). Under a progressive-ratio schedule, the self-administered amount of cocaine increases with the dose of cocaine obtained in each infusion (Depoortere *et al.*, 1993). Peltier and Schenk (1993) reported a decrease in cocaine self-administration after pre-treatment with the 5-HT_{1A} receptor agonist 8-OH-DPAT. Homberg *et al.* (2004) investigated the effects of 8-OH-DPAT and buspirone on cocaine self-administration in a progressive-ratio and a fixed-ratio schedule of reinforcement. In the progressive-ratio schedule, 8-OH-DPAT increased absolute responding for cocaine, while it significantly reduced responding in a fixed-ratio schedule. Buspirone, in contrast, partially reduced responding for cocaine in a progressive-ratio schedule, but enhanced it in a fixed-ratio schedule. The authors interpreted their findings as a preferential decrease in cocaine self-administration by 5-HT_{1A} receptor activation (Homberg *et al.*, 2004). This interpretation was supported by findings in mice showing an inhibitory effect of 8-OH-DPAT (0.3 mg/kg) in a progressive-ratio schedule of cocaine self-administration (Parsons *et al.*, 1998). One study investigated the effects of the 5-HT_{1A} receptor agonist ipsapirone on oral cocaine self-administration in rats. In that study, cocaine and ipsapirone were mixed in the drinking fluid with unrestricted access. The study, however, did not find an effect of ipsapirone on voluntary oral cocaine consumption (Mosner *et al.*, 1997). Several studies investigated the effects of 5-HT_{1A} receptor agonists on cocaine self-administration in non-human primates. Gold and Balster (1992) compared the effects of the two partial 5-HT_{1A} receptor agonists buspirone and gepirone on the self-administration of cocaine in a fixed-ratio schedule in rhesus monkeys. At medium doses, buspirone increased self-administration of cocaine; at the highest dose of buspirone, a significant decrease was observed. Gepirone did not affect self-administration of cocaine at any dose tested. The highest doses of both drugs also disrupted the consumption of food. When buspirone and gepirone administration was repeated over 10 days, no effect on the self-administration of cocaine was evident (Gold and Balster, 1992). Since buspirone has a much higher affinity to D2 dopamine (DA) receptors than gepirone (Van Wijngaarden *et al.*, 1990), the effects of buspirone on the self-administration of cocaine may have been mediated by a dopaminergic mechanism (Gold and Balster, 1992). A study in cynomolgus monkeys, where animals could respond for different doses of cocaine or food in a fixed-ratio schedule of reinforcement, found that 8-OH-DPAT increased the choice of cocaine over food, but only when a low dose of cocaine was tested, which was originally not preferred over food. In this condition, 8-OH-DPAT seemed to increase the reinforcing strength of cocaine. At cocaine doses which were originally preferred to the food, 8-OH-DPAT pre-treatment did not modify responding (Czoty *et al.*, 2005). Also, 5-HT_{1A} receptor KO mice have been tested regarding cocaine-induced behavior. These mice acquired cocaine self-administration in the same rate as their wild-type littermates, and showed the same regular within-session responding (Rocha *et al.*, 1998). In summary, the majority of studies in rodents and monkeys suggest that activation of 5-HT_{1A} receptors attenuates the self-administration of cocaine. Importantly, the inhibitory action was often effective only at high doses of the agonists. The dose–response curves suggest that the stimulation of postsynaptic 5-HT_{1A} receptors is required for the inhibitory effect.

It was found that cocaine, as well as cocaine-associated cues, can reinstate self-administration in rats (De Wit and Stewart, 1981; Shaham *et al.*, 2003). Schenk (2000) investigated the effects of the 5-HT_{1A} receptor antagonist WAY 100635 on cocaine-induced reinstatement of cocaine-seeking behavior in rats. Pre-treatment with WAY 100635 blocked the reinstatement of cocaine-seeking by an injection of cocaine. The same dose of WAY 100635 did not affect oral self-administration of the artificial sweetener saccharin, thus indicating that WAY 100635 did not induce a generalized anhedonia (Schenk, 2000). In contrast to the attenuating effects on cocaine-induced reinstatement of cocaine-seeking behavior in rats, WAY 100635 (0.1–1.0 mg/kg, s.c.) did not affect the reinstatement elicited by a cocaine-associated cue (Cervo *et al.*, 2003). Both findings were replicated in a study by Burmeister *et al.* (2004). The authors reported a failure of WAY 100635 to block cue-induced reinstatement of cocaine-seeking, but an attenuating effect of the same doses on the cocaine-induced reinstatement of cocaine-seeking behavior (Burmeister *et al.*, 2004). In summary, the available data suggest a facilitatory role for 5-HT_{1A} receptors in the reinstatement of cocaine-seeking behavior induced by cocaine, but not when induced by sensory cues.

5-HT_{1B} receptor

The 5-HT_{1B} receptor was shown to facilitate cocaine-induced hyperlocomotion in rats and mice (Sershen *et al.*, 1996; Castanon *et al.*, 2000; Przegalinski *et al.*, 2001a; Hoplight *et al.*, 2005). However, using 5-HT_{1B} receptor knockout mice, suppressory effects were found for cocaine-induced hyperlocomotion (Rocha *et al.*, 1998). At the local level, fewer regional dissociations and deviations from systemic effects were observed for the 5-HT_{1B} receptor compared with the 5-HT_{1A} receptor. Pharmacological evidence suggests a facilitatory role of 5-HT_{1B} receptors in the Nac/shell in cocaine-induced hyperlocomotion (Przegalinski *et al.*, 2002), which mimics systemic effects. In contrast, no effect was found when Nac/core 5-HT_{1B} receptors were activated (Przegalinski *et al.*, 2002). Przegalinski *et al.* (2004) observed a potentiation of cocaine-induced hyperlocomotion by intra-VTA pre-treatment with the 5-HT_{1B} receptor agonist CP 93129. In rats with a viral vector-mediated local over-expression of 5-HT_{1B} receptors in the VTA, cocaine-induced hyperlocomotion was also increased, which might suggest a facilitatory role for VTA 5-HT_{1B} receptors (Neumaier *et al.*, 2002).

The 5-HT_{1B} receptor contributed to neither the establishment nor the expression of locomotor sensitization in rats (Przegalinski *et al.*, 2001a). In 5-HT_{1B} receptor knockout mice, however, suppressory effects on the establishment of sensitization were reported (Rocha *et al.*, 1998). 5-HT_{1B} receptors in the Nac/shell facilitate the establishment as well as the expression of locomotor sensitization, while in the VTA 5-HT_{1B} receptors only contribute to the establishment but not to the expression. Nac/core 5-HT_{1B} receptors are not involved in locomotor sensitization (Przegalinski *et al.*, 2002, 2004).

Systemic 5-HT_{1B} receptor stimulation with the receptor agonist RU 24969 partially substitutes for cocaine (Callahan and Cunningham, 1995), although with a temporal restriction (Parsons *et al.*, 1998). 5-HT_{1B} receptor stimulation with RU 24969 or CP 94,253 was also found to contribute to the discriminative stimulus properties of cocaine at the systemic level (Callahan and Cunningham 1997; Filip *et al.*, 2001a). In contrast to the systemic effects, neither VTA nor Nac/core or Nac/shell 5-HT_{1B} receptor stimulation with CP 93129 substituted for cocaine. This suggests that the partial substitution observed at the systemic level is mediated by other 5-HT_{1B} receptor populations. VTA and Nac/core, but not Nac/shell, 5-HT_{1B} receptors contribute to the discriminative stimulus properties of cocaine (Filip *et al.*, 2002, 2003).

Pharmacological and transgenic approaches support a facilitatory role for 5-HT_{1B} receptors in the establishment of cocaine-induced CPP in rats as well as in mice (Belzung *et al.*, 2000; Cervo *et al.*, 2002). However, a potentiation of cocaine-induced CPP was only found with the selective 5-HT_{1B} receptor agonist CP 94,253 with a low dose of cocaine (Cervo *et al.*, 2002). In rats with a viral vector-mediated local over-expression of 5-HT_{1B} receptors in the VTA the dose–response curve for cocaine was shifted to the left in the establishment of a CPP, suggesting a facilitatory role for VTA 5-HT_{1B} receptors (Neumaier *et al.*, 2002). Over-expression of 5-HT_{1B} receptors in the Nac/shell also enhanced cocaine-induced CPP, but also the delayed aversive effects of cocaine (Barot *et al.*, 2007).

The self-administration of cocaine in rats was facilitated by pre-treatment with the 5-HT_{1B} receptor agonist RU 24969 (Parsons *et al.*, 1998) but inhibited by the agonist CP94253 (Przegalinski *et al.*, 2007). In this study, the 5-HT_{1B} receptor antagonist SB 216641 did not affect cocaine self-administration. Pre-treatment with the selective 5-HT_{1B} receptor antagonist GR 127,935 blocked self-administration in mice when cocaine was administered into the VTA (David *et al.*, 2004). In contrast, in 5-HT_{1B} receptor KO mice a potentiated cocaine self-administration was observed compared to wild-type mice (Rocha *et al.*, 1998). Cue- as well as cocaine-induced reinstatement of cocaine self-administration was blocked by 5-HT_{1B} receptor activation (Acosta *et al.*, 2005).

5-HT_{2A} receptor

Several studies, using selective 5-HT_{2A} receptor agonists and antagonists, suggest a facilitatory role of the 5-HT_{2A} receptor in cocaine-induced hyperlocomotion at a systemic level (O'Neill *et al.*, 1999; Fletcher *et al.*, 2002a). A study on local 5-HT_{2A} receptor effects applied the selective antagonist M 100,907 into the VTA or Nac/shell, and provided evidence for a facilitatory contribution of VTA but not Nac/shell 5-HT_{2A} receptors to cocaine-induced hyperlocomotion (McMahon *et al.*, 2001). Thus the VTA but not Nac/shell 5-HT_{2A} receptor effects concur with the systemic contribution of this receptor to cocaine-induced hyperlocomotion.

5-HT_{2A} receptors also facilitate the expression of locomotor sensitization, while they are not involved in its establishment (Davidson *et al.*, 2002; Filip *et al.*, 2004). Neither 5-HT_{2A} receptor agonism nor antagonism substitutes for cocaine. However, several studies have demonstrated a contribution to the discriminative stimulus properties of cocaine in rats and monkeys (Schama *et al.*, 1997; Munzar *et al.*, 2002; Filip *et al.*, 2006). There is no evidence for a role of 5-HT_{2A} receptors in cocaine self-administration in rats (Howell and Byrd, 1995; Fletcher *et al.*, 2002a). However, high doses of the 5-HT_{2A} receptor antagonist MDL100907 might produce a downward shift in the dose–effect curve of cocaine (Fantegrossi *et al.*, 2002). Interestingly, several 5-HT_{2A} receptor antagonists can attenuate the reinstatement of cocaine self-administration by cues or cocaine, which might suggest a facilitatory role of the 5-HT_{2A} receptor in this behavior (Fletcher *et al.*, 2002a; Burmeister *et al.*, 2004; Filip *et al.*, 2005). At the systemic level, 5-HT_{2A} receptors clearly regulate many cocaine-addiction related behaviors.

5-HT_{2B} receptor

There is little evidence to suggest a 5-HT_{2B} receptor contribution to addiction-related behavior in animal studies. Neither 5-HT_{2B} receptor agonism nor antagonism affected cocaine-induced hyperlocomotion or its sensitization at the systemic level (Fletcher *et al.*, 2002b; Filip *et al.*, 2004). Also, 5-HT_{2B} receptor antagonism neither substituted for cocaine nor affected its discriminative stimulus properties (Filip *et al.*, 2006). So far, there is no evidence for a contribution of the 5-HT_{2B} receptor to any of the cocaine effects on behavior.

5-HT_{2C} receptor

The 5-HT_{2C} receptor is the only presently known 5-HT receptor subtype which predominantly attenuates several core addiction-related behaviors of cocaine at the systemic level. The pharmacological approach in rats as well as the 5-HT_{2C} receptor KO approach in mice provide convincing evidence for a sustained inhibitory role of 5-HT_{2C} receptors at the systemic level in cocaine-induced hyperlocomotion (Fletcher *et al.*, 2006; Neisewander and Acosta, 2007), cocaine self-administration, and the cue-, context-, stress- and cocaine-induced reinstatement of cocaine self-administration (Grottick *et al.*, 2000; Fletcher *et al.*, 2002a; Rocha *et al.*, 2002; Burmeister *et al.*, 2004; Filip *et al.*, 2004; Neisewander and Acosta, 2007; Burbassi and Cervo, 2008; Fletcher *et al.*, 2008). In contrast, no evidence for a role in the establishment or expression of locomotor sensitization has been reported (Filip *et al.*, 2004). However, 5-HT_{2C} receptors play an inhibitory role in the expression of cocaine-induced conditioned hyperactivity, which was inhibited by pre-treatment with a 5-HT_{2C} receptor agonist and enhanced by an antagonist (Liu and Cunningham, 2006).

Neither systemic 5-HT_{2C} receptor agonists nor antagonists substituted for cocaine (Callahan and Cunningham, 1997; Frankel and Cunningham, 2004; Filip *et al.*, 2006). Nevertheless, there is evidence for an attenuation of cocaine's discriminative stimulus properties by 5-HT_{2C} receptor activation (Callahan and Cunningham, 1997; Frankel and Cunningham, 2004), and evidence for an enhancement by 5-HT_{2C} receptor antagonism (Filip *et al.*, 2006).

Moreover, for the 5-HT_{2C} receptor considerable local dissociations in the contribution to cocaine-related behaviors have been discovered. Thus, Nac/shell 5-HT_{2C} receptors differ in their effects largely from the systemic and other local 5-HT_{2C} receptor effects (McMahon *et al.*, 2001). Nac/shell 5-HT_{2C} receptor stimulation with the agonists MK 212 and Ro 60-0175 facilitated cocaine-induced hyperlocomotion (Filip and Cunningham, 2003). In contrast, 5-HT_{2C} receptor stimulation in the PFC or the VTA blocked cocaine-induced hyperlocomotion, and thus worked in the direction of the systemic effects (Filip and Cunningham, 2003; Fletcher *et al.*, 2004b). In line with the systemic effects, VTA 5-HT_{2C} receptors attenuated cocaine self-administration behavior (Fletcher *et al.*, 2004b). Neither Nac/shell nor PFC 5-HT_{2C} receptor stimulation substituted for cocaine. However, Nac/shell 5-HT_{2C} receptors facilitate the discriminative stimulus properties of cocaine, which is also in opposition to the systemic effects. The PFC 5-HT_{2C} receptors seems to work in the other direction, and attenuate them (Filip and Cunningham, 2002, 2003).

5-HT_3 receptor

Several pharmacological studies in rats and mice, using a wide range of 5-HT_3 receptor antagonists, provide evidence for a facilitatory role of the 5-HT_3 receptor in cocaine-induced hyperlocomotion (Reith, 1990; Kankaanpää *et al.*, 2002). Such a role has been confirmed by a transgenic approach, which used 5-HT_3 receptor over-expressing mice (Allan *et al.*, 2001). However, 5-HT_3 receptor KO mice did not differ in their acute hyperlocomotor effects from wild types after different doses of cocaine (Hodge *et al.*, 2008). At the local level of the Nac, application of the 5-HT_3 receptor antagonist ondansetron attenuated cocaine-induced hyperactivity (Herges and Taylor, 2000), which suggests a permissive role for Nac 5-HT_3 receptors in this cocaine effect. Such a role would be largely congruent with the systemic effects.

In rats, 5-HT_3 receptors were found to contribute to the expression (Davidson *et al.* 2002, 2004, 2007; Zhang *et al.*, 2007) but not to the establishment (Szumlinski *et al.*, 2003) of locomotor sensitization. A study in 5-HT_3 receptor KO mice reported an attenuated establishment of cocaine sensitization, but no differences in conditioned hyperlocomotion, following a saline challenge after cocaine sensitization (Hodge *et al.*, 2008).

5-HT_3 receptor stimulation with the receptor agonist mCPBG partially substitutes for cocaine, but only minimally modulated its discriminative stimulus properties (Koetzner *et al.*, 1995; De La Garza *et al.*, 1996), although various 5-HT_3 receptor antagonists failed to modulate either parameter (Paris and Cunningham, 1991; Lane *et al.*, 1992).

Pharmacological studies with the 5-HT_3 receptor antagonists MDL 72222 and ICS 205-930 suggest a facilitatory role in the establishment of CPP (Suzuki *et al.*, 1992; Kankaanpää *et al.*, 2002), but not in CPP expression (Cervo *et al.*, 1996; Kankaanpää *et al.*, 2002). A study using 5-HT_3 receptor over-expressing mice, however, suggested an inhibitory role (Allan *et al.*, 2001).

The 5-HT_3 receptor seems not to contribute to cocaine self-administration and its reinstatement by cues (Lane *et al.*, 1992; Davidson *et al.*, 2002, 2004). However, when the 5-HT_3 receptor was blocked with ondansetron after an oral self-administration session, it reduced cocaine intake on the following day (Davidson *et al.*, 2004). Likewise, it reduced the resumption of i.v. cocaine self-administration after a withdrawal period in rats (Davidson *et al.*, 2002). Interestingly, VTA 5-HT_3 receptor antagonism with ICS 205-930 attenuated cocaine self-administration into the VTA (Rodd *et al.*, 2005), which is in contrast to the systemic effects.

5-HT_4 receptor

While systemic application of the 5-HT_4 receptor partial agonist RS 67333 did not affect cocaine-induced hyperlocomotion, the receptor antagonist SDZ 205,557 attenuated this cocaine effect (McMahon and Cunningham, 1999). These findings might suggest a permissive, but not a potentiating, role for 5-HT_4 receptors at a systemic level in the locomotor effects of cocaine. In the same study, the role of Nac/shell 5-HT_4 receptors was investigated. Agonism as well as antagonism of the Nac/shell 5-HT_4 receptors attenuated cocaine-induced hyperlocomotion (McMahon and Cunningham, 1999).

5-HT_6 receptor

A study on the role of the 5-HT_6 receptor did not provide evidence for a contribution at the systemic level to cocaine-induced hyperlocomotion or self-administration (Frantz *et al.*, 2002). In line with this are findings which show that the 5-HT_6 receptor antagonist MS-245 did not substitute for cocaine, or modulate its discriminative stimulus properties (Pullagurla *et al.*, 2004). However, a recent study using viral-mediated 5-HT_6 receptor overexpression in the Nac/shell found an inhibitory effect on the establishment of cocaine-induced CPP, but no effect on cocaine-induced hyperlocomotion or the sensitization of hyperlocomotion (Ferguson *et al.*, 2008). In line with this were pharmacological findings with the 5-HT_6 receptor antagonist Ro4368545, which facilitated cocaine-induced CPP (Ferguson *et al.*, 2008). Clearly, the role of the 5-HT_6 receptor in cocaine addiction requires further research.

Human studies

The effects of 5-HT depletion on acute subjective effects of cocaine were investigated in humans (Aronson *et al.*, 1995). Each of 12 subjects, who were all DSM-III-R diagnosed for cocaine dependence, were given two low-tryptophan diet pre-treatments, a placebo and a sham pre-treatment, each followed by an intranasal test-dose of cocaine. The subjective rating of the 'high' after cocaine was 20 percent lower during tryptophan depletion than during sham depletion. The depletion itself did not have significant effects on mood in this study. These results suggest a role for 5-HT in the acute euphoric effects of cocaine. However, it is not clear how far the effects are related to cocaine craving or self-administration. Satel *et al.* (1995) found, in a study of cocaine-dependent inpatients, an attenuation of cue-induced cocaine craving after drinking an amino acid cocktail that was devoid of tryptophan.

Walsh *et al.* (1994) investigated the effects of the SSRI, fluoxetine, on the subjective effects of a cocaine challenge in volunteers with a history of cocaine abuse. Fluoxetine or placebo was administered for 4 weeks before subjects were challenged with either cocaine (i.v.) or saline. It was found that fluoxetine dose-dependently attenuated the cocaine effects on subjective measures, such as 'liking', 'rush' and 'good effects'. Although fluoxetine reduced the hedonic effects of cocaine, its application as a treatment for cocaine addiction yielded inconsistent results. In a study with cocaine-dependent patients, Grabowski *et al.* (1995) reported a transient reduction of the cocaine consumption in a fluoxetine-treated group, while another study in crack cocaine-dependent outpatients did not find reduced cocaine use or cocaine craving in a fluoxetine-treated group compared to a placebo group (Batki *et al.*, 1996). Covi *et al.* (1995) even found increased cocaine consumption in patients treated with a high dose of fluoxetine in a 12-week trial compared to placebo-treated patients. Moeller *et al.* (2007) investigated the effects of the SSRI citalopram in combination with behavioral therapy and contingency management in 76 cocaine-dependent patients. In this study, citalopram-treated subjects showed a significant increase in abstinence rates as measured by a reduction of cocaine-positive urine samples. Increased 5-HT availability induced by a diet enriched in tryptophan and tyrosine had no effect on cocaine craving in humans. In a double-blind study, Chadwick *et al.* (1990) used a supplementary tryptophan and tyrosine diet in cocaine-dependent subjects in order to investigate the effects of an increased 5-HT level on drug craving. They did not find significant differences in drug-craving and other subjective syndromes compared to a placebo-treated control group.

No compelling evidence for association of serotonin receptors in cocaine abuse and dependence has been presented in the few studies published to date. No association of cocaine abuse with seven SNPs in the 5-HT_{1B} gene (Cigler *et al.*, 2001), the 5-HT_{1B} gene in Han Chinese, the serotonin transporter (5-HTT) (Patkar *et al.*, 2002) in African Americans and the trytophane hydroxylase 2 (TPH2) gene in African Americans (Dahl *et al.*, 2006) could be observed. By today's standards, however, these studies were underpowered, and SNP selection procedures were less rigorous and systematic than current procedures, thus indicating the need for larger, well-designed studies in this area.

The investigation of single receptor subtypes in terms of their contribution to the hedonic effects of cocaine is restricted by the limited availability of selective 5-HT receptor ligands that can be applied in humans in a therapy setting. However, there are some preliminary reports that suggest a reduction of cocaine's subjective effects in humans by the 5-HT_2 antagonist ritanserin and the 5-HT_3 antagonist ondansetron (Walsh and Cunningham, 1997). At present, very little is known about the contribution of other 5-HT receptor subtypes to the subjective effects of cocaine. Only very few 5-HT receptor ligands have been tested as complementary treatments for cocaine addiction. Ehrman *et al.* (1996) investigated the 5-HT_2 antagonist ritanserin in a double-blind study in patients on cue-elicited cocaine craving. Ritanserin did not affect the cue-induced 'high' or craving. These results were confirmed by Cornish *et al.* (2001), who used ritanserin or placebo as an adjunct therapy in cocaine-dependent outpatients for a 4-week period. Also in this study, ritanserin failed to reduce craving in a therapeutic application in cocaine addicts.

The 5-HT system in cocaine addiction

The role of 5-HT in cocaine addiction has been investigated with great effort. While basic 5-HT activity is required for all sorts of behavior (including addiction-related), global manipulation of the 5-HT system has shown that a strong increase in extracellular 5-HT activity quite uniformly inhibits crucial addiction-related behaviors, while 5-HT depletion enhances them. Does this mean that the 5-HT increase observed in virtually all brain areas measured so far is a purely limiting factor for cocaine addiction that only counteracts, for example, the prominent DA effects (Di Chiara and Imperato, 1988)? The answer is probably 'no', with a more sophisticated approach lying in the decomposition of local 5-HT effects. It appears that the 5-HT response has to be in balance with the DA (and possibly noradrenergic) response to fully express cocaine addiction-related behaviors. If this balance is changed towards or away from 5-HT, it can modify the behavioral effects of the drug in either way (see, for example, Rothman *et al.*, 2006; Carey *et al.*, 2008). Most of the investigated 5-HT receptors contribute to cocaine addiction-related behaviors. Interestingly, for most receptors this contribution is facilitating for one addiction-related behavior but inhibitory for others. Exceptions to this 'rule' seem to be the 5-HT_{2A}- and 5-HT_{2C} receptors. While 5-HT_{2A} receptors tend to facilitate addiction-related behaviors, 5-HT_{2C} receptor stimulation constantly inhibits many of them. This would make 5-HT_{2C} agonism a desirable treatment approach for cocaine addiction. However, the prominent anxiogenic effects of 5-HT_{2C} receptor agonists are still limiting this approach in humans (however, for an alternative approach see Ji *et al.*, 2006; Müller and Carey, 2006).

Using the preclinical findings for pharmacological approaches in human cocaine addiction, therapy has not proven successful yet. The available evidence suggests that the relatively crude approach of increasing 5-HT activity with either precursors or SSRI treatment does not work. Also, unspecific 5-HT_2- and 5-HT_3-receptor ligands did not yield significant improvement. Considering the pre-clinical evidence, the most promising target within the 5-HT system is probably the 5-HT_{2C} receptor, given that mechanisms can be found to avoid the aversive effects of the agonists that quite certainly would limit patient compliance.

Amphetamine

The neurochemical effects of amphetamine on 5-HT

The synthetic substance amphetamine occurs as stereo isomers d- and l-amphetamine, or as the racemic mixture. Notably, the d- or (+) isomer is usually more potent at the behavioral and neurochemical level, and is more often used for scientific investigation. Amphetamine increases extracellular 5-HT levels in the Nac and the striatum, parallel to the expression of hyperlocomotion and behavioral stereotypies (Hernandez *et al.*, 1987; Kuczenski and Segal, 1989, 1997; Kuczenski *et al.*, 1995; Kankaanpää *et al.*, 1998; Millan *et al.*, 1999). An amphetamine-induced 5-HT increase was also observed in the frontal cortex (FC) (Millan *et al.*, 1999), in the mPFC (Kuroki *et al.*, 1996; Pum *et al.*, 2007) and in the entorhinal and perirhinal cortices parallel to locomotor activation

(Pum *et al.*, 2007). Increased extracellular 5-HT levels were reported in the DRN during local amphetamine application (Ferre *et al.*, 1994), but not in the striatum (Dawson *et al.*, 2003). Local application of amphetamine by reverse dialysis into the infralimbic and anterior cingular subregions of the PFC also increases extracellular 5-HT levels in the respective regions (Hedou *et al.*, 2000). Also at the neurochemical level, d-amphetamine appears to be more potent than l-amphetamine (Kuczenski *et al.*, 1995; Seiden *et al.*, 1993). As for cocaine, amphetamine's acute effects on 5-HT suggest significant modulation of the behavioral and subjective effects.

Reducing and potentiating 5-HT activity

Studies using global 5,7-DHT lesions found attenuation (Lipska *et al.*, 1992) or no effect (Gately *et al.*, 1985) on amphetamine-induced hyperactivity. Local lesion studies found no effect when 5,7-DHT was injected into amygdala regions (Kusljic and van den Buuse, 2006), the ventral hippocampus (Adams *et al.*, 2008) or the MRN and/or DRN (Lehmann *et al.*, 2000; Kusljic *et al.*, 2003), but significant attenuation when injected in the fimbria–fornix/cingular bundle (Lehmann *et al.*, 2000). Pre-treatment with the SSRIs fluoxetine and citalopram had either no effect (Maj *et al.*, 1996) or, with fluoxetine and sertraline, enhanced the acute hyperlocomotor effects of amphetamine in rats (Sills *et al.*, 1999a, b, 2000). However, SSRI effects must be interpreted with some caution, since they were also observed after combined 5,7-DHT lesions of the MRN and DRN. Both SSRIs also enhanced CNS concentrations of amphetamine, possibly by decreasing the amphetamine metabolism. These findings suggest a non-serotonergic mechanism in the potentiation of the amphetamine effects (Sills *et al.*, 1999a, b, 2000).

Reducing brain 5-HT availability by 5,7-DHT injection into the medial forebrain bundle or i.c.v. did not affect the establishment of amphetamine self-administration behavior, but resulted in higher drug-taking per session (Lyness *et al.*, 1980; Leccese and Lyness, 1984). Other studies, however, did not find any changes in d-amphetamine self-administration after a profound reduction of brain serotonin following DRN and MRN lesions with 5,7-DHT (Fletcher *et al.*, 1999), or after local lesion in the Nac (Lyness *et al.*, 1980). Enhancing the 5-HT availability by pre-treatment with the precursor L-tryptophan or the SSRI fluoxetine reduced responding for amphetamine and amphetamine self-administration in rats (Leccese and Lyness, 1984; Smith *et al.*, 1986; Yu *et al.*, 1986; Porrino *et al.*, 1989). Also, the direct injection of 5-HT into the Nac reduced amphetamine self-administration in rats (Fletcher *et al.*, 2002b), which might at least suggest a restrictive function of 5-HT when activated.

The SSRIs fluoxetine and paroxetine did not affect the increase in brain-stimulation reward threshold after withdrawal from repeated amphetamine administration in rats. Nevertheless, a high dose of fluoxetine could reduce the duration of the withdrawal-induced elevation of ICSS thresholds (Harrison *et al.*, 2001; Markou *et al.*, 2005).

5-HT receptors

5-HT$_{1A}$ receptor

Amphetamine can increase locomotor activity at low to medium doses, but causes behavioral stereotypies at higher doses (Segal, 1975; Geyer *et al.*, 1987; Paulus *et al.*, 1993; Kuczenski and Segal, 1999). Partial and full 5-HT$_{1A}$ receptor agonists blocked the hyperlocomotion induced by d-amphetamine in mice and rats (Maj *et al.*, 1987; Przegalinski and Filip, 1997; Millan *et al.*, 1998). While the non-selective 5-HT$_{1A}$ receptor agonist buspirone blocked the hyperlocomotor effects of amphetamine in one study, the application of the selective agonist 8-OH-DPAT significantly potentiated the hyperlocomotor effects of amphetamine (Jackson *et al.*, 1994). The selective 5-HT$_{1A}$ receptor antagonist WAY 100135 failed to influence hyperlocomotion induced by amphetamine (Przegalinski and Filip, 1997). Layer *et al.* (1992), in contrast, reported a marked inhibition of d-amphetamine stimulated locomotor activity by the 5-HT$_{1A}$ receptor antagonist NAN-190 in rats. A local application study found a facilitatory role of the 5-HT$_{1A}$ receptor in the Nac in d-amphetamine-induced hyperlocomotion (Layer *et al.*, 1992). Overall, the available studies support the view that 5-HT$_{1A}$ receptors are essentially involved in acute amphetamine-induced hyperlocomotion, while the direction of effect might depend on 5-HT$_{1A}$ receptor subtype activation.

Przegalinski *et al.* (2000) found that the 5-HT$_{1A}$ receptor agonist 8-OH-DPAT blocked the induction as well as the expression of amphetamine-induced sensitization. While these effects were reversed by WAY 100135, the antagonist alone did not have an influence.

Amphetamine can be discriminated from saline in operant tasks (Snoddy and Tessel, 1983). The discriminative stimulus properties of amphetamine were not modulated by 8-OH-DPAT in rats (Przegalinski and Filip, 1997). 8-OH-DPAT neither generalized to d-amphetamine when given alone, nor antagonized its stimulus effects when given in combination (Young *et al.*, 2006a). However, pre-treatment with 8-OH-DPAT appeared to potentiate the discriminative stimulus properties of a sub-threshold dose of d-amphetamine. The authors concluded that d-amphetamine is a more effective discriminative stimulus in the presence of 8-OH-DPAT (Young *et al.*, 2006a). A study in rhesus monkeys found no substitution for the effects of intra-gastrically applied amphetamine by the 5-HT_{1A} receptor agonists buspirone, gepirone or 8-OH-DPAT. Nevertheless, this study reported inhibition of the discriminative stimulus properties of amphetamine (Nader and Woolverton, 1994). The potency to block the amphetamine discriminative stimulus effects was 8-OH-DPAT > buspirone > gepirone. Considering the high effectiveness of 8-OH-DPAT in this study, the ED_{50} suggests that the observed effects may be primarily mediated by 5-HT_{1A} autoreceptor activation rather than by affecting all 5-HT_{1A} receptor populations of the brain. In contrast, the lack of an amphetamine substitution was evident at much higher doses of 8-OH-DPAT, which are known to affect 5-HT_{1A} autoreceptors and postsynaptic 5-HT_{1A} receptors (Nader and Woolverton, 1994). Also, the studies that failed to find an inhibitory effect for 8-OH-DPAT used concentrations in this range. Thus, it may be speculated that some of the seemingly contradictory findings might be due to dose differences, which result in the preferential activation of 5-HT_{1A} autoreceptors at low doses, or in the activation of 5-HT_{1A} autoreceptors and postsynaptic receptors at higher doses. Accordingly, the present data suggest that 5-HT_{1A} autoreceptors in the raphe nuclei might inhibit the discriminative stimulus properties of amphetamine. This effect, however, may be cancelled out by the additional activation of (some) postsynaptic 5-HT_{1A} receptor populations. Most studies show that 5-HT_{1A} receptor agonists do not substitute for amphetamine, and 5-HT_{1A} receptor antagonists do not modulate the discriminative stimulus properties.

Amphetamine induces a profound CPP in animals (Bardo and Bevins, 2000; Tzschentke, 2007). A CPP study in mice showed that the 5-HT_{1A} receptor antagonist WAY 100635 (2 mg/kg, i.p.) did not affect the establishment of amphetamine-induced CPP (Budygin *et al.*, 2004).

Several studies have shown drug-seeking and self-administration of amphetamine in rodents (Yokel and Pickens, 1973) and monkeys (Johanson *et al.*, 1976; Ritz and Kuhar, 1989) under different schedules of reinforcement. The role of 5-HT_{1A} receptors in the self-administration of amphetamine and its derivates, however, has received little attention. A study by Fletcher *et al.* (2002b) showed that intra-Nac application of 8-OH-DPAT did not affect the self-administration of d-amphetamine in rats in a progressive-ratio schedule of reinforcement.

During withdrawal, thresholds to maintain intracranial self-stimulation were increased in rats. This indicates a higher reward threshold in the brain consistent with a decreased hedonic tone (Markou *et al.*, 2005). The 5-HT_{1A} receptor antagonist p-MPPI did not affect the increase in the brain-stimulation reward threshold after withdrawal from repeated amphetamine administration in rats. However, co-administration of the SSRIs paroxetine or fluoxetine with p-MPPI reduced the duration of the reward deficits significantly (Harrison *et al.*, 2001; Markou *et al.*, 2005). Although very little is known about the role of 5-HT_{1A} receptors in psychostimulant withdrawal effects, the present data suggest a possible beneficial effect when administered in combination with a treatment that increases 5-HT activity.

5-HT_{1B} receptor

5-HT_{1B} receptor stimulation was found to enhance d-amphetamine-induced locomotor stimulation (Fletcher and Korth, 1999; Przegalinski *et al.*, 2001b; but see also Layer *et al.*, 1992), while 5-HT_{1B} receptor antagonism inhibited it (Przegalinski *et al.*, 2001b) or had no significant effect (Chaouloff *et al.*, 1999).

The establishment, but not the expression, of amphetamine-induced sensitization was blocked by 5-HT_{1B} receptor antagonism and facilitated by the agonism (Przegalinski *et al.*, 2001b). Although not necessary for the hyperlocomotor effects of amphetamine, 5-HT_{1B} receptors in the VTA were able to enhance the amphetamine effects when pharmacologically stimulated (Papla *et al.*, 2002). Surprisingly, 5-HT_{1B} receptor KO mice showed enhanced acute effects of amphetamine and a more pronounced establishment of the sensitization than wild types (Bronsert *et al.*, 2001). Altogether, the majority of pharmacological studies suggest a facilitatory role for 5-HT_{1B} receptor stimulation in the acute hyperlocomotor effects of amphetamine as well as in the establishment of their sensitization.

A partial generalization for the amphetamine stimulus by two 5-HT_{1B} receptor agonists was found in a conditioned taste aversion paradigm in rats. Agonist pre-treatment also enhanced the discriminative stimulus properties of amphetamine. 5-HT_{1B} receptor antagonism neither generalized to amphetamine, nor affected its discriminative stimulus properties (Miranda *et al.*, 2007). A local stimulation study did not find evidence for a role of 5-HT_{1B} receptors in the Nac/shell in the discriminative stimulus effects of amphetamine (Filip *et al.*, 2001b), supporting the view that 5-HT_{1B} receptors do not play a major role in this amphetamine effect.

The 5-$HT_{1B/1A}$ receptor agonist RU-24969 reduced d-amphetamine self-administration. This effect was reversed by a 5-$HT_{1B/1D}$ antagonist, but not by a selective 5-HT_{1A} receptor antagonist (Fletcher and Korth, 1999). Fletcher *et al.* (2002b) identified the Nac as one site of 5-HT_{1B} receptor action. Local application of a 5-HT_{1B}

receptor agonist into the Nac inhibited amphetamine self-administration in this study.

5-HT$_{2A}$ receptor

A number of studies showed that the 5-HT$_{2A}$ receptor antagonist MDL 100907 inhibited the locomotor stimulant effects of amphetamine in rats and mice (Sorensen *et al.*, 1993; Moser *et al.*, 1996; O'Neill *et al.*, 1999; but see also Carlsson *et al.*, 1999), but did not affect its discriminative stimulus properties in rats (Moser *et al.*, 1996). The acute hyperlocomotor effect of amphetamine, but not its sensitization, was potentiated in 5-HT$_{2A}$ receptor KO mice. This paradoxical effect, however, was accompanied by disruption of the 5-HT$_{2A}$ receptor regulation of noradrenergic activity (Salomon *et al.*, 2007). Altogether, there is accumulating evidence for a permissive role of 5-HT$_{2A}$ receptors in the locomotor stimulant effects of amphetamine in normal animals. There is no evidence for a role in the sensitization or the discriminative stimulus properties of amphetamine.

5-HT$_{2C}$ receptor

The 5-HT$_{2C}$ receptor antagonist SB242084 potentiated amphetamine-induced hyperlocomotion in rats (Fletcher *et al.*, 2006), while the selective 5-HT$_{2C}$ receptor agonist WAY-163909 inhibited it (Marquis *et al.*, 2007). These findings suggest an inhibitory role for the 5-HT$_{2C}$ receptor in amphetamine-induced hyperlocomotion. No evidence for an effect of 5-HT$_{2C}$ receptor antagonism on the establishment of amphetamine-induced hyperlocomotor sensitization was found (Lanteri *et al.*, 2008).

5-HT$_3$ receptor

Pharmacological stimulation of the 5-HT$_3$ receptor was found to enhance the hyperlocomotor effects of d-amphetamine (Layer *et al.*, 1992), while systemic inhibition reduced it in rats (Shankar *et al.*, 2000). Systemic as well as intra-Nac administration of the 5-HT$_3$ receptor antagonist GR38032F blocked the hyperlocomotor effects of intra-Nac d-amphetamine in rats (Costall *et al.*, 1987). Local application studies suggest a mainly facilitatory role of 5-HT$_3$ receptors in the VTA, but not in the Nac, in d-amphetamine-induced hyperlocomotion (Gillies *et al.*, 1996). Overall, evidence points to a facilitatory role of 5-HT$_3$ receptors in amphetamine-induced hyperlocomotion.

A series of 5-HT$_3$ receptor antagonists was tested to modulate the discriminative stimulus effects of amphetamine in rats. However, none of them blocked the amphetamine effects in this test (Moser, 1992). Another study, though, found a potentiating effect of the 5-HT$_3$ receptor antagonist ICS 205-930 on the discriminative stimulus properties of low, but not high, doses of d-amphetamine in rats (West *et al.*, 1995). The partial 5-HT$_3$ receptor agonist MD-354 did not substitute for d-amphetamine or modify its discriminative stimulus effects. However, it enhanced the discriminative stimulus effects when low doses of d-amphetamine were used (Dukat *et al.*, 2007). There was no effect of 5-HT$_3$ receptor antagonism on amphetamine-induced CPP (Acquas *et al.*, 1988; Carboni *et al.*, 1988).

5-HT$_4$ and 5-HT$_6$ receptors

The 5-HT$_4$ receptor antagonist SB-204070-A did not affect amphetamine-induced hyperlocomotion in rats, suggesting no contribution of the 5-HT$_4$ receptor to this amphetamine-induced behavior (Reavill *et al.*, 1998). The 5-HT$_6$ receptor antagonist MS-245 did not substitute for d-amphetamine, but enhanced the discriminative stimulus properties of a low dose of d-amphetamine (Pullagurla *et al.*, 2004).

Human studies

The efficacy of the 5-HT$_3$ receptor antagonist ondansetron to reduce the subjective effects of amphetamine was investigated in nine healthy human volunteers in a double-blind, placebo-controlled, balanced-crossover study. In this study, amphetamine did not change the self-rating of mood but increased self-rating for overall subjective state. This effect was attenuated by ondansetron (Silverstone *et al.*, 1992). Another placebo-controlled study in 10 healthy human volunteers confirmed an inhibitory effect of ondansetron (i.v.) on amphetamine-induced activation-euphoria ratings, although this seemed to depend on the subjects' susceptibility to the amphetamine effects in the first place (Grady *et al.*, 1996). Altogether, there is evidence suggesting a facilitatory contribution of 5-HT$_3$ receptors to the subjective effects of amphetamine in humans.

The 5-HT system in amphetamine addiction

Compared to cocaine, relatively little is known about the role of 5-HT in amphetamine addiction. However, at the neurochemical level, amphetamine also induces a profound activation of the 5-HT system, thus suggesting mechanisms similar to those of cocaine. Global 5-HT manipulations have similar effects on amphetamine- and cocaine-related behaviors: 5-HT enrichment inhibits some important addiction-related behaviors. However, this strategy has not been tested in a clinical setting. The knowledge about single receptor contributions is much more limited for amphetamine. We still lack information on crucial paradigms such as CPP and self-administration, and their reinstatement after withdrawal, as well as on local effects of these receptors. Although inhibitory effects of a 5-HT_3 receptor antagonist on the acute subjective effects of amphetamine were repeatedly found in humans, there is as yet not sufficient direct evidence that suggests the use of any 5-HT receptor ligand for the treatment of amphetamine addiction in humans.

Methamphetamine

The neurochemical effects of methamphetamine on 5-HT

The synthetic amphetamine derivative methamphetamine occurs as stereo-isomers and as a racemic mixture. An increase of extracellular 5-HT activity in the Nac and dorsal striatum was reported after methamphetamine application (Kuczenski *et al.*, 1995; Segal and Kuczenski, 1997), with the d-isomer being neurochemically more active than the l-isomer (Kuczenski *et al.*, 1995). d-methamphetamine increased 5-HT levels in the ventral hippocampus of freely moving rats (Rocher and Gardier, 2001). In mice, a 5-HT increase was reported in the PFC after methamphetamine administration (Ago *et al.*, 2006b).

Reducing and potentiating 5-HT activity

Methamphetamine increases locomotor activity with an inverted U-shaped dose–response curve (Witkin *et al.*, 1999). The SSRI fluoxetine did not affect acute methamphetamine-induced hyperlocomotion in mice, but significantly reduced the establishment of methamphetamine sensitization (Takamatsu *et al.*, 2006). Increasing 5-HT availability in the brain by pre-treatment with the monoamine oxidase (MAO)-B inhibitor clorgyline attenuated methamphetamine-induced hyperlocomotion in mice (Kitanaka *et al.*, 2005a). Repeated treatment with the MAO-B inhibitor also blocked the expression of methamphetamine sensitization in mice (Kitanaka *et al.*, 2005b).

The SSRI fluoxetine did not substitute for methamphetamine in rats in a fixed-ratio schedule of food presentation, but shifted the dose–response curve markedly to the left. This effect was blocked by the 5-$HT_{2A/2C}$ antagonist ketanserin (Munzar *et al.*, 1999a). In a drug discrimination study in pigeons, the 5-HT releaser fenfluramine, but not the SSRI fluoxetine, produced methamphetamine-appropriate responding (Sasaki *et al.*, 1995).

Fluoxetine independently attenuated the establishment as well as expression of methamphetamine-induced CPP in mice (Takamatsu *et al.*, 2006). Fenfluramine attenuated the self-administration of methamphetamine in rats. However, this effect developed tolerance after only a few days of treatment (Munzar *et al.*, 1999b).

5-HT receptors

5-HT_{1A} receptor

A recent study in mice reported an inhibitory effect of the 5-HT_{1A} receptor agonist osemozotan on methamphetamine-induced hyperlocomotion (Ago *et al.*, 2006a). Repeated methamphetamine application for 7 days induced a profound sensitization of the hyperlocomotor response in mice. The maintenance of the sensitized response was blocked by twice-daily treatment with osemozotan, as revealed by a challenge injection with methamphetamine 8 days later. The osemozotan effect was antagonized by WAY 100635 (Ago *et al.*, 2006a).

Methamphetamine is discriminated from saline by pigeons and rodents (Sasaki *et al.*, 1995; Witkin *et al.*, 1999). A study in pigeons showed that 8-OH-DPAT at high doses substituted for methamphetamine in at least 50 percent of the animals tested. Low doses of 8-OH-DPAT blocked the discriminative stimulus properties of methamphetamine in pigeons, which might be mediated by presynaptic 5-HT_{1A} autoreceptors (Sasaki *et al.*, 1995). However, since only four pigeons were tested, it is difficult to draw sound conclusions from this study concerning effective dose ranges. The systemic application of high doses of 8-OH-DPAT and buspirone partially generalized to a training dose of methamphetamine under a fixed-ratio schedule of food presentation in rats. Effective doses, however, markedly decreased general responding. The partial generalization was antagonized by the 5-HT_{1A} receptor antagonist WAY 100635 (Munzar *et al.*, 1999b). At all doses tested, neither 8-OH-DPAT nor buspirone altered the discriminative stimulus properties of methamphetamine in rats. WAY 100635 also failed to affect the methamphetamine discriminative stimulus properties (Munzar *et al.*, 1999b).

5-HT_2 receptors

The unselective 5-$HT_{2A/2C}$ agonist DOI showed a partial generalization for methamphetamine at high doses in rats in a fixed-ratio schedule of food presentation, and shifted the dose–response curve to the left when injected 120 minutes before the test session. The unselective 5-$HT_{2A/2C}$ antagonist ketanserin shifted the dose–response curve to the right (Munzar *et al.*, 1999b).

5-HT_3 receptor

The acute hyperlocomotor effects of methamphetamine were blocked by pre-treatment with the 5-HT_3 receptor antagonist MDL 72222 in mice (Yoo *et al.*, 2006). Overexpression of 5-HT_3 receptors in frontal areas of the brain of mice went along with an increased sensitivity to the hyperlocomotor effects of methamphetamine (Allan *et al.*, 2001). These findings suggest a facilitatory role of 5-HT_3 receptors in acute methamphetamine-induced hyperlocomotion. 5-HT_3 receptor antagonism reduced the establishment as well as the expression of methamphetamine-induced locomotor sensitization in mice (Yoo *et al.*, 2006). Neither the 5-HT_3 receptor agonist m-CPBG nor the antagonist tropisetron substituted for the methamphetamine discriminative stimulus in rats. However, tropisetrone shifted the dose–response curve significantly to the left (Munzar *et al.*, 1999b). In a drug discrimination study in pigeons, the 5-HT_3 receptor antagonist MDL 72222 partially substituted for methamphetamine, but had no effect when administered together with methamphetamine (Sasaki *et al.*, 1995).

The establishment of methamphetamine-induced CPP was blocked by the 5-HT_3 receptor antagonists MDL 72222 and ICS 205-930, which suggests a facilitatory role of the 5-HT_3 receptor in the motivational properties of methamphetamine (Suzuki *et al.*, 1992).

Human studies

Based on the findings in animals, Johnson and colleagues (2008) investigated whether the 5-HT_3 receptor antagonist ondansetron would be superior to placebo in reducing methamphetamine use in 150 methamphetamine-dependent men and women. In this multi-site randomized, double-blind, placebo-controlled study, ondansetron (three doses) or placebo (p.o.) were given for 8 weeks in combination with cognitive behavioral therapy. The primary outcome measure was the number of methamphetamine-free study weeks assessed by methamphetamine-free urine samples. Although ondansetron was tolerated well, there was no difference between the ondansetron and placebo groups in the primary outcome measure. Also, secondary measures such as methamphetamine urine levels and self-reported non-use or achievement were not improved by ondansetron (Johnson *et al.*, 2008).

The 5-HT system in methamphetamine addiction

Methamphetamine reliably increases 5-HT activity in the brain. Similar to cocaine and amphetamine, increasing the 5-HT activity before drug administration inhibited several addiction-related behaviors. However, crucial behaviors still need to be investigated. The knowledge regarding 5-HT receptor contribution to the methamphetamine behavioral effects is still very limited, basically being restricted to the 5-HT_{1A} and 5-HT_3 receptors. While preclinical findings with the 5-HT_3 receptor suggested an application of the antagonist for addiction treatment, a recent clinical trial did not succeed in finding beneficial effects in human methamphetamine addicts. In the light of the findings with cocaine, investigating the role of the 5-HT_{2A} and 5-HT_{2C} receptors in methamphetamine addiction could be a promising future goal.

MDMA

The neurochemical effects of MDMA on 5-HT

Methylenedioxymethamphetamine (MDMA, ecstasy) is a synthetic amphetamine derivative with the stereo isomers (+)MDMA and (−)MDMA (Green *et al.*, 2003). MDMA increases extracellular 5-HT activity in the Nac (Kankaanpää *et al.*, 1998; Trigo *et al.*, 2007; Baumann *et al.*, 2008; Doly *et al.*, 2008), striatum (Gough *et al.*, 1991; Sabol and Seiden, 1998; Baumann *et al.*, 2008), VTA (Doly *et al.*, 2008), SN (Yamamoto *et al.*, 1995), hippocampus (Gartside *et al.*, 1997; Mechan *et al.*, 2002) and FC (Gartside *et al.*, 1997; Trigo *et al.*, 2007; Baumann *et al.*, 2008), as measured by *in vivo* microdialysis studies in rats, or in superfused rat brain slices (Schmidt *et al.*, 1987). Microdialysis measurements in slice preparations also revealed an increase of extracellular 5-HT levels in the striatum (Koch and Galloway, 1997) and in the DRN (Sprouse *et al.*, 1989, 1990). Altogether, MDMA effects on the 5-HT system suggest a contribution of 5-HT to the acute effects (Baumann *et al.*, 2008).

Interestingly, basal extracellular 5-HT levels in the FC and hippocampus and the 5-HT neuronal firing pattern were not changed after repeated MDMA treatment in rats, when measured 10–12 days after the end of the treatment. However, 5-HT release induced by d-fenfluramine in the FC (Series *et al.*, 1994), and stimulation-evoked 5-HT release in the FC but not in the ventral hippocampus, was markedly decreased after MDMA treatment (Gartside *et al.*, 1996). MDMA exerts potent neurotoxic effects on the 5-HT system (Schmidt, 1987; Schmidt and Taylor, 1988) which are associated with a loss of 5-HT neurons and projections. The 5-HT neurotoxic effects of MDMA are only partially reversible, and may contribute to the long-term behavioral impairments in MDMA users (Quednow *et al.*, 2006, 2007).

Reducing and potentiating 5-HT activity

An increase in locomotor activity was observed for MDMA (Gold *et al.*, 1988; Callaway *et al.*, 1990;

Bankson and Cunningham, 2001). Both stereoisomers (+)MDMA and (−)MDMA, as well as the racemic mixture (+/−)MDMA, induce hyperlocomotion, with the racemic mixture being the most potent (Fantegrossi *et al.*, 2003). A global lesion of the 5-HT system by 5,7-DHT attenuated (+/−)MDMA-induced locomotor stimulation (Kehne *et al.*, 1996). Pre-treatment with the SSRI fluoxetine also attenuated the hyperlocomotor effect of (+/−)MDMA, but potentiated that of (+)MDMA (Fantegrossi *et al.*, 2003). However, Callaway *et al.* (1990) found that (+)MDMA-induced hyperlocomotion was inhibited by fluoxetine and other SSRIs. Reducing 5-HT activity with the synthesis inhibitor pCPA attenuated the discriminative stimulus properties of MDMA in rats (Schechter, 1991). Strong support for a role of 5-HT in MDMA self-administration was found in mice. SERT KO mice did not self-administer MDMA at any doses tested. MDMA increased DA in the Nac in SERT KO mice and wild types to an equal extent. However, MDMA effects on 5-HT in the PFC were significantly attenuated in the SERT-deficient mice (Trigo *et al.*, 2007).

5-HT receptors

5-HT$_{1A}$ receptor

Early evidence suggested an inhibitory role of 5-HT$_{1A}$ receptor antagonist pre-treatment with propranolol and pindolol on (+)MDMA-induced hyperlocomotion (Callaway *et al.*, 1992; Kehne *et al.*, 1996). However, propranolol and pindolol also display a β-adrenoceptor antagonist profile, which restricts the interpretation of these studies. Subchronic 5-HT$_{1A}$ receptor activation with the agonist 8-OH-DPAT for 3 days reduced the hyperlocomotor effects of (+)MDMA when tested 35–37 hours after the last pre-treatment (Callaway and Geyer, 1992). When the effects of the selective 5-HT$_{1A}$ receptor antagonist WAY 100635 on (+)MDMA-induced hyperlocomotion were investigated in rats, however, no effect could be found (McCreary *et al.*, 1999). The authors concluded that the 5-HT$_{1A}$ receptor is not involved in (+)MDMA-induced locomotor activation (Bankson and Cunningham, 2001).

MDMA can be discriminated from saline by animals in operant tasks (Schechter, 1988; Baker *et al.*, 1997). However, only a minor role for 5-HT$_{1A}$ receptors was found regarding the discriminative stimulus properties of MDMA. Stimulation of 5-HT$_{1A}$ receptors with 8-OH-DPAT did not substitute for MDMA in rats in a study by Schechter (1988). Another study in rats, however, found a substitution for the MDMA stimulus. A detailed investigation revealed an ED_{50} value for racemic 8-OH-DPAT of 0.3 mg/kg, and for the isomers R(+)8-OH-DPAT and S(−)8OH-DPAT of 0.2 and 0.4 mg/kg, respectively (Glennon and Young, 2000). Racemic 8-OH-DPAT, in contrast, did not change the discriminative stimulus properties of MDMA (Glennon and Young, 2000). Pre-treatment with the 5-HT$_{1A}$ receptor antagonist NAN-190 slightly attenuated the discriminative stimulus effects of MDMA in rats, but this effect failed to reach statistical significance (Glennon *et al.*, 1992). WAY 1001635 also had little effect on the discrimination of either optical isomer of MDMA (Baker *et al.*, 1997).

5-HT$_{1B}$ receptor

5-HT$_{1B/1D}$ receptors were shown to play a facilitatory role in MDMA-induced hyperlocomotion (McCreary *et al.*, 1999; Fletcher *et al.*, 2002c). The 5-HT$_{1B/1D}$ receptor antagonist GR 127935 blocked the acute hyperlocomotor effects of MDMA in mice. These findings were supported by 5-HT$_{1B}$ receptor knockout mice, which showed a reduced hyperlocomotor response to MDMA compared to wild-type mice (Scearce-Levie *et al.*, 1999). Subchronic 5-HT$_{1B}$ receptor activation with the agonist RU24969 for 3 days, however, reduced the hyperlocomotor effects of (+)MDMA when tested 35–37 hours after the last pre-treatment (Callaway and Geyer, 1992). 5HT$_{1B/1D}$ receptors were also found to contribute to the sensitization of the MDMA-induced hyperlocomotion (McCreary *et al.*, 1999).

5-HT$_{2A}$ receptor

5-HT$_{2A}$ receptors were shown to play a facilitatory role in MDMA-induced hyperlocomotion (McCreary *et al.*, 1999; Fletcher *et al.*, 2002c). MDMA-induced hyperlocomotion was significantly changed by pre-treatment with the selective 5-HT$_{2A}$ receptor antagonist MDL100907. 5-HT$_{2A}$ receptor antagonism significantly attenuated the hyperlocomotor effects of (+/−)MDMA (Kehne *et al.*, 1996; Fantegrossi *et al.*, 2003; Ball and Rebec, 2005). The role of the 5-HT$_{2A}$ receptor in (+)MDMA-induced hyperlocomotion, however, is still unclear. Fantegrossi *et al.* (2003) reported potentiating effects of the 5-HT$_{2A}$ receptor antagonist MDL100907, while Herin *et al.* (2005) found an attenuation of (+)MDMA-induced hyperlocomotion. (+)MDMA and (−)MDMA, as well as the racemic mixture (+/−)MDMA, are self-administered by rodents and primates (Fantegrossi *et al.*, 2002, 2003). 5-HT$_{2A}$ receptor antagonism with MDL100907 reduced the self-administration of (+)MDMA and (−)MDMA in monkeys (Fantegrossi *et al.*, 2002).

5-HT$_{2B}$ receptor

A recent study that reported an essential involvement of 5-HT$_{2B}$ receptors in the MDMA-induced increase in extracellular 5-HT levels in the Nac and VTA also reported an essential requirement for these receptors for the hyperlocomotor effects of acute MDMA treatment in mice. This study showed that a selective 5-HT_{2B} receptor antagonist RS 127445 blocked the hyperlocomotor effects of MDMA dose-dependently, while MDMA had no hyperlocomotor effects in 5-HT_{2B} receptor (−/−) knockout mice (Doly *et al.*, 2008). In rats, however, the unselective 5-$HT_{2B/2C}$ receptor antagonist SB-206553 did not affect MDMA-induced hyperlocomotion (Ball and Rebec, 2005).

5-HT_{2C} receptor

The 5-HT_{2C} receptor antagonist SB242084 potentiated MDMA-induced hyperlocomotion in rats (Fletcher *et al.*, 2006). This confirmed an earlier study in rats that used more unselective receptor ligands (Bankson and Cunningham, 2002), together suggesting an inhibitory role for the 5-HT_{2C} receptor in MDMA-induced hyperlocomotion. Local application of the 5-HT_{2C} receptor agonist MK212 into the mPFC blocked the expression of MDMA-induced sensitization, while the antagonist RS 102221 had no effect (Ramos *et al.*, 2005).

5-HT_3 receptor

As other psychostimulant drugs do, MDMA induces a profound CPP in rodents (Tzschentke, 2007). The establishment of MDMA-induced CPP could be blocked in rats by pre-treatment with the 5-HT_3 receptor antagonists MDL 72222 and tropisetron (Bilsky and Reid, 1991; Braida *et al.*, 2005).

Human studies

Systematic studies in humans showed that MDMA significantly enhances mood, characterized by the feeling of happiness, euphoria, and sometimes mania-like experience. It also increases wellbeing and extroversion. MDMA leads to a moderate derealization, depersonalization and intensified sensory perception (Peroutka *et al.*, 1988; Vollenweider *et al.*, 1998). A case study in four ecstasy users suggested that the SSRI fluoxetine (p.o.), at a relatively low dose, did not block the reinforcing effects of MDMA (McCann and Ricaurte, 1993). A systematic study in 16 healthy volunteers showed, in contrast, that pre-treatment with the SSRI citalopram (i.v.) could indeed inhibit most of the subjective effects of MDMA. However, citalopram had some side effects, such as fatigue, headache and nausea, which might account for the loss of the effects on mood scales (Liechti *et al.*, 2000a).

The mixed 5-HT_{1A}/β-adrenoceptor antagonist pindolol (p.o.) only marginally altered the effects of MDMA on subjective experience in a double-blind, placebo-controlled study in healthy volunteers. Pindolol affected mainly the MDMA effects on the scales 'positive basic mood' and 'mania-like experience', which overall suggests only a minor contribution of 5-HT_{1A} receptors to the reinforcing effects of MDMA. In healthy volunteers, the 5-$HT_{2A/2C}$ antagonist ketanserin (p.o.) also reduced MDMA-induced changes in perception and emotional excitation, and the acute adverse effects. However, ketanserin had little impact on the positive mood and wellbeing effects of MDMA (Liechti *et al.*, 2000b).

The 5-HT system in MDMA addiction

As an amphetamine derivative, MDMA increases 5-HT activity in the brain after acute application, but also has profound 5-HT neurotoxic effects. Pre-clinical evidence clearly suggests involvement of the 5-HT system in the addiction-related effects of MDMA, but with little focus on core behaviors and with many receptors not investigated. Only 5-HT_{2A} and 5-HT_3 receptors were shown to have a facilitatory role in CPP and self-administration, respectively. Based on this pre-clinical knowledge, it might appear hard to design a 5-HT-based pharmacotherapy for MDMA addiction. As for the other psychostimulant drugs, increasing 5-HT levels before drug administration might limit drug effects, but was also found to have some unwanted side effects. For a better treatment design, more basic research into the serotonergic mechanisms of MDMA abuse and addiction is required.

Morphine and heroin

The neurochemical effects of morphine and heroin on the 5-HT system

Following the application of opioid drugs of abuse, an increase of extracellular levels of 5-HT is usually observed in *in vivo* microdialysis experiments. The most consistent effects were reported in the dorsal striatum, where an increase of extracellular 5-HT was found (Tao and Auerbach, 1994, 1995; Fadda *et al.*, 2005). In the diencephalon, too, the systemic application of morphine consistently results in the elevation of extracellular 5-HT levels (Grauer *et al.*, 1992; Tao and Auerbach, 1994, 1995; Munn and Borszcz, 2002). For the Nac, an increase of extracellular 5-HT was reported following both systemic (Tao and Auerbach, 1995, 2002; Fadda *et al.*, 2005) as well as local application of morphine

(Tao and Auerbach, 1995). However, this neurochemical effect seems to be modulated by genetic factors, as in one study a decrease of 5-HT in the nucleus accumbens was found in DBA/2J mice, but an increase of 5-HT activity was evident in CB57BL/6J mice (Fadda *et al.*, 2005). Mixed results were obtained when measurements of 5-HT were taken in the mPFC. While both Sudakov *et al.* (2007) and Tao and Auerbach (1995) reported an increase of 5-HT induced by self-administration or systemic application of morphine, respectively, Bland *et al.* (2003) found such a morphine response only in animals that had been exposed to inescapable stress. For the hippocampus, the available results suggest a differentiation between its dorsal and ventral part. Thus, a morphine-induced increase of extracellular 5-HT was found in the ventral hippocampus (Tao and Auerbach, 1994, 1995), while no effect was seen in the dorsal hippocampus (Tao and Auerbach, 1995). Increases of 5-HT were also seen in the amygdala and hypothalamus, while in the medial septum no effects were evident following systemic or local application of morphine (Tao and Auerbach, 1995). Furthermore, an increase of 5-HT activity following morphine could be observed in the rostroventral medulla (Taylor and Basbaum, 2003). For the cell nucleus region of the serotonergic system there seems to be a dissociation between the DRN and MRN regarding the neurochemical effects of morphine. While morphine induced an increase of extracellular 5-HT in the DRN following systemic (Tao and Auerbach, 1995, 2002) or local (Tao and Auerbach, 1995) application, no effects could be found in the MRN (Tao and Auerbach, 1995).

Continuous delivery of morphine via subcutaneously implanted morphine pellets was associated with up-regulation of 5-HT_2 receptors and no change in 5-HT_1 receptors in the cortex, whereas the withdrawal of morphine by removal of these pellets induced up-regulation of 5-HT_1 receptors and no change in 5-HT_2 receptors (Gulati and Bhargava, 1988). The findings regarding the 5-HT_2 receptors were confirmed by a subsequent study in which increased 5-HT_2 receptor binding was evident in the amygdala, midbrain and brainstem of rats with implanted morphine pellets (Gulati and Bhargava, 1989). A study with a 5-HT_{1A} selective ligand found down-regulation of these receptors in the hypothalamus, while no effects were found in limbic brain areas, the cortex, the brainstem, the midbrain or the spinal cord (Gulati and Bhargava, 1990).

Chronic morphine treatment was also found to enhance the stimulatory effect of morphine on 5-HT synthesis, and the inhibitory effect of a 5-HT_{1A} receptor agonist on 5-HT synthesis, indicating the development of super-sensitivity of somatic 5-HT autoreceptors during morphine-dependence (Sastre-Coll *et al.*, 2002). This finding corroborated previous results reporting an enhanced hypothermic response to a 5-HT_{1A} receptor agonist during morphine withdrawal (Gulati and Bhargava, 1990).

Withdrawal following chronic consumption of a morphine solution resulted in a decrease of 5-HT turnover rates (Ahtee *et al.*, 1989). Furthermore, a challenge injection of morphine following 1 day of withdrawal elicited an attenuated increase of extracellular 5-HT in the DRN. A decrease of 5-HT in the DRN was found following naloxone-precipitated withdrawal (Tao *et al.*, 1998).

Reducing and potentiating 5-HT activity

Morphine-induced hyperactivity and sensitization of the behavioral response following repeated treatment with morphine were blocked by the SSRI fluoxetine (Sills and Fletcher, 1997), suggesting an inhibitory role of 5-HT.

The effectiveness of morphine as a discriminative stimulus was not influenced by pre-treatment with a 5-HT-synthesis inhibitor or an unspecific 5-HT receptor antagonist (Winter, 1977). No effects could be found on the discriminative stimulus properties of heroin or morphine following application of the 5-HT releaser dexfenfluramine or the SSRI fluoxetine (Hynes *et al.*, 1985; Higgins *et al.*, 1993). Fluoxetine also did not substitute for morphine (Platt *et al.*, 1999).

Opiate drugs induce a profound CPP in animals (Tzschentke, 2007). Morphine-induced CPP was attenuated following depletion of 5-HT from the Nac (Spyraki *et al.*, 1988). Both the acquisition (Subhan *et al.*, 2000) and the expression (Maleki *et al.*, 2008) of morphine-induced CPP were enhanced by the application of SSRIs. Contrasting findings were reported when zimelidine was co-administered during the acquisition of morphine-induced CPP, as no effects of the 5-HT uptake blocker were evident (Kruszewska *et al.*, 1986). However, recent findings support an inhibitory role of elevated 5-HT levels on CPP, as the application of fluoxetine or the 5-HT precursor 5-hydroxytryptophan (5-HTP) in morphine-trained rats blocked the expression of morphine-induced CPP. This effect was mimicked by local application of fluoxetine into the Nac (Harris and Aston-Jones, 2001). Notably, this effect was reversed when animals that were not trained on morphine were treated with systemic fluoxetine – i.e., CPP for a morphine-paired environment was enhanced (Harris and Aston-Jones, 2001).

Intravenous self-administration of morphine was enhanced after 5,7-DHT lesions of the Nac (Smith *et al.*, 1987). The self-administration of heroin was reduced by dexfenfluramine (Higgins *et al.*, 1993, 1994; Wang *et al*, 1995).

5,7-DHT lesions of the DRN and MRN did not have effects on spontaneous or naloxone-precipitated withdrawal (Caillé *et al.*, 2002). The inhibition of 5-HT synthesis was also reported to have only minor effects on morphine withdrawal (Berthold *et al.*, 1989), but unspecific 5-HT receptor antagonists were reported to attenuate symptoms induced by naloxone-precipitated withdrawal (el-Kadi and Sharif, 1995). Moreover, the global increase in 5-HT levels with the application of SSRIs was effective in attenuating the signs of opiate withdrawal induced by naloxone (Gray, 2002; Singh *et al.*, 2003) and by spontaneous morphine withdrawal (Harris and Aston-Jones, 2001).

5-HT receptors

5-HT$_1$ receptors

The discriminative stimulus properties of morphine were attenuated by a selective 5-HT$_{1A}$ receptor agonist, but not by a partial 5-HT$_{1A}$ agonist (Powell *et al.*, 1994). Agonism of the 5-HT$_{1A}$ receptor was found to impair acquisition of morphine-induced CPP (Colpaert *et al.*, 2006). After infusion into the DRN, it also blocked the stress-induced potentiation of morphine-induced CPP (Will *et al.*, 2004). 5-HT$_1$ receptors seem to play a role in morphine withdrawal, as agonists of the 5-HT$_{1A}$ and the 5-HT$_{1B}$ receptor were shown to reduce signs of naloxone-induced withdrawal in morphine-dependent mice (Berthold *et al.*, 1989).

5-HT$_2$ receptors

Locomotor activity evoked by morphine was attenuated by the application of a 5-HT$_{2A}$ receptor antagonist (Auclair *et al.*, 2004), while a 5-HT$_{2C}$ receptor antagonist had the opposite effect and potentiated morphine-induced hyperactivity (Fletcher *et al.*, 2006). The systemic application of an unselective 5-HT$_2$ receptor antagonist reduced the acquisition of morphine-induced CPP (Nomikos and Spyraki, 1988). Involvement of 5-HT$_2$ receptors in opiate withdrawal was suggested by the finding that naloxone-precipitated morphine withdrawal was attenuated by non-selective 5-HT$_2$ receptor antagonists (Neal and Sparber, 1986a, b).

5-HT$_3$ receptor

Morphine-evoked hyperactivity is attenuated by blocking 5-HT$_3$ receptors (Pei *et al.*, 1993). The discriminative stimulus properties of morphine, however, were not influenced by treatment with 5-HT$_3$ receptor antagonists (Joharchi *et al.*, 1993). Morphine-induced CPP was blocked by application of 5-HT$_3$ antagonists during the conditioning phase (Carboni *et al.*, 1989; Higgins *et al.*, 1992; Bisaga *at al.*, 1993). Application of the 5-HT$_3$ receptor antagonist ondansetron reduced oral self-administration of morphine when administered chronically, but not following acute application (Hui *et al.*, 1993). No effects of ondansetron on the intravenous self-administration of heroin were evident (Higgins *et al.*, 1994). Also, the 5-HT$_3$ receptor antagonists MDL 72222 and tropisetron did not influence morphine or heroin self-administration (Hui *et al.*, 1993; Higgins *et al.*, 1994). Tropisetron, however, reduced the preference for a morphine solution (Hui *et al.*, 1993).

There is evidence of involvement of 5-HT$_3$ receptors in opiate withdrawal. It was reported that application of a 5-HT$_3$ receptor antagonist attenuated the morphine-withdrawal syndrome induced by a naloxone injection given to morphine-dependent animals (Roychoudhury and Kulkarni, 1996; Pinelli *et al.*, 1997). However, no effects on naloxone-precipitated withdrawal were found when rats were made dependent on morphine by the implantation of morphine pellets. Ondansetron and MDL 72222 were still able to counteract the CPA induced by an injection of a low dose of naloxone to morphine-dependent rats (Higgins *et al.*, 1991).

5-HT$_4$ receptor

Regarding the role of 5-HT$_4$ receptors, there is only one study, applying an antagonist with affinity for 5-HT$_4$ and 5-HT$_3$ receptors, which led to a reduction of morphine-induced CPP (Bisaga *et al.*, 1993). However, due to the established role of 5-HT$_3$ receptors in morphine-induced CPP, further studies, using selective ligands at the 5-HT$_4$ receptor, are needed.

Human studies

Studies analyzing the association between polymorphisms at the 5-HT transporter gene and opiate addiction reported mixed results. One study in Asian subjects did not find evidence supporting an association between addiction to opiates and polymorphisms of the serotonin transporter gene (Li *et al.*, 2002), but another study did find such an association in Asian subjects (Tan *et al.*, 1999). The same was the case when Caucasian subjects were investigated: while one study found an association with a 5-HT transporter gene polymorphism (Gerra *et al.*, 2004), another did not (Kotler *et al.*, 1999). In addition to ethnic diversity, small sample size or different sample selection across the studies may underlie these conflicting results. Further evidence for involvement of serotonergic genes in heroin dependence has been provided by Nielsen *et al.*

(2008), who reported an association of genetic variations of an enzyme catalyzing serotonin synthesis, tryptonphane hydroxylase 1 (TPH1), with heroin dependence in Hispanics, and a haplotype of TPH2 in African-American individuals. Also, an association between heroin dependence and a polymorphism of the 5-HT_{2A} receptor gene was reported (Sáiz *et al.*, 2008) and awaits replication. An association between a polymorphism of the 5-HT_{1B} receptor gene and a protective effect from heroin addiction was suggested for a Caucasian population (Proudnikov *et al.*, 2006).

The partial 5-HT_{1A} receptor agonist buspirone was effective in reducing the subjective and objective signs of opiate withdrawal (Rose *et al.*, 2003). Treatment of opiate addicts with a 5-HT uptake inhibitor resulted in a greater probability of being abstinent only when they lived in a non-adverse environment (Carpenter *et al.*, 2004). Adjunct therapy with a 5-HT uptake inhibitor enhanced the effectiveness of naltrexone treatment in heroin-addicted patients (Landabaso *et al.*, 1998). Fluoxetine was also shown to be effective in the treatment of premature ejaculation after opioid detoxification (Abdollahian *et al.*, 2006).

5-HT in morphine and heroin addiction

Several studies reported a genetic predisposition to opiate addiction in humans, which is located in genes coding for parts of the 5-HT system. Unfortunately, the replication rate for these findings is somewhat disappointing and does not yet allow sound conclusions. In general, the application of opiates leads to an increase of 5-HT in certain brain areas. Interestingly, pharmacologically increased 5-HT activity in the brain also attenuated opiate addiction-related behaviors with some success in a clinical setting. Receptor research showed a predominantly inhibitory role for 5-HT_{1A} receptors and a mainly facilitatory role for 5-HT_3 receptors in opiate-addiction-related behaviors. However, further studies are needed that investigate the role of other 5-HT receptors, and apply local manipulations to account for the specificity of the functions of local 5-HT in different brain areas.

Nicotine

The neurochemical effects of nicotine on the 5-HT system

The effects of nicotine on extracellular levels of 5-HT are not consistent. Following the systemic application of nicotine, there is usually an increase of 5-HT in the FC (Ribeiro *et al.*, 1993; Summers and Giacobini, 1995; Rossi *et al.*, 2005; Liang *et al.*, 2008), but a lack of effect has also been reported (Singer *et al.*, 2004). An increase of 5-HT levels in the FC could be also induced by local application of nicotine (Toth *et al.*, 1992), but again in one study no effect was reported, although this may have been due to the low nicotine concentration applied (Shearman *et al.*, 2005). Systemic nicotine also increases 5-HT in the medial temporal cortex and the VTA (Singer *et al.*, 2004; Rossi *et al.*, 2005), whereas local administration of a low dose of nicotine had no effect in the medial temporal cortex and decreased 5-HT levels in the VTA (Shearman *et al.*, 2005). In the Nac, an increase in extracellular 5-HT was evident following systemic administration of nicotine (Schiffer *et al.*, 2001; Liang *et al.*, 2008). For the dorsal hippocampus, both an increase (Singer *et al.*, 2004) and a decrease (Rossi *et al.*, 2005) in extracellular 5-HT levels were reported. The latter result was replicated with the local application of nicotine through the microdialysis probe (Shearman *et al.*, 2005). In the ventral hippocampus, nicotine either had no effect (Singer *et al.*, 2004) or decreased 5-HT levels (Rossi *et al.*, 2005). The same occurred when nicotine was applied locally to the ventral hippocampus. There was either no effect (Toth *et al.*, 1992) or a decrease in 5-HT levels (Shearman *et al.*, 2005). Furthermore, locally administered nicotine increased 5-HT in the basolateral amygdala (Shearman *et al.*, 2005), the lateral hypothalamus (Yang *et al.*, 1999) and the DRN (Ma *et al.*, 2005). No effects were evident in the dorsal striatum, the SN, the cerebellum and the pontine nucleus (Toth *et al.*, 1992).

Studies on the serotonergic mechanisms of nicotine withdrawal indicate that chronic application of nicotine may enhance the sensitivity of the serotonin system. For example, an increase of the sensitivity of 5-HT_{1A} autoreceptors in the DRN was reported after 2 days of withdrawal (Rasmussen and Czachura, 1997). Consistent with this finding, increased sensitivity of DRN neurons to the application of the SSRI citalopram could be observed 48 hours after chronic nicotine treatment (Touiki *et al.*, 2008). However, there was no effect of nicotine withdrawal on serotonin transporter mRNA expression in the DRN (Semba and Wakuta, 2008). Also, no alterations in 5-HT_{1A} receptor mRNA could be found during nicotine withdrawal in the cortex, hippocampus or lateral septum (Kenny *et al.*, 2001).

Repeated injections with nicotine were found to enhance the behavioral response to a 5-HT_2 receptor agonist between 24 hours and 24 days after discontinuation of nicotine treatment (Suemaru *et al.*, 2001; Yasuda *et al.*, 2002), indicating the sensitization of 5-HT_2 receptors during nicotine withdrawal. Opposing findings were reported during early nicotine withdrawal (16–18 hours after the last injection), where down-regulation of 5-HT_{2A}

receptors in the cortex (Hayslett and Tizabi, 2005) was evident. Accordingly, both chronic and acute treatment with nicotine attenuated the response to 5-HT_2 receptor stimulation when there was no withdrawal phase before the application of the 5-HT_2 receptor agonist (Tizabi *et al.*, 2001).

Reducing and potentiating 5-HT activity

Global 5-HT depletion was shown to promote sensitization of the locomotor response to nicotine (Olausson *et al.*, 2001b). Correspondingly, co-administration of nicotine and the SSRI citalopram blocked the expression of behavioral sensitization to nicotine (Olausson *et al.*, 1999). No effect on the discriminative stimulus properties of nicotine was found following the application of a non-specific 5-HT receptor antagonist (Stolerman *et al.*, 1983) or a 5-HT releaser (Mansbach *et al.*, 1998). Elevation of brain 5-HT levels also blocked the dysphoric component of nicotine withdrawal as measured by the reduction of brain-stimulation reward thresholds, without affecting the somatic signs of nicotine withdrawal (Harrison *et al.*, 2001).

5-HT receptors

5-HT_{1A} receptors

There is evidence indicating involvement of 5-HT_{1A} receptors in the long-term behavioral effects of nicotine. Thus, it was found that the enhancement of the auditory startle response following withdrawal from nicotine was blocked by the application of 5-HT_{1A} receptor antagonists (Rasmussen *et al.*, 1997, 2000). Furthermore, behavioral disinhibition on the elevated plus maze following repeated nicotine treatment was counteracted by 5-HT_{1A} receptor stimulation (Olausson *et al.*, 2001b).

5-HT_2 receptors

There are a number of studies suggesting a role for 5-HT_2 receptors in nicotine-induced locomotion and reward. The acute locomotor effects of nicotine were attenuated by selective stimulation of 5-HT_{2C} receptors (Grottick *et al.*, 2001). Pharmacological antagonism of these receptors potentiated the locomotor effects of nicotine (Fletcher *et al.*, 2006). The sensitization of nicotine-induced locomotion was blocked by the administration of an unselective 5-HT_2 receptor agonist (Olausson *et al.*, 2001a) or a selective 5-HT_{2C} receptor agonist (Grottick *et al.*, 2001). A role for 5-HT_2 receptors in the mediation of nicotine's effects is corroborated by experiments testing the effects of 5-HT_2 receptor agonists on the discriminative stimulus properties of nicotine. Thus, drugs activating 5-$HT_{2A/2C}$ receptors as well as drugs selectively activating 5-HT_{2C} receptors reduced nicotine-lever responding in operant tasks (Batman *et al.*, 2005; Quarta *et al.*, 2007; Zaniewska *et al.*, 2007). 5-HT_{2C} receptor agonism did not substitute for nicotine (Quarta *et al.*, 2007). Nicotine-induced CPP was attenuated by 5-HT_{2C} receptor agonists (Ji *et al.*, 2006; Hayes *et al.*, 2008). The unselective 5-HT_2 receptor antagonist ketanserin was reported to reduce intravenous nicotine self-administration (Levin *et al.*, 2009). Also, selective stimulation of the 5-HT_{2C} receptor reduced nicotine self-administration (Grottick *et al.*, 2001). Altogether, 5-HT_{2C} receptor activation reliably reduces virtually all nicotine-addiction related behaviors tested so far.

5-HT_3 receptor

5-HT_3 receptors do not seem to play a role in the mediation of the acute locomotor effects of nicotine, as the administration of 5-HT_3 receptor antagonists did not block nicotine-induced hyperactivity (Corrigall and Coen, 1994; Arnold *et al.*, 1995). Antagonists at the 5-HT_3 receptor were found to block nicotine-induced CPP (Carboni *et al.*, 1988, 1989). In contrast, neither intravenous nicotine self-administration (Corrigall and Coen, 1994) nor the effects of nicotine on intracranial self-stimulation (Ivanova and Greenshaw, 1997) were influenced by 5-HT_3 receptor antagonists.

Withdrawal from chronic nicotine treatment induces behavioral depression, which was attenuated by the co-administration of a 5-HT_3 receptor antagonist (Costall *et al.*, 1990a). This effect could be replicated by direct injection of ondansetron into the amygdala or the DRN, but not by injection of the 5-HT_3 receptor antagonist into the MRN, Nac or dorsal striatum (Costall *et al.*, 1990b). Furthermore, the CPA caused by application of the nicotinic receptor antagonist mecamylamine to animals that had been chronically treated with nicotine was attenuated by pre-treatment with the 5-HT_3 receptor antagonist ondansetron (Suzuki *et al.*, 1997).

5-HT_4 and 5-HT_6 receptor

The 5-HT_4 receptor does not seem to be involved in the acute hyperlocomotor effects of nicotine (Reavill *et al.*, 1998). Antagonism of 5-HT_6 receptors potentiated the discriminative stimulus effects of nicotine and also the locomotor effects of this drug (Young *et al.*, 2006b). However, further studies are needed to confirm the role of these receptors in the mediation of nicotine's effects.

Human studies

A study testing the relationship between tryptophan hydroxylase gene polymorphisms and smoking behavior did not find associations between gene variants and smoking status, but rather an association between genotype and the age of smoking initiation, indicating a non-specific genetic effect (Lerman *et al.*, 2001). Concerning the role of 5-HT transporter gene polymorphism, contrasting results have been published. One study did not find an association between 5-HT transporter polymorphism and smoking behavior or abstinence from smoking (Brody *et al.*, 2005), while another report did suggest an association with motives for smoking and nicotine dependence (Lerman *et al.*, 2000).

Mixed results were obtained when serotonergic drugs were used as a treatment of nicotine addiction. Treatment with the SSRI sertraline, in combination with the partial 5-HT_{1A} receptor agonist buspirone, increased abstinence from smoking (Carrao *et al.*, 2007). When the SSRI fluoxetine was used as pharmacological treatment to support smoking cessation, no effect on nicotine abstinence was evident, irrespective whether the treated subjects had a history of a depressive disorder or not (Saules *et al.*, 2004). In this study, there were only beneficial effects on weight gain following discontinuation of smoking (Saules *et al.*, 2004). This was consistent with other studies using drugs increasing serotonergic neurotransmission (Spring *et al.*, 1991, 1995). These results support an earlier study which applied the SSRI sertraline as treatment, and did not find effects on abstinence from nicotine (Covey *et al.*, 2002). However, sertraline did reduce withdrawal symptoms (Covey *et al.*, 2002). In contrast to the above reports, a multicenter study in which the effects of fluoxetine treatment applied in addition to behavioral therapy were investigated found positive effects of the SSRI on abstinence rates (Niaura *et al.*, 2002). When tested in patients with depression, paroxetine was effective in reducing cigarette consumption (Miyamoto *et al.*, 2007); however, in another study in depressive smokers it was shown that treatment with fluoxetine only initially reduced smoking, but 6 months after discontinuation of smoking the probability of relapse was higher in the treatment-group (Spring *et al.*, 2007). The selective 5-HT_3 receptor antagonist ondansetron was also tested for its potential to support smoking cessation, but did not prove to be effective (Cropp and Goraharper, 1995; West and Hajek, 1996).

The 5-HT system in nicotine addiction

The effects of nicotine on 5-HT levels in the brain are complex, and seem to depend on the specific brain area. The relatively scarce pre-clinical evidence suggests that a general enrichment of 5-HT in the brain might also limit nicotine addiction-related behaviors. Some clinical trials with SSRIs already support this view when short-term success is considered. In the long run the SSRI treatment might increase relapse risk, one element of addiction that is not sufficiently covered by animal studies. At single receptor level there is accumulating evidence for an inhibitory role of 5-HT_{2C} receptors in many nicotine-addiction related behaviors in animals. There is also some evidence for an inhibitory role of the 5-HT_6 and a facilitating role of the 5-HT_3 receptors. However, 5-HT_3 receptor antagonism did not prove to be an effective treatment strategy in human smokers. Many other 5-HT receptors still await investigation for their role in nicotine addiction. Given the indications that serotonergic drugs may have some efficacy in supporting discontinuation of smoking, it may be of interest to discover whether genetic polymorphisms within the 5-HT_{2C} receptor gene and other components of the 5-HT system contribute to the development and maintenance of nicotine addiction and/or influence the efficacy of the pharmacotherapy of nicotine addiction.

Conclusion

All of the discussed drugs have a strong effect on 5-HT activity in the brain, either by direct interaction with components of the 5-HT system (cocaine and psychostimulants) or by indirect interaction via other transmitter systems (opiates, nicotine). Early addiction research has tried to link acute and chronic neurochemical effects causally to the observed behavioral effects in a 'proactive way' – i.e., assuming that the increase in neurochemical activity is indeed triggering behavior. As research on the relevance of the 5-HT increase for addiction-related behaviors developed, it became obvious that the 5-HT increase does not uniformly have such a proactive effect, but might rather (or in addition) serve a 'counteractive function' and thus limit the expression of behavior, which is driven by other neurochemical effects of the drug (e.g., by DA and glutamate). At the current state of knowledge, it may well be that some of the 5-HT responses are 'neurochemical epiphenomena' – i.e., they are without any causal effect on addiction-related behavior. A predominant counteractive function is suggested by the well-supported observation that an induced increase in 5-HT extracellular activity prior to addiction-related behavior (e.g., by and SSRI) almost uniformly attenuates subsequent drug-related behaviors for virtually all of the reviewed drugs in this chapter. Complementary to this, a reduction in 5-HT activity can enhance these behaviors. However, a number of studies have shown that for the full expression of addiction-related behaviors, the ratio of the 5-HT and DA activation is important. Therefore, single manipulation of the 5-HT activity might serve as a potential therapeutic intervention, but may mask the real function of 5-HT in drug addiction. A better view can possibly be obtained when looking at the contribution of single receptors. Given the large number of different 5-HT receptors and their splice variants, it may not be surprising that these receptors play very different roles for normal as well as addiction-related behavior. These multiple functional dissociations are taken to another level of complexity when local receptor populations are considered. A great number of studies revealed that one 5-HT receptor can have very different effects on addiction-related (and normal) behavior depending on its brain area, cell type and synaptic localization (Müller and Huston, 2006). Thus, some receptor populations did appear to have proactive effects (i.e., are translating the drug-induced 5-HT increase into addiction-related behaviors) while others counteracted them, or were not involved at all. While acknowledging this complexity will be important in understanding the 5-HT role in drug addiction, it might also obscure the use of 5-HT system's components for addiction pharmacotherapy in humans. So far, serotonergic approaches alone and in combination with other, non-pharmacological therapies have not proved successful in the long term. However, there are new targets emerging, such as the 5-HT_{2C} receptor, for which the search for a suitable intervention has started. Possibly, personalized treatment approaches using, for example, 5-HT system gene polymorphisms as biological markers might also enhance the use of the 5-HT system to treat drug addiction in the future.

Acknowledgement

This work was supported by the research grant HU306/23-5 from the Deutsche Forschungsgemeinschaft, by IMAGEN and the BRC.

References

Abdollahian, E., Javanbakht, A., Javidi, K., Samari, A.A., Shakiba, M. and Sargolzaee, M.R. (2006) Study of the efficacy of fluoxetine and clomipramine in the treatment of premature ejaculation after opioid detoxification. Am. J. Addict., 15: 100–104.

Acosta, J.I., Boynton, F.A., Kirschner, K.F. and Neisewander, J.L. (2005) Stimulation of 5-HT1B receptors decreases cocaine- and sucrose-seeking behavior. Pharmacol. Biochem. Behav., 80: 297–307.

Acquas, E., Carboni, E., Leone, P. and Di, C.G. (1988) 5-HT3 receptors antagonists block morphine- and nicotine- but not amphetamine-induced place-preference conditioning. Pharmacol. Res. Commun., 20: 1113–1114.

Adams, W., Kusljic, S. and van den, B.M. (2008) Serotonin depletion in the dorsal and ventral hippocampus: effects on locomotor hyperactivity, prepulse inhibition and learning and memory. Neuropharmacology, 55: 1048–1055.

Ago, Y., Nakamura, S., Hayashi, A., Itho, S., Baba, A. and Matsuda, T. (2006a) Effects of osemozotan, ritanserin and azasetron on cocaine-induced behavioral sensitization in mice. Pharmacol. Biochem. Behav., 85: 198–205.

Ago, Y., Nakamura, S., Uda, M., Kajii, Y., Abe, M., Baba, A. and Matsuda, T. (2006b) Attenuation by the 5-HT1A receptor agonist osemozotan of the behavioral effects of single and repeated methamphetamine in mice. Neuropharmacology, 51: 914–922.

Ahtee, L., Attila, L.M.J., Carlson, K.R. and Haikala, H. (1989) Changes in brain monoamine metabolism during withdrawal from chronic oral self-administration of morphine and in response to a morphine challenge in the withdrawn state. J. Pharmacol. Exp. Therap., 249: 303–310.

Ali, I. and Kelly, M.E. (1996) 8-OH-DPAT enhances the expression of cocaine-induced place preference. Br. J. Pharmacol. Suppl., 119: 207P.

Ali, I. and Kelly, M.E. (1997) Buspirone fails to affect cocaine-induced conditioned place preference in the mouse. Pharmacol. Biochem. Behav., 58: 311–315.

Allan, A.M., Galindo, R., Chynoweth, J., Engel, S.R. and Savage, D.D. (2001) Conditioned place preference for cocaine is attenuated in mice over-expressing the 5-HT(3) receptor. Psychopharmacology (Berl.), 158: 18–27.

American Psychiatric Association (1994) Diagnostic Statistical Manual of Mental Disorders. 4th ed. DSM IV, APA, Washington DC.

Andrews, C.M. and Lucki, I. (2001) Effects of cocaine on extracellular dopamine and serotonin levels in the nucleus accumbens. Psychopharmacology (Berl.), 155(3): 221–229.

Arnold, B., Allison, K., Ivanova, S., Paetsch, P.R., Paslawski, T. and Greenshaw, A.J. (1995) 5HT3 receptor antagonists do not block nicotine induced hyperactivity in rats. Psychopharmacology, 119: 213–221.

Aronson, S.C., Black, J.E., McDougle, C.J., Scanley, B.E., Jatlow, P., Kosten, T.R., Heninger, G.R. and Price, L.H. (1995) Serotonergic mechanisms of cocaine effects in humans. Psychopharmacology, 119: 179–185.

Arroyo, M., Markou, A., Robbins, T.W. and Everitt, B.J. (1998) Acquisition, maintenance and reinstatement of intravenous cocaine self-administration under a second-order schedule of reinforcement in rats: effects of conditioned cues and continuous access to cocaine. Psychopharmacology, 140: 331–344.

Auclair, A., Drouin, C., Cotecchia, S., Glowinski, J. and Tassin, J.P. (2004) 5-HT2A and alpha 1b-adrenergic receptors entirely mediate dopamine release, locomotor response and behavioral sensitization to opiates and psychostimulants. Eur. J. Neurosci., 20: 3073–3084.

Baker, L.E., Riddle, E.E., Saunders, R.B. and Appel, J.B. (1993) The role of monoamine uptake in the discriminative stimulus effects of cocaine and related compounds. Behav. Pharmacol., 4: 69–79.

Baker, L.E., Virden, T.B., Miller, M.E. and Sullivan, C.L. (1997) Time course analysis of the discriminative stimulus effects of the optical isomers of 3,4-methylenedioxymethamphetamine (MDMA). Pharmacol. Biochem. Behav., 58: 505–516.

Ball, K.T. and Rebec, G.V. (2005) Role of 5-HT2A and 5-HT2C/B receptors in the acute effects of 3,4-methylenedioxymethamphetamine (MDMA) on striatal single-unit activity and

locomotion in freely moving rats. Psychopharmacology (Berl.), 181: 676–687.
Bankson, M.G. and Cunningham, K.A. (2001) 3,4-methylenedioxymethamphetamine (MDMA) as a unique model of serotonin receptor function and serotonin–dopamine interactions. J. Pharmacol. Exp. Therap., 297: 846–852.
Bankson, M.G. and Cunningham, K.A. (2002) Pharmacological studies of the acute effects of (+)-3,4-methylenedioxymethamphetamine on locomotor activity: role of 5-HT(1B/1D) and 5-HT(2) receptors. Neuropsychopharmacology, 26: 40–52.
Bardo, M.T. and Bevins, R.A. (2000) Conditioned place preference: what does it add to our preclinical understanding of drug reward?. Psychopharmacology (Berl.), 153: 31–43.
Barot, S.K., Ferguson, S.M. and Neumaier, J.F. (2007) 5-HT(1B) receptors in nucleus accumbens efferents enhance both rewarding and aversive effects of cocaine. Eur. J. Neurosci., 25: 3125–3131.
Batki, S.L., Washburn, A.M., Delucchi, K. and Jones, R.T. (1996) A controlled trial of fluoxetine in crack cocaine dependence. Drug Alcohol Depend., 41: 137–142.
Batman, A.M., Munzar, P. and Beardsley, P.M. (2005) Attenuation of nicotine's discriminative stimulus effects in rats and its locomotor activity effects in mice by serotonergic 5-HT2A/2C receptor agonists. Psychopharmacology, 179: 393–401.
Baumann, M.H. and Rothman, R.B. (1998) Alterations in serotonergic responsiveness during cocaine withdrawal in rats: similarities to major depression in humans. Biol. Psychiat., 44: 578–591.
Baumann, M.H., Clark, R.D. and Rothman, R.B. (2008) Locomotor stimulation produced by 3,4-methylenedioxymethamphetamine (MDMA) is correlated with dialysate levels of serotonin and dopamine in rat brain. Pharmacol. Biochem. Behav., 90: 208–217.
Bedford, J.A., Borne, R.F. and Wilson, M.C. (1980) Comparative behavioral profile of cocaine and morcocaine in rats and monkeys. Pharmacol. Biochem. Behav., 13: 69–75.
Belzung, C., Scearce Levie, K., Barreau, S. and Hen, R. (2000) Absence of cocaine-induced place conditioning in serotonin 1B receptor knockout mice. Pharmacol. Biochem. Behav., 66: 221–225.
Berthold, H., Fozard, J.R. and Engel, G. (1989) 5-HT1 receptor agonists attenuate the naloxone-induced jumping behavior in morphine-depe ndent mice. Eur. J. Pharmacol., 162: 19–27.
Bilsky, E.J. and Reid, L.D. (1991) MDL72222, a serotonin 5-HT3 receptor antagonist, blocks MDMA's ability to establish a conditioned place preference. Pharmacol. Biochem. Behav., 39: 509–512.
Bisaga, A., Sikora, J. and Kostowski, W. (1993) The effect of drugs interacting with serotonergic 5-HT3 and 5-HT4 receptors on morphine place conditioning. Pol. J. Pharmacol., 45: 513–519.
Bland, S.T., Hargrave, D., Pepin, J.L., Amat, J., Watkins, L.R. and Maier, S.F. (2003) Stressor controllability modulates stress-induced dopamine and serotonin efflux and morphine-induced serotonin efflux in the medial prefrontal cortex. Neuropsychopharmacology, 28: 1589–1596.
Bradberry, C.W., Nobiletti, J.B., Elsworth, J.D., Murphy, B., Jatlow, P. and Roth, R.H. (1993) Cocaine and cocaethylene: microdialysis comparison of brain drug levels and effects on dopamine and serotonin. J. Neurochem., 60: 1429–1435.
Braida, D., Iosue, S., Pegorini, S. and Sala, M. (2005) 3,4 Met hylenedioxymethamphetamine-induced conditioned place preference (CPP) is mediated by endocannabinoid system. Pharmacol. Res., 51: 177–182.
Broderick, P.A., Kornak-EP, J., Eng, F. and Wechsler, R. (1993) Real time detection of acute (IP) cocaine-enhanced dopamine and serotonin release in ventrolateral nucleus accumbens of the behaving Norway rat. Pharmacol. Biochem. Behav., 46: 715–722.
Brody, C.L., Hamer, D.H. and Haaga, D.A.F. (2005) Depression vulnerability, cigarette smoking, and the serotonin transporter gene. Addict. Behav., 30: 557–566.
Bronsert, M.R., Mead, A.N., Hen, R. and Rocha, B.A. (2001) Amphetamine-induced locomotor activation in 5-HT(1B) knockout mice: effects of injection route on acute and sensitized responses. Behav. Pharmacol., 12: 549–555.
Bubar, M.J., McMahon, L.R., De Deurwaerdere, P., Spampinato, U. and Cunningham, K.A. (2003) Selective serotonin reuptake inhibitors enhance cocaine-induced locomotor activity and dopamine release in the nucleus accumbens. Neuropharmacology, 44: 342–353.
Budygin, E.A., Brodie, M.S., Sotnikova, T.D., Mateo, Y., John, C.E., Cyr, M., Gainetdinov, R.R. and Jones, S.R. (2004) Dissociation of rewarding and dopamine transporter-mediated properties of amphetamine. Proc. Natl. Acad. Sci. USA, 101: 7781–7786.
Burbassi, S. and Cervo, L. (2008) Stimulation of serotonin2C receptors influences cocaine-seeking behavior in response to drug-associated stimuli in rats. Psychopharmacology (Berl.), 196: 15–27.
Burmeister, J.J., Lungren, E.M., Kirschner, K.F. and Neisewander, J.L. (2004) Differential roles of 5-HT receptor subtypes in cue and cocaine reinstatement of cocaine-seeking behavior in rats. Neuropsychopharmacology, 29: 660–668.
Caillé, S., Espejo, E.F., Koob, G.F. and Stinus, L. (2002) Dorsal and median raphe serotonergic system lesion does not alter the opiate withdrawal syndrome. Pharmacol. Biochem. Behav., 72: 979–986.
Callahan, P.M. and Cunningham, K.A. (1995) Modulation of the discriminative stimulus properties of cocaine by 5-HT1B and 5-HT2C receptors. J. Pharmacol. Exp. Ther., 274: 1414–1424.
Callahan, P.M. and Cunningham, K.A. (1997) Modulation of the discriminative stimulus properties of cocaine: comparison of the effects of fluoxetine with 5-HT1A and 5-HT1B receptor agonists. Neuropharmacology, 36: 373–381.
Callaway, C.W. and Geyer, M.A. (1992) Tolerance and cross-tolerance to the activating effects of 3,4-methylenedioxymethamphetamine and a 5-hydroxytryptamine1B agonist. J. Pharmacol. Exp. Ther., 263: 318–326.
Callaway, C.W., Wing, L.L. and Geyer, M.A. (1990) Serotonin release contributes to the locomotor stimulant effects of 3,4-methylenedioxymethamphetamine in rats. J. Pharmacol. Exp. Ther., 254: 456–464.
Callaway, C.W., Rempel, N., Peng, R.Y. and Geyer, M.A. (1992) Serotonin 5-HT1-like receptors mediate hyperactivity in rats induced by 3,4-methylenedioxymethamphetamine. Neuropsychopharmacology, 7: 113–127.
Camp, D.M., Browman, K.E. and Robinson, T.E. (1994) The effects of methamphetamine and cocaine on motor behavior and extracellular dopamine in the ventral striatum of Lewis versus Fischer-344 rats. Brain Res., 668: 180–193.
Carboni, E., Acquas, E., Leone, P., Perezzani, L. and Di, C.G. (1988) 5-HT3 receptor antagonists block morphine- and nicotine-induced place-preference conditioning. Eur. J. Pharmacol., 151: 159–160.

Carboni, E., Acquas, E., Leone, P. and Di Chiara, G. (1989) 5HT3 receptor antagonists block morphine- and nicotine- but not amphetamine-induced reward. Psychopharmacology, 97: 175–178.
Carey, R.J., DePalma, G. and Damianopoulos, E. (2001) Cocaine and serotonin: a role for the 5-HT(1A) receptor site in the mediation of cocaine stimulant effects. Behav. Brain Res., 126: 127–133.
Carey, R.J., DePalma, G. and Damianopoulos, E. (2002) 5-HT1A agonist/antagonist modification of cocaine stimulant effects: implications for cocaine mechanisms. Behav. Brain Res., 132: 37–46.
Carey, R.J., DePalma, G., Damianopoulos, E., Müller, C.P. and Huston, J.P. (2004a) The 5-HT(1A) receptor and behavioral stimulation in the rat: effects of 8-OHDPAT on spontaneous and cocaine-induced behavior. Psychopharmacology (Berl.), 177: 46–54.
Carey, R.J., DePalma, G., Damianopoulos, E., Hopkins, A., Shanahan, A., Müller, C.P. and Huston, J.P. (2004b) Dopaminergic and serotonergic autoreceptor stimulation effects are equivalent and additive in the suppression of spontaneous and cocaine induced locomotor activity. Brain Res., 1019: 134–143.
Carey, R.J., DePalma, G., Damianopoulos, E., Shanahan, A., Müller, C.P. and Huston, J.P. (2005a) Evidence that the 5-HT1A autoreceptor is an important pharmacological target for the modulation of cocaine behavioral stimulant effects. Brain Res., 1034: 162–171.
Carey, R.J., DePalma, G., Damianopoulos, E., Shanahan, A., Müller, C.P. and Huston, J.P. (2005b) Pharmacological inhibition of DA- and 5-HT activity blocks spontaneous and cocaine-activated behavior: reversal by chronic cocaine treatment. Brain Res., 1047: 194–204.
Carey, R.J., DePalma, G., Damianopoulos, E. and Shanahan, A. (2005c) Stimulus gated cocaine sensitization: Interoceptive drug cue control of cocaine locomotor sensitization. Pharmacol. Biochem. Behav., 82: 353–360.
Carey, R.J., Huston, J.P. and Müller, C.P. (2008) Pharmacological inhibition of dopamine and serotonin activity blocks spontaneous and cocaine-activated behavior. Prog. Brain Res., 172: 347–360.
Carlsson, M.L., Martin, P., Nilsson, M., Sorensen, S.M., Carlsson, A., Waters, S. and Waters, N. (1999) The 5-HT2A receptor antagonist M100907 is more effective in counteracting NMDA antagonist- than dopamine agonist-induced hyperactivity in mice. J. Neural Transm., 106: 123–129.
Carpenter, K.M., Brooks, A.C., Vosburg, S.K. and Nunes, E.V. (2004) The effect of sertraline and environmental context on treating depression and illicit substance use among methadone maintained opiate dependent patients: a controlled clinical trial. Drug Alcohol Depend., 74: 123–134.
Carrao, J.L., Moreira, L.B. and Fuchs, F.D. (2007) The efficacy of the combination of sertraline with buspirone for smoking cessation – a randomized clinical trial in nondepressed smokers. Eur. Arch. Psychiat. Clin. Neurosci., 257: 383–388.
Castanon, N., Scearce-Levie, K., Lucas, J.J., Rocha, B. and Hen, R. (2000) Modulation of the effects of cocaine by 5-HT1B receptors: a comparison of knockouts and antagonists. Pharmacol. Biochem. Behav., 67: 559–566.
Cervo, L., Pozzi, L. and Samanin, R. (1996) 5-HT3 receptor antagonists do not modify cocaine place conditioning or the rise in extracellular dopamine in the nucleus accumbens of rats. Pharmacol. Biochem. Behav., 55: 33–37.
Cervo, L., Rozio, M., Ekalle Soppo, C.B., Carnovali, F., Santangelo, E. and Samanin, R. (2002) Stimulation of serotonin1B receptors induces conditioned place aversion and facilitates cocaine place conditioning in rats. Psychopharmacology (Berl.), 163: 142–150.
Cervo, L., Carnovali, F., Stark, J.A. and Mennini, T. (2003) Cocaine-seeking behavior in response to drug-associated stimuli in rats: involvement receptors of D-3 and D-2 dopamine receptors. Neuropsychopharmacology, 28: 1150–1159.
Chadwick, M.J., Gregory, D.L. and Wendling, G. (1990) A double-blind amino acids, L-tryptophan and L-tyrosine, and placebo study with cocaine-dependent subjects in an inpatient chemical dependency treatment center. Am. J. Drug Alcohol Abuse, 16: 275–286.
Chaouloff, F., Courvoisier, H., Moisan, M.P. and Mormede, P. (1999) GR 127935 reduces basal locomotor activity and prevents RU 24969-, but not D-amphetamine-induced hyperlocomotion, in the Wistar-Kyoto hyperactive (WKHA) rat. Psychopharmacology (Berl.), 141: 326–331.
Chen, N.H. and Reith, M.E. (1994) Autoregulation and monoamine interactions in the ventral tegmental area in the absence and presence of cocaine: a microdialysis study in freely moving rats. J. Pharmacol. Exp. Ther., 271: 1597–1610.
Cigler, T., LaForge, K.S., McHugh, P.F., Kapadia, S.U., Leal, S.M. and Kreek, M.J.. (2001) Novel and previously reported single-nucleotide polymorphisms in the human 5-HT(1B) receptor gene: no association with cocaine or alcohol abuse or dependence. Am. J. Med. Genet., 105: 489–497.
Colpaert, F.C., Deseure, K., Stinus, L. and Adriaensen, H. (2006) High-efficacy 5-hydroxytryptamine 1A receptor activation counteracts opioid hyperallodynia and affective conditioning. J. Pharmacol. Exp. Ther., 316: 892–899.
Cornish, J.W., Maany, I., Fudala, P.J., Ehrman, R.N., Robbins, S.J. and O'Brien, C.P. (2001) A randomized, double-blind, placebo-controlled study of ritanserin pharmacotherapy for cocaine dependence. Drug Alcohol Depend., 61: 183–189.
Corrigall, W.A. and Coen, K.M. (1994) Nicotine self-administration and locomotor activity are not modified by the 5-HT3 antagonist ICS 205-930 and MDL 72222. Pharmacol. Biochem. Behav., 49: 67–71.
Costall, B., Domeney, A.M., Naylor, R.J. and Tyers, M.B. (1987) Effects of the 5-HT3 receptor antagonist, GR38032F, on raised dopaminergic activity in the mesolimbic system of the rat and marmoset brain. Br. J. Pharmacol., 92: 881–894.
Costall, B., Jones, B.J., Kelly, M.E., Naylor, R.J., Onaivi, E.S. and Tyers, M.B. (1990a) Ondansetron inhibits a behavioral consequence of withdrawing from drugs of abuse. Pharmacol. Biochem. Behav., 36: 339–344.
Costall, B., Jones, B.J., Kelly, M.E., Naylor, R.J., Onaivi, E.S. and Tyers, M.B. (1990b) Sites of Action of Ondansetron to Inhibit Withdrawal from Drugs of Abuse. Pharmacol. Biochem. Behav., 36: 97–104.
Covey, L.S., Glassman, A.H., Stetner, F., Rivelli, S. and Stage, K. (2002) A randomized trial of sertraline as a cessation aid for smokers with a history of major depression. Am. J. Psychiatry, 159: 1731–1737.
Covi, L., Hess, J.M., Kreiter, N.A. and Haertzen, C.A. (1995) Effects of combined fluoxetine and counseling in the outpatient treatment of cocaine abusers. Am. J. Drug Alcohol Abuse, 21: 327–344.
Cropp, C.D. and Goraharper, M.L. (1995) Ondansetron use for smoking cessation. Ann. Pharmacotherapy, 29: 1041–1042.

Czoty, P.W., Ginsburg, B.C. and Howell, L.L. (2002) Serotonergic attenuation of the reinforcing and neurochemical effects of cocaine in squirrel monkeys. J. Pharmacol. Exp. Ther., 300: 831–837.

Czoty, P.W., McCabe, C. and Nader, M.A. (2005) Effects of the 5-HT1A agonist (+/−)-8-hydroxy-2-(di-n-propylamino)tetralin (8-OH-DPAT) on cocaine choice in cynomolgus monkeys. Behav. Pharmacol., 16: 187–191.

Dahl, J.P., Cubells, J.F., Ray, R., Weller, A.E., Lohoff, F.W., Ferraro, T.N., Oslin, D.W., Kampman, K.M., Dackis, C., Tang, Y., Gelernter, J., Kranzler, H.R., O'Brien, C.P. and Berrettini, W.H. (2006) Analysis of variations in the tryptophan hydroxylase-2 (TPH2) gene in cocaine dependence. Addict. Biol., 1(1): 76–83.

David, V., Segu, L., Buhot, M.C., Ichaye, M. and Cazala, P. (2004) Rewarding effects elicited by cocaine microinjections into the ventral tegmental area of C57BL/6 mice: involvement of dopamine D1 and serotonin1B receptors. Psychopharmacology (Berl.), 174: 367–375.

Davidson, C., Lee, T.H., Xiong, Z. and Ellinwood, E.H. (2002) Ondansetron given in the acute withdrawal from a repeated cocaine sensitization dosing regimen reverses the expression of sensitization and inhibits self-administration. Neuropsychopharmacology, 27: 542–553.

Davidson, C., Lazarus, C., Lee, T.H. and Ellinwood, E.H. (2004) Ondansetron, given during the acute cocaine withdrawal, attenuates oral cocaine self-administration. Eur. J. Pharmacol., 503: 99–102.

Davidson, C., Gopalan, R., Ahn, C., Chen, Q., Mannelli, P., Patkar, A.A., Weese, G.D., Lee, T.H. and Ellinwood, E.H. (2007) Reduction in methamphetamine induced sensitization and reinstatement after combined pergolide plus ondansetron treatment during withdrawal. Eur. J. Pharmacol., 565: 113–118.

Dawson, L.A., Nguyen, H.Q. and Li, P. (2003) Potentiation of amphetamine-induced changes in dopamine and 5-HT by a 5-HT6 receptor antagonist. Brain Res. Bull., 59: 513–521.

De La Garza, R., Callahan, P.M. and Cunningham, K.A. (1996) Detailed investigations of 5-HT3 compounds in a drug discrimination model. Pharmacol. Biochem. Behav., 54: 533–540.

De La Garza, R., Callahan, P.M. and Cunningham, K.A. (1998) The discriminative stimulus properties of cocaine: effects of microinfusion of cocaine, a 5-HT1A agonist or antagonist, into the ventral tegmental area. Psychopharmacology (Berl.), 137: 1–6.

De La Garza, R.D. and Cunningham, K.A. (2000) The effects of the 5-hydroxytryptamine(1A) agonist 8-hydroxy-2-(di-n-propylamino)tetralin on spontaneous activity, cocaine-induced hyperactivity and behavioral sensitization: a microanalysis of locomotor activity. J. Pharmacol. Exp. Ther., 292: 610–617.

De Wit, H. and Stewart, J. (1981) Reinstatement of cocaine-reinforced responding in the rat. Psychopharmacology (Berl.), 75: 134–143.

Depoortere, R.Y., Li, D.H., Lane, J.D. and Emmett Oglesby, M.W. (1993) Parameters of self-administration of cocaine in rats under a progressive-ratio schedule. Pharmacol. Biochem. Behav., 45: 539–548.

Deroche Gamonet, V., Belin, D. and Piazza, P.V. (2004) Evidence for addiction-like behavior in the rat. Science, 305: 1014–1017.

Di Chiara, G. (1995) The role of dopamine in drug abuse viewed from the perspective of its role in motivation. Drug Alcohol Depend., 38: 95–137.

Di Chiara, G. and Imperato, A. (1988) Drugs abused by humans preferentially increase synaptic dopamine concentrations in the mesolimbic system of freely moving rats. Proc. Natl. Acad. Sci. USA, 85: 5274–5278.

Doly, S., Valjent, E., Setola, V., Callebert, J., Herve, D., Launay, J.M. and Maroteaux, L. (2008) Serotonin 5-HT2B receptors are required for 3,4-methylenedioxymethamphetamine-induced hyperlocomotion and 5-HT release *in vivo* and *in vitro*. J. Neurosci., 28: 2933–2940.

Dukat, M., Wesolowska, A., Young, R. and Glennon, R.A. (2007) The 5-HT3 receptor partial agonist MD-354 (meta-chlorophenylguanidine) enhances the discriminative stimulus actions of (+)amphetamine in rats. Pharmacol. Biochem. Behav., 87: 203–207.

Durazzo, T.C., Gauvin, D.V., Goulden, K.L., Briscoe, R.J. and Holloway, F.A. (1994) Cocaine-induced conditioned place approach in rats: the role of dose and route of administration. Pharmacol. Biochem. Behav., 49: 1001–1005.

Ehrman, R.N., Robbins, S.J., Cornish, J.W., Childress, A.R. and O'Brien, C.P. (1996) Failure of ritanserin to block cocaine cue reactivity in humans. Drug Alcohol Depend., 42: 167–174.

El-Kadi, A.O.S. and Sharif, S.I. (1995) The role of 5-HT in the expression of morphine-withdrawal in mice. Life Sci., 57: 511–516.

Ettenberg, A. and Bernardi, R.E. (2007) Effects of buspirone on the immediate positive and delayed negative properties of intravenous cocaine as measured in the conditioned place preference test. Pharmacol. Biochem. Behav., 87: 171–178.

Fadda, P., Scherma, M., Fresu, A., Collu, M. and Fratta, W. (2005) Dopamine and serotonin release in dorsal striatum and nucleus accumbens is differentially modulated by morphine in DBA/2J and C57BL/6J mice. Synapse, 56: 29–38.

Fantegrossi, W.E., Ullrich, T., Rice, K.C., Woods, J.H. and Winger, G. (2002) 3,4-Methylenedioxymethamphetamine (MDMA, 'ecstasy') and its stereoisomers as reinforcers in rhesus monkeys: serotonergic involvement. Psychopharmacology (Berl.), 161: 356–364.

Fantegrossi, W.E., Godlewski, T., Karabenick, R.L., Stephens, J.M., Ullrich, T., Rice, K.C. and Woods, J.H. (2003) Pharmacological characterization of the effects of 3,4-methylenedioxymethamphetamine ('ecstasy') and its enantiomers on lethality, core temperature, and locomotor activity in singly housed and crowded mice. Psychopharmacology (Berl.), 166: 202–211.

Ferguson, S.M., Mitchell, E.S. and Neumaier, J.F. (2008) Increased expression of 5-HT6 receptors in the nucleus accumbens blocks the rewarding but not psychomotor activating properties of cocaine. Biol. Psychiat., 63: 207–213.

Ferre, S., Cortes, R. and Artigas, F. (1994) Dopaminergic regulation of the serotonergic raphe–striatal pathway – microdialysis studies in freely moving rats. J. Neurosci., 14: 4839–4846.

Filip, M. and Cunningham, K.A. (2002) Serotonin 5-HT2C receptors in nucleus accumbens regulate expression of the hyperlocomotive and discriminative stimulus effecs of cocaine. Pharmacol. Biochem. Behav., 71: 745–756.

Filip, M. and Cunningham, K.A. (2003) Hyperlocomotive and discriminative stimulus effects of cocaine are under the control of serotonin(2C) (5-HT(2C)) receptors in rat prefrontal cortex. J. Pharmacol. Exp. Ther., 306: 734–743.

Filip, M., Nowak, E., Papla, I. and Przegalinski, E. (2001a) Role of 5-hydroxytryptamine1B receptors and 5-hydroxytryptamine uptake inhibition in the cocaine-evoked discriminative stimulus effects in rats. J. Physiol. Pharmacol., 52: 249–263.

Filip, M., Nowak, E., Baran, L. and Przegalinski, E. (2001b) Serotonin1B receptor ligands in the nucleus accumbens shell do not affect the discriminative stimulus effects of amphetamine in rats. Pol. J. Pharmacol., 53: 449–457.

Filip, M., Papla, I., Nowak, E., Jungersmith, K. and Przegalinski, E. (2002) Effects of serotonin (5-HT)(1B) receptor ligands, microinjected into accumbens subregions, on cocaine discrimination in rats. Naunyn Schmiedebergs Arch. Pharmacol., 366: 226–234.

Filip, M., Papla, I., Nowak, E., Czepiel, K. and Przegalinski, E. (2003) Effects of 5-HT1B receptor ligands microinjected into the ventral tegmental area on cocaine discrimination in rats. Eur. J. Pharmacol., 459: 239–245.

Filip, M., Bubar, M.J. and Cunningham, K.A. (2004) Contribution of serotonin (5-hydroxytryptamine; 5-HT) 5-HT2 receptor subtypes to the hyperlocomotor effects of cocaine: acute and chronic pharmacological analyses. J. Pharmacol. Exp. Ther., 310: 1246–1254.

Filip, M., Frankowska, M., Zaniewska, M., Golda, A. and Przegalinski, E. (2005) The serotonergic system and its role in cocaine addiction. Pharmacol. Rep., 57: 685–700.

Filip, M., Bubar, M.J. and Cunningham, K.A. (2006) Contribution of serotonin (5-HT) 5-HT2 receptor subtypes to the discriminative stimulus effects of cocaine in rats. Psychopharmacology (Berl.), 183: 482–489.

Fletcher, P.J. and Korth, K.M. (1999) RU-24969 disrupts d-amphetamine self-administration and responding for conditioned reward via stimulation of 5-HT1B receptors. Behav. Pharmacol., 10: 183–193.

Fletcher, P.J., Korth, K.M. and Chambers, J.W. (1999) Depletion of brain serotonin following intra-raphe injections of 5,7-dihydroxytryptamine does not alter d-amphetamine self-administration across different schedule and access conditions. Psychopharmacology (Berl.), 146: 185–193.

Fletcher, P., Grottick, A. and Higgins, G. (2002a) Differential effects of the 5-HT(2A) receptor antagonist M100,907 and the 5-HT(2C) receptor antagonist SB242,084 on cocaine-induced locomotor activity, cocaine self-administration and cocaine-induced reinstatement of responding. Neuropsychopharmacology, 27: 576–586.

Fletcher, P.J., Azampanah, A. and Korth, K.M. (2002b) Activation of 5-HT1B receptors in the nucleus accumbens reduces self-administration of amphetamine on a progressive ratio schedule. Pharmacol. Biochem. Behav., 71: 717–725.

Fletcher, P.J., Korth, K.M., Robinson, S.R. and Baker, G.B. (2002c) Multiple 5-HT receptors are involved in the effects of acute MDMA treatment: studies on locomotor activity and responding for conditioned reinforcement. Psychopharmacology, 162: 282–291.

Fletcher, P.J., Sinyard, J., Salsali, M. and Baker, G.B. (2004a) Fluoxetine, but not sertraline or citalopram, potentiates the locomotor stimulant effect of cocaine: possible pharmacokinetic effects. Psychopharmacology (Berl.), 174: 406–413.

Fletcher, P.J., Chintoh, A.F., Sinyard, J. and Higgins, G.A. (2004b) Injection of the 5-HT2C receptor agonist Ro60-0175 into the ventral tegmental area reduces cocaine-induced locomotor activity and cocaine self-administration. Neuropsychopharmacology, 29: 308–318.

Fletcher, P.J., Sinyard, J. and Higgins, G.A. (2006) The effects of the 5-HT2C receptor antagonist SB242084 on locomotor activity induced by selective, or mixed, indirect serotonergic and dopaminergic agonists. Psychopharmacology, 187: 515–525.

Fletcher, P.J., Rizos, Z., Sinyard, J., Tampakeras, M. and Higgins, G.A. (2008) The 5-HT2C receptor agonist Ro60-0175 reduces cocaine self-administration and reinstatement induced by the stressor yohimbine, and contextual cues. Neuropsychopharmacology, 33: 1402–1412.

Frankel, P.S. and Cunningham, K.A. (2004) m-Chlorophenylpiperazine (mCPP) modulates the discriminative stimulus effects of cocaine through actions at the 5-HT2C receptor. Behav. Neurosci., 118: 157–162.

Frantz, K.J., Hansson, K.J., Stouffer, D.G. and Parsons, L.H. (2002) 5-HT(6) receptor antagonism potentiates the behavioral and neurochemical effects of amphetamine but not cocaine. Neuropharmacology, 42: 170–180.

Gartside, S.E., McQuade, R. and Sharp, T. (1996) Effects of repeated administration of 3,4-methylenedioxymethamphetamine on 5-hydroxytryptamine neuronal activity and release in the rat brain *in vivo*. J. Pharmacol. Exp. Ther., 279: 277–283.

Gartside, S.E., McQuade, R. and Sharp, T. (1997) Acute effects of 3,4-methylenedioxymethamphetamine (MDMA) on 5-HT cell firing and release: comparison between dorsal and median raphe 5-HT systems. Neuropharmacology, 36: 1697–1703.

Gately, P.F., Poon, S.L., Segal, D.S. and Geyer, M.A. (1985) Depletion of brain serotonin by 5,7-dihydroxytryptamine alters the response to amphetamine and the habituation of locomotor activity in rats. Psychopharmacology (Berl.), 87: 400–405.

Gawin, F.H. (1991) Cocaine addiction: psychology and neurophysiology. Science, 251: 1580–1586.

Gerra, G., Garofano, L., Santoro, G., Bosari, S., Pellegrini, C., Zaimovic, Z., Moi, G., Bussandri, M., Moi, A., Brambilla, F. and Donnini, C. (2004) Association between low-activity serotonin transporter genotype and heroin dependence: behavioral and personality correlates. Am. J. Med. Gen. Part B-Neuropsychiatr. Genet., 126B: 37–42.

Geyer, M.A., Russo, P.V., Segal, D.S. and Kuczenski, R. (1987) Effects of apomorphine and amphetamine on patterns of locomotor and investigatory behavior in rats. Pharmacol. Biochem. Behav., 28: 393–399.

Gillies, D.M., Mylecharane, E.J. and Jackson, D.M. (1996) Effects of 5-HT3 receptor-selective agents on locomotor activity in rats following injection into the nucleus accumbens and the ventral tegmental area. Eur. J. Pharmacol., 303: 1–12.

Glennon, R.A. and Young, R. (2000) MDMA stimulus generalization to the 5-HT1A serotonin agonist 8-hydroxy-2-(di-n-propylamino)tetralin. Pharmacol. Biochem. Behav., 66: 483–488.

Glennon, R.A., Higgs, R., Young, R. and Issa, H. (1992) Further-studies on N-methyl-1(3, 4-methylenedioxyphenyl)-2-aminopropane as a discriminative stimulus – antagonism by 5-hydroxytryptamine3 antagonists. Pharmacol. Biochem. Behav., 43: 1099–1106.

Gold, L.H. and Balster, R.L. (1992) Effects of buspirone and gepirone on i, v. cocaine self-administration in rhesus monkeys. Psychopharmacology (Berl.), 108: 289–294.

Gold, L.H., Koob, G.F. and Geyer, M.A. (1988) Stimulant and hallucinogenic behavioral profiles of 3,4-methylenedioxymethamphetamine and N-ethyl-3,4-methylenedioxyamphetamine in rats. J. Pharmacol. Exp. Ther., 247: 547–555.

Gough, B., Ali, S.F., Slikker, W. and Holson, R.R. (1991) Acute effects of 3,4-methylenedioxymethamphetamine (MDMA) on monoamines in rat caudate. Pharmacol. Biochem. Behav., 39: 619–623.

Grabowski, J., Rhoades, H., Elk, R., Schmitz, J., Davis, C., Creson, D. and Kirby, K. (1995) Fluoxetine is ineffective for treatment of cocaine dependence or concurrent opiate and cocaine dependence: two placebo-controlled double-blind trials. J. Clin. Psychopharmacol., 15: 163–174.

Grady, T.A., Broocks, A., Canter, S.K., Pigott, T.A., Dubbert, B., Hill, J.L. and Murphy, D.L. (1996) Biological and behavioral responses to D-amphetamine, alone and in combination with the serotonin3 receptor antagonist ondansetron, in healthy volunteers. Psychiat. Res., 64: 1–10.

Grauer, S.M., Tao, R. and Auerbach, S.B. (1992) Morphine induces an increase in extracellular serotonin in the rat diencephalon. Brain Res., 599: 277–282.

Gray, A.M. (2002) The effects of fluvoxamine and sertraline on the opioid withdrawal syndrome: a combined *in vivo* cerebral microdialysis and behavioral study. Eur. Neuropsychopharmacol., 12: 245–254.

Green, A.R., Mechan, A.O., Elliott, J.M., O'Shea, E. and Colado, M.I. (2003) The pharmacology and clinical pharmacology of 3,4-methylenedioxymethamphetamine (MDMA, 'ecstasy'). Pharmacol. Rev., 55: 463–508.

Grottick, A.J., Fletcher, P.J. and Higgins, G.A. (2000) Studies to investigate the role of 5-HT(2C) receptors on cocaine- and food-maintained behavior. J. Pharmacol. Exp. Ther., 295: 1183–1191.

Grottick, A.J., Corrigal, W.A. and Higgins, G.A. (2001) Activation of 5-HT(2C) receptors reduces the locomotor and rewarding effects of nicotine. Psychopharmacology, 157: 292–298.

Gulati, A. and Bhargava, H.N. (1988) Cerebral cortical 5-Ht1 and 5-Ht2 receptors of morphine tolerant-dependent rats. Neuropharmacology, 27: 1231–1237.

Gulati, A. and Bhargava, H.N. (1989) Brain and spinal-cord 5-Ht2 receptors of morphine-tolerant-dependent and morphine-abstinent rats. Eur. J. Pharmacol., 167: 185–192.

Gulati, A. and Bhargava, H.N. (1990) Down-regulation of hypothalamic 5-HT1A receptors in morphine-abstinent rats. Eur. J. Pharmacol., 182: 253–259.

Harris, G.C. and Aston-Jones, G. (2001) Augmented accumbal serotonin levels decrease the preference for a morphine associated environment during withdrawal. Neuropsychopharmacology, 24: 75–85.

Harrison, A.A., Liem, Y.T.B. and Markou, A. (2001) Fluoxetine combined with a serotonin-1A receptor antagonist reversed reward deficits observed during nicotine and amphetamine withdrawal in rats. Neuropsychopharmacology, 25: 55–71.

Hayes, D.J., Mosher, T.M. and Greenshaw, A.J. (2008) Differential effects of 5-HT(2C) receptor activation by WAY 161503 on nicotine-induced place conditioning and locomotor activity in rats. Behav. Brain Res., in press, doi:10.1016/j.bbr.2008.08.034.

Hayslett, R.L. and Tizabi, Y. (2005) Effects of donepezil, nicotine and haloperidol on the central serotonergic system in mice: Implications for Tourette's syndrome. Pharmacol. Biochem. Behav., 81: 879–886.

Hedou, G., Homberg, J., Martin, S., Wirth, K., Feldon, J. and Heidbreder, C.A. (2000) Effect of amphetamine on extracellular acetylcholine and monoamine levels in subterritories of the rat medial prefrontal cortex. Eur. J. Pharmacol., 390: 127–136.

Herges, S. and Taylor, D.A. (1998) Involvement of serotonin in the modulation of cocaine-induced locomotor activity in the rat. Pharmacol. Biochem. Behav., 59: 595–611.

Herges, S. and Taylor, D.A. (1999) Modulation of cocaine-induced locomotor activity, rears and head bobs by application of WAY100635 into the dorsal and median raphe nuclei of the rat. Naunyn Schmiedebergs Arch. Pharmacol., 360: 129–134.

Herges, S. and Taylor, D.A. (2000) Involvement of 5-HT(3) receptors in the nucleus accumbens in the potentiation of cocaine-induced behaviors in the rat. Br. J. Pharmacol., 131: 1294–1302.

Herin, D.V., Liu, S., Ullrich, T., Rice, K.C. and Cunningham, K.A. (2005) Role of the serotonin 5-HT2A receptor in the hyperlocomotive and hyperthermic effects of (+)-3,4-methylenedioxymethamphetamine. Psychopharmacology (Berl.), 178: 505–513.

Hernandez, L., Lee, F. and Hoebel, B.G. (1987) Simultaneous microdialysis and amphetamine infusion in the nucleus accumbens and striatum of freely moving rats – increase in extracellular dopamine and serotonin. Brain Res. Bull., 19: 623–628.

Higgins, G.A., Nguyen, P., Joharchi, N. and Sellers, E.M. (1991) Effects of 5-Ht3 receptor antagonists on behavioral measures of naloxone-precipitated opioid withdrawal. Psychopharmacology, 105: 322–328.

Higgins, G.A., Joharchi, N., Nguyen, P. and Sellers, E.M. (1992) Effect of the 5-Ht3 receptor antagonists, Mdl72222 and ondansetron on morphine place conditioning. Psychopharmacology, 106: 315–320.

Higgins, G.A., Wang, Y.H. and Sellers, E.M. (1993) Preliminary findings with the indirect 5-Ht agonist dexfenfluramine on heroin discrimination and self-administration in rats. Pharmacol. Biochem. Behav., 45: 963–966.

Higgins, G.A., Wang, Y.P., Corrigall, W.A. and Sellers, E.M. (1994) Influence of 5-Ht3 receptor antagonists and the indirect 5-Ht agonist, dexfenfluramine, on heroin self-administration in rats. Psychopharmacology, 114: 611–619.

Hodge, C.W., Bratt, A.M. and Kelley, S.P. (2008) Deletion of the 5-HT(3A) receptor subunit blunts the induction of cocaine sensitization. Genes Brain Behav., 7: 96–102.

Homberg, J.R., Arends, B., Wardeh, G., Raaso, H.S., Schoffelmeer, A.N.M. and De Vries, T.J. (2004) Individual differences in the effects of serotonergic anxiolytic drugs on the motivation to self-administer cocaine. Neuroscience, 128: 121–130.

Homberg, J.R., De Boer, S.F., Raaso, H.S., Olivier, J.D., Verheul, M., Ronken, E., Cools, A.R., Ellenbroek, B.A., Schoffelmeer, A.N., Vanderschuren, L.J., De Vries, T.J. and Cuppen, E. (2008) Adaptations in pre- and postsynaptic 5-HT1A receptor function and cocaine supersensitivity in serotonin transporter knockout rats. Psychopharmacology (Berl.), 200: 367–380.

Hoplight, B.J., Vincow, E.S. and Neumaier, J.F. (2005) The effects of SB 224289 on anxiety and cocaine-related behaviors in a novel object task. Physiol. Behav., 84: 707–714.

Howell, L.L. and Byrd, L.D. (1995) Serotonergic modulation of the behavioral effects of cocaine in the squirrel monkey. J. Pharmacol. Exp. Ther., 275: 1551–1559.

Hui, S.C., Sevilla, E.L. and Ogle, C.W. (1993) 5-HT3 antagonists reduce morphine self-administration in rats. Br. J. Pharmacol., 110: 1341–1346.

Hyman, S.E., Malenka, R.C. and Nestler, E.J. (2006) Neural mechanisms of addiction: the role of reward-related learning and memory. Ann. Rev. Neurosci., 29: 565–598.

Hynes, M.D., Lochner, M.A., Bemis, K.G. and Hymson, D.L. (1985) Fluoxetine, a selective inhibitor of serotonin uptake, potentiates morphine analgesia without altering its discriminative stimulus properties or affinity for opioid receptors. Life Sci., 36: 2317–2323.

Ivanova, S. and Greenshaw, A.J. (1997) Nicotine-induced decreases in VTA electrical self-stimulation thresholds: blockade by haloperidol and mecamylamine but not scopolamine or ondansetron. Psychopharmacology, 134: 187–192.

Jackson, D.M., Johansson, C., Lindgren, L.M. and Bengtsson, A. (1994) Dopamine receptor antagonists block amphetamine and phencyclidine-induced motor stimulation in rats. Pharmacol. Biochem. Behav., 48: 465–471.

Ji, S.P., Zhang, Y., Van Cleemput, J., Jiang, W., Liao, M., Li, L., Wan, Q., Backstrom, J.R. and Zhang, X. (2006) Disruption of PTEN coupling with 5-HT2C receptors suppresses behavioral responses induced by drugs of abuse. Nat. Med., 12: 324–329.

Johanson, C.E., Balster, R.L. and Bonese, K. (1976) Self-administration of psychomotor stimulant drugs: the effects of unlimited access. Pharmacol. Biochem. Behav., 4: 45–51.

Joharchi, N., Sellers, E.M. and Higgins, G.A. (1993) Effect of 5-Ht(3) receptor antagonists on the discriminative stimulus properties of morphine in rats. Psychopharmacology, 112: 111–115.

Johnson, B.A., it-Daoud, N., Elkashef, A.M., Smith, E.V., Kahn, R., Vocci, F., Li, S.H. and Bloch, D.A. (2008) A preliminary randomized, double-blind, placebo-controlled study of the safety and efficacy of ondansetron in the treatment of methamphetamine dependence. Intl. J. Neuropsychopharmacol., 11: 1–14.

Kalivas, P.W. (2005) How we determine which drug-induced neuroplastic changes are important?. Nat. Neurosci., 8: 1440–1441.

Kalivas, P.W. and Volkow, N.D. (2005) The neural basis of addiction: a pathology of motivation and choice. Am. J. Psychiat., 162: 1403–1413.

Kalivas, P.W., Duffy, P., Dumars, L.A. and Skinner, C. (1988) Behavioral and neurochemical effects of acute and daily cocaine administration in rats. J. Pharmacol. Exp. Ther., 245: 485–492.

Kamien, J.B. and Woolverton, W.L. (1990) Buspirone blocks the discriminative stimulus effects of apomorphine in monkeys. Pharmacol. Biochem. Behav., 35: 117–120.

Kankaanpää, A., Meririnne, E., Lillsunde, P. and Seppala, T. (1998) The acute effects of amphetamine derivatives on extracellular serotonin and dopamine levels in rat nucleus accumbens. Pharmacol. Biochem. Behav., 59: 1003–1009.

Kankaanpää, A., Meririnne, E. and Seppala, T. (2002) 5-HT3 receptor antagonist MDL 72222 attenuates cocaine- and mazindol-, but not methylphenidate-induced neurochemical and behavioral effects in the rat. Psychopharmacology (Berl.), 159: 341–350.

Kehne, J.H., Ketteler, H.J., McCloskey, T.C., Sullivan, C.K., Dudley, M.W. and Schmidt, C.J. (1996) Effects of the selective 5-HT2A receptor antagonist MDL 100,907 on MDMA-induced locomotor stimulation in rats. Neuropsychopharmacology, 15: 116–124.

Kelley, A.E. (2004) Memory and addiction: shared neural circuitry and molecular mechanisms. Neuron, 44: 161–179.

Kelley, A.E. and Berridge, K.C. (2002) The neuroscience of natural reward: relevance to addictive drugs. J. Neurosci., 22: 3306–3311.

Kenny, P.J., File, S.E. and Rattray, M. (2001) Nicotine regulates 5-HT(1A) receptor gene expression in the cerebral cortex and dorsal hippocampus. Eur. J. Neurosci., 13: 1267–1271.

King, G.R., Joyner, C., Lee, T.H. and Ellinwood, E.H. (1993) Withdrawal from continuous or intermittent cocaine – effects of Nan-190 on cocaine-induced locomotion. Pharmacol. Biochem. Behav., 44: 253–262.

Kitanaka, N., Kitanaka, J. and Takemura, M. (2005a) Inhibition of methamphetamine-induced hyperlocomotion in mice by clorgyline, a monoamine oxidase-a inhibitor, through alteration of the 5-hydroxytriptamine turnover in the striatum. Neuroscience, 130: 295–308.

Kitanaka, N., Kitanaka, J. and Takemura, M. (2005b) Repeated clorgyline treatment inhibits methamphetamine-induced behavioral sensitization in mice. Neurochem. Res., 30: 445–451.

Klein, D.A. and Gulley, J.M. (2009) Reduced sensitivity to the locomotor-stimulant effects of cocaine is associated with increased sensitivity to its discriminative stimulus properties. Behav. Pharmacol., in press.

Kleven, M.S. and Koek, W. (1998) Discriminative stimulus properties of cocaine: enhancement by monoamine reuptake blockers. J. Pharmacol. Exp. Ther., 284: 1015–1025.

Koch, S. and Galloway, M.P. (1997) MDMA induced dopamine release in vivo: role of endogenous serotonin. J. Neural Transm., 104: 135–146.

Koetzner, L., Mastropaolo, J. and Deutsch, S.I. (1995) Facilitating role of m-chlorophenylbiguanide in a cocaine discrimination. Pharmacol. Biochem. Behav., 52: 815–818.

Kolb, B., Gorny, G., Li, Y.L., Samaha, A.N. and Robinson, T.E. (2003) Amphetamine or cocaine limits the ability of later experience to promote structural plasticity in the neocortex and nucleus accumbens. Proc. Natl. Acad. Sci. USA, 100: 10523–10528.

Koob, G.F. (1992) Drugs of abuse: anatomy, pharmacology and function of reward pathways. Trends Pharmacol. Sci., 13: 177–184.

Koob, G.F. and Le, M.M. (2008) Addiction and the brain antireward system. Ann. Rev. Psychol., 59: 29–53.

Kotler, M., Cohen, H., Kremer, I., Mel, H., Horowitz, R., Ohel, N., Gritsenko, I., Nemanov, L., Katz, M. and Ebstein, R. (1999) No association between the serotonin transporter promoter region (5-HTTLPR) and the dopamine D3 receptor (Ball D2DR) polymorphisms and heroin addiction. Mol. Psychiatry, 4: 313–314.

Kramer, J.C., Fischman, V.S. and Littlefield, D.C. (1967) Amphetamine abuse – pattern and effects of high doses taken intravenously. J. Am. Med. Assoc., 201: 89–93.

Kruszewska, A., Romandini, S. and Samanin, R. (1986) Differents effects of zimelidine on the reinforcing effects of d-amphetamine and morphine on conditioned place preference in rats. Eur. J. Pharmacol., 125: 283–286.

Kuczenski, R. and Segal, D. (1989) Concomitant characterization of behavioral and striatal neurotransmitter response to amphetamine using *in vivo* microdialysis. J. Neurosci., 9: 2051–2065.

Kuczenski, R. and Segal, D.S. (1997) Effects of methylphenidate on extracellular dopamine, serotonin, and norepinephrine: comparison with amphetamine. J. Neurochem., 68: 2032–2037.

Kuczenski, R. and Segal, D.S. (1999) Sensitization of amphetamine-induced stereotyped behaviors during the acute response. J. Pharmacol. Exp. Ther., 288: 699–709.

Kuczenski, R., Segal, D.S., Cho, A.K. and Melega, W. (1995) Hippocampus norepinephrine, caudate dopamine and serotonin,

and behavioral responses to the stereoisomers of amphetamine and methamphetamine. J. Neurosci., 15: 1308–1317.
Kuroki, T., Ichikawa, J., Dai, J. and Meltzer, H.Y. (1996) R(+)-8-OH-DPAT, a 5-HT1A receptor agonist, inhibits amphetamine-induced serotonin and dopamine release in rat medial prefrontal cortex. Brain Res., 743: 357–361.
Kusljic, S. and van den Buuse, M. (2006) Differential involvement of 5-HT projections within the amygdala in prepulse inhibition but not in psychotomimetic drug-induced hyperlocomotion. Behav. Brain Res., 168: 74–82.
Kusljic, S., Copolov, D.L. and van den Buuse, M. (2003) Differential role of serotonergic projections arising from the dorsal and median raphe nuclei in locomotor hyperactivity and prepulse inhibition. Neuropsychopharmacology, 28: 2138–2147.
Landabaso, M.A., Iraurgi, I., Jiménez-Lerma, J.M., Sanz, J., Fernádez de Corres, B., Araluce, K., Calle, R. and Gutiérrez-Fraile, M. (1998) A randomized trial of adding fluoxetine to a naltrexone treatment programm for heroin addicts. Addiction, 93: 739–744.
Lane, J.D., Pickering, C.L., Hooper, M.L., Fagan, K., Tyers, M.B. and Emmett Oglesby, M.W. (1992) Failure of ondansetron to block the discriminative or reinforcing stimulus effects of cocaine in the rat. Drug Alcohol Depend., 30: 151–162.
Lanteri, C., Salomon, L., Torrens, Y., Glowinski, J. and Tassin, J.P. (2008) Drugs of abuse specifically sensitize noradrenergic and serotonergic neurons via a non-dopaminergic mechanism. Neuropsychopharmacology, 33: 1724–1734.
Layer, R.T., Uretsky, N.J. and Wallace, L.J. (1992) Effect of serotonergic agonists in the nucleus-accumbens on D-amphetamine-stimulated locomotion. Life Sci., 50: 813–820.
Leccese, A.P. and Lyness, W.H. (1984) The effects of putative 5-hydroxytryptamine receptor active agents on D-amphetamine self-administration in controls and rats with 5,7-dihydroxytryptamine median forebrain bundle lesions. Brain Res., 303: 153–162.
Lehmann, O., Jeltsch, H., Bertrand, F., Lazarus, C., Will, B. and Cassel, J. (2000) When injected into the fimbria–fornix/cingular bundle, not in the raphe, 5,7-dihydroxytryptamine prevents amphetamine-induced hyperlocomotion. Behav. Brain Res., 114: 213–217.
Lerman, C., Caporaso, N.E., Audrain, J., Main, D., Boyd, N.R. and Shields, P.G. (2000) Interacting effects of the serotonin transporter gene and neuroticism in smoking practices and nicotine dependence. Mol. Psychiatry, 5: 189–192.
Lerman, C., Caporaso, N.E., Bush, A., Zheng, Y.L., Audrain, J., Main, D. and Shields, P.G. (2001) Tryptophan hydroxylase gene variant and smoking behavior. Am. J. Med. Gen., 105: 518–520.
Levin, E.D., Slade, S., Johnson, M., Petro, A., Horton, K., Williams, P., Rezvani, A.H. and Rose, J.E. (2009) Ketanserin, a 5-HT2 receptor antagonist, decreases nicotine self-administration in rats. Eur. J. Pharmacol., 600: 93–97.
Li, T., Liu, X.H., Zhao, J.H., Hu, X., Ball, D.M., Loh, E.W., Sham, P.C. and Collier, D.A. (2002) Allelic association analysis of the dopamine D2, D3, 5-HT2A, and GABA(A) gamma 2 receptors and serotonin transporter genes with heroin abuse in Chinese subjects. Am. J. Med. Gen., 114: 329–335.
Liang, Y., Boules, M., Shaw, A.M., Williams, K., Fredrickson, P. and Richelson, E. (2008) Effect of novel neurotensin analog, NT69L, on nicotine-induced alteration in monoamine levels in rat brain. Brain Res., 1231: 6–15.
Liechti, M.E., Baumann, C., Gamma, A. and Vollenweider, F.X. (2000a) Acute psychological effects of 3,4-methylenedioxymethamphetamine (MDMA, 'Ecstasy') are attenuated by the serotonin uptake inhibitor citalopram. Neuropsychopharmacology, 22: 513–521.
Liechti, M.E., Saur, M.R., Gamma, A., Hell, D. and Vollenweider, F.X. (2000b) Psychological and physiological effects of MDMA ('Ecstasy') after pretreatment with the 5-HT(2) antagonist ketanserin in healthy humans. Neuropsychopharmacology, 23: 396–404.
Lipska, B.K., Jaskiw, G.E., Arya, A. and Weinberger, D.R. (1992) Serotonin depletion causes long-term reduction of exploration in the rat. Pharmacol. Biochem. Behav., 43: 1247–1252.
Liu, S. and Cunningham, K.A. (2006) Serotonin2C receptors (5-HT2C R) control expression of cocaine-induced conditioned hyperactivity. Drug Alcohol Depend., 81: 275–282.
Loh, E.A. and Roberts, D.C. (1990) Break-points on a progressive ratio schedule reinforced by intravenous cocaine increase following depletion of forebrain serotonin. Psychopharmacology, 101: 262–266.
Lyness, W.H., Friedle, N.M. and Moore, K.E. (1980) Increased self-administration of d-amphetamine after destruction of 5-hydroxytryptaminergic neurons. Pharmacol. Biochem. Behav., 12: 937–941.
Ma, Z., Strecker, R.E., McKenna, J.T., Thakkar, M.M., McCarley, R.W. and Tao, R. (2005) Effects on serotonin of (−)nicotine and dimethylphenylpiperazinium in the dorsal raphe and nucleus accumbens of freely behaving rats. Neuroscience, 135: 949–958.
Maj, J., Chojnackawojcik, E., Tatarczynska, E. and Klodzinska, A. (1987) Central action of ipsapirone, a new anxiolytic drug, on serotoninergic, noradrenergic and dopaminergic functions. J. Neural Transm., 70: 1–17.
Maj, J., Rogoz, Z., Skuza, G. and Wedzony, K. (1996) The synergistic effect of fluoxetine on the locomotor hyperactivity induced by MK-801, a non-competitive NMDA receptor antagonist. J. Neural Transm., 103: 131–146.
Maleki, S.A., Samini, M., Babapour, V., Mehr, S.E., Cheraghiyan, S. and Nouri, M.H.K. (2008) Potentiation of morphine-induced conditioned place preference with concurrent use of amantadine and fluvoxamine by the intraperitoneal and intracerebroventricular injection in rat. Behav. Brain Res., 190: 189–192.
Mangiavacchi, S., Masi, F., Scheggi, S., Leggio, B., De Montis, M.G. and Gambarana, C. (2001) Long-term behavioral and neurochemical effects of chronic stress exposure in rats. J. Neurochem., 79: 1113–1121.
Mansbach, R.S., Rovetti, C.C. and Freedland, C.S. (1998) The role of monoamine neurotransmitter systems in the nicotine discriminative stimulus. Drug Alcohol Depend., 52: 125–134.
Markou, A. and Koob, G.F. (1992) Construct-validity of a self-stimulation threshold paradigm – effects of reward and performance manipulations. Physiol. Behav., 51: 111–119.
Markou, A., Kosten, T.R. and Koob, G.F. (1998) Neurobiological similarities in depression and drug dependence: a self-medication hypothesis. Neuropsychopharmacology, 18: 135–174.
Markou, A., Harrison, A.A., Chevrette, J. and Hoyer, D. (2005) Paroxetine combined with a 5-HT1A receptor antagonist reversed reward deficits observed during amphetamine withdrawal in rats. Psychopharmacology, 178: 133–142.
Marquis, K.L., Sabb, A.L., Logue, S.F., Brennan, J.A., Piesla, M.J., Comery, T.A., Grauer, S.M., Ashby, C.R. Jr., Nguyen,

H.Q., Dawson, L.A., Barrett, J.E., Stack, G., Meltzer, H.Y., Harrison, B.L. and Rosenzweig-Lipson, S. (2007) WAY-163909 [(7bR,10aR)-1,2,3,4,8,9,10,10a-octahydro-7bH-cyclopenta-[b][1,4]diazepino[6,7,1hi]indole]: a novel 5-hydroxytryptamine 2C receptor-selective agonist with preclinical antipsychotic-like activity. J. Pharmacol. Exp. Ther., 320: 486–496.

McCann, U.D. and Ricaurte, G.A. (1993) Reinforcing subjective effects of (+/−) 3,4-methylenedioxymethamphetamine ('ecstasy') may be separable from its neurotoxic actions: clinical evidence. J. Clin. Psychopharmacol., 13: 214–217.

McCreary, A.C., Bankson, M.G. and Cunningham, K.A. (1999) Pharmacological studies of the acute and chronic effects of (+)-3,4-methylenedioxymethamphetamine on locomotor activity: role of 5-hydroxytryptamine(1A) and 5-hydroxytryptamine(1B/1D) receptors. J. Pharmacol. Exp. Ther., 290: 965–973.

McGregor, A., Lacosta, S. and Roberts, D.C. (1993) L-tryptophan decreases the breaking point under a progressive ratio schedule of intravenous cocaine reinforcement in the rat. Pharmacol. Biochem. Behav., 44: 651–655.

McMahon, L.R. and Cunningham, K.A. (1999) Antagonism of 5-hydroxytryptamine(4) receptors attenuates hyperactivity induced by cocaine: putative role for 5-hydroxytryptamine(4) receptors in the nucleus accumbens shell. J. Pharmacol. Exp. Ther., 291: 300–307.

McMahon, L.R., Filip, M. and Cunningham, K.A. (2001) Differential regulation of the mesoaccumbens circuit by serotonin 5-hydroxytryptamine (5-HT)2A and 5-HT2C receptors. J. Neurosci., 21: 7781–7787.

Mechan, A.O., Esteban, B., O'Shea, E., Elliott, J.M., Colado, M.I. and Green, A.R. (2002) The pharmacology of the acute hyperthermic response that follows administration of 3,4-methylenedioxymethamphetamine (MDMA, 'ecstasy') to rats. Br. J. Pharmacol., 135: 170–180.

Miczek, K.A. and Yoshimura, H. (1982) Disruption of primate social-behavior by D-amphetamine and cocaine – differential antagonism by antipsychotics. Psychopharmacology, 76: 163–171.

Millan, M.J., Schreiber, R., Dekeyne, A., Rivet, J.M., Bervoets, K., Mavridis, M., Sebban, C., Maurel-Remy, S., Newman-Tancredi, A., Spedding, M., Muller, O., Lavielle, G. and Brocco, M. (1998) S 16924 ((R)-2-{1-[2-(2,3-dihydro-benzo[1,4]dioxin-5-yloxy)-ethyl]-pyrrolidin-3yl}-1-(4-fluoro-phenyl)-ethanone), a novel, potential antipsychotic with marked serotonin (5-HT)(1A) agonist properties: II. Functional profile in comparison to clozapine and haloperidol. J. Pharmacol. Exp. Ther., 286: 1356–1373.

Millan, M.J., Brocco, M., Gobert, A., Joly, F., Bervoets, K., Rivet, J.M., Newman-Tancredi, A., Audinot, V. and Maurel, S. (1999) Contrasting mechanisms of action and sensitivity to antipsychotics of phencyclidine versus amphetamine: importance of nucleus accumbens 5-HT2A sites for PCP-induced locomotion in the rat. Eur. J. Neurosci., 11: 4419–4432.

Miranda, F., Sandoval-Sanchez, A., Cedillo, L.N., Jimenez, J.C., Millan-Mejia, P. and Velazquez-Martinez, D.N. (2007) Modulatory role of 5-HT1B receptors in the discriminative signal of amphetamine in the conditioned taste aversion paradigm. Pharmacol. Rep., 59: 517–524.

Miyamoto, K., Yoshimura, R., Ueda, N., Sugita, A., Umene, W., Hori, H., Mitoma, M. and Nakamura, J. (2007) Effects of acute paroxetine treatment on the consumption of cigarette smoking and caffeine in depressed patients. Hum. Psychopharmacol. Clin. Exp., 22: 483–490.

Moeller, F.G., Schmitz, J.M., Steinberg, J.L., Green, C.M., Reist, C., Lai, L.Y., Swann, A.C. and Grabowski, J. (2007) Citalopram combined with behavioral therapy reduces cocaine use: a double-blind, placebo-controlled trial. Am. J. Drug Alcohol Abuse, 33: 367–378.

Molina, P.E., Ahmed, N., Gatley, J., Volkow, N.D. and Abumrad, N.N. (2001) L-tryptophan attenuation of the dopaminergic and behavioral responses to cocaine. Life Sci., 69: 1897–1906.

Morrow, B.A. and Roth, R.H. (1996) Serotonergic lesions alter cocaine-induced locomotor behavior and stress-activation of the mesocorticolimbic dopamine system. Synapse, 23: 174–181.

Moser, P.C. (1992) The effect of 5-HT3 receptor antagonists on the discriminative stimulus effects of amphetamine. Eur. J. Pharmacol., 212: 271–274.

Moser, P.C., Moran, P.M., Frank, R.A. and Kehne, J.H. (1996) Reversal of amphetamine-induced behaviors by MDL 100,907, a selective 5-HT2A antagonist. Behav. Brain Res., 73: 163–167.

Mosner, A., Kuhlman, G., Roehm, C. and Vogel, W.H. (1997) Serotonergic receptors modify the voluntary intake of alcohol and morphine but not of cocaine and nicotine by rats. Pharmacology, 54: 186–192.

Müller, C.P. and Carey, R.J. (2006) Intracellular 5-HT 2C receptor dephosphorylation: a new target for treating drug addiction. Trends Pharmacol. Sci., 27: 455–458.

Müller, C.P. and Huston, J.P. (2006) Determining the region-specific contributions of 5-HT receptors to the psychostimulant effects of cocaine. Trends Pharmacol. Sci., 27: 105–112.

Müller, C.P., De Souza Silva, M.A., DePalma, G., Tomaz, C., Carey, R.J. and Huston, J.P. (2002a) The selective serotonin(1A) receptor antagonist WAY 100635 blocks behavioral stimulating effects of cocaine but not ventral striatal dopamine increase. Behav. Brain Res., 134: 337–346.

Müller, C.P., Carey, R.J., De Souza Silva, M.A., Jocham, G. and Huston, J.P. (2002b) Cocaine increases serotonergic activity in the hippocampus and nucleus accumbens *in vivo*: 5-HT1a receptor antagonism blocks behavioral but potentiates serotonergic activation. Synapse, 45: 67–77.

Müller, C.P., Carey, R.J., Salloum, J.B. and Huston, J.P. (2003) Serotonin1A receptor agonism attenuates the cocaine-induced increase in serotonin levels in the hippocampus and nucleus accumbens but potentiates hyperlocomotion: an *in vivo* microdialysis study. Neuropharmacology, 44: 592–603.

Müller, C.P., Thonnessen, H., Barros, M., Tomaz, C., Carey, R.J. and Huston, J.P. (2004a) Hippocampus 5-HT(1A) receptors attenuate cocaine-induced hyperlocomotion and the increase in hippocampal but not nucleus accumbens 5-HT. Hippocampus, 14: 710–721.

Müller, C.P., Thonnessen, H., De Souza Silva, M.A., Fink, H., Bert, B., Carey, R.J. and Huston, J.P. (2004b) Nucleus accumbens serotonin1A receptors control cocaine-induced hyperactivity but not local serotonin increase: an *in vivo* microdialysis study. Neuropharmacology, 47: 205–215.

Müller, C.P., Carey, R.J., Huston, J.P. and De Souza Silva, M.A. (2007a) Serotonin and psychostimulant addiction: focus on 5-HT1A receptors. Prog. Neurobiol., 81: 133–178.

Müller, C.P., De Souza Silva, M.A. and Huston, J.P. (2007b) Double dissociating effects of sensory stimulation and cocaine on serotonin activity in the occipital and temporal cortices. Neuropharmacology, 52: 854–862.

Munn, E.M. and Borszcz, G.S. (2002) Increases in the release and metabolism of serotonin in nucleus parafascicularis thalami following systemically administered morphine in the rat. Neurosci. Lett., 332: 151–154.

Munzar, P., Baumann, M.H., Shoaib, M. and Goldberg, S.R. (1999a) Effects of dopamine and serotonin-releasing agents on methamphetamine discrimination and self-administration in rats. Psychopharmacology (Berl.), 141: 287–296.

Munzar, P., Laufert, M.D., Kutkat, S.W., Novakova, J. and Goldberg, S.R. (1999b) Effects of various serotonin agonists, antagonists, and uptake inhibitors on the discriminative stimulus effects of methamphetamine in rats. J. Pharmacol. Exp. Ther., 291: 239–250.

Munzar, P., Justinova, Z., Kutkat, S.W. and Goldberg, S.R. (2002) Differential involvement of 5-HT(2A) receptors in the discriminative-stimulus effects of cocaine and methamphetamine. Eur. J. Pharmacol., 436: 75–82.

Nader, M.A. and Woolverton, W.L. (1994) Blockade of the discriminative stimulus effects of D-amphetamine in rhesus monkeys with serotonin 5-Ht1A agonists. Behav. Pharmacol., 5: 591–598.

Nakamura, S., Ago, Y., Hayashi, A., Itho, S., Kakuda, M., Hashimoto, H., Baba, A. and Matsuda, T. (2006) Modification of cocaine-induced behavioral and neurochemical effects by serotonin1A receptor agonist/antagonst in mice. Synapse, 60: 479–484.

Neal, B.S. and Sparber, S.B. (1986a) Mianserin attenuates naloxone-precipitated withdrawal signs in rats acutely or chronically dependent upon morphine. J. Pharmacol. Exp. Ther., 236: 157–165.

Neal, B.S. and Sparber, S.B. (1986b) Ketanserin and pirenperone attenuate acute morphine withdrawal in rats. Eur. J. Pharmacol., 132: 299–304.

Neisewander, J.L. and Acosta, J.I. (2007) Stimulation of 5-HT2C receptors attenuates cue and cocaine-primed reinstate ment of cocaine-seeking behavior in rats. Behav. Pharmacol., 18: 791–800.

Nestler, E.J. (2002) Common molecular and cellular substrates of addiction and memory. Neurobiol. Learn. Mem., 78: 637–647.

Neumaier, J.F., Vincow, E.S., Arvanitogiannis, A., Wise, R.A. and Carlezon, W.A. (2002) Elevated expression of 5-HT1B receptors in nucleus accumbens efferents sensitizes animals to cocaine. J. Neurosci., 22: 10856–10863.

Niaura, R., Spring, B., Borrelli, B., Hedeker, D., Goldstein, M.G., Keuthen, N., Depue, J., Kristeller, J., Ockene, J., Prochazka, A., Chiles, J.A. and Abrams, D.B. (2002) Multicenter trial of fluoxetine as an adjunct to behavioral smoking cessation treatment. J. Cons. Clin. Psychol., 70: 887–896.

Nielsen, D.A., Barral, S., Proudnikov, D., Kellogg, S., Ho, A., Ott, J. and Kreek, M.J. (2008) TPH2 and TPH1: association of variants and interactions with heroin addiction. Behav. Genet., 38: 133–150.

Nomikos, G.G. and Spyraki, C. (1988) Effects of ritanserine on the rewarding properties of d-amphetamine, morphine and diazepam revealed by conditioned place preference in rats. Pharmacol. Biochem. Behav., 30: 853–858.

Nutt, D., King, L.A., Saulsbury, W. and Blakemore, C. (2007) Development of a rational scale to assess the harm of drugs of potential misuse. Lancet, 369: 1047–1053.

O'Neill, M.F., Heron-Maxwell, C.L. and Shaw, G. (1999) 5-HT2 receptor antagonism reduces hyperactivity induced by amphetamine, cocaine, and MK-801 but not D1 agonist C-APB. Pharmacol. Biochem. Behav., 63: 237–243.

Olausson, P., Engel, J.A. and Soderpalm, B. (1999) Behavioral sensitization to nicotine is associated with behavioral disinhibition; counteraction by citalopram. Psychopharmacology, 142: 111–119.

Olausson, P., Akesson, P., Engel, J.A. and Soderpalm, B. (2001a) Effects of 5-HT1A and 5-HT2 receptor agonists on the behavioral and neurochemical consequences of repeated nicotine treatment. Eur. J. Pharmacol., 420: 45–54.

Olausson, P., Akesson, P., Petersson, A., Engel, J.A. and Soderpalm, B. (2001b) Behavioral and neurochemical consequences of repeated nicotine treatment in the serotonin-depleted rat. Psychopharmacology, 155: 348–361.

Olds, J. and Milner, P. (1954) Positive reinforcement produced by electrical stimulation of septal area and other regions of rat brain. J. Comp. Physiol. Psychol., 47: 419–427.

Olmstead, M.C. (2006) Animal models of drug addiction: where do we go from here? Q. J. Exp. Psychol., Colchester, 59: 625–653.

Papla, I., Filip, M., Przegalinski, E., Filip, M., Nowak, E., Baran, L. and Przegalinski, E. (2002) Effect of intra-tegmental microinjections of 5-HT1B receptor ligands on the amphetamine-induced locomotor hyperactivity in rats. Pol. J. Pharmacol., 54: 351–357.

Paris, J.M. and Cunningham, K.A. (1991) Serotonin 5-HT3 antagonists do not alter the discriminative stimulus properties of cocaine. Psychopharmacology (Berl.), 104: 475–478.

Parsons, L.H. and Justice, J.-B.J. (1993) Serotonin and dopamine sensitization in the nucleus accumbens, ventral tegmental area, and dorsal raphe nucleus following repeated cocaine administration. J. Neurochem., 61: 1611–1619.

Parsons, L.H., Koob, G.F. and Weiss, F. (1995) Serotonin dysfunction in the nucleus accumbens of rats during withdrawal after unlimited access to intravenous cocaine. J. Pharmacol. Exp. Ther., 274: 1182–1191.

Parsons, L.H., Weiss, F. and Koob, G.F. (1998) Serotonin1B receptor stimulation enhances cocaine reinforcement. J. Neurosci., 18: 10078–10089.

Patkar, A.A., Berrettini, W.H., Hoehe, M., Hill, K.P., Gottheil, E., Thornton, C.C. and Weinstein, S.P. (2002) No association between polymorphisms in the serotonin transporter gene and susceptibility to cocaine dependence among African-American individuals. Psychiatr. Genet., 12: 161–164.

Paulus, M.P., Callaway, C.W. and Geyer, M.A. (1993) Quantitative assessment of the microstructure of rat behavior. 2. Distinctive effects of dopamine releasers and uptake inhibitors. Psychopharmacology, 113: 187–198.

Pei, Q., Zetterstrom, T., Leslie, R.A. and Grahamesmith, D.G. (1993) 5-Ht3 receptor antagonists inhibit morphine-induced stimulation of mesolimbic dopamine release and function in the rat. Eur. J. Pharmacol., 230: 63–68.

Peltier, R. and Schenk, S. (1993) Effects of serotonergic manipulations on cocaine self-administration in rats. Psychopharmacology, 110: 390–394.

Peroutka, S.J., Newman, H. and Harris, H. (1988) Subjective effects of 3,4-methylenedioxymethamphetamine in recreational users. Neuropsychopharmacology, 1: 273–277.

Pinelli, A., Trivulzio, S. and Tomasoni, L. (1997) Effects of ondansetron administration on opioid withdrawal syndrome observed in rats. Eur. J. Pharmacol., 340: 111–119.

Platt, D.M., Grech, D.M., Rowlett, J.K. and Spealman, R.D. (1999) Discriminative stimulus effects of morphine in squirrel monkeys: stimulants, opioids, and stimulant-opioid combinations. J. Pharmacol. Exp. Ther., 290: 1092–1100.

Porrino, L.J., Ritz, M.C., Goodman, N.L., Sharpe, L.G., Kuhar, M.J. and Goldberg, S.R. (1989) Differential effects of the pharmacological manipulation of serotonin systems on cocaine and amphetamine self-administration in rats. Life Sci., 45: 1529–1535.

Powell, K.R., Picker, M.J. and Dykstra, L.A. (1994) Serotonin involvement in the discriminative stimulus effects of mu-opioids and kappa-opioids in rats. Behav. Pharmacol., 5: 255–264.

Pradhan, S., Roy, S.N. and Pradhan, S.N. (1978) Correlation of behavioral and neurochemical effects of acute administration of cocaine in rats. Life Sci., 22: 1737–1743.

Proudnikov, D., LaForge, K.S., Hofflich, H., Levenstien, M., Gordon, D., Barral, S., Ott, J. and Kreek, M.J. (2006) Association analysis of polymorphisms in serotonin 1B receptor (HTR1B) gene with heroin addiction: a comparison of molecular and statistically estimated haplotypes. Pharmacogenet. Genomics, 16: 25–36.

Przegalinski, E. and Filip, M. (1997) Stimulation of serotonin (5-HT)1A receptors attenuates the locomotor, but not the discriminative, effects of amphetamine and cocaine in rats. Behav. Pharmacol., 8: 699–706.

Przegalinski, E., Siwanowicz, J., Baran, L. and Filip, M. (2000) Activation of serotonin (5-HT)(1A) receptors inhibits amphetamine sensitization in mice. Life Sci., 66: 1011–1019.

Przegalinski, E., Filip, M., Papla, I. and Siwanowicz, J. (2001a) Effect of serotonin (5-HT)1B receptor ligands on cocaine sensitization in rats. Behav. Pharmacol., 12: 109–116.

Przegalinski, E., Siwanowicz, J., Nowak, E., Papla, I. and Filip, M. (2001b) Role of 5-HT(1B) receptors in the sensitization to amphetamine in mice. Eur. J. Pharmacol., 422: 91–99.

Przegalinski, E., Filip, M., Papla, I. and Czepiel, K. (2002) Effects of 5-HT1B receptor ligands microinjected into the accumbal shell or core on the cocaine-induced locomotor hyperactivity in rats. J. Physiol. Pharmacol., 53: 383–394.

Przegalinski, E., Papla, I., Siwanowicz, J. and Filip, M. (2004) Effects of 5-HT1B receptor ligands microinjected into the ventral tegmental area on the locomotor and sensitizating effects of cocaine in rats. Eur. Neuropsychopharmacol., 14: 217–225.

Przegalinski, E., Golda, A., Frankowska, M., Zaniewska, M. and Filip, M. (2007) Effects of serotonin 5-HT1B receptor ligands on the cocaine- and food-maintained self-administration in rats. Eur. J. Pharmacol., 559: 165–172.

Pullagurla, M., Bondareva, T., Young, R. and Glennon, R.A. (2004) Modulation of the stimulus effects of (+)amphetamine by the 5-HT6 antagonist MS-245. Pharmacol. Biochem. Behav., 78: 263–268.

Pum, M., Carey, R.J., Huston, J.P. and Müller, C.P. (2007) Dissociating effects of cocaine and d-amphetamine on dopamine and serotonin in the perirhinal, entorhinal, and prefrontal cortex of freely moving rats. Psychopharmacology (Berl.), 193: 375–390.

Pum, M.E., Carey, R.J., Huston, J.P. and Müller, C.P. (2008) Role of medial prefrontal, entorhinal, and occipital 5-HT in cocaine-induced place preference and hyperlocomotion: evidence for multiple dissociations. Psychopharmacology (Berl.), 201: 391–403.

Quarta, D., Naylor, C.G. and Stolerman, I.P. (2007) The serotonin 2C receptor agonist Ro-60-0175 attenuates effects of nicotine in the five-choice serial reaction time task and in drug discrimination. Psychopharmacology (Berl.), 193: 391–402.

Quednow, B.B., Jessen, F., Kuhn, K.U., Maier, W., Daum, I. and Wagner, M. (2006) Memory deficits in abstinent MDMA (ecstasy) users: neuropsychological evidence of frontal dysfunction. J. Psychopharmacol., 20: 373–384.

Quednow, B.B., Kuhn, K.U., Hoppe, C., Westheide, J., Maier, W., Daum, I. and Wagner, M. (2007) Elevated impulsivity and impaired decision-making cognition in heavy users of MDMA ('Ecstasy'). Psychopharmacology (Berl.), 189: 517–530.

Ramos, M., Goni-Allo, B. and Aguirre, N. (2005) Administration of SCH 23390 into the medial prefrontal cortex blocks the expression of MDMA-induced behavioral sensitization in rats: an effect mediated by 5-HT2C receptor stimulation and not by D1 receptor blockade. Neuropsychopharmacology, 30: 2180–2191.

Rapoza, D. (1993) Buspirone fails to affect the discriminative stimulus effects of cocaine. Pharmacol. Biochem. Behav., 45: 179–183.

Rasmussen, K. and Czachura, J.F. (1997) Nicotine withdrawal leads to increased sensitivity of serotonergic neurons to the 5-HT1A agonist 8-OH-DPAT. Psychopharmacology, 133: 343–346.

Rasmussen, K., Kallman, M.J. and Helton, D.R. (1997) Serotonin-1A antagonists attenuate the effects of nicotine withdrawal on the auditory startle response. Synapse, 27: 145–152.

Rasmussen, K., Calligaro, D.O., Czachura, J.F., Dreshfield-Ahmad, L.J., Evans, D.C., Hemrick-Luecke, S.K., Kallman, M.J., Kendrick, W.T., Leander, J.D., Nelson, D.L., Overshiner, C.D., Wainscott, D.B., Wolff, M.C., Wong, D.T., Branchek, T.A., Zgombick, J.M. and Xu, Y.C. (2000) The novel 5-hydroxytryptamine(1A) antagonist LY426965: effects on nicotine withdrawal and interactions with fluoxetine. J. Pharmacol. Exp. Ther., 294: 688–700.

Reavill, C., Hatcher, J.P., Lewis, V.A., Sanger, G.J. and Hagan, J. (1998) 5-HT4 receptor antagonism does not affect motor and reward mechanisms in the rat. Eur. J. Pharmacol., 357: 115–120.

Reith, M.E. (1990) 5-HT3 receptor antagonists attenuate cocaine-induced locomotion in mice. Eur. J. Pharmacol., 186: 327–330.

Reith, M.E., Li, M.Y. and Yan, Q.S. (1997) Extracellular dopamine, norepinephrine, and serotonin in the ventral tegmental area and nucleus accumbens of freely moving rats during intracerebral dialysis following systemic administration of cocaine and other uptake blockers. Psychopharmacology, 134: 309–317.

Ribeiro, E.B., Bettiker, R.L., Bogdanov, M. and Wurtman, R.J. (1993) Effects of systemic nicotine on serotonin release in rat brain. Brain Res., 621: 311–318.

Richardson, N.R. and Roberts, D.C.S. (1996) Progressive ratio schedules in drug self-administration studies in rats: a method to evaluate reinforcing efficacy. J. Neurosci. Meth., 66: 1–11.

Rijnders, H.J. and Slangen, J.L. (1993) The discriminative stimulus properties of buspirone involve dopamine-2 receptor antagonist activity. Psychopharmacology, 111: 55–61.

Risner, M.E. and Jones, B.E. (1980) Intravenous self-administration of cocaine and norcocaine by dogs. Psychopharmacology, 71: 83–89.

Ritz, M.C. and Kuhar, M.J. (1989) Relationship between self-administration of amphetamine and monoamine receptors in brain – comparison with cocaine. J. Pharmacol. Exp. Ther., 248: 1010–1017.

Robbins, T.W. and Everitt, B.J. (1996) Neurobehavioral mechanisms of reward and motivation. Curr. Opin. Neurobiol., 6: 228–236.

Roberts, D.C.S., Brebner, K., Vincler, M. and Lynch, W.J. (2002) Patterns of cocaine self-administration in rats produced by various access conditions under a discrete trials procedure. Drug Alcohol Depend., 67: 291–299.

Robinson, T.E. and Berridge, K.C. (1993) The neural basis of drug craving: an incentive-sensitization theory of addiction. Brain Res. Rev., 18: 247–291.

Robinson, T.E. and Berridge, K.C. (2003) Addiction. Ann. Rev. Psychol., 54: 25–53.

Robinson, T.E. and Kolb, B. (2004) Structural plasticity associated with exposure to drugs of abuse. Neuropharmacology, 47: 33–46.

Rocha, B.A., Scearce, L.K., Lucas, J.J., Hiroi, N., Castanon, N., Crabbe, J.C., Nestler, E.J. and Hen, R. (1998) Increased vulnerability to cocaine in mice lacking the serotonin-1B receptor [see comments]. Nature, 393: 175–178.

Rocha, B.A., Goulding, E.H., O'Dell, L.E., Mead, A.N., Coufal, N.G., Parsons, L.H. and Tecott, L.H. (2002) Enhanced locomotor, reinforcing, and neurochemical effects of cocaine in serotonin 5-hydroxytryptamine 2C receptor mutant mice. J. Neurosci., 22: 10039–10045.

Rocher, C. and Gardier, A.M. (2001) Effects of repeated systemic administration of d-Fenfluramine on serotonin and glutamate release in rat ventral hippocampus: comparison with methamphetamine using in vivo microdialysis. Naunyn Schmiedebergs Arch. Pharmacol., 363: 422–428.

Rodd, Z.A., Bell, R.L., Kuc, K.A., Zhang, Y., Murphy, J.M. and McBride, W.J. (2005) Intracranial self-administration of cocaine within the posterior ventral tegmental area of Wistar rats: evidence for involvement of serotonin-3 receptors and dopamine neurons. J. Pharmacol. Exp. Ther., 313: 134–145.

Rose, J.S., Branchey, M., Wallach, L. and Buydens-Branchey, L. (2003) Effects of buspirone in withdrawal from opiates. Am. J. Addict., 12: 253–259.

Rossi, S., Singer, S., Shearman, E., Sershen, H. and Lajtha, A. (2005) The effects of cholinergic and dopaminergic antagonists on nicotine-induced cerebral neurotransmitter changes. Neurochem. Res., 30: 541–558.

Rothman, R.B., Blough, B.E. and Baumann, M.H. (2006) Dual dopamine-5-HT releasers: potential treatment agents for cocaine addiction. Trends Pharmacol. Sci., 27: 612–618.

Roychoudhury, M. and Kulkarni, S.K. (1996) Prevention of morphine discontinuation phenomenon in mice by ondansetron, a selective 5-HT3 antagonist. Meth. Find. Exp. Clin. Pharmacol., 18: 677–683.

Rueter, L.E., Fornal, C.A. and Jacobs, B.L. (1997) A critical review of 5-HT brain microdialysis and behavior. Rev. Neurosci., 8: 117–137.

Rutter, J.J., Baumann, M.H. and Waterhouse, B.D. (1998) Systemically administered cocaine alters stimulus-evoked responses of thalamic somatosensory neurons to perithreshold vibrissae stimulation. Brain Res., 798: 7–17.

Sabol, K.E. and Seiden, L.S. (1998) Reserpine attenuates D-amphetamine and MDMA-induced transmitter release *in vivo*: a consideration of dose, core temperature and dopamine synthesis. Brain Res., 806: 69–78.

Sáiz, P.A., Garcia-Portilla, M.P., Arango, C., Morales, B., Martinez-Barrondo, S., Alvarez, C., Narciso, G.S., Carreno, E., Alvarez, V., Coto, E. and Bobes, J. (2008) Association between heroin dependence and 5-HT2A receptor gene polymorphisms. Eur. Addict. Res., 14: 47–52.

Salomon, L., Lanteri, C., Godeheu, G., Blanc, G., Gingrich, J. and Tassin, J.P. (2007) Paradoxical constitutive behavioral sensitization to amphetamine in mice lacking 5-HT2A receptors. Psychopharmacology (Berl.), 194: 11–20.

Sanchis-Segura, C. and Spanagel, R. (2006) Behavioral assessment of drug reinforcement and addictive features in rodents: an overview. Addict. Biol., 11: 2–38.

Sasaki, J.E., Tatham, T.A. and Barrett, J.E. (1995) The discriminative stimulus effects of methamphetamine in pigeons. Psychopharmacology, 120: 303–310.

Sastre-Coll, A., Esteban, S. and García-Sevilla, J.A. (2002) Supersensitivity of 5-HT1A autoreceptors and alpha2-adrenoceptors regulating monoamine synthesis in the brain of morphine-dependent rats. Naunyn Schmiedebergs Arch. Pharmacol., 365: 210–219.

Satel, S.L., Krystal, J.H., Delgado, P.L., Kosten, T.R. and Charney, D.S. (1995) Tryptophan depletion and attenuation of cue-induced craving for cocaine. Am. J. Psychiatry, 152: 778–783.

Saules, K.K., Schuh, L.M., Arfken, C.L., Reed, K., Kilbey, M.M. and Schuster, C.R. (2004) Double-blind placebo-controlled trial of fluoxetine in smoking cessation treatment including nicotine patch and cognitive-behavioral group therapy. Am. J. Addict., 13: 438–446.

Scearce-Levie, K., Viswanathan, S.S. and Hen, R. (1999) Locomotor response to MDMA is attenuated in knockout mice lacking the 5-HT1B receptor. Psychopharmacology (Berl.), 141: 154–161.

Schama, K.F., Howell, L.L. and Byrd, L.D. (1997) Serotonergic modulation of the discriminative-stimulus effects of cocaine in squirrel monkeys. Psychopharmacology (Berl.), 132: 27–34.

Schechter, M.D. (1988) Serotonergic–dopaminergic mediation of 3,4-methylenedioxymethamphetamine (MDMA, Ecstasy). Pharmacol. Biochem. Behav., 31: 817–824.

Schechter, M.D. (1991) Effect of serotonin depletion by p-chlorophenylalanine upon discriminative behaviors. Gen. Pharmacol., 22: 889–893.

Scheel-Krüger, J., Braestrup, C., Nielson, M., Golembiowska, K. and Mogilincka, E. (1977) Cocaine: discussion of the role of dopamine in the biochemical mechanism of action. In: Ellinwood, E.H. and Kilbey, M.M. (Eds.), Advances in Behavioral Biology: Cocaine and other Stimulants. Plenum Press, New York, NY, pp. 373–407.

Schenk, S. (2000) Effects of the serotonin 5-HT(2) antagonist, ritanserin, and the serotonin 5-HT(1A) antagonist, WAY 100635, on cocaine-seeking in rats. Pharmacol. Biochem. Behav., 67: 363–369.

Schiffer, W.K., Gerasimov, M.R., Marsteller, D.A., Geiger, J., Barnett, C., Alexoff, D.L. and Dewey, S.L. (2001) Topiramate selectively attenuates nicotine-induced increases in monoamine release. Synapse, 42: 196–198.

Schmidt, C.J. (1987) Acute administration of methylenedioxymethamphetamine – comparison with the neurochemical effects of its N-desmethyl and N-ethyl analogs. Eur. J. Pharmacol., 136: 81–88.

Schmidt, C.J. and Taylor, V.L. (1988) Direct central effects of acute methylenedioxymethamphetamine on serotonergic neurons. Eur. J. Pharmacol., 156: 121–131.

Schmidt, C.J., Levin, J.A. and Lovenberg, W. (1987) *In vitro* and *in vivo* neurochemical effects of methylenedioxymethamphetamine on striatal monoaminergic systems in the rat brain. Biochem. Pharmacol., 36: 747–755.

Segal, D.S. (1975) Behavioral characterization of D-amphetamine and L-amphetamine – neurochemical implications. Science, 190: 475–477.

Segal, D.S. and Kuczenski, R. (1997) Repeated binge exposures to amphetamine and methamphetamine: behavioral and neurochemical characterization. J. Pharmacol. Exp. Ther., 282: 561–573.

Seiden, L.S., Sabol, K.E. and Ricaurte, G.A. (1993) Amphetamine – effects on catecholamine systems and behavior. Ann. Rev. Pharmacol. Toxicol., 33: 639–677.

Semba, J. and Wakuta, M. (2008) Chronic effect of nicotine on serotonin transporter mRNA in the raphe nucleus of rats: reversal by co-administration of bupropion. Psychiat. Clin. Neurosci., 62: 435–441.

Series, H.G., Cowen, P.J. and Sharp, T. (1994) P-Chloroamphetamine (Pca), 3,4-methylenedioxymethamphetamine (MDMA) and D-fenfluramine pretreatment attenuates D-fenfluramine-evoked release of 5-Ht *in vivo*. Psychopharmacology, 116: 508–514.

Sershen, H., Hashim, A. and Lajtha, A. (1996) Effect of ibogaine on cocaine-induced efflux of [3H] dopamine and [3H] serotonin from mouse striatum. Pharmacol. Biochem. Behav., 53: 863–869.

Shaham, Y., Shalev, U., Lu, L., de Wit, H. and Stewart, J. (2003) The reinstatement model of drug relapse: history, methodology and major findings. Psychopharmacology, 168: 3–20.

Shankar, R.P., Karan, R.S., Handu, S.S. and Bhargava, V.K. (2000) Effect of the 5-HT3 receptor antagonist ondansetron on amphetamine-induced hyperactivity and stereotypy in rats. Indian J. Physiol. Pharmacol., 44: 355–358.

Shearman, E., Rossi, S., Sershen, H., Hashim, A. and Lajtha, A. (2005) Locally administered loc nicotine-induced neurotransmitter changes in areas of cognitive function. Neurochem. Res., 30: 1055–1066.

Shimizu, N., Take, S., Hori, T. and Oomura, Y. (1992) *In vivo* measurement of hypothalamic serotonin release by intracerebral microdialysis: significant enhancement by immobilization stress in rats. Brain Res. Bull., 28: 727–734.

Shuster, L., Yu, G. and Bates, A. (1977) Sensitization to cocaine stimulation in mice. Psychopharmacology, 52: 185–190.

Sills, T.L. and Fletcher, P.J. (1997) Fluoxetine attenuated morphine-induced locomotion and blocks morphine-sensitization. Eur. J. Pharmacol., 337: 161–164.

Sills, T.L., Greenshaw, A.J., Baker, G.B. and Fletcher, P.J. (1999a) Acute fluoxetine treatment potentiates amphetamine hyperactivity and amphetamine-induced nucleus accumbens dopamine release: possible pharmacokinetic interaction. Psychopharmacology (Berl.), 141: 421–427.

Sills, T.L., Greenshaw, A.J., Baker, G.B. and Fletcher, P.J. (1999b) The potentiating effect of sertraline and fluoxetine on amphetamine-induced locomotor activity is not mediated by serotonin. Psychopharmacology (Berl.), 143: 426–432.

Sills, T.L., Greenshaw, A.J., Baker, G.B. and Fletcher, P.J. (2000) Subchronic fluoxetine treatment induces a transient potentiation of amphetamine-induced hyperlocomotion: possible pharmacokinetic interaction. Behav. Pharmacol., 11: 109–116.

Silverstone, P.H., Oldman, D., Johnson, B. and Cowen, P.J. (1992) Ondansetron, a 5-HT3 receptor antagonist, partially attenuates the effects of amphetamine: a pilot study in healthy volunteers. Intl. Clin. Psychopharmacol., 7: 37–43.

Singer, S., Rossi, S., Verzosa, S., Hashim, A., Lonow, R., Cooper, T., Sershen, H. and Lajtha, A. (2004) Nicotine-induced changes in neurotransmitter levels in brain areas associated with cognitive function. Neurochem. Res., 29: 1779–1792.

Singh, V.P., Jain, N.K. and Kulkarni, S.K. (2003) Fluoxetine suppresses morphine tolerance and dependence: modulation of NO-cGMP/DA/serotonergic pathways. Meth. Find. Exp. Clin. Pharmacol., 25: 273–280.

Sizemore, G.M., Co, C. and Smith, J.E. (2000) Ventral pallidal extracellular fluid levels of dopamine, serotonin, gamma amino-butyric acid, and glutamate during cocaine self-administration in rats. Psychopharmacology (Berl.), 150: 391–398.

Smith, F.L., Yu, D.S., Smith, D.G., Leccese, A.P. and Lyness, W.H. (1986) Dietary tryptophan supplements attenuate amphetamine self-administration in the rat. Pharmacol. Biochem. Behav., 25: 849–855.

Smith, J.E., Shultz, K., Co, C., Goeders, N.E. and Dworkin, S.I. (1987) Effects of 5,7-dihydroxytryptamine lesions of the nucleus accumbens on rat intravenous morphine self-administration. Pharmacol. Biochem. Behav., 26: 607–612.

Snoddy, A.M. and Tessel, R.E. (1983) Nisoxetine and amphetamine share discriminative stimulus properties in mice. Pharmacol. Biochem. Behav., 19: 205–210.

Solinas, M., Chauvet, C., Thiriet, N., El, R.R. and Jaber, M. (2008) Reversal of cocaine addiction by environmental enrichment. Proc. Natl. Acad. Sci. USA, 105: 17145–17150.

Solomon, R.L. (1980) The opponent-process theory of acquired motivation: the costs of pleasure and the benefits of pain. Am. Psychol., 35: 691–712.

Sora, I., Hall, F.S., Andrews, A.M., Itokawa, M., Li, X.F., Wei, H.B., Wichems, C., Lesch, K.P., Murphy, D.L. and Uhl, G.R. (2001) Molecular mechanisms of cocaine reward: combined dopamine and serotonin transporter knockouts eliminate cocaine place preference. Proc. Natl. Acad. Sci. USA, 98: 5300–5305.

Sorensen, S.M., Kehne, J.H., Fadayel, G.M., Humphreys, T.M., Ketteler, H.J., Sullivan, C.K., Taylor, V.L. and Schmidt, C.J. (1993) Characterization of the 5-HT2 receptor antagonist MDL 100907 as a putative atypical antipsychotic: behavioral, electrophysiological and neurochemical studies. J. Pharmacol. Exp. Ther., 266: 684–691.

Spealman, R.D. (1993) Modification of behavioral effects of cocaine by selective serotonin and dopamine uptake inhibitors in squirrel monkeys. Psychopharmacology, 112: 93–99.

Spealman, R.D. and Goldberg, S.R. (1978) Drug self-administration by laboratory animals: control by schedules of reinforcement. Ann. Rev. Pharmacol. Toxicol., 18: 313–339.

Spring, B., Wurtman, J., Gleason, R., Wurtman, R. and Kessler, K. (1991) Weight-gain and withdrawal symptoms after smoking cessation – a preventive intervention using D-fenfluramine. Health Psychol, 10: 216–223.

Spring, B., Wurtman, J., Wurtman, R., Elkhoury, A., Goldberg, H., Mcdermott, J. and Pingitore, R. (1995) Efficacies of dexfenfluramine and fluoxetine in preventing weight-gain after smoking cessation. Am. J. Clin. Nutr., 62: 1181–1187.

Spring, B., Doran, N., Pagoto, S., McChargue, D., Cook, J.W., Bailey, K., Crayton, J. and Hedeker, D. (2007) Fluoxetine, smoking, and history of major depression: A randomized controlled trial. J. Cons. Clin. Psychol., 75: 85–94.

Sprouse, J.S., Bradberry, C.W., Roth, R.H. and Aghajanian, G.K. (1989) MDMA (3,4-methylenedioxymethamphetamine) inhibits the firing of dorsal raphe neurons in brain slices via release of serotonin. Eur. J. Pharmacol., 167: 375–383.

Sprouse, J.S., Bradberry, C.W., Roth, R.H. and Aghajanian, G.K. (1990) 3,4-methylenedioxymethamphetamine-induced release of serotonin and inhibition of dorsal raphe cell firing – potentiation by L-tryptophan. Eur. J. Pharmacol., 178: 313–320.

Spyraki, C., Nomikos, G.G. and Varonos, D.D. (1987) Intravenous cocaine-induced place preference: attenuation by haloperidol. Behav. Brain Res., 26: 57–62.

Spyraki, C., Nomikos, G.G., Galanopoulou, P. and Daifotis, Z. (1988) Drug-induced place preference in rats with 5,7-dihydroxytryptamine lesions of the nucleus accumbens. Behav. Brain Res., 29: 127–134.

Stolerman, I.P., Pratt, J.A., Garcha, H.S., Giardini, V. and Kumar, R. (1983) Nicotine cue in rats analysed with drugs acting on cholinergic and 5-hydroxytryptamine mechanisms. Neuropharmacology, 22: 1029–1037.

Subhan, F., Pache, D.M. and Sewell, R.D.E. (2000) Potentiation of opioid-induced conditioned place preference by the selective serotonin reuptake inhibitor fluoxetine. Eur. J. Pharmacol., 390: 137–143.

Sudakov, S.K., Rusakova, I.V., Trigub, M.M., Kudrin, V.S. and Klodt, P.M. (2007) Self-administration of morphine by rats causes monoamine release in the anterior cingulate cortex. Bull. Exp. Biol. Med., 144: 210–213.

Suemaru, K., Araki, H., Kitamura, Y., Yasuda, K. and Gomita, Y. (2001) Cessation of chronic nicotine administration enhances wet-dog shake responses to 5-HT2 receptor stimulation in rats. Psychopharmacology, 159: 38–41.

Summers, K.L. and Giacobini, E. (1995) Effects of local and repeated systemic administration of (−)nicotine on extracellular levels of acetylcholine, norepinephrine, dopamine, and serotonin in rat cortex. Neurochem. Res., 20: 753–759.

Suzuki, T., Shiozaki, Y., Masukawa, Y. and Misawa, M. (1992) 5-HT3 receptor antagonists block cocaine- and methamphetamine-induced place preference. Yakubutsu Seishin Kodo, 12: 33–38.

Suzuki, T., Ise, Y., Mori, T. and Misawa, M. (1997) Attenuation of mecamylamine-precipitated nicotine-withdrawal aversion by the 5-HT3 receptor antagonist ondansetron. Life Sci., 61: L249–L254.

Svingos, A.L. and Hitzemann, R. (1992) 5-HT3 receptor antagonists block cocaine-induced locomotion via a PCPA-sensitive mechanism. Pharmacol. Biochem. Behav., 43: 871–879.

Szumlinski, K.K., Frys, K.A. and Kalivas, P.W. (2003) Pretreatment with serotonin 5-HT(3) receptor antagonists produces no observable blockade of long-term motor sensitization to cocaine in rats. Psychopharmacology (Berl.), 165: 329–336.

Szumlinski, K.K., Frys, K.A. and Kalivas, P.W. (2004) Dissociable roles for the dorsal and median raphe in the faciliatory effect of 5-HT1A receptor stimulation upon cocaine-induced locomotion and sensitization. Neuropsychopharmacology, 29: 1675–1687.

Takamatsu, Y., Yamamoto, H., Ogai, Y., Hagino, Y., Markou, A. and Ikeda, K. (2006) Fluoxetine as a potential pharmacotherapy for methamphetamine dependence: studies in mice. Ann. N.Y. Acad. Sci., 1074: 295–302.

Tan, E.C., Yeo, B.K., Ho, B.K., Tay, A.H. and Tan, C.H. (1999) Evidence for an association between heroin dependence and a VNTR polymorphism at the serotonin transporter locus. Mol. Psychiatry, 4: 215–217.

Tao, R. and Auerbach, S.B. (1994) Increased extracellular serotonin in rat brain after systemic or intraraphe administration of morphine. J. Neurochem., 63: 517–524.

Tao, R. and Auerbach, S.B. (1995) Involvement of the dorsal raphe but not median raphe nucleus in morphine-induced increases in serotonin release in the rat forebrain. Neuroscience, 68: 553–561.

Tao, R. and Auerbach, S.B. (2002) GABAergic and glutamatergic afferents in the dorsal raphe nucleus mediate morphine-induced increases in serotonin efflux in the rat central nervous system. J. Pharmacol. Exp. Ther., 303: 704–710.

Tao, R., Ma, Z.Y. and Auerbach, S.B. (1998) Alteration in regulation of serotonin release in rat dorsal raphe nucleus after prolonged exposure to morphine. J. Pharmacol. Exp. Ther., 286: 481–488.

Taylor, B.K. and Basbaum, A.I. (2003) Systemic morphine-induced release of serotonin in the rostroventral medulla is not mimicked by morphine microinjection into the periaqueuctal gray. J. Neurochem., 86: 1129–1141.

Teneud, L.M., Baptista, T., Murzi, E., Hoebel, B.G. and Hernandez, L. (1996) Systemic and local cocaine increase extracellular serotonin in the nucleus accumbens. Pharmacol. Biochem. Behav., 53: 747–752.

Thomsen, M., Hall, F.S., Uhl, G.R. and Caine, S.B. (2009) Dramatically decreased cocaine self-administration in dopamine but not serotonin transporter knockout mice. J. Neurosci., 29: 1087–1092.

Tizabi, Y., Russell, L.T., Johnson, M. and Darmani, N.A. (2001) Nicotine attenuates DOI-induced head-twitch response in mice: implications for Tourette syndrome. Prog. Neuro-Psychopharm. Biol. Psychiat., 25: 1445–1457.

Toth, E., Sershen, H., Hashim, A., Vizi, E.S. and Lajtha, A. (1992) Effect of nicotine on extracellular levels of neurotransmitters assessed by microdialysis in various brain regions: role of glutamic acid. Neurochem. Res., 17: 265–271.

Touiki, K., Rat, P., Arib, O., Molimard, R., Chait, A. and de Beaurepaire, R. (2008) Chronic intermittent nicotine treatment dose-dependently alters serotonergic neurons response to citalopram in the rat. Pharmacol. Biochem. Behav., 89: 338–344.

Tran-Nguyen, L.T.L., Baker, D.A., Grote, K.A., Solano, J. and Neisewander, J.L. (1999) Serotonin depletion attenuates cocaine-seeking behavior in rats. Psychopharmacology, 146: 60–66.

Tran-Nguyen, L.T., Bellew, J.G., Grote, K.A. and Neisewander, J.L. (2001) Serotonin depletion attenuates cocaine seeking but enhances sucrose seeking and the effects of cocaine priming on reinstatement of cocaine seeking in rats. Psychopharmacology (Berl.), 157: 340–348.

Trigo, J.M., Renoir, T., Lanfumey, L., Hamon, M., Lesch, K.P., Robledo, P. and Maldonado, R. (2007) 3,4-methylenedioxymethamphetamine self-administration is abolished in serotonin transporter knockout mice. Biol. Psychiat., 62: 669–679.

Tzschentke, T.M. (2007) Measuring reward with the conditioned place preference (CPP) paradigm: update of the last decade. Addict. Biol., 12: 227–462.

Ungless, M.A., Whistler, J.L., Malenka, R.C. and Bonci, A. (2001) Single cocaine exposure *in vivo* induces long-term potentiation in dopamine neurons. Nature, 411: 583–587.

Vanderschuren, L.J. and Kalivas, P.W. (2000) Alterations in dopaminergic and glutamatergic transmission in the induction and expression of behavioral sensitization: a critical review of preclinical studies. Psychopharmacology (Berl.), 151: 99–120.

Vanderschuren, L.J.M.J. and Everitt, B.J. (2004) Drug seeking becomes compulsive after prolonged cocaine self-administration. Science, 305: 1017–1019.

Van Wijngaarden, I., Tulp, M.T.M. and Soudijn, W. (1990) The concept of selectivity in 5-Ht receptor research. Eur. J. Pharmacol. Mol. Pharmacol. Sect., 188: 301–312.

Vollenweider, F.X., Gamma, A., Liechti, M. and Huber, T. (1998) Psychological and cardiovascular effects and short-term sequelae of MDMA ('ecstasy') in MDMA-naive healthy volunteers. Neuropsychopharmacology, 19: 241–251.

Walsh, S.L. and Cunningham, K.A. (1997) Serotonergic mechanisms involved in the discriminative stimulus, reinforcing and subjective effects of cocaine. Psychopharmacology, 130: 41–58.

Walsh, S.L., Preston, K.L., Sullivan, J.T., Fromme, R. and Bigelow, G.E. (1994) Fluoxetine alters the effects of intravenous cocaine in humans. J. Clin. Psychopharmacol., 14: 396–407.

Wang, Y., Joharchi, N., Fletcher, P.J., Sellers, E.M. and Higgins, G.A. (1995) Further studies to examine the nature of dexfenfluramine-induced suppression of heroin self-administration. Psychopharmacology, 120: 134–141.

West, R. and Hajek, P. (1996) Randomised controlled trial of odansetron in smoking cessation. Psychopharmacology, 126: 95–96.

West, W.B., Van Groll, B.J. and Appel, J.B. (1995) Stimulus effects of d-amphetamine II: DA, NE, and 5-HT mechanisms. Pharmacol. Biochem. Behav., 51: 69–76.

Will, M.J., Der-Avakian, A., Bland, S.T., Grahn, R.E., Hammack, S.E., Sparks, P.D., Pepin, J.L., Watkins, L.R. and Maier, S.F. (2004) Electrolytic lesions and pharmacological inhibition of the dorsal raphe nucleus prevent stressor potentiation of morphine conditioned place preference in rats. Psychopharmacology, 171: 191–198.

Winter, J.C. (1977) Morphine and ethanol as discriminative stimuli – absence of antagonism by para-chlorophenylalanine methyl ester, cinanserin, or Bc-105. Psychopharmacology, 53: 159–163.

Wise, R.A. (1980) The dopamine synapse and the notion of 'pleasure centers' in the brain. Trends Neurosci.: 91–95.

Wise, R.A. (2002) Brain reward circuitry: insights from unsensed incentives. Neuron, 36: 229–240.

Witkin, J.M., Savtchenko, N., Mashkovsky, M., Beekman, M., Munzar, P., Gasior, M., Goldberg, S.R., Ungard, J.T., Kim, J., Shippenberg, T. and Chefer, V. (1999) Behavioral, toxic, and neurochemical effects of sydnocarb, a novel psychomotor stimulant: comparisons with methamphetamine. J. Pharmacol. Exp. Ther., 288: 1298–1310.

Yamamoto, B.K., Nash, J.F. and Gudelsky, G.A. (1995) Modulation of methylenedioxymethamphetamine-induced striatal dopamine release by the interaction between serotonin and gamma aminobutyric acid in the substantia nigra. J. Pharmacol. Exp. Ther., 273: 1063–1070.

Yang, Z.J., Blaha, V., Meguid, M.M., Oler, A. and Miyata, G. (1999) Infusion of nicotine into the LHA enhances dopamine and 5-HT release and suppresses food intake. Pharmacol. Biochem. Behav., 64: 155–159.

Yasuda, K., Suemaru, K., Araki, H. and Gomita, Y. (2002) Effect of nicotine cessation on the central serotonergic systems in mice: involvement of 5-HT2 receptors. Naunyn Schmiedebergs Arch. Pharmacol., 366: 276–281.

Yeh, S.Y. and Haertzen, C.A. (1991) Cocaine-induced locomotor activity in rats. Pharmacol. Biochem. Behav., 39: 723–727.

Yokel, R.A. and Pickens, R. (1973) Self-administration of optical isomers of amphetamine and methylamphetamine by rats. J. Pharmacol. Exp. Ther., 187: 27–33.

Yoo, J.H., Cho, J.H., Yu, H.S., Lee, K.W., Lee, B.H., Jeong, S.M., Nah, S.Y., Kim, H.C., Lee, S.Y. and Jang, C.G. (2006) Involvement of 5-HT receptors in the development and expression of methamphetamine-induced behavioral sensitization: 5-HT receptor channel and binding study. J. Neurochem., 99: 976–988.

Young, R., Bondareva, T., Wesolowska, A. and Glennon, R.A. (2006a) Modulation of a (+)amphetamine discriminative stimulus in rats by 8-hydroxy-2-(N,N-di-n-propylamino)tetralin (8-OH DPAT). Pharmacol. Biochem. Behav., 83: 612–617.

Young, R., Bondareva, T., Wesolowska, A., Young, S. and Glennon, R.A. (2006b) Effect of the 5-HT(6) serotonin antagonist MS-245 on the actions of (-)nicotine. Pharmacol. Biochem. Behav., 85: 170–177.

Yu, D.S., Smith, F.L., Smith, D.G. and Lyness, W.H. (1986) Fluoxetine-induced attenuation of amphetamine self-administration in rats. Life Sci., 39: 1383–1388.

Zaniewska, M., McCreary, A.C., Przegalinski, E. and Filip, M. (2007) Effects of the serotonin 5-HT2A and 5-HT2C receptor ligands on the discriminative stimulus effects of nicotine in rats. Eur. J. Pharmacol., 571: 156–165.

Zhang, X., Lee, T.H., Davidson, C., Lazarus, C., Wetsel, W.C. and Ellinwood, E.H. (2007) Reversal of cocaine-induced behavioral sensitization and associated phosphorylation of the NR2B and GluR1 subunits of the NMDA and AMPA receptors. Neuropsychopharmacology, 32: 377–387.

CHAPTER 4.4

The Serotonergic System in Obsessive-Compulsive Disorder

Addy van Dijk, Andre Klompmakers and Damiaan Denys*

Department of Psychiatry, Academic Medical Center, University of Amsterdam, Amsterdam, The Netherlands

Abstract: Obsessive-compulsive disorder (OCD) is a psychiatric disorder consisting of obsessions and compulsions. Over the past two decades it has been suggested that OCD might be related to the functioning of brain serotonin systems, mainly because of the anti-obsessional efficacy of selective serotonin reuptake inhibitors (SSRIs). Though the efficacy of SSRIs suggests a role of the serotonergic system in OCD, the exact function of serotonin (5-HT) is still unclear. Is the serotonergic system implicated in the pathophysiology of OCD, or is it implicated in the treatment effect in OCD? Do SSRIs compensate for a fundamental abnormality of the serotonergic system, or do SSRIs modulate an intact serotonergic system to compensate for another neurotransmitter mechanism? This chapter reviews five lines of evidence that have been cited in support of the 5-HT hypothesis: (1) pharmacotherapy; (2) pharmacologic challenge studies; (3) receptor binding studies; (4) genetic association studies; and (5) animal models.

Keywords: obsessive-compulsive disorder, anxiety disorder, pharmacotherapy, neurobiology, serotonin, dopamine.

Introduction

Obsessive-compulsive disorder (OCD) is a psychiatric disorder which consists of obsessions and compulsions. Obsessions are unwanted ideas or impulses that repeatedly well up in the mind, such as persistent fears to harm a loved one, an unreasonable concern with becoming contaminated or an excessive need to do things correctly or perfectly. Compulsions are repetitive behaviors or rituals such as washing and checking, counting, mentally repeating phrases, list-making or endlessly rearranging objects.

Over the past two decades it has been suggested that OCD might be related to the functioning of brain serotonin systems, mainly because of the anti-obsessional efficacy of SSRIs. The 5-HT hypothesis originated from the observed efficacy of SSRIs in OCD, compared to inefficacy of selective noradrenergic and dopaminergic reuptake inhibitors, suggesting a specific abnormality of the serotonergic system in OCD (Baumgarten and Grozdanovic, 1998). However, the exact function of serotonin (5-HT) in OCD is still unclear. This chapter provides an overview of the evidence supporting a role for the serotonergic system in OCD. It reviews five lines of evidence that have been cited in support of the 5-HT hypothesis: (1) pharmacotherapeutic studies; (2) pharmacotherapy; (3) receptor binding studies; (4) genetic association studies; and (5) animal models.

Pharmacotherapy

A common method to unravel the pathophysiology of a disorder in psychiatry is to study the mechanism of action of effective drug therapy. The cornerstone of pharmacotherapy for OCD is serotonin reuptake inhibition, either with clomipramine or with selective serotonin reuptake inhibitors. Over the past 25 years, a number of double-blind, randomized, placebo controlled studies in OCD have confirmed the efficacy of clomipramine and the selective serotonin reuptake inhibitors fluvoxamine, paroxetine, sertraline, fluoxetine and citalopram. Efficacy of drug trials in OCD is commonly expressed in absolute changes of Yale-Brown Obsessive-Compulsive Scale (Y-BOCS) scores. The Y-BOCS is a clinician rated 10-item scale with a total range of 0 to 40 that measures severity of obsessions and compulsions (Goodman *et al.*, 1989a, b). In many cases, patients are classified as responders when their scores on the Y-BOCS decrease by 25–35 percent.

The efficacy of clomipramine was established in more than 10 placebo-controlled trials. In a recent meta-analysis, clomipramine shows a net improvement (difference) compared with placebo of 8.20 points on the Y-BOCS

*Corresponding author
E-mail: d.denys@amc.nl

Christian Müller & Barry Jacobs (Eds.)
Handbook of Behavioral Neurobiology of Serotonin
ISBN 978-0-12-374634-4

DOI: 10.1016/B978-0-12-374634-4.00032-0

(Ackerman and Greenland, 2002). Responder rates to clomipramine vary between 40 and 50 percent of patients. Though in some cases low doses of clomipramine over a short period of time may result in significant improvement of symptoms, clomipramine treatment should last at least 10–12 weeks with a dose of 250–300 mg/day to demonstrate a full effect. Clomipramine has never been tested in a controlled fixed-dose study to assess optimal doses. Fluvoxamine was superior to placebo in four placebo-controlled trials and equipotent to clomipramine in five comparison trials (Dell'osso *et al.*, 2005). The pooled difference of fluvoxamine compared with placebo in these trials was 4.8 points on the Y-BOCS. Fluvoxamine has never been examined in a fixed-dose trial, but appears to be efficacious from 150 to 300 mg/day. Fluoxetine was effective in three placebo-controlled trials and equipotent to clomipramine in two comparison trials. The pooled difference on the Y-BOCS of fluoxetine compared with placebo was 1.6 points (Ackerman and Greenland, 2002). Responder rates to fluoxetine vary between 25 and 30 percent of patients. Sertraline was effective in four placebo-controlled trials and equipotent to clomipramine in one comparison trial. The pooled difference of sertraline compared with placebo in these trials was 2.5 points on the Y-BOCS (Ackerman and Greenland, 2002). Although no correlation was found of sertraline plasma levels with treatment outcome, in a large multicenter trial, sertraline 50 mg/day and 200 mg/day were superior to placebo, whereas 100 mg/day was not.

For a long time, paroxetine was shown to be effective in just one published placebo-controlled trial and equipotent to clomipramine in the same trial (Zohar and Judge, 1996; Ackerman and Greenland, 2002; Ninan, 2003). The difference of paroxetine compared with placebo was 3.1 points on the Y-BOCS. Doses of 40 and 60 mg/day were significantly better than placebo, whereas 20 mg was not. The efficacy of citalopram in OCD was recently established in a placebo-controlled study including 401 patients who were randomized to receive citalopram 20, 40 or 60 mg/day or placebo for 12 weeks (Montgomery *et al.*, 2001). The difference of citalopram compared with placebo was 3.6 points on the Y-BOCS. The highest responder rate (65 percent), defined as 25 percent improvement in Y-BOCS entry score, was observed in the 60 mg group compared with 52 percent and 57.4 percent in the 40 mg and 20 mg groups, respectively. Surprisingly, the responder rate on placebo was 36.6 percent with a mean decrease on the Y-BOCS of 5.6 points. This high improvement rate on placebo, uncommon in OCD, has been explained as a result of inclusion of milder, atypical cases, in which spontaneous remission is more frequent. Patients with longer duration of OCD, more severe OCD symptoms or previous SSRI use were less likely to be responders to citalopram, whereas patients who received adequate medication doses for sufficient periods of time were more likely to be responders (Stein *et al.*, 2001). In a small study, clinical response did not appear to be related to citalopram plasma concentrations (Bareggi *et al.*, 2004).

The efficacy of clomipramine and five SSRIs has been established with placebo-controlled trials, but are clomipramine, fluvoxamine, paroxetine, sertraline, citalopram and fluoxetine equally effective (Soomro *et al.*, 2008)? Meta-analyses using statistical methods to pool samples from different studies may answer this question, offering a precise estimate of the treatment effect from separate drug trials. Consistent with previous meta-analyses, Ackerman and colleagues recently confirmed that clomipramine is more effective than SSRIs in placebo-controlled trials (Ackerman and Greenland, 2002). For seven clomipramine trials the net Y-BOCS decrease compared with placebo was 8.19, whereas for four fluvoxamine trials the net Y-BOCS decrease was 4.84, for three fluoxetine trials 1.61, and for four sertraline trials 2.47. It is worth noting, however, that the largest effect values for clomipramine were found in early studies and have not been as clearly replicated in more recent studies comparing it with SSRIs. Moreover, in the long term, drop-out rates in clomipramine-treated patients are higher than in SSRI-treated patients. In another meta-analysis which included 32 studies covering a total of 3588 OCD patients, again, clomipramine had the largest effect value of 1.55, versus 0.81 for fluoxetine and 1.36 for sertraline (Eddy *et al.*, 2004). This meta-analysis also demonstrated that two-thirds of the patients who completed a medication trial improved. Across all active treatments, the mean Y-BOCS decrease was 7.1, resulting in a mean post-treatment Y-BOCS score of 17.9, whereas for placebo conditions the Y-BOCS decrease was 1.8.

Clearly, in effectiveness, clomipramine is superior to all the SSRIs. In practice, however, the choice of a drug also depends on its side-effect profile and potential for drug interactions. Clomipramine has the greatest anticholinergic side-effect profile, potential cardiotoxicity, and is the most noxious in an overdose. In contrast, SSRIs may be associated with more complaints of headaches, nausea, insomnia or agitation, but are safer and less prone to drug interactions. Considering side effects, toxicity and potential drug interactions, SSRIs are usually the first treatment of choice.

The 5-HT hypothesis of OCD has emerged from the observation that only antidepressants that preferentially block the serotonin transporter (5-HTT) are efficacious in OCD. A comparison of the clinical efficacy of clomipramine, fluvoxamine and sertraline with that of desipramine, an antidepressant that selectively blocks the uptake of norepinephrine, supports the notion that inhibition of the 5-HTT is required for antidepressants to be efficacious in OCD (Zohar and Insel, 1987; Leonard *et al.*, 1989; Goodman *et al.*, 1990; Hoehn-Saric *et al.*, 2000). Denys and colleagues have shown that venlafaxine and paroxetine are equally efficacious in OCD, suggesting that inhibition of the norepinephrine uptake does not contribute to the effect of antidepressants as well (Denys *et al.*, 2003). As mentioned previously, this finding does not necessarily point to an abnormal serotonergic system in OCD. On the contrary, several observations lend support to the idea that inhibition of the 5-HTT, although required for antidepressants to work in OCD, is not causally related to the genesis of the disorder. First, nearly half of OCD patients do not respond to SSRI treatment (Denys, 2006). Even with adequate treatment, the mean symptom decrease is 30–50 percent at most (Denys *et al.*, 2002). Second, the doses of SSRIs required for treatment of OCD are substantially higher than necessary to completely block the 5-HTT (Kent *et al.*, 2002; Meyer, 2007; Voineskos *et al.*, 2007). Third, depletion studies of tryptophan, necessary for the synthesis of 5-HT, did not show a worsening of symptoms in OCD patients (Berney *et al.*, 2006). Fourth, the therapeutic effects in OCD are usually not seen within 8 weeks of treatment, which is much later than in patients with depression. Fifth, there is no relation between the selectivity and binding potential of the different SSRIs and clinical effect size (Baumgarten and Grozdanovic, 1998).

In conclusion, the 5-HT hypothesis was initially motivated by the observed differential efficacy of SSRIs in alleviating OCD symptoms. However, these findings, although attesting to the therapeutic versatility of serotonin transporter inhibition in OCD, do not necessarily reflect the existence of a neurobiological abnormality in the central serotonergic system in OCD. SSRIs may modulate serotonin via an intact serotonergic system to compensate for the underlying pathogenesis.

Pharmacologic challenge studies

Another approach to assessing the function of the serotonergic system in OCD is by administration of indirect or direct serotonin receptor agonists or antagonists. Pharmacological challenges with meta-chlorophenylpiperazine (mCPP) have shown to exacerbate obsessive-compulsive (OC) symptoms. mCPP is a non-selective 5-HT receptor agonist acting mainly at the 5-HT_{2C} receptors, but with affinity also for the 5-$HT_{1B/1D}$ and 5-HT_{1A} receptors (Westenberg *et al.*, 2007). Administration of psilocybin, a psychedelic agent with potent 5-$HT_{2A/2C}$ and 5-HT_{1A} agonist properties, in nine subjects resulted in acute reductions of the Y-BOCS score in a controlled clinical environment (Moreno *et al.*, 2006). A challenge with the selective 5-HT_{1A} agonist ipsapirone did not induce exacerbation of OC symptoms in OCD patients (Lesch *et al.*, 1991a, b). Pindolol, on the other hand, a β-blocker and 5-HT_{1A} receptor antagonist, was shown to augment the therapeutic effect of paroxetine by increasing serotonergic transmission (Dannon *et al.*, 2000). Pharmacological challenge studies with 5-HT_{1D} receptor agonists in OCD are inconsistent. Some studies show an exacerbation of OC symptoms with sumatriptan, though zolmitriptan, with a better penetration of the blood–brain barrier than sumatriptan, failed to increase OC symptoms (Pian *et al.*, 1998; Stein *et al.*, 1999; Boshuisen and den Boer, 2000; Koran *et al.*, 2001; Gross-Isseroff *et al.*, 2004). Atypical antipsychotics with potent 5-HT_{2A} and D_2 antagonistic properties have anti-obsessional qualities in addition to SSRI treatment. Risperidone, an atypical antipsychotic, is the most potent and selective 5-HT_{2A} antagonist available to clinicians (Marek *et al.*, 2003). Multiple double-blind studies have found risperidone augmentation of ongoing SRI treatment in OCD to be effective (McDougle *et al.*, 2000; Hollander *et al.*, 2003; Erzegovesi *et al.*, 2005). Hewlett and colleagues initiated a pilot study with the 5-HT_{3A} receptor antagonist ondansetron. They found that of the eight OCD patients, six completed the trial and three subjects achieved a clinically significant response of ≥35 percent reduction in Y-BOCS score (Hewlett *et al.*, 2003). On the contrary, Broocks and colleagues found that pre-treatment with ondansetron did not affect any of the self-rated behavioral symptoms induced by an mCPP challenge with 11 OCD-patients (Broocks *et al.*, 1998).

In conclusion, pharmacological challenges implicate a role for the 5-HT_{2A} receptor and/or 5-HT_{2C} receptor, and possibly the $5HT_{1B/1D}$, receptor in the pathophysiology of OCD.

Receptor binding studies

Neuroimaging with positron emission tomography (PET) or single photon emission computer tomography (SPECT) offers a unique tool to probe the serotonergic system *in vivo*. For a summary of the different receptor binding studies discussed in this review, see Table 1.

5-HTT binding studies at baseline

Pogarell and colleagues reported a 25 percent higher 5-HTT availability in the midbrain region of nine unmedicated OCD patients using the radiotracer $[^{123}I]$-2β-carbomethoxy-3β-(4-iodophenyl)tropane ($[^{123}I]$β-CIT) and SPECT imaging (Pogarell *et al.*, 2003). Stratification of the patients according to onset of the disorder revealed that the significant difference was only present in patients with an early age of onset compared to control.

Table 1 Receptor binding studies

Research group	Transporter, receptor of interest	Patients in study	Results	Y-BOCS (mean)	Co-morbidity	Control groups	Acquisition time (hours)	Radiotracer	Imaging technique
Pogarell *et al.*, 2003	5-HTT	9, heterogeneous, drug-free (7 drug-naïve)	25% higher 5-HTT availability in the midbrain region	23.0 ± 8.2	Depression	—	24	[^{123}I]β-CIT	SPECT
Stengler-Wenzke *et al.*, 2004	5-HTT	10, heterogeneous, drug-free	Significant reduction of 5-HTT availability in the midbrain and upper brainstem	30 ± 2.5	—	Age matched	24	[^{123}I]β-CIT	SPECT
Hesse *et al.*, 2005	5-HTT/DAT	15, heterogeneous, drug-naïve	Reduced availability of striatal DAT and of thalamic/hypothalamic, midbrain and brainstem 5-HTT	25.3 ± 8.8	—	—	24	[^{123}I]β-CIT	SPECT
Zitterl *et al.*, 2007	5-HTT	24, homogeneous OC-checkers, drug-free	18% reduction in 5-HTT availability in the thalamus and hypothalamus	25.1 ± 5.0	—	Age and gender matched	4	[^{123}I]β-CIT	SPECT
Hasselbalch *et al.*, 2007	5-HTT	9, heterogeneous, drug-free	13% reduction in midbrain-pons 5-HTT binding	22	Anxiety, depression	—	10 min–7 hours	[^{123}I]β-CIT	SPECT
Reimold *et al.*, 2007	5-HTT	9, heterogeneous, drug-free	Significant reduction in 5-HTT availability in thalamus and midbrain	21.4 ± 7.7	—	Age matched	0–80 min	[^{11}C]DASB	PET
van der Wee *et al.*, 2004	5-HTT	15, heterogeneous, psychotropic-naïve	No difference in 5-HTT binding density in the midbrain, thalamus and pons of OCD patients compared to controls	23.4 ± 4.2 (cut off score of 16)	—	Age and gender matched	4	[^{123}I]β-CIT	SPECT
Simpson *et al.*, 2003	5-HTT	11, heterogeneous, drug-free	No significant difference was found between OCD patients and control subjects	20 ± 4	—	Age and gender matched	130 min	[^{11}C]McN 5652	PET
Stengler-Wenzke *et al.*, 2006	5-HTT	10, heterogeneous, drug-free at baseline; 5, after 1 year 60-mg citalopram treatment. Patients on stable treatment during 2^{e} scan	The availability of 5-HTT in the thalamus, midbrain and brainstem decreased significantly after treatment with 60 mg citalopram for 1 year	32.0 ± 2.5 (pre-treatment)	—	Age matched	24	[^{123}I]β-CIT	SPECT
Zitterl *et al.*, 2008	5-HTT	24, OC-checkers, drug-free at baseline; 12-week clomipramine treatment. Patients on stable 150-mg treatment during 2^{e} scan	A 48% reduced brain SERT availability in the thalamus–hypothalamus after 12-week treatment. Significantly negative associations between SERT availability and Y-BOCS both at baseline and after treatment	26.2 ± 4.9 (pre-treatment) 18.9 ± 6.7 (after treatment)	—	—	4	[^{123}I]β-CIT	SPECT

Pogarell *et al.*, 2005	5-HTT/DAT	2, heterogeneous, drug-naïve; 12 weeks 40-mg citalopram treatment. Patients on stable treatment during 2ᵉ scan	A 36.5% decrease in 5-HTT availability after 12 weeks' therapy with 40 mg citalopram. An increase of 40% in the availability of DAT	?	—	No control group	20–24	[123I]β-CIT	SPECT
Kim *et al.*, 2007	DAT	10, heterogeneous, drug-free; 16-week treatment 73.3 mg fluoxetine (n = 6) 53.3 mg paroxetine (n = 3) 250 mg clomipramine (n = 1). Patients on stable treatment during 2ᵉ scan	Decreased DAT binding ratio in the right basal ganglia after treatment with SRIs	33.10 ± 6.38 (pre-treatment)	Depression, tic disorder	No control group	2	[(123)I]IPT	SPECT
Moresco *et al.*, 2007	D_2	9, heterogeneous, drug-naïve; 12-week treatment 150–300 mg fluvoxamine. Patients on stable treatment during 2ᵉ scan	Significant increase between 6.9 and 9.7% in striatal [(11)C]Rac binding potential after chronic treatment with fluvoxamine; the mean values after treatment were closer to those observed in normal subjects	29 ± 5	—	—	0–60 min	[(11)C]Rac	PET
Adams *et al.*, 2005	5-HT_{2A}	15, heterogeneous, drug-free 12-week treatment 60–80 mg paroxetine 50–150 mg sertraline 60–80 mg fluoxetine 60–80 mg citalopram. Patients on stable treatment during 2ᵉ scan	9% higher 5-HT_{2A} receptor binding potential in the caudate nucleus before treatment; no significant change after SSRI treatment	30 ± 6.8	Depression, anorexia, anxiety, single phobia	Age and gender matched	90 min	[18F]altanserin	PET
Perani *et al.*, 2008	5-HT_{2A}	9, heterogeneous, drug naïve	Significant reduction of 5-HT_{2A} receptor availability in frontal cortex, cingulated cortex, parietal and temporal associative cortices; significant inverse correlation between 5-HT_{2A} availability and Y-BOCS	29.44 ± 4.42	—	Age matched	0 min	[11C]MDL-BP	PET

Abbreviations: [123I]β-CIT, [123I]-2β-carbomethoxy-3β-4-iodophenyltropane; SPECT, single photon emission computer tomography; DAT, dopamine transporter; 5-HTT, serotonin transporter; OCD, obsessive-compulsive disorder; PET, positron emission tomography; [123I]IPT, iodine-123-labeled N-3-iodopropen-2-yl-2 beta-carbomethoxy-3beta-4-chlorophenyl tropane; [11C]Rac, [11C]raclopride.

Patients and controls in this study were not matched by age and gender (Pogarell *et al.*, 2003).

In contrast to this finding, Stengler-Wenzke and colleagues described a significant reduction of 5-HTT availability in the midbrain and upper brainstem in 10 drug-free OCD patients compared to age-matched controls. However, they did not report the age of onset of OCD symptoms (Stengler-Wenzke *et al.*, 2004).

In agreement with this, Hesse and colleagues reported a significantly reduced 5-HTT density in the midbrain and the thalamus in 15 drug-naïve OCD patients compared to controls (Hesse *et al.*, 2005). Comparing 24 drug-free OC checkers to controls, Zitterl and colleagues found an 18 percent reduction in 5-HTT availability in the thalamus and hypothalamus. Reduced 5-HTT availability correlated with increased severity of OC symptomatology and short duration of illness, although not with age and age at onset (Zitterl *et al.*, 2007). Hasselbalch and colleagues found a 13 percent reduced midbrain-pons 5-HTT binding in nine drug-free OCD patients. No correlation between 5-HTT binding and any clinical variables (age at onset, disease duration and Y-BOCS score) was shown (Hasselbalch *et al.*, 2007). A [^{11}C] DASB PET study reported significantly reduced 5-HTT availability in the thalamus and in the midbrain in nine drug-free OCD patients (Reimold *et al.*, 2007).

Finally, van der Wee and colleagues found no difference in 5-HTT binding density in the midbrain, thalamus and pons of drug-naïve OCD patients compared to controls using [^{123}I]β-CIT and SPECT (van der Wee *et al.*, 2004). Simspon and colleagues, using [^{11}C]McN 5652 as a radioligand in combination with PET, found no significant difference between OCD patients and control subjects (Simpson *et al.*, 2003).

5-HTT binding studies after SSRI treatment

Treatment with 60 mg citalopram for a year in five patients resulted in a decrease of 5-HTT availability in the thalamus, midbrain and brainstem (Stengler-Wenzke *et al.*, 2006). In a group of 24 OC checkers, 12 weeks of treatment with clomipramine resulted in a 48 percent decrease in brain SERT availability in the thalamus-hypothalamus. A significantly negative association between SERT availability and Y-BOCS both at baseline and after treatment was also found in this group of 24 OC checkers (Zitterl *et al.*, 2008). A 12-week period of treatment with 40 mg citalopram in two patients led to a 36.5 percent decrease in 5-HTT availability and an increase of 40 percent in the availability of dopamine transporter (DAT) in the midbrain-pons region (Pogarell *et al.*, 2005). A SPECT study with iodine-123-labeled N-(3-iodopropen-2-yl)-2β-carbomethoxy-3β-(4-chlorophenyl)tropane ([(123)I]IPT) as radiotracer found a significantly decreased DAT binding ratio in the right basal ganglia after 16 weeks of treatment with SSRIs in 10 drug-free patients (Kim *et al.*, 2007). Moresco and colleagues found, in a PET study with [(11)C]raclopride ([(11)C]Rac) as radiotracer, a slight but significant increase in striatal [(11)C]Rac binding potential, varying from 6.9 percent to 9.7 percent, after a 12-week course of treatment with 150–300 mg fluvoxamine (Moresco *et al.*, 2007).

5-HT$_{2A}$ receptor binding studies

A PET study with [^{18}F]altanserin in naïve OCD patients showed a significantly higher 5-HT$_{2A}$ receptor binding potential in the caudate nucleus compared to age- and gender-matched healthy controls. This increase in 5-HT$_{2A}$ receptor binding was not influenced by subsequent treatment with SSRIs (Adams *et al.*, 2005). The increased 5-HT$_{2A}$ binding density might result from a lack of 5-HT in the basal ganglia. However, a PET study with the radioligand [^{11}C]MDL-BP found a significant reduction of 5-HT$_{2A}$ receptor availability in frontal cortex and cingulated cortex, as well as in parietal and temporal associative cortices, in drug-naïve OCD patients compared to age-matched controls. The reduced 5-HT$_{2A}$ receptor availability in the orbitofrontal and dorsolateral frontal cortex correlated with clinical severity (Perani *et al.*, 2008).

In summary, though some of the neuroimaging studies suggest that an impaired 5-HTT function might play a role in the pathogenesis of OCD, others indicate that successful SSRI treatment is associated with normalization of the dopamine (DA) function. There are mixed results regarding the possible role of the 5-HT$_{2A}$ receptor in the neuroimaging studies.

Genetic association studies

Genetic studies investigating the possible role for polymorphism in the promoter region of the 5-HTT gene (5-HTTPR) have generated conflicting results (Table 2). Some studies reported an association between the S allele of 5-HTTPR and OCD (Denys *et al.*, 2006; Perez *et al.*, 2006; Grados *et al.*, 2007; Lin, 2007). The S allele reduces 5-HTT mRNA expression, and a higher frequency of this allele would suggest a lower density of 5-HTT in OCD patients. Others found an association between the two copies of the long allele (L) in the 5-HTTLPR region and OCD (McDougle *et al.*, 1998; Bengel *et al.*, 1999). Kim and colleagues investigated the 5-HTTPR polymorphism in association with the phenotypic characteristics of OCD, and found that patients with the L-genotype had

Table 2 Genetic association studies

Research group	Transporter, receptor of interest	Patients in study	Origin	Results
Grados *et al.*, 2007	5-HTTPR	149 OCD patients, 397 controls	Caucasians (>90%)	A trend for both the S/s Genotype and the S allele frequency was found
Lin *et al.*, 2007	5-HTTPR	Meta-analysis of 13 independent case-control association studies 3000 subjects	—	An association between OCD and the SS homozygous genotype was found; it is inversely associated with the LS heterozygous genotype and has no association with the LL homozygous genotype or the allelic distribution was found
Perez *et al.*, 2006	5-HTTPR	26 OCD patients, 87 controls	Caucasians	A significant association between the S/S genotype and OCD
Denys *et al.*, 2006	5-HTTPR/ 5-HT_{1B}/5-HT_{2A}	156 OCD patients, 134 controls	Caucasians	No significant differences in genotype distribution and allele frequency for polymorphisms investigated were found relative to controls; however, there was an indication towards an association of the 5-HTTLPR S allele with female OCD patients, and the 5-HT2A G-allele and GG genotype with patients with a positive family history of OCD and an early onset of disease
Bengel *et al.*, 1999	5-HTTPR	75 OCD patients, 397 ethnically matched controls	Caucasians	A statistically significant association between the L/L genotype and OCD was found
McDougle *et al.*, 1998	5-HTTPR	34 family trios	European-Americans	An association between the L allele in the 5-HTTPR region and OCD was found
Kim *et al.*, 2005	5-HTTPR	124 OCD patients, 171 controls	Koreans	No significant association between OCD and 5-HTTPR was found; however, patients with the L-genotype had higher scores for the religious and somatic obsessions
Cavallini *et al.*, 2002	5-HTTPR	180 OCD patients, 112 controls	Italians	Patients who were homozygote for the L-allele could more often be classified as symmetry/ordering subtype OCD patients
Denys *et al.*, 2007	5-HTT, 5-HT_{2A}	91 OCD patients	Dutch	The S/L genotype of the 5-HTTLPR polymorphism was amongst the majority of OCD patients who responded with a Y-BOCS decrease of more than 25% to a 12-week treatment period with 300 mg venlafaxine; the G/G genotype of the 5-HT_{2A} polymorphism was amongst the majority of OCD patients who were responders to a 12-week treatment period with 60 mg paroxetine
Camarena *et al.*, 2001	5-HTTPR	115 OCD patients, 136 controls. Family-based design: 43 trios	Mexicans	No significant association between l allele and OCD was found; no preferential transmission of L allele to OCD probands was found
Chabane *et al.*, 2004	5-HTTPR	Case-control study: 106 OCD patients, 171 controls. Family association study: 116 trios including an OCD patient	French Caucasians	There was no association between the 5-HTTLPR polymorphism and OCD in either the case-control study or the family study
Di Bella *et al.*, 2002	5-HTTPR	181 OCD patients, 191 controls	Italians	No significant differences in allele/genotype distribution of the 5-HTTLPR was found

(Continued)

Table 2 (*Continued*)

Research group	Transporter, receptor of interest	Patients in study	Origin	Results
Dickel *et al.*, 2007	5-HTTPR/ 5-HT_{2A}	21 parent–child trios from singleplex families; 7 parent–child trios from multiplex families; 26 probands with early-onset OCD	Americans	There was no evidence for association detected at any of the polymorphisms in the entire set of subjects
Frisch *et al.*, 2000	5-HTTPR/ 5-HT_{2A}/ 5-HT2C	75 patients, 172 ethnically matched controls	Jews	There were no statistically significant differences between patients and control for the genotypic and allelic distribution of all polymorphisms tested
Meira-Lima *et al.*, 2004	5-HTTPR/ 5-HT_{2A}	79 OCD patients, 202 controls	Brazilians, Caucasian (90.5%)	A statistically significant difference in the genotypic distribution and in the allelic frequencies for the C516T 5HT2A gene polymorphism was found; however, no significant difference for the 5HTTLPR and the T102C 5HT_{2A} gene polymorphisms was found
Wendland *et al.*, 2006	5-HTTPR	279 OCD-probands, 390 controls	Caucasian	Allelic frequencies for G56A did not differ significantly between probands and controls
Saiz *et al.*, 2008	5-HTTPR/ 5-HT_{2A}	99 OCD patients, 56 non-OCD psychiatric patients, 420 controls	Spanish Caucasian	No significant differences were found with respect to 5-HTTLPR and 5-HT_{2A} polymorphisms
Mundo *et al.*, 2002	$5\text{-HT}_{1B/1D}$	Family-based design: 121 families, 157 OCD patients	Canadians, Caucasian (96.7%)	A significant association between OCD and the G861 polymorphism of the 5-HT_{1D} gene was found; the G-allele was associated with OCD and predicted OC severity
Camarena *et al.*, 2004	$5\text{-HT}_{1B/1D}$	72 trios	Mexicans	No significant association between the G-allele and OCD was found; however, subjects with a preferential transmission of the G861 variant showed higher Y-BOCS obsession scores compared to patients carrying the C861 allele
Levitan *et al.*, 2006	$5\text{-HT}_{1B/1D}$	165 female probands with bulimia nervosa	Canadians	There was no significant association between the G-allele and OCD; the G861C polymorphism did strongly differentiate between full syndrome vs partial syndrome OCD
Di Bella *et al.*, 2002	$5\text{-HT}_{1B/1D}$	Family-based design: 79 trios	Italians	No preferential transmission of either allele of the 5-HT1D gene was observed
Enoch *et al.*, 2001	5-HT_{2A}	101 OCD patients, 138 controls	Caucasian	A higher frequency of the −1438G/A allele was found in female OCD patients compared to healthy women; however, there was no difference in allele frequencies between male OCD patients and healthy men
Walitza *et al.*, 2002	5-HT_{2A}	55 OCD patients, 223 controls	Germans	A statistically significant association between the A-allele of the 5-HT_{2A} receptor gene with OCD was found
Kim *et al.*, 2000	5-HT_{2B}	7 early-onset probands, 10 autism probands, 10 controls	?	No evidence for functional mutation was found in the sequenced regions of HTR2B
Cavallini *et al.*, 1998	5-HT_{2C}	109 OCD patients, 107 controls	Italian	No allelic or genotypic association of OCD with the 5HT2C receptor gene mutation was found
Mossner *et al.*, 2007	5-HT_{3A}	Family-based design: ±75 trios	Germans, Caucasian	No evidence for a preferential transmission of either allele of the 5-HT_{3A} was found

higher scores for the religious and somatic obsessions (Kim *et al.*, 2005). A principal component analysis on the Y-BOCS checklist performed by Cavallini and colleagues demonstrated that patients who were homozygote for the L-allele more often could be classified as symmetry/ordering subtype OCD patients (Cavallini *et al.*, 2002). The S/L genotype of the 5-HTTLPR polymorphism was reported among the majority of OCD patients who were responders to a 12-week treatment period with 300 mg venlafaxine (Denys *et al.*, 2007). However, eight studies reported no association at all between 5-HTTPR and OCD (Frisch *et al.*, 2000; Camarena *et al.*, 2001; Di Bella *et al.*, 2002b; Chabane *et al.*, 2004; Meira-Lima *et al.*, 2004; Wendland *et al.*, 2006; Dickel *et al.*, 2007; Saiz *et al.*, 2008).

Polymorphisms of the 5-HT_{1B} receptor were studied by Mundo and colleagues in a large, Canadian family-based sample. They found a significant association between OCD and the G861 polymorphism of this gene. The G-allele was associated with OCD and predicted OC severity (Mundo *et al.*, 2002). Two subsequent studies also reported an association between OC severity and the G861 variant, although they did not find a significant association between G861 polymorphism and OCD (Camarena *et al.*, 2004; Levitan *et al.*, 2006). In an Italian family based study, no preferential transmission of either allele of the 5-$HT_{1B/1D}$ was observed (Di Bella *et al.*, 2002a).

The −1438G/A polymorphism of the 5-HT_{2A} receptor gene has also been linked to OCD. Enoch and colleagues reported a higher frequency of the −1438G/A allele in female OCD patients compared to healthy women (Enoch *et al.*, 2001). A statistically significant difference in the frequency of the A-allele was also reported in a study with 55 children and adolescents with OCD compared to control subjects (Walitza *et al.*, 2002). Denys and colleagues found an association between the 5-HT_{2A} G-allele in 156 OCD patients and a positive family history and an early onset of the disease compared to 134 control individuals (Denys *et al.*, 2006).

The G/G genotype of the 5-HT_{2A} polymorphism was also reported amongst the majority of OCD patients who responded with a Y-BOCS decrease of more than 25 percent to a 12-week treatment period with 60 mg paroxetine (Denys *et al.*, 2007). Difference in OCD phenotype may explain these discrepant findings. However, two studies found no association between 5-HT_{2A} polymorphism and OCD (Dickel *et al.*, 2007; Saiz *et al.*, 2008).

A genetic study in several early-onset OCD probands found no evidence for functional mutations in the sequenced regions of 5-HTR_{2B} (Kim *et al.*, 2000).

Cavallini and colleagues found no association between OCD and the CYS23Ser mutation of the 5-HT_{2C} receptor gene in a genetic study comparing 109 OCD patients with 107 healthy control subjects (Cavallini *et al.*, 1998).

The C178T variant of the 5-HT_{3A} receptor influences, among others, 5-HT_3 receptor expression. In a family-based approach with 75 children and adolescents with OCD as well as their biological parents, no evidence for a preferential transmission of either allele of the 5-HT_{3A} receptor was found. This is contrary to the involvement of the 5-HT_{3A} receptor in OCD (Mossner *et al.*, 2007).

The results of genetic association studies are, as yet, inconsistent and, except for the 5-HT_{2A} receptor, mainly negative.

Animal models

In the past decades, OCD-like animal models have been developed using chemical, genetic and behavioral induction. Table 3 gives a summary of the different animal models for OCD available.

Suppression of spontaneous burying in the marble-burying behavior model was originally used as a measure of anxiolytic drug action. Analysis of the marble-burying behavior later led to the suggestion that it was more related to compulsive behavior (Joel, 2006).

The 5-HT_{1A} agonists perospirone, MKC-242 and aripiprazole were tested in the marble-burying behavior model. All three inhibited marble-burying behavior. Inhibition of the behavior may be antagonized with WAY100135, a selective 5-HT_{1A} antagonist (Abe *et al.*, 1998; Matsushita *et al.*, 2005; Egashira *et al.*, 2008). Ichumaru and colleagues demonstrated that the suppressive effect of fluvoxamine on marble-burying behavior may be inhibited by the 5-HT_{1A} antagonist NAN-190 (Ichimaru *et al.*, 1995). Ketanserin, a 5-HT_{2A} antagonist, had no effect on marble-burying behavior (Egashira *et al.*, 2008). However, YM992, a compound with selective 5-HT reuptake inhibition and 5-HT_{2A} receptor antagonistic activity, significantly inhibited marble-burying behavior (Takeuchi *et al.*, 2002).

An extensive study examined the effects of 36 compounds, including typical, atypical and novel antipsychotics, on marble-burying behavior in mice. It concluded that the inhibition of marble-burying behavior may result from the interplay of several receptor systems, such as 5-HT_2 receptor blockade, 5-HT_{1A} agonism and dopamine D_2 partial agonism (Bruins Slot *et al.*, 2008).

The adjunctive drinking in the scheduled-induced polydipsia (SIP) model decreased after acute administration of WAY-181187, a novel and selective 5-HT_6 receptor agonist, and several 5-HT_{2C} receptor agonists (Bos *et al.*, 1997; Martin *et al.*, 1998; Dunlop *et al.*, 2006; Rosenzweig-Lipson *et al.*, 2007; Schechter *et al.*, 2007).

Table 3 Summary of the animal models available for OCD

	Inductor	Face validity symptoms	Predictive validity	Reference
Pharmacological models				
Spontaneous alternation behavior	8-OHDPAT 5-HT1A agonist	Indecision (T-maze)	—	Seibell *et al.*, 2003
Spontaneous alternation behavior	8-OHDPAT 5-HT1A agonist	Perseveration (T-maze)	SSRI	Yadin *et al.*, 1991
Spontaneous alternation behavior	Quinpirole (D_2 agonist)	Perseveration (T-maze)	SSRI	Einat and Szechtman, 1995
Reinforced spatial alternation	m-CPP (5-HT1A/1B/2C agonist)	Persistence (T-maze)	SSRI	Tsaltas *et al.*, 2005
Singular behavior	DOI 5-HT2A/2C agonist	Head twitch	SSRI	Rojas-Corrales *et al.*, 2007
Singular behavior	m-CPP (5-HT1A/1B/2C agonist)	Self-grooming	5-HT depletion caused more grooming	Graf, 2006
Singular behavior	Oxytocin	Hyper-grooming	—	Marroni *et al.*, 2007
Motor behavior	Quinpirole (D_2 agonist)	Compulsive checking	SSRI	Szechtman *et al.*, 1998
Behavioral models				
Marble burying	—	Compulsive burying	SSRI	Njung'e and Handley, 1991
Scheduled induced polydipsia (SIP)	Food deprived/fixed time feeding	Excessive drinking	SSRI	Woods *et al.*, 1993
Signal attenuation	Food deprivation	Compulsive lever pressing	SSRI	Joel, 2006
Head dipping	—	Head dipping	—	Chou-Green *et al.*, 2003
Plastic-mesh screen chewing	—	Plastic-mesh screen chewing	—	Chou-Green *et al.*, 2003
Genetic models				
Sapap3-mutant mice		Compulsive grooming behaviour	SSRI	Welch *et al.*, 2007
5-HT7 receptor knockout mice		None	—	Hedlund and Sutcliffe 2007
5-HT2C receptor knockout mice		Compulsive like behaviour: chewing of non-nutritive clay, chewing patterns on plastic-mesh screens, head dipping	—	Chou-Green *et al.*, 2003
Hoxb8 mutant mice		Excessive grooming	—	Greer and Capecchi, 2002
DAT KD mice		Rigid syntactic grooming chain patterns	—	Berridge *et al.*, 2005
DICT-7 mice		Compulsive behavior	—	Campbell *et al.*, 1999

In the signal attenuation rat model, the 5-HT_{2A} antagonist MDL as well as the $5\text{-HT}_{2A/2C}$ agonist DOI had no effect on compulsive lever pressing (Flaisher-Grinberg *et al.*, 2008). However, the 5-HT_{2C} antagonist RS 102221 decreased compulsive lever pressing after systemic administration, and after administration directly into the orbitofrontal cortex (Flaisher-Grinberg *et al.*, 2008).

Naratriptan, a selective 5-HT_{1D} receptor agonist, did not exacerbate compulsive behavior in the reinforced spatial alternation animal model. The fact that mCPP, a non-selective 5-HT receptor agonist acting at the 5-HT_{2C}, $5\text{-HT}_{1B/1D}$ and 5-HT_{1A} receptors, did increase compulsive behavior in this animal model speaks against the involvement of the $5\text{-HT}_{1B/1D}$ receptor in the OCD pathophysiology (Tsaltas *et al.*, 2005).

The mCPP-induced self-grooming in rats was reversed by a subtype-selective 5-HT_{2C} antagonist SB-242084, but not by the subtype-selective 5-HT_{2B} receptor antagonist SB-215505 (Graf *et al.*, 2003).

Several animal models for OCD originated from genetic modification of the serotonergic system. The 5-HT_{2C} receptor knockout mouse exhibited compulsive behavior. It chewed more clay, produced a distinct pattern of 'neat' chewing of plastic screens, and exhibited reduced habituation of head-dipping activity. This suggests that the 5-HT_{2C} receptor knockout mouse could be a promising model of compulsive behavior (Chou-Green *et al.*, 2003). Hedlund and colleagues investigated mice lacking the 5-HT_7 receptor compared to wild-type mice in three behavioral models for OCD. The 5-$HT_7^{-/-}$ mice buried fewer marbles than 5-$HT_7^{+/+}$ mice in the marble-burying behavioral model. Furthermore when the 5-$HT_7^{+/+}$ mice were treated with SB-269970, a selective 5-HT_7 receptor antagonist, they also buried fewer marbles than the 5-$HT_7^{+/+}$ mice treated with vehicle. In other words, inactivation as well as blockade of the 5-HT_7 receptor had a positive effect on OCD activity. However, in the two other behavioral models, head dipping and plastic-mesh screen chewing, no difference was found between the 5-$HT_7^{-/-}$ mice and wild-type mice (Hedlund and Sutcliffe, 2007).

Preclinical research into 5-HTT function has progressed rapidly since 5-HTT knockout animal models became available. However, as yet, research on the 5-HTT knockout rat or mouse behavior paradigms has not directly linked any behavioral deviations to OCD (Mathews *et al.*, 2004; Homberg *et al.*, 2007; Olivier *et al.*, 2008).

In summary, preclinical research gives conflicting results. It suggests involvement of the 5-HT_{2C} receptor and provides interesting preliminary results in animal models for the 5-HT_6 and 5-HT_7 receptors.

Conclusion

There is compelling evidence that the serotonergic system plays a major role in the treatment of OCD. Pharmacotherapeutic studies have shown that SSRIs are more effective in achieving clinical response compared to placebo, and that atypical antipsychotics, which have a high affinity for 5-HT_{2A} and 5-HT_{2C} receptors may augment the effect of SSRIs. However, these findings do not necessarily reflect the existence of a neurobiological abnormality in the central serotonergic system in OCD. SSRIs may modulate serotonin via an intact serotonergic system to compensate for the underlying pathogenesis that is related to another neurotransmitter system.

A general survey of the specific role of 5-HT transporters and 5-HT receptors in OCD arrives at inconsistent conclusions. Though an impaired 5-HTT function is suggested by some neuroimaging studies, others indicate that successful SSRI treatment is associated with normalization of the dopamine function. Genetic and animal studies have provided insufficient evidence to implicate a fundamental role for 5-HTT in the pathophysiology of OCD. As yet the role of 5-HT_{1A} receptors in OCD is unclear, since 5-HT_{1A} agonists as well as 5-HT_{1A} antagonists improve OC symptoms in pharmacologic challenge studies. Assuming that chronic SSRI treatment leads to a desensitization of the terminal $5HT_{1B/1D}$ receptor, some authors have hypothesized that $5HT_{1B/1D}$ receptors are supersensitive in OCD, resulting in chronic reductions in synaptic levels of 5-HT (Moret and Briley, 2000). There is circumstantial evidence that the 5-HT_{2A} receptor may be of particular relevance to OCD. Genetic studies linked the −1438G/A polymorphism of the 5-HT_{2A} receptor gene to OCD, but non-selective 5-HT_{2A} agonists as well as antagonists have a positive effect on OC symptoms. Reports on involvement of the 5-HT_{2A} receptor in neuroimaging and animal research are inconsistent. Though pharmacological challenges with mCPP suggest a role for the 5-HT_{2C} receptor, animal models report conflicting results, since 5-HT_{2C} agonists as well as antagonists reduce or induce obsessive-compulsive behavior. At this moment, there is no evidence to underscore a role for the 5-HT_{2B}, 5-HT_3, 5-HT_6, or 5-HT_7 receptors in OCD.

In summary, direct evidence for an abnormality in the serotonergic system in the pathophysiology of OCD is still scarce. Hence, it is open to discussion whether the serotonergic system plays a primary role in the pathophysiology OCD or is secondarily involved due to the efficacy of SSRIs. Reasonably, SSRIs modulate an intact serotonergic system to compensate for another neurotransmitter mechanism. Several observations lend support to the assumption that inhibition of 5-HTT is not causally related to the genesis of the disorder (Baumgarten and Grozdanovic, 1998).

First, the mean response rate in OCD is 50 percent or less (Denys *et al.*, 2002), indicating that for a substantial number of patients 5-HTT inhibition is insufficient to alleviate OC symptoms. Phenotypic heterogeneity may partly explain this phenomenon, as, for example, patients with contamination fear respond more favorably to SSRIs than do patients from the subtype symmetry/perfectionism/hoarding (Denys *et al.*, 2004a). Second, the doses of antidepressants necessary for OCD are substantially higher than those for depression or anxiety disorders, and also substantially higher than necessary to completely block the 5-HTT (Kent *et al.*, 2002). Third, the synthesis of 5-HT depends on the availability of tryptophan in the brain. Depletion studies in depression have shown deterioration in patients on antidepressants in remission (Neumeister, 2003), but no worsening of symptoms was seen in patients with OCD who underwent a tryptophan-depletion paradigm (Berney *et al.*, 2006). Fourth, the therapeutic effects in OCD are usually not seen within 8 weeks of treatment, which is much later than in patients with depression. El Mansari and colleagues have argued, on the basis of their electrophysiological work, that this difference in onset of action between depression and OCD can be accounted for by a greater delay in effect of SSRIs on 5-HT release in the orbitofrontal cortex (OFC), a brain region supposedly implicated in OCD, as compared to other brain regions (El Mansari and Blier, 2006). According to these investigators, this delay in effect in the OFC might be explained by a slower desensitization of the 5-HT_{1B} autoreceptors in this region. They also used this finding as an argument to explain why, in OCD, larger doses of SSRIs are needed. In line with this finding, Dannon and colleagues have reported pindolol, a non-selective 5-HT_{1A} receptor antagonist, to hasten the effect of SSRIs in OCD patients (Dannon *et al.*, 2000). They also suggest that the effect of SSRIs in OCD might be explained by the delayed stimulation of the postsynaptic 5-HT_{2A} receptors in the OFC. If this were to be true, one would expect mirtazapine (which among others is a 5-HT_{2A} receptor antagonist) and atypical antipsychotics that also have antagonistic effects at this receptor to attenuate the effect of SSRIs. Clinical studies with these drugs, however, have shown the opposite. Mirtazapine, although not effective by itself, has been shown to hasten the effect of paroxetine (Pallanti *et al.*, 2004), and several studies have shown that atypical antipsychotics augment the effects of SSRIs in refractory OCD patients (Bloch *et al.*, 2006). Moreover, mCPP, a non-selective 5-HT_2 receptor agonist, is either not effective or causes a worsening of OC symptoms after acute administration (Charney *et al.*, 1988; Goodman *et al.*, 1995; Hollander *et al.*, 1992; Ho Pian *et al.*, 1998; Gross-Isseroff *et al.*, 2004).

In conclusion, future research should focus much more on the interaction of the serotonergic system with other neurochemical systems, such as dopamine and glutamate. There is some evidence that dopamine might be involved in OCD. Atypical antipsychotics with potent 5-HT_{2A}, 5-HT_{2C} and D_2 antagonistic properties have anti-obsessional properties as additional therapy to SSRIs. Denys and colleagues found that the combination of quetiapine and fluvoxamine may cause a synergistic dopamine increase in the prefrontal cortex and thalamus (Denys *et al.*, 2004b). Moreover, SPECT and PET studies hint at decreased D_2 receptor binding in the ventral striatum (Denys *et al.*, 2004c), and indicate that successful SSRI treatment in OCD is associated with normalization of the dopamine function. Additionally, more selective agents need to be developed to investigate the specific role of the different receptors of the 5-HT system. In the pipeline are a series of 5-HT_{2C} agonists which are being investigated by Hoffman-La Roche, while Janssen Pharmaceutica is studying an azeprine derivative with mixed 5-$HT_{2A/2C}$ antagonism for use in OCD. Furthermore, a 5-$HT_{1B/1D}$ antagonist is being investigated by GlaxoSmithKline, and Vanderbilt University is looking at a selective 5-HT_3 antagonist (Davidson and Bjorgvinsson, 2003). The development of better animal models will be a tool to further enrich our understanding of OCD. OCD is a complex psychiatric disorder affecting cognition, affect and volition. It is therefore unlikely that one single neurotransmitter is involved in its pathophysiology.

References

Abe, M., Nakai, H., Tabata, R., Saito, K. and Egawa, M. (1998) Effect of 5-[3-[((2S)-1,4-benzodioxan-2-ylmethyl) amino]propoxy]-1,3-benzodioxole HCl (MKC-242), a novel 5-HT1A-receptor agonist, on aggressive behavior and marble burying behavior in mice. Jpn. J. Pharmacol., 76: 297–304.

Ackerman, D.L. and Greenland, S. (2002) Multivariate meta-analysis of controlled drug studies for obsessive-compulsive disorder. J. Clin. Psychopharmacol., 22: 309–317.

Adams, K.H., Hansen, E.S., Pinborg, L.H., Hasselbalch, S.G., Svarer, C., Holm, S., Bolwig, T.G. and Knudsen, G.M. (2005) Patients with obsessive-compulsive disorder have increased 5-HT2A receptor binding in the caudate nuclei. Int. J. Neuropsychopharmacol., 8: 391–401.

Bareggi, S.R., Bianchi, L., Cavallaro, R., Gervasoni, M., Siliprandi, F. and Bellodi, L. (2004) Citalopram concentrations and response in obsessive-compulsive disorder. Preliminary results. CNS Drugs, 18: 329–335.

Baumgarten, H.G. and Grozdanovic, Z. (1998) Role of serotonin in obsessive-compulsive disorder. Br. J. Psychiatry Suppl.: 13–20.

Bengel, D., Greenberg, B.D., Cora-Locatelli, G., Altemus, M., Heils, A., Li, Q. and Murphy, D.L. (1999) Association of the serotonin transporter promoter regulatory region polymorphism and obsessive-compulsive disorder. Mol. Psychiatry, 4: 463–466.

Berney, A., Sookman, D., Leyton, M., Young, S.N. and Benkelfat, C. (2006) Lack of effects on core obsessive-compulsive symptoms of tryptophan depletion during symptom provocation in remitted obsessive-compulsive disorder patients. Biol. Psychiatry, 59: 853–857.

Berridge, K.C., Aldridge, J.W., Houchard, K.R. and Zhuang, X. (2005) Sequential super-stereotypy of an instinctive fixed action pattern in hyper-dopaminergic mutant mice: a model of obsessive-compulsive disorder and Tourette's. BMC Biol., 3: 4.

Bloch, M.H., Landeros-Weisenberger, A., Kelmendi, B., Coric, V., Bracken, M.B. and Leckman, J.F. (2006) A systematic review: antipsychotic augmentation with treatment refractory obsessive-compulsive disorder. Mol. Psychiatry, 11: 622–632.

Bos, M., Jenck, F., Martin, J.R., Moreau, J.L., Sleight, A.J., Wichmann, J. and Widmer, U. (1997) Novel agonists of 5HT2C receptors. Synthesis and biological evaluation of substituted 2-(indol-1-yl)-1-methylethylamines and 2-(indeno[1,2-b]pyrrol-1-yl)-1-methylethylamines. Improved therapeutics for obsessive-compulsive disorder. J. Med. Chem., 40: 2762–2769.

Boshuisen, M.L. and den Boer, J.A. (2000) Zolmitriptan (a 5-HT1B/1D receptor agonist with central action) does not increase symptoms in obsessive-compulsive disorder. Psychopharmacology (Berl.), 152: 74–79.

Broocks, A., Pigott, T.A., Hill, J.L., Canter, S., Grady, T.A., L'Heureux, F. and Murphy, D.L. (1998) Acute intravenous administration of ondansetron and m-CPP, alone and in combination, in patients with obsessive-compulsive disorder (OCD): behavioral and biological results. Psychiatry Res., 79: 11–20.

Bruins Slot, L.A., Bardin, L., Auclair, A.L., Depoortere, R. and Newman-Tancredi, A. (2008) Effects of antipsychotics and reference monoaminergic ligands on marble burying behavior in mice. Behav. Pharmacol., 19: 145–152.

Camarena, B., Rinetti, G., Cruz, C., Hernandez, S., de, L.F. Jr. and Nicolini, H. (2001) Association study of the serotonin transporter gene polymorphism in obsessive-compulsive disorder. Int. J. Neuropsychopharmacol., 4: 269–272.

Camarena, B., Aguilar, A., Loyzaga, C. and Nicolini, H. (2004) A family-based association study of the 5-HT-1Dbeta receptor gene in obsessive-compulsive disorder. Intl. J. Neuropsychopharmacol., 7: 49–53.

Campbell, K.M., de Lecea, L., Severynse, D.M., Caron, M.G., McGrath, M.J., Sparber, S.B., Sun, L.Y. and Burton, F.H. (1999) OCD-Like behaviors caused by a neuropotentiating transgene targeted to cortical and limbic D1+ neurons. J. Neurosci., 19: 5044–5053.

Cavallini, M.C., Di Bella, D., Pasquale, L., Henin, M. and Bellodi, L. (1998) 5HT2C CYS23/SER23 polymorphism is not associated with obsessive-compulsive disorder. Psychiatry Res., 77: 97–104.

Cavallini, M.C., Di Bella, D., Siliprandi, F., Malchiodi, F. and Bellodi, L. (2002) Exploratory factor analysis of obsessive-compulsive patients and association with 5-HTTLPR polymorphism. Am. J. Med. Genet., 114: 347–353.

Chabane, N., Millet, B., Delorme, R., Lichtermann, D., Mathieu, F., Laplanche, J.L., Roy, I., Mouren, M.C., Hankard, R., Maier, W., Launay, J.M. and Leboyer, M. (2004) Lack of evidence for association between serotonin transporter gene (5-HTTLPR) and obsessive-compulsive disorder by case control and family association study in humans. Neurosci. Lett., 363: 154–156.

Charney, D.S., Goodman, W.K., Price, L.H., Woods, S.W., Rasmussen, S.A. and Heninger, G.R. (1988) Serotonin function in obsessive-compulsive disorder. A comparison of the effects of tryptophan and m-chlorophenylpiperazine in patients and healthy subjects. Arch. Gen. Psychiatry, 45: 177–185.

Chou-Green, J.M., Holscher, T.D., Dallman, M.F. and Akana, S.F. (2003) Compulsive behavior in the 5-HT2C receptor knockout mouse. Physiol. Behav., 78: 641–649.

Dannon, P.N., Sasson, Y., Hirschmann, S., Iancu, I., Grunhaus, L.J. and Zohar, J. (2000) Pindolol augmentation in treatment-resistant obsessive-compulsive disorder: a double-blind placebo controlled trial. Eur. Neuropsychopharmacol., 10: 165–169.

Davidson, J. and Bjorgvinsson, T. (2003) Current and potential pharmacological treatments for obsessive-compulsive disorder. Expert. Opin. Investig. Drugs, 12: 993–1001.

Dell'osso, B., Allen, A. and Hollander, E. (2005) Fluvoxamine: a selective serotonin re-uptake inhibitor for the treatment of obsessive-compulsive disorder. Expert. Opin. Pharmacother., 6: 2727–2740.

Denys, D. (2006) Pharmacotherapy of obsessive-compulsive disorder and obsessive-compulsive spectrum disorders. Psychiatr. Clin. North Am., 29: 553–584, xi

Denys, D., van Megan, H. and Westenberg, H. (2002) The adequacy of pharmacotherapy in outpatients with obsessive-compulsive disorder. Intl. Clin. Psychopharmacol., 17: 109–114.

Denys, D., van der Wee, N., van Megen, H.J. and Westenberg, H.G. (2003) A double blind comparison of venlafaxine and paroxetine in obsessive-compulsive disorder. J. Clin. Psychopharmacol., 23: 568–575.

Denys, D., de Geus, F., van Megen, H.J. and Westenberg, H.G. (2004a) Use of factor analysis to detect potential phenotypes in obsessive-compulsive disorder. Psychiatry Res., 128: 273–280.

Denys, D., Klompmakers, A.A. and Westenberg, H.G. (2004b) Synergistic dopamine increase in the rat prefrontal cortex with the combination of quetiapine and fluvoxamine. Psychopharmacology (Berl.), 176: 195–203.

Denys, D., van der Wee, N., Janssen, J., de Geus, F. and Westenberg, H.G. (2004c) Low level of dopaminergic D2 receptor binding in obsessive-compulsive disorder. Biol. Psychiatry, 55: 1041–1045.

Denys, D., van Nieuwerburgh, F., Deforce, D. and Westenberg, H.G. (2006) Association between serotonergic candidate genes and specific phenotypes of obsessive-compulsive disorder. J. Affect Disord., 91: 39–44.

Denys, D., van Nieuwerburgh, F., Deforce, D. and Westenberg, H.G. (2007) Prediction of response to paroxetine and venlafaxine by serotonin-related genes in obsessive-compulsive disorder in a randomized, double-blind trial. J. Clin. Psychiatry, 68: 747–753.

Di Bella, D., Cavallini, M.C. and Bellodi, L. (2002a) No association between obsessive-compulsive disorder and the 5-HT(1Dbeta) receptor gene. Am. J. Psychiatry, 159: 1783–1785.

Di Bella, D., Erzegovesi, S., Cavallini, M.C. and Bellodi, L. (2002b) Obsessive-compulsive disorder, 5-HTTLPR polymorphism and treatment response. Pharmacogenomics J., 2: 176–181.

Dickel, D.E., Veenstra-VanderWeele, J., Bivens, N.C., Wu, X., Fischer, D.J., van Etten-Lee, M., Himle, J.A., Leventhal, B.L., Cook, E.H. Jr. and Hanna, G.L. (2007) Association studies of serotonin system candidate genes in early-onset obsessive-compulsive disorder. Biol. Psychiatry, 61: 322–329.

Dunlop, J., Marquis, K.L., Lim, H.K., Leung, L., Kao, J., Cheesman, C. and Rosenzweig-Lipson, S. (2006) Pharmacological profile of the 5-HT(2C) receptor agonist WAY-163909; therapeutic potential in multiple indications. CNS Drug Rev., 12: 167–177.

Eddy, K.T., Dutra, L., Bradley, R. and Westen, D. (2004) A multidimensional meta-analysis of psychotherapy and pharmacotherapy for obsessive-compulsive disorder. Clin. Psychol. Rev., 24: 1011–1030.

Egashira, N., Okuno, R., Matsushita, M., Abe, M., Mishima, K., Iwasaki, K., Nishimura, R., Oishi, R. and Fujiwara, M. (2008) Aripiprazole inhibits marble-burying behavior via 5-hydroxytryptamine (5-HT)(1A) receptor-independent mechanisms. Eur. J. Pharmacol., 592: 103–108.

Einat, H. and Szechtman, H. (1995) Perseveration without hyperlocomotion in a spontaneous alternation task in rats sensitized to the dopamine agonist quinpirole. Physiol. Behav., 57: 55–59.

El Mansari, M. and Blier, P. (2006) Mechanisms of action of current and potential pharmacotherapies of obsessive-compulsive disorder. Prog. Neuropsychopharmacol. Biol. Psychiatry, 30: 362–373.

Enoch, M.A., Greenberg, B.D., Murphy, D.L. and Goldman, D. (2001) Sexually dimorphic relationship of a 5-HT2A promoter polymorphism with obsessive-compulsive disorder. Biol. Psychiatry, 49: 385–388.

Erzegovesi, S., Guglielmo, E., Siliprandi, F. and Bellodi, L. (2005) Low-dose risperidone augmentation of fluvoxamine treatment in obsessive-compulsive disorder: a double-blind, placebo-controlled study. Eur. Neuropsychopharmacol., 1: 69–74.

Flaisher-Grinberg, S., Klavir, O. and Joel, D. (2008) The role of 5-HT2A and 5-HT2C receptors in the signal attenuation rat model of obsessive-compulsive disorder. Int. J. Neuropsychopharmacol., 14: 1–15.

Frisch, A., Michaelovsky, E., Rockah, R., Amir, I., Hermesh, H., Laor, N., Fuchs, C., Zohar, J., Lerer, B., Buniak, S.F., Landa, S., Poyurovsky, M., Shapira, B. and Weizman, R. (2000) Association between obsessive-compulsive disorder and polymorphisms of genes encoding components of the serotonergic and dopaminergic pathways. Eur. Neuropsychopharmacol., 10: 205–209.

Goodman, W.K., Price, L.H., Rasmussen, S.A., Mazure, C., Fleischmann, R.L., Hill, C.L., Heninger, G.R. and Charney, D.S. (1989a) The Yale-Brown Obsessive-compulsive Scale. I. Development, use, and reliability. Arch. Gen. Psychiatry, 46: 1006–1011.

Goodman, W.K., Price, L.H., Rasmussen, S.A., Mazure, C., Delgado, P., Heninger, G.R. and Charney, D.S. (1989b) The Yale-Brown Obsessive-compulsive Scale. II. Validity. Arch. Gen. Psychiatry, 46: 1012–1016.

Goodman, W.K., Price, L.H., Delgado, P.L., Palumbo, J., Krystal, J.H., Nagy, L.M., Rasmussen, S.A., Heninger, G.R. and Charney, D.S. (1990) Specificity of serotonin reuptake inhibitors in the treatment of obsessive-compulsive disorder. Comparison of fluvoxamine and desipramine. Arch. Gen. Psychiatry, 47: 577–585.

Goodman, W.K., McDougle, C.J., Price, L.H., Barr, L.C., Hills, O.F., Caplik, J.F., Charney, D.S. and Heninger, G.R. (1995) m-Chlorophenylpiperazine in patients with obsessive-compulsive disorder: absence of symptom exacerbation. Biol. Psychiatry, 38: 138–149.

Grados, M.A., Samuels, J., Shugart, Y.Y., Willour, V.L., Wang, Y., Cullen, B., Bienvenu, O.J., Hoehn-Saric, R., Valle, D., Liang, K.Y., Riddle, M.A., Wendland, J.R., Murphy, D.L., Nestadt, G. and tera-Wadleigh, S. (2007) Rare plus common SERT variants in obsessive-compulsive disorder. Mol. Psychiatry, 12: 422–423.

Graf, M. (2006) 5-HT2c receptor activation induces grooming behavior in rats: possible correlations with obsessive-compulsive disorder. Neuropsychopharmacol. Hung, 8: 23–28.

Graf, M., Kantor, S., Anheuer, Z.E., Modos, E.A. and Bagdy, G. (2003) m-CPP-induced self-grooming is mediated by 5-HT2C receptors. Behav. Brain Res., 142: 175–179.

Greer, J.M. and Capecchi, M.R. (2002) Hoxb8 is required for normal grooming behavior in mice. Neuron, 33: 23–34.

Gross-Isseroff, R., Cohen, R., Sasson, Y., Voet, H. and Zohar, J. (2004) Serotonergic dissection of obsessive-compulsive symptoms: a challenge study with m-chlorophenylpiperazine and sumatriptan. Neuropsychobiology, 50: 200–205.

Hasselbalch, S.G., Hansen, E.S., Jakobsen, T.B., Pinborg, L.H., Lonborg, J.H. and Bolwig, T.G. (2007) Reduced midbrain-pons serotonin transporter binding in patients with obsessive-compulsive disorder. Acta Psychiatr. Scand., 115: 388–394.

Hedlund, P.B. and Sutcliffe, J.G. (2007) The 5-HT7 receptor influences stereotypic behavior in a model of obsessive-compulsive disorder. Neurosci. Lett., 414: 247–251.

Hesse, S., Muller, U., Lincke, T., Barthel, H., Villmann, T., Angermeyer, M.C., Sabri, O. and Stengler-Wenzke, K. (2005) Serotonin and dopamine transporter imaging in patients with obsessive-compulsive disorder. Psychiatry Res., 140: 63–72.

Hewlett, W.A., Schmid, S.P. and Salomon, R.M. (2003) Pilot trial of ondansetron in the treatment of 8 patients with obsessive-compulsive disorder. J. Clin. Psychiatry, 64: 1025–1030.

Hoehn-Saric, R., Ninan, P., Black, D.W., Stahl, S., Greist, J.H., Lydiard, B., McElroy, S., Zajecka, J., Chapman, D., Clary, C. and Harrison, W. (2000) Multicenter double-blind comparison of sertraline and desipramine for concurrent obsessive-compulsive and major depressive disorders. Arch. Gen. Psychiatry, 57: 76–82.

Hollander, E., DeCaria, C.M., Nitescu, A., Gully, R., Suckow, R.F., Cooper, T.B., Gorman, J.M., Klein, D.F. and Liebowitz, M.R. (1992) Serotonergic function in obsessive-compulsive disorder. Behavioral and neuroendocrine responses to oral m-chlorophenylpiperazine and fenfluramine in patients and healthy volunteers. Arch. Gen. Psychiatry, 49: 21–28.

Hollander, E., Baldini, R.N., Sood, E. and Pallanti, S. (2003) Risperidone augmentation in treatment-resistant obsessive-compulsive disorder: a double-blind, placebo-controlled study. Int. J. Neuropsychopharmacol., 6: 397–401.

Homberg, J.R., Olivier, J.D., Smits, B.M., Mul, J.D., Mudde, J., Verheul, M., Nieuwenhuizen, O.F., Cools, A.R., Ronken, E., Cremers, T., Schoffelmeer, A.N., Ellenbroek, B.A. and Cuppen, E. (2007) Characterization of the serotonin transporter knockout rat: a selective change in the functioning of the serotonergic system. Neuroscience, 146: 1662–1676.

Ho Pian, K.L., Westenberg, H.G., den Boer, J.A., de Bruin, W.I. and van Rijk, P.P. (1998) Effects of meta-chlorophenylpiperazine on cerebral blood flow in obsessive-compulsive disorder and controls. Biol. Psychiatry, 44: 367–370.

Ichimaru, Y., Egawa, T. and Sawa, A. (1995) 5-HT1A-receptor subtype mediates the effect of fluvoxamine, a selective serotonin reuptake inhibitor, on marble-burying behavior in mice. Jpn. J. Pharmacol., 68: 65–70.

Joel, D. (2006) The signal attenuation rat model of obsessive-compulsive disorder: a review. Psychopharmacology (Berl.), 186: 487–503.

Kent, J.M., Coplan, J.D., Lombardo, I., Hwang, D.R., Huang, Y., Mawlawi, O., Van Heertum, R.L., Slifstein, M., bi-Dargham, A., Gorman, J.M. and Laruelle, M. (2002) Occupancy of brain serotonin transporters during treatment with paroxetine in patients with social phobia: a positron emission tomography study with 11C McN 5652. Psychopharmacology (Berl.), 164: 341–348.

Kim, C.H., Cheon, K.A., Koo, M.S., Ryu, Y.H., Lee, J.D., Chang, J.W. and Lee, H.S. (2007) Dopamine transporter density in the basal ganglia in obsessive-compulsive disorder, measured with [123I]IPT SPECT before and after treatment with serotonin reuptake inhibitors. Neuropsychobiology, 55: 156–162.

Kim, S.J., Veenstra-VanderWeele, J., Hanna, G.L., Gonen, D., Leventhal, B.L. and Cook, E.H. Jr. (2000) Mutation screening of human 5-HT(2B)receptor gene in early-onset obsessive-compulsive disorder. Mol. Cell Probes, 14: 47–52.

Kim, S.J., Lee, H.S. and Kim, C.H. (2005) Obsessive-compulsive disorder, factor-analyzed symptom dimensions and serotonin transporter polymorphism. Neuropsychobiology, 52: 176–182.

Koran, L.M., Pallanti, S. and Quercioli, L. (2001) Sumatriptan, 5-HT(1D) receptors and obsessive-compulsive disorder. Eur. Neuropsychopharmacol., 11: 169–172.

Leonard, H.L., Swedo, S.E., Rapoport, J.L., Koby, E.V., Lenane, M.C., Cheslow, D.L. and Hamburger, S.D. (1989) Treatment of obsessive-compulsive disorder with clomipramine and

desipramine in children and adolescents. A double-blind crossover comparison. Arch. Gen. Psychiatry, 46: 1088–1092.
Lesch, K.P., Hoh, A., Schulte, H.M., Osterheider, M. and Muller, T. (1991a) Long-term fluoxetine treatment decreases 5-HT1A receptor responsivity in obsessive-compulsive disorder. Psychopharmacology (Berl.), 105: 415–420.
Lesch, K.P., Hoh, A., Sselkamp-Tietze, J., Wiesmann, M., Osterheider, M. and Schulte, H.M. (1991b) 5-Hydroxytryptamine1A receptor responsivity in obsessive-compulsive disorder. Comparison of patients and controls. Arch. Gen. Psychiatry, 48: 540–547.
Levitan, R.D., Kaplan, A.S., Masellis, M., Basile, V.S., Richter, M.A. and Kennedy, J.L. (2006) The serotonin-1Dbeta receptor gene and severity of obsessive-compulsive disorder in women with bulimia nervosa. Eur. Neuropsychopharmacol., 16: 1–6.
Lin, P.Y. (2007) Meta-analysis of the association of serotonin transporter gene polymorphism with obsessive-compulsive disorder. Prog. Neuropsychopharmacol. Biol. Psychiatry, 31: 683–689.
Marek, G.J., Carpenter, L.L., McDougle, C.J. and Price, L.H. (2003) Synergistic action of 5-HT2A antagonists and selective serotonin reuptake inhibitors in neuropsychiatric disorders. Neuropsychopharmacology, 28: 402–412.
Marroni, S.S., Nakano, F.N., Gati, C.D., Oliveira, J.A., Ntunes-Rodrigues, J. and Garcia-Cairasco, N. (2007) Neuroanatomical and cellular substrates of hypergrooming induced by microinjection of oxytocin in central nucleus of amygdala, an experimental model of compulsive behavior. Mol. Psychiatry, 12: 1103–1117.
Martin, J.R., Bos, M., Jenck, F., Moreau, J., Mutel, V., Sleight, A.J., Wichmann, J., Andrews, J.S., Berendsen, H.H., Broekkamp, C.L., Ruigt, G.S., Kohler, C. and Delft, A.M. (1998) 5-HT2C receptor agonists: pharmacological characteristics and therapeutic potential. J. Pharmacol. Exp. Ther., 286: 913–924.
Mathews, T.A., Fedele, D.E., Coppelli, F.M., Avila, A.M., Murphy, D.L. and Andrews, A.M. (2004) Gene dose-dependent alterations in extraneuronal serotonin but not dopamine in mice with reduced serotonin transporter expression. J. Neurosci. Methods, 140: 169–181.
Matsushita, M., Egashira, N., Harada, S., Okuno, R., Mishima, K., Iwasaki, K., Nishimura, R. and Fujiwara, M. (2005) Perospirone, a novel antipsychotic drug, inhibits marble-burying behavior via 5-HT1A receptor in mice: implications for obsessive-compulsive disorder. J. Pharmacol. Sci., 99: 154–159.
McDougle, C.J., Epperson, C.N., Price, L.H. and Gelernter, J. (1998) Evidence for linkage disequilibrium between serotonin transporter protein gene (SLC6A4) and obsessive-compulsive disorder. Mol. Psychiatry, 3: 270–273.
McDougle, C.J., Epperson, C.N., Pelton, G.H., Wasylink, S. and Price, L.H. (2000) A double-blind, placebo-controlled study of risperidone addition in serotonin reuptake inhibitor-refractory obsessive-compulsive disorder. Arch. Gen. Psychiatry, 57: 794–801.
Meira-Lima, I., Shavitt, R.G., Miguita, K., Ikenaga, E., Miguel, E.C. and Vallada, H. (2004) Association analysis of the catechol-o-methyltransferase (COMT), serotonin transporter (5-HTT) and serotonin 2A receptor (5HT2A) gene polymorphisms with obsessive-compulsive disorder. Genes Brain Behav., 3: 75–79.
Meyer, J.H. (2007) Imaging the serotonin transporter during major depressive disorder and antidepressant treatment. J. Psychiatry Neurosci., 32: 86–102.
Montgomery, S.A., Kasper, S., Stein, D.J., Bang, H.K. and Lemming, O.M. (2001) Citalopram 20 mg, 40 mg and 60 mg are all effective and well tolerated compared with placebo in obsessive-compulsive disorder. Intl. Clin. Psychopharmacol., 16: 75–86.
Moreno, F.A., Wiegand, C.B., Taitano, E.K. and Delgado, P.L. (2006) Safety, tolerability, and efficacy of psilocybin in 9 patients with obsessive-compulsive disorder. J. Clin. Psychiatry, 67: 1735–1740.
Moresco, R.M., Pietra, L., Henin, M., Panzacchi, A., Locatelli, M., Bonaldi, L., Carpinelli, A., Gobbo, C., Bellodi, L., Perani, D. and Fazio, F. (2007) Fluvoxamine treatment and D2 receptors: a pet study on OCD drug-naive patients. Neuropsychopharmacology, 32: 197–205.
Moret, C. and Briley, M. (2000) The possible role of 5-HT(1B/D) receptors in psychiatric disorders and their potential as a target for therapy. Eur. J. Pharmacol., 404: 1–12.
Mossner, R., Doring, N., Scherag, A., Schafer, H., Herpertz-Dahlmann, B., Remschmidt, H., Schulz, E., Renner, T., Wewetzer, C., Warnke, A., Lesch, K.P. and Walitza, S. (2007) Transmission disequilibrium analysis of the functional 5-HT3A receptor variant C178T in early-onset obsessive-compulsive-disorder. J. Psychopharmacol., 21: 833–836.
Mundo, E., Richter, M.A., Zai, G., Sam, F., McBride, J., Macciardi, F. and Kennedy, J.L. (2002) 5HT1Dbeta Receptor gene implicated in the pathogenesis of obsessive-compulsive disorder: further evidence from a family-based association study. Mol. Psychiatry, 7: 805–809.
Neumeister, A. (2003) Tryptophan depletion, serotonin, and depression: where do we stand? Psychopharmacol. Bull., 37: 99–115.
Ninan, P.T. (2003) Obsessive-compulsive disorder: implications of the efficacy of an SSRI, paroxetine. Psychopharmacol. Bull., 37(Suppl. 1): 89–96.
Njung'e, K. and Handley, S.L. (1991) Evaluation of marble-burying behavior as a model of anxiety. Pharmacol. Biochem. Behav., 38: 63–67.
Olivier, J.D., Van Der Hart, M.G., Van Swelm, R.P., Dederen, P. J., Homberg, J.R., Cremers, T., Deen, P.M., Cuppen, E., Cools, A.R. and Ellenbroek, B.A. (2008) A study in male and female 5-HT transporter knockout rats: an animal model for anxiety and depression disorders. Neuroscience, 152: 573–584.
Pallanti, S., Quercioli, L. and Bruscoli, M. (2004) Response acceleration with mirtazapine augmentation of citalopram in obsessive-compulsive disorder patients without comorbid depression: a pilot study. J. Clin. Psychiatry, 65: 1394–1399.
Perani, D., Garibotto, V., Gorini, A., Moresco, R.M., Henin, M., Panzacchi, A., Matarrese, M., Carpinelli, A., Bellodi, L. and Fazio, F. (2008) In vivo PET study of 5HT(2A) serotonin and D(2) dopamine dysfunction in drug-naive obsessive-compulsive disorder. Neuroimage, 42: 306–314.
Perez, M., Brown, J.S., Vrshek-Schallhorn, S., Johnson, F. and Joiner, T.E. Jr. (2006) Differentiation of obsessive-compulsive-, panic-, obsessive-compulsive personality-, and non-disordered individuals by variation in the promoter region of the serotonin transporter gene. J. Anxiety Disord, 20: 794–806.
Pian, K.L., Westenberg, H.G., van Megen, H.J. and den Boer, J.A. (1998) Sumatriptan (5-HT1D receptor agonist) does not exacerbate symptoms in obsessive-compulsive disorder. Psychopharmacology (Berl.), 140: 365–370.
Pogarell, O., Hamann, C., Popperl, G., Juckel, G., Chouker, M., Zaudig, M., Riedel, M., Moller, H.J., Hegerl, U. and Tatsch, K. (2003) Elevated brain serotonin transporter availability in

patients with obsessive-compulsive disorder. Biol. Psychiatry, 54: 1406–1413.
Pogarell, O., Poepperl, G., Mulert, C., Hamann, C., Sadowsky, N., Riedel, M., Moeller, H.J., Hegerl, U. and Tatsch, K. (2005) SERT and DAT availabilities under citalopram treatment in obsessive-compulsive disorder (OCD). Eur. Neuropsychopharmacol., 15: 521–524.
Reimold, M., Smolka, M.N., Zimmer, A., Batra, A., Knobel, A., Solbach, C., Mundt, A., Smoltczyk, H.U., Goldman, D., Mann, K., Reischl, G., Machulla, H.J., Bares, R. and Heinz, A. (2007) Reduced availability of serotonin transporters in obsessive-compulsive disorder correlates with symptom severity – a [11C]DASB PET study. J. Neural Transm., 114: 1603–1609.
Rojas-Corrales, M.O., Gibert-Rahola, J. and Mico, J.A. (2007) Role of atypical opiates in OCD. Experimental approach through the study of 5-HT(2A/C) receptor-mediated behavior. Psychopharmacology (Berl.), 190: 221–231.
Rosenzweig-Lipson, S., Sabb, A., Stack, G., Mitchell, P., Lucki, I., Malberg, J.E., Grauer, S., Brennan, J., Cryan, J.F., Sukoff Rizzo, S.J., Dunlop, J., Barrett, J.E. and Marquis, K. L. (2007) Antidepressant-like effects of the novel, selective, 5-HT2C receptor agonist WAY-163909 in rodents. Psychopharmacology (Berl.), 192: 159–170.
Saiz, P.A., Garcia-Portilla, M.P., Arango, C., Morales, B., Bascaran, M.T., Martinez-Barrondo, S., Florez, G., Sotomayor, E., Paredes, B., Alvarez, C., San, N.G., Carreno, E., Bombin, I., Alvarez, V., Coto, E., Fernandez, J.M., Bousono, M. and Bobes, J. (2008) Association study between obsessive-compulsive disorder and serotonergic candidate genes. Prog. Neuropsychopharmacol. Biol. Psychiatry, 32: 765–770.
Schechter, L.E., Lin, Q., Smith, D.L., Zhang, G., Shan, Q., Platt, B., Brandt, M.R., Dawson, L.A., Cole, D., Bernotas, R., Robichaud, A., Rosenzweig-Lipson, S. and Beyer, C.E. (2007) Neuropharmacological profile of novel and selective 5-HT(6) receptor agonists: WAY-181187 and WAY-208466. Neuropsychopharmacology, 33(6): 132–335.
Seibell, P.J., Demarest, J. and Rhoads, D.E. (2003) 5-HT1A receptor activity disrupts spontaneous alternation behavior in rats. Pharmacol. Biochem. Behav., 74: 559–564.
Simpson, H.B., Lombardo, I., Slifstein, M., Huang, H.Y., Hwang, D.R., bi-Dargham, A., Liebowitz, M.R. and Laruelle, M. (2003) Serotonin transporters in obsessive-compulsive disorder: a positron emission tomography study with [(11)C]McN 5652. Biol. Psychiatry, 54: 1414–1421.
Soomro, G.M., Altman, D., Rajagopal, S. and Oakley-Browne, M. (2008) Selective serotonin re-uptake inhibitors (SSRIs) versus placebo for obsessive-compulsive disorder (OCD). Cochrane Database Syst. Rev.: CD001765.
Stein, D.J., Van Heerden, B., Wessels, C.J., Van Kradenburg, J., Warwick, J. and Wasserman, H.J. (1999) Single photon emission computed tomography of the brain with Tc-99m HMPAO during sumatriptan challenge in obsessive-compulsive disorder: investigating the functional role of the serotonin autoreceptor. Prog. Neuropsychopharmacol. Biol. Psychiatry, 23: 1079–1099.
Stein, D.J., Montgomery, S.A., Kasper, S. and Tanghoj, P. (2001) Predictors of response to pharmacotherapy with citalopram in obsessive-compulsive disorder. Intl. Clin. Psychopharmacol., 16: 357–361.
Stengler-Wenzke, K., Muller, U., Angermeyer, M.C., Sabri, O. and Hesse, S. (2004) Reduced serotonin transporter-availability in obsessive-compulsive disorder (OCD). Eur. Arch. Psychiatry Clin. Neurosci., 254: 252–255.
Stengler-Wenzke, K., Muller, U., Barthel, H., Angermeyer, M. C., Sabri, O. and Hesse, S. (2006) Serotonin transporter imaging with [123I]beta-CIT SPECT before and after one year of citalopram treatment of obsessive-compulsive disorder. Neuropsychobiology, 53: 40–45.
Szechtman, H., Sulis, W. and Eilam, D. (1998) Quinpirole induces compulsive checking behavior in rats: a potential animal model of obsessive-compulsive disorder (OCD). Behav. Neurosci., 112: 1475–1485.
Takeuchi, H., Yatsugi, S. and Yamaguchi, T. (2002) Effect of YM992, a novel antidepressant with selective serotonin reuptake inhibitory and 5-HT 2A receptor antagonistic activity, on a marble-burying behavior test as an obsessive-compulsive disorder model. Jpn. J. Pharmacol., 90: 197–200.
Tsaltas, E., Kontis, D., Chrysikakou, S., Giannou, H., Biba, A., Pallidi, S., Christodoulou, A., Maillis, A. and Rabavilas, A. (2005) Reinforced spatial alternation as an animal model of obsessive-compulsive disorder (OCD): investigation of 5-HT2C and 5-HT1D receptor involvement in OCD pathophysiology. Biol. Psychiatry, 57: 1176–1185.
van der Wee, N.J., Stevens, H., Hardeman, J.A., Mandl, R.C., Denys, D.A., van Megen, H.J., Kahn, R.S. and Westenberg, H.M. (2004) Enhanced dopamine transporter density in psychotropic-naive patients with obsessive-compulsive disorder shown by [123I]{beta}-CIT SPECT. Am. J. Psychiatry, 161: 2201–2206.
Voineskos, A.N., Wilson, A.A., Boovariwala, A., Sagrati, S., Houle, S., Rusjan, P., Sokolov, S., Spencer, E.P., Ginovart, N. and Meyer, J.H. (2007) Serotonin transporter occupancy of high-dose selective serotonin reuptake inhibitors during major depressive disorder measured with [11C]DASB positron emission tomography. Psychopharmacology (Berl.), 193: 539–545.
Walitza, S., Wewetzer, C., Warnke, A., Gerlach, M., Geller, F., Gerber, G., Gorg, T., Herpertz-Dahlmann, B., Schulz, E., Remschmidt, H., Hebebrand, J. and Hinney, A. (2002) 5-HT2A promoter polymorphism -1438G/A in children and adolescents with obsessive-compulsive disorders. Mol. Psychiatry, 7: 1054–1057.
Welch, J.M., Lu, J., Rodriguiz, R.M., Trotta, N.C., Peca, J., Ding, J.D., Feliciano, C., Chen, M., Adams, J.P., Luo, J., Dudek, S.M., Weinberg, R.J., Calakos, N., Wetsel, W.C. and Feng, G. (2007) Cortico-striatal synaptic defects and OCD-like behaviors in Sapap3-mutant mice. Nature, 448: 894–900.
Wendland, J.R., Kruse, M.R. and Murphy, D.L. (2006) Functional SLITRK1 var321, varCDfs and SLC6A4 G56A variants and susceptibility to obsessive-compulsive disorder. Mol. Psychiatry, 11: 802–804.
Westenberg, H.G., Fineberg, N.A. and Denys, D. (2007) Neurobiology of obsessive-compulsive disorder: serotonin and beyond. CNS Spectr., 12: 14–27.
Woods, A., Smith, C., Szewczak, M., Dunn, R.W., Cornfeldt, M. and Corbett, R. (1993) Selective serotonin re-uptake inhibitors decrease schedule-induced polydipsia in rats: a potential model for obsessive-compulsive disorder. Psychopharmacology (Berl.), 112: 195–198.
Yadin, E., Friedman, E. and Bridger, W.H. (1991) Spontaneous alternation behavior: an animal model for obsessive-compulsive disorder? Pharmacol. Biochem. Behav., 40: 3113–3115.
Zitterl, W., Aigner, M., Stompe, T., Zitterl-Eglseer, K., Gutierrez-Lobos, K., Schmidl-Mohl, B., Wenzel, T., Demal, U., Zettinig, G., Hornik, K. and Thau, K. (2007)

[123I]-beta-CIT SPECT imaging shows reduced thalamus-hypothalamus serotonin transporter availability in 24 drug-free obsessive-compulsive checkers. Neuropsychopharmacology, 32: 1661–1668.

Zitterl, W., Aigner, M., Stompe, T., Zitterl-Eglseer, K., Gutierrez-Lobos, K., Wenzel, T., Zettinig, G., Hornik, K., Pirker, W. and Thau, K. (2008) Changes in thalamus–hypothalamus serotonin transporter availability during clomipramine administration in patients with obsessive-compulsive disorder. Neuropsychopharmacology, 33: 3126–3134.

Zohar, J. and Insel, T.R. (1987) Obsessive-compulsive disorder: psychobiological approaches to diagnosis, treatment, and pathophysiology. Biol. Psychiatry, 22: 667–687.

Zohar, J. and Judge, R. (1996) Paroxetine versus clomipramine in the treatment of obsessive-compulsive disorder. OCD Paroxetine Study Investigators. Br. J. Psychiatry, 169: 468–474.

CHAPTER 4.5

The Role of Serotonin in Attention-Deficit Hyperactivity Disorder (ADHD)

Robert D. Oades*

Clinic for Child and Adolescent Psychiatry and Psychotherapy, University of Duisburg-Essen, Essen, Germany

Abstract: The symptoms of childhood attention-deficit hyperactivity disorder (ADHD) cover the domains of cognition, impulsivity and motor control, and often embody motivational/emotional features. Prevalence of the combined-type disorder may decrease with age, but the former features predominate increasingly through adolescence to adulthood. The widespread innervation through the CNS by fibers containing serotonin (5-HT) means that, as cause or effect, it is likely that 5-HT is involved where function in these domains is impaired. The relative success of treatment with catecholaminergic agents has overshadowed investigation of the reasons for the partial response of half of these patients. This chapter suggests that anomalous 5-HT function may provide one potential explanation. The best evidence lies with the high heritability of the disorder, where several genes with small effect contribute to the availability of 5-HT and its metabolism. Extra-neuronal availability is under the control of enzymes responsible for 5-HT synthesis, and its removal depends on breakdown or reuptake. Alleles are expressed in cases with ADHD that disturb this availability; they can be over- or under-expressed, depending on phenotype (e.g., impulsivity). The effect of altered levels of 5-HT is mediated by the expression of pre- and postsynaptic receptors of the 5-HT_1 and 5-HT_2 families. Neuropsychological and neurophysiological studies show that the processes affected range from perception, through attention to motivationally guided learning and mood. While the function of a range of brain regions may be involved, the thalamus and frontal lobes are taken as potentially influential examples. Study in this field is in its infancy, and many putative interactions (e.g., between genes, gene/environment, co-morbid problems, development and motor function) require closer investigation.

Keywords: ADHD, attention, co-morbidity, genetics, impulsivity, monoamine oxidase, serotonin, transporter, tryptophan hydroxylase.

Abbreviations: ADHD, attention-deficit hyperactivity disorder: ct – combined type, ia – inattentive type, hi – hyperactive-impulsive type; *DSM, Diagnostic and Statistical Manual of the American Psychiatric Association*; EEG, electroencephalogram; ERN, event-related negativity; ERP, event-related potential; (f)-MRI, (functional)-magnetic resonance imaging; 5-HIAA, 5-hydroxyindole-acetic acid; 5-HT, serotonin/5-hydroxy-tryptamine; 5-$HT_{1/2\ A/B}$, serotonin receptor subtypes; HTR1/2 A/B, genes for serotonin receptor subtypes; 5-HTTLPR, 5-HT transporter linked promoter region; MAOA/B, monoamine oxidase (type A, type B); PET, positron emission tomography; SNP, single nucleotide polymorphism; TPH, tryptophan hydroxylase (genes TPH1, TPH2); VNTR, variable number of tandem repeats.

Introduction

ADHD is a developmental disorder with a childhood onset, where symptoms usually set in before the age of 6 years. The diagnosis with a functional impairment is made worldwide in about 2–7 percent of youngsters (Faraone *et al.*, 2003). Rates at the lower end of this range are often reported outside the USA using the criteria of the World Health Organization (ICD-10) for the similar hyperactivity disorder. Higher rates have often been reported from the USA, where the research diagnostic criteria of the American Psychiatric Association (*DSM-IV*) constitute the preferred diagnostic instrument. To a large extent, the variability may reflect the assessment of a clinically significant functional impairment. For those with the childhood disorder, this persists in about two from three cases into adolescence. In about one in three cases the childhood condition continues on into adulthood, where it is often overshadowed by co-morbid problems (Cumyn *et al.*,

*E-mail: robert.oades@uni-due.de

Christian Müller & Barry Jacobs (Eds.)
Handbook of Behavioral Neurobiology of Serotonin
ISBN 978-0-12-374634-4

DOI: 10.1016/B978-0-12-374634-4.00034-4

2007). ADHD presents more often in male than female children (*ca.* 3–5:1; Buitelaar *et al.*, 2006), but the ratio tends to even out among adults (Biederman *et al.*, 2004).

As the name denotes, problems in ADHD of the combined type (ADHDct) cover cognitive (attention-related) and motor (restlessness) functions, and impulsivity in both of these domains. Smaller numbers of patients can present with problems in only one of these domains ('inattentive', ADHDia; and 'hyperactive-impulsive', ADHDhi). The characteristic of impulsivity may be extended to cognitive function (e.g., rapid and risky decisions; White *et al.*, 1994; Oades, 2008), but also to motivation, where the term typically reflects a preference for immediate reward. This feature may also reflect a further specific subtype or endophenotype within the mantle diagnosis of ADHD (Sonuga-Barke, 2005). Arguably also falling within the motivational domain may be problems concerning the control of affect that can have internalizing features (e.g., lack of self-esteem, anxiety) or an externalizing nature (e.g., aggressive outbursts). Both can lead to poor social networking, and both can occur in one and the same patient (Boylan *et al.*, 2008). However, it should also be emphasized, as noted in the *DSM*, that these features are '*often*' present; they are not necessarily evident all the time. Many current studies of ADHD are focusing on the frequent finding of an increased moment-to-moment variability of the laboratory measure under study. Taking into consideration the range of cognitive and motor functions, discussed elsewhere in this volume, in which serotonin (5-HT) plays an important role, then one can reasonably suspect 5-HT activity to be altered in ADHD: be this as an effect of changes in the function of other neural systems or itself a cause of the anomalous manifestation of cognitive, motivational and motor function (Oades, 2007, 2008).

A significant confound for the interpretation of many studies of patients with ADHD involves the role of co-morbid problems. This is particularly marked in adult cases, where more than 80 percent may exhibit co-morbid psychiatric problems and over half express symptoms satisfying two or more co-morbid disorders (Cumyn *et al.*, 2007). In children and adolescents, more than one in five may have a co-morbid anxiety, mood or depressive disorder, and a similar proportion are reported to have a conduct or oppositional defiant disorder (*DSM-IIIR*; August *et al.*, 1996). It is widely acknowledged that children with ADHD can show autistic symptoms that may reflect a genetic overlap between the two disorders (Ronald *et al.*, 2008). A recent study showed that as many as one in four autistic children showed enough clinical features to be diagnosed with ADHD (Goldstein and Schwebach, 2004). Lastly, around 20 percent of children with ADHD may also show co-morbid tic disorders (Banaschewski *et al.*, 2007). The relevance of this is that each of these adjunctive problems has been associated with dysfunctional 5-HT systems – for example, mood, depression and internalizing problems (Stahl, 1997; Schmidt *et al.*, 2007), conduct and antisocial disorders (Stadler *et al.*, 2004; van Goozen *et al.*, 2007), autistic spectrum (Whitaker-Azmitia, 2005; Murphy *et al.*, 2006) and tic or Tourette syndrome (Haugbol *et al.*, 2007; Wong *et al.*, 2008).

With these issues in mind, this chapter discusses first the more direct categories of evidence for a contribution of 5-HT to the ADHD phenotype from genetics studies, followed by the often less direct evidence of the role of 5-HT activity in the impairments in the dimensions of impulsivity and cognition-related processes. The undoubted contributions of anomalous noradrenergic and dopaminergic function to the full expression of ADHD are not the focus of this chapter.

Genetics

In psychiatry, as in other fields of medicine, there is a current explosion of genetics studies purporting to find an association between a marker for an unusual physiological function in the brain (e.g., a single nucleotide polymorphism, SNP) and the feature encoded (e.g., a form of transmitter receptor or growth modulator). However, determining the criteria for a reliable result is a process still at an early stage of development, where most mental disorders are polygenic, with many alterations of small effect. Examples of methods that are still developing include the determination of an adequate sample size (the need for hundreds or, better, thousands of cases for statistical power), the nature of data cleaning and the conservatism appropriate to the performance of hundreds of statistical tests. Nonetheless, with these caveats in mind, the importance of 5-HT synthesis, breakdown, transport and receptor type for ADHD phenotypes will now be considered.

5-HT synthesis

The rate-limiting enzyme for the synthesis of 5-HT is tryptophan hydroxylase. In brain tissue, it is encoded by TPH2 on chromosome 12Q21.1. Less likely to be relevant to ADHD is TPH1, which encodes the enzyme more usually found peripherally (Walther *et al.*, 2003). Eight SNPs that influence TPH2 activity and the supply of 5-HT have been examined. Initially, an Irish study of 179 families reported an association with ADHD for three alleles that was enhanced by paternal transmission (Sheehan *et al.*, 2005). Although this group could not replicate their results in a UK sample (Sheehan *et al.*, 2007), the IMAGE study, based on 12 European centers (674 families, 776 children), partially confirmed the influence of at least one of these SNPs (Brookes *et al.*, 2006), albeit the significance of a genome-wide association did not survive correction (Lasky-Su *et al.*, 2008a). Nevertheless, nominally significant associations have been reported in the IMAGE sample, specifically with impulsive symptoms (Oades *et al.*, 2008). Three SNPs were over-expressed in association with cognitive impulsivity; a further SNP was under-expressed in association with impulsive aggressive symptoms. There is modest evidence suggestive of a paternal influence on TPH2 transmission (Anney *et al.*, 2008a). A report based on 103 German families (225 children) also found a trend for preferential transmission of two SNPs in the regulatory region for TPH2 (Walitza *et al.*, 2005). One of these markers is of particular interest, as it lies in the promoter region for TPH2 and leads to a reduction of TPH2 transcriptional activity (Scheuch *et al.*, 2007).

From a functional point of view, TPH alleles can moderate working memory performance (Reuter *et al.*, 2008) and the topography of potentials elicited by stimuli requiring response inhibition (Baehne *et al.*, 2008). Both features are typically altered in patients with ADHD (Nigg *et al.*, 2005). An examination of 8 SNPs, overlapping with those studied above (Manor *et al.*, 2008), showed borderline associations in an Israeli sample of 344 patients. The associations with the subjects' accuracy (errors of omission not commission) and reaction times, and their variability while performing a sustained attention task, were striking. These are behavioral measures that characterize impaired ADHD performance in most studies (Willcutt *et al.*, 2005). Initially surprising was that the improved performance after methylphenidate treatment was also related to TPH2, as methylphenidate has no direct influence on 5-HT function (Leonard *et al.*, 2004). However, this result would be compatible with the impaired regulatory activity for the enzyme described above, and a reduced inhibitory control of 5-HT in its well-known interactions with DA (Oades, 2008).

Lastly, it may be instructive to point to potential gene–environmental interactions that are gaining increased attention. Work with rodent stress models has shown that increasing glucocorticoid levels can reduce the expression of TPH2 and 5-HT synthesis (Clark *et al.*, 2008). In the IMAGE study, TPH genes were among a number of monoaminergic candidate genes that were associated with measures of the maternal expressions of criticism toward their offspring and the conduct problems arising (Sonuga-Barke *et al.*, 2008). Indeed, a subgroup of this patient sample was exposed to experimenter-determined degrees of parental warmth or negative hostile remarks. In contrast to those receiving encouraging remarks, patients with increased levels of perceived criticism failed to show the normal circadian decrease of salivary cortisol levels: warmth was associated with the protective effect (Christiansen *et al.*, 2010). Adverse social environments (e.g., maternal hostility) may 'switch off' or socially benign environments 'switch on' genetic effects, whereby 5-HT availability plays a role. Separate phenotypes may be associated with increases and decreases of levels of 5-HT made available through the activity of the enzyme responsible for its synthesis.

5-HT breakdown

The availability of 5-HT in and around the synapse depends on the rate of synthesis and release, reuptake and catabolism. The enzyme responsible for the catabolic breakdown (oxidative deamination) of dopamine, noradenaline and 5-HT is for the most part monoamine oxidase A (MAOA: chromosome locus Xp11.23). Thus, it should be borne in mind that any findings relating to MAOA may influence monoamines other than 5-HT. However, studies of knockout mice lacking MAOA confirm larger alterations in the 5-HT than catecholamine systems (Cases *et al.*, 1995; Evrard *et al.*, 2002). In detail, 5-HT levels were decreased in the hippocampus, frontal cortex and dorsal raphe, the transporter was down-regulated, and in the dorsal raphe (the source of many 5-HT projections) the 5-HT$_{1A}$ autoreceptors were desensitized and reduced in number. Consistent with this, an over-expression of MAOA (at least in healthy women) has been associated with greater 5-HT$_{1A}$ availability (Mickey *et al.*, 2008). Perhaps regard for gender is crucial. A study of girls and boys with ADHD reported that one particular haplotype (ATT) was associated with working memory performance in the former but motor control in the latter (Rommelse *et al.*, 2008). This could reflect influences on separate transmitter systems.

Among a number of polymorphisms examined in studies of ADHD are those with a variable number of tandem repeats (VNTR) in the promoter region of the MAOA gene, in which there may be shorter or longer repeat sequences that give rise to forms with lower or higher enzymatic activity, respectively. Modest associations for several *long* alleles giving rise to high-activity forms of MAOA have been reported to be preferentially transmitted in ADHD (Jiang *et al.*, 2001; Payton *et al.*, 2001; Manor *et al.*, 2002; Domschke *et al.*, 2005) in samples of 82, 110 and 179 cases, respectively. However, Domschke and colleagues were unable to replicate the results for the alleles studied in the other laboratories. They went on to suggest it is one (or more) of the *shorter* alleles that may be preferentially transmitted. This is supported by other reports with Caucasian and Asian samples (Brookes *et al.*, 2006; Das *et al.*, 2006). Yet while Brookes and colleagues found gene-wide significance for a marker in this region, Ribases *et al.* (2007) could not confirm this. Rather, they emphasized a marker in the overlapping gene for MAOB. Indeed, one small study (Lung *et al.*, 2006) could find no evidence at all for long or short alleles being associated with ADHD.

Contributing to these disparate results may be the uncontrolled co-morbid problems of the cases examined. In a normal population, low-activity alleles were associated with impulsivity and aggressive tendencies (Manuck *et al.*, 2000). Indeed, trends for association of the low-activity allele with ADHD boys showing disruptive behavior have been reported (Lawson *et al.*, 2003; Malmberg *et al.*, 2008). It appears that the associations of MAOA markers have less to do with the diagnosis of ADHD than with antisocial behavior seen in the context of ADHD (Thapar *et al.*, 2007).

An alternative explanation should be considered; namely, that both high- and low-activity markers may be relevant. For example, an fMRI study of subjects under dietary control for tryptophan, a 5-HT precursor, showed that decisions about smaller and larger rewards after shorter and longer delays were under the differential control of 5-HT levels (Tanaka *et al.*, 2008). Reward prediction at short timescales (under low levels of 5-HT) was associated with activity in the ventral striatum, but prediction at long timescales was correlated (under high levels of 5-HT) with activity in the dorsal striatum (Schweighofer *et al.*, 2008). The results represent essentially a gradient where low and high levels both contribute to components of the decision and the potential to be impulsive. However, care must be taken in the interpretation of these results. Manor *et al.* (2002) found that an association in children with ADHD between the long MAOA allele and impulsive errors of commission on a test of sustained attention was strongly attenuated following methylphenidate treatment. As this medication inhibits catecholamine reuptake, the apparent association with impulsive decisions may not reflect 5-HT activity.

The interest in MAO here lies with demonstrations of the influence that an interference with 5-HT breakdown can have in development and neurotransmission: this could be crucial for understanding the origins of developmental disorders such as ADHD. In animal models, treatment of pregnant rats with MAO inhibitors severely affected 5-HT (not dopamine) neurotransmission and led to hyperactivity and impulsivity in the offspring (Whitaker-Azmitia *et al.*, 1994). In the case of mice, interference with MAO early in development rather than in the adult leads to enhanced aggressive responses later in the maturation process (Mejia *et al.*, 2002). Genetically speaking, it is the low-expression variant that is associated with aggression in humans (Meyer-Lindenberg *et al.*, 2006). There is strong evidence for an association of MAOA expression leading to externalizing disorders such as ADHD and conduct disorder (Baler *et al.*, 2008). This may reflect on an allele that itself induces high activity, or, as Baler and colleagues emphasize, a genetic interaction with environmental risk. Baler *et al.* (2008) show that a genotype that leads to reduced fetal MAOA activity can modulate development and induce changes of morphology in limbic and cognitive areas of the brain (reductions in the amygdala and cingulate gyrus, increases in the lateral orbito-frontal cortex; Meyer-Lindenberg *et al.*, 2006). Alone, the behavioral outcome may remain within the normal range, but neural differentiation will also be influenced by prenatal exposure to gonadal steroid hormones. The extent to which this experience facilitates aggressive or impulsive tendencies also depends on exposure to stress: the threshold for expressing disruptive behavior may be lower if the foregoing conditions of genotype and brain circuitry are met. Some claim that the risk of developing conduct disorder is even greater if the fetus is exposed to the environmental feature of a mother who smokes tobacco (components of tobacco smoke can lower fetal MAO activity and influence several neurodevelopmental processes, including the control of emotions). The message is that environmental constellations may regulate the outcome, whether the expression of the genotype alone facilitates more or less activity.

Thus, in the present case, the low-activity form of MAO may be involved only where the children are at risk and, for example, may have already been maltreated (Caspi *et al.*, 2002). Although ADHD is often grouped among the externalizing disorders, the evidence of association with MAOA influence on 5-HT activity is strongest where conduct disorder is diagnosed at least as a primary disorder (Kruesi *et al.*, 1992; van Goozen and Fairchild, 2006), but perhaps also as a co-morbid trait. In many cases, ADHD with conduct problems should be considered a separate diagnostic entity (Christiansen *et al.*, 2008).

Clearly, gene–environment associations merit closer attention. It may also be appropriate not to neglect MAOB, which has a considerable structural and genetic overlap with MAOA, and is an indicator of 5-HT as well as dopamine activity (Oreland and Hallman, 1995). MAOB polymorphisms have been nominally implicated in contributing to ADHD (Ribases *et al.*, 2007; Li *et al.*, 2008). A recent twin study (Malmberg *et al.*, 2008) showed that low MAOB activity could relate to impulsive tendencies and monotony avoidance. As with 5-HT in the previous section, genetic variants reflecting high and low rates of breakdown may be associated with different phenotypes, and their effect may be gender sensitive. Possibly, shorter sequences in the promoter region are associated with delayed discounting, hyperactivity and conduct problems (e.g., ADHDct and ADHDhi), and cognitive difficulties arise in the presence of the longer variant (e.g., ADHDia), with expression depending on developmental experience.

5-HT transport and uptake

The uptake efficiency of the 5-HT transporter (5-HTT), with a genetic locus on chromosome 17q11.1–12, has a major influence on the perisynaptic availability of 5-HT and the likelihood of postsynaptic neuromodulation by 5-HT. Most studies relating to ADHD concern variants of three polymorphisms of this locus (SLC6A4): (1) an in/del promoter polymorphism (5-HTTLPR); (2) a VNTR polymorphism in intron 2 (STin2): and (3) an SNP in the 3′-untranslated region (3′UTR). The 5-HTTLPR has short (14) and long (16) repeat sequences, but the STin2 polymorphism also has shorter and longer repeat sequences (9, 10, 11, and 12). Homozygotes with the 5-HTTLPR long allele (in contrast to homo- or heterozygotes for the short form) are associated with increased 5-HT uptake, transporter expression, and efficiency of transcription (Lesch *et al.*, 1996; Bradley *et al.*, 2005).

Initially, based on several relatively small samples, it seemed that ADHD cases were characterized by either a significantly low frequency of the short allele, or a high frequency of the long (l/l) allele in the 5-HTTLPR (Manor *et al.*, 2001; Seeger *et al.*, 2001; Kent *et al.*, 2002; Retz *et al.*, 2002; Zoroglu *et al.*, 2002). By 2005, a pooled odds ratio of 1.31 had been calculated (95% CI 1.09–1.59) for the long allele (Faraone *et al.*, 2005). This result was supported by studies of 250 children showing the most ADHD features in a normal population of 3819 children (Curran *et al.*, 2005) and an association with ADHD features in 184 delinquents (Retz *et al.*, 2008).

However, this was then followed by a sequence of studies yielding negative results in children and adults, from Caucasian as well as Asian samples (Langley *et al.*, 2003; Xu *et al.*, 2005; Banerjee *et al.*, 2006; Wigg *et al.*, 2006; Grevet *et al.*, 2007; Guimaraes *et al.*, 2007; Heiser *et al.*, 2007). Most recently, no association for 5-HTTLPR or STin2-VNTR was found in the IMAGE sample of 1166 ADHDct cases or in a sub-sample of 199 cases with co-morbid mood disorder (Xu *et al.*, 2008). However, demonstrating the potential merit of looking for intermediate phenotypes, separate analyses using a family-based association test on repeat polymorphisms (microsatellites) of the STin2-VNTR provided a modestly significant association with symptoms associated with cognitive (not aggressive) impulsiveness for the 12 *vs* 9 and 10 repeat alleles and not for the short or long repeats of the promoter region for the 5-HTTLPR (Oades *et al.*, 2008).

Fewer groups have studied the STin2 polymorphism, and in general their results lead to equivocal conclusions. Initially, Zoroglu *et al.* (2002) claimed that the absence of the 12 repeat variant characterized their ADHD sample. However, in India a small group of subjects showed a maternally biased transmission (Banerjee *et al.*, 2006), and borderline support for an association with attentional features was described for a Korean sample (Kim *et al.*, 2005). Currently, negative findings predominate (Langley *et al.*, 2003; Curran *et al.*, 2005; Xu *et al.*, 2005, 2008; Heiser *et al.*, 2007). Despite an early description of preferential transmission for the T allele of the 3'UTR (Kent *et al.*, 2002), this has not been confirmed (Curran *et al.*, 2005; Xu *et al.*, 2005; Wigg *et al.*, 2006; Heiser *et al.*, 2007).

It is unfortunate that both a series of 'false positive' reports as well as failed replications have failed to look for associations with other genes or gene–environment interactions. In contrast to the above reports on the long allele of the 5-HTTLPR, but consistent with an early report that ADHD cases may have fewer uptake binding sites (Stoff *et al.*, 1987), some trends for a preferential transmission of the short allele for the 5-HTTLPR have been described (Heiser *et al.*, 2007; Li *et al.*, 2007; Nyman *et al.*, 2007). Indeed, Schmidt and colleagues (2007), in a study of 107 normal 7-year-olds, found that if the short form is expressed along with a long allele from the DRD4 gene, then there is a marked association with increased internalizing and externalizing symptoms. The presence of the long allele of the promoter seemed to be protective. The association of the over-expression of 5-HT markers with cognitive impulsivity has been mentioned above: under-expressed, in this study, were SNPs for the dopamine transporter and D1 receptor (Oades *et al.*, 2008). Such results argue for more intensive study of subtypes of patient defined by endophenotype, and with particular regard for 5-HT/dopamine interactions.

A further example from outside the field of ADHD studies lies with the dramatic relationship between the volume of parts of the thalamus, the 5-HTT short genotype and various mood disorders (Young *et al.*, 2008). This may be pertinent and illustrative, as similarly dramatic and much neglected is the extensive concentration of dopamine uptake sites in the human thalamus (Garcia-Cabezas *et al.*, 2008), especially in nuclei where some of the highest levels of 5-HT innervation and of 5-HTT in the brain are to be found (Lavoie and Parent, 1991; Frankle *et al.*, 2004; Figure 1). Of particular interest are some of the 'aspecific' nuclei, such as the reticular and lateral habenula nuclei. These have recently been shown, from neurophysiological studies of non-human primates, to exert prominent roles in selective attention to ascending sensory information in the geniculate (McAlonan *et al.*, 2008), and in responses to the valence of these stimuli and reinforcement prediction (Hong and Hikosaka, 2008), respectively. Levels of 5-HT (reuptake inhibition, as well as mRNA expression of

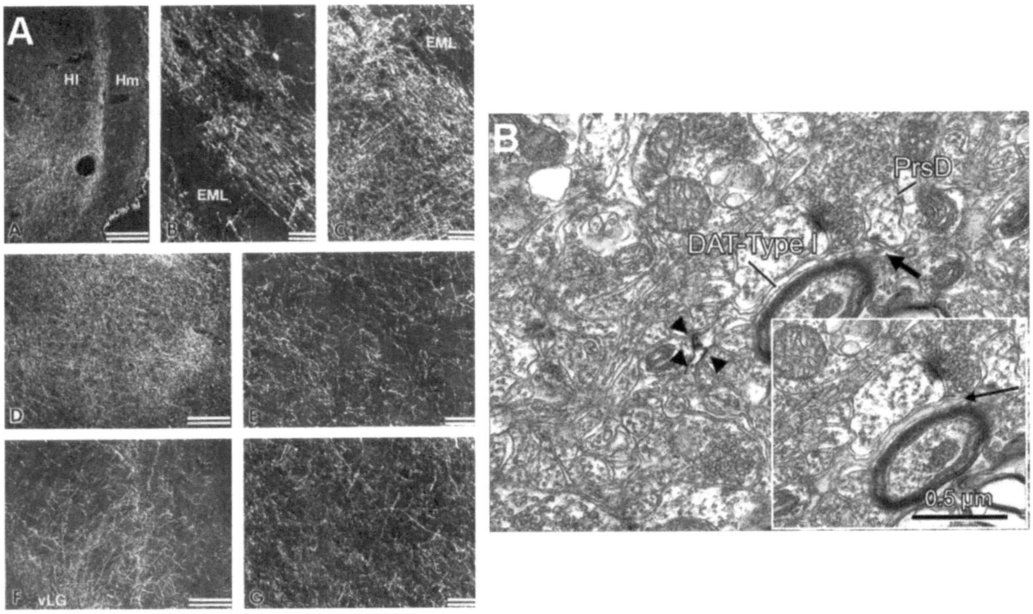

Figure 1 (Left) Darkfield photomicrographs depicting 5-HT innervation in different thalamic nuclei of the squirrel monkey. (A) Relatively weak innervation of the habenula. (B, C) Distribution of 5-HT fibers, isolated varicosities and some clusters along the external medullary lamina in the reticular nucleus and lateral posterior nucleus, respectively. (D) Two dense clusters of 5-HT fibers and isolated varicosities in the caudal-most portion of the mediodorsal nucleus. (E, G) 5-HT innervation in the ventral anterior and anteromedial nuclei, respectively. (F) Fibers coursing from the reticular nucleus dorsally to the lateral geniculate nucleus ventrally. Scale bars: (A, D, F) 200μm; (B, C, E, G) 50μm. Reprinted with permission from Lavoie and Parent (1991) and John Wiley & Sons, Inc. (Right) Thin unmyelinated dopamine transporter-immunoreactive (DAT-ir) axon cut longitudinally in the macaque medio-dorsal thalamus (see arrow in the inset). The DAT reaction product is electron-dense, and rims the plasma membrane in a varicosity and adjacent intervaricose segment (black arrowheads). The DAT-ir axon establishes a synaptic contact (black arrow) located far from the DAT immunoperoxidase product. Note the presence of vesicles in the postsynaptic element, thus identified as a presynaptic dendrite (PrsD), typical of thalamic interneurons. Photomicrograph courtesy of Drs P. Martinez-Sanchez and C. Cavada, Universidad Autónoma de Madrid, Spain.

5-HT$_{1B}$ sites, and 5-HT$_7$ receptors) are also notably high in posterior regions such as the intralaminar as well as the mediodorsal nuclei (Alelu-Paz and Gimenez-Amaya, 2008). The latter part of the thalamus is well known for projecting to the same frontal regions innervated by the mesocortical dopamine system, in which anomalous activity can contribute to ADHD. In turn, these thalamic regions are intimately involved in functions ranging from cortical activation to maintaining the frontal involvement in working memory (Monckton and McCormick, 2002). Against a background of major dopamine/5-HT interactions in the brain (Oades, 2008), polymorphisms with small effect in these two systems may together in the thalamus (among other regions) have considerable repercussions for cerebral function relevant to ADHD.

Consideration of the importance of interactions between genes has already given rise to some investigation of the potential for subgroups of ADHD patients. Cadoret *et al.* (2003) observed, in a group of adoptees at high risk for externalizing disorders, that the males developing these conduct problems more often carried the short allele than the females in the group. Indeed, in a population of twins assessed for ADHD ($n = 247$), the heterozygote was predictive for conduct disorder in the boys (Malmberg *et al.*, 2008). In their study of delinquents, Retz and colleagues (2008) show that in the presence of a history of childhood adversity, the presence of the short allele is associated with the presence of ADHD features. In turn, there is also evidence that childhood aggression is related to the expression of the low-activity variant (Beitchman *et al.*, 2006, 2008; Haberstick *et al.*, 2006).

To sum up, it seems more than likely that changes, up and down, in the level and function of 5-HT around the synapse, as controlled by the uptake site, are related to subgroups of ADHD patients. These forms may be defined by co-morbidity, or endophenotype defined by temperament (e.g., opposition; Malmberg *et al.*, 2008) or neuropsychological control (e.g., cognitive impulsivity; Oades *et al.*, 2008). As in previous sections, interactions with gender and environment must be taken into consideration, but while the high-activity long form of the promoter may be associated with inappropriate cognitive processes, the shorter form may permit poor control of externalizing tendencies.

Either way, it should not be overlooked that persistent high or low levels of 5-HT, controlled by reuptake efficacy, will have effects on the numbers of pre- and postsynaptic 5-HT receptors. Thus, for example, knockout mice without the transporter show reduced 5-HT_{1A} binding in the brainstem, but increases in forebrain regions (Fabre *et al.*, 2000). This provides substance for the next section on 5-HT receptors.

Pre- and postsynaptic 5-HT receptors

To date, most work has been directed towards the 5-HT_1 and 5-HT_2 families of receptors, rather than the 5-HT_{3-7}–related binding sites. Thus, a concentration here on studies of the former should not preclude a serious consideration of potential contributions of the latter to ADHD in the future. Relevant studies have mostly concerned the 5-$HT_{1A,\ 1B,\ 1E}$ and 5-$HT_{2A,\ 2C}$ sites (i.e., HTR1A, 1B, 1E; HTR2A, 2C) that map to widely separated loci in the genome (chromosomes 5q11.2-q13, 6q13, 6q14, 13q14, and Xp24, respectively).

5-HT1A sites are numerous in limbic and frontal regions, but less abundant in the basal ganglia. While in the forebrain 5-HT_{1A} sites are often postsynaptic, in the midbrain they more usually have a presynaptic location (Sharma and Shapiro, 1996). In contrast, 5-HT_{1B} sites have a marked fronto-striatal distribution and are usually located presynaptically. Stimulation of either site can have disinhibitory effects that can, for example, enhance the release of dopamine (Millan *et al.*, 2007). 5-HT_{2A} binding sites at pre- and postsynaptic loci are widely but differentially distributed through forebrain regions, with the highest density in frontal areas (Rosel *et al.*, 2002). They can exert a phasic, excitatory influence on catecholamine release from terminal regions, whereas 5HT_{2C} receptors exert a tonic, inhibitory influence, usually at somatic loci (Millan *et al.*, 2007). Thus, the 5-HT_{1B} and 5-HT_{2A} receptors tend to be 'stimulatory' while the 5-HT_{1A} and 5HT_{2C} sites are more often 'inhibitory' (Martin *et al.*, 1998). Surprisingly little is known about 5-HT_{1E} receptors, which make up a quarter of those binding 5-HT in the forebrain in *post-mortem* studies of cortical and hippocampal regions (especially the CA4 division; Barone *et al.*, 1994).

Historically, genetics studies of ADHD concentrated first on the HTR1B gene. As might be expected from the last paragraph, early studies with mice showed that activation of this site was associated with aggression and motor activity (Sandou *et al.*, 1994). However, it is not surprising (continuing the above line of discussion) that it was soon shown that knockout of this gene in mice affected attention-related (sensory gating) and reward-related behavior (Dulawa *et al.*, 1997; Barot *et al.*, 2007). Thus, HTR1B seems to be implicated in three dimensions of responsiveness relevant to ADHD (attention, reinforcement and motor organization). Work with ADHD patients implicated more or less significantly sites in or around the HTR1B locus (e.g., the 861G allele) in the disorder (Hawi *et al.*, 2002; Quist *et al.*, 2003; Ogdie *et al.*, 2004; Smoller *et al.*, 2006; Ribases *et al.*, 2007). Despite this combined effort on some 1188 families, four research groups were then unable to replicate the results with a further 953 cases (Bobb *et al.*, 2005; Li *et al.*, 2005; Heiser *et al.*, 2007; Ickowicz *et al.*, 2007). Indeed, the IMAGE study of 776 ADHDct cases that used 11 different SNPs from the HTR1B gene could report no significant association (Brookes *et al.*, 2006). It is easy to qualify the results of each study, be they positive or negative, in a way that is suggestive of a subgroup of ADHD cases that may usually show an association with HTR1B. For example, the report of Smoller *et al.* (2006) emphasized paternal transmission, especially in cases of the inattentive subtype. The study of Li *et al.* (2005) was restricted to Han Chinese subjects. However, the hypotheses that can be developed regarding the diagnostic subtype or demographic stratification from such qualifications still await testing.

Despite an early report that HTR1A, along with TPH1 and 5-HTT, contributed to more than 3 percent of the variance in ADHD (Comings *et al.*, 2000), markers for this receptor have received little attention. Ribases *et al.* (2007) included SNPs covering this gene, but found no association with childhood or adult ADHD. However, it appears that HTR1E, which sits adjacent to HTR1B on chromosome 6, may soon elicit more concerted study. Although Ribases' group could report no significant findings, a nominally significant association was a novel finding in the IMAGE study of 674 families (Brookes *et al.*, 2006). Nominal significance also appeared in a genome-wide scan of the full IMAGE sample of 909 family trios (Lasky-Su *et al.*, 2008a). Separate analyses for phenotypes showing symptoms of aggressive impulsivity and cognitive impulsiveness in 1160 offspring from 590 families with an ADHD child revealed nominal associations for both phenotypes with an HTR1E SNP (Oades *et al.*, 2008). This SNP was under-expressed in behavioral impulsivity, but over-expressed in cognitive impulsiveness.

A link with aggressive behavioral impulsivity is more firmly established for increased 5-HT_{2A}, decreased 5-HT_{1A} receptor binding and low levels of 5-HT activity (Coccaro *et al.*, 1997; Parsey *et al.*, 2002; Carver and Miller, 2006). An early report of preferential transmission of the 452Tyr rather than the T102C polymorphism of HTR2A (Quist *et al.*, 2000) has not been confirmed using various methods (Hawi *et al.*, 2002; Galili-Weisstub and Segman,

2003; Bobb *et al.*, 2005; Guimaraes *et al.*, 2007; Heiser *et al.*, 2007). However, one case–control study of the T102 polymorphism did describe a highly frequent transmission of the T and a far less frequent transmission of the C allele (Su *et al.*, 2004). Yet in healthy subjects it has been suggested that the C allele is associated with *hyperactivity and impulsivity* (Reuter *et al.*, 2006). This complements the study of Ribases *et al.* (2007), above, that described an association for an SNP in HTR1D with *inattentive* cases of ADHD. Intriguingly, symptoms of hyperactivity and impulsivity were markedly inversely related to the age of onset in the IMAGE sample, which, in the genome-wide scan, was nominally associated with two HTR2A polymorphisms (Lasky-Su *et al.*, 2008b). While it is difficult to weigh up the evidence for and against an involvement of the determinants of 5-HT_{2A} receptor in ADHD, one should not overlook that these reports made use of a variety of methods and are not always directly comparable. Further, having noted the involvement of 5-HT_{2A} activity with impulsive aggression, it must be emphasized that recruitment rarely considers the nature and degree of co-morbidity. It is therefore salutary to see that, in a family-based association test of three measures of conduct problems, a categorical diagnosis was quite strongly associated ($P = 0.0008$) with a marker for HTR2A on chromosome 13 in the IMAGE sample (Anney *et al.*, 2008b).

There have been few attempts to see if other 5-HT receptors could be involved in ADHD. Interest in the 5-HT_{2C} site derives from its usually inhibitory interactions with the mesostriatal and mesolimbic dopamine systems: blockade of this interaction in rodents can lead to hyperlocomotion (Stiedl *et al.*, 2007) and cognitive impulsivity (Robinson *et al.*, 2008). An examination of several HTR2C haplotypes in a Chinese sample was able to demonstrate both over- and under-expression in different cases, with biased transmission occurring in those with ADHDct rather than ADHDia (Li *et al.*, 2006a). This research group has also reported equivocal results favoring the involvement of HTR4 polymorphisms, but a marked absence of any effects for HTR5A or HTR6 markers (Li *et al.*, 2006b, 2006c). Markers for the 5-HT_4 receptor deserve further attention, as activation in the rodent model facilitates aspects of cortical and limbic processing where knockout animals show impaired responses to novelty and cognitive challenge, potentially modeling aspects of the ADHD phenotype (Micale *et al.*, 2006).

It seems that receptors of both the 5-HT_1 and 5-HT_2 families can be associated with the disorder in some patients exhibiting ADHD, especially where the emphasis of the disorder is on externalizing features. Together these reports may be taken to point towards the role of co-morbid conduct problems, but also that the variability of the results may reflect that the availability of 5-HT through the activity of synthesis and breakdown may influence the expression of some variants for the 5-HT receptors. Further study of the subtypes of 5-HT receptors is imperative.

Neurotransmitter activity and functional associations

Biochemical signs of activity

A number of laboratories have reported on peripheral measures of 5-HT or its metabolites, even though there is wide agreement that most of what is measured derives from somatic sources. Thus, the actual levels recorded are distant reflections of CNS activity. Nonetheless, if there were any deleterious influences on the control of 5-HT metabolism, these could affect both central and peripheral systems and potentially incur psychopathological consequences. Perhaps a more serious argument against the general usefulness of such measures is the great sensitivity of monoamine systems to the state and to the stage of development of the child (Oades *et al.*, 1996a). These features, along with the numerous co-morbid conditions of the patients, all prove very difficult to control adequately (discussion: Oades, 2007)

Reports on the levels of 5-HT and its major metabolite 5-hydroxy-indoleacetic acid (5-HIAA) in the urine, blood (plasma, serum, platelets) and cerebrospinal fluid (CSF) of healthy children and those with ADHD have been reviewed in detail (Oades, 2005). While some larger samples reporting decreased levels of 5-HT, many smaller studies found no differences. 5-HIAA levels have been more rarely studied. Perhaps surprisingly, even using measures from the CSF of children with ADHD, with healthy and conduct disordered children for comparison, no group differences were found (Castellanos *et al.*, 1994). The only reports on both 5-HT and 5-HIAA described an increased 5-HT metabolism in 24-hour urine samples taken from children with ADHD (Oades and Müller, 1997; Oades *et al.*, 1998), which is consistent with the reports (above) of decreased 5-HT excretion. Also in accord with this finding, it has been reported that treatment with a psychostimulant resulted in marked decreases of catecholamine metabolites, 5-HIAA and platelet MAO activity (Uzbekov, 2006).

5-HT activity and functional associations

Studies of groups of ADHD children may fail to provide unequivocal associations with the general levels of 5-HT activity measured. Nonetheless, there may remain dimensional relationships for these measures with the expression of various traits across individuals that reflect some of the genetic features of 5-HT discussed above. For example, an association of an MAOA allele with the degree of cognitive impulsivity shown by ADHD children was described in the previous section (Manor *et al.*, 2002). Thus, while no gender or group differences in the 5-HT uptake by blood platelets could be found in a sample of 84 girls and boys with ADHD aged 4–14 years (Novkovic *et al.*, 2009), the 5-HT levels in the platelets were much higher in those with marked impulsive symptoms. The same result has been reported for a small group of impulsive adolescents (Askenazy *et al.*, 2000).

Measures of the uptake of 5-HT by blood platelets model well the transport mechanisms in brain tissue (Lesch *et al.*, 1993) and thus merit attention in the study of ADHD. The affinity of this uptake site in 20 children with ADHD was measured with paroxetine and shown to correlate with different aspects of impulsivity (Oades *et al.*, 2002). Increases of affinity (but not of the B_{max} measure of density) related to ratings of impulsive aggressive behavior. In contrast, though, decreases of affinity were related to the ability on a stop-task to attend to stimuli requiring the children to withhold response after they had received a 'Go' stimulus. This attention-related ability in turn correlated well with ratings of distractibility and impulsivity. The context here of two sorts of impulsivity with different associations with 5-HTT affinity supports the description above of modest small effects of alleles influencing synthesis (TPH2), uptake (5-HTT) and synaptic effect (HTR1E) in ADHD that was also dependent on the predominant type of impulsiveness (Oades *et al.*, 2008).

An alternative method for registering the status of the 5-HT system (in particular receptors belonging to the 5-HT_2 family) is to register the responsiveness of prolactin to a challenge dose of a drug. In the context of the control of behavioral impulsivity, it is of interest to note that poor responsivity to a challenge with fenfluramine as a child (aged 7–11 years) was reported to be associated with aggression and antisocial behavior 6–7 years later as an adolescent (Halperin *et al.*, 2006; Flory *et al.*, 2007). In contrast to the role of 5-HT_2 receptors in facilitating aggressive traits, animal work has clearly emphasized the role of receptors of the 5-HT_1 family in non-aggressive impulsivity. For example, administration of 5-HT_{1A} agonists to rats was noted to enhance impulsiveness in delayed discounting tasks (Winstanley *et al.*, 2005). While there is unfortunately no directly comparable study with ADHD children who have problems in awaiting reinforcement, there is evidence from the use of a challenge dose of a 5-HT_{1D} agonist with oppositional children showing ADHD symptoms (Snoek *et al.*, 2002). In terms of the amount of growth hormone release elicited by this drug, receptors in these children were twice as sensitive as those in a healthy comparison group. This could provide a basis and consequence for the cognitive impulsivity, described in the genetics section, where there is an indication for over-expression of the HTR1E site. Animal studies indicate the potential effect in practice. On choice reaction time tasks, the impulsive errors of commission made by rodents were reported to correlate with 5-HT release in frontal regions, as measured by microdialysis (Dalley *et al.*, 2002). Interestingly, this over-responsiveness could be blocked by administering 5-HT_{2A} antagonists, thus indicating the way the 5-HT system can exert opposing influences on two forms of impulsivity.

Potentially, the anomalous activity in the 5-HT system in those with ADHD may affect information processing from sensory input and perception through attention to learning and memory (see Chapter 3.9 of this volume). For example, in the auditory system 5-HT innervation and activity is marked in the cochlea (Doleviczenyi *et al.*, 2008) and the primary auditory cortex (Jacobs and Azmitia, 1992). It should not surprise, then, that d-prime, a signal detection marker of perception measured on a continuous performance task, was associated – indeed was correlated negatively with – excretion of 5-HIAA in children with ADHD (Oades, 2000). Later in processing, the selective aspects of attention documented by conditioned blocking were also influenced. Here, the normal ability to restrict attention and not learn about redundant conditioned stimuli was correlated positively with 5-HT metabolism (Oades and Müller, 1997). This association was the opposite to that obtained for children with the tic syndrome. These data suggest how 5-HT metabolism is associated with enhanced target processing and restricted distractor processing in ADHD.

5-HT activity and neurophysiological recording

In this section, some examples of the potential interaction of 5-HT activity with such stimulus processing, namely 'event-related processing', are considered from records of the EEG. Salient stimuli elicit a negatively-directed event-related potential (ERP) in the EEG (the N1) after about 100 ms and shortly thereafter a positively-directed component (the P2) that reduces the influence of other stimuli competing for processing. The combined N1–P2 amplitude can depend on the loudness of an auditory stimulus. In 'augmenters' it increases with loudness and the slope of the amplitude–intensity relationship becomes steeper; 'reducers' it decreases and the slope becomes flatter. Early studies of those with 'ADHD', using earlier diagnostic criteria, showed that putative combined types were more likely augmenters than putative inattentive types or healthy controls (Buchsbaum and Wender, 1973; Dykman *et al.*, 1983). The augmenting response correlates with the sort of behavioral impulsivity that reflects decreased 5-HT activity (Hegerl *et al.*, 1992).

Studies of the ERP and its tangential dipole in the primary auditory cortex, which receives a dense 5-HT innervation (Azmitia and Gannon, 1986), indicate that a steep slope correlates with low 5-HT and high 5-HIAA levels and *vice versa*. This was deduced from the effects of challenges with drugs that have agonistic or antagonistic effects on the 5-HT system (e.g., lithium, alcohol, zimelidine, fenfluramine; Hegerl and Juckel, 1993, 1994). Studies of the similar phenomenon in rats following administration of the uptake inhibitor citalopram (Wutzler *et al.*, 2008) or in cats after treatment with a 5-HT_{1A} agonist (8-OH-DPAT) described decreased intensity dependence, while there was increased intensity dependence after administering the 5-HT_2 antagonist ketanserin (Juckel *et al.*, 1997, 1999). In humans, support for the locus of the effect (primary auditory cortex) has come from fMRI studies (Mulert *et al.*, 2005), and for the 5-HT involvement from genetics. In this instance, homozygotes for the high-activity, long allele of the 5-HTT promoter exhibit weaker intensity dependence than heterozygotes (Gallinat *et al.*, 2003). The problems are that (1) others have found that the long allele related more to a stronger intensity dependence (Strobel *et al.*, 2003; Hensch *et al.*, 2006); (2) acute depletion of tryptophan with an amino-acid drink did not clearly influence intensity dependence (Dierks *et al.*, 2000; Debener *et al.*, 2002; O'Neill *et al.*, 2008a); and (3) intensity dependence is not sensitive to challenge from all drugs influencing 5-HT metabolism (O'Neill *et al.*, 2008b). Resolution of the controversy is not simple, because (1) the long allele of the 5-HTTLPR consists of two alleles (A and G) with stronger and weaker effects in this paradigm (Gallinat *et al.*, 2007); (2) the various studies report using different stimulus intensities and degrees of 5-HT depletion after tryptophan drinks; and (3) extensive drug testing has usefully highlighted the additional contribution of other neurotransmitters (e.g., dopamine) to loudness dependency (Nathan *et al.*, 2005). Direct investigation of these parameters in those with a diagnosis of ADHD would be useful. Nonetheless, an influence of 5-HT activity on loudness dependency and in particular on early stimulus processing in those with ADHD seems likely.

If one accepts that the N1–P2 peak to peak amplitude is influenced by 5-HT activity, it is a small step to propose that the same activity underlies the frequent appearance of a larger P2 amplitude in oddball discriminations in those with ADHD than is normal (Oades, 2006). This has been reported for the childhood (Satterfield *et al.*, 1984; Oades *et al.*, 1996b; Johnstone *et al.*, 2001) and the adult condition (Barry *et al.*, 2009). The next inhibitory ERP after the P2 is the P3, which reflects the need to update a working-memory-like template following stimulation and the effort required to do this. This component is usually recorded from posterior scalp regions, and may have its origin in several brain regions. These sources apparently differ between straightforward discriminations and a go/no-go task where withholding a response elicits a marked 'P3' that is normally more anteriorly located in frontal regions. The presence of alleles for the TPH2 gene that are considered a risk for ADHD alters the topography of this component, and thus implies an important role for 5-HT modulation of the response (Baehne *et al.*, 2008). This was reported to hold for adult carriers of the respective SNPs independent of whether they had a diagnosis of ADHD or were considered healthy.

Finally, there is a sequence of ERPs that can be recorded if the 'event' is the response rather than the stimulus. This is of special interest where the response is an error. In this case it is followed by a component called the error-related negativity (ERN). Studies of ERN in ADHD patients have yielded disparate results in the eight studies published so far, largely due to the broad developmental age range of the subjects, the presence of variable co-morbid problems and the use of different tasks to produce errors. However, half the reports with young patients which usually used a modified Eriksen flanker task found the ERN to be more or less reduced (Liotti *et al.*, 2005; Van Meel *et al.*, 2007; Albrecht *et al.*, 2008; Groen *et al.*, 2008; Wild-Wall *et al.*, 2009). Independent of the proposed involvement of the dopaminergic system in modulating the ERN (Holroyd and Coles, 2002), 5-HT activity likely also plays a role (Jocham and Ullsperger, 2008). Healthy young adults with one or two copies of the low-activity, short allele variant of the 5-HTT promoter showed much larger ERN amplitudes than age- and gender-matched subjects homozygous for the long allele (Fallgatter *et al.*, 2004). Where ADHD cases carry the long allele (Seeger *et al.*, 2001; Curran *et al.*, 2005), a smaller ERN may be anticipated (Figure 2).

In conclusion, there is evidence from neurophysiological recording studies, as with the previous section, that 5-HT activity can influence processes ranging from perception, through attention to learning. There is evidence at each level of processing for anomalies relating to 5-HT responsivity in ADHD. If the evidence is sometimes circumstantial, this section points to the sorts of study needed to validate these claims.

5-HT activity and neuroimaging

One direct way to demonstrate 5-HT involvement in the problems of ADHD children would be to examine ligand

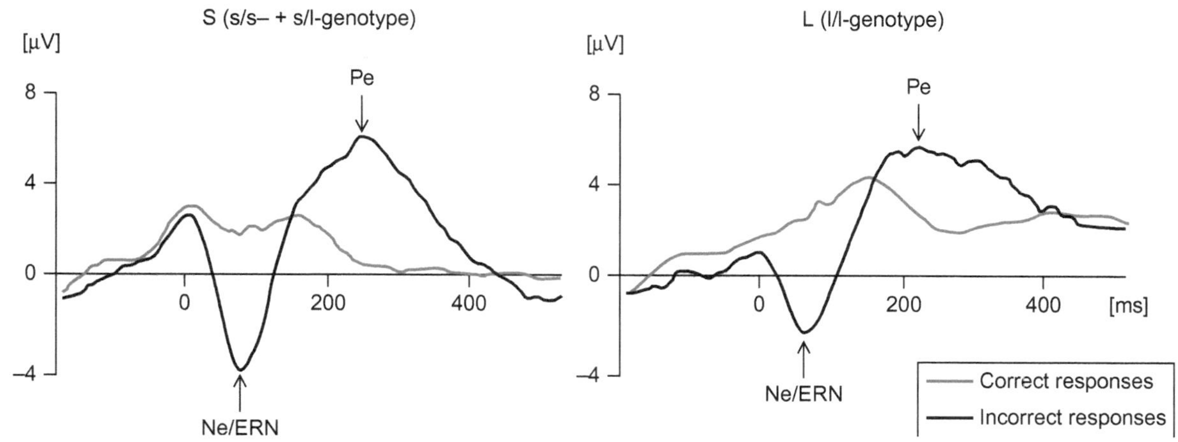

Figure 2 Response-locked grand-averaged event related potentials (ERP). (Left) Comparison between the ERPs of correct (thin line) and incorrect (bold line) responses of 11 subjects with one or two copies of the low-activity short 5-HTTLPR variant. (Right) Comparison between the ERPs to correct (thin line) and incorrect (bold line) responses of 11 subjects homozygous for the long allele. Adapted by permission from Macmillan Publishers Ltd, (*Neuropsychopharmacology*, Fallgatter *et al.* 2004).

binding with PET. This form of study is difficult to carry out with children for ethical reasons, and hence there are only two reports with some relevance. 5-HTT binding (^{123}I-FP-CIT) in the midbrain and brainstem of adult ADHD patients was examined using SPECT methods (Hesse *et al.*, 2006). The authors were unable to report any unusual binding characteristics. However, another study in children with fetal alcohol syndrome, who all showed ADHD, used the ^{123}I-labeled nor-CIT as a specific label for 5-HTT (Riikonen *et al.*, 2005). The report described significantly less binding (25 percent) in the anterior cingulate cortex, but no alteration in the temporal lobe or midbrain. It would be useful to seek direct evidence for the role of 5-HT either with a pharmacological fMRI paradigm (the uptake inhibitor venlafaxine has been successfully used with adult patients) or with subjects expressing genotypes with known effects on 5-HT metabolism (e.g., 5-HTT or MAO alleles).

Let us briefly consider indirect evidence for proposing that parts of the frontal cortex (as indicated above) would provide interesting target areas for study, and first look at the more ventral regions. Gray- and white-matter reductions in the orbito-frontal cortex have been reported in adolescents with or at high risk for ADHD (van t'Ent *et al.*, 2007; Huebner *et al.*, 2008) as well as adults with ADHD (Hesslinger *et al.*, 2002). The data emphasize impairment in the left hemisphere. In contrast, prominent conduct problems were associated more with anomalies in limbic regions. However, the connectivity of orbito-frontal with limbic regions (Plessen *et al.*, 2006) or between the hemispheres (over the rostral corpus callosum) may also be compromised (Giedd *et al.*, 1994). Although an earlier report on children with ADHD was not supportive (Hill *et al.*, 2003), half the subjects showed an oppositional disorder that may have reflected developing conduct problems.

The orbito-frontal cortex receives a marked 5-HT innervation, and genetic variants of MAOA and 5-HTT exert a clear influence on 5-HT activity in orbito-frontal regions (Carver and Miller, 2006; Meyer-Lindenberg *et al.*, 2006; Passamonti *et al.*, 2006; Way *et al.*, 2007). Activation in MRI scans of orbito-frontal regions has been related to features typical of ADHD patients, albeit in non-ADHD subjects. These features include response inhibition abilities and ratings of impulsivity from the Eysenck and Barratt scales (Horn *et al.*, 2003). Damage to these regions impairs delayed discounting as a measure of motivationally influenced impulsivity (Kalenscher *et al.*, 2006). Under-activation of orbito-frontal (and anterior cingulate) regions, illustrated by fMRI, is associated with errors of commission in a continuous performance task, a measure of cognitive impulsivity (Casey *et al.*, 1997). Reduced bilateral orbito-frontal activation during sustained attention was confirmed as specific to ADHD children without co-morbidity, and contrasted with a sample exhibiting pure conduct disorder who were only impaired in a second condition associated with reward (Rubia *et al.*, 2008a).

Not only ventral, but also more dorsal and lateral frontal regions are prominent in the circuitry that has been reported to be underactive in ADHD children performing a go/no-go task (Durston *et al.*, 2006). In addition, the study found that siblings of these cases also showed reduced activity in these regions, implying a genetic influence on their ability to activate prefrontal regions when appropriate. Similarly, inferior- and mid-frontal regions were reported to be under-activated in ADHD cases during errors or even successful inhibition of responses on go/no-go and stop-tasks (Rubia *et al.*, 2005, 2008b; Smith and Taylor, 2006). The effects were seen bilaterally, with an emphasis on the right hemisphere, and correlated with symptom ratings. This lack of appropriate activation will be reflected in processing impairments in connected brain regions, as was shown by impaired fractional anisotropy in fiber tracts in right prefrontal areas (Casey *et al.*, 2007). These results were obtained with ADHD children, but were reported to be tightly associated with findings from their parents, again implying a familial if not a genetic link. The potential involvement of 5-HT activity here is indicated by a study of the effect of acute tryptophan depletion that likely suppressed 5-HT synthesis in the healthy young adults studied (Rubia *et al.*, 2004). Regional frontal activation shown with fMRI during a go/no-go task was specifically reduced in the right inferior frontal and orbito-frontal regions, as recorded for ADHD patients (above).

This section has taken only the frontal cortices for consideration of their role in ADHD, and described the likelihood that this role is influenced by 5-HT activity. Earlier sections focused on 5-HT in the thalamus and the temporal lobe, which reflect important afferent and efferent connections of the frontal regions discussed here. It is not difficult to point out additionally the importance of function (and apparent dysfunction in ADHD) of fronto-striatal and fronto-parietal processing to planning and execution of responses in which 5-HT exerts marked interactions with the catecholamine innervation and glutamatergic mediation of activity.

Conclusions

The genetic contributions to the high heritability (76 percent; Faraone *et al.*, 2005) of ADHD are under active investigation. Part of the variance is explained by modest contributions of a range of genes with small effect. Some of these influence catecholaminergic neuromodulators, growth factors and development, and among these, some affect the 5-HT system. The initial evidence so far is controversial in detail, but points to a combined role for genes influencing the availability of extra-neuronal 5-HT (i.e., primarily synthesis, breakdown and re-uptake and secondarily the receptor mediating and controlling function). A differential or combined role is evident for subtypes of the disorder, especially with respect to the variable expression of impulsivity and conduct problems. The evidence does not speak for a single process for the full disorder. The normal functions affected range from perception, to attention to learning and related emotional/motivational moderation of function. Memorial and motor mechanisms still require study. The problems that those with ADHD have in the laboratory, the clinic, at home and in the world at large are only partly ameliorated by current treatments that concentrate on catecholaminergic function, and some cases do not respond to such treatments. More attention to the widespread involvement of 5-HT in all the domains of function affected in ADHD should lead to an understanding of the disorder that can only benefit the development of even more effective treatments.

References

Albrecht, B., Brandeis, D., Uebel, H., Heinrich, H., Müller, U.C., Hasselhorn, M., Steinhausen, H.-C., Rothenberger, A. and Banaschewski, T. (2008) Action monitoring in boys with attention-deficit/hyperactivity disorder, their nonaffected siblings, and normal control subjects: evidence for an endophenotype. Biol. Psychiat., 64: 615–625.

Alelu-Paz, R. and Gimenez-Amaya, J.M. (2008) The mediodorsal thalamic nucleus and schizophrenia. J. Psychiat. Neurosci., 33: 489–498.

Anney, R.J.L., Hawi, Z., Sheehan, K.A., Mulligan, A., Pinto, C., Brookes, K.-J., Xu, X., Zhou, K., Franke, B., Buitelaar, J. K., Vermeulen, S., Banaschewski, T., Sonuga-Barke, E.J.S., Ebstein, R.P., Manor, I., Miranda, A., Mulas, F., Oades, R.D., Rothenberger, A., Sergeant, J.A., Steinhausen, H.-C., Taylor, E. A., Thompson, M.J., Asherson, P., Faraone, S.V. and Gill, M. (2008a) Parent of origin effects in attention/deficit hyperactivity disorder (ADHD): analysis of data from the International Multicentre ADHD Genetics (IMAGE) programme. Am. J. Med. Genet. Part B, 147B: 1495–1500.

Anney, R.J.L., Lasky-Su, J., O'Dushlaine, C., Kenny, E., Neale, B.M., Mulligan, A., Franke, B., Zhou, K., Chen, W., Christiansen, H., Arias-Vasquez, A., Banaschewski, T., Buitelaar, J.K., Ebstein, R.P., Miranda, A., Mulas, F., Oades, R.D., Roeyers, H., Rothenberger, A., Sergeant, J.A., Sonuga-Barke, E.J.S., Steinhausen, H.-C., Asherson, P., Faraone, S.V. and Gill, M. (2008b) Conduct disorder and ADHD: evaluation of conduct problems as a categorical and quantitative trait in the International Multicentre ADHD Genetics Study. Am. J. Med. Genet. Part B, 147B: 1369–1378.

Askenazy, F., Caci, H., Myquel, M., Darcourt, G. and Lecrubier, Y. (2000) Relationship between impulsivity and platelet serotonin content in adolescents. Psychiat. Res., 94: 19–28.

August, G.J., Realmuto, G.M., MacDonald, A.W., Nugent, S.M. and Crosby, R. (1996) Prevalence of ADHD and co-morbid disorders among elementary school children screened for disruptive behavior. J. Abnorm. Child Psychol., 24: 571–595.

Azmitia, E.C. and Gannon, P.J. (1986) The primate serotonergic system: a review of human and animal studies and a report on *Macaca fascicularis*. Adv. Neurol., 43: 407–468.

Baehne, C.G., Ehlis, A.-C., Plichta, M.M., Conzelmann, A., Pauli, P., Jacob, C., Gutknecht, L., Lesch, K.-P. and Fallgatter, A.J. (2008) TPH2 gene variants modulate response control processes in adult ADHD patients and healthy individuals. Mol. Psychiatry doi:10.1038/mp.2008.39.

Baler, R.D., Volkow, N.D., Fowler, J.S. and Benveniste, H. (2008) Is fetal brain monoamine oxidase inhibition the missing link between maternal smoking and conduct disorders? J. Psychiat. Neurosci., 33: 187–195.

Banaschewski, T., Neale, B.M., Rothenberger, A. and Roessner, V. (2007) Co-morbidity of tic disorders & ADHD conceptual and methodological considerations. Eur. Child Adolesc. Psychiat., 16: 5–14.

Banerjee, E., Sinha, S., Chatterjee, A., Gangopdhyay, P.K., Singh, M. and Nandogopal, K. (2006) A family-based study of Indian subjects from Kolkata reveals allelic association of the serotonin transporter intron-2 (STin2) polymorphism and attention-deficit-hyperactivity disorder (ADHD). Am. J. Med. Genet. Part B, 141: 361–366.

Barone, P., Jordan, D., Atger, F., Kopp, N. and Fillion, G. (1994) Quantitative autoradiography of 5-HT1D and 5-HT1E binding sites labelled by [3H]5-HT, in frontal cortex and the hippocampal region of the human brain. Brain Res., 638: 85–94.

Barot, S.K., Ferguson, S.M. and Neumaier, J.F. (2007) 5-HT_{1B} receptors in nucleus accumbens efferents enhance both rewarding and aversive effects of cocaine. Eur. J. Neurosci., 25: 3125–3131.

Barry, R.J., Clarke, A.R., McCarthy, R., Selikowitz, M., Brown, C.R. and Heaven, P. (2009) Event-related potentials in adults with attention-deficit/hyperactivity disorder: an investigation using an inter-modal auditory/visual oddball task. Intl. J. Psychophysiol., 71:124–131.

Beitchman, J.H., Baldasarra, L., Mik, H., De Luca, V., King, N., Bender, D., Ehtesham, S. and Kennedy, J.L. (2006) Serotonin Transporter polymorphisms and persistent, pervasive childhood aggression. Am. J. Psychiatry, 163: 1103–1105.

Beitchman, J.H., Davidge, K.M., Kennedy, J.L., Atkinson, L., Lee, V., Shapiro, S. and Douglas, L. (2008) The serotonin transporter gene in aggressive children with and without ADHD and nonaggressive matched controls. Ann. N.Y. Acad. Sci., 1008: 248–251.

Biederman, J., Faraone, S.V., Monuteaux, M.C., Bober, M. and Cadogen, E. (2004) Gender effects on attention-deficit/hyperactivity disorder in adults, revisited. Biol. Psychiatry, 55: 692–700.

Bobb, A.J., Addington, A.M., Sidransky, E., Gornick, M.C., Lerch, J.P., Greenstein, D.K., Clasen, L.S., Sharp, W.S., Inoff-Germain, G., Wavrant-De Vrie'ze, F., Arcos-Burgos, M., Straub, R.E., Hardy, J.A., Castellanos, F.X. and Rapoport, J.L. (2005) Support for Association between ADHD and two candidate genes: NET1 and DRD1. Am. J. Med. Genet. Part B, 134B: 67–72.

Boylan, K., Vaillancourt, T., Boyle, M. and Szatmari, P. (2008) Co-morbidity of internalizing disorders in children with oppositional defiant disorder. Eur. Child Adolesc. Psychiatry, 16: 484–494.

Bradley, S.L., Dodelzon, K., Sandu, H.K. and Philibert, R.A. (2005) Relationship of serotonin transporter gene polymorphisms and haplotypes to mRNA transcription. Am. J. Med. Genet. Part B, 136: 58–61.

Brookes, K.-J., Xu, X., Chen, W., Zhou, K., Neale, B.M., Lowe, N., Aneley, R., Franke, B., Gill, M., Ebstein, R.P., Buitelaar, J.K., Sham, P., Cambell, D., Knight, J., Andreou, P., Altink, M., Arnold, R., Boer, F., Buschgens, C.J.M., Butler, L., Christiansen, H., Feldman, L., Fleischman, K., Fliers, E., Howe-Forbes, R., Goldfarb, A., Heise, A., Gabriels, I., Lubetzki, I., Marco, R., Medad, S., Minderaa, R.B., Mulas, F., Müller, U.C., Mulligan, A., Rabin, K., Rommelse, N.N.J., Sethna, V., Sorohan, J., Uebel, H., Psychogiou, L., Weeks, A., Barrett, R., Craig, I., Banaschewski, T., Sonuga-Barke, E.J.S., Eisenberg, J., Kuntsi, J., Manor, I., McGuffin, P., Miranda, A., Oades, R.D., Plomin, R., Roeyers, H., Rothenberger, A., Sergeant, J.A., Steinhausen, H.-C., Taylor, E.A., Thompson, M.J., Faraone, S.V., Asherson, P. and Johansson, L. (2006) Analysis of 51 candidate genes in DSM-IV combined subtype attention deficit hyperactivity disorder: association signals in DRD4, DAT1 and 16 other genes. Mol. Psychiatry, 11: 934–953.

Buchsbaum, M.S. and Wender, P.H. (1973) Averaged evoked responses in normal and minimally brain dysfunctioned children treated with amphetamine. A preliminary report. Arch. Gen. Psychiatry, 29: 764–770.

Buitelaar, J.K., Barton, J., Danckaerts, M., Friedrichs, E., Gillberg, C., Hazell, P.L., Hellemans, H., Johnson, M., Kalverdijk, L.J., Masi, G., Michelson, D., Revol, O., San Sebastian, J., Zhang, S. and Zuddas, A. (2006) A comparison of North American versus non-North American ADHD study populations. Eur. Child Adolesc. Psychiatry, 15: 177–181.

Cadoret, R.J., Langbehn, D., Caspers, K., Troughton, E.P., Yucuis, R., Sandhu, H.K. and Philibert, R. (2003) Associations of the serotonin transporter promoter polymorphism with aggressivity, attention deficit, and conduct disorder in an adoptee population. Compr. Psychiatry, 44: 88–101.

Carver, C.S. and Miller, C.J. (2006) Relations of serotonin function to personality: current views and a key methodological issue. Psychiatry Res., 144: 1–15.

Cases, O., Seif, I., Grimsby, J., Gaspar, P., Chen, K., Pournin, S., Mueller, U., Aguet, M., Babinet, C., Shih, J.C. and De Maeyer, E. (1995) Aggressive behavior and altered amounts of brain serotonin and norepinephrine in mice lacking MAOA. Science, 268: 1763–1766.

Casey, B.J., Trainor, R., Giedd, J.N., Vauss, Y., Vaituzis, C.K., Hamburger, S.D., Kozuch, P. and Rapoport, J.L. (1997) The role of the anterior cingulate in automatic and controlled processes: a developmental neuroanatomical study. Dev. Psychobiol., 30: 61–69.

Casey, B.J., Epstein, J.N., Buhle, J., Liston, C., Davidson, M.C., Tonev, S.T., Spicer, J., Niogi, S., Millner, A.J., Reiss, A.L., Garrett, A., Hinshaw, S.P., Greenhill, L.L., Shafritz, K.M., Vitolo, A., Kotler, L.A., Jarrett, M.A. and Glover, G. (2007) Frontostriatal connectivity and its role in cognitive control in parent–child dyads with ADHD. Am. J. Psychiatry, 164: 1729–1736.

Caspi, A., McClay, J., Moffitt, T.E., Mill, J., Martin, J., Craig, I.W. et al. (2002) Role of genotype in the cycle of violence in maltreated children. Science, 297: 851–854.

Castellanos, F.X., Elia, J., Kruesi, M.J.P., Gulotta, C.S., Mefford, I.N., Potter, W.Z., Ritchie, G.F. and Rapoport, J.L. (1994) Cerebrospinal fluid monoamine metabolites in boys with attention-deficit hyperactivity disorder. Psychiat. Res., 52: 305–316.

Christiansen, H., Chen, W., Oades, R.D., Asherson, P., Taylor, E.A., Lasky-Su, J., Zhou, K., Banaschewski, T., Buschgens, C.J.M., Franke, B., Gabriels, I., Manor, I., Marco, R., Müller, U., Mulligan, A., Psychogiou, L., Rommelse, N.N.J., Uebel, H., Buitelaar, J.K., Ebstein, R.P., Eisenberg, J., Gill, M., Miranda, A., Mulas, F., Gill, M., Miranda, A., Rothenberger, A., Sergeant, J.A., Sonuga-Barke, E.J.S., Steinhausen, H.-C., Thompson, M.J. and Faraone, S.V. (2008) Co-transmission of conduct problems with attention deficit hyperactivity disorder: familial evidence for a distinct disorder. J. Neural Transm., 115: 163–175.

Christiansen, H., Oades, R.D., Psychogiou, Hauffa B.P. and L., Sonuga-Barke, E.J.S. (2010) Does stress reactivity mediate the link between expressed emotion and oppositional behavior in attention deficit/hyperactivity disorder (ADHD)? J. Abnom. psycol., submitted.

Clark, J.A., Flick, R.B., Pai, L.-Y., Szalayova, I., Key, S., Conley, R.K., Deutch, A.Y., Hutson, P.H. and Mezey, E. (2008) Glucocorticoid modulation of tryptophan hydroxylase-2 protein in raphe nuclei and 5-hydroxytryptophan concentrations in frontal cortex of C57/Bl6 mice. Mol. Psychiatry, 13: 498–506.

Coccaro, E.F., Kavoussi, R.J., Sheline, Y.I., Berman, M.E. and Csernansky, J.G. (1997) Impulsive aggression in personality disorder correlates with platelet 5-HT_{2A} receptor binding. Neuropsychopharmacology, 16: 211–216.

Comings, D.E., Gade-Andavolu, R., Gonzalez, N., Wu, S., Muhleman, D., Blake, H., Wang, E., Farwell, K., Darakjy, S., Baker, R., Dietz, G., Saucier, G. and MacMurray, J.P. (2000) Multivariate analysis of associations of 42 genes in ADHD, ODD and conduct disorder. Clin. Genet., 58: 31–40.

Cumyn, L., Kolar, D., Keller, A. and Hechtman, L. (2007) Current issues and trends in the diagnosis and treatment of adults with ADHD. Expert. Rev. Neurother., 7: 1375–1390.

Curran, S., Purcell, S., Craig, I., Asherson, P. and Sham, P. (2005) The serotonin transporter gene as a QTL for ADHD. Am. J. Med. Genet. Part B, 134B: 42–47.

Dalley, J.W., Theobald, D.E., Eagle, D.M., Passetti, F. and Robbins, T.W. (2002) Deficits in impulse control associated with tonically-elevated serotonergic function in rat prefrontal cortex. Neuropsychopharmacology, 26: 716–728.

Das, M., Bhowmik, A.D., Sinha, S., Chattopadhyay, A., Chaudhuri, K., Singh, M. and Mukhopahyay, K. (2006) MAOA Promoter polymorphism and attention deficit hyperactivity disorder (ADHD) in Indian children. Am. J. Med. Genet. Part B, 141B: 637–642.

Debener, S.I., Strobel, A., Kurschner, K., Kranczioch, C., Hebenstreit, J., Maercker, A., Beauducel, A. and Brocke, B. (2002) Is auditory evoked potential augmenting/reducing affected by acute tryptophan depletion? Biol. Psychol., 59: 121–133.

Dierks, T., Barta, S., Demisch, L., Schmeck, K., Englert, E., Kewitz, A., Maurer, K. and Poustka, F. (2000) Intensity dependence of auditory evoked potentials (AEPs) as biological marker for cerebral serotonin levels: effects of tryptophan depletion in healthy subjects. Psychopharmacology, 146: 101–107.

Doleviczenyi, Z., Vizi, E.S., Gacsalyi, I., Pallagi, K., Volk, B., Harsing, L.G., Halmos, G., Lendvai, B. and Zelles, T. (2008) 5-$HT_{6/7}$ receptor antagonists facilitate dopamine release in the cochlea via a GABAergic disinhibitory mechanism. Neurochem. Res., 33: 2364–2372.

Domschke, K., Sheehan, K., Lowe, N., Kirley, A., Mullins, C., O'Sullivan, R., Freitag, C., Becker, T., Conroy, J., Fitzgerald, M., Gill, M. and Hawi, Z. (2005) Association analysis of the monoamine oxidase A and B genes with attention deficit hyperactivity disorder (ADHD) in an Irish sample: preferential transmission of the MAOA 941G allele to affected children. Am. J. Med. Genet. Part B, 134B: 110–114.

Dulawa, S.C., Hen, R., Scearce-Levie, K. and Geyer, M.A. (1997) $Serotonin_{1B}$ receptor modulation of startle reactivity, habituation and prepulse inhibition in wild-type and $serotonin_{1B}$ knockout mice. Psychopharmacology, 132: 125–134.

Durston, S., Mulder, M., Casey, B.J., Ziemans, T. and van Engeland, H. (2006) Activation in ventral prefrontal cortex is sensitive to genetic vulnerability for attention-deficit hyperactivity disorder. Biol. Psychiatry, 60: 1062–1070.

Dykman, R.A., Holcomb, P.J., Ackerman, P.T. and McCray, D.S. (1983) Auditory ERP augmentation-reduction and methylphenidate dosage needs in attention and reading disordered children. Psychiat. Res., 9: 255–269.

Evrard, A., Malagie, I., Laporte, A.M., Boni, C., Hanoun, N., Trillat, A.C., Seif, I., De Maeyer, E., Gardier, A., Hamon, M. and Adrien, J. (2002) Altered regulation of the 5-HT system in the brain of MAOA knockout mice. Eur. J. Neurosci., 15: 841–851.

Fabre, V., Beaufour, C., Evrard, A., Rioux, A., Hanoun, N., Lesch, K.-P., Lanfumey, L., Hamon, M. and Martres, M.-P. (2000) Altered expression and functions of serotonin 5-HT_{1A} and 5-HT_{1B} receptors in knockout mice lacking the 5-HT transporter. Eur. J. Neurosci., 12: 2299–2310.

Fallgatter, A.J., Herrmann, M.J., Roemmler, J., Ehlis, A.-C., Wagener, A., Heidrich, A., Ortega, G., Zeng, Y. and Lesch, K.-P. (2004) Allelic variation of serotonin transporter function modulates the brain electrical response for error processing. Neuropsychopharmacology, 29: 1506–1511.

Faraone, S.V., Sergeant, J.A., Gillberg, C. and Biederman, J. (2003) The worldwide prevalence of ADHD: is it an American condition? World Psychiatry, 2: 104–113.

Faraone, S.V., Perlis, R.H., Doyle, A.E., Smoller, J.W., Goralnick, J.J., Holmgren, M.A. and Sklar, P. (2005) Molecular genetics of attention deficit hyperactivity disorder. Biol. Psychiatry, 57: 1313–1323.

Flory, J.D., Newcorn, J.H., Miller, C., Harty, S. and Halperin, J.M. (2007) Serotonergic function in children with attention-deficit hyperactivity disorder: relationship to later antisocial personality disorder. Br. J. Psychiatry, 190: 410–414.

Frankle, W.G., Huang, Y., Hwang, D.R., Talbot, P.S., Slifstein, M., Van Heertum, R., Abi-Dargham, A. and Laruelle, M. (2004) Comparative evaluation of serotonin transporter radioligands 11C-DASB and 11C-McN 5652 in healthy humans. J. Nucl. Med., 45: 682–694.

Galili-Weisstub, E. and Segman, R.H. (2003) Attention deficit and hyperactivity disorder: review of genetic association studies. Isr. J. Psychiatr. Relat. Sci., 40: 57–66.

Gallinat, J., Senkowski, D., Wernicke, C., Juckel, G., Becker, I., Sander, T., Smolka, M., Hegerl, U., Rommelspacher, H., Winterer, G. and Herrmann, W.M. (2003) Allelic variants of the functional promoter polymorphism of the human serotonin transporter gene is associated with auditory cortical stimulus processing. Neuropsychopharmacology, 28: 530–532.

Gallinat, J., Müller, D.J., Bierbrauer, J., Rommelspacher, H., Juckel, G. and Wernicke, C. (2007) Functional cortical effects of novel allelic variants of the serotonin transporter gene-linked polymorphic region (5-HTTLPR) in humans. Pharmacopsychiat, 40: 191–195.

Garcia-Cabezas, M.A., Martinez-Sanchez, P., Sanchez-Gonzalez, M.A., Garzon, M. and Cavada, C. (2008) Dopamine innervation in the thalamus: monkey versus rat. Cereb. Cortex doi:10.1093/cercor/bhn093.

Giedd, J.N., Castellanos, F.X., Casey, B.J., Kozuch, P., King, A.C., Hamburger, S.D. and Rapoport, J.L. (1994) Quantitative morphology of the corpus callosum in attention deficit hyperactivity disorder. Am. J. Psychiatry, 151: 665–669.

Goldstein, S. and Schwebach, A.J. (2004) The co-morbidity of pervasive developmental disorder and attention deficit hyperactivity disorder: results of a retrospective chart review. J. Aut. Devel. Disord., 34: 329–339.

Grevet, E.H., Marques, F.Z.C., Salgado, C.A.I., Fischer, A.G., Kalil, K.L., Victor, M.M., Garcia, C., Sousa, N.O., Belmonte-de-Abreu, P. and Bau, C.H.D. (2007) Serotonin transporter gene polymorphism and the phenotypic heterogeneity of adult ADHD. J. Neural Transm., 114: 1631–1636.

Groen, Y., Wijers, A.A., Mulder, L.J.M., Waggeveld, B., Minderaa, R.B. and Althaus, M. (2008) Error and feedback processing in children with ADHD and children with Autistic Spectrum Disorder: an EEG event-related potential study. Clin. Neurophysiol., 119: 2476–2493.

Guimaraes, A.P.M., Zeni, C., Polanczyk, G.V., Genro, J.P., Roman, T., Rohde, L.A. and Hutz, M.H. (2007) Serotonin genes and attention deficit/hyperactivity disorder in a Brazilian sample: preferential transmission of the HTR2A 452His allele to affected boys. Am. J. Med. Genet. Part B, 144B: 69–73.

Haberstick, B.C., Smolen, A. and Hewitt, J.K. (2006) Family-based association test of the 5HTTLPR and aggressive behavior in a general population sample of children. Biol. Psychiatry, 59: 836–843.

Halperin, J.M., Kalmar, J., Schulz, K.P., Marks, D.J., Sharma, V. and Newcorn, J.H. (2006) Elevated childhood serotonergic function protects against adolescent aggression in disruptive boys. J. Am. Acad. Child. Adolesc. Psychiatry, 45: 833–840.

Haugbol, S., Pinborg, L.H., Regeur, L., Hansen, E.S., Bolwig, T.G., Nielsen, F.A., Svarer, C., Skovgaard, L.T. and Knudsen, G.M. (2007) Cerebral 5-HT_{2A} receptor binding is increased in patients with Tourette's syndrome. Intl. J. Neuropsychopharmacol., 10: 245–252.

Hawi, Z., Dring, M., Kirley, A., Foley, D., Kent, L., Craddock, N., Asherson, P., Curran, S., Gould, A., Richards, S., Lawson, D., Pay, H., Turic, D., Langley, K., Owen, M., O'Donovan, M., Thapar, A., Fitzgerald, M. and Gill, M. (2002) Serotonergic system and attention deficit hyperactivity disorder (ADHD): a potential susceptibility locus at the 5-HT_{1B} receptor gene in 273 nuclear families from a multi-centre sample. Mol. Psychiatry, 7: 718–725.

Hegerl, U. and Juckel, G. (1993) Intensity dependence of auditory evoked potentials as an indicator of central serotonergic neurotransmission: a new hypothesis. Biol. Psychiatry, 33: 173–187.

Hegerl, U. and Juckel, G. (1994) Auditory evoked dipole source activity: indicator of central serotonergic dysfunction in psychiatric patients? Pharmacopsychiat, 27: 75–78.

Hegerl, U., Wulff, H. and Muller-Oerlinghausen, B. (1992) Intensity dependence of auditory evoked potentials and clinical response to prophlactic lithium medication: a replication study. Psychiatry Res., 44: 181–191.

Heiser, P., Dempfle, A., Friedel, S., Konrad, K., Hinney, A., Kiefl, H., Walitaz, S., Bettecken, T., Saar, K., Linder, M., Warnke, A., Herpetz-Dahlmann, B., Schäfer Remschmidt, H. and Hebebrand, J. (2007) Family-based association study of serotonergic candidate genes and attention-deficit/hyperactivity disorder in a German sample. J. Neural Transm., 114: 513–521.

Hensch, T., Wargelius, H.-L., Herold, U., Lesch, K.-P., Oreland, L. and Brocke, B. (2006) Further evidence for an association of 5-HTTLPR with intensity dependence of auditory-evoked potentials. Neuropsychopharmacology, 31: 2047–2054.

Hesse, S., Ballaschkle, O., Barthel, H., von Cramon, D.Y. and Sabri, O. (2006) The striatal dopamine transporter availability is reduced in adults with attention-deficit/hyperactivity disorder. J. Nucl. Med., 47: 142P.

Hesslinger, B., Tebartz van Elst, L., Thiel, T., Haegele, K., Hennig, J. and Ebert, D. (2002) Fronto-orbital volume reductions in adult patients with attention deficit hyperactivity disorder. Neurosci. Lett., 328: 319–321.

Hill, D.E., Yeo, R.A., Campbell, R.A., Hart, B., Vigil, J. and Brooks, W. (2003) Magnetic resonance imaging correlates of attention-deficit/hyperactivity disorder in children. Neuropsychology, 17: 496–506.

Holroyd, C.B. and Coles, M.G.H. (2002) The neural basis of human error processing: reinforcement learning, dopamine and the error-related negativity. Psychol. Rev., 109: 679–709.

Hong, S. and Hikosaka, O. (2008) The globus pallidus sends reward-related signals to the lateral habenula. Neuron, 60: 720–729.

Horn, N.R., Dolan, M., Elliott, R., Deakin, J.F. and Woodruff, P.W.R. (2003) Response inhibition and impulsivity: an fMRI study. Neuropsychologia, 41: 1959–1966.

Huebner, T., Vloet, T., Marx, I., Kondrad, K., Fink, G.R., Herpertz, S.C. and Herpertz-Dahlmann, B. (2008) Morphometric brain abnormalities in boys with conduct disorder. J. Am. Acad. Child Adolesc. Psychiatry, 47: 540–547.

Ickowicz, A., Feng, Y., Wigg, K., Quist, J.F., Pathare, J., Roberts, W., Malone, M., Schachar, R.J., Tannock, R., Kennedy, J.L. and Barr, C.L. (2007) The serotonin receptor HTR1B: gene polymorphisms in attention deficit hyperactivity disorder. Am. J. Med. Genet. Part B, 144B: 121–125.

Jacobs, B.L. and Azmitia, E.C. (1992) Structure and function of the brain serotonin system. Physiol. Rev., 72: 165–229.

Jiang, S., Xin, R., Lin, S., Qian, Y., Tang, G., Wang, D. and Wu, X. (2001) Linkage studies between attention-deficit/hyperactivity disorder and the monoamine oxidase genes. Am. J. Med. Genet. Part B, 105: 783–788.

Jocham, G. and Ullsperger, M. (2008) Neuropharmacologyogy of performance monitoring. Neurosci. Biobehav. Rev., 33: 48–60.

Johnstone, S.J., Barry, R.J. and Anderson, J.W. (2001) Topographic distribution and developmental timecourse of auditory event-related potentials in two subtypes of attention-deficit hyperactivity disorder. Intl. J. Psychophysiol., 42: 73–94.

Juckel, G., Molnar, M., Hegerl, U., Csepe, V. and Karmos, G. (1997) Auditory evoked potentials as indicator of brain serotonergic activity – first evidence in behaving cats. Biol. Psychiatry, 41: 1181–1195.

Juckel, G., Hegerl, U., Molnar, M., Csepe, V. and Karmos, G. (1999) Auditory evoked potentials reflect serotonergic neuronal activity – a study in behaving cats administered drugs acting on 5-HT1a autoreceptors in the dorsal raphe nucleus. Neuropsychopharmacology, 21: 710–716.

Kalenscher, T., Ohmann, T. and Güntürkün, O. (2006) The neuroscience of impulsive and self-controlled decisions. Intl. J. Psychophysiol., 62: 203–211.

Kent, L., Doerry, U., Hardy, E., Parmar, R., Gingell, K., Hawi, Z., Kirley, A., Lowe, N., Fitzgerald, M., Gill, M. and Craddock, N. (2002) Evidence that variation at the serotonin transporter gene influences susceptibility to attention deficit hyperactivity disorder (ADHD): analysis and pooled analysis. Mol. Psychiatry, 7: 908–912.

Kim, S.J., Badner, J., Cheon, K.A., Kim, B.N., Yoo, H.J., Cook, E., Leventhal, B.L. and Kim, Y.S. (2005) Family-based association study of the serotonin transporter gene polymorphisms in Korean ADHD trios. Am. J. Med. Genet. Part B, 139: 14–18.

Kruesi, M.J.P., Hibbs, E.D., Zahn, T.P., Keysor, C.C., Hamburger, S.D., Bartko, J.J. and Rapoport, J.L. (1992) A 2-year prospective follow-up study of children and adolescents with disruptive behavior disorders. Arch. Gen. Psychiatry, 49: 429–435.

Langley, K., Payton, A., Hamshere, M.L., Pay, H.M., Lawson, D.C., Turic, D., Ollier, W., Worthington, J., Owen, M.J., O'Donovan, M.C. and Thapar, A. (2003) No evidence of association of two 5HT transporter gene polymorphisms and attention deficit hyperactivity disorder. Psychiatry Genet., 13: 107–110.

Lasky-Su, J., Neale, B.M., Franke, B., Anney, R.J.L., Zhou, K., Maller, J.B., Vasquez, A.A., Chen, W., Asherson, P., Buitelaar, J.K., Banaschewski, T., Ebstein, R.P., Gill, M., Miranda, A., Mulas, F., Oades, R.D., Roeyers, H., Rothenberger, A., Sergeant, J.A., Sonuga-Barke, E.J.S., Steinhausen, H.-C., Taylor, E.A., Daly, M., Laird, N., Lange, C. and Faraone, S.V. (2008a) Genome-wide association scan of quantitative traits for Attention Deficit Hyperactivity Disorder identifies novel associations and confirms candidate gene associations. Am. J. Med. Genet. Part B, 147B: 1345–1354.

Lasky-Su, J., Anney, R.J.L., Neale, B.M., Franke, B., Zhou, K., Maller, J.B., Vasquez, A.A., Chen, W., Asherson, P., Buitelaar, J.K., Banaschewski, T., Ebstein, R.P., Gill, M., Miranda, A., Mulas, F., Oades, R.D., Roeyers, H., Rothenberger, A., Sergeant, J.A., Sonuga-Barke, E.J.S., Steinhausen, H.-C., Taylor, E.A., Daly, M., Laird, N., Lange, C. and Faraone, S.V. (2008b) Genome-wide association scan of the time to onset of attention deficit hyperactivity disorder. Am. J. Med. Genet. Part B, 147B: 1355–1358.

Lavoie, B. and Parent, A. (1991) Serotoninergic innervation of the thalamus in the primate: an immunohistochemical study. J. Comp. Neurol., 312: 1–18.

Lawson, D.C., Turie, D., Langley, K., Pay, H.M., Govan, C.F., Norton, N., Hamshere, M.L., Owen, M.J., O'Donovan, M.C. and Thapar, A. (2003) Association analysis of monoamine oxidase A and attention deficit hyperactivity disorder. Am. J. Med. Genet. Part B, 116B: 84–89.

Leonard, B.E., McCartan, D., White, J. and King, D.J. (2004) Methylphenidate: a review of its Neuropharmacologyogical, neuropsychological and adverse clinical effects. Hum. Psychopharmacol. Clin. Exp., 19: 151–180.

Lesch, K.-P., Wolzin, B., Murphy, D. and Riederer, P. (1993) Primary structure of the human platelet serotonin uptake site: identity with the brain serotonin transporter. J. Neurochem., 60: 2319–2322.

Lesch, K.-P., Bengel, D., Heils, A., Sabol, S.Z., Greenberg, B.D., Petri, S., Benjamin, J., Muller, C.R., Hamer, D.H. and Murphy, D.L. (1996) Association of anxiety-related traits with a polymorphism in the serotonin transporter gene regulatory region. Science, 274: 1527–1531.

Li, J., Wang, Y., Zhu, R., Zhang, H., Yang, L., Wang, B., Khan, S. and Faraone, S.V. (2005) Serotonin 5-HT1B receptor gene and attention deficit hyperactivity disorder in Chinese Han subjects. Am. J. Med. Genet. Part B, 132B: 59–63.

Li, J., Wang, Y., Zhou, R., Zhang, H., Yang, L., Wang, B. and Faraone, S.V. (2006a) Association between polymorphisms in serotonin 2C receptor gene and attention-deficit/hyperactivity disorder in Han Chinese subjects. Neurosci. Lett., 407: 107–111.

Li, J., Wang, Y., Zhou, R., Wang, B., Zhang, H., Yang, L. and Faraone, S.V. (2006b) Association of attention-deficit/hyperactivity disorder with serotonin 4 receptor gene polymorphisms in Han Chinese subjects. Neurosci. Lett., 401: 6–9.

Li, J., Wang, Y., Zhou, R., Wang, B., Zhang, H., Yang, L. and Faraone, S.V. (2006c) No association of attention-deficit/hyperactivity disorder with genes of the serotonergic pathway in Han Chinese subjects. Neurosci. Lett., 403: 172–175.

Li, J., Wang, Y., Zhou, R., Zhang, H., Yang, H., Yang, L., Wang, B. and Faraone, S.V. (2007) Association between polymorphisms in serotonin transporter gene and attention deficit hyperactivity disorder in Chinese Han subjects. Am. J. Med. Genet. Part B, 144B: 14–19.

Li, J., Wang, Y., Hu, S., Zhou, R., Yu, X., Wang, B., Guan, L., Yang, L., Zhang, F. and Faraone, S.V. (2008) The monoamine oxidase B gene exhibits significant association to ADHD. Am. J. Med. Genet. Part B, 147B: 370–374.

Liotti, M., Pliszka, S.R., Perez, R., Kothmann, D. and Woldorff, M.G. (2005) Abnormal brain activity related to performance monitoring and error detection in children with ADHD. Cortex, 41: 377–388.

Lung, F.-W., Yang, P., Cheng, T.-S. and Kao, W.-T. (2006) No allele variation of the MAOA gene promoter in male Chinese subjects with attention deficit hyperactivity disorder. Neuropsychobiology, 54: 147–151.

Malmberg, K., Wargelius, H.-L., Lichtenstein, P., Oreland, L. and Larsson, J.-O. (2008) ADHD and disruptive behavior scores-associations with MAOA and 5-HTT genes and with platelet MAOB activity in adolescents. BMC Psychiatry, 8: 28-doi:10.1186/1471-244X-8-28.

Manor, I., Eisenberg, J., Tyano, S., Sever, Y., Cohen, H., Ebstein, R.P. and Kotler, M. (2001) Family-based association study of the serotonin transporter promoter region polymorphism (5-HTTLPR) in attention deficit hyperactivity disorder. Am. J. Med Genet. Part B, 105: 91–95.

Manor, I., Tyano, S., Mel, E., Eisenberg, J., Bachner-Melman, R., Kotler, M. and Ebstein, R.P. (2002) Family-based and association studies of monoamine oxidase A and attention deficit hyperactivity disorder (ADHD): preferential transmission of the long promoter-region repeat and its association with impaired performance on a continuous performance test (TOVA). Mol. Psychiatry, 7: 626–632.

Manor, I., Laiba, E., Eisenberg, J., Meidad, S., Lerer, E., Israel, S., Gritsenko, I., Tyano, S., Faraone, S.V. and Ebstein, R.P. (2008) Association between tryptophan hydroxylase 2 (*TPH2*) SNPs, performance on a continuance performance test and response to methylphenidate in ADHD participants. Am. J. Med. Genet. Part B, 147B: 1501–1508.

Manuck, S.B., Flory, J.D., Ferrell, R.E., Mann, J.J. and Muldoon, M.F. (2000) A regulatory polymorphism of the monoamine oxidase-A gene may be associated with variability in aggression, impulsivity, and central nervous system serotonergic responsivity. Psychiatry Res., 95: 9–23.

Martin, P., Waters, N., Schmidt, C.J., Carlsson, A. and Carlsson, M.L. (1998) Rodent data and general hypothesis: antipsychotic action exerted through 5-HT2A receptor antagonism is dependent on increased serotonergic tone. J. Neural. Transm., 105: 365–396.

McAlonan, K., Cavenaugh, J. and Wurtz, R.H. (2008) Guarding the gateway to cortex with attention in visual thalamus. Nature, 456: 391–395.

Mejia, JM., Ervin, F.R., Baker, G.B. and Palmour, R.M. (2002) Monoamine oxidase inhibition during brain development induces pathological aggressive behavior in mice. Biol. Psychiatry, 52: 811–821.

Meyer-Lindenberg, A.S., Buckholtz, J.W., Kolachana, B., Hariri, A.R., Pezawas, L., Blasi, G., Wabnitz, A., Honea, R., Verchinski, B., Callicott, J.H., Egan, M.F., Mattay, V.S. and Weinberger, D.R. (2006) Neural mechanisms of genetic risk for impulsivity and violence in humans. Proc. Natl. Acad. Sci. (USA), 103: 6269–6274.

Micale, V., Leggio, G.M., Mazzola, C. and Drago, F. (2006) Cognitive effects of SL65.0155, a serotonin 5-HT_4 receptor partial agonist, in animal models of amnesia. Brain Res., 1121: 207–215.

Mickey, B.J., Ducci, F., Hodgkinson, C.A., Langenecker, S.A., Goldman, D. and Zubieta, J.-K. (2008) Monoamine oxidase a genotype predicts human Serotonin 1A receptor availability *in vivo*. J. Neurosci., 28: 11354–11359.

Millan, M.J., Lejeune, F. and Gobert, A. (2007) Reciprocal autoreceptor and heteroreceptor control of serotonergic, dopaminergic and noradrenergic transmission in the frontal cortex: relevance to the actions of antidepressant agents. J. Psychopharmacol., 14: 114–138.

Monckton, J.E. and McCormick, D.A. (2002) Neuromodulatory role of serotonin in the ferret thalamus. J. Neurophysiol., 87: 2124–2136.

Mulert, C., Jäger, L., Propp, S., Karch, S., Störmann, S., Pogarell, O., Möller, H.-J., Juckel, G. and Hegerl, U. (2005) Sound level dependence of the primary auditory cortex: Simultaneous measurement with 61-channel EEG and fMRI. Neuroimage, 28: 49–58.

Murphy, D.G.M., Daly, E., Schmitz, N., Toal, F., Murphy, K., Curran, S., Erlandsson, K., Kerwin, R., Ell, P. and Travis, M. (2006) Cortical Serotonin 5-HT_{2A} receptor binding and social communication in adults with Asperger's syndrome: an *in vivo* SPECT study. Am. J. Psychiatry, 163: 934–936.

Nathan, P.J., O'Neill, B. and Croft, R.J. (2005) Is the loudness dependence of the auditory evoked potential a sensitive and selective *in vivo* marker of central serotonergic function? Neuropsychopharmacology, 30: 1584–1585.

Nigg, J.T., Willcutt, E.G., Doyle, A.E. and Sonuga-Barke, E.J.S. (2005) Causal heterogeneity in attention-deficit/hyperactivity disorder: do we need neuropsychologically impaired subtypes? Biol. Psychiatry, 57: 1224–1230.

Novkovic, V.H., Rudan, V., Pivac, N., Nedic, G. and Muck-Seler, D. (2009). Platelet serotonin concentration in children with attention-deficit/hyperactivity disorder. Neuropsychobiology, in press.

Nyman, E.S., Ogdie, M.N., Loukola, A., Varilo, T., Taanila, A., Hurtig, T., Moilanen, I.K., Loo, S.K., McGough, J.J., Järvelin, M.-R., Smalley, S.L., Nelson, S.F. and Peltonen, L. (2007) ADHD candidate gene study in a population-based birth cohort: association with DBH and DRD2. J. Am. Acad. Child Psychiatry, 46: 1614–1621.

Oades, R.D. (2000) Differential measures of sustained attention in children with attention-deficit/hyperactivity or tic disorders: relationship to monoamine metabolism. Psychiatry Res., 93: 165–178.

Oades, R.D. (2005) The roles of norepinephrine and serotonin in ADHD. In: Gozal, D. and Molfese, D.L. (Eds.), Attention Deficit Hyperactivity Disorder: From Genes to Animal Models to Patients. Humana Press, Totawa, NY, pp. 97–130.

Oades, R.D. (2006) Function and dysfunction of monoamine interactions in children and adolescents with AD/HD. In: Levin, E.D. (Ed.), Neurotransmitter Interactions and Cognitive Function. Birkhäuser Verlag, Basel, pp. 207–244.

Oades, R.D. (2007) The role of the serotonin system in ADHD: treatment implications. Expert. Rev. Neurother., 7: 1357–1374.

Oades, R.D. (2008) Dopamine-serotonin interactions in attention-deficit/hyperactivity disorder (ADHD). In: Di Giovanni, G., Di Matteo, V. and Esposito, E. (Eds.), Serotonin-Dopamine Interaction: Experimental Evidence and Therapeutic Relevance. Prog. Brain Res, vol. 172, Elsevier, Amsterdam, pp. 543–565.

Oades, R.D. and Müller, B.W. (1997) The development of conditioned blocking and monoamine metabolism in children with attention-deficit-hyperactivity disorder or complex tics and healthy controls: an exploratory analysis. Behav. Brain Res., 88: 95–102.

Oades, R.D., Röpcke, B. and Schepker, R. (1996a) A test of conditioned blocking and its development in childhood and adolescence: relationship to personality and monoamines metabolism. Dev. Neuropsychol., 12: 207–230.

Oades, R.D., Dittmann-Balcar, A., Schepker, R. and Eggers, C. (1996b) Auditory event-related potentials and mismatch negativity in healthy children and those with attention-deficit- or Tourette-like symptoms. Biol. Psychol., 43: 163–185.

Oades, R.D., Daniels, R. and Rascher, W. (1998) Plasma neuropeptide Y levels, monoamine metabolism, electrolyte excretion, and drinking behavior in children with attention-deficit hyperactivity-disorder (ADHD). Psychiat. Res., 80: 177–186.

Oades, R.D., Slusarek, M., Velling, S. and Bondy, B. (2002) Serotonin platelet-transporter measures in childhood attention-deficit/hyperactivity disorder (ADHD): clinical versus experimental measures of impulsivity. World J. Biol. Psychiatry, 3: 96–100.

Oades, R.D., Lasky-Su, J., Christiansen, H., Faraone, S.V., Sonuga-Barke, E.J.S., Banaschewski, T., Chen, W., Anney, R.J.L., Buitelaar, J.K., Ebstein, R.P., Franke, B., Gill, M., Miranda, A., Rothenberger, A., Sergeant, J.A., Steinhausen, H.-C., Taylor, E.A., Thompson, M. and Asherson, P. (2008) The influence of serotonin and other genes on impulsive behavioral aggression and cognitive impulsivity in children with attention-deficit/hyperactivity disorder (ADHD): Findings from a family-based association test (FBAT) analysis. Behav. Brain Funct., 4: 48.

Ogdie, M.N., Fisher, S.E., Yang, M., Ishii, J., Francks, C., Loo, S.K., Cantor, R.M., McCracken, J., McGough, J.J., Smalley, S.L. and Nelson, S.F. (2004) Attention deficit hyperactivity disorder: fine mapping supports linkage to 5p13, 6q12, 16p13 and 17p11. Am. J. Hum. Genet., 75: 661–668.

O'Neill, B.V., Croft, R.J. and Nathan, P.J. (2008a) The loudness dependence of the auditory evoked potential (LDAEP) as an *in vivo* biomarker of central serotonergic function in humans: rationale, evaluation and review of findings. Hum. Psychopharmacol. Clin. Exp., 23: 355–370.

O'Neill, B.V., Guille, V., Croft, R.J., Leung, S.M.S., Scholes, K.E., Phan, K.L. and Nathan, P.J. (2008b) Effects of selective and combined serotonin and dopamine depletion on the loudness dependence of the auditory evoked potential (LDAEP) in humans. Hum. Psychopharmacol. Clin. Exp., 23: 301–312.

Oreland, L. and Hallman, J. (1995) The correlation between platelet MAO activity and personality: short review of findings and a discussion on possible mechanisms. Prog. Brain Res., 106: 77–82.

Parsey, R.V., Oquendo, M.A., Simpson, N.R., Ogden, R.T., Van Heertum, R., Arango, V. and Mann, J.J. (2002) Effects of sex, age, and aggressive traits in man on brain serotonin 5-HT(1A) receptor binding potential measured by PET using [C-11]WAY-100635. Brain Res., 954: 173–182.

Passamonti, L., Fera, F., Magariello, A., Cerasa, A., Gioia, M.C., Muglia, M., Nicoletti, G., Gallo, O., Provinciali, L. and Quattrone, A. (2006) Monoamine oxidase-A genetic variations influence brain activity associated with inhibitory control: new-insight into the neural correlates of impulsivity. Biol. Psychiatry, 59: 334–340.

Payton, A., Holmes, J., Barrett, J.H., Hever, T., Fitzpatrick, H., Trumper, A.L., Harrington, R., McGuffin, P., O'Donovan, M., Owen, M., Ollier, W., Worthington, J. and Thapar, A. (2001) Examining for association between candidate gene polymorphisms in the dopamine pathway and attention-deficit hyperactivity disorder: a family based study. Am. J. Med. Genet. Part B, 105: 464–470.

Plessen, K.J., Bansal, R., Zhu, H., Whiteman, R., Amat, J., Quackenbush, G.A., Martin, L., Durkin, K., Blair, C., Royal, J.,

Hugdahl, K. and Peterson, B.S. (2006) Hippocampus and amygdala morphology in attention-deficit/hyperactivity disorder. Arch. Gen. Psychiatry, 63: 795–807.

Quist, J.F., Barr, C.L., Schachar, R.J., Roberts, W., Malone, M., Tannock, R., Basile, V.S., Beitchman, J.H. and Kennedy, J.L. (2000) Evidence for the serotonin HTR2A receptor gene as a susceptibility factor in attention deficit hyperactivity disorder (ADHD). Mol. Psychiatry, 5: 537–541.

Quist, J.F., Barr, C.L., Schachar, R.J., Roberts, W., Malone, M., Tannock, R., Basile, V.S., Beitchman, J.H. and Kennedy, J.L. (2003) The serotonin 5-HT1B receptor gene and attention deficit hyperactivity disorder. Mol. Psychiatry, 8: 98–102.

Retz, W., Thome, J., Blücher, D., Baader, M. and Rösler, M. (2002) Association of attention deficit hyperactivity disorder-related psychopathology and personality traits with the serotonin transporter promoter region polymorphism. Neurosci. Lett., 319: 133–136.

Retz, W., Freitag, C.M., Retz-Junginger, P., Wenzler, D., Schneider, M., Kissling, C., Thome, J. and Rösler, M. (2008) A functional serotonin transporter promoter gene polymorphism increases ADHD symptoms in delinquents: Interaction with adverse childhood environment. Psychiatry Res., 158: 123–131.

Reuter, M., Kirsch, P. and Hennig, J. (2006) Inferring candidate genes for attention deficit hyperactivity disorder (ADHD) assessed by the world health organization adult ADHD self-report scale (ASRS). J. Neural Transm., 113: 929–938.

Reuter, M., Esslinger, C., Montag, C., Lis, S., Gallhofer, B. and Kirsch, P. (2008) A functional variant of the tryptophan hydroxylase 2 gene impacts working memory: a genetic imaging study. Biol. Psychol., 79: 111–117.

Ribases, M., Ramos-Quiroga, J.A., Hervas, A., Bosch, R., Bielsa, A., Gastaminza, X., Artigas, J., Rodriguez-Ben, S., Estivill, X., Casas, M., Cormand, B. and Bayes, M. (2009) Exploration of 19 serotoninergic candidate genes in adults and children with attention-deficit/hyperactivity disorder identifies association for 5HT2A, DDC and MAOB. Mol. Psychiatry, 14: 71–85.

Riikonen, R.S., Nokelainen, P., Valkonen, K., Kolemainen, A.I., Kupulainen, K.I., Koenoenen, M., Vanninen, R.-L.S. and Kuikka, J.T. (2005) Deep serotonergic and dopaminergic structures in fetal alcoholic syndrome: a study with nor-β-CIT-single-photon emission computed tomography and magnetic resonance imaging volumetry. Biol. Psychiatry, 57: 1565–1572.

Robinson, E.S.J., Dalley, J.W., Theobald, D.E.H., Glennon, J. C., Pezze, M.A., Murphy, E.R. and Robbins, T.W. (2008) Opposing Roles for 5-HT_{2A} and 5-HT_{2C} receptors in the nucleus accumbens on inhibitory response control in the 5-choice serial reaction time task. Neuropsychopharmacology, 33: 2398–2406.

Rommelse, N.N.J., Altink, M.E., Arias-Vasquez, A., Buschgens, C.J.M., Fliers, E., Faraone, S.V., Buitelaar, J.K., Sergeant, J.A., Oosterlaan, J. and Franke, B. (2008) Differential association between *MAOA*, ADHD and neuropsychological functioning in boys and girls. Am. J. Med. Genet. Part B, 147B: 1524–1530.

Ronald, A., Simonoff, E., Kuntsi, J., Asherson, P. and Plomin, R. (2008) Evidence for overlapping genetic influences on autistic and ADHD behaviours in a community twin sample. J. Child. Psychol. Psychiatry, 49: 535–542.

Rosel, P., Arranz, B., Urretavizcaya, M., Oros, M., San, L., Vallejo, J. and Navarro, M.A. (2002) Different distributions of the 5-HT reuptake complex and the postsynaptic 5-HT_{2A} receptors in Brodmann areas and brain hemispheres. Psychiatry Res., 111: 105–117.

Rubia, K., Lee, F., Cleare, A.J., Tunstall, N., Fu, C.H.Y., Brammer, M.J. and McGuire, P.K. (2004) Tryptophan depletion reduces right inferior prefrontal activation during no-go trials in fast, event-related fMRI. Psychopharmacology, 179: 791–803.

Rubia, K., Smith, A.B., Brammer, M.J., Toone, B. and Taylor, E.A. (2005) Abnormal brain activation during inhibition and error detection in medication-naive adolescents with ADHD. Am. J. Psychiatry, 162: 1067–1075.

Rubia, K., Halari, R., Smith, A.B., Mohammed, M., Scott, S., Giampietro, V., Taylor, E.A. and Brammer, M.J. (2008a) Dissociated functional brain abnormalities of inhibition in boys with pure conduct disorder and in boys with pure attention deficit hyperactivity disorder. Am. J. Psychiatry, 165: 889–897.

Rubia, K., Smith, A. B., Halari, R., Matsukura, F., Mohammad, M., Taylor, E.A. and Brammer, M.J. (2008b) Disorder-specific dissociation of orbitofrontal dysfunction in boys with pure conduct disorder during reward and ventrolateral prefrontal dysfunction in boys with pure ADHD during sustained attention. Am. J. Psychiatry 166: 83–94.

Sandou, F., Amara, D.A., Dierich, A., Le Meur, M., Ramboz, S., Segu, L., Buhot, M.C. and Hen, R. (1994) Enhanced aggressive behavior in mice lacking 5-HT1b receptor. Science, 265: 1875–1878.

Satterfield, J.H., Schell, A.M., Backs, R.W. and Hidaka, K.C. (1984) A cross-sectional and longitudinal study of age effects of electrophysiological measures in hyperactive and normal children. Biol. Psychiatry, 19: 973–990.

Scheuch, K., Lautenschlager, M., Grohmann, M., Stahlberg, S., Kirchheiner, J., Zill, P., Heint, A., Walther, D.J. and Priller, J. (2007) Characterization of a functional promoter polymorphism of the human tryptophan hydroxylase 2 gene in serotonergic raphe neurons. Biol. Psychiatry, 62: 1288–1294.

Schmidt, L.A., Fox, N.A. and Hamer, D.H. (2007) Evidence for a gene–gene interaction in predicting children's behavior problems: Association of serotonin transporter short and dopamine receptor D4 long genotypes with internalizing and externalizing behaviors in typically developing 7-year-olds. Dev. Psychopathol., 19: 1105–1116.

Schweighofer, N., Bertin, M., Shishida, K., Okamoto, Y., Tanaka, S.C., Yamawaki, S. and Doya, K. (2008) Low-serotonin levels increase delayed reward discounting in humans. J. Neurosci., 28: 4528–4532.

Seeger, G., Schloss, P. and Schmidt, M.H. (2001) Marker gene polymorphisms in hyperkinetic disorder – predictors of clinical response to treatment with methylphenidate? Neurosci. Lett., 313: 45–48.

Sharma, R.P. and Shapiro, L.E. (1996) The 5HT1a receptor system: possible implications for schizophrenic negative symptomatology. Psychiatry Ann., 26: 88–92.

Sheehan, K., Lowe, N., Kirley, A., Mullins, C., Fitzgerald, M., Gill, M. and Hawi, Z. (2005) Tryptophan hydroxylase 2 (TPH2) gene variants associated with ADHD. Mol. Psychiatry, 10: 944–949.

Sheehan, K., Hawi, Z., Gill, M. and Kent, L. (2007) No association between TPH2 gene polymorphisms and ADHD in a UK sample. Neurosci. Lett., 412: 105–107.

Smith, A.B. and Taylor, E.A. (2006) Response inhibition and hyperactivity in clinical and non-clinical populations:

a meta-analysis using the stop task. In: Oades, R.D. (Ed.), Attention-Deficit/Hyperactivity Disorder and the Hyperkinetic Syndrome: Current Ideas and Ways Forward. Nova Science Publishing Inc., Hauppauge, New York, pp. 203–225.

Smoller, J.W., Biederman, J., Arbeitman, L., Doyle, A.E., Fagerness, J., Perlis, R.H., Sklar, P. and Faraone, S.V. (2006) Association between the 5HT1B receptor gene (*HTR1B*) and the inattentive subtype of ADHD. Biol. Psychiatry, 59: 460–467.

Snoek, H., van Goozen, S.H.M., Matthys, W., Sigling, H.O., Koppeschaar, H.P.F., Westenberg, H.G.M. and van Engeland, H. (2002) Serotonergic functioning in children with oppositional defiant disorder: a sumatriptan challenge study. Biol. Psychiatry, 51: 319–325.

Sonuga-Barke, E.J.S. (2005) Causal models of attention-deficit/hyperactivity disorder: from common simple deficits to multiple developmental pathways. Biol. Psychiatry, 57: 1231–1238.

Sonuga-Barke, E.J.S., Lasky-Su, J., Neale, B.M., Oades, R.D., Chen, W., Franke, B., Buitelaar, J.K., Banaschewski, T., Ebstein, R.P., Gill, M., Anney, R.J.L., Miranda, A., Mulas, F., Roeyers, H., Rothenberger, A., Sergeant, J.A., Steinhausen, H.-C., Thompson, M., Asherson, P. and Faraone, S.V. (2008) Does parental expressed emotion moderate genetic effects in ADHD? An exploration using a genome wide association scan. Am. J. Med. Genet. Part B, 147B: 1359–1368.

Stadler, C., Schmeck, K., Nowraty, I., Müller, W.E. and Poustka, F. (2004) Platelet 5-HT uptake in boys with conduct disorder. Neuropsychobiology, 50: 244–251.

Stahl, S.M. (1997) Mixed depression and anxiety: serotonin (1A) receptors as a common phamacologic link. J. Clin. Psychiatry, 58 (Suppl. 8): 20–26.

Stiedl, O., Misane, I., Koch, M., Pattij, T., Meyer, M. and Ögren, S.O. (2007) Activation of the brain 5-HT_{2C} receptors causes hypolocomotion without anxiogenic-like cardiovascular adjustments in mice. Neuropharmacology, 52: 949–957.

Stoff, D.M., Pollock, L., Vitiello, B., Behar, D. and Bridger, W.H. (1987) Reduction of (3H)-imipramine binding sites on platelets of conduct-disordered children. Neuropsychopharmacology, 1: 55–62.

Strobel, A., Debener, S., Schmidt, D., Hünnerkopf, R., Lesch, K.-P. and Brocke, B. (2003) Allelic variation in serotonin transporter function associated with the intensity dependence of the auditory evoked potential. Am. J. Med. Genet. Part B, 118B: 41–47.

Su, L., Cheng, D. and Gao, X. (2004) Association of the 5-HT2A receptor polymorphism and attention deficit hyperactivity disorder (ADHD). Proceedings of the 16th World Congress of the IACAPAP, 22–26 Aug, Berlin 169.

Tanaka, S.C., Schweighofer, N., Asahi, S., Shishida, K., Okamoto, Y., Yamawaki, S. and Doya, K. (2008) Serotonin differentially regulates short- and long-term prediction of rewards in the ventral and dorsal striatum. PLoS One, 2.

Thapar, A., Langley, K., Asherson, P. and Gill, M. (2007) Gene–environment interplay in attention-deficit hyperactivity disorder and the importance of a developmental perspective. Br. J. Psychiatry, 190: 1–3.

Uzbekov, M.G. (2006) Hyperkinetic syndrome as a manifestation of a disturbance of metabolism and mental development. In: Oades, R.D. (Ed.), Attention-Deficit/Hyperactivity Disorder and the Hyperkinetic Syndrome: Current Ideas and Ways Forward. Nova Science Publishers, Inc., Hauppauge, NewYork, pp. 133–154.

van Goozen, S.H.M. and Fairchild, G. (2006) Neuroendocrine and neurotransmitter correlates in children with antisocial behavior. Horm. Behav., 50: 647–654.

van Goozen, S.H.M., Fairchild, G., Snoek, H. and Harold, G.T. (2007) The evidence for a neurobiological model of childhood antisocial behavior. Psychol. Bull., 133: 149–182.

Van Meel, C.S., Heslenfeld, D.J., Oosterlaan, J. and Sergeant, J.A. (2007) Adaptive control deficits in attention-deficit/hyperactivity disorder (ADHD): the role of error processing. Psychiatry Res., 151: 211–220.

van t'Ent, D., Lehn, H., Derks, E.M., Hudziak, J.J., Van Strien, N.M., Veltman, D.J., de Geus, E.J.C., Todd, R.D. and Boomsma, D.I. (2007) A structural MRI study in monozygotic twins concordant or discordant for attention/hyperactivity problems: evidence for genetic and environmental heterogeneity in the developing brain. Neuroimage, 35: 1004–1020.

Walitza, S., Renner, T.J., Dempfle, A., Konrad, K., Wewetzer, C., Halbach, A., Herpetz-Dahlmann, B., Remschmidt, H., Smidt, J., Linder, M., Flier, L., Knöllker, U., Friedel, S., Schäfer, H., Gross, C., Hebebrand, J., Warnke, A. and Lesch, K.-P. (2005) Transmission disequilibrium of polymorphic variants in the tryptophan hydroxylase-2 gene in attention-deficit/hyperactivity disorder. Mol. Psychiatry, 10: 1126–1132.

Walther, D.J., Peter, J.U., Bashammakh, S., Hörtnagl, H., Voits, M., Fink, H. and Bader, M. (2003) Synthesis of serotonin by a second tryptophan hydroxylase isoform. Science, 299: 76.

Way, B.M., Lacan, G., Fairbanks, LA. and Melega, W.P. (2007) Architectonic distribution of the serotonin transporter within the orbitofrontal cortex of the vervet monkey. Neurosci., 148: 937–948.

Whitaker-Azmitia, P.M. (2005) Behavioral and cellular consequences of increasing serotonergic activity during brain development: a role in autism? Intl. J. Dev. Neurosci., 23: 75–83.

Whitaker-Azmitia, P.M., Zhang, X. and Clarke, C. (1994) Effects of gestational exposure to monoamine oxidase inhibitors in rats: preliminary behavioral and neurochemical studies. Neuropsychopharmacology, 11: 125–132.

White, J.L., Moffitt, T.E., Caspi, A., Bartsch, DJ., Needles, D.J. and Stouthamer-Loeber, M. (1994) Measuring impulsivity and examining its relationship to delinquency. J. Abnorm. Psychol., 103: 192–205.

Wigg, K.G., Takhar, A., Ickowicz, A., Tannock, R., Kennedy, J.L., Pathare, T., Malone, M., Schachar, R.J. and Barr, C.L. (2006) Gene for the serotonin transporter and ADHD: no association with two functional polymorphisms. Am. J. Med. Genet. Part B, 141B: 566–570.

Wild-Wall, N., Oades, R.D., Schmidt-Wessels, M., Christiansen, H. and Falkenstein, M. (2009) Neural activity associated with conflict and performance monitoring in attention-deficit/hyperactivity disorder (ADHD). Intl. J. Psychophysiol., 74: 19–27.

Willcutt, E.G., Doyle, A.E., Nigg, J.T., Faraone, S.V. and Pennington, B.F. (2005) Validity of the executive function theory of attention-deficit/hyperactivity disorder: a meta-analytic review. Biol. Psychiatry, 57: 1336–1346.

Winstanley, C.A., Theobald, D.E.H., Dalley, J.W. and Robbins, T.W. (2005) Interactions between serotonin and dopamine in the control of impulsive choice in rats: therapeutic implications for impulse control disorders. Neuropsychopharmacology, 30: 669–682.

Wong, D.F., Brasic, J.R., Singer, H.S., Schretlen, D.J., Kuwabara, H., Zhou, Y., Nandi, A., Maris, M.A., Alexander, M., Ye, W., Rousset, O., Kumar, A., Szabo, Z., Gjedde, A. and Grace, A.A. (2008) Mechanisms of dopaminergic and serotonergic neurotransmission in Tourette syndrome: clues from an *in vivo* neurochemistry study with PET. Neuropsychopharmacology, 33: 1239–1251.

Wutzler, A., Winter, C., Kitzrow, W., Uhl, I., Wolf, R.J., Heinz, A. and Juckel, G. (2008) Loudness dependence of auditory evoked potentials as indicator of central serotonergic neurotransmission: simultaneous electrophysiological recordings and *in vivo* microdialysis in the rat primary auditory cortex. Neuropsychopharmacology, 33: 3182–3189.

Xu, X., Mill, J., Chen, C.K., Brookes, K.-J., Taylor, E.A. and Asherson, P. (2005) Family-based association study of serotonin transporter gene polymorphisms in attention deficit hyperactivity disorder: no evidence for association in UK and Taiwanese samples. Am. J. Med. Genet. Part B, 139: 11–13.

Xu, X., Aysimi, E., Anney, R.J.L., Brookes, K.-J., Franke, B., Zhou, K., Bitsakou, P., Baeyens, D., Buschgens, C.J.M., Chen, W., Christiansen, H., Eisenberg, J., Manor, I., Marco, R., Gabriels, I., Mulligan, A., Rommelse, N.N.J., Müller, U.C., Uebel, H., Fleischman, K., Banaschewski, T., Buitelaar, J.K., Ebstein, R.P., Gill, M., Miranda, A., Oades, R.D., Roeyers, H., Rothenberger, A., Sergeant, J.A., Sonuga-Barke, E.J.S., Steinhausen, H.-C., Taylor, E.A., Faraone, S.V. and Asherson, P. (2008) No association between two polymorphisms of the serotonin transporter gene and combined type attention deficit hyperactivity disorder. Am. J. Med. Genet. Part B, 147B: 1306–1309.

Young, K.A., Bonkale, W.L., Holcomb, L.A., Hicks, P.B. and German, D.C. (2008) Major depression, 5HTTLPR genotype, suicide and antidepressant influences on thalamic volume. Br. J. Psychiatry, 192: 284–289.

Zoroglu, S.S., Erdal, M.E., Alasehirli, B., Erdal, N., Sivasli, E., Tutkun, H., Savas, H.A. and Herken, H. (2002) Significance of serotonin transporter gene 5-HTTLPR and variable number of tandem repeat polymorphism in attention deficit hyperactivity disorder. Neuropsychobiology, 45: 176–181.

CHAPTER 4.6

Serotonin and Schizophrenia

Boris B. Quednow*,[1], Mark A. Geyer[2] and Adam L. Halberstadt[2]

[1]*University Hospital of Psychiatry, University of Zurich, Switzerland*
[2]*Department of Psychiatry, University of California at San Diego, California*

Abstract: Although the serotonin hypothesis of schizophrenia is one of the oldest neurochemical hypotheses on the pathogenesis of this disease, it is still highly topical. The concept of how the serotonin system is involved in the origin and progress of schizophrenia has considerably changed over the past decades. Therefore, the present work will give an overview about the development and the current directions of the serotonin hypothesis of schizophrenia. In this regard, we will discuss the phenomenology of hallucinogenic drug action, model psychosis and translational research, *post mortem* studies on receptors and transporters, imaging studies, antipsychotic drug action, neuroendocrine challenge studies, platelet and cerebrospinal fluid data, genetic association studies, developmental aspects, and the cross-talk between the glutamate and the serotonin system. In sum, there are several lines of evidence suggesting that the serotonin system plays a major role in the pathogenesis of at least a subpopulation of schizophrenia patients. Further studies are needed to better characterize patients whose psychotic symptoms are suspected to have a serotonergic origin.

Keywords: serotonin, 5-HT, schizophrenia, model psychosis, LSD, psilocybin, atypical antipsychotics, 5-HT_{2A} receptor, platelets, positron emission tomography, single nucleotide polymorphism, developmental disorder, dopamine, glutamate, cognition.

Introduction

The symptoms of schizophrenia can be divided into three major domains: (1) positive symptoms such as hallucinations, perceptual disturbances, delusional phenomena and formal thought disorder; (2) cognitive dysfunction, which includes motivational and executive function deficits; and (3) negative symptoms, including flat affect, poverty of speech, avolition and inappropriate emotional responses (Tamminga and Holcomb, 2005). Presentation of symptoms from these three domains is heterogeneous, making the illness difficult to diagnose and treat. The highest risk period for developing schizophrenia is during young adulthood. Both sexes are equally affected by the disorder, although the age of onset of symptoms is typically younger for men than women (Goldstein *et al.*, 1989; Faraone *et al.*, 1994; Bromet and Fennig, 1999). Although incidence figures vary depending on the diagnostic criteria, schizophrenia affects approximately 1 percent of the general population. Individuals with schizophrenic parents or siblings have an increased risk for developing the illness (8–12 percent). For monozygotic twins, the concordance rate is approximately 50 percent (Holzman and Matthysse, 1990; Gottesman, 1991). The elevated familial incidence of schizophrenia strongly indicates that there is a genetic contribution to the disorder, although the fact that concordance rates for monozygotic twins are lower than 100 percent suggests that environmental factors are also involved. It is therefore likely that a combination of genetic susceptibility and environmental factors is required for the illness to develop (Gottesman, 1991). Linkage studies of schizophrenia have identified several chromosomal regions and candidate genes that are associated with the disorder (reviewed by Harrison and Owen, 2003; Harrison and Weinberger, 2005).

Although there is evidence for enlarged ventricles and decreased cerebral (cortical and hippocampal) volume associated with schizophrenia, there is not a distinct 'diagnostic' neuropathology associated with the disease (reviewed by Harrison, 1999a, 2004; Harrison and Owen, 2003). Misplaced and clustered neurons, particularly in the entorhinal cortex, indicate problems of neuronal migration, and suggest an early developmental anomaly (Jakob and Beckmann, 1986; Arnold *et al.*, 1991; Falkai *et al.*, 2000). Pyramidal neurons in the hippocampus and neocortex have been shown to have smaller cell bodies and fewer dendritic spines and dendritic

*Corresponding author
E-mail: quednow@bli.uzh.ch

Christian Müller & Barry Jacobs (Eds.)
Handbook of Behavioral Neurobiology of Serotonin
ISBN 978-0-12-374634-4

DOI: 10.1016/B978-0-12-374634-4.00034-4

arborizations (reviewed by Harrison and Weinberger, 2005). Additionally, decreased presynaptic proteins such as synaptophysin, SNAP-25 and complexin II have been observed in schizophrenia brains (Harrison and Eastwood, 2001; Honer and Young, 2004), as well as decreased density of interneurons (e.g., parvalbumin-immunoreactive cells; Lewis, 2000; Reynolds *et al.*, 2002a). There are also reports of decreases in cell numbers in the thalamus and a decreased number of oligodendrocytes. Neuroimaging data and *post mortem* studies have shown that N-acetylaspartate (NAA), a marker of neuronal integrity, is decreased in first-episode and never-medicated patients (Bertolino and Weinberger, 1999; Nudmamud *et al.*, 2003). Based on these neuropathological changes, investigators have conceptualized schizophrenia as a disease of functional 'dysconnectivity' (Weinberger *et al.*, 1992; Friston and Frith, 1995; McGlashan and Hoffman, 2000), or a 'disorder of the synapse' (Mirnics *et al.*, 2001; Frankle *et al.*, 2003) affecting the machinery of the synapse (Harrison and Eastwood, 2001; Honer and Young, 2004).

Not only structural alterations but also neurochemical changes have been proposed to play a role in the etiopathogenesis of schizophrenia. In the following sections, we give an overview on the serotonin hypothesis of schizophrenia. Although it is one of the oldest neurochemical hypotheses on the pathogenesis of this disease, it is still highly topical, as will be shown in the following sections.

History of the serotonin hypothesis of schizophrenia

The first step in the direction of the idea that the serotonin system may contribute to schizophrenia was probably made by the German psychiatrist Kurt Beringer (1923). He was the first to propose the use of the hallucinogen mescaline as an experimental model of psychosis, despite the fact that he had no knowledge of serotonin receptors or the principles of neurotransmission. Previously, on the eve of World War I, Knauer and Maloney (1913) had already recommended the mescaline self-experience for psychiatrists to gain better insights into the psychotic states of their patients. Subsequently, we have come to understand that mescaline is a selective serotonin-2A (5-HT_{2A}) receptor agonist that played an important role in the development of the transmethylation hypothesis of schizophrenia. In 1943, Albert Hofmann identified the impressive psychotomimetic effects of d-lysergic acid diethylamid (LSD) during an unintentional self-intoxication in his laboratory at Sandoz Pharmaceutical Company (Stoll, 1947). During subsequent repeated self-experiments, Hofmann noted that the necessary dose of LSD to cause psychological effects was very small, strongly suggesting that there must be a receptor or some other specific site of action for the LSD molecule. Mescaline, in contrast, had to be given in hundreds of milligrams to produce psychotomimetic effects that were comparable to the effects of several micrograms of LSD (Stoll, 1947). Hofmann gave LSD to Walter Stoll, a psychiatrist at the University Hospital of Psychiatry Zurich 'Burghölzli', and the son of Hofmann's supervisor Arthur Stoll at Sandoz. The younger Stoll explored the psychopathological effects of LSD in 16 healthy volunteers and found that the LSD effects were strikingly similar to the symptoms of schizophrenia (Stoll, 1947).[1] Subsequently, both Stoll and his colleague Condrau administered LSD to patients with schizophrenia, hoping that the LSD 'shock' may have some therapeutic benefits. They noted that LSD is much less potent in schizophrenia patients than in normal controls, and therefore concluded that a toxic substance similar to LSD may cause schizophrenic psychoses (Stoll, 1947, 1949; Condrau, 1949). With this perception, they paved the way for the transmethylation hypothesis. Moreover, both authors noted that LSD may prove to be a valuable tool to induce psychotic states experimentally in the laboratory.

While searching for a vasoconstrictive substance in platelets, Rapport and colleagues (1948) discovered serotonin; soon thereafter, the structure of serotonin was deduced (Rapport, 1949). Betty Twarog and Page (1953) subsequently demonstrated that serotonin could be found in the mammalian brain. Initially it was thought that serotonin was simply a residue of blood in the brain, but the structural similarities between LSD and serotonin led to the suggestion that serotonin may act directly in the brain (Healy, 2002). Gaddum (1953) quickly determined that the oxytocic effects of serotonin could be antagonized by LSD. As was fashionable at the time among pharmacologists, Gaddum took LSD himself. The intense experience encouraged him to propose that serotonin in the brain may play a role in preserving sanity (Gaddum and Hameed, 1954; Healy, 2002). At the same time, Woolley and Shaw (1953) independently discovered that other centrally acting indoleamines (yohimbine, ergot alkaloids, harmine) also antagonize the vasoconstrictive action of serotonin, and they also concluded that serotonin may play a role in nervous disorders (Woolley and Shaw, 1954). Gaddum and Hameed (1954) and Woolley and Shaw (1954) proposed that serotonin activity might be decreased in the brain of schizophrenia patients. Subsequent evidence indicating that LSD is an agonist rather than an antagonist

[1] Interestingly, Stoll (1947) had already suggested radioactive labeling of LSD to investigate, in animals, which brain regions LSD acts upon.

questioned this hypothesis (Baumeister and Hawkins, 2004). Later, Woolley (1962) revoked his initial suggestion and stated that schizophrenia may result from an excess of brain serotonin.

Shortly before the discoveries of Gaddum, Woolley and Shaw, another serotonin-related hypothesis of schizophrenia had also appeared. As early as 1932, Henk de Jong noted that mescaline is chemically related to epinephrine. He therefore supposed that a disturbance of epinephrine metabolism might lead to the synthesis of a mescaline-like substance that causes catatonia, one of the primary forms of schizophrenia at that time (de Jong, 1932). Twenty years later, Osmond and Smythies (1952) reinvented this idea and proposed the influential transmethylation hypothesis of schizophrenia. Osmond and Smythies observed that an asthmatic patient developed psychotic symptoms after he had taken old (and therefore oxidized) epinephrine during an asthmatic attack. In a self-experiment, Osmond and his director Abram Hoffer then took adrenochrome – a breakdown product of epinephrine, pink in color – and reported that it produced hallucinogenic responses (Healy, 2002). These observations led to their assumption that schizophrenia results from an endogenous neurotoxin that is formed by aberrant metabolic processes during the biosynthesis of catecholamines. The last step of the biosynthesis of epinephrine is methylation of the amino group of norepinephrine. If the phenolic hydroxyl groups were irregularly methylated instead, then a mescaline-like compound would be produced. Later Hoffer, Osmond and Smythies (1991) expanded the transmethylation hypothesis by proposing the possibility of an aberrant endogenous biosynthesis of methylated indolamine hallucinogens such as LSD. In the following years, many researchers tried to find the 'pink spot' of adrenochrome and other suspected endogenous neurotoxins in the brain, blood or urine of schizophrenia patients; however, it was never convincingly found. Moreover, Hoffer and Osmond brought their theory directly to the clinic and treated schizophrenia patients with large doses of nicotinic acid because it acts to trap methyl donors, and thus the aberrant transmethylation of catecholamines or indolamines should be decreased. The authors reported that nicotinic acid alone, as well as in combination with chlorpromazine, had some beneficial effects in the treatment of schizophrenia, but these results could not be replicated in later studies performed by the Canadian Association of Mental Health (Healy, 2002). Although the transmethylation hypothesis still has strong face validity, it fell out of favor after the 1960s for two reasons: first, the schizophrenogenic substances could not be isolated; and secondly, a new influential theory targeting another neurotransmitter commandeered the focus of schizophrenia research. For the time being, the serotonin hypotheses were superseded by the influential dopamine hypothesis of schizophrenia.

The dopamine hypothesis of schizophrenia

Based on the finding of Brodie *et al.* (1955, 1956) that reserpine acutely releases brain serotonin while post-acutely depeleting it, the group of Arvid Carlsson demonstrated that reserpine has the same effect on catecholamines (Bertler *et al.*, 1956). These results suggested that serotonin and catecholamines may play a role in the sedative and motor-depressant effects of reserpine. Carlsson *et al.* (1957) tested this hypothesis by administering the precursors L-dopa and 5-hydroxytryptophan to animals after a pre-treatment with reserpine. Only L-dopa attenuated the behavioral effects of reserpine, whereas 5-hydroxytryptophan had no effect. Subsequently, it was shown that L-dopa increases brain dopamine but not norepinephrine (Carlsson, 1959). These results suggested an important role of dopamine in brain function. In 1963, Carlsson and Lindquist reported that chlorpromazine and haloperidol reduced catecholamine activity through a postsynaptic action (Carlsson and Lindqvist, 1963). Later, van Rossum (1966) explicated that blockade of postsynaptic dopamine receptors is responsible for the behavioral effects of these neuroleptic drugs. The dopamine hypothesis of schizophrenia – which is actually a hypothesis of neuroleptic drug action – was born. For more than three decades, the dopamine hypothesis has dominated biological research on the etiopathogenesis of schizophrenia. The assumption that schizophrenia is caused by a significant disturbance of dopamine transmission (or metabolism) that results in an increase of dopamine function was initially supported by the following data (Bleich *et al.*, 1988):

1. All (admitted) antipsychotic drugs are dopamine-D_2 (D_2) receptor antagonists, and before the advent of 'atypical' antipsychotics 20 years ago it was shown that antipsychotic potency of the neuroleptics was directly correlated with D_2 receptor binding (Meltzer and Stahl, 1976; Seeman, 1987). However, the latter is not true for clozapine, which is still the gold standard of antipsychotic drug action, because it has only a moderate affinity for D_2 receptors but higher affinity for 5-HT_{2A}, 5-HT_{2C}, 5-HT_6 and 5-HT_7 receptors, as well as for D_4, histamine H_1, muscarinic M_1, α_1, and α_2 receptors (Arnt and Skarsfeldt, 1998; Abi-Dargham and Krystal, 2000).
2. Sustained or high-dose exposure to indirect dopamine agonists (e.g., L-dopa, cocaine, amphetamine) may cause psychotic symptoms in healthy subjects that

are similar to those of paranoid schizophrenia (Segal *et al.*, 1981). Moreover, indirect dopamine agonists provoke exacerbation of symptoms in schizophrenia patients. Amphetamine is known to release presynaptic dopamine and norepinephrine, and it was shown that antipsychotics could improve the acute symptoms of amphetamine psychosis (Snyder, 1973; Carlsson, 1988). Nevertheless, psychotic states induced by indirect dopamine agonists mimic only the positive symptoms of schizophrenia; thus, maybe only the positive symptoms might be due to an increased dopaminergic activity (Angrist and Gershon, 1970).

3. *Post-mortem* studies and imaging studies with positron emission tomography (PET) initially indicate an increase of striatal D_2 receptor levels in the brains of schizophrenia patients (Wong *et al.*, 1986; Seeman, 1987). However, up-regulation of D_2 receptor expression could be the result of adaptation to antipsychotic drug treatment rather than a pathological abnormality inherent to schizophrenia, and in fact many *post-mortem* and PET studies could not replicate the finding of increased striatal D_2 receptor density in drug-naïve schizophrenia patients (Harrison, 1999a; Weinberger and Laruelle, 2002). However, there is accumulating evidence for a presynaptic dopaminergic abnormality in schizophrenia, implying disturbances in presynaptic storage, vesicular transport, release, reuptake and metabolic mechanisms in mesolimbic dopamine systems (Laruelle *et al.*, 1999; Weinberger and Laruelle, 2002).

The current view on the role of dopamine in schizophrenia is that subcortical mesolimbic dopamine projections might be hyperactive (causing productive symptoms) and that the mesocortical dopamine projections to the prefrontal cortex (PFC) and the anterior cingulate are hypoactive (causing negative symptoms and cognitive impairment). These two dysfunctions might be linked, as the cortical dopamine system generally inhibits the subcortical dopamine system (Weinberger and Laruelle, 2002).

Addressing the psychopathological heterogeneity of schizophrenia, Timothy Crow (1980a, 1980b) proposed that schizophrenia be divided into two syndromes: type I is characterized by positive symptoms that reflect an increase in striatal dopamine function, and responds well to antipsychotics; while type II is more characterized by negative symptoms, structural brain abnormalities (cortical atrophy and/or ventricular enlargement) and limited response to (typical) antipsychotics. Bleich *et al.* (1988) suggested that the type II syndrome might respond better to serotonin antagonistic compounds, and thus he proposed a dopaminergic and serotonergic form of schizophrenia. This view is supported by the fact that some atypical substances having a strong 5-HT_{2A} receptor antagonistic component may be superior in the treatment of negative symptoms when compared to typical neuroleptics without a strong serotonin antagonistic action (Meltzer, 1999). The fact that amisulpride, a pure D_2/D_3 antagonist, has nevertheless a strong impact on not only positive but also negative symptoms may be due to its unique pharmacokinetic properties (Leucht, 2004).

Phenomenology of hallucinogenic drug action

Effects of hallucinogens in human subjects

Serotonergic hallucinogens produce profound alterations in thought, mood, affect and sensory perception. The effects of these drugs are often characterized by visual illusions and elementary hallucinations, altered sense of time and space, and depersonalization. Hallucinogen-induced Altered States of Consciousness (ASCs) are highly subjective, and are typically assessed using self-reports. Various rating scales have been used to assess the effects of hallucinogens (reviewed by Strassman, 1995). The Addiction Research Center Inventory (Haertzen *et al.*, 1963) is an older instrument that emphasized the unpleasant effects of hallucinogens. The Hallucinogen Rating Scale (HRS) was designed specifically to detect the effects of intravenous N,N-dimethyltryptamine (Strassman *et al.*, 1994), and has now been validated for other hallucinogens (Gouzoulis-Mayfrank *et al.*, 1999). Another rating scale, the Altered States of Consciousness Questionnaire (APZ), was developed by Dittrich to assess various types of ASCs, independent of their etiology (Dittrich, 1998). The original APZ includes three dimensions that have been labeled: Oceanic Boundlessness (OB), Anxious Ego Dissolution (AED) and Visionary Restructuralization (VR). The OB dimension measures states that resemble mystical experiences, the AED dimension reflects 'bad trip'-like experiences, and the VR dimension refers to altered visual perception. An updated version of the APZ, the 5D-ASC, includes two additional dimensions: Reduction of Vigilance (RV) and Auditory Alterations (AA). For a detailed description of the APZ and 5D-ASC core dimensions, see Table 1.

Clinical studies have demonstrated that psilocybin, DMT and mescaline increase scores in the OB, AED and VR dimensions of the APZ (Hermle *et al.*, 1992; Vollenweider *et al.*, 1997a; Dittrich, 1998;

Table 1 Core dimensions of the 5D-ASC (Dittrich, 1998)

Dimension	Symptoms assessed
Oceanic Boundlessness (OB)	*Positive derealization*
	Positive depersonalization
	Altered sense of time
	Positive mood
	Mania-like experience
Anxious Ego Dissolution (AED)	*Anxious derealization*
	Thought disorder
	Delusion
	Fear of loss of control
Visionary Restructuralization (VR)	*Elementary hallucinations*
	Visual pseudohallucinations
	Synesthesia
	Changed meaning of percepts
	Facilitated recollection
	Facilitated imagination
Auditory Alterations (AA)	*Auditory illusions*
	Auditory pseudohallucinations
Reduction of Vigilance (RV)	*Drowsiness*
	Decreased alertness
	Impaired cognitive function

Gouzoulis-Mayfrank *et al.*, 1999). Additional studies have shown that psilocybin produces a dose-dependent increase of scores in the five core dimensions of the 5D-ASC rating scale (Hasler *et al.*, 2004). However, AED and AA scores are increased significantly only after administration of a high dose of psilocybin (0.315 mg/kg, p.o.), and are relatively unaffected by lower doses (0.045–0.215 mg/kg).

A large amount of preclinical evidence indicates that the 5-HT_{2A} receptor mediates most of the behavioral effects of hallucinogens. Pre-treatment with the 5-HT_{2A} antagonist ketanserin blocks the effects of psilocybin on the APZ in human volunteers (Vollenweider *et al.*, 1998), confirming the involvement of the 5-HT_{2A} receptor. According to a recent PET study with [^{18}F]altanserin, the ability of psilocybin to increase 5D-ASC scores is directly correlated with the level of 5-HT_{2A} receptor occupation in the anterior cingulate cortex and medial PFC (Hasler, Quednow, Vollenweider, unpublished data, Figure 1). These findings are consistent with those of an [^{18}F]fluorodeoxyglucose PET study (Vollenweider *et al.*, 1997a), which found that the effects of psilocybin on the APZ are correlated with increases in PFC and anterior cingulate metabolic activity.

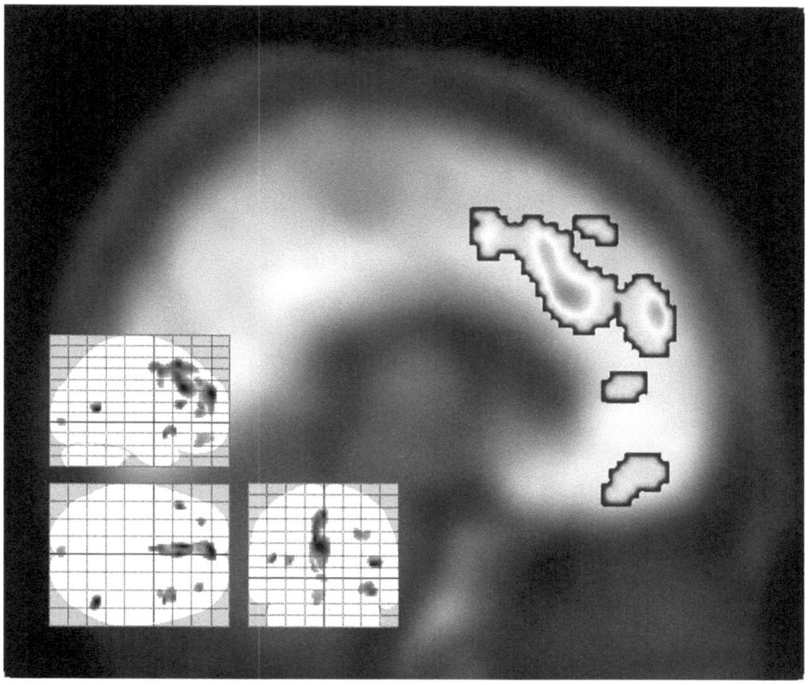

Figure 1 Inverse Correlation of 5D-ASC Global Scale scores and [^{18}F]altanserin apparent distribution volume [DV′]. Results of a voxel based correlation analysis (Δ 5D-ASC global vs ΔDV′, threshold $P < 0.005$, uncorrected) using Statistical Parametric Mapping (SPM2) (Hasler, Quednow, Vollenweider, unpublished data). To see the full color version of this figure, please refer to the color plate in the back of the book. Copies produced via our print on demand service do not contain color plates. If your copy does not have the color plate, please go to this website to view the figure in color www.elsevierdirect.com/companions/9780123746344

Comparison of hallucinogen effects and endogenous psychoses

As noted earlier, Beringer was the first investigator to propose that hallucinogens could be used to produce a 'model psychosis' in healthy humans (Beringer, 1923). Subsequent investigations confirmed that administration of mescaline, psilocybin and LSD induces states that resemble the symptoms of the earliest phases of schizophrenia (Rinkel *et al.*, 1952, 1955; Keeler, 1965; Bowers and Freedman, 1966). Indeed, the loss of control over thought processes that occurs after ingestion of psilocybin (Vollenweider *et al.*, 1997a) closely parallels acute psychotic decompensation (Keeler, 1965; Bowers and Freedman, 1966). Despite these similarities, Hollister (1962) and other clinicians have argued that there are notable differences between the effects of hallucinogens and the symptomatology of schizophrenia, leading them to question whether hallucinogen-induced psychedelic phenomena is a valid model for endogenous psychotic states. For example, Hollister noted that auditory but not visual hallucinations are prominent in schizophrenia, whereas changes of visual perception are a characteristic effect of hallucinogens. However, disturbances in visual perception, including hallucinations and synesthesias, do occur during the acute phase of schizophrenia (McCabe *et al.*, 1972; Freedman and Chapman, 1973). Hollister (1962) also argued that schizophrenics often display social and emotional withdrawal, but this effect is rarely observed after administration of serotonergic hallucinogens. There is evidence, however, that administration of hallucinogens, especially at high doses, can sometimes induce withdrawal and states resembling catatonia (Gouzoulis-Mayfrank *et al.*, 1998a).

In a study conducted by Gouzoulis-Mayfrank and colleagues (1998b), the symptoms of schizophrenia were assessed using the APZ rating scale. The goal of that investigation was to determine, using objective criteria, whether psychotic patients experience hallucinogen-like psychedelic effects. The study compared APZ scores from 50 healthy controls and 93 patients with acute schizophrenia, schizophreniform disorder or schizoaffective disorder. The APZ scores of psychotic patients were found to be significantly higher than those of controls. The study also examined whether the APZ scores correlate with scores on the Brief Psychiatric Rating Scale (BPRS), which measures positive symptoms and general psychopathology. Correlation analysis revealed that the OB subscale of the APZ correlates with BPRS factor 3 (reflecting most of the typical positive symptoms of schizophrenia), whereas the AED subscale correlates with BPRS factor 1 (reflecting anxiety and depression). These findings confirm that patients with acute schizophrenia experience hallucinogen-like effects, indicating that the syndrome induced by hallucinogens is a valid model of acute schizophrenia.

Animal models of hallucinogen effects relevant to schizophrenia

In laboratory animals, serotonergic hallucinogens have been shown to (1) potentiate neophobia (Tilson *et al.*, 1975; Adams and Geyer, 1982, 1985); (2) increase the responsiveness to sensory stimulation (Key, 1964; Geyer *et al.*, 1978; Geyer, 1998); and (3) retard habituation in a variety of input modalities and response output systems (Key, 1964; Geyer *et al.*, 1978; Geyer, 1998; Dulawa and Geyer, 2000; Geyer and Moghaddam, 2002). Given the similarities between the psychedelic state induced by hallucinogens and the symptoms of acute schizophrenia, there has been substantial interest in developing animal models of schizophrenia based on the acute behavioral effects of hallucinogens (Geyer and Vollenweider, 2008). Unfortunately, many of the unconditioned behaviors induced by hallucinogens in animals (e.g., head-twitch response, ear scratch) have no human counterpart, and thus it is not clear how these behaviors relate to the subjective effects of hallucinogens. However, hallucinogens produce effects on habituation and prepulse inhibition (PPI) of startle in animals that are analogous to hallucinogen effects in humans. Based partially on these cross-species similarities, the effects of hallucinogens on habituation and PPI have been proposed as potential behavioral models of schizophrenia (reviewed by Powell and Geyer, 2007). A brief description of these two behavioral models is provided below.

Habituation

Repeated presentation of irrelevant stimuli leads to a marked response decrement, a process known as habituation. Habituation is the simplest form of learning, and is necessary for selective attention. Deficits of attention and information-processing are core features of schizophrenia (Braff, 1985; Braff and Geyer, 1990). Patients with schizophrenia are often unable to filter out extraneous stimuli, leading to distractibility, sensory flooding and impaired cognition (McGhie and Chapman, 1961). Several studies have found that schizophrenic patients show deficits of startle reflex habituation, potentially contributing to the sensory overload and disorganized cognitive processes that occur in the disorder (e.g., Geyer and Braff, 1982, 1987; Bolino *et al.*, 1994; Parwani *et al.*, 2000; Ludewig *et al.*, 2003; Quednow *et al.*, 2006). An advantage of using habituation as a behavioral model is that similar testing procedures can be used to assess habituation in experimental animals and humans. For example, LSD and mescaline have been shown to decrease habituation to startling tactile stimuli in rats (Geyer *et al.*, 1978; Braff and Geyer, 1980), similar to the finding in patients with schizophrenia.

Prepulse inhibition

The PPI paradigm has been multiply applied in order to assess the loss of sensorimotor gating functions in schizophrenia. PPI refers to the fact that weak pre-stimuli presented at brief intervals (30–500 ms) prior to a startle-eliciting stimulus reduce (or gate) the amplitude of the startle response. Studies have consistently detected robust PPI deficits in schizophrenia patients (see, for example, Braff *et al.*, 1978; Braff and Geyer, 1990; Bolino *et al.*, 1994; Parwani *et al.*, 2000; Ludewig *et al.*, 2003; Quednow *et al.*, 2006). It was proposed that the mechanism underlying PPI regulates sensory input by filtering out irrelevant or distracting stimuli in order to prevent sensory information overflow and to allow for selective and efficient processing of relevant information (Swerdlow and Geyer, 1998). The consistently reported PPI deficits in schizophrenia patients contributed to the view that schizophrenia could be seen as gating- or filter-deficit disorder (Carlsson, 1995). As detailed in Chapter 4.7, hallucinogens such as LSD and DOI also disrupt PPI. Thus, the hallucinogen-treated animals tested in the PPI paradigm exhibit an increased or unfiltered responsiveness to sensory stimuli – that is, they fail to exhibit the gating or inhibition of the response normally produced by the prepulse stimulus. As reviewed elsewhere (Geyer *et al.*, 2001; Swerdlow *et al.*, 2001), this cross-species phenomenon of PPI is very robust, unlearned and ubiquitous. Indeed, depending on the testing parameters used, the hallucinogen psilocybin has been shown to produce PPI deficits in normal human volunteers (Vollenweider *et al.*, 2007). Hence, the ability of hallucinogens to alter PPI has been considered to be a useful model to study the positive symptoms of schizophrenia.

Serotonin receptor and transporter changes *in vivo* and *post-mortem* in schizophrenia

Early *post-mortem* studies with schizophrenia patients revealed that 5-HT and 5-hydroxyindoleacetic acid (5-HIAA) levels were increased in subcortical brain regions such as the putamen, nucleus accumbens and globus pallidus (Crow *et al.*, 1979; Farley *et al.*, 1980), and that 5-HIAA levels are decreased in cortical regions, including cingulate and frontal areas (Winblad *et al.*, 1979). Many subsequent studies investigated altered serotonin receptor and transporter expression in schizophrenia patients *in vivo* and *post-mortem*, especially with radiolabeled compounds. Most of these receptor investigations explored the 5-HT_{1A} or 5-HT_{2A} receptor density, usually with highly divergent results.

5-HT_{1A} receptors

Among the most consistent alterations of 5-HT parameters in schizophrenia, as identified in *post-mortem* studies, is an increase in the density of 5-HT_{1A} receptors in the PFC (Bantick *et al.*, 2001; Weinberger and Laruelle, 2002). Seven of ten studies – using either [^{3}H]8-hydroxy-2,3-(dipropylamino)-tetralin ([^{3}H]8-OH-DPAT) or the more specific compound [^{3}H]WAY-100653 for 5-HT_{1A} receptor binding and audioradiography, and/or analysis of receptor mRNA – have reported a 15–80 percent increase in 5-HT_{1A} receptor levels in the dorsolateral or orbital PFC, whereas other brain regions such as the anterior cingulate cortex and the temporal cortex have shown less consistent results, including possible increases (for review and citations, see Bantick *et al.*, 2001; Gray *et al.*, 2006). Interestingly, the increase in prefrontal 5-HT_{1A} receptor density was not necessarily accompanied by a change in 5-HT_{1A} receptor mRNA (Burnet *et al.*, 1996a). Moreover, the only study that stained for 5-HT_{1A}-like immunoreactivity did not find differences between schizophrenia patients and controls (Cruz *et al.*, 2004). Since receptors located at other cellular locations could not be visualized with the antibody used, changes in the overall density of the 5-HT_{1A} receptor could not be excluded by this study.

In contrast to the consistent findings in *post-mortem* studies, recent investigations of 5-HT_{1A} receptor distribution using [^{11}C]WAY-100653 PET in schizophrenia patients are contradictory. One study reported increased binding only in the medial temporal lobe (Tauscher *et al.*, 2002), another study described decreased binding in the amygdala (Yasuno *et al.*, 2004), whereas two further studies found no alterations in cortical 5-HT_{1A} receptor binding (Bantick *et al.*, 2004; Frankle *et al.*, 2006). Moreover, Bantick *et al.* (2004) found no differences in 5-HT_{1A} receptor binding between clozapine-treated patients, patients medicated with antipsychotics with low 5-HT_{1A} affinity, and healthy human volunteers. The authors concluded that clozapine did not occupy 5-HT_{1A} receptors at clinical doses. With respect to the inconsistencies between PET and *post-mortem* studies, Frankle *et al.* (2006) suggested that the alterations found in *post-mortem* studies cannot be reliably detected in PET studies, which puts into question whether 5-HT_{1A} receptors play a major role in the pathophysiology of schizophrenia. Due to the fact that schizophrenia patients included in *post-mortem* studies are rarely antipsychotic-naïve, whereas PET studies have mostly assessed in drug-naïve or unmedicated patients, it is likely that the 5-HT_{1A} receptor changes found in the majority of *post-mortem* studies are probably the result of chronic medication with antipsychotics or other psychotropics. However, in two of the *post-mortem* studies, 5-HT_{1A} receptor increases were seen also in drug-free patients (Hashimoto *et al.*, 1991; Sumiyoshi *et al.*, 1996).

In sum, findings with respect to 5-HT_{1A} receptor changes are highly contradictory. Whereas *post-mortem* studies consistently suggest an increase of 5-HT_{1A} receptor especially in the PFC, PET studies did not find changes of prefrontal receptor binding. Effects of chronic antipsychotic medication may contribute to these different results. When schizophrenia patients actually display frontal up-regulation in 5-HT_{1A} receptors, this might reflect an abnormal glutamatergic network because in the neocortex these receptors are mainly located on pyramidal cells (Bantick *et al.*, 2001).

5-HT_{2A} receptors

The 5-HT_{2A} receptor is the most intensively investigated 5-HT receptor in *post-mortem* schizophrenia studies in the last 30 years. Of 18 *post-mortem* studies, 14 reported finding decreased 5-HT_{2A} receptor binding/densities (or decreased 5-HT_{2A} receptor mRNA expression) in cortical areas, especially in the frontal cortex, of schizophrenia patients (for references and details, see Table 2). Two of the studies reported an increase in several brain regions, whereas the two remaining studies did not find 5-HT_{2A} receptor changes. Moreover, only five investigations explored 5-HT_{2A} receptors in the basal ganglia, but just one report suggested increased 5-HT_{2A} levels, whereas the other four studies found no changes. It should be noted that the radioligands that were used in these studies have high affinity for 5-HT_{2A} receptors, but they also label other receptor types. For example, ketanserin additionally labels α-adrenoreceptors, histamine H_1 receptors and vesicular monoamine transporters; LSD binds to 5-HT_{1A}, 5-HT_{1E}, 5-HT_{2C}, 5-HT_6, 5-HT_7 and dopamine-D_1 receptors; whereas spiperone also has high affinity for D_2 receptors (Harrison, 1999b). This lack of specificity must be taken into account when these studies are interpreted.

Legitimately, the question has been raised whether these receptor changes are simply the result of chronic drug treatment, because most of the patients studied were treated with antipsychotics for many years. Indeed, it was shown that long-term treatment with clozapine decreases 5-HT_{2A} receptor binding and mRNA expression in the cingulate and frontal cortex of rats. In contrast, haloperidol did not alter cortical 5-HT_{2A} receptor density or expression in the frontal cortex of rats (Reynolds *et al.*, 1983a; Wilmot and Szczepanik, 1989; O'Dell *et al.*, 1990; Burnet *et al.*, 1996b). Other atypical antipsychotics that are 5-HT_{2A} antagonists may also reduce cortical 5-HT_{2A} receptors when given chronically (Mikuni and Meltzer, 1984; Andree *et al.*, 1986; Padin *et al.*, 2006). However, particularly in the early studies, only a very small number of patients were treated with clozapine or other atypical substances. Additionally, antipsychotic medication may increase rather than decrease 5-HT_{2A} receptor expression (Hernandez and Sokolov, 2000), and many studies also found decreased 5-HT_{2A} receptor densities in unmedicated subjects, or did not detect dose effects of previous antipsychotic drug treatment (Table 2). Thus, the decrease of 5-HT_{2A} receptors especially in the dorsolateral PFC could not be explained only by chronic drug treatment; a pathological process also has to be involved (Dean, 2003).

PET studies applying 5-HT_{2A} receptor tracers to schizophrenia patients show controversial results. Three studies using $[^{18}F]$septoperone and one study using $[^{11}C]$*N*-methylspiperone did not show any significant differences in 5-HT_{2A} receptor densities between schizophrenia patients and controls, either with regions-of-interest (ROI)-based or voxel-based analyses (Trichard *et al.*, 1998; Lewis *et al.*, 1999; Okubo *et al.*, 2000; Verhoeff *et al.*, 2000). However, both tracers suffer from a relatively low affinity for 5-HT_{2A} receptors, and thus they have an insufficient signal-to-noise ratio in subcortical areas (Erritzoe *et al.*, 2008). In contrast, one study using $[^{18}F]$septoperone found decreased frontal 5-HT_{2A} receptor densities in antipsychotic-naïve schizophrenia patients (-16.3 percent; Ngan *et al.*, 2000), while a recently published study using the more selective 5-HT_{2A} antagonist $[^{18}F]$altanserin could not demonstrate a frontal decrease but rather an increase of 5-HT_{2A} receptors in the caudate in a similar patient sample (Erritzoe *et al.*, 2008). Two further studies investigated 5-HT_{2A} receptor density with $[^{18}F]$altanserin PET in subjects supposed to be in a prodromal state of schizophrenia and both reported decreased binding of the radio tracer in the PFC (Hurlemann *et al.*, 2005, 2008). In the later study, Hurlemann *et al.* (2008) additionally detected decreased 5-HT_{2A} receptor binding in the right insular cortex, the left amygdala, both hippocampi, the right caudate and the left putamen in never-medicated subjects in a late prodromal stage. Interestingly, a low 5-HT_{2A} receptor density in the right caudate predicted later conversion to full-blown psychosis, a finding that is highly discrepant with the results of Erritzoe *et al.* (2008). Taken together, in contrast to the consistency of the *post-mortem* findings,

Table 2 *Post-mortem* studies investigating 5-HT_{2A} receptor density in schizophrenia (modified and updated according to Harrison, 1999a)

Study	Method[1]	Brain region[2]	Cases/controls	Medicated cases	Main findings
Decrease in cortical binding					
Bennett *et al.* (1979)	HB with [^{3}H]LSD	BA 6, 8–11, 44–47	26/25[3]	18	↓ 40–50%, no effect of medication
Mita *et al.* (1986)	HB with [^{3}H]ketanserin	BA 9	11/9	7	↓ 36%, no effect of medication
Arora and Meltzer (1991)	HB with [^{3}H]spiperone	BA 8/9	11/11	11	↓ 33%, no effect of medication
Laruelle *et al.* (1993a)	HB with [^{3}H]ketanserin	BA 10, 17/18	10/12[4]	6	↓ 21% in BA 10, no effect of medication
Burnet *et al.* (1993b)	a) RA with [^{3}H]ketanserin	BA 17, 22, 46, MTL, AC	13/15	12	↓ 27% in BA 46, ↓ 38% MTL, similar trend in AC
	b) mRNA using ISH				↓ 49–63% in BA 17, 22, 46, AC, ↔ in MTL
Dean and Hayes (1996)	RA with [^{3}H]ketanserin	BA 8, 9, 10	20/20	19	↓ 25–33% in all frontal regions
Gurevich and Joyce (1997)	RA with [^{125}I]LSD	BA 1-3, 4, 6, 8, 9, 31, 32, 40, 44–46, AC, PC	10/12	5	↓ ~60% in BA 6, 24 in drug-free cases, ↓ ~70–90% in all brain regions in medicated cases
Kouzmenko *et al.* (1997)	RA with [^{3}H]ketanserin	BA 9/46	63/62[5]	60	↓ 33%
Dean *et al.* (1998)	RA with [^{3}H]ketanserin	BA 9	55/55	55	↓ 33%
Dean *et al.* (1999a)	RA with [^{3}H]ketanserin	BA 9	19/19	17	↓ 35%
Hernandez and Sokolov (2000)	mRNA using ISH	BA 9	21/14	18	↓ 60% in patients being drug free for > 26 weeks, antipsychotic treatment increased 5-HT2A mRNA
Pralong *et al.* (2000)	a) RA with [^{3}H]ketanserin	BA 22 (planum temporale)	20/20	17	↓ 32%
	b) HB with [^{3}H]ketanserin	BA 22 (planum temporale)	10/10	10	↓ 34% B_{max}, ↑ 119%, changes in affinity (K_d) but not density (B_{max}) explained by medication effects
Scarr *et al.* (2004)	RA with [^{3}H]ketanserin	MTL (only hippocampus)	20/20	20	↓ ~29–47% across different regions of hippocampus
Matsumoto *et al.* (2005)	RA with [^{3}H]ketanserin	BA 9, MTL	6/6	6	↓ 39% in BA 9, ↔ in MTL
Increase in cortical binding					
Whitaker *et al.* (1981)	HB with [^{3}H]LSD	BA 4, 10, 11	13/8	8	↔, ↑ 55% in unmedicated cases
Joyce *et al.* (1993)	RA with [^{125}I]LSD	BA 4, 9, 21, AC, PC, MTL	8/10	4	↑ ~50–100% only in MTL, BA 21, PC
No changes in cortical binding					
Reynolds *et al.* (1983b)	HB with [^{3}H]ketanserin	BA 10	11/10	11	↔

(Continued)

Table 2 (*Continued*)

Study	Method[1]	Brain region[2]	Cases/controls	Medicated cases	Main findings
Dean *et al.* (1996)	HB with [^{3}H]ketanserin	BA 9	20/20	19	↔
Increase in basal ganglia					
Joyce *et al.* (1993)	RA with [^{3}H]ketanserin	Caudate, putamen, NAC	8/10	4	↑ ~30–75%
No changes in basal ganglia					
Mackay *et al.* (1978)	HB with [^{3}H]spiperone	NAC	26/17	?	↔
Owen *et al.* (1981)	HB with [^{3}H]LSD, [^{3}H]5-HT	Caudate, putamen	19/20	?12	↔
Seeman *et al.* (1993)	HB with [^{3}H]ketanserin	Striatum	9/4	6	↔
Matsumoto *et al.* (2005)	RA with [^{3}H]ketanserin	Caudate, putamen	6/6	6	Not significant but strong trend for decrease (↓ 34%)

Notes:
[1] HB, homogenate binding; ISH, *in situ* hybridization; RA, receptor autoradiography.
[2] BA, Brodmann area; BA 4, motor cortex; BA 6, 8, 9, 10, 11, 44–47, prefrontal cortex; BA 17/18, occipital cortex; BA 21, 22, temporal cortex; AC, anterior cingulate cortex; PC, posterior cingulate cortex; MTL, mediotemporal lobe including hippocampus, amygdala, uncus, parahippocampal gyrus, entorhinal cortex; NAC, nucleus accumbens.
[3] Sum of three separate case–control groups. The decrease in [^{3}H]LSD binding was demonstrated in all three comparisons.
[4] Includes six subjects with schizoaffective disorder. Significant differences remained when these subjects were excluded.
[5] Included cases of Burnet *et al.* (1996b) and Dean and Hayes (1996).

the PET results are highly contradictory. Given that the methodological differences between the PET studies are not really obvious, further studies are needed to clarify whether 5-HT$_{2A}$ receptor changes could also be detected with an *in vivo* imaging approach. The new and highly selective 5-HT$_{2A}$ receptor radioligand [^{11}C]MDL 100,907 may be a promising tool to further investigate 5-HT$_{2A}$ receptor alterations in schizophrenia (Ito *et al.*, 1998).

Other serotonin receptors

Other serotonin receptor types have been investigated in *post-mortem* studies: two studies using [^{3}H]GR113808 autoradiography showed that the density of 5-HT$_4$ receptors is unaltered in either the dorsolateral PFC or the hippocampus of deceased schizophrenia patients when compared to control subjects (Dean *et al.*, 1999b; Scarr *et al.*, 2004). A *post-mortem* study investigating the concentration of 5-HT$_3$ receptors (which is the only ion channel in the 5-HT receptor family) in the amygdala of schizophrenia patients and controls with [^{3}H]LY278584 autoradiography did not find group differences (Abi-Dargham *et al.*, 1993). The 5-HT$_6$ receptor density measured with [^{125}I]SB-258585 in the frontal cortex was not changed in 20 schizophrenia patients compared to 17 control subjects (East *et al.*, 2002). Recently, two studies investigated the densities of 5-HT$_{1D}$ and 5-HT$_{1F}$ receptors in the dorsolateral PFC and the hippocampus, respectively, of schizophrenia patients using methiothepin-sensitive and -insensitive [^{3}H]sumatriptane autoradiography (Scarr *et al.*, 2004; Dean *et al.*, 2006). While Scarr *et al.* (2004) reported a decrease of 5-HT$_{1F}$ but unaltered 5-HT$_{1D}$ receptors in the hippocampus, Dean and colleagues did not find any changes in HT$_{1F}$ and 5-HT$_{1D}$ receptors in the dorsolateral PFC of the patients. In the same study, however, Dean *et al.* (2006) found decreased 5-HT$_7$ receptor levels in dorsolateral PFC of schizophrenia patients using [^{3}H]SB-269970. By contrast, haloperidol treatment increased the number of 5-HT$_7$ receptors in the cortex of rats (Dean *et al.*, 2006). The authors therefore concluded that 5-HT$_7$ receptors are possibly involved in the pathological processes of schizophrenia and that appropriate 5-HT$_7$ receptor levels may be critical for normal cortical development. These recent findings regarding alterations of HT$_{1F}$ and 5-HT$_7$ receptors in schizophrenia need confirmation by further *post-mortem* and, if possible, PET studies.

Serotonin transporter (SERT)

Serotonin transporters (SERT) are located presynaptically on serotonergic axon terminals and are believed to serve as an index of serotonergic innervation (Abi-Dargham and Krystal, 2000). Two *post-mortem* studies applying [^{3}H]cyano-imipramine and [^{3}H]paroxetine initially showed that the density of SERT was decreased in the frontal cortex (Joyce *et al.*, 1993; Laruelle *et al.*, 1993a). In the study of Joyce and colleagues (1993), SERT was also decreased in the anterior and posterior cingulate cortex of schizophrenia patients, but increased in the striatum. On the contrary, later studies using radiolabeled serotonin reuptake inhibitors (SRIs), such as [^{3}H]paroxetine, [^{3}H]citalopram or [^{125}I]RTI-55, did not demonstrate any alterations in the SERT density in several brain regions, including the PFC or the cingulate cortex, in schizophrenia patients (Dean *et al.*, 1995, 1999b; Naylor *et al.*, 1996; Gurevich and Joyce, 1997). In fact, three of these studies did not report altered SERT densities but did find a decrease in the affinity of [^{3}H]paroxetine for SERT in hippocampal membranes, whereas the affinity of [^{3}H]paroxetine for SERT binding in the frontal cortex was unaltered (Dean *et al.*, 1995, 1996; Naylor *et al.*, 1996). Gurevich and Joyce (1997) concluded that the initial positive findings were probably confounded by the large number of schizophrenia patients in the samples who committed suicide. The same may be true for the finding of decreased SRI affinity in hippocampal SERT, as Dean and colleagues showed that the effect was more pronounced in schizophrenia patients who committed suicide (Dean *et al.*, 2006).

Examining the expression of SERT mRNA, Hernandez and Sokolov (1997) found a four-fold increase in the level of SERT mRNA in the dorsolateral PFC, but a two-fold decrease in the temporolateral cortex, of schizophrenics. However, since these changes were strongly correlated with previous antipsychotic drug treatment, they cannot be attributed to the illness process.

A SPECT study using [^{123}I]RTI-55 could not detect any differences in SERT concentration in midbrain areas of schizophrenia patients (Laruelle *et al.*, 2000). However, [^{123}I]RTI-55 is not specific for SERT but also labels the dopamine transporter (DAT) (Neumeyer *et al.*, 1991). In addition, [^{123}I]RTI-55 does not permit measurement of SERT availability in regions other than the midbrain (Laruelle *et al.*, 1993b). Recently, Frankle *et al.* (2005) also failed to detect any differences in SERT binding between schizophrenia patients and controls when using the more specific radiotracer [^{11}C]DASB. However, [^{11}C]DASB does not have a good signal-to-noise ratio when assessing regions with low SERT density such as the neocortex (Frankle *et al.*, 2005), complicating a possible detection of group differences in, for example, the frontal cortex. Thus, taking all these findings into account, it is unlikely that SERT plays an important role in the pathophysiology of schizophrenia.

Genetic association studies regarding schizophrenia and serotonin

Given that, to date, more than 1400 association studies searching for potential genetic risk factors have been published with largely inconsistent results, a regularly updated online database ('SzGene') including meta-analyses of all published genetic studies for schizophrenia has recently been established (Allen *et al.*, 2008; www.schizophreniaforum.org). Single nucleotide polymorphisms (SNPs) having genotype data available in at least four independent case–control samples were included in random-effect meta-analyses using allelic contrasts. In the ranking of the meta-analyses showing the strongest effect sizes, two serotonin-related SNPs are currently placed in the top 20 (status 30 September, 2009): the tryptophan hydroxylase 1 (TPH1) A218C polymorphism ranked eleventh (odds ratio (OR) = 1.25), and the 5-HT$_{2A}$ A-1438G polymorphism ranked sixteenth (OR = 1.16). Table 3 displays all serotonin SNPs for which meta-analyses were done. For comparison, the strongest effect so far was shown for the A2897G polymorphism of the disrupted-in-schizophrenia gene 1 (DISC1; OR = 1.80, confidence interval (CI) = 1.2–2.68). The second ranking belongs to the A277C polymorphism of the vesicular monoamine transporter 1 (VMAT1; OR = 1.63, CI = 1.03–2.57). VMAT1 is involved in the intracellular transport of all monoamines including serotonin. The rs6556547 SNP of the GABA-A beta-2 receptor (OR = 0.70, CI = 0.52–0.95), is ranked third. These ORs reported for schizophrenia susceptibility genes are comparable with those found in other genetically complex neuropsychiatric diseases such as Alzheimer's disease (Allen *et al.*, 2008).

TPH1 and TPH2 are rate-limiting enzymes in 5-HT synthesis, but while TPH1 is primarily expressed in peripheral regions such as the pineal gland and enterochromaffin cells of the gut, TPH2 is expressed predominantly in serotonergic neurons of the raphe nuclei (Zhang *et al.*, 2006). Moreover, the TPH1 A779C and A218C SNPs are intronic (non-coding), and alternative mechanisms probably providing gene expression from intronic sequences such as splicing and exon skipping have been ruled out (Shaltiel *et al.*, 2005). For these reasons, the positive associations between the TPH1 A218C polymorphism and schizophrenia were strongly criticized (Reuter *et al.*, 2007). However, in a *post-mortem* study, Zill *et al.* (2007) recently demonstrated the expression of TPH1 mRNA in several brain regions. Therefore, it seems to be largely unclear to date which role TPH1 plays in cerebral function and schizophrenia. Most likely, the intronic TPH1 A218C SNP is in linkage disequilibrium with other functional as yet unidentified gene variations. Variations of the TPH2 gene have not been positively linked to

Table 3 Meta-analyses of association studies between polymorphisms of the serotonin system and schizophrenia*

Gene	Substrate	SNP	Chromosome (location)	Synonymy	Number of studies	Number of patients	Number of controls	Minor allele (% frequency in controls)	Risk allele	Odds ratio (all studies)	95% confidence interval
5-HT1B	Receptor	rs6296 (G861C)	6 (6q13)	Synonymous	4	763	1123	C (38%)	C	0.95	0.75–1.20
5-HT2A	Receptor	rs6311 (A-1438G)	13 (13q14-q21)	Synonymous	8	2678	2964	A (42%)	G	1.16	1.01–1.33
		rs6313 (T102C)	13 (13q14-q21)	Synonymous	46	9369	10076	T (47%)	T	0.96	0.89–1.04
		rs6314 (His452Tyr)	13 (13q14-q21)	Non-synonymous	5	2706	2878	T (9%)	T	0.96	0.79–1.17
5-HT6	Receptor	rs1805054 (C267T)	1 (1p36-p35)	Synonymous	4	530	519	T (25%)	–	1.00	0.73–1.37
SERT	Transporter	5-HTTVNTR	17 (17q11.1-q12)	Synonymous	13	2629	3042	10 (30%)	10	0.90	0.78–1.03
		5-HTTLPR	17 (17q11.1-q12)	Synonymous	23	3861	4998	L (48%)	L	1.03	0.97–1.10
TPH1	Enzyme	rs1800532 (A218C)	11 (11p13.3-p14)	Synonymous	6	1239	1708	A (45%)	C	1.25	1.08–1.44
		rs1799913 (A779C)	11 (11p13.3-p14)	Synonymous	5	653	994	A (48%)	C	0.91	0.78–1.06

Published on the SchizophreniaGene database (www.schizophreniaforum.org/res/sczgene/default.asp, status 30 September 2009.
Meta-analyses were done for SNPs with a minor allele frequency >1%, and when more than three independent case–control samples were available (Allen *et al.*, 2008).
Abbreviations: SNP, single nucleotide polymorphisms; SERT, serotonin transporter; TPH1, tryptophan hydroxylase 1.

schizophrenia so far (De Luca *et al.*, 2005; Higashi *et al.*, 2007).

The positive association findings regarding the 5-HT_{2A} A-1438G polymorphism once more underscore the significance of this receptor for schizophrenia (Table 3). The A-1438G polymorphism is silent and does not result in an alteration of the amino acid sequence of the 5-HT_{2A} receptor, but is located within the promoter region of the 5-HT_{2A} receptor gene (Spurlock *et al.*, 1998), and thus it has been proposed that the A-1438G polymorphism alters promoter activity and expression of 5-HT_{2A} receptors (Parsons *et al.*, 2004). The 5-HT_{2A} A-1438G and T102G receptor polymorphisms are usually in perfect linkage disequilibrium. We have recently shown that the sensorimotor gating deficits of schizophrenia patients – which are seen as a promising endophenotype of schizophrenia (Gottesman and Gould, 2003) – are strongly modulated by the 5-HT_{2A} A-1438G and T102C receptor SNPs that were completely linked in our sample (Quednow *et al.*, 2008). In accordance with the genetic association studies, carriers of the high-risk G and C alleles displayed diminished sensorimotor gating. We have just replicated this finding in a sample of 94 normal subjects (Quednow *et al.*, 2009). In conclusion, the C allele of the T102C variation and the G allele of the A-1438G variation may cause lower 5-HT_{2A} receptor densities in some brain areas, which may lead to a less flexible serotonin system and worse dopaminergic modulation (Serretti *et al.*, 2007).

The SzGene meta-analyses regarding 5-HT_{1B} G861C, 5-HT_{2A} His452Tyr and 5-HT_6 C267T receptor SNPs, and the well-known SERT polymorphisms, suggest rather weak or no associations with schizophrenia (Table 3).

Currently, there are also some new interesting data coming from single association studies that still have to be replicated. Huang *et al.* (2004) reported an increased frequency of the G allele of the 5-HT_{1A} C-1019G promoter polymorphism in schizophrenia patients. The occurrence of the G allele is associated with enhanced gene expression (Lemonde *et al.*, 2003), which would fit with the *post-mortem* data on increased 5-HT_{1A} receptor levels in the PFC of schizophrenics. However, in a large *post-mortem* autoradiographic study, Huang *et al.* (2004) could not demonstrate differences in 5-HT_{1A} receptor binding between 5-HT_{1A} C-1019G genotypes in suicidal, depressed and healthy subjects. Moreover, significant associations were reported for the 5-HT_{2C} G68C (Segman *et al.*, 2000), a 5-HT_4 haplotype (Suzuki *et al.*, 2003), and several 5-HT_{5A} (Birkett *et al.*, 2000; Iwata *et al.*, 2001;

Dubertret *et al.*, 2004) and two 5-HT$_7$ receptor polymorphisms (Ikeda *et al.*, 2006). There have also, however, been some negative reports for 5-HT$_{2C}$ G68C (Segman *et al.*, 1997; Semwal *et al.*, 2002) and the 5-HT$_7$ receptor polymorphisms (Fallin *et al.*, 2005). Further studies are needed to finally assess the sustainability of these initial findings.

Not only schizophrenia itself, but also drug response or risk for developing side effects such as tardive dyskinesia or weight gain, were the subject of pharmacogenetic studies of mutations in the serotonin system (for a comprehensive review, see Arranz and de Leon, 2007). The aim of these studies was the development of genetic predictors for treatment response and side effects to preselect and therefore improve antipsychotic treatment. The functional C-1019G variant of the 5-HT$_{1A}$ receptor was recently demonstrated to influence the response to atypical antipsychotics (Reynolds *et al.*, 2006; Wang *et al.*, 2008; Mossner *et al.*, 2009). These studies have consistently shown that carriers of the C allele show greater improvement, especially in regard to negative symptoms. A number of studies reported significant associations between the linked 5-HT$_{2A}$ A-1438G and T102C receptor SNPs and response to atypical antipsychotics (mostly clozapine), but there are some discrepant results with respect to the risk alleles for non-responding (Arranz *et al.*, 1995, 1998a, 1998b; Joober *et al.*, 1999; Yu *et al.*, 2001; Lane *et al.*, 2002; Hamdani *et al.*, 2005; see also Arranz and de Leon, 2007). In addition, there are several studies reporting that there is no association between the T102C substitution and the therapeutic response to clozapine and other antipsychotics (Masellis *et al.*, 1995; Nothen *et al.*, 1995; Jonsson *et al.*, 1996; Malhotra *et al.*, 1996; Lin *et al.*, 1999). A more consistent picture comes from studies showing an influence of the functional His452Tyr polymorphism of the 5-HT$_{2A}$ receptor on the clozapine response (Arranz *et al.*, 1996; Masellis *et al.*, 1998). Although several other studies failed to detect an association of this SNP and treatment response (Masellis *et al.*, 1995; Nothen *et al.*, 1995; Jonsson *et al.*, 1996; Malhotra *et al.*, 1996; Lin *et al.*, 1999), a recent meta-analysis has shown a relatively strong association of the Tyr variant with poor response to clozapine (Arranz *et al.*, 1998b). Mutations in the promoter (VNTR, T-759C and G-995A) and coding region (Cys23Ser) of the 5-HT$_{2C}$ receptor have also been associated with clozapine response and improvement in negative symptoms (Sodhi *et al.*, 1995; Arranz *et al.*, 2000a; Reynolds *et al.*, 2005). A further study was unable to detect an effect of the Cys23Ser SNP on treatment response (Rietschel *et al.*, 1997). In sum, the contribution of the variants to general drug response is relatively moderate, possibly indicating contribution to specific symptoms or side effects that need further investigation (Arranz and de Leon, 2007). There are also some indices for an association of the 5-HT$_6$ T-267C receptor variant with the response to treatment with clozapine and risperidone in Chinese patients (Yu *et al.*, 1999; Lane *et al.*, 2004), although this association has not been replicated in US patients (Masellis *et al.*, 2001).

Additionally, SERT polymorphisms have been investigated in respect to drug response but, with the exception of an initial positive finding (Arranz *et al.*, 2000b), all further reports did not find significant associations (Arranz *et al.*, 2000a; Tsai *et al.*, 2000; Kaiser *et al.*, 2001). Through combining the data of gene variants previously associated with clozapine response, Arranz *et al.* (2000b) found that a combination of SNPs in the genes coding for 5-HT$_{2A}$, 5-HT$_{2C}$, histamine H_2 receptors and the SERT resulted in the correct prediction of response in 76 percent of cases. However, so far this finding has not replicated (Schumacher *et al.*, 2000).

The findings with respect to the association of 5-HT$_{2A}$ polymorphisms and tardive dyskinesia are controversial (Basile *et al.*, 2001; Segman *et al.*, 2001; Tan *et al.*, 2001; Lattuada *et al.*, 2004; Deshpande *et al.*, 2005; Lerer *et al.*, 2005). The same is true for 5-HT$_{2C}$ polymorphisms (Rietschel *et al.*, 1997; Zhang *et al.*, 2002; Deshpande *et al.*, 2005). Two recent studies reported promising results regarding a possible interaction of the 5-HT$_{2A}$, 5-HT$_{2C}$ and D_3 receptor genotype with respect to the risk of developing tardive dyskinesia under antipsychotic treatment (Segman *et al.*, 2000; Segman and Lerer, 2002).

The most significant results associated the 5-HT$_{2C}$ T-759C receptor variant with antipsychotic-induced weight gain (Arranz and de Leon, 2007). Although a number of studies failed to replicate this finding (Hong *et al.*, 2001; Basile *et al.*, 2002; Tsai *et al.*, 2002; Theisen *et al.*, 2004), the evidence for a protective effect of the T allele is convincing (Reynolds *et al.*, 2002b, 2003; Buckland *et al.*, 2005; Ellingrod *et al.*, 2005; Miller *et al.*, 2005; Templeman *et al.*, 2005). Given the strength of the reported associations, this could be a discovery with a useful clinical application as a predictor of drug-induced weight gain (Arranz and de Leon, 2007).

In summary, genetic studies imply a critical role of mainly 5-HT$_{1A}$ and 5-HT$_{2A}$ receptors in the pathophysiology of schizophrenia and drug-response to antipsychotics.

Serotonergic mechanisms of atypical antipsychotics

5-HT$_{2A}$ receptor antagonism

For a long time, one of the most important arguments for an involvement of the serotonin system in the etiology of schizophrenia was the serotonergic action of most of the so-called atypical antipsychotics. Due to clinical experiences with the early antipsychotics, neuropharmacologists initially believed that extrapyramidal side effects (EPS) are an essential part of the antipsychotic effectiveness. The antiquated term 'neuroleptics' ('seize the neuron') still refers to this association (Lidow, 2000). The reason for the positive correlation of antipsychotic effectiveness and EPS is that the antipsychotic potency of the early neuroleptics is proportional to their ability to block striatal D_2 receptors, which is also the cause for EPS (Seeman *et al.*, 1976). As a consequence, D_2 receptor blockade was proposed to be the principal mechanism of action of neuroleptics known until then (Creese *et al.*, 1976), such as the phenothiazines chlorpromazine, perphenazine, fluphenazine and thioridazine, the thioxanthenes thiothixene and flupentixol, and the butyrophenone haloperidol (which is still the most widely used neuroleptic drug). However, the dibenzodiazepine clozapine broke these rules, because its therapeutic effectiveness was not paired with notable EPS. Therefore, clozapine was described as an 'atypical' antipsychotic.[2] Unfortunately, the conditions for atypicality are not well defined. The narrowest definition is that atypical drugs produce lower EPS than typical drugs. However, in the past two decades several further prerequisites have been proposed – that atypical drugs should have (1) a lower capacity to elevate prolactin levels and more strongly ameliorate negative and cognitive symptoms of schizophrenia compared to typical substances; (2) a multi-receptor profile, and higher *in vivo* selectivity for corticolimbic D_2 receptors compared to striatal D_2 receptors; and (3) a higher D_4 receptor affinity, and a serotonergic component or a higher affinity for 5-HT_{2A} receptors than for D_2 receptors (Meltzer, 1991, 1999; Blin, 1999; Lidow, 2000; Seeman, 2002). It becomes clear that, in the end, all these definitions are derived from the multiple mechanisms of action of clozapine, and thus only clozapine itself matches all of these criteria, reducing the concept of atypicality to absurdity.

Although clozapine is still the gold standard regarding antipsychotic effectiveness, it has a not especially rare (0.5–2 percent) and potentially life-threatening side effect: agranulocytosis (Buchanan, 1995). Therefore, scientists have aimed to develop novel antipsychotics having the antipsychotic potency but not the dangerous side effects of clozapine. Given that the superior efficacy of clozapine had been attributed to its high 5-HT_{2A} receptor selectivity relative to the D_2 receptor (Meltzer *et al.*, 1989; Meltzer, 1991), the development of 'balanced' $5\text{-HT}_{2A}/D_2$ antagonists as potential antipsychotics was initiated in the late 1980s (Abi-Dargham and Krystal, 2000). This approach led to the discovery of novel antipsychotic substances, such as risperidone, olanzapine, quetiapine, ziprasidone and sertindole. All these compounds have higher affinity for the 5-HT_{2A} receptor than for the D_2 receptor, even if none of them shows as high a $D_2/5\text{-HT}_{2A}$-binding ratio as clozapine (only the dibenzoxazepine amoxapine has a higher ratio than clozapine) (Seeman, 2002). As a consequence, Meltzer (1999) proposed that atypical antipsychotics with a high $D_2/5\text{-HT}_{2A}$-binding ratio are more effective against negative symptoms, show a stronger improvement of cognitive functions, and cause less EPS than typical antipsychotics. Several clinical trials have shown that atypical antipsychotics with strong 5-HT_{2A} antagonism – first and foremost clozapine – improve negative symptoms more efficaciously than typical compounds (see, for example, Kane *et al.*, 1988; Marder and Meibach, 1994; Moller *et al.*, 1995; Tollefson and Sanger, 1997). However, meta-analyses revealed rather moderate advantages of atypical antipsychotics in the treatment of negative symptoms (Carman *et al.*, 1995; Leucht *et al.*, 1999, 2009). Some scientists argued that these beneficial effects are only related to the improvement of secondary negative symptoms, which are correlated with the improvement of positive symptoms, depressive symptoms, EPS or environmental deprivation, but that primary negative symptoms (also called as the 'deficit syndrome') are still unaffected by atypicals (Carpenter *et al.*, 1995; Buchanan *et al.*, 1998; Lidow, 2000). Moreover, the view that a 5-HT_{2A} receptor blockade is probably not necessary to improve negative symptoms is supported by large meta-analyses showing that amisulpride – an atypical antipsychotic that is a selective D_2/D_3 receptor antagonist – has a comparable efficacy to clozapine with regard to negative symptoms (Leucht *et al.*, 2002, 2009).

Cognitive dysfunctions constitute core symptoms of schizophrenia, and improvement of cognitive function is highly relevant for functional outcome such as social and occupational functioning (Green, 1996; Liddle, 2000). Many studies have shown that when compared to haloperidol, the atypicals clozapine, risperidone and olanzapine differently improved functioning in several cognitive domains, including semantic memory, verbal learning and memory, sustained attention, and working memory (Kern *et al.*, 1999; Meltzer and McGurk, 1999; Purdon *et al.*, 2000; Bilder *et al.*, 2002). However, most of these clinical trials did not use a control group or did not measure the control groups repeatedly. Meanwhile, recent data suggest that the measured cognitive improvements are only in the range of the expected test–retest enhancement (Goldberg *et al.*, 2007; Quednow and Wagner, unpublished data).

[2] Second- or new-generation antipsychotics, multireceptor antipsychotics, or modern antipsychotics are often used (but not necessarily better) synonyms for atypical antipsychotics. Typical antipsychotics are also termed as classical or first-generation antipsychotics or neuroleptics, respectively.

Additionally, the large ($n = 817$) CATIE (Clinical Antipsychotic Trials of Intervention Effectiveness) schizophrenia trial funded by the National Institute of Mental Health (NIMH) has recently shown only small effects of several atypical drugs on neurocognitive composite scores after 2, 6 and 18 months of continued treatment (Keefe *et al.*, 2007). After 2 months, treatment with the atypicals ziprasidone ($z = 0.12$), olanzapine (0.13), quetiapine (0.18) and risperidone (0.26), as well as the typical antipsychotic perphenazine (0.25) (where z is the least-squares mean improvement in the neurocognitive composite score), resulted in only small but significant neurocognitive improvements, with no significant differences between treatment groups. In contrast, after 18 months of treatment, neurocognitive enhancement was significantly greater in the perphenazine group than in the olanzapine and risperidone groups, despite the fact that perphenazine is a stronger D_2 receptor than a 5-HT_{2A} receptor blocker. This is in line with our previous data that treatment with the selective D_2/D_3 blocker amisulpride resulted in greater improvement of all cognitive domains (attention, executive function, working memory and declarative memory) in schizophrenia patients compared to the clozapine-copy olanzapine (Wagner *et al.*, 2005). These data strongly call the following two hypotheses into question: (1) that atypical antipsychotics improve cognitive deficits beyond simple test–retest effects; and (2) that 5-HT_{2A} receptor blockade is necessary for the cognition enhancing effects of atypical substances.

A recent meta-analysis showed that clozapine is still the antipsychotic drug with the lowest risk to produce EPS (measured by the amount of antiparkinsonian medication), followed by sertindole and olanzapine (Leucht *et al.*, 2009). Several suggestions have been made to explain the low probability of EPS under clozapine treatment. The anticholinergic properties, the lack of ability to increase acetylcholine in the striatum, D_1 and D_4 receptor blockade, α_1- or α_2-adrenoreceptor antagonism, and the 5-HT_{2A} receptor antagonism of clozapine have been proposed to reduce the risk of EPS. Data from animal models of schizophrenia as well as clinical data suggest that a high 5-HT_{2A} receptor blockade in combination with a low D_2 receptor blockade may help to avoid EPS, whereas the D_1 receptor did not play a meaningful role (Meltzer, 1999; Roth and Meltzer, 2000). Given that many atypical compounds still induce EPS if higher doses are given, 5-HT_{2A} blockade may not be sufficient for the reduction of EPS in the presence of complete or near-complete D_2 blockade. However, 5-HT_{2A} antagonism may reduce the risk for EPS when D_2 receptors are not completely saturated (Abi-Dargham and Krystal, 2000).

Animal studies first indicated that selective 5-HT_{2A} receptor antagonists lacking a dopaminergic component may have antipsychotic properties (Geyer *et al.*, 2001). The selective 5-HT_{2A} receptor blocker MDL 100,907 was the first compound whose antipsychotic activity was exclusively predicted by preclinical animal models (Varty *et al.*, 1999). In a subsequent clinical trial MDL 100,907 was not sufficiently more effective than haloperidol in the treatment of schizophrenia, although it was more effective than placebo in reducing psychotic symptoms (de Paulis, 2001). Nevertheless, so far there is no efficacious and approved antipsychotic without a dopaminergic mechanism of action. Moreover, the mechanism underlying the therapeutic superiority of clozapine is still unclear. One alternative hypothesis for atypicality focused on the special kinetics of interaction with the D_2 receptor displayed by atypical drugs and, therefore, negating the role of 5-HT_{2A} receptors (Kapur and Seeman, 2001). Data reported in Seeman (2002) are interpreted to suggest that most of the atypical drugs dissociate much faster from D_2 receptors than do typical compounds. The dibenzapines clozapine and quetiapine and the benzamides amisulpride and remoxipride show the fastest dissociation from the D_2 receptor. Seeman (2002) concluded that transient occupation of D_2 receptors allows relatively normal dopamine neurotransmission, which is likely to be a prerequisite for normal prolactin levels, intact cognition and avoidance of EPS. This 'fast-off-D_2' theory was strongly criticized because it applies only to clozapine and quetiapine and is inconsistent with the relatively slow dissociation of several atypicals, including olanzapine, risperidone, ziprasidone and sertindole (Meltzer *et al.*, 2003). However, so far there is no other theory that explains the high antipsychotic efficacy of both clozapine and amisulpride.

Schizophrenia is most likely not a homogeneous entity of an illness, but rather a cluster of diverse schizophreniform diseases with different pathogeneses. Thus, some patients may have more benefit from a serotonergic compound than others. However, to date there are no criteria to safely predict the response to treatment with either antipsychotic.

Role of other 5-HT receptors

Most of the atypical antipsychotics have affinities for multiple 5-HT receptors (Table 4). This section briefly discusses the interaction of antipsychotics with 5-HT_{1A}, 5-HT_{2C}, 5-HT_3, 5-HT_4, 5-HT_6 and 5-HT_7 receptors.

Numerous antipsychotics display activity at human 5-HT_{1A} receptors – aripiprazole, clozapine, quetiapine and ziprasidone display marked affinity and act as agonists or partial agonists, whereas risperidone and sertindole display low affinity and act as antagonists. Moreover, several of the typical compounds, including haloperidol and chlorpromazine, also exhibit relatively low affinity and antagonistic properties at 5-HT_{1A} receptors (Newman-Tancredi *et al.*, 1998; Shapiro *et al.*, 2003). Thus, a specific 5-HT_{1A} action is likely not necessary for antipsychotic activity. However, a 5-HT_{1A} agonist activity was proposed to enhance memory and cognition in schizophrenia because it was shown that (1) 5-HT_{1A} receptors are concentrated in brain regions thought to mediate several cognitive functions (e.g., hippocampus, thalamus, cingulate cortex and PFC) (Roth *et al.*, 2004), and (2) clozapine increases dopamine release in the PFC via its 5-HT_{1A} agonism (Rollema *et al.*, 1997). In support of this hypothesis, Sumiyoshi and colleagues reported that chronic administration of the selective 5-HT_{1A} receptor agonist tandospirone as a co-therapy with typical antipsychotics enhanced verbal memory and executive functions in schizophrenia patients (Sumiyoshi *et al.*, 2001a, 2001b). In contrast, chronic co-administration of the 5-HT_{1A} receptor partial agonist buspirone with atypical antipsychotics improved psychomotor speed but not memory or executive function in schizophrenia patients (Sumiyoshi *et al.*, 2007). On the other hand, tandospirone exerted negative effects on memory function in demented patients (Yasuno *et al.*, 2003), and the potent 5-HT_{1A} agonist NAE-086 induced hallucinations and nightmares in healthy volunteers after repeated doses (Renyi *et al.*, 2001). Thus, augmentation with a 5-HT_{1A} partial agonist for cognitive enhancement in schizophrenia seems only to be effective in combination with antipsychotics that lack 5-HT_{1A} activity. Contrarily, atypical antipsychotics with a 5-HT_{1A} agonistic action should not be combined with tandospirone or buspirone because this may worsen psychotic symptoms and has no additional effects on cognition (Roth *et al.*, 2004).

In addition, clozapine has higher affinity for the 5-HT_{2C} receptor than for the 5-HT_{2A} receptor. Animal studies first suggested that activation of 5-HT_{2C} receptors is inhibitory, while activation of 5-HT_{2A} receptors is stimulatory (Martin *et al.*, 1997, 1998). This led to the conclusion that 5-HT_{2C} receptor agonists might be antipsychotic (Abi-Dargham and Krystal, 2000). Newer data have shown that 5-HT_{2C} receptor antagonists can directly increase dopamine release in the nucleus accumbens and PFC (Di Matteo *et al.*, 1998), while 5-HT_{2C} receptor agonists markedly decrease dopamine and noradrenalin levels in the frontal cortex of rats (Millan *et al.*, 1998). Administration of the $5\text{-HT}_{2C/2A}$ agonist m-CPP caused deterioration of positive psychotic symptoms in schizophrenia, an effect that could be prevented by the $5\text{-HT}_{2C/2A}$ blocker ritanserin (Abi-Saab *et al.*, 2002). Moreover, ritanserin in combination with risperidone showed significant superiority over risperidone alone in decreasing negative symptoms in schizophrenia patients (Akhondzadeh *et al.*, 2008). These results suggest that 5-HT_{2C} blockade may actually have beneficial effects on positive, negative and cognitive symptoms in schizophrenia (Meltzer *et al.*, 2003). In contrast, earlier work demonstrated that affinity to the 5-HT_{2C} receptor did not distinguish typical from atypical antipsychotics (Roth *et al.*, 1992), and Meltzer *et al.* (2003) concluded that the high 5-HT_{2C} receptor affinity of some atypical substances (e.g., clozapine, olanzapine, sertindole) roughly corresponds with their potential to produce weight gain rather than with potential antipsychotic activity. However, some

Table 4 Affinities of selected antipsychotic drugs for 5-HT receptors expressed as pK_i (the negative logarithm to base 10 of the equilibrium dissociation constant, K_i, in molar concentration units)*

Drug	5-HT_{1A}	5-HT_{1B}	5-HT_{1D}	5-HT_{1E}	5-HT_{1F}	5-HT_{2A}	5-HT_{2B}	5-HT_{2C}	5-HT_{5A}	5-HT_{6}	5-HT_{7}
Aripiprazol	8.2^{ag}	6.1^{ag}	7.2^{ag}			$7.5–8.1^{ag}$		7.6^{ag}			
Chlorpromazine	6.2^{ant}					8.1^{iag}		$7.6–8.2^{ant}$		$7.7–7.8^{iag}$	7.6^{iag}
Clozapine	$6.8–6.9^{ag}$	6.2^{ag}	6.4^{ag}	6.4^{ag}	6.9^{ag}	$7.6–9.0^{iag}$	$8.0–8.8^{ant}$	$7.4–8.7^{iag}$	$6.0–6.5^{ant}$	$7.8–8.1^{iag}$	$7.2–7.8^{iag}$
Haloperidol	$5.7–5.8^{ant}$		6.6^{ant}			$6.7–7.3^{ant}$	$5.8–6.4^{ant}$				$6.3–6.6^{ant}$
Olanzapine	$5.6–5.8^{ag}$	6.3^{ag}	6.2^{ag}	5.7^{ag}	6.5^{ag}	$8.6–8.7^{ant}$		$8.1–8.2^{iag}$		8^{iag}	6.5^{ant}
Perphenazine						8.2^{ant}		6.9^{ant}		7.1^{iag}	7.2^{iag}
Quetiapine	$6.5–6.6^{ag}$		5.7^{ag}	5.9^{ag}	5.6^{ag}	6.4–7.0					
Risperidone	$6.4–6.5^{ant}$	$6.6–7.0^{ant}$	$7.8 – 8.0^{ant}$	5.9^{ant}	5.9^{ant}	$9.3–10.0^{iag}$		$7.5–7.6^{iag}$		5.6^{ant}	$8.3–8.7^{iag}$
Sertindole	$6.4–6.6^{ant}$	7^{ant}	7.2^{ant}	6.4^{ant}	6.4^{ant}	$9.2–9.4^{ant}$		$9.0–9.2^{iag}$			
Ziprasidone	$7.9–8.9^{pag}$	8.3^{ag}	9^{ag}	6.4^{ag}		$8.8–9.5^{ant}$		$7.9–8.4^{iag}$			8.4^{iag}

*Higher values imply a higher affinity. No value is shown if no data were available or if the pK_i was below 3. All data were drawn from the IUPHAR database (Harmar *et al.*, 2009; www.iuphar-db.org).

Abbreviations: ag, agonist; ant, antagonist; pag, partial agonist; iag, inverse agonist.

clinical data did suggest that augmentation with the $5\text{-HT}_{2C/2A}$ antagonists ritanserin and mianserin may have some beneficial effects, especially on negative and cognitive symptoms in schizophrenia (Lieberman *et al.*, 1998; Meltzer *et al.*, 2003; Akhondzadeh *et al.*, 2008).

5-HT_3 receptor antagonists have also been investigated as potential antipsychotics because clozapine has a moderate affinity for this receptor, and preclinical animal studies indicated possible antipsychotic efficacy (Lieberman *et al.*, 1998). Although an open-label and uncontrolled clinical trial demonstrated a moderate antipsychotic activity of the selective 5-HT_3 receptor antagonist ondansetron (DeVeaugh-Geiss *et al.*, 1992), these results could not be replicated in a double-blind study (Gaster and King, 1997). Also, the 5-HT_3 antagonist zacopride was not effective in the treatment of schizophrenia (Newcomer *et al.*, 1992), suggesting that the 5-HT_3 receptor is not a promising drug target in the treatment of schizophrenia.

Given that 5-HT_4 receptors modulate acetylcholine and GABA release, and that 5-HT_4 receptors are found in high densities in the frontal cortex and the hippocampus, it was suggested that modification of 5-HT_4 receptor activity could be helpful in improving cognition in schizophrenia. Several animal studies support this assumption, but studies in healthy human volunteers and schizophrenia patients are lacking so far (Roth *et al.*, 2004; Gray and Roth, 2007). Since atypical antipsychotic drugs are devoid of major 5-HT_4 receptor actions, Roth *et al.* (2004) recommended that a 5-HT_4 partial agonist would be potentially beneficial as add-on therapy for improving cognition in schizophrenia.

On the basis of animal studies, the $5HT_6$ receptor was suggested to be a promising drug target to specifically improve cognition in schizophrenia as well (Meltzer *et al.*, 2003; Roth *et al.*, 2004; Gray and Roth, 2007). The 5-HT_6-selective antagonist SB-271046 is currently undergoing preclinical testing as a cognitive enhancer in schizophrenia (Hatcher *et al.*, 2005; Marcos *et al.*, 2008; Da Silva Costa *et al.*, 2009). However, several typical (e.g., chlorpromazine, fluphenazine) and atypical (e.g., clozapine, olanzapine, ziprasidone and quetiapine) antipsychotics have high affinity for the 5-HT_6 receptor, making it unlikely that addition of a 5-HT_6 antagonistic drug would further improve cognition in schizophrenia patients treated with these antipsychotics (Roth *et al.*, 1994, 2004). Moreover, both 5-HT_6 agonists and antagonists have shown pro-cognitive properties in preclinical animal studies, but an explanation for these paradoxical effects is currently missing (Fone, 2008). Thus, further studies are needed to further understand the role of the 5-HT_6 receptor in the modulation of cognition and to develop 5-HT_6 antagonist compounds for the treatment of cognitive deficits in schizophrenia.

Both clozapine and risperidone, as well as the typical drugs chlorpromazine, fluphenazine and pimozide, have high affinity for the 5-HT_7 receptor (Roth *et al.*, 1994); this suggests that a 5-HT_7 action is not a feature of atypicality (Abi-Dargham and Krystal, 2000). Evidence primarily drawn from knockout studies in mice indicates that the 5-HT_7 receptor plays an important role in hippocampus-dependent functions, including learning and memory (Gray and Roth, 2007). These data warrant further investigation into the potential use of 5-HT_7 receptor antagonist compounds in the treatment of memory dysfunction in schizophrenia (Gray and Roth, 2007).

Antipsychotic drug action and serotonin receptor occupancy

Most of the molecular imaging studies investigating the role of receptor occupancy in antipsychotic activity by PET or SPECT have traditionally focused on the dopamine system. Here it was consistently shown that typical antipsychotics usually produce higher striatal D_2 receptor occupancy rates (>70 percent) than atypical antipsychotics (<70 percent) at mean therapeutic doses (Lieberman *et al.*, 1998; Kasper *et al.*, 1999; Weinberger and Laruelle, 2002). Given that the atypicals clozapine and quetiapine display the lowest rates of D_2 occupancy (20–67 percent) at clinically effective doses and that most of the studies could not demonstrate a linear correlation between striatal D_2 binding and therapeutic efficacy, striatal D_2 receptor occupancy rates alone cannot sufficiently explain antipsychotic activity (Kasper *et al.*, 1999; Weinberger and Laruelle, 2002). On the contrary, several studies consistently found a clear correlation between EPS and striatal D_2 receptor occupancy, indicating a high likelihood of EPS when D_2 occupancy exceeds a threshold of 80 percent (Kasper *et al.*, 1999; Weinberger and Laruelle, 2002; Zipursky *et al.*, 2007). Since at least 50–60 percent D_2 receptor occupancy is required to observe rapid clinical response with typical antipsychotics such as haloperidol, an optimal antipsychotic dose range resulting in 70–80 percent D_2 occupancy was suggested (Nordstrom *et al.*, 1993; Nyberg *et al.*, 1999; Kapur *et al.*, 2000). However, this rule does not apply to clozapine and quetiapine.

In line with the *in vivo* data (see above), most of the atypical drugs display higher 5-HT_{2A} than D_2 occupancy rates when dual-tracer approaches are used (Kapur *et al.*, 1998; Nyberg *et al.*, 1999; Gefvert *et al.*, 2001; Mamo *et al.*, 2004). However, although it was suggested that the predominant 5-HT_{2A} receptor antagonism of atypical drugs protects against EPS (Meltzer, 1999), even atypical substances such as olanzapine or risperidone cause

EPS when given in higher doses that lead to D_2 receptor occupancy of more than 80 percent (Kapur *et al.*, 1998; Nyberg *et al.*, 1999). Thus, occupancy of 5-HT_{2A} receptors does not confer protection against EPS, because the threshold of D_2 receptor occupancy associated with EPS is not markedly reduced for atypical substances with a balanced 5-HT_{2A}/D_2 receptor profile (Weinberger and Laruelle, 2002). Compared to the other atypicals, aripiprazole is an interesting exception regarding D_2, 5-HT_{2A} and 5-HT_{1A} receptor occupancy. A recent study has shown that aripiprazole exhibits very high striatal D_2 occupancy (81–94 percent), lower occupancy of frontal and temporal 5-HT_{2A} receptors (31–84 percent), and even lower occupancy at frontal and temporal 5-HT_{1A} receptors (−2 percent to 44 percent) at doses between 10 and 30 mg in schizophrenia patients. EPS was seen only in two of four subjects with D_2 occupancies exceeding 90 percent (Mamo *et al.*, 2007). In accordance with the study of Bantick *et al.* (2004), who showed that clozapine did not occupy the 5-HT_{1A} receptor at clinical doses, these data do not support an important role of the 5-HT_{1A} receptor regarding antipsychotic activity. In sum, molecular imaging studies do not support the view that the 5-HT_{2A} or 5-HT_{1A} mechanism of several atypical drugs contributes significantly to their clinical superiority.

Serotonergic challenge studies

Given that the release of several hormones, such as cortisol, prolactin and growth hormone (GH), is under monoaminergic control, the neuroendocrine challenge paradigm is suitable to investigate the functional state of central monoaminergic systems. In a hypersensitive system the stimulation of 5-HT receptors will induce augmented hormonal release, whereas in a hypoactive system increased release would be expected. If 5-HT receptors are antagonized, the reverse results are anticipated (Murphy *et al.*, 1986).

Early neuroendocrine challenge studies investigating small samples of schizophrenia patients and employing the 5-HT precursors tryptophan and 5-hydroxytryptophan (5-HTP) reported inconsistent results. Two studies reported an increased prolactin response and a blunted GH release (Cowen *et al.*, 1985; Kolakowska *et al.*, 1987). One study found decreased prolactin responses and decreased GH response only in long-term haloperidol-treated patients, whereas short-term treated patients did not differ from controls in both measures (Hoshino *et al.*, 1985). However, precursor effects are relatively muted because of their 'upstream' (and therefore secondary) actions on synaptic function, making these studies hard to interpret (Breier, 1995).

Challenge studies with the serotonin releaser fenfluramine (racemate or d-form) provided some conflicting results as well. Whereas an initial study reported a decreased prolactin release in chronic patients (Lerer *et al.*, 1988), two later studies found prolactin hyperresponsivity in drug-free patients (Abel *et al.*, 1996; Monteleone *et al.*, 1999). However, in a study by Monteleone *et al.* (1999), the elevated prolactin response was restricted to patients who were refractory to typical neuroleptics. In line with that, Mohr *et al.* (1998) reported that a higher prolactin response to d-fenfluramine, and therefore a higher responsiveness of the 5-HT system, was associated with poorer treatment response to haloperidol in unmedicated first-episode patients. Additionally, Sharma *et al.* (1999) found that a higher prolactin response to dl-fenfluramine was correlated with more negative symptoms. These studies also varied with regard to psychotic symptom provocation after fenfluramine: some reported no changes, while others described exacerbation of positive symptoms.

The serotonin and noradrenaline reuptake inhibitor clomipramine, which also acts as a 5-HT_2 receptor antagonist, provoked an increased prolactin response in drug-naïve schizophrenia patients – an effect that was positively correlated with duration of illness and negatively correlated with treatment response (Angelopoulos *et al.*, 2002). However, another study did not find changes in the prolactin release after clomipramine in patients treated with typical antipsychotics (Markianos *et al.*, 2001).

The drug m-chlorophenylpiperazine (mCPP) preferentially acts as a partial agonist at 5-HT_{2C} receptors and as an antagonist at 5-HT_{2A} receptors, but also binds to several other 5-HT receptor subtypes (Kahn and Wetzler, 1991). mCPP increases anxiety, body temperature and plasma levels of prolactin, cortisol, GH and ACTH, but does not provoke psychotic symptoms in healthy human volunteers (Breier, 1995). Schizophrenia patients show either blunted (Iqbal *et al.*, 1991; Maes and Meltzer, 1996) or normal prolactin response to mCPP (Kahn *et al.*, 1992; Krystal *et al.*, 1993). Moreover, mCPP has been reported to exacerbate (Iqbal *et al.*, 1991; Krystal *et al.*, 1993; Abi-Saab *et al.*, 2002), reduce (Kahn *et al.*, 1992) or have no effect on psychotic symptoms (Breier *et al.*, 1993; Owen *et al.*, 1993; Koreen *et al.*, 1997). Clozapine has been reported to block the symptom-worsening and hormone-releasing effects of mCPP; this was attributed to 5-HT_{2C} antagonistic effects of clozapine (Breier *et al.*, 1993; Kahn *et al.*, 1993a; Krystal *et al.*, 1993; Owen *et al.*, 1993). Similar effects were shown for olanzapine (Abi-Saab *et al.*, 2002) and the 5-HT_2 antagonist ritanserin (Scheepers *et al.*, 2001a).

In general, the contradictory results across the different serotonergic challenge studies point to heterogeneity in central serotonergic sensitivity within different subpopulations of schizophrenia patients. This assumption is also supported by the consistent observation that a hypersensitive 5-HT system is associated with poor treatment response to mostly typical antipsychotics. Serotonergic challenge studies might therefore be useful for tailoring individual antipsychotic pharmacotherapy.

Serotonin metabolites in the cerebrospinal fluid

Many studies have measured monoamine metabolite concentrations in the cerebrospinal fluid (CSF) of schizophrenia patients in order to investigate central 5-HT and dopamine turnover. Most of the early studies did not find changes in the CSF concentration of the major 5-HT metabolite 5-HIAA, but some reported decreased 5-HIAA CSF levels in schizophrenia (for review and references, see Bleich *et al.*, 1991). A more recent meta-analysis and a recent study with a large sample of schizophrenia patients have supported the view that mean 5-HIAA concentrations in the CSF are generally relatively unaltered in schizophrenia patients (Tuckwell and Koziol, 1996; Wieselgren and Lindstrom, 1998). In contrast, another meta-analysis has indicated that CSF levels of the main metabolite of dopamine, homovanillic acid (HVA), are lowered in schizophrenia patients (Tuckwell and Koziol, 1993); this finding was confirmed by a more recent study investigating 90 schizophrenia patients and 47 healthy controls (Wieselgren and Lindstrom, 1998). Studies linking specific characteristics of the illness with 5-HIAA CSF levels have shown that low 5-HIAA concentrations are associated with advanced brain atrophy (Nyback *et al.*, 1983; Potkin *et al.*, 1983; Jennings *et al.*, 1985; Losonczy *et al.*, 1986), more prominent negative symptoms (Pickar *et al.*, 1986; Csernansky *et al.*, 1990), and failure to activate the PFC during the Wisconsin Card Sorting Test (Weinberger *et al.*, 1988). However, all of these measures have been found to be associated with decreased HVA CSF levels as well (Nyback *et al.*, 1983; Potkin *et al.*, 1983; Jennings *et al.*, 1985; Losonczy *et al.*, 1986; Pickar *et al.*, 1986; Csernansky *et al.*, 1990; Weinberger *et al.*, 1988; Scheepers *et al.*, 2001b). However, one of the best replicated findings in biological psychiatry is the strong intercorrelation of monoamine metabolites in the CSF, which possibly could be explained by similar transport mechanisms of all monoamines (Hsiao *et al.*, 1993). This idea has led to the approach of calculating HVA/5-HIAA concentration ratios to investigate the relation of serotonergic and dopaminergic activity in schizophrenia (Hsiao *et al.*, 1993). Lewine *et al.* (1991) demonstrated, for example, that the HVA/5-HIAA ratio was a better predictor of the extent of brain atrophy than HVA or 5-HIAA CSF levels alone (see also Nyback *et al.*, 1983). Additionally, while 5-HIAA and HVA levels alone could not predict treatment outcome, a low HVA/5-HIAA CSF ratio was significantly associated with better response to clozapine and typical antipsychotics in several studies (Pickar *et al.*, 1992; Kahn *et al.*, 1993b; Risch and Lewine, 1993; Szymanski *et al.*, 1993; Lieberman *et al.*, 1994; Risch, 1995). These results suggested that the antipsychotic effect is associated with changing dopamine function relative to 5-HT function, rather than changing dopamine or 5-HT function *per se* (Scheepers *et al.*, 2001b). However, at least two studies failed to find a predictive value of the HVA/5-HIAA CSF ratios regarding treatment response to olanzapine, clozapine or haloperidol treatment (Jacobsen *et al.*, 1997; Scheepers *et al.*, 2001b), while one study reported a worse long-term outcome in patients with a low HVA/5-HIAA CSF ratio (Wieselgren and Lindstrom, 1998). These discrepancies may be explained due to differences in patient populations, duration of treatment, method of analysis or criteria of response.

Surprisingly, several investigations demonstrated that neither typical nor atypical antipsychotics changed 5-HIAA CSF levels in the course of treatment, although many of these substances strongly affect the 5-HT system (van Kammen *et al.*, 1986; Kahn *et al.*, 1994; Jacobsen *et al.*, 1997; Wieselgren and Lindstrom, 1998; Scheepers *et al.*, 2001b). These results put the idea that 5-HIAA CSF concentrations are a valid marker of the central 5-HT turnover somewhat into question. Moreover, it was suggested that 5-HIAA concentrations may not mirror 5-HT metabolism in the whole brain, but rather reflect turnover in specific brain regions such as frontal cortices and the striatum (Scheepers *et al.*, 2001b). On the contrary, typical antipsychotics seemed to consistently elevate HVA CSF levels and HVA/5-HIAA CSF ratios, while atypical substances did not (Hsiao *et al.*, 1993; Kahn *et al.*, 1993b; Wieselgren and Lindstrom, 1998; Scheepers *et al.*, 2001b).

In sum, investigations on 5-HT metabolite levels in the CSF in schizophrenia are hard to interpret because it is not clear what the specific neuronal substrate of 5-HIAA CSF levels is. However, there is some consistency in the data showing that at least a subpopulation of patients display changes in global 5-HT and dopamine turnover, and these patients may respond differentially to antipsychotics than other subpopulations.

Platelet studies

Human blood platelets have been proposed as a peripheral model of central 5-HT function because platelets are neuroectodermal derivatives that share several biochemical and morphological characteristics with 5-HT synapses (Bleich *et al.*, 1991).

Most of the studies investigating platelet or whole blood 5-HT concentrations in schizophrenia patients found elevated values, although there have also been some contradictory results (for review, see Bleich *et al.*, 1991; Iqbal and van Praag, 1995). The increase in peripheral 5-HT concentrations reported in the early studies was apparently not an artifact of medication, as no *in vivo* effect of antipsychotics on platelet 5-HT could be demonstrated (Bleich *et al.*, 1991). On the contrary, accumulating evidence suggests that treatment with clozapine and other atypical and typical antipsychotics increases 5-HT plasma levels in schizophrenia patients (Joseph *et al.*, 1977; Schulz *et al.*, 1997; Fleischhaker *et al.*, 1998; van der Heijden *et al.*, 2004; Ertugrul *et al.*, 2007). These findings suggest that antipsychotics still have an impact on peripheral 5-HT concentrations, and indicate that medication may have indeed influenced previous results.

The findings regarding platelet 5-HT uptake are less consistent. The numbers of studies reporting reduced or unchanged platelet 5-HT uptake are more or less equal (for review, see Bleich *et al.*, 1991; Iqbal and van Praag, 1995). However, Arora and Meltzer (1983) have convincingly demonstrated that a 2-week treatment with chlorpromazine significantly decreased platelet 5-HT uptake in schizophrenia patients and healthy controls. Thus, previous findings of reduced platelet 5-HT uptake in schizophrenia patients are likely explained by acute or residual antipsychotic treatment effects. Moreover, several studies investigating [^{3}H]imipramine binding sites on platelets, which have been suggested as another measure of 5-HT uptake or transport, predominantly yielded no differences between normals and schizophrenia patients (for review, see Bleich *et al.*, 1991; Iqbal and van Praag, 1995).

Platelet 5-HT_{2A} receptors are identical to brain 5-HT_{2A} receptors in terms of their pharmacological properties (Ostrowitzki *et al.*, 1993). Although Arora and Meltzer (1983) detected an increased number of 5-HT_{2A} receptors on platelets from suicidal schizophrenia patients, a newer study reported increased platelet 5-HT_{2A} receptor density in chronic, medication-free patients with schizophrenia (Arranz *et al.*, 2003). Given that treatment with risperidone strongly increased platelet 5-HT_{2A} receptor density, Arranz *et al.* (2003) concluded that the increased platelet 5-HT_{2A} receptor density in their drug-free sample was a residual drug effect caused by previous antipsychotic treatment. Additionally, these authors reported recently that low baseline platelet 5-HT_{2A} receptor levels may predict clinical response to olanzapine in a group of antipsychotic-naïve schizophrenia patients (Arranz *et al.*, 2007).

The activity of platelet monoamine oxidase (MAO) has also been studied in schizophrenia, demonstrating results similar to platelet 5-HT_{2A} receptor density. Although there are some indications of decreased platelet MAO activity in at least some subgroups of schizophrenia patients (Zureick and Meltzer, 1988), it could not be excluded that this effect is primarily caused by antipsychotic treatment (DeLisi *et al.*, 1981; Ohuoha *et al.*, 1993; Ertugrul *et al.*, 2007).

It should be noted that the changes of serotonergic markers found in platelets are largely in the opposite direction to the alterations that have been found in more centrally relevant 5-HT measures in schizophrenia patients (e.g., decreased 5-HT in CSF and brain tissue vs increased 5-HT in blood and platelets; decreased 5-HT receptors in several brain regions vs increased 5-HT_{2A} receptor density in platelets). In addition, treatment with antipsychotics has also had mostly opposite effects on platelet and brain 5-HT markers, respectively. These facts suggest that platelets are not an ideal model for brain 5-HT function (Roth and Meltzer, 2000).

Neurotrophic role of serotonin in the developmental disorder schizophrenia

As reviewed by Whitaker-Azmitia in Chapter 3.1 of this volume, serotonin plays a major role at several stages of neuroplasticity. During embryogenesis, the serotonin system is one of the first neurotransmitter systems that innervates brain structures and demonstrates functional activity. In this phase, serotonin acts as a growth factor that influences neuronal and glial morphology, and connectivity. Some of these effects are direct, whereas some others are mediated by the interaction with further chemical messengers (such as brain-derived neurotrophic factor (BDNF) or S100β) and other neurotransmitter systems (such as dopamine, GABA and glutamate). However, postnatal serotonin also influences the formation and degradation of synapses and axon terminals, indicating that serotonin is important not only for neuronal development but also for the preservation and maintenance of normal function in the adult brain (see also Sodhi and Sanders-Bush, 2004).

Accumulating evidence from several domains suggests that schizophrenia could be a neurodevelopmental disorder that is – at least in part – caused by aberrant early brain development:

1. Many schizophrenia patients exhibit delayed developmental milestones in childhood, including cognitive, motor and behavioral abnormalities, which indicates abnormal brain function prior to diagnosis of schizophrenia.

2. Obstetric complications and prenatal infections increase the risk for schizophrenia.
3. *Post-mortem* studies have not found indicators for neurodegenerative processes such as gliosis or loss of neurons in the brain of schizophrenia patients.
4. Several anatomical and functional disruptions are associated with exacerbation of schizophrenia in adulthood, and these disruptions can be simulated in developmental animal models (Marenco and Weinberger, 2000; Miyamoto *et al.*, 2003).

As suggested by Murray *et al.* (1992), aberrant developmental processes may play a major role, especially in the congenital subform of schizophrenia that shows a gradual increase in behavioral disturbances until the disorder is diagnosed in adolescence or early adulthood. Maynard and colleagues (2001) have proposed a two-hit hypothesis of schizophrenia. According to their suggestion, a lesion occurring in early neurodevelopment (first hit) and caused by a genetic load or adverse embryonic and perinatal events, in combination with a second hit arising from hormonal events, excitotoxicity, psychosocial stress or oxygen radical formation, may cause schizophrenia.

Immunocytochemical and ultrastructural *post-mortem* studies have demonstrated neurocellular alterations in schizophrenia, such as decreased neuronal size, increased cellular packing density, fewer dendritic spines and synapses, and distortions in neuronal orientation (for review, see Arnold, 1999). The abnormalities in the cytoarchitecture, such as neuronal disarray, heterotopias and malpositioning, indicate disruption of proliferation or migration during the gestational period (Miyamoto *et al.*, 2003). In accordance, it was consistently shown that the expression of reelin, a glycoprotein that regulates neuronal migration, is strongly decreased in schizophrenia patients (Impagnatiello *et al.*, 1998; Guidotti *et al.*, 2000). Moreover, anatomical studies found enlargements of the lateral and third ventricles in conjunction with a decrease in cortical volume, especially within the hippocampal formation and the amygdala; additionally, subcortical structures appear to be reduced in size, including the thalamus and striatum (for review, see Sodhi and Sanders-Bush, 2004). It is unlikely that these macrostructural alterations are simply caused by neurodegenerative processes, because some of them have also been shown at a prodromal state of schizophrenia (Wood *et al.*, 2003; Morey *et al.*, 2005; Jessen *et al.*, 2006), and *post-mortem* studies did not find gliosis and neuronal cell loss. Thus, these anatomical and cytoarchitectual changes are likely to arise during brain maturation.

Several lines of evidence suggest that abnormalities in brain development may contribute to the pathogenesis of schizophrenia in a subset of patients. Moreover, we know that serotonin plays an important role in neurogenesis and neuronal plasticity. However, future studies will have to determine whether genetic or early developmental insults could alter the serotonin system in a manner that leads to sustained neuronal changes during brain development, which consequently induces the symptoms of schizophrenia.

Serotonin–glutamate interactions

NMDA antagonists such as phencyclidine (PCP) and ketamine produce effects in humans that closely mimic the symptoms of schizophrenia (Javitt and Zukin, 1991; Krystal *et al.*, 1994). Microdialysis studies have demonstrated that ketamine and PCP increase glutamate outflow in PFC (Moghaddam *et al.*, 1997; Adams and Moghaddam, 1998). Potentially related to this effect is evidence that increases in glutamatergic activity may contribute to the psychotomimetic and behavioral effects of these drugs. Indeed, diminution of PCP-induced glutamate release by activation of metabotropic glutamate 2/3 ($mGlu_{2/3}$) receptors attenuates the effects of PCP on locomotor activity and stereotypy (Moghaddam and Adams, 1998). Other agents that decrease glutamate release also reduce the behavioral effects of PCP and ketamine (Anand *et al.*, 2000; Idris *et al.*, 2005). In each of these cases the actions of the released glutamate would presumably be on non-NMDA glutamate receptors – AMPA, kainate, or metabotropic – since PCP and ketamine block NMDA receptor functions. The involvement of glutamate release in the psychotomimetic effects of NMDA antagonists is consistent with the hypothesis that dysfunction of glutamatergic systems underlies the psychopathology of schizophrenia (Javitt and Zukin, 1991; Halberstadt, 1995; Jentsch and Roth, 1999).

Electrophysiological evidence demonstrates that LSD and other serotonergic hallucinogens can modulate cellular responses to glutamate (Rahman and Neuman, 1993; Arvanov *et al.*, 1999). Recent studies indicate that hallucinogens increase the release of glutamate in neocortex (Scruggs *et al.*, 2003; Muschamp *et al.*, 2004). Activation of 5-HT_{2A} receptors by 5-HT and the hallucinogen 2,5-dimethoxy-4-iodoamphetamine (DOI) produces an enhancement of the frequency and amplitude of spontaneous excitatory postsynaptic potentials/currents (EPSPs/EPSCs) in most layer V pyramidal cells of PFC (Aghajanian and Marek, 1997; Lambe *et al.*, 2000; Klodzinska *et al.*, 2002; Benneyworth *et al.*, 2007); this effect is mediated by increased glutamate efflux and subsequent activation of AMPA receptors (Zhang and Marek, 2008). There is also evidence that 5-HT- and DOI-induced EPSCs are suppressed by activation of $mGlu_{2/3}$ receptors, and augmented by $mGlu_{2/3}$ receptor blockade (Marek *et al.*, 2000; Klodzinska *et al.*, 2002; Benneyworth *et al.*, 2007). Although it is generally accepted that $5\text{-}HT_{2A}$ receptor activation increases the terminal release of glutamate in PFC, there has been some controversy regarding the source of these glutamatergic terminals. Based on evidence that lesions of the medial thalamus attenuate 5-HT-induced EPSCs, Marek and colleagues have argued that thalamocortical afferents are involved (Marek *et al.*, 2001). However, Béïque *et al.* (2007) recently identified a subpopulation of pyramidal cells in the deep layers of PFC that are excited by $5\text{-}HT_{2A}$ receptor activation, indicating that the spontaneous EPSCs evoked by 5-HT may be a product of PFC recurrent network activity.

As was found with PCP, the behavioral effects of serotonergic hallucinogens are attenuated by activation of $mGlu_{2/3}$ receptors. The ability of DOI to induce the head-twitch response in mice and rats is suppressed by the selective $mGlu_{2/3}$ agonists LY354740 and LY379268; conversely, the selective $mGlu_{2/3}$ antagonist LY341495 enhances the frequency of DOI-induced head-twitch (Gewirtz and Marek, 2000; Klodzinska *et al.*, 2002). Likewise, the $mGlu_2$ positive allosteric modulator biphenyl-indanone A inhibits the head-twitch response induced by the hallucinogen (−)-2,5-dimethoxy-4-bromoamphetamine (DOB) (Benneyworth *et al.*, 2007). It has also been shown that the discriminative stimulus effects of LSD are potentiated by LY341495 and partially antagonized by LY379268 (Winter *et al.*, 2004). The ability of $mGlu_{2/3}$ receptor ligands to alter the behavioral response to DOI, DOB and LSD indicates that the behavioral effects of hallucinogens are linked to their ability to increase glutamate release.

Taken together, the aforementioned findings demonstrate that NMDA receptor antagonists and serotonergic hallucinogens increase glutamate release, and it has been suggested that the glutamatergic system may represent a common final pathway for their psychotomimetic effects (Vollenweider and Geyer, 2001). This view is consistent with the fact that both ketamine and psilocybin produce metabolic hyperfrontality (Vollenweider *et al.*, 1997a, 1997b), and have somewhat similar effects on perception and cognition (Vollenweider and Geyer, 2001). Additional support for the convergence of serotonergic and glutamatergic systems is derived from the finding that the behavioral effects of hallucinogens are potentiated by co-administration of NMDA antagonists (Dall'Olio *et al.*, 1999; Winter *et al.*, 2000, 2004; Zhang and Marek, 2008). Recently, evidence has emerged that $mGlu_2$ and $5\text{-}HT_{2A}$ receptors are co-localized in cortical neurons, where they may form functional complexes (Gonzalez-Maeso *et al.*, 2008). The existence of a $mGlu_2/5\text{-}HT_{2A}$ receptor complex is intriguing in light of a recent report that a prodrug for a selective $mGlu_{2/3}$ agonist possesses significant antipsychotic efficacy in schizophrenic patients (Patil *et al.*, 2007).

Conclusions and future directions

As reviewed above, considerable evidence derived from converging methods suggests that schizophrenia patients display abnormalities in serotonergic function. Nevertheless, different approaches intended to measure identical biological markers have frequently produced contradictory results (e.g., autoradiographic *post-mortem* studies vs PET studies). In particular, results from peripheral measures (CSF, platelets, blood, hormone response) have often not matched findings based upon more central parameters of serotonin function (receptor density, brain levels of 5-HT and metabolites). Moreover, it has repeatedly been shown that some alterations of the 5-HT system reported in schizophrenia patients could be explained by chronic treatment with antipsychotic drugs. Despite some methodological reservations and the many contradictory results, there is accumulating evidence that the $5\text{-}HT_{1A}$ and $5\text{-}HT_{2A}$ receptor subtypes play an especially important role in schizophrenia. *Post-mortem* studies and some PET data suggest that schizophrenia patients display an increase of $5\text{-}HT_{1A}$ and a decrease of $5\text{-}HT_{2A}$ receptors, especially in the PFC. Genetic variations of the $5\text{-}HT_{2A}$ receptor (and perhaps also of the $5\text{-}HT_{1A}$ receptor) appear to contribute to the risk of developing schizophrenia and the response to antipsychotic treatment. These genetic variations also appear to be associated with endophenotypic markers of schizophrenia, such as sensorimotor gating. Hallucinogenic $5\text{-}HT_{2A}$ agonists produce some schizophrenia-like symptoms, and also mimic several endophenotypes of schizophrenia. In contrast, the hypothesis that a serotonergic mechanism is necessary for the claimed therapeutic superiority of the so-called atypical antipsychotics is not well supported by the data so far, because a 5-HT antagonistic action seems to be not sufficient for an antipsychotic effect (at least on the level of large and heterogenous populations of schizophrenia patients). Nevertheless, $5\text{-}HT_{1A}$ agonists and $5\text{-}HT_{2C}$ antagonists may have some beneficial effects, particularly on cognition and negative symptoms. Additionally, agents acting at other 5-HT receptor subtypes ($5\text{-}HT_4$, $5\text{-}HT_6$, $5\text{-}HT_7$) may have some pro-cognitive effects in schizophrenia patients.

The highly contradictory results regarding serotonergic alterations in schizophrenia might have two origins:

1. Alterations of the serotonin system are not sufficient to explain the full picture of schizophrenia. This view is supported by the fact that other transmitter systems

(e.g., dopamine, GABA, glutamate, acetylcholine) and biochemical substrates (e.g., reelin, BDNF, synaptophysin, SNAP-25 and complexin II) are also affected in schizophrenia patients.

2. Not all but only a subpopulation of the patients within the broad disease cluster schizophrenia display changes in serotonin function. This assumption is supported by several studies showing that some patients better respond to serotonergic antipsychotic drugs than other patients, that some alterations of the 5-HT systems at baseline could predict treatment response, and that serotonergic challenges induce a broad range of reactions ranging from improvement to worsening of symptoms, pointing to substantial heterogeneity of central serotonergic activity.

The 5-HT system is probably only one piece of the enigmatic mosaic of the multifactorial causation of the group of schizophrenia spectrum disorders. Specific polymorphisms within the 5-HT system might influence – for example, the expression of serotonin receptors during neurogenesis – and these changes could have an impact on later brain maturation and 5-HT function. However, only in combination with further neurodevelopmental 'hits' (such as pre- and postnatal infections, stressful events or drug use during pregnancy, obstetric complications, a stressful adolescence or further critical life events) and other genetic variations (DISC1, VMAT1, GRIN2) could the symptom pattern of a schizophreniform disorder arise.

Future studies should devote more attention to the demarcation of subpopulations of schizophrenia patients exhibiting specific changes of the 5-HT system, who could then be successfully treated with specific serotonergic drugs. These subpopulations should be characterized not only by distinct biological markers but also by a more precise psychopathological description. Moreover, the behavioral consequences of genetic variations within the 5-HT system or of pharmacological manipulations of the system might help us to better understand disturbed brain functions of schizophrenia patients. Finally, recent preclinical data suggest that also alterations in the interaction between the serotonin and the glutamate system might have an influence on the development and the symptoms of schizophrenia. These interactions should be further investigated in healthy humans and schizophrenia patients.

Acknowledgements

Mark Geyer and Adam Halberstadt were supported by grants from the National Institute on Drug Abuse (DA002925, DA025412) and the National Institute of Mental Health (MH52885, MH42228) and by the Veterans Affairs VISN 22 Mental Illness Research, Education, and Clinical Center. Boris Quednow was supported by the Deutsche Forschungsgemeinschaft (DFG, grant QU 218/1-1) and by the Nachwuchsförderungskredit of the University of Zurich.

References

Abel, K.M., O'Keane, V. and Murray, R.M. (1996) Enhancement of the prolactin response to d-fenfluramine in drug-naive schizophrenic patients. Br. J. Psychiatry, 168: 57–60.

Abi-Dargham, A. and Krystal, J. (2000) Serotonin receptors as targets of antipsychotic medication. In: Lidow, M.S. (Ed.), Neurotransmitter Receptors in Actions of Antipsychotic Medications. CRC Press LLC, Boca Raton, FL, pp. 79–107.

Abi-Dargham, A., Laruelle, M., Lipska, B., Jaskiw, G.E., Wong, D.T., Robertson, D.W., Weinberger, D.R. and Kleinman, J.E. (1993) Serotonin 5-HT3 receptors in schizophrenia: a postmortem study of the amygdala. Brain Res., 616: 53–57.

Abi-Saab, W., Seibyl, J.P., D'Souza, D.C., Karper, L.P., Gueorgueva, R., Abi-Dargham, A., Wong, M.L., Rajhans, S., Erdos, J.P., Heninger, G.R., Charney, D.S. and Krystal, J.H. (2002) Ritanserin antagonism of m-chlorophenylpiperazine effects in neuroleptic-free schizophrenics patients: support for serotonin-2 receptor modulation of schizophrenia symptoms. Psychopharmacology (Berl.), 162: 55–62.

Adams, L.M. and Geyer, M.A. (1982) LSD-induced alterations of locomotor patterns and exploration in rats. Psychopharmacology (Berl.), 77: 179–185.

Adams, L.M. and Geyer, M.A. (1985) A proposed animal model for hallucinogens based on LSD's effects on patterns of exploration in rats. Behav. Neurosci., 99: 881–900.

Adams, B. and Moghaddam, B. (1998) Corticolimbic dopamine neurotransmission is temporally dissociated from the cognitive and locomotor effects of phencyclidine. J. Neurosci., 18: 5545–5554.

Aghajanian, G.K. and Marek, G.J. (1997) Serotonin induces excitatory postsynaptic potentials in apical dendrites of neocortical pyramidal cells. Neuropharmacology, 36: 589–599.

Akhondzadeh, S., Malek-Hosseini, M., Ghoreishi, A., Raznahan, M. and Rezazadeh, S.A. (2008) Effect of ritanserin, a 5HT2A/2C antagonist, on negative symptoms of schizophrenia: a double-blind randomized placebo-controlled study. Prog. Neuropsychopharmacol. Biol. Psychiatry, 32: 1879–1883.

Allen, N.C., Bagade, S., McQueen, M.B., Ioannidis, J.P., Kavvoura, F.K., Khoury, M.J., Tanzi, R.E. and Bertram, L. (2008) Systematic meta-analyses and field synopsis of genetic association studies in schizophrenia: the SzGene database. Nat. Genet., 40: 827–834.

Anand, A., Charney, D.S., Oren, D.A., Berman, R.M., Hu, X.S., Cappiello, A. and Krystal, J.H. (2000) Attenuation of the neuropsychiatric effects of ketamine with lamotrigine: support for hyperglutamatergic effects of N-methyl-D-aspartate receptor antagonists. Arch. Gen. Psychiatry, 57: 270–276.

Andree, T.H., Mikuni, M., Tong, C.Y., Koenig, J.I. and Meltzer, H.Y. (1986) Differential effect of subchronic treatment with various neuroleptic agents on serotonin2 receptors in rat cerebral cortex. J. Neurochem., 46: 191–197.

Angelopoulos, E.K., Markianos, M., Daskalopoulou, E.G., Hatzimanolis, J. and Tzemos, J. (2002) Changes in central serotonergic function as a correlate of duration of illness in paranoid schizophrenia. Psychiatry Res., 110: 9–17.
Angrist, B.M. and Gershon, S. (1970) The phenomenology of experimentally induced amphetamine psychosis – preliminary observations. Biol. Psychiatry, 2: 95–107.
Arnold, S.E. (1999) Neurodevelopmental abnormalities in schizophrenia: insights from neuropathology. Dev. Psychopathol., 11: 439–456.
Arnold, S.E., Hyman, B.T., Van Hoesen, G.W. and Damasio, A.R. (1991) Some cytoarchitectural abnormalities of the entorhinal cortex in schizophrenia. Arch. Gen. Psychiatry, 48: 625–632.
Arnt, J. and Skarsfeldt, T. (1998) Do novel antipsychotics have similar pharmacological characteristics? A review of the evidence. Neuropsychopharmacology, 18: 63–101.
Arora, R.C. and Meltzer, H.Y. (1983) Effects of chlorpromazine on serotonin uptake in blood platelets. Psychiatry Res., 9: 23–28.
Arora, R.C. and Meltzer, H.Y. (1991) Serotonin2 (5-HT2) receptor binding in the frontal cortex of schizophrenic patients. J. Neural Transm. Gen. Sect., 85: 19–29.
Arranz, M., Collier, D., Sodhi, M., Ball, D., Roberts, G., Price, J., Sham, P. and Kerwin, R. (1995) Association between clozapine response and allelic variation in 5-HT2A receptor gene. Lancet, 346: 281–282.
Arranz, M.J., Collier, D.A., Munro, J., Sham, P., Kirov, G., Sodhi, M., Roberts, G., Price, J. and Kerwin, R.W. (1996) Analysis of a structural polymorphism in the 5-HT2A receptor and clinical response to clozapine. Neurosci. Lett., 217: 177–178.
Arranz, M.J. and de Leon, J. (2007) Pharmacogenetics and pharmacogenomics of schizophrenia: a review of last decade of research. Mol. Psychiatry, 12: 707–747.
Arranz, M.J., Munro, J., Owen, M.J., Spurlock, G., Sham, P.C., Zhao, J., Kirov, G., Collier, D.A. and Kerwin, R.W. (1998a) Evidence for association between polymorphisms in the promoter and coding regions of the 5-HT2A receptor gene and response to clozapine. Mol. Psychiatry, 3: 61–66.
Arranz, M.J., Munro, J., Sham, P., Kirov, G., Murray, R.M., Collier, D.A. and Kerwin, R.W. (1998b) Meta-analysis of studies on genetic variation in 5-HT2A receptors and clozapine response. Schizophr Res., 32: 93–99.
Arranz, M.J., Munro, J., Birkett, J., Bolonna, A., Mancama, D., Sodhi, M., Lesch, K.P., Meyer, J.F., Sham, P., Collier, D.A., Murray, R.M. and Kerwin, R.W. (2000a) Pharmacogenetic prediction of clozapine response. Lancet, 355: 1615–1616.
Arranz, M.J., Bolonna, A.A., Munro, J., Curtis, C.J., Collier, D.A. and Kerwin, R.W. (2000b) The serotonin transporter and clozapine response. Mol. Psychiatry, 5: 124–125.
Arranz, B., Rosel, P., Sarro, S., Ramirez, N., Duenas, R., Cano, R., Maria Sanchez, J. and San, L. (2003) Altered platelet serotonin 5-HT2A receptor density but not second messenger inositol trisphosphate levels in drug-free schizophrenic patients. Psychiatry Res., 118: 165–174.
Arranz, B., Rosel, P., San, L., Ramirez, N., Duenas, R.M., Salavert, J., Centeno, M. and del Moral, E. (2007) Low baseline serotonin-2A receptors predict clinical response to olanzapine in first-episode schizophrenia patients. Psychiatry Res., 153: 103–109.
Arvanov, V.L., Liang, X., Russo, A. and Wang, R.Y. (1999) LSD and DOB: Interaction with 5-HT2A receptors to inhibit NMDA receptor-mediated transmission in the rat prefrontal cortex. Eur. J. Neurosci., 11: 3064–3072.
Bantick, R.A., Deakin, J.F. and Grasby, P.M. (2001) The 5-HT1A receptor in schizophrenia: a promising target for novel atypical neuroleptics? J. Psychopharmacol., 15: 37–46.
Bantick, R.A., Montgomery, A.J., Bench, C.J., Choudhry, T., Malek, N., McKenna, P.J., Quested, D.J., Deakin, J.F. and Grasby, P.M. (2004) A positron emission tomography study of the 5-HT1A receptor in schizophrenia and during clozapine treatment. J. Psychopharmacol., 18: 346–354.
Basile, V.S., Ozdemir, V., Masellis, M., Meltzer, H.Y., Lieberman, J.A., Potkin, S.G., Macciardi, F.M., Petronis, A. and Kennedy, J.L. (2001) Lack of association between serotonin-2A receptor gene (HTR2A) polymorphisms and tardive dyskinesia in schizophrenia. Mol. Psychiatry, 6: 230–234.
Basile, V.S., Masellis, M., De Luca, V., Meltzer, H.Y. and Kennedy, J.L. (2002) 759C/T genetic variation of 5HT(2C) receptor and clozapine-induced weight gain. Lancet, 360: 1790–1791.
Baumeister, A.A. and Hawkins, M.F. (2004) The serotonin hypothesis of schizophrenia: a historical case study on the heuristic value of theory in clinical neuroscience. J. Hist. Neurosci., 13: 277–291.
Béïque, J.C., Imad, M., Mladenovic, L., Gingrich, J.A. and Andrade, R. (2007) Mechanism of the 5-hydroxytryptamine 2A receptor-mediated facilitation of synaptic activity in prefrontal cortex. Proc. Natl. Acad. Sci. USA, 104: 9870–9875.
Bennett, J.P. Jr., Enna, S.J., Bylund, D.B., Gillin, J.C., Wyatt, R.J. and Snyder, S.H. (1979) Neurotransmitter receptors in frontal cortex of schizophrenics. Arch. Gen. Psychiatry, 36: 927–934.
Benneyworth, M.A., Xiang, Z., Smith, R.L., Garcia, E.E., Conn, P.J. and Sanders-Bush, E. (2007) A selective positive allosteric modulator of metabotropic glutamate receptor subtype 2 blocks a hallucinogenic drug model of psychosis. Mol. Pharmacol., 72: 477–484.
Beringer, K. (1923) Experimentelle Psychosen durch Mescalin. Z. Ges. Neurol. Psychiatry, 84: 426–433.
Bertler, A., Carlsson, A. and Rosengren, E. (1956) Release by reserpine of catechol amines from rabbits' hearts. Naturwissenschaften, 43: 521.
Bertolino, A. and Weinberger, D.R. (1999) Proton magnetic resonance spectroscopy in schizophrenia. Eur. J. Radiol., 30: 132–141.
Bilder, R.M., Goldman, R.S., Volavka, J., Czobor, P., Hoptman, M., Sheitman, B., Lindenmayer, J.P., Citrome, L., McEvoy, J., Kunz, M., Chakos, M., Cooper, T.B., Horowitz, T.L. and Lieberman, J.A. (2002) Neurocognitive effects of clozapine, olanzapine, risperidone, and haloperidol in patients with chronic schizophrenia or schizoaffective disorder. Am. J. Psychiatry, 159: 1018–1028.
Birkett, J.T., Arranz, M.J., Munro, J., Osbourn, S., Kerwin, R.W. and Collier, D.A. (2000) Association analysis of the 5-HT5A gene in depression, psychosis and antipsychotic response. Neuroreport, 11: 2017–2020.
Bleich, A., Brown, S.L., Kahn, R. and van Praag, H.M. (1988) The role of serotonin in schizophrenia. Schizophr. Bull., 14: 297–315.
Bleich, A., Brown, S.L. and van Praag, H.M. (1991) A serotonergic theory of schizophrenia. In: Brown, S.L. and van Praag, H.M. (Eds.), The Role of Serotonin in Psychiatric Disorders. Brunner/Mazel, Publishers, New York, NY, pp. 183–214.
Blin, O. (1999) A comparative review of new antipsychotics. Can. J. Psychiatry, 44: 235–244.
Bolino, F., Di Michele, V., Di Cicco, L., Manna, V., Daneluzzo, E. and Casacchia, M. (1994) Sensorimotor gating and habituation

evoked by electro-cutaneous stimulation in schizophrenia. Biol. Psychiatry, 36: 670–679.
Bowers, M.B. Jr. and Freedman, D.X. (1966) 'Psychedelic' experiences in acute psychoses. Arch. Gen. Psychiatry, 15: 240–248.
Braff, D. (1985) Attention, habituation and information processing in psychiatric disorders. In: Micheks, B., Cavenar, J., Brodie, H., Cooper, A., Guze, S. and Judd, L. (Eds.), Psychiatry. Lippincott, Philadelphia, PA, pp. 1–12.
Braff, D.L. and Geyer, M.A. (1980) Acute and chronic LSD effects on rat startle: data supporting an LSD–rat model of schizophrenia. Biol. Psychiatry, 15: 909–916.
Braff, D.L. and Geyer, M.A. (1990) Sensorimotor gating and schizophrenia. Human and animal model studies. Arch. Gen. Psychiatry, 47: 181–188.
Braff, D.L., Stone, C., Callaway, E., Geyer, M., Glick, I. and Bali, L. (1978) Prestimulus effects on human startle reflex in normals and schizophrenics. Psychophysiology, 15: 339–343.
Breier, A. (1995) Serotonin, schizophrenia and antipsychotic drug action. Schizophr. Res., 14: 187–202.
Breier, A., Kirkpatrick, B. and Buchanan, R.W. (1993) Clozapine attenuates meta-chlorophenylpiperazine (mCPP)-induced plasma cortisol increases in schizophrenia. Biol. Psychiatry, 34: 492–494.
Brodie, B.B., Pletscher, A. and Shore, P.A. (1955) Evidence that serotonin has a role in brain function. Science, 122: 968.
Brodie, B.A., Shore, P.A. and Pletscher, A. (1956) Serotonin-releasing activity limited to Rauwolfia alkaloids with tranquilizing action. Science, 123: 992–993.
Bromet, E.J. and Fennig, S. (1999) Epidemiology and natural history of schizophrenia. Biol. Psychiatry, 46: 871–881.
Buchanan, R.W. (1995) Clozapine: efficacy and safety. Schizophr. Bull., 21: 579–591.
Buchanan, R.W., Breier, A., Kirkpatrick, B., Ball, P. and Carpenter, W.T. Jr. (1998) Positive and negative symptom response to clozapine in schizophrenic patients with and without the deficit syndrome. Am. J. Psychiatry, 155: 751–760.
Buckland, P.R., Hoogendoorn, B., Guy, C.A., Smith, S.K., Coleman, S.L. and O'Donovan, M.C. (2005) Low gene expression conferred by association of an allele of the 5-HT2C receptor gene with antipsychotic-induced weight gain. Am. J. Psychiatry, 162: 613–615.
Burnet, P.W., Eastwood, S.L. and Harrison, P.J. (1996a) 5-HT1A and 5-HT2A receptor mRNAs and binding site densities are differentially altered in schizophrenia. Neuropsychopharmacology, 15: 442–455.
Burnet, P.W., Chen, C.P., McGowan, S., Franklin, M. and Harrison, P.J. (1996b) The effects of clozapine and haloperidol on serotonin-1A, -2A and -2C receptor gene expression and serotonin metabolism in the rat forebrain. Neuroscience, 73: 531–540.
Carlsson, A. (1959) The occurrence, distribution and physiological role of catecholamines in the nervous system. Pharmacol. Rev., 11: 490–493.
Carlsson, A. (1988) The current status of the dopamine hypothesis of schizophrenia. Neuropsychopharmacology, 1: 179–186.
Carlsson, A. (1995) Neurocircuitries and neurotransmitter interactions in schizophrenia. Intl. Clin. Psychopharmacol., 10(Suppl. 3): 21–28.
Carlsson, A. and Lindqvist, M. (1963) Effect of chlorpromazine or haloperidol on formation of 3-methoxytyramine and nor-metanephrine in mouse brain. Acta Pharmacol. Toxicol. (Copenh), 20: 140–144.
Carlsson, A., Lindqvist, M. and Magnusson, T. (1957) 3,4-Dihydroxyphenylalanine and 5-hydroxytryptophan as reserpine antagonists. Nature, 180: 1200.
Carman, J., Peuskens, J. and Vangeneugden, A. (1995) Risperidone in the treatment of negative symptoms of schizophrenia: a meta-analysis. Intl. Clin. Psychopharmacol., 10: 207–213.
Carpenter, W.T. Jr., Conley, R.R., Buchanan, R.W., Breier, A. and Tamminga, C.A. (1995) Patient response and resource management: another view of clozapine treatment of schizophrenia. Am. J. Psychiatry, 152: 827–832.
Condrau, G. (1949) Klinische Erfahrungen an Geisteskranken mit Lysergsäure-diäthylamid. Acta Psychiatr. Neurol., 24: 9–32.
Cowen, P.J., Gadhvi, H., Gosden, B. and Kolakowska, T. (1985) Responses of prolactin and growth hormone to L-tryptophan infusion: effects in normal subjects and schizophrenic patients receiving neuroleptics. Psychopharmacology (Berl.), 86: 164–169.
Creese, I., Burt, D.R. and Snyder, S.H. (1976) Dopamine receptor binding predicts clinical and pharmacological potencies of antischizophrenic drugs. Science, 192: 481–483.
Crow, T.J. (1980a) Molecular pathology of schizophrenia: more than one disease process? Br. Med. J., 280: 66–68.
Crow, T.J. (1980b) Positive and negative schizophrenic symptoms and the role of dopamine. Br. J. Psychiatry, 137: 383–386.
Crow, T.J., Baker, H.F., Cross, A.J., Joseph, M.H., Lofthouse, R., Longden, A., Owen, F., Riley, G.J., Glover, V. and Killpack, W.S. (1979) Monoamine mechanisms in chronic schizophrenia: post-mortem neurochemical findings. Br. J. Psychiatry, 134: 249–256.
Cruz, D.A., Eggan, S.M., Azmitia, E.C. and Lewis, D.A. (2004) Serotonin1A receptors at the axon initial segment of prefrontal pyramidal neurons in schizophrenia. Am. J. Psychiatry, 161: 739–742.
Csernansky, J.G., King, R.J., Faustman, W.O., Moses, J.A. Jr., Poscher, M.E. and Faull, K.F. (1990) 5-HIAA in cerebrospinal fluid and deficit schizophrenic characteristics. Br. J. Psychiatry, 156: 501–507.
Da Silva Costa, V., Duchatelle, P., Boulouard, M. and Dauphin, F. (2009) Selective 5-HT6 receptor blockade improves spatial recognition memory and reverses age-related deficits in spatial recognition memory in the mouse. Neuropsychopharmacology, 34: 488–500.
Dall'Olio, R., Gaggi, R., Bonfante, V. and Gandolfi, O. (1999) The non-competitive NMDA receptor blocker dizocilpine potentiates serotonergic function. Behav. Pharmacol., 10: 63–71.
de Jong, H. (1932) Die experimentelle Katatonie als vielfach vorkommende Reaktionsform des Zentralnervensystems. Z. Ges. Neurol. Psychiatry, 139: 468–499.
De Luca, V., Likhodi, O., Van Tol, H.H., Kennedy, J.L. and Wong, A.H. (2005) Tryptophan hydroxylase 2 gene expression and promoter polymorphisms in bipolar disorder and schizophrenia. Psychopharmacology (Berl.), 183: 378–382.
de Paulis, T. (2001) M-100907 (Aventis). Curr. Opin. Investig. Drugs, 2: 123–132.
Dean, B. (2003) The cortical serotonin2A receptor and the pathology of schizophrenia: a likely accomplice. J. Neurochem., 85: 1–13.

Dean, B. and Hayes, W. (1996) Decreased frontal cortical serotonin2A receptors in schizophrenia. Schizophr. Res., 21: 133–139.

Dean, B., Opeskin, K., Pavey, G., Naylor, L., Hill, C., Keks, N. and Copolov, D.L. (1995) [3H]paroxetine binding is altered in the hippocampus but not the frontal cortex or caudate nucleus from subjects with schizophrenia. J. Neurochem., 64: 1197–1202.

Dean, B., Hayes, W., Opeskin, K., Naylor, L., Pavey, G., Hill, C., Keks, N. and Copolov, D.L. (1996) Serotonin2 receptors and the serotonin transporter in the schizophrenic brain. Behav. Brain Res., 73: 169–175.

Dean, B., Hayes, W., Hill, C. and Copolov, D. (1998) Decreased serotonin2A receptors in Brodmann's area 9 from schizophrenic subjects. A pathological or pharmacological phenomenon? Mol. Chem. Neuropathol., 34: 133–145.

Dean, B., Hussain, T., Hayes, W., Scarr, E., Kitsoulis, S., Hill, C., Opeskin, K. and Copolov, D.L. (1999a) Changes in serotonin2A and GABA(A) receptors in schizophrenia: studies on the human dorsolateral prefrontal cortex. J. Neurochem., 72: 1593–1599.

Dean, B., Tomaskovic-Crook, E., Opeskin, K., Keks, N. and Copolov, D. (1999b) No change in the density of the serotonin1A receptor, the serotonin4 receptor or the serotonin transporter in the dorsolateral prefrontal cortex from subjects with schizophrenia. Neurochem. Intl., 34: 109–115.

Dean, B., Pavey, G., Thomas, D. and Scarr, E. (2006) Cortical serotonin7, 1D and 1F receptors: effects of schizophrenia, suicide and antipsychotic drug treatment. Schizophr. Res., 88: 265–274.

DeLisi, L.E., Wise, C.D., Bridge, T.P., Rosenblatt, J.E., Wagner, R.L., Morihisa, J., Karson, C., Potkin, S.G. and Wyatt, R.J. (1981) A probable neuroleptic effect on platelet monoamine oxidase in chronic schizophrenic patients. Psychiatry Res., 4: 95–107.

Deshpande, S.N., Varma, P.G., Semwal, P., Rao, A.R., Bhatia, T., Nimgaonkar, V.L., Lerer, B. and Thelma, B.K. (2005) II. Serotonin receptor gene polymorphisms and their association with tardive dyskinesia among schizophrenia patients from North India. Psychiatr. Genet., 15: 157–158.

DeVeaugh-Geiss, J., McBain, S., Cooksey, P. and Bell, J.M. (1992) The effects of a novel 5-HT3 anatgonist, ondansetron, in schizophrenia. In: Meltzer, H.Y. (Ed.), Novel Antipsychotic Drugs. Raven Press, New York, NY, pp. 225–232.

Di Matteo, V., Di Giovanni, G., Di Mascio, M. and Esposito, E. (1998) Selective blockade of serotonin2C/2B receptors enhances dopamine release in the rat nucleus accumbens. Neuropharmacology, 37: 265–272.

Dittrich, A. (1998) The standardized psychometric assessment of altered states of consciousness (ASCs) in humans. Pharmacopsychiatry, 31(Suppl 2): 80–84.

Dubertret, C., Hanoun, N., Ades, J., Hamon, M. and Gorwood, P. (2004) Family-based association studies between 5-HT5A receptor gene and schizophrenia. J. Psychiatr. Res., 38: 371–376.

Dulawa, S.C. and Geyer, M.A. (2000) Effects of strain and serotonergic agents on prepulse inhibition and habituation in mice. Neuropharmacology, 39: 2170–2179.

East, S.Z., Burnet, P.W., Leslie, R.A., Roberts, J.C. and Harrison, P.J. (2002) 5-HT6 receptor binding sites in schizophrenia and following antipsychotic drug administration: autoradiographic studies with [125I]SB-258585. Synapse, 45: 191–199.

Ellingrod, V.L., Perry, P.J., Ringold, J.C., Lund, B.C., Bever-Stille, K., Fleming, F., Holman, T.L. and Miller, D. (2005) Weight gain associated with the −759C/T polymorphism of the 5HT2C receptor and olanzapine. Am. J. Med. Genet. B Neuropsychiatr. Genet., 134B: 76–78.

Erritzoe, D., Rasmussen, H., Kristiansen, K.T., Frokjaer, V.G., Haugbol, S., Pinborg, L., Baare, W., Svarer, C., Madsen, J., Lublin, H., Knudsen, G.M. and Glenthoj, B.Y. (2008) Cortical and subcortical 5-HT2A receptor binding in neuroleptic-naive first-episode schizophrenic patients. Neuropsychopharmacology, 33: 2435–2441.

Ertugrul, A., Ucar, G., Basar, K., Demir, B., Yabanoglu, S. and Ulug, B. (2007) Influence of clozapine on platelet serotonin, monoamine oxidase and plasma serotonin levels. Psychiatry Res., 149: 49–57.

Falkai, P., Schneider-Axmann, T. and Honer, W.G. (2000) Entorhinal cortex pre-alpha cell clusters in schizophrenia: quantitative evidence of a developmental abnormality. Biol. Psychiatry, 47: 937–943.

Fallin, M.D., Lasseter, V.K., Avramopoulos, D., Nicodemus, K.K., Wolyniec, P.S., McGrath, J.A., Steel, G., Nestadt, G., Liang, K.Y., Huganir, R.L., Valle, D. and Pulver, A.E. (2005) Bipolar I disorder and schizophrenia: a 440-single-nucleotide polymorphism screen of 64 candidate genes among Ashkenazi Jewish case-parent trios. Am. J. Hum. Genet., 77: 918–936.

Faraone, S.V., Chen, W.J., Goldstein, J.M. and Tsuang, M.T. (1994) Gender differences in age at onset of schizophrenia. Br. J. Psychiatry, 164: 625–629.

Farley, I.J., Shannak, K.S. and Hornykiewicz, O. (1980) Brain monoamine changes in chronic paranoid schizophrenia and their possible relation to increased receptor sensitivity. In: Pepeu, G., Kuhar, M.J. and Enna, S.J. (Eds.), Receptors for Neurotransmitters and Peptides. Raven Press, New York, NY, pp. 427–433.

Fleischhaker, C., Schulz, E. and Remschmidt, H. (1998) Biogenic amines as predictors of response to clozapine treatment in early-onset schizophrenia. J. Psychiatr. Res., 32: 325–333.

Fone, K.C. (2008) An update on the role of the 5-hydroxytryptamine6 receptor in cognitive function. Neuropharmacology, 55: 1015–1022.

Frankle, W.G., Lerma, J. and Laruelle, M. (2003) The synaptic hypothesis of schizophrenia. Neuron, 39: 205–216.

Frankle, W.G., Narendran, R., Huang, Y., Hwang, D.R., Lombardo, I., Cangiano, C., Gil, R., Laruelle, M. and Abi-Dargham, A. (2005) Serotonin transporter availability in patients with schizophrenia: a positron emission tomography imaging study with [11C]DASB. Biol. Psychiatry, 57: 1510–1516.

Frankle, W.G., Lombardo, I., Kegeles, L.S., Slifstein, M., Martin, J.H., Huang, Y., Hwang, D.R., Reich, E., Cangiano, C., Gil, R., Laruelle, M. and Abi-Dargham, A. (2006) Serotonin 1A receptor availability in patients with schizophrenia and schizo-affective disorder: a positron emission tomography imaging study with [11C]WAY 100635. Psychopharmacology (Berl.), 189: 155–164.

Freedman, B. and Chapman, L.J. (1973) Early subjective experience in schizophrenic episodes. J. Abnorm. Psychol., 82: 46–54.

Friston, K.J. and Frith, C.D. (1995) Schizophrenia: a disconnection syndrome? Clin. Neurosci., 3: 89–97.

Gaddum, J.H. (1953) Antagonism between lysergic acid diethylamide and 5-hydroxytryptamine. J. Physiol., 121: 15P.

Gaddum, J.H. and Hameed, K.A. (1954) Drugs which antagonize 5-hydroxytryptamine. Br. J. Pharmacol. Chemother., 9: 240–248.

Gaster, L.M. and King, F.D. (1997) Serotonin 5-HT3 and 5-HT4 receptor antagonists. Med. Res. Rev., 17: 163–214.

Gefvert, O., Lundberg, T., Wieselgren, I.M., Bergstrom, M., Langstrom, B., Wiesel, F. and Lindstrom, L. (2001) D(2) and 5HT(2A) receptor occupancy of different doses of quetiapine in schizophrenia: a PET study. Eur. Neuropsychopharmacol., 11: 105–110.

Gewirtz, J.C. and Marek, G.J. (2000) Behavioral evidence for interactions between a hallucinogenic drug and group II metabotropic glutamate receptors. Neuropsychopharmacology, 23: 569–576.

Geyer, M.A. (1998) Behavioral studies of hallucinogenic drugs in animals: implications for schizophrenia research. Pharmacopsychiatry, 31(Suppl. 2): 73–79.

Geyer, M.A. and Braff, D.L. (1982) Habituation of the Blink reflex in normals and schizophrenic patients. Psychophysiology, 19: 1–6.

Geyer, M.A. and Braff, D.L. (1987) Startle habituation and sensorimotor gating in schizophrenia and related animal models. Schizophr. Bull., 13: 643–668.

Geyer, M. and Moghaddam, B. (2002) Animal models relevant to schizophrenia disorders. In: Davis, K.L., Charney, C., Coyle, J.T. and Nemeroff, C. (Eds.), Neuropsychopharmacology: The Fifth Generation of Progress. Lippincott Williams and Wilkins, Philadelphia PA, pp. 689–701.

Geyer, M.A. and Vollenweider, F.X. (2008) Serotonin research: contributions to understanding psychoses. Trends Pharmacol. Sci., 29: 445–453.

Geyer, M.A., Petersen, L.R., Rose, G.J., Horwitt, D.D., Light, R.K., Adams, L.M., Zook, J.A., Hawkins, R.L. and Mandell, A.J. (1978) The effects of lysergic acid diethylamide and mescaline-derived hallucinogens on sensory-integrative function: tactile startle. J. Pharmacol. Exp. Ther., 207: 837–847.

Geyer, M.A., Krebs-Thomson, K., Braff, D.L. and Swerdlow, N.R. (2001) Pharmacological studies of prepulse inhibition models of sensorimotor gating deficits in schizophrenia: a decade in review. Psychopharmacology (Berl.), 156: 117–154.

Goldberg, T.E., Goldman, R.S., Burdick, K.E., Malhotra, A.K., Lencz, T., Patel, R.C., Woerner, M.G., Schooler, N.R., Kane, J.M. and Robinson, D.G. (2007) Cognitive improvement after treatment with second-generation antipsychotic medications in first-episode schizophrenia: is it a practice effect? Arch. Gen. Psychiatry, 64: 1115–1122.

Goldstein, J.M., Tsuang, M.T. and Faraone, S.V. (1989) Gender and schizophrenia: implications for understanding the heterogeneity of the illness. Psychiatry Res., 28: 243–253.

Gonzalez-Maeso, J., Ang, R.L., Yuen, T., Chan, P., Weisstaub, N.V., Lopez-Gimenez, J.F., Zhou, M., Okawa, Y., Callado, L.F., Milligan, G., Gingrich, J.A., Filizola, M., Meana, J.J. and Sealfon, S.C. (2008) Identification of a serotonin/glutamate receptor complex implicated in psychosis. Nature, 452: 93–97.

Gottesman, I. (1991) Schizophrenia Genesis: The Origin of Madness. W. H. Freeman, New York, NY.

Gottesman, I. and Gould, T.D. (2003) The endophenotype concept in psychiatry: etymology and strategic Intentions. Am. J. Psychiatry, 160: 636–645.

Gouzoulis-Mayfrank, E., Hermle, L., Thelen, B. and Sass, H. (1998a) History, rationale and potential of human experimental hallucinogenic drug research in psychiatry. Pharmacopsychiatry, 31(Suppl. 2): 63–68.

Gouzoulis-Mayfrank, E., Habermeyer, E., Hermle, L., Steinmeyer, A.M., Kunert, H.J. and Sass, H. (1998b) Hallucinogenic drug induced states resemble acute endogenous psychoses: Results of an empirical study. Eur. Psychiatry, 13: 399–406.

Gouzoulis-Mayfrank, E., Thelen, B., Habermeyer, E., Kunert, H.J., Kovar, K.A., Lindenblatt, H., Hermle, L., Spitzer, M. and Sass, H. (1999) Psychopathological, neuroendocrine and autonomic effects of 3,4-methylenedioxyethylamphetamine (MDE), psilocybin and d-methamphetamine in healthy volunteers. Results of an experimental double-blind placebo-controlled study. Psychopharmacology (Berl.), 142: 41–50.

Gray, J.A. and Roth, B.L. (2007) Molecular targets for treating cognitive dysfunction in schizophrenia. Schizophr. Bull., 33: 1100–1119.

Gray, L., Scarr, E. and Dean, B. (2006) Serotonin 1a receptor and associated G-protein activation in schizophrenia and bipolar disorder. Psychiatry Res., 143: 111–120.

Green, M.F. (1996) What are the functional consequences of neurocognitive deficits in schizophrenia? Am. J. Psychiatry, 153: 321–330.

Guidotti, A., Pesold, C. and Costa, E. (2000) New neurochemical markers for psychosis: a working hypothesis of their operation. Neurochem. Res., 25: 1207–1218.

Gurevich, E.V. and Joyce, J.N. (1997) Alterations in the cortical serotonergic system in schizophrenia: a postmortem study. Biol. Psychiatry, 42: 529–545.

Haertzen, C.A., Hill, H.E. and Belleville, R.E. (1963) Development of the Addiction Research Center Inventory (ARCI): selection of items that are sensitive to the effects of various drugs. Psychopharmacologia, 4: 155–166.

Halberstadt, A.L. (1995) The phencyclidine–glutamate model of schizophrenia. Clin. Neuropharmacol., 18: 237–249.

Hamdani, N., Bonniere, M., Ades, J., Hamon, M., Boni, C. and Gorwood, P. (2005) Negative symptoms of schizophrenia could explain discrepant data on the association between the 5-HT2A receptor gene and response to antipsychotics. Neurosci. Lett., 377: 69–74.

Harmar, A.J., Hills, R.A., Rosser, E.M., Jones, M., Buneman, O.P., Dunbar, D.R., Greenhill, S.D., Hale, V.A., Sharman, J.L., Bonner, T.I., Catterall, W.A., Davenport, A.P., Delagrange, P., Dollery, C.T., Foord, S.M., Gutman, G.A., Laudet, V., Neubig, R.R., Ohlstein, E.H., Olsen, R.W., Peters, J., Pin, J.P., Ruffolo, R.R., Searls, D.B., Wright, M.W. and Spedding, M. (2009) IUPHAR-DB: the IUPHAR database of G protein-coupled receptors and ion channels. Nucleic Acids Res., 37: D680–D685.

Harrison, P.J. (1999a) The neuropathology of schizophrenia. A critical review of the data and their interpretation. Brain, 122(Pt 4): 593–624.

Harrison, P.J. (1999b) Neurochemical alterations in schizophrenia affecting the putative receptor targets of atypical antipsychotics. Focus on dopamine (D1, D3, D4) and 5-HT2a receptors. Br. J. Psychiatry Suppl.: 12–22.

Harrison, P.J. (2004) The hippocampus in schizophrenia: a review of the neuropathological evidence and its pathophysiological implications. Psychopharmacology (Berl.), 174: 151–162.

Harrison, P.J. and Eastwood, S.L. (2001) Neuropathological studies of synaptic connectivity in the hippocampal formation in schizophrenia. Hippocampus, 11: 508–519.

Harrison, P.J. and Owen, M.J. (2003) Genes for schizophrenia? Recent findings and their pathophysiological implications. Lancet, 361: 417–419.

Harrison, P.J. and Weinberger, D.R. (2005) Schizophrenia genes, gene expression, and neuropathology: on the matter of their convergence. Mol. Psychiatry, 10: 40–68; image 45
Hashimoto, T., Nishino, N., Nakai, H. and Tanaka, C. (1991) Increase in serotonin 5-HT1A receptors in prefrontal and temporal cortices of brains from patients with chronic schizophrenia. Life Sci., 48: 355–363.
Hasler, F., Grimberg, U., Benz, M.A., Huber, T. and Vollenweider, F.X. (2004) Acute psychological and physiological effects of psilocybin in healthy humans: a double-blind, placebo-controlled dose–effect study. Psychopharmacology (Berl.), 172: 145–156.
Hatcher, P.D., Brown, V.J., Tait, D.S., Bate, S., Overend, P., Hagan, J.J. and Jones, D.N. (2005) 5-HT6 receptor antagonists improve performance in an attentional set shifting task in rats. Psychopharmacology (Berl.), 181: 253–259.
Healy, D. (2002) The Creation of Psychopharmacology. Harvard University Press, Cambridge, MA.
Hermle, L., Funfgeld, M., Oepen, G., Botsch, H., Borchardt, D., Gouzoulis, E., Fehrenbach, R.A. and Spitzer, M. (1992) Mescaline-induced psychopathological, neuropsychological, and neurometabolic effects in normal subjects: experimental psychosis as a tool for psychiatric research. Biol. Psychiatry, 32: 976–991.
Hernandez, I. and Sokolov, B.P. (1997) Abnormal expression of serotonin transporter mRNA in the frontal and temporal cortex of schizophrenics. Mol. Psychiatry, 2: 57–64.
Hernandez, I. and Sokolov, B.P. (2000) Abnormalities in 5-HT2A receptor mRNA expression in frontal cortex of chronic elderly schizophrenics with varying histories of neuroleptic treatment. J. Neurosci. Res., 59: 218–225.
Higashi, S., Ohnuma, T., Shibata, N., Higashi, M., Matsubara, Y. and Arai, H. (2007) No genetic association between tryptophan hydroxylase 2 gene polymorphisms and Japanese schizophrenia. Psychiatr. Genet., 17: 123.
Hollister, L.E. (1962) Drug-induced psychoses and schizophrenic reactions: a critical comparison. Ann. N.Y. Acad. Sci., 96: 80–92.
Holzman, P. and Matthysse, S. (1990) The genetics of schizophrenia: a review. Psychol. Sci., 1: 279–285.
Honer, W. and Young, C. (2004) Presynaptic proteins and schizophrenia. In: Smythies, J. (Ed.), Disorders of Synaptic Plasticity and Schizophrenia. Elsevier, Amsterdam, pp. 175–201.
Hong, C.J., Lin, C.H., Yu, Y.W., Yang, K.H. and Tsai, S.J. (2001) Genetic variants of the serotonin system and weight change during clozapine treatment. Pharmacogenetics, 11: 265–268.
Hoshino, Y., Kaneko, M., Kumashiro, H. and Tachibana, R. (1985) Endocrinological function in schizophrenic patients under haloperidol treatment: plasma PRL, HGH and 5HT levels after L-5HTP loading. Folia Psychiatr. Neurol. Jpn, 39: 25–31.
Hsiao, J.K., Potter, W.Z., Agren, H., Owen, R.R. and Pickar, D. (1993) Clinical investigation of monoamine neurotransmitter interactions. Psychopharmacology (Berl.), 112: S76–S84.
Huang, Y.Y., Battistuzzi, C., Oquendo, M.A., Harkavy-Friedman, J., Greenhill, L., Zalsman, G., Brodsky, B., Arango, V., Brent, D.A. and Mann, J.J. (2004) Human 5-HT1A receptor C(−1019)G polymorphism and psychopathology. Intl. J. Neuropsychopharmacol., 7: 441–451.
Hurlemann, R., Boy, C., Meyer, P.T., Scherk, H., Wagner, M., Herzog, H., Coenen, H.H., Vogeley, K., Falkai, P., Zilles, K., Maier, W. and Bauer, A. (2005) Decreased prefrontal 5-HT2A receptor binding in subjects at enhanced risk for schizophrenia. Anat. Embryol. (Berl.), 210: 519–523.
Hurlemann, R., Matusch, A., Kuhn, K.U., Berning, J., Elmenhorst, D., Winz, O., Kolsch, H., Zilles, K., Wagner, M., Maier, W. and Bauer, A. (2008) 5-HT2A receptor density is decreased in the at-risk mental state. Psychopharmacology (Berl.), 195: 579–590.
Idris, N.F., Repeto, P., Neill, J.C. and Large, C.H. (2005) Investigation of the effects of lamotrigine and clozapine in improving reversal-learning impairments induced by acute phencyclidine and D-amphetamine in the rat. Psychopharmacology (Berl.), 179: 336–348.
Ikeda, M., Iwata, N., Kitajima, T., Suzuki, T., Yamanouchi, Y., Kinoshita, Y. and Ozaki, N. (2006) Positive association of the serotonin 5-HT7 receptor gene with schizophrenia in a Japanese population. Neuropsychopharmacology, 31: 866–871.
Impagnatiello, F., Guidotti, A.R., Pesold, C., Dwivedi, Y., Caruncho, H., Pisu, M.G., Uzunov, D.P., Smalheiser, N.R., Davis, J.M., Pandey, G.N., Pappas, G.D., Tueting, P., Sharma, R.P. and Costa, E. (1998) A decrease of reelin expression as a putative vulnerability factor in schizophrenia. Proc. Natl. Acad. Sci. USA, 95: 15718–15723.
Iqbal, N. and van Praag, H.M. (1995) The role of serotonin in schizophrenia. Eur. Neuropsychopharmacol., 5(Suppl.): 11–23.
Iqbal, N., Asnis, G.M., Wetzler, S., Kahn, R.S., Kay, S.R. and van Praag, H.M. (1991) The MCPP challenge test in schizophrenia: hormonal and behavioral responses. Biol. Psychiatry, 30: 770–778.
Ito, H., Nyberg, S., Halldin, C., Lundkvist, C. and Farde, L. (1998) PET imaging of central 5-HT2A receptors with carbon-11-MDL 100,907. J. Nucl. Med., 39: 208–214.
Iwata, N., Ozaki, N., Inada, T. and Goldman, D. (2001) Association of a 5-HT(5A) receptor polymorphism, Pro15Ser, to schizophrenia. Mol. Psychiatry, 6: 217–219.
Jacobsen, L.K., Frazier, J.A., Malhotra, A.K., Karoum, F., McKenna, K., Gordon, C.T., Hamburger, S.D., Lenane, M.C., Pickar, D., Potter, W.Z. and Rapoport, J.L. (1997) Cerebrospinal fluid monoamine metabolites in childhood-onset schizophrenia. Am. J. Psychiatry, 154: 69–74.
Jakob, H. and Beckmann, H. (1986) Prenatal developmental disturbances in the limbic allocortex in schizophrenics. J. Neural Transm., 65: 303–326.
Javitt, D.C. and Zukin, S.R. (1991) Recent advances in the phencyclidine model of schizophrenia. Am. J. Psychiatry, 148: 1301–1308.
Jennings, W.S. Jr., Schulz, S.C., Narasimhachari, N., Hamer, R.M. and Friedel, R.O. (1985) Brain ventricular size and CSF monoamine metabolites in an adolescent inpatient population. Psychiatry Res., 16: 87–94.
Jentsch, J.D. and Roth, R.H. (1999) The neuropsychopharmacology of phencyclidine: from NMDA receptor hypofunction to the dopamine hypothesis of schizophrenia. Neuropsychopharmacology, 20: 201–225.
Jessen, F., Scherk, H., Traber, F., Theyson, S., Berning, J., Tepest, R., Falkai, P., Schild, H.H., Maier, W., Wagner, M. and Block, W. (2006) Proton magnetic resonance spectroscopy in subjects at risk for schizophrenia. Schizophr. Res., 87: 81–88.
Jonsson, E., Nothen, M.M., Bunzel, R., Propping, P. and Sedvall, G. (1996) 5HT 2a receptor T102C polymorphism and schizophrenia. Lancet, 347: 1831.
Joober, R., Benkelfat, C., Brisebois, K., Toulouse, A., Turecki, G., Lal, S., Bloom, D., Labelle, A., Lalonde, P., Fortin, D., Alda, M., Palmour, R. and Rouleau, G.A. (1999) T102C polymorphism in

the 5HT2A gene and schizophrenia: relation to phenotype and drug response variability. J. Psychiatry Neurosci., 24: 141–146.
Joseph, M.H., Owen, F., Baker, H.F. and Bourne, R.C. (1977) Platelet serotonin concentration and monoamine oxidase activity in unmedicated chronic schizophrenic and in schizo-affective patients. Psychol. Med., 7: 159–162.
Joyce, J.N., Shane, A., Lexow, N., Winokur, A., Casanova, M.F. and Kleinman, J.E. (1993) Serotonin uptake sites and serotonin receptors are altered in the limbic system of schizophrenics. Neuropsychopharmacology, 8: 315–336.
Kahn, R.S. and Wetzler, S. (1991) m-Chlorophenylpiperazine as a probe of serotonin function. Biol. Psychiatry, 30: 1139–1166.
Kahn, R.S., Siever, L.J., Gabriel, S., Amin, F., Stern, R.G., DuMont, K., Apter, S. and Davidson, M. (1992) Serotonin function in schizophrenia: effects of meta-chlorophenylpiperazine in schizophrenic patients and healthy subjects. Psychiatry Res., 43: 1–12.
Kahn, R.S., Davidson, M., Siever, L., Gabriel, S., Apter, S. and Davis, K.L. (1993a) Serotonin function and treatment response to clozapine in schizophrenic patients. Am. J. Psychiatry, 150: 1337–1342.
Kahn, R.S., Davidson, M., Knott, P., Stern, R.G., Apter, S. and Davis, K.L. (1993b) Effect of neuroleptic medication on cerebrospinal fluid monoamine metabolite concentrations in schizophrenia. Serotonin–dopamine interactions as a target for treatment. Arch. Gen. Psychiatry, 50: 599–605.
Kahn, R.S., Davidson, M., Siever, L.J., Sevy, S. and Davis, K.L. (1994) Clozapine treatment and its effect on neuroendocrine responses induced by the serotonin agonist, m-chlorophenylpiperazine. Biol. Psychiatry, 35: 909–912.
Kaiser, R., Tremblay, P.B., Schmider, J., Henneken, M., Dettling, M., Muller-Oerlinghausen, B., Uebelhack, R., Roots, I. and Brockmoller, J. (2001) Serotonin transporter polymorphisms: no association with response to antipsychotic treatment, but associations with the schizoparanoid and residual subtypes of schizophrenia. Mol. Psychiatry, 6: 179–185.
Kane, J., Honigfeld, G., Singer, J. and Meltzer, H. (1988) Clozapine for the treatment-resistant schizophrenic. A double-blind comparison with chlorpromazine. Arch. Gen. Psychiatry, 45: 789–796.
Kapur, S. and Seeman, P. (2001) Does fast dissociation from the dopamine d(2) receptor explain the action of atypical antipsychotics?: a new hypothesis. Am. J. Psychiatry, 158: 360–369.
Kapur, S., Zipursky, R.B., Remington, G., Jones, C., DaSilva, J., Wilson, A.A. and Houle, S. (1998) 5-HT2 and D2 receptor occupancy of olanzapine in schizophrenia: a PET investigation. Am. J. Psychiatry, 155: 921–928.
Kapur, S., Zipursky, R., Jones, C., Remington, G. and Houle, S. (2000) Relationship between dopamine D(2) occupancy, clinical response, and side effects: a double-blind PET study of first-episode schizophrenia. Am. J. Psychiatry, 157: 514–520.
Kasper, S., Tauscher, J., Kufferle, B., Barnas, C., Pezawas, L. and Quiner, S. (1999) Dopamine- and serotonin-receptors in schizophrenia: results of imaging-studies and implications for pharmacotherapy in schizophrenia. Eur. Arch. Psychiatry Clin. Neurosci., 249(Suppl. 4): 83–89.
Keefe, R.S., Bilder, R.M., Davis, S.M., Harvey, P.D., Palmer, B.W., Gold, J.M., Meltzer, H.Y., Green, M.F., Capuano, G., Stroup, T.S., McEvoy, J.P., Swartz, M.S., Rosenheck, R.A., Perkins, D.O., Davis, C.E., Hsiao, J.K. and Lieberman, J.A. (2007) Neurocognitive effects of antipsychotic medications in patients with chronic schizophrenia in the CATIE Trial. Arch. Gen. Psychiatry, 64: 633–647.
Keeler, M.H. (1965) Similarity of schizophrenia and the psilocybin syndrome as determined by objective methods. Intl. J. Neuropsychiatry, 1: 630–634.
Kern, R.S., Green, M.F., Marshall, B.D. Jr., Wirshing, W.C., Wirshing, D., McGurk, S.R., Marder, S.R. and Mintz, J. (1999) Risperidone versus haloperidol on secondary memory: can newer medications aid learning? Schizophr. Bull., 25: 223–232.
Key, B.J. (1964) The effect of LSD-25 on the interaction between conditioned and non-conditioned stimuli in a simple avoidance situation. Psychopharmacologia, 6: 319–326.
Klodzinska, A., Bijak, M., Tokarski, K. and Pilc, A. (2002) Group II mGlu receptor agonists inhibit behavioural and electrophysiological effects of DOI in mice. Pharmacol. Biochem. Behav., 73: 327–332.
Knauer, A. and Maloney, W.J.M.A. (1913) A prelimnary note on the psychic action of mescaline, with special reference to the mechanism of visual hallucinations. J. Nerv. Ment. Dis., 40: 425–436.
Kolakowska, T., Cowen, P.J. and Murdock, P. (1987) Endocrine responses to tryptophan infusion in schizophrenic patients treated with neuroleptics. Psychoneuroendocrinology, 12: 193–202.
Koreen, A.R., Lieberman, J.A., Alvir, J. and Chakos, M. (1997) The behavioral effect of m-chlorophenylpiperazine (mCPP) and methylphenidate in first-episode schizophrenia and normal controls. Neuropsychopharmacology, 16: 61–68.
Kouzmenko, A.P., Hayes, W.L., Pereira, A.M., Dean, B., Burnet, P.W. and Harrison, P.J. (1997) 5-HT2A receptor polymorphism and steady state receptor expression in schizophrenia. Lancet, 349: 1815.
Krystal, J.H., Seibyl, J.P., Price, L.H., Woods, S.W., Heninger, G.R., Aghajanian, G.K. and Charney, D.S. (1993) m-Chlorophenylpiperazine effects in neuroleptic-free schizophrenic patients. Evidence implicating serotonergic systems in the positive symptoms of schizophrenia. Arch. Gen. Psychiatry, 50: 624–635.
Krystal, J.H., Karper, L.P., Seibyl, J.P., Freeman, G.K., Delaney, R., Bremner, J.D., Heninger, G.R., Bowers, M.B. Jr. and Charney, D.S. (1994) Subanesthetic effects of the noncompetitive NMDA antagonist, ketamine, in humans. Psychotomimetic, perceptual, cognitive, and neuroendocrine responses. Arch. Gen. Psychiatry, 51: 199–214.
Lambe, E.K., Goldman-Rakic, P.S. and Aghajanian, G.K. (2000) Serotonin induces EPSCs preferentially in layer V pyramidal neurons of the frontal cortex in the rat. Cereb. Cortex, 10: 974–980.
Lane, H.Y., Chang, Y.C., Chiu, C.C., Chen, M.L., Hsieh, M.H. and Chang, W.H. (2002) Association of risperidone treatment response with a polymorphism in the 5-HT(2A) receptor gene. Am. J. Psychiatry, 159: 1593–1595.
Lane, H.Y., Lin, C.C., Huang, C.H., Chang, Y.C., Hsu, S.K. and Chang, W.H. (2004) Risperidone response and 5-HT6 receptor gene variance: genetic association analysis with adjustment for nongenetic confounders. Schizophr. Res., 67: 63–70.
Laruelle, M., Abi-Dargham, A., Casanova, M.F., Toti, R., Weinberger, D.R. and Kleinman, J.E. (1993a) Selective abnormalities of prefrontal serotonergic receptors in schizophrenia. A postmortem study. Arch. Gen. Psychiatry, 50: 810–818.
Laruelle, M., Baldwin, R.M., Malison, R.T., Zea-Ponce, Y., Zoghbi, S.S., al-Tikriti, M.S., Sybirska, E.H., Zimmermann, R.C., Wisniewski, G., Neumeyer, J.L. et al. (1993b) SPECT imaging of dopamine and serotonin transporters with

[123I]beta-CIT: pharmacological characterization of brain uptake in nonhuman primates. Synapse, 13: 295–309.
Laruelle, M., Abi-Dargham, A., Gil, R., Kegeles, L. and Innis, R. (1999) Increased dopamine transmission in schizophrenia: relationship to illness phases. Biol. Psychiatry, 46: 56–72.
Laruelle, M., Abi-Dargham, A., van Dyck, C., Gil, R., D'Souza, D.C., Krystal, J., Seibyl, J., Baldwin, R. and Innis, R. (2000) Dopamine and serotonin transporters in patients with schizophrenia: an imaging study with [(123)I]beta-CIT. Biol. Psychiatry, 47: 371–379.
Lattuada, E., Cavallaro, R., Serretti, A., Lorenzi, C. and Smeraldi, E. (2004) Tardive dyskinesia and DRD2, DRD3, DRD4, 5-HT2A variants in schizophrenia: an association study with repeated assessment. Intl. J. Neuropsychopharmacol., 7: 489–493.
Lemonde, S., Turecki, G., Bakish, D., Du, L., Hrdina, P.D., Bown, C.D., Sequeira, A., Kushwaha, N., Morris, S.J., Basak, A., Ou, X.M. and Albert, P.R. (2003) Impaired repression at a 5-hydroxytryptamine 1A receptor gene polymorphism associated with major depression and suicide. J. Neurosci., 23: 8788–8799.
Lerer, B., Ran, A., Blacker, M., Silver, H., Weller, M.P., Drummer, D., Ebstein, B. and Calev, A. (1988) Neuroendocrine responses in chronic schizophrenia. Evidence for serotonergic dysfunction. Schizophr. Res., 1: 405–410.
Lerer, B., Segman, R.H., Tan, E.C., Basile, V.S., Cavallaro, R., Aschauer, H.N., Strous, R., Chong, S.A., Heresco-Levy, U., Verga, M., Scharfetter, J., Meltzer, H.Y., Kennedy, J.L. and Macciardi, F. (2005) Combined analysis of 635 patients confirms an age-related association of the serotonin 2A receptor gene with tardive dyskinesia and specificity for the non-orofacial subtype. Int. J. Neuropsychopharmacol, 8: 411–425.
Leucht, S. (2004) Amisulpride a selective dopamine antagonist and atypical antipsychotic: results of a meta-analysis of randomized controlled trials. Int. J. Neuropsychopharmacol., 7(Suppl 1): S15–S20.
Leucht, S., Pitschel-Walz, G., Abraham, D. and Kissling, W. (1999) Efficacy and extrapyramidal side-effects of the new antipsychotics olanzapine, quetiapine, risperidone, and sertindole compared to conventional antipsychotics and placebo. A meta-analysis of randomized controlled trials. Schizophr. Res., 35: 51–68.
Leucht, S., Pitschel-Walz, G., Engel, R.R. and Kissling, W. (2002) Amisulpride, an unusual 'atypical' antipsychotic: a meta-analysis of randomized controlled trials. Am. J. Psychiatry, 159: 180–190.
Leucht, S., Corves, C., Arbter, D., Engel, R.R., Li, C. and Davis, J.M. (2009) Second-generation versus first-generation antipsychotic drugs for schizophrenia: a meta-analysis. Lancet, 373: 31–41.
Lewine, R.R., Risch, S.C., Risby, E., Stipetic, M., Jewart, R.D., Eccard, M., Caudle, J. and Pollard, W. (1991) Lateral ventricle–brain ratio and balance between CSF HVA and 5-HIAA in schizophrenia. Am. J. Psychiatry, 148: 1189–1194.
Lewis, D.A. (2000) GABAergic local circuit neurons and prefrontal cortical dysfunction in schizophrenia. Brain Res. Brain Res. Rev., 31: 270–276.
Lewis, R., Kapur, S., Jones, C., DaSilva, J., Brown, G.M., Wilson, A.A., Houle, S. and Zipursky, R.B. (1999) Serotonin 5-HT2 receptors in schizophrenia: a PET study using [18F]setoperone in neuroleptic-naive patients and normal subjects. Am. J. Psychiatry, 156: 72–78.
Liddle, P.F. (2000) Cognitive impairment in schizophrenia: its impact on social functioning. Acta Psychiatr. Scand. Suppl., 400: 11–16.
Lidow, M.S. (2000) General overview of contemporary antipsychotic medications. In: Lidow, M.S. (Ed.), Neurotransmitter Receptors in Actions of Antipsychotic Medications. CRC Press LLC, Boca Raton, FL, pp. 17–29.
Lieberman, J.A., Kane, J.M., Safferman, A.Z., Pollack, S., Howard, A., Szymanski, S., Masiar, S.J., Kronig, M.H., Cooper, T. and Novacenko, H. (1994) Predictors of response to clozapine. J. Clin. Psychiatry, 55(Suppl. B): 126–128.
Lieberman, J.A., Mailman, R.B., Duncan, G., Sikich, L., Chakos, M., Nichols, D.E. and Kraus, J.E. (1998) Serotonergic basis of antipsychotic drug effects in schizophrenia. Biol. Psychiatry, 44: 1099–1117.
Lin, C.H., Tsai, S.J., Yu, Y.W., Song, H.L., Tu, P.C., Sim, C.B., Hsu, C.P., Yang, K.H. and Hong, C.J. (1999) No evidence for association of serotonin-2A receptor variant (102T/C) with schizophrenia or clozapine response in a Chinese population. Neuroreport, 10: 57–60.
Losonczy, M.F., Song, I.S., Mohs, R.C., Small, N.A., Davidson, M., Johns, C.A. and Davis, K.L. (1986) Correlates of lateral ventricular size in chronic schizophrenia, I: behavioral and treatment response measures. Am. J. Psychiatry, 143: 976–981.
Ludewig, K., Geyer, M.A. and Vollenweider, F.X. (2003) Deficits in prepulse inhibition and habituation in never-medicated, first-episode schizophrenia. Biol. Psychiatry, 54: 121–128.
Mackay, A.V., Doble, A., Bird, E.D., Spokes, E.G., Quik, M. and Iversen, L.L. (1978) 3H-Spiperone binding in normal and schizophrenic post-mortem human brain. Life Sci., 23: 527–532.
Maes, M. and Meltzer, H.Y. (1996) Effects of meta-chlorophenylpiperazine on neuroendocrine and behavioral responses in male schizophrenic patients and normal volunteers. Psychiatry Res., 64: 147–159.
Malhotra, A.K., Goldman, D., Ozaki, N., Breier, A., Buchanan, R. and Pickar, D. (1996) Lack of association between polymorphisms in the 5-HT2A receptor gene and the antipsychotic response to clozapine. Am. J. Psychiatry, 153: 1092–1094.
Mamo, D., Kapur, S., Shammi, C.M., Papatheodorou, G., Mann, S., Therrien, F. and Remington, G. (2004) A PET study of dopamine D2 and serotonin 5-HT2 receptor occupancy in patients with schizophrenia treated with therapeutic doses of ziprasidone. Am. J. Psychiatry, 161: 818–825.
Mamo, D., Graff, A., Mizrahi, R., Shammi, C.M., Romeyer, F. and Kapur, S. (2007) Differential effects of aripiprazole on D(2), 5-HT(2), and 5-HT(1A) receptor occupancy in patients with schizophrenia: a triple tracer PET study. Am. J. Psychiatry, 164: 1411–1417.
Marcos, B., Chuang, T.T., Gil-Bea, F.J. and Ramirez, M.J. (2008) Effects of 5-HT6 receptor antagonism and cholinesterase inhibition in models of cognitive impairment in the rat. Br. J. Pharmacol., 155: 434–440.
Marder, S.R. and Meibach, R.C. (1994) Risperidone in the treatment of schizophrenia. Am. J. Psychiatry, 151: 825–835.
Marek, G.J., Wright, R.A., Schoepp, D.D., Monn, J.A. and Aghajanian, G.K. (2000) Physiological antagonism between 5-hydroxytryptamine(2A) and group II metabotropic glutamate receptors in prefrontal cortex. J. Pharmacol. Exp. Ther., 292: 76–87.
Marek, G.J., Wright, R.A., Gewirtz, J.C. and Schoepp, D.D. (2001) A major role for thalamocortical afferents in

serotonergic hallucinogen receptor function in the rat neocortex. Neuroscience, 105: 379–392.

Marenco, S. and Weinberger, D.R. (2000) The neurodevelopmental hypothesis of schizophrenia: following a trail of evidence from cradle to grave. Dev. Psychopathol., 12: 501–527.

Markianos, M., Hatzimanolis, J. and Lykouras, L. (2001) Neuroendocrine serotonergic and dopaminergic responsivity in male schizophrenic patients during treatment with neuroleptics and after switch to risperidone. Psychopharmacology (Berl.), 157: 55–59.

Martin, P., Waters, N., Carlsson, A. and Carlsson, M.L. (1997) The apparent antipsychotic action of the 5-HT2a receptor antagonist M100907 in a mouse model of schizophrenia is counteracted by ritanserin. (Rapid communication.) J. Neural Transm., 104: 561–564.

Martin, P., Waters, N., Schmidt, C.J., Carlsson, A. and Carlsson, M.L. (1998) Rodent data and general hypothesis: antipsychotic action exerted through 5-Ht2A receptor antagonism is dependent on increased serotonergic tone. J. Neural Transm., 105: 365–396.

Masellis, M., Paterson, A.D., Badri, F., Lieberman, J.A., Meltzer, H.Y., Cavazzoni, P. and Kennedy, J.L. (1995) Genetic variation of 5-HT2A receptor and response to clozapine. Lancet, 346: 1108.

Masellis, M., Basile, V., Meltzer, H.Y., Lieberman, J.A., Sevy, S., Macciardi, F.M., Cola, P., Howard, A., Badri, F., Nothen, M.M., Kalow, W. and Kennedy, J.L. (1998) Serotonin subtype 2 receptor genes and clinical response to clozapine in schizophrenia patients. Neuropsychopharmacology, 19: 123–132.

Masellis, M., Basile, V.S., Meltzer, H.Y., Lieberman, J.A., Sevy, S., Goldman, D.A., Hamblin, M.W., Macciardi, F.M. and Kennedy, J.L. (2001) Lack of association between the T→C 267 serotonin 5-HT6 receptor gene (HTR6) polymorphism and prediction of response to clozapine in schizophrenia. Schizophr. Res., 47: 49–58.

Matsumoto, I., Inoue, Y., Iwazaki, T., Pavey, G. and Dean, B. (2005) 5-HT2A and muscarinic receptors in schizophrenia: a postmortem study. Neurosci. Lett., 379: 164–168.

Maynard, T.M., Sikich, L., Lieberman, J.A. and LaMantia, A.S. (2001) Neural development, cell–cell signaling, and the 'two-hit' hypothesis of schizophrenia. Schizophr. Bull., 27: 457–476.

McCabe, M.S., Fowler, R.C., Cadoret, R.J. and Winokur, G. (1972) Symptom differences in schizophrenia with good and poor prognosis. Am. J. Psychiatry, 128: 1239–1243.

McGhie, A. and Chapman, J. (1961) Disorders of attention and perception in early schizophrenia. Br. J. Med. Psychol., 34: 103–116.

McGlashan, T.H. and Hoffman, R.E. (2000) Schizophrenia as a disorder of developmentally reduced synaptic connectivity. Arch. Gen. Psychiatry, 57: 637–648.

Meltzer, H.Y. (1991) The mechanism of action of novel antipsychotic drugs. Schizophr. Bull., 17: 263–287.

Meltzer, H.Y. (1999) The role of serotonin in antipsychotic drug action. Neuropsychopharmacology, 21: 106S–115S.

Meltzer, H.Y. and McGurk, S.R. (1999) The effects of clozapine, risperidone, and olanzapine on cognitive function in schizophrenia. Schizophr. Bull., 25: 233–255.

Meltzer, H.Y. and Stahl, S.M. (1976) The dopamine hypothesis of schizophrenia: a review. Schizophr. Bull., 2: 19–76.

Meltzer, H.Y., Bastani, B., Ramirez, L. and Matsubara, S. (1989) Clozapine: new research on efficacy and mechanism of action. Eur. Arch. Psychiatry Neurol. Sci., 238: 332–339.

Meltzer, H.Y., Li, Z., Kaneda, Y. and Ichikawa, J. (2003) Serotonin receptors: their key role in drugs to treat schizophrenia. Prog. Neuropsychopharmacol. Biol. Psychiatry, 27: 1159–1172.

Mikuni, M. and Meltzer, H.Y. (1984) Reduction of serotonin-2 receptors in rat cerebral cortex after subchronic administration of imipramine, chlorpromazine, and the combination thereof. Life Sci., 34: 87–92.

Millan, M.J., Dekeyne, A. and Gobert, A. (1998) Serotonin (5-HT)2C receptors tonically inhibit dopamine (DA) and noradrenaline (NA), but not 5-HT, release in the frontal cortex in vivo. Neuropharmacology, 37: 953–955.

Miller, D.D., Ellingrod, V.L., Holman, T.L., Buckley, P.F. and Arndt, S. (2005) Clozapine-induced weight gain associated with the 5HT2C receptor -759C/T polymorphism. Am. J. Med. Genet. B Neuropsychiatr. Genet., 133B: 97–100.

Mirnics, K., Middleton, F.A., Lewis, D.A. and Levitt, P. (2001) Analysis of complex brain disorders with gene expression microarrays: schizophrenia as a disease of the synapse. Trends Neurosci., 24: 479–486.

Mita, T., Hanada, S., Nishino, N., Kuno, T., Nakai, H., Yamadori, T., Mizoi, Y. and Tanaka, C. (1986) Decreased serotonin S2 and increased dopamine D2 receptors in chronic schizophrenics. Biol. Psychiatry, 21: 1407–1414.

Miyamoto, S., LaMantia, A.S., Duncan, G.E., Sullivan, P., Gilmore, J.H. and Lieberman, J.A. (2003) Recent advances in the neurobiology of schizophrenia. Mol. Interv., 3: 27–39.

Moghaddam, B. and Adams, B.W. (1998) Reversal of phencyclidine effects by a group II metabotropic glutamate receptor agonist in rats. Science, 281: 1349–1352.

Moghaddam, B., Adams, B., Verma, A. and Daly, D. (1997) Activation of glutamatergic neurotransmission by ketamine: a novel step in the pathway from NMDA receptor blockade to dopaminergic and cognitive disruptions associated with the prefrontal cortex. J. Neurosci., 17: 2921–2927.

Mohr, P., Horacek, J., Motlova, L., Libiger, J. and Czobor, P. (1998) Prolactin response to D-fenfluramine challenge test as a predictor of treatment response to haloperidol in acute schizophrenia. Schizophr. Res., 30: 91–99.

Moller, H.J., Muller, H., Borison, R.L., Schooler, N.R. and Chouinard, G. (1995) A path-analytical approach to differentiate between direct and indirect drug effects on negative symptoms in schizophrenic patients. A re-evaluation of the North American risperidone study. Eur. Arch. Psychiatry Clin. Neurosci., 245: 45–49.

Monteleone, P., Tortorella, A., Borriello, R., Cassandro, P. and Maj, M. (1999) Prolactin hyperresponsiveness to D-fenfluramine in drug-free schizophrenic patients: a placebo-controlled study. Biol. Psychiatry, 45: 1606–1611.

Morey, R.A., Inan, S., Mitchell, T.V., Perkins, D.O., Lieberman, J.A. and Belger, A. (2005) Imaging frontostriatal function in ultra-high-risk, early, and chronic schizophrenia during executive processing. Arch. Gen. Psychiatry, 62: 254–262.

Mossner, R., Schuhmacher, A., Kuhn, K.U., Cvetanovska, G., Rujescu, D., Zill, P., Quednow, B.B., Rietschel, M., Wolwer, W., Gaebel, W., Wagner, M. and Maier, W. (2009) Functional serotonin 1A receptor variant influences treatment response to atypical antipsychotics in schizophrenia. Pharmacogenet. Genomics, 19: 91–94.

Murphy, D.L., Mueller, E.A., Garrick, N.A. and Aulakh, C.S. (1986) Use of serotonergic agents in the clinical assessment of central serotonin function. J. Clin. Psychiatry, 47(Suppl.): 9–15.

Murray, R.M., O'Callaghan, E., Castle, D.J. and Lewis, S.W. (1992) A neurodevelopmental approach to the classification of schizophrenia. Schizophr. Bull., 18: 319–332.

Muschamp, J.W., Regina, M.J., Hull, E.M., Winter, J.C. and Rabin, R.A. (2004) Lysergic acid diethylamide and [-]-2,5-dimethoxy-4-methylamphetamine increase extracellular glutamate in rat prefrontal cortex. Brain Res., 1023: 134–140.

Naylor, L., Dean, B., Opeskin, K., Pavey, G., Hill, C., Keks, N. and Copolov, D. (1996) Changes in the serotonin transporter in the hippocampus of subjects with schizophrenia identified using [3H]paroxetine. J. Neural Transm., 103: 749–757.

Neumeyer, J.L., Wang, S.Y., Milius, R.A., Baldwin, R.M., Zea-Ponce, Y., Hoffer, P.B., Sybirska, E., al-Tikriti, M., Charney, D.S., Malison, R.T. et al. (1991) [123I]-2 beta-carbomethoxy-3 beta-(4-iodophenyl)tropane: high-affinity SPECT radiotracer of monoamine reuptake sites in brain. J. Med. Chem., 34: 3144–3146.

Newcomer, J.W., Faustman, W.O., Zipursky, R.B. and Csernansky, J.G. (1992) Zacopride in schizophrenia: a single-blind serotonin type 3 antagonist trial. Arch. Gen. Psychiatry, 49: 751–752.

Newman-Tancredi, A., Gavaudan, S., Conte, C., Chaput, C., Touzard, M., Verriele, L., Audinot, V. and Millan, M.J. (1998) Agonist and antagonist actions of antipsychotic agents at 5-HT1A receptors: a [35S]GTPgammaS binding study. Eur. J. Pharmacol., 355: 245–256.

Ngan, E.T., Yatham, L.N., Ruth, T.J. and Liddle, P.F. (2000) Decreased serotonin 2A receptor densities in neuroleptic-naive patients with schizophrenia: a PET study using [(18)F]setoperone. Am. J. Psychiatry, 157: 1016–1018.

Nordstrom, A.L., Farde, L., Wiesel, F.A., Forslund, K., Pauli, S., Halldin, C. and Uppfeldt, G. (1993) Central D2-dopamine receptor occupancy in relation to antipsychotic drug effects: a double-blind PET study of schizophrenic patients. Biol. Psychiatry, 33: 227–235.

Nothen, M.M., Rietschel, M., Erdmann, J., Oberlander, H., Moller, H.J., Nober, D. and Propping, P. (1995) Genetic variation of the 5-HT2A receptor and response to clozapine. Lancet, 346: 908–909.

Nudmamud, S., Reynolds, L.M. and Reynolds, G.P. (2003) N-acetylaspartate and N-acetylaspartylglutamate deficits in superior temporal cortex in schizophrenia and bipolar disorder: a postmortem study. Biol. Psychiatry, 53: 1138–1141.

Nyback, H., Berggren, B.M., Hindmarsh, T., Sedvall, G. and Wiesel, F.A. (1983) Cerebroventricular size and cerebrospinal fluid monoamine metabolites in schizophrenic patients and healthy volunteers. Psychiatry Res., 9: 301–308.

Nyberg, S., Eriksson, B., Oxenstierna, G., Halldin, C. and Farde, L. (1999) Suggested minimal effective dose of risperidone based on PET-measured D2 and 5-HT2A receptor occupancy in schizophrenic patients. Am. J. Psychiatry, 156: 869–875.

O'Dell, S.J., La Hoste, G.J., Widmark, C.B., Shapiro, R.M., Potkin, S.G. and Marshall, J.F. (1990) Chronic treatment with clozapine or haloperidol differentially regulates dopamine and serotonin receptors in rat brain. Synapse, 6: 146–153.

Ohuoha, D.C., Hyde, T.M. and Kleinman, J.E. (1993) The role of serotonin in schizophrenia: an overview of the nomenclature, distribution and alterations of serotonin receptors in the central nervous system. Psychopharmacology (Berl.), 112: S5–S15.

Okubo, Y., Suhara, T., Suzuki, K., Kobayashi, K., Inoue, O., Terasaki, O., Someya, Y., Sassa, T., Sudo, Y., Matsushima, E., Iyo, M., Tateno, Y. and Toru, M. (2000) Serotonin 5-HT2 receptors in schizophrenic patients studied by positron emission tomography. Life Sci., 66: 2455–2464.

Osmond, H. and Smythies, J. (1952) Schizophrenia: a new approach. J Ment. Sci., 98: 309–315.

Ostrowitzki, S., Rao, M.L., Redei, J. and Andres, A.H. (1993) Concurrence of cortex and platelet serotonin2 receptor binding characteristics in the individual and the putative regulation by serotonin. J. Neural Transm. Gen. Sect., 93: 27–35.

Owen, F., Cross, A.J., Crow, T.J., Lofthouse, R. and Poulter, M. (1981) Neurotransmitter receptors in brain in schizophrenia. Acta Psychiatr. Scand. Suppl., 291: 20–28.

Owen, R.R. Jr., Gutierrez-Esteinou, R., Hsiao, J., Hadd, K., Benkelfat, C., Lawlor, B.A., Murphy, D.L. and Pickar, D. (1993) Effects of clozapine and fluphenazine treatment on responses to m-chlorophenylpiperazine infusions in schizophrenia. Arch. Gen. Psychiatry, 50: 636–644.

Padin, J.F., Rodriguez, M.A., Dominguez, E., Dopeso-Reyes, I.G., Buceta, M., Cano, E., Sotelo, E., Brea, J., Caruncho, H.J., Isabel Cadavid, M., Castro, M. and Isabel Loza, M. (2006) Parallel regulation by olanzapine of the patterns of expression of 5-HT2A and D3 receptors in rat central nervous system and blood cells. Neuropharmacology, 51: 923–932.

Parsons, M.J., D'Souza, U.M., Arranz, M.J., Kerwin, R.W. and Makoff, A.J. (2004) The −1438A/G polymorphism in the 5-hydroxytryptamine type 2A receptor gene affects promoter activity. Biol. Psychiatry, 56: 406–410.

Parwani, A., Duncan, E.J., Bartlett, E., Madonick, S.H., Efferen, T.R., Rajan, R., Sanfilipo, M., Chappell, P.B., Chakravorty, S., Gonzenbach, S., Ko, G.N. and Rotrosen, J.P. (2000) Impaired prepulse inhibition of acoustic startle in schizophrenia. Biol. Psychiatry, 47: 662–669.

Patil, S.T., Zhang, L., Martenyi, F., Lowe, S.L., Jackson, K.A., Andreev, B.V., Avedisova, A.S., Bardenstein, L.M., Gurovich, I.Y., Morozova, M.A., Mosolov, S.N., Neznanov, N.G., Reznik, A.M., Smulevich, A.B., Tochilov, V.A., Johnson, B.G., Monn, J.A. and Schoepp, D.D. (2007) Activation of mGlu2/3 receptors as a new approach to treat schizophrenia: a randomized Phase 2 clinical trial. Nat. Med., 13: 1102–1107.

Pickar, D., Roy, A., Breier, A., Doran, A., Wolkowitz, O., Colison, J. and Agren, H. (1986) Suicide and aggression in schizophrenia. Neurobiologic correlates. Ann. N. Y. Acad. Sci., 487: 189–196.

Pickar, D., Owen, R.R., Litman, R.E., Konicki, E., Gutierrez, R. and Rapaport, M.H. (1992) Clinical and biologic response to clozapine in patients with schizophrenia. Crossover comparison with fluphenazine. Arch. Gen. Psychiatry, 49: 345–353.

Potkin, S.G., Weinberger, D.R., Linnoila, M. and Wyatt, R.J. (1983) Low CSF 5-hydroxyindoleacetic acid in schizophrenic patients with enlarged cerebral ventricles. Am. J. Psychiatry, 140: 21–25.

Powell, S.B. and Geyer, M.A. (2007) Overview of animal models of schizophrenia. Curr. Protoc. Neurosci., Chapter 9, Unit 9 24.

Pralong, D., Tomaskovic-Crook, E., Opeskin, K., Copolov, D. and Dean, B. (2000) Serotonin(2A) receptors are reduced in the planum temporale from subjects with schizophrenia. Schizophr. Res., 44: 35–45.

Purdon, S.E., Jones, B.D., Stip, E., Labelle, A., Addington, D., David, S.R., Breier, A. and Tollefson, G.D. (2000) Neuropsychological change in early phase schizophrenia during 12 months of treatment with olanzapine, risperidone, or haloperidol. The Canadian Collaborative Group for research in schizophrenia. Arch. Gen. Psychiatry, 57: 249–258.

Quednow, B.B., Wagner, M., Westheide, J., Beckmann, K., Bliesener, N., Maier, W. and Kuhn, K.U. (2006) Sensorimotor gating and habituation of the startle response in schizophrenic

patients randomly treated with amisulpride or olanzapine. Biol. Psychiatry, 59: 536–545.
Quednow, B.B., Kuhn, K.U., Mossner, R., Schwab, S. G., Schuhmacher, A., Maier, W. and Wagner, M. (2008) Sensorimotor gating of schizophrenia patients is influenced by 5-HT2A receptor polymorphisms. Biol. Psychiatry, 64: 434–437.
Quednow, B.B., Schmechtig, A., Ettinger, U., Petrovsky, N., Collier, D.A., Vollenweider, F.X., Wagner, M. and Kumari, V. (2009) Sensorimotor gating depends on polymorphisms of the serotonin-2A receptor and catechol-O-methyltransferase, but not on neuregulin-1 Arg38Gln genotype: a replication study. Biol. Psychiatry, 66: 614–620.
Rahman, S. and Neuman, R.S. (1993) Activation of 5-HT2 receptors facilitates depolarization of neocortical neurons by N-methyl-D-aspartate. Eur. J. Pharmacol., 231: 347–354.
Rapport, M.M. (1949) Serum vasoconstrictor (serotonin) the presence of creatinine in the complex; a proposed structure of the vasoconstrictor principle. J. Biol. Chem., 180: 961–969.
Rapport, M.M., Green, A.A. and Page, I.H. (1948) Serum vasoconstrictor, serotonin; isolation and characterization. J. Biol. Chem., 176: 1243–1251.
Renyi, L., Evenden, J.L., Fowler, C.J., Jerning, E., Kelder, D., Lake-Bakaar, D., Larsson, L.G., Mohell, N., Sallemark, M. and Ross, S.B. (2001) The pharmacological profile of (R)-3, 4-dihydro-N-isopropyl-3-(N-isopropyl-N-propylamino)-2H-1-benzopyran-5-carboxamide, a selective 5-hydroxytryptamine(1A) receptor agonist. J. Pharmacol. Exp. Ther., 299: 883–893.
Reuter, M., Ott, U., Vaitl, D. and Hennig, J. (2007) Impaired executive control is associated with a variation in the promoter region of the tryptophan hydroxylase 2 gene. J. Cogn. Neurosci., 19: 401–408.
Reynolds, G.P., Garrett, N.J., Rupniak, N., Jenner, P. and Marsden, C.D. (1983a) Chronic clozapine treatment of rats down-regulates cortical 5-HT2 receptors. Eur. J. Pharmacol., 89: 325–326.
Reynolds, G.P., Rossor, M.N. and Iversen, L.L. (1983b) Preliminary studies of human cortical 5-HT2 receptors and their involvement in schizophrenia and neuroleptic drug action. J. Neural Transm. Suppl., 18: 273–277.
Reynolds, G.P., Beasley, C.L. and Zhang, Z.J. (2002a) Understanding the neurotransmitter pathology of schizophrenia: selective deficits of subtypes of cortical GABAergic neurons. J. Neural Transm., 109: 881–889.
Reynolds, G.P., Zhang, Z.J. and Zhang, X.B. (2002b) Association of antipsychotic drug-induced weight gain with a 5-HT2C receptor gene polymorphism. Lancet, 359: 2086–2087.
Reynolds, G.P., Zhang, Z. and Zhang, X. (2003) Polymorphism of the promoter region of the serotonin 5-HT(2C) receptor gene and clozapine-induced weight gain. Am. J. Psychiatry, 160: 677–679.
Reynolds, G.P., Yao, Z., Zhang, X., Sun, J. and Zhang, Z. (2005) Pharmacogenetics of treatment in first-episode schizophrenia: D3 and 5-HT2C receptor polymorphisms separately associate with positive and negative symptom response. Eur. Neuropsychopharmacol., 15: 143–151.
Reynolds, G.P., Arranz, B., Templeman, L.A., Fertuzinhos, S. and San, L. (2006) Effect of 5-HT1A receptor gene polymorphism on negative and depressive symptom response to antipsychotic treatment of drug-naive psychotic patients. Am. J. Psychiatry, 163: 1826–1829.
Rietschel, M., Naber, D., Fimmers, R., Moller, H.J., Propping, P. and Nothen, M.M. (1997) Efficacy and side-effects of clozapine not associated with variation in the 5-HT2C receptor. Neuroreport, 8: 1999–2003.
Rinkel, M., DeShon, H.J., Hyde, R.W. and Solomon, H.C. (1952) Experimental schizophrenia-like symptoms. Am. J. Psychiatry, 108: 572–578.
Rinkel, M., Hyde, R.W., Solomon, H.C. and Hoagland, H. (1955) Experimental psychiatry. II. Clinical and physio-chemical observations in experimental psychosis. Am. J. Psychiatry,, 111: 881–895.
Risch, S.C. (1995) Clozapine therapy and increases in homovanillic acid. Arch. Gen. Psychiatry, 52: 244.
Risch, S.C. and Lewine, R.R. (1993) Low cerebrospinal fluid homovanillic acid-5-hydroxyindoleacetic acid ratio predicts clozapine efficacy: a replication. Arch. Gen. Psychiatry, 50: 670.
Rollema, H., Lu, Y., Schmidt, A.W. and Zorn, S.H. (1997) Clozapine increases dopamine release in prefrontal cortex by 5-HT1A receptor activation. Eur. J. Pharmacol., 338: R3–R5.
Roth, B. and Meltzer, H.Y. (2000) The role of serotonin in schizophrenia. In: Bloom, F.E. and Kupfer, D.J. (Eds.), Psychopharmacology: The Fourth Generation of Progress. American College of Neuropsychopharmacology, Nashville, TN, pp. 1215–1228.
Roth, B.L., Ciaranello, R.D. and Meltzer, H.Y. (1992) Binding of typical and atypical antipsychotic agents to transiently expressed 5-HT1C receptors. J. Pharmacol. Exp. Ther., 260: 1361–1365.
Roth, B.L., Craigo, S.C., Choudhary, M.S., Uluer, A., Monsma, F.J. Jr., Shen, Y., Meltzer, H.Y. and Sibley, D.R. (1994) Binding of typical and atypical antipsychotic agents to 5-hydroxytryptamine-6 and 5-hydroxytryptamine-7 receptors. J. Pharmacol. Exp. Ther., 268: 1403–1410.
Roth, B.L., Hanizavareh, S.M. and Blum, A.E. (2004) Serotonin receptors represent highly favorable molecular targets for cognitive enhancement in schizophrenia and other disorders. Psychopharmacology (Berl.), 174: 17–24.
Scarr, E., Pavey, G., Copolov, D. and Dean, B. (2004) Hippocampal 5-hydroxytryptamine receptors: abnormalities in postmortem brain from schizophrenic subjects. Schizophr. Res., 71: 383–392.
Scheepers, F.E., Gespen de Wied, C.C. and Kahn, R.S. (2001a) The effect of olanzapine treatment on m-chlorophenylpiperazine-induced hormone release in schizophrenia. J. Clin. Psychopharmacol., 21: 575–582.
Scheepers, F.E., Gispen-de Wied, C.C., Westenberg, H.G. and Kahn, R.S. (2001b) The effect of olanzapine treatment on monoamine metabolite concentrations in the cerebrospinal fluid of schizophrenic patients. Neuropsychopharmacology, 25: 468–475.
Schulz, E., Fleischhaker, C., Clement, H.W. and Remschmidt, H. (1997) Blood biogenic amines during clozapine treatment of early-onset schizophrenia. J. Neural Transm., 104: 1077–1089.
Schumacher, J., Schulze, T.G., Wienker, T.F., Rietschel, M. and Nothen, M.M. (2000) Pharmacogenetics of the clozapine response. Lancet, 356: 506–507.
Scruggs, J.L., Schmidt, D. and Deutch, A.Y. (2003) The hallucinogen 1-[2,5-dimethoxy-4-iodophenyl]-2-aminopropane (DOI) increases cortical extracellular glutamate levels in rats. Neurosci. Lett., 346: 137–140.
Seeman, P. (1987) Dopamine receptors and the dopamine hypothesis of schizophrenia. Synapse, 1: 133–152.

Seeman, P. (2002) Atypical antipsychotics: mechanism of action. Can. J. Psychiatry, 47: 27–38.
Seeman, P., Lee, T., Chau-Wong, M. and Wong, K. (1976) Antipsychotic drug doses and neuroleptic/dopamine receptors. Nature, 261: 717–719.
Seeman, P., Guan, H.C. and Van Tol, H.H. (1993) Dopamine D4 receptors elevated in schizophrenia. Nature, 365: 441–445.
Segal, D.S., Geyer, M.A. and Schuckit, M.A. (1981) Stimulant-induced psychosis: An evaluation of animal models. In: Youdim, M.B.H., Lovenberg, W., Sharman, D.F. and Lagnado, J.R. (Eds.), Essays in Neurochemistry and Neuropharmacology. John Wiley & Sons, New York, NY, pp. 95–130.
Segman, R.H. and Lerer, B. (2002) Age and the relationship of dopamine D3, serotonin 2C and serotonin 2A receptor genes to abnormal involuntary movements in chronic schizophrenia. Mol. Psychiatry, 7: 137–139.
Segman, R.H., Ebstein, R.P., Heresco-Levy, U., Gorfine, M., Avnon, M., Gur, E., Nemanov, L. and Lerer, B. (1997) Schizophrenia, chronic hospitalization and the 5-HT2C receptor gene. Psychiatr. Genet., 7: 75–78.
Segman, R.H., Heresco-Levy, U., Finkel, B., Inbar, R., Neeman, T., Schlafman, M., Dorevitch, A., Yakir, A., Lerner, A., Goltser, T., Shelevoy, A. and Lerer, B. (2000) Association between the serotonin 2C receptor gene and tardive dyskinesia in chronic schizophrenia: additive contribution of 5-HT2Cser and DRD3gly alleles to susceptibility. Psychopharmacology (Berl.), 152: 408–413.
Segman, R.H., Heresco-Levy, U., Finkel, B., Goltser, T., Shalem, R., Schlafman, M., Dorevitch, A., Yakir, A., Greenberg, D., Lerner, A. and Lerer, B. (2001) Association between the serotonin 2A receptor gene and tardive dyskinesia in chronic schizophrenia. Mol. Psychiatry, 6: 225–229.
Semwal, P., Prasad, S., Varma, P.G., Bhagwat, A.M., Deshpande, S.N. and Thelma, B.K. (2002) Candidate gene polymorphisms among North Indians and their association with schizophrenia in a case–control study. J. Genet., 81: 65–71.
Serretti, A., Drago, A. and De Ronchi, D. (2007) HTR2A gene variants and psychiatric disorders: a review of current literature and selection of SNPs for future studies. Curr. Med. Chem., 14: 2053–2069.
Shaltiel, G., Shamir, A., Agam, G. and Belmaker, R.H. (2005) Only tryptophan hydroxylase (TPH)-2 is relevant to the CNS. Am. J. Med. Genet. B Neuropsychiatr. Genet., 136B: 106.
Shapiro, D.A., Renock, S., Arrington, E., Chiodo, L.A., Liu, L.X., Sibley, D.R., Roth, B.L. and Mailman, R. (2003) Aripiprazole, a novel atypical antipsychotic drug with a unique and robust pharmacology. Neuropsychopharmacology, 28: 1400–1411.
Sharma, R.P., Singh, V., Janicak, P.G., Javaid, J.I. and Pandey, G.N. (1999) The prolactin response to fenfluramine in schizophrenia is associated with negative symptoms. Schizophr. Res., 39: 85–89.
Snyder, S.H. (1973) Amphetamine psychosis: a 'model' schizophrenia mediated by catecholamines. Am. J. Psychiatry, 130: 61–67.
Sodhi, M.S. and Sanders-Bush, E. (2004) Serotonin and brain development. Intl. Rev. Neurobiol., 59: 111–174.
Sodhi, M.S., Arranz, M.J., Curtis, D., Ball, D.M., Sham, P., Roberts, G.W., Price, J., Collier, D.A. and Kerwin, R.W. (1995) Association between clozapine response and allelic variation in the 5-HT2C receptor gene. Neuroreport, 7: 169–172.
Spurlock, G., Heils, A., Holmans, P., Williams, J., D'Souza, U.M., Cardno, A., Murphy, K.C., Jones, L., Buckland, P.R., McGuffin, P., Lesch, K.P. and Owen, M.J. (1998) A family based association study of T102C polymorphism in 5HT2A and schizophrenia plus identification of new polymorphisms in the promoter. Mol. Psychiatry, 3: 42–49.
Stoll, W.A. (1947) Lysergsäure-diäthylamid, ein Phantastikum aus der Mutterkorngruppe. Schweiz Arch. Neurol. Psychiatr., 60: 279–323.
Stoll, W.A. (1949) Ein neues, in sehr kleinen Mengen wirksames Phantastikum. Schweiz Arch. Neurol. Psychiatr., 64: 483–484.
Strassman, R.J. (1995) Hallucinogenic drugs in psychiatric research and treatment. Perspectives and prospects. J. Nerv. Ment. Dis., 183: 127–138.
Strassman, R.J., Qualls, C.R., Uhlenhuth, E.H. and Kellner, R. (1994) Dose–response study of N,N-dimethyltryptamine in humans. II. Subjective effects and preliminary results of a new rating scale. Arch. Gen. Psychiatry, 51: 98–108.
Sumiyoshi, T., Stockmeier, C.A., Overholser, J.C., Dilley, G.E. and Meltzer, H.Y. (1996) Serotonin1A receptors are increased in postmortem prefrontal cortex in schizophrenia. Brain Res., 708: 209–214.
Sumiyoshi, T., Matsui, M., Nohara, S., Yamashita, I., Kurachi, M., Sumiyoshi, C., Jayathilake, K. and Meltzer, H.Y. (2001a) Enhancement of cognitive performance in schizophrenia by addition of tandospirone to neuroleptic treatment. Am. J. Psychiatry, 158: 1722–1725.
Sumiyoshi, T., Matsui, M., Yamashita, I., Nohara, S., Kurachi, M., Uehara, T., Sumiyoshi, S., Sumiyoshi, C. and Meltzer, H.Y. (2001b) The effect of tandospirone, a serotonin(1A) agonist, on memory function in schizophrenia. Biol. Psychiatry, 49: 861–868.
Sumiyoshi, T., Park, S., Jayathilake, K., Roy, A., Ertugrul, A. and Meltzer, H.Y. (2007) Effect of buspirone, a serotonin1A partial agonist, on cognitive function in schizophrenia: a randomized, double-blind, placebo-controlled study. Schizophr. Res., 95: 158–168.
Suzuki, T., Iwata, N., Kitamura, Y., Kitajima, T., Yamanouchi, Y., Ikeda, M., Nishiyama, T., Kamatani, N. and Ozaki, N. (2003) Association of a haplotype in the serotonin 5-HT4 receptor gene (HTR4) with Japanese schizophrenia. Am. J. Med. Genet. B Neuropsychiatr. Genet., 121B: 7–13.
Swerdlow, N.R. and Geyer, M.A. (1998) Using an animal model of deficient sensorimotor gating to study the pathophysiology and new treatments of schizophrenia. Schizophr. Bull., 24: 285–301.
Swerdlow, N.R., Geyer, M.A. and Braff, D.L. (2001) Neural circuit regulation of prepulse inhibition of startle in the rat: current knowledge and future challenges. Psychopharmacology (Berl.), 156: 194–215.
Szymanski, S., Lieberman, J., Pollack, S., Munne, R., Safferman, A., Kane, J., Kronig, M. and Cooper, T. (1993) The dopamine–serotonin relationship in clozapine response. Psychopharmacology (Berl.), 112: S85–S89.
Tamminga, C.A. and Holcomb, H.H. (2005) Phenotype of schizophrenia: a review and formulation. Mol. Psychiatry, 10: 27–39.
Tan, E.C., Chong, S.A., Mahendran, R., Dong, F. and Tan, C.H. (2001) Susceptibility to neuroleptic-induced tardive dyskinesia and the T102C polymorphism in the serotonin type 2A receptor. Biol. Psychiatry, 50: 144–147.
Tauscher, J., Kapur, S., Verhoeff, N.P., Hussey, D.F., Daskalakis, Z.J., Tauscher-Wisniewski, S., Wilson, A.A., Houle, S., Kasper, S. and Zipursky, R.B. (2002) Brain serotonin 5-HT(1A) receptor

binding in schizophrenia measured by positron emission tomography and [11C]WAY-100635. Arch. Gen. Psychiatry, 59: 514–520.

Templeman, L.A., Reynolds, G.P., Arranz, B. and San, L. (2005) Polymorphisms of the 5-HT2C receptor and leptin genes are associated with antipsychotic drug-induced weight gain in Caucasian subjects with a first-episode psychosis. Pharmacogenet. Genomics, 15: 195–200.

Theisen, F.M., Hinney, A., Bromel, T., Heinzel-Gutenbrunner, M., Martin, M., Krieg, J.C., Remschmidt, H. and Hebebrand, J. (2004) Lack of association between the -759C/T polymorphism of the 5-HT2C receptor gene and clozapine-induced weight gain among German schizophrenic individuals. Psychiatr. Genet., 14: 139–142.

Tilson, H.A., Baker, T.G. and Chamberlain, J.H. (1975) Behavioral and neuropharmacological analysis of amphetamine and 2,5-dimethoxy-4-methylamphetamine in rats. Psychopharmacologia, 44: 229–239.

Tollefson, G.D. and Sanger, T.M. (1997) Negative symptoms: a path analytic approach to a double-blind, placebo- and haloperidol-controlled clinical trial with olanzapine. Am. J. Psychiatry, 154: 466–474.

Trichard, C., Paillere-Martinot, M.L., Attar-Levy, D., Blin, J., Feline, A. and Martinot, J.L. (1998) No serotonin 5-HT2A receptor density abnormality in the cortex of schizophrenic patients studied with PET. Schizophr. Res., 31: 13–17.

Tsai, S.J., Hong, C.J., Yu, Y.W., Lin, C.H., Song, H.L., Lai, H.C. and Yang, K.H. (2000) Association study of a functional serotonin transporter gene polymorphism with schizophrenia, psychopathology and clozapine response. Schizophr. Res., 44: 177–181.

Tsai, S.J., Hong, C.J., Yu, Y.W. and Lin, C.H. (2002) − 759C/T genetic variation of 5HT(2C) receptor and clozapine-induced weight gain. Lancet, 360: 1790.

Tuckwell, H.C. and Koziol, J.A. (1993) A meta-analysis of homovanillic acid concentrations in schizophrenia. Int. J. Neurosci., 73: 109–114.

Tuckwell, H.C. and Koziol, J.A. (1996) On the concentration of 5-hydroxyindoleacetic acid in schizophrenia: a meta-analysis. Psychiatry Res., 59: 239–244.

Twarog, B.M. and Page, I.H. (1953) Serotonin content of some mammalian tissues and urine and a method for its determination. Am. J. Physiol., 175: 157–161.

van der Heijden, F.M., Tuinier, S., Fekkes, D., Sijben, A.E., Kahn, R.S. and Verhoeven, W.M. (2004) Atypical antipsychotics and the relevance of glutamate and serotonin. Eur. Neuropsychopharmacol., 14: 259–265.

van Kammen, D.P., Peters, J. and van Kammen, W.B. (1986) Cerebrospinal fluid studies of monoamine metabolism in schizophrenia. Psychiatr. Clin. North Am., 9: 81–97.

van Rossum, J.M. (1966) The significance of dopamine-receptor blockade for the mechanism of action of neuroleptic drugs. Arch. Int. Pharmacodyn. Ther., 160: 492–494.

Varty, G.B., Bakshi, V.P. and Geyer, M.A. (1999) M100907, a serotonin 5-HT2A receptor antagonist and putative antipsychotic, blocks dizocilpine-induced prepulse inhibition deficits in Sprague-Dawley and Wistar rats. Neuropsychopharmacology, 20: 311–321.

Verhoeff, N.P., Meyer, J.H., Kecojevic, A., Hussey, D., Lewis, R., Tauscher, J., Zipursky, R.B. and Kapur, S. (2000) A voxel-by-voxel analysis of [18F]setoperone PET data shows no substantial serotonin 5-HT(2A) receptor changes in schizophrenia. Psychiatry Res., 99: 123–135.

Vollenweider, F.X. and Geyer, M.A. (2001) A systems model of altered consciousness: integrating natural and drug-induced psychoses. Brain Res. Bull., 56: 495–507.

Vollenweider, F.X., Leenders, K.L., Scharfetter, C., Maguire, P., Stadelmann, O. and Angst, J. (1997a) Positron emission tomography and fluorodeoxyglucose studies of metabolic hyperfrontality and psychopathology in the psilocybin model of psychosis. Neuropsychopharmacology, 16: 357–372.

Vollenweider, F.X., Leenders, K.L., Scharfetter, C., Antonini, A., Maguire, P., Missimer, J. and Angst, J. (1997b) Metabolic hyperfrontality and psychopathology in the ketamine model of psychosis using positron emission tomography (PET) and [18F]fluorodeoxyglucose (FDG). Eur. Neuropsychopharmacol., 7: 9–24.

Vollenweider, F.X., Vollenweider-Scherpenhuyzen, M.F., Babler, A., Vogel, H. and Hell, D. (1998) Psilocybin induces schizophrenia-like psychosis in humans via a serotonin-2 agonist action. Neuroreport, 9: 3897–3902.

Vollenweider, F.X., Csomor, P.A., Knappe, B., Geyer, M.A. and Quednow, B.B. (2007) The effects of the preferential 5-HT2A agonist psilocybin on prepulse inhibition of startle in healthy human volunteers depend on interstimulus interval. Neuropsychopharmacology, 32: 1876–1887.

Wagner, M., Quednow, B.B., Westheide, J., Schlaepfer, T.E., Maier, W. and Kuhn, K.U. (2005) Cognitive improvement in schizophrenic patients does not require a serotonergic mechanism: randomized controlled trial of olanzapine vs amisulpride. Neuropsychopharmacology, 30: 381–390.

Wang, L., Fang, C., Zhang, A., Du, J., Yu, L., Ma, J., Feng, G., Xing, Q. and He, L. (2008) The −1019 C/G polymorphism of the 5-HT(1)A receptor gene is associated with negative symptom response to risperidone treatment in schizophrenia patients. J. Psychopharmacol., 22: 904–909.

Weinberger, R.D. and Laruelle, M. (2002) Neurochemical and neuropharmacological imaging in schizophrenia. In: Davis, K.L., Charney, D.S., Coyle, J.T. and Nemeroff, C.B. (Eds.), Neuropsychopharmacology: The Fifth Generation of Progress. Lippincott, Williams and Wilkins, Philadelphia, PA, pp. 833–855.

Weinberger, D.R., Berman, K.F. and Illowsky, B.P. (1988) Physiological dysfunction of dorsolateral prefrontal cortex in schizophrenia. III. A new cohort and evidence for a monoaminergic mechanism. Arch. Gen. Psychiatry, 45: 609–615.

Weinberger, D.R., Berman, K.F., Suddath, R. and Torrey, E.F. (1992) Evidence of dysfunction of a prefrontal-limbic network in schizophrenia: a magnetic resonance imaging and regional cerebral blood flow study of discordant monozygotic twins. Am. J. Psychiatry, 149: 890–897.

Whitaker, P.M., Crow, T.J. and Ferrier, I.N. (1981) Tritiated LSD binding in frontal cortex in schizophrenia. Arch. Gen. Psychiatry, 38: 278–280.

Wieselgren, I.M. and Lindstrom, L.H. (1998) CSF levels of HVA and 5-HIAA in drug-free schizophrenic patients and healthy controls: a prospective study focused on their predictive value for outcome in schizophrenia. Psychiatry Res., 81: 101–110.

Wilmot, C.A. and Szczepanik, A.M. (1989) Effects of acute and chronic treatments with clozapine and haloperidol on serotonin (5-HT2) and dopamine (D2) receptors in the rat brain. Brain Res., 487: 288–298.

Winblad, B., Bucht, G., Gottfries, C.G. and Roos, B.E. (1979) Monoamines and monoamine metabolites in brains from demented schizophrenics. Acta Psychiatr. Scand., 60: 17–28.

Winter, J.C., Doat, M. and Rabin, R.A. (2000) Potentiation of DOM-induced stimulus control by non-competitive NMDA antagonists: a link between the glutamatergic and serotonergic hypotheses of schizophrenia. Life Sci., 68: 337–344.

Winter, J.C., Eckler, J.R. and Rabin, R.A. (2004) Serotonergic/glutamatergic interactions: the effects of mGlu2/3 receptor ligands in rats trained with LSD and PCP as discriminative stimuli. Psychopharmacology (Berl.), 172: 233–240.

Wong, D.F., Wagner, H.N. Jr., Tune, L.E., Dannals, R.F., Pearlson, G.D., Links, J.M., Tamminga, C.A., Broussolle, E.P., Ravert, H.T., Wilson, A.A., Toung, J.K., Malat, J., Williams, J.A., O'Tuama, L.A., Snyder, S.H., Kuhar, M.J. and Gjedde, A. (1986) Positron emission tomography reveals elevated D2 dopamine receptors in drug-naive schizophrenics. Science, 234: 1558–1563.

Wood, S.J., Berger, G., Velakoulis, D., Phillips, L.J., McGorry, P.D., Yung, A.R., Desmond, P. and Pantelis, C. (2003) Proton magnetic resonance spectroscopy in first episode psychosis and ultra high-risk individuals. Schizophr. Bull., 29: 831–843.

Woolley, D.W. (1962) The Biochemical Bases of Psychosis or the Serotonin Hypothesis about Mental Disease. John Wiley & Sons, New York, NY.

Woolley, D.W. and Shaw, E. (1953) Antimetabolites of serotonin. J. Biol. Chem., 203: 69–79.

Woolley, D.W. and Shaw, E. (1954) Some neurophysiological aspects of serotonin. Br. Med. J., 2: 122–126.

Yasuno, F., Suhara, T., Nakayama, T., Ichimiya, T., Okubo, Y., Takano, A., Ando, T., Inoue, M., Maeda, J. and Suzuki, K. (2003) Inhibitory effect of hippocampal 5-HT1A receptors on human explicit memory. Am. J. Psychiatry, 160: 334–340.

Yasuno, F., Suhara, T., Ichimiya, T., Takano, A., Ando, T. and Okubo, Y. (2004) Decreased 5-HT1A receptor binding in amygdala of schizophrenia. Biol. Psychiatry, 55: 439–444.

Yu, Y.W., Tsai, S.J., Lin, C.H., Hsu, C.P., Yang, K.H. and Hong, C.J. (1999) Serotonin-6 receptor variant (C267T) and clinical response to clozapine. Neuroreport, 10: 1231–1233.

Yu, Y.W., Tsai, S.J., Yang, K.H., Lin, C.H., Chen, M.C. and Hong, C.J. (2001) Evidence for an association between polymorphism in the serotonin-2A receptor variant (102T/C) and increment of N100 amplitude in schizophrenics treated with clozapine. Neuropsychobiology, 43: 79–82.

Zhang, C. and Marek, G.J. (2008) AMPA receptor involvement in 5-hydroxytryptamine2A receptor-mediated pre-frontal cortical excitatory synaptic currents and DOI-induced head shakes. Prog. Neuropsychopharmacol. Biol. Psychiatry, 32: 62–71.

Zhang, X., Beaulieu, J.M., Gainetdinov, R.R. and Caron, M.G. (2006) Functional polymorphisms of the brain serotonin synthesizing enzyme tryptophan hydroxylase-2. Cell. Mol. Life Sci., 63: 6–11.

Zhang, Z.J., Zhang, X.B., Sha, W.W., Zhang, X.B. and Reynolds, G.P. (2002) Association of a polymorphism in the promoter region of the serotonin 5-HT2C receptor gene with tardive dyskinesia in patients with schizophrenia. Mol. Psychiatry, 7: 670–671.

Zill, P., Buttner, A., Eisenmenger, W., Moller, H.J., Ackenheil, M. and Bondy, B. (2007) Analysis of tryptophan hydroxylase I and II mRNA expression in the human brain: a post-mortem study. J. Psychiatr. Res., 41: 168–173.

Zipursky, R.B., Meyer, J.H. and Verhoeff, N.P. (2007) PET and SPECT imaging in psychiatric disorders. Can. J. Psychiatry, 52: 146–157.

Zureick, J.L. and Meltzer, H.Y. (1988) Platelet MAO activity in hallucinating and paranoid schizophrenics: a review and meta-analysis. Biol. Psychiatry, 24: 63–78.

CHAPTER 4.7

Serotonin and Serotonin Receptors in Hallucinogen Action

Adam L. Halberstadt[1] and David E. Nichols[*,2]

[1]Department of Psychiatry, University of California, San Diego, La Jolla, California
[2]Department of Medicinal Chemistry and Molecular Pharmacology, School of Pharmacy – RHPH, Purdue University, West Lafayette, Indiana

Abstract: Hallucinogens are a class of substances that induce profound changes in perception and cognition. A closely related drug, 3,4-methylenedioxymethamphetamine (MDMA), produces euphoria and a feeling of empathy, with minimal sensory distortion. Both of these classes of substances produce their effects by interacting with the serotonergic system. This chapter will review the receptor interactions that contribute to the behavioral effects of serotonergic hallucinogens and MDMA. In rodents, the behavioral effects of hallucinogens such as lysergic acid diethylamide (LSD), psilocybin and mescaline are primarily mediated by activation of 5-HT$_{2A}$ receptors. There is evidence, however, that 5-HT$_{1A}$ receptors, 5-HT$_{2C}$ receptors and dopamine receptors may play a secondary role. The molecular requirements for interaction of hallucinogens with the 5-HT$_{2A}$ receptor are well-defined on the basis of structure–activity relationships. By contrast with the hallucinogens, MDMA is a potent releaser of monoamines that has complex effects on serotonergic, dopaminergic and noradrenergic systems. In recent years, psilocybin and MDMA have been administered to human volunteers in controlled clinical trials. Human studies confirm that the 5-HT$_{2A}$ receptor plays a primary role in mediating the subjective effects of psilocybin, whereas the effects of MDMA are largely attributable to carrier-mediated release of serotonin. These findings emphasize the importance of clinical investigation of hallucinogenic drugs. Additionally, there is a growing consensus that these drugs are likely to show therapeutic efficacy in the treatment of certain psychiatric disorders.

Keywords: 5-HT$_{2A}$ receptor, behavior, drug discrimination, entactogen, hallucinogen, head twitch, locomotor activity, MDMA, psychedelic, prepulse inhibition.

Abbreviations: 5-HT, serotonin; 5-MeO-DMT, 5-methoxy-*N,N*-dimethyltryptamine; AED, Anxious Ego Dissolution; AMPH, amphetamine; APZ, Altered States of Consciousness rating scale; BPM, Behavioral Pattern Monitor; DMT, *N,N*-dimethyltryptamine; DPT, *N,N*-dipropyltryptamine; DOB, 2,5-dimethoxy-4-bromoamphetamine; DOET, 2,5-dimethoxy-4-ethylamphetamine; DOI, 2,5-dimethoxy-4-iodoamphetamine; DOM, 2,5-dimethoxy-4-methylamphetamine; DRN, dorsal raphe nucleus; EL2, extracellular loop 2; HTR, head-twitch response; LSD, lysergic acid diethylamide; MBDB, 1-(1,3-benzodioxol-5-yl)-N-methylbutan-2-amine; MDE, N-ethyl-3,4-methylenedioxyamphetamine; MDMA, 3,4-methylenedioxymethamphetamine; OB, Oceanic Boundlessness; PPI, prepulse inhibition; RV, Reduction of Vigilance; SERT, 5-HT transporter; VR, Visionary Restructuralization.

Introduction

Hallucinogenic drugs have been used by humans for thousands of years to diagnose and cure disease, induce mystical and spiritual states, and produce euphoria and inebriation (Schultes and Hofmann, 1980). Typically, these drugs have been derived from botanical sources. Some examples include the mescaline-containing peyote cactus (*Lophophora williamsii*), which is legally used in the United States by members of the Native American Church; *teonanácatl* mushrooms (a species of *Psilocybe* containing psilocybin and psilocin), which have a history of ceremonial use in Mexico; and *ayahuasca*, an infusion or decoction prepared from the bark of β-carboline-containing species of *Banisteriopsis* to which admixture plants containing *N,N*-dimethyltryptamine (DMT) are typically added. Modern scientific investigation of the hallucinogens began with the identification of mescaline as the active principal of peyote by Arthur Heffter in 1897.

* Corresponding author
E-mail: drdave@pharmacy.purdue.edu

Christian Müller & Barry Jacobs (Eds.)
Handbook of Behavioral Neurobiology of Serotonin
ISBN 978-0-12-374634-4

DOI: 10.1016/B978-0-12-374634-4.00035-6

The discovery of the potent hallucinogenic effects of lysergic acid diethylamide (LSD) by Albert Hoffmann in 1943 markedly increased scientific interest in these substances.

The publication of the first description of the hallucinogenic effects of LSD (Stoll, 1947) coincided with the initial isolation and characterization of serotonin (5-HT) (Rapport *et al.*, 1948). In 1953, Gaddum reported that LSD antagonized the contractile effect of 5-HT on the isolated uterine smooth muscle (Gaddum, 1953). It was soon recognized, however, that LSD acts as a 5-HT agonist in some assay systems (Shaw and Woolley, 1956). It is now widely accepted that LSD and other hallucinogens are agonists or partial agonists at 5-HT receptors (Glennon, 1990).

The psychological effects produced by hallucinogens are highly subjective, usually including changes in thought and mood, depersonalization, and perceptual alterations that may include visual hallucinations, synesthesias and tactile sensations. The effects of serotonergic hallucinogens are distinguishable from those of other drug classes such as dissociative anesthetics (e.g., phencyclidine, ketamine), κ-agonists (e.g., salvinorin A), and cannabinoids, which produce some hallucinogen-like effects.

Soon after the discovery of LSD it was recognized that there are similarities between the effects of hallucinogens and the symptomatology of schizophrenia, and LSD and mescaline were used experimentally to produce a model psychosis. There is now renewed interest in the possibility that hallucinogens may be useful therapeutic drugs, and the efficacy of psilocybin in the treatment of obsessive-compulsive disorder and as an anxiolytic in terminal cancer patients has been investigated in clinical trials (Biello, 2006; Moreno *et al.*, 2006). It also has been claimed that 3,4-methylenedioxymethamphetamine (MDMA, 'Ecstasy') has value as an adjunct to psychotherapy. MDMA produces mood elevation and an increased sense of empathy with little sensory disruption. The effects of MDMA are distinct from those typically produced by hallucinogens, however, and it has been proposed that MDMA is an entactogen, a distinct class of drugs (Nichols, 1986).

The mechanism of action of hallucinogens and entactogens such as MDMA has been the subject of intense investigation. This chapter will review the receptor interactions that contribute to the behavioral effects of hallucinogens and MDMA.

Receptors mediating the behavioral effects of hallucinogens in animals

5-HT$_{2A}$ receptor

The 5-HT$_{2A}$ receptor is a $G_{q/11}$-coupled receptor that is linked to the phosphoinositide hydrolysis signaling cascade (Nichols and Nichols, 2008). Members of both the indoleamine and phenylalkylamine classes of hallucinogens bind to 5-HT$_{2A}$ receptors with high affinity. Phenylalkylamine hallucinogens such as 2,5-dimethoxy-4-methylamphetamine (DOM), 2,5-dimethoxy-4-iodoamphetamine (DOI), and 2,5-dimethoxy-4-bromoamphetamine (DOB) are highly selective for 5-HT$_2$ receptor subtypes (Titeler *et al.*, 1988; Pierce and Peroutka, 1989), and there is a general consensus in the literature that the behavioral effects of hallucinogens are primarily mediated by the 5-HT$_{2A}$ receptor (Nichols, 2004).

Drug discrimination

Much of what is known about the mechanism of action of hallucinogenic drugs has been derived from drug discrimination studies. Rats can be trained to discriminate hallucinogens (including LSD, mescaline, DOM, DOI, DOB, DPT, psilocybin and 5-methoxy-DMT (5-MeO-DMT) from saline using two-lever operant procedures (Hirschhorn and Winter, 1971; Glennon *et al.*, 1982). Cross-generalization occurs between these training drugs, indicating that they produce similar interoceptive stimulus effects. By contrast, members of other drug classes reliably fail to evoke hallucinogen-like stimulus effects (Glennon *et al.*, 1982; Appel and Cunningham, 1986).

Glennon and colleagues first proposed in 1983 that the behavioral effects of hallucinogens are mediated by activation of 5-HT$_2$ receptors (Glennon *et al.*, 1983). This proposal was based on the finding that the high-affinity, selective 5-HT$_2$ antagonists pirenperone and ketanserin are highly effective antagonists of the discriminative stimulus properties of DOM and LSD (Colpaert *et al.*, 1982; Colpaert and Janssen, 1983), and of DOM-stimulus generalization to LSD, mescaline, and 5-MeO-DMT (Glennon *et al.*, 1983). It was subsequently demonstrated that a robust linear correlation ($r = 0.938$) exists between the 5-HT$_2$ affinities of a series of phenylalkylamine hallucinogens and their ED_{50} values for substitution in animals trained with DOM (Glennon *et al.*, 1984). An antagonist correlation analysis confirmed that there is a strong correlation between the 5-HT$_{2A}$ affinity of a series of 5-HT antagonists and their potencies for blocking LSD stimulus generalization to *R*-(−)-DOM (Fiorella *et al.*, 1995). Additionally, stimulus control in animals trained with LSD (Winter *et al.*, 2004), DOI (Smith *et al.*, 1999; Schreiber *et al.*, 1994) and DOM (Li *et al.*, 2007) is blocked by M100907, a highly selective 5-HT$_{2A}$ antagonist. Taken together, these findings demonstrate that interactions with the 5-HT$_{2A}$ receptor are required for hallucinogens to induce discriminative stimulus effects.

Head-twitch response

Hallucinogens evoke a variety of stereotypical motor responses in laboratory animals, including ear scratch in mice, head twitch in mice and rats, reciprocal hindleg body scratch in gerbils, limb flicks and abortive grooming in cats, and limb jerks in primates. The head-twitch response (HTR), which in mice consists of a paroxysmal rotational movement of the head (Corne and Pickering, 1967), is one of the best characterized of the hallucinogen-induced stereotypies. Administration of hallucinogens to rats induces an analogous behavior ('wet-dog shakes') that typically involves not only the head but also the neck and trunk. There is extensive evidence that the 5-HT_{2A} receptor mediates the hallucinogen-induced HTR. The first evidence for a linkage between 5-HT_2 receptors and the HTR induced by hallucinogens was the finding that the potencies of 19 5-HT antagonists to block the mescaline HTR is significantly correlated ($r = 0.88$) with their 5-HT_2 affinity (Leysen *et al.*, 1982). Similar findings were later reported for the HTR induced by DOI (Schreiber *et al.*, 1995). It was also demonstrated that ketanserin blocks the HTR to DOI (Darmani *et al.*, 1990). Although ketanserin is an antagonist at 5-HT_{2A} and 5-HT_{2C} receptors, the fact that the HTR to DOI is completely blocked by M100907, but not by the highly selective 5-HT_{2C} antagonist SB 242,084 confirmed that the behavior is mediated by the 5-HT_{2A} receptor and not by the 5-HT_{2C} receptor (Vickers *et al.*, 2001). Further evidence for a link between 5-HT_{2A} receptors and head twitch was provided by the recent finding that the ability of LSD, DOI, DOM, DOB, mescaline and psilocin to induce the HTR is abolished in 5-$HT_{2A}{}^{-/-}$ knockout mice (González-Maeso *et al.*, 2007).

Exploratory and investigatory behavior

Measures of unconditioned locomotor activity are frequently used to assess compounds for stimulant or depressant effects. Tests of hallucinogens using conventional open-field measures, however, produce inconsistent results and fail to distinguish the effects of hallucinogens from those of non-hallucinogenic agents (Brimblecombe, 1963; Cohen and Wakeley, 1968; Kabes *et al.*, 1972; Silva and Calil, 1975). Given the complexity of the effects of hallucinogens, it is not surprising that the locomotor paradigm produces inconclusive findings because it does not provide a qualitative assessment of behavior and does not measure whether there are changes in sensitivity to environmental stimuli. The Behavioral Pattern Monitor (BPM), a combination of activity and holeboard chambers, was designed to overcome the shortcomings of open-field measurements (Geyer, 1990). The BPM provides both quantitative and qualitative measures of unconditioned locomotor and investigatory activity in rats, and can be used to assess animals for changes in responsivity to environmental stimuli.

Hallucinogens produce a characteristic profile of behavioral effects in the BPM. When phenylalkylamine (mescaline, DOM, DOI and DOET) and indolealkylamine (psilocin, DMT and 5-MeO-DMT) hallucinogens are tested in a novel environment, they decrease locomotor activity and investigatory behaviors (rearings and holepokes) and increase avoidance of the center of the chamber (Geyer *et al.*, 1979; Adams and Geyer, 1985a; Wing *et al.*, 1990). These effects are not observed when animals are tested in a familiar environment, and have been attributed to exacerbation of the neophobia and agoraphobia normally exhibited by rats. LSD has similar effects on investigatory behavior and center entries (Adams and Geyer, 1985b), but it produces a biphasic locomotor pattern with activity initially suppressed and then increasing as time progresses (Mittman and Geyer, 1991).

Most of the effects of DOI in the BPM paradigm are blocked by M100907 but not by the 5-$HT_{2C/2B}$ antagonist SER-082, and are therefore likely mediated by activation of 5-HT_{2A} receptors (Krebs-Thomson *et al.*, 1998). Mechanistically, the effect of the indoleamines in the BPM is more complex. For example, both the mixed 5-HT_1/β-adrenergic antagonist propranolol (Mittman and Geyer, 1991) and the selective 5-HT_{1A} antagonist WAY-100635 (Krebs-Thomson and Geyer, 1996) block the initial suppression of locomotor activity induced by LSD, whereas LSD-induced hyperactivity is blocked by M100907 (Ouagazzal *et al.*, 2001) and the 5-HT_2 antagonist ritanserin (Mittman and Geyer, 1991). The effects of 5-MeO-DMT are antagonized by WAY-100635 but not by M100907 (Krebs-Thomson *et al.*, 2006), indicating that they are mediated by 5-HT_{1A} receptors. However, when 5-MeO-DMT is administered in combination with a MAO-A inhibitor such as harmaline or clorgyline, it produces LSD-like late hyperactivity that is completely blocked by the 5-HT_{2A}-selective antagonist MDL 11,939 (Halberstadt *et al.*, 2008). In summary, the effects of LSD and 5-MeO-DMT in the BPM are mediated by both 5-HT_{1A} and 5-HT_{2A} receptors.

Prepulse inhibition of startle

Acoustic startle responses are attenuated when preceded by a weak prestimulus, a phenomenon known as prepulse inhibition (PPI). PPI serves as an operational measure of sensorimotor gating and has been found to be deficient in patients with a variety of psychiatric illnesses, including schizophrenia. Hallucinogens, including LSD, DOI, DOB and 5-MeO-DMT, disrupt PPI in rats (Rigdon and Weatherspoon, 1992; Sipes and Geyer, 1994; Johansson *et al.*, 1995; Ouagazzal *et al.*, 2001). The decrease in PPI induced by DOI and LSD is blocked by M100907 but not by SB 242,084 or SER-082 (Sipes and Geyer, 1995; Ouagazzal *et al.*, 2001), demonstrating the involvement of 5-HT_{2A} receptors. Compared with DOI and LSD, the mechanism for the effect of 5-MeO-DMT on PPI is more complex and involves interactions with both 5-HT_{1A} and 5-HT_{2C} receptors (Krebs-Thomson *et al.*, 2006).

5-HT_{2C} receptor

Phenylalkylamine and indoleamine hallucinogens bind to 5-HT_{2C} receptors with high affinity, and are relatively non-selective for 5-HT_{2A} versus 5-HT_{2C} receptors (Nelson *et al.*, 1999; Porter *et al.*, 1999). The discovery that hallucinogens interact with 5-HT_{2C} receptors confounded the hypothesis that activation of 5-HT_{2A} receptors is the primary mechanism by which LSD-like drugs act (Glennon *et al.*, 1984). This hypothesis was proposed prior to the discovery of 5-HT_{2C} sites, and was partially based on evidence demonstrating that ketanserin and other classical 5-HT_{2A} antagonists block the behavioral effects of hallucinogens in various animal paradigms. Unfortunately, those serotonin antagonists are now known to act at both 5-HT_{2A} and 5-HT_{2C} sites. The potent action of hallucinogens at 5-HT_{2A} and 5-HT_{2C} receptors has led authors to speculate that hallucinogenesis may be mediated by interactions with both of these receptor populations (Titeler *et al.*, 1988; Glennon, 1990). In fact, Sanders-Bush and others have gone so far as to suggest that 5-HT_{2C} receptor occupation plays a primary mechanistic role in the action of hallucinogenic drugs (Burris *et al.*, 1991).

There is a significant correlation between the affinities of phenylalkylamine hallucinogens for 5-HT_{2C} receptors and their potencies to evoke psychoactive effects in humans and to substitute in DOM-trained animals (Titeler *et al.*, 1988). Nevertheless, 5-HT_{2A} and 5-HT_{2C} receptors display parallel structure–affinity relationships for ligand binding (Nelson *et al.*, 1999), so it is possible that the observed correlation between 5-HT_{2C} receptor affinity and hallucinogen potency merely reflects the relationship between 5-HT_{2A} and 5-HT_{2C} affinities. Importantly, for a series of structurally diverse 5-HT_2 antagonists, the rank order of potencies for blocking LSD-like stimulus effects (Fiorella *et al.*, 1995) and DOI-induced HTR (Schreiber *et al.*, 1995) does not parallel their 5-HT_{2C} affinity. These findings strongly indicate that interactions with 5-HT_{2C} receptors are not primarily responsible for mediating the behavioral effects of hallucinogens. Further, 5-$HT_{2C/2B}$ antagonists (SB 200,646A, SB 206,553, and SER-082) and the selective 5-HT_{2C} antagonist SB 242,084 consistently fail to block the effects of hallucinogens in a variety of behavioral paradigms (Schreiber *et al.*, 1994; Sipes and Geyer, 1995; Krebs-Thomson *et al.*, 1998; Wettstein *et al.*, 1999; Smith *et al.*, 1999; Ouagazzal *et al.*, 2001; Winter *et al.*, 2007). Although findings suggesting that 5-HT_{2C} receptor interactions may contribute to or modify the effects of some hallucinogens have occasionally been reported (Smith *et al.*, 2003; Krebs-Thomson *et al.*, 1998, 2006), the vast majority of evidence demonstrates that the behavioral effects of these drugs are not dependent on the 5-HT_{2C} receptor.

5-HT_{1A} receptor

LSD and other indoleamine hallucinogens are agonists at 5-HT_{1A} autoreceptors, and the indoleamines potently inhibit the firing of serotonergic neurons in the dorsal raphe nucleus (DRN) (Aghajanian *et al.*, 1968, 1970). Indoleamine hallucinogens also directly inhibit firing in many DRN target regions, but at low doses they preferentially inhibit DRN cells while leaving downstream neurons virtually unaffected (Haigler and Aghajanian, 1974; DeMontigny and Aghajanian, 1977). The finding that hallucinogens are more potent at presynaptic than postsynaptic sites led to the proposal that by depressing the activity of neurons in the DRN and thereby decreasing 5-HT outflow, hallucinogenic drugs remove the tonic inhibition of downstream neurons mediated by serotonin. Further research, however, identified several problems with this 'presynaptic hypothesis' of hallucinogen action. One of the most serious discrepancies was the finding that phenylalkylamine hallucinogens such as mescaline and DOM display negligible 5-HT_{1A} binding affinity and do not consistently depress DRN firing (Haigler and Aghajanian, 1973; Penington and Reiffenstein, 1986). Furthermore, selective 5-HT_{1A} agonists such as buspirone, ipsapirone and flesinoxan inhibit DRN cell firing but are not hallucinogenic.

The ability to inhibit DRN cell firing is clearly an epiphenomenon unrelated to hallucinogenesis. Nonetheless, the 5-HT_{1A} agonist activity of indoleamines does appear to contribute to their behavioral effects. As was noted earlier, the effects of LSD and 5-MeO-DMT in the BPM are partially mediated by the 5-HT_{1A} receptor. There is also considerable evidence that the 5-HT_{1A} receptor plays a role in the discriminative stimulus effects of LSD and 5-MeO-DMT. LSD elicits intermediate levels of drug-lever selection in rats trained with the 5-HT_{1A} agonist 8-OH-DPAT, and the 5-HT_{1A} agonist ipsapirone elicits partial substitution in LSD-trained animals (Arnt, 1989). Likewise, the LSD stimulus completely generalizes to the mixed 5-HT_{1A} agonist/α_2-adrenoceptor antagonist yohimbine (Colpaert, 1984; Marona-Lewicka and Nichols, 1995). 8-OH-DPAT and ipsapirone produce full substitution in animals trained with 5-MeO-DMT (Spencer *et al.*, 1987; Schreiber and DeVry, 1993; Winter *et al.*, 2000). In fact, it has been reported that WAY-100635 and the mixed 5-HT_1/β-adrenergic antagonist pindolol are more effective than pirenperone and ketanserin at blocking stimulus control by 5-MeO-DMT, indicating that the 5-MeO-DMT discriminative stimulus involves a substantial 5-HT_{1A} receptor-mediated component (Spencer *et al.*, 1987; Winter *et al.*, 2000). These observations are consistent with evidence that the behavioral response to 5-MeO-DMT in rats is predominantly mediated by the 5-HT_{1A} receptor (Tricklebank *et al.*, 1985; Krebs-Thomson *et al.*, 2006).

There is also evidence that activation of the 5-HT_{1A} receptor can inhibit 5-HT_{2A} receptor-induced behavioral effects. 8-OH-DPAT pretreatment dose-dependently attenuates the head-twitch response to DOI in rats and mice (Arnt and Hyttel, 1989; Berendsen and Broekkamp, 1990; Darmani *et al.*, 1990), and the reciprocal hindleg body scratch induced by DOI in the gerbil (Eison and Wright, 1992). Furthermore, although both 8-OH-DPAT and DOI reduce locomotor activity in rats, there is an infra-additive interaction when the two drugs are co-administered, suggesting that they act as functional antagonists (Krebs-Thomson and Geyer, 1998).

$5\text{-HT}_{5,6,7}$ receptors

Indoleamine hallucinogens such as LSD, 5-MeO-DMT and psilocin bind with relatively high affinity to 5-HT_5, 5-HT_6, and 5-HT_7 receptors. Until recently, however, there were few pharmacological tools available to probe 5-HT_5, 5-HT_6, and 5-HT_7 sites, and therefore there has been no systematic investigation to determine whether these receptors contribute to the behavioral effects of the indoleamines. Experiments conducted in $5\text{-HT}_{5A}{}^{-/-}$ knockout mice indicate that the ability of LSD to increase exploratory behavior is partially dependent on the 5-HT_{5A} receptor (Grailhe *et al.*, 1999). It has been reported that the 5-HT_6 receptor is not involved in the LSD-induced decrease of PPI in rats (Ouagazzal *et al.*, 2001). Although interactions with 5-HT_5, 5-HT_6, and 5-HT_7 receptors may modulate the effects of the indoleamines, given the low affinity of the phenylalkylamines for these receptors it is unlikely that any of these sites are responsible for mediating the characteristic behavioral effects of hallucinogens.

Dopamine receptors

Based on evident similarities between the symptoms of schizophrenia and those of the so-called 'model psychosis' induced by hallucinogenic drugs, and because dopamine antagonists such as chlorpromazine and haloperidol have been used to counteract hallucinogen-induced psychotic episodes and panic reactions, it has been proposed that interactions with dopaminergic systems are involved in the mechanism of action of hallucinogens. It is now well established that indolealkylamine and phenylalkylamine hallucinogens lack appreciable affinity for dopamine receptors; consequently, it is unlikely that direct interactions with dopamine receptors are responsible for the behavioral effects of those agents. Conversely, like many other ergot alkaloids, LSD binds to D_1, D_2, D_3, D_4 and D_5 receptors (Nichols *et al.*, 2002), and acts as a partial agonist at D_1 receptors (Watts *et al.*, 1995).

There is evidence that dopaminergic mechanisms play a role in the discriminative stimulus properties of LSD. Although the $5\text{-HT}_{2A}/D_2$ antagonist risperidone and the 5-HT_2 antagonist ritanserin bind to the 5-HT_{2A} receptor with comparable affinities (Leysen, 1990), risperidone blocks LSD discrimination with 414 times the potency of ritanserin (Koek *et al.*, 1992). A 0.63-mg/kg dose of ritanserin produces full occupation of 5-HT_{2A} sites in rats, yet fails to attenuate the LSD stimulus; it is not until a much higher dose of 40 mg/kg is administered, where ritanserin is capable of interacting with catecholamine receptors, that it can completely antagonize the LSD cue (Meert *et al.*, 1990; Koek *et al.*, 1992). Conversely, risperidone and ritanserin antagonize the DOM cue with similar potencies (ED_{50} values of 0.024 mg/kg and 0.15 mg/kg, respectively; Meert, 1996). Taken together, these findings argue that both 5-HT_{2A} and D_2 receptor interactions contribute to the interoceptive state governing LSD discrimination, whereas DOM discrimination involves only the 5-HT_{2A} receptor. In support of this conclusion, it was observed that the ED_{50} of ritanserin for blocking LSD stimulus control is reduced 53-fold when administered in combination with 0.04 mg/kg haloperidol (Meert and Awouters, 1991). It has also recently been reported that the dopaminergic component of the LSD discriminative stimulus is time-dependent (Marona-Lewicka *et al.*, 2005). Studies that have trained rats to discriminate LSD have typically used pretreatment times of 15–30 minutes, and have consistently shown that the LSD cue is blocked by 5-HT_{2A} antagonists. When a longer pretreatment time of 90 minutes is used, however, the resulting LSD cue is apparently mediated by D_4 receptors rather than by 5-HT_{2A} receptors (Marona-Lewicka *et al.*, 2008).

Chemistry and structure–activity relationships

As discussed above, the serotonin 5-HT_{2A} receptor is now generally considered to be the key target for hallucinogens (Nichols, 2004). As such, one might therefore expect the

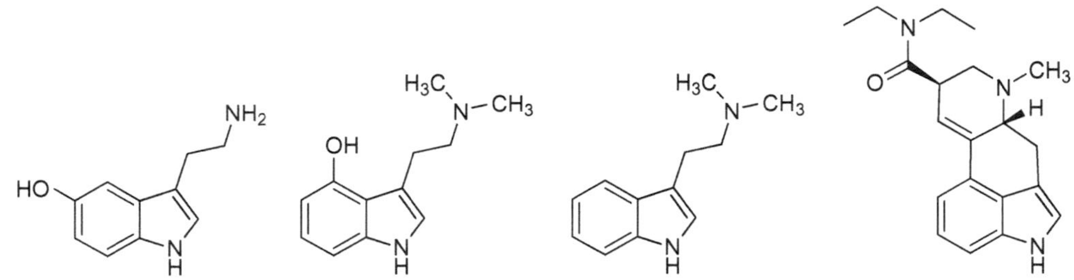

Serotonin   Psilocin   N,N-Dimethyltryptamine (DMT)   LSD-25

Figure 1 The structures of the natural transmitter serotonin and the tryptamine-type hallucinogens psilocin, DMT and LSD.

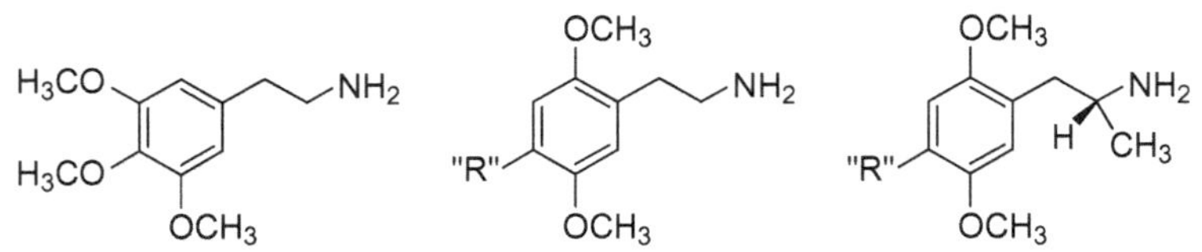

Mescaline   "2C" compounds   "Amphetamines"

R = CH_3; DOM
R = Br; DOB
R = I, DOI
R = CF_3; DOTFM

Figure 2 The structures of mescaline, the prototypical phenethylamine hallucinogen, and general formulas for synthetic substituted hallucinogenic phenethylamines and 'amphetamines'. For the alpha-substituted amphetamines, the stereoisomer with the *R*-(−)- absolute configuration (shown) is more potent.

hallucinogens to bear some structural resemblance to serotonin. This relationship is quite evident with the simple tryptamines and with the tetracyclic ergoline LSD. Indeed, it was the recognition that the tryptamine fragment within the structure of LSD also was the core feature of serotonin itself that first led to the realization that serotonin might play a role in behavior. A comparison of the structures of serotonin and the hallucinogens DMT, 5-methoxytryptamine, psilocin and LSD is presented in Figure 1. It is not difficult to understand why all of these structures interact with serotonin receptors, because of their close structural resemblance. Although LSD is a more complex molecule, the tryptamine core fragment is still readily apparent.

What is less clear, though, is how the phenethylamine type hallucinogens (Figure 2) interact with serotonin receptors. The only apparent similarity is a basic nitrogen atom two carbon atoms removed from an aromatic system. Recent mutagenesis studies of the 5-HT_{2A} receptor have, however, begun to reveal the basis for the requirement of certain structural features of the phenethylamines to interact with the receptor (Figure 3).

Studies of the mutant S239A 5-HT_{2A} receptor have demonstrated that it is this serine residue, in transmembrane helix 5, that likely hydrogen bonds to the 5-hydroxy group of serotonin, as well as to the 5-methoxy group of the 2,5-dimethoxy-substituted phenethylamines (Braden and Nichols, 2007). Figure 4 suggests possible binding modes for the 2,5-dimethoxy-substituted phenethylamines and the tryptamines, based on those studies.

This view could explain why a primary amine is most potent for the phenethylamines, with *N*-alkyl or *N,N*-dialkyl derivatives being essentially inert, whereas tertiary amines are potent in the tryptamine series. That is, one might envision that the binding site must 'shrink' or compress to accommodate a phenethylamine. That is most readily achieved by rotation of the side-chain of Asp155 in transmembrane helix 3. If this aspartate engages the amino group in an 'end on' conformation for the phenethylamines, it might be speculated that *N*-alkyl groups would hinder that interaction. By contrast, aspartate residue could engage the amino group of the tryptamines through an approach more perpendicular to the plane of the molecule. In this event, *N*-alkyl groups would be expected to have less of an effect on the ligand interaction. The argument can likely be extended to understand the binding of LSD, because its structure is so rigid that interaction of Asp155 with the protonated electron pair of the amino group can only occur through an 'axial' approach to the amine moiety.

Substitution of an alpha-methyl group on the phenethylamine side-chain gives more potent compounds.

Phenethylamine SAR	Tryptamine SAR	Lysergamide SAR
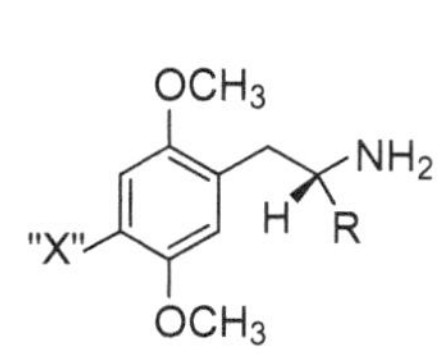	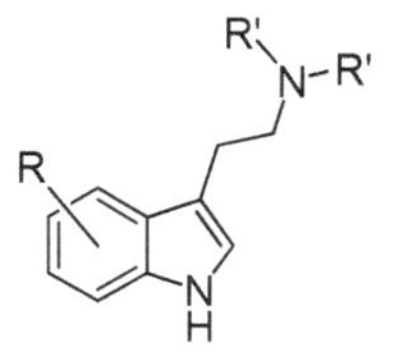	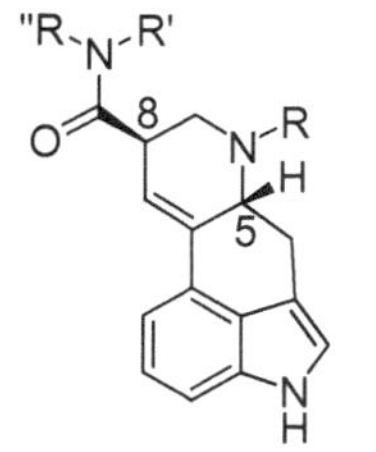
• 2,5-dimethoxy substitution is optimal • Hydrophobic X substituent must be at the 4-position • *R*-stereochemistry at the alpha-carbon; no larger than methyl • Only the primary amines are active • Orally active	• Unsubstituted, 4-OH, or 5-methoxy-substituted are active • If substituted at the alpha side-chain carbon, the *S*-stereochemistry is most potent • Tertiary amines are most potent • 4-OH, or bulky *N,N*-dialkyls are orally active	• No data for A-ring substituted lysergamides • 5*R*,8*R* stereochemistry required for activity • Tertiary amine required; ethyl, propyl > methyl • *N,N*-diethylamide has unique potency; other amides are much less potent

Figure 3 Summary of structure–activity relationships (SAR) for classes of hallucinogens.

There are at least three possible explanations, all of which probably make some contribution. First of all, the alpha methyl will retard metabolic deamination, giving better oral availability and increased *in vivo* stability. Second, the alpha methyl adds significant hydrophobicity, leading to better brain penetration. Finally, *R*-alpha-methyl-substituted phenethylamines generally are more efficacious in activating second messenger systems than their non-alkylated congeners (Parrish *et al.*, 2005). At least for the phenethylamines ('amphetamines') the alpha methyl group enhances the intrinsic activity, possibly through an interaction with a critical phenylalanine residue in transmembrane helix 6. The steric limitation in this region is very strict, however, as any larger alkyl group, even an ethyl, destroys activity.

Interestingly, the stereochemistry is reversed for alpha-substituted tryptamines. One may speculate that the two classes of ligands bind in such a way as to project the alpha-methyl groups into the same general region of the receptor, but there are no data to support that supposition. The different lengths of the ligands also may be a key here. For example, Figure 4 illustrates *S*-(+)-alpha-methyl serotonin and the more potent *R*-(−)- isomer of a rigid phenethylamine. In these orientations, representing potential receptor-bound conformations, the distance from the side-chain amino nitrogen to the 5-oxygen atom is 6.1 and 6.6 Ångstroms for the phenethylamine and the tryptamine, respectively.

Incorporating the 2- and 5-methoxy groups into furan or dihydrofuran rings gives compounds that are more potent, but, more importantly, establishes the binding orientations for the two methoxys. Simulating docking experiments suggest that the 2-methoxy group of the phenethylamines engages Ser159 through a hydrogen bond (Braden, 2007). Attempts to test that hypothesis with the S159A mutant receptor gave a protein that had such drastically reduced function that it was not possible to examine the effect of the presence or absence of the 2-methoxy of the ligand.

Although substitution at the 6-indole position of the tryptamines renders them inactive (Blair *et al.*, 2000), substituting the phenethylamines at the 4-position is essential to provide good activity. Again, the different sizes of the phenethylamines and tryptamines are likely at the heart of this observation. Simulating docking experiments suggest that the 4-substituent of the phenethylamines projects into a hydrophobic area of the receptor between transmembrane helices 3, 4 and 5 created by residues Phe234, Ile206 and Val156. It may be that the larger indole ring of the tryptamines projects into this space, and additional bulk attached to the ligand in this region creates too large a steric footprint for the receptor to accommodate.

3,4,5-substituted phenethylamines (e.g., mescaline) do not bind in the same way as the 2,4,5-substituted compounds, because site specific mutations of the receptor lead to different effects on 2,4,5- vs the 3,4,5- substituted

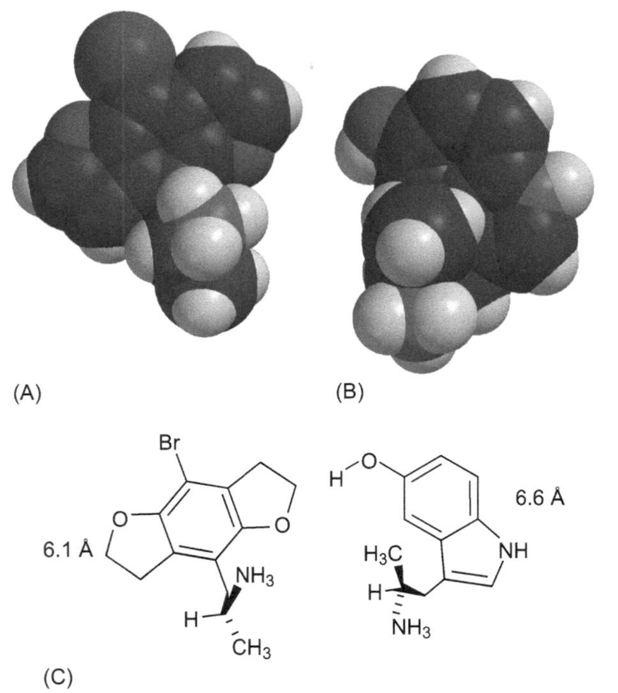

Figure 4 Comparison of the *R*-(−)- isomer of a potent difuranobenzene phenethylamine type hallucinogen (A) with *S*-(+)-α-methylserotonin (B). Although the stereochemistry at the carbon bearing the alpha-methyl is reversed for the phenethylamines and the tryptamines, the lengths of the ligands, differing by about 0.5 Ångstroms (C), could be the underlying reason for this reversal. The bicyclic aromatic indole system of the tryptamines is longer than the phenyl ring of the phenethylamines, and thus when the side-chain is extended, the tryptamines will occupy a larger space in the receptor. Because the receptor evolved to accept serotonin, a tryptamine, the perspective must be one that the phenethylamines somehow adapt themselves to a binding site that prefers tryptamines. To see the full color version of this figure please refer to the color plate at the back of the book. Copies produced via our print on demand service do not contain color plates. If your copy does not have the color plate, please go to this website to view the figure in color www.elsevierdirect.com/companions/9780123746344

compounds (Braden, 2007). Nevertheless, a hydrophobic 4-alkoxy substituent in the 3,4,5-substituted compounds also gives highest activity in this series (Nichols and Dyer, 1977; Shulgin and Shulgin, 1991), and simulated docking into homology models of the receptor suggests that this substituent projects into the same receptor space as the 4-substituent in the 2,4,5-substituted series.

The rigid nature of LSD limits its conformational flexibility upon binding. Mutagenesis studies suggest that the indole NH hydrogen bonds to Ser242 in transmembrane helix 5 (Braden and Nichols, 2007), and the electron pair on the amine nitrogen is probably directed in an axial orientation toward the conserved aspartate in helix 3.

Very little structural modification can be tolerated by ergolines related to LSD. The stereochemistry at both the 5- and 8-positions must be of the *R* configuration, shown in Figures 1 and 4. Inversion at C(8) affords iso-LSD, which is inactive. Halogenation at the 2-indole of LSD position affords antagonists such as BOL that lack hallucinogenic activity. Although N(1)-acetylation leads to an active compound, this molecule is actually a prodrug and the *N*-acetyl is readily cleaved *in vivo* to afford LSD itself.

The bulky diethylamide moiety is relatively limited in its conformational flexibility (Pfaff *et al.*, 1994), and virtually any modification of the diethylamide gives about an order of magnitude decrease in potency. Constraining the diethylamide into a 2,4-dimethylazetidide, however, did lead to a compound with potency comparable to LSD in the rat drug discrimination paradigm (Nichols *et al.*, 2002) The stereoisomer with the 2*S*,4*S*-absolute configuration in the azetidine ring gave the most potent lysergamide, demonstrating that the receptor binding site for the amide moiety of LSD is stereospecific, and uniquely complementary to the diethylamide function. Simulating docking experiments show that the diethylamide moiety is forced to project toward extracellular loop 2 (EL2), suggesting that EL2 may play an important role in LSD binding in the 5-HT_{2A} receptor, as is now known for many other GPCRs.

The only structural modification known to enhance activity is homologation of the N(6)-alkyl group. That is, replacing the N(6)-methyl with an ethyl, n-propyl, or n-allyl gives compounds slightly more active than LSD in rat behavioral models (Hoffman and Nichols, 1985).

Receptors mediating the behavioral effects of MDMA in animals

MDMA (Figure 5) is a potent monoamine-releasing drug (Johnson *et al.*, 1986; Nichols *et al.*, 1986; Baumann *et al.*, 2008). Neurochemical studies have demonstrated that MDMA binds to monoamine transporters, is a substrate for the transporters, and increases the non-exocytotic release of serotonin, dopamine and norepinephrine. MDMA also displays moderate affinity for 5-HT_{2A} receptors (Lyon *et al.*, 1986).

As contrasted with the hallucinogenic amphetamines, where the stereochemistry at the alpha side-chain carbon bearing the methyl group is *R*, the stereochemistry in the side-chain of the more potent enantiomer of MDMA is *S*, which corresponds to that of the psychostimulant amphetamines. The alpha-methyl can be homologated to an ethyl

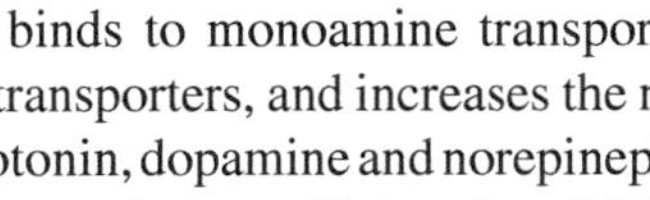

Figure 5 The structure of 3,4-methylenedioxymethamphetamine (MDMA).

group with retention of activity, and the *N*-methyl that is present in MDMA is not tolerated in the hallucinogenic amphetamines. These structural features are strong evidence that MDMA is not simply a hallucinogenic amphetamine, but belongs to a distinct class of psychoactive agents called entactogens (Nichols *et al.*, 1986; Nichols and Oberlender, 1990).

Drug discrimination

Rats can be trained to discriminate MDMA from vehicle (Schechter, 1987; Oberlender and Nichols, 1988). The MDMA stimulus completely generalizes to drugs that produce MDMA-like subjective effects in humans, including N-ethyl-3,4-methylenedioxyamphetamine (MDE) and 1-(1,3-benzodioxol-5-yl)-N-methylbutan-2-amine (MBDB) (Oberlender and Nichols, 1988; Glennon and Misenheimer, 1989). By contrast, the MDMA stimulus does not generalize to LSD or DOM (Oberlender and Nichols, 1988), and MDMA does not substitute in rats trained with LSD (Nichols *et al.*, 1986). These findings demonstrate that the stimulus cue evoked by MDMA is clearly distinct from that produced by hallucinogens.

The neurochemical basis for the discriminative stimulus effects of MDMA is complex, and is incompletely characterized. The fact that MDMA generalizes to the indirect 5-HT agonists norfenfluramine (Schechter, 1988) and *p*-methoxyamphetamine (Nichols and Oberlender, 1989) indicates that 5-HT release plays a role in the stimulus effects of MDMA. The finding that the 5-HT antagonist pizotyline completely blocks MDMA discrimination (Young *et al.*, 2005) is consistent with 5-HT receptor mediation of MDMA stimulus control, but the identity of the specific 5-HT receptor subtype(s) involved remains to be elucidated. There is some disagreement in the literature regarding the extent to which the MDMA stimulus generalizes to the selective 5-HT_{1A} receptor agonist 8-OH-DPAT (Schechter, 1988; Glennon and Young, 2000). Nonetheless, the finding that the 5-HT_{1A} antagonist NAN-190 fails to block MDMA discrimination completely (Glennon *et al.*, 1992) suggests that 5-HT_{1A} receptors do not play a primary role in MDMA discrimination. Although the 5-HT_2 antagonists ketanserin and pirenperone partially attenuate MDMA stimulus control (Glennon *et al.*, 1992), it is unlikely that 5-HT_2 receptor subtypes contribute appreciably to the MDMA discriminative stimulus because DOM does not substitute in animals trained with MDMA. In light of the structural similarities between MDMA and amphetamine, and given the fact that both drugs increase dopamine release, it does not seem unreasonable to assume that there may be a psychostimulant-like component to the MDMA interoceptive stimulus. Indeed, the MDMA stimulus completely generalizes to *S*-(+)-amphetamine (AMPH) and cocaine (Oberlender and Nichols, 1988; Khorana *et al.*, 2004). At the same time, however, MDMA does not substitute in animals trained with either AMPH or cocaine (Oberlender and Nichols, 1988; Khorana *et al.*, 2004). The asymmetric generalization between MDMA and AMPH indicates that MDMA produces a compound stimulus, with one element of the MDMA cue being AMPH-like. It has been reported that AMPH does not substitute in animals trained with (+)-MBDB (Oberlender and Nichols, 1990), a compound that produces symmetrical cross-generalization with MDMA. (+)-MBDB increases 5-HT release but has little effect on dopamine release (Johnson *et al.*, 1986). Taken together, these findings indicate that the most salient component of the MDMA stimulus (i.e., the (+)-MBDB-like effect) is transduced via 5-HT release, with dopamine release mediating a secondary nonessential AMPH-like component. The fact that dopamine antagonists have little or no effect on MDMA discrimination (Schechter, 1988; Glennon *et al.*, 1992) provides further support for the conclusion that dopamine release is not essential for the induction of MDMA stimulus control.

Locomotor activity

Administration of MDMA to rats and mice produces an increase in locomotor activity that is accompanied by thigmotaxis (Gold *et al.*, 1988). MDMA does not induce hyperactivity in 5-HT transporter $(SERT)^{-/-}$ knockout mice (Bengel *et al.*, 1998) or in rats pretreated with 5-HT uptake inhibitors (Callaway *et al.*, 1990), indicating that the effect of MDMA is dependent on carrier-mediated release of 5-HT. There is considerable evidence that 5-HT_{1B} and 5-HT_{2A} receptors mediate the locomotor response to MDMA (McCreary *et al.*, 1999; Scearce-Levie *et al.*, 1999; Fletcher *et al.*, 2002; Ball and Rebec, 2005). However, there is also a dopaminergic component to the MDMA locomotor response (Gold *et al.*, 1989; Baumann *et al.*, 2008). MDMA is a potent releaser of dopamine, and there is a significant correlation between MDMA-induced hyperlocomotion and extracellular dopamine levels in the nucleus accumbens (Baumann *et al.*, 2008). Antagonist and gene deletion studies have confirmed that MDMA acts partially via D_1 and D_2 receptors to increase locomotor activity (Ball *et al.*, 2003; Bubar *et al.*, 2004; Risbrough *et al.*, 2006). Recently, it was reported that the selective α_1-adrenoceptor antagonist prazosin can block the locomotor response to MDMA (Selken and Nichols, 2007), demonstrating that MDMA-induced norepinephrine release also contributes to the locomotor effects of the drug.

Prepulse inhibition of startle

PPI is attenuated by MDMA and MDE in rats and mice (Mansbach *et al.*, 1989; Dulawa and Geyer, 1996; Kehne *et al.*, 1996). Padich *et al.* (1996) have shown that the ability of MDMA to disrupt PPI is completely blocked by M100907, indicating 5-HT_{2A} receptor involvement. Pretreatment with haloperidol did not block the effect of MDMA on PPI. Several other indirect 5-HT agonists, including *p*-chloroamphetamine and fenfluramine, have also been shown to disrupt PPI (Kehne *et al.*, 1996; Padich *et al.*, 1996). As was found with MDMA, the effect of fenfluramine on PPI is dependent on 5-HT_{2A} receptors (Padich *et al.*, 1996).

Clinical studies of hallucinogens in humans

Assessment of hallucinogen-induced altered states of consciousness

The 'Altered States of Consciousness' (APZ) rating scale is a self-administered questionnaire that was developed by Dittrich to assess alterations of consciousness (Dittrich, 1998). Numerous clinical trials have confirmed that the APZ rating scale is sensitive to the effects of hallucinogens, including DMT, psilocybin, *ayahuasca* and mescaline (Hermle *et al.*, 1992; Vollenweider *et al.*, 1997; Dittrich, 1998; Vollenweider *et al.*, 1998a; Gouzoulis-Mayfrank *et al.*, 1999, 2005; Riba *et al.*, 2002; Hasler *et al.*, 2004; Griffiths *et al.*, 2006). Dittrich has proposed that the subjective effects of hallucinogens can be divided into three major dimensions: (1) 'Oceanic Boundlessness' (OB), (2) 'Anxious Ego Dissolution' (AED), and (3) 'Visionary Restructuralization' (VR). The OB dimension is similar to classical mystical experiences, and involves a pleasant state of depersonalization and derealization. The AED dimension includes dysphoric effects such as anxiety, thought disorder, delusions, fear of losing control, and ego disintegration. The VR dimension includes perceptual phenomena such as visual hallucinations and illusions, synesthesia, and changes in the meaning of percepts. Studies using an updated version of the APZ, the 5D-ASC, have shown that psilocybin also increases scores in a dimension known as 'Reduction of Vigilance' (RV), which includes measures of drowsiness, cognitive impairment and decreased alertness (Hasler *et al.*, 2004; Carter *et al.*, 2007; Vollenweider *et al.*, 2007). Psilocybin can produce auditory effects, but only when high doses are administered (Hasler *et al.*, 2004). An [^{18}F]FDG PET study has demonstrated that psilocybin increases cerebral glucose metabolic activity in the prefrontal cortex, anterior cingulate, temporomedial cortex and putamen, and the increased metabolic rate within those regions was found to be significantly correlated with subjective effects in the OB, VUS and AED dimensions (Vollenweider *et al.*, 1997).

Mechanism of hallucinogen effects in humans

One focus of investigation has been to determine how interactions with specific neurotransmitter systems contribute to the effects of hallucinogens in humans. A few early studies addressed this issue by attempting to block the effects of LSD with serotonin, dopamine and norepinephrine receptor antagonists (Isbell and Logan, 1957; Isbell *et al.*, 1959). Unfortunately, due to the fact that high-affinity receptor-selective antagonists were not available at the time, those studies failed to link the action of hallucinogens to any particular transmitter system. The recent resumption of human hallucinogen studies has coincided with the development of selective antagonist ligands, and these compounds have been used to assess how interactions with serotonergic and dopaminergic receptors contribute to the effects of hallucinogens.

DMT

Meltzer and colleagues (Tueting *et al.*, 1992) examined whether the moderately selective 5-$HT_{2A/2C}$ antagonist cyproheptadine and the D_2 antagonist haloperidol could block the psychological effects induced by DMT in normal human volunteers. Neither drug effectively blocked the effects of DMT, and in some subjects the effects of DMT were actually intensified by pretreatment with cyproheptadine. The finding that cyproheptadine intensified the effects of DMT is intriguing in light of reports that high doses of cyproheptadine produce LSD-like behavioral effects in rats (Colpaert *et al.*, 1982). Nonetheless, it is not clear that the dose of cyproheptadine used in the study (4 mg, p.o.) produces significant occupation of 5-HT_2 sites. The mixed 5-$HT_{1A/1B}$/β-adrenergic antagonist pindolol has been used to assess the involvement of 5-HT_{1A} receptors in the response to DMT (Strassman, 1996). Pretreatment with pindolol markedly intensified the subjective effects of DMT, suggesting that activation of 5-HT_{1A} receptors by DMT acts to attenuate the hallucinogenic effects of the drug.

Psilocybin

As part of a series of studies of the effects of psilocybin in human volunteers, Vollenweider and colleagues have assessed the effect of pretreatment with the 5-HT_{2A} antagonist ketanserin, the mixed D_2/5-HT_{2A} antagonist risperidone, and haloperidol. The effects of psilocybin on the OB, AED and VR dimensions of the APZ were reduced by pretreatment with 20 mg ketanserin (p.o.) and almost completely blocked by pretreatment with either 40 or 50 mg ketanserin (Vollenweider *et al.*, 1998a; Carter *et al.*, 2007). Likewise, pretreatment with 0.5 and 1.0 mg risperidone (p.o.) produced partial and complete blockade, respectively, of the subjective effects of psilocybin (Vollenweider *et al.*, 1998a). Conversely, haloperidol was much less effective, producing only partial blockade of psilocybin effects in the OB dimension, while actually increasing scores in the AED dimension. Taken together, these findings confirm that the hallucinogenic effects of psilocybin in humans are primarily mediated by actions at the 5-HT_{2A} receptor, with D_2 receptor interactions playing only a minor role.

Effect of MDMA in humans

The effects of MDMA and the related homolog MDE in humans have been examined in clinical trials. Treatment with either MDMA or MDE produced effects such as heightened mood, a sense of relaxation, empathy, feelings of closeness to others, stimulation, and minor perceptual and cognitive disturbances (Hermle *et al.*, 1993; Vollenweider *et al.*, 1998b). Although MDMA and MDE did produce some psychostimulant- and hallucinogen-like effects, a double-blind placebo controlled study confirmed that the subjective effects of the entactogens are distinct from those of *S*-(+)-methamphetamine and psilocybin (Gouzoulis-Mayfrank *et al.*, 1999). The latter study also found that the effects of MDE on core APZ dimensions were intermediate between those of psilocybin and *S*-(+)-methamphetamine, providing additional confirmation that differences exist between the subjective effects of these three drugs. Interestingly, in trials comparing the two enantiomers of MDE, the *S*-enantiomer produced entactogenic effects whereas the *R*-enantiomer elicited mainly dysphoric effects (Spitzer *et al.*, 2001). This finding distinguishes MDE from phenylisopropylamine hallucinogens such as DOM, DOB and DOET, for which the *R*-enantiomer is the more active stereoisomer (Shulgin, 1973; Shulgin *et al.*, 1971; Snyder *et al.*, 1974).

Mechanism of MDMA effect in humans

Pretreatment with the selective 5-HT reuptake inhibitor citalopram significantly attenuated most of the subjective effects of MDMA, consistent with the proposed involvement of carrier-mediated 5-HT release in the effects of MDMA (Liechti *et al.*, 2000a; see also this volume). Citalopram also blocked the cardiovascular response to MDMA (Liechti and Vollenweider, 2000a) and inhibited the effect of MDMA on PPI (Liechti *et al.*, 2001). Pretreatment with ketanserin produced some attenuation of the subjective effects of MDMA (Liechti *et al.*, 2000b). However, the most significant reduction of MDMA effects by ketanserin occurred in the VR dimension of the APZ, indicating that 5-HT_{2A} receptors are responsible for the effects of MDMA on perception. Haloperidol pretreatment reduced the euphoric effects of MDMA but increased MDMA-induced dysphoria (i.e., scores in the AED dimension) (Liechti and Vollenweider, 2000b). Neither ketanserin nor haloperidol blocked the cardiovascular effects of MDMA (Liechti and Vollenweider, 2000b; Liechti *et al.*, 2000b) or the ability of the drug to increase PPI (Liechti *et al.*, 2001). Pretreatment with pindolol did not substantially alter either the subjective effects of MDMA or its effect on measures of cognitive performance (Hasler *et al.*, 2008). The fact that interactions with the 5-HT_{1A} receptor do not appear to contribute to the effects of MDMA in humans stands in contrast to the preclinical evidence in rodents indicating that the 5-HT_{1A} receptor plays a role in the behavioral effects of MDMA (Millan and Colpaert, 1991; Glennon and Young, 2000; Morley *et al.*, 2005). In summary, these findings demonstrate that the subjective effects of MDMA are largely attributable to increases in 5-HT release, with D_2 and 5-HT_{2A} receptor stimulation contributing to the mood-elevating and hallucinogen-like perceptual effects, respectively, of MDMA.

Summary

Over the past three decades, substantial progress has been made toward understanding how hallucinogens exert their complex behavioral effects. It is now well established that serotonergic hallucinogens bind to the 5-HT_{2A} receptor, and a large body of preclinical evidence demonstrates that the behavioral effects of the hallucinogens are primarily mediated by interactions with the 5-HT_{2A} receptor. Importantly, due to the recent resumption of human clinical studies with hallucinogens, it has been possible to confirm that the 5-HT_{2A} receptor also plays a primary role in the effects of hallucinogens in humans.

The drug MDMA has been a topic of intense study in recent years. Although there is some overlap between the effects of entactogens and hallucinogens, there are considerable pharmacological differences between these two classes of compounds. The effects of MDMA are largely attributable to increases in the release of serotonin, a fact that has been confirmed in clinical studies with human subjects. Studies of MDMA and the hallucinogens have proven to be extremely informative regarding the pharmacology of these agents. These investigations have also provided insight into how the central serotonergic system acts to modulate behavior.

Most recently, in a study where psilocybin was administered to normal human volunteers, effects were characterized by a majority of the subjects as among the most meaningful experiences of their lives. Positive effects on mood and behavior were manifest, and at 14-month follow-up were still significantly above values prior to the study (Griffiths *et al.*, 2006, 2008). Several clinical studies are now under way to determine whether these effects may provide value in reduction of stress and anxiety in the treatment of cancer patients. It is anticipated that future investigations with hallucinogens, especially research conducted in humans, will yield additional insight into the psychopharmacology of hallucinogens and the behavioral influence of serotonin.

Acknowledgements

This work was supported by National Institute on Drug Abuse (NIDA) Awards DA025412 and DA002925 (A. L. Halberstadt) and DA002189 (D.E. Nichols).

References

Adams, L. and Geyer, M.A. (1985a) Effects of DOM and DMT in a proposed animal model of hallucinogenic activity. Prog. Neuropsychopharmacol. Biol. Psychiatry, 9: 121–132.

Adams, L. and Geyer, M.A. (1985b) A proposed animal model for hallucinogens based on LSD's effects on patterns of exploration in rats. Behav. Neurosci., 99: 881–900.

Aghajanian, GK., Foote, W.E. and Sheard, ME. (1968) Lysergic acid diethylamide: sensitive neuronal units in the midbrain raphe. Science, 161: 706–707.

Aghajanian, G.K., Foote, W.E. and Sheard, ME. (1970) Action of psychotogenic drugs on midbrain raphe neurons. J. Pharmacol. Exp. Ther., 171: 178–187.

Appel, J.B. and Cunningham, KA. (1986) The use of drug discrimination procedures to characterize hallucinogenic drug actions. Psychopharmacol. Bull., 22: 959–967.

Arnt, J. (1989) Characterization of the discriminative stimulus properties induced by 5-HT_1 and 5-HT_2 agonists in rats. Pharmacol. Toxicol., 64: 165–172.

Arnt, J. and Hyttel, J. (1989) Facilitation of 8-OH DPAT-induced forepaw treading of rats by the 5-HT_2 agonist DOI. Eur. J. Pharmacol., 161: 45–51.

Ball, K.T. and Rebec, GV. (2005) Role of 5-HT2A and 5-HT2C/B receptors in the acute effects of 3,4-methylenedioxymethamphetamine (MDMA) on striatal single-unit activity and locomotion in freely moving rats. Psychopharmacology, 181: 676–687.

Ball, K.T., Budreau, D. and Rebec, GV. (2003) Acute effects of 3,4-methylenedioxymethamphetamine on striatal single-unit activity and behavior in freely moving rats: differential involvement of dopamine D_1 and D_2 receptors. Brain Res., 994: 203–215.

Baumann, M.H., Clark, R.D. and Rothman, R.B. (2008) Locomotor stimulation produced by 3,4-methylenedioxymethamphetamine (MDMA) is correlated with dialysate levels of serotonin and dopamine in rat brain. Pharmacol. Biochem. Behav., 90: 208–217.

Bengel, D., Murphy, D.L., Andrews, A.M., Wichems, C.H., Feltner, D., Heils A Mössner, R., Westphal, H. and Lesch, K.P. (1998) Altered brain serotonin homeostasis and locomotor insensitivity to 3,4-methylenedioxymethamphetamine ('Ecstasy') in serotonin transporter-deficient mice. Mol. Pharmacol., 53: 649–655.

Berendsen, H.H. and Broekkamp, C.L. (1990) Behavioural evidence for functional interactions between 5-HT-receptor subtypes in rats and mice. Br. J. Pharmacol., 101: 667–673.

Biello, D. (2006) Not imagining it. Research into hallucinogens cautiously resumes. Sci. Am., 295: 33–35.

Blair, J.B., Kurrasch-Orbaugh, D., Marona-Lewicka, D., Cumbay, M.G., Watts, V.J. and Barker EL and Nichols, D.E. (2000) Effect of ring fluorination on the pharmacology of hallucinogenic tryptamines. J. Med. Chem., 43: 4701–4710.

Braden, M.R. (2007) Towards a biophysical understanding of hallucinogen action. PhD Dissertation. West Lafayette, IN, Purdue University.

Braden, M.R. and Nichols, D.E. (2007) Assessment of the roles of serines 5.43(239) and 5.46(242) for binding and potency of agonist ligands at the human serotonin 5-HT2A receptor. Mol. Pharmacol., 72: 1200–1209.

Brimblecombe, R.W. (1963) Effects of psychotropic drugs on open-field behavior in rats. Psychopharmacologie, 4: 139–147.

Bubar, M.J., Pack, K.M., Frankel, P.S. and Cunningham, K.A. (2004) Effects of dopamine D_1- or D_2-like receptor antagonists on the hypermotive and discriminative stimulus effects of (+)-MDMA. Psychopharmacology, 173: 326–336.

Burris, K.D., Breeding, M. and Sanders-Bush, E. (1991) (+)Lysergic acid diethylamide, but not its nonhallucinogenic congeners, is a potent serotonin 5-HT_{1C} receptor agonist. J. Pharmacol. Exp. Ther., 258: 891–896.

Callaway, C.W., Wing, L.L. and Geyer, M.A. (1990) Serotonin release contributes to the locomotor stimulant effects of 3,4-methylenedioxymethamphetamine in rats. J. Pharmacol. Exp. Ther., 254: 456–464.

Carter, O.L., Hasler, F., Pettigrew, J.D., Wallis, G.M., Liu, G. B. and Vollenweider, F.X. (2007) Psilocybin links binocular rivalry switch rate to attention and subjective arousal levels in humans. Psychopharmacology, 195: 415–424.

Cohen, M. and Wakeley, H. (1968) A comparative behavioral study of ditran and LSD in mice, rats, and dogs. Arch. Int. Pharmacodyn., 173: 316–325.

Colpaert, F.C. (1984) Cross generalization with LSD and yohimbine in the rat. Eur. J. Pharmacol., 102: 73–77.

Colpaert, F.C. and Janssen, P.A.J. (1983) A characterization of the LSD-antagonist effects of pirenperone in the rat. Neuropharmacology, 22: 1001–1003.

Colpaert, F.C., Niemegeers, G.J.E. and Janssen, P.A.J. (1982) A drug discrimination analysis of lysergic acid diethylamide (LSD): *in vivo* agonist and antagonist effects of purported 5-hydroxytryptamine antagonists and of pirenperone, a LSD-antagonist. J. Pharmacol. Exp. Ther., 221: 206–214.

Corne, S.J. and Pickering, R.W. (1967) A possible correlation between drug-induced hallucinations in man and a behavioural response in mice. Psychopharmacologia, 11: 65–78.

Darmani, N.A., Martin, B.R., Pandey, U. and Glennon, R.A. (1990) Do functional relationships exist between 5-HT$_{1A}$ and 5-HT$_2$ receptors?. Pharmacol. Biochem. Behav., 36: 901–906.

DeMontigny, C. and Aghajanian, G.K. (1977) Preferential action of 5-methoxytryptamine and 5-methoxydimethyltryptamine on presynaptic serotonin receptors: a comparative iontophoretic study with LSD and serotonin. Neuropharmacology, 16: 811–818.

Dittrich, A. (1998) The standardized psychometric assessment of altered states of consciousness (ASCs) in humans. Pharmacopsychiatry, 31(Suppl. 2): 80–84.

Dulawa, S.C. and Geyer, M.A. (1996) Psychopharmacology of prepulse inhibition in mice. Chin. J. Physiol., 39: 139–146.

Eison, A.S. and Wright, R.N. (1992) 5-HT1A and 5-HT2 receptors mediate discrete behaviors in the Mongolian gerbil. Pharmacol. Biochem. Behav., 43: 131–137.

Fiorella, D., Rabin, R.A. and Winter, J.C. (1995) The role of the 5-HT$_{2A}$ and 5-HT$_{2C}$ receptors in the stimulus effects of hallucinogenic drugs I: Antagonist correlation analysis. Psychopharmacology, 121: 347–356.

Fletcher, P.J., Korth, K.M., Robinson, S.R. and Baker, G.B. (2002) Multiple 5-HT receptors are involved in the effects of acute MDMA treatment: studies on locomotor activity and responding for conditioned reinforcement. Psychopharmacology, 162: 282–291.

Gaddum, J.H. (1953) Antagonism between lysergic acid diethylamide and 5-hydroxytryptamine. J. Physiol. (Lond.), 121: 15P.

Geyer, M.A. (1990) Approaches to the characterization of drug effects on locomotor activity in rodents. In: Adler, M.W. and Cowan, A. (Eds.), Modern Methods in Pharmacology: Testing and Evaluation of Drugs of Abuse. Wiley-Liss, New York, NY, pp. 81–99.

Geyer, M.A., Light, R.K., Rose, G.J., Petersen, L.R., Horwitt, D.D., Adams, L.M. and Hawkins, R.L. (1979) A characteristic effect of hallucinogens on investigatory responding in rats. Psychopharmacology, 65: 35–40.

Glennon, R.A. (1990) Do classical hallucinogens act as 5-HT$_2$ agonists or antagonists? Neuropsychopharmacology, 3: 509–517.

Glennon, R.A. and Misenheimer, B.R. (1989) Stimulus effects of N-monoethyl-1-(3,4-methylenedioxyphenyl)-2-aminopropane (MDE) and N-hydroxy-1-(3,4-methylenedioxyphenyl)-2-aminopropane (N-OH MDA) in rats trained to discriminate MDMA from saline. Pharmacol. Biochem. Behav., 33: 909–912.

Glennon, R.A. and Young, R. (2000) MDMA stimulus generalization to the 5-HT$_{1A}$ serotonin agonist 8-hydroxy-2-(di-n-propylamino)tetralin. Pharmacol. Biochem. Behav., 66: 483–488.

Glennon, R.A., Rosecrans, J.A. and Young, R. (1982) The use of the drug discrimination paradigm for studying hallucinogenic agents. A review. In: Colpaert, F.C. and Slangen, J.L. (Eds.), Drug Discrimination: Applications in CNS Pharmacology. Elsevier Biomedical Press, Amsterdam, pp. 69–96.

Glennon, R.A., Young, R. and Rosecrans, J.A. (1983) Antagonism of the effects of the hallucinogen DOM and the puported 5-HT agonist quipazine by 5-HT$_2$ antagonists. Eur. J. Pharmacol., 91: 189–196.

Glennon, R.A., Titeler, M. and McKenney, J.D. (1984) Evidence for 5-HT$_2$ involvement in the mechanisms of action of hallucinogenic agents. Life Sci., 35: 2505–2511.

Glennon, R.A., Higgs, R., Young, R. and Issa, H. (1992) Further studies on N-methyl-1(3,4-methylenedioxyphenyl)-2-aminopropane as a discriminative stimulus: antagonism by 5-hydroxytryptamine3 antagonists. Pharmacol. Biochem. Behav., 43: 1099–1106.

Gold, L.H., Koob, G.F. and Geyer, M.A. (1988) Stimulant and hallucinogenic behavioral profiles of 3,4-methylenedioxymethamphetamine and N-ethyl-3,4-methylenedioxyamphetamine in rats. J. Pharmacol. Exp. Ther., 247: 547–555.

Gold, L.H., Hubner, C.B. and Koob, G.F. (1989) A role for the mesolimbic dopamine system in the psychostimulant actions of MDMA. Psychopharmacology, 99: 40–47.

González-Maeso, J., Weisstaub, N.V., Zhou, M., Chan, P., Ivic, L., Ang, R., Lira, A., Bradley-Moore, M., Ge, Y., Zhou, Q., Sealfon, S.C. and Gingrich, J.A. (2007) Hallucinogens recruit specific cortical 5-HT$_{2A}$ receptor-mediated signaling pathways to affect behavior. Neuron, 53: 439–452.

Gouzoulis-Mayfrank, E., Thelen, B., Habermeyer, E., Kunert, H.J., Kovar, K.A., Lindenblatt, H., Hermle, L., Spitzer, M. and Sass, H. (1999) Psychopathological, neuroendocrine and autonomic effects of 3,4-methylenedioxyethylamphetamine (MDE), psilocybin and d-methamphetamine in healthy volunteers. Results of an experimental double-blind placebo-controlled study. Psychopharmacology, 142: 41–50.

Gouzoulis-Mayfrank, E., Heekeren, K., Neukirch, A., Stoll, M., Stock, C., Obradovic, M. and Kovar, K.A. (2005) Psychological effects of (S)-ketamine and, N,N-dimethyltryptamine (DMT): a double-blind, cross-over study in healthy volunteers. Pharmacopsychiatry, 38: 301–311.

Grailhe, R., Waeber, C., Dulawa, S.C., Hornung, J.P., Zhuang, X., Brunner, D., Geyer, M.A. and Hen, R. (1999) Increased exploratory activity and altered response to LSD in mice lacking the 5-HT$_{5A}$ receptor. Neuron, 22: 581–591.

Griffiths, R., Richards, W. and McCann U and Jesse, R. (2006) Psilocybin can occasion mystical-type experiences having substantial and sustained personal meaning and spiritual significance. Psychopharmacology, 187: 268–283.

Griffiths, R., Richards, W., Johnson, M. and McCann U and Jesse, R. (2008) Mystical-type experiences occasioned by psilocybin mediate the attribution of personal meaning and spiritual significance 14 months later. J. Psychopharmacol., 22: 621–632.

Haigler, H.J. and Aghajanian, G.K. (1973) Mescaline and LSD: direct and indirect effects on serotonin-containing neurons in brain. Eur. J. Pharmacol., 21: 53–60.

Haigler, H.J. and Aghajanian, G.K. (1974) Lysergic acid diethylamide and serotonin: a comparision of effects on serotonergic neurons and neurons receiving a serotonergic input. J. Pharmacol. Exp. Ther., 188: 688–699.

Halberstadt, A.L., Buell, M.R., Masten, V.L., Risbrough, V. B. and Geyer, M.A. (2008) Modification of the effects of 5-methoxy-N,N-dimethyltryptamine on exploratory behavior in rats by monoamine oxidase inhibitors. Psychopharmacology, 201: 55–66.

Hasler, F., Grimberg, U., Benz, M.A., Huber, T. and Vollenweider, F.X. (2004) Acute psychological and physiological effects of psilocybin in healthy humans: a double-blind, placebo-controlled dose-effect study. Psychopharmacology, 172: 145–156.

Hasler, F., Studerus, E., Lindner, K., Ludewig, S. and Vollenweider, F. (2008) Investigation of serotonin-1A receptor function in the human psychopharmacology of MDMA. J. Psychopharmacol. doi:10.1177/0269881108094650.

Hermle, L., Fünfgeld, M., Oepen, G., Botsch, H., Borchardt, D., Gouzoulis, E., Fehrenbach, R.A. and Spitzer, M. (1992) Mescaline-induced psychopathological, neuropsychological, and neurometabolic effects in normal subjects: experimental psychosis as a tool for psychiatric research. Biol. Psychiatry, 32: 976–991.

Hermle, L., Spitzer, M., Borchardt, D., Kovar, K.A. and Gouzoulis, E. (1993) Psychological effects of MDE in normal subjects. Are entactogens a new class of psychoactive agents? Neuropsychopharmacology, 8: 171–176.

Hirschhorn, I.D. and Winter, J.C. (1971) Mescaline and lysergic acid diethylamide (LSD) as discriminative stimuli. Psychopharmacologia, 22: 64–71.

Hoffman, A.J. and Nichols, D.E. (1985) Synthesis and LSD-like discriminative stimulus properties in a series of N(6)-alkyl norlysergic acid, N,N-diethylamide derivatives. J. Med. Chem., 28: 1252–1255.

Isbell, H. and Logan, C.R. (1957) Studies on the diethylamide of lysergic acid (LSD-25). II. Effects of chlorpromazine, azacyclonol, and reserpine on the intensity of the LSD-reaction. AMA Arch. Neurol. Psychiatry, 77: 350–358.

Isbell, H., Logan, C.R. and Miner, E.J. (1959) Studies on lysergic acid diethylamide (LSD-25). III. Attempts to attenuate the LSD-reaction in man by pretreatment with neurohumoral blocking agents. AMA Arch. Neurol. Psychiatry, 81: 20–27.

Johansson, C., Jackson, D.M., Zhang, J. and Svensson, L. (1995) Prepulse inhibition of acoustic startle, a measure of sensorimotor gating: effects of antipsychotics and other agents in rats. Pharmacol. Biochem. Behav., 52: 649–654.

Johnson, M.P., Hoffman, A.J. and Nichols, D.E. (1986) Effects of the enantiomers of MDA, MDMA and related analogues on [^{3}H]serotonin and [^{3}H]dopamine release from superfused rat brain slices. Eur. J. Pharmacol., 132: 269–276.

Kabes, J., Fink, Z. and Roth, Z. (1972) A new device for measuring spontaneous motor activity-effects of lysergic acid diethylamide in rats. Psychopharmacologia, 23: 75–85.

Kehne, J.H., Padich, R.A., McCloskey, T.C., Taylor, V.L. and Schmidt, C.J. (1996) 5-HT modulation of auditory and visual sensorimotor gating: I. Effects of 5-HT releasers on sound and light prepulse inhibition in Wistar rats. Psychopharmacology, 124: 95–106.

Khorana, N., Pullagurla, M.R., Young, R. and Glennon, R.A. (2004) Comparison of the discriminative stimulus effects of 3,4-methylenedioxymethamphetamine (MDMA) and cocaine: asymmetric generalization. Drug Alcohol Depend., 74: 281–287.

Koek, W., Jackson, A. and Colpaert, F.C. (1992) Behavioral pharmacology of antagonists at 5-HT$_2$/5-HT$_{1C}$ receptors. Neurosci. Biobehav. Rev., 16: 95–105.

Krebs-Thomson, K. and Geyer, M.A. (1996) The role of 5-HT$_{1A}$ receptors in the locomotor-suppressant effects of LSD: WAY-100635 studies of 8-OH-DPAT, DOI and LSD in rats. Behav. Pharmacol., 7: 551–559.

Krebs-Thomson, K. and Geyer, M.A. (1998) Evidence for a functional interaction between 5-HT1A and 5-HT2 receptors in rats. Psychopharmacology, 140: 69–74.

Krebs-Thomson, K., Paulus, M.P. and Geyer, M.A. (1998) Effects of hallucinogens on locomotor and investigatory activity and patterns: influence of 5-HT$_{2A}$ and 5-HT$_{2C}$ receptors. Neuropsychopharmacology, 18: 339–351.

Krebs-Thomson, K., Ruiz, E.M., Masten, V., Buell, M. and Geyer, M.A. (2006) The roles of 5-HT$_{1A}$ and 5-HT$_2$ receptors in the effects of 5-MeO-DMT on locomotor activity and prepulse inhibition in rats. Psychopharmacology, 189: 319–329.

Leysen, J.E. (1990) Gaps and peculiarities in 5-HT$_2$ receptor studies. Neuropsychopharmacology, 3: 361–369.

Li, J.-X., Rice, K.C. and France, C.P. (2007) Behavioral effects of dipropyltryptamine in rats: evidence for 5-HT$_{1A}$ and 5-HT$_{2A}$ agonist activity. Behav. Pharmacol., 18: 283–288.

Liechti, M.E. and Vollenweider, F.X. (2000a) The serotonin uptake inhibitor citalopram reduces acute cardiovascular and vegetative effects of 3,4-methylenedioxymethamphetamine ('Ecstasy') in healthy volunteers. J. Psychopharmacol., 14: 269–274.

Liechti, M.E. and Vollenweider, F.X. (2000b) Acute psychological and physiological effects of MDMA ('Ecstasy') after haloperidol pretreatment in humans. Eur. Neuropsychopharmacol., 10: 289–295.

Liechti, M.E., Baumann, C., Gamma, A. and Vollenweider, F.X. (2000a) Acute psychological effects of 3,4-methylenedioxymethamphetamine (MDMA, 'Ecstasy') are attenuated by the serotonin uptake inhibitor citalopram. Neuropsychopharmacology, 22: 513–521.

Liechti, M.E., Saur, M.R., Gamma, A., Hell, D. and Vollenweider, F.X. (2000b) Psychological and physiological effects of MDMA ('Ecstasy') after pretreatment with the 5-HT$_2$ antagonist ketanserin in healthy humans. Neuropsychopharmacology, 23: 396–404.

Liechti, M.E., Geyer, M.A., Hell, D. and Vollenweider, F.X. (2001) Effects of MDMA (ecstasy) on prepulse inhibition and habituation of startle in humans after pretreatment with citalopram, haloperidol, or ketanserin. Neuropsychopharmacology, 24: 240–252.

Lyon, R.A., Glennon, R.A. and Titeler, M. (1986) 3,4-Methylene dioxymethamphetamine (MDMA): stereoselective interactions at brain 5-HT1 and 5-HT2 receptors. Psychopharmacology, 188: 525–526.

Mansbach, R.S., Braff, D.L. and Geyer, M.A. (1989) Prepulse inhibition of the acoustic startle response is disrupted by N-ethyl-3,4-methylenedioxyamphetamine (MDEA) in the rat. Eur. J. Pharmacol., 167: 49–55.

Marona-Lewicka, D. and Nichols, D.E. (1995) Complex stimulus properties of LSD: a drug discrimination study with α_2-adrenoceptor agonists and antagonists. Psychopharmacology, 120: 384–391.

Marona-Lewicka, D., Thisted, R.A. and Nichols, D.E. (2005) Distinct temporal phases in the behavioral pharmacology of LSD: dopamine D2 receptor-mediated effects in the rat and implications for psychosis. Psychopharmacology, 180: 427–435.

Marona-Lewicka, D., Chemel, B.R. and Nichols, D.E. (2008) Dopamine D$_4$ receptor involvement in the discriminative stimulus effects in rats of LSD, but not the phenethylamine hallucinogen DOI. Psychopharmacology. doi:10.1007/s00213-008-1238-0.

McCreary, A.C., Bankson, M.G. and Cunningham, K.A. (1999) Pharmacological studies of the acute and chronic effects of (+)-3,4-methylenedioxymethamphetamine on locomotor activity: role of 5-hydroxytryptamine$_{1A}$ and 5-hydroxytryptamine$_{1B/1D}$ receptors. J. Pharmacol. Exp. Ther., 290: 965–973.

Meert, T.F. (1996) Serotonin mechanisms in antipsychotic treatment: evidence from drug discrimination studies. In: Kane,

J.M., Möller, H.J. and Awouters, F. (Eds.), Serotonin in Antipsychotic Treatment. Mechanisms and Clinical Practice. Marcel Dekker, Inc., New York, NY, pp. 109–130.

Meert, T.F. and Awouters, F. (1991) Serotonin 5-HT_2 antagonists: a preclinical evaluation of possible therapeutic effects. In: Idzikowski, C. and Cowen, P.J. (Eds.), Serotonin, Sleep and Mental Disorders. Washington Biomedical Publishing, pp. 65–76.

Meert, T.F., DeHaes, P.L.A.J., Vermote, P.C.M. and Janssen, P. A.J. (1990) Pharmacological validation of ritanserin and risperidone in the drug discrimination test procedure in the rat. Drug Dev. Res., 19: 353–373.

Millan, M.J. and Colpaert, F.C. (1991) Methylenedioxymeth amphetamine induces spontaneous tail-flicks in the rat via 5-HT1A receptors. Eur. J. Pharmacol., 193: 145–152.

Mittman, S.M. and Geyer, M.A. (1991) Dissociation of multiple effects of acute LSD on exploratory behavior in rats by ritanserin and propranolol. Psychopharmacology, 105: 69–76.

Moreno, F.A., Wiegand, C.B., Taitano, E.K. and Delgado, P. L. (2006) Safety, tolerability, and efficacy of psilocybin in 9 patients with obsessive-compulsive disorder. J. Clin. Psychiatry, 67: 1735–1740.

Morley, K.C., Arnold, J.C. and McGregor, I.S. (2005) $Serotonin_{1A}$ receptor involvement in acute 3,4-methylenedioxymethamphetamine (MDMA) facilitation of social interaction in the rat. Prog. Neuropsychopharmacol. Biol. Psychiatry, 29: 648–657.

Nelson, D.L., Lucaites, V.L., Wainscott, D.B. and Glennon, R.A. (1999) Comparisions of hallucinogenic phenylisopropylamine binding affinities at cloned human 5-HT_{2A}, 5-HT_{2B} and 5-HT_{2C} receptors. Naunyn-Schmiedeberg's Arch. Pharmacol., 359: 1–6.

Nichols, D.E. (1986) Differences between the mechanism of action of MDMA, MBDB, and the classic hallucinogens. Identification of a new therapeutic class: entactogens. J. Psychoactive Drugs, 18: 305–313.

Nichols, D.E. (2004) Hallucinogens. Pharmacol. Ther., 101: 131–181.

Nichols, D.E. and Dyer, D.C. (1977) Lipophilicity and serotonin agonist activity in a series of 4- substituted mescaline analogues. J. Med. Chem., 20: 299–301.

Nichols, D.E. and Nichols, C.D. (2008) Serotonin receptors. Chem. Rev., 108: 1614–1641.

Nichols, D.E. and Oberlender, R. (1989) Structure-activity relationships of MDMA-like substances. NIDA Res. Monogr., 94: 1–29.

Nichols, D.E. and Oberlender, R. (1990) Structure-activity relationships of MDMA and related compounds: a new class of psychoactive drugs?. Ann. N. Y. Acad. Sci., 600: 613–623.

Nichols, D.E., Hoffman, A.J., Oberlender, R.A., Jacob, P. 3rd and Shulgin, A.T. (1986) Derivatives of 1-(1,3-benzodioxol-5-yl)-2-butanamine: representatives of a novel therapeutic class. J. Med. Chem., 29: 2009–2015.

Nichols, D.E., Frescas, S., Marona-Lewicka, D. and Kurrasch-Orbaugh, D.M. (2002) Lysergamides of isomeric 2,4-dimethylazetidines map the binding orientation of the diethylamide moiety in the potent hallucinogenic agent, N,N-diethyllysergamide (LSD). J. Med. Chem., 45: 4344–4349.

Oberlender, R. and Nichols, D.E. (1988) Drug discrimination studies with MDMA and amphetamine. Psychopharmacology, 95: 71–76.

Oberlender, R. and Nichols, D.E. (1990) (+)-N-methyl-1-(1,3-benzodioxol-5-yl)-2-butanamine as a discriminative stimulus in studies of 3,4-methylenedioxy-methamphetamine-like behavioral activity. J. Pharmacol. Exp. Ther., 255: 1098–1106.

Ouagazzal, A., Grottick, A.J., Moreau, J. and Higgins, G.A. (2001) Effect of LSD on prepulse inhibition and spontaneous behavior in the rat. A pharmacological analysis and comparison between two rat strains. Neuropsychopharmacology, 25: 565–575.

Padich, R.A., McCloskey, T.C. and Kehne, J.H. (1996) 5-HT modulation of auditory and visual sensorimotor gating: II. Effects of the 5-HT2A antagonist MDL 100,907 on disruption of sound and light prepulse inhibition produced by 5-HT agonists in Wistar rats. Psychopharmacology, 124: 107–116.

Parrish, J.C., Braden, M.R. and Gundy E and Nichols, D. E. (2005) Differential phospholipase C activation by phenylalkylamine serotonin 5-HT 2A receptor agonists. J. Neurochem., 95: 1575–1584.

Penington, N.J. and Reiffenstein, R.J. (1986) Direct comparision of hallucinogenic phenethylamines and D-amphetamine on dorsal raphe neurons. Eur. J. Pharmacol., 122: 373–377.

Pfaff, R.C., Huang, X., Marona-Lewicka, D., Oberlender R. and Nichols, D.E. (1994) Lysergamides revisited. NIDA Res. Monogr., 146: 52–73.

Pierce, P.A. and Peroutka, S.J. (1989) Hallucinogenic drug interactions with neurotransmitter receptor binding sites in human cortex. Psychopharmacology, 97: 118–122.

Porter, R.H.P., Benwell, K.R., Lamb, H., Malcolm, C.S., Allen, N.H., Revell, D.F., Adams, D.R. and Sheardown, M.J. (1999) Functional characterization of agonists at recombinant human 5-HT_{2A}, 5-HT_{2B} and 5-HT_{2C} receptors in CHO-K1 cells. Br. J. Pharmacol., 128: 13–20.

Rapport, M.M., Green, A.A. and Paige, I.H. (1948) Serum vasoconstrictor (serotonin) IV. Isolation and characterization. J. Biol. Chem., 176: 1243–1251.

Riba, J., Rodriguez-Fornells, A. and Barbanog, M.J. (2002) Effects of ayahuasca on sensory and sensorimotor gating in humans as measured by P50 suppression and prepulse inhibition of the startle reflex, respectively. Psychopharmacology, 165: 18–28.

Rigdon, G.C. and Weatherspoon, J.K. (1992) 5-$Hydroxytryptamine_{1a}$ receptor agonists block prepulse inhibition of acoustic startle reflex. J. Pharmacol. Exp. Ther., 263: 486–493.

Risbrough, V.B., Masten, V.L., Caldwell, S., Paulus, M.P., Low, M.J. and Geyer, M.A. (2006) Differential contributions of dopamine D1, D2, and D3 receptors to MDMA-induced effects on locomotor behavior patterns in mice. Neuropsychopharmacology, 31: 2349–2358.

Scearce-Levie, K., Viswanathan, S.S. and Hen, R. (1999) Locomotor response to MDMA is attenuated in knockout mice lacking the 5-HT_{1B} receptor. Psychopharmacology, 141: 154–161.

Schechter, M.D. (1987) MDMA as a discriminative stimulus: isomeric comparisons. Pharmacol. Biochem. Behav., 27: 41–44.

Schechter, M.D. (1988) Serotonergic-dopaminergic mediation of 3,4-methylenedioxymethamphetamine (MDMA, 'ecstasy'). Pharmacol. Biochem. Behav., 31: 817–824.

Schreiber, R. and DeVry, J. (1993) Studies on the neuronal circuits involved in the discriminative stimulus effects of 5-$hydroxytryptamine_{1A}$ receptor agonists in the rat. J. Pharmacol. Exp. Ther., 265: 572–579.

Schreiber, R., Brocco, M. and Millan, M.J. (1994) Blockade of the discriminative stimulus effect of DOI by MDL 100,907 and the 'atypical' antipsychotics, clozapine and risperidone. Eur. J. Pharmacol., 264: 99–102.

Schreiber, R., Brocco, M., Audinot, V., Gobert, A., Veiga, S. and Millan, M.J. (1995) (1-(2,5-Dimethoxy-4 iodophenyl)-2-aminopropane)-induced head-twitches in the rat are mediated by 5-hydroxytryptamine (5-HT) $_{2A}$ receptors: modulation by novel 5-$HT_{2A/2C}$ antagonists, D_1 antagonists and 5-HT_{1A} agonists. J. Pharmacol. Exp. Ther., 273: 101–112.

Schultes, R.E. and Hofmann, A. (1980) The Botany and Chemistry of Hallucinogens. Charles C. Thomas, Springfield, IL.

Selken, J. and Nichols, D.E. (2007) Alpha1-adrenergic receptors mediate the locomotor response to systemic administration of (+/−)-3,4-methylenedioxymethamphetamine (MDMA) in rats. Pharmacol. Biochem. Behav., 86: 622–630.

Shaw, E. and Woolley, D.W. (1956) Some serotoninlike activities of lysergic acid diethylamide. Science, 124: 121–122.

Shulgin, A.T. (1973) Stereospecific requirements for hallucinogenesis. J. Pharm. Pharmacol., 25: 271–272.

Shulgin, A. and Shulgin, A. (1991) PIHKAL A Chemical Love Story. Transform Press, Berkeley, CA.

Shulgin, A.T., Sargent, T. and Naranjo, C. (1971) 4-Bromo-2,5-dimethoxyphenylisopropylamine, a new centrally active amphetamine analog. Pharmacology, 5: 103–107.

Silva, M.T.A. and Calil, H.M. (1975) Screening hallucinogenic drugs: systematic study of three behavioral tests. Psychopharmacologia, 42: 163–171.

Sipes, T.E. and Geyer, M.A. (1994) Multiple serotonin receptor subtypes modulate prepulse inhibition of the startle response in rats. Neuropharmacology, 33: 441–448.

Sipes, T.E. and Geyer, M.A. (1995) DOI disruption of prepulse inhibition of startle in the rat is mediated by 5-HT_{2A} and not by 5-HT_{2C} receptors. Behav. Pharmacol., 6: 839–842.

Smith, R.L., Barrett, R.J. and Sanders-Bush, E. (1999) Mechanism of tolerance development to 2,5-dimethoxy-4-iodoamphetamine in rats: down-regulation of the 5-HT_{2A}, but not 5-HT_{2C}, receptor. Psychopharmacology, 144: 248–254.

Smith, R.L., Barrett, R.J. and Sanders-Bush, E. (2003) Discriminative stimulus properties of 1-(2,5-dimethoxy-4-iodophenyl)-2-aminopropane [(+/−)DOI] in C57BL/6J mice. Psychopharmacology, 166: 61–68.

Snyder, S.H., Unger, S., Blatchley, R. and Barfknecht, C.F. (1974) Stereospecific actions of DOET (2,5-dimethoxy-4-ethylamphetamine) in man. Arch. Gen. Psychiatry, 31: 103–106.

Spencer, D.G., Glaser, T. and Traber, J. (1987) Serotonin receptor subtype mediation of the interoceptive discriminative stimuli induced by 5-methoxy-N,N-dimethyltryptamine. Psychopharmacology, 93: 158–166.

Spitzer, M., Franke, B., Walter, H., Buechler, J., Wunderlich, A.P., Schwab, M., Kovar, K.A., Hermle, L. and Grön, G. (2001) Enantio-selective cognitive and brain activation effects of N-ethyl-3,4-methylenedioxyamphetamine in humans. Neuropharmacology, 41: 263–271.

Stoll, W.A. (1947) Lysergsäure-diathylamid, ein Phantastikum aus der Mutterkorngruppe. Schweiz Arch. Neurol. Psychiatry, 60: 279–323.

Strassman, R.J. (1996) Human psychopharmacology of, N,N-dimethyltryptamine. Behav. Brain Res., 73: 121–124.

Titeler, M., Lyon, L.A. and Glennon, R.A. (1988) Radioligand binding evidence implicates the brain 5-HT_2 receptor as a site of action for LSD and phenylisopropyl amine hallucinogens. Psychopharmacology, 94: 213–216.

Tricklebank, M.D., Forler, C., Middlemiss, D.N. and Fozard, J. R. (1985) Subtypes of the 5-HT receptor mediating the behavioral responses to 5-methoxy-N,N-dimethyltryptamine in the rat. Eur. J. Pharmacol., 117: 15–24.

Tueting, P.A., Metz, J., Rhoades, B.K. and Boutros, N.N. (1992) Pharmacologic challenge in ERP research. Ann. N.Y. Acad. Sci., 658: 223–255.

Vickers, S.P., Easton, N., Malcolm, C.S., Allen, N.H., Porter, R.H., Bickerdike, M.J. and Kennett, G.A. (2001) Modulation of 5-HT_{2A} receptor-mediated head-twitch behaviour in the rat by 5-HT_{2C} receptor agonists. Pharmacol. Biochem. Behav., 69: 643–652.

Vollenweider, F.X., Leenders, K.L., Scharfetter, C., Maguire, P., Stadelmann, O. and Angst, J. (1997) Positron emission tomography and fluorodeoxyglucose studies of metabolic hyperfrontality and psychopathology in the psilocybin model of psychosis. Neuropsychopharmacology, 16: 357–372.

Vollenweider, F.X., Vollenweider-Scherpenhuyzen, M.F., Bäbler, A., Vogel, H. and Hell, D. (1998a) Psilocybin induces schizophrenia-like psychosis in humans via a serotonin-2 agonist action. Neuroreport, 9: 3897–3902.

Vollenweider, F.X., Gamma, A., Liechti, M. and Huber, T. (1998b) Psychological and cardiovascular effects and short-term sequelae of MDMA ('ecstasy') in MDMA-naïve healthy volunteers. Neuropsychopharmacology, 19: 241–251.

Vollenweider, F.X., Csomor, P.A., Knappe, B., Geyer, M.A. and Quednow, B.B. (2007) The effects of the preferential 5-HT2A agonist psilocybin on prepulse inhibition of startle in healthy human volunteers depend on interstimulus interval. Neuropsychopharmacology, 32: 1876–1887.

Watts, V.J., Lawler, C.P., Rox, D.R., Neve, K.A., Nichols, D.E. and Mailman, R.B. (1995) LSD and structural analogs: pharmacological evaluation at D_1 dopamine receptors. Psychopharmacology, 118: 401–409.

Wettstein, J.G., Host, M. and Hitchcock, J.M. (1999) Selectivity of action of typical and atypical antipsychotic drugs as antagonists of the behavioral effects of 1-[2,5-dimethoxy-4-iodophenyl]-2-aminopropane (DOI). Prog. Neuro-Psychopharmacol. Biol. Psychiatry, 23: 533–544.

Wing, L.L., Tapson, G.S. and Geyer, M.A. (1990) 5HT-2 mediation of acute behavioral effects of hallucinogens in rats. Psychopharmacology, 100: 417–425.

Winter, J.C., Filipink, R.A., Timineri, D., Helsley, S.E. and Rabin, R.A. (2000) The paradox of 5-methoxy-*N,N*-dimethyltryptamine: an indoleamine hallucinogen that induces stimulus control via 5-HT_{1A} receptors. Pharmacol. Biochem. Behav., 65: 75–82.

Winter, J.C., Eckler, J.R. and Rabin, R.A. (2004) Serotonergic/glutamatergic interactions: the effects of mGlu2/3 receptor ligands in rats trained with LSD and PCP as discriminative stimuli. Psychopharmacology, 172: 233–240.

Winter, J.C., Rice, K.C., Amorosi, D.J. and Rabin, R.A. (2007) Psilocybin-induced stimulus control in the rat. Pharmacol. Biochem. Behav., 87: 472–480.

Young, R., Khorana, N., Bondareva, T. and Glennon, R.A. (2005) Pizotyline effectively attenuates the stimulus effects of N-methyl-3,4-methylenedioxyamphetamine (MDMA). Pharmacol. Biochem. Behav., 82: 404–410.

CHAPTER 4.8

The Role of Serotonin in Cortical Development: Implications for Autism Spectrum Disorder

Christine F. Hohmann[*,1] and Mary E. Blue[2]

[1]*Department of Biology, Morgan State University, Baltimore, Maryland*
[2]*Department of Neurology and Neuroscience, Johns Hopkins University School of Medicine, Hugo W. Moser Research Institute at Kennedy Krieger, Neuroscience Laboratory, Kennedy Krieger Institute, Baltimore, Maryland*

Abstract: This chapter provides a review of basic research as well as clinical literature in support of the idea that altered serotonergic homeostasis is prominently involved in the pathogenesis of Autism Spectrum Disorder (ASD). Animal models of cortical serotonergic effects in development and plasticity are reviewed and juxtaposed to neuropathological and clinical findings in ASD. Based on the emerging conceptualization of ASD as a developmental disorder of cortical connectivity, we present a hypothesis about the role of serotonin in altered cortical network formation, as a substrate for altered cognitive function. Investigations into the role of serotonin in immune regulation, and particularly neuroimmune regulation, support a role for serotonin in a variety of homeostatic functions that may be impaired in ASD and can contribute to altered brain development and function. A unified hypothesis such as proposed here could provide an avenue towards early identification of biomarkers as well as therapeutic strategies. We hope to provide an impetus for new investigations into a 'serotonergic hypothesis' for ASD.

Keywords: serotonin, cerebral cortex, development and plasticity, autism spectrum disorder, neuroinflammation, cytokines, BDNF.

Introduction

Attention on serotonin as modulator of cortical development and plasticity has come relatively late compared to research on the role of other afferent neuromodulator systems such as norepinephrine, dopamine and acetylcholine. In the mid-1990s, when Hohmann co-authored a review on neuromodulatory influences in cortical development, little direct evidence was available for serotonergic effects during critical periods of cortical development (Berger-Sweeney and Hohmann, 1997). This has dramatically changed in the decade since. Such interest likely has been precipitated by the increasing awareness of serotonergic involvement in a spectrum of mental health disorders with developmental origins and the rising prominence of serotonergically acting drugs in the treatment of mental health disorders (Stokes and Holtz, 1997; Murrin *et al.*, 2007; Thase and Denko, 2008). This, in turn, has led to rising concerns about the potential effects of such drug treatments on the immature nervous system (Bonati and Clavenna, 2005; Gentile, 2007).

Our goal in this chapter is to present an overview of the experimental literature concerning serotonergic influences on the development of the cerebral cortical structure and, to a lesser extent, function within the context of serotonin involvement in the pathologies of Autism Spectrum Disorder [ASD]. We will largely concentrate on the rat/mouse somatosensory cortex and rat and cat visual cortex as model systems to illustrate current thinking concerning the role of 5-HT in the modulation of cortical development and plasticity. These two different paradigms, long used to study developmental plasticity in the cortex, are often contrasted in terms of the established contributions of synaptic activity on thalamocortical afferent pattern formation. It is generally presumed that the development of visual connections in cortex is more fundamentally under the control of patterned synaptic activity and that the somatosensory cortex is more strictly controlled by genetic information (Rebsam *et al.*, 2005). Both structures illustrate the importance of patterned sensory input on structural and functional development of the neocortex, and both

*Corresponding author
E-mail: christine.hohmann@morgan.edu

Christian Müller & Barry Jacobs (Eds.)
Handbook of Behavioral Neurobiology of Serotonin
ISBN 978-0-12-374634-4

DOI: 10.1016/B978-0-12-374634-4.00036-8

paradigms do have a 'critical period' for the proper development of functional sensory maps.

Knowing that other authors in this volume are likely to give some detail to general serotonergic effects in morphogenesis as well as to cognitive social and emotional behavioral consequences of altered serotonergic innervation in maturity, we will discuss behavioral effects only to the extent that they are specific to developmental serotonergic manipulations and differ from what is seen in the mature nervous system. Such behavioral studies will serve as a bridge to understanding the potential relevance of serotonin in the etiology of ASD, a disorder with early developmental onset. We will review the clinical evidence that points towards a developmental serotonergic imbalance in ASD along with the neuropathology that supports a serotonergic role, based on insights from the experimental literature. We will conclude by presenting a model for the role of 5-HT in the pathogenesis of ASD, and discuss this within the emerging conceptual framework of understanding that ASD is a disorder of neural network development, particularly in the cerebral cortex. There are many avenues, genetic and environmental, that can lead to altered serotonergic innervation during brain development, and, depending on the time that the innervation is disrupted, different endophenotypes might emerge and thus account for the spectrum of behavioral and neurobiological sequelae in ASD and, beyond that, other mental health disorders.

Serotonin in cortical development

Development of the cortical serotonergic innervation

In the mammalian cerebral cortex, serotonergic fibers typically appear perinatally and their maturation, as well as the maturation of their postsynaptic receptors, brackets the period of maximal developmental plasticity (Goldman-Rakic and Brown, 1982; Jonsson and Kasamatsu, 1983; Foote and Morrison, 1987; Gu *et al.*, 1990; Dyck and Cynader, 1993; Dori *et al.*, 1998; Harper and Wallace, 1995). In rat and mouse, 5-HT immunoreactive fibers enter the cortex at approximately embryonic days 18–19 and do not reach adult distribution patterns until 3–4 weeks postnatally (Lidov and Molliver, 1982; D'Amato *et al.*, 1987; Dori *et al.*, 1998; Janusonis *et al.*, 2004). Serotonin immunoreactive fibers from the medial forebrain bundle appear first in frontal cortical areas, and thereafter spread across parietal and occipital areas. Studies by Janusonis and colleagues suggest that serotonergic afferents first contact Cajal-Retzius neurons in the marginal zone, prior to birth, before entering the cortical plate (Janusonis *et al.*, 2004). The subsequent cortical innervation pattern does not necessarily represent a gradual increase of fiber density or transmitter levels. 5-HT immunoreactivity in rat cortex is distributed in dense patches in sensory areas during the first postnatal week. After postnatal day 10 (P10) the patches of 5-HT immunoreactivity are no longer apparent, but the density of 5-HT axons gradually increases until an adult pattern is present at approximately 4 weeks of age (D'Amato *et al.*, 1987). Using electronmicroscopy, Dinopoulos and colleagues have observed transient accumulation of serotonin synapses during the most plastic period of development in rat visual and somatosensory cortex (Dinopoulos *et al.*, 1997). Rodent somatosensory cortex also displays a transient, patchy appearance when labeled with antibodies against the serotonin transporter (SERT) (Fujimiya *et al.*, 1986), and these transient immunoreactive patches are associated with clusters of thalamocortical afferents in their cortical targets, the barrel field (Blue *et al.*, 1991; Bruning and Liangos, 1997; Hansson *et al.*, 1998; Boylan *et al.*, 2000a). Serotonergic receptors (5-HT_{2A}) also are transiently over-expressed in layer IV of barrel-field cortex (Mansour-Robaey *et al.*, 1998). Our early studies of monoamine levels in cortex, during the first few postnatal weeks, confirmed a peak of 5-HT and its metabolites within the first 5 days after birth, which is followed by a more gradual incline over the next 2 weeks (Hohmann *et al.*, 1988). We have since observed that the early postnatal peak is present in male mice but not apparent in female mice (Connell *et al.*, 2004). Different studies have confirmed, to date, that thalamocortical afferents actively sequester serotonin during this critical period of synaptic maturation in sensory cortical areas (Bennett-Clarke *et al.*, 1991; Bruning and Liangos, 1997). These data strongly suggest a role for the serotonergic innervation in the formation and perhaps plasticity of thalamocortical sensory afferent connections. We will further discuss the implication of these findings below.

Data from non-rodents suggest that a transient peak in serotonergic neurotransmission, associated with the development of thalamocortical afferent connections, may be a universal feature of mammalian cortical development (Murrin *et al.*, 2007). In cat visual cortex, a patchy distribution of 5-HT receptors has been reported in association with thalamic input columns, and most receptors have their peak expression during the 'critical period' of development (Dyck and Cynader, 1993; Kojic *et al.*, 2000). Receptors (particularly 5-HT_{2A} receptors) in cat as well as rat visual and auditory cortex are increased several-fold during the first postnatal month compared to adult levels (Jonsson and Kasamatsu, 1983; Dyck and Cynader, 1993; Basura *et al.*, 2008). Similar observations have been made for 5-HT_1 and 5-HT_2 receptor binding autoradiography in prefrontal, somatosensory and visual cortex of rhesus monkeys, where levels at 2 months of age were twice that in the adult (Lidow and Rakic, 1992). In kittens, an early peak of adult 5-HT levels is seen between 3 and 5 weeks postnatally, followed by a decline in levels and then a gradual recovery (Jonsson and Kasamatsu, 1983). High 5-HT and metabolite levels were also recently reported for rodent auditory cortex during the second postnatal week (Basura *et al.*, 2008). Last but not least, studies by Diane Chugani's laboratory, using imaging approaches, show a transient elevation of cortical serotonergic synthesis during the pre-school years in humans (Chugani *et al.*, 1999). These recent studies were preceded by earlier findings that serotonin receptor binding in human brain tissue increases postnatally in infant hippocampus and frontal cortex to levels as high as in the adult (Marcusson *et al.*, 1984a, 1984b).

The barrel field as a model system for the study of serotonergic impact on morphogenesis

The barrel field of the rodent somatosensory cortex has long served as a model system to study cortical patterns of ascending sensory innervation in development and their plastic responses to altered afferent sensory information (Jeanmonod *et al.*, 1981; Killackey, 1982; Killackey *et al.*, 1995; Erzurumlu and Kind, 2001). Sensory information travels from the whisker pad via the trigeminal nucleus in the brainstem and the ventrobasal nucleus of the thalamus to form an anatomically distinct area in parietal somatosensory cortex (see Figure 1). Individual whiskers have 'barrel-shaped' cellular clusters in layer IV of the somatosensory cortex. The clustering of neurons in barrels first becomes visible around postnatal days 5–7 in mouse and rat, and only fully develops if their afferent sensory inputs are properly developed and maintained (Jeanmonod *et al.*, 1981; Killackey, 1982; Schlaggar and O'Leary, 1994; Killackey *et al.*, 1995; Erzurumlu and Kind, 2001). For instance, if input from a row of whiskers is removed early postnatally (see Figure 1), then adjacent, 'active' whiskers will take over the cortical 'barrel' space vacated by the inactive thalamocortical afferents in the deafferented row.

The current literature leaves little doubt that a normal 5-HT innervation and synaptic tone serve an important role in establishing the proper thalamocortical connectivity and cortical cytoarchitecture that supports the barrel

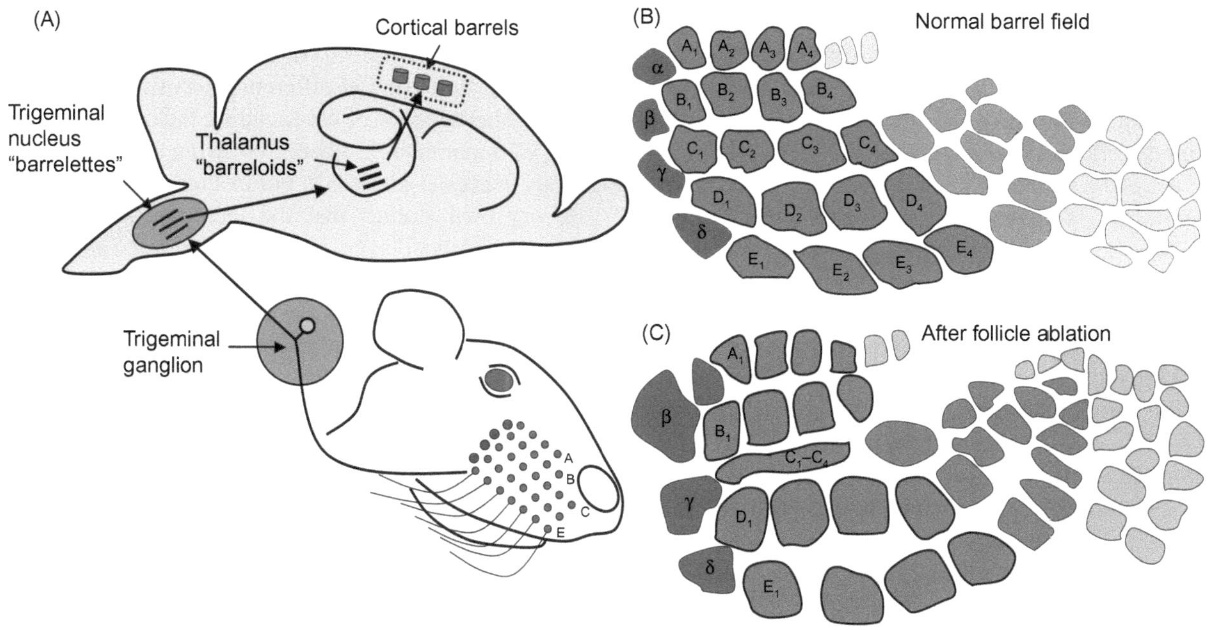

Figure 1 Schematic of the whisker-to-barrel pathway. Sensory information from the whisker pad on the mouse snout is conveyed to the cerebral cortex via the trigeminal nerve to the trigeminal nucleus, then to the ventrobasal thalamus and to the barrel field in layer IV of somatosensory cortex (left-hand panel). All along this pathway, the precise somatotopy is preserved such that 'whisker maps' are present in the trigeminal nucleus (barrelettes), the thalamus (barreloids) and cortex (barrels). The cortical representations were named barrels due to their 'barrel' shape when viewed in sections cut in the coronal plane. In flattened sections that are cut through layer IV, barrels are formed by a clustering of neurons around a zone with fewer cells or 'hollows', where the thalamocortical axons travel. The upper-right schematic shows the positions of straddler ($\alpha-\gamma$) and the large barrels in rows A–E (shown in lighter gray), and was drawn from cytochrome-oxidase stained barrel field. This barrel map is altered after neonatal whisker follicle ablation of whiskers C_1-C_4 (lower-right panel). As a result of this, the barrels in row C shrink, while those in adjacent rows B and D expand.

field in rat and mouse cortex. Moreover, too much or too little serotonin disrupts thalamocortical terminal fields and barrel formation, albeit with somewhat different results. Studies by Blue and colleagues, in rat, first demonstrated that systemic serotonergic *depletions* at birth, using injections of the substituted amphetamine parachloroamphetamine (PCA), delays the development of barrel formation in cortex (Blue *et al.*, 1991). Others subsequently showed that neonatal systemic 5,7-DHT injections reduce cortical serotonin by 75–80 percent in rat and reduce the size and activity of thalamocortical terminal fields without reducing the overall size of cortex or the barrel field (Bennett-Clarke *et al.*, 1994; Turlejski *et al.*, 1997; Rhoades *et al.*, 1998).

Increases in serotonin availability, likewise, alter barrel formation. Several groups have used monoamine oxidase A (MAOA) inhibition via chlorgyline and/or $MAOA^{-/-}$ mice to increase 5-HT availability. Both paradigms show increased barrel size as well as poorly differentiated barrels (Vitalis *et al.*, 1998, 2002; Boylan *et al.*, 2000b; Alvarez *et al.*, 2002). Another study reported the absence of barrel formation and thalamic afferent patterns, altogether, in $MAOA^{-/-}$ mice (Cases *et al.*, 1996). Instead of forming clusters that innervate selected 'barrels' in layer IV of somatosensory cortex, thalamocortical afferents maintain a uniform distribution in $MAO^{-/-}$ mice (Rebsam *et al.*, 2002). Similar effects are seen using the serotonin uptake inhibitor (SSRI) paroxetine (Xu *et al.*, 2004) or using genetically altered mice lacking the 5-HT uptake pump (Salinchon *et al.*, 2001; Esaki *et al.*, 2005). Importantly, effects of increased cortical 5-HT, via inhibition of reuptake or metabolism, can be reversed by lowering cortical 5-HT availability, pharmacologically or via double knockout mutations, thus indicating that alterations in barrel-field formation and plasticity are indeed specific to altered serotonergic tone (Young-Davies *et al.*, 2000; Salinchon *et al.*, 2001; Rebsam *et al.*, 2002, 2005). The 5-HT_{1B} receptor, in particular, has emerged as mediator of the developmental serotonergic effects. Crossing either 5-HTT or MAOA knockout mice with $5\text{-HT}_{1B}{}^{-/-}$ mice normalizes afferent thalamic branching patterns and barrel-field formation (Salinchon *et al.*, 2001; Rebsam *et al.*, 2002). Pharmacological stimulation of the 5-HT_{1B} receptor, conversely, elicits disruptions of barrel-field formation reminiscent of increased 5-HT availability (Young-Davies *et al.*, 2000).

The current concept of how 5-HT levels and 5-HT_{1B} receptor stimulation modulate thalamocortical target formation in cortex is based on serotonergic modulation of glutamate receptors. 5-HT_{1B} receptors inhibit glutamate activity or release in thalamocortical afferents (TCAs) *in vivo* and *in vitro* (Rhoades *et al.*, 1994; Laurent *et al.*, 2002). Thus, if 5-HT_{1B} receptors are present, glutamate receptors are inhibited by the *excessive* 5-HT caused by the SERT or MAO KOs. When 5-HT_{1B} receptors are inhibited or knocked out, along with the others, although 5-HT levels remain high, TCA activity is increased and barrel boundaries are maintained.

Modulation of glutamatergic postsynatic effects is also compatible with a number of reports of altered cortical neuronal differentiation and cytoarchitectonic maturation. Tryptophan restriction, which reduces cortical 5-HT synthesis, appears to increase dendritic branching in cortical pyramidal cells and increase dendritic spines as well (González-Burgos *et al.*, 1996). The latter findings are compatible with ours showing increases in cortical width after focal serotonergic depletions with 5,7-DHT injections into the medial forebrain bundle in Balb/CbyJ mice (Hohmann *et al.*, 2000).

However, other studies present conflicting findings. For example, Altamura and colleagues reported increases in cortical width in 5-HTT knockout mice, on a C57Bl6 background, with increased levels of 5-HT (Altamura *et al.*, 2007). Likewise, Vitalis and colleagues recently showed that late-embryonic, PCA-mediated 5-HT depletions lead to reductions in the dendritic arborization of cortical layer III and V pyramidal cells (Vitalis *et al.*, 2007). An earlier study reported delays in the differentiation of cortical layers following neonatal 5,7-DHT injections (Osterheld-Haas and Hornung, 1996). Such discrepancies in effect on dendritic growth may be the result of measuring at different intervals or ages after the 5-HT alteration, or of confounding factors such as sprouting of dopaminergic fibers, following serotonergic depletions (Benes *et al.*, 2000). For example, we have observed in our own studies that the increases in the width of selected cortical layers in young adult mice attenuate with increased age (Hohmann *et al.*, 2000, 2007).

Direct 5-HT effects on thalamocortical efferent outgrowth have also been observed *in vitro*. 5-HT stimulation enhances thalamic neuron process outgrowth (Lieske *et al.*, 1999; Lotto *et al.*, 1999), and this effect is sensitive to impulse inhibition via tetrodotoxin (TTX) and mimicked by 5-HT_{1B} agonists (Lotto *et al.*, 1999). Extrapolating from these observations, absence of 5-HT_{1B} receptor activation would decrease thalamocortical fiber growth into cortex. Thus, depending on the availability of serotonin and on 5-HT_{1B} receptor occupancy, thalamocortical afferent ingrowth into cortex is either accelerated or slowed. Such a mechanism would be compatible with a delay in barrel formation after neonatal 5-HT depletion (Blue and Molliver, 1989) and reduction of barrel size (Bennett-Clarke *et al.*, 1994; Rhoades *et al.*, 1998). It remains to be established how postsynaptic glutamatergic and presynaptic growth regulatory mechanisms, modulated by 5-HT, interact to effect the normal formation of the barrel field.

There also are studies of serotonin depletion that point towards a postsynaptic mediation of cortical changes that may be independent of afferent thalamocortical activity. Studies by Rhodes and colleagues suggest that the decrease in barrel size after systemic 5,7-DHT lesions at birth is not dependent on afferent activity, since treatment with TTX does not alter it (Rhoades *et al.*, 1998). Boylan *et al.* (2001) did not detect effects of NMDA receptor blockage on barrel-field plasticity, but that stands in stark contrast to studies by Schlaggar *et al.* (1993). *In vitro* studies have shown that serotonin can directly affect the survival and differentiation of cortical neurons (Dooley *et al.*, 1997; Lavdas *et al.*, 1997). Such effects are more likely mediated by 5-HT_2- and 5-HT_3-type receptors. A confound of comparing *in vitro* with *in vivo* studies of the developing cortex is, of course, knowing whether added 5-HT concentrations *in vitro* represent reduced, normal or elevated levels *in vivo*. Thus, it is noteworthy that Lavdas and colleagues saw increased cortical neuron survival in organotypic cultures with co-cultured nucleus raphe neurons (Lavdas *et al.*, 1997). Decreased apoptosis of cortical neurons has also been described, furthermore, in 5-HTT KAO mice where 5-HT is increased (Persico *et al.*, 2003). Increased apoptosis occurs in genetic constructs missing the vesicular 5-HT transporter, which renders 5-HT nearly absent and is blocked by MAOA inhibition (Stankovski *et al.*, 2007).

Visual cortex

For the past 50 years, kitten visual cortex also has been used to study neurotransmitter modulation of morphological and physiological correlates of use-dependent cortical plasticity. There have been many excellent reviews over the years describing this paradigm, which allows for the morphological and electrophysiological assessment of peripheral visual influences on the patterning of visual inputs (Blakemore and Mitchell, 1973; Wiesel, 1999; Daw and Beaver, 2001; Taha and Stryker, 2005). Basically, thalamocortical afferents from the lateral geniculate nucleus segregate into columns with differential responsiveness to visual input from the left and right eye over the course of the first 2–3 months after birth, also named the 'critical period'. Monocular deprivation of vision from either eye during this critical period allows cortical responsiveness to drift towards the eye that is receiving visual input, effectively disconnecting the occluded eye from cortex. The extent of this 'ocular dominance shift' depends on the duration and timing of eye occlusion. This process is clearly dependent on excitatory glutamatergic neurotransmission, particularly NMDA receptor activation (Gu *et al.*, 1989; Bear *et al.*, 1990; Rauschecker, 1991a). The development of the 'ocular dominance columns' in cat visual cortex has been regarded as a classical model of cortical learning and memory formation in that it is susceptible to both long-term potentiation (LTP) and long-term depression (LTD) (Rauschecker, 1991a; Kirkwood *et al.*, 1995; Johansen-Berg and Walsh, 2001).

The effect of monoaminergic and cholinergic modulation of developmental plasticity became an intense focus of research during the 1980s (Singer and Rauschecker, 1982; Sillito *et al.*, 1985; Gordon *et al.*, 1988). Based on the similarities in timing and adult modulatory properties on cortical synapses, Gu and Singer conducted a comprehensive assessment of 5-HT effects on ocular dominance plasticity (Gu and Singer, 1995). They showed that infusion of 5,7-DHT into visual cortex at 1 month postnatally, combined with 1 week of eye occlusion in the kittens, substantially inhibited ocular dominance shift to the open (not occluded) eye. This was specific to 5-HT depletion, since comparable effects were elicited with comprehensive 5-HT receptor blockage. Although neither 5-HT_1 nor 5-HT_2 receptor blockage alone inhibited plasticity, combined cortical application of the antagonists methysurgide and ketanserin did effectively block the drift of responsive cortical cells towards the open eye. Other groups have since confirmed that serotonergic antagonists can interfere with ocular dominance plasticity (Wang *et al.*, 1997). The 5-HT_{2C} receptor has become most closely associated with the regulation of OD plasticity in cat, as well as with the mediation of, curiously, both LTP and LTD (Dyck and Cynader, 1993; Wang *et al.*, 1997; Kojic *et al.*, 2000, 2001). In his 2000 Commentary in *PNAS*, Kirkwood elucidates this apparent conundrum. Not only are serotonergic receptors distributed in a highly laminar and developmentally changing manner; 5-HT_{2C} receptors actually also have a patterned, columnar appearance even within the developing layer IV in kitten occipital cortex (Kirkwood, 2000). Thus, serotonin has the ability to transiently modulate plasticity during the critical period in kitten cortex in a manner that interacts with and shapes the microcolumnar arrangement of the cortex (Kirkwood, 2000). Modulation of plasticity via 5-HT_{2C} receptors has also been reported in the rat hippocampus (Tecott *et al.*, 1998), and 5-HT attenuates LTP in slice preparations of immature rat visual cortex via receptors that aren't clearly classified (Edagawa *et al.*, 2001).

5-HT_1 receptors also appear to play a role in the shaping of cat visual cortex. Mower (1991) showed a selective up-regulation of 5-HT_1 receptors in the visual cortex of dark-reared cats that spared other cortical areas (Mower, 1991). In rat visual cortex, neonatal removal of visual input via enucleation (eye removal) also increases the number of 5-HT_{1B} receptors, and neonatal 5,7-DHT injections lead to a transient increase in 5-HT_{1B} receptors (Sari *et al.*, 1999).

It has been shown that the SSRI fluoxetine, which is used to treat depression in humans, can augment visual plasticity in rats, even in the adult (Maya Vetencourt *et al.*, 2008). Besides blocking serotonin uptake, fluoxetine is also known to interact directly with 5-HT_{2C} receptors (Ni and Miledi, 1997).

So what is it that we have learned from the study of serotonergic regulation of cortical plasticity in the rodent barrel field and cat visual system? Cortical plasticity is clearly a highly complex process that cannot easily be comprehended by observing single receptor related effects in isolation. 5-HT_1-type and 5-HT_2-type receptor families both influence this process in the barrel field and visual cortex, and 5-HT_{1B} and 5-HT_{2C} receptors have emerged as the two most likely candidates to mediate serotonergic responses. These receptors apparently have different locations in the developing nervous system. The 5-HT_{1B} receptor is located on thalamocortical afferents in the barrel field, at least in mouse and rat (Bennett-Clarke *et al.*, 1993; Laurent *et al.*, 2002). This is consistent with unrelated observations that this receptor apparently localizes predominantly to axonal terminals (Sari, 2004). On the other hand, in the somatosensory and visual cortex of rodents and felines, the 5-HT_{2C} receptor is clearly localized postsynaptically, on cortical neurons, and influences LTP and LTD in these neurons (Kirkwood, 2000; Edagawa *et al.*, 2001). Both model systems also show that cortical serotonergic modulation is mediated via the NMDA receptor mechanisms. Thus, the emerging picture is one of a dual serotonergic influence on both afferent thalamocortical fiber ingrowth into cortex (i.e., fasciculation, target selection) and the ability of the arriving fibers to activate their cortical targets (i.e., LTP/LTD induction, synapse formation and elimination). Removing serotonergic afferents during critical periods of cortical plasticity in visual cortex clearly arrests the ability of cortex to reorganize itself in response to altered sensory stimulation, hence inappropriate map formation. While changes in serotonergic tone affect the timing and ultimate shape of the barrel field, it is less clear in this paradigm how they affect cortical reorganization in response to altered sensory input. However, these barrel-field studies were performed after whole body changes in 5-HT or receptors, opening up the possibility that other modulatory systems were affected as well, perhaps compensating for plasticity disruptions. In contrast, 5-HT levels and receptor function were manipulated with direct infusions into the visual cortex, limiting the possibilities of compensatory effects within the timeframe of study. Thus, the proof of concept, in rodent, might be a localized depletion of the relevant cortical areas, such as our focused lesions of the afferent 5-HT innervation to cortex, described below.

Serotonin in ASD

Brief characterization of ASD

ASD encompasses a spectrum of developmental brain disorders characterized by deficits in social and communication behavior, perseveration or stereotyped behavior, altered sensory–motor performance and a variety of aspects of altered cognitive performance and emotional responsiveness (DeMyer, 1975; Rapin, 1991; Gillberg, 1999; Lord *et al.*, 2000; Dawson *et al.*, 2002a; Baron-Cohen, 2004; DiCicco-Bloom *et al.*, 2006; Landa *et al.*, 2007). ASD is approximately four times as prevalent in males compared to females (Murcia *et al.*, 2005; Landa, 2008). In the absence of a clear neurobiological understanding of the disorder and without any biomarkers, an ASD diagnosis is still entirely based on a pattern of behavioral deficits and alterations compared to typically developing children. ASD is currently regarded as predominantly of genetic origin, based on a 70 percent or larger concordance among monozygotic twins (Abramson *et al.*, 1992; Buitelaar and Willemsen-Swinkels, 2000; Baron-Cohen, 2004), yet sporadic occurrences of autism still outnumber autism in multiplex families where a pedigree can be established (Trottier *et al.*, 1999). Part of the problem is that it is nearly impossible to investigate environmental influences on ASD in the human population in the absence of a clearly defined etiology and neurobiology of the disorder (Serajee *et al.*, 2004; Arndt *et al.*, 2005; Santangelo and Tsatsanis, 2005; Bello, 2007; Newschaffer *et al.*, 2007; Pardo and Eberhart, 2007). Meanwhile, hundreds of genes and restriction enzyme polymorphisms have been linked to ASD in the human population (Wassink and Piven, 2000; Cook, 2001; Muhle *et al.*, 2004; Polleux and Lauder, 2004), and the list is growing steadily. Many of these candidate genes have become the focus of investigation in animal models, *in vitro* and *in vivo*, and a number of them have been shown to influence morphological or behavioral endophenotypes shared with ASD (see, for example, Wassink *et al.*, 2001, 2004; Andres, 2002; Bernardet and Crusio, 2006; Nishijima *et al.*, 2006; Paylor *et al.*, 2006; Kuemerle *et al.*, 2007; Jamain *et al.*, 2008). It is clear, however, that no one single gene can be linked to ASD as causative; this is a polygenetic disorder with, at best, a final common neurobiological phenotype (Gillberg, 1999; Arndt *et al.*, 2005; Santangelo and Tsatsanis, 2005). It is also increasingly becoming accepted that environmental influences likely serve as triggers for the disorder in individuals with a vulnerable genotype (Trottier *et al.*, 1999; Herbert *et al.*, 2006; Bauer *et al.*, 2007; Pardo and Eberhart, 2007). We will begin, below, with a description of the most prevalent neurobiological deficits in ASD, particularly as they pertain to cerebral cortex.

Developmental neurobiology of ASD

Although some specialists are now able to diagnose ASD with some certainly as early as 12 to 18 months postnatally, the disorder is still typically diagnosed between 2 and 3 years of age, when social and communication deficits are becoming overwhelmingly obvious to caregivers (Osterling *et al.*, 2002; Landa, 2007, 2008). At this time in development, there is already a significant increase of cortical volume in children with ASD, which in the white matter volume is about a 10 percent change (Courchesne *et al.*, 2001; Courchesne, 2002, 2004; Sparks *et al.*, 2002; Carper and Courchesne, 2005; Hazlett *et al.*, 2005). Volume increases have been confirmed by many additional studies in older children and adults, yet comparisons of such studies suggest that brain volume decreases with age, leaving the margin of difference closer to 1 percent (Aylward *et al.*, 2002; Herbert *et al.*, 2004; Redcay and Courchesne, 2005; Mostofsky *et al.*, 2007; Wassink *et al.*, 2007). Head circumference is also increased in the toddler years and attenuates with age (Aylward *et al.*, 2002). The cortical areas that show the most prominent increase in volume of both white and gray matter are the frontal and temporal lobes, while the least change is seen in occipital cortex. Frontal and temporal cortical areas display altered event-related potentials and fMRI activation to social stimuli as early as ages 2–4 years, demonstrating the functional significance of the anatomical findings (Boddaert and Zilbovicius, 2002; Dawson *et al.*, 2002b; Kuhl *et al.*, 2005; Courchesne *et al.*, 2007; Knaus *et al.*, 2008; Muller, 2008). Volume changes in parietal areas have not received as much attention, but significant evidence of increased gray and white matter volume is present there as well (Courchesne *et al.*, 1993; Saitoh and Courchesne, 1998; Mostofsky *et al.*, 2007).

Cortical gray matter cytoarchitecture is also disturbed in autism. *Post-mortem* microscopic evaluation of brain tissue from adults with autism demonstrates abnormalities in cerebral cortical development that include increased cortical thickness and neuronal density, neuronal disorganization with some suggestions of abnormal migration, poorly differentiated gray–white matter boundaries, and increased number of neurons in the white matter (Bailey *et al.*, 1998; Kemper and Bauman, 1998). While the former changes are not easily reconciled with any particular functional network in cortex, Casanova and colleagues have reported changes in minicolumns in frontal, temporal and, most recently, auditory cortex that point towards specific information-processing deficits. Minicolumns in ASD are smaller, more numerous and less compact in adolescent subjects compared to those of control subjects (Casanova *et al.*, 2002; Buxhoeveden *et al.*, 2006). Alterations in the size and cellular distribution within cortical minicolumns in autism likely reflect disturbances in the processing of thalamic inputs to cortex that are rooted in the ontogeny of these connections (Buxhoeveden and Casanova, 2002). In normal cortical development, cortical minicolumns are generally regarded as the smallest cortical processing unit that self-assembles according to Hebbian principles (Shenoy *et al.*, 1993; Favorov and Kelly, 1994). In the sensory and motor cortex of the mature organism, minicolumns are responsible for generating basic receptive field properties, and combine to constitute the columnar processing arrangements originally described by Hubel and Wiesel, Mountcastle and colleagues (see Favorov and Kelly, 1994; Amirikian and Georgopoulos, 2003). The functional significance of minicolumns in fronto-cortical areas has been demonstrated in a modeling study (Koene and Hasselmo, 2005). Recent psychometric studies in ASD corroborate observations of altered minicolumn anatomy by demonstrating impairments in sensory perceptual changes in adults with ASD (Tommerdahl *et al.*, 2008). Furthermore, altered minicolumns are consistent with evidence for abnormal sensory cortical maps in autism, including modifications in cortical sulcal maps (Levitt *et al.*, 2003) and visual–motor maps (Muller *et al.*, 2003), deficits in the development of cortical sound processing (Gage *et al.*, 2003), and problems with face recognition (Pierce *et al.*, 2001; Dawson *et al.*, 2002b). Studies by Janusonis and colleagues have suggested that 5-HT innervation of reelin containing Cajal Retzius (CR) cells is involved in appropriate maturation of the columnar arrangement in cortex (Janusonis *et al.*, 2004).

As recently pointed out by Courchesne, a fundamental problem in understanding the developmental pathology of ASD is that the vast majority of current brain studies have been conducted in older children or adults (Courchesne *et al.*, 2007). Thus, with the exception of the abovementioned measurements of cortical volume and head circumference in toddlers, at the age of first diagnosis most current neuropathology is looking at a brain well after the initial developmental disruptions – a cortex that has already made adjustments to altered structure and function. In comparison with the experimental model systems discussed in the first section of this review, this also would be a cortex past the ‘critical period’ for developmental plasticity. To better illustrate this, a brief review of human cortical development is merited.

The processes of neocortical neuronal differentiation, dendritic elaboration and axon outgrowth substantially occur between late gestation and school age, but extend into adolescence in frontal cortical areas (for reviews, see Huttenlocher, 1990; Mrzljak *et al.*, 1992; Kostovic and Judas, 2002; Courchesne *et al.*, 2004; Courchesne and Pierce, 2005; Kostovic and Jovanov-Milosevic, 2006; Bystron *et al.*, 2008). The cortical plate forms in humans around gestational weeks 6–7 (Marin-Padilla, 1970, 1978; Mrzljak *et al.*, 1992), and neuronal migration into the cortex is concluded by gestational weeks 24–26 when synaptogenesis commences (Rakic, 1988; Bystron *et al.*, 2008). The bulk of neuronal differentiation and synapse formation occurs postnatally, and shows regional variations along a posterior to anterior gradient (Huttenlocher, 1979, 1984; Huttenlocher *et al.*, 1982; Huttenlocher and Dabholkar, 1997). For example, according to electron microscopic (EM) quantifications of synaptic profiles, peak synapse formation in visual cortex occurs roughly around 1 year of age, while in frontal cortex, synapse formation continues to increase for another couple of years (Huttenlocher and Dabholkar, 1997). As recently illustrated by Courchesne and colleagues, and according to Golgi data from earlier studies, the dendritic expansion of cortical pyramidal neurons closely mirrors the increase in synapse formation (Courchesne *et al.*, 2007). In subsequent postnatal years, synapses are pruned in cortex to fine-tune functional connectivity, a process that in frontal areas lasts into the second decade of life (Huttenlocher, 1984; Huttenlocher and Dabholkar, 1997; Courchesne *et al.*, 2007). Studies of molecular synaptic markers have essentially confirmed these EM studies but contributed some additional information about synapse type. Synapse numbers increase into late childhood (10 years of age), as indicated by synaptophysin quantification, while the peak number of presumed glutamatergic synapses, as labeled by PSD-95, is reached earlier but then decreases (Glantz *et al.*, 2007). Western blot analysis of complexin 1 (CX1) and complexin 2 (CX2), proteins associated predominantly with inhibitory (GABAergic) of excitatory (glutamatergic) synapses respectively, substantiates the impression that excitatory synapses in cortical development peak at around 10 years of age, while inhibitory synapses continue to increase into adulthood (Salimi *et al.*, 2008). Moreover, glucose utilization, which also is regarded as an indicator of cortical synaptic activity, increases between birth and 3 years of age to nearly three times adult levels, followed by a decrease after age 8 years (Chugani, 1998; Muzik *et al.*, 1999). Chugani has likened this glucose utilization plateau between the ages of 3 and 8 years to the critical period in animal models (Chugani, 1998).

The temporal and regional differences in cortical synaptogenesis can seriously confound the analysis of volume measurements in ASD, compared to typical development, and may have led, in past studies, to an underestimation of quantitative gray-matter changes. Several groups have recently used MRI technology to assess regional brain growth in typically developing children pre- and postnatally (Sowell *et al.*, 2001; Kostovic and Judas, 2002). Kostovic demonstrated a 50 percent increase in cortical gray matter, between gestational week 29 and term, which he attributed to afferent ingrowths of subcortical and cortico-cortical axons, concomitantly with the expansion of cortical neuropil, as dendritic branches and synapses expanded (Kostovic and Judas, 2002). Sowell and colleagues analyzed cortical thickness in the same group of typically developing children at 5 and again at 11 years of age (Sowell *et al.*, 2001). This study revealed a complex pattern of intermingled areas of gray-matter expansions and contractions in development, presumably as a consequence of fine-tuning and pruning of cortical connections along regionally specific timelines.

An emerging neurobiological concept is that ASD represents a failure in the development of networks (Rubenstein and Merzenich, 2003; Belmonte *et al.*, 2004; Dong and Greenough, 2004; Courchesne *et al.*, 2007; Minshew and Williams, 2007). Neuropathologies such as altered minicolumns and cortical map formation, in conjunction with transient increases in cortical volume, all point towards developmental disruptions of cortical circuitry formation and pruning. As illustrated in the preceding paragraph, connections between cortical neurons and their thalamic afferents begin to form in the late gestational period, after neuronal generation has largely concluded in cortex, and they continue after birth. Connections between cortical regions commence to form and mature for years postnatally, with networks in frontal cortical areas being the latest to reach maturity, in adolescence. Thus, based on the currently available neuropathology data, cortical development appears to proceed normally until the late gestational period when afferent innervation and synapse formation begins to be disrupted, as suggested by altered minicolumn formation. Such cortical neuropathologies in ASD are compatible with the disruptions of serotonergic modulatory influences in cortical network formation, as outlined in the first section of this chapter.

Evidence for serotonergic imbalances in ASD

Substantial evidence supports altered serotonergic homeostasis in ASD. A longstanding literature has shown *increased* platelet serotonin levels in both children and adults with ASD (Anderson *et al.*, 1990; Cook, 1990; Burgess *et al.*, 2006; Lam *et al.*, 2006). In contrast, cortical serotonergic innervation appears to be transiently *decreased* during early postnatal development, and perhaps also prenatally. Chugani and colleagues have shown an altered developmental trajectory of serotonergic innervation to the cerebral cortex in ASD (Chugani *et al.*, 1999). Using positron emission tomography (PET) imaging of a tryptophan analog, they found that young children with autism do not display the developmental peak in brain 5-HT synthesis capacity seen in typically developing children; this attenuation correlated with altered language development (Chugani *et al.*, 1999; Chandana *et al.*, 2005). Boys with autism also showed specific decreases in 5-HT synthesis in the dentato-thalamocortical pathway, with simultaneous increases in the contralateral dentate deep cerebellar nucleus (Chugani *et al.*, 1997). Interestingly, in adults with ASD, serotonin synthesis has normalized or may even show increases (Chugani *et al.*, 1999; Chandana *et al.*, 2005). The significance of these findings has been highlighted recently by the demonstration of decreased SERT binding, particularly in frontal cortex, in young children with ASD, using single photon emission computed tomography (SPECT) (Makkonen *et al.*, 2008). Lowered SERT is suggestive of decreased serotonergic fiber densities in the cortex of these young children. Furthermore, studies in adults with Asperger's syndrome show a significant decrease in cortical 5-HT_{2A} receptors (Murphy *et al.*, 2006). Thus, while 5-HT levels may normalize or even increase in adults with ASD, their postsynaptic receptive mechanisms still may be down-regulated.

There is substantial genetic evidence for altered serotonin homeostasis in ASD and family members. This includes promoter region polymorphisms (5-HTTLPR) of the 5-HTT gene SLC6A4 (Cook *et al.*, 1997; Yirmiya *et al.*, 2001; Conroy *et al.*, 2004; Devlin *et al.*, 2005; Sutcliffe *et al.*, 2005; Guhathakurta *et al.*, 2006; Koishi *et al.*, 2006) that can correlate significantly with cortical gray-matter volume in young children with ASD (Wassink *et al.*, 2007), or with a rigid-compulsive endophenotype (McCauley *et al.*, 2004). Other genotype–phenotype interactions for the 5-HTTLPR also are observed in ASD (Brune *et al.*, 2006). Promotor region polymorphisms also may be associated with peripheral hyperserotonemia in a subset of patients (Anderson *et al.*, 2002; Coutinho *et al.*, 2004). Linkage disequilibrium involving alleles for the 5-HT_{2A} receptor has been described as well (Veenstra-VanderWeele *et al.*, 2002; Cho *et al.*, 2007). Moreover, Hu and colleagues demonstrated increased SERT expression in peripheral lymphocytes is associated with ASD phenotype in monozygotic twins who are discordant for autism (Hu *et al.*, 2006).

Observation of elevated peripheral serotonin in individuals with ASD and early central-developmental decreases may not be as contradictory as they seem. As described in more detail below, exposure of rats to a serotonergic agonist from mid-gestation to adolescence, leading to elevations in pre- and postnatal peripheral 5-HT levels, actually resulted in a loss of central serotonergic innervation in the pups, presumably due to autoregulation via 5-HT_{1A} receptors (Whitaker-Azmitia, 2005). Additionally, 5-HT_{1A} receptor knockout mice present with platelet hyperserotonemia in development and adulthood, but show transient, early, postnatal serotonin decreases in the brain (Janusonis *et al.*, 2006). Moreover, studies of gestational tryptophan hydroxylase inhibition and, more recently, studies involving tryptophan-1 deficient (tph-$1^{-/-}$) mice have demonstrated that peripheral serotonin levels and normal brain morphogenesis in offspring are a direct consequence of peripheral and maternal serotonin availability during gestation (Lauder *et al.*, 1981; Cote *et al.*, 2007). The mechanisms that regulate the relationship between CNS and peripheral serotonergic synthesis and control are not understood, and clearly much work is needed here.

Likewise, information regarding the development of the serotonergic innervation in typically developing human cortex is limited at present. Using the serotonin transporter (SERT) and vesicular transporter (VMAT2) as markers for 5-HT, fibers from the raphe are shown to enter the immature cortex around gestational week 13 (Verney *et al.*, 2002). A transient gestational presence of SERT immunoreactive fibers in the internal capsule suggests transient SERT expression on thalamocortical afferents in humans, as observed for other mammals (Verney *et al.*, 2002). Similarly, a peak in 5-HT_{1A} receptor expression is observed in the human cortex during gestational weeks 16–22, when thalamocortical fibers first approach and then enter the cortical plate (Bar-Peled *et al.*, 1991). Thus, current data strongly suggest that serotonergic afferents in humans exert similar morphogenetic functions in cortex as seen for several mammalian animal models.

Experimental approaches to testing a serotonergic hypothesis of ASD

Only a few laboratories have devised experimental paradigms that directly address the 'serotonergic hypothesis' that is, the role of serotonin in the etiology of ASD. We have developed a mouse model based on the hypothesis that selective neonatal serotonergic depletions of cortex will produce morphological and behavioral pathologies consistent with ASD. In this model, we have deliberately limited 5-HT manipulations to cortical areas of the forebrain in order to isolate resulting behavioral deficits. Our goal has been to distinguish alterations that are due directly to serotonergic depletion, from changes that might be secondary effects of 5-HT depletion on other afferent neuromodulatory systems, such as (for example) catecholaminergic or cholinergic afferents, or the HPA axes (Berger-Sweeney *et al.*, 1998; Hohmann *et al.*, 2000, 2007; Boylan *et al.*, 2007). As shown in Figure 2A, we neonatally inject the neurotoxin 5,7-DHT into the medial forebrain bundle to provide a selective, albeit transient, serotonergic denervation of cortex and hippocampus (Figures 2C, 2D). Our earliest studies (Hohmann *et al.*, 2000) demonstrated significant, quantifiable alterations in the cortical cytoarchitecture of adult mice following the neoanatal serotonergic depletion. These changes were dimorphic, and more substantial in males. Subsequent work (Boylan *et al.*, 2007; Hohmann *et al.*, 2007; Blue *et al.*, 2008) revealed a significant increase in early cortical volume that is no longer apparent by adulthood (see Figure 2B). Altered cortical morphogenesis is also accompanied by changes in cortical excitatory amino acid neurotransmission consistent with altered development of cortical networks (Boylan *et al.*, 2007). While we are still in the process of characterizing the full range of morphological and molecular changes that account for altered cortical development, we have observed increased immunoreactivity for glial fibrillary protein, possibly an indicator of a neuro-inflammatory response, in the cortex of 5-HT-depleted mice at the time of the transient volume increase (Figures 2E, 2F; unpublished observations). Morphological alterations in our mouse model are accompanied by substantial changes in social and cognitive behaviors, and these behavioral abnormalities are also more prevalent in males than females. As adults, lesioned mice show learning deficits in a socially motivated food preference task, along with altered social interactions with conspecifics and increased stereotypical behaviors (Boylan *et al.*, 2007). Preliminary data suggest that social behaviors are altered as early as the first week postnatally, when cortical volume increases are measured (unpublished observations). The neonatally 5-HT-depleted mice also show, as adults, substantial neophobia within a spatial exploratory context that occurs in the absence of generalized anxiety (Hohmann *et al.*, 2007). Additional tests also indicate sensory processing abnormalities and problems with impulse inhibition in the lesioned mice (Berger-Sweeney *et al.*, 1998; Boylan *et al.*, 2007; Hohmann *et al.*, 2007).

Whittaker-Azmitia and colleagues devised a mouse model aimed at testing the effects of hyperserotonemia on brain and behavioral development, by treating pregnant rats with the serotonin agonist 5-methoxytryptamine (5-MT) between gestational day 12 and various postnatal time periods (Kahne *et al.*, 2002). As already stated above, while such treatment did increase peripheral 5-HT levels in the offspring of such treated dams, brain 5-HT was actually decreased throughout the critical periods of cortical differentiation and synaptogenesis. These rats displayed behavioral alterations compatible with the ones seen in our neonatally 5-HT depleted mice, such as altered sensory responsiveness and stereotopies (Whitaker-Azmitia, 2005). In addition, the Whitaker-Azmitia group has described substantially altered mother–pup interactions, post-injection seizures and metabolic disturbances, as indicated by magnetic resonance spectroscopy (MRS) studies in the animals as adults (Kahne *et al.*, 2002; Whitaker-Azmitia, 2005; McNamara *et al.*, 2008). Unlike our neonatal serotonin depletion model, the perinatal 5-MT rats show altered serotonergic homeostasis throughout the brain, and neurochemical changes were observed in the hypothalamus and amygdala, including a decrease in oxytocin in the paraventricular nucleus (Whitaker-Azmitia, 2005; McNamara *et al.*, 2008). Thus, the more encompassing behavioral and neurobiological changes in the perinatally treated rats are compatible with a more severe serotonergic disruption, suggesting that the developing cortex is not the sole target of serotonergic imbalance in ASD.

Genetic models for serotonergic depletion have, to date, not been studied extensively within the context of ASD. Serotonin transporter gene (SERT) knockout mouse models have shown systemically altered 5-HT homeostasis, and have been linked to altered gut function, inflammatory responses, stress responsiveness and anxiety responses (Murphy and Lesch, 2008). Moy and colleagues recently found decreased sociability in mice with such targeted disruptions (Moy *et al.*, 2009), but little can be deduced about brain pathologies in these mice that presumably have substantial increases of 5-HT availability in development. The Pet-1 ETS transcription factor appears to be responsible for the specification of brainstem 5-HT neurons (Hendricks *et al.*, 1999; Hendricks *et al.*, 2003; Scott *et al.*, 2005). Mice with this mutation have an approximately 80 percent reduction of CNS serotonergic innervation; their neurological effects appear relatively mild, although postsynaptic receptors or SERT content were not assessed, making it difficult to deduct the level of functional neurotransmission (Hendricks *et al.*, 2003). Cortical cytoarchitecture appears qualitatively normal, and, behaviorally, increased anxiety and aggression were the most notable observations (Hendricks *et al.*, 2003). Clearly, these models merit further scrutiny.

In addition, a variety of rodent models, designed to test for the effects of specific environmental or genetic manipulations on the development of ASD endophenotypes, coincidentally include altered serotonergic homestasis. One of the most longstanding ASD models, gestational exposure to valproic acid (VPA), shows altered development of afferent serotonergic innervation to the forebrain, along with behavioral manifestations such as increased stereotopies, altered sensory responsiveness and altered eye-blink reflexes (Miyazaki *et al.*, 2005; Schneider and Przewlocki, 2005; Stanton *et al.*, 2007; Murawski *et al.*, 2009). The engrailed 2 (ENG-2) mouse mutant, studied as a model for homebox gene involvement in ASD, also displays altered serotonergic innervation to the forebrain

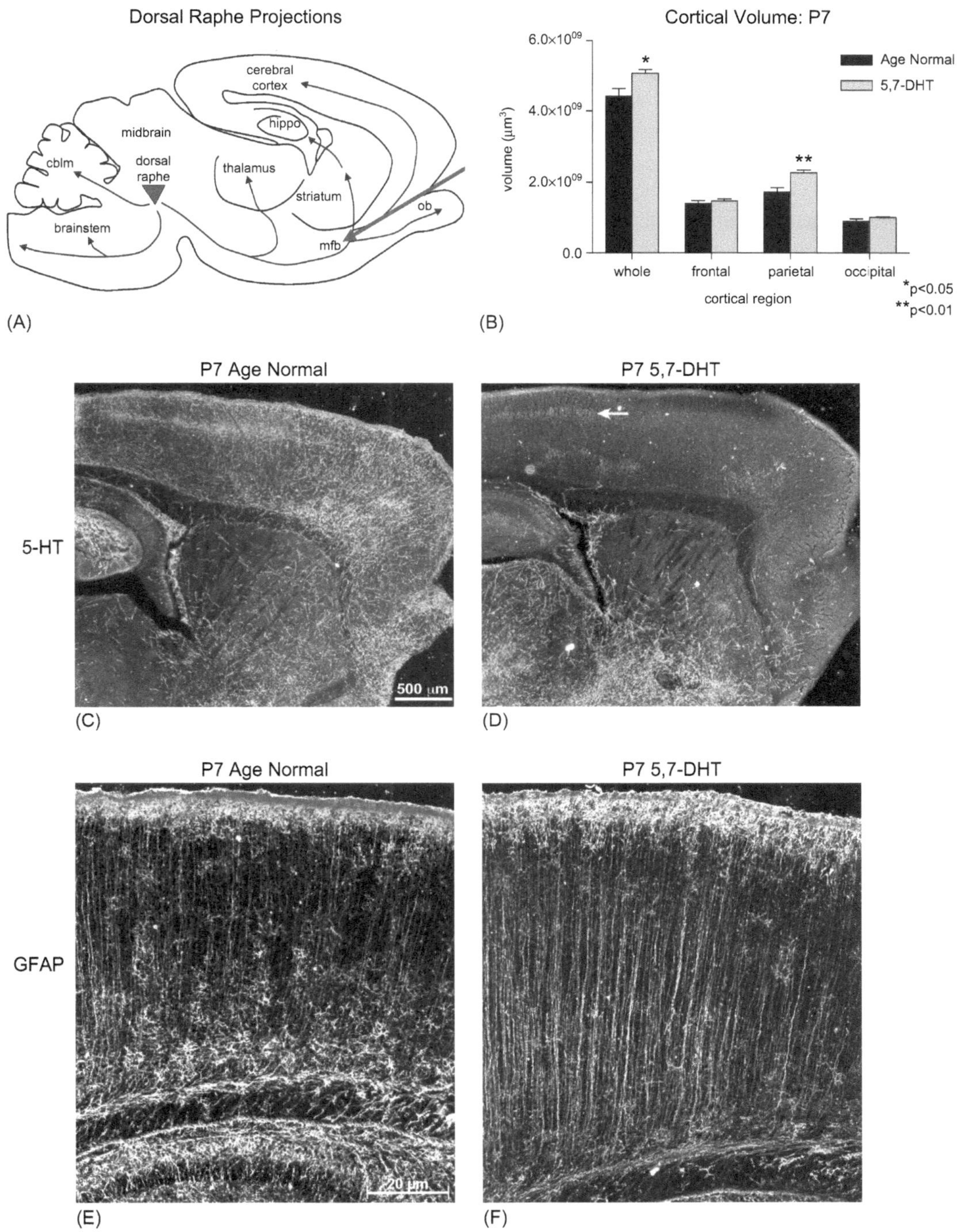

Figure 2 5,7-DHT mouse model. (A) Schematic drawing of a parasagittal section through a P7 mouse brain showing the location of the serotonergic neurons in the dorsal raphe and the projections from the raphe to multiple brain regions. Serotonergic axons innervate the hippocampus (hippo) and cerebral cortex via the medial forebrain bundle (mfb). The red arrow indicates the path of the 5,7-DHT injections. cblm, cerebellum; ob, olfactory bulb. (B) The volume of the whole cerebral cortex and parietal regions was greater in 5,7-DHT lesioned mice than in Age Normal controls. (C, D) Dark field photomicrographs show the distribution of serotonergic axons (5-HT) in parasagittal sections of (C) P7 Age Normal and (D) 5,7-DHT lesioned mouse brains. In lesioned mice, there are many fewer afferent serotonergic axons in the cortex and the hippocampus. However, some serotonin remains in thalamocortical terminals due to their transient expression of the serotonin transporter (white arrow). (E, F) The pattern of immunostaining for glial fibrillary acetic protein (GFAP) is shown in the dark field photomicrographs of the cortex in (E) Age Normal and (F) 5,7-DHT lesioned mouse brains. At P7, many of the radial glial fibers remain present, and their density appears higher in the lesioned mice. To see the full color version of this figure please refer to the color plate at the back of the book. Copies produced via our print on demand service do not contain color plates. If your copy does not have the color plate, please go this website to view the figure in color www.elsevierdirect.com/companions/9780123746344

and cerebellum, suggesting a role of this brainstem segmentation gene in the formation of the raphe nucleus (Cheh *et al.*, 2006; Kuemerle *et al.*, 2007). Mouse models for Fragile X and Smith-Lemli-Opitz, both regarded as disorders on the autism spectrum, show altered 5-HT innervation (Gruss and Braun, 2001; Waage-Baudet *et al.*, 2003; Moy and Nadler, 2008). Interestingly, serotonergic innervation is up- rather than down-regulated in the brain of several of the aforementioned mouse models, confirming the assessment in basic research studies (see first section of this chapter) that too much as well as too little serotonin can be detrimental to forebrain development and function.

Factors that can interact with serotonin homeostasis in ASD

Neuroinflammatory/neuroimmune interactions

As Herbert and others have pointed out, ASD is not solely a disorder of altered brain and behavior (Lipkin and Hornig, 2003; Herbert *et al.*, 2006; Herbert and Anderson, 2008; Pessah *et al.*, 2008). Thus, a 'whole body' approach may offer further conceptualizations of core deficits and intervention strategies for ASD. Reports of altered immune and digestive function in ASD abound, suggesting that up to 60 percent of patients with ASD have some type of systemic immune dysfunction as part of their cellular or humoral immune response (Posey and McDougle, 2001; Horvath and Perman, 2002; Korvatska *et al.*, 2002; Krause *et al.*, 2002; Licinio *et al.*, 2002; Torres, 2003; Cohly and Panja, 2005; Ashwood *et al.*, 2006; Wills *et al.*, 2007). The most prevalent observations concern altered innate immune responses and an increase in pro-inflammatory cytokines, particularly interleukin-1 (IL-1) and IL-6 and tumor necrosis factor (Gupta *et al.*, 1998; Croonenberghs *et al.*, 2002; Chez *et al.*, 2007; Grigorenko *et al.*, 2008; Jyonouchi *et al.*, 2008). However, there are also reports of an altered inducible immune response and other cytokine changes (Ashwood *et al.*, 2006; Molloy *et al.*, 2006). Since the cited reports concern a mix of measurements in serum, in cerebrospinal fluid or *in vitro* blood cell explants, some contradictory observations might be the product of assessing different compartments as well as differently affected individuals. Recent studies also have provided evidence for a neuroinflammatory response directly in the brain of ASD. Reports show lymphocyte infiltration as well as astroglial and microglial infiltration (Pardo *et al.*, 2005; Vargas *et al.*, 2005). Cytokines/chemokines predominantly of the pro-inflammatory variety, such as MCP-1, IL-6 and TGFβ1, which are derived from activated neuroglia, are up-regulated in brain tissues in ASD (Cohly and Panja, 2005; Pardo *et al.*, 2005; Vargas *et al.*, 2005; Molloy *et al.*, 2006). Gastrointestinal problems in individuals with ASD are linked to an altered ability to absorb nutrients, and are generally of an inflammatory nature (Horvath and Perman, 2002; Wakefield, 2002; Jyonouchi *et al.*, 2005). Thus, there may be a direct connection between digestive dysfunction and immune disruption.

Interestingly, serotonin emerges as a common denominator in both systemic as well as CNS pathologies associated with ASD. Peripherally, serotonin is synthesized by the enzyme Th1 in the enterochromaffin cells of the gut and in enteric neurons, and enters the bloodstream from there to be selectively taken up by platelets (Kiba, 2006; Schroecksnadel *et al.*, 2006; Grundy, 2008). Serotonergic release is involved in the modulation of gut motility and the secretion of intestinal fluids (Ormsbee and Fondacaro, 1985; Kiba, 2006; Li, 2007; Hansen and Witte, 2008. These processes are under vagus nerve control, which in turn receives input from serotonin neurons in the raphe nucleus (Tache *et al.*, 1993; Li, 2007; Kiba, 2006). As mentioned above, mouse constructs with altered systemic serotonergic neurotransmission, such as SERT-null mice, have altered intestinal functions, along with inflammatory responses, stress responsiveness and anxiety response (Murphy and Lesch, 2008). It has been postulated that, under inflammatory conditions, serotonergic modulation of gut activity may become disrupted (Grundy, 2008). However, given that SERT-null mice also show altered inflammatory responses, serotonergic disruptions may also be primary to the occurrence of inflammation. 5-HT is a known immune system regulator (Burgess *et al.*, 2006); it directly binds to receptors on both T- and B-lymphocytes to activate or suppress their function (Meredith *et al.*, 2005; Soga *et al.*, 2007; Levite, 2008). The complete mechanisms for these actions are not currently understood, but the expression of a plethora of different 5-HT receptors on lymphocytes, in addition to the SERT transporter, strongly suggest a complex regulatory network (Meredith *et al.*, 2005). Recent studies suggest that peripheral 5-HT promotes pro-inflammatory responses, along with cellular stimulation and cytokine release (Idzko *et al.*, 2004; Katoh *et al.*, 2006; Kushnir-Sukhov *et al.*, 2006). In particular, 5-HT appears to affect IL6 and TNFα release from a variety of different cell types (Basterzi *et al.*, 2005; Durk *et al.*, 2005; Lieb *et al.*, 2005; Cloez-Tayarani and Changeux, 2007; Menard *et al.*, 2007; Yu *et al.*, 2008). Central serotonin levels are also able to modulate inflammatory conditions in the periphery (Harbuz *et al.*, 1998).

Likewise, cytokine release can regulate the availability of tryptophan, the precursor amino acid used by Th1 in the periphery and Th2 in the CNS to generate serotonin (for extensive review, see Wirleitner *et al.*, 2003;

Schroecksnadel *et al.*, 2006a). In particular, interferon-γ (INFγ) and IL-2 stimulate the expression of the enzyme indolamine-2,3-dioxygenase [IDO], which degrades tryptophan to generate, among its end-products, kynurenic acid (Wirleitner *et al.*, 2003; Schroecksnadel *et al.*, 2006a, b). The physiological 'purpose' of this accelerated catabolism presumably is to reduce protein availability and 'starve' infectious agents while similarly providing regulatory feedback to the immune-response (Schroecksnadel *et al.*, 2006a, b). A 'side effect' is limitation of 5-HT synthesis, particularly in the CNS, since tryptophan competes with similar amino acids at the uptake pump across the blood–brain barrier. It has been shown that the resulting serotonin deficit in the brain is related to a central depression that has also been described as 'sickness behavior', since it is commonly associated with infections, and particularly chronic inflammatory problems as well as autoimmune conditions (Russo *et al.*, 2003; Wirleitner *et al.*, 2003; Capuron and Miller, 2004; Spalletta *et al.*, 2006).

On the other hand, there are many indications that pro-inflammatory cytokines such as TNFα, IL-1 and IL-6 increase serotonin levels throughout the brain and also in the periphery. Dunn and others demonstrated increases in the 5-HT metabolite 5-HIAA throughout the brain following IL-1 injections (Dunn, 2006). Both TNFα and cytokine IL-1 activate the expression of SERT (Zhu *et al.*, 2006), and IL-6 increases peripheral serotonin levels in a rat model for pulmonary hypertension (Miyata *et al.*, 2001). Peripheral IL-6 injection leads to increased striatal 5-HT levels (Zhang *et al.*, 2001). Treatment of colonic mucosa cells *in vitro* with pro-inflammatory cytokines, including TNFα, reduced 5-HT function (Foley *et al.*, 2007). Moreover, a selective serotonin increase was observed in frontal cortical regions following infectious immune-stimulation in the lungs (Lowry *et al.*, 2007). Thus, 5-HT levels appear to be modulated in a push–pull fashion by different types of cytokines and peripheral immune responses.

The pro-inflammatory cytokines IL-6 and TNFα, along with macrophage inflammatory protein (MIP-1) and interferon-α (IF-α), are also described as pyrogenes – substances that cause and modulate fever (Blatteis, 1990; Morrow *et al.*, 1996; Oka *et al.*, 2001). Fever represents an alteration of the body's temperature set-point, and occurs in response to both infectious and inflammatory processes, as well as to stress, particularly psychogenic stress (Blatteis, 1990; Morrow *et al.*, 1996; Oka *et al.*, 2001). The site for such temperature regulation appears to be the pre-optic area (POA) of the hypothalamus (Blatteis, 1990; Morrow *et al.*, 1996). Along with pro-inflammatory cytokines, serotonin itself is reported to modulate fever production via the POA (Blatteis, 1990). Thus, fever may represent a state of increased serotonin availability in the brain. This could be meaningful within the context of observations that fever improves behavioral performance in children with ASD (Curran *et al.*, 2007).

Environmental modulators

Another intriguing aspect of the 5-HT hypothesis for ASD is the susceptibility of the serotonergic system to environmental toxins. Environmental toxins such as organophosphates, organomercury compounds and polychlorinated biphenyls have long been known to alter brain development and cognitive behavior in animal models (Chang, 1984; Trask and Kosofsky, 2000; Faroon *et al.*, 2001; Ricceri *et al.*, 2003; Venerosi *et al.*, 2006; Timofeeva *et al.*, 2008), and recently have been implicated in the etiology of ASD (D'Amelio *et al.*, 2005; Cohly and Panja, 2005; Bello, 2007; Pessah *et al.*, 2008). Although many, particularly organophosphates, were originally linked to the disruption of other modulatory neurotransmitter pathways, substantial serotonergic effects are documented in the literature as well (Seegal *et al.*, 1986; Oudar *et al.*, 1989; Raulf and Konig, 1991; Morse *et al.*, 1996; Marcusson *et al.*, 2000; Mariussen and Fonnum, 2001; Khan and Thomas, 2004; Aldridge *et al.*, 2005a, b, c; Slotkin and Seidler, 2007).

Environmental stressors, ranging from psychosocial to traumatic, shape the development of serotonergic neurotransmission, and are in turn regulated by serotonergic innervation (Herrenkohl *et al.*, 1988; Peters, 1988; Ogawa *et al.*, 1994; Baumann *et al.*, 1998; Schneider *et al.*, 1998; Meaney *et al.*, 2000; Vazquez *et al.*, 2000; Carrasco and Van de Kar, 2003; Hayley *et al.*, 2005; Matsumoto *et al.*, 2005). For example, male rodent pups, maternally separated for periods of hours over the course of infancy, show significantly decreased 5-HT in hippocampus and frontal cortex (Resnicov and Nosenko, 1996; Vazquez *et al.*, 2000; Mathews *et al.*, 2001). At the same time, both 5-HT_1 and 5-HT_{2A} receptors are up-regulated following such treatment (Vazquez *et al.*, 2000; Mathews *et al.*, 2001). Several experimental studies have now demonstrated that the effects of perinatal stress can become epigenetically hard-wired during brain development, and that the serotonergic innervation to the forebrain is instrumental in this process (Meaney *et al.*, 2000; Seckl and Meaney, 2004; Weaver *et al.*, 2004). A recent review describes a frontocortical–amygdala–dorsal raphe feedback loop involved in the serotonergic modulation of stress responsiveness in the adult (Holmes, 2008). It is presumably this circuitry that becomes epigenetically altered as a consequence of early developmental stress or trauma, as suggested by findings in mice with genetically or pharmacologically altered 5-HT_1- and 5-HT_{2A}-type receptor availability (Holmes, 2008).

BDNF

Towards the beginning of this chapter, we discussed basic research evidence for the morphogenetic role of serotonin in cortical network formation. A molecule that is well positioned to serve as an intermediary in this process is Brain Derived Nerve Growth Factor (BDNF). Altered BDNF levels have consistently been found in ASD. Most studies have been based on analyzing serum or blood levels, and both show significant increases (Miyazaki *et al.*, 2004; Connolly *et al.*, 2006; Nelson *et al.*, 2006; Nishimura *et al.*, 2007) or decreases (Hashimoto *et al.*, 2006; Katoh-Semba *et al.*, 2007) of BDNF in ASD. Recent assessments of normal developmental BDNF trajectories in blood show that BDNF protein is increased in childhood, and subsequently attenuates towards adult levels (Nelson *et al.*, 2006). Thus, reports of increased vs. decreased levels might depend on developmental age at the time of testing or, alternatively, the widely divergent methods employed to measure peripheral BDNF. In addition, BDNF levels are subject to diet, exercise, stress and neuroinflammation (Kerschensteiner *et al.*, 2003; Tabakman *et al.*, 2004; Ferris *et al.*, 2007; Araya *et al.*, 2008; Tang *et al.*, 2008), suggesting that these variables need to be considered, in addition to age and sex (Trajkovska *et al.*, 2007), when assessing patient cohorts.

BDNF is among a small group of CNS neurotrophins that are instrumental in regulating neuronal survival, differentiation and synapse formation, and are necessary for the plastic adjustment of the developing brain to the environmental cues as well as for normal cognitive function in the adult (Huang *et al.*, 1999; McAllister, 2001; Branchi *et al.*, 2004; Garoflos *et al.*, 2005a; Kuipers and Bramham, 2006; Gunstad *et al.*, 2008). Neurotrophins and their receptors are expressed in the neocortex and hippocampus (Cellerino and Maffei, 1996; Marty *et al.*, 1996; Huang *et al.*, 1999; Lein *et al.*, 2000; McAllister, 2002) and the patterns of neurotrophin expression are activity-dependent and regulated by sensory inputs, electrical activity and stimulation (Aloyz *et al.*, 1999; Huang *et al.*, 1999; Balkowiec and Katz, 2000; McAllister, 2001). BDNF and its receptor, trkB, are densely expressed on cortical and hippocampal neurons, and influence both axonal and dendritic growth in a highly neuron-specific and age-dependent manner (Huang *et al.*, 1999; Xu *et al.*, 2000; McAllister, 2001). The expression of the trkB receptor peaks in the first 2 weeks postnatally in rodents, near the peak age of synapse formation, but BDNF action on cortical plasticity continues into adulthood (McAllister, 2002; Gomes *et al.*, 2006; Kuipers and Bramham, 2006). With maturation, trkB becomes enriched at the site of glutamatergic synapses and is uniquely able to modulate experience-dependent plasticity in cortical areas (Gomes *et al.*, 2006). Based on this neurobiology, it is somewhat surprising to find BDNF expression in serum and peripheral tissues. However, recent studies suggest an immune-modulatory role for BDNF (Kerschensteiner *et al.*, 2003; Schulte-Herbruggen *et al.*, 2007). Although neurotrophins have primarily been localized to neurons, subsets of cells from the immune system (e.g., macrophages, lymphocytes, mast cells) synthesize and produce neurotrophins and associated receptors in response to injury or inflammation (Kerschensteiner *et al.*, 1999; Tabakman *et al.*, 2004; Nassenstein *et al.*, 2005; Nockher and Renz, 2006). There are even indications that BDNF can be introduced to the brain by immune cells (Correale and Villa, 2004; Tabakman *et al.*, 2004).

BDNF and serotonin show reciprocal regulation and respond to the same environmental factors (Branchi *et al.*, 2004; Mattson *et al.*, 2004). For example, mice heterozygous for BDNF expression ($BDNF^{+/-}$) display premature, age-associated loss in forebrain serotonergic innervation (Lyons *et al.*, 1999) and impaired SERT function (Daws *et al.*, 2007). Localized increases in BDNF expression promote 5-HT fiber sprouting after injury (Mamounas *et al.*, 1995; Grider *et al.*, 2005). In turn, 5-HT depletion, via inhibition of synthesis, is accompanied by decreases in BDNF levels in the mature hippocampus (Zetterstrom *et al.*, 1999). During brain development, factors such as perinatal stress or environmental enrichment lead to long-term alterations in BDNF expression in brain and blood plasma (Vaidya *et al.*, 1997; Branchi *et al.*, 2004; Fumagalli *et al.*, 2004; Roceri *et al.*, 2004; Garoflos *et al.*, 2005b). In rat, maternal infection can cause long-term increases in BDNF within the cerebral cortex and other brain areas (Gilmore *et al.*, 2003). Interestingly, within the context of gender disparity in ASD and sex differences observed in our mouse model, there are indications of differential BDNF regulation in males and females (Cavus and Duman, 2003; Szapacs *et al.*, 2004; Sohrabji and Lewis, 2006).

A conceptual framework for the etiology of ASD behavioral endophenotypes

As indicated in the introduction, the particular focus of this chapter is the role of serotonin in the cortical pathology of ASD, largely because this is an area where a richness of basic research can inform translational hypotheses. We do not regard ASD as a purely cortical brain disorder, as there are well-known neuropathologies in the cerebellum and brainstem, among other regions (Kemper and Bauman, 1998; Trottier *et al.*, 1999; Keller and Persico, 2003; Courchesne *et al.*, 2005; Santangelo and Tsatsanis, 2005). Some brainstem neuropathologies (Rodier *et al.*, 1996; Courchesne, 1997; Bailey *et al.*, 1998), along with alterations in segmentation genes in ASD (Hashimoto *et al.*, 1992; Ingram *et al.*, 2000; Conciatori *et al.*, 2004; Tischfield *et al.*, 2005), are actually consistent with serotonergic abnormalities. The serotonergic neurons in the raphe nuclei form within regions of the brainstem reported to be affected in ASD (Lauder *et al.*, 1982; Wallace *et al.*, 1982). Moreover, the cerebellum receives a modulatory 5-HT innervation (Dieudonne, 2001).

We do hypothesize that most behavioral indicators of ASD are the consequence of late developmental and early postnatal alterations in brain development and plasticity. Moreover, data from our animal model suggest that cerebral cortical network formation can be implicated in substantial aspects of the behavioral pathologies. This view has important therapeutic implications for designing postnatal interventions that may be effective in substantially improving behavioral and cognitive outcomes in ASD (see Bethea and Sikich, 2007). This has to some extent already been borne out by the effectiveness of some early behavioral interventions (Landa, 2007; Dawson, 2008).

With many others, we are proponents of the theory that ASD is not a single disorder, but in fact a spectrum of disorders with a similar 'final common pathway' of altered brain network assembly and fine-tuning. The hypothesis that serotonergic tone may be a major factor in this 'final common pathway' is compatible with the available data as reviewed in this chapter. Figure 3 illustrates the pervasive role of serotonergic tone in brain and body homeostasis and development.

In essence, there are two different serotonergic systems, one each in the PNS and CNS, but these two 'branches' are linked within an intricate web of mutually regulatory mechanisms. It is most significant that the main difference between the PNS and CNS 'branches' is the location of 5-HT synthesis and the synthesizing enzyme, but beyond that most molecular components of signal transduction (receptors, uptake pumps) are shared. Thus, a genetic or environmental insult to the function of most gene products involved in serotonergic communication will affect both peripheral and central functions. At the same time, dysregulation of just the PNS serotonergic system by way of inflammatory or infectious insults, or via genetic deficits in aspects of immune response or digestive function, can derail central serotonergic innervation and modulation of brain development. Research has only scratched the surface of identifying the molecular regulators, such as neurotrophins (BDNF) or transcription factors (Pet-1, homebox genes), that might be intermediates in the CNS serotonergic modulation of brain development. Likewise, the study of neuro-immune modulation of brain morphogenesis has just begun. It is not hard to imagine, with this scenario, how dozens or even hundreds of different genes could be implicated in derailing serotonergic tone. A recent review illustrates, on the basis of genetically engineered mice for various serotonergic genes, the plethora of epistatic and/or additive effects that might result from just combining a handful of altered genes (Murphy *et al.*, 2003).

It would make sense that altered function of multiple genes or gene-products is necessary to imbalance serotonergic homeostasis/neurotransmission sufficiently to generate a disease state. In some instances, inherited

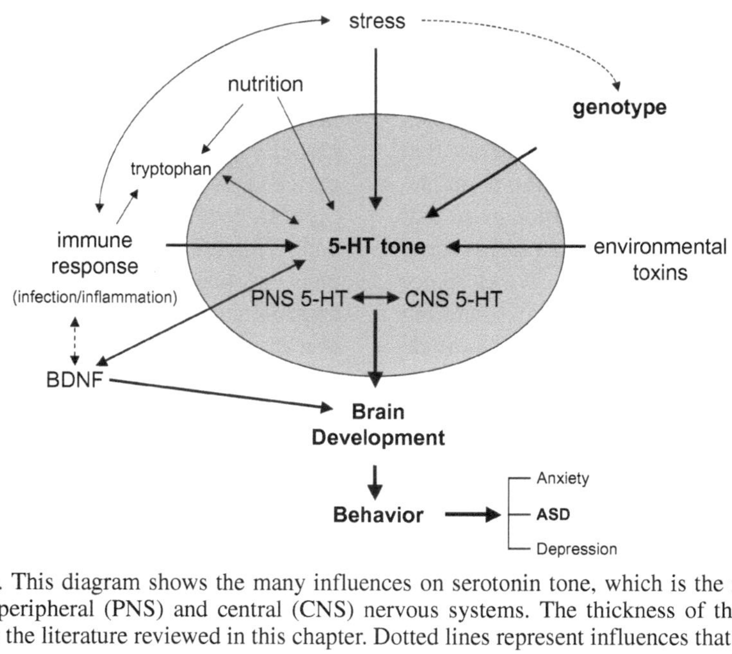

Figure 3 Summary diagram. This diagram shows the many influences on serotonin tone, which is the result of interactions between serotonergic systems in the peripheral (PNS) and central (CNS) nervous systems. The thickness of the lines denotes the estimated degree of influence, based on the literature reviewed in this chapter. Dotted lines represent influences that are understood less.

multi-gene vulnerabilities might be sufficient to precipitate ASD; in the majority of instances, multiple gene vulnerabilities might require one or more environmental factors, such as stress, infection or toxicants, to result in a functional breakdown and ASD. Furthermore, the severity of imbalances and developmental timing of a serotonergic dysfunction might determine whether the resulting disease state will resemble full-blown autism, Asperger's syndrome, anxiety disorder, or perhaps even depression or bipolar illness. Additional phenotype differences are the likely consequence of 5-HT imbalances interacting with other background genes.

How, in turn, do we envision a CNS serotonin imbalance to affect cortical and perhaps also limbic and cerebellar network formation? As the data in the first section of this review show, the cortical serotonergic innervation is instrumental in shaping afferent thalamocortical fiber ingrowth. It also influences the ability of the arriving fibers to appropriately activate their cortical targets and subsequently reorganize in a plastic way to accommodate afferent sensory input in the process of cortical map formation. The effects of the serotonergic innervation to frontal cortical association areas has not received much scrutiny in animal models, likely because in animals such as mouse, rat and cat it presents a much harder paradigm to study. Unlike parietal and occipital sensory cortex, frontal association cortex receives its input predominantly from other cortical areas (Goldman and Nauta, 1977; Jacobson and Trojanowski, 1977). Yet, as the data reviewed in the second section of this chapter show, frontal cortex undergoes a period of synaptic development, followed by pruning and fine-tuning of connectivity that is likely to benefit from serotonergic modulation in similar ways as primary sensory cortical areas. Moreover, altered minicolumns are observed in the frontal cortex as well as in parietal areas (Casanova *et al.*, 2002). Finally, the frontal cortical circuitry is intimately linked to limbic system control, and constitutes a network that is tightly controlled by serotonergic innervation in the mature brain (Holmes, 2008). Thus, we hypothesize that altered serotonergic tone in development can alter proper cortical network formation in both sensory and motor areas, as well as in association areas of cortex. Altering sensorimotor network function, in itself, would provide frontal association cortex with 'scrambled' information about the peripheral world, and impede both executive functions and proper social–emotional responses. Altered maturation of associative cortical networks, on top of that, would surely exacerbate the effects. The extent of effects on cortical regions may be related to the timing of serotonergic inputs. As reviewed above, CNS serotonergic abnormalities in ASD appear to be restricted to the early postnatal years and attenuate later. Since frontal cortex matures much later than the primary sensory and motor areas, it might be spared the brunt of the central serotonergic imbalances. However, PNS effects on serotonergic tone, including neuro-inflammatory events, could alter this time-line.

The molecular mechanism involved in serotonergic modulation of cortical network formation, without doubt, involves both glutamatergic and GABA-ergic synaptic connections. Thalamocortical afferents use glutamate as their excitatory neurotransmitter (Salt and Hill, 1983). Experimental evidence shows that glutamate receptors (GluRs) play roles in the activity-dependent refinement of synaptic connectivity (Constantine-Paton and Law, 1982; Daw *et al.*, 1993; Fox and Daw, 1993). GluRs are classified broadly into two groups: ionotropic sites, linked to ion channels; and metabotropic sites, linked to second messengers. The ionotropic sites include those activated by the exogenous agonists *N*-methyl-D-aspartate (NMDA), amino-3-hydroxy-5-methyl-4-isoxazole propionic acid (AMPA) and kainate (KA). NMDA receptors influence both the retraction of incorrectly placed axon arbors and synapses and the elaboration of correctly positioned terminals, and are instrumental in learning and memory functions in immature as well as mature cortex (Daw *et al.*, 1993; Fox and Daw, 1993; Rocha and Sur, 1995; Bear, 1996; Iwasato *et al.*, 1997). In addition, metabotropic glutamate receptors are involved in fine-tuning synaptic connectivity (Anwyl, 1991; Reid and Romano, 2001; Catania *et al.*, 2007).

Although they are excitatory during early cortical development, GABAergic interneurons in cortex eventually serve inhibitory functions and are essential in receptive field and thus proper map formation (Herlenius and Lagercrantz, 2001). Several different recent papers have pursued the hypothesis that ASD might be a disease of reduced GABAergic inhibition (Rubenstein and Merzenich, 2003; Minshew and Williams, 2007). This has partially been based on the concept that the morphological changes associated with minicolums in ASD might suggest reductions in GABAergic neurons (Minshew and Williams, 2007). Additional support for this argument has been provided by the occurrence of seizures in approximately one-third of ASD cases (Rubenstein and Merzenich, 2003).

Our concept is compatible with both these theoretical frameworks. Altered cortical network formation, such as would occur with altered serotonergic modulatory tone, creates 'noisy' connectivity by altering both excitatory and inhibitory synapses. In keeping with the idea that brainstem modulators to cortex, including serotonergic mechanisms, affect signal-to-noise ratios in their cortical targets, these afferents would be unable properly to 'drive' postsynaptic cortical neurons, whether these are excitatory (glutamatergic) or GABA-ergic (Rauschecker, 1991b; Herlenius and Lagercrantz, 2001; Gu, 2002; Siri *et al.*, 2007). This idea is supported by the literature reviewed in the first section of this chapter, concerning serotonergic modulation of thalamocortical glutamatergic tone as well as direct modulation of cortical neuronal activity. Activity regulated neurotrophins, such as BDNF, would clearly play a prominent role in mediating these modulatory events. The likely result would be an initial overproduction of weak synapses in cortex, followed by a later phase of pruning that is somewhat haphazard rather than selective for the most appropriately 'tuned' connections. Such a cortex would be more vulnerable to seizure activity. This scenario also is compatible with the increased cortical volume early in development that is followed by decreased connections later, as indicated by diffusion tensor imaging studies (see Minshew and Williams, 2007). As discussed above, serotonergic disruptions are also compatible with the findings of early brainstem changes and the well-established alterations in cerebellar development. In some cases of ASD, brainstem abnormalities, whether the consequence of genetic or toxic influences, may be the primary deficit. Segmentation genes such as Engrailed [Eng] alter the generation and/or location of dorsal raphe serotonergic neurons along with cerebellar histogenesis (Gharani *et al.*, 2004; Cheh *et al.*, 2006). Alternatively, the cerebellum might be directly impacted by an abnormal serotonergic innervation. While the serotonergic innervation to the cerebellum is very well characterized, little is known about its developmental function in mammals.

Conclusion

In this chapter we have presented a review of a large body of basic science and ASD-related literature that supports the idea that altered serotonergic homeostasis in the CNS and PNS may be prominently involved in the pathogenesis of ASD. In the CNS, experimental animal models have provided some proof of concept for this. However, investigations into the role of serotonin in immune regulation, and particularly neuroimmune regulation, are very much in their infancy. A unified hypothesis such as proposed here could provide an avenue towards early identification of biomarkers as well as therapeutic strategies. We hope to have provided an impetus for new investigations into a 'serotonergic hypothesis' for ASD.

References

Abramson, R.K., Wright, H.H., Cuccaro, M.L., Lawrence, L.G., Babb, S., Pencarinha, D., Marsteller, F. and Harris, E.C. (1992) Biological liability in families with autism. J. Am. Acad. Child. Adolesc Psychiatry, 31: 370–371.

Aldridge, J.E., Levin, E.D., Seidler, F.J. and Slotkin, T.A. (2005a) Developmental exposure of rats to chlorpyrifos leads to behavioral alterations in adulthood, involving serotonergic mechanisms and resembling animal models of depression. Environ. Health Perspect., 113: 527–531.

Aldridge, J.E., Meyer, A., Seidler, F.J. and Slotkin, T.A. (2005b) Alterations in central nervous system serotonergic and dopaminergic synaptic activity in adulthood after prenatal or neonatal chlorpyrifos exposure. Environ. Health Perspect., 113: 1027–1031.

Aldridge, J.E., Meyer, A., Seidler, F.J. and Slotkin, T.A. (2005c) Developmental exposure to terbutaline and chlorpyrifos: pharmacotherapy of preterm labor and an environmental neurotoxicant converge on serotonergic systems in neonatal rat brain regions. Toxicol. Appl. Pharmacol., 203: 132–144.

Aloyz, R., Fawcett, J.P., Kaplan, D.R., Murphy, R.A. and Miller, F.D. (1999) Activity-dependent activation of TrkB neurotrophin receptors in the adult CNS. Learn Mem., 6: 216–231.

Altamura, C., Dell'acqua, M.L., Moessner, R., Murphy, D.L., Lesch, K.P. and Persico, A.M. (2007) Altered neocortical cell density and layer thickness in serotonin transporter knockout mice: a quantitation study. Cereb. Cortex, 17: 1394–1401.

Alvarez, C., Vitalis, T., Fon, E.A., Hanoun, N., Hamon, M., Seif, I., Edwards, R., Gaspar, P. and Cases, O. (2002) Effects of genetic depletion of monoamines on somatosensory cortical development. Neuroscience, 115: 753–764.

Amirikian, B. and Georgopoulos, A.P. (2003) Modular organization of directionally tuned cells in the motor cortex: is there a short-range order? Proc. Natl. Acad. Sci. USA, 100: 12474–12479.

Anderson, G., Horne, W., Chatterjee, D. and Cohen, D. (1990) The hyperserotonemia of autism. Ann. NY. Acad. Sci., 660: 331.

Anderson, G.M., Gutknecht, L., Cohen, D.J., Brailly-Tabard, S., Cohen, J.H., Ferrari, P., Roubertoux, P.L. and Tordjman, S. (2002) Serotonin transporter promoter variants in autism: functional effects and relationship to platelet hyperserotonemia. Mol. Psychiatry, 7: 831–836.

Andres, C. (2002) Molecular genetics and animal models in autistic disorder. Brain Res. Bull., 57: 109–119.

Anwyl, R. (1991) The role of the metabotropic receptor in synaptic plasticity. Trends Pharmacol. Sci., 12: 324–326.

Araya, A.V., Orellana, X. and Espinoza, J. (2008) Evaluation of the effect of caloric restriction on serum BDNF in overweight and obese subjects: preliminary evidences. Endocrine, 33(3): 300–304.

Arndt, T.L., Stodgell, C.J. and Rodier, P.M. (2005) The teratology of autism. Intl. J. Dev. Neurosci., 23: 189–199.

Ashwood, P., Wills, S. and Van de Water, J. (2006) The immune response in autism: a new frontier for autism research. J. Leukoc. Biol., 80: 1–15.

Aylward, E., Minshew, N., Field, K., Sparks, B. and Singh, N. (2002) Effects of age on brain volume and head circumference in autism. Neurology, 59: 175–183.

Bailey, A., Luthert, P., Dean, A., Harding, B., Janota, I., Montgomery, M., Rutter, M. and Lantos, P. (1998) A clinicopathological study of autism. Brain, 121(Pt 5): 889–905.

Balkowiec, A. and Katz, D.M. (2000) Activity-dependent release of endogenous brain-derived neurotrophic factor from primary sensory neurons detected by ELISA *in situ*. J. Neurosci., 20: 7417–7423.

Bar-Peled, O., Gross-Isseroff, R., Ben-Hur, H., Hoskins, I., Groner, Y. and Biegon, A. (1991) Fetal human brain exhibits a prenatal peak in the density of serotonin 5-HT1A receptors. Neurosci. Lett., 127: 173–176.

Baron-Cohen, S. (2004) Autism: research into causes and intervention. Pediatr. Rehabil., 7: 73–78.

Basterzi, A.D., Aydemir, C., Kisa, C., Aksaray, S., Tuzer, V., Yazici, K. and Goka, E. (2005) IL-6 levels decrease with SSRI treatment in patients with major depression. Hum. Psychopharmacol., 20: 473–476.

Basura, G.J., Abbas, A.I., O'Donohue, H., Lauder, J.M., Roth, B.L., Walker, P.D. and Manis, P.B. (2008) Ontogeny of serotonin and serotonin2A receptors in rat auditory cortex. Hear Res., 244: 45–50.

Bauer, S., Kerr, B.J. and Patterson, P.H. (2007) The neuropoietic cytokine family in development, plasticity, disease and injury. Nat. Rev. Neurosci., 8: 221–232.

Baumann, M.H., Ayestas, M.A. and Rothman, R.B. (1998) Functional consequences of central serotonin depletion produced by repeated fenfluramine administration in rats. J. Neurosci., 18: 9069–9077.

Bear, M.F. (1996) A synaptic basis for memory storage in the cerebral cortex. Proc. Natl. Acad. Sci. USA, 93: 13453–13459.

Bear, M.F., Kleinschmidt, A., Gu, Q.A. and Singer, W. (1990) Disruption of experience-dependent synaptic modifications in striate cortex by infusion of an NMDA receptor antagonist. J. Neurosci., 10: 909–925.

Bello, S.C. (2007) Autism and environmental influences: review and commentary. Rev. Environ. Health, 22: 139–156.

Belmonte, M.K., Cook, E.H. Jr., Anderson, G.M., Rubenstein, J.L., Greenough, W.T., Beckel-Mitchener, A., Courchesne, E., Boulanger, L.M., Powell, S.B., Levitt, P.R., Perry, E.K., Jiang, Y.H., DeLorey, T.M. and Tierney, E. (2004) Autism as a disorder of neural information processing: directions for research and targets for therapy. Mol. Psychiatry, 9: 646–663.

Benes, F.M., Taylor, J.B. and Cunningham, M.C. (2000) Convergence and plasticity of monoaminergic systems in the medial prefrontal cortex during the postnatal period: implications for the development of psychopathology. Cereb. Cortex, 10: 1014–1027.

Bennett-Clarke, C.A., Chiaia, N.L., Crissman, R.S. and Rhoades, R.W. (1991) The source of the transient serotoninergic input to the developing visual and somatosensory cortices in rat. Neuroscience, 43: 163–183.

Bennett-Clarke, C.A., Leslie, M.J., Chiaia, N.L. and Rhoades, R.W. (1993) Serotonin 1B receptors in the developing somatosensory and visual cortices are located on thalamocortical axons. Proc. Natl. Acad. Sci. USA, 90: 153–157.

Bennett-Clarke, C.A., Leslie, M.J., Lane, R.D. and Rhoades, R.W. (1994) Effect of serotonin depletion on vibrissa-related patterns of thalamic afferents in the rat's somatosensory cortex. J. Neurosci., 14: 7594–7607.

Berger-Sweeney, J. and Hohmann, C.F. (1997) Behavioral consequences of abnormal cortical development: insights into developmental disabilities. Behav. Brain Res., 86: 121–142.

Berger-Sweeney, J., Libbey, M., Arters, J., Junagadhwalla, M. and Hohmann, C.F. (1998) Neonatal monoaminergic depletion in mice (*Mus musculus*) improves performance of a novel odor discrimination task. Behav. Neurosci., 112: 1318–1326.

Bernardet, M. and Crusio, W.E. (2006) Fmr1 KO mice as a possible model of autistic features. ScientificWorld J., 6: 1164–1176.

Bethea, T.C. and Sikich, L. (2007) Early pharmacological treatment of autism: a rationale for developmental treatment. Biol. Psychiatry, 61: 521–537.

Blakemore, C. and Mitchell, D.E. (1973) Environmental modification of the visual cortex and the neural basis of learning and memory. Nature, 241: 467–468.

Blatteis, C.M. (1990) Neuromodulative actions of cytokines. Yale J. Biol. Med., 63: 133–146.

Blue, M.E. and Molliver, M.E. (1989) Serotonin influences barrel formation in developing somatosensory cortex of the rat. Soc. Neurosci. Abstr., 15: 1214.

Blue, M.E., Erzurumlu, R.S. and Jhaveri, S. (1991) A comparison of pattern formation by thalamocortical and serotonergic afferents in the rat barrel field cortex. Cereb. Cortex, 1: 380–389.

Blue, M.E., Johnston, M., Moloney, C. and Hohmann, C.F. (2008) Serotonin dysfunction in autism. In: Zimmerman, A.W. (Ed.), Autism: Current Theories and Evidence. Humana Press, Totowa NJ, pp. 111–132.

Boddaert, N. and Zilbovicius, M. (2002) Functional neuroimaging and childhood autism. Pediatr. Radiol., 32: 1–7.

Bonati, M. and Clavenna, A. (2005) The epidemiology of psychotropic drug use in children and adolescents. Intl. Rev. Psychiatry, 17: 181–188.

Boylan, C.B., Bennett-Clarke, C.A., Chiaia, N.L. and Rhoades, R.W. (2000a) Time course of expression and function of the serotonin transporter in the neonatal rat's primary somatosensory cortex. Somatosens Mot. Res., 17: 52–60.

Boylan, C.B., Bennett-Clarke, C.A., Crissman, R.S., Mooney, R.D. and Rhoades, R.W. (2000b) Clorgyline treatment elevates cortical serotonin and temporarily disrupts the vibrissae-related pattern in rat somatosensory cortex. J.Comp. Neurol. 427:139–149.

Boylan, C.B., Kesterson, K.L., Bennett-Clarke, C.A., Chiaia, N.L. and Rhoades, R.W. (2001) Neither peripheral nerve input nor cortical NMDA receptor activity are necessary for recovery of a disrupted barrel pattern in rat somatosensory cortex. Brain Res. Dev. Brain Res., 129: 95–106.

Boylan, C.B., Blue, M.E. and Hohmann, C.F. (2007) Modeling early cortical serotonergic deficits in autism. Behav. Brain Res., 176: 94–108.

Branchi, I., Francia, N. and Alleva, E. (2004) Epigenetic control of neurobehavioural plasticity: the role of neurotrophins. Behav. Pharmacol., 15: 353–362.

Brune, C.W., Kim, S.J., Salt, J., Leventhal, B.L., Lord, C. and Cook, E.H. Jr. (2006) 5-HTTLPR genotype-specific phenotype in children and adolescents with autism. Am. J. Psychiatry, 163: 2148–2156.

Bruning, G. and Liangos, O. (1997) Transient expression of the serotonin transporter in the developing mouse thalamocortical system. Acta Histochem., 99: 117–121.

Buitelaar, J.K. and Willemsen-Swinkels, S.H. (2000) Autism: current theories regarding its pathogenesis and implications for rational pharmacotherapy. Paediatr. Drugs, 2: 67–81.

Burgess, N.K., Sweeten, T.L., McMahon, W.M. and Fujinami, R.S. (2006) Hyperserotoninemia and altered immunity in autism. J. Autism Dev. Disord., 36: 697–704.

Buxhoeveden, D.P. and Casanova, M.F. (2002) The minicolumn hypothesis in neuroscience. Brain, 125: 935–951.

Buxhoeveden, D.P., Semendeferi, K., Buckwalter, J., Schenker, N., Switzer, R. and Courchesne, E. (2006) Reduced minicolumns in the frontal cortex of patients with autism. Neuropathol. Appl. Neurobiol., 32: 483–491.

Bystron, I., Blakemore, C. and Rakic, P. (2008) Development of the human cerebral cortex: boulder committee revisited. Nat. Rev. Neurosci., 9: 110–122.

Capuron, L. and Miller, A.H. (2004) Cytokines and psychopathology: lessons from interferon-alpha. Biol. Psychiatry, 56: 819–824.

Carper, R.A. and Courchesne, E. (2005) Localized enlargement of the frontal cortex in early autism. Biol. Psychiatry, 57: 126–133.

Carrasco, G.A. and Van de Kar, L.D. (2003) Neuroendocrine pharmacology of stress. Eur. J. Pharmacol., 463: 235–272.

Casanova, M.F., Buxhoeveden, D.P., Switala, A.E. and Roy, E. (2002) Minicolumnar pathology in autism. Neurology, 58: 428–432.

Cases, O., Vitalis, T., Seif, I., De Maeyer, E., Sotelo, C. and Gaspar, P. (1996) Lack of barrels in the somatosensory cortex of monoamine oxidase A-deficient mice: role of a serotonin excess during the critical period. Neuron, 16: 297–307.

Catania, M.V., D'Antoni, S., Bonaccorso, C.M., Aronica, E., Bear, M.F. and Nicoletti, F. (2007) Group I metabotropic glutamate receptors: a role in neurodevelopmental disorders? Mol. Neurobiol., 35: 298–307.

Cavus, I. and Duman, R.S. (2003) Influence of estradiol, stress, and 5-HT2A agonist treatment on brain-derived neurotrophic factor expression in female rats. Biol. Psychiatry, 54: 59–69.

Cellerino, A. and Maffei, L. (1996) The action of neurotrophins in the development and plasticity of the visual cortex. Prog. Neurobiol., 49: 53–71.

Chandana, S.R., Behen, M.E., Juhasz, C., Muzik, O., Rothermel, R.D., Mangner, T.J., Chakraborty, P.K., Chugani, H.T. and Chugani, D.C. (2005) Significance of abnormalities in developmental trajectory and asymmetry of cortical serotonin synthesis in autism. Intl. J. Dev. Neurosci., 23: 171–182.

Chang, L.W. (Ed.), (1984) Developmental Toxicology of Methylmercury. Elsevier Science Publishers, Amsterdam.

Cheh, M.A., Millonig, J.H., Roselli, L.M., Ming, X., Jacobsen, E., Kamdar, S. and Wagner, G.C. (2006) En2 knockout mice display neurobehavioral and neurochemical alterations relevant to autism spectrum disorder. Brain Res., 1116: 166–176.

Chez, M.G., Dowling, T., Patel, P.B., Khanna, P. and Kominsky, M. (2007) Elevation of tumor necrosis factor-alpha in cerebrospinal fluid of autistic children. Pediatr. Neurol., 36: 361–365.

Cho, I.H., Yoo, H.J., Park, M., Lee, Y.S. and Kim, S.A. (2007) Family-based association study of 5-HTTLPR and the 5-HT2A receptor gene polymorphisms with autism spectrum disorder in Korean trios. Brain Res., 1139: 34–41.

Chugani, H.T. (1998) A critical period of brain development: studies of cerebral glucose utilization with PET. Prev. Med., 27: 184–188.

Chugani, D.C., Muzik, O., Rothermel, R., Behen, M., Chakraborty, P., Mangner, T., da Silva, E.A. and Chugani, H.T. (1997) Altered serotonin synthesis in the dentatothalamocortical pathway in autistic boys. Ann. Neurol., 42: 666–669.

Chugani, D.C., Muzik, O., Behen, M., Rothermel, R., Janisse, J.J., Lee, J. and Chugani, H.T. (1999) Developmental changes in brain serotonin synthesis capacity in autistic and nonautistic children. Ann. Neurol., 45: 287–295.

Cloez-Tayarani, I. and Changeux, J.P. (2007) Nicotine and serotonin in immune regulation and inflammatory processes: a perspective. J. Leukoc Biol., 81: 599–606.

Cohly, H.H. and Panja, A. (2005) Immunological findings in autism. Intl. Rev. Neurobiol., 71: 317–341.

Conciatori, M., Stodgell, C.J., Hyman, S.L., O'Bara, M., Militerni, R., Bravaccio, C., Trillo, S., Montecchi, F., Schneider, C., Melmed, R., Elia, M., Crawford, L., Spence, S.J., Muscarella, L., Guarnieri, V., D'Agruma, L., Quattrone, A., Zelante, L., Rabinowitz, D., Pascucci, T., Puglisi-Allegra, S., Reichelt, K.L., Rodier, P.M. and Persico, A.M. (2004) Association between the HOXA1 A218G polymorphism and increased head circumference in patients with autism. Biol. Psychiatry, 55: 413–419.

Connell, S., Karikari, C. and Hohmann, C.F. (2004) Sex-specific development of cortical monoamine levels in mouse. Brain Res. Dev. Brain Res., 151: 187–191.

Connolly, A.M., Chez, M., Streif, E.M., Keeling, R.M., Golumbek, P.T., Kwon, J.M., Riviello, J.J., Robinson, R.G., Neuman, R.J. and Deuel, R.M. (2006) Brain-derived neurotrophic factor and autoantibodies to neural antigens in sera of children with autistic spectrum disorders, Landau-Kleffner syndrome, and epilepsy. Biol. Psychiatry, 59: 354–363.

Conroy, J., Meally, E., Kearney, G., Fitzgerald, M., Gill, M. and Gallagher, L. (2004) Serotonin transporter gene and autism: a haplotype analysis in an Irish autistic population. Mol. Psychiatry, 9: 587–593.

Constantine-Paton, M. and Law, M.I. (1982) The development of maps and stripes in the brain. Sci. Am., 247: 62–70.

Cook, E.H. (1990) Autism: review of neurochemical investigation. Synapse, 6: 292–308.

Cook, E.H. Jr. (2001) Genetics of autism. Child Adolesc Psychiatr. Clin. N. Am., 10: 333–350.

Cook, E.H. Jr., Courchesne, R., Lord, C., Cox, N.J., Yan, S., Lincoln, A., Haas, R., Courchesne, E. and Leventhal, B.L. (1997) Evidence of linkage between the serotonin transporter and autistic disorder. Mol. Psychiatry, 2: 247–250.

Correale, J. and Villa, A. (2004) The neuroprotective role of inflammation in nervous system injuries. J. Neurol., 251: 1304–1316.

Cote, F., Fligny, C., Bayard, E., Launay, J.M., Gershon, M.D., Mallet, J. and Vodjdani, G. (2007) Maternal serotonin is crucial for murine embryonic development. Proc. Natl. Acad. Sci. USA, 104: 329–334.

Courchesne, E. (1997) Brainstem, cerebellar and limbic neuroanatomical abnormalities in autism. Curr. Opin. Neurobiol., 7: 269–278.

Courchesne, E. (2002) Abnormal early brain development in autism. Mol. Psychiatry, 7(Suppl. 2): S21–S23.

Courchesne, E. (2004) Brain development in autism: early overgrowth followed by premature arrest of growth. Ment. Retard. Dev. Disabil. Res. Rev., 10: 106–111.

Courchesne, E. and Pierce, K. (2005) Brain overgrowth in autism during a critical time in development: implications for frontal pyramidal neuron and interneuron development and connectivity. Intl. J. Dev. Neurosci., 23: 153–170.

Courchesne, E., Press, G.A. and Yeung-Courchesne, R. (1993) Parietal lobe abnormalities detected with MR in patients with infantile autism. AJR Am. J. Roentgenol., 160: 387–393.

Courchesne, E., Karns, C.M., Davis, H.R., Ziccardi, R., Carper, R.A., Tigue, Z.D., Chisum, H.J., Moses, P., Pierce, K., Lord, C., Lincoln, A.J., Pizzo, S., Schreibman, L., Haas, R.H., Akshoomoff, N.A. and Courchesne, R.Y. (2001) Unusual brain growth patterns in early life in patients with autistic disorder: an MRI study. Neurology, 57: 245–254.

Courchesne, E., Redcay, E. and Kennedy, D.P. (2004) The autistic brain: birth through adulthood. Curr. Opin. Neurol., 17: 489–496.

Courchesne, E., Redcay, E., Morgan, J.T. and Kennedy, D.P. (2005) Autism at the beginning: microstructural and growth

abnormalities underlying the cognitive and behavioral phenotype of autism. Dev. Psychopathol., 17: 577–597.
Courchesne, E., Pierce, K., Schumann, C.M., Redcay, E., Buckwalter, J.A., Kennedy, D.P. and Morgan, J. (2007) Mapping early brain development in autism. Neuron, 56: 399–413.
Coutinho, A.M., Oliveira, G., Morgadinho, T., Fesel, C., Macedo, T.R., Bento, C., Marques, C., Ataide, A., Miguel, T., Borges, L. and Vicente, A.M. (2004) Variants of the serotonin transporter gene (SLC6A4) significantly contribute to hyperserotonemia in autism. Mol. Psychiatry, 9: 264–271.
Croonenberghs, J., Bosmans, E., Deboutte, D., Kenis, G. and Maes, M. (2002) Activation of the inflammatory response system in autism. Neuropsychobiology, 45: 1–6.
Curran, L.K., Newschaffer, C.J., Lee, L.C., Crawford, S.O., Johnston, M.V. and Zimmerman, A.W. (2007) Behaviors associated with fever in children with autism spectrum disorders. Pediatrics, 120: e1386–e1392.
D'Amato, R.J., Blue, M.E., Largent, B.L., Lynch, D.R., Ledbetter, D.J., Molliver, M.E. and Snyder, S.H. (1987) Ontogeny of the serotonergic projection to rat neocortex: transient expression of a dense innervation to primary sensory areas. Proc. Natl. Acad. Sci. USA, 84: 4322–4326.
D'Amelio, M., Ricci, I., Sacco, R., Liu, X., D'Agruma, L., Muscarella, L.A., Guarnieri, V., Militerni, R., Bravaccio, C., Elia, M., Schneider, C., Melmed, R., Trillo, S., Pascucci, T., Puglisi-Allegra, S., Reichelt, K.L., Macciardi, F., Holden, J.J. and Persico, A.M. (2005) Paraoxonase gene variants are associated with autism in North America, but not in Italy: possible regional specificity in gene-environment interactions. Mol. Psychiatry, 10: 1006–1016.
Daw, N.W. and Beaver, C.J. (2001) Developmental changes and ocular dominance plasticity in the visual cortex. Keio J. Med., 50: 192–197.
Daw, N.W., Stein, P.S. and Fox, K. (1993) The role of NMDA receptors in information processing. Annu. Rev. Neurosci., 16: 207–222.
Daws, L.C., Munn, J.L., Valdez, M.F., Frosto-Burke, T. and Hensler, J.G. (2007) Serotonin transporter function, but not expression, is dependent on brain-derived neurotrophic factor (BDNF): *in vivo* studies in BDNF-deficient mice. J. Neurochem., 101: 641–651.
Dawson, G. (2008) Early behavioral intervention, brain plasticity, and the prevention of autism spectrum disorder. Dev. Psychopathol., 20: 775–803.
Dawson, G., Webb, S., Schellenberg, G.D., Dager, S., Friedman, S., Aylward, E. and Richards, T. (2002a) Defining the broader phenotype of autism: genetic, brain, and behavioral perspectives. Dev. Psychopathol., 14: 581–611.
Dawson, G., Carver, L., Meltzoff, A.N., Panagiotides, H., McPartland, J. and Webb, S.J. (2002b) Neural correlates of face and object recognition in young children with autism spectrum disorder, developmental delay, and typical development. Child Dev., 73: 700–717.
DeMyer, M.K. (1975) The nature of the neuropsychological disability in autistic children. J. Autism Child Schizophr., 5: 109–128.
Devlin, B., Cook, E.H. Jr., Coon, H., Dawson, G., Grigorenko, E.L., McMahon, W., Minshew, N., Pauls, D., Smith, M., Spence, M. A., Rodier, P.M., Stodgell, C. and Schellenberg, G.D. (2005) Autism and the serotonin transporter: the long and short of it. Mol. Psychiatry, 10: 1110–1116.
DiCicco-Bloom, E., Lord, C., Zwaigenbaum, L., Courchesne, E., Dager, S.R., Schmitz, C., Schultz, R.T., Crawley, J. and Young, L.J. (2006) The developmental neurobiology of autism spectrum disorder. J. Neurosci., 26: 6897–6906.
Dieudonne, S. (2001) Serotonergic neuromodulation in the cerebellar cortex: cellular, synaptic, and molecular basis. Neuroscientist, 7: 207–219.
Dinopoulos, A., Dori, I. and Parnavelas, J.G. (1997) The serotonin innervation of the basal forebrain shows a transient phase during development. Brain Res. Dev. Brain Res., 99: 38–52.
Dong, W.K. and Greenough, W.T. (2004) Plasticity of nonneuronal brain tissue: roles in developmental disorders. Ment. Retard. Dev. Disabil. Res. Rev., 10: 85–90.
Dooley, A.E., Pappas, I.S. and Parnavelas, J.G. (1997) Serotonin promotes the survival of cortical glutamatergic neurons *in vitro*. Exp. Neurol., 148: 205–214.
Dori, I.E., Dinopoulos, A. and Parnavelas, J.G. (1998) The development of the synaptic organization of the serotonergic system differs in brain areas with different functions. Exp. Neurol., 154: 113–125.
Dunn, A.J. (2006) Effects of cytokines and infections on brain neurochemistry. Clin. Neurosci. Res., 6: 52–68.
Durk, T., Panther, E., Muller, T., Sorichter, S., Ferrari, D., Pizzirani, C., Di Virgilio, F., Myrtek, D., Norgauer, J. and Idzko, M. (2005) 5-Hydroxytryptamine modulates cytokine and chemokine production in LPS-primed human monocytes via stimulation of different 5-HTR subtypes. Intl. Immunol., 17: 599–606.
Dyck, R.H. and Cynader, M.S. (1993) Autoradiographic localization of serotonin receptor subtypes in cat visual cortex: transient regional, laminar, and columnar distributions during postnatal development. J. Neurosci., 13: 4316–4338.
Edagawa, Y., Saito, H. and Abe, K. (2001) Endogenous serotonin contributes to a developmental decrease in long-term potentiation in the rat visual cortex. J. Neurosci., 21: 1532–1537.
Erzurumlu, R.S. and Kind, P.C. (2001) Neural activity: sculptor of 'barrels' in the neocortex. Trends Neurosci., 24: 589–595.
Esaki, T., Cook, M., Shimoji, K., Murphy, D.L., Sokoloff, L. and Holmes, A. (2005) Developmental disruption of serotonin transporter function impairs cerebral responses to whisker stimulation in mice. Proc. Natl. Acad. Sci. USA, 102: 5582–5587.
Faroon, O., Jones, D. and de Rosa, C. (2001) Effects of polychlorinated biphenyls on the nervous system. Toxicol. Ind. Health, 16: 305–333.
Favorov, O.V. and Kelly, D.G. (1994) Minicolumnar organization within somatosensory cortical segregates: I. Development of afferent connections. Cereb. Cortex, 4: 408–427.
Ferris, L.T., Williams, J.S. and Shen, C.L. (2007) The effect of acute exercise on serum brain-derived neurotrophic factor levels and cognitive function. Med. Sci. Sports Exerc., 39: 728–734.
Foley, K.F., Pantano, C., Ciolino, A. and Mawe, G.M. (2007) IFN-{gamma} and TNF-{alpha} decrease serotonin transporter function and expression in Caco2 cells. Am. J. Physiol. Gastrointest. Liver Physiol., 292: G779–G784.
Foote, S.L. and Morrison, J.H. (1987) Development of the noradrenergic, serotonergic, and dopaminergic innervation of neocortex. Curr. Top. Dev. Biol., 21: 391–423.
Fox, K. and Daw, N.W. (1993) Do NMDA receptors have a critical function in visual cortical plasticity? Trends Neurosci., 16: 116–122.
Fujimiya, M., Hosoda, S., Kitahama, K., Kimura, H. and Maeda, T. (1986) Early development of serotonin neuron in the rat brain

as studied by immunohistochemistry combined with tryptophan administration. Brain Dev., 8: 335–342.

Fumagalli, F., Bedogni, F., Perez, J., Racagni, G. and Riva, M.A. (2004) Corticostriatal brain-derived neurotrophic factor dysregulation in adult rats following prenatal stress. Eur. J. Neurosci., 20: 1348–1354.

Gage, N.M., Siegel, B., Callen, M. and Roberts, T.P. (2003) Cortical sound processing in children with autism disorder: an MEG investigation. Neuroreport, 14: 2047–2051.

Garoflos, E., Panagiotaropoulos, T., Pondiki, S., Stamatakis, A., Philippidis, E. and Stylianopoulou, F. (2005a) Cellular mechanisms underlying the effects of an early experience on cognitive abilities and affective states. Ann. Gen. Psychiatry, 4: 8.

Garoflos, E., Stamatakis, A., Mantelas, A., Philippidis, H. and Stylianopoulou, F. (2005b) Cellular mechanisms underlying an effect of 'early handling' on pCREB and BDNF in the neonatal rat hippocampus. Brain Res., 1052: 187–195.

Gentile, S. (2007) Serotonin reuptake inhibitor-induced perinatal complications. Paediatr. Drugs, 9: 97–106.

Gharani, N., Benayed, R., Mancuso, V., Brzustowicz, L.M. and Millonig, J.H. (2004) Association of the homeobox transcription factor, ENGRAILED 2, 3, with autism spectrum disorder. Mol. Psychiatry, 9: 474–484.

Gillberg, C. (1999) Neurodevelopmental processes and psychological functioning in autism. Dev. Psychopathol., 11: 567–587.

Gilmore, J.H., Jarskog, L.F. and Vadlamudi, S. (2003) Maternal infection regulates BDNF and NGF expression in fetal and neonatal brain and maternal-fetal unit of the rat. J. Neuroimmunol., 138: 49–55.

Glantz, L.A., Gilmore, J.H., Hamer, R.M., Lieberman, J.A. and Jarskog, L.F. (2007) Synaptophysin and postsynaptic density protein 95 in the human prefrontal cortex from mid-gestation into early adulthood. Neuroscience, 149: 582–591.

Goldman, P.S. and Nauta, W.J. (1977) Columnar distribution of cortico-cortical fibers in the frontal association, limbic, and motor cortex of the developing rhesus monkey. Brain Res., 122: 393–413.

Goldman-Rakic, P.S. and Brown, R.M. (1982) Postnatal development of monoamine content and synthesis in the cerebral cortex of rhesus monkeys. Brain Res., 256: 339–349.

Gomes, R.A., Hampton, C., El-Sabeawy, F., Sabo, S.L. and McAllister, A.K. (2006) The dynamic distribution of TrkB receptors before, during, and after synapse formation between cortical neurons. J. Neurosci., 26: 11487–11500.

González-Burgos, I., del Angel-Meza, A.R., Barajas-López, G. and Feria-Velasco, A. (1996) Tryptophan restriction causes long–term plastic changes in corticofrontal pyramidal neurons. Intl. J. Dev. Neurosci., 14: 673–679.

Gordon, B., Allen, E.E. and Trombley, P.Q. (1988) The role of norepinephrine in plasticity of visual cortex. Prog. Neurobiol., 30: 171–191.

Grider, M.H., Mamounas, L.A., Le, W. and Shine, H.D. (2005) *In situ* expression of brain-derived neurotrophic factor or neurotrophin-3 promotes sprouting of cortical serotonergic axons following a neurotoxic lesion. J. Neurosci. Res., 82: 404–412.

Grigorenko, E.L., Han, S.S., Yrigollen, C.M., Leng, L., Mizue, Y., Anderson, G.M., Mulder, E.J., de Bildt, A., Minderaa, R.B., Volkmar, F.R., Chang, J.T. and Bucala, R. (2008) Macrophage migration inhibitory factor and autism spectrum disorders. Pediatrics, 122: e438–e445.

Grundy, D. (2008) 5-HT system in the gut: roles in the regulation of visceral sensitivity and motor functions. Eur. Rev. Med. Pharmacol. Sci., 12(Suppl 1): 63–67.

Gruss, M. and Braun, K. (2001) Alterations of amino acids and monoamine metabolism in male Fmr1 knockout mice: a putative animal model of the human Fragile X mental retardation syndrome. Neural Plast., 8: 285–298.

Gu, Q. (2002) Neuromodulatory transmitter systems in the cortex and their role in cortical plasticity. Neuroscience, 111: 815–835.

Gu, Q. and Singer, W. (1995) Involvement of serotonin in developmental plasticity of kitten visual cortex. Eur. J. Neurosci., 7: 1146–1153.

Gu, Q., Patel, B. and Singer, W. (1990) The laminar distribution and postnatal development of serotonin- immunoreactive axons in the cat primary visual cortex. Exp. Brain Res., 81: 257–266.

Gu, Q.A., Bear, M.F. and Singer, W. (1989) Blockade of NMDA-receptors prevents ocularity changes in kitten visual cortex after reversed monocular deprivation. Brain Res. Dev. Brain Res., 47: 281–288.

Guhathakurta, S., Ghosh, S., Sinha, S., Chatterjee, A., Ahmed, S., Chowdhury, S.R., Gangopadhyay, P.K., Ghosh, S., Singh, M. and Usha, R. (2006) Serotonin transporter promoter variants: analysis in Indian autistic and control population. Brain Res., 1092: 28–35.

Gunstad, J., Benitez, A., Smith, J., Glickman, E., Spitznagel, M.B., Alexander, T., Juvancic-Heltzel, J. and Murray, L. (2008) Serum brain-derived neurotrophic factor is associated with cognitive function in healthy older adults. J. Geriatr. Psychiatry Neurol., 21: 166–170.

Gupta, S., Aggarwal, S., Rashanravan, B. and Lee, T. (1998) Th1- and Th2-like cytokines in CD4+ and CD8+ T cells in autism. J. Neuroimmunol., 85: 106–109.

Hansen, M.B. and Witte, A.B. (2008) The role of serotonin in intestinal luminal sensing and secretion. Acta Physiol. (Oxf.), 193: 311–323.

Hansson, S.R., Mezey, E. and Hoffman, B.J. (1998) Serotonin transporter messenger RNA in the developing rat brain: early expression in serotonergic neurons and transient expression in non-serotonergic neurons. Neuroscience, 83: 1185–1201.

Harbuz, M.S., Marti, O., Lightman, S.L. and Jessop, D.S. (1998) Alteration of central serotonin modifies onset and severity of adjuvanT-induced arthritis in the rat. Br. J. Rheumatol., 37: 1077–1083.

Harper, M.S. and Wallace, M.N. (1995) Changes in density of brainstem afferents in ferret primary auditory cortex (AI) during postnatal development. J. Anat., 186(Pt 2): 373–382.

Hashimoto, K., Iwata, Y., Nakamura, K., Tsujii, M., Tsuchiya, K.J., Sekine, Y., Suzuki, K., Minabe, Y., Takei, N., Iyo, M. and Mori, N. (2006) Reduced serum levels of brain-derived neurotrophic factor in adult male patients with autism. Prog. Neuropsychopharmacol. Biol. Psychiatry, 30: 1529–1531.

Hashimoto, T., Tayama, M., Miyazaki, M., Sakurama, N., Yoshimoto, T., Murakawa, K. and Kuroda, Y. (1992) Reduced brainstem size in children with autism. Brain Dev., 14: 94–97.

Hayley, S., Poulter, M.O., Merali, Z. and Anisman, H. (2005) The pathogenesis of clinical depression: stressor- and cytokine-induced alterations of neuroplasticity. Neuroscience, 135: 659–678.

Hazlett, H.C., Poe, M., Gerig, G., Smith, R.G., Provenzale, J., Ross, A., Gilmore, J. and Piven, J. (2005) Magnetic resonance imaging and head circumference study of brain size in autism: birth through age 2 years. Arch. Gen. Psychiatry, 62: 1366–1376.

Hendricks, T., Francis, N., Fyodorov, D. and Deneris, E.S. (1999) The ETS domain factor PeT-1 is an early and precise marker of central serotonin neurons and interacts with a conserved element in serotonergic genes. J. Neurosci., 19: 10348–10356.

Hendricks, T.J., Fyodorov, D.V., Wegman, L.J., Lelutiu, N.B., Pehek, E.A., Yamamoto, B., Silver, J., Weeber, E.J., Sweatt, J.D. and Deneris, E.S. (2003) PeT-1 ETS gene plays a critical role in 5-HT neuron development and is required for normal anxiety-like and aggressive behavior. Neuron, 37: 233–247.

Herbert, M. and Anderson, M.P. (2008) An expanding spectrum of austism models. In: Zimmerman, A.W. (Ed.), Autism: Current Theories and Evidence. Humana Press, Totowa, NJ, pp. 429–464.

Herbert, M.R., Ziegler, D.A., Makris, N., Filipek, P.A., Kemper, T.L., Normandin, J.J., Sanders, H.A., Kennedy, D.N. and Caviness, V.S. Jr. (2004) Localization of white matter volume increase in autism and developmental language disorder. Ann. Neurol., 55: 530–540.

Herbert, M.R., Russo, J.P., Yang, S., Roohi, J., Blaxill, M., Kahler, S.G., Cremer, L. and Hatchwell, E. (2006) Autism and environmental genomics. Neurotoxicology, 27: 671–684.

Herlenius, E. and Lagercrantz, H. (2001) Neurotransmitters and neuromodulators during early human development. Early Hum. Dev., 65: 21–37.

Herrenkohl, L.R., Ribary, U., Schlumpf, M. and Lichtensteiger, W. (1988) Maternal stress alters monoamine metabolites in fetal and neonatal rat brain. Experientia, 44: 457–459.

Hohmann, C.F., Hamon, R., Batshaw, M.L. and Coyle, J.T. (1988) Transient postnatal elevation of serotonin levels in mouse neocortex. Brain Res., 471: 163–166.

Hohmann, C.F., Richardson, C., Pitts, E. and Berger-Sweeney, J. (2000) Neonatal 5,7-DHT lesions cause sex-specific changes in mouse cortical morphogenesis. Neural Plast., 7: 213–232.

Hohmann, C.F., Walker, E.M., Boylan, C.B. and Blue, M.E. (2007) Neonatal serotonin depletion alters behavioral responses to spatial change and novelty. Brain Res., 1139: 163–177.

Holmes, A. (2008) Genetic variation in cortico-amygdala serotonin function and risk for stress-related disease. Neurosci. Biobehav. Rev., 32: 1293–1314.

Horvath, K. and Perman, J.A. (2002) Autistic disorder and gastrointestinal disease. Curr. Opin. Pediatr., 14: 583–587.

Hu, V.W., Frank, B.C., Heine, S., Lee, N.H. and Quackenbush, J. (2006) Gene expression profiling of lymphoblastoid cell lines from monozygotic twins discordant in severity of autism reveals differential regulation of neurologically relevant genes. BMC Genomics, 7: 118.

Huang, Z.J., Kirkwood, A., Pizzorusso, T., Porciatti, V., Morales, B., Bear, M.F., Maffei, L. and Tonegawa, S. (1999) BDNF regulates the maturation of inhibition and the critical period of plasticity in mouse visual cortex. Cell, 98: 739–755.

Huttenlocher, P.R. (1979) Synaptic density in human frontal cortex – developmental changes and effects of aging. Brain Res., 163: 195–205.

Huttenlocher, P.R. (1984) Synapse elimination and plasticity in developing human cerebral cortex. Am. J. Ment Defic., 88: 488–496.

Huttenlocher, P.R. (1990) Morphometric study of human cerebral cortex development. Neuropsychologia, 28: 517–527.

Huttenlocher, P.R. and Dabholkar, A.S. (1997) Regional differences in synaptogenesis in human cerebral cortex. J. Comp. Neurol., 387: 167–178.

Huttenlocher, P.R., de Courten, C., Garey, L.J. and Van der Loos, H. (1982) Synaptogenesis in human visual cortex–evidence for synapse elimination during normal development. Neurosci. Lett., 33: 247–252.

Idzko, M., Panther, E., Stratz, C., Muller, T., Bayer, H., Zissel, G., Durk, T., Sorichter, S., Di Virgilio, F., Geissler, M., Fiebich, B., Herouy, Y., Elsner, P., Norgauer, J. and Ferrari, D. (2004) The serotoninergic receptors of human dendritic cells: identification and coupling to cytokine release. J. Immunol., 172: 6011–6019.

Ingram, J.L., Stodgell, C.J., Hyman, S.L., Figlewicz, D.A., Weitkamp, L.R. and Rodier, P.M. (2000) Discovery of allelic variants of HOXA1 and HOXB1: genetic susceptibility to autism spectrum disorders. Teratology, 62: 393–405.

Iwasato, T., Erzurumlu, R.S., Huerta, P.T., Chen, D.F., Sasaoka, T., Ulupinar, E. and Tonegawa, S. (1997) NMDA receptor-dependent refinement of somatotopic maps. Neuron, 19: 1201–1210.

Jacobson, S. and Trojanowski, J.Q. (1977) Prefrontal granular cortex of the rhesus monkey. I. Intrahemispheric cortical afferents. Brain Res., 132: 209–233.

Jamain, S., Radyushkin, K., Hammerschmidt, K., Granon, S., Boretius, S., Varoqueaux, F., Ramanantsoa, N., Gallego, J., Ronnenberg, A., Winter, D., Frahm, J., Fischer, J., Bourgeron, T., Ehrenreich, H. and Brose, N. (2008) Reduced social interaction and ultrasonic communication in a mouse model of monogenic heritable autism. Proc. Natl. Acad. Sci. USA, 105: 1710–1715.

Janusonis, S., Gluncic, V. and Rakic, P. (2004) Early serotonergic projections to Cajal-Retzius cells: relevance for cortical development. J. Neurosci., 24: 1652–1659.

Janusonis, S., Anderson, G.M., Shifrovich, I. and Rakic, P. (2006) Ontogeny of brain and blood serotonin levels in 5-HT receptor knockout mice: potential relevance to the neurobiology of autism. J. Neurochem., 99: 1019–1031.

Jeanmonod, D., Rice, F.L. and Van der Loos, H. (1981) Mouse somatosensory cortex: alterations in the barrelfield following receptor injury at different early postnatal ages. Neuroscience, 6: 1503–1535.

Johansen-Berg, H. and Walsh, V. (2001) Cognitive neuroscience: early learning centres. Curr. Biol., 11: R613–R615.

Jonsson, G. and Kasamatsu, T. (1983) Maturation of monoamine neurotransmitters and receptors in cat occipital cortex during postnatal critical period. Exp. Brain Res., 50: 449–458.

Jyonouchi, H., Geng, L., Ruby, A., Reddy, C. and Zimmerman-Bier, B. (2005) Evaluation of an association between gastrointestinal symptoms and cytokine production against common dietary proteins in children with autism spectrum disorders. J. Pediatr., 146: 605–610.

Jyonouchi, H., Geng, L., Cushing-Ruby, A. and Quraishi, H. (2008) Impact of innate immunity in a subset of children with autism spectrum disorders: a case control study. J. Neuroinflammation, 5: 52.

Kahne, D., Tudorica, A., Borella, A., Shapiro, L., Johnstone, F., Huang, W. and Whitaker-Azmitia, P.M. (2002) Behavioral and magnetic resonance spectroscopic studies in the rat hyperserotonemic model of autism. Physiol. Behav., 75: 403–410.

Katoh, N., Soga, F., Nara, T., Tamagawa-Mineoka, R., Nin, M., Kotani, H., Masuda, K. and Kishimoto, S. (2006) Effect of serotonin on the differentiation of human monocytes into dendritic cells. Clin. Exp. Immunol., 146: 354–361.

Katoh-Semba, R., Wakako, R., Komori, T., Shigemi, H., Miyazaki, N., Ito, H., Kumagai, T., Tsuzuki, M., Shigemi, K., Yoshida, F. and Nakayama, A. (2007) Age-related changes in BDNF protein levels in human serum: differences between autism cases and normal controls. Intl. J. Dev. Neurosci., 25: 367–372.

Keller, F. and Persico, A.M. (2003) The neurobiological context of autism. Mol. Neurobiol., 28: 1–22.

Kemper, T.L. and Bauman, M. (1998) Neuropathology of infantile autism. J. Neuropathol. Exp. Neurol., 57: 645–652.

Kerschensteiner, M., Gallmeier, E., Behrens, L., Leal, V.V., Misgeld, T., Klinkert, W.E., Kolbeck, R., Hoppe, E., Oropeza-Wekerle, R.L., Bartke, I., Stadelmann, C., Lassmann, H., Wekerle, H. and Hohlfeld, R. (1999) Activated human T cells, B cells, and monocytes produce brain-derived neurotrophic factor *in vitro* and in inflammatory brain lesions: a neuroprotective role of inflammation? J. Exp. Med., 189: 865–870.

Kerschensteiner, M., Stadelmann, C., Dechant, G., Wekerle, H. and Hohlfeld, R. (2003) Neurotrophic cross-talk between the nervous and immune systems: implications for neurological diseases. Ann. Neurol., 53: 292–304.

Khan, I.A. and Thomas, P. (2004) Aroclor 1254 inhibits tryptophan hydroxylase activity in rat brain. Arch. Toxicol., 78: 316–320.

Kiba, T. (2006) Relationships between the autonomic nervous system, humoral factors and immune functions in the intestine. Digestion, 74: 215–227.

Killackey, H.P. (1982) Development and plasticity of somatosensory cortex. Neurosci. Res. Program Bull., 20: 507–513.

Killackey, H.P., R.W., R. and Bennett-Clarke, C.A. (1995) The formation of a cortical somatotopic map. Trends Neurosci., 18: 402–407.

Kirkwood, A. (2000) Serotonergic control of developmental plasticity. Proc. Natl. Acad. Sci., 97: 1951–1952.

Kirkwood, A., Lee, H.K. and Bear, M.F. (1995) Co-regulation of long-term potentiation and experience-dependent synaptic plasticity in visual cortex by age and experience. Nature, 375: 328–331.

Knaus, T.A., Silver, A.M., Lindgren, K.A., Hadjikhani, N. and Tager-Flusberg, H. (2008) fMRI activation during a language task in adolescents with ASD. J. Intl. Neuropsychol. Soc., 14: 967–979.

Koene, R.A. and Hasselmo, M.E. (2005) An integrate-and-fire model of prefrontal cortex neuronal activity during performance of goal-directed decision making. Cereb. Cortex, 15: 1964–1981.

Koishi, S., Yamamoto, K., Matsumoto, H., Koishi, S., Enseki, Y., Oya, A., Asakura, A., Aoki, Y., Atsumi, M., Iga, T., Inomata, J., Inoko, H., Sasaki, T., Nanba, E., Kato, N., Ishii, T. and Yamazaki, K. (2006) Serotonin transporter gene promoter polymorphism and autism: a family-based genetic association study in Japanese population. Brain Dev., 28: 257–260.

Kojic, L., Dyck, R.H., Gu, Q., Douglas, R.M., Matsubara, J. and Cynader, M.S. (2000) Columnar distribution of serotonin-dependent plasticity within kitten striate cortex. Proc. Natl. Acad. Sci. USA, 97: 1841–1844.

Kojic, L., Gu, Q., Douglas, R.M. and Cynader, M.S. (2001) Laminar distribution of cholinergic- and serotonergic-dependent plasticity within kitten visual cortex. Brain Res. Dev. Brain Res., 126: 157–162.

Korvatska, E., Van de Water, J., Anders, T.F. and Gershwin, M.E. (2002) Genetic and immunologic considerations in autism. Neurobiol. Dis., 9: 107–125.

Kostovic, I. and Jovanov-Milosevic, N. (2006) The development of cerebral connections during the first 20-45 weeks' gestation. Semin. Fetal Neonatal Med., 11: 415–422.

Kostovic, I. and Judas, M. (2002) Correlation between the sequential ingrowth of afferents and transient patterns of cortical lamination in preterm infants. Anat. Rec., 267: 1–6.

Krause, I., He, X.S., Gershwin, M.E. and Shoenfeld, Y. (2002) Brief report: immune factors in autism: a critical review. J. Autism Dev. Disord., 32: 337–345.

Kuemerle, B., Gulden, F., Cherosky, N., Williams, E. and Herrup, K. (2007) The mouse Engrailed genes: a window into autism. Behav. Brain Res., 176: 121–132.

Kuhl, P.K., Coffey-Corina, S., Padden, D. and Dawson, G. (2005) Links between social and linguistic processing of speech in preschool children with autism: behavioral and electrophysiological measures. Dev. Sci., 8: F1–F12.

Kuipers, S.D. and Bramham, C.R. (2006) Brain-derived neurotrophic factor mechanisms and function in adult synaptic plasticity: new insights and implications for therapy. Curr. Opin. Drug Discov. Devel., 9: 580–586.

Kushnir-Sukhov, N.M., Gilfillan, A.M., Coleman, J.W., Brown, J.M., Bruening, S., Toth, M. and Metcalfe, D.D. (2006) 5-hydroxytryptamine induces mast cell adhesion and migration. J. Immunol., 177: 6422–6432.

Lam, K.S., Aman, M.G. and Arnold, L.E. (2006) Neurochemical correlates of autistic disorder: a review of the literature. Res. Dev. Disabil., 27: 254–289.

Landa, R. (2007) Early communication development and intervention for children with autism. Ment. Retard Dev. Disabil. Res. Rev., 13: 16–25.

Landa, R.J. (2008) Diagnosis of autism spectrum disorders in the first 3 years of life. Nat. Clin. Pract. Neurol., 4: 138–147.

Landa, R.J., Holman, K.C. and Garrett-Mayer, E. (2007) Social and communication development in toddlers with early and later diagnosis of autism spectrum disorders. Arch. Gen. Psychiatry, 64: 853–864.

Lauder, J.M., Sze, P.Y. and Krebs, H. (1981) Maternal influences on tryptophan hydroxylase activity in embryonic rat brain. Dev. Neurosci., 4: 291–295.

Lauder, J.M., Wallace, J.A., Krebs, H., Petrusz, P. and McCarthy, K. (1982) *In vivo* and *in vitro* development of serotonergic neurons. Brain Res. Bull., 9: 605–625.

Laurent, A., Goaillard, J.M., Cases, O., Lebrand, C., Gaspar, P. and Ropert, N. (2002) Activity-dependent presynaptic effect of serotonin 1B receptors on the somatosensory thalamocortical transmission in neonatal mice. J. Neurosci., 22: 886–900.

Lavdas, A.A., Blue, M.E., Lincoln, J. and Parnavelas, J.G. (1997) Serotonin promotes the differentiation of glutamate neurons in organotypic slice cultures of the developing cerebral cortex. J. Neurosci., 17: 7872–7880.

Lein, E.S., Hohn, A. and Shatz, C.J. (2000) Dynamic regulation of BDNF and NT-3 expression during visual system development. J. Comp. Neurol., 420: 1–18.

Levite, M. (2008) Neurotransmitters activate T-cells and elicit crucial functions via neurotransmitter receptors. Curr. Opin. Pharmacol., 8: 460–471.

Levitt, J.G., Blanton, R.E., Smalley, S., Thompson, P.M., Guthrie, D., McCracken, J.T., Sadoun, T., Heinichen, L. and Toga, A.W. (2003) Cortical sulcal maps in autism. Cereb. Cortex, 13: 728–735.

Li, Y. (2007) Sensory signal transduction in the vagal primary afferent neurons. Curr. Med. Chem., 14: 2554–2563.

Licinio, J., Alvarado, I. and Wong, M.L. (2002) Autoimmunity in autism. Mol. Psychiatry, 7: 329.

Lidov, H.G. and Molliver, M.E. (1982) Immunohistochemical study of the development of serotonergic neurons in the rat CNS. Brain Res. Bull., 9: 559–604.

Lidow, M.S. and Rakic, P. (1992) Scheduling of monoaminergic neurotransmitter receptor expression in the primate neocortex during postnatal development. Cereb. Cortex, 2: 401–416.

Lieb, K., Biersack, L., Waschbisch, A., Orlikowski, S., Akundi, R.S., Candelario-Jalil, E., Hull, M. and Fiebich, B.L. (2005) Serotonin via 5-HT7 receptors activates p38 mitogen-activated protein kinase and protein kinase C epsilon resulting in interleukin-6 synthesis in human U373 MG astrocytoma cells. J. Neurochem., 93: 549–559.

Lieske, V., Bennett-Clarke, C.A. and Rhoades, R.W. (1999) Effects of serotonin on neurite outgrowth from thalamic neurons *in vitro*. Neuroscience, 90: 967–974.

Lipkin, W.I. and Hornig, M. (2003) Microbiology and immunology of autism spectrum disorders. Novartis Found Symp., 251: 129–143, discussion 144–128, 281–197.

Lord, C., Cook, E.H., Leventhal, B.L. and Amaral, D.G. (2000) Autism spectrum disorders. Neuron, 28: 355–363.

Lotto, B., Upton, L., Price, D.J. and Gaspar, P. (1999) Serotonin receptor activation enhances neurite outgrowth of thalamic neurones in rodents. Neurosci. Lett., 269: 87–90.

Lowry, C.A., Hollis, J.H., de Vries, A., Pan, B., Brunet, L.R., Hunt, J.R., Paton, J.F., van Kampen, E., Knight, D.M., Evans, A.K., Rook, G.A. and Lightman, S.L. (2007) Identification of an immune-responsive mesolimbocortical serotonergic system: potential role in regulation of emotional behavior. Neuroscience, 146: 756–772.

Lyons, W.E., Mamounas, L.A., Ricaurte, G.A., Coppola, V., Reid, S.W., Bora, S.H., Wihler, C., Koliatsos, V.E. and Tessarollo, L. (1999) Brain-derived neurotrophic factor-deficient mice develop aggressiveness and hyperphagia in conjunction with brain serotonergic abnormalities. Proc. Natl. Acad. Sci. USA, 96: 15239–15244.

Makkonen, I., Riikonen, R., Kokki, H., Airaksinen, M.M. and Kuikka, J.T. (2008) Serotonin and dopamine transporter binding in children with autism determined by SPECT. Dev. Med. Child Neurol., 50: 593–597.

Mamounas, L.A., Blue, M.E., Siuciak, J.A. and Altar, C.A. (1995) Brain-derived neurotrophic factor promotes the survival and sprouting of serotonergic axons in rat brain. J. Neurosci., 15: 7929–7939.

Mansour-Robaey, S., Mechawar, N., Radja, F., Beaulieu, C. and Descarries, L. (1998) Quantified distribution of serotonin transporter and receptors during the postnatal development of the rat barrel field cortex. Brain Res. Dev. Brain Res., 107: 159–163.

Marcusson, J., Oreland, L. and Winblad, B. (1984a) Effect of age on human brain serotonin (S-1) binding sites. J. Neurochem., 43: 1699–1705.

Marcusson, J.O., Morgan, D.G., Winblad, B. and Finch, C.E. (1984b) Serotonin-2 binding sites in human frontal cortex and hippocampus. Selective loss of S-2A sites with age. Brain Res., 311: 51–56.

Marcusson, J.A., Cederbrant, K. and Gunnarsson, L.G. (2000) Serotonin production in lymphocytes and mercury intolerance. Toxicol. In Vitro, 14: 133–137.

Marin-Padilla, M. (1970) Prenatal and early postnatal ontogenesis of the human motor cortex: a golgi study. I. The sequential development of the cortical layers. Brain Res., 23: 167–183.

Marin-Padilla, M. (1978) Dual origin of the mammalian neocortex and evolution of the cortical plate. Anat. Embryol. (Berl.), 152: 109–126.

Mariussen, E. and Fonnum, F. (2001) The effect of polychlorinated biphenyls on the high affinity uptake of the neurotransmitters, dopamine, serotonin, glutamate and GABA, into rat brain synaptosomes. Toxicology, 159: 11–21.

Marty, S., Carroll, P., Cellerino, A., Castren, E., Staiger, V., Thoenen, H. and Lindholm, D. (1996) Brain-derived neurotrophic factor promotes the differentiation of various hippocampal nonpyramidal neurons, including Cajal-Retzius cells, in organotypic slice cultures. J. Neurosci., 16: 675–687.

Mathews, K., Dalley, J.W., Mathews, C., Tsai, T.H. and Robbins, T.W. (2001) Periodic maternal separation of neonatal rats produces region and gender specific effects on biogenic amine content in adult brain. Synapse, 40: 1–10.

Matsumoto, M., Higuchi, K., Togashi, H., Koseki, H., Yamagichi, T., Kanno, M. and Yoshioka, M. (2005) early postnatal stress alters the 5-HTergic modulation to emotional stress at post-adolescent periods of rat. Hippocampus, 15: 775–781.

Mattson, M.P., Maudsley, S. and Martin, B. (2004) BDNF and 5-HT: a dynamic duo in age-related neuronal plasticity and neurodegenerative disorders. Trends Neurosci., 27: 589–594.

Maya Vetencourt, J.F., Sale, A., Viegi, A., Baroncelli, L., De Pasquale, R., O'Leary, O.F., Castren, E. and Maffei, L. (2008) The antidepressant fluoxetine restores plasticity in the adult visual cortex. Science, 320: 385–388.

McAllister, A.K. (2001) Neurotrophins and neuronal differentiation in the central nervous system. Cell Mol. Life Sci., 58: 1054–1060.

McAllister, A.K. (2002) Neurotrophins and cortical development. Results Probl. Cell Differ, 39: 89–112.

McCauley, J.L., Olson, L.M., Dowd, M., Amin, T., Steele, A., Blakely, R.D., Folstein, S.E., Haines, J.L. and Sutcliffe, J.S. (2004) Linkage and association analysis at the serotonin transporter (SLC6A4) locus in a rigid-compulsive subset of autism. Am. J. Med. Genet. B Neuropsychiatr. Genet., 127: 104–112.

McNamara, I.M., Borella, A.W., Bialowas, L.A. and Whitaker-Azmitia, P.M. (2008) Further studies in the developmental hyperserotonemia model (DHS) of autism: social, behavioral and peptide changes. Brain Res., 1189: 203–214.

Meaney, J.M., Diorio, J., Francis, D., Weaver, S., Yau, J., Chapman, K. and Seckl, J.R. (2000) postnatal handling increases the expression of cAMP-inducible transcription factors in the rat hippocampus: the effects of thyroid hormones and serotonin. J. Neurosci., 20: 3926–3935.

Menard, G., Turmel, V. and Bissonnette, E.Y. (2007) Serotonin modulates the cytokine network in the lung: involvement of prostaglandin E2. Clin. Exp. Immunol., 150: 340–348.

Meredith, E.J., Chamba, A., Holder, M.J., Barnes, N.M. and Gordon, J. (2005) Close encounters of the monoamine kind: immune cells betray their nervous disposition. Immunology, 115: 289–295.

Minshew, N.J. and Williams, D.L. (2007) The new neurobiology of autism: cortex, connectivity, and neuronal organization. Arch. Neurol., 64: 945–950.

Miyata, M., Ito, M., Sasajima, T., Ohira, H. and Kasukawa, R. (2001) Effect of a serotonin receptor antagonist on interleukin-6-induced pulmonary hypertension in rats. Chest, 119: 554–561.

Miyazaki, K., Narita, N., Sakuta, R., Miyahara, T., Naruse, H., Okado, N. and Narita, M. (2004) Serum neurotrophin concentrations in autism and mental retardation: a pilot study. Brain Dev., 26: 292–295.

Miyazaki, K., Narita, N. and Narita, M. (2005) Maternal administration of thalidomide or valproic acid causes abnormal

serotonergic neurons in the offspring: implication for pathogenesis of autism. Intl. J. Dev. Neurosci., 23: 287–297.
Molloy, C.A., Morrow, A.L., Meinzen-Derr, J., Schleifer, K., Dienger, K., Manning-Courtney, P., Altaye, M. and Wills-Karp, M. (2006) Elevated cytokine levels in children with autism spectrum disorder. J. Neuroimmunol., 172: 198–205.
Morrow, L.E., McClellan, J.L., Klir, J.J. and Kluger, M.J. (1996) The CNS site of glucocorticoid negative feedback during LPS- and psychological stress-induced fevers. Am. J. Physiol., 271: R732–R737.
Morse, D.C., Seegal, R.F., Borsch, K.O. and Brouwer, A. (1996) Long-term alterations in regional brain serotonin metabolism following maternal polychlorinated biphenyl exposure in the rat. Neurotoxicology, 17: 631–638.
Mostofsky, S.H., Burgess, M.P. and Gidley Larson, J.C. (2007) Increased motor cortex white matter volume predicts motor impairment in autism. Brain, 130: 2117–2122.
Mower, G.D. (1991) Comparison of serotonin 5-HT1 receptors and innervation in the visual cortex of normal and dark-reared cats. J. Comp. Neurol., 312: 223–230.
Moy, S.S. and Nadler, J.J. (2008) Advances in behavioral genetics: mouse models of autism. Mol. Psychiatry, 13: 4–26.
Moy, S.S., Nadler, J.J., Young, N.B., Nonneman, R.J., Grossman, A.W., Murphy, D.L., D'Ercole, A.J., Crawley, J.N., Magnuson, T.R. and Lauder, J.M. (2009) Social approach in genetically-engineered mouse lines relevant to autism. Genes Brain Behav, 8: 129–142.
Mrzljak, L., Uylings, H.B., Kostovic, I. and van Eden, C.G. (1992) Prenatal development of neurons in the human prefrontal cortex. II. A quantitative Golgi study. J. Comp. Neurol., 316: 485–496.
Muhle, R., Trentacoste, S.V. and Rapin, I. (2004) The genetics of autism. Pediatrics, 113: e472–e486.
Muller, R.A. (2008) From loci to networks and back again: anomalies in the study of autism. Ann. N.Y. Acad. Sci., 1145: 300–315.
Muller, R.A., Kleinhans, N., Kemmotsu, N., Pierce, K. and Courchesne, E. (2003) Abnormal variability and distribution of functional maps in autism: an fMRI study of visuomotor learning. Am. J. Psychiatry, 160: 1847–1862.
Murawski, N.J., Brown, K.L. and Stanton, M.E. (2009) Interstimulus interval (ISI) discrimination of the conditioned eyeblink response in a rodent model of autism. Behav. Brain Res., 196: 297–303.
Murcia, C.L., Gulden, F. and Herrup, K. (2005) A question of balance: a proposal for new mouse models of autism. Intl. J. Dev. Neurosci., 23: 265–275.
Murphy, D.G., Daly, E., Schmitz, N., Toal, F., Murphy, K., Curran, S., Erlandsson, K., Eersels, J., Kerwin, R., Ell, P. and Travis, M. (2006) Cortical serotonin 5-HT2A receptor binding and social communication in adults with Asperger's syndrome: an in vivo SPECT study. Am. J. Psychiatry, 163: 934–936.
Murphy, D.L. and Lesch, K.P. (2008) Targeting the murine serotonin transporter: insights into human neurobiology. Nat. Rev. Neurosci., 9: 85–96.
Murphy, D.L., Uhl, G.R., Holmes, A., Ren-Patterson, R., Hall, F.S., Sora, I., Detera-Wadleigh, S. and Lesch, K.P. (2003) Experimental gene interaction studies with SERT mutant mice as models for human polygenic and epistatic traits and disorders. Genes Brain Behav., 2: 350–364.
Murrin, L.C., Sanders, J.D. and Bylund, D.B. (2007) Comparison of the maturation of the adrenergic and serotonergic neurotransmitter systems in the brain: implications for differential drug effects on juveniles and adults. Biochem. Pharmacol., 73: 1225–1236.
Muzik, O., Janisse, J., Ager, J., Shen, C., Chugani, D.C. and Chugani, H.T. (1999) A mathematical model for the analysis of cross-sectional brain glucose metabolism data in children. Prog. Neuropsychopharmacol. Biol. Psychiatry, 23: 589–600.
Nassenstein, C., Braun, A., Nockher, W.A. and Renz, H. (2005) Neurotrophin effects on eosinophils in allergic inflammation. Curr. Allergy Asthma Rep., 5: 204–211.
Nelson, P.G., Kuddo, T., Song, E.Y., Dambrosia, J.M., Kohler, S., Satyanarayana, G., Vandunk, C., Grether, J.K. and Nelson, K.B. (2006) Selected neurotrophins, neuropeptides, and cytokines: developmental trajectory and concentrations in neonatal blood of children with autism or Down syndrome. Intl. J. Dev. Neurosci., 24: 73–80.
Newschaffer, C.J., Croen, L.A., Daniels, J., Giarelli, E., Grether, J.K., Levy, S.E., Mandell, D.S., Miller, L.A. and Pinto-Martin, J. (2007) The epidemiology of autism spectrum disorders. Annu. Rev. Public Health, 28: 235–258.
Ni, Y.G. and Miledi, R. (1997) Blockage of 5HT2C serotonin receptors by fluoxetine (Prozac). Proc. Natl. Acad. Sci. USA, 94: 2036–2040.
Nishijima, I., Yamagata, T., Spencer, C.M., Weeber, E.J., Alekseyenko, O., Sweatt, J.D., Momoi, M.Y., Ito, M., Armstrong, D.L., Nelson, D.L., Paylor, R. and Bradley, A. (2006) Secretin receptor-deficient mice exhibit impaired synaptic plasticity and social behavior. Hum. Mol. Genet., 15: 3241–3250.
Nishimura, K., Nakamura, K., Anitha, A., Yamada, K., Tsujii, M., Iwayama, Y., Hattori, E., Toyota, T., Takei, N., Miyachi, T., Iwata, Y., Suzuki, K., Matsuzaki, H., Kawai, M., Sekine, Y., Tsuchiya, K., Sugihara, G., Suda, S., Ouchi, Y., Sugiyama, T., Yoshikawa, T. and Mori, N. (2007) Genetic analyses of the brain-derived neurotrophic factor (BDNF) gene in autism. Biochem. Biophys. Res. Commun., 356: 200–206.
Nockher, W.A. and Renz, H. (2006) Neurotrophins and asthma: novel insight into neuroimmune interaction. J. Allergy Clin. Immunol., 117: 67–71.
Ogawa, T., Mikuni, M., Kuroda, Y., Muneoka, K., Mori, K.J. and Takahashi, K. (1994) Effects of the altered serotonergic signalling by neonatal treatment with 5,7-dihydroxytryptamine, ritanserin or clomipramine on the adrenocortical stress response and the glucocorticoid receptor binding in the hippocampus in adult rats. J. Neural Transm. Gen. Sect., 96: 113–123.
Oka, T., Oka, K. and Hori, T. (2001) Mechanisms and mediators of psychological stress-induced rise in core temperature. Psychosom. Med., 63: 476–486.
Ormsbee, H.S. III and Fondacaro, J.D. (1985) Action of serotonin on the gastrointestinal tract. Proc. Soc. Exp. Biol. Med., 178: 333–338.
Osterheld-Haas, M.C. and Hornung, J.P. (1996) Laminar development of the mouse barrel cortex: effects of neurotoxins against monoamines. Exp. Brain Res., 110: 183–195.
Osterling, J.A., Dawson, G. and Munson, J.A. (2002) Early recognition of 1-year-old infants with autism spectrum disorder versus mental retardation. Dev. Psychopathol., 14: 239–251.
Oudar, P., Caillard, L. and Fillion, G. (1989) *In vitro* effect of organic and inorganic mercury on the serotonergic system. Pharmacol. Toxicol., 65: 245–248.
Pardo, C.A. and Eberhart, C.G. (2007) The neurobiology of autism. Brain Pathol., 17: 434–447.

Pardo, C.A., Vargas, D.L. and Zimmerman, A.W. (2005) Immunity, neuroglia and neuroinflammation in autism. Intl. Rev. Psychiatry, 17: 485–495.

Paylor, R., Glaser, B., Mupo, A., Ataliotis, P., Spencer, C., Sobotka, A., Sparks, C., Choi, C.H., Oghalai, J., Curran, S., Murphy, K.C., Monks, S., Williams, N., O'Donovan, M.C., Owen, M.J., Scambler, P.J. and Lindsay, E. (2006) Tbx1 haploinsufficiency is linked to behavioral disorders in mice and humans: implications for 22q11 deletion syndrome. Proc. Natl. Acad. Sci. USA, 103: 7729–7734.

Persico, A.M., Baldi, A., Dell'Acqua, M.L., Moessner, R., Murphy, D.L., Lesch, K.P. and Keller, F. (2003) Reduced programmed cell death in brains of serotonin transporter knockout mice. Neuroreport, 14: 341–344.

Pessah, I.N., Seegal, R.F., Lein, P.J., LaSalle, J., Yee, B.K., Van De Water, J. and Berman, R.F. (2008) Immunologic and neurodevelopmental susceptibilities of autism. Neurotoxicology, 29: 532–545.

Peters, D.A. (1988) Effects of maternal stress during different gestational periods on the serotonergic system in adult rat offspring. Pharmacol. Biochem. Behav., 31: 839–843.

Pierce, K., Muller, R.A., Ambrose, J., Allen, G. and Courchesne, E. (2001) Face processing occurs outside the fusiform 'face area' in autism: evidence from functional MRI. Brain, 124: 2059–2073.

Polleux, F. and Lauder, J.M. (2004) Towards a developmental neurobiology of autism. Ment. Retard, Dev. Disabil. Res. Rev., 10: 303–317.

Posey, D.J. and McDougle, C.J. (2001) The pathophysiology and treatment of autism. Curr. Psychiatry Rep., 3: 101–108.

Rakic, P. (1988) Defects of neuronal migration and the pathogenesis of cortical malformations. Prog. Brain Res., 73: 15–37.

Rapin, I. (1991) Autistic children: diagnosis and clinical features. Pediatrics, 87: 751–760.

Raulf, M. and Konig, W. (1991) *In vitro* effects of polychlorinated biphenyls on human platelets. Immunology, 72: 287–291.

Rauschecker, J.P. (1991a) Mechanisms of visual cortical plasticity: Hebb synapses, NMDA receptors and beyond. Physiol. Rev., 71: 587–615.

Rauschecker, J.P. (1991b) Mechanisms of visual plasticity: Hebb synapses, NMDA receptors, and beyond. Physiol. Rev., 71: 587–615.

Rebsam, A., Seif, I. and Gaspar, P. (2002) Refinement of thalamocortical arbors and emergence of barrel domains in the primary somatosensory cortex: a study of normal and monoamine oxidase a knockout mice. J. Neurosci., 22: 8541–8552.

Rebsam, A., Seif, I. and Gaspar, P. (2005) Dissociating barrel development and lesion-induced plasticity in the mouse somatosensory cortex. J. Neurosci., 25: 706–710.

Redcay, E. and Courchesne, E. (2005) When is the brain enlarged in autism? A meta-analysis of all brain size reports. Biol. Psychiatry, 58: 1–9.

Reid, S.N. and Romano, C. (2001) Developmental and sensory-dependent changes of group II metabotropic glutamate receptors. J. Comp. Neurol., 429: 270–276.

Resnicov, A.G. and Nosenko, N.D. (1996) Early postnatal changes in sexual dimorphism of catecholamine and indolamine content in the brain of prenatally stressed rats. Neuroscience, 70: 547–551.

Rhoades, R.W., Bennett-Clarke, C.A., Shi, M.Y. and Mooney, R.D. (1994) Effects of 5-HT on thalamocortical synaptic transmission in the developing rat. J. Neurophysiol., 72: 2438–2450.

Rhoades, R.W., Chiaia, N.L., Lane, R.D. and Bennett-Clarke, C.A. (1998) Effect of activity blockade on changes in vibrissae-related patterns in the rat's primary somatosensory cortex induced by serotonin depletion. J. Comp. Neurol., 402: 276–283.

Ricceri, L., Markina, N., Valanzano, A., Fortuna, S., Cometa, M.F., Meneguz, A. and Calamandrei, G. (2003) Developmental exposure to chlorpyrifos alters reactivity to environmental and social cues in adolescent mice. Toxicol. Appl. Pharmacol., 191: 189–201.

Roceri, M., Cirulli, F., Pessina, C., Peretto, P., Racagni, G. and Riva, M.A. (2004) Postnatal repeated maternal deprivation produces age-dependent changes of brain-derived neurotrophic factor expression in selected rat brain regions. Biol. Psychiatry, 55: 708–714.

Rocha, M. and Sur, M. (1995) Rapid acquisition of dendritic spines by visual thalamic neurons after blockade of N-methyl-D-aspartate receptors. Proc. Natl. Acad. Sci. USA, 92: 8026–8030.

Rodier, P.M., Ingram, J.L., Tisdale, B., Nelson, S. and Romano, J. (1996) Embryological origin for autism: developmental anomalies of the cranial nerve motor nuclei. J. Comp. Neurol., 370: 247–261.

Rubenstein, J.L. and Merzenich, M.M. (2003) Model of autism: increased ratio of excitation/inhibition in key neural systems. Genes Brain Behav., 2: 255–267.

Russo, S., Kema, I.P., Fokkema, M.R., Boon, J.C., Willemse, P.H., de Vries, E.G., den Boer, J.A. and Korf, J. (2003) Tryptophan as a link between psychopathology and somatic states. Psychosom. Med., 65: 665–671.

Saitoh, O. and Courchesne, E. (1998) Magnetic resonance imaging study of the brain in autism. Psychiatry Clin. Neurosci., 52(Suppl.): S219–S222.

Salimi, K., Glantz, L.A., Hamer, R.M., German, T.T., Gilmore, J.H. and Jarskog, L.F. (2008) Regulation of complexin 1 and complexin 2 in the developing human prefrontal cortex. Synapse, 62: 273–282.

Salinchon, N., Gaspar, P., Upton, A.L., Picaud, S., Hanoun, N., Hamon, M., DeMaeyer, E., Murphy, D.L., Mossner, R., Lesch, K.P., Hen, R. and Seif, L. (2001) Excessive activation of serotonin (5-HT) 1B receptors disrupts the formation of sensory maps in monoamine oxidase and 5-HT transporter knockout mice. J. Neurosci., 21: 884–896.

Salt, T.E. and Hill, R.G. (1983) Neurotransmitter candidates of somatosensory primary afferent fibres. Neuroscience, 10: 1083–1103.

Santangelo, S.L. and Tsatsanis, K. (2005) What is known about autism: genes, brain, and behavior. Am. J. Pharmacogenomics, 5: 71–92.

Sari, Y. (2004) Serotonin1B receptors: from protein to physiological function and behavior. Neurosci. Biobehav. Rev., 28: 565–582.

Sari, Y., Miquel, M.C., Brisorgueil, M.J., Ruiz, G., Doucet, E., Hamon, M. and Verge, D. (1999) Cellular and subcellular localization of 5-hydroxytryptamine1B receptors in the rat central nervous system: immunocytochemical, autoradiographic and lesion studies. Neuroscience, 88: 899–915.

Schlaggar, B.L. and O'Leary, D.D. (1994) Early development of the somatotopic map and barrel patterning in rat somatosensory cortex. J. Comp. Neurol., 346: 80–96.

Schlaggar, B.L., Fox, K. and O'Leary, D.D. (1993) Postsynaptic control of plasticity in developing somatosensory cortex. Nature, 364: 623–626.

Schneider, M.L., Clarke, A.S., Kraemer, G.W., Roughton, E.C., Lubach, G.R., Rimm-Kaufman, S., Schmidt, D. and Ebert, M. (1998) Prenatal stress aters brain amine levels in primates. Dev. Psychopathol., 10: 427–440.

Schneider, T. and Przewlocki, R. (2005) Behavioral alterations in rats prenatally exposed to valproic acid: animal model of autism. Neuropsychopharmacology, 30: 80–89.

Schrocksnadel, K., Wirleitner, B., Winkler, C. and Fuchs, D. (2006a) Monitoring tryptophan metabolism in chronic immune activation. Clin. Chim. Acta, 364: 82–90.

Schroecksnadel, K., Winkler, C., Duftner, C., Wirleitner, B., Schirmer, M. and Fuchs, D. (2006b) Tryptophan degradation increases with stage in patients with rheumatoid arthritis. Clin. Rheumatol., 25: 334–337.

Schulte-Herbruggen, O., Hellweg, R., Chourbaji, S., Ridder, S., Brandwein, C., Gass, P. and Hortnagl, H. (2007) Differential regulation of neurotrophins and serotonergic function in mice with genetically reduced glucocorticoid receptor expression. Exp. Neurol., 204: 307–316.

Scott, M.M., Wylie, C.J., Lerch, J.K., Murphy, R., Lobur, K., Herlitze, S., Jiang, W., Conlon, R.A., Strowbridge, B.W. and Deneris, E.S. (2005) A genetic approach to access serotonin neurons for *in vivo* and *in vitro* studies. Proc. Natl. Acad. Sci. USA, 102: 16472–16477.

Seckl, J.R. and Meaney, M.J. (2004) Glucocorticoid programming. Ann. N. Y. Acad. Sci., 1032: 63–84.

Seegal, R.F., Brosch, K.O. and Bush, B. (1986) Regional alterations in serotonin metabolism induced by oral exposure of rats to polychlorinated biphenyls. Neurotoxicology, 7: 155–165.

Serajee, F.J., Nabi, R., Zhong, H. and Huq, M. (2004) Polymorphisms in xenobiotic metabolism genes and autism. J. Child Neurol., 19: 413–417.

Shenoy, K.V., Kaufman, J., McGrann, J.V. and Shaw, G.L. (1993) Learning by selection in the trion model of cortical organization. Cereb. Cortex, 3: 239–248.

Sillito, A.M., Salt, T.E. and Kemp, J.A. (1985) Modulatory and inhibitory processes in the visual cortex. Vision Res., 25: 375–381.

Singer, W. and Rauschecker, J.P. (1982) Central core control of developmental plasticity in the kitten visual cortex: II. Electrical activation of mesencephalic and diencephalic projections. Exp. Brain Res., 47: 223–233.

Siri, B., Quoy, M., Delord, B., Cessac, B. and Berry, H. (2007) Effects of Hebbian learning on the dynamics and structure of random networks with inhibitory and excitatory neurons. J. Physiol. Paris, 101: 136–148.

Slotkin, T.A. and Seidler, F.J. (2007) Prenatal chlorpyrifos exposure elicits presynaptic serotonergic and dopaminergic hyperactivity at adolescence: critical periods for regional and sex-selective effects. Reprod. Toxicol., 23: 421–427.

Soga, F., Katoh, N., Inoue, T. and Kishimoto, S. (2007) Serotonin activates human monocytes and prevents apoptosis. J. Invest. Dermatol., 127: 1947–1955.

Sohrabji, F. and Lewis, D.K. (2006) Estrogen-BDNF interactions: implications for neurodegenerative diseases. Front. Neuroendocrinol., 27: 404–414.

Sowell, E.R., Thompson, P.M., Tessner, K.D. and Toga, A.W. (2001) Mapping continued brain growth and gray matter density reduction in dorsal frontal cortex: Inverse relationships during postadolescent brain maturation. J. Neurosci., 21: 8819–8829.

Spalletta, G., Bossu, P., Ciaramella, A., Bria, P., Caltagirone, C. and Robinson, R.G. (2006) The etiology of poststroke depression: a review of the literature and a new hypothesis involving inflammatory cytokines. Mol. Psychiatry, 11: 984–991.

Sparks, B., Friedman, S., Shaw, D., Aylward, E., Echelard, D., Artru, A., Maravilla, K., Giedd, J., Munson, J., G, D. and SR., D. (2002) Brain structural abnormalities in young children with autism spectrum disorder. Neurology, 59: 184–192.

Stankovski, L., Alvarez, C., Ouimet, T., Vitalis, T., El-Hachimi, K.H., Price, D., Deneris, E., Gaspar, P. and Cases, O. (2007) Developmental cell death is enhanced in the cerebral cortex of mice lacking the brain vesicular monoamine transporter. J. Neurosci., 27: 1315–1324.

Stanton, M.E., Peloso, E., Brown, K.L. and Rodier, P. (2007) Discrimination learning and reversal of the conditioned eyeblink reflex in a rodent model of autism. Behav. Brain Res., 176: 133–140.

Stokes, P.E. and Holtz, A. (1997) Fluoxetine tenth anniversary update: the progress continues. Clin. Ther., 19: 1135–1250.

Sutcliffe, J.S., Delahanty, R.J., Prasad, H.C., McCauley, J.L., Han, Q., Jiang, L., Li, C., Folstein, S.E. and Blakely, R.D. (2005) Allelic heterogeneity at the serotonin transporter locus (SLC6A4) confers susceptibility to autism and rigid-compulsive behaviors. Am. J. Hum. Genet., 77: 265–279.

Szapacs, M.E., Mathews, T.A., Tessarollo, L., Ernest Lyons, W., Mamounas, L.A. and Andrews, A.M. (2004) Exploring the relationship between serotonin and brain-derived neurotrophic factor: analysis of BDNF protein and extraneuronal 5-HT in mice with reduced serotonin transporter or BDNF expression. J. Neurosci. Methods, 140: 81–92.

Tabakman, R., Lecht, S., Sephanova, S., Arien-Zakay, H. and Lazarovici, P. (2004) Interactions between the cells of the immune and nervous system: neurotrophins as neuroprotection mediators in CNS injury. Prog. Brain Res., 146: 387–401.

Tache, Y., Yang, H. and Yoneda, M. (1993) Vagal regulation of gastric function involves thyrotropin-releasing hormone in the medullary raphe nuclei and dorsal vagal complex. Digestion, 54: 65–72.

Taha, S.A. and Stryker, M.P. (2005) Molecular substrates of plasticity in the developing visual cortex. Prog. Brain Res., 147: 103–114.

Tang, S.W., Chu, E., Hui, T., Helmeste, D. and Law, C. (2008) Influence of exercise on serum brain-derived neurotrophic factor concentrations in healthy human subjects. Neurosci. Lett., 431: 62–65.

Tecott, L.H., Logue, S.F., Wehner, J.M. and Kauer, J.A. (1998) Perturbed dentate gyrus function in serotonin 5-HT2C receptor mutant mice. Proc. Natl. Acad. Sci. USA, 95: 15026–15031.

Thase, M.E. and Denko, T. (2008) Pharmacotherapy of mood disorders. Annu. Rev. Clin. Psychol., 4: 53–91.

Timofeeva, O.A., Roegge, C.S., Seidler, F.J., Slotkin, T.A. and Levin, E.D. (2008) Persistent cognitive alterations in rats after early postnatal exposure to low doses of the organophosphate pesticide, diazinon. Neurotoxicol. Teratol., 30: 38–45.

Tischfield, M.A., Bosley, T.M., Salih, M.A., Alorainy, I.A., Sener, E.C., Nester, M.J., Oystreck, D.T., Chan, W.M., Andrews, C., Erickson, R.P. and Engle, E.C. (2005) Homozygous HOXA1 mutations disrupt human brainstem, inner ear, cardiovascular and cognitive development. Nat. Genet., 37: 1035–1037.

Tommerdahl, M., Tannan, V., Holden, J.K. and Baranek, G.T. (2008) Absence of stimulus-driven synchronization effects on sensory perception in autism: Evidence for local underconnectivity? Behav. Brain Funct., 4: 19.

Torres, A.R. (2003) Is fever suppression involved in the etiology of autism and neurodevelopmental disorders? BMC Pediatr., 3: 9.

Trajkovska, V., Marcussen, A.B., Vinberg, M., Hartvig, P., Aznar, S. and Knudsen, G.M. (2007) Measurements of brain-derived neurotrophic factor: methodological aspects and demographical data. Brain Res. Bull., 73: 143–149.

Trask, C.L. and Kosofsky, B.E. (2000) Developmental considerations of neurotoxic exposures. Neurol. Clin., 18: 541–562.

Trottier, G., Srivastava, L. and Walker, C.D. (1999) Etiology of infantile autism: a review of recent advances in genetic and neurobiological research. J. Psychiatry Neurosci., 24: 103–115.

Turlejski, K., Djavadian, R.L. and Kossut, M. (1997) Neonatal serotonin depletion modifies development but not plasticity in rat barrel cortex. Neuroreport, 8: 1823–1828.

Vaidya, V.A., Marek, G.J., Aghajanian, G.K. and Duman, R.S. (1997) 5-HT2A receptor-mediated regulation of brain-derived neurotrophic factor mRNA in the hippocampus and the neocortex. J. Neurosci., 17: 2785–2795.

Vargas, D.L., Nascimbene, C., Krishnan, C., Zimmerman, A.W. and Pardo, C.A. (2005) Neuroglial activation and neuroinflammation in the brain of patients with autism. Ann. Neurol., 57: 67–81.

Vazquez, D.M., Lopez, J.F., Van Hoers, H., Watson, S.J. and Levine, S. (2000) Maternal deprivation regulates serotonin 1A and 2A receptors in the infant rat. Brain Res., 855: 76–82.

Veenstra-VanderWeele, J., Kim, S.J., Lord, C., Courchesne, R., Akshoomoff, N., Leventhal, B.L., Courchesne, E. and Cook, E.H. Jr. (2002) Transmission disequilibrium studies of the serotonin 5-HT2A receptor gene (HTR2A) in autism. Am. J. Med. Genet., 114: 277–283.

Venerosi, A., Calamandrei, G. and Ricceri, L. (2006) A social recognition test for female mice reveals behavioral effects of developmental chlorpyrifos exposure. Neurotoxicol. Teratol., 28: 466–471.

Verney, C., Lebrand, C. and Gaspar, P. (2002) Changing distribution of monoaminergic markers in the developing human cerebral cortex with special emphasis on the serotonin transporter. Anat. Rec., 267: 87–93.

Vitalis, T., Cases, O., Callebert, J., Launay, J.M., Price, D.J., Seif, I. and Gaspar, P. (1998) Effects of monoamine oxidase A inhibition on barrel formation in the mouse somatosensory cortex: determination of a sensitive developmental period. J. Comp. Neurol., 393: 169–184.

Vitalis, T., Cases, O., Gillies, K., Hanoun, N., Hamon, M., Seif, I., Gaspar, P., Kind, P. and Price, D.J. (2002) Interactions between TrkB signaling and serotonin excess in the developing murine somatosensory cortex: a role in tangential and radial organization of thalamocortical axons. J. Neurosci., 22: 4987–5000.

Vitalis, T., Cases, O., Passemard, S., Callebert, J. and Parnavelas, J.G. (2007) Embryonic depletion of serotonin affects cortical development. Eur. J. Neurosci., 26: 331–344.

Waage-Baudet, H., Lauder, J.M., Dehart, D.B., Kluckman, K., Hiller, S., Tint, G.S. and Sulik, K.K. (2003) Abnormal serotonergic development in a mouse model for the Smith-Lemli-Opitz syndrome: implications for autism. Intl. J. Dev. Neurosci., 21: 451–459.

Wakefield, A.J. (2002) The gut–brain axis in childhood developmental disorders. J. Pediatr. Gastroenterol. Nutr., 34(Suppl. 1): S14–S17.

Wallace, J.A., Petrusz, P. and Lauder, J.M. (1982) Serotonin immunocytochemistry in the adult and developing rat brain: methodological and pharmacological considerations. Brain Res. Bull., 9: 117–129.

Wang, Y., Gu, Q. and Cynader, M.S. (1997) Blockade of serotonin-2C receptors by mesulergine reduces ocular dominance plasticity in kitten visual cortex. Exp. Brain Res., 114: 321–328.

Wassink, T.H. and Piven, J. (2000) The molecular genetics of autism. Curr. Psychiatry Rep., 2: 170–175.

Wassink, T.H., Piven, J., Vieland, V.J., Huang, J., Swiderski, R.E., Pietila, J., Braun, T., Beck, G., Folstein, S.E., Haines, J.L. and Sheffield, V.C. (2001) Evidence supporting WNT2 as an autism susceptibility gene. Am. J. Med. Genet., 105: 406–413.

Wassink, T.H., Piven, J., Vieland, V.J., Pietila, J., Goedken, R.J., Folstein, S.E. and Sheffield, V.C. (2004) Examination of AVPR1a as an autism susceptibility gene. Mol. Psychiatry, 9: 968–972.

Wassink, T.H., Hazlett, H.C., Epping, E.A., Arndt, S., Dager, S.R., Schellenberg, G.D., Dawson, G. and Piven, J. (2007) Cerebral cortical gray matter overgrowth and functional variation of the serotonin transporter gene in autism. Arch. Gen. Psychiatry, 64: 709–717.

Weaver, I.C., Cervoni, N., Champagne, F.A., D'Alessio, A.C., Sharma, S., Seckl, J.R., Dymov, S., Szyf, M. and Meaney, M.J. (2004) Epigenetic programming by maternal behavior. Nat. Neurosci., 7: 847–854.

Whitaker-Azmitia, P.M. (2005) Behavioral and cellular consequences of increasing serotonergic activity during brain development: a role in autism? Intl. J. Dev. Neurosci., 23: 75–83.

Wiesel, T.N. (1999) Early explorations of the development and plasticity of the visual cortex: a personal view. J. Neurobiol., 41: 7–9.

Wills, S., Cabanlit, M., Bennett, J., Ashwood, P., Amaral, D. and Van de Water, J. (2007) Autoantibodies in autism spectrum disorders (ASD). Ann. N. Y. Acad. Sci., 1107: 79–91.

Wirleitner, B., Neurauter, G., Schrocksnadel, K., Frick, B. and Fuchs, D. (2003) Interferon-gamma-induced conversion of tryptophan: immunologic and neuropsychiatric aspects. Curr. Med. Chem., 10: 1581–1591.

Xu, B., Gottschalk, W., Chow, A., Wilson, R.I., Schnell, E., Zang, K., Wang, D., Nicoll, R.A., Lu, B. and Reichardt, L.F. (2000) The role of brain-derived neurotrophic factor receptors in the mature hippocampus: modulation of long-term potentiation through a presynaptic mechanism involving TrkB. J. Neurosci., 20: 6888–6897.

Xu, Y., Sari, Y. and Zhou, F.C. (2004) Selective serotonin reuptake inhibitor disrupts organization of thalamocortical somatosensory barrels during development. Brain Res. Dev. Brain Res., 150: 151–161.

Yirmiya, N., Pilowsky, T., Nemanov, L., Arbelle, S., Feinsilver, T., Fried, I. and Ebstein, R.P. (2001) Evidence for an association with the serotonin transporter promoter region polymorphism and autism. Am. J. Med. Genet., 105: 381–386.

Young-Davies, C.L., Bennett-Clarke, C.A., Lane, R.D. and Rhoades, R.W. (2000) Selective facilitation of the serotonin(1B) receptor causes disorganization of thalamic afferents and barrels in somatosensory cortex of rat. J. Comp. Neurol., 425: 130–138.

Yu, B., Becnel, J., Zerfaoui, M., Rohatgi, R., Boulares, A.H. and Nichols, C.D. (2008) Serotonin 5-hydroxytryptamine(2A) receptor activation suppresses tumor necrosis factor-alpha-induced inflammation with extraordinary potency. J. Pharmacol. Exp. Ther., 327: 316–323.

Zetterstrom, T.S., Pei, Q., Madhav, T.R., Coppell, A.L., Lewis, L. and Grahame-Smith, D.G. (1999) Manipulations of brain 5-HT levels affect gene expression for BDNF in rat brain. Neuropharmacology, 38: 1063–1073.

Zhang, J., Terreni, L., De Simoni, M.G. and Dunn, A.J. (2001) Peripheral interleukin-6 administration increases extracellular concentrations of serotonin and the evoked release of serotonin in the rat striatum. Neurochem. Intl., 38: 303–308.

Zhu, C.B., Blakely, R.D. and Hewlett, W.A. (2006) The proinflammatory cytokines interleukin-1beta and tumor necrosis factor-alpha activate serotonin transporters. Neuropsychopharmacology, 31: 2121–2131.

CHAPTER 4.9

Serotonin in Panic and Anxiety Disorders

Francisco S. Guimarães*[,1], Hélio Zangrossi Jr.[1], Cristina M. Del Ben[2] and Frederico G. Graeff[2]

[1]*Department of Pharmacology, Faculdade de Medicina de Ribeirão Preto, Universidade de São Paulo, Ribeirão Preto, Brazil*
[2]*Department of Neuroscience and Behavioral Sciences, Hospital das Clínicas da Faculdade de Medicina de Ribeirão Preto, Universidade de São Paulo, Ribeirão Preto, Brazil*

Abstract: Selective serotonin reuptake inhibitors (SSRIs) are largely used to treat several anxiety disorders. Nevertheless, despite being the focus of intensive research interest over the past 35 years, the precise role of serotonin (5-HT) in anxiety remains incompletely understood. More recently, it has been suggested that SSRI effects depend on facilitation of 5-HT-mediated neurotransmission in two parallel, longitudinally organized defensive systems that would be distinctly engaged by different threatening stimuli. 5-HT would play a dual role in these systems, facilitating anxiety but inhibiting panic. As reviewed in the present chapter, this proposal has received support from clinical and basic studies. Recent evidence has also suggested that disturbances in 5-HT-mediated neurotransmission in these systems could be involved in the genesis of pathological anxiety in humans, particularly panic disorder.

Keywords: serotonin, anxiety, panic, anxiolytics, antidepressants, periaqueductal gray, amygdala, hippocampus, raphe nuclei, prefrontal cortex.

Introduction

The first suggestions regarding the role of serotonin (5-HT) in anxiety were based on experimental evidence obtained in laboratory animals. Among the early animal models of anxiety, the so-called punishment or conflict test had shown the highest predictive value for the clinical response in anxious patients for the drugs available in the 1960s and 1970s, mainly barbiturates and benzodiazepines. Typically, drugs that relieved anxiety in humans were able to release behavior maintained by response-contingent presentation of a reinforcer (such as food or water) which was also suppressed by the delivery of an aversive stimulus (such as electric shock) following the same response. In this way, an approach–avoidance conflict is generated, the avoidance tendency being decreased by anxiolytics.

As a consequence, the observation that drugs that acted on 5-HT changed punished behavior has implicated this neurotransmitter in anxiety. Using the rat conflict test designed by Geller and Seifter (1960), Robichaud and Sledge (1969) showed that decreasing brain 5-HT by injecting the selective inhibitor of 5-HT synthesis para-chlorophenylalanine (PCPA) releases bar-pressing simultaneously rewarded by sweetened milk presentation and punished by foot-shock. In the same direction, Graeff and Schoenfeld (1970) reported that the non-selective 5-HT receptor antagonists methysergide and bromolysergic acid (BOL) increase key-pecking rates maintained by food presentation, concurrently suppressed by response-contingent electric shocks, whereas the receptor agonist α-methyltryptamine has the opposite effect. On the basis of these results, they suggested that tryptaminergic mechanisms in the brain mediate the response suppression determined by punishment.

Shortly thereafter, Wise *et al.* (1972) showed that the benzodiazepine anxiolytic oxazepam reduces 5-HT turnover in the rat midbrain, at the same dose that released punished responding in the Geller-Seifter conflict test. Since it was already known that midbrain raphe nuclei contained the cell bodies of 5-HT neurons that project to the forebrain and brainstem, they proposed that ascending 5-HT pathways would mediate the effects of punishment by acting on structures that suppress ongoing behavior localized in both the forebrain and in the periaqueductal gray matter (PAG) of the midbrain. Benzodiazepine anxiolytics would reduce anxiety by decreasing 5-HT release in these brain areas. Therefore, in the theoretical model designed by Wise *et al.* (1972), 5-HT is supposed to enhance anxiety by acting both in the forebrain and in the PAG (see Figure 1).

The first tenet of the above hypothesis was supported by the results obtained by Tye *et al.* (1977) with the

*Corresponding author
E-mail: fsguimar@fmrp.usp.br

Christian Müller & Barry Jacobs (Eds.)
Handbook of Behavioral Neurobiology of Serotonin
ISBN 978-0-12-374634-4

DOI: 10.1016/B978-0-12-374634-4.00037-X

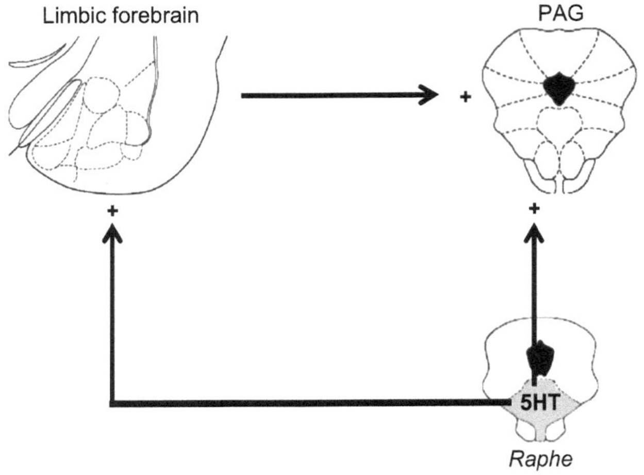

Figure 1 Classical view on the role of 5-HT in anxiety. 5-HT was proposed by Wise *et al.* (1972) to mediate the effects of punishment by acting on structures that suppress ongoing behavior localized in both the forebrain and in the periaqueductal gray matter (PAG) of the midbrain. Benzodiazepine anxiolytics would reduce anxiety by decreasing 5-HT release in these brain areas. +, facilitation. Drawings based on *The Rat Brain in Stereotaxic Coordinates – The New Coronal Set*, 5th edn, by George Paxinos and Charles Watson, 2005.

neurotoxin 5,7-dihydroxytryptamine (5,7-DHT) , which selectively damages 5-HT neurons. Their results showed that microinjection of 5,7-DHT into the ventromedial tegmentum of the rat midbrain causes 70 percent depletion of cortical 5-HT, indicating a major lesion of ascending 5-HT pathways, and prevents the acquisition of response suppression induced by foot-shock in a modified Geller-Seifter procedure. In the same vein, Schoenfeld (1976) reported that the hallucinogens lysergic acid diethylamide (LSD) and mescaline release punished licking in the rat, and suggested that this effect may be due to decreased activity of ascending serotonergic neurons. As complementary evidence, Graeff and Silveira Filho (1978) found that electrical stimulation of the median raphe nucleus (MRN) inhibits ongoing lever-pressing behavior maintained by a variable-interval schedule of water presentation. In addition, such brain stimulation induces defecation, urination, piloerection, teeth clattering and exophthalmos, mimicking the effect of a conditioned aversive stimulus in the conditioned suppression and conditioned emotional response procedures (Estes and Skinner, 1941; Millenson and Leslie, 1974). Moreover, pre-treatment with PCPA was shown to reduce the effect of MRN electrical stimulation, suggesting a 5-HT mediation of the response suppression. Since MRN 5-HT neurons project mainly to the dorsal hippocampus (Azmitia and Segal, 1978), the latter results agree with Gray's (1982) proposal that anxiety is due to activation of the septo-hippocampal system. In addition to the septo-hippocampal system, 5-HT may also act in the amygdala to mediate the effects of punishment. For instance, microinjection of 5-HT antagonists into the basolateral amygdala was been shown to release water licking suppressed by electric shock punishment (Petersen and Scheel-Krüger, 1984). These results add to a large body of evidence implicating the amygdala in the learning and expression of conditioned fear and anxiety (Davis, 1992; LeDoux, 1993).

In summary, these initial studies clearly suggested that facilitation of 5-HT-mediated neurotransmission facilitates anxiety. As will be seen below, a large body of clinical and animal studies, although clearly implicating 5-HT in the neurobiology of anxiety, indicates that this classical view can not be sustained any longer.

5-HT related drugs and the treatment of anxiety disorders

The clinical link between serotonin and anxiety could be traced back to the key observation by Klein and Flink (1962) that chronic treatment with imipramine has a beneficial effect in panic patients. Although it was subsequently recognized that imipramine is also effective in generalized anxiety disorder (GAD) (Kahn *et al.*, 1986), this initial finding was critical for the recognition of panic as a distinct anxiety disorder. Imipramine, however, is able to block the reuptake of both noradrenaline and serotonin, and the involvement of the latter neurotransmitter in its anxiolytic effects was not immediately recognized. The link between 5-HT and clinical anxiolytic effects became clear during the late 1970s–1980s, with the clinical introduction of buspirone, a partial 5-HT_{1A} receptor agonist effective in GAD patients (Goldberg and Finnerty, 1979), and the selective serotonin reuptake inhibitors (SSRIs). The latter group, being usually well tolerated and having a broad spectrum anxiolytic efficacy, is nowadays considered as firsT-line pharmacological treatment for several anxiety disorders, including GAD, panic (PD), social anxiety disorder (SAD) and post-traumatic stress disorder (PTSD) (Baldwin *et al.*, 2005).

Although other 5-HT-related compounds have also been tested in patients with anxiety disorders, results are far from encouraging. For example, whereas ritanserin, a 5-$HT_{2A/2C}$ receptor antagonist, seems to be effective in GAD patients (Ceulemans *et al.*, 1985), it has no effect or even aggravates PD (den Boer and Westenberg, 1990). Mixed results have also been observed with 5-HT_3 receptor antagonists (Lecrubier *et al.*, 1999; Romach *et al.*, 1998; Hewlett *et al.*, 2003) and 5-HT_{1A} agonists (van Vliet *et al.*, 1996). Open studies have also suggested that d-fenfluramine, a 5-HT releaser, could improve panic patients resistant to traditional treatments (Solyom, 1994;

Hetem, 1996). This drug was, however, withdrawn from the market due to side-effects.

How do SSRIs and buspirone work?

The acute pharmacological effects of SSRIs and buspirone are well-known. However, it needs at least 2–3 weeks before the beneficial effects of these compounds become evident in patients with anxiety disorders. Several proposals have been made to try to explain the delay in therapeutic effects. Electrophysiological studies indicate that repeated administration of SSRIs or buspirone can attenuate the function of inhibitory 5-HT_{1A} autosomic receptors, without affecting 5-HT_{1A} postsynaptic function (Blier and Abbott, 2001; Blier and Ward, 2003). As a consequence, there would be an increased 5-HT-mediated neurotransmission in projection areas. Although imipramine does not seem to share this effect with SSRIs, it can sensitize 5-HT_{1A} postsynaptic receptors located in areas such as the hippocampus and amygdala (Gordon and Hen, 2004). Since buspirone is also a partial 5-HT_{1A} agonist, stimulation of these receptors located postsynaptically in limbic areas has been proposed to mediate at least part of the anxiolytic effects of these drugs (Gordon and Hen, 2004; Guimarães *et al.*, 2008). Postsynaptic changes in 5-HT_2 receptors could also be involved. Long-term treatment with several classes of antidepressants desensitizes 5-HT_{2C} receptors in limbic areas (Millan, 2006a). These receptors have been proposed to exert a negative action over mood that would be opposed by postsynaptic 5-HT_{1A} receptors (Millan, 2006b). Therefore, antidepressants would cause anxiolytic effects by both down-regulating 5-HT_{2C} and facilitating effects mediated by postsynaptic 5-HT_{1A} receptors (Millan, 2006a, b). Since 5-HT_{2C} receptors may be constitutively active, whereas stimulation of postsynaptic 5-HT_{1A} receptors requires spontaneous release of 5-HT, a general decrease in 5-HT-mediated neurotransmission would shift the balance towards 5-HT_{2C}-mediated anxiogenic effects Milan (2006b). In addition to 5-HT_{2C}, down-regulation of 5-HT_{2A}-mediated neurotransmission by repeated treatment with buspirone or antidepressants could also be involved in the anxiolytic effects of these drugs (Millan, 2003; Weisstaub *et al.*, 2006).

A strong argument relating 5-HT to the anxiolytic effects of SSRIs came from controlled clinical trials using tryptophan depletion to decrease brain 5-HT level. These studies showed that the therapeutic effect of SSRIs is reverted by 5-HT depletion in PD and SAD (Bell *et al.*, 2002; Argyropoulos *et al.*, 2004). The 5-HT precursor l-5-hydroxytryptophan (5-HTP), on the other hand, was able to attenuate panic symptoms induced by CCK4 or CO_2 in PD patients (Schruers *et al.*, 2002; Maron *et al.*, 2004a), whereas tryptophan depletion and metergoline facilitated these symptoms (Ben-Zion *et al.*, 1999; Miller *et al.*, 2000). Results regarding OCD are not so clear. Only mood, but not OCD, symptoms were sensitive to tryptophan depletion. This suggests that the therapeutic effects of SSRIs over the latter cluster of symptoms do not depend on short-term availability of 5-HT (Kulz *et al.*, 2007).

In addition to these effects, long-term treatment with SSRIs, probably by facilitating 5-HT-mediated neurotransmission, has been shown to induce several plastic changes and increase neurogenesis in the hippocampus (Santarelli *et al.*, 2003).Chronic treatment with these compounds facilitates neurogenesis in the dentate gyrus, probably by activation of hippocampal 5-HT_{1A} receptors. Hippocampal neurogenesis inhibition has been shown to prevent anxiolytic effects of these drugs in mice (Santarelli *et al.*, 2003). In addition to neurogenesis, hippocampal levels of the brain-derived neurotrophic factor (BDNF) have also been related to stress-induced behavioral changes. Studies investigating the effects of SSRIs on these levels have produced conflicting results (Greenwood *et al.*, 2007). However, the behavioral effects of chronically administered SSRIs were shown to depend on BDNF signaling in some studies (Saarelainen *et al.*, 2003). Despite these results, the acute reappearance of anxiety symptoms in SSRI-treated patients induced by tryptophan depletion (see above) indicates that caution must be exerted when considering these neuroplastic proposals as unitary explanations of the long-term effects of these drugs.

Are anxiety disorders caused by disturbances in 5-HT-mediated neurotransmission?

Although the therapeutic efficacy of 5-HT-interacting compounds clearly points to an involvement of this neurotransmitter in anxiety disorders (Gordon and Hen, 2004), the question remains of whether these disorders are caused by disturbances in 5-HT-mediated neurotransmission.

Impairment of 5-HT_{1A}-mediated neurotransmission in PD has been shown by several neuroendocrine studies (Lesch *et al.*, 1992; Broocks *et al.*, 2000; Mortimore and Anderson, 2000). These studies also suggest an increased 5-HT_{2C} sensitivity in PD and post-traumatic stress disorder (PTSD), but a decreased sensitivity in OCD. Since blunted responses to the 5-HT releaser d-fenfluramine have been found in patients with PD and OCD, the increased sensitivity of 5-HT_{2C} receptors in these disorders may be reflecting a compensatory up-regulation of 5-HT_{2C} receptors due to a decrease in 5-HT neurotransmission (Guimarães *et al.*, 2008).

Genetic association studies also suggest that gene variants of the 5-HT system may contribute to anxiety susceptibility, particularly in PD (for review, see Maron and Shlik, 2006). Neuroimaging studies have not only supported this proposal but also been helpful in elucidating genetic influences on these disorders (Hariri *et al.*, 2005; Iidaka *et al.*, 2005).

An inverse correlation has been found between 5-HT_{1A} binding and anxiety in healthy subjects and in patients with other neuropsychiatric disorders (Tauscher *et al.*, 2001; Savic *et al.*, 2004; Yasuno *et al.*, 2004). There is also a report of association between a single polymorphism in the regulatory region of the 5-HT_3 gene and greater reactivity in the amygdala and in the dorsal and medial PFC (Iidaka *et al.*, 2005). In patients with PD, fewer 5-HT_{1A} receptors in the forebrain and raphe nuclei have been reported (Neumeister *et al.*, 2004; Nash *et al.*, 2008). A reduction of these receptors in limbic areas, particularly the amygdala, has also been shown in patients with SAD (Lanzenberger *et al.*, 2007), but not in patients with PSTD or GAD (Bonne *et al.*, 2005). In these patients, successful treatment with SSRIs antagonized activity changes during symptom provocation in the insula, and in prefrontal and inferior frontal cortices, suggesting the involvement of 5-HT (Fernandez *et al.*, 2001; Hoehn-Saric *et al.*, 2004). This same treatment has been shown to attenuate the increased activation of the anterolateral orbitofrontal cortex, caudate nucleus, thalamus and temporal regions in OCD patients (Carey *et al.*, 2004).

In addition to 5-HT receptors, several studies have investigated changes in the serotonin transporter (5-HTT). An increased 5-HTT binding has been reported in patients with GAD (van der Wee *et al.*, 2008), but an inverse correlation between 5-HTT binding and anxiety has also been reported (Maron *et al.*, 2004b). The presence of one or two copies of the short allele of the 5-HTT promoter predicts greater amygdala neuronal activity in response to fearful and angry facial expressions, compared with individuals homozygous for the long allele, suggesting a bias toward increased reactivity to stressful life experiences (Hariri *et al.*, 2005). However, studies comparing the presence of the short allele in anxiety disorders, such as OCD or GAD, have produced mainly negative results (Hesse *et al.*, 2004; Maron *et al.*, 2004b). Nevertheless, SAD patients with one or two copies of the short allele in the promoter region of the 5-HTT gene show increased anxiety-related traits, state anxiety, and enhanced right amygdala responding to anxiety provocation (Furmark *et al.*, 2004). Studies with PD patients have not found an association between the presence of polymorphisms in the 5-HTT promoter gene and the disorder (Blaya *et al.*, 2007; Strug *et al.*, 2008). However, an association between a polymorphism in the serotonin transporter gene and PD has been reported (Strug *et al.*, 2008).

Another point to be considered in these studies is that although several results support the presence of an association between changes in 5-HTT expression and emotional processing in the amygdala, the functional consequences of these changes on 5-HT-mediated neurotransmission are still not clear. A decrease in the expression of 5-HTT would, theoretically, increase synaptic 5-HT. There is evidence, however, that this polymorphism is associated with blunted 5-HT function (Reist *et al.*, 2001). Moreover, it is probable that this genetic trait interferes with neuronal development, causing long-term effects that go far beyond a simple change in 5-HT availability (see bellow).

Recent studies have also focused on the brain-specific tryptophan hydroxylase-2 (TPH2) gene, the rate-limiting enzyme in 5-HT synthesis. Although no association between polymorphisms in this gene and PD was found (Maron and Shlik, 2006), some polymorphisms have been shown to modulate amygdala responsiveness to emotional arousal. In addition, there is an additive effect on emotional processing in individuals carrying the combination of the T variant of TPH2 with the short variant (l/l) of the serotonin transporter gene (5-HTT) (Canli *et al.*, 2008).

In conclusion, several results from the clinical literature indicate that anxiety disorders, particularly PD, are associated with disturbances of 5-HT-mediated neurotransmission. How these disturbances bring about the complex and sometimes distinct symptomatology found in these disorders is still unclear. Therefore, data from studies with laboratory animals could bring a valuable help.

The dual role of 5-HT in anxiety: the Deakin and Graeff theory

The complexity of the 5-HT system greatly contributes to the difficulty in understanding its role in anxiety. 5-HT can act on more than 14 receptor subtypes, located either pre- or postsynaptically, and in some cases in both locations. Moreover, their activation could produce opposite effects (Hoyer *et al.*, 2002). In addition to this intricate neurochemistry, the system also has a complex neuroanatomy. The widespread distribution of 5-HT fibers in the forebrain comes from two distinct raphe nuclei, the dorsal and median raphe nuclei (DRN and MRN, respectively). These nuclei innervate partially distinct structures (Pazos and Palacios, 1985; Barnes and Sharp, 1999; Hensler, 2006) and may present a high degree of heterogeneity (Clark *et al.*, 2006). For example, it has recently been shown that the projections to cortical and limbic structures arise from neurons located in a specific part of the caudal DRN that is particularly sensitive to stressful stimuli (for review, see Lowry *et al.*, 2005).

Studies with animal models of anxiety can potentially overcome some limitations of clinical investigations, and have been extensively used to investigate the role of 5-HT in anxiety (for comprehensive reviews, see Griebel, 1995; Blanchard *et al.*, 2003; Bourin and Hascoet, 2003; Millan, 2003; Olivier *et al.*, 2003; Prut and Belzung, 2003; Sanchez, 2003). Although the evidence reviewed so far using these models supports the hypothesis that 5-HT enhances anxiety by acting upon forebrain limbic structures, contrary results have been obtained in the PAG. Electrical stimulation of the dorsal PAG in laboratory animals induces defensive reactions, such as vigorous flight or defensive aggression (Hunsperger, 1956). These defense strategies are expressed in natural conditions when a predator is very close to or in direct contact with the prey (Blanchard *et al.*, 2001). Testifying to the aversive character of such stimulation, laboratory animals easily learn to switch off electrical stimulation of the dorsal PAG (Delgado *et al.*, 1954; Hunsperger, 1956). According to the above theoretical model, 5-HT is expected to facilitate such escape behavior. However, reported results with a procedure in which rats are trained to lever-press in order to decrease the intensity of dorsal PAG electrical stimulation suggest the opposite action. Indeed, using this decremental escape procedure, Kiser and coworkers showed that PCPA markedly increased the rate of bar-pressing (Kiser and Lebovitz, 1975; Kiser *et al.*, 1978a). Further results have shown that the precursor of 5-HT synthesis 5-hydroxytryptophan (5-HTP) reduces the same bar-pressing, and that the 5-HT reuptake inhibitor clomipramine has the same effect (Kiser *et al.*, 1978b). Finally, electrical stimulation of the dorsal raphe nucleus (DRN), which sends 5-HT nerve fibers to the PAG, has also been reported to depress responding (Kiser *et al.*, 1980). Together, these results show that manipulations that increase 5-HT activity in the dorsal PAG have an anti-aversive effect, whereas 5-HT depletion results in facilitation of escape from PAG electrical stimulation.

Further supporting an inhibitory function of 5-HT on aversion generated in the dorsal PAG, Schenberg and Graeff (1978) have shown that the 5-HT receptor antagonists cyproheptadine and methysergide decrease responding that switches off electrical stimulation of the dorsal PAG, whereas non-sedative doses of the benzodiazepine chlordiazepoxide facilitate switch-off responding (Schenberg and Graeff, 1978). Such opposing effects of a benzodiazepine anxiolytic and the 5-HT antagonist contrast with the similar anti-punishment effects of the two classes of drugs verified in conflict tests. Furthermore, using electrical stimulation of the dorsal PAG as a punishing stimulus, Morato de Carvalho and de Aguiar, (1981) have shown that, as expected, the anxiolytics chlordiazepoxide and pentobarbital released responding maintained by water reinforcement, suppressed by the response-contingent delivery of electrical stimuli in the dorsal PAG. In contrast, the 5-HT receptor antagonist cyproheptadine did not increase punished responding at doses that had been shown to markedly release behavior punished by foot-shock (Graeff, 1974), and methysergide was also ineffective.

To explore the function of 5-HT in the PAG more directly, rats bearing a chemitrode – thus allowing microinjection of drugs associated with electrical stimulation – implanted into the dorsal PAG were placed inside a shuttle box, and the intensity of a sinusoidal current was gradually increased until the rat ran towards the opposite compartment of the shuttle box. This response was made to switch off the brain stimulation. The same procedure was repeated three times to determine the basal aversive threshold. Soon after, a drug microinjection was administered and 10 minutes later the aversive threshold was determined again. The difference between the magnitude of the post-drug and that of the basal threshold measured the drug effect on aversion. In the first study with this method, conducted by Schütz *et al.* (1985), microinjection of 5-HT as well as of the 5-HT receptor agonist 5-methoxy-N,N-dimethyltryptamine (5-MeODMT) into the dorsal PAG raised the aversive threshold dose-dependently. Local pretreatment with the 5-HT receptor antagonists metergoline or ketanserin blocked the anti-aversive effect of 5-HT. Since ketanserin is a relatively selective 5-HT_{2A}-receptor antagonist, these results suggest that 5-HT inhibits aversion in the dorsal PAG by stimulating such 5-HT receptor subtype. This view has been further supported by results showing that the 5-$HT_{2A/2C}$ agonist 2,5-dimethoxy-4-iodophenyl-2-aminopropane (DOI) increases the aversive threshold (Nogueira and Graeff, 1995). The latter study has also shown that the 5-HT_{1A} agonists, 8-OH-DPAT and ipsapirone, have a similar effect, while the 5-HT_{2C} preferential agonist 1-(m-chlorophenyl) piperazine (mCPP) was ineffective. Using chemical instead of electrical stimulation, Marsden and coworkers showed that microinjection of 5-HT_{1A} agonists into the PAG attenuated running behavior induced by microinjection of an excitatory amino acid into the dorsal PAG; the selective 5-HT_{1A}-receptor blocker WAY-100635 antagonized this effect, whereas the preferential 5-HT_{2C}-receptor agonist mCPP facilitated running (Beckett *et al.*, 1992; Beckett and Marsden, 1997). Therefore, stimulation of either the 5-HT_{1A} or the 5-HT_{2A} receptor in the dorsal PAG seems to inhibit escape.

It is worth remarking that in the results reported by Schütz *et al.* (1985), microinjection of a selective inhibitor of 5-HT reuptake into the dorsal PAG not only potentiated the anti-aversive effect of the following administration of 5-HT, but also had an anti-aversive effect of its own. This implies the existence of 5-HT nerve fibers in the dorsal PAG that are physiologically regulating defense. This view has been strengthened by further experimental evidence showing that blockade of presynaptic 5-HT_{1B} receptors that inhibit 5-HT release with isamoltane, microinjected into the dorsal PAG, also has an anti-aversive effect, which is blocked by local pre-treatment with the 5-HT_2-receptor blockers ketanserin and ritanserin (Nogueira and Graeff, 1991).

Interestingly, administration of 5-HT receptor antagonists alone into the dorsal PAG has no effect on the aversive threshold (Schütz *et al.*, 1985; Nogueira and Graeff, 1991). This finding contrasts with the major aversive effects caused by compounds like bicuculline, which block γ-aminobutyric acid type A (GABAA) receptors in the dorsal PAG (Brandão *et al.*, 1982). It may be concluded that while GABAergic terminals tonically inhibit the neurons of the dPAG that control defensive behavior, serotonergic fibers seem to exert a phasic inhibition. As a consequence, it has been suggested that the modulatory action of 5-HT would only appear under conditions, such as stressful situations, that engage 5-HT systems (Deakin and Graeff, 1991).

Overall, the results reviewed so far indicate that 5-HT facilitates inhibitory avoidance by acting on forebrain structures such as the septo-hippocampal system and the amygdala, while inhibiting escape in the dorsal PAG. The amygdala and the PAG, plus the medial hypothalamus, constitute a set of interrelated structures, often referred to as the brain defense system (BDS), which controls defensive strategies and elaborates the accompanying emotional and motivational states. Although the three components of the BDS would work together to generate different kinds of defensive behavior, each structure is likely to exert specific functions. In this regard, Fanselow (1991) has suggested that the amygdala synthesizes the stimulus input from the environment and then signals to the PAG the degree of threat that this represents to the organism, whereas the PAG is in charge of selecting, organizing and executing the appropriate behavioral and neurovegetative defensive reactions.

On the basis of the systematic study of anti-predator defense in wild rats, Blanchard and Blanchard (1988) formulated the concept of defense levels. Three such levels have been identified as a function of the threat being potential (uncertain), distal or proximal, each evoking a peculiar defense reaction. Thus, when danger is uncertain, like in novel environments, rats perform cautious exploration aimed at risk assessment; when the predator is perceived at a safe distance, tense and attentive immobility (freezing) ensues; and when the predator is near or in actual contact with the rat, the animal rapidly flees, whenever possible, or otherwise threatens or even attacks the predator defensively. Comparative studies led to the conclusion that these basic defense strategies can be found in other animals, including non-mammalian species (Blanchard *et al.*, 2001). Attempts to relate each level of defense with critical brain structures and to particular emotions (e.g., Graeff, 1994) led to the suggestion that the septo-hippocampal system and the amygdala would be the key structures for risk-assessment behavior and inhibitory avoidance, the related emotion being anxiety. The septo-hippocampal system would provide the cognitive component of anxiety, while the affective component would be integrated in the amygdala (Gray and McNaughton, 2000). At the other extreme, the vigorous undirected flight elicited by proximal danger would be related to panic, the critical structure being the dorsal PAG. At the intermediate level of distal threat, well-directed escape related to fear would be elaborated by the medial hypothalamus. Concerning psychopathology, the first level of defense has been related to generalized anxiety disorder (GAD) and the third to panic disorder (PD), the second being implicated in specific phobias (Deakin and Graeff, 1991; Gray and McNaughton, 2000).

Both the amygdala and the PAG receive serotonergic input mainly from the DRN. The axons that project onto the amygdala follow the DRN–forebrain tract, while those that go to the PAG run via the DRN–periventricular tract (Azmitia and Segal, 1978). Most of the nerve fibers that originate in the DRN are thin and have small varicosities that make preferential contact with 5-HT_2 receptors (Mamounas *et al.*, 1991). Given this background, a working hypothesis has been conceived to reconcile the experimental evidence on the role of 5-HT in defense, reviewed above (Graeff, 1991). According to this view, activation of the DRN results in facilitation of defense strategies that are mainly integrated at the amygdala and other forebrain areas, such as risk assessment and inhibitory avoidance. At the same time, proximal defense reactions that are chiefly organized in the dorsal PAG, such as escape and fight, are inhibited. The adaptive function of this disposition would be to inhibit extreme defense patterns, like flight or fight, in situations where predatory threat is uncertain or far from the prey. In these circumstances, such reactions are inappropriate, because they enhance the probability of detection by the predator. Instead, more flexible, largely learned, responses are likely to lead to successful defense. This construct has been extended to psychopathology, relating inhibitory avoidance and risk assessment to GAD and proximal defense to PD (Deakin and Graeff, 1991, Figure 2).

Experimental tests of the Deakin and Graeff theory: clinical studies

Two experimental models of anxiety in humans have been used to test this theoretical model: the simulated public-speaking (SPS) and the skin conductance response (CSCR) tests (review of Graeff *et al.*, 2003). The SPS test consists, basically, in the preparation and performance of a speech in front of a videocamera. During the procedure, the participant can see his or her own image on a TV screen, and anxiety and other subjective measures are taken before, during and after the speech. Since fear of speaking is highly prevalent and generates anxiety in healthy persons irrespective of trait anxiety level (Palma *et al.*, 1994), it is supposed to be a species-specific fear. Pharmacological studies have shown that benzodiazepines decrease anxiety – measured by the anxiety factor of the Visual Analog Mood Scale (VAMS) – before and after the speech, but do not attenuate the increase of VAMS anxiety during speech preparation or performance (speaking fear). In contrast, and in agreement with the predictions of Deakin and Graeff's proposal, drugs that facilitate 5-HT-mediated neurotransmission decrease, whereas drugs that impair 5-HT function increase, speaking fear (Graeff *et al.*, 2003).

The CSCR test consists of the presentation, through earphones, of 10 neutral tones (habituation phase), followed by a neutral tone paired with loud white noise (acquisition phase) and then by the representation of 10 neutral tones (extinction phase). During the procedure, measures of skin conductance (level, SCL; response, SCR and spontaneous fluctuations, SF) are taken. Anxiolytic drugs, such as diazepam and the 5-HT_{1A} partial agonist buspirone, impaired conditioning, while anxiogenic drugs, such as the 5-HT agonist methyl-chlorphenylpiperidine (mCPP) and the 5-HT releaser d-fenfluramine, tend to facilitate conditioning (Graeff *et al.*, 2003).

The pharmacological profile of the above experimental human tests has some similarities with the pharmacological response observed in clinical practice in patients with the diagnoses of anxiety disorders. The anxiolytic drugs diazepam and buspirone are effective in the treatment of GAD (Baldwin and Polkinghorn, 2005), but not of PD (Bakker *et al.*, 2005). Both drugs had an anxiolytic effect in the CSCR test, without decreasing speaking fear in the SPS test.

Some antidepressants largely used for the treatment of anxiety disorders have also been tested in these models of human experimental anxiety. The acute administration

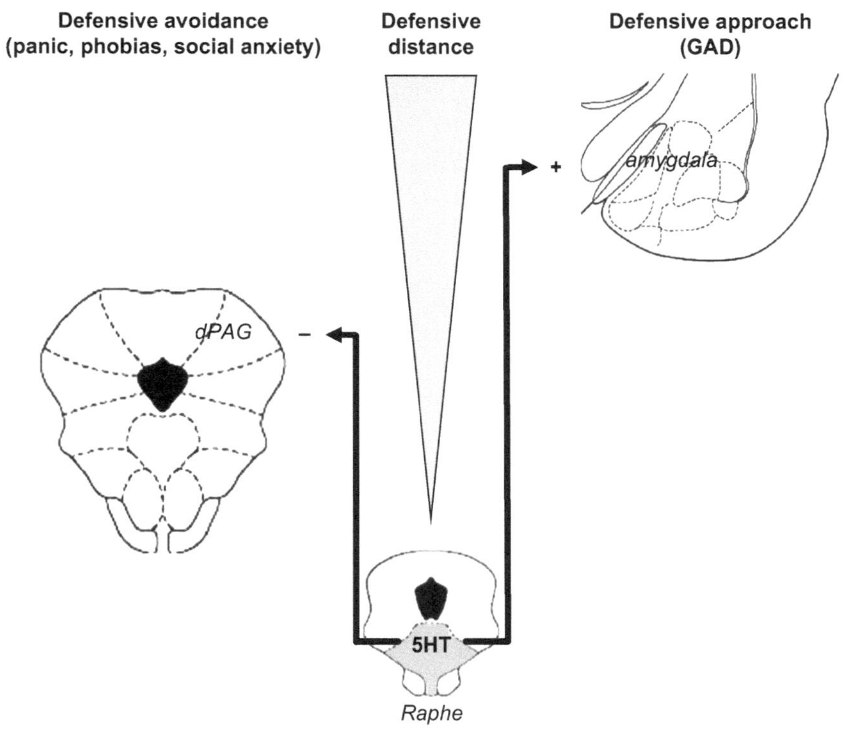

Figure 2 The dual role of 5-HT on defensive behaviors proposed by Deakin and Graeff (1991). By activating the dorsal raphe nucleus, threatening stimuli would facilitate defense strategies that are mainly integrated at the amygdala and other forebrain areas, such as risk assessment and inhibitory avoidance. At the same time, proximal defense reactions that are chiefly organized in the dorsal periaqueductal gray (dPAG), such as escape and fight, would be inhibited. In humans these strategies have been associated, respectively, with generalized anxiety (GAD) and with panic, phobias and social anxiety. +, facilitation; –: inhibition. Drawings modified from *The Rat Brain in Stereotaxic Coordinates – The New Coronal Set*, 5th edn, by George Paxinos and Charles Watson, 2005.

of clomipramine (Guimarães *et al.*, 1987) and nefazodone (Silva *et al.*, 2001) increased speaking fear in the SPS test, and this effect has been related to the worsening of anxiety symptoms frequently observed at the beginning of the treatment with antidepressants. It has been proposed that the fear-enhancing effect of a single dose of antidepressants in SPS is due to a lack of 5-HT inhibition of brain systems that generate panic attacks, likely to be localized in the dorsal PAG. Although some microdialysis studies in animals have shown increases in cortical extracellular 5-HT following acute SSRI administration (Moret and Bridley, 1996; David *et al.*, 2003; Felton *et al.*, 2003) others have shown a greater increase of extracellular 5-HT in the raphe nuclei than in the neocortex (Bel and Artigas, 1992). Therefore, it is possible that acute doses of antidepressants preferentially increase 5-HT concentration near the cell bodies of serotonergic neurons, reducing their firing rate due to the activation of somatodendritic 5-HT_{1A} autoreceptors (Gartside *et al.*, 1995), decreasing 5-HT release and lowering the postsynaptic concentration of 5-HT.

Taken together, the pharmacological data obtained so far with human experimental models are compatible with the hypothesis that 5-HT enhances anxiety, evaluated by the CSCR test, whereas it inhibits fear, accessed by the SPS test.

If the predictions derived from pharmacological tests discussed above are correct, it is expected that panic patients would perform differently from healthy volunteers in the SPS but not in the CSCR test, since the former would engage the neural substrates that would be implicated in the neurobiology of PD but the latter would not. It is important to highlight that the SPS test is not considered as a model of panic attack or a probe aimed at provoking panic attacks in susceptible individuals. The rationale of the relationship between the model and the anxiety disorder is that the public speaking would engage the brain mechanisms involved in the process of innate fear, which would be impaired in PD.

As expected, our studies have shown that panic patients' response to CSCR is similar to that of controls, except for the occurrence of more spontaneous fluctuations of skin conductance along the experimental session in panic patients – a finding that is consistent with their high baseline level of subjective anxiety. In the SPS test, panic patients showed higher levels of anxiety than normal controls along the whole session, but were less responsive to the speaking challenge (Del-Ben *et al.*, 2001; Garcia-Leal *et al.*, 2005; Parente *et al.*, 2005). The profile of panic patients during the SPS test resembles the effect of the non-selective 5-HT-receptor blocker metergoline in healthy volunteers, which increased anxiety before and after the speech, but not during the speech preparation or performance (Graeff *et al.*, 1985). These results support the suggestion that 5-HT deficiency, and therefore less inhibition of PAG, may be present in PD. Interestingly, the above SPS task did not increase cortisol secretion (Garcia-Leal *et al.*, 2005). Accordingly, neither panic attacks (reviewed in Graeff *et al.*, 2005) nor the electrical stimulation of the dorsal PAG of the rat (Schenberg *et al.*, 2008) seem to activate the hypothalamic–pituitary–adrenal axis.

Further evidence that panic patients process innate fear responses abnormally has come from a study carried out in our laboratory in patients with social anxiety disorder (SAD) submitted to the SPS test (M.C. Freitas, A. Santos Filho, F. Osório, S.R. Loureiro, C.M. Del-Ben, A.W. Zuardi, F.G. Graeff and J.A.S. Crippa, unpublished results). Although SAD and PD are different anxiety disorders they have some similarities, such as the response to the treatment with antidepressants that act on 5-HT function. Nevertheless, differently from panic patients, SAD patients have shown a greater increase in the fear induced by the SPS test, in comparison to healthy controls. Therefore, low sensitivity to the SPS test seems to be a specific feature of PD.

Experimental tests of the Deakin and Graeff theory in laboratory animals: the elevated T-maze

Since the initial findings showing that impairment of 5-HT-mediated neurotransmission causes anti-conflict effects (Robichaud and Sledge, 1969; Graeff and Schoenfeld, 1970), a large number of studies aimed at investigating the role of 5-HT in anxiety has been performed. The results, however, have been far from conclusive (Griebel, 1995; Belzung, 2001; Blanchard *et al.*, 2003; Bourin and Hascoet, 2003; Millan, 2003; Olivier *et al.*, 2003; Prut and Belzung, 2003; Sanchez, 2003). This fact is easily illustrated with results obtained in the elevated plus-maze (EPM) (Handley *et al.*, 1993; Griebel, 1995; for review, see Pinheiro *et al.*, 2007). Trying to explain these contradictory results, the British psychologist Sheila Handley proposed that the EPM engages different defense strategies, which could be influenced in opposing directions by 5-HT (Handley *et al.*, 1993). These strategies would include avoidance of the open arms (when the rat is in one of the enclosed arms) and escape from these same arms to enter a safer, enclosed arm. Based on this proposal, and trying to develop an experimental approach to test the predictions of the Deakin and Graeff (1991) theory, Graeff and co-workers developed the elevated T-maze (ETM) (Graeff *et al.*, 1993, 1998). The apparatus was built by closing the entrance of one of the enclosed arms of the elevated plus-maze. The test measures both inhibitory avoidance (the latency to withdraw from the enclosed arm) and one-way escape (the withdrawal latency from the open arm) in the same animal. Pharmacological results have shown that diazepam impairs inhibitory avoidance but does not affect one-way escape, whereas chronic treatment with imipramine and other antidepressants increases escape latency (Graeff *et al.*, 1993; Teixeira *et al.*, 2000; Poltronieri *et al.*, 2003). The latter parameter was enhanced by the panicogenic agent cholecystokinin (Zanoveli *et al.*, 2004). These results indicate that inhibitory avoidance and escape measurements in the ETM may be modeling two different anxiety disorders, respectively GAD and PD.

In the first test of the dual role of 5-HT in the defense hypothesis, it was shown that systemic administration of d-fenfluramine, a drug that selectively releases 5-HT from the thin serotonergic fibers that come mainly from the DRN, impairs escape, whilst tending to increase inhibitory avoidance (Graeff *et al.*, 1997). These results not only fulfill the predictions from the above hypothesis, but also correlate with the results described above with the same compound administered to healthy subjects submitted to the SPS and CSCR models. Moreover, they agree with clinical reports on antipanic effects of this drug in PD patients (Solyom, 1994; Hetem, 1996).

To test the dual 5-HT hypothesis more directly, a series of experiments has been conducted to explore the effects of intracerebral injection of 5-HT-acting compounds in rats submitted to the ETM (Table 1; Figure 3). Results employing microinjection of the 5-HT_{1A} receptor antagonist WAY-100635 and the 5-HT_{1A} agonist 8-OH-DPAT into the dorsal raphe nucleus have produced consistent results. Whereas the former drug, supposedly enhancing neuronal activity of 5-HT neurons, impaired escape and facilitated inhibitory avoidance, the latter did exactly the opposite (Sena *et al.*, 2003; Pobbe and Zangrossi, 2005).

As shown in Table 1 and Figure 3, results obtained with direct injections into projecting areas of 5-HT neurons have also produced results that are, in general, compatible with the proposal of a dual role of 5-HT (Deakin and Graeff, 1991). There are, though, some inconsistencies. Intra-dorsal PAG injection of the endogenous agonist 5-HT not only inhibited escape performance, as expected, but also caused an anxiogenic-like effect on inhibitory

Table 1 Effects in the elevated T-maze of 5-HT-related drugs injected into brain areas involved in defensive responses

Region	Drug	Avoidance	Escape	Reference(s)
Dorsal hippocampus	8-OHDPAT	+	0	dos Santos *et al.*, 2008
	WAY100635	0	0	
Lateral septum	8-OHDPAT	+	0	Viana *et al.*, 2008
	WAY100635	0	0	
BLAmyg	5-HT	+	0	M.A. Vicente, C.V.A.
	MK212	+	0	Strauss and H. Zangrossi
	SB242084	−	0	Jr, personal communication
	8-OHDPAT	−	−	
	DOI	−	−	
MeAmyg	8-OHDPAT	−	0	
	DOI	0	0	C.V.A. Strauss and H. Zangrossi Jr, personal communication
dPAG	5-HT	+	−	
	mCPP	+	−	de Paula Soares *et al.*,
	8-OHDPAT	−	−	2004; Zanoveli *et al.*, 2003
	DOI	0	−	
	WAY100635	0	0	
	Ketanserine	0	0	
	SDZ SER082	0	0	
vlPAG	5-HT	−	0	
	8-OHDPAT	−	0	de Paula Soares *et al.*,
	DOI	−	0	2008.
	WAY100635	−	0	
	Ketanserine	0	0	

+, facilitation; −, inhibition; 0, no effect.

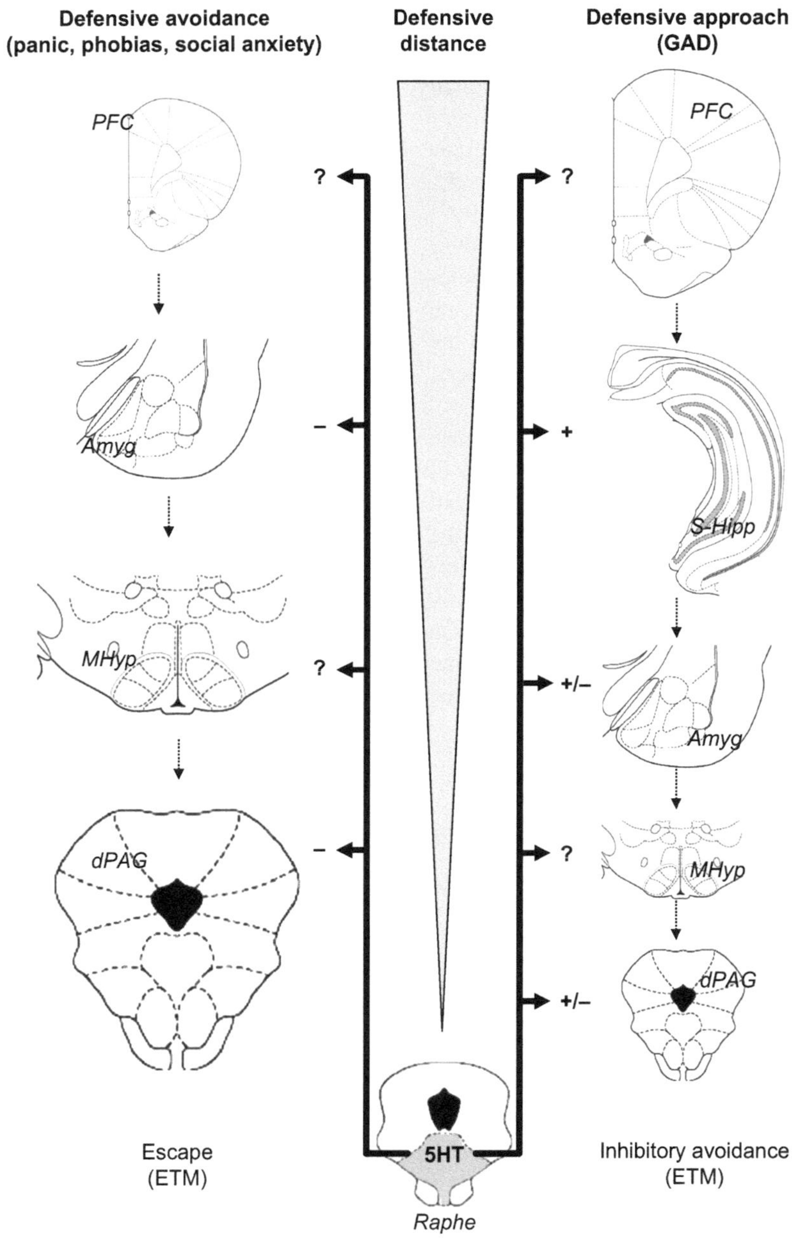

Figure 3 An alternative view of the proposal by Deakin and Graeff (1991). Instead of the rostro-caudal hierarchy originally suggested, 5-HT would play a dual role in two parallel, longitudinally organized systems (McNaughton and Corr, 2004). The approach defense system would be engaged mainly by approach-avoidance conflict. Its main neural representation would involve the septum-hippocampus and other forebrain structures. The avoidance defense system, on the other hand, would be activated when there is no tendency to approach the source of danger. In this case, the experienced emotion is fear, not anxiety. According to this proposal, fear or panic would be elicited depending on the defensive distance. Distal threat would activate forebrain structures and elicit fear, while proximal threat would activate the PAG and elicit panic. The figure also illustrates the results obtained after intracerebral injections of 5-HT agonists (see Table 1) into the dorsal periaqueductal gray (dPAG), medial hypothalamus (MHyp), septum-hippocampus (S-Hipp), amygdaloid nuclei (Amyg) or medial prefrontal cortex (PFC) on inhibitory avoidance and escape latencies of rats submitted to the elevated T-maze (ETM). +, facilitation; −, inhibition. ? indicates areas where the effects of these drugs have not yet been tested. Drawings modified from *The Rat Brain in Stereotaxic Coordinates – The New Coronal Set*, 5th edn, by George Paxinos and Charles Watson, 2005.

avoidance (Zanoveli *et al.*, 2003; de Paula Soares and Zangrossi, 2004). Other 5-HT receptor agonists, such as 8-OH-DPAT and DOI, similarly impaired escape, but with a differential effect on inhibitory avoidance – i.e., while the former compound was anxiolytic, the latter had no effect (Zanoveli *et al.*, 2003; de Paula Soares and Zangrossi, 2004). Intra-dorsal PAG administration of either 5-HT_{1A} (WAY-100635) or 5-$HT_{2A/2C}$ (ketanserin, SDZ SER082) receptor antagonists, was ineffective, corroborating previous evidence indicating that in this brain region 5-HT exerts a phasic rather than tonic regulatory influence on defensive behaviors.

Similar to the results obtained in the dorsal PAG, the role of different amygdaloid 5-HT receptor subtypes in the modulation of the two tasks generated by the elevated T-maze seems to be more complex than originally thought. Thus, in the basolateral nucleus (BLA), stimulation of 5-HT_{1A} or 5-HT_{2A} receptors by 8-OH-DPAT and DOI, respectively, impaired rather than facilitated inhibitory avoidance acquisition, indicating an anxiolytic effect. Also unexpectedly, these compounds equally inhibited escape behavior (C. Villela de Andrade Strauss and H. Zangrossi Jr, unpublished results). However, in agreement with Deakin and Graeff's proposal, local injection of 5-HT itself or of the 5-HT_{2C} receptor agonist MK212 caused an anxiogenic effect upon inhibitory avoidance, without changing escape performance. On the other hand, microinjection of the 5-HT_{2C} receptor antagonist SB242084 had the opposite effect on inhibitory avoidance, also without interfering with escape (M.A. Vicente and H. Zangrossi Jr, unpublished results). Since antidepressants have been shown to down-regulate 5-HT_{2C} after repeated administration, this finding may be related to their therapeutic effects in GAD patients.

Together, these results indicate that 5-HT, through the activation of BLA 5-HT_{2C} receptors, plays a tonic and selective role in the mediation of inhibitory avoidance. Considering that the BLA has been proposed to assign emotional salience to both rewarding and aversive stimuli, the contradictory findings found after direct injections of 5-HT_{1A}-, 5-HT_{2A}- and 5-HT_{2C}-related compounds may involve complex interference in both approach and avoidance behaviors (Lowry *et al.*, 2005).

Regarding the septo-hippocampal system, only studies assessing the involvement of 5-HT_{1A} receptors have been published using the ETM, the results being compatible with the proposal that 5-HT is mainly implicated in the regulation of anxiety. Accordingly, 8-OH-DPAT injection into either the lateral septum or the dorsal hippocampus has an anxiogenic effect upon inhibitory avoidance, without affecting escape. In the lateral septum, the effect of 8-OH-DPAT on anxiety was opposed to that caused by the benzodiazepine midazolam. Microinjection of WAY-100635 into either the septum or the hippocampus had no effect on the two elevated T-maze defensive tasks (dos Santos *et al.*, 2008; Viana *et al.*, 2008). However, several conflicting results suggest that the role of 5-HT in defensive responses that involve the septo-hippocampal system is more complex than originally believed. Although effects on different 5-HT receptor subtypes could help to explain these results (Gordon and Hen, 2004), most studies so far have targeted the dorsal hippocampus, an area that receives 5-HT fibers from the median rather than the dorsal raphe nucleus. The latter structure innervates mainly the ventral hippocampus, a region that has been particularly related to anxiety (Bannerman *et al.*, 2004).

As a final remark, results obtained with the administration of 5-HT-acting drugs in the ventrolateral column of the PAG point to the selective involvement of this area in anxiety modulation. In this respect, it is noteworthy that whereas electrical or chemical stimulation of the dorsal PAG evokes fight and escape responses accompanied by tachycardia and hypertension, as mentioned before, the same procedure in the ventrolateral PAG leads to behavioral inhibition (e.g., hyporeactive immobility), accompanied by bradycardia and hypotension (Carrive and Bandler, 1991). In agreement, intra-ventrolateral PAG injection of 5-HT, 8-OH-DPAT and DOI has been shown consistently to impair inhibitory avoidance acquisition, without interfering with one-way escape. Curiously, while both ketanserin and WAY-100635 were also ineffective in changing escape performance, the latter drug impaired avoidance acquisition – an effect that has been attributed to the drug interference with non-serotonergic mechanisms. Ketanserin did not affect inhibitory avoidance (de Paula Soares and Zangrossi, 2008).

Studies in the ETM have also been employed to unveil the mechanism of action of therapeutic drugs. As discussed above, several pieces of evidence suggest that the anti-panic effects of SSRIs depend on facilitation of 5-HT-mediated neurotransmission in the PAG. If this is the case, however, a general 5-HT_{2C} receptor down-regulation induced by chronic SSRIs treatment would be expected to facilitate panic symptoms. On the other hand, the several described isoforms of 5-HT_{2C} receptors may show distinct pharmacological characteristics and coupling mechanisms in, for example, the amygdala and the PAG (Millan, 2006a). In addition, other 5-HT receptors (5-HT_{2A}, 5-HT_{1A}) could play a more important role in the PAG. Recent results obtained in one of our laboratories (HZJ) support this view. Repeated treatment with imipramine, sertraline or fluoxetine enhanced the inhibitory effect of intra-dPAG injection of 5-HT_{1A} or 5-HT_2 agonists on escape responses either induced by local electrical stimulation or measured in the elevated T-maze (Zanoveli *et al.*, 2005; de Bortoli *et al.*, 2006). The same effect is also reported after long- but not short-term treatment with alprazolam, a high-potency benzodiazepine receptor agonist successfully used in the treatment of PD. Curiously, chronic treatment with alprazolam facilitates 5-HT_{1A}- or 5-HT_{2A}-mediated neurotransmission in the dPAG, without affecting the responsiveness of benzodiazepine receptors in the same brain area (de Bortoli *et al.*, 2008). Differently from SSRIs, buspirone failed to enhance the inhibitory effects of 5-HT_{1A} or 5-HT_2 agonists on escape responses mediated by the dPAG (Zanoveli *et al.*, 2005; de Bortoli *et al.*, 2006), which could explain its lack of therapeutic efficacy in PD.

Overall, the results obtained in the ETM, particularly those in the dorsal PAG and BLA, agree with a proposed dual role of 5-HT in defensive responses. They question, however, the rostro-caudal hierarchy originally suggested by Deakin and Graeff (1991) (Figure 2). More recently, McNaughton and Corr (2004) proposed that defensive responses are organized in two parallel, longitudinally organized systems. In this view, the 'approach defense system' would deal with anxiety and be engaged mainly by approach-avoidance conflict. Its main neural representation involves the septum-hippocampus and other forebrain structures. The 'avoidance defense system', on the other hand, would be activated when there is no tendency to approach the source of danger. In this case, the experienced emotion is fear, not anxiety (Figure 3). According to this proposal, fear or panic would be elicited depending on the defensive distance. Distal threat would activate forebrain structures and elicit fear, while proximal threat would activate the PAG and elicit panic. This proposal has recently received strong support from a study in healthy volunteers using functional magnetic resonance imaging (MRI) (Mobbs *et al.*, 2007), the results of which have shown that brain activity shifts from the prefrontal cortex (PFC) to the midbrain PAG as a virtual predator capable of inflicting pain grows nearer to the virtual prey. In addition, the activity of the dorsal raphe nucleus correlated with reported subjective degree of dread and decreased confidence of escape.

Genetic manipulations of the 5-HT system

Gene targeting manipulation developed in the last decade is a new and promising approach to investigate the biological vulnerability that underlies 'traits' of fear/anxiety (Gottesman and Gould, 2003; Holmes *et al.*, 2005; Lesch, 2005; Hariri and Holmes, 2006; Leonardo and Hen, 2006; Wrase *et al.*, 2006). Using this approach, several studies have investigated the 5-HT system. By focusing mainly on the deletion of genes coding for 5-HT receptors, 5-HT-related enzymes and the 5-HT transporter (5-HTT), these studies have produced a wealth of new evidence relating 5-HT to anxiety. They show that impairment of 5-HT_{1A} neurotransmission results in a state of chronic fearfulness and higher responsivity to stressful stimuli. For example, 5-HT_{1A}-knockout mice display increased anxiety-related behavior in the open field, elevated zero-maze, EPM and novelty-suppressed feeding tests (Heisler *et al.*, 1998; Parks *et al.*, 1998; Ramboz *et al.*, 1998). These effects seem to depend on postsynaptic, rather than inhibitory, autosomic receptors (Gross *et al.*, 2002). Also, mice with genetically-induced depletion of brain serotonin exhibit reduced anxiety-like behaviors and enhanced contextual fear memory. These animals also showed electrophysiological changes in hippocampal slices that can be attenuated by 5-HT or a 5-HT_{1A} agonist (Dai *et al.*, 2008).

The anxious-like behavior displayed by 5-HT_{1A} knockout mice in tests that involve conflict contrasts with the predictions from the Deakin–Graeff theory (see Figures 2 and 3) and with the increased inhibitory avoidance found in the ETM after direct intra-dorsal hippocampus or amygdala drug injection (Table 1). However, developmental changes induced by this genetic manipulation could be an additional factor involved in this phenotype. The presence of forebrain 5-HT_{1A} receptors seems to be necessary, between the days 5 and 21 postnatally, for the development of mechanisms underlying normal adult defensive behavior (Gross *et al.*, 2002; Leonardo and Hen, 2005). In addition, a recent study has shown that, surprisingly, these mice perform normally in the Vogel conflict test, an animal model also based on approach-avoidance conflict that is sensitive to anxiolytic benzodiazepines. Different from the open field, elevated zero-maze or EPM, however, this test involves a discrete, non-spatial threatening stimulus (Klemenhagen *et al.*, 2006). Therefore, rather than being more 'anxious', 5-HT_{1A} knockout mice may show a specific subtype of anxiety-related behavior, characterized by excessive behavioral inhibition to ambiguous and complex partial cues (Klemenhagen *et al.*, 2006). Recent data showing that 5-HT_{1A} knockout mice exhibit enhanced fear conditioning to ambiguous conditioned stimuli are compatible with this proposal (Tsetsenis *et al.*, 2007). The latter study has also revealed that pharmacogenetic inhibition of the dentate gyrus blocked this effect, suggesting that the hippocampal formation plays a crucial role in the anxiety-related changes observed in these knockout animals (Klemenhagen *et al.*, 2006).

Since generalization of fear responses to ambiguous environments has been proposed to be characteristic of PD and PTSD, impairments in hippocampal 5-HT_{1A}-receptor mediated function have been related to the pathophysiology of these disorders (Klemenhagen *et al.*, 2006).

This proposal agrees with clinical data showing hippocampal abnormalities and decreased 5-HT_{1A}-receptor binding in these patients (van Pragg, 2004; Klemenhagen *et al.*, 2006).

Contrasting with 5-HT_{1A} knockout mice, and corroborating the Deakin and Graeff (1991) proposal, animals with disruption of 5-HT_{2A} show decreased anxiety in animal models (Cohen, 2005; Weisstaub *et al.*, 2006). This change disappears after selective restoration of 5-HT_{2A} signaling in the cortex (Weisstaub *et al.*, 2006).

Compatible with the clinical reports in subjects carrying the short-allele 5-HTT gene polymorphism (see above), 5-HTT knockout mice also show increased anxiety-like behavior (Holmes *et al.*, 2003a, b) and a failure in the capacity to cope with stress (Adamec *et al.*, 2006; Hariri and Holmes, 2006). As discussed above, however, this genetic change could produce neuroadaptive compensatory effects. Similar increased anxiety-like behavior was found in wild-type mice that were treated with fluoxetine between postnatal days 4 and 21 (Ansorge *et al.*, 2004). Neuronal changes observed in these knockout animals include enhanced 5-HT synthesis and turnover, increased extracellular and decreased intracellular levels of 5-HT (Bengel *et al.*, 1998; Mathews *et al.*, 2004; Kim *et al.*, 2005), down-regulation of 5-HT_{1A}-receptors in the dorsal raphe and at some forebrain targets (Li *et al.*, 1999, 2000; Gobbi *et al.*, 2001), down-regulation of 5-HT_{1B} receptors (Fabre *et al.*, 2000) and up-regulation of 5-HT_{2A}, 5-HT_{2C} and 5-HT_3 receptors in forebrain regions (Rioux *et al.*, 1999; Li *et al.*, 2003; Mössner *et al.*, 2004; Hariri and Holmes, 2006).

Conclusions

Although the precise role of 5-HT in anxiety is not yet clear, converging evidence gathered along these more than 35 years that followed the original proposal by Stein and co-workers (Wise *et al.*, 1972) has clearly shown that 5-HT (1) modulates normal defensive responses; (2) is altered in anxiety disorders; and (3) is involved in the therapeutic effects of drugs, particularly SSRIs. It has also indicated that this role is much more complex than initially thought. The modulatory effects of 5-HT on anxiety-related behavior seem to depend on several factors, including the nature of the threatening stimulus, the defensive strategies available and the brain structures engaged. In addition, these effects can suffer great interference from genetic and neurodevelopmental factors.

There is now good evidence that 5-HT inhibits a defense system engaged by proximal threats that is akin to fear/panic symptoms in humans. In this sense, facilitation of 5-HT-mediated neurotransmission in brain structures related to this system, such as the dorsal PAG, could be part of the mechanism of the therapeutic effects of SSRIs in panic and perhaps other anxiety disorders that share some neurobiological features with panic, such as SAD and PTSD. Clinical studies have also started relating impairments in 5-HT neurotransmission with the pathophysiology of PD. In this way, a recently published study has investigated if 5-HT neurons in the DRN would be activated in rats submitted to a model of PD. The results show that rats submitted to chronic inhibition of GABA synthesis in the dorsal hypothalamus present, in contrast to controls, marked behavioral and autonomic changes when injected with sodium lactate, a well-known panicogenic agent in panic patients. The same animals fail to activate a subset of serotonergic neurons in the DRN that project to the dorsal PAG and the rostral ventromedial medulla, which is activated by lactate in control animals and is likely to inhibit both the behavioral and autonomic effects of lactate (Johnson *et al.*, 2008).

The role of 5-HT in generalized anxiety, however, is not so clear. Although several studies have been compatible with the Deakin and Graeff proposal that 5-HT facilitates anxiety while inhibiting panic, these effects may involve interference in longitudinally organized systems instead of the rostral–caudal hierarchy originally proposed. Even so, some contradictory data still persist. Despite several pieces of evidence suggesting that facilitation of 5-HT_{2C}-mediated neurotransmission in forebrain areas facilitates anxiety, the results regarding the involvement of forebrain 5-HT_{1A} receptors are much less clear. Studies using genetically modified animals indicate that impairment of forebrain 5-HT_{1A}-mediated neurotransmission facilitates, rather than decreases, anxiety. This proposal has received strong support from data originating in neuroimaging studies. Although recent results suggest that this picture is not so clear (Tsetsenis *et al.*, 2007), understanding the precise role of forebrain 5-HT_{1A} receptors in anxiety is still a challenge. In addition, we also need to understand the role of serotonin in mechanisms involved in PTSD and OCD. Although the involvement of 5-HT in neurobiological mechanism of the latter disorder is yet poorly understood, PTSD could involve 5-HT modulation of extinction processes, perhaps mediated by 5-HT_{1A} receptors in the hippocampus or medial prefrontal cortex (van Pragg, 2004).

It is hoped that future studies combining the traditional approaches (behavioral, pharmacological, anatomical and electrophysiological) with the new molecular genetic and functional neuroimaging techniques will help us to further unveil the complex role of serotonin in normal and pathological anxiety.

References

Adamec, R., Burton, P., Blundell, J., Murphy, D.L. and Holmes, A. (2006) Vulnerability to mild predator stress in serotonin transporter knockout mice. Behav. Brain Res., 170: 126–140.

Ansorge, M.S., Zhou, M., Lira, A., Hen, R. and Gingrich, J.A. (2004) Early-life blockade of the 5-HT transporter alters emotional behavior in adult mice. Science, 306: 879–881.

Argyropoulos, S.V., Hood, S.D., Adrover, M., Bell, C.J., Rich, A.S., Nash, J.R., Rich, N.C., Witchel, H.J. and Nutt, D.J. (2004) Tryptophan depletion reverses the therapeutic effect of selective serotonin reuptake inhibitors in social anxiety disorder. Biol. Psychiatry, 56: 503–509.

Azmitia, E.C. and Segal, M. (1978) An autodiographic analysis of the differential ascending projections of the dorsal and median raphe nuclei in the rat. J. Comp. Neurol., 179: 641–688.

Bakker, A., van Balkom, A.J. and Stein, D.J. (2005) Evidence-based pharmacotherapy of panic disorder. Intl. J. Neuropsychopharmacol., 8: 473–482.

Baldwin, D.S. and Polkinghorn, C. (2005) Evidence-based pharmacotherapy of Generalized Anxiety Disorder. Intl. J. Neuropsychopharmacol., 8: 293–302.

Baldwin, D.S., Anderson, I.M., Nutt, D.J., Bandelow, B., Bond, A., Davidson, J.R.T., den Boer, J.A., Fineberg, N.A., Knapp, M., Scott, J. and Wittchen, H.-U. (2005) Evidence-based guidelines for the pharmacological treatment of anxiety disorders: recommendations from the British Association for Psychopharmacology. J. Psychopharmacology, 19: 567–596.

Bannerman, D.M., Rawlins, J.N., McHugh, S.B., Deacon, R.M., Yee, B.K., Bast, T., Zhang, W.N., Pothuizen, H.H. and Feldon, J. (2004) Regional dissociations within the hippocampus –memory and anxiety. Neurosci. Biobehav. Rev., 28: 273–283.

Barnes, N. and Sharp, T. (1999) A review of central 5-HT receptors and their function. Neuropharmacology, 38: 1083–1152.

Beckett, S. and Marsden, C.A. (1997) The effect of central and systemic injection of the 5-HT1A receptor agonist 8-OH-DPAT and the 5-HT1A receptors antagonist WAY 100635 on periaqueductal grey-induced defence behaviour. J. Psychopharmacology, 11: 35–40.

Beckett, S.R.G., Lawrence, A.J., Marsden, C.A. and Marshal, P.W. (1992) Attenuation of chemically induced defence response by 5-HT1 receptor agonists administered into the periaqueductal gray. Psychopharmacology, 108: 110–114.

Bel, N. and Artigas, F. (1992) Fluvoxamine preferentially increases extracellular 5-hydroxytryptamine in the raphe nuclei: an in vivo microdialysis study, Eur. J. Pharmacology, 229: 101–103.

Bell, C., Forshall, S., Adrover, M., Nash, J., Hood, S., Argyropoulos, S., Rich, A. and Nutt, D.J. (2002) Does 5-HT restrain panic? A tryptophan depletion study in panic disorder patients recovered on paroxetine. J. Psychopharmacology, 16: 5–14.

Belzung, C. (2001) Rodent models of anxiety-like behaviors: are they predictive for compounds acting via non-benzodiazepine mechanisms? Curr. Opin. Investig. Drugs, 2: 1108–1111.

Bengel, D., Murphy, D.L., Andrews, A.M., Wichems, C.H., Feltner, D., Heils, A., Mossner, R., Westphal, H. and Lesch, K.P. (1998) Altered brain serotonin homeostasis and locomotor insensitivity to 3, 4-methylenedioxymethamphetamine ('Ecstasy') in serotonin transporter-deficient mice. Mol. Pharmacol., 53: 649–655.

Ben-Zion, I.Z., Meiri, G., Greenberg, B.D., Murphy, D.L. and Benjamin, J. (1999) Enhancement of CO2-induced anxiety in healthy volunteers with the serotonin antagonist metergoline. Am. J. Psychiatry, 156: 1635–1637.

Blanchard, D.C. and Blanchard, R.J. (1988) Ethoexperimental approaches to the biology of emotion. Annu. Rev. Psychol., 39: 43–68.

Blanchard, D.C., Hynd, A.L., Minke, K.A., Minemoto, T. and Blanchard, R.J. (2001) Human defensive behaviors to threat scenarios show parallels to fear- and anxiety-related defense patterns on non-human mammals. Neurosci. Biobehav. Rev., 25: 761–770.

Blanchard, D.C., Griebel, G. and Blanchard, R.J. (2003) The mouse defense test battery: pharmacological and behavioral assay for anxiety and panic. Eur. J. Pharmacol., 463: 97–116.

Blaya, C., Salum, G.A., Lima, M.S., Leistner-Segal, S. and Manfro, G.G. (2007) Lack of association between the serotonin transporter promoter polymorphism (5-HTTLPR) and panic disorder: a systematic review and meta-analysis. Behav. Brain Funct., 3: 41.

Blier, P. and Abbott, F.V. (2001) Putative mechanisms of action of antidepressant drugs in affective and anxiety disorders and pain. J. Psychiatry Neurosci., 26: 37–43.

Blier, P. and Ward, N.M. (2003) Is there a role for 5-HT1A agonists in the treatment of depression? Biol. Psychiatry, 53: 193–203.

Bonne, O., Bain, E., Neumeister, A., Nugent, A.C., Vythilingam, M., Carson, R.E., Luckenbaugh, D.A., Eckelman, W., Herscovitch, P., Drevets, W.C. and Charney, D.S. (2005) No change in serotonin type 1A receptor binding in patients with posttraumatic stress disorder. Am. J. Psychiatry, 162: 383–385.

Bourin, M. and Hascoet, M. (2003) The mouse light/dark box test. Eur. J. Pharmacol., 463: 55–65.

Brandão, M.L., de Aguiar, J.C. and Graeff, F.G. (1982) GABA mediation of the anti-aversive action of minor tranquilizers. Pharmacol. Biochem. Behav., 16: 397–402.

Broocks, A., Bandelow, B., George, A., Jestrabeck, C., Opitz, M., Bartmann, U., Gleiter, C.H., Meineke, I., Roed, I.S., Ruther, E. and Hajak, G. (2000) Increased psychological responses and divergent neuroendocrine responses to m-CPP and ipsapirone in patients with panic disorder. Int. Clin. Psychopharmacol., 15: 153–161.

Canli, T., Congdon, E., Todd Constable, R. and Lesch, K.P. (2008) Additive effects of serotonin transporter and tryptophan hydroxylase-2 gene variation on neural correlates of affective processing. Biol. Psychol., 79: 118–125.

Carey, P.D., Warwick, J., Niehaus, D.J., van der Linden, G., van Heerden, B.B., Harvey, B.H., Seedat, S. and Stein, D.J. (2004) Single photon emission computed tomography (SPECT). BMC. Psychiatry, 4: 30.

Carrive, P. and Bandler, R. (1991) Viscerotopic organization of neurons subserving hypotensive reactions within the midbrain periaqueductal grey: a correlative functional and anatomical study. Brain Res., 541: 206–215.

Ceulemans, D.L.S., Hoppenbrowers, M.L.J.A., Gelders, Y.G. and Reyntjens, A.J.M. (1985) The influence of ritanserin, a serotonin antagonist, in anxiety disorders: a double-blind, placebo-controlled study versus lorazepam. Pharmacopsychiatry, 18: 303–305.

Clark, M.S., McDevitt, R.A. and Neumaier, J.F. (2006) Quantitative mapping of tryptophan hydroxylase-2, 5-HT1A, 5-HT1B, and serotonin transporter expression across the anteroposterior axis of the rat dorsal and median raphe nuclei. J. Comp. Neurol., 498: 611–623.

Cohen, H. (2005) Anxiolytic effect and memory improvement in rats by antisense oligodeoxynucleotide to 5-hydroxytryptamine-2A precursor protein. Depress. Anxiety, 22: 84–93.

Dai, J.-X., Han, H.-L., Cao, J., Xiu, J.-B., Song, N.-N., Huan, Y., Xu, T.-L., Ding, Y.-Q. and Xu, L. (2008) Enhanced contextual fear memory in central serotonin-deficient mice. Proc. Natl. Acad. Sci. USA, 105: 11981–11986.

David, D.J., Bourin, M., Jego, G., Przybylski, C., Jolliet, P. and Gardier, A.M. (2003) Effects of acute treatment with paroxetine, citalopram and venlafaxine in vivo on noradrenaline and serotonin outflow: a microdialysis study in Swiss mice. Br. J. Pharmacology, 140: 1128–1136.

Davis, M. (1992) The role of the amygdala in fear and anxiety. Annu. Rev. Neurosci., 15: 353–375.

Deakin, J.F.W. and Graeff, F.G. (1991) 5-HT and mechanisms of defence. J. Psychopharmacol., 5: 305–315.

de Bortoli, V.C., Nogueira, R.L. and Zangrossi, H. Jr. (2006) Effects of fluoxetine and buspirone on the panicolytic-like response induced by the activation of 5-HT1A and 5-HT2A receptors in the rat dorsal periaqueductal gray. Psychopharmacology, 183: 422–428.

de Bortoli, V.C., Nogueira, R.L. and Zangrossi, H. Jr. (2008) Alprazolam potentiates the antiaversive effect induced by the activation of 5-HT(1A) and 5-HT (2A) receptors in the rat dorsal periaqueductal gray. Psychopharmacology (Berl.), 198: 341–349.

Del-Ben, C.M., Vilela, J.A.A., Hetem, L.A.B., Guimarães, F.S., Graeff, F.G. and Zuardi, A.W. (2001) Do panic patients process unconditioned fear versus conditioned anxiety differently from normal subjects? Psychiatry Res., 104: 227–237.

Delgado, J.M.R., Roberts, W.W. and Miller, N.E. (1954) Learning motivated by electrical stimulation of the brain. Am. J. Physiol., 179: 587–593.

den Boer, J.A. and Westenberg, H.G.M. (1990) Serotonin function in panic disorder: a double blind placebo controlled study with fluvoxamine and ritanserin. Psychopharmacology, 102: 85–94.

de Paula Soares, V. and Zangrossi, H. Jr. (2004) Involvement of 5-HT1A and 5-HT2 receptors of the dorsal periaqueductal gray in the regulation of the defensive behaviors generated by the elevated T-maze. Brain Res. Bull., 64: 181–188.

de Paula Soares, V. and Zangrossi, H. Jr. (2008) Stimulation of 5-HT1A or 5-HT2A receptors in the ventrolateral periaqueductal gray causes anxiolytic-, but not panicolytic-like effect in rats. Behav. Brain Res., in press.

dos Santos, L., de Andrade, T.G. and Zangrossi, H. Jr. (2008) 5-HT1A receptors in the dorsal hippocampus mediate the anxiogenic effect induced by the stimulation of 5-HT neurons in the median raphe nucleus. Eur. Neuropsychopharmacol., 18: 286–294.

Estes, W.K. and Skinner, F.B. (1941) Some quantitative properties of anxiety. J. Exp. Psychol., 29: 390–400.

Fabre, V., Beaufour, C., Evrard, A., Rioux, A., Hanoun, N., Lesch, K.P., Murphy, D.L., Lanfumey, L., Hamon, M. and Martres, M.P. (2000) Altered expression and functions of serotonin 5-HT1A and 5-HT1B receptors in knockout mice lacking the 5-HT transporter. Eur. J. Neurosci., 12: 2299–2310.

Fanselow, M.S. (1991) The midbrain periaqueductal gray as a coordinator of action in response to fear and anxiety. In: Depaulis, A. and Bandler, R. (Eds.), The Midbrain Periaqueductal Gray Matter: Functional, Anatomical and Immunohistochemical Organization. Plenum, New York, NY, pp. 151–173.

Felton, T.M., Kang, T.B., Hjorth, S. and Auerbach, S.B. (2003) Effects of selective serotonin and serotonin/noradrenaline reuptake inhibitors on extracellular serotonin in rat diencephalon and frontal cortex, Naunyn Schmiedebergs Arch. Pharmacology, 367: 297–305.

Fernandez, M., Pissiota, A., Frans, O., von Knorring, L., Fischer, H. and Fredrikson, M. (2001) Brain function in a patient with torture related post-traumatic stress disorder before and after fluoxetine treatment: a positron emission tomography provocation study. Neurosci. Lett., 297: 101–104.

Furmark, T., Tillfors, M., Garpenstrand, H., Marteinsdottir, I., Langstrom, B., Oreland, L. and Fredrikson, M. (2004) Serotonin transporter polymorphism related to amygdala excitability and symptom severity in patients with social phobia. Neurosci. Lett., 362: 189–192.

Garcia-Leal, C., Parente, A.C.B.V., Del-Bem, C.M., Guimarães, F.S., Moreira, A.C., Elias, L.L.K. and Graeff, F.G. (2005) Anxiety and salivary cortisol in symptomatic and nonsymptomatic panic patients and healthy volunteers performing simulated public speaking. Psychiatry Res., 133: 239–252.

Gartside, S.E., Umbers, V., Hajos, M. and Sharp, T. (1995) Interaction between a selective 5-HT1A receptor antagonist and an SSRI *in vivo*: effects on 5-HT cell firing and extracellular 5-HT, Br. J. Pharmacology, 115: 1064–1070.

Geller, I. and Seifter, J. (1960) The effects of meprobamate, barbiturates, d-amphetamine and promazine on experimentally induced conflict in the rat. Psychopharmacologia, 1: 482–492.

Gobbi, G., Murphy, D.L., Lesch, K. and Blier, P. (2001) Modifications of the serotonergic system in mice lacking serotonin transporters: an in vivo electrophysiological study. J. Pharmacol. Exp. Ther., 296: 987–995.

Goldberg, H.L. and Finnerty, R.J. (1979) The comparative efficacy of buspirone and diazepam in the treatment of anxiety. Am. J. Psychiatry, 136: 1184–1187.

Gordon, J.A. and Hen, R. (2004) The serotonergic system and anxiety. NeuroMolecular Med., 5: 27–40.

Gottesman, I.I. and Gould, D. (2003) The endophenotype concept in psychiatry: etymology and strategic intentions. Am. J. Psychiatry., 160: 636–645.

Graeff, F.G. (1974) Tryptamine antagonists and punished behavior. J. Pharmacol. Exp. Ther., 189: 344–350.

Graeff, F.G. (1991) Neurotransmitters in the dorsal periaqueductal gray and animal models of panic anxiety. In: Briley, M. and File, S.E. (Eds.), New Concepts in Anxiety. Macmillan Press, London, pp. 288–312.

Graeff, F.G. (1994) Neuroanatomy and neurotransmitter regulation of defensive behaviors and related emotions in mammals. Braz. J. Med. Biol. Res., 27: 811–829.

Graeff, F.G. and Schoenfeld, R.I. (1970) Tryptaminergic mechanisms in punished and nonpunished behavior. J. Pharmacol. Exp. Ther., 173: 277–283.

Graeff, F.G. and Silveira Filho, N.G. (1978) Behavioral inhibition induced by electrical stimulation of the median raphe nucleus of the rat. Physiol. Behav., 21: 477–484.

Graeff, F.G., Zuardi, A.W., Giglio, J.S., Lima Filho, E.C. and Karniol, I.G. (1985) Effect of metergoline on human anxiety. Psychopharmacology, 86: 334–338.

Graeff, F.G., Viana, M.B. and Tomaz, C. (1993) The elevated T maze, a new experimental model of anxiety and memory: effect of diazepam. Braz. J. Med. Biol. Res., 26: 67–70.

Graeff, F.G., Viana, M.B. and Mora, P. (1997) Dual role of 5-HT in defense and anxiety. Neurosci. Biobehav. Rev., 21: 791–799.

Graeff, F.G., Netto, C.F. and Zangrossi, H. Jr. (1998) The elevated T-maze as an experimental model of anxiety. Neurosci. Biobehav. Rev., 23: 237–246.
Graeff, F.G., Parente, A., Del-Ben, C.M. and Guimarães, F.S. (2003) Pharmacology of human experimental anxiety. Brazilian J. Med. Biol. Res., 36: 421–432.
Graeff, F.G., Garcia-Leal, C., Del-Ben, C.M. and Guimarães, F.S. (2005) Does the panic attack activate the hypothalamic–pituitary–adrenal axis? An. Acad. Brasil. Ciências, 77: 1–15.
Gray, J.A. (1982) The Neuropsychology of Anxiety. Oxford University Press, New York. NY.
Gray, J.A. and McNaughton, N. (2000) The Neuropsychology of Anxiety. 2nd edn. Oxford University Press, New York, NY.
Greenwood, B.N., Strong, P.V., Foley, T.E., Thompson, R.S. and Fleshner, M. (2007) Learned helplessness is independent of levels of brain-derived neurotrophic factor in the hippocampus. Neuroscience, 144: 1193–1208.
Griebel, G. (1995) 5-Hydroxytryptamine-interacting drugs in animal models of anxiety disorders: more than 30 years of research. Pharmacol. Ther., 65: 319–395.
Gross, C., Zhuang, X., Stark, K., Ramboz, S., Oosting, R., Kirby, L., Santarelli, L., Beck, S. and Hen, R. (2002) Serotonin1A receptor acts during development to establish normal anxiety-like behaviour in the adult. Nature, 416: 396–400.
Guimarães, F.S., Zuardi, A.W. and Graeff, F.G. (1987) Effect of chlorimipramine and maprotiline on experimental anxiety in humans. J. Psychopharmacol., 1: 184–192.
Guimarães, F.S., Carobrez, A.P. and Graeff, F.G. (2008) Modulation of anxiety disorders by 5-HT-interacting drugs. In: Blanchard, R.J., Blanchard, D.C., Griebel, G. and Nutt, D. (Eds.), Handbook of Anxiety and Fear. Elesivier, Amsterdam, pp. 241–268.
Handley, S.L., McBlane, J.W., Critchley, M.A.E. and Njung'e, K. (1993) Multiple serotonin mechanisms in animal models of anxiety: environmental, emotional and cognitive factors. Behav. Brain Res., 58: 203–210.
Hariri, A.R. and Holmes, A. (2006) Genetics of emotional regulation: the role of the serotonin transporter in neural function. Trends Cogn. Sci., 10(4): 182–191.
Hariri, A.R., Drabant, E.M., Munoz, K.E., Kolachana, B.S., Mattay, V.S., Egan, M.F. and Weinberger, D.R. (2005) A susceptibility gene for affective disorders and the response of the human amygdala. Arch. Gen. Psychiatry, 62: 146–152.
Heisler, L.K., Chu, H.M., Brennan, T.J., Danao, J.A., Bajwa, P., Parsons, L.H. and Tecott, L.H. (1998) Elevated anxiety and antidepressant-like responses in serotonin 5-HT1A receptor mutant mice. Proc. Natl. Acad. Sci. USA, 95: 15049–15054.
Hensler, J.G. (2006) Serotonergic modulation of the limbic system. Neurosci. Biobehav. Rev., 30: 203–214.
Hesse, S., Barthel, H., Schwarz, J., Sabri, O. and Müller, U. (2004) Advances in in vivo imaging of serotonergic neurons in neuropsychiatric disorders. Neurosci. Biobehav. Rev., 28: 547–563.
Hetem, L.A. (1996) Addition of d-fenfluramine to benzodiazepines produces a marked improvement in refractory panic disorder – a case report. J. Clin. Psychopharmacology, 16: 77–78.
Hewlett, W.A., Schmid, S.P. and Salomon, R.M. (2003) Pilot trial of ondansetron in the treatment of 8 patients with obsessive-compulsive disorder. J. Clin. Psychiatry, 64: 1025–1030.
Hoehn-Saric, R., Schlund, M.W. and Wong, S.H. (2004) Effects of citalopram on worry and brain activation in patients with generalized anxiety disorder. Psychiatry Res., 131: 11–21.
Holmes, A., Lit, Q., Murphy, D.L., Gold, E. and Crawley, J.N. (2003a) Abnormal anxiety-related behavior in serotonin transporter null mutant mice: the influence of genetic background. Genes Brain Behav., 2: 365–380.
Holmes, A., Yang, R.J., Lesch, K.P., Crawley, J.N. and Murphy, D.L. (2003b) Mice lacking the serotonin transporter exhibit 5-HT(1A) receptor-mediated abnormalities in tests for anxiety-like behavior. Neuropsychopharmacology, 28: 2077–2088.
Holmes, A., le Guisquet, A.M., Vogel, E., Millstein, R.A., Leman, S. and Belzung, C. (2005) Early life genetic, epigenetic and environmental factors shaping emotionality in rodents. Neurosci. Biobehav. Rev., 29: 1335–1346.
Hoyer, D., Hannon, J.P. and Martin, G.R. (2002) Molecular, pharmacological and functional diversity of 5-HT receptors. Pharmacol. Biochem. Behav., 71: 533–554.
Hunsperger, R.W. (1956) Role of substantia grisea centralis mesencephali in electrically-induced rage reactions. Prog. Neurobiol., 2: 289–294.
Iidaka, T., Ozaki, N., Matsumoto, A., Nogawa, J., Kinoshita, Y., Suzuki, T., Iwata, N., Yamamoto, Y., Okada, T. and Sadato, N. (2005) A variant C178T in the regulatory region of the serotonin receptor gene HTR3A modulates neural activation in the human amygdala. J. Neurosci., 25: 4646–6460.
Johnson, P.L., Truitt, W.A., Fitz, S.D., Lowry, C.A. and Shekhar, A. (2008) Neural pathways underlying lactate-induced panic. Neuropsychopharmacology, 33: 2093–2107.
Kahn, R.J., McNair, D.M., Lipman, R.S., Covi, L., Rickels, K., Downing, R., Fisher, S. and Frankenthaler, L.M. (1986) Imipramine and chlordiazepoxide in depressive and anxiety disorders. II Efficacy in anxious outpatients. Arch. Gen. Psychiatry, 43: 79–85.
Kim, D.K., Tolliver, T.J., Huang, S.J., Martin, B.J., Andrews, A.M., Wichems, C., Holmes, A., Lesch, K.P. and Murphy, D.L. (2005) Altered serotonin synthesis, turnover and dynamic regulation in multiple brain regions of mice lacking the serotonin transporter. Neuropharmacology, 49: 798–810.
Kiser, R.S. Jr. and Lebovitz, R.M. (1975) Monoaminergic mechanisms in aversive brain stimulation. Physiol. Behav., 15: 47–53.
Kiser, R.S., Lebovitz, R.M. and German, D.C. (1978a) Anatomic and pharmacologic differences between two types of aversive midbrain stimulation. Brain Res., 155: 331–342.
Kiser, R.S., German, D.C. and Lebovitz, R.M. (1978b) Serotonergic reduction of dorsal central gray area stimulation-produced aversion. Pharmacol. Biochem. Behav., 9: 27–31.
Kiser, R.S., Brown, C.A., Sanghera, M.K. and German, D.C. (1980) Dorsal raphe nucleus stimulation reduces centrally-elicited fearlike behavior. Brain Res., 191: 265–272.
Klein, F.F. and Flink, M. (1962) Psychiatric reaction patterns to imipramine. Am. J. Psychiatry, 119: 432–438.
Klemenhagen, K.C., Gordon, J.A., David, D.J., Hen, R. and Gross, C.T. (2006) Increased fear response to contextual cues in mice lacking the 5-HT1A receptor. Neuropsychopharmacology, 31: 101–111.
Kulz, A.K., Meinzer, S., Kopasz, M. and Voderholzer, U. (2007) Effects of tryptophan depletion on cognitive functioning, obsessive-compulsive symptoms and mood in obsessive-compulsive disorder: preliminary results. Neuropsychobiology, 56: 127–131.
Lanzenberger, R.R., Mitterhauser, M., Spindelegger, C., Wadsak, W., Klein, N., Mien, L.-K., Holik, A., Attarbaschi, T., Mossaheb, N., Sacher, J., Geiss-Granadia, T., Kletter, K., Kasper, S. and Tauscher, J. (2007) Reduced serotonin-1A receptor binding in social anxiety disorder. Biol. Psychiatry, 61: 1081–1089.

Lecrubier, Y., Pucch, A.J., Azeona, A., Bailey, P.E. and Lataste, X. (1999) A randomized double-blind place-controlled study of tropisetron in the treatment of outpatients with generalized anxiety disorder. Psychopharmacology, 112: 129–133.
LeDoux, J.E. (1993) Emotional memory systems in the brain. Behav. Brain Res., 58: 69–79.
Leonardo, E.D. and Hen, R. (2006) Genetics of affective and anxiety disorders. Ann. Rev. Psychol., 57: 117–137.
Lesch, K.P. (2005) Genetic alterations of the murine serotonergic gene pathway: the neurodevelopmental basis of anxiety. Handb. Exp. Pharmacol., 169: 71–112.
Lesch, K.P., Wiesmann, M., Hoh, A., Muller, T., Disselkamp-Tietze, J., Osterheider, M. and Schulte, H.M. (1992) 5-HT1A receptor–effector system responsivity in panic disorder. Psychopharmacology, 106: 111–117.
Li, Q., Wichems, C., Heils, A., Van De Kar, L.D., Lesch, K.P. and Murphy, D.L. (1999) Reduction of 5-hydroxytryptamine (5-HT1A)-mediated temperature and neuroendocrine responses and 5-HT(1A) binding sites in 5-HT transporter knockout mice. J. Pharmacol. Exp. Ther., 291: 999–1007.
Li, Q., Wichems, C., Heils, A., Lesch, K.P. and Murphy, D.L. (2000) Reduction in the density and expression, but not G-protein coupling, of serotonin receptors (5-HT1A) in 5-HT transporter knockout mice: gender and brain region differences. J. Neurosci., 20: 7888–7895.
Li, Q., Wichems, C., Ma, L., Van de Kar, L.D., Garcia, F. and Murphy, D.L. (2003) Brain region-specific alterations of 5-HT2A and 5-HT2C receptors in serotonin transporter knockout mice. J. Neurochem., 84: 1256–1265.
Lowry, C.A., Johnson, P.L., Hay-Schmidt, A., Mikkelsen, J. and Shekhar, A. (2005) Modulation of anxiety circuits by serotonergic systems. Stress, 8: 233–246.
Mamounas, L.A. and Mullen, C.A. *et al.* (1991) Dual serotoninergic projections to forebrain in the rat: morphologically distinct 5-HT axon terminals exhibit differential vulnerability to neurotoxic amphetamine derivatives. J. Comp. Neurol., 314: 558–586.
Maron, E. and Shlik, J. (2006) Serotonin function in panic disorder: important, but why? Neuropsychopharmacology, 31: 1–11.
Maron, E., Toru, I., Vasar, V. and Shlik, J. (2004a) The effect of 5-hydroxytryptophan on cholecystokinin-4-induced panic attacks in healthy volunteers. J. Psychopharmacol., 18: 194–199.
Maron, E., Kuikka, J.T., Ulst, K., Tiihonen, J., Vasar, V. and Shlik, J. (2004b) SPECT imaging of serotonin transporter binding in patients with generalized anxiety disorder. Eur. Arch. Psychiatry Clin. Neurosci., 254: 392–396.
Mathews, T.A., Fedele, D.A., Coppelli, F.M., Avila, A.M., Murphy, D.L. and Andrews, A.M. (2004) Gene dose-dependent alterations in extraneuronal serotonin but not dopamine in mice with reduced serotonin transporter expression. J. Neurosci. Meth., 140: 169–181.
McNaughton, N. and Corr, P.J. (2004) A two-dimensional neuropsychology of defense: fear/anxiety and defensive distance. Neurosci. Biobehav. Rev., 28: 285–305.
Millan, M.J. (2003) The neurobiology and control of anxious states. Prog. Neurobiol., 70: 53–244.
Millan, M.J. (2006a) Serotonin 5-HT2C receptors as a target for the treatment of depressive and anxious states: focus on novel therapeutic strategies. Therapie, 60: 441–460.
Millan, M.J. (2006b) Multi-target strategies for the improved treatment of depressive states: conceptual foundations and neuronal substrates, drug discovery and therapeutic application. Pharmacol. Ther., 110: 135–370.
Millenson, J.R. and Leslie, J. (1974) The conditioned emotional response (CER) as a baseline for the study of anti-anxiety drugs. Neuropharmacology, 13: 1–9.
Miller, H.E., Deakin, J.F. and Anderson, I.M. (2000) Effect of acute tryptophan depletion on CO2-induced anxiety in patients with panic disorder and normal volunteers. Br. J. Psychiatry, 176: 182–188.
Mobbs, D., Petrovic, P., Marchant, J.L., Hassabis, D., Weiskopf, N., Seymour, B., Dolan, R.J. and Frith, C.D. (2007) When fear is near: threat imminence elicits prefrontal-periaqueductal gray shifts in humans. Science, 317: 1079–1083.
Morato de Carvalho, S. and de Aguiar, J.C.etal. (1981) Effect of minor tranquilizers, tryptamine antagonists and amphetamine on behavior punished by brain stimulation. Pharmacol. Biochem. Behav., 15: 351–356.
Moret, C. and Bridley, M. (1996) Effects of acute and repeated administration of citalopram on extracellular levels of serotonin in rat brain, Eur. J. Pharmacology, 295: 189–197.
Mortimore, C. and Anderson, I.M. (2000) d-Fenfluramine in panic disorder: a dual role for 5-hydroxytryptamine. Psychopharmacology, 149: 251–258.
Mössner, R., Schmitt, A., Hennig, T., Benninghoff, J., Gerlach, M., Riederer, P., Deckert, J. and Lesch, K.P. (2004) Quantification of 5-HT3 receptors in forebrain of serotonin transporter deficient mice. J. Neural Transm., 111: 27–35.
Nash, J.R., Sargent, P.A., Rabiner, E.A., Hood, S.D., Argyropoulos, S.V., Potokar, J.P., Grasby, P.M. and Nutt, D.J. (2008) Serotonin 5-HT1A receptor binding in people with panic disorder: positron emission tomography study. Br. J. Psychiatry, 193(3): 229–234.
Neumeister, A., Bain, E., Nugent, A.C., Carson, R.E., Bonne, O., Luckenbaugh, D.A., Eckelman, W., Herscovitch, P., Charney, D.S. and Drevets, W.C. (2004) Reduced serotonin type 1A receptor binding in panic disorder. J. Neurosci., 24: 589–591.
Nogueira, R.L. and Graeff, F.G. (1991) Mediation of the anti-aversive effect of isamoltane injected into the dorsal periaqueductal grey. Behav. Pharmacol., 2: 73–77.
Nogueira, R.L. and Graeff, F.G. (1995) Role of 5-HT receptor subtypes in the modulation of aversion generated in the dorsal periaqueductal gray. Pharmacol. Biochem. Behav., 52: 1–6.
Olivier, B., Zethof, T., Parttij, T., van Boogaert, M., van Oorschot, R., Leahy, C., Oosting, R., Bouwknecht, A., Veening, J., van der Gugten, J. and Groenink, L. (2003) Stress-induced hyperthermia and anxiety: pharmacological validation. Eur. J. Pharmacol., 463: 117–132.
Palma, S.M., Guimarães, F.S. and Zuardi, A.W. (1994) Anxiety induced by simulated public speaking and stroop color word test in healthy subjects: effects of different trait-anxiety levels, Braz. J. Med. Biol. Res., 27: 2895–2902.
Parente, A.C.B.V., Garcia-Leal, C., Del-Ben, C.M., Guimarães, F.S. and Graeff, F.G. (2005) Subjective and neurovegetative changes in healthy volunteers and panic patients performing simulated public speaking, Eur. Neuropsychopharmacology, 15: 663–671.
Parks, C.L., Robinson, P.S., Sibille, E., Shenk, T. and Toth, M. (1998) Increased anxiety of mice lacking the serotonin1A receptor. Proc. Natl. Acad. Sci. USA, 95: 10734–10739.
Paxinos, G. and Watson, C. (2005) The Rat Brain in Stereotaxic Coordinates. 5th edn. Elsevier, Amsterdam.
Pazos, A. and Palacios, J.M. (1985) Quantitative autoradiographic mapping of serotonin receptors in the rat brain. I. Serotonin-1 receptors. Brain Res., 346: 205–230.

Petersen, E.N. and Scheel-Krüger, J. (1984) Anticonflict effects of 5-HT antagonists by intraamygdaloid injection. Abstracts of the 14th CINP Congress, p. 654.

Pinheiro, S.N., Zangrossi, H. Jr., Del-Ben, C.M. and Graeff, F.G. (2007) Elevated mazes as animal models of anxiety: Effects of serotonergic agents. An. Acad. Bras. Cienc., 79: 71–85.

Pobbe, R.L. and Zangrossi, H. Jr. (2005) 5-HT(1A) and 5-HT(2A) receptors in the rat dorsal periaqueductal gray mediate the antipanic-like effect induced by the stimulation of serotonergic neurons in the dorsal raphe nucleus. Psychopharmacology, 183: 314–321.

Poltronieri, S.C., Zangrossi, H. Jr. and Viana, M.B. (2003) Antipanic-like effect of serotonin reuptake inhibitors in the elevated T-maze. Behav. Brain Res., 147: 185–192.

Prut, L. and Belzung, C. (2003) The open field as a paradigm to measure the effects of drugs on anxiety-like behaviors: a review. Eur. J. Pharmacol., 463: 3–33.

Ramboz, S., Oosting, R., Amara, D.A., Kung, H.F., Blier, P., Mendelsohn, M., Mann, J.J., Brunner, D. and Hen, R. (1998) Serotonin receptor 1A knockout: an animal model of anxiety-related disorder. Proc. Natl. Acad. Sci. USA, 95: 14476–14481.

Reist, C., Mazzanti, C., Vu, R., Tran, D. and Goldman, D. (2001) Serotonin transporter promoter polymorphism is associated with attenuated prolactin response to fenfluramine. Am. J. Med. Genetics, 105: 363–368.

Rioux, A., Fabre, V., Lesch, K.P., Moessner, R., Murphy, D.L., Lanfumey, L., Hamon, M. and Martres, M.P. (1999) Adaptive changes of serotonin 5-HT2A receptors in mice lacking the serotonin transporter. Neurosci. Lett., 262: 113–116.

Robichaud, R.C. and Sledge, K.L. (1969) The effects of p-chlorophenylalanine on experimentally induced conflict in the rat. Life Sci., 8: 965–969.

Romach, M.K., Kaplan, H.L., Busto, U.E., Somer, G. and Sellers, E.M. (1998) A controlled trial of ondansetran, a 5-HT3 antagonist, in benzodiazepine discontinuation. J. Clin. Psychopharmacol., 18: 121–131.

Saarelainen, T., Hendolin, P. *et al.* (2003) Activation of the TrkB neurotrophin receptor is induced by antidepressant drugs and is required for antidepressant-induced behavioral effects. J. Neurosci., 23: 349–357.

Sanchez, C. (2003) Stress-induced vocalization in adult animals. A valid model of anxiety? Eur. J. Pharmacol., 463: 133–143.

Santarelli, L., Saxe, M., Gross, C., Surget, A., Battaglia, F., Dulawa, S., Weisstaub, N., Lee, J., Duman, R., Arancio, O., Belzung, C. and Hen, R. (2003) Requirement of hippocampal neurogenesis for the behavioral effects of antidepressants. Science, 301: 805–809.

Savic, I., Lindstrom, P., Gulyas, B., Halldin, C., Andree, B. and Farde, L. (2004) Limbic reductions of 5-HT1A receptor binding in human temporal lobe epilepsy. Neurology, 62: 1343–1351.

Schenberg, L.C. and Graeff, F.G. (1978) Role of the periaqueductal gray substance in the antianxiety action of benzodiazepines. Pharmacol. Biochem. Behav., 9: 287–295.

Schenberg, L.C., Dos Reis, A.M., Ferreira Povoa, R.M., Tufik, S. and Silva, S.R. (2008) A panic attack-like unusual stress reaction. Horm. Behav., 55: 584–591.

Schoenfeld, R.I. (1976) Lysergic acid diethylamide- and mescaline-induced attenuation of the effect of punishment in the rat. Science, 192: 801–803.

Schruers, K., van Diest, R., Overbeek, T. and Griez, E. (2002) Acute L-5-hydroxytryptophan administration inhibits carbon dioxide-induced panic in panic disorder patients. Psychiatry Res., 113: 237–243.

Schütz, M.T.B., de Aguiar, J.C. and Graeff, F.G. (1985) Anti-aversive role of serotonin on the dorsal periaqueductal grey matter. Psychopharmacology, 85: 340–345.

Sena, L.M., Bueno, C., Pobbe, R.L., Andrade, T.G., Zangrossi, H. and Viana, M.B. (2003) The dorsal raphe nucleus exerts opposed control on generalized anxiety and panic-related defensive responses in rats. Behav. Brain Res., 142: 125–133.

Silva, M., Hetem, L.A., Guimarães, F.S. and Graeff, F.G. (2001) Opposed effects of nefazodone in two human models of anxiety. Psychopharmacology, 156: 454–460.

Solyom, L. (1994) Controlling panic attacks with fenfluramine. Am. J. Psychiatry, 151: 621–622.

Strug, L.J., Suresh, R., Fyer, A.J., Talati, A., Adams, P.B., Li, W., Hodge, S.E., Gilliam, T.C. and Weissman, M.M. (2008) Panic disorder is associated with the serotonin transporter gene (SLC6A4) but not the promoter region (5-HTTLPR). Mol. Psychiatry, in press.

Tauscher, J., Bagby, R.M., Javanmard, M., Christensen, B.K., Kasper, S. and Kapur, S. (2001) Inverse relationship between serotonin 5-HT(1A) receptor binding and anxiety: a[(11)C]WAY-100635 PET investigation in healthy volunteers. Am. J. Psychiatry, 158: 1326–1328.

Teixeira, R.C., Zangrossi, H. Jr. and Graeff, F.G. (2000) Behavioral effects of acute and chronic imipramine in the elevated T-maze model of anxiety. Pharmacol. Biochem. Behav., 65: 571–576.

Tsetsenis, T., Ma, X.-H., Iacono, L.L., Beck, S.G. and Gross, C. (2007) Suppression of conditioning to ambiguous cues by pharmacogenetic inhibition of the dentate gyrus. Nat. Neurosci., 10: 896–902.

Tye, N.C., Everitt, B.J. and Iversen, S.D. (1977) 5-Hydroxytryptamine and punishment. Nature, 268: 741–742.

van der Wee, N.J., van Veen, J.F., Stevens, H., van Vliet, I.M., van Rijk, P.P. and Westenberg, H.G. (2008) Increased serotonin and dopamine transporter binding in psychotropic medication-naive patients with generalized social anxiety disorder shown by 123I-beta-(4-iodophenyl)-tropane SPECT. J. Nucl. Med., 49: 757–763.

van Pragg, H.M. (2004) The cognitive paradox in posttraumatic stress disorder: a hypothesis. Prog. Neuropsychopharmacol. Biol. Psychiatry, 28: 923–935.

van Vliet, I.M., Westenberg, H.G. and den Boer, J.A. (1996) Effects of the 5-HT1A receptor agonist flesinoxan in panic disorder. Psychopharmacology, 127: 174–180.

Viana, M.B., Zangrossi, H. Jr. and Onusic, G.M. (2008) 5-HT1A receptors of the lateral septum regulate inhibitory avoidance but not escape behavior in rats. Pharmacol. Biochem. Behav., 89: 360–366.

Weisstaub, N.V., Zhou, M., Lira, A., Gonzalez-Maeso, J., Hornung, J.-P., Sibille, E., Underwood, M., Itohara, S., Dauer, W.T., Ansorge, M.S., Morelli, E., Mann, J.J., Toth, M., Aghajanian, G., Sealfon, S.C., Hen, R. and Gingrich, J.A. (2006) Cortical 5-HT2A receptor signaling modulates anxiety-like behaviors in mice. Science, 313: 536–540.

Wise, C.D., Berger, B.D. and Stein, L. (1972) Benzodiazepines: anxiety-reducing activity by reduction of serotonin turnover in the brain. Science, 177: 180–183.

Wrase, J., Reimold, M., Puls, I., Kienast, T. and Heinz, A. (2006) Serotonergic dysfunction: brain imaging and behavioral correlates. Cogn. Affect. Behav. Neurosci., 6: 53–61.

Yasuno, F., Suhara, T., Ichimiya, T., Takano, A., Ando, T. and Okubo, Y. (2004) Decreased 5-HT1A receptor binding in amygdala of schizophrenia. Biol. Psychiatry, 55: 439–444.

Zanoveli, J.M., Nogueira, R.L. and Zangrossi, H. Jr. (2003) Serotonin in the dorsal periaqueductal gray modulates inhibitory avoidance and one-way escape behaviors in the elevated T-maze. Eur. J. Pharmacol., 473: 153–161.

Zanoveli, J.M., Netto, C.F., Guimarães, F.S. and Zangrossi, H. Jr. (2004) Systemic and intra-dorsal periaqueductal gray injections of cholecystokinin sulfated octapeptide (CCK-8s) induce a panic-like response in rats submitted to the elevated T-maze. Peptides, 25: 1935–1941.

Zanoveli, J.M., Nogueira, R.L. and Zangrossi, H. Jr. (2005) Chronic imipramine treatment sensitizes 5-HT1A and 5-HT2A receptors in the dorsal periaqueductal gray matter: evidence from the elevated T-maze test of anxiety. Behav. Pharmacol., 16: 543–552.

CHAPTER 4.10

Serotonin and Aggression

Isabel M. Quadros, Aki Takahashi and Klaus A. Miczek*

Tufts University, Medford and Boston, Massachusetts

Abstract: Traditional clinical research on the neurobiology of aggressive behavior focuses on individuals who are characterized by their impulsive, hostile, antisocial and violent traits and who show some deficiency in brain serotonin (5-HT) activity relative to those who have a propensity to engage in premeditated, calculating and instrumental aggressive acts. Preclinical research has focused on territorial, dominant or maternal aggressive behavior patterns, chiefly for reproductive purposes. Recent efforts study individuals who engage in very high levels of aggressive behavior due to genetic selection, consumption of a moderate dose of alcohol, social instigation or pharmacological insults. The early proposal that lower CSF levels of 5-HT and/or 5-HIAA were associated with a greater propensity for violent outbursts (the 'serotonin deficiency hypothesis') has been challenged by the evidence implicating receptor subtypes of the 5-HT_1 and 5-HT_2 families, the serotonin transporter (5-HTT) and metabolic enzyme monoamine oxidase A (MAOA), acting at different levels of the neuroaxis. *In vivo* techniques revealed a significant role of cortical 5-HT release during the termination of an aggressive confrontation and its consequences, rather than its initiation. Considerable clinical and preclinical evidence implicates 5-HT_{1A}, 5-HT_{1B} and 5-HT_{2A} as targets for pharmacotherapeutic management of escalated aggressive behavior, which are accompanied to a varying extent by non-specific effects, whereas the genetic data for linking these receptors to specific types of aggression are less consistent. The prevalent anti-aggressive effects emerge with drugs targeting the 5-HTT, and the genetic data from human and non-human primates suggest the short allele 5-HTT polymorphism as a risk factor for violent traits, particularly in combination with environmental stress. Contrary to pharmacological inhibition of MAO, mutation of the MAOA gene in humans and mice suggests that chronically elevated 5-HT levels may promote escalated aggressive behavior, most prominently after early life maltreatment. Interactions between 5-HTT and MAOA polymorphisms suggest that complex epistatic interactions may ultimately reveal strong effects on aggressive phenotypes.

Keywords: violence, aggression, agonistic behavior, serotonin transporter (5-HTT, SERT), monoamine oxidase A (MAOA), antidepressants, selective serotonin reuptake inhibitors (SSRI); mutants.

Introduction

No other transmitter substance in the brain has been linked more often to aggressive behavior than the phylogenetically old indolamine serotonin, as is well illustrated by the large number of reviews during the past four decades (see, for example, Valzelli and Garattini, 1968; Brown *et al.*, 1979a; Pucilowski and Kostowski, 1983; Miczek and Donat, 1989; Linnoila and Virkkunen, 1992; Olivier and Mos, 1992; Edwards and Kravitz, 1997; Higley and Bennett, 1999; Nelson and Chiavegatto, 2001; Miczek *et al.*, 2002, 2007). Ever since its discovery in the brain, its role in mediating physiological and behavioral inhibition has been highlighted (Brodie and Shore, 1957). Currently, it has become evident that serotonin contributes to the neural mechanisms mediating aggressive behavior in multiple ways at different levels of the neuroaxis, and certain themes have emerged and guide the agenda for the current research efforts.

Clinical and preclinical research of the neurobiology of aggression fundamentally differ in that the former focuses most often on aggressive *traits*, and particularly maladaptive expressions of such traits, whereas the latter examines *state* changes in adaptive species-typical aggressive behavior. The most common research strategy for clinical research relates biological assays of amines or their acid metabolites or neuroimages from human subjects to their lifelong history of aggressive behavior, but these behavioral events are divorced in time from the biological

*Corresponding author
E-mail: Klaus.miczek@tufts.edu

Christian Müller & Barry Jacobs (Eds.)
Handbook of Behavioral Neurobiology of Serotonin
ISBN 978-0-12-374634-4

DOI: 10.1016/B978-0-12-374634-4.00038-1

assay. In addition to the most widely used scales for assessing aggressive and impulsive traits (see McKinley *et al.*, 1948; Buss and Durkee, 1957; Beck *et al.*, 1961; Brown *et al.*, 1979b; Barratt *et al.*, 1997; DeMoja and Spielberger, 1997), it has been possible to validate behavioral challenge methods for provoking aggressive responses under controlled laboratory conditions in parolees of violent crimes (Cherek *et al.*, 1999). Behavioral challenges, in combination with pharmaco-challenges, reveal distinctive sensitivity of serotonergic receptors in violence-prone individuals (Coccaro *et al.*, 1997a, 1997b). Individuals who are characterized by their impulsive, hostile, antisocial and violent outbursts are more likely to show some deficiency in brain serotonin activity relative to those who engage in premeditated, calculating and instrumental aggressive acts (Stoff and Vitiello, 1996; Vitiello and Stoff, 1997).

One of the most important areas in clinical research during the past decade is the epigenetic analysis of the interactions between allelic variants for one of the genes that encode for serotonergic receptors, transporters or enzymes, with life experiences during critical developmental periods as determinants for adult antisocial and violent behavior. As discussed below, only after maltreatment in childhood do boys with the low monoamineoxidase A (MAOA) activity allele develop violent behavior as adults, whereas the functional polymorphism for high MAOA activity and no childhood maltreatment do not engender such adult problem behavior (Caspi *et al.*, 2002).

Animal models of aggression rely on confrontations between a territorial resident and an intruder, as is typical for mice (*M. musculus*) or between a dominant male in a colony and a challenger, as is characteristic of socially organized rats (*R. norvegicus*) or troops of monkeys (e.g., *M. mulatta*, *C. aethiops*, *S. sciureus*) (Miczek *et al.*, 2002). The resident or dominant individual engages in a well-scripted sequence of offensive acts, postures and displays, to which the intruder reacts with defensive responses and eventually submits or, if possible, flees. These patterns of agonistic interactions have evolved in a species-typical manner (as illustrated in Figure 1). Animal species that disperse during their reproductively active phase of life, and those that live cohesively, engage in aggressive behavior in order to secure the resources for reproduction or to protect the offspring. Recent efforts aim to model pathologically violent behavior in humans by a focus on genetic and environmental conditions that engender escalated aggressive behavior in mice and rats relative to the species-normative patterns (Table 1). These efforts focus on subsets of individuals that engage in very high levels of aggressive behavior due to genetic selection, consumption of a moderate dose of alcohol, social instigation, or pharmacological insults (Miczek *et al.*, 1998a, 2002, 2007; Miczek and Fish, 2005; Haller and Kruk, 2006; Natarajan *et al.*, 2009).

Figure 1 Mouse agonistic behavior. Behaviors of resident and intruder mice engaged in an aggressive confrontation: (A) the resident leaps and bites the intruder as the intruder attempts to escape; (B) the resident (*right*) threatens as the intruder (*left*) holds a defensive upright posture; (C) the resident investigates the intruder's anogenital region; (D) the resident pursues the fleeing intruder; (E) both resident and intruder engage in a mutual upright defensive posture. Copyright 1978 by Springer-Verlag. Reprinted with permission from Miczek and O'Donnell (1978).

Table 1 Types of aggressive behavior

	Situational or experimental variable	Agonistic behavioral measurements	References
(A) Species-typical aggressive behavior			
Dominance behavior, mainly in primates and rats	In the stable colony where dominance hierarchy to be established and maintained.	Frequency and duration of agonistic acts, postures, and displays including supplants, threat, pursue, and fight.	Mehlman *et al.*, 1994 Higley *et al.*, 1996a Fairbanks *et al.*, 1999 Bennett *et al.*, 2002 Bernstein *et al.*, 1974 Steiniger, 1950 Vandenbergh, 1967
Territorial aggression (*resident–intruder* test), mainly in mice and hamsters	In the established territory. In the laboratory, home-cage of experimental male (*resident*) where it is pair-housed with a female. A male stimulus animal (*intruder*) that is group housed with other males is introduced into resident's cage.	Frequency of attack bite, sideways threat, tail-rattle, pursue, upright posture. Latency to the first bite.	Van Oortmerssen and Bakker, 1981 Eibl-Eibesfeldt, 1950 Miczek and O'Donnell, 1978 Crawley *et al.*, 1975
Maternal aggression, mainly in rats, mice, and hamsters	Home-cage of lactating females from postpartum days 1 to 7. Either male or female intruder is introduced into dam's cage.	Frequency of attack bite (especially directed at the snout and the face), sideways threat, tail-rattle, pursue, upright posture. Latency to the first bite	Hurst, 1987 Sgoifo *et al.*, 1992 Lonstein and De Vries, 2000 Noirot *et al.*, 1975 Haney *et al.*, 1989
Female aggression, mainly in primates and rodents.	Dominant hierarchy among female monkeys. In the laboratory settings, female rodent pair-housed with a breeding male. Sexually matured female is introduced as an intruder.	Harrassing attacks by dominant female. Frequency of attack bite, sideways threat, tail-rattle, pursue, upright posture. Latency to the first bite	Smuts, 1986 Palanza *et al.*, 2005 DeBold and Miczek, 1981 Zitzman *et al.*, 2004
Isolation-induced aggression (similar to territorial aggression)	Male isolated for some time, ranging from 24 hours to 8 weeks prior to *resident–intruder* encounter.	Frequency of attack bite, sideways threat, tail-rattle, pursue, upright posture. Latency to the first bite.	Malick, 1979 Valzelli and Bernasconi, 1979 Cairns and Nakelski, 1971 Yen *et al.*, 1959
(B) Escalated aggressive behavior			
Alcohol-heightened aggression, mainly in rats and mice	Animals receive ethanol (1.0 g/kg) intraperitoneally or orally before the *resident–intruder* encounter.	(These three methods measure aggressive behavior using the *resident-intruder* procedure) Frequency of attack bite, sideways threat, tail-rattle, pursue, upright posture. Latency to the first bite.	Peeke and Figler 1981 Cherek *et al.*, 1984 Blanchard *et al.*, 1987 Miczek *et al.*, 1992, 1998 Miczek and de Almeida, 2001
Social provocations (instigations), mainly in hamsters, mice, and rats	A resident male pre-exposed to another breeding male in his home-cage without direct agonistic interaction (stimulus animals are behind protective screen), followed by *resident–intruder* encounter.		Heiligenberg, 1974 Potegal and Tenbrink, 1984 Potegal, 1991 Fish *et al.*, 1999
Frustration-heightened aggression	A resident male trained to obtain rewards. Before the *resident–intruder* encounter, the reward is omitted.		Berkowitz, 1993 De Almeida and Miczek, 2002 Cherek *et al.*, 1997
Affective defense ('rage'), mainly in cats	Electrical stimulation (0.2–0.8 mA, 63 Hz, 1 ms per half cycle duration) delivered in medial hypothalamus or midbrain periaqueductal gray.	Hissing, arching of the back, retraction of the ears, piloerection, unsheathing of the claws, papillary dilatation and paw striking	Leyhausen, 1979 Siegel *et al.*, 1999 Hess, 1954

The emergence of the serotonin–aggression link

Given the prominent role of serotonin in the neurobiology of aggression, it is not surprising that 5-HT signaling mechanisms are promising pharmacotherapeutic targets in the management of impulsive, aggressive and violent behavior, as discussed in the next section. The initial evidence suggesting that 5-HT might have an inhibitory action on aggressive behavior stemmed from a study in isolated mice. *Post-mortem* assays of the brains of isolated aggressive mice revealed lower brain 5-HT and 5-HIAA contents and turnover when compared with group-housed control mice (Giacalone *et al.*, 1968). In humans, concordant evidence of the involvement of 5-HT function in violent behavior originated from US marines with a life history of repeated aggressive behavior. In this population of marines, the levels of the 5-HT metabolite 5-HIAA detected in the cerebrospinal fluid (CSF) were inversely correlated with the life history of aggression (Brown *et al.*, 1979a, 1979b). The evidence on suppressed CSF 5-HIAA as a biomarker for increased aggressive behavior is, however, divided. On the one hand, the proposal that lower CSF levels of 5-HT and/or 5-HIAA were associated with a greater propensity for violent outbursts (the 'serotonin deficiency hypothesis') was further supported by studies in patients with antisocial personality disorder and others showing impulsive aggression and type-II alcoholism (Linnoila *et al.*, 1983; Kruesi *et al.*, 1990; Virkkunen and Linnoila, 1993a; Pihl and LeMarquand, 1998; Lesch and Merschdorf, 2000; Placidi *et al.*, 2001). Likewise, negative correlations between CSF 5-HIAA levels and a history of impulsive aggression were observed in non-human primates (Higley *et al.*, 1992, 1996b; Westergaard *et al.*, 1999; Manuck *et al.*, 2003; Fairbanks *et al.*, 2004). On the other hand, many studies in humans, non-human primates and rodents failed to find such inverse correlations between 5-HT or 5-HIAA and levels of aggressive behavior (see, for example, Yodyingyuad *et al.*, 1985; Miczek and Donat, 1989; Gardner *et al.*, 1990; Lidberg *et al.*, 2000; van der Vegt *et al.*, 2003a). This observation raises the possibility that the importance of serotonergic deficiency for trait aggression could be restricted to groups of individuals who engage in aggressive behavior in an impulsive, escalated manner, rather than in individuals with other types of aggressive behavior (Coccaro, 1989; Virkkunen and Linnoila, 1993b; Mehlman *et al.*, 1994). For example, in male rats showing species-typical aggression, 5-HT and 5-HIAA levels in the CSF were actually positively correlated with trait aggression, indexed as percentage of time spent in aggressive behavior towards a conspecific male in a resident–intruder confrontation (van der Vegt *et al.*, 2003a). In that study, when 5-HT terminals were lesioned with the neurotoxin 5,7-dihydroxytryptamine (5,7-DHT), producing a reduction in CSF 5-HT levels, aggressive behavior was not affected (van der Vegt *et al.*, 2003a). Thus, it seems likely that when aggression occurs in a more species-typical social context, with an ethological function and purpose (i.e., establishment of territory and dominance), the association with 5-HT function is different (if not opposite) from what is observed in pathological, escalated types of aggression.

Similar patterns are observed using the pharmaco-challenge approach to assess 5-HT function in individuals with a history of aggressive outbursts, administering a 5-HT agonist and assaying an endocrine response in the blood. In these studies, patients with personality and mood disorders (Coccaro *et al.*, 1989, 1997b), violent alcoholics (Moss *et al.*, 1990; Handelsman *et al.*, 1996) and violent offenders (O'Keane *et al.*, 1992) showed a blunted plasma prolactin response after receiving administration of 5-HT agonists (such as fenfluramine or buspirone). These results supported the proposal of an inverse correlation between 5-HT function and a history of impulsive aggressive behavior, and implicated 5-HT projections to the hypothalamus as an important pathway associated with aggression (Coccaro, 1992). However, similarly to studies of CSF 5-HT levels, some studies using the pharmaco-challenge approach also failed to reveal an association between prolactin responses and the propensity to engage in aggressive acts (Halperin *et al.*, 1994; Pine *et al.*, 1997; Handelsman *et al.*, 1998).

Due to the prolonged period that separates the time when 5-HT, 5-HIAA or prolactin levels are assayed from the actual aggressive events, the results supporting a serotonin deficiency hypothesis reflect a *trait* for impulsive aggression in individuals with lower basal 5-HT/5-HIAA levels, but offer little explanation regarding the role for 5-HT in the preparation, execution and recovery from the actual aggressive acts and behaviors (Coccaro, 1989; Miczek *et al.*, 1994). Furthermore, analysis of transmitter and metabolites in the CSF lacks neuroanatomical resolution. It is not possible to determine which 5-HT brain pathways contribute primarily to the 5-HT/5-HIAA contents that are measured in the CSF, which can generate a confounding factor in the interpretation of these data.

Using *in vivo* microdialysis techniques, transient changes in 5-HT extracellular levels were monitored before, during, after, and in anticipation of an aggressive encounter in rats. In one study, reduced 5-HT levels in the prefrontal cortex were revealed during the aggressive confrontation, while no changes in 5-HT were detected in another terminal region, the nucleus accumbens (Van Erp and Miczek, 2000). Interestingly, reduced 5-HT in the prefrontal cortex was detected for at least 1 hour after the termination of the fight, and after the fight dopamine levels were increased in the nucleus accumbens (Figure 2; Van Erp and Miczek, 2000). On the other hand, in rats that

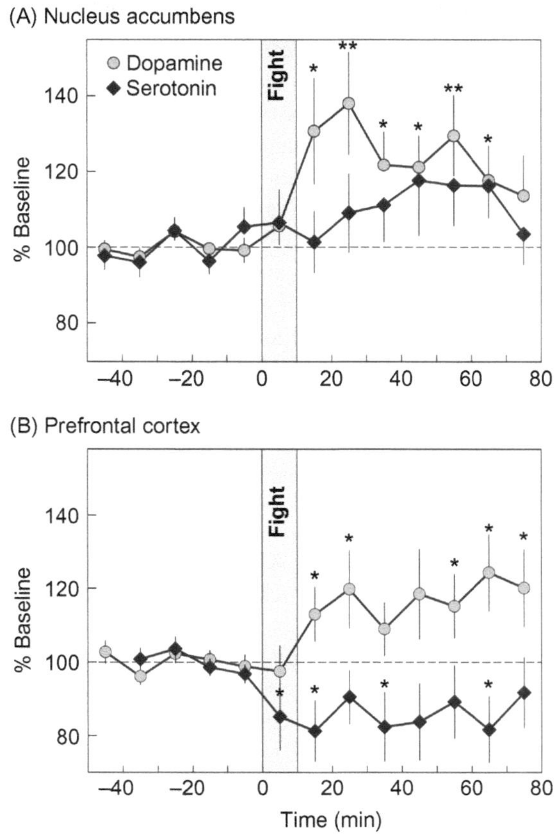

Figure 2 Dopamine and serotonin during aggression. Measurements of extracellular dopamine and serotonin via *in vivo* microdialysis in resident male rats before, during and after a confrontation with an intruder. (A) In the nucleus accumbens (*top panel*), dopamine levels (gray circles) rise and remain elevated after the confrontation, while serotonin levels (black diamonds) do not significantly change. (B) In the prefrontal cortex (*bottom panel*), dopamine levels rise after the confrontation, while serotonin levels decline and remain lower after the confrontation. Samples were collected every 10 minutes and levels are expressed as mean (SEM, vertical lines) percent of baseline. Baseline was measured for 50 minutes before the fight. The vertical light-gray bar indicates the occurrence of the 10-minute fight. * and ** represent significance from baseline (dashed line) at the $P < 0.05$ and $P < 0.01$ levels, respectively. Copyright 2000 by the Society for Neuroscience. Reprinted with permission from Van Erp and Miczek (2000).

were expecting an aggressive confrontation (which had previously occurred every day at the same time), 5-HT levels were reduced in the nucleus accumbens even in the absence of the actual confrontation (Ferrari *et al.*, 2003). Thus, mesocorticolimbic 5-HT activity is transiently modified during the aggressive confrontation, and in the phases that precede and follow the aggressive encounter (Van Erp and Miczek, 2000; Ferrari *et al.*, 2003).

Adding another degree of complexity to the serotonin–aggression link is the consideration that the effects of 5-HT are mediated by a wide variety of receptors, and 5-HT action is modulated by other synaptic and non-synaptic components (e.g., the 5-HT transporter or the synthetic enzyme, tryptophan hydroxylase). Monitoring the extracellular 5-HT content in one particular brain region reveals the summation of 5-HT release, uptake and presynaptic gating processes in that region. Pharmacological manipulations of pre- and postsynaptic components of the 5-HT system, particularly the 5-HT_1 and 5-HT_2 receptor families as well as 5-HT transporter, markedly affect aggressive behavior, and will be discussed below. Furthermore, there are prominent genetic influences on aggressive behavior and temperament. Heritability of antisocial behavior has been estimated to range from 28 percent to 47 percent (Coccaro *et al.*, 1997c). In animal models, selective breeding has been successful for aggressive behavior in mice and rats (Lagerspetz *et al.*, 1968; Cairns and Nakelski, 1971; van Oortmerssen and Bakker, 1981), which also suggests that genetics can affect aggression as a trait. It is of interest to discuss the genes that modulate aggression by their direct or indirect regulation of 5-HT pathways. Gene targeting techniques are useful to elucidate the effects of one particular gene on behavior, and analysis of gene polymorphisms can uncover a genetic basis for individual differences in aggressive behavior. Here, we will further discuss the serotonin–aggression link, focusing on recent developments and tools for pharmacological and genetic assessment of 5-HT function.

Modulation of 5-HT synthesis and aggression

Tryptophan depletion or supplementation and aggression

Removing the aromatic acid precursor L-tryptophan from the diet causes depletion of plasma tryptophan and brain serotonin (Delgado *et al.*, 1990; Nishizawa *et al.*, 1997). Acute tryptophan depletion in healthy male volunteers promotes increased aggressive responding in experimental models of human aggression (Pihl *et al.*, 1995; Moeller *et al.*, 1996; Bjork *et al.*, 1999), suggesting that acutely decreased levels of 5-HT could be associated with increased display of aggressive behavior. In one of these studies, men received a low-tryptophan diet for 24 hours, followed by a tryptophan-free amino acid mixture drink on the test day, and aggressive responding was assessed using the Point Subtraction Aggression Paradigm. Aggression responding was increased 5–6 hours after drinking the tryptophan-free mixture, and at the same time corresponding low levels of plasma tryptophan were detected (Moeller *et al.*, 1996). Conversely, supplementation of tryptophan in the diet has been shown to increase 5-HT synthesis in the brain, resulting in decreased irritability and aggression in experimental human settings (see Young and Leyton, 2002, for a review). According to these results, a proximal and direct role for 5-HT is proposed whereby acute reductions in 5-HT would promote increased risk to engage in violent behavior, whereas chronic maintenance of higher 5-HT concentrations would have protective anti-aggressive effects.

In rodents, depletion of tryptophan from the diet reduces 5-HT synthesis in the brain (Gessa *et al.*, 1974), and further supports the negative correlation between 5-HT levels and aggressive behavior. While a tryptophan-free diet resulted in augmented isolation-induced aggression in mice and shock-induced defensive reactions in rats, tryptophan supplementation reduced some types of aggressive behavior (Kantak *et al.*, 1980a, 1980b; Lasley and Thurmond, 1985). In humans and lower animals, however, it remains unclear whether or not manipulation of tryptophan availability, which presumably regulates 5-HT synthesis, will also result in augmented or decreased brain 5-HT release and function (Young and Leyton, 2002). Overall, results with tryptophan (or 5-HTP) depletion or supplementation should be interpreted in light of some limitations. These manipulations produce effects that are not specifically related to 5-HT, including alteration of other amines (e.g., melatonin and tryptamine), alterations in protein synthesis, and changes in mood and temperament (Young and Leyton, 2002).

Tryptophan hydroxylase and aggression

Tryptophan hydroxylase is the rate-limiting enzyme in the synthesis of serotonin, catalyzing the formation of 5-hydroxytryptophan (5-HTP) from the dietary precursor l-tryptophan.

Pharmacology of tryptophan hydroxylase and aggression

In preclinical studies, para-chlorophenylalanine (pCPA), an irreversible inhibitor of tryptophan hydroxylase, is used as a pharmacological tool to deplete 5-HT (Koe and Weissman, 1966; Jequier *et al.*, 1967). Significant depletion of 5-HT levels is found for approximately 2 weeks after pCPA administration, with peak depletion (approximately 80–90 percent depletion) 2–3 days after pCPA treatment (Aghajanian *et al.*, 1973; Vergnes *et al.*, 1986). Behaviorally, pCPA treatment increased defensive reactions to pulses of electric shock to the feet of a pair of rats, and these effects could be attributed to altered pain sensitivity after serotonin depletion (Ellison and Bresler, 1974; Knutson *et al.*, 1979). Inhibition of tryptophan hydroxylase also escalated offensive, but not defensive, aggression in a resident–intruder encounter in rats (see, for example, Vergnes *et al.*, 1986; Keele, 2001). These results suggest that depletion of 5-HT would facilitate and/or promote escalated levels of aggressive behavior in rodents. Interpretation of these findings is complicated due to the pervasive effects of depletion of whole brain 5-HT that extend from changes in perception, sensory functions and motor routines that are necessary for the display of different types of aggression.

Genetics of tryptophan hydroxylase and aggression

Variations in tryptophan hydroxylase (TPH) activity can directly affect 5-HT neurotransmission. Several studies have shown that polymorphisms in the TPH1 or TPH2 gene change the enzyme activity and 5-HT level (Jonsson *et al.*, 1997 for the TPH1 gene; Zhang *et al.*, 2004, 2005; Lin *et al.*, 2007; and Cichon *et al.*, 2008 for the TPH2 gene). Two single nucleotide polymorphisms (SNPs) in tight linkage disequilibrium, A218C and A779C located in the intron of TPH1 gene (which are often referred to as the U (upper) and L (lower) alleles), have been shown to be associated with state and trait anger as assessed by interview and self report questionnaires (Manuck *et al.*, 1999; Rujescu *et al.*, 2002). Lower CSF 5-HIAA levels have been observed in healthy men, but not women, with the U allele (Jonsson *et al.*, 1997), whereas the lowest CSF 5-HIAA was observed in LL carriers diagnosed with antisocial alcoholism (Nielsen *et al.*, 1994). Apparently, both U and L alleles may relate to different types of aggression. Hennig *et al.* (2005) performed factor analysis on the Buss-Durkee Hostility Inventory (BDHI) and found two factors, named as ‘neurotic hostility’ and ‘aggressive hostility’. Neurotic hostility is characterized by irritability, resentment, verbal hostility and guilt, whereas aggressive hostility represents a more ‘cold’ aggression, including assault and negativism but low guilt. Individuals with the UU allele showed higher aggressive hostility but slightly lower neurotic hostility (Hennig *et al.*, 2005), whereas LL homozygote showed a higher level of impulsivity in the measurements that focused more on neurotic hostility (New *et al.*, 1998). This may explain controversial results about the association between the polymorphisms and aggression; depending on which dimensions of aggressive behavior are measured, the association pattern may change.

In mouse models, the focus has been on the Tph2 gene, because Tph2 is preferentially expressed in the brain whereas Tph1 is more widely distributed throughout the body (Walther *et al.*, 2003). Mice with an R439H point mutation in the Tph2 gene have strongly reduced 5-HTP, 5-HT and its acidic metabolite levels in brain (Beaulieu *et al.*, 2008). This mutant mouse showed increased aggressive behaviors in the social interaction test in a neutral arena where, typically, very low levels of aggression are seen (Beaulieu *et al.*, 2008). In contrast, another polymorphism in Tph2, C1473G, which inhibits Tph2 activity in the midbrain, reduced the intensity of aggression in mice (Osipova *et al.*, 2009). The mouse data on the R439H mutation in the Tph2 gene may be relevant to a large association study in children with attention-deficit hyperactivity disorder (ADHD), which identified a TPH2 gene polymorphism associated with impulsivity (Oades *et al.*, 2008).

Summary for 5-HT synthesis and aggression

Evidence from pharmacological manipulations suggests that reduction of 5-HT synthesis (tryptophan depletion and pCPA treatment) increases aggressive behaviors. However, those effects depend on the type of aggressive behavior examined. Genetic studies indicate that both alleles, L and U, of polymorphism in the TPH1 gene are involved in human aggression, but each allele is associated with a different type of aggression. There is support to suggest that reduced central 5-HT activity is implicated in abnormal 'irritable' impulsive aggression but not in species-typical aggressive behaviors (Mehlman *et al.*, 1994; Coccaro *et al.*, 1989; de Boer and Koolhaas, 2005).

5-HT receptors and aggression

Considerable evidence implicates 5-HT receptors in the neural mechanisms for aggressive behaviors in various animal species. In particular, receptors of the 5-HT_1 and 5-HT_2 families are of significance in the control of aggressive behavior, based on initial pharmacological studies (Miczek *et al.*, 2002; Olivier, 2004). Other receptor subtypes (e.g., $5\text{-HT}_{5,6,7,1e,1f}$) have not been explored as of yet, primarily because of a lack of adequate pharmacological tools. There is some suggestion of the involvement of 5-HT_3 receptors in aggression (Rudissaar *et al.*, 1999; Ricci *et al.*, 2004; McKenzie-Quirk *et al.*, 2005); however, further investigation will be needed. Here we will discuss chiefly the data implicating 5-HT_1 and 5-HT_2 receptors in aggressive behaviors, since abundant data are available for these receptors.

5-HT_1 receptors and aggression

Pharmacology of 5-HT_1 receptors and aggression

Clinically, the 5-HT_{1A} receptor partial agonist buspirone reduces aggressive behavior in mentally retarded patients (Ratey *et al.*, 1991; Kavoussi *et al.*, 1997). Preclinically, systemic injections of 5-HT_{1A} receptor agonists reduce aggressive behavior across several species, including fish, amphibian, birds, rodents, guinea pigs and non-human primates (Tompkins *et al.*, 1980; Dompert *et al.*, 1985; Lindgren and Kantak, 1987; Blanchard *et al.*, 1988; McMillen *et al.*, 1988; Haug *et al.*, 1990; Nikulina *et al.*, 1992; Olivier and Mos, 1992; Sanchez *et al.*, 1993; Bell and Hobson, 1994; Muehlenkamp *et al.*, 1995; Joppa *et al.*, 1997; Miczek *et al.*, 1998b; de Boer *et al.*, 1999, 2000; Sperry *et al.*, 2003; de Boer and Koolhaas, 2005; Clotfelter *et al.*, 2007; Ten Eyck, 2008a). Only one exception was observed, in fruit flies (*Drosophila melanogaster*), where 8-OH-DPAT treatment increased aggressive behavior (Johnson *et al.*, 2009). Although 5-HT_{1A} agonists have consistent anti-aggressive effect in vertebrates, most of those compounds have an unfavorable profile of side-effects, including sedation, slow motor routines, reduced social interest or stereotypic behavior in the dose range that produces effective reduction of aggressive behaviors (Olivier *et al.*, 1995; Joppa *et al.*, 1997; Miczek *et al.*, 1998b; de Boer *et al.*, 2005). Some 5-HT_{1A} agonists, at least in specific rat strains, can reduce aggressive behavior specifically, without changing the other non-aggressive behaviors (i.e., alnespirone and S-15535; de Boer *et al.*, 1999, 2000; de Boer and Koolhaas, 2005). It is possible that those compounds act on a subpopulation of 5-HT_{1A} receptors to exert this anti-aggressive effect, and thereby achieve higher behavioral specificity. Antagonists with preferential action at 5-HT_{1A} receptors, such as WAY-100635, blocked the anti-aggressive effect of 5-HT_{1A} agonists, while having no effect by themselves on aggression (Miczek *et al.*, 1998; Mendoza *et al.*, 1999; de Boer and Koolhaas, 2005).

5-HT_{1B} receptor agonists seem more specific in their anti-aggressive effects than 5-HT_{1A} agonists. Systemic injection of 5-HT_{1B} agonists reduced aggressive behavior without sedation or motor or sensory impairment in mice and rats (Olivier *et al.*, 1990; Fish *et al.*, 1999; de Almeida *et al.*, 2001; de Almeida and Miczek, 2002; Miczek *et al.*, 2002, 2004; Olivier, 2004; de Boer *et al.*, 2005). Figure 3 shows that 5-HT_{1B} agonists effectively decreased escalated level of aggressive behavior (Fish *et al.*, 1999; de Almeida and Miczek, 2002). These effects were antagonized by the $5\text{-HT}_{1B/1D}$ antagonist GR-127935, which confirmed that the anti-aggressive effects of these compounds were mediated by 5-HT_{1B} receptors (de Boer and Koolhaas, 2005). However, it is known that there is an amino acid difference in the binding domain of 5-HT_{1B} receptors between humans and rodents, and therefore pharmacological sensitivity and specificity of 5-HT_{1B} agonists in rodents may differ from those in humans (Olivier, 2004).

It remains to be resolved exactly how and where 5-HT_{1A} and 5-HT_{1B} receptor agonists modulate 5-HT neurotransmission and reduce aggressive behaviors. One site of action for 5-HT_1 receptors is the autoreceptors,

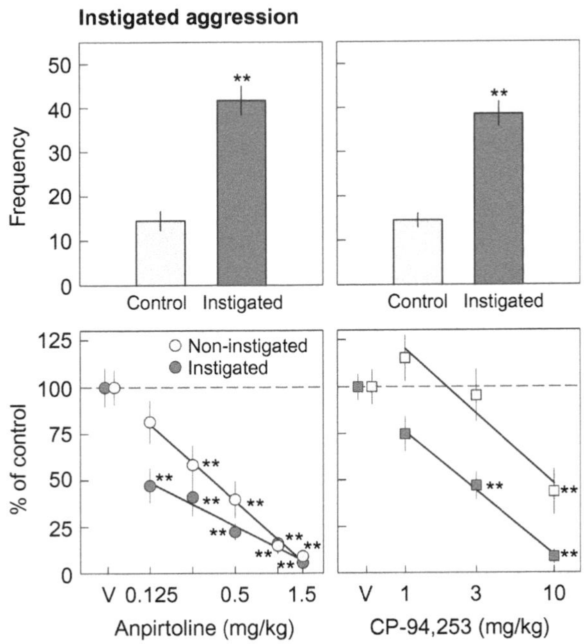

Figure 3 (*Top panels*) Effects of social instigation on aggressive behavior by a resident mouse toward a male intruder. Bars represent the mean frequency ± SEM (vertical lines) of attack bites under control (light-gray) and instigated (dark-gray) conditions. Asterisks denote statistical significance from control ($^{**}P < 0.01$). (*Bottom panels*) Preferential reduction of instigated aggressive behavior by the 5-HT_{1B} agonist anpirtoline (left panel, filled circles) and CP-94,253 (right panel, filled squares). Symbols represent the mean frequency of attack bites, expressed as a percentage of vehicle (*V*) baseline, ±SEM. Light-gray symbols represent non-instigated fighting and dark-gray symbols represent instigated levels of fighting. Asterisks denote significance from vehicle baseline ($P < 0.05$).Copyright 2002 by Springer-Verlag. Reprinted with permission from Miczek *et al.* (2002).

which can be localized at either the somata or the synaptic terminals of 5-HT neurons. Stimulation of 5-HT_1 autoreceptors can inhibit 5-HT neuronal activity and 5-HT impulse flow. On the other hand, agonists that act postsynaptically on projection sites of 5-HT_1 receptors (heteroreceptors) may actually be based on increased 5-HT neurotransmission. To identify neural sites of action of the anti-aggressive effect by 5-HT_{1A} and 5-HT_{1B} receptor agonists, intracerebral microinjection techniques have been used.

5-HT in the forebrain derives mainly from ascending projections that originate in the dorsal and median raphe nuclei (DRN and MRN, respectively) in rodents. Microinfusion of 5-HT_{1A} agonists into the DRN consistently reduced aggressive behavior in rats and mice (Mos *et al.*, 1993; van der Vegt *et al.*, 2003b; Faccidomo *et al.*, 2008). 5-HT_{1A} agonist injection into the MRN also reduced aggressive behavior of female rats (de Almeida and Lucion, 1997). Moreover, intra-DRN microinjection of 5-HT_{1B} agonists reduced aggressive behavior in mice (Bannai *et al.*, 2007; Faccidomo *et al.*, 2008). However, intra-raphe injection of either 5-HT_{1A} or 5-HT_{1B} agonists concomitantly reduced activity and social interest (Mos *et al.*, 1993; Bannai *et al.*, 2007; Faccidomo *et al.*, 2008). Microinjection of 5-HT_{1A} and 5-HT_{1B} receptor agonists reduced 5-HT neuronal activity and 5-HT release in the projection sites (Sprouse and Aghajanian, 1987; Bonvento *et al.*, 1992; Adell *et al.*, 2001); therefore, these results suggest that reduction of 5-HT neurotransmission reduces aggressive behaviors, which challenges the 5-HT deficiency hypothesis (de Boer and Koolhaas, 2005).

On the other hand, the importance of 5-HT_{1A} and 5-HT_{1B} receptors on projection sites has been emphasized in studies using lesion or depletion of raphe 5-HT neurons either by systemic administration of the tryptophan hydroxylase inhibitor pCPA or by intracerebral injection of the 5-HT neurotoxin, 5,7-dihydroxytryptamine (5,7-DHT). The logic is that by depleting 5-HT neurons, one can observe the effect of 5-HT_1 agonists on postsynaptic receptors. If only somata and presynaptic autoreceptors are the site of action of 5-HT_{1A} and 5-HT_{1B} agonists, pCPA or 5,7-DHT treatment should eliminate the anti-aggressive effect of 5-HT_{1A} and 5-HT_{1B} agonists. However, lesions or depletion of 5-HT neurons did not affect anti-aggressive effects of 5-HT_{1A} and 5-HT_{1B} agonists (Sijbesma *et al.*, 1991; Sanchez and Hyttel, 1994; Miczek *et al.*, 1998b; de Almeida *et al.*, 2001), suggesting postsynaptic 5-HT receptors as critical sites of action. This interpretation needs to be qualified, since these neurotoxic lesions are typically incomplete and spare a considerable percentage of presynaptic receptors (de Boer and Koolhaas, 2005). In the projection sites of 5-HT neurons, it is often hypothesized that 5-HT_{1B} receptors modulate 5-HT release from synaptic terminal as autoreceptors, whereas both 5-HT_{1A} and 5-HT_{1B} receptors change activation of postsynaptic neurons (Olivier and Mos, 1992). Figure 4 shows a summary of the effects of regional 5-HT_{1A} and 5-HT_{1B} receptor populations on different kinds of aggressive behaviors. A reduction of territorial aggression was observed when 5-HT_{1A} and 5-HT_{1B} agonists are microinjected into the medial pre-optic area, lateral septum or orbitofrontal cortex in mice (Cologer-Clifford *et al.*, 1997; de Almeida *et al.*, 2006). Vasopressin-escalated aggression in hamsters was reduced by microinjection of a 5-HT_{1A}, but not 5-HT_{1B}, agonist into the anterior hypothalamus (Ferris *et al.*, 1999). Interestingly, a 5-HT_{1B} receptor agonist increased alcohol-heightened aggressive behavior when it was microinjected into the medial prefrontal cortex (Faccidomo *et al.*, 2008). This escalation may be due to reduction of 5-HT neurotransmission by acting on presynaptic terminals of 5-HT neurons. However, a 5-HT_{1A} agonist increased maternal aggression when it

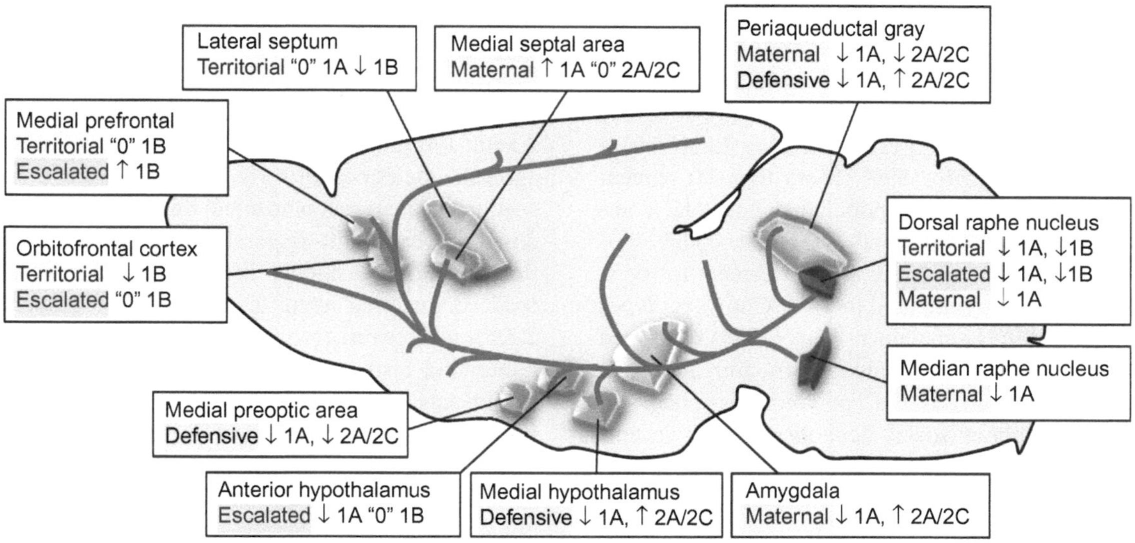

Figure 4 Modulation of aggressive behaviors by microinjections of $5HT_{1A}$, $5\text{-}HT_{1B}$, and $5\text{-}HT_{2A/2C}$ receptor agonists. Text boxes show that local injection of $5\text{-}HT_{1A}$ receptor (1A), $5\text{-}HT_{1B}$ receptor (1B), or $5\text{-}HT_{2A/2C}$ receptor (2A/2C) agonist increases (↑), decreases (↓) or has no effect ('0') on territorial, escalated, maternal, and defensive aggressive behaviors. Serotonergic neurons originated from raphe nuclei project to several brain areas. To see the full color version of this figure please refer to the colour plate in the back fo the book. Copies produced via our print on demand service do not contain color plate. If your copy does not have the color plate, please go to this website to view the figure in color www.elsevierdirect.com/companions/9780123746344

was microinjected into the medial septal area (de Almeida and Lucion, 1997), and this result is inconsistent with the 5-HT deficiency hypothesis. Regarding defensive aggression, $5\text{-}HT_{1A}$ agonist in the periaqueductal gray (PAG) and medial hypothalamus reduced defensive hissing in cats (Shaikh *et al.*, 1997; Hassanain *et al.*, 2003). Most of the evidence suggests that $5\text{-}HT_{1A}$ and $5\text{-}HT_{1B}$ agonists reduce territorial aggression by acting at somatodendritic presynaptic and postsynaptic receptor sites. In contrast, in maternal and escalated types of aggression, increased aggressive behavior was observed after a $5\text{-}HT_{1A}$ agonist was injected into medial septal area and after a $5\text{-}HT_{1B}$ agonist was injected into medial prefrontal cortex. Further studies are required to delineate the mechanisms for those pro-aggressive effects.

Genetics of 5-HT$_1$ receptors and aggression

In contrast to the strong implication of $5\text{-}HT_{1A}$ receptor in aggressive behavior from pharmacological research, no prominent linkage has been reported with the polymorphisms in the gene for $5\text{-}HT_{1A}$ and aggression so far. However, there is evidence for a correlation between $5\text{-}HT_{1A}$ receptor expression and aggression. Human PET studies found a higher $5\text{-}HT_{1A}$ receptor distribution in the prefrontal cortex of subjects with higher aggression scores in self-report questionnaires (Witte *et al.*, 2008). Also, rats selected by higher defensive reactions showed reduced $5\text{-}HT_{1A}$ receptor expression in several brain areas (Popova *et al.*, 1998). It is possible that the polymorphisms which directly or indirectly affect $5\text{-}HT_{1A}$ receptor transcription may be associated with aggressive or defensive responses. $5\text{-}HT_{1A}$ receptor knockout mice engaged in less aggressive behavior relative to wild-type controls, which also implicates possible involvement of $5\text{-}HT_{1A}$ gene in aggressive behavior (Zhuang *et al.*, 1999). Given the consistent pharmacological data on anti-aggressive effects of $5\text{-}HT_{1A}$ agonists, lack of complementary genetic data is surprising.

On the other hand, data from several genetics studies have shown the involvement of $5\text{-}HT_{1B}$ receptor in aggressive behavior. Male mice with disrupted $5\text{-}HT_{1B}$ receptor expression ($Htr1b^{-/-}$) increased aggressive behavior in the resident–intruder test after a month of isolation (Saudou *et al.*, 1994; Bouwknecht *et al.*, 2001). However, due to very low (close to zero) levels of aggressive behavior in the wild-type mice (129/Sv-ter), the number of attacks in $Htr1b^{-/-}$ was low compared to other strains of mice. These mice displayed behavioral disinhibition in other behavioral tests, including hyperlocomotor activity (Ramboz *et al.*, 1995; Brunner *et al.*, 1999), drug intake (Crabbe *et al.*, 1996; Rocha *et al.*, 1998), anxiety (Brunner *et al.*, 1999; Malleret *et al.*, 1999) and autonomic hyperreactivity to novelty (Bouwknecht *et al.*, 2001). Female $Htr1b^{-/-}$ mice also showed increased aggressive behavior during the postpartum period (Brunner and Hen, 1997). These results suggest that $5\text{-}HT_{1B}$ receptor may have a role in inhibiting aggressive and impulsive behaviors, although the genetic deletion and pharmacological blockade of $5\text{-}HT_{1B}$ result in opposite outcomes.

Several polymorphisms have been reported in the 5-HT_{1B} gene; SNPs consisting of G861C and C129T have been the most commonly investigated. Both SNPs are non-synonymous transversions, which do not affect the amino acid structure of the 5-HT_{1B} receptor (Lappalainen *et al.*, 1995; Huang *et al.*, 1999). More than 80 percent of individuals in the general population have the combination of C129 and G861, and 5-HT_{1B} receptor binding in *post-mortem* human brain was 20 percent lower in C129/G861 genotypes compared to T129/C861 genotypes (Huang *et al.*, 1999). Lappalainen *et al.* (1998) examined the linkage of 5-HT_{1B} with alcoholism and antisocial personality disorder, and G861C polymorphism had significant linkage with antisocial alcoholism in two groups. Specifically, C861, the SNP with higher 5-HT_{1B} receptor expression, was related to antisocial behavior in alcoholics. However, these associations between the 5-HT_{1B} polymorphisms (G861C, G261T or C129T) and aggression/antisocial behavior have not been seen in other studies in humans (Huang *et al.*, 1999; Kranzler *et al.*, 2002; Sinha *et al.*, 2003) or in dogs (van den Berg *et al.*, 2008). Linkage analysis of the other SNP in the 5′UTR region of the 5-HT_{1B} gene, A161T, found a significant relationship between this SNP and the history of aggression in subjects who committed suicide (Zouk *et al.*, 2007). Individuals with the T161 locus had higher levels of lifetime aggressive behaviors. T161 polymorphism had reduced transcriptional activity (Sun *et al.*, 2002), and thus lower 5-HT_{1B} receptor expression may be related to lifetime aggression in suicidal victims.

5-HT_2 receptors and aggression

Pharmacology of 5-HT_2 receptors and aggression

Atypical antipsychotic agents (e.g., risperidone) with significant antagonist action at 5-HT_{2A} receptors have been successfully used to reduce aggressive behaviors in patients of varying ages and with a wide range of diagnoses, such as depression, schizophrenia, dementia and mental retardation, and post-traumatic stress disorder (Czobor *et al.*, 1995; Buckley *et al.*, 1997; Fava, 1997; De Deyn *et al.*, 1999; Keck *et al.*, 2000; Buitelaar *et al.*, 2001; Zarcone *et al.*, 2001). In animal models, risperidone decreases aggressive behaviors but has been shown to concomitantly reduce mobility (Rodriguez-Arias *et al.*, 1998). 5-HT_{2A} specific antagonists, such as ketanserin, ritanserin and MDL 100907, reduced aggressive behavior in monoamine oxidase A (MAOA)-deficient mice (Shih *et al.*, 1999) and in socially isolated mice (White *et al.*, 1991; Sakaue *et al.*, 2002).

DOI and other substituted phenylisopropylamines, agonists that act at both 5-HT_{2A} and 5-HT_{2C} receptors, also reduce aggressive behavior in several species, including flies, amphibians, mice and rats (Sanchez *et al.*, 1993; Bonson *et al.*, 1994; de Almeida and Lucion, 1994; Muehlenkamp *et al.*, 1995; Olivier *et al.*, 1995; Ten Eyck, 2008b; Johnson *et al.*, 2009). However, similar to 5-HT_{1A} agonists, the effects of 5-HT_2 ligands are not behaviorally specific and are accompanied by sedative effects at the doses that exert anti-aggressive effects. Local administration of $5\text{-HT}_{2A/2C}$ agonist into periaqueductal gray (PAG) reduced maternal aggression in rats (de Almeida *et al.*, 2005), whereas microinjections into the medial hypothalamus and also into the PAG increased defensive hissing in cats (Shaikh *et al.*, 1997; Hassanain *et al.*, 2003; Figure 4). This latter effect is likely related to the role of $5\text{-HT}_{2A/2C}$ receptors in anxiety-like behavior (Lucki and Wieland, 1990; Nogueira and Graeff, 1995). The development of more selectively acting pharmacological tools will allow a more adequate differentiation of 5-HT_2 receptor subtypes. These novel tools promise to dissociate the anti-aggressive and sedative effects.

Genetics of 5-HT_2 receptors and aggression

Platelet 5-HT_{2A} receptor binding is increased in personality disorder patients and in a psychiatric population with greater lifetime aggression scores (McBride *et al.*, 1994; Coccaro *et al.*, 1997d). In *post-mortem* brains, lifetime aggression was positively correlated with prefrontal 5-HT_{2A} receptor binding in suicide victims (Oquendo *et al.*, 2006). Therefore, it is possible that polymorphisms that affect the level of expression of 5-HT_{2A} receptors can be associated with self-directed aggression. In some samples, significant linkage was found between polymorphisms in the 5-HT_{2A} receptors, T102C, A1438G and His452Tyr, and aggressive-impulsive trait or adolescent-onset antisocial behavior in humans (Bjork *et al.*, 2002; Assal *et al.*, 2004; Nomura *et al.*, 2006; Burt and Mikolaiewski, 2008). However, others have reported no such relation between aggression and 5-HT_{2A} polymorphisms (Khait *et al.*, 2005; Van den Berg *et al.*, 2008). Again, the success with pharmacotherapeutic management of aggressive patients using compounds with affinity for 5-HT_{2A} receptors would suggest that violence-prone individuals may be characterized by distinctive 5-HT polymorphisms.

Summary for 5-HT receptors and aggression

Genetic studies have shown that associations between aggressive behavior and polymorphisms of 5-HT_{1A}, 5-HT_{1B} and 5-HT_{2A} receptor genes failed to have a clear-cut pattern of results compared to relatively consistent pharmacological data. It may be important to determine where in the brain the gene expression is changed by the polymorphisms. Behavioral pharmacological studies reveal consistent anti-aggressive effects of 5-HT_{1A}, 5-HT_{1B} and 5-HT_{2A} receptor agonists, which are accompanied, to different extents, by non-specific effects. Novel pharmacological tools promote 5-HT_{1B} as a promising target for serenic drugs, despite the fact that these drugs are not currently used in clinical settings. Agonist microinjection experiments further reveal that the role of these receptors in aggressive behavior differs depending on the brain areas targeted. It should be kept in mind that pharmacological findings modulate *phasic* 5-HT activity, while genetic findings from gene knockout or polymorphisms reflect *trait* characteristics of 5-HT function.

5-HT transporters (5-HTT) and aggression

Serotonin transporters are uptake molecules that are responsible for tuning tonic and phasic levels of synaptic 5-HT in the brain in addition to peripheral 5-HT activity.

Pharmacology of 5-HTT and aggression

Blocking serotonin transporter molecules is effective in reducing and preventing aggressive behavior in humans and animals, due to increased brain 5-HT levels. Clinically, blocking 5-HT transporters with the administration of selective serotonin reuptake inhibitors (SSRIs) reduces aggressive outbursts and violent behavior in psychiatric patients (Coccaro and Kavoussi, 1997; Reist *et al.*, 2003; Bond, 2005; Barkan *et al.*, 2006; Blader, 2006). Similarly to the observed delay in therapeutic effects of SSRIs in the treatment of mood disorders, aggression-reducing effects of SSRIs are usually observed after chronic treatment (>3 weeks). For example, New *et al.* (2004) reported that chronic treatment with fluoxetine significantly reduced aggressive behavior in patients with borderline personality disorders, and normalized prefrontal cortical metabolic activity as assessed by PET (positron emission tomography). Overall, most evidence shows that SSRIs have a beneficial impact in clinical populations that present violent outbursts (Walsh and Dinan, 2001). However, there are occasional reports that SSRIs may facilitate aggressivity and suicidal behavior, and the causes for these unusual outcomes remain to be determined (Troisi *et al.*, 1995; Spigset, 1999).

In animal models, both acute and chronic treatment with SSRIs can dose-dependently reduce aggressive behavior (Olivier *et al.*, 1989; Delville *et al.*, 1996; Pinna *et al.*, 2003; Carrillo *et al.*, 2009). Acute administration of several SSRIs, such as fluoxetine, fluvoxamine and sertraline, reduces aggression in different contexts and species, including mice, rats and non-human primates (Delville *et al.*, 1996; Fuller, 1996; Cutler *et al.*, 1997; Ferris *et al.*, 1997; Sanchez and Meier, 1997; Fairbanks *et al.*, 2001; Ho *et al.*, 2001). Fewer studies used chronic administration of SSRIs in rodents, which would translate more readily to the clinical treatment that produces anti-aggressive effects in humans (see, for example, Mitchell and Redfern, 1997; Delini-Stula and Vassout, 1981). Interestingly, in placid laboratory rats that do not engage in any aggressive behavior, chronic SSRI treatment restores competent agonistic behavior (Mitchell *et al.*, 1991; Mitchell and Redfern, 1992; Mitchell, 2005). On the other hand, escalated levels of aggression induced by a moderate dose of alcohol were completely prevented during the course of 3 weeks of daily treatment with a highly selective SSRI, citalopram (Figure 5; Caldwell and Miczek, 2008). Baseline levels of aggression were also modestly reduced in those conditions. Thus, the outcomes of chronic SSRI treatment seem to vary depending on the type of aggression (species-typical vs escalated levels) and social context of the animals. Anti-aggressive effects of SSRIs are more prominent in conditions of augmented,

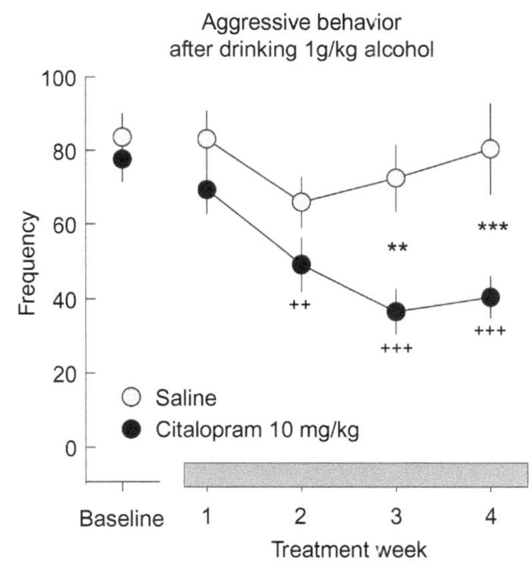

Figure 5 Effects of repeated, twice daily administration of the SSRI citalopram (10 mg/kg, i.p.) on aggressive behavior in mice after drinking 1.0 g/kg alcohol in operant self-administration panels. Frequency of aggressive acts (±SEM) is defined as sum of attack bites, threats, pursuits and tail rattles, and was analyzed in 5-minute confrontations against a male intruder, during the course of 4 weeks of citalopram (or saline control) treatment. + symbols represent differences from baseline (+ + $P < 0.01$; + + + $P < 0.001$); * symbols represent group (citalopram vs saline-control) differences ($^{**}P < 0.01$; $^{***}P < 0.001$). Copyright 2008 by Springer-Verlag. Adapted from Caldwell and Miczek (2008).

exaggerated aggressive behavior such as those promoted by alcohol (Caldwell and Miczek, 2008).

In support of the 5-HT deficiency hypothesis, acute administration, as well as chronic infusion, of citalopram (or the more potent and selective isomer escitalopram) both elevate extracellular levels of 5-HT in the prefrontal cortex of rats, suggesting increased cortical 5-HT as a putative therapeutic mechanism for SSRIs' effects on aggression and other mood disorders (Ceglia *et al.*, 2004). It is also possible that the anti-aggressive effects of another SSRI, fluoxetine, are primarily mediated by actions on neurosteroids and GABA transmission, and only secondarily via 5-HT (Pinna *et al.*, 2003, 2006). Several neuroadaptations are reported to occur in the 5-HT system as a result of chronic treatment with SSRIs. These neuroadaptations may include changes in serotonin transporter gene expression, binding and reuptake function in the hippocampus and raphe regions (Piñeyro *et al.*, 1994; Benmansour *et al.*, 1999), and 5-HT_{1A} autoreceptor desensitization (Blier and de Montigny, 1998; Ceglia *et al.*, 2004). These observations suggest that long-term effects of SSRIs possibly recruit other pre- and postsynaptic downstream mechanisms (e.g., 5-HT receptor-mediated signaling) for their therapeutic effects, rather than just transiently increasing 5-HT synaptic availability.

Genetics of 5-HTT and aggressive behavior

A variation in the length of the 5′-flanking transcriptional control region (promoter) of the 5-HTT gene (the serotonin-transporter-gene-linked polymorphic region; 5-HTTLPR) has been identified in humans (Heils *et al.*, 1996), great apes and rhesus monkeys (Lesch *et al.*, 1997). This variation affects the transcriptional activity of the 5-HTT gene, and the short length (*s*) allele reduces 5-HTT expression *in vitro* and lowers prolactin response to clomipramine in humans, which reflects reduced 5-HT function, compared to the long-length (*l/l*) homozygote (Heils *et al.*, 1995; Lesch *et al.*, 1996; Whale *et al.*, 2000). An association study in humans showed that individuals with one or two copies of the *s* allele (*s/s*, *s/l*) were characterized by higher anxiety, depression, hostility and aggression, and lower agreeableness, than *l/l* homozygotes in both sexes (Lesch *et al.*, 2000). Higher frequency of the *s* allele was observed in alcoholics accompanied by impulsivity and antisocial behaviors (type 2 alcoholism) compared to alcoholics without antisocial behavior (type 1) or healthy controls (Hallikainen *et al.*, 1999). Consistent with the human polymorphism, rhesus monkeys that possess the *s* allele engaged in higher rates of aggressive behaviors compared to *l/l* individuals (Lesch and Merschdorf, 2000; Jarrell *et al.*, 2008).

Again, a genotype–environment interaction can be found in 5-HTT polymorphism. Rhesus monkeys with the *s* allele had lower levels of 5-hydroxyindoleacetic acid (5HIAA) in CSF than *l/l* individuals when they were reared without their mother (peer-reared). This difference disappeared under the mother-rearing condition (Bennett *et al.*, 2002). Peer-reared monkeys showed increased aggression-related behavior as well as altered CSF 5HIAA levels (Kraemer *et al.*, 1989; Higley *et al.*, 1991). In humans, higher suicide ideations or attempts were observed in the individuals carrying the *s* allele than in *l/l* homozygotes when they encountered a number of stressful life events, but not in less stressful situations (Caspi *et al.*, 2003). Therefore, it is possible that individuals with the *s* allele are more vulnerable to the stressful challenges, and subsequently escalate their aggressive behaviors towards others and themselves. However, the effect of the *s* allele on aggression differed among sexes (Cadoret *et al.*, 2003) and even cultures (Baca-Garcia *et al.*, 2004).

Seemingly contrary results were observed in mice with a deletion of the 5-HTT gene (Slc6a4). Homozygote and heterozygote 5-HTT knockout mice on a C57BL/6J background showed fewer attack bites and longer latencies to start fighting relative to wild-type mice in the resident–intruder test (Holmes *et al.*, 2002). 5-HTT knockout mice have lower 5-HT uptake and higher extracellular 5-HT concentrations in the forebrain compared to wild types (Mathews, 2004). 5-HTT knockout mice underwent changes in more than 50 phenotypes, including morphological, physiological, sensory and behavioral functions (Murphy and Lesch, 2008), and thus those pleiotropic changes of other phenotypes may contribute to the reduction of aggressive behaviors in these mice.

Summary for 5-HTT and aggression

Consistent anti-aggressive effects of SSRI treatment among several species strongly suggest that the serotonin transporter is a promising therapeutic target for the management of impulsive, escalated aggression (for review, see Carrillo *et al.*, 2009). More studies are required to determine the precise neurobiological mechanisms recruited for the anti-aggressive effects of SSRIs to emerge, especially as a result of chronic treatment that induces pre- and postsynaptic neuroadaptive processes. The genetic studies in human and non-human primates also suggest 5-HTT polymorphisms (short allele) as a risk factor for violent traits, which seem to be particularly relevant in combination with environmental stress. Results from transgenic mice lacking 5-HTT are more difficult to interpret due to the wide variety of behavioral functions that are affected by the genetic manipulation.

Monoamine oxidase A (MAOA) and aggression

MAOA, an enzyme that metabolizes monoamine neurotransmitters including norepinephrine (NE), dopamine (DA) and 5-HT, plays a major role in modulating the synaptic availability of 5-HT and other amines.

Pharmacology of MAOA and aggression

Inhibition of MAOA reduces the oxidative metabolism of monoamines, thus presumably increasing the availability of 5-HT and other monoamines in the brain. Inhibitors of MAOA have been clinically used as first-generation antidepressants, especially in cases that are resistant to treatment with tricyclic antidepressants or SSRIs. Clorgyline was one of the initial selective MAOA inhibitors, but its irreversible inhibition of the enzyme posed serious side effects. More recent developments generated drugs such as moclobemide, a reversible inhibitor of clinical use. Despite the early importance of MAO inhibitors as antidepressants, there have been only a few preclinical studies that systematically evaluated the effects of MAO inhibitors on aggression (Miczek, 1987). For the most part, non-selective inhibitors of both MAOA and MAOB (e.g., phenelzine, isocarboxazid, tranylcypromine) show acute anti-aggressive effects in doses that also promote alterations in motor and other non-aggressive behaviors (DaVanzo *et al.*, 1966; Valzelli *et al.*, 1967; Welch and Welch, 1968; Sofia, 1969). Clinically, non-selective MAO inhibitors or selective MAOB inhibitors can be useful in the pharmacological management of personality disorders that include impulsive aggression and suicidal tendencies as important symptoms, but they are accompanied by an unfavorable profile of side effects (Hollander, 1999; Raj, 2004).

Genetics of MAOA and aggression

The gene for MAOA was the first candidate identified as a determinant in the susceptibility for aggression in humans. Brunner and colleagues (1993a) identified a large Dutch family with a syndrome of borderline mental retardation and dysregulation of impulsive aggression. All affected males showed aggressive outbursts, and some exhibited sexually aberrant behavior, attempted murder and arson. Linkage and sequence analyses showed that all affected males in this family possessed one missense mutation in the MAOA gene on the X chromosome, so that MAOA function was completely disturbed (Brunner *et al.*, 1993a). The affected males had higher serotonin and lower metabolites of NE, DA and 5-HT in the urine, whereas they had normal MAOB activity (Brunner *et al.*, 1993b). MAOA also is of significance in the probability of fighting in animals. Male mice with a disrupted MAOA gene on either C3H/He or 129Sv background expression showed escalated aggressive behaviors compared to wild types, as is evident by skin wounds among the cage-mates and a short latency to initiating attacks in the resident–intruder test (Cases *et al.*, 1995; Scott *et al.*, 2008). MAOA-deficient mice also showed a large increase in 5-HT and NE, and a subtle DA elevation, in the brain and liver (Cases *et al.*, 1995; Kim *et al.*, 1997). It is likely that the change of 5-HT function is the cause of the behavioral changes in the MAOA-deficient mice. Ketanserin and MDL100907, antagonists that preferentially bind to 5-HT_{2A} receptors, blocked the escalated aggression in the MAOA mutant mice (Shih *et al.*, 1999). Depletion of 5-HT by pCPA during the early developmental stage improved some behavioral and brain structural abnormality in the MAOA-deficient mice (Cases *et al.*, 1995, 1996).

Polymorphisms in the MAOA gene can alter MAOA activity and thus modulate catecholaminergic and serotonergic function. Variable-number tandem repeat (VNTR) polymorphism, which exists on the upstream region of the MAOA gene, regulates MAOA expression depending on the number of repeats: alleles with 3.5 or 4 repeats have 2- to 10-times higher transcription than 3 or 5 repeat alleles *in vitro* (Sabol *et al.*, 1998; Denney *et al.*, 1999). A prominent interaction between MAOA genotype and environment (maltreatment) on aggressive behavior has been reported (Figure 6). Under a stressful rearing environment, such as abuse or neglect, or with traumatic life events in the first 15 years of their lives, individuals with low MAOA expression (MAOA-L) polymorphisms showed a higher propensity for criminal arrest, a violent history and adolescent conduct disorder, and also higher aggressive disposition, in the self-report questionnaire compared to individuals with higher MAOA expression (MAOA-H) or MAOA-L individuals without abuse (Caspi *et al.*, 2002; Foley *et al.*, 2004; Kim-Cohen *et al.*, 2006; Widom and Brzustowicz, 2006; Frazzetto *et al.*, 2007; Weder *et al.*, 2009). If rearing environments were lumped together, the effect of the MAOA genotype disappeared (Fresan *et al.*, 2007) or sometimes MAOA-H individuals reported higher aggression using interviews and questionnaires (Manuck *et al.*, 2000, 2002). Therefore, individuals with the MAOA-L allele are vulnerable to environmental factors and show a high propensity to engage in aggressive behaviors only when they are in a stressful environment. These findings are consistent in males, but not in females (Sjoberg *et al.*, 2007). Rhesus monkeys also have a repeat length variation polymorphism (rhMAOA-LPR)

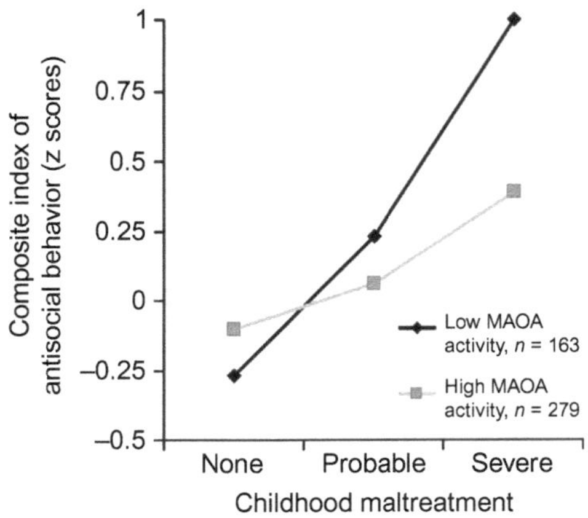

Figure 6 Means on the composite index of antisocial behavior as a function of MAOA activity and a childhood history of maltreatment. MAOA activity is the gene expression level associated with allelic variants of the functional promoter polymorphism, grouped into low and high activity; childhood maltreatment is grouped into three categories of increasing severity. The antisocial behavior composite is standardized (z score) to M = 0 and SD = 1; group differences are interpretable in SD unit differences (d). Copyright 2002 by The American Association for the Advancement of Science. Reprinted with permission from Caspi *et al.* (2002).

in the MAOA gene. Monkeys with a low-activity allele exhibited higher aggressive behavior and tended to attain higher dominance rank when they were reared by their mother. In contrast, when they were reared separately from their parents (peer-reared), monkeys with the low-activity allele engaged in less aggressive behavior (Newman *et al.*, 2005). This inhibition of aggression has been attributed to increased fear and anxiety in peer-reared monkeys (Higley and Suomi, 1986).

Neuroimaging studies have indicated pronounced differences in the volume and activity of limbic system and neocortical areas between individuals with MAOA-L and MAOA-H (see Buckholtz and Meyer-Lindenberg, 2008, for review). fMRI analysis in healthy human volunteers showed that MAOA-L males had smaller limbic and orbitofrontal volumes, and higher activity in amygdala and hippocampus during aversive recall (Meyer-Lindenberg *et al.*, 2006), which may be related to violent behavior in these individuals. Alia-Klein *et al.* (2008) reported that lower MAOA activity in cortical and subcortical brain areas is associated with high aggression, measured by self-report questionnaire, independent from MAOA polymorphism.

Summary for MAOA and aggression

Strong evidence for a role of MAOA in aggressive behaviors is supported by genetic studies of a deleterious mutation of the MAOA gene in humans and knockout mice. In contrast to the serotonin deficiency hypothesis, these results suggest that chronically increased 5-HT levels due to reduced MAOA function (*trait-like* change) may promote escalated aggressive displays. Gene polymorphism studies further revealed critical gene–environment interactions. Individuals with low MAOA activity and a history of early life maltreatment are particularly prone to engage in violent behavior. Conversely, pharmacological inhibition of MAO promotes phasic changes of 5-HT (*state-like* change), and generally reduces aggressive behaviors in doses that produce non-specific behavioral effects. It would be of interest to study more selective MAOA inhibitors and how they affect different types of aggression in rodents.

Other molecules that directly or indirectly affect 5-HT pathways and aggression

Here we will briefly discuss selected molecules that indirectly affect 5-HTergic pathways and modulate aggressive behaviors, based mainly on findings from gene knockout mouse studies, as summarized in Figure 7.

Pet-1 (also known as Fev)

Pet-1, one of the transcription factors, is specifically expressed in the raphe 5-HT neurons, and has a critical role in 5-HT neuronal development. Deletion of Pet-1 expression reduced 5-HT levels in the forebrain, and also depleted expression of TPH, 5-HTT and the vesicular monoamine transporter 2 (Vmat2). In the resident–intruder test, Pet-1 knockout mice (Pet-1$^{-/-}$) on a mixed C57BL/6J and 129Sv background engage in higher frequency and intensity of attacks toward a conspecific male (Hendricks *et al.*, 2003). These increases in aggressive behavior in Pet-1$^{-/-}$ mice are embedded in broad behavioral disruptions that extend to maternal and anxiety-like behaviors (Hendricks *et al.*, 2003; Lerch-Haner *et al.*, 2008), and it is possible that those other behavioral changes promote aggressive behaviors. Because general gene knockouts can affect an animal's development and induce compensatory changes of other gene expressions, investigation using conditional knockout of Pet-1 expression at discrete developmental stages promises to give more insight into the role of this gene in more specific behaviors.

Brain-derived neurotrophic factor (BDNF)

BDNF has several important roles in the neuron, including neuronal survival, development, differentiation and plasticity. Mice with decreased BDNF expression, including

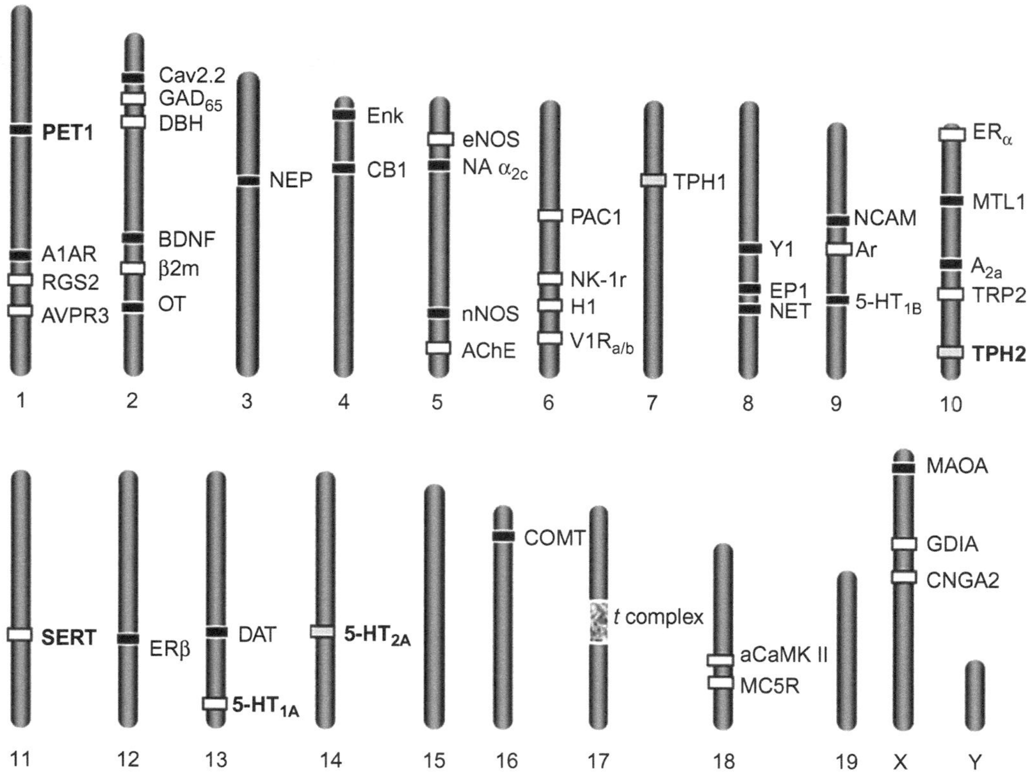

Figure 7 Genes/alleles implicated in mouse aggression.This figure shows representations of mouse chromosomes (19 autosomes + sex chromosomes) and the approximate location of genes related to aggressive behaviors. Genes related to 5-HT pathways are highlighted as bold. Genes that have been knocked out in mice and engendered heightened (black-filled bars) or suppressed (open bars) levels of aggressive behaviors are also indicated. No knockout mice data were available for some 5-HT related genes (gray bars). The *t* region and Y chromosome are also indicated because of the prominent effect on male aggressive behavior (see Miczek *et al.*, 2001, for review). Abbreviations and references for knockout studies: 5-HT_{1A}, serotonin receptor 1A (Zhuang *et al.*, 1999); 5-HT_{1B}, serotonin receptor 1B (Saudou *et al.*, 1994); A1AR, adenosine receptor A1 (Gimenez-Llort *et al.*, 2002); A_{2a}, adenosine receptor 2A (Ledent *et al.*, 1997); AchE, acetylcholinesterase (Duysen *et al.*, 2002); αCaMK II, alpha-calcium-calmodulin kinase II (Chen *et al.*, 1994); Ar, aromatase (Matsumoto *et al.*, 2003); AVPR3, arginine vasopressin 1B receptor (Wersinger *et al.*, 2002); β2m, beta2-microglobulin (Loconto *et al.*, 2003); Cav2.2, α1B subunit (Young *et al.*, 2002) of N-type Ca^{2+} channels (Kim *et al.*, 2009); CB1, cannabinoid receptor 1 (Martin *et al.*, 2002); CNGA2, cyclic nucleotide gated channel alpha 2 (Mandiyan *et al.*, 2005); COMT, catechol-O-methyltransferase (Gogos *et al.*, 1998); DAT, dopamine transporter (Rodriguiz *et al.*, 2004); DBH, dopamine beta hydroxylase (Marino *et al.*, 2005); eNOS, endothelial nitric oxide synthase (Demas *et al.*, 1999); ENK, enkephalin (Konig *et al.*, 1996); EP1, prostaglandin E receptor 1 (Matsuoka *et al.*, 2005); ERα, estrogen receptor alpha (Ogawa *et al.*, 1998); ERβ, estrogen receptor beta (Ogawa *et al.*, 1999); GAD65, glutamic acid decarboxylase (65 amino acids) (Stork *et al.*, 2000); GDIA, guanosine diphosphate (GDP) dissociation inhibitor 1 (D'Adamo *et al.*, 2002); H1, histamine receptor 1 (Yanai *et al.*, 1998); MAOA, monoamine oxidase A (Cases *et al.*, 1995); MC5R, melanocortin-5 receptor (Morgan and Cone, 2006); MTL1, nuclear receptor subfamily 2, group E, member 1 (aka 'FIERCE') (Young *et al.*, 2002); $Na\alpha_{2C}$, adrenergic alpha receptor 2C (Sallinen *et al.*, 1998); NCAM, neural cell adhesion molecule (Stork *et al.*, 1997); NEP, neutral endopeptidase (Fischer *et al.*, 2000); NET, norepinephrine transporter (Haller *et al.*, 2002); NK-1r, neurokinin receptor 1 (De Felipe *et al.*, 1998); nNOS, neuronal nitric oxide synthase (Nelson *et al.*, 1995); OT, oxytocin (Winslow *et al.*, 2000); PAC1, adenylate cyclase activating polypeptide 1 receptor 1 (Nicot *et al.*, 2004); PET-1, ETS (E26 transformation specific) domain transcription factor (Hendricks *et al.*, 2003); RGS2, regulator of G protein signaling (Oliveira-dos-Santos *et al.*, 2000); SERT, serotonin transporter (Holmes *et al.*, 2002); Trp2, transient receptor potential family 2 (Stowers *et al.*, 2002); $V1R_{a/b}$, a cluster of vomeronasal receptor genes located on chromosome six, V1ra1–9 and V1rb1–4, 7–9 (Del Punta *et al.*, 2002); Y1, neuropeptide Y receptor 1 (Karl *et al.*, 2004).

knockout ($BDNF^{+/-}$) and conditional knockout ($BDNF^{2L/1LNes-cre}$ and $BDNF^{2L/2LCk-Cre}$) mice with a mixed background all showed increased inter-male aggression (Lyons *et al.*, 1999; Chan *et al.*, 2006). Fluoxetine reduced their heightened aggressive behavior (Lyons *et al.*, 1999). All mutants changed 5-HT_{2A} receptor expression; however, $BDNF^{+/-}$ mice showed increased 5-HT_{2A} expression in the lateral frontal cortex and hypothalamus (Lyons *et al.*, 1999), whereas $BDNF^{2L/1LNes-Cre}$ and $BDNF^{2L/2LCk-Cre}$ mice exhibited reduced 5-HT_{2A} receptor expression in the prefrontal cortex (Chan *et al.*, 2006; Rios *et al.*, 2006). An SNP in the BDNF gene, Val66Met, has attracted strong interest because of its association with mood disorders and hippocampal function in humans (Neves-Pereira *et al.*, 2002; Egan *et al.*, 2003). Although this polymorphism is found only in humans, an animal model was established by inserting a $BDNF^{Met}$ knock-in allele in the mouse (Chen *et al.*, 2006). This animal also showed increased aggressive behavior, and changed the response to SSRI treatment (Chen *et al.*, 2006). In contrast to the consistent results on aggressive behavior among BDNF mutant mice, studies on polymorphisms of the BDNF gene in aggressive behavior in humans remain to be resolved. No association was observed between Val66Met polymorphism and proneness to violence in a Chinese male sample (Tsai *et al.*, 2005). Other SNPs in the BDNF gene may be associated with high impulsivity in children with ADHD (Oades *et al.*, 2008).

Nitric oxide

Nitric oxide, a free radical gas which diffuses across membranes, is involved in several cellular functions (for review, see Calabrese *et al.*, 2007). Mice lacking neuronal nitric oxide synthase ($nNOS^{-/-}$) on a mixed C57BL/6J and 129Sv background show various deficits in their physiological development and also behavior (Huang *et al.*, 1993). $nNOS^{-/-}$ males, but not females, showed a higher duration of aggressive behavior, and also displayed much fewer submissive postures compared to wild types (Nelson *et al.*, 1995). Serotonergic dysfunction was observed in the $nNOS^{-/--}$ mice, specifically reduced 5-HT turnover in the brain and deficient 5-HT_{1A} and 5-HT_{1B} receptor function (Chiavegatto *et al.*, 2001). Escalated aggression in the $nNOS^{-/-/}$ mice was rescued by 5-HTP treatment, which increased 5-HT level and turnover. These findings point to an important role of nitric oxide for the normal 5-HT function, and thus increased aggression in $nNOS^{-/-}$ mice may be induced by changing 5-HT activity.

Neuropeptide Y (NPY)

NPY primarily controls food intake, energy balance, and metabolic regulation (Herzog, 2003). This molecule, which is critical for energy homeostasis, is also implicated in aggressive behavior (Emeson and Morabito, 2005). Male mice with deleted expression of Y1-receptor ($Y1^{-/-}$) on a mixed C57BL/6-129SvJ background showed obesity and reduced energy homeostasis (Kushi *et al.*, 1998), and also exhibited increased aggressive behaviors in the resident–intruder test (Karl *et al.*, 2004). However, this escalated aggression in $Y1^{-/-}$ mice was observed only in the home-cage but not in the novel environment. This result suggests a specific increase in territorial aggression. Karl *et al.* (2004) provided evidence that altered 5-HT function underlies the aggressive behavior of $Y1^{-/-}$ mice. TPH mRNA expression in the raphe nuclei was reduced in the $Y1^{-/-}$ mice. In addition, 5-HT_{1A} agonist treatment reduced escalated aggression in $Y1^{-/-}$ mice.

α-Calcium-calmodulin kinase II (α-CaMKII)

α-CaMKII is a neural-specific enzyme, and has been shown to be involved in long-term potentiation (LTP) (Silva *et al.*, 1992). Heterozygotes of α-CaMK II knockout mice on a mixed C57BL/6-129/OU-BALB/c background showed escalated defensive aggression but not offensive aggression (Chen *et al.*, 1994). In the resident–intruder test, resident α-CaMKII heterozygotes showed similar aggressive behavior to that of wild-type mice. In contrast, when the mutant was tested as an intruder, it exhibited high defensive reactions toward the resident. Reduced 5-HT release was observed in the dorsal raphe of the α-CaMKII mutant *in vitro*, and thus the changed 5-HT function may be associated with defensive aggression in this mouse.

General conclusions

Across different species, a critical role for serotonin in aggressive, violent behaviors is supported by a wide variety of observations and experimental approaches. Starting with the 'serotonin deficiency hypothesis', which stated that lower 5-HT function was inversely correlated with an increased risk to engage in impulsive aggressive outbursts, new developments have revealed the complexity of the serotonin–aggression link. Such complexity can be reflected by the intrinsically different types of aggression, particularly when comparing normative, species-typical patterns of aggressive behavior with fundamentally different types of escalated, pathological aggression. From clinical and preclinical studies, most evidence suggests that a serotonin dysfunction is characteristic of individuals presenting impulsive, hostile and violent behavior. These individuals may particularly benefit from pharmacological treatments aimed at inhibiting 5-HT transporters (using SSRIs such as fluoxetine or citalopram), or activating 5-HT_{1A} (buspirone) or 5-HT_{2A} receptors (risperidone). Acutely, these drugs induce *phasic* changes in 5-HT function that are associated to their anti-aggressive effects (i.e., *state* changes). On the other hand, chronic treatment with these anti-aggressive compounds may promote neuroadaptive changes in 5-HT function that are associated with the emergence of therapeutic effects (i.e., autoreceptor desensitization). The critical anti-aggressive effects of 5-HT receptor agonists appear to be mediated by both pre- and post-synaptic sites of action. The use of drug microinjection techniques into discrete brain areas (terminals vs. soma), together with the development of novel pharmacological tools that specifically target receptor subpopulations, may help to define these mechanisms more adequately. Furthermore, promising and reasonably selective anti-aggressive effects have been obtained in preclinical studies using 5-HT_{1B} receptor agonists, and are awaiting clinical validation.

While it is clear that multiple genes contribute in the determination of *traits* for an aggressive phenotype, it is remarkable that in several cases a gene–environment interaction is required for increased propensity to engage in violent outbursts, as observed with MAOA and 5-HTT polymorphisms. In a recent study, Passamonti *et al.* (2008) showed interactions between 5-HTT and MAOA polymorphisms, and those interactions exerted stronger effects on the activity of the anterior cingulate cortex, one of the brain areas implicated in impulsivity and aggression. Many other genes may have subtle effects on aggressive phenotypes, and it is possible that those genes have complex epistatic interactions from which emerge stronger effects (Miczek *et al.*, 2001). In rodents, most genetic studies on aggression make use of conventional knockout techniques in which the expression of a gene is generally deleted in the whole body, affecting all developmental stages and inducing compensatory changes in other genes (trait-like changes). Novel tools, including conditional knockout, viral vector microinfusion, or drug-induceable knockout techniques, can produce transient and local changes in gene expression, enabling the examination of more 'phasic' changes in gene expression and how they affect aggressive behavior. The use of these techniques may reduce some discrepancies in the results from genetic and pharmacological studies of 5-HT function in aggression.

In summary, understanding the complex role of 5-HT in aggression requires consideration of multiple factors, including (1) the type of aggressive behavior, its topography and function; (2) the genetic background of the individual and the typical phenotype for this background; (3) the trait characteristics of the human subject or the mouse (e.g., *Mus musculus* is a pugnacious species and many recombinant inbred strains are quite placid); and (4) the situational conditions under which aggressive behaviors have been engaged. By defining the specific characteristics of the individual, the environment and the type of social interaction, the serotonin–aggression link can be further explored, providing new pharmacological and molecular targets for interventions.

References

Adell, A., Celada, P. and Artigas, F. (2001) The role of 5-HT1B receptors in the regulation of serotonin cell firing and release in the rat brain. J. Neurochem., 79: 172–182.

Aghajanian, G.K., Kuhar, M.J. and Roth, R.H. (1973) Serotonin-containing neuronal perikarya and terminals: differential effects of P-chlorophenylalanine. Brain Res., 54: 85–101.

Alia-Klein, N., Goldstein, R.Z., Kriplani, A., Logan, J., Tomasi, D., Williams, B. *et al.* (2008) Brain monoamine oxidase a activity predicts trait aggression. J. Neurosci., 28: 5099–5104.

Assal, F., Alarcon, M., Solomon, E.C., Masterman, D., Geschwind, D.H. and Cummings, J.L. (2004) Association of the serotonin transporter and receptor gene polymorphisms in neuropsychiatric symptoms in Alzheimer disease. Arch. Neurol., 61: 1249–1253.

Baca-Garcia, E., Vaquero, C., Diaz-Sastre, C., Garcia-Resa, E., Saiz-Ruiz, J., Fernandez-Piqueras, J. *et al.* (2004) Lack of association between the serotonin transporter promoter gene polymorphism and impulsivity or aggressive behavior among suicide attempters and healthy volunteers. Psychiatry Res., 126: 99–106.

Bannai, M., Fish, E.W., Faccidomo, S. and Miczek, K.A. (2007) Anti-aggressive effects of agonists at 5-HT(1B) receptors in the dorsal raphé nucleus of mice. Psychopharmacology, 193: 295–304.

Barkan, T., Peled, A., Modai, I., Weizman, A. and Rehavi, M. (2006) Characterization of the serotonin transporter in lymphocytes and platelets of schizophrenia patients treated with atypical or typical antipsychotics compared to healthy individuals. Eur. Neuropsychopharmacol., 16: 429–436.

Barratt, E.S., Stanford, M.S., Kent, T.A. and Felthous, A. (1997) Neuropsychological and cognitive psychophysiological substrates of impulsive aggression. Biol. Psychiatry, 41: 1045–1061.

Beaulieu, J.M., Zhang, X., Rodriguiz, R.M., Sotnikova, T.D., Cools, M.J., Wetsel, W.C. *et al.* (2008) Role of GSK3 beta in behavioral abnormalities induced by serotonin deficiency. Proc. Natl. Acad. Sci. USA, 105: 1333–1338.

Beck, A.T., Erbaugh, J., Ward, C.A., Mock, J. and Mendelsohn, M. (1961) An inventory for measuring depression. Arch. Gen. Psychiatry, 4: 561–571.

Bell, R. and Hobson, H. (1994) 5-HT(1A) receptor influences on rodent social and agonistic behavior: a review and empirical study. Neurosci. Biobehav. Rev., 18: 325–338.

Benmansour, S., Cecchi, M., Morilak, D.A., Gerhardt, G.A., Javors, M.A., Gould, G.G. *et al.* (1999) Effects of chronic antidepressant treatments on serotonin transporter function, density, and mRNA level. J. Neurosci., 19: 10494–10501.

Bennett, A.J., Lesch, K.P., Heils, A., Long, J.C., Lorenz, J.G., Shoaf, S.E. *et al.* (2002) Early experience and serotonin transporter gene variation interact to influence primate CNS function. Mol. Psychiatry, 7: 118–122.

Berkowitz, L. (1993) Aggression. Its Causes, Consequences and Control. McGraw Hill, New York, NY.

Bernstein, I.S., Gordon, T.P. and Rose, R.M. (1974) Aggression and social controls in rhesus monkey (Macaca mulatta) groups revealed in group formation studies. Folia Primatologica, 21: 81–107.

Bjork, J.M., Dougherty, D.M., Moeller, F.G., Cherek, D.R. and Swann, A.C. (1999) The effects of tryptophan depletion and loading on laboratory aggression in men: time course and a food-restricted control. Psychopharmacology, 142: 24–30.

Bjork, J.M., Moeller, F.G., Dougherty, D.M., Swann, A.C., Machado, M.A. and Hanis, C.L. (2002) Serotonin 2a receptor T102C polymorphism and impaired impulse control. Am. J. Med. Genet., 114: 336–339.

Blader, J.C. (2006) Pharmacotherapy and postdischarge outcomes of child inpatients admitted for aggressive behavior. J. Clin. Psychopharmacol., 26: 419–425.

Blanchard, R.J., Hori, K., Blanchard, D.C. and Hall, J. (1987) Ethanol effects on aggression of rats selected for different levels of aggressiveness. Pharmacol. Biochem. Behav., 27: 641–644.

Blanchard, D.C., Rodgers, R.J., Hendrie, C.A. and Hori, K. (1988) 'Taming' of wild rats (Rattus rattus) by 5HT1A agonists buspirone and gepirone. Pharmacol. Biochem. Behav., 31: 269–278.

Blier, P. and de Montigny, C. (1998) Possible serotonergic mechanisms underlying the antidepressant and anti-obsessive-compulsive disorder responses. Biol. Psychiatry, 44: 313–323.

Bond, A.J. (2005) Antidepressant treatments and human aggression. Eur. J. Pharmacol., 526: 218–225.

Bonson, K.R., Johnson, R.G., Fiorella, D., Rabin, R.A. and Winter, J.C. (1994) Serotonergic control of androgen-induced dominance. Pharmacol. Biochem. Behav., 49: 313–322.

Bonvento, G., Scatton, B., Claustre, Y. and Rouquier, L. (1992) Effect of local injection of 8-OH-DPAT into the dorsal or median raphé nuclei on extracellular levels of serotonin in serotonergic projection areas in the rat-brain. Neurosci. Lett., 137: 101–104.

Bouwknecht, J.A., Hijzen, T.H., van der Gugten., J., Maes, R.A., Hen, R. and Olivier, B. (2001) Absence of 5-HT(1B) receptors is associated with impaired impulse control in male 5-HT(1B) knockout mice. Biol. Psychiatry, 49: 557–568.

Brodie, B.B. and Shore, P.A. (1957) A concept for a role of serotonin and norepinephrine as chemical mediators in the brain. Analgesia, 66: 631–642.

Brown, G.L., Ballanger, J.C., Minichiello, M.D. and Goodwin, F.K. (1979a) Human aggression and its relationship to cerebrospinal fluid 5- hydroxyindoleacetic acid, 3-methoxy-4-hydroxyphenylglycol, and homovanillic acid. In: Sandler, M. (Ed.), Psychopharmacology of Aggression. Raven Press, New York, NY, pp. 131–147.

Brown, G.L., Goodwin, F.K., Ballenger, J.C., Goyer, P.F. and Major, L.F. (1979b) Aggression in humans correlates with cerebrospinal fluid amine metabolites. Psychiatry Res., 1: 131–139.

Brunner, D. and Hen, R. (1997) Insights into the neurology of impulsive behavior from serotonin receptor knockout mice. Ann. N.Y. Acad. Sci., 836: 81–105.

Brunner, H.G., Nelen, M.R., Vanzandvoort, P., Abeling, N. G.G.M., Vangennip, A.H., Wolters, E.C. *et al.* (1993a) X-linked borderline mental retardation with prominent behavioral disturbance – phenotype, genetic localization, and evidence for disturbed monoamine metabolism. Am. J. Hum. Genet., 52: 1032–1039.

Brunner, H.G., Nelen, M., Breakefield, X.O., Ropers, H.H. and Vanoost, B.A. (1993b) Abnormal behavior associated with a point mutation in the structural gene for monoamine oxidase-A. Science, 262: 578–580.

Brunner, D., Buhot, M.C., Hen, R. and Hofer, M. (1999) Anxiety, motor activation, and maternal-infant interactions in 5HT(1B) knockout mice. Behav. Neurosci., 113: 587–601.

Buckholtz, J.W. and Meyer-Lindenberg, A. (2008) MAOA and the neurogenetic architecture of human aggression. Trends Neurosci., 31: 120–129.

Buckley, P.F., Ibrahim, Z.Y., Singer, B., Orr, B., Donenwirth, K. and Brar, P.S. (1997) Aggression and schizophrenia: Efficacy of risperidone. J. Am. Acad. Psychiatry Law, 25: 173–181.

Buitelaar, J.K., van der Gaag, R.J., Cohen-Kettenis, P. and Melman, T.M. (2001) A randomized controlled trial of risperidone in the treatment of aggression in hospitalized adolescents with subaverage cognitive abilities. J. Clin. Psychiatry, 62: 239–248.

Burt, S.A. and Mikolaiewski, A.J. (2008) Preliminary evidence that specific candidate genes are associated with adolescent-onset antisocial behavior. Aggress. Behav., 34: 437–445.

Buss, A.H. and Durkee, A. (1957) An inventory for assessing different kinds of hostility. J. Consult. Psychol., 21: 343–349.

Cadoret, R.J., Langbehn, D., Caspers, K., Troughton, E.P., Yucuis, R., Sandhu, H.K. *et al.* (2003) Associations of the serotonin transporter promoter polymorphism with aggressivity, attention deficit, and conduct disorder in an adoptee population. Compr. Psychiatry, 44: 88–101.

Cairns, R.B. and Nakelski, J.S. (1971) On fighting in mice: ontogenetic and experiential determinants. J. Comp. Physiol. Psychol., 74: 354–364.

Calabrese, V., Mancuso, C., Calvani, M., Rizzarelli, E., Butterfield, D.A. and Stella, A.M.G. (2007) Nitric oxide in the central nervous system: neuroprotection versus neurotoxicity. Nat. Rev. Neurosci., 8: 766–775.

Caldwell, E.E. and Miczek, K.A. (2008) Long-term citalopram maintenance in mice: selective reduction of alcohol-heightened aggression. Psychopharmacology (Berl.), 196: 407–416.

Carrillo, M., Ricci, L.A., Coppersmith, G.A. and Melloni, R.H. Jr. (2009). The effect of increased serotonergic neurotransmission on aggression: a critical meta-analytical review of preclinical studies. Psychopharmacology (Berl.), in press.

Cases, O., Seif, I., Grimsby, J., Gaspar, P., Chen, K., Pournin, S. *et al.* (1995) Aggressive behavior and altered amounts of brain serotonin and norepinephrine in mice lacking MAOA. Science, 268: 1763–1766.

Cases, O., Vitalis, T., Seif, I., DeMaeyer, E., Sotelo, C. and Gaspar, P. (1996) Lack of barrels in the somatosensory cortex of monoamine oxidase A-deficient mice: role of a serotonin excess during the critical period. Neuron, 16: 297–307.

Caspi, A., McClay, J., Moffitt, T.E., Mill, J., Martin, J., Craig, I. W. *et al.* (2002) Role of genotype in the cycle of violence in maltreated children. Science, 297: 851–854.

Caspi, A., Sugden, K., Moffitt, T.E., Taylor, A., Craig, I.W., Harrington, H., McClay, J., Mill, J., Martin, J., Braithwaite, A. and Poulton, R. (2003) Influence of life stress on depression: moderation by a polymorphism in the 5-HTT gene. Science, 301: 386–389.

Ceglia, I., Acconcia, S., Fracasso, C., Colovic, M., Caccia, S. and Invernizzi, R.W. (2004) Effects of chronic treatment with escitalopram or citalopram on extracellular 5-HT in the prefrontal cortex of rats: role of 5-HT1A receptors. Br. J. Pharmacol., 142: 469–478.

Chan, J.P., Unger, T.J., Byrnesa, J. and Rios, M. (2006) Examination of behavioral deficits triggered by targeting Bdnf in fetal or postnatal brains of mice. Neuroscience, 142: 49–58.

Chen, C., Rainnie, D.G., Greene, R.W. and Tonegawa, S. (1994) Abnormal fear response and aggressive behavior in mutant mice deficient for alpha-calcium-calmodulin kinase II. Science, 265: 291–294.

Chen, Z.Y., Jing, D.Q., Bath, K.G., Ieraci, A., Khan, T., Siao, C. J. *et al.* (2006) Genetic variant BDNF (Val66Met) polymorphism alters anxiety-related behavior. Science, 314: 140–143.

Cherek, D.R., Steinberg, J.L. and Vines, R.V. (1984) Low doses of alcohol affect human aggressive responses. Biol. Psychiatry, 19: 263–267.
Cherek, D.R., Moeller, F.G., Schnapp, W. and Dougherty, D.M. (1997) Studies of violent and nonviolent male parolees: I. Laboratory and Psychometric Measurements of Aggression. Biol. Psychiatry, 41: 514–522.
Cherek, D.R., Moeller, F.G., Khan-Dawood, F., Swann, A. and Lane, S.D. (1999) Prolactin response to buspirone was reduced in violent compared to nonviolent parolees. Psychopharmacology, 142: 144–148.
Chiavegatto, S., Dawsons, V.L., Mamounas, L.A., Koliatsos, V.E., Dawson, T.M. and Nelson, R.J. (2001) Brain serotonin dysfunction accounts for aggression in male mice lacking neuronal nitric oxide synthase. Proc. Natl. Acad. Sci. USA, 98: 1277–1281.
Cichon, S., Winge, I., Mattheisen, M., Georgi, A., Karpushova, A., Freudenberg, J. *et al.* (2008) Brain-specific tryptophan hydroxylase 2 (TPH2): a functional Pro206Ser substitution and variation in the 5′-region are associated with bipolar affective disorder. Hum. Mol. Genet., 17: 87–97.
Clotfelter, E.D., O'Hare, E.P., McNitt, M.M., Carpenter, R.E. and Summers, C.H. (2007) Serotonin decreases aggression via 5-HT1A receptors in the fighting fish Betta splendens. Pharmacol. Biochem. Behav., 87: 222–231.
Coccaro, E.F. (1989) Central serotonin and impulsive aggression. Br. J. Psychiatry, 155: 52–62.
Coccaro, E.F. (1992) Impulsive aggression and central serotonergic system function in humans — an example of a dimensional brain-behavior relationship. Int. Clin. Psychopharmacol., 7: 3–12.
Coccaro, E.F. and Kavoussi, R.J. (1997) Fluoxetine and impulsive aggressive behavior in personality-disordered subjects. Arch. Gen. Psychiatry, 54: 1081–1088.
Coccaro, E.F., Siever, L.J., Klar, H.M., Maurer, G., Cochrane, K., Cooper, T.B. *et al.* (1989) Serotonergic studies in patients with affective and personality disorders. Arch. Gen. Psychiatry, 46: 587–599.
Coccaro, E.F., Kavoussi, R.J., Trestman, R.L., Gabriel, S.M., Cooper, T.B. and Siever, L.J. (1997a) Serotonin function in human subjects: Intercorrelations among central 5-HT indices and aggressiveness. Psychiatry Res., 73: 1–14.
Coccaro, E.F., Kavoussi, R.J., Cooper, T.B. and Hauger, R.L. (1997b) Central serotonin activity and aggression: inverse relationship with prolactin response to d-Fenfluramine, but not CSF 5-HIAA concentration, in human subjects. Am. J. Psychiatry, 154: 1430–1435.
Coccaro, E.F., Bergeman, C.S., Kavoussi, R.J. and Seroczynski, A.D. (1997c) Heritability of aggression and irritability: a twin study of the Buss-Durkee aggression scales in adult male subjects. Biol. Psychiatry, 41: 273–284.
Coccaro, E.F., Kavoussi, R.J., Sheline, Y.I., Berman, M.E. and Csernansky, J.G. (1997d) Impulsive aggression in personality disorder correlates with platelet 5- HT-2A receptor binding. Neuropsychopharmacology, 16: 211–216.
Cologer-Clifford, A., Simon, N.G., Lu, S.F. and Smoluk, S.A. (1997) Serotonin agonist-induced decreases in intermale aggression are dependent on brain region and receptor subtype. Pharmacol. Biochem. Behavior., 58: 425–430.
Crabbe, J.C., Phillips, T.J., Feller, D.J., Hen, R., Wenger, C.D., Lessov, C.N. *et al.* (1996) Elevated alcohol consumption in null mutant mice lacking 5-HT1B serotonin receptors. Nat. Genet., 14: 98–101.
Crawley, J.N., Schleidt, W.M. and Contrera, J.F. (1975) Does social environment decrease propensity to fight in male mice? Behav. Biol., 15: 73–83.
Cutler, M.G., Rodgers, R.J. and Jackson, J.E. (1997) Behavioural effects in mice of subchronic buspirone, ondansetron, and tianeptine. I. Social interactions. Pharmacol. Biochem. Behav., 56: 287–293.
Czobor, P., Volavka, J. and Meibach, R.C. (1995) Effect of risperidone on hostility in schizophrenia. J. Clin. Psychopharmacol., 15: 243–249.
D'Adamo, P., Welzl, H., Papadimitriou, S., Raffaele, D.B., Tiveron, C., Tatangelo, L. *et al.* (2002) Deletion of the mental retardation gene Gdi1 impairs associative memory and alters social behavior in mice. Hum. Mol. Genet., 11: 2567–2580.
DaVanzo, J.P., Daugherty, M., Ruckart, R. and Kang, L. (1966) Pharmacological and biochemical studies in isolation-induced fighting mice. Psychopharmacologia, 9: 210–219.
de Almeida, R.M.M. and Lucion, A. (1994) Effects of intracerebroventricular administration of 5-HT receptor agonists on the maternal aggression of rats. Eur. J. Neurosci., 264: 445–448.
de Almeida, R.M.M. and Lucion, A.B. (1997) 8-OH-DPAT in the median raphé, dorsal periaqueductal gray and corticomedial amygdala nucleus decreases, but the medial septal area it can increase maternal aggressive behavior in rats. Psychopharmacology, 134: 392–400.
de Almeida, R.M.M. and Miczek, K.A. (2002) Aggression escalated by social instigation or by discontinuation of reinforcement ('frustration') in mice: inhibition by anpirtoline – a 5-HT1B receptor agonist. Neuropsychopharmacology, 27: 171–181.
de Almeida, R.M.M., Nikulina, E.M., Faccidomo, S., Fish, E.W. and Miczek, K.A. (2001) Zolmitriptan – a 5-HT1B/D agonist, alcohol, and aggression in mice. Psychopharmacology (Berl.), 157: 131–141.
de Almeida, R.M.M., Giovenardi, M., da Silva, S.P., de Oliveira, V.P. and Stein, D.J. (2005) Maternal aggression in Wistar rats: effect of 5-HT2A/2C receptor agonist and antagonist microinjected into the dorsal periaqueductal gray matter and medial septum. Braz. J. Med. Biol. Res., 38: 597–602.
de Almeida, R.M., Rosa, M.M., Santos, D.M., Saft, D.M., Benini, Q. and Miczek, K.A. (2006) 5-HT(1B) receptors, ventral orbitofrontal cortex, and aggressive behavior in mice. Psychopharmacology (Berl.), 185: 441–450.
de Boer, S.F. and Koolhaas, J.M. (2005) 5-HT1A and 5-HT1B receptor agonists and aggression: A pharmacological challenge of the serotonin deficiency hypothesis. Eur. J. Pharmacol., 526: 125–139.
de Boer, S.F., Lesourd, M., Mocaer, E. and Koolhaas, J.M. (1999) Selective antiaggressive effects of alnespirone in resident–intruder test are mediated via (5-hydroxytryptamine)1A receptors: A comparative pharmacological study with 8-hydroxy-2-dipropylaminotetralin, ipsapirone, buspirone, eltoprazine, and WAY-100635. J. Pharmacol. Exp. Ther., 288: 1125–1133.
de Boer, S.F., Lesourd, M., Mocaer, E. and Koolhaas, J.M. (2000) Somatodendritic 5-HT(1A) autoreceptors mediate the anti-aggressive actions of 5-HT(1A) receptor agonists in rats: An ethopharmacological study with S-15535, alnespirone, and WAY-100635. Neuropsychopharmacology, 23: 20–33.
DeBold, J.F. and Miczek, K.A. (1981) Sexual dimorphism in the hormonal control of aggressive behavior of rats. Pharmacol. Biochem. Behav., 14: 89–93.

De Deyn, P.P., Rabheru, K., Rasmussen, A., Bocksberger, J.P., Dautzenberg, P.L., Eriksson, S. *et al.* (1999) A randomized trial of risperidone, placebo, and haloperidol for behavioral symptoms of dementia. Neurology, 53: 946–955.

De Felipe, C., Herrero, J.F., O'Brien, J.A., Palmer, J.A., Doyle, C.A., Smith, A.J.H. *et al.* (1998) Altered nocieption, analgesia and aggression in mice lacking the receptor for substance P. Nature, 392: 394–397.

Del Punta, K., Leinders-Zufall, T., Rodriguez, I., Jukam, D., Wysocki, C.J., Ogawa, S. *et al.* (2002) Deficient pheromone responses in mice lacking a cluster of vomeronasal receptor genes. Nature, 419: 70–74.

Delgado, P.L., Charney, D.S., Price, L.H., Aghajanian, G.K., Landis, H. and Heninger, G.R. (1990) Serotonin function and the mechanism of antidepressant action – reversal of antidepressant-induced remission by rapid depletion of plasma tryptophan. Arch. Gen. Psychiatry, 47: 411–418.

Delini-Stula, A. and Vassout, A. (1981) The effects of antidepressants on aggressiveness induced by social deprivation in mice. Pharmacol. Biochem. Behav., 14(Suppl. 1): 33–41.

Delville, Y., Mansour, K.M. and Ferris, C.F. (1996) Serotonin blocks vasopressin-facilitated offensive aggression: Interactions within the ventrolateral hypothalamus of golden hamsters. Physiol. Behav., 59: 813–816.

Demas, G.E., Kriegsfeld, L.J., Blackshaw, S., Huang, P., Gammie, S.C., Nelson, R.J. *et al.* (1999) Elimination of aggressive behavior in male mice lacking endothelial nitric oxide synthase. J. Neurosci., 19: 1–5.

DeMoja, C.A. and Spielberger, C.D. (1997) Anger and drug addiction. Psychol. Rep., 81: 152–154.

Denney, R.M., Koch, H. and Craig, I.W. (1999) Association between monoamine oxidase A activity in human male shin fibroblasts and genotype of the MAOA promoter-associated variable number tandem repeat. Hum. Genet., 105: 542–551.

Dompert, W.U., Glaser, T. and Traber, J. (1985) 3H-TVX Q 7821: Identification of 5-HT1 binding sites as target for a novel putative anxiolytic. Naunyn-Schmiedeberg's Arch. Pharmacol., 328: 467–470.

Duysen, E.G., Stribley, J.A., Fry, D.L., Hinrichs, S.H. and Lockridge, O. (2002) Rescue of the acetylcholinesterase knockout mouse by feeding a liquid diet; phenotype of the adult acetylcholinesterase deficient mouse. Brain Res. Dev. Brain Res., 137: 43–54.

Edwards, D.H. and Kravitz, E.A. (1997) Serotonin, social status and aggression. Curr. Opin. Neurobiol., 7: 812–819.

Egan, M.F., Kojima, M., Callicott, J.H., Goldberg, T.E., Kolachana, B.S., Bertolino, A. *et al.* (2003) The BDNF val66met polymorphism affects activity-dependent secretion of BDNF and human memory and hippocampal function. Cell, 112: 257–269.

Eibl-Eibesfeldt, I. (1950) Beiträge zur Biologie der Haus- und der Ährenmaus nebst einigen Beobachtungen an anderen Nagern. Z Tierpsychol., 7: 558–587.

Ellison, G.D. and Bresler, D.E. (1974) Tests of emotional behavior in rats following depletion of norepinephrine, of serotonin, or of both. Psychopharmacologia, 34: 275–288.

Emeson, R.B. and Morabito, M.V. (2005) Food fight: the NPY–serotonin link between aggression and feeding behavior. Sci. STKE, 2005: e12.

Faccidomo, S., Bannai, M. and Miczek, K.A. (2008) Escalated aggression after alcohol drinking in male mice: dorsal raphé and prefrontal cortex serotonin and 5-HT(1B) receptors. Neuropsychopharmacology, in press.

Fairbanks, L.A., Fontenot, M.B., Phillips-Conroy, J.E., Jolly, C.J., Kaplan, J.R. and Mann, J.J. (1999) CSF monoamines, age and impulsivity in wild grivet monkeys (*Cercopithecus aethiops aethiops*). Brain Behav. Evol., 53: 305–312.

Fairbanks, L.A., Melega, W.P., Jorgensen, M.J., Kaplan, J.R. and McGuire, M.T. (2001) Social impulsivity inversely associated with CSF 5-HIAA and fluoxetine exposure in vervet monkeys. Neuropsychopharmacology, 24: 370–378.

Fairbanks, L.A., Jorgensen, M.J., Huff, A., Blau, K., Hung, Y.Y. and Mann, J.J. (2004) Adolescent impulsivity predicts adult dominance attainment in male. Am. J. Primatol., 64: 1–17.

Fava, M. (1997) Psychopharmacologic treatment of pathologic aggression. Anger, Aggression, Violence, 20: 427–451.

Ferrari, P.F., Van Erp, A.M.M., Tornatzky, W. and Miczek, K.A. (2003) Accumbal dopamine and serotonin in anticipation of the next aggressive episode in rats. Eur. J. Neurosci., 17: 371–378.

Ferris, C.F., Melloni, R.H., Koppel, G., Perry, K.W., Fuller, R.W. and Delville, Y. (1997) Vasopressin/serotonin interactions in the anterior hypothalamus control aggressive behavior in golden hamsters. J. Neurosci., 17: 4331–4340.

Ferris, C.F., Stolberg, T. and Delville, Y. (1999) Serotonin regulation of aggressive behavior in male golden hamsters (*Mesocricetus auratus*). Behav. Neurosci., 113: 804–815.

Fischer, H.S., Zernig, G., Schuligoi, R., Miczek, K.A., Hauser, K.F., Gerard, C. *et al.* (2000) Alterations within the endogenous opioid system in mice with targeted deletion of the neutral endopeptidase ('enkephalinase') gene. Regulatory Peptides, 96: 53–58.

Fish, E.W., Faccidomo, S. and Miczek, K.A. (1999) Aggression heightened by alcohol or social instigation in mice: reduction by the 5-HT1B receptor agonist CP-94,253. Psychopharmacology, 146: 391–399.

Foley, D.L., Eaves, L.J., Wormley, B., Silberg, J.L., Maes, H.H., Kuhn, J. *et al.* (2004) Childhood adversity, monoamine oxidase A genotype, and risk for conduct disorder. Arch. Gen. Psychiatry, 61: 738–744.

Frazzetto, G., Di Lorenzo, G., Carola, V., Proietti, L., Sokolowska, E., Siracusano, A. *et al.* (2007) Early trauma and increased risk for physical aggression during adulthood: the moderating role of MAOA genotype. PLoS ONE., 2: e486.

Fresan, A., Camarena, B., Apiquian, R., Aguilar, A., Urraca, N. and Nicolini, H. (2007) Association study of MAO-A and DRD4 genes in schizophrenic patients with aggressive behavior. Neuropsychobiology, 55: 171–175.

Fuller, R.W. (1996) The influence of fluoxetine on aggressive behavior. Neuropsychopharmacology, 14: 77–81.

Gardner, D.L., Lucas, P.B. and Cowdry, R.W. (1990) Csf metabolites in borderline personality-disorder compared with normal controls. Biol. Psychiatry, 28: 247–254.

Gessa, G.L., Biggio, G., Fadda, F., Corsini, G.U. and Tagliamonte, A. (1974) Effect of the oral administration of tryptophan-free amino acid mixtures on serum tryptophan, brain tryptophan and serotonin metabolism. J Neurochem., 22: 869–870.

Giacalone, E., Tansella, M., Valzelli, L. and Garattini, S. (1968) Brain serotonin metabolism in isolated aggressive mice. Biochemical Pharmacology, 17: 1315–1327.

Gimenez-Llort, L., Fernandez-Teruel, A., Escorihuela, R.M., Fredholm, B.B., Tobena, A., Pekny, M. *et al.* (2002) Mice lacking the adenosine A1 receptor are anxious and aggressive, but are normal learners with reduced muscle strength and survival rate. Eur. J. Neurosci., 16: 547–550.

Gogos, J.A., Morgan, M., Luine, V., Santha, M., Ogawa, S., Pfaff, D. *et al.* (1998) Catechol-O-methyltransferase-deficient

mice exhibit sexually dimorphic changes in catecholamine levels and behavior. Proc. Natl. Acad. Sci. USA, 95: 9991–9996.
Haller, J. and Kruk, M.R. (2006) Normal and abnormal aggression: human disorders and novel laboratory models. Neurosci. Biobehav. Rev., 30: 292–303.
Haller, J., Bakos, N., Rodriguiz, R.M., Caron, M.G., Wetsel, W.C. and Liposits, Z. (2002) Behavioral responses to social stress in noradrenaline transporter knockout mice: effects on social behavior and depression. Brain Res. Bull., 58: 279–284.
Hallikainen, T., Saito, T., Lachman, H.M., Volavka, J., Pohjalainen, T., Ryynanen, O.P. *et al.* (1999) Association between low activity serotonin transporter promoter genotype and early onset alcoholism with habitual impulsive violent behavior. Mol. Psychiatry, 4: 385–388.
Halperin, J.M., Sharma, V., Siever, L.J., Schwartz, S.T., Matier, K., Wornell, G. *et al.* (1994) Serotonergic function in aggressive and nonaggressive boys with attention deficit hyperactivity disorder. Am. J. Psychiatry, 151: 243–248.
Handelsman, L., Holloway, K., Kahn, R.S., Sturiano, C., Rinaldi, P.J., Bernstein, D.P. *et al.* (1996) Hostility is associated with a low prolactin response to meta-chlorophenylpiperazine in abstinent alcoholics. Alcohol. Clin. Exp. Res., 20: 824–829.
Handelsman, L., Kahn, R.S., Sturiano, C., Rinaldi, P.J., Gabriel, S., Schmeidler, J.P. *et al.* (1998) Hostility is associated with a heightened prolactin response to meta-chlorophenylpiperazine in abstinent cocaine addicts. Psychiatry Res., 80: 1–12.
Haney, M., DeBold, J.F. and Miczek, K.A. (1989) Maternal aggression in mice and rats towards male and female conspecifics. Aggress. Behav., 15: 443–453.
Hassanain, M., Bhatt, S. and Siegel, A. (2003) Differential modulation of feline defensive rage behavior in the medial hypothalamus by 5-HT1A and 5-HT2 receptors. Brain Res., 981: 201–209.
Haug, M., Wallian, L. and Brain, P.F. (1990) Effects of 8-OH-DPAT and fluoxetine on activity and attack by female mice towards lactating intruders. Gen. Pharmacol., 21: 845–849.
Heiligenberg, W. (1974) Processes governing behavioral states of readiness. Adv. Study Behav., 5: 173–200.
Heils, A., Teufel, A., Petri, S., Seemann, M., Bengel, D., Balling, U. *et al.* (1995) Functional promoter and polyadenylation site mapping of the human serotonin (5-HT) transporter gene. J. Neural Transm. Gen. Sect., 102: 247–254.
Heils, A., Teufel, A., Petri, S., Stober, G., Riederer, P., Bengel, D. *et al.* (1996) Allelic variation of human serotonin transporter gene expression. J. Neurochem., 66: 2621–2624.
Hendricks, T.J., Fyodorov, D.V., Wegman, L.J., Lelutiu, N.B., Pehek, E.A., Yamamoto, B. *et al.* (2003) Pet-1 ETS gene plays a critical role in 5-HT neuron development and is required for normal anxiety-like and aggressive behavior. Neuron, 37: 233–247.
Hennig, J., Reuter, M., Netter, P., Burk, C. and Landt, O. (2005) Two types of aggression are differentially related to serotonergic activity and the A779C TPH polymorphism. Behav. Neurosci., 119: 16–25.
Herzog, H. (2003) Neuropeptide Y and energy homeostasis: insights from Y receptor knockout models. Eur. J. Pharmacol., 480: 21–29.
Hess, W.R. (1954) Das Zwischenhirn. Benno Schwabe and Co., Basel.
Higley, J.D. and Bennett, A.J. (1999) Central nervous system serotonin and personality as variables contributing to excessive alcohol consumption in non-human primates. Alcohol. Alcohol., 34: 402–418.
Higley, J.D. and Suomi, S.J. (1986) Parental behavior in primates. In: Sluckin, W. (Ed.), Parental Behavior in Animals and Humans. Blackwell Press, Oxford, pp. 152–207.
Higley, J.D., Suomi, S.J. and Linnoila, M. (1991) CSF monoamine metabolite concentrations vary according to age, rearing, and sex, and are influenced by the stressor of social separation in rhesus monkeys. Psychopharmacology, 103: 551–556.
Higley, J.D., Mehlman, P.T., Taub, D.M., Higley, S.B., Suomi, S.J., Vickers, J.H. *et al.* (1992) Cerebrospinal fluid monoamine and adrenal correlates of aggression in free-ranging rhesus monkeys. Arch. Gen. Psychiatry, 49: 436–441.
Higley, J.D., King, S.T., Hasert, M.F., Champoux, M., Suomi, S.J. and Linnoila, M. (1996a) Stability of interindividual differences in serotonin function and its relationship to severe aggression and competent social behavior in rhesus macaque females. Neuropsychopharmacology, 14: 67–76.
Higley, J.D., Mehlman, P.T., Poland, R.E., Taub, D.M., Vickers, J., Suomi, S.J. *et al.* (1996b) CSF testosterone and 5-HIAA correlate with different types of aggressive behaviors. Biol. Psychiatry, 40: 1067–1082.
Ho, H.P., Olsson, M., Westberg, L., Melke, J. and Eriksson, E. (2001) The serotonin reuptake inhibitor fluoxetine reduces sex steroid-related aggression in female rats: an animal model of premenstrual irritability? Neuropsychopharmacology, 24: 502–510.
Hollander, E. (1999) Managing aggressive behavior in patients with obsessive-compulsive disorder and borderline personality disorder. J. Clin. Psychiatry, 60(Suppl. 15): 38–44.
Holmes, A., Murphy, D.L. and Crawley, J.N. (2002) Reduced aggression in mice lacking the serotonin transporter. Psychopharmacology, 161: 160–167.
Huang, P.L., Dawson, T.M., Bredt, D.S., Snyder, S.H. and Fishman, M.C. (1993) Targeted disruption of the neuronal nitric oxide synthase gene. Cell, 75: 1273–1286.
Huang, Y.Y., Grailhe, R., Arango, V., Hen, R. and Mann, J.J. (1999) Relantionship of psychopatology to the human serotonin (1B) genotype and receptor binding kinectics in postmortem brain tissue. Neuropsychopharmacology, 21(2): 238–246.
Hurst, J.L. (1987) Behavioral variation in wild house mice Mus domesticus Rutty: a quantitative assessment of female social organization. Anim. Behav., 35: 1846–1857.
Jarrell, H., Hoffman, J.B., Kaplan, J.R., Berga, S., Kinkead, B. and Wilson, M.E. (2008) Polymorphisms in the serotonin reuptake transporter gene modify the consequences of social status on metabolic health in female rhesus monkeys. Physiol. Behav., 93: 807–819.
Jequier, E., Lovenberg, W. and Sjoerdsma, A. (1967) Tryptophan hydroxylase inhibition: the mechanism by which p-chlorophenylalanine depletes rat brain serotonin. Mol. Pharmacol., 3: 274–278.
Johnson, O., Becnel, J. and Nichols, C.D. (2009) Serotonin 5-Ht2 and 5-Ht1A-like receptors differentially modulate aggressive behaviors in drosophila melanogaster. Neuroscience, 158: 1292–1300.
Jonsson, E.G., Goldman, D., Spurlock, G., Gustavsson, J.P., Nielsen, D.A., Linnoila, M. *et al.* (1997) Tryptophan hydroxylase and catechol-O-methyltransferase gene polymorphisms: relationships to monoamine metabolite concentrations in CSF of healthy volunteers. Eur. Arch. Psychiatry Clin. Neurosci., 247: 297–302.
Joppa, M.A., Rowe, R.K. and Meisel, R.L. (1997) Effects of serotonin 1A or 1B receptor agonists on social aggression in male and female Syrian hamsters. Pharmacol. Biochem. Behav., 58: 349–353.
Kantak, K.M., Hegstrand, L.R. and Eichelman, B. (1980a) Dietary tryptophan modulation and aggressive behavior in mice. Pharmacol. Biochem. Behav., 12: 675–679.

Kantak, K.M., Hegstrand, L.R., Whitman, J. and Eichelman, B. (1980b) Effects of dietary supplements and a tryptophan-free diet on aggressive behavior in rats. Pharmacol. Biochem. Behav., 12: 173–179.
Karl, T., Lin, S., Schwarzer, C., Sainsbury, A., Couzens, M., Wittmann, W. *et al.* (2004) Y1 receptors regulate aggressive behavior by modulating serotonin pathways. Proc. Natl. Acad. Sci. USA, 101: 12742–12747.
Kavoussi, R., Armstead, P. and Coccaro, E. (1997) The neurobiology of impulsive aggression. Psychiatr. Clin. N Am, 20: 395–403.
Keck, P.E. Jr., Strakowski, S.M. and McElroy, S.L. (2000) The efficacy of atypical antipsychotics in the treatment of depressive symptoms, hostility, and suicidality in patients with schizophrenia. J. Clin. Psychiatry, 61(Suppl. 3): 4–9.
Keele, N.B. (2001) Phenytoin inhibits isolation-induced aggression specifically in rats with low serotonin. NeuroReport, 12: 1107–1112.
Khait, V.D., Huang, Y.Y., Zalsman, G., Oquendo, M.A., Brent, D.A., Harkavy-Friedman, J.M. *et al.* (2005) Association of serotonin 5-HT2A receptor binding and the T102C polymorphism in depressed and healthy Caucasian subjects. Neuropsychopharmacology, 30: 166–172.
Kim, C., Jeon, D., Kim, Y.H., Lee, C.J., Kim, H. and Shin, H.S. (2009) Deletion of N-type Ca2+ channel Cav2.2 results in hyperaggressive behaviors in Mice. J. Biol. Chem., 284: 2738–2745.
Kim, J.J., Shih, J.C., Chen, K., Chen, L., Bao, S.W., Maren, S. *et al.* (1997) Selective enhancement of emotional, but not motor, learning in monoamine oxidase A-deficient mice. Proc. Natl. Acad. Sci. USA, 94: 5929–5933.
Kim-Cohen, J., Caspi, A., Taylor, A., Williams, B., Newcombe, R., Craig, I.W. *et al.* (2006) MAOA, maltreatment, and gene-environment interaction predicting children's mental health: new evidence and a meta-analysis. Mol. Psychiatry, 11: 903–913.
Knutson, J.F., Kane, N.L., Schlosberg, A.J., Fordyce, D.J. and Simansky, K.J. (1979) Influence of PCPA, shock level, and home-cage conditions on shock-induced aggression. Physiol. Behav., 23: 897–907.
Koe, B.K. and Weissman, A. (1966) p-Chlorophenylalanine: a specific depletor of brain serotonin. J. Pharmacol. Exp. Ther., 154: 499–516.
Konig, M., Zimmer, A.M., Steiner, H., Holmes, P.V., Crawley, J.N., Brownstein, M.J. *et al.* (1996) Pain responses, anxiety and aggression in mice deficient in pre-proenkephalin. Nature, 383: 535–538.
Kraemer, G.W., Ebert, M.H., Schmidt, D.E. and McKinney, W.T. (1989) A longitudinal study of the effect of different social rearing conditions on cerebrospinal fluid norepinephrine and biogenic amine metabolites in rhesus monkeys. Neuropsychopharmacology, 2: 175–189.
Kranzler, H.R., Hernandez-Avila, C.A. and Gelernter, J. (2002) Polymorphism of the 5-HT1B receptor gene (HTR1B): Strong within-locus linkage disequilibrium without association to antisocial substance dependence. Neuropsychopharmacology, 26: 115–122.
Kruesi, M.J.P., Rapoport, J.L., Hamburger, S., Hibbs, E., Potter, W.Z., Lenane, M. *et al.* (1990) Cerebrospinal fluid monoamine metabolites, aggression, and impulsivity in disruptive behavior disorders of children and adolescents. Arch. Gen. Psychiatry, 47: 419–426.
Kushi, A., Edamura, K., Noguchi, M., Akiyama, K., Nishi, Y. and Sasai, H. (1998) Generation of mutant mice with large chromosomal deletion by use of irradiated ES cells – analysis of large deletion around hprt locus of ES cell. Mamm. Genome, 9: 269–273.
Lagerspetz, K.Y.H., Tirri, R. and Lagerspetz, K.M.J. (1968) Neurochemical and endocrinological studies of mice selectively bred for aggressiveness. Scand. J. Psychol., 9: 157–160.
Lappalainen, J., Dean, M., Charbonneau, L., Virkkunen, M., Linnoila, M. and Goldman, D. (1995) Mapping of the Serotonin 5-Ht1D-Beta Autoreceptor Gene on Chromosome-6 and Direct Analysis for Sequence Variants. Am. J. Med. Genet., 60: 157–161.
Lappalainen, J., Long, J.C., Eggert, M., Ozaki, N., Robin, R.W., Brown, G.L. *et al.* (1998) Linkage of antisocial alcoholism to the serotonin 5-HT1B receptor gene in 2 populations. Arch. Gen. Psychiatry, 55: 989–994.
Lasley, S.M. and Thurmond, J.B. (1985) Interaction of dietary tryptophan and social isolation on territorial aggression, motor activity, and neurochemistry in mice. Psychopharmacology, 87: 313–321.
Ledent, C., Vaugeois, J.M., Schiffmann, S.N., Pedrazzini, T., El Yacoubi, M., Vanderhaeghen, J.J. *et al.* (1997) Aggressiveness, hypoalgesia and high blood pressure in mice lacking the adenosine A(2a) receptor. Nature, 388: 674–678.
Lerch-Haner, J.K., Frierson, D., Crawford, L.K., Beck, S.G. and Deneris, E.S. (2008) Serotonergic transcriptional programming determines maternal behavior and offspring survival. Nat. Neurosci., 11: 1001–1003.
Lesch, K.P. and Merschdorf, U. (2000) Impulsivity, aggression, and serotonin: a molecular psychobiological perspective. Behav. Sci. Law, 18: 581–604.
Lesch, K.P., Bengel, D., Heils, A., Sabol, S.Z., Greenberg, B.D., Petri, S. *et al.* (1996) Association of anxiety-related traits with a polymorphism in the serotonin transporter gene regulatory region. Science, 264: 1537–1551.
Lesch, K.P., Meyer, J., Glatz, K., Flugge, G., Hinney, A., Hebebrand, J. *et al.* (1997) The 5-HT transporter gene-linked polymorphic region (5-HTTLPR) in evolutionary perspective: alternative biallelic variation in rhesus monkeys. J. Neural Transm., 104: 1259–1266.
Leyhausen, P. (1979) Cat Behavior: Predatory and Social Behavior of Domestic and Wild Cats. New York, NY: Garland STPM Press.
Lidberg, L., Belfrage, H., Bertilsson, L., Evenden, M.M. and Asberg, M. (2000) Suicide attempts and impulse control disorder are related to low cerebrospinal fluid 5-HIAA in mentally disordered violent offenders. Acta Psychiatr. Scand., 101: 395–402.
Lin, Y.M.J., Chao, S.C., Chen, T.M., Lai, T.J., Chen, J.S. and Sun, H.S. (2007) Association of functional polymorphisms of the human tryptophan hydroxylase 2 gene with risk for bipolar disorder in Han Chinese. Arch. Gen. Psychiatry, 64: 1015–1024.
Lindgren, T. and Kantak, K.M. (1987) Effects of serotonin receptor agonists and antagonists on offensive aggression in mice. Aggress. Behav., 13: 87–96.
Linnoila, V.M.I. and Virkkunen, M. (1992) Aggression, suicidality, and serotonin. J. Clin. Psychiatry, 53: 46–51.
Linnoila, M., Virkkunen, M., Scheinin, M., Nuutila, A., Rimon, R. and Goodwin, F.K. (1983) Low cerebrospinal fluid 5-hydroxyindoleacetic acid concentration differentiates impulsive from nonimpulsive violent behavior. Life Sci., 33: 2609–2614.
Loconto, J., Papes, F., Chang, E., Stowers, L., Jones, E.P., Takada, T. *et al.* (2003) Functional expression of murine V2R

pheromone receptors involves selective association with the M10 and M1 families of MHC class Ib molecules. Cell, 112: 607–618.
Lonstein, J.S. and De Vries, G.J. (2000) Maternal behaviour in lactating rats stimulates c-*fos* in glutamate decarboxylase-synthesizing neurons of the medial preoptic area, ventral bed nucleus of the stria terminalis, and ventrocaudal periaqueductal gray. Neuroscience, 100: 557–568.
Lucki, I. and Wieland, S. (1990) 5-Hydroxytryptamine-1A receptors and behavioral responses. Neuropsychopharmacology, 3: 481–493.
Lyons, W.E., Mamounas, L.A., Ricaurte, G.A., Coppola, V., Reid, S.W., Bora, S.H. *et al.* (1999) Brain-derived neurotrophic factor-deficient mice develop aggressiveness and hyperphagia in conjunction with brain serotonergic abnormalities. Proc. Natl. Acad. Sci. USA, 96: 15239–15244.
Malick, J.B. (1979) The pharmacology of isolation-induced aggressive behavior in mice. In: Essman, W.B. (Ed.), Current Developments in Psychopharmacology. 5 ed. SP Medical and Scientific Books, New York, pp. 1–27.
Malleret, G., Hen, R., Guillou, J.L., Segu, L. and Buhot, M.C. (1999) 5-HT1B receptor knockout mice exhibit increased exploratory activity and enhanced spatial memory performance in the Morris water maze. J. Neurosci., 19: 6157–6168.
Mandiyan, V.S., Coats, J.K. and Shah, N.M. (2005) Deficits in sexual and aggressive behaviors in Cnga2 mutant mice. Nat. Neurosci., 8: 1660–1662.
Manuck, S.B., Flory, J.D., Ferrell, R.E., Dent, K.M., Mann, J.J. and Muldoon, M.F. (1999) Aggression and anger-related traits associated with a polymorphism of the tryptophan hydroxylase gene. Biol. Psychiatry, 45: 603–614.
Manuck, S.B., Flory, J.D., Ferrell, R.E., Mann, J.J. and Muldoon, M.F. (2000) A regulatory polymorphism of the monoamine oxidase-A gene may be associated with variability in aggression, impulsivity, and central nervous system serotonergic responsivity. Psychiatry Res., 95: 9–23.
Manuck, S.B., Flory, J.D., Muldoon, M.F. and Ferrell, R.E. (2002) Central nervous system serotonergic responsivity and aggressive disposition in men. Physiol. Behav., 77: 705–709.
Manuck, S.B., Kaplan, J.R., Rymeski, B.A., Fairbanks, L.A. and Wilson, M.E. (2003) Approach to a social stranger is associated with low central nervous system serotonergic responsivity in female cynomolgus monkeys (Macaca fascicularis). Am. J. Primatol., 61: 187–194.
Marino, M.D., Bourdelat-Parks, B.N., Liles, L.C. and Weinshenker, D. (2005) Genetic reduction of noradrenergic function alters social memory and reduces aggression in mice. Behav. Brain Res., 161: 197–203.
Martin, M., Ledent, C., Parmentier, M., Maldonado, R. and Valverde, O. (2002) Involvement of CB1 cannabinoid receptors in emotional behaviour. Psychopharmacology (Berl.), 159: 379–387.
Mathews, T.A. (2004) Gene dose-dependent alterations in extraneuronal serotonin but not dopamine in mice with reduced serotonin transporter expression. J. Neurosci. Methods, 140: 169–181.
Matsumoto, T., Honda, S. and Harada, N. (2003) Neurological effects of aromatase deficiency in the mouse. J. Steroid Biochem. Mol. Biol., 86: 357–365.
Matsuoka, Y., Furuyashiki, T., Yamada, K., Nagai, T., Bito, H., Tanaka, Y. *et al.* (2005) Prostaglandin E receptor EP1 controls impulsive behavior under stress. Proc. Natl. Acad. Sci. USA, 102: 16066–16071.
McBride, P.A., Brown, R.P., Demeo, M., Keilp, J., Mieczkowski, T. and Mann, J.J. (1994) The relationship of platelet 5-Ht(2) receptor indexes to major depressive disorder, personality-traits, and suicidal-behavior. Biol. Psychiatry, 35: 295–308.
McKenzie-Quirk, S.D., Girasa, K.A., Allan, A.M. and Miczek, K.A. (2005) 5-HT(3) receptors, alcohol and aggressive behavior in mice. Behav. Pharmacol., 16: 163–170.
McKinley, J.C., Hathaway, S.R. and Meehl, P.E. (1948) The Minesota multiphasic personality inventory the k-scale. J. Consult. Psychol., 12: 20–31.
McMillen, B.A., DaVanzo, E.A., Scott, S.M. and Song, A.H. (1988) N-alkyl-substituted aryl-piperazine drugs: relationship between affinity for serotonin receptors and inhibition of aggression. Drug Dev. Res., 12: 53–62.
Mehlman, P.T., Higley, J.D., Faucher, I., Lilly, A.A., Taub, D.M., Vickers, J. *et al.* (1994) Low CSF 5-HIAA concentrations and severe aggression and impaired impulse control in non-human primates. Am. J. Psychiatry, 151: 1485–1491.
Mendoza, D.L., Bravo, H.A. and Swanson, H.H. (1999) Antiaggresive and anxiolytic effects of gepirone in mice, and their attenuation by WAY 100635. Pharmacol. Biochem. Behav., 62: 499–509.
Meyer-Lindenberg, A., Buckholtz, J.W., Kolachana, B., Hariri, A.R., Pezawas, L., Blasi, G. *et al.* (2006) Neural mechanisms of genetic risk for impulsivity and violence in humans. Proc. Natl. Acad. Sci. USA, 103: 6269–6274.
Miczek, K.A. (1987) The psychopharmacology of aggression. In: Iversen, L.L. (Ed.), Handbook of Psychopharmacology, Volume 19: New Directions in Behavioral Pharmacology. Plenum, New York, pp. 183–328.
Miczek, K.A. and de Almeida, R.M.M. (2001) Oral drug self-administration in the home cage of mice: alcohol-heightened aggression and inhibition by the 5-HT1B agonist anpirtoline. Psychopharmacology (Berl.), 157: 421–429.
Miczek, K.A. and Donat, P. (1989) Brain 5-HT systems and inhibition of aggressive behavior. In: Bevan, P., Cools, A.R. and Archer, T. (Eds.), Behavioural Pharmacology of 5-HT. Lawrence Erlbaum Associates, Hillsdale, NJ, pp. 117–144.
Miczek, K.A. and Fish, E.W. (2005) Dopamine, glutamate and aggression. In: Schmidt, W.J. and Reith, M.E.A. (Eds.), Dopamine and Glutamate in Psychiatric Disorders. Humana Press, Totowa, New Jersey, pp. 237–266.
Miczek, K.A. and O'Donnell, J.M. (1978) Intruder-evoked aggression in isolated and nonisolated mice: Effects of psychomotor stimulants and l-dopa. Psychopharmacology, 57: 47–55.
Miczek, K.A., Weerts, E.M., Tornatzky, W., DeBold, J.F. and Vatne, T.M. (1992) Alcohol and 'bursts' of aggressive behavior: ethological analysis of individual differences in rats. Psychopharmacology, 107: 551–563.
Miczek, K.A., Weerts, E.M., Haney, M. and Tidey, J. (1994) Neurobiological mechanisms controlling aggression: Preclinical developments for pharmacotherapeutic interventions. Neurosci. Biobehav. Rev., 18: 97–110.
Miczek, K.A., Barros, H.M., Sakoda, L. and Weerts, E.M. (1998a) Alcohol and heightened aggression in individual mice. Alcohol. Clin. Exp. Res., 22: 1698–1705.
Miczek, K.A., Hussain, S. and Faccidomo, S. (1998b) Alcohol-heightened aggression in mice: attenuation by 5-HT1A receptor agonists. Psychopharmacology, 139: 160–168.
Miczek, K.A., Maxson, S.C., Fish, E.W. and Faccidomo, S. (2001) Aggressive behavioral phenotypes in mice. Behav. Brain Res., 125: 167–181.

Miczek, K.A., Fish, E.W., DeBold, J.F. and de Almeida, R. M.M. (2002) Social and neural determinants of aggressive behavior: pharmacotherapeutic targets at serotonin, dopamine and γ-aminobutyric acid systems. Psychopharmacology, 163: 434–458.

Miczek, K.A., Faccidomo, S., de Almeida, R.M.M., Bannai, M., Fish, E.W. and DeBold, J.F. (2004) Escalated aggressive behavior: new pharmacotherapeutic approaches and opportunities. Ann. N.Y. Acad. Sci., 1036: 336–355.

Miczek, K.A., Faccidomo, S.P., Fish, E.W. and DeBold, J.F. (2007) Neurochemistry and Molecular Neurobiology of Aggressive Behavior. In: Blaustein, J. (Ed.), Behavioral Neurochemistry, Neuroendocrinology and Molecular Neurobiology. 3rd ed., Springer, New York, pp. 285–336.

Mitchell, P.J. (2005) Antidepressant treatment and rodent aggressive behaviour. Eur. J. Pharmacol., 526: 147–162.

Mitchell, P.J. and Redfern, P.H. (1992) Acute and chronic antidepressant drug treatments induce opposite effects in the social behavior of rats. J. Psychopharmacol., 6: 241–257.

Mitchell, P.J. and Redfern, P.H. (1997) Potentiation of the time-dependent, antidepressant-induced changes in the agonistic behaviour of resident rats by the 5-HT(1A) receptor antagonist, WAY-100635. Behav. Pharmacol., 8: 585–606.

Mitchell, P.J., Fletcher, A. and Redfern, P.H. (1991) Is antidepressant efficacy revealed by drug-induced changes in rat behaviour exhibited during social interaction? Neurosci. Biobehav. Rev., 15: 539–544.

Moeller, F.G., Dougherty, D.M., Swann, A.C., Collins, D., Davis, C.M. and Cherek, D.R. (1996) Tryptophan depletion and aggressive responding in healthy males. Psychopharmacology, 126: 97–103.

Morgan, C. and Cone, R.D. (2006) Melanocortin-5 receptor deficiency in mice blocks a novel pathway influencing Pheromone-Induced Aggression. Behav. Genet.: 1–10.

Mos, J., Olivier, B., Poth, M., Van Oorschot, R. and Van Aken, H. (1993) The effects of dorsal raphé administration of eltoprazine, TFMPP and 8-OH-DPAT on resident intruder aggression in the rat. Eur. J. Pharmacol., 238: 411–415.

Moss, H.B., Yao, J.K. and Panzak, G.L. (1990) Serotonergic Responsivity and Behavioral Dimensions in Antisocial Personality Disorder with Substance Abuse. Biol. Psychiatry, 28: 325–338.

Muehlenkamp, F., Lucion, A. and Vogel, W.H. (1995) Effects of selective serotenergic agonists on aggressive behavior in rats. Pharmacol. Biochem. Behav., 50: 671–674.

Murphy, D.L. and Lesch, K.P. (2008) Targeting the murine serotonin transporter: insights into human neurobiology. Nat. Rev. Neurosci., 9: 85–96.

Natarajan, D., de Vries, H., Saaltink, D.J., de Boer, S.F. and Koolhaas, J.M. (2009) Delineation of violence from functional aggression in mice: an ethological approach. Behav Genet., 39: 73–90.

Nelson, R.J. and Chiavegatto, S. (2001) Molecular basis of aggression. Trends Neurosci., 24: 713–719.

Nelson, R.J., Demas, G.E., Huang, P.L., Fishman, M.C., Dawson, V.L., Dawson, T.M. *et al.* (1995) Behavioural abnormalities in male mice lacking neuronal nitric oxide synthase. Nature, 378: 383–386.

Neves-Pereira, M., Mundo, E., Muglia, P., King, N., Macciardi, F. and Kennedy, J.L. (2002) The brain-derived neurotrophic factor gene confers susceptibility to bipolar disorder: evidence from a family-based association study. Am. J. Hum. Genet., 71: 651–655.

New, A.S., Gelernter, J., Yovell, Y., Trestman, R.L., Nielsen, D. A., Silverman, J. *et al.* (1998) Tryptophan hydroxylase genotype is associated with impulsive-aggression measures: a preliminary study. Am. J. Med. Genet., 81: 13–17.

New, A.S., Buchsbaum, M.S., Hazlett, E.A., Goodman, M., Koenigsberg, H.W., Lo, J. *et al.* (2004) Fluoxetine increases relative metabolic rate in prefrontal cortex in impulsive aggression. Psychopharmacology, 176: 451–458.

Newman, T.K., Syagailo, Y.V., Barr, C.S., Wendland, J.R., Champoux, M., Graessle, M. *et al.* (2005) Monoamine oxidase A gene promoter variation and rearing experience influences aggressive behavior in rhesus monkeys. Biol. Psychiatry, 57: 167–172.

Nicot, A., Otto, T., Brabet, P. and Dicicco-Bloom, E.M. (2004) Altered social behavior in pituitary adenylate cyclase-activating polypeptide type I receptor-deficient mice. J. Neurosci., 24: 8786–8795.

Nielsen, D.A., Goldman, D., Virkkunen, M., Tokola, R., Rawlings, R. and Linnoila, M. (1994) Suicidality and 5-hydroxyindoleacetic acid concentration associated with a tryptophan-hydroxylase polymorphism. Arch. Gen. Psychiatry, 51: 34–38.

Nikulina, E.M., Avgustinovich, D.F. and Popova, N.K. (1992) Role of 5HT1A receptors in a variety of kinds of aggressive behavior in wild rats and counterparts selected for low defensiveness towards man. Aggress. Behav., 18: 357–364.

Nishizawa, S., Benkelfat, C., Young, S.N., Leyton, M., Mzengeza, S., de Montigny, C. *et al.* (1997) Differences between males and females in rates of serotonin synthesis in human brain. Proc. Natl. Acad. Sci. USA, 94: 5308–5313.

Nogueira, R.L. and Graeff, F.G. (1995) Role of 5HT receptor subtypes in the modulation of dorsal periaqueductal gray generated aversion. Pharmacol. Biochem. Behav., 52: 1–6.

Noirot, E., Goyens, J. and Buhot, M.C. (1975) Aggressive behavior of pregnant mice towards males. Horm. Behav., 6: 9–17.

Nomura, M., Kusumi, I., Kaneko, M., Masui, T., Daiguji, M., Ueno, T. *et al.* (2006) Involvement of a polymorphism in the 5-HT2A receptor gene in impulsive behavior. Psychopharmacology, 187: 30–35.

Oades, R.D., Lasky-Su, J., Christiansen, H., Faraone, S.V., Sonuga-Barke, E.J.S., Banaschewski, T. *et al.* (2008) The influence of serotonin- and other genes on impulsive behavioral aggression and cognitive impulsivity in children with attention-deficit/hyperactivity disorder (ADHD): findings from a family-based association test (FBAT) analysis. Behav. Brain Funct., 4: 48.

Ogawa, S., Washburn, T.F., Taylor, J., Lubahn, D.B., Korach, K.S. and Pfaff, D.W. (1998) Modifications of testosterone-dependent behaviors by estrogen receptor-alpha gene disruption in male mice. Endocrinology, 139: 5058–5069.

Ogawa, S., Chan, J., Chester, A.E., Gustafsson, J.A., Korach, K. S. and Pfaff, D.W. (1999) Survival of reproductive behaviors in estrogen receptor alpha gene-deficient (alphaERKO) male and female mice. Proc. Natl. Acad. Sci., 96: 12887–12892.

O'Keane, V., Moloney, E., O'Neill, H., O'Connor, A., Smith, C. and Dinan, T.G. (1992) Blunted prolactin responses to d-fenfluramine in sociopathy – evidence for subsensitivity of central serotonergic function. Br. J. Psychiatry, 160: 643–646.

Oliveira-dos-Santos, A.J., Matsumoto, G., Snow, B.E., Bai, D., Houston, F.P., Whishaw, I.Q. *et al.* (2000) Regulation of T cell activation, anxiety, and male aggression by RGS2. Proc. Natl. Acad. Sci., 97: 12272–12277.

Olivier, B. (2004) Serotonin and aggression. Ann. N.Y. Acad. Sci., 1036: 382–392.
Olivier, B. and Mos, J. (1992) Rodent models of aggressive behavior and serotonergic drugs. Prog. Neuropsychopharmacol. Biol. Psychiatry, 16: 847–870.
Olivier, B., Mos, J., Van der Heyden, J. and Hartog, J. (1989) Serotonergic modulation of social interactions in isolated male mice. Psychopharmacology, 97: 154–156.
Olivier, B., Mos, J. and Rasmussen, D. (1990) Behavioural pharmacology of the serenic, eltoprazine. Rev. Drug Metab. Drug Interact., 8: 31–83.
Olivier, B., Mos, J., Van Oorschot, R. and Hen, R. (1995) Serotonin receptors and animal models of aggressive behavior. Pharmacopsychiatry, 28: 80–90.
Oquendo, M.A., Russo, S.A., Underwood, M.D., Kassir, S.A., Ellis, S.P., Mann, J.J. *et al.* (2006) Higher postmortem prefrontal 5-HT2A receptor binding correlates with lifetime aggression in suicide. Biol. Psychiatry, 59: 235–243.
Osipova, D.V., Kulikov, A.V. and Popova, N.K. (2009) C1473G polymorphism in mouse tph2 gene is linked to tryptophan hydroxylase-2 activity in the brain, intermale aggression, and depressive-like behavior in the forced swim test. J. Neurosci. Res., 87: 1168–1174.
Palanza, P., Della Seta, D., Ferrari, P.F. and Parmigiani, S. (2005) Female competition in wild house mice depends upon timing of female/male settlement and kinship between females. Anim. Behav., 69: 1259–1271.
Passamonti, L., Cerasa, A., Gioia, M.C., Magariello, A., Muglia, M., Quattrone, A. *et al.* (2008) Genetically dependent modulation of serotonergic inactivation in the human prefrontal cortex. Neuroimage, 40: 1264–1273.
Peeke, H.V.S. and Figler, M.H. (1981) Modulation of aggressive behavior in fish by alcohol and congeners. Pharmacol. Biochem. Behav., 14(Suppl. 1): 79–84.
Pihl, R.O. and LeMarquand, D. (1998) Serotonin and aggression and the alcohol-aggression relationship. Alcohol. Alcohol., 33: 55–65.
Pihl, R.O., Young, S.N., Harden, P., Plotnick, S., Chamberlain, B. and Ervin, F.R. (1995) Acute effect of altered tryptophan levels and alcohol on aggression in normal human males. Psychopharmacology, 119: 353–360.
Pine, D.S., Coplan, J.D., Wasserman, G.A., Miller, L.S., Fried, J.E., Davies, M. *et al.* (1997) Neuroendocrine response to fenfluramine challenge in boys – associations with aggressive behavior and adverse rearing. Arch. Gen. Psychiatry, 54: 839–846.
Piñeyro, G., Blier, P., Dennis, T. and de Montigny, C. (1994) Desensitization of the neuronal 5-HT carrier following its long-term blockade. J. Neurosci., 14: 3036–3047.
Pinna, G., Dong, E., Matsumoto, K., Costa, E. and Guidotti, A. (2003) In socially isolated mice, the reversal of brain allopregnanolone down-regulation mediates the anti-aggressive action of fluoxetine. Proc. Natl. Acad. Sci USA, 100: 2035–2040.
Pinna, G., Costa, E. and Guidotti, A. (2006) Fluoxetine and norfluoxetine stereospecifically and selectively increase brain neurosteroid content at doses that are inactive on 5-HT reuptake. Psychopharmacology (Berl.), 186: 362–372.
Placidi, G.P.A., Oquendo, M.A., Malone, K.M., Huang, Y. Y., Ellis, S.P. and Mann, J.J. (2001) Aggressivity, suicide attempts, and depression: relationship to cerebrospinal fluid monoamine metabolite levels. Biol. Psychiatry, 50: 783–791.
Popova, N.K., Avgustinovich, D.F., Kolpakov, V.G. and Plyusnina, I.Z. (1998) Specific [H-3]8-OH-DPAT binding in brain regions of rats genetically predisposed to various defense behavior strategies. Pharmacol. Biochem. Behav., 59: 793–797.
Potegal, M. (1991) Attack priming and satiation in female golden hamsters: tests of some alternatives to the aggression arousal interpretation. Aggress. Behav., 17: 327–335.
Potegal, M. and Tenbrink, L. (1984) Behavior of attack-primed and attack-satiated female golden hamsters (*Mesocricetus auratus*). J. Comp. Psychol., 98: 66–75.
Pucilowski, O. and Kostowski, W. (1983) Aggressive behaviour and the central serotonergic systems. Behav. Brain Res., 9: 33–48.
Raj, Y.P. (2004) Psychopharmacology of borderline personality disorder. Curr. Psychiatry Rep., 6: 225–231.
Ramboz, S., Saudou, F., Amara, D.A., Belzung, C., Segu, L., Misslin, R. *et al.* (1995) 5-HT1B receptor knock out – behavioral consequences. Behav. Brain Res., 73: 305–312.
Ratey, J., Sovner, R., Parks, A. and Rogentine, K. (1991) Buspirone treatment of aggression and anxiety in mentally retarded patients: a multiple-baseline, placebo lead-in study. J. Clin. Psychiatry, 52: 159–162.
Reist, C., Nakamura, K., Sagart, E., Sokolski, K.N. and Fujimoto, K.A. (2003) Impulsive aggressive behavior: open-label treatment with citalopram. J. Clin. Psychiatry, 64: 81–85.
Ricci, L.A., Grimes, J.M. and Melloni, R.H. Jr. (2004) Serotonin type 3 receptors modulate the aggression-stimulating effects of adolescent cocaine exposure in Syrian hamsters (*Mesocricetus auratus*). Behav. Neurosci., 118: 1097–1110.
Rios, M., Lambe, E.K., Liu, R.J., Teillon, S., Liu, J.H., Akbarian, S. *et al.* (2006) Severe deficits in 5-HT2A-mediated neurotransmission in BDNF conditional mutant mice. J. Neurobiol., 66: 408–420.
Rocha, B.A., Scearce-Levie, K., Lucas, J.J., Hiroi, N., Castanon, N., Crabbe, J.C. *et al.* (1998) Increased vulnerability to cocaine in mice lacking the serotonin-1B receptor. Nature, 393: 175–178.
Rodriguiz, R.M., Chu, R., Caron, M.G. and Wetsel, W.C. (2004) Aberrant responses in social interaction of dopamine transporter knockout mice. Behav. Brain Res., 148: 185–198.
Rodriguez-Arias, M., Minarro, J., Aguilar, M.A., Pinazo, J. and Simon, V.M. (1998) Effects of risperidone and SCH 23390 on isolation-induced aggression in male mice. Eur. Neuropsychopharmacol., 8: 95–103.
Rudissaar, R., Pruus, K., Skrebuhhova, T., Allikmets, L. and Matto, V. (1999) Modulatory role of 5- HT3 receptors in mediation of apomorphine-induced aggressive behaviour in male rats. Behav. Brain Res., 106: 91–96.
Rujescu, D., Giegling, I., Bondy, B., Gietl, A., Zill, P. and Moller, H.J. (2002) Association of anger-related traits with SNPs in the TPH gene. Mol. Psychiatry, 7: 1023–1029.
Sabol, S.Z., Hu, S. and Hamer, D. (1998) A functional polymorphism in the monoamine oxidase A gene promoter. Hum. Genet., 103: 273–279.
Sakaue, M., Ago, Y., Sowa, C., Sakamoto, Y., Nishihara, B., Koyama, Y. *et al.* (2002) Modulation by 5-HT2A receptors of aggressive behavior in isolated mice. Jpn. J. Pharmacol., 89: 89–92.
Sallinen, J., Haapalinna, A., Viitamaa, T., Kobilka, B.K. and Scheinin, M. (1998) Adrenergic alpha2c receptors modulate the acoustic startle reflex, prepulse inhibition, and aggression in mice. J. Neurosci., 18: 3035–3042.
Sanchez, C. and Hyttel, J. (1994) Isolation-induced aggression in mice: effects of 5-hydroxytryptamine uptake inhibitors and involvement of postsynaptic 5-HT(1A) receptors. Eur. J. Pharmacol., 264: 241–247.

Sanchez, C. and Meier, E. (1997) Behavioral profiles of SSRIs in animal models of depression, anxiety and aggression. Psychopharmacology, 129: 197–205.

Sanchez, C., Arnt, J., Hyttel, J. and Moltzen, E.K. (1993) The role of serotonergic mechanisms in inhibition of isolation-induced aggression in male mice. Psychopharmacology, 110: 53–59.

Saudou, F., Amara, D.A., Dierich, A., Lemeur, M., Ramboz, S., Segu, L. *et al.* (1994) Enhanced aggressive behavior in mice lacking 5-HT1B receptor. Science, 265: 1875–1878.

Scott, A.L., Bortolato, M., Chen, K. and Shih, J.C. (2008) Novel monoamine oxidase A knock out mice with human-like spontaneous mutation. Neuroreport, 19: 739–743.

Sgoifo, A., Stilli, D., Musso, E., Mainardi, D. and Parmigiani, S. (1992) Offensive and defensive bite-target topographies in attacks by lactating rats. Aggress. Behav., 17: 47–52.

Shaikh, M.B., De Lanerolle, N.C. and Siegel, A. (1997) Serotonin 5-HT1A and 5-HT2/1C receptors in the midbrain periaqueductal gray differentially modulate defensive rage behavior elicited from the medial hypothalamus of the cat. Brain Res., 765: 198–207.

Shih, J.C., Ridd, M.J., Chen, K., Meehan, W.P., Kung, M.P., Seif, I. *et al.* (1999) Ketanserin and tetrabenazine aggression in mice lacking monoamine oxidase A. Brain Res., 835: 104–112.

Siegel, A., Roeling, T.A.P., Gregg, T.R. and Kruk, M.R. (1999) Neuropharmacology of brain-stimulation-evoked aggression. Neurosci. Biobehav. Rev., 23: 359–389.

Sijbesma, H., Schipper, J., De Kloet, E.R., Mos, J., Van Aken, H. and Olivier, B. (1991) Postsynaptic 5-HT1 receptors and offensive aggression in rats: a combined behavioural and autoradiographic study with eltoprazine. Pharmacol. Biochem. Behav., 38: 447–458.

Silva, A.J., Paylor, R., Wehner, J.M. and Tonegawa, S. (1992) Impaired spatial-learning in alpha-calcium-calmodulin kinase II mutant mice. Science, 257: 206–211.

Sinha, R., Cloninger, C.R. and Parsian, A. (2003) Linkage disequilibrium and haplotype analysis between serotonin receptor 1B gene variations and subtypes of alcoholism. Am. J. Med. Genet. B Neuropsychiatr. Genet., 121B: 83–88.

Sjoberg, R.L., Nilsson, K.W., Wargelius, H.L., Leppert, J., Lindstrom, L. and Oreland, L. (2007) Adolescent girls and criminal activity: role of MAOA-LPR genotype and psychosicial factors. Am. J. Med. Genet. B Neuropsychiatr. Genet., 144B: 159–164.

Smuts, B.B. (1986) Sexual competition and mate choice. In: Smuts, B.B., Cheney, D.L., Seyfarth, R.M., Wrangham, R.W. and Struhsaker, T.T. (Eds.), Primate Societies. University of Chicago Press, Chicago, IL, pp. 385–399.

Sofia, R.D. (1969) Structural relationship and potency of agents which selectively block mouse killing (muricide) behavior in rats. Life Sci., 8: 1201–1210.

Sperry, T.S., Thompson, C.K. and Wingfield, J.C. (2003) Effects of acute treatment with 8-OH-DPAT and fluoxetine on aggressive behaviour in male song sparrows (*Melospiza melodia* morphna). J. Neuroendocrinol., 15: 150–160.

Spigset, O. (1999) Adverse reactions of selective serotonin reuptake inhibitors – reports from a spontaneous reporting system. Drug Saf., 20: 277–287.

Sprouse, J.S. and Aghajanian, G.K. (1987) Electrophysiological responses of serotoninergic dorsal raphé neurons to 5-HT1A and 5-HT1B agonists. Synapse, 1: 3–9.

Steiniger, F. (1950) Beitrag zur Soziologie und sonstigen Biologie der Wanderratte. Zeitschrift fuer Tierpsychologie, 7: 356–379.

Stoff, D.M. and Vitiello, B. (1996) Role of serotonin in aggression of children and adolesecents: biochemical and phamacological studies. In: Stoff, D.M. (Ed.), Aggression and Violence: Genetic, Neurobiological and Biosocial Perspectives. Lawrence Erlbaum Associates, Mahwah, NJ, pp. 101–123.

Stork, O., Welzl, H., Cremer, H. and Schachner, M. (1997) Increased intermale aggression and neuroendocrine response in mice deficient for the neural cell adhesion molecule (NCAM). Eur. J. Neurosci., 9: 1117–1125.

Stork, O., Ji, F.Y., Kaneko, K., Stork, S., Yoshinobu, Y., Moriya, T. *et al.* (2000) Postnatal development of a GABA deficit and disturbance of neural functions in mice lacking GAD65. Brain Res., 865: 45–58.

Stowers, L., Holy, T.E., Meister, M., Dulac, C. and Koentges, G. (2002) Loss of sex discrimination and male-male aggression in mice deficient for TRP2. Science, 295: 1493–1500.

Sun, H.F., Chang, Y.T., Fann, C.S., Chang, C.J., Chen, Y.H., Hsu, Y.P. *et al.* (2002) Association study of novel human serotonin 5-HT(1B) polymorphisms with alcohol dependence in Taiwanese Han. Biol. Psychiatry, 51: 896–901.

Ten Eyck, G.R. (2008a) Serotonin modulates vocalizations and territorial behavior in an amphibian. Behav. Brain Res., 193: 144–147.

Ten Eyck, G.R. (2008b) Serotonin modulates vocalizations and territorial behavior in an amphibian. Behav. Brain Res., 193: 144–147.

Tompkins, E.C., Clemento, A.J., Taylor, D.P. and Perhach, J.L. Jr. (1980) Inhibition of aggressive behavior in rhesus monkeys by buspirone. Res. Commun. Psychol. Psychiatr. Behav., 5: 337–352.

Troisi, A., Vicario, E., Nuccetelli, F., Ciani, N. and Pasini, A. (1995) Effects of fluoxetine on aggressive behavior of adult inpatients with mental retardation and epilepsy. Pharmacopsychiatry, 28: 1–4.

Tsai, S.J., Liao, D.L., Yu, Y.W.Y., Chen, T.J., Wu, H.C., Lin, C.H. *et al.* (2005) A study of the association of (Val66Met) polymorphism in the brain-derived neurotrophic factor gene with alcohol dependence an extreme violence in Chinese males. Neurosci. Lett., 381: 340–343.

Valzelli, L. and Bernasconi, S. (1979) Aggressiveness by isolation and brain serotonin turnover changes in different strains of mice. Neuropsychobiology, 5: 129–135.

Valzelli, L. and Garattini, S. (1968) Behavioral changes and 5-hydroxytryptamine turnover in animals. Adv. Pharmacol., 6B: 249–260.

Valzelli, L., Giacalone, E. and Garattini, S. (1967) Pharmacological control of aggressive behavior in mice. Eur. J. Pharmacol., 2: 144–146.

van den Berg, L., Vos-Loohuis, M., Schilder, M.B.H., van Oost, B.A., Hazewinkel, H.A.W., Wade, C.M. *et al.* (2008) Evaluation of the serotonergic genes htr1A, htr1B, htr2A, and slc6A4 in aggressive behavior of Golden Retriever dogs. Behav. Genet., 38: 55–66.

Vandenbergh, J.G. (1967) The development of social structure in free-ranging rhesus monkeys. Behaviour, 29: 179–194.

van der Vegt, B.J., Lieuwes, N., Cremers, T.I.F.H., de Boer, S.F. and Koolhaas, J.M. (2003a) Cerebrospinal fluid monoamine and metabolite concentrations and aggression in rats. Horm. Behav., 44: 199–208.

van der Vegt, B.J., Lieuwes, N., van de Wall, E.H., Kato, K., Moya-Albiol, L., Martinez-Sanchis, S. *et al.* (2003b) Activation of serotonergic neurotransmission during the performance of aggressive behavior in rats. Behav. Neurosci., 117: 667–674.

Van Erp, A.M.M. and Miczek, K.A. (2000) Aggressive behavior, increased accumbal dopamine, and decreased cortical serotonin in rats. J. Neurosci., 20: 9320–9325.

van Oortmerssen, G.A. and Bakker, T.C.M. (1981) Artificial selection for short and long attack latencies in wild mus (musculus domesticus). Behav. Genet., 11: 115–126.

Vergnes, M., Depaulis, A. and Boehrer, A. (1986) Parachlorophenylalanine-induced serotonin depletion increases offensive but not defensive aggression in male rats. Physiol. Behav., 36: 653–658.

Virkkunen, M. and Linnoila, M. (1993a) Brain serotonin, type II alcoholism and impulsive violence. J. Stud. Alcohol. Suppl., 11: 163–169.

Virkkunen, M. and Linnoila, M. (1993b) Serotonin in personality disorders with habitual violence and impulsivity. In: Hodgins, S. (Ed.), Mental Disorder and Crime. Sage, Newbury Park, CA, pp. 227–243.

Vitiello, B. and Stoff, D.M. (1997) Subtypes of aggression and their relevance to child psychiatry. J. Am. Acad. Child Adolesc. Psychiatry, 36: 307–315.

Walsh, M.T. and Dinan, T.G. (2001) Selective serotonin reuptake inhibitors and violence: a review of the available evidence. Acta Psychiatr. Scand., 104: 84–91.

Walther, D.J., Peter, J.U., Bashammakh, S., Hortnagl, H., Voits, M., Fink, H. *et al.* (2003) Synthesis of serotonin by a second tryptophan hydroxylase isoform. Science, 299: 76.

Weder, N., Yang, B.Z., Douglas-Palumberi, H., Massey, J., Krystal, J.H., Gelernter, J. *et al.* (2009) MAOA genotype, maltreatment, and aggressive behavior: the changing impact of genotype at varying levels of trauma. Biol. Psychiatry, 65: 417–424.

Welch, B.L. and Welch, A.S. (1968) Rapid modification of isolation-induced aggressive behavior and elevation of brain catecholamines and serotonin by the quick-acting monoamine- oxidase inhibitor pargyline. Commun. Behav. Biol., 1: 347–351.

Wersinger, S.R., Ginns, E.I., O'Carroll, A.M., Lolait, S.J. and Young, W.S. III (2002) Vasopressin V1b receptor knockout reduces aggressive behavior in male mice. Mol. Psychiatry, 7: 975–984.

Westergaard, G.C., Suomi, S.J., Higley, J.D. and Mehlman, P.T. (1999) CSF 5-HIAA and aggression in female macaque monkeys: species and interindividual differences. Psychopharmacology, 146: 440–446.

Whale, R., Quested, D.J., Laver, D., Harrison, P.J. and Cowen, P.J. (2000) Serotonin transporter (5-HTT) promoter genotype may influence the prolactin response to clomipramine. Psychopharmacology, 150: 120–122.

White, S.M., Kucharik, R.F. and Moyer, J.A. (1991) Effects of serotonergic agents on isolation-induced aggression. Pharmacol. Biochem. Behav., 39: 729–736.

Widom, C.S. and Brzustowicz, L.M. (2006) MAOA and the 'Cycle of violence': childhood abuse and neglect, MAOA genotype, and risk for violent and antisocial behavior. Biol. Psychiatry, 60: 684–689.

Winslow, J.T., Hearn, E.F., Ferguson, J., Young, L.J., Matzuk, M.M. and Insel, T.R. (2000) Infant vocalization, adult aggression, and fear behavior of an oxytocin null mutant mouse. Horm. Behav., 37: 145–155.

Witte, A.V., Floel, A., Stein, P., Savli, M., Mien, L.K., Wadsak, W., *et al.* (2008). Aggression is related to frontal serotonin-1A receptor distribution as revealed by PET in healthy subjects. Hum. Brain Mapp., in press.

Yanai, K., Son, L.Z., Endou, M., Sakurai, E., Nakagawasai, O., Tadano, T. *et al.* (1998) Behavioral characterization and amounts of brain monoamines and their metabolites in mice lacking histamine H{-1} receptors. Neuroscience, 87: 479–487.

Yen, C.Y., Stanger, R.L. and Millman, N. (1959) Ataractic suppression of isolation-induced aggressive behavior. Archives Internationales de Pharmacodynamie et de Therapie, 123: 179–185.

Yodyingyuad, U., de la Riva, C., Abbott, D.H., Herbert, J. and Keverne, E.B. (1985) Relationship between dominance hierarchy, cerebrospinal fluid levels of amine transmitter metabolites (5-hydroxyindole acetic acid and homovanillic acid) and plasma cortisol in monkeys. Neuroscience, 16: 851–858.

Young, S.N. and Leyton, M. (2002) The role of serotonin in human mood and social interaction-Insight from altered tryptophan levels. Pharmacol. Biochem. Behav., 71: 857–865.

Young, K.A., Berry, M.L., Mahaffey, C.L., Saionz, J.R., Hawes, N.L., Chang, B. *et al.* (2002) Fierce: a new mouse deletion of Nr2e1; violent behaviour and ocular abnormalities are background-dependent. Behav. Brain Res., 132: 145–158.

Zarcone, J.R., Hellings, J.A., Crandall, K., Reese, R.M., Marquis, J., Fleming, K. *et al.* (2001) Effects of risperidone on aberrant behavior of persons with developmental disabilities: I. A double-blind crossover study using multiple measures. Ame. J. Ment. Retard., 106: 525–538.

Zhang, X.D., Beaulieu, J.M., Sotnikova, T.D., Gainetdinov, R.R. and Caron, M.G. (2004) Tryptophan hydroxylase-2 controls brain serotonin synthesis. Science, 305: 217.

Zhang, X.D., Gainetdinov, R.R., Beaulieu, J.M., Sotnikova, T.D., Burch, L.H., Williams, R.B. *et al.* (2005) Loss-of-function mutation in tryptophan hydroxylase-2 identified in unipolar major depression. Neuron, 45: 11–16.

Zhuang, X., Gross, C., Santarelli, L., Compan, V., Trillat, A.C. and Hen, R. (1999) Altered emotional states in knockout mice lacking 5-HT1A or 5-HT1B receptors. Neuropsychopharmacology, 21: S52–S60.

Zouk, H., McGirr, A., Lebel, V., Benkelfat, C., Rouleau, G. and Turecki, G. (2007) The effect of genetic variation of the serotonin 1B receptor gene on impulsive aggressive Behavior and suicide. Am. J. Med. Genet. B Neuropsychiatr. Genet., 144B: 996–1002.

CHAPTER 4.11

Altered Serotonin Function in Anorexia and Bulimia Nervosa

Vikas Duvvuri[1], Ursula F. Bailer[1,2] and Walter H. Kaye[*,1]

[1]*University of California San Diego, School of Medicine, Department of Psychiatry, La Jolla, California*
[2]*Medical University of Vienna, Department of Psychiatry and Psychotherapy, Division of Biological Psychiatry, Vienna, Austria*

Abstract: Anorexia nervosa (AN) and bulimia nervosa (BN) are related disorders of unknown etiology that most commonly begin during adolescence in women. They are characterized by aberrant patterns of feeding behavior and weight regulation, and deviant attitudes and perceptions toward body weight and shape. In AN, an inexplicable fear of weight gain and unrelenting obsession with being fat, even in the face of increasing emaciation, accounts for a protracted course. BN usually emerges after a period of food restriction, which may or may not have been associated with weight loss. Binge eating is followed either by self-induced vomiting, or by other means of compensation for the excess of ingested food. There is growing acknowledgement that neurobiological vulnerabilities make a substantial contribution to the pathogenesis of AN and BN, suggesting that altered brain serotonin (5-HT) function contributes to dysregulation of appetite, mood, and impulse control. Brain imaging studies, using 5-HT specific ligands, show that disturbances of 5-HT function occur when people are ill, and persist after recovery. It is possible that a trait-related disturbance of 5-HT neuronal modulation predates the onset of AN, and contributes to premorbid symptoms of anxiety, obsessionality and inhibition. This dysphoric temperament may involve an inherent dysregulation of emotional and reward pathways which also mediates the hedonic aspects of feeding, thus making these individuals vulnerable to disturbed appetitive behaviors. Individuals with AN may discover that reduced dietary intake, by reducing plasma tryptophan availability, is a means by which they can modulate brain 5-HT functional activity and anxious mood. They enter a vicious cycle, because caloric restriction results in a brief respite from dysphoric mood. However, malnutrition and weight loss, in turn, produce alterations in many neuropeptides and monoamine function, which also exaggerates dysphoric mood. Thus, those with AN pursue starvation in an attempt to avoid the dysphoric consequences of eating.

Keywords: serotonin, anorexia, bulimia, anxiety, neuroimaging, harm avoidance, impulsivity.

Introduction

Anorexia nervosa (AN) and bulimia nervosa (BN) are related disorders of unknown etiology that most commonly begin during adolescence in women (*DSM-IV*; Table 1). They are frequently chronic and often disabling conditions that are characterized by aberrant patterns of feeding behavior and weight regulation, and deviant attitudes and perceptions toward body weight and shape. In AN, an inexplicable fear of weight gain and unrelenting obsession with being fat, even in the face of increasing emaciation, accounts for a protracted course. Sequelae include extreme medical and psychological morbidity, and standardized mortality rates exceeding those of all other psychiatric disorders. BN usually emerges after a period of food restriction, which may or may not have been associated with weight loss. Binge eating is followed by either self-induced vomiting, or by some other means of compensation for the excess of food that is ingested. Although abnormally low body weight is an exclusion criterion for the diagnosis of BN, some 25–30 percent of individuals with BN have a prior history of AN.

Because AN and BN present most often during adolescence in women, they are often theorized to be caused by cultural pressures for thinness (Kaye *et al.*, 2009), since dieting and the pursuit of thinness are common in industrialized countries. Still, AN and BN affect only an estimated 0.3–0.7 percent and 1.5–2.5 percent, respectively, of females in the general population (Hoek *et al.*, 1995), suggesting that an underlying biological predisposition might

*Corresponding author
E-mail: wkaye@ucsd.edu

ISBN 978-0-12-374634-4
DOI: 10.1016/B978-0-12-374634-4.00039-3

Table 1 *DSM IV* diagnostic criteria for anorexia nervosa and bulimia nervosa

DSM IV, Diagnostic criteria for Anorexia Nervosa	
A	Refusal to maintain body weight at or above a minimally normal weight for age and height (e.g., weight loss leading to maintenance of body weight less than 85% of that expected; or failure to make expected weight gain during period of growth, leading to body weight less than 85% of that expected).
B	Intense fear of gaining weight or becoming fat, even though underweight.
C	Disturbance in the way in which one's body weight or shape is experienced, undue influence of body weight or shape on self-evaluation, or denial of the seriousness of the current low body weight.
D	In postmenarcheal females, amenorrhea, i.e., the absence of at least three consecutive menstrual cycles. (A woman is considered to have amenorrhea if her periods occur only following hormone, e.g., estrogen, administration.)
	Specify type:
	Restricting type: during the current episode of anorexia nervosa, the person has not regularly engaged in binge-eating or purging behavior (i.e., self-induced vomiting or the misuse of laxatives, diuretics, or enemas)
	Binge-eating/purging type: during the current episode of anorexia nervosa, the person has regularly engaged in binge-eating or purging behavior (i.e., self-induced vomiting or the misuse of laxatives, diuretics, or enemas)
DSM IV, Diagnostic criteria for Bulimia Nervosa	
A	Recurrent episodes of binge-eating. An episode of binge-eating is characterized by both of the following (1) Eating, in a discrete period of time (e.g., within any 2-hour period), an amount of food that is definitely larger than most people would eat during a similar period of time and under similar circumstances. (2) A sense of lack of control over eating during the episode (e.g., a feeling that one cannot stop eating or control what or how much one is eating)
B	Recurrent inappropriate compensatory behavior in order to prevent weight gain, such as self-induced vomiting, misuse of laxatives, diuretics, enemas, or other medications; fasting, or excessive exercise.
C	The binge-eating and inappropriate compensatory behaviors occur, on average, at least twice a week for three months.
D	Self-evaluation is unduly influenced by body shape and weight.
E	The disturbance does not occur exclusively during episodes of Anorexia Nervosa.
	Type: purging type vs. non-purging type (exercise and fasting to compensate)
	Specify type
	Purging type: during the current episode of Bulimia Nervosa, the person has regularly engaged in self-induced vomiting or the misuse of laxatives, diuretics, or enemas
	Non-purging type: during the current episode of Bulimia Nervosa, the person has used other inappropriate compensatory behaviors, such as fasting or excessive exercise, but has not regularly engaged in self-induced vomiting or the misuse of laxatives, diuretics, or enemas

play an important role. Indeed, there is clear evidence of AN occurring at least several centuries ago (Treasure and Campbell, 1994). Its stereotypic presentation, substantial heritability, and a developmentally specific age-of-onset distribution further underscore the possibility of contributing biological vulnerabilities.

Considering that transitions between syndromes occur in many individuals, it has been argued that AN and BN share at least some risk and liability factors (Kendler *et al.*, 1991; Schweiger and Fichter, 1997). In fact, AN and BN are cross-transmitted in families (Lilenfeld *et al.*, 1998; Strober *et al.*, 2000). Moreover, there is an increased prevalence of AN and BN as well as subthreshold forms of eating disorders (EDs) in relatives, consistent with the possibility of a continuum of transmitted liability in at-risk families manifesting a broad spectrum of eating disorder phenotypes. Twin studies can differentiate genetic from environmental effects by comparing concordance for a trait, or disorder, between identical (monozygotic, MZ) and fraternal (dizygotic, DZ) twins. Twin studies of AN and BN suggest that there is an approximately 50–80 percent genetic contribution to liability (Kendler *et al.*, 1991; Bulik *et al.*, 1998, 2006; Klump *et al.*, 2001; Wade *et al.*, 2000) accounted for by additive genetic factors. These heritability estimates are similar to those found in schizophrenia and bipolar disorder, suggesting that AN and BN may be as genetically influenced as disorders traditionally viewed to be biological in nature.

Clinical symptoms and puzzling behaviors

The *DSM-IV* diagnostic criteria for AN and BN focus on eating behavior and body image distortions. Because of their unusual and prominent nature, these symptoms tend to capture much attention. The pathogenesis of the disturbed eating behaviors is poorly understood (Vitousek and

Manke, 1994; Schweiger and Fichter, 1997). Individuals with AN rarely have complete suppression of appetite, but rather exhibit an ego-syntonic resistance to feeding drives while simultaneously being preoccupied with food and eating rituals to the point of obsession. Individuals with AN severely restrict food intake, particularly fats and carbohydrates, but rarely stop eating completely; instead, they restrict their caloric intake to a few hundred calories a day. They tend to be vegetarians, have monotonous choices in the food they eat, select unusual combinations of foods and flavors, and have ritualized eating behaviors. Similarly, BN is not associated with a primary, pathological increase in appetite; rather, like individuals with AN, individuals with BN have a seemingly relentless drive to restrain their food intake, an extreme fear of weight gain, and often have a distorted view of their actual body shape. Loss of control with overeating in individuals with BN usually occurs intermittently, and typically only some time after the onset of dieting behavior. Restrained eating behavior and dysfunctional cognition relating weight and shape to self-concept are shared by all types of patients with EDs.

AN and BN individuals commonly have clusters of other puzzling symptoms. Excessive exercise and motor restlessness are common in AN (Shroff *et al.*, 2006). While not well studied, excessive exercise is thought to be associated particularly with the purging subtype of AN as well as with a constellation of anxious/obsessional temperaments. Individuals with AN often have resistance to treatment (Halmi *et al.*, 2005). This is in part due to the ego-syntonic nature of the disorder, which is demonstrated by the patient's denial of being underweight and refusal to accept the seriousness of the medical consequences of the disorder. Consequently, it has been difficult to enlist the cooperation of individuals with AN for treatment studies. This, and the fact that psychological/pharmacological strategies effective in other disorders are less so for AN, likely explains the dearth of controlled treatment trials of any kind.

Mood and impulse control

Individuals with AN and BN have elevated rates of lifetime diagnoses of anxiety and depressive disorders, and obsessive-compulsive disorder (Lilenfeld *et al.*, 1998; Godart *et al.*, 2002, 2007; Kaye *et al.*, 2004a). In addition, individuals with AN and BN are consistently characterized by perfectionism, obsessive-compulsiveness, neuroticism, negative emotionality, harm avoidance, low self-directedness, low cooperativeness, and traits associated with avoidant personality disorder (PD). There are consistent differences that emerge between groups of individuals with ED. Those with restricting-type AN exhibit high constraint and persistence, low novelty-seeking, constriction of affect and emotional expressiveness, anhedonia and asceticism, and reduced social spontaneity. Individuals with BN are more likely to be of high impulsivity, sensation-seeking, novelty-seeking, and to have traits associated with borderline personality disorder and substance use disorders (Cassin and von Ranson, 2005). Thus, several domains, such as eating, affect and impulse control, are involved in systematic ways in EDs. Specifically, AN is associated with over-control and BN is associated with switches between over-control and under-control, which raises the question of whether there is a concerted disturbance in the modulation of multiple systems.

Neurocognition

Individuals with AN have an obsessive, perseverative and rigid personality style, and have difficulty shifting sets. While those with AN do well on goal-directed behavior, they have difficulties incorporating feedback and modifying their behavior. For example, they often feel that they should be able to do things perfectly without making mistakes, and they have little appreciation for the fact that mistakes are a normal learning experience. Moreover, they often fail to accurately recognize and incorporate affective and social stimuli in the environment, as confirmed by laboratory tests (Strupp *et al.*, 1986; Kingston *et al.*, 1996). Those who are ill with (ILL) AN and recovered from (REC) AN tend to have delayed set-shifting (Tchanturia *et al.*, 2005), an ability that normally allows for the adaptation of behavior in line with the changing demands of one's environment. Furthermore, individuals with AN have an enhanced ability to pay attention to detail or use a logical/analytic approach, but exhibit worse performance with global strategies (Strupp *et al.*, 1986; Lopez *et al.*, 2007).

State and trait

It has long been debated whether symptoms in individuals with AN and BN are the cause or consequence of malnutrition. Confounding this understanding is the issue that most studies of symptoms have been done when individuals are ill with an ED. Recent studies have shown that the majority of people with AN and BN exhibit childhood perfectionism, obsessive-compulsive personality patterns, and anxiety, all of which predate the onset of AN and BN (Fairburn *et al.*, 1997; Godart *et al.*, 2002; Anderluh *et al.*, 2003). Moreover, studies done on three continents (Table 2) have shown that in AN and BN individuals with a lifetime history of an anxiety-disorder diagnosis, the anxiety disorder most often began in childhood before the onset of the ED (Deep *et al.*, 1995; Bulik *et al.*, 1997b, 1997c; Godart *et al.*, 2000). The most common premorbid childhood disorders were OCD and social phobia (Kaye *et al.*, 2004a). In summary, such symptoms may be susceptibility factors that make people vulnerable to developing an ED. Malnutrition tends to exaggerate premorbid behavioral traits, not cause them (Pollice *et al.*, 1997).

It is important to note that such temperament and personality traits persist after recovery from an ED. While a definition of recovery has not been formalized, investigators tend to include people once they have been at a stable and healthy body weight for months or years, are not malnourished, and have not engaged in pathological eating behavior during that period of recovery. Some investigators include a criterion of a normal menstrual cycle, and a minimal duration of recovery, such as 1 year. Investigators (Strober, 1980; Casper, 1990; Srinivasagam *et al.*, 1995; Kaye *et al.*, 1998) have found that women who are long-term recovered from AN and BN have a persistence of anxiety, perfectionism and obsessional behaviors (particularly symmetry, exactness and order). Determining cause and effect is a major methodological confound in this illness. Thus, this chapter will focus on studies that have been done in people who have recovered from AN and BN. The persistence of many alterations after recovery suggests that these may be traits that create a vulnerability to develop an ED, although it is possible these are 'scars' caused by the ill state.

Neurobiology

There is growing acknowledgement that neurobiological vulnerabilities make a substantial contribution to the pathogenesis of AN and BN (Treasure and Campbell, 1994); however, we have little understanding of how such vulnerabilities result in disturbances of brain pathways or which systems are primarily involved. For example, are there disturbances of pathways related to the modulation of feeding behaviors, mood, temperament, obsessionality or impulse control? Are there primary disturbances of pathways that may modulate some factors related to body proprioception, and thus result in body image distortions? In the past, answering these questions has been thwarted by the inaccessibility of the brain. Technologies capable of characterizing the complexity of brain circuits in humans, such as imaging or genetics, have only recently become available. Still, past studies have been useful in terms of grossly identifying systems that may be involved in ED. Because these technologies could only characterize one molecule at a time, they weren't able to provide answers to questions about complex interactions and functions.

Serotonin

Much technology has focused on characterizing the function of monoamines, such as serotonin (5-HT), dopamine (DA) and norepinephrine (NE). In part, this is because many of the medications used to treat psychiatric disorders act on these systems. Although neurotransmission occurs via distinct receptors, these systems have reciprocal interactions, so it is virtually impossible to act on a specific neuronal element without a cascade effect on the two other systems (Tremblay and Blier, 2006).

Investigations of serotonin metabolism
Theoretically, 5-HT disturbances could contribute to appetite dysregulation (Blundell, 1984; Leibowitz and Shor-Posner, 1986), anxious and obsessive behaviors, and extremes of impulse control (Soubrie, 1986; Cloninger,

Table 2 Lifetime and premorbid rates of anxiety disorders

Study	ED	N	Lifetime D	AD before ED
Deep *et al.*, 1995	AN	24	68%	58%
Bulik *et al.*, 1997a	AN	68	60%	54%
Bulik *et al.*, 1997c	BN	116	57%	54%
Godart *et al.*, 2000	AN	29	83%	62%
Godart *et al.*, 2000	BN	34	71%	62%
Kaye *et al.*, 2004a	AN, BN	672	64%	61%
				23% OCD
				13% social phobia

Abbreviations: AD, anxiety disorder; AN, anorexia nervosa; BN, bulimia nervosa; ED, eating disorder; Lifetime D, lifetime disorder.

1987; Higley and Linnoila, 1997; Lucki, 1998; Mann, 1999). Considerable evidence suggests that disturbances of monoamine function occur when people are ill with ED, and persist after recovery from AN and BN (Brewerton *et al.*, 1990; Jimerson *et al.*, 1997; Walsh and Devlin, 1998; Steiger *et al.*, 2001a; Kaye *et al.*, 2004b).

ILL AN subjects have a significant reduction in 5-hydroxyindoleacetic acid (5-HIAA) in their cerebrospinal fluid (CSF) compared to control subjects (CW), whereas CSF 5-HIAA levels are normal in ILL BN (Jimerson *et al.*, 1992; Kaye *et al.*, 1984, 1990, 1991a, 1998). In comparison, REC AN and BN subjects have higher than normal concentrations of CSF 5-HIAA, which is about 50 percent greater than CSF 5-HIAA levels found in the ILL state. While CSF levels have been found to reflect monoamine function in the brains of humans (Stanley *et al.*, 1985), they are an indirect measure of 5-HT function.

ILL AN and BN patients have abnormal levels of prolactin and other hormones in response to 5-HT specific challenges in most studies (Goodwin *et al.*, 1989; Brewerton *et al.*, 1990; Goldbloom *et al.*, 1990; Hadigan *et al.*, 1995; Jimerson *et al.*, 1997; Levitan *et al.*, 1997). After various degrees of recovery, hormonal responses tend to normalize in AN and BN, but some abnormal hormonal responses may persist (Wolfe *et al.*, 1994, 2000; Hadigan *et al.*, 1995; O'Dwyer *et al.*, 1996; Kaye *et al.*, 1998; Ward *et al.*, 1998). While hormones have been used as a reflection of CNS 5-HT, hormones are altered by malnutrition (Ishizuka *et al.*, 1983; Anderson *et al.*, 1990; O'Keane *et al.*, 1991) or may be relatively independent of brain 5-HT activity (McCann *et al.*, 1996).

Clinically, women with AN and BN appear to have opposing relationships between feeding and behavior. BN women are dysphoric when starved, whereas binge behavior appears temporarily to reduce dysphoric mood (Kaye *et al.*, 1986). In contrast, AN women describe feeling better when they starve. There is some evidence that these subgroups show different responses to 5-HT challenges. Studies that have examined the behavioral response to 5-HT challenges have found that depletion of tryptophan (TRP), the precursor of 5-HT (Fernstrom and Wurtman, 1971a, 1972) or serotonergic agents, such as mCPP, reduced dysphoric mood in ILL and REC AN subjects (O'Dwyer *et al.*, 1996; Gendall *et al.*, 1998; Ward *et al.*, 1998; Frank *et al.*, 2001; Kaye *et al.*, 2003). In contrast, in ILL and REC BN women, TRP depletion or mCPP is associated with increased caloric intake and dysphoric mood (Weltzin *et al.*, 1995; Kaye *et al.*, 1998, 2000; Smith *et al.*, 1999). In addition, BN individuals show an abnormal response to other measures, such as platelet binding of the selective serotonin reuptake inhibitor (SSRI) paroxetine (Steiger *et al.*, 2005a), which has also been found to be altered in non-affected first-degree relatives (Steiger *et al.*, 2006).

Diet and brain 5-HT neurotransmission

Tryptophan (TRP), an essential amino acid only available in the diet, is the precursor of 5-HT. Meal consumption, depending on the proportion of carbohydrate and protein, can enhance brain 5-HT release (Fernstrom and Wurtman, 1971b, 1972); thereby affecting appetite regulation. In brief, carbohydrate consumption causes an insulin-mediated fall in plasma levels of the large neutral amino acids (LNAA, tyrosine, phenylalanine, valine, leucine, isoleucine) which compete with TRP for uptake into the brain. This elevates the plasma TRP/large neutral amino acid ratio (TRP/LNAA), and thus brain TRP, which rapidly accelerates brain 5-HT synthesis and release. Dietary proteins tend to block these effects by contributing large amounts of LNAA to the bloodstream. Considerable evidence in animals and healthy humans (Fernstrom and Wurtman, 1971b, 1972; Biggio *et al.*, 1974; Messing *et al.*, 1976; Gibbons *et al.*, 1979; Young and Gauthier, 1981; Anderson *et al.*, 1990) shows that a restricted diet significantly lowers plasma TRP, resulting in a decreased plasma ratio of TRP to neutral amino acids and, in turn, a reduction in the availability of TRP to the brain. Thus, restricted diet (and experimentally reduced TRP) decreases brain 5-HT synthesis, down-regulates the density of 5-HT transporters (Huether *et al.*, 1997), and produces a compensatory supersensitivity of post-synaptic receptors in response to reduced 5-HT turnover (Goodwin *et al.*, 1987a, 1987b). Limited data show that malnourished and emaciated AN women have a reduction of plasma TRP availability (Schweiger *et al.*, 1986). In addition, these alterations in post-meal amino acid metabolism are only partly reversed by nutritional rehabilitation (Schweiger *et al.*, 1986). It has been speculated (Vitousek and Manke, 1994; Strober, 1995) that dietary restraint may reduce anxiety in people with AN. In fact, we found that the anxiolytic effects of dieting in AN were related to a reduction in 5-HT neurotransmission (Kaye *et al.*, 2003). Furthermore, administration of meta-chlorophenylpiperazine (mCPP), a relatively selective serotonergic agent (Hoyer *et al.*, 1989; Kahn and Wetzler, 1991; Graeff *et al.*, 1996; Thomas *et al.*, 1996; Willins and Meltzer, 1997) was associated with a significant reduction in dysphoric mood states in one study (Frank *et al.*, 2001) but not in another (Hadigan *et al.*, 1995).

Serotonin medication

While BN individuals show a response to higher doses of fluoxetine, an SSRI (Fluoxetine Bulimia Nervosa Collaborative Study Group, 1992), the efficacy of such medication has been questioned since relatively few individuals abstain from binge and purge behaviors, and relapse during treatment is common (Walsh and Devlin, 1995). Despite the abundance of data implicating 5-HT dysregulation in AN, it remains controversial whether SSRIs are effective in restricting-type AN (RAN) individuals (Kaye *et al.*, 2001; Walsh *et al.*, 2006). Our clinical experience (Kaye *et al.*, 1991b) suggests that RAN respond better to fluoxetine than do binge eating–purging type AN (BAN) individual, and that some BN individuals can be relatively insensitive to high doses of SSRIs. For severely emaciated patients, hospitalization for supportive medical care and weight restoration may be useful or necessary. Still, relapse is common after discharge.

Brain imaging

The past decade has seen the introduction of tools, such as brain imaging, which hold the promise of characterizing complex neurocircuits and their relationship to behavior in living humans. In fact, these tools have rapidly advanced knowledge to the point where we can begin to make educated guesses about the pathophysiology of AN and BN and start to develop testable hypotheses about mechanisms.

Brain-imaging studies in AN and BN can be divided into several categories. First, there is a substantial literature, using computerized tomography (CT) and, more recently, magnetic resonance imaging (MRI), that has sought to determine whether there are brain structural alterations in individuals with ED. Second, there are imaging studies, such as positron emission tomography (PET) and single photon emission computed tomography (SPECT), that employ a radioligand. These studies, which may use fluorodeoxyglucose (FDG) to study glucose metabolism, or a ligand that is specific for a serotonin receptor, provide information that is specific for the system being studied, such as the 5-HT_{2A} receptor. Third, more recent studies have used fMRI or other technologies to assess blood-flow responses to some stimuli, such as pictures of food. Overall, imaging studies have been relatively consistent, in that many studies show differences in ILL and REC ED individuals in frontal, cingulate, temporal and/or parietal regions compared to control subjects (CW). However, it should be noted that these studies have not consistently identified regions, pathways or behavioral correlates. Sample sizes have been small, and imaging technologies and methods vary widely. Moreover, investigations have tended to assess relatively large regions of the brain that vary widely between studies. Published studies to date indicate gross alterations of brain function. Because brain pathways are highly complex, the neuroanatomy of AN and BN has only begun to be characterized.

Imaging studies of 5-HT function

The development of selective tracers for the 5-HT system has made *in vivo* study of 5-HT receptor function possible using PET brain imaging. In turn, this offers the possibility of better understanding of 5-HT neurotransmitter activity and dynamic relationships to behavior.

5-HT1A receptor Our group used PET imaging with the radioligand [^{11}C]WAY100635 to assess the binding potential (BP) of the 5-HT_{1A} receptor. The 5-HT_{1A} autoreceptor is located presynaptically on serotonergic somatodendritic cells in the raphe nucleus, where it functions to decrease 5-HT neurotransmission (Staley *et al.*, 1998). High densities of postsynaptic 5-HT_{1A} exist in the hippocampus, septum, amygdale, and entorhinal and frontal cortex, where they serve to mediate the effects of released 5-HT. Although the molecular organization for the receptor transduction seems to be identical in all of the areas where 5-HT_{1A} receptors are expressed, some differences in both functional and regulatory properties have been reported from area to area (Lanfumey and Hamon, 2000). Studies in animals and humans implicate the 5-HT_{1A} receptor in anxiety (Cervo *et al.*, 2000; File *et al.*, 2000; Olivier *et al.*, 2001) and depression and/or suicide (Matsubara *et al.*, 1991; Arango *et al.*, 1995; Mann, 1999).

Bailer and colleagues (Bailer *et al.*, 2007a) reported that ILL AN individuals had a 50–70 percent increase in [^{11}C]WAY100635 BP in subgenual, mesial temporal, orbital frontal and raphe brain regions, as well as prefrontal, lateral temporal, anterior cingulate and parietal regions. Similarly, REC BAN and BN subjects (Bailer *et al.*, 2005; Kaye, unpublished data; Figure 1) had a significant 20–40 percent increase in [^{11}C]WAY100635 BP in these same regions compared to CW (Bailer *et al.*, 2005). In contrast, REC RAN women showed no difference in [^{11}C]WAY100635 BP compared to controls (Bailer *et al.*, 2005). Increased 5-HT_{1A} postsynaptic activity has also been reported in ILL BN subjects (Tiihonen *et al.*, 2004).

The role of the 5-HT_{1A} receptor in behavior is not certain. 5-HT_{1A} receptor activity has been reported to play a role in anxiety (File *et al.*, 2000). For example, some, but not all, studies show that 5-HT_{1A} knockout mice have increased anxiety (Groenink *et al.*, 2003). REC RAN subjects showed positive relationships between harm avoidance, a trait characterized by anticipatory worry and fear of uncertainty, and postsynaptic [^{11}C]WAY100635 BP in subgenual cingulate, mesial temporal, lateral temporal, medial orbital frontal and parietal cortex. It is not clear why this animal model and human phenomena appear to have opposite effects.

As noted above, EDs are frequently co-morbid with depression and anxiety disorders. Reduced [^{11}C]WAY100635 BP has been found in ILL (Drevets *et al.*, 1999; Sargent

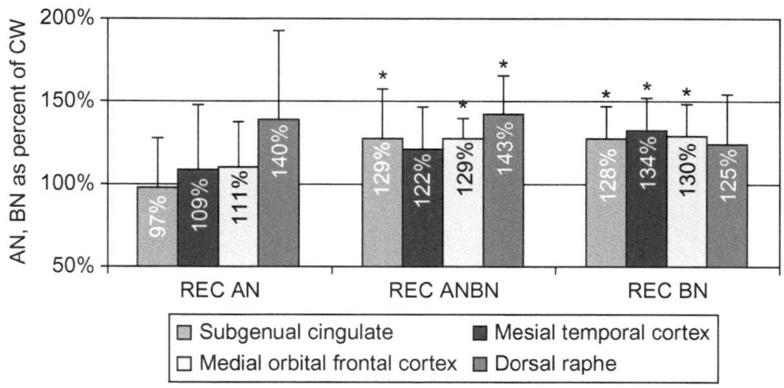

Figure 1 Results of PET/[^{11}C]WAY100635 BP (5-HT$_{1A}$ receptor) where individuals recovered from eating disorders (REC ED) are shown as percent of control subject (CW) values (mean CW values set at 100%). $^{*}P < 0.05$ REC ED vs. CW.

et al., 2000) and REC (Bhagwagar *et al.*, 2004) depressed subjects, as well as in a primate model for depression (Shively *et al.*, 2006). Parsey and colleagues found no difference in carbonyl-[^{11}C]WAY100635 BP in major depressive disorder, although a subgroup of never medicated subjects had elevated carbonyl-[^{11}C]WAY100635 BP (Parsey *et al.*, 2005). Recent studies have found reduced [^{11}C]WAY100635 BP in social phobia (Lanzenberger *et al.*, 2007) and panic disorder (Neumeister *et al.*, 2004). These findings suggest that ED, mood and depression share disturbances of common systems but are etiologically different.

There is an extensive literature associating the serotonergic systems and fundamental aspects of behavioral inhibition (Soubrie, 1986; Geyer, 1996). Reduced CSF 5-HIAA levels are associated with increased impulsivity and aggression in humans and non-human primates, whereas increased CSF 5-HIAA levels are related to behavioral inhibition (Fairbanks *et al.*, 2001; Westergaard *et al.*, 2003). Brainstem 5-HT$_{1A}$ receptors inhibit stress-induced sympathetic activity and inhibit fight-or-flight behavioral responses, supporting a role for this receptor in behavioral inhibition and self-control (Johnson *et al.*, 2004). Furthermore, recent animal studies also support modulation of impulse control via 5-HT$_{1A}$ receptors through effects on catecholamine systems (Winstanley *et al.*, 2005). Other studies have shown that blunted 5-HT$_{1A}$ receptor number or function is associated with increased aggression (Coccaro *et al.*, 1990; Cleare and Bond, 2000). A recent study (Fischer *et al.*, 2007) found a significant inverse relationship between dorsal raphe 5-HT$_{1A}$ autoreceptor BP and bilateral amygdala reactivity. 5-HT$_{1A}$ receptor function could contribute to behavioral inhibition in BN, although there is no direct evidence for this conjecture. Still other studies show that various measures of 5-HT activity are related to measures of affective instability and impulsivity in ill BN subjects (Steiger *et al.*, 2001a, 2001b, 2001c).

5-HT2A receptor Our group used PET imaging with the radioligand [^{18}F]altanserin to assess BP of 5-HT$_{2A}$ receptors (Figure 2). Postsynaptic 5-HT$_{2A}$ receptors are present in high densities in the cerebral cortex and other brain regions of rodents and humans (Saudou and Hen, 1994; Burnet *et al.*, 1997). The 5-HT$_{2A}$ receptor is of interest in ED because it has been implicated in the modulation of feeding and mood, as well as SSRI response (Simansky, 1996; Stockmeier, 1997; Bonhomme and Esposito, 1998; De Vry and Schreiber, 2000; Bailer *et al.*, 2004).

ILL AN subjects had normal [^{18}F]altanserin BP values (Bailer *et al.*, 2007a). In comparison, REC RAN individuals (Frank *et al.*, 2002) had reduced [^{18}F]altanserin BP in mesial temporal and parietal cortical areas, as well as in subgenual and pregenual cingulate cortex. Similarly, REC BAN (Bailer *et al.*, 2004) women had reduced [^{18}F]altanserin BP relative to controls in the left subgenual cingulate, left parietal, and right occipital cortex. Moreover, REC BN women only had reduced [^{18}F]altanserin BP relative to controls in the orbitofrontal region. Audenaert *et al.* (2003) used SPECT and 123I-5-I-R91159 and found that ILL AN subjects had reduced binding of postsynaptic 5-HT$_{2A}$ receptors in the left frontal, bilateral parietal and occipital cortex, with bulimic type BAN having reduced 5-HT$_{2A}$ binding in the parietal cortex in comparison to RANs (Goethals *et al.*, 2007).

Brain regions/pathways enervated by 5HT$_{1A/2A}$ receptors In REC subjects, altered 5-HT$_{1A}$ and 5-HT$_{2A}$ receptor BP shows persistent alterations in frontal, subgenual cingulate

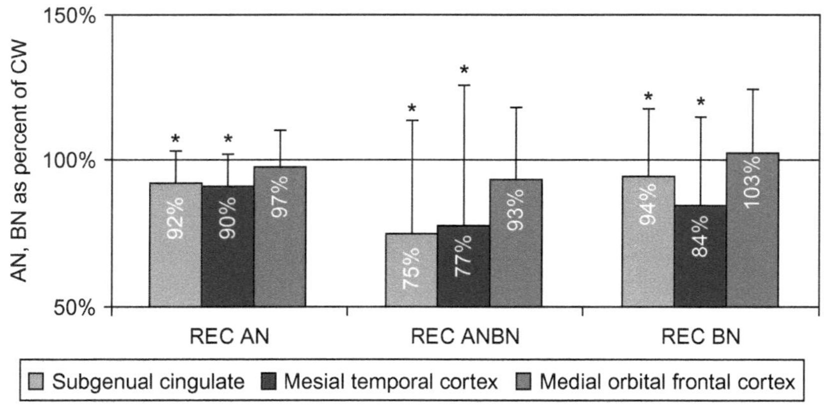

Figure 2 Results of PET/[18F]altanserin binding potential (5-HT_{2A} receptor) where individuals recovered from eating disorders (REC ED) are shown as percent of control subject (CW) values (mean CW values set at 100%). $^{*}P < 0.05$ REC ED vs. CW.

and mesial temporal regions that are part of the ventral limbic system. The subcaudal cingulate regions play a role in emotion ('affect component') and have extensive connections with the amygdala, periaqueductal gray, frontal lobes and ventral striatum. They are involved in conditioned emotional learning, vocalizations associated with expressing internal states, and assigning emotional valence to internal and external stimuli (Freedman *et al.*, 2000). Mesial temporal regions include the amygdala and related regions that play a pivotal role in anxiety and fear (Charney and Deutch, 1996) as well in the modulation and integration of cognition and mood. The amygdala may enable the individual to initiate adaptive behaviors to threat, based upon the nature of the threat and prior experience.

Several lines of evidence show that 5-HT_{1A} and 5-HT_{2A} receptors interact in the brain. In rats, 5-HT_{1A} and 5-HT_{2A} receptors interact robustly to regulate the inhibition of exploration of novel environments produced by either 5-HT_{1A} or 5-HT_{2A} receptor agonists (Krebs-Thomson and Geyer, 1998). 5-HT_{2A} and 5-HT_{1A} receptors are highly co-localized in rodent frontal cortex (Amargos-Bosch *et al.*, 2004). Postsynaptic 5-HT_{1A} and 5-HT_{2A} receptors mediate, respectively, the direct hyperpolarizing and depolarizing actions of 5-HT on prefrontal neurons (Santana *et al.*, 2004), which in turn project to numerous cortical and subcortical areas. Thus, a balance between postsynaptic 5-HT_{1A} and 5-HT_{2A} receptor activity on neurons may modulate the descending excitatory input into limbic and motor structures. These data raise speculation that postsynaptic 5-HT_{1A} and 5-HT_{2A} receptors fine-tune cortical systems that modulate behavioral inhibition and self-control. Mixed 5-$HT_{2A/1A}$ agonists, such as psilocybin, seem to disrupt the 5-$HT_{1A/2A}$ balance (Vollenweider *et al.*, 1999) by driving 5-HT_{2A} activity, thus resulting in excessive neuronal output that contributes to extremes of disinhibition, disorganization, and loss of self-control. In our studies, REC ED subjects had a relative increase in 5-HT_{1A} receptor activity compared to 5-HT_{2A} receptor binding. While speculative, this possible imbalance could contribute to behavioral inhibition and over-control commonly seen in ED. As discussed below, we found considerable correlations between binding of these two receptors and harm avoidance. Taken together, these findings raise the possibility that mesial temporal (amygdala)–cingulate $5HT_{1A/2A}$ imbalance may also be a trait shared by AN subgroups related to behavioral inhibition, anticipatory anxiety, or integration of cognition and mood.

5-HT transporter (5-HTT)

Studies with SPECT and [123I] beta-CIT showed that ILL BN subjects had a 17 percent reduced 5-HTT availability in the hypothalamus and thalamus (Tauscher *et al.*, 2001). Our group (Bailer *et al.*, 2007b) used PET imaging with [^{11}C]McN5652 to determine if alterations of 5-HTT persist after recovery from AN and BN. We compared 11 subjects recovered (>1 year normal weight, regular menstrual cycles, no bingeing or purging) from restricting type AN (REC RAN), 7 REC from bulimia-type AN (REC BAN), and 10 healthy CW. After correcting for multiple comparisons, we found that the REC RAN had significantly increased [^{11}C]McN5652 BP compared to REC BAN for the dorsal raphe and antero-ventral striatum. A SPECT study compared 5-HTT availability in the midbrain and thalamus in 13 female twins with BN (9 with purging and 4 with non-purging) versus 25 CW, using a different radiologand for 5-HTT, [123I]ADAM (Koskela *et al.*, 2007). They found that purging bulimics had increased midbrain [123I]ADAM binding compared to CW, supporting a serotonin-based distinction between those with purging and non-purging behaviors, across both studies. Differences in sample selection, imaging resolution of the methods, the regional binding of the radioligands and the data analysis approaches likely contributed to whether 5-HTT binding was relatively elevated or diminished in those with purging behaviors. It remains controversial whether SSRIs are effective in RAN individuals. Our clinical experience and data (Kaye *et al.*, 1991b, 2001; Walsh *et al.*, 2006) suggest that individuals with RAN respond better to fluoxetine than do those with BAN. While highly speculative, our findings raise the provocative possibility that decreased 5-HTT function may be related to poor response to SSRI medication, whereas individuals with increased 5-HTT activity may respond to higher SSRI doses. In general, the REC RAN individuals had elevated 5-HTT binding, suggesting that they have relatively greater 5-HT uptake, and reduced extracellular 5-HT, compared to REC BAN. In support of this possibility, the REC BAN individuals tend to have higher binding of 5-HT_{1A} postsynaptic receptors and autoreceptors (Bailer *et al.*, 2005), which may be a compensatory means of downregulating raphe activity (Cooper, 1996; Hajos *et al.*, 2003). Moreover, reduced 5-HTT activity, resulting from functional polymorphisms (Steiger *et al.*, 2005b), has been associated with affect dysregulation, which tends to be more common in the bulimic subgroups. Traits such as sensation-seeking and insecure attachment are elevated in bulimic syndromes carrying low function alleles of the 5-HTT promoter polymorphism, who report prior physical or sexual maltreatment (Steiger *et al.*, 2007). Furthermore, in people with impulsive aggression, reduced 5-HTT binding was found in the anterior cingulate cortex – a region involved in affect regulation (Frankle *et al.*, 2005).

5-HT and harm avoidance

The PET imaging studies in ILL and REC AN and BN subjects described above have found significant correlations between harm avoidance and binding for the 5-HT_{1A} and 5-HT_{2A} receptors in mesial temporal and other limbic regions. Bailer and colleagues have found that REC AN–BN subjects showed a positive relationship between [^{18}F]altanserin BP in the left subgenual cingulate and mesial temporal cortex and harm avoidance (Bailer *et al.*, 2004). For ILL AN subjects, [^{18}F]altanserin BP was positively related to harm avoidance in the supra-pragenual cingulate, frontal and parietal regions. 5-HT_{2A} receptor binding and harm avoidance were shown to be negatively correlated in the frontal cortex in healthy subjects (Moresco *et al.*, 2002), and in the prefrontal cortex in patients who attempted suicide (van Heeringen *et al.*, 2003). Furthermore, genetic studies have shown that the low 5-HTT activity polymorphism (125 women with BN and 94 CW (Monteleone *et al.*, 2006)) as well as the A218C polymorphism of the biosynthetic tryptophan hydroxylase-1 gene (91 women with BN and 89 CW (Monteleone *et al.*, 2007)) predispose BN individuals to increased harm avoidance.

Clinical and epidemiological studies have consistently shown that one or more anxiety disorders occur in the majority of people with AN or BN (Kendler *et al.*, 1995; Walters and Kendler, 1995; Godart *et al.*, 2002; Kaye *et al.*, 2004a). Silberg and Bulik, using twins, found a unique genetic effect that influences liability to early anxiety and eating disorder symptoms (Silberg and Bulik, 2005). When a lifetime anxiety disorder is present, the anxiety most commonly occurs first in childhood, preceding the onset of AN or BN (Deep *et al.*, 1995; Bulik *et al.*, 1997a; Godart *et al.*, 2000). Anxiety and harm avoidance remain elevated after recovery from AN, AN–BN and BN (Wagner *et al.*, 2006), even if individuals had never had a lifetime anxiety disorder diagnosis (Kaye *et al.*, 2004a). Finally, anxiety (Spielberger *et al.*, 1970) and harm avoidance from the Cloninger Temperament and Character Inventory (TCI) (Cloninger *et al.*, 1994) have been robust signals in our genetic studies (Bacanu *et al.*, 2005). In summary, the premorbid onset and the persistence of anxiety and harm avoidance symptoms after recovery suggest that these are traits that contribute to the pathogenesis of AN and BN. The PET imaging data suggest that such behaviors are related to disturbances of 5-HT neurotransmitter function in limbic and executive pathways.

Implications

Phillips and colleagues (2003) have described a ventral limbic system, containing the amygdala, insula, ventral striatum and ventral regions of the anterior cingulate gyrus and prefrontal cortex, which identifies the emotional significance of a stimulus and the production of an affective state in response to that stimulus. In addition, these regions are important for automatic regulation and mediation of autonomic responses to emotional stimuli and contexts accompanying the production of affective states.

The findings described above offer evidence that individuals with ED have 5-HT dysregulation within brain regions that constitute limbic circuits. In general, such alterations tend to be present in the ill state and persist after recovery. Patterns of 5-HT receptor binding vary by subtype in the REC subjects, raising the possibility that each ED subtype has a unique pathophysiology (Table 3). Similar patterns of binding (elevated 5-HT_{1A} and reduced 5-HT_{2A}) have also been found in other temporal, cingulate and parietal regions, suggesting a widespread distribution involving more than just limbic function.

Table 3 PET findings with serotonin radioligands in ED

	AN	AN–BN	BN
ROI	Medial orbital frontal, subgenual cingulate, mesial temporal		
5-HT_{1A} BP	—	↑	↑
5-HT_{2A} BP	↓	↓	—
ROI	Anterio-ventral striatum		
5-HTT BP	↑	↓	—

Abbreviations: AN, anorexia nervosa; BN, bulimia nervosa; BP, binding potential; ROI, region of interest.

Drawing inferences about behavior from these 5-HT findings is speculative. Moreover, no receptor works in isolation in the brain. Monoamine enervation of limbic pathways is complex, and involves many pathways, receptors, enzymes, intracellular transcription messengers, other neurotransmitters and molecules. It can be hypothesized that these receptor findings might reflect the 'health' of the 5-HT function within limbic circuits, but not necessarily be the specific cause of these disorders. The genetic literature suggests that behavioral disorders are complex and consist of multiple factors, each of small effect. Finally, altered 5-HT receptor binding is often found in depression and anxiety, although patterns of binding tend to be different. Disturbances of affect and impulse regulation may involve similar pathways, but with very different patterns of molecular disturbances.

Conclusion

We hypothesize that a trait-related disturbance of 5-HT neuronal modulation predates the onset of AN and contributes to premorbid symptoms of anxiety and inhibition. This dysphoric temperament may involve an inherent dysregulation of emotional and reward pathways (Kelley, 2004), which also mediate the hedonic aspects of feeding, thus making these individuals vulnerable to disturbed appetitive behaviors. This 5-HT disturbance contributes to a vulnerability for restricted eating and dysphoric mood states such as increased harm avoidance. Most importantly, we think that restricting food intake is powerfully reinforcing because it provides a temporary respite from dysphoric mood. Several factors may act on these vulnerabilities to cause AN to start in adolescence. First, puberty-related female gonadal steroids or age-related changes may exacerbate 5-HT dysregulation. Second, stress and/or cultural and societal pressures may contribute by increasing anxious and obsessional temperament. We hypothesize that people with AN may discover that reduced dietary intake, by reducing plasma TRP availability (Schweiger *et al.*, 1986), is a means by which they can crudely modulate brain 5-HT functional activity and anxious mood (Kaye *et al.*, 2003). People with AN enter a vicious cycle, which accounts for the chronicity of this disorder, because caloric restriction results in a brief respite from dysphoric mood. However, malnutrition and weight loss, in turn, produce alterations in many neuropeptides and monoamine function, perhaps in the service of conserving energy, but which also exaggerate dysphoric mood. Thus, those with AN pursue starvation in an attempt to avoid the dysphoric consequences of eating. SSRI administration does not appear to be effective in counteracting 5-HT disturbances in patients ill with AN, perhaps because of the extreme changes induced by malnutrition in the 5-HT_{1A}-receptor and extracellular 5-HT concentrations.

References

Amargos-Bosch, M., Bortolozzi, A., Puig, M.J.S., Adell, A., Celada, P., Toth, M., Mengod, G. and Artigas, F. (2004) Co-expression and *in vivo* interaction of $serotonin_{1A}$ and $serotonin_{2A}$ receptors in pyramidal neurons of prefrontal cortex. Cereb. Cortex, 14: 281–299.

Anderluh, M.B., Tchanturia, K., Rabe-Hesketh, S. and Treasure, J. (2003) Childhood obsessive-compulsive personalitiy traits in adult women with eating disorders: defining a broader eating disorder phenotype. Am. J. Psychiatry, 160: 242–247.

Anderson, I.M., Parry-Billings, M., Newsholme, E.A., Fairburn, C.G. and Cowen, P.J. (1990) Dieting reduces plasma tryptophan and alters brain 5-HT function in women. Psychol. Med., 20: 785–791.

Arango, V., Underwood, M.D., Gubbi, A.V. and Mann, J.J. (1995) Localized alterations in pre- and postsynaptic serotonin binding sites in the ventrolateral prefrontal cortex of suicide victims. Brain Res., 688: 121–133.

Audenaert, K., Van Laere, K., Dumont, F., Vervaet, M., Goethals, I., Slegers, G., Mertens, J., van Heeringen, C. and Dierckx, R. (2003) Decreased 5-HT2a receptor binding in patients with anorexia nervosa. J. Nucl. Med., 44: 163–169.

Bacanu, S., Bulik, C., Klump, K., Fichter, M., Halmi, K., Keel, P., Kaplan, A., Strober, M., Treasure, J., Woodside, D.B., Berrettini, W. and Kaye, W. (2005) Linkage analysis of anorexia and bulimia nervosa cohorts using selected behavioral phenotypes as quantitative traits or covariates. Am. J. Med. Genet. B. Neuropsychiatr. Genet., 139: 61–68.

Bailer, U.F., Price, J.C., Meltzer, C.C., Mathis, C.A., Frank, G.K., Weissfeld, L., McConaha, C.W., Henry, S.E., Brooks-Achenbach, S., Barbarich, N.C. and Kaye, W.H. (2004) Altered 5-HT_{2A} receptor binding after recovery from bulimia-type anorexia nervosa: relationships to harm avoidance and drive for thinness. Neuropsychopharmacology, 29: 1143–1155.

Bailer, U.F., Frank, G.K., Henry, S.E., Price, J.C., Meltzer, C.C., Weissfeld, L., Mathis, C.A., Drevets, W.C., Wagner, A., Hoge, J., Ziolko, S.K., McConana, C.W. and Kaye, W.H. (2005) Altered brain serotonin 5-HT_{1A} receptor binding after recovery from anorexia nervosa measured by positron emission tomography and [^{11}C]WAY100635. Arch. Gen. Psychiatry, 62: 1032.

Bailer, U.F., Frank, G., Henry, S., Price, J., Meltzer, C., Mathis, C., Wagner, A., Thornton, L., Hoge, J., Ziolko, S.K., Becker, C., McConaha, C. and Kaye, W.H. (2007a) Exaggerated 5-HT_{1A} but normal 5-HT_{2A} receptor activity in individuals ill with anorexia nervosa. Biol. Psychiatry, 61: 1090–1099.

Bailer, U.F., Frank, G., Henry, S., Price, J., Meltzer, C.C., Becker, C., Ziolko, S.K., Mathis, C., Wagner, A., Barbarich-Marsteller, N., Putman, K. and Kaye, W.H. (2007b) Serotonin transporter binding after recovery from eating disorders. Psychopharmacology, 195: 315–324.

Bhagwagar, Z., Rabiner, E., Sargent, P., Grasby, P. and Cowen, P. (2004) Persistent reduction in brain serotonin$_{1A}$ receptor binding in recovered depressed men mesured by positron emission tomography with [^{11}C]WAY-100635. Mol. Psychiatry, 9: 386–392.

Biggio, G., Fadda, F., Fanni, P., Tagliamonte, A. and Gessa, G.L. (1974) Rapid depletion of serum tryptophan, brain tryptophan, serotonin and 5-hydroxyindoleacetic acid by a tryptophan-free diet. Life Sci., 14: 1321–1329.

Blundell, J.E. (1984) Serotonin and appetite. Neuropharmacology, 23: 1537–1551.

Bonhomme, N. and Esposito, E. (1998) Involvement of serotonin and dopamine in the mechanism of action of novel antidepressant drugs: a review. J. Clin. Psychopharmacol., 18: 447–454.

Brewerton, T.D., Brandt, H.A., Lessem, M.D., Murphy, D.L. and Jimerson, D.C. (1990) Serotonin in eating disorders. In: Coccaro, E.F. and Murphy, D.L. (Eds.), Serotonin in Major Psychiatric Disorders. Progress in Psychiatry. American Psychiatric Press, Inc, Washington, DC, Vol. 21, pp. 155–184.

Bulik, C., Sullivan, P.F., Tozzi, F., Furberg, H., Lichtenstein, P. and Pedersen, N.L. (2006) Prevalence, heritability and prospective risk factors for anorexia nervosa. Arch. Gen. Psychiatry, 63: 305–312.

Bulik, C.M., Sullivan, P.F., Fear, J.L. and Joyce, P.R. (1997a) Eating disorders and antecedent anxiety disorders: a controlled study. Acta Psychiatr. Scand., 96: 101–107.

Bulik, C.M., Sullivan, P.F., Fear, J. and Pickering, A. (1997b) Predictors of the development of bulimia nervosa in women with anorexia nervosa. J. Nerv. Ment. Dis., 185: 704–707.

Bulik, C.M., Sullivan, P.F., Carter, F.A. and Joyce, P.R. (1997c) Lifetime comorbidity of alcohol dependence in women with bulimia nervosa. Addict. Behav., 22: 437–446.

Bulik, C.M., Sullivan, P.F. and Kendler, K.S. (1998) Heritability of binge-eating and broadly defined bulimia nervosa. Biol. Psychiatry, 44: 1210–1218.

Burnet, P.W., Eastwood, S.L. and Harrison, P.J. (1997) [3H]WAY-100635 for 5-HT1A receptor autoradiography in human brain: a comparison with [3H]8-OH-DPAT and demonstration of increased binding in the frontal cortex in schizophrenia. Neurochem. Intl., 30: 565–574.

Casper, R.C. (1990) Personality features of women with good outcome from restricting anorexia nervosa. Psychosom. Med., 52: 156–170.

Cassin, S. and von Ranson, K. (2005) Personality and eating disorders: a decade in review. Clin. Psychol. Rev., 25: 895–916.

Cervo, L., Mocaer, E., Bertaglia, A. and Samanin, R. (2000) Roles of 5-HT_{1A} receptors in the dorsal raphe and dorsal hippocampus in anxiety assessed by the behavioral effects of 8-OH-DPAT and S 15535 in a modified Geller-Seifter conflict model. Neuropharmacology, 39: 1037–1043.

Charney, D.S. and Deutch, A. (1996) A functional neuroanatomy of anxiety and fear: implications for the pathophysiology and treatment of anxiety disorders. Crit. Rev. Neurobiol., 10: 419–446.

Cleare, A. and Bond, A.J. (2000) Ipaspirone challenge in aggressive men shows an inverse correlation between 5HT1A receptor function and aggression. Psychopharmacology (Berl.), 148: 344.

Cloninger, C.R. (1987) A systematic method for clinical description and classification of personality variants. A proposal. Arch. Gen. Psychiatry, 44: 573–588.

Cloninger, C.R., Przybeck, T.R., Svrakic, D.M. and Wetzel, R.D. (1994) The Temperament and Character Inventory (TCI): A Guide to its Development and Use. Center for Psychobiology of Personality, Washington University, St Louis, MO.

Coccaro, E.F., Gabriel, S. and Siever, L.J. (1990) Buspirone challenge: preliminary evidence for a role of 5HT1A receptors in behavioral irritability in personality disordered patients. Psychopharmacol. Bull., 26: 285.

Cooper, S.J. (1996) Cholecystokinin modulation of serotonergic control of feeding behavior. Ann. N.Y. Acad. Sci., 780: 213–222.

De Vry, J. and Schreiber, R. (2000) Effects of selected serotonin 5-HT(1) and 5-HT(2) receptor agonists on feeding behavior: possible mechanisms of action. Neurosci. Biobehav. Rev., 24: 341–353.

Deep, A.L., Nagy, L.M., Weltzin, T.E., Rao, R. and Kaye, W.H. (1995) Premorbid onset of psychopathology in long-term recovered anorexia nervosa. Intl. J. Eat. Disord., 17: 291–297.

Drevets, W.C., Frank, E., Price, J.C., Kupfer, D.J., Holt, D., Greer, P.J., Huang, Y., Gautier, C. and Mathis, C. (1999) PET imaging of serotonin 1A receptor binding in depression. Biol. Psychiatry, 46: 1375–1387.

Fairbanks, L., Melega, W., Jorgensen, M., Kaplan, J. and McGuire, M. (2001) Social impulsivity inversely associated with CSF 5-HIAA and fluoxetine exposure in vermet monkeys. Neuropschopharmacology, 24: 370–378.

Fairburn, C.G., Welch, S.L., Doll, H.A., Davies, B.A. and O'Connor, M.E. (1997) Risk factors for bulimia nervosa. A community-based case-control study. Arch. Gen. Psychiatry, 54: 509–517.

Fernstrom, J.D. and Wurtman, R.J. (1971a) Brain serotonin content: physiological dependence on plasma tryptophan levels. Science, 173: 149–152.

Fernstrom, J.D. and Wurtman, R.J. (1971b) Brain serotonin content: increase following ingestion of carbohydrate diet. Science, 174: 1023–1025.

Fernstrom, J.D. and Wurtman, R.J. (1972) Brain serotonin content: physiological regulation by plasma neutral amino acids. Science, 178: 414–416.

File, S.E., Kenny, P.J. and Cheeta, S. (2000) The role of the dorsal hippocampal serotonergic and cholinergic systems in the modulation of anxiety. Pharmacol. Biochem. Behav., 66: 65.

Fischer, P., Meltzer, C., Ziolko, S., Price, J. and Hariri, A.R. (2007) Capacity for 5-HT1A-mediated autoregulation predicts amygdala reactivity. Nat. Neurosci., 9: 1362.

Fluoxetine Bulimia Nervosa Collaborative Study Group (1992) Fluoxetine in the treatment of bulimia nervosa, A multicenter, placebo-controlled, double-blind trial. Arch. Gen. Psychiatry, 49: 139–147.

Frank, G.K., Kaye, W.H., Weltzin, T.E., Perel, J., Moss, H., McConaha, C. and Pollice, C. (2001) Altered response to meta-chlorophenylpiperazine in anorexia nervosa: support for a persistent alteration of serotonin activity after short-term weight restoration. Intl. J. Eat. Disord., 30: 57–68.

Frank, G.K., Kaye, W.H., Meltzer, C.C., Price, J.C., Greer, P., McConaha, C. and Skovira, K. (2002) Reduced 5-HT2A receptor binding after recovery from anorexia nervosa. Biol. Psychiatry, 52: 896–906.

Frankle, W., Lombardo, I., New, A.S., Goodman, M., Talbot, P., Huang, Y., Hwang, D., Slifstein, M., Curry, S., Abi-Dargham, A., Lauruelle, M. and Siever, L. (2005) Brain serotonin transporter distribution in subjects with impulsive aggressivity: a positron emission study with [11C]McN 5652. Am. J. Psychiatry, 162: 915–923.

Freedman, L.J., Insel, T.R. and Smith, Y. (2000) Subcortical projections of area 25 (subgenual cortex) of the macaque monkey. J. Comp. Neurol., 421: 172–188.

Gendall, K.A., Kaye, W.H. and McConaha, C. (1998). Investigation of serotonin activity in anorexia nervosa by administration of d-fenfluramine and acute tryptophan depletion. Annual Meeting of the Eating Disorders Research Society, Boston, MA

Geyer, M.A. (1996) Serotonergic functions in arousal and motor activity. Behav. Brain Res., 73: 31.

Gibbons, J.L., Barr, G.A., Bridger, W.H. and Leibowitz, S.F. (1979) Manipulations of dietary tryptophan: effects on mouse killing and brain serotonin in the rat. Brain Res., 169: 139–153.

Godart, N.T., Flament, M.F., Lecrubier, Y. and Jeammet, P. (2000) Anxiety disorders in anorexia nervosa and bulimia nervosa: co-morbidity and chronology of appearance. Eur. Psychiatry, 15: 38–45.

Godart, N.T., Flament, M.F., Perdereau, F. and Jeammet, P. (2002) Comorbidity between eating disorders and anxiety disorders: a review. Intl. J. Eat. Disord., 32: 253–270.

Godart, N., Perdereau, F., Rein, Z., Berthoz, S., Wallier, J., Jeammet, P. and Flament, M. (2007) Comorbidity studies of eating disorders and mood disorders. Critical review of the literature. J. Affect. Disord., 97: 37–49.

Goethals, I., Vervaet, M., Audenaert, K., Jacobs, F., Ham, H. and Van Heeringen, C. (2007) Does regional brain perfusion correlate with eating disorder symptoms in anorexia and bulimia nervosa. J. Psychiatr. Res., 41: 1005–1011.

Goldbloom, D.S., Hicks, L.K. and Garfinkel, P.E. (1990) Platelet serotonin uptake in bulimia nervosa. Biol. Psychiatry, 28: 644–647.

Goodwin, G.M., Fairburn, C.G. and Cowen, P.J. (1987a) Dieting changes serotonergic function in women, not men: implications for the aetiology of anorexia nervosa? Psychol. Med., 17: 839–842.

Goodwin, G.M., Fairburn, C.G. and Cowen, P.J. (1987b) The effects of dieting and weight loss on neuroendocrine responses to tryptophan, clonidine, and apomorphine in volunteers. Important implications for neuroendocrine investigations in depression. Arch. Gen. Psychiatry, 44: 952–957.

Goodwin, G.M., Shapiro, C.M., Bennie, J., Dick, H., Carroll, S. and Fink, G. (1989) The neuroendocrine responses and psychological effects of infusion of L-tryptophan in anorexia nervosa. Psychol. Med., 19: 857–864.

Graeff, F.G., Guimaraes, F.S., De Andrade, T.G. and Deakin, J.F. (1996) Role of 5-HT in stress, anxiety, and depression. Pharmacol. Biochem. Behav., 54: 129–141.

Groenink, L., van Bogaert, M., van der Gugten, J. and Oosting, R. (2003) 5-HT_{1A} receptor and 5-HT_{1B} receptor knockout mice in stress and anxiety paradigms. Behav. Pharmacol., 14: 369–383.

Hadigan, C.M., Walsh, B.T., Buttinger, C. and Hollander, E. (1995) Behavioral and neuroendocrine responses to *meta*CPP in anorexia nervosa. Biol. Psychiatry, 37: 504–511.

Hajos, M., Gartside, S.E., Varga, V. and Sharp, T. (2003) *In vivo* inhibition of neuronal activity in the rat ventromedial prefrontal cortex by midbrain-raphe nuclei: role of 5-HT_{1A} receptors. Neuropharmacology, 45: 72–81.

Halmi, K., Agras, W.S., Crow, S., Mitchell, J., Wilson, G., Bryson, S. and Kraemer, H.C. (2005) Predictors of treatment acceptance and completion in anorexia nervosa. Arch. Gen. Psychiatry, 62: 776–781.

Higley, J.D. and Linnoila, M. (1997) Low central nervous system serotonergic activity is traitlike and correlates with impulsive behavior. A nonhuman primate model investigating genetic and environmental influences on neurotransmission. Ann. NY Acad. Sci., 836: 39–56.

Hoek, H.W., Bartelds, A.I., Bosveld, J.J., van der Graaf, Y., Limpens, V.E., Maiwald, M. and Spaaij, C.J. (1995) Impact of urbanization on detection rates of eating disorders. Am. J. Psychiatry, 152: 1272–1278.

Hoyer, D., Neijt, H.C. and Karpf, A. (1989) Competitive interaction of agonists and antagonists with 5-HT3 recognition sites in membranes of neuroblastoma cells labelled with [3H]ICS 205-930. J. Recept. Res., 9: 65–79.

Huether, G., Zhou, D. and Ruther, E. (1997) Long-term modulation of presynaptic 5-HT-output: experimentally induced changes in cortical 5-HT-transporter density, tryptophan hydroxylase content and 5-HT innervation density. J. Neural. Transm. Gen. Sect., 104: 993–1004.

Ishizuka, B., Quigley, M.E. and Yen, S.S. (1983) Pituitary hormone release in response to food ingestion: evidence for neuroendocrine signals from gut to brain. J. Clin. Endocrinol. Metab., 57: 1111–1116.

Jimerson, D.C., Lesem, M.D., Kaye, W.H. and Brewerton, T.D. (1992) Low serotonin and dopamine metabolite concentrations in cerebrospinal fluid from bulimic patients with frequent binge episodes. Arch. Gen. Psychiatry, 49: 132–138.

Jimerson, D.C., Wolfe, B.E., Metzger, E.D., Finkelstein, D.M., Cooper, T.B. and Levine, J.M. (1997) Decreased serotonin function in bulimia nervosa. Arch. Gen. Psychiatry, 54: 529–534.

Johnson, P., Lightman, S. and Lowry, C. (2004) A functional subset of serotonergic neruons in the rat ventrolateral periaqueductal gray implicated in the inhibition of sympathoexcitation and panic. Ann. N.Y. Acad. Sci., 1018: 58.

Kahn, R.S. and Wetzler, S. (1991) m-Chlorophenylpiperazine as a probe of serotonin function. Biol. Psychiatry, 30: 1139–1166.

Kaye, W., Bulik, C., Thornton, L., Barbarich, N., Masters, K., Fichter, M., Halmi, K., Kaplan, A., Strober, M., Woodside, D.B., Bergen, A., Crow, S., Mitchell, J., Rotondo, A., Mauri, M., Cassano, G., Keel, P.K., Plotnicov, K., Pollice, C., Klump, K., Lilenfeld, L.R., Devlin, B., Quadflieg, R. and Berrettini, W.H. (2004a) Comorbidity of anxiety disorders with anorexia and bulimia nervosa. Am. J. Psychiatry, 161: 2215–2221.

Kaye, W., Strober, M. and Jimerson, D. (2004b) The neurobiology of eating disorders. In: Charney, D.S. and Nestler, E.J. (Eds.), The Neurobiology of Mental Illness. Oxford Press, New York, NY, pp. 1112–1128.

Kaye, W., Strober, M. and Jimerson, D. (2009) The neurobiology of eating disorders. In: Charney, D. and Nestler, E. (Eds.), The Neurobiology of Mental Illness. 3rd edn. Oxford Press, New York, NY.

Kaye, W.H., Ebert, M.H., Raleigh, M. and Lake, R. (1984) Abnormalities in CNS monoamine metabolism in anorexia nervosa. Arch. Gen. Psychiatry, 41: 350–355.

Kaye, W.H., Gwirtsman, H.E., George, D.T., Weiss, S.R. and Jimerson, D.C. (1986) Relationship of mood alterations to bingeing behaviour in bulimia. Br. J. Psychiatry, 149: 479–485.

Kaye, W.H., Gwirtsman, H.E., George, D.T., Jimerson, D.C., Ebert, M.H. and Lake, C.R. (1990) Disturbances of noradrenergic systems in normal weight bulimia: relationship to diet and menses. Biol. Psychiatry, 27: 4–21.

Kaye, W.H., Gwirtsman, H.E., George, D.T. and Ebert, M.H. (1991a) Altered serotonin activity in anorexia nervosa after long-term weight restoration. Does elevated cerebrospinal fluid 5-hydroxyindoleacetic acid level correlate with rigid and obsessive behavior? Arch. Gen. Psychiatry, 48: 556–562.

Kaye, W.H., Weltzin, T.E., Hsu, L.K. and Bulik, C.M. (1991b) An open trial of fluoxetine in patients with anorexia nervosa. J. Clin. Psychiatry, 52: 464–471.

Kaye, W.H., Greeno, C.G., Moss, H., Fernstrom, J., Fernstrom, M., Lilenfeld, L.R., Weltzin, T.E. and Mann, J.J. (1998) Alterations in serotonin activity and psychiatric symptomatology after recovery from bulimia nervosa. Arch. Gen. Psychiatry, 55: 927–935.

Kaye, W.H., Gendall, K.A., Fernstrom, M.H., Fernstrom, J.D., McConaha, C.W. and Weltzin, T.E. (2000) Effects of acute tryptophan depletion on mood in bulimia nervosa. Biol. Psychiatry, 47: 151–157.

Kaye, W.H., Nagata, T., Weltzin, T.E., Hsu, L.K., Sokol, M.S., McConaha, C., Plotnicov, K.H., Weise, J. and Deep, D. (2001) Double-blind placebo-controlled administration of fluoxetine in restricting- and restricting-purging-type anorexia nervosa. Biol. Psychiatry, 49: 644–652.

Kaye, W.H., Barbarich, N.C., Putnam, K., Gendall, K.A., Fernstrom, J., Fernstrom, M., McConaha, C.W. and Kishore, A. (2003) Anxiolytic effects of acute tryptophan depletion in anorexia nervosa. Intl. J. Eat. Disord., 33: 257–267.

Kelley, A.E. (2004) Ventral striatal control of appetite motivation: role in ingestive behavior and reward-related learning. Neurosci. Biobehav. Rev., 27: 765–776.

Kendler, K.S., MacLean, C., Neale, M., Kessler, R., Heath, A. and Eaves, L. (1991) The genetic epidemiology of bulimia nervosa. Am. J. Psychiatry, 148: 1627–1637.

Kendler, K.S., Walters, E.E., Neale, M.C., Kessler, R.C., Heath, A.C. and Eaves, L.J. (1995) The structure of the genetic and environmental risk factors for six major psychiatric disorders in women. Phobia, generalized anxiety disorder, panic disorder, bulimia, major depression, and alcoholism. Arch. Gen. Psychiatry, 52: 374–383.

Kingston, K., Szmukler, G., Andrewes, D., Tress, B. and Desmond, P. (1996) Neuropsychological and structural brain changes in anorexia nervosa before and after refeeding. Psychol. Med., 26: 15–28.

Klump, K.L., McGue, M. and Iacono, W.G. (2001) Genetic and environmental influences on anorexia nervosa syndromes in a population-based sample of twins. Psychol. Med., 31: 737–740.

Koskela, A.K.-R.A., Sihvola, E., Kauppinen, T., Kaprio, J.A.A. and Rissanen, A. (2007) Serotonin transporter binding of [123I]ADAM in bulimic women, their healthy twin sisters, and healthy women: a SPET study. BMC Psychiatry, 21: 7–9.

Krebs-Thomson, K. and Geyer, M.A. (1998) Evidence for a functional interaction between 5-HT_{1A} and 5-HT_{2A} receptors in rats. Psychopharmacology (Berl.), 140: 69–74.

Lanfumey, L. and Hamon, M. (2000) Central 5-HT_{1A} receptors: regional distribution and functional characteristics. Nucl. Med. Biol., 27: 429–435.

Lanzenberger, R., Mitterhauser, M., Spindelegger, C., Wadsak, W., Klein, N., Mien, L., Holik, A., Attarbaschi, T., Mossaheb, N., Sacher, J., Geiss-Granadia, T., Keletter, K., Kasper, S. and Tauscher, J. (2007) Reduced serotonin-1A receptor binding in social anxiety disorder. Biol. Psychiatry, 61: 1081–1089.

Leibowitz, S.F. and Shor-Posner, G. (1986) Brain serotonin and eating behavior. Appetite, 7: 1–14.

Levitan, R.D., Kaplan, A.S., Joffe, R.T., Levitt, A.J. and Brown, G.M. (1997) Hormonal and subjective responses to intravenous *meta*-chlorophenylpiperazine in bulimia nervosa. Arch. Gen. Psychiatry, 54: 521–527.

Lilenfeld, L.R., Kaye, W.H., Greeno, C.G., Merikangas, K.R., Plotnicov, K., Pollice, C., Rao, R., Strober, M., Bulik, C.M. and Nagy, L. (1998) A controlled family study of anorexia nervosa and bulimia nervosa: psychiatric disorders in first-degree relatives and effects of proband comorbidity. Arch. Gen. Psychiatry, 55: 603–610.

Lopez, C., Tchanturia, K., Stahl, D., Booth, R., Holliday, J. and Treasure, J. (2007). An examination of central coherence in women with anorexia nervosa. Intl. J. Eat. Disord., Epub ahead of print.

Lucki, I. (1998) The spectrum of behaviors influenced by serotonin. Biol. Psychiatry, 44: 151–162.

Mann, J.J. (1999) Role of the serotonergic system in the pathogenesis of major depression and suicidal behavior. Neuropsychopharmacology, 21: 99S–105S.

Matsubara, S., Arora, R.C. and Meltzer, H.Y. (1991) Serotonergic measures in suicide brain: 5-HT1A binding sites in frontal cortex of suicide victims. J. Neural. Transm. Gen. Section,, 85: 181–194.

McCann, U.D., Hatzidimitriou, G. and Ricaurte, G.A. (1996) Prolactin response to fenfluramine is independent of serotonin release. Eur. J. Pharmacol., 312: R1–R2.

Messing, R.B., Fisher, L.A., Phebus, L. and Lytle, L.D. (1976) Interaction of diet and drugs in the regulation of brain 5-hydroxyindoles and the response to painful electric shock. Life Sci., 18: 707–714.

Monteleone, P., Santonastaso, P., Mauri, M., Bellodi, L., Erzegovesi, S., Fuschino, A., Favaro, A., rotondo, A., Castaldo, E. and Maj, M. (2006) Investigation of serotonin transporter regulatory region polymorphism in bulimia nervosa: relationships to harm avoidance, nutritional parameters, and psychiatric comorbidity. Psychosom. Med., 68: 99–103.

Monteleone, P., Tortorella, A., Martiadis, V., Serino, I., Di Filippo, C. and Maj, M. (2007) Association between A218C polymorphism of the tryptophan-hydroxylase-1 gene, harm avoidance and binge eating behavior in bulimia nervosa. Neurosci. Lett., 421: 42–46.

Moresco, F.M., Dieci, M., Vita, A., Messa, C., Gobbo, C., Galli, L., Rizzo, G., Panzacchi, A., De Peri, L., Ivernizzi, G. and Fazio, F. (2002) *In vivo* serotonin $5HT_{2A}$ receptor binding and personality traits in healthy subjects: a positron emission tomography study. NeuroImage, 17: 1470–1478.

Neumeister, A., Brain, E., Nugent, A., Carson, R., Bonne, O., Lucnekbaugh, D., Eckelman, W., Herschovitch, P., Charney, D.

and Drevets, W. (2004) Reduced serotinin type 1_A receptor binding in panic disorder. J. Neurosci., 24: 589–591.

O'Dwyer, A.M., Lucey, J.V. and Russell, G.F. (1996) Serotonin activity in anorexia nervosa after long-term weight restoration: response to D-fenfluramine challenge. Psychol. Med., 26: 353–359.

O'Keane, V., O'Hanlon, M., Webb, M. and Dinan, T. (1991) d-fenfluramine/prolactin response throughout the menstrual cycle: evidence for an oestrogen-induced alteration. Clin. Endocrinol. (Oxf.), 34: 289–292.

Olivier, B., Pattij, T., Wood, S., Oosting, R., Sarnyai, Z. and Toth, M. (2001) The 5-HT_{1A} receptor knockout mouse and anxiety. Behav. Pharmacol., 12: 439–450.

Parsey, R.V., Oquendo, M.A., Ogden, R.T., Olvet, D., Simpson, N., Huang, Y., Van Heertum, R., Arango, V. and Mann, J.J. (2005) Altered serotonin 1A binding in major depression: a [carbonyl-C-11]WAY100635 positron emission tomography study. Biol. Psychiatry, 59: 106–113.

Phillips, M.L., Drevets, W.C., Rauch, S.L. and Lane, R. (2003) Neurobiology of emotion perception I: the neural basis of normal emotion perception. Biol. Psychiatry, 54: 504–514.

Pollice, C., Kaye, W.H., Greeno, C.G. and Weltzin, T.E. (1997) Relationship of depression, anxiety, and obsessionality to state of illness in anorexia nervosa. Intl. J. Eat. Disord., 21: 367–376.

Santana, N., Bortolozzi, A., Serrats, J., Mengod, G. and Artigas, F. (2004) Expression of serotoinin$_{1A}$ and serotonin$_{2A}$ receptor in pyramidal and GABAergic neurons of the rat prefrontal cortex. Cereb. Cortex., 14: 1100–1109.

Sargent, P.A., Kjaer, K.H., Bench, C.J., Rabiner, E.A., Messa, C., Meyer, J., Gunn, R.N., Grasby, P.M. and Cowen, P.J. (2000) Brain serotonin$_{1A}$ receptor binding measured by positron emission tomography with [^{11}C]WAY-100635: effects of depression and antidepressant treatment. Arch. Gen. Psychiatry, 57: 174–180.

Saudou, F. and Hen, R. (1994) 5-Hydroxytryptamine receptor subtypes in vertebrates and invertebrates. Neurochem. Intl., 25: 503–532.

Schweiger, U. and Fichter, M. (1997) Eating disorders: clinical presentation, classification and etiologic models. In: Jimerson, D.C. and Kaye, W.H. (Eds.), Balliere's Clinical Psychiatry. Balliere's Tindall, London, pp. 199–216.

Schweiger, U., Warnhoff, M., Pahl, J. and Pirke, K.M. (1986) Effects of carbohydrate and protein meals on plasma large neutral amino acids, glucose, and insulin plasma levels of anorectic patients. Metabolism, 35: 938–943.

Shively, C., Friedman, D., Gage, H., Bounds, M., Brown-Proctor, C., Blair, J., Henderson, J., Smith, M. and Buchheimer, N. (2006) Behavioral depression and positron emission tomography-determined serotonin 1A receptor binding potential in cynomolgus monkeys. Arch. Gen. Psychiatry, 63: 396–403.

Shroff, H., Reba, L., Thornton, L., Tozzi, F., Klump, K., Berrettini, W., Brandt, H., Crawford, S., Crow, S., Fichter, M., Goldman, D., Halmi, K., Johnson, C., Kaplan, A., Keel, P., LaVia, M., Mitchell, J., Rotondo, A., Strober, M., Treasure, J., Woodside, D., Kaye, W.H. and Bulik, C. (2006) Features associated with excessive exercise in women with eating disorders. Intl. J. Eat. Disord., 39: 454–461.

Silberg, J. and Bulik, C. (2005) Developmental association between eating disorders symptoms and symptoms of depression and anxiety in juvenile twin girls. J. Child. Psychol. Psychiat., 46: 1317–1326.

Simansky, K.J. (1996) Serotonergic control of the organization of feeding and satiety. Behav. Brain Res., 73: 37–42.

Smith, K.A., Fairburn, C.G. and Cowen, P.J. (1999) Symptomatic relapse in bulimia nervosa following acute tryptophan depletion. Arch. Gen. Psychiatry, 56: 171–176.

Soubrie, P. (1986) Reconciling the role of central serotonin neurons in human and animal behavior. Behav. Brain Sci., 9: 319.

Spielberger, C.D., Gorsuch, R.L. and Lushene, R.E. (1970) STAI Manual for the State Trait Anxiety Inventory. Consulting Psychologists Press, Palo Alto, CA.

Srinivasagam, N.M., Kaye, W.H., Plotnicov, K.H., Greeno, C., Weltzin, T.E. and Rao, R. (1995) Persistent perfectionism, symmetry, and exactness after long-term recovery from anorexia nervosa. Am. J. Psychiatry, 152: 1630–1634.

Staley, J., Malison, R. and Innis, R. (1998) Imaging of the serotonergic system: interactions of neuroanatomical and functional abnormalities of depression. Biol. Psychiatry, 44: 534–549.

Stanley, M., Traskman-Bendz, L. and Dorovini-Zis, K. (1985) Correlations between aminergic metabolites simultaneously obtained from human CSF and brain. Life Sci., 37: 1279–1286.

Steiger, H., Young, S.N., Kin, N.M., Koerner, N., Israel, M., Lageix, P. and Paris, J. (2001a) Implications of impulsive and affective symptoms for serotonin function in bulimia nervosa. Psychol. Med., 31: 85–95.

Steiger, H., Koerner, N., Engelberg, M.J., Israel, M., Ng Ying Kin, N.M. and Young, S.N. (2001b) Self-destructiveness and serotonin function in bulimia nervosa. Psychiatry Res., 103: 15–26.

Steiger, H., Gauvin, L., Israel, M., LKoerner, N., Ng Ying Kin, N.M., Paris, J. and Young, S.N. (2001c) Association of serotonin and cortisol indices with childhood abuse in bulimia nervosa. Arch. Gen. Psychiatry, 58: 837–843.

Steiger, H., Richardson, J., Israel, M., Ng Ying Kin, N., Bruce, K., Mansour, S. and Marie Parent, A. (2005a) Reduced density of platelet-binding sites for [3H]paroxetine in remitted bulimic women. Neuropsychopharmacology, 30: 1028–1032.

Steiger, H., Joober, R., Israel, M., Young, S., Ng Ying Kin, N.M., Gauvin, L., Bruce, K., Joncas, J. and Torkaman-Zehi, A. (2005b) The 5HTTLPR polymorphism, psychopathological symptoms, and platelet [3H-] paroxetine binding in bulimic syndromes. Intl. J. Eat. Disord., 37: 57–60.

Steiger, H., Gauvin, L., Joober, R., Israel, M., Ng Ying Kin, N., Bruce, K., Richardson, J., Young, S. and Hakim, J. (2006) Intrafamilial correspondences on platelet [3H-]paroxetine-binding indices in bulimic probands and their unaffected first-degree relatives. Neuropsychopharmacology, 31: 1785–1792.

Steiger, H., Richardson, J., Joober, R., Gauvin, L., Israel, M., Bruce, K., Ying Kin, N., Howard, H. and Young, S. (2007) The 5HTTLPR polymorphism, prior maltreatment and dramatic–erratic personality manifestations in women with blimic syndromes. J. Psychiatr. Res., 32: 354–362.

Stockmeier, C.A. (1997) Neurobiology of serotonin in depression and suicide. Ann. NY Acad. Sci., 836: 220–232.

Strober, M. (1980) Personality and symptomatological features in young, nonchronic anorexia nervosa patients. J. Psychosom. Res., 24: 353–359.

Strober, M. (1995) Family-genetic perspectives on anorexia nervosa and bulimia nervosa. In: Brownell, K. and Fairburn, C. (Eds.), Eating Disorders and Obesity – A Comprehensive Handbook. The Guilford Press, New York, NY, pp. 212–218.

Strober, M., Freeman, R., Lampert, C., Diamond, J. and Kaye, W. (2000) Controlled family study of anorexia nervosa and bulimia nervosa: evidence of shared liability and transmission of partial syndromes. Am. J. Psychiatry, 157: 393–401.

Strupp, B.J., Weingartner, H., Kaye, W. and Gwirtsman, H. (1986) Cognitive processing in anorexia nervosa. A disturbance in automatic information processing. Neuropsychobiology, 15: 89–94.

Tauscher, J., Pirker, W., Willeit, M., de Zwaan, M., Bailer, U., Neumeister, A., Asenbaum, S., Lennkh, C., Praschak-Rieder, N., Brücke, T. and Kasper, S. (2001) [^{123}I]beta-CIT and single photon emission computed tomography reveal reduced brain serotonin transporter availability in bulimia nervosa. Biol. Psychiatry, 49: 326–332.

Tchanturia, K., Campbell, I., Morris, R. and Treasure, J. (2005) Neuropsychological studies in anorexia nervosa. Intl. J. Eat. Disord., 37: S72–S76.

Thomas, D.R., Gager, T.L., Holland, V., Brown, A.M. and Wood, M.D. (1996) m-Chlorophenylpiperazine (mCPP) is an antagonist at the cloned human 5- HT2B receptor. Neuroreport, 7: 1457–1460.

Tiihonen, J., Keski-Rahkonen, A., Lopponen, M., Muhonen, M., Kajander, J., Allonen, T., Nagren, K., Hietala, J. and Rissanen, A. (2004) Brain serotonin 1A receptor binding in bulimia nervosa. Biol. Psychiatry, 55: 871.

Treasure, J. and Campbell, I. (1994) The case for biology in the aetiology of anorexia nervosa. Psychol. Med., 24: 3–8.

Tremblay, P. and Blier, P. (2006) Catecholaminergic strategies for the treatment of major depression. Curr. Drug Targets, 7: 149–158.

van Heeringen, C., Audenaert, K., Van Laere, K., Dumont, F., Slegers, G., Mertens, J. and Dierckx, R.A. (2003) Prefrontal 5-HT_{2a} receptor binding index, hopelessness and personality characteristics in attempted suicide. J. Affect. Disord., 74: 149–158.

Vitousek, K. and Manke, F. (1994) Personality variables and disorders in anorexia nervosa and bulimia nervosa. J. Abnorm. Psychol., 103: 137–147.

Vollenweider, F., Vontobel, P., Hell, D. and Leenders, K. (1999) 5-HT modulation of dopamine release in basal ganglia in psilocybin-induced psychosis in man – a PET study with [^{11}C]raclopride. Neuropsychopharmacology, 20: 424–433.

Wade, T.D., Bulik, C.M., Neale, M. and Kendler, K.S. (2000) Anorexia nervosa and major depression: Shared genetic and environmental risk factors. Am. J. Psychiatry, 157: 469–471.

Wagner, A., Barbarich, N., Frank, G., Bailer, U.F., Weissfeld, L., Henry, S., Achenbach, S., Vogel, V., Plotnicov, K., McConaha, C., Kaye, W. and Wonderlich, S. (2006) Personality traits after recovery from eating disorders: Do subtypes differ? Intl. J. Eat. Disord., 39: 276–284.

Walsh, B., Kaplan, A., Attia, E., Olmsted, M., Parides, M., Carter, J., Pike, K., Devlin, M., Woodside, B., Roberto, C. and Rocket, W. (2006) Fluoxetine after weight restoration in anorexia nervosa: a randomized controlled trial. J. Am. Med. Assoc., 295: 2605–2612.

Walsh, B.T. and Devlin, M.J. (1995) Pharmacotherapy of bulimia nervosa and binge eating disorder. Addict. Behav., 20: 757–764.

Walsh, B.T. and Devlin, M.J. (1998) Eating disorders: progress and problems. Science, 280: 1387–1390.

Walters, E.E. and Kendler, K.S. (1995) Anorexia nervosa and anorexic-like syndromes in a population-based female twin sample. Am. J. Psychiatry, 152: 64–71.

Ward, A., Brown, N., Lightman, S., Campbell, I.C. and Treasure, J. (1998) Neuroendocrine, appetitive and behavioural responses to d-fenfluramine in women recovered from anorexia nervosa. Br. J. Psychiatry, 172: 351–358.

Weltzin, T.E., Fernstrom, M.H., Fernstrom, J.D., Neuberger, S.K. and Kaye, W.H. (1995) Acute tryptophan depletion and increased food intake and irritability in bulimia nervosa. Am. J. Psychiatry, 152: 1668–1671.

Westergaard, G., Suomi, S., Chavanne, T., Houser, L., Hurley, A., Cleveland, A., Snoy, P. and Higley, J. (2003) Physiological correlates of aggression and impulsivity in free-ranging female primates. Neuropsychopharmacology, 28: 1045–1055.

Willins, D.L. and Meltzer, H.Y. (1997) Direct injection of 5-HT2A receptor agonists into the medial prefrontal cortex produces a head-twitch response in rats. J. Pharmacol. Exp. Ther., 282: 699–706.

Winstanley, C.A., Theobald, D.E., Dalley, J.W. and Robbins, T.W. (2005) Interactions between serotonin and dopamine in the control of impulsive choice in rats: Therapeutic implications for impulse control disorders. Neuropsychopharmacology, 30: 669.

Wolfe, B., Metzger, E., Levine, J., Finkelstein, D., Cooper, T. and Jimerson, D. (2000) Serotonin function following remission from bulimia nervosa. Neuropsychopharmacology, 22: 257–263.

Wolfe, B.E., Jimerson, D.C. and Levine, J.M. (1994) Impulsivity ratings in bulimia nervosa: relationship to binge eating behaviors. Intl. J. Eat. Disord., 15: 289–292.

Young, S.N. and Gauthier, S. (1981) Effect of tryptophan administration on tryptophan, 5- hydroxyindoleacetic acid and indoleacetic acid in human lumbar and cisternal cerebrospinal fluid. J. Neurol. Neurosurg. Psychiatry, 44: 323–327. 2010

CHAPTER 4.12

Genetic, Epigenetic and Environmental Factors in Serotonin Associated Disease Condition

Eva Friedel and Andreas Heinz*

Department of Psychiatry and Psychotherapy, Charité - University Medicine Berlin, Germany

Abstract: Genetic, epigenetic and environmental factors have been shown to influence serotonergic neurotransmission implicated in psychiatric and neurologic disorders. Due to the prominent effects of serotonin reuptake inhibition on negative mood states, a series of studies examined central serotonin transporter (5-HTT) availability and turnover in patients suffering from major depression, anxiety, obsessive-compulsive disorder and alcohol dependence. Human studies and animal experiments in rodents and non-human primates identified effects of genetic, epigenetic and environmental factors that modulate 5-HTT mRNA transcription and *in vivo* availability. In this chapter we review the literature and discuss studies in non-human primates that underwent social stress and displayed a reduction of the serotonin turnover rate among carriers of one or two short alleles of a functional polymorphism of the 5-HTT regulatory region. Further studies in these primates suggested specific effects of stress hormone responses on serotonin transporter availability depending on 5-HTT genotype and gender. In humans suffering from negative mood states such as depression or alcoholism, reduced availability of brainstem and thalamic serotonin transporters was associated with anxiety and may reflect stress exposure mediated by epigenetic interactions between stress hormone levels, gender and 5-HTT genotype that bias central processing of emotionally salient information.

Keywords: serotonin transporters, SLC6A4, anxiety, impulsivity, negative mood states, reversal learning.

Introduction

Serotonergic neurons are located in the brainstem raphe nuclei, from where they project to a multitude of brain areas and stimulate a wide range of serotonin receptor subtypes (Baumgarten and Grozdanovic, 1997). Serotonergic dyfsunction was implicated in a variety of psychiatric disorders, and may primarily contribute to negative affect such as anxiety and depression (Cloninger, 1987; Owens and Nemeroff, 1994; Grove *et al.*, 1997). Uptake of intrasynaptic serotonin is regulated by the function and availability of serotonin transporters (5-HTT). A two-fold difference in serotonin reuptake rates was associated with a functional polymorphism in the regulatory region of the 5-HTT gene, and may contribute to the risk of developing negative mood states (Lesch *et al.*, 1996; Caspi *et al.*, 2003). Besides negative mood states such as anxiety and depression (Praag, 1977; Traskman-Bendz *et al.*, 1984; Artigas, 1995; Mann *et al.*, 1996; Barr *et al.*, 2003), serotonergic dysfunction may also contribute to the pathogenesis and maintenance of impulsive behavior, aggressiveness and excessive alcohol intake (LeMarquand *et al.*, 1994; FilsAime *et al.*, 1996). Serotonergic neurotransmission modulates the acute response to alcohol; a low response to the acute effects of alcohol intake was found both in humans with a 5-HTT genotype that is associated with high *in vivo* 5-HTT availability, and in non-human primates displaying a comparable increase in brainstem 5-HTT availability in association with a stress-induced decrease in serotonin turnover rates (Doudet *et al.*, 1995; Schuckit and Smith, 1996; Heinz *et al.*, 1998a; Schuckit *et al.*, 1999; Barr *et al.*, 2003; Hu *et al.*, 2005). These observations appear to be of clinical relevance, because a low level of response to acute alcohol intake is more common in relatively alcohol-naïve children of alcoholics and is predictive of subsequent alcohol abuse and dependence (Rodriguez *et al.*, 1993; Schuckit *et al.*, 1999; Schuckit and Smith, 1996; Volavka *et al.*, 1996). In this chapter, we will discuss behavioral correlates of central serotonergic neurotransmission and focus on negative mood states and a low response to alcohol as behaviorally relevant correlates of serotonin function.

*Corresponding author
E-mail: andreas.heinz@charite.de

Christian Müller & Barry Jacobs (Eds.)
Handbook of Behavioral Neurobiology of Serotonin
ISBN 978-0-12-374634-4
DOI: 10.1016/B978-0-12-374634-4.00040-X

Genetic, epigenetic and environmental effects that modulate serotonergic neurotransmission

Genetic, epigenetic and environmental factors contribute to central serotonin turnover as assessed by the serotonin metabolite 5-hydroxyindoleacetic acid (5-HIAA) in the cerebrospinal fluid (CSF) (Oxenstierna *et al.*, 1986; Higley *et al.*, 1991, 1993; Clarke *et al.*, 1996). Serotonin turnover rates can be measured *in vivo* in non-human primates, and heritability estimates of serotonin turnover accounted for 42 percent of the variance (Kaplan, 2000) – a number that is rather close to heritability estimates in human twin studies, which accounted for about 35 percent of the variance in CSF 5-HIAA concentrations; therefore, epigenetic and environmental factors appear to play a predominant role in the regulation of central serotonin turnover (Beck *et al.*, 1984; Oxenstierna *et al.*, 1986). Environmental factors are of special interest if they have long-lasting effects on serotonergic neurotransmission, and such lasting effects were observed following developmentally early social separation stress (Higley *et al.*, 1991; Jones *et al.*, 1992; Clarke *et al.*, 1996).

Intrasynaptic serotonin concentrations are modulated by the availability and functional capacity of serotonin transporters (5-HTT), which regulate reuptake of extracellular serotonin. A variable number of tandem repeat (VNTR) polymorphic sites in the regulatory unit of the serotonin transporter gene (SLC6A4) has been shown to influence serotonin transporter expression: homozygous carriers of a long allele (*ll*-genotype) express about twice as many serotonin transporters as carriers of one or two short alleles (*s*-carriers) (Lesch *et al.*, 1996; Little *et al.*, 1998; Heinz *et al.*, 2000). A subsequently discovered $A > G$ exchange in the 5-HTT regulatory region was associated with differential transcriptional activation of 5-HTT mRNA in lymphoblastoid cell lines (Hu *et al.*, 2006), and *in vivo* imaging studies suggested that carriers of two long (*ll*) alleles of the 5-HTT regulator region express elevated serotonin transporter availability only if they do not carry the $A > G$ exchange ($l_A l_A$ homozygotes) (Praschak-Rieder *et al.*, 2007; Reimold *et al.*, 2007). One previous study that reported increased radioligand binding in *ll* homozygotes was reanalyzed, and revealed that all *ll* indeed carried the $l_A l_A$ genotype.

Stress exposure is one environmental factor that was examined in the context of gene–environment interactions and their effect on central serotonin function. A brain-imaging study in non-human primates used single photon emission tomography (SPECT) and a radioligand that binds with high affinity to central serotonin transporters to assess *in vivo* serotonin transporter availability and its interaction with serotonin turnover rates. The adult rhesus macaques examined in this study had experienced social isolation stress during their early development (Heinz *et al.*, 1998a). In previous studies, concentrations of the serotonin metabolite 5-HIAA in the CSF were decreased in primates that had been separated from their mothers after birth and peer-raised as compared with mother-reared animals, and CSF 5-HIAA levels decreased further when the primates were completely isolated from their peers (Clarke *et al.*, 1996; Higley *et al.*, 1996). Such reductions in CSF 5-HIAA concentrations were trait-like, and persisted during adulthood (Higley *et al.*, 1996; Heinz *et al.*, 1998a). The rhesus macaques displayed increased anxiety-like behaviors during early social isolation stress exposure (Higley *et al.*, 1991); when the male monkeys reached adulthood, they displayed an increased impulsive aggressiveness (Higley *et al.*, 1996), which was correlated with low CSF 5-HIAA concentrations and an increased availability of serotonin transporters in the brainstem (raphe) area with low CSF 5-HIAA availability (Heinz *et al.*, 1998a). Low CSF 5-HIAA concentrations and a high availability of brainstem 5-HTT were also associated with an increased frequency of self-initiated aggressive acts, and reduced time spent in social contacts (Heinz *et al.*, 1998a). Gender differences may play a role, because it was observed that in female rhesus monkeys low CSF 5-HIAA levels were correlated with low-intensity, restrained aggression, a behavior pattern that is found in matrilineal defense of social status (Westergaard *et al.*, 2003).

The effects of early social isolation stress on serotonin turnover rates (measured as CSF 5-HIAA concentrations) interacted with genetic variance in the structure of the regulatory region of the 5-HTT gene: only carriers of one or two short (*s*) alleles of the 5-HTT regulatory region displayed a reduction in serotonin turnover rates following social isolation, while this was not the case in homozygote carriers of two long (*l*) alleles (Bennett *et al.*, 2002). When the rhesus monkeys were further isolated from their peers – i.e., experienced complete social isolation during the first 6 months of life – carriers of one or two *s* alleles of the 5-HTT regulatory region displayed increased levels of adrenocorticotropin hormone (ACTH) and cortisol compared to *ll* homozygotes (Barr *et al.*, 2004). Animal experiments suggest that stress-associated activation of the hypothalamic–pituitary–adrenal (HPA) axis may further interfere with serotonergic neurotransmission via down-regulation of serotonin transporters (Slotkin *et al.*, 1997). Also in human alcohol-dependent patients and healthy control subjects, plasma cortisol concentrations were negatively correlated with the availability of brainstem 5-HTT (Heinz *et al.*, 2002; Figure 1).

In non-human primate imaging studies, a high availability of raphe serotonin transporters was associated with low CSF 5-HIAA concentrations – a finding that may be due to a real increase in serotonin transporter

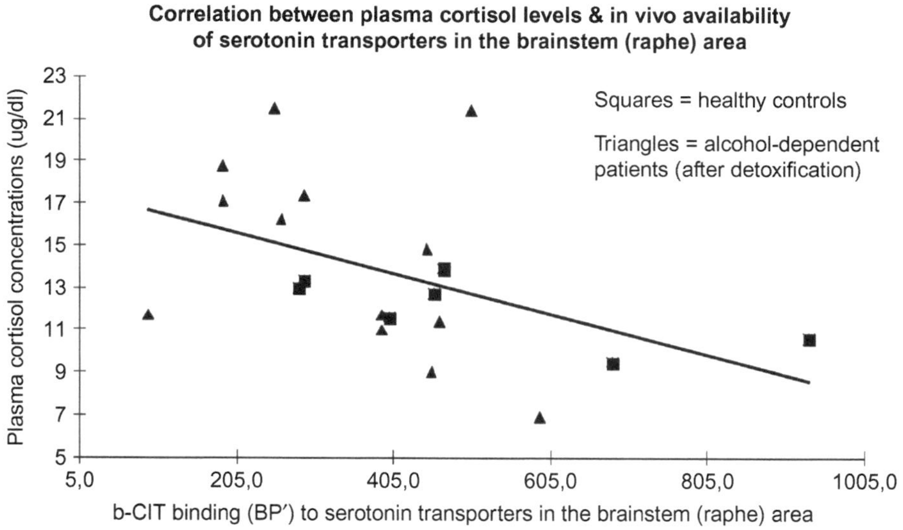

Figure 1 In male healthy control subjects (squares) and alcohol-dependent patients during early abstinence, high cortisol plasma concentrations (as observed during detoxification) are associated with low binding of the radioligand iodine-123-β-CIT to brainstem (raphe) serotonin transporters. Modified from Heinz *et al.*, 2002, by permission from Georg Thieme Verlag KG.

density. Such an increase in functional reuptake capacity could cause a reduction in extracellular concentrations of monoamine neurotransmitters and their metabolites. A *post-mortem* study in non-human primates supported this hypothesis, and observed a negative correlation between striatal dopamine transporter density and individual concentrations of the dopamine metabolite homovanillic acid (HVA) (Mash *et al.*, 1996). This observation was further confirmed in a combined microdialysis and SPECT study, which showed that the availability of monoamine transporters negatively correlated with extracellular neurotransmitter concentrations (Heinz, 1999). Conversely, blockade or loss of serotonin or dopamine transporters induces a dramatic increase in extracellular neurotransmitter concentrations in, for example, knockout animal models (Kreiss and Lucki, 1995; Giros *et al.*, 1996). These observations support the hypothesis that extracellular monoamine concentrations are regulated by monoamine reuptake capacity. However, an alternative explanation is also possible: low extracellular 5-HIAA concentrations may be correlated with increased binding of the radiotracers to monoamine reuptake sites (transporters) because of decreased competition between endogenenous serotonin and the applied radioligand (Jones *et al.*, 1998; Figure 2).

Indeed, when synaptic serotonin (5-HT) levels were restored after serotonin depletion in non-human primates, this was accompanied by a significant reduction in radioligand binding raphe 5-HTT (Heinz *et al.*, 2004). Independent of the underlying nature of the neurobiological process, *in vivo* and *in vitro* studies suggest inverse correlations between serotonin transporter availability and both extracellular neurotransmitter concentrations and serotonin metabolite levels in CSF (Heinz *et al.*, 1998a, 2002).

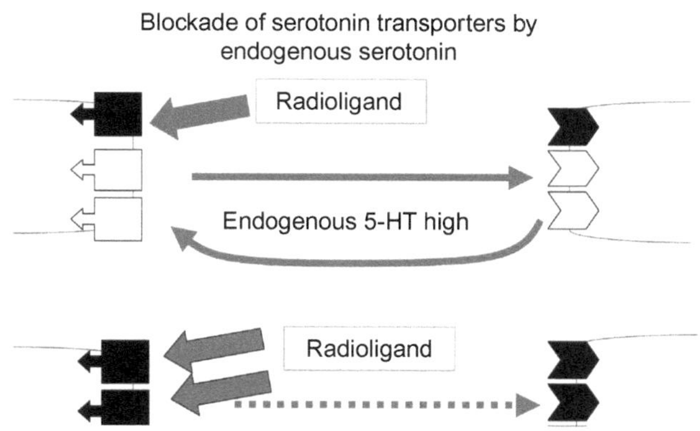

Figure 2 Competition between endogenous extracellular serotonin levels and radioligands such as iodine-123-β-CIT that bind to presynaptic (left) serotonin transporters (*Upper panel*): when extracellular serotonin concentrations are high, presynaptic serotonin transporters are occupied to a considerable degree by endogenous serotonin and radioligand binding is relatively low. (*Lower panel*): low extracellular serotonin concentrations occupy less serotonin reuptake sites and radioligand binding is relatively high.

Genetic factors, epigenetic programming and environmental factors have been implicated in the etiology of depression and anxiety. Nevertheless, it remains unclear what exactly causes the observed neurochemical changes that are considered crucial, like serotonin and catecholamine variations, elevated levels of CRH and nerve growth factor quantities. Previous stress exposure, especially when occurring early in life, has great impact on the reactivity to stressors. These stressor effects might reflect sensitization of neuronal functioning, phenotypic changes of processes that lead to neurochemical release or receptor sensitivity, or epigenetic processes that modify the expression of specific genes associated with stressor reactivity (Anisman *et al.*, 2008). Experiences of stressful life events in early pregnancy or during postnatal development are able to induce epigenetic effects such as alterations in DNA methylation and transcription (Weaver *et al.*, 2007; Mueller and Bale, 2008; Oberlander *et al.*, 2008a, 2008b). Moreover, such stress exposure can lead to maladaptive stress responses and anhedonia later in life (Mueller and Bale, 2008). Long-term alterations in central corticotropin-releasing factor (CRF) and glucocorticoid receptor (GR) expression, as well as dysregulated hypothalamic–pituitary–adrenal (HPA) axis responsivity and altered serotonin transporter availability, are probable factors contributing to elevated stress sensitivity later in life (Oberlander *et al.*, 2008a).

Sex-specific differences in prenatal stress vulnerability were described in animal experiments (Mueller and Bale, 2008), with an increased vulnerability of male offspring to maternal stress in early gestation due to stress effects on placental function in male but not female placentas. In this study changes in CRF glucocorticoid receptor (GR) gene methylation were associated with altered gene expression, and provide evidence of epigenetic programming during early pregnancy. Human studies confirm that prenatal (Oberlander *et al.*, 2008b) and postnatal (Oberlander *et al.*, 2008a) stress exposure is associated with increased stress reactivity in early infancy via gene modulation. For example, methylation status of a CpG-rich region in the promoter and exon 1F of the human glucocorticoid receptor (GR) gene (NR3C1) in newborns is sensitive to prenatal maternal mood, and may offer a potential epigenetic process that links prenatal maternal mood and altered hypothalamic–pituitary–adrenal (HPA) axis stress reactivity during infancy (Oberlander *et al.*, 2008a).

Postnatal differences in nurturing behavior influence glucocrticoid receptor (GR) gene expression in the hippocampus of mice offspring (Weaver *et al.*, 2007): increased postnatal licking/grooming behavior enhanced the binding of nerve growth factor inducible protein A (NGFI-A) transcription factor to the exon 1(7) GR promotor within the hippocampus of the offspring. Knock-down experiments regarding NGFI-A in hippocampal primary cell culture showed that NGFI-A is required for serotonin-induced DNA demethylation and increased exon 1(7) GR promoter expression, which may contribute to protection from stress exposure. These findings suggest a protective mechanism that directly links the effect of maternal care on both gene expression and stress sensitivity. Conversely, NGFI-A binding to its response element is inhibited by DNA methylation. Human data (Oberlander *et al.*, 2008a) describe similar influences of maternal care (here, breast-feeding) on the developing HPA regulation system in 3-month-old infants.

In adults, stress experiences have been linked to the 5-HTT genotype (Caspi *et al.*, 2003), postulating a 'risk' genotype associated with the short allele of the 5-HTT gene, leading to increased levels of depression following stressful life events compared to carriers of the long allele. Recent studies offer a potential account of this mechanism. Philibert and coworkers described a CpG island in the 5′ region of the 5-HTT gene that contains n alternative exon 1 and possible promoter for 5-HTT. An increase of CpG methylation was associated with decreased transcription of 5-HTT mRNA (Philibert *et al.*, 2008). Women displayed increased levels of CpG methylation and, respectively, low levels of 5-HTT mRNA (Philibert *et al.*, 2008), implicating a sex-specific vulnerability for developing mood disorders, whereas male subjects showed higher 5-HTT mRNA expression. Carrying a short allele of the 5-HTT regulatory region was also associated with lower amounts of 5-HTT mRNA transcription, explaining up to 10 percent of the observed variance (Philibert *et al.*, 2008). Altogether, these findings suggest that prenatal and postnatal stress exposure can induce long-lasting alterations in stress hormone axis activation, and may impact on serotonergic neurotransmission. Early human imaging studies directly linked acute stress hormone elevations with reductions in serotonin transporter availability and with negative mood states (Heinz *et al.*, 2002), thus providing additional evidence for interactions between stress exposure and serotonin dysfunction.

Aggression and impulsivity as behavioral manifestations of central serotonergic dysfunction

Impulsive and aggressive behavior may be a primary or secondary correlate of reduced central serotonin turnover. Gray (1982) hypothesized that serotonergic activation of the septo-hippocampal behavior inhibition system (BIS) is unpleasant and therefore experienced as punishment, which induces passive avoidance (Gray, 1982). Serotonergic dysfunction may thus interfere with the function of the behavior inhibition system, and this lack of behavioral control would then manifest as impulsiveness and uncontrolled aggressive behavior. However, further studies suggested that impulsive aggressiveness is not a specific correlate of central serotonergic dysfunction, which instead is primarily associated with negative mood states such as anxiety or depression (Owens and Nemeroff, 1994; Young *et al.*, 1994; Artigas, 1995; Knutson *et al.*, 1998). This hypothesis is supported by the primarily pleasant effects of methylenedioxymethamphetamine (MDMA, 'ecstasy'), which have been associated with acute increases in serotonin release (Huether *et al.*, 1997). In a human study, medication with selective serotonin reuptake inhibitors (SSRIs) has been associated with improvement of negative mood states due to an *increase* rather than a decrease in synaptic serotonin concentrations, and this observation is difficult to reconcile with the notion of serotonergic neurotransmission stimulating a punishment system (Limberger *et al.*, 1990; Artigas, 1995; Kreiss and Lucki, 1995; Muck-Seler *et al.*, 1996). Human CSF studies observed a significant association between the clinical remission of depressive symptoms and an *increase* in primarily low CSF 5-HIAA concentrations (Praag, 1977; Traskman-Bendz *et al.*, 1984). Serotonin depletion studies pointed in the same direction by showing that reduced serotonin content induced negative mood states in patients suffering from obsessive-compulsive disorder, and among patients with major depression who previously displayed an improvement of their depressive symptoms following SSRI medication (Delgado *et al.*, 1990; Barr *et al.*, 1994). Altogether, these studies suggest that an increase in synaptic serotonin concentrations induces a *reduction* in negative mood states such as anxiety and depressiveness, rather than representing the neurobiological correlate of punishment.

So how can the association between low serotonin turnover rates and impulsive aggression be explained? The correlation between reduced central serotonergic dysfunction and both impulsive aggression and negative mood states may derive from common ground: individuals who are overwhelmed by feelings of insecurity and threat may response with impulsively aggressive behavior (Heinz *et al.*, 2001). This hypothesis is based on animal experiments and human studies. In rodents, increased serotonin turnover was associated with higher social competence in competitive games (Knutson *et al.*, 1996a), while serotonin depletion induced anxious and insecure behavior patterns (Knutson *et al.*, 1996b). A human study of Knutson *et al.* (1998) demonstrated that SSRI application primarily induced a reduction in negative mood among individuals engaging in a competitive game. Specifically, Knutson and coworkers (1998) observed that the reduction in aggressive behavior which occurred following SSRI application was statistically explained by the decrease in negative emotions. The reduction in aggressiveness may thus be due to a decrease in perceptions of threat and insecurity.

On the other hand, a high serotonin turnover rate may protect against the impact of threat-related environmental cues: serotonergic neurotransmission modulates thalamo-cortical circuits and may thus support a 'protective filter effect' that reduces the impact of sensory inputs (McCormick, 1992; Baumgarten and Grozdanovic, 1997). Indeed, genetic variation in serotonin transporter expression and function has been associated with limbic, particularly amygdala, responses to aversive environmental stimuli (Hariri *et al.*, 2002; Heinz *et al.*, 2005). Exacerbation of central serotonergic dysfunction during clinically relevant negative mood states may lead to over-activation of the amygdala by aversive environmental stimuli, a hypothesis that was supported by the observation of increased glucose turnover rates in the amygdala of patients suffering from major depression (Drevets, 2000).

Altogether, these observations suggest that serotonergic neurotransmission promotes feelings of security and tranquility (Raleigh *et al.*, 1988; Knutson, 1996b), while subjects suffering from a deficit in serotonergic neurotransmission may feel threatened, insecure and anxious (Higley *et al.*, 1991; Jones *et al.*, 1992; Clarke *et al.*, 1996). In this view, the manifestation of clinically relevant syndromes such as clinical depression or impulsive aggression after serotonin depletion are secondary consequences from general feelings of insecurity and tension that manifest as negative mood states (Knutson *et al.*, 1998; Heinz *et al.*, 2001). Whether behavioral responses are more tuned towards depression or aggression may depend upon the manifestation of other causal factors, such as learned behavior in social contexts, which is modulated by, for example, socially constructed roles of men and women (Kraemer and McKinney, 1979; Raleigh, 1991; Heinz *et al.*, 2001). *In vivo* interactions between serotonergic dysfunction, amygdala activation and specific negative mood states will have to be explored in future studies.

Serotonergic neurotransmission and the response to acute alcohol intake

It has been suggested that early onset of alcoholism (i.e., before 25 years of age) is characterized by increased impulsiveness, aggressive behavior and serotonin dysfunction (FilsAime, 1996). The suggested association between dysfunction of central serotonergic neurotransmission, impulsive aggression and excessive alcohol intake has been explored in a series of animal experiments and human studies. In humans and non-human primates, individuals who show a low response to the acute effects of alcohol may fail to experience the negative consequences of excessive alcohol intake, and thus lack a 'warning signal' that limits the amount of alcohol consumed. Studies among adolescents and young adults confirmed that a low response to acute alcohol intake was indeed associated with excessive drinking during longer observation periods, and with an increased risk to develop alcohol dependence in prospective studies (Schuckit and Smith, 1996; Schuckit *et al.*, 1999; Hinckers *et al.*, 2006). Reduced serotonin turnover rates and a high availability of serotonin transporters appear to represent one factor that contributes to a low alcohol response: following alcohol intake, reduced sedation and ataxia were observed in non-human primates with a genetic disposition to elevated serotonin transporter availability and function (i.e., carriers of two long alleles of the 5-HTT regulatory region compared with carriers of one or two short alleles) (Barr *et al.*, 2003). A decrease in serotonin turnover rates in association with elevated raphe serotonin transporter and a low response to alcohol can also result from environmental factors such as early social isolation stress, because it was observed in non-human primates separated from their mothers after birth compared with animals that were raised by their mothers (Heinz *et al.*, 1998a). Low serotonin turnover rates appear to contribute to a low alcohol response by reducing the sedative, GABAergic effects of acute alcohol intake. The sedative effects of ethanol are mediated by activation of $GABA_A$ receptors (Liljequist and Engel, 1982; Silveri and Spear, 2002) and by blockade of glutamatergic NMDA receptors, particularly after consumption of high doses of alcohol (Tsai *et al.*, 1995; Krystal *et al.*, 1998). Dysfunction of serotonergic neurotransmission in the prefrontal cortex modulates both GABAergic and glutamatergic neurotransmission via $5\text{-}HT_{1A}$ and $5\text{-}HT_4$ receptors (Krystal, 1999; Feng *et al.*, 2001; Cai *et al.*, 2002a, 2002b; Figure 3).

Therefore, low serotonin turnover rates may reduce the sedative effects of acute alcohol intake via modulation of prefrontal GABAergic and glutamatergic alcohol effects, and indeed Doudet and coworkers observed that low CSF 5-HIAA concentrations following social isolation stress in non-human primates were associated with reduced effects of GABAergic stimulation on prefrontal glucose utilization rates (Doudet *et al.*, 1995). In further studies with non-human primates that suffered from social isolation during their early development, low CSF 5-HIAA levels and a high availability of serotonin transporters were associated with both a low acute response to alcohol and an increased alcohol intake in a 3-month observation period when they had access to alcohol in a free choice paradigm (Heinz *et al.*, 2003).

Disposition to excessive alcohol intake: gene–environment interactions

Gene–environment interactions modulate the disposition to excessive alcohol intake. Environmental factors are implicated by the observation that early social isolation stress was followed by (1) low levels of CSF 5-HIAA concentrations, (2) a high availability of brainstem serotonin transporters, (3) reduced drug-elicited GABAergic sedation, and (4) increased alcohol intake in a free choice paradigm (Doudet *et al.*, 1995; Higley *et al.*, 1996; Heinz *et al.*, 1998a). Variation in the genetic constitution of 5-HTT transporters and $GABA_A$ receptors appears to result in similar interactions between serotonergic and GABAergic neurotransmission, which mimic stress effects on a phenotypical level: human studies observed that a low response to acute alcohol intake was associated with a 5-HTT genotype that predisposes to high serotonin

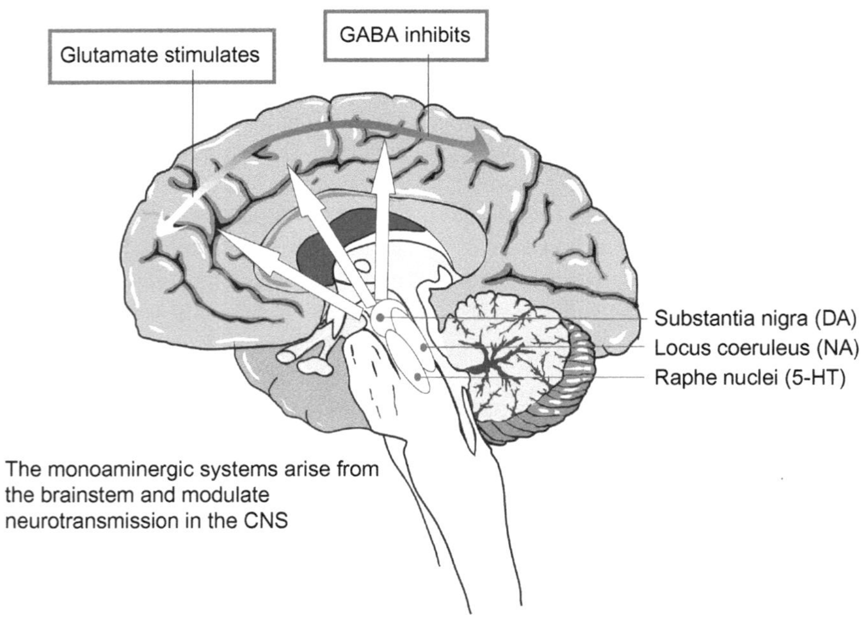

Figure 3 The ascending monoaminergic neurotransmitter systems (serotonin, dopamine and norepinephrine) arise from brainstem nuclei and modulate fast excitatory glutamatergic and inhibitory GABAergic neurotransmission.

transporter availability (Schuckit *et al.*, 1999; Hu *et al.*, 2005; Hinckers *et al.*, 2006). In one of the human studies, a low acute response to alcohol was also associated with an allelic variant, Ser385, of the $GABA_{A\alpha6}$ receptor polymorphism Pro385Ser (Schuckit *et al.*, 1999). Animal experiments have further elucidated potential effects of such genetic variation on acute alcohol effects: in rodents; a single nucleotide mutation in the $GABA_{A\alpha6}$ receptor subunit was associated with the effects of moderate alcohol and benzodiazepine intake on motor impairment (Korpi *et al.*, 1993). Schuckit *et al.* (1999) observed in a pilot study that all human subjects who carried a 5-HTT genotype associated with elevated 5-HTT expression plus the Pro/Ser genotype of the $GABA_{A\alpha6}$ receptor subunit developed alcohol dependence during the decade-long follow-up period. In summary, these animal experiments and human studies suggest that both genetic and environmental (stress) factors interact with central serotonergic neurotransmission and thus affect the balance between GABAergic inhibition and glutamatergic excitation such that low serotonin turnover and high transporter availability reduce the sedative and ataxic effects of acute alcohol intake (Higley *et al.*, 1996; Heinz *et al.*, 1998a; Schuckit *et al.*, 1999).

Studies that assessed the potential association of serotonin transporter genotype and the risk to develop alcoholism gave inconsistent results (Gelernter *et al.*, 1997; Edenberg *et al.*, 1998; Sander *et al.*, 1998). These results may be explained by phenocopies of genetic effects on serotonergic neurotransmission – i.e., environmental stress factors appear to mimic genetic effects on serotonin transporters and GABAergic neurotransmission. Bennett and coworkers observed that exposure to early social isolation stress reduced CSF 5-HIAA concentrations in non-human primates only if they carried one or two short alleles of the 5-HTT regulatory unit, while it did not affect CSF 5-HIAA concentrations in carriers of two long alleles (*ll*) of the 5-HTT gene (Bennett *et al.*, 2002). All but one of the primates examined in the imaging study of Heinz *et al.* (1998a) were carriers of this 'stress-vulnerable' 5-HTT genotype – i.e., they carried one or two short alleles of the 5-HTT regulatory unit. In this genotype a high availability of brainstem serotonin transporters was found, and can result from stress-induced decreases in central serotonin turnover due to reduced competition between endogenous serotonin and the radioligand for binding at 5-HTT sites (Heinz *et al.*, 1998a, 2004). In the *ll*-genotype, on the other hand, a high availability of central serotonin transporters may be mainly due to genetic factors directly affecting gene expression and function (Lesch *et al.*, 1996). It appears that a common phenotype, high availability of 5-HTT transporters *in vivo*, contributes to a low response to acute alcohol intake, and that this association is independent of its genetic or environmental causation (Heinz *et al.*, 1998a; Barr *et al.*, 2003).

Altogether, animal experiments suggest that 5-HTT genotype mediates the impact of early social separation stress on the central serotonin turnover rate, and this effect may be mediated by differences in the stress hormone axis (Bennett *et al.*, 2002; Barr *et al.*, 2004). In non-human primates, social isolation stress was associated with a low serotonin turnover rate and associated changes in the availability of brainstem (raphe) serotonin transporters (Higley *et al.*, 1996; Heinz *et al.*, 1998a). Behaviorally, non-human primates with a low serotonin turnover rate and a high availability of brainstem serotonin transporters displayed high levels of anxiety, impulsive aggression, and a low response to alcohol (Higley *et al.*, 1996; Heinz *et al.*, 1998a). Increased aggressiveness and a low response to alcohol were also found in mice lacking the $5\text{-}HT_{1B}$ receptor (Coccaro *et al.*, 1989; Saudou *et al.*, 1994). Human studies also suggest that a low serotonin turnover rate is associated with anxious behavior and perceptions of insecurity and threat, which may secondarily contribute to impulsive aggression (Praag, 1977; Meltzer *et al.*, 1994; Knutson *et al.*, 1998). Serotonin function may contribute to negative mood states via modulation of amygdala responses to emotionally negative stimuli. Indeed, in healthy human volunteers amygdala activation elicited by aversive stimuli was modulated by 5-HTT genotype, with carriers of a short allele of the 5-HTT gene displaying more amygdala activation; healthy volunteers carrying this genotype may be able to maintain limited levels of negative mood states because aversive stimuli also elicit increased activation of the medial prefrontal cortex, which can contribute to emotion regulation (Heinz *et al.*, 2005; Pezawas *et al.*, 2005; Figure 4).

However, stress effects may interfere with this prefrontal–limbic connectivity. In human studies, 5-HTT *s*-carriers who had experienced a high number of traumatizing events displayed a higher risk of developing negative mood states such as major depression (Caspi *et al.*, 2003). Also in primate studies, 5HTT *s*-carriers were specifically vulnerable to social isolation stress, which induced significant reductions in the central serotonin turnover rate (Bennett *et al.*, 2002).

Gender-specific effects on the interaction between early social isolation stress, serotonin transporter genotype and HPA axis were implicated in studies among non-human primates (Barr *et al.*, 2004). Female and male rhesus monkeys were either raised without their mothers by their peers (peer-raised, PR), or grew up with their mothers (mother-raised, MR). When subjected to complete social isolation, female primates displayed an increase in cortisol concentrations in *s*-carriers compared with *ll*-homozygotes only if they were previously raised without

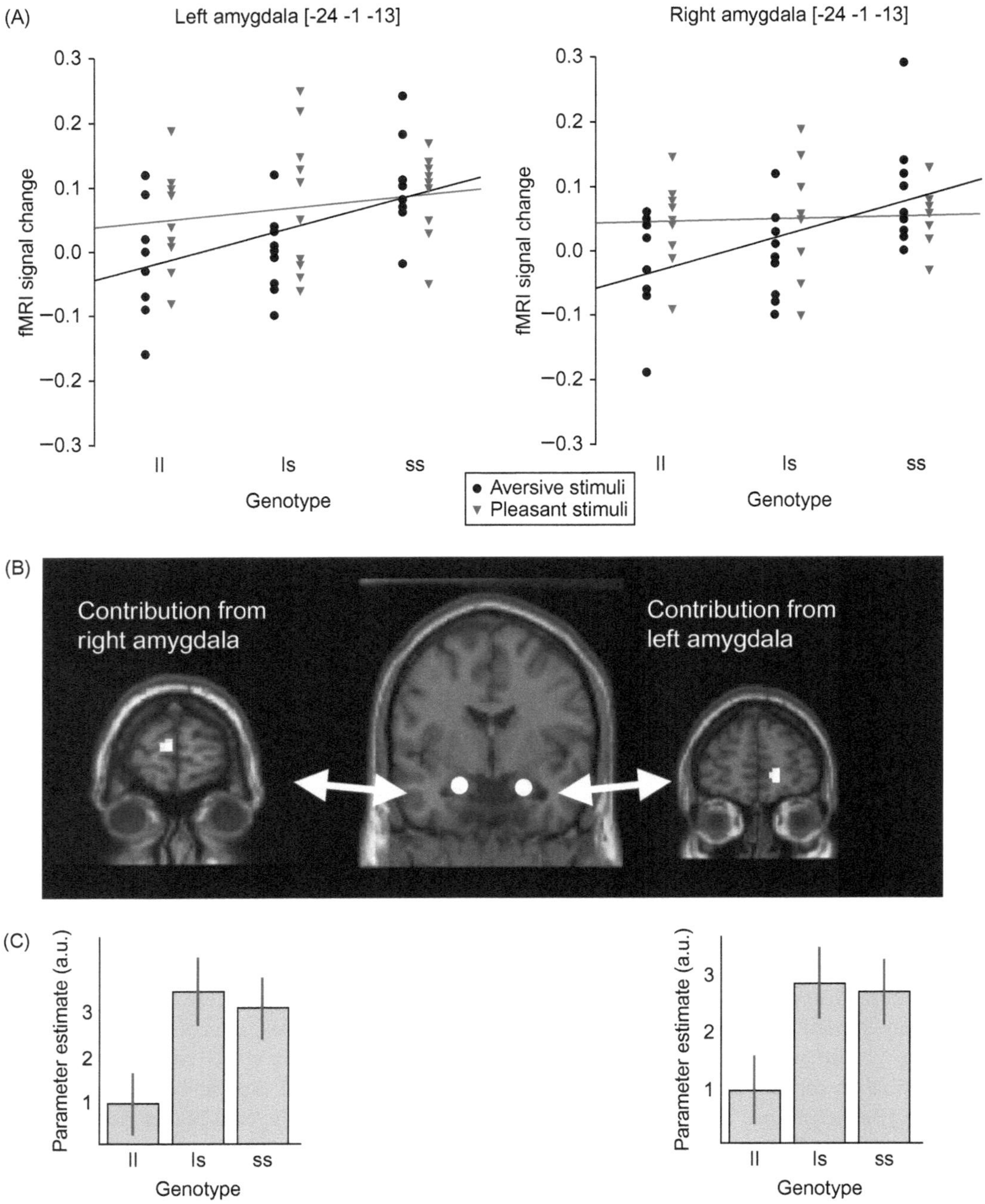

Figure 4 (A) 5-HTT genotype modulates amygdala activation elicited by aversive versus neutral (black) but not by pleasant versus neutral (red) pictures. (B) Assessment of bilateral functional connectivity between the amygdala and medial prefrontal cortex (PFC). In healthy control subjects, 5-HTT genotypes with increased amygdala activation elicited by aversive cues also showed increased functional connectivity between the amygdalae and medial prefrontal cortex (Brodman's area: BA10), which may limit effects of limbic stimulation and facilitate cortical control. (C) Parameter estimates of connectivity between the amygdala and medial PFC modulated by 5-HTT genotype: risk allele (s-allele) carriers display increased parameter estimates for functional connectivity. (B) Modified by permission from Macmillan Publishers Ltd (*Psychiatry Research*, Heinz *et al.*, ©2005).

their mothers (PR) – i.e., if they had experienced early adversity in the form of social separation stress – while male *s*-carriers displayed increased ACTH concentrations during social isolation independent of whether they were PR or MR.

Altogether, it appears that 5-HTT genotype modulates stress exposure and stress hormone effects on 5-HTT expression and function. Specifically, stress-related alterations in serotonin turnover and transporter availability may interfere with the 5-HTT genotype-driven interaction between the medial prefrontal cortex and the amygdala, thus impairing prefrontal emotional control. Studies in humans and non-human primates suggest that gender effects may influence whether serotonin dysfunction results in negative mood states or impulsive aggression – an observation that is particularly relevant for clinical depression, which is more frequently observed in women than in men (Heinz *et al.*, 1998a; Barr *et al.*, 2004).

Serotonin and major depression

Serotonin dysfunction has long been the focus of research on the neurobiological correlates of major depression. Asberg *et al.* (1976, 1984) showed that serotonin turnover rates are reduced in major depression, and increase with recovery. A series of studies measured serotonin transporter availability and function, because several antidepressive drugs block serotonin reuptake and thus alleviate negative mood states (Pirker *et al.*, 1995; Meltzer *et al.*, 1997; Meyer *et al.*, 2004a). Single photon emission computed tomography (SPECT) imaging studies in major depression (Malison *et al.*, 1998; Staley *et al.*, 2006) and seasonal affective disorder (Willeit *et al.*, 2000) revealed a reduction in binding of the radioligand [I^{123}]β-CIT to 5-HTT in the diencephalon and/or brainstem. However, these first studies used the radioligand β-CIT, which competes with endogenous serotonin for binding to central serotonin transporters (Laruelle *et al.*, 1993; Heinz *et al.*, 2004). Therefore, reduced binding of this radioligand to serotonin transporters could either reflect a true reduction in 5-HTT densitiy or increased extracellular serotonin levels that compete with binding at serotonin reuptake sites. In accordance with this latter hypothesis, a high serotonin turnover rate (as indicated by elevated CSF 5-HIAA concentrations) was negatively correlated with serotonin transporter availability measured with radioligand β-CIT in humans and non-human primates (Heinz *et al.*, 1998a, 2002). Interpretation of imaging studies with β-CIT is further complicated by the fact that this radioligand also binds to dopamine and norepinephrine transporters (Laruelle *et al.*, 1993; Seibyl *et al.*, 1994; Rinne *et al.*, 1995); therefore, β-CIT binding to 5-HTT can be assessed only in brain areas such as the brainstem and thalamus, where dopamine transporters are rare and where the radioligand is displaced exclusively by selective serotonin reuptake inhibitors (Laruelle *et al.*, 1993; Pirker *et al.*, 1995, 2000; Heinz *et al.*, 2004).

Parsey *et al.* (2006a, 2006b) used positron emission tomography (PET) and the radioligand [^{11}C]McN, which displays selective affinity for 5-HTT and is assumed to be not displaceable by endogenous serotonin (Hummerich *et al.*, 2006). They observed reduced midbrain and amygdala 5-HTT availability in a group of 25 patients with major depression. However, this finding was not replicated by Ichimiya *et al.* (2002), who observed unaltered [^{11}C]McN binding in the midbrain and *increased* binding in the thalamus in a small sample of seven patients with major depression (Ichimiya *et al.*, 2002). Another PET ligand that selectively binds to serotonin transporters is [^{11}C]DASB (Houle *et al.*, 2000). DASB has a better signal-to-noise ratio than [^{11}C]McN (Frankle *et al.*, 2004), and is assumed to measure 5-HTT availability independent from synaptic levels of endogenous serotonin (Hummerich *et al.*, 2004). Using DASB and PET in unmedicated patients with major depression and healthy controls, Meyer and coworkers (2004a) observed unaltered binding in various subcortical and cortical regions, including the thalamus and midbrain, with the exception of a subgroup of patients with highly negativistic attitudes that displayed *increased* 5-HTT binding potential. It was suggested that the previously observed reductions in 5-HTT availability in major depression may depend on comorbidity such as anxiety (Meyer, 2007), and on factors like age, gender, smoking status (Staley *et al.*, 2001, 2006) and the genetic constitution of the 5-HTT regulatory region (5-HTTLPR) (Lesch *et al.*, 1996; Heinz *et al.*, 2000; van Dyck *et al.*, 2004; Praschak-Rieder *et al.*, 2007; Reimold *et al.*, 2007). Therefore, discrepancies in the literature may in part be explained by interactions between these covariates (Staley *et al.*, 2006). In a study that compared patients suffering from unipolar major depression with healthy control subjects individually matched for age, smoking status, 5-HTT genotype and gender, the unmedicated patients displayed a significant reduction in their thalamic serotonin transporter (5-HTT) availability (Reimold *et al.*, 2008). In accordance with the hypothesis that serotonin dysfunction is primarily associated with negative mood states, a significant negative correlation was found between serotonin transporter availability in the thalamus and the severity of anxiety (STAI), and, to a lower degree, with the severity of depressed mood (BDI). Furthermore, low serotonin transporter availability in the thalamus was associated with low levels of novelty-seeking (Reimold *et al.*, 2008). These results suggest that reduced serotonin reuptake capacity in the thalamus may interfere with the activity and excitability of thalamic and cortical neurons (McCormick, 1992), and thus contribute to negative mood states such as anxiety, and to a low tendency for seeking new experiences. Therefore, discrepancies between studies using DASB to assess 5-HTT availability in major depression may be explained by different severity of anxiety: Meyer and coworkers (2004b), who did not observe a significant difference in DASB binding in patients suffering from major depression compared with healthy control subjects, had excluded patients suffering from anxiety disorders, while in the study of Reimold and coworkers (2008) the severity of anxiety rather than depressive mood *per se* was associated with low 5-HTT availability in the thalamus. This interpretation is in accordance with the study of Malison *et al.* (1998), who also reported a negative correlation between anxiety (Hamilton Anxiety Rating Scale) and brainstem β-CIT binding ($r = -0.32$); however, this correlation was not statistically significant ($n = 15$) and 5-HTT availability in other regions was not investigated.

Is has been suggested that interactions between gender, age and depression can further help to explain discrepancies in the literature: Staley *et al.* (2006) observed that β-CIT binding was reduced particularly in the diencephalon of younger female patients. Furthermore, several studies did not scan medication-naïve patients, but rather patients suffering from major depression who had been medication-free long enough to avoid direct blockade of serotonin transporters by SSRIs. However, serotonin transporters appear to remain down-regulated after previous exposure to antidepressive medication (Meyer *et al.*, 2001; Benmansour *et al.*, 2002). Therefore, an imaging study in medication-naïve patients matched for gender, 5-HTT genotype and smoking status may help to elucidate interactions between these factors.

In the amygdala, increased activation elicited by aversive stimuli was observed among subjects who carried a 5-HTT genotype that was associated with low serotonin transporter expression (Hariri *et al.*, 2002, 2005; Heinz *et al.*, 2005). Therefore, a low availability of amygdala 5-HTT may directly contribute to negative mood states. However, it is currently not clear whether the assessed 5-HTT genotype exerts its effect via developmental or current expression of 5-HTT in the amygdala or in some other serotonergic brain area (Hansen *et al.*, 1997; Pezawas *et al.*, 2005). Published PET and SPECT studies that revealed reductions in 5-HTT availability in major depression differ with respect to the described anatomical pattern. Reimold *et al.* (2008) and Staley *et al.* (2006) observed 5-HTT reductions in the diencephalons, including the thalamus, but not in the brainstem, while Parsey *et al.* (2006a) observed a significant reduction in SERT availability in midbrain and amygdala, but not in the thalamus. Malison *et al.* (1998) also observed reductions in the brainstem, but did not report on 5-HTT availability in other regions. The study of Reimold and coworkers observed correlations between 5-HTT availability and severity of anxiety in all these regions (Reimold *et al.*, 2008), though group differences may not reach statistical significance due to signal-to-noise issues. The latter may include methodological aspects such as ROI definitions and tracer properties (Frankle *et al.*, 2004), as well as genotype or smoking status. Reimold and coworkers (2008) showed that explorative voxelwise analysis indeed showed further regions with reduced 5-HTT availability in patients suffering from major depression compared with healthy controls, namely in the midbrain (a cluster that extended into the bilateral thalamus), the left putamen, the right insula, the anterior cingulate and the cingulate gyrus. Altogether, these brain-imaging studies in major depression support the hypothesis that serotonin dysfunction is associated with negative mood states (Heinz *et al.*, 1998a; Knutson *et al.*, 1998; Spillmann *et al.*, 2001; Hasler *et al.*, 2004). Some studies suggest that anxiety is a primary correlate of reduced central 5-HTT availability in major depression (Malison *et al.*, 1998), particularly when patients are matched with healthy control subjects for age, 5-HTT genotype, gender and smoking status. Anxiety may be mediated by 5-HTT effects on limbic processing of emotionally aversive stimuli (Hariri *et al.*, 2002, 2005; Heinz *et al.*, 2005), and clinically relevant depression may manifest when prefrontal control of limbic activation is impaired (Hariri *et al.*, 2002, 2005; Heinz *et al.*, 2005). However, so far no study has directly assessed the interaction between *in vivo* central 5-HTT availability, negative mood states and the processing of affective stimuli in limbic brain areas in health and in major depression.

Depression and anxiety are increased in women with stress hormone dysregulation following severe childhood abuse (Heim *et al.*, 2000). Among patients with anxiety and major depression, dysregulation of the stress hormone axis is a common finding, and appears to be caused at least in part by elevated levels of the hypothalamic neuropeptide corticotrophin releasing hormone (CRH) (Holsboer and Ising, 2008). In alcohol-dependent patients, elevated cortisol levels following detoxification were correlated with a reduction in serotonin transporter availability, which was in turn associated with high levels of anxiety during withdrawal (Heinz *et al.*, 1998a, 1998b, 2002). However, if stress hormone axis activation correlates with negative mood states, it appears counterintuitive that some studies in humans and non-human primates showed a blunted rather than an increased stress hormone response (Barr *et al.*, 2004; Smith *et al.*, 2004) in individuals carrying a serotonin transporter (5-HTT) genotype that is associated with reduced 5-HTT expression, anxiety and dysfunctional amygdala activation elicited by fearful faces (Lesch *et al.*, 1996; Hariri *et al.*, 2005). Moreover, human subjects with this genotype displayed an increased risk to develop clinical depression in cases where they had experienced several severe stressful life events (Caspi *et al.*, 2003). Together, these findings indicate that stress hormones may have differential effects on serotonin transporter availability depending on 5-HTT genotype.

Serotonin dysfunction and obsessive-compulsive disorder

Several lines of evidence link central serotonergic dysfunction with the manifestation of obsessive-compulsive disorder (OCD) (Swedo and Leonard, 1994). Studies measuring monoamine metabolites in the cerebrospinal fluid (CSF) observed elevated concentrations of the serotonin metabolite 5-hydroxyindoleacetic acid (5-HIAA) in OCD patients (Asberg, 1982; Insel *et al.*, 1985). CSF 5-HIAA concentrations decreased during successful drug treatment of OCD with clomipramine (Thoren *et al*, 1980), while acute stimulation of 5-HT_2 receptors with m-chloro-phenylpiperazine exacerbated obsessive-compulsive symptoms (Zohar, 1987; Brooks *et al.*, 1998). The important role of serotonergic neurotransmission in OCD is further emphasized by the observation that OCD symptoms are more efficiently reduced when serotonin reuptake is inhibited by clomipramine or selective serotonin reuptake inhibitors (SSRIs) than by the application of norepinephrine reuptake inhibitors (Zohar and Insel, 1987; Goodman *et al.*, 1990).

Dysfunction of fronto-striato-thalamic circuits is implicated in the pathogenesis of both OCD and complex tics (Alexander, 1986; Cummings, 1993). Increased glucose metabolism was found in the orbitofrontal cortex, the head of the caudate and the thalamus of OCD patients compared to healthy controls (Baxter, 1987; Perani *et al.*, 1995). After successful treatment of OCD with behavior therapy or selective serotonin reuptake inhibitors (SSRIs), glucose metabolism decreased in the head of the caudate (Baxter *et al.*, 1992). Ascending fibers from the brainstem, including the serotonergic system, control the firing mode of thalamocortical neurons (McCormick, 1992; Monckton and McCormick, 2002). Therefore, altered serotonin reuptake capacity, for example in the thalamus of OCD patients, may interfere with the role of the thalamus as a 'noise filter' and contribute to overactivity of fronto-striatal-thalamic neurocircuits in OCD. However, fronto-striatal-thalamic overactivity in OCD also appears to involve dopaminergic dysfunction. The dopaminergic and serotonergic systems closely interact, and both animal and human studies support a role for dopamine dysfunction in OCD (van der Wee *et al.*, 2004; Berridge *et al.*, 2005; Hesse *et al.*, 2005). However, studies on dopamine transporter availability in OCD provided disparate results: Hesse *et al.* (2005) described decreased *in vivo* availability of striatal dopamine transporter (DATs) in OCD, while Van der Wee *et al.* (2004) observed an increased DAT availability among patients suffering from OCD. These discrepancies may be due to variation in the DAT genotype that interacts with dopamine transporter availability *in vivo* (van Dyck *et al.*, 2005), comorbid tic disorder (Heinz, 1998c) or other unknown factors. In patients suffering from Tourette's syndrome (TS), who often display both tics and OCD symptoms, the severity of tics has been positively correlated with dopamine D2 receptor availability in the caudate nucleus (Wolf *et al.*, 1996). Dopaminergic dysfunction may thus primarily contribute to the motor/compulsive aspects of OCD, whereas serotonergic dysfunction may be associated more closely with obsessive ideation (Heinz, 1999).

Several studies measured the *in vivo* availability of serotonin transporters in the brainstem (raphe) area, the center of origin of serotonergic projections, among medication-free OCD patients with positron emission tomography (PET) or single photon emission computed tomography (SPECT) and radioligands that show high affinity for serotonin transporters. However, results have so far been inconsistent: two groups used SPECT and the radioligand $[^{123}I]\beta$-CIT and observed a significant reduction in serotonin transporter availability in the thalamus/hypothalamus region (Hesse *et al.*, 2005; Zitterl *et al.*, 2007), which was associated with symptom severity. Hesse and coworkers also observed a significant 5-HTT reduction in the midbrain and brainstem of OCD patients (Hesse *et al.*, 2005). On the other hand, Pogarell and colleagues (2003) reported a 25 percent increase in 5-HTT availability in the midbrain using the same ligand, while no significant differences between OCD patients and healthy control subjects were found in two further studies that used either $[^{11}C]$McN 5652 and PET (Simpson *et al.*, 2003) or $[^{123}I]\beta$-CIT and SPECT (van der Wee *et al.*, 2004). The disparate results may in part be due to the fact that the SPECT radioligand β-CIT competes with endogenous serotonin for binding to central serotonin transporters (Laruelle *et al.*, 1993; Heinz *et al.*, 2004), which resulted in lower radioligand binding in humans and non-human primates with high serotonin turnover as indicated by elevated CSF 5-HIAA levels (Heinz *et al.*, 1998a, 2002). This interaction may be particularly relevant in OCD, where CSF studies suggested significant increases in 5-HIAA concentrations (Asberg, 1982; Insel *et al.*, 1985). On the other hand, *in vivo* studies showed that 5-HTT binding potential measured with $[^{11}C]$DASB was not affected by depletion of the serotonin precursor tryptophan (Praschak-Rieder *et al.*, 2005; Talbot *et al.*, 2005), suggesting that DASB PET can measure 5-HTT independently of synaptic levels of serotonin. Further factors that can modify 5-HTT availability are age, 5-HTT genotype (Lesch *et al.*, 1996; Heinz *et al.*, 2000; Praschak-Rieder *et al.*, 2007; Reimold *et al.*, 2007), gender and smoking status (Staley *et al.*, 2001). One study that accounted for the *individual* combination of gender, genotype and smoking status in OCD patients and healthy controls observed significantly reduced 5-HTT availability in the midbrain and thalamus of OCD patients (Reimold *et al.*, 2007).

Together, these findings suggest that unmedicated OCD patients display reduced thalamus 5-HTT availability when matched for gender, age, 5-HTT genotype and smoking status, compared with healthy control subjects. The imaging studies thus support the hypothesis of serotonergic dysfunction in OCD (Zohar, 1987; Brooks *et al.*, 1998), which may contribute to dysfunctional information processing in fronto-striato-thalamic neurocircuits. Animal experiments and human studies suggest serotonergic dysfunction in the orbitofrontal cortex directly interferes with behavioral flexibility: perseverative responding to a previously rewarded stimulus was provoked by selective serotonin depletion of the prefrontal cortex in non-human primates (Clarke *et al.*, 2004). Serotonin depletion also interfered with further task performance associated with the orbitofrontal cortex in non-human primates (Walker *et al.*, 2006). Studies in humans confirmed that acute serotonin depletion interfered with the function of the dorsomedial prefrontal cortex during reversal learning and negative feedback (Evers *et al.*, 2005). Disruption of dopaminergic or noradrenergic neurotransmission, on the other hand, had no significant effects on reveral learning (Robbins and Roberts, 2007).

Altogether, these findings confirm a prominent role for serotonergic dysfunction in OCD. Unlike in major depression, CSF studies measuring the serotonin metabolite 5-HIAA pointed to an increased rather than a decreased serotonin turnover rate (Asberg and Bertilsson, 1982; Insel *et al.*, 1985). In the brainstem and thalamus of unmedicated OCD patients, serotonin transporters appear to be reduced when compared with healthy controls matched for gender, 5-HTT genotype and smoking status (Reimold *et al.*, 2007). Direct effects of such an alteration in brainstem (raphe) serotonin reuptake capacity have not yet been explored; however, serotonin depletion studies suggest that dysfunction of serotonergic neurotransmission in the orbitofrontal cortex interferes with flexible responding (Robbins and Roberts, 2007), and may thus contribute to obsessive ideation and perhaps also compulsive behavior patterns in OCD.

Summary and outlook

This review suggests that central serotonin dysfunction is associated with increased anxiety and feelings of threat and insecurity across a wide range of clinical disorders. Anxiety was a direct correlate of reduced serotonin transporter availability in patients suffering from major depression (Malison *et al.*, 1998; Reimold *et al.*, 2008) and alcohol dependence (Heinz *et al.*, 1998a). Impulsive aggression may be a secondary correlate of serotonergic dysfunction due to feelings of threat and insecurity interacting with sex-specific behavior patterns and hormonal regulation (Knutson *et al.*, 1998; Heinz *et al.*, 2001; Barr *et al.*, 2004). In OCD, an increased rather than decreased serotonin turnover rate may interfere with a thalamic 'noise filter' and with orbitofrontal function during reversal learning, thus contributing to overactivation of fronto-striato-thalamic neurocircuits and behavioral inflexibility (Baxter *et al.*, 1987; Heinz, 1999; Clarke *et al.*, 2008).

Serotonin dysfunction can result from developmentally early social stress factors, and appears to modulate responses to stress and sedative drugs such as alcohol in adolescent and adult humans and non-human primates (Higley *et al.*, 1996; Heinz *et al.*, 1998a; Schuckit *et al.*, 1999; Hinckers *et al.*, 2006). Early epigenetic studies suggest that alterations in DNA methylation may contribute to decreased transcription of 5-HTT mRNA and hence reduced expression of central serotonin transporters (Philibert *et al.*, 2008). Variation in the 5-HTT gene has been shown to correlate with 5-HTT expression and function (Lesch *et al.*, 1996; Heinz *et al.*, 2000) and with limbic and prefrontal responses to aversive stimuli (Heinz *et al.*, 2005; Pezawas *et al.*, 2005; Smolka *et al.*, 2007), and appears to modulate HPA axis activation and the stress-associated risk to develop affective disorders (Caspi *et al.*, 2003; Smith *et al.*, 2004; Jabbi *et al.*, 2007). However, so far we understand only a small part of the picture, and in the future the complex interaction between genetic, epigenetic and environmental factors on serotonin function and dysfunction in specific brain areas can be further elucidated by combined (epi)genetic, endocrinological and imaging studies that assess monoaminergic neurotransmission, sex and stress hormone responses plus functional activation during specific tasks. Such studies may help to develop an individualized therapeutic approach that combines specific behavioral and pharmacological interventions.

Acknowledgements

This section was supported in part by NGFN O1GSO8159 and DFG 2597/7-3, 4-3.

References

Alexander, G., DeLong, M.R. and Strick, P.L. (1986) Parallel organization of functionally segregated circuits linking basal ganglia and cortex. Ann. Rev. Neurosci., 9: 357–381.

Anisman, H., Merali, Z. and Stead, J.D. (2008) Experiential and genetic contributions to depressive- and anxiety-like disorders: clinical and experimental studies. Neurosci. Biobehav. Rev., 32: 1185–1206.

Artigas, F. (1995) Pindolol, 5-hydroxytryptamine, and antidepressant augmentation. Arch. Gen. Psychiatry, 52: 969–971.

Asberg, M.T.L. and Bertilsson, L. (1982) Clomipramine treatment of obsessive-compulsive disorder: biochemical and clinical aspects. Psychopharmacol. Bull., 18: 13–21.

Asberg, M., Bertilsson, L., Martensson, B., Scalia-Tomba, G.P., Thoren, P. and Traskman-Bendz, L. (1984) CSF monoamine metabolites in melancholia. Acta Psychiatrica Scandinavica, 69: 201–219.

Asberg, M., Thoren, P., Traskman, L., Bertilsson, L. and Ringberger, V. (1976) 'Serotonin depression' – a biochemical subgroup within the affective disorders? Science (NY), 191: 478–480.

Barr, C.S., Newman, T.K., Becker, M.L., Champoux, M., Lesch, K.P., Suomi, S.J., Goldman, D. and Higley, J.D. (2003) Serotonin transporter gene variation is associated with alcohol sensitivity in rhesus macaques exposed to early-life stress. Alcohol. Clin. Exp. Res., 27: 812–817.

Barr, C.S., Newman, T.K., Shannon, C., Parker, C., Dvoskin, R.L., Becker, M.L., Schwandt, M., Champoux, M., Lesch, K.P., Goldman, D., Suomi, S.J. and Higley, J.D. (2004) Rearing condition and rh5-HTTLPR interact to influence limbic–hypothalamic–pituitary–adrenal axis response to stress in infant macaques. Biol. Psychiatry, 55: 733–738.

Barr, L.C., Goodman, W.K., McDougle, C.J., Delgado, P. L., Heninger, G.R., Charney, D.S. and Price, L.H. (1994) Tryptophan depletion in patients with obsessive-compulsive disorder who respond to serotonin reuptake inhibitors. Arch. Gen. Psychiatry, 51: 309–317.

Baumgarten, H.G. and Grozdanovic, Z. (1997). Anatomy of central serotoninergic projection systems. In: Handbook of Experimental Pharmacology. Springer, Berlin, pp. 41–89.

Baxter, L.R., Phleps, M.E., Mazziotta, J.C., Guze, B.H., Schwartz, J.M. and Selin, C.E. (1987) Local cerebral glucose metabolic rates in obsessive-compulsive disorder. Arch. Gen. Psychiatry, 44: 211–218.

Baxter, L.R. Jr., Schwartz, J.M., Bergman, K.S., Szuba, M.P., Guze, B.H., Mazziotta, J.C., Alazraki, A., Selin, C.E., Ferng, H.K., Munford, P. et al. (1992) Caudate glucose metabolic rate changes with both drug and behavior therapy for obsessive-compulsive disorder. Arch. Gen. Psychiatry, 49: 681–689.

Beck, O., Borg, S., Edman, G., Fyro, B., Oxenstierna, G. and Sedvall, G. (1984) 5-hydroxytryptophol in human cerebrospinal fluid: conjugation, concentration gradient, relationship to 5-hydroxyindoleacetic acid, and influence of hereditary factors. J. Neurochem., 43: 58–61.

Benmansour, S., Owens, W.A., Cecchi, M., Morilak, D.A. and Frazer, A. (2002) Serotonin clearance *in vivo* is altered to a greater extent by antidepressant-induced downregulation of the serotonin transporter than by acute blockade of this transporter. J. Neurosci., 22: 6766–6772.

Bennett, A.J., Lesch, K.P., Heils, A., Long, J.C., Lorenz, J.G., Shoaf, S.E., Champoux, M., Suomi, S.J., Linnoila, M.V. and Higley, J.D. (2002) Early experience and serotonin transporter gene variation interact to influence primate CNS function. Mol. Psychiatry, 7: 118–122.

Berridge, K.C., Aldridge, J.W., Houchard, K.R. and Zhuang, X. (2005) Sequential super-stereotypy of an instinctive fixed action pattern in hyper-dopaminergic mutant mice: a model of obsessive compulsive disorder and Tourette's. BMC Biol., 3: 4.

Brooks, A., Pigott, T.A., Hill, J.L. et al. (1998) Acute intravenous administration of ondansetron and m-CPP, alone and in combination, in patients with obsessive-compulsive disorder (OCD): behavioural and biological results. Psychiatry Res., 79: 2077–2082.

Cai, X., Flores-Hernandez, J., Feng, J. and Yan, Z. (2002a) Activity-dependent bidirectional regulation of GABA(A) receptor channels by the 5-HT(4) receptor-mediated signalling in rat prefrontal cortical pyramidal neurons. J. Physiol., 540: 743–759.

Cai, X., Gu, Z., Zhong, P., Ren, Y. and Yan, Z. (2002b) Serotonin 5-HT1A receptors regulate AMPA receptor channels through inhibiting Ca^{2+}/calmodulin-dependent kinase II in prefrontal cortical pyramidal neurons. J. Biol. Chem., 277: 36553–36562.

Caspi, A., Sugden, K., Moffitt, T.E., Taylor, A., Craig, I.W., Harrington, H., McClay, J., Mill, J., Martin, J., Braithwaite, A. and Poulton, R. (2003) Influence of life stress on depression: moderation by a polymorphism in the 5-HTT gene. Science (NY), 301: 386–389.

Clarke, A.S., Hedeker, D.R., Ebert, M.H., Schmidt, D.E., McKinney, W.T. and Kraemer, G.W. (1996) Rearing experience and biogenic amine activity in infant rhesus monkeys. Biol. Psychiatry, 40: 338–352.

Clarke, H.F., Dalley, J.W., Crofts, H.S., Robbins, T.W. and Roberts, A.C. (2004) Cognitive inflexibility after prefrontal serotonin depletion. Science (NY), 304: 878–880.

Clarke, H.F., Robbins, T.W. and Roberts, A.C. (2008) Lesions of the medial striatum in monkeys produce perseverative impairments during reversal learning similar to those produced by lesions of the orbitofrontal cortex. J. Neurosci., 28: 10972–10982.

Cloninger, C.R. (1987) A systematic method for clinical description and classification of personality variants. A proposal. Arch. Gen. Psychiatry, 44: 573–588.

Coccaro, E.F., Siever, L.J., Klar, H.M., Maurer, G., Cochrane, K., Cooper, T.B., Mohs, R.C. and Davis, K.L. (1989) Serotonergic studies in patients with affective and personality disorders. Correlates with suicidal and impulsive aggressive behavior. Arch. Gen. Psychiatry, 46: 587–599.

Cummings, J. (1993) Frontal-subcortical circuits and motor behavior. Arch. Neurol., 50: 873–880.

Delgado, P.L., Charney, D.S., Price, L.H., Aghajanian, G.K., Landis, H. and Heninger, G.R. (1990) Serotonin function and the mechanism of antidepressant action. Reversal of antidepressant-induced remission by rapid depletion of plasma tryptophan. Arch. Gen. Psychiatry, 47: 411–418.

Doudet, D., Hommer, D., Higley, J.D., Andreason, P.J., Moneman, R., Suomi, S.J. and Linnoila, M. (1995) Cerebral glucose metabolism, CSF 5-HIAA levels, and aggressive behavior in rhesus monkeys. Am. J. Psychiatry, 152: 1782–1787.

Drevets, W.C. (2000) Neuroimaging studies of mood disorders. Biol. Psychiatry, 48: 813–829.

Edenberg, H.J., Reynolds, J., Koller, D.L., Begleiter, H., Bucholz, K.K., Conneally, P.M., Crowe, R., Goate, A., Hesselbrock, V., Li, T.K., Nurnberger, J.I. Jr., Porjesz, B., Reich, T., Rice, J.P., Schuckit, M., Tischfield, J.A. and Foroud, T. (1998) A family-based analysis of whether the functional promoter alleles of the serotonin transporter gene HTT affect the risk for alcohol dependence. Alcohol. Clin. Exp. Res., 22: 1080–1085.

Evers, E.A., Cools, R., Clark, L., van der Veen, F.M., Jolles, J., Sahakian, B.J. and Robbins, T.W. (2005) Serotonergic modulation of prefrontal cortex during negative feedback in probabilistic reversal learning. Neuropsychopharmacology, 30: 1138–1147.

Feng, J., Cai, X., Zhao, J. and Yan, Z. (2001) Serotonin receptors modulate GABA(A) receptor channels through activation of anchored protein kinase C in prefrontal cortical neurons. J. Neurosci., 21: 6502–6511.

FilsAime, M.L., Eckardt, M.J., George, D.T., Brown, G.L., Mefford, I. and Linnoila, M. (1996) Early-onse*t al*coholics have lower cerebrospinal fluid 5-hydroxyindoleacetic acid levels than late-onse*t al*coholics. Arch. Gen. Psychiatry, 53: 211–216.

Frankle, W.G., Huang, Y., Hwang, D.R., Talbot, P.S., Slifstein, M., Van Heertum, R., Abi-Dargham, A. and Laruelle, M. (2004) Comparative evaluation of serotonin transporter radioligands 11C-DASB and 11C-McN 5652 in healthy humans. J. Nucl. Med., 45: 682–694.

Gelernter, J., Kranzler, H. and Cubells, J.F. (1997) Serotonin transporter protein (SLC6A4) allele and haplotype frequencies and linkage disequilibria in African- and European-American and Japanese populations and in alcohol-dependent subjects. Hum. Genet., 101: 243–246.

Giros, B., Jaber, M., Jones, S.R., Wightman, R.M. and Caron, M.G. (1996) Hyperlocomotion and indifference to cocaine and amphetamine in mice lacking the dopamine transporter. Nature, 379: 606–612.

Goodman, W.K., Price, L.H., Delgado, P.L., Palumbo, J., Krystal, J.H., Nagy, L.M., Rasmussen, S.A., Heninger, G.R. and Charney, D.S. (1990) Specificity of serotonin reuptake inhibitors in the treatment of obsessive-compulsive disorder. Comparison of fluvoxamine and desipramine. Arch. Gen. Psychiatry, 47: 577–585.

Gray, J. (1982) The Neuropsychology of Anxiety. An Inquiry into the Function of the Septo-Hippocampal System. Oxford University Press, New York, NY.

Grove, G., Coplan, J.D. and Hollander, E. (1997) The neuroanatomy of 5-HT dysregulation and panic disorder. J. Neuropsychiatry Clin. Neurosci., 9: 198–207.

Hansen, H.H., Sanchez, C. and Meier, E. (1997) Neonatal administration of the selective serotonin reuptake inhibitor Lu 10-134-C increases forced swimming-induced immobility in adult rats: a putative animal model of depression? J. Pharmacol. Exp. Ther., 283: 1333–1341.

Hariri, A.R., Mattay, V.S., Tessitore, A., Kolachana, B., Fera, F., Goldman, D., Egan, M.F. and Weinberger, D.R. (2002) Serotonin transporter genetic variation and the response of the human amygdala. Science (NY), 297: 400–403.

Hariri, A.R., Drabant, E.M., Munoz, K.E., Kolachana, B.S., Mattay, V.S., Egan, M.F. and Weinberger, D.R. (2005) A susceptibility gene for affective disorders and the response of the human amygdala. Arch. Gen. Psychiatry, 62: 146–152.

Hasler, G., Drevets, W.C., Manji, H.K. and Charney, D.S. (2004) Discovering endophenotypes for major depression. Neuropsychopharmacology, 29: 1765–1781.

Heim, C., Newport, D.J., Miller, A.H. and Nemeroff, C.B. (2000) Long-term neuroendocrine effects of childhood maltreatment. J. Am. Med. Assoc., 284: 2321.

Heinz, A. (1999) Neurobiological and anthropological aspects of compulsions and rituals. Pharmacopsychiatry, 32: 223–229.

Heinz, A., Braus, D.F., Smolka, M.N., Wrase, J., Puls, I., Hermann, D., Klein, S., Grusser, S.M., Flor, H., Schumann, G., Mann, K. and Buchel, C. (2005) Amygdala–prefrontal coupling depends on a genetic variation of the serotonin transporter. Nat. Neurosci., 8: 20–21.

Heinz, A., Higley, J.D., Gorey, J.G., Saunders, R.C., Jones, D.W., Hommer, D., Zajicek, K., Suomi, S.J., Lesch, K.P., Weinberger, D.R. and Linnoila, M. (1998a) *In vivo* association between alcohol intoxication, aggression, and serotonin transporter availability in nonhuman primates. Am. J. Psychiatry, 155: 1023–1028.

Heinz, A., Ragan, P., Jones, D.W., Hommer, D., Williams, W., Knable, M.B., Gorey, J.G., Doty, L., Geyer, C., Lee, K. S., Coppola, R., Weinberger, D.R. and Linnoila, M. (1998b) Reduced central serotonin transporters in alcoholism. Am. J. Psychiatry, 155: 1544–1549.

Heinz, A., Knable, M.B., Wolf, S.S. et al. (1998c) Tourette's syndrome: [123I]h-CIT SPECT correlates of vocal tic severity. Neurology, 51: 1069–1074.

Heinz, A., Jones, D.W., Mazzanti, C., Goldman, D., Ragan, P., Hommer, D., Linnoila, M. and Weinberger, D.R. (2000) A relationship between serotonin transporter genotype and *in vivo* protein expression and alcohol neurotoxicity. Biol. Psychiatry, 47: 643–649.

Heinz, A., Mann, K., Weinberger, D.R. and Goldman, D. (2001) Serotonergic dysfunction, negative mood states, and response to alcohol. Alcohol. Clin. Exp. Res., 25: 487–495.

Heinz, A., Jones, D.W., Bissette, G., Hommer, D., Ragan, P., Knable, M., Wellek, S., Linnoila, M. and Weinberger, D.R. (2002) Relationship between cortisol and serotonin metabolites and transporters in alcoholism. Pharmacopsychiatry, 35: 127–134.

Heinz, A., Schafer, M., Higley, J.D., Krystal, J.H. and Goldman, D. (2003) Neurobiological correlates of the disposition and maintenance of alcoholism. Pharmacopsychiatry, 36(Suppl. 3): S255–S258.

Heinz, A., Jones, D.W., Zajicek, K., Gorey, J.G., Juckel, G., Higley, J.D. and Weinberger, D.R. (2004) Depletion and restoration of endogenous monoamines affects beta-CIT binding to serotonin but not dopamine transporters in non-human primates. J. Neural Transm.: 29–38.

Hesse, S., Muller, U., Lincke, T., Barthel, H., Villmann, T., Angermeyer, M.C., Sabri, O. and Stengler-Wenzke, K. (2005) Serotonin and dopamine transporter imaging in patients with obsessive-compulsive disorder. Psychiatry Res., 140: 63–72.

Higley, J.D., Suomi, S.J. and Linnoila, M. (1991) CSF monoamine metabolite concentrations vary according to age, rearing, and sex, and are influenced by the stressor of social separation in rhesus monkeys. Psychopharmacology, 103: 551–556.

Higley, J.D., Thompson, W.W., Champoux, M., Goldman, D., Hasert, M.F., Kraemer, G.W., Scanlan, J.M., Suomi, S.J. and Linnoila, M. (1993) Paternal and maternal genetic and environmental contributions to cerebrospinal fluid monoamine metabolites in rhesus monkeys (*Macaca mulatta*). Arch. Gen. Psychiatry, 50: 615–623.

Higley, J.D., Suomi, S.J. and Linnoila, M. (1996) A nonhuman primate model of type II excessive alcohol consumption? Part 1. Low cerebrospinal fluid 5-hydroxyindoleacetic acid concentrations and diminished social competence correlate with excessive alcohol consumption. Alcohol. Clin. Exp. Res., 20: 629–642.

Hinckers, A.S., Laucht, M., Schmidt, M.H., Mann, K.F., Schumann, G., Schuckit, M.A. and Heinz, A. (2006) Low level of response to alcohol as associated with serotonin transporter genotype and high alcohol intake in adolescents. Biol. Psychiatry, 60: 282–287.

Holsboer, F. and Ising, M. (2008) Central CRH system in depression and anxiety – evidence from clinical studies with CRH1 receptor antagonists. Eur. J. Pharmacol., 583: 350–357.

Houle, S., Ginovart, N., Hussey, D., Meyer, J.H. and Wilson, A.A. (2000) Imaging the serotonin transporter with positron emission tomography: initial human studies with [11C]DAPP and [11C]DASB. Eur. J. Nucl. Med., 27: 1719–1722.

Hu, X., Oroszi, G., Chun, J., Smith, T.L., Goldman, D. and Schuckit, M.A. (2005) An expanded evaluation of the relationship of four alleles to the level of response to alcohol and the alcoholism risk. Alcohol. Clin. Exp. Res., 29: 8–16.

Hu, X.Z., Lipsky, R.H., Zhu, G., Akhtar, L.A., Taubman, J., Greenberg, B.D., Xu, K., Arnold, P.D., Richter, M. A., Kennedy, J.L., Murphy, D.L. and Goldman, D. (2006) Serotonin transporter promoter gain-of-function genotypes are linked to obsessive-compulsive disorder. Am. J. Hum. Genet., 78: 815–826.

Huether, G., Zhou, D. and Ruther, E. (1997) Causes and consequences of the loss of serotonergic presynapses elicited by the consumption of 3,4-methylenedioxymethamphetamine (MDMA, 'ecstasy') and its congeners. J. Neural. Transm., 104: 771–794.

Hummerich, R., Reischl, G., Ehrlichmann, W., Machulla, H. J., Heinz, A. and Schloss, P. (2004) DASB – *in vitro* binding characteristics on human recombinant monoamine transporters with regard to its potential as positron emission tomography (PET) tracer. J. Neurochem., 90: 1218–1226.

Hummerich, R., Schulze, O., Radler, T., Mikecz, P., Reimold, M., Brenner, W., Clausen, M., Schloss, P. and Buchert, R. (2006) Inhibition of serotonin transport by (+)McN5652 is noncompetitive. Nucl. Med. Biol., 33: 317–323.

Ichimiya, T., Suhara, T., Sudo, Y., Okubo, Y., Nakayama, K., Nankai, M., Inoue, M., Yasuno, F., Takano, A., Maeda, J. and Shibuya, H. (2002) Serotonin transporter binding in patients with mood disorders: a PET study with [11C](+)McN5652. Biol. Psychiatry, 51: 715–722.

Insel, T.R., Mueller, E.A., Alterman, I., Linnoila, M. and Murphy, D.L. (1985) Obsessive-compulsive disorder and serotonin: is there a connection? Biol. Psychiatry, 20: 1174–1188.

Jabbi, M., Korf, J., Kema, I.P., Hartman, C., van der Pompe, G., Minderaa, R.B., Ormel, J. and den Boer, J.A. (2007) Convergent genetic modulation of the endocrine stress response involves polymorphic variations of 5-HTT, COMT and MAOA. Mol. Psychiatry, 12: 483–490.

Jones, D., Gorey, J., Zajicek, K., Das, S., Urbina, R., Lee, K., Heinz, A., Knable, M., Higley, D., Weinberger, D. and Linnoila, M. (1998) Depletion-restoration studies reveal the impact of endogenous dopamine and serotonin on [I123]?CIT SPECT imaging in primate brain. J. Nucl. Med., 35.

Jones, G.H., Hernandez, T.D., Kendall, D.A., Marsden, C.A. and Robbins, T.W. (1992) Dopaminergic and serotonergic function following isolation rearing in rats: study of behavioural responses and postmortem and *invivo* neurochemistry. Pharmacol. Biochem. Behav., 43: 17–35.

Kaplan, J., Martin, L., Comuzzie, A., Manuck, S., Mann, J. and Rogers, J. (2000) Heritability of monoaminergic metabolites measured in the cerebrospinal fluid of baboons. Social Neurosci. Abstract, 26.

Knutson, B., Panksepp, J. and Pruitt, D. (1996a) Effects of fluoxetine on play dominance in juvenile rats. Aggress. Behav., 22: 297–307.

Knutson, B., Panksepp, J., Narayanana, T. and Rossi, J. (1996b) Early central serotonin damage increases 'anxious' behaviors in juvenile rats. Social Neurosci., 22: 446.

Knutson, B., Wolkowitz, O.M., Cole, S.W., Chan, T., Moore, E.A., Johnson, R.C., Terpstra, J., Turner, R.A. and Reus, V.I. (1998) Selective alteration of personality and social behavior by serotonergic intervention. Am. J. Psychiatry, 155: 373–379.

Korpi, E.R., Kleingoor, C., Kettenmann, H. and Seeburg, P.H. (1993) Benzodiazepine-induced motor impairment linked to point mutation in cerebellar GABAA receptor. Nature, 361: 356–359.

Kraemer, G.W. and McKinney, W.T. (1979) Interactions of pharmacological agents which alter biogenic amine metabolism and depression – an analysis of contributing factors within a primate model of depression. J. Affect. Disord., 1: 33–54.

Kreiss, D.S. and Lucki, I. (1995) Effects of acute and repeated administration of antidepressant drugs on extracellular levels of 5-hydroxytryptamine measured *in vivo*. J. Pharmacol. Exp. Ther., 274: 866–876.

Krystal, J.H., Petrakis, I.L., Webb, E., Cooney, N.L., Karper, L. P., Namanworth, S., Stetson, P., Trevisan, L.A. and Charney, D.S. (1998) Dose-related ethanol-like effects of the NMDA antagonist, ketamine, in recently detoxified alcoholics. Arch. Gen. Psychiatry, 55: 354–360.

Krystal, J., Abi-Dargham, A., Laruelle, M. and Moghaddam, B. (1999) Pharmacological models of psychosis. In: Charney, E. N.D. and Bunney, B. (Eds.), Neurobiology of Mental Illness. Oxford University Press, New York, NY, pp. 214–224.

Laruelle, M., Baldwin, R.M., Malison, R.T., Zea-Ponce, Y., Zoghbi, S.S., al-Tikriti, M.S., Sybirska, E.H., Zimmermann, R.C., Wisniewski, G., Neumeyer, J.L. et al. (1993) SPECT imaging of dopamine and serotonin transporters with [123I]beta-CIT: pharmacological characterization of brain uptake in nonhuman primates. Synapse (NY), 13: 295–309.

LeMarquand, D., Pihl, R.O. and Benkelfat, C. (1994) Serotonin and alcohol intake, abuse, and dependence: findings of animal studies. Biol. Psychiatry, 36: 395–421.

Lesch, K.P., Bengel, D., Heils, A., Sabol, S.Z., Greenberg, B. D., Petri, S., Benjamin, J., Muller, C.R., Hamer, D.H. and Murphy, D.L. (1996) Association of anxiety-related traits with a polymorphism in the serotonin transporter gene regulatory region. Science (NY), 274: 1527–1531.

Liljequist, S. and Engel, J. (1982) Effects of GABAergic agonists and antagonists on various ethanol-induced behavioral changes. Psychopharmacology, 78: 71–75.

Limberger, N., Starke, K. and Singer, E.A. (1990) Serotonin uptake blockers influence serotonin autoreceptors by increasing the biophase concentration of serotonin and not through a 'molecular link'. Naunyn-Schmiedeberg's Arch. Pharmacol., 342: 363–370.

Little, K.Y., McLaughlin, D.P., Zhang, L., Livermore, C.S., Dalack, G.W., McFinton, P.R., DelProposto, Z.S., Hill, E., Cassin, B.J., Watson, S.J. and Cook, E.H. (1998) Cocaine, ethanol, and genotype effects on human midbrain serotonin transporter binding sites and mRNA levels. Am. J. Psychiatry, 155: 207–213.

Malison, R.T., Price, L.H., Berman, R., van Dyck, C.H., Pelton, G.H., Carpenter, L., Sanacora, G., Owens, M.J., Nemeroff, C. B., Rajeevan, N., Baldwin, R.M., Seibyl, J.P., Innis, R.B. and Charney, D.S. (1998) Reduced brain serotonin transporter availability in major depression as measured by [123I]-2 beta-carbomethoxy-3 beta-(4-iodophenyl)tropane and single photon emission computed tomography. Biol. Psychiatry, 44: 1090–1098.

Mann, J.J., Malone, K.M., Psych, M.R., Sweeney, J.A., Brown, R.P., Linnoila, M., Stanley, B. and Stanley, M. (1996) Attempted suicide characteristics and cerebrospinal fluid amine metabolites in depressed inpatients. Neuropsychopharmacology, 15: 576–586.

Mash, D.C., Staley, J.K., Doepel, F.M., Young, S.N., Ervin, F. R. and Palmour, R.M. (1996) Altered dopamine transporter densities in alcohol-preferring vervet monkeys. Neuroreport, 7: 457–462.

McCormick, D.A. (1992) Neurotransmitter actions in the thalamus and cerebral cortex. J. Clin. Neurophysiol., 9: 212–223.

Meltzer, H., Bastani, B., Jayathilake, K. and Maes, M. (1997) Fluoxetine, but not tricyclic antidepressants, potentiates the 5-hydroxytryptophan-mediated increase in plasma cortisol and prolactin secretion in subjects with major depression or with obsessive compulsive disorder. Neuropsychopharmacology, 17: 1–11.

Meltzer, H., Maes, M. and Elkis, H. (1994) The biological basis of refractory depression. In: Nolen, J.Z.W., Roose, S. and Amsterdam, J. (Eds.), Refractory Depression: Current Strategies and Future Directions. John Wiley & Sons, Chichester, pp. 177–198.

Meyer, J.H. (2007) Imaging the serotonin transporter during major depressive disorder and antidepressant treatment. J. Psychiatry Neurosci., 32: 86–102.

Meyer, J.H., Wilson, A.A., Ginovart, N., Goulding, V., Hussey, D., Hood, K. and Houle, S. (2001) Occupancy of serotonin transporters by paroxetine and citalopram during treatment of depression: a [(11)C]DASB PET imaging study. Am. J. Psychiatry, 158: 1843–1849.

Meyer, J.H., Houle, S., Sagrati, S., Carella, A., Hussey, D. F., Ginovart, N., Goulding, V., Kennedy, J. and Wilson, A. A. (2004a) Brain serotonin transporter binding potential measured with carbon 11-labeled DASB positron emission tomography: effects of major depressive episodes and severity of dysfunctional attitudes. Arch. Gen. Psychiatry, 61: 1271–1279.

Meyer, J.H., Wilson, A.A., Sagrati, S., Hussey, D., Carella, A., Potter, W.Z., Ginovart, N., Spencer, E.P., Cheok, A. and Houle, S. (2004b) Serotonin transporter occupancy of five selective serotonin reuptake inhibitors at different doses: an [11C]DASB positron emission tomography study. Am. J. Psychiatry, 161: 826–835.

Monckton, J.E. and McCormick, D.A. (2002) Neuromodulatory role of serotonin in the ferret thalamus. J. Neurophysiol., 87: 2124–2136.

Muck-Seler, D., Jevric-Causevic, A. and Diksic, M. (1996) Influence of fluoxetine on regional serotonin synthesis in the rat brain. J. Neurochem., 67: 2434–2442.

Mueller, B.R. and Bale, T.L. (2008) Sex-specific programming of offspring emotionality after stress early in pregnancy. J. Neurosci., 28: 9055–9065.

Oberlander, T.F., Grunau, R., Mayes, L., Riggs, W., Rurak, D., Papsdorf, M., Misri, S. and Weinberg, J. (2008a) Hypothalamic–pituitary–adrenal (HPA) axis function in 3-month-old infants with prenatal selective serotonin reuptake inhibitor (SSRI) antidepressant exposure. Early Hum. Dev., 84: 689–697.

Oberlander, T.F., Weinberg, J., Papsdorf, M., Grunau, R., Misri, S. and Devlin, A.M. (2008b) Prenatal exposure to maternal depression, neonatal methylation of human glucocorticoid receptor gene (NR3C1) and infant cortisol stress responses. Epigenetics, 3: 97–106.

Owens, M.J. and Nemeroff, C.B. (1994) Role of serotonin in the pathophysiology of depression: focus on the serotonin transporter. Clin. Chem., 40: 288–295.

Oxenstierna, G., Edman, G., Iselius, L., Oreland, L., Ross, S.B. and Sedvall, G. (1986) Concentrations of monoamine metabolites in the cerebrospinal fluid of twins and unrelated individuals – a genetic study. J. Psychiatr. Res., 20: 19–29.

Parsey, R.V., Hastings, R.S., Oquendo, M.A., Huang, Y.Y., Simpson, N., Arcement, J., Huang, Y., Ogden, R.T., Van Heertum, R.L., Arango, V. and Mann, J.J. (2006a) Lower serotonin transporter binding potential in the human brain during major depressive episodes. Am. J. Psychiatry, 163: 52–58.

Parsey, R.V., Kent, J.M., Oquendo, M.A., Richards, M.C., Pratap, M., Cooper, T.B., Arango, V. and Mann, J.J. (2006b) Acute occupancy of brain serotonin transporter by sertraline as measured by [11C]DASB and positron emission tomography. Biol. Psychiatry, 59: 821–828.

Perani, D., Colombo, C., Bressi, S., Bonfanti, A., Grassi, F., Scarone, S., Bellodi, L., Smeraldi, E. and Fazio, F. (1995) [18F]FDG PET study in obsessive-compulsive disorder. A clinical/metabolic correlation study after treatment. Br. J. Psychiatry, 166: 244–250.

Pezawas, L., Meyer-Lindenberg, A., Drabant, E.M., Verchinski, B.A., Munoz, K.E., Kolachana, B.S., Egan, M.F., Mattay, V.S., Hariri, A.R. and Weinberger, D.R. (2005) 5-HTTLPR polymorphism impacts human cingulate–amygdala interactions: a genetic susceptibility mechanism for depression. Nat. Neurosci., 8: 828–834.

Philibert, R.A., Sandhu, H., Hollenbeck, N., Gunter, T., Adams, W. and Madan, A. (2008) The relationship of 5HTT (SLC6A4) methylation and genotype on mRNA expression and liability to major depression and alcohol dependence in subjects from the Iowa Adoption Studies. Am. J. Med. Genet. B Neuropsychiatr. Genet., 147B: 543–549.

Pirker, W., Asenbaum, S., Kasper, S., Walter, H., Angelberger, P., Koch, G., Pozzera, A., Deecke, L., Podreka, I. and Brucke, T. (1995) beta-CIT SPECT demonstrates blockade of 5HT-uptake sites by citalopram in the human brain *in vivo*. J. Neural Transm., 100: 247–256.

Pirker, W., Asenbaum, S., Hauk, M., Kandlhofer, S., Tauscher, J., Willeit, M., Neumeister, A., Praschak-Rieder, N., Angelberger, P. and Brucke, T. (2000) Imaging serotonin and dopamine transporters with 123I-beta-CIT SPECT: binding kinetics and effects of normal aging. J. Nucl. Med., 41: 36–44.

Pogarell, O., Hamann, C., Popperl, G., Juckel, G., Chouker, M., Zaudig, M., Riedel, M., Moller, H.J., Hegerl, U. and Tatsch, K. (2003) Elevated brain serotonin transporter availability in patients with obsessive-compulsive disorder. Biol. Psychiatry, 54: 1406–1413.

Praag, H.M. (1977) The Harold E. Himwich Memorial Lecture. Significance of biochemical parameters in the diagnosis, treatment, and prevention of depressive disorders. Biol. Psychiatry, 12: 101–131.

Praschak-Rieder, N., Wilson, A.A., Hussey, D., Carella, A., Wei, C., Ginovart, N., Schwarz, M.J., Zach, J., Houle, S. and Meyer, J.H. (2005) Effects of tryptophan depletion on the serotonin transporter in healthy humans. Biol. Psychiatry, 58: 825–830.

Praschak-Rieder, N., Kennedy, J., Wilson, A.A., Hussey, D., Boovariwala, A., Willeit, M., Ginovart, N., Tharmalingam, S., Masellis, M., Houle, S. and Meyer, J.H. (2007) Novel 5-HTTLPR allele associates with higher serotonin transporter binding in putamen: a [(11)C] DASB positron emission tomography study. Biol. Psychiatry, 62: 327–331.

Raleigh, M., McGuire, M. and Brammer, G. (1988) Behavioral and cognitive effects of altered tryptophan and tyrosine supply. In: Huether, G. (Ed.), Amino Acid Availability and Brain Function in Health and Disease. Springer, Berlin, pp. 299–308.

Raleigh, M.M. (1991) Bidirectional relationship between tryptophan and social behavior in vervet monkeys. In: Schwarcz, S.Y.R. and Brown, R. (Eds.), Kynurenine and Serotonin Pathways. Progress in Tryptophan Research. Plenum Press, New York, NY, pp. 289–298.

Reimold, M., Smolka, M.N., Schumann, G., Zimmer, A., Wrase, J., Mann, K., Hu, X.Z., Goldman, D., Reischl, G., Solbach, C., Machulla, H.J., Bares, R. and Heinz, A. (2007) Midbrain serotonin transporter binding potential measured with [11C]DASB is affected by serotonin transporter genotype. J. Neural. Transm., 114: 635–639.

Reimold, M., Batra, A., Knobel, A., Smolka, M.N., Zimmer, A., Mann, K., Solbach, C., Reischl, G., Schwarzler, F., Grunder, G., Machulla, H.J., Bares, R. and Heinz, A. (2008) Anxiety is associated with reduced central serotonin transporter availability in unmedicated patients with unipolar major depression: a [11C]DASB PET study. Mol. Psychiatry, 13: 606–613, 557

Rinne, J.O., Laihinen, A., Nagren, K., Ruottinen, H., Ruotsalainen, U. and Rinne, U.K. (1995) PET examination of the monoamine transporter with [11C]beta-CIT and [11C]beta-CFT in early Parkinson's disease. Synapse (NY), 21: 97–103.

Robbins, T.W. and Roberts, A.C. (2007) Differential regulation of fronto-executive function by the monoamines and acetylcholine. Cereb. Cortex., 17(Suppl 1): i151–i160.

Rodriguez, L.A., Wilson, J.R. and Nagoshi, C.T. (1993) Does psychomotor sensitivity to alcohol predict subsequent alcohol use? Alcohol. Clin. Exp. Res., 17: 155–161.

Sander, T., Harms, H., Dufeu, P., Kuhn, S., Hoehe, M., Lesch, K.P., Rommelspacher, H. and Schmidt, L.G. (1998) Serotonin transporter gene variants in alcohol-dependent subjects with dissocial personality disorder. Biol. Psychiatry, 43: 908–912.

Saudou, F., Amara, D.A., Dierich, A., LeMeur, M., Ramboz, S., Segu, L., Buhot, M.C. and Hen, R. (1994) Enhanced aggressive behavior in mice lacking 5-HT1B receptor. Science (NY), 265: 1875–1878.

Schuckit, M.A. and Smith, T.L. (1996) An 8-year follow-up of 450 sons of alcoholic and control subjects. Arch. Gen. Psychiatry, 53: 202–210.

Schuckit, M.A., Mazzanti, C., Smith, T.L., Ahmed, U., Radel, M., Iwata, N. and Goldman, D. (1999) Selective genotyping for the role of 5-HT2A, 5-HT2C, and GABA alpha 6 receptors and the serotonin transporter in the level of response to alcohol: a pilot study. Biol. Psychiatry, 45: 647–651.

Seibyl, J.P., Wallace, E., Smith, E.O., Stabin, M., Baldwin, R.M., Zoghbi, S., Zea-Ponce, Y., Gao, Y., Zhang, W.Y., Neumeyer, J.L. et al. (1994) Whole-body biodistribution, radiation absorbed dose and brain SPECT imaging with iodine-123-beta-CIT in healthy human subjects. J. Nucl. Med., 35: 764–770.

Silveri, M.M. and Spear, L.P. (2002) The effects of NMDA and GABAA pharmacological manipulations on ethanol sensitivity in immature and mature animals. Alcohol. Clin. Exp. Res., 26: 449–456.

Simpson, H.B., Lombardo, I., Slifstein, M., Huang, H.Y., Hwang, D.R., Abi-Dargham, A., Liebowitz, M.R. and Laruelle, M. (2003) Serotonin transporters in obsessive-compulsive disorder: a positron emission tomography study with [(11)C]McN 5652. Biol. Psychiatry, 54: 1414–1421.

Slotkin, T.A., McCook, E.C., Ritchie, J.C., Carroll, B.J. and Seidler, F.J. (1997) Serotonin transporter expression in rat brain regions and blood platelets: aging and glucocorticoid effects. Biol. Psychiatry, 41: 172–183.

Smith, G.S., Lotrich, F.E., Malhotra, A.K., Lee, A.T., Ma, Y., Kramer, E., Gregersen, P.K., Eidelberg, D. and Pollock, B.G. (2004) Effects of serotonin transporter promoter polymorphisms on serotonin function. Neuropsychopharmacology, 29: 2226–2234.

Smolka, M.N., Buhler, M., Schumann, G., Klein, S., Hu, X.Z., Moayer, M., Zimmer, A., Wrase, J., Flor, H., Mann, K., Braus, D.F., Goldman, D. and Heinz, A. (2007) Gene–gene effects on central processing of aversive stimuli. Mol. Psychiatry, 12: 307–317.

Spillmann, M.K., Van der Does, A.J., Rankin, M.A., Vuolo, R.D., Alpert, J.E., Nierenberg, A.A., Rosenbaum, J.F., Hayden, D., Schoenfeld, D. and Fava, M. (2001) Tryptophan depletion in SSRI-recovered depressed outpatients. Psychopharmacology, 155: 123–127.

Staley, J.K., Krishnan-Sarin, S., Zoghbi, S., Tamagnan, G., Fujita, M., Seibyl, J.P., Maciejewski, P.K., O'Malley, S. and Innis, R.B. (2001) Sex differences in [123I]beta-CIT SPECT measures of dopamine and serotonin transporter availability in healthy smokers and nonsmokers. Synapse (NY), 41: 275–284.

Staley, J.K., Sanacora, G., Tamagnan, G., Maciejewski, P.K., Malison, R.T., Berman, R.M., Vythilingam, M., Kugaya, A., Baldwin, R.M., Seibyl, J.P., Charney, D. and Innis, R.B. (2006) Sex differences in diencephalon serotonin transporter availability in major depression. Biol. Psychiatry, 59: 40–47.

Swedo, S.E. and Leonard, H.L. (1994) Childhood movement disorders and obsessive compulsive disorder. J. Clin. Psychiatry, 55(Suppl.): 32–37.

Talbot, P.S., Frankle, W.G., Hwang, D.R., Huang, Y., Suckow, R.F., Slifstein, M., Abi-Dargham, A. and Laruelle, M. (2005) Effects of reduced endogenous 5-HT on the *in vivo* binding of the serotonin transporter radioligand 11C-DASB in healthy humans. Synapse (NY), 55: 164–175.

Thoren, P., Asberg, M., Bertilsson, B., Mellstrom, B., Sjoqvist, F. and Traskman, L. (1980) Clomipramine treatment of obsessive-compulsive disorder: II. Biochemical aspects. Arch. Gen. Psychiatry, 37: 1289–1294.

Traskman-Bendz, L., Asberg, M., Bertilsson, L. and Thoren, P. (1984) CSF monoamine metabolites of depressed patients during illness and after recovery. Acta Psychiatrica Scandinavica, 69: 333–342.

Tsai, G., Gastfriend, D.R. and Coyle, J.T. (1995) The glutamatergic basis of human alcoholism. Am. J. Psychiatry, 152: 332–340.

van der Wee, N.J., Stevens, H., Hardeman, J.A., Mandl, R.C., Denys, D.A., van Megen, H.J., Kahn, R.S. and Westenberg, H.M. (2004) Enhanced dopamine transporter density in psychotropic-naive patients with obsessive-compulsive disorder shown by [123I]{beta}-CIT SPECT. Am. J. Psychiatry, 161: 2201–2206.

van Dyck, C.H., Malison, R.T., Staley, J.K., Jacobsen, L.K., Seibyl, J.P., Laruelle, M., Baldwin, R.M., Innis, R.B. and Gelernter, J. (2004) Central serotonin transporter availability measured with [123I]beta-CIT SPECT in relation to serotonin transporter genotype. Am. J. Psychiatry, 161: 525–531.

van Dyck, C.H., Malison, R.T., Jacobsen, L.K., Seibyl, J.P., Staley, J.K., Laruelle, M., Baldwin, R.M., Innis, R.B. and Gelernter, J. (2005) Increased dopamine transporter availability associated with the 9-repeat allele of the SLC6A3 gene. J. Nucl. Med., 46: 745–751.

Volavka, J., Czobor, P., Goodwin, D.W., Gabrielli, W.F., Penick, E.C., Mednick, S.A., Jensen, P., Knop, J. and Schulsinger, F.

(1996) The electroencephalogram after alcohol administration in high-risk men and the development of alcohol use disorders 10 years later – preliminary findings. Arch. Gen. Psychiatry, 53: 258–263.
Walker, S.C., Mikheenko, Y.P., Argyle, L.D., Robbins, T.W. and Roberts, A.C. (2006) Selective prefrontal serotonin depletion impairs acquisition of a detour-reaching task. Eur. J. Neurosci., 23: 3119–3123.
Weaver, I.C., D'Alessio, A.C., Brown, S.E., Hellstrom, I.C., Dymov, S., Sharma, S., Szyf, M. and Meaney, M.J. (2007) The transcription factor nerve growth factor-inducible protein a mediates epigenetic programming: altering epigenetic marks by immediate-early genes. J. Neurosci., 27: 1756–1768.
Westergaard, G.C., Suomi, S.J., Chavanne, T.J., Houser, L., Hurley, A., Cleveland, A., Snoy, P.J. and Higley, J.D. (2003) Physiological correlates of aggression and impulsivity in free-ranging female primates. Neuropsychopharmacology, 28: 1045–1055.
Willeit, M., Praschak-Rieder, N., Neumeister, A., Pirker, W., Asenbaum, S., Vitouch, O., Tauscher, J., Hilger, E., Stastny, J., Brucke, T. and Kasper, S. (2000) [123I]-beta-CIT SPECT imaging shows reduced brain serotonin transporter availability in drug-free depressed patients with seasonal affective disorder. Biol. Psychiatry, 47: 482–489.
Wolf, S.S., Jones, D.W., Knable, M.B., Gorey, J.G., Lee, K.S., Hyde, T.M., Coppola, R. and Weinberger, D.R. (1996) Tourette syndrome: prediction of phenotypic variation in monozygotic twins by caudate nucleus D2 receptor binding. Science (NY), 273: 1225–1227.
Young, L.T., Warsh, J.J., Kish, S.J., Shannak, K. and Hornykeiwicz, O. (1994) Reduced brain 5-HT and elevated NE turnover and metabolites in bipolar affective disorder. Biol. Psychiatry, 35: 121–127.
Zitterl, W., Aigner, M., Stompe, T., Zitterl-Eglseer, K., Gutierrez-Lobos, K., Schmidl-Mohl, B., Wenzel, T., Demal, U., Zettinig, G., Hornik, K. and Thau, K. (2007) [123I]-beta-CIT SPECT imaging shows reduced thalamus-hypothalamus serotonin transporter availability in 24 drug-free obsessive-compulsive checkers. Neuropsychopharmacology, 32: 1661–1668.
Zohar, J. and Insel, T.R. (1987) Obsessive-compulsive dirorder: psychobiological approaches to diagnosis, treatment and pathophysiology. Biol. Psychiatry, 22: 667–687.

CHAPTER 4.13

The Behavioral Genetics of Serotonin: Relevance to Anxiety and Depression

Olivia F. O'Leary[1,2] and John F. Cryan[*,1,2,3]

[1]*School of Pharmacy, University College Cork, Ireland*
[2]*Department of Pharmacology and Therapeutics, University College Cork, Ireland*
[3]*Alimentary Pharmabiotic Centre, University College Cork, Ireland*

Abstract: 5-hydroxytryptamine (5-HT, serotonin) is a major neurotransmitter involved in the modulation of behavior, the manifestation of various psychiatric disorders, and is a pharmacological target in the treatment of depression and anxiety disorders. The physiological effects of serotonin are modulated by a variety of proteins that regulate its synthesis, storage, release, uptake and degradation. In addition, serotonin signaling is mediated by at least 14 distinct receptors. Alterations in the expression of genes that regulate the biological effects of serotonin in the brain could alter serotonergic signaling, and thus could ultimately alter behaviors where serotonin has been implicated. The purpose of this chapter is to describe the behavioral consequences of manipulation of genes that regulate serotonergic signaling in rodents and to review analogous genetic association studies in humans that relate to psychiatric disorders, with a particular focus on depression and anxiety disorders. Many of these studies provide supportive evidence of a role for the serotonergic system in various behavioral responses related to depression and anxiety, as well as other psychiatric disorders. In particular, dysfunction of tryptophan hydroxylase, the serotonin transporter (SERT) or 5-HT_{1A} receptors can induce behaviors in rodents that are associated with anxiety and depression, and can also alter behavioral responses to antidepressant treatments. Moreover, many of these findings are supported by human genetic association studies. Studies in rodents also suggest that interference with SERT or 5-HT_{1A} receptor function specifically during brain development can program anxiety levels and depression in adulthood. Finally, while some evidence suggests that other components of the serotonergic system might also play important roles in these disorders, such findings remain to be refined using genetically modified mice, selective pharmacological tools or human genetic association studies.

Keywords: 5-HT, antidepressant, genetically modified mouse, knockout, single-nucleotide polymorphism, tryptophan, SERT, TPH2.

Introduction

5-hydroxytryptamine (5-HT), or serotonin, is a major neurotransmitter that plays an important role in the modulation of behavior. Serotonin-producing neurons originating in the dorsal and median raphe nuclei of the midbrain (Dahlstrom and Fuxe, 1964) innervate most of the brain, including structures that play important roles in mood regulation as well as the response to stress, which is a precipitating factor of depression and some anxiety disorders (Azmitia and Segal, 1978; Vertes, 1991; Vertes *et al.*, 1999; Sapolsky, 2000; Pittenger and Duman, 2008).

The physiological effects of serotonin are modulated by a variety of proteins that regulate its synthesis, storage, release, uptake and degradation. In addition, serotonin signaling is mediated by at least 14 distinct receptors that are classified into seven groups, 5-HT_{1-7} (Figure 1) (Barnes and Sharp, 1999; Hoyer *et al.*, 2002). Considering the multitude of serotonergic receptor subtypes, it is no surprise that serotonin is involved in a diverse array of behavioral functions (Lucki, 1998). Furthermore, serotonin plays a significant role in a wide range of psychiatric disorders, including depression, anxiety disorders, post-traumatic stress disorder, schizophrenia and anorexia nervosa, as well as a range of impulse-related disorders, such as aggression, substance abuse, obsessive-compulsive disorder, gambling and attention-deficit disorder (Lucki, 1998; Gingrich and

*Corresponding author
E-mail: j.cryan@ucc.ie

Christian Müller & Barry Jacobs (Eds.)
Handbook of Behavioral Neurobiology of Serotonin
ISBN 978-0-12-374634-4

DOI: 10.1016/B978-0-12-374634-4.00041-1

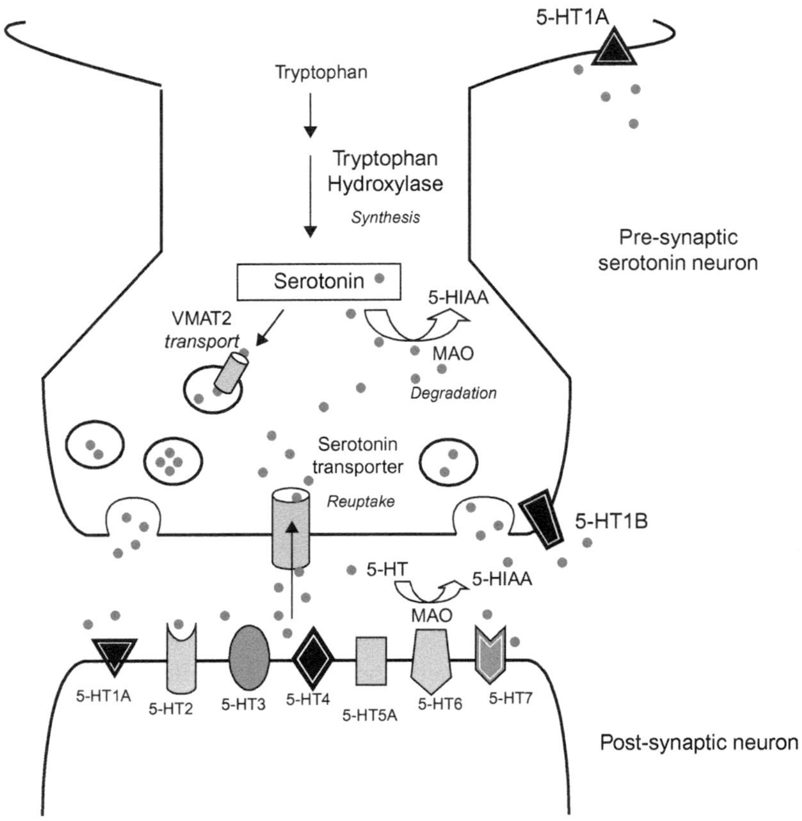

Figure 1 A model of the serotonergic synapse. The biosynthesis of serotonin involves hydroxylation of L-tryptophan by the enzyme tryptophan hydroxylase, to form L-5-hydroxytryptophan (L-5-HTP). L-5-HTP is then decarboxylated by aromatic L-amino acid decarboxylase to form 5-hydroxytryptamine (5-HT; serotonin). Serotonin is packaged into storage vesicles by the vesicular monoamine transporter 2 (VMAT2). Serotonin is removed from the synaptic cleft by the serotonin transporter (SERT). The metabolism of serotonin involves its conversion to 5-hydroxyindoleacetic acid (5-HIAA) by monoamine oxidase (MAO). A variety of presynaptic and postsynaptic serotonin receptors mediate its signaling (5-HT_{1-7}).

Hen, 2001). Moreover, serotonin is an important pharmacological target in the treatment of depression and some anxiety disorders (Delgado *et al.*, 1991; Heninger *et al.*, 1996; Vaswani *et al.*, 2003; Lucki and O'Leary, 2004).

Alterations in the expression of genes that regulate the biological effects of serotonin in the brain could alter serotonergic signaling, and thus might ultimately alter behaviors where 5-HT has been implicated. The purpose of this chapter is to describe the behavioral consequences of manipulation of genes that regulate serotonergic signaling in rodents, and to review analogous genetic association studies in humans that relate to psychiatric disorders with a particular focus on depression and anxiety disorders.

5-HT synthesis, storage, release, uptake and degradation

Alterations in the expression of genes that regulate synthesis, storage, release, uptake and degradation of 5-HT in the brain would ultimately result in changes in the synaptic availability of 5-HT. Such changes in gene expression could have important consequences for signaling at various 5-HT receptors, and thus might alter behaviors that are thought to be influenced by the serotonergic system. The following section, which is summarized in Table 1, describes the behavioral effects of genetic manipulation of these genes in mice as well as analogous genetic association studies in humans that relate to anxiety disorders and depression.

5-HT synthesis: tryptophan hydroxylase, Pet-1 and LIM homeobox transcription factor 1β

Preclinical studies Tryptophan hydroxylase (TPH) is the rate-limiting enzyme for the synthesis of serotonin from L-tryptophan (Grahame-Smith, 1964; Fitzpatrick, 1999; Ruddick *et al.*, 2006). Until recently, it was thought that only one gene (now called *Tph1*) encoded TPH in

Table 1 The effect of genetic mutations of proteins that regulate serotonin synthesis, storage, release, uptake and degradation on behaviors associated with depression and anxiety disorders

	Affective disorders and antidepressant response	Anxiety
5-HT SYNTHESIS		
Animal studies:		
Tph 1		
Tph 1 knockout		No change (Walther *et al.*, 2003)
NZW and SWR mice: Tph1 promoter and 3′UTR polymorphisms which result in low 5-HT during development but normal levels in adulthood	Increased immobility in TST and FST (Nakamura *et al.*, 2006) No response to the SSRI, paroxetine in the FST (Nakamura *et al.*, 2006)	
Tph 2		
R439H (equivalent of R441H in human Tph2)	Increased immobility in the TST (Beaulieu *et al.*, 2008).	Increased anxiety in dark–light emergence test (Beaulieu *et al.*, 2008)
C1473G	Mice homozygous for G allele do not respond to citalopram in FST (Cervo *et al.*, 2005) No association with behavioral response to citalopram in TST (Crowley *et al.*, 2005) Congenic C57BL/6 mice carrying either the C or G allele exhibit similar behavior in TST (Tenner *et al.*, 2008)	Congenic C57BL/6 mice carrying either the C or G allele exhibit similar behavior in the EPM (Tenner *et al.*, 2008)
Pet-1		
Pet-1 knockout		Increased anxiety in the elevated plus maze and open field (Hendricks *et al.*, 2003)
Human studies:		
Tph 1		
A218C	No association with major depression (Kunugi *et al.*, 1999a; Du *et al.*, 2001; Souery *et al.*, 2001; Serretti *et al.*, 2002) Moderates the influence of social support in depressive symptoms (Jokela *et al.*, 2007) AA genotype is associated with poor clinical response to SSRIs (Serretti *et al.*, 2001; Ham *et al.*, 2007) Generally no association with bipolar disorder (Bellivier *et al.*, 1998; Kunugi *et al.*, 1999a; McQuillin *et al.*, 1999; Reitschel *et al.*, 2000; Souery *et al.*, 2001; Serretti *et al.*, 2002) Association with the clinical response to the mood stabilizer lithium (Serretti *et al.*, 1999) No association with seasonal affective disorder (Johansson *et al.*, 2001) No association with premenstrual dysphoric disorder (Magnay *et al.*, 2006)	No association with panic disorder (Fehr *et al.*, 2001; Kim *et al.*, 2006; Maron *et al.*, 2005) No association with generalized anxiety disorder (Fehr *et al.*, 2001; You *et al.*, 2005)
T1095C		No association with obsessive compulsive disorder (Han *et al.*, 1999)
G-6526A	No association with premenstrual dysphoric disorder (Magnay *et al.*, 2006)	
G-5806T	No association with premenstrual dysphoric disorder (Magnay *et al.*, 2006)	

(*Continued*)

Table 1 *(Continued)*

	Affective disorders and antidepressant response	Anxiety
rs1799913	A allele associated with major depression (Gizatullin *et al.*, 2006)	
Tph 2		
G1463A (R441H)	Associated with severe refractory depression in elderly patients (Zhang *et al.*, 2005a) No association with major depression (Glatt *et al.*, 2005; Van Den Bogart *et al.*, 2005; Zhou *et al.*, 2005a; Delorme *et al.*, 2006) No association with bipolar disorder (Zhang *et al.*, 2005a)	No association with panic disorder (Henningsson *et al.*, 2007) No association with social phobia (Henningsson *et al.*, 2007)
rs10748185	No association with unipolar depression (Gizatullin *et al.*, 2008)	
rs2129575	No association with unipolar depression (Gizatullin *et al.*, 2008)	
rs1386495	No association with unipolar depression (Gizatullin *et al.*, 2008)	
rs1386494 A/G	No association with unipolar depression (Gizatullin *et al.*, 2008)	No association with panic disorder (Maron *et al.*, 2007).
rs7305115	No association with unipolar depression (Gizatullin *et al.*, 2008)	
rs4131347 (-C8347G)	No association with unipolar depression (Mann *et al.*, 2008) No association with bipolar disorder (Mann *et al.*, 2008)	
rs17110563 (Pro206Ser)	Associated with bipolar disorder (Cichon *et al.*, 2008)	
rs4290270	Associated with bipolar disorder (Harvey *et al.*, 2007)	
rs1386483 C/T		No association with panic disorder (Maron *et al.*, 2007)
Various haplotyes	Association with bipolar disorder (Van den Bogaert *et al.*, 2006; Lin *et al.*, 2007) Association with unipolar depression (Van den Bogaert *et al.*, 2006)	Association with early onset OCD (Mossner *et al.*, 2006)
5-HT STORAGE AND RELEASE		
Animal studies:		
VMAT2		
VMAT $2^{+/-}$ mice	Increased immobility in the FST and TST (Fukui *et al.*, 2007) Anhedonia (decreased preference for sweet solutions) (Fukui *et al.*, 2007) Increased learned helplessness (Fukui *et al.*, 2007)	No change in open field, light–dark exploration, novelty-suppressed feeding and zero maze (Fukui *et al.*, 2007)
Human studies:		
VMAT2		
T440G	No association with bipolar disorder (Gutierrez *et al.*, 2007)	
C1368T	No association with bipolar disorder (Gutierrez *et al.*, 2007)	
T2666C	No association with bipolar disorder (Gutierrez *et al.*, 2007)	
A2683C	No association with bipolar disorder (Gutierrez *et al.*, 2007)	

Table 1 *(Continued)*

	Affective disorders and antidepressant response	**Anxiety**
A745G	No association with bipolar disorder (Gutierrez *et al.*, 2007)	
Haplotype: T allele of rs363399 with C allele of rs4752045	Reduced risk of depression (Christiansen *et al.*, 2007)	
Haplotypes: 2666T-2683A-745G (TAG) and 2666C-2683C-745A (CCA)	Higher frequency in bipolar disorder (Gutierrez *et al.*, 2007)	
Haplotypes: 666T-2683C-745A (TCA) and 2666C-2683A-745G (CAG)	Less common in bipolar disorder patients than controls (Gutierrez *et al.*, 2007)	
5-HT REUPTAKE		
Animal studies:		
SERT		
SERT knockout mice on 129S6 background	Antidepressant-like reduction in immobility in the TST (Holmes *et al.*, 2002a; Lira *et al.*, 2003)	No change in anxiety in the open field or in the elevated plus maze (Lira *et al.*, 2003)
	Increased immobility in the FST (Holmes *et al.*, 2002a; Lira *et al.*, 2003)	
	Increased latency to feed in novelty-suppressed feeding test and to escape from shock environment (Lira *et al.*, 2003)	
SERT knockout mice on C57BL/6 background	No change in immobility in the TST or FST upon first exposure (Holmes *et al.*, 2002a)	Increased anxiety in the open field, elevated plus maze and light–dark test (Holmes *et al.*, 2003; Zhao *et al.*, 2006b; Caroll *et al.*, 2007; Kalueff *et al.*, 2007a, 2007b)
	Increased immobility in the TST and FST upon repeated testing (Zhao *et al.*, 2006b; Carroll *et al.*, 2007; Wellman *et al.*, 2007)	
	No change in anhedonia as measured in the sucrose preference test (Kalueff *et al.*, 2006)	
	No response to the SSRI, fluoxetine in the TST (Holmes *et al.*, 2002a)	
Mutation at C-terminus of SERT in mice on C57BL/6 background	Increased immobility in the TST upon first exposure (Zhao *et al.*, 2006b)	
SERT-targeting siRNA in adult mouse brain (2 weeks)	Decreased immobility in the FST (Thakker *et al.*, 2005)	
SERT blockade during early postnatal development and behavioral testing in adulthood	Increased latency to feed in a novel environment (Ansorge *et al.*, 2008)	Increased latency to feed in a novel environment (Ansorge *et al.*, 2008)
	Increased latency to drink a sweet solution in a novel environment (Ansorge *et al.*, 2004, 2008)	Increased latency to drink a sweet solution in a novel environment (Ansorge *et al.*, 2004, 2008)
	Increased latency to escape from an environment associated with electric shock (Ansorge *et al.*, 2004, 2008)	
SERT over-expressing mice		Reduced anxiety in elevated plus maze and hyponeophagia (Jennings *et al.*, 2006)
SERT knockout rat	Increased immobility in the FST (Olivier *et al.*, 2008)	Higher latency to feed in a novel environment (Olivier *et al.*, 2008)
	Anhedonia as demonstrated by reduced preference for a sweet solution (Olivier *et al.*, 2008)	Higher latency to escape from home cage (Olivier *et al.*, 2008)
	Higher latency to feed in a novel environment (Olivier *et al.*, 2008)	Increased anxiety in the open field and elevated plus maze (Olivier *et al.*, 2008)
	Higher latency to escape from home cage (Olivier *et al.*, 2008)	

(Continued)

Table 1 *(Continued)*

	Affective disorders and antidepressant response	Anxiety
Human studies:		
SERT		
Insertion/deletion polymorphism in the promoter of the SERT gene	Short variant is a risk factor for major depression (Collier *et al.*, 1996; Gutierrez *et al.*, 1998a; Dorado *et al.*, 2007)	Short variant associated with obsessive compulsive disorder in females (Denys *et al.*, 2006)
	No association with depression (Kunugi *et al.*, 1997; Mendlewicz *et al.*, 2004; Lasky-Su *et al.*, 2005; Bozina *et al.*, 2006; Christiansen *et al.*, 2007; Zaboli *et al.*, 2008)	Short variant associated with obsessive compulsive disorder in females (Grados *et al.*, 2007)
	Short variant is a risk factor for major depression in response to stressful life events (Caspi *et al.*, 2003; Kendler *et al.*, 2005; Brummet *et al.*, 2008)	Short variant associated with anxiety-related personality traits (Lesch *et al.*, 1996; Melke *et al.*, 2001)
	Short variant confers poor clinical response to SSRI antidepressant drugs (Smeraldi *et al.*, 1998; Rausch *et al.*, 2002; Yu *et al.*, 2002; Arias *et al.*, 2005; Perna *et al.*, 2005; Bozina *et al.*, 2007)	Short variant associated with generalized anxiety disorder (You *et al.*, 2005)
	Short variant confers poor clinical response to light therapy with sleep deprivation (Benedetti *et al.*, 2003)	No association with obsessive compulsive disorder (Billett *et al.*, 1997; Camarena *et al.*, 2001; Meira-Lima *et al.*, 2004)
	No association with bipolar disorder (Kunugi *et al.*, 1997; Ewald *et al.*, 1998; Gutierrez *et al.*, 1998b; Mendlewicz *et al.*, 2004; Meira-Lima *et al.*, 2005)	No association with panic disorder (Hamilton *et al.*, 1999; Samochowiec *et al.*, 2004)
	Short allele associated with bipolar disorder (Collier *et al.*, 1996; Mynett-Johnson *et al.*, 2000; Rotondo *et al.*, 2002; Anguelova *et al.*, 2003; Cho *et al.*, 2005; Lasky-Su *et al.*, 2005)	No association with anxiety-related personality traits (Lang *et al.*, 2004; Schinka *et al.*, 2004)
		No association with generalized anxiety disorder (Samochowiec *et al.*, 2004)
		No association with social phobia (Samochowiec *et al.*, 2004)
Variable number of tandem repeats (VNTR) polymorphism in intron 2 of the SERT gene	No association with unipolar depression (Hoehe *et al.*, 1998; Kunugi *et al.*, 1997; Lasky-Su., 2005)	12 allele was associated with obsessive compulsive disorder (Baca-Garcia *et al.*, 2007)
	Association with unipolar depression (Bozina *et al.*, 2006; Liu *et al.*, 1999)	No association with generalized anxiety disorder (You *et al.*, 2005)
	Low number of repeats associated with a better response to the SSRI paroxetine (Bozina *et al.*, 2007)	
	No association with bipolar disorder (Esterling *et al.*, 1998; Hoehe *et al.*, 1998; Bochchetta *et al.*, 1999; Liu *et al.*, 1999; Saleem *et al.*, 2000; Dimitrova *et al.*, 2002; Lasky-Su *et al.*, 2005)	
A to G at 1273 (ile425Val in the 9th exon (Gain of function mutation))		Associated with obsessive compulsive disorder (Ozaki *et al.*, 2003)
5-HT METABOLISM		
Animal studies:		
MAOA		
MAOA (Tg8) (transgenic insertional mutation which inhibits enzyme function)	Reduced basal immobility in the TST (Cases *et al.*, 1995)	Enhanced cued and contextual fear conditioning (Kim *et al.*, 1997)
		Decreased extinction of a conditioned passive avoidance response (Dubrovina *et al.*, 2006)
		Reduced nose-poking in the hole board test and reduced exploration in light–dark test (Popova *et al.*, 2000)

Table 1 *(Continued)*

	Affective disorders and antidepressant response	Anxiety
		Reduced social investigation (Vishnivetskaya *et al.*, 2007)
		Decreased anxiety in the open field (Cases *et al.*, 1995)
		No change in time spent in light side of light–dark box (Popova *et al.*, 2000)
MAOA (A863T) mice (Spontaneous mutation)	No change in immobility in the TST (Scott *et al.*, 2008)	Decreased activity in the open field (Scott *et al.*, 2008)
Human studies:		
MAOA		
C-A repeat in intron 2	No association with bipolar disorder (Craddock *et al.*, 1995; Muramatsu *et al.*, 1997; Furlong *et al.*, 1999)	
	No association with unipolar depression (Furlong *et al.*, 1999)	
	No association with unipolar depression (Muramatsu *et al.*, 1997)	
VNTR in intron 1	No association with bipolar disorder (Craddock *et al.*, 1995; Muramatsu *et al.*, 1997)	
Fnu4HI RFLP G/T silent polymorphism at position 941	No association with bipolar disorder (Craddock *et al.*, 1995; Muramatsu *et al.*, 1997; Sasaki *et al.*, 1998; Furlong *et al.*, 1999)	T allele is over-represented in generalized anxiety disorder (Tadic *et al.*, 2003)
	No association with unipolar depression (Furlong *et al.*, 1999)	No association with panic disorder (Tadic *et al.*, 2003)
	No association with unipolar depression (Muramatsu *et al.*, 1997)	Association with OCD (Karayiorgou *et al.*, 1999)
	No association with unipolar depression (Sasaki *et al.*, 1998)	
	No association with unipolar depression (Tadic *et al.*, 2003)	
VNTR in promoter	No association with bipolar disorder (Furlong *et al.*, 1999; Kunugi *et al.*, 1999b)	Higher activity allele is associated with Panic disorder (Deckert *et al.*, 1999)
	Less active VNTR associated with unipolar depression (Brummett *et al.*, 2007)	
	High-activity MAOA gene promoter alleles associated with major depressive disorder in females (Schulze *et al.*, 2000)	
	Increased frequency of 4-repeat (4R) allele in MDD patients, especially in the female population (Yu *et al.*, 2005)	
	MDD female patients who were 3R homozygotes had a better response to fluoxetine treatment when compared to 4R carriers (Yu *et al.*, 2006)	
	No association with unipolar depression (Furlong *et al.*, 1999)	
	No association with unipolar depression (Gutierrez *et al.*, 2004)	
	No association with unipolar depression (Kunugi *et al.*, 1999b)	
	No association with unipolar depression (Muramatsu *et al.*, 1997)	
30-bp repeat in the MAOA gene promoter (uMAOA)	No association with bipolar disorder (Gutierrez *et al.*, 2004)	
	No association with depression (Syagailo *et al.*, 2001)	

vertebrates. However, in 2003 Walther and colleagues reported that two genes, *Tph1* and *Tph2*, which are found on different chromosomes, encode TPH in the mouse. Specifically, they reported that *Tph1* encoded Tph predominantly in the periphery and the pineal gland, while *Tph2* was expressed exclusively in the brain (Walther *et al.*, 2003). This discovery arose following the finding that despite a dramatic reduction of serotonin in the periphery, midbrain serotonin levels were largely intact in *Tph* knockout mice (Walther *et al.*, 2003). These *Tph* (*Tph1*) knockout mice also exhibited anxiety levels similar to those of the wild-type controls (Walther *et al.*, 2003). This breakthrough shifted attention away from the *Tph1* gene to characterizing the role of the *Tph2* gene in behavior and brain disorders.

Recently, a knockin mouse was generated which expresses a rare human variant of *Tph2* (R441H) that is found in some individuals with unipolar depression (Beaulieu *et al.*, 2008). Mice heterozygous or homozygous for this knockin mutation exhibit between 40 and 80 percent reductions in forebrain 5-HT content, increased immobility in the tail suspension test (TST), and a higher latency to enter the bright compartment in the dark–light emergence test, thus suggesting that the mice have depression- and anxiety-like phenotypes, respectively (Beaulieu *et al.*, 2008). Mice homozygous for this mutation also exhibited increased aggression, while neither genotype differed from wild-type controls in a social interaction test or in basal locomotor activity (Beaulieu *et al.*, 2008). Moreover, it was reported that the R441H polymorphism occurred at a higher frequency in a cohort of patients with unipolar major depression compared to control subjects (Zhang *et al.*, 2005a, but see also Glatt *et al.*, 2005; Van Den Bogaert *et al.*, 2005; Zhou *et al.*, 2005a; Delorme *et al.*, 2006). Prior to the generation of this knockin mouse, a single nucleotide polymorphism which replaces C with G at the 1473 position of *Tph2* and thus reduces 5-HT synthesis *in vitro* was reported (Zhang *et al.*, 2004). Mouse strains homozygous for the 1473G allele, namely the BALB/cJ and DBA/2 mouse strains, had reduced serotonin synthesis and tissue content compared to mouse strains homozygous for the 1473C allele, including the 129×1/SvJ and C57BL/6 mouse strains. This finding generated a lot of interest because of reports that BALB/cJ mice exhibited higher levels of anxiety-like behavior compared with other mouse strains (Griebel *et al.*, 2000; Belzung and Griebel, 2001). Following up on these findings, the role of the C1473G *Tph2* polymorphism in the behavioral response to antidepressants was investigated. In contrast to mouse strains homozygous for the C allele, both BALB/cJ and DBA/2 mice (which are homozygous for the G allele) failed to respond to the selective serotonin reuptake inhibitor (SSRI), citalopram, in the forced swimming test (FST; Cervo *et al.*, 2005). However, this effect appears to be dependent upon the behavioral test employed, because no correlation between this polymorphism and the behavioral response to citalopram was observed in the TST (Crowley *et al.*, 2005). Finally, a role for the C1473G polymorphism has also been described in aggression, with mouse strains homozygous for the G allele being more aggressive than those homozygous for the C allele (Kulikov *et al.*, 2005). Such correlations between the C1473G polymorphism and various behaviors should be interpreted with caution, however, because the reported strain-dependent behaviors could also be a result of other variations in the genetic background. Indeed, it was just recently reported that congenic C57BL/6 mice carrying either the C1473 or G1473 polymorphism have the same levels of TPH activity and brain 5-HT tissue content, as well as similar behaviors in tests of depression- and anxiety-like behavior, thus suggesting that distinct behavioral characteristics between mouse strains may be due to other variations in the genetic background rather than the C1473G *Tph2* polymorphism alone (Tenner *et al.*, 2008).

Despite initial reports that *Tph1* encoded TPH predominantly in the periphery while *Tph2* was expressed exclusively in the brain, emerging reports suggest that *Tph1* may still play an important role in behavior and the brain. *Tph1* mRNA expression has been reported to be equal to that of *Tph2* mRNA expression in many regions of the human brain (Zill *et al.*, 2007). Moreover, while *Tph2* mRNA expression is four-fold higher than *Tph1* mRNA expression in the raphe nuclei, *Tph1* mRNA levels are higher than those of *Tph2* in the human hypothalamus and amygdala, brain regions which are important in the stress response (Zill *et al.*, 2007). While the extent of *Tph1* mRNA translation in the human brain has yet to be confirmed, various human genetic association studies have already established an association of *Tph1* polymorphisms with various psychiatric conditions as well as the response to antidepressant drugs (Serretti *et al.*, 2001; Gizatullin *et al.*, 2006; Ham *et al.*, 2007). Recent studies also suggest that *Tph1* may be important during brain development. *Tph1* is preferentially expressed during mouse brain development, and the resulting protein exhibits higher enzymatic activity than that encoded by *Tph2* when tested in a model of the developing brainstem (Nakamura *et al.*, 2006). New Zealand white and SWR mice, which have common polymorphisms in both the promoter and 3′ untranslated region (UTR) of *Tph1*, exhibit reduced 5-HT levels in the raphe nuclei during development but normal levels in adulthood (Nakamura *et al.*, 2006). Interestingly, both of these mouse strains exhibited depression-like behavior in adulthood, as demonstrated by increases in immobility in both the FST and the TST, and were insensitive to the behavioral effects of the SSRI antidepressant, paroxetine, in the FST (Nakamura *et al.*, 2006). Moreover, peripheral serotonin synthesis by the TPH encoded by *Tph1* in the mouse dam is a critical determinant of normal brain development in her offspring (Cote *et al.*, 2007). Both $Tph1^{+/-}$ and $Tph1^{-/-}$ embryos from $Tph1^{-/-}$ mothers had morphological brain abnormalities compared to embryos of the same genotype which were born to *Tph1* mothers (Cote *et al.*, 2007). Taken together, it appears that *Tph1* may also play an important role in the development of 5-HT neurons and could thus ultimately affect behavior in adulthood.

The behavioral genetics of 5-HT during development is also strongly influenced by specific transcription factors that are required to establish the enzymatic machinery that is necessary for the production and metabolism of 5-HT (Gaspar *et al.*, 2003). These transcription factors are expressed in postmitotic cells, and include the Lim homeobox transcription factor 1β (*Lmx1b*) (Ding *et al.*, 2003) and plasmacytoma expressed transcript 1 (Pet-1). Pet-1 has a unique expression pattern: it is strictly limited to the raphe nuclei, and appears 1 day before serotonergic neurons can be identified. This factor could directly activate the transcription of *Tph* in addition to other genes involved in 5-HT neurotransmission, such as aromatic amino acid decarboxylase (AADC), the 5-HT transporter (SERT) and the vesicular monoamine transporter (VMAT) (Hendricks *et al.*, 1999; Pfaar *et al.*, 2002; Gaspar *et al.*, 2003). The role of developmental serotonin synthesis in anxiety and aggression are further supported by a recent study of mice lacking Pet-1. Adult Pet-1 knockout mice exhibit a 70 percent loss of serotonergic neurons and an 80–85 percent reduction in 5-HT tissue content in multiple brain regions. Moreover, in the serotonergic neurons that do remain, dramatic reductions in tryptophan hydroxylase, SERT and VMAT are observed (Hendricks *et al.*, 2003). These mice exhibit increased aggression as well as increased anxiety in the open field and elevated plus maze (Hendricks *et al.*, 2003). A later study generated Lmx1b conditional knockout mice (Lmx1bf/f/p) in which Lmx1b was only deleted in Pet-1-expressing 5-HT neurons. While the initial generation of central 5-HT neurons appeared normal in these mice, 5-HT-specific markers including *Tph2* mRNA were absent at later stages of development, thus suggesting that Lmx1b maintains Pet-1 expression. These Lmx1bf/f/p mice survive to adulthood and exhibit normal locomotor activity, and thus may prove to be a useful animal model for examining the roles of central 5-HT in a variety of animal behaviors (Zhao *et al.*, 2006a). In support, it has already been reported that inactivation of Lmx1b in serotonergic neurons results in enhanced contextual fear conditioning, a behavior thought to be related to cognitive deficits associated with anxiety (Dai *et al.*, 2008).

Human studies Human genetic association studies of both *Tph1* and *Tph2* have been conducted, but the contribution of the *Tph1* gene to brain disorders has been investigated more extensively. Most human genetic associations of the *Tph1* gene have focused on the 218A/C polymorphism. While this polymorphism has not been associated with major depressive disorder (Kunugi *et al.*, 1999a; Du *et al.*, 2001; Souery *et al.*, 2001), other polymorphisms (Gizatullin *et al.*, 2006) and a microsatellite of the *Tph1* gene (Nash *et al.*, 2005) have been reported to be associated with this disorder. Furthermore, it appears that the 218A/C and 779A/C polymorphisms moderate the influence of social support on depressive symptoms (Jokela *et al.*, 2007). Moreover, the 218A/C polymorphism appears to play an important role in the response to antidepressant drugs, with the A/A genotype being associated with a poor clinical response to various SSRI antidepressant drugs (Serretti *et al.*, 2001; Ham *et al.*, 2007). The contribution of the 218A/C polymorphism to susceptibility to bipolar disorder has also been investigated, with the majority of studies reporting a lack of association (Bellivier *et al.*, 1998; Kunugi *et al.*, 1999a; McQuillin *et al.*, 1999; Rietschel *et al.*, 2000; Souery *et al.*, 2001; Serretti *et al.*, 2002) but interestingly an association with the clinical response to the mood-stabilizer lithium has been reported (Serretti *et al.*, 1999). Similarly, a lack of association of various other SNPs of the *Tph1* gene has also been reported in bipolar disorder (Lai *et al.*, 2005).

The contribution of the 218A/C polymorphism to risk for suicide has also been investigated. While some studies including two meta-analyses reported an over-representation of the A allele in suicide attempters or victims (Abbar *et al.*, 2001; Rujescu *et al.*, 2003; Bellivier *et al.*, 2004), many original investigations have reported a lack of association of suicide with this polymorphism (Kunugi *et al.*, 1999a; Bennett *et al.*, 2000; Ono *et al.*, 2000; Zalsman *et al.*, 2001; Lalovic and Turecki, 2002; Ohtani *et al.*, 2004; Stefulj *et al.*, 2005).

For the most part, studies have reported a lack of association of the 218A/C polymorphism and other *Tph1* polymorphisms with various anxiety disorders, including panic disorder (Fehr *et al.*, 2001; Maron *et al.*, 2005; Kim *et al.*, 2006), generalized anxiety disorder (Fehr *et al.*, 2001; You *et al.*, 2005) and obsessive-compulsive disorder (Han *et al.*, 1999). Similarly, a lack of association of various *Tph1* gene polymorphisms has been reported for many other disorders, including anorexia nervosa (Han *et al.*, 1999), seasonal affective disorder (Johansson *et al.*, 2001), premenstrual dysphoric disorder (Magnay *et al.*, 2006), bulimia (Monteleone *et al.*, 2007), autism (Ramoz *et al.*, 2006) and narcolepsy (Fehr *et al.*, 2001).

Given the recent discovery of the *Tph2* gene, only a limited number of genetic association studies of this gene have been reported. In 2005, a loss-of function SNP (G1463A) in the *Tph2* gene which replaces the arginine441 with histidine was identified and reported to occur more frequently in subjects with unipolar major depression compared to controls (Zhang *et al.*, 2005a), a finding further supported by the depression-like phenotype of knockin mice expressing this mutation (Beaulieu *et al.*, 2008). However, the validity and significance of this study has been cast in doubt by the findings of other groups, who failed to replicate it in much larger cohorts (Glatt *et al.*, 2005; Van Den Bogaert *et al.*, 2005; Zhou *et al.*, 2005a; Delorme *et al.*, 2006). This discrepancy was explained by the original authors as being due to differences in the study groups. All patients involved in the initial study were over 60 years of age, mostly severely ill with many refractory to treatment, while these criteria did not apply to the patients involved in subsequent investigations (Zhang *et al.*, 2005b). This suggests that the G1463A polymorphism in humans may be linked to a very small subset of elderly, severely ill and refractory depressed patients (Delorme *et al.*, 2006; Henningsson *et al.*, 2007). Other studies have also reported a lack of association between unipolar depression and various other *Tph2* polymorphisms (Gizatullin *et al.*, 2008; Mann *et al.*, 2008). However, many other studies do support a possible *Tph2* involvement in depression (and depression-related suicide) through various SNPs in the *Tph2* gene and its upstream promoter region (Zhou *et al.*, 2005b; Ke *et al.*, 2006; Van Den Bogaert *et al.*, 2006; Lopez de Lara *et al.*, 2007). These studies, in addition to findings in animal models, suggest that *Tph2* may be one of many candidate genes that play a contributory role in the development of depression.

Multiple studies have reported that many *Tph2* polymorphisms and haplotypes are associated with bipolar disorder (Van Den Bogaert *et al.*, 2006; Harvey *et al.*, 2007; Lin *et al.*, 2007; Lopez *et al.*, 2007; Cichon *et al.*, 2008), but many other polymorphisms are not associated with this disorder (Zhang *et al.*, 2005a; Mann *et al.*, 2008). Some polymorphisms of the *Tph2* gene have also been associated with anxiety disorders such as obsessive-compulsive disorder (Mossner *et al.*, 2006), while a lack of association of other polymorphisms with panic disorder and social phobia has been reported (Henningsson *et al.*, 2007; Maron *et al.*, 2007). Finally, various *Tph2* polymorphisms have also been associated with aggression (Hennig *et al.*, 2005).

Taken together, it appears that in both human and preclinical rodent studies the *Tph2* gene plays a much larger role in behavior and brain disorders than *Tph1*. Although, animal studies suggest that *Tph1* contributes to brain development and sometimes to behavior, human genetic association studies have for the most part reported a lack of association with many disorders. Nevertheless, a role for *Tph1* in brain and behavior cannot be entirely ruled out because the 218A/C polymorphism is important in the clinical response to SSRI antidepressant drugs. Although the number of human and animal studies investigating the role of *Tph2* in behavior and brain disorders is limited, current evidence suggests that this gene may play a role in unipolar and bipolar depression as well as anxiety disorders.

5-HT storage and release: vesicular monoamine transporter 2 (VMAT 2)

Following 5-HT synthesis, an integral membrane protein of secretory vesicles called the vesicular monoamine transporter 2 (VMAT2) removes serotonin from the cytoplasm and packages it into secretory vesicles which later transport serotonin to the presynaptic terminal for release (Schuldiner *et al.*, 1995). However, VMAT2 is also responsible for the vesicular packaging of other monoamine neurotransmitters as well as serotonin, thus making it difficult to ascertain precisely whether the effects of genetic alterations of VMAT2 on behavior would be specifically due to the loss of vesicular stores of serotonin. Nevertheless, the phenotype of VMAT2 knockout mice and genetic association studies in humans will be described in this chapter.

VMAT2$^{-/-}$ mice are unable to package monoamines into synaptic vesicles, and as a result all monoamines are rapidly degraded, the consequence of which is severe depletion of 5-HT, dopamine and noradrenaline. These mice are hypoactive, prone to hypothermia, exhibit severe growth retardation, and most die within a few days after birth (Wang *et al.*, 1997). However, heterozygous *VMAT2* mice are viable, and show changes in depression-like behavior in adulthood. Specifically, these mutant mice exhibit a modest increase in immobility in both the TST and FST, and this immobility is reversed by antidepressant treatment (Fukui *et al.*, 2007). *VMAT2*$^{+/-}$ mice are anhedonic, as indicated by a loss of sucrose preference, and also demonstrate increased learned helplessness and increased stress-induced release of corticosterone, thus further supporting a depression-like phenotype (Fukui *et al.*, 2007). In tests of anxiety including the open field, light–dark exploration, novelty-suppressed feeding and zero maze, mutant mice behaved similarly to their wild-type counterparts. Taken together, these findings suggest that *VMAT2*$^{+/-}$ mice demonstrate a depressive-like phenotype which is devoid of anxiety-like behaviors.

Few human genetic association studies on *VMAT2* have been published. However, several haplotypes of *VMAT2* polymorphisms have been associated with depression

(Christiansen *et al.*, 2007) and bipolar disorder (Gutierrez *et al.*, 2007), and some of the behaviors related to these disorders have been reported to be altered in *VMAT2* mice as described above. Further characterization of *VMAT2* in human psychiatric genetic studies and animal behavioral studies is eagerly awaited.

5-HT reuptake from the synaptic cleft: the serotonin transporter (SERT)

The serotonin transporter (SERT) is the principal mechanism for the removal of serotonin from the synaptic cleft (Murphy *et al.*, 2004; Blakely *et al.*, 2005). Studies on rodents either over-expressing or lacking functional serotonin transporters have largely focused on behaviors associated with anxiety and depression. This is due in large part to the clinical use of drugs that selectively block the serotonin transporter (SSRIs) in the treatment of depression and some anxiety disorders.

Preclinical studies It is now very clear that the genetic background strain of a rodent can markedly influence behavior (Crawley *et al.*, 1997; Jacobson and Cryan, 2007). The SERT knockout mouse has been bred on two different inbred background strains, and this has had some important consequences in defining some of the phenotypes of SERT knockout mice. SERT knockout mice bred onto a 129S6 background exhibit an antidepressant-like reduction in baseline immobility when tested in the TST, but demonstrate increased immobility in the FST (Holmes *et al.*, 2002a; Lira *et al.*, 2003). The increase in FST-induced immobility in SERT knockout mice of the 129S6 background has been suggested to be a result of compromised neuromuscular strength, which likely interferes with their capacity to swim (Holmes *et al.*, 2002a). To further clarify the phenotype of these animals in tests of antidepressant-like activity, these mice have also been tested in the novelty-suppressed feeding test and a shock escape paradigm. In these tests, the SERT knockout mice exhibited an increased latency to feed in a novel environment and increased latency to escape from the shock environment, thus supporting a depression-like phenotype for these animals (Lira *et al.*, 2003).

In contrast to the SERT knockout mice bred on a 129S6 background, SERT knockout mice on a C57BL/6J background do not exhibit changes in baseline immobility upon first exposure to the TST or FST (Holmes *et al.*, 2002a), and do not exhibit changes in a measure of anhedonia, the sucrose preference test (Kalueff *et al.*, 2006). However, they do exhibit higher baseline immobility values in the TST and FST upon repeated testing (Zhao *et al.*, 2006b; Carroll *et al.*, 2007; Wellman *et al.*, 2007), and when the mutation is restricted to the C-terminus of SERT these mice on a C57BL/6 background exhibit higher baseline immobility upon first exposure to the TST (Zhao *et al.*, 2006b). Furthermore, when tested in the TST, SERT knockout mice on a C57BL/6 background fail to respond to the acute effects of the SSRI antidepressant drug fluoxetine but not the norepinephrine reuptake inhibitor desipramine or the mixed serotonin/norepinephrine reuptake inhibitor imipramine (Holmes *et al.*, 2002a), thus confirming that SERT is a primary target for the acute behavioral effects of fluoxetine. Although the depression-like phenotypes of the SERT knockout mice on a C57BL/6 background are somewhat subtle compared to those on a 129S6 background, recent studies using a SERT knockout rat suggest that constitutive knockout of SERT does indeed result in a depressive-like phenotype. The SERT knockout rat exhibits increased immobility in the FST, anhedonia as demonstrated by reduced preference for a sweet sucrose solution, a higher latency to feed in a novel environment, and a higher latency to escape from their home cage (Olivier *et al.*, 2008).

Overall, data from experiments with SERT knockout rodents suggest that constitutive knockout of the SERT gene results in depression-like behavior. This finding was initially surprising, because chronic treatment with SSRI antidepressants downregulates SERT in both humans and animals and therefore deletion of SERT would be expected to induce an antidepressant-like phenotype. Recently, a number of investigators tackled this conundrum by suggesting that life-long blockade of SERT, particularly during early brain development, could have different consequences to those of prolonged periods of pharmacological SERT blockade during adulthood. Indeed, down-regulation of SERT by intracerebroventricular infusion of SERT-targeting siRNA into the adult mouse brain for 2 weeks decreases immobility in the FST in a manner similar to chronic SSRI treatment, and thus has an antidepressant-like effect (Thakker *et al.*, 2005). In contrast, when blockade of the serotonin transporter (by treatment with various SSRIs) is restricted to early development (postnatal days 4–21), depression-like behaviors, including increased latency to feed in a novel environment (novelty suppressed feeding), increased latency to drink a sweet solution in a novel environment (novelty-induced hypophagia) and an increased latency to escape from an environment associated with an electrical shock, are observed in adulthood (Ansorge *et al.*, 2004, 2008). However, if adult mice are chronically treated with an SSRI and behaviorally tested 2.5 months following the last drug treatment, changes in depression-associated behaviors are not observed (Ansorge *et al.*, 2008). Taken together, these studies suggest that the depression-like phenotype of SERT knockout mice is a functional consequence of the loss of SERT early in brain development.

SERT knockout mice have also been tested in behavioral tests of anxiety, but once again strain-dependent effects have been reported. SERT knockout mice on a C57BL/6 background exhibit increased anxiety in the open field, elevated plus maze and light–dark test (Holmes *et al.*, 2003; Carroll *et al.*, 2007; Kalueff *et al.*, 2007a, 2007b; Zhao *et al.*, 2006b). In contrast, SERT knockout mice on a 129S6 background do not exhibit any changes in anxiety-like behavior in these tests (Lira *et al.*, 2003). However, SERT knockout rats exhibit increased anxiety levels in the open field and elevated plus maze, thus supporting the hypothesis that constitutive loss of SERT results in an anxious phenotype (Olivier *et al.*, 2008). Moreover, transgenic mice over-expressing human SERT exhibit reduced anxiety, and this effect is reversed by acute administration of the SSRI paroxetine (Jennings *et al.*, 2006). Finally, a recent study suggests that, like the depression phenotype, the anxious phenotype of SERT knockout mice is also a direct consequence of SERT blockade during early postnatal development but not during adulthood (Ansorge *et al.*, 2008). Taken together, animal studies suggest that during early postnatal development, SERT plays an important role in programming anxiety levels in adulthood.

In addition to altered emotional behaviors, SERT knockout mice also exhibit a number of other key behavioral and physiological phenotypes, including late onset obesity (Holmes *et al.*, 2002a; Murphy and Lesch, 2008), reduced basal corticosterone levels (Murphy and Lesch, 2008), enhanced stress-induced ACTH release (Li *et al.*, 1999), sleep abnormalities such as increased duration and number of REM sleep episodes (Wisor *et al.*, 2003; Alexandre *et al.*, 2006), reduced social interaction (Kalueff *et al.*, 2007a), decreased compulsive activity in the marble burying test (Zhao *et al.*, 2006b) and reduced basal locomotor activity in C57BL/6 SERT knockout mice (Holmes *et al.*, 2002b; Zhao *et al.*, 2006b; Kalueff *et al.*, 2007b) but not in 129S6 SERT knockout mice (Lira *et al.*, 2003). Finally, both SERT knockout mice and rats exhibit reduced aggression (Holmes *et al.*, 2002b; Homberg *et al.*, 2007).

Human studies Genetic association studies of the serotonin transporter gene in humans have primarily focused on the role of two SERT-function-modifying gene variants, the first of which is an insertion/deletion polymorphism in the promoter region of *SERT* (the serotonin-transporter-gene-linked polymorphic region, 5-HTTLPR) which results in a long (L) and a short (S) allele. The short (S) allele of this insertion/deletion polymorphism confers lower transcriptional activity relative to the long (L) allele. The second variant is a variable number of tandem repeats (VNTR) polymorphism in functional intron 2, which consists of 9, 10 or 12 repeats (SERTin2:9, 10 or 12) and can display functionality both *in vitro* (Fiskerstrand *et al.*, 1999; Lovejoy *et al.*, 2003; Klenova *et al.*, 2004) and *in vivo* (MacKenzie and Quinn, 1999).

The short variant of the 5-HTTLPR is a contributory risk factor for the development of major depression, an effect which is only clearly observed when the S allele occurs in combination with other genetic or environmental risk factors. While an association between the S allele and unipolar depression has been reported by some studies (Collier *et al.*, 1996; Gutierrez *et al.*, 1998a; Dorado *et al.*, 2007), many other studies, including a meta-analysis, have reported a lack of association (Kunugi *et al.*, 1997; Mendlewicz *et al.*, 2004; Lasky-Su *et al.*, 2005; Bozina *et al.*, 2006; Christiansen *et al.*, 2007; Zaboli *et al.*, 2008). Nevertheless, it appears that the S allele of this polymorphism does confer a risk to the development of depressive symptoms in response to stressful life events (Caspi *et al.*, 2003; Kendler *et al.*, 2005; Brummett *et al.*, 2008). Moreover, the 5-HTTLPR polymorphism also plays a critical role in the clinical response to antidepressant treatments because carriers of the S allele are more likely to be poor responders to SSRIs as well as some other antidepressant treatments (Smeraldi *et al.*, 1998; Rausch *et al.*, 2002; Yu *et al.*, 2002; Benedetti *et al.*, 2003; Arias *et al.*, 2005; Perna *et al.*, 2005; Bozina *et al.*, 2007).

The contribution of the SERTin2 polymorphism to the risk of unipolar depression is unclear, with some studies reporting a lack of association (Kunugi *et al.*, 1997; Hoehe *et al.*, 1998; Lasky-Su *et al.*, 2005) and others reporting an association (Liu *et al.*, 1999; Bozina *et al.*, 2006) particularly if it is part of a haplotype with the 5-HTTLPR polymorphism (Gutierrez *et al.*, 1998a). It has also been reported that carriers of alleles with a low number of repeats of the SERTin2 VNTR polymorphism showed better responses to paroxetine (Bozina *et al.*, 2007). Taken together, studies suggest that some polymorphisms of the SERT gene confer a risk to unipolar depression, particularly when in association with other risk factors for the disorder.

Many studies have investigated the contribution of *SERT* polymorphisms to risk for anxiety disorders. While the short allele of the 5-HTTLPR variant has been associated with anxiety-related traits and anxiety disorders in many studies (Lesch *et al.*, 1996; Melke *et al.*, 2001; You *et al.*, 2005; Denys *et al.*, 2006; Grados *et al.*, 2007), numerous other studies have also reported a lack of association (Billett *et al.*, 1997; Hamilton *et al.*, 1999; Camarena *et al.*, 2001; Lang *et al.*, 2004; Meira-Lima *et al.*, 2004; Samochowiec *et al.*, 2004; Schinka *et al.*,

2004). However, it is important to note that such discrepancies may well be a function of which anxiety traits were measured in a particular study (Schinka *et al.*, 2004). Finally, a neuroimaging study of a brain area associated with anxiety, the amygdala, has suggested that S allele carriers exhibited hyperactivity of this brain region in response to fearful stimuli (Hariri *et al.*, 2002), thus further supporting a role for the S allele of the 5-HTTLPR polymorphism in anxiety disorders. The SERTin2 polymorphism has been associated with obsessive-compulsive disorder (Baca-Garcia *et al.*, 2007), and anxiety-related personality traits (Vormfelde *et al.*, 2006), but not generalized anxiety disorder (You *et al.*, 2005). Moreover, a rare polymorphism in the ninth exon, Ile425Val, has been found in unrelated families with obsessive-compulsive disorder (Ozaki *et al.*, 2003; Wendland *et al.*, 2008). Taken together, genetic association studies suggest an association of *SERT* gene variants with risk for anxiety disorders in humans.

The contribution of the 5-HTTLPR and SERTin2 polymorphisms to the risk for bipolar disorder has been investigated, but conflicting results have been reported. Although many individual studies have reported no association between the 5-HTTLPR polymorphism and bipolar disorder (Kunugi *et al.*, 1997; Ewald *et al.*, 1998; Gutierrez *et al.*, 1998b; Meira-Lima *et al.*, 2005; Mendlewicz *et al.*, 2004), several meta-analysis studies and original investigations have revealed an association with bipolar disorder (Collier *et al.*, 1996; Mynett-Johnson *et al.*, 2000; Rotondo *et al.*, 2002; Anguelova *et al.*, 2003; Cho *et al.*, 2005; Lasky-Su *et al.*, 2005). Conflicting results for the SERTin2 polymorphism in bipolar disorder have also been published with the majority reporting a lack of association (Esterling *et al.*, 1998; Hoehe *et al.*, 1998; Bocchetta *et al.*, 1999; Liu *et al.*, 1999; Saleem *et al.*, 2000; Dimitrova *et al.*, 2002; Lasky-Su *et al.*, 2005).

Taken together, studies from both animals and humans suggest a clear role for SERT in anxiety disorders and the mechanism of action of SSRI antidepressant drugs. While some *SERT* polymorphisms may also contribute to risk for unipolar depression, their contribution may only become apparent when combined with other risk factors for the disorder – for example, stressful life events. Finally, studies in rodents suggest that interference with SERT function during brain development can program anxiety levels and depression in adulthood.

5-HT degradation: monoamine oxidase A (MAOA)

Following SERT-mediated 5-HT reuptake into the cell, monoamine oxidase A (MAOA) catalyzes the oxidative deamination of serotonin (Youdim *et al.*, 2006). However, MAOA also metabolizes monoamines other than serotonin, including norepinephrine and dopamine, thus making it difficult to ascertain the exact role of MAOA-induced serotonin metabolism in behavioral phenotypes. Nevertheless, mice lacking functional MAOA exhibit behavioral traits that would be expected by the increased availability of serotonin. Moreover, these mice exhibit a nine-fold increase in brain 5-HT levels and lowered concentrations of its metabolite 5-HIAA, but the only a two-fold increase in brain norepinephrine levels (Cases *et al.*, 1995; Popova *et al.*, 2001). The predominant behavioral feature of MAOA-deficient mice is increased aggression, and this has been observed in two different MAOA knockout mouse lines (Cases *et al.*, 1995; Vishnivetskaya *et al.*, 2007; Scott *et al.*, 2008). Other phenotypic features include enhanced emotional learning, as evidenced by increased cued and contextual fear conditioning (Kim *et al.*, 1997); decreased memory extinction (Dubrovina *et al.*, 2006); decreased exploratory activity in the open field, light–dark and hole board tests (Popova *et al.*, 2000; Scott *et al.*, 2008); reduced social investigation (Vishnivetskaya *et al.*, 2007); sleep apnoea (Real *et al.*, 2007); and attenuated stress-induced corticosterone release (Popova *et al.*, 2006).

Studies examining behaviors related to anxiety and depression have yielded conflicting results in these mice. While one group reported decreased anxiety in the open field (Cases *et al.*, 1995), another reported no change in the light–dark test (Popova *et al.*, 2000). Similarly, while reduced baseline immobility in the FST has been reported in one MAOA knockout strain (Cases *et al.*, 1995), no changes in baseline immobility in the TST were observed in another MAOA knockout strain (Scott *et al.*, 2008). The ambiguity of these results might be a function of disruptions to other monoaminergic systems such as norepinephrine or dopamine, which are also thought to play a role in these psychiatric disorders (Cryan *et al.*, 2004; O'Leary *et al.*, 2007).

The contribution of *MAOA* gene polymorphisms in humans to susceptibility for mood disorders has been investigated extensively. The majority of studies have reported a lack of association for *MAOA* with bipolar disorder (Craddock *et al.*, 1995; Kawada *et al.*, 1995; Nothen *et al.*, 1995; Muramatsu *et al.*, 1997; Parsian and Todd, 1997; Sasaki *et al.*, 1998; Furlong *et al.*, 1999; Kunugi *et al.*, 1999b; Serretti *et al.*, 2002; Gutierrez *et al.*, 2004). On the other hand, the role of *MAOA* polymorphisms in conferring risk for unipolar depression is less clear, with some studies reporting an association particularly in females (Schulze *et al.*, 2000; Yu *et al.*, 2005; Brummett *et al.*, 2007), and others reporting a lack of association with various polymorphisms (Muramatsu *et al.*, 1997;

Sasaki *et al.*, 1998; Furlong *et al.*, 1999; Kunugi *et al.*, 1999b; Syagailo *et al.*, 2001; Tadic *et al.*, 2003; Gutierrez *et al.*, 2004). An association of *MAOA* with generalized anxiety disorder and obsessive-compulsive disorder has been reported (Karayiorgou *et al.*, 1999; Tadic *et al.*, 2003), while a lack of association has been predominantly reported for panic disorder (Deckert *et al.*, 1999; Hamilton *et al.*, 2000; Tadic *et al.*, 2003). Finally, similarly to the mouse model, *MAOA* polymorphisms have been associated with aggression (Brunner *et al.*, 1993; Manuck *et al.*, 2000).

Taken together, both preclinical and human studies suggest a clear role for MAOA in aggression, but the contribution to depression and anxiety disorders remains ambiguous.

5-HT receptors

The biological effects of serotonin are mediated by at least 14 distinct receptors that have been classified into seven groups, 5-HT_{1-7}, based on their structural, functional and biochemical characteristics (Barnes and Sharp, 1999; Hoyer *et al.*, 2002). The heterogeneity of these receptors is further complicated by various processes such as post-translational modifications and heterodimerization (Hoyer *et al.*, 2002; Bockaert *et al.*, 2006), thus generating further complexity when attempting to identify the role of a given 5-HT receptor in behavior. The following section, which is summarized in Table 2, describes the behavioral effects of genetic manipulation of genes encoding serotonergic receptors in mice, as well as analogous genetic association studies in humans that relate to psychiatric disorders.

5-HT_1 receptors

The 5-HT_1 receptor class is comprised of five receptor subtypes, 5-HT_{1A}, 5-HT_{1B}, 5-HT_{1D}, 5-HT_{1E} and 5-HT_{1F} (Cryan and Leonard, 2000; Hoyer *et al.*, 2002). To date, few studies have examined the roles of 5-HT_{1E} and 5-HT_{1F} receptors on behavior; therefore, the function of these receptors is not discussed in this chapter.

5-HT_{1A} receptors

5-HT_{1A} receptors are largely distributed throughout the CNS. 5-HT_{1A} receptors localized in the raphe nuclei are somatodendritic autoreceptors that inhibit cell firing. 5-HT_{1A} receptors are also found postsynaptically in a number of limbic brain regions important in the regulation of emotion, such as the hippocampus (Cryan and Leonard, 2000; Hoyer *et al.*, 2002), thus complicating attempts to unravel the function of 5-HT_{1A} receptors in behavior. However, activation of pre- versus postsynaptic 5-HT_{1A} receptors can be distinguished by discrete behavioral changes. Selective activation of postsynaptic 5-HT_{1A} receptors induces a behavioral syndrome that comprises a flat body posture, reciprocal forepaw treading and head weaving (Hoyer *et al.*, 2002; Lucki, 1992), while activation of presynaptic receptors induces hyperphagia (Simansky, 1996).

Preclinical studies In 1998, three independent groups reported the generation of 5-HT_{1A} receptor knockout mice on different genetic backgrounds (Heisler *et al.*, 1998; Parks *et al.*, 1998; Ramboz *et al.*, 1998). The predominant phenotype of these animals is anxiety. 5-HT_{1A} knockout mice on various genetic backgrounds exhibit elevated anxiety levels when tested in the open field, elevated zero maze, elevated plus maze, novel-object tests, novelty-suppressed feeding test and the light–dark test (Heisler *et al.*, 1998; Parks *et al.*, 1998; Ramboz *et al.*, 1998; Santarelli *et al.*, 2003; Klemenhagen *et al.*, 2006); in some of these tests, *5-HT_{1A}* mice demonstrated anxiety levels intermediate to those of wild-type and *5-HT_{1A}*$^{-/-}$ mice (Heisler *et al.*, 1998; Ramboz *et al.*, 1998). This anxious phenotype is a direct consequence of the loss of 5-HT_{1A} receptors specifically in the forebrain during early postnatal development but not during adulthood, thus suggesting that early postnatal processes are critical determinants of anxiety in adulthood (Gross *et al.*, 2002).

Using fear conditioning paradigms, other studies also support an anxious phenotype in 5-HT_{1A} receptor knockout mice. When placed in an environment identical to that where the animal previously received a foot shock, mice exhibit freezing behavior. When placed in this identical environment, 5-HT_{1A} receptor knockout mice demonstrated more freezing behavior than wild-type mice (Klemenhagen *et al.*, 2006). Moreover, when the mice were placed in an environment that was similar to the original one but also contained novel cues, the freezing behavior of wild-type but not 5-HT_{1A} receptor knockout mice decreased (Klemenhagen *et al.*, 2006). This suggests that 5-HT_{1A} receptor knockout mice have an innate inappropriate bias towards threatening cues even in a relatively safe environment – a trait often associated with subsets of anxiety disorders. Furthermore, it was recently demonstrated that selective suppression of neural activity of granule cells in the dentate gyrus of the hippocampus suppresses the freezing response of 5-HT_{1A} receptor knockout mice to such ambiguous cues (Tsetsenis *et al.*, 2007). However, it is important to note that the enhanced freezing response observed in the 5-HT_{1A} receptor knockout mice in contextual fear conditioning may be dependent upon the background strain of the animal (Groenink

Table 2 The effect of genetic mutations of serotonin receptors on behaviors associated with depression and anxiety disorders

5-HT receptors	Affective disorders and antidepressant response	Anxiety
5-HT$_{1A}$ receptors		
Animal studies:		
5-HT$_{1A}$ knockout mice	Decreased immobility in the TST (Heisler *et al.*, 1998; Mayorga *et al.*, 2001) Decreased immobility in the FST (Parks *et al.*, 1998; Ramboz *et al.*, 1998) Do not respond to SSRI antidepressant drugs in behavioral tests (Mayorga *et al.*, 2001; Santarelli *et al.*, 2003)	Increased anxiety in the open field, elevated plus maze, elevated zero maze, novel object test, novelty suppressed feeding and light–dark test (Heisler *et al.*, 1998; Parks *et al.*, 1998; Ramboz *et al.*, 1998; Gross *et al.*, 2002; Santarelli *et al.*, 2003; Klemenhagen *et al.*, 2006)
	Enhanced stress-induced corticosterone release (Gross *et al.*, 2000)	Increased freezing behavior in contextual fear conditioning (Klemenhagen *et al.*, 2006; Tsetsenis *et al.*, 2007)
		Resistance to anxiolytic effects of benzodiazepines (Sibille *et al.*, 2000)
5-HT$_{1A}$ over-expression in the forebrain during early development		Reduced anxiety in elevated plus maze (Kusserow *et al.*, 2004)
Human studies:		
C/G at −1019	C/C genotype carriers showed a better response to fluvoxamine (Serretti *et al.*, 2004)	
	C/C genotype female patients showed a better antidepressant response to fluoxetine (Yu *et al.*, 2006)	
	C/C patients showed a better response to TMS (Zanardi *et al.*, 2007)	
	G/G genotpye associated with depression (Lemonde *et al.*, 2003)	
	No association with risk for depression (Anttila *et al.*, 2007)	
	G allele carriers associated with personality traits that are linked to depression and anxiety such as neuroticism (Strobel *et al.*, 2003)	
	Increased frequency in G/G genotype in suicide victims (Lemonde *et al.*, 2003)	
Gly272Asp	272Asp carriers showed enhanced response to fluvoxamine (Suzuki *et al.*, 2004)	
	No association with response to fluoxetine (Yu *et al.*, 2006)	
rs10042486C/C	Showed a better response to fluoxetine, paroxetine and milnacipran (Kato *et al.*, 2009)	
rs6295G/G	Showed a better response to fluoxetine, paroxetine and milnacipran (Kato *et al.*, 2009)	
rs1364043T/T	Showed a better response to fluvoxamine, paroxetine and milnacipran (Kato *et al.*, 2009)	
5-HT$_{1B}$ receptors		
Animal studies:		
5-HT$_{1B}$ knockout mice	Enhanced behavioral responses to fluoxetine in the TST (Mayorga *et al.*, 2001)	No change in anxiety in the elevated plus maze (Parks *et al.*, 1998; Brunner *et al.*, 1999; Malleret *et al.*, 1999; Sibille *et al.*, 2007)
	Female knockouts have lower immobility (antidepressant-like phenotype) in the TST and the FST (Jones and Lucki, 2005)	No change in anxiety in the open field (Sibille *et al.*, 2007)

(Continued)

Table 2 (*Continued*)

5-HT receptors	Affective disorders and antidepressant response	Anxiety
		Decreased anxiety in the open field (Zhuang *et al.*, 1999)
		Decreased anxiety as measured by increased social activity (Bouwknecht *et al.*, 2001a)
		Decreased anxiety as indicated by increased exploration of a novel object and a novel environment (Malleret *et al.*, 1999; Zhuang *et al.*, 1999)
Viral-mediated over-expression of 5-HT_{1B} receptor gene in the rat dorsal raphe nucleus	No changes in the FST (Clark *et al.*, 2002)	Increased anxiety in the open field and elevated plus maze following inescapable stress (Clark *et al.*, 2002)
P11 over-expressing mice (enhanced 5-HT_{1B} receptor function)	Decreased immobility in the TST (Svenningsson *et al.*, 2006)	Decreased anxiety as represented by reduced thigmotaxis in the open field (Svenningsson *et al.*, 2006)
P11 knockout mice (decreased 5-HT_{1B} receptors at cell surface)	Increased immobility in the TST (Svenningsson *et al.*, 2006)	Increased anxiety as represented by increased thigmotaxis in the open field (Svenningsson *et al.*, 2006)
	Reduced sucrose consumption (Svenningsson *et al.*, 2006)	
	Decreased response to imipramine in the TST (Svenningsson *et al.*, 2006)	
Human studies:		
861 G/C	No association with bipolar disorder (Huang *et al.*, 2003)	
161 A/T	No association with major depression (Tsai *et al.*, 2004)	
	No association with response to fluoxetine (Tsai *et al.*, 2004)	
129 C/T	No association with major depression (Huang *et al.*, 1999)	
5-HT_2 receptors		
5-HT_{2A} receptors		
Animal studies:		
5-HT_{2A} knockout mice	No change in immobility in the FST or TST (Weisstaub *et al.*, 2006)	Decreased anxiety in the open-field, dark–light box, elevated plus maze and novelty-suppressed feeding (Weisstaub *et al.*, 2006)
Human studies:		
G-1438A	No association with bipolar disorder (Etain *et al.*, 2004)	G allele associated with obsessive compulsive disorder (Denys *et al.*, 2006)
	G allele associated with major depression (Choi *et al.*, 2004)	A allele associated with obsessive compulsive disorder (Enoch *et al.*, 2001; Walitza *et al.*, 2002)
	A allele associated with unipolar depression (Christiansen *et al.*, 2007)	A allele associated with severity of OCD symptoms (Tot *et al.*, 2003)
	G allele associated with better clinical response to citalopram (Choi *et al.*, 2005)	
T102C	No association with bipolar disorder (Arranz *et al.*, 1997; Gutierrez *et al.*, 1997; Mahieu *et al.*, 1997; Massat *et al.*, 2000; Blairy *et al.*, 2000; Tut *et al.*, 2000; Ni *et al.*, 2002; Ranade *et al.*, 2003)	Association with panic disorder (Inada *et al.*, 2003)
	C allele associated with major depression (Zhang *et al.*, 1997; Du *et al.*, 2000)	No association with panic disorder (Fehr *et al.*, 2001; Rothe *et al.*, 2004)
	T allele associated with higher depression scores (Eley *et al.*, 2004)	No association with generalized anxiety disorder (Fehr *et al.*, 2001)

Table 2 (*Continued*)

5-HT receptors	Affective disorders and antidepressant response	Anxiety
	No association with unipolar depression (Anguelova *et al.*, 2003; Arias *et al.*, 2001a, 2001b; Minov *et al.*, 2001; Oswald *et al.*, 2003)	
−1438A/G	A allele more frequent in bipolar disorder (Bonnier *et al.*, 2002)	
	G/G genotype associated with a good therapeutic response to SSRIs (Kato *et al.*, 2006)	
1354C/T (His452Tyr)	Associated with bipolar disorder (Ranade *et al.*, 2003)	
	No association with bipolar disorder (Arranz *et al.*, 1997; Etain *et al.*, 2004; Gutierrez *et al.*, 1997; Ni *et al.*, 2002)	
	No association with unipolar depression (Minov *et al.*, 2001)	
516C/T (Asp172Asp)	Associated with bipolar disorder (Ranade *et al.*, 2003)	
	No association with bipolar disorder (Arranz *et al.*, 1997; Gutierrez *et al.*, 1997)	
Thr25Asn	No association with bipolar disorder (Arranz *et al.*, 1997; Gutierrez *et al.*, 1997)	
rs7997012	Associated with therapeutic response to citalopram (McMahon *et al.*, 2006)	
5-HT$_{2C}$ receptors		
Animal studies:		
5-HT$_{2C}$ knockout mice	Enhanced response to fluoxetine in the TST (Cremers *et al.*, 2004)	Decreased anxiety in dark-light choice test (Tecott *et al.*, 1998)
	No change in baseline immobility in the TST (Cremers *et al.*, 2004)	
Human studies:		
Cys23Ser	Ser allele associated with subgroups of bipolar disorder patients including female and early onset patients (Gutierrez *et al.*, 1996, 2001; Oruc *et al.*, 1997; Lerer *et al.*, 2001; Massat *et al.*, 2007)	
	Ser allele associated with unipolar depression (Lerer *et al.*, 2001)	
	No association with unipolar depression (Frisch *et al.*, 1999)	
	No association with seasonal affective disorder (Johansson *et al.*, 2001)	
(CT)4 ± 5 resulting in a short or long variant		No association with panic disorder (Deckert *et al.*, 2000)
5-HT$_3$ receptors		
Animal studies:		
5-HT$_3$ over-expressing mice on B6SJL/F1 background		Increased anxiety or emotional memory as indicated by enhanced contextual fear conditioning (Harrell and Allan, 2003)
		Decreased anxiety in the elevated plus maze (Harrell and Allan, 2003)
		Increased exploratory behavior in response to novel environmental stimuli (Harrell and Allan, 2003)

(*Continued*)

Table 2 *(Continued)*

5-HT receptors	Affective disorders and antidepressant response	Anxiety
$5\text{-}HT_3$ over-expressing mice on C57BL/6J background		Increased anxiety or emotional memory as indicated by enhanced contextual fear conditioning (Metz *et al.*, 2006) Decreased anxiety in the elevated plus maze (Metz *et al.*, 2006) Increased exploratory behavior in response to novel environmental stimuli (Metz *et al.*, 2006)
$5\text{-}HT_3$ over-expressing mice on DBA/2J background		No change in contextual fear conditioning (Metz *et al.*, 2006) No change in anxiety in the elevated plus maze (Metz *et al.*, 2006) Increased exploratory behavior in response to novel environmental stimuli (Metz *et al.*, 2006)
$5\text{-}HT_{3A}$ knockout mice	Increased immobility in female $5\text{-}HT_{3A}$ knockout mice in the FST (Bhatnagar *et al.*, 2004a)	Decreased anxiety in the elevated plus maze (Kelley *et al.*, 2003; Bhatnagar *et al.*, 2004b) Decreased anxiety in the light–dark box (Kelley *et al.*, 2003) Decreased anxiety in the novel object tests (Kelley *et al.*, 2003) No effect on anxiety levels in the open field (Bhatnagar *et al.*, 2004b) No effect on anxiety levels in the light–dark box (Bhatnagar *et al.*, 2004b) Enhanced fear conditioning (Bhatnagar *et al.*, 2004b)
Human studies:		
$5\text{-}HT_{3A}$		
178-C/T	Associated with bipolar disorder (Niesler *et al.*, 2001) No association with bipolar disorder (Yamada *et al.*, 2006) No association with unipolar depression (Yamada *et al.*, 2006)	No association with OCD (Mossner *et al.*, 2007)
$5\text{-}HT_{3B}$		
3-bp deletion −100-102delAAG	Under-represented in bipolar disorder (Frank *et al.*, 2004)	
$5\text{-}HT_4$ receptors		
Animal studies:		
$5\text{-}HT_4$ knockout mice		Increased anxiety in the open field (Compan *et al.*, 2004)
Human studies:		
exon d g.83097C/T	Significant association with bipolar disorder (Ohtsuki *et al.*, 2002)	
exon d g.83159G/A	Significant association with bipolar disorder (Ohtsuki *et al.*, 2002)	
exon d g.83164 (T)9-10	Significant association with bipolar disorder (Ohtsuki *et al.*, 2002)	
exon d g.83198A/G	Significant association with bipolar disorder (Ohtsuki *et al.*, 2002)	
$5\text{-}HT_{5A}$ receptors		
Animal studies:		
$5\text{-}HT_{5A}$ knockout mice		No changes in anxiety levels in the elevated plus maze, electric probe burying or acoustic startle (Grailhe *et al.*, 1999)

Table 2 *(Continued)*

5-HT receptors	Affective disorders and antidepressant response	Anxiety
Human studies:		
A12T	T-T genotype associated with unipolar depression (Birkett *et al.*, 2000)	
	No association with unipolar depression (Arias *et al.*, 2001c)	
	No association with bipolar disorder (Arias *et al.*, 2001c)	
−19G/C	C allele associated with unipolar depression (Arias *et al.*, 2001c)	
	No association with unipolar depression (Arias *et al.*, 2001c)	
	C allele associated with bipolar disorder (Birkett *et al.*, 2000)	
	No association with bipolar disorder (Arias *et al.*, 2001c)	
5-HT$_6$ receptors		
Animal studies:		
i.c.v. administration of 5-HT$_6$ antisense oligonucleotides in the rat		Increased anxiety in the elevated plus maze and the social interaction test (Hamon *et al.*, 1999)
Human studies:		
C267T	C allele associated with bipolar disorder (Vogt *et al.*, 2000)	
	No association with unipolar depression (Lee *et al.*, 2005)	
5-HT$_7$ receptors		
Animal studies:		
5-HT$_7$ knockout mice	Decreased immobility in the TST and the FST (Guscott *et al.*, 2005; Hedlund *et al.*, 2005)	Reduced marble burying (Hedlund and Sutcliffe, 2007)
		No changes in anxiety in the light–dark test (Roberts *et al.*, 2004)
		Decreased anxiety as demonstrated by impaired (less freezing) contextual fear conditioning (Roberts *et al.*, 2004)
Human studies:		
Pro279Leu	No association with bipolar disorder (Erdmann *et al.*, 1996)	

et al., 2003). Nevertheless, the majority of studies support an anxiety-like phenotype for 5-HT$_{1A}$ receptor knockout mice.

In addition to anxious behaviors, 5-HT$_{1A}$ receptor knockout mice also exhibit enhanced autonomic changes in response to stresses such as heart rate and body temperature (Pattij *et al.*, 2002) and augmented stress-induced corticosterone release (Gross *et al.*, 2000), as well as a resistance to the anxiolytic effects of benzodiazepines (Sibille *et al.*, 2000). Curiously, these increases in anxiety-like behaviors and autonomic stress responses are accompanied by an antidepressant-like phenotype. 5-HT$_{1A}$ receptor knockout mice exhibit decreased immobility in both the TST and FST (Heisler *et al.*, 1998; Parks *et al.*, 1998; Ramboz *et al.*, 1998; Mayorga *et al.*, 2001) without changes in general locomotor activity (Parks *et al.*, 1998; Ramboz *et al.*, 1998). The reduced immobility in the TST was reversed by pharmacologically-induced depletion of endogenous catecholamines but not serotonin (Mayorga *et al.*, 2001), thus suggesting that the excess serotonin levels induced by the loss of 5-HT$_{1A}$ autoreceptors does not mediate the antidepressant-like phenotype. Moreover, 5-HT$_{1A}$ receptor knockout mice fail to respond behaviorally to acute or chronic treatment with the SSRIs fluoxetine and paroxetine but respond equally well to some other classes of antidepressant drugs, thus suggesting that 5-HT$_{1A}$ receptors play an important role in the behavioral response to at least serotonin-selective antidepressant drugs (Mayorga *et al.*, 2001; Santarelli *et al.*, 2003). Other phenotypic changes observed in 5-HT$_{1A}$ receptor knockout mice include sleep abnormalities (Boutrel *et al.*, 2002; Monaca *et al.*, 2003) and impairments in hippocampal-dependent learning and memory tasks but not in hippocampal-independent or motor memory tasks (Parks *et al.*, 1998; Sarnyai *et al.*, 2000).

The importance of serotonin during brain development in shaping behavioral phenotypes in adulthood has been further demonstrated using mice which temporally over-express the 5-HT_{1A} receptor during embryonic and early postnatal development (up to postnatal day 1.5; Kusserow *et al.*, 2004; Bert *et al.*, 2005, 2006). These mice, which over-express 5-HT_{1A} receptors predominantly in the dentate gyrus of the hippocampus and outer cortical layers, exhibit reductions in anxiety levels in adulthood, a phenotype opposite to that described previously in the 5-HT_{1A} receptor knockout mice (Kusserow *et al.*, 2004). However, these mice also exhibit phenotypes which fail to oppose those of the 5-HT_{1A} receptor knockout mice, including impairments in the Morris water maze memory task (Bert *et al.*, 2005), decreased locomotor activity and reduced body temperature (Bert *et al.*, 2006).

Human studies Despite intense investigation of the 5-HT_{1A} receptor knockout mouse, relatively few studies have examined the association of 5-HT_{1A} receptor gene polymorphisms with human behavior and brain disorders. Of the studies conducted, however, the majority have primarily focused on the contribution of 5-HT_{1A} receptor gene polymorphisms to the clinical response to antidepressant treatments. In this regard, the main polymorphisms examined include the −1019C/G and Gly272Asp variants. There is convincing evidence for a more favorable response to transcranial magnetic stimulation, fluoxetine and various other chemical antidepressants in patients of the −1019C/C genotype (Lemonde *et al.*, 2004; Serretti *et al.*, 2004; Yu *et al.*, 2006; Zanardi *et al.*, 2007). However, one recent study did report that carriers of the −1019C/C genotype demonstrated a poor response to a variety of antidepressant treatments (Baune *et al.*, 2008). Overall, however, it appears that the −1019C/C genotype is associated with a favorable response to antidepressant drugs. The role of the Gly272Asp variant in the therapeutic effects of antidepressant drugs has also been investigated, with some reporting enhanced response in 272Asp carriers (Suzuki *et al.*, 2004) and others reporting no effect (Yu *et al.*, 2006). Other polymorphisms, including rs10042486C/C, rs6295G/G and rs1364043T/T, have also been associated with a favorable clinical response to antidepressant drugs (Kato *et al.*, 2009). From both animal and clinical studies, it is clear that the 5-HT_{1A} receptor plays an important role in the mechanism of antidepressant action.

The contribution of 5-HT_{1A} receptor gene polymorphisms to depression, anxiety and suicide has also been investigated, but conflicting results have been reported. In a sample of depressed patients, the −1019G/G genotype was higher than in controls (Lemonde *et al.*, 2003), and carriers of the G allele were more likely to have personality traits associated with depression and anxiety, such as neuroticism (Strobel *et al.*, 2003). However, others have reported a lack of association between the −1019C/G polymorphism and the risk for depression (Anttila *et al.*, 2007). Finally, an increased frequency of the −1019G/G genotype has been reported in suicide victims (Lemonde *et al.*, 2003), but no other variants have been associated with suicide (Nishiguchi *et al.*, 2002; Ohtani *et al.*, 2004; Videtic *et al.*, 2006).

Taken together, both human and preclinical studies have revealed a role for the 5-HT_{1A} receptor in the mechanism of action of several antidepressant treatments. Although animal studies suggest that the 5-HT_{1A} receptor is an important modulator of anxiety and depression-like behavior, human studies examining the contribution of polymorphisms in this receptor to mood disorders is lacking and warrants further investigation. Nevertheless, human studies suggest that the 5-HT_{1A} receptor can modulate at least some behavioral traits associated with depression and anxiety disorders.

5-HT_{1B} receptor

5-HT_{1B} receptors are primarily expressed in the basal ganglia, striatum and frontal cortex. They are localized on presynaptic 5-HT terminals, where they serve as inhibitory autoreceptors of 5-HT release, as well as on other nerve terminals, where they act as heteroreceptors controlling the release of various neurotransmitters including acetylcholine, glutamate, dopamine, norepinephrine and gamma-aminobutyric acid (Barnes and Sharp, 1999; Cryan and Leonard, 2000; Hoyer *et al.*, 2002).

Preclinical studies The 5-HT_{1B} receptor knockout mouse has been predominantly characterized for the acquisition and sensitivity to drugs of abuse, antidepressant-like and anxiety behaviors, as well as learning.

In the FST and the TST, male 5-HT_{1B} receptor knockout mice exhibit baseline immobility values similar to wild-type mice (Mayorga *et al.*, 2001; Jones and Lucki, 2005), but their behavioral responses to low doses of fluoxetine are enhanced when compared with wild-types in the TST (Mayorga *et al.*, 2001). In contrast, female 5-HT_{1B} receptor knockout mice exhibit lower immobility in both the TST and FST when compared to female wild-types, and this effect is mediated by endogenous serotonin levels (Jones and Lucki, 2005). Recently, it was discovered that the cellular localization of 5-HT_{1B} receptors is regulated by their interaction with the protein p11 (Svenningsson *et al.*, 2006). Mice that over-express p11 and thus exhibit enhanced 5-HT_{1B} receptor function have an antidepressant-like phenotype (Svenningsson *et al.*, 2006). In contrast, p11 knockout mice exhibit a depression-like phenotype and have a diminished response to antidepressant treatments (Svenningsson *et al.*, 2006). Such findings are conflicting with the behavioral phenotype of the 5-HT_{1B} receptor knockout mice (Mayorga *et al.*, 2001; Jones and Lucki, 2005), perhaps suggesting that other mechanisms may also exist through which p11 exerts such effects.

Conflicting results on anxiety levels in 5-HT_{1B} receptor knockout mice have been reported, with some studies reporting no change in standard measures of anxiety in the elevated plus maze (Parks *et al.*, 1998; Brunner *et al.*, 1999; Malleret *et al.*, 1999; Sibille *et al.*, 2007) or the open field (Sibille *et al.*, 2007) and others reporting reduced anxiety in the open field and the social interaction test (Zhuang *et al.*, 1999; Bouwknecht *et al.*, 2001a). These conflicting results may be due to dual functions of 5-HT_{1B} receptors as either autoreceptors at serotonergic terminals or heteroreceptors on non-serotonergic neurons. To investigate whether the 5-HT_{1B} autoreceptor mediates anxiety, Clark and colleagues (2002, 2004) used viral-mediated gene transfer to selectively over-express 5-HT_{1B} receptors on serotonergic neurons in the rat dorsal raphe nucleus. These rats exhibited increased anxiety in the open field and elevated plus maze, but only when tested following inescapable stress (Clark *et al.*, 2002). These rats also exhibited reduced fear-potentiated startle (Clark *et al.*, 2004) but no changes in the FST (Clark *et al.*, 2002). Together with the reduced anxiety levels reported in some studies of the 5-HT_{1B} receptor knockout mice (Zhuang *et al.*, 1999; Bouwknecht *et al.*, 2001a), these results suggest that it is the 5-HT_{1B} autoreceptor on serotonergic neurons, and not heteroreceptors, which play the predominant role in anxiety-associated behavior.

The behavior of 5-HT_{1B} receptor knockout mice has also been examined in learning and memory tests. In a test of spatial memory, the Morris water maze, 5-HT_{1B} receptor knockout mice exhibited superior performance (Malleret *et al.*, 1999), and this facilitation of learning performance was selectively enhanced when the complexity of the task was increased (Buhot *et al.*, 2003a; Wolff *et al.*, 2003). Moreover, it was reported that 5-HT_{1B} receptor knockout mice were more resistant to age-related deterioration in spatial memory than wild-types (Buhot *et al.*, 2003b). Taken together, experimental data suggest that the 5-HT_{1B} receptor plays an important role in learning and memory.

Other phenotypes of the 5-HT_{1B} receptor knockout mice include increased aggression (Saudou *et al.*, 1994; Ramboz *et al.*, 1996), increased body weight (Bouwknecht *et al.*, 2000, 2001b), reduced longevity (Sibille *et al.*, 2007), early onset of age-related motor decline (Sibille *et al.*, 2007), circadian rhythm and sleep abnormalities (Boutrel *et al.*, 1999; Sollars *et al.*, 2006), and increased exploration of a novel object and a novel environment (Malleret *et al.*, 1999; Zhuang *et al.*, 1999).

Human studies The most commonly investigated polymorphisms of the human 5-HT_{1B} receptor gene include the 861G/C and −161A/T variants. Although the role of these variants has been predominantly investigated in attention deficit hyperactivity disorder (ADHD), suicide substance abuse and dependence, the role of 5-HT_{1B} receptor gene polymorphisms in behaviors associated with abnormal emotion has also been investigated. Conflicting reports on the role of the 861G/C polymorphism in aggression have been described, with one group reporting a trend for an increased frequency of the C allele in aggression (Davidge *et al.*, 2004) and another reporting no change (Huang *et al.*, 1999). In general, there was no association between suicide, obsessive-compulsive disorder, bipolar disorder, unipolar depression or antidepressant response and various 5-HT_{1B} receptor gene polymorphisms including the 861G/C and −161A/T variants (Nishiguchi *et al.*, 2001; Huang *et al.*, 2003; Rujescu *et al.*, 2003; Hong *et al.*, 2004; Stefulj *et al.*, 2004; Tsai *et al.*, 2004; Videtic *et al.*, 2006; Dickel *et al.*, 2007; Kia-Keating *et al.*, 2007; Shaikh *et al.*, 2008).

Taken together, it is clear from preclinical and human studies that the 5-HT_{1B} receptor plays an important role in aggression. However, while animal studies suggest that the 5-HT_{1B} receptor may also be an important modulator of depression-like and anxiety behaviors, such findings have yet to be substantiated in human studies.

5-HT_2 receptors

This class comprises the 5-HT_{2A}, 5-HT_{2B} and 5-HT_{2C} receptors, which preferentially couple to Gq/11.

5-HT_{2A} receptor

5-HT_{2A} receptors are primarily located in the cortex, claustrum and basal ganglia (Hoyer *et al.*, 2002).

Preclinical studies Constitutive knockout of the 5-HT_{2A} receptor inhibits anxiety in a number of conflict-based anxiety paradigms (Weisstaub *et al.*, 2006). When compared to their wild-type counterparts, 5-HT_{2A} receptor knockout mice spend more time in the center of a brightly-lit open field and exhibit increased rearing, which reflects risk-type behavior. Furthermore, when tested in the dark–light choice test or elevated plus maze, 5-HT_{2A} knockout mice spend more time than the wild-type controls in risky or anxiogenic areas of the tests – namely, the brightly-illuminated compartment of the dark–light choice test or the open arms of the elevated plus maze. Moreover, when compared with their wild-type counterparts, 5-HT_{2A} receptor knockout mice demonstrate a decreased latency to begin feeding in a novel environment (the novelty-suppressed feeding test). These reductions in anxiety are not accompanied by changes in fear-conditioned behavior or antidepressant-like behaviors in either the TST or FST. Importantly, the reduction in anxiety-like behaviors is not a result of alterations in HPA axis activity, locomotor activity or feeding behavior, and the anxiety phenotype can be reversed by specific restoration of cortical 5-HT_{2A} receptor expression (Weisstaub *et al.*, 2006). In addition to reduced anxiety, 5-HT_{2A} receptor knockout mice also exhibit reduced non-rapid eye movement sleep compared to wild-type controls (Popa *et al.*, 2005). Together, the data suggest that the 5-HT_{2A} receptor plays important roles in anxiety and sleep.

Human studies Data from the rodent studies suggest that the 5-HT_{2A} receptor may play an important role in anxiety; however, the contribution of polymorphisms in this gene to anxiety disorders in humans is less clear. For example, conflicting reports of the association of the −1438G/A polymorphism with obsessive-compulsive disorder have been published, with some studies reporting an association of the G allele (Denys *et al.*, 2006), others suggesting an association of the A allele (Enoch *et al.*, 2001; Walitza *et al.*, 2002), and others still suggesting a lack of association of either allele with the disorder but an association of the A allele with severity of symptoms (Tot *et al.*, 2003). Similarly, conflicting results of the association of the T102C polymorphism with panic disorder have been reported, with some suggesting an association (Inada *et al.*, 2003) and others suggesting a lack of association (Fehr *et al.*, 2001; Rothe *et al.*, 2004). Finally, a lack of association of the T102C polymorphism with generalized anxiety disorder has been reported (Fehr *et al.*, 2001).

In addition to anxiety, the frequency of 5-HT_{2A} receptor gene polymorphisms has also been investigated in mood disorders. Although a few studies have reported an association between bipolar disorder and various 5-HT_{2A} receptor gene polymorphisms (Chee *et al.*, 2001; Bonnier *et al.*, 2002; Ranade *et al.*, 2003), the majority of studies suggest that there is no association (Arranz *et al.*, 1997; Gutierrez *et al.*, 1997; Mahieu *et al.*, 1997; Blairy *et al.*, 2000; Massat *et al.*, 2000; Tut *et al.*, 2000; Murphy *et al.*, 2001; Ni *et al.*, 2002; Etain *et al.*, 2004). The contribution of 5-HT_{2A} receptor gene polymorphisms to unipolar depression is less clear, with some studies reporting an association (Zhang *et al.*, 1997; Du *et al.*, 2000; Choi *et al.*, 2004; Eley *et al.*, 2004; Christiansen *et al.*, 2007) and others reporting a lack of association (Ohara *et al.*, 1998; Arias *et al.*, 2001a, 2001b; Minov *et al.*, 2001; Anguelova *et al.*, 2003; Oswald *et al.*, 2003). However, many studies have reported that 5-HT_{2A} receptor gene polymorphisms are associated with the clinical efficacy of SSRI antidepressant drugs (Choi *et al.*, 2005: Kato *et al.*, 2006; McMahon *et al.*, 2006; Denys *et al.*, 2007).

Although the predominant phenotype of the 5-HT_{2A} receptor knockout mice is reduced anxiety, the role of the 5-HT_{2A} receptor in anxiety disorders has not been substantiated in human genetics association studies. The most consistent finding of the human genetic association studies was that the 5-HT_{2A} receptor plays an important role in the efficacy of antidepressant drugs. However, such studies have not yet been conducted in the 5-HT_{2A} receptor knockout mice, thus making it difficult to define the precise role of this receptor in the mechanism of antidepressant action.

5-HT_{2B} receptor

The function of 5-HT_{2B} receptors in the brain remains largely unexplored. This is primarily due to a lack of pharmacological agents that show selectivity for the 5-HT_{2B} receptor subtype (Doly *et al.*, 2008). However, there are several pharmacological studies which suggest that this receptor may play a role in anxiety (Kennett *et al.*, 1998) and sleep regulation (Popa *et al.*, 2005). A 5-HT_{2B} receptor knockout mouse has also been generated (Nebigil *et al.*, 2000), and this mouse is insensitive to MDMA-induced hyperlocomotion and 5-HT release in the nucleus accumbens and ventral tegmental area (Doly *et al.*, 2008). Whether this mouse exhibits behavioral changes related to psychiatric disorders such as anxiety and depression has not yet been reported.

5-HT_{2C} receptor

The 5-HT_{2C} receptor (formerly 5-HT_{1C}) (Baxter *et al.*, 1995) is a postsynaptically located, seven-transmembrane spanning receptor present in highest concentrations in the choroid plexus, but significant densities are also found in the subthalamic nucleus, hypothalamus, hippocampus and amygdala (Mengod *et al.*, 1990). The 5-HT_{2C} receptor is believed to have an integral function in the control of many physiological and behavioral responses, including feeding, anxiety, temperature regulation, locomotion, sexual behavior and the occurrence of seizures, in addition to a role in mood disorders (Lucki *et al.*, 1989; Baxter *et al.*, 1995; Cryan and Lucki, 2000; Gurevich *et al.*, 2002; Englander *et al.*, 2005).

Preclinical studies One of the most intensely investigated phenotypes of the 5-HT_{2C} receptor knockout mouse is its obesity. 5-HT_{2C} knockout mice consume more food despite normal satiety responses to leptin administration, and this chronic hyperphagia leads to late-onset obesity (Tecott *et al.*, 1995; Nonogaki *et al.*, 1998). However, it was later reported that this obesity only develops after 6 months of age and without further increases in the level of hyperphagia (Nonogaki *et al.*, 2003). This obesity has been associated with a partial leptin resistance, insulin resistance and impaired glucose tolerance, thus suggesting that disruption of 5-HT_{2C} receptor function might play a role in the predisposition to type 2 diabetes (Nonogaki *et al.*, 1998). Furthermore, it has been suggested that an age-dependent decline in the energy cost of physical activity could contribute to the development of late-onset obesity in these mice (Nonogaki *et al.*, 2003).

The role of the 5-HT_{2C} receptor in the antidepressant response has also been investigated. While pharmacological studies have reported that 5-HT_{2C} receptor antagonists have antidepressant-like effects in a variety of behavioral tests, including the rat FST and the olfactory bulbectomy model (Cryan and Lucki, 2000; Rosenzweig-Lipson *et al.*, 2007), 5-HT_{2C} receptor knockout mice do not exhibit such behavior in the TST (Cremers *et al.*, 2004). However, these mice do exhibit enhanced behavioral responses to the antidepressant fluoxetine when tested in the TST (Cremers *et al.*, 2004). Taken together, data suggest that the 5-HT_{2C} receptor may be an important target in antidepressant drug development, but more studies are required to elucidate its precise contribution to depression and antidepressant efficacy.

Other phenotypes reported in the 5-HT_{2C} receptor knockout mice include reduced aversion to a novel environment (Tecott *et al.*, 1998), alterations in HPA axis activity (Chou-Green *et al.*, 2003a; Heisler *et al.*, 2007), impaired spatial learning in the Morris water maze (Tecott *et al.*, 1998), compulsive-like behavior (Chou-Green *et al.*, 2003b), sleep abnormalities including more wakefulness and enhanced responses to sleep deprivation (Frank *et al.*, 2002), and enhanced susceptibility to spontaneous and chemically- or electrically-induced seizures (Tecott *et al.*, 1995; Applegate and Tecott, 1998).

Human studies Similarly to preclinical studies, genetic association studies of the 5-HT_{2C} receptor gene have revealed a link between this receptor and body weight in humans. Specifically, the frequency of the Ser allele of the Cys23Ser variant of the 5-HT_{2C} receptor gene has been reported to be higher in patients with anorexia nervosa compared to controls (Westberg *et al.*, 2002; Hu *et al.*, 2003). Moreover, the −759C/T polymorphism has been implicated in the weight gain induced by antipsychotic drug treatment. It is in general agreement that the T allele of this polymorphism is protective against antipsychotic drug-induced weight gain (Reynolds *et al.*, 2002; Ellingrod *et al.*, 2005; Miller *et al.*, 2005; Templeman *et al.*, 2005; De Luca *et al.*, 2007; Ryu *et al.*, 2007), although a lack of association has also been reported (Tsai *et al.*, 2002). In addition, various other polymorphisms of the 5-HT_{2C} receptor gene have also been associated with antipsychotic drug-induced weight gain (Mulder *et al.*, 2007). Furthermore, the frequency of the T allele has been reported to be lower in obese women (Pooley *et al.*, 2004), and various other polymorphisms of the 5-HT_{2C} receptor gene have been associated with obesity in men (Yuan *et al.*, 2000). Moreover, a recent study reported that cell lines with haplotypes containing the -759T allele demonstrated lower expression of the 5-HT_{2C} gene (Buckland *et al.*, 2005), thus suggesting that antipsychotic drug-induced weight gain may be a result of reduced 5-HT_{2C} receptor expression. In support, constitutive knockout of the 5-HT_{2C} receptor results in hyperphagia and obesity in mice (see above).

The role of 5-HT_{2C} receptor gene polymorphisms in mood disorders has also been examined. A significant association between the Ser23 allele and bipolar disorder has been reported, but this is generally restricted to female patients and patients with early onset (Gutierrez *et al.*, 1996, 2001; Oruc *et al.*, 1997; Lerer *et al.*, 2001; Massat *et al.*, 2007). While one study reported an excess of Ser23 carriers in patients with major depressive disorder (Lerer *et al.*, 2001), another reported a lack of association of 5-HT_{2C} polymorphisms with depression (Frisch *et al.*, 1999). Finally, a lack of association between suicide and the Cys23Ser polymorphism has been reported (Serretti *et al.*, 2007).

The role of 5-HT_{2C} receptor gene polymorphisms has been examined in various other disorders. No association was found with panic disorder (Deckert *et al.*, 2000), sleep apnoea (Sakai *et al.*, 2005) or seasonal affective disorder (Johansson *et al.*, 2001). An association of the-759C/T polymorphism with the clinical response to antipsychotic drugs has been reported (Reynolds *et al.*, 2005), while conflicting reports on the role of the Cys23Ser variant have been published (Sodhi *et al.*, 1995; Malhotra *et al.*, 1996; Rietschel *et al.*, 1997).

Taken together, both preclinical studies and human genetic studies clearly implicate a role for the 5-HT_{2C} receptor in eating disorders. While pharmacological and genetic studies in animals suggest that functional 5-HT_{2C} receptors might be important in the mechanism of antidepressant action, such findings must be further explored particularly in human studies.

5-HT$_3$ receptor

5-HT$_3$ receptors differ from the other 5-HT receptors in that they are ligand-gated ion channels. They are found either as homomeric receptors composed of five 5-HT$_{3A}$ receptor subunits or as complex heteromers composed of at least one 5-HT$_{3A}$ receptor subunit in combination with one or more of the 5-HT$_{3B}$, 5-HT$_{3C}$, 5-HT$_{3D}$ and 5-HT$_{3E}$ receptor subunits (Barnes and Sharp, 1999; Hoyer *et al.*, 2002; Jensen *et al.*, 2008).

Both 5-HT$_{3A}$ receptor knockout mice and 5-HT$_3$-over-expressing mice have been generated. However, the phenotype of mice over-expressing the 5-HT$_3$ receptor has been characterized more extensively. The predominant phenotype of 5-HT$_3$ receptor-over-expression in mice is enhanced sensitivity to drugs of abuse (Engel *et al.*, 1998; Engel and Allan, 1999; Allan *et al.*, 2001). In addition to changes in sensitivity to drugs of abuse, 5-HT$_3$ receptor-over-expressing transgenic mice demonstrate a number of other behavioral phenotypes, including enhanced contextual fear conditioning, enhanced exploratory behavior to novel environmental stimuli and decreased anxiety in the elevated plus maze (Harrell and Allan, 2003). Importantly, however, it should be noted that some of the described behavioral characteristics of 5-HT$_3$ receptor-over-expressing mice are dependent upon the background strain. For example, while the reduced anxiety and enhanced contextual conditioning phenotype of 5-HT$_3$ receptor-over-expression on a B6SJL/F2 background were recapitulated by 5-HT$_3$ receptor-over-expression on a C57BL/6J background, no such phenotype was apparent on a DBA/2J background (Metz *et al.*, 2006).

The 5-HT$_{3A}$ receptor knockout mice have been characterized in anxiety tests, but conflicting results have been reported. Kelley and colleagues described reductions in anxiety-like behavior in 5-HT$_{3A}$ receptor knockout mice when tested in the elevated plus maze, light–dark box and novel object test (Kelley *et al.*, 2003). Similarly, decreased anxiety in the elevated plus maze was reported by Bhatagnar and colleagues, but the same study did not find any differences in anxiety-like behavior in the light–dark test or the open field (Bhatnagar *et al.*, 2004b). Since both over-expression and deletion of the 5-HT$_3$ receptor generally reduces anxiety, the exact role of the 5-HT$_3$ receptor in anxiety remains ambiguous.

5-HT$_{3A}$ receptor knockout mice also exhibit attenuated stress-induced increases in ACTH when compared to wild-type littermates (Bhatnagar *et al.*, 2004b). Enhanced fear conditioning has been reported in 5-HT$_{3A}$ receptor knockout mice (Bhatnagar *et al.*, 2004b), but enhanced fear conditioning has also been reported in 5-HT$_3$ receptor-over-expressing mice (Harrell and Allan, 2003), thus making it difficult to decipher the exact contribution of the 5-HT$_3$ receptor to this form of learning and memory. Finally, deletion of the 5-HT$_3$ receptor also increases immobility in female but not male mice when tested in the FST, thus suggesting a potential role for this receptor in gender differences in antidepressant-like behaviors (Bhatnagar *et al.*, 2004a).

Human genetic studies of the 5-HT$_3$ receptor have largely focused on its role in the efficacy of anti-emetic drugs, and few studies have examined its role in human behavior and brain disorders. A lack of association of the C178T 5-HT$_{3A}$ receptor gene variant with early onset obsessive-compulsive disorder has been reported (Mossner *et al.*, 2007). However, associations with mood disorders have been reported, with one study reporting differences in the frequency of the C178T variant in bipolar disorder patients compared to controls (Niesler *et al.*, 2001), and more recently polymorphisms in the 5-HT3B receptor gene have been associated with major depression (Yamada *et al.*, 2006) and bipolar disorder (Frank *et al.*, 2004).

The results from rodent studies suggest that the 5-HT$_3$ receptor may play a role in the neurobiology of anxiety; however, as far as we are aware human genetic studies of the 5-HT$_3$ receptor gene have not yet thoroughly investigated anxiety disorders, although one study did report a lack of association with OCD (Mossner *et al.*, 2007). Finally, while there is some evidence to suggest that the 5-HT$_3$ receptor may play a role in mood disorders, more exhaustive animal and human studies are required.

5-HT$_4$ receptor

The 5-HT$_4$ receptor is a postsynaptically located seven-transmembrane spanning receptor present at highest densities in limbic brain areas such as the olfactory tubercules, septum, hippocampus and amygdala as well as the basal ganglia (Bockaert *et al.*, 1997).

Despite the generation of 5-HT$_4$ receptor knockout mice a number of years ago, little has been reported regarding their behavioral phenotype. However, it has been demonstrated that the 5-HT$_4$ receptor is required for both stress- and MDMA-induced hypophagia (Compan *et al.*, 2004; Jean *et al.*, 2007), and the absence of MDMA-induced anorexia in 5-HT$_4$ knockout mice has been replicated by both pharmacological blockade and siRNA-mediated 5-HT$_4$ receptor knockdown in the nucleus accumbens (Jean *et al.*, 2007). In addition to changes in feeding behavior, decreased reactivity to novelty, increased anxiety and enhanced pentylene-induced seizures have been reported in these mice (Compan *et al.*, 2004). Although the behavior of these mice in behavioral tests of antidepressant activity has yet to be reported, a recent pharmacological study reported that treatment with 5-HT_4 receptor agonists has antidepressant effects in the FST, olfactory bulbectomy model and the chronic mild stress paradigm (Lucas *et al.*, 2007).

Similarly to animal studies, few studies have examined the role of polymorphisms in the human 5-HT_4 receptor gene in human behavior and brain disorders. Nevertheless, it was recently reported that the 5-HT_4 receptor gene may contribute to genetic predisposition to ADHD and schizophrenia (Suzuki *et al.*, 2003; Li *et al.*, 2006). Finally, several polymorphisms of the 5-HT_4 receptor gene have been associated with bipolar disorder (Ohtsuki *et al.*, 2002).

Since extensive preclinical and human genetic studies of the 5-HT_4 receptor are lacking, the role of this receptor in brain and behavior remains to be more extensively characterized.

5-HT_{5A} receptor

The 5-HT_5 receptor family consists of two members designated 5-HT_{5A} and 5-HT_{5B}. To date, the 5-HT_{5A} receptor has been identified in the mouse, rat and human. The 5-HT_{5B} receptor is also expressed in the mouse and rat, but not in the human where the coding sequence is interrupted by stop codons (Nelson, 2004; Thomas, 2006). Within the CNS, the 5-HT_{5A} receptor shows a relatively broad distribution in mouse (Plassat *et al.*, 1992; Matthes *et al.*, 1993), rat (Erlander *et al.*, 1993) and human (Rees *et al.*, 1994; Pasqualetti *et al.*, 1998) brain, including the cerebral cortex, hippocampus, thalamus, hypothalamus, habenula, cerebellum and spinal cord. This widespread mRNA localization pattern contrasts with that reported for the 5-HT_{5B} receptor, the mRNA for which appears to be restricted to the hippocampus, habenula and dorsal raphe nucleus in mouse and rat (Erlander *et al.*, 1993; Matthes *et al.*, 1993; Serrats *et al.*, 2004). Pharmacologically, both receptors show their highest affinity for lysergic acid (LSD), which appears to act as a partial agonist at the 5-HT_{5A} receptor (Nelson, 2004). In terms of animal behavior, 5-HT_{5A} receptor knockout mice exhibit increased exploratory activity in a novel environment but no change in home-cage locomotor activity or anxiety-related behaviors (Grailhe *et al.*, 1999). These mice also demonstrate a reduced locomotor response to LSD, a drug which can produce a psychotic state in humans (Grailhe *et al.*, 1999).

The contribution of 5-HT_{5A} receptor polymorphisms to the risk of bipolar and unipolar depression has been investigated. Birkett *et al.*, (2000) reported that the frequency of the T/T genotype of the 12A/T polymorphism and the C allele of the −19G/C polymorphism was higher in unipolar depression; however, a later study did not replicate these findings (Arias *et al.*, 2001c). Similarly, a protective effect of the G allele of the −19G/C polymorphism was reported in bipolar disorder, but this was not replicated by a later study (Birkett *et al.*, 2000; Arias *et al.*, 2001c). Since wide-scale preclinical and human genetic studies of the 5-HT_{5A} receptor are lacking, the role of this receptor in brain and behavior remains to be elucidated.

5-HT_6 receptor

5-HT_6 receptors are receiving increasing attention as a potential therapeutic target in mood disorders (Branchek and Blackburn, 2000; Schechter *et al.*, 2005). 5-HT_6 receptors are abundant in limbic and cortical regions, and are positively coupled via G_s to the AC5 subtype of adenylate cyclase (Baker *et al.*, 1998; Hamblin *et al.*, 1998). Stimulation of 5-HT_6 receptors in the striatum increases phosphorylation of DARPP-32 via the protein kinase A phosphorylation site (Thr34; Svenningsson *et al.*, 2002), and it is conceivable that some of the biochemical actions of 5-HT_6 receptor activation occur via DARPP-32 phosphorylation. As far as we are aware, 5-HT_6 receptor knockout mice have been generated by the Tecott laboratory (US Patent 6060642) and have been reported to have an anxiety phenotype, but this has not been published in a peer-reviewed journal to date. However, the expression of the receptor in the rat brain has been manipulated by using antisense oligonucleotides or viral-mediated gene transfer. Continuous intracerebroventricular administration of an antisense oligonucleotide against the 5-HT_6 receptor for 3–4 days increased anxiety-like behaviors in rats when tested in the social interaction test and the elevated plus maze (Hamon *et al.*, 1999). However, some pharmacological studies have reported that a 5-HT_6 receptor antagonist reduces anxiety and also affects antidepressant-like behaviors (Svenningsson *et al.*, 2007; Wesolowska and Nikiforuk, 2007, 2008; Wesolowska *et al.*, 2007). 5-HT_6 receptor agonism has also been shown to induce antidepressant-like behavior (Svenningsson *et al.*, 2007). In a more recent study, viral-mediated overexpression of the 5-HT_6 receptor gene in the rat nucleus accumbens prevented conditioned place preference for cocaine, but had no effect on either the acute locomotor response to cocaine or the development of cocaine-induced locomotor sensitization (Ferguson *et al.*, 2008). Other than these studies, the role of the 5-HT_6 receptor remains largely uncharacterized.

In human genetic studies, the role of the C267T polymorphism in the 5-HT_6 receptor gene has been predominantly investigated for association with schizophrenia and response to antipsychotic drugs. While one study reported that the T allele of this polymorphism was associated with an increased risk for schizophrenia (Tsai *et al.*, 1999), other studies reported a lack of association between this polymorphism as well as other polymorphisms with schizophrenia (Shinkai *et al.*, 1999; Vogt *et al.*, 2000; Ohmori *et al.*, 2001). Similarly, while one study reported that the T/T genotype confers better therapeutic efficacy of the antipsychotic drug risperidone (Lane *et al.*, 2004), this polymorphism was not associated with the efficacy of another antipsychotic drug, clozapine (Masellis *et al.*, 2001). The association of the C267T polymorphism with mood disorders has also been investigated, and reports of increased frequency of the C allele in bipolar disorder (Vogt *et al.*, 2000) and of no association with major depressive disorder (Lee *et al.*, 2005) have been published. Finally, a lack of association with suicide has been reported (Okamura *et al.*, 2005).

Taken together, the use of genetic studies to unravel the role of the 5-HT$_6$ receptor in behavior remains largely unexplored. The availability of 5-HT$_6$ receptor knockout animals would significantly aid in the characterization of the role this receptor in behavior and brain disorders.

5-HT$_7$ receptor

The 5-HT$_7$ receptor is the most recently described 5-HT receptor member (Bard *et al.*, 1993; Lovenberg *et al.*, 1993; Hedlund and Sutcliffe, 2004). Functionally, 5-HT$_7$ receptor activation increases cyclic AMP formation. In the rodent and human brain, the highest receptor densities are found in the thalamus, hypothalamus (including the suprachiasmatic nucleus), amygdala, hippocampus, cortex and dorsal raphe (To *et al.*, 1995; Thomas *et al.*, 2002; Varnas *et al.*, 2004). In the CNS, important physiological roles for the 5-HT$_7$ receptor have been established in thermoregulation and circadian rhythmicity (Lovenberg *et al.*, 1993; Hagan *et al.*, 2000; Glass *et al.*, 2003; Guscott *et al.*, 2003; Hedlund *et al.*, 2003). Moreover, recent pharmacological data suggest a role for 5-HT$_7$ in antidepressant-related behaviors (Wesolowska *et al.*, 2006; Bonaventure *et al.*, 2007). Such assertions are bolstered by a number of genetic studies. Despite its only recent generation, the 5-HT$_7$ receptor knockout mouse has been well-characterized. These mice exhibit reduced baseline immobility in both the TST and the FST (Guscott *et al.*, 2005; Hedlund *et al.*, 2005). Irrespective of the already reduced baseline immobility, 5-HT$_7$ receptor knockout mice still respond to the SSRI citalopram in these tests, thus suggesting that the 5-HT$_7$ receptor does not play a role in acute behavioral effects of at least this particular SSRI (Hedlund *et al.*, 2005). Consistent with the antidepressant-like phenotype in the TST and the FST, 5-HT$_7$ receptor knockout mice also exhibit reductions in the frequency and time spent in REM sleep – an effect often induced by antidepressant treatment (Hedlund *et al.*, 2005). In extracellular recordings from hypothalamic slices of 5-HT$_7$ receptor knockout mice, 8-OH-DPAT-induced circadian rhythm phase shifts are attenuated, thus suggesting a role for the 5-HT$_7$ receptor in the regulation of circadian rhythms (Guscott *et al.*, 2005). Disruptions in circadian rhythms are often reported in depression. In addition to their antidepressant-like phenotype, 5-HT$_7$ receptor knockout mice exhibit reduced stereotypic or compulsive behavior in the marble burying test (Hedlund and Sutcliffe, 2007), impaired contextual fear conditioning and hippocampal LTP (Roberts *et al.*, 2004), but no changes in other forms of hippocampus-dependent learning or motor learning (Roberts *et al.*, 2004; Guscott *et al.*, 2005). Finally, no changes in anxiety or locomotor activity were reported in the 5-HT$_7$ receptor knockout mice (Roberts *et al.*, 2004).

The association of 5-HT$_7$ receptor gene polymorphisms with human behavior and brain disorders remains largely unexplored. Early studies did not find any association of the proline279 to leucine amino acid substitution with alcoholism (Pesonen *et al.*, 1998), obesity (Hinney *et al.*, 1999), anorexia nervosa (Hinney *et al.*, 1999), bipolar disorder (Erdmann *et al.*, 1996) or schizophrenia (Erdmann *et al.*, 1996). However, a more recent study reported the association of two SNPs with susceptibility to schizophrenia in a Japanese population (Ikeda *et al.*, 2006).

While animal studies suggest a clear association of the 5-HT$_7$ receptor in depression- and antidepressant-like behaviors, these phenotypes have yet to be fully investigated in human genetic association studies.

Summary and concluding remarks

A vast number of both preclinical and human studies provide supportive evidence of a role for the serotonergic system in various behavioral responses related to depression and anxiety, as well as many other psychiatric disorders. In particular, dysfunction of TPH, SERT or 5-HT$_{1A}$ receptors can induce behaviors in rodents that are associated with anxiety and depression, and can also alter behavioral responses to antidepressant treatments. Moreover, many of these findings are supported by human genetic association studies. Studies in mice also suggest that interference with SERT or 5-HT$_{1A}$ receptor function specifically during brain development can programme anxiety levels and depression in adulthood. Finally, while there is some evidence to suggest that other components of the serotonergic system might also play important roles in such disorders, more extensive investigations using genetically modified mice, selective pharmacological tools or human genetic association studies are warranted.

Acknowledgement

Olivia O'Leary holds a postdoctoral fellowship from the Health Research Board, Ireland (PD/208/26).

References

Abbar, M., Courtet, P., Bellivier, F., Leboyer, M., Boulenger, J.P., Castelhau, D., Ferreira, M., Lambercy, C., Mouthon, D., Paoloni-Giacobino, A., Vessaz, M., Malafosse, A. and Buresi, C. (2001) Suicide attempts and the tryptophan hydroxylase gene. Mol. Psychiatry, 6: 268–273.

Alexandre, C., Popa, D., Fabre, V., Bouali, S., Venault, P., Lesch, K.P., Hamon, M. and Adrien, J. (2006) Early life blockade of 5-hydroxytryptamine 1A receptors normalizes sleep and depression-like behavior in adult knockout mice lacking the serotonin transporter. J. Neurosci., 26: 5554–5564.

Allan, A.M., Galindo, R., Chynoweth, J., Engel, S.R. and Savage, D.D. (2001) Conditioned place preference for cocaine is attenuated in mice over-expressing the 5-HT(3) receptor. Psychopharmacology (Berl.), 158: 18–27.

Anguelova, M., Benkelfat, C. and Turecki, G. (2003) A systematic review of association studies investigating genes coding for serotonin receptors and the serotonin transporter: I. Affective disorders. Mol. Psychiatry, 8: 574–591.

Ansorge, M.S., Morelli, E. and Gingrich, J.A. (2008) Inhibition of serotonin but not norepinephrine transport during development produces delayed, persistent perturbations of emotional behaviors in mice. J. Neurosci., 28: 199–207.

Ansorge, M.S., Zhou, M., Lira, A., Hen, R. and Gingrich, J.A. (2004) Early-life blockade of the 5-HT transporter alters emotional behavior in adult mice. Science, 306: 879–881.

Anttila, S., Huuhka, K., Huuhka, M., Rontu, R., Hurme, M., Leinonen, E. and Lehtimaki, T. (2007) Interaction between 5-HT1A and BDNF genotypes increases the risk of treatment-resistant depression. J. Neural. Transm., 114: 1065–1068.

Applegate, C.D. and Tecott, L.H. (1998) Global increases in seizure susceptibility in mice lacking 5-HT2C receptors: a behavioral analysis. Exp. Neurol., 154: 522–530.

Arias, B., Gutierrez, B., Pintor, L., Gasto, C. and Fananas, L. (2001a) Variability in the 5-HT(2A) receptor gene is associated with seasonal pattern in major depression. Mol. Psychiatry, 6: 239–242.

Arias, B., Gasto, C., Catalan, R., Gutierrez, B., Pintor, L. and Fananas, L. (2001b) The 5-HT(2A) receptor gene 102T/C polymorphism is associated with suicidal behavior in depressed patients. Am. J. Med. Genet., 105: 801–804.

Arias, B., Collier, D.A., Gasto, C., Pintor, L., Gutierrez, B., Valles, V. and Fananas, L. (2001c) Genetic variation in the 5-HT5A receptor gene in patients with bipolar disorder and major depression. Neurosci. Lett., 303: 111–114.

Arias, B., Catalan, R., Gasto, C., Gutierrez, B. and Fananas, L. (2005) Evidence for a combined genetic effect of the 5-HT(1A) receptor and serotonin transporter genes in the clinical outcome of major depressive patients treated with citalopram. J. Psychopharmacol., 19: 166–172.

Arranz, M.J., Erdmann, J., Kirov, G., Rietschel, M., Sodhi, M., Albus, M., Ball, D., Maier, W., Davies, N., Franzek, E., Abusaad, I., Weigelt, B., Murray, R., Shimron-Abarbanell, D., Kerwin, R., Propping, P., Sham, P., Nothen, M.M. and Collier, D.A. (1997) 5-HT2A receptor and bipolar affective disorder: association studies in affected patients. Neurosci. Lett., 224: 95–98.

Azmitia, E.C. and Segal, M. (1978) An autoradiographic analysis of the differential ascending projections of the dorsal and median raphe nuclei in the rat. J. Comp. Neurol., 179: 641–667.

Baca-Garcia, E., Vaquero-Lorenzo, C., Diaz-Hernandez, M., Rodriguez-Salgado, B., Dolengevich-Segal, H., Arrojo-Romero, M., Botillo-Martin, C., Ceverino, A., Piqueras, J.F., Perez-Rodriguez, M.M. and Saiz-Ruiz, J. (2007) Association between obsessive-compulsive disorder and a variable number of tandem repeats polymorphism in intron 2 of the serotonin transporter gene. Prog. Neuropsychopharmacol. Biol. Psychiatry, 31: 416–420.

Baker, L.P., Nielsen, M.D., Impey, S., Metcalf, M.A., Poser, S.W., Chan, G., Obrietan, K., Hamblin, M.W. and Storm, D.R. (1998) Stimulation of type 1 and type 8 Ca^{2+}-sensitive adenylyl cyclases by the Gs-coupled 5-hydroxytryptamine subtype 5-HT7A receptor. J. Biol. Chem., 273: 17469–17476.

Bard, J.A., Zgombick, J., Adham, N., Vaysse, P., Branchek, T.A. and Weinshank, R.L. (1993) Cloning of a novel human serotonin receptor (5-HT7) positively linked to adenylate cyclase. J. Biol. Chem., 268: 23422–23426.

Barnes, N.M. and Sharp, T. (1999) A review of central 5-HT receptors and their function. Neuropharmacology, 38: 1083–1152.

Baune, B.T., Hohoff, C., Roehrs, T., Deckert, J., Arolt, V. and Domschke, K. (2008) Serotonin receptor 1A-1019C/G variant: impact on antidepressant pharmacoresponse in melancholic depression? Neurosci. Lett., 436: 111–115.

Baxter, G., Kennett, G., Blaney, F. and Blackburn, T. (1995) 5-HT2 receptor subtypes: a family re-united? Trends Pharmacol. Sci., 16: 105–110.

Beaulieu, J.M., Zhang, X., Rodriguiz, R.M., Sotnikova, T.D., Cools, M.J., Wetsel, W.C., Gainetdinov, R.R. and Caron, M.G. (2008) Role of GSK3 beta in behavioral abnormalities induced by serotonin deficiency. Proc. Natl. Acad. Sci. USA, 105: 1333–1338.

Bellivier, F., Leboyer, M., Courtet, P., Buresi, C., Beaufils, B., Samolyk, D., Allilaire, J.F., Feingold, J., Mallet, J. and Malafosse, A. (1998) Association between the tryptophan hydroxylase gene and manic-depressive illness. Arch. Gen. Psychiatry, 55: 33–37.

Bellivier, F., Chaste, P. and Malafosse, A. (2004) Association between the TPH gene A218C polymorphism and suicidal behavior: a meta-analysis. Am. J. Med. Genet. B Neuropsychiatr. Genet., 124B: 87–91.

Belzung, C. and Griebel, G. (2001) Measuring normal and pathological anxiety-like behavior in mice: a review. Behav. Brain Res., 125: 141–149.

Benedetti, F., Colombo, C., Serretti, A., Lorenzi, C., Pontiggia, A., Barbini, B. and Smeraldi, E. (2003) Antidepressant effects of light therapy combined with sleep deprivation are influenced by a functional polymorphism within the promoter of the serotonin transporter gene. Biol. Psychiatry, 54: 687–692.

Bennett, P.J., McMahon, W.M., Watabe, J., Achilles, J., Bacon, M., Coon, H., Grey, T., Keller, T., Tate, D., Tcaciuc, I., Workman, J. and Gray, D. (2000) Tryptophan hydroxylase polymorphisms in suicide victims. Psychiatr. Genet., 10: 13–17.

Bert, B., Dere, E., Wilhelmi, N., Kusserow, H., Theuring, F., Huston, J.P. and Fink, H. (2005) Transient overexpression of

the 5-HT1A receptor impairs water-maze but not hole-board performance. Neurobiol. Learn. Mem., 84: 57–68.
Bert, B., Fink, H., Hortnagl, H., Veh, R.W., Davies, B., Theuring, F. and Kusserow, H. (2006) Mice over-expressing the 5-HT(1A) receptor in cortex and dentate gyrus display exaggerated locomotor and hypothermic response to 8-OH-DPAT. Behav. Brain Res., 167: 328–341.
Bhatnagar, S., Nowak, N., Babich, L. and Bok, L. (2004a) Deletion of the 5-HT3 receptor differentially affects behavior of males and females in the Porsolt forced swim and defensive withdrawal tests. Behav. Brain Res., 153: 527–535.
Bhatnagar, S., Sun, L.M., Raber, J., Maren, S., Julius, D. and Dallman, M.F. (2004b) Changes in anxiety-related behaviors and hypothalamic–pituitary–adrenal activity in mice lacking the 5-HT-3A receptor. Physiol. Behav., 81: 545–555.
Billett, E.A., Richter, M.A., King, N., Heils, A., Lesch, K.P. and Kennedy, J.L. (1997) Obsessive-compulsive disorder, response to serotonin reuptake inhibitors and the serotonin transporter gene. Mol. Psychiatry, 2: 403–406.
Birkett, J.T., Arranz, M.J., Munro, J., Osbourn, S., Kerwin, R.W. and Collier, D.A. (2000) Association analysis of the 5-HT5A gene in depression, psychosis and antipsychotic response. Neuroreport, 11: 2017–2020.
Blairy, S., Massat, I., Staner, L., Le Bon, O., Van Gestel, S., Van Broeckhoven, C., Hilger, C., Hentges, F., Souery, D. and Mendlewicz, J. (2000) 5-HT2a receptor polymorphism gene in bipolar disorder and harm avoidance personality trait. Am. J. Med. Genet., 96: 360–364.
Blakely, R.D., Defelice, L.J. and Galli, A. (2005) Biogenic amine neurotransmitter transporters: just when you thought you knew them. Physiology (Beth MD), 20: 225–231.
Bocchetta, A., Piccardi, M.P., Palmas, M.A., Chillotti, C., Oi, A. and Del Zompo, M. (1999) Family-based association study between bipolar disorder and DRD2, DRD4, DAT, and SERT in Sardinia. Am. J. Med. Genet., 88: 522–526.
Bockaert, J., Fagni, L. and Dumuis, A. (1997) 5-HT4 receptors: an update. In: Baumgarten, H.G. and Gothert, M. (Eds.), Serotoninergic Neurons and 5-HT Receptors in the CNS. Springer, Berlin, pp. 239–268.
Bockaert, J., Claeysen, S., Becamel, C., Dumuis, A. and Marin, P. (2006) Neuronal 5-HT metabotropic receptors: fine-tuning of their structure, signaling, and roles in synaptic modulation. Cell Tissue Res., 326: 553–572.
Bonaventure, P., Kelly, L., Aluisio, L., Shelton, J., Lord, B., Galici, R., Miller, K., Atack, J., Lovenberg, T.W. and Dugovic, C. (2007) Selective blockade of 5-hydroxytryptamine (5-HT)7 receptors enhances 5-HT transmission, antidepressant-like behavior, and rapid eye movement sleep suppression induced by citalopram in rodents. J. Pharmacol. Exp. Ther., 321: 690–698.
Bonnier, B., Gorwood, P., Hamon, M., Sarfati, Y., Boni, C. and Hardy-Bayle, M.C. (2002) Association of 5-HT(2A) receptor gene polymorphism with major affective disorders: the case of a subgroup of bipolar disorder with low suicide risk. Biol. Psychiatry, 51: 762–765.
Boutrel, B., Franc, B., Hen, R., Hamon, M. and Adrien, J. (1999) Key role of 5-HT1B receptors in the regulation of paradoxical sleep as evidenced in 5-HT1B knockout mice. J. Neurosci., 19: 3204–3212.
Boutrel, B., Monaca, C., Hen, R., Hamon, M. and Adrien, J. (2002) Involvement of 5-HT1A receptors in homeostatic and stress-induced adaptive regulations of paradoxical sleep: studies in 5-HT1A knockout mice. J. Neurosci., 22: 4686–4692.
Bouwknecht, J.A., Hijzen, T.H., van der Gugten, J., Maes, R.A., Hen, R. and Olivier, B. (2000) Ethanol intake is not elevated in male 5-HT(1B) receptor knockout mice. Eur. J. Pharmacol., 403: 95–98.
Bouwknecht, J.A., Hijzen, T.H., van der Gugten, J., Maes, R.A., Hen, R. and Olivier, B. (2001a) Absence of 5-HT(1B) receptors is associated with impaired impulse control in male 5-HT(1B) knockout mice. Biol. Psychiatry, 49: 557–568.
Bouwknecht, J.A., van der Gugten, J., Hijzen, T.H., Maes, R.A., Hen, R. and Olivier, B. (2001b) Male and female 5-HT(1B) receptor knockout mice have higher body weights than wild-types. Physiol. Behav., 74: 507–516.
Bozina, N., Mihaljevic-Peles, A., Sagud, M., Jakovljevic, M. and Sertic, J. (2006) Serotonin transporter polymorphism in Croatian patients with major depressive disorder. Psychiatr. Danub., 18: 83–89.
Bozina, N., Peles, A.M., Sagud, M., Bilusic, H. and Jakovljevic, M. (2007) Association study of paroxetine therapeutic response with SERT gene polymorphisms in patients with major depressive disorder. World J. Biol. Psychiatry: 1–8.
Branchek, T.A. and Blackburn, T.P. (2000) 5-ht6 receptors as emerging targets for drug discovery. Annu. Rev. Pharmacol. Toxicol., 40: 319–334.
Brummett, B.H., Krystal, A.D., Siegler, I.C., Kuhn, C., Surwit, R.S., Zuchner, S., Ashley-Koch, A., Barefoot, J.C. and Williams, R.B. (2007) Associations of a regulatory polymorphism of monoamine oxidase-A gene promoter (MAOA-uVNTR) with symptoms of depression and sleep quality. Psychosom. Med., 69: 396–401.
Brummett, B.H., Boyle, S.H., Siegler, I.C., Kuhn, C.M., Ashley-Koch, A., Jonassaint, C.R., Zuchner, S., Collins, A. and Williams, R.B. (2008) Effects of environmental stress and gender on associations among symptoms of depression and the serotonin transporter gene linked polymorphic region (5-HTTLPR). Behav. Genet., 38: 34–43.
Brunner, D., Buhot, M.C., Hen, R. and Hofer, M. (1999) Anxiety, motor activation, and maternal–infant interactions in 5HT1B knockout mice. Behav. Neurosci., 113: 587–601.
Brunner, H.G., Nelen, M., Breakefield, X.O., Ropers, H.H. and van Oost, B.A. (1993) Abnormal behavior associated with a point mutation in the structural gene for monoamine oxidase A. Science, 262: 578–580.
Buckland, P.R., Hoogendoorn, B., Guy, C.A., Smith, S.K., Coleman, S.L. and O'Donovan, M.C. (2005) Low gene expression conferred by association of an allele of the 5-HT2C receptor gene with antipsychotic-induced weight gain. Am. J. Psychiatry, 162: 613–615.
Buhot, M.C., Wolff, M., Benhassine, N., Costet, P., Hen, R. and Segu, L. (2003a) Spatial learning in the 5-HT1B receptor knockout mouse: selective facilitation/impairment depending on the cognitive demand. Learn. Mem., 10: 466–477.
Buhot, M.C., Wolff, M., Savova, M., Malleret, G., Hen, R. and Segu, L. (2003b) Protective effect of 5-HT1B receptor gene deletion on the age-related decline in spatial learning abilities in mice. Behav. Brain Res., 142: 135–142.
Camarena, B., Rinetti, G., Cruz, C., Hernandez, S., de la Fuente, J.R. and Nicolini, H. (2001) Association study of the serotonin transporter gene polymorphism in obsessive-compulsive disorder. Intl. J. Neuropsychopharmacol., 4: 269–272.
Carroll, J.C., Boyce-Rustay, J.M., Millstein, R., Yang, R., Wiedholz, L.M., Murphy, D.L. and Holmes, A. (2007) Effects of mild early life stress on abnormal emotion-related behaviors in 5-HTT knockout mice. Behav. Genet., 37: 214–222.

Cases, O., Seif, I., Grimsby, J., Gaspar, P., Chen, K., Pournin, S., Muller, U., Aguet, M., Babinet, C., Shih, J.C. et al. (1995) Aggressive behavior and altered amounts of brain serotonin and norepinephrine in mice lacking MAOA. Science, 268: 1763–1766.

Caspi, A., Sugden, K., Moffitt, T.E., Taylor, A., Craig, I.W., Harrington, H., McClay, J., Mill, J., Martin, J., Braithwaite, A. and Poulton, R. (2003) Influence of life stress on depression: moderation by a polymorphism in the 5-HTT gene. Science, 301: 386–389.

Cervo, L., Canetta, A., Calcagno, E., Burbassi, S., Sacchetti, G., Caccia, S., Fracasso, C., Albani, D., Forloni, G. and Invernizzi, R.W. (2005) Genotype-dependent activity of tryptophan hydroxylase-2 determines the response to citalopram in a mouse model of depression. J. Neurosci., 25: 8165–8172.

Chee, I.S., Lee, S.W., Kim, J.L., Wang, S.K., Shin, Y.O., Shin, S.C., Lee, Y.H., Hwang, H.M. and Lim, M.R. (2001) 5-HT2A receptor gene promoter polymorphism -1438A/G and bipolar disorder. Psychiatr. Genet., 11: 111–114.

Cho, H.J., Meira-Lima, I., Cordeiro, Q., Michelon, L., Sham, P., Vallada, H. and Collier, D.A. (2005) Population-based and family-based studies on the serotonin transporter gene polymorphisms and bipolar disorder: a systematic review and meta-analysis. Mol. Psychiatry, 10: 771–781.

Choi, M.J., Lee, H.J., Ham, B.J., Cha, J.H., Ryu, S.H. and Lee, M.S. (2004) Association between major depressive disorder and the -1438A/G polymorphism of the serotonin 2A receptor gene. Neuropsychobiology, 49: 38–41.

Choi, M.J., Kang, R.H., Ham, B.J., Jeong, H.Y. and Lee, M.S. (2005) Serotonin receptor 2A gene polymorphism (-1438A/G) and short-term treatment response to citalopram. Neuropsychobiology, 52: 155–162.

Chou-Green, J.M., Holscher, T.D., Dallman, M.F. and Akana, S.F. (2003a) Repeated stress in young and old 5-HT(2C) receptor knockout mice. Physiol. Behav., 79: 217–226.

Chou-Green, J.M., Holscher, T.D., Dallman, M.F. and Akana, S.F. (2003b) Compulsive behavior in the 5-HT2C receptor knockout mouse. Physiol. Behav., 78: 641–649.

Christiansen, L., Tan, Q., Iachina, M., Bathum, L., Kruse, T.A., McGue, M. and Christensen, K. (2007) Candidate gene polymorphisms in the serotonergic pathway: influence on depression symptomatology in an elderly population. Biol. Psychiatry, 61: 223–230.

Cichon, S., Winge, I., Mattheisen, M., Georgi, A., Karpushova, A., Freudenberg, J., Freudenberg-Hua, Y., Babadjanova, G., Van Den Bogaert, A., Abramova, L.I., Kapiletti, S., Knappskog, P.M., McKinney, J., Maier, W., Jamra, R.A., Schulze, T.G., Schumacher, J., Propping, P., Rietschel, M., Haavik, J. and Nothen, M.M. (2008) Brain-specific tryptophan hydroxylase 2 (TPH2): a functional Pro206Ser substitution and variation in the 5′-region are associated with bipolar affective disorder. Hum. Mol. Genet., 17: 87–97.

Clark, M.S., Sexton, T.J., McClain, M., Root, D., Kohen, R. and Neumaier, J.F. (2002) Overexpression of 5-HT1B receptor in dorsal raphe nucleus using Herpes Simplex Virus gene transfer increases anxiety behavior after inescapable stress. J. Neurosci., 22: 4550–4562.

Clark, M.S., Vincow, E.S., Sexton, T.J. and Neumaier, J.F. (2004) Increased expression of 5-HT1B receptor in dorsal raphe nucleus decreases fear-potentiated startle in a stress dependent manner. Brain Res., 1007: 86–97.

Collier, D.A., Stober, G., Li, T., Heils, A., Catalano, M., Di Bella, D., Arranz, M.J., Murray, R.M., Vallada, H.P., Bengel, D., Muller, C.R., Roberts, G.W., Smeraldi, E., Kirov, G., Sham, P. and Lesch, K.P. (1996) A novel functional polymorphism within the promoter of the serotonin transporter gene: possible role in susceptibility to affective disorders. Mol. Psychiatry, 1: 453–460.

Compan, V., Zhou, M., Grailhe, R., Gazzara, R.A., Martin, R., Gingrich, J., Dumuis, A., Brunner, D., Bockaert, J. and Hen, R. (2004) Attenuated response to stress and novelty and hypersensitivity to seizures in 5-HT4 receptor knockout mice. J. Neurosci., 24: 412–419.

Cote, F., Fligny, C., Bayard, E., Launay, J.M., Gershon, M.D., Mallet, J. and Vodjdani, G. (2007) Maternal serotonin is crucial for murine embryonic development. Proc. Natl. Acad. Sci. USA, 104: 329–334.

Craddock, N., Daniels, J., Roberts, E., Rees, M., McGuffin, P. and Owen, M.J. (1995) No evidence for allelic association between bipolar disorder and monoamine oxidase A gene polymorphisms. Am. J. Med. Genet., 60: 322–324.

Crawley, J.N., Belknap, J.K., Collins, A., Crabbe, J.C., Frankel, W., Henderson, N., Hitzemann, R.J., Maxson, S.C., Miner, L.L., Silva, A.J., Wehner, J.M., Wynshaw-Boris, A. and Paylor, R. (1997) Behavioral phenotypes of inbred mouse strains: implications and recommendations for molecular studies. Psychopharmacology (Berl.), 132: 107–124.

Cremers, T.I., Giorgetti, M., Bosker, F.J., Hogg, S., Arnt, J., Mork, A., Honig, G., Bogeso, K.P., Westerink, B.H., den Boer, H., Wikstrom, H.V. and Tecott, L.H. (2004) Inactivation of 5-HT(2C) receptors potentiates consequences of serotonin reuptake blockade. Neuropsychopharmacology, 29: 1782–1789.

Crowley, J.J., Blendy, J.A. and Lucki, I. (2005) Strain-dependent antidepressant-like effects of citalopram in the mouse tail suspension test. Psychopharmacology (Berl.), 183: 257–264.

Cryan, J.F. and Leonard, B.E. (2000) 5-HT1A and beyond: the role of serotonin and its receptors in depression and the antidepressant response. Hum. Psychopharmacol., 15: 113–135.

Cryan, J.F. and Lucki, I. (2000) Antidepressant-like behavioral effects mediated by 5-Hydroxytryptamine(2C) receptors. J. Pharmacol. Exp. Ther., 295: 1120–1126.

Cryan, J.F., O'Leary, O.F., Jin, S.H., Friedland, J.C., Ouyang, M., Hirsch, B.R., Page, M.E., Dalvi, A., Thomas, S.A. and Lucki, I. (2004) Norepinephrine-deficient mice lack responses to antidepressant drugs, including selective serotonin reuptake inhibitors. Proc. Natl. Acad. Sci. USA, 101: 8186–8191.

Dahlstrom, A. and Fuxe, K. (1964) Localization of monoamines in the lower brain stem. Experientia, 20: 398–399.

Dai, J.X., Han, H.L., Tian, M., Cao, J., Xiu, J.B., Song, N.N., Huang, Y., Xu, T.L., Ding, Y.Q. and Xu, L. (2008) Enhanced contextual fear memory in central serotonin-deficient mice. Proc. Natl. Acad. Sci. USA, 105: 11981–11986.

Davidge, K.M., Atkinson, L., Douglas, L., Lee, V., Shapiro, S., Kennedy, J.L. and Beitchman, J.H. (2004) Association of the serotonin transporter and 5HT1Dbeta receptor genes with extreme, persistent and pervasive aggressive behavior in children. Psychiatr. Genet., 14: 143–146.

De Luca, V., Mueller, D.J., de Bartolomeis, A. and Kennedy, J.L. (2007) Association of the HTR2C gene and antipsychotic induced weight gain: a meta-analysis. Intl. J. Neuropsychopharmacol., 10: 697–704.

Deckert, J., Catalano, M., Syagailo, Y.V., Bosi, M., Okladnova, O., Di Bella, D., Nothen, M.M., Maffei, P., Franke, P., Fritze, J., Maier, W., Propping, P., Beckmann, H., Bellodi, L. and Lesch, K.P. (1999) Excess of high activity monoamine oxidase A gene promoter alleles in female patients with panic disorder. Hum. Mol. Genet., 8: 621–624.

Deckert, J., Meyer, J., Catalano, M., Bosi, M., Sand, P., DiBella, D., Ortega, G., Stober, G., Franke, P., Nothen, M.M., Fritze, J., Maier, W., Beckmann, H., Propping, P., Bellodi, L. and Lesch, K.P. (2000) Novel 5′-regulatory region polymorphisms of the 5-HT2C receptor gene: association study with panic disorder. Intl. J. Neuropsychopharmacol., 3: 321–325.

Delgado, P.L., Price, L.H., Miller, H.L., Salomon, R.M., Licinio, J., Krystal, J.H., Heninger, G.R. and Charney, D.S. (1991) Rapid serotonin depletion as a provocative challenge test for patients with major depression: relevance to antidepressant action and the neurobiology of depression. Psychopharmacol. Bull., 27: 321–330.

Delorme, R., Durand, C.M., Betancur, C., Wagner, M., Ruhrmann, S., Grabe, H.J., Nygren, G., Gillberg, C., Leboyer, M., Bourgeron, T., Courtet, P., Jollant, F., Buresi, C., Aubry, J.M., Baud, P., Bondolfi, G., Bertschy, G., Perroud, N. and Malafosse, A. (2006) No human tryptophan hydroxylase-2 gene R441H mutation in a large cohort of psychiatric patients and control subjects. Biol. Psychiatry, 60: 202–203.

Denys, D., Van Nieuwerburgh, F., Deforce, D. and Westenberg, H.G. (2006) Association between serotonergic candidate genes and specific phenotypes of obsessive-compulsive disorder. J. Affect. Disord., 91: 39–44.

Denys, D., Van Nieuwerburgh, F., Deforce, D. and Westenberg, H.G. (2007) Prediction of response to paroxetine and venlafaxine by serotonin-related genes in obsessive-compulsive disorder in a randomized, double-blind trial. J. Clin. Psychiatry, 68: 747–753.

Dickel, D.E., Veenstra-VanderWeele, J., Bivens, N.C., Wu, X., Fischer, D.J., Van Etten-Lee, M., Himle, J.A., Leventhal, B.L., Cook, E.H. Jr. and Hanna, G.L. (2007) Association studies of serotonin system candidate genes in early-onset obsessive-compulsive disorder. Biol. Psychiatry, 61: 322–329.

Dimitrova, A., Georgieva, L., Nikolov, I., Poriazova, N., Krastev, S., Toncheva, D., Owen, M.J. and Kirov, G. (2002) Major psychiatric disorders and the serotonin transporter gene (SLC6A4): family-based association studies. Psychiatr. Genet., 12: 137–141.

Ding, Y.Q., Marklund, U., Yuan, W., Yin, J., Wegman, L., Ericson, J., Deneris, E., Johnson, R.L. and Chen, Z.F. (2003) Lmx1b is essential for the development of serotonergic neurons. Nat. Neurosci., 6: 933–938.

Doly, S., Valjent, E., Setola, V., Callebert, J., Herve, D., Launay, J.M. and Maroteaux, L. (2008) Serotonin 5-HT2B receptors are required for 3,4-methylenedioxymethamphetamine-induced hyperlocomotion and 5-HT release *in vivo* and *in vitro*. J. Neurosci., 28: 2933–2940.

Dorado, P., Penas-Lledo, E.M., Gonzalez, A.P., Caceres, M.C., Cobaleda, J. and Llerena, A. (2007) Increased risk for major depression associated with the short allele of the serotonin transporter promoter region (5-HTTLPR-S) and the CYP2C9*3 allele. Fundam. Clin. Pharmacol., 21: 451–453.

Du, L., Bakish, D., Lapierre, Y.D., Ravindran, A.V. and Hrdina, P.D. (2000) Association of polymorphism of serotonin 2A receptor gene with suicidal ideation in major depressive disorder. Am. J. Med. Genet., 96: 56–60.

Du, L., Bakish, D. and Hrdina, P.D. (2001) Tryptophan hydroxylase gene 218A/C polymorphism is associated with somatic anxiety in major depressive disorder. J. Affect. Disord., 65: 37–44.

Dubrovina, N.I., Popova, N.K., Gilinskii, M.A., Tomilenko, R.A. and Seif, I. (2006) Acquisition and extinction of a conditioned passive avoidance reflex in mice with genetic knockout of monoamine oxidase A. Neurosci. Behav. Physiol., 36: 335–339.

Eley, T.C., Sugden, K., Corsico, A., Gregory, A.M., Sham, P., McGuffin, P., Plomin, R. and Craig, I.W. (2004) Gene–environment interaction analysis of serotonin system markers with adolescent depression. Mol. Psychiatry, 9: 908–915.

Ellingrod, V.L., Perry, P.J., Ringold, J.C., Lund, B.C., Bever-Stille, K., Fleming, F., Holman, T.L. and Miller, D. (2005) Weight gain associated with the -759C/T polymorphism of the 5HT2C receptor and olanzapine. Am. J. Med. Genet. B Neuropsychiatr. Genet., 134B: 76–78.

Engel, S.R. and Allan, A.M. (1999) 5-HT3 receptor over-expression enhances ethanol sensitivity in mice. Psychopharmacology (Berl.), 144: 411–415.

Engel, S.R., Lyons, C.R. and Allan, A.M. (1998) 5-HT3 receptor over-expression decreases ethanol self administration in transgenic mice. Psychopharmacology (Berl.), 140: 243–248.

Englander, M.T., Dulawa, S.C., Bhansali, P. and Schmauss, C. (2005) How stress and fluoxetine modulate serotonin 2C receptor pre-mRNA editing. J. Neurosci., 25: 648–651.

Enoch, M.A., Greenberg, B.D., Murphy, D.L. and Goldman, D. (2001) Sexually dimorphic relationship of a 5-HT2A promoter polymorphism with obsessive-compulsive disorder. Biol. Psychiatry, 49: 385–388.

Erdmann, J., Nothen, M.M., Shimron-Abarbanell, D., Rietschel, M., Albus, M., Borrmann, M., Maier, W., Franzek, E., Korner, J., Weigelt, B., Fimmers, R. and Propping, P. (1996) The human serotonin 7 (5-HT7) receptor gene: genomic organization and systematic mutation screening in schizophrenia and bipolar affective disorder. Mol. Psychiatry, 1: 392–397.

Erlander, M.G., Lovenberg, T.W., Baron, B.M., de Lecea, L., Danielson, P.E., Racke, M., Slone, A.L., Siegel, B.W., Foye, P.E., Cannon, K. et al. (1993) Two members of a distinct subfamily of 5-hydroxytryptamine receptors differentially expressed in rat brain. Proc. Natl. Acad. Sci. USA, 90: 3452–3456.

Esterling, L.E., Yoshikawa, T., Turner, G., Badner, J.A., Bengel, D., Gershon, E.S., Berrettini, W.H. and Detera-Wadleigh, S.D. (1998) Serotonin transporter (5-HTT) gene and bipolar affective disorder. Am. J. Med. Genet., 81: 37–40.

Etain, B., Rousseva, A., Roy, I., Henry, C., Malafosse, A., Buresi, C., Preisig, M., Rayah, F., Leboyer, M. and Bellivier, F. (2004) Lack of association between 5HT2A receptor gene haplotype, bipolar disorder and its clinical subtypes in a West European sample. Am. J. Med. Genet. B Neuropsychiatr. Genet., 129B: 29–33.

Ewald, H., Flint, T., Degn, B., Mors, O. and Kruse, T.A. (1998) A functional variant of the serotonin transporter gene in families with bipolar affective disorder. J. Affect. Disord., 48: 135–144.

Fehr, C., Schleicher, A., Szegedi, A., Anghelescu, I., Klawe, C., Hiemke, C. and Dahmen, N. (2001) Serotonergic polymorphisms in patients suffering from alcoholism, anxiety disorders and narcolepsy. Prog. Neuropsychopharmacol. Biol. Psychiatry, 25: 965–982.

Ferguson, S.M., Mitchell, E.S. and Neumaier, J.F. (2008) Increased expression of 5-HT6 receptors in the nucleus accumbens blocks the rewarding but not psychomotor activating properties of cocaine. Biol. Psychiatry, 63: 207–213.

Fiskerstrand, C.E., Lovejoy, E.A. and Quinn, J.P. (1999) An intronic polymorphic domain often associated with susceptibility to affective disorders has allele dependent differential enhancer activity in embryonic stem cells. FEBS Lett., 458: 171–174.

Fitzpatrick, P.F. (1999) Tetrahydropterin-dependent amino acid hydroxylases. Annu. Rev. Biochem., 68: 355–381.

Frank, B., Niesler, B., Nothen, M.M., Neidt, H., Propping, P., Bondy, B., Rietschel, M., Maier, W., Albus, M. and Rappold, G. (2004) Investigation of the human serotonin receptor gene HTR3B in bipolar affective and schizophrenic patients. Am. J. Med. Genet. B Neuropsychiatr. Genet., 131B: 1–5.

Frank, M.G., Stryker, M.P. and Tecott, L.H. (2002) Sleep and sleep homeostasis in mice lacking the 5-HT2c receptor. Neuropsychopharmacology, 27: 869–873.

Frisch, A., Postilnick, D., Rockah, R., Michaelovsky, E., Postilnick, S., Birman, E., Laor, N., Rauchverger, B., Kreinin, A., Poyurovsky, M., Schneidman, M., Modai, I. and Weizman, R. (1999) Association of unipolar major depressive disorder with genes of the serotonergic and dopaminergic pathways. Mol. Psychiatry, 4: 389–392.

Fukui, M., Rodriguiz, R.M., Zhou, J., Jiang, S.X., Phillips, L.E., Caron, M.G. and Wetsel, W.C. (2007) Vmat2 heterozygous mutant mice display a depressive-like phenotype. J. Neurosci., 27: 10520–10529.

Furlong, R.A., Ho, L., Rubinsztein, J.S., Walsh, C., Paykel, E.S. and Rubinsztein, D.C. (1999) Analysis of the monoamine oxidase A (MAOA) gene in bipolar affective disorder by association studies, meta-analyses, and sequencing of the promoter. Am. J. Med. Genet., 88: 398–406.

Gaspar, P., Cases, O. and Maroteaux, L. (2003) The developmental role of serotonin: news from mouse molecular genetics. Nat. Rev. Neurosci., 4: 1002–1012.

Gingrich, J.A. and Hen, R. (2001) Dissecting the role of the serotonin system in neuropsychiatric disorders using knockout mice. Psychopharmacology (Berl.), 155: 1–10.

Gizatullin, R., Zaboli, G., Jonsson, E.G., Asberg, M. and Leopardi, R. (2006) Haplotype analysis reveals tryptophan hydroxylase (TPH) 1 gene variants associated with major depression. Biol. Psychiatry, 59: 295–300.

Gizatullin, R., Zaboli, G., Jonsson, E.G., Asberg, M. and Leopardi, R. (2008) The tryptophan hydroxylase (TPH) 2 gene unlike TPH-1 exhibits no association with stress-induced depression. J. Affect. Disord., 107: 175–179.

Glass, J.D., Grossman, G.H., Farnbauch, L. and DiNardo, L. (2003) Midbrain raphe modulation of nonphotic circadian clock resetting and 5-HT release in the mammalian suprachiasmatic nucleus. J. Neurosci., 23: 7451–7460.

Glatt, C.E., Carlson, E., Taylor, T.R., Risch, N., Reus, V.I. and Schaefer, C.A. (2005) Response to Zhang *et al.* (2005): loss-of-function mutation in tryptophan hydroxylase-2 identified in unipolar major depression. Neuron, 45: 11–16, Neuron **48,** 704–705; author reply 705–706.

Grados, M.A., Samuels, J., Shugart, Y.Y., Willour, V.L., Wang, Y., Cullen, B., Bienvenu, O.J., Hoehn-Saric, R., Valle, D., Liang, K.Y., Riddle, M.A., Wendland, J.R., Murphy, D.L., Nestadt, G. and Detera-Wadleigh, S. (2007) Rare plus common SERT variants in obsessive-compulsive disorder. Mol. Psychiatry, 12: 422–423.

Grahame-Smith, D.G. (1964) Tryptophan hydroxylation in brain. Biochem. Biophys. Res. Commun., 16: 586–592.

Grailhe, R., Waeber, C., Dulawa, S.C., Hornung, J.P., Zhuang, X., Brunner, D., Geyer, M.A. and Hen, R. (1999) Increased exploratory activity and altered response to LSD in mice lacking the 5-HT(5A) receptor. Neuron, 22: 581–591.

Griebel, G., Belzung, C., Perrault, G. and Sanger, D.J. (2000) Differences in anxiety-related behaviors and in sensitivity to diazepam in inbred and outbred strains of mice. Psychopharmacology (Berl.), 148: 164–170.

Groenink, L., Pattij, T., De Jongh, R., Van der Gugten, J., Oosting, R.S., Dirks, A. and Olivier, B. (2003) 5-HT1A receptor knockout mice and mice overexpressing corticotropin-releasing hormone in models of anxiety. Eur. J. Pharmacol., 463: 185–197.

Gross, C., Santarelli, L., Brunner, D., Zhuang, X. and Hen, R. (2000) Altered fear circuits in 5-HT(1A) receptor KO mice. Biol. Psychiatry, 48: 1157–1163.

Gross, C., Zhuang, X., Stark, K., Ramboz, S., Oosting, R., Kirby, L., Santarelli, L., Beck, S. and Hen, R. (2002) Serotonin1A receptor acts during development to establish normal anxiety-like behavior in the adult. Nature, 416: 396–400.

Gurevich, I., Tamir, H., Arango, V., Dwork, A.J., Mann, J.J. and Schmauss, C. (2002) Altered editing of serotonin 2C receptor pre-mRNA in the prefrontal cortex of depressed suicide victims. Neuron, 34: 349–356.

Guscott, M.R., Egan, E., Cook, G.P., Stanton, J.A., Beer, M.S., Rosahl, T.W., Hartmann, S., Kulagowski, J., McAllister, G., Fone, K.C. and Hutson, P.H. (2003) The hypothermic effect of 5-CT in mice is mediated through the 5-HT7 receptor. Neuropharmacology, 44: 1031–1037.

Guscott, M., Bristow, L.J., Hadingham, K., Rosahl, T.W., Beer, M.S., Stanton, J.A., Bromidge, F., Owens, A.P., Huscroft, I., Myers, J., Rupniak, N.M., Patel, S., Whiting, P.J., Hutson, P.H., Fone, K. C., Biello, S.M., Kulagowski, J.J. and McAllister, G. (2005) Genetic knockout and pharmacological blockade studies of the 5-HT7 receptor suggest therapeutic potential in depression. Neuropharmacology, 48: 492–502.

Gutierrez, B., Fananas, L., Arranz, M.J., Valles, V., Guillamat, R., van Os, J. and Collier, D. (1996) Allelic association analysis of the 5-HT2C receptor gene in bipolar affective disorder. Neurosci. Lett., 212: 65–67.

Gutierrez, B., Bertranpetit, J., Collier, D., Arranz, M.J., Valles, V., Guillamat, R., Van Os, J. and Fananas, L. (1997) Genetic variation of the 5-HT2A receptor gene and bipolar affective disorder. Hum. Genet., 100: 582–584.

Gutierrez, B., Pintor, L., Gasto, C., Rosa, A., Bertranpetit, J., Vieta, E. and Fananas, L. (1998a) Variability in the serotonin transporter gene and increased risk for major depression with melancholia. Hum. Genet., 103: 319–322.

Gutierrez, B., Arranz, M.J., Collier, D.A., Valles, V., Guillamat, R., Bertranpetit, J., Murray, R.M. and Fanas, L. (1998b) Serotonin transporter gene and risk for bipolar affective disorder: an association study in Spanish population. Biol. Psychiatry, 43: 843–847.

Gutierrez, B., Arias, B., Papiol, S., Rosa, A. and Fananas, L. (2001) Association study between novel promoter variants at the 5-HT2C receptor gene and human patients with bipolar affective disorder. Neurosci. Lett., 309: 135–137.

Gutierrez, B., Arias, B., Gasto, C., Catalan, R., Papiol, S., Pintor, L. and Fananas, L. (2004) Association analysis between a functional polymorphism in the monoamine oxidase A gene promoter and severe mood disorders. Psychiatr. Genet., 14: 203–208.

Gutierrez, B., Rosa, A., Papiol, S., Arrufat, F.J., Catalan, R., Salgado, P., Peralta, V., Cuesta, M.J. and Fananas, L. (2007) Identification of two risk haplotypes for schizophrenia and bipolar disorder in the synaptic vesicle monoamine transporter gene (SVMT). Am. J. Med. Genet. B Neuropsychiatr. Genet., 144B: 502–507.

Hagan, J.J., Price, G.W., Jeffrey, P., Deeks, N.J., Stean, T., Piper, D., Smith, M.I., Upton, N., Medhurst, A.D., Middlemiss, D.N., Riley, G.J., Lovell, P.J., Bromidge, S.M. and Thomas, D.R. (2000) Characterization of SB-269970-A, a selective 5-HT(7) receptor antagonist. Br. J. Pharmacol., 130: 539–548.

Ham, B.J., Lee, B.C., Paik, J.W., Kang, R.H., Choi, M.J., Choi, I.G. and Lee, M.S. (2007) Association between the tryptophan hydroxylase-1 gene A218C polymorphism and citalopram antidepressant response in a Korean population. Prog. Neuropsychopharmacol. Biol. Psychiatry, 31: 104–107.

Hamblin, M.W., Guthrie, C.R., Kohen, R. and Heidmann, D.E. (1998) Gs protein-coupled serotonin receptors: receptor isoforms and functional differences. Ann. NY Acad. Sci., 861: 31–37.

Hamilton, S.P., Heiman, G.A., Haghighi, F., Mick, S., Klein, D.F., Hodge, S.E., Weissman, M.M., Fyer, A.J. and Knowles, J.A. (1999) Lack of genetic linkage or association between a functional serotonin transporter polymorphism and panic disorder. Psychiatr. Genet., 9: 1–6.

Hamilton, S.P., Slager, S.L., Heiman, G.A., Haghighi, F., Klein, D.F., Hodge, S.E., Weissman, M.M., Fyer, A.J. and Knowles, J.A. (2000) No genetic linkage or association between a functional promoter polymorphism in the monoamine oxidase-A gene and panic disorder. Mol. Psychiatry, 5: 465–466.

Hamon, M., Doucet, E., Lefevre, K., Miquel, M.C., Lanfumey, L., Insausti, R., Frechilla, D., Del Rio, J. and Verge, D. (1999) Antibodies and antisense oligonucleotide for probing the distribution and putative functions of central 5-HT6 receptors. Neuropsychopharmacology, 21: 68S–76S.

Han, L., Nielsen, D.A., Rosenthal, N.E., Jefferson, K., Kaye, W., Murphy, D., Altemus, M., Humphries, J., Cassano, G., Rotondo, A., Virkkunen, M., Linnoila, M. and Goldman, D. (1999) No coding variant of the tryptophan hydroxylase gene detected in seasonal affective disorder, obsessive-compulsive disorder, anorexia nervosa, and alcoholism. Biol. Psychiatry, 45: 615–619.

Hariri, A.R., Mattay, V.S., Tessitore, A., Kolachana, B., Fera, F., Goldman, D., Egan, M.F. and Weinberger, D.R. (2002) Serotonin transporter genetic variation and the response of the human amygdala. Science, 297: 400–403.

Harrell, A.V. and Allan, A.M. (2003) Improvements in hippocampal-dependent learning and decremental attention in 5-HT(3) receptor overexpressing mice. Learn. Mem., 10: 410–419.

Harvey, M., Gagne, B., Labbe, M. and Barden, N. (2007) Polymorphisms in the neuronal isoform of tryptophan hydroxylase 2 are associated with bipolar disorder in French Canadian pedigrees. Psychiatr. Genet., 17: 17–22.

Hedlund, P.B. and Sutcliffe, J.G. (2004) Functional, molecular and pharmacological advances in 5-HT7 receptor research. Trends Pharmacol. Sci., 25: 481–486.

Hedlund, P.B. and Sutcliffe, J.G. (2007) The 5-HT7 receptor influences stereotypic behavior in a model of obsessive-compulsive disorder. Neurosci. Lett., 414: 247–251.

Hedlund, P.B., Danielson, P.E., Thomas, E.A., Slanina, K., Carson, M.J. and Sutcliffe, J.G. (2003) No hypothermic response to serotonin in 5-HT7 receptor knockout mice. Proc. Natl. Acad. Sci. USA, 100: 1375–1380.

Hedlund, P.B., Huitron-Resendiz, S., Henriksen, S.J. and Sutcliffe, J.G. (2005) 5-HT7 receptor inhibition and inactivation induce antidepressantlike behavior and sleep pattern. Biol. Psychiatry, 58: 831–837.

Heisler, L.K., Chu, H.M., Brennan, T.J., Danao, J.A., Bajwa, P., Parsons, L.H. and Tecott, L.H. (1998) Elevated anxiety and antidepressant-like responses in serotonin 5-HT1A receptor mutant mice. Proc. Natl. Acad. Sci. USA, 95: 15049–15054.

Heisler, L.K., Pronchuk, N., Nonogaki, K., Zhou, L., Raber, J., Tung, L., Yeo, G.S., O'Rahilly, S., Colmers, W.F., Elmquist, J.K. and Tecott, L.H. (2007) Serotonin activates the hypothalamic–pituitary–adrenal axis via serotonin 2C receptor stimulation. J. Neurosci., 27: 6956–6964.

Hendricks, T., Francis, N., Fyodorov, D. and Deneris, E.S. (1999) The ETS domain factor Pet-1 is an early and precise marker of central serotonin neurons and interacts with a conserved element in serotonergic genes. J. Neurosci., 19: 10348–10356.

Hendricks, T.J., Fyodorov, D.V., Wegman, L.J., Lelutiu, N.B., Pehek, E.A., Yamamoto, B., Silver, J., Weeber, E.J., Sweatt, J.D. and Deneris, E.S. (2003) Pet-1 ETS gene plays a critical role in 5-HT neuron development and is required for normal anxiety-like and aggressive behavior. Neuron, 37: 233–247.

Heninger, G.R., Delgado, P.L. and Charney, D.S. (1996) The revised monoamine theory of depression: a modulatory role for monoamines, based on new findings from monoamine depletion experiments in humans. Pharmacopsychiatry, 29: 2–11.

Hennig, J., Reuter, M., Netter, P., Burk, C. and Landt, O. (2005) Two types of aggression are differentially related to serotonergic activity and the A779C TPH polymorphism. Behav. Neurosci., 119: 16–25.

Henningsson, S., Annerbrink, K., Olsson, M., Allgulander, C., Andersch, S., Sjodin, I., Gustafson, D., Skoog, I., Eriksson, E. and Westberg, L. (2007) Absence of the Arg441His polymorphism in the tryptophan hydroxylase 2 gene in adults with anxiety disorders and depression. Am. J. Med. Genet. B Neuropsychiatr. Genet., 144B: 816–817.

Hinney, A., Herrmann, H., Lohr, T., Rosenkranz, K., Ziegler, A., Lehmkuhl, G., Poustka, F., Schmidt, M.H., Mayer, H., Siegfried, W., Remschmidt, H. and Hebebrand, J. (1999) No evidence for an involvement of alleles of polymorphisms in the serotonin1Dbeta and 7 receptor genes in obesity, underweight or anorexia nervosa. Intl. J. Obes. Relat. Metab. Disord., 23: 760–763.

Hoehe, M.R., Wendel, B., Grunewald, I., Chiaroni, P., Levy, N., Morris-Rosendahl, D., Macher, J.P., Sander, T. and Crocq, M. A. (1998) Serotonin transporter (5-HTT) gene polymorphisms are not associated with susceptibility to mood disorders. Am. J. Med. Genet., 81: 1–3.

Holmes, A., Yang, R.J., Murphy, D.L. and Crawley, J.N. (2002a) Evaluation of antidepressant-related behavioral responses in mice lacking the serotonin transporter. Neuropsychopharmacology, 27: 914–923.

Holmes, A., Murphy, D.L. and Crawley, J.N. (2002b) Reduced aggression in mice lacking the serotonin transporter. Psychopharmacology (Berl.), 161: 160–167.

Holmes, A., Yang, R.J., Lesch, K.P., Crawley, J.N. and Murphy, D.L. (2003) Mice lacking the serotonin transporter exhibit 5-HT(1A) receptor-mediated abnormalities in tests for anxiety-like behavior. Neuropsychopharmacology, 28: 2077–2088.

Homberg, J.R., Pattij, T., Janssen, M.C., Ronken, E., De Boer, S.F., Schoffelmeer, A.N. and Cuppen, E. (2007) Serotonin transporter deficiency in rats improves inhibitory control but not behavioral flexibility. Eur. J. Neurosci., 26: 2066–2073.

Hong, C.J., Pan, G.M. and Tsai, S.J. (2004) Association study of onset age, attempted suicide, aggressive behavior, and schizophrenia with a serotonin 1B receptor (A-161T) genetic polymorphism. Neuropsychobiology, 49: 1–4.

Hoyer, D., Hannon, J.P. and Martin, G.R. (2002) Molecular, pharmacological and functional diversity of 5-HT receptors. Pharmacol. Biochem. Behav., 71: 533–554.

Hu, X., Giotakis, O., Li, T., Karwautz, A., Treasure, J. and Collier, D.A. (2003) Association of the 5-HT2c gene with susceptibility and minimum body mass index in anorexia nervosa. Neuroreport, 14: 781–783.

Huang, Y.Y., Grailhe, R., Arango, V., Hen, R. and Mann, J.J. (1999) Relationship of psychopathology to the human serotonin1B genotype and receptor binding kinetics in postmortem brain tissue. Neuropsychopharmacology, 21: 238–246.

Huang, Y.Y., Oquendo, M.A., Friedman, J.M., Greenhill, L.L., Brodsky, B., Malone, K.M., Khait, V. and Mann, J.J. (2003) Substance abuse disorder and major depression are associated with the human 5-HT1B receptor gene (HTR1B) G861C polymorphism. Neuropsychopharmacology, 28: 163–169.

Ikeda, M., Iwata, N., Kitajima, T., Suzuki, T., Yamanouchi, Y., Kinoshita, Y. and Ozaki, N. (2006) Positive association of the serotonin 5-HT7 receptor gene with schizophrenia in a Japanese population. Neuropsychopharmacology, 31: 866–871.

Inada, Y., Yoneda, H., Koh, J., Sakai, J., Himei, A., Kinoshita, Y., Akabame, K., Hiraoka, Y. and Sakai, T. (2003) Positive association between panic disorder and polymorphism of the serotonin 2A receptor gene. Psychiatry Res., 118: 25–31.

Jacobson, L.H. and Cryan, J.F. (2007) Feeling strained? Influence of genetic background on depression-related behavior in mice: a review. Behav. Genet., 37: 171–213.

Jean, A., Conductier, G., Manrique, C., Bouras, C., Berta, P., Hen, R., Charnay, Y., Bockaert, J. and Compan, V. (2007) Anorexia induced by activation of serotonin 5-HT4 receptors is mediated by increases in CART in the nucleus accumbens. Proc. Natl. Acad. Sci. USA, 104: 16335–16340.

Jennings, K.A., Loder, M.K., Sheward, W.J., Pei, Q., Deacon, R.M., Benson, M.A., Olverman, H.J., Hastie, N.D., Harmar, A.J., Shen, S. and Sharp, T. (2006) Increased expression of the 5-HT transporter confers a low-anxiety phenotype linked to decreased 5-HT transmission. J. Neurosci., 26: 8955–8964.

Jensen, A.A., Davies, P.A., Brauner-Osborne, H. and Krzywkowski, K. (2008) 3B but which 3B and that's just one of the questions: the heterogeneity of human 5-HT3 receptors. Trends Pharmacol. Sci., 29: 437–444.

Johansson, C., Smedh, C., Partonen, T., Pekkarinen, P., Paunio, T., Ekholm, J., Peltonen, L., Lichtermann, D., Palmgren, J., Adolfsson, R. and Schalling, M. (2001) Seasonal affective disorder and serotonin-related polymorphisms. Neurobiol. Dis., 8: 351–357.

Jokela, M., Raikkonen, K., Lehtimaki, T., Rontu, R. and Keltikangas-Jarvinen, L. (2007) Tryptophan hydroxylase 1 gene (TPH1) moderates the influence of social support on depressive symptoms in adults. J. Affect. Disord., 100: 191–197.

Jones, M.D. and Lucki, I. (2005) Sex differences in the regulation of serotonergic transmission and behavior in 5-HT receptor knockout mice. Neuropsychopharmacology, 30: 1039–1047.

Kalueff, A.V., Gallagher, P.S. and Murphy, D.L. (2006) Are serotonin transporter knockout mice 'depressed'?: hypoactivity but no anhedonia. Neuroreport, 17: 1347–1351.

Kalueff, A.V., Fox, M.A., Gallagher, P.S. and Murphy, D.L. (2007a) Hypolocomotion, anxiety and serotonin syndrome-like behavior contribute to the complex phenotype of serotonin transporter knockout mice. Genes Brain Behav., 6: 389–400.

Kalueff, A.V., Jensen, C.L. and Murphy, D.L. (2007b) Locomotory patterns, spatiotemporal organization of exploration and spatial memory in serotonin transporter knockout mice. Brain Res., 1169: 87–97.

Karayiorgou, M., Sobin, C., Blundell, M.L., Galke, B.L., Malinova, L., Goldberg, P., Ott, J. and Gogos, J.A. (1999) Family-based association studies support a sexually dimorphic effect of COMT and MAOA on genetic susceptibility to obsessive-compulsive disorder. Biol. Psychiatry, 45: 1178–1189.

Kato, M., Fukuda, T., Wakeno, M., Fukuda, K., Okugawa, G., Ikenaga, Y., Yamashita, M., Takekita, Y., Nobuhara, K., Azuma, J. and Kinoshita, T. (2006) Effects of the serotonin type 2A, 3A and 3B receptor and the serotonin transporter genes on paroxetine and fluvoxamine efficacy and adverse drug reactions in depressed Japanese patients. Neuropsychobiology, 53: 186–195.

Kato, M., Fukuda, T., Wakeno, M., Okugawa, G., Takekita, Y., Watanabe, S., Yamashita, M., Hosoi, Y., Azuma, J., Kinoshita, T. and Serretti, A. (2009) Effect of 5-HT1A gene polymorphisms on antidepressant response in major depressive disorder. Am. J. Med. Genet. B Neuropsychiatr. Genet., 150B: 115–123.

Kawada, Y., Hattori, M., Dai, X.Y. and Nanko, S. (1995) Possible association between monoamine oxidase A gene and bipolar affective disorder. Am. J. Hum. Genet., 56: 335–336.

Ke, L., Qi, Z.Y., Ping, Y. and Ren, C.Y. (2006) Effect of SNP at position 40237 in exon 7 of the TPH2 gene on susceptibility to suicide. Brain Res., 1122: 24–26.

Kelley, S.P., Bratt, A.M. and Hodge, C.W. (2003) Targeted gene deletion of the 5-HT3A receptor subunit produces an anxiolytic phenotype in mice. Eur. J. Pharmacol., 461: 19–25.

Kendler, K.S., Kuhn, J.W., Vittum, J., Prescott, C.A. and Riley, B. (2005) The interaction of stressful life events and a serotonin transporter polymorphism in the prediction of episodes of major depression: a replication. Arch. Gen. Psychiatry, 62: 529–535.

Kennett, G.A., Trail, B. and Bright, F. (1998) Anxiolytic-like actions of BW 723C86 in the rat Vogel conflict test are 5-HT2B receptor mediated. Neuropharmacology, 37: 1603–1610.

Kia-Keating, B.M., Glatt, S.J. and Tsuang, M.T. (2007) Meta-analyses suggest association between COMT, but not HTR1B, alleles, and suicidal behavior. Am. J. Med. Genet. B Neuropsychiatr. Genet., 144B: 1048–1053.

Kim, J.J., Shih, J.C., Chen, K., Chen, L., Bao, S., Maren, S., Anagnostaras, S.G., Fanselow, M.S., De Maeyer, E., Seif, I. and Thompson, R.F. (1997) Selective enhancement of emotional, but not motor, learning in monoamine oxidase A-deficient mice. Proc. Natl. Acad. Sci. USA, 94: 5929–5933.

Kim, W., Choi, Y.H., Yoon, K.S., Cho, D.Y., Pae, C.U. and Woo, J.M. (2006) Tryptophan hydroxylase and serotonin transporter gene polymorphism does not affect the diagnosis, clinical features and treatment outcome of panic disorder in the Korean population. Prog. Neuropsychopharmacol. Biol. Psychiatry, 30: 1413–1418.

Klemenhagen, K.C., Gordon, J.A., David, D.J., Hen, R. and Gross, C.T. (2006) Increased fear response to contextual cues in mice lacking the 5-HT1A receptor. Neuropsychopharmacology, 31: 101–111.

Klenova, E., Scott, A.C., Roberts, J., Shamsuddin, S., Lovejoy, E.A., Bergmann, S., Bubb, V.J., Royer, H.D. and Quinn, J.P. (2004) YB-1 and CTCF differentially regulate the 5-HTT polymorphic intron 2 enhancer which predisposes to a variety of neurological disorders. J. Neurosci., 24: 5966–5973.

Kulikov, A.V., Osipova, D.V., Naumenko, V.S. and Popova, N.K. (2005) Association between Tph2 gene polymorphism, brain tryptophan hydroxylase activity and aggressiveness in mouse strains. Genes Brain Behav., 4: 482–485.

Kunugi, H., Hattori, M., Kato, T., Tatsumi, M., Sakai, T., Sasaki, T., Hirose, T. and Nanko, S. (1997) Serotonin transporter gene polymorphisms: ethnic difference and possible association with bipolar affective disorder. Mol. Psychiatry, 2: 457–462.

Kunugi, H., Ishida, S., Kato, T., Sakai, T., Tatsumi, M., Hirose, T. and Nanko, S. (1999a) No evidence for an association of polymorphisms of the tryptophan hydroxylase gene with affective disorders or attempted suicide among Japanese patients. Am. J. Psychiatry, 156: 774–776.

Kunugi, H., Ishida, S., Kato, T., Tatsumi, M., Sakai, T., Hattori, M., Hirose, T. and Nanko, S. (1999b) A functional polymorphism in the promoter region of monoamine oxidase-A gene and mood disorders. Mol. Psychiatry, 4: 393–395.

Kusserow, H., Davies, B., Hortnagl, H., Voigt, I., Stroh, T., Bert, B., Deng, D.R., Fink, H., Veh, R.W. and Theuring, F. (2004) Reduced anxiety-related behavior in transgenic mice overexpressing serotonin 1A receptors. Brain Res. Mol. Brain Res., 129: 104–116.

Lai, T.J., Wu, C.Y., Tsai, H.W., Lin, Y.M. and Sun, H.S. (2005) Polymorphism screening and haplotype analysis of the tryptophan hydroxylase gene (TPH1) and association with bipolar affective disorder in Taiwan. BMC Med. Genet., 6: 14.

Lalovic, A. and Turecki, G. (2002) Meta-analysis of the association between tryptophan hydroxylase and suicidal behavior. Am. J. Med. Genet., 114: 533–540.

Lane, H.Y., Lin, C.C., Huang, C.H., Chang, Y.C., Hsu, S.K. and Chang, W.H. (2004) Risperidone response and 5-HT6 receptor gene variance: genetic association analysis with adjustment for nongenetic confounders. Schizophr Res., 67: 63–70.

Lang, U.E., Bajbouj, M., Wernicke, C., Rommelspacher, H., Danker-Hopfe, H. and Gallinat, J. (2004) No association of a functional polymorphism in the serotonin transporter gene promoter and anxiety-related personality traits. Neuropsychobiology, 49: 182–184.

Lasky-Su, J.A., Faraone, S.V., Glatt, S.J. and Tsuang, M.T. (2005) Meta-analysis of the association between two polymorphisms in the serotonin transporter gene and affective disorders. Am. J. Med. Genet. B Neuropsychiatr. Genet., 133B: 110–115.

Lee, S.H., Lee, K.J., Lee, H.J., Ham, B.J., Ryu, S.H. and Lee, M.S. (2005) Association between the 5-HT6 receptor C267T polymorphism and response to antidepressant treatment in major depressive disorder. Psychiatry Clin. Neurosci., 59: 140–145.

Lemonde, S., Turecki, G., Bakish, D., Du, L., Hrdina, P.D., Bown, C.D., Sequeira, A., Kushwaha, N., Morris, S.J., Basak, A., Ou, X.M. and Albert, P.R. (2003) Impaired repression at a 5-hydroxytryptamine 1A receptor gene polymorphism associated with major depression and suicide. J. Neurosci., 23: 8788–8799.

Lemonde, S., Du, L., Bakish, D., Hrdina, P. and Albert, P.R. (2004) Association of the C(-1019)G 5-HT1A functional promoter polymorphism with antidepressant response. Intl. J. Neuropsychopharmacol., 7: 501–506.

Lerer, B., Macciardi, F., Segman, R.H., Adolfsson, R., Blackwood, D., Blairy, S., Del Favero, J., Dikeos, D.G., Kaneva, R., Lilli, R., Massat, I., Milanova, V., Muir, W., Noethen, M., Oruc, L., Petrova, T., Papadimitriou, G.N., Rietschel, M., Serretti, A., Souery, D., Van Gestel, S., Van Broeckhoven, C. and Mendlewicz, J. (2001) Variability of 5-HT2C receptor cys23ser polymorphism among European populations and vulnerability to affective disorder. Mol. Psychiatry, 6: 579–585.

Lesch, K.P., Bengel, D., Heils, A., Sabol, S.Z., Greenberg, B.D., Petri, S., Benjamin, J., Muller, C.R., Hamer, D.H. and Murphy, D.L. (1996) Association of anxiety-related traits with a polymorphism in the serotonin transporter gene regulatory region. Science, 274: 1527–1531.

Li, J., Wang, Y., Zhou, R., Wang, B., Zhang, H., Yang, L. and Faraone, S.V. (2006) Association of attention-deficit/hyperactivity disorder with serotonin 4 receptor gene polymorphisms in Han Chinese subjects. Neurosci Lett, 401: 6–9.

Li, Q., Wichems, C., Heils, A., Van De Kar, L.D., Lesch, K.P. and Murphy, D.L. (1999) Reduction of 5-hydroxytryptamine (5-HT)(1A)-mediated temperature and neuroendocrine responses and 5-HT(1A) binding sites in 5-HT transporter knockout mice. J. Pharmacol. Exp. Ther., 291: 999–1007.

Lin, Y.M., Chao, S.C., Chen, T.M., Lai, T.J., Chen, J.S. and Sun, H.S. (2007) Association of functional polymorphisms of the human tryptophan hydroxylase 2 gene with risk for bipolar disorder in Han Chinese. Arch. Gen. Psychiatry, 64: 1015–1024.

Lira, A., Zhou, M., Castanon, N., Ansorge, M.S., Gordon, J.A., Francis, J.H., Bradley-Moore, M., Lira, J., Underwood, M.D., Arango, V., Kung, H.F., Hofer, M.A., Hen, R. and Gingrich, J.A. (2003) Altered depression-related behaviors and functional changes in the dorsal raphe nucleus of serotonin transporter-deficient mice. Biol. Psychiatry, 54: 960–971.

Liu, W., Gu, N., Feng, G., Li, S., Bai, S., Zhang, J., Shen, T., Xue, H., Breen, G., St Clair, D. and He, L. (1999) Tentative association of the serotonin transporter with schizophrenia and unipolar depression but not with bipolar disorder in Han Chinese. Pharmacogenetics, 9: 491–495.

Lopez, V.A., Detera-Wadleigh, S., Cardona, I., Kassem, L. and McMahon, F.J. (2007) Nested association between genetic variation in tryptophan hydroxylase II, bipolar affective disorder, and suicide attempts. Biol. Psychiatry, 61: 181–186.

Lopez de Lara, C., Brezo, J., Rouleau, G., Lesage, A., Dumont, M., Alda, M., Benkelfat, C. and Turecki, G. (2007) Effect of tryptophan hydroxylase-2 gene variants on suicide risk in major depression. Biol. Psychiatry, 62: 72–80.

Lovejoy, E.A., Scott, A.C., Fiskerstrand, C.E., Bubb, V.J. and Quinn, J.P. (2003) The serotonin transporter intronic VNTR enhancer correlated with a predisposition to affective disorders has distinct regulatory elements within the domain based on the primary DNA sequence of the repeat unit. Eur. J. Neurosci., 17: 417–420.

Lovenberg, T.W., Baron, B.M., de Lecea, L., Miller, J.D., Prosser, R.A., Rea, M.A., Foye, P.E., Racke, M., Slone, A.L., Siegel, B.W. et al. (1993) A novel adenylyl cyclase-activating serotonin receptor (5-HT7) implicated in the regulation of mammalian circadian rhythms. Neuron, 11: 449–458.

Lucas, G., Rymar, V.V., Du, J., Mnie-Filali, O., Bisgaard, C., Manta, S., Lambas-Senas, L., Wiborg, O., Haddjeri, N., Pineyro, G., Sadikot, A.F. and Debonnel, G. (2007)

Serotonin(4) (5-HT(4)) receptor agonists are putative antidepressants with a rapid onset of action. Neuron, 55: 712–725.

Lucki, I. (1992) 5-HT1 receptors and behavior. Neurosci. Biobehav. Rev., 16: 83–93.

Lucki, I. (1998) The spectrum of behaviors influenced by serotonin. Biol. Psychiatry, 44: 151–162.

Lucki, I. and O'Leary, O.F. (2004) Distinguishing roles for norepinephrine and serotonin in the behavioral effects of antidepressant drugs. J. Clin. Psychiatry, 65(Suppl 4): 11–24.

Lucki, I., Ward, H.R. and Frazer, A. (1989) Effect of 1-(m-chlorophenyl)piperazine and 1-(m-trifluoromethylphenyl) piperazine on locomotor activity. J. Pharmacol. Exp. Ther., 249: 155–164.

MacKenzie, A. and Quinn, J. (1999) A serotonin transporter gene intron 2 polymorphic region, correlated with affective disorders, has allele-dependent differential enhancer-like properties in the mouse embryo. Proc. Natl. Acad. Sci. USA, 96: 15251–15255.

Magnay, J.L., Ismail, K.M., Chapman, G., Cioni, L., Jones, P.W. and O'Brien, S. (2006) Serotonin transporter, tryptophan hydroxylase, and monoamine oxidase A gene polymorphisms in premenstrual dysphoric disorder. Am. J. Obstet. Gynecol., 195: 1254–1259.

Mahieu, B., Souery, D., Lipp, O., Mendelbaum, K., Verheyen, G., De Maertelaer, V., Van Broeckhoven, C. and Mendlewicz, J. (1997) No association between bipolar affective disorder and a serotonin receptor (5-HT2A) polymorphism. Psychiatry Res., 70: 65–69.

Malhotra, A.K., Goldman, D., Ozaki, N., Rooney, W., Clifton, A., Buchanan, R.W., Breier, A. and Pickar, D. (1996) Clozapine response and the 5HT2C Cys23Ser polymorphism. Neuroreport, 7: 2100–2102.

Malleret, G., Hen, R., Guillou, J.L., Segu, L. and Buhot, M.C. (1999) 5-HT1B receptor knockout mice exhibit increased exploratory activity and enhanced spatial memory performance in the Morris water maze. J. Neurosci., 19: 6157–6168.

Mann, J.J., Currier, D., Murphy, L., Huang, Y.Y., Galfalvy, H., Brent, D., Greenhill, L. and Oquendo, M. (2008) No association between a TPH2 promoter polymorphism and mood disorders or monoamine turnover. J. Affect. Disord., 106: 117–121.

Manuck, S.B., Flory, J.D., Ferrell, R.E., Mann, J.J. and Muldoon, M.F. (2000) A regulatory polymorphism of the monoamine oxidase-A gene may be associated with variability in aggression, impulsivity, and central nervous system serotonergic responsivity. Psychiatry Res., 95: 9–23.

Maron, E., Lang, A., Tasa, G., Liivlaid, L., Toru, I., Must, A., Vasar, V. and Shlik, J. (2005) Associations between serotonin-related gene polymorphisms and panic disorder. Intl. J. Neuropsychopharmacol., 8: 261–266.

Maron, E., Toru, I., Must, A., Tasa, G., Toover, E., Vasar, V., Lang, A. and Shlik, J. (2007) Association study of tryptophan hydroxylase 2 gene polymorphisms in panic disorder. Neurosci. Lett., 411: 180–184.

Massat, I., Souery, D., Lipp, O., Blairy, S., Papadimitriou, G., Dikeos, D., Ackenheil, M., Fuchshuber, S., Hilger, C., Kaneva, R., Milanova, V., Verheyen, G., Raeymaekers, P., Staner, L., Oruc, L., Jakovljevic, M., Serretti, A., Macciardi, F., Van Broeckhoven, C. and Mendlewicz, J. (2000) A European multicenter association study of HTR2A receptor polymorphism in bipolar affective disorder. Am. J. Med. Genet., 96: 136–140.

Massat, I., Lerer, B., Souery, D., Blackwood, D., Muir, W., Kaneva, R., Nothen, M.M., Oruc, L., Papadimitriou, G.N., Dikeos, D., Serretti, A., Bellivier, F., Golmard, J.L., Milanova, V., Del-Favero, J., Van Broeckhoven, C. and Mendlewicz, J. (2007) HTR2C (cys23ser) polymorphism influences early onset in bipolar patients in a large European multicenter association study. Mol. Psychiatry, 12: 797–798.

Masellis, M., Basile, V.S., Meltzer, H.Y., Lieberman, J.A., Sevy, S., Goldman, D.A., Hamblin, M.W., Macciardi, F.M. and Kennedy, J.L. (2001) Lack of association between the T→C 267 serotonin 5-HT6 receptor gene (HTR6) polymorphism and prediction of response to clozapine in schizophrenia. Schizophr Res., 47: 49–58.

Matthes, H., Boschert, U., Amlaiky, N., Grailhe, R., Plassat, J.L., Muscatelli, F., Mattei, M.G. and Hen, R. (1993) Mouse 5-hydroxytryptamine5A and 5-hydroxytryptamine5B receptors define a new family of serotonin receptors: cloning, functional expression, and chromosomal localization. Mol. Pharmacol., 43: 313–319.

Mayorga, A.J., Dalvi, A., Page, M.E., Zimov-Levinson, S., Hen, R. and Lucki, I. (2001) Antidepressant-like behavioral effects in 5-hydroxytryptamine(1A) and 5-hydroxytryptamine(1B) receptor mutant mice. J. Pharmacol. Exp. Ther., 298: 1101–1107.

McMahon, F.J., Buervenich, S., Charney, D., Lipsky, R., Rush, A.J., Wilson, A.F., Sorant, A.J., Papanicolaou, G.J., Laje, G., Fava, M., Trivedi, M.H., Wisniewski, S.R. and Manji, H. (2006) Variation in the gene encoding the serotonin 2A receptor is associated with outcome of antidepressant treatment. Am. J. Hum. Genet., 78: 804–814.

McQuillin, A., Lawrence, J., Kalsi, G., Chen, A., Gurling, H. and Curtis, D. (1999) No allelic association between bipolar affective disorder and the tryptophan hydroxylase gene. Arch. Gen. Psychiatry, 56: 99–101.

Meira-Lima, I., Shavitt, R.G., Miguita, K., Ikenaga, E., Miguel, E.C. and Vallada, H. (2004) Association analysis of the catechol-o-methyltransferase (COMT), serotonin transporter (5-HTT) and serotonin 2A receptor (5HT2A) gene polymorphisms with obsessive-compulsive disorder. Genes Brain Behav., 3: 75–79.

Meira-Lima, I., Michelon, L., Cordeiro, Q., Cho, H.J. and Vallada, H. (2005) Allelic association analysis of the functional insertion/deletion polymorphism in the promoter region of the serotonin transporter gene in bipolar affective disorder. J. Mol. Neurosci., 27: 219–224.

Melke, J., Landen, M., Baghei, F., Rosmond, R., Holm, G., Bjorntorp, P., Westberg, L., Hellstrand, M. and Eriksson, E. (2001) Serotonin transporter gene polymorphisms are associated with anxiety-related personality traits in women. Am. J. Med. Genet., 105: 458–463.

Mendlewicz, J., Massat, I., Souery, D., Del-Favero, J., Oruc, L., Nothen, M.M., Blackwood, D., Muir, W., Battersby, S., Lerer, B., Segman, R.H., Kaneva, R., Serretti, A., Lilli, R., Lorenzi, C., Jakovljevic, M., Ivezic, S., Rietschel, M., Milanova, V. and Van Broeckhoven, C. (2004) Serotonin transporter 5HTTLPR polymorphism and affective disorders: no evidence of association in a large European multicenter study. Eur. J. Hum. Genet., 12: 377–382.

Mengod, G., Nguyen, H., Le, H., Waeber, C., Lubbert, H. and Palacios, J.M. (1990) The distribution and cellular localization of the serotonin 1C receptor mRNA in the rodent brain examined by *in situ* hybridization histochemistry. Comparison with receptor binding distribution. Neuroscience, 35: 577–591.

Metz, A.V., Chynoweth, J. and Allan, A.M. (2006) Influence of genetic background on alcohol drinking and behavioral phenotypes of 5-HT3 receptor over-expressing mice. Pharmacol. Biochem. Behav., 84: 120–127.

Miller, D.D., Ellingrod, V.L., Holman, T.L., Buckley, P.F. and Arndt, S. (2005) Clozapine-induced weight gain associated with the 5HT2C receptor -759C/T polymorphism. Am. J. Med. Genet. B Neuropsychiatr. Genet., 133B: 97–100.

Minov, C., Baghai, T.C., Schule, C., Zwanzger, P., Schwarz, M.J., Zill, P., Rupprecht, R. and Bondy, B. (2001) Serotonin-2A-receptor and -transporter polymorphisms: lack of association in patients with major depression. Neurosci. Lett., 303: 119–122.

Monaca, C., Boutrel, B., Hen, R., Hamon, M. and Adrien, J. (2003) 5-HT 1A/1B receptor-mediated effects of the selective serotonin reuptake inhibitor, citalopram, on sleep: studies in 5-HT 1A and 5-HT 1B knockout mice. Neuropsychopharmacology, 28: 850–856.

Monteleone, P., Tortorella, A., Martiadis, V., Serino, I., Di Filippo, C. and Maj, M. (2007) Association between A218C polymorphism of the tryptophan-hydroxylase-1 gene, harm avoidance and binge eating behavior in bulimia nervosa. Neurosci. Lett., 421: 42–46.

Mossner, R., Walitza, S., Geller, F., Scherag, A., Gutknecht, L., Jacob, C., Bogusch, L., Remschmidt, H., Simons, M., Herpertz-Dahlmann, B., Fleischhaker, C., Schulz, E., Warnke, A., Hinney, A., Wewetzer, C. and Lesch, K.P. (2006) Transmission disequilibrium of polymorphic variants in the tryptophan hydroxylase-2 gene in children and adolescents with obsessive-compulsive disorder. Intl. J. Neuropsychopharmacol., 9: 437–442.

Mossner, R., Doring, N., Scherag, A., Schafer, H., Herpertz-Dahlmann, B., Remschmidt, H., Schulz, E., Renner, T., Wewetzer, C., Warnke, A., Lesch, K.P. and Walitza, S. (2007) Transmission disequilibrium analysis of the functional 5-HT3A receptor variant C178T in early-onset obsessive-compulsive-disorder. J. Psychopharmacol., 21: 833–836.

Mulder, H., Franke, B., van der-Beek van der, A.A., Arends, J., Wilmink, F.W., Egberts, A.C. and Scheffer, H. (2007) The association between HTR2C polymorphisms and obesity in psychiatric patients using antipsychotics: a cross-sectional study. Pharmacogenomics J., 7: 318–324.

Muramatsu, T., Matsushita, S., Kanba, S., Higuchi, S., Manki, H., Suzuki, E. and Asai, M. (1997) Monoamine oxidase genes polymorphisms and mood disorder. Am. J. Med. Genet., 74: 494–496.

Murphy, D.L. and Lesch, K.P. (2008) Targeting the murine serotonin transporter: insights into human neurobiology. Nat. Rev. Neurosci., 9: 85–96.

Murphy, D.L., Lerner, A., Rudnick, G. and Lesch, K.P. (2004) Serotonin transporter: gene, genetic disorders, and pharmacogenetics. Molecular interventions, 4: 109–123.

Murphy, V.E., Mynett-Johnson, L.A., Claffey, E., Shields, D.C. and McKeon, P. (2001) No association between 5HT-2A and bipolar disorder irrespective of genomic imprinting. Am. J. Med. Genet., 105: 422–425.

Mynett-Johnson, L., Kealey, C., Claffey, E., Curtis, D., Bouchier-Hayes, L., Powell, C. and McKeon, P. (2000) Multimarkerhaplotypes within the serotonin transporter gene suggest evidence of an association with bipolar disorder. Am. J. Med. Genet., 96: 845–849.

Nakamura, K., Sugawara, Y., Sawabe, K., Ohashi, A., Tsurui, H., Xiu, Y., Ohtsuji, M., Lin, Q.S., Nishimura, H., Hasegawa, H. and Hirose, S. (2006) Late developmental stage-specific role of tryptophan hydroxylase 1 in brain serotonin levels. J. Neurosci., 26: 530–534.

Nash, M.W., Sugden, K., Huezo-Diaz, P., Williamson, R., Sterne, A., Purcell, S., Sham, P.C. and Craig, I.W. (2005) Association analysis of monoamine genes with measures of depression and anxiety in a selected community sample of siblings. Am. J. Med. Genet. B Neuropsychiatr. Genet., 135B: 33–37.

Nebigil, C.G., Choi, D.S., Dierich, A., Hickel, P., Le Meur, M., Messaddeq, N., Launay, J.M. and Maroteaux, L. (2000) Serotonin 2B receptor is required for heart development. Proc. Natl. Acad. Sci. USA, 97: 9508–9513.

Nelson, D.L. (2004) 5-HT5 receptors. Curr. Drug Targets, 3: 53–58.

Ni, X., Trakalo, J.M., Mundo, E., Lee, L., Parikh, S. and Kennedy, J.L. (2002) Family-based association study of the serotonin-2A receptor gene (5-HT2A) and bipolar disorder. Neuromolecular Med., 2: 251–259.

Niesler, B., Flohr, T., Nothen, M.M., Fischer, C., Rietschel, M., Franzek, E., Albus, M., Propping, P. and Rappold, G.A. (2001) Association between the 5′UTR variant C178T of the serotonin receptor gene HTR3A and bipolar affective disorder. Pharmacogenetics, 11: 471–475.

Nishiguchi, N., Shirakawa, O., Ono, H., Nishimura, A., Nushida, H., Ueno, Y. and Maeda, K. (2001) No evidence of an association between 5HT1B receptor gene polymorphism and suicide victims in a Japanese population. Am. J. Med. Genet., 105: 343–345.

Nishiguchi, N., Shirakawa, O., Ono, H., Nishimura, A., Nushida, H., Ueno, Y. and Maeda, K. (2002) Lack of an association between 5-HT1A receptor gene structural polymorphisms and suicide victims. Am. J. Med. Genet., 114: 423–425.

Nonogaki, K., Strack, A.M., Dallman, M.F. and Tecott, L.H. (1998) Leptin-independent hyperphagia and type 2 diabetes in mice with a mutated serotonin 5-HT2C receptor gene. Nat. Med., 4: 1152–1156.

Nonogaki, K., Abdallah, L., Goulding, E.H., Bonasera, S.J. and Tecott, L.H. (2003) Hyperactivity and reduced energy cost of physical activity in serotonin 5-HT(2C) receptor mutant mice. Diabetes, 52: 315–320.

Nothen, M.M., Eggermann, K., Albus, M., Borrmann, M., Rietschel, M., Korner, J., Maier, W., Minges, J., Lichtermann, D., Franzek, E. et al. (1995) Association analysis of the monoamine oxidase A gene in bipolar affective disorder by using family-based internal controls. Am. J. Hum. Genet., 57: 975–978.

Ohara, K., Nagai, M., Tsukamoto, T., Tani, K. and Suzuki, Y. (1998) 5-HT2A receptor gene promoter polymorphism – 1438G/A and mood disorders. Neuroreport, 9: 1139–1141.

Ohmori, O., Shinkai, T., Hori, H. and Nakamura, J. (2001) Novel polymorphism in the 5′-upstream region of the human 5-HT6 receptor gene and schizophrenia. Neurosci. Lett., 310: 17–20.

Ohtani, M., Shindo, S. and Yoshioka, N. (2004) Polymorphisms of the tryptophan hydroxylase gene and serotonin 1A receptor gene in suicide victims among Japanese. Tohoku J. Exp. Med., 202: 123–133.

Ohtsuki, T., Ishiguro, H., Detera-Wadleigh, S.D., Toyota, T., Shimizu, H., Yamada, K., Yoshitsugu, K., Hattori, E., Yoshikawa, T. and Arinami, T. (2002) Association between serotonin 4 receptor gene polymorphisms and bipolar disorder in Japanese case-control samples and the NIMH Genetics Initiative Bipolar Pedigrees. Mol. Psychiatry, 7: 954–961.

Okamura, K., Shirakawa, O., Nishiguchi, N., Ono, H., Nushida, H., Ueno, Y. and Maeda, K. (2005) Lack of an association between 5-HT receptor gene polymorphisms and suicide victims. Psychiatry Clin. Neurosci., 59: 345–349.

O'Leary, O.F., Bechtholt, A.J., Crowley, J.J., Hill, T.E., Page, M.E. and Lucki, I. (2007) Depletion of serotonin and catecholamines block the acute behavioral response to different classes of antidepressant drugs in the mouse tail suspension test. Psychopharmacology (Berl.), 192: 357–371.

Olivier, J.D., Van Der Hart, M.G., Van Swelm, R.P., Dederen, P.J., Homberg, J.R., Cremers, T., Deen, P.M., Cuppen, E., Cools, A.R. and Ellenbroek, B.A. (2008) A study in male and female 5-HT transporter knockout rats: an animal model for anxiety and depression disorders. Neuroscience, 152: 573–584.

Ono, H., Shirakawa, O., Nishiguchi, N., Nishimura, A., Nushida, H., Ueno, Y. and Maeda, K. (2000) Tryptophan hydroxylase gene polymorphisms are not associated with suicide. Am. J. Med. Genet., 96: 861–863.

Oruc, L., Verheyen, G.R., Furac, I., Jakovljevic, M., Ivezic, S., Raeymaekers, P. and Van Broeckhoven, C. (1997) Association analysis of the 5-HT2C receptor and 5-HT transporter genes in bipolar disorder. Am. J. Med. Genet., 74: 504–506.

Oswald, P., Souery, D., Massat, I., Del-Favero, J., Linotte, S., Papadimitriou, G., Dikeos, D., Kaneva, R., Milanova, V., Oruc, L., Ivezic, S., Serretti, A., Lilli, R., Van Broeckhoven, C. and Mendlewicz, J. (2003) Lack of association between the 5HT2A receptor polymorphism (T102C) and unipolar affective disorder in a multicentric European study. Eur. Neuropsychopharmacol., 13: 365–368.

Ozaki, N., Goldman, D., Kaye, W.H., Plotnicov, K., Greenberg, B.D., Lappalainen, J., Rudnick, G. and Murphy, D.L. (2003) Serotonin transporter missense mutation associated with a complex neuropsychiatric phenotype. Mol. Psychiatry, 8: 933–936.

Parks, C.L., Robinson, P.S., Sibille, E., Shenk, T. and Toth, M. (1998) Increased anxiety of mice lacking the serotonin1A receptor. Proc. Natl. Acad. Sci. USA, 95: 10734–10739.

Parsian, A. and Todd, R.D. (1997) Genetic association between monoamine oxidase and manic-depressive illness: comparison of relative risk and haplotype relative risk data. Am. J. Med. Genet., 74: 475–479.

Pasqualetti, M., Ori, M., Nardi, I., Castagna, M., Cassano, G.B. and Marazziti, D. (1998) Distribution of the 5-HT5A serotonin receptor mRNA in the human brain. Brain Res. Mol. Brain Res., 56: 1–8.

Pattij, T., Groenink, L., Hijzen, T.H., Oosting, R.S., Maes, R.A., van der Gugten, J. and Olivier, B. (2002) Autonomic changes associated with enhanced anxiety in 5-HT(1A) receptor knockout mice. Neuropsychopharmacology, 27: 380–390.

Perna, G., Favaron, E., Di Bella, D., Bussi, R. and Bellodi, L. (2005) Antipanic efficacy of paroxetine and polymorphism within the promoter of the serotonin transporter gene. Neuropsychopharmacology, 30: 2230–2235.

Pesonen, U., Koulu, M., Bergen, A., Eggert, M., Naukkarinen, H., Virkkunen, M., Linnoila, M. and Goldman, D. (1998) Mutation screening of the 5-hydroxytryptamine7 receptor gene among Finnish alcoholics and controls. Psychiatry Res., 77: 139–145.

Pfaar, H., von Holst, A., Vogt Weisenhorn, D.M., Brodski, C., Guimera, J. and Wurst, W. (2002) mPet-1, a mouse ETS-domain transcription factor, is expressed in central serotonergic neurons. Dev. Genes Evolution., 212: 43–46.

Pittenger, C. and Duman, R.S. (2008) Stress, depression, and neuroplasticity: a convergence of mechanisms. Neuropsychopharmacology, 33: 88–109.

Plassat, J.L., Boschert, U., Amlaiky, N. and Hen, R. (1992) The mouse 5HT5 receptor reveals a remarkable heterogeneity within the 5HT1D receptor family. EMBO J., 11: 4779–4786.

Pooley, E.C., Fairburn, C.G., Cooper, Z., Sodhi, M.S., Cowen, P.J. and Harrison, P.J. (2004) A 5-HT2C receptor promoter polymorphism (HTR2C - 759C/T) is associated with obesity in women, and with resistance to weight loss in heterozygotes. Am. J. Med. Genet. B Neuropsychiatr. Genet., 126B: 124–127.

Popa, D., Lena, C., Fabre, V., Prenat, C., Gingrich, J., Escourrou, P., Hamon, M. and Adrien, J. (2005) Contribution of 5-HT2 receptor subtypes to sleep-wakefulness and respiratory control, and functional adaptations in knockout mice lacking 5-HT2A receptors. J. Neurosci., 25: 11231–11238.

Popova, N.K., Vishnivetskaya, G.B., Ivanova, E.A., Skrinskaya, J.A. and Seif, I. (2000) Altered behavior and alcohol tolerance in transgenic mice lacking MAO A inhibitor clorgyline. Pharmacol. Biochem. Behav., 67: 719–727.

Popova, N.K., Gilinsky, M.A., Amstislavskaya, T.G., Morosova, E.A., Seif, I. and De Maeyer, E. (2001) Regional serotonin metabolism in the brain of transgenic mice lacking monoamine oxidase A. J. Neurosci. Res., 66: 423–427.

Popova, N.K., Maslova, L.N., Morosova, E.A., Bulygina, V.V. and Seif, I. (2006) MAO A knockout attenuates adrenocortical response to various kinds of stress. Psychoneuroendocrino logy, 31: 179–186.

Ramboz, S., Saudou, F., Amara, D.A., Belzung, C., Segu, L., Misslin, R., Buhot, M.C. and Hen, R. (1996) 5-HT1B receptor knock out – behavioral consequences. Behav. Brain Res., 73: 305–312.

Ramboz, S., Oosting, R., Amara, D.A., Kung, H.F., Blier, P., Mendelsohn, M., Mann, J.J., Brunner, D. and Hen, R. (1998) Serotonin receptor 1A knockout: an animal model of anxiety-related disorder. Proc. Natl. Acad. Sci. USA, 95: 14476–14481.

Ramoz, N., Cai, G., Reichert, J.G., Corwin, T.E., Kryzak, L.A., Smith, C.J., Silverman, J.M., Hollander, E. and Buxbaum, J.D. (2006) Family-based association study of TPH1 and TPH2 polymorphisms in autism. Am. J. Med. Genet. B Neuropsychiatr. Genet., 141B: 861–867.

Ranade, S.S., Mansour, H., Wood, J., Chowdari, K.V., Brar, L.K., Kupfer, D.J. and Nimgaonkar, V.L. (2003) Linkage and association between serotonin 2A receptor gene polymorphisms and bipolar I disorder. Am. J. Med. Genet. B Neuropsychiatr. Genet., 121B: 28–34.

Rausch, J.L., Johnson, M.E., Fei, Y.J., Li, J.Q., Shendarkar, N., Hobby, H.M., Ganapathy, V. and Leibach, F.H. (2002) Initial conditions of serotonin transporter kinetics and genotype: influence on SSRI treatment trial outcome. Biol. Psychiatry, 51: 723–732.

Real, C., Popa, D., Seif, I., Callebert, J., Launay, J.M., Adrien, J. and Escourrou, P. (2007) Sleep apneas are increased in mice lacking monoamine oxidase A. Sleep, 30: 1295–1302.

Rees, S., den Daas, I., Foord, S., Goodson, S., Bull, D., Kilpatrick, G. and Lee, M. (1994) Cloning and characterisation of the human 5-HT5A serotonin receptor. FEBS Lett., 355: 242–246.

Reynolds, G.P., Zhang, Z.J. and Zhang, X.B. (2002) Association of antipsychotic drug-induced weight gain with a 5-HT2C receptor gene polymorphism. Lancet, 359: 2086–2087.

Reynolds, G.P., Yao, Z., Zhang, X., Sun, J. and Zhang, Z. (2005) Pharmacogenetics of treatment in first-episode schizophrenia: D3 and 5-HT2C receptor polymorphisms separately

associate with positive and negative symptom response. Eur. Neuropsychopharmacol., 15: 143–151.
Rietschel, M., Naber, D., Fimmers, R., Moller, H.J., Propping, P. and Nothen, M.M. (1997) Efficacy and side-effects of clozapine not associated with variation in the 5-HT2C receptor. Neuroreport, 8: 1999–2003.
Rietschel, M., Schorr, A., Albus, M., Franzek, E., Kreiner, R., Held, T., Knapp, M., Muller, D.J., Schulze, T.G., Propping, P., Maier, W. and Nothen, M.M. (2000) Association study of the tryptophan hydroxylase gene and bipolar affective disorder using family-based internal controls. Am. J. Med. Genet., 96: 310–311.
Roberts, A.J., Krucker, T., Levy, C.L., Slanina, K.A., Sutcliffe, J.G. and Hedlund, P.B. (2004) Mice lacking 5-HT receptors show specific impairments in contextual learning. Eur. J. Neurosci., 19: 1913–1922.
Rosenzweig-Lipson, S., Sabb, A., Stack, G., Mitchell, P., Lucki, I., Malberg, J.E., Grauer, S., Brennan, J., Cryan, J.F., Sukoff Rizzo, S.J., Dunlop, J., Barrett, J.E. and Marquis, K.L. (2007) Antidepressant-like effects of the novel, selective, 5-HT2C receptor agonist WAY-163909 in rodents. Psychopharmacology (Berl.), 192: 159–170.
Rothe, C., Koszycki, D., Bradwejn, J., King, N., De Luca, V., Shaikh, S., Franke, P., Garritsen, H., Fritze, J., Deckert, J. and Kennedy, J.L. (2004) Association study of serotonin-2A receptor gene polymorphism and panic disorder in patients from Canada and Germany. Neurosci. Lett., 363: 276–279.
Rotondo, A., Mazzanti, C., Dell'Osso, L., Rucci, P., Sullivan, P., Bouanani, S., Gonnelli, C., Goldman, D. and Cassano, G.B. (2002) Catechol o-methyltransferase, serotonin transporter, and tryptophan hydroxylase gene polymorphisms in bipolar disorder patients with and without comorbid panic disorder. Am. J. Psychiatry, 159: 23–29.
Ruddick, J.P., Evans, A.K., Nutt, D.J., Lightman, S.L., Rook, G.A. and Lowry, C.A. (2006) Tryptophan metabolism in the central nervous system: medical implications. Expert. Rev. Mol. Med., 8: 1–27.
Rujescu, D., Giegling, I., Sato, T., Hartmann, A.M. and Moller, H.J. (2003) Genetic variations in tryptophan hydroxylase in suicidal behavior: analysis and meta-analysis. Biol. Psychiatry, 54: 465–473.
Ryu, S., Cho, E.Y., Park, T., Oh, S., Jang, W.S., Kim, S.K., Lee, D. and Hong, K.S. (2007) -759 C/T polymorphism of 5-HT2C receptor gene and early phase weight gain associated with antipsychotic drug treatment. Prog. Neuropsychopharmacol. Biol. Psychiatry, 31: 673–677.
Sakai, K., Takada, T., Nakayama, H., Kubota, Y., Nakamata, M., Satoh, M., Suzuki, E., Akazawa, K. and Gejyo, F. (2005) Serotonin-2A and 2C receptor gene polymorphisms in Japanese patients with obstructive sleep apnea. Intern. Med., 44: 928–933.
Saleem, Q., Ganesh, S., Vijaykumar, M., Reddy, Y.C., Brahmachari, S.K. and Jain, S. (2000) Association analysis of 5HT transporter gene in bipolar disorder in the Indian population. Am. J. Med. Genet., 96: 170–172.
Samochowiec, J., Hajduk, A., Samochowiec, A., Horodnicki, J., Stepien, G., Grzywacz, A. and Kucharska-Mazur, J. (2004) Association studies of MAO-A, COMT, and 5-HTT genes polymorphisms in patients with anxiety disorders of the phobic spectrum. Psychiatry Res., 128: 21–26.
Santarelli, L., Saxe, M., Gross, C., Surget, A., Battaglia, F., Dulawa, S., Weisstaub, N., Lee, J., Duman, R., Arancio, O., Belzung, C. and Hen, R. (2003) Requirement of hippocampal neurogenesis for the behavioral effects of antidepressants. Science, 301: 805–809.
Sapolsky, R.M. (2000) The possibility of neurotoxicity in the hippocampus in major depression: a primer on neuron death. Biol. Psychiatry, 48: 755–765.
Sarnyai, Z., Sibille, E.L., Pavlides, C., Fenster, R.J., McEwen, B.S. and Toth, M. (2000) Impaired hippocampal-dependent learning and functional abnormalities in the hippocampus in mice lacking serotonin(1A) receptors. Proc. Natl. Acad. Sci. USA, 97: 14731–14736.
Sasaki, T., Hattori, M., Sakai, T., Kato, T., Kunugi, H., Hirose, T. and Nanko, S. (1998) The monoamine oxidase-A gene and major psychosis in Japanese subjects. Biol. Psychiatry, 44: 922–924.
Saudou, F., Amara, D.A., Dierich, A., LeMeur, M., Ramboz, S., Segu, L., Buhot, M.C. and Hen, R. (1994) Enhanced aggressive behavior in mice lacking 5-HT1B receptor. Science, 265: 1875–1878.
Schechter, L.E., Ring, R.H., Beyer, C.E., Hughes, Z.A., Khawaja, X., Malberg, J.E. and Rosenzweig-Lipson, S. (2005) Innovative approaches for the development of antidepressant drugs: current and future strategies. NeuroRx, 2: 590–611.
Schinka, J.A., Busch, R.M. and Robichaux-Keene, N. (2004) A meta-analysis of the association between the serotonin transporter gene polymorphism (5-HTTLPR) and trait anxiety. Mol. Psychiatry, 9: 197–202.
Schuldiner, S., Shirvan, A. and Linial, M. (1995) Vesicular neurotransmitter transporters: from bacteria to humans. Physiol. Rev., 75: 369–392.
Schulze, T.G., Muller, D.J., Krauss, H., Scherk, H., Ohlraun, S., Syagailo, Y.V., Windemuth, C., Neidt, H., Grassle, M., Papassotiropoulos, A., Heun, R., Nothen, M.M., Maier, W., Lesch, K.P. and Rietschel, M. (2000) Association between a functional polymorphism in the monoamine oxidase A gene promoter and major depressive disorder. Am. J. Med. Genet., 96: 801–803.
Scott, A.L., Bortolato, M., Chen, K. and Shih, J.C. (2008) Novel monoamine oxidase A knock out mice with human-like spontaneous mutation. Neuroreport, 19: 739–743.
Serrats, J., Raurich, A., Vilaro, M.T., Mengod, G. and Cortes, R. (2004) 5-ht5B receptor mRNA in the raphe nuclei: coexpression with serotonin transporter. Synapse (NY), 51: 102–111.
Serretti, A., Lilli, R., Lorenzi, C., Gasperini, M. and Smeraldi, E. (1999) Tryptophan hydroxylase gene and response to lithium prophylaxis in mood disorders. J. Psychiatr. Res., 33: 371–377.
Serretti, A., Zanardi, R., Rossini, D., Cusin, C., Lilli, R. and Smeraldi, E. (2001) Influence of tryptophan hydroxylase and serotonin transporter genes on fluvoxamine antidepressant activity. Mol. Psychiatry, 6: 586–592.
Serretti, A., Cristina, S., Lilli, R., Cusin, C., Lattuada, E., Lorenzi, C., Corradi, B., Grieco, G., Costa, A., Santorelli, F., Barale, F., Nappi, G. and Smeraldi, E. (2002) Family-based association study of 5-HTTLPR, TPH, MAO-A, and DRD4 polymorphisms in mood disorders. Am. J. Med. Genet., 114: 361–369.
Serretti, A., Artioli, P., Lorenzi, C., Pirovano, A., Tubazio, V. and Zanardi, R. (2004) The C(-1019)G polymorphism of the 5-HT1A gene promoter and antidepressant response in mood disorders: preliminary findings. Intl. J. Neuropsychopharmacol., 7: 453–460.
Serretti, A., Mandelli, L., Giegling, I., Schneider, B., Hartmann, A.M., Schnabel, A., Maurer, K., Moller, H.J. and Rujescu, D. (2007)

HTR2C and HTR1A gene variants in German and Italian suicide attempters and completers. Am. J. Med. Genet. B Neuropsychiatr. Genet., 144B: 291–299.

Shaikh, S.A., Strauss, J., King, N., Bulgin, N.L., Vetro, A., Kiss, E., George, C.J., Kovacs, M., Barr, C.L. and Kennedy, J.L. (2008) Association study of serotonin system genes in childhood-onset mood disorder. Psychiatr. Genet., 18: 47–52.

Shinkai, T., Ohmori, O., Kojima, H., Terao, T., Suzuki, T. and Abe, K. (1999) Association study of the 5-HT6 receptor gene in schizophrenia. Am. J. Med. Genet., 88: 120–122.

Sibille, E., Pavlides, C., Benke, D. and Toth, M. (2000) Genetic inactivation of the Serotonin(1A) receptor in mice results in downregulation of major GABA(A) receptor alpha subunits, reduction of GABA(A) receptor binding, and benzodiazepine-resistant anxiety. J. Neurosci., 20: 2758–2765.

Sibille, E., Su, J., Leman, S., Le Guisquet, A.M., Ibarguen-Vargas, Y., Joeyen-Waldorf, J., Glorioso, C., Tseng, G.C., Pezzone, M., Hen, R. and Belzung, C. (2007) Lack of serotonin1B receptor expression leads to age-related motor dysfunction, early onset of brain molecular aging and reduced longevity. Mol. Psychiatry, 12: 1042–1056., 1975.

Simansky, K.J. (1996) Serotonergic control of the organization of feeding and satiety. Behav. Brain Res., 73: 37–42.

Smeraldi, E., Zanardi, R., Benedetti, F., Di Bella, D., Perez, J. and Catalano, M. (1998) Polymorphism within the promoter of the serotonin transporter gene and antidepressant efficacy of fluvoxamine. Mol. Psychiatry, 3: 508–511.

Sodhi, M.S., Arranz, M.J., Curtis, D., Ball, D.M., Sham, P., Roberts, G.W., Price, J., Collier, D.A. and Kerwin, R.W. (1995) Association between clozapine response and allelic variation in the 5-HT2C receptor gene. Neuroreport, 7: 169–172.

Sollars, P.J., Ogilvie, M.D., Simpson, A.M. and Pickard, G.E. (2006) Photic entrainment is altered in the 5-HT1B receptor knockout mouse. J. Biol. Rhythms, 21: 21–32.

Souery, D., Van Gestel, S., Massat, I., Blairy, S., Adolfsson, R., Blackwood, D., Del-Favero, J., Dikeos, D., Jakovljevic, M., Kaneva, R., Lattuada, E., Lerer, B., Lilli, R., Milanova, V., Muir, W., Nothen, M., Oruc, L., Papadimitriou, G., Propping, P., Schulze, T., Serretti, A., Shapira, B., Smeraldi, E., Stefanis, C., Thomson, M., Van Broeckhoven, C. and Mendlewicz, J. (2001) Tryptophan hydroxylase polymorphism and suicidality in unipolar and bipolar affective disorders: a multicenter association study. Biol. Psychiatry, 49: 405–409.

Stefulj, J., Buttner, A., Skavic, J., Zill, P., Balija, M., Eisenmenger, W., Bondy, B. and Jernej, B. (2004) Serotonin 1B (5HT-1B) receptor polymorphism (G861C) in suicide victims: association studies in German and Slavic population. Am. J. Med. Genet. B Neuropsychiatr. Genet., 127B: 48–50.

Stefulj, J., Kubat, M., Balija, M., Skavic, J. and Jernej, B. (2005) Variability of the tryptophan hydroxylase gene: study in victims of violent suicide. Psychiatry Res., 134: 67–73.

Strobel, A., Gutknecht, L., Rothe, C., Reif, A., Mossner, R., Zeng, Y., Brocke, B. and Lesch, K.P. (2003) Allelic variation in 5-HT1A receptor expression is associated with anxiety- and depression-related personality traits. J. Neural. Transm., 110: 1445–1453.

Suzuki, T., Iwata, N., Kitamura, Y., Kitajima, T., Yamanouchi, Y., Ikeda, M., Nishiyama, T., Kamatani, N. and Ozaki, N. (2003) Association of a haplotype in the serotonin 5-HT4 receptor gene (HTR4) with Japanese schizophrenia. Am. J. Med. Genet. B Neuropsychiatr. Genet., 121B: 7–13.

Suzuki, Y., Sawamura, K. and Someya, T. (2004) The effects of a 5-hydroxytryptamine 1A receptor gene polymorphism on the clinical response to fluvoxamine in depressed patients. Pharmacogenomics J., 4: 283–286.

Svenningsson, P., Tzavara, E.T., Liu, F., Fienberg, A.A., Nomikos, G.G. and Greengard, P. (2002) DARPP-32 mediates serotonergic neurotransmission in the forebrain. Proc. Natl. Acad. Sci. USA, 99: 3188–3193.

Svenningsson, P., Chergui, K., Rachleff, I., Flajolet, M., Zhang, X., El Yacoubi, M., Vaugeois, J.M., Nomikos, G.G. and Greengard, P. (2006) Alterations in 5-HT1B receptor function by p11 in depression-like states. Science, 311: 77–80.

Svenningsson, P., Tzavara, E.T., Qi, H., Carruthers, R., Witkin, J.M., Nomikos, G.G. and Greengard, P. (2007) Biochemical and behavioral evidence for antidepressant-like effects of 5-HT6 receptor stimulation. J. Neurosci., 27: 4201–4209.

Syagailo, Y.V., Stober, G., Grassle, M., Reimer, E., Knapp, M., Jungkunz, G., Okladnova, O., Meyer, J. and Lesch, K.P. (2001) Association analysis of the functional monoamine oxidase A gene promoter polymorphism in psychiatric disorders. Am. J. Med. Genet., 105: 168–171.

Tadic, A., Rujescu, D., Szegedi, A., Giegling, I., Singer, P., Moller, H.J. and Dahmen, N. (2003) Association of a MAOA gene variant with generalized anxiety disorder, but not with panic disorder or major depression. Am. J. Med. Genet. B Neuropsychiatr. Genet., 117B: 1–6.

Tecott, L.H., Sun, L.M., Akana, S.F., Strack, A.M., Lowenstein, D.H., Dallman, M.F. and Julius, D. (1995) Eating disorder and epilepsy in mice lacking 5-HT2c serotonin receptors. Nature, 374: 542–546.

Tecott, L.H., Logue, S.F., Wehner, J.M. and Kauer, J.A. (1998) Perturbed dentate gyrus function in serotonin 5-HT2C receptor mutant mice. Proc. Natl. Acad. Sci. USA, 95: 15026–15031.

Templeman, L.A., Reynolds, G.P., Arranz, B. and San, L. (2005) Polymorphisms of the 5-HT2C receptor and leptin genes are associated with antipsychotic drug-induced weight gain in Caucasian subjects with a first-episode psychosis. Pharmacogenet. Genomics, 15: 195–200.

Tenner, K., Qadri, F., Bert, B., Voigt, J.P. and Bader, M. (2008) The mTPH2 C1473G single nucleotide polymorphism is not responsible for behavioral differences between mouse strains. Neurosci. Lett., 431: 21–25.

Thakker, D.R., Natt, F., Husken, D., van der Putten, H., Maier, R., Hoyer, D. and Cryan, J.F. (2005) siRNA-mediated knockdown of the serotonin transporter in the adult mouse brain. Mol. Psychiatry, 10: 782–789, 714

Thomas, D.R. (2006) 5-ht5A receptors as a therapeutic target. Pharmacol. Ther., 111: 707–714.

Thomas, D.R., Atkinson, P.J., Hastie, P.G., Roberts, J.C., Middlemiss, D.N. and Price, G.W. (2002) [3H]-SB-269970 radiolabels 5-HT7 receptors in rodent, pig and primate brain tissues. Neuropharmacology, 42: 74–81.

To, Z.P., Bonhaus, D.W., Eglen, R.M. and Jakeman, L.B. (1995) Characterization and distribution of putative 5-ht7 receptors in guinea-pig brain. Br. J. Pharmacol., 115: 107–116.

Tot, S., Erdal, M.E., Yazici, K., Yazici, A.E. and Metin, O. (2003) T102C and -1438 G/A polymorphisms of the 5-HT2A receptor gene in Turkish patients with obsessive-compulsive disorder. Eur. Psychiatry, 18: 249–254.

Tsai, S.J., Chiu, H.J., Wang, Y.C. and Hong, C.J. (1999) Association study of serotonin-6 receptor variant (C267T) with schizophrenia and aggressive behavior. Neurosci. Lett., 271: 135–137.

Tsai, S.J., Hong, C.J., Yu, Y.W. and Lin, C.H. (2002) -759C/T genetic variation of 5HT(2C) receptor and clozapine-induced weight gain. Lancet, 360: 1790.

Tsai, S.J., Hong, C.J., Yu, Y.W., Chen, T.J., Wang, Y.C. and Lin, W.K. (2004) Association study of serotonin 1B receptor (A-161T) genetic polymorphism and suicidal behaviors and response to fluoxetine in major depressive disorder. Neuropsychobiology, 50: 235–238.

Tsetsenis, T., Ma, X.H., Lo Iacono, L., Beck, S.G. and Gross, C. (2007) Suppression of conditioning to ambiguous cues by pharmacogenetic inhibition of the dentate gyrus. Nat. Neurosci., 10: 896–902.

Tut, T.G., Wang, J.L. and Lim, C.C. (2000) Negative association between T102C polymorphism at the 5-HT2A receptor gene and bipolar affective disorders in Singaporean Chinese. J. Affect. Disord., 58: 211–214.

Van Den Bogaert, A., De Zutter, S., Heyrman, L., Mendlewicz, J., Adolfsson, R., Van Broeckhoven, C. and Del-Favero, J. (2005) Response to Zhang *et al* (2005): loss-of-function mutation in tryptophan hydroxylase-2 identified in unipolar major depression. Neuron, 45: 11–16. , Neuron **48,** 704; author reply 705–706.

Van Den Bogaert, A., Sleegers, K., De Zutter, S., Heyrman, L., Norrback, K.F., Adolfsson, R., Van Broeckhoven, C. and Del-Favero, J. (2006) Association of brain-specific tryptophan hydroxylase, TPH2, with unipolar and bipolar disorder in a Northern Swedish, isolated population. Arch. Gen. Psychiatry, 63: 1103–1110.

Varnas, K., Thomas, D.R., Tupala, E., Tiihonen, J. and Hall, H. (2004) Distribution of 5-HT7 receptors in the human brain: a preliminary autoradiographic study using [3H]SB-269970. Neurosci. Lett., 367: 313–316.

Vaswani, M., Linda, F.K. and Ramesh, S. (2003) Role of selective serotonin reuptake inhibitors in psychiatric disorders: a comprehensive review. Prog. Neuropsychopharmacol. Biol. Psychiatry, 27: 85–102.

Vertes, R.P. (1991) A PHA-L analysis of ascending projections of the dorsal raphe nucleus in the rat. J. Comp. Neurol., 313: 643–668.

Vertes, R.P., Fortin, W.J. and Crane, A.M. (1999) Projections of the median raphe nucleus in the rat. J. Comp. Neurol., 407: 555–582.

Videtic, A., Pungercic, G., Pajnic, I.Z., Zupanc, T., Balazic, J., Tomori, M. and Komel, R. (2006) Association study of seven polymorphisms in four serotonin receptor genes on suicide victims. Am. J. Med. Genet. B Neuropsychiatr. Genet., 141B: 669–672.

Vishnivetskaya, G.B., Skrinskaya, J.A., Seif, I. and Popova, N.K. (2007) Effect of MAO A deficiency on different kinds of aggression and social investigation in mice. Aggress Behav., 33: 1–6.

Vogt, I.R., Shimron-Abarbanell, D., Neidt, H., Erdmann, J., Cichon, S., Schulze, T.G., Muller, D.J., Maier, W., Albus, M., Borrmann-Hassenbach, M., Knapp, M., Rietschel, M., Propping, P. and Nothen, M.M. (2000) Investigation of the human serotonin 6 [5-HT6] receptor gene in bipolar affective disorder and schizophrenia. Am. J. Med. Genet., 96: 217–221.

Vormfelde, S.V., Hoell, I., Tzvetkov, M., Jamrozinski, K., Sehrt, D., Brockmoller, J. and Leibing, E. (2006) Anxiety- and novelty seeking-related personality traits and serotonin transporter gene polymorphisms. J. Psychiatr. Res., 40: 568–576.

Walitza, S., Wewetzer, C., Warnke, A., Gerlach, M., Geller, F., Gerber, G., Gorg, T., Herpertz-Dahlmann, B., Schulz, E., Remschmidt, H., Hebebrand, J. and Hinney, A. (2002) 5-HT2A promoter polymorphism -1438G/A in children and adolescents with obsessive-compulsive disorders. Mol. Psychiatry, 7: 1054–1057.

Walther, D.J., Peter, J.U., Bashammakh, S., Hortnagl, H., Voits, M., Fink, H. and Bader, M. (2003) Synthesis of serotonin by a second tryptophan hydroxylase isoform. Science, 299: 76.

Wang, Y.M., Gainetdinov, R.R., Fumagalli, F., Xu, F., Jones, S.R., Bock, C.B., Miller, G.W., Wightman, R.M. and Caron, M.G. (1997) Knockout of the vesicular monoamine transporter 2 gene results in neonatal death and supersensitivity to cocaine and amphetamine. Neuron, 19: 1285–1296.

Weisstaub, N.V., Zhou, M., Lira, A., Lambe, E., Gonzalez-Maeso, J., Hornung, J.P., Sibille, E., Underwood, M., Itohara, S., Dauer, W.T., Ansorge, M.S., Morelli, E., Mann, J.J., Toth, M., Aghajanian, G., Sealfon, S.C., Hen, R. and Gingrich, J.A. (2006) Cortical 5-HT2A receptor signaling modulates anxiety-like behaviors in mice. Science, 313: 536–540.

Wellman, C.L., Izquierdo, A., Garrett, J.E., Martin, K.P., Carroll, J., Millstein, R., Lesch, K.P., Murphy, D.L. and Holmes, A. (2007) Impaired stress-coping and fear extinction and abnormal corticolimbic morphology in serotonin transporter knockout mice. J. Neurosci., 27: 684–691.

Wendland, J.R., DeGuzman, T.B., McMahon, F., Rudnick, G., Detera-Wadleigh, S.D. and Murphy, D.L. (2008) SERT Ileu425Val in autism, Asperger syndrome and obsessive-compulsive disorder. Psychiatr. Genet., 18: 31–39.

Wesolowska, A. and Nikiforuk, A. (2007) Effects of the brain-penetrant and selective 5-HT6 receptor antagonist SB-399885 in animal models of anxiety and depression. Neuropharmacology, 52: 1274–1283.

Wesolowska, A. and Nikiforuk, A. (2008) The selective 5-HT(6) receptor antagonist SB-399885 enhances anti-immobility action of antidepressants in rats. Eur. J. Pharmacol., 582: 88–93.

Wesolowska, A., Nikiforuk, A., Stachowicz, K. and Tatarczynska, E. (2006) Effect of the selective 5-HT7 receptor antagonist SB 269970 in animal models of anxiety and depression. Neuropharmacology, 51: 578–586.

Wesolowska, A., Nikiforuk, A. and Stachowicz, K. (2007) Anxiolytic-like and antidepressant-like effects produced by the selective 5-HT6 receptor antagonist SB-258585 after intrahippocampal administration to rats. Behav. Pharmacol., 18: 439–446.

Westberg, L., Bah, J., Rastam, M., Gillberg, C., Wentz, E., Melke, J., Hellstrand, M. and Eriksson, E. (2002) Association between a polymorphism of the 5-HT2C receptor and weight loss in teenage girls. Neuropsychopharmacology, 26: 789–793.

Wisor, J.P., Wurts, S.W., Hall, F.S., Lesch, K.P., Murphy, D.L., Uhl, G.R. and Edgar, D.M. (2003) Altered rapid eye movement sleep timing in serotonin transporter knockout mice. Neuroreport, 14: 233–238.

Wolff, M., Savova, M., Malleret, G., Hen, R., Segu, L. and Buhot, M.C. (2003) Serotonin 1B knockout mice exhibit a task-dependent selective learning facilitation. Neurosci. Lett., 338: 1–4.

Yamada, K., Hattori, E., Iwayama, Y., Ohnishi, T., Ohba, H., Toyota, T., Takao, H., Minabe, Y., Nakatani, N., Higuchi, T., Detera-Wadleigh, S.D. and Yoshikawa, T. (2006) Distinguishable haplotype blocks in the HTR3A and HTR3B region in the Japanese reveal evidence of association of HTR3B with female major depression. Biol. Psychiatry, 60: 192–201.

You, J.S., Hu, S.Y., Chen, B. and Zhang, H.G. (2005) Serotonin transporter and tryptophan hydroxylase gene polymorphisms in Chinese patients with generalized anxiety disorder. Psychiatr. Genet., 15: 7–11.

Youdim, M.B., Edmondson, D. and Tipton, K.F. (2006) The therapeutic potential of monoamine oxidase inhibitors. Nat. Rev. Neurosci., 7: 295–309.

Yu, Y.W., Tsai, S.J., Chen, T.J., Lin, C.H. and Hong, C.J. (2002) Association study of the serotonin transporter promoter polymorphism and symptomatology and antidepressant response in major depressive disorders. Mol. Psychiatry, 7: 1115–1119.

Yu, Y.W., Tsai, S.J., Hong, C.J., Chen, T.J., Chen, M.C. and Yang, C.W. (2005) Association study of a monoamine oxidase a gene promoter polymorphism with major depressive disorder and antidepressant response. Neuropsychopharmacology, 30: 1719–1723.

Yu, Y.W., Tsai, S.J., Liou, Y.J., Hong, C.J. and Chen, T.J. (2006) Association study of two serotonin 1A receptor gene polymorphisms and fluoxetine treatment response in Chinese major depressive disorders. Eur. Neuropsychopharmacol., 16: 498–503.

Yuan, X., Yamada, K., Ishiyama-Shigemoto, S., Koyama, W. and Nonaka, K. (2000) Identification of polymorphic loci in the promoter region of the serotonin 5-HT2C receptor gene and their association with obesity and type II diabetes. Diabetologia, 43: 373–376.

Zaboli, G., Jonsson, E.G., Gizatullin, R., De Franciscis, A., Asberg, M. and Leopardi, R. (2008) Haplotype analysis confirms association of the serotonin transporter (5-HTT) gene with schizophrenia but not with major depression. Am. J. Med. Genet. B Neuropsychiatr. Genet., 147: 301–307.

Zalsman, G., Frisch, A., King, R.A., Pauls, D.L., Grice, D.E., Gelernter, J., Alsobrook, J., Michaelovsky, E., Apter, A., Tyano, S., Weizman, A. and Leckman, J.F. (2001) Case control and family-based studies of tryptophan hydroxylase gene A218C polymorphism and suicidality in adolescents. Am. J. Med. Genet., 105: 451–457.

Zanardi, R., Magri, L., Rossini, D., Malaguti, A., Giordani, S., Lorenzi, C., Pirovano, A., Smeraldi, E. and Lucca, A. (2007) Role of serotonergic gene polymorphisms on response to transcranial magnetic stimulation in depression. Eur. Neuropsychopharmacol., 17: 651–657.

Zhang, H.Y., Ishigaki, T., Tani, K., Chen, K., Shih, J.C., Miyasato, K. and Ohara, K. (1997) Serotonin2A receptor gene polymorphism in mood disorders. Biol. Psychiatry, 41: 768–773.

Zhang, X., Beaulieu, J.M., Sotnikova, T.D., Gainetdinov, R.R. and Caron, M.G. (2004) Tryptophan hydroxylase-2 controls brain serotonin synthesis. Science, 305: 217.

Zhang, X., Gainetdinov, R.R., Beaulieu, J.M., Sotnikova, T.D., Burch, L.H., Williams, R.B., Schwartz, D.A., Krishnan, K.R. and Caron, M.G. (2005a) Loss-of-function mutation in tryptophan hydroxylase-2 identified in unipolar major depression. Neuron, 45: 11–16.

Zhang, X.D., Gainetdinov, R.R., Beaulieu, J.M., Sotnikova, T.D., Burch, L.H., Williams, R.B., Schwartz, D.A., Krishnan, K.R.R. and Caron, M.G. (2005b) Response to correspondence: Loss-of-function mutation in tryptophan hydroxylase-2 identified in unipolar major depression. Neuron, 48: 705–706.

Zhao, Z.Q., Scott, M., Chiechio, S., Wang, J.S., Renner, K.J., Gereau, R.W.t., Johnson, R.L., Deneris, E.S. and Chen, Z.F. (2006a) Lmx1b is required for maintenance of central serotonergic neurons and mice lacking central serotonergic system exhibit normal locomotor activity. J. Neurosci., 26: 12781–12788.

Zhao, S., Edwards, J., Carroll, J., Wiedholz, L., Millstein, R.A., Jaing, C., Murphy, D.L., Lanthorn, T.H. and Holmes, A. (2006b) Insertion mutation at the C-terminus of the serotonin transporter disrupts brain serotonin function and emotion-related behaviors in mice. Neuroscience, 140: 321–334.

Zhou, Z., Peters, E.J., Hamilton, S.P., McMahon, F., Thomas, C., McGrath, P.J., Rush, J., Trivedi, M.H., Charney, D.S., Roy, A., Wisniewski, S., Lipsky, R. and Goldman, D. (2005a) Response to Zhang *et al.* (2005): loss-of-function mutation in tryptophan hydroxylase-2 identified in unipolar major depression. Neuron, 45: 11–16. , Neuron **48,** 702–703; author reply 705–706.

Zhou, Z., Roy, A., Lipsky, R., Kuchipudi, K., Zhu, G., Taubman, J., Enoch, M.A., Virkkunen, M. and Goldman, D. (2005b) Haplotype-based linkage of tryptophan hydroxylase 2 to suicide attempt, major depression, and cerebrospinal fluid 5-hydroxyindoleacetic acid in 4 populations. Arch. Gen. Psychiatry, 62: 1109–1118.

Zhuang, X., Gross, C., Santarelli, L., Compan, V., Trillat, A.C. and Hen, R. (1999) Altered emotional states in knockout mice lacking 5-HT1A or 5-HT1B receptors. Neuropsychopharmacology, 21: 52S–60S.

Zill, P., Buttner, A., Eisenmenger, W., Moller, H.J., Ackenheil, M. and Bondy, B. (2007) Analysis of tryptophan hydroxylase I and II mRNA expression in the human brain: a post-mortem study. J. Psychiatr. Res., 41: 168–173.

CHAPTER 4.14

Functional Pharmacogenetics of Serotonin Receptors in Psychiatric Drug Action

Olga O. Yevtushenko[1] and Gavin P. Reynolds[*,2]

[1]*Department of Neuropharmacology, Institute of Pharmacology and Toxicology AMS Ukraine, Kyiv, Ukraine*
[2]*Division of Psychiatry and Neuroscience, Queen's University Belfast, Whitla Medical Building, Belfast, UK*

Abstract: There is substantial evidence for the importance of serotonin system dysfunction in the biology of mood, anxiety disorders and psychotic illness, as well as in the behavioral disturbances associated with neurodegenerative disorders. Interactions with serotonin receptors are likely to contribute to the mechanisms of action of many psychiatric drugs. There is substantial individual difference in response to drug treatment – not all patients show benefit, the treatment is often inadequate, and adverse effects of drugs may reduce their tolerability. Genetic factors are likely to contribute substantially to this individual variability; variability in serotonin receptor genes can influence both response to drug treatment and the emergence of adverse effects. This chapter concentrates on the functional pharmacogenetics of the main serotonin receptors involved in psychiatric drug action. The 5-HT_{1A}, 5-HT_{2A} and 5-HT_{2C} receptors are the main focus; however, 5-HT_{1B}, 5-HT_4, 5-HT_6 and 5-HT_7 receptors are also discussed. The genes for these receptors exhibit polymorphisms that in some cases have functional effects on the expression or structure of the gene product. The potential involvement of these functional polymorphisms of serotonin receptor genes in the pharmacological mechanism of psychiatric drug action, notably treatment response and development of side effects, is discussed.

Keywords: serotonin receptors, pharmacogenetics, antipsychotics, antidepressants, treatment response, side effects.

Introduction

It would be difficult to over-emphasize the importance of serotonin systems in the pharmacotherapy of psychiatric disorders. Serotonin appears to be central to the biology of affective and anxiety disorders, and implicated in psychotic disease, as well as having a role in the behavioral disturbances associated with various neurodegenerative disorders. The convergence of pharmacological mechanisms of the antidepressant drugs on serotoninergic activity has underpinned the long-established serotonin hypothesis of depression, and the efficacy of many antidepressant drugs in anxiety disorders has also implicated serotonin in these clinical problems. Furthermore, there is a role for serotonin in the effects of antipsychotic drugs. The majority of the second-generation, 'atypical' antipsychotics have a high affinity for 5-HT_{2A} receptors; this is considered to contribute to a diminution of certain side effects as well as, possibly, to their symptom response profile. These drugs are also being used in the treatment of bipolar disorder; again, serotonergic mechanisms are thought to contribute.

Thus, many of the drugs used in the treatment of psychiatric disorders act, directly or indirectly, via serotonin receptors. These receptors include 5-HT_{1A}, 5-HT_{2A} and 5-HT_{2C}; effects at certain other receptors, including 5-HT_{1B}, 5-HT_4 and 5-HT_6, may also be important. This chapter will address the functional pharmacogenetics of these receptors in the context of psychiatric drug action. It will not attempt an in-depth review of the pharmacogenetics of the serotonin receptors – the interested reader will be guided to some articles that do this far more comprehensively – but will attempt to bring together interesting findings that not only contribute to understanding the individual differences in drug response, but also provide clues that give us some insight into underlying mechanisms.

The treatment of psychiatric disorders is often unsatisfactory; the response to treatment is often inadequate, and adverse effects of drugs may reduce their tolerability. These limitations are made worse by the fact they are unpredictable; psychiatric treatment regimes are often a process of trial and error. What underlies this individual

*Corresponding author
E-mail: g.reynolds@qub.ac.uk

ISBN 978-0-12-374634-4
DOI: 10.1016/B978-0-12-374634-4.00042-3

variability in response to drug treatment is multifactorial; environmental factors – lifestyle, social/home environment, personal history (e.g. development, medical illness, trauma) – may play a part, but it is likely that much of the variability is genetic.

Pharmacogenetics – the study of genetic factors underlying individual variability in effects of drug treatment – has undergone a rapid growth in its application to psychiatric pharmacotherapy research over the past 15 years. A large proportion, if not the majority, of pharmacogenetic studies have addressed genes governing, or influencing, serotonin neurotransmission. The most studied of these is the 5-HT transporter (5-HTT) gene, which has a common polymorphism (the HTTLPR) resulting in a sequence deletion within the promoter region. This is considered to influence the activity of the transporter and, perhaps not surprisingly, is implicated in antidepressant action, as well as being associated with depression (Yu *et al.*, 2002; Caspi *et al.*, 2003; Gonda *et al.*, 2005; Ott *et al.*, 2007) We do not intend to discuss the pharmacogenetics of 5-HTT at any length, it having been reviewed frequently and extensively by others (Alessandro and Kato, 2008; Gerretsen and Pollock, 2008; Kronenberg *et al.*, 2008; Serretti *et al.*, 2005), although as an important and functional polymorphism the HTTLPR will inevitably be mentioned; studies of the psychiatric pharmacogenetics of 5-HT receptors may also include investigation of this polymorphism. In addition, there is substantial regulatory interaction between the various components of synaptic 5-HT neurotransmission, and this can result in effects of a polymorphism within one gene (for example, that for 5-HTT) on other gene products (such as 5-HT receptors). This provides a mechanism that might underlie some gene–gene and gene–environment interactions.

However, this chapter will concentrate on some key 5-HT receptors by discussing their genetic variability and how these genetic polymorphisms may have effects on the function of the gene product. Most importantly, we shall discuss how the pharmacogenetics of 5-HT receptors may provide some further understanding of the mechanisms that involve these receptors in the various effects of drugs used in the treatment of psychiatric disorders.

Genetic polymorphisms and pharmacogenetic mechanisms

Individual variability of 5-HT receptor genes can influence the gene product, the receptor protein, in several ways. The easiest to understand is, of course, the non-synonymous single nucleotide polymorphism (SNP) in a coding region of DNA that results in an amino acid substitution in the protein. This will influence the structure of the receptor and is likely to affect its function, perhaps in binding of receptor ligands or in interaction with G proteins. Sequence deletions, SNPs or variations in the length of tandem repeat sequences can occur in the promoter region of the gene and affect binding of transcription factors regulating gene expression; many of the more interesting polymorphisms discussed here are known, or likely, to influence promoter activity. Other effects on gene transcription and translation, perhaps via influences on DNA folding, production of splice variants, RNA stability and editing, etc., can potentially occur from polymorphisms elsewhere in the gene sequence.

The mechanisms underlying genetic association with drug effects can result from a variety of gene–drug interactions. It has recently been suggested that there are four main levels of pharmacogenetic associations (Reynolds, 2007), of which three pharmacodynamic interactions are relevant here. Thus, the genetic polymorphism can affect the primary drug target directly, thereby influencing the immediate physiological response to treatment. Alternatively, it can affect a protein that is secondary to this initial interaction but necessary for the effect of drug on the organism. Finally, it may influence disease pathology or other processes, which may then be more or less responsive to the effects of drug treatment. Understanding which of these processes is important can shed light on the mechanisms of both symptom response to drug treatment and the adverse effects of that treatment.

In the following sections, we discuss some of the serotonin receptors that are involved, or potentially involved, in psychiatric disorders and their treatment. Of these, most effort has focused on the 5-HT_{1A}, 5-HT_{2A} and 5-HT_{2C} receptors, and this chapter will reflect that emphasis. Nevertheless, other subtypes have been studied, although not always with any notable pharmacogenetic finds to date; these too will be briefly reviewed. In each case, we briefly assess their known or potential involvement in psychopharmacological mechanisms and identify some of the more important genetic polymorphisms, with an emphasis on those demonstrating some functionality – i.e., polymorphisms known to influence protein structure (non-synonymous coding region SNPs) or gene transcription. Of course, as implied above, there may be functional effects of particular polymorphisms that are as yet unknown, or there may be SNPs that, while not imparting functional effects themselves, may be in linkage disequilibrium with others that do. This may be apparent from genetic association studies, and some potentially relevant case–control studies in disease will also be mentioned prior to briefly reviewing the pharmacogenetics.

5-HT_{1A} receptor

The 5-HT_{1A} receptor is an inhibitory G-protein-coupled receptor found in the region of origin of 5-HT projections in the CNS, the raphe nuclei of the brainstem. It acts as an autoreceptor on these serotonergic neurons, where it imparts an inhibitory control over neuronal activity. It is also found postsynaptically in terminal regions such as the hippocampus, cortex and basal ganglia. Its actions are mediated via inhibition of adenylate cyclase activity.

The 5-HT_{1A} receptor is the site of action of some anxiolytic drugs, and several other drugs, including some antipsychotics, also have effects at this site. However, its greatest relevance in pharmacotherapeutics is its proposed involvement in the antidepressant efficacy of the selective serotonin reuptake inhibitors (SSRIs). These drugs block reuptake of 5-HT into neurons via the 5-HT transporter, resulting in an increase in extraneuronal 5-HT that can interact with the regulatory somatodendritic 5-HT_{1A} receptors in the raphe nuclei. The binding of transmitter to these inhibitory autoreceptors results initially in a decrease in neuronal firing, followed by a compensatory down-regulation of receptor density and a return to normal, or elevated, neuronal activity. Thus, SSRIs have been shown to desensitize somatodendritic autoreceptor sites within a few days of administration (Le Poul *et al.*, 1995) and induce a chronic decrease in hippocampal and cortical (postsynaptic) receptors (Spindelegger *et al.*, 2008). The time taken for these regulatory changes may account for the delay in clinical response to antidepressant treatment.

Genetic variability

There are reports of polymorphisms in the 5-HT_{1A} gene that have been investigated in the context of psychiatric disorders and their treatment. There are multiple polymorphisms in the promoter (5′) and 3′ regions of the gene, as well as several non-synonymous coding region SNPs, including Ile28Val, Asp272Gly and Pro16Leu. The pharmacological effect of these three, and other amino acid substitutions, has been investigated by Del Tredici *et al.* (2004), and their importance in psychiatric disorders by Drago *et al.* (2008). However, little in the way of meaningful results has come from association studies of 5-HT_{1A} receptor coding region polymorphisms, which are generally found at low frequencies, with the minor allele consistently less, and sometimes much less, than 5 percent of the sample. Although the non-exonic polymorphisms have higher frequencies of variability, the only one studied to any great extent is −1019C/G, one of several promoter region polymorphisms. The finding that the 5-HT_{1A} receptor gene has a common variant in its promoter sequence (Wu and Comings, 1999) which has yielded a clear molecular mechanism potentially explaining its associations with disease and symptom response to treatment was a major advance in functional psychiatric pharmacogenetics. This −1019C/G SNP occurs at a binding site for transcription factors; the G allele prevents the binding of inhibitory transcription factors, including the repressor Deaf-1, theoretically resulting in an increase in the expression of 5-HT_{1A} autoreceptors with a consequent reduction in 5-HT neuronal transmission (Lemonde *et al.*, 2003). Consistent with this, an apparent, but not significant, increase in 5-HT_{1A} receptor PET binding potential associated with the G allele has been reported (Parsey *et al.*, 2006). Interestingly, the effect of the transcription factor is considered to be neuron-specific – contrary to the observation at the presynaptic site, receptor expression in postsynaptic regions is reportedly activated by Deaf-1 and hence diminished by the G allele (Czesak *et al.*, 2006), although the PET findings are not consistent with this *in vitro* observation. Nevertheless, the genotypic differences in receptor expression appear to predispose for depression (Lemonde *et al.*, 2003), as well as anxiety and depression-related personality traits (Strobel *et al.*, 2003). A comprehensive review of associations of this SNP in psychiatric disorders is provided by Drago *et al.* (2008).

We find that this polymorphism can influence receptor expression determined in *post-mortem* human brain tissue (C. L. Dalton and colleagues, unpublished data). Interestingly, a human neuroimaging study did not manage to identify this relationship, but did report an association between the 5-HT transporter HTTLPR polymorphism and 5-HT_{1A} receptor binding in the brain.

Pharmacogenetics

Genetically determined differences in 5-HT_{1A} receptor density or function may thus influence the response to drugs which act directly on 5-HT_{1A} receptors or otherwise influence 5-HT neurotransmission, as well as the response to treatment of any neuro- or psychopathology that might involve 5-HT systems. Clearly, the prime candidate for involvement in 5-HT_{1A} receptor pharmacogenetics is treatment of depression; here, Lemonde *et al.* (2004) and subsequently several other studies have, fairly consistently, demonstrated that the −1019G allele is associated with a poorer response to treatment with SSRIs and, in some studies, other antidepressant drugs (reviewed in Drago *et al.*, 2008).

The effect on SSRI response is not restricted to depression; our recent study on the initial response of panic disorder to SSRI treatment has also shown a beneficial effect of the C allele (Yevtushenko *et al.*, 2009).

The 5-HT_{1A} receptor has also been implicated in the action of the atypical antipsychotic drugs. The effects these drugs have in promoting frontal cortical dopamine release, considered to be a mechanism contributing to improvement of negative symptoms and mediated primarily via 5-HT_{2A} receptor antagonism, requires intact 5-HT_{1A} receptors (Díaz-Mataix *et al.*, 2005). Reynolds *et al.* (2006a) have shown that the −1019C/G polymorphism of the 5-HT_{1A} receptor gene is strongly associated with changes in both negative features and depressive symptoms after 3 months of initial treatment of first-episode psychosis. The G allele was associated with poorer response of the patients in this study, and this finding was replicated by Wang *et al.* (2008) in an Asian population receiving risperidone treatment.

Of the other SNPs in the 5-HT_{1A} receptor gene, the only notable pharmacogenetic finding was that of Gly272Asp associated with response to fluvoxamine with the (few) Asp allele carriers having a significantly poorer response in HAM-D-17 score in a Japanese population (Suzuki *et al.*, 2004). This was not replicated in a study with fluoxetine (Yu *et al.*, 2006). The low frequency of the minor allele in this study limits its value, and the possibility of linkage with another functional polymorphism, such as −1019C/G, cannot be excluded.

5-HT_{2A} receptor

5-HT_2 receptors are excitatory, primarily postsynaptic, G-protein-coupled receptors, bringing about their effects via stimulation of phospholipase C. Of the two major subtypes, 5-HT_{2A} receptors are found in many regions of the brain, where they are located predominantly in the prefrontal cortex, claustrum, olfactory nuclei and basal ganglia; here, they modulate the activity of other neurotransmitters, including glutamate, dopamine, acetylcholine and noradrenaline. Thus, the 5-HT_{2A} receptors have been shown to modulate the activation of dopaminergic neurons, enhancing transmitter release in the prefrontal cortex (Bortolozzi *et al.*, 2005) with presumed effects on cognitive function, and in the striatum, where they influence motor behavior.

Thus, 5-HT_{2A} receptor inhibition is thought to be the mechanism involved in some of the beneficial effects of atypical antipsychotic drugs on cognitive and negative features of psychotic illness, and in minimizing extrapyramidal side effects. There is also some evidence for their role in antidepressant action; a common consequence of antidepressant drug administration is the down-regulation of 5-HT_{2A} receptors. Interestingly, these receptors decrease their expression not only in response to elevated agonist activity, but also following exposure to antagonists such as clozapine (Reynolds *et al.*, 1983). This may relate to the inverse agonism that clozapine and many other 5-HT_{2A} antagonists demonstrate; the receptor has a high level of intrinsic activity which can be elevated by agonists, while these inverse agonists not only, as antagonists, block agonist action, but also decrease the inherent activity of the receptor in the absence of agonist effects (Egan *et al.*, 1998).

Genetic variability

The substantial sequence variability in the 5-HT_{2A} receptor gene has been much studied, particularly in association studies. Of these, the 102T/C polymorphism has been the most intensively investigated. It is located in exon 1 near the promoter region, but is a synonymous polymorphism, and hence does not result in alteration of the amino acid sequence of the receptor protein. The 102T/C is in complete linkage disequilibrium with another polymorphic site of the gene referred to as −1438A/G, which in turn has also been shown to be in linkage disequilibrium with a third polymorphism, −783A/G (Myers *et al.*, 2007). There are other SNPs that have been investigated in association with clinical measures – a silent 516C/T and two non-synonymous SNPs, Thr25Asn and His452Tyr. However, the minor allele frequencies in the European population are 2 percent for 516C/T and Thr25Asn, and 8 percent for His452Tyr (Erdmann *et al.*, 1996), while that for the common −1438 SNP is 45 percent (Cargill *et al.*, 1999). No significant associations of these rarer polymorphisms with psychiatric drug treatment response have been reported.

Parsons *et al.* (2004) reported that the presence of the A allele of −1438A/G significantly increased promoter activity in a reporter gene assay compared to the G allele, although whether this translates to effects in the brain *in vivo* is unclear. A *post-mortem* brain study found that the C allele of 102T/C was associated with lower mRNA and lower protein expression than the T allele (Polesskaya and Sokolov, 2002), while a study by Bray *et al.* (2004) using the highly accurate quantitative allele-specific primer extension assay found no evidence for differences in mRNA expression between the 102T and C allelic forms of the 5-HT_{2A} receptor mRNA in *post-mortem* brain tissue derived from frontal, parietal or temporal cortex of the individuals heterozygous for this polymorphism. These findings would be consistent with the functionality residing with the −1438 SNP and apparent effects of the 102T/C site due to linkage disequilibrium with this site.

Myers *et al.* (2007) reported that variation in gene expression was associated with the −1438A/G and −783A/G polymorphisms, the minor G allele at −783 when expressed with the G allele at −1438 significantly decreased promoter activity of the gene. In the same study, mRNA expression in human fibroblast cell lines confirmed that −783A/G polymorphism significantly modified the effects of the −1438A/G SNP. Despite the assumption that the promoter polymorphisms are primarily responsible for functional differences in 5-HT_{2A} receptor expression, there is some evidence that the 102C/T site SNPs have an influence. DNA methylation can be responsible for expression differences between alleles; Polesskaya *et al.* (2006) studied methylation in two CpG sites, which are specific to allele C (102C/T polymorphism), and found that levels of methylation in the promoter correlated significantly with expression of the 5-HT_{2A} receptor.

Polymorphisms of the 5-HT_{2A} receptor gene have been associated with many psychiatric diagnoses, including psychoses, affective disorders and anxiety disorders, although these genetic association studies with disease report generally weak and conflicting results. Meta-analysis of the many studies in schizophrenia suggested an association of the 102C/T SNP (Abdolmaleky *et al.*, 2004), not confirmed in a later study (Li *et al.*, 2006a); there was also no significant effect in meta-analyses of association with suicidal behavior (Anguelova *et al.*, 2003; Li *et al.*, 2006a), despite positive findings in each case. However, in addition to the many association studies with psychiatric diagnoses, there are strong indications of association with response to treatment, particularly relating to antidepressants and antipsychotics (as reviewed by Lane *et al.*, 2005; Serretti *et al.*, 2007).

Pharmacogenetics

Some examples of these associations provide an indication of the difficulties associated with drawing definitive conclusions regarding the value of these pharmacogenetic studies. Lane *et al.* (2002) indicated that the 5-HT_{2A} receptor 102C/C genotype is related to better clinical response to risperidone in Chinese patients. However, Arranz *et al.* (1998) showed that the 102C/C genotype is more frequent among clozapine non-responders than responders in Europeans. The fact that the clozapine study was performed with previously treatment-resistant patients may contribute to such discrepancies in results.

Joober *et al.* (1999) suggest that the 102C/C genotype tended to be more frequent in patients with schizophrenia with poor long-term outcome and poor response to typical antipsychotic drugs. This difference was significant in men but not in women. In addition, the age at first contact with psychiatric care was significantly lower in the patients with schizophrenia with genotype C/C than in patients with genotype TT.

In a large prospective study of patients with major depressive disorder who were treated with citalopram, McMahon *et al.* (2006) showed that participants who were homozygous for the −1438A/G allele had an 18 percent reduction in absolute risk of having no response to treatment, compared with those homozygous for the G allele. However, there may be racial differences in outcomes of antidepressant treatment. Choi *et al.* (2005) reported a better response to citalopram treatment in Korean patients with major depressive disorder who had the G/G genotype.

There also have been studies trying to identify genetic markers for intolerance to antidepressant medication in the light of 5-HT_{2A} functionality. Murphy *et al.* (2003) reported that discontinuation of paroxetine treatment in elderly patients with major depression due to side effects was strongly associated with the 102C/T polymorphism. There have been several studies investigating possible associations with drug-induced weight gain, although no consistently replicated evidence supports a role for 5-HT_{2A} receptor polymorphisms in this important side effect associated with several antipsychotic drugs.

Some studies concentrated on investigating the association of 5-HT_{2A} receptor gene polymorphisms with tardive dyskinesia as a side effect of long-term antipsychotic treatment. A recent study by Boke *et al.* (2007) showed that the −1438A/A genotype is a risk factor for developing tardive dyskinesia in Turkish schizophrenia patients. A large, multicenter study by Lerer *et al.* (2005) found that the 102C/T genotype was significantly associated with tardive dyskinesia in older (median age 47) but not younger patients, and in patients with non-orofacial but not orofacial tardive dyskinesia. An earlier study by Segman *et al.* (2001) also suggested that the 5-HT_{2A} receptor gene is significantly associated with susceptibility to tardive dyskinesia, with the 102C/C and the −1438G/G genotypes being associated with significantly higher abnormal involuntary movement scores. Association of the 102C/T genotype with tardive diskinesia was also shown in the study by Tan *et al.* (2001), with a significant difference remaining after adjustment for age and antipsychotic dosage.

However, there are some studies with negative association results. Thus, Basile *et al.* (2001) studied three polymorphisms in the 5-HT_{2A} receptor gene without demonstrating any association with the development of tardive dyskinesia, and a more recent study came to the same conclusion with the 102C/T allele (Güzey *et al.*, 2007). Racial differences, as well as differences between the types of disorder (orofacial or limb/trunk dyskinesias), may account for some of the differences regarding what is at most a weak pharmacogenetic effect.

5-HT_{2C} receptor

Like the 5-HT_{2A} receptor, the 5-HT_{2C} receptor is a primarily postsynaptic excitatory G-protein-coupled receptor mediating many aspects of 5-HT function. 5-HT_{2C}

receptors have their highest density in the choroid plexus, but are found in basal ganglia and limbic regions of the brain, where they may also be involved in the effects of some of antipsychotic and antidepressant drugs. These receptors appear to inhibit dopaminergic activity, influencing motor function (Alex and Pehek, 2007), and are implicated in the motor effect of antipsychotic drugs (Reavill *et al.*, 1999) and the emergence of dyskinesias (Eberle-Wang *et al.*, 1996).

5-HT_{2C} receptors also have effects on food intake and other endocrine-influenced behaviors, due to their presence in the hypothalamus. Thus, transgenic mice without a functional 5-HT_{2C} receptor gene demonstrate late-onset obesity (Tecott *et al.*, 1995). These 5-HT_{2C} knockout mice are hyperphagic and exhibit metabolic hormone changes, including the development of obesity, hyperleptinemia and hyperinsulinemia. Blockade of this receptor also leads to weight gain. Thus, some atypical antipsychotics and antidepressants with potent 5-HT_{2C} antagonism may induce weight gain in susceptible individuals (Reynolds *et al.*, 2006a). However, this pharmacological action may also contribute to beneficial effects, including symptom amelioration and protection from extra-pyramidal side effects. Like 5-HT_{2A} receptors, they are down-regulated following antidepressant treatment – an effect that, again like 5-HT_{2A} sites, may also be brought about by antipsychotic drugs that have antagonist or inverse agonist effects at the 5-HT_{2C} receptor.

Genetic variability

There are several polymorphisms of the 5-HT_{2C} receptor gene, with initial pharmacogenetic evaluation focusing on a non-synonymous coding polymorphism: Cys23Ser. Although frequent in Caucasian populations, this polymorphism is rare in Asians. Okada *et al.* (2004) showed functional differences in the activity of the expressed receptor, indicating that the Ser23 allele may be constitutively more active than Cys23. The other important 5-HT_{2C} receptor polymorphisms studied are found within the promoter region and are present in both Caucasian and Asian populations, with minor allele frequencies of between 15 and 30 percent. Yuan *et al.* (2000) identified three single nucleotide polymorphisms (−995, now −997G/A, −759C/T and −697G/C) and a tandem repeat sequence (−1027GT), all in linkage disequilibrium. The functionality of these promoter region polymorphisms has been studied, with somewhat conflicting results. Thus, while initial studies showed the −759T and −697C alleles to demonstrate increased promoter activity, Castensson *et al.* (2005) were unable to replicate these findings fully, reporting a significant reduction in promoter activity only in haplotypes containing −697C and −759C. More recently, it was found that in a physiologically relevant human neuroblastoma cell line, the minor −759T and −697C alleles demonstrated relatively reduced promoter activity (Hill and Reynolds, 2007). In a small study in *post-mortem* tissue reported in Reynolds *et al.* (2005a), a significant association between the −759C/T genotype and receptor density in human brain could not be identified.

Pharmacogenetics

There are several studies of 5-HT_{2C} polymorphisms and their association with drug response. An early study related the Cys23Ser SNP to hospitalization of patients with schizophrenia (Segman *et al.*, 1997), indicating a potential relationship to treatment response. Sodhi *et al.* (1995) found that the presence of at least one Ser23 allele was more common in patients who responded to clozapine than in those who did not, but the association of Cys23Ser polymorphism and antipsychotic drug treatment response was not confirmed in further studies (Malhotra *et al.*, 1996; Masellis *et al.*, 1998). The −759C/T polymorphism is reported to be associated with improvement of negative- and general-symptom Chinese first-episode drug-naïve patients with schizophrenia, the C allele being associated with improvement of symptoms (Reynolds *et al.*, 2005b).

Reflecting the involvement of 5-HT_{2C} receptors in both motor function and dopaminergic control, pharmacogenetic studies have investigated association with tardive dyskinesia, the chronic consequence of treatment with several (particularly older) antipsychotic drugs. Segman *et al.* (2000) found an association of the Cys23Ser SNP with orofacial tardive dyskinesia in females, while Zhang *et al.* (2002) also found a relationship with tardive dyskinesia in Chinese patients (in whom the Cys23Ser SNP is rare) of the −697C/G promoter SNP. A recent study by Gunes *et al.* (2008) looked at the association of −997 G/A, −759 C/T, −697 G/C and Cys23Ser polymorphisms with extrapyramidal side effects in male schizophrenia patients. The 23Ser allele was associated with significantly higher frequency of extrapyramidal pathology.

The majority of pharmacogenetic studies of the 5-HT_{2C} receptor gene have investigated drug-induced weight gain (reviewed by Reynolds *et al.*, 2005a, 2006b), reflecting the potential role of the receptor in this process, as well as the topicality of concern over iatrogenic metabolic disease. There are several functional SNPs in the 5-HT_{2C} receptor gene which have been studied in relation to association with weight gain in patients treated with antipsychotics. The strongest evidence lies with the −759C/T polymorphism of the promoter region in the 5-HT_{2C} receptor gene, although this may only relate to the fact that this is the most studied of the several promoter region polymorphisms.

An association between the −759C/T polymorphism and antipsychotic drug-induced weight gain was demonstrated in a Chinese first-episode population, the T allele protecting against initial weight gain (Reynolds *et al.*, 2002). The finding was replicated in a Caucasian first-episode psychosis series following a longer-term (up to 9 months) treatment with antipsychotics (Templeman *et al.*, 2005), and has also been found in chronically-treated patients receiving olanzapine (Ellingrod *et al.*, 2005). It has been observed, too, in patients receiving clozapine (Reynolds *et al.* 2003; Miller *et al.*, 2005), although there are some failed replications (see, for example, De Luca *et al.*, 2007). This protective effect of the T allele on antipsychotic-induced weight gain has also been shown over a short treatment period (4 weeks) in Korean patients (Ryu *et al.*, 2007), which is interesting in respect of the suggestion that the 5-HT_{2C} receptor polymorphisms are more related to short-term effects on body weight, while other genes, including leptin, may impart greater control over the longer-term effects (Templeman *et al.*, 2005). Interactions and synergy with other genes seem likely; a haplotype study carried out on patients treated with olanzapine showed that the 102T allele of HTR2A and the 23Cys allele of HTR2C, in combination with polymorphisms in two other risk genes, were significantly associated with olanzapine-induced weight gain, with significant additive effects (Ujike *et al.*, 2008).

5-HT_{2C} genotypes are associated with increased risk of metabolic syndrome in patients taking antipsychotics. In a cross-sectional study (Mulder *et al.*, 2007), the association of 5-HT_{2C} receptor polymorphisms with the presence of metabolic syndrome in patients using antipsychotics was investigated. A similar study demonstrated an association with metabolic syndrome of a functional leptin promoter genotype previously found to be associated with drug-induced weight gain, but no significant effect of the −759C/T 5-HT_{2C} polymorphism (Yevtushenko *et al.*, 2008).

A haplotype study carried out on patients treated with olanzapine showed that the 102T allele of HTR2A and the 23Cys allele of HTR2C, in combination with polymorphisms of the genes for GNB3 (G protein) and the beta3 adrenoceptor, were significantly associated with olanzapine-induced weight gain, with significant additive effects (Ujike *et al.*, 2008).

5-HT_{1B} and 5-HT_{1D} receptors

These receptors share many similarities in pharmacology, and have only been clearly and definitively distinguished following the identification of their respective genes; there is also substantial species variability which has complicated extrapolation of pharmacological findings from rat to human (Hartig *et al.*, 1996). Past studies have often failed to distinguish the two receptor subtypes, both of which act presynaptically to regulate presynaptic release of 5-HT and other neurotransmitters.

5-HT_{1B} receptors are widely distributed in the human brain. The highest levels of mRNA expression are found in the striatum, cortex, lateral geniculate nucleus and raphe nuclei (Varnäs *et al.*, 2005). 5-HT_{1B} receptors are located on the axon terminals of both serotonergic and non-serotonergic neurons. These receptors inhibit the release of serotonin and its biosynthesis, as well as release of other neurotransmitters, including GABA, acetylcholine, glutamate and dopamine.

The results of a number of studies suggest that reduced 5-HT_{1B} heteroreceptor activity may increase impulsive behaviors, whereas reduced 5-HT_{1B} autoreceptor activity may have an antidepressant-like effect (reviewed by Clark and Neumaier, 2001). Thus 5-HT_{1B} receptors are implicated in antidepressant action and reinforcement of substance abuse. 5-HT_{1B} receptor gene knockout mice demonstrate increased aggressive behavior and impulsivity, and are also characterized by increased susceptibility to cocaine and ethanol (Saudou *et al.*, 1994; Crabbe *et al.*, 1996; Rocha *et al.*, 1998). Over-expression of the receptor in the rat dorsal raphe nucleus has been shown to increase anxiety behavior after inescapable stress (Clark *et al.*, 2002). Further preclinical research has also shown that the 5-HT_{1B} receptor has important modulatory effects on feeding behavior and thus body weight. In the hypophagic effect induced by 5-HT, activation of 5-HT_{1B} receptors is required (Mancilla-Diaz *et al.*, 2005). The role of the 5-HT_{1B} receptor in a variety of behaviors has made it the target for much neuropsychiatric and neuropharmacological research.

The same potential may be argued for the 5-HT_{1D} site, although research here is still at an early stage. In the human brain this receptor is far less abundant than 5-HT_{1B}, being found mainly in the ventral pallidum (Varnäs *et al.*, 2001). While the 5-HT_{1D} receptor is likely to have a presynaptic role similar to that of the 5-HT_{1B} site, little else about its functional role is understood.

Genetic variability and pharmacogenetics

Investigations into the 5-HT_{1B} receptor gene have identified a number of polymorphisms in the coding sequence and surrounding 5′- and 3′-untranslated regions (Sanders *et al.*, 2002; Duan *et al.*, 2003). These include two coding mutations, a silent 861G/C, and 371T/G, which leads to an amino acid substitution: Phe124Cys. This substitution significantly affects the pharmacological properties of the 5-HT_{1B} receptor, with 124Cys exhibiting higher affinity for various receptor ligands (Bruss *et al.*, 1999). Huang *et al.* (1999) described another silent 129C/T polymorphism in the coding region that was in complete linkage disequilibrium with 861C/G. The 5′ SNPs appear to have functional effects on gene expression: −261T/G and −161A/T influence the binding of transcription factors (Duan *et al.*, 2003). Both of these SNPs appear to be in linkage disequilibrium with 861C/G, suggesting that the associations ascribed to the 861 SNP (see below) may be due to effects of the promoter region polymorphisms.

An allelic association of the 861 SNP with 5-HT_{1B} binding kinetics has been reported, where the C allele was associated with decreased 5-HT_{1B} binding (Huang *et al.*, 1999).

This SNP, first identified by Sidenberg *et al.* (1993), has attracted the most attention for a range of different hypothesis-driven associational studies, including alcoholism, suicide, eating disorders and obsessive-compulsive disorder; however, no consistent positive findings have emerged. Reports by Mundo *et al.* (2000, 2002) showed association of the 861C variant with obsessive-compulsive disorder; the absence of an association of 5-HT_{1B} polymorphisms with schizophrenia has also been reported (Cordeiro and Vallada, 2005; Duan *et al.*, 2005). However, presence of the 861C allele has been associated with significantly lower minimum lifetime body mass indices in women with bulimia nervosa (Levitan *et al.*, 2001). Despite the apparent potential in relation to, for example, antidepressant action and weight gain, there appear to be no substantial reports relating to the psychiatric pharmacogenetics of this gene.

The 5-HT_{1D} gene lies in a region of chromosome 1 that has been linked to anorexia; polymorphisms in the 5-HT_{1D} receptor gene have been identified and an association with anorexia has been found and replicated (Bergen *et al.*, 2003; Brown *et al.*, 2007), although there have been studies in ADHD (Li *et al.*, 2006b) and other behavioral disorders that are negative. The functionality of the 5-HT_{1D} SNPs has not apparently been explored. The association with anorexia would certainly indicate the potential for further investigation in association with drug-induced weight gain. However, as yet no pharmacogenetic studies have been reported.

5-HT_3 receptor

The 5-HT_3 receptor is unique among the 5-HT receptor family, being the only ionotropic receptor, defining a new class of excitatory ligand-gated channels. The 5-HT_3 receptor has five subunits (A–E) that surround an ion channel (Boess *et al.*, 1995), although to date only homomeric (A only) or heteromeric (A and B) subunit complexes have been functionally characterized (Davies *et al.*, 1999). Within the central nervous system, 5-HT_3 receptors are highly expressed in the area postrema, hippocampus and amygdala, as well as in spinal cord and peripherally (Tecott *et al.*, 1993).

Fast excitatory neurotransmission mediated by 5-HT_3 receptors results in neurotransmitter release, especially of dopamine in mesolimbic and nigrostriatal pathways. 5-HT_3 receptor antagonists exhibit anti-emetic (Sugita *et al.*, 1992; Kazemi-Kjellberg *et al.*, 2001), anxiolytic (Nakagawa *et al.*, 1998) and atypical antipsychotic actions (Warburton *et al.*, 1994). Different classes of antidepressants act as non-competitive antagonists at the human 5-HT_{3A} receptor (Eisensamer *et al.*, 2003), as well as neuroleptics (Rammes *et al.*, 2004).

Gonadal steroids might interact allosterically with the 5-HT_3 receptor at the receptor–membrane interface (Wetzel *et al.*, 1998). The functional antagonism of these sex hormones at the 5-HT_3 receptor may play a role in the development and pathophysiology of psychiatric disorders (Wetzel *et al.*, 1998).

Genetic variability and pharmacogenetics

Several mutations of 5-HT_{3A} gene have been described (Niesler *et al.*, 2001a). One of higher frequency (approximately 25 percent) is the 178C/T polymorphism, which encodes a Pro16Ser substitution (Niesler *et al.*, 2001b). In this study, the less common T allele was associated with an increase in the expression of the protein *in vitro* compared with the more common C allele. Association studies showed that the 178C/T polymorphism affected the susceptibility to bipolar disorder (Niesler *et al.*, 2001b) and was associated with anxiety- and aggression-related traits in women (Melke *et al.*, 2003).

Ji *et al.* (2008) reported the association of 5-HT_{3A} receptor gene 178C/T polymorphism with the development of treatment-resistant schizophrenia. In this study, patients with the T/T genotype were receiving significantly higher neuroleptic dosage than C allele carriers. However, an associational study by Gutierrez *et al.* (2002) failed to find a difference in 178C/T genotype distributions among clozapine responders and non-responders.

Another polymorphism of the 5-HT_{3A} receptor (1596A/G) was implicated in an association with score reductions of negative and general PANSS subscales after 8 weeks of risperidone monotherapy in Chinese drug-naïve schizophrenia patients (Gu *et al.*, 2008), although there is little evidence of or theoretical basis for consistent effects of 5-HT_3 subunit polymorphisms on antipsychotic drug response.

5-HT_4 receptor

5-HT_4 receptors are expressed in wide variety of tissues. According to Varnäs *et al.* (2003), the highest density of 5-HT_4 receptors is in regions of the basal ganglia, the hippocampal formation and superficial layers in the neocortex. 5-HT_4 receptors are also located in the periphery, including on the enteric neurons and smooth muscle cells of the gastrointestinal tract, where they are implicated in the control of motility and secretion (Crowell, 2001). 5-HT_4 receptors contribute to the control of dopamine secretion (Bonhomme *et al.*, 1995) and are involved in learning and memory (Letty *et al.*, 1997), providing a target for the development of drugs addressing cognitive deficits in neurodegenerative and psychiatric disorders (King *et al.*, 2008).

Genetic variability and pharmacogenetics

A series of eight polymorphisms of the 5-HT_4 receptor gene has been described (Ohtsuki *et al.*, 2002), although their functionality has yet to be explored. While associations with these polymorphisms, and/or their haplotypes, within the 5-HT_4 receptor gene have been observed for schizophrenia (Suzuki *et al.*, 2003), bipolar disorder (Ohtsuki *et al.*, 2002) and attention deficit hyperactivity disorder (Li *et al.*, 2006c), there have been no reports of pharmacogenetic association studies with these polymorphisms. Equally surprising is that there have been no reports of association with Alzheimer's disease or its treatment outcome, given the action of the receptor on amyloid processing (Cho and Hua, 2007).

5-HT_6 receptor

The 5-HT_6 receptor is a G-protein-linked receptor, and is expressed primarily in the striatum, olfactory tubercle, frontal and entorhinal cortex, nucleus accumbens, hippocampus and molecular layer of the cerebellum (Gérard *et al.*, 1997; Monsma *et al.*, 1993).

5-HT_6 receptors appear to regulate cholinergic and glutamatergic neuronal neurotransmission in the brain. Along with their pharmacology and anatomic distribution, this suggests possible roles for 5-HT_6 receptor antagonists in the treatment of learning, memory disorders, obesity and eating disorders, and, possibly, affective state and seizures (Woolley *et al.*, 2004). 5-HT_6 receptor antagonists improved learning consolidation in rats, improved retention in the water maze and produced a significant performance improvement in aged rats in an operant-delayed alternation task (Meneses *et al.*, 2001). A number of other studies using different animals models have shown that 5-HT_6 antagonists can improve memory consolidation (King *et al.*, 2004; Mitchell and Neumaier, 2005; Mitchell *et al.*, 2007), indicating their potential in the treatment of cognitive disorders (King *et al.*, 2008; Liu and Robichaud, 2009).

This receptor exhibits high affinity for a number of tricyclic antidepressant and antipsychotic drugs, which, if this pharmacology were contributing to clinical effects, supports the view that 5-HT_6 receptors are a target for treating affective disorders, schizophrenia and anxiety (Monsma *et al.*, 1993; Roth *et al.*, 1994, 2004).

Genetic variability and pharmacogenetics

Vogt *et al.* (2000) identified six polymorphisms in the gene of the 5-HT_6 receptor, of which two (a silent 267C/T in exon 1 and a trinucleotide repeat polymorphism ([GCC](2/3)) in the 5′-upstream region of the gene) have been studied more extensively. An association study of these polymorphisms with susceptibility to suicide was negative (Okamura *et al.*, 2005), and a significant association between the 5-HT_6 receptor 267C/T polymorphism with Alzheimer's disease (Tsai *et al.*, 1999) has not been replicated (Thome *et al.*, 2001).

Pharmacogenetic studies of 267C/T polymorphism and clozapine response in schizophrenia have given contradictory results. The study by Yu *et al.* (1999) showed that patients homogeneous for 267T/T genotype had a better response to clozapine treatment; however, Masellis *et al.* (2001) did not replicate this. Absence of a positive association of −267C/T with treatment response to antidepressants in major depressive disorder was also reported by Wu *et al.* (2001).

Lee *et al.* (2005) did not find any difference in 5-HT_6 267C/T genotype and allele distribution between patients with major depressive disorder and a control group; however, C/T carriers showed significantly better antidepressant treatment response than homozygote counterparts. Lane *et al.* (2004) found an association between 267C/T polymorphism and response to risperidone in acutely ill schizophrenia inpatients. T/T carriers showed less severe positive symptoms and general psychopathology compared with C allele carriers.

5-HT_7 receptor

The 5-HT_7 receptor is a G-protein-linked receptor, detected in high densities in layers I–III of the cortex, and several subcortical structures. The limbic system (centromedial amygdala, anterior hippocampus, hypothalamus) is particularly well represented, indicating a potential role for the 5-HT_7 receptor in affective processes (Gustafson *et al.*, 1996). The 5-HT_7 receptor is also known to be involved in thermoregulation, circadian rhythm, learning and memory, sleep and hippocampal signaling. Hedlund and Sutcliffe (2004) hypothesized that this receptor might be involved in mood regulation, suggesting that the 5-HT_7 receptor is a putative target in the treatment of depression, and hence a candidate gene for antidepressant pharmacogenetics.

Several studies indicate changes in 5-HT_7 receptor density following antidepressant treatment. Mullins *et al.* (1999) suggested that the number of postsynaptic 5-HT_7 receptors in the hypothalamus is decreased in response to chronic antidepressant treatment, consistent with an earlier report by Sleight *et al.* (1995).

The 5-HT_7 receptor has high affinity for some atypical and typical antipsychotics (Roth *et al.*, 1994). Zhukovskaya and Neumaier (2000) reported that clozapine induced an increase in 5-HT_7 receptor density, as has been shown for haloperidol in rat cortex, while a significant decrease in receptor binding has been reported in schizophrenia patients compared with controls (Dean *et al.*, 2006).

Genetic variability and pharmacogenetics

A point mutation causing amino acid substitution Pro279Leu in the 5-HT_7 receptor gene has been described (Pesonen *et al.*, 1998). Although the Leu279 variant is approximately three times more common among alcoholic offenders than among healthy controls, it was not significantly associated with alcoholism or impulsivity in this study.

No association between this polymorphism and obesity and eating disorders was found (Hinney *et al.*, 1999). A large case–control association study identified association of some 5-HT_7 SNPs with schizophrenia, notably including one in the gene promoter region. However, in a functional assay of the effect of the promoter SNP on gene expression, the authors were unable to identify an underlying functional mechanism (Ikeda *et al.*, 2006).

Despite evidence implicating 5-HT_7 receptors in antidepressant action, there appear to be no pharmacogenetic association studies investigating this hypothesis, or apparently assessing association with antipsychotic drug effects.

Summary and synthesis

There is a substantial body of evidence demonstrating the involvement of genetic polymorphisms in serotonin receptor genes in the symptom response to, and other effects of, treatment with psychiatric drugs. Many of the studies behind these findings have been strongly grounded in hypotheses that postulate association of clinical effects with known functional polymorphisms which, directly or indirectly, are likely to influence receptor density, activity and/or regulatory response. Some of the best examples here involve the −1019C/G SNP in the 5-HT_{1A} receptor gene, in which a polymorphism influences the binding of a known transcription factor, resulting in an effect on gene expression and regulation, and consequent effects on response to treatment with drugs influencing the activity of serotonin neurons. Although this is one of the best understood processes linking a gene to therapeutics in psychiatric genetics, there are still substantial gaps in our understanding of all the mechanisms behind the effect of the SNP on 5-HT_{1A} pharmacology. So much more remains unclear in terms of the mechanisms underlying other pharmacogenetic associations.

Much past work has inevitably investigated pharmacogenetic association with polymorphisms that have not been shown to have functional activity. Some SNPs have emerged as having functional effects on gene expression or, for missense coding region SNPs, resulting in changes in protein structure. However, many have not been investigated in this respect, or have not yielded indications of functionality, reliance being placed on their likelihood of being in linkage disequilibrium with other SNPs which influence the eventual activity of the gene product – in this case, receptor density or pharmacology.

There remains great potential in further pharmacogenetic investigation with 5-HT receptor genes. In particular, the 5-HT_{1B} receptor, with effects on the synaptic activity of serotonin and other neurotransmitters, provides a valuable candidate gene, all the more so given the reports of functional effects of several 5-HT_{1B} receptor gene polymorphisms. This gene, and perhaps also the 5-HT_{1D} receptor gene, therefore provides a strong basis for hypothesis-driven association studies into, for example, drug response in the treatment of affective disorder. Drug-induced weight gain and eating disorders also provide potential studies of interest with these candidate genes. Similarly, there is a strong theoretical basis for studies on the pharmacogenetics of 5-HT_4 receptor polymorphisms, and particularly their role in the effects of drugs enhancing cognition in, for example, dementing disorders, schizophrenia and depression.

Of course, the pharmacogenetics of serotonin systems needs also to take into account a variety of other important genes, most notably those for the 5-HTT, tryptophan hydroxylase 2 (TPH2) and monoamine oxidase type A (MAOA), all of which demonstrate functionally important polymorphisms and which are likely to influence synaptic activity of serotonin. These are beyond the scope of the current chapter, although the reader needs to bear in mind their potential for interaction with the serotonin receptor pharmacogenetics described here. Thus it can be seen that effects on regulatory changes in receptors, a frequent consequence of changes in synaptic neurotransmitter (and, of course, receptor ligand drugs), can be brought about by polymorphisms in these genes. This will result in effects whereby one such polymorphism may have different consequences depending on the receptor genotype present – i.e., a gene–gene interaction. Few studies have either looked for, or been powered to detect, such interactions, although they are an inevitable consequence of current pharmacogenetic findings.

References

Abdolmaleky, H.M., Faraone, S.V., Glatt, S.J. and Tsuang, M.T. (2004) Meta-analysis of association between the T102C polymorphism of the 5HT2a receptor gene and schizophrenia. Schizophr. Res., 67: 53–62.

Alessandro, S. and Kato, M. (2008) The serotonin transporter gene and effectiveness of SSRIs. Expert. Rev. Neurother., 8: 111–120.

Alex, K.D. and Pehek, E.A. (2007) Pharmacologic mechanisms of serotonergic regulation of dopamine neurotransmission. Pharmacol. Ther., 113: 296–320.

Anguelova, M., Benkelfat, C. and Turecki, G. (2003) A systematic review of association studies investigating genes coding for serotonin receptors and the serotonin transporter: II. Suicidal behavior. Mol. Psychiatry, 8: 646–653.

Arranz, M.J., Munro, J., Sham, P., Kirov, G., Murray, R.M., Collier, D.A. and Kerwin, R.W. (1998) Meta-analysis of studies on genetic variation in 5-HT_{2A} receptors and clozapine response. Schizophr. Res., 32: 93–99.

Basile, V.S., Ozdemir, V., Masellis, M., Meltzer, H.Y., Lieberman, J.A., Potkin, S.G., Macciardi, F.M., Petronis, A. and Kennedy, J.L. (2001) Lack of association between serotonin-2A receptor gene (HTR2A) polymorphisms and tardive dyskinesia in schizophrenia. Mol. Psychiatry, 6: 230–234.

Bergen, A.W., van den Bree, M.B., Yeager, M., Welch, R., Ganjei, J.K., Haque, K., Bacanu, S., Berrettini, W.H., Grice, D.E., Goldman, D., Bulik, C.M., Klump, K., Fichter, M., Halmi, K., Kaplan, A., Strober, M., Treasure, J., Woodside, B. and Kaye, W.H. (2003) Candidate genes for anorexia nervosa in the 1p33-36 linkage region: serotonin 1D and delta opioid receptor loci exhibit significant association to anorexia nervosa. Mol. Psychiatry, 8: 397–406.

Boess, F.G., Beroukhim, R. and Martin, I.L. (1995) Ultrastructure of the 5-hydroxytryptamine3 receptor. J. Neurochem., 64: 1401–1405.

Boke, O., Gunes, S., Kara, N., Aker, S., Sahin, A.R., Basar, Y. and Bagci, H. (2007) Association of serotonin 2A receptor and lack of association of CYP1A2 gene polymorphism with tardive dyskinesia in a Turkish population. DNA Cell Biol., 26: 527–531.

Bonhomme, N., De Deurwaerdere, P., Le Moal, M. and Spampinato, U. (1995) Evidence for 5–HT4 receptor subtype involvement in the enhancement of striatal dopamine release induced by serotonin: a microdialysis study in the halothane-anesthetized rat. Neuropharmacology, 34: 269–279.

Bortolozzi, A., Díaz-Mataix, L., Scorza, M.C., Celada, P. and Artigas, F. (2005) The activation of 5-HT receptors in prefrontal cortex enhances dopaminergic activity. J. Neurochem., 95: 1597–1607.

Bray, N.J., Buckland, P.R., Hall, H., Owen, M.J. and O'Donovan, M.C. (2004) The serotonin-2A receptor gene locus does not contain common polymorphism affecting mRNA levels in adult brain. Mol. Psychiatry, 9: 109–114.

Brown, K.M., Bujac, S.R., Mann, E.T., Campbell, D.A., Stubbins, M.J. and Blundell, J.E. (2007) Further evidence of association of OPRD1 & HTR1D polymorphisms with susceptibility to anorexia nervosa. Biol. Psychiatry, 61: 367–373.

Bruss, M., Bonisch, H., Bühlen, M., Nöthen, M.M., Propping, P. and Göthert, M. (1999) Modified ligand binding to the naturally occurring Cys-124 variant of the human serotonin 5-HT1B receptor. Pharmacogenetics, 9: 95–102.

Cargill, M., Altshuler, D., Ireland, J., Sklar, P., Ardlie, K., Patil, N., Shaw, N., Lane, C.R., Lim, E.P., Kalyanaraman, N., Nemesh, J., Ziaugra, L., Friedland, L., Rolfe, A., Warrington, J., Lipshutz, R., Daley, G.Q. and Lander, E.S. (1999) Characterization of single-nucleotide polymorphisms in coding regions of human genes. Nat. Genet., 22: 231–238.

Caspi, A., Sugden, K., Moffitt, T.E., Taylor, A., Craig, I.W., Harrington, H., McClay, J., Mill, J., Martin, J., Braithwaite, A. and Poulton, R. (2003) Influence of life stress on depression: moderation by a polymorphism in the 5-HTT gene. Science, 301: 386–389.

Castensson, A., Aberg, K., McCarthy, S., Saetre, P., Andersson, B. and Jazin, E. (2005) Serotonin receptor 2C (HTR2C) and schizophrenia: examination of possible medication and genetic influences on expression levels. Am. J. Med. Genet. B. Neuropsychiatr. Genet., 134B: 84–89.

Cho, S. and Hua, Y. (2007) Activation of 5-HT4 receptors inhibits secretion of ß-amyloid peptides and increases neuronal survival. Exp. Neurol., 203: 274–278.

Choi, M.J., Kang, R.H., Ham, B.J., Jeong, H.Y. and Lee, M.S. (2005) Serotonin receptor 2A gene polymorphism (-1438A/G) and short-term treatment response to citalopram. Neuropsychobiology, 52: 155–162.

Clark, M.S. and Neumaier, J.F. (2001) The 5-HT1B receptor: behavioral implications. Psychopharmacol. Bull., 35: 170–185.

Clark, M.S., Sexton, T.J., McClain, M., Root, D., Kohen, R. and Neumaier, J.F. (2002) Overexpression of 5-HT1B receptor in dorsal raphe nucleus using herpes simplex virus gene transfer increases anxiety behavior after inescapable stress. J. Neurosci., 22: 4550–4562.

Cordeiro, Q. and Vallada, H. (2005) Lack of association between the G681C polimorphism in the 5-HT1D(beta) autoreceptor gene and schizophrenia. Arq. Neuropsiquiatr., 63: 380–382.

Crabbe, J.C., Phillips, T.J., Feller, D.J., Hen, R., Wenger, C.D., Lessov, C.N. and Schafer, G.L. (1996) Elevated alcohol consumption in null mutant mice lacking 5-HT1B serotonin receptors. Nat. Genet., 14: 98–101.

Crowell, M.D. (2001) The role of serotonin in the pathophysiology of irritable bowel syndrome. Am. J. Manag. Care, 7(8 Suppl): S252–S260.

Czesak, M., Lemonde, S., Peterson, E.A., Rogaeva, A. and Albert, P.R. (2006) Cell-specific repressor or enhancer activities of Deaf-1 at a serotonin 1A receptor gene polymorphism. J. Neurosci., 26: 1864–1871.

Davies, P.A., Pistis, M., Hanna, M.C., Peters, J.A., Lambert, J.J., Hales, T.G. and Kirkness, E.F. (1999) The 5-HT3B subunit is a major determinant of serotonin-receptor function. Nature, 397: 359–363.

Dean, B., Pavey, G., Thomas, D. and Scarr, E. (2006) Cortical serotonin7, 1D and 1F receptors: effects of schizophrenia, suicide and antipsychotic drug treatment. Schizophr. Res., 88: 265–274.

De Luca, V., Müller, D.J., Hwang, R., Lieberman, J.A., Volavka, J., Meltzer, H.Y. and Kennedy, J.L. (2007) HTR2C haplotypes and antipsychotics-induced weight gain: X-linked multimarker analysis. Hum. Psychopharmacol., 22: 463–467.

Del Tredici, A.L., Schiffer, H.H., Burstein, E.S., Lameh, J., Mohell, N., Hacksell, U., Brann, M.R. and Weiner, D.M. (2004) Pharmacology of polymorphic variants of the human 5-HT1A receptor. Biochem. Pharmacol., 67: 479–490.

Díaz-Mataix, L., Scorza, M.C., Bortolozzi, A., Toth, M., Celada, P. and Artigas, F. (2005) Involvement of 5-HT1A receptors in prefrontal cortex in the modulation of dopaminergic activity: role in atypical antipsychotic action. J. Neurosci., 25: 10831–10843.

Drago, A., Ronchi, D.D. and Serretti, A. (2008) 5-HT1A gene variants and psychiatric disorders: a review of current literature and selection of SNPs for future studies. Int. J. Neuropsychopharmacol., 11: 701–721.

Duan, J., Sanders, A.R., Molen, J.E., Martinolich, L., Mowry, B. J., Levinson, D.F., Crowe, R.R., Silverman, J.M. and Gejman, P.V. (2003) Polymorphisms in the 5'-untranslated region of the human serotonin receptor 1B (HTR1B) gene affect gene expression. Mol. Psychiatry, 8: 901–910.

Duan, S., Yin, H., Chen, W., Xing, Q., Chen, Q., Guo, T., Gao, J., Li, X., Gao, R., Liu, Z., Gu, N., Feng, G. and He, L. (2005) No association between the serotonin 1B receptor gene and schizophrenia in a case–control and family-based association study. Neurosci. Lett., 376: 93–97.

Egan, C.T., Herrick-Davis, K. and Teitler, M. (1998) Creation of a constitutively activated state of the 5-hydroxytryptamine2A receptor by site-directed mutagenesis: inverse agonist activity of antipsychotic drugs. J. Pharmacol. Exp. Ther., 286: 85–90.

Eberle-Wang, K., Lucki, I. and Chesselet, M.F. (1996) A role for the subthalamic nucleus in 5-HT2C-induced oral dyskinesia. Neuroscience, 72: 117–128.

Eisensamer, B., Rammes, G., Gimpl, G., Shapa, M., Ferrari, U., Hapfelmeier, G., Bondy, B., Parsons, C., Gilling, K., Zieglgänsberger, W. and Holsboer, F. (2003) Rupprecht R. Antidepressants are functional antagonists at the serotonin type 3 (5-HT3) receptor. Mol. Psychiatry, 8: 994–1007.

Ellingrod, V.L., Perry, P.J., Ringold, J.C., Lund, B.C., Bever-Stille, K., Fleming, F., Holman, T.L. and Miller, D. (2005) Weight gain associated with the -759C/T polymorphism of the 5HT2C receptor and olanzapine. Am. J. Med. Genet. B. Neuropsychiatr. Genet., 134B: 76–78.

Erdmann, J., Shimron-Abarbanell, D., Rietschel, M., Albus, M., Maier, W., Körner, J., Bondy, B., Chen, K., Shih, J.C., Knapp, M., Propping, P. and Nöthen, M.M. (1996) Systematic screening for mutations in the human serotonin-2A (5-HT2A) receptor gene: identification of two naturally occurring receptor variants and association analysis in schizophrenia. Hum. Genet., 97: 614–619.

Gérard, C., Martres, M.P., Lefèvre, K., Miquel, M.C., Vergé, D., Lanfumey, L., Doucet, E., Hamon, M. and el Mestikawy, S. (1997) Immuno-localization of serotonin 5-HT6 receptor-like material in the rat central nervous system. Brain Res., 746: 207–219.

Gerretsen, P. and Pollock, B.G. (2008) Pharmacogenetics and the serotonin transporter in late-life depression. Expert. Opin. Drug Metab. Toxicol., 4: 1465–1478.

Gonda, X., Juhasz, G., Laszik, A., Rihmer, Z. and Bagdy, G. (2005) Subthreshold depression is linked to the functional polymorphism of the 5HT transporter gene. J. Affect. Disord., 87: 291–297.

Gu, B., Wang, L., Zhang, A.P., Ma, G., Zhao, X.Z., Li, H.F., Feng, G.Y., He, L. and Xing, Q.H. (2008) Association between a polymorphism of the HTR3A gene and therapeutic response to risperidone treatment in drug-naïve Chinese schizophrenia patients. Pharmacogenet. Genomics, 18: 721–727.

Gunes, A., Dahl, M.L., Spina, E. and Scordo, M.G. (2008) Further evidence for the association between 5-HT2C receptor gene polymorphisms and extrapyramidal side effects in male schizophrenic patients. Eur. J. Clin. Pharmacol., 64: 477–482.

Gustafson, E.L., Durkin, M.M., Bard, J.A., Zgombick, J. and Branchek, T.A. (1996) A receptor autoradiographic and *in situ* hybridization analysis of the distribution of the 5-ht7 receptor in rat brain. Br. J. Pharmacol., 117: 657–666.

Gutierrez, B., Arranz, M.J., Huezo-Diaz, P., Dempster, D., Matthiasson, P., Travis, M., Munro, J., Osborne, S. and Kerwin, R.W. (2002) Novel mutations in 5-HT3A and 5-HT3B receptor genes not associated with clozapine response. Schizophr. Res., 58: 93–97.

Güzey, C., Scordo, M.G., Spina, E., Landsem, V.M. and Spigset, O. (2007) Antipsychotic-induced extrapyramidal symptoms in patients with schizophrenia: associations with dopamine and serotonin receptor and transporter polymorphisms. Eur. J. Clin. Pharmacol., 63: 233–241.

Hartig, P.R., Hoyer, D., Humphrey, P.P. and Martin, G.R. (1996) Alignment of receptor nomenclature with the human genome: classification of 5-HT1B and 5-HT1D receptor subtypes. Trends Pharmacol. Sci., 17: 103–105.

Hedlund, P.B. and Sutcliffe, J.G. (2004) Functional, molecular and pharmacological advances in 5-HT7 receptor research. Trends Pharmacol. Sci., 25: 481–486.

Hill, M.J. and Reynolds, G.P. (2007) 5-HT2C receptor gene polymorphisms associated with antipsychotic drug action alter promoter activity. Brain Res., 1149: 14–17.

Hinney, A., Herrmann, H., Löhr, T., Rosenkranz, K., Ziegler, A., Lehmkuhl, G., Poustka, F., Schmidt, M.H., Mayer, H., Siegfried, W., Remschmidt, H. and Hebebrand, J. (1999) No evidence for an involvement of alleles of polymorphisms in the serotonin1Dbeta and 7 receptor genes in obesity, underweight or anorexia nervosa. Intl J. Obes. Relat. Metab. Disord., 23: 760–763.

Huang, Y.Y., Grailhe, R., Arango, V., Hen, R. and Mann, J.J. (1999) Relationship of psychopathology to the human serotonin1B genotype and receptor binding kinetics in postmortem brain tissue. Neuropsychopharmacology, 21: 238–246.

Ikeda, M., Iwata, N., Kitajima, T., Suzuki, T., Yamanouchi, Y., Kinoshita, Y. and Ozaki, N. (2006) Positive association of the serotonin 5-HT7 receptor gene with schizophrenia in a Japanese population. Neuropsychopharmacology, 31: 866–871.

Ji, X., Takahashi, N., Saito, S., Ishihara, R., Maeno, N., Inada, T. and Ozaki, N. (2008) Relationship between three serotonin receptor subtypes (HTR3A, HTR2A and HTR4) and treatment-resistant schizophrenia in the Japanese population. Neurosci. Lett., 435: 95–98.

Joober, R., Benkelfat, C., Brisebois, K., Toulouse, A., Turecki, G., Lal, S., Bloom, D., Labelle, A., Lalonde, P., Fortin, D., Alda, M., Palmour, R. and Rouleau, G.A. (1999) T102C polymorphism in the 5HT2A gene and schizophrenia: relation to phenotype and drug response variability. J. Psychiatry Neurosci., 24: 141–146.

Kazemi-Kjellberg, F., Henzi, I. and Tramer, M. (2001) Treatment of established postoperative nausea and vomiting: a quantitative systematic review. BMC Anesthesiol., 1: 2.

King, M.V., Sleight, A.J., Woolley, M.L., Topham, I.A., Marsden, C.A. and Fone, K.C. (2004) 5-HT6 receptor antagonists reverse delay-dependent deficits in novel object discrimination by enhancing consolidation – an effect sensitive to NMDA receptor antagonism. Neuropharmacology, 47: 195–204.

King, M.V., Marsden, C.A. and Fone, K.C. (2008) A role for the 5-HT(1A), 5-HT4 and 5-HT6 receptors in learning and memory. Trends Pharmacol., 29: 482–492.

Kronenberg, S., Frisch, A., Rotberg, B., Carmelm, M., Apterm, A. and Weizmanm, A. (2008) Pharmacogenetics of selective serotonin reuptake inhibitors in pediatric depression and anxiety. Pharmacogenomics, 9: 1725–1736.

Lane, H.Y., Chang, Y.C., Chiu, C.C., Chen, M.L., Hsieh, M.H. and Chang, W.H. (2002) Association of risperidone treatment response with a polymorphism in the 5-HT(2A) receptor gene. Am. J. Psychiatry, 159: 1593–1595.

Lane, H.Y., Lin, C.C., Huang, C.H., Chang, Y.C., Hsu, S.K. and Chang, W.H. (2004) Risperidone response and 5-HT6 receptor gene variance: genetic association analysis with adjustment for nongenetic confounders. Schizophr. Res., 67: 63–70.

Lane, H.Y., Lee, C.C., Liu, Y.C. and Chang, W.H. (2005) Pharmacogenetic studies of response to risperidone and other newer atypical antipsychotics. Pharmacogenomics, 6: 139–149.

Lee, S., Lee, K., Lee, H., Ham, B., Ryu, S. and Lee, M. (2005) Association between the 5-HT6 receptor C267T polymorphism and response to antidepressant treatment in major depressive disorder. Psychiatry Clin. Neurosci., 59: 140–145.

Lemonde, S., Turecki, G., Bakish, D., Du, L., Hrdina, P.D., Bown, C.D., Sequeira, A., Kushwaha, N., Morris, S.J., Basak, A., Ou, X.M. and Albert, P.R. (2003) Impaired repression at a 5-hydroxytryptamine 1A receptor gene polymorphism associated with major depression and suicide. J. Neurosci., 23: 8788–8799.

Lemonde, S., Du, L., Bakish, D., Hrdina, P. and Albert, P. R. (2004) Association of the C(-1019)G 5-HT1A functional promoter polymorphism with antidepressant response. Intl. J. Neuropsychopharmacol., 7: 501–506.

Le Poul, E., Laaris, N., Doucet, E., Laporte, A.M., Hamon, M. and Lanfumey, L. (1995) Early desensitization of somato-dendritic 5-HT1A autoreceptors in rats treated with fluoxetine or paroxetine. Naunyn Schmiedebergs Arch. Pharmacol., 352: 141–148.

Lerer, B., Segman, R.H., Tan, E.C., Basile, V.S., Cavallaro, R., Aschauer, H.N., Strous, R., Chong, S.A., Heresco-Levy, U., Verga, M., Scharfetter, J., Meltzer, H.Y., Kennedy, J.L. and Macciardi, F. (2005) Combined analysis of 635 patients confirms an age-related association of the serotonin 2A receptor gene with tardive dyskinesia and specificity for the non-orofacial subtype. Intl J. Neuropsychopharmacol., 8: 411–425.

Letty, S., Child, R., Dumuis, A., Pantaloni, A., Bockaert, J. and Rondouin, G. (1997) 5-HT4 receptors improve social olfactory memory in the rat. Neuropharmacology, 36: 681–687.

Levitan, R.D., Kaplan, A.S., Masellis, M., Basile, V.S., Walker, M.L., Lipson, N., Siegel, G.I., Woodside, D.B., Macciardi, F.M., Kennedy, S.H. and Kennedy, J.L. (2001) Polymorphism of the serotonin 5-HT1B receptor gene (HTR1B) associated with minimum lifetime body mass index in women with bulimia nervosa. Biol. Psychiatry, 50: 640–643.

Li, D., Duan, Y. and He, L. (2006a) Association study of serotonin 2A receptor (5-HT2A) gene with schizophrenia and suicidal behavior using systematic meta-analysis. Biochem. Biophys. Res. Commun., 340: 1006–1015.

Li, J., Zhang, X., Wang, Y., Zhou, R., Zhang, H., Yang, L., Wang, B. and Faraone, S.V. (2006b) The serotonin 5-HT1D receptor gene and attention-deficit hyperactivity disorder in Chinese Han subjects. Am. J. Med. Genet. B. Neuropsychiatr., 141B: 874–876.

Li, J., Wang, Y., Zhou, R., Wang, B., Zhang, H., Yang, L. and Faraone, S.V. (2006c) Association of attention-deficit/hyperactivity disorder with serotonin 4 receptor gene polymorphisms in Han Chinese subjects. Neurosci. Lett., 401: 6–9.

Liu, K.G. and Robichaud, A.J. (2009) 5-HT6 antagonists as potential treatment for cognitive dysfunction. Drug Dev. Res., 70: 145–168.

Malhotra, A.K., Goldman, D., Ozaki, N., Rooney, W., Clifton, A., Buchanan, R.W., Breier, A. and Pickar, D. (1996) Clozapine response and the 5-HT2C Cys23Ser polymorphism. Neuroreport, 7: 2100–2102.

Mancilla-Diaz, J.M., Escartin-Perez, R.E., Lopez-Alonso, V.E., Floran-Garduño, B. and Romano-Camacho, J.B. (2005) Role of 5-HT1A and 5-HT1B receptors in the hypophagic effect of 5-HT on the structure of feeding behavior. Med. Sci. Monit., 11: BR74–BR79.

Masellis, M., Basile, V., Meltzer, H.Y., Lieberman, J.A., Sevy, S., Macciardi, F.M., Cola, P., Howard, A., Badri, F., Nothen, M.M., Kalow, W. and Kennedy, J.L. (1998) Serotonin subtype 2 receptor genes and clinical response to clozapine in schizophrenia patients. Neuropsychopharmacology, 19: 123–132.

Masellis, M., Basile, V.S., Meltzer, H.Y., Lieberman, J.A., Sevy, S., Goldman, D.A., Hamblin, M.W., Macciardi, F.M. and Kennedy, J.L. (2001) Lack of association between the T→C 267 serotonin 5-HT6 receptor gene (HTR6) polymorphism and prediction of response to clozapine in schizophrenia. Schizophr. Res., 47: 49–58.

McMahon, F.J., Buervenich, S., Charney, D., Lipsky, R., Rush, A.J., Wilson, A.F., Sorant, A.J., Papanicolaou, G.J., Laje, G., Fava, M., Trivedi, M.H., Wisniewski, S.R. and Manji, H. (2006) Variation in the gene encoding the serotonin 2A receptor is associated with outcome of antidepressant treatment. Am. J. Hum. Genet., 78: 804–814.

Melke, J., Westberg, L., Nilsson, S., Landen, M., Soderstrom, H., Baghaei, F., Rosmond, R., Holm, G., Björntorp, P., Nilsson, L.G., Adolfsson, R. and Eriksson, E. (2003) A polymorphism in the serotonin receptor 3A (HTR3A) gene and its association with harm avoidance in women. Arch. Gen. Psychiatry, 60: 1017–1023.

Meneses, A. (2001) Role of 5-HT6 receptors in memory formation. Drug News Perspect., 14: 396–400.

Miller, D.D., Ellingrod, V.L., Holman, T.L., Buckley, P.F. and Arndt, S. (2005) Clozapine-induced weight gain associated with the 5HT2C receptor -759C/T polymorphism. Am. J. Med. Genet. B. Neuropsychiatr. Genet., 133B: 97–100.

Mitchell, E.S. and Neumaier, J.F. (2005) Neuropharmacology. 5-HT6 receptors: a novel target for cognitive enhancement. Pharmacol. Ther., 108: 320–333.

Mitchell, E.S., Sexton, T. and Neumaier, J.F. (2007) Increased expression of 5-HT6 receptors in the rat dorsomedial striatum impairs instrumental learning. Neuropsychopharmacology, 32: 1520–1530.

Monsma, F.J. Jr., Shen, Y., Ward, R.P., Hamblin., M.W. and Sibley, D.R. (1993) Cloning and expression of a novel serotonin receptor with high affinity for tricyclic psychotropic drugs. Mol. Pharmacol., 43: 320–327.

Mulder, H., Franke, B., van der-Beek van der, A.A., Arends, J., Wilmink, F.W., Scheffer, H. and Egberts, A.C. (2007) The association between HTR2C gene polymorphisms and the metabolic syndrome in patients with schizophrenia. J. Clin. Psychopharmacol., 27: 338–343.

Mullins, U.L., Gianutsos, G. and Eison, A.S. (1999) Effects of antidepressants on 5-HT7 receptor regulation in the rat hypothalamus. Neuropsychopharmacology, 21: 352–367.

Mundo, E., Richter, M.A., Sam, F., Macciardi, F. and Kennedy, J.L. (2000) Is the 5-HT1Dβ receptor gene implicated in the pathogenesis of obsessive-compulsive disorder? Am. J. Psychiatry, 157: 1160–1161.

Mundo, E., Richter, M.A., Zai, G., Sam, F., McBride, J., Macciardi, F. and Kennedy, J.L. (2002) 5HT1Dbeta receptor gene implicated in the pathogenesis of obsessive-compulsive disorder: further evidence from a family-based association study. Mol. Psychiatry, 7: 805–809.

Murphy, G.M. Jr., Kremer, C., Rodrigues, H.E. and Schatzberg, A.F. (2003) Pharmacogenetics of antidepressant medication intolerance. Am. J. Psychiatry, 160: 1830–1835.

Myers, R.L., Airey, D.C., Manier, D.H., Shelton, R.C. and Sanders-Bush, E. (2007) Polymorphisms in the regulatory region of the human serotonin 5-HT2A receptor gene (HTR2A) influence gene expression. Biol. Psychiatry, 61: 167–173.

Nakagawa, Y., Ishima, T. and Takashima, T. (1998) The 5-HT_3 receptor agonist attenuates the action of antidepressants in the forced swim test in rats. Brain Res., 786: 189–193.

Niesler, B., Weiss, B., Fischer, C., Nothen, M.M., Propping, P., Bondy, B., Rietschel, M., Maier, W., Albus, M., Franzek, E. and Rappold, G.A. (2001a) Serotonin receptor gene *HTR3A* variants in schizophrenic and bipolar affective patients. Pharmacogenetics, 11: 21–27.

Niesler, B., Flohr, T., Nothen, M.M., Fischer, C., Rietschel, M., Franzek, E., Albus, M., Propping, P. and Rappold, G.A. (2001b) Association between the 5_ UTR variant C178T of the serotonin receptor gene HTR3A and bipolar affective disorder. Pharmacogenetics, 11: 471–475.

Ohtsuki, T., Ishiguro, H., Detera-Wadleigh, S.D., Toyota, T., Shimizu, H., Yamada, K., Yoshitsugu, K., Hattori, E., Yoshikawa, T. and Arinami, T. (2002) Association between serotonin 4 receptor gene polymorphisms and bipolar disorder in Japanese case–control samples and the NIMH Genetics Initiative Bipolar Pedigrees. Mol. Psychiatry, 7: 954–961.

Okada, M., Northup, J.K., Ozaki, N., Russell, J.T., Linnoila, M. and Goldman, D. (2004) Modification of human 5-HT(2C) receptor function by Cys23Ser, an abundant, naturally occurring amino-acid substitution. Mol. Psychiatry, 9: 55–64.

Okamura, K., Shirakawa, O., Nishiguchi, N., Ono, H., Nushida, H., Ueno, Y. and Maeda, K. (2005) Lack of an association between 5-HT receptor gene polymorphisms and suicide victims. Psychiatry Clin. Neurosci., 59: 345–349.

Ott, C., McCaffery, J., Ali, S. and Whooley, M.A. (2007) Association of a serotonin transporter polymorphism (5-HTTLPR) with depression, perceived stress, and norepinephrine in patients with coronary disease: the Heart and Soul Study. Am. J. Psychiatry, 164: 1379–1384.

Parsey, R.V., Oquendo, M.A., Ogden, R.T., Olvet, D.M., Simpson, N., Huang, Y.Y., Van Heertum, R.L., Arango, V. and Mann, J.J. (2006) Altered serotonin 1A binding in major depression: a [carbonyl-C-11]WAY100635 positron emission tomography study. Biol. Psychiatry, 59: 106–113.

Parsons, M.J., D'Souza, U.M., Arranz, M.J., Kerwin, R.W. and Makoff, A.J. (2004) The -1438A/G polymorphism in the 5-hydroxytryptamine type 2A receptor gene affects promoter activity. Biol. Psychiatry, 56: 406–410.

Pesonen, U., Koulu, M., Bergen, A., Eggert, M., Naukkarinen, H., Virkkunen, M., Linnoila, M. and Goldman, D. (1998) Mutation screening of the 5-hydroxytryptamine7 receptor gene among Finnish alcoholics and controls. Psychiatry Res., 77: 139–145.

Polesskaya, O.O. and Sokolov, B.P. (2002) Differential expression of the 'C' and 'T' alleles of the 5-HT_{2A} receptor gene in the temporal cortex of normal individuals and schizophrenics. J. Neurosci. Res., 67: 812–822.

Polesskaya, O.O., Aston, C. and Sokolov, B.P. (2006) Allele C-specific methylation of the 5-HT2A receptor gene: evidence for correlation with its expression and expression of DNA methylase DNMT1. J. Neurosci. Res., 83: 362–373.

Rammes, G., Eisensamer, B., Ferrari, U., Shapa, M., Gimpl, G., Gilling, K., Parsons, C., Riering, K., Hapfelmeier, G., Bondy, B., Zieglgänsberger, W., Holsboer, F. and Rupprecht, R. (2004) Antipsychotic drugs antagonize human serotonin type 3 receptor currents in a noncompetitive manner. Mol. Psychiatry, 9: 846–858.

Reavill, C., Kettle, A., Holland, V., Riley, G. and Blackburn, T.P. (1999) Attenuation of haloperidol-induced catalepsy by a 5-HT2C receptor antagonist. Br. J. Pharmacol., 126: 572–574.

Reynolds, G.P. (2007) The impact of pharmacogenetics on the development and use of antipsychotic drugs. Drug Discov. Today, 12: 953–959.

Reynolds, G.P., Garrett, N.J., Rupniak, N., Jenner, P. and Marsden, C.D. (1983) Chronic clozapine treatment of rats down-regulates cortical 5-HT2 receptors. Eur. J. Pharmacol., 89: 325–326.

Reynolds, G.P., Zhang, Z.J. and Zhang, X.B. (2002) Association of antipsychotic drug-induced weight gain with a 5-HT2C receptor gene polymorphism. Lancet, 359: 2086–2087.

Reynolds, G.P., Zhang, Z. and Zhang, X. (2003) Polymorphism of the promoter region of the serotonin 5-HT(2C) receptor gene and clozapine-induced weight gain. Am. J. Psychiatry, 160: 677–679.

Reynolds, G.P., Templeman, L.A. and Zhang, Z.J. (2005a) The role of 5-HT2C receptor polymorphisms in the pharmacogenetics of antipsychotic drug treatment. Prog. Neuropsychopharmacol. Biol. Psychiatry, 29: 1021–1028.

Reynolds, G.P., Yao, Z.J., Zhang, X.B., Sun, J. and Zhang, Z.J. (2005b) Pharmacogenetics of treatment in first-episode schizophrenia: D3 and 5-HT2C receptor polymorphisms separately associate with positive and negative symptom response. Eur. Neuropsychopharmacol., 15: 143–151.

Reynolds, G.P., Hill, M.J. and Kirk, S.L. (2006a) The 5-HT2C receptor and antipsychotic-induced weight gain – mechanisms and genetics. J. Psychopharmacol., 20(4 Suppl): 15–18.

Reynolds, G.P., Arranz, B., Templeman, L.A., Fertuzinhos, S. and San, L. (2006b) Effect of 5-HT1A receptor gene polymorphism on negative and depressive symptom response to antipsychotic treatment of drug-naive psychotic patients. Am. J. Psychiatry, 163: 1826–1829.

Rocha, B.A., Scearce-Levie, K., Lucas, J.J., Hiroi, N., Castanon, N., Crabbe, J.C., Nestler, E.J. and Hen, R. (1998) Increased vulnerability to cocaine in mice lacking the serotonin-1B receptor. Nature, 393: 175–178.

Roth, B.L., Craigo, S.C., Choudhary, M.S., Uluer, A., Monsma, F.J. Jr., Shen, Y., Meltzer, H.Y. and Sibley, D.R. (1994) Binding of typical and atypical antipsychotic agents to 5-hydroxytryptamine-6 and 5-hydroxytryptamine-7 receptors. J. Pharmacol. Exp. Ther., 268: 1403–1410.

Roth, B.L., Hanizavareh, S.M. and Blum, A.E. (2004) Serotonin receptors represent highly favorable molecular targets for cognitive enhancement in schizophrenia and other disorders. Psychopharmacology (Berl.), 174: 17–24.

Ryu, S., Cho, E.Y., Park, T., Oh, S., Jang, W.S., Kim, S.K., Lee, D. and Hong, K.S. (2007) -759 C/T polymorphism of 5-HT2C receptor gene and early phase weight gain associated with antipsychotic drug treatment. Prog. Neuropsychopharmacol. Biol. Psychiatry, 31: 673–677.

Sanders, A.R., Duan, J. and Gejman, P.V. (2002) DNA variation and psychopharmacology of the human serotonin receptor 1B (HTR1B) gene. Pharmacogenomics, 3: 745–762.

Saudou, F., Amara, D.A., Dierich, A., LeMeur, M., Ramboz, S., Segu, L., Buhot, M.C. and Hen, R. (1994) Enhanced aggressive behavior in mice lacking 5-HT1B receptor. Science, 265: 1875–1878.

Segman, R.H., Ebstein, R.P., Heresco-Levy, U., Gorfine, M., Avnon, M., Gur, E., Nemanov, L. and Lerer, B. (1997) Schizophrenia, chronic hospitalization and the 5-HT2C receptor gene. Psychiatr. Genet., 7: 75–78.

Segman, R.H., Heresco-Levy, U., Finkel, B., Inbar, R., Neeman, T., Schlafman, M., Dorevitch, A., Yakir, A., Lerner, A., Goltser, T., Shelevoy, A. and Lerer, B. (2000) Association between the serotonin 2C receptor gene and tardive dyskinesia in chronic schizophrenia: additive contribution of 5-HT2CSer and DRD3Gly alleles to susceptibility. Psychopharmacology, 152: 408–413.

Segman, R.H., Heresco-Levy, U., Finkel, B., Goltser, T., Shalem, R., Schlafman, M., Dorevitch, A., Yakir, A., Greenberg, D., Lerner, A. and Lerer, B. (2001) Association between the serotonin 2A receptor gene and tardive dyskinesia in chronic schizophrenia. Mol. Psychiatry, 6: 225–229.

Serretti, A., Benedetti, F., Zanardi, R. and Smeraldi, E. (2005) The influence of Serotonin Transporter Promoter Polymorphism (SERTPR) and other polymorphisms of the serotonin pathway on the efficacy of antidepressant treatments. Prog. Neuropsychopharmacol. Biol. Psychiatry, 29: 1074–1084.

Serretti, A., Drago, A. and De Ronchi, D. (2007) HTR2A gene variants and psychiatric disorders: a review of current literature and selection of SNPs for future studies. Curr. Med. Chem., 14: 2053–2069.

Sidenberg, D.G., Bassett, A.S., Demchyshyn, L., Niznik, H.B., Macciardi, F., Kamble, A.B., Honer, W.G. and Kennedy, J.L. (1993) New polymorphism for the human serotonin 1D receptor variant (5-HT1D beta) not linked to schizophrenia in five Canadian pedigrees. Hum. Hered., 43: 315–318.

Sleight, A.J., Carolo, C., Petit, N., Zwingelstein, C. and Bourson, A. (1995) Identification of 5-hydroxytrytamine$_7$ receptor binding sites in rat hypothalamus: sensitivity to chronic antidepressant treatment. Mol. Pharm., 47: 99–103.

Sodhi, M.S., Arranz, M.J., Curtis, D., Ball, D.M., Sham, P., Roberts, G.W., Price, J., Collier, D.A. and Kerwin, R.W. (1995) Association between clozapine response and allelic variation in the 5-HT2C receptor gene. Neuroreport, 7: 169–172.

Spindelegger, C., Lanzenberger, R., Wadsak, W., Mien, L.K., Stein, P., Mitterhauser, M., Moser, U., Holik, A., Pezawas, L., Kletter, K. and Kasper, S. (2008). Influence of escitalopram treatment on 5-HT(1A) receptor binding in limbic regions in patients with anxiety disorders. Mol. Psychiatry, published online: 25 Mar 2008. PMID: 18362913

Strobel, A., Gutknecht, L., Rothe, C., Reif, A., Mossner, R., Zeng, Y., Brocke, B. and Lesch, K.P. (2003) Allelic variation in 5-HT1A receptor expression is associated with anxiety- and depression-related personality traits. J. Neural. Transm., 110: 1445–1453.

Sugita, S., Shen, K.Z. and North, R.A. (1992) 5-Hydroxytryptamine is a fast excitatory transmitter at 5-HT$_3$ receptors in rat amygdala. Neuron, 8: 199–203.

Suzuki, T., Iwata, N., Kitamura, Y., Kitajima, T., Yamanouchi, Y., Ikeda, M., Nishiyama, T., Kamatani, N. and Ozaki, N. (2003) Association of a haplotype in the serotonin 5-HT4 receptor gene (HTR4) with Japanese schizophrenia. Am. J. Med. Genet. B. Neuropsychiatr. Genet., 121B: 7–13.

Suzuki, Y., Sawamura, K. and Someya, T. (2004) The effects of a 5-hydroxytryptamine 1A receptor gene polymorphism on the clinical response to fluvoxamine in depressed patients. Pharmacogenomics J., 4: 283–286.

Tan, E.C., Chong, S.A., Mahendran, R., Dong, F. and Tan, C.H. (2001) Susceptibility to neuroleptic-induced tardive dyskinesia and the T102C polymorphism in the serotonin type 2A receptor. Biol. Psychiatry, 50: 144–147.

Tecott, L.H., Maricq, A.V. and Julius, D. (1993) Nervous system distribution of the serotonin 5-HT3 receptor mRNA. Proc. Natl. Acad. Sci. USA, 90: 1430–1434.

Tecott, L.H., Sun, L.M., Akana, S.F., Strack, A.M., Lowenstein, D.H., Dallman, M.F. and Julius, D. (1995) Eating disorder and epilepsy in mice lacking 5-HT2c serotonin receptors. Nature, 374: 542–546.

Templeman, L.A., Reynolds, G.P., Arranz, B. and San, L. (2005) Polymorphisms of the 5-HT2C receptor and leptin genes are associated with antipsychotic drug-induced weight gain in Caucasian subjects with a first-episode psychosis. Pharmacogenetics, 15: 195–200.

Thome, J., Retz, W., Baader, M., Pesold, B., Hu, M., Cowen, M., Durany, N., Adler, G., Henn, F.A. and Rösler, M. (2001) Association analysis of HTR6 and HTR2A polymorphisms in sporadic Alzheimer's disease. J. Neural. Transm., 108: 1175–1180.

Tsai, S.J., Liu, H.C., Liu, T.Y., Wang, Y.C. and Hong, C.J. (1999) Association analysis of the 5-HT6 receptor polymorphism C267T in Alzheimer's disease. Neurosci. Lett., 276: 138–139.

Ujike, H., Nomura, A., Morita, Y., Morio, A., Okahisa, Y., Kotaka, T., Kodama, M., Ishihara, T. and Kuroda, S. (2008) Multiple genetic factors in olanzapine-induced weight gain in schizophrenia patients: a cohort study. J. Clin. Psychiatry, 69: 1416–1422.

Varnäs, K., Hall, H., Bonaventure, P. and Sedvall, G. (2001) Autoradiographic mapping of 5-HT(1B) and 5-HT(1D) receptors in the post mortem human brain using [(3)H]GR 125743. Brain Res., 915: 47–57.

Varnäs, K., Halldin, C., Pike, V.W. and Hall, H. (2003) Distribution of 5-HT4 receptors in the postmortem human brain – an autoradiographic study using [125I]SB 207710. Eur. Neuropsychopharmacol., 13: 228–234.

Varnäs, K., Hurd, Y.L. and Hall, H. (2005) Regional expression of 5-HT1B receptor mRNA in the human brain. Synapse, 56: 21–28.

Vogt, I.R., Shimron-Abarbanell, D., Neidt, H., Erdmann, J., Cichon, S., Schulze, T.G., Müller, D.J., Maier, W., Albus, M., Borrmann-Hassenbach, M., Knapp, M., Rietschel, M., Propping, P. and Nöthen, M.M. (2000) Investigation of the human serotonin 6 [5-HT6] receptor gene in bipolar affective disorder and schizophrenia. Am. J. Med. Genet., 96: 217–221.

Wang, L., Fang, C., Zhang, A., Du, J., Yu, L., Ma, J., Feng, G., Xing, Q. and He, L. (2008) The -1019 C/G polymorphism of the 5-HT(1)A receptor gene is associated with negative symptom response to risperidone treatment in schizophrenia patients. J. Psychopharmacol., 22: 904–909.

Warburton, A.C., Cole, J.C. and Tredwell, J.M. (1994) Antagonism of amphetamine-induced disruption of latent inhibition in rats by haloperidol and ondansetron: implications for a possible antipsychotic action of ondansetron. Psychopharmacology, 114: 657–664.

Wetzel, C.H.R., Hermann, B., Behl, C., Pestel, E., Rammes, G., Ziegelgänsberger, W., Holsboer, F. and Rupprecht, R. (1998) Functional antagonism of gonadal steroids at the 5-hydroxytryptamine type 3 receptor. Mol. Endocrinol., 12: 1441–1451.

Woolley, M.L., Marsden, C.A. and Fone, K.C. (2004) 5-ht6 receptors. Curr. Drug Targets CNS Neurol. Disord., 3: 59–79.

Wu, S. and Comings, D.E. (1999) A common C-1018G polymorphism in the human 5-HT1A receptor gene. Psychiatr. Genet., 9: 105–106.

Wu, W.H., Huo, S.J., Cheng, C.Y., Hong, C.J. and Tsai, S. J. (2001) Association study of the 5-HT(6) receptor polymorphism (C267T) and symptomatology and antidepressant response in major depressive disorders. Neuropsychobiology, 44: 172–175.

Yevtushenko, O.O., Cooper, S.J., O'Neill, R., Doherty, J.K., Woodside, J.V. and Reynolds, G.P. (2008) Influence of 5-HT2C receptor and leptin gene polymorphisms, smoking and drug treatment on metabolic disturbances in patients with schizophrenia. Br. J. Psychiatry, 192: 424–428.

Yevtushenko, O.O. et al. (2009) Early response to selective serotonin reuptake inhibitors in panic disorder is associated with a functional 5-HT1A receptor gene polymorphism. J. Affect. Disord., doi:10.1016/j.jad.2009.09.007. [AUQ

Yu, Y.W., Tsai, S.J., Lin, C.H., Hsu, C.P., Yang, K.H. and Hong, C.J. (1999) Serotonin-6 receptor variant (C267T) and clinical response to clozapine. Neuroreport, 10: 1231–1233.

Yu, Y.W., Tsai, S.J., Chen, T.J., Lin, C.H. and Hong, C.J. (2002) Association study of the serotonin transporter promoter polymorphism and symptomatology and antidepressant response in major depressive disorders. Mol. Psychiatry, 7: 1115–1119.

Yu, Y.W., Tsai, S.J., Liou, Y.J., Hong, C.J. and Chen, T.J. (2006) Association study of two serotonin 1A receptor gene polymorphisms and fluoxetine treatment response in Chinese major depressive disorders. Eur. Neuropsychopharmacol., 16: 498–503.

Yuan, X., Yamada, K., Ishiyama-Shigemoto, S., Koyama, W. and Nonaka, K. (2000) Identification of polymorphic loci in the promoter region of the serotonin 5-HT2C receptor gene and their association with obesity and type II diabetes. Diabetologia, 43: 373–376.

Zhang, Z.J., Zhang, X.B., Sha, W.W. and Zhang, X.B. (2002) Association of a polymorphism of the promoter region of the serotonin 5-HT2C receptor gene with tardive dyskinesia in patients with schizophrenia. Mol. Psychiatry, 7: 670–671.

Zhukovskaya, N.L. and Neumaier, J.F. (2000) Clozapine downregulates 5-hydroxytryptamine6 (5-HT6) and upregulates 5-HT7 receptors in HeLa cells. Neurosci. Lett., 288: 236–240.

Index

B

E

I

R

S

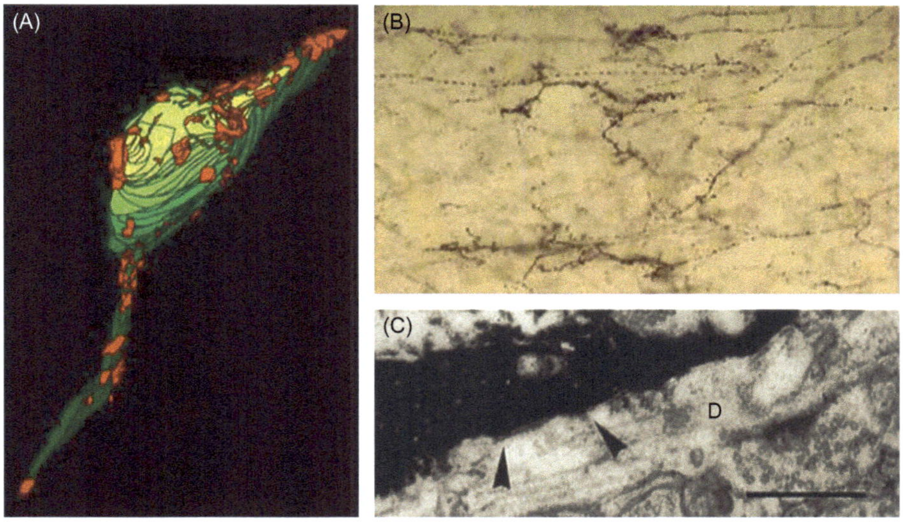

Plate 1 (A) Three-dimensional reconstruction of a cortical interneuron in layer II of the cat primary auditory cortex (green) contacted by varicose 5-HT-immunoreactive axons (red) at the surface of the soma and primary dendrites. (B) Micrograph of a coronal section in the marmoset frontal cortex containing 5-HT-immunoreactive thin and varicose axons, the latter forming a conspicuous pericellular array surrounding tightly the cell body and proximal dendrites of individual neurons. (C) Electron micrograph of an unstained post-synaptic dendrite of a cat auditory cortex interneuron (D) forming a synaptic contact with a serotoninergic large varicose immunoreactive axon terminal (arrowheads). Scale bar: for (A), 8 microns; for (B), 40 microns; for (C), 1 micron.

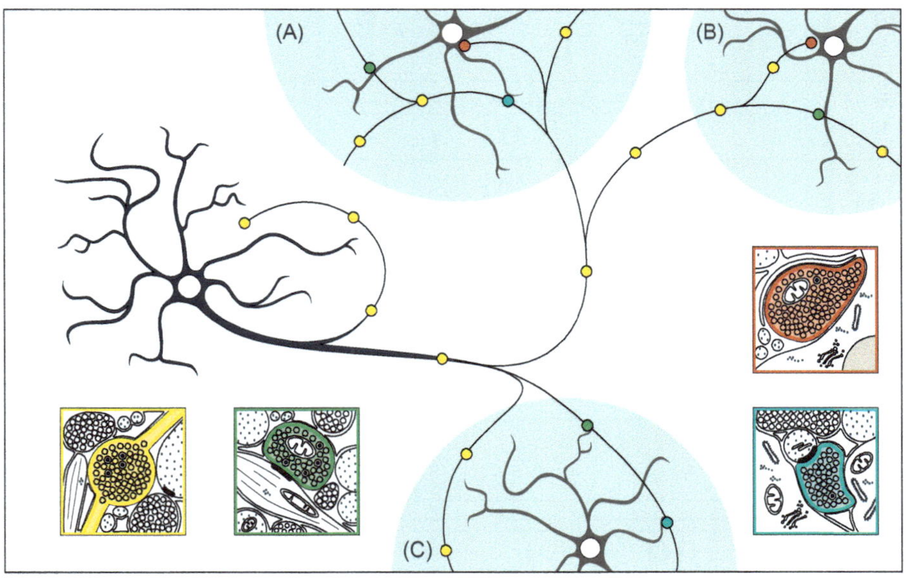

Plate 2 Schematic representation of a 5-HT neuron. As depicted here, a single 5-HT neuron may innervate different anatomical regions (A, B, C), including its nucleus of origin (through recurrent collaterals of its axon). In the territories of innervation, many of the 5-HT varicosities (small colored dots), distributed as rosary beads on distal branches of the axon, do not make synaptic contact (yellow dots). Others are endowed with junctional complexes, the small zone of membrane specialization, either symmetrical or asymmetrical, that is the hallmark of synapses, and make such a contact with dendritic branches (green dots), and, more rarely, dendritic spines (blue dots). Juxtapositions to cell bodies are also seen (red dots). These various configurations, schematized in the boxes, may be found within a given anatomical territory, and probably on a single axonal branch. The vesicular content of 5-HT varicosities may also vary from one anatomical region to another, and in some regions, such as the cerebellum with its mossy 5-HT fibers, or the locus coeruleus and its 5-HT varicosities filled with microvesicles and canaliculi, 5-HT terminals of a unique morphology may be found. It has been calculated that some 5-HT neurons may possess as many as hundreds of thousands of axon varicosities (Descarries *et al.*, 1990). The demonstration of the existence of asynaptic 5-HT varicosities in mammalian CNS was at the origin of the concept of diffuse (volume) transmission (for review, see Descarries and Mechawar, 2000).

Gene	Gene structure	Gene products	Cds
HTR1A 5q11.2-q13		5-HT_{1A}: 422aa	1269b
HTR1B 6q13		5-HT_{1B}: 390aa	1173b
HTR1D 1p36.3-p34.3		5-HT_{1D}: 377aa	1134b
HTR1E 6q14-q15		5-HT_{1E}: 365aa	1098b
HTR1F 3p12		5-HT_{1F}: 366aa	1101b
HTR2A 6q14-q21		5-HT_{2A}: 471aa	>60kb
HTR2B 2q36.3-q37.1		5-HT_{2B}: 481aa	>15kb
HTR2C Xq24		5-HT_{2C}: 458aa	>183kb
HTR3A 11q23.1		5-$HT_{3A(a)}$: 510aa 5-$HT_{3A(b)}$: 478aa	>15kb
HTR3B 11q23.1		5-HT_{3B}: 441aa	>41kb
HTR3C 3q27.1		5-HT_{3C}: 447aa	>7kb
HTR3D 3q27.1		5-HT_{3D}: 279aa	>7kb
HTR3E 3q27.1		5-HT_{3E}: 471aa	>6kb
HTR4 5q31-q33		5-$HT_{4(a)}$: 387aa $_{(b)}$: 388aa, $_{(c)}$: 411aa, $_{(d)}$: 360aa, $_{(e)}$: 371aa, $_{(f)}$: 363aa, $_{(g)}$: 378aa, $_{(hb)}$: 402aa, $_{(i)}$: 428aa, $_{(n)}$: 359aa,	>185kb
HTR5A 7q36.1		5-HT_{5A} : 357aa	>14kb
HTR5B 2q14.1		Pseudogene	none
HTR6 1p36-p35		5-HT_6 : 440aa	>15kb
HTR7 10q21-q24		5-$HT_{7(a)}$: 445aa 5-$HT_{7(b)}$: 432aa 5-$HT_{7(d)}$: 479aa	>116kb

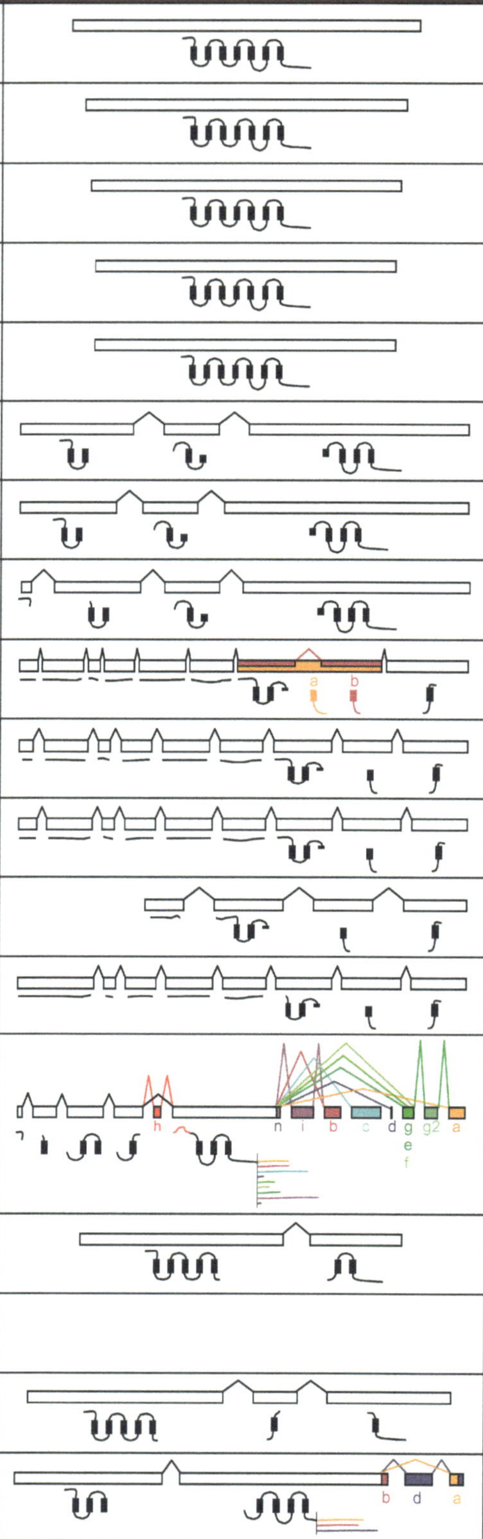

Plate 3 Human gene structures of 5-HT receptor coding sequences. This figure was generated using released data of the human genome project available on the NCBI site (http://www.ncbi.nlm.nih.gov) using Map Viewer interface, Gene Database and Consensus CDS Database. Builds 36.2–36.3 were used to localize the exons, except for HTR5B which was only annotated on build 35. The first column presents gene names and cytogenetic localizations. The second column schematizes the exon composition of the coding sequences and the part of the receptor encoded by each exon: open boxes, exons, bridges, splicing events, black boxes, transmembrane domains, curved lines, loops or N- or C-terminal domains. Colors indicate alternative exons and the receptor part encoded in the splice variants. The third column indicates the name and size of the gene products and splice variants (aa, amino acids). The fourth column presents size of the complete coding sequence on the human genome (b, bases; kb, kilobases).

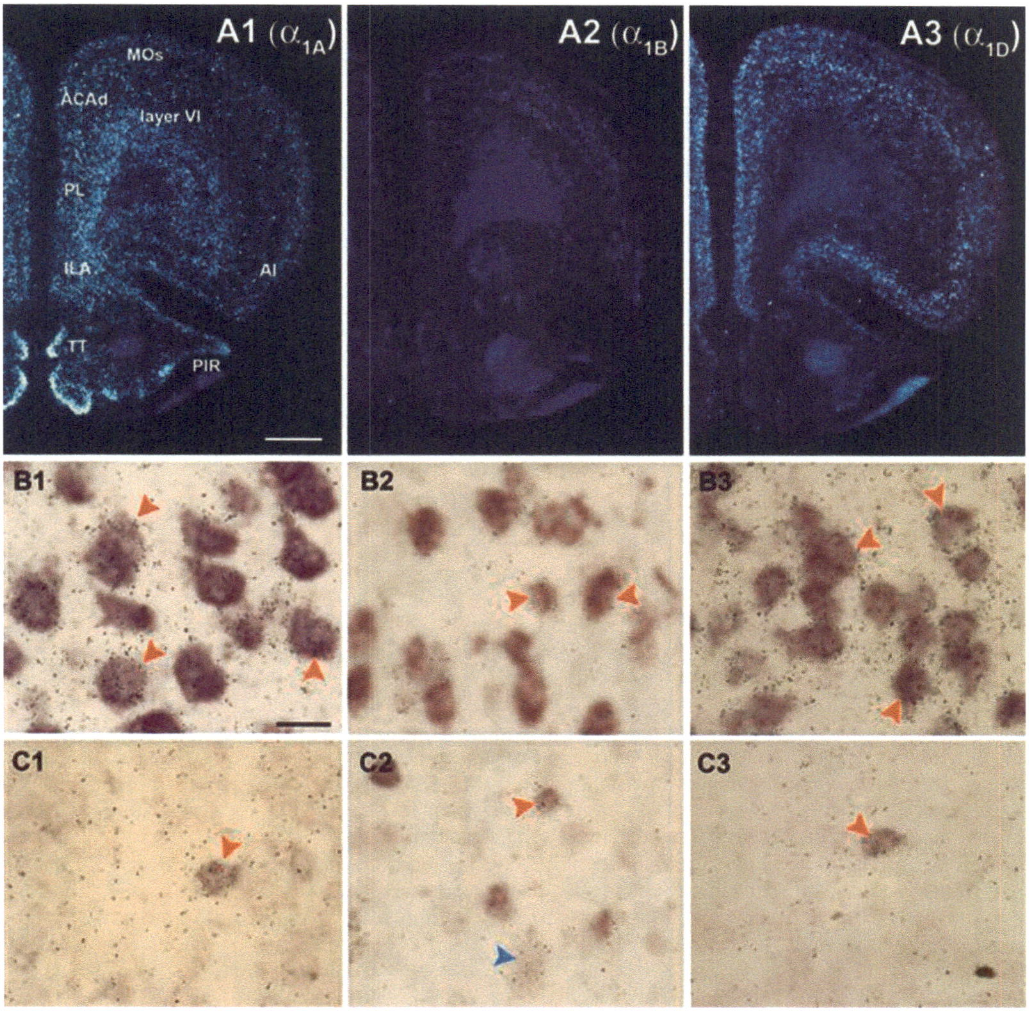

Plate 4 (A) Low-magnification dark-field photomicrographs showing the localization of α_{1A}-adrenoceptor (A1), α_{1B}-adrenoceptor (A2) and α_{1D}-adrenoceptor (A3) mRNAs in the rat prefrontal cortex using *in situ* hybridization histochemistry. The sections correspond approximately to AP + 3.7 mm (Paxinos and Watson, 2005). Each receptor transcript was detected with ^{33}P-labeled oligonucleotides. (B) High-magnification photomicrographs showing the presence of: α_{1A}-adrenoceptor (B1), α_{1B}-adrenoceptor (B2) and α_{1D}-adrenoceptor (B3) mRNA (^{33}P-labeled oligonucleotides, seen as silver grains) in pyramidal cells of the prelimbic area. Pyramidal neurons were identified by the presence of vGluT1 mRNA (Dig-labeled oligonucleotides). (C) High-magnification photomicrographs showing the presence of: α_{1A}-adrenoceptor (C1), α_{1B}-adrenoceptor (C2) and α_{1D}-adrenoceptor (C3) mRNA (^{33}P-labeled oligonucleotides, seen as silver grains) in GABAergic cells of the prelimbic area. GABAergic neurons were identified by the presence of $GAD_{65/67}$ mRNA (Dig-labeled oligonucleotides). Red arrowheads mark cells positive for both transcripts, blue arrowheads mark cells only positive for the receptor mRNA. For the sake of simplicity, only a few cells of each type are marked. Abbreviations: ACAd, dorsal anterior cingulate cortex; PL, prelimbic area; ILA infralimbic area, TT, tenia tecta, PIR, piriform cortex; AI, agranular insular cortex; MOs, secondary motor area (nomenclature from Swanson, 1998). Bar size: 1 mm (A1), 20 μm (B1).

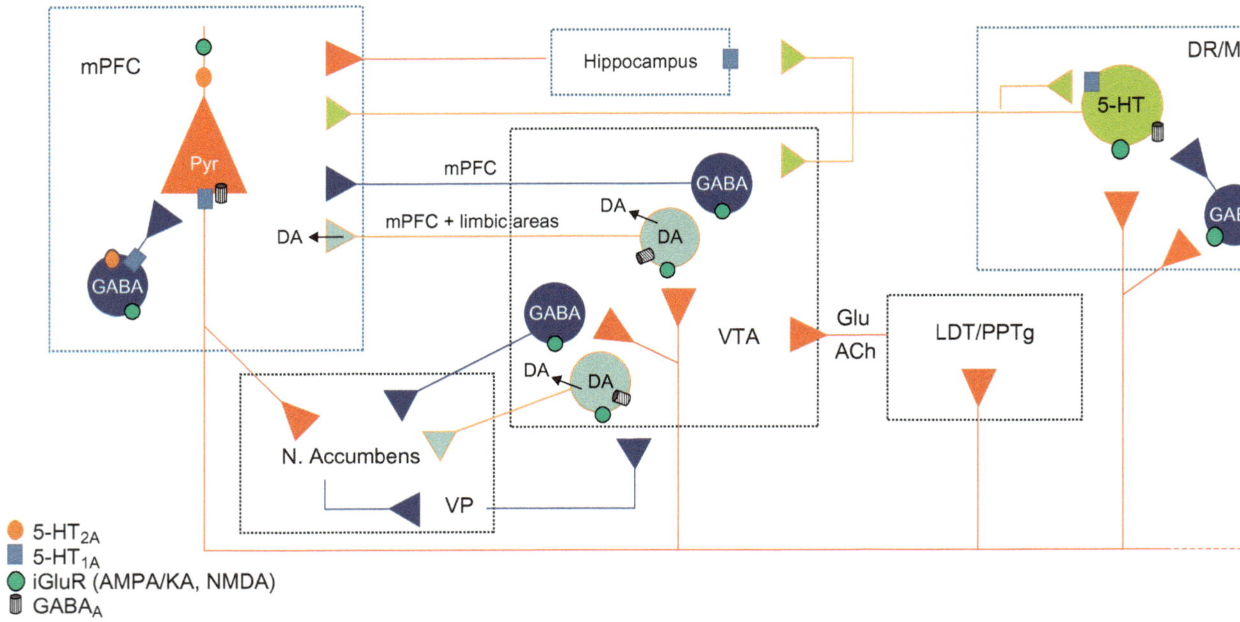

Plate 5 Anatomical and functional relationship between the mPFC, the VTA, and the dorsal and median raphe nuclei (DR/MnR). Pyramidal neurons of the mPFC project directly to mesocortical (but not mesoaccumbal) dopamine (DA) neurons, closing a mPFC-VTA circuit. GABAergic cells in the VTA project to mPFC and limbic areas as well. The mPFC may also modulate the activity of VTA neurons indirectly, through the basal ganglia circuit (e.g., mPFC nucleus accumbens ventral pallidum (VP) pathway) or through afferents to the LDT/PPTg. Likewise, the mPFC is reciprocally connected with the DR/MnR. Pyramidal neurons of the mPFC project to raphe 5-HT and GABA neurons and modulate their activity. In turn, 5-HT neurons modulate the activity of pyramidal cells in the mPFC through various receptors, in particular 5-HT$_{1A}$ and 5-HT$_{2A}$ receptors, which are expressed by pyramidal and GABAergic neurons. The pharmacological activation of these receptors in the mPFC has been shown to modulate 5-HT neuron activity and terminal 5-HT release via descending afferents to the midbrain. In addition, to the mPFC, 5-HT$_{1A}$ receptors are densely expressed in the DR/MnR (autoreceptors) and in areas projecting to the mPFC, such as the hippocampal formation. Direct 5-HT afferents to the VTA may also be involved in the control of DA neurons by other 5-HT receptors, notably the 5-HT$_{2C}$ subtype, present in GABAergic neurons of the VTA. Abbreviations: Glu, glutamate; Pyr, pyramidal neuron; Ach, acetylcholine; N. Accumbens, nucleus accumbens; iGluR, Ionotropic glutamate receptor; KA, kainic acid.

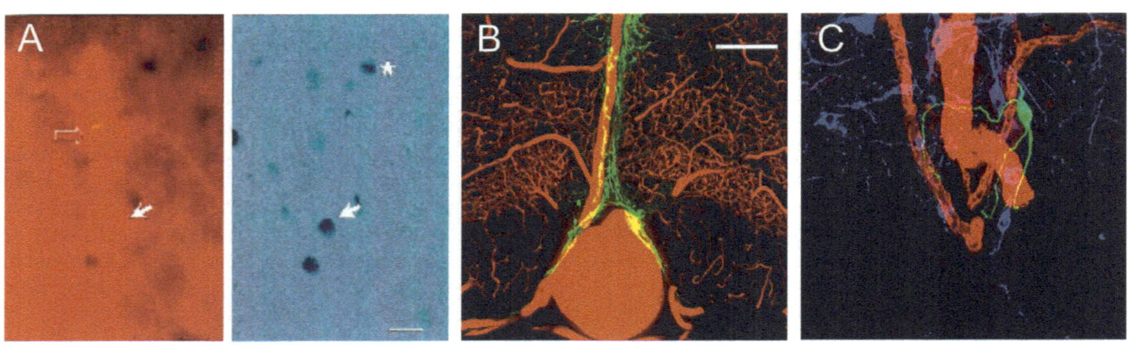

Plate 6 5-HT neurons are chemosensitive *in vivo* and are closely associated with large blood vessels. (A) A subset of neurons in the raphe pallidus express c-Fos when exposed to hypercapnia (right panel). Some of these neurons are immunoreactive for 5-HT (left panel). Open arrow: 5-HT neuron that does not express c-Fos. Closed arrows: 5-HT neuron that is immunoreactive for c-Fos. Asterisk: C-Fos positive neuron that does not express 5-HT. Scale bar: 20μm. (B) Medullary raphe neurons are closely related to the basilar artery and its large penetrating branches. Shown is a transverse section of the medulla. 5-HT neurons are immunostained with an antibody against tryptophan hydroxylase (green). Blood vessels are filled with fluorescent albumin (red). Scale bar: 200μm. (C). Shown is a neuron in the medullary raphe that was filled with biocytin (green) after patch-clamp recording. This neuron was stimulated by acidosis and was immunoreactive for tryptophan hydroxylase (blue). It had processes that were closely associated with large blood vessels (red). (A) Reproduced with permission from Elsevier: *Respiration Physiology* (Haxhiu *et al.*, 2001) 129: 191–209, ©2001. (B, C) Reproduced with permission from Nature Publishing Group: *Nature Neuroscience* (Bradley *et al.*, 2002) 5: 401–2, ©2002.

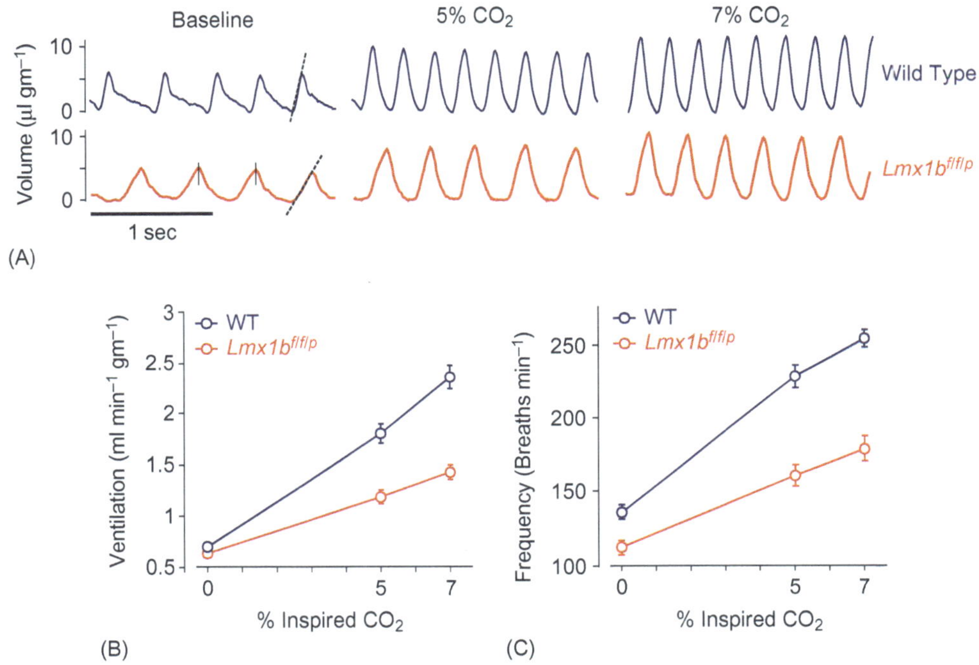

Plate 7 Mice with genetic deletion of 5-HT neurons have a defect in central respiratory chemoreception. (A) Plethysmograph recordings from WT (blue) and Lmx1b$^{f/f/p}$ (red) mice. Lmx1b$^{f/f/p}$ breathe with a slower frequency at baseline, and do not have as large a response to 5% and 7% CO_2. (B, C) Lmx1b$^{f/f/p}$ have a smaller response to an increase in ambient CO_2 and this is due to a smaller increase in breathing frequency. Reproduced with permission from the Society for Neuroscience: *Journal of Neuroscience* (Hodges *et al.*., 2009) 28: 2495–505, ©2009.

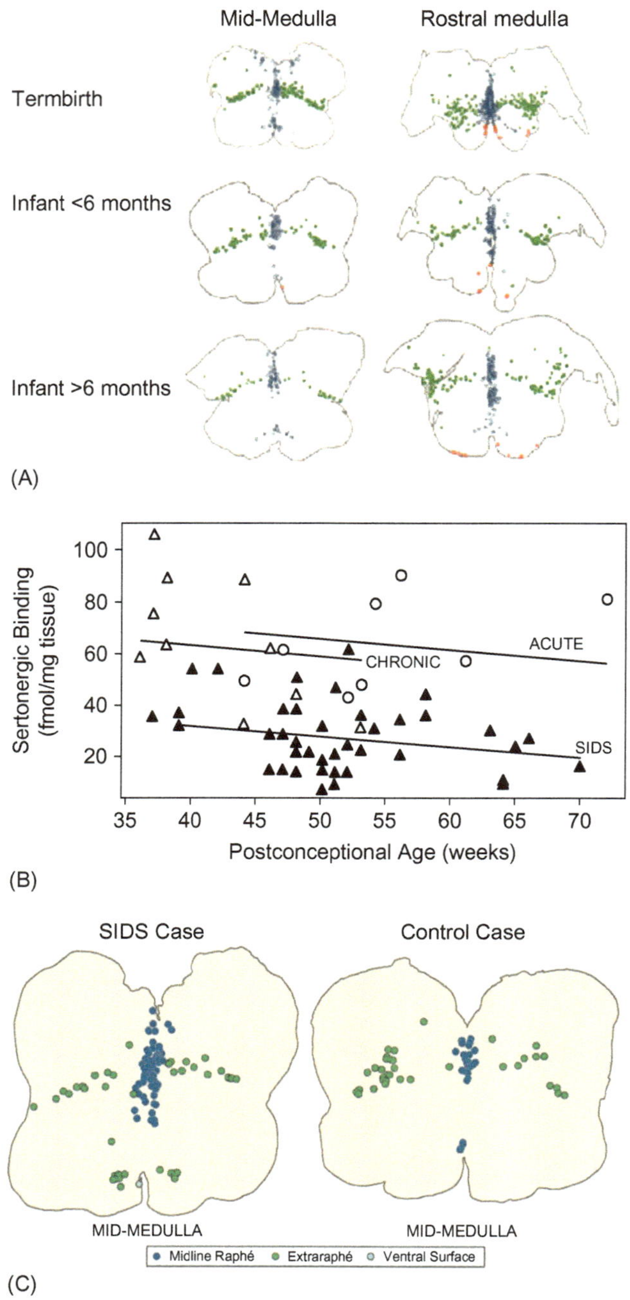

Plate 8 Sudden infant death syndrome (SIDS) is associated with abnormalities of the 5-HT system. (A) Developmental changes in the human medullary raphe nuclei at the mid- and rostral medullary levels. Blue dots, raphe 5-HT neurons; green dots, extra-raphe 5-HT neurons; red dots, 5-HT neurons on the ventral surface including the arcuate nucleus. Scale bar: 1 cm. (B) Tritiated LSD binding to 5-HT receptors is decreased in the raphe obscurus of infants that have died of SIDS. The scatter plot shows ^{3}H-LSD binding as measured with quantitative tissue receptor autoradiograhy in the raphe obscurus relative to postconceptional age for SIDS cases (solid triangles) compared to control infants that died of acute causes (open circles) or diseases that caused chronic hypoxia (closed circles). (C) Comparison of distribution of 5-HT neurons of a SIDS case and a control case. In the midline raphe of the SIDS infant there is an increased number of 5-HT neurons. (A) Reproduced with permission from Elsevier: *Autonomic Neuroscience* (Kinney, 2007) 132: 81–102, ©2007. (B) Reproduced with permission from Lippincott, Williams & Wilkins, Inc.: *Journal of Neuropathology and Experimental Neurology* (Panigrahy *et al.*, 2000) 59: 377–87. (C) Reproduced with permission from the American Medical Association: *Journal of the American Medical Association* (Paterson *et al.*, 2006) 296: 2124–2132 ©2006.

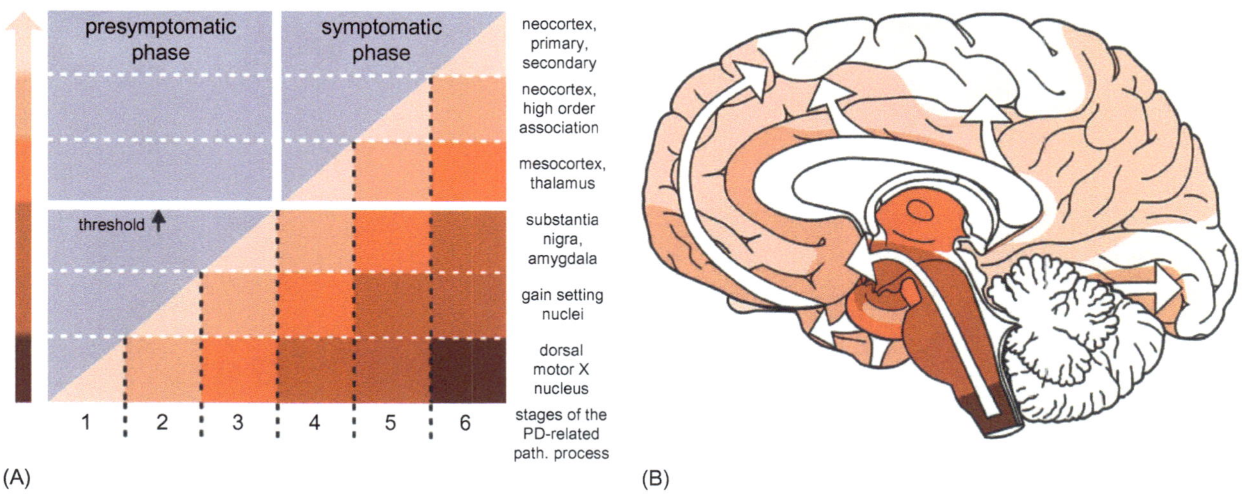

Plate 9 Parkinson's disease involves the 5-HT system early and many patients have abnormalities of breathing and chemoreception. (A) Braak staging of neurodegeneration in idiopathic PD. Note that 'gain-setting nuclei', which includes the raphe nuclei, are involved earlier in the disease process than dopaminergic neurons of the sustantia nigra. (B) Progression of disease in PD begins in the lower brainstem and progresses rostrally. (A) Reproduced with permission from Springer: *Cell Tissue Research* (Braak *et al.*, 2004) 318: 121–34, ©2004.

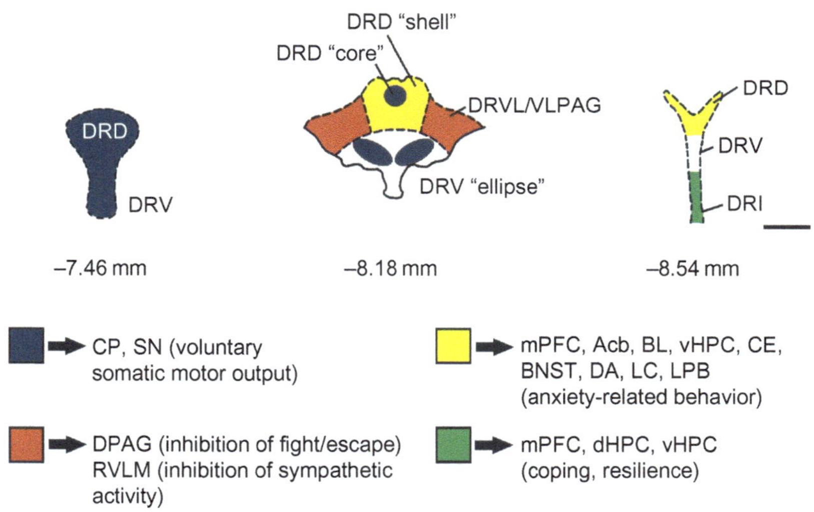

Plate 10 Diagrammatic illustration of the topographical distribution of serotonergic neurons projecting to functionally related forebrain and hindbrain targets in the rat brain. Serotonergic neurons projecting to the caudate putamen and substantia nigra (blue) are located predominantly in the rostral part of the DR and are concentrated within clusters in the DRN 'core' and DRV 'ellipse' regions of the mid-rostrocaudal parts of the DR. Serotonergic neurons projecting to brain structures implicated in the regulation of anxious states and anxiety-related behaviors (yellow) are concentrated in the mid-rostrocaudal DRN 'shell' region and caudal DR. Populations of serotonergic neurons in the DRVL/VLPAG region (red) project to the DPAG and RVLM and have been implicated in the inhibition of panic-like responses. Serotonergic neurons in the DRI (green) have projections to the prefrontal cortex, hippocampus, and midline thalamus, are thought to be activated by peripheral immune activation, and may play an important role in coping and resilience. Abbreviations: Acb, nucleus accumbens; BL, basolateral amygdala, BNST, bed nucleus of the stria terminalis; CE, central nucleus of the amygdala; CP, caudate putamen; DA, dorsal hypothalamic area; dHPC, dorsal hippocampus; DRN, dorsal raphe nucleus, dorsal part; DRI, dorsal raphe nucleus, interfascicular part, DRV, dorsal raphe nucleus, ventral part; DRVL/VLPAG, dorsal raphe nucleus, ventrolateral part/ventrolateral periaqueductal gray; DPAG, dorsal periaqueductal gray; LC, locus coeruleus; LPB, lateral parabrachial nucleus; mPFC, medial prefrontal cortex; RVLM, rostral ventrolateral medulla; SN, substantia nigra; vHPC, ventral hippocampus. Scale bar, 500 μm. Reproduced from *Stress: The Biology of Stress*, 12 January 2005, Christopher A. Lowry, Philip L. Johnson, Anders Hay-Schmidt *et al.* 'Modulation of anxiety circuits by serotonergic systems' 8:4, pp. 233–246, reprinted by permission of Taylor & Francis Ltd, http://www.tandf.co.uk/journals.

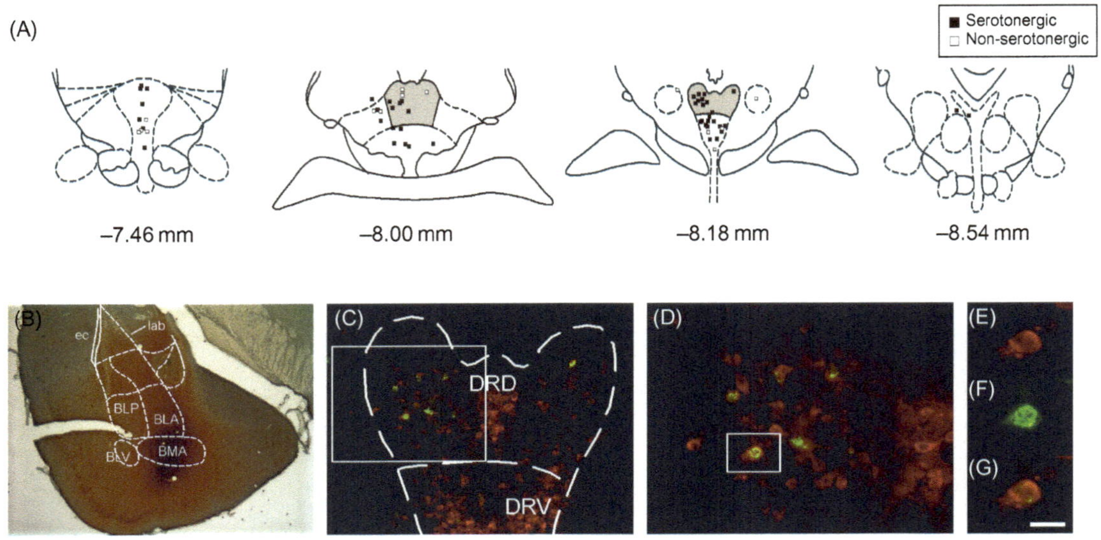

Plate 11 Illustrations and photomicrographs showing retrogradely labeled neurons in the dorsal raphe nucleus following injections of Cholera Toxin B-subunit (CTb) into the basolateral nucleus of the amygdala. (A) Illustrations of the distribution of basolateral amygdala-projecting serotonergic (black squares) and non-serotonergic (white squares) neurons within subdivisions of the dorsal raphe nucleus. Retrogradely labeled neurons are concentrated in the mid-rostrocaudal dorsal raphe nucleus (gray shaded areas at −8.00 mm and −8.18 mm Bregma) (Paxinos and Watson, 1998). Most but not all BL-projecting neurons in the dorsal raphe nucleus are serotonergic. (B) Photomicrograph illustrating the CTb injection site restricted to the BLA and BMA. (C) Double-labeled immunofluorescence photomicrograph showing the mid-rostrocaudal dorsal raphe nucleus (corresponding to −8.18 mm Bregma) illustration in (A). White box in (C) indicates region shown at higher magnification in (D). White box in (D) indicates region shown at higher magnification in E, F and G. Tryptophan hydroxylase-immunoreactive (TrpOH-ir; serotonergic) neurons appear red (E), while CTb-ir (BL-projecting) neuron appears green (F). Double TrpOH/CTb immunofluorescent neuron appears yellow. Abbreviations: BLA, basolateral nucleus of the amygdala, anterior part; BLP, basolateral nucleus of the amygdala, posterior part; BLV, basolateral nucleus of the amygdala, ventral part; BMA, basomedial nucleus of the amygdala, anterior part; DRN, dorsal raphe nucleus, dorsal part; DRV, dorsal raphe nucleus, ventral part; ec, external capsule; lab, longitudinal association bundle. Scale bar: $448\,\mu m$ (A); $500\,\mu m$ (B); $100\,\mu m$ (C); $50\,\mu m$ (D), $25\,\mu m$ (E, F, G).

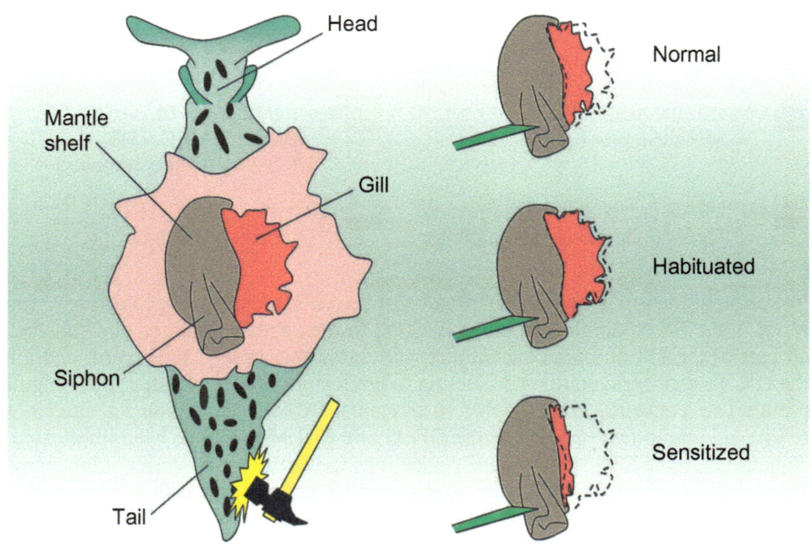

Plate 12 A dorsal view of an *Aplysia* showing the animal's respiratory organ: the gill. When the siphon is stimulated by the application of a tactile stimulus, the siphon is contracted and the gill is withdrawn under the mantle shelf (right, top). This withdrawal reflex is a protection reflex that undergoes habituation (right, middle) when the stimulation is repeated, or which may show sensitization after a noxious stimulus is applied to another part of the body (e.g., a hurt on the tail; right bottom). Dishabituation (not-illustrated) is a term used to describe what happens once habituation has been established and a noxious stimulus is applied to another part of the body; in this case, the habituation of the withdrawal reflex is relieved. The scattered line indicates the limits of the non-retracted gill, as shown on the *Aplysia*, left.

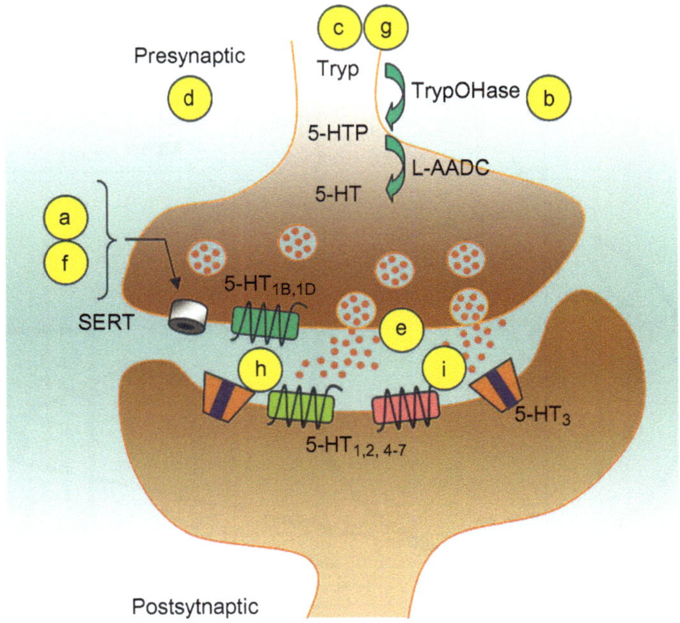

Plate 13 Alterations of the serotonergic system/synapses can be achieved by different methods. Depletion may be obtained by irreversible damage to the neurons by the means of intracerebroventricular 5,7-dihydroxytryptamine (5,7-DHT), which is taken up by the serotonin transporter (a), by pCPA-induced blockade of the 5-HT-synthesizing enzyme, tryptophane hydroxylase (b), by placing subjects under tryptophan-depleted food regimens (c), or by pCA- or MDMA-induced alterations of 5-HT terminals (d). Increased 5-HT functions can be obtained by acute pCA-induced release of 5-HT (e), by blockade of the serotonin transporter (SERT; f), or the administration of tryptophan-enriched food regimens (g). Another way to interfere with serotonergic function is based on the administration of ligands that may more or less selectively block (antagonists) or activate (agonists) the receptors (h), which are of the metabotropic (5-HT_1, 5-HT_2, 5-HT_4, 5-HT_7) or inotropic (5-HT_3) type; depending on the species, the auto-receptor is of the 5-HT_{1B} or 5-HT_{1D} subtype. As regards the receptors, it is also possible to genetically delete them in mice (i). In general, (d), (g) and (h) can be used both in experimental animals and human subjects, but all other approaches are restricted to animal models (most often to mice, rats, monkeys, less frequently to guinea pigs, as concerns the study of CNS functions).

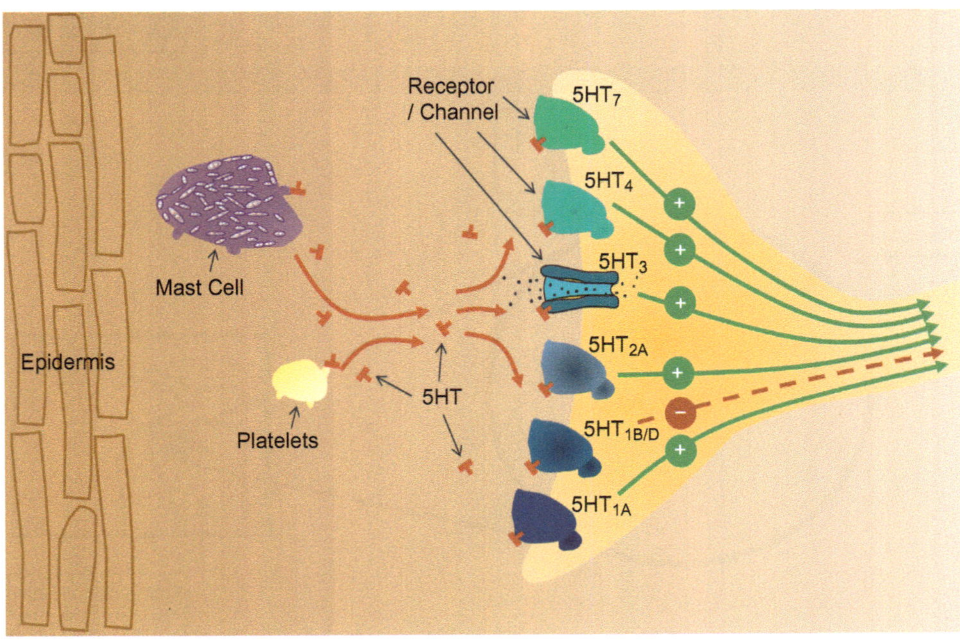

Plate 14 Simplified schematic drawing of 5-HT action in the periphery.

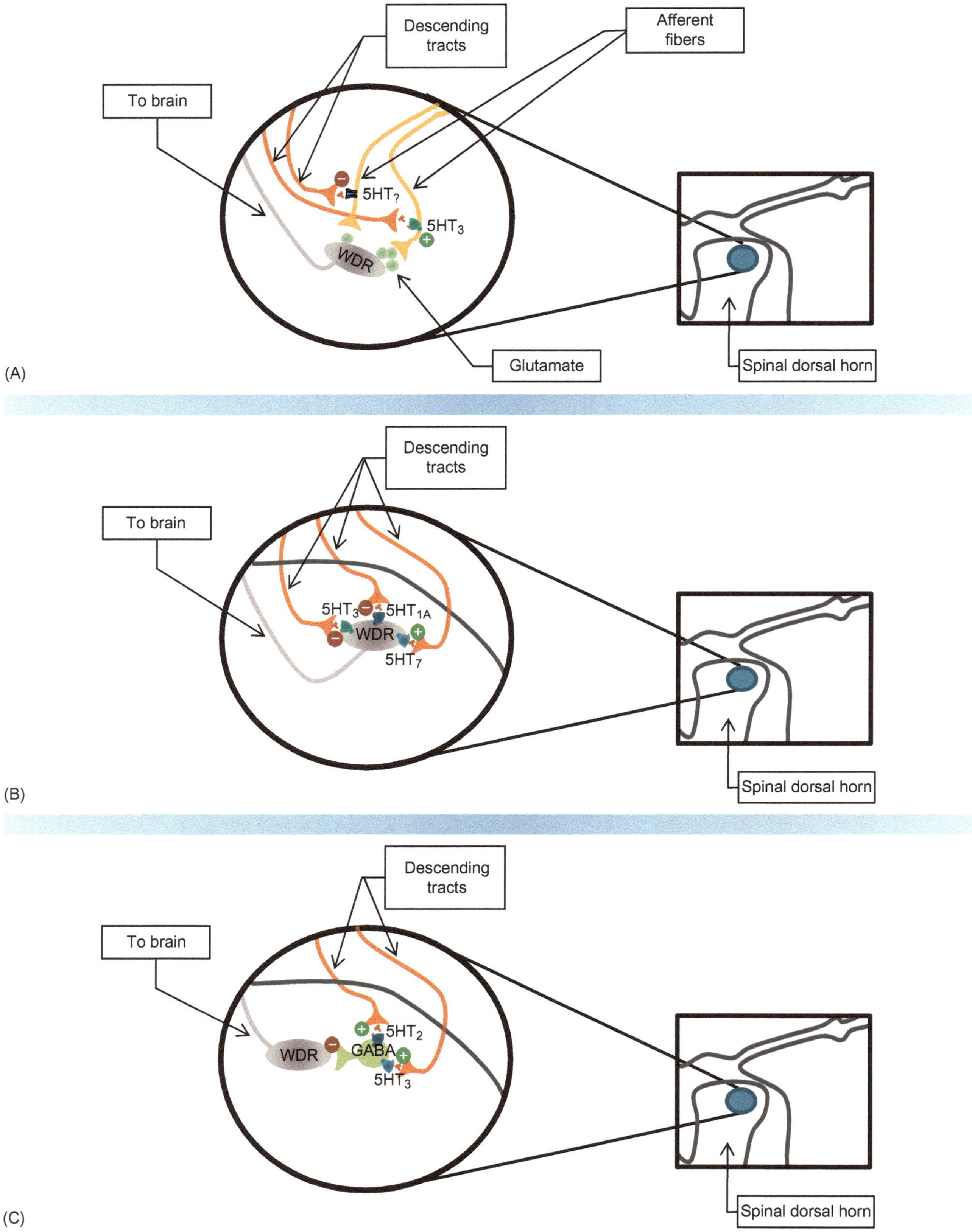

Plate 15 Simplified schematic drawing of the 5-HT descending system in the spinal cord.

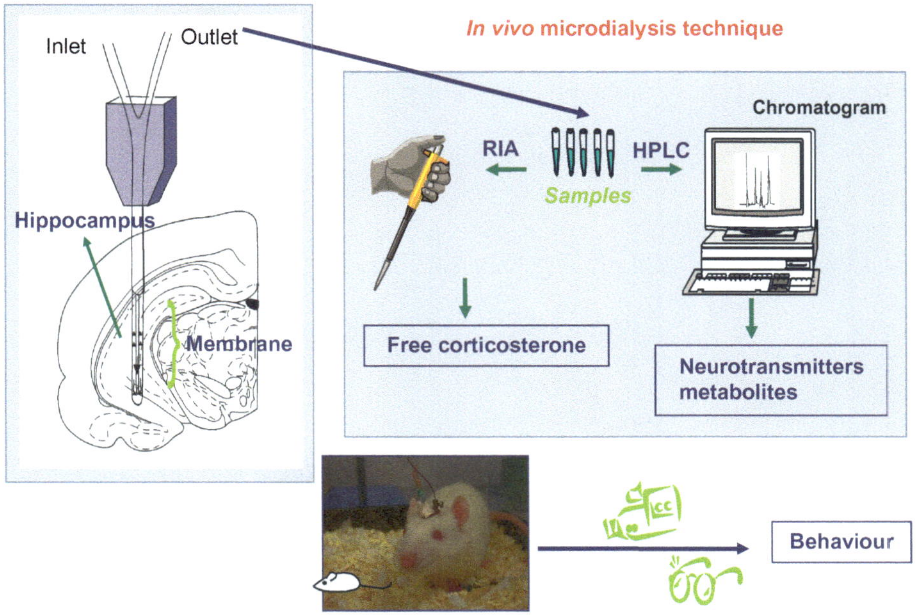

Plate 16 Schematic overview of the microdialysis technique as applied in our laboratories. A microdialysis probe is inserted through a guide cannula (not depicted, implanted 7–10 days earlier; the dialysis probe depicted is the CMA12 of CMA Microdialysis/AB, Stockholm, Sweden) exposing the dialysis membrane to the brain tissue of interest (shown here is the hippocampus). The microdialysis probe is perfused with Ringer solution via a microinfusion pump (not depicted). Drugs can be delivered to the target tissue by inclusion in the perfusion fluid (so-called retrodialysis or reverse dialysis). Animals are connected to a swivel system and counterbalance arm via a little peg anchored to the dental cement on their skull (see picture) giving them the opportunity to move freely in all directions. Samples are collected using automated microsamplers, and are subsequently used to measure various molecules in the dialysates – e.g., neurotransmitters and their metabolites – by high-pressure liquid chromatography (HPLC) with electrochemical detection or free corticosterone levels via radioimmunoassay (RIA). Behavior of the rats or mice is continuously scored during the microdialysis experiments by visual observation or is subsequently analyzed from videotape. Experiments are performed under baseline conditions, and during and following stressful challenges or behavioral testing. Note that this method allows the simultaneous collection of information on the status and responsivity of various neurotransmitter systems in the brain, on the activity of the HPA axis and on behavior, providing strong, integrative data and representing a clear example of refinement and reduction (Russell and Burch, 1959) in animal research.

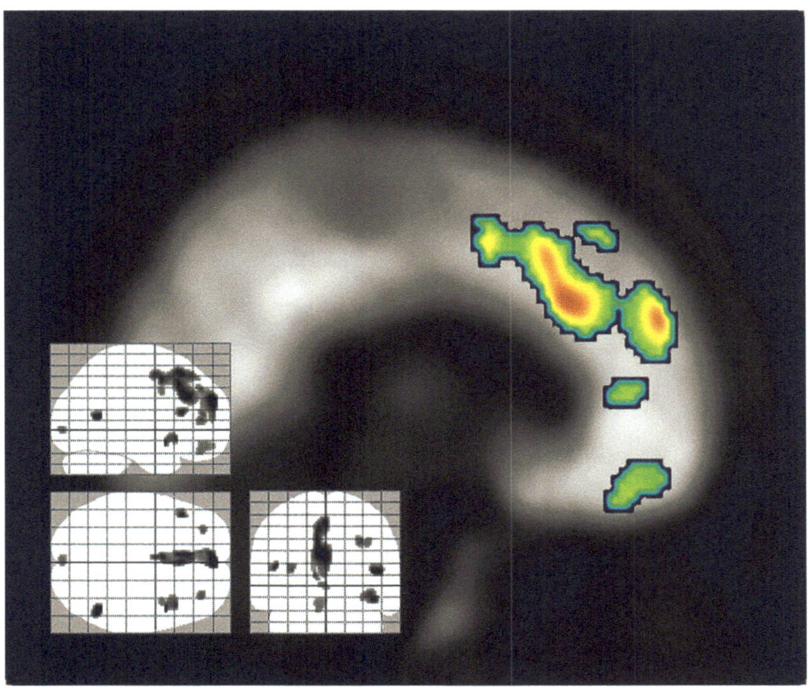

Plate 17 Inverse Correlation of 5D-ASC Global Scale scores and [^{18}F]altanserin apparent distribution volume [DV′]. Results of a voxel based correlation analysis (? 5D-ASC global *vs* ΔDV′, threshold $P < 0.005$, uncorrected) using Statistical Parametric Mapping (SPM2) (Hasler, Quednow, Vollenweider, unpublished data).

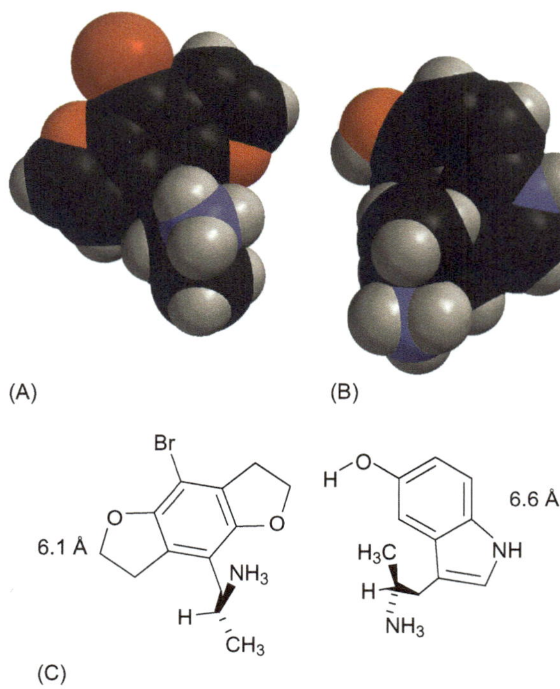

Plate 18 Comparison of the R-(_)- isomer of a potent difuranobenzene phenethylamine type hallucinogen (A) with S-(_)-α-methylserotonin (B). Although the stereochemistry at the carbon bearing the alpha-methyl is reversed for the phenethylamines and the tryptamines, the lengths of the ligands, differing by about 0.5 Ångstorms (C), could be the underlying reason for this reversal. The bicyclic aromatic indole system of the tryptamines is longer than the phenyl ring of the phenethylamines, and thus when the side-chain is extended, the tryptamines will occupy a larger space in the receptor. Because the receptor evolved to accept serotonin, a tryptamine, the perspective must be one that the phenethylamines somehow adapt themselves to a binding site that prefers tryptamines.

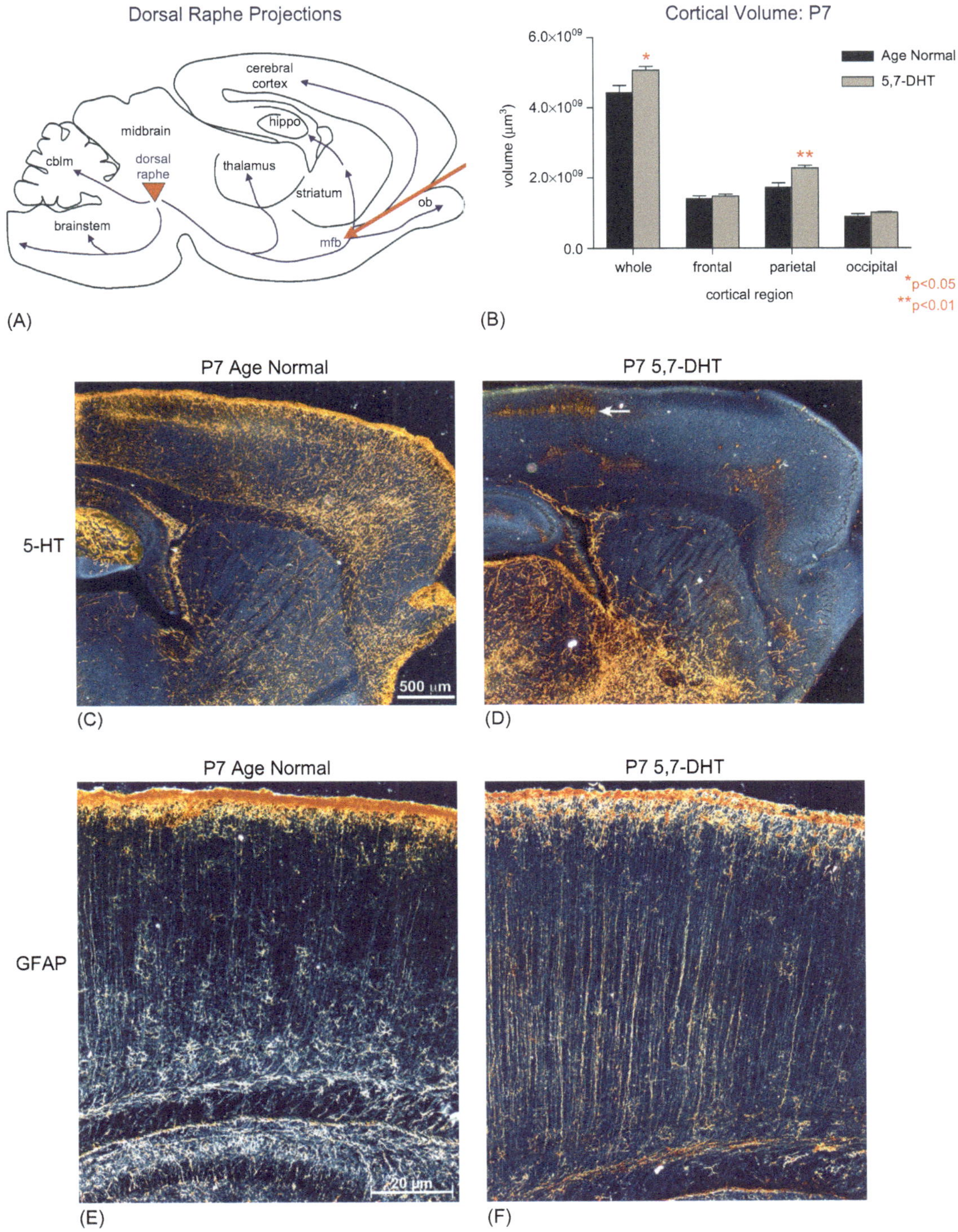

Plate 19 5,7-DHT mouse model. (A) Schematic drawing of a parasagittal section through a P7 mouse brain showing the location of the serotonergic neurons in the dorsal raphe and the projections from the raphe to multiple brain regions. Serotonergic axons innervate the hippocampus (hippo) and cerebral cortex via the medial forebrain bundle (mfb). The red arrow indicates the path of the 5,7-DHT injections. cblm, cerebellum; ob, olfactory bulb. (B) The volume of the whole cerebral cortex and parietal regions was greater in 5,7-DHT lesioned mice than in Age Normal controls. (C, D) Dark field photomicrographs show the distribution of serotonergic axons (5-HT) in parasagittal sections of (C) P7 Age Normal and (D) 5,7-DHT lesioned mouse brains. In lesioned mice, there are many fewer afferent serotonergic axons in the cortex and the hippocampus. However, some serotonin remains in thalamocortical terminals due to their transient expression of the serotonin transporter (white arrow). (E, F) The pattern of immunostaining for glial fibrillary acetic protein (GFAP) is shown in the dark field photomicrographs of the cortex in (E) Age Normal and (F) 5,7-DHT lesioned mouse brains. At P7, many of the radial glial fibers remain present, and their density appears higher in the lesioned mice.

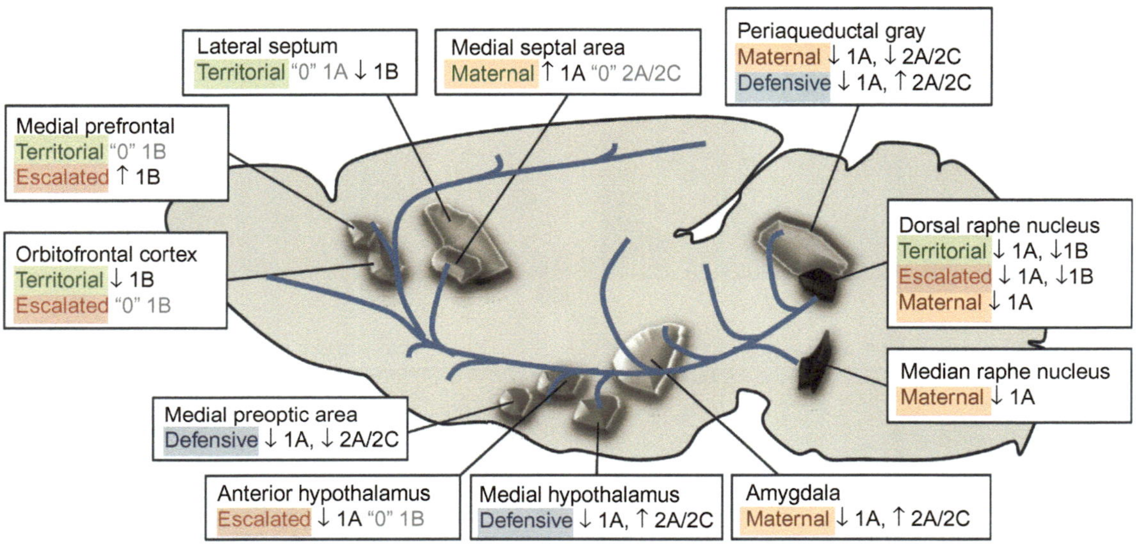

Plate 20 Modulation of aggressive behaviors by microinjections of $5HT_{1A}$, $5\text{-}HT_{1B}$, and $5\text{-}HT_{2A/2C}$ receptor agonists. Text boxes show that local injection of $5\text{-}HT_{1A}$ receptor (1 A), $5\text{-}HT_{1B}$ receptor (1B), or $5\text{-}HT_{2A/2C}$ receptor (2A/2C) agonist increases (↑), decreases (↓) or has no effect ('0') on territorial, escalated, maternal, and defensive aggressive behaviors. Serotonergic neurons originated from raphe nuclei project to several brain areas.

FSC
www.fsc.org
MIX
Papier | Fördert gute Waldnutzung
FSC® C083411

Kolibri 360 GmbH
Ferdinand-Jühlke-Straße 7
99095 Erfurt, Deutschland
produktsicherheit@kolibri360.de